Families of Functions

Linear Functions

$y = mx + b, m > 0$

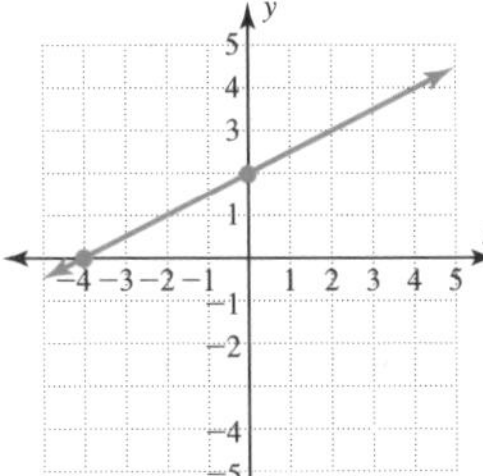

$y = mx + b, m < 0$

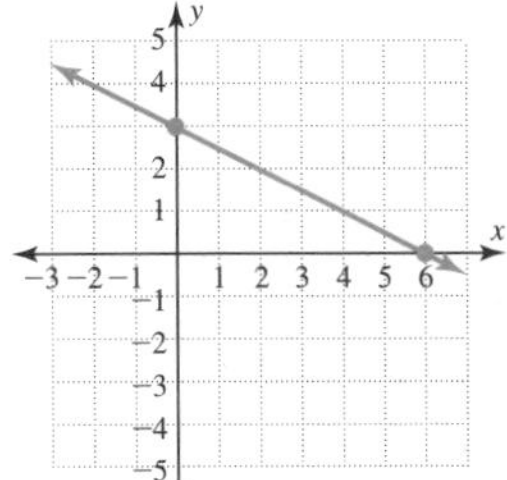

$y = mx + b, m = 0$
(a constant function)

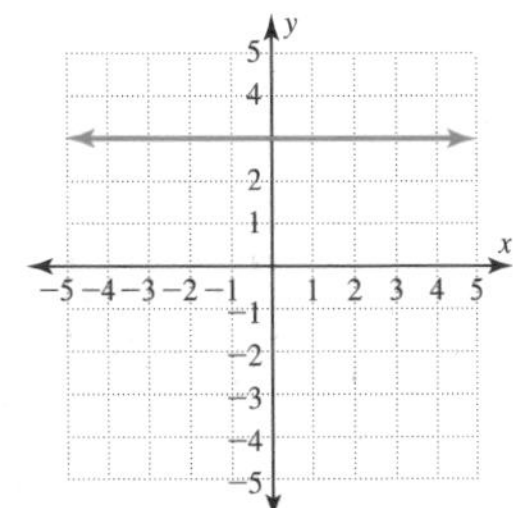

Absolute Value Function

$y = |x|$

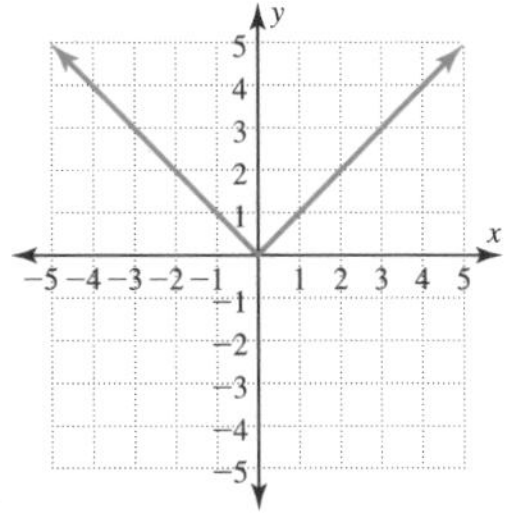

Quadratic Functions

$y = x^2$

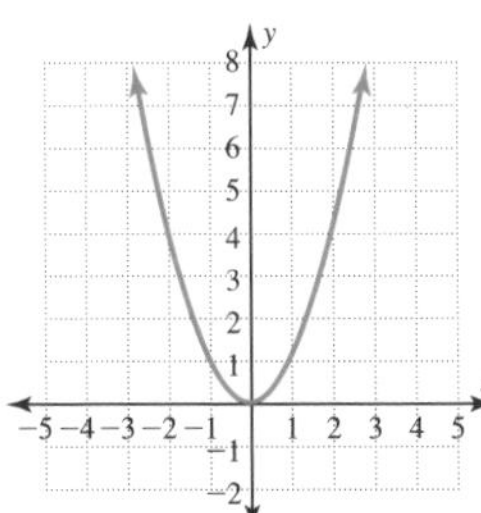

$y = ax^2 + bx + c, a > 0$

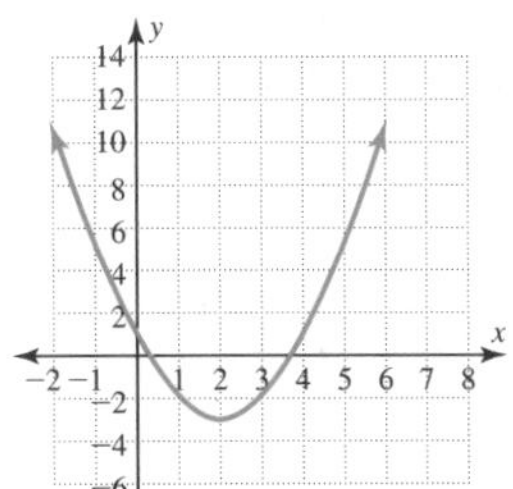

$y = ax^2 + bx + c, a < 0$

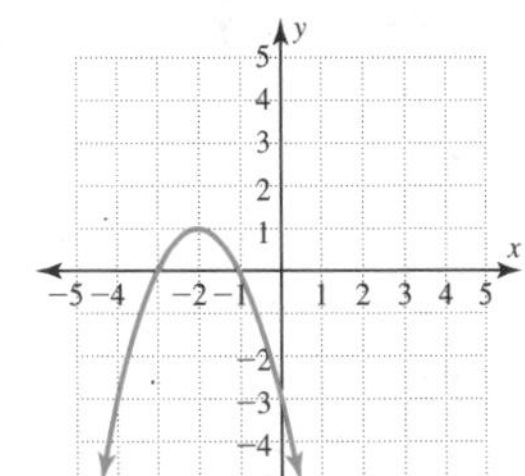

Square Root Function

$y = \sqrt{x}$

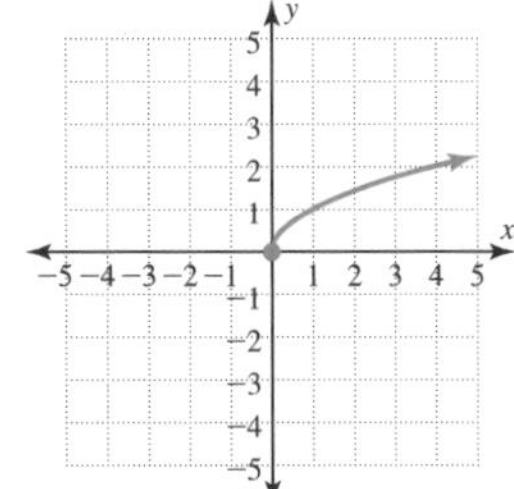

Cubic Functions

$y = x^3$

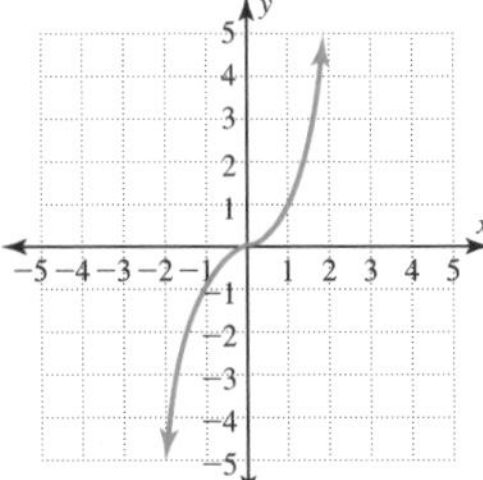

$y = ax^3 + bx^2 + cx + d,$
$a > 0$

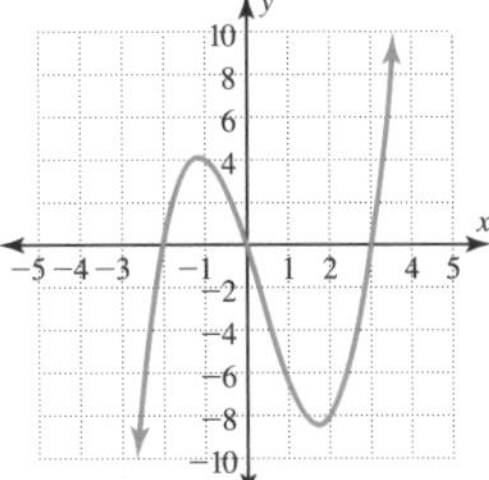

$y = ax^3 + bx^2 + cx + d,$
$a < 0$

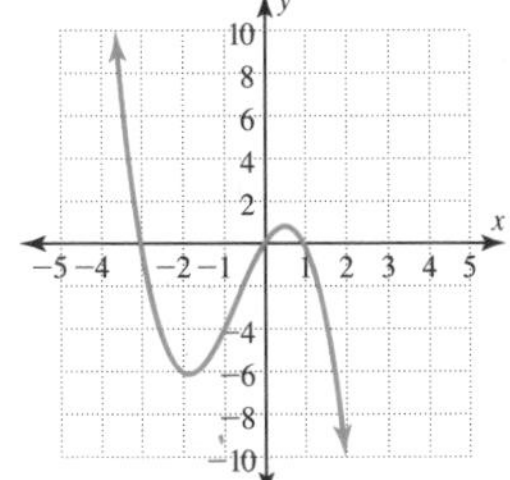

Cube Root Function

$y = \sqrt[3]{x}$

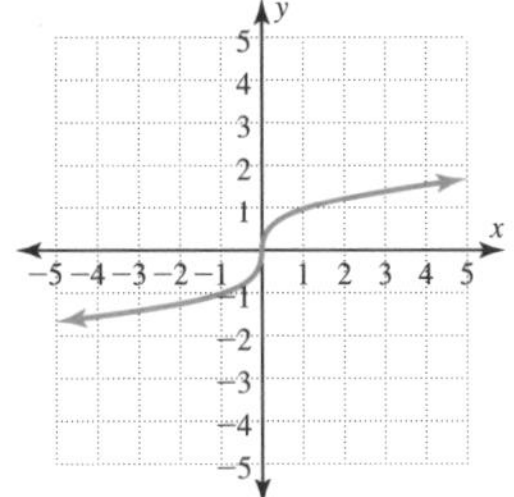

Rational Function

$y = \dfrac{1}{x}$

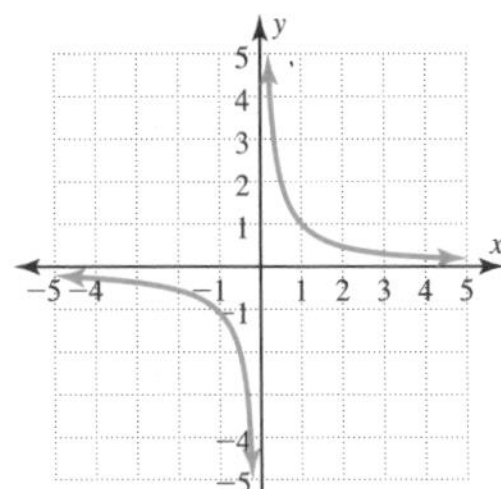

Exponential Growth Function

$y = b^x, b > 1$

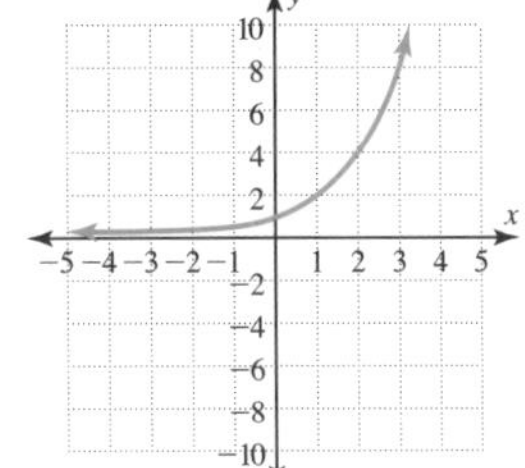

Exponential Decay Function

$y = b^x, 0 < b < 1$

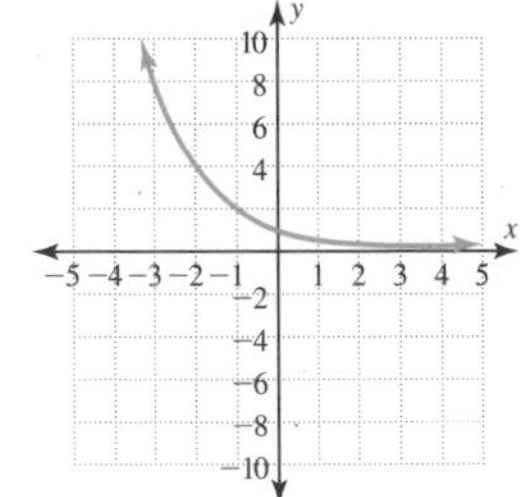

Logarithmic Function

$y = \log_b x, b > 0, b \neq 1$

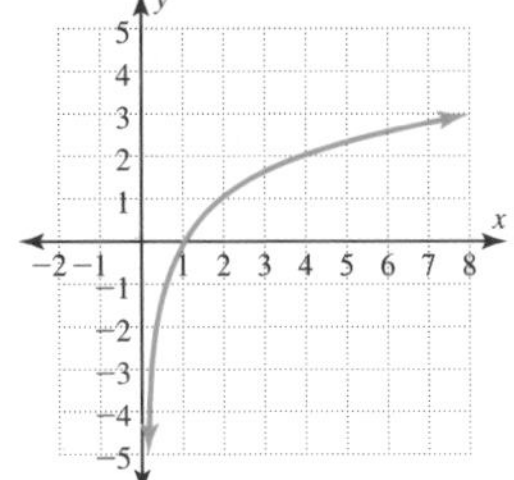

Subsets of the Real Numbers

Natural numbers: $\{1, 2, 3, 4, 5, 6, \ldots\}$
Whole numbers: $\{0, 1, 2, 3, 4, 5, 6, \ldots\}$
Integers: $\{\ldots -3, -2, -1, 0, 1, 2, 3, \ldots\}$
Rational: $\left\{\frac{a}{b}: a, b \text{ are integers and } b \neq 0\right\}$
Irrational: The set of real numbers that are not rational.

CONSTANTS: $\pi \approx 3.14159$, $e \approx 2.71828$

Operations with Fractions

Addition: $\frac{a}{b} + \frac{c}{b} = \frac{a + c}{b}$ for $b \neq 0$

Subtraction: $\frac{a}{b} - \frac{c}{b} = \frac{a - c}{b}$ for $b \neq 0$

Reducing fractions: $\frac{ac}{bc} = \frac{a}{b}$ for $b \neq 0$ and $c \neq 0$

Multiplication: $\frac{a}{b} \cdot \frac{c}{d} = \frac{a \cdot c}{b \cdot d}$ for $b \neq 0$ and $d \neq 0$

Division: $\frac{a}{b} \div \frac{c}{d} = \frac{a}{b} \cdot \frac{d}{c} = \frac{a \cdot d}{b \cdot c}$ for $b \neq 0$, $c \neq 0$, and $d \neq 0$

Properties of the Real Numbers

Commutative properties:

of addition	$a + b = b + a$
of multiplication	$ab = ba$

Associative properties:

of addition	$(a + b) + c = a + (b + c)$
of multiplication	$(ab)(c) = (a)(bc)$

Distributive properties of multiplication over addition:

$a(b + c) = ab + ac$

$(b + c)a = ba + ca$

Identities:

additive	$a + 0 = 0 + a = a$
multiplicative	$a \cdot 1 = 1 \cdot a = a$

Inverses:

additive	$a + (-a) = (-a) + a = 0$
multiplicative	$a \cdot \frac{1}{a} = \frac{1}{a} \cdot a = 1$ for a $\neq 0$

Order of Operations

Step 1. Start with the expression within the innermost pair of grouping symbols.
Step 2. Perform all exponentiations.
Step 3. Perform all multiplications and divisions as they appear from left to right.
Step 4. Perform all additions and subtractions as they appear from left to right.

Interval Notation

Inequality Notation	Verbal Meaning	Graph	Interval Notation
$x > a$	x is greater than a	a	(a, ∞)
$x \geq a$	x is greater than or equal to a	a	$[a, \infty)$
$x < a$	x is less than a	a	$(-\infty, a)$
$x \leq a$	x is less than or equal to a	a	$(-\infty, a]$
$a < x < b$	x is greater than a and less than b	a b	(a, b)
$a < x \leq b$	x is greater than a and less than or equal to b	a b	$(a, b]$
$a \leq x < b$	x is greater than or equal to a and less than b	a b	$[a, b)$
$a \leq x \leq b$	x is greater than or equal to a and less than or equal to b	a b	$[a, b]$
$-\infty < x < \infty$	x is any real number		$(-\infty, \infty)$

Beginning and Intermediate ALGEBRA

THE LANGUAGE AND SYMBOLISM OF MATHEMATICS

Beginning and Intermediate ALGEBRA

THE LANGUAGE AND SYMBOLISM OF MATHEMATICS

Second Edition

James W. Hall
Parkland College

Brian A. Mercer
Parkland College

Boston Burr Ridge, IL Dubuque, IA New York San Francisco St. Louis
Bangkok Bogotá Caracas Kuala Lumpur Lisbon London Madrid Mexico City
Milan Montreal New Delhi Santiago Seoul Singapore Sydney Taipei Toronto

Higher Education

BEGINNING AND INTERMEDIATE ALGEBRA: THE LANGUAGE AND SYMBOLISM OF MATHEMATICS, SECOND EDITION

Published by McGraw-Hill, a business unit of The McGraw-Hill Companies, Inc., 1221 Avenue of the Americas, New York, NY 10020.

 This book is printed on acid-free paper.

3 4 5 6 7 8 9 0 DOW/DOW 0 9 8

ISBN 978–0–07–293322–2
MHID 0–07–293322–4

Publisher: *Elizabeth J. Haefele*
Sponsoring Editor: *David Millage*
Developmental Editor: *Suzanne Alley*
Marketing Manager: *Barbara Owca*
Project Manager: *April R. Southwood*
Senior Production Supervisor: *Laura Fuller*
Lead Media Project Manager: *Stacy A. Patch*
Media Producer: *Amber M. Huebner*
Designer: *Laurie B. Janssen*
Cover Designer: *Ron Bissell*
(USE) Cover Image: ©*Daryl Benson/Masterfile*
Senior Photo Research Coordinator: *John C. Leland*
Photo Research: *Mary Reeg*
Supplement Producer: *Melissa M. Leick*
Compositor: *Techbooks*
Typeface: 10.5/12 *Times*
Printer: *R. R. Donnelley, Willard, OH*

The credits section for this book begins on page 928 and is considered an extension of the copyright page.

Library of Congress Cataloging-in-Publication Data

Hall, James W.
Beginning and intermediate algebra : the language and symbolism of mathematics / James W. Hall, Brian A. Mercer.–2nd ed.
p. cm.
Includes index.
ISBN 978–0–07–293322–2 — ISBN 0–07–293322–4 (hard copy : alk. paper)
1. Algebra–Textbooks. I. Mercer, Brian A. II. Title.
QA152.3.H28 2008
512–dc22

2006033785

www.mhhe.com

Dedication

To our wives, Peggy Hall and Nikki Mercer, for their support and encouragement.

About the Authors

JAMES W. HALL is a retired professor and chair from the mathematics department at Parkland College in Champaign, Illinois. He started teaching mathematics in 1969 at Northern Arizona University and also taught at Clayton State College in Georgia prior to joining Parkland College in 1975. From 1989 to 1990 he taught at Dandenong College in Victoria, Australia. He received a B.S. and an M.A. in mathematics from Eastern Illinois University and an Ed.D. from Oklahoma State University. He was Midwest Regional Vice President of AMATYC (American Mathematical Association of Two-Year Colleges) from 1987 to 1989 and President of IMACC (Illinois Mathematics Association of Community Colleges) from 1995 to 1996. In 1978 he edited the "Report on Microcomputers in the Classroom" for ICTM (Illinois Council of Teachers of Mathematics), and from 1991 to 1995 he was chairperson of the editorial review committee for AMATYC. He was the writing team chair for Chapter 6 on Curriculum and Program Development of "Beyond Crossroads". This is his eighteenth mathematics textbook.

BRIAN A. MERCER is an associate professor of mathematics at Parkland College in Champaign, Illinois. Before starting at Parkland College in 1998, he taught at Neoga High School from 1994 to 1995 and at Lakeland College from 1997 to 1998. He received a B.S. in mathematics from Eastern Illinois University and an M.S. in mathematics from Southern Illinois University. He is a member of AMATYC, NADE (National Association for Developmental Education), and has served as a board member of IMACC.

Contents

Preface

The Universe is a grand book which cannot be read until one first learns to comprehend the language and become familiar with the characters in which it is composed. It is written in the language of mathematics.
—Galileo

Beginning and Intermediate Algebra, the Language and Symbolism of Mathematics, covers the topics from both Beginning and Intermediate Algebra. It is fully integrated rather than the combination of two separate texts.

Our primary goal is to implement the AMATYC standards outlined in ***Beyond Crossroads,*** and to give strong support to faculty members to teach this material. These standards were used as guiding principles to organize the topics. This organization was designed to work for students with a variety of learning styles and for teachers with a variety of experiences and backgrounds. An example of this organization is an early presentation of function notation and the graphing of linear equations in two variables.

The inclusion of multiple perspectives—verbal, numerical, algebraic, and graphical—has proven popular with a broad cross section of students. Calculator Perspectives help students see the relationship between mathematics and technology. The TI-84 Plus keystroke information provided in these Calculator Perspectives eliminates the need for instructors to create separate keystroke handouts.

The Beginning Algebra portion of this text concentrates primarily on material related to linear equations. (Most of the nonlinear material is reserved for the Intermediate Algebra portion of the text.) The review material in Chapter 1 is presented through the evaluation of algebraic expressions, the checking of solutions to equations, and other contexts. This nontraditional approach motivates students through real-life applications to review background concepts. It also gives students who have already had this material in high school a fresh approach and helps them to connect previously separated topics.

TEACHING APPROACH

- **Emphasis on the Rule of Four and Multiple Perspectives**
  The "rule of four" suggests that concepts should be examined algebraically (symbolically), numerically, graphically, and verbally. Users of the first edition were very pleased that we integrated multiple perspectives throughout the book. We use multiple perspectives not only in examples and exercises, but also in definitions and exposition. Our experience leads us to believe that students who use the rule of four develop a deeper understanding of the concepts they study. They are less likely to memorize steps, they are more likely to retain the material they understand, and they are more likely to apply mathematics outside the classroom. (See AMATYC Standard for Intellectual Development: Linking Multiple Representations.)

- **Technology Is Built-In, Not Added-On**
  Topics that once were postponed until many manipulative skills had been developed can now be considered earlier by using technology to focus on concepts instead of computation. The use of the TI-84 Plus graphing calculator is demonstrated throughout the book. The students' use of technology enables them to examine realistic problems such as producing the payment schedule for a car loan. Together, realistic applications and the use of calculators facilitate the development of modeling skills by the students. Technology is woven throughout the text—it is not simply inserted into a standard presentation. (See AMATYC Standard for Intellectual Development: Using Technology.)

- **Functions**
The Beginning Algebra portion of the book introduces function notation in Chapter 2 and gives the student many opportunities to become familiar with this notation and with the input-output concept. The Intermediate Algebra portion of the textbook contains a formal definition of a function and various notations used to represent functions. Chapter 7 through 10 examine linear, absolute value, quadratic, rational, radical, and exponential families of functions. For those wanting to apply these functions to realistic data, curve fitting for linear, quadratic, and exponential functions is also covered. Properties of functions are examined further in Chapter 11 in a preview of college algebra. (See AMATYC Standard for Content: Function.)

- **Mathematical Modeling and Word Equations**
The residual value of mathematics—the mathematics that students can still use four or more years after taking a course—is not a collection of tricks or memorized steps. What endures is an understanding that allows students to see mathematics as useful in improving their daily lives. Most people encounter mathematics through words, either orally or in writing, not through equations. Word equations help students bridge the gap between the statement of a word problem and the formation of an algebraic equation that models the problem. Students must model real problems in a course if we expect them to use mathematics on their own. To that end, the text presents many realistic examples and exercises involving data (see the Index of Applications). (See AMATYC Standards for Intellectual Development: Modeling and for Content: Symbolism and Algebra.)

- **Using Systems of Equations**
Word problems that involve two unknowns are solved in Chapter 3 using two variables rather than one variable. This approach has been received well by the students who often have more trouble identifying two unknowns using one variable than using a separate variable for each unknown. This approach also received favorable feedback from users of the first edition. Later in the book we examine alternate approaches that build on creating functional models and the relation of one variable to another.

- **Using Discrete Data**
The book includes problems that give the students experience with discrete data. This experience will help students develop a better perspective on mathematical models, especially those who will use Intermediate Algebra as their prerequisite to an Introductory Statistics course. (See AMATYC Standards for Content: Continuous and Discrete Models and Data Analysis, Statistics, and Probability.)

- **Using the Language and Symbolism of Mathematics**
Each exercise set starts with a few questions on the language and symbolism of mathematics. One benefit of assigning these exercises that we have noted in our classes is that the students spend more time reading the book before starting the other exercises. (See AMATYC Standards for Intellectual Development: Communicating, and Standards for Content: Symbolism and Algebra.)

TEXTBOOK FEATURES

- **Geometrical Problems**
Examples and exercises based on geometrical shapes are placed throughout the textbook. Many exercise sets have problems involving perimeter, area, and volume. (See AMATYC Standards for Content: Geometry and Measurement.)

- **Design of Exercises**
Many exercises are composed of multiple parts to help the students address common misconceptions about the language and symbolism of mathematics. A few examples are:

Exercise 1.7, #28	**a.** Simplify $(5 + 7)^2$.
	b. Simplify $5^2 + 7^2$.
Exercise 6.1, #29	**a.** Expand $x(x + 4) + 2(x + 4)$.
	b. Factor $x(x + 4) + 2(x + 4)$.

Exercise 6.6, #55	**a.** Solve $(5m - 3)(m - 2) = 0$.
	b. Simplify $(5m - 3)(m - 2)$.
Exercise 9.1, #21	**a.** Mentally estimate $\sqrt{66}$.
	b. Use a calculator to approximate $\sqrt{66}$.

- **Mathematical Notes**
  Mathematical Notes are placed throughout the book to give the students some sense of historical perspective and to help connect mathematics to other disciplines. These short vignettes give the origin of some of the symbols and terms that we now use and provide brief glimpses into the lives of some of the men and women of mathematics. (See AMATYC Standards for Pedagogy: Making Connections.)
- **Self-Checks**
  Self-Checks and the answers to these questions are placed in each section to help the students become active learners and to monitor their own progress. (See AMATYC Standards for Intellectual Development: Developing Mathematical Power.)
- **Example Format**
  The format of the examples provides the students a clear model that they can use to work the exercises. The side-bar explanations can be used by the students as needed. This format keeps the text from being wordy and lets students with different ability levels use the examples in different ways. Many examples are worked using multiple perspectives so students can compare the algebraic, numerical, graphical, and verbal approaches to the same problem. (See AMATYC Standards for Pedagogy: Using Multiple Approaches.)
- **Group Discussion Questions and Group Projects**
  Each exercise set has group discussion questions and there is a group project at the end of each of the first ten chapters. Some of these exercises build bridges to past or future material. Some of them are more challenging for the students. Many of these problems seek to engage the students in communicating with mathematics orally and in writing and to involve the students in interactive and collaborative learning. (See AMATYC Standards for Pedagogy: Active Learning.)
- **Key Concepts, Chapter Review, and Mastery Tests**
  Each chapter ends with these features. The Key Concepts outline the main points covered in that chapter. The Chapter Review contains a selection of exercises designed to help the students review the material from the chapter and to gauge their readiness for an exam. The Chapter Review is longer than an hour exam, and the order of the questions may not parallel the order of each topic within the chapter. The Mastery Test is more limited in its purpose. Each question matches an objective stated at the beginning of one of the sections in the chapter. The students can use this Mastery Test in a diagnostic way to determine which sections and objectives that they have mastered and which may still need more work.

WHAT'S NEW IN THE SECOND EDITION?

Three features that the first edition introduced have been very well received and are continued in the second edition. 1. Multiple Perspectives are still used to state definitions and properties. 2. Full keystroke examples for the TI-84 Plus are given in the text. 3. A class-tested lecture guide is available for students and instructors. Some of the changes in the second edition include:

- Some group exercises have been revised due to user input.
- There are more cumulative reviews in this edition.
- The presentation on factoring has been revised due to user input. The strong emphasis on the role of the distributive property and the connection between multiplying polynomials and factoring polynomials has been continued.
- More one-variable applications are included in Chapter 2.
- The exercises sets in the intermediate algebra portion of the book start with a quick review of some concepts from earlier chapters.

- More chapters start with a realistic problem or scenario to focus student attention on the chapter.
- Chapters 7 through 10 have more of a function emphasis and start by discussing different families of functions.
- Material on the vertex of a parabola and related applications is included in Chapter 7.

SUPPLEMENTS FOR THE INSTRUCTOR

Instructor's Edition *ISBN-13: 978-0-07-330350-5 (ISBN-10: 0-07-330350-X)*

The Instructor's Edition (IE) contains answers to problems and exercises in the text, including answers to all Language and Symbolism of Mathematics vocabulary questions, all end-of-section exercises, all end-of-chapter review exercises, and all end-of-chapter mastery tests.

Instructor's Solutions Manual *ISBN-13: 978-0-07-330352-9 (ISBN-10: 0-07-330352-6)*

The Instructor's Solutions Manual, prepared by Mark Smith of College of Lake County, provides comprehensive **worked-out solutions** to exercises in the text.

Lecture Guide *ISBN-13: 978-0-07-293323-9 (ISBN-10: 0-07-293323-2)*

This supplement by Brian Mercer and James Hall, with the assistance of Mark Stevenson of Oakland Community College, provides instructors with the framework of day-by-day class activities for each section in the book. Each lecture guide can help instructors make more efficient use of class time and can help keep students focused on active learning. Students who use the lecture guides have the framework of well-organized notes that can be completed with the instructor in class.

Instructor's Testing and Resource CD-ROM *ISBN-13: 978-0-07-330351-2 (ISBN-10: 0-07-330351-8)*

This cross-platform CD-ROM provides a wealth of resources for the instructor. Among the supplements featured on the CD-ROM is a **computerized test bank** utilizing Brownstone Diploma® algorithm-based testing software to quickly create customized exams. This user-friendly program enables instructors to search for questions by topic, format, or difficulty level; to edit existing questions or to add new ones; and to scramble questions and answer keys for multiple versions of a single test. Hundreds of text-specific open-ended and multiple-choice questions are included in the question bank. Sample chapter tests, midterms, and final exams in Microsoft Word® and PDF formats are also provided.

Video Lectures on Digital Video Disk (DVD) *ISBN-13: 978-0-07-293325-3 (ISBN-10: 0-07-293325-9)*

In the videos, the authors, James Hall and Brian Mercer, work through selected problems from the textbook, following the solution methodology employed in the text. The video series is available on DVD or online as an assignable element of MathZone (see next). The DVDs are closed-captioned for the hearing impaired, subtitled in Spanish, and meet the Americans with Disabilities Act Standards for Accessible Design. Instructors may use them as resources in a learning center, for online courses, and/or to provide additional help to students who require extra practice.

MathZone—www.mathzone.com

McGraw-Hill's **MathZone 3.0** is a complete **web-based tutorial and course-management system** for mathematics and statistics, designed for greater ease of use than any other system available. Free upon adoption of a McGraw-Hill textbook, the system enables instructors to **create and share courses and assignments** with colleagues, adjunct faculty members, and teaching assistants with only a few mouse clicks. All **assignments, exercises, "e-Professor" multimedia tutorials, video lectures, and NetTutor® live tutors** follow the textbook's learning objectives and problem-solving style and notation. Using MathZone's **assignment builder,** instructors can **edit questions and algorithms, import their own content,** and **create announcements and due-dates** for homework and quizzes. MathZone's **automated grading function** reports the results of easy-to-assign algorithmically-generated homework, quizzes, and tests. All student activity within MathZone is recorded and available through a **fully integrated gradebook** that can be downloaded to Microsoft Excel®. MathZone also is available on CD-ROM. (See "Supplements for the Student" for descriptions of the elements of MathZone.)

ALEKS

ALEKS (**A**ssessment and **LE**arning in **K**nowledge **S**paces) is an artificial intelligence-based system for mathematics learning, available over the Web 24/7. Using unique adaptive questioning, ALEKS accurately assesses what topics each student knows and then determines exactly what each student is ready to learn next. ALEKS interacts with the students much as a skilled human tutor would, moving between explanation and practice as needed, correcting and analyzing errors, defining terms and changing topics on request, and helping them master the course content more quickly and easily. Moreover, the new ALEKS 3.0 now links to text-specific videos, multimedia tutorials, and textbook pages in PDF format. ALEKS also offers a robust classroom management system that allows instructors to monitor and direct student progress toward mastery of curricular goals. See www.aleks.com.

SUPPLEMENTS FOR THE STUDENT

Student's Solutions Manual *ISBN-13: 978-0-07-293324-6 (ISBN-10: 0-07-293324-0)*

The Student's Solutions Manual, provided by Mark Smith of College of Lake County, provides comprehensive, **worked-out solutions** to odd-numbered exercises. The steps shown in solutions match the style of solved examples in the textbook.

Lecture Guide *ISBN-13: 978-0-07-293323-9 (ISBN-10: 0-07-293323-2)*

This supplement by Brian Mercer and James Hall, with the assistance of Mark Stevenson of Oakland Community College, provides instructors with the framework of day-by-day class activities for each section in the book. Each lecture guide can help instructors make more efficient use of class time and can help keep students focused on active learning. Students who use the lecture guides have the framework of well-organized notes that can be completed with the instructor in class.

MathZone—www.mathzone.com

McGraw-Hill's MathZone is a powerful web-based tutorial for homework, quizzing, testing, and multimedia instruction. Also available in CD-ROM format, MathZone offers:

Practice exercises based on the text and generated in an unlimited quantity for as much practice as needed to master any objective.

Video clips of classroom instructors showing step-by-step how to solve exercises from the text.

e-Professor animations that take the student through step-by-step instructions, delivered on-screen and narrated by a teacher on audio, for solving exercises from the textbook; the user controls the pace of the explanations and can review as needed.

NetTutor offers personalized instruction by live tutors familiar with the textbook's objectives and problem-solving methods.

Every assignment, exercise, video lecture, and e-Professor is derived from the textbook.

Video Lectures on Digital Video Disk (DVD) *ISBN-13: 978-0-07-293325-3 (ISBN-10: 0-07-293325-9)*

The video series is based on exercises from the textbook. The authors, James Hall and Brian Mercer, work through selected problems, following the solution methodology employed in the text. The video series is available on DVD or online as part of MathZone. The DVDs are closed-captioned for the hearing impaired, subtitled in Spanish, and meet the Americans with Disabilities Act Standards for Accessible Design.

NetTutor

Available through MathZone, NetTutor is a revolutionary system that enables students to interact with a live tutor over the Web. NetTutor's web-based, graphical chat capabilities enable students and tutors to use mathematical notation and even to draw graphs as they work through a problem together. Students can also submit questions and receive answers, browse previously answered questions, and view previous sessions. Tutors are familiar with the textbook's objectives and problem-solving styles.

ACKNOWLEDGMENTS

We would like to thank students who used the preliminary versions of this text and users of the first edition for their suggestions and encouragement. We are also very appreciative of the support from McGraw-Hill. Elizabeth Haefele, publisher; David Dietz, director of development; and David Millage, sponsoring editor, gave us the support to create a book that incorporates technology and embraces the AMATYC standards. We are also very appreciative of the support of Suzanne Alley, developmental editor; April Southwood, project manager; and Patti Scott, copyeditor, for helping us accomplish all the miracles that put text, art, and answers into a quality textbook. We would also like to thank Amber Huebner, media producer, and Stacy Patch, lead media project manager, who did an outstanding job creating the MathZone components.

Special thanks are due to: Chris Nord and Rick Mercer for their accuracy-check of the text; Mark Stevenson for helping with the preparation of the lecture guide that accompanies this book; Brian Smith of Parkland College for creating the online Spreadsheet Activities; Mark Smith of College of Lake County, who not only authored the Instructor's and Student's Solutions Manuals but also prepared the appendix of answers; Link Systems International for their work on MathZone; David Sze for his authorship of the test bank included on the Instructor's Testing and Resource CD-ROM; and David Winter and Igor Torgeson at Publishers Media Group for their innovative work on the text's video series.

Commentary from the following reviewers was indispensable during the development of the text:

Mary Bergs, *Mercy College of Northwest Ohio*
Barbara Burke, *Hawaii Pacific University*
Susan Caldiero, *Consumnes River College*
Jackie Coomes, *Eastern Washington University*
Cheryl Davids, *Central Carolina Technical College*
Carol Dawson, *Middle Tennessee State University*
Cynthia Gubitose, *University of Southern Connecticut*
Annette Hawkins, *Wayne Community College*
Elias Jureidini, *Lamar State College—Orange*
Tamie McCabe, *Redlands Community College*
David Otts, *Middle Tennessee State University*
Mohsen Shirani, *Tennessee State University—Nashville*
Shirley Smith, *Tomball College*
Doug Smucker, *Columbia College*
Erin Wall, *College of the Redwoods*

Walk-Through

Tools for Learning

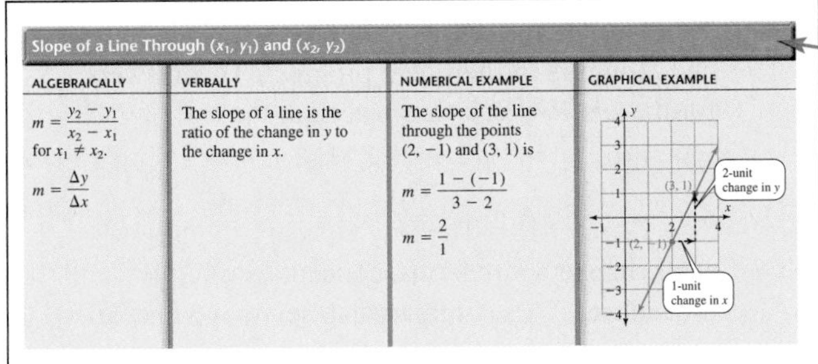

Slope of a Line Through (x_1, y_1) and (x_2, y_2)

ALGEBRAICALLY	VERBALLY	NUMERICAL EXAMPLE	GRAPHICAL EXAMPLE
$m = \frac{y_2 - y_1}{x_2 - x_1}$ for $x_1 \neq x_2$. $m = \frac{\Delta y}{\Delta x}$	The slope of a line is the ratio of the change in y to the change in x.	The slope of the line through the points $(2, -1)$ and $(3, 1)$ is $m = \frac{1 - (-1)}{3 - 2}$ $m = \frac{2}{1}$	

Multiple Perspective boxes—Demonstrating concepts in multiple perspectives—algebraically (symbolically), numerically, graphically, and verbally—helps students develop a deeper understanding of mathematics.

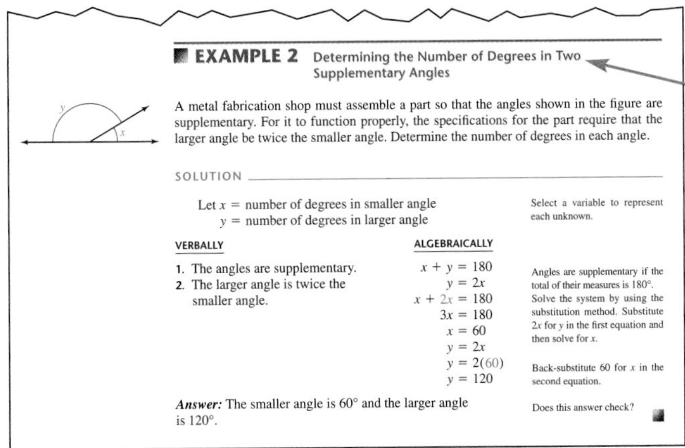

EXAMPLE 2 Determining the Number of Degrees in Two Supplementary Angles

A metal fabrication shop must assemble a part so that the angles shown in the figure are supplementary. For it to function properly, the specifications for the part require that the larger angle be twice the smaller angle. Determine the number of degrees in each angle.

SOLUTION

Let x = number of degrees in smaller angle
y = number of degrees in larger angle

Select a variable to represent each unknown.

VERBALLY	ALGEBRAICALLY	
1. The angles are supplementary. 2. The larger angle is twice the smaller angle.	$x + y = 180$ $y = 2x$	Angles are supplementary if the total of their measures is 180°.
	$x + 2x = 180$ $3x = 180$ $x = 60$	Solve the system by using the substitution method. Substitute $2x$ for y in the first equation and then solve for x.
	$y = 2x$ $y = 2(60)$ $y = 120$	Back-substitute 60 for x in the second equation.

Answer: The smaller angle is 60° and the larger angle is 120°.

Does this answer check?

Examples—Each chapter includes many worked examples. These examples advance skills, develop concepts, connect concepts, and show application of concepts. Many examples are presented using multiple perspectives. These problems are worked in the same format students should use and have explanations horizontally aligned with the steps.

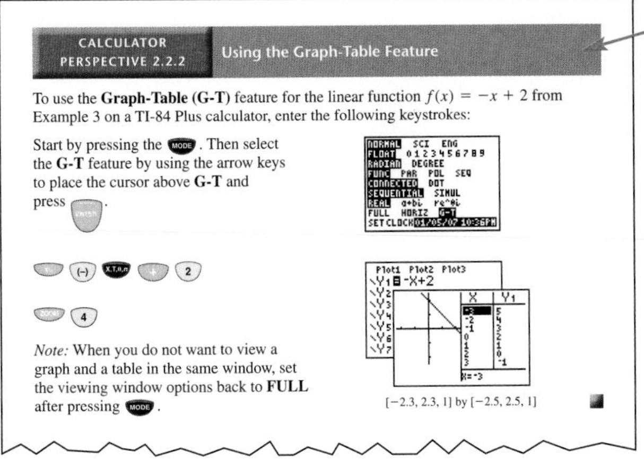

CALCULATOR PERSPECTIVE 2.2.2 Using the Graph-Table Feature

To use the **Graph-Table (G-T)** feature for the linear function $f(x) = -x + 2$ from Example 3 on a TI-84 Plus calculator, enter the following keystrokes:

Start by pressing the MODE. Then select the **G-T** feature by using the arrow keys to place the cursor above **G-T** and press ENTER.

Note: When you do not want to view a graph and a table in the same window, set the viewing window options back to **FULL** after pressing MODE.

[−2.3, 2.3, 1] by [−2.5, 2.5, 1]

Calculator Perspective boxes—The TI-84 Plus™ calculator is used to work sample problems as concepts are developed. The location of the Calculator Perspectives within the textbook is not only convenient for the student, it also helps to keep the focus on the mathematical concepts while showing the appropriate use of technology for exploration and for computation.

$$y - (-4) = \frac{3}{2}[x - (-6)]$$ Substitute $(-6, -4)$ for (x_1, y_1) and $\frac{3}{2}$ for m.

$$y + 4 = \frac{3}{2}(x + 6)$$ Actually, either point can be used in this step.

$$2(y + 4) = 3(x + 6)$$ Multiply both sides by the LCD, 2.

$$2y + 8 = 3x + 18$$ Simplify, and write the equation in the slope-intercept form, $y = mx + b$.

$$2y = 3x + 10$$

Answer: $y = \frac{3}{2}x + 5$ Can you use a graphing calculator to check that this is the correct equation?

SELF-CHECK 3.2.5

1. The equation $y - 1 = \frac{1}{3}(x - 1)$ can also be written as $y - 2 = \frac{1}{3}(x - 4)$. Show that these equations are equivalent by converting both equations to general form.
2. Determine by inspection the slope of $y + 7 = -\frac{3}{4}(x - 4)$ and one point on this line.
3. Write in slope-intercept form the equation of a line through (0, 4) and (5, 2).
4. Graph the line through $(-2, 4)$ with slope $-\frac{3}{4}$.

Self-Checks—Self-Checks are located in each section to help students to become active learners and monitors of their own progress. The answers for these self-checks are placed at the end of each section.

Key Concepts—The Key Concepts highlight each chapter's key terms, concepts, principles, and procedures. This summary can serve as a checklist of key items for students to review.

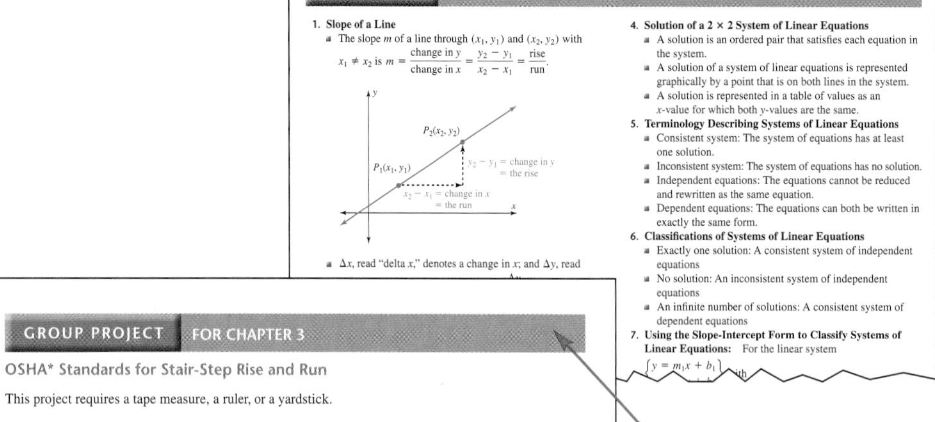

GROUP PROJECT FOR CHAPTER 3

OSHA* Standards for Stair-Step Rise and Run

This project requires a tape measure, a ruler, or a yardstick.

The OSHA standards for stairs for single-family residences vary from the standards for other buildings. The standards for single-family residences set a maximum on the rise of 8 in and a minimum on the run of 9 in.

OSHA Standard 1910.24

Definitions:

Rise: *The vertical distance from the top of a tread to the top of the next-higher tread.*

Run: *The horizontal distance from the leading edge of a tread to the leading edge of an adjacent tread.*

Group Project—Chapters end with a Group Project that encourages collaborative problem-solving while reinforcing connections among mathematical concepts.

126 (2-40) Chapter 2 Linear Equations and Patterns

The general strategy for solving a linear equation, regardless of its complexity, is to isolate the variable on one side of the equation and place all other terms on the other side.

Example 4 contains constants and variables on both sides of the equation. We will use the addition-subtraction principle of equality to isolate the variables on the left side of the equation and the constant terms on the right side of the equation.

EXAMPLE 4 Solving a Linear Equation with Variable and Constant Terms on Both Sides

A Mathematical Note

Emmy Noether (1882–1935) was called by Albert Einstein the greatest of all the women who were creative mathematical geniuses. Emmy completed her doctoral dissertation in algebra at the University of Erlangen in 1907. She taught at the University of Göttingen in Germany from 1915 until 1933. She then moved to the United States and taught at Bryn Mawr for 2 years until her death in 1935.

EXAMPLE 1 Solving a Proportion

Solve $\frac{x}{3} = \frac{5}{4}$.

SOLUTION

$$\frac{x}{3} = \frac{5}{4}$$

$$12\left(\frac{x}{3}\right) = 12\left(\frac{5}{4}\right)$$ LCD = 3 · 4 = 12. Multiply both sides of the equation by the LCD of 12.

$$4x = 3(5)$$

$$4x = 15$$

$$\frac{4x}{4} = \frac{15}{4}$$ Divide both sides of the equation by 4.

$$x = \frac{15}{4}$$

Commentary and Guidance

Marginal Notes—The margins of the text are populated with brief hints and explanations of nearby material that will help teachers teach and students learn.

Mathematical Notes—These short vignettes describe the origin of some of the symbols and terms used in the text and provide brief glimpses into the lives of some of the men and women of mathematics.

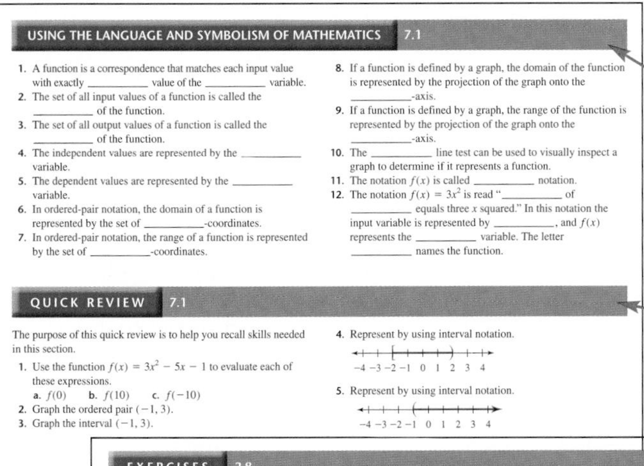

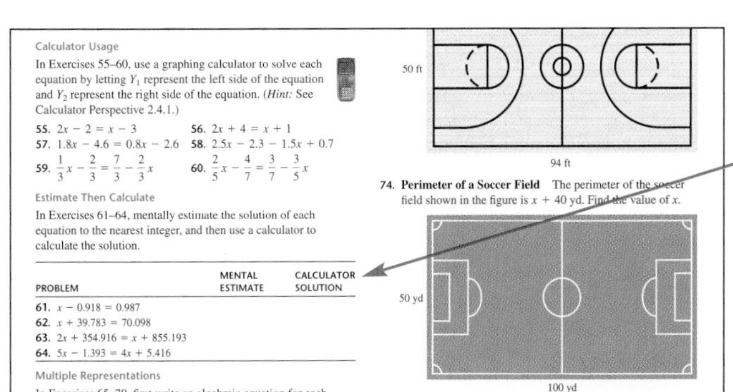

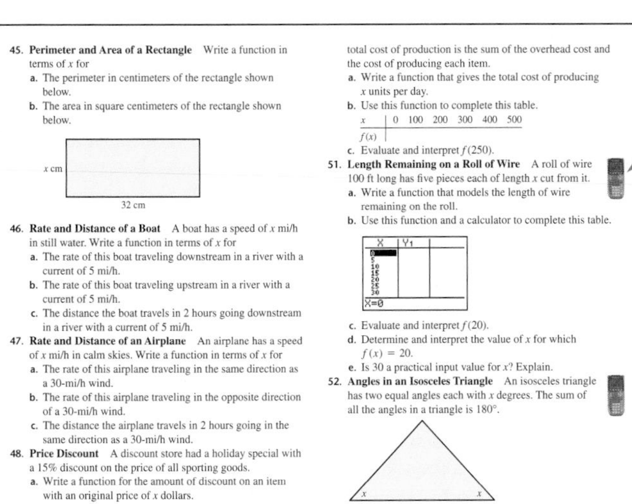

Conceptual Understanding Through Practice

In every section of the text:

Using the Language and Symbolism of Mathematics—Every section includes a series of fill-in-the-blank questions to help students gain fluency in the language and symbolism of mathematics.

Quick Review—Every section in the Intermediate Algebra portion of the book includes five questions that review concepts or notation that will be used in this section.

Exercises—Each end-of-section exercise set is carefully constructed to develop and to reinforce the skills and concepts of algebra, and to provide an appropriate review of the section. Exercise sets include:

Estimation Skills and Error Analysis Exercises—Estimation skills, concern for reasonable answers, and the ability to detect calculator errors are critical elements of students' mathematical knowledge. There are examples and exercises in the book specifically designed to help the students develop these skills.

Calculator Exercises—When the use of a calculator is appropriate to the solution of an exercise, it is indicated in the text by a calculator icon, right next to the exercise.

Group Discussion Questions—These exercises involve students in interactive and collaborative learning and encourage them to communicate mathematics both orally and in writing.

Spreadsheet Exploration—Some sections include spreadsheet exercises. These exercises show how algebra is used to construct spreadsheets and also how spreadsheets can be used to explore algebraic concepts.

In every chapter of the text:

Chapter Review Exercises—These comprehensive exercise sets provide ample and well-distributed practice on the topics of the chapter.

Chapter Mastery Test—These tests are written specifically to cover each objective presented in the chapter.

Cumulative Review—To ensure students have mastered the concepts of the previous chapters, cumulative reviews can be found after Chapters 3, 6, and 9. Direct references to relevant examples placed alongside the answers immediately follow each review.

Accuracy Statement

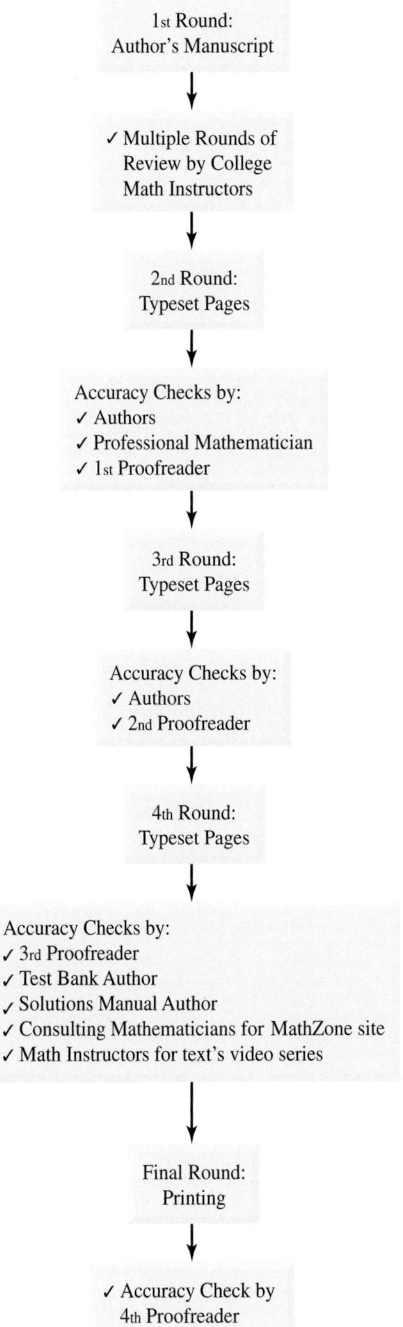

A COMMITMENT TO ACCURACY

You have a right to expect an accurate textbook, and McGraw-Hill invests considerable time and effort to make sure that we deliver one. Listed below are the many steps we take to make sure this happens.

OUR ACCURACY VERIFICATION PROCESS

First Round

Step 1: Numerous **college math instructors** review the manuscript and report on any errors that they may find, and the authors make these corrections in their final manuscript.

Second Round

Step 2: Once the manuscript has been typeset, the **authors** check their manuscript against the first page proofs to ensure that all illustrations, graphs, examples, exercises, solutions, and answers have been correctly laid out on the pages, and that all notation is correctly used.

Step 3: An outside, **professional mathematician** works through every example and exercise in the page proofs to verify the accuracy of the answers.

Step 4: A **proofreader** adds a triple layer of accuracy assurance in the first pages by hunting for errors, then a second, corrected round of page proofs is produced.

Third Round

Step 5: The **author team** reviews the second round of page proofs for two reasons: 1) to make certain that any previous corrections were properly made, and 2) to look for any errors they might have missed on the first round.

Step 6: A **second proofreader** is added to the project to examine the new round of page proofs to double check the author team's work and to lend a fresh, critical eye to the book before the third round of paging.

Fourth Round

Step 7: A **third proofreader** inspects the third round of page proofs to verify that all previous corrections have been properly made and that there are no new or remaining errors.

Step 8: Meanwhile, in partnership with **independent mathematicians,** the text accuracy is verified from a variety of fresh perspectives:

- The **test bank author** checks for consistency and accuracy as they prepare the computerized test item file.
- The **solutions manual author** works every single exercise and verifies their answers, reporting any errors to the publisher.
- A **consulting group of mathematicians,** who write material for the text's MathZone site, notifies the publisher of any errors they encounter in the page proofs.
- A video production company employing **expert math instructors** for the text's videos will alert the publisher of any errors they might find in the page proofs.

Final Round

Step 9: The **project manager,** who has overseen the book from the beginning, performs a **fourth proofread** of the textbook during the printing process, providing a final accuracy review.

⇒ What results is a mathematics textbook that is as accurate and error-free as is humanly possible, and our authors and publishing staff are confident that our many layers of quality assurance have produced textbooks that are the leaders of the industry for their integrity and correctness.

Index of Applications

Index of Group Projects

Index of Applications

BIOLOGY/HEALTH/LIFE SCIENCES

BUSINESS

CHEMISTRY

CONSTRUCTION

CONSUMER APPLICATIONS

SCHOOL

SCIENCE

SPORTS

STATISTICS/DEMOGRAPHICS

chapter 1

Operations with Real Numbers

CHAPTER OUTLINE

Many of our most important decisions, especially financial decisions, are based upon judgments about mathematical information. Mathematics provides many valuable tools, such as graphs, tables, and formulas for making these decisions.

As illustrated by the graph, both the interest rate and the number of years of a mortgage are important factors when one is comparing mortgage options. Sometimes changing the interest rate by just 1% and keeping the same monthly payment can result in saving 5 or more years of payments. A shorter-term mortgage will mean larger monthly payments, a steeper climb, but a financial journey that will end sooner. A longer-term mortgage will result in smaller monthly payments, an easier hill to climb, but a journey that takes longer to finish. It pays to shop before you sign!

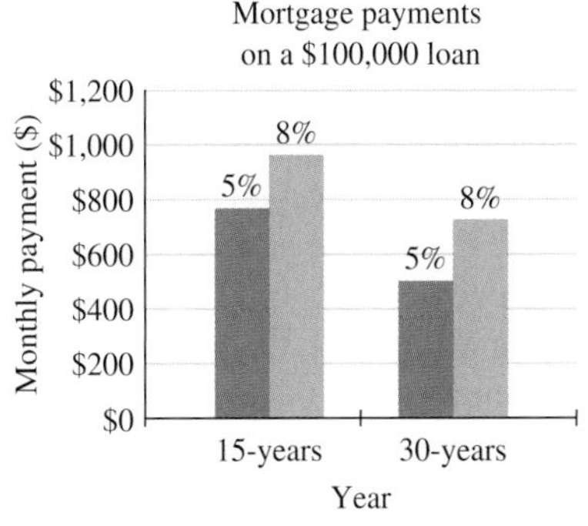

REALITY CHECK

The reality check at the end of this chapter compares the results of a 5% mortgage for 15 years with an 8% mortgage for 30 years—a choice that can have expensive consequences. From the graph shown, which monthly payment option is less, a 5% mortgage for 15 years or an 8% mortgage for 30 years?

Section 1.1 Preparing for an Algebra Class

Objective: 1. Understand your class syllabus and textbook features.

You control the most important factors in your success in any class. Your positive attitude, determination, and work ethic will be the primary factors in your success in this algebra class. Your instructor is also a key source of information and assistance. It is important to read the syllabus and materials provided by your instructor. These materials may contain information on office hours or other supplemental help that is available on your campus. Be sure to consult with your instructor if you have any questions about placement, tutoring services, or supplements to this textbook.

The success tips and study hints given below are based on many years of experience and research by the authors. We are sure that some of these hints will be familiar to you, and we also expect that others will be new things for you to try. We urge you to consider all these tips and use them in this course and in other courses.

Success Tips and Study Hints

Your education can never be taken from you. Your ability to know how to learn will be with you in all future economic circumstances.

Set Goals: Students with clearly defined academic goals outperform students without goals by a wide margin. Students with goals often overcome the conflicts of work and other time commitments to reach their goals. Goals provide a direction and a motivation for the work and time necessary to meet the challenge presented by a mathematics class. If you have not already established some academic goals for the term, it would help you to set some now. Clearly establish a realistic goal, even if it is tentative or short term, and routinely remind yourself of this goal to help filter out distractions. Check with your faculty adviser to be sure that you select courses that will help you meet your goals and that you understand the requirements.

Time Management: The most important factor in your success will be time on task. The more quality time you spend, the more you will gain. Two places to use your time are in attending class daily and doing homework daily. The importance of these two elements to your learning cannot be overstated. To do both consistently requires time management. Perhaps the most frequent mistake made by beginning college students is the underestimation of the time required to study and learn.

It is possible to shortchange yourself in sleep, study, or any area for a short time. However, frequently shortchanging any of these areas can have serious consequences. As a college student, you must budget enough time for studying in addition to attending classes. A good rule of thumb is to allow about 2 hours of study for each hour of class. Therefore if you are enrolled in 16 credit hours, you should plan to study another 32 hours per week. For academics, this represents a total of 48 hours a week or 8 hours per day for 6 days of the week. This leaves only 16 hours of each of these 6 days for family, sleep, work, and recreation. To decide the amount of time to allow for each activity, you must *set priorities*.

How important is an education to you? If it is one of your top priorities, then you must allow adequate time to study. You may need to decrease the time spent on other commitments to gain sufficient time to study. If you cannot do this, then maybe you should consider taking fewer courses this term. Keep in mind that your education is of long-term benefit. What you invest in it now will determine how qualified and competitive you will be in the marketplace. If you have any concerns about your schedule, it is advisable to consult with your college adviser or counselor for assistance.

Organizing Your Work: A major mistake made by many mathematics students is that they overestimate the value of their answer to a problem and underestimate the importance of their organization and overall work. Teachers want to ensure that students have mastered the concepts and procedures for solving all problems of a given type, and therefore they will often examine the steps and logic of a student's work to be sure that the problems have been worked correctly. It will help you both in class and in your future career if you organize your work so that others can understand it easily. A paper that is well organized gives the impression that its writer understands the material. A messy or disorganized paper gives the reader many negative impressions about the writer, whether they are true or not. A well-organized paper makes it easier for you to proof your own work. The samples below illustrate some pointers for organizing mathematics problems. The problem to be worked is

Simplify $$\frac{-3-\sqrt{3^2-4(-2)(2)}}{2(-2)}$$

Here is a sample of a well-organized problem:

$$\begin{aligned}\frac{-3-\sqrt{3^2-4(-2)(2)}}{2(-2)} &= \frac{-3-\sqrt{9+16}}{-4}\\ &= \frac{-3-\sqrt{25}}{-4}\\ &= \frac{-3-5}{-4}\\ &= \frac{-8}{-4}\\ &= 2\end{aligned}$$

The problem is recopied so that the starting point is clear.

The equals signs are aligned to help the reader see the connection between the steps.

Each expression following an equals symbol is equal to the previous expression and thus to the original expression.

The location of the last step makes the answer clear.

In contrast, here is a sample of a poorly organized problem:

$$-3-\sqrt{3^2-4(-2)(2)} = 9+16 = 25 = \sqrt{25} = 5$$
$$2(-2) = -4$$
$$-3-5 = \frac{-8}{-4} = 2$$

The problem is not recopied so the relation of the work to the question is not clear. The equals signs are not aligned and are used incorrectly. Note that

$$25 \neq \sqrt{25} \quad \text{and} \quad -3-5 \neq \frac{-8}{-4}$$

The expressions written are just pieces of the original problem. Confusion is likely for the reader regarding the relationship of these pieces.

The answer to this problem is not clear with this organization.

You should check all your homework to verify that your practice is correct. Any problems with wrong answers should be circled so that you can ask questions about them in class. It is important to review and then review some more. As you review, look over the problems you have circled, and be sure that you have mastered any troublesome points. By keeping track of the sources of your errors, you can be self-correcting on both homework and tests. If you have extra time on a test, double-check your work, especially for any of the errors that you are prone to make.

Many teachers recommend a three-ring binder for your homework, quizzes, tests, and other papers. Bring this binder, your textbook, and your calculator to each class.

Getting the Most out of Lectures: Attending every class is very important. To maximize your gain from lectures, we suggest that you preview, participate, and review. Students who preview the material before the lecture have a big advantage over students who wait until after the lecture to examine the material. If you have read the material, you can focus your attention on any points you find troublesome. You can also concentrate on listening for concepts and the connections between different ideas rather than worrying about details. Participate by listening actively rather than passively. So that your mind does not wander, try to predict ahead of time the point your instructor is going to make, and keep pace with the lecture. Watch for your instructor's nonverbal signals, which will clarify meanings or intent. Ask questions; your instructor will appreciate your interest, and other students are likely to have the same questions you do. Review; all review is beneficial, especially an early review of your lecture notes. Before you start your homework, make sure that all the examples covered in class make sense. Clarify the organization of your notes, fill in any steps that you may have omitted, and be sure to highlight any hints that your instructor may have given regarding possible test questions.

Study in a Group: Research projects have shown that studying and doing some of your homework with two or three of your classmates can improve your grade. Organizing your questions or your responses to questions well enough to verbalize them to others can help you clarify your understanding of the material. Exchange phone numbers and e-mail addresses with members of your study group, and consider meeting them to discuss the major purposes of an assignment or the best way to approach certain problems. You can even practice testing each other prior to an exam. Having other points of view to consider is itself part of the educational process; few people in business, education, or industry work alone. Learning in a group situation will help prepare you for working in a group environment.

Calculator and Spreadsheet Usage: Learning how to take full advantage of your knowledge of mathematics requires a familiarity with modern technology. Students who integrate calculators and spreadsheets into their course work will be better prepared for their careers. Do not try to substitute calculator usage for understanding the material, but do use a calculator or a spreadsheet to expedite your work and to experiment and consider alternatives.

Some general suggestions on calculator usage follow. For specific instructions consult this textbook or your calculator's operating manual. If you run into problems, ask your instructor for help.

1. Use your calculator to work a problem with simple values whose result is known before you undertake the calculation of similar problems whose results are unknown.
2. For calculations, use as many digits of accuracy as your calculator will allow. Since most calculators store more digits than they display, you can increase accuracy by observing the following guidelines.
   a. Leave intermediate values in the calculator rather than copying down the display digits and then reentering these values. Learn to use the memory and parentheses keys.
   b. Enter e and π by using the special keys on your calculator. Such entries will typically be accurate to two or three more digits than are shown on the display.

Preparing For Tests: You will benefit most if the majority of your test preparation occurs well before the test. However, near test time you may find it helpful to take some practice tests to gauge your level of preparation. You can use the *Key Concepts* and the *Mastery Tests* at the end of each chapter for this purpose. There are also sample quizzes available on the website for this text. If you work with a study group, you can gain by testing each other. One way to overcome test anxiety is to test yourself frequently so that a test is no longer a big deal to you.

Another strategy for testing yourself is as follows: Each day, copy a few representative problems from your homework assignment on the front of 3×5 cards. Put the solution and/or a page reference on the back of each card. Review this stack of cards for a few minutes after each assignment. You don't have to actually rework the problems. Just shuffle the cards, and then mentally ask yourself what steps you would follow if this problem were given on a test. You will be practicing the discrimination skills that are required on tests and that are easily overlooked on the day-to-day homework. Many students have used this strategy to raise their grade by 10% to 20%.

Solving Word Problems: Maintain a positive attitude; it really does help. Many students are harsher critics of themselves than anyone else is. No one understands every topic immediately; basketball, cooking, golf, mathematics, and piano all take practice. Work steadily, and don't expect to master all the information presented the first time through it. By working in groups you can help one another, and you will observe how others must also practice to gain their understanding.

Review your progress and note your gains, as well as problems that may still require further practice. Take pride in what you do understand, and ask yourself how this material is similar to (or different from) your new material. Try to understand ideas and build connections between different ideas—avoid memorizing without understanding.

Almost all the mathematics problems that you will encounter outside the classroom will be stated in words. Thus to profit from your algebra, you must be able to work with word problems.

Almost all the mathematics problems that you will encounter outside the classroom will be stated in words. Thus to profit from your algebra, you must be able to work with word problems. There are several suggestions that accompany the word problems worked in the textbook. Many of these suggestions are based on research articles describing successful study techniques.

Using Text Supplements: The most important factor in your success in this class is the time that you spend in class and on your homework. However, realize that no one understands each new topic immediately. If one topic should cause you particular concern, consult the many features of this text and its supplements that are designed to help you master the material and obtain additional practice. Be sure to start by going over the examples and key points that are boxed in the text. You may also want to consider using the end of chapter materials including the *Key Concepts,* the *Chapter Review*, and the *Mastery Test,* all of which can be used to polish your earlier work. The *MathZone* website provides multiple resources, including video lectures, linked to this text. Also feel free to go to your instructor with well-organized questions. Instructors enjoy students who work hard and are usually willing to provide these students with additional help during office hours. Instructors may also have recommendations regarding useful text supplements. The main point is to obtain this help as soon as you need it.

Preparing For a Final Exam: Your preparation for the final exam started on the first day of class. Rarely is a student's performance significantly different on the final exam from what it is on homework and other hour exams. To maximize the benefit of your review for the final, you should spread this review over several days. It should definitely not consist of one long marathon session. Start by spending a couple of hours going over your first hour exam with particular emphasis on correcting any careless errors that you may have made. On the next day do the same thing with the second hour exam. By the end of the week you will have gone over all your hour exams. Then review the key concepts at the end of each chapter, and practice on some of the exercises in the cumulative reviews in the text. It would be wise to go through your notes and look over the troublesome problems you circled earlier. However, it is generally unwise to try to cram in new material that could leave you confused and cause you to miss reviewing concepts that you have mastered. Just review what you already know, and don't miss any problems that you can work. The final exam will then give an accurate measure of your knowledge.

A significant part of a college education is learning how to learn.

Preparing For Your Next Term: A significant part of a college education is learning how to learn. Many of the job titles that are common today did not exist 25 years ago, and many people must change jobs at least once during their careers. Although you may have a specific job in mind now, you must educate yourself as broadly as possible rather than focus yourself narrowly. One of the reasons algebra is required of so many students is that it does more than prepare you for more advanced mathematical topics. It also helps you see how mathematics can be applied to real-life problems, and it aids you in improving your reasoning and critical-thinking skills. As you prepare for your next term, try to keep in mind your long-term goal of improving yourself through education. Review all the other study hints given in this book, and reflect on the learning strategies that you have developed during this term. Then try to apply these strategies right from the start of your next term.

Building Your Professional Library: As you advance in your education, it is important to begin to build your own professional library of reference books. In addition to a good dictionary, thesaurus, and college handbook, you should include basic textbooks that you have studied in detail. You are already familiar with their style, language, and format, so when you refer to these books in later sequential courses, it will be like working with an old friend. Any financial gain that you may receive by selling these books now may cost you much time and effort later if you have to adjust to the new style and format of another book. This *Beginning and Intermediate Algebra* book has been designed with several goals in mind. One important goal is to prepare you for subsequent mathematics courses and to serve as a resource for you in these courses. We hope that you find this book a valuable reference and addition to your professional library.

EXERCISES 1.1

1. What is the name of your instructor? __________
2. Your instructor's office is in __________.
3. Your instructor's office hours are __________.
4. Your instructor's phone number is __________.
5. Your instructor's e-mail address is __________.
6. Is there a mathematics tutoring lab on your campus? __________ If so, it is located in __________.
7. I realize that to succeed in this course I will need to work about __________ hours per day outside of class.
8. The date and time for the final exam (if known) are __________. Mark this date on your calendar. Also record the dates of the hour exams if these are known.
9. The calculator used in this class is the __________.
10. In addition to the textbook, calculator, and notebook, the supplies recommended by your instructor are __________.
11. The textbook has __________-Check Exercises by the examples in the text for you to check your own understanding as you read through the text.
12. The *Student's Solutions Manual* for this book contains the worked-out solutions to the __________-numbered problems. (See the Preface.)
13. The __________ *Guide* for this book has the key language and symbolism, definitions, and examples structured in the framework of a class lecture. (See the Preface.)
14. The video series for this book has a separate video for each __________. (See the Preface.)
15. The __________ website provides multiple resources, including video lectures, linked to this text. (See the Preface.)
16. NetTutor provides a live tutor that can interact with you over the __________. (See the Preface.)
17. ALEKS is an artificial intelligence-based system that can interact with you much as a human __________ would. (See the Preface.)
18. After rereading the tips on time management, fill out your schedule for the term including your class time, your study time, and work or other time commitments.

	SUNDAY	MONDAY	TUESDAY	WEDNESDAY	THURSDAY	FRIDAY	SATURDAY
7:00 A.M.							
8:00 A.M.							
9:00 A.M.							
10:00 A.M.							
11:00 A.M.							
12:00 P.M.							
1:00 P.M.							
2:00 P.M.							
3:00 P.M.							
4:00 P.M.							
5:00 P.M.							
6:00 P.M.							
7:00 P.M.							
8:00 P.M.							
9:00 P.M.							

19. When you form your study group, record the names, phone numbers, and e-mail addresses of your study group here.

NAME	PHONE NUMBER	E-MAIL ADDRESS

20. One hint given under the heading "Preparing for Tests" suggests that you copy a few representative problems on a __________ for each assignment. One advantage of this format is that the questions can be shuffled and you can review and give yourself a quick practice test each day.

Group Discussion Questions

21. A quote attributed to Henry Ford states, "If you think you can, you can. If you think you can't, you can't. In either case you are right." Discuss what you think is meant by this quote.
22. A quote from Samuel Johnson states, "The future is purchased by the present." Discuss what you think is meant by this quote. Write two things you think this might mean about your attendance, homework, and test preparation for this class.
23. Two very common comments by students are "I knew this material, my mind just went blank" and "I can do all the problems, I just don't test well." The purpose of this group exercise is to remind you of test-taking strategies that you may have used successfully in the past, to share new test-taking strategies, and to improve your test scores.
    a. In your group, discuss and list two kinds of test-taking strategies.
        i. Strategies to use before the test. (For example: Do the chapter review a couple of days before the test.)
        ii. Strategies to use during the test. (For example: Never change an answer unless you are very confident that you have a better answer.)

b. In class, share two strategies to use before the test and two strategies to use during the test that your group likes best. Perhaps the teacher can record the contributions from each group and give each student a copy of all these strategies.
c. Decide on at least one strategy that you will use before the test and one strategy that you will use during the test, and then use these strategies.
d. After the exam have each member of your group share her or his perceptions of the results that each experienced while using these strategies. What worked and what did not work? What will you do differently for the next exam?

Section 1.2 The Real Number Line

Objectives:
2. Identify additive inverses.
3. Evaluate absolute value expressions.
4. Use interval notation.
5. Estimate and approximate square roots.
6. Identify natural numbers, whole numbers, integers, rational numbers, and irrational numbers.

The word *algebra* comes from the book *Hisab al-jabr*, written by an Arab mathematician in A.D. 830; translations of this text on solving equations became widely known in Europe as *al-jabr*. Algebra, however, is concerned with more than solving equations; it is a generalization of arithmetic. Algebra is used to explore, explain, and to model life's questions from a quantitative viewpoint. We will examine mathematical modeling throughout this book.

One of the primary uses of algebra is to manipulate mathematical expressions into a more desirable form. Three of the reasons that we rewrite an algebraic expression follow.

1. To put the expression in simplest form.
2. To simplify an equation in order to solve the equation.
3. To place a function in a form that reveals more information about the graph of the function.

This book will also focus on the relationship between algebra and geometry, including various graphs. Graphs are used widely in newspapers, magazines, and mathematics texts. Graphs can help us to understand data, to quickly note patterns and trends, or to visualize abstract concepts. Common graphs are histograms, bar graphs, pie charts, line graphs, the real number line, and the Cartesian plane.

To interpret or create graphs, we need to understand the terminology and concepts reviewed in this section. The number line is one of the most used graphs in mathematics. It provides a model of the real numbers and is used when students are first introduced to a concept. In arithmetic we use only constants such as $-13, 0, \frac{2}{3}, \sqrt{2}$, and π, which have a fixed value. Algebra uses not only constants but also variables such as a, b, P, W, x, y, and z. A **variable** is a letter that can be used to represent different numbers. The constants and variables that we use most often will represent real numbers.

Variables also are used to identify cells in a spreadsheet.

	A	B
1	1	3
2	4	5
3	2	7
4		

The real number line

Origin

Negative numbers | Positive numbers

−5 −4 −3 −2 −1 0 1 2 3 4 5

The symbol ℝ is used to represent the set of all real numbers.

There are an infinite number of points on the number line, and every point is associated with a **real number.** The set of all real numbers can be denoted by ℝ. The numbers to the right of 0 are called the **positive numbers,** and those to the left of 0 are called the **negative numbers.** The number 0 is neither positive nor negative. The set of numbers $\{1, 2, 3, 4, \ldots\}$ associated with the points marked off to the right of the origin is called the set of **natural numbers.** The three dots inside this set notation indicate that these numbers continue without end and thus the set is infinite. The set $\{0, 1, 2, 3, 4, \ldots\}$ is called the set of **whole numbers.**

Each real number represents a directed distance from the origin. For example, $+2$ represents a distance 2 units to the right, while -2 represents a distance 2 units to the left of the origin.

Real numbers the same distance from the origin, but on opposite sides of the origin, are called **opposites** of each other, or **additive inverses.** For example, -2 (negative two) and 2 are additive inverses of each other.

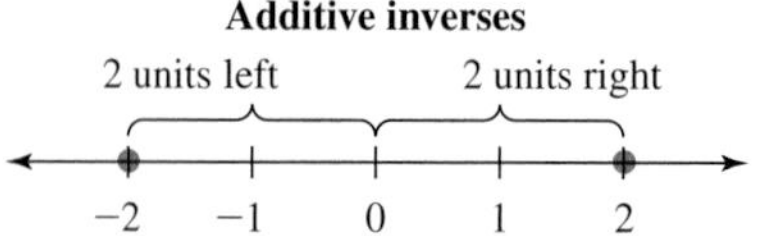

The sum of a number and its additive inverse is zero. For example, $-2 + 2 = 0$ and $5 + (-5) = 0$. The additive inverse of 0 is 0. Zero is called the **additive identity** because 0 is the only real number with the property that $a + 0 = a$ and $0 + a = a$ for every real number a.

Definitions and the solution of problems will be presented by using multiple representations throughout this book. The definition of additive inverses is given algebraically, verbally, numerically, and graphically. Comparing these representations can help you increase your understanding of mathematics.

Opposites or Additive Inverses

ALGEBRAICALLY	VERBALLY	NUMERICAL EXAMPLE	GRAPHICAL EXAMPLE
1. If a is a real number, the *opposite* of a is $-a$.	Except for zero, the additive inverse of a real number is formed by changing the sign of the number.	-5 is the opposite of 5 5 is the opposite of -5 0 is the opposite of 0	Opposites (number line with −5 and 5 marked; labels −5, 0, 5)
2. $a + (-a) = 0$ $-a + a = 0$	The sum of a real number and its additive inverse is zero.	$5 + (-5) = 0$ $-5 + 5 = 0$ $0 + 0 = 0$	

The set $\{\ldots, -4, -3, -2, -1, 0, 1, 2, 3, 4, \ldots\}$ is called the set of **integers.** The set of integers contains all the whole numbers and their opposites, or additive inverses. As illustrated in the figure, the integers are spaced 1 unit apart on the number line.

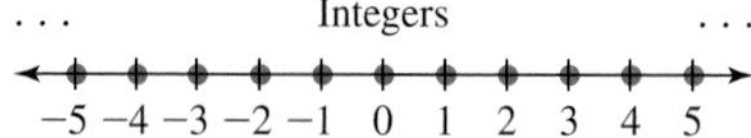

The opposite of any positive number is a negative number; for example, the opposite of 4 is -4. The opposite of any negative number is a positive number; for example, the opposite of -4 is 4. The opposite of x is $-x$. If x is -4, then $-x$ is 4. Thus $-(-4) = 4$, and in general $-(-x) = x$.

Double Negative Rule

ALGEBRAICALLY	VERBALLY	NUMERICAL EXAMPLE
For any real number a, $-(-a) = a$.	The opposite of the additive inverse of a is a.	$-(-5) = 5$

A Mathematical Note

π is defined as the ratio of the circumference of a circle to its diameter. $\pi \approx 3.14159265$. ($\approx$ means "approximately equal to.") However, the decimal form of π does not terminate or repeat. In 1897, House Bill #246 was introduced in the Indiana legislature to make 3.2 the value of π in that state. Fortunately, better judgment prevailed, and this foolish bill was defeated.

EXAMPLE 1 Writing Additive Inverses

Write the additive inverse of each of the following real numbers.

SOLUTION

NUMBER	ADDITIVE INVERSE	
(a) 7	-7	Except for zero, the additive inverse is formed by changing the sign of the number.
(b) -3	$-(-3) = 3$	The opposite of negative three is three.
(c) 0	0	Zero is its own additive inverse.
(d) $\frac{4}{17}$	$-\frac{4}{17}$	
(e) π	$-\pi$	The opposite of pi is negative pi.
(f) $3x$	$-3x$	The opposite of $3x$ is represented by $-3x$.

Graphing calculators have special keys to represent some of the frequently used constants such as π. In Calculator Perspective 1.2.1 the calculator screen shows an excellent approximation of π, but this approximation is not the exact value of π. Calculator perspectives will occur throughout the text and will integrate mathematical concepts with the appropriate use of technology. Note that we use the (-) key to form the opposite of a number and the − key to subtract one number from another.

CALCULATOR PERSPECTIVE 1.2.1 Using the Pi Key

To approximate π, -2π, and $-\pi + \pi$ on a TI-84 Plus calculator, enter the following keystrokes:

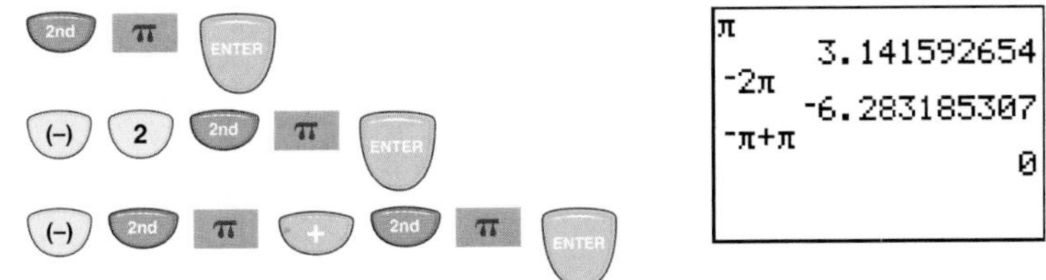

Note: The π key is the secondary feature of the ^ key. The change of sign key (-), not the subtraction key −, must be used to enter the additive inverse of a number.

Example 2 illustrates that the sum of a number and its additive inverse is 0.

EXAMPLE 2 Adding Additive Inverses and the Additive Identity

Evaluate each of the following sums.

SOLUTION

(a) $7 + (-7)$	$7 + (-7) = 0$	The sum of a number and its additive inverse is zero.
(b) $-7 + 7$	$-7 + 7 = 0$	
(c) $-\pi + \pi$	$-\pi + \pi = 0$	This sum was also examined in Calculator Perspective 1.2.1.

(d) $3x + (-3x)$ $\quad$ $3x + (-3x) = 0$

(e) $19 + 0$ $\quad$ $19 + 0 = 19$ $\quad$ Zero is the additive identity; $a + 0 = a$ for all real numbers a.

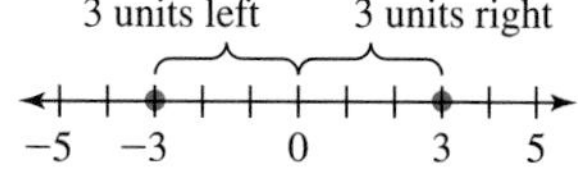

Additive inverses are the same distance from the origin, so these numbers are said to have the same magnitude, or absolute value. The **absolute value** of a real number x, denoted by $|x|$, is the distance on the number line between 0 and x. Distance is never negative, so $|x|$ is never negative: $|0| = 0$, $|3| = 3$, and $|-3| = 3$. Both -3 and 3 are a distance of 3 units from the origin. If x is positive, $|x| = x$, as in $|3| = 3$. If x is negative, however, $|x| = -x$, as in $|-3| = -(-3) = 3$. Thus the definition of absolute value is often stated by using the notation given in the following box.

Absolute Value

ALGEBRAICALLY	VERBALLY	NUMERICAL EXAMPLE	GRAPHICAL EXAMPLE
$|x| = \begin{cases} x & \text{if } x \text{ is nonnegative} \\ -x & \text{if } x \text{ is negative} \end{cases}$	The absolute value of x is the distance between 0 and x on the number line.	$|2| = 2$ $|-2| = 2$	2 units left, 2 units right (number line: −2 −1 0 1 2)

We use this definition of absolute value in Example 3 to evaluate four absolute value expressions.

EXAMPLE 3 Evaluating Absolute Value Expressions

Evaluate each of the following absolute value expressions.

SOLUTION

(a) $|26|$ $\quad$ $|26| = 26$ $\quad$ $|x| = x$ if x is nonnegative. 26 is 26 units from 0 on the number line.

(b) $|-4.29|$ $\quad$ $|-4.29| = 4.29$ $\quad$ $|x|$ is the opposite of x if x is negative. -4.29 is 4.29 units from 0 on the number line.

(c) $-|5.8|$ $\quad$ $-|5.8| = -5.8$ $\quad$ The additive inverse symbol is outside the absolute value symbols.

(d) $|0|$ $\quad$ $|0| = 0$ $\quad$ 0 is 0 units from 0 on the number line.

CALCULATOR PERSPECTIVE 1.2.2 Evaluating Absolute Value Expressions

To evaluate $|26|$, $|-4.29|$, and $-|5.8|$ from Example 3 on a TI-84 Plus calculator, enter the following keystrokes:

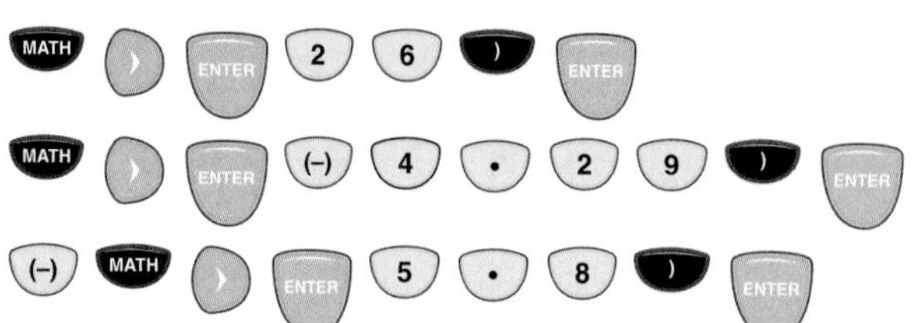


```
abs(26)
                26
abs(-4.29)
              4.29
-abs(5.8)
              -5.8
```

Note: The first option under **NUM** within the **MATH** menu is **abs(**. This includes the left parenthesis. The expression within the absolute value is completed by entering the right parenthesis key.

SELF-CHECK 1.2.1

1. Write the additive inverse of -45, 0, and $\frac{17}{43}$.
2. Evaluate $-45 + 0$.
3. Evaluate $-45 + 45$.
4. Evaluate $|-45|$.
5. Evaluate $|45|$.

The number line also provides an excellent model for examining the inequalities *less than* and *greater than*. Because these inequalities describe the order of numbers on the number line, they are also known as the **order relations.** Table 1.2.1 illustrates all possible orders for arranging two numbers x and y on the number line.

Table 1.2.1 Equality and Inequality Symbols

ALGEBRAIC NOTATION	VERBAL MEANING	GRAPHICAL RELATIONSHIP ON THE NUMBER LINE
$x = y$	x equals y	x and y are the same point.
$x \approx y$	x is approximately equal to y	x and y are "close" but are not the same point.
$x \neq y$	x is not equal to y	x and y are different points.
$x < y$	x is less than y	Point x is to the left of point y.
$x \leq y$	x is less than or equal to y	Point x is on or to the left of point y.
$x > y$	x is greater than y	Point x is to the right of point y.
$x \geq y$	x is greater than or equal to y	Point x is on or to the right of point y.

The statements $x < y$ and $y > x$ are equivalent because they both specify that point x is to the left of point y.

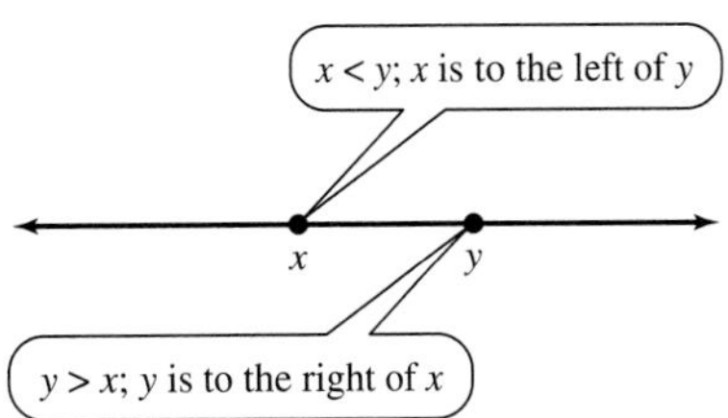

EXAMPLE 4 Showing Order Relationships on the Number Line

Plot each pair of numbers on the line, and determine the order relationship between the numbers.

SOLUTION

	Number line	
(a) -4 and -2	-5 -4 -3 -2 -1 0 1 2 3 4 5	-4 is to the left of -2; thus -4 is less than -2; $-4 < -2$.
(b) π and 3	2.9 3 3.1 π 3.2 3.3 3.4	$\pi \approx 3.14159$; thus π is to the right of 3 and is greater than 3; $\pi > 3$.
(c) 0 and -3	-5 -4 -3 -2 -1 0 1 2 3 4 5	0 is to the right of -3; thus $0 > -3$.

SELF-CHECK 1.2.2

Determine whether each statement is true (T) or false (F).

1. $9 \geq 9$ **2.** $9 > 9$ **3.** $-7 \leq -3$

4. $-5 \geq 0$ **5.** $4\frac{1}{3} > 4\frac{1}{2}$ **6.** $\frac{1}{3} > 0.3$

A Mathematical Note

The symbol ∞ was used to represent infinity by John Wallis in *Arithmetica Infinitorum* in 1655. The Romans had commonly used this symbol to represent 1,000. Likewise, we now use the word *myriad* to mean any large number, although the Greeks used it to mean 10,000.

The real number line extends infinitely to both the left and the right. The **infinity symbol** ∞ is not a real number; rather it signifies that the values continue through extremely large values without any end or bound. The symbol $-\infty$ indicates values unbounded to the left, and $+\infty$ (or just ∞) indicates values unbounded to the right. The interval notation $(3, \infty)$ is a compact notation to represent all real numbers greater than 3.

INEQUALITY NOTATION	VERBAL MEANING	GRAPH	INTERVAL NOTATION
$x > 3$	x is greater than 3	3	$(3, \infty)$
$x \leq 5$	x is less than or equal to 5	5	$(-\infty, 5]$

Note in the interval notation defined below that a parenthesis indicates that a value is not included in the interval and a bracket indicates that a value is included in the interval. An infinity symbol at either end of an interval is enclosed with parentheses since this denotes an unbounded interval that continues without any ending point.

Interval Notation

INEQUALITY NOTATION	VERBAL MEANING	GRAPH	INTERVAL NOTATION
$x > a$	x is greater than a	a	(a, ∞)
$x \geq a$	x is greater than or equal to a	a	$[a, \infty)$
$x < a$	x is less than a	a	$(-\infty, a)$
$x \leq a$	x is less than or equal to a	a	$(-\infty, a]$
$a < x < b$	x is greater than a and less than b	a b	(a, b)
$a < x \leq b$	x is greater than a and less than or equal to b	a b	$(a, b]$
$a \leq x < b$	x is greater than or equal to a and less than b	a b	$[a, b)$
$a \leq x \leq b$	x is greater than or equal to a and less than or equal to b	a b	$[a, b]$
$-\infty < x < \infty$	x is any real number		$(-\infty, \infty)$

The interval $(-\infty, \infty)$ can also be represented by $\mathbb{R}$. In the interval $[a, b]$, a is called the left endpoint and b is called the right endpoint. The smaller value is always the left endpoint, and the larger value is always the right endpoint. For example, $[2, 5]$ is proper notation while $[5, 2]$ is improper notation because the larger value is listed first. The interval $(-1, 3)$ contains all points between -1 and 3.

EXAMPLE 5 Writing Multiple Representations for Intervals

Express each interval in inequality notation, in verbal form, and graphically.

SOLUTION

INTERVAL NOTATION	INEQUALITY NOTATION	VERBALLY	GRAPHICALLY
(a) $[2, \infty)$	$x \geq 2$	x is greater than or equal to 2	−1 0 1 2 3 4 5 6
(b) $(-\infty, \pi)$	$x < \pi$	x is less than π	π −1 0 1 2 3 4 5 6
(c) $(-2, 1]$	$-2 < x \leq 1$	x is greater than -2 and less than or equal to 1	−4 −3 −2 −1 0 1 2 3
(d) $[-3, 2]$	$-3 \leq x \leq 2$	x is greater than or equal to -3 and less than or equal to 2	−4 −3 −2 −1 0 1 2 3

SELF-CHECK 1.2.3

Express each of these intervals by using interval notation.

1. All real numbers greater than or equal to -2
2. $x < 4$
3. $-5 < x \leq 3$
4. (number line) $-\pi$ 1

We use $-\sqrt{4}$ to represent -2, a negative number that is also a square root of 4. The key point is that $\sqrt{x}$ denotes only one of the square roots of x and $-\sqrt{x}$ denotes the other.

Although calculators are used frequently to approximate square roots, it is important to be able to make rough mental estimates of square roots. For $x \geq 0$, the **principal square root of x**, denoted by $\sqrt{x}$, is a nonnegative real number r so that $r^2 = x$. For example, $\sqrt{4} = 2$ since $2^2 = 4$, and $\sqrt{9} = 3$ since $3^2 = 9$. To estimate $\sqrt{5}$, note that $\sqrt{4} < \sqrt{5} < \sqrt{9}$. Thus $2 < \sqrt{5} < 3$. A calculator approximation yields $\sqrt{5} \approx 2.2361$, a value consistent with our mental estimation.

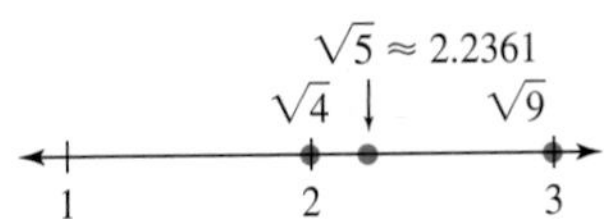

CALCULATOR PERSPECTIVE 1.2.3 **Evaluating Square Roots**

To approximate $\sqrt{9}$, $\sqrt{5}$, and $\sqrt{47}$ on a TI-84 Plus calculator, enter the following keystrokes:

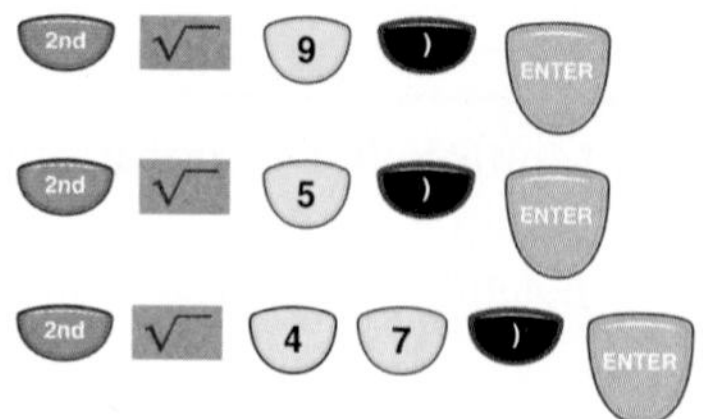

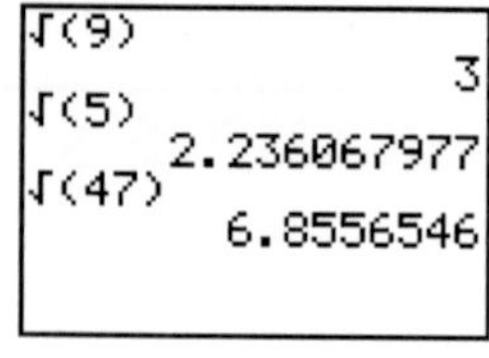

Note: The square root √ is the secondary feature of the x^2 key. Most graphing calculators include the left parenthesis when the square root button is pressed. Then the expression under the square root is entered and completed by pressing the right parenthesis key. The value shown for $\sqrt{9}$ is exact while the values shown for $\sqrt{5}$ and $\sqrt{47}$ are approximations.

EXAMPLE 6 Estimate Then Calculate

Estimate each of the square roots to the nearest integer, and then use a calculator to approximate each value to the nearest thousandth. Is the calculator value a reasonable answer?

SOLUTION

	ESTIMATED VALUE	CALCULATOR APPROXIMATION
(a) $\sqrt{47}$	$47 \approx 49$ $\sqrt{47} \approx \sqrt{49} = 7$ Because $\sqrt{47} \approx 7$, 6.856 seems to be a reasonable approximation of $\sqrt{47}$. ***Answer:*** $\sqrt{47} \approx 6.856$	√(47) 6.8556546 √(10100) 100.4987562
(b) $\sqrt{10{,}100}$	$10{,}100 \approx 10{,}000$ $\sqrt{10{,}100} \approx \sqrt{10{,}000} = 100$ $\sqrt{10{,}100} \approx 100.499$ seems to be a reasonable approximation of $\sqrt{10{,}100}$ because $100^2 = 10{,}000$. ***Answer:*** $\sqrt{10{,}100} \approx 100.499$	

If you pick up a mathematics or science textbook from 1970 or earlier, you will likely find several tables in the back of the book. One of the most common tables is a square root table because these values were needed frequently. Advances in technology have made the inclusion of these tables unnecessary as we can quickly use a calculator or computer to determine square roots. We can also use this technology to create custom tables.

To create a custom table of values, first enter the appropriate algebraic expression under the **Y** = screen. The table setup is used to fix the initial value of x and to input the increment (change) in the x values. This setup determines the x-values in the table. The y-values are calculated based on the expression entered in the **Y** = screen. Calculator Perspective 1.2.4 illustrates how to create a custom table of square roots.

To create a table of square roots for input values of $x = 10, 15, 20, 25, 30, 35, 40$ on a TI-84 calculator, enter the following keystrokes.

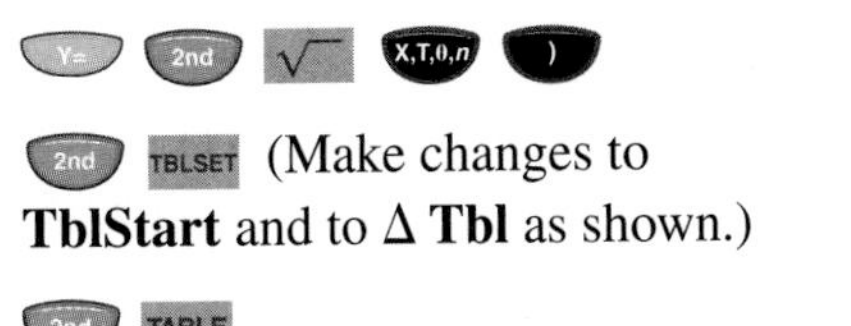

2nd TBLSET (Make changes to **TblStart** and to **Δ Tbl** as shown.)

2nd TABLE

Plot1 Plot2 Plot3
\Y1■√(X)

TABLE SETUP
TblStart=10
ΔTbl=5
Indpnt: Auto Ask
Depend: Auto Ask

X	Y1
10	3.1623
15	3.873
20	4.4721
25	5
30	5.4772
35	5.9161
40	6.3246

X=10

Note: **Δ Tbl** identifies the change in the x-values in the table. Both **Indpnt** and **Depend** should be set to **Auto.** The square roots are displayed in the Y_1 column. The TBLSET feature is the secondary function of the WINDOW key, and the TABLE feature is the secondary function of the GRAPH key.

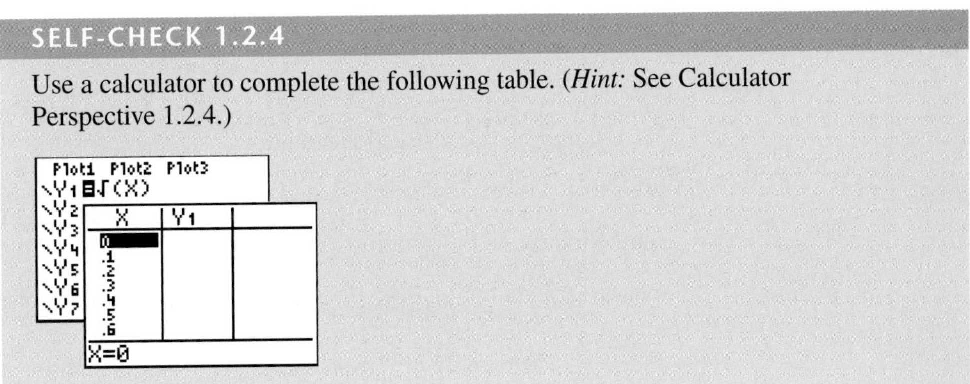

SELF-CHECK 1.2.4

Use a calculator to complete the following table. (*Hint:* See Calculator Perspective 1.2.4.)

The real numbers are classified as either rational or irrational. A real number that can be written as a ratio of two integers is called a **rational number.** (Notice that the first five letters of the word *rational* spell *ratio*.) The other real numbers, which cannot be written as the ratio of two integers, are called **irrational numbers.**

Rational and Irrational Numbers

ALGEBRAICALLY	NUMERICALLY	NUMERICAL EXAMPLES	VERBAL EXAMPLES
Rational: A real number x is rational if $x = \frac{a}{b}$ for integers a and b, with $b \neq 0$.	In decimal form, a rational number is either a terminating decimal or an infinite repeating decimal.	$\frac{1}{2} = 0.5$	$\frac{1}{2}$ in decimal form is a terminating decimal.
		$\frac{1}{3} = 0.333 \cdots = 0.\overline{3}$	$\frac{1}{3}$ in decimal form is a repeating decimal.
		$\frac{5}{33} = 0.151515 \cdots = 0.\overline{15}$	$\frac{5}{33}$ in decimal form is a repeating decimal.
Irrational: A real number x is irrational if it cannot be written as $x = \frac{a}{b}$ for integers a and b.	In decimal form, an irrational number is an infinite nonrepeating decimal.	$\sqrt{2} \approx 1.414214$	$\sqrt{2}$ cannot be written as a rational fraction—it is an infinite nonrepeating decimal.
		$\pi \approx 3.141593$	π cannot be written as a rational fraction—it is an infinite nonrepeating decimal.
		$0.1010010001 \cdots$	This irrational number does exhibit a pattern, but it does not terminate and it does not repeat.

All integers are also rational numbers, since every integer can be written as the ratio of itself to 1. For example, -3, 0, and 7 can be written as $\frac{-3}{1}, \frac{0}{1}$, and $\frac{7}{1}$. Another form that can be interpreted as a ratio is percent notation. For example, 17% read as "17 percent" means 17 parts out of 100; $17\% = \frac{17}{100}$. Other examples of rational numbers are $\frac{1}{7}, \frac{-2}{13}$, and $\frac{479}{-22}$. The numbers $4\frac{2}{5}$, 0.17, and $0.333 \cdots$ are also rational since each of these numbers can be written as a ratio of two integers: $4\frac{2}{5} = \frac{22}{5}$, $0.17 = \frac{17}{100}$, and $0.333 \cdots = \frac{1}{3}$. The repeating decimal form for $\frac{1}{3}$ can be written as either $0.\overline{3}$ or as $0.333 \cdots$. Numbers such as π, $\sqrt{2}$, $\sqrt{3}$, and $-\sqrt{5}$ are irrational since they cannot be written as the ratio of two integers.

A number is not automatically irrational just because it has a square root symbol. For example, $\sqrt{4}$ is another notation for the rational number 2. All numbers with rational numbers as their square roots are called **perfect squares.** The integers 1, 4, 9, 16, and 25 are perfect squares since they have as their square roots 1, 2, 3, 4, and 5, respectively. The number $\frac{4}{81}$ is also a perfect square since $\sqrt{\frac{4}{81}} = \frac{2}{9}$. However, $\sqrt{2}$ is irrational, since it can be shown that 2 is not a perfect square of any rational number.

EXAMPLE 7 Identifying Rational and Irrational Numbers

Classify each of the following real numbers as either rational or irrational.

SOLUTION

(a) $1\frac{3}{5}$	Rational	$1\frac{3}{5}$ can be written as the fraction $\frac{8}{5}$.
(b) 1.35	Rational	1.35 is a terminating decimal.
(c) $0.666\cdots$	Rational	All repeating decimals are rational numbers. This number can also be written as $\frac{2}{3}$.
(d) $0.101101110\cdots$	Irrational	Although there is a pattern (the number of 1s is increasing between the 0s), this is not a repeating decimal.
(e) $\sqrt{\frac{36}{49}}$	Rational	$\sqrt{\frac{36}{49}} = \frac{6}{7}$ since $\left(\frac{6}{7}\right)^2 = \frac{36}{49}$.
(e) $\sqrt{0}$	Rational	$\sqrt{0} = 0$ since $0^2 = 0$.
(f) $\sqrt{5}$	Irrational	5 is not the square of any rational number.
(g) $\sqrt{0.25}$	Rational	$\sqrt{0.25} = 0.5$ since $0.5^2 = 0.25$.

The relationship of the important subsets of the real numbers that have been defined in this section is summarized by the tree diagram in the following figure.

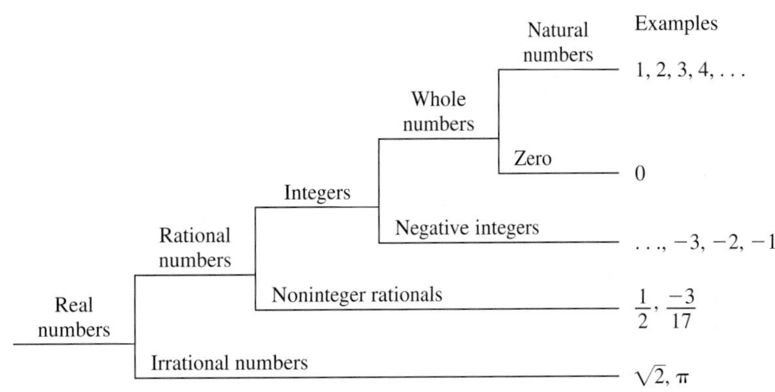

EXAMPLE 8 Classifying Real Numbers

Given the set $\left\{-5, -3.2, -\sqrt{3}, 0, \frac{2}{3}, 2, \pi, 4\frac{1}{3}, 6\right\}$, list all the elements in this set of real numbers that are

SOLUTION

(a) Natural numbers	2, 6	The natural numbers consist of 1, 2, 3, . . .
(b) Whole numbers	0, 2, 6	The whole numbers consist of 0, 1, 2, . . .
(c) Integers	$-5, 0, 2, 6$	The integers consist of . . ., $-2, -1, 0, 1, 2,$ . . .
(d) Rational numbers	$-5, -3.2, 0, \frac{2}{3}, 2, 4\frac{1}{3}, 6$	The rational numbers consist of the integers, fractions, and terminating or repeating decimals.
(e) Irrational numbers	$-\sqrt{3}, \pi$	Irrational numbers consist of the real numbers that are not rational.

SELF-CHECK 1.2.5

Given the set $\left\{-4, -\sqrt{2}, -1, -\frac{3}{7}, 0, 4, 5\frac{3}{8}, 7\right\}$, list all the elements in this set of real numbers that are

1. Natural numbers
2. Whole numbers
3. Integers
4. Rational numbers
5. Irrational numbers

SELF-CHECK ANSWERS

1.2.1

1. $45, 0, -\frac{17}{43}$ 2. -45
3. 0 4. 45 5. 45

1.2.2

1. T 2. F 3. T
4. F 5. F 6. T

1.2.3

1. $[-2, \infty)$ 2. $(-\infty, 4)$
3. $(-5, 3]$ 4. $[-\pi, 1)$

1.2.4

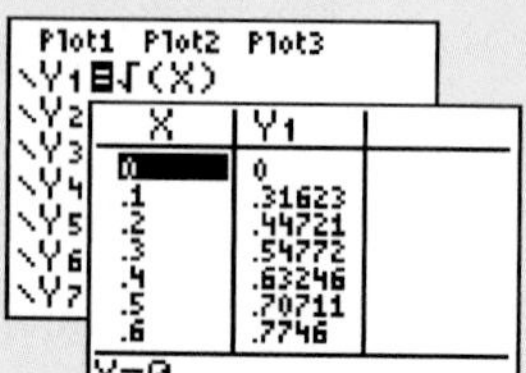

1.2.5

1. 4, 7 2. 0, 4, 7
3. $-4, -1, 0, 4, 7$
4. $-4, -1, -\frac{3}{7}, 0, 4, 5\frac{3}{8}, 7$
5. $-\sqrt{2}$

Each exercise set will start with exercises on "Using the Language and Symbolism of Mathematics." It is highly recommended that you complete these before your instructor lectures on each section. These language and symbolism exercises are repeated in the lecture guide that accompanies this text.

USING THE LANGUAGE AND SYMBOLISM OF MATHEMATICS 1.2

1. A __________ is a letter that can be used to represent different numbers.
2. The notation $-a$ is read "the opposite of a" or the "__________ __________ of a."
3. The additive identity is __________.
4. The additive inverse of $-a$ is __________.
5. If one number is 7 units to the right of the origin, its additive inverse is __________ units to the __________ of the origin.
6. The notation $|x|$ is read "the __________ __________ of x." $|x|$ represents the __________ between 0 and x on the number line.
7. The notation $x \geq y$ is read "x is __________ than or __________ to y."
8. The notation $x < y$ is read "x is __________ than y." On the number line, x is to the __________ of y.
9. The notation $x \neq y$ is read "x is not __________ to y."
10. The notation $x \approx y$ is read "x is __________ __________ to y."
11. The notation $\sqrt{x}$ is read "the principal __________ __________ of x."
12. A real number that is a terminating decimal is a __________ number.
13. A real number that is a repeating decimal is a __________ number.
14. A real number that is an infinite nonrepeating decimal is an __________ number.
15. The symbol π is the symbol for __________.
16. The symbol ∞ is the symbol for __________.
17. In interval notation, a parenthesis indicates that a value **is/is not** included in the interval. (Select the correct choice.)
18. In interval notation, a bracket indicates that a value **is/is not** included in the interval. (Select the correct choice.)
19. The set of all real numbers can be represented by the interval notation $(-\infty, \infty)$ or by __________.
20. The notation $[2, -3]$ is incorrect notation because the larger value is listed first. The correct interval notation for $-3 \leq x \leq 2$ is __________.
21. The interval notation to represent $a < x \leq b$ is __________.

EXERCISES 1.2

In Exercises 1 and 2, write the additive inverse of each number.

1. a. 9 b. -13 c. $-\frac{3}{5}$
   d. 7.23 e. $-\sqrt{13}$ f. 0
2. a. 8 b. -30 c. $\frac{5}{7}$
   d. -8.94 e. $-\pi$ f. 0.03
3. Simplify each expression to either 7 or -7.
   a. $-(-7)$ b. $-[-(-7)]$ c. $0 + 7$
4. Simplify each expression to either 9 or -9.
   a. $-(-9)$ b. $-[-(-9)]$ c. $-9 + 0$

In Exercises 5–8, simplify each expression.

5. a. $17 + (-17)$ b. $-17 + 17$
   c. $-\pi + \pi$ d. $x + (-x)$
6. a. $2.3 + (-2.3)$ b. $-2.3 + 2.3$
   c. $e + (-e)$ d. $-y + y$
7. a. $-6x + 6x$ b. $6x + 0$
   c. $6x + (-6x)$ d. $-6x + 0$
8. a. $2y + (-2y)$ b. $0 + 2y$
   c. $-2y + 2y$ d. $0 + (-2y)$

In Exercises 9–12, evaluate each expression.

9. a. $|29|$ b. $|-29|$
   c. $-|29|$ d. $-|-29|$
10. a. $|37|$ b. $|-37|$
    c. $-|37|$ d. $-|-37|$
11. a. $\sqrt{25}$ b. $-\sqrt{25}$
    c. $\sqrt{0.25}$ d. $\sqrt{2{,}500}$
12. a. $\sqrt{36}$ b. $-\sqrt{36}$
    c. $\sqrt{0.36}$ d. $\sqrt{3{,}600}$

Estimation Skills

In Exercises 13–16, mentally estimate the value of each square root and then select the choice that is closest to your estimate.

13. $\sqrt{17}$ a. 3.88 b. 4.12 c. 4.92 d. 9.24
14. $\sqrt{24}$ a. 4.90 b. 5.10 c. 12.01 d. 12.99
15. $\sqrt{62.41}$ a. 6.7 b. 7.1 c. 7.9 d. 8.9
16. $\sqrt{123.21}$ a. 61.6 b. 58.4 c. 12.3 d. 11.1

Estimate Then Calculate

In Exercises 17–22, complete the following table by

a. Estimating each square root to the nearest integer;
b. Determining whether this integer estimate is less than or greater than the actual value;
b. Using a calculator to approximate each expression to the nearest thousandth. (*Hint:* See Calculator Perspective 1.2.3.)

	INTEGER ESTIMATE	INEQUALITY	APPROXIMATION
Example $\sqrt{5}$	2	$2 < \sqrt{5}$	2.236
17. $\sqrt{50.6}$			
18. $\sqrt{8.92}$			
19. $\sqrt{0.976}$			
20. $\sqrt{65.3}$			
21. $\sqrt{15.1}$			
22. $\sqrt{146.6}$			

23. What value of x makes the statement $x < 7$ false but makes the statement $x \le 7$ true?
24. What value of x makes the statement $x > 9$ false but makes the statement $x \ge 9$ true?
25. What value of x makes both $x \le 3$ and $x \ge 3$ true statements?
26. What value of x makes both $x \le -5$ and $x \ge -5$ true statements?

In Exercises 27–38, insert $<$, $=$, or $>$ in the blank to make each statement true.

27. a. 83 ____ 38 b. -83 ____ -38
28. a. -9.4 ____ 0 b. 0 ____ 0.009
29. a. -8.4 ____ -4.8 b. 8.4 ____ 4.8
30. a. 0 ____ $\frac{1}{3}$ b. 0 ____ $-\frac{2}{5}$
31. a. $-\frac{1}{2}$ ____ $-\frac{1}{5}$ b. $8\frac{3}{4}$ ____ $8\frac{3}{5}$
32. a. $-\frac{5}{8}$ ____ $-\frac{5}{9}$ b. $7\frac{1}{11}$ ____ $7\frac{1}{15}$
33. a. $\frac{1}{2}$ ____ 0.5 b. $|21|$ ____ $|-21|$
34. a. $-\frac{1}{4}$ ____ -0.25 b. $|-37|$ ____ $|37|$
35. a. $|-5|$ ____ $-|5|$ b. $|-5|$ ____ $-|-5|$
36. a. $-|7|$ ____ $-|-7|$ b. $-|-7|$ ____ $|7|$
37. a. $|7|$ ____ $\sqrt{7}$ b. $|0.49|$ ____ $\sqrt{0.49}$
38. a. $|3|$ ____ $\sqrt{3}$ b. $|0.16|$ ____ $\sqrt{0.16}$
39. List all the elements of $\left\{-11, -4.8, -\sqrt{9}, 0, 1\frac{3}{5}, \sqrt{5}, 15\right\}$ that are
    a. Natural numbers b. Whole numbers
    c. Integers d. Rational numbers
    e. Irrational numbers
40. List all the elements of $\left\{-9.3, -5, 0, \sqrt{\frac{25}{36}}, 1, 2, \sqrt{7}, 5\frac{1}{9}\right\}$ that are
    a. Natural numbers b. Whole numbers
    c. Integers d. Rational numbers
    e. Irrational numbers

Multiple Representations

In Exercises 41–48, use the given information to complete each row as illustrated by the example in the first row.

	INEQUALITY NOTATION	VERBALLY	GRAPHICALLY	INTERVAL NOTATION
Example	$x > 2$	x is greater than 2.	2	$(2, \infty)$
41.			−3 3	
42.			6	
43.	$x > 1$			
44.	$1 \le x \le 5$			
45.		x is greater than or equal to -3.		
46.		x is greater than or equal to 0 and less than 2.		
47.				$(-\infty, 4]$
48.				$(-3, 5)$

In Exercises 49–56, list the set or sets to which the given real number belongs. The choices are natural numbers, whole numbers, integers, rational numbers, and irrational numbers.

49. -18 **50.** $-\frac{1}{8}$ **51.** 81 **52.** $\sqrt{81}$

53. $\sqrt{8}$ **54.** 8 **55.** $5\frac{3}{7}$ **56.** 0

57. Plot these numbers on a real number line:
$-5, -3.5, 0, 1\frac{3}{4}, 4$

58. Plot these numbers on a real number line:
$-4, -2, -\frac{1}{2}, 0, 1, 2\frac{1}{2}, 5$

59. Plot these numbers on a real number line:
$-2.5, \frac{5}{4}, |-3|, -|4|$

60. Plot these numbers on a real number line:
$-\frac{3}{2}, -|-2|, |-2|, 2\frac{1}{3}$

In Exercises 61–66, identify each real number as either rational or irrational.

61. **a.** $\sqrt{9}$ **b.** $\sqrt{10}$

62. **a.** $\sqrt{20}$ **b.** $\sqrt{25}$

63. **a.** $0.171717\cdots$ **b.** $0.171771777\cdots$

64. **a.** $0.232232223\cdots$ **b.** $-0.133133133\cdots$

65. **a.** $\sqrt{0}$ **b.** $\sqrt{\frac{49}{16}}$

66. **a.** 0.13562 **b.** $0.101001000\cdots$

Multiple Representations

In Exercises 67–72, write each verbal statement in algebraic form.

67. The absolute value of x is equal to y.

68. The square root of x is equal to y.

69. π is greater than 3.14 and less than 3.15.

70. The square root of two is greater than one and less than two.

71. The opposite of x is less than or equal to negative two.

72. The absolute value of negative eleven equals eleven.

73. **a.** Give an example of a number x so that $\sqrt{x} < x$.
b. Give an example of a number x so that $\sqrt{x} > x$.
c. Give an example of a number x so that $\sqrt{x} = x$.

74. **a.** What whole number is not a natural number?
b. List two integers that are not natural numbers.
c. List two negative rational numbers that are not integers.
d. List two rational numbers between 1 and 2.

75. **a.** What number is its own additive inverse?
b. List two positive rational numbers that are not integers.
c. List two real numbers that are not rational numbers.
d. List two rational numbers between -2 and -3.

76. Use a calculator to complete the following table.

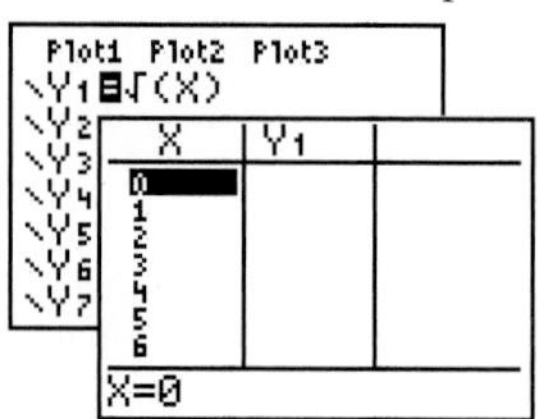

77. Use a calculator to complete the following table.

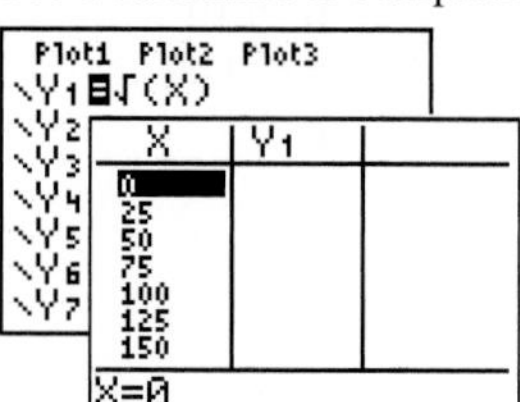

78. **Stock Price Change** The bar graph below shows the change in the price of a stock for each day of a given week. Select the data that this bar graph best represents.
a. What was the change in price of this stock on Thursday?
b. On which day did the price of the stock drop by \$0.75?

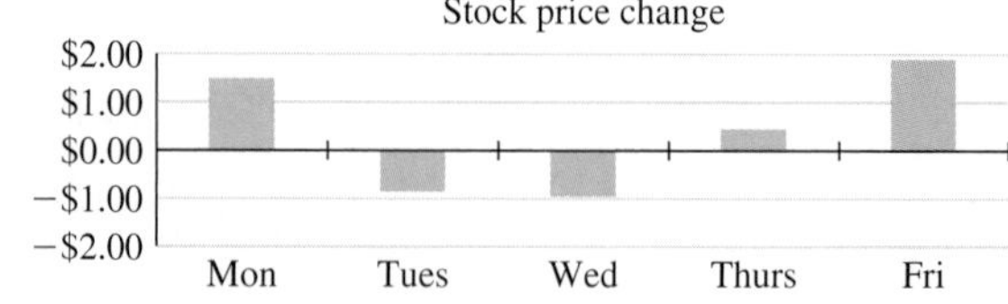

79. **Home Runs** The bar graph below shows the total number of home runs hit by Albert Pujols of the St. Louis Cardinals for four consecutive seasons.
   a. How many home runs did he hit in 2005?
   b. During which year did he hit 37 home runs?
   c. In which year did he hit the most home runs?

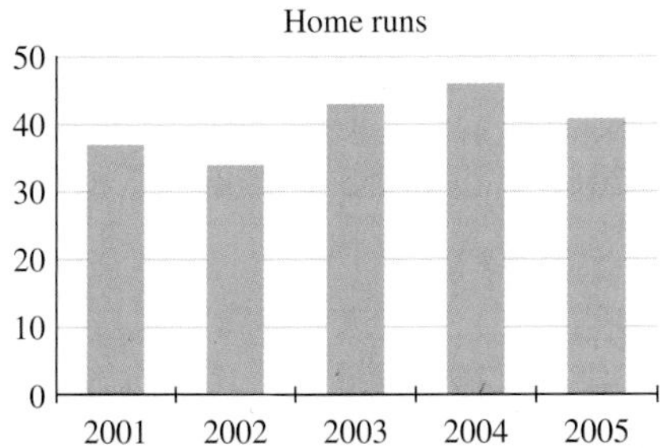

80. **Asset Allocation** The pie chart below gives the breakdown for a family's assets. Are the assets in stocks less than or more than 25%?

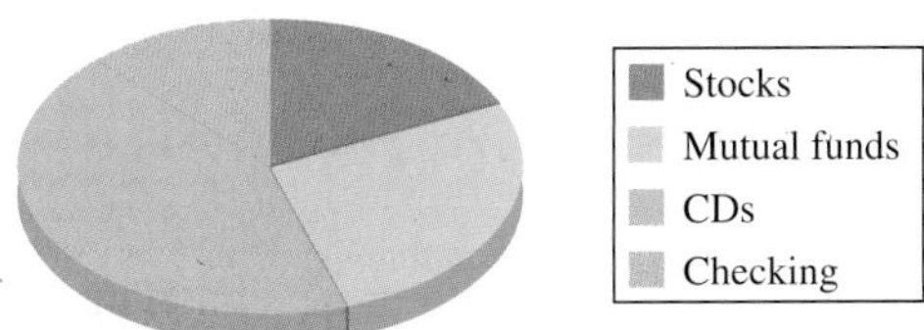

81. **Home Energy Usage** According to the U.S. Department of Energy, the average U.S. home uses energy in four main areas. Which uses more energy, heating and cooling or lights and appliances?

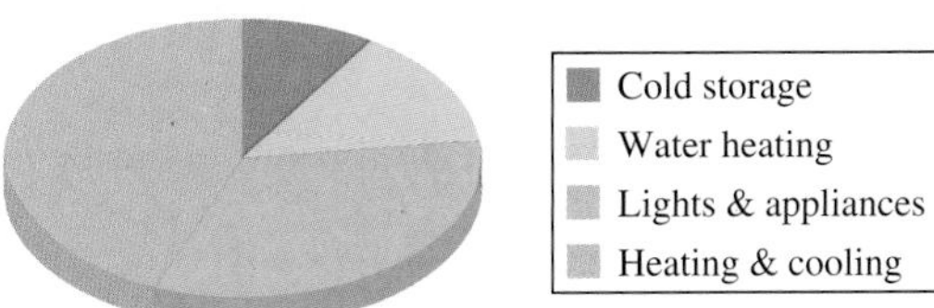

Group Discussion Questions

82. **Multiple Representations** A variety of notations can be used to represent the same real number. For example, 0.25, $\frac{1}{4}$, and 25% all represent the same real number.
   a. Plot 0.25 on this number line:

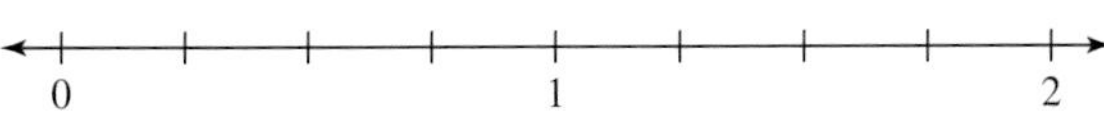

   b. Shade $\frac{1}{4}$ of this region:

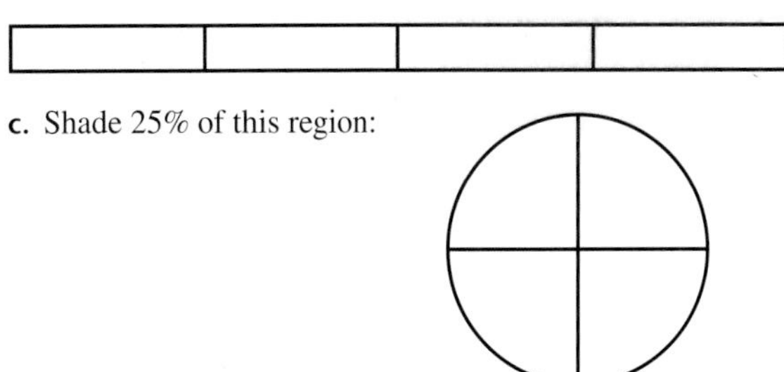

   c. Shade 25% of this region:

83. **Multiple Representations** Complete the following table to represent each of these rational numbers in fractional form, in decimal form, and as a percent.

	FRACTION	DECIMAL	PERCENT
Example	$\frac{1}{2}$	0.50	50%
a.	$\frac{1}{4}$		
b.		0.10	
c.			5%

84. **Communicating Mathematically** There are two square roots of 19. Write an algebraic representation for both of these square roots.

85. **Communicating Mathematically**
   a. Using 3.14 as an approximation for π, one student reports the area of this circle as 12.56 cm^2. Is this student's answer a rational number or an irrational number? Give a verbal justification for your answer.

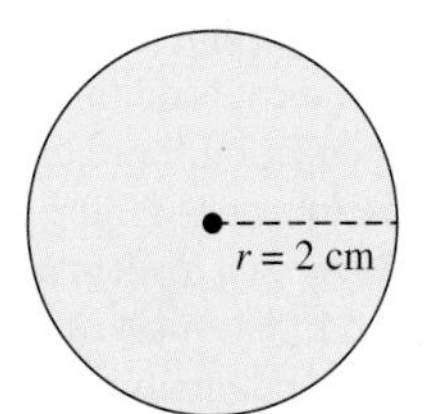

   b. Another student reports the area of this circle as 4π cm^2. Is this student's answer a rational number or an irrational number? Give a verbal justification for your answer.

86. **Communicating Mathematically** Describe in your own words how to evaluate the absolute value of a real number. Is there any real number whose absolute value is not a positive number?

87. **Communicating Mathematically** Write a paragraph describing the relationship of the natural numbers, whole numbers, integers, rationals, irrationals, and real numbers.

Spreadsheet Exploration

88. One of the first things to understand about a spreadsheet is that information is contained in **cells.** Each cell is given a unique name by first indicating the column and then the row. For example, cell D3 contains the value 252. Notice that the 252 is found in column D, row 3.
   a. What value is in cell B4? ______
   b. What value is in cell C5? ______

	A	B	C	D
1	Dimensions for Jones Home			
2	Room	Length	Width	Area
3	Bedroom 1	14	18	252
4	Bedroom 2	12	14	168
5	Den	12	16	192
6				

Cell D3

Section 1.3 Addition of Real Numbers

Objectives:
7. Add positive and negative real numbers.
8. Use the commutative and associative properties of addition.
9. Evaluate an algebraic expression for given values of the variables.

Positive and negative numbers both occur in many real-life applications: the price changes of a stock, the profit or loss of a business, the temperature on a cold day, the elevation of an object, the net yardage on a football play, and the directed distance as shown on a number line.

SYMBOL	LAST	CHANGE
BRCM	41.15	−0.35
DELL	25.68	+0.09
DJIA	11,577.74	+138.88
INTC	19.57	−0.01
NASDAQ	2,342.57	+18.67
MSFT	23.84	−0.54

A Mathematical Note

The symbols + and −, used to indicate addition and subtraction, were introduced by Johannes Widman, a Bohemian mathematician, in the 15th century. The words *plus* and *minus* are from the Latin for "more" and "less," respectively.

In Example 1 below, we will begin to consider operations with these signed numbers. We will start by building on the skills that we already have with addition. Recall the terminology for the addition $5 + 8 = 13$; the numbers 5 and 8 are called **terms** or **addends,** and the result 13 is called the **sum.**

EXAMPLE 1 Adding Real Numbers with Like Signs

Complete the following table, using a positive number to indicate a profit and a negative number to indicate a loss.

Quarterly Profit/Loss Statement for H&M Printers

	QUARTER 1	QUARTER 2	QUARTER 3	QUARTER 4
2004	+\$25,000 (Profit)	+\$30,000 (Profit)	−\$5,000 (Loss)	−\$3,000 (Loss)
2005	+\$40,000 (Profit)	+\$50,000 (Profit)	−\$7,000 (Loss)	−\$4,000 (Loss)
Total				

SOLUTION

The answer +\$65,000 is usually just written as \$65,000 and is understood to represent a positive sixty-five thousand dollars.

Total for quarter 1:	\$25,000 + \$40,000 = +\$65,000	A positive number represents a profit.
Total for quarter 2:	\$30,000 + \$50,000 = +\$80,000	A positive number represents a profit.
Total for quarter 3:	−\$5,000 + (−\$7,000) = −\$12,000	A negative number represents a loss.
Total for quarter 4:	−\$3,000 + (−\$4,000) = −\$7,000	A negative number represents a loss.

Example 1 illustrates that when both terms are positive, the sum is positive and its magnitude is found by adding the magnitudes or absolute values of the terms. When both terms are negative, the sum is negative and its magnitude is found by adding the magnitudes or absolute values of the terms. As Example 2 illustrates, when one term is positive and the other is negative, the result is that one term tends to diminish the magnitude of the other term.

EXAMPLE 2 Adding Real Numbers with Unlike Signs

Complete the following table, using a positive number to indicate a profit and a negative number to indicate a loss.

Branch Results for H&M Productions

	QUARTER 1	QUARTER 2	QUARTER 3	QUARTER 4
Texas branch	+\$35,000	−\$35,000	+\$5,000	−\$10,000
California branch	−\$10,000	+\$50,000	−\$10,000	+\$7,000
Total				

SOLUTION

Total for quarter 1: \$35,000 + (−\$10,000) = +\$25,000, a profit
Total for quarter 2: −\$35,000 + \$50,000 = +\$15,000, a profit
Total for quarter 3: \$5,000 + (−\$10,000) = −\$5,000, a loss
Total for quarter 4: −\$10,000 + \$7,000 = −\$3,000, a loss

The + symbol has two distinct meanings. One meaning indicates that the direction of a signed number on the number line is to the right of the origin. The second meaning indicates the operation of addition, which results in movement on the number line.

Another application of signed numbers is directed distance. The addition of real numbers can be represented by combining directed distances on the number line. View each term as a trip in a specific direction and addition as the combining of these individual trips. This is illustrated in the following box.

Addition of Real Numbers

	VERBALLY	NUMERICAL EXAMPLE	GRAPHICAL EXAMPLE
Like signs	First add the absolute values of the terms. Then use the same sign as the terms.	$(+3) + (+4) = +7$	(+3) + (+4) = +7 SUM; +3; +4; −2 −1 0 1 2 3 4 5 6 7 8
		$(-3) + (-4) = -7$	(−3) + (−4) = −7 SUM; −4; −3; −8 −7 −6 −5 −4 −3 −2 −1 0 1 2
Unlike signs	First subtract the smaller absolute value from the larger absolute value. Then use the sign of the term with larger absolute value.	$-3 + (+4) = +1$	(−3) + (+4) = +1 SUM; +4; −3; −5 −4 −3 −2 −1 0 1 2 3 4 5
		$(+3) + (-4) = -1$	(+3) + (−4) = −1 SUM; −4; +3; −5 −4 −3 −2 −1 0 1 2 3 4 5

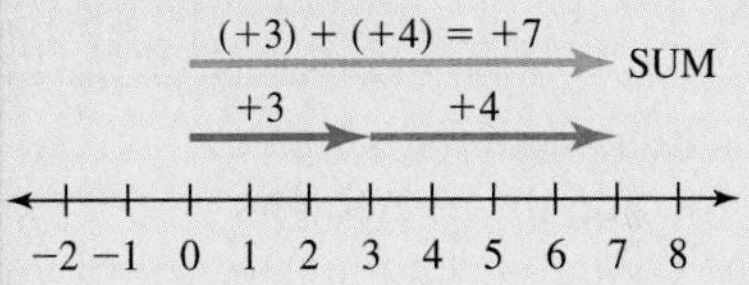

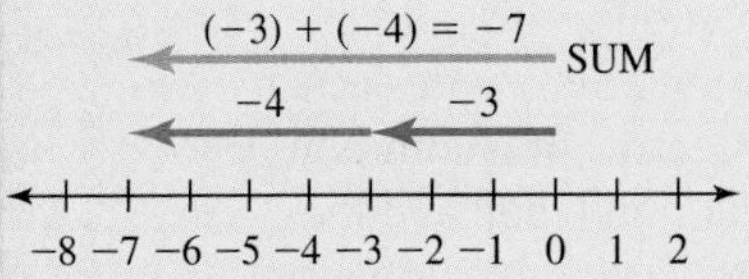

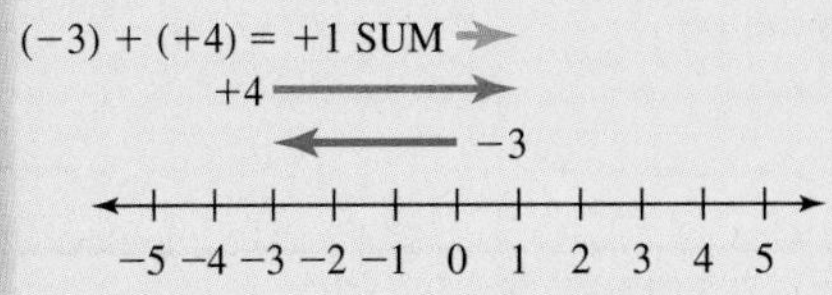

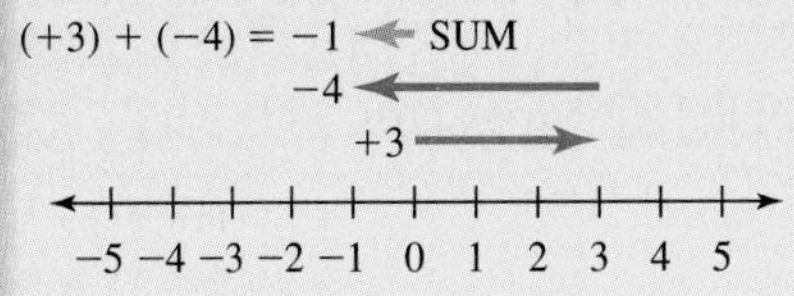

Example 3 illustrates the addition of terms with both like signs and unlike signs.

EXAMPLE 3 Adding Terms with Like and Unlike Signs

Calculate the following sums.

SOLUTION

(a) $4.7 + 5.3$ — $4.7 + 5.3 = 10.0$ — Both terms and the sum are positive.

(b) $-2.9 + (-5.1)$ — $-2.9 + (-5.1) = -8.0$ — Both terms and the sum are negative.

(c) $-4.7 + 5.3$ — $-4.7 + 5.3 = 0.6$ — The terms have unlike signs. The difference of the absolute values is $5.3 - 4.7 = 0.6$. The sum is positive as is the term 5.3.

(d) $2.9 + (-5.1)$ — $2.9 + (-5.1) = -2.2$ — The terms have unlike signs. The difference of the absolute values is 2.2. The sum is negative as is the term -5.1.

SELF-CHECK 1.3.1

Calculate each sum.

1. $11.1 + 22.2$ **2.** $-55.5 + (-44.4)$ **3.** $11.1 + (-22.2)$ **4.** $55.5 + (-44.4)$

The procedure for adding fractions is summarized in the following box. The group exercises at the end of this section contain a graphical perspective on adding fractions.

Addition of Fractions

VERBALLY	ALGEBRAICALLY	NUMERICAL EXAMPLE
To add fractions with the same denominator, add the numerators and use the common denominator.	$\frac{a}{b} + \frac{c}{b} = \frac{a + c}{b}$ for $b \neq 0$	$\frac{1}{5} + \frac{3}{5} = \frac{1 + 3}{5} = \frac{4}{5}$
To add fractions with unlike denominators, first express each fraction in terms of a common denominator and then add the numerators, using this common denominator.	$\frac{a}{b} + \frac{c}{d} = \frac{ad}{bd} + \frac{bc}{bd} = \frac{ad + bc}{bd}$ for $b \neq 0$ and $d \neq 0$	$\frac{1}{4} + \frac{2}{3} = \frac{3}{12} + \frac{8}{12} = \frac{3 + 8}{12} = \frac{11}{12}$

EXAMPLE 4 Adding Fractions with Like and Unlike Denominators

Calculate the following sums.

SOLUTION

(a) $\frac{2}{9} + \frac{5}{9}$

$$\frac{2}{9} + \frac{5}{9} = \frac{2 + 5}{9}$$
$$= \frac{7}{9}$$

$\frac{a}{b} + \frac{c}{b} = \frac{a + c}{b}$ for $b \neq 0$. To add fractions with the same denominator, add the numerators and use the common denominator.

(b) $-\frac{5}{13} + \frac{7}{13}$

$$-\frac{5}{13} + \frac{7}{13} = \frac{-5+7}{13} = \frac{2}{13}$$

$\frac{a}{b} + \frac{c}{b} = \frac{a+c}{b}$ for $b \neq 0$. To add fractions with the same denominator, add the numerators and use the common denominator.

(c) $-\frac{1}{4} + \left(-\frac{1}{3}\right)$

$$-\frac{1}{4} + \left(-\frac{1}{3}\right) = -\frac{3}{12} + \left(-\frac{4}{12}\right) = \frac{-3+(-4)}{12} = \frac{-7}{12} = -\frac{7}{12}$$

To add fractions with different denominators, first express each fraction in terms of a common denominator. In this case, express each term in terms of the least common denominator of 12. Then add these like terms.

Both $\frac{-7}{12}$ and $-\frac{7}{12}$ represent the same number. This is discussed further in Section 1.6.

Most graphing calculators have a limited ability (for fractions with only a few digits in the denominator) to convert the decimal form of a rational number to its fractional form. This means that we also can use a calculator to add some fractions and obtain the sum in fractional form. To convert a fraction to a decimal, enter the fraction, using the division symbol for the fraction bar. Pressing ENTER will then display the fraction in decimal form. To convert a decimal answer to a fraction, use the **Ans ▶ Frac** feature. By pressing   at the end of the expression with fractions, we could directly convert the answer to a fraction.

CALCULATOR PERSPECTIVE 1.3.1 Converting Between Decimals and Fractions

To convert $\frac{3}{8}$ to a decimal and to evaluate $-\frac{1}{4} + \left(-\frac{1}{3}\right)$ from Example 4(c) on a TI-84 Plus calculator, enter the following keystrokes:

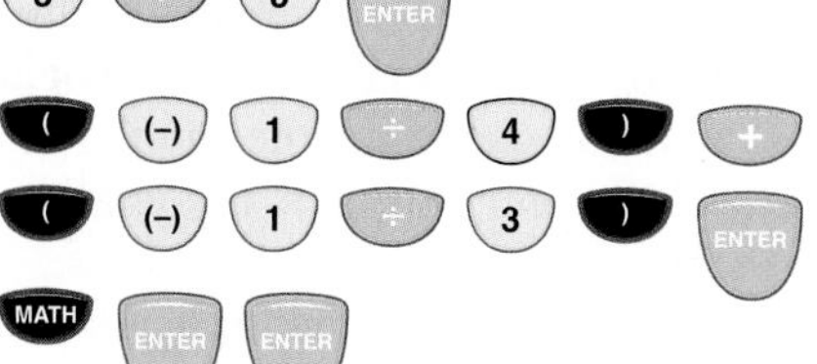

```
3/8
                .375
(-1/4)+(-1/3)
       -.5833333333
Ans▸Frac
              -7/12
```

Note: Pressing the MATH key pulls up the **Frac** conversion as one option. Pressing  once selects this option. Pressing ENTER a second time executes this conversion to a fraction. Although the parentheses enclosing the fractions are not required in this example, their usage makes input more readable and ensures the correct interpretation.

SELF-CHECK 1.3.2

Calculate each sum.

1. $\frac{2}{7} + \left(-\frac{5}{7}\right)$ **2.** $-\frac{5}{13} + \left(-\frac{7}{13}\right)$ **3.** $\frac{1}{4} + \left(-\frac{1}{3}\right)$ **4.** $-\frac{1}{4} + \frac{1}{3}$

Use a calculator to work these problems.

5. Convert $\frac{5}{8}$ to a decimal. **6.** Add $-\frac{5}{13} + \frac{7}{13}$. **7.** Add $\frac{1}{4} + \left(-\frac{1}{3}\right)$.

Use only parentheses when you enter expressions into a calculator, as brackets have a special meaning.

When grouping symbols are used, the implied order of operations is to perform the operations within the innermost grouping symbols first. The most commonly used grouping symbols for printed copy are parentheses ( ) and brackets [ ]. We will examine order of operations in greater detail in Section 1.7.

EXAMPLE 5 Simplifying Expressions with Grouping Symbols

Determine the following sums.

SOLUTION

(a) $-7 + [-8 + (+5)]$

$-7 + [-8 + (+5)] = -7 + (-3)$
$= -10$

The indicated order of operations is to first add the terms inside the brackets.

(b) $[-9 + (-7 + 3)] + (-11)$

$[-9 + (-7 + 3)] + (-11)$
$= [-9 + (-4)] + (-11)$
$= -13 + (-11)$
$= -24$

Add the terms inside the inner parentheses first; then add the terms inside the brackets.

SELF-CHECK 1.3.3

1. Use a calculator to check the answer to $[-9 + (-7 + 3)] + (-11)$ from Example 5(b). Use only parentheses when you enter the expression.

When we purchase two items at a store, the total will be the same no matter which item is charged first. (Can you imagine the confusion that would result if the order did affect the total price?) For example, $9 + 5 = 5 + 9$. We say that addition is **commutative** to describe the fact that the order of the terms does not change the sum. That is, $a + b = b + a$.

When we purchase several items at a store, we may cluster them together in different ways. For example, all canned goods and all fresh produce might be totaled separately. Again if we subtotal these groups separately, we are confident that the total price of all items will not change. We say that addition is **associative** to describe the fact that terms can be regrouped without changing the sum. For example, $(3 + 4) + 5$ and $3 + (4 + 5)$ both equal 12. In general, the associative property of addition says $(a + b) + c = a + (b + c)$.

The properties of addition, subtraction, multiplication, and division are presented as needed. These properties are summarized in the Key Concepts for Chapter 1 after we have reviewed all these operations.

Properties of Addition

	ALGEBRAICALLY	VERBALLY	NUMERICAL EXAMPLE
Commutative property of addition	$a + b = b + a$	The sum of two terms in either order is the same.	$4 + 5 = 5 + 4$
Associative property of addition	$(a + b) + c = a + (b + c)$	Terms can be regrouped without changing the sum.	$(3 + 4) + 5 = 3 + (4 + 5)$

Together the commutative and associative properties allow us to add terms in any order we desire. People who are very skilled at mental arithmetic often take advantage of this to add terms producing multiples of 10 or other convenient values. For expressions with several terms, it is often convenient to first add all terms with like signs.

EXAMPLE 6 Using the Properties of Addition

Calculate these sums.

SOLUTION

(a) $47 + 19 + 53 + 31 + 7$

$$\begin{aligned} &47 + 19 + 53 + 31 + 7 \\ &= (47 + 53) + (19 + 31) + 7 \\ &= 100 + 50 + 7 \\ &= 157 \end{aligned}$$

For convenience, we use the commutative and associative properties to regroup the terms in order to pair terms that produce multiples of 10. With practice you may be able to do this entire computation mentally. For most problems writing some of these steps is usually a good idea.

(b) $3 + (-4) + 5 + (-6) + 7$

$$\begin{aligned} &3 + (-4) + 5 + (-6) + 7 \\ &= (3 + 5 + 7) + [(-4) + (-6)] \\ &= 15 + (-10) \\ &= 5 \end{aligned}$$

For convenience, we first regroup terms with like signs together.

SELF-CHECK 1.3.4

Calculate these sums.

1. $-11 + [-5 + (+9)]$
2. $[-2 + (-8 + 5)] + (-12)$
3. $-9 + 8 + (-13) + (-11) + 22$

The ability to mentally estimate answers in an important skill that can be used in a variety of real-world applications. Example 7 illustrates how we can mentally estimate sums.

EXAMPLE 7 Estimate Then Calculate

Mentally estimate the value of each expression and then calculate the exact value.

SOLUTION

		ESTIMATED VALUE	ACTUAL VALUE
(a) $-24.937 + (-15.054)$	$-24.937 \approx -25$ $-15.054 \approx -15$	-25 $(+)\ -15$ -40	-24.937 $(+)\ -15.054$ -39.991
(b) $-9.823 + 61.150$	$61.150 \approx 61$ $-9.823 \approx -10$	61 $(+)\ -10$ 51	61.150 $(+)\ -9.823$ 51.327

We rounded each term to the nearest integer to simplify the mental estimation.

Algebraic expressions can be as simple as a single constant or a variable or can be more involved when we combine constants and variables with operations. Examples of algebraic expressions are $-\frac{7}{12}$, π, x, $x + y$, $2x + y + (-11)$, $|x| + |y|$, and $\sqrt{x + y}$.

To **evaluate an algebraic expression** for given values of the variables means to replace each variable by the specific value given for that variable and then to simplify this expression. When a constant is substituted for a variable, it is often wise to use parentheses to avoid a careless error in sign or an incorrect order of operations.

EXAMPLE 8 Evaluating an Algebraic Expression

Evaluate these expressions for $x = 7$ and $y = -5$.

SOLUTION

(a) $x + y + 4$

$$\begin{aligned} x + y + 4 &= 7 + (-5) + 4 \\ &= (7 + 4) + (-5) \\ &= 11 + (-5) \\ &= 6 \end{aligned}$$

Replace x by 7 and y by -5 and then use the commutative property of addition to reorder the terms. Add the terms with like signs before combining the terms with unlike signs. Note the use of parentheses when -5 is substituted for y.

(b) $x + (-y) + (-11)$

$$\begin{aligned} x + (-y) + (-11) &= 7 + [-(-5)] + (-11) \\ &= 7 + 5 + (-11) \\ &= 12 + (-11) \\ &= 1 \end{aligned}$$

Replace x by 7 and y by -5 and then add the terms. Note the use of parentheses when -5 is substituted for y.

(c) $|x| + |y|$

$$\begin{aligned} |x| + |y| &= |7| + |-5| \\ &= 7 + 5 \\ &= 12 \end{aligned}$$

Replace x by 7 and y by -5 and then evaluate each absolute value before adding the two terms.

Graphing calculators allow us to input both constants and variables. Most of the constants can be entered by using the numeric keys. The TI-84 Plus calculator also has variables that can store different real numbers at different times. This is illustrated in Calculator Perspective 1.3.2. To store values under any variable on a TI-84 Plus calculator, use the STO▸ and ALPHA keys, followed by any letter you need. Once the values have been stored under each variable, you may enter the expression to be evaluated.

CALCULATOR PERSPECTIVE 1.3.2 Storing Values Under Variable Names

To store $x = 7$, $y = -5$, and evaluate $x + (-y) + (-11)$ from Example 8(b) on a TI-84 Plus calculator, enter the following keystrokes:

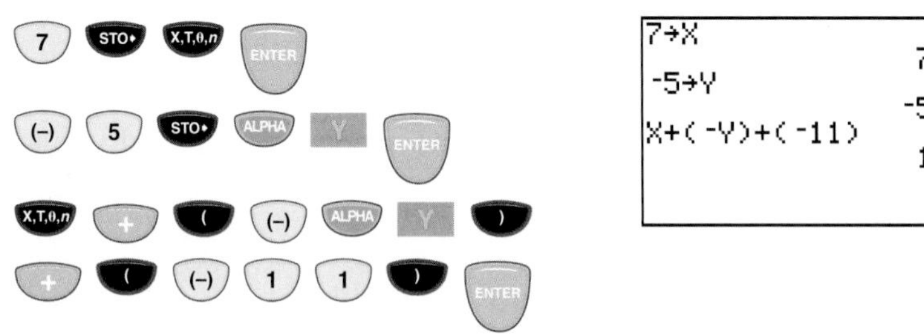

Notes: The change of sign key (-), not the subtraction key −, must be used to enter the additive inverse of a number.

The variable x is used so frequently that the X,T,θ,n key is provided to enter this variable with a single keystroke.

Many different words and phrases are commonly used in word problems and applications to indicate addition. Table 1.3.1 lists some common phrases and corresponding algebraic expressions.

It is customary to convert percents to decimal form before performing operations; $0.5\% = 0.005$.

Table 1.3.1 Phrases Used to Indicate Addition

KEY PHRASE	VERBAL EXAMPLE	ALGEBRAIC EXAMPLE
Plus	"Twelve plus eight"	$12 + 8$
Total	"The total of \$25 and \$40"	\$25 + \$40
Sum	"The sum of x and y"	$x + y$
Increased by	"An interest rate r is increased by 0.5%"	$r + 0.005$
More than	"Seven more than x"	$x + 7$

SELF-CHECK 1.3.5

1. Write an algebraic expression for "11 more than y."
2. Write an algebraic expression for "x increased by 2."
3. Evaluate $x + y + (-5)$ for $x = 6$ and $y = -9$.

You will encounter algebraic concepts in a variety of contexts in the real world. Bar graphs are frequently used in newspapers and magazines to display both positive and negative results. Sometimes it is useful to take results like those displayed in Example 9 and calculate additional information.

EXAMPLE 9 Totaling the Results of a Bar Graph

Use the midyear results shown in this bar graph to determine the total profit of the company for this 6-month period.

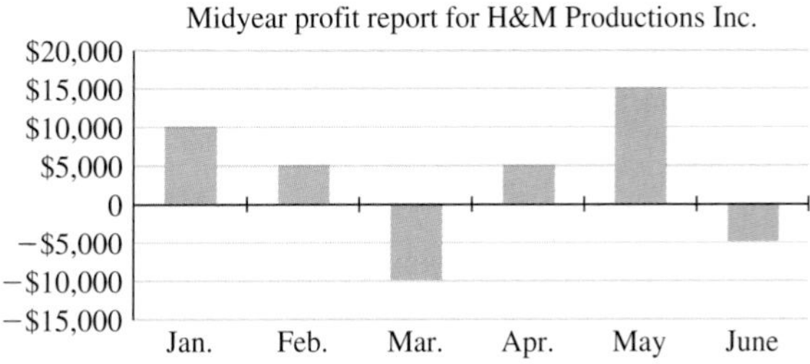

SOLUTION

Read each of the six monthly results from the table and then add these terms to obtain the 6-month total.

$10{,}000 + 5{,}000 + (-10{,}000) + 5{,}000 + 15{,}000 + (-5{,}000)$
$= (10{,}000 + 5{,}000 + 5{,}000 + 15{,}000) + [(-10{,}000) + (-5{,}000)]$
$= 35{,}000 + (-15{,}000)$
$= 20{,}000$

Answer: The total profit for the company for the first 6 months of the year was \$20,000.

SELF-CHECK ANSWERS

1.3.1

1. 33.3 2. −99.9
3. −11.1 4. 11.1

1.3.2

1. $-\frac{3}{7}$ 2. $-\frac{12}{13}$
3. $-\frac{1}{12}$ 4. $\frac{1}{12}$

```
5/8
                .625
-5/13+7/13▶Frac
                2/13
1/4+(-1/3)▶Frac
               -1/12
```

5. 0.625 6. $\frac{2}{13}$
7. $-\frac{1}{12}$

1.3.3

1. −24

```
(-9+(-7+3))+(-11
)
                 -24
```

1.3.4

1. −7 2. −17
3. −3

1.3.5

1. $y + 11$ 2. $x + 2$
3. −8

USING THE LANGUAGE AND SYMBOLISM OF MATHEMATICS 1.3

1. In the addition $27 + (-13) = 14$, 27 and −13 are called __________ or __________, and 14 is called the __________.
2. The property that says $a + b = b + a$ for all real numbers a and b is called the __________ property of addition.
3. The property that says $(a + b) + c = a + (b + c)$ for all real numbers a, b, and c is called the __________ property of addition.
4. If x and y have the same sign, then the sum $x + y$ will have the same sign as __________ terms.
5. If x and y have unlike signs, then the sum $x + y$ will have the same sign as the term with the __________ absolute value.
6. Together the __________ and __________ properties of addition allow us to add terms in any order and to obtain the same sum.
7. The sum $\frac{w}{x} + \frac{y}{x} =$ __________ for $x \neq$ __________.
8. To __________ an algebraic expression for given values of the variables means to replace each variable by the specific value given for that variable.
9. The phrase "x increased by 5" is represented algebraically by __________.
10. The phrase "the sum of x, y, and z" is represented algebraically by __________.

EXERCISES 1.3

In Exercises 1–22, calculate each sum without using a calculator.

1. a. $8 + (+5)$ b. $-8 + (+5)$
   c. $8 + (-5)$ d. $-8 + (-5)$
2. a. $9 + (+11)$ b. $-9 + (+11)$
   c. $9 + (-11)$ d. $-9 + (-11)$
3. a. $-45 + 45$ b. $31 + (-47)$
   c. $-58 + 67$ d. $-19 + (-24)$
4. a. $97 + (-97)$ b. $-50 + 23$
   c. $70 + (-58)$ d. $-31 + (-17)$
5. a. $-8 + (-7) + (-6)$
   b. $-17 + [(-11) + 9]$
6. a. $-3 + (-9) + (-2)$
   b. $[(-12) + (-9)] + 8$
7. a. $[11 + (-17)] + [(-8) + (-4)]$
   b. $(-19 + 15) + [17 + (-30)]$
8. a. $(-19 + 7) + [13 + (-19)]$
   b. $[23 + (-19)] + [41 + (-51)]$
9. $[-5 + (-6) + (-7)] + (8 + 4 + 3)$
10. $(11 + 12 + 13) + [(-10) + (-15) + (-16)]$
11. $-5 + 7 + (-4) + 8 + (-9) + 2$
12. $-3 + 5 + (-6) + (-7) + 4 + 11$
13. a. $-85.6 + 85.6$ b. $1.2 + 0 + (-1.5)$
    c. $-4.8 + 0 + 6.09$
14. a. $-77.9 + 77.9$ b. $-5.4 + 0 + 8.6$
    c. $-8.6 + 0 + 6.09$
15. a. $|-17| + |17|$ b. $-|17| + |-17|$
16. a. $|19| + |-19|$ b. $-|19| + |-19|$
17. $\frac{7}{11} + \left(-\frac{3}{11}\right)$
18. $-\frac{11}{17} + \frac{5}{17}$
19. $-\frac{2}{5} + \frac{3}{4}$
20. $-\frac{2}{3} + \frac{3}{5}$
21. $-\frac{1}{2} + \left(-\frac{1}{3}\right) + \left(-\frac{1}{4}\right)$
22. $-\frac{2}{5} + \left(-\frac{2}{4}\right) + \frac{2}{3}$
23. The property that says $(5 + 7) + 9 = (7 + 5) + 9$ is the ________ property of addition.
24. The property that says $(5 + 7) + 9 = 5 + (7 + 9)$ is the ________ property of addition.
25. Use the associative property of addition to rewrite $11 + (12 + 13)$.
26. Use the commutative property of addition to rewrite $17 + 13$.

Multiple Representations

In Exercises 27–30, represent the addition numerically, verbally, or graphically as described in each problem. (*Hint:* See the following example.)

NUMERICALLY	VERBALLY	GRAPHICALLY*
$4 + 2 = 6$	The sum of four and two is six.	+6 SUM; +4 +2; −3 −2 −1 0 1 2 3 4 5 6 7

*An arrow 4 units long is drawn to the right, starting at the origin. From the tip of this arrow, draw another arrow 2 units to the right. The sum, 6, is represented by an arrow from the origin to the tip of the last arrow.

27. Give the numerical and verbal representations for the addition shown below.

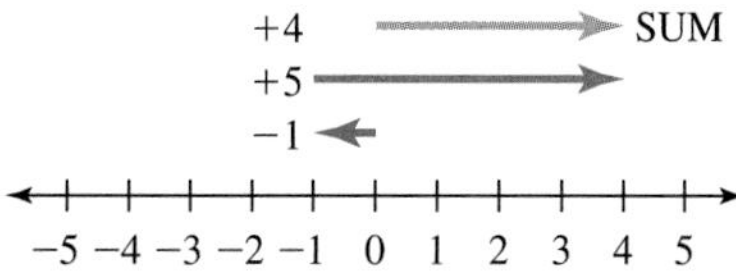

28. Represent the addition $-3 + 5 = 2$ both verbally and graphically.

29. Represent the addition $5 + (-2) = 3$ both verbally and graphically.

30. Represent the addition "The sum of two and three is five" both numerically and graphically.

In Exercises 31–34, evaluate each expression for $w = 7$, $x = -3$, and $y = -6$.

31. **a.** $|x|$ **b.** $-|x|$ **c.** $|-x|$

32. **a.** $|y|$ **b.** $-|y|$ **c.** $|-y|$

33. **a.** $w + x + y$ **b.** $-(w + x + y)$

34. **a.** $-w + (-x) + y$ **b.** $w + (-x) + (-y)$

In Exercises 35–38, use a calculator to evaluate each expression for $w = 17.5$, $x = -153.76$, and $y = -72.25$.

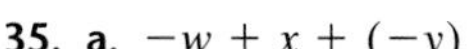

35. **a.** $-w + x + (-y)$ **b.** $\sqrt{-x}$

36. **a.** $-w + (-x) + y$ **b.** $\sqrt{-y}$

37. **a.** $w + |x|$ **b.** $-w + |y|$

38. **a.** $x + |y|$ **b.** $-y + |x|$

Perimeter

The perimeter of a geometric figure is the distance around the figure. For a rectangle or triangle, the perimeter is the sum of the lengths of all sides. In Exercises 39–42, calculate the perimeter of each figure. All units are centimeters (cm).

39. **40.**

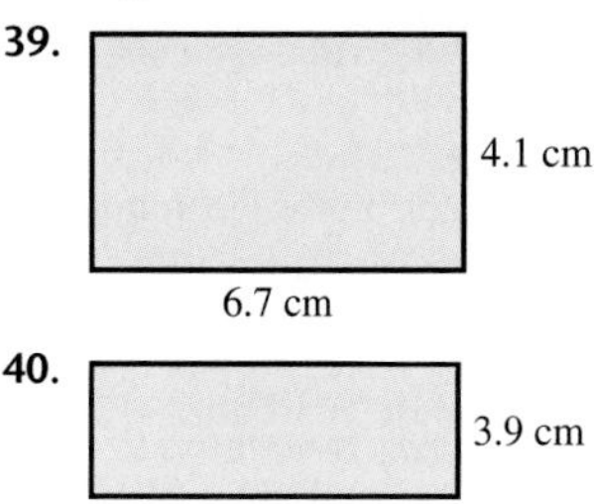

41. **42.**

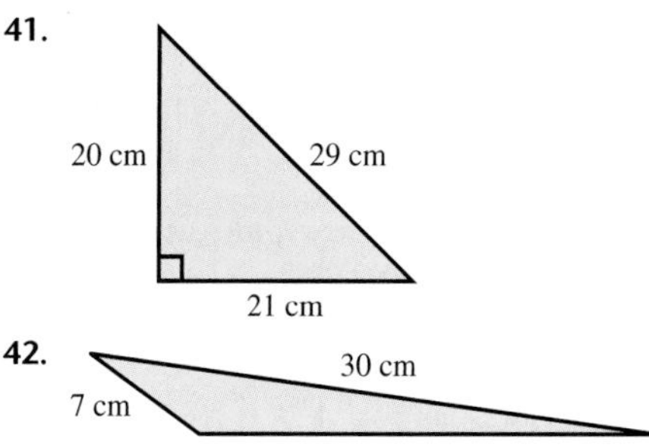

Tuition Bill

The table below gives the 2005–2006 tuition bill for an Indiana Purdue Fort Wayne University business administration student.

	HOURS	TUITION
Fall	12	\$2,251.80
Spring	14	\$2,627.10
Summer	6	\$1,125.90

43. Calculate the total number of semester hours taken by this student in 2005–2006.

44. Calculate the total tuition billed to this student for 2005–2006.

iPod Sales

45. Use this bar graph to determine the total sales of iPods for 2004.

46. Use this bar graph to determine the total sales of iPods for 2005.

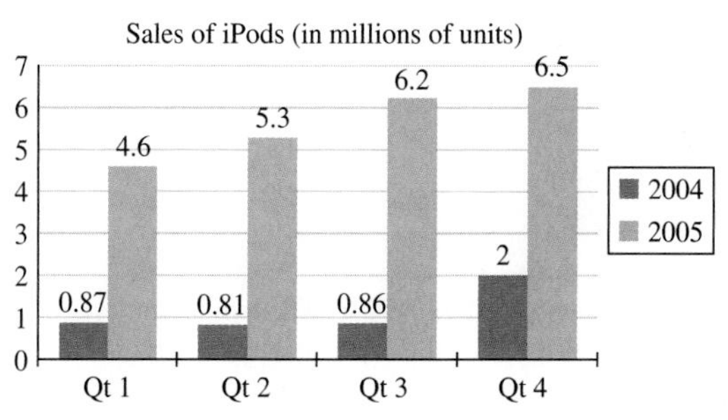

Source: www.apple.com

Land Usage in the United States

According to the U.S. Department of Agriculture, the pie chart describes how land is used in the United States in 2000. In Exercises 47–49, use the pie chart to answer each question.

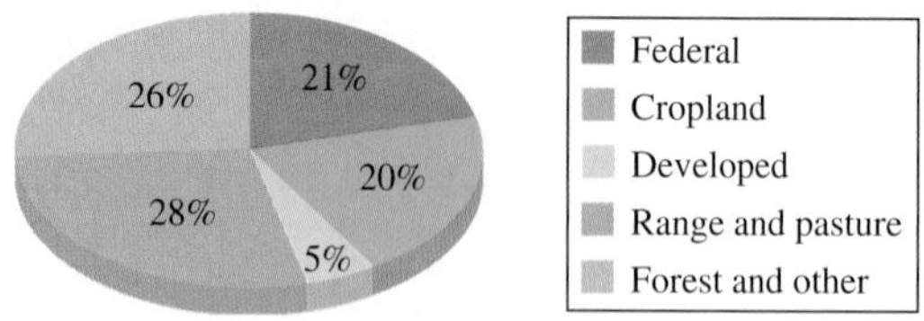

47. What total percent of the land in the United States is used for crops, pasture, or range?

48. What total percent of the land in the United States is developed or designated as federal land?

49. What percent of the land in the United States is not developed?

Dietary Guidelines

In Exercises 50–52, use the pie chart to answer the questions regarding the percent of calories from a 2,000-calorie diet that should come from each of the given sources, as recommended by the USDA Dietary Guidelines.

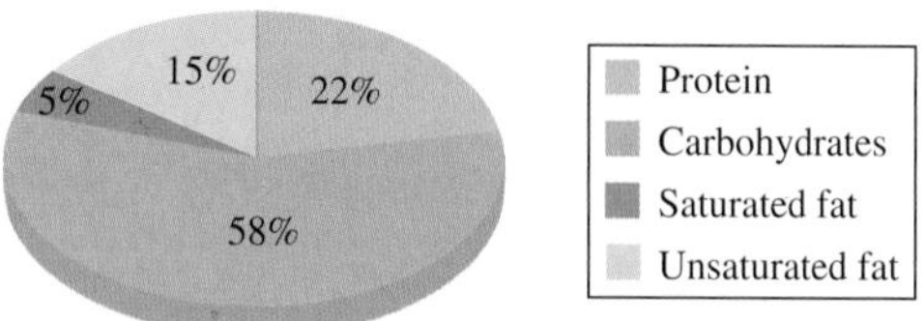

Source: *USDA 2005 Dietary Guidelines for Americans.*

50. What percent of one's diet should come from carbohydrates?

51. What percent of one's diet should come from fats, saturated and unsaturated?

52. What percent of one's diet should come from sources other than fats?

Thermometer Reading

53. A thermometer at Harper College in Palatine, Illinois, registers −12 degrees Fahrenheit (−12°F) on one February morning. The temperature rises 21 degrees that afternoon. What does the thermometer then register?

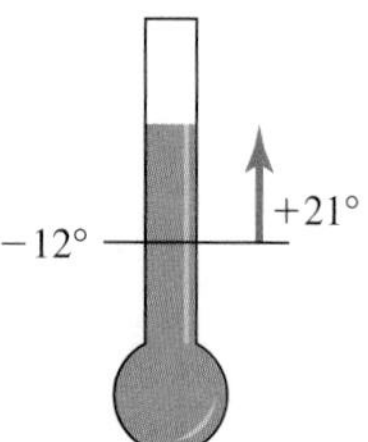

54. A thermometer registers −32°F. The temperature rises 30 degrees. What does the thermometer then register?

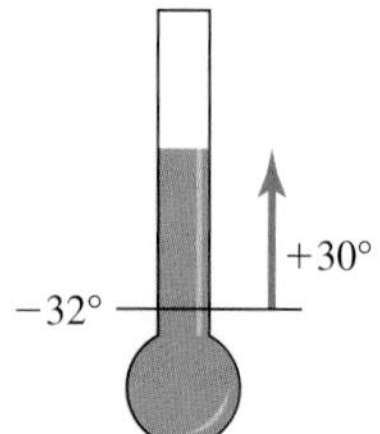

Perimeter of a Room

To order the crown molding trim for a basement recreation room, a contractor must calculate the total length of trim needed. To do this the contractor calculates the perimeter of the room. In Exercises 55 and 56, calculate the perimeter of each room. All units are given in meters (m). (*Hint:* The perimeter can be calculated by using the given measurements even though the measurements of some sides are not given.)

55.

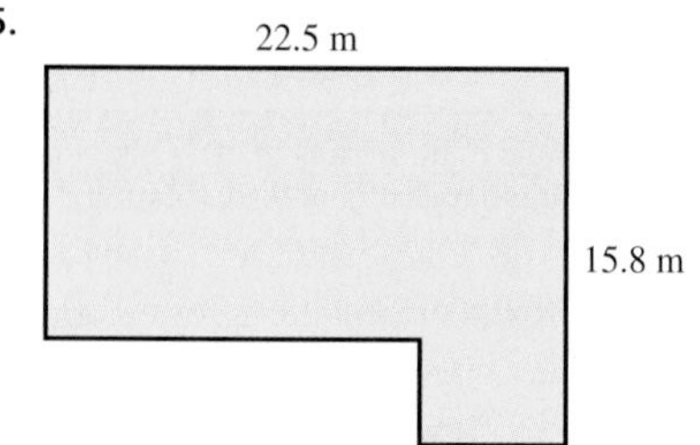

56.

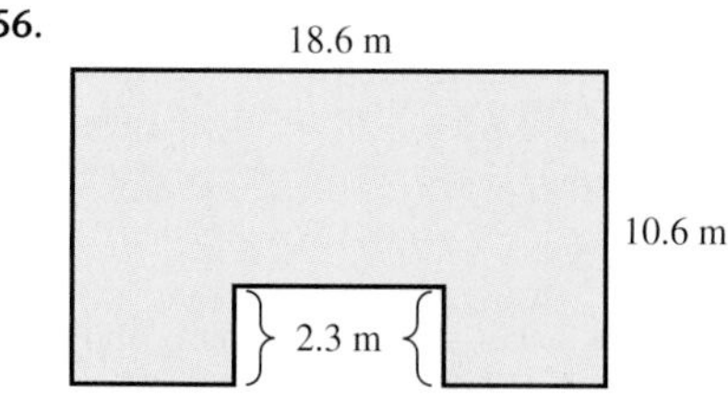

Perimeter of a Nickel Blank

57. Some of the U.S. coinage is based on the metric system, which was legalized in 1866. The nickel is 2 centimeters in diameter and has a mass of 5 grams. Nickels are stamped from a blank metal sheet. In the example shown here, one nickel is stamped from a surrounding square. Calculate the perimeter of the square.

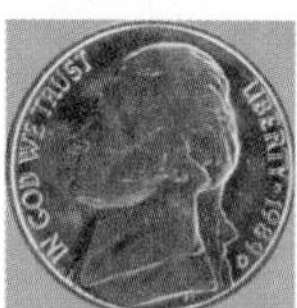

58. The blank metal sheet shown below has four nickels stamped from it. Calculate the perimeter of this rectangular sheet. (*Hint:* See Exercise 57.)

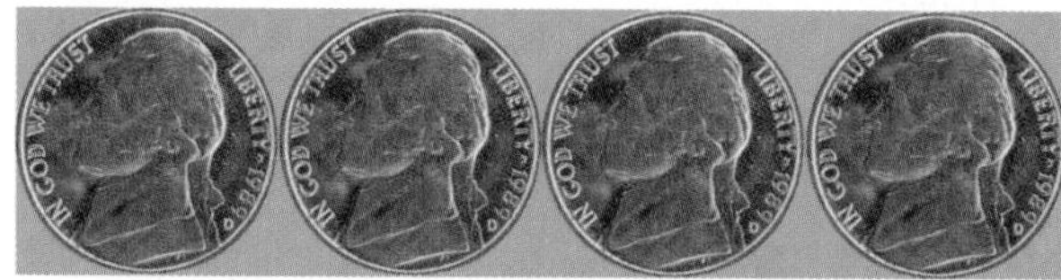

Business Profit or Loss

59. Complete the last row of the table, using a positive number to represent a total profit for the two stores and a negative number to represent a total that is a loss.

Quarterly Profit/Loss Statement for Emmett's Barbeque

	QUARTER 1	QUARTER 2	QUARTER 3	QUARTER 4
Springfield store	+$7,000	+$7,000	−$7,000	−$7,000
Washington store	+$5,000	−$5,000	+$5,000	−$5,000
Total				

Football Yardage

60. A football team started a series on its own 20-yd line. The results of five plays are given in the following table. Complete this table by giving the position of the football at the end of each play.

PLAY	RESULT	NEW YARD LINE POSITION
a. 1	6-yd gain	
b. 2	3-yd loss	
c. 3	12-yd gain	
d. 4	0 yd	
e. 5	7-yd loss	

Estimate Then Calculate

In Exercises 61–64, first mentally determine the correct sign of the sum and then use a calculator to calculate this sum. See Calculator Perspective 1.3.1 for Exercises 63 and 64.

PROBLEM	SIGN OF SUM WITHOUT CALCULATOR	SUM WITH CALCULATOR
61. $-47.6 + 53.4$		
62. $47.6 + (-53.4)$		
63. $-\frac{8}{5} + \left(-\frac{3}{4}\right)$		
64. $\frac{8}{5} + \left(-\frac{3}{4}\right)$		

Estimation Skills

In Exercises 65–72, mentally estimate each sum, and then select the most appropriate answer.

65. $6.73296 + (-7.5186)$
 a. Less than -1 b. Between -1 and 0
 c. Between 0 and 1 d. Greater than 1
66. $-13.54462 + 14.92857$
 a. Less than -1 b. Between -1 and 0
 c. Between 0 and 1 d. Greater than 1
67. $-5.40073 + (-1.29875) + (-2.4463)$
 a. Less than -10 b. Between -10 and 0
 c. Between 0 and 10 d. Greater than 10
68. $12.438 + (-13.794) + (-6.214) + (-2.847)$
 a. Less than -10 b. Between -10 and 0
 c. Between 0 and 10 d. Greater than 10
69. $\pi + (-3.15)$
 a. Less than -1 b. Between -1 and 0
 c. Between 0 and 1 d. Greater than 1
70. $\frac{1}{3} + (-0.3333)$
 a. Less than -1 b. Between -1 and 0
 c. Between 0 and 1 d. Greater than 1
71. $-35.764 + 66.056$
 a. -100 b. $+100$
 c. -30 d. 30
72. $-35.764 + (-66.056)$
 a. -100 b. $+100$
 c. -30 d. 30

Calculator Usage

In Exercises 73–76, complete each row of the table to express each rational number in both fractional form and decimal form. See Calculator Perspective 1.3.1.

	FRACTIONAL FORM	DECIMAL FORM
Example	$\frac{1}{2}$	0.5
73.		0.4
74.		0.125
75.	$\frac{3}{8}$	
76.	$\frac{7}{16}$	

Multiple Representations

In Exercises 77–80, write each verbal statement in algebraic form.

77. a. The sum of a and b equals the sum of b and a.
 b. The sum of a real number x and its additive inverse is zero.
78. a. The sum of x and three is less than eleven.
 b. The sum of negative four and y is greater than or equal to twelve.
79. a. π is not equal to 3.14.
 b. π is approximately equal to 3.14.
80. a. 0.333 is not equal to $\frac{1}{3}$.
 b. 0.333 is approximately equal to $\frac{1}{3}$.
81. If r is the rate in miles per hour (mi/h) of one airplane, write an algebraic expression for the rate of an airplane that is traveling 50 mi/h faster.
82. An investor placed part of her portfolio in a savings bond and the rest in a CD. She invested $3,000 more in the CD than in the savings bond. If P is the principal invested in the savings bond, write an algebraic expression for the amount invested in a CD.
83. If n represents an even integer, write an algebraic expression for the next integer.
84. If n represents an even integer, write an algebraic expression for the next even integer.

Group Discussion Questions

85. **Challenge Question** Determine the following sums and explain your reasoning for part **c**.
 a. $1 + (-2) + 3 + (-4)$
 b. $1 + (-2) + 3 + (-4) + 5 + (-6)$
 c. $1 + (-2) + 3 + (-4) + \cdots + 99 + (-100)$
86. **Multiple Representations** A fraction represents a part of something. The addition of fractions can be represented graphically by combining these parts as illustrated by the two examples below.

Example: $\frac{1}{5} + \frac{3}{5} = \frac{4}{5}$

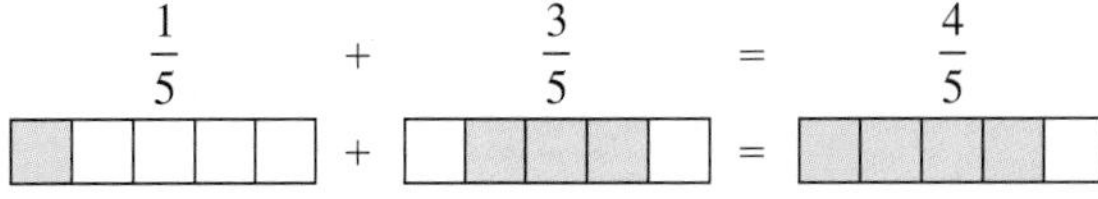

Example: $\frac{1}{3} + \frac{1}{2} = \frac{5}{6}$

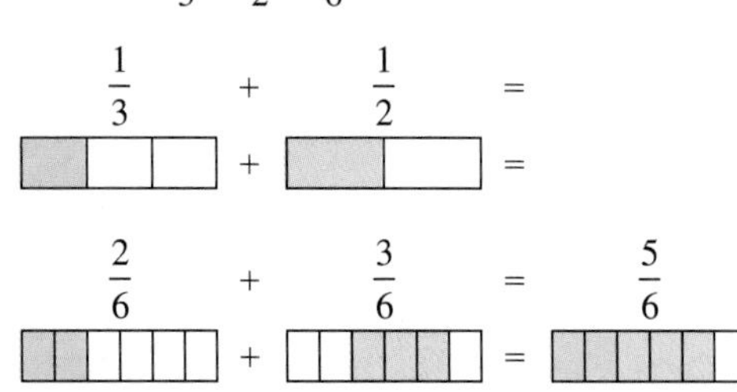

Using the format illustrated above, represent the addition of the fractions in parts **a–d.**

a. $\frac{2}{9} + \frac{5}{9}$ **b.** $\frac{1}{4} + \frac{1}{3}$

c. $\frac{1}{2} + \frac{1}{6}$ **d.** $\frac{3}{10} + \frac{7}{15}$

87. Error Analysis If you use a TI-84 Plus calculator to evaluate the expression $-7 + (-8 + (+5))$ from Example 5(a) as shown below, this will produce an error. What is the source of the syntax error?

Hint: Consider a similar expression $-7 + (-8 + (-5))$ that has -5 instead of $+5$.

88. Spreadsheet Exploration After an interview for a new position, a candidate is offered a job with a starting salary of $48,000 per year paid monthly. She is given two options:

I: An annual increase of $1,200 per year (paid at $100 per month effective at the end of each year)

II: Increases of $300 every 6 months (paid at $50 per month effective at the end of each 6-month period)

a. Complete the entries in this spreadsheet.

b. Use the information from part **a** to select the option that would pay the candidate the greater amount.

	A	B	C	D	E	F
1			Comparing Salary Options			
2		YEARLY INCREASES OF $1,200			SEMIANNUAL INCREASES OF $300	
3	MONTH	MONTHLY PAY	TOTAL PAY		MONTHLY PAY	TOTAL PAY
4	1	$4,000	$4,000		$4,000	$4,000
5	2	$4,000	$8,000		$4,000	$8,000
6	3	$4,000	$12,000		$4,000	$12,000
7	4	$4,000	$16,000		$4,000	$16,000
8	5	$4,000	$20,000		$4,000	$20,000
9	6	$4,000	$24,000		$4,000	$24,000
10	7	$4,000	$28,000		$4,050	$28,050
11	8	$4,000	$32,000		$4,050	
12	9	$4,000	$36,000		$4,050	
13	10	$4,000	$40,000		$4,050	
14	11	$4,000	$44,000		$4,050	
15	12	$4,000	$48,000		$4,050	
16	13	$4,100	$52,100		$4,100	
17	14	$4,100			$4,100	
18	15	$4,100			$4,100	
19	16	$4,100			$4,100	
20	17	$4,100			$4,100	
21	18	$4,100			$4,100	
22	19	$4,100				
23	20	$4,100				
24	21	$4,100				
25	22	$4,100				
26	23	$4,100				
27	24	$4,100				
28						

Section 1.4 Subtraction of Real Numbers

Objectives:

10. Subtract positive and negative real numbers.
11. Calculate the terms of a sequence.
12. Check a possible solution of an equation.

In this section, we will continue our presentation of operations with signed numbers by considering subtraction. Recall the terminology for the subtraction $13 - 5 = 8$; the number 13 is called the **minuend,** the number 5 is called the **subtrahend,** and 8 is called the **difference.** Table 1.4.1 lists some of the common words and phrases used in word problems and applications to indicate subtraction.

Many mathematical concepts such as slope (Section 3.1) are an examination or comparison of different changes.

Phrases Used to Indicate Subtraction

KEY PHRASE	VERBAL EXAMPLE	ALGEBRAIC EXAMPLE
Minus	"x minus y"	$x - y$
Difference	"The difference between \$12 and \$8"	\$12 − \$8
Decreased by	"An interest rate r is decreased by 0.5%"	$r - 0.005$
Less than	"Seven less than x"	$x - 7$
Change	"The change from 70° to 96°"	$96° - 70°$

Table 1.4.1

The **change** from a to b is determined by the subtraction $b - a$. In Example 1 we will build on our previous skills to examine temperature changes and to develop the rules for subtracting signed numbers.

EXAMPLE 1 Subtracting Positive and Negative Real Numbers

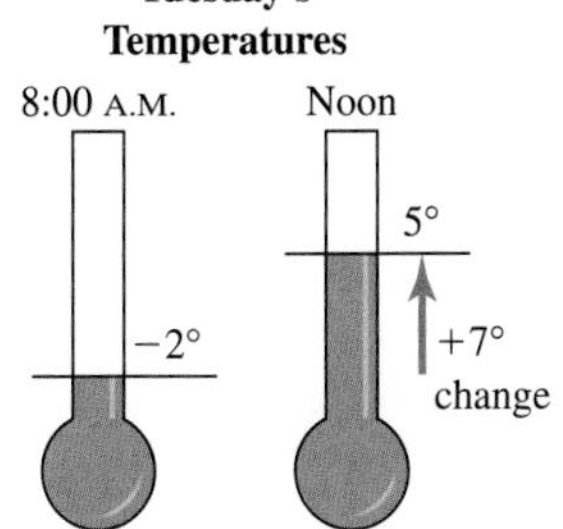

The following table gives the Celsius temperatures at 8 A.M. and noon for four consecutive days in January in Fort Wayne, Indiana. Use subtraction to calculate the change from the 8 A.M. temperature to the noon temperature.

Temperatures at Indiana University–Purdue University in Fort Wayne

	MONDAY	TUESDAY	WEDNESDAY	THURSDAY
Noon	+5°	+5°	−5°	−5°
8 A.M.	+2°	−2°	−2°	+2°
Change				

SOLUTION

The change from the 8:00 A.M. temperature to the noon temperature is

[Noon temperature] [−] [8:00 A.M. temperature]

The change from a to b is represented algebraically by $b - a$.

	MONDAY	TUESDAY	WEDNESDAY	THURSDAY
Change	$+5° - (+2°) = +3°$	$+5° - (-2°) = +7°$	$-5° - (-2°) = -3°$	$-5° - (+2°) = -7°$

We can check the answer to any subtraction problem by using addition. In Example 1:

$5 - 2 = 3$	checks because	$3 + 2 = 5$
$5 - (-2) = 7$	checks because	$7 + (-2) = 5$
$-5 - (-2) = -3$	checks because	$-3 + (-2) = -5$
$-5 - 2 = -7$	checks because	$-7 + 2 = -5$

SELF-CHECK 1.4.1

1. Determine the change in temperature from 60° to 74°.
2. Determine the change in temperature from 74° to 60°.
3. Write an algebraic expression for "x decreased by 2."
4. Write an algebraic expression for "3 less than y."

We can say that subtraction is the inverse of addition and can define subtraction as the addition of an additive inverse.

Definition of Subtraction

ALGEBRAICALLY	VERBALLY	NUMERICAL EXAMPLE
For any real numbers x and y, $x - y = x + (-y)$.	To subtract y from x, add the opposite of y to x.	$5 - 2 = 5 + (-2)$ $= 3$

When you use your calculator to perform a subtraction, there is no need to rewrite the problem as an addition problem (the calculator takes care of that for you). For pencil-and-paper calculations or mental computations, you may find it wise to either write or think of the subtraction in terms of addition, as illustrated in Example 2.

EXAMPLE 2 Subtracting Two Real Numbers

Determine the following differences.

SOLUTION

(a) $9 - 13$

$9 - 13 = 9 + (-13)$
$= -4$

Change the operation to addition, and change 13 to its additive inverse, −13. Do not change the sign of 9.

(b) $17 - (-5)$

$17 - (-5) = 17 + (5)$
$= 22$

Change the operation to addition, and change −5 to its additive inverse, 5. Do not change the sign of 17.

(c) $-8 - 4$

$-8 - 4 = -8 + (-4)$
$= -12$

Change the operation to addition, and change 4 to its additive inverse, −4. Do not change the sign of −8.

(d) $-6 - (-11)$

$-6 - (-11) = -6 + (11)$
$= 5$

Change the operation to addition, and change −11 to its additive inverse, 11. Do not change the sign of −6.

For subtraction problems it is wise to write the conversion step until your proficiency level is very high. Nonetheless, it is natural to want to use shortcuts. This can be done mentally, as illustrated in Example 3.

EXAMPLE 3 Subtracting Two Real Numbers

Mentally calculate these differences.

SOLUTION

(a) $-6 - 5$	$-6 - 5 = -11$	Think "-6 *plus* -5."
(b) $-8 - (-3)$	$-8 - (-3) = -5$	Think "-8 *plus* $+3$."
(c) $11 - (-5)$	$11 - (-5) = 16$	Think "11 *plus* $+5$."

SELF-CHECK 1.4.2

Determine the following differences.

1. $-7 - 8$ 2. $-7 - (-8)$ 3. $11 - 17$ 4. $11 - (-17)$

The procedure for subtracting fractions is summarized in the following box.

Subtracting Fractions

VERBALLY	ALGEBRAICALLY	NUMERICAL EXAMPLE
To subtract fractions with the same denominator, subtract the numerators and use the common denominator.	$\frac{a}{b} - \frac{c}{b} = \frac{a - c}{b}$ for $b \neq 0$	$\frac{5}{7} - \frac{3}{7} = \frac{5 - 3}{7} = \frac{2}{7}$
To subtract fractions with different denominators, first express each fraction in terms of a common denominator, and then subtract the numerators, using this common denominator.	$\frac{a}{b} - \frac{c}{d} = \frac{ad}{bd} - \frac{bc}{bd}$ $= \frac{ad - bc}{bd}$ for $b \neq 0$ and $d \neq 0$	$\frac{3}{5} - \frac{1}{3} = \frac{9}{15} - \frac{5}{15}$ $= \frac{9 - 5}{15}$ $= \frac{4}{15}$

Example 4 illustrates the subtraction of fractions. The group exercises at the end of this section contain a graphical perspective on subtracting fractions.

EXAMPLE 4 Subtracting Decimals and Fractions

Determine the following differences.

SOLUTION

(a) $7.7 - 4.05$ $\quad 7.7 - 4.05 = 3.65$

(b) $7.7 - (-4.05)$ $\quad 7.7 - (-4.05) = 7.7 + 4.05$
$= 11.75$

Change the operation to addition, and change -4.05 to its additive inverse 4.05. Do not change the sign of 7.7.

(c) $-\frac{7}{13} - \frac{3}{13}$

$$-\frac{7}{13} - \frac{3}{13} = \frac{-7 - 3}{13}$$
$$= \frac{-7 + (-3)}{13}$$
$$= \frac{-10}{13}$$
$$= -\frac{10}{13}$$

$\frac{a}{b} - \frac{c}{b} = \frac{a - c}{b}$ for $b \neq 0$. To subtract fractions with the same denominator, subtract the numerators and use the common denominator.

$\frac{-10}{13}$ can also be written as $-\frac{10}{13}$.

(d) $\frac{7}{10} - \frac{4}{15}$

$$\frac{7}{10} - \frac{4}{15} = \frac{7}{10} \cdot \frac{3}{3} - \frac{4}{15} \cdot \frac{2}{2}$$
$$= \frac{21}{30} - \frac{8}{30}$$
$$= \frac{21 - 8}{30}$$
$$= \frac{13}{30}$$

To subtract fractions with different denominators, first express each fraction in terms of a common denominator. In this case, express each fraction in terms of the least common denominator of 30. Then subtract like terms.

Two calculator keys often confused are the change in sign key and the subtraction key. To subtract one number from another, use the [−] key. To form the opposite of a number, use the [(−)] key. The correct use of these keys is illustrated in Calculator Perspective 1.4.1.

CALCULATOR PERSPECTIVE 1.4.1 Using Both the Change in Sign Key and the Subtraction Key

To enter 14,495 − (−282) from Example 5 on a TI-84 Plus calculator, enter the following keystrokes:

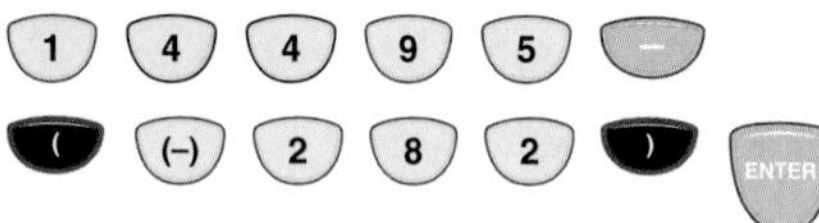

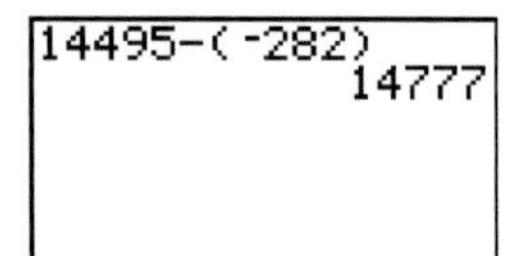

SELF-CHECK 1.4.3

Calculate these differences and then check your answers on a calculator.

1. $-37.4 - 87.8$ **2.** $-37.4 - (-87.8)$ **3.** $\frac{5}{8} - \frac{7}{9}$ **4.** $-\frac{3}{4} - \left(-\frac{7}{8}\right)$

There are many words used in the English language that imply either addition or subtraction. Many of these words are antonyms since addition and subtraction are opposites. For example, we sometimes use the word *increase* to indicate addition and the word *decrease* to indicate subtraction. In Example 5 the word *change* implies that subtraction is the appropriate operation.

EXAMPLE 5 Determining Changes in Altitude

Death Valley and Mt. Whitney

Death Valley, the lowest spot in the continental United States, is 282 ft below sea level. Mt. Whitney has an altitude of 14,495 ft above sea level. What is the change in altitude of a hiker who goes from Death Valley to the top of Mt. Whitney?

SOLUTION

The change in altitude from Death Valley to Mt. Whitney is

[Change in altitude] = [Mt. Whitney altitude] − [Death Valley altitude]

Subtract the altitude of Death Valley from the altitude of Mt. Whitney to determine the change in altitude.

$$= 14{,}495 - (-282)$$
$$= 14{,}495 + 282$$
$$= 14{,}777$$

Answer: The change in altitude from the bottom of Death Valley to the top of Mt. Whitney is 14,777 ft. If the hiker then reversed his trip, going from the top of Mt. Whitney to the bottom of Death Valley, the subtraction would be $-282 - 14{,}495 = -14{,}777$, a drop in altitude.

In Example 6 we use the data from the bar graph and subtraction to obtain information related to the company's change in performance.

EXAMPLE 6 Determining Changes from a Bar Graph

The bar graph below displays the quarterly dividends for two years for Chevron-Texaco. Use these values to determine the change in dividends for each quarter from year 1 to year 2.

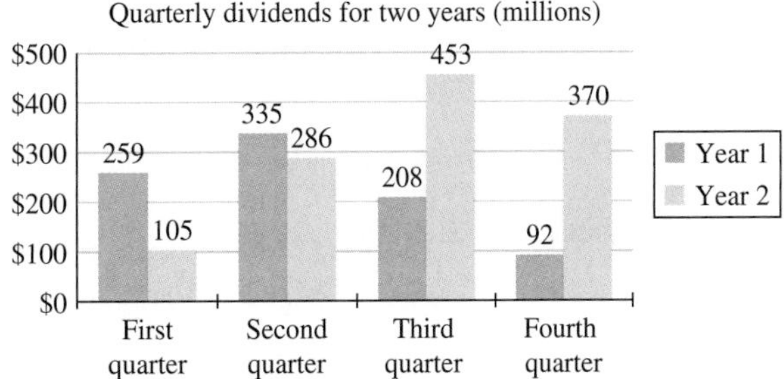

SOLUTION

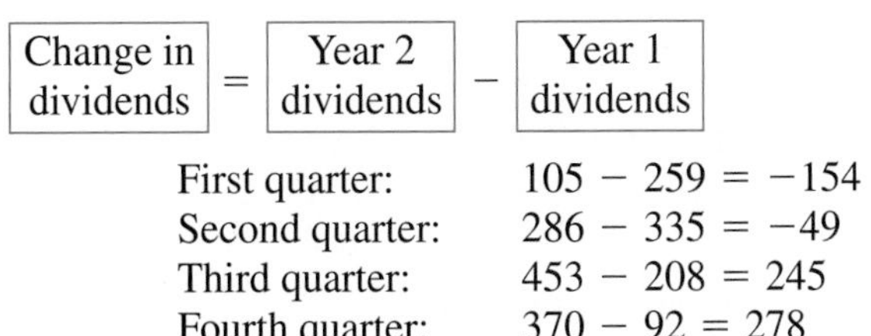

[Change in dividends] = [Year 2 dividends] − [Year 1 dividends]

First quarter: $105 - 259 = -154$
Second quarter: $286 - 335 = -49$
Third quarter: $453 - 208 = 245$
Fourth quarter: $370 - 92 = 278$

Answer: The dividends decreased \$154 million in the first quarter.
The dividends decreased \$49 million in the second quarter.
The dividends increased \$245 million in the third quarter.
The dividends increased \$278 million in the fourth quarter.

The algebraic expressions in Example 7 involve both addition and subtraction. We perform the operations as they occur from left to right.

EXAMPLE 7 Evaluating an Algebraic Expression

Evaluate these expressions for $w = -11$, $x = -9$, and $y = -5$.

SOLUTION

(a) $w + x - y$

$$\begin{aligned} w + x - y &= -11 + (-9) - (-5) \\ &= -20 + 5 \\ &= -15 \end{aligned}$$

Substitute for w, x, and y. Then perform the operations as they occur from left to right. Note the use of parentheses to enclose the values of x and y.

(b) $x + y - w$

$$\begin{aligned} x + y - w &= -9 + (-5) - (-11) \\ &= -14 + 11 \\ &= -3 \end{aligned}$$

Substitute for w, x, and y. Then perform the operations as they occur from left to right. Note the use of parentheses to enclose the values of y and w.

A Mathematical Note

The first recorded use of $=$ as the equals symbol was by Robert Recorde, an English author, in 1557. His justification for this notation was "No two things could be more equal than two straight lines."

We evaluate algebraic expressions routinely when we work with equations. An equation that contains a variable may or may not be a true statement when a value is substituted for the variable. A value of a variable that makes the equation true is called a **solution** of the equation, and this solution is said to **satisfy the equation.** The collection of all solutions is called the **solution set.**

For example, 3 is a solution of $x + 2 = 5$ because $3 + 2 = 5$; 3 satisfies the equation. To substitute a value into the equation to determine whether it satisfies the equation is referred to as a **check** of the value or as **checking a possible solution.**

At this point we are not solving equations. We are only checking possible solutions and establishing what it means to be a solution.

EXAMPLE 8 Checking Possible Solutions of an Equation

Substitute 2 and 3 for x to determine whether either is a solution of $x + 5 = x + x + 2$.

SOLUTION

We are using the notation $\stackrel{?}{=}$ as we check to determine whether the two sides of the equation are equal.

Check $x = 2$:

$$\begin{aligned} x + 5 &= x + x + 2 \\ 2 + 5 &\stackrel{?}{=} 2 + 2 + 2 \\ 7 &\stackrel{?}{=} 4 + 2 \\ 7 &\stackrel{?}{=} 6 \text{ does not check.} \end{aligned}$$

Evaluate each side of the equation for $x = 2$.

Because the statement $7 = 6$ is false, 2 does not satisfy the equation.

Answer: 2 is not a solution.

Check $x = 3$:

$$\begin{aligned} x + 5 &= x + x + 2 \\ 3 + 5 &\stackrel{?}{=} 3 + 3 + 2 \\ 8 &\stackrel{?}{=} 6 + 2 \\ 8 &\stackrel{?}{=} 8 \text{ checks.} \end{aligned}$$

Evaluate each side of the equation for $x = 3$.

Because the statement $8 = 8$ is true, 3 satisfies the equation.

Answer: 3 is a solution.

SELF-CHECK 1.4.4

1. Substitute 5 for x to determine whether it is a solution of $x - 1 - x = x - 6$.
2. Substitute 6 for x to determine whether it is a solution of $x - 1 - x = x - 6$.

We can use a calculator to assist us in checking possible solutions of an equation. One way to do this is to store the value being checked as the variable x. Then evaluate the expression on the left side of the equation and the expression on the right side of the equation. If the stored value of x causes both expressions to have the same value, then that stored value is a solution. This is illustrated in Calculator Perspective 1.4.2.

CALCULATOR PERSPECTIVE 1.4.2 Checking Possible Solutions of an Equation

To check both 2 and 3 in the equation $x + 5 = x + x + 2$ from Example 8 on a TI-84 Plus calculator, enter the following keystrokes:

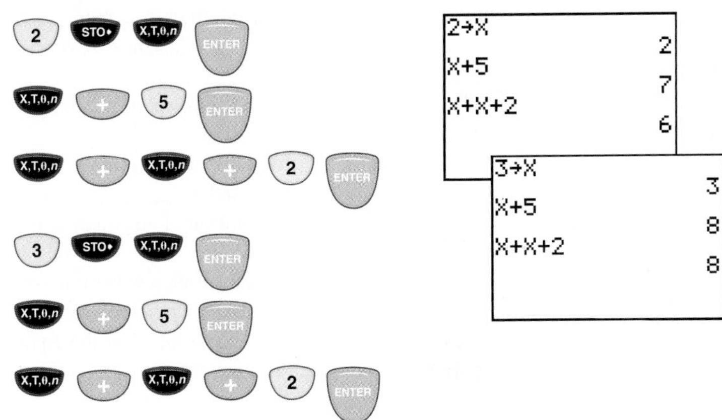

Note: The value $x = 2$ does not check because the two expressions are not equal. Both expressions are equal for $x = 3$, so this value checks.

a_1 denotes the first term and a_2 denotes the second term. Note the close similarity of this subscript notation to the notation adopted by spreadsheets where A1 denotes the first cell in column A and A2 denotes the second cell in column A.

A basic human characteristic is curiosity and the search for patterns. In everyday language we use the term *sequence* to indicate a succession of events. Mathematically, a **sequence** is an ordered set of numbers with a first number, a second number, a third number, etc. **Subscript notation** often is used to denote the terms of a sequence: a_1, a_2, and a_n. These terms are read *a sub one*, *a sub two*, and *a sub n*, respectively. If a sequence follows a predictable pattern, then we may be able to describe this pattern with a formula for a_n. This is illustrated in Example 9.

EXAMPLE 9 Calculating the Terms of a Sequence

Use the formula $a_n = n - 3$ to calculate the first five terms of this sequence: a_1, a_2, a_3, a_4, and a_5.

SOLUTION

$a_n = n - 3$

$a_1 = 1 - 3 = -2$ Replace n by 1 to calculate a_1.

$a_2 = 2 - 3 = -1$ Replace n by 2 to calculate a_2.

$a_3 = 3 - 3 = 0$ Replace n by 3 to calculate a_3.

$a_4 = 4 - 3 = 1$ Replace n by 4 to calculate a_4.

$a_5 = 5 - 3 = 2$ Replace n by 5 to calculate a_5.

Answer: The first five terms are $-2, -1, 0, 1$, and 2.

SELF-CHECK 1.4.5

1. Calculate the first five terms of the sequence with $a_n = n - 9$.

SELF-CHECK ANSWERS

1.4.1
1. 14° 2. −14°
3. $x - 2$ 4. $y - 3$

1.4.2
1. −15 2. 1
3. −6 4. 28

1.4.3
1. −125.2 2. 50.4
3. $-\frac{11}{72}$ 4. $\frac{1}{8}$

1.4.4
1. 5 is a solution
2. 6 is not a solution

1.4.5
1. −8, −7, −6, −5, and −4

USING THE LANGUAGE AND SYMBOLISM OF MATHEMATICS 1.4

1. In the subtraction $11 - 4 = 7$, 11 is called the minuend, 4 is called the ___________, and 7 is called the ___________.
2. The phrase "x decreased by 5" is represented algebraically by ___________.
3. The phrase "17 less than w" is represented algebraically by ___________.
4. The difference $\frac{w}{x} - \frac{y}{x} =$ ___________ for $x \neq$ ___________.
5. The change from x_1 to x_2 is represented algebraically by ___________.
6. The change from y_1 to y_2 is represented algebraically by ___________.
7. The notation a_1 is read ___________ ___________ ___________, and a_n is read ___________ ___________ ___________.
8. An ordered set of numbers with a first term, a second term, and so on is called a ___________.
9. A value of a variable that makes an equation a true statement is called a ___________ of an equation.
10. A solution of an equation is said to ___________ the equation.
11. To substitute a value into an equation to determine whether it makes the equation true is referred to as ___________ a possible solution of the equation.
12. The collection of all solutions of an equation is called the ___________ set.

EXERCISES 1.4

In Exercises 1–16, calculate the value of each expression without using a calculator.

1. a. $7 - 13$ b. $7 - (-13)$
   c. $-7 - 13$ d. $-7 - (-13)$
2. a. $-15 - (-8)$ b. $15 - 8$
   c. $-15 - 8$ d. $15 - (-8)$
3. a. $-22 - (-15)$ b. $22 - 15$
   c. $-22 - 15$ d. $22 - (-15)$
4. a. $-11 - (-17)$ b. $-11 - 17$
   c. $11 - 17$ d. $11 - (-17)$
5. a. $-\frac{2}{3} - \left(-\frac{2}{3}\right)$ b. $-\frac{2}{3} - \frac{5}{3}$
6. a. $-\frac{3}{5} - \left(-\frac{3}{5}\right)$ b. $-\frac{3}{7} - \frac{2}{7}$
7. a. $\frac{3}{5} - \left(-\frac{5}{6}\right)$ b. $-\frac{3}{5} - \left(-\frac{5}{6}\right)$
8. a. $\frac{4}{7} - \left(-\frac{1}{2}\right)$ b. $-\frac{4}{7} - \left(-\frac{1}{2}\right)$
9. a. $-18 - 18 - 21$ b. $18 - 18 - (-21)$
10. a. $-15 - 17 - (-4)$ b. $15 - 17 - 4$
11. a. $-7 - 8 - 9 - 10$ b. $-7 - (-8) - (-9) - 10$
12. a. $-6 - 7 - 13 - 14$ b. $-6 - 7 - (-13) - (-14)$
13. a. $43.8 - 43.8$ b. $-12{,}348.56 - 0$
14. a. $27.9 - 27.9$ b. $5.4794 - 0$
15. a. $\sqrt{100} - \sqrt{64}$ b. $|100| + |-64|$
16. a. $\sqrt{169} - \sqrt{144}$ b. $|169| + |-144|$

In Exercises 17–20, evaluate each expression without using a calculator. Use the values of $w = -3$, $x = -5$, and $y = -8$.

17. a. $w + x$ b. $w - y$
18. a. $w - x$ b. $w + y$
19. a. $w + x - y$ b. $w - x - y$
20. a. $w - x + y$ b. $-w + x - y$

In Exercises 21–26, use a calculator to evaluate each expression. Use the values of $w = 28.09$, $x = -50.41$, and $y = -19.36$.

21. $w + x + y$
22. $w - x + y$
23. $\sqrt{w} - x$
24. $\sqrt{-x} - y$
25. $\sqrt{-y} + |x| + 75.12$
26. $|x| + \sqrt{w} - 17.45$

In Exercises 27–30, check both $x = 4$ and $x = 5$ to determine whether either is a solution of the given equations.

27. a. $x + 7 = 12$ b. $-x - 7 = -11$
28. a. $x - 10 = 6$ b. $-x + 3 = -2$

29. **a.** $x - 1 = 7 - x$ **b.** $x + x - 3 = x + 2$

30. **a.** $x - 2 = 8 - x$ **b.** $x - 2 = 6 - x$

In Exercises 31–34, calculate the first five terms of each sequence.

31. $a_n = n - 4$ **32.** $a_n = n - 7$

33. $a_n = 3 - n$ **34.** $a_n = 11 - n$

35. **Length of a Fence** A fencing company has been contracted to construct a fence on three sides of the rectangular lot shown below. The contractor has 120 ft of chain-link fencing available for this job. If the width of the fenced area is 25 ft, determine the amount of fencing available for the length of this rectangle.

36. **Material Wasted in Coin Production** The area of a nickel is π cm^2 (square centimeters). Two nickels are stamped from a blank sheet with an area of 8 cm^2 as shown in the figure. Determine the area of the metal sheet that is wasted by this stamping process.

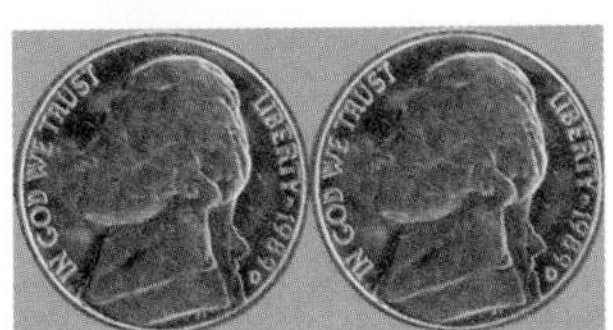

Tuition Bill

The table below gives 2001–2002 tuition comparisons for Austin Community College and the University of Texas at Austin.

Semester Hours	Tuition at ACC	Tuition at UTA
6	$192	$ 504
12	384	1,008
18	576	1,512

37. How much can a student save by taking 12 semester hours at ACC rather than at UTA?

38. How much can a student save by taking 18 semester hours at ACC rather than at UTA?

Earnings Reports

39. Determine the change in net income for Ford Motor Company from 2001 to 2003.

Annual net income for Ford Motor Company
(in millions of dollars)

$1,000 | $0 | −$1,000 | −$2,000 | −$3,000 | −$4,000 | −$5,000 | −$6,000

2001: −5,453 | 2002: −980 | 2003: 495

Source: http://www.ford.com

40. Determine the change in sales for the fourth quarter from 1997 to 1998.

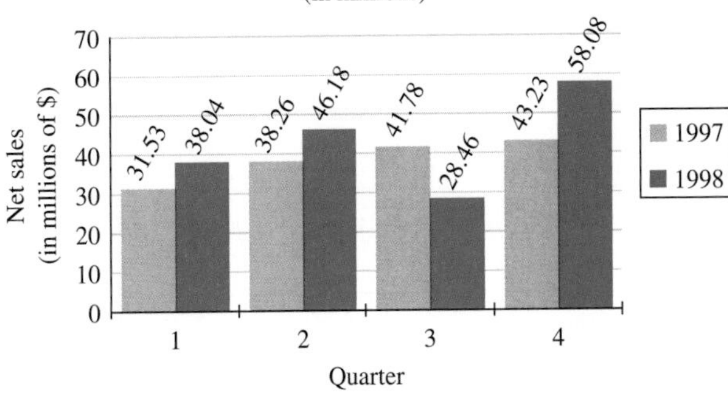

In Exercises 41 and 42, use the results shown in the pie charts to determine the value of the missing entry.

41. **U.S. Auto Sales**

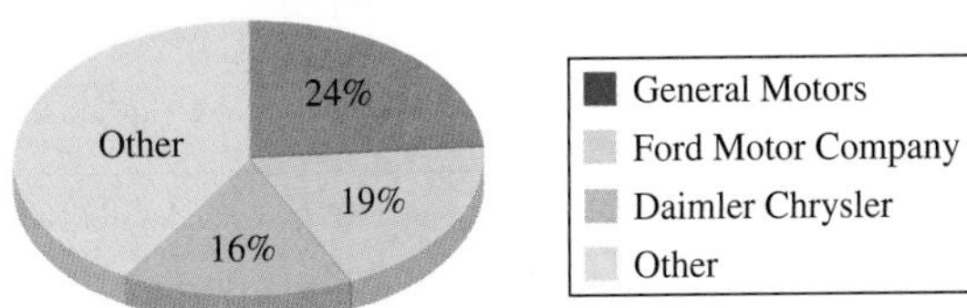

Source: *USA Today*, 3/1/2006.

42. **Snack Food Sales**

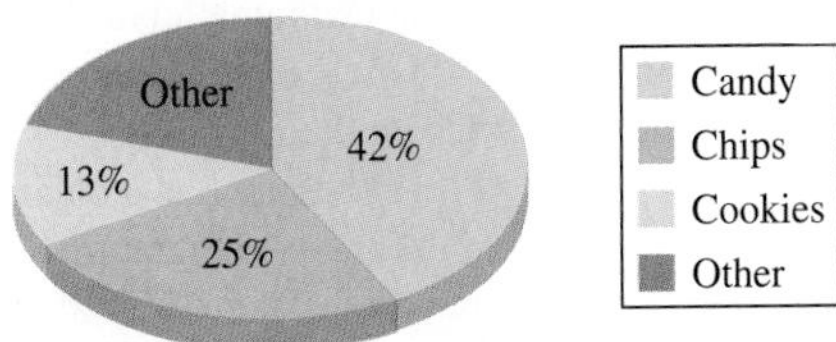

Source: *USA Today*, 5/11/2004.

Change in Thermometer Readings

In Exercises 43 and 44, determine the change in Fahrenheit temperature from 8 A.M. to noon.

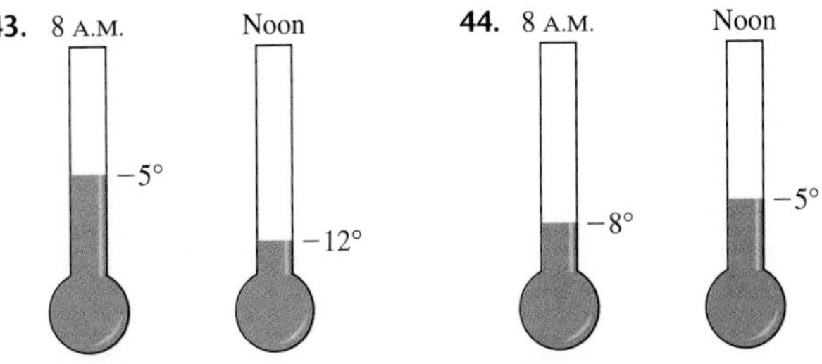

In Exercises 45–50, find each temperature change.

45. **a.** Find the change in temperature from 2° to 8°.
b. Find the change in temperature from 8° to 2°.

46. **a.** Find the change in temperature from 7° to 3°.
b. Find the change in temperature from 3° to 7°.

47. **a.** Find the change in temperature from −3° to 6°.
b. Find the change in temperature from 6° to −3°.

48. **a.** Find the change in temperature from −2° to 3°.
b. Find the change in temperature from 3° to −2°.

49. **a.** Find the change in temperature from −5° to −9°.
b. Find the change in temperature from −9° to −5°.

50. a. Find the change in temperature from $-7°$ to $-1°$.
b. Find the change in temperature from $-1°$ to $-7°$.

51. **College Graduation Rates** Complete the last row of this table to determine the change in graduation rates from 1990 to 2002.

College Graduation Rates

	SELECTIVE PUBLIC COLLEGES	OPEN-ADMISSION COLLEGES
2002	74.6%	25.8%
1990	61.0%	42.0%
Change		

Source: *USA Today*, 2/3/2005.

52. **Television Viewing Habits** Complete the last row of this table to determine the change in viewing habits from 2000 to 2004.

News Source for Ages 18 to 29

	NETWORK NEWS	LOCAL NEWS	TV NEWS MAGAZINES	COMEDY TV SHOWS
2004	22%	29%	26%	21%
2000	39%	42%	18%	4%
Change				

Source: The Pew Research Center for the People & the Press, Jan. 11, 2004.

53. **Football Yardage** A football team lost 7 yd on their first play, gained 10 yd on the second play, and gained $2\frac{1}{2}$ yd on the third play. What was the net yardage for the three plays?

54. **Age Difference** Two of the greatest mathematicians in history were Archimedes, born in 287 B.C., and Newton, born in 1642 A.D. How many years separate their births?

55. **Credit Card Balance** A credit card account had a balance of \$143.12 at the start of July. On the July statement there was a credit of \$37.16 for items returned. There were also purchases of \$14.83, \$17.89, and \$34.11. The payment made in July was \$75. What was the new balance?

56. **Submarine Depth** A submarine was submerged to a depth of 85 ft. It ascended 40 ft, descended 25 ft, and then descended another 50 ft. What was its new depth?

57. **Length of a Steel Rod** The length of a steel rod was 79.4506 m when first measured. As the temperature rose, the bar expanded by 0.0036 m. Then the temperature dropped, and the bar contracted by 0.0052 m. What is the resulting length of the bar?

58. **Water Storage** The reservoir for Charleston started the month with 15,000 acre feet of water (An acre foot of water is a volume of water determined by one acre covered to a depth of one foot). Due to evaporation and usage, the water supply decreased 3,000 acre feet during the month. On the last day of the month the runoff from a 3-in rain increased the supply by 1,500 acre feet. What is the volume of water in the reservoir at the end of the month?

Estimate Then Calculate

In Exercises 59–64, mentally determine the correct sign of the difference and then use a calculator to calculate this difference. Approximate 63 and 64 to the nearest thousandth.

PROBLEM	SIGN OF DIFFERENCE (WITHOUT CALCULATOR)	DIFFERENCE (WITH CALCULATOR)
59. $-78.43 - 97.5$		
60. $-78.43 - (-97.5)$		
61. $\frac{7}{3} - \frac{3}{8}$		
62. $-\frac{7}{3} - \left(-\frac{3}{8}\right)$		
63. $\sqrt{71} - \sqrt{17}$		
64. $-\sqrt{83} - (-\sqrt{77})$		

In Exercises 65–70, mentally estimate the value of each expression and then select the most appropriate answer.

65. $12.9854 - 13.0168$
a. Less than -1
b. Between -1 and 0
c. Between 0 and 1
d. Greater than 1

66. $-0.65438 - 0.360583$
a. Less than -1
b. Between -1 and 0
c. Between 0 and 1
d. Greater than 1

67. $0.666 - \frac{2}{3}$
a. Less than -1
b. Between -1 and 0
c. Between 0 and 1
d. Greater than 1

68. $3.14 - \pi$
a. Less than -1
b. Between -1 and 0
c. Between 0 and 1
d. Greater than 1

69. $1 - 17.89 - 82.416$
a. -100
b. 100
c. -65
d. 65

70. $-40.23 - (60.28 - 70.41)$
a. -50
b. -30
c. -10
d. 20

Multiple Representations

In Exercises 71–76, write each verbal statement in algebraic form.

71. x decreased by eight equals z.
72. The change from y to z equals x.
73. The difference of x minus y equals seven.
74. The difference of negative seven minus w equals x plus five.
75. The absolute value of x minus the absolute value of the opposite of y equals eleven.
76. The square root of x minus the square root of y is less than six.
77. An investor placed part of her portfolio in a savings bond and the rest in a CD. She invested \$2,000 less in the CD than in the savings bond. If P is the principal invested in the savings bond, write an expression for the amount invested in a CD.
78. An investor has \$10,000 to invest. If A represents the amount invested in stocks, write an expression for the amount remaining for other investments.
79. If n represents an odd integer, write an expression for
    a. the previous odd integer
    b. the next odd integer
80. If n represents an even integer, write an expression for
    a. the previous even integer
    b. the next even integer
81. If r is the rate in miles per hour of one airplane, write an expression for the rate of an airplane that is traveling
    a. 40 mi/h slower
    b. 40 mi/h faster
82. If r is the speed in miles per hour of an airplane flying in still air, write an expression for the groundspeed of this airplane when it encounters a
    a. 20 mi/h headwind
    b. 20 mi/h tailwind

Group Discussion Questions

83. **Communicating Mathematically**
    a. Explain the difference between the meanings of the symbol + in the expressions $-7-(+4)$ and $-7+(-4)$.
    b. Explain the difference between the meanings of the symbol − in the expressions $8-3$ and $8+(-3)$.
    c. Explain the difference between the (−) key and the (−) key on a calculator.
84. **Communicating Mathematically**
    a. Make a list of words or phrases that are often used to indicate the operation of addition.
    b. Make a list of words or phrases that are often used to indicate the operation of subtraction.
85. **Multiple Representations** A fraction represents a part of something. The subtraction of fractions can be represented graphically by removing some of these parts as illustrated by the two examples below.

Example: $\frac{5}{7}-\frac{3}{7}=\frac{2}{7}$

$\frac{5}{7} - \frac{3}{7} = \frac{2}{7}$

Example: $\frac{1}{2}-\frac{1}{3}=\frac{1}{6}$

$\frac{1}{2} - \frac{1}{3} =$

$\frac{3}{6} - \frac{2}{6} = \frac{1}{6}$

Using the format illustrated above, represent the subtraction of the fractions in parts **a–c**.

a. $\frac{7}{9}-\frac{2}{9}$
b. $\frac{7}{12}-\frac{1}{6}$
c. $\frac{2}{3}-\frac{1}{2}$

86. **Error Analysis** If you use a TI-84 Plus calculator to evaluate the expression $-7+(-8+5)$ as shown below, this will produce an error. What is the source of the syntax error?

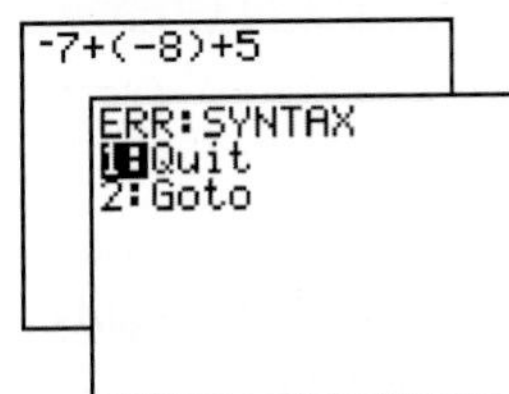

87. **Spreadsheet Exploration** The perimeter of a triangular truss is based on the lengths of the sides A, B, and C. Complete the missing entries in this spreadsheet which gives the dimensions in feet for three different trusses.

	A	B	C	D
1	Perimeter of Triangular Truss			
2	A	B	C	Perimeter
3	3	4	6	13
4	4	5	8	
5	7	6		18
6				

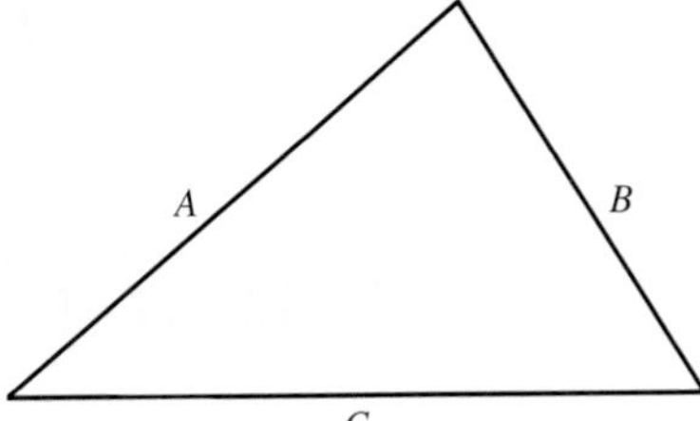

Section 1.5 Multiplication of Real Numbers and Natural Number Exponents

Objectives:

13. Multiply positive and negative real numbers.
14. Use the commutative and associative properties of multiplication.
15. Use natural number exponents.
16. Use algebraic formulas.

This section will examine the multiplication of signed numbers and the use of natural number exponents. Recall the terminology for the multiplication $5 \times 8 = 40$; both 5 and 8 are called **factors** of the **product** 40. The operation of multiplication is seldom represented by the times sign $\times$ in algebra, in part, due to the possible confusion with the variable x. Instead, we usually use implied multiplication, the dot, or parentheses as shown in the following box. The symbol * is used more on calculator displays and in computer programs than in written mathematics.

A Mathematical Note

The symbol $\times$ for multiplication was introduced by an English mathematician William Oughtred (1574–1660).

Notations for the Product of the Factors x and y

xy $\quad x \cdot y$ $\quad (x)(y)$ $\quad x(y)$ $\quad (x)y$ $\quad x * y$

Some of the words and phrases used in word problems and applications to indicate multiplication are given in Table 1.5.1.

Phrases Used to Indicate Multiplication

KEY PHRASE	VERBAL EXAMPLE	ALGEBRAIC EXAMPLE
Times	"x times y"	xy
Product	"The product of 5 and 7"	$5 \cdot 7$
Multiplied by	"The rate r is multiplied by the time t"	rt
Twenty percent of	"Twenty percent of x"	$0.20x$
Twice	"Twice y"	$2y$
Double	"Double the price P"	$2P$
Triple	"Triple the coupon value V"	$3V$

Table 1.5.1

EXAMPLE 1 Evaluating an Algebraic Expression

Evaluate each of these products for $x = 4$, $y = 5$, and $z = 6$.

SOLUTION

(a) xy $\quad xy = (4)(5) = 20$ $\quad$ Substitute 4 for x and 5 for y in the implied product of x and y.

(b) yx $\quad yx = (5)(4) = 20$ $\quad$ Compare the product of yx to xy in part (a).

(c) $(xy)(z)$ $\quad (xy)(z) = (4 \cdot 5)(6)$ $\quad$ Substitute 4 for x, 5 for y, and 6 for z in this product of three factors.

$$= 20(6)$$
$$= 120$$

(d) $x(yz)$ $\quad x(yz) = (4)(5 \cdot 6)$ $\quad$ Compare this result to the result in part (c).

$$= 4(30)$$
$$= 120$$

Using the correct mathematical vocabulary gives you an advantage throughout your course work. Remember that factors are multiplied and terms are added.

Example 1 illustrates that multiplication is both commutative and associative. Together, the commutative and associative properties of multiplication allow us to multiply factors in any order we desire.

Properties of Multiplication

	ALGEBRAICALLY	VERBALLY	NUMERICAL EXAMPLE
Commutative property of multiplication	$ab = ba$	The product of two factors in either order is the same.	$4 \cdot 5 = 5 \cdot 4$
Associative property of multiplication	$(ab)(c) = (a)(bc)$	Factors can be regrouped without changing the product.	$(4 \cdot 5)(6) = (4)(5 \cdot 6)$

The rules for multiplying signed numbers can be developed by considering the following four cases. Also recall that the product of 0 and any other factor is 0.

Case 1: *A Positive Factor Times a Positive Factor*
We are already familiar with this case. The product of two positive factors is positive. For example, $3 \cdot 4 = 12$.

Case 2: *A Positive Factor Times a Negative Factor*
Because multiplication can be interpreted as repeated addition, $3(-4)$ can be interpreted as $(-4) + (-4) + (-4) = -12$. Thus the product of a positive factor and a negative factor is negative.

Case 3: *A Negative Factor Times a Positive Factor*
Because multiplication is commutative, $(-4)(3)$ equals $3(-4) = -12$. Thus the product of a negative factor and a positive factor is negative.

Case 4: *A Negative Factor Times a Negative Factor*
Examine the pattern illustrated by the following products:

$$3(-3) = -9$$
$$2(-3) = -6$$
$$1(-3) = -3$$
$$0(-3) = 0$$
$$-1(-3) = ?$$

As the factor to the left decreases by 1, the product to the right increases by 3.

To continue this pattern, $(-1)(-3) = 3$. This suggests that the product of two negative factors is positive. For example, $(-3)(-4) = 12$.

These four cases can be summarized by examining the product of like signs and unlike signs.

Multiplication of Two Real Numbers

	VERBALLY	NUMERICAL EXAMPLE
Like signs:	Multiply the absolute values of the two factors and use a positive sign for the product.	$(+3)(+6) = +18$ $(-3)(-6) = +18$
Unlike signs:	Multiply the absolute values of the two factors and use a negative sign for the product.	$(+3)(-6) = -18$ $(-3)(+6) = -18$
Zero factor:	The product of 0 and any other factor is 0.	$(3)(0) = 0$ $(0)(-6) = 0$

Example 2 illustrates the multiplication of two real numbers, including the multiplication of fractions. The procedure for multiplying fractions is summarized in the following box.

Multiplication of Fractions

VERBALLY	ALGEBRAICALLY	NUMERICAL EXAMPLE
To multiply two fractions, multiply the numerators and multiply the denominators.	$\frac{a}{b} \cdot \frac{c}{d} = \frac{ac}{bd}$ for $b \neq 0$ and $d \neq 0$	$\frac{2}{7} \cdot \frac{3}{5} = \frac{2 \cdot 3}{7 \cdot 5} = \frac{6}{35}$

EXAMPLE 2 Multiplying Two Real Numbers

Determine the following products.

SOLUTION

(a) $5(-9)$

$5(-9) = -45$

The product of unlike signs is negative.

(b) $-6(7)$

$-6(7) = -42$

The product of unlike signs is negative.

(c) $\left(\frac{-2}{3}\right)\left(\frac{-4}{5}\right)$

$$\left(\frac{-2}{3}\right)\left(\frac{-4}{5}\right) = \frac{(-2)(-4)}{3 \cdot 5} = \frac{8}{15}$$

$\frac{a}{b} \cdot \frac{c}{d} = \frac{a \cdot c}{b \cdot d}$ for $b \neq 0$ and $d \neq 0$. To multiply two fractions, multiply the numerators and multiply the denominators. The product of like signs is positive.

(d) $\left(\frac{-5}{6}\right)\left(\frac{7}{11}\right)$

$$\left(\frac{-5}{6}\right)\left(\frac{7}{11}\right) = \frac{-5 \cdot 7}{6 \cdot 11} = \frac{-35}{66} = -\frac{35}{66}$$

The product of unlike signs is negative. The answer is usually written in the form $-\frac{35}{66}$.

(e) $(-17)(0)$

$(-17)(0) = 0$

The product of 0 and any other factor is 0.

SELF-CHECK 1.5.1

1. Calculate the product of -5 and 17.
2. Multiply -11 times -9.
3. Calculate 25% of 80.
4. Write an algebraic expression for "four times w."
5. The radius r of a circle is tripled. Write an algebraic expression for the new radius.

Two notations for indicating the product of two factors are illustrated in Calculator Perspective 1.5.1.

CALCULATOR PERSPECTIVE 1.5.1 Using the Multiplication Symbol on a Calculator

To evaluate $-6(7)$ from Example 2(b) on a TI-84 Plus calculator, enter the following keystrokes:

Note: In the first expression the $*$ symbol indicates multiplication. In the second expression the multiplication is implied.

SELF-CHECK 1.5.2

Mentally determine the sign of each product, and then use a calculator to calculate the product.

1. $(-45.7)(0.13)$ **2.** $17.6(-4.8)$

3. $\left(-\frac{2}{3}\right)\left(-\frac{5}{7}\right)$ **4.** $0\left(-\frac{8}{19}\right)$

The product of several nonzero factors is either positive or negative, as illustrated in Table 1.5.2 and as summarized in the box following the table. When a product is positive, it is customary not to write the plus symbol. For example, in the table $+2$ is usually written as just 2.

Table 1.5.2 Sign Pattern for Negative Factors

NUMBER OF NEGATIVE FACTORS	PRODUCT	SIGN OF PRODUCT
2 (even)	$(-1)(-2) = +2$	Positive
3 (odd)	$(-1)(-2)(-3) = -6$	Negative
4 (even)	$(-1)(-2)(-3)(-4) = +24$	Positive
5 (odd)	$(-1)(-2)(-3)(-4)(-5) = -120$	Negative
6 (even)	$(-1)(-2)(-3)(-4)(-5)(-6) = +720$	Positive

Product of Negative Factors

Even number: The product is positive if the number of negative factors is even.

Odd number: The product is negative if the number of negative factors is odd.

Thus to multiply several signed factors, it may be most efficient to first determine the sign of the product and then multiply the absolute values of the factors. The sign pattern of negative factors is reexamined when exponents are covered.

EXAMPLE 3 Determining Products of Negative Factors

Determine the following products.

SOLUTION

(a) $(-2)(3)(-5)(-6)$

$(-2)(3)(-5)(-6) = -180$

The product is negative because the number of negative factors is odd (3).

(b) $(-17)(33)(0)(-45)$

$(-17)(33)(0)(-45) = 0$

Since one factor is 0, the product is 0.

(c) $(-8)(-5)(2)(-2)(-10)$

$(-8)(-5)(2)(-2)(-10) = 1{,}600$

The product is positive because the number of negative factors is even (4). The product is written as 1,600, which represents +1,600.

The following examples illustrate the effect of a factor of -1:

$-1(-5) = +5 \qquad (-1)(+5) = -5$

Thus the product of -1 and a number results in the opposite of that number.

A Factor of −1

ALGEBRAICALLY	VERBALLY	NUMERICAL EXAMPLE
For any real number a, $-1 \cdot a = -a$	The product of negative one and any real number is the opposite of that real number.	$-1 \cdot 3 = -3$ and $-1 \cdot (-4) = 4$

We will use this property in Example 4 to write $-x$ as $(-1)(x)$. One of the properties that we use in this example is $\sqrt{a}\sqrt{b} = \sqrt{ab}$ for both $a > 0$ and $b > 0$.

EXAMPLE 4 Evaluating Algebraic Expressions

Evaluate these expressions for $x = -5$, $y = 3$, and $z = 12$.

SOLUTION

(a) $-xy$

$$\begin{aligned} -xy &= (-1)(x)(y) \\ &= (-1)(-5)(3) \\ &= (5)(3) \\ &= 15 \end{aligned}$$

Replace $-x$ by $(-1)(x)$ and substitute -5 for x and 3 for y. The product of two negative factors yields a positive product.

(b) $\sqrt{y}\sqrt{z}$

$$\begin{aligned} \sqrt{y}\sqrt{z} &= \sqrt{3}\sqrt{12} \\ &= \sqrt{36} \\ &= 6 \end{aligned}$$

Substitute 3 for y and 12 for z. Then simplify, using the fact that $\sqrt{a}\sqrt{b} = \sqrt{ab}$ for both $a > 0$ and $b > 0$. The principal square root of 36 is 6.

The **reciprocal** or **multiplicative inverse** of any nonzero real number a can be represented by $\frac{1}{a}$. The product of a number and its multiplicative inverse is $+1$. For example, $3\left(\frac{1}{3}\right) = 1$ and $-\frac{3}{5}\left(-\frac{5}{3}\right) = 1$.

One is called the **multiplicative identity** because 1 is the only real number with the property that $1 \cdot a = a$ and $a \cdot 1 = a$ for every real number a.

Reciprocals or Multiplicative Inverses

ALGEBRAICALLY	VERBALLY	NUMERICAL EXAMPLE
For any real number $a \neq 0$, $a \cdot \frac{1}{a} = 1$	For any real number a other than zero, the product of the number a and its multiplicative inverse $\frac{1}{a}$ is one. Zero has no multiplicative inverse.	$-4 \cdot \left(-\frac{1}{4}\right) = 1$ and $\frac{4}{3} \cdot \frac{3}{4} = 1$

The product of zero and any real number is always zero; this product can never be 1. Thus zero has no multiplicative inverse.

EXAMPLE 5 Determining Multiplicative Inverses

Determine the multiplicative inverse of each number, and then multiply each number by its multiplicative inverse.

SOLUTION

(a) $\frac{2}{7}$ $\left(\frac{2}{7}\right)\left(\frac{7}{2}\right) = 1$ The reciprocal or multiplicative inverse of $\frac{2}{7}$ is $\frac{7}{2}$.

(b) $-\frac{1}{5}$ $\left(-\frac{1}{5}\right)(-5) = 1$

(c) y (for $y \neq 0$) $y\left(\frac{1}{y}\right) = 1$

(d) 0 0 has no multiplicative inverse.

SELF-CHECK 1.5.3

Write the additive inverse and the multiplicative inverse of each number.

1. 8 **2.** $-\frac{3}{8}$ **3.** 1 **4.** -1

The multiplicative inverse of an expression can be entered into a calculator by pressing the x^{-1} key after the expression. This is illustrated in Calculator Perspective 1.5.2.

CALCULATOR PERSPECTIVE 1.5.2 Finding the Multiplicative Inverse

To determine the multiplicative inverse of $-\frac{1}{5}$ from Example 5(b) on a TI-84 Plus calculator, enter the following keystrokes:

```
(-1/5)-1
                -5
```

Note: The notation x^{-1} will be discussed in Section 5.3.

It is easy to make keystroke errors when we enter values into a calculator or computer. You often can spot serious keystroke errors by performing a quick mental estimate of the answer. Example 6 illustrates how we can mentally estimate a product.

EXAMPLE 6 Estimate Then Calculate

Note that multiplication is implied by the parentheses between the factors in this example.

Estimate $-25.32(-3.89)$ and then use a calculator to perform this multiplication.

SOLUTION

ESTIMATED VALUE

$$-25.32 \approx -25 \qquad -3.89 \approx -4 \qquad \begin{array}{r} -25 \\ \times \quad (-4) \\ \hline +100 \end{array}$$

CALCULATOR VALUE

```
-25.32(-3.89)
         98.4948
```

Answer: 98.4948 seems reasonable based on the estimated value of 100.

To denote repeated multiplication by the same factor, we use exponential notation. The expression 5^2 means $5 \cdot 5$ and is read as "five squared." In the expression 5^2, 5 is called the **base** and 2 is called the **exponent.** The compact use of exponential notation to indicate repeated multiplication is defined in the box below. The expression b^n indicates that b is used as a factor n times, where the exponent n can be 1, 2, 3, 4, or any other natural number.

Exponential Notation

ALGEBRAICALLY	VERBALLY	NUMERICAL EXAMPLE
For any natural number n, $b^n = \underbrace{b \cdot b \cdots b}_{n \text{ factors of } b}$ with base b and exponent n.	For any natural number n, b^n is the product of b used as a factor n times. The expression b^n is read as "b to the nth power."	$7^4 = 7 \cdot 7 \cdot 7 \cdot 7$ $(-3)^2 = (-3)(-3)$

A Mathematical Note

x^2, which can be read as "x to the second power," is usually read as "x squared" because this is the area of a square with sides of length x.

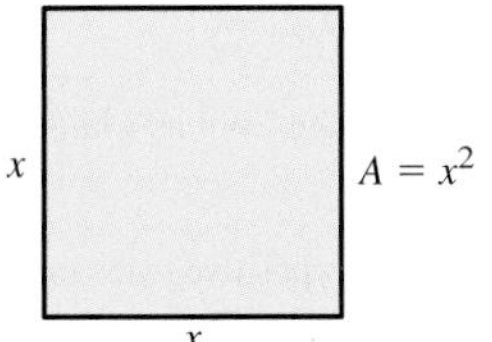

x^3, which can be read as "x to the third power," is usually read as "x cubed" because this is the volume of a cube with sides of length x.

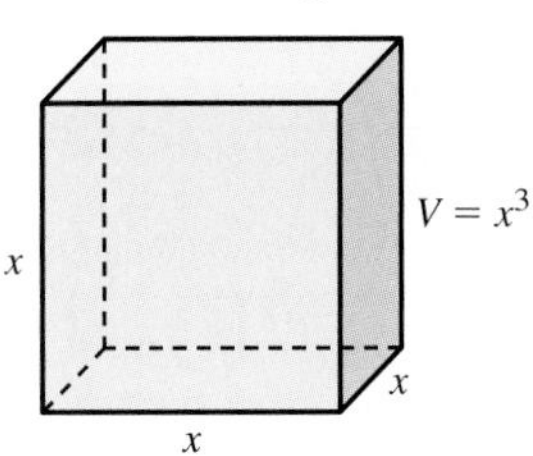

Some of the words and phrases used in word problems and applications to indicate exponentiation are given in Table 1.5.3.

Phrases Used to Indicate Exponentiation

KEY PHRASE	VERBAL EXAMPLE	ALGEBRAIC EXAMPLE
To a power	"Three to the sixth power"	3^6
Raised to	"y raised to the fifth power"	y^5
Squared	"Four squared"	4^2
Cubed	"x cubed"	x^3

Table 1.5.3

EXAMPLE 7 Evaluating Exponential Expressions

Evaluate each of the following exponential expressions.

SOLUTION

(a) 3^4

$3^4 = 3 \cdot 3 \cdot 3 \cdot 3$
$= 81$

The base is 3. The exponent of 4 indicates that 3 is used as a factor 4 times. This is read as "3 to the fourth power equals 81."

(b) 2^5

$2^5 = 2 \cdot 2 \cdot 2 \cdot 2 \cdot 2$
$= 32$

"2 to the fifth power equals 32." The expanded form of 2^5 is $2 \cdot 2 \cdot 2 \cdot 2 \cdot 2$.

(c) 1^6

$1^6 = 1 \cdot 1 \cdot 1 \cdot 1 \cdot 1 \cdot 1$
$= 1$

"1 to the sixth power equals 1."

(d) $\left(\frac{2}{5}\right)^3$

$\left(\frac{2}{5}\right)^3 = \frac{2}{5} \cdot \frac{2}{5} \cdot \frac{2}{5}$
$= \frac{8}{125}$

The base of $\frac{2}{5}$ is used as a factor 3 times. This is read, "Two-fifths cubed equals eight one hundred twenty fifths."

(e) 0.12^2

$0.12^2 = (0.12) \cdot (0.12)$
$= 0.0144$

"Twelve hundredths squared equals one hundred forty-four ten-thousandths."

SELF-CHECK 1.5.4

Write each expression in exponential form.

1. $4 \cdot 4 \cdot 4 \cdot 4 \cdot 4$ **2.** $(-3)(-3)(-3)(-3)$
3. $x \cdot x \cdot x \cdot x \cdot x \cdot x$ **4.** $(2w)(2w)(2w)$

Caution: -3^2 is not the same as $(-3)^2$.

An exponent is understood to refer only to the constant or variable that immediately precedes it. This can lead to misconceptions when expressions such as $(-3)^2$ and -3^2 are considered. Note the distinction between $(-3)^2$ and -3^2 in Example 8.

EXAMPLE 8 Evaluating Exponential Expressions

Evaluate each of the following exponential expressions.

SOLUTION

(a) $(-3)^2$ $\quad (-3)^2 = (-3)(-3)$
$= 9$

The base is -3. The product is positive since the number of negative factors is even (2).

(b) -3^2 $\quad -3^2 = -(3)(3)$
$= -9$

The base is 3. Square 3, and then form its additive inverse.

(c) $(-2)^4$ $\quad (-2)^4 = (-2)(-2)(-2)(-2)$
$= 16$

The base is -2. The product is positive since the number of negative factors is even (4).

(d) -2^4 $\quad -2^4 = -(2 \cdot 2 \cdot 2 \cdot 2)$
$= -16$

The base is 2. Raise 2 to the fourth power, and then form its additive inverse. Contrast this to part (c) where the base is -2.

(e) $(-1)^{117}$ $\quad (-1)^{117} = -1$

The base of -1 is used as a factor an odd number of times (117), so the result is negative.

SELF-CHECK 1.5.5

Evaluate each of the following exponential expressions.

1. 3^4 **2.** 4^3 **3.** $(-3)^4$ **4.** -3^4

Graphing calculators have a key for squaring a value and generally have a menu for cubing a value. The squaring key is very useful because this is the most common power that you will need. For other powers we recommend using the caret key ^.

To square a value, press the x^2 button after entering the expression to be squared. For exponents other than 2, use ^ followed by the desired exponent. The expression x^4 is sometimes read "x raised to the fourth power" and can be written as x ^ 4. These notations for raising a quantity to a power are illustrated in Calculator Perspective 1.5.3.

CALCULATOR PERSPECTIVE 1.5.3 Raising Quantities to a Power

To evaluate $(-3)^2$ and -2^4 from Example 8 on a TI-84 Plus calculator, enter the following keystrokes:

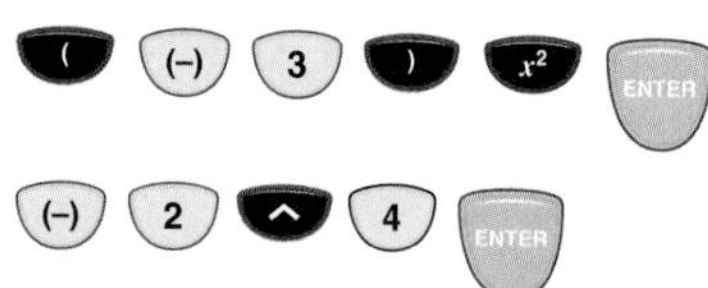

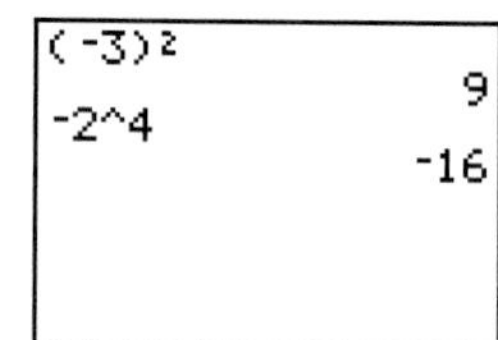

One important use of equations is to state formulas. A **formula** is an equation that states the relationship between different quantities. For example, the formula $A = l \cdot w$ relates the area of a rectangle A to its length l and its width w. Each variable used in a formula has a selected meaning and represents every possible value to which it applies. It is helpful to use representative letters for variables, such as V for volume, P for perimeter, r for radius, and t for time. A list of common formulas, including many formulas from geometry, is given on the inside back cover of this book.

EXAMPLE 9 Calculating the Interest on an Investment

Use the formula $I = PRT$ to calculate the interest on an investment of \$5,000 at a rate of 8% per year for 1 year.

SOLUTION

$I = PRT$

$I = (5{,}000)(0.08)(1)$ Substitute into the interest formula \$5,000 for the principal P, 0.08 for the interest rate R of 8%, and 1 for the time T of 1 year.

$I = 400$

Answer: The interest on this investment for 1 year is \$400.

SELF-CHECK 1.5.6

Use a calculator to answer these exercises.

1. Use the formula $A = lw$ to determine the area of a rectangle with a width of 7 cm and a length of 11 cm.
2. Use the formula $I = PRT$ to determine the interest on an investment of \$13,500 at 8.5% for 1 year.
3. Determine the first five terms of the sequence defined by $a_n = -6n$.

SELF-CHECK ANSWERS

1.5.1
1. -85 2. 99 3. 20
4. $4w$ 5. $3r$

1.5.2
1. Negative, -5.941
2. Negative, -84.48
3. Positive, $\frac{10}{21}$
4. 0 is neither negative nor positive; 0

1.5.3
1. $-8, \frac{1}{8}$ 2. $\frac{3}{8}, -\frac{8}{3}$
3. $-1, 1$ 4. $1, -1$

1.5.4
1. 4^5 2. $(-3)^4$
3. x^6 4. $(2w)^3$

1.5.5
1. 81 2. 64
3. 81 4. -81

1.5.6
1. 77 cm^2 2. \$1,147.50
3. $-6, -12, -18, -24$, and -30

USING THE LANGUAGE AND SYMBOLISM OF MATHEMATICS 1.5

1. In the multiplication $9 \cdot 11 = 99$, 9 and 11 are called __________ and 99 is called the __________.
2. The phrase "the product of P times R times T" is represented algebraically by __________.
3. The phrase "fourteen percent of C" is represented algebraically by __________.
4. The property that says $ab = ba$ for all real numbers is called the __________ property of multiplication.
5. The property that says $(ab)(c) = (a)(bc)$ for all real numbers is called the __________ property of multiplication.
6. If either x or y is zero, the product $xy =$ __________.
7. If x and y have the same sign, then the sign of the product xy is __________.
8. If x and y have unlike signs, then the sign of the product xy is __________.
9. If an even number of negative factors are multiplied, the product will be __________.
10. If an odd number of negative factors are multiplied, the product will be __________.
11. If 12 * 5 is shown on a calculator display, then the result will be __________.
12. The product $\frac{w}{x} \cdot \frac{y}{z} =$ __________ for $x \neq$ __________ and $z \neq$ __________.
13. The real numbers $-\frac{4}{7}$ and $-\frac{7}{4}$ are __________ or multiplicative __________ of each other.
14. The product of a number and its reciprocal is __________.
15. The only real number that does not have a multiplicative inverse is __________.
16. The multiplicative identity is __________.
17. For both $a > 0$ and $b > 0$, $\sqrt{a}\sqrt{b} =$ __________.
18. In the expression w^4, the base is __________ and the exponent is __________.

19. __________ is read "x to the seventh power."
20. y^5 is read "y to the __________ power."
21. a^8 means that a is used as a __________ 8 times.
22. In the expression $(-x)^2$, the base is __________.
23. In the expression $-x^2$, the base is __________.

EXERCISES 1.5

In Exercises 1–14, calculate each product without using your calculator.

1. a. $7(-11)$ b. $-7(11)$
   c. $(-7)(-11)$ d. $-(-7)(-11)$
2. a. $-5(12)$ b. $-(5)(12)$
   c. $(5)(-12)$ d. $(-5)(-12)$
3. a. $2(-3)(10)$ b. $2(-3)(-10)$
   c. $(-2)(-3)(-10)$ d. $-(-2)(-3)(-10)$
4. a. $2(5)(-20)$ b. $2(-5)(-20)$
   c. $(-2)(-5)(-20)$ d. $-(-2)(-5)(-20)$
5. a. $(-1)(-1)(-1)(-8)$ b. $(-1)(1)(1)(-8)$
   c. $(-1)(1)(-1)(-8)$ d. $(-1)(1)(0)(-8)$
6. a. $(-1)(2)(-3)(5)$ b. $(-1)(-2)(-3)(-5)$
   c. $(-1)(-2)(3)(-5)$ d. $(-1)(2)(0)(5)$
7. a. $-0.1(1{,}234)$ b. $-100(1{,}234)$
   c. $-0.001(1{,}234)$ d. $-1{,}000(1{,}234)$
8. a. $-0.01(-45.96)$ b. $-10(-45.96)$
   c. $-100(45.96)$ d. $-0.001(45.96)$
9. a. $\frac{1}{2} \cdot \frac{3}{5}$ b. $-\frac{1}{2} \cdot \frac{3}{5}$
10. a. $\frac{1}{3} \cdot \frac{4}{5}$ b. $-\frac{1}{3} \cdot \frac{4}{5}$
11. a. $\left(\frac{1}{5}\right)(5)$ b. $\left(\frac{1}{5}\right)(-1)$
12. a. $\left(-\frac{3}{2}\right)\left(-\frac{2}{3}\right)$ b. $\left(-\frac{2}{3}\right)\left(-\frac{3}{2}\right)$
13. a. $\sqrt{2}\sqrt{18}$ b. $\sqrt{5}\sqrt{20}$
14. a. $\sqrt{2}\sqrt{50}$ b. $\sqrt{3}\sqrt{27}$

In Exercises 15 and 16, write each expression in exponential form.

15. a. $5 \cdot 5 \cdot 5 \cdot 5$ b. $(-4)(-4)(-4)$
    c. $y \cdot y \cdot y \cdot y \cdot y$ d. $(3z)(3z)(3z)(3z)(3z)(3z)$
16. a. $7 \cdot 7 \cdot 7$ b. $(-6)(-6)(-6)(-6)$
    c. $w \cdot w \cdot w \cdot w \cdot w \cdot w$ d. $(-5z)(-5z)$

In Exercises 17 and 18, write each exponential expression in expanded form.

17. a. -3^2 b. $(-3)^2$
    c. $3x^2$ d. $(3x)^2$
18. a. -5^4 b. $(-5)^4$
    c. $5x^4$ d. $(5x)^4$

In Exercises 19–26, evaluate each expression without using a calculator.

19. a. 2^3 b. 3^2 c. $(-3)^2$ d. -3^2
20. a. 2^5 b. 5^2 c. $(-5)^2$ d. -5^2
21. a. 0^8 b. 1^8 c. $(-1)^8$ d. -1^8
22. a. 0^{10} b. 1^{10} c. $(-1)^{10}$ d. -1^{10}
23. a. -10^2 b. $(-10)^2$ c. $(0.1)^3$ d. $(-10)^3$
24. a. -10^3 b. $(-0.1)^2$ c. -0.1^2 d. $(-10)^4$
25. a. $\left(\frac{1}{2}\right)^3$ b. $\left(-\frac{1}{2}\right)^3$ c. $\left(\frac{1}{3}\right)^2$ d. $\left(-\frac{1}{3}\right)^2$
26. a. $\left(\frac{1}{2}\right)^5$ b. $\left(-\frac{1}{2}\right)^5$ c. $\left(\frac{1}{5}\right)^2$ d. $\left(-\frac{1}{5}\right)^2$

In Exercises 27–30, evaluate each expression for $w = -2$, $x = -5$, and $y = -30$.

27. a. $(wx)(y)$ b. $w(xy)$
28. a. $(wx)(-y)$ b. $w[(x)(-y)]$
29. a. $-3w(xy)$ b. $4w(xy)$
30. a. $-10w(xy)$ b. $(-10wx)(-y)$

In Exercises 31 and 32, check both $x = -2$ and $x = 2$ to determine whether either is a solution of these equations. (*Hint:* See Example 8 in Section 1.4.)

31. a. $-4x = x + 10$ b. $-x = x - 4$
32. a. $-5x = x - 12$ b. $-x = x + 4$

In Exercises 33 and 34, calculate the first five terms of each sequence. (*Hint:* See Example 9 in Section 1.4.)

33. a. $a_n = 2n$ b. $a_n = n^2$
34. a. $a_n = 3n$ b. $a_n = n^3$

Comparing Addition and Multiplication

In Exercises 35–40, first compute the sum of x and y and then compute the product of x and y.

	X	Y	$X + Y$	XY
Example	3	4	7	12
35.	5	8		
36.	−5	8		
37.	5	−8		
38.	−5	−8		
39.	0	π		
40.	π	−1		

Multiple Representations

In Exercises 41 and 42, write each verbal statement in algebraic form.

41. a. The product of eight and x
    b. x to the eighth power
42. a. Twice the value of y
    b. y squared
43. The initial value of an investment is P. Write an expression for the new value of this investment if this investment
    a. triples in value b. decreases by \$300
44. A board is cut into two pieces so that the length of the shorter piece is s cm. Write an expression for the length of the longer piece if the longer piece is
    a. 26 cm longer than the shorter piece
    b. 4 times as long as the shorter piece

In Exercises 45–50, fill in each blank.

45. a. The result of the addition "a and b" is called their __________.

b. The result of the subtraction "a minus b" is called their __________.

c. The result of the multiplication "a times b" is called their __________.

46. a. The sum of a number and its additive inverse is ______.
    b. The product of a nonzero number and its multiplicative inverse is ______.
47. a. The property that says $5 \cdot (7 \cdot 9) = 5 \cdot (9 \cdot 7)$ is the __________ property of multiplication.
    b. The property that says $5 \cdot (7 \cdot 9) = (5 \cdot 7) \cdot 9$ is the __________ property of multiplication.
48. a. The property that says $wx + yz = yz + wx$ is the commutative property of __________.
    b. The property that says $wx + yz = wx + zy$ is the commutative property of __________.
49. a. The property that says $w + x(yz) = w + (xy)(z)$ is the associative property of __________.
    b. The property that says $w + (x + yz) = (w + x) + yz$ is the associative property of __________.
50. a. The number 1 is its own multiplicative inverse. The only other number that is its own multiplicative inverse is __________.
    b. __________ is the only real number that does not have a multiplicative inverse.

Area of a Triangle

In Exercises 51 and 52, use the formula $A = \frac{1}{2}bh$ to calculate the area of each triangle.

51.

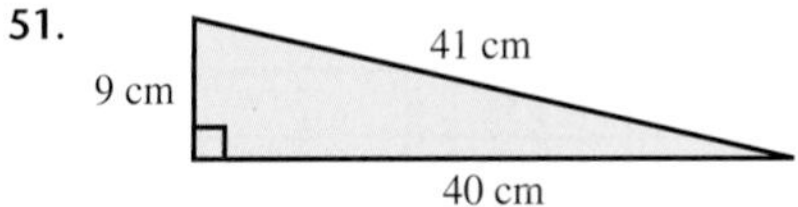

52.

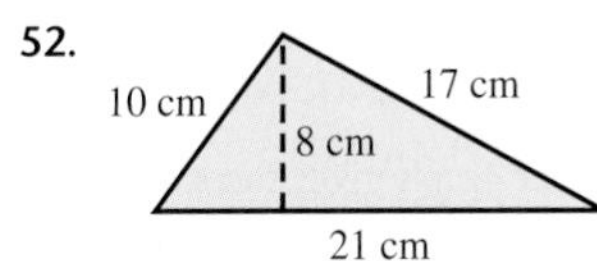

Volume of a Box

In Exercises 53 and 54, use the formula $V = l \cdot w \cdot h$ to calculate the volume of each object.

53. The mini refrigerator shown below has interior dimensions of 14 in by 15 in by 16 in. Determine the interior volume of this refrigerator in cubic inches (in^3).

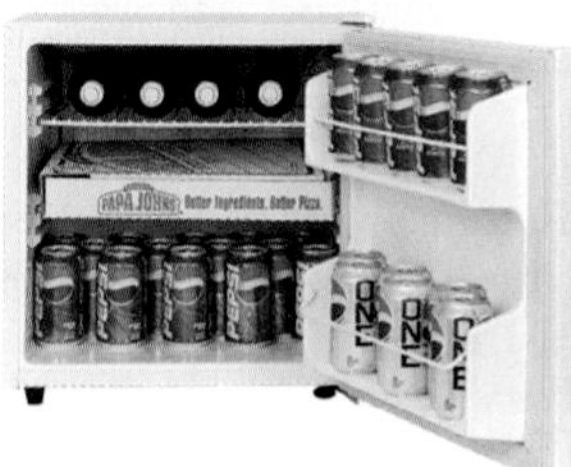

54. A box that contains 10 reams of printer paper has dimensions of 9 in by 11.5 in by 18 in. Determine the volume of this box in cubic inches (in^3).

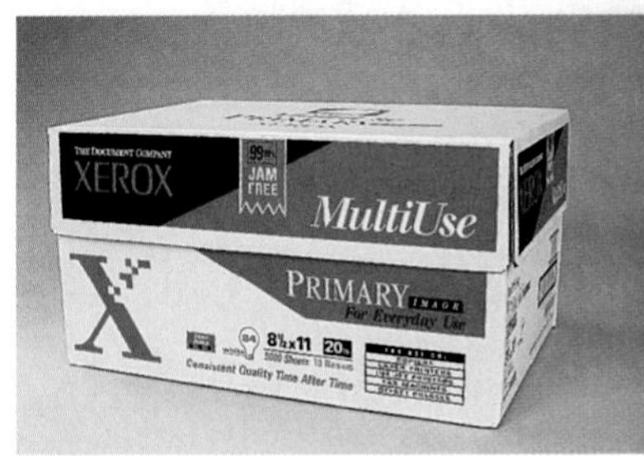

55. **Perimeter of a Hexagon** Each side of the regular hexagon (six equal sides) shown is 2.47 cm long. Determine the perimeter of this hexagon.

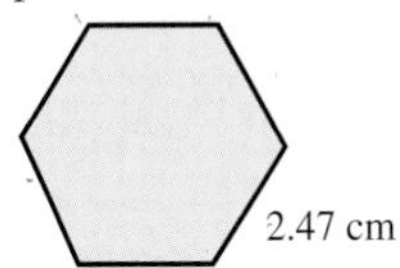

56. **Perimeter of a Pentagon** Each side of the regular pentagon (five equal sides) shown is 11.56 cm long. Determine the perimeter of this pentagon.

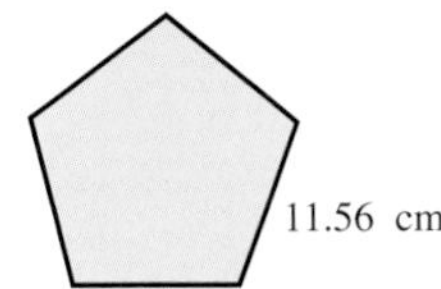

Simple Interest

In Exercises 57 and 58, use a calculator and the formula $I = P \cdot R \cdot T$ to calculate the interest for each investment.

57. a. An investment of $4,000 at a rate of 7% for 1 year
    b. An investment of $4,000 at a rate of 7.25% for 1 year
58. a. An investment of $7,000 at a rate of 8% for 1 year
    b. An investment of $7,000 at a rate of 8.5% for 1 year

Active Ingredient

The amount of active ingredient in a solution can be calculated by multiplying the percent of concentration of the active ingredient by the volume of the solution. In 4 gal of 65% antifreeze solution there would be $(0.65)(4)$, or 2.6, gal of pure antifreeze, and the rest is presumed to be water. In Exercises 59–62, use a calculator and the formula $A = RB$ to calculate the amount of active ingredient in each solution.

59. 500 milliliters (mL) of a 15% juice solution
60. 200 gal of a 3.8% insecticide solution
61. 3 liters (L) of a 22.5% hydrochloric acid solution
62. 1.5 gal of a 70% antifreeze solution

Distance Traveled

The distance an object travels can be calculated by finding the product of its rate and the time it travels. In Exercises 63–66, use the formula $D = RT$ compute the distance each object travels.

63. A plane flies 420 mi/h for 3 h.
64. An explorer walks 3 days and averages 15 mi/day.
65. An ant crawls 2 m/min for 6.5 min.
66. A spacecraft flies 6,200 ft/s for 120 s.

67. Volume of a Cube The volume of a cube is given by $V = s^3$. Calculate the volume of the cube in the figure below.

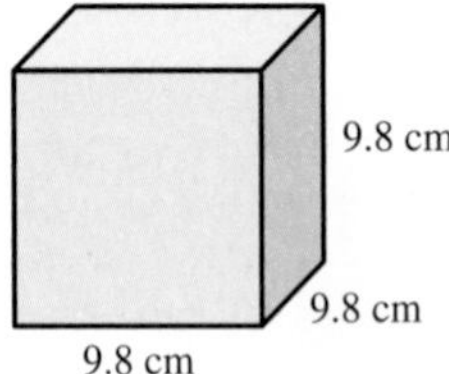

68. Circumference and Area of a Circle

a. Use the given circle and the formula $C = 2\pi r$ to approximate to the nearest tenth of a centimeter the circumference of the circle.

b. Use the given circle and the formula $A = \pi r^2$ to approximate to the nearest tenth of a square centimeter the area of the circle.

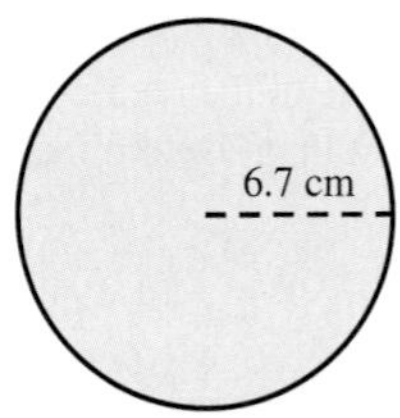

69. Surface Area and Volume of a Sphere

a. Use the given sphere and the formula $S = 4\pi r^2$ to approximate to the nearest tenth of a square centimeter the surface area of the sphere.

b. Use the given sphere and the formula $V = \frac{4}{3}\pi r^3$ to approximate to the nearest tenth of a cubic centimeter the volume of the sphere.

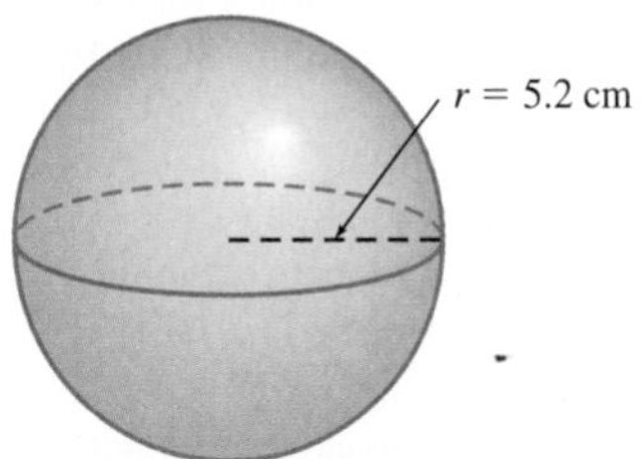

70. Mass of Four Nickels A nickel is 2 cm in diameter and has a mass of 5 grams(g). Four nickels are stamped from a rectangular sheet as shown below.

a. Determine the total mass of the four nickels.

b. Determine the area of this rectangular sheet.

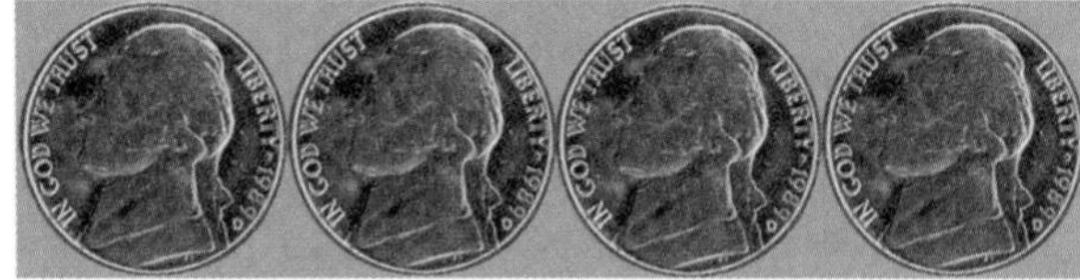

71. Sales Tax on a Television Determine the Texas sales tax on a television set that costs \$499.99. The sales tax rate in Texas is 6.25%.

72. Sales Tax on a Microwave Oven Determine the Minnesota sales tax on a microwave oven that costs \$229.99. The sales tax rate in Minnesota is 6.5%.

73. Weekly Wages A student worked 52 hours during one week. If she earned \$5.25 per hour for the first 40 hours and \$7.88 per hour for overtime, how much did she earn that week?

74. Cost of Bolts The wholesale cost of a bolt was reduced from \$0.027 to \$0.023.

a. What is the cost of 1,000 bolts at the reduced price?

b. How much will be saved if these 1,000 bolts are bought at the lower price rather than at the higher price?

75. Basketball Ticket Payments The spreadsheet below is a partially completed order form for college basketball tickets. Complete the last column of this spreadsheet by calculating the cost for each game and then determine the total cost for all of these tickets.

	A	B	C	D
1	GAME	NUMBER OF TICKETS	PRICE PER TICKET	COST
2	1	4	\$16	
3	2	2	\$16	
4	3	3	\$25	
5	4	6	\$25	
6				

76. Concert Ticket Payments The spreadsheet below is a partially completed order form for concert tickets. Complete the last column of this spreadsheet by calculating the cost for each concert, and then determine the total cost for all these tickets.

	A	B	C	D
1	CONCERT	NUMBER OF TICKETS	PRICE PER TICKET	COST
2	1	5	\$40	
3	2	3	\$35	
4	3	2	\$50	
5	4	1	\$80	
6				

77. Income From Sales The first bar graph gives the sales price per item for each quarter for a business. The second bar graph gives the number of items sold each quarter. Calculate the sales revenue for this item for each quarter.

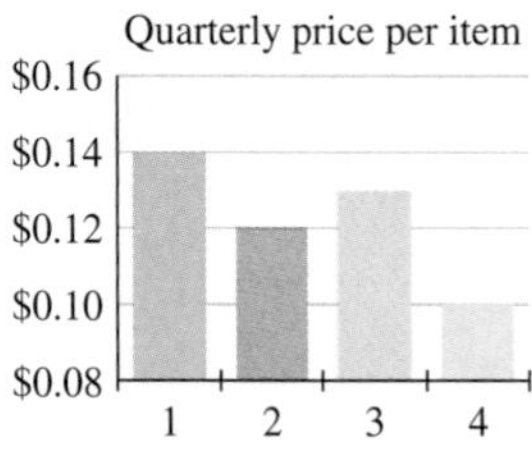

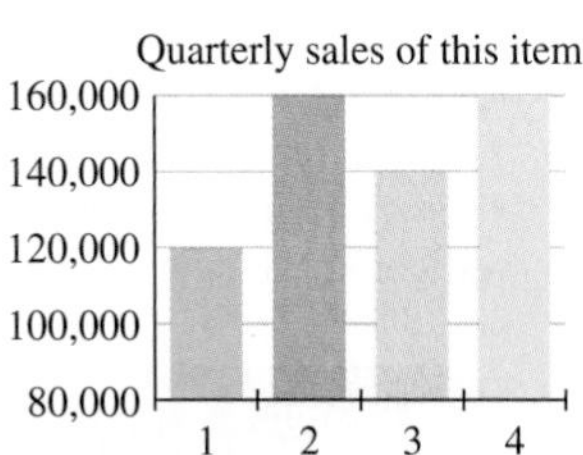

78. Profit From Sales The first bar graph gives the profit per item for each quarter for a business. The second bar graph gives the number of items sold in each quarter. Calculate the sales profit for this item for each quarter.

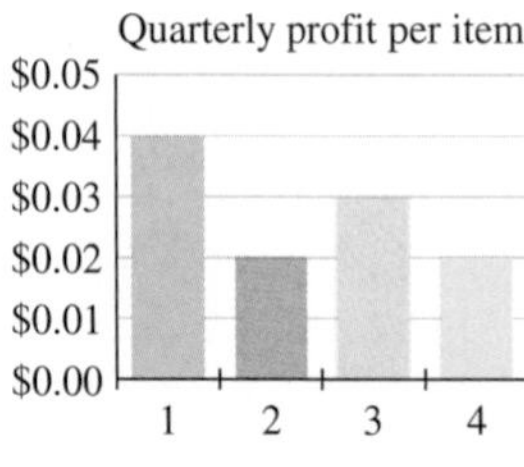

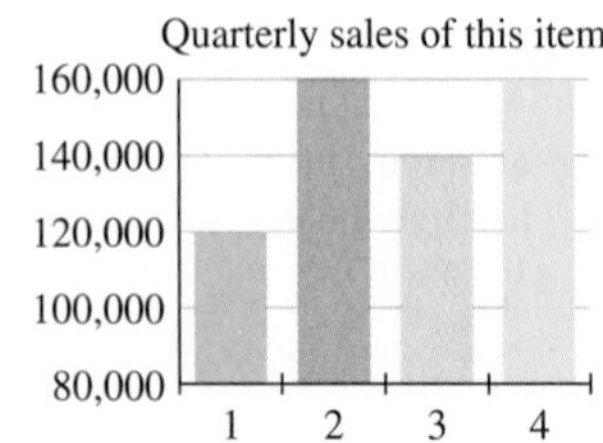

Estimate Then Calculate

In Exercises 79–84, mentally determine the correct sign of the product and then use a calculator to calculate the product. Approximate the product in Exercises 83 and 84 to the nearest thousandth.

PROBLEM	SIGN OF PRODUCT (WITHOUT CALCULATOR)	PRODUCT (WITH CALCULATOR)
79. $-8.7(4.5)(-3.2)$		
80. $-5.6(-7.8)(-9.3)$		
81. $\left(-\frac{4}{5}\right)\left(-\frac{7}{13}\right)$		
82. $\left(-\frac{3}{5}\right)\left(\frac{4}{7}\right)\left(-\frac{6}{11}\right)$		
83. $(-\sqrt{2})(-\sqrt{3})$		
84. $\sqrt{7}(-\sqrt{11})$		

Estimation Skills

In Exercises 85–88, mentally estimate the value of each expression and then select the most appropriate answer.

85. $(-9.9)(15.12)$
a. -149.688 b. -249.688 c. 0
d. 149.688 e. 249.688

86. $(-159.63)(-5.1)$
a. -814.113 b. -514.113 c. 51.413
d. 81.4113 e. 814.113

87. $(93.45)(187.8)(0)(-35.3)$
a. $-620{,}000$ b. $-260{,}000$ c. 0
d. 260,000 e. 260,000

88. $(-0.09)(-311.83)$
a. -28.0647 b. -58.0647 c. 0
d. 28.0647 e. 58.0647

Group Discussion Questions

89. Risks and Choices Being a good citizen involves making personal decisions based on facts. Consider the following situation, which involves alcohol consumption.

Three workers each had a drink after work. One drank a 12-oz bottle of beer that was 5% alcohol; the second had a 4-oz glass of wine that was 15% alcohol; and the third had a 1.5-oz shot that was 40% alcohol. Which worker consumed the most alcohol?

90. Risks and Choices A Web search for the "Karvonen formula" will produce information about the target heart rate for a person who is exercising. In this formula a is the age of the person, r is the resting heart rate, and T is the target heart rate for fitness level training. Use the formula $T = r + 0.65(220 - a - r)$ to compute the fitness zone target heart rate for three or four people in your group. You can determine your heart rate by placing your index finger on the side of your neck between the middle of your collar bone and your jaw line. Count the heart beats from the carotid artery for 60 seconds.

91. Communicating Mathematically The cost of one medium soft drink at a fast-food service is \$1.13. However, if you buy two at the same time, the cost is \$2.27. Can you give a mathematical explanation for this?

92. Communicating Mathematically Discuss the reason that the * symbol was used to denote multiplication on computers rather than using the symbols × and · , which were already common conventions for multiplication.

93. Spreadsheet Exploration Complete the missing entries in the spreadsheet shown. This carpet estimate is based upon lengths and widths given in feet and area in square feet. The costs for padding and carpet were originally quoted to the customer in dollars per square foot.

	A	B	C	D	E	F	G
1			Carpet Estimate for the Smith Home				
2	ROOM	LENGTH (ft)	WIDTH (ft)	AREA (ft²)	PADDING	CARPET	ROOM TOTAL
3	Family room	20	14	280	\$84.00	\$616.00	\$700.00
4	Bedroom 1	14	12	168	50.40	369.60	
5	Bedroom 2	12		120	36.00	264.00	300.00
6	Den	15	12	180		396.00	
7							

Section 1.6 Division of Real Numbers

Objectives:
17. Divide positive and negative real numbers.
18. Express ratios in lowest terms.

This section will examine the division of signed numbers. Recall the terminology for the division $16 \div 8 = 2$; 16 is called the **dividend,** 8 is called the **divisor,** and 2 is called the **quotient.**

The notation x/y can easily be misinterpreted, especially when you enter expressions such as $\frac{x}{y+z}$ on a calculator. We will consider this again in Section 1.7 when the order of operations is covered.

Notations for the Quotient of x Divided by y for $y \neq 0$

$x \div y$ $\frac{x}{y}$ x/y $x : y$

Some of the words and phrases used in word problems and applications to indicate division are given in Table 1.6.1.

Caution: x divided *by* y is written as $\frac{x}{y}$, but x divided *into* y is written as $\frac{y}{x}$.

Phrases Used to Indicate Division

KEY PHRASE	VERBAL EXAMPLE	ALGEBRAIC EXAMPLE
Divided by	"x divided by y"	$\frac{x}{y}$
Quotient	"The quotient of five and three"	$5 \div 3$
Ratio	"The ratio of x to two"	$x : 2$ or $\frac{x}{2}$

Table 1.6.1

A Mathematical Note

The symbol ÷ for division is an imitation of fractional division; the dots indicate the numerator and the denominator of a fraction. This symbol was invented by an English mathematician John Pell (1611–1685).

If two people agree to share a \$50 restaurant bill equally, we can divide the bill by 2; $\$50 \div 2 = \$25$ for each person. Equivalently, we could say that each person gets one-half of the bill; $\frac{1}{2}(\$50) = \$25$ for each person. Dividing by 2 produces the same result as multiplying by $\frac{1}{2}$.

The division $50 \div 2 = 25$ checks because $2 \cdot 25 = 50$. We can say that division is the inverse of multiplication, and we can define division as the multiplication of a multiplicative inverse.

Definition of Division

ALGEBRAICALLY	VERBALLY	NUMERICAL EXAMPLE
For any real numbers x and y with $y \neq 0$, $x \div y = x\left(\frac{1}{y}\right)$	Dividing two real numbers is the same as multiplying the first number by the multiplicative inverse of the second number.	$3 \div 2 = 3\left(\frac{1}{2}\right)$

Since division is defined in terms of multiplication, the rules for dividing signed numbers are derived from the rules for multiplying signed numbers.

Division of Two Real Numbers

Like signs: Divide the absolute values of the two numbers and use a positive sign for the quotient.

Unlike signs: Divide the absolute values of the two numbers and use a negative sign for the quotient.

Zero dividend: $\frac{0}{x} = 0 \quad \text{for} \quad x \neq 0$

Zero divisor: $\frac{x}{0}$ is undefined for every real number x.

Example 1 illustrates the division of two real numbers, including the division of fractions. The procedure for dividing fractions is summarized in the following box. The reciprocal or the multiplicative inverse of $\frac{c}{d}$ is $\frac{d}{c}$ for c and d not equal to zero.

Division of Fractions

VERBALLY	ALGEBRAICALLY	NUMERICAL EXAMPLE
To divide two fractions, multiply the first fraction by the reciprocal of the second fraction.	$\frac{a}{b} \div \frac{c}{d} = \frac{a}{b} \cdot \frac{d}{c}$ $= \frac{ad}{bc}$ for $b \neq 0$, $c \neq 0$, and $d \neq 0$	$\frac{3}{4} \div \frac{4}{5} = \frac{3}{4} \cdot \frac{5}{4}$ $= \frac{15}{16}$

EXAMPLE 1 Dividing Real Numbers

Determine the following quotients.

SOLUTION

(a) $21 \div (-7)$ $\quad 21 \div (-7) = -3$ The quotient of unlike signs is negative.

(b) $-36 \div (+4)$ $\quad -36 \div (+4) = -9$ The quotient of unlike signs is negative.

(c) $-45 \div (-9)$ $\quad -45 \div (-9) = 5$ The quotient of like signs is positive.

(d) $\frac{1}{2} \div \frac{1}{3}$

$$\frac{1}{2} \div \frac{1}{3} = \frac{1}{2} \cdot \frac{3}{1} = \frac{3}{2}$$

$\frac{a}{b} \div \frac{c}{d} = \frac{a}{b} \cdot \frac{d}{c} = \frac{a \cdot d}{b \cdot c}$ for $b \neq 0$, $c \neq 0$, and $d \neq 0$. To divide by $\frac{1}{3}$, multiply by its multiplicative inverse $\frac{3}{1}$. The product of like signs is positive.

(e) $\frac{-3}{5} \div \frac{4}{7}$

$$\frac{-3}{5} \div \frac{4}{7} = \frac{-3}{5} \cdot \frac{7}{4}$$

To divide by $\frac{4}{7}$, multiply by its multiplicative inverse $\frac{7}{4}$. Then multiply the numerators and multiply the denominators.

$$= \frac{-3 \cdot 7}{5 \cdot 4}$$

$$= \frac{-21}{20}$$

The product of unlike signs is negative.

$$= -\frac{21}{20}$$

Since $1 \div (-2) = -\frac{1}{2}$ and $(-1) \div 2 = -\frac{1}{2}$, we can write $\frac{1}{-2} = \frac{-1}{2} = -\frac{1}{2}$. The relationship of the sign of the numerator, the sign of the denominator, and the sign of an algebraic fraction is given in the following box. This relationship is given not just for rational numbers such as $\frac{3}{5}$, but for all algebraic fractions $\frac{a}{b}$ where a and b are algebraic expressions representing real numbers and $b \neq 0$.

Three Signs of a Fraction

ALGEBRAICALLY	VERBALLY	NUMERICAL EXAMPLE
For all algebraic expressions a and b with $b \neq 0$, $-\frac{a}{b} = \frac{-a}{b} = \frac{a}{-b} = -\frac{-a}{-b}$ and $\frac{a}{b} = \frac{-a}{-b} = -\frac{-a}{b} = -\frac{a}{-b}$	Each fraction has three signs associated with it. Any two of these signs can be changed, and the value of the fraction will stay the same.	$-\frac{2}{3} = \frac{-2}{3} = \frac{2}{-3} = -\frac{-2}{-3}$ and $\frac{2}{3} = \frac{-2}{-3} = -\frac{-2}{3} = -\frac{2}{-3}$

SELF-CHECK 1.6.1

1. Calculate the quotient of 46 and -23.
2. Write the ratio of 11 to 17 in fractional form.
3. Write an equation to express "The ratio of the circumference C of a circle to its diameter d is the constant π."
4. Calculate the quotient $\left(\frac{2}{3}\right) \div \left(-\frac{7}{11}\right)$.

Fractions are generally expressed in **reduced form,** a form in which the numerator and denominator have no common factor other than -1 or 1. To multiply fractions, it is usually easier first to divide both the numerator and the denominator by any reducing factor and then to simplify the numerator and the denominator.

EXAMPLE 2 Simplifying Fractions

Express each fraction in reduced form.

SOLUTION

(a) $\frac{30}{66}$

$$\frac{30}{66} = \frac{5 \cdot \overset{1}{\cancel{6}}}{\underset{1}{\cancel{6}} \cdot 11} = \frac{5}{11}$$

For $b \neq 0$ and $c \neq 0$, $\frac{ac}{bc} = \frac{a}{b}$ with c the common reducing factor. In this example, divide both the numerator and the denominator by the reducing factor 6.

(b) $\frac{-33}{35} \cdot \frac{49}{22}$

$$\frac{-33}{35} \cdot \frac{49}{22} = \frac{-3 \cdot \overset{1}{\cancel{11}}}{5 \cdot \underset{1}{\cancel{7}}} \cdot \frac{\overset{1}{\cancel{7}} \cdot 7}{2 \cdot \underset{1}{\cancel{11}}} = \frac{-3 \cdot 7}{5 \cdot 2} = \frac{-21}{10} = -\frac{21}{10}$$

Factor the numerator and denominator and divide them by their common factors, 7 and 11.

The product of unlike signs is negative.

(c) $\frac{27}{50} \div \frac{12}{10}$

$$\frac{27}{50} \div \frac{12}{10} = \frac{\overset{9}{\cancel{27}}}{\underset{5}{\cancel{50}}} \cdot \frac{\overset{1}{\cancel{10}}}{\underset{4}{\cancel{12}}} = \frac{9}{20}$$

Invert the divisor and multiply. The reducing factors are 3 and 10.

Division by zero must be undefined. To define division by zero would create inconsistencies in our algebraic system.

Division by Zero

The division $50 \div 2 = 25$ checks because $2 \cdot 25 = 50$. Likewise if $3 \div 0 = a$, then $0 \cdot a = 3$ must check. However, this is impossible because $0 \cdot a = 0$ for all values of a. Thus this division by zero is undefined. Next, we will examine $0 \div 0 = b$ which will check if $0 \cdot b = 0$. However, this is true for all real numbers b. We say that $\frac{0}{0}$ is **indeterminate** since there is no reason to select or determine one value of b as preferable to any other value of b. Thus, this division by zero is also undefined.

EXAMPLE 3 Dividing Expressions Involving Zero

Determine the following quotients.

SOLUTION

(a) $\frac{10}{5}$	$\frac{10}{5} = 2$	*Check:* $2 \cdot 5 = 10$.
(b) $\frac{0}{5}$	$\frac{0}{5} = 0$	*Check:* $0 \cdot 5 = 0$.
(c) $\frac{5}{0}$	$\frac{5}{0}$ is undefined	Division by zero is undefined. There is no value to multiply times 0 that will produce 5.
(d) $\frac{0}{0}$	$\frac{0}{0}$ is undefined	Division by zero is undefined. All values multiplied times 0 will yield 0, thus there is no unique quotient.

SELF-CHECK 1.6.2

Calculate each quotient, using a calculator.

1. $\frac{25}{25}$ **2.** $\frac{25}{-1}$ **3.** $\frac{25}{0}$ **4.** $\frac{0}{25}$

Entering ÷ on a calculator produces a slash symbol on the calculator screen. This is illustrated in Calculator Perspective 1.6.1. Also note what happens when we try to divide by 0.

CALCULATOR PERSPECTIVE 1.6.1 Using the Division Symbol on a Calculator

To evaluate $\frac{10}{5}$, $\frac{0}{5}$, and $\frac{5}{0}$ from Example 3 on a TI-84 Plus calculator, enter the following keystrokes:

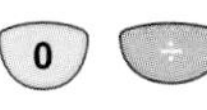

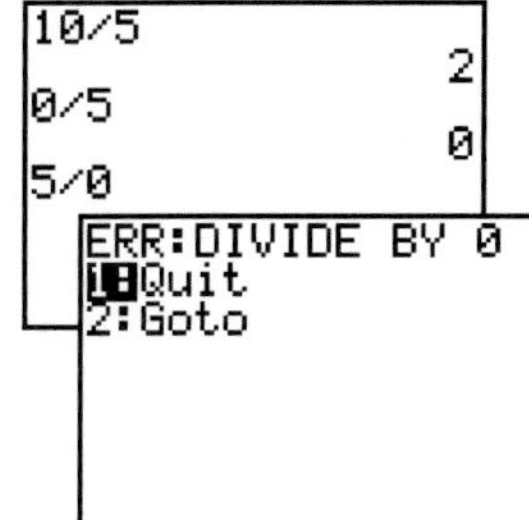

Note: Dividing any real number by zero is undefined.

SELF-CHECK 1.6.3

Mentally determine the sign of each quotient, and then use a calculator to calculate the quotient.

1. $-8.05 \div 2.3$ **2.** $0.728 \div (-0.014)$ **3.** $\left(-\frac{2}{3}\right) \div \left(-\frac{5}{7}\right)$ **4.** $-\frac{3}{7} \div 0$

The algebraic expressions evaluated in Example 4 both involve division. One of the properties that we use in this example is that $\frac{\sqrt{a}}{\sqrt{b}} = \sqrt{\frac{a}{b}}$ for both $a > 0$ and $b > 0$.

EXAMPLE 4 Evaluating Algebraic Expressions

Evaluate the following expressions for $x = 48$, $y = -4$, and $z = 3$.

SOLUTION

(a) $\frac{-x}{y}$

$$\frac{-x}{y} = \frac{-(48)}{(-4)}$$

Substitute 48 for x and -4 for y.

$$= \frac{12(\cancel{-4})}{\cancel{-4}}$$

Divide both the numerator and the denominator by the reducing factor -4. Note that $\frac{-4}{-4}$ equals 1; this is often understood and not written. The quotient of like signs is positive.

$$= 12$$

(b) $\frac{\sqrt{x}}{\sqrt{z}}$

$$\frac{\sqrt{x}}{\sqrt{z}} = \frac{\sqrt{48}}{\sqrt{3}}$$

$$= \sqrt{\frac{48}{3}}$$

Substitute for x and z. Then simplify, using the fact that $\frac{\sqrt{a}}{\sqrt{b}} = \sqrt{\frac{a}{b}}$ for both $a > 0$ and $b > 0$.

$$= \sqrt{16}$$
$$= 4$$

Reduce the fraction to lowest terms and then take the square root of 16.

The ability to estimate answers is such an important skill that we examine this topic in a variety of contexts. In Example 5 we estimate a quotient.

EXAMPLE 5 Estimate Then Calculate

Estimate $682.5 \div 2.1$ and then use a calculator to perform this division.

SOLUTION

ESTIMATED VALUE

$682.5 \approx 680$ $\quad 680 \div 2 = 340$
$2.1 \approx 2$

Answer: 325 seems reasonable based on the estimated value of 340.

CALCULATOR VALUE

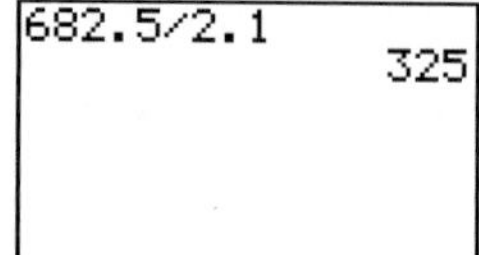

In Example 5, the estimated value of 340 is 15 more than the actual value of 325. To judge how good this estimate is, we should examine not only our error of estimate +15, but also the relative size of this error compared to the magnitude of the number being estimated. The **error,** or **error of estimate,** is the estimated value minus the actual value. **Relative error** is the error divided by the actual value. Relative error is often expressed as a percent and referred to as **percent of error.** For Example 5 the relative error is $\frac{15}{325} \approx 0.0462$, and the percent of error is 4.62%. Thus, our mental estimate in this example was within 5% of the actual answer. Many calculator errors are the result of keystrokes that produce incorrectly placed decimal points, missing digits, etc. These errors generate large relative errors and thus often can be detected by simply thinking about our results and asking if the results seem reasonable based upon our mental estimate.

EXAMPLE 6 Calculating Percent of Error

44.1 m

58.7 m

A contractor estimated the dimensions of the region shown to the left as 45 m by 60 m and calculated the approximate area. What was the percent of error of this estimate?

SOLUTION

Actual area:

$A = l \cdot w$

$A = (44.1 \text{ m})(58.7 \text{ m})$

$A = 2{,}588.67 \text{ m}^2$

Use the formula for the area of a rectangle with the actual dimensions. The calculator value gives the area in square meters.

Estimated area:

$A = l \cdot w$

$A \approx (45 \text{ m})(60 \text{ m})$

$A \approx 2{,}700 \text{ m}^2$

Use the estimated values in the same formula.

$$\boxed{\text{Error of estimate}} = \boxed{\text{Estimated value}} - \boxed{\text{Actual value}}$$

$$= 2{,}700 \text{ m}^2 - 2{,}588.67 \text{ m}^2$$

$$= +111.33 \text{ m}^2$$

The estimated value is larger than the actual value.

$$\boxed{\text{Percent of error}} = \boxed{\text{Error of estimate}} \div \boxed{\text{Actual value}}$$

$$= 111.33 \text{ m}^2 \div 2{,}588.67 \text{ m}^2$$

$$\approx 0.043$$

$$\approx 4.3\%$$

First use a calculator to determine the percent of error as a decimal, and then convert to a percent.

Answer: The contractor's percent of error was +4.3%.

Many contractors prefer to be a little high on their estimate so that they do not run short on materials.

SELF-CHECK 1.6.4

1. Rework Example 6, using estimated dimensions of 44 m by 59 m to calculate the area of the region. Then calculate the percent of error for this estimate.

Two statistics that are commonly used to analyze a set of numbers are the range and the mean. The **range** of a set of numerical scores measures the number of units that separate the smallest score from the largest score and is calculated by subtracting the smallest score from the largest score. The **mean** of a set of numerical scores is an average calculated by dividing the sum of scores by the number of scores. This mean is the number most commonly referred to when the term *average* is used. Other common uses of the term *average* include the average grade of a student, average number of miles per gallon of gasoline an automobile will achieve, and average cost per ounce of a product. In each case, as illustrated in the following examples, division is used to obtain the desired average.

EXAMPLE 7 Finding the Mean and Range of Test Scores

A student made five grades of 75, 88, 94, 78, and 91 on 100-point tests. Find the mean and the range of these test scores.

SOLUTION

$$\boxed{\text{Mean test score}} = \boxed{\text{Total points}} \div \boxed{\text{Number of tests}}$$

$$\text{Mean} = \frac{75 + 88 + 94 + 78 + 91}{5}$$

$$= \frac{426}{5}$$

$$= 85.2$$

Substitute the given values into the word equation for the mean. The numerator is the total number of points, and the denominator is the number of tests.

$$\boxed{\text{Range}} = \boxed{\text{Highest score}} - \boxed{\text{Lowest score}}$$

$$= 94 - 75$$

$$= 19$$

Substitute the highest score of 94 and the lowest score into the word equation for the range.

Answer: The student's mean score is 85.2, and the range of these scores is 19.

A Mathematical Note

The colon has been used by the French to indicate division. Thus the use of the colon in the ratio $a : b$ indicates $a \div b$.

The ratio of a to b is the quotient $a \div b$. This ratio is sometimes written as $a : b$ and read as "the ratio of a to b." Since $a \div b = \frac{a}{b}$, ratios are frequently written as fractions in reduced form.

Ratio

VERBALLY	ALGEBRAICALLY	NUMERICAL EXAMPLE
The ratio of a to b is the quotient of a divided by b.	The ratio a to b can be denoted by either $a : b$ or $\frac{a}{b}$.	The ratio 3 to 7 can be denoted by either 3 : 7 or $\frac{3}{7}$.

Sometimes both the numerator and the denominator of a ratio can be expressed in terms of the same unit. In this case the reduced form of the ratio is a number free of units. This is illustrated in Example 8.

EXAMPLE 8 Determining the Ratio of Two Measurements

A flagpole 8 m tall casts a 2-m shadow (figure is on next page). Determine the ratio of the height of the flagpole to the length of its shadow.

SOLUTION

$$\frac{\boxed{\text{Height of flagpole}}}{\boxed{\text{Length of shadow}}} = \frac{8 \text{ m}}{2 \text{ m}}$$ Divide both the numerator and denominator of this ratio by 2 m.

$$= \frac{4}{1}$$ The reduced form of the ratio is a number free of units.

$$= 4$$ The flagpole is taller than the shadow by a factor of 4.

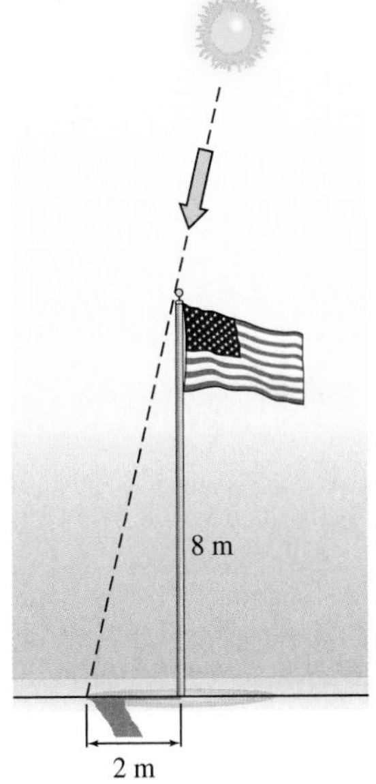

Answer: The ratio of the height of the flagpole to the length of its shadow is $\frac{4}{1} = 4:1$.

SELF-CHECK 1.6.5

1. The front gear of a bicycle has 52 teeth, and the rear gear has 16 teeth. What is the ratio of the teeth on the front gear to those on the rear gear?

The best price for a commodity (such as laundry detergent) that is used in various quantities over time is often determined by calculating the lowest unit price. The **unit price** of an item is the ratio of the price to the number of units.

EXAMPLE 9 Determining a Unit Price

A brand-name liquid laundry detergent is packaged in a 6-lb 4-oz container that sells for \$7.89 and a 4-lb container that sells for \$4.80. Which size is the better buy?

SOLUTION

The size with the lower unit price is the better buy.

$\boxed{\text{Unit price}} = \boxed{\text{Price}} \div \boxed{\text{Number of ounces}}$

The unit price is the average price per ounce. Calculate the unit price of each box, and then determine the one with the lower unit price.

SMALLER SIZE

$$\text{Unit Price} = \frac{\$4.80}{4 \text{ lb}} = \frac{\$4.80}{64 \text{ oz}} = \$0.075/\text{oz}$$

LARGER SIZE

$$\text{Unit price} = \frac{\$7.89}{6 \text{ lb } 4 \text{ oz}} = \frac{\$7.89}{100 \text{ oz}} = \$0.0789/\text{oz}$$

$4 \text{ lb} = 4(16) \text{ oz} = 64 \text{ oz}$

$6 \text{ lb } 4 \text{ oz} = [6(16) + 4] \text{ oz} = 100 \text{ oz}$

Answer: Since the smaller size has the lower price per ounce, it is the better buy.

Although many people think larger sizes are always a better buy, this is not always true.

SELF-CHECK ANSWERS

1.6.1

1. -2
2. $\frac{11}{17}$
3. $\frac{C}{d} = \pi$
4. $-\frac{22}{21}$

1.6.2

1. 1
2. -25
3. Undefined
4. 0

1.6.3

1. Negative, -3.5
2. Negative, -52
3. Positive, $\frac{14}{15}$
4. An error message on a calculator since dividing by zero is undefined.

1.6.4

1. 2,596 m^2 is 0.3% greater than the actual value.

1.6.5

1. The ratio of the teeth on the front gear to those on the rear gear is 13 : 4.

USING THE LANGUAGE AND SYMBOLISM OF MATHEMATICS 1.6

1. In the division $24 \div 3 = 8$, 24 is called the dividend, 3 is called the __________, and 8 is called the __________.
2. The equation "The quotient of D divided by R equals T" is represented algebraically by __________.
3. The equation "The ratio of men m to women w is three-fifths" is represented algebraically by __________.
4. If x and y ($y \neq 0$) have the same sign, then the sign of the quotient $\frac{x}{y}$ is __________.
5. If x and y ($y \neq 0$) have unlike signs, then the sign of the quotient $\frac{x}{y}$ is __________.
6. To divide $\frac{5}{6}$ by $\frac{2}{3}$, we can multiply $\frac{5}{6}$ by $\frac{3}{2}$, the __________ inverse or the __________ of $\frac{2}{3}$.
7. The quotient $\frac{w}{x} \div \frac{y}{z} =$ __________ for $x \neq 0$, __________ $\neq 0$, and $z \neq$ __________.
8. The notation $a : b$ is read the __________ of a to b.
9. Division by zero is __________.
10. For both $a > 0$ and $b > 0$, $\frac{\sqrt{a}}{\sqrt{b}} =$ __________.
11. The __________ of a set of numbers is the largest number minus the smallest number.
12. The __________ of a set of numerical scores is an average calculated by dividing the sum of the scores by the number of scores.
13. The difference between an estimate and the actual value is the error of the estimate. The __________ error is the error of the estimate divided by the actual value.
14. The unit price of an item is the ratio of the price to the number of __________.

EXERCISES 1.6

In Exercises 1–12, calculate each quotient without using a calculator.

1. a. $48 \div (-6)$ b. $-48 \div (-6)$ c. $-48 \div 6$ d. $0 \div 6$
2. a. $56 \div (-8)$ b. $-56 \div (-8)$ c. $-56 \div 8$ d. $0 \div (-8)$
3. a. $-48 \div \left(\frac{1}{2}\right)$ b. $-48 \div 2$ c. $-48 \div \left(-\frac{1}{3}\right)$ d. $-48 \div (-3)$
4. a. $48 \div \left(-\frac{1}{4}\right)$ b. $48 \div (-4)$ c. $-48 \div \left(-\frac{1}{8}\right)$ d. $-48 \div (-8)$
5. a. $123 \div (0.1)$ b. $123 \div (-0.01)$ c. $123 \div (-1{,}000)$ d. $-123 \div (-10)$
6. a. $-123 \div (0.01)$ b. $-123 \div (100)$ c. $123 \div (-0.001)$ d. $123 \div (-10{,}000)$
7. a. $\frac{2}{3} \div \frac{1}{6}$ b. $-\frac{2}{3} \div \frac{1}{6}$ c. $-\frac{2}{3} \div (-6)$ d. $\frac{2}{3} \div (-1)$
8. a. $\frac{4}{5} \div 20$ b. $-\frac{4}{5} \div \frac{1}{20}$ c. $-\frac{4}{5} \div (-20)$ d. $\frac{1}{20} \div \left(-\frac{4}{5}\right)$
9. a. $0 \div (-7)$ b. $-7 \div 0$ c. $7 \div 0$ d. $0 \div 0$
10. a. $0 \div (-11)$ b. $-11 \div 0$ c. $11 \div 0$ d. $-0.01 \div 0$
11. a. $\frac{\sqrt{18}}{\sqrt{2}}$ b. $\frac{\sqrt{75}}{\sqrt{3}}$
12. a. $\frac{\sqrt{44}}{\sqrt{11}}$ b. $\frac{\sqrt{45}}{\sqrt{5}}$

In Exercises 13 and 14, evaluate each expression for $w = 15$, $x = 5$, and $y = 10$.

13. a. $\dfrac{w}{x}$ **b.** $\dfrac{-w}{x}$

c. $\dfrac{w}{-x}$ **d.** $\dfrac{-w}{-x}$

14. a. $\dfrac{x}{y}$ **b.** $\dfrac{-x}{y}$

c. $\dfrac{x}{-y}$ **d.** $\dfrac{-x}{-y}$

In Exercises 15 and 16, check both −6 and 6 to determine whether either is a solution of these equations. (*Hint:* See Example 8 in Section 1.4.)

15. a. $-\dfrac{1}{3}x = x - 8$ **b.** $\dfrac{x}{3} = -8 - x$

16. a. $7 + x = \dfrac{-x}{6}$ **b.** $7 - x = \dfrac{1}{6}x$

In Exercises 17 and 18, calculate the first five terms of each sequence. (*Hint:* See Example 9 in Section 1.4.)

17. a. $a_n = \dfrac{120}{n}$ **b.** $a_n = \dfrac{-n}{2}$

18. a. $a_n = \dfrac{600}{-n}$ **b.** $a_n = \dfrac{n^2}{2}$

Sum of Two Factors

In Exercises 19–24, complete the following table. First determine the missing factor of 36, and then compute the sum of x and y.

	x	y	$x + y$	xy
Example	1	36	37	36
19.	2			36
20.	3			36
21.	−4			36
22.	−6			36
23.	6			36
24.	12			36

25. Ratio of Diamonds Thirteen of 52 cards in a deck of cards are diamonds. What is the ratio of diamonds to all cards in the deck?

26. Defective Computer Chips An inspection of 1,056 experimental computer chips found 132 defective. What fractional portion of the chips was defective?

27. Tree Shadow A tree 24 m tall casts an 8-m shadow. Determine the ratio of the height of the tree to the length of its shadow.

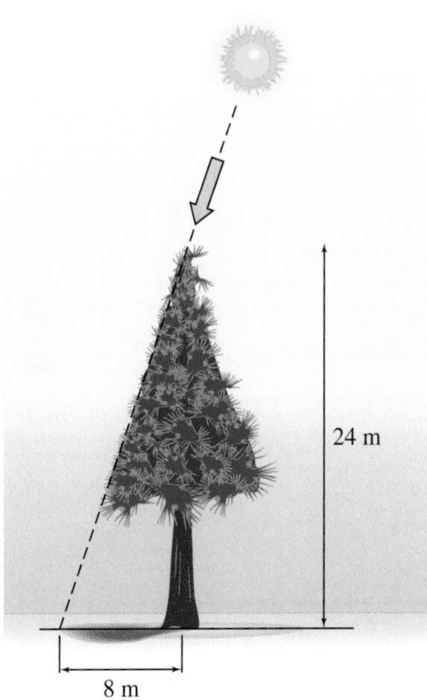

28. Shadow of a Man A young man 2 m tall casts a 6-m shadow. Determine the ratio of his height to the length of his shadow.

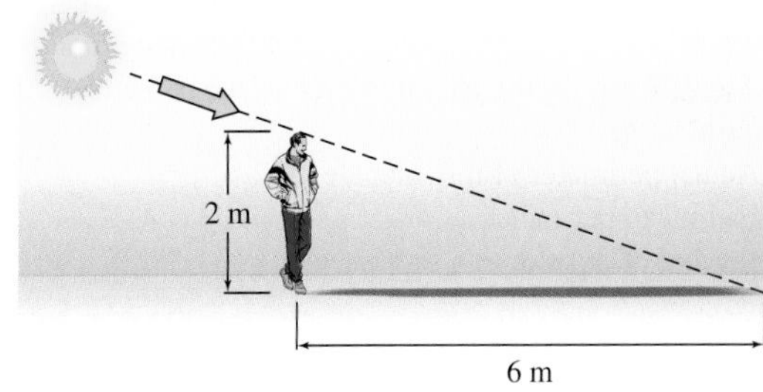

29. Banded Ducks A wildlife study involved banding ducks and then studying those birds when they were recaptured. Of the 85 captured ducks, 17 had already been banded.

a. What is the ratio of banded ducks to the total captured?

b. What is the ratio of banded ducks to those not banded?

30. Banded Rabbits A wildlife study involved banding rabbits' ears and then studying those banded rabbits when they were recaptured. Of the 117 recaptured rabbits, 26 had already been banded.

a. What is the ratio of banded rabbits to the total recaptured?

b. What is the ratio of banded rabbits to those not banded?

31. Comparison of Two Pulleys The radius of one pulley is 24 cm, and the radius of a second pulley is 15 cm. Determine the ratio of the radius of the larger pulley to that of the smaller pulley.

32. Gear Ratio The front gear of a go-cart has 48 teeth, and the rear gear has 16 teeth. What is the ratio of the teeth on the front gear to those on the rear gear?

33. Best Buy for Ketchup One brand of ketchup is sold in two popular sizes. Which is the better buy, a 40-oz bottle that is priced at $2.52 or a 64-oz bottle that is priced at $3.68?

34. Best Buy for Corn Flakes One brand of corn flakes is sold in several sizes. Which is the best buy, a 7-oz box that is priced at \$1.22, a 12-oz box priced at \$2.04, or an 18-oz box priced at \$3.15?

35. **Test Average** A student scored 78, 85, 93, and 72 on four history exams. What is the student's average for these four exams?

36. **Test Average** A student scored 86, 78, 82, and 80 on four algebra exams. What is the student's average for these four exams?

37. **Percent of Error** The perimeter of the rectangle in the figure is estimated by using 30 cm for the width and 50 cm for the length. Determine the
   a. Actual perimeter of this rectangle
   b. Estimated perimeter
   c. Percent of error of this estimate

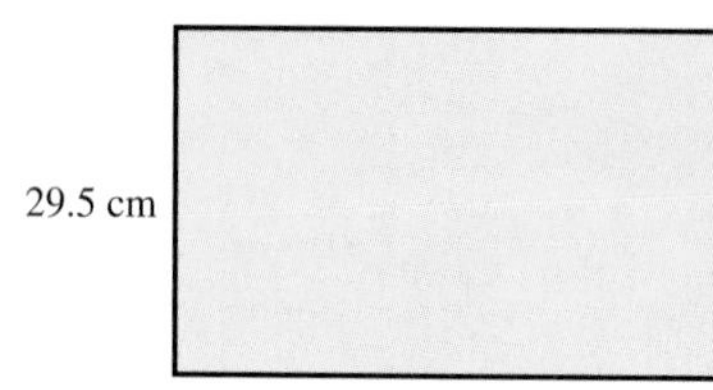

38. **Percent of Error** Using the estimates for the width and length of the rectangle in Exercise 37, determine the
   a. Actual area of this rectangle
   b. Estimated area
   c. Percent of error of this estimate

39. **Percent of Error** The perimeter of the isosceles triangle in the figure is estimated by using 20 cm for the length of the base and 16 cm for the length of each of the two equal sides. Determine the
   a. Actual perimeter of this triangle
   b. Estimated perimeter
   c. Percent of error of this estimate

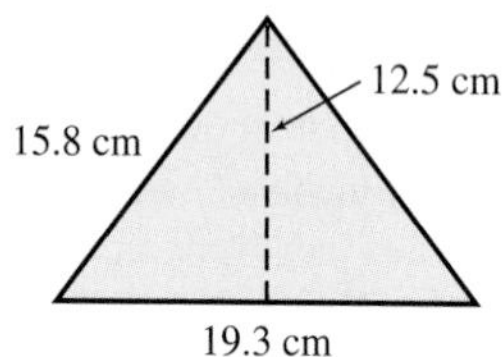

40. **Percent of Error** Using the 20-cm estimate for the base and a 12-cm estimate for the height of the triangle in Exercise 39, determine the
   a. Actual area of this triangle
   b. Estimated area
   c. Percent of error of this estimate

41. **Range and Mean** Use a calculator and the utility bills for the months January through June to determine the
   a. Range of these bills b. Mean of these bills

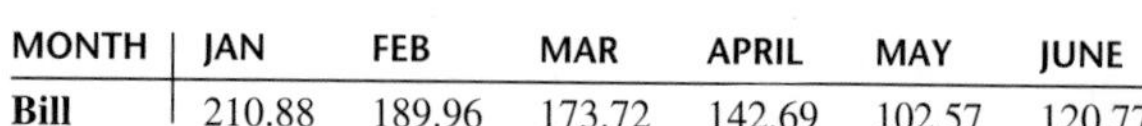

MONTH	JAN	FEB	MAR	APRIL	MAY	JUNE
Bill	210.88	189.96	173.72	142.69	102.57	120.77

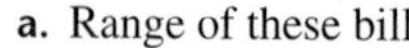

42. **Range and Mean** Use a calculator and the telephone bills for the months January through June to determine the
   a. Range of these bills b. Mean of these bills

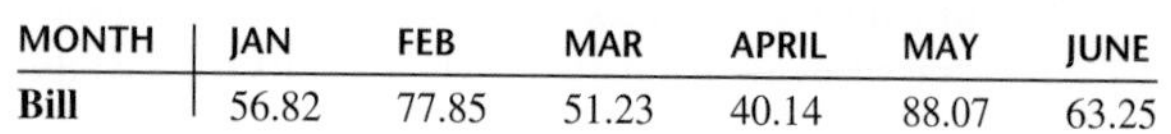

MONTH	JAN	FEB	MAR	APRIL	MAY	JUNE
Bill	56.82	77.85	51.23	40.14	88.07	63.25

Rate of Travel

The average rate at which an object travels can be calculated by dividing the distance it travels by the time it takes to travel that distance. A car that drives 300 miles in 5 hours travels at an average rate of $\frac{300 \text{ miles}}{5 \text{ hours}} = 60$ mi/h. In Exercises 43–46, use the formula $R = \frac{D}{T}$ to compute the rate at which each object travels.

43. A plane flies 870 mi in 3 h.
44. A hiker walks 40 mi in 2 days.
45. A snail crawls 240 cm in 10 min.
46. A boat sails 1,200 km in 5 weeks.

Rate of Work

The rate at which a task is completed depends on the time required to complete the task. If a painter can paint 1 room in 3 hours, then the rate of work is $\frac{1}{3}$ room per hour. In Exercises 47–50, use the formula $R = \frac{W}{T}$ to compute the rate of work for each situation.

47. An assembly line produces 2,400 lightbulbs in 8 hours.
48. A bricklayer lays 500 bricks in 10 hours.
49. A hose fills 1 swimming pool in 5 hours.
50. A roofer shingles 1 roof in 12 hours.

Estimate Then Calculate

In Exercises 51–56, mentally determine the correct sign of the quotient, and then use a calculator to calculate the quotient. Approximate the quotient in Exercises 55 and 56 to the nearest thousandth. (*Hint:* See Calculator Perspective 1.6.1.)

PROBLEM	SIGN OF QUOTIENT (WITHOUT CALCULATOR)	QUOTIENT (WITH CALCULATOR)
51. $-8.2 \div 2.5$		
52. $-5.7 \div (-0.02)$		
53. $-\frac{12}{25} \div \left(\frac{-16}{15}\right)$		
54. $\frac{-42}{55} \div \frac{-14}{-66}$		
55. $\sqrt{37} \div (-\sqrt{17})$		
56. $-\sqrt{111} \div \sqrt{11}$		

Estimation Skills

In Exercises 57–60 mentally estimate the value of each expression, and then select the most appropriate answer.

57. $-28.5 \div 11.4$
   a. -17.5 b. 17.5 c. -2.5
   d. 2.5 e. -5.2

58. $-97.68 \div (-4.4)$
   a. -22.2 b. 22.2 c. -32.2
   d. 32.2 e. 52.2

59. $45.1 \div (-0.11)$
   a. -610 b. 610 c. -410
   d. 410 e. 31

60. $7.8 \div (-0.12)$
   a. -95 b. 95 c. -65
   d. 65 e. -35

In Exercises 61 and 62, mentally estimate the mean of each set of data, and then select the answer closest to your estimate.

61. 8.4, 9.7, and 10.4

a. 8.5 b. 9.5 c. 10.1 d. 10.5

62. 2.98, 7.05, and 5.75

a. 3.26 b. 4.26 c. 5.26 d. 6.26

63. College Tuition The table below gives the tuition costs for a student enrolling in 15 semester hours at two different colleges. Complete the rightmost column of this table by calculating the cost per credit hour at each college.

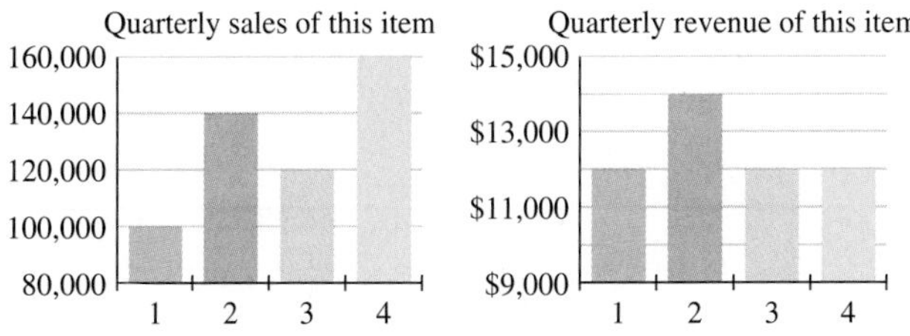

	A	B	C
1	COLLEGE	TUITION ($)	COST PER CREDIT HOUR ($)
2	University of Florida	1547	
3	North Florida Community College	930	
4			

64. Tuition Comparison Use the table in Exercise 63 to determine the ratio of the cost of tuition at the University of Florida to the cost of tuition at the North Florida Community College for a student enrolling in 15 semester hours. Interpret this result.

65. The first bar graph gives the number of an item sold each quarter. The second bar graph gives the sales revenue for this item for each quarter. Calculate the price per item for each quarter.

Quarterly sales of this item

160,000
140,000
120,000
100,000
80,000
1 2 3 4

Quarterly revenue of this item

$15,000
$13,000
$11,000
$9,000
1 2 3 4

66. The first bar graph gives the number of an item sold each quarter. The second bar graph gives the profit generated by the sale of this item for each quarter. Calculate the profit per item for each quarter.

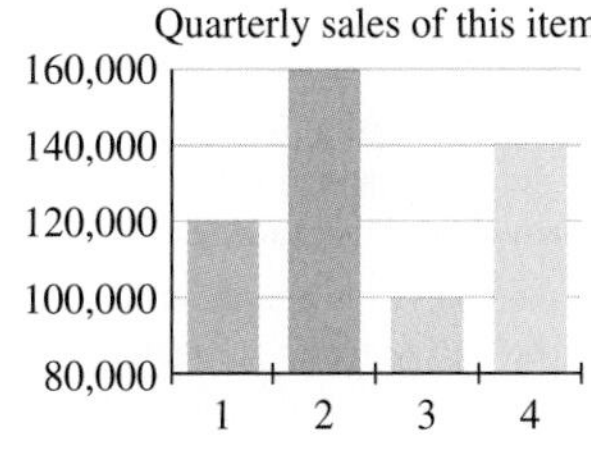

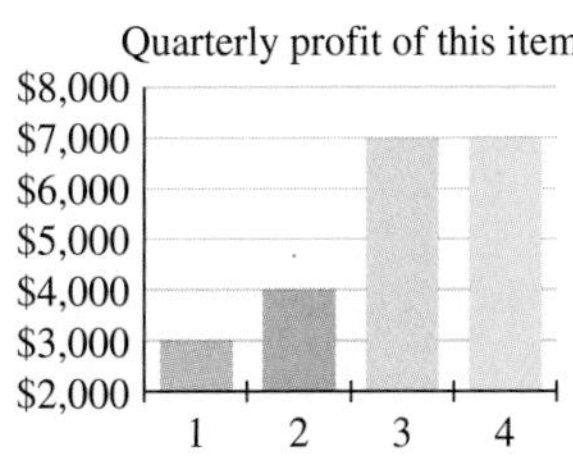

67. Height of a Box The volume of the box shown below is 2,552 cm^3 (cubic centimeters). The width of the box is 8 cm and the length is 22 cm. Find the height h of this box.

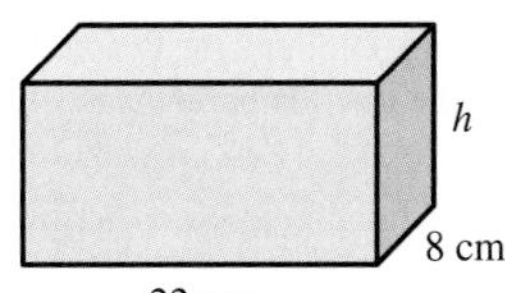

68. Width of a Box The volume of the box shown below is 1,512 cm^3. The length of the box is 18 cm and the height is 7 cm. Find the width w of this box.

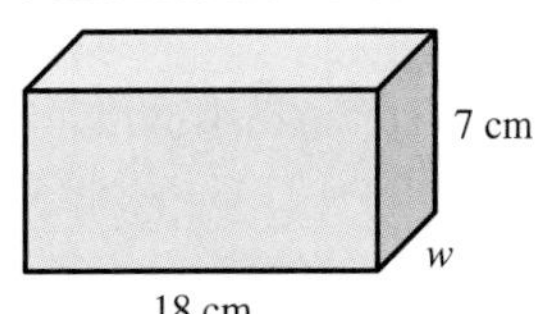

Multiple Representations

In Exercises 69–73 write each verbal statement in algebraic form.

69. a. The quotient of a divided by five.

b. The quotient of d divided by t is seven.

70. a. The ratio of h to r.

b. The ratio of V to R is I.

71. a. The quotient of x and y is eight.

b. The product of x and y is eight.

72. a. Twelve divided by x is approximately equal to y.

b. Nine divided by z is not equal to four.

73. a. Five is greater than x divided by three.

b. Negative two is less than or equal to eight divided by w.

In Exercises 74 and 75, fill in each blank.

74. The only real number that is neither negative nor positive is __________.

75. The quotient of a nonzero number and its additive inverse is __________.

Group Discussion Questions

76. Challenge Question

a. Write four numbers with a mean of 100 and a range of 0.

b. Write four numbers with a mean of 100 and a range of 100.

77. Challenge Question Simplify

$$\frac{11}{12}\cdot\frac{12}{13}\cdot\frac{13}{14}\cdot\frac{14}{15}\cdot\frac{15}{16}\cdot\frac{16}{17}\cdot\frac{17}{18}\cdot\frac{18}{19}\cdot\frac{19}{20}\cdot\frac{20}{21}\cdot\frac{21}{22}$$

78. Error Analysis In trying to evaluate the expression $\frac{2}{5} \div \frac{3}{8}$ on a TI-84 Plus calculator, a student enters the expression shown below. What is the error and how can it be corrected?

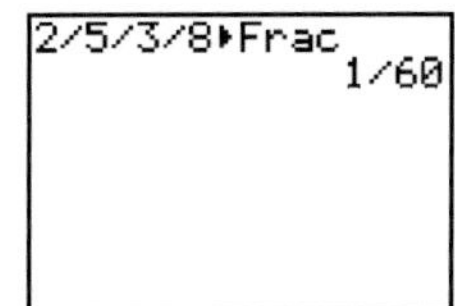

79. Discovery Question

a. Have each person in your group mentally estimate the appropriate tip for each of the restaurant bills in the table below. Then compare how you made your estimates. Which method do you think is best?

BILL	10% TIP	15% TIP	20% TIP
$13.86			
$25.43			
$36.17			
$45.82			
$68.14			

b. What does your group consider to be the appropriate amount to tip, 10%, 15%, or 20%?

Section 1.7 Order of Operations

Objectives:
19. Use the correct order of operations.
20. Use the distributive property.
21. Simplify expressions by adding like terms.

Symbols of grouping are used in mathematical expressions to separate signs, to enclose an expression to be treated as one quantity, and to indicate the order of operations within an expression. The common grouping symbols are the fraction bar —, which separates the numerator from the denominator; parentheses ( ); brackets [ ]; and braces { }. For problems involving more than one operation, mathematicians have developed the following order of operations so that each expression will have a unique interpretation and value.

Order of Operations

Step 1 Start with the expression within the innermost pair of grouping symbols.
Step 2 Perform all exponentiations.
Step 3 Perform all multiplications and divisions as they appear from left to right.
Step 4 Perform all additions and subtractions as they appear from left to right.

This order of operations has been adopted by the manufacturers of most calculators and incorporated into many computer languages. You should verify this hierarchy on a calculator before undertaking any important calculation.

EXAMPLE 1 Using the Correct Order of Operations

Evaluate each of the following expressions, using the correct order of operations.

SOLUTION

(a) $3 + 4 \cdot 7$

$$3 + 4 \cdot 7 = 3 + 28$$
$$= 31$$

Multiplication has priority over addition.

(b) $(3 + 4)(7)$

$$(3 + 4)(7) = 7(7)$$
$$= 49$$

First simplify the expression within the parentheses. Note the distinction between part (a) and part (b) of this example.

(c) $5 \cdot 8 - 12 \div 3 + 1$

$$5 \cdot 8 - 12 \div 3 + 1 = 40 - 12 \div 3 + 1$$
$$= 40 - 4 + 1$$
$$= 36 + 1$$
$$= 37$$

Multiplication and division are performed as they appear from left to right. Then addition and subtraction are performed as they appear from left to right.

(d) $24 \div 6 \cdot 2$

$$24 \div 6 \cdot 2 = 4 \cdot 2$$
$$= 8$$

Multiplication and division are performed as they appear from left to right.

(e) $\dfrac{12 - 3(2)}{(12 - 3)(2)}$

$$\frac{12 - 3(2)}{(12 - 3)(2)} = \frac{12 - 6}{9(2)}$$
$$= \frac{6}{18}$$
$$= \frac{1}{3}$$

Simplify the numerator and the denominator separately, as the fraction bar indicates this grouping.
Then reduce the fraction to lowest terms.

When you are entering an expression containing fractions, it is usually wise to include a set of parentheses around the entire numerator and the entire denominator. This is illustrated in Calculator Perspective 1.7.1.

CALCULATOR PERSPECTIVE 1.7.1 **Using Implied Parentheses**

To evaluate $\dfrac{12 - 3(2)}{(12 - 3)(2)}$ from Example 1(e) on a TI-84 Plus calculator, enter the following keystrokes:

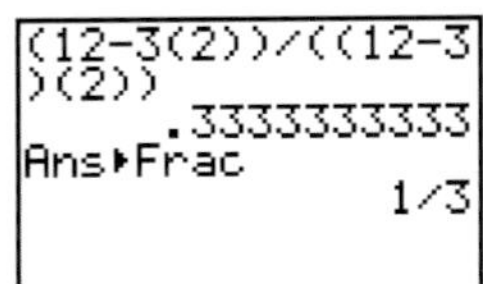

Note: Failure to use this additional set of parentheses on both the numerator and the denominator will result in an error.

Whenever more than one set of grouping symbols appears in an expression, you should start from the innermost pair of symbols and work toward the outermost pair of symbols. In Example 2 we simplify first within the parentheses and then within the brackets.

EXAMPLE 2 Evaluating Expressions with Grouping Symbols

Evaluate each of the following expressions.

SOLUTION

(a) $20 + 3[(7 + 8) - (3 + 6)]$

$20 + 3[(7 + 8) - (3 + 6)]$
$= 20 + 3[15 - 9]$ Simplify first within the parentheses and then within the brackets.
$= 20 + 3[6]$ Multiplication has priority over addition.
$= 20 + 18$
$= 38$

(b) $149 - 2[7 + 3(11 + 2^2) - 5]$

$149 - 2[7 + 3(11 + 2^2) - 5]$
$= 149 - 2[7 + 3(11 + 4) - 5]$ Start with the expression within the innermost pair of grouping symbols and work outward.
$= 149 - 2[7 + 3(15) - 5]$ Multiplication has priority over addition and subtraction.
$= 149 - 2[7 + 45 - 5]$
$= 149 - 2[47]$
$= 149 - 94$
$= 55$

SELF-CHECK 1.7.1

Evaluate each of the following expressions

1. $18 - 4 \cdot 3$ **2.** $(18 - 4)(3)$ **3.** $\dfrac{20 - 45 \div 3}{8 + 6 \cdot 2}$ **4.** $5 \cdot 6 - 8 \div 2$

5. $[14 - 5(13 - 11)][6 + 2(14 - 3)]$

In Example 2 both brackets and parentheses are used in the same expression. This notation is usually used in printed mathematics because it is less confusing than using multiple levels of parentheses. However, $20 + 3[(7 + 8) - (3 + 6)]$ can also be denoted by $20 + 3((7 + 8) - (3 + 6))$. In fact, multiple levels of parentheses must be used with many calculators and in some computer program statements. To be interpreted correctly, these statements must have the parentheses paired properly and must be evaluated by starting with the innermost pair of parentheses.

When you are entering expressions containing parentheses, be careful to keep track of pairs of left and right parentheses. Note the nested parentheses in Calculator Perspective 1.7.2.

CALCULATOR PERSPECTIVE 1.7.2 **Using Parentheses**

To evaluate $20 + 3[(7 + 8) - (3 + 6)]$ from Example 2(a) on a TI-84 Plus calculator, enter the following keystrokes:

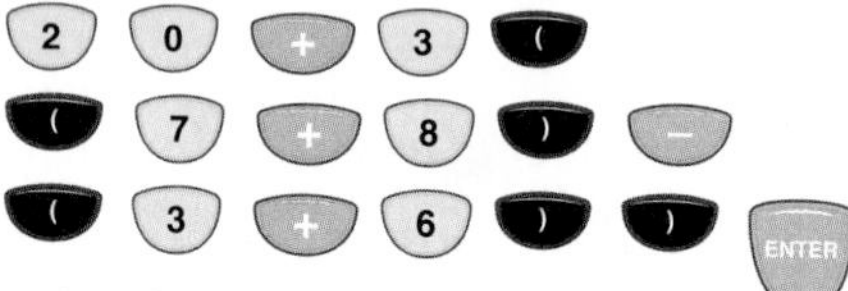

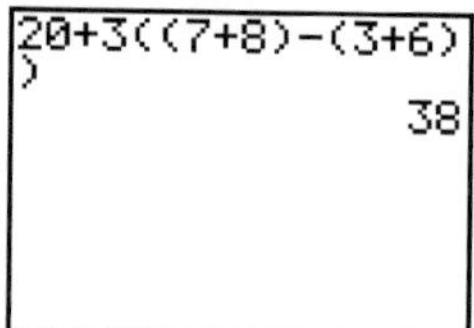

Note: Do not use the brackets on a calculator as they are reserved for a special purpose and are not used as general-purpose grouping symbols.

Order-of-operations errors can occur subtly in expressions involving exponents. One common error involves confusing the notations in parts (a) and (b) in Example 3. Note carefully the distinct meanings of each of these two expressions.

EXAMPLE 3 Using the Correct Order of Operations

Evaluate each of the following expressions, using the correct order of operations.

SOLUTION

(a) $2^2 + 3^2$

$2^2 + 3^2 = 4 + 9$
$= 13$

Exponentiation has priority over addition.

(b) $(2 + 3)^2$

$(2 + 3)^2 = 5^2$
$= 25$

First simplify within the parentheses, then square. Notice that this is a different answer from that in part (a) of this example.

(c) $2 \cdot 5^3 - 6^2 \div 3$

$2 \cdot 5^3 - 6^2 \div 3$
$= 2(125) - 36 \div 3$
$= 250 - 12$
$= 238$

Exponentiation has the highest priority, followed by multiplication and division, and lastly subtraction.

Absolute value symbols and square root symbols are also used to group terms. It is important to note the significant differences in the order of operations illustrated in Example 4.

EXAMPLE 4 Using the Correct Order of Operations

Evaluate each of the following expressions, using the correct order of operations.

SOLUTION

(a) $\sqrt{25} + \sqrt{144}$

$\sqrt{25} + \sqrt{144} = 5 + 12$
$= 17$

First evaluate each separate square root, and then add these two terms.

(b) $\sqrt{25 + 144}$

$\sqrt{25 + 144} = \sqrt{169}$
$= 13$

First add the terms underneath the square root symbol, and then take the square root of this result. Notice that this is a different answer from that in part (a) of this example.

(c) $|-20| + |14|$

$|-20| + |14| = 20 + 14$
$= 34$

First evaluate each absolute value, and then add these two terms.

(d) $|-20 + 14|$

$|-20 + 14| = |-6|$
$= 6$

First add the terms within the absolute value symbols, and then take the absolute value of this result. Notice that this is a different answer from that in part (c) of this example.

SELF-CHECK 1.7.2

Evaluate each of the following expressions.

1. $-4^2 + 5^2$
2. $(-4)^2 + 5^2$
3. $(-4 + 5)^2$
4. $|-4 + 5|$
5. $|-4| + |5|$
6. $\sqrt{25 - 9}$
7. $\sqrt{25} - \sqrt{9}$

To be interpreted correctly, expressions involving different operations, such as addition and multiplication, often require grouping symbols. However, expressions involving only the operation of addition or only the operation of multiplication are usually written without grouping symbols. In these circumstances, grouping symbols are not needed, as all possible groupings yield the same value. For example, we can write $3 + 4 + 5$ without parentheses since $(3 + 4) + 5$ and $3 + (4 + 5)$ are equal. Recall that it is the associative property of addition that allows terms to be regrouped in different ways. Similarly, $3 \cdot 4 \cdot 5$ can be written as either $(3 \cdot 4)(5)$ or $(3)(4 \cdot 5)$ because of the associative property of multiplication.

There is also one property that relates multiplication and addition. Compare the following products.

$$5(2 + 6) = 5 \cdot 8 = 40 \qquad 5 \cdot 2 + 5 \cdot 6 = 10 + 30 = 40$$

Thus $5(2 + 6) = 5 \cdot 2 + 5 \cdot 6$. In general, the **distributive property** of multiplication over addition says $a(b + c) = ab + ac$.

A Mathematical Note

The French mathematician Servois (ca. 1814) introduced the terms *commutative* and *distributive*. The term *associative* is attributed to the Irish mathematician William R. Hamilton (1805–1865).

Distributive Property of Multiplication Over Addition

ALGEBRAICALLY	VERBALLY	NUMERICAL EXAMPLES
For all real numbers a, b, and c, $a(b + c) = ab + ac$ and $(b + c)a = ba + ca$	Multiplication distributes over addition.	$5(2 + 6) = 5 \cdot 2 + 5 \cdot 6$ $5 \cdot 8 = 10 + 30$ $40 = 40$ $(2 + 6)5 = 2 \cdot 5 + 6 \cdot 5$ $8 \cdot 5 = 10 + 30$ $40 = 40$

Consider both $5(6 + 7)$ and $5(x + 7)$. We can use the distributive property to rewrite $5(6 + 7)$ as $5 \cdot 6 + 5 \cdot 7$ and rewrite $5(x + 7)$ as $5x + 35$. Using the distributive property is not crucial in $5(6 + 7)$, but it is necessary to expand $5(x + 7)$ as $5x + 35$. We use the distributive property extensively both to expand expressions such as $5(x + 7)$ to $5x + 35$ and to factor expressions such as $5x + 35$ to $5(x + 7)$.

EXAMPLE 5 Using the Distributive Property to Expand an Expression

Use the distributive property to expand each expression.

SOLUTION

(a) $9(x + 4)$

$$9(x + 4) = 9 \cdot x + 9 \cdot 4$$
$$= 9x + 36$$

Distribute the factor of 9 to both the x term and the 4 term.

(b) $11(2y - 3)$

$$11(2y - 3) = 11(2y) + 11(-3)$$
$$= 22y + (-33)$$
$$= 22y - 33$$

Multiplication distributes over both addition and subtraction. We can first think of this expression as $11[2y + (-3)]$.

(c) $-(7x - 3y)$

$$-(7x - 3y) = -1(7x - 3y)$$
$$= (-1)(7x) + (-1)(-3y)$$
$$= -7x + 3y$$

The factor of -1 is understood. Distribute the factor of -1 to both terms.

In Example 6 we use the distributive property to factor two expressions. This is the same property used in Example 5, just in the opposite direction.

EXAMPLE 6 Using the Distributive Property to Factor an Expression

Use the distributive property to factor each expression.

SOLUTION

(a) $5x - 5y$

$$5x - 5y = 5(x - y)$$

Five is a factor of both terms. Use the distributive property to factor out this common factor.

(b) $-4x + 8y$

$$-4x + 8y = -4(x) + (-4)(-2y)$$
$$= -4(x - 2y)$$

Factor out the common factor of -4. You can check this factorization by multiplying these two factors.

The distributive property also plays a key role in adding like terms and simplifying algebraic expressions. **Like** terms have exactly the same variable factors. Using the distributive property, we can rewrite $5x + 7x$ as $(5 + 7)x$ or $12x$. The constant factor in a term is called the **numerical coefficient** of the term. For example, the numerical coefficient of $5x$ is 5, and the numerical coefficient of $7x$ is 7. The coefficient of x is understood to be 1; that is, $1x = x$. The process of adding like terms is sometimes called **collecting like terms** or **combining like terms.**

EXAMPLE 7 Simplifying an Expression by Adding Like Terms

Use the distributive property to add the like terms.

SOLUTION

(a) $13x + 8x$

$$13x + 8x = (13 + 8)x$$
$$= 21x$$

(b) $13x - 8x$

$$13x - 8x = (13 - 8)x$$
$$= 5x$$

Usually we skip writing the middle step and perform this addition mentally. It is the distributive property that justifies what we are doing.

(c) $-7w + 4w$ $\quad$ $-7w + 4w = (-7 + 4)w$
$= -3w$

(d) $2a + (3b + 4a)$ $\quad$ $2a + (3b + 4a) = 2a + (4a + 3b)$
$= (2a + 4a) + 3b$
$= (2 + 4)a + 3b$
$= 6a + 3b$

Reorder by using the commutative property of addition. Then regroup by using the associative property of addition. Then add like terms, using the distributive property. Again, it is common to perform this addition mentally.

(e) $3\sqrt{2} + 4\sqrt{2}$ $\quad$ $3\sqrt{2} + 4\sqrt{2} = (3 + 4)\sqrt{2}$
$= 7\sqrt{2}$

Use the distributive property to factor out the common factor of $\sqrt{2}$. Then add the coefficients 3 and 4.

SELF-CHECK 1.7.3

1. Expand $5(x + 3y)$.
2. Factor $2a - 10b$.
3. Simplify $-8y + 17y$.
4. Simplify $5x + (2x - 3)$.

The estimation skill that we practice in Example 8 includes recognizing the correct order of operations in expressions involving exponents.

EXAMPLE 8 Estimate Then Calculate

Estimate $(6.07 + 3.98)^2 - (6.07^2 + 3.98^2)$, and then use a calculator to evaluate this expression.

SOLUTION

ESTIMATED VALUE

$$(6.07 + 3.98)^2 - (6.07^2 + 3.98^2) \approx 10^2 - (6^2 + 4^2)$$
$$\approx 100 - (36 + 16)$$
$$\approx 100 - 52$$
$$\approx 48$$

CALCULATOR VALUE

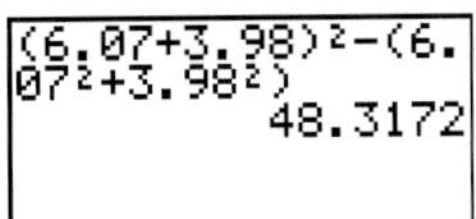

Answer: 48.3172 seems reasonable based on the estimated value of 48.

When we use the term **quantity** in the verbal statement of an algebraic expression, we are indicating that parentheses group the sum or difference that follows. This is significant because the expression $2 \cdot 3x + 1$ does not have the same meaning as $2(3x + 1)$. The statement "two times three x plus one" refers to $2 \cdot 3x + 1$, and the statement "two times the quantity three x plus one" refers to $2(3x + 1)$. We now illustrate this in Example 9.

EXAMPLE 9 Using the Language and Symbolism of Mathematics for Expressions Involving Parentheses

Give a verbal statement of each algebraic expression.

SOLUTION

ALGEBRAICALLY	VERBALLY
(a) $5(2x + 3)$	Five times the quantity two x plus three.
(b) $4(5x - 6)$	Four times the quantity five x minus six.
(c) $(4x + 5y)(x - 3)$	The quantity four x plus five y times the quantity x minus three.

SELF-CHECK ANSWERS

1.7.1
1. 6
2. 42
3. $\frac{1}{4}$
4. 26
5. 112

1.7.2
1. 9
2. 41
3. 1
4. 1
5. 9
6. 4
7. 2

1.7.3
1. $5x + 15y$
2. $2(a - 5b)$
3. $9y$
4. $7x - 3$

USING THE LANGUAGE AND SYMBOLISM OF MATHEMATICS 1.7

1. In the expression $2 - 5 \cdot 7$, the first operation to perform is __________.
2. In the expression $(4 + 9) \cdot 8$, the first operation to perform is __________.
3. In the expression $2 + 5^2 \cdot 3$, the first operation to perform is __________.
4. In the expression $3x^5$, the base is __________ and the exponent is __________.
5. In the expression $(3x)^5$, the base is __________ and the exponent is __________.
6. In the expression $(-x)^4$, the base is __________ and the exponent is __________.
7. In the expression $-x^4$, the base is __________ and the exponent is __________.
8. Considering the commutative, associative, and distributive properties, the property that involves two operations is the __________ property of __________ over __________.
9. In the expression $7v$, 7 is the __________ of v.
10. The __________ property is used to add like terms.
11. The process of adding like terms is sometimes called collecting like terms or __________ like terms.

EXERCISES 1.7

In Exercises 1–44, calculate each expression without using a calculator.

1. a. $5 + 2 \cdot 8$ b. $(5 + 2) \cdot 8$
2. a. $11 + 3 \cdot 9$ b. $(11 + 3) \cdot 9$
3. a. $17 - 3 \cdot 5$ b. $(17 - 3)(5)$
4. a. $41 - 6 \cdot 4$ b. $(41 - 6)(4)$
5. a. $(6 + 4)^2$ b. $6^2 + 4^2$
6. a. $(6 - 4)^2$ b. $6^2 - 4^2$
7. a. $25 \div 5 \cdot 5$ b. $25 \div (5 \cdot 5)$
8. a. $25 \div 5 \div 5$ b. $25 \div (5 \div 5)$
9. a. $19 - 19 \cdot 2$ b. $(19 - 19) \cdot 2$
10. a. $10 - 10 \cdot 3$ b. $(10 - 10) \cdot 3$
11. a. $4 - 7^2$ b. $(4 - 7)^2$
12. a. $1 - 9^2$ b. $(1 - 9)^2$
13. a. $3 \cdot 5 - 6 \cdot 7$ b. $3 \cdot (5 - 6) \cdot 7$
14. a. $4 \cdot 9 - 8 \cdot 11$ b. $4 \cdot (9 - 8) \cdot 11$
15. a. $(-6 + 2 \cdot 4)^2$ b. $-6 + 2 \cdot 4^2$
16. a. $(-7 + 5) \cdot 2^4$ b. $-7 + 5 \cdot 2^4$
17. a. $15 - 3^2 - 6$ b. $15 - (3^2 - 6)$
18. a. $(15 - 3)^2 - 6$ b. $(15 - 3 - 6)^2$
19. a. $2^3 - 5^3$ b. $(2 - 5)^3$
20. a. $5^3 - 10^3$ b. $(5 - 10)^3$
21. a. $4{,}000 - 5 \cdot 10^3$ b. $4{,}000 \div 5 \cdot 10^3$
22. a. $36 - 24 \div 3 + 5$ b. $36 - (24 \div 3 + 5)$
23. a. $16 \div 4 \cdot 3 + 5$ b. $16 \div (4 \cdot 3 + 5)$
24. a. $24 \div 8 \cdot 2 - 4$ b. $24 \div 8 \cdot (2 - 4)$
25. a. $\dfrac{15 - 3(4)}{(15 - 3)(4)}$ b. $\dfrac{15 \div 3(4)}{(15 - 3)(4)}$
26. a. $\dfrac{18 + 2(5)}{(18 + 2)(5)}$ b. $\dfrac{18 + 2^5}{(18 + 2)(5)}$
27. a. $(3 + 4)^2$ b. $3^2 + 4^2$
28. a. $(5 + 7)^2$ b. $5^2 + 7^2$
29. a. $\sqrt{9} + \sqrt{16}$ b. $\sqrt{9 + 16}$
30. a. $\sqrt{169} - \sqrt{144}$ b. $\sqrt{169 - 144}$
31. a. $\sqrt{169} - \sqrt{144} - \sqrt{16}$ b. $\sqrt{169 - 144 - 16}$
32. a. $\sqrt{81} - \sqrt{64} - \sqrt{16}$ b. $\sqrt{81 - 64 - 16}$
33. a. $|-23| + |17|$ b. $|-23 + 17|$
34. a. $|36| + |-22|$ b. $|36 - 22|$
35. $12 + 2[8 - 3(7 - 5)]$
36. $15 + 5[19 - 4(11 - 9)]$
37. $(5 - 3)^3 + (5^3 - 3^3)$
38. $(2 + 3)^3 + (2^3 + 3^3)$
39. $113 + 5[8 + 2(13 - 3^2) - 7]$
40. $217 + 6[11 - 3(50 - 7^2)]$
41. $18 + 2[(41 - 8) - (5 + 3^2)]$
42. $125 - 3[(38 - 27) - (31 - 5^2)]$
43. $\dfrac{5}{16} + \dfrac{1}{2} \cdot \dfrac{3}{8}$
44. $\dfrac{6}{5} - \dfrac{1}{5}\left(\dfrac{7}{8} + \dfrac{3}{8}\right)$

In Exercises 45–52, use the distributive property to expand the first expression and to factor the second expression. In part (b) of 51 and 52, factor -1 out of each expression.

	Expand	Factor
45.	a. $7(x + 5)$	b. $9x + 36$
46.	a. $3(y + 9)$	b. $4y + 20$
47.	a. $-2(3x - 7)$	b. $-5a - 5b$
48.	a. $-4(5b - 9)$	b. $-7a - 21$
49.	a. $(2x - 3y)(5)$	b. $22x - 33y$
50.	a. $(4w - 5z)(6)$	b. $36w - 48z$
51.	a. $-(b - 3)$	b. $-b + 5$
52.	a. $-(y + 5)$	b. $-y - 9$

In Exercises 53–60, use the distributive property to add like terms.

53. a. $2x + 5x$ b. $2\sqrt{3} + 5\sqrt{3}$
54. a. $3y - 8y$ b. $3\sqrt{7} - 8\sqrt{7}$
55. a. $2w - 3w + 7w$ b. $2w - 3z + 7w$
56. a. $-4y + 5y - 9y$ b. $-4y + 5z - 9y$
57. a. $(2a + 3b) + (4a - 5b)$ b. $(2a + 3b) - (4a - 5b)$
58. a. $(3v - 4w) + (5v - 6w)$ b. $(3v - 4w) - (5v - 6w)$
59. $(2\sqrt{3} + 3\sqrt{5}) + (4\sqrt{3} - 5\sqrt{5})$
60. $(3\sqrt{7} - 4\sqrt{11}) - (5\sqrt{7} - 6\sqrt{11})$

In Exercises 61–66, check each equation for $x = -2$ and $x = 5$.

61. $3(x + 4) = x + 8$
62. $(x + 2)(x - 5) = 0$
63. $4(x - 4) = x - 1$
64. $(x + 2)(x - 5) = 1$
65. $2x - 3(x + 5) = -4x$
66. $-3x + 2(5x - 1) = 8x$

In Exercises 67–70, insert the correct inequality symbol, $<$ or $>$, between each pair of expressions.

67. 2^5 ___ 5^2
68. 2^3 ___ 3^2
69. $\sqrt{9 + 16}$ ___ $\sqrt{9} + \sqrt{16}$
70. $\sqrt{25 + 144}$ ___ $\sqrt{25} + \sqrt{144}$

Estimate Then Calculate

In Exercises 71–74, mentally estimate the value of each expression, and then use a calculator to approximate each value to the nearest thousandth.

PROBLEM	MENTAL ESTIMATE	CALCULATOR APPROXIMATION
71. $(4.738 + 4.229)^2$		
72. $5.112^2 + 3.987^2$		
73. $7.129^2 - 2.978^2$		
74. $(7.129 + 2.978)^2$		

75. **Heating Space** A warehouse with 500,000 ft^3 of storage space is to be used for storing crates. For the purposes of planning the heating of this warehouse, the designer must know the volume of air that is to be heated. Each of the 40,000 crates planned for storage in the warehouse is a cube 2 ft on a side. After the crates are stored, what volume of air will remain in the warehouse?

76. **Surface Area of a Water Tank** Allied Manufacturing produces water tanks using a design with a hemisphere (half-sphere) on top of a right circular cylinder. To plan for the painting process, the company determines the surface area of each tank. The surface area S is given by the formula $S = 2\pi r^2 + 2\pi rh$, where r is the radius and h is the height of the cylinder. Determine the surface area in square feet of the tank shown below.

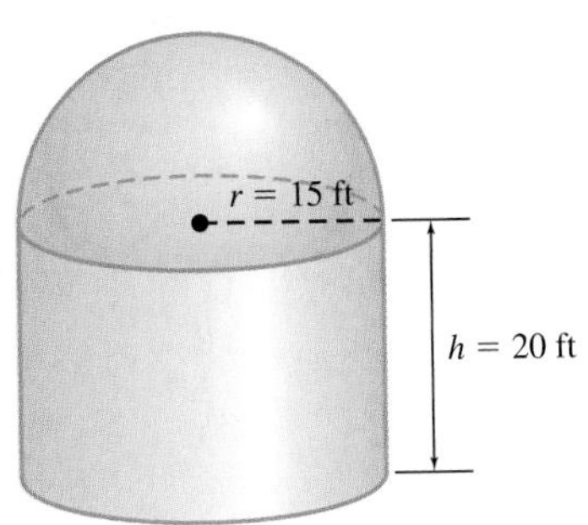

77. Use the formula $V = \frac{2}{3}\pi r^3 + \pi r^2 h$ to calculate the volume in cubic feet of the tank in Exercise 76.

Multiple Representations

In Exercises 78–83, write each verbal statement in algebraic form.

78. a. Seven times x minus three
b. Seven times the quantity x minus three
79. a. Two times x plus five y
b. Two times the quantity x plus five y
80. a. y is equal to the quotient of k divided by the cube of x.
b. y is equal to the cube of the quotient of k divided by x.
81. a. c is equal to the square root of the quantity a plus b.
b. c is equal to the square root of a plus the square root of b.
82. a. The sum of x squared plus y squared.
b. The square of the quantity x plus y.
83. a. The ratio of the quantity y_2 minus y_1 to the quantity x_2 minus x_1.
b. m equals the ratio of the quantity y_2 minus y_1 to the quantity x_2 minus x_1.

The rate of a boat in still water is r mi/h. In Exercises 84–87, write an algebraic expression for each quantity.

84. The rate of the boat going downstream in an 8 mi/h current
85. The rate of the boat going upstream in an 8 mi/h current
86. The distance the boat goes downstream in an 8 mi/h current in 3 h
87. The distance the boat goes upstream in an 8 mi/h current in 3 h

Group Discussion Questions

88. **Error Analysis** In trying to evaluate the expression $\frac{16}{2 + 8}$ on a TI-84 Plus calculator, a student enters the expression shown below. What is the error and how can it be corrected?

89. **Calculator Discovery** Calculators and computer languages such as C++, Java, and FORTRAN require that algebraic expressions be entered on one line. Expressions with fractions must be entered by using grouping symbols to clearly indicate the numerator and denominator. Write each expression by hand as it would appear on a calculator or in a computer program. Then enter the expressions in a calculator. Compare your results with those of others in your group.

a. $\frac{2(4 - 7)}{9 + 16}$
b. $\frac{3 + 1}{2} \cdot 8$
c. $\frac{4 - 3}{9} + 16$
d. $\frac{1 + 13}{2(7)}$

90. Spreadsheet Exploration Investment A yielded a dividend of \$5,000 at the end of the first year; and the dividend at the end of each of the next 4 years doubled. How much did this investment yield at the end of the fifth year? Investment B yielded a dividend of \$25,000 at the end of the first year; and the dividend at the end of each of the next 4 years increased by \$5,000 per year. How much did this investment yield at the end of the fifth year? Which investment produced the greater total yield over the 5-year period? Would your answer change if the pattern continued for 1 more year? (*Hint:* Complete the table below.)

	A	B	C	D	E	F
1		INVESTMENT A			INVESTMENT B	
2	YEAR	DIVIDEND	TOTAL		DIVIDEND	TOTAL
3	1	\$5,000.00	\$5,000.00		\$25,000.00	\$25,000.00
4	2	\$10,000.00	\$15,000.00		\$30,000.00	\$55,000.00
5	3	\$20,000.00			\$35,000.00	
6	4					
7	5					
8	6					
9						

KEY CONCEPTS FOR CHAPTER 1

1. **Real Numbers:** The real numbers consist of the rational numbers and the irrational numbers.
   Rational Numbers: Numbers that can be written as a ratio of two integers are rational numbers. Expressed decimally, rational numbers are either terminating decimals or repeating decimals.
   Irrational Numbers: Real numbers, such as $\sqrt{2}$ or π, that cannot be written as a ratio of two integers are irrational numbers. Expressed decimally, irrational numbers are infinite nonrepeating decimals.
   Subsets of the rational numbers:
   - **Natural Numbers:** $1, 2, 3, 4, 5, 6, \ldots$
   - **Whole Numbers:** $0, 1, 2, 3, 4, 5, 6, \ldots$
   - **Integers:** $\cdots, -3, -2, -1, 0, 1, 2, 3, \cdots$
2. **Additive Inverses:** Real numbers that are the same distance from the origin but on opposite sides of the origin are called opposites of each other or additive inverses. If a is a real number, the opposite of a is $-a$. The opposite of $-a$ is a. The numbers -2 and 2 are opposites of each other. The sum of a number and its opposite is zero. Both $2 + (-2) = 0$ and $-2 + 2 = 0$.
3. **Additive Identity:** Zero is called the additive identity because 0 is the only real number with the property that $a + 0 = a$ and $0 + a = a$ for every real number a.
4. **Inequality and Interval Notation:**

INEQUALITY NOTATION	MEANING	INTERVAL NOTATION
$x > a$	x is greater than a	(a, ∞)
$x \ge a$	x is greater than or equal to a	$[a, \infty)$
$x < a$	x is less than a	$(-\infty, a)$
$x \le a$	x is less than or equal to a	$(-\infty, a]$
$a < x < b$	x is greater than a and less than b	(a, b)
$a < x \le b$	x is greater than a and less than or equal to b	$(a, b]$
$a \le x < b$	x is greater than or equal to a and less than b	$[a, b)$
$a \le x \le b$	x is greater than or equal to a and less than or equal to b	$[a, b]$

5. **Absolute Value:** The absolute value of a real number is the distance between this number and the origin on the real number line.

$$|x| = \begin{cases} x & \text{if } x \text{ is nonnegative} \\ -x & \text{if } x \text{ is negative} \end{cases}$$

6. **Square Roots:** If x is a positive real number, then $\sqrt{x}$ denotes a positive number r such that $r^2 = x$. For example, $\sqrt{9} = +3$ since $3^2 = 9$.
7. **Evaluating an Algebraic Expression:** To evaluate an algebraic expression for given values of the variables means to replace each variable by the specific value given for that variable and then to simplify this expression. When a constant is substituted for a variable, it is often wise to use parentheses to avoid careless errors with either pencil and paper or your calculator.
8. **Addition of Real Numbers**
   - **Like signs:** Add the absolute values of the numbers, and use the common sign of these terms.
   - **Unlike signs:** Find the difference of the absolute values of the numbers, and use the sign of the term that has the larger absolute value.

9. **Subtraction of Real Numbers:**
   Step 1 Rewrite the problem as an addition problem by changing the operation to addition and by changing the sign of the subtrahend (the number being subtracted).
   Step 2 Add, using the rule for adding signed numbers.

10. **Multiplication of Real Numbers**
   - **Like signs:** Multiply the absolute values of the two factors, and use a positive sign for the product.
   - **Unlike signs:** Multiply the absolute values of the two factors, and use a negative sign for the product.
   - **Zero factor:** The product of 0 and any other factor is 0.

11. **Multiplicative Inverse:** The reciprocal or multiplicative inverse of any nonzero real number a can be represented by $\frac{1}{a}$. The product of a number and its multiplicative inverse is $+1$.

12. **Multiplicative Identity:** One is called the multiplicative identity because 1 is the only real number with the property that $1 \cdot a = a$ and $a \cdot 1 = a$ for every real number a.

13. **Product of Negative Factors**
   - The product is positive if the number of negative factors is even.
   - The product is negative if the number of negative factors is odd.

14. **Division of Real Numbers**
   - **Like signs:** Divide the absolute values of the two numbers, and use a positive sign for the quotient.
   - **Unlike signs:** Divide the absolute values of the two numbers, and use a negative sign for the quotient.
   - **Zero dividend:** $\frac{0}{x} = 0$ for $x \neq 0$
   - **Zero divisor:** $\frac{x}{0}$ is undefined for every real number x.

15. **Ratio** The ratio a to b can be denoted by either $a : b$ or $\frac{a}{b}$.

16. **Exponential Notation:** For any natural number n, $b^n = \underbrace{b \cdot b \cdots b}_{n \text{ factors of } b}$ with base b and exponent n.

17. **Sequence:** A sequence is an ordered set of numbers. Subscript notation ($a_1, a_2, \ldots, a_n, \ldots$, read a sub one, a sub two, and a sub n, respectively) is often used to denote the terms of a sequence. The spreadsheet notation A1, A2, . . . is used to denote the cells in column A of a spreadsheet.

18. **Operations with Fractions**
   - **Addition:** $\frac{a}{b} + \frac{c}{b} = \frac{a+c}{b}$ for $b \neq 0$
   - **Subtraction:** $\frac{a}{b} - \frac{c}{b} = \frac{a-c}{b}$ for $b \neq 0$
   - **Multiplication:** $\frac{a}{b} \cdot \frac{c}{d} = \frac{a \cdot c}{b \cdot d}$ for $b \neq 0$ and $d \neq 0$
   - **Division:** $\frac{a}{b} \div \frac{c}{d} = \frac{a}{b} \cdot \frac{d}{c} = \frac{a \cdot d}{b \cdot c}$ for $b \neq 0$, $c \neq 0$, and $d \neq 0$
   - **Reducing Fractions:** $\frac{ac}{bc} = \frac{a}{b}$ for $b \neq 0$ and $c \neq 0$

19. **Order of Operations**
   Step 1 Start with the expression within the innermost pair of grouping symbols.
   Step 2 Perform all exponentiations.
   Step 3 Perform all multiplications and divisions as they appear from left to right.
   Step 4 Perform all additions and subtractions as they appear from left to right.

20. **Operations with Radicals**
   Multiplication: $\sqrt{a}\sqrt{b} = \sqrt{ab}$ for $a > 0$ and $b > 0$
   Division: $\frac{\sqrt{a}}{\sqrt{b}} = \sqrt{\frac{a}{b}}$ for $a > 0$ and $b > 0$

21. **Properties of Addition and Multiplication**
   Commutative Properties: Properties dealing with order of operations:
   - $a + b = b + a$ Addition is commutative
   - $ab = ba$ Multiplication is
   Associative Properties: Properti
   - $(a + b) + c = a + (b + c)$
   - $(ab)(c) = a(bc)$ Multipl
   Distributive Property of Multi
   Property describing the relations
   and addition:
   - $a(b + c) = ab + ac$ Mu
   - $(b + c)a = ba + ca$ adc

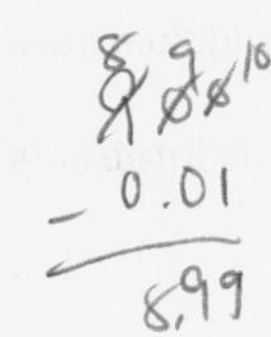

REVIEW EXERCISES FOR CHAPTER 1

In Exercises 1–28, perform the indicated operations.

1. a. $-16 + 4$ b. $-16 + (-4)$
   c. $-16 - 4$ d. $-16 - (-4)$
2. a. $-16(4)$ b. $-16(-4)$
   c. $-16 \div 4$ d. $-16 \div (-4)$
3. a. $-7 + 0$ b. $-7(0)$
   c. $\frac{0}{-7}$ d. $\frac{-7}{0}$
4. a. $24 + (-6)$ b. $24 - (-6)$
   c. $24(-6)$ d. $24 \div (-6)$
5. a. $9 + 0.01$ b. $9 - 0.01$
   c. $9(0.01)$ d. $9 \div 0.01$
6. a. $-4.5 + 1{,}000$ b. $-4.5 - 1{,}000$
   c. $-4.5(1{,}000)$ d. $-4.5 \div 1{,}000$
7. a. $-7 - 8 + 9$ b. $-7 - (8 + 9)$
   c. $-(7 - 8) + 9$ d. $-(7 - 8 + 9)$
8. a. $-6 + 8 - 11 - 15$ b. $(-6 + 8) - (11 - 15)$
   c. $-6 + (8 - 11 - 15)$ d. $-(6 + 8) - (11 - 15)$
9. a. $(15 - 6)(9 - 11)$ b. $15 - 6(9 - 11)$
   c. $15 - 6 \cdot 9 - 11$ d. $(15 - 6)9 - 11$
10. a. $-36 \div 4 \cdot 3$ b. $-36 \div (4 \cdot 3)$
    c. $-(36 \div 4 \cdot 3)$ d. $-(36 \div 4) \cdot 3$
11. a. 25% of 36 b. 35% of 15
    c. 10% of 12 d. 18% of 100

12. a. $(-3)^2$ b. -3^2
c. $(-3)^3$ d. -3^3

13. a. $(3 + 4)^2$ b. $3^2 + 4^2$
c. $(3 - 4)^2$ d. $3^2 - 4^2$

14. a. 5^2 b. 2^5
c. $(-5)^2$ d. -2^5

15. a. $|3 - 11|$ b. $|3| - |11|$
c. $-|3 - 11|$ d. $|3| + |-11|$

16. a. $\sqrt{64} + \sqrt{36}$ b. $\sqrt{64 + 36}$
c. $\sqrt{1} + \sqrt{1} + \sqrt{1} + \sqrt{1}$ d. $\sqrt{1 + 1 + 1 + 1}$

17. a. $(-1)(-2)(-3)(-4)$
b. $(-1)(-2)(-3)(-4)(-5)$
c. $(0)(-1)(-2)(-3)(-4)$
d. $(-1)(-2)(-3)(-4)(-5)(-6)$

18. a. 0^{37} b. 0^{38}
c. $(-1)^{37}$ d. $(-1)^{38}$

19. a. $\frac{2}{3} + \frac{3}{4}$ b. $\frac{2}{3} - \frac{3}{4}$
c. $\left(\frac{2}{3}\right)\left(\frac{3}{4}\right)$ d. $\left(\frac{2}{3}\right) \div \left(\frac{3}{4}\right)$

20. a. $-\frac{14}{15} + \frac{21}{25}$ b. $-\frac{14}{15} - \frac{21}{25}$
c. $-\frac{14}{15} \cdot \frac{21}{25}$ d. $-\frac{14}{15} \div \frac{21}{25}$

21. a. $-48 + 12 \div 6 + 3 - 1$ b. $-48 - 12 \cdot 6 \div 3 + 1$
c. $-48 + 12 - 6 \div 3 \cdot 1$ d. $-48 \div 12 \cdot 6 - 3 + 1$

22. $14 - 2[11 - 5(13 - 10)]$

23. $19 + 3[(40 - 7) - (4 + 3^2)]$

24. $\frac{-3 + 7}{8 - 10}$ 25. $\frac{14 - 7(-5)}{-3(9) - 2(-10)}$

26. $\sqrt{3}\sqrt{12}$ 27. $\sqrt{5}\sqrt{45}$

28. $\sqrt{2}\sqrt{5}\sqrt{10}$

In Exercises 29–32, match each number with the description given to the right.

29. 0
30. $\frac{3}{4}$
31. $\sqrt{4}$
32. $\sqrt{5}$

A. A natural number
B. An integer that is not a natural number
C. An irrational number
D. A rational number that is not an integer

Estimate Then Calculate

In Exercises 33–38, mentally determine the sign of each expression, and then use a calculator to compute the value.

PROBLEM	SIGN	CALCULATOR VALUE
33. $-312 + (-221)$		
34. $-41 - (-72)$		
35. $(53)(-37)(-21)$		
36. $(-3)^2 \cdot 5 \cdot (-7)$		
37. $-|361| - |-196|$		
38. $\sqrt{196} - \sqrt{361}$		

In Exercises 39–44, mentally estimate the value of each expression, and then use a calculator to compute the exact value.

PROBLEM	MENTAL ESTIMATE	CALCULATOR VALUE
39. $8.903 + 7.146$		
40. $1{,}049.8768 - 250.4132$		
41. $(301.06)(9.8)$		
42. $43.5969 \div 7.23$		
43. $(5.01)^3$		
44. $\sqrt{50.41}$		

In Exercises 45–50, fill in the missing columns.

Inequality	Graph	Interval
45.	(number line, parenthesis at −2, shaded to the right) −2	
46.		$(-\infty, 4]$
47. $x < 3$		
48.		$(-3, 0)$
49. $-7 \le x \le -1$		
50.	(number line, bracket at 4, parenthesis at 10) 4 10	

Multiple Representations

In Exercises 51–59, write each verbal statement in algebraic form.

51. The opposite of x equals eleven.
52. The absolute value of x is less than or equal to seven.
53. The square root of twenty-six is greater than five.
54. The sum of x and five is equal to four.
55. The difference of x minus three is less than y.
56. The product of negative three and y is negative one.
57. The quotient of x and three is twelve.
58. The ratio of x to y equals three-fourths.
59. Five times the quantity three x minus four equals thirteen.

In Exercises 60 and 61, calculate the first five terms of each sequence.

60. $a_n = 10n - 3$ 61. $a_n = n^2 - 10$

In Exercises 62–66, match each algebraic equation with the property it illustrates.

62. $(xy)z = x(yz)$
63. $(x + y) + z = x + (y + z)$
64. $w(xy + z) = w(yx + z)$
65. $w(xy + z) = wxy + wz$
66. $w(xy + z) = w(z + xy)$

A. Associative property of addition
B. Associative property of multiplication
C. Commutative property of addition
D. Commutative property of multiplication
E. Distributive property of multiplication over addition

In Exercises 67 and 68, evaluate the expression $\frac{y_2 - y_1}{x_2 - x_1}$ for the given values of x_1, x_2, y_1, and y_2.

67. $x_1 = 3, x_2 = -2, y_1 = 6$, and $y_2 = -4$
68. $x_1 = 5, x_2 = 4, y_1 = -7$, and $y_2 = -3$

In Exercises 69 and 70, check $x = 2$ and $x = 3$ as possible solutions of the given equation.

69. $3x + 5 = 4x + 2$
70. $3(x + 5) = 4(x + 3) + 1$
71. **Surface Area and Volume** The crate in the photograph has a length of 6 ft, a width of 5 ft, and a height of 4.5 ft.

a. Use the formula $V = l \cdot w \cdot h$ to determine the volume of this crate.
b. Use the formula $S = 2lw + 2hl + 2hw$ to determine the surface area of this crate.

72. **Perimeter and Area**
a. Determine the perimeter of the volleyball court shown here.
b. Determine the area of the volleyball court shown here.

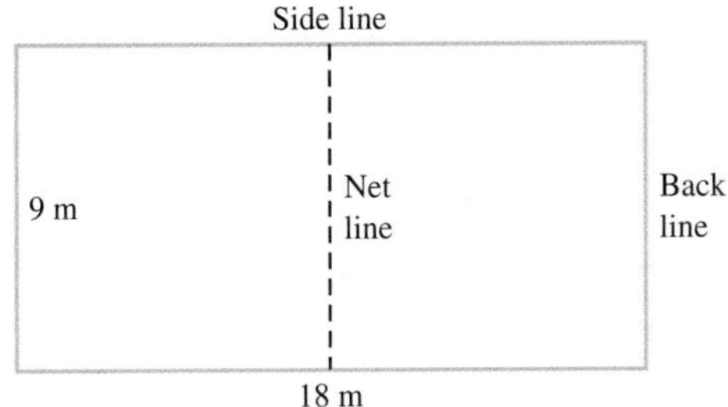

73. a. Determine the error in the calculation of the perimeter if a painter measured the volleyball court in Exercise 72 as 18.1 m by 9.1 m.
b. Determine the relative error of this calculation.

74. **Best Buy for Sports Drink** One brand of sport drink is sold in two popular sizes. Which is the better buy; a 20-ounce bottle that is priced at \$0.99 or a 32-ounce bottle that is priced at \$1.69?
75. **Heights of a Daughter and Father** A 20-year-old student is 68 in tall, and her father is 72 in tall. Determine the ratio of the height of the daughter to that of her father.
76. **Computer Sale** To make room for a new computer model, the prices of all older models have been reduced by 20%. What is the sale price of a computer that normally sells for \$1,299?
77. **Active Ingredient** Calculate the amount of the active ingredient in each mixture.
a. 400 L of a 10% insecticide mixture
b. 600 gal of a 5% ethanol mixture
78. **Rate of Work** Use the formula $R = \frac{W}{T}$ to compute the rate of work for each situation.
a. An assembly line produces 1,600 golf balls in 8 hours.
b. A pipe can fill 1 brewery vat in 4 hours.
79. **Rates and Distances** An airplane has an airspeed of r mi/h. Write an algebraic expression for
a. The distance the plane will travel in 2 hours in still air
b. The rate of the plane if it is traveling with a 30 mi/h tailwind
c. The distance the plane will travel in 2 hours with a 30 mi/h tailwind
d. The rate of the plane if it is traveling against a 30 mi/h headwind
e. The distance the plane will travel in 2 hours against a 30 mi/h headwind
80. **Monthly Sales** The bar graph below gives the number of homes sold each month by one realtor in Chandler, Arizona.
a. Determine the total sales for this 6-month period.
b. Calculate the mean number of homes sold for this 6-month period.
c. Calculate the range in the number of homes sold for this 6-month period.

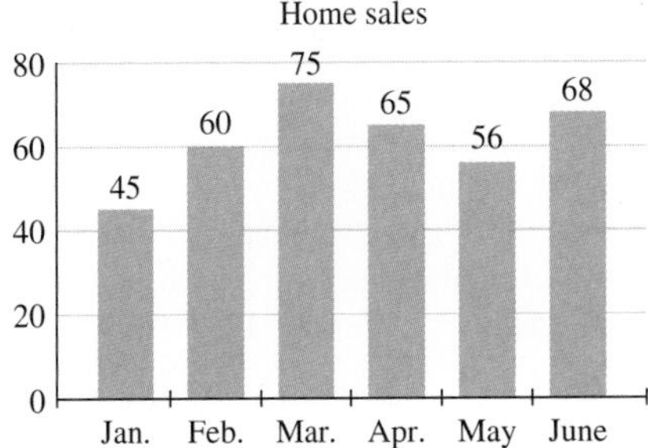

MASTERY TEST FOR CHAPTER 1

[1.1] 1. a. The name of your instructor is __________.
b. Your instructor's office is in __________.
c. Your instructor's e-mail address is __________.
d. The back of the book contains the answers to the __________ numbered exercises.

[1.2] 2. Write the additive inverse of each of the real numbers graphed below.
a. (number line: −2, 0)
b. (number line: 0)
c. (number line: 0, 1.5)
d. (number line: 0, 25)

[1.2] 3. Evaluate each of these absolute value expressions.
a. $|23|$ b. $|-23|$
c. $|23 - 23|$ d. $|23| + |-23|$

[1.2] 4. Express each of these intervals by using interval notation.
a. (number line: −2, 3)
b. (number line: −1)
c. $x < 4$ d. $5 \leq x < 9$

[1.2] 5. Mentally estimate, to the nearest integer, the value of each of the following square roots; then use your calculator to approximate each value to the nearest hundredth.
a. $\sqrt{26}$ b. $\sqrt{99}$
c. $\sqrt{4+4}$ d. $\sqrt{4} + \sqrt{4}$

[1.2] 6. Match each number with the most appropriate description.
a. -5 A. A whole number that is not a natural number
b. $-\frac{2}{3}$ B. An integer that is not a whole number
c. 0 C. A rational number that is not an integer
d. $\sqrt{9}$ D. A positive number that is an irrational number
e. $\sqrt{11}$ E. A natural number

[1.3] 7. Calculate each sum without using a calculator.
a. $17 + (-11)$ b. $-17 + (-11)$
c. $-17 + 11$ d. $\frac{3}{8} + \frac{1}{8}$
e. $-\frac{3}{8} + \frac{1}{8}$ f. $-\frac{2}{5} + \left(-\frac{3}{8}\right)$
g. $57.3 + (-57.3)$
h. $(-5 + 3) + [-8 + (-2)]$

[1.3] 8. a. The property illustrated by $5(6 + 7) = 5(7 + 6)$ is the __________ property of __________.
b. The property illustrated by $5 + (6 + 7) = (5 + 6) + 7$ is the __________ property of __________.
c. Use the associative property of addition to rewrite $(x + y) + 5$.
d. Use the commutative property of addition to rewrite $5(x + y)$.

[1.3] 9. Evaluate each expression for $x = -1$, $y = -2$, and $z = -3$.
a. $x + y + z$ b. $-x + y + z$
c. $-x + yz$ d. $2x + |y| + |z|$

[1.4] 10. Calculate each difference without using a calculator.
a. $15 - 8$ b. $-15 - 8$
c. $-15 - (-8)$ d. $15 - (-8)$
e. $\frac{5}{6} - \frac{1}{6}$ f. $-\frac{5}{6} - \frac{1}{6}$
g. $\frac{1}{2} - \left(-\frac{1}{3}\right)$ h. $-\frac{3}{5} - \left(-\frac{5}{6}\right)$

[1.4] 11. Calculate the first five terms of each sequence.
a. $a_n = n + 2$ b. $a_n = 12 - n$
c. $a_n = 2n + 1$ d. $a_n = n^2 - 5$

[1.4] 12. Check $x = 3$ to determine whether it is a solution of each equation.
a. $x + 2 = 8 - x$ b. $2(x + 1) = x + 4$
c. $2x + 1 = x + 4$ d. $3x + 1 = 4x - 7$

[1.5] 13. Calculate each product without using a calculator.
a. $5(-6)$ b. $-5(-6)$
c. $-5(6)$ d. $-5(0)$
e. $\frac{1}{3} \cdot \frac{1}{5}$ f. $\left(-\frac{3}{7}\right)\left(-\frac{7}{9}\right)$
g. $-\frac{2}{5}\left(\frac{15}{14}\right)$ h. $(-2)(-3)(-4)$

[1.4] 14. a. The property illustrated by $5 \cdot (6 + 7) = (6 + 7) \cdot 5$ is the __________ property of __________.
b. The property illustrated by $5 \cdot (6 \cdot 7) = (5 \cdot 6) \cdot 7$ is the __________ property of __________.
c. Use the associative property of multiplication to rewrite $(5x)(y)$.
d. Use the commutative property of multiplication to rewrite $x(a + b)$.

[1.5] 15. Calculate the value of each expression without using a calculator.
a. 5^2 b. 2^5
c. $(-5)^2$ d. -5^2
e. 0^7 f. $\left(\frac{3}{7}\right)^2$
g. 219^1 h. $(-1)^{219}$

[1.5] 16. a. Use the formula $A = l \cdot w$ to calculate the area of the rectangle below.
b. Use the formula $P = 2l + 2w$ to calculate the perimeter of this rectangle.

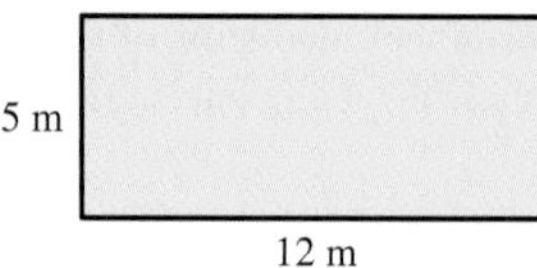

[1.6] 17. Calculate each quotient without using a calculator.
a. $-12 \div 4$ b. $-12 \div (-4)$
c. $12 \div (-4)$ d. $0 \div 4$
e. $-4 \div 0$ f. $-\frac{1}{2} \div \frac{1}{3}$
g. $-\frac{5}{6} \div \left(-\frac{2}{3}\right)$ h. $\frac{12}{35} \div \left(-\frac{14}{11}\right)$

[1.6] 18. a. **Defective Keyboard Trays** The computer support services at Apollo College reported that 21 of 105 keyboard trays in an open computer lab had to be repaired one year. What fractional portion of the trays was repaired?

b. **Best Buy for Salad Dressing** One brand of salad dressing is sold in two popular sizes. Which is the better buy, a 16-oz bottle that sells for $2.59 or a 12-oz bottle that sells for $1.92?

[1.7] 19. Calculate the value of each expression without using a calculator.

a. $15 - 2(8 - 5)$

b. $4 \cdot 7 - 2 \cdot 3 + 5$

c. $15 - 6 \div 3 + 11$

d. $\dfrac{14 - 3 \cdot 6}{(14 - 3) \cdot 6}$

e. $(3 + 5)^2$

f. $3^2 + 5^2$

g. $-8 - 3[5 - 4(6 - 9)]$

h. $\sqrt{25 - 16} - (\sqrt{25} - \sqrt{16})$

[1.7] 20. a. The __________ property of __________ over __________ states that $a(b + c) = ab + ac$ and $(b + c)\,a = ba + ca$.

b. Expand $11(3x + 4)$.

c. Factor $33x + 44$.

d. Expand $-7(2x - 5)$.

e. Factor $16x - 24$.

[1.7] 21. Simplify each expression by adding like terms.

a. $20x - 11x + 3x$

b. $20x - 11(x + 3)$

c. $3(x - 1) - 2(x + 2)$

d. $2a + 3b - 5 + 2(a - 2b + 4)$

REALITY CHECK FOR CHAPTER 1

A shorter-term mortgage means larger monthly payments with a quicker payout time. A longer-term mortgage means smaller monthly payments but with an extended payout time. Examine the given spreadsheets which display some key information for two mortgage options;

A. $790 per month payment for a $100,000 mortgage at 5% for 15 years

B. $733 per month payment for a $100,000 mortgage at 8% for 30 years

	A	B	C	D
1	**Rate**	**Principal**	**Term**	**Payment**
2	5%	$100,000.00	15 years	$790.79
3				
4	**Years**	**Total Paid**	**Total Interest**	**Balance**
5	5	$47,447.40	$22,004.72	$74,557.09
6	10	$94,894.80	$36,799.98	$41,904.70
7	15	$142,342.18	$42,342.18	$0.00
8				

	A	B	C	D
1	**Rate**	**Principal**	**Term**	**Payment**
2	8%	$100,000.00	30 years	$733.76
3				
4	**Years**	**Total Paid**	**Total Interest**	**Balance**
5	5	$44,025.60	$39,095.75	$95,069.83
6	10	$88,051.20	$75,776.48	$87,724.66
7	15	$132,076.80	$108,859.23	$76,781.49
8	20	$176,102.40	$136,581.51	$60,477.90
9	25	$220,128.00	$156,317.51	$36,188.04
10	30	$264,153.47	$164,153.47	$0.00
11				

1. How much can you save in interest costs by choosing the 15-year option?
2. Why might someone select the 15-year option?
3. Why might someone select the 30-year option?

GROUP PROJECT FOR CHAPTER 1

Risks and Choices: Shopping for Interest Rates on a Home Loan

The formula for determining the monthly payment for a loan is

$$P_n = \frac{AR\left(1 + \frac{R}{12}\right)^{12n}}{12\left(1 + \frac{R}{12}\right)^{12n} - 12}$$

where A represents the amount of the loan, R represents the interest rate, n represents the number of years, and P_n represents the monthly payment required to pay off the loan in n years.

Use this formula and a calculator to complete the following tables.

Consider a \$100,000 loan at several different interest rates.

INTEREST RATE (%)	LOAN AMOUNT (\$)	NUMBER OF YEARS	MONTHLY PAYMENT (\$)	TOTAL OF ALL PAYMENTS (\$)
5.50	100,000	30		
6.00	100,000	30		
6.50	100,000	30		
7.00	100,000	30		
7.50	100,000	30		
8.00	100,000	30		

Consider a \$50,000 loan at 7.00% interest for several different time periods.

INTEREST RATE (%)	LOAN AMOUNT (\$)	NUMBER OF YEARS	MONTHLY PAYMENT (\$)	TOTAL OF ALL PAYMENTS (\$)
7.00	50,000	10		
7.00	50,000	15		
7.00	50,000	20		
7.00	50,000	25		
7.00	50,000	30		

Use these facts to discuss the advantages and disadvantages for the borrower for each of these options.

chapter 2

Linear Equations and Patterns

CHAPTER OUTLINE

The holiday season in New York is a dazzling display of lights and decorations in a city that can get very cold and windy in December. The wintry combination of wind and temperature can combine to produce frostbite on unprotected tissues in just minutes. Weather reports describe these conditions by using the windchill factor.

The weather forecast for New York City on December 24 is for steady winds blowing at 10 mi/h with dropping temperatures. The graph gives the windchill factor for this wind speed in Fahrenheit temperatures from −15° to 15°.

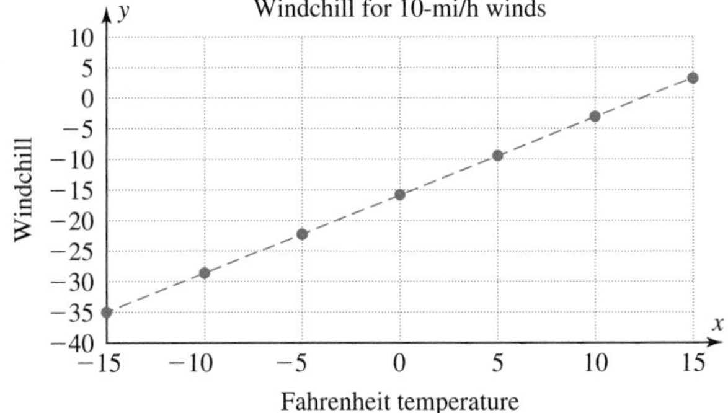

Source: For more information on this topic go to the Additional Resources link on the MathZone website at **www.mhhe.com/hallmercer.**

REALITY CHECK

Before venturing out on a walk to see the light display at Rockefeller Center, you should check on the expected weather conditions including the windchill. If the temperature drops to −5°F, what will the windchill in a 10-mi/h wind be?

Section 2.1 The Rectangular Coordinate System and Arithmetic Sequences

Objectives:

1. Plot ordered pairs on a rectangular coordinate system.
2. Draw a scatter diagram of a set of points.
3. Identify an arithmetic sequence.

The primary focus of this section is on plotting data points and creating and interpreting graphs. Data reported periodically by businesses often form a sequence of data values which can be examined graphically for a pattern. The repayment schedule in the next paragraph illustrates this and provides a context for the topics examined in this section.

A portion of the repayment schedule for a car loan of $3,300 at 8.5% interest for 2 years is shown. This payment schedule contains the sequences 1, 2, 3, 4, 5 in the leftmost column and 150, 300, 450, 600, 750 in the rightmost column. Both of these sequences have a constant change from term to term. In the first sequence, the constant change is 1, and in the second sequence, the constant change is 150. As a prelude to the graphing material to be covered in this section, consider the graph of the sequence of totals paid from the payment schedule shown. Note the linear pattern exhibited by the points on this graph.

Each point visually pairs the payment number with the total paid after this payment is made. The dashes between the points emphasize the linear pattern formed by these points.

PAYMENT NUMBER	PAYMENT AMOUNT, $	TOTAL PAID, $
1	150	150
2	150	300
3	150	450
4	150	600
5	150	750

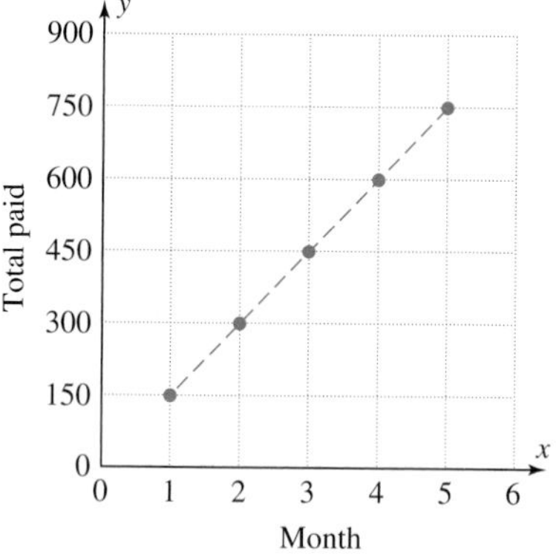

A Mathematical Note

Rene Descartes (1596–1650), born to a noble French family, was known for his studies in anatomy, astronomy, chemistry, physics, and philosophy as well as mathematics. Prior to Descartes, algebra was concerned with numbers and calculations, and geometry was concerned with figures and shapes. Descartes merged the power of these two areas into analytic geometry—his most famous discovery.

This graph is shown on a **rectangular coordinate system,** also known as the **Cartesian coordinate system,** or the **coordinate plane.** On the rectangular coordinate system, the horizontal number line is called the ***x*-axis,** the vertical number line is called the ***y*-axis,** and the point where they cross is called the **origin.** These axes divide the plane into four **quadrants,** labeled counterclockwise as I, II, III, and IV, as illustrated in Figure 2.1.1. The points on the axis are not considered to be in any of the quadrants.

On the x-axis, points to the right of the origin are positive and those to the left of the origin are negative. On the y-axis, points above the origin are positive and those below are negative. Any point in the plane can be uniquely identified by specifying its horizontal and vertical location with respect to the origin. We identify a particular point by giving an ordered pair (x, y) with **coordinates** x and y. The first coordinate is called the ***x*-coordinate,** and the second coordinate is called the ***y*-coordinate.** The ordered pair (0, 0) identifies the origin.

The point in Figure 2.1.1(c) is identified by the ordered pair (5, 3) in quadrant I. The x-coordinate is 5 and the y-coordinate is 3. The identification of this point can be associated with the horizontal and vertical segments of the shaded rectangle—hence the name *rectangular coordinate system.*

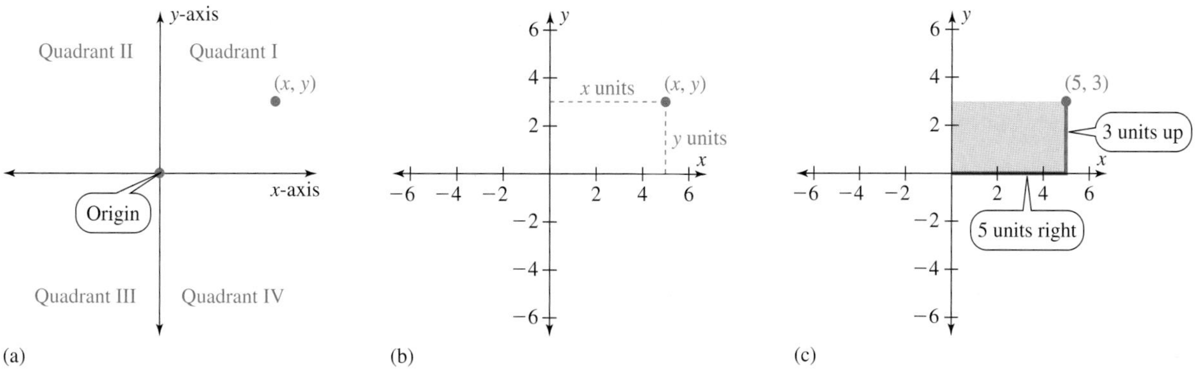

Figure 2.1.1 Cartesian coordinate system.

EXAMPLE 1 Plotting Points on a Rectangular Coordinate System

Plot $(-3, 4)$ and $(2, -5)$ on a rectangular coordinate system.

SOLUTION

Parentheses are used to indicate both ordered pairs as in the point $(-3, 4)$ and intervals as in the interval $(-3, 4)$:

These meanings are not easily confused because the context in which they occur will make each meaning clear.

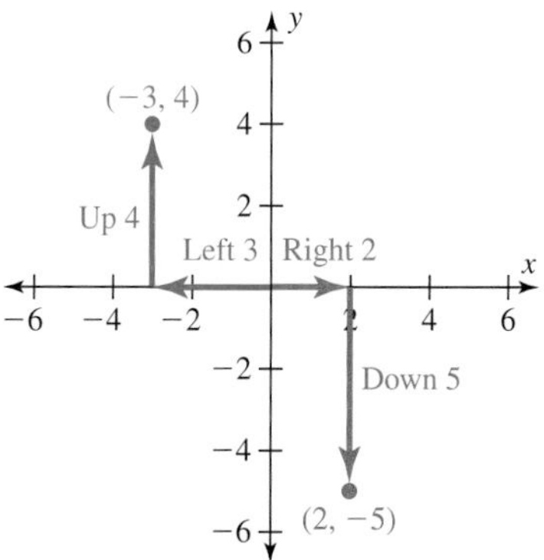

To plot $(-3, 4)$, start at the origin and move 3 units left and then 4 units up.

To plot $(2, -5)$, start at the origin and move 2 units right and then 5 units down.

EXAMPLE 2 Identifying Coordinates of Points

Identify the coordinates of each of the points A through E in the following figure, and give the quadrant in which each point is located.

SOLUTION

A: $(5, 1)$; quadrant I
B: $(-2, 5)$; quadrant II
C: $(-4, -2)$; quadrant III
D: $(3, -6)$; quadrant IV
E: $(4, 0)$ is on the x-axis and thus is not in any quadrant.

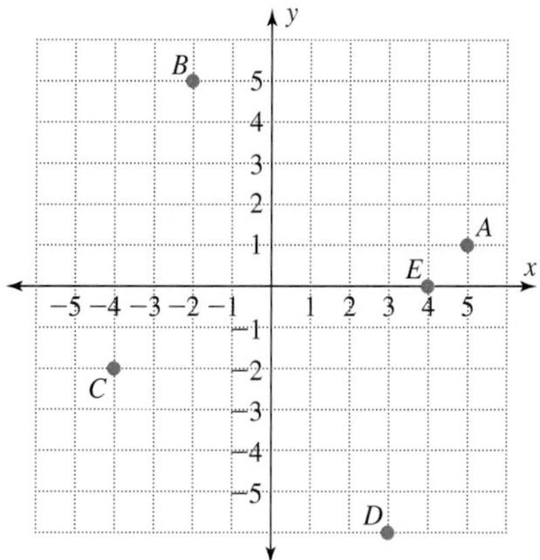

The integer setting on a graphing calculator can be a useful tool to examine the rectangular coordinate system. This is illustrated in Calculator Perspective 2.1.1. The notation below the graph indicates the minimum and maximum x and y values used as well as the scale in the calculator window. The notation $[-47, 47, 10]$ means the x-values extend from -47 to 47 with each mark on the x-axis representing 10 units. Likewise, $[-31, 31, 10]$ means the y-values extend from -31 to 31 with each mark on the y-axis representing 10 units. This decimal viewing window on a TI-84 calculator is obtained by pressing ZOOM 6 to access the standard viewing window and then ZOOM 8 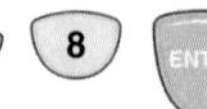 to access the integer window centered at the origin.

CALCULATOR PERSPECTIVE 2.1.1 **Using the Integer Setting to Examine the Rectangular Coordinate System**

To examine the rectangular coordinate system by using a TI-84 Plus calculator, enter the following keystrokes:

Y= (Press to make sure this screen is clear.)

ZOOM 6

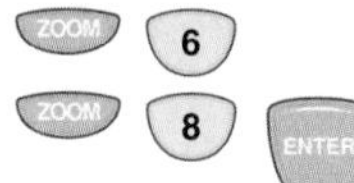

Then use the arrows

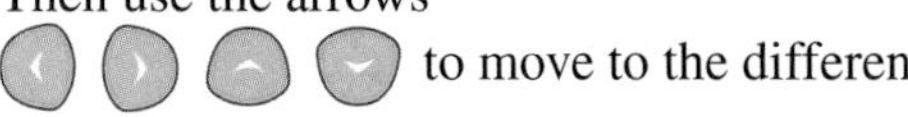 to move to the different quadrants. Make note of the signs of the x- and y-values in each quadrant.

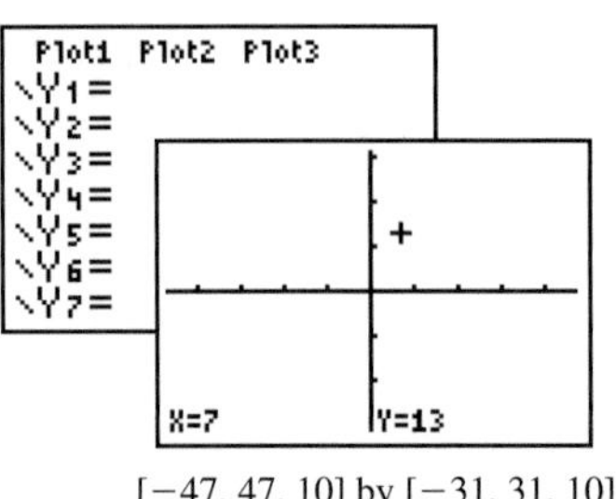

$[-47, 47, 10]$ by $[-31, 31, 10]$

Note: What do you notice about all points (a) on the x-axis, (b) on the y-axis, and (c) in each quadrant?

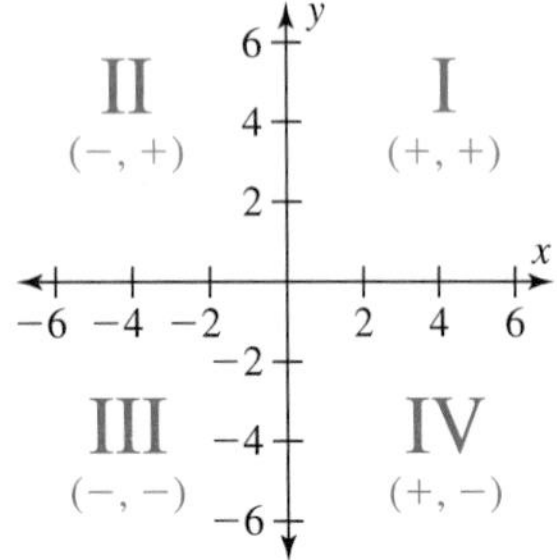

Figure 2.1.2 Quadrant sign pattern.

All the points within the same quadrant have the same sign pattern. For example, in quadrant I both coordinates are positive, which we can denote by $(+, +)$. The sign pattern for each quadrant is shown in Figure 2.1.2. Knowing the sign pattern in each quadrant is useful in trigonometry. These patterns can also be helpful in analyzing scatter diagrams in statistics.

A **scatter diagram** for a set of data points is simply a graph of these points. It allows us to examine the data for some type of visual pattern. For example, if the data points all lie near a line, then the pattern exhibited is called a **linear relationship.**

EXAMPLE 3 Drawing a Scatter Diagram

Draw a scatter diagram for the data points given in the following table.

SOLUTION

x	y
−15	−12
−10	−4
−5	−1
0	6
5	8
9	15
15	18

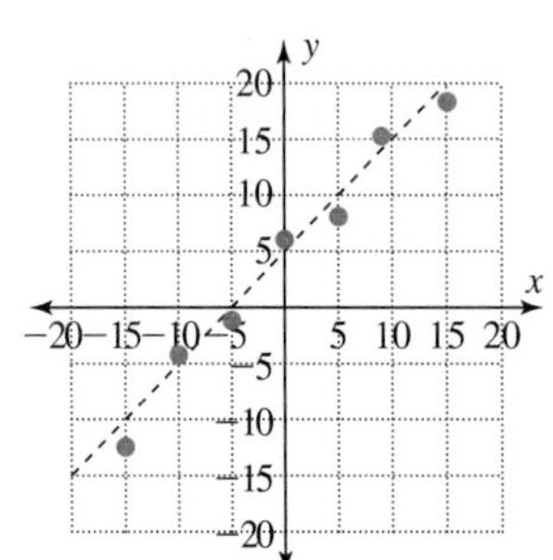

The table format is often used to give ordered pairs. The first ordered pair from this table is $(-15, -12)$.

The dashed line is placed on the graph to show that the data points all lie near this line. Therefore, the relationship between x and y is approximately a linear relationship.

SELF-CHECK 2.1.1

1. Give the quadrant for $(1, -3)$, $(-1, -3)$, $(-1, 3)$, $(1, 3)$.
2. The origin is the point ___________.
3. Every point on the x-axis has a y-coordinate of ___________.
4. Every point on the y-axis has 0 for the ___________-coordinate.

The size of the coordinates is used to determine the scale on each axis. If the coordinates differ significantly in size, then a different scale can be used on each axis. This practice is common in statistics, where the x- and y-variables often represent quantities measured in different units. This is illustrated in Example 4, which also shows that we sometimes start the input with a gap between the origin and the first value.

EXAMPLE 4 Drawing a Scatter Diagram

Draw a scatter diagram for the data in these financial reports from Harley-Davidson. (For more information on this topic, go to the Additional Resources link on the MathZone website at **www.mhhe.com/hallmercer.**)

SOLUTION

YEAR x	NET INCOME y, MILLIONS OF \$
2001	438
2002	580
2003	761
2004	914
2005	785

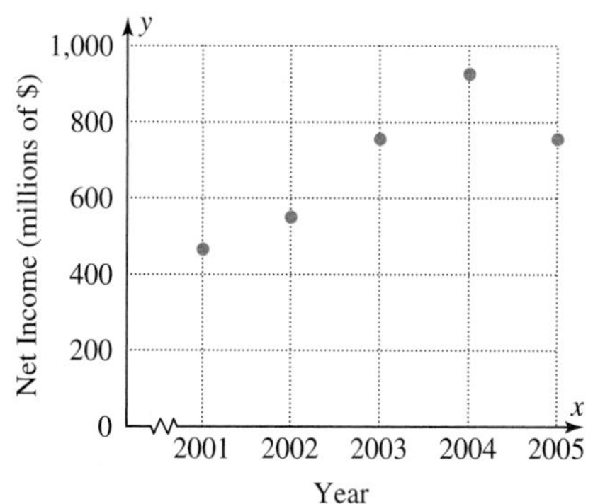

Business results often defy perfect patterns and have a nasty habit of changing patterns just when this information is used to make investment decisions. Nonetheless, the pattern given here by this very limited set of data shows a business that is growing and doing well.

We first examined sequences and subscript notation in Section 1.4. Recall that a sequence is an ordered set of numbers with a first number, a second number, a third number, etc. We will now examine some other ways to represent sequences. The sequence 3, 5, 7, 9, 11 can also be written as $a_1 = 3$, $a_2 = 5$, $a_3 = 7$, $a_4 = 9$, and $a_5 = 11$, or as the set of ordered pairs $\{(1, 3), (2, 5), (3, 7), (4, 9), (5, 11)\}$. Note that in ordered-pair notation the term number is the x-value and the sequence value is the y-value. A scatter diagram of these points is given in Example 5.

EXAMPLE 5 Drawing a Scatter Diagram of a Sequence

Draw a scatter diagram for the sequence 3, 5, 7, 9, 11.

SOLUTION

The term number is the x-coordinate of each point. The sequence value is the y-coordinate.

n	a_n
1	3
2	5
3	7
4	9
5	11

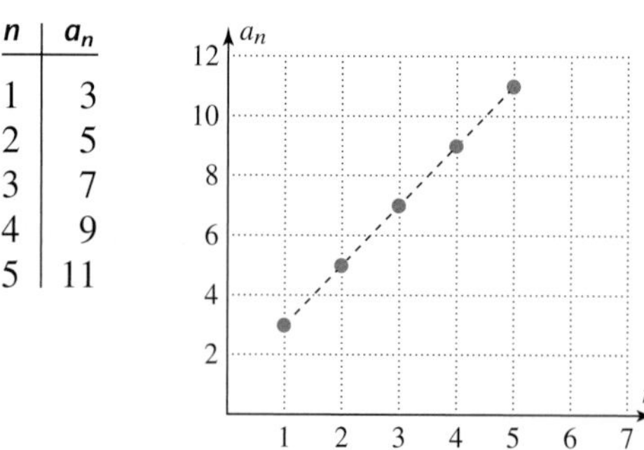

Use the table to express the first five terms as ordered pairs. Then plot these ordered pairs and note that these points all lie on a straight line.

SELF-CHECK 2.1.2

1. Represent the sequence −5, −2, 1, 4, 7 by using subscript notation.
2. Form a table of values for this sequence.
3. Draw a scatter diagram for this sequence.

The sequence 3, 5, 7, 9, 11 in Example 5 is one example of an arithmetic sequence. An **arithmetic sequence** is a sequence with a constant change from term to term. The constant change is called the **common difference** and is often denoted by ***d.*** The common difference d can be positive, negative, or zero. The common difference for the sequence 3, 5, 7, 9, 11 is +2 since the change from one term to the next is +2. The graph of the terms of an arithmetic sequence will consist of individual disconnected points that lie on a straight line.

We can identify an arithmetic sequence either by calculating the difference from term to term or by observing the visual pattern formed by graphing the sequence. This is illustrated in Example 6.

EXAMPLE 6 Identifying Arithmetic Sequences

Determine whether each sequence is an arithmetic sequence. If the sequence is arithmetic, determine the common difference d.

(a) −4, −1, 2, 5, 8 **(b)** 2, 4, 7, 8, 6

SOLUTION

(a) NUMERICAL DIFFERENCE BETWEEN TERMS

$$-1 - (-4) = 3$$
$$2 - (-1) = 3$$
$$5 - 2 = 3$$
$$8 - 5 = 3$$

$$d = 3$$

VERBALLY

This is an arithmetic sequence because there is a common difference of 3. The points establish a linear pattern with the points rising 3 units from one term to the next.

GRAPHICALLY

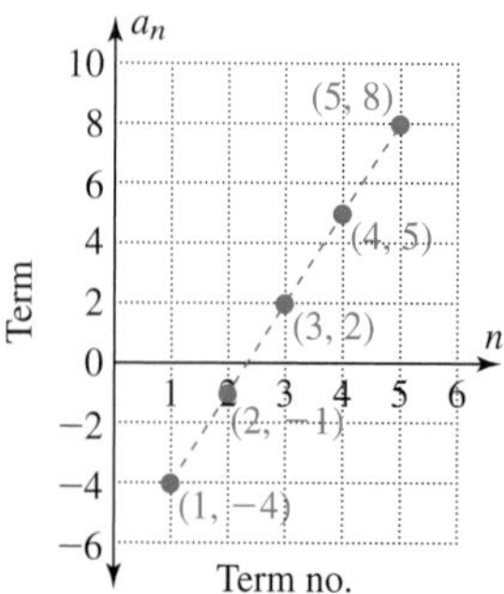

(b) NUMERICAL DIFFERENCE BETWEEN TERMS	VERBALLY	GRAPHICALLY
$4 - 2 = 2$ $7 - 4 = 3$ $2 \neq 3$	This is not an arithmetic sequence because the change from term to term is not constant. The points do not form a linear pattern. The height change between consecutive terms is not constant.	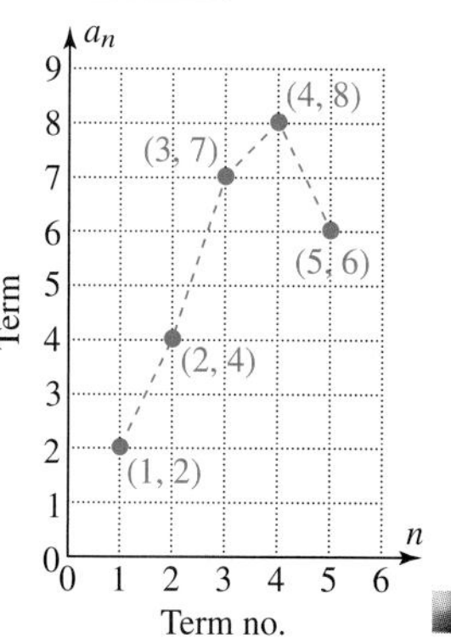

SELF-CHECK 2.1.3

Write the terms of each sequence and determine whether the sequence is arithmetic.

1. 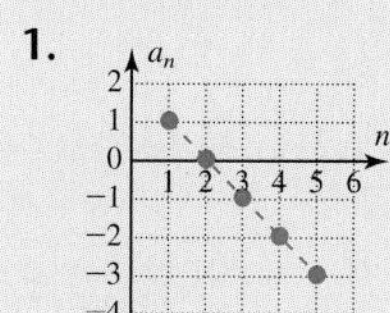

2. 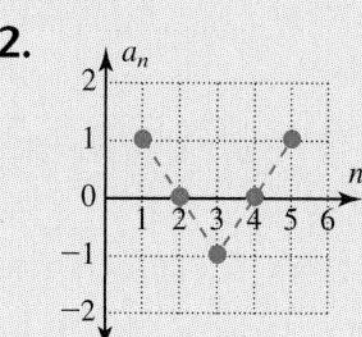

The sequence of total payments \$150, \$300, \$450, . . . in the payment schedule at the beginning of the chapter can also be represented by the algebraic equation $a_n = 150n$, where n gives the number of the term and a_n represents the value of the nth term. One of the advantages of giving an equation for a sequence is its generality. In Example 7 we do not need to complete the entire payment schedule in order to determine a specific monthly result. The formula allows us to directly compute any monthly result we want.

Consider a new car lease, which requires a \$1,200 initial payment and then a \$300 payment at the first of each month for 2 years. We will again let a_n represent the total paid by the nth month.

Writing out a word equation for each word problem is well worth your time. Research has shown that writing this step can produce a significant improvement in understanding how to set up word problems.

Verbally

Total of all payments = Total of monthly payments + Initial payment

Multiply the number of payments by the monthly amount to determine the total of the monthly payments.

Algebraically

$$a_n = 300n + 1{,}200$$

EXAMPLE 7 Using an Equation for an Arithmetic Sequence

Use the equation $a_n = 300n + 1{,}200$ to complete a payments table, which will give a snapshot of the total of all payments after every 6-month period on a car lease with a \$1,200 initial payment followed by monthly payments of \$300 for 2 years. Then graph this table of points.

SOLUTION

Often a_0 denotes the initial conditions of a problem, the value at a time zero.

ALGEBRAICALLY

$$a_n = 300n + 1200$$
$$a_0 = 300(0) + 1200 = 1200$$
$$a_6 = 300(6) + 1200 = 3000$$
$$a_{12} = 300(12) + 1200 = 4800$$
$$a_{18} = 300(18) + 1200 = 6600$$
$$a_{24} = 300(24) + 1200 = 8400$$

NUMERICALLY

n	a_n
0	1200
6	3000
12	4800
18	6600
24	8400

GRAPHICALLY

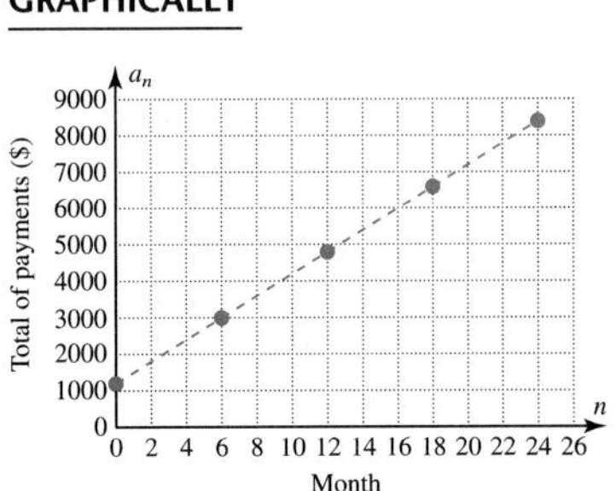

VERBALLY

The cost of leasing the car is increasing by $1,800 every 6 months or equivalently by $300 per month. The graph illustrates the linear pattern exhibited by this arithmetic sequence. Note that the total of the payments for leasing this car for 2 years (24 months) is $8,400.

SELF-CHECK 2.1.4

1. Write the first six terms of the sequence defined by $a_n = 2n - 5$.
2. Is the sequence in question 1 an arithmetic sequence? If so, what is the common difference?

The graph of the points from an arithmetic sequence forms a linear pattern. However, these points do not form an entire line; they consist only of distinct unconnected points corresponding to the input values 1, 2, 3, 4, ... We will examine these graphs and their equations further in Section 2.3. The following box summarizes what we have observed regarding arithmetic sequences.

Arithmetic Sequences

Numerically: An arithmetic sequence has a constant change d from term to term.

Graphically: The distinct points of the graph of an arithmetic sequence all lie on a straight line. There is a constant change in height between consecutive points.

Many people, including the authors, understand and relate to mathematics better when they can visualize the material. Graphs can display not only a table of values but also a variety of information about the relationship of the x-y data pairs. This information can also be obtained by inspecting the table of values, but it may be easier to see by inspecting the graph. Example 8 illustrates how to use a graph to determine a maximum value and to determine trends in the data.

EXAMPLE 8 Reading Information from a Graph

The graph shown gives the distance a taxi is from a dispatching office at a given time. The time x is given in minutes from the start of the shift of one driver, and the distance y is given in miles away from the dispatching office. Answer each of these questions by examining this graph.

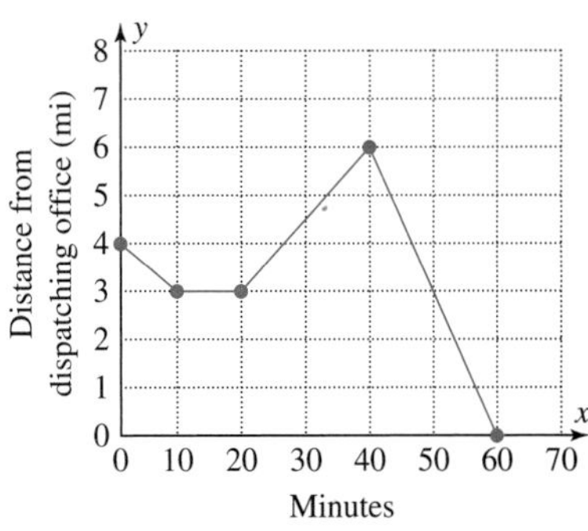

SOLUTION

(a) At a time of 0 min, how far is the taxi from the dispatching office?
The taxi is 4 mi from the office at a time of 0 min.
The point (0, 4) indicates that when $x = 0$, $y = 4$.

(b) When did the taxi arrive at the dispatching office?
The taxi arrives at the dispatching office after 60 min.
The point (60, 0) indicates after 60 min the distance from the office is 0 mi.

(c) What was the maximum distance the taxi was from the office?
The maximum distance the taxi was from the dispatch office was 6 mi.
The highest point on the graph is the point (40, 6). Thus the maximum y-value is $y = 6$.

(d) The taxi was parked for a period of time. During what time was the taxi parked?
The taxi was parked from 10 to 20 min after the start of the driver's shift.
From $x = 10$ to $x = 20$ the y-value is a constant 3 mi. This corresponds to the taxi being parked and not changing its distance.

SELF-CHECK ANSWERS

2.1.1

1. IV, III, II, I
2. (0, 0)
3. 0
4. x

2.1.2

1. $a_1 = -5, a_2 = -2, a_3 = 1, a_4 = 4, a_5 = 7$
2.

n	a_n
1	−5
2	−2
3	1
4	4
5	7

3.

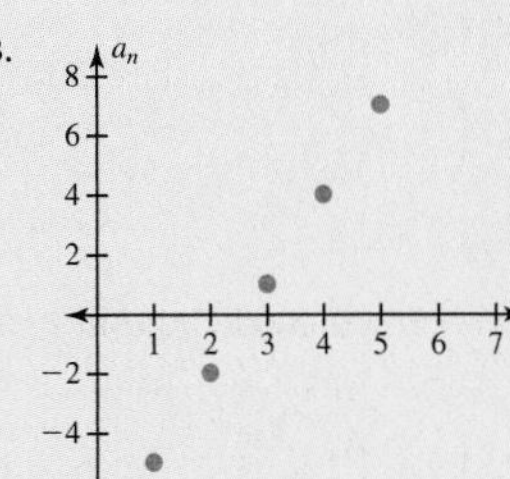

2.1.3

1. 1, 0, −1, −2, −3 is an arithmetic sequence with $d = -1$.
2. 1, 0, −1, 0, 1 is not an arithmetic sequence.

2.1.4

1. −3, −1, 1, 3, 5, 7
2. Arithmetic sequence with $d = 2$

USING THE LANGUAGE AND SYMBOLISM OF MATHEMATICS 2.1

1. The rectangular coordinate system is also called the __________ coordinate system.
2. On the rectangular coordinate system, the horizontal axis is called the __________-axis, and the vertical axis is called the __________-axis.
3. The point where the horizontal and vertical axes cross is called the __________ and has coordinates __________.
4. All points with an x-coordinate of 0 are on the __________-axis.
5. All points with a y-coordinate of 0 are on the __________-axis.
6. The four quadrants are numbered I, II, III, and IV in a **clockwise/counterclockwise** direction. (Select the correct choice.)
7. In quadrant __________ both coordinates are positive.
8. In quadrant __________ both coordinates are negative.
9. A __________ diagram for a set of points is a graph of these points.
10. A sequence with a constant change from term to term is called an __________ sequence.
11. In an arithmetic sequence, d represents the common __________.
12. The graph of an arithmetic sequence consists of distinct unconnected points that form a __________ pattern.

EXERCISES 2.1

In Exercises 1 and 2, identify the coordinates of points A, B, C, and D. Also give the quadrant in which each point is located.

1.

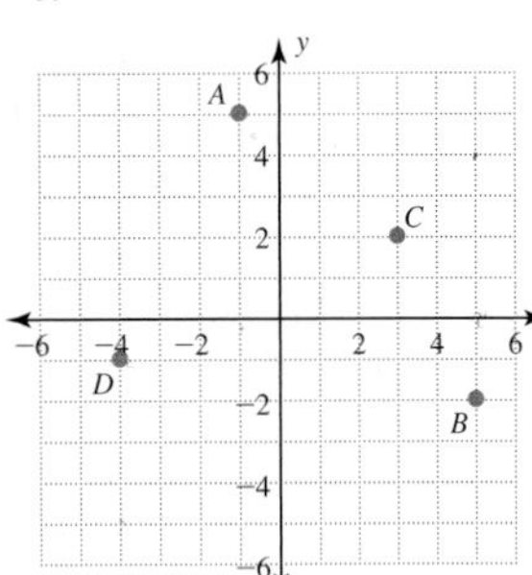

2.

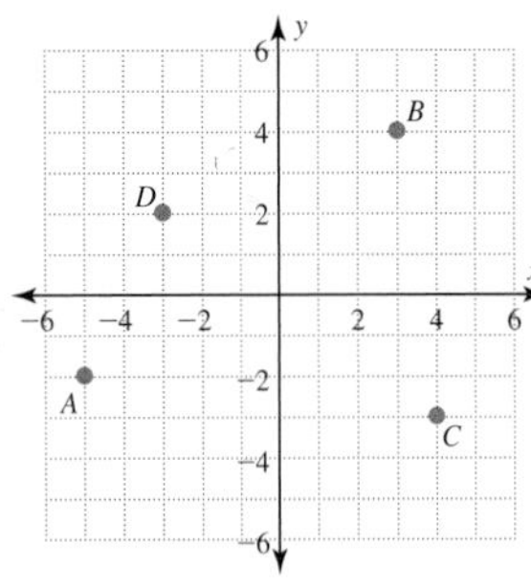

In Exercises 3 and 4, plot and label the points whose coordinates are given. Plot these points on the same coordinate system.

3. $A(5, 4)$ $B(-4, -5)$ $C(3, -4)$ $D(-6, 1)$ $E(4, 0)$ $F(0, -5)$

4. $A(2, 6)$ $B(-3, -6)$ $C(4, -2)$ $D(-5, 3)$ $E(-3, 0)$ $F(0, 6)$

In Exercises 5 and 6, use the sign pattern of the coordinates to determine the quadrant in which each point is located. Do not plot these points.

5. a. $(-1.3, 3.7)$ **b.** $(-5.8, -9.3)$ **c.** $(6.7, -2.1)$ **d.** $(0.4, 0.1)$

6. a. $(-193, -217)$ **b.** $\left(\frac{2}{3}, -\frac{3}{5}\right)$ **c.** $(801, 913)$ **d.** $(-\pi, \sqrt{2})$

In Exercises 7 and 8, determine the axis on which each point lies. Do not plot these points.

7. a. $(0, 8)$ **b.** $(-8, 0)$ **c.** $(-\sqrt{3}, 0)$ **d.** $(0, \sqrt{2})$

8. a. $\left(\frac{1}{8}, 0\right)$ **b.** $(0, 81)$ **c.** $(0, -\pi)$ **d.** $(-\sqrt{5}, 0)$

In Exercises 9 and 10, draw a scatter diagram for each set of data. Do these points lie approximately on a straight line?

9.

x	−2	−1	0	1	2	3
y	−10	−9	−4	1	6	8

10.

x	−2	−1	0	1	2	3
y	8	6	4	4	2	−2

In Exercises 11 and 12, use the given sequence to **(a)** represent the sequence by using subscript notation and **(b)** form a table of x-y values for this sequence.

11. 5, 13, 19, 8, 6 **12.** 2, 9, 15, 24, 4

In Exercises 13–16, use the given sequence to complete each part of the problem.

a. Graph each sequence.
b. Do these points form a linear pattern?
c. Is the sequence an arithmetic sequence? If the sequence is arithmetic, what is the common difference?

13. $a_1 = -4, a_2 = -2, a_3 = 0, a_4 = 2, a_5 = 4$

14. $a_1 = 5, a_2 = 3, a_3 = 1, a_4 = -1, a_5 = -3$

15. $8, 5, 2, -1, -4$ **16.** $-5, -2, 1, 4, 7$

In Exercises 17–20, write the first five terms of these arithmetic sequences.

17. $a_n = 5n - 1$ **18.** $a_n = -4n + 6$

19. $a_n = 1 - 5n$ **20.** $a_n = 4n - 6$

21. Use the equation $a_n = 3n + 7$ to evaluate each expression.
a. a_0 **b.** a_1 **c.** a_4 **d.** a_{10}

22. Use the equation $a_n = 5n - 6$ to evaluate each expression.
a. a_0 **b.** a_2 **c.** a_{20} **d.** a_{100}

23. Use the equation $a_n = -6n + 4$ to evaluate each expression.
a. a_0 **b.** a_1 **c.** a_4 **d.** a_{40}

24. Use the equation $a_n = -8n + 5$ to evaluate each expression.
a. a_0 **b.** a_1 **c.** a_5 **d.** a_{50}

25. Use the equation
$$a_n = \frac{n + 1}{2}$$
to complete this table.

n	a_n
1	
2	
3	
4	
5	

26. Use the equation
$$a_n = \frac{n - 1}{2}$$
to complete this table.

n	a_n
1	
2	
3	
4	
5	

In Exercises 27 and 28, use the graph of each arithmetic sequence to write the first five terms of each sequence.

27.

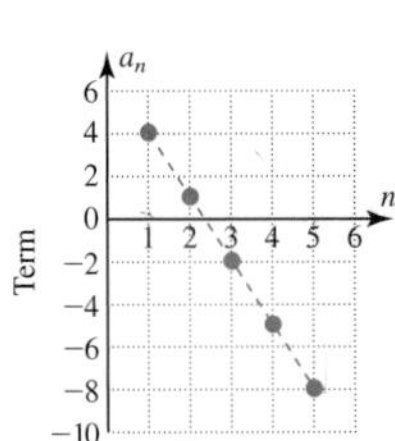

28.

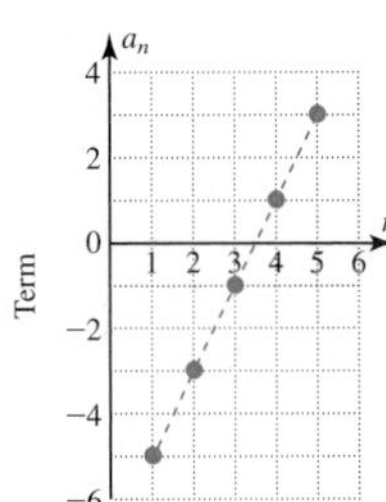

In Exercises 29–36, determine whether the sequence is arithmetic. If the sequence is arithmetic, write the common difference d.

29. $-3, 1, 5, 9, \ldots$ **30.** $-11, -6, -1, 4, 9, \ldots$

31. 40, 30, 20, 10, 0 **32.** 20, 17, 14, 11, 8, 5

33. 3, 5, 8, 12, 17 **34.** 2, 4, 8, 16, 32, 64

35. a. $a_n = 2n$ **b.** $a_n = 2^n$ **36. a.** $a_n = \frac{n}{2}$ **b.** $a_n = \frac{60}{n}$

In Exercises 37–44, determine whether the sequence shown in each graph is an arithmetic sequence. For those that are arithmetic sequences, find the common difference d.

37.

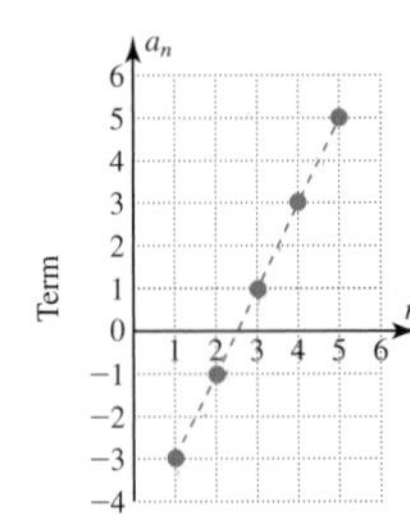

38.

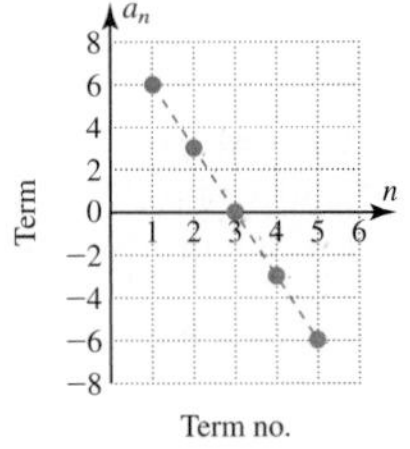

39.

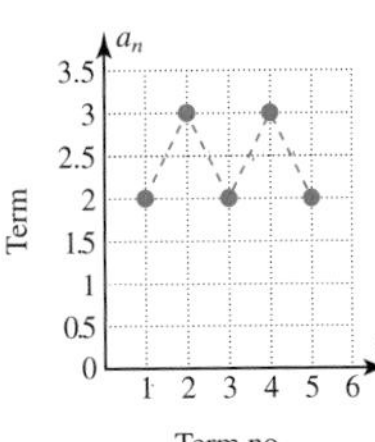

40.

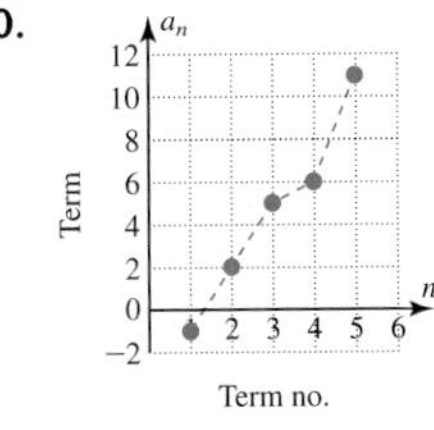

41.

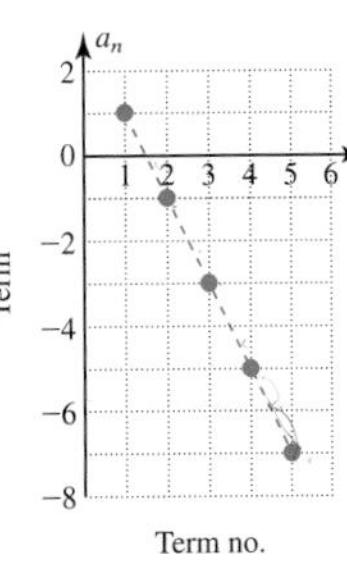

42.

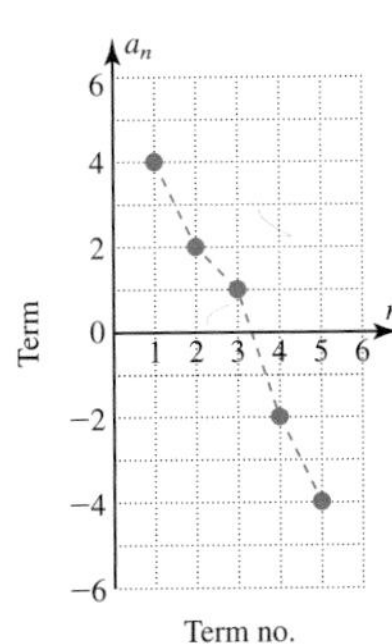

43.

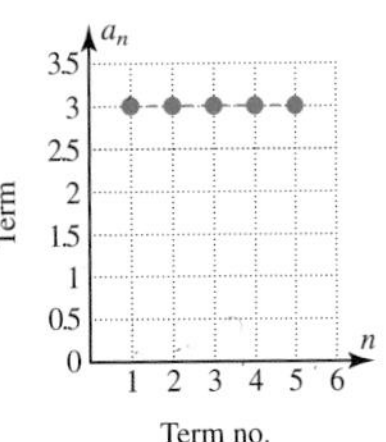

44.

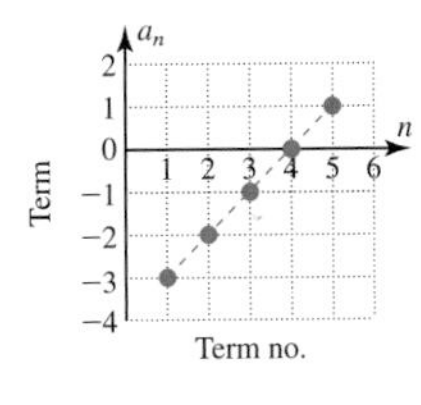

In Exercises 45–50, use the graph to complete the table. Then determine whether the sequence in the table is an arithmetic sequence.

45. Graph

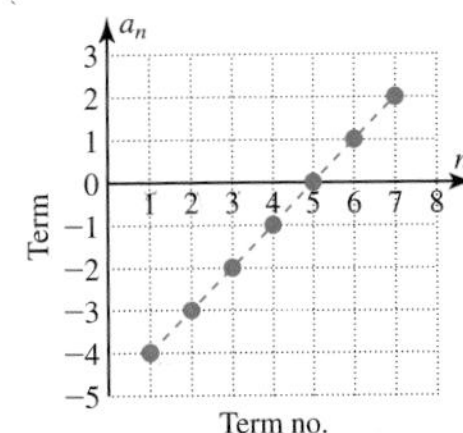

Table

n	a_n
1	-4
2	-3
3	-2
4	-1
5	0
6	1
7	2

46. Graph

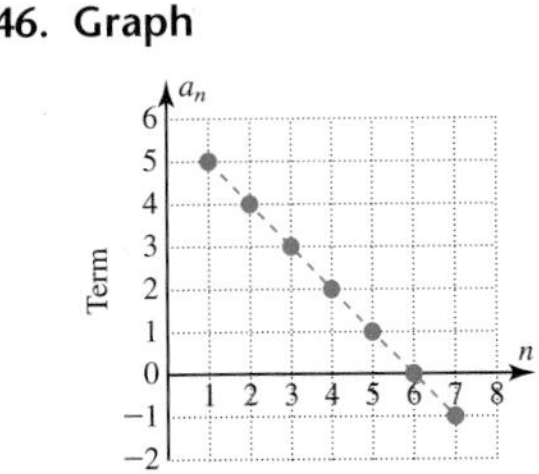

Table

n	a_n
1	
2	
3	
4	
5	
6	
7	

47. Graph

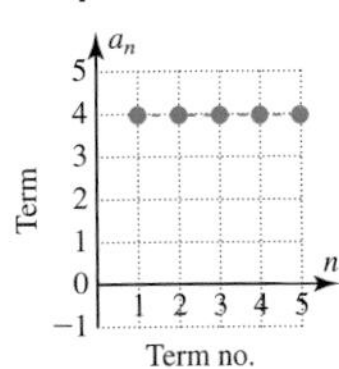

Table

n	a_n
1	
2	
3	
4	
5	

48. Graph

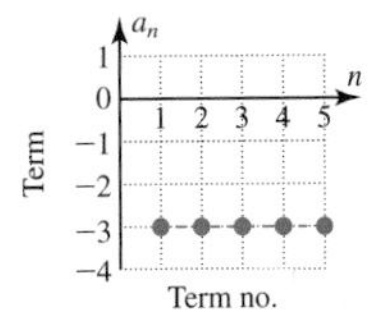

Table

n	a_n
1	
2	
3	
4	
5	

49. Graph

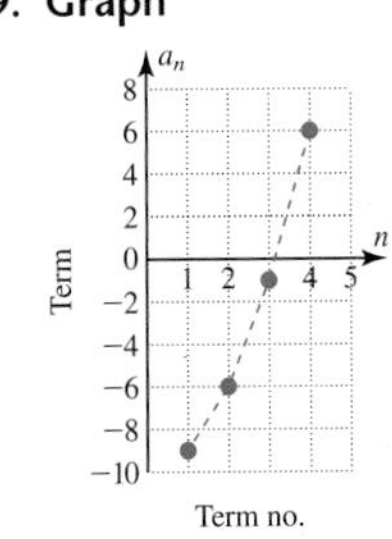

Table

n	a_n
1	
2	
3	
4	

50. Graph

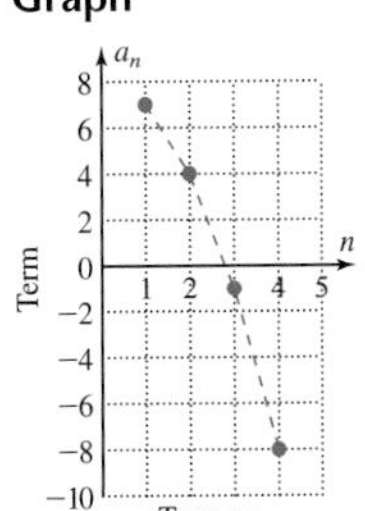

Table

n	a_n
1	
2	
3	
4	

Automobile Loan

In Exercises 51 and 52, the equation $a_n = 425n$ gives the total dollars paid on a car loan after n months. Calculate and interpret each value of a_n.

51. a. a_0 **b.** a_6 **c.** a_{12}

52. a. a_1 **b.** a_{18} **c.** a_{24}

Automobile Lease

In Exercises 53 and 54, the equation $a_n = 450n + 2{,}400$ gives the total payments in dollars after n months on the lease for a luxury sports utility vehicle. Calculate and interpret each value of a_n.

53. a. a_1 **b.** a_{18} **c.** a_{24}

54. a. a_0 **b.** a_6 **c.** a_{12}

In Exercises 55–58, complete each inequality by filling in each blank with either $<$ or $>$.

55. If a point (x, y) is in quadrant I, then x __>__ 0 and y __>__ 0.

56. If a point (x, y) is in quadrant II, then x ____ 0 and y ____ 0.

57. If a point (x, y) is in quadrant III, then x ____ 0 and y ____ 0.

58. If a point (x, y) is in quadrant IV, then x ____ 0 and y ____ 0.

59. If $xy > 0$, then the point (x, y) is in either quadrant ____ or quadrant ____.

60. If $xy < 0$, then the point (x, y) is in either quadrant ____ or quadrant ____.

61. Consider an arithmetic sequence that has a positive common difference. Does the line through the graph of this sequence go up or down as the points move to the right?

62. Consider an arithmetic sequence that has a negative common difference. Does the line through the graph of this sequence go up or down as the points move to the right?

63. Are the even integers an arithmetic sequence? If so, what is the common difference d?

64. Are the odd integers an arithmetic sequence? If so, what is the common difference d?

Area and Perimeter

65. Determine the area and perimeter of a rectangle whose corners are located at (0, 0), (5, 0), (5, 4), and (0, 4).

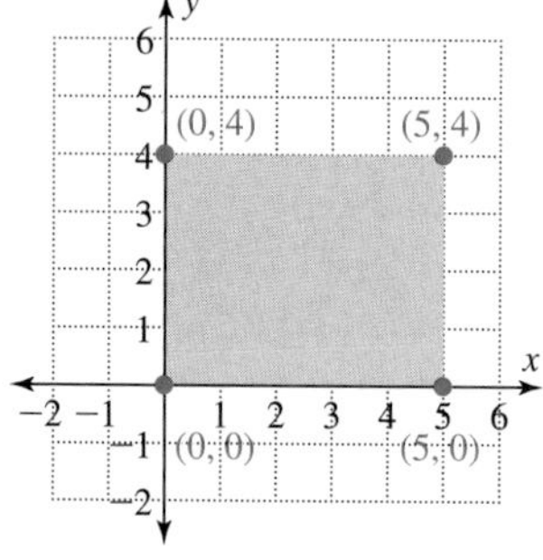

66. Determine the area and perimeter of a rectangle whose corners are located at (0, −3), (2, −3), (2, 3), and (0, 3).

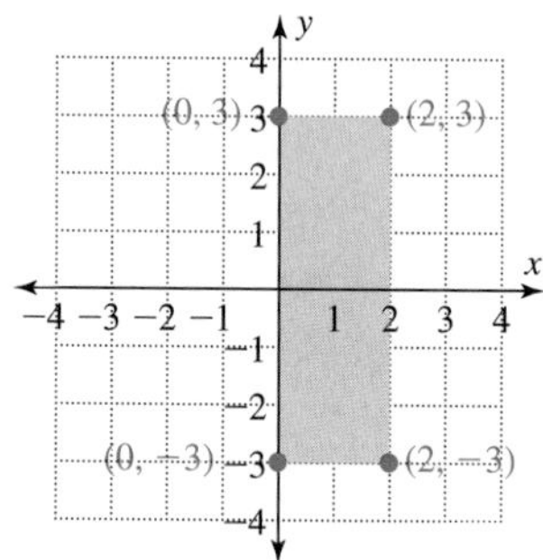

Exercises 67–69 refer to the given payment schedule.

PAYMENT NUMBER	PAYMENT AMOUNT, $	TOTAL PAID, $
1	150	150
2	150	300
3	150	450
4	150	600
5	150	750

67. Is the sequence of payment numbers 1, 2, 3, 4, 5 an arithmetic sequence? If so, what is the common difference?

68. Is the sequence of monthly payments 150, 150, 150, 150, 150 an arithmetic sequence? If so, what is the common difference?

69. Is the sequence of total amount paid 150, 300, 450, 600, 750 an arithmetic sequence? If so, what is the common difference?

70. **Depth of Water in a Reservoir** The depth of water in a city reservoir is carefully recorded for a 6-month period. The results are displayed in the given graph. The time x is given in months, and the depth y is given in feet. Use this graph to answer the following questions.

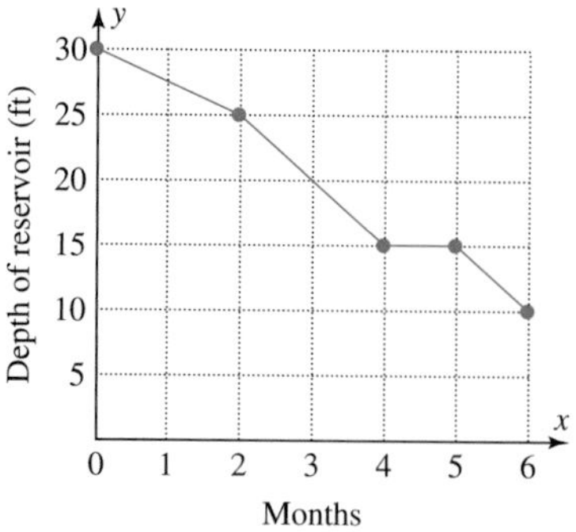

a. At the start of the recording time, a time of 0 months, what is the depth of the water in the reservoir?

b. At the end of 5 months, what is the depth of the water in the reservoir?

c. When did the depth of the water reach 25 ft?

d. What was the minimum depth of the water during this 6-month period?

71. **Weight of a Patient** The weight of a patient is recorded for an 8-week period by a physical therapist. The results are displayed in the given graph. The time x is given in weeks, and the weight y is given in pounds. Use this graph to answer the following questions.

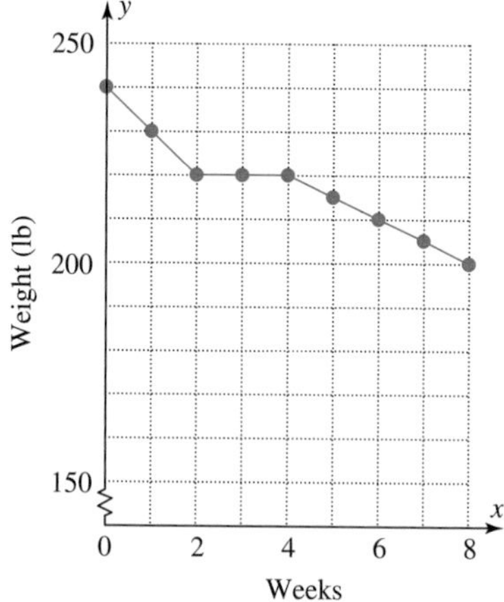

a. At the start of the recording time, a time of 0 weeks, what is the weight of the patient?

b. At the end of 1 week, what is the weight of the patient?

c. When did the weight of the patient reach 210 lb?

d. The patient maintained a steady weight of 220 lb for a period of time. During what time period did the patient weigh 220 lb?

72. **Price of a Stock** Each graph corresponds to the price of a stock over a 6-month period. The time x is given in months, and the price y is given in dollars. Match each of these descriptions with the corresponding graph.

a. The price is increasing for the entire 6 months.

b. The price is decreasing for the entire 6 months.

c. The price is constant for the entire 6 months.

d. The price has a maximum after 3 months that is higher than the price at any other time.

A.

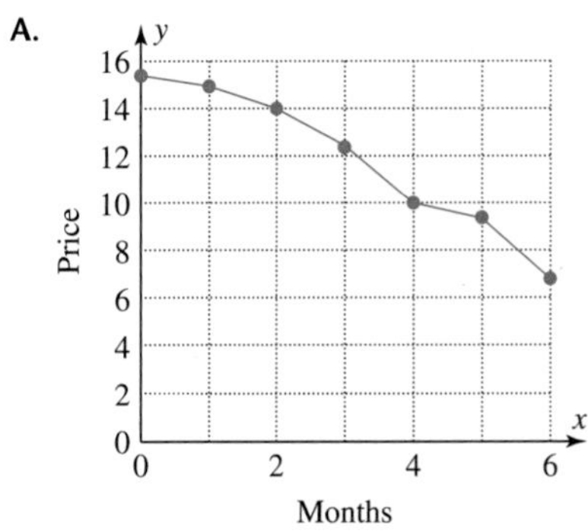

B.

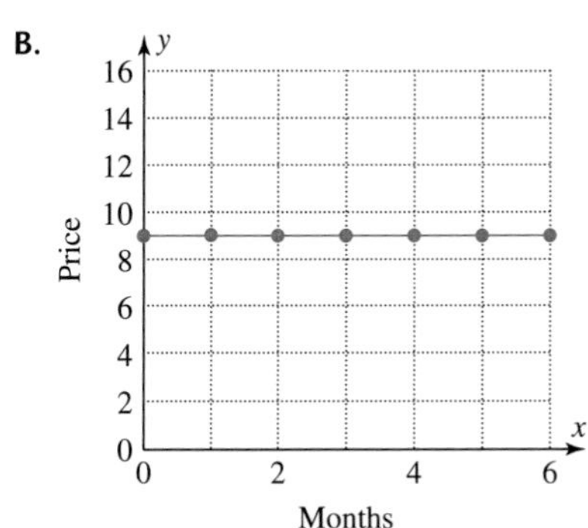

C.

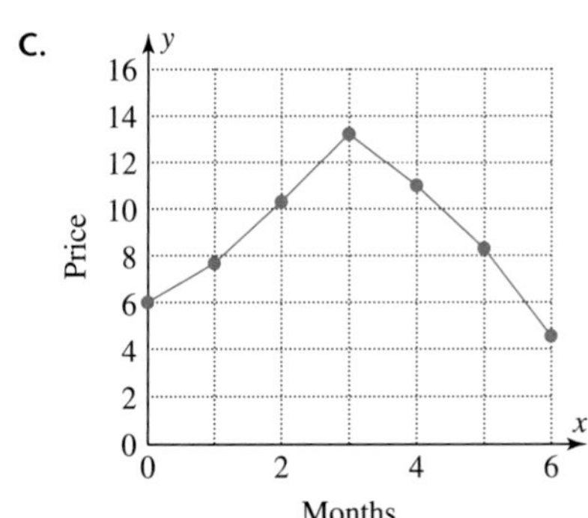

D.

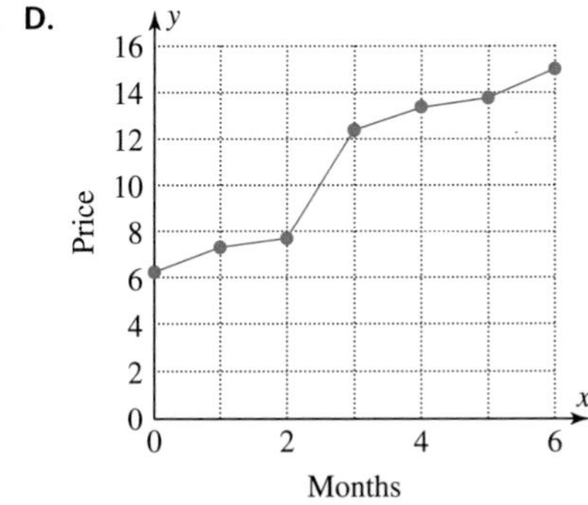

73. Temperature of a Child Each graph corresponds to the temperature of a sick child over a 6-hour period. The time x is given in hours, and the temperature y is given in degrees Fahrenheit. Match each of these descriptions with the corresponding graph.

a. The temperature is increasing for the entire 6 hours.

b. The temperature is decreasing for the entire 6 hours.

c. The temperature is constant for the entire 6 hours.

d. The temperature reaches a minimum after 3 hours and then increases again.

A.

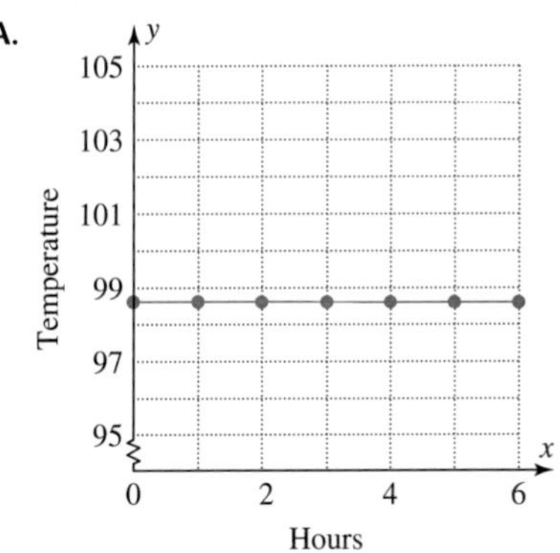

B.

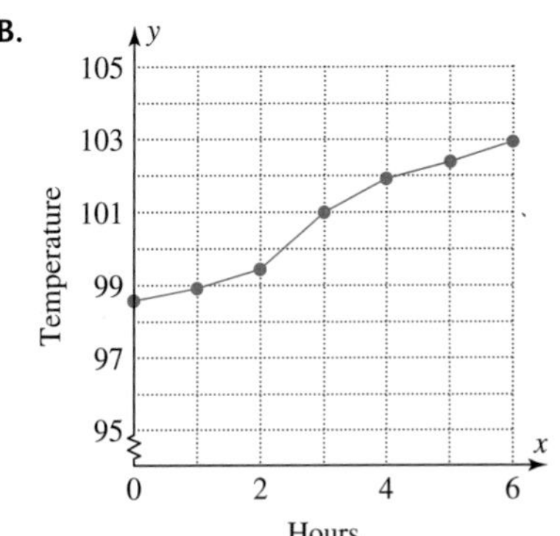

C.

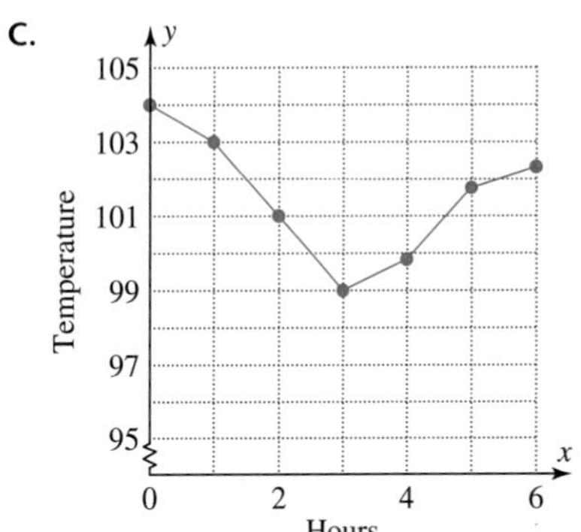

D.

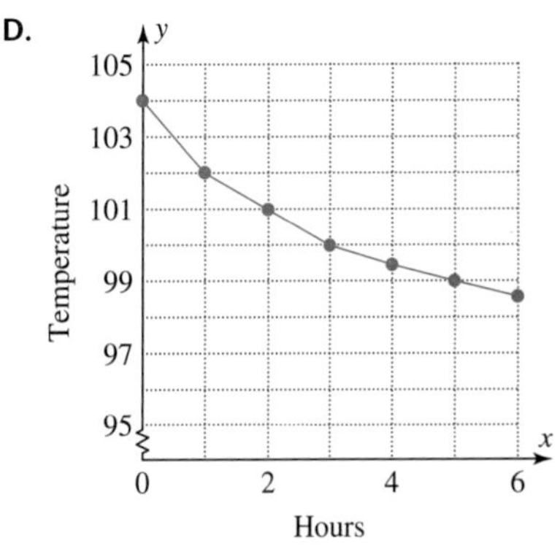

74. **Water Bills** The water bills of a homeowner are recorded under a variety of situations over a 6-month period. The time x is given in months, and the water bill y is given in dollars. Sketch a separate graph to match each of the following verbal descriptions.
   a. The homeowner has limited water usage and each month has the same minimum usage charge.
   b. The homeowner has had family members leave the house and has gradually reduced her water bill each month.
   c. The homeowner has increased her water bill each month as she has watered the yard during an extended dry period.
   d. The homeowner has increased her water bill for 3 months as she has watered the yard during a dry period, and then a series of rains allowed the homeowner to gradually reduce the water usage to the minimal level by the end of the sixth month.

75. **Blood Pressure** The diastolic blood pressure (for more information on blood pressure, go to the Additional Resources link on the MathZone website at **www.mhhe.com/hallmercer**) of a man is recorded under a variety of situations over a 5-minute period. The time x is given in minutes, and the blood pressure y is given in millimeters (mm) of mercury. Assuming the man starts the period with a diastolic blood pressure of 90 (representing 90 mm of mercury), sketch a separate graph to match each of the following verbal descriptions.
   a. The man is inactive, and the blood pressure remains steady.
   b. The man has just completed a physical activity, and as he rests, his blood pressure steadily drops.
   c. The man has just initiated an exercise on a treadmill, and his blood pressure is steadily increasing.
   d. The man is watching a movie, and at first a calm scene helps to lower his blood pressure, but then an intense scene causes his blood pressure to rise.

Group Discussion Questions

76. **Risks and Choices** Suppose an employer offered you a $30,000 starting salary for your first year of employment with raises for the first 5 years determined by either option A or option B.

   Option A: A $1,500 raise for each of the next 5 years.
   Option B: A 5% raise for each of the next 5 years.

   a. Can you determine without any calculations which option will produce an arithmetic sequence? Explain your reasoning.
   b. Write the sequence of salaries for each option.
   c. Use the changes from term to term in each sequence to explain which option you would prefer.

77. **Writing Mathematically** A basketball player scored 5, 10, and 15 points in her first three games. Do you think an arithmetic sequence would be a good model of her scores in the next eight games? Explain your answer.

78. **Challenge Question** The given table shows a payment schedule for a $5,000 car loan at 9%. Use a calculator and the interest formula $I = P \cdot R \cdot T$ to complete the table. Can you adjust the last payment so that the ending balance is $0? (*Hint:* One month equals $\frac{1}{12}$ of a year.)

MONTHLY PAYMENT NO.	STARTING BALANCE, $	MONTHLY PAYMENT, $	TOTAL PAID, $	NEW BALANCE, $
1	5,000.00	500.00	500.00	4,537.50
2	4,537.50	500.00	1,000.00	4,071.53
3	4,071.53			
4				
.				
.				
.				

79. **Writing Mathematically** Write a story to accompany the given graph, and label each axis to match your story. Then use this story to ask four questions that the members of your group can answer by using this graph. (*Hint:* See Example 8.)

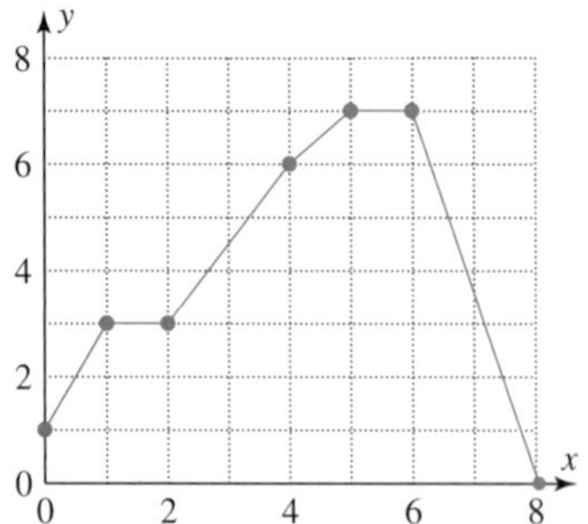

80. Spreadsheet Exploration The given graph was created from an Excel spreadsheet that displays the cost of operating a paint sprayer for different time periods. Use the graph to complete the missing entries in this spreadsheet.

	A	B
1	Hours	Cost
2	1	$125.00
3	2	$175.00
4		$225.00
5	4	$275.00
6	5	
7		

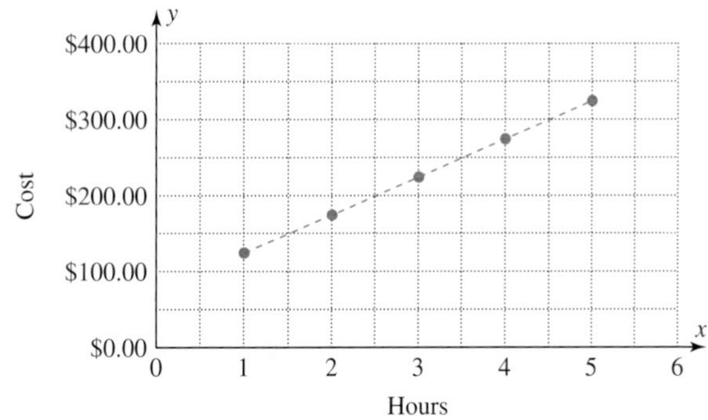

Section 2.2 Function Notation and Linear Functions

Objectives:

4. Use function notation.
5. Use a linear equation to form a table of values and to graph a linear equation.
6. Write a function to model an application.

A Mathematical Note

The history of mathematics has many noted women mathematicians. One of the earliest women about whom anything is written is Hypatia (c. 360–415). She studied in Athens and then taught geometry and astronomy at Alexandria. She also wrote on algebra. History records she was killed in a religious riot. For more information on women mathematicians please reference **www.mhhe.com/hallmercer.**

The primary purpose of this section is to introduce function notation and to illustrate how linear functions can be used to model applications. This introduction will enable us to use function notation throughout this text. We will wait until Chapter 7 to examine the function concept more formally. For now, we will lay the foundation for this work by becoming familiar with function notation.

A table of values and a graph of a few points both give us valuable representations for showing the relationship between a set of x-y pairs. For many x-y pairs it may be inconvenient or impossible to put all these points in a table. Thus it is useful to be able to express the relationship between x and y by using an equation. One notation for writing an equation relating x-y pairs is called function notation.

The notation $f(x)$ is referred to as **function notation** and is read "f **of** x" or "$f(x)$ **is the output value for an input value of** x." Function notation is used to give an equation that describes a unique output that we can calculate for each input of x. To evaluate $f(x)$ for a specific value of x, replace x on both sides of the equation by this specific value, as illustrated in Example 1.

Caution: $f(x)$ does not mean f times x in this context. It represents an output value for a specific input value of x.

EXAMPLE 1 Using Function Notation to Evaluate a Function

Evaluate each expression for $f(x) = 3x + 5$

SOLUTION

(a) $f(0)$

$$f(x) = 3x + 5$$
$$f(0) = 3(0) + 5$$
$$f(0) = 0 + 5$$

Substitute each input value of x into the equation $f(x) = 3x + 5$.

Answer: $f(0) = 5$

For an input of 0 the output is 5.

(b) $f(4)$

$$f(x) = 3x + 5$$
$$f(4) = 3(4) + 5$$
$$f(4) = 12 + 5$$

Substitute 4 for x in the equation.

Answer: $f(4) = 17$

For an input of 4 the output is 17.

(c) $f(-4)$

$$f(x) = 3x + 5$$
$$f(-4) = 3(-4) + 5 \quad \text{Substitute } -4 \text{ for } x \text{ in the equation.}$$
$$f(-4) = -12 + 5$$

Answer: $f(-4) = -7$ For an input of -4 the output is -7.

The AMATYC standards in "Beyond Crossroads" strongly recommend that functions be examined verbally, numerically, graphically, and symbolically. This book has been written to follow this guideline and to use technology to enhance your learning. Nonetheless, it is important that you understand how to produce these tables and sketch these graphs by hand.

In Example 2 we will use the function $f(x) = 2x + 1$ to form a table and to sketch a graph. A major difference between the function $f(x) = 2x + 1$ and the sequence $a_n = 2n + 1$ is that x can be any real number while n is restricted to the natural numbers 1, 2, 3, 4, ... Thus the graph of a sequence will consist of individual discrete points, whereas the graph of $f(x) = 2x + 1$ will be the solid line shown in Example 2. Because the input values of x can be any real number we arbitrarily selected the input values used in Example 2. The graph is the continuous line that is drawn through all the points in the table.

EXAMPLE 2 Graphing a Function

Use the function $f(x) = 2x + 1$ to complete a table for x-values of $-2, -1, 0, 1$, and 2. Then use these input-output pairs to graph the line through these points.

SOLUTION

ALGEBRAICALLY

$$f(x) = 2x + 1$$
$$f(-2) = 2(-2) + 1 = -3$$
$$f(-1) = 2(-1) + 1 = -1$$
$$f(0) = 2(0) + 1 = 1$$
$$f(1) = 2(1) + 1 = 3$$
$$f(2) = 2(2) + 1 = 5$$

NUMERICALLY

x	f(x)
−2	−3
−1	−1
0	1
1	3
2	5

GRAPHICALLY

VERBALLY

The function $f(x) = 2x + 1$ defines a continuous line (no gaps) that contains the points in this table.

SELF-CHECK 2.2.1

Given $f(x) = 4x + 11$, evaluate each expression.

1. $f(0)$ 2. $f(1)$ 3. $f(10)$ 4. $f(-10)$
5. Use the function $f(x) = -x + 3$ to complete this table

x	−2	−1	0	1	2
y					

6. Graph this function.
7. Graph the line through the points (0, 2) and (5, 0).

The equation in Example 2, $f(x) = 2x + 1$, is an excellent example of a function of the form $f(x) = mx + b$. A function of this form is called a **linear function** because its graph is a straight line. We will examine linear functions in detail in Section 3.2. The linear function $f(x) = mx + b$ can also be represented by $y = mx + b$. Using the subscript notation introduced in Section 1.4, graphing calculators can denote several functions using Y_1, Y_2, Y_3, and so on.

Once a function is entered into a graphing calculator by using the [Y=] key, there are many ways this function can be examined. In Calculator Perspective 1.2.4 we showed how to generate a table. It is very useful to compare the table of values to the graph of these points. Calculator Perspective 2.2.1 illustrates how to produce a table and a graph for the equation given in Example 2. The notation below the graph indicates the minimum and maximum x- and y-values used as well as the scale in the calculator window. The notation $[-4.7, 4.7, 1]$ means the x-values extend from -4.7 to 4.7 with each mark on the x-axis representing 1 unit. Likewise, $[-3.1, 3.1, 1]$ means the y-values extend from -3.1 to 3.1 with each mark on the y-axis representing 1 unit. The values shown here represent the decimal viewing window on a TI-84 Plus calculator. This window is obtained by pressing [ZOOM] [4]. We will examine other viewing windows in Section 2.3.

CALCULATOR PERSPECTIVE 2.2.1 Generating a Table and a Graph

To generate a table and a graph for the equation $y = 2x + 1$ from Example 2 on a TI-84 Plus calculator, enter the following keystrokes:

[Y=]
[2] [X,T,θ,n] [+] [1]
[2nd] TABLE
[ZOOM] [4]

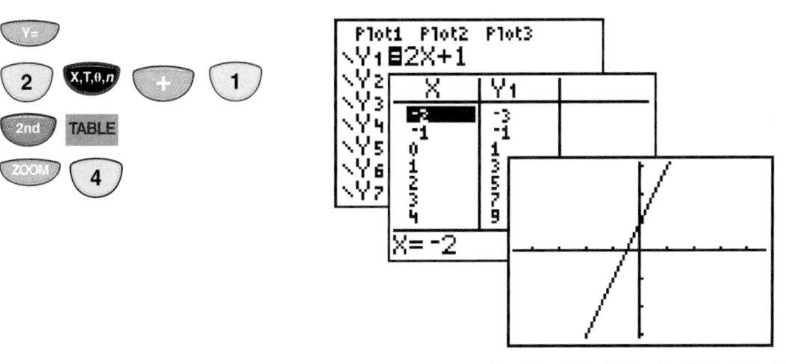

$[-4.7, 4.7, 1]$ by $[-3.1, 3.1, 1]$

Note: The line is a solid connected line containing all the points in the table as well as an infinite number of other points.

EXAMPLE 3 Using Multiple Perspectives to Examine a Linear Function

Use a graphing calculator to examine the linear function $f(x) = -x + 2$ numerically and graphically. Then verbally describe the relationship given by this algebraic equation. Use the x-values of -3, -2, -1, 0, 1, 2, and 3 to form the table of values.

SOLUTION

ALGEBRAICALLY

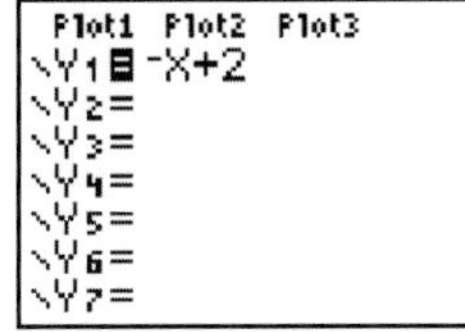

NUMERICALLY

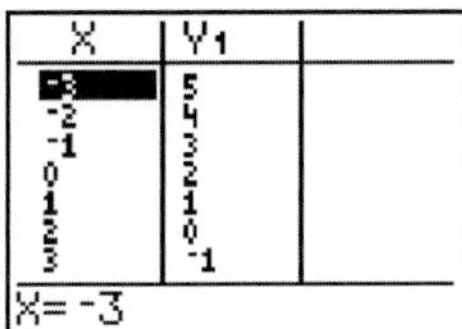

GRAPHICALLY

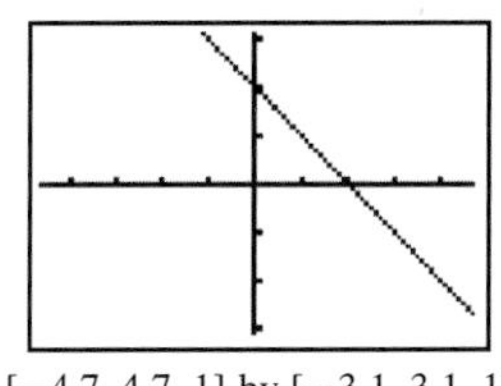

$[-4.7, 4.7, 1]$ by $[-3.1, 3.1, 1]$

VERBALLY

The function defines a table of points that all lie on a straight line. Note that the line drops 1 unit for every 1-unit move to the right.

In Example 3, we used the equation $f(x) = -x + 2$ to generate input-output pairs. We can also use graphs and tables to identify input and output values. There are two types of questions that we can ask regarding these input-output pairs. The first type of question uses a given input value of x to determine the output value $f(x)$. The second type of question uses a given output value $f(x)$ to determine input value(s) of x that produce this output. This is illustrated in Example 4.

EXAMPLE 4 Evaluating a Function From a Table and a Graph

The graph and table to the right both define the same function, one whose algebraic form is $f(x) = mx + b$. Use this graph and table to determine the following input and output values.

GRAPH

TABLE

x	$f(x)$
-2	-3
0	-2
2	-1
4	0
6	1
8	2

SOLUTION

(a) $f(-2) = \underline{?}$ $\quad f(-2) = -3$ Both the table and the graph contain the point $(-2, -3)$. The function pairs the input value of -2 with the output value of -3.

(b) $f(4) = \underline{?}$ $\quad f(4) = 0$ Both the table and the graph contain the point (4, 0). The function pairs the input value of 4 with the output value of 0.

(c) $f(x) = -1; x = \underline{?}$ $\quad f(x) = -1; x = 2$ Both the table and the graph contain the point $(2, -1)$. The function pairs the input value of 2 with the output value of -1.

(d) $f(x) = 2; x = \underline{?}$ $\quad f(x) = 2; x = 8$ Both the table and the graph contain the point (8, 2). The function pairs the input value of 8 with the output value of 2.

SELF-CHECK 2.2.2

Use a graphing calculator and $y = 1.75x - 2.25$ to do the following:

1. Form a table of values for the input values of x of $-1, 0, 1, 2, 3, 4$ and 5.
2. Graph the line defined by this equation, using the window $[-4.7, 4.7, 1]$ by $[-3.1, 3.1, 1]$.

The given graph and table both define the same function, one whose algebraic form is $f(x) = mx + b$. Use this graph and table to determine the following input and output values.

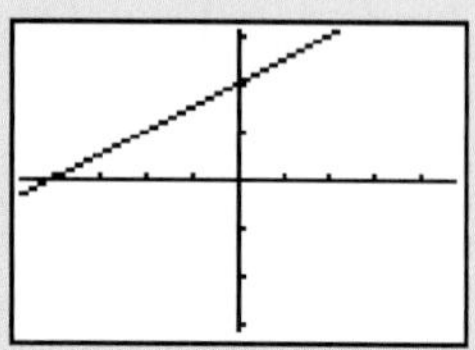

[−4.7, 4.7, 1] by [−3.1, 3.1, 1]

X	Y1
-5	-.5
-4	0
-3	.5
-2	1
-1	1.5
0	2
1	2.5

X= -5

3. $f(0) = \underline{?}$ 4. $f(1) = \underline{?}$ 5. $f(x) = 0; x = \underline{?}$ 6. $f(x) = 1; x = \underline{?}$

Key point: A graph of discrete data contains only selected discrete values whereas continuous data will produce a graph that is connected. The real world has many applications of both discrete and continuous data. This book has a goal of giving you experience with both these data types, as recommended in AMATYC's "Beyond Crossroads."

One of the connections that you may have observed is that a linear function produces an arithmetic sequence for input values of 1, 2, 3, ... The graph of an arithmetic sequence consists of discrete points that lie on a line. The graph of a linear function $f(x) = mx + b$ consists of the entire line. Thus the graph of a linear function is a solid connected line.

EXAMPLE 5 Comparing the Graphs of an Arithmetic Sequence and a Linear Function

Use the points $\{(1, -1), (2, 1), (3, 3), (4, 5), (5, 7)\}$ to graph both the arithmetic sequence $a_n = 2n - 3$ and the linear function $f(x) = 2x - 3$.

SOLUTION

ARITHMETIC SEQUENCE

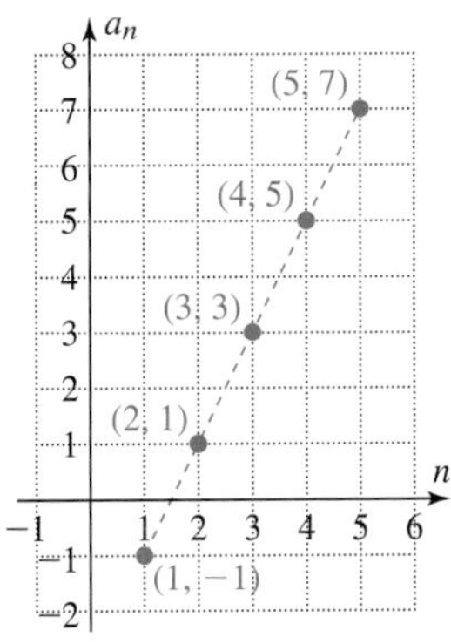

LINEAR FUNCTION

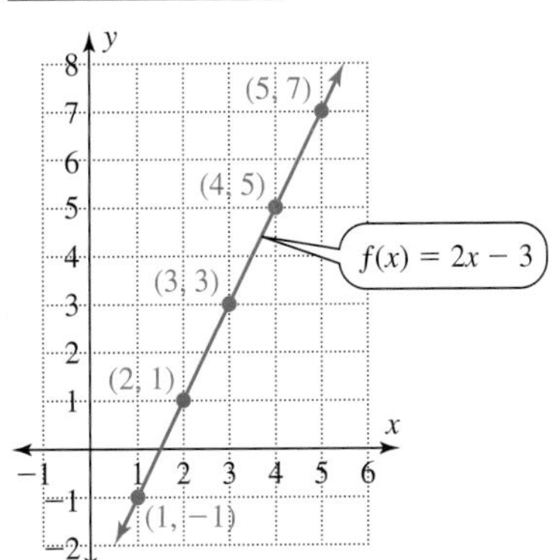

Note the common difference of the arithmetic sequence is 2. Also note the graph of the line goes up 2 units for each 1-unit movement to the right.

Functions are used frequently to describe the relationship between two variables. The function concept is useful because we can observe the changes to the output value of the function as we change the input value. Creating an equation or a function to describe an application is called mathematical modeling. Once the functional model has been created in Example 6, we can create a table and examine various options. The mathematical models presented in the examples and exercises in this section help develop skills needed for word problems that occur later in this book.

EXAMPLE 6 Using a Function to Model the Length of a Board

A board 12 ft long has a piece x ft long cut off. Write a function for the length of the remaining piece in terms of x, and examine this function numerically and verbally.

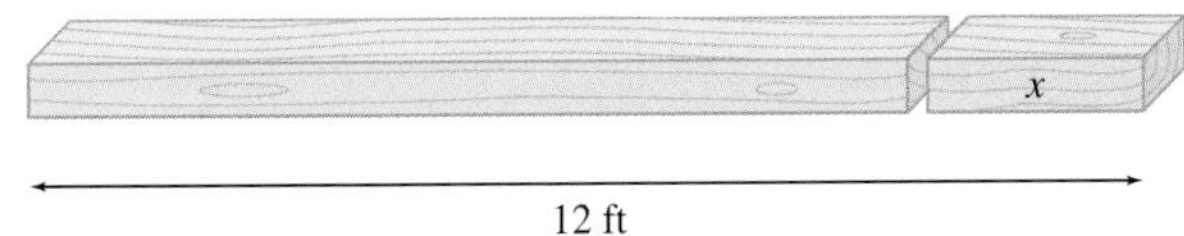

SOLUTION

ALGEBRAICALLY

$f(x) = 12 - x$

If x ft is cut off, then $12 - x$ ft remains.

NUMERICALLY

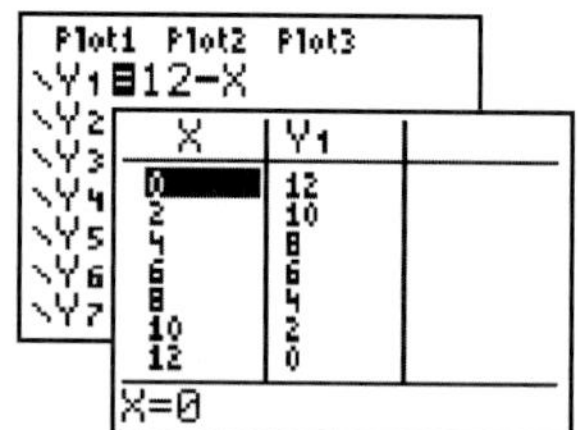

Let Y_1 represent the function $f(x) = 12 - x$. Enter $Y_1 = 12 - x$.

By changing the table setup we can examine other options.

VERBALLY

From the table we observe that if the first piece is 2 ft, then the remaining piece is 10 ft. If the first piece is 4 ft, then the remaining piece is 8 ft. The input-output pairs in the table can be used to examine other possibilities.

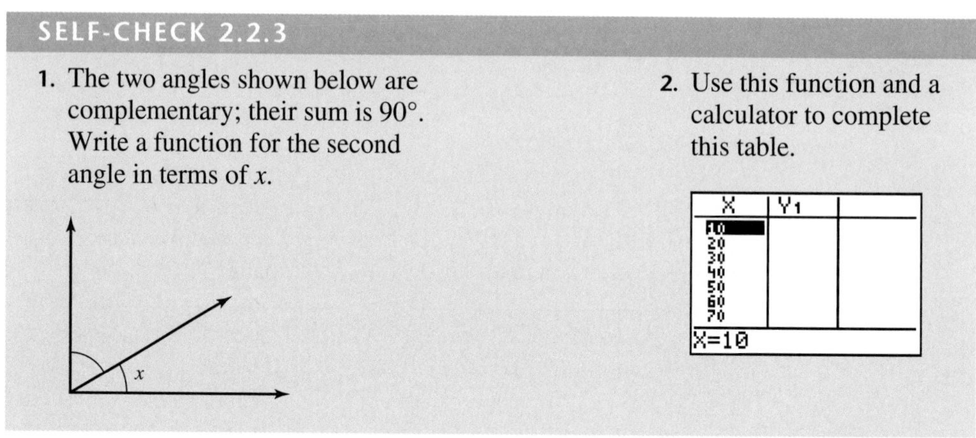

SELF-CHECK 2.2.3

1. The two angles shown below are complementary; their sum is 90°. Write a function for the second angle in terms of x.

2. Use this function and a calculator to complete this table.

In Example 7 note that the length x cannot be less than 0 ft or more than 60 ft. Clearly negative lengths are not possible. Also note that two sides of 60 ft each will use up all the available fencing. Thus the input values for this function have practical limitations. We will explore this concept more throughout the book. In this table note the output values correspond to x-values from 0 to 60.

EXAMPLE 7 Using a Function to Model the Length of a Rectangle

A fencing company has been contracted to construct a fence on three sides of the rectangular lot shown below. The perimeter of this fenced area is 120 ft. If the width is x, write a function for the length in terms of x and examine this function numerically and verbally.

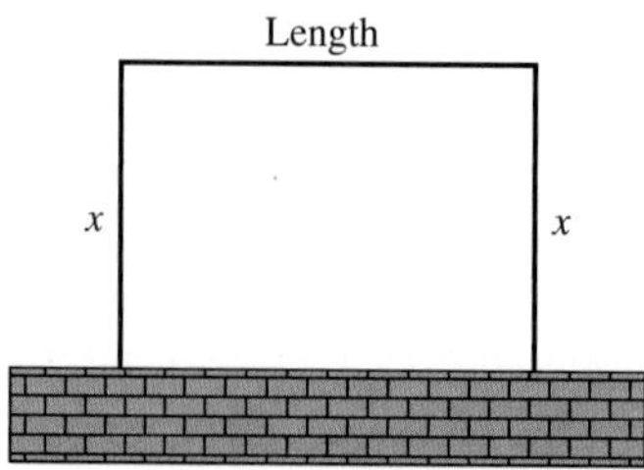

SOLUTION

ALGEBRAICALLY

$f(x) = 120 - 2x$

Of the 120 ft, x ft is used on each of the two sides. This leaves $120 - 2x$ ft for the third side to be fenced.

NUMERICALLY

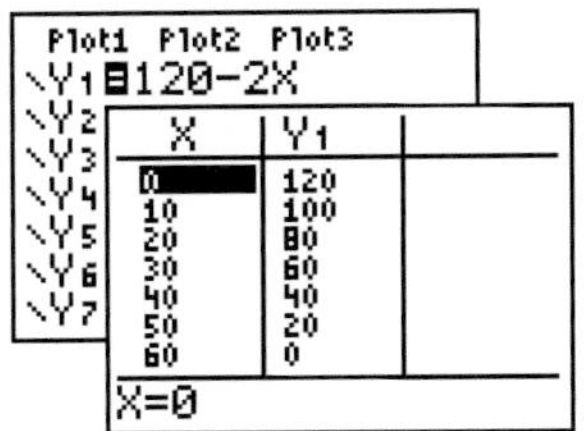

Let Y_1 represent the function $f(x) = 120 - 2x$.

VERBALLY

From the table we observe that if the width is 20 ft, then the length is 80 ft. If the width is 30 ft, then the length is 60 ft. The input-output pairs in the table can be used to examine other possibilities. ■

Note in Examples 2 and 3 in this section, and in many other examples throughout the book, that graphs and tables are placed side by side so that you can relate each piece of numerical information to the overall pattern shown by the graph formed by the data. There is a graphing calculator feature that some students like that displays a graph and table side by side. This **Graph-Table** feature does have the advantage of displaying a graph and table simultaneously, but it also has the disadvantage that both the graph and the table are smaller and somewhat harder to read. This feature is shown in Calculator Perspective 2.2.2 so you make your own judgment on its appropriateness. The **ZDecimal** window is shown here. This window gives nice decimal values and is obtained by pressing [ZOOM] [4] in the **Graph-Table** mode on a TI-84 calculator.

CALCULATOR PERSPECTIVE 2.2.2 Using the Graph-Table Feature

To use the **Graph-Table (G-T)** feature for the linear function $f(x) = -x + 2$ from Example 3 on a TI-84 Plus calculator, enter the following keystrokes:

Start by pressing the [MODE]. Then select the **G-T** feature by using the arrow keys to place the cursor above **G-T** and press .

 [(−)]  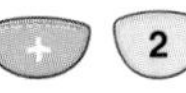 [2]

 [4]

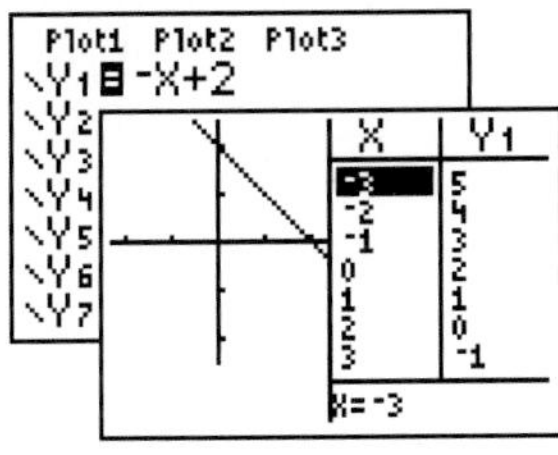

[−2.3, 2.3, 1] by [−2.5, 2.5, 1]

Note: When you do not want to view a graph and a table in the same window, set the viewing window options back to **FULL** after pressing [MODE]. ■

SELF-CHECK 2.2.4

1. Use the function $f(x) = 2x + 1$ and the **Graph-Table** feature on a graphing calculator to complete this graph and table.

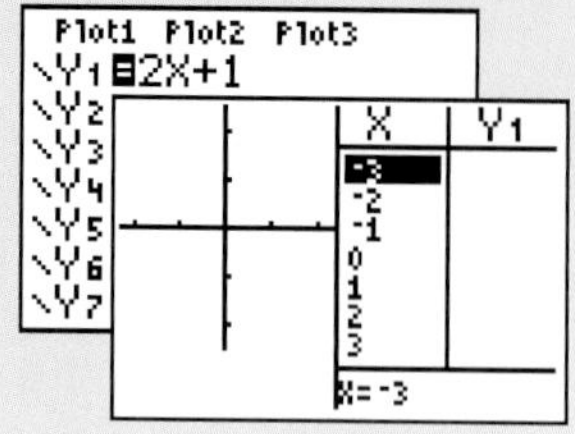

$[-2.3, 2.3, 1]$ by $[-2.5, 2.5, 1]$

SELF-CHECK ANSWERS

2.2.1

1. $f(0) = 11$ 2. $f(1) = 15$
3. $f(10) = 51$ 4. $f(-10) = -29$
5.

x	-2	-1	0	1	2
y	5	4	3	2	1

6.

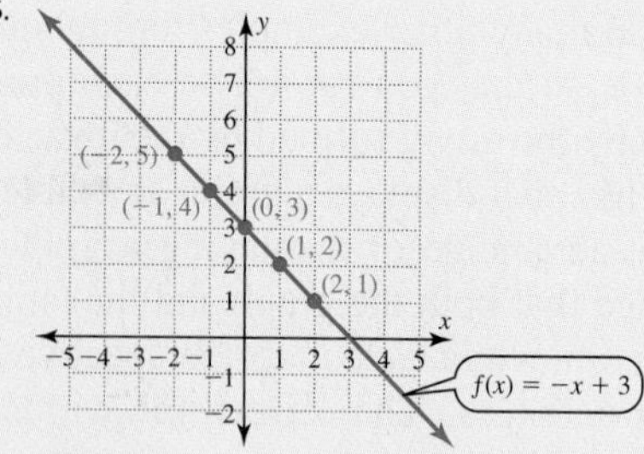

7.

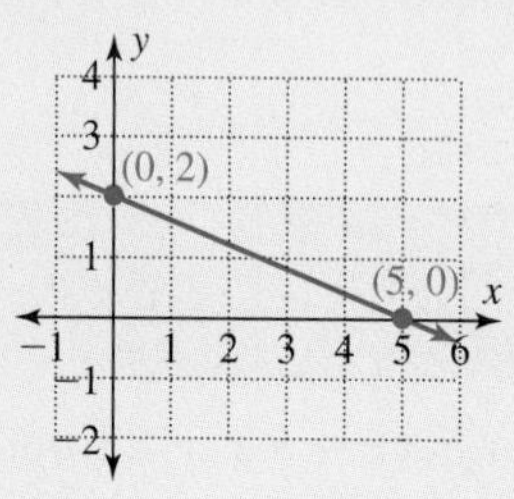

2.2.2

1.

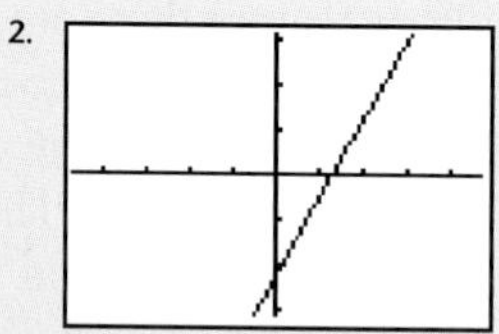

2.

$[-4.7, 4.7, 1]$ by $[-3.1, 3.1, 1]$

3. 2 4. 2.5 5. -4 6. -2

2.2.3

1. $f(x) = 90 - x$

2.

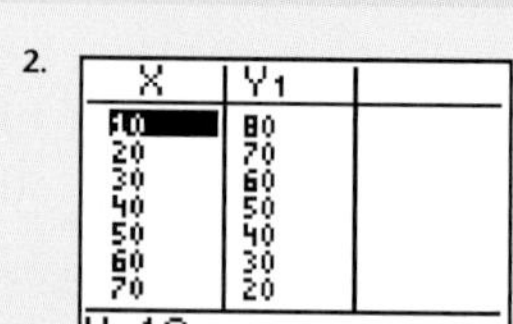

2.2.4

1.

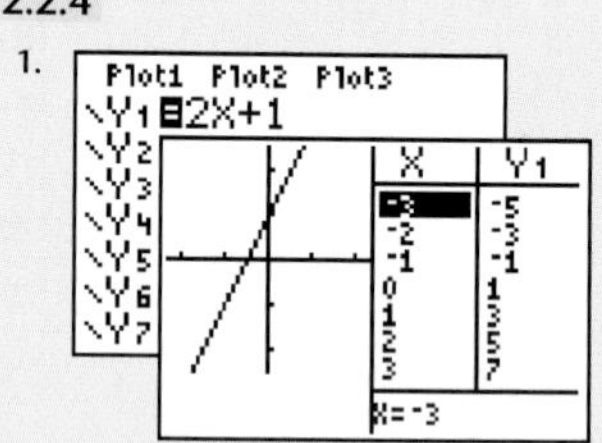

$[-2.3, 2.3, 1]$ by $[-2.5, 2.5, 1]$

USING THE LANGUAGE AND SYMBOLISM OF MATHEMATICS 2.2

1. The notation $f(x)$ is called __________ notation.
2. The notation $f(x) = 8x - 2$ is read "__________ of __________ equals eight x minus two."
3. In the notation $f(x) = 8x - 2$, the input variable is represented by __________ and $f(x)$ represents the __________ variable.
4. In the notation $f(5) = 9$, the input value is __________ and the output value is __________.
5. The graph of $f(x) = 8x - 2$ is a straight __________.
6. The function $f(x) = mx + b$ is called a __________ function.
7. Creating an equation or a function to describe an application is called mathematical __________.

EXERCISES 2.2

In Exercises 1–4, use the linear function $f(x) = 3x + 7$ to evaluate each expression.

1. $f(0)$
2. $f(1)$
3. $f(-1)$
4. $f(5)$

In Exercises 5–8, use the linear function $f(x) = 5x - 6$ to evaluate each expression.

5. $f(2)$
6. $f(-2)$
7. $f(-10)$
8. $f(10)$

In Exercises 9–12, use the linear function $f(x) = -6x + 4$ to evaluate each expression.

9. $f(-4)$
10. $f(4)$
11. $f\left(\frac{1}{6}\right)$
12. $f\left(-\frac{1}{6}\right)$

In Exercises 13–16, use the linear function $f(x) = -8x + 5$ to evaluate each expression.

13. $f\left(\frac{1}{4}\right)$
14. $f\left(-\frac{1}{2}\right)$
15. $f\left(-\frac{1}{8}\right)$
16. $f(0)$

17. Use the linear function $f(x) = -2x + 4$ to complete this table.

x	f(x)
−2	
−1	
0	
1	
2	

18. Use the linear function $f(x) = \frac{x - 3}{2}$ to complete this table.

x	f(x)
−2	
−1	
0	
1	
2	

19. Use the linear function $f(x) = \frac{x + 3}{2}$ to complete this table and to graph this function.

x	f(x)
0	
1	
2	
3	

20. Use the linear function $f(x) = -\frac{1}{2}x + 1$ to complete this table and to graph this function.

x	f(x)
−2	
−1	
0	
1	
2	

21. Graph the line through (3, 0) and (0, 2).
22. Graph the line through (−4, 0) and (0, 2).

In Exercises 23–26, use the given function and a graphing calculator to complete each table.

23. $f(x) = 2x + 3$

a. 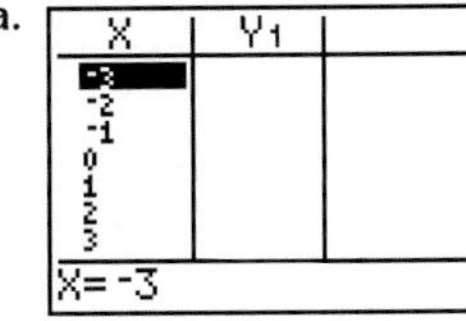

b. 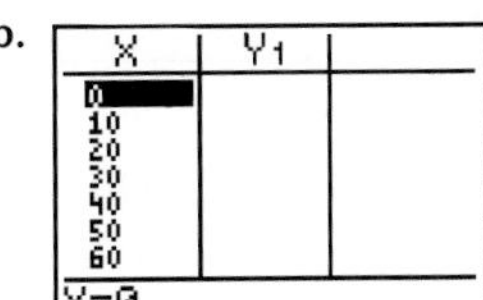

24. $f(x) = 4x - 5$

a. 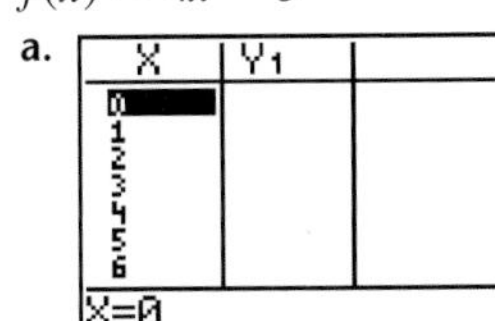

b. 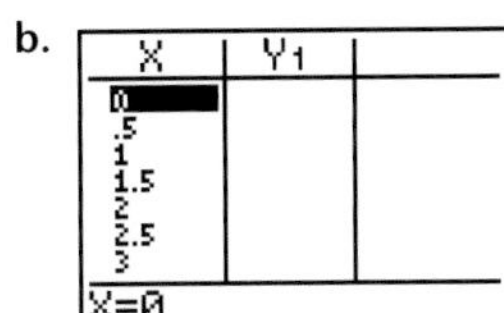

25. $f(x) = \frac{1}{4}x - 2$

a. 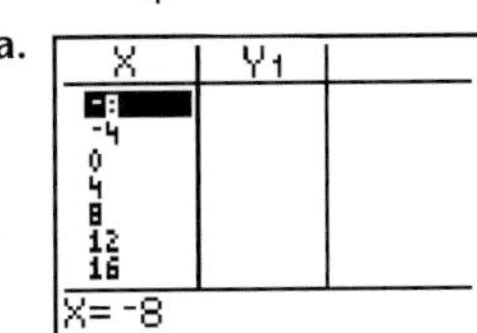

b. 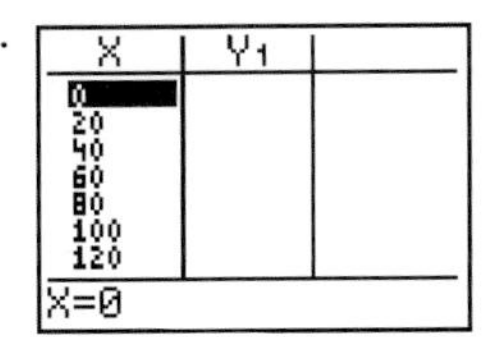

26. $f(x) = -\frac{1}{2}x + 4$

a. 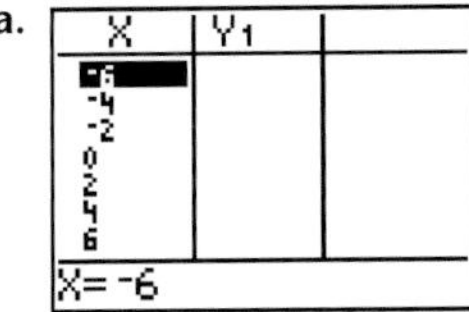

b. 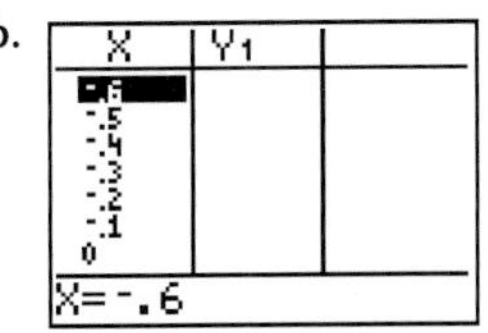

27. Use the graph shown below to complete this table.

x	f(x)
−2	
−1	
0	
1	
2	

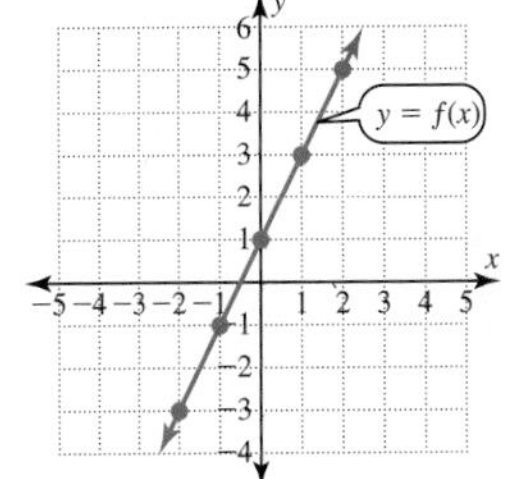

28. Use the graph shown below to complete this table.

x	f(x)
−2	
−1	
0	
1	
2	

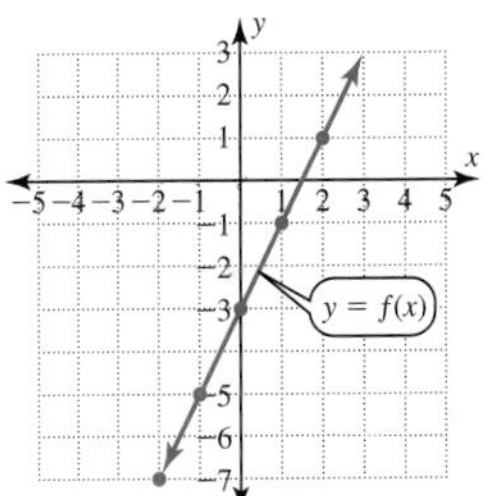

29. Use the graph shown below to complete this table.

x	f(x)
	−5
	−3
	0
	2

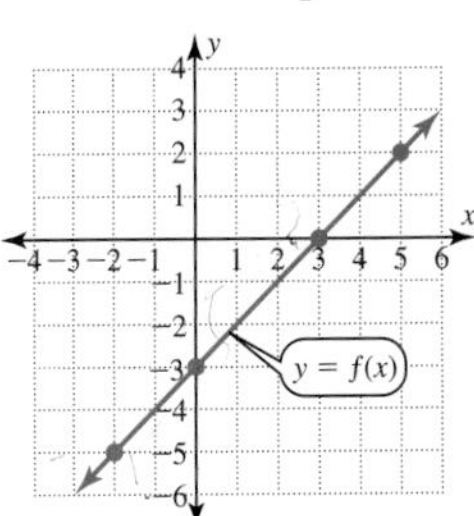

30. Use the graph shown below to complete this table.

x	f(x)
	−3
	−1
	0
	3

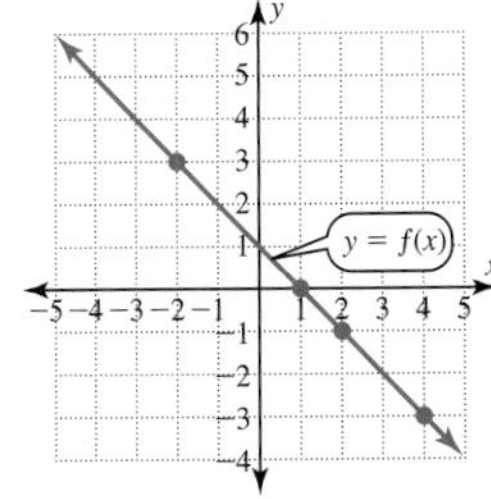

31. Use the given table of values to determine the missing input and output values.

x	f(x)
0	−10
1	−8
6	2
8	6

a. $f(1) =$ _____
b. $f(6) =$ _____
c. $f(x) = 6$; $x =$ _____
d. $f(x) = -10$; $x =$ _____

32. Use the given table of values to determine the missing input and output values.

x	f(x)
0	2
2	3
6	5
8	6

a. $f(0) =$ _____
b. $f(8) =$ _____
c. $f(x) = 3$; $x =$ _____
d. $f(x) = 5$; $x =$ _____

33. Use the given graph to determine the missing input and output values.

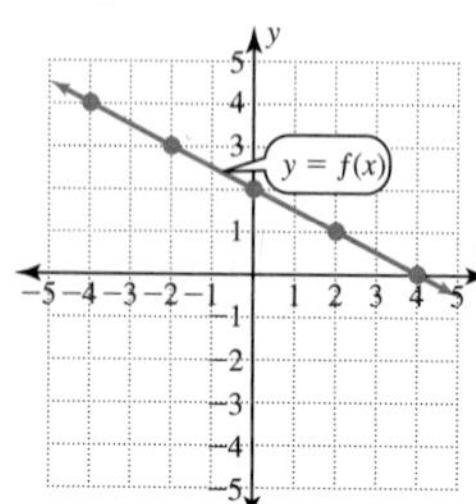

a. $f(-2) =$ _____
b. $f(2) =$ _____
c. $f(x) = 2$; $x =$ _____
d. $f(x) = 0$; $x =$ _____

34. Use the given graph to determine the missing input and output values.

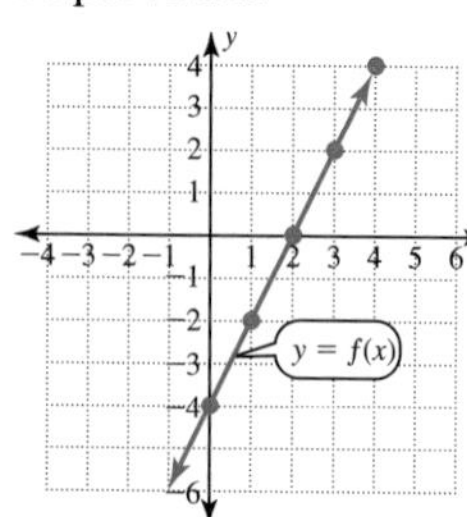

a. $f(0) =$ _____
b. $f(2) =$ _____
c. $f(x) = 0$; $x =$ _____
d. $f(x) = -2$; $x =$ _____

Comparing an Arithmetic Sequence and a Linear Function

35. a. Graph the first five terms of $a_n = -n + 5$.
b. For the linear function $f(x) = -x + 5$, plot the points with x-coordinates of 1, 2, 3, 4, and 5, and then sketch the line through these points.
c. Compare the graphs in parts **a** and **b**.

36. a. Graph the first five terms of $a_n = n - 4$.
b. For the linear function $f(x) = x - 4$, plot the points with x-coordinates of 1, 2, 3, 4, and 5, and then sketch the line through these points.
c. Compare the graphs in parts **a** and **b**.

Calculator Exercises

In Exercises 37 and 38, use a calculator and the given linear equation to **(a)** complete the table and **(b)** graph the equation on the given window. (*Hint:* See Calculator Perspective 2.2.1.)

37. $y = -\frac{x}{2}$

a.
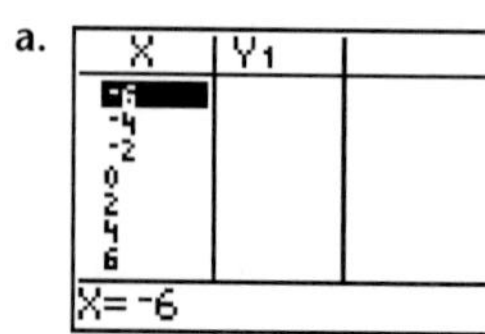

b.
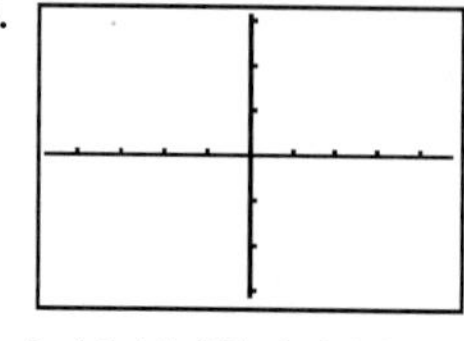

[−4.7, 4.7, 1] by [−3.1, 3.1, 1]

38. $y = \frac{x}{3}$

a.
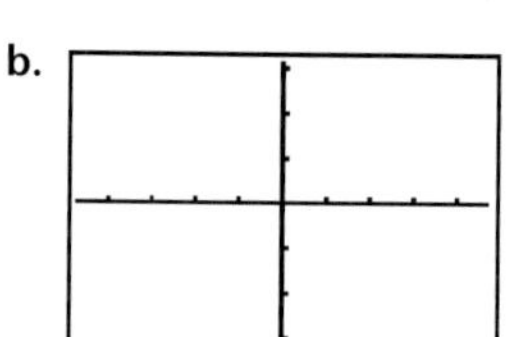

b.

[−4.7, 4.7, 1] by [−3.1, 3.1, 1]

In Exercises 39 and 40, use a calculator, the **G-T** feature, and the given linear equation to complete the table and graph the equation on the following window. (*Hint:* See Calculator Perspective 2.2.2.)

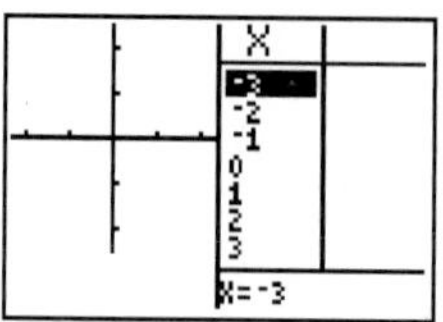

[−2.3, 2.3, 1] by [−2.5, 2.5, 1]

39. $y = 2x - 1$

40. $y = -\frac{x}{2} - \frac{1}{2}$

Using Functions to Create Mathematical Models

41. **Length of a Board** A board 18 ft long has three pieces each x ft long cut off. Write a function for the length of the remaining piece in terms of x.

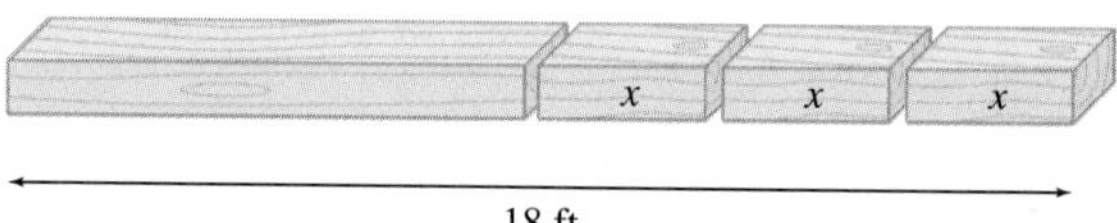

42. **Supplementary Angles** The two angles shown below are supplementary; their sum is 180°. Write a function for the second angle in terms of x.

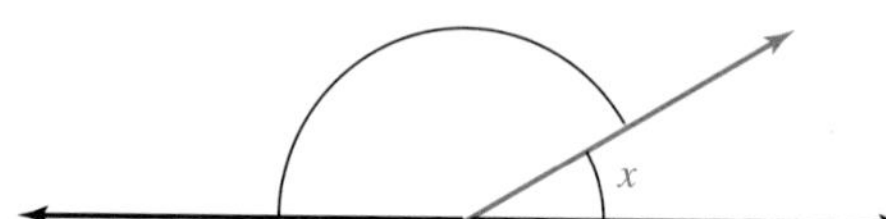

43. **Perimeter of an Isosceles Triangle** Write a function in terms of x for the perimeter (in centimeters) of the isosceles triangle shown below.

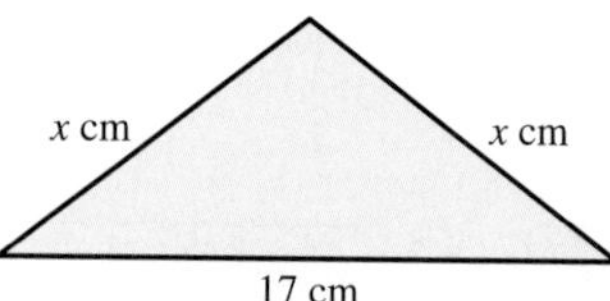

44. **Perimeter of an Equilateral Triangle** Write a function in terms of x for the perimeter in centimeters of the equilateral triangle shown below.

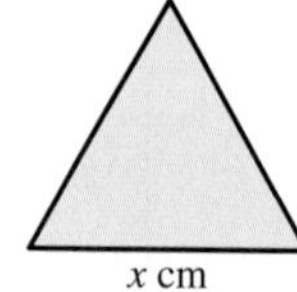

45. **Perimeter and Area of a Rectangle** Write a function in terms of x for
  a. The perimeter in centimeters of the rectangle shown below.
  b. The area in square centimeters of the rectangle shown below.

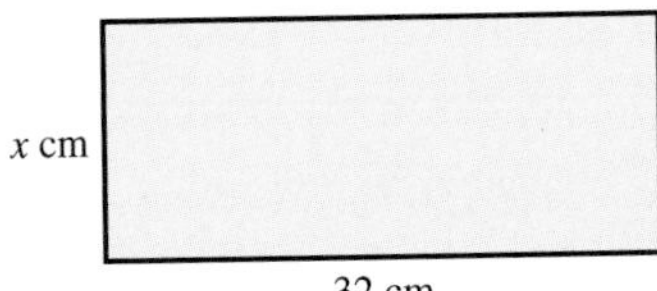

46. **Rate and Distance of a Boat** A boat has a speed of x mi/h in still water. Write a function in terms of x for
  a. The rate of this boat traveling downstream in a river with a current of 5 mi/h.
  b. The rate of this boat traveling upstream in a river with a current of 5 mi/h.
  c. The distance the boat travels in 2 hours going downstream in a river with a current of 5 mi/h.

47. **Rate and Distance of an Airplane** An airplane has a speed of x mi/h in calm skies. Write a function in terms of x for
  a. The rate of this airplane traveling in the same direction as a 30-mi/h wind.
  b. The rate of this airplane traveling in the opposite direction of a 30-mi/h wind.
  c. The distance the airplane travels in 2 hours going in the same direction as a 30-mi/h wind.

48. **Price Discount** A discount store had a holiday special with a 15% discount on the price of all sporting goods.
  a. Write a function for the amount of discount on an item with an original price of x dollars.
  b. Write a function for the new price of an item with an original price of x dollars.
  c. Complete the following table for the new price of each item whose original price is given.

x	$f(x)$
20	17
44	
60	
68	
90	

49. **Price Increase** A wholesaler is increasing the price of all items 2% to cover increased costs.
  a. Write a function for the amount of the increase on an item with an original price of x dollars.
  b. Write a function for the new price on an item with an original price of x dollars.
  c. Complete the following table for the new price of each item whose original price is given.

x	$f(x)$
20	20.40
35	
44	
80	
90	

50. **Cost of Production** The overhead cost for a company is \$500 per day. The cost of producing each item is \$15. The total cost of production is the sum of the overhead cost and the cost of producing each item.
  a. Write a function that gives the total cost of producing x units per day.
  b. Use this function to complete this table.

x	0	100	200	300	400	500
$f(x)$						

  c. Evaluate and interpret $f(250)$.

51. **Length Remaining on a Roll of Wire** A roll of wire 100 ft long has five pieces each of length x cut from it.
  a. Write a function that models the length of wire remaining on the roll.
  b. Use this function and a calculator to complete this table.

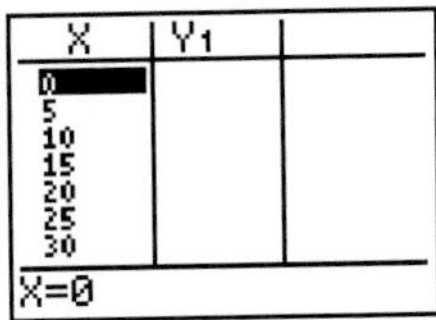

  c. Evaluate and interpret $f(20)$.
  d. Determine and interpret the value of x for which $f(x) = 20$.
  e. Is 30 a practical input value for x? Explain.

52. **Angles in an Isosceles Triangle** An isosceles triangle has two equal angles each with x degrees. The sum of all the angles in a triangle is 180°.

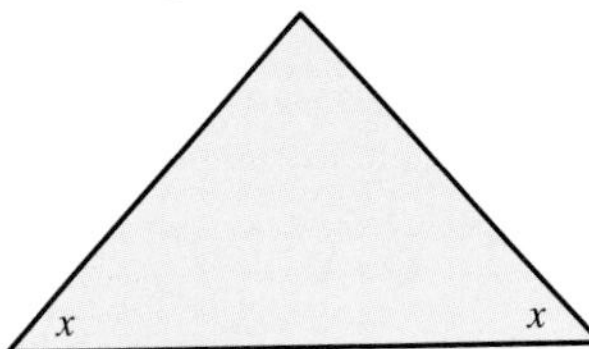

  a. Write a function that models the number of degrees in the third angle.
  b. Use this function and a calculator to complete this table.

X	Y1
10	
20	
30	
40	
50	
60	
70	

X=10

  c. Evaluate and interpret $f(40)$.
  d. Determine and interpret the value of x for which $f(x) = 40$.
  e. Is 100 a practical input value for x? Explain.

Group Discussion Questions

53. **Discovery Question** Use a graphing calculator and the equation $y = \frac{x}{2}$ to do the following:

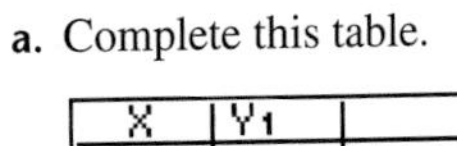

  a. Complete this table.

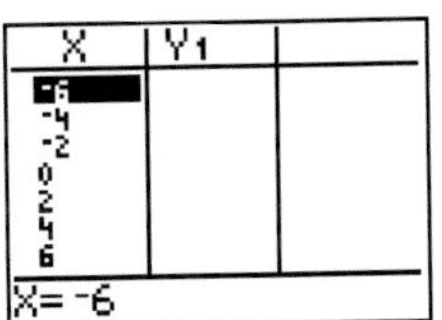

  b. Complete this graph.

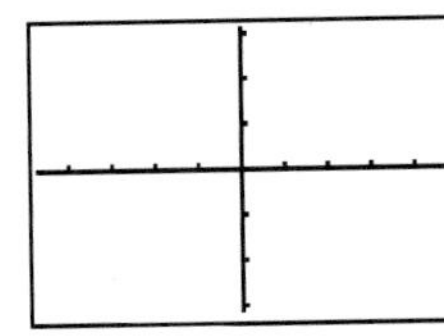

[−4.7, 4.7, 1] by [−3.1, 3.1, 1]

  c. For each increase of 2 units in x in the table, what is the change in y?

d. For each 2-unit movement to the right on the graph, what is the change in the y-coordinate?

54. **Discovery Question** Use a graphing calculator to graph $f(x) = \frac{x}{2}$, $f(x) = \frac{x}{2} + 1$, $f(x) = \frac{x}{2} + 3$, and $f(x) = \frac{x}{2} - 3$. What relationship do you observe among these graphs?

55. **Discovery Question** Use a graphing calculator to graph $f(x) = x^2$ and $f(x) = x^2 - 3$. Describe the shape of each of these graphs.

56. **Discovery Question** Use a graphing calculator to graph $f(x) = |x| - 4$ and $f(x) = |x|$. Describe the shape of each of these graphs.

Section 2.3 Graphs of Linear Equations in Two Variables

Objectives:

7. Check possible solutions of a linear equation.
8. Determine the intercepts from a graph.
9. Determine the point where two lines intersect.

The two lease options given here are summaries of two advertisements by two different automobile agencies both trying to lease the same automobile.

Lease option A: A \$2,000 initial down payment followed by 36 monthly payments of \$250.

Lease option B: A \$2,500 initial down payment followed by 36 monthly payments of \$200.

In this section, we will examine a method for comparing these two options. In Example 8 we will examine which of these two options is the more costly. To prepare for this we will examine linear equations in greater detail.

Key point: A solution of $y = mx + b$ is an ordered pair—not just an x-value or a y-value.

A **solution** of a linear equation of the form $y = mx + b$ is an ordered pair (x, y) that makes the equation a true statement. A solution (x, y) is said to **satisfy** the equation. Example 1 illustrates the fact that an ordered pair (x, y) satisfies an equation if and only if that point lies on the graph of that equation.

EXAMPLE 1 Checking Ordered Pairs in an Equation

Determine algebraically whether each ordered pair is on the graph of $y = 2x - 3$.

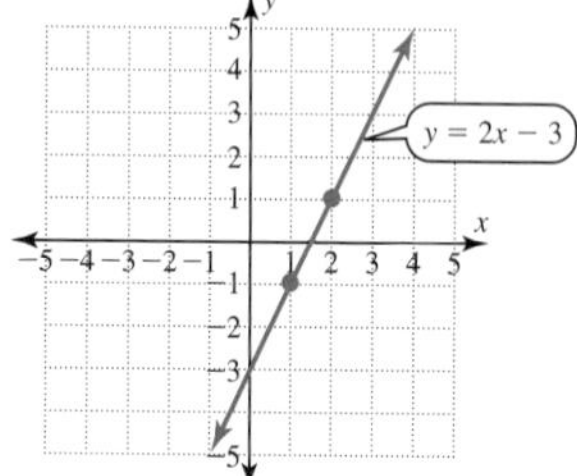

SOLUTION

(a) $(0, -3)$

$y = 2x - 3$

$-3 \stackrel{?}{=} 2(0) - 3$

$-3 \stackrel{?}{=} 0 - 3$

$-3 \stackrel{?}{=} -3$ checks

Substitute -3 for y and 0 for x and then simplify.

$(0, -3)$ is on the graph of $y = 2x - 3$.

Examine the graph of $y = 2x - 3$ to note that $(0, -3)$ is a point on this graph.

(b) (2, 3)

$y = 2x - 3$
$3 \stackrel{?}{=} 2(2) - 3$ Substitute 3 for y and 2 for x and then simplify.
$-3 \stackrel{?}{=} 4 - 3$
$-3 \stackrel{?}{=} 1$ *is false*

(2, 3) is not on the graph of $y = 2x - 3$. Examine the graph of $y = 2x - 3$ to note that (2, 3) is not a point on this graph.

Key point: Both y and $f(x)$ are used to denote the output associated with an input x.

In the linear equation $y = mx + b$, x represents the input variable and y represents the output variable. Linear equations are also frequently given using function notation in the form $f(x) = mx + b$. In this notation, x represents the input variable and $f(x)$ represents the output variable. Both y and $f(x)$ are used to denote the output value. In Example 2 we check an ordered pair to determine if this point is on the graph defined by the function.

EXAMPLE 2 Checking Ordered Pairs in a Function

Determine algebraically whether each ordered pair is on the graph of $f(x) = -2x + 8$.

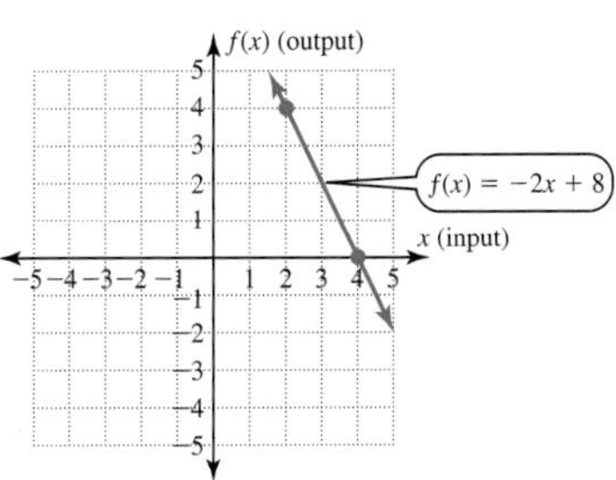

SOLUTION

(a) (4, 0)

$f(x) = -2x + 8$
$f(4) = -2(4) + 8$ Substitute 4 for the input value of x to determine the output value $f(4)$.
$f(4) = -8 + 8$
$f(4) = 0$

(4, 0) is on the graph of $f(x) = -2x + 8$. Because the output value is 0, (4, 0) is a point on the graph of $f(x) = -2x + 8$.

(b) (3, 5)

$f(x) = -2x + 8$
$f(3) = -2(3) + 8$ Substitute 3 for the input value of x to determine the output value $f(3)$.
$f(3) = -6 + 8$
$f(3) = 2$

(3, 5) is not on the graph of $f(x) = -2x + 8$. Because the output value is 2, not 5, (3, 5) is not a point on the graph of $f(x) = -2x + 8$.

SELF-CHECK 2.3.1

Determine whether each ordered pair is a solution of $y = -3x + 4$.

1. $(2, -2)$ **2.** $(1, 5)$ **3.** $\left(\frac{4}{3}, 0\right)$ **4.** $(0, 4)$

A linear equation in two variables has an infinite set of ordered pairs that are solutions of the equation. Graphically all the solutions lie on a line, and all points not on the line are not solutions. In Example 3 we check selected points on the line defined by $y = -2x + 4$ to confirm that they satisfy the equation.

EXAMPLE 3 Checking the Coordinates of a Point on a Line

Determine whether the points A, B, and C on this graph satisfy the equation $y = -2x + 4$.

SOLUTION

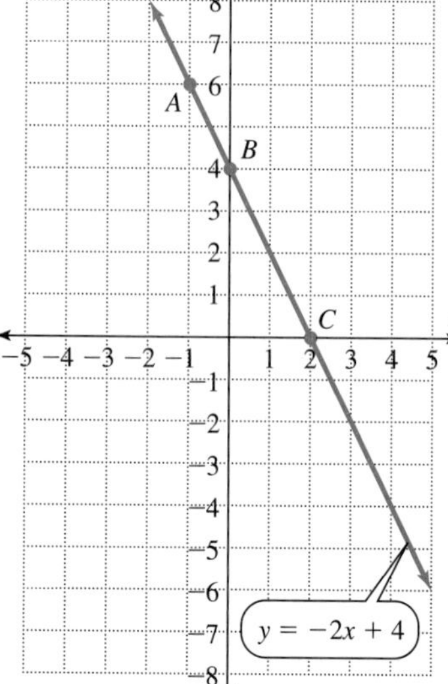

(a) Point A has coordinates $(-1, 6)$.

$y = -2x + 4$
$6 \stackrel{?}{=} -2(-1) + 4$
$6 \stackrel{?}{=} 2 + 4$
$6 \stackrel{?}{=} 6$ checks

Substitute -1 for the input value of x and 6 for the output value of y.

Answer: $(-1, 6)$ is a solution of $y = -2x + 4$.

(b) Point B has coordinates $(0, 4)$.

$y = -2x + 4$
$4 \stackrel{?}{=} -2(0) + 4$
$4 \stackrel{?}{=} 0 + 4$
$4 \stackrel{?}{=} 4$ checks

The point $(0, 4)$ lies on the y-axis and is called the y-intercept of the graph. We will examine intercepts more in Examples 4 and 5.

Answer: $(0, 4)$ is a solution of $y = -2x + 4$.

(c) Point C has coordinates $(2, 0)$.

$y \stackrel{?}{=} -2x + 4$
$0 \stackrel{?}{=} -2(2) + 4$
$0 \stackrel{?}{=} -4 + 4$
$0 \stackrel{?}{=} 0$ checks

The point $(2, 0)$ lies on the x-axis and is called the x-intercept of the graph. We will examine intercepts more in Examples 4 and 5.

Answer: $(2, 0)$ is a solution of $y = -2x + 4$.

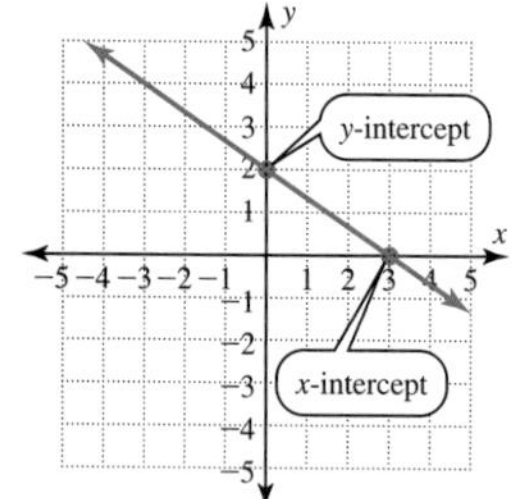

The intercepts are points with both x- and y-coordinates.

The line defined by a linear equation is completely determined by any two points on the line. To graph the line, we will often select points that are easy to calculate or that have a special significance. The points where a graph intersects, or crosses, the axes are called the **intercepts.** The ***x*-intercept** is often denoted by $(a, 0)$ and the ***y*-intercept** by $(0, b)$. These points frequently have special significance in applications, as illustrated in Example 4.

EXAMPLE 4 Interpreting the Intercepts of a Graph: Overhead Costs and Break-even Values

In the graph to the right, the input value of x is the number of units a machine produces. The output y is the profit generated by the sale of these units when they are produced. Determine the intercepts and interpret the meaning of these points.

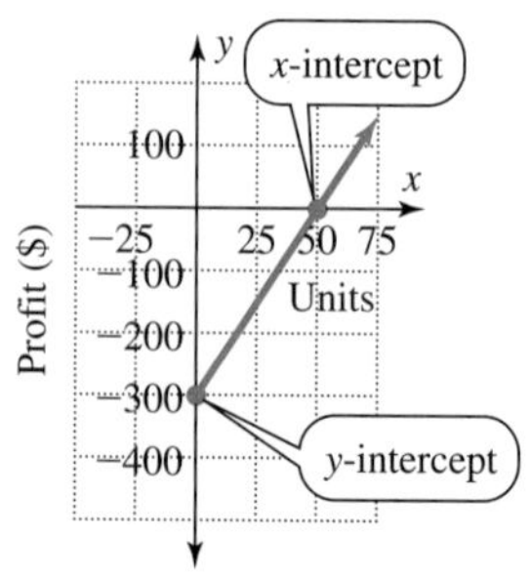

SOLUTION

The y-intercept is $(0, -300)$. If 0 units are produced, then \$300 will be lost. (Often there are overhead costs of rent, electricity, etc., even if no units are produced.)

The x-intercept is $(50, 0)$. If 50 units are produced, then \$0 will be made. (This is the break-even value for this machine—the number of units the machine must make before a profit can be realized.)

The graph of a linear function $f(x) = mx + b$ can be obtained by using any two points that lie on the line. One particularly easy point to determine is the y-intercept, which can be found by evaluating $f(0)$. Another point may be obtained by substituting any other value for x. It is also wise to plot a third point as a double-check on our work. Since all points should be on the line, a third point not on the line is a sure indication of a computation error or an error in plotting the points.

EXAMPLE 5 Graphing a Linear Equation

Use the y-intercept and another point to graph the line defined by $f(x) = \frac{1}{2}x - 2$. Plot a third point as a check on your work.

SOLUTION

y-INTERCEPT

$f(x) = \frac{1}{2}x - 2$

$f(0) = \frac{1}{2}(0) - 2$ To find the y-intercept, substitute 0 for x.

$f(0) = 0 - 2$

$f(0) = -2$

$(0, -2)$ is the y-intercept.

SECOND POINT

$f(x) = \frac{1}{2}x - 2$

$f(4) = \frac{1}{2}(4) - 2$ To find a second point, find y when x is 4.

$f(4) = 2 - 2$

$f(4) = 0$

Note that $(4, 0)$ is the x-intercept.

THIRD POINT

$f(x) = \frac{1}{2}x - 2$

$f(6) = \frac{1}{2}(6) - 2$ To find a third point to double-check our work, find y when x is 6.

$f(6) = 3 - 2$

$f(6) = 1$

$(6, 1)$ is also on the line.

TABLE

x	y
0	−2
4	0
6	1

GRAPH

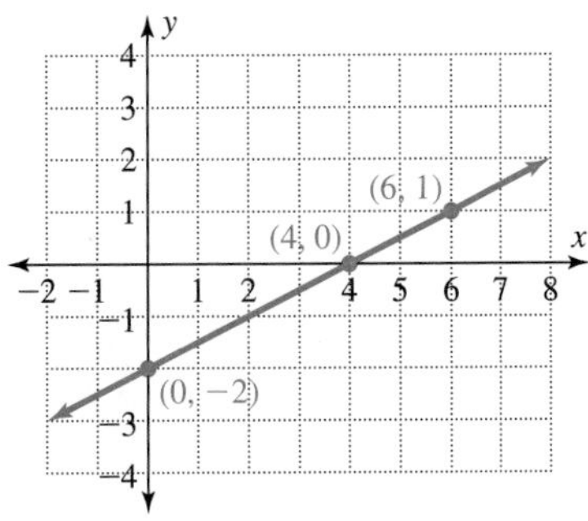

Plot these points and sketch the line through the points.

Sometimes we can determine both the x- and y-intercepts by inspecting either the graph or the table associated with the linear function. To find the x-intercept, look for the point with a y-coordinate of zero, and to find the y-intercept, look for the point with an x-coordinate of zero. The purpose of Calculator Perspective 2.3.1 is to illustrate how we can use a table and a graph to determine the x- and y-intercepts by inspection. For the linear equation $y = -2x + 3$, the y-intercept is highlighted in the table and the x-intercept is shown on the graph. We will reexamine how to find the x-intercept of a linear equation in Section 2.7 after we have solved linear equations.

CALCULATOR PERSPECTIVE 2.3.1 Finding the Intercepts of a Graph by Using the Table or Trace Features

To create a table of values for $y = -2x + 3$ and to use the **Trace** feature on a TI-84 Plus calculator to find the x-intercept of the graph of this function, enter the following keystrokes:

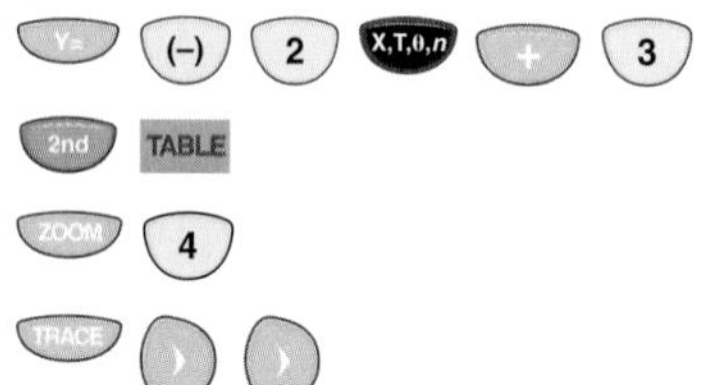

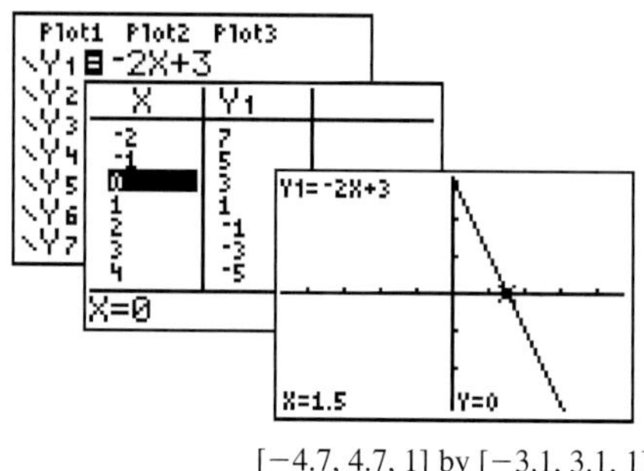

(Press the right arrow key until the x-y values shown here appear.)

[−4.7, 4.7, 1] by [−3.1, 3.1, 1]

Note: The y-intercept (0, 3) is highlighted in the table, and the x-intercept (1.5, 0) is shown on the graph.

SELF-CHECK 2.3.2

1. Determine the x- and y-intercepts of this graph.

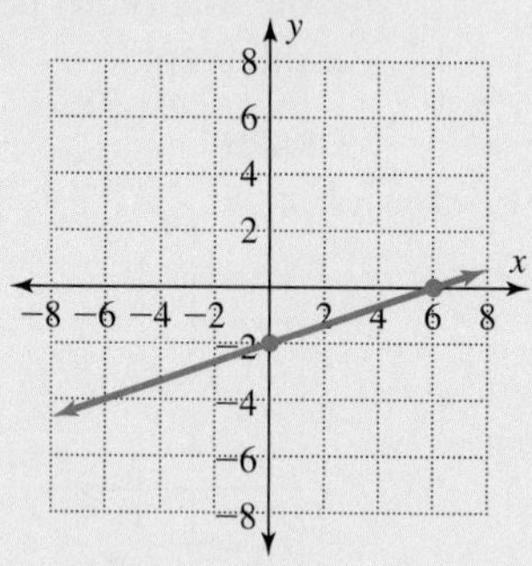

2. Use a graphing calculator to determine the x- and y-intercepts of the graph of $f(x) = -2x + 1$.

Sometimes it is useful to examine two graphs simultaneously or to compare two different options. When we refer to two or more equations at the same time, we refer to this as a **system of equations.** A **solution of a system of linear equations in two variables** is an ordered pair that satisfies each equation in the system. If there is a unique solution to a system of linear equations, it is represented graphically by their point of intersection, the point that is on both graphs.

Solution of a System of Two Linear Equations

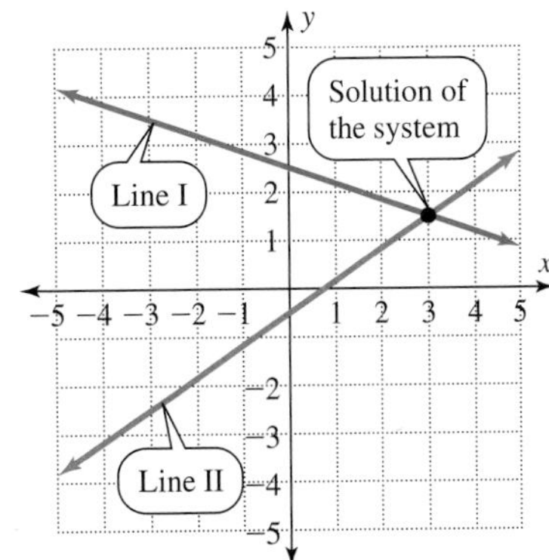

To solve a system of equations graphically, graph each equation and then estimate the coordinates of the point of intersection. Because serious errors of estimation can occur, it is wise to check an estimated solution by substituting it into each equation of the system.

EXAMPLE 6 Solving a System of Linear Equations Graphically

The graphs of $y = x - 1$ and $y = \frac{x}{3} + 1$ are shown to the right. Determine their point of intersection by inspecting the graph. Then check this point in both equations.

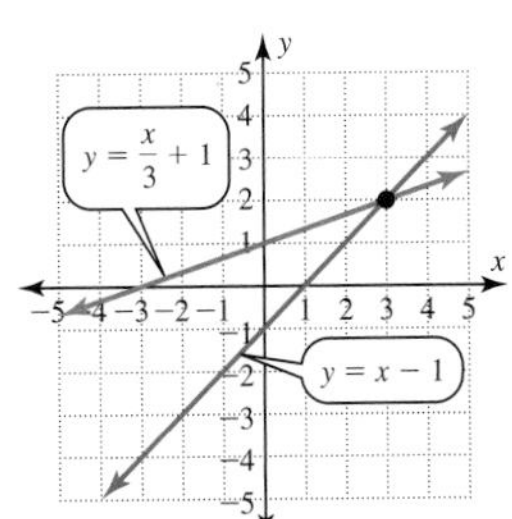

SOLUTION

An estimate of the solution is (3, 2).

Check:

FIRST EQUATION	SECOND EQUATION
$y = x - 1$	$y = \frac{x}{3} + 1$
$2 \stackrel{?}{=} 3 - 1$	$2 \stackrel{?}{=} \frac{3}{3} + 1$
$2 \stackrel{?}{=} 2$ checks.	$2 \stackrel{?}{=} 1 + 1$
	$2 \stackrel{?}{=} 2$ checks.

Substitute (3, 2) into both equations to check this point.

Answer: (3, 2) is a solution of this system of linear equations because it checks in both equations.

As you can verify by checking any other point, this is the only point that checks in both equations.

SELF-CHECK 2.3.3

Determine whether each point is a solution of the linear system $\begin{Bmatrix} y = 3x - 5 \\ y = -x + 3 \end{Bmatrix}$.

1. $(0, -5)$ **2.** $(3, 0)$ **3.** $(2, 1)$

In Example 6 we were given the graphs of two linear equations. If instead we are given a system of two linear equations, we can find the solution by first graphing these equations. In Example 7 we start by forming a table of points so we can graph the lines.

EXAMPLE 7 Solving a System of Linear Equations Graphically

Solve the system of linear equations $\begin{Bmatrix} y = 2x + 5 \\ y = -2x - 3 \end{Bmatrix}$ graphically.

SOLUTION

First graph both lines on the same coordinate system.

$y = 2x + 5$

x	y
0	5
$-\frac{5}{2}$	0
−1	3

$y = -2x - 3$

x	y
0	−3
$-\frac{3}{2}$	0
−1	−1

Set $x = 0$ to find the y-intercept. Then find the coordinates of two other points. Note that we also selected the x-intercept. A third point was calculated by letting $x = -1$.

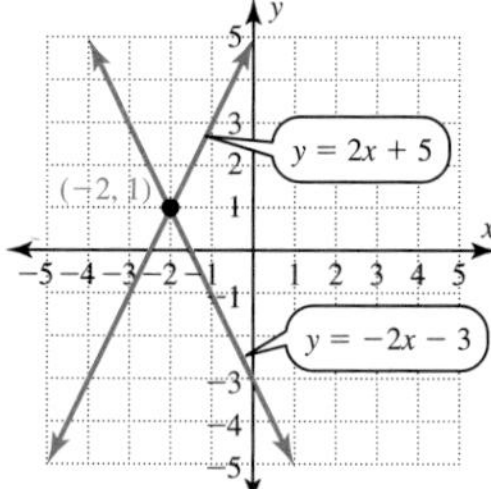

Then estimate the coordinates of the point of intersection.
The point of intersection appears to be $(-2, 1)$.

Check:

FIRST EQUATION	SECOND EQUATION
$y = 2x + 5$	$y = -2x - 3$
$1 \stackrel{?}{=} 2(-2) + 5$	$1 \stackrel{?}{=} -2(-2) - 3$
$1 \stackrel{?}{=} -4 + 5$	$1 \stackrel{?}{=} 4 - 3$
$1 \stackrel{?}{=} 1$ checks.	$1 \stackrel{?}{=} 1$ checks.

Substitute $(-2, 1)$ into both equations to check this point.

Answer: $(-2, 1)$ is a solution of this system of linear equations because it checks in both equations.

Calculator Perspective 2.3.2 shows how we can use a graphing calculator to assist in problems such as Example 7. To solve a system of two linear equations on a graphing calculator, enter each equation and then use the **Intersect** feature to approximate the point of intersection.

CALCULATOR PERSPECTIVE 2.3.2 Approximating a Point of Intersection

To solve the system of linear equations $\begin{cases} y = 2x + 5 \\ y = -2x - 3 \end{cases}$ from Example 7 on a TI-84 Plus calculator, enter the following keystrokes:

2nd TABLE

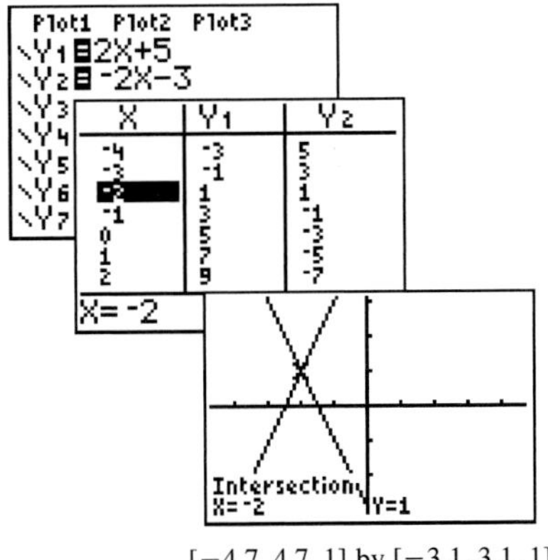

[−4.7, 4.7, 1] by [−3.1, 3.1, 1]

Note: In the table, both Y_1 and Y_2 have the value of 1 when $x = -2$. On the graph the two lines intersect at the point $(-2, 1)$. Thus both the table and the graph show that the simultaneous solution of this system of linear equations is the ordered pair $(-2, 1)$.

SELF-CHECK 2.3.4

1. Complete the following table to solve the system of linear equations $\begin{cases} y = -2x + 3 \\ y = -x - 1 \end{cases}$.

X	Y1	Y2
-4		
-2		
0		
2		
4		
6		
8		

X=-4

2. Use the **Intersect** feature on a graphing calculator and the window shown below to solve the system of linear equations $\begin{cases} y = -x + 0.3 \\ y = x - 1.3 \end{cases}$.

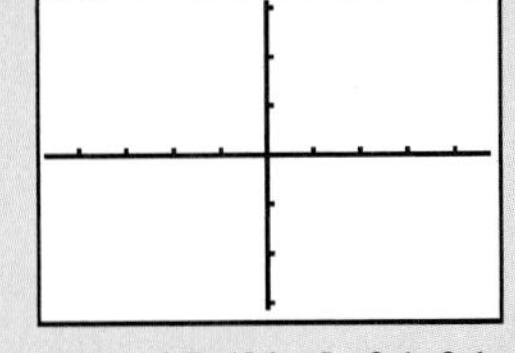

[−4.7, 4.7, 1] by [−3.1, 3.1, 1]

The important features of a graph may not always display in the first window you select on your calculator. We have already illustrated that we can obtain the window [−4.7, 4.7, 1] by [−3.1, 3.1, 1] on a TI-84 Plus calculator by using the **ZOOM 4** option. We can also obtain the window [−10, 10, 1] by [−10, 10, 1] by using the **ZOOM 6** option. Calculator Perspective 2.3.3 illustrates another way to change the display window.

CALCULATOR PERSPECTIVE 2.3.3 Changing Window Settings

To change the window settings on a TI-84 Plus calculator to [0, 20, 5] by [0, 7000, 1000] for Example 8, enter the following keystrokes:

WINDOW

Note: The values in the window setting are entered by using the numeric keys and the ENTER key.

GRAPH

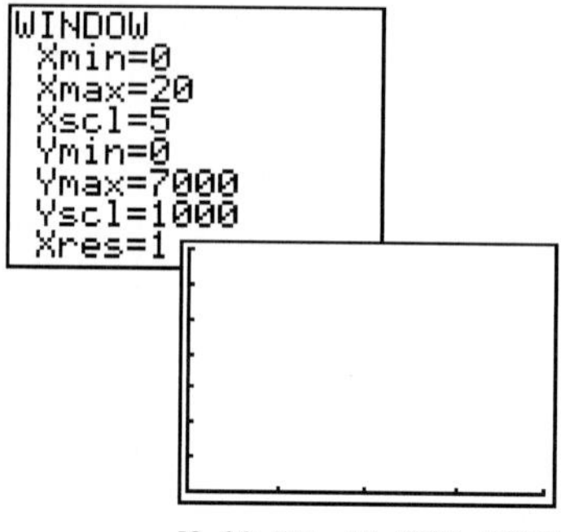

[0, 20, 5] by [0, 7000, 1000]

Note: If the Y= screen has been cleared, then you should see the blank window shown here.

Comparing Car Lease Options

The two lease options given at the beginning of this section will be examined now. Assume there are no factors other than the given costs that will affect the selection between these two options.

EXAMPLE 8 Determining Equivalent Costs

Examine these lease options numerically and graphically, and determine if these costs will ever be the same.

Lease option A: A $2,000 initial down payment followed by 36 monthly payments of $250.
Lease option B: A $2,500 initial down payment followed by 36 monthly payments of $200.

SOLUTION

We will let y represent the total cost incurred by the xth month.

VERBALLY Total cost = Total of monthly payments + Initial payment

The total of the monthly payments is calculated by multiplying the monthly payment by the number of months this payment has been made.

ALGEBRAICALLY

Option A: $y = 250x + 2{,}000$
Option B: $y = 200x + 2{,}500$

NUMERICALLY

Plot1 Plot2 Plot3
\Y1=250X+2000
\Y2=200X+2500

X	Y1	Y2
0	2000	2500
5	3250	3500
10	4500	4500
15	5750	5500
20	7000	6500
25	8250	7500
30	9500	8500

X=10

GRAPHICALLY

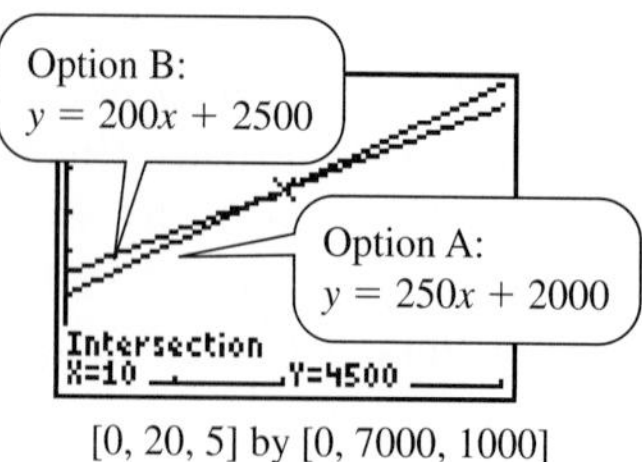

[0, 20, 5] by [0, 7000, 1000]

A graphical approximation of the coordinates of the point of intersection is (10, 4,500); this value is supported by the numerical table.

Note from both the table and the graph that option A starts with the lower initial cost, and the output values (the cost) are lower until the tenth month. After the tenth month, the higher monthly cost for option A will produce output values (the cost) that are higher than the output values for option B.

Answer: After 10 months, both options will have cost $4,500.

SELF-CHECK ANSWERS

2.3.1

1. Solution 2. Not a solution
3. Solution 4. Solution

2.3.2

1. The x-intercept is $(6, 0)$, and the y-intercept is $(0, -2)$.
2. The x-intercept is $(0.5, 0)$, and the y-intercept is $(0, 1)$.

2.3.3

1. Not a solution
2. Not a solution
3. Solution

2.3.4

1. $(4, -5)$
2. $(0.8, -0.5)$

USING THE LANGUAGE AND SYMBOLISM OF MATHEMATICS 2.3

1. The point $(a, 0)$ on the graph of a line is called the ________.
2. The point $(0, b)$ on the graph of a line is called the ________.
3. A solution of a linear equation $y = mx + b$ is an ordered pair (x, y) that makes the equation a __________ statement.
4. A linear equation of the form $y = mx + b$ has an ________ number of solutions.
5. An ordered pair that satisfies each equation in a system of linear equations is called a __________ of the system of linear equations.
6. A unique solution to a system of two linear equations is represented graphically by their point of __________.
7. If a point is the x-intercept of a graph, then the _____-coordinate is zero.
8. If a point is the y-intercept of a graph, then the _____-coordinate is zero.
9. If a point satisfies $f(x) = mx + b$ and $x = 0$, then this point is the __________-intercept of the graph of this linear function.
10. If a point satisfies $f(x) = mx + b$ and $f(x) = 0$, then this point is the __________-intercept of the graph of this linear function.
11. Even when no units are produced there are costs for rent, electricity, and so on. These costs are called __________ costs.
12. The number of units a machine must make before a profit can be realized is called the __________ value for this machine.

EXERCISES 2.3

In Exercises 1–8, determine whether each ordered pair is a solution of $y = 6x - 1$.

1. $(1, 5)$ 2. $(2, 4)$ 3. $\left(\frac{1}{6}, 6\right)$ 4. $(0, -1)$

5. $(2, 11)$ 6. $(1, -1)$ 7. $\left(\frac{1}{6}, 0\right)$ 8. $(0, 3)$

In Exercises 9–16, determine whether each point is on the graph $f(x) = -\frac{x}{2} + 5$.

9. $(0, 6)$ 10. $(10, 0)$ 11. $(-6, -8)$ 12. $(4, 3)$
13. $(0, 5)$ 14. $(6, 0)$ 15. $(12, -1)$ 16. $(-8, -9)$

In Exercises 17–20, determine whether the points A, B, C, and D on the graph to the right are solutions of the given equation.

17. $y = \frac{x}{4} + 1$
18. $y = -2x + 10$
19. $y = 3$
20. $x = -4$

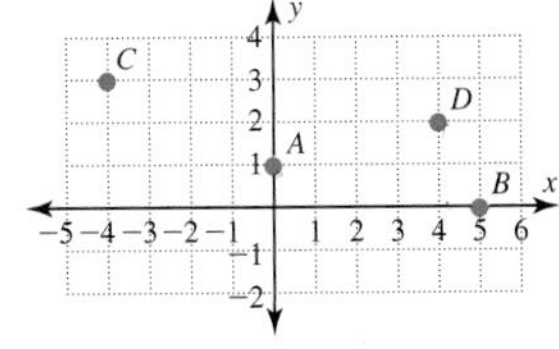

In Exercises 21 and 22, identify the x- and y-intercepts on the given graph.

21.

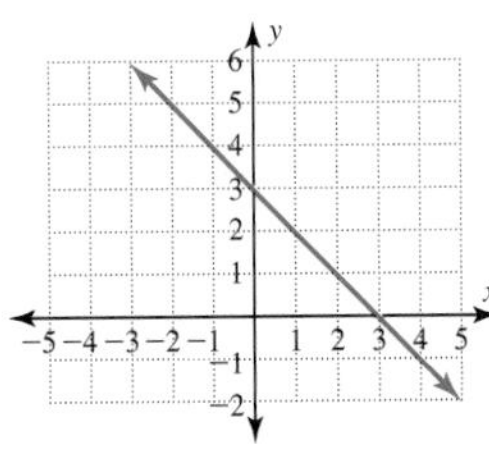

22.

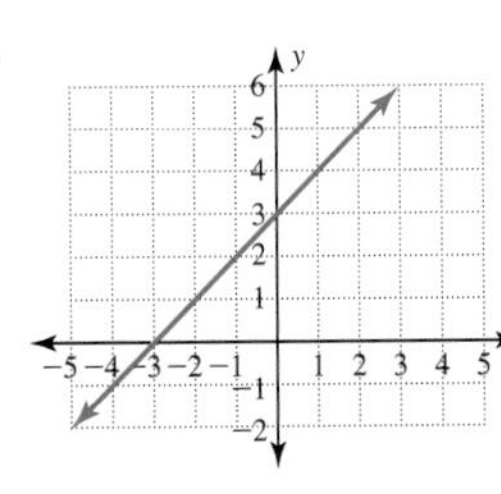

In Exercises 23 and 24, use the table for the linear equation $y = mx + b$ to identify the x- and y-intercepts of the graph of the line through these points.

23.

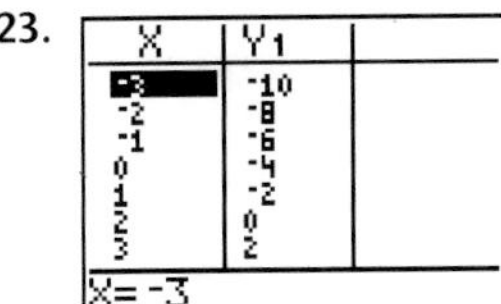

X	Y1
-3	-10
-2	-8
-1	-6
0	-4
1	-2
2	0
3	2

X= -3

24.

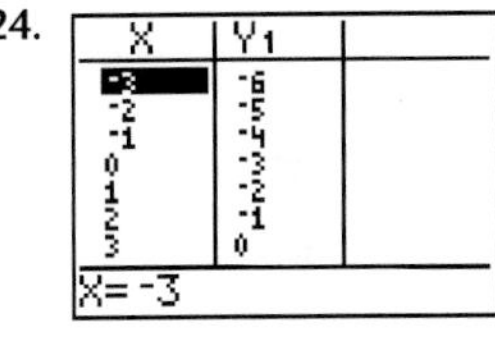

X	Y1
-3	-6
-2	-5
-1	-4
0	-3
1	-2
2	-1
3	0

X= -3

25. Graph the line with x-intercept $(3, 0)$ and y-intercept $(0, 2)$.
26. Graph the line with x-intercept $(-2, 0)$ and y-intercept $(0, -3)$.
27. Graph the line if every point on the line has an x-coordinate of -4.
28. Graph the line if every point on the line has an x-coordinate of 3.
29. Graph the line if every point on the line has a y-coordinate of 4.
30. Graph the line if every point on the line has a y-coordinate of -3.

In Exercises 31–38, plot the y-intercept of each line and one other point to graph the line. Select a third point to double-check your work.

31. $y = x + 2$ 32. $y = x - 3$
33. $y = \frac{x}{3} - 1$ 34. $y = \frac{x}{4} + 1$
35. $f(x) = -x + 4$ 36. $f(x) - x - 2$
37. $f(x) = 2x + 3$ 38. $f(x) = -\frac{x}{2} + 1$

Calculator Exercises

In Exercises 39 and 40, use a calculator to complete the table of values and then identify the x- and y-intercepts of the graph of the line through these points. (*Hint:* See Calculator Perspective 2.3.1.)

39.

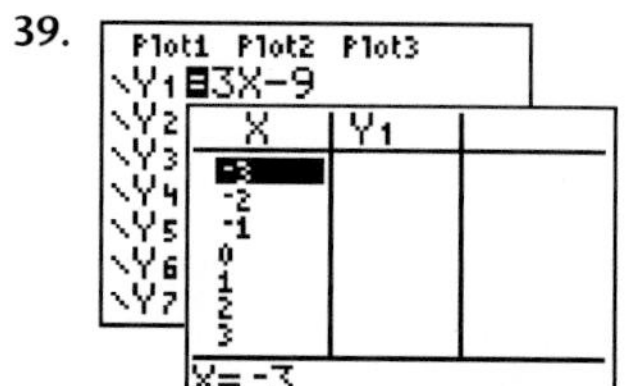

40.

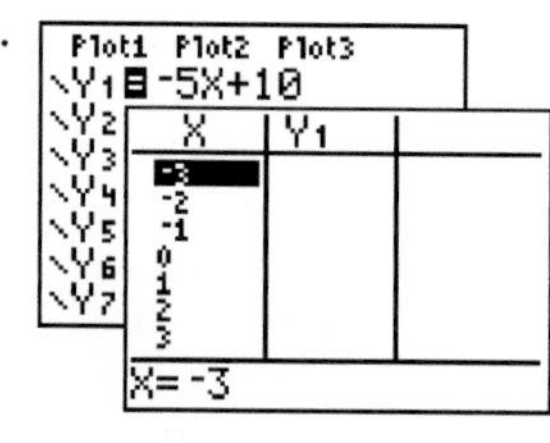

In Exercises 41 and 42, use a calculator with a **Trace** feature to graph the linear function on the window shown and then identify the x- and y-intercepts of the line. (*Hint:* See Calculator Perspective 2.3.1.)

41.

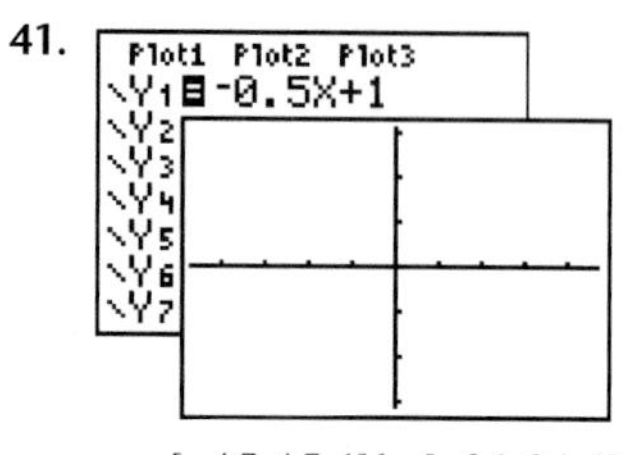

$[-4.7, 4.7, 1]$ by $[-3.1, 3.1, 1]$

42.

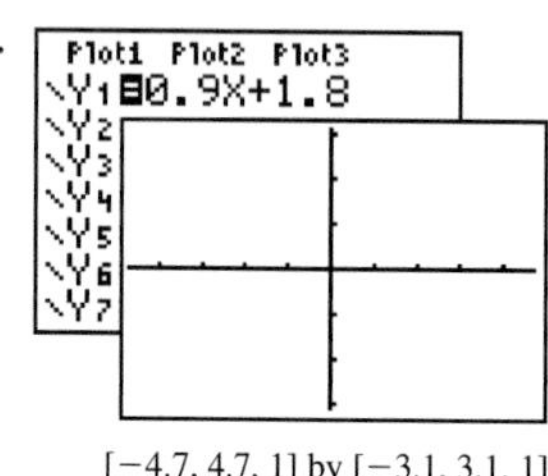

$[-4.7, 4.7, 1]$ by $[-3.1, 3.1, 1]$

In Exercises 43–48, determine the point of intersection of the two lines.

43.

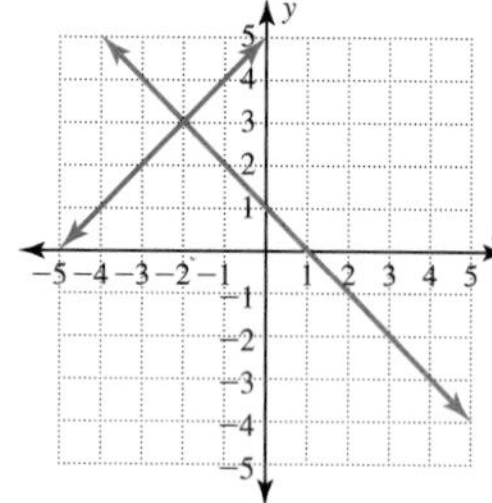

44.

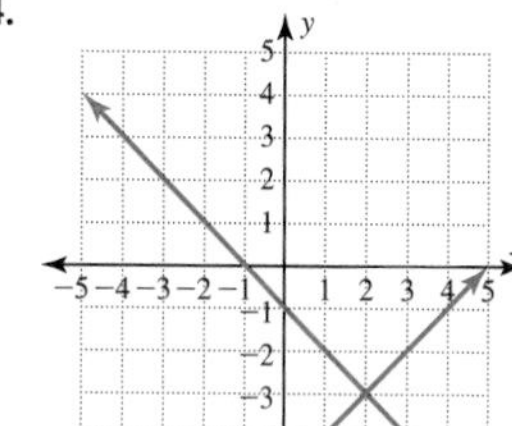

45.

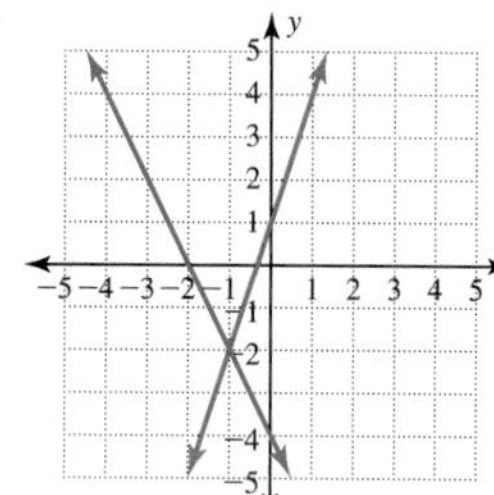

46.

47.

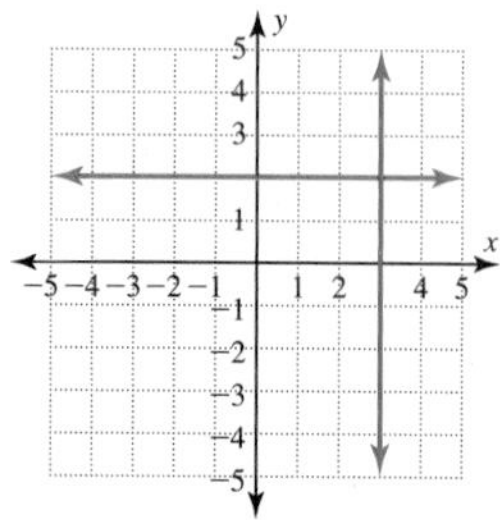

48.

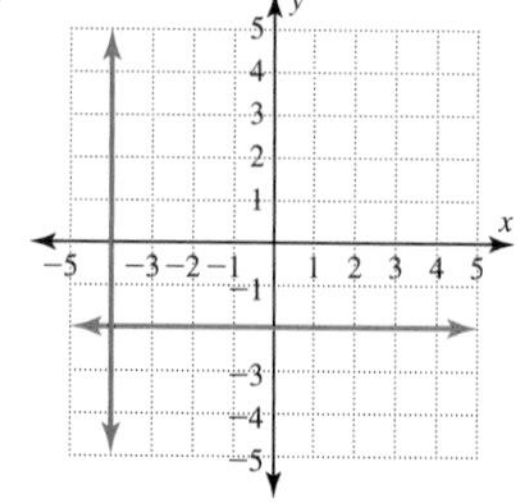

Calculator Exercises

In Exercises 49 and 50, use a calculator to complete the table of values and then identify the point of intersection of the lines through these two sets of points. (*Hint:* See Calculator Perspective 2.3.2.)

49.

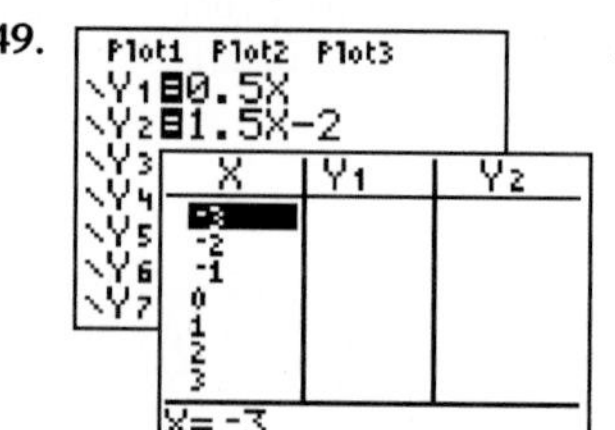

50.

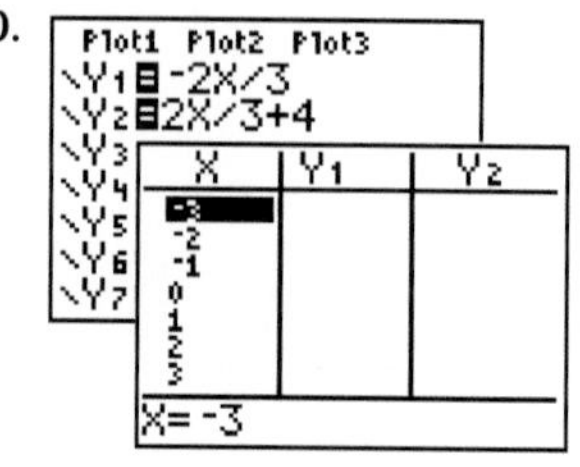

In Exercises 51–54, solve each system of linear equations by using a calculator to graph each equation on the same coordinate system, and determine the point of intersection. Check the coordinates of this point in both of the linear equations. (*Hint:* See Calculator Perspective 2.3.2.)

51. $y = 2x - 5$
$y = -2x - 1$

52. $y = x - 5$
$y = -x + 3$

53. $y = 2x - 1$
$y = \dfrac{x}{2} + 2$

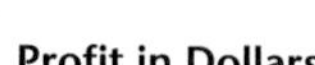

54. $y = -x + 5$
$y = \dfrac{x}{3} + 1$

Interpreting the Intercepts of a Graph

In Exercises 55 and 56, determine the intercepts of each line and interpret the meaning of these points. In each graph the input value of x is the number of units of production by one assembly line at a factory, and the output value y is the profit in dollars generated by the sale of these units when they are produced. (*Hint:* See Example 4.)

55.

Profit in Dollars

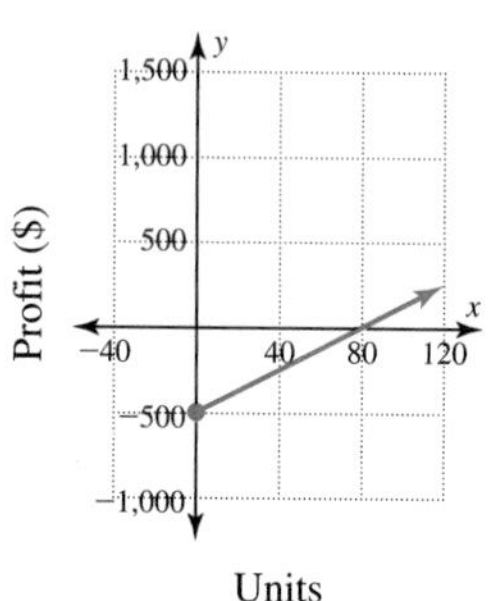

56.

Profit in Dollars

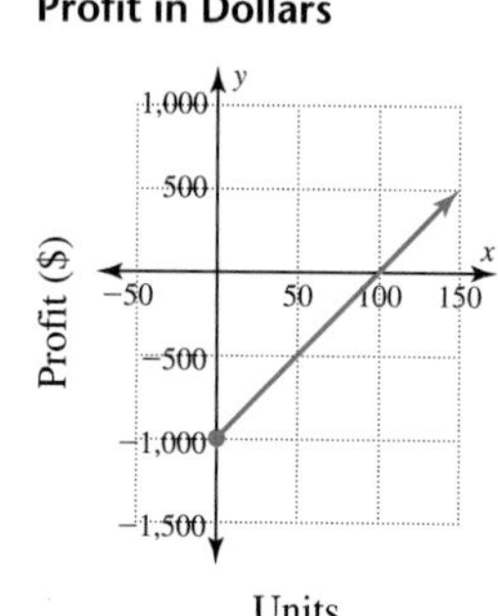

Car Payments

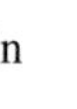

In Exercises 57 and 58, use a calculator and the given equation to prepare a table of the total payments for each of months 1 through 12. In the equation, x represents the number of months, and $f(x)$ represents the total payment in dollars at the end of x months.

57. $f(x) = 333.33x + 457.50$

58. $f(x) = 287.67x + 393.80$

Equivalent Costs

In Exercises 59–62, determine when both options will have the same cost; also determine what this cost will be.

59. The input variable x represents the number of units produced, and the output variable $f(x)$ represents the cost of this production. (*Hint:* Use a calculator window of $[-10, 10, 1]$ by $[-10, 15, 1]$.)
Option A: $f(x) = x + 7$
Option B: $f(x) = 2x + 3$

60. The input variable x represents the number of units produced, and the output variable $f(x)$ represents the cost of this production. (*Hint:* Use a calculator window of $[-5, 10, 1]$ by $[-5, 20, 1]$.)
Option A: $f(x) = 2x + 9$
Option B: $f(x) = 3x + 6$

61. The input variable x represents the number of months an apartment has been rented. The output variable $f(x)$ represents the rental cost by the end of the xth month including an initial nonrefundable deposit. (*Hint:* Use a calculator window of $[0, 5, 1]$ by $[0, 1500, 500]$.)
Option A: $f(x) = 250x + 50$
Option B: $f(x) = 225x + 100$

62. The input variable x represents the number of months an apartment has been rented. The output variable $f(x)$ represents the rental cost by the end of the xth month including an initial nonrefundable deposit. (*Hint:* Use a calculator window of $[0, 4, 1]$ by $[0, 1000, 200]$.)
Option A: $f(x) = 400x + 100$
Option B: $f(x) = 450x$

Group Discussion Questions

63. Discovery Question Graph $y = x + 1$, $y = x + 5$, and $y = x + 9$ all on the same coordinate system. Compare these graphs. What is the same? What is different? What can you predict about the graph from the form $y = mx + b$?

64. Discovery Question Graph $y = x + 1$, $y = \frac{x}{2} + 1$, and $y = 2x + 1$ all on the same coordinate system. Compare these graphs. What is the same? What is different? What can you predict about the graph from the form $y = mx + b$?

65. Discovery Question Graph $y = -2$, $y = 1$, and $y = 3$ all on the same coordinate system. Compare these graphs. What is the same? What is different? What can you predict about the graph from the form $y = mx + b$?

66. Error Analysis A student is trying to graph $y = x + 20$ on a TI-84 Plus calculator. However, after pressing **ZOOM 6**, the student does not see the graph on the display shown below. What is the error and how can it be corrected?

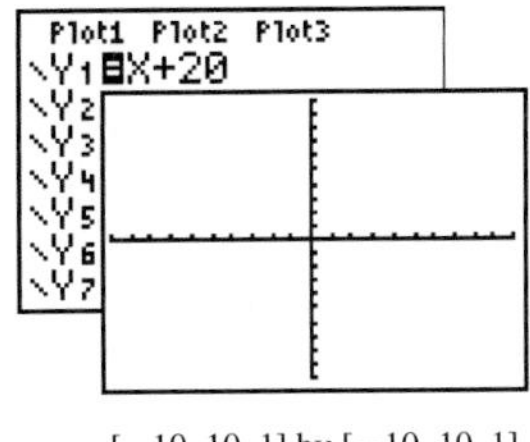

$[-10, 10, 1]$ by $[-10, 10, 1]$

Section 2.4 Solving Linear Equations in One Variable by Using the Addition-Subtraction Principle

Objectives:

10. Solve linear equations in one variable by using the addition-subtraction principle.
11. Identify a linear equation as a conditional equation, an identity, or a contradiction.
12. Use tables and graphs to solve a linear equation in one variable.

One purpose of Sections 2.1 through 2.3 was to develop the concepts and skills to examine linear equations by using both tables and graphs. We will now use these skills to solve linear equations containing only one variable. We will concentrate on linear equations in one variable for the rest of this chapter and return to linear equations in two variables in Chapter 3.

In this section we will concentrate on developing algebraic skills for solving linear equations. We have already noted that $y = mx + b$, an equation with two variables, has a graph that is a straight line. Thus, we refer to $y = mx + b$ as a linear equation. Note that the exponent on both x and y is understood to be 1; we refer to this by saying that the equation is **first degree in both x and y.** If each variable in an equation is first degree, then we extend the terminology and call the equation a **linear equation.**

A Mathematical Note

Francois Viete (1540–1603) is considered by many to be the father of algebra, as we know it today. Prior to Viete, algebra was expressed rhetorically. Viete promoted the use of variables to represent unknowns, which led to the algebraic notation we now use.

Linear Equation in One Variable

ALGEBRAICALLY	VERBALLY	ALGEBRAIC EXAMPLE
A **linear equation in one variable** x is an equation that can be written in the form $Ax = B$, where A and B are real constants and $A \neq 0$.	A linear equation in one variable is first degree in this variable.	$2x = 24$

EXAMPLE 1 Identifying Linear Equations

Determine which of the following choices are linear equations in one variable.

SOLUTION

(a) $3x - 1 = 23$ Linear equation in one variable

The exponent of the variable x is understood to be 1. This is a first-degree equation.

(b) $x^2 = 24$ Not a linear equation

The exponent on variable x is 2.

(c) $3(y - 5) = 4y + 7$ Linear equation in one variable

The only variable is y, and the exponent on y is 1 on both sides of the equation. This equation can be simplified to the form $y = -22$.

(d) $7x - 2 = 5y + 1$ Not a linear equation in one variable

This is a linear equation, but it contains two variables, x and y.

(e) $5x - 3 + 4(x - 1)$ Not a linear equation

This is an algebraic expression, but it is not an equation since it has no symbol of equality.

SELF-CHECK 2.4.1

Determine which of the following choices are linear equations in one variable.

1. $w + 6 = 4$ **2.** $2(w + 6) - 4$ **3.** $5(3x - 1) = 2(7x + 5)$ **4.** $(x - 1)^2 = 4$

To solve an equation whose solution is not obvious, we form simpler equivalent equations until we obtain an equation whose solution is obvious. **Equivalent equations** have the same solution set.

Think of a Balance Scale When You Solve Equations

When thinking of an equation, you may find it helpful to use the concept of a balance scale that has the left side in balance with the right side. We always must perform the same operation on both sides to preserve this balance.

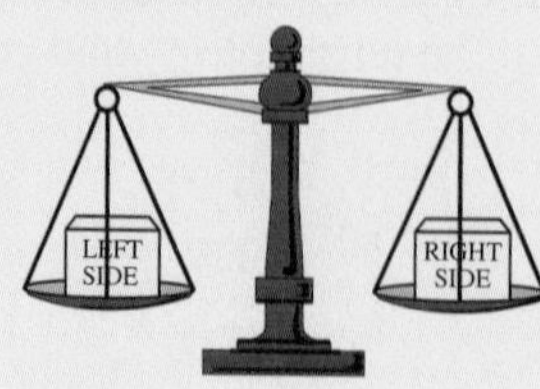

The general strategy for solving a linear equation, regardless of its complexity, is to isolate the variable whose value is to be determined on one side of the equation and to place all other terms on the other side. One of the principles used to accomplish this is given in the following box. Another principle, the Multiplication-Division Principle of Equality, is covered in the Section 2.5. Many students benefit by practicing on the material in this section before using the multiplication-division principle.

Addition-Subtraction Principle of Equality

VERBALLY	ALGEBRAICALLY	NUMERICAL EXAMPLE
If the same number is added to or subtracted from both sides of an equation, the result is an equivalent equation.	If a, b, and c are real numbers, then $a = b$ is equivalent to $a + c = b + c$ and to $a - c = b - c$.	$x + 2 = 5$ is equivalent to $x + 2 - 2 = 5 - 2$ and to $x = 3$

EXAMPLE 2 Solving a Linear Equation with One Variable Term

Solve $x - 3 = 8$.

SOLUTION

To remove a constant or variable from one side of an equation, we add its additive inverse to both sides of the equation.

$$x - 3 = 8$$
$$x - 3 + 3 = 8 + 3$$
$$x = 11$$

Use the addition-subtraction principle of equality to add 3 to both sides of the equation in order to isolate the variable term on the left side and the constant terms on the right side.

Check: $x - 3 = 8$

$11 - 3 \stackrel{?}{=} 8$

$8 \stackrel{?}{=} 8$ checks

Answer: $x = 11$

EXAMPLE 3 Solving a Linear Equation with Variable Terms on Both Sides

Solve $3y = 2y - 5$.

SOLUTION

$$3y = 2y - 5$$
$$3y - 2y = 2y - 5 - 2y$$
$$y = 2y - 2y - 5$$
$$y = -5$$

Subtract $2y$ from both sides of the equation to isolate the variable term on the left side of the equation. Then simplify both sides by combining like terms.

Check: $3y = 2y - 5$

$3(-5) \stackrel{?}{=} 2(-5) - 5$

$-15 \stackrel{?}{=} -10 - 5$

$-15 \stackrel{?}{=} -15$ checks.

Answer: $y = -5$

The general strategy for solving a linear equation, regardless of its complexity, is to isolate the variable on one side of the equation and place all other terms on the other side.

Example 4 contains constants and variables on both sides of the equation. We will use the addition-subtraction principle of equality to isolate the variables on the left side of the equation and the constant terms on the right side of the equation.

EXAMPLE 4 Solving a Linear Equation with Variable and Constant Terms on Both Sides

Solve $6v - 7 = 5v - 3$.

SOLUTION

$$6v - 7 = 5v - 3$$
$$6v - 7 - 5v = 5v - 3 - 5v$$
$$v - 7 = -3$$
$$v - 7 + 7 = -3 + 7$$
$$v = 4$$

Subtract $5v$ from both sides of the equation to isolate the variable term on the left side of the equation.

Add 7 to both sides of the equation to isolate the constant term on the right side of the equation.

Check:
$$6v - 7 = 5v - 3$$
$$6(4) - 7 \stackrel{?}{=} 5(4) - 3$$
$$24 - 7 \stackrel{?}{=} 20 - 3$$
$$17 \stackrel{?}{=} 17 \text{ checks}$$

Answer: $v = 4$

Example 5 illustrates an equation that contains grouping symbols on both sides of the equation. The strategy is first to simplify the left and right sides of the equation and then to isolate the variable terms on one side of the equation and the constant terms on the other side. To solve a linear equation, try to make each step produce a simpler equation than in the previous step.

EXAMPLE 5 Solving a Linear Equation Containing Parentheses

Solve $4(2b - 3) = 7(b - 2)$.

SOLUTION

$$4(2b - 3) = 7(b - 2)$$
$$4(2b) + 4(-3) = 7b + 7(-2)$$
$$8b - 12 = 7b - 14$$
$$8b - 12 - 7b = 7b - 14 - 7b$$
$$b - 12 = -14$$
$$b - 12 + 12 = -14 + 12$$
$$b = -2$$

Use the distributive property $a(b + c) = ab + ac$ to remove the parentheses, and then simplify both sides of the equation.

Subtract $7b$ from both sides of the equation.

Add 12 to both sides of the equation.

Answer: $b = -2$

Does this answer check?

SELF-CHECK 2.4.2

Solve each linear equation.

1. $w + 6 = 4$ **2.** $8a + 9 = 7a + 16$ **3.** $5(3x - 1) = 2(7x + 5)$

The linear equations we have examined thus far have all had exactly one solution. These equations are examples of conditional equations. A **conditional equation** is true for some values of the variable and false for other values.

To complete our discussion we must consider two other types of equations. An equation that is true for all values of the variable is called an **identity.** An equation that is false for all values of the variable is called a **contradiction.** These two types of equations are considered in Examples 6 and 7, respectively.

Conditional Equation, Identity, and Contradiction

	VERBALLY	ALGEBRAIC EXAMPLE	NUMERICAL EXAMPLE
Conditional equation	A conditional equation is true for some values of the variable and false for other values.	$2x = x + 3$ *Answer:* $x = 3$ The only value of x that checks is $x = 3$.	Plot1 Plot2 Plot3 \Y1■2X \Y2■X+3 X: -1, 0, 1, 2, 3, 4, 5; Y1: -2, 0, 2, 4, 6, 8, 10; Y2: 2, 3, 4, 5, 6, 7, 8 X=3
Identity	An identity is an equation that is true for all values of the variable.	$2x = x + x$ *Answer:* All real numbers. All real numbers will check. $x + x$ is always $2x$.	Plot1 Plot2 Plot3 \Y1■2X \Y2■X+X X: -1, 0, 1, 2, 3, 4, 5; Y1: -2, 0, 2, 4, 6, 8, 10; Y2: -2, 0, 2, 4, 6, 8, 10 X=-1
Contradiction	A contradiction is an equation that is false for all values of the variable.	$x = x + 3$ *Answer:* No solution. No real numbers will check because no real number is 3 greater than its own value.	Plot1 Plot2 Plot3 \Y1■X \Y2■X+3 X: -1, 0, 1, 2, 3, 4, 5; Y1: -1, 0, 1, 2, 3, 4, 5; Y2: 2, 3, 4, 5, 6, 7, 8 X=-1

EXAMPLE 6 Solving an Identity

Solve $4x + 1 - x = 2x + 1 + x$.

SOLUTION

$$4x + 1 - x = 2x + 1 + x$$

$$3x + 1 = 3x + 1$$ First simplify both sides of the equation.

$$3x + 1 - 3x = 3x + 1 - 3x$$ Then subtract $3x$ from both sides of the equation.

$$1 = 1$$ is a true statement. The last equation is an identity, so the original equation is also an identity.

Answer: Every real number is a solution since the equation is an identity. Test a couple of values to verify that all real numbers will check.

EXAMPLE 7 Solving a Contradiction

Solve $4v - 2 = 4v + 1$.

SOLUTION

$$4v - 2 = 4v + 1$$

$$4v - 2 - 4v = 4v + 1 - 4v$$ Subtract $4v$ from both sides of the equation.

$-2 = 1$ is a false statement. This last equation is a contradiction, so the original equation is also a contradiction.

Answer: Because the equation is a contradiction, there is no solution. No matter what values you may test, none of them will check.

In Examples 6 and 7, all variables were eliminated in the process of solving the equation. If all the variables are eliminated and an identity results, then every real number is a solution of the equation. If all the variables are eliminated and a contradiction results, then the equation has no solution.

SELF-CHECK 2.4.3

Solve each equation.

1. $v + 1 = 1$ **2.** $v + 1 = 1 + v$ **3.** $v + 1 = 2 + v$

We now use the material that we examined earlier for linear equations in two variables to solve linear equations in one variable. In Calculator Perspective 2.4.1 we use x as the variable regardless of the actual variable given in the equation. Then we let Y_1 equal the left side of the equation and Y_2 equal the right side of the equation. The solution of the equation is the x-coordinate that causes Y_1 to equal Y_2. It is important to remember since we are solving a linear equation with only one variable that the answer we are looking for is not an ordered pair but only the x-coordinate.

CALCULATOR PERSPECTIVE 2.4.1 Using a Table or a Graph to Solve a Linear Equation in One Variable

To solve $2x - 5 = x - 7$ by using a TI-84 Plus calculator, enter the following keystrokes:

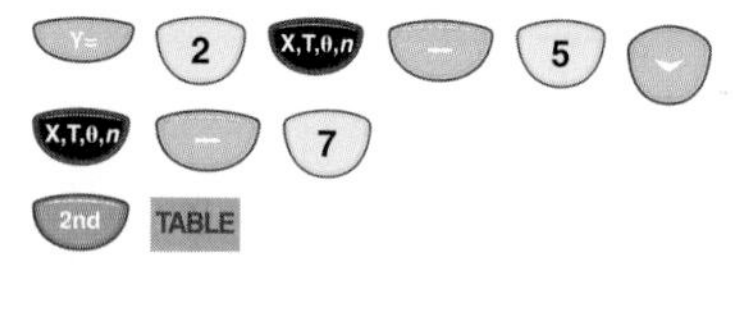

WINDOW

(reset the window to the values shown)

GRAPH

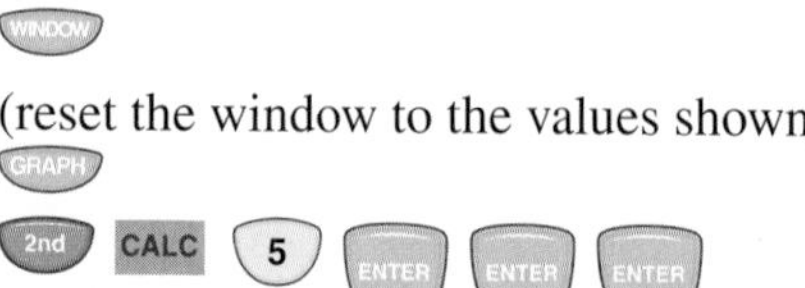

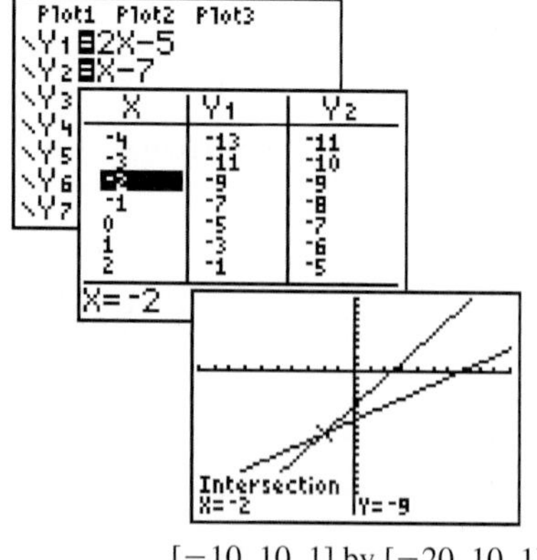

[−10, 10, 1] by [−20, 10, 1]

Answer: $x = -2$

Note: In the table, both Y_1 and Y_2 have the value -9 when $x = -2$. In the graph, -2 is the x-coordinate where the two lines intersect.

In Example 8 we use multiple representations to solve a linear equation. We examine this equation algebraically, graphically, numerically, and verbally.

EXAMPLE 8 Using Multiple Representations to Solve a Linear Equation

Solve $5x - 5 = 4x - 3$.

SOLUTION

ALGEBRAICALLY

$$\begin{aligned} 5x - 5 &= 4x - 3 \\ 5x - 5 - 4x &= 4x - 3 - 4x \\ x - 5 &= -3 \\ x - 5 + 5 &= -3 + 5 \\ x &= 2 \end{aligned}$$

GRAPHICALLY

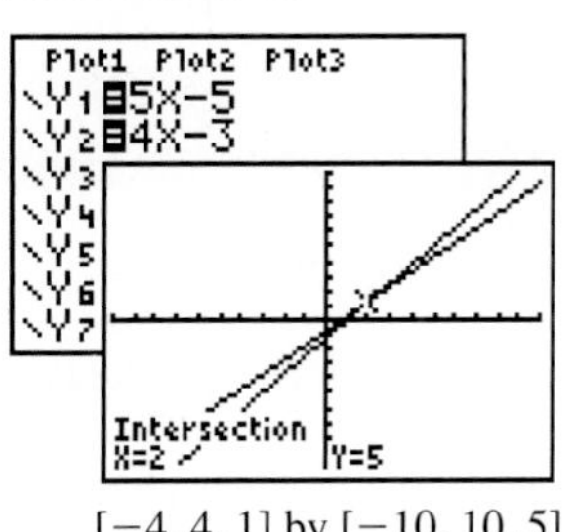

[−4, 4, 1] by [−10, 10, 5]

NUMERICALLY

X	Y1	Y2
0	-5	-3
1	0	1
2	5	5
3	10	9
4	15	13
5	20	17
6	25	21

X=2

VERBALLY

The value of x that satisfies the equation is $x = 2$. From the table, $Y_1 = Y_2$ when $x = 2$. From the graph, the x-coordinate of the point of intersection is $x = 2$.

SELF-CHECK 2.4.4

1. Use a graphing calculator and a table to solve $4.7x - 1.5 = 3.7x + 1.5$.

In Example 9 we solve a verbally stated problem by first rephrasing the problem into a word equation and then translating this word equation into algebraic form.

EXAMPLE 9 Solving a Verbally Stated Equation

If twice a number is decreased by seven, the result is the same as four more than the number. Find this number.

SOLUTION

[Twice a number decreased by 7] [is] [4 more than the number]

$$2n - 7 = n + 4$$

Let n represent the number and translate the **word equation** into an algebraic equation. Subtract n from both sides of the equation.

$$\begin{aligned} 2n - 7 - n &= n + 4 - n \\ n - 7 &= 4 \end{aligned}$$

Add 7 to both sides of the equation.

$$n - 7 + 7 = 4 + 7$$

$$n = 11$$

Does this value check?

Answer: The number is 11.

SELF-CHECK ANSWERS

2.4.1

1. Linear equation in one variable
2. Not a linear equation
3. Linear equation in one variable
4. Not a linear equation

2.4.2

1. $w = -2$ 2. $a = 7$
3. $x = 15$

2.4.3

1. $v = 0$
2. Every real number is a solution.
3. There is no solution

2.4.4

1. Answer: $x = 3$

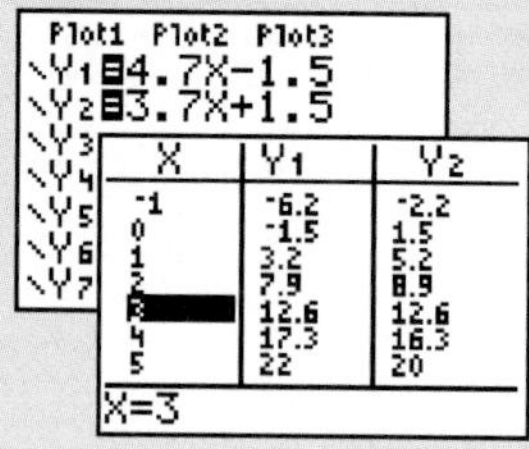

USING THE LANGUAGE AND SYMBOLISM OF MATHEMATICS 2.4

1. An equation is first degree in x if the exponent on x is __________.
2. An equation that is first degree in each variable is called a __________ equation.
3. The equation $y = mx + b$ is a __________ equation in two variables.
4. The equation $Ax = B$ is a __________ equation in one variable.
5. Equations with the same solution are called __________ equations.
6. By the __________-subtraction principle of equality, $a = b$ is equivalent to $a + c = b + c$.
7. By the addition-subtraction principle of equality, $a = b$ is equivalent to $a - c =$ __________.
8. A __________ equation is true for some values of the variable and false for other values.
9. An equation that is true for all values of the variable is called an __________.
10. An equation that is false for all values of the variable is called a __________.

EXERCISES 2.4

In Exercises 1 and 2, determine which of the following are linear equations in one variable.

1. a. $5x + 7$ b. $5x + 7 = 4x$
   c. $y = 5x + 7$ d. $x^2 = 25$
2. a. $x^2 = 4$ b. $4x + 5$
   c. $y = 4x + 5$ d. $4x + 5 = 3x$

In Exercises 3–12, solve each equation, and then check your solution.

3. $x - 11 = 13$
4. $x + 11 = 13$
5. $v + 6 = 2$
6. $v - 6 = -9$
7. $2y = y - 1$
8. $3y = 2y + 4$
9. $3z + 7 = 2z - 4$
10. $5z - 1 = 4z + 3$
11. $10y + 17 = 9y + 21$
12. $8y + 17 = 7y - 21$

In Exercises 13–40, solve each equation.

13. $6x + 2 = 5x + 18$
14. $7x - 4 = 6x + 28$
15. $9y - 11 = 8y + 52$
16. $14y + 21 = 13y - 6$
17. $5v + 7 = 4v + 9$
18. $7v - 8 = 6v + 11$
19. $12y + 2 = 11y + 2$
20. $24y - 8 = 23y - 8$
21. $5m - 7 + 3m = 19 + 7m$
22. $9m - 7 = 13m - 11 - 5m$
23. $12n - 103 = 9n - 4 + 2n$
24. $17n - 18 - 7n = 9n + 18$
25. $3t + 3t + 3t = 4t + 4t$
26. $5t + 5t + 5t = 7t + 7t$
27. $10v - 8v = 9v - 8v$
28. $12v - 10v = 14v - 13v$
29. $3x - 4 + 5x = 6x + 3 + x$
30. $2x + 7 + 9x = 4x - 3 + 6x$
31. $3(v - 5) = 2(v + 5)$
32. $5(v + 2) = 4(v - 6)$
33. $5(w - 2) = 4(w + 3)$
34. $9(w + 4) = 8(w - 1)$
35. $8(m - 2) = 7(m - 3)$
36. $7(m + 5) = 6(m + 3)$
37. $4(n - 6) = 3(n - 5)$
38. $8(n - 2) = 7(n - 4)$
39. $3(5y + 2) = 2(7y - 3)$
40. $7(3y + 2) = 5(4y - 3)$

In Exercises 41–44, each equation is a conditional equation, an identity, or a contradiction. Identify the type of equation and solve it.

41. a. $2x = x$ b. $x = x + 2$ c. $x + 2 = x + 2$
42. a. $2x + 1 = x$ b. $2x = x + x$ c. $x + 3 = 3 + x$
43. a. $v = v + 4$ b. $4 + v = v + 4$ c. $4v = 3v + 4$
44. a. $x + x = 2x + 1$ b. $2(x + 1) = 2x + 2$
    c. $x + x = x + 2$

In Exercises 45–50, simplify the expression in the first column by adding like terms, and solve the equation in the second column.

	Simplify	Solve
45.	a. $5x + 1 + 4x - 6$	b. $5x + 1 = 4x - 6$
46.	a. $8x + 5 + (7x + 2)$	b. $8x + 5 = 7x + 2$
47.	a. $6x - 4 - (5x + 3)$	b. $6x - 4 = 5x + 3$
48.	a. $12x - 5 - (11x + 1)$	b. $12x - 5 = 11x + 1$
49.	a. $3.4x - 1.7 + 2.4x + 2.3$	b. $3.4x - 1.7 = 2.4x + 2.3$
50.	a. $2.6x - 1.9 - (1.6x + 4.1)$	b. $2.6x - 1.9 = 1.6x + 4.1$

In Exercises 51–54, Y_1 represents the left side of the equation and Y_2 represents the right side. Use the table shown from a graphing calculator to solve the equation.

51. $4.8x + 6.3 = 3.8x + 11.3$

X	Y_1	Y_2
0	6.3	11.3
1	11.1	15.1
2	15.9	18.9
3	20.7	22.7
4	25.5	26.5
5	30.3	30.3
6	35.1	34.1
X=0		

52. $12.4x - 15.8 = 11.4x - 13.8$

X	Y_1	Y_2
-2	-40.6	-36.6
-1	-28.2	-25.2
0	-15.8	-13.8
1	-3.4	-2.4
2	9	9
3	21.4	20.4
4	33.8	31.8
X= -2		

53. $3.5x - 4.2 = 2.5x - 1.9$

X	Y_1	Y_2
2	2.8	3.1
2.1	3.15	3.35
2.2	3.5	3.6
2.3	3.85	3.85
2.4	4.2	4.1
2.5	4.55	4.35
2.6	4.9	4.6
X=2		

54. $6.3x + 7.4 = 5.3x + 5.8$

X	Y_1	Y_2
-2	-5.2	-4.8
-1.9	-4.57	-4.27
-1.8	-3.94	-3.74
-1.7	-3.31	-3.21
-1.6	-2.68	-2.68
-1.5	-2.05	-2.15
-1.4	-1.42	-1.62
X= -2		

Calculator Usage

In Exercises 55–60, use a graphing calculator to solve each equation by letting Y_1 represent the left side of the equation and Y_2 represent the right side of the equation. (*Hint:* See Calculator Perspective 2.4.1.)

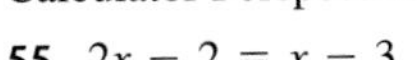

55. $2x - 2 = x - 3$

56. $2x + 4 = x + 1$

57. $1.8x - 4.6 = 0.8x - 2.6$

58. $2.5x - 2.3 = 1.5x + 0.7$

59. $\frac{1}{3}x - \frac{2}{3} = \frac{7}{3} - \frac{2}{3}x$

60. $\frac{2}{5}x - \frac{4}{7} = \frac{3}{7} - \frac{3}{5}x$

Estimate Then Calculate

In Exercises 61–64, mentally estimate the solution of each equation to the nearest integer, and then use a calculator to calculate the solution.

PROBLEM	MENTAL ESTIMATE	CALCULATOR SOLUTION
61. $x - 0.918 = 0.987$		
62. $x + 39.783 = 70.098$		
63. $2x + 354.916 = x + 855.193$		
64. $5x - 1.393 = 4x + 5.416$		

Multiple Representations

In Exercises 65–70, first write an algebraic equation for each verbal statement, using the variable m to represent the number, and then solve for m.

65. Seven more than three times a number equals eight less than twice the number.

66. If three is subtracted from five times a number, the result equals eight more than four times the number.

67. Twelve minus nine times a number is the same as two minus ten times the number.

68. Seven minus six times a number is the same as twelve minus seven times the number.

69. Twice the quantity three times a number minus nine is the same as five times the sum of the number and thirteen.

70. Four times the sum of a number and eleven is equal to three times the difference of the number and five.

Perimeter of a Triangle (Exercises 71 and 72)

71. The perimeter of the triangle shown in the figure is 28 cm. Find the value of a.

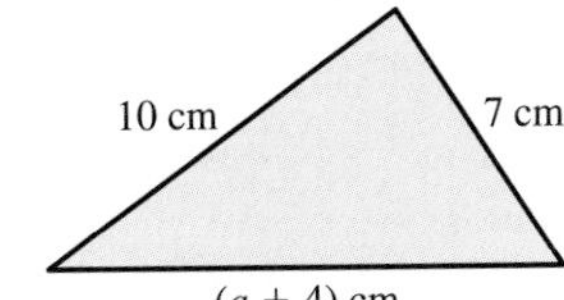

72. The perimeter of the triangle shown in the figure is 31 cm. Find the value of a.

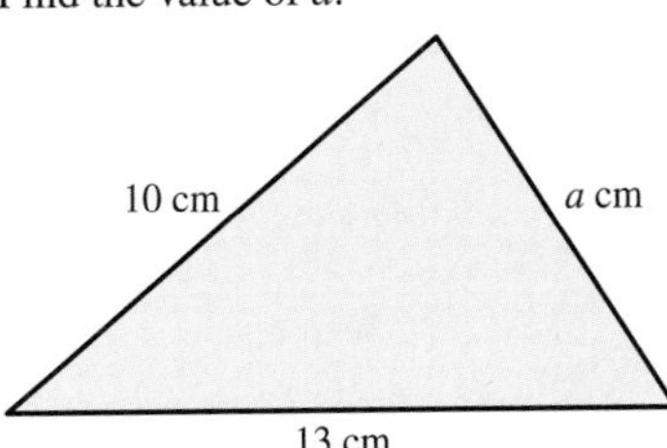

73. Perimeter of a Basketball Court The perimeter of the basketball court shown in the figure is $x + 35$ ft. Find the value of x.

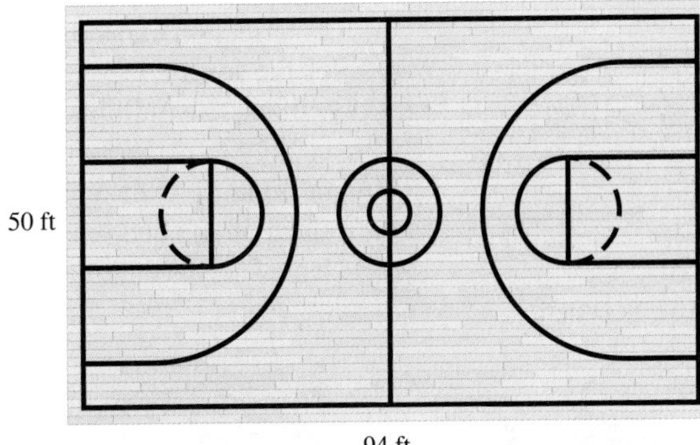

74. Perimeter of a Soccer Field The perimeter of the soccer field shown in the figure is $x + 40$ yd. Find the value of x.

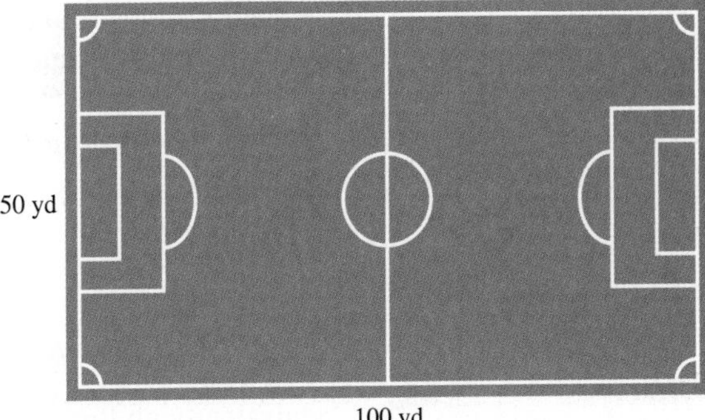

Group Discussion Questions

75. Writing Mathematically Write a paragraph using the analogy of a balance scale to describe the addition-subtraction principle.

76. Challenge Question

a. Complete the equation $8x + 7 = 7x + \underline{?}$ so that the solution is $x = 4$.

b. Complete the equation $12x - 3 = 11x + \underline{?}$ so that the solution is $x = 8$.

77. Challenge Question

a. Complete the equation $3x + 5 = 2x + \underline{?}$ so that the solution is $x = 7$.

b. Complete the equation $3x + 5 = 2x + \underline{?}$ so that the equation is a contradiction.

c. Complete the equation $3x + 5 = 2x + \underline{?}$ so that the equation is an identity.

78. **Discovery Question** Let Y_1 represent the left side of $x^2 = 5x - 6$ and Y_2 represent the right side. Then use a graphing calculator to explore this equation by using the same strategy used for linear equations. Record your observations and explanation of these observations.

Section 2.5 Solving Linear Equations in One Variable by Using the Multiplication-Division Principle

Objective: 13. Solve linear equations in one variable by using the multiplication-division principle.

This section expands our ability to solve linear equations in one variable. We solve equations of the form $Ax = B$ by obtaining a coefficient of 1 for x. To obtain the coefficient of 1, we will use the multiplication-division principle of equality. Students sometimes conclude that it is okay to perform any operation on both sides of an equation. This is incorrect. Multiplying or dividing both sides of an equation by zero is an exception. This is especially tricky if both sides of an equation are multiplied by a variable. If this variable is zero, this can cause problems such as extraneous values. We will examine this situation later in the book.

Key point: Do not multiply or divide both sides of an equation by zero.

Multiplication-Division Principle of Equality

VERBALLY	ALGEBRAICALLY	NUMERICAL EXAMPLE
If both sides of an equation are multiplied or divided by the same nonzero number, the result is an equivalent equation.	If a, b, and c are real numbers and $c \neq 0$, then $a = b$ is equivalent to $ac = bc$ and to $\frac{a}{c} = \frac{b}{c}$.	$\frac{x}{2} = 5$ is equivalent to $2\left(\frac{x}{2}\right) = 2(5)$; and $3x = 12$ is equivalent to $\frac{3x}{3} = \frac{12}{3}$.

To solve $Ax = B$ with $A \neq 0$, we will either divide both sides of the equation by A or equivalently multiply by $\frac{1}{A}$, the multiplicative inverse of A. If A is a rational fraction, then we usually multiply by the reciprocal of A. Otherwise, we usually use division.

EXAMPLE 1 Solving a Linear Equation with an Integer Coefficient

Solve $7y = 91$ and check the solution.

SOLUTION

$7y = 91$

$\frac{7y}{7} = \frac{91}{7}$ Divide both sides of the equation by 7, the coefficient of y.

$y = 13$ This produces a coefficient on y of 1. Note that $1 \cdot y$ is usually written as just y.

Check: $7y = 91$

$7(13) \stackrel{?}{=} 91$

$91 \stackrel{?}{=} 91$ checks.

Answer: $y = 13$

A linear equation is not solved until the coefficient of the variable is 1. If the coefficient is -1, the equation is not yet solved.

EXAMPLE 2 Solving a Linear Equation with a Coefficient of −1.

Solve $-t = 17$.

SOLUTION

$-t = 17$	The coefficient of t is -1.
$-t(-1) = 17(-1)$	Multiply both sides of the equation by -1. (The multiplicative inverse of -1 is -1.)
$t = -17$	Does this value check?

The ability to estimate answers is an important skill not only in recognizing incorrect answers from keystroke errors on a calculator, but also in making routine consumer decisions. Families make many choices from the dimensions of bedspreads and carpet to room additions that can be facilitated by good estimation skills. Example 3 illustrates how to estimate the solution of a linear equation.

EXAMPLE 3 Estimate Then Calculate

Mentally estimate the solution of $-7.15m = 42.042$ and then calculate the exact solution.

SOLUTION

Estimate by considering the equation $-7m = 42$.

ESTIMATED SOLUTION

$$-7m = 42$$
$$\frac{-7m}{-7} = \frac{42}{-7}$$
$$m = -6$$

CALCULATOR SOLUTION

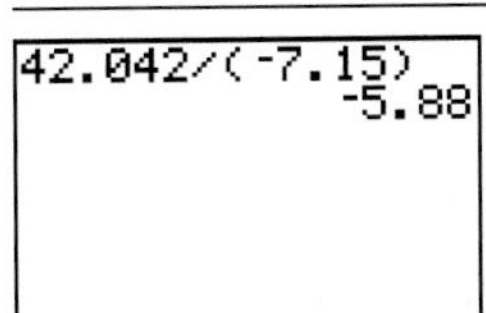

Answer: The solution $m = -5.88$ seems reasonable based upon the estimated solution of $m \approx -6$.

SELF-CHECK 2.5.1

Solve each equation.

1. $6v = -42$
2. $-v = 4$
3. $-8x = -48$

Mentally estimate to the nearest integer the solution of each equation.

4. $x + 6.12 = 18.29$
5. $6.12x = 18.29$

Linear equations often contain variables and constants on both sides of the equation. To solve these linear equations, first isolate the variable terms on one side of the equation by using the addition-subtraction principle. Once the equation has been simplified to the form $Ax = B$, use the multiplication-division principle to solve for the variable. This is illustrated in Example 4.

EXAMPLE 4 Solving a Linear Equation with Constant Terms on Both Sides

Solve $3x + 2 = 14$ and check the solution.

SOLUTION

$$3x + 2 = 14$$

$$3x + 2 - 2 = 14 - 2$$

Subtract 2 from both sides of the equation to isolate the constants on the right side of the equation.

$$3x = 12$$

$$\frac{3x}{3} = \frac{12}{3}$$

Divide both sides of the equation by 3 to obtain a coefficient of 1 for x.

$$x = 4$$

Check:

$$3x + 2 = 14$$

$$3(4) + 2 \stackrel{?}{=} 14$$

$$12 + 2 \stackrel{?}{=} 14$$

$$14 \stackrel{?}{=} 14 \text{ checks.}$$

Answer: $x = 4$

Example 5 has constant and variable terms on both sides of the equation. We also use Calculator Perspective 1.4.2 to check our solution.

EXAMPLE 5 Solving a Linear Equation with Variables on Both Sides

Solve $5v - 4 = 7v + 10$ and check the solution on a calculator.

SOLUTION

$$5v - 4 = 7v + 10$$

$$5v - 4 - 7v = 7v + 10 - 7v$$

Subtract $7v$ from both sides of the equation to isolate the variables on the left side.

$$-2v - 4 = 10$$

$$-2v - 4 + 4 = 10 + 4$$

Add 4 to both sides of the equation to isolate the constants on the right side.

$$-2v = 14$$

$$\frac{-2v}{-2} = \frac{14}{-2}$$

Divide both sides of the equation by -2.

$$v = -7$$

Check:

```
-7→X
                -7
5X-4
               -39
7X+10
               -39
```

Using Calculator Perspective 1.4.2, store the value to be checked under the variable x. Then calculate the value of both sides of the equation.

If we use -7 for the value of the variable, both sides of the equation have the same value, so this solution checks.

Answer: $v = -7$

The strategy used in the previous examples is described in the following box. We will then use this strategy to solve equations containing parentheses and fractions.

A good rule of thumb to use when solving a linear equation is to try to produce simpler expressions with each step.

Strategy for Solving Linear Equations

STEP 1. Simplify each side of the equation.

a. If the equation contains fractions, simplify by multiplying both sides of the equation by the least common denominator (LCD) of all the fractions.

b. If the equation contains grouping symbols, simplify by using the distributive property to remove the grouping symbols and then combine like terms.

STEP 2. Using the addition-subtraction principle of equality, isolate the variable terms on one side of the equation and the constant terms on the other side.

STEP 3. Using the multiplication-division principle of equality, solve the equation produced in step 2.

EXAMPLE 6 Solving a Linear Equation Containing Parentheses

Solve $2(3x - 5) = 7(2x + 2)$.

SOLUTION

$$2(3x - 5) = 7(2x + 2)$$

Remove parentheses by using the distributive property.

$$6x - 10 = 14x + 14$$

$$6x - 10 - 14x = 14x + 14 - 14x$$

$$-8x - 10 + 10 = 14 + 10$$

$$-8x = 24$$

Isolate the variable terms on the left side of the equation and the constant terms on the right side of the equation. Do this by subtracting $14x$ from both sides of the equation and adding 10 to both sides of the equation.

$$\frac{-8x}{-8} = \frac{24}{-8}$$

Solve for x by dividing both sides of the equation by -8.

$$x = -3$$

Check:

$$2(3x - 5) = 7(2x + 2)$$

$$2[3(-3) - 5] \stackrel{?}{=} 7[2(-3) + 2]$$

$$2(-9 - 5) \stackrel{?}{=} 7(-6 + 2)$$

$$2(-14) \stackrel{?}{=} 7(-4)$$

$$-28 \stackrel{?}{=} -28 \text{ checks.}$$

Note the use of parentheses when we substitute -3 for x. The use of these parentheses is highly recommended to avoid errors in the order of operations.

Answer $x = -3$

SELF-CHECK 2.5.2

Solve each equation.

1. $5x - 3 = 27$ **2.** $3(v - 3) = 2(5v - 11) - 1$

EXAMPLE 7 Solving a Linear Equation Containing Parentheses

Solve $4y + 3(y - 2) = 2(y + 4) - (2y - 7)$.

SOLUTION

$$\begin{aligned} 4y + 3(y - 2) &= 2(y + 4) - (2y - 7) \\ 4y + 3y - 6 &= 2y + 8 - 2y + 7 \\ 7y - 6 &= 15 \\ 7y - 6 + 6 &= 15 + 6 \\ 7y &= 21 \\ \frac{7y}{7} &= \frac{21}{7} \end{aligned}$$

Remove parentheses by using the distributive property, and then combine like terms.

Isolate the constant terms on the right side of the equation by adding 6 to both sides.

Solve for y by dividing both sides of the equation by 7.

Answer: $y = 3$

Does this value check?

If a linear equation contains fractions, then we can simplify it by converting it to an equivalent equation that does not involve fractions. To do this, multiply both sides of the equation by the LCD of all the terms. In Example 8 the LCD is 12.

EXAMPLE 8 Solving a Linear Equation Containing Fractions

Solve $\frac{z}{6} + 2 = \frac{z}{4}$.

SOLUTION

$$\begin{aligned} \frac{z}{6} + 2 &= \frac{z}{4} \\ 12\left(\frac{z}{6} + 2\right) &= 12\left(\frac{z}{4}\right) \\ 12\left(\frac{z}{6}\right) + 12(2) &= 12\left(\frac{z}{4}\right) \\ 2z + 24 &= 3z \\ 2z + 24 - 2z &= 3z - 2z \\ 24 &= z \text{ or} \end{aligned}$$

$6 = 2 \cdot 3$
$4 = 2 \cdot 2$
$\text{LCD} = 2 \cdot 2 \cdot 3 = 12$

Multiply both sides of the equation by the LCD, 12.

Use the distributive property to remove the parentheses, and then simplify.

Subtract $2z$ from both sides of the equation.

Answer: $z = 24$

Does this solution check?

Observe in Examples 8 and 9 the execution of our basic strategy—at each step of the solution process we use operations that will produce simpler expressions.

EXAMPLE 9 Solving a Linear Equation Containing Fractions

Solve $\frac{3v-3}{6} = \frac{4v+1}{15} + 2$.

SOLUTION

$$\frac{3v-3}{6} = \frac{4v+1}{15} + 2$$

$6 = 2 \cdot 3$
$15 = 3 \cdot 5$
$\text{LCD} = 2 \cdot 3 \cdot 5 = 30$

$$30\left(\frac{3v-3}{6}\right) = 30\left(\frac{4v+1}{15} + 2\right)$$

Multiply both sides of the equation by the LCD, 30.

$$\frac{30}{6}(3v-3) = \frac{30}{15}(4v+1) + 30(2)$$

Use the distributive property to remove parentheses and then combine like terms.

$$5(3v-3) = 2(4v+1) + 30(2)$$
$$15v - 15 = 8v + 2 + 60$$
$$15v - 15 = 8v + 62$$
$$15v - 15 - 8v = 8v + 62 - 8v$$

Subtract $8v$ from both sides of the equation.

$$7v - 15 = 62$$
$$7v - 15 + 15 = 62 + 15$$

Then add 15 to both sides of the equation.

$$7v = 77$$
$$\frac{7v}{7} = \frac{77}{7}$$

Divide both sides of the equation by 7.

$$v = 11$$

Does this solution check?

SELF-CHECK 2.5.3

Solve each equation.

1. $\frac{y}{3} - \frac{y}{12} = -\frac{1}{2}$
2. $\frac{4x+2}{5} - 1 = \frac{7x-3}{2}$

SELF-CHECK ANSWERS

2.5.1
1. $v = -7$ 2. $v = -4$
3. $x = 6$ 4. $x \approx 12$
5. $x \approx 3$

2.5.2
1. $x = 6$ 2. $v = 2$

2.5.3
1. $y = -2$ 2. $x = \frac{1}{3}$

USING THE LANGUAGE AND SYMBOLISM OF MATHEMATICS 2.5

1. Equations with the same solution are called __________ equations.
2. By the __________-division principle, $a = b$ is equivalent to $ac = bc$ for $c \neq 0$.
3. By the multiplication-division principle of equality, $a = b$ is equivalent to $\frac{a}{c} =$ __________ for $c \neq 0$.
4. The equation $-x = 9$ is equivalent to $x =$ __________.
5. A __________ equation is true for some values of the variable and false for other values.
6. The statement "You can always multiply both sides of an equation by the same number to produce an equivalent equation" is false because we do not obtain equivalent equations if we multiply both sides of the equation by __________.
7. The __________ property often is used to help us remove parentheses from expressions in an equation.
8. To clear an equation of fractions, we can multiply both sides of the equation by the __________ __________ __________ of all the terms in the equation.
9. The multiplicative inverse of $\frac{1}{4}$ is __________.
10. The multiplicative inverse of $-\frac{1}{4}$ is __________.

EXERCISES 2.5

In Exercises 1–6, solve each equation, and then check your solution.

1. a. $7x = 42$ b. $x + 7 = 42$
2. a. $-5y = 45$ b. $y - 5 = 45$
3. a. $-v = 8$ b. $v - 1 = 8$
4. a. $-m = -6$ b. $m - 1 = -6$
5. a. $\frac{3}{4}t = -36$ b. $\frac{4}{3}t = -36$
6. a. $-\frac{2}{3}m = 48$ b. $-\frac{3}{2}m = 48$

In Exercises 7–14, solve each equation and then check your solution by using a graphing calculator. (*Hint:* See Calculator Perspective 1.4.2.)

7. $3(2t - 1) = 7t + 1$
8. $-3(4t - 1) = -(t - 14)$
9. $4(2 - 3x) = 3 - 13x$
10. $-2(5x + 4) = -(x - 10)$
11. $\frac{y}{3} + \frac{2}{3} = \frac{y}{2} + \frac{3}{2}$
12. $\frac{v}{4} - \frac{2}{5} = \frac{v}{10} + \frac{1}{2}$
13. $2 - 6(y + 1) = 4(2 - 3y) + 6$
14. $13 + 3(2v - 5) = 1 - 2(3 - 6v)$

In Exercises 15–40, solve each equation.

15. $24 = -8z$
16. $-15 = 25z$
17. $-1 = 9m$
18. $1 = -7m$
19. $-47w = 0$
20. $0 = 31w$
21. $\frac{-5v}{3} = 25$
22. $-\frac{2v}{11} = 66$
23. $8x - 1 = 13x - 1$
24. $-17x + 5 = 17x + 5$
25. $4(3y - 5) = 5(4y + 4)$
26. $7(2y - 3) = 3(6y + 5)$
27. $0.12a = 13.2$
28. $0.07a = -3.5$
29. $2.3x + 29.3 = 1.2(4 - x)$
30. $5.7x + 35.7 = 1.2(x - 4)$
31. $6 - 3(2v + 2) = 4(8 - v)$
32. $2(1 - v) + 9 = 3(2v + 1)$
33. $4(x + 1) + 3(x + 2) = 9(x + 1) - 5$
34. $11(x + 3) + 4(2x - 1) = 5(3x - 2) + 13$
35. $\frac{w + 1}{3} = \frac{w - 5}{5}$
36. $\frac{24w - 67}{60} = \frac{3w - 8}{12}$
37. $\frac{x + 2}{3} - 4 = -\frac{x}{2}$
38. $\frac{2x}{7} = 1 - \frac{2x + 1}{3}$
39. $7(x - 1) - 4(2x + 3) = 2(x + 1) - 3(4 - x)$
40. $5(2x - 1) + 3(x - 3) = -4(x - 6) + 2(13 - 3x)$

In Exercises 41 and 42, each equation is a conditional equation, an identity, or a contradiction. Identify the type of each equation and then solve it.

41. a. $3(x + 1) = 3x$ b. $3(x + 1) = 3x + 3$
    c. $3(x + 2) = 2(x + 3)$
42. a. $4(x + 5) = 4x$ b. $4(x + 5) = 4x + 20$
    c. $4(x + 5) = 5(x + 4)$

In Exercises 43–46, simplify the expression in the first column, and solve the equation in the second column.

	Simplify	Solve
43.	a. $3(2x - 4) - 5(x - 2)$	b. $3(2x - 4) = 5(x - 2)$
44.	a. $3(2x - 4) + 5(x - 2)$	b. $3(2x - 4) = -5(x - 2)$
45.	a. $1.5(4x - 6) + 2.5(6x - 4)$	b. $1.5(4x - 6) = -2.5(6x - 4)$
46.	a. $1.5(4x - 6) - 2.5(6x - 4)$	b. $1.5(4x - 6) = 2.5(6x - 4)$

Drug Prescriptions for Children

Children are often prescribed the same drugs used for adults. Two commonly used formulas for adjusting the dosage to account for the age of the child are Cowling's formula and Young's formula.

47. Cowling's formula for a 10-year old child is $y = 0.8x$, where x is the adult dosage and y is the child dosage. What is the adult dosage if the child dosage of a medication is 5 mg?
48. Young's formula for a 12-year-old child is $y = 0.5x$, where x is the adult dosage and y is the child dosage. What is the adult dosage if the child dosage of a medication is 4 mg?

Estimation Skills (Exercises 49–52)

49. **Perimeter of a Square** ($P = 4s$) The perimeter of a square is 99.837 cm. Mentally estimate to the nearest centimeter the length of each side.

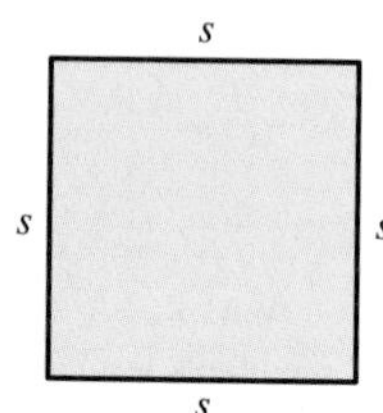

50. **Perimeter of an Equilateral Triangle** ($P = 3s$) The perimeter of an equilateral triangle is 99.378 m. Mentally estimate to the nearest meter the length of each side.

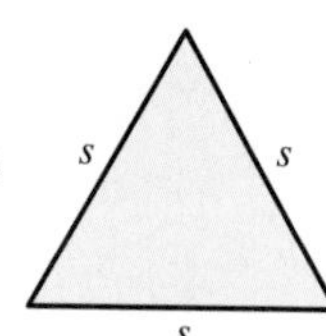

51. **Hotel Room Tax** ($T = 0.10C$) A hotel tax on a bill is 10% of the room charges. Mentally estimate the pretax room charges for a room with a tax charge of \$7.58.

52. **Gratuity** ($G = 0.20B$) A customer added a gratuity of $13 to the restaurant charge. If the gratuity was approximately 20% of the restaurant bill, mentally estimate the bill before the gratuity was added.

Estimate Then Calculate

In Exercises 53–56, mentally estimate the solution of each equation to the nearest integer, and then use a calculator to calculate the solution.

PROBLEM	MENTAL ESTIMATE	CALCULATOR SOLUTION
53. $-2.1x = 8.82$		
54. $4.9x = -14.945$		
55. $-0.49x = -2.009$		
56. $-0.24x = 2.01$		

Calculator Usage

In Exercises 57–60, Y_1 represents the left side of the equation and Y_2 represents the right side. Use the table shown from a graphics calculator to solve the equation.

57. $5.8x - 1.71 = 2.7x + 2.94$

X	Y_1	Y_2
0	−1.71	2.94
.5	1.19	4.29
1	4.09	5.64
1.5	6.99	6.99
2	9.89	8.34
2.5	12.79	9.69
3	15.69	11.04

X=0

58. $7.2x + 4.2 = 4.7x + 0.95$

X	Y_1	Y_2
−1.5	−6.6	−6.1
−1.4	−5.88	−5.63
−1.3	−5.16	−5.16
−1.2	−4.44	−4.69
−1.1	−3.72	−4.22
−1	−3	−3.75
−.9	−2.28	−3.28

X=−1.5

59. $2.4(4x - 1) = 1.8(2x + 1)$

X	Y_1	Y_2
.2	−.48	2.52
.3	.48	2.88
.4	1.44	3.24
.5	2.4	3.6
.6	3.36	3.96
.7	4.32	4.32
.8	5.28	4.68

X=.2

60. $2.5(6x + 1) = 12x + 10.9$

X	Y_1	Y_2
2	32.5	34.9
2.2	35.5	37.3
2.4	38.5	39.7
2.6	41.5	42.1
2.8	44.5	44.5
3	47.5	46.9
3.2	50.5	49.3

X=2

Multiple Representations

In Exercises 61–68, write an algebraic equation for each verbal statement, and then solve for the variable.

61. Five times the sum of x and two is the same as seven times the quantity x minus three.
62. Three times the sum of x and six equals eleven times the quantity x minus two.
63. Twice the quantity three v minus two equals four times the quantity v plus nine.
64. Six times the quantity seven minus v is the same as five times the quantity two v plus nine.
65. Twice the difference of three m minus five is four less than three times the sum of m and six.
66. Five times the sum of two m plus eleven is three more than two times the quantity m minus nine.
67. One-third of the quantity two x plus five is the same as the quantity four x plus two.
68. One-half of the quantity three x plus seven equals the quantity six x plus two.
69. **Perimeter of a Wrestling Mat** The perimeter of each square wrestling mat shown in the figure is 168 ft. The length of one side of a mat is $4x + 18$ ft. Find the value of x.

70. **Perimeter of a Yield Sign** The perimeter of the yield sign shown in the figure is 60 in. Assuming this sign is approximately an equilateral triangle, find the value of x.

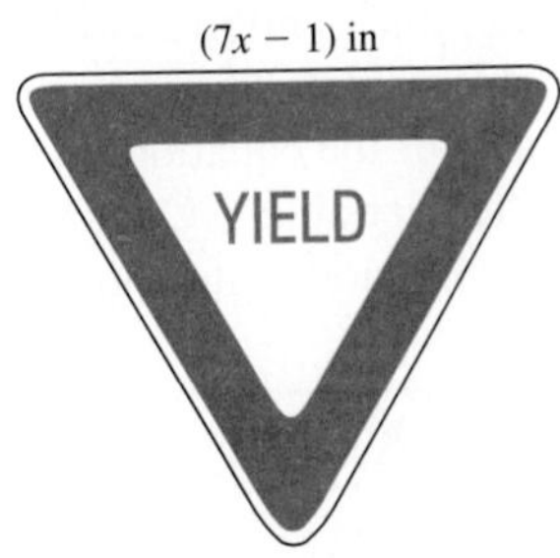

71. **Perimeter of a Rectangle** The perimeter of the rectangle shown in the figure is 17 ft. Find the value of a.

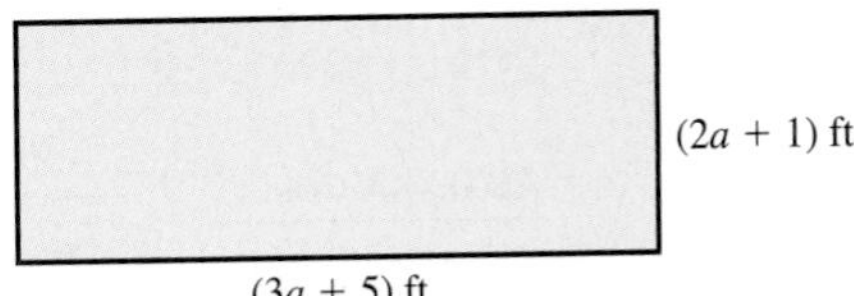

72. **Perimeter of a Parallelogram** The perimeter of the parallelogram shown in the figure is 18 cm. Find the value of a.

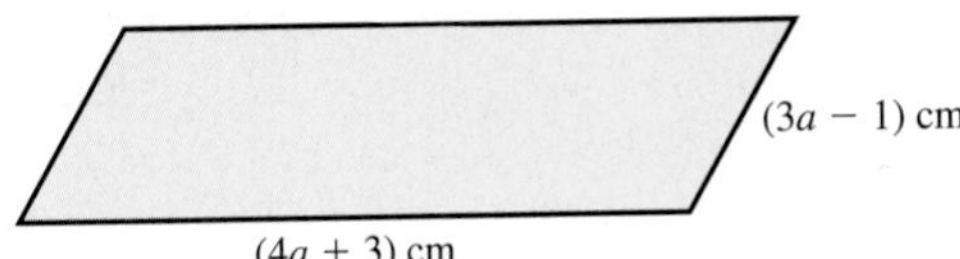

73. **Area of a Triangle** The area of the triangle shown in the figure is 51 cm^2 (square centimeters). Find the value of x. $\left(A = \frac{1}{2}bh\right)$

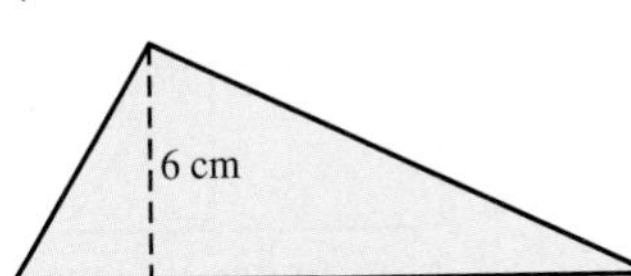

74. Area of a Rectangle The area of the rectangle shown in the figure is 102 cm^2. Find the value of x. ($A = lw$)

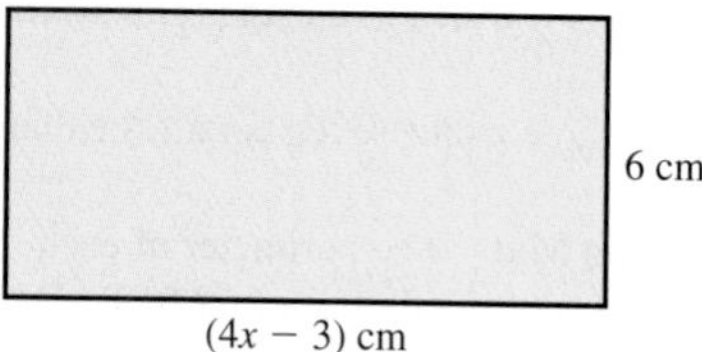

Group Discussion Questions

75. Writing Mathematically Solve $5(x - 3) + 4 = 4(2x + 1) - 9$, and then write an explanation of each step of the solution. Also explain why you performed the steps in the order in which you have them listed rather than in some other order.

76. Challenge Question

a. Complete the equation $\frac{4}{5}x + 11 = \underline{?}$ so that the solution is $x = 31$.

b. Complete the equation $5(x - 3) + 1 = 2(x + 4) + \underline{?}$ so that the solution is $x = 3$.

77. Error Analysis Examine the following argument and explain the error in the reasoning. If $x = 0$, then $2x = 0$; thus $x = 2x$. Dividing both sides of this equation by x, we conclude that $1 = 2$.

Section 2.6 Using and Rearranging Formulas

Objectives:

14. Solve an equation for a specified variable.
15. Rewrite a linear equation in the form $y = mx + b$ and calculate the x- and y-intercepts of the graph of a linear equation.

To enter information into calculators or computers, we often must rearrange equations so that the information we have fits the required format. We will develop this ability in this section by using the addition-subtraction and the multiplication-division principles of equality.

If an equation contains more than one variable, we can specify the variable that we wish to solve for. Then we can rearrange the equality to write it in the desired form. This variable is referred to as the **specified variable.** To solve an equation for a specified variable, isolate this variable on one side of the equation with all other variables on the other side of the equation. In Example 1 we solve the linear equation $4x + 2y = 6$ for the specified variable y. Once we have the equation written in the form $y = mx + b$, we can enter the equation into a graphing calculator.

EXAMPLE 1 Solving a Linear Equation for y

Key point: A first-degree equation $Ax + By = C$ can always be rewritten in the form $y = mx + b$. We will examine this form more in Section 3.2 where we examine the meaning of both m and b.

Solve the linear equation $4x + 2y = 6$ for y, and then use a graphing calculator to graph this equation.

SOLUTION

ALGEBRAICALLY

$$4x + 2y = 6$$
$$4x + 2y - 4x = 6 - 4x$$

Use the addition principle of equality to subtract $4x$ from both sides of the equation to isolate the y-term on the left side of the equation.

$$2y = -4x + 6$$
$$2y = 2(-2x + 3)$$

Use the distributive property $ab + ac = a(b + c)$ to rewrite the right side of the equation.

$$\frac{2y}{2} = \frac{2(-2x + 3)}{2}$$
$$y = -2x + 3$$

Then use the division principle of equality to divide both sides of the equation by 2 to solve for y.

GRAPHICALLY

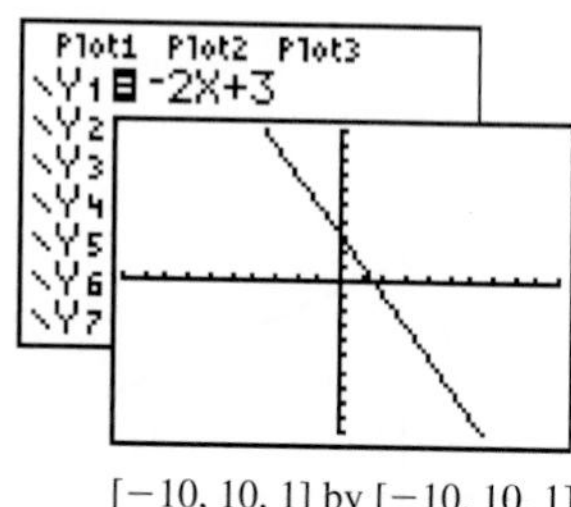

[−10, 10, 1] by [−10, 10, 1]

Now that the linear equation is in the form $y = mx + b$, enter this equation into a graphing calculator and graph this equation.

SELF-CHECK 2.6.1

1. Solve $x - 2y = 2$ for y.
2. Graph this equation on a graphing calculator.

Once a linear function is rewritten in the form $y = mx + b$, we can use a calculator to examine the function both numerically and graphically. In Example 2 we will create a table of values for the given linear function.

EXAMPLE 2 Solving a Linear Equation for y

Solve the linear equation $3x - y = 3(2x - y) - 1$ for y, and then use a graphing calculator to complete a table of values for the x-values $-3, -2, -1, 0, 1, 2, 3$.

SOLUTION

ALGEBRAICALLY

$$3x - y = 3(2x - y) - 1$$

$$3x - y = 6x - 3y - 1$$

First use the distributive property to remove the parentheses from the right side.

$$3x - 3x - y = 6x - 3x - 3y - 1$$

Subtract $3x$ from both sides to move all x-terms to the right side.

$$-y = 3x - 3y - 1$$

$$-y + 3y = 3x - 3y - 1 + 3y$$

Add $3y$ to both sides to isolate all y-terms on the left side.

$$2y = 3x - 1$$

$$\frac{2y}{2} = \frac{3x - 1}{2}$$

Then divide both sides of the equation by 2 to solve for y.

Answer: $y = \frac{3}{2}x - \frac{1}{2}$

NUMERICALLY

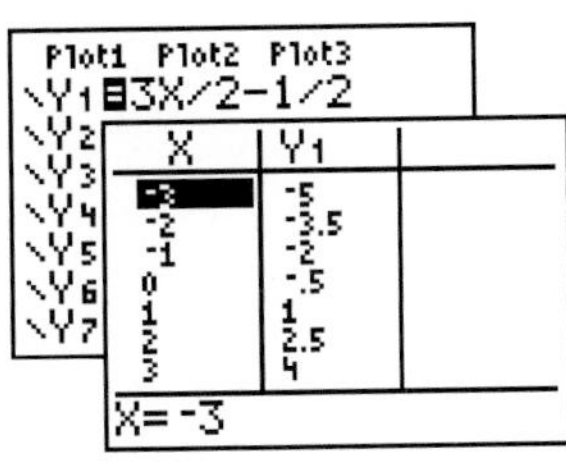

Enter the given equation into a graphing calculator to create the table shown.

The ability to solve an equation for either x or y is a skill that we use frequently when working with linear equations. In Section 2.3 we determined the x- and y-intercepts of a line by examining either a table of values or a graph of the linear function. We will now examine how to determine the exact values of these intercepts algebraically. In Example 3 we use the fact that the x-intercept is of the form $(a, 0)$ with a y-coordinate of zero, and the y-intercept is of the form $(0, b)$ with an x-coordinate of zero.

EXAMPLE 3 Calculating the Intercepts of a Line

Calculate the x- and y-intercepts of $3x - 2y = 6$ and use these intercepts to sketch the graph of this equation.

SOLUTION

CALCULATION OF THE x-INTERCEPT

$$3x - 2y = 6$$
$$3x - 2(0) = 6$$
$$3x - 0 = 6$$
$$3x = 6$$
$$x = 2$$

To calculate the x-intercept, set $y = 0$ and solve for x.

Simplify the left side of the equation and then divide both sides by 3.

(2, 0) is the x-intercept.

Intercepts are points on the line and should be written as ordered pairs. The x-intercept is not 2, but the point (2, 0). The y-intercept is not -3, but the point $(0, -3)$.

CALCULATION OF THE y-INTERCEPT

$$3x - 2y = 6$$
$$3(0) - 2y = 6$$
$$0 - 2y = 6$$
$$-2y = 6$$
$$y = -3$$

To calculate the y-intercept, set $x = 0$ and solve for y.

Simplify the left side of the equation and then divide both sides of the equation by -2.

$(0, -3)$ is the y-intercept.

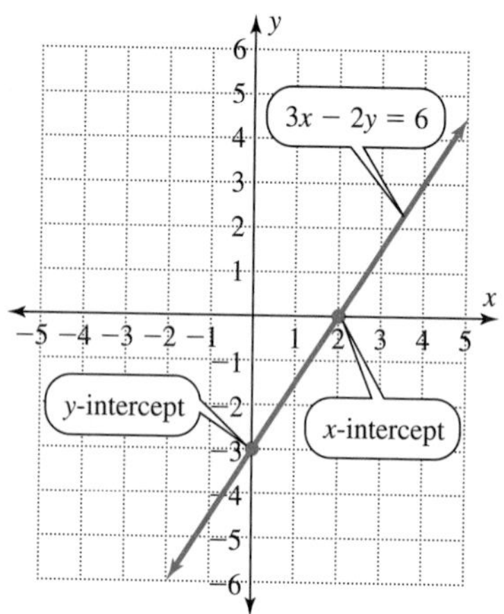

To sketch the graph of $3x - 2y = 6$, plot both the x- and y-intercepts and then draw the line through these two points.

The x- and y-intercepts are key points on the graph of an equation with important interpretations in real-world problems that are modeled by these equations. Thus it is useful to be able to calculate the coordinates of these points exactly rather than to approximate them from a graph. One example of this is the examination of the overhead cost and the break-even value of a profit function. The **fixed costs** or the **overhead costs** for a company can include insurance, rent, electricity, and other expenses that must be paid even when 0 units are produced. The **break-even value** is the number of units that must be produced and sold to pay for the overhead costs and production costs and to produce a profit of \$0. Example 4 examines both the overhead cost and the break-even value for the production at a pizza parlor.

EXAMPLE 4 Profit on Pizzas

A pizza parlor in a small town has daily overhead costs of \$150. They cover production costs and make a profit of \$2 on each pizza that they sell.

(a) Write an equation that gives the profit for this business when they sell x pizzas.

(b) Using the equation from part (a), determine the x-intercept of the graph of this equation and interpret this point.

(c) Using the equation from part (a), determine the y-intercept of the graph of this equation and interpret this point.

SOLUTION

Let x = number of pizzas sold
y = profit made by selling these pizzas

(a) VERBALLY

[Profit] = [Profit made by selling x pizzas] − [Overhead costs]

We first concisely state the word equation that we will use to from the algebraic equation.

ALGEBRAICALLY

$$y = 2x - 150$$

The profit per pizza is \$2, and the overhead costs are \$150.

(b) CALCULATION OF THE x-INTERCEPT

$$y = 2x - 150$$
$$0 = 2x - 150$$

To calculate the x-intercept, set $y = 0$ and solve for x.

$$0 + 150 = 2x - 150 + 150$$
$$150 = 2x$$
$$75 = x \quad \text{or}$$
$$x = 75$$

First add 150 to both sides of the equation, and then divide both sides of the equation by 2.

The x-intercept is (75, 0). Producing 75 pizzas will result in a profit of \$0. Thus 75 pizzas is the break-even value for this business.

The break-even value is the number of units that must be produced to pay for the overhead costs and production costs and produce a profit of \$0.

(c) CALCULATION OF THE y-INTERCEPT

$$y = 2x - 150$$
$$y = 2(0) - 150$$
$$y = 0 - 150$$
$$y = -150$$

To calculate the y-intercept, set $x = 0$ and solve for y.

The y-intercept is $(0, -150)$. Producing 0 pizzas will result in a loss of \$150. The given overhead costs for this business are \$150. ■

SELF-CHECK 2.6.2

Given the linear equation $2(x - y) = 4(5x + y) - 12$, calculate the

1. x-intercept **2.** y-intercept

The profit y for a manufacturer is given by $y = 3x - 420$, where x is the number of units produced.

3. Determine the overhead costs for this manufacturer.

4. Determine the break-even value for this manufacturer.

Many common formulas are given in a standard form and then rearranged depending on the variable that is needed for a particular application. In Example 5 we rearrange the formula for the perimeter of a trapezoid.

EXAMPLE 5 Solving the Formula for the Perimeter of an Isosceles Trapezoid for One Side

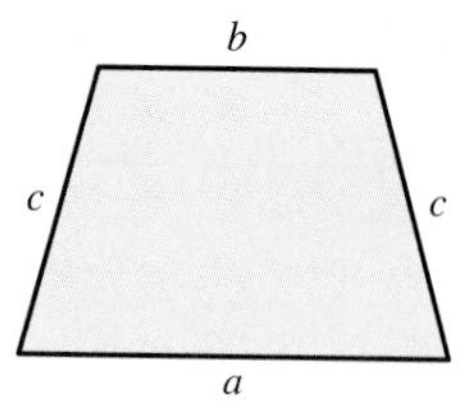

Solve $P = a + b + 2c$ for c.

SOLUTION

$$P = a + b + 2c$$

$$P - a - b = a + b + 2c - a - b$$

$$P - a - b = 2c$$

$$\frac{P - a - b}{2} = \frac{2c}{2}$$

$$\frac{P - a - b}{2} = c \quad \text{or}$$

$$c = \frac{P - a - b}{2}$$

An isosceles trapezoid (as shown to the right) has two parallel sides and two equal nonparallel sides.

Subtract a and b from both sides to isolate the term containing c on one side of the equation.

Then divide both sides of the equation by 2, the coefficient of c.

In Example 6 there are four variables: S, n, a_1, and a_n. We will solve this equation for a_1. This is an equation that we will revisit in Section 11.6.

EXAMPLE 6 Solving the Formula for the Sum of an Arithmetic Sequence for the First Term

Solve $S = \frac{n}{2}(a_1 + a_n)$ for a_1.

SOLUTION

$$S = \frac{n}{2}(a_1 + a_n)$$

$$\left(\frac{2}{n}\right)S = \left(\frac{2}{n}\right)\left(\frac{n}{2}\right)(a_1 + a_n)$$

Multiply both sides of the equation by $\frac{2}{n}$, the reciprocal of $\frac{n}{2}$.

$$\frac{2S}{n} = a_1 + a_n$$

$$\frac{2S}{n} - a_n = a_1 + a_n - a_n$$

To isolate a_1 on the right side, subtract a_n (read as "a sub n") from both sides of the equation.

$$\frac{2S}{n} - a_n = a_1 \quad \text{or}$$

$$a_1 = \frac{2S}{n} - a_n$$

SELF-CHECK 2.6.3

1. Solve $3(2x - 4y) = 10(2x - y) + 8$ for y.
2. Solve $P = a + b + c + 2d$ for b.
3. Solve $6ab + 3 = 3cd$ for d. Assume $c \neq 0$.

The equation $I = PRT$ allows us to directly calculate the simple interest I on a principal P at an interest rate R for a time T. In Example 7 we solve this equation for P. If we have a budgeted amount to spend on interest, then this form of the equation allows us to directly calculate the principal corresponding to that interest amount.

EXAMPLE 7 Solving the Simple Interest Formula for *P*

Solve $I = PRT$ for P. Then use a graphing calculator to prepare a table of principal amounts that can be borrowed when the monthly interest is \$33. Use interest rates of 7%, 7.5%, 8%, 8.5%, 9%, 9.5%, and 10%.

SOLUTION

If you are trying to solve for a specified variable, you may find it useful to highlight this variable. In more complicated equations this will help you clearly focus on the variable that needs to be isolated on the left side of the equation.

$$I = PRT$$

$$\frac{I}{RT} = \frac{PRT}{RT}$$ Divide both sides of the equation by RT to isolate the specified variable P on the right side of the equation.

$$\frac{I}{RT} = P \quad \text{or}$$

$$P = \frac{I}{RT}$$ This is an alternate form of the equation that has been solved for P.

$$P = \frac{33}{R\left(\frac{1}{12}\right)}$$ To determine the principal that can be paid for with a monthly interest payment of \$33, substitute \$33 for the interest I and $\frac{1}{12}$ for T for a time of $\frac{1}{12}$ year.

$$P = \frac{33(12)}{R\left(\frac{1}{12}\right)(12)}$$ Then simplify by multiplying the numerator and denominator by 12.

$$P = \frac{396}{R}$$

$$y = \frac{396}{x}$$ To use a graphing calculator, use x for the input variable instead of R and y for the specified output variable instead of P.

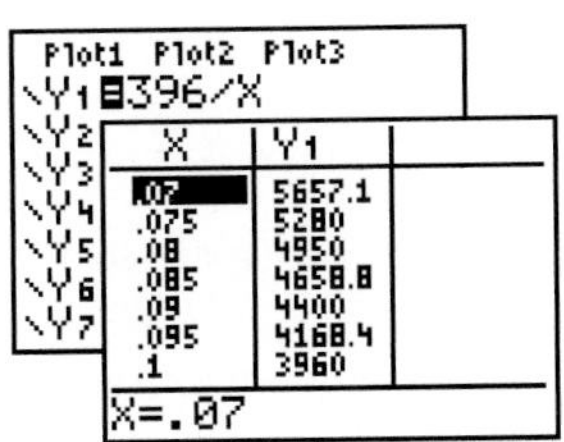

Enter the equation $y = \frac{396}{x}$ into a graphing calculator and then generate the table with an initial value of 0.07 for x with an increment of 0.005.

Answer: The amount of principal that can be borrowed for a \$33 monthly interest payment varies from \$5,657.10 at 7% to \$3,960.00 at 10%. The greater the interest rate, the less one can borrow on a fixed budget.

SELF-CHECK ANSWERS

2.6.1

1. $y = \frac{1}{2}x - 1$
2. 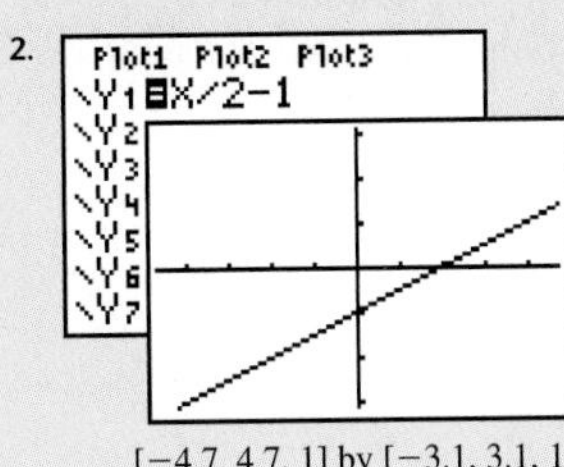

[−4.7, 4.7, 1] by [−3.1, 3.1, 1]

2.6.2

1. $\left(\frac{2}{3}, 0\right)$
2. (0, 2)
3. \$420
4. 140 units

2.6.3

1. $y = -7x - 4$
2. $b = P - a - c - 2d$
3. $d = \frac{2ab + 1}{c}$

USING THE LANGUAGE AND SYMBOLISM OF MATHEMATICS 2.6

1. Rewriting an equation to solve for one particular variable is called solving for a __________ variable.
2. The __________ property of multiplication over addition allows us to rewrite $2(3x + y)$ as $6x + 2y$.
3. In a profit function, the __________-__________ value produces a profit of $0.
4. The __________ costs for a company can include insurance, rent, electricity, and other fixed expenses when 0 units are produced.
5. The point on the graph of a linear equation with a y-coordinate of zero is the __________-intercept.
6. The point on the graph of a linear equation with an x-coordinate of zero is the __________-intercept.

EXERCISES 2.6

In Exercises 1–14, solve each equation for y.

1. $2x + y = 7$
2. $5x + y = 8$
3. $3x - y = 2$
4. $4x - y = 3$
5. $-6x + 3y = -9$
6. $-30x - 5y = 20$
7. $\frac{x}{2} - \frac{y}{4} = -1$
8. $\frac{x}{3} + \frac{y}{12} = -\frac{1}{6}$
9. $-0.3x - 0.1y = 0.2$
10. $-0.4x + 0.8y = -0.2$
11. $5x - 2y = x - 3y + 4$
12. $-6x + 5y = 8x - 2y - 21$
13. $2(3x - y + 1) = 3(4x - y - 2)$
14. $4(2x + 3y - 5) = 3(2x + 5y + 4)$

In Exercises 15–18, solve each equation for y. Then use a graphing calculator to complete a table of values for the x-values $-3, -2, -1, 0, 1, 2, 3$ and to graph the equation by using the window $[-4.7, 4.7, 1]$ by $[-3.1, 3.1, 1]$. (*Hint:* See Examples 1 and 2.)

15. $2x + 3y = 4x + 2y - 3$
16. $7x - 5y = 6(x - y) + 1$
17. $-5x + 4y = -2(2x - 3y) + 4$
18. $-4x + 7y = -5(x - y) + 3$

In Exercises 19–42, solve each literal equation for the variable specified. The context in which you may encounter these formulas is indicated in parentheses. Assume all variables are nonzero.

19. $A = lw$ for l (area of a rectangle)
20. $A = \frac{1}{2}bh$ for h (area of a triangle)
21. $C = 2\pi r$ for r (circumference of a circle)
22. $V = lwh$ for w (volume of a parallelepiped)
23. $V_1T_2 = V_2T_1$ for V_1 (Charles's law in chemistry)
24. $P_1V_1 = P_2V_2$ for P_2 (Boyle's law in chemistry)
25. $V = \frac{1}{3}\pi r^2 h$ for h (volume of a cone)

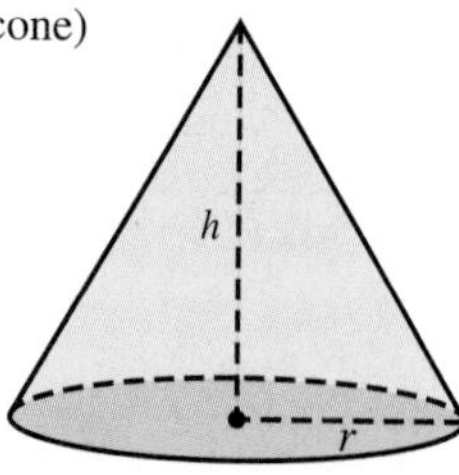

26. $P = a + b + c$ for a (perimeter of a triangle)
27. $V - E + F = 2$ for E (Euler's theorem)
28. $F = \frac{9}{5}C + 32$ for C (Fahrenheit and Celsius temperatures)
29. $A = \frac{1}{2}h(a + b)$ for b (area of trapezoid)
30. $A = \frac{1}{2}h(a + b)$ for a (area of trapezoid)
31. $C = \frac{5}{9}(F - 32)$ for F (Fahrenheit and Celsius temperatures)
32. $y = mx + b$ for x (slope-intercept form of a line)
33. $l = a + (n - 1)d$ for a (last term of an arithmetic sequence)
34. $l = a + (n - 1)d$ for d; $n \neq 1$ (last term of an arithmetic sequence)
35. $l = a + (n - 1)d$ for n (last term of an arithmetic sequence)
36. $S = 2\pi r^2 + 2\pi rh$ for h (surface area of a cylinder)

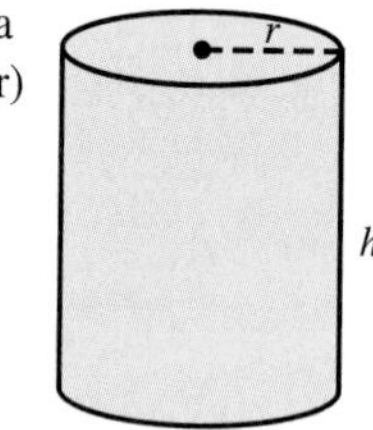

37. $S = \frac{a}{1 - r}$ for a (sum of an infinite geometric sequence)
38. $S = \frac{a}{1 - r}$ for r (sum of an infinite geometric sequence)
39. $y = mx + b$ for m (slope-intercept form of a line)
40. $S = \frac{n}{2}(a + l)$ for n (sum of an arithmetic sequence)
41. $P = 2l + 2w$ for w (perimeter of a rectangle)
42. $E = mc^2$ for m (Einstein's formula)
43. **Sales Tax** The sales tax T on a purchase amount P at a sales tax rate R is given by the formula $T = PR$.
    a. Solve this formula for P.
    b. Write the specific formula for P that is used for a store in a state with a sales tax rate of 8%.
    c. The following table gives various sales tax amounts collected by the store. Use a calculator to prepare a table of purchase amounts that correspond to these sales tax amounts.

SALES TAX T, $	PURCHASE AMOUNT P, $
100	
200	
300	
400	
500	
600	

44. **Dimensions of a Building** The area of a rectangular building is given by the formula $A = LW$.

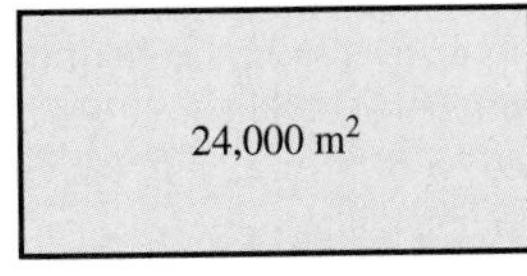

a. Solve this formula for W.
b. Write the specific formula that is used for a rectangular building that will have an area of 24,000 m^2.
c. The following table gives various possible lengths for this building. Use a calculator to prepare a table of widths that correspond to these lengths.

LENGTH L, m	WIDTH W, m
100	
110	
120	
130	
140	
150	

45. **Dimensions of a Building** The perimeter of a building is given by the formula $P = 2W + 2L$.
a. Solve this formula for W.
b. Write the specific formula that is used for a rectangular building that will have a perimeter of 800 m.
c. The following table gives various possible lengths for this building. Use a calculator to prepare a table of widths that correspond to these lengths.

LENGTH L, m	WIDTH W, m
100	
120	
140	
160	
180	
200	

46. **Isosceles Triangle** An isosceles triangle has two equal sides and two equal angles. In the isosceles triangle shown. $x + 2y = 180$ (degrees).

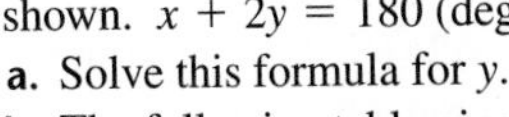

a. Solve this formula for y.
b. The following table gives the number of degrees for several possible values for angle x. Use a calculator to prepare a table for corresponding values of y.

x, DEGREES	y, DEGREES
5	
20	
30	
40	
60	
85	

47. **Profit on Ice Cream** An ice cream shop in a small town has daily overhead costs of \$120. It makes \$0.75 on each ice cream product that it sells.
a. Write an equation that gives the profit for this business when it sells x orders for its ice cream products.
b. Using the equation from part **a,** determine the x-intercept of the graph of this equation and interpret this point.
c. Using the equation from part **a,** determine the y-intercept of the graph of this equation and interpret this point.

48. **Profit on Coffee** A small drive-through coffee franchise has daily overhead costs of \$75. They make \$0.50 on each cup of coffee that they sell.
a. Write an equation that gives the profit for this business when they sell x cups of coffee.
b. Using the equation from part **a,** determine the x-intercept of the graph of this equation and interpret this point.
c. Using the equation from part **a,** determine the y-intercept of the graph of this equation and interpret this point.

49. **Simple Interest Formula** A bank uses an annual percent rate (APR) of R and a time of $\frac{1}{12}$ year to determine the monthly interest it pays customers on their certificates of deposit. Solve $I = PRT$ for R. Then use a graphing calculator to prepare a table of interest rates that will result in a monthly interest of \$50 for CDs in these amounts: \$10,000; \$12,000; \$14,000; \$16,000; \$18,000; \$20,000; and \$22,000 (*Hint:* See Example 7.)

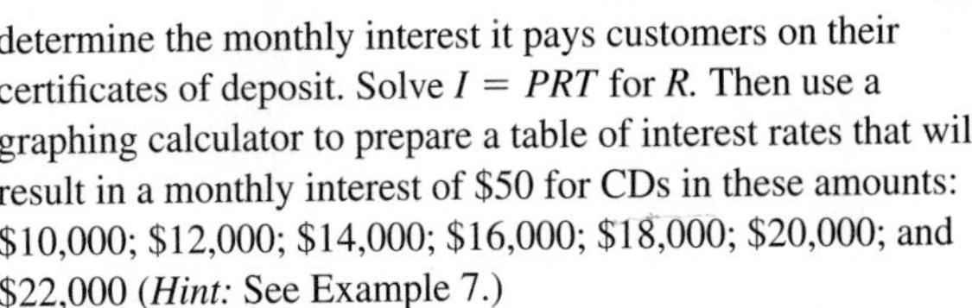

50. **Determining the Length of a Copper Wire** A copper wire of a fixed radius is really a very long cylinder whose volume is given by the formula $V = \pi r^2 h$, where r is the radius of the wire and the height of the cylinder is the length of the wire. Solve this formula for h and then complete this table for a wire of radius 0.8 cm. For each volume of copper delivered to a wire extrusion machine the table will give the length of wire that can be produced.

VOLUME OF COPPER V, cm^2	LENGTH OF WIRE L, cm	LENGTH OF WIRE L, m
1,000		
2,000		
3,000		
4,000		
5,000		

Group Discussion Questions

51. **Writing Mathematically** Write a paragraph describing why it is handier to have the formula $F = \frac{9}{5}C + 32$ for some applications, whereas the transposed equation $C = \frac{5}{9}(F - 32)$ is handier for other applications. Then complete the following table.

Celsius	Fahrenheit	
100°		The temperature at which water boils.
94°		Is this hot?
	98.6°	Normal body temperature.
	72°	A comfortable room temperature.
0°		Water freezes.
	0°	Zero on Fahrenheit scale.
		Both temperatures are equal.

52. **Challenge Questions** It is approximately 5 ft between the centers of the front tires on an SUV.

a. A driver of this SUV drives around so that the inside tire traces out a circle with a radius of 40 ft. How much farther does the outside tire travel than the inside tire?
b. If the radius of the circle formed by the inside tire is 400 ft, predict without calculating how much farther the outside tire travels.
c. Check the prediction made for part **b** by calculating this distance.
d. Explain the connection between the result in parts **a** and **c** by analyzing the formula for the circumference of a circle.

53. **Challenge Question** A sheet metal worker had two identical rectangular sheets of metal that he used to form two cylindrical heating ducts. One of the cylindrical ducts is 50 cm long with a circumference of 80 cm. The second cylindrical duct is 80 cm long with a circumference of 50 cm. If you unroll each cylinder as illustrated below, you can determine the dimensions of each original rectangular sheet of metal.

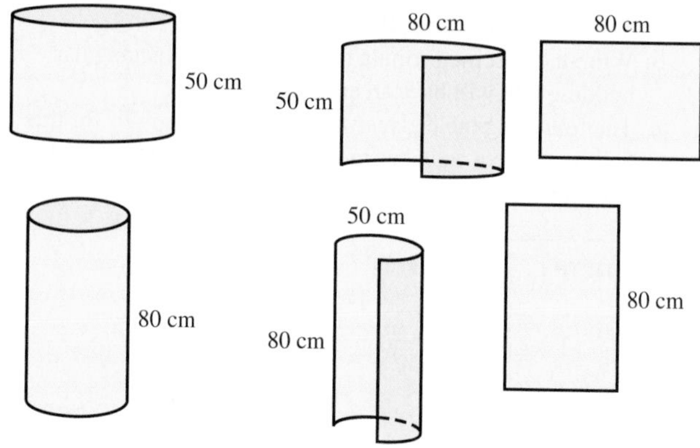

a. Calculate the volume of the cylinder that is 50 cm long. $(V = \pi r^2 h)$
b. Calculate the volume of the cylinder that is 80 cm long.
c. Are the volumes the same? If not, use the formula for volume to describe why one of these has a larger volume.
d. Determine the dimensions of each original rectangular sheet of metal.
e. Calculate the area of each sheet of metal.
f. Is the area of each sheet of metal the same? Use the formula $A = lw$ to explain your answer.

54. **Spreadsheet Exploration** In Example 7 we solved the simple interest formula $I = PRT$ for P. We did this because we wanted to calculate the principal we could borrow based on several different interests rates. Time was fixed at 1 month$\left(\frac{1}{12} \text{ year}\right)$, and the interest amount was fixed at \$33.00. Note that the values in column D were entered as $\frac{1}{12}$, but the decimal 0.0833 is displayed. Calculations involving column D will use $\frac{1}{12}$ rather than the decimal value that is displayed.

a. The formula in cell B2 is = **A2/(C2*D2)**. Determine the formula and value for each of the remaining cells in column B.

	A	B	C	D
1	Interest	Principal	Rate	Time
2	$33.00	$5,657.14	0.070	0.0833
3	$33.00		0.075	0.0833
4	$33.00		0.080	0.0833
5	$33.00		0.085	0.0833
6	$33.00		0.090	0.0833
7	$33.00		0.095	0.0833
8	$33.00		0.100	0.0833
9				

Cell D2 = **1/12**

b. Create a table similar to the one given, using the same values in columns C and D but with a fixed interest payment of \$50.00. Complete column B to determine the principal that can be borrowed under these conditions.

Section 2.7 Proportions and Direct Variation

Objectives:

16. Use proportions to solve word problems.
17. Solve problems involving direct variation.

The ability to solve real-world problems often requires us to create a mathematical model of the problem. Linear equations are used to model many problems. In this section we will use proportions to form linear equations to model some applications.

A classic type of problem is to determine the height of one object by comparing it to the height of a second object. We will use proportions to accomplish this. A **proportion** is an equation that states that two ratios are equal. For example, the proportion $\frac{a}{b} = \frac{c}{d}$ is read "a is to b as c is to d." The four numbers a, b, c, and d are called the **terms of the proportion:** a is the first term, b is the second term, c is the third term, and d is the fourth term. The first and fourth terms are called the **extremes,** and the second and third terms are called the **means.**

The word *mean* has now been used in two contexts: as an average of a set of values and as one of the middle terms of a proportion.

Proportion

ALGEBRAICALLY	VERBALLY	NUMERICAL EXAMPLE
$\frac{a}{b} = \frac{c}{d}$ or $a:b = c:d$	This proportion is read "a is to b as c is to d." The extremes are a and d, and the means are b and c.	$\frac{2}{5} = \frac{40}{100}$ or $2:5 = 40:100$

Since proportions are equations, we can solve them by using the same rules and procedures that apply to all equations.

A Mathematical Note

Emmy Noether (1882–1935) was called by Albert Einstein the greatest of all the women who were creative mathematical geniuses. Emmy completed her doctoral dissertation in algebra at the University of Erlangen in 1907. She taught at the University of Göttingen in Germany from 1915 until 1933. She then moved to the United States and taught at Bryn Mawr for 2 years until her death in 1935.

EXAMPLE 1 Solving a Proportion

Solve $\frac{x}{3} = \frac{5}{4}$.

SOLUTION

$$\frac{x}{3} = \frac{5}{4}$$

$$12\left(\frac{x}{3}\right) = 12\left(\frac{5}{4}\right)$$ LCD = 3 · 4 = 12
Multiply both sides of the equation by the LCD of 12.

$$4x = 3(5)$$

$$4x = 15$$

$$\frac{4x}{4} = \frac{15}{4}$$ Divide both sides of the equation by 4.

$$x = \frac{15}{4}$$

Answer: $x = \frac{15}{4}$ Does this solution check?

EXAMPLE 2 Solving a Proportion

Solve $\frac{35}{6m} = \frac{5}{9}$.

SOLUTION

$$\frac{35}{6m} = \frac{5}{9}$$

$6m = 2 \cdot 3 \cdot m$
$9 = 3 \cdot 3$
LCD $= 2 \cdot 3 \cdot 3 \cdot m = 18m$

$$18m\left(\frac{35}{6m}\right) = 18m\left(\frac{5}{9}\right)$$

Multiply both sides by the LCD of $18m$.

$$\frac{\overset{3}{\cancel{18m}}}{\underset{1}{\cancel{6m}}}(35) = \frac{\overset{2m}{\cancel{18m}}}{\underset{1}{\cancel{9}}}(5)$$

Then simplify both sides of the equation.

$$3(35) = (2m)(5)$$
$$105 = 10m$$
$$\frac{105}{10} = \frac{10m}{10}$$

To solve for m, divide both sides by 10.

$$10.5 = m \quad \text{or}$$

Answer: $m = 10.5$

Does this value check?

SELF-CHECK 2.7.1

1. Solve $\frac{11}{6} = \frac{x}{3}$.
2. Solve $\frac{10}{x} = \frac{8}{12}$.
3. Solve $2 : 7 = x : 35$.

Ratios and proportions are so useful that many employment tests, including civil service tests, have problems that test your ability to use proportions. A typical problem involves a constant ratio under two different situations; that is,

$$\boxed{\text{Ratio for first situation}} = \boxed{\text{Ratio for second situation}}.$$

This type of problem is illustrated in Example 3, which compares the ratio of distances on a first map reading to distances on a second map reading.

EXAMPLE 3 Distance Using a Map

On a map 2 cm represents a distance of 230 km. What distance corresponds to 6.5 cm on the map?

SOLUTION

Let d = distance in kilometers corresponding to 6.5 cm.

VERBALLY

$$\frac{\text{First map reading}}{\text{First distance}} = \frac{\text{second map reading}}{\text{second distance}}$$

First stating the strategy as a verbal equation is an excellent way to start a word problem.

A common question regarding Example 3 is to inquire if it is OK to form the ratio $\frac{\text{First map reading}}{\text{Second map reading}} = \frac{\text{first distance}}{\text{second distance}}$. This is a good question and the answer is yes. More generally, the equation $\frac{a}{b} = \frac{c}{d}$ is equivalent to $\frac{a}{c} = \frac{b}{d}$.

ALGEBRAICALLY

$$\frac{2}{230} = \frac{6.5}{d}$$

Use the given values to translate the verbal equation into an algebraic model of problem.

$$230d\left(\frac{2}{230}\right) = 230d\left(\frac{6.5}{d}\right)$$

Multiply both sides of the equation by the LCD of $230d$.

$$2d = 230(6.5)$$

$$\frac{2d}{2} = \frac{1{,}495}{2}$$

Divide both sides by 2.

$$d = 747.5$$

Answer: On the map 6.5 cm represents 747.5 km.

Does this answer seem reasonable?

SELF-CHECK 2.7.2

1. A recipe for old-fashioned vegetable soup suggests using 4 cups of water to prepare a serving for 6 people. How much water should be used to prepare a serving for 10?

Describing the relationship among variables is a very important part of mathematics. Some relationships are so common that it is not surprising that we have developed multiple ways to describe the relationships. The concept of direct variation, which we examine now, is closely related to arithmetic sequences, linear equations, and ratios and proportions.

Direct Variation

If x and y are real variables and k is a real constant with $k \neq 0$, then:

VERBALLY	ALGEBRAICALLY	NUMERICAL EXAMPLE
y varies directly as x with the constant of variation k.	$y = kx$ *Example:* $y = 3x$	see table below

x	$y = 3x$
1	3
2	6
3	9
4	12
5	15

EXAMPLE 4 Translating Statements of Variation

SOLUTION

(a) Translate $C = \pi d$ into a verbal statement of variation.

C varies directly as d with the constant of variation π.

The circumference of a circle varies directly as the diameter with π being the constant of variation.

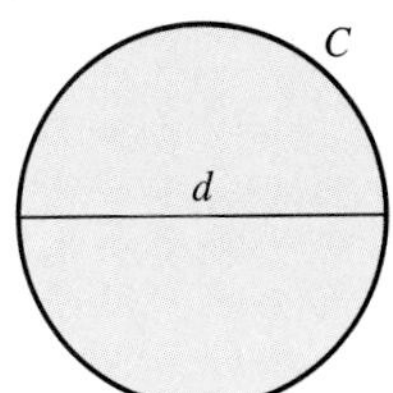

(b) As a result of Hooke's law, we know that the distance d a spring stretches varies directly as the mass m attached to the spring. Translate this statement of variation into an algebraic equation.

$d = km$

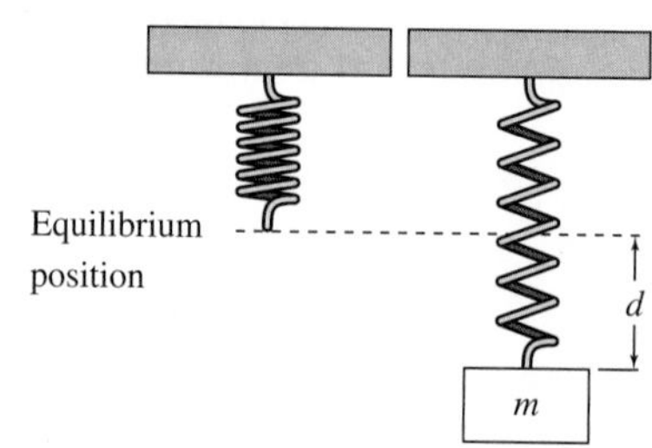

We will revisit direct variation in Section 8.6 when we examine inverse and joint variation.

Sometimes the constant of variation is so important that this constant becomes well known or has a special symbol assigned to it. An example of this is π, which is the ratio of the circumference of a circle to its diameter. In Example 5 we solve for the constant of variation that represents the exchange rate between two currencies.

As we shall see in Example 6, problems involving direct variation can be solved by using proportions. However, Example 5 illustrates that using direct variation is more useful for situations where the constant of variation has a significant meaning or where we want to generate a table of values.

EXAMPLE 5 Determining a Currency Exchange Rate

Traveling in New Zealand during 2007, one could exchange 800 USD (U.S. dollars) for 1,176 NZD (New Zealand dollars). The number of NZD received varies directly as the number of USD exchanged. The constant of variation is called the exchange rate.

(a) What was the exchange rate at this time?

(b) Use the exchange rate to write an equation relating the number of NZD received to the number of USD exchanged.

(c) Use this equation and a graphing calculator to create a table of values displaying exchange amounts in \$50 increments starting at \$50 USD.

SOLUTION

Let U = number of U.S. dollars
N = number of New Zealand dollars
k = constant of variation

VERBALLY

The number of NZD varies directly as the number of USD.

Verbally state a precise word equation that is easily translated into algebraic form.

ALGEBRAICALLY

$$N = kU$$

Translate the word equation into algebraic form by using the variables identified above.

$$1{,}176 \text{ NZD} = k(800 \text{ USD})$$

Substitute in the given values for N and U.

$$\frac{1{,}176 \text{ NZD}}{800 \text{ USD}} = k$$

Then solve for k, the exchange rate.

$$k \approx 1.47$$

This means we would receive 1.47 NZD for each 1 USD exchanged.

(a) The exchange rate was 1.47 NZD for 1 USD.

(b) Equation: $N = 1.47U$

(c)

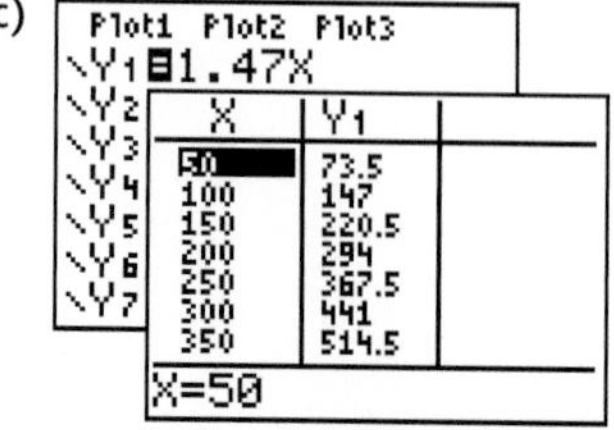

To generate a table of values, enter the equation $N = 1.47U$ as $y = 1.47x$. Then use 50 for the **TblStart** and Δ**Tbl** values.

SELF-CHECK 2.7.3

1. A traveler exchanged 350 USD (U.S. dollars) for 402.5 CAD (Canadian dollars). The number of CAD received varies directly as the number of USD exchanged.
   (a) What was the exchange rate at this time?
   (b) Use the exchange rate to write an equation relating the number of CAD received to the number of USD exchanged.
   (c) Use this equation to create a table of values displaying exchange amounts in \$50 increments starting at 50 USD.

(x_1, y_1) represents one input-output pair, and (x_2, y_2) represents a second input-output pair.

Example 6 shows how to solve a problem involving direct variation without calculating the constant of variation. If $y = kx$, then $k = \frac{y}{x}$. When y varies directly as x, k is a constant while x and y vary. For two sets of values we can denote the two x-values by x_1 (read "x sub 1") and x_2 (read "x sub 2"). Likewise the two y-values can be denoted by y_1 and y_2. If $k = \frac{y_1}{x_1}$ and $k = \frac{y_2}{x_2}$ for two sets of values, then $\frac{y_1}{x_1} = \frac{y_2}{x_2}$. This can be described by saying that *x and y are directly proportional when y varies directly as x.*

EXAMPLE 6 Using Direct Variation to Form a Proportion

If y varies directly as x, and y is 12 when x is 9, find y when x is 15.

SOLUTION

$$\frac{y_1}{x_1} = \frac{y_2}{x_2}$$ x and y are directly proportional when y varies directly as x.

$$\frac{12}{9} = \frac{y_2}{15}$$ Substitute 12 for y_1, 9 for x_1, and 15 for x_2.

$$\frac{4}{3} = \frac{y_2}{15}$$ Reduce the left side of the equation by dividing both the numerator and the denominator by 3.

$$15\left(\frac{4}{3}\right) = 15\left(\frac{y_2}{15}\right)$$ Multiply both sides by the LCD of 15.

$$5(4) = y_2$$
$$20 = y_2 \quad \text{or}$$
$$y_2 = 20$$ Then simplify both sides of the equation.

Check: $\frac{12}{9} \stackrel{?}{=} \frac{20}{15}$

$\frac{4}{3} \stackrel{?}{=} \frac{4}{3}$ checks. Since $\frac{y_1}{x_1} = \frac{y_2}{x_2}$, x and y are directly proportional.

Answer: $y = 20$ when $x = 15$.

Ratios and proportions are the basis for many indirect measurements. Example 7 illustrates how to determine the height of a tree by measuring the length of a shadow. This same technique has been used to determine the heights of mountains on Mars.

EXAMPLE 7 Modeling the Height of a Tree by Using Its Shadow

At a fixed time of the day the length of a shadow varies directly as the height of the object casting the shadow. If a person 2 m tall casts a 5-m shadow, how tall is a fir tree that casts a 40-m shadow?

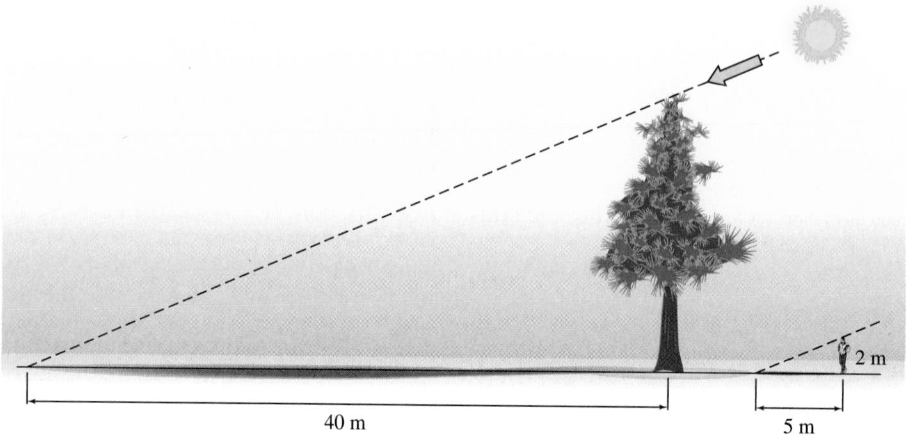

SOLUTION

Let t = height of tree in meters.

When y varies directly as x, increasing magnitudes of x result in increasing magnitudes of y. Likewise, decreasing magnitudes of x result in decreasing magnitudes of y. For example, a taller tree produces a longer shadow, and a shorter tree produces a shorter shadow.

VERBALLY

$$\frac{\text{Height of person}}{\text{Length of person's shadow}} = \frac{\text{height of tree}}{\text{length of tree's shadow}}$$

Verbally we form an equation based on the proportion given by this statement of direct variation. Also note that this proportion could be formed by using corresponding sides of similar right triangles.

ALGEBRAICALLY

$$\frac{2}{5} = \frac{t}{40}$$

Substitute the given values to translate the word equation into an algebraic model of the problem.

$$40\left(\frac{2}{5}\right) = 40\left(\frac{t}{40}\right)$$

Multiply both sides of the equation by the LCD of 40.

$$8(2) = t$$
$$16 = t \quad \text{or}$$
$$t = 16$$

Then simplify both sides of the equation.

Does this answer seem reasonable?

Answer: The tree is 16 m tall.

SELF-CHECK 2.7.4

1. If y varies directly as x and y is 16 when x is 20, find y when x is 25.
2. Rework Example 7, assuming that the tree casts a 35-m shadow.

Example 8 examines the concept of direct variation from multiple perspectives. Note the connection of this concept to linear equations, arithmetic sequences, and straight lines. The sequence of y-values shown in this table forms an arithmetic sequence with a common difference that is equal to the constant of variation.

EXAMPLE 8 Using Multiple Perspectives to Examine Direct Variation

Examine the following statement of variation verbally, algebraically, numerically, and graphically: y varies directly as x with constant of variation 2.

SOLUTION

VERBALLY

y varies directly as x with constant of variation 2.

ALGEBRAICALLY

$y = 2x$

NUMERICALLY

x	$y = 2x$
1	2
2	4
3	6
4	8
5	10

GRAPHICALLY

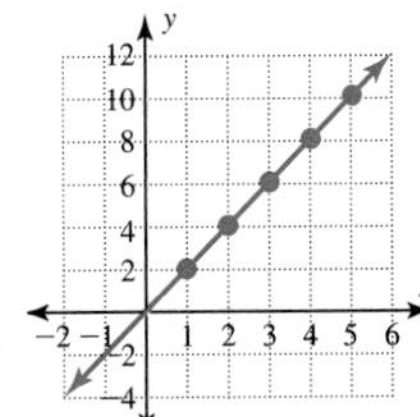

Note that the y-values in this table form an arithmetic sequence with a common difference of 2. The graph of $y = 2x$ is a line that goes up 2 units for every 1-unit move to the right. ■

SELF-CHECK ANSWERS

2.7.1

1. $x = \frac{11}{2}$ 2. $x = 15$
3. $x = 10$

2.7.2

1. $6\frac{2}{3}$ cups

2.7.3

1. (a) 1.15 CAD for 1 USD
(b) $C = 1.15U$

(c)

```
Plot1 Plot2 Plot3
\Y1=1.15X
\Y2=
\Y3=
\Y4=
\Y5=
\Y6=
\Y7=

  X   | Y1
  50  | 57.5
  100 | 115
  150 | 172.5
  200 | 230
  250 | 287.5
  300 | 345
  350 | 402.5
X=50
```

2.7.4

1. $y = 20$
2. The tree is 14 m tall.

USING THE LANGUAGE AND SYMBOLISM OF MATHEMATICS 2.7

1. An equation that states two ratios are equal is called a __________.
2. In the proportion $\frac{a}{b} = \frac{c}{d}$, a, b, c, and d are called the __________ of the proportion.
3. The means of $\frac{a}{b} = \frac{c}{d}$ are __________ and __________.
4. The extremes of $\frac{a}{b} = \frac{c}{d}$ are __________ and __________.
5. If x and y are directly proportional, then y varies __________ as x.
6. If y varies directly as x, then the equation expressing this is a __________ equation.
7. If y varies directly as x and x increases in magnitude, then y __________ in magnitude.
8. If y varies directly as x and x decreases in magnitude, then y __________ in magnitude.
9. If y varies directly as x, then inputting consecutive natural numbers for x will produce output y-values that form an __________ sequence.

EXERCISES 2.7

In Exercises 1–22, solve each proportion.

1. $\frac{x}{20} = \frac{3}{5}$
2. $\frac{x}{16} = \frac{3}{2}$
3. $\frac{3}{5} = \frac{6}{x}$
4. $\frac{2}{7} = \frac{10}{x}$
5. $\frac{5x}{2} = \frac{45}{6}$
6. $\frac{11x}{3} = \frac{154}{21}$
7. $\frac{9}{2y} = \frac{27}{66}$
8. $\frac{6}{3y} = \frac{5}{6}$

9. $\frac{5}{6} = \frac{40}{3y}$

10. $\frac{13}{5} = \frac{52}{10y}$

11. $a : 3 = 4 : 6$

12. $7z : 10 = 7 : 2$

13. $5 : 2v = 3 : 4$

14. $5 : 11 = 15 : b$

15. $\frac{x + 1}{5} = \frac{8}{10}$

16. $\frac{x + 1}{20} = \frac{3}{5}$

17. $\frac{z + 2}{3} = \frac{z}{4}$

18. $\frac{t - 1}{5} = \frac{t}{2}$

19. $\frac{x}{3} = \frac{2x - 3}{5}$

20. $\frac{3w}{4} = \frac{w - 5}{1}$

21. $\frac{2v - 1}{5} = \frac{2v - 5}{1}$

22. $\frac{x - 1}{6} = \frac{2x - 4}{9}$

In Exercises 23 and 24, write an equation for each statement of variation, using k as the constant of variation.

23. At a fixed speed the distance d that a car travels varies directly as the time t it travels.

24. At a fixed pressure the volume V of a gas varies directly as the absolute temperature T.

In Exercises 25 and 26, write a statement of variation for each equation, assuming k is the constant of variation.

25. $v = kw$

26. $b = ka$

In Exercises 27–32, solve each of these problems involving direct variation.

27. If v varies directly as w, and $v = 22$ when $w = 77$, find v when $w = 35$.

28. If m varies directly as n, and $m = 12$ when $n = 20$, find n when $m = 18$.

29. If p varies directly as g, and p is 0.375 when g is 1, find g when p is 1.5.

30. If b varies directly as d, and b is 0.3125 when d is 1, find b when d is 0.4.

31. If v varies directly as w, and $v = 12$ when $w = 15$, find the constant of variation k.

32. If b varies directly as a, and $b = \frac{2}{3}$ when $a = \frac{4}{5}$, find the constant of variation k.

33. **Sales Commission Rate** The commission earned by a salesman varies directly as his gross sales. In one month, he earned a commission of \$2,700 on \$18,000 in gross sales.
   a. What was the sales commission rate?
   b. Use the sales commission rate to write an equation relating the commission earned to the amount of gross sales for that month.
   c. Use this equation to complete the table of values.

	A	B
1	Gross Sales	Commission
2	\$16,000	
3	\$18,000	
4	\$20,000	
5	\$22,000	
6	\$24,000	
7	\$26,000	
8	\$28,000	
9		

34. **Running Rate** The distance run by a runner varies directly as the amount of time she runs. She runs 3 mi in 30 minutes.
   a. What was her running rate in miles per hour?
   b. Use the running rate to write an equation relating the distance she ran to her running time.
   c. Use this equation to complete the table of values.

	A	B
1	Time (h)	Distance (mi)
2	0.5	
3	1.0	
4	1.5	
5	2.0	
6	2.5	
7	3.0	
8	3.5	
9		

35. **Currency Conversion Rate** The number of EUR (euros) received varies directly as the number of USD (U.S. dollars) exchanged. In 2006, one could exchange 500 USD for 410 EUR.
   a. What was the exchange rate at that time?
   b. Use the exchange rate to write an equation relating the number of EUR received to the number of USD exchanged.
   c. Use this equation and a graphing calculator to create a table of values displaying exchange amounts in \$50 increments starting at 50 USD.

36. **Currency Conversion Rate** The number of AUD (Australian dollars) received varies directly as the number of USD (U.S. dollars) exchanged. In 2006, one could exchange 700 USD for 938 AUD.
   a. What was the exchange rate at that time?
   b. Use the exchange rate to write an equation relating the number of AUD received to the number of USD exchanged.
   c. Use this equation and a graphing calculator to create a table of values displaying exchange amounts in \$50 increments starting at 50 USD.

37. **Weight of Jet Fuel** The weight of jet fuel is approximately 6.7 lb/gal. The amount of jet fuel used at cruising altitude by an MD80 passenger jet is approximately 11.4 gal/min. The weight of fuel used varies directly as the time.
   a. Write an equation that gives W, the pounds of fuel used, in terms of t, the time in minutes.
   b. What does the constant of variation represent?

38. **Snow Cone Revenue** There are 128 one-ounce servings of snow cone juice in 1 gal. Each snow cone serving produces revenue of \$1.75. The revenue generated by selling snow cones varies directly as the number of gallons of juice used in a day.
   a. Write an equation that gives D, the dollars of revenue generated, in terms of G, the gallons of snow cone juice served.
   b. What does the constant of variation represent?

Multiple Representations

In Exercises 39–42, represent each direct variation algebraically, numerically, verbally, or graphically as described in each problem. (*Hint:* See the following example.)

ALGEBRAICALLY	NUMERICALLY	GRAPHICALLY	VERBALLY
$y = -2x$	x | y 1 | -2 2 | -4 3 | -6 4 | -8 5 | -10		y varies directly as x with constant of variation -2.

39. Represent the direct variation "y varies directly as x with constant of variation $\frac{1}{2}$" algebraically, numerically, and graphically.

40. Represent the direct variation $y = -\frac{x}{2}$ numerically, graphically, and verbally.

41. Represent the direct variation $y = 3x$ numerically, graphically, and verbally.

42. Represent the direct variation "y varies directly as x with constant of variation -1" algebraically, numerically, and graphically.

Recipe Proportions (Exercises 43 and 44)

43. A recipe for 50 people used 3 cups of sugar. How many cups of sugar are needed for 75 people?

44. A recipe for six adults called for $\frac{3}{4}$ tsp of salt. If this recipe is used to cook for four people, how much salt is required?

Similar Geometric Figures (Exercises 45–48)

The lengths of corresponding sides of similar geometric figures are directly proportional.

45. The triangles shown in the figure are similar. Find the length of the side labeled a.

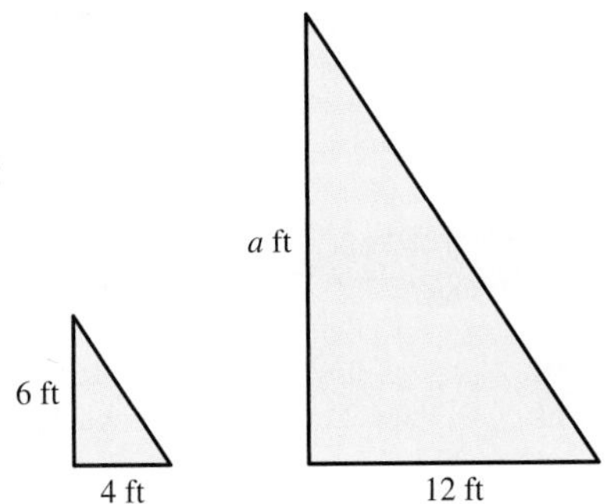

46. The two rectangles shown in the figure are similar. Find the length of the side labeled w.

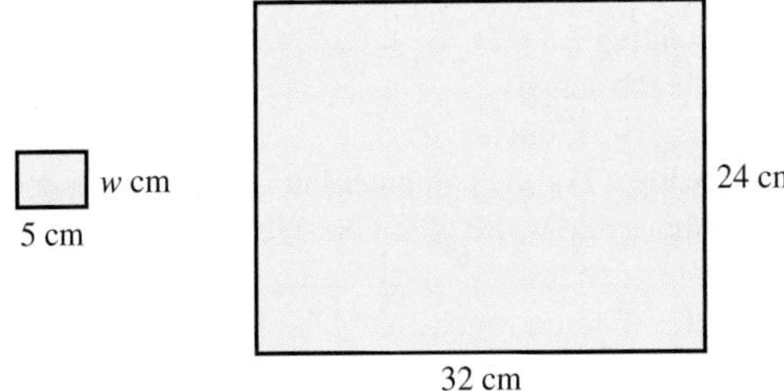

47. Use similar triangles and proportional sides to find the value of x.

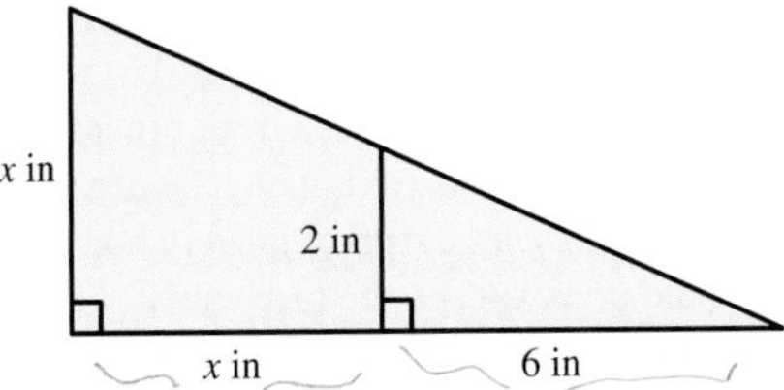

48. Use similar triangles and proportional sides to find the value of x.

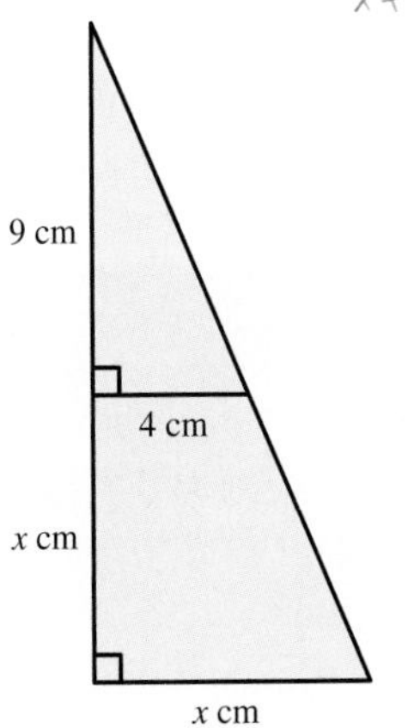

49. Microscope Magnification Under a microscope an insect antenna that measures 1.5 mm appears to be 7.5 mm. Find the apparent length under this magnification of an object that is 2 mm long.

50. Distance on a Map If 6 cm on a map represents 300 km, how much does 1 cm represent?

51. Defective Lightbulbs A factory that produces lightbulbs finds that 2.5 out of every 500 bulbs are defective. In a group of 10,000 bulbs, how many would be expected to be defective?

52. **Gold Ore** A mining company was able to recover 3 oz of gold from 10 tons of ore. How many tons would be needed to recover 12 oz of gold?
53. **Bricks for a Wall** If 118 bricks are used in constructing a 2-ft section of a wall, how many bricks will be needed for a similar wall that is 24 ft long?
54. **Sand in Concrete** If 11 ft^3 of sand is needed to make 32 ft^3 of concrete, how much sand is needed to make 224 ft^3 of concrete?
55. **Scale Drawing** An architect's sketch of one room in a new computer lab is shown here. The dimensions of the sketch are given in centimeters. The architect used a constant scale factor of 1 cm representing 2.5 m. What will be the actual dimensions of the room?

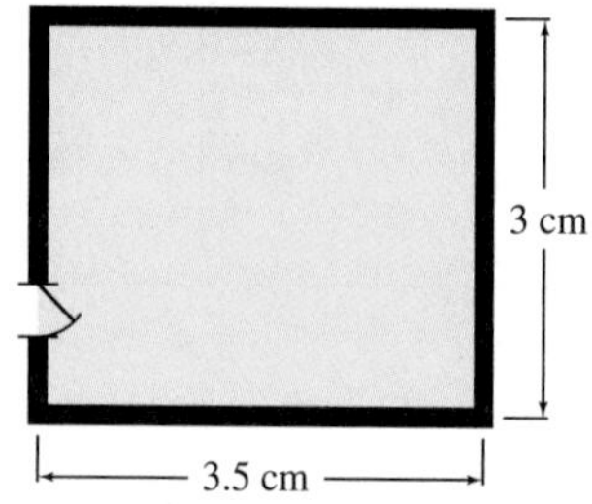

56. **Scale Drawing** For the computer lab sketch in Exercise 55 determine the actual width of the door if the door in the sketch is 0.4 cm wide.
57. **Grass Seed** A homeowner used 3 lb of grass seed for 100 ft^2. How many pounds would be needed for 450 ft^2 if the amount of seed used varies directly as the area covered?
58. **Quality Control** A quality control inspector found 4 defective computer chips in the 250 that were tested. How many defective chips would be expected in a shipment of 10,000 chips?
59. **Face Cards in Blackjack** A blackjack dealer in a casino dealt 39 cards. Each of the decks from which he dealt had 12 face cards for every 52 cards. Approximately how many of these 39 cards would you expect to be face cards?
60. **Maglev Train** Plans for an operational scale model of a maglev (magnetically levitating) train call for a scale of 1 to 50. The two-car train will be about 28 in long and will sell to hobbyists for \$2,000 to \$5,000. What is the length of the maglev train on which this scale model is based?
61. **Width of a River** Use the dimensions shown in the following figure to determine the width of the river. Triangles AB_1C_1 and AB_2C_2 are similar.

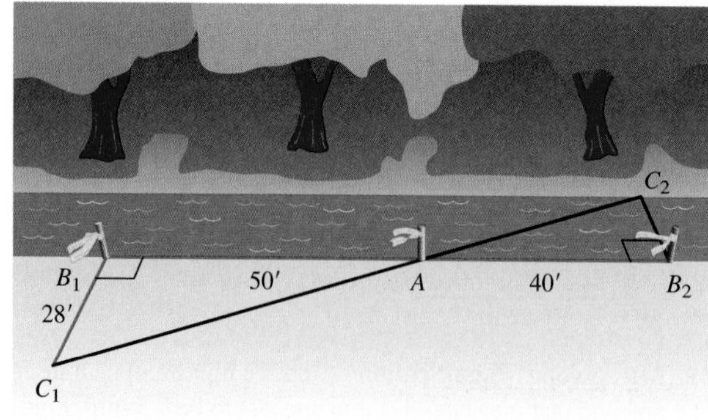

62. **Dosage Instructions** A pharmaceutical company testing a new drug finds that the ideal dosages appear to be 292.5 mg for a patient weighing 45 kg; 390 mg for a patient weighing 60 kg; and 487.5 mg for a patient weighing 75 kg. Determine the correct dosage instructions for this drug.
63. **Photograph Enlargement** A photograph is 7 cm wide and 10 cm long. What would be the width of the photograph if it were enlarged so that the length was 25 cm?
64. **Photocopier Reduction** A photocopier is to be used to reduce the printed material on a page 8.5 in wide and 11 in high so that it will occupy a page 4.25 in wide. How high will be the reduced page of material?
65. **Height of a Flagpole** The length of a shadow varies directly as the length of the object casting the shadow. A boy 5 ft tall casts a 6-ft shadow. How high is the flagpole beside him that casts a 21-ft shadow? (See the figure.)

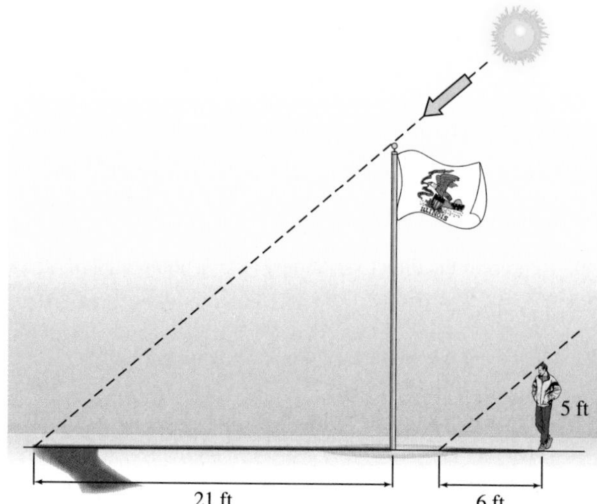

66. **Height of a Building** An architect who is designing a building is concerned with the shadow that will be created by this building during a certain time of the day. If a yardstick (3 ft) casts a 10-ft shadow, what height of building would cast a 400-ft shadow?
67. **Height of a Television Tower** The slope of a cable is the ratio of the rise to the run. A woman wished to determine the height of a television tower. The distance (or run) from the bottom of the cable to the bottom of the tower measured 40 ft. She then measured a rise of 8 ft and found a run of 6 ft. What is the height of the tower whose tip is at the top end of the cable?
68. **Scale Drawing of a House** The scale drawing of the house in the figure reveals that the roof rises 3 in over a run of 12 in. When the roof is actually built, how far will the roof rise over a run of 16 ft?

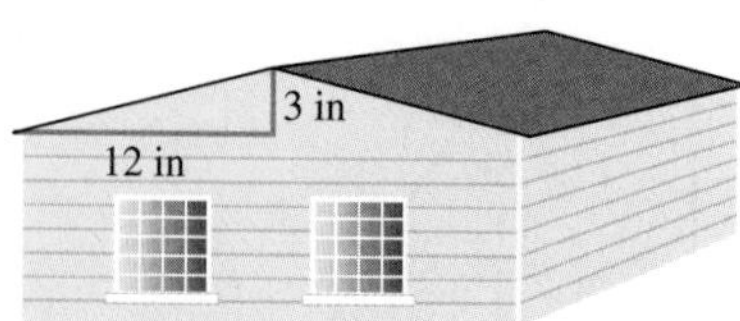

69. **Baseball ERA** A baseball pitcher's ERA (earned run average) is computed by determining the number of earned runs he allowed while recording 27 outs (one game). If a pitcher allowed two earned runs while recording only six outs in his first game, what would his ERA be?
70. **Baseball ERA** In his first major league game a pitcher allowed five earned runs while recording nine outs. What was his ERA at the end of this game? (See Exercise 69.)

71. **Doses of Medicine** The child dosage of a certain medicine is $\frac{2}{3}$ oz. A nurse has a bottle that contains 16 oz of this medicine. How many children's doses of this medicine does this bottle contain?

72. **Distance Traveled by a Truck Tire** The radius of a truck tire is 50 cm.
   a. To the nearest tenth of a centimeter, how far will the truck travel when the tire makes 1 revolution?
   b. To the nearest hundred thousandth of a kilometer, how far will the truck travel when the tire makes 1 revolution.
   c. Let x represent the number of revolutions the tire turns. Write an equation that gives the distance that the truck travels in kilometers.
   d. Determine the distance traveled when the tire makes 500 revolutions.
   e. Determine the number of revolutions necessary to travel 1 km.

73. **Similar Triangles** Similar triangles have the same shape, and the corresponding angles have the same measure. The lengths of the corresponding sides are proportional. The sum of the measures of the interior angles of any triangle is 180°. Triangles ABC and $A'B'C'$ are similar.

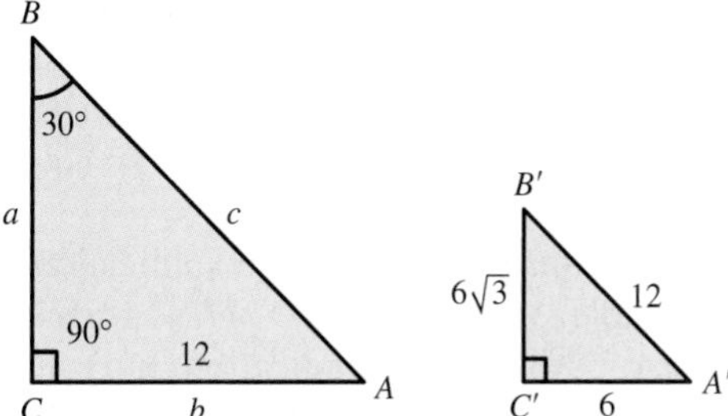

   a. Determine the measure of angle A.
   b. Determine the measure of angle A'.
   c. Determine the measure of angle B'.
   d. Determine the length of side c in triangle ABC.
   e. Determine the length of side a in triangle ABC.

74. **Health Care Costs** A patient has an insurance policy that requires a 10% copayment by the patient for all medical expenses. Thus, the patient's responsibility varies directly with the medical bill with a constant variation of 0.10. Let x represent the dollar cost of a medical bill.
   a. Write an equation that gives the patient's responsibility for this medical bill.
   b. Determine the patient's responsibility for a $500 medical bill.
   c. Determine the medical bill if the patient's responsibility is $500.

75. **Health Care Costs** A patient has an insurance policy that requires a 20% copayment by the patient for all medical expenses. Thus, the insurance company's responsibility varies directly with the medical bill with a constant of variation of 0.80. Let x represent the dollar cost of a medical bill.
   a. Write an equation that gives the insurance company's responsibility for this patient.
   b. Determine the insurance company's responsibility for a $500 medical bill.
   c. Determine the medical bill if the insurance company's responsibility is $500.

In Exercises 76 and 77, y varies directly as x. Use the given table of values to determine the constant of variation.

76.

x	3	4	5	6	7
y	4.5	6	7.5	9	10.5

77.

x	3	4	5	6	7
y	2.4	3.2	4	4.8	5.6

Group Discussion Questions

78. **Solve Then Explain** Use the proportion $\frac{x}{a} = \frac{b}{x}$ from the semicircle shown to answer the following question. An architect is planning a semicircular arch over a hallway as illustrated. The minimum height of the hallway is at the sides. The architect has set this measurement from building codes to be 8 ft. To design this arch, the architect has selected an imaginary point 4 ft from the bottom of the wall for this semicircle to pass through, if the arch were to continue.

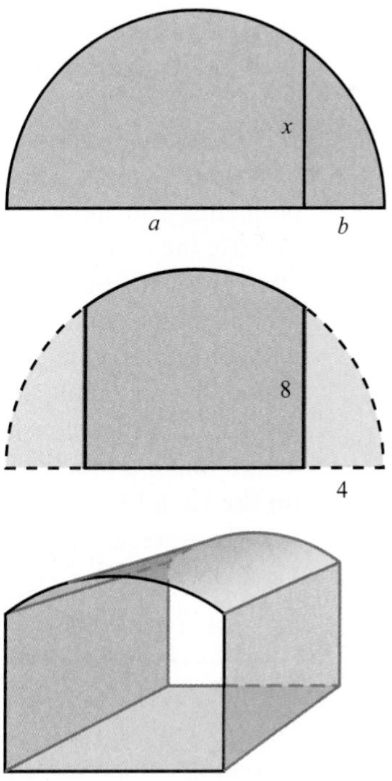

   a. Determine the height of the hallway at the center.
   b. Determine the width of this hallway.
   c. Do these height and width values seem reasonable?
   d. Prepare an explanation of the logic you used to solve this problem that you can present to other groups in your class.

79. Challenge Question Complete the equation $\frac{2}{m-4} = \frac{4}{2m-?}$ so that it is a proportion for all values of $m \neq 4$.

80. Challenge Question An observer located at point A has a clear line of sight over level ground to points B and C. Describe a procedure using linear measurements that could be employed to determine indirectly the distance from point B to point C. (*Hint:* Use similar triangles.)

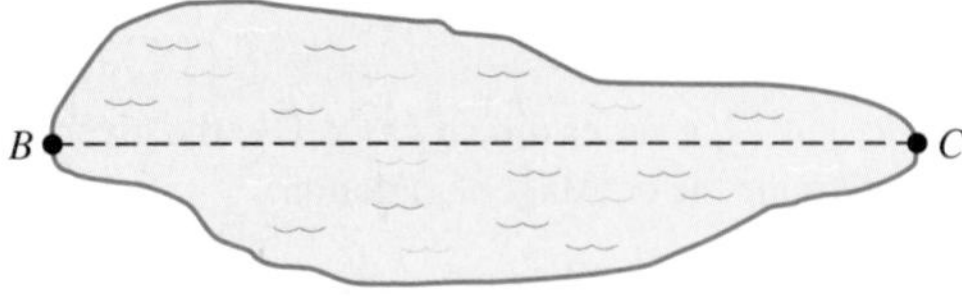

81. Spreadsheet Exploration Example 5 examined the exchange rate of U.S. dollars to Australian dollars. A spreadsheet can be a useful tool for creating a quick reference sheet to use when making purchases in a foreign country. As of May 2007, the exchange rate for U.S. dollars to Mexican pesos was 1 U.S. dollar = 11.1329 Mexican pesos. (Source: For more information on this topic go to the Additional Resources link on the MathZone website at **www.mhhe.com/hallmercer.**) The following table shows the number of pesos a traveler could get in exchange for several U.S. dollar amounts in Mexico.

	A	B
1	**U.S. Dollars**	**Mexican Pesos**
2	10	111.33
3	20	
4	50	
5	100	
6		

a. Complete the missing entries in the spreadsheet by writing the formula for cells B3, B4, and B5 and then giving the value for each cell.

b. Use the referenced website to determine today's exchange rate for U.S. dollars to euros.

c. Then create a table similar to the one shown. You may want to consider creating the given table and using column C to display the conversions to euros.

Section 2.8 More Applications of Linear Equations

Objectives:

18. Determine the restrictions on a variable in an application.
19. Use the mixture principle to create a mathematical model.

It is likely that almost all the mathematics you will encounter outside the classroom will be stated in words. Thus to profit from your algebra, you must be able to work with word problems. The applications in Section 2.7 involved proportions. The problems in this section use the mixture principle and are carefully selected to gradually develop and broaden your problem-solving skills. Some problems are selected for the same purpose as drill exercises on a piano or in basketball—to hone the skills needed for a real performance that will come later. One of the important goals of this book is to enable you to write equations that model real applications.

An important consideration when we write an equation to model an application is the values that are practical to use for the input variable. We will use the phrase **restrictions on the variable** to refer to the values that are permissible in an application. In Example 1 we illustrate an application with the input values restricted to the natural numbers $\{1, 2, 3, \ldots, 20\}$.

EXAMPLE 1 Amount of Guttering Remaining on a Roll

A roofing company installs seamless guttering from a roll of aluminum 900 ft long. Working in a new subdivision, an installer cuts several 45-ft pieces of aluminum from the roll.

(a) Identify an input variable x so that we can write a function f so that $f(x)$ represents the length remaining on the roll.

(b) Write a function f so that $f(x)$ represents the length of aluminum remaining on the roll after the pieces have been cut from the roll.

(c) What restrictions should be placed on the variable x?

SOLUTION

(a) Let x = number of 45-ft pieces of aluminum cut from roll
$f(x)$ = length in feet remaining after x pieces are cut off

(b) VERBAL EQUATION

| Length remaining | = | Total length | − | Length cut off |

ALGEBRAIC EQUATION

$f(x) = 900 - 45x$

The starting length is 900 ft, and 45 ft is used for each length cut off.

(c) The variable x must be restricted to the values $\{0, 1, 2, 3, \ldots, 20\}$.

The number of pieces cut off cannot be less than 0 pieces. If 20 pieces of length 45 ft are cut from the roll, this will use all 900 ft. Thus x cannot exceed 20. The value of x must be one of the following whole numbers: $\{0, 1, 2, \ldots, 20\}$. ■

In many applications, restrictions on the variable are implied from the given problem rather than stated explicitly. Example 2 illustrates some possibilities. These restrictions are important to recognize so that we do not present impractical solutions in the workplace. Knowing the restrictions on the variable can also help us to select an appropriate viewing window on a graphing calculator. We will illustrate this in Section 3.3.

EXAMPLE 2 Identifying the Restrictions on a Variable

Identify the restrictions on the variable for each of these applications.

SOLUTION

(a) The function $L(x) = 100 - 4x$ models the length of wire remaining on a 100-ft roll after an electrician has cut off four equal pieces of length x.

The value of x is any real number in the interval $[0, 25]$.

The length x that is cut off cannot be less than 0 ft. Also this length cannot be more than 25 ft because four pieces of 25 ft will use all the available wire. Thus x can be any real number from 0 to 25.

(b) The function $C(x) = 225x + 200$ models the total rental cost of an apartment for a 1-year lease, where x is the number of months after the lease started.

The value of x can be any of these integers: $\{0, 1, 2, \ldots, 11, 12\}$.

The input values of x can be only the discrete values 0, 1, 2, 3, 4, 5, 6, 7, 8, 9, 10, 11, or 12 because the rent is paid only once a month. The value 0 is included to represent the time when the lease is signed. Because the lease is only for 1 year, the x-values stop at 12. If the lease is extended, then a new model would apply and the values could be extended beyond 12.

(c) The function $D(x) = 55x$ models the distance in miles that a car can travel in 1 day averaging 55 mi/h. The input variable x is the number of hours the car travels.

The value of x is any real number in the interval $[0, 24]$.

The car could travel up to 24 hours in the day, but x cannot assume any negative value or any value more than 24 for this specific application. ■

SELF-CHECK 2.8.1

1. The function $f(x) = 10 - 2x$ models the length of board after x pieces each 2 ft long have been cut from this board. Identify the restrictions on the variable for this application.
2. The function $f(x) = 10 - 2x$ models the length of board after two pieces each x ft long have been cut from this board. Identify the restrictions on the variable for this application.

A basic strategy for solving word problems is outlined in the following box. Until you become quite proficient in solving word problems, we suggest you use the steps in this box as a checklist. Learning how to break down a seemingly complicated problem into a series of manageable steps is an important lifetime skill that extends well beyond word problems. When you start the first step, read actively—not passively. Circle key words, underline key phrases, make sketches, or jot down key facts as you read.

A Mathematical Note

The problem-solving strategy given here and used throughout this book is not new or unique. George Polya (1887–1985) is well known for teaching problem-solving techniques. His bestselling book *How to Solve It* includes four steps to problem solving:

- **Step 1** Understand the problem.
- **Step 2** Devise a plan.
- **Step 3** Carry out the plan.
- **Step 4** Check back.

Compare these steps to those given in the box.

Strategy for Solving Word Problems

STEP 1. Read the problem carefully to determine what you are being asked to find.

STEP 2. Select a variable to represent each unknown quantity. Specify precisely what each variable represents and note any restrictions on each variable.

STEP 3. If necessary, make a sketch and translate the problem into a word equation or a system of word equations. Then translate each word equation into an algebraic equation.

STEP 4. Solve the equation or system of equations, and answer the question asked by the problem.

STEP 5. Check the reasonableness of your answer.

The mixture principle is a great example of how one general mathematical principle can be used to form equations for seemingly unrelated problems. One of the main goals of this book is to help you see these connections. Mathematics can be difficult if you try to memorize it bit by bit; it is much easier when you understand how to use its general principles.

As you solve problems throughout this book, try to recognize how problems share general principles of mathematics. Students who recognize general principles can profit from their mathematics more than those who memorize individual problems.

Mixture Principle for Two Ingredients
Amount in first + Amount in second = Amount in mixture

Applications of the Mixture Principle

1. Amount of product A + Amount of product B = Total amount of mixture
2. Variable cost + Fixed cost = Total cost
3. Interest on bonds + Interest on CDs = Total interest
4. Distance by first plane + Distance by second plane = Total distance
5. Antifreeze in first solution + Antifreeze in second solution = Total amount of antifreeze

The overhead cost for a company includes such expenses as rent, salaries, and insurance. This cost does not change with the number of items produced and is commonly referred to as the **fixed cost** or **overhead cost.** The cost involved in making each unit of a product is commonly called the **variable cost** because this cost varies with the number of items produced. The variable cost is determined by multiplying the cost per item by the number of items produced.

EXAMPLE 3 Fixed and Variable Costs for a Company

For July 2007, a snow cone business had a fixed monthly cost of \$120 and a variable cost of \$0.30 per snow cone. How many snow cones did it sell in July 2007 if the total cost was \$405 for that month?

SOLUTION

DEFINE THE VARIABLE

Let $x =$ number of snow cones sold in July 2007.

RESTRICTIONS ON THE VARIABLE

The variable x must take on values from the set of whole numbers.

Assuming the business does not sell partial snow cones, the variable x is restricted to the whole numbers $\{0, 1, 2, 3, \ldots\}$. Although there is clearly some upper limit to the actual number of snow cones that can be made, this upper limit is unknown, so an upper limit was not placed on this set of values.

VERBAL EQUATION

| Variable cost | + | Fixed cost | = | Total cost |

ALGEBRAIC EQUATION

$$\begin{aligned} 0.30x + 120 &= 405 \\ 0.30x &= 285 \\ x &= 950 \end{aligned}$$

The variable cost for the month is $0.30x$, the cost per item times the number of items. The fixed cost for the month is \$120. The total cost for the month is \$405.

Answer: The business sold 950 snow cones in July 2007.

This answer seems reasonable. Does it check?

SELF-CHECK 2.8.2

1. Rework Example 3, assuming that the total cost for the month was \$450. Practice using each of the steps illustrated in Example 3. Then write your answer as a full sentence.

The distance formula $D = R \cdot T$ allows us to calculate the distance an object has traveled by multiplying its rate by the time it has traveled. In Example 4 the distance formula is used to determine the distance flown by each plane.

EXAMPLE 4 Time It Takes Two Planes to Fly a Given Distance

Two passenger airplanes depart from an airport simultaneously. One flies east at 300 mi/h, and the other flies west at 450 mi/h. Determine how long it will take the planes to be 2,250 mi apart. (Fuel capacity limits the flight time to a maximum of 5 hours for each plane.)

SOLUTION

DEFINE THE VARIABLE

Let $x =$ number of hours each plane has been flying.

RESTRICTIONS ON THE VARIABLE

The variable x is restricted to the set of real numbers $[0, 5]$.

The planes can fly from 0 to their 5-hour maximum. In many cases it is not clear exactly what the upper limit of the variable may be, but in most physical applications there is some upper limit.

VERBAL EQUATION

| Distance flown by first plane | + | Distance flown by second plane | = | Total distance between planes |

Note the verbal equation is based on the mixture principle.

ALGEBRAIC EQUATION

$$\begin{aligned} 300x + 450x &= 2{,}250 \\ 750x &= 2{,}250 \\ x &= 3 \end{aligned}$$

If we use $D = RT$ and a rate of 300 mi/h, the distance flown by the eastbound plane after x hours is $300x$. If we use $D = RT$ and a rate of 450 mi/h, the distance flown by the westbound plane after x hours is $450x$.

Answer: After 3 hours, the two planes will be 2,250 mi apart.

This answer seems reasonable and it checks. In 3 hours, one plane flies $3(300) = 900$ mi and the other plane flies $3(450) = 1{,}350$ mi for a total of 2,250 mi.

It is very important to check answers to be sure they meet the restrictions on the variable. If the planes in Example 4 were far enough apart to require 5.5 hours of flight, this would exceed the fuel capacity of the planes and this would not be an acceptable answer for a nonstop flight. Another situation in which it is important to check the restrictions on the variable is given in Example 5. The answer to the algebraic equation will check in the equation, but it is not practical in this application.

EXAMPLE 5 Amount of a Pesticide in a Solution

A farmer needs to apply a pesticide mixture that is 0.2% pesticide. From an earlier application she has 200 gal of a more concentrated mixture that is 0.6% pesticide in a 500-gal tank. Rather than discard the existing mixture, she decides to dilute it with water. How much water does she need to add to the 500-gal tank to have the desired concentration? Does she have enough room in this tank?

SOLUTION

DEFINE THE VARIABLE

Let $x =$ number of gallons of water that should be added to tank.

RESTRICTIONS ON THE VARIABLE

The variable x could take on any real number from 0 to 300.

The smallest amount of water the farmer could add is 0 gal. She could add up to 300 gal of water—the amount of empty space remaining in the tank.

VERBAL EQUATION

[Pesticide in existing mixture] + [Pesticide in water added] = [Total pesticide in desired mixture]

Note the verbal equation is based on the mixture principle.

ALGEBRAIC EQUATION

$$\begin{aligned} 0.006(200) + 0.0(x) &= 0.002(200 + x) \\ 1.2 &= 0.4 + 0.002x \\ 0.8 &= 0.002x \\ 400 &= x \\ x &= 400 \end{aligned}$$

The volume of pesticide in 200 gal of 0.6% pesticide solution is 0.006(200), the product of the concentration and the volume of the mixture. There is 0.0% pesticide in the water added, so the volume of pesticide in x gal of water is $0.0(x)$. Note that the new volume of the mixture would be $(200 + x)$ gal. The volume of pesticide in $(200 + x)$ gal of 0.2% pesticide solution is $0.002(200 + x)$.

Answer: The farmer would need to add 400 gal of water to dilute the mixture to 0.2% insecticide. Since she started with 200 gal of the 0.6% mixture, adding 400 gal of water would give a total volume of 600 gal in the tank. She does not have enough room in the 500-gal tank.

SELF-CHECK 2.8.3

A chemist needs to decrease the concentration of a salt solution. She has 4 liters (L) of a 30% salt solution in 6-L container.

1. How much water should she add to the container to decrease the concentration to 20%?
2. How much water should she add to the container to decrease the concentration to 15%?

The interest formula $I = PRT$ can be used to calculate the interest earned by an investment P at an annual rate R for a time of T years. In Example 6 we use this formula to determine the income from two different investments.

EXAMPLE 6 Interest Earned on an Investment

A recent college graduate is planning a budget. She has a yearly maximum of $4,800 for investment purposes. She has already committed $1,440 for an insurance investment earning 2%. How much would she have to invest in a CD earning 5% to yield a total gain on her two investments of 4%?

SOLUTION

DEFINE THE VARIABLE

Let x = number of dollars she needs to invest in a CD earning 5% to yield a total gain on her two investments of 4%.

Note she does not have to invest the entire $4,800. She wants to invest enough so that she will earn a 4% return overall.

RESTRICTIONS ON THE VARIABLE

The variable x may take on values from $0 to $3,360.

The minimum she could invest in a CD is $0. The maximum she could invest is $3,360, the amount remaining from her $4,800 after the insurance investment of $1,440.

VERBAL EQUATION

Interest from insurance	+	Interest from CD investment	=	Total interest earned

Note the verbal equation is based on the mixture principle.

ALGEBRAIC EQUATION

$$\begin{aligned} 0.02(1{,}440) + 0.05(x) &= 0.04(1{,}440 + x) \\ 28.8 + 0.05x &= 57.6 + 0.04x \\ 0.01x &= 28.8 \\ x &= 2{,}880 \end{aligned}$$

The interest earned from $1,440 invested at 2% is 0.02(1,440), the product of the interest rate and the amount invested at that rate. The interest earned from x dollars invested at 5% is $0.05(x)$. She has a goal of earning a combined return of 4% on the two investments. For this to be true, the total interest earned would need to be $0.04(1{,}440 + x)$.

Answer: If she invests $2,880 in a CD earning 5%, she will yield a total gain on her two investments of 4%.

Note the value of $2,880 falls within the stated restrictions of $0 to $3,360.

SELF-CHECK 2.8.4

1. A retiree receives a yearly income of $20,000 from her IRA to help fund her retirement. She has placed $40,000 of this IRA in a secure Treasury bond earning 5% yearly interest. If she earns an 8% rate of return on the rest of this IRA through an investment in bank stocks, how much of the IRA is invested in bank stocks?

SELF-CHECK ANSWERS

2.8.1

1. {0, 1, 2, 3, 4, 5}
2. [0, 5]

2.8.2

1. The business sold 1,100 snow cones in July 2007.

2.8.3

1. Add 2 L of water.
2. There is not enough room in the 6-L container to add 4 L of water.

2.8.4

1. $225,000 is invested in bank stocks.

USING THE LANGUAGE AND SYMBOLISM OF MATHEMATICS 2.8

1. We use the phrase __________ *on the variable* to refer to the values of the variable that are permissible in an application.
2. The first step in the word problem strategy given in this book is to read the problem carefully to determine what you are being asked to __________.
3. The second step in the word problem strategy given in this book is to select a __________ to represent each unknown quantity.
4. The third step in the word problem strategy given in this book is to model the problem verbally with word equations and then to translate these word equations into __________ equations.
5. The fourth step in the word problem strategy given in this book is to __________ the equation or system of equations and answer the question asked by the problem.
6. The fifth step in the word problem strategy given in this book is to check your answer to make sure the answer is __________.
7. The __________ principle states that the amount obtained by combining two parts is equal to the amount obtained from the first part plus the amount obtained from the second part.
8. In the formula $D = RT$, D represents distance. R represents __________, and T represents __________.
9. In the formula $I = PRT$, I represents interest, P represents __________, R represents __________, and T represents __________.
10. The cost for a company that does not change with the number of units produced is called the __________ cost or the overhead cost.
11. The cost for a company that depends on how many units the company produces is called the __________ cost.

EXERCISES 2.8

In Exercises 1–4, match each problem with the restricted values for the variable in this application.

1. The function $f(x) = 500 - 50x$ models the length of optical fiber remaining on a spool after x pieces each 50 ft long have been removed from this spool.
2. The function $f(x) = 500 - 50x$ models the length of optical fiber remaining on a spool after 50 pieces each of length x have been removed from this spool.
3. The function $f(x) = 0.05(70) + 0.08x$ models the amount of acid in a 100-gal tank that has x gal of an 8% acidic solution poured into a tank that already contains 70 gal of a 5% acidic solution.
4. The function $f(x) = 100x + 250$ models the cost in dollars to lease a highway messaging system for x days during the month of June.

A. $\{0, 1, 2, 3, \ldots, 30\}$

B. $\{0, 1, 2, 3, \ldots, 10\}$

C. $[0, 10]$

D. $[0, 30]$

In Exercises 5–8, determine the restrictions on the variable for each application.

5. **Quick Pass Account** The function $V(x) = 100 - 2x$ models the number of dollars remaining on a \$100 toll road Quick Pass account. Each trip on this toll road costs \$2, and x represents the number of trips on this toll road.
6. **Cost of Gasoline** The function $C(x) = 3.10x$ models the total cost of filling a 20-gal tank with x gal of gasoline costing \$3.10 per gallon.
7. **Acidic Solution** The function $f(x) = 0.07(175) + 0.12x$ models the number of gallons of acid in a 400-gal tank. The 400-gal tank originally contained 175 gal of a 7% acidic solution, and then x gal of a stronger solution was added to the tank.
8. **Cost of Renting a Tent** The function $C(x) = 250x$ models the cost of renting a tent for x days in one week. The rental company will not rent the tent for partial days.

In Exercises 9 and 10, solve each problem by completing each of the given steps.

9. **Width of a Brake Pad** The brake pad on a new truck is 11.2 mm thick. The maintenance shop that services this truck estimates that about 0.2 mm will wear from the pad each month. How many months can the truck be used before the thickness of the brake pad is reduced to 4.0 mm?
   a. Select a variable to represent the unknown quantity, and identify this variable.
   b. Write a word equation that models this problem.
   c. Translate this word equation into an algebraic equation.
   d. Solve this equation.
   e. Is this solution within the restrictions on the variable, and does it seem reasonable?
   f. Write a sentence that answers the problem.
10. **Depth of a Tire Tread** The depth of the tread on a new tire on a taxi is 9.8 mm. It is estimated that the tire will wear about 0.3 mm per month. Determine the number of months until the depth of the tread on these tires is 2.6 mm.
    a. Select a variable to represent the unknown quantity and identify this variable.
    b. Write a word equation that models this problem.
    c. Translate this word equation into an algebraic equation.
    d. Solve this equation.
    e. Is this solution within the restrictions on the variable, and does it seem reasonable?
    f. Write a sentence that answers the problem.

In Exercises 11–30, solve each problem by using each step of the word problem strategy outlined in this section.

11. **Telephone Costs** A long-distance phone package charges a fixed amount of \$24.95 per month plus a variable charge of

\$0.05 per minute. How many minutes were billed if the long-distance phone charges were \$42.20 for a month of service?

12. **Taxi Costs** The cost of a taxi ride involves a fixed cost of \$3.00 and a variable cost of \$4.00 per mile. If a taxi fare was \$25.40, how many miles did the taxi cover in this trip?
13. **Fixed and Variable Costs of Production** The manufacturer of custom sofa cushions has fixed daily costs of \$150 and a variable cost of \$6 per cushion. Determine how many cushions the manufacturer needs to make per day to meet the goal of an average cost of \$9 per cushion. Note that the total cost can also be calculated by multiplying the average cost per cushion by the number of cushions produced.
14. **Fixed and Variable Costs of Production** The manufacturer of custom bricks places names of individuals on donor bricks sold as fund-raisers. The manufacturer has fixed daily costs of \$180 and a variable cost of \$1.50 per brick. Determine how many bricks the company needs to make per day to meet its goal of an average cost of \$2 per brick. Note that the total cost can also be calculated by multiplying the average cost per brick by the number of bricks produced.
15. **Income Earned on an IRA Investment** A retiree needs a yearly income of \$10,200 from his \$150,000 IRA to help fund his retirement. He has placed \$60,000 of this account in a secure Treasury bond earning 5% yearly interest. What rate of return must he earn on the rest of this investment to reach his \$10,200 income goal?
16. **Income Earned on an IRA Investment** A retiree receives a yearly income of \$13,000 from her IRA to help fund her retirement. She has placed \$50,000 of this IRA in a secure Treasury bond earning 5% yearly interest. If she earns a 7% rate of return on the rest of this IRA through an insurance annuity, what is the principal that is invested in the insurance annuity?
17. **Income Earned on an Educational Account** A student receives a yearly income of \$4,000 from an account set up by her grandparents to help fund her education. From this educational account \$10,000 is invested in a secure Treasury bond earning 5.5% yearly interest. If she earns a 6% rate of return on the rest of this educational account through a corporate bond, what is the principal that is invested in the corporate bond?
18. **Income Earned on an Educational Account** A student receives a yearly income of \$4,900 from a \$70,000 account set up by his grandparents to help fund his education. From this educational account \$20,000 is invested in a secure Treasury bond earning 4.5% yearly interest. What rate of return must he earn on the rest of this investment to produce the yearly income of \$4,900?
19. **Travel Time** Two boats leave a dock at the same time. One travels downstream at a rate of 12 mi/h, and the other travels upstream 4 mi/h slower than the first boat. Determine the number of hours it will take the boats to be 64 mi apart.
20. **Exercise Time** Two friends go to a park to exercise. One starts walking at a rate of 4 mi/h on the 7.5-mi path around the park, and the other starts from the same point, jogging at a rate of 6 mi/h in the opposite direction on this path. How many minutes will it be before they meet on this path?

21. **Travel Time** A car and a truck simultaneously leave two towns 364 mi apart and head toward each other. The car travels at a rate of 70 mi/h, and the truck travels 10 mi/h slower. How long will it take the two to meet?
22. **Flight Time** At the time a refueling request is made by a jet pilot, his plane is 175 mi from a refueling tanker. The two planes head toward each other so the jet can refuel. The jet flies 450 mi/h, and the tanker flies 200 mi/h slower. How long will it take the two to meet?
23. **Mixture of Gasohol** A gasoline distributor has a 20,000-gal tank to store a gasohol mix. The tank already contains 12,000 gal of gasohol that contains 4% ethanol by volume. How many gallons of ethanol must be added to the tank to produce a mix that is 10% ethanol?
24. **Amount of a Pesticide in a Solution** A crop-dusting service needs to prepare a 0.4% pesticide solution. It has an airplane with a 200-gal tank that contains 75 gal of a 1% solution. How much water should be added to the tank to decrease the concentration to the desired 0.4% level?
25. **Amount of a Pesticide in a Solution** A crop-dusting service needs to prepare a 0.8% pesticide solution. It has an airplane with a 300-gal tank that contains 100 gal of a 0.5% solution. How much of a previously mixed 1% pesticide solution should be added to the tank to increase the concentration to the desired 0.8% level?
26. **Amount of Acid in a Solution** A chemist needs to create a 20% acid solution. She has 4 L of a 30% acid solution in a 20-L container. How much water should she mix with this solution to decrease the concentration to the desired 20%?
27. **Electricity Generated by Windmills** One growing community in California currently uses 2 MW of electricity per year. All this electricity is generated from coal. The community plans to add x MW of windmill generation capacity to its system to meet increasing demand. How many megawatts of windmill generation capacity must be built in order that 20% of the community's total electricity usage comes from windmills?

28. **Electricity Generated by Windmills** One small island community in the Pacific currently uses 1.8 MW of electricity per year. All this electricity is generated from plants fired by natural gas. The island community plans to add x MW of offshore windmill generation capacity to its system. How many megawatts of windmill generation capacity must be built in order that 25% of the island community's total electricity usage comes from windmills?

29. **Amount of Sand in a Concrete Mix** A company that mixes and sells concrete has a contract to deliver several loads of concrete containing an aggregate that is 20% sand by weight. The company has a pile that contains 360,000 lb of aggregate that is 15% sand. How much sand will be required to mix with the aggregate in this pile to produce a mixture that is 20% sand?

30. **Amount of Lean Beef in Hamburger** The meat department in a grocery store purchased 800 lb of hamburger that is 90% lean beef. How many pounds of hamburger that is 75% lean must be mixed with this 90% lean mix to produce hamburger that is 85% lean beef?

In Exercises 31–36, use the given information to solve each part of the problem. (*Hint:* Do your proposed answers meet the restrictions placed on the variable?)

31. **Cost of Manufacturing Ceramic Tile** A manufacturer of custom ceramic tiles has the capacity to produce 8,000 tiles per day. It has a fixed daily cost of $2,000 plus a variable cost of $0.50 per tile.
   a. How many tiles were created in a batch that had a total daily cost of $5,750?
   b. The total cost of the ceramic tiles can also be calculated by multiplying the average cost per tile by the number of tiles produced. If the company needs to reach an average cost of $0.70 per tile, how many tiles must it produce per day?

32. **Cost of Manufacturing Mugs** A manufacturer of custom drinking mugs has the capacity to produce 1,800 mugs per day. It has a fixed daily cost of $1,000 plus a variable cost of $1.50 per mug.
   a. How many mugs were created in a batch that had a total daily cost of $2,800?
   b. The total cost of the mugs can also be calculated by multiplying the average cost per mug by the number of mugs produced. If the company needs to reach an average cost of $2.00 per mug, how many mugs must it produce per day?

33. **Mixture of Gasohol** A gasoline distributor has a 20,000-gal tank to store a gasohol mix. The tank already contains 19,000 gal of gasohol that contains 4% ethanol by volume.
   a. How many gallons of ethanol must be added to the tank to produce a mix that is 8% ethanol? (Round to the nearest gallon.)
   b. How many gallons of ethanol must be added to the tank to produce a mix that is 10% ethanol? (Round to the nearest gallon.)

34. **Amount of a Pesticide in a Solution** A crop-dusting service needs to decrease the concentration of the pesticide in a tank. It has an airplane with a 200-gal tank that contains 75 gal of a 1% solution.
   a. How much water should be added to the tank to decrease the concentration to 0.3%?
   b. How much water should be added to the tank to decrease the concentration to 0.4%?

35. **Amount of a Pesticide in a Solution** A crop-dusting service needs to increase the concentration of the pesticide in a tank. It has an airplane with a 300-gal tank that contains 100 gal of a 0.5% solution.
   a. How much of a previously mixed 1% pesticide solution should be added to the tank to increase the concentration to 0.9%?
   b. How much of a previously mixed 1% pesticide solution should be added to the tank to increase the concentration to 0.8%?

36. **Amount of Acid in a Solution** A chemist needs to decrease the concentration of an acid solution. She has 4 L of a 30% acid solution on hand. She will store the resulting solution in a 5-L container.
   a. How much water should she add to the mixture to decrease the concentration to 20%?
   b. How much water should she add to the mixture to decrease the concentration to 25%?

37. **Cost of Grass Seed** A landscaping company has two popular types of grass seed mixes for its area, a premium mix that sells for $1.25 per pound and a standard mix that sells for $0.95 pound.
   a. If the store manager dumps a 20-lb bag of each mix into a large barrel, what is the total value of the grass seed in the barrel?
   b. What price per pound should the company sell this new mix for to produce the same income as selling the two mixes separately?
   c. If the store manager started with 20 lb of the premium mix in a large barrel, how many pounds of the standard mix should be added to create a mix that could be sold for $1.00 per pound?
   d. How many pounds of grass seed must the barrel hold in order to contain all the mixture created in part **c**?
   e. How many pounds of the standard mix should be put in a barrel with 20 lb of the premium mix to create a mix worth $1.05 per pound?

38. **Cost of Paving Bricks** A construction company used two different types of paving bricks to create a design around a water fountain. The company charges $1.50 each for the rectangular bricks and $2.00 each for the hexagonal bricks. These bricks are stored on pallets with each pallet of the rectangular bricks containing 144 bricks and each pallet of the hexagonal bricks containing 120 bricks.
   a. If a forklift operator places one pallet of each type of bricks on a truck, what is the total value of the bricks on this truck?
   b. What is the average price per brick for all the bricks on this load?
   c. If the design for the fountain calls for 2,000 hexagonal bricks, how many rectangular bricks should be added to create a mix that costs an average of $1.60 per brick?
   d. If the design for the fountain calls for 2,000 rectangular bricks, how many hexagonal bricks should be added to create a mix that costs an average of $1.60 per brick?
   e. If the design for the fountain calls for 2,000 hexagonal bricks and the budget for the cost of the bricks will allow for an expense of $9,250, how many rectangular bricks can be used?

Group Discussion Questions

39. **Challenge Question** A farmer raising watermelons for a contest produced a 200-lb watermelon that was 98% water. After the watermelon was moved, weighed, and left in the sun, some of the moisture evaporated. The result was that the watermelon was now only 95% water. What is the new weight of the watermelon?

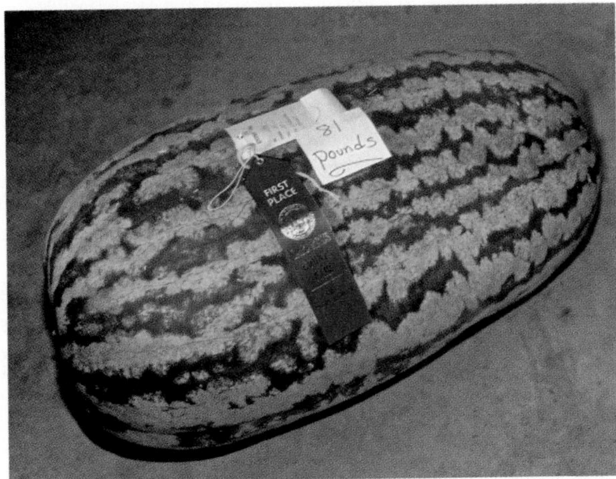

40. **Challenge Question** The antifreeze mixture in a car radiator in Michigan should be kept between 40% and 60%. The cooling system of a car holds 6 L of a 40% antifreeze mixture. To raise the antifreeze concentration to 60%, how much of this mixture should be drained from the radiator and replaced with pure antifreeze?
41. **Communicating Mathematically** Write a word problem that can be solved by using the equation $0.06(4{,}000) + 0.10x = 0.07(4{,}000 + x)$ and that models
    a. The income earned from an investment.
    b. The amount of ethanol in a gasohol mixture.
42. **Communicating Mathematically** Write a word problem that can be solved by using the equation $50{,}000 - 8{,}500x = 16{,}000$ and that models
    a. The number of meters of transmission cable on an installation truck.
    b. The number of pounds of jet fuel on an airplane.

KEY CONCEPTS FOR CHAPTER 2

1. **Cartesian Coordinate System:** The sign pattern for the coordinates in each quadrant of the rectangular coordinate system is shown in the figure.

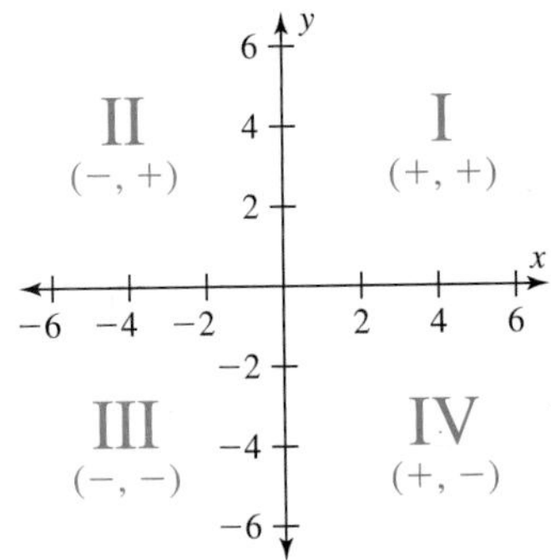

2. **Scatter Diagram:** A scatter diagram for a set of data points is a graph of these points that can be used to examine the data for some type of visual pattern.
3. **Arithmetic Sequences and Linear Equations**
   - A sequence is an ordered set of numbers.
   - An arithmetic sequence is a sequence with a constant change from term to term.
   - The graph of the terms of an arithmetic sequence consists of discrete points that lie on a straight line.
   - The linear equation $y = mx + b$ will produce an arithmetic sequence of output values y for input values of x of $1, 2, 3, \ldots$.
4. **Function Notation:** The function notation $f(x)$ is read "f of x." $f(x)$ represents a unique output value for each input value of x.
5. **Solution of a Linear Equation:** A solution of a linear equation of the form $y = mx + b$ is an ordered pair (x, y) that makes the equation a true statement.
6. **Intercepts of a Line**
   - The x-intercept, often denoted by $(a, 0)$, is the point where the line crosses the x-axis.
   - The y-intercept, often denoted by $(0, b)$, is the point where the line crosses the y-axis.
7. **Solution of a System of Linear Equations**
   - A solution of a system of linear equations is an ordered pair that satisfies each equation in the system.
   - A solution of a system of linear equations is represented graphically by a point that is on both lines in the system.
   - A solution is represented in a table of values as an x-value for which both y-values are the same.
8. **Linear Equation in One Variable:** A linear equation in one variable x is an equation that can be written in the form $Ax = B$, where A and B are real constants and $A \neq 0$.
9. **Equivalent Equations:** Equivalent equations have the same solution set.
10. **Addition-Subtraction Principle of Equality:** If the same number is added to or subtracted from both sides of an equation, the result is an equivalent equation.
    - $a = b$ is equivalent to $a + c = b + c$.
    - $a = b$ is equivalent to $a - c = b - c$.
11. **Multiplication-Division Principle of Equality:** If both sides of an equation are multiplied or divided by the same nonzero number, the result is an equivalent equation.
    - $a = b$ is equivalent to $ac = bc$ for $c \neq 0$.
    - $a = b$ is equivalent to $\frac{a}{c} = \frac{b}{c}$ for $c \neq 0$.
12. **Classification of Equations**
    - **Conditional Equation:** A conditional equation is true for some values of the variable and false for other values.
    - **Contradiction:** An equation that is false for all values of the variable is a contradiction.

- **Identify:** An equation that is true for all values of the variable is an identity.

13. **Strategy for Solving Linear Equations Algebraically**

    Step 1. Simplify each side of the equation.

    a. If the equation contains fractions, simplify by multiplying both sides of the equation by the least common denominator (LCD) of all the fractions.

    b. If the equation contains grouping symbols, simplify by using the distributive property to remove the grouping symbols and then combine like terms.

    Step 2. Using the addition-subtraction principle of equality, isolate the variable terms on one side of the equation and the constant terms on the other side.

    Step 3. Using the multiplication-division principle of equality, solve the equation produced in step 2.

14. **Solving Linear Equations by Using Tables:** Enter the left side of the equation as Y_1 and the right side of the equation as Y_2. Then select x-values and form a table displaying the values for x, Y_1, and Y_2. The x-coordinate that causes Y_1 to equal Y_2 is the solution of the original linear equation in x. (*Caution:* This value of x may not appear on the first table that you select.)

15. **Solving for a Specified Variable:** To solve for a specified variable, isolate this variable on one side of the equation with all other variables on the other side of the equation.

16. **Proportions**

    - A proportion is an equation that states two ratios are equal.
    - In the proportion $\frac{a}{b} = \frac{c}{d}$, the terms a and d are called the *extremes* and the terms b and c are called the *means*.

17. **Direct Variation**

    - If x and y are variables and k is a real constant with $k \neq 0$, then stating "y varies directly as x with constant of variation k" means $y = kx$.
    - If y varies directly as x, then the graph of the (x, y) pairs will lie on a straight line.

18. **Restrictions on a Variable:** The restrictions on a variable refer to the values that are permissible in an application.

19. **Strategy for Solving Word Problems**

    Step 1. Read the problem carefully to determine what you are being asked to find.

    Step 2. Select a variable to represent each unknown quantity. Specify precisely what each variable represents, and note any restrictions on each variable.

    Step 3. If necessary, make a sketch and translate the problem into a word equation or a system of word equations. Then translate each word equation into an algebraic equation.

    Step 4. Solve the equation or system of equations, and answer the question asked by the problem.

    Step 5. Check the reasonableness of your answer.

20. **Mixture Principle:**

    Amount in first + Amount in second = Amount in mixture

REVIEW EXERCISES FOR CHAPTER 2

1. Identify the coordinates of the points A through D in the figure, and give the quadrant in which each point is located.

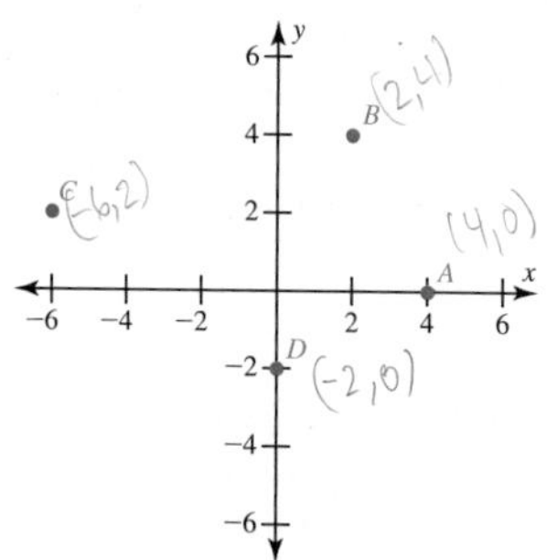

2. Draw a scatter diagram for the data points in the given table.

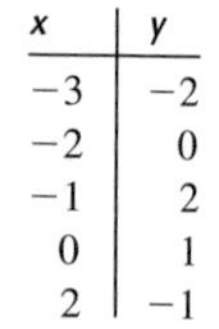
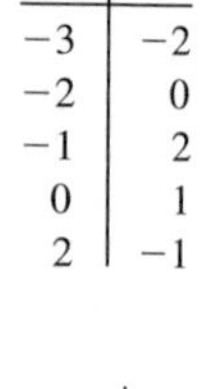

x	y
−3	−2
−2	0
−1	2
0	1
2	−1

3. For the arithmetic sequence defined by $a_n = -2n + 3$, complete the given table and graph the points.

n	a_n
1	
2	
3	
4	
5	

4. For the arithmetic sequence defined by $y = 0.5x + 1.5$, complete the given table and graph the points.

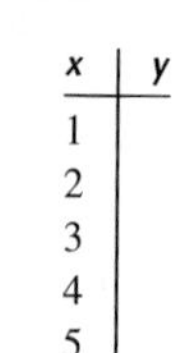

x	y
1	
2	
3	
4	
5	

In Exercises 5–10, determine which of these sequences are arithmetic sequences. For those that are arithmetic, find the common difference d.

5. 4, 7, 10, 13, 16
6. 0, 1, 3, 6, 10
7. $a_n = -2n + 5$
8. $a_n = n^2 - 5$
9. 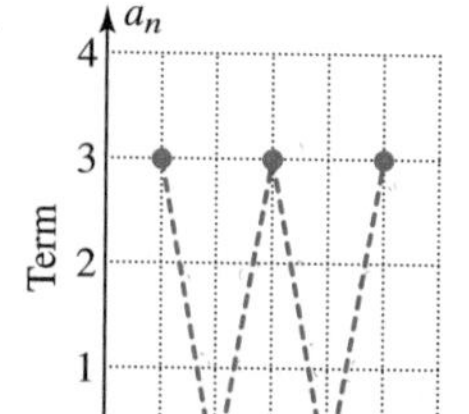
10. 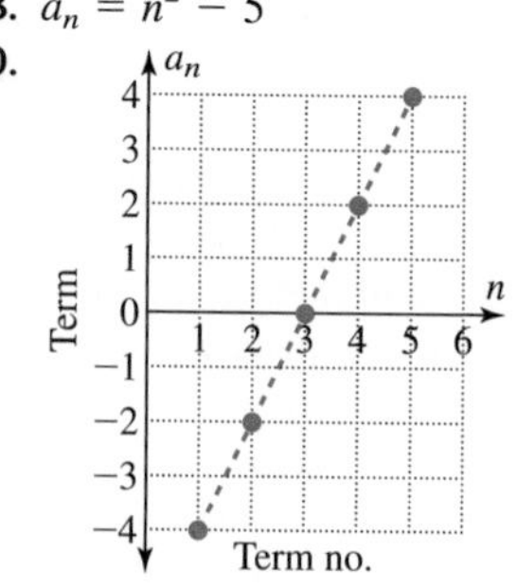

In Exercises 11–14, determine whether either $(3, -2)$ or $(-3, 2)$ is a solution of each equation.

11. $y - 3 = 4(x + 2)$
12. $y = x + 5$

13.

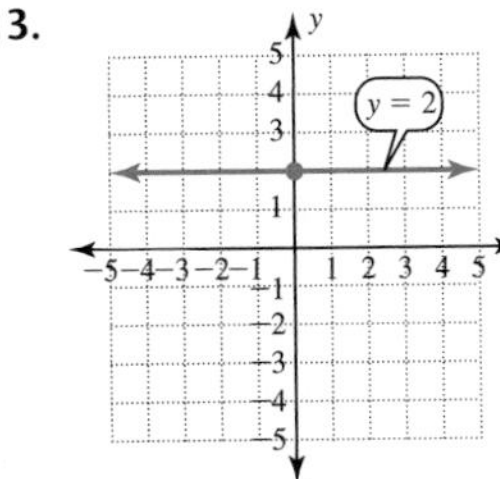

14.

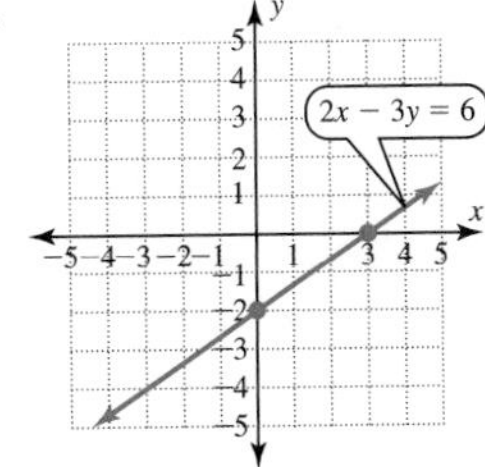

15. Determine the x- and y-intercepts of the line in Exercise 14.

16. Determine the x- and y-intercepts of $y = x + 5$.

17. Determine the x- and y-intercepts of $y - 3 = 4(x + 2)$.

In Exercises 18–21, match each exercise with the most appropriate choice.

18. $5x + 3 = 2(x + 4)$

19. $5x^2 = 3$

20. $y = 2x + 4$

21. $5x + 3 - 2(x + 4)$

A. A first-degree expression in x but not a linear equation

B. A linear equation in x

C. Not a linear equation

D. A linear equation in two variables

In Exercises 22 and 23, determine the point of intersection of the two lines.

22.

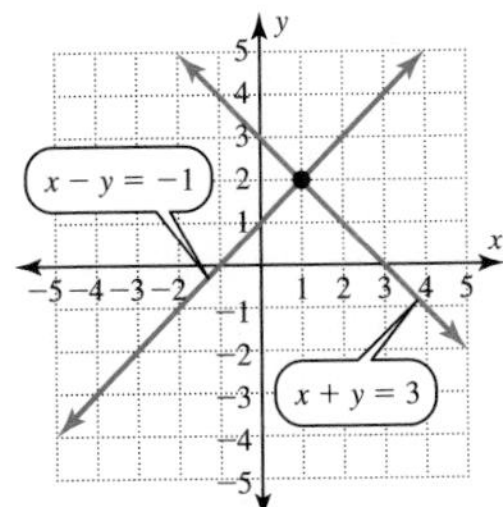

23.

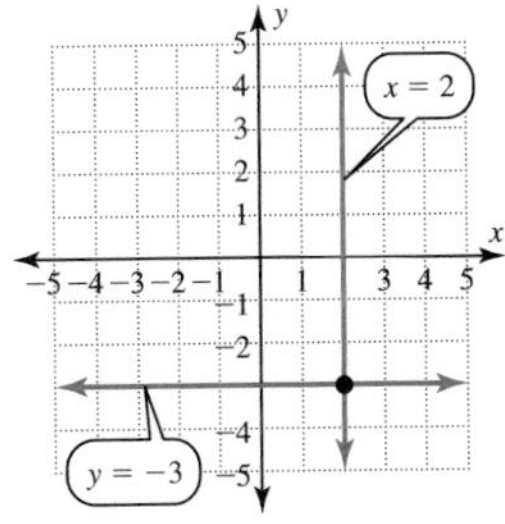

In Exercises 24 and 25, Y_1 represents the left side of the equation and Y_2 represents the right side of the equation. Use the table shown from a graphing calculator to solve this equation.

24. $3.6x + 1.2 = 1.8x + 5.7$

X	Y1	Y2
0	1.2	5.7
.5	3	6.6
1	4.8	7.5
1.5	6.6	8.4
2	8.4	9.3
2.5	10.2	10.2
3	12	11.1

X=0

25. $8.5x - 55 = 10.5x - 103$

X	Y1	Y2
20	115	107
21	123.5	117.5
22	132	128
23	140.5	138.5
24	149	149
25	157.5	159.5
26	166	170

X=20

In Exercises 26 and 27, check $x = 2$ and $x = 3$ as possible solutions of the given equation.

26. $3x + 5 = 4x + 2$

27. $3(x + 5) = 4(x + 3) + 1$

In Exercises 28–52, solve each equation.

28. $x + 2 = 17$

29. $5v + 7 = 4v + 9$

30. $7m = 6m$

31. $-x = 3$

32. $-17m = -34$

33. $\frac{n}{3} = -12$

34. $-\frac{2y}{7} = 28$

35. $0.045w = -0.18$

36. $117 = 0.01t$

37. $0 = -147b$

38. $5x - 8 = 27$

39. $-8z + 7 - 2z = 5z + 4 + z$

40. $-r - r - r - r - r = r$

41. $3(2x + 7) = x + 1$

42. $(6x - 9) - (3x + 8) = 4(x - 11)$

43. $7y - 5(2 - y) = 3(2y + 1) + 5$

44. $9(2y - 3) - 13(5y - 1) = 14(3y - 1)$

45. $0.55(w - 10) = 0.80(w + 3)$

46. $\frac{t}{8} - 5 = \frac{t}{12} - 6$

47. $\frac{7m + 27}{6} = \frac{4m + 6}{5}$

48. $\frac{167 - 3n}{10} = 6 - \frac{5n - 1}{4}$

49. $1{,}293y = 1{,}294y + 1{,}295$

50. $0.289x - 5 = 3.1 - 0.78(2 - x)$

51. $\frac{3v}{10} - \frac{5v}{6} = -\frac{8v}{15}$

52. $1.8a - 7.8a + 1.97a - 8.3a = 0$

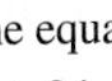

In Exercises 53 and 54, use a graphing calculator to solve each equation by letting Y_1 represent the left side of the equation and Y_2 represent the right side of the equation.

53. $1.5x - 1.2 = 2.7x - 3.0$

54. $8.5x + 24 = 12.5x + 33.6$

In Exercises 55–58, each equation is a conditional equation, an identity, or a contradiction. Identify the type of equation, and then solve it.

55. $7(y - 3) - 4(y + 2) = 3(y - 7)$

56. $3(2v + 1) = 2(3v + 1)$

57. $4(w - 1) + (7 - 2w) = 2(w + 1) + 1$

58. $2v + 2v + 2v = 5v$

59. Use the function $f(x) = 7x - 11$ to evaluate the following.
a. $f(-10)$ **b.** $f(0)$ **c.** $f(9)$ **d.** $f(\pi)$

60. Use the function $f(x) = 5(4x - 3) + 1$ to evaluate the following.
a. $f(-3)$ **b.** $f(0)$ **c.** $f(6)$ **d.** $f(100)$

In Exercises 61–64, solve each equation for x.

61. $v + w = x + y$

62. $vw = xy$

63. $3x - 5y = 7z$

64. $2(x - y) + 3(2x + y) = 38$

In Exercises 65–68, simplify the expression in the first column by adding like terms, and solve the equation in the second column.

	Simplify	Solve
65.	**a.** $3x + 4 + 5x + 6$	**b.** $3x + 4 = 5x + 6$
66.	**a.** $3x + 4 - (5x + 6)$	**b.** $3x + 4 = -5x + 6$
67.	**a.** $-2x + 3 + 4(x - 1)$	**b.** $-2x + 3 = -4(x - 1)$
68.	**a.** $5(x + 1) - 6(x + 2)$	**b.** $5(x + 1) = 6(x + 2)$

Estimate Then Calculate

In Exercises 69 and 70, mentally estimate the solution of each equation to the nearest integer, and then use a calculator to calculate the solution.

PROBLEM	MENTAL ESTIMATE	CALCULATOR SOLUTION
69. $0.99x - 5.1 = 4.899$		
70. $11x - 20 = 9x + 19.9$		

Multiple Representations

In Exercises 71–74, first write an algebraic equation for each verbal statement, using the variable m to represent the number, and then solve for m.

71. Four more than five times a number is forty-nine.

72. One-third of the sum of a number and seven equals twelve.

73. Twice the quantity of two more than a number is the opposite of the quantity of three less than the number.

74. The sum of a number, one more than the number, and two more than the number is ninety-three.

75. **Perimeter of a Rectangle**
   a. Write an expression for the perimeter of the rectangle shown in terms of the variable w.
   b. Determine the perimeter of this rectangle when $w = 12$.
   c. Write an equation that represents a perimeter of 48 cm for this rectangle.
   d. Solve for the equation in part **c** for w.

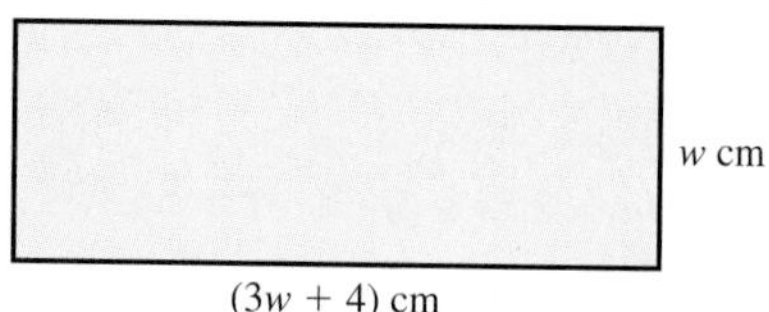

76. **Fixed and Variable Costs** At the end of the summer, the business records for Cones-R-Us revealed that it had a fixed cost of $500 for running the business (utilities, insurance, etc.). The accountant also calculated that each snow cone cost 25 cents for ice, juice, and a cup.
   a. Write a function that gives the total cost of selling x snow cones in a summer.
   b. Use this function to complete this table.

x	0	1,000	2,000	3,000	4,000	5,000
$f(x)$						

   c. Evaluate and interpret $f(2{,}500)$.

77. **Fixed and Variable Costs** A small business makes custom ornamental pottery. The fixed cost is $160 per week, and the variable cost is $26 per piece of pottery made. The company's sales are seasonal, so the company has a weekly budget of $1,200 for making the pottery pieces. It then stores this pottery until orders are received. Let x represent the number of pieces of pottery made per week.
   a. Write an expression for the cost of making x pieces of pottery per week.
   b. Determine the weekly cost of making 60 pieces of pottery per week.
   c. Write an equation that represents a total weekly cost of $1,200.
   d. Solve the equation in part **c**.

78. **Distance on a Map** On a map of the United States 1 in represents 20 mi. If the distance on the map between Louisville, Kentucky, and Columbus, Ohio, is 10.5 in, find the distance in miles between these two cities.

79. **Similar Polygons** The two figures shown are similar. Thus, their corresponding parts are proportional. Find the lengths of sides a, b, and c.

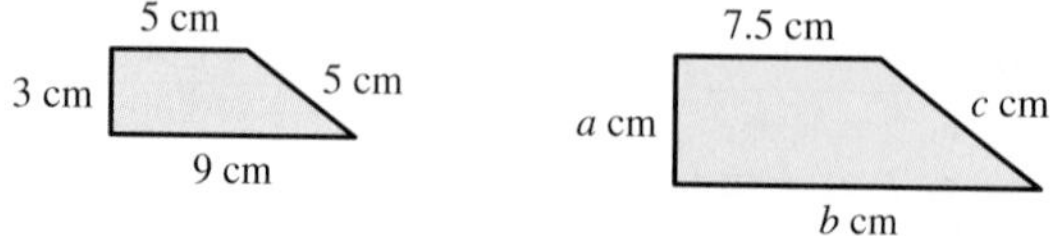

80. **Height of a Tree** If a shrub 0.64 m tall casts a 6.72-m shadow, how tall is a tree that casts a 46.2-m shadow? (See the figure.)

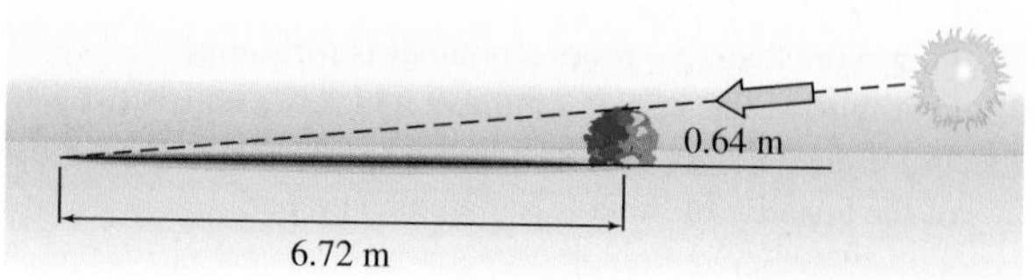

81. **Direct Variation**
   a. If v varies directly as w, and $v = 22$ when $w = 77$, find v when $w = 35$.
   b. If v varies directly as w, and $v = 12$ when $w = 15$, determine the constant of variation.

82. **Quality Control** A quality control inspector discovered 3 defective display panels in the 75 handheld computing devices that he tested. How many defective display panels would be expected in a shipment of 500 of these items?

83. **Airspeed of a Plane** Two ground observers are 20 mi apart. They communicate with each other by cell phone and note that a passenger plane has passed over each of them exactly 5 minutes apart. If this plane is flying into a 50-mi/h headwind, what is the airspeed of this plane?

84. **Car Payments** At the beginning of July, a student purchased a used car with an initial payment of $680. She then owed $280 on the first of August and on the first of the month for each of the next 24 months. Let y represent the total of all payments and x represent the number of monthly $280 payments.
   a. **Algebraically:** Write an equation to express y in terms of x.
   b. **Numerically:** Use a graphing calculator to form a table of y-values for $x = 0, 1, 2, 3, 4, 5, 6$.
   c. **Graphically:** Use a graphing calculator to graph this equation.
   d. Is the sequence of monthly payment totals an arithmetic sequence? If so, what is the common difference d?

In Exercises 85–88, match each problem with the appropriate restrictions on the variable for this application.

A. $\{0, 1, 2, 3, \ldots, 365\}$
B. $\{0, 1, 2, 3, \ldots, 20\}$
C. $[0, 20]$
D. $[0, 365]$

85. **Newsprint** The function $f(x) = 1{,}000 - 50x$ models the length of paper remaining on a spool of newsprint after x pieces each 50 ft long have been removed from this spool.

86. **Newsprint** The function $f(x) = 1{,}000 - 50x$ models the length of paper remaining on a spool of newsprint after 50 pieces each of length x have been removed from this spool.

87. **Gasohol** A 1,000-gal tank contains 635 gal of gasohol that is 8% ethanol by volume. The function $f(x) = 0.08(100) + 0.12x$ models the amount of ethanol that will be in the tank after x gal of gasohol that is 12% by volume is pumped into this tank.

88. **Sign Rental** The function $f(x) = 100x + 250$ models the cost in dollars to lease a highway messaging system for x days during the year 2007.

89. **Time Traveled** Two passenger airplanes depart from an airport simultaneously. One flies east at 400 mi/h, and the other flies west at 450 mi/h. Determine how long it will take the planes to be 2,975 mi apart. (Fuel capacity limits the flight time to a maximum of 5 hours for each plane.)

90. **Amount of a Pesticide in a Solution** A crop-dusting company needs to apply a pesticide mixture that is 0.2% pesticide. From an earlier application, they already have 300 gal of a more concentrated mixture that is 0.3% pesticide. This mixture is in a 500-gal tank on the plane that they plan to use. Rather than discard the existing mixture, they decide to dilute the existing mixture with water. How much water do they need to add to the 500-gal tank to have the desired concentration? Do they have enough room in this tank?

MASTERY TEST FOR CHAPTER 2

[2.1] 1. Identify the coordinates of the points A through D in the figure and give the quadrant in which each point is located.

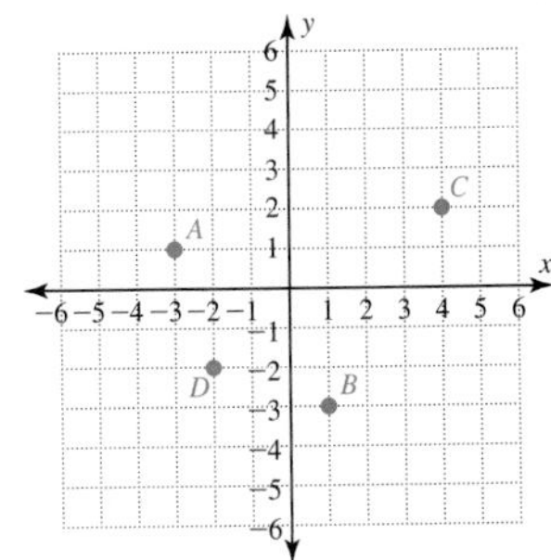

[2.1] 2. Draw a scatter diagram for the data points in the table to the right.

x	y
−2	3
0	−2
1	2
3	1
4	−3

[2.1] 3. Determine whether each sequence is an arithmetic sequence. If the sequence is arithmetic, write the common difference d.

a. 0, 2, 4, 6, 8 b. 1, 2, 4, 8, 16
c. 4, 4, 4, 4, 4 d. $a_n = 15 - 3n$

[2.2] 4. Use the function $f(x) = 11x - 7$ to evaluate each expression.

a. $f(0)$ b. $f(-1)$
c. $f(7)$ d. $f(10)$

[2.2] 5. Use each equation to complete a table with input values of 1, 2, 3, 4, and 5; then graph these points and the line through these points.

a. $y = x + 1$ b. $y = -x + 2$
c. $y = 2x - 5$ d. $y = -2x + 4$

[2.2] 6. **Car Loan** A new car loan requires an $800 down payment and a monthly payment of $375.

a. Write a function that gives the total paid by the end of the xth month.
b. Use this function to complete this table.

x	0	6	12	18	24	30
$f(x)$						

c. Evaluate and interpret $f(25)$.

[2.3] 7. Determine whether $(-2, 3)$ is a solution to each equation.

a. $y = 2x + 7$ b. $y = -x - 1$
c. $y = 3$ d. $x = -2$

[2.3] 8. Determine the x- and y-intercepts of these lines.

a.

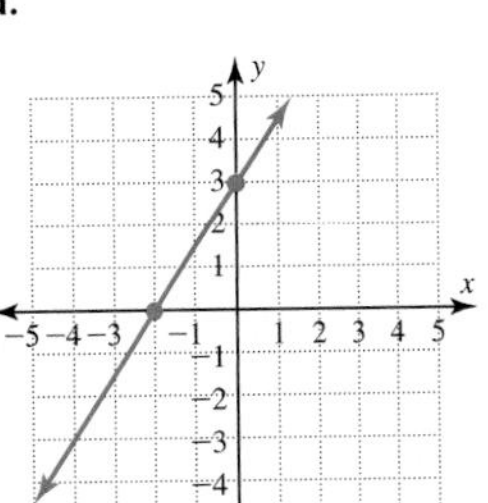

b.

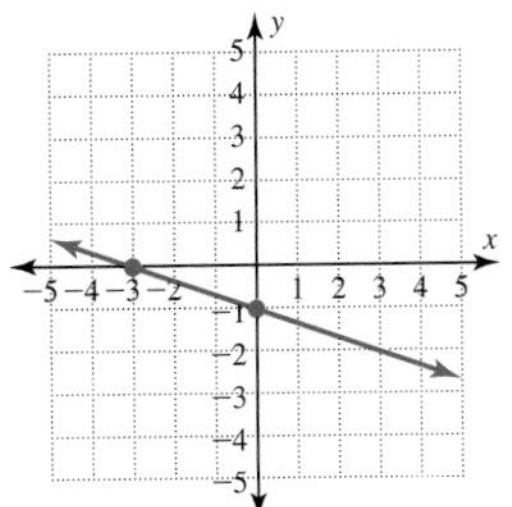

c.

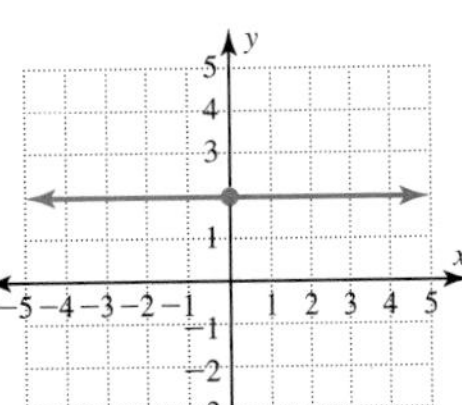

b.

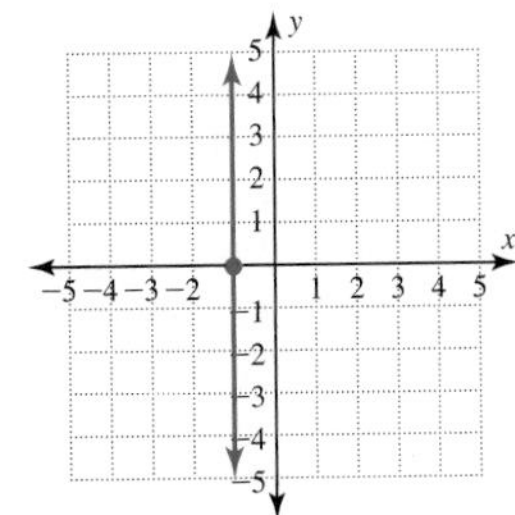

[2.3] 9. Determine the point of intersection of the lines in each graph, and then check this point in both equations.

a.

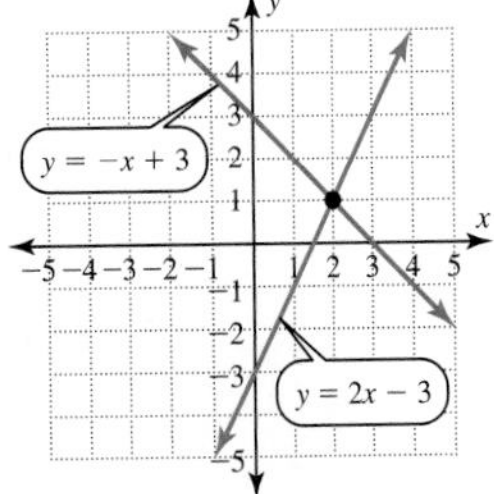

b.

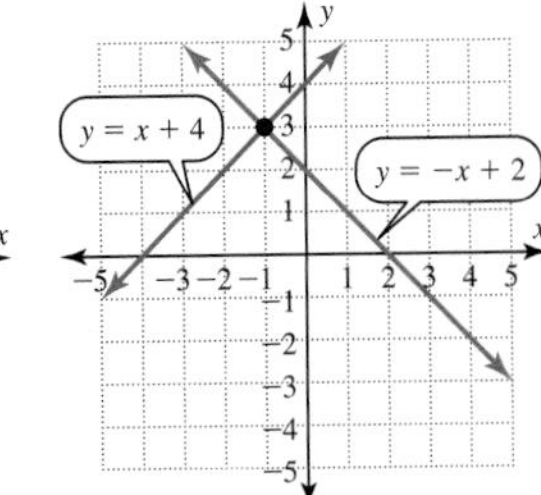

[2.4] 10. Solve each linear equation.

a. $x + 7 = 11$ b. $2x + 7 = x + 21$
c. $7x + 1 = 6x + 1$ d. $x + 2(x + 1) = 4x - 4$

[2.4] 11. Each of these equations is a conditional equation, an identity or a contradiction. Identify the type of each equation and then solve it.

a. $4(x + 2) = 4x + 8$ b. $4(x + 2) = 4x + 2$
c. $4(x + 2) = 2(x + 4)$

[2.4] 12. Let Y_1 equal the left side of each equation and Y_2 equal the right side. Then use a graphing calculator to solve each equation by examining graphs and tables for Y_1 and Y_2.

a. $\frac{x - 4}{2} = \frac{-3x + 4}{2}$ b. $2.3x + 1.6 = 1.7x + 0.4$

[2.5] **13.** Solve each linear equation.

a. $-x = 3x - 12$ **b.** $-\frac{5}{21} = \frac{3y}{7}$

c. $3(2v + 4) = 4(v + 3) + 3$ **d.** $\frac{w - 1}{6} = \frac{w - 3}{5}$

[2.6] **14.** Solve the linear equation $2x - y = 2(3x - y - 1) + 1$ for y, and then use a graphing calculator to complete this table and to graph the equation, using a window of $[-10, 10, 1]$ by $[-10, 10, 1]$.

X	Y_1
-3	
-2	
-1	
0	
1	
2	
3	

X=-3

[2.6] **15.** Calculate the x- and y-intercepts of the graphs of these equations.

a. $y = 2x - 6$ **b.** $y = 4$

c. $2x - 5y = 10$ **d.** $3(2x - 1) = 4(3y - 2) - 7$

[2.7] **16.** Solve the proportions in parts **a** and **b**.

a. $\frac{m - 1}{m - 7} = \frac{2}{5}$ **b.** $\frac{2x - 1}{7} = \frac{x + 1}{4}$

c. On a map 2 cm corresponds to a distance of 75 km. What distance corresponds to a distance of 5 cm on the map?

[2.7] **17. a.** If y varies directly as x, and y is 33 when x is 22, find y when x is 8.

b. The number of MXP (Mexican pesos) varies directly as the number of USD (U.S. dollars) exchanged. In 2007, one could exchange 50 USD for 475 MXP. What was the constant of variation (the exchange rate) at that time?

[2.8] **18.** Determine the restrictions on the variable for each problem.

a. The function $C(x) = 17.50x$ models the cost of filling a 30-gal sprayer tank with x gal of insecticide for a pest control service.

b. The function $C(x) = 17.50x$ models the cost of renting a paint sprayer for x days during the month of June. The rental company will not rent the sprayer for partial days.

c. The function $f(x) = 0.06(200) + 0.10x$ models the number of gallons of acid in a 500-gal tank. The 500-gal tank originally contained 200 gal of a 6% acidic solution, and then x gal of a stronger solution was added to the tank.

d. The function $f(x) = 250 - 25x$ models the length of speaker wire remaining on a spool after x pieces each 25 ft long have been removed from this spool.

[2.8] **19. Mixture of Gasohol** A gasoline distributor has 12,000-gal tank to store a gasohol mix. The tank already contains 5,000 gal of gasohol that contains 5% ethanol by volume.

a. If 1,000 gal of ethanol is added to the tank, what percent of alcohol will be in the resulting mixture? (Round to the nearest tenth of a percent.)

b. How many gallons of ethanol must be added to the tank to produce a mix that is 8% ethanol? (Round to the nearest gallon.)

c. How many gallons of ethanol must be added to the tank to produce a mix that is 10% ethanol? (Round to the nearest gallon.)

REALITY CHECK FOR CHAPTER 2

The windchill chart shown covers Fahrenheit temperatures from $-15°$ to $15°$ and wind speeds ranging from 0 to 40 mi/h. The shaded entries in this chart indicate conditions with a windchill of $-18°$ or lower. These hazardous conditions can produce frostbite in less than 15 minutes, especially in children waiting for school buses. Use this chart to compare the conditions of 5° with a 30-mi/h wind to that of 0° with a 5-mi/h wind. Which conditions would you prefer? Why?

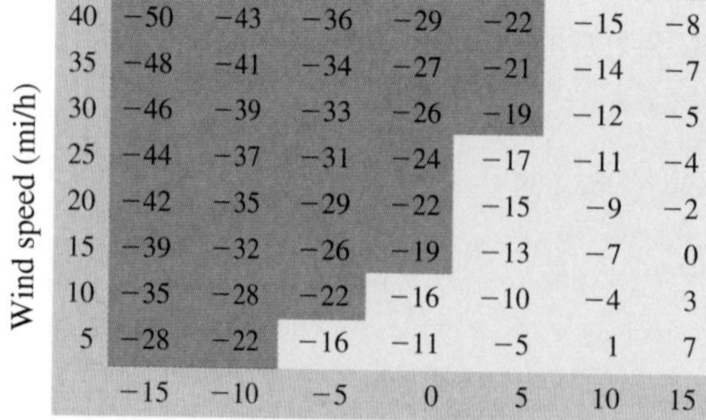
Windchill

Wind speed (mi/h)							
40	−50	−43	−36	−29	−22	−15	−8
35	−48	−41	−34	−27	−21	−14	−7
30	−46	−39	−33	−26	−19	−12	−5
25	−44	−37	−31	−24	−17	−11	−4
20	−42	−35	−29	−22	−15	−9	−2
15	−39	−32	−26	−19	−13	−7	0
10	−35	−28	−22	−16	−10	−4	3
5	−28	−22	−16	−11	−5	1	7
	−15	−10	−5	0	5	10	15

Fahrenheit temperature

GROUP PROJECT FOR CHAPTER 2

Risks and Choice: Comparing the Cost of Leasing Versus Purchasing an Automobile

This project requires the group to collect current data from a local automobile dealership.

Select a specific model of a new car that can be either leased or purchased. Assume there is no trade-in for either the lease or the purchase. To facilitate a comparison, get a price for both options over a 4-year period. For each option determine the total of all initial costs (the required down payment, licenses, taxes, etc.) and the required monthly payment for the 4-year period. Use the information you have collected to complete the following items.

1.

	INITIAL COST	MONTHLY COST
Lease option:	__________	__________
Purchase option:	__________	__________

2. **Lease option:** Write a function $L(x)$ for the total monthly cost of the lease option at the end of the xth month.
   $L(x) =$ ________
3. **Purchase option:** Write a function $P(x)$ for the total monthly cost of the purchase option at the end of the xth month.
   $P(x) =$ ________
4. Graph $y = L(x)$ and $y = P(x)$ on the same coordinate system.
5. Verbally compare the costs of the two options, using the graph from part 4.
6. Determine the estimated value of the car at the end of the 4-year period if you purchase the car.
7. Which option would you select? Why?

chapter 3

Lines and Systems of Linear Equations in Two Variables

CHAPTER OUTLINE

Fossil fuels cannot sustain our energy needs forever, and they produce pollution. Wind power doesn't produce pollutants and indirectly captures heat energy radiating from our sun. Thus this energy source won't run out or even diminish for billions of years. Wind power is one of the fastest-growing energy resources. Since 1995, global wind-generating capacity has increased by almost 500%.

Mathematics is used to design more efficient towers and wind turbines and to reduce the power loss over transmission lines. Increases of efficiency by even 1% can produce tremendous savings and make this energy source more competitive with other energy sources. The graph shows how the production of wind energy has grown in Denmark and Spain since 1996.

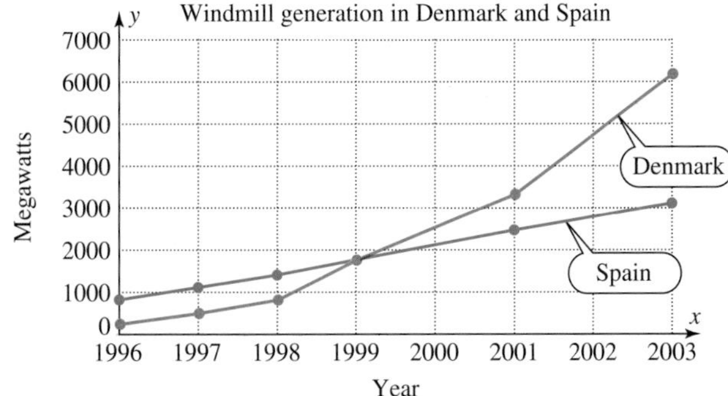

Source: European Wind Energy Association.

REALITY CHECK

In what year did the windmill generation capacity in Denmark first equal or exceed the windmill generation capacity in Spain?

Section 3.1 Slope of a Line and Applications of Slope

Objectives:

1. Determine the slope of a line.
2. Calculate and interpret rates of change.
3. Use the slope to determine whether two lines are parallel, perpendicular, or neither.

Highway engineers must consider safety, construction difficulty, cost, and many other factors when designing new highways. The slope or grade of a highway influences all the decisions related to these factors. In Example 7 we will examine a section of highway that cannot exceed a 6% grade. An important part of mathematics involves analyzing trends and change. This chapter will examine the slope of a line and other key skills needed to study trends and change.

Question: Which of the following lines is the steepest or represents the greatest change in share price per year?

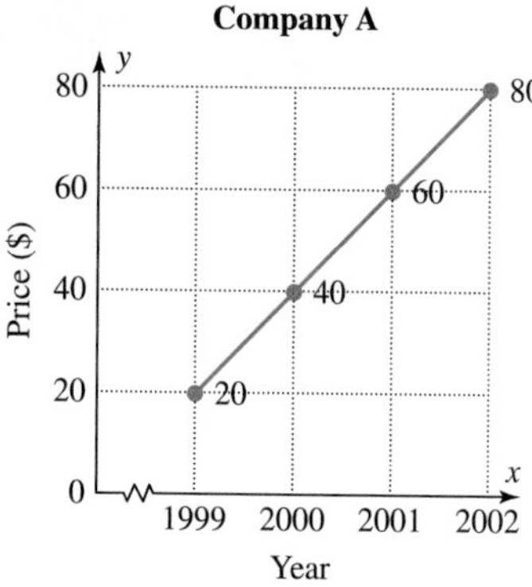

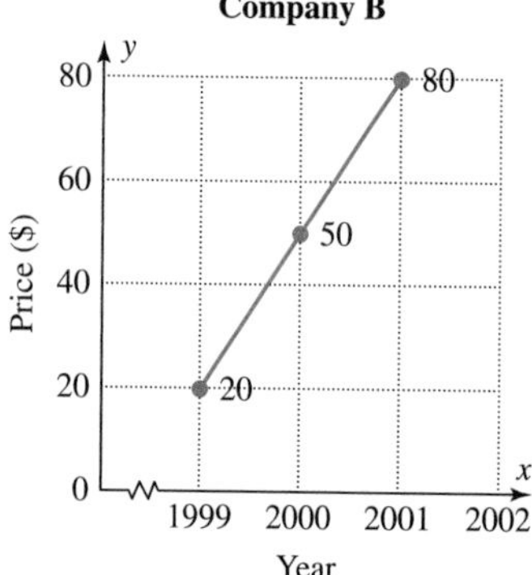

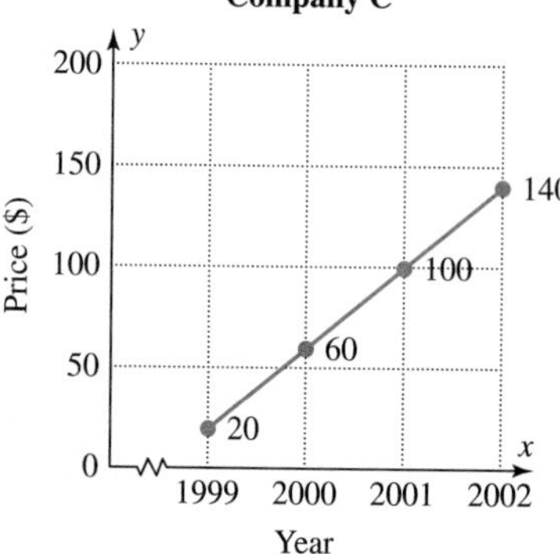

Answer: The graph for company B appears steeper than the graph for company A and accurately reflects that the share price of company B is increasing faster than that of company A ($30 per year for B versus $20 per year for A). Although the graph for company C does not appear as steep as the other two graphs, it is company C whose share price is increasing most rapidly—at $40 per year. The reason this graph does not appear as steep is that a different scale was used on the vertical axis for this graph.

This example illustrates the importance of looking beyond the picture portion of a graph and also examining the numerical information or the scale of the graph. Since our eyes can be misled when examining the steepness, we will also look at this concept algebraically.

The **slope of a line** is a measure of the steepness of the line. In Figure 3.1.1, consider the steepness of the line connecting the point (x_1, y_1) and the point (x_2, y_2). The slope of this line is defined to be the ratio of the change in y (the rise) to the change in x (the run). Slope is usually represented by the letter m.

> **A Mathematical Note**
>
> The origin of the use of m to designate slope is unknown. In his book *Mathematical Circles Revisited,* Howard Eves says that m might have been used because slopes were first studied with respect to mountains. Some suggest that m may have been derived from the French word *monter,* which means "to mount, to climb, or to slope up."

$$m = \frac{\text{change in } y}{\text{change in } x}$$

$$m = \frac{y_2 - y_1}{x_2 - x_1}$$

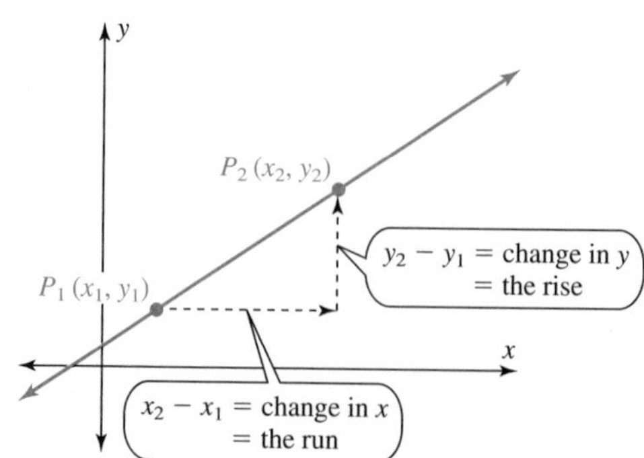

Figure 3.1.1

The delta symbol Δ is often used to designate change. In Calculator Perspective 1.2.4 the **ΔTbl** notation was used to note the change from one input value to the next in a table setup on a TI-84 Plus calculator. Many books also use Δx, read "delta *x*," to denote a change in the *x*-variable and Δy, read "delta *y*," to denote a change in the *y*-variable. Using this notation, we can represent the slope of line by $m = \frac{\Delta y}{\Delta x}$.

Slope of a Line Through (x_1, y_1) and (x_2, y_2)

ALGEBRAICALLY	VERBALLY	NUMERICAL EXAMPLE	GRAPHICAL EXAMPLE
$m = \frac{y_2 - y_1}{x_2 - x_1}$ for $x_1 \neq x_2$. $m = \frac{\Delta y}{\Delta x}$	The slope of a line is the ratio of the change in *y* to the change in *x*.	x: 1, 2, 3, 4; y: −3, −1, 1, 3. A 1-unit change in *x* produces a 2-unit change in *y*. For the points (2, −1) and (3, 1), $m = \frac{1-(-1)}{3-2}$ $m = \frac{2}{1}$	(3, 1); (2, −1); 2-unit change in *y*; 1-unit change in *x*

A key point is that the slope measures the steepness of a line and compares the rate of change of *y* with respect to *x*. Example 1 gives the slope in decimal form, thus making it easy to determine the change in *y* for each 1-unit change in *x*.

EXAMPLE 1 Calculating the Slope of a Line Through Two Points

Calculate the slope of the line through the given points.

SOLUTION

(a) (−5, 8) and (3, −2)

$$m = \frac{y_2 - y_1}{x_2 - x_1}$$

Substitute the given points into the formula for slope with $x_1 = -5$, $y_1 = 8$, $x_2 = 3$, and $y_2 = -2$. Note the use of parentheses in the denominator to prevent an error in sign.

$$m = \frac{-2 - 8}{3 - (-5)}$$

$$m = \frac{-10}{8}$$

Express the slope as a fraction in reduced form.

$$m = -\frac{5}{4} \quad \text{or}$$

y decreases 5 units for each 4-unit increase in *x*, or *y* decreases 1.25 units for every 1-unit increase in *x*.

$$m = -1.25$$

Answer: $m = -1.25$

(b) (3, −2) and (−5, 8)

$$m = \frac{y_2 - y_1}{x_2 - x_1}$$

Although the order of these two points is different, these are the same points as those in part **a** and the slope is also the same. Substitute $x_1 = 3$, $y_1 = -2$, $x_2 = -5$, and $y_2 = 8$.

$$m = \frac{8 - (-2)}{-5 - 3}$$

$$m = \frac{10}{-8}$$

$$m = -\frac{5}{4} \quad \text{or}$$

$$m = -1.25$$

Answer: $m = -1.25$

The slope of a line is the same no matter which two points on the line are used to calculate the slope. As shown in Example 1, this slope is also the same no matter which point is taken first. This is true because the slope is the ratio of the change in y to the corresponding change in x. However, it is crucial that the x- and y-values be kept in the correct pairings. Do not match the x-coordinate of one point with the y-coordinate of another point.

Keep the x- and y-values paired in the same order.

$$m = \frac{y_2 - y_1}{x_2 - x_1} \text{ or } m = \frac{y_1 - y_2}{x_1 - x_2} \qquad m \neq \frac{y_2 - y_1}{x_1 - x_2}$$

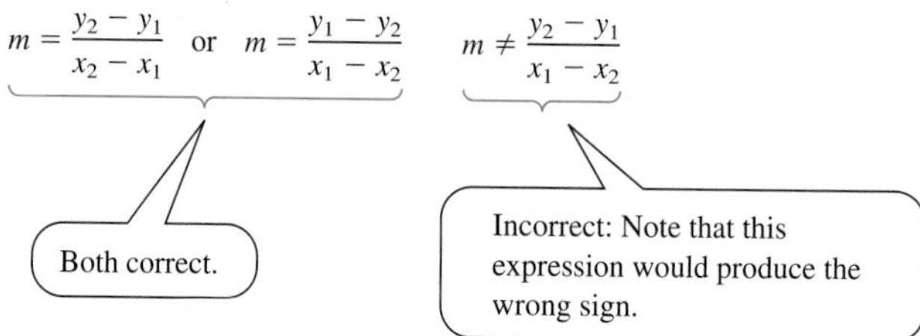

EXAMPLE 2 Calculating the Slope of a Line Through Given Points

Calculate the slope of each of the following lines, using the points labeled on the line.

SOLUTION

(a)

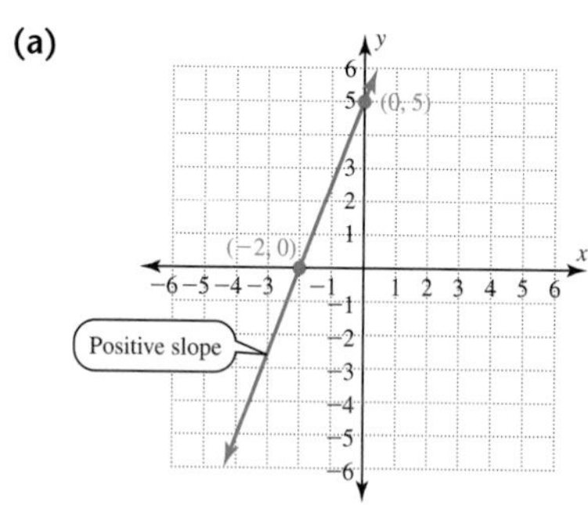

$$m = \frac{y_2 - y_1}{x_2 - x_1}$$

The line slopes upward to the right; its slope is a positive number.

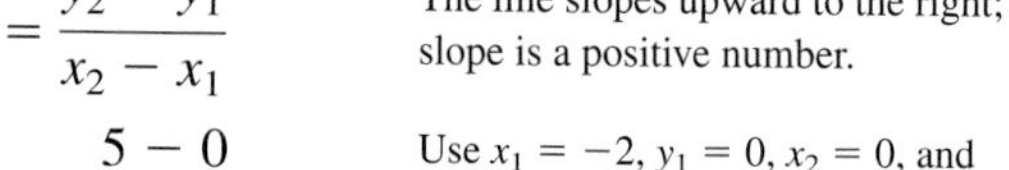

$$m = \frac{5 - 0}{0 - (-2)}$$

Use $x_1 = -2$, $y_1 = 0$, $x_2 = 0$, and $y_2 = 5$.

$$m = \frac{5}{2} \quad \text{or}$$

y increases 5 units for each 2-unit increase in x, or y increases 2.5 units for each 1-unit increase in x.

$$m = 2.5$$

Answer: $m = 2.5$

(b)

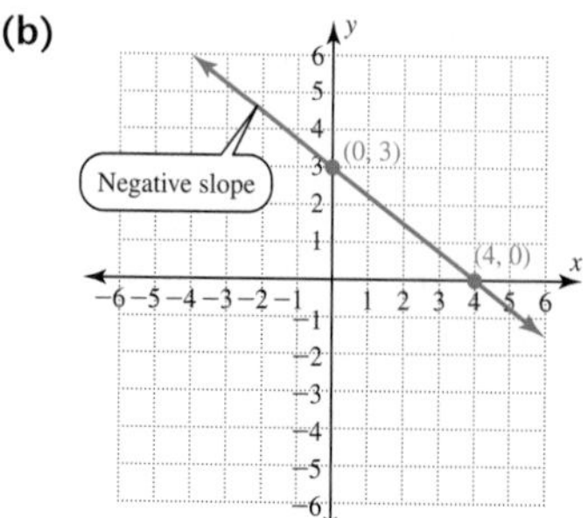

$$m = \frac{y_2 - y_1}{x_2 - x_1}$$

The line slopes downward to the right; its slope is a negative number.

$$m = \frac{0 - 3}{4 - 0}$$

Use $x_1 = 0$, $y_1 = 3$, $x_2 = 4$, and $y_2 = 0$.

$$m = -\frac{3}{4} \quad \text{or}$$

y decreases 3 units for each 4-unit increase in x, or y decreases 0.75 units for each 1-unit increase in x.

$$m = -0.75$$

Answer: $m = -0.75$

Although only two points are needed to determine the slope of a line, it is sometimes easy to determine the slope of a line by examining a table of values of points on the line. This is illustrated in Example 3. If you are creating a table of values for a linear equation with slope $m = \dfrac{\Delta y}{\Delta x}$, often it is convenient to let the change in x in the table be the same as Δx in the fraction $\dfrac{\Delta y}{\Delta x}$.

EXAMPLE 3 Determining Slope From a Table of Values

Calculate the slope of the line containing the points in the table.

SOLUTION

(a)

X	Y1
0	-4
5	-2
10	0
15	2
20	4
25	6
30	8

X=0

$m = \dfrac{2}{5}$

For each increase of 5 units for x, y increases by 2 units. Thus $\dfrac{\Delta y}{\Delta x} = \dfrac{2}{5}$.

(b)

X	Y1
0	7
4	4
8	1
12	-2
16	-5
20	-8
24	-11

X=0

$m = -\dfrac{3}{4}$

For each increase of 4 units for x, y decreases by 3 units. Thus $\dfrac{\Delta y}{\Delta x} = -\dfrac{3}{4}$.

(c)

X	Y1
-3	.5
-2	1
-1	1.5
0	2
1	2.5
2	3
3	3.5

X= -3

$m = 0.5$

For each increase of 1 unit for x, y increases by 0.5 unit. Thus $\dfrac{\Delta y}{\Delta x} = \dfrac{0.5}{1}$.

SELF-CHECK 3.1.1

1. Calculate the slope of the line through $(5, -9)$ and $(-4, -21)$.
2. Calculate the slope of the line, using the points labeled on the graph.

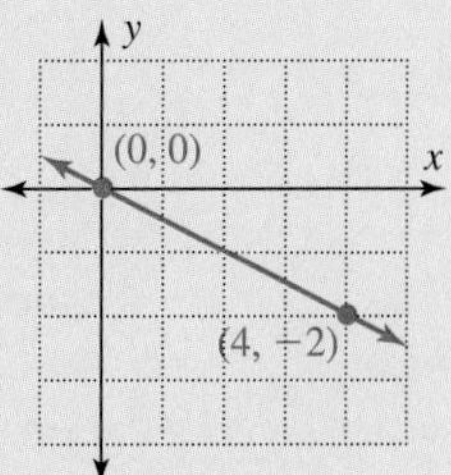

3. Complete the y-values in this table so that the line through these points will have a slope of $m = \dfrac{2}{3}$.

x	y
0	-8
3	
6	
9	
12	

Given any two points on a line, we can graph the line. We can also graph a line if we are given one point and the slope of the line. Start by plotting the given point, and then use the changes for x and y from $\frac{\Delta y}{\Delta x}$ to move horizontally and vertically to a second point on the line. Finally draw the line as illustrated in Example 4. We encourage you to practice graphing lines by this method.

EXAMPLE 4 Using the Slope and y-Intercept to Graph a Line

A line has a y-intercept of (0, 1) and a slope of $\frac{5}{3}$. Use this information to determine another point on the line and to graph this line.

SOLUTION

y-Intercept: (0, 1)

Second point: $(0 + 3, 1 + 5) = (3, 6)$

Start with the y-intercept and use the slope of $\frac{5}{3}$ to determine a second point. Do this by increasing y by 5 units for a 3-unit increase in x.

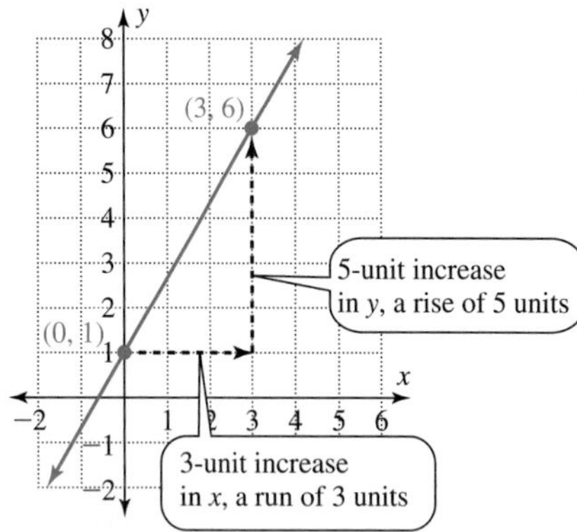

Plot both (0, 1) and (3, 6), and sketch the line through these points. There is a 5-unit rise for each run of 3 units.

Check: $m = \frac{y_2 - y_1}{x_2 - x_1}$

$m = \frac{6 - 1}{3 - 0}$

$m = \frac{5}{3}$ checks.

SELF-CHECK 3.1.2

The slope of a line is $-\frac{2}{5}$.

1. How much change in y will a 5-unit increase in x produce?
2. How much change in y will a 1-unit increase in x produce?
3. Graph a line with this slope and a y-intercept of (0, 3).

All lines with positive slope go upward to the right since the x- and y-coordinates increase together. All lines with negative slope go downward to the right since the y-coordinate decreases as the x-coordinate increases. If we compare two lines graphed on the same coordinate system, the steeper line will have the slope with the larger absolute value.

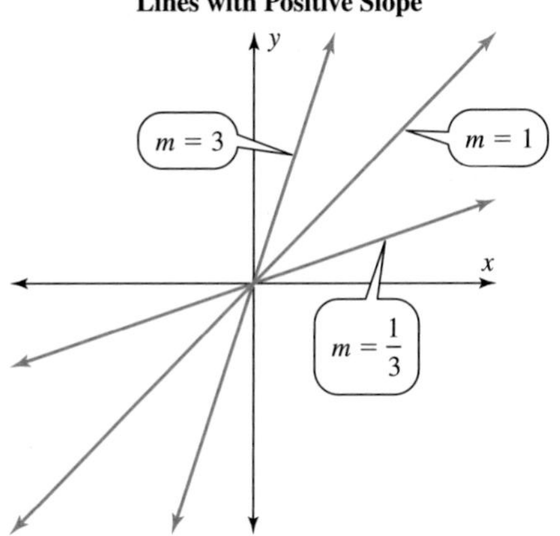

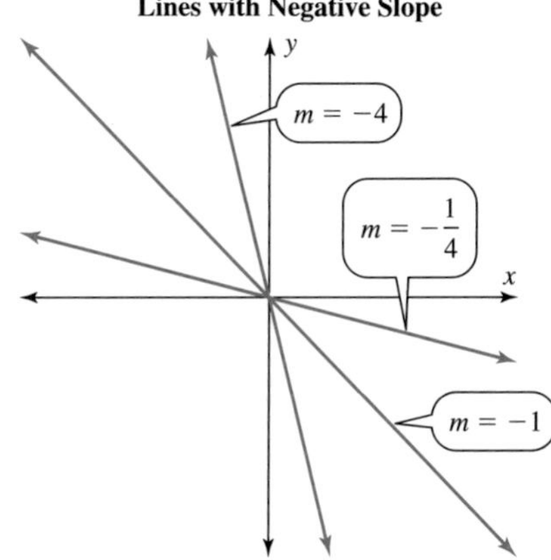

Example 5 shows that the slope of a horizontal line is 0 and the slope of a vertical line is undefined.

EXAMPLE 5 Calculating the Slopes of Horizontal and Vertical Lines

Calculate the slope of each of the following lines, using the points labeled on the line.

SOLUTION

(a)

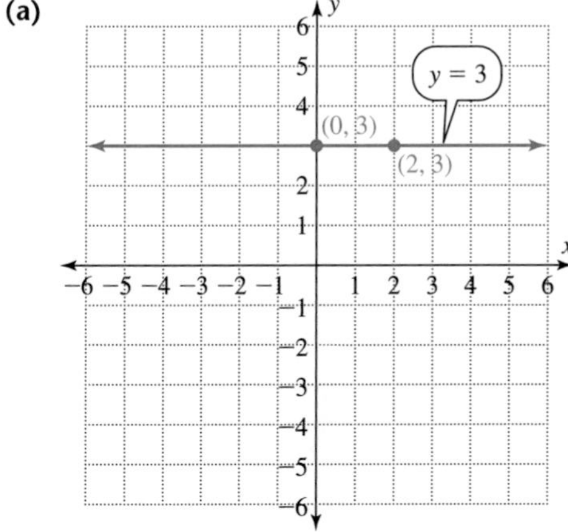

$$m = \frac{y_2 - y_1}{x_2 - x_1}$$

$$m = \frac{3 - 3}{2 - 0}$$ Use $x_1 = 0$, $y_1 = 3$, $x_2 = 2$, and $y_2 = 3$.

$$m = \frac{0}{2}$$ For any change in x, the change in y is 0.

Answer: $m = 0$ This line is horizontal, and its slope is 0.

(b)

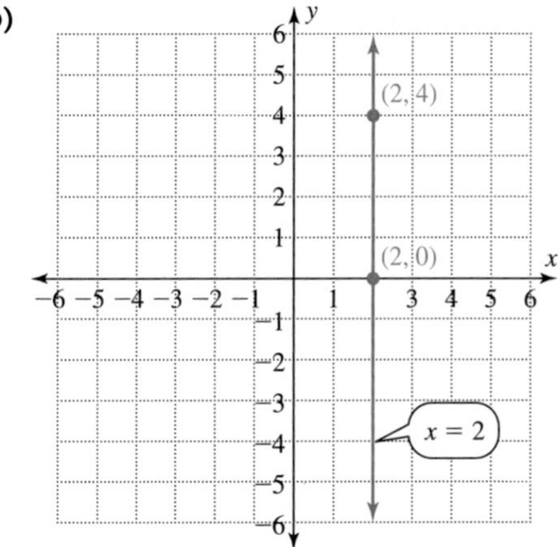

$$m = \frac{y_2 - y_1}{x_2 - x_1}$$

$$m = \frac{4 - 0}{2 - 2}$$ Use $x_1 = 2$, $y_1 = 0$, $x_2 = 2$, and $y_2 = 4$.

$$m = \frac{4}{0}$$

Answer: m is undefined. This line is vertical, and its slope is undefined since division by 0 is undefined.

Classifying Lines by Their Slopes

NUMERICALLY	VERBALLY	GRAPHICALLY
m is positive	The line slopes upward to the right.	
m is negative	The line slopes downward to the right.	
m is zero	The line is horizontal.	
m is undefined	The line is vertical.	

The slope of a line can also be determined from its equation, as illustrated in Example 6. One approach is to determine two points on the line and then use these points in the definition of the slope. In Section 3.2 we will examine how to determine the slope directly from a linear equation.

EXAMPLE 6 Calculating the Slope of a Line from Its Equation

Determine the slope of each of the following lines.

SOLUTION

(a) $2x - 9y = 18$

$$\begin{aligned} 2x - 9y &= 18 \\ 2(0) - 9y &= 18 \\ -9y &= 18 \\ y &= -2 \end{aligned}$$

$$\begin{aligned} 2x - 9(0) &= 18 \\ 2x &= 18 \\ x &= 9 \end{aligned}$$

First determine the intercepts so that we can have two points on the line.

x	y
0	−2
9	0

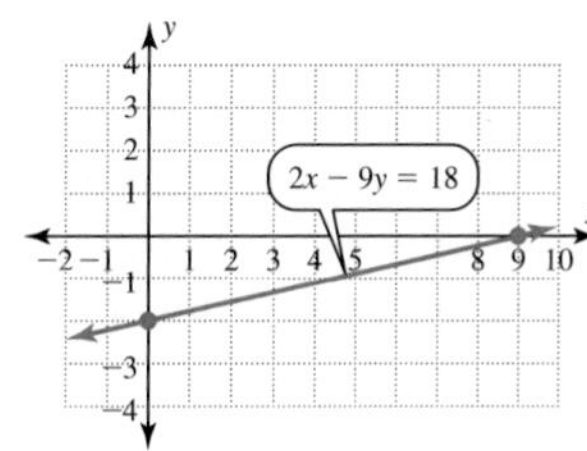

$m = \dfrac{y_2 - y_1}{x_2 - x_1}$

$m = \dfrac{0 - (-2)}{9 - 0}$ Substitute $(0, -2)$ for (x_1, y_1) and $(9, 0)$ for (x_2, y_2).

Answer: $m = \dfrac{2}{9}$

(b) $y = 4$

x	y
0	4
3	4

All points on this line have a y-coordinate of 4.

This is a horizontal line with a slope of 0. The y-coordinate does not change when the x-coordinate changes.

The equation $y = 4$ can also be written as $0x + 1y = 4$. Check the points $(0, 4)$ and $(3, 4)$ in the equation $0x + 1y = 4$.

$m = \dfrac{y_2 - y_1}{x_2 - x_1}$

$m = \dfrac{4 - 4}{3 - 0}$

$m = \dfrac{0}{3}$

Answer: $m = 0$

(c) $x = -3$

x	y
−3	0
−3	2

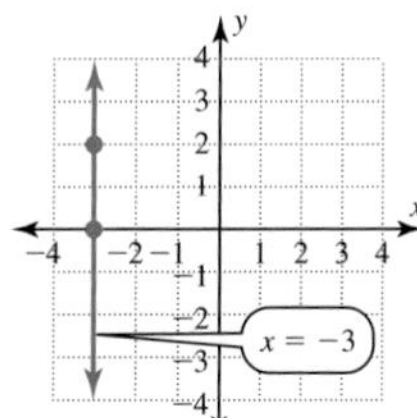

All points on this line have an x-coordinate of -3.

This is a vertical line whose slope is undefined. The x-coordinate does not change.

The equation $x = -3$ can also be written as $1x + 0y = -3$. Check the points $(-3, 0)$ and $(-3, 2)$ in the equation $1x + 0y = -3$.

$m = \dfrac{y_2 - y_1}{x_2 - x_1}$

$m = \dfrac{2 - 0}{-3 - (-3)}$

$m = \dfrac{2}{0}$

Answer: m is undefined

SELF-CHECK 3.1.3

Calculate the slope of the lines defined by these equations.

1. $x + y = 5$ **2.** $y = 5x$
3. $y = 5$ **4.** $x = 5$

A Mathematical Note

The concept of slope as a rate of change is one of the most important concepts in algebra. The term *slope* is widely used in many disciplines. A quick search on the Internet can reveal many uses of the term *slope* from the "glide slope" of an airplane to the "bunny slope" at a ski resort.

A key concept is that slope represents a rate of change. Slope is used in many applications to compare how one quantity is changing with respect to another quantity. Some of these applications use the terms *rise* and *run*. Other applications involving slope may use the terms *angle of elevation* or *grade*. Example 7 involves the grade of a highway.

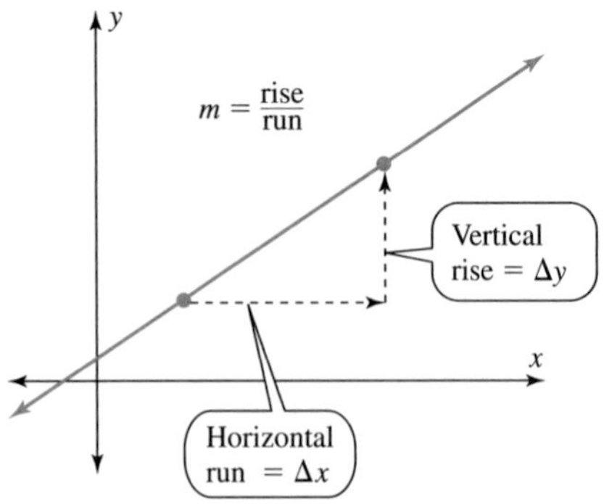

EXAMPLE 7 Modeling the Slope of a Highway

For several reasons, including highway safety, an engineer has determined that the slope of a particular section of highway should not exceed a 6% grade (slope). If this maximum grade is allowed on a section covering a horizontal distance of 2,000 m, how much change in elevation is permitted on this section? (The change of elevation can be controlled by topping hills and filling low places.)

SOLUTION

Let y = change in elevation. Sketch the problem, using a convenient placement of the origin to simplify the computations (see the figure).

$$m = \frac{y_2 - y_1}{x_2 - x_1}$$ Use the formula for slope.

$$\frac{6}{100} = \frac{y - 0}{2{,}000 - 0}$$ Substitute in the changes for y and x, and write 6% as $\frac{6}{100}$.

$$\frac{6}{100} = \frac{y}{2{,}000}$$

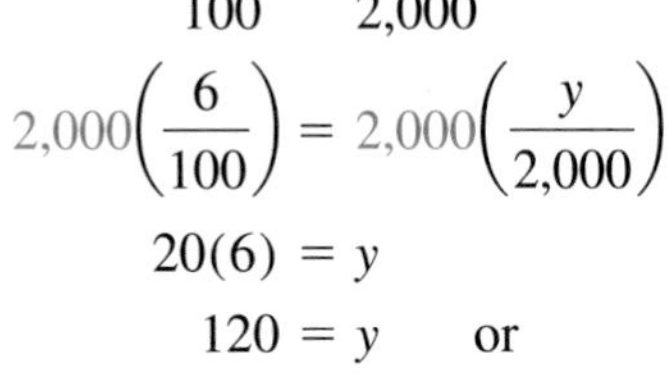

$$2{,}000\left(\frac{6}{100}\right) = 2{,}000\left(\frac{y}{2{,}000}\right)$$ To solve this proportion for y, multiply both sides by the LCD of 2,000.

$$20(6) = y$$

$$120 = y \quad \text{or}$$

$$y = 120$$

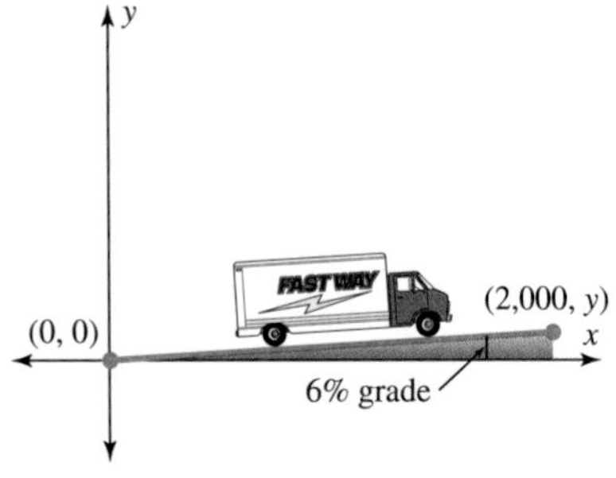

Answer: This section of road could change 120 m in elevation.

In Example 7 the slope represents a rate of change of elevation. In Example 8 the slope represents a rate of change in the depth of water of 3 in/h. Also note that the points in this example form an arithmetic sequence with a common difference of $d = 3$.

EXAMPLE 8 Rate of Change of the Depth of Water in a Pool

A pool is being filled with water from a hose flowing at a constant rate. The depth of the water is measured every hour. These depths recorded in inches are shown in the following sequence: 20, 23, 26, 29, 32, 35, 38.

(a) Graph the points in this sequence.
(b) Determine the slope of the line through these points.

(c) Determine whether this is an arithmetic sequence.
(d) Interpret the slope of the line and the common difference of the arithmetic sequence.

SOLUTION

(a)

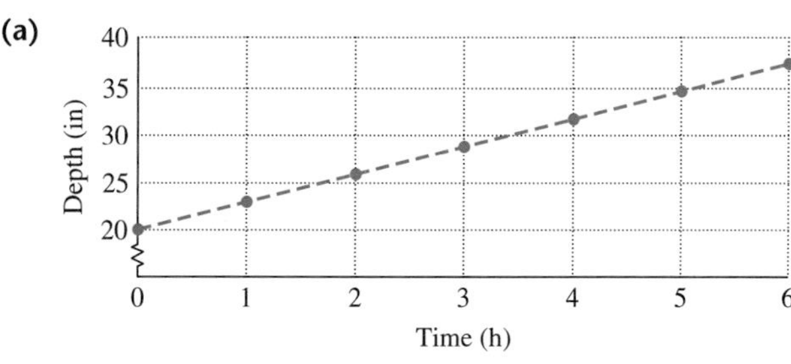

Each of the terms of the sequence corresponds to a 1-hour time period. The dashed line through these points illustrates these points form a linear pattern.

(b) $m = \frac{3}{1} = 3$

The line rises 3 units for every 1-unit move to the right.

(c) $a_0 = 20$
$a_1 = 20 + 3 = 23$
$a_2 = 23 + 3 = 26$
$a_3 = 26 + 3 = 29$
$a_4 = 29 + 3 = 32$
$a_5 = 32 + 3 = 35$
$a_6 = 35 + 3 = 38$

The starting depth of the water a_0 is 20 in, and the depth increases by 3 in by the end of each hour.

This is an arithmetic sequence because there is a common difference $d = 3$.

The linear pattern formed by the points in the graph also confirms that this is an arithmetic sequence.

(d) Both the slope $m = 3$ and the common difference of $d = 3$ represent that the depth of the water in the pool is increasing 3 in/h.

The common difference of 3 between each term of the sequence represents the change in depth of the water during each hour.

SELF-CHECK 3.1.4

1. In Example 7 calculate the elevation change that is permitted if the highway is restricted to a 5% grade.
2. The slope of a line is $m = 27$. In the equation of this line, the variable x represents the number of hours a concrete mixer has been rented and y represents the rental cost of the customer in dollars. Interpret the meaning of the slope of this line.

Another use of slope is to determine whether two lines are parallel or perpendicular. **Parallel lines** are lines that lie in the same plane but never intersect. Parallel lines have the same slope because they rise or fall at the same rate.

Perpendicular lines form a 90° angle when they intersect. (See Figure 3.1.2.) If one line slopes upward to the right, then a line perpendicular to it will slope downward to the right. Perpendicular lines have slopes that are opposite reciprocals. Opposite reciprocals, such as $\frac{2}{3}$ and $-\frac{3}{2}$, have a product of -1, as in $\left(\frac{2}{3}\right)\left(-\frac{3}{2}\right) = -1$. Thus lines that are perpendicular to one another have slopes whose product is -1. All vertical lines are parallel to one another, and all vertical lines are perpendicular to all horizontal lines.

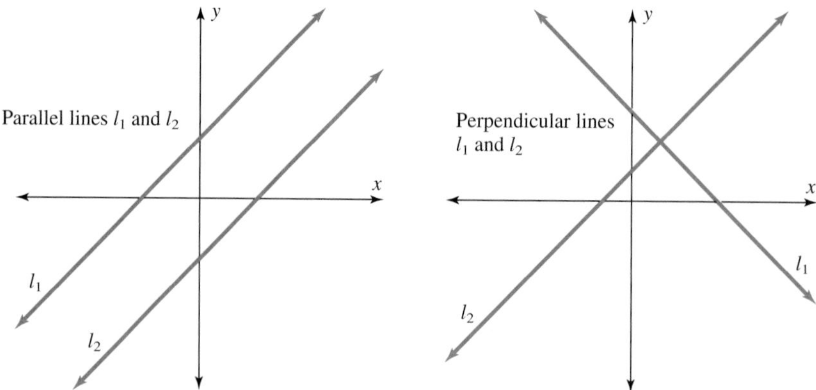

Figure 3.1.2 Parallel and perpendicular lines.

Parallel and Perpendicular Lines

If l_1 and l_2 are distinct nonvertical* lines with slopes m_1 and m_2, respectively, then:

ALGEBRAICALLY	VERBALLY	GRAPHICALLY
$m_1 = m_2$	Lines l_1 and l_2 are parallel because they have the same slope.	y, x, l_1, l_2
$m_1 = -\frac{1}{m_2}$ or $m_1 m_2 = -1$	Lines l_1 and l_2 are perpendicular because their slopes are opposite reciprocals.	y, x, l_1, l_2

*Also note all vertical lines are parallel to one another, and all vertical lines are perpendicular to all horizontal lines.

EXAMPLE 9 Determining Whether Two Lines Are Parallel or Perpendicular

Determine whether the line through the first pair of points is parallel to, perpendicular to, or neither parallel nor perpendicular to the line through the second pair of points.

SOLUTION

(a) (1, 3) and (−1, −1)
(2, 1) and (3, 3)

$$m = \frac{y_2 - y_1}{x_2 - x_1} \qquad m = \frac{y_2 - y_1}{x_2 - x_1}$$

$$m = \frac{-1 - 3}{-1 - 1} \qquad m = \frac{3 - 1}{3 - 2}$$

Calculate the slope of each line by substituting in the given points.

$$m = \frac{-4}{-2} \qquad m = \frac{2}{1}$$

$$m = 2 \qquad m = 2$$

Since the slopes are equal, the lines are parallel.

Answer: The two lines are parallel.

(b) (3, 1) and (6, 3)
(2, −1) and (4, −4)

$m = \frac{y_2 - y_1}{x_2 - x_1}$ $m = \frac{y_2 - y_1}{x_2 - x_1}$

$m = \frac{3 - 1}{6 - 3}$ $m = \frac{-4 - (-1)}{4 - 2}$

$m = \frac{2}{3}$ $m = -\frac{3}{2}$

Since $\left(\frac{2}{3}\right)\left(-\frac{3}{2}\right) = -1$, the two lines are perpendicular.

Answer: The two lines are perpendicular.

(c) (1, 7) and (−1, −1)
(1, 1) and (−1, 9)

$m = \frac{y_2 - y_1}{x_2 - x_1}$ $m = \frac{y_2 - y_1}{x_2 - x_1}$

$m = \frac{-1 - 7}{-1 - 1}$ $m = \frac{9 - 1}{-1 - 1}$

$m = \frac{-8}{-2}$ $m = \frac{8}{-2}$

$m = 4$ $m = -4$

The slopes are not equal, and $4(-4) \neq -1$. Thus the lines are neither parallel nor perpendicular.

Answer: The lines are neither parallel nor perpendicular.

Two features of linear functions that are key to characterizing members of the family of linear functions have now been examined: the intercepts of a line and the slope of a line. In Section 3.2 we will investigate how to observe these two key features directly from the algebraic equation defining the line. This will also allow us to examine parallel and perpendicular lines further and to look at examples involving horizontal and vertical lines.

SELF-CHECK 3.1.5

1. Determine whether the line through (5, 3) and (−5, −1) is parallel to, perpendicular to, or neither parallel nor perpendicular to the line through (2, −3) and (−2, 7).

SELF-CHECK ANSWERS

3.1.1

1. $m = \frac{4}{3}$; y increases $1\frac{1}{3}$ units for each 1-unit increase in x.
2. $m = -\frac{1}{2}$; y decreases 0.5 unit for each 1-unit increase in x.
3.

x	y
0	−8
3	−6
6	−4
9	−2
12	0

3.1.2

1. 2-unit decrease in y
2. −0.4-unit decrease in y
3.

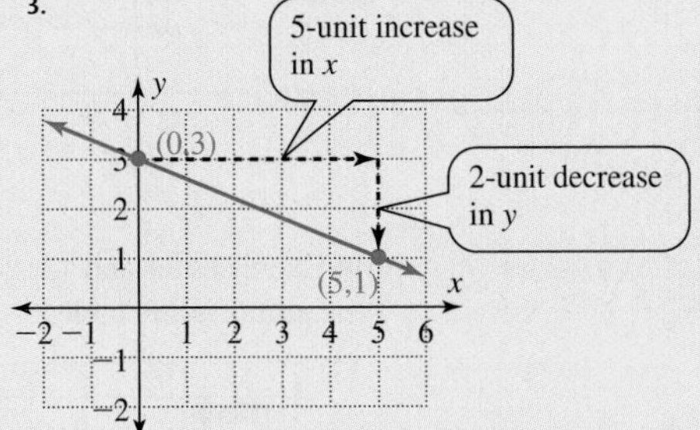

3.1.3

1. $m = -1$; this line slopes downward to the right.
2. $m = 5$; this line slopes upward to the right.
3. $m = 0$; this is a horizontal line.
4. m is undefined; this is a vertical line.

3.1.4

1. This section of road could change 100 m in elevation.
2. The concrete mixer costs $27/h to rent.

3.1.5

1. The lines are perpendicular.

USING THE LANGUAGE AND SYMBOLISM OF MATHEMATICS 3.1

1. The rise between two points on a line refers to the change in the __________ variable.
2. The run between two points on a line refers to the change in the __________ variable.
3. The letter that is used to represent the slope of a line is __________.
4. The formula for the slope of a line through (x_1, y_1) and (x_2, y_2) is __________.
5. The slope of a line gives the change in y for each __________ unit change in x.
6. The slope of a horizontal line is __________.
7. The slope of a vertical line is __________.
8. The symbol Δx represents the __________ in x.
9. The symbol Δy represents the __________ in y.
10. Slope represents a __________ of change.
11. If the slope of a line is positive and the x-coordinate increases, then the y-coordinate __________.
12. If the slope of a line is negative and the x-coordinate increases, then the y-coordinate __________.
13. If two lines have the same slope, the lines are __________.
14. If the slopes of two lines are opposite reciprocals, the lines are __________.
15. The common difference of an arithmetic sequence is the same as the __________ of the line through the points of the sequence.
16. The slope of a highway is sometimes referred to as the __________ of the highway.

EXERCISES 3.1

In Exercises 1–8, calculate the slope of the line through the given points.

1. a. (4, 3) and (3, 1) b. (3, 1) and (4, 3)
2. a. (5, 8) and (2, 2) b. (2, 2) and (5, 8)
3. a. $(2, -7)$ and $(-7, 4)$ b. $(2, -7)$ and $(4, -7)$
4. a. $(-6, 2)$ and $(2, 9)$ b. $(-6, 2)$ and $(9, 2)$
5. a. $(2, -7)$ and $(-10, 2)$ b. $(2, -7)$ and $(2, -10)$
6. a. $(-2, -1)$ and $(-10, -2)$ b. $(-2, -1)$ and $(-2, -10)$
7. a. (1.40, 0.56) and (0, 0) b. (0, 0) and $\left(\frac{1}{3}, \frac{1}{5}\right)$
8. a. $\left(\frac{1}{2}, \frac{1}{4}\right)$ and (0, 0) b. (0, 0) and (0.51, 0.34)

In Exercises 9 and 10, calculate the slope of each line.

9. a.

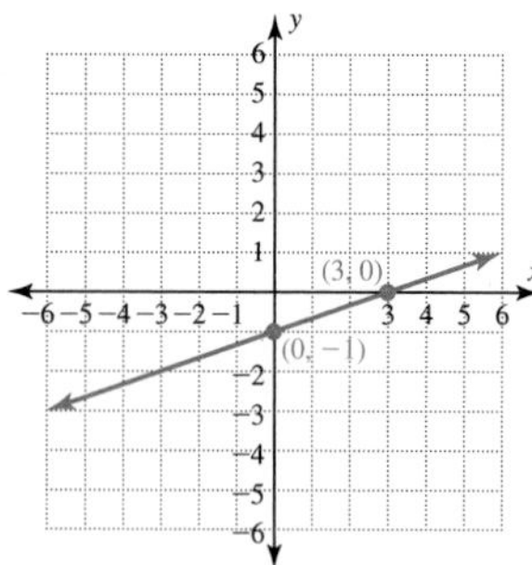

b.

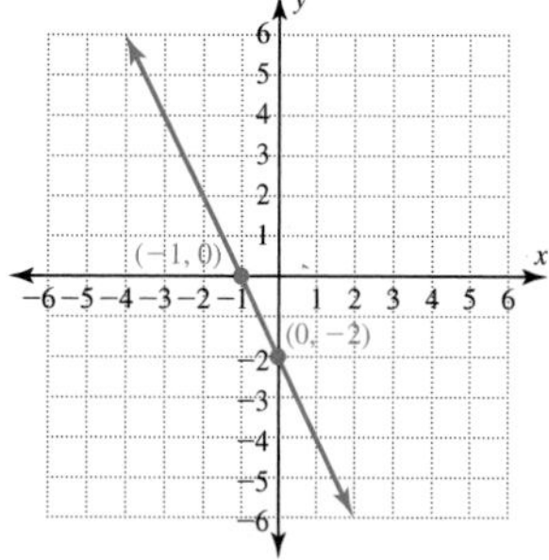

c.

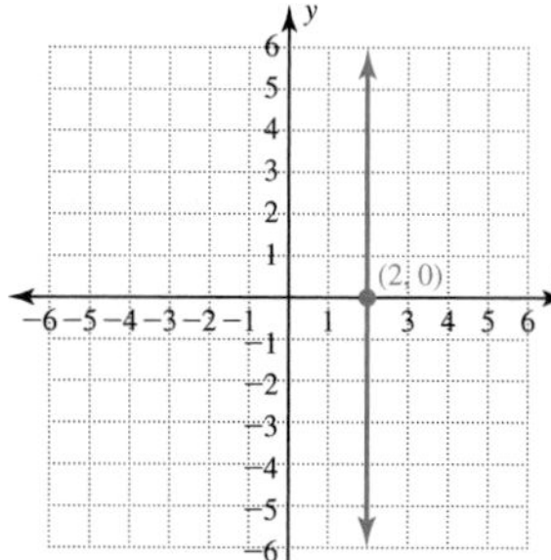

d.

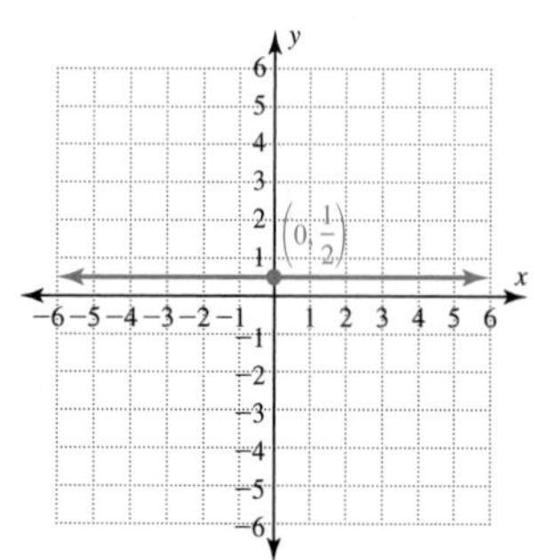

10. a.

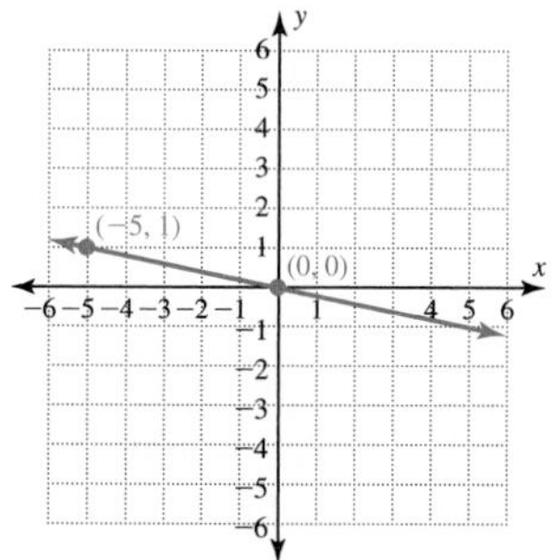

b.

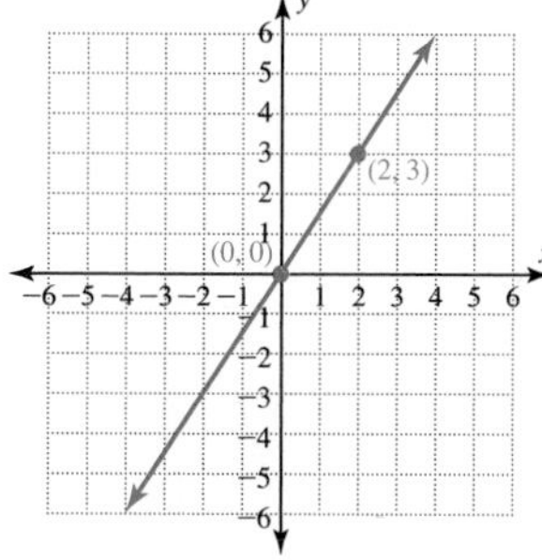

c.

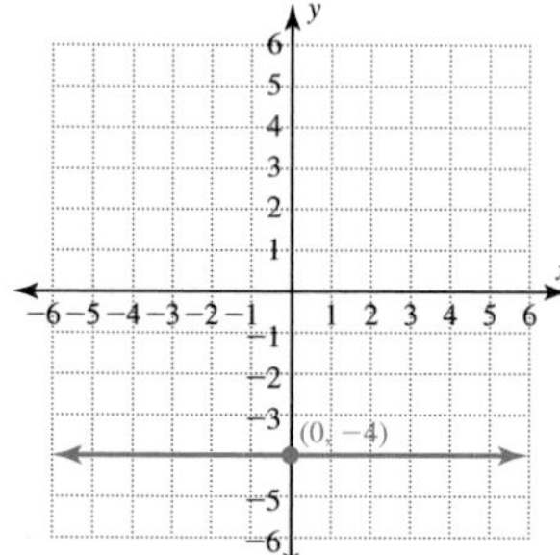

d.

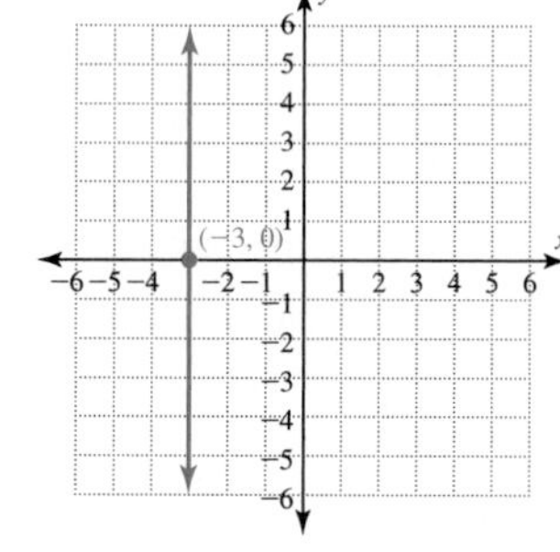

In Exercises 11 and 12, calculate the slope of the line containing the points in the table.

11.

X	Y1
0	1
2	4
4	7
6	10
8	13
10	16
12	19

X=0

12.

X	Y1
0	11
7	9
14	7
21	5
28	3
35	1
42	-1

X=0

In Exercises 13–16, complete each table so that the points all lie on a line with the given slope. Complete each table without using a calculator.

13. $m = \frac{3}{2}$

x	y
0	5
2	
4	
6	
8	

14. $m = -\frac{3}{2}$

x	y
0	5
2	
4	
6	
8	

15. $m = -\frac{2}{3}$

a.

x	y
0	5
3	
6	
9	
12	

b.

x	y
0	5
6	
12	
18	
24	

16. $m = \frac{2}{3}$

a.

x	y
0	5
3	
6	
9	
12	

b.

x	y
0	5
9	
18	
27	
36	

17. Given the linear equation $y = \frac{5}{3}x - 2$:

a. Use this equation and a calculator to complete the table of values shown.

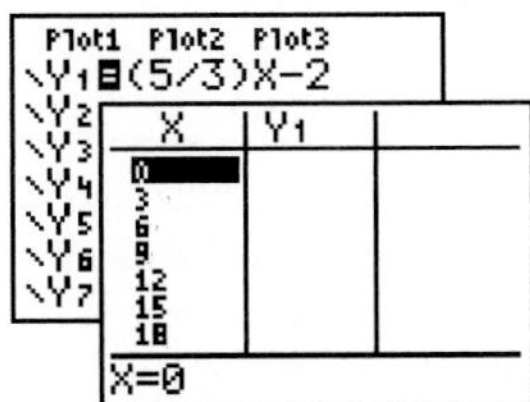

b. What is the value of Δx used in this table?

c. What is the value of Δy that is produced in this table?

d. What is the slope of the line through these points?

18. Given the linear equation $y = -\frac{2}{5}x + 1$:

a. Use this equation and a calculator to complete the table of values shown.

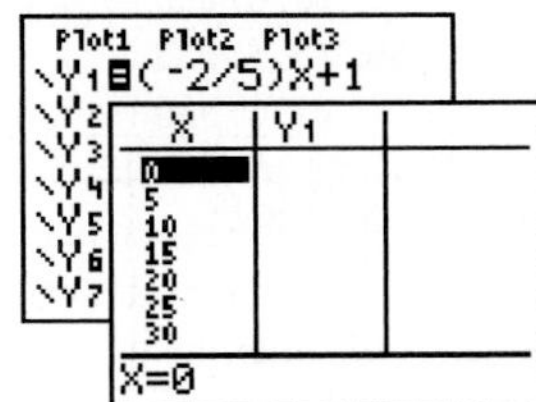

b. What is the value of Δx used in this table?

c. What is the value of Δy that is produced in this table?

d. What is the slope of the line through these points?

In Exercises 19 and 20, determine two points on each line and then calculate the slope of the line.

19. a. $3x + 5y = 15$ **b.** $5x = 8y$
c. $x = 11$ **d.** $y = -11$

20. a. $4x + 3y = 24$ **b.** $7x = 4y$
c. $x = -7$ **d.** $y = 7$

In Exercises 21–32, complete the following table involving the change in x, the change in y, and the slope of the line defined by $y = mx + b$.

	CHANGE IN x	CHANGE IN y	SLOPE
21.	−5	8	
22.	−7	−2	
23.	3		$\frac{2}{3}$
24.	3		$-\frac{2}{3}$
25.	1		$\frac{2}{3}$
26.	1		$-\frac{2}{3}$
27.		2	$\frac{2}{3}$
28.		2	$-\frac{2}{3}$
29.		6	$-\frac{2}{3}$
30.		−6	$-\frac{2}{3}$
31.	6		0
32.	−6		0

In Exercises 33–36, a line has the given y-intercept and slope. Use this information to determine another point on the line and to graph the line.

33. $(0, 2); m = \frac{3}{4}$ **34.** $(0, 2); m = -\frac{3}{4}$

35. $(0, -3); m = -2$ **36.** $(0, -3); m = 2$

In Exercises 37–42, draw a line through the point (1, 3) that has the given slope.

37. $m = 0$
38. $m = 1$
39. $m = 2$
40. $m = -1$
41. $m = -2$
42. m is undefined

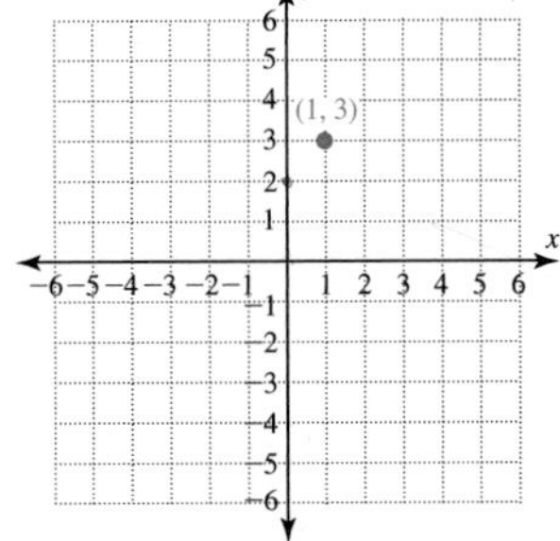

In Exercises 43 and 44, the table gives x and y values for a linear equation $y = mx + b$. Determine

a. The x-intercept of this line
b. The y-intercept of this line
c. The slope of this line

43.

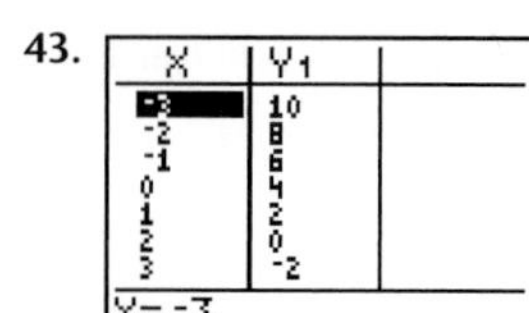

X	Y1
-3	10
-2	8
-1	6
0	4
1	2
2	0
3	-2

X=-3

44.

X	Y1
-3	-6
-2	-3
-1	0
0	3
1	6
2	9
3	12

X=-3

45. Distance Traveled The following graph gives the distance an automobile has traveled after different periods of time. Determine the slope of this line, and then interpret the meaning of this rate of change.

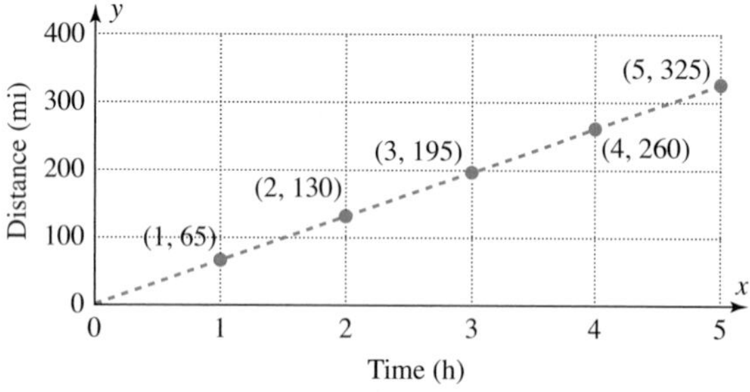

46. Cost of Shirts The following graph gives the total order and shipping costs for an order of shirts from an online vendor. Determine the slope of this line, and then interpret the meaning of this rate of change.

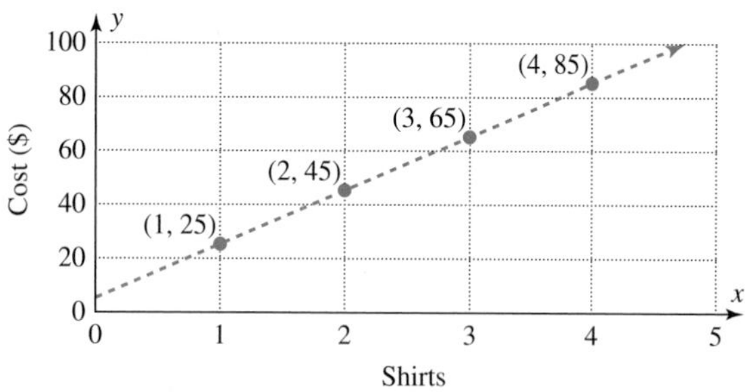

Estimate Then Calculate

In Exercises 47–50, mentally estimate the slope of each line. Then use a calculator to approximate the slope to the nearest thousandth.

	GRAPH	MENTAL ESTIMATE OF m	CALCULATOR APPROXIMATION OF m
47.	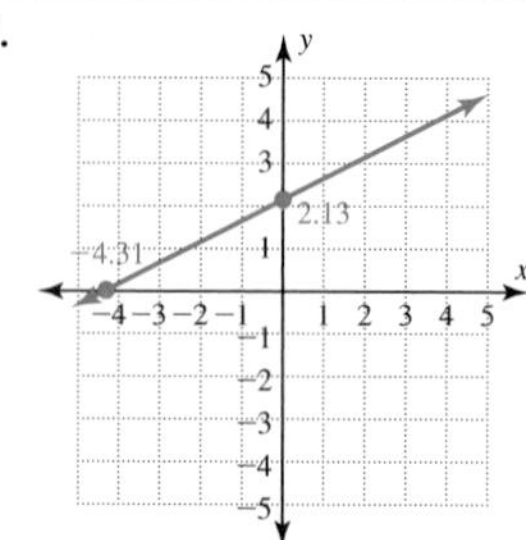		
48.	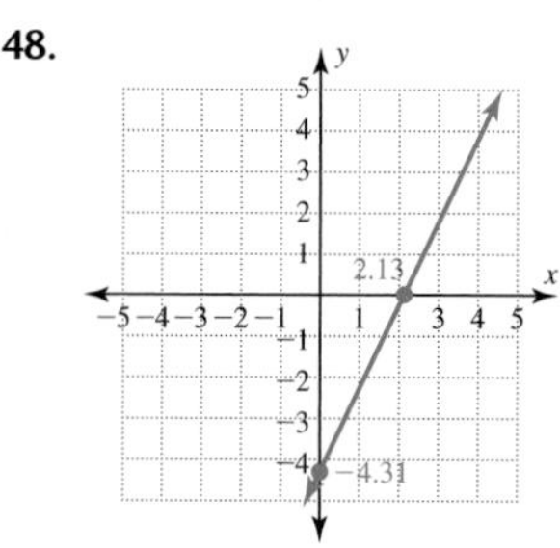		
49.	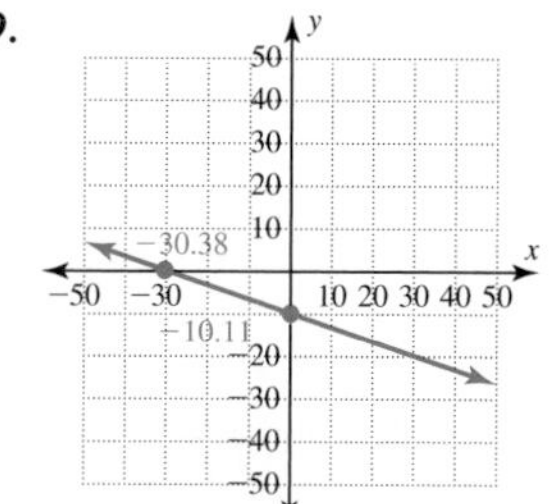		
50.	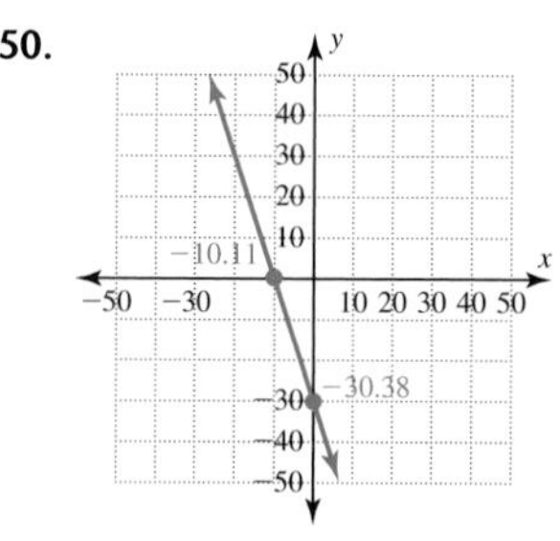		

In Exercises 51–56, m_1 is the slope of line l_1; m_2 is the slope of line l_2; and m_3 is the slope of line l_3. Line l_1 is parallel to l_2, and l_1 is perpendicular to l_3. Complete the following table.

	m_1	m_2	m_3
51.	$\frac{3}{7}$		
52.	$-\frac{3}{7}$		
53.	0		
54.	Undefined		
55.		$\frac{8}{5}$	
56.			$-\frac{3}{11}$

In Exercises 57–66, determine whether the line through the first pair of points is parallel to, perpendicular to, or neither parallel nor perpendicular to the line through the second pair of points. Assume that no line passes through both pairs of points.

57. $(-3, 5)$ and $(-6, 3)$
$(3, -3)$ and $(6, -1)$

58. $(4, 4)$ and $(-4, -2)$
$(3, 1)$ and $(-3, 9)$

59. $(5, 4)$ and $(10, 1)$
$(6, 8)$ and $(3, 3)$

60. $(7, -1)$ and $(-7, -7)$
$(7, 5)$ and $(-7, -1)$

61. $(0, 6)$ and $(8, 0)$
$(-5, 0)$ and $(0, -7)$

62. $(0, -11)$ and $(9, 0)$
$(-6, 0)$ and $(0, 6)$

63. $(0, 6)$ and $(4, 6)$
$(0, 6)$ and $(0, 0)$

64. $(0, 0)$ and $(0, -3)$
$(0, 0)$ and $(3, 0)$

65. $(5, 3)$ and $(5, -3)$
$(7, 6)$ and $(7, -6)$

66. $(5, 3)$ and $(-5, 3)$
$(7, 6)$ and $(-7, 6)$

67. Rate of Flow The water tank on a firetruck holds 500 gal of water. This water is used so the firefighters can begin pumping water as soon as they arrive at a fire. The volume of water remaining in the tank x seconds after the pump has been turned on is displayed in the table.

	A	B
1	Time (s)	Volume (gal)
2	0	500
3	20	440
4	40	380
5	60	320
6		

a. Determine the rate of change of the volume with respect to time.
b. Interpret the meaning of this value.
c. At this rate, how long do the firefighters have to connect to a hydrant before the water in the tank runs out?

68. **Rate of Descent** At a local flight school, the air traffic control tower began recording the altitude of an airplane flown by a student pilot x seconds after the plane began its descent. These data are displayed in the given table.

	A	B
1	Time (s)	Altitude (ft)
2	0	2800
3	15	2740
4	30	2680
5	45	2620
6		

a. Determine the rate of change of the altitude with respect to time.
b. Interpret the meaning of this value.
c. At this rate, how long after the plane begins its descent will it land?

69. **Using a Sequence That Models House Payments** The equation $a_n = 1{,}250n + 5{,}000$ gives the total payment in dollars for a house for n months.
a. Determine the sequence of total payments for the first 4 months.
b. Graph this sequence.
c. Determine the slope of the line through these points. Interpret the meaning of the rate of change in this application.
d. Calculate a_0, and interpret the meaning of a_0.

70. **Using a Sequence That Models House Payments** The equation $a_n = 500n + 1{,}500$ gives the total payment in dollars for a house for n months.
a. Determine the sequence of total payments for the first 4 months.
b. Graph this sequence.
c. Determine the slope of the line through these points. Interpret the meaning of the rate of change in this application.
d. Calculate a_0, and interpret the meaning of a_0.

71. **Slope of a Wheelchair Ramp** Determine the slope of the wheelchair ramp shown in the figure.

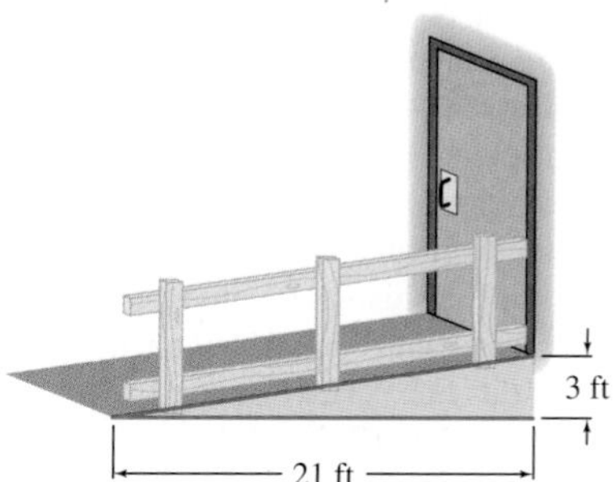

72. **Slope of a Roof** Determine the slope of the roof on the gable of the house whose cross section is sketched.

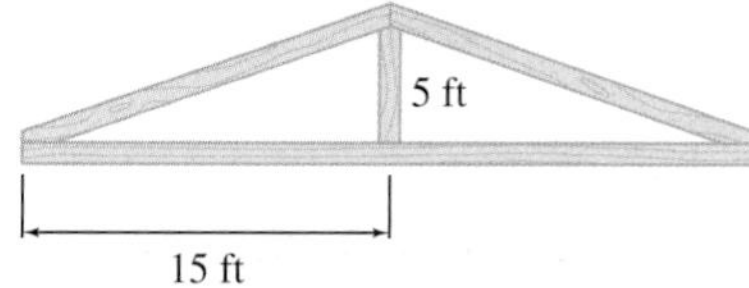

73. **Modeling the Grade of a Highway** For several reasons, including highway safety, an engineer has determined that the slope of a particular section of highway should not exceed a 5% grade (slope). If this maximum grade is allowed on a section covering a horizontal distance of 1,800 m, how much change in elevation is permitted on this section? (The change of elevation can be controlled by topping hills and filling in low places. See the given figure.)

74. **Modeling the Grade of a Highway** How much change in elevation would be permitted on a section covering a horizontal distance of 2,400 m on the highway in Exercise 73?

75. **Modeling the Height of a Roof Brace** The roof of the mountain cabin in the figure rises 14 ft over a run of 12 ft. Determine the height h of the brace placed 3 ft from the side of the house.

76. **Modeling the Height of a Tree** A man who is 6 ft tall is standing so that the tip of his head is exactly in the shadow line of a tree. His shadow is 4 ft long, and the shadow of the tree is 34 ft long. How tall is the tree?
77. What is the slope of the x-axis?
78. What is the slope of the y-axis?
79. A line passes through quadrants I, II, and III but not quadrant IV. Is the slope of this line positive or negative?
80. A line passes through quadrants I, II, and IV but not quadrant III. Is the slope of this line positive or negative?
81. A line passes through quadrants II, III, and IV but not quadrant I. Is the slope of this line positive or negative?
82. A line passes through quadrants I, III, and IV but not quadrant II. Is the slope of this line positive or negative?
83. A line passes through quadrants I and III but not quadrants II and IV. Is the slope of this line positive or negative?
84. A line passes through quadrants II and IV but not quadrants I and III. Is the slope of this line positive or negative?
85. Variable y varies directly as x with a constant of variation 2. What is the slope of the line connecting these (x, y) data points?
86. Variable y varies directly as x with a constant of variation -3. What is the slope of the line connecting these (x, y) data points?

Group Discussion Questions

87. **Communicating Mathematically** Use the language of slope to compare the difficulty in pedaling for a bicycle rider in these situations. Assume the rider is moving from left to right as you present your descriptions. (Thanks to Fred Worth of Henderson State University for suggesting this problem.)
    a. A flat road
    b. A road that goes up 10 ft over 1 mi of roadway
    c. A road that goes up 10 ft over 50 ft of roadway
    d. A road that goes up 20 ft over 200 ft of roadway
    e. A road that goes down 10 ft over 1 mi of roadway
88. **Discovery Question** The equation of a line is $f(x) = mx + b$.
    a. Calculate the y-intercept of this line.
    b. Calculate the x-intercept of this line.
    c. Use the intercepts and the formula for slope to calculate the slope of this line.
    d. Use the results obtained in parts **a** to **c** to determine by inspection the slope and y-intercept of $f(x) = 5x - 3$.
89. **Discovery Question** The equation of a line is $y - y_1 = m(x - x_1)$.
    a. Is (x_1, y_1) a point on this line? Justify your answer.
    b. Determine the slope of this line and justify your answer.
    c. Use the results obtained in parts **a** and **b** to determine by inspection the slope and one point on the line defined by $y - 2 = 4(x - 5)$.
90. **Communicating Mathematically** Can you calculate the slope between $(-2, -10)$ and $(-2, -10)$? Explain why or why not.
91. **Error Analysis** A student examined the graph shown on the calculator display below and concluded the line was vertical and therefore the slope was undefined. Describe the error the student has made.

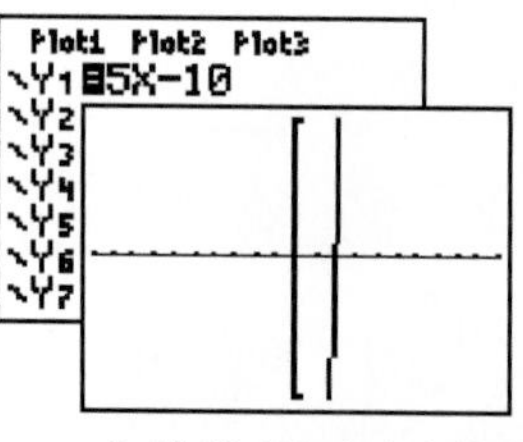

$[-10, 10, 1]$ by $[-1, 1, 1]$

92. **Error Analysis** A student examined the graphs shown on the calculator display below and concluded that the lines were parallel. Describe the error the student has made.

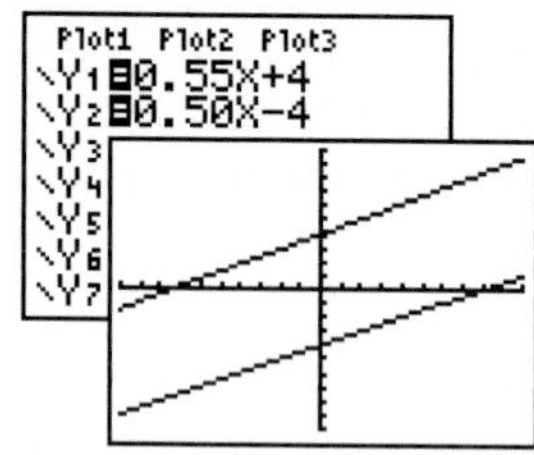

$[-10, 10, 1]$ by $[-10, 10, 1]$

Section 3.2 Special Forms of Linear Equations in Two Variables

Objectives:
4. Use the slope-intercept and point-slope forms of a linear equation.
5. Use the special forms of equations for horizontal and vertical lines.
6. Graph a line, given one point and its slope.

We use a variety of graphs—bar graphs, pie graphs, graphs in the Cartesian plane, etc.—to examine mathematical concepts from a graphical perspective. Each type of graph has a context where it is more informative than other graphs. Bar graphs are good for quick comparisons of data; pie charts are good for comparing parts of a whole; and line graphs are good for showing trends.

We also use a variety of algebraic forms of linear equations. Again, the context will determine which form is the most useful. Three forms of linear equations that we will examine in this section are $y = mx + b$, $y - y_1 = m(x - x_1)$, and $Ax + By = C$.

A Mathematical Note

W. W. Sawyer makes the point in his 1943 book *Mathematician's Delight* that the first mathematicians were practical men who built or made things. Some of our terminology can be traced to this source. For example, the word *straight* comes from Old English for "stretched," whereas the word *line* is the same as that for "linen thread." Thus a straight line is literally a stretched linen thread—as anyone who is planting potatoes or laying bricks knows.

Slope-Intercept Form

One of the most useful forms of a linear equation is the form $y = mx + b$. We have used this form frequently to enter equations into a graphing calculator. This form is also the function form for a linear equation and can be written as $f(x) = mx + b$ to stress the relationship between the x-y input-output pairs.

The form $y = mx + b$ is called the slope-intercept form because this form displays the slope m and the y-intercept $(0, b)$ directly from the equation. This form can be developed by using the slope m, the y-intercept $(0, b)$, and an arbitrary point (x, y) in the formula for slope.

$$m = \frac{y_2 - y_1}{x_2 - x_1}$$ Start with the formula for slope and substitute in m for the slope and $(0, b)$ and (x, y) for the two points.

$$m = \frac{y - b}{x - 0}$$

$$m = \frac{y - b}{x}$$

$$mx = y - b$$ Then multiply both sides by the LCD x.

$$mx + b = y \quad \text{or}$$ Add b to both sides.

$$y = mx + b$$ This form of a linear equation is referred to as the slope-intercept form.

$$f(x) = mx + b$$ This is the function form for a linear function.

Slope-Intercept Form

ALGEBRAICALLY	ALGEBRAIC EXAMPLE	VERBAL EXAMPLE	GRAPHICAL EXAMPLE
$y = mx + b$ is the equation of a line with slope m and y-intercept $(0, b)$.	$y = \frac{1}{2}x + 3$	This line has slope $\frac{1}{2}$ and a y-intercept of $(0, 3)$.	

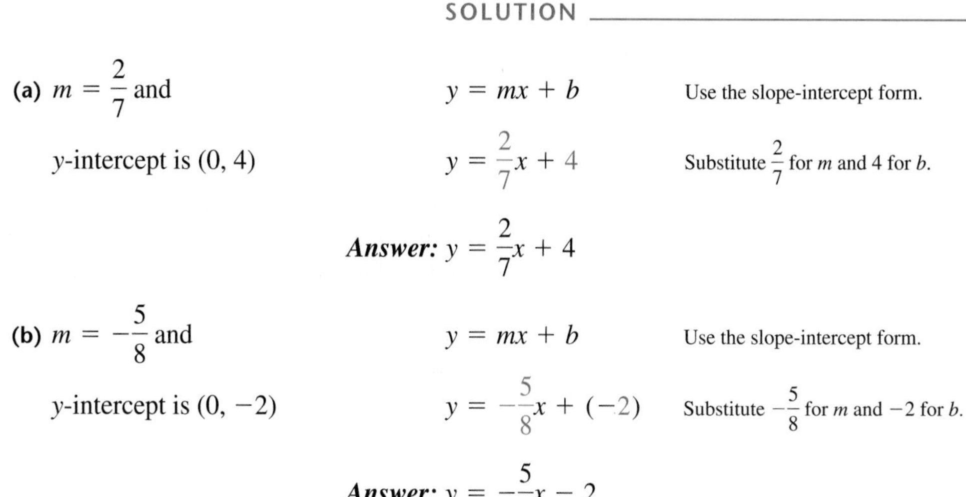

EXAMPLE 1 Writing the Equation of a Line, Given Its Slope and y-Intercept

Write the equation of a line satisfying the given conditions.

SOLUTION

(a) $m = \frac{2}{7}$ and y-intercept is $(0, 4)$

$$y = mx + b$$ Use the slope-intercept form.

$$y = \frac{2}{7}x + 4$$ Substitute $\frac{2}{7}$ for m and 4 for b.

Answer: $y = \frac{2}{7}x + 4$

(b) $m = -\frac{5}{8}$ and y-intercept is $(0, -2)$

$$y = mx + b$$ Use the slope-intercept form.

$$y = -\frac{5}{8}x + (-2)$$ Substitute $-\frac{5}{8}$ for m and -2 for b.

Answer: $y = -\frac{5}{8}x - 2$

(c) $m = 0$, and y-intercept is $(0, 3)$

$$y = mx + b$$
Use the slope-intercept form.

$$y = 0x + 3$$
Substitute 0 for m and 3 for b.

Answer: $y = 3$

This is the equation of a horizontal line with slope 0.

One of the advantages of the slope-intercept form of a line is that we can determine by inspection two key pieces of information about the line—both the slope and the y-intercept. This is illustrated in Example 2.

EXAMPLE 2 Using the Slope-Intercept Form

Determine by inspection the slope and y-intercept of the following lines.

SOLUTION

(a) $f(x) = -\dfrac{5}{13}x + \dfrac{1}{3}$

$$f(x) = -\frac{5}{13}x + \frac{1}{3}$$

This equation is in the slope-intercept form, $f(x) = mx + b$, with $m = -\dfrac{5}{13}$ and $b = \dfrac{1}{3}$.

Answer: $m = -\dfrac{5}{13}$, and y-intercept is $\left(0, \dfrac{1}{3}\right)$

(b) $f(x) = -6x$

$$f(x) = -6x$$
$$f(x) = -6x + 0$$

This equation is in the slope-intercept form, with b understood to be 0.

Answer: $m = -6$, and y-intercept is $(0, 0)$

(c) $2x + 5y = -6$

$$2x + 5y = -6$$
$$5y = -2x - 6$$
$$y = -\frac{2}{5}x - \frac{6}{5}$$

To put the equation in slope-intercept form, solve for y. Subtract $2x$ from both sides and then divide both sides by 5.

Answer: $m = -\dfrac{2}{5}$, and y-intercept is $\left(0, -\dfrac{6}{5}\right)$

SELF-CHECK 3.2.1

1. Write the equation of a line with slope $\dfrac{3}{4}$ and y-intercept $(0, -1)$.
2. Determine the slope and y-intercept of $y = \dfrac{7}{11}x - 8$.
3. Determine the slope and y-intercept of $3x + 4y = 12$.

We can sketch the graph of the line directly from the slope-intercept form without forming a table of values. In Example 3 we do this by plotting the y-intercept and then using the slope to move over horizontally and vertically. This is the skill that we introduced in Example 4 of Section 3.1.

EXAMPLE 3 Graphing a Line by Using Slope-Intercept Form

Graph the line defined by $f(x) = \frac{5}{3}x - 2$.

SOLUTION

$f(x) = \frac{5}{3}x - 2$

Slope: $m = \frac{5}{3}$

y-Intercept: $(0, -2)$

Second point: $(0 + 3, -2 + 5) = (3, 3)$

Use the slope-intercept form $f(x) = mx + b$ to determine the slope and y-intercept.

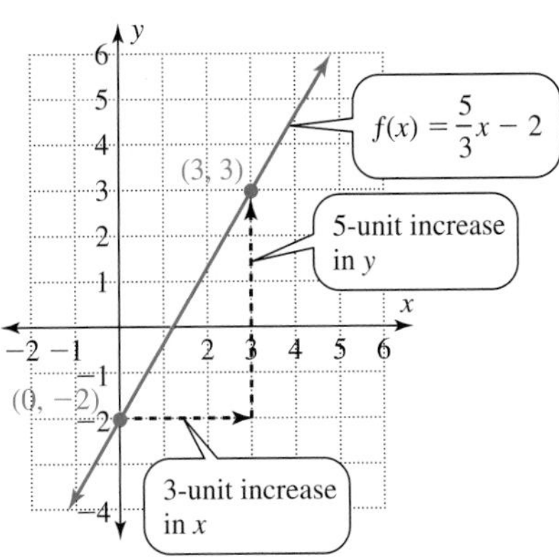

Plot the y-intercept $(0, -2)$ and then move 3 units to the right and 5 units up to plot a second point. Sketch the line through these two points.

In Example 4 we use the slope-intercept form $y = mx + b$ and the table of values to write the equation of the line containing the points in the table. This is easily done by using inspection to determine the slope and y-intercept of this line. This is a very useful skill.

EXAMPLE 4 Using a Table of Values to Write the Equation of a Line

Use this table of values for a linear function to do the following:

(a) Write the y-intercept of this line.
(b) Find the slope of this line.
(c) Write the equation of this line in slope-intercept form.
(d) Use a graphing calculator to check this result.

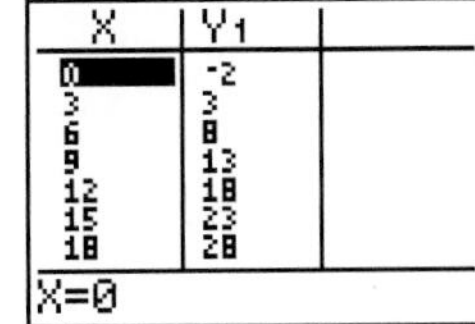

X	Y1
0	-2
3	3
6	8
9	13
12	18
15	23
18	28

X=0

SOLUTION

(a) The y-intercept is $(0, -2)$.

From the first row in the table, the y-intercept is $(0, -2)$.

(b) $m = \frac{5}{3}$

From one row to the next in the table, x increases by 3 units and y increases by 5 units. $m = \frac{\Delta y}{\Delta x}$ so the slope is $\frac{5}{3}$.

(c) $y = \frac{5}{3}x - 2$

Substitute this y-intercept and the slope into the slope-intercept form $y = mx + b$.

(d)

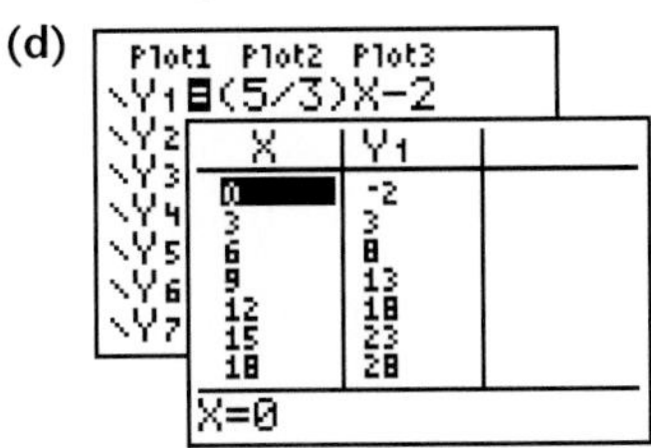

X	Y1
0	-2
3	3
6	8
9	13
12	18
15	23
18	28

X=0

After entering this equation into a graphing calculator, we can produce the given table.

Parallel and Perpendicular Lines

In Section 3.1 we noted that parallel lines have the same slope and that perpendicular lines have slopes that are opposite reciprocals. If the equations of lines are written in slope-intercept form, it is easy to identify their slopes and to determine whether the lines are parallel or perpendicular.

EXAMPLE 5 Determining Whether Two Lines Are Parallel or Perpendicular

Determine whether the first line is parallel to, perpendicular to, or neither parallel nor perpendicular to the second line.

SOLUTION

(a) $f(x) = \frac{2}{3}x + 7$

$f(x) = -\frac{3}{2}x + 5$

FIRST EQUATION	SECOND EQUATION
$f(x) = \frac{2}{3}x + 7$	$f(x) = -\frac{3}{2}x + 5$
Slope: $m = \frac{2}{3}$	Slope: $m = -\frac{3}{2}$

Each slope can be determined by inspection since the equations are in slope-intercept form.

Answer: Since $\left(\frac{2}{3}\right)\left(-\frac{3}{2}\right) = -1$, these lines are perpendicular.

Lines are perpendicular if the product of their slopes is -1.

(b) $5x - 3y = 12$

$5x = 3y + 8$

FIRST EQUATION	SECOND EQUATION
$5x - 3y = 12$	$5x = 3y + 8$
$-3y = -5x + 12$	$5x - 8 = 3y$
$y = \frac{5}{3}x - 4$	$\frac{5}{3}x - \frac{8}{3} = y$
	$y = \frac{5}{3}x - \frac{8}{3}$
Slope: $m = \frac{5}{3}$	Slope: $m = \frac{5}{3}$

Write each equation in slope-intercept form so the slope can be determined by inspection.

The lines are parallel, but they are not the same line because the y-intercepts are different.

Answer: Since the slopes are equal, the lines are parallel.

SELF-CHECK 3.2.2

Use this table of values for a linear function to do the following.

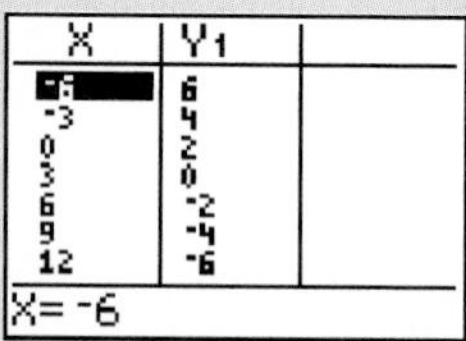

X	Y1
-6	6
-3	4
0	2
3	0
6	-2
9	-4
12	-6

X= -6

1. Write the y-intercept of this line.
2. Find the slope of this line.
3. Write the equation of this line in slope-intercept form.
4. Use a graphing calculator to graph this line, using a window of $[-4.7, 4.7, 1]$ by $[-3.1, 3.1, 1]$.

Determine whether the first line is parallel to, perpendicular to, or neither parallel nor perpendicular to the second line.

5. $f(x) = \frac{2}{5}x - 8$ $\quad f(x) = \frac{5}{2}x + 8$
6. $f(x) = \frac{2}{5}x - 8$ $\quad f(x) = \frac{2}{5}x + 8$
7. $4x - 3y = 9$ $\quad 3x + 4y = 8$

Horizontal and Vertical Lines

The y-coordinate of every point on a horizontal line is the same—thus y is a constant.

We examined horizontal and vertical lines in Section 3.1. Now we will examine their equations. Recall that all horizontal lines have a slope of zero—there is no rise between two points on a horizontal line. The equation of a horizontal line is developed next by substituting $m = 0$ and a y-intercept of $(0, b)$ into the slope-intercept form $y = mx + b$.

$y = mx + b$ Start with slope-intercept form.

$y = 0x + b$ Substitute in 0 for m.

$y = b$

The x-coordinate of every point on a vertical line is the same—thus x is a constant.

The equation of a horizontal line can always be written in the form $y = b$ or $f(x) = b$. The equation of a vertical line has a similar form. A vertical line with an x-intercept $(a, 0)$ must have the same x-coordinate for every point. The equation that specifies this is $x = a$.

Horizontal and Vertical Lines

ALGEBRAICALLY	NUMERICAL EXAMPLE	GRAPHICAL EXAMPLE	VERBALLY
$y = b$ is the equation of a horizontal line with y-intercept $(0, b)$. ***Example:*** $y = 3$	x | y −2 | 3 −1 | 3 0 | 3 1 | 3 2 | 3	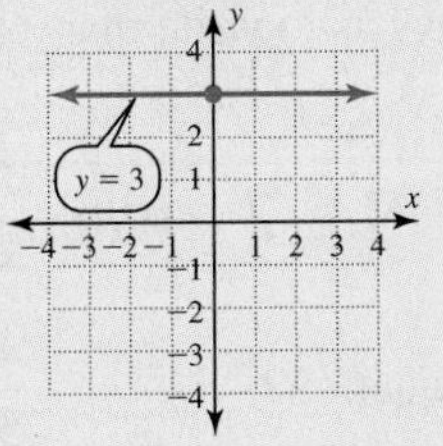 y = 3	This horizontal line has a y-intercept of $(0, 3)$ and a slope of 0.
$x = a$ is the equation of a vertical line with x-intercept $(a, 0)$. ***Example:*** $x = -2$	x | y −2 | −2 −2 | −1 −2 | 0 −2 | 1 −2 | 2	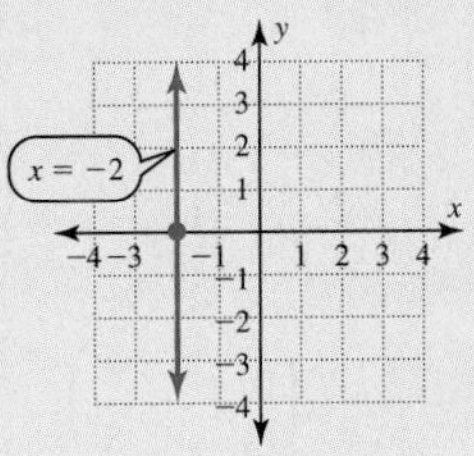 x = −2	This vertical line has an x-intercept of $(-2, 0)$, and its slope is undefined.

EXAMPLE 6 Writing the Equations of Horizontal and Vertical Lines

Write the equation of each of these lines.

SOLUTION

(a)

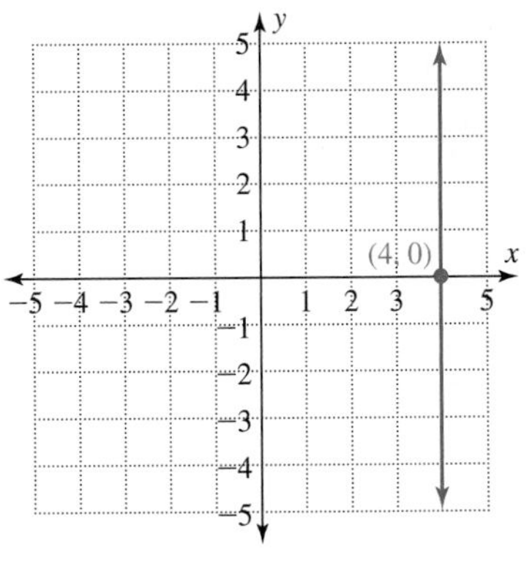

(4, 0)

Answer: $x = 4$

This is a vertical line with an x-intercept of $(4, 0)$. Use the form $x = a$ with 4 as the value of a.

(b)

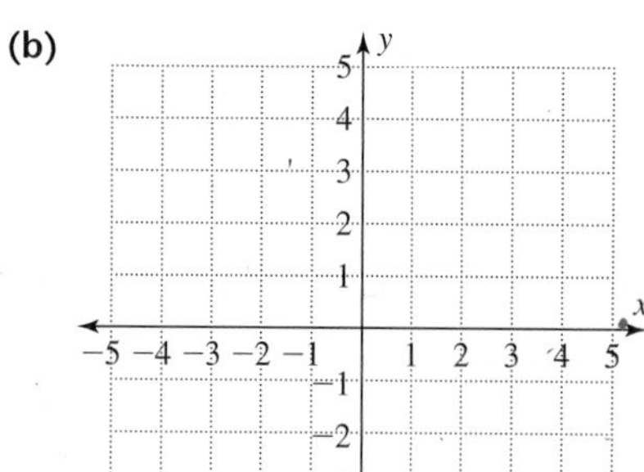

Answer: $y = -4$

This is a horizontal line with a y-intercept of $(0, -4)$. Use the form $y = b$ with -4 as the value of b.

We now examine two other forms of linear equations—the general form and the point-slope form.

General Form

A first-degree equation in both x and y means that the exponent on both of these variables is 1. Sometimes the form $Ax + By + C = 0$ is also referred to as general form.

Every linear equation with variables x and y is a first-degree equation in both x and y that can be written in the form $Ax + By = C$, where A, B, and C are real number constants. The **general form** $Ax + By = C$ is used to represent linear equations in many disciplines because this form allows us to write many equations in a nice, clean form that is free of fractions.

EXAMPLE 7 Writing a Linear Equation in General Form

Rewrite the equation $y = \frac{2}{3}x + \frac{4}{5}$ in the general form $Ax + By = C$.

SOLUTION

The general form is generally written so that the coefficient of x is positive.

$$y = \frac{2}{3}x + \frac{4}{5}$$

$$15y = 15\left(\frac{2}{3}x + \frac{4}{5}\right)$$ Multiply both sides of the equation by 15, the LCD of all these terms.

$$15y = 15\left(\frac{2}{3}x\right) + 15\left(\frac{4}{5}\right)$$ Distribute the factor of 15 and then simplify.

$$15y = 5(2x) + 3(4)$$

$$15y = 10x + 12$$

$$-12 = 10x - 15y \quad \text{or}$$ Subtract $15y$ from both sides of the equation, and subtract 12 from both sides of the equation.

Answer: $10x - 15y = -12$

SELF-CHECK 3.2.3

1. Write the equation of a horizontal line through $(6, -1)$.
2. Write the equation of a vertical line through $(6, -1)$.
3. Rewrite $y = -\frac{1}{2}x + \frac{3}{7}$ in general form with integer coefficients.

Point-Slope Form

The point-slope form that is developed on the following page is often used to write the equation of a line, given either a point on the line and its slope or two points on the line. Suppose that a line passes through the point (x_1, y_1) and has slope m. To determine the equation that relates x and y for any other point (x, y) on this line, we substitute (x, y) for (x_2, y_2) into the formula for slope to obtain

$$m = \frac{y_2 - y_1}{x_2 - x_1}$$

Start with the formula for slope. Then substitute m and the two points (x_1, y_1) and (x, y) into this formula.

$$m = \frac{y - y_1}{x - x_1}$$

$$m(x - x_1) = y - y_1 \quad \text{or}$$
$$y - y_1 = m(x - x_1)$$

To produce the point-slope form, multiply both sides of this equation by $x - x_1$.

Point-Slope Form $y - y_1 = m(x - x_1)$

ALGEBRAICALLY	ALGEBRAIC EXAMPLE	GRAPHICAL EXAMPLE	VERBAL EXAMPLE
$y - y_1 = m(x - x_1)$ is the equation of a line through (x_1, y_1) with slope m.	$y - 1 = \frac{1}{3}(x - 1)$		This line passes through the point (1, 1) with slope $\frac{1}{3}$.

It is often convenient to be able to sketch the graph of the line directly from the point-slope form. This method can sometimes avoid fractional values that may result from alternate methods, and it also can save the step of rewriting the equation in another form or forming a table of values. Compare Self-Check 3.2.4 with Example 8 to observe the advantage of this method.

EXAMPLE 8 Graphing a Line by Using Point-Slope Form

Graph the line defined by $y + 2 = \frac{3}{4}(x - 1)$.

SOLUTION

$$y + 2 = \frac{3}{4}(x - 1)$$

Point: $(1, -2)$

Slope: $m = \frac{3}{4}$

Use the point-slope form $y - y_1 = m(x - x_1)$ to determine a point and the slope.

Second point: $(1 + 4, -2 + 3) = (5, 1)$

Plot the point $(1, -2)$ and then move over 4 units to the right and 3 units up to plot a second point.

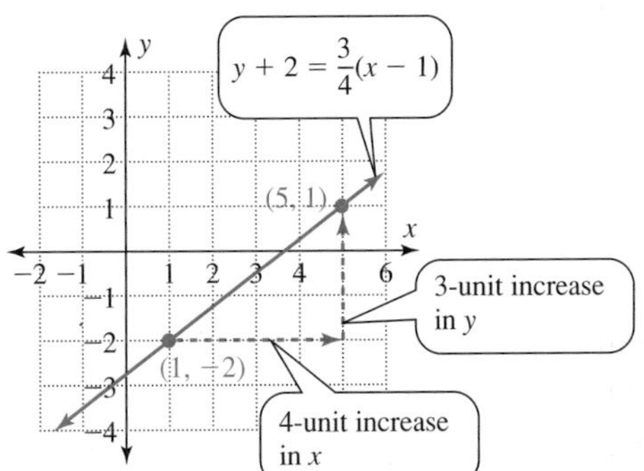

Sketch the line through these two points.

SELF-CHECK 3.2.4

1. Rewrite the equation $y + 2 = \frac{3}{4}(x - 1)$ in slope-intercept form.
2. Use this slope-intercept form to graph the line.
3. Compare this result with Example 8.

EXAMPLE 9 Writing the Equation of a Line Perpendicular to a Given Line

Write in slope-intercept form the equation of a line passing through $(-5, 6)$ and perpendicular to the line $y = -\frac{5}{3}x - 6$. Then use a graphing calculator to check that your equation is correct.

SOLUTION

$y = -\frac{5}{3}x - 6$ From the slope-intercept form of the given line, its slope is $-\frac{5}{3}$.

Slope of given line $= -\frac{5}{3}$ Because the new line is perpendicular to the given line, its slope is $\frac{3}{5}$.

$m = \frac{3}{5}$

$y - y_1 = m(x - x_1)$ Use the point-slope form to determine the equation of the new line. Substitute $(-5, 6)$ for (x_1, y_1) and $\frac{3}{5}$ for m.

$y - 6 = \frac{3}{5}[x - (-5)]$ Multiply both sides by the LCD, 5.

$5(y - 6) = 3(x + 5)$

$5y - 30 = 3x + 15$

$5y = 3x + 45$ Simplify, and write the equation in slope-intercept form, $y = mx + b$.

Answer: $y = \frac{3}{5}x + 9$

Check:

NUMERICALLY

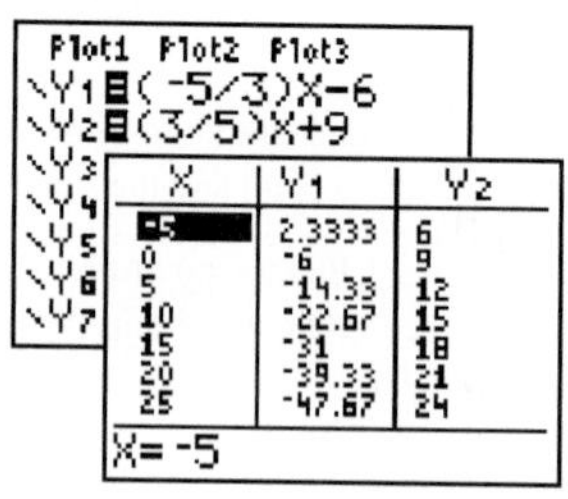

The point $(-5, 6)$ and the y-intercept $(0, 9)$ both appear in this table for Y_2.
Also, note that a 5-unit increase in x results in a 3-unit increase in Y_2; a slope of $\frac{3}{5}$.

GRAPHICALLY

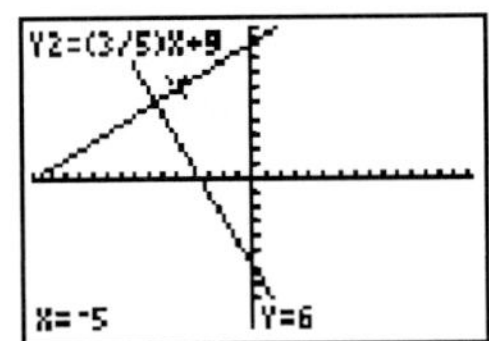

$[-15.2, 15.2, 1]$ by $[-10, 10, 1]$

This graph visually displays a slope of $\frac{3}{5}$ and passes through the point $(-5, 6)$ and the Y_2-intercept $(0, 9)$.

On a TI-84 Plus calculator, using ZOOM **ZStandard** followed by **ZSquare** creates a screen that squares the rectangular display and makes the perpendicular lines really appear perpendicular on the display. ■

A line can be graphed if any two points on the line are known. Similarly if we know any two points on the line, then we can write the equation of the line. This is illustrated in Example 10.

EXAMPLE 10 Writing the Equation of the Line Through Two Points

Write in slope-intercept form the equation of a line through $(-6, -4)$ and $(-2, 2)$.

SOLUTION

$$m = \frac{y_2 - y_1}{x_2 - x_1} = \frac{2 - (-4)}{-2 - (-6)} = \frac{6}{4} = \frac{3}{2}$$

Step 1: Since the slope is not given, calculate it by using the two given points.

$$y - y_1 = m(x - x_1)$$

Step 2: Use the point-slope form.

$$y - (-4) = \frac{3}{2}[x - (-6)]$$

Substitute $(-6, -4)$ for (x_1, y_1) and $\frac{3}{2}$ for m.

$$y + 4 = \frac{3}{2}(x + 6)$$

Actually, either point can be used in this step.

$$2(y + 4) = 3(x + 6)$$

Multiply both sides by the LCD, 2.

$$2y + 8 = 3x + 18$$
$$2y = 3x + 10$$

Simplify, and write the equation in the slope-intercept form, $y = mx + b$.

Answer: $y = \frac{3}{2}x + 5$

Can you use a graphing calculator to check that this is the correct equation? ■

In Example 10, the same equation can be obtained by substituting in the point $(-2, 2)$ instead of the point $(-6, -4)$. It is worthwhile to note that both points were used to calculate the slope of the line. Thus either point can be used to select the specific line that has this slope.

SELF-CHECK 3.2.5

1. The equation $y - 1 = \frac{1}{3}(x - 1)$ can also be written as $y - 2 = \frac{1}{3}(x - 4)$. Show that these equations are equivalent by converting both equations to general form.
2. Determine by inspection the slope of $y + 7 = -\frac{3}{4}(x - 4)$ and one point on this line.
3. Write in slope-intercept form the equation of a line through $(0, 4)$ and $(5, 2)$.
4. Graph the line through $(-2, 4)$ with slope $-\frac{3}{4}$.

SELF-CHECK ANSWERS

3.2.1

1. $y = \frac{3}{4}x - 1$
2. $m = \frac{7}{11}$, y-intercept $(0, -8)$
3. $m = -\frac{3}{4}$, y-intercept $(0, 3)$

3.2.2

1. $(0, 2)$ 2. $m = -\frac{2}{3}$
3. $y = -\frac{2}{3}x + 2$
4.

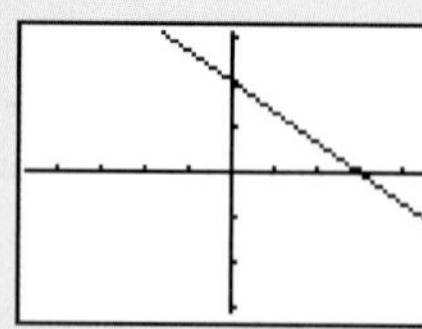

$[-4.7, 4.7, 1]$ by $[-3.1, 3.1, 1]$

5. Neither parallel nor perpendicular
6. Parallel
7. Perpendicular

3.2.3

1. $y = -1$ 2. $x = 6$
3. $7x + 14y = 6$

3.2.4

1. $y = \frac{3}{4}x - \frac{11}{4}$
2.

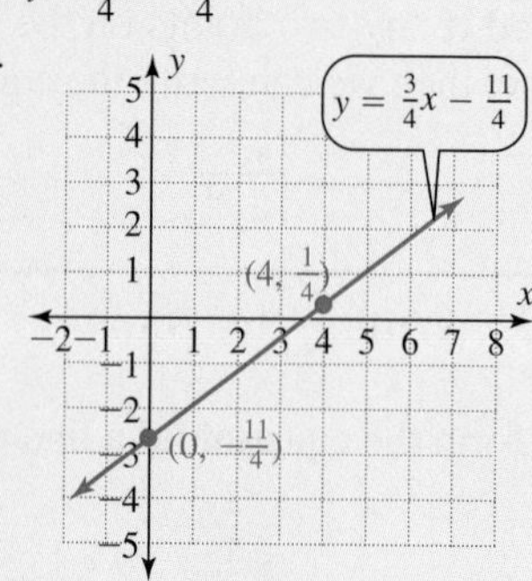

3. The answers are the same, but the slope-intercept form requires plotting a point with a fractional y-intercept.

3.2.5

1. The general form for both equations is $x - 3y = -2$.
2. $m = -\frac{3}{4}$; $(4, -7)$
3. $y = -\frac{2}{5}x + 4$
4.

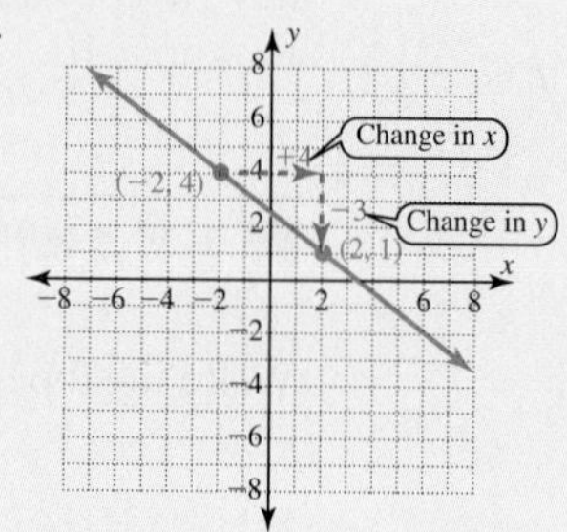

USING THE LANGUAGE AND SYMBOLISM OF MATHEMATICS 3.2

1. The slope-intercept form of the equation of a line is __________.
2. Lines with the same slope are __________.
3. Lines whose slopes are opposite reciprocals of each other are __________.
4. The line defined by $x = 5$ is a __________ line whose x-intercept is __________.
5. The line defined by $y = -5$ is a __________ line whose y-intercept is __________.
6. The general form of the equation of a line is __________.
7. When we say that a linear equation in x and y is a first-degree equation, we mean that the exponent on both x and y is __________.
8. The equation $y - y_1 = m(x - x_1)$ is the __________-__________ form of a linear equation with (x_1, y_1) representing a __________ on the line and m representing the __________ of the line.

EXERCISES 3.2

In Exercises 1–4, determine by inspection the slope and y-intercept of each line.

1. a. $f(x) = 2x + 5$ b. $f(x) = -\frac{3}{11}x - \frac{4}{5}$
   c. $f(x) = 6x$
2. a. $f(x) = -3x + 7$ b. $f(x) = \frac{4}{9}x - \frac{2}{3}$
   c. $f(x) = -6x$
3. a. $y = 2$ b. $x = -7$
   c. $y = -x$
4. a. $y = -2$ b. $x = 7$
   c. $y = x$

In Exercises 5–10, write in the slope-intercept form $y = mx + b$ the equation of the line with the given slope and y-intercept.

5. $m = 4$; $(0, 7)$ 6. $m = 3$; $(0, 5)$
7. $m = -\frac{2}{11}$; $(0, 5)$ 8. $m = -\frac{5}{7}$; $(0, -2)$
9. $m = 0$; $(0, -6)$ 10. $m = 0$; $(0, 4)$

In Exercises 11–16, sketch the graph of each line without using a calculator. (*Hint:* See Example 3.)

11. $f(x) = \frac{2}{3}x - 4$ 12. $f(x) = \frac{3}{4}x - 5$
13. $f(x) = -\frac{5}{3}x + 2$ 14. $f(x) = -\frac{5}{2}x + 3$
15. $f(x) = 2$ 16. $f(x) = -3$

In Exercises 17–20, write the equation for each line in the slope-intercept form $f(x) = mx + b$.

17.

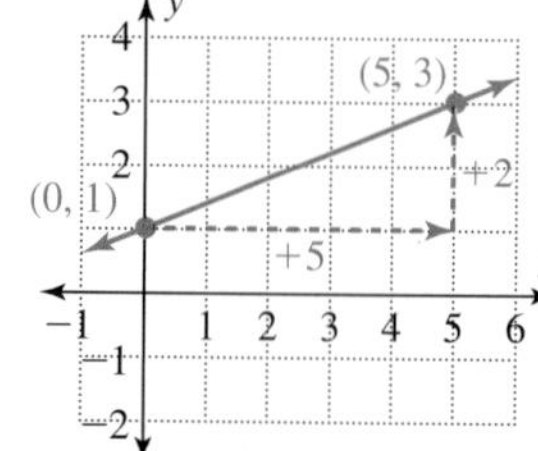

18.

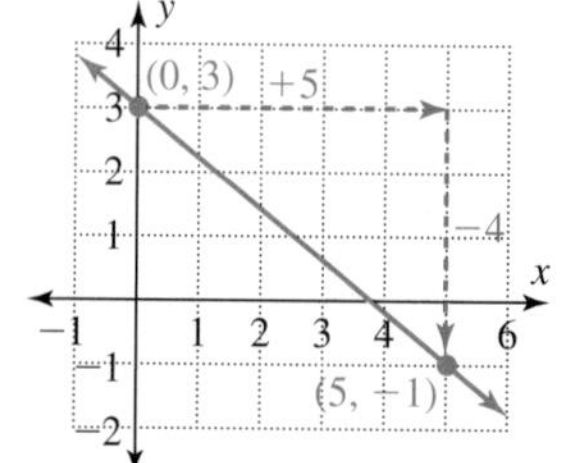

19.

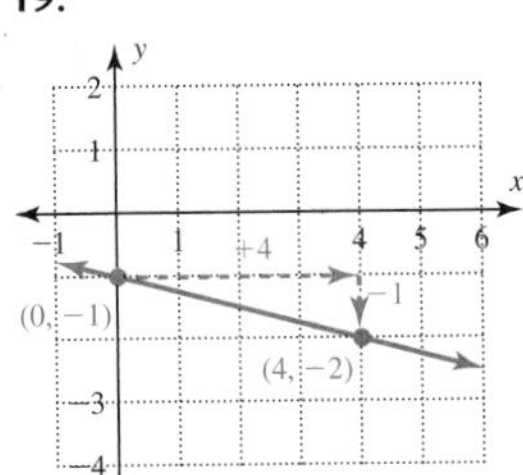

20.

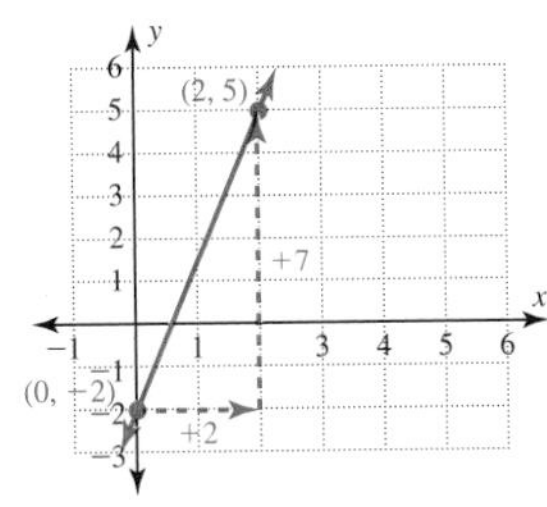

21. Distance Traveled This graph gives the distance from the dispatch center for a truck at different times. Use this graph to determine the linear equation $f(x) = mx + b$ of this line. Then interpret the meaning of m and b in this problem.

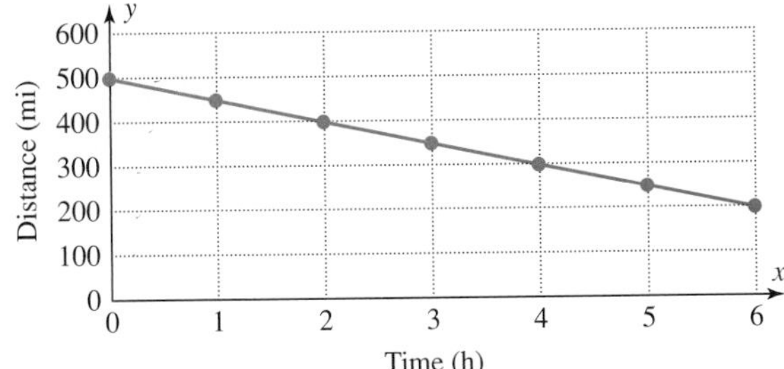

22. Cost of Telephone Minutes The graph below gives the monthly cost for a cell phone based upon the number of minutes used. Use this graph to determine the linear equation $f(x) = mx + b$ of this line. Then interpret the meaning of m and b in this problem.

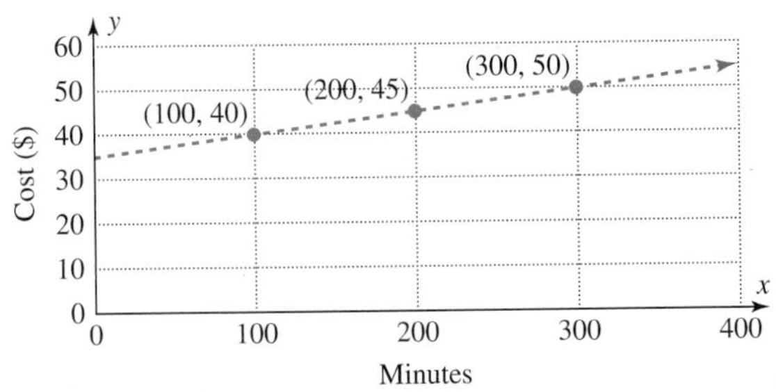

23. The slope of a line is $m = \frac{3}{4}$, and its y-intercept is $(0, -2)$. Use this information to complete this table.

x	y
0	
4	
8	
12	
16	

24. The slope of a line is $m = -\frac{4}{3}$, and its y-intercept is $(0, 2)$. Use this information to complete this table.

x	y
0	
3	
6	
9	
12	

Calculator Usage

In Exercises 25–28, the graphing calculator display shows a table of values for a linear equation $f(x) = mx + b$. Determine the following:

a. Value of Δx shown in the table
b. Value of Δy produced in the table
c. Slope of this line
d. The y-intercept of this line
e. Equation of this line

25.

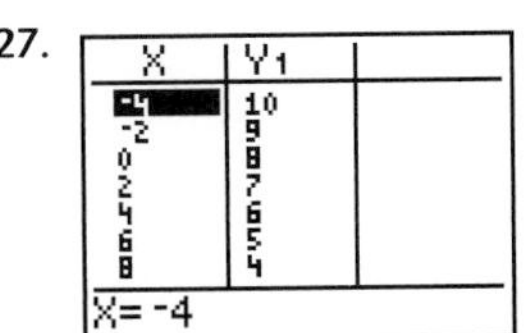

X	Y1
-3	-20
-2	-13
-1	-6
0	1
1	8
2	15
3	22

X=-3

26.

X	Y1
-3	18
-2	13
-1	8
0	3
1	-2
2	-7
3	-12

X=-3

27.

X	Y1
-4	10
-2	9
0	8
2	7
4	6
6	5
8	4

X=-4

28.

X	Y1
-10	-6
-5	-5
0	-4
5	-3
10	-2
15	-1
20	0

X=-10

29. Jet Fuel Usage The main fuel tank on one aircraft contains 3,500 gal of jet fuel when fuel begins to be used from this tank. The volume of fuel remaining in the tank x minutes after the pump has been turned on is displayed in the given table.

	A	B
1	Time (min)	Volume (gal)
2	0	3500
3	15	3320
4	30	3140
5	45	2960
6		

a. Determine the slope of the line containing these points.
b. Interpret the meaning of the slope from part **a.**
c. Write a function f so that $f(x)$ gives the number of gallons of fuel left after x minutes.
d. Is 300 a practical input value for x? Explain.

30. Postage Meter Usage A postage meter has had \$100 of credit placed into it at a post office. The number of dollars of credit remaining in the postage meter after x letters have had postage applied by this meter is displayed in the given table.

	A	B
1	Letters	Credit ($)
2	0	$100.00
3	25	$90.25
4	50	$80.50
5	75	$70.75
6		

a. Determine the slope of the line containing these points.
b. Interpret the meaning of the slope from part **a.**
c. Write a function f so that $f(x)$ gives the number of dollars of credit left after x letters have had postage applied.
d. Is 300 a practical input value for x? Explain.

31. Cost of a Service Call The given table displays the dollar cost of a service call by Reliable Heating and Air Conditioning, based on the number of hours the service person is at the customer's location. Use this table to determine the linear equation $f(x) = mx + b$ for these data points. Then interpret the meaning of m and b in this problem.

HOURS x	COST y ($)
0	75
0.5	100
1.0	125
1.5	150
2.0	175
2.5	200

32. Salary of a Salesperson The given table displays the total monthly salary of a salesperson at an automobile dealership. This salary consists of a base salary and a commission based upon this person's sales for the month. Use this table to determine the linear equation $f(x) = mx + b$ for these data points. Then interpret the meaning of m and b in this problem.

SALES x (\$)	SALARY y (\$)
0	1,200
20,000	1,400
40,000	1,600
60,000	1,800
80,000	2,000
100,000	2,200

In Exercises 33–36, write the equation of each line.

33.

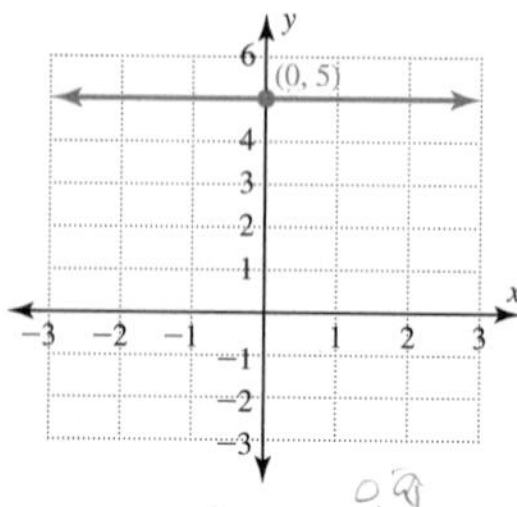

34.

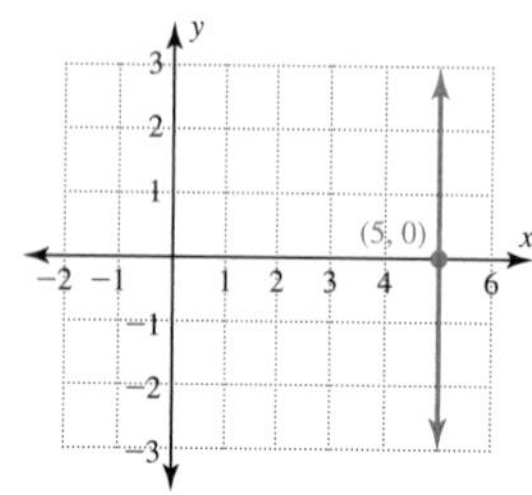

35.

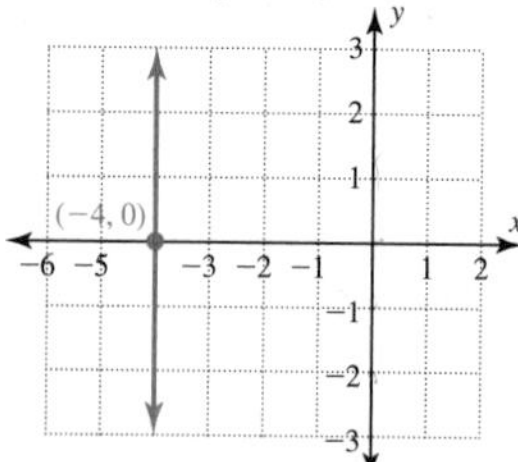

36.

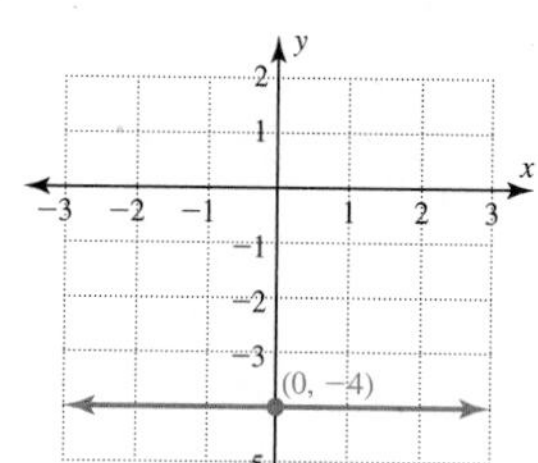

In Exercises 37–44, determine whether the first line is parallel to, perpendicular to, or neither parallel nor perpendicular to the second line.

37. $y = 5x + 11$
$y = -5x + 11$

38. $y = 5x + 11$
$y = 5x - 11$

39. $y = \frac{2}{3}x + 7$
$y = -\frac{3}{2}x - 2$

40. $y = \frac{3}{13}x - 4$
$y = \frac{13}{3}x + 4$

41. $2x + 5y = 8$
$6x + 15y - 7 = 0$

42. $3x - 4y + 8 = 0$
$4x + 3y = -15$

43. $y - 8 = 4(x + 5)$
$y + 4 = 5(x - 3)$

44. $y + 6 = 2(x - 1)$
$y - 1 = 2(x + 3)$

In Exercises 45 and 46, determine by inspection the slope of the line and one point on the line.

45. a. $y - 4 = 5(x - 3)$
b. $y + 5 = -7(x - 2)$
c. $y = \frac{1}{2}(x + 6)$

46. a. $y - 7 = x - 2$
b. $y - 9 = -4(x + 11)$
c. $y = \frac{2}{3}(x + 7)$

In Exercises 47–54, graph the line through the given point with the slope specified. (*Hint:* See Example 3.)

47. $(3, -2)$; $m = -\frac{3}{2}$

48. $(4, 2)$; $m = -\frac{5}{3}$

49. $(-6, 0)$; $m = 2$

50. $(0, -4)$; $m = \frac{1}{4}$

51. $(0, 0)$; $m = \frac{1}{4}$

52. $(0, 0)$; $m = -\frac{1}{3}$

53. $(3, 4)$; $m = 0$

54. $(3, 4)$; slope is undefined

In Exercises 55 and 56, write in slope-intercept form the equation of a line passing through the given point with the slope specified.

55. $(2, 3)$; $m = -4$

56. $(5, 2)$; $m = -2$

In Exercises 57–60, determine the slope and one point on each line; then write the equation of the line in point-slope form.

57.

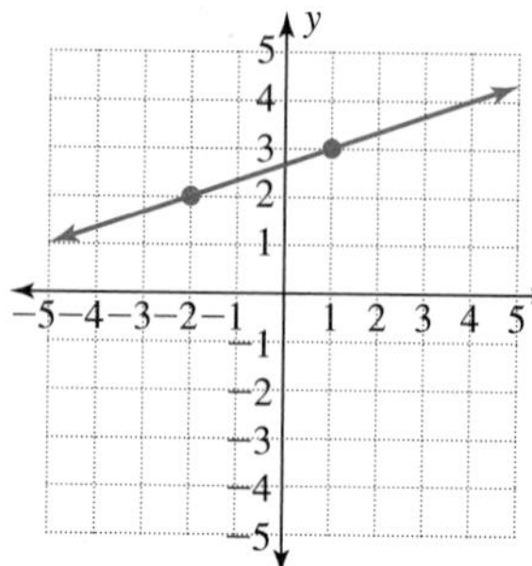

58.

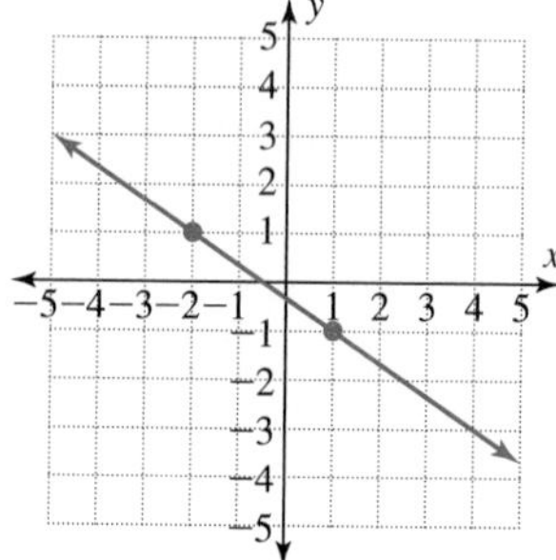

59.

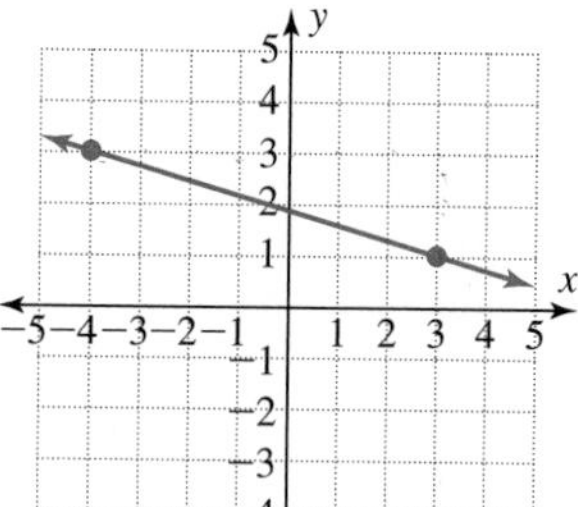

60.

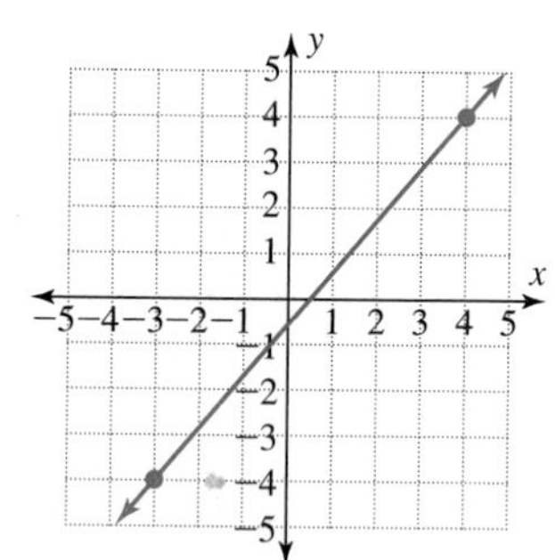

In Exercises 61–64, write in slope-intercept form the equation of a line passing through the given points.

61. $(-4, 2), (4, 4)$

62. $(2, 3), (-2, 1)$

63. $(0, 6), (-3, 0)$

64. $(5, 0), (-4, 3)$

In Exercises 65 and 66, write the equation of a line passing through the given points.

65. a. $(2, 7), (2, 4)$
b. $(6, 3), (1, 3)$

66. a. $(-1, 6), (2, 6)$
b. $(-1, 6), (-1, -3)$

67. Write in slope-intercept form the equation of a line passing through $(2, 3)$ and parallel to $y = \frac{3}{7}x - 1$.

68. Write in slope-intercept form the equation of a line passing through $(2, 3)$ and perpendicular to $y = \frac{3}{7}x - 1$.

69. Write in slope-intercept form the equation of a line passing through $(-5, 1)$ and perpendicular to $y = -\frac{4}{3}x + 7$.

70. Write in slope-intercept form the equation of a line passing through $(-5, 1)$ and parallel to $y = -\frac{4}{3}x + 7$.

71. Write the equation of a line passing through $(4, -8)$ and parallel to the x-axis.

72. Write the equation of a line passing through $(4, -8)$ and perpendicular to the x-axis.

73. **Taxi Fares** The given table displays the dollar cost of a taxi ride based upon the number of miles traveled. Write the linear equation $f(x) = mx + b$ for the line that contains these data points. Then interpret the meaning of m and b in this problem. (*Hint:* Although you would not pay for a ride of 0 mi, the y-intercept still has a meaningful interpretation.)

MILES x	COST y ($)
1.5	4.45
4	6.45
7	8.85
12.5	13.25

74. **Cost of Books** The given table displays the dollar cost of a shipment of math books from a publisher to a campus bookstore. Write the linear equation $f(x) = mx + b$ for the line that contains these data points. Then interpret the meaning of m and b in this problem. (*Hint:* Although you would not pay for a shipment of zero books, the y-intercept still has a meaningful interpretation.)

BOOKS x	COST y ($)
70	4,210
140	8,410
250	15,010
400	24,010

In Exercises 75–78, complete the slope-intercept form and the general form for each linear equation.

POINT-SLOPE FORM	SLOPE-INTERCEPT FORM	GENERAL FORM
Example: $y - 1 = 2(x - 3)$	$y = 2x - 5$	$2x - y = 5$
75. $y - 3 = 4(x + 1)$		
76. $y + 2 = -3(x - 1)$		
77. $y + 4 = -\frac{2}{3}(x - 5)$		
78. $y - 5 = \frac{7}{2}(x + 3)$		

79. **Displacement of a Spring** A spring is 8 cm long. The spring stretches 2 cm for each kilogram attached. Express the new length of the spring y in terms of the mass x. Write this equation in slope-intercept form. What is the significance of the slope and the y-intercept?

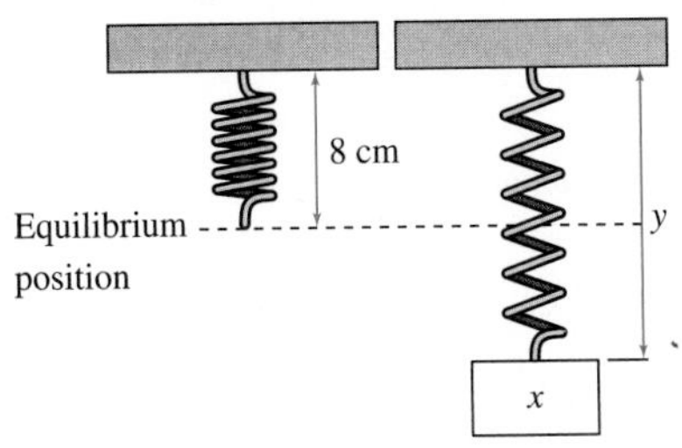

80. **Fixed and Variable Production Costs** The monthly cost y for producing toner cartridges at a factory is \$17,500 plus \$5 for each of the x cartridges produced. Write a linear equation in slope-intercept form that expresses the relationship between x and y. What is the significance of the slope and the y-intercept?

81. **Duke Energy Dividends** An investor had the annual dividend results for Duke Energy for the years 2002 and 2005. In 2002 the annual dividend per share was \$1.10, and in 2005 the annual dividend per share was \$1.17.
   a. Using x for the number of years after 2000 and y as the annual dividend, write a linear equation to relate the dividend to the year.
   b. Use this equation to estimate the dividend per share that the investor might receive in 2007. (Source for more information on this topic is the Additional Resources link on the MathZone website at **www.mhhe.com/hallmercer.**)

Group Discussion Questions

82. **Communicating Mathematically** The profit y of a company over a period of time x is graphed on a rectangular coordinate system. Write a paragraph describing your interpretation when the slope is **(a)** negative, **(b)** zero, and **(c)** positive.

83. **Challenge Question**
   a. Determine by inspection which point is not on the same line as all the other points in this table.
   b. Explain your reasoning, using the notation $m = \frac{\Delta y}{\Delta x}$.

x	y
−22	−3
−11	4
0	11
11	18
22	21

84. **Challenge Question** Write in general form the equation of the line satisfying the conditions given.
   a. Parallel to x-axis through (2, 3)
   b. Parallel to y-axis through (2, 3)
   c. Perpendicular to x-axis through (−5, 8)
   d. Perpendicular to y-axis through (−5, 8)
   e. Parallel to $y = \frac{3}{7}x + 7$ through (2,−3)
   f. Parallel to $y = \frac{2}{9}x - 5$ through (2,−3)
   g. Perpendicular to $y - 1 = \frac{1}{2}(x + 3)$ through (3, 5)
   h. Perpendicular to $y + 2 = -\frac{1}{3}(x - 7)$ through (4,−9)

85. **Challenge Question** Write in general form the equation of the line satisfying the conditions given.
   a. y-intercept (0, −5) with $m = \frac{1}{5}$
   b. y-intercept (0, 5) with $m = -\frac{1}{5}$
   c. x-intercept (3, 0) with $m = 7$
   d. x-intercept (2, 0) with $m = -4$
   e. Through (−7, −2) with $m = 0$
   f. Through (−7, 5) with slope undefined

86. **Spreadsheet Exploration** Use Spreadsheet Exploration 3.2 from the website **www.mhhe.com/hallmercer** to explore the changes to the forms $y = mx + b$ and $y - y_1 = m(x - x_1)$. Use the sliders to explore the patterns in the tables of values for m and b and the graph of the line defined by $y = mx + b$. Also use the sliders to observe the patterns in the table of values for m and (x_1, y_1) and the graph defined by $y - y_1 = m(x - x_1)$.

Section 3.3 Solving Systems of Linear Equations in Two Variables Graphically and Numerically

Objectives:

7. Solve a system of linear equations by using graphs and tables.
8. Identify inconsistent systems and systems of dependent linear equations.

It is often useful for a business to examine two different options or situations simultaneously and to be able to compare these options. We will illustrate this in Example 3 when we compare two different options for producing pasta. If we use equations to model both options, then we can examine both equations simultaneously. When we consider two or more equations at the same time, we refer to this as a **system of equations.** A **solution of a system of linear equations** in two variables is an ordered pair that is a solution of each equation in the system. Each solution of a linear equation in two variables is represented by a point on the line. If there is a unique solution to a system of linear equations, it is represented by their point of intersection, the point that is on both graphs.

The graphical method for solving systems of linear equations, first covered in Section 2.3, is an excellent way to develop a good understanding of what the solution to a system of linear equations represents. It is easy to "see" the point of intersection of two lines, the ordered pair that satisfies the equation of both lines. Although it may be easy to approximate the coordinates of this point of intersection, you may need to use the algebraic methods covered in Sections 3.4 and 3.5 to obtain the exact values of these coordinates.

Solution of a system of two linear equations

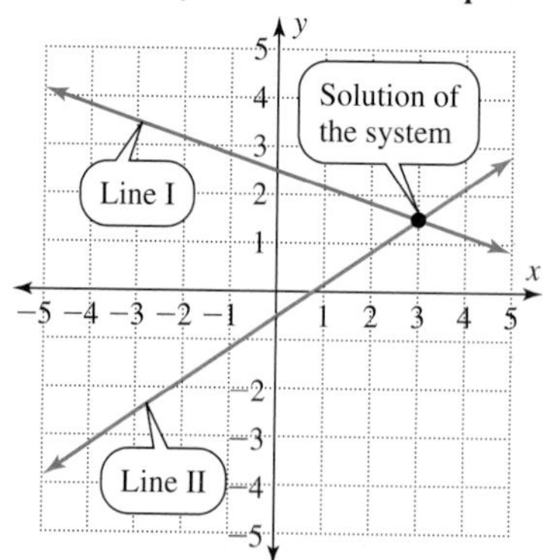

To solve a system of equations graphically, graph each equation and then estimate the coordinates of the point of intersection. Because serious errors of estimation can occur, it is wise to check an estimated solution by substituting it into each equation of the system. One way to check these coördinates is to use a table of values. In Example 1 we use a table to check the point (2, 4) in both equations.

EXAMPLE 1 Solving a System of Linear Equations Graphically and Numerically

Solve the linear system $\begin{Bmatrix} 7x - 3y = 2 \\ 9x - 5y = -2 \end{Bmatrix}$ graphically and numerically. (For help with keystrokes, see Calculator Perspective 2.3.2.)

SOLUTION

FIRST EQUATION

$$\begin{aligned} 7x - 3y &= 2 \\ -3y &= -7x + 2 \\ y &= \frac{7}{3}x - \frac{2}{3} \end{aligned}$$

SECOND EQUATION

$$\begin{aligned} 9x - 5y &= -2 \\ -5y &= -9x - 2 \\ y &= \frac{9}{5}x + \frac{2}{5} \end{aligned}$$

If you plan to use a graphing calculator, you will first need to rewrite the equations in slope-intercept form as shown here.

GRAPHICALLY

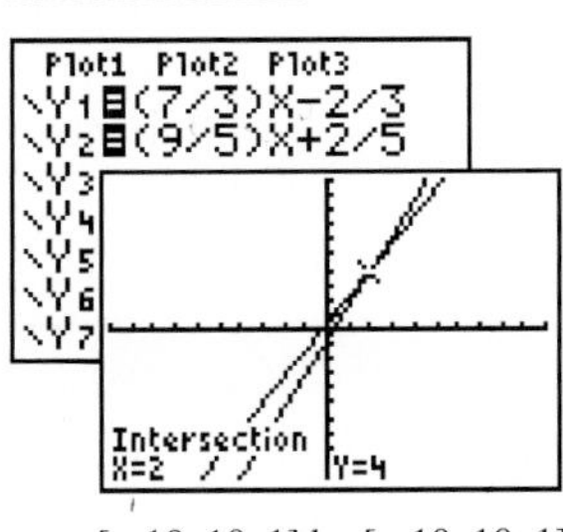

[−10, 10, 1] by [−10, 10, 1]

NUMERICALLY

X	Y1	Y2
-2	-5.333	-3.2
-1	-3	-1.4
0	-.6667	.4
1	1.6667	2.2
2	4	4
3	6.3333	5.8
4	8.6667	7.6

X=2

Observing the point of intersection on the graph can help us select an appropriate table of values to examine.

The answer to a system of two linear equations should be written in ordered-pair notation. Writing this solution as (2, 4) emphasizes there is only one point of intersection, which has two coordinates.

The graph yields an approximate solution of (2, 4). This is confirmed by using a table. Both lines have a y-coordinate of 4 for an x-value of 2.

Answer: (2, 4)

SELF-CHECK 3.3.1

1. Use a graphing calculator to solve the system of linear equations $\begin{Bmatrix} 4x - 3y = -15 \\ x + 3y = -6 \end{Bmatrix}$.

If the solution for a system of linear equations is not an ordered pair of small integers, then the graphical method may require some effort to obtain the solution. We will examine a couple of techniques that you can use to facilitate graphical solutions. In Calculator Perspective 1.3.1 we showed how to convert a decimal to fractional form. We use this feature in Calculator Perspective 3.3.1 to express both coordinates of the solution in fractional form. Note that the QUIT feature is the secondary function of the MODE key.

CALCULATOR PERSPECTIVE 3.3.1 Determining Rational Solutions of a System of Linear Equations

To solve the system of equations $\begin{cases} y = 2x - 3 \\ y = -5x + 6 \end{cases}$ on a TI-84 Plus calculator and convert the decimal coordinates of x and y to fractions, enter the following keystrokes:

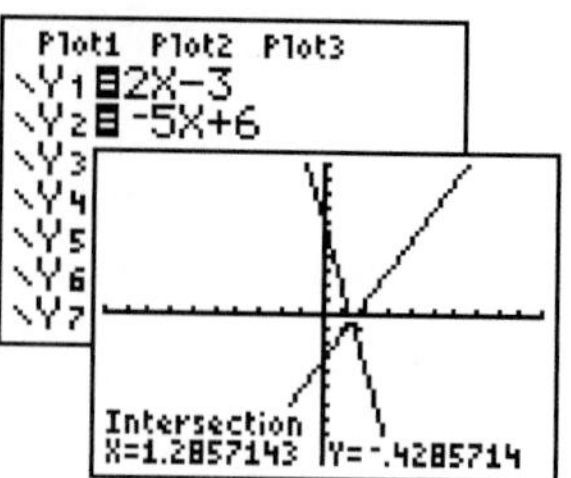

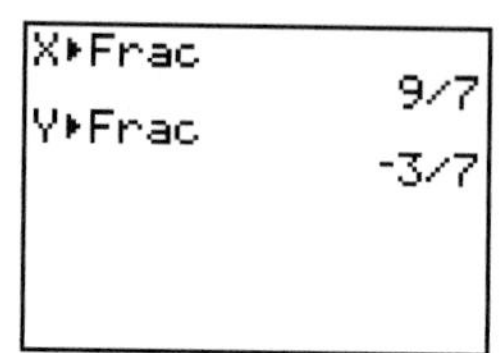

Note: The point of intersection was first determined and displayed in decimal form. Then individually the x- and y-coordinates were converted to fractional form.

Answer: The solution of the system $\begin{cases} y = 2x - 3 \\ y = -5x + 6 \end{cases}$ is the ordered pair $\left(\frac{9}{7}, -\frac{3}{7}\right)$.

SELF-CHECK 3.3.2

1. Use a graphing calculator to determine the fractional form of the solution of $\begin{cases} 5x - y = 6 \\ 3x + 2y = 4 \end{cases}$.

One of the prominent uses of systems of linear equations is to solve problems involving two quantities that are both changing. Often changing one variable (the input variable) will control the change in the other variable (the output variable). Example 2 examines the relationship between two angles.

EXAMPLE 2 Modeling the Angles in a Metal Part

A metal fabrication shop must assemble a part so that the angles shown in the figure are supplementary. To function properly, the specifications for the part require that the larger angle be twice the smaller angle. Determine the number of degrees in each angle.

SOLUTION

Let x = number of degrees in smaller angle
y = number of degrees in larger angle

Select a variable to represent each unknown.

VERBALLY	ALGEBRAICALLY
1. The angles are supplementary.	$x + y = 180$ or $y = 180 - x$
2. The larger angle is twice the smaller angle.	$y = 2x$

Angles are supplementary if the total of their measures is 180°.

Specifications require that the larger angle be twice the smaller angle.

NUMERICALLY

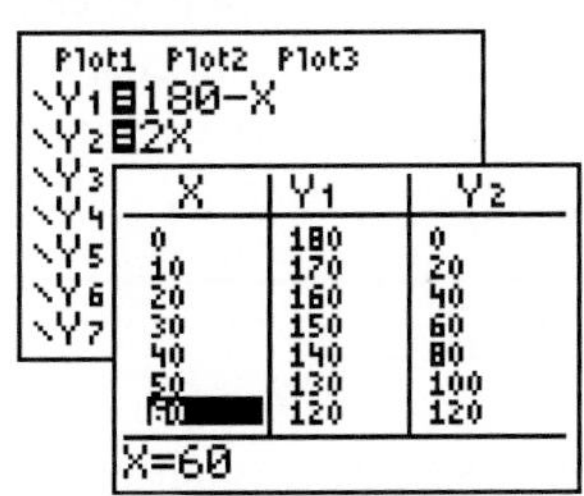

Enter both equations into a graphing calculator, and examine a table of values to determine the solution of the system.

It is not practical in this application for either angle to be negative. Also their total is 180°. Thus the x-variable is restricted to the values 0 to 180. You can scroll down the table on your calculator to 180 if this is needed.

Answer: The smaller angle is 60° and the larger angle is 120°.

Does this answer check?

SELF-CHECK 3.3.3

1. Rework Example 2 with the same conditions except that the angles must be complementary (sum is 90°).

One of the disadvantages of solving a system of linear equations by using a table or a graph is that it is sometimes difficult to select the appropriate table setup or the appropriate viewing window on a calculator. Fortunately our work in Section 2.8 concerning the restricted values on a variable in an application can assist us in selecting an appropriate table or viewing rectangle. Calculator Perspective 3.3.2 explores an option that can help us create a viewing rectangle to display the point of intersection of two lines. In this perspective we enter the restrictions from 0 to 180 for x and then use **ZoomFit** to select the y-values for the viewing window. The **ZoomFit** option is obtained by pressing ZOOM 0.

CALCULATOR PERSPECTIVE 3.3.2 **Using the ZoomFit Option to Select an Appropriate Viewing Window**

To solve the system of linear equations $\begin{Bmatrix} x + y = 180 \\ y = 2x \end{Bmatrix}$ from Example 2 by restricting the x-values from 0 to 180, enter the following keystrokes:

Y= 1 8 0 − X,T,θ,n ▼

2 X,T,θ,n

WINDOW (Set **Xmin** and **Xmax** based on the restrictions on the x-values.)

ZOOM 0

2nd CALC 5 ENTER ENTER ENTER

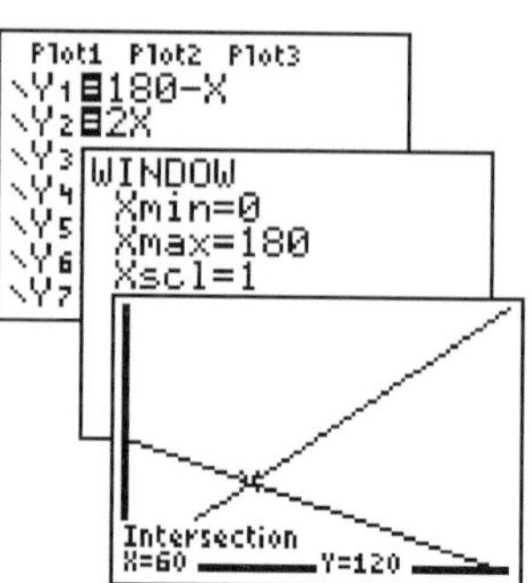

Answer: The solution of this system of equations is (60, 120).

Note: The window shown to the right displays the y-values obtained by using the **ZoomFit** option for this problem.

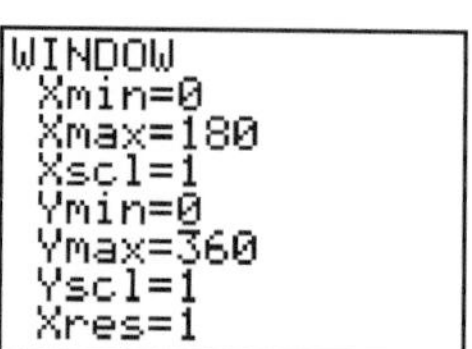

Example 3 examines two options that a pizza business can use to prepare its pasta. The mixture principle is used to form the equation: Total cost = Variable cost + Fixed cost. Then the **ZoomFit** feature on a calculator is used to examine these two options.

EXAMPLE 3 Business Options with Fixed and Variable Costs

A small family restaurant needs to be able to prepare up to 20 batches of its specialty pasta per day. There are two different machines used to make this pasta. For the first machine there is a fixed daily cost of \$46, plus a variable cost of \$5 for each order of pasta produced. The second machine has a fixed daily cost of \$22 and a variable cost of \$8 for each order produced. Using x to represent the number of orders of pasta produced and $C(x)$ to represent the total daily cost of producing these orders, determine the number of orders for which the daily costs will be the same on the two machines.

SOLUTION

Let x = number of orders of pasta produced
$C(x)$ = total daily cost of producing x orders of pasta

VERBALLY	ALGEBRAICALLY
Machine 1: Total daily cost equals variable cost plus fixed cost.	$C(x) = 5x + 46$
Machine 2: Total daily cost equals variable cost plus fixed cost.	$C(x) = 8x + 22$

The function notation $C(x)$ stresses that for the input x, the number of pasta orders, we can calculate the output $C(x)$, the daily cost.

GRAPHICALLY

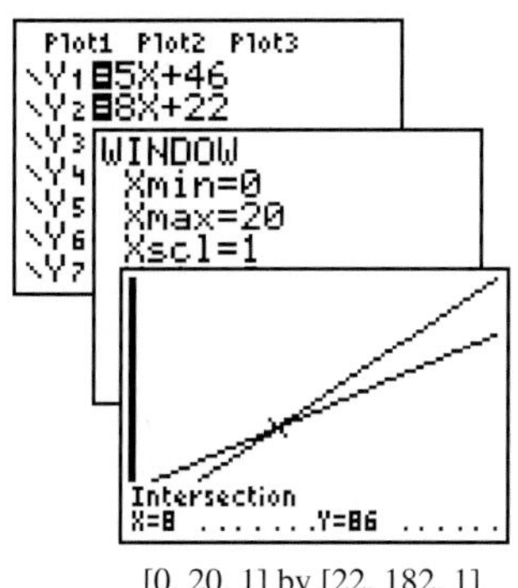

[0, 20, 1] by [22, 182, 1]

NUMERICALLY

X	Y1	Y2
3	61	46
4	66	54
5	71	62
6	76	70
7	81	78
8	86	86
9	91	94

X=8

From the problem statement, the variable x is restricted to the values from 0 to 20. Use the WINDOW key to enter these values for x, and then use the **ZoomFit** option to select the window shown here. The **Intersect** feature is used with this window to determine the point of intersection. The calculator window can be used to suggest an appropriate table of values to examine. This table confirms the costs are the same when eight batches are produced.

Answer: The total daily costs are the same on the two machines when eight batches are produced.

Write the answers to word problems as full sentences. This helps reinforce that you understand the question, and it increases your ability to use mathematics in the workforce.

SELF-CHECK 3.3.4

1. One machine has a fixed daily cost of \$120 and a variable cost of \$4 per item produced, whereas a second machine has a fixed daily cost of \$80 and a variable cost of \$4.50 per item produced. Using y to represent the total daily costs of these items, determine the number of items x for which the total daily costs will be the same. Assume that the number of units that can be produced per day is limited to 100 units.

One Solution

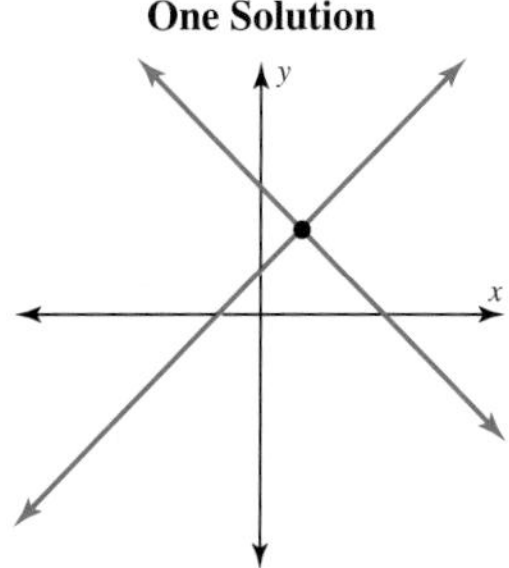

No Solution

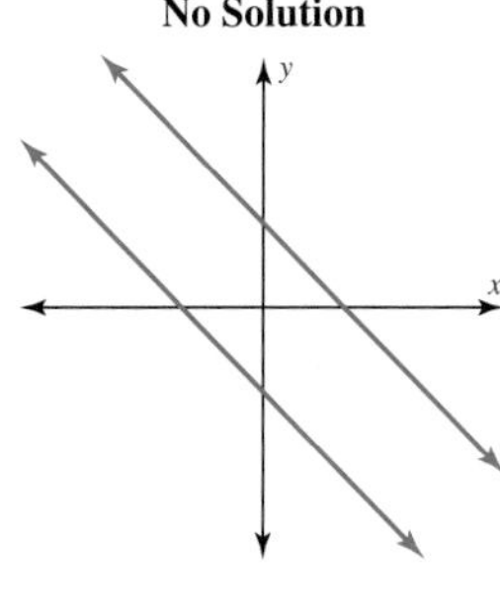

An Infinite Number of Solutions

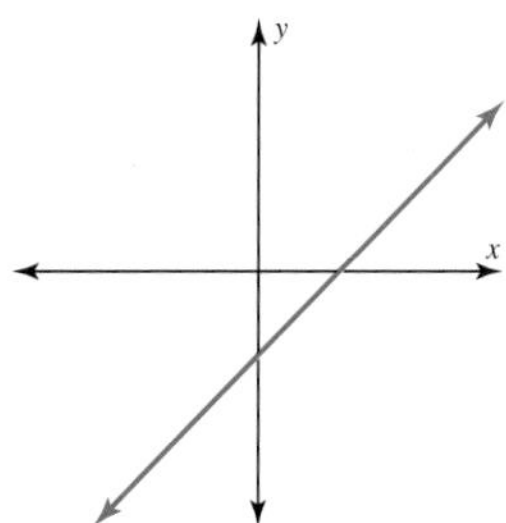

Examples 1 to 3 in this section have illustrated systems of linear equations with exactly one simultaneous solution. There are two other possibilities that we examine now.

Classification of Linear Systems

Two lines in a plane can be related in three ways:

1. **One solution:** The lines are distinct and intersect at a single point. This point represents the only simultaneous solution.
2. **No solution:** The lines are parallel and distinct. There is no point of intersection and no simultaneous solution.
3. **An infinite number of solutions:** The lines coincide (both equations represent the same line). There are an infinite number of points on this shared line, and each represents a solution. Only these points are solutions; the points not on the line are not solutions.

If a system of two equations in two variables has a solution, the system is called **consistent;** otherwise it is called **inconsistent.** If the equations have distinct graphs, the equations are called **independent;** if the graphs coincide, the equations are called **dependent.** Example 4 presents an inconsistent system of equations.

EXAMPLE 4 Solving an Inconsistent System

Solve the linear system $\begin{cases} x + 2y = 2 \\ 2x + 4y = 8 \end{cases}$ numerically and graphically.

SOLUTION

FIRST EQUATION

$$x + 2y = 2$$
$$2y = -x + 2$$
$$y = -\frac{1}{2}x + 1$$

The slope is $-\frac{1}{2}$ and the y-intercept is (0, 1).

SECOND EQUATION

$$2x + 4y = 8$$
$$4y = -2x + 8$$
$$y = -\frac{1}{2}x + 2$$

The slope is $-\frac{1}{2}$ and the y-intercept is (0, 2).

We are writing each equation in slope-intercept form both to facilitate an algebraic comparison and to prepare the equations for graphing either by hand or by using a graphing calculator.

Thus the two lines are parallel and distinct.

NUMERICALLY

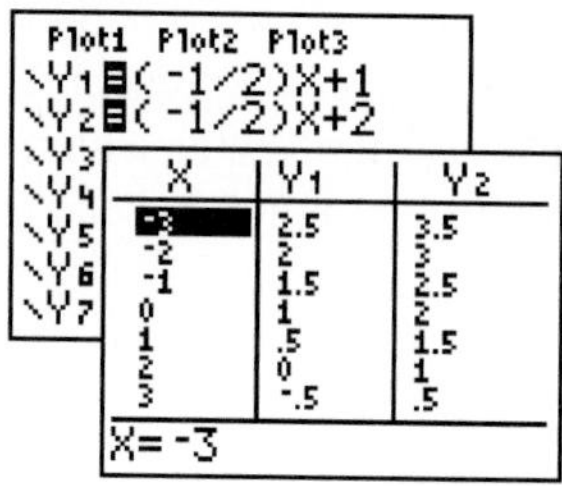

For the same x-value, the Y_1 and Y_2 values are always 1 unit apart.

GRAPHICALLY

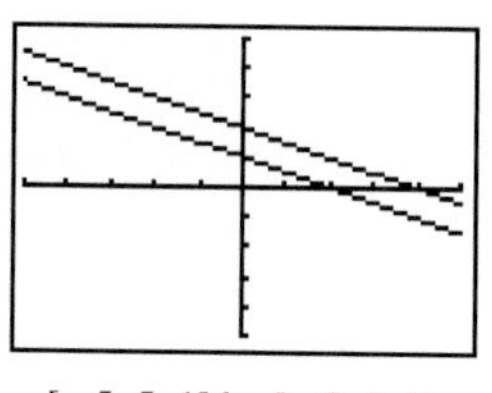

[−5, 5, 1] by [−5, 5, 1]

The lines appear parallel and are always the same distance apart.

These parallel lines are always the same distance apart. This is shown both by the table and by the graph. Since these parallel lines never meet, the system has no solution and is classified as inconsistent.

Answer: No solution; this is an inconsistent system.

Example 5 presents the third possibility that can occur with a system of linear equations. This is a system of dependent linear equations.

EXAMPLE 5 Solving a System of Dependent Equations

Solve the linear system $\begin{cases} 3x - 4y = 12 \\ 6x - 8y = 24 \end{cases}$ numerically and graphically.

SOLUTION

FIRST EQUATION

$$3x - 4y = 12$$
$$-4y = -3x + 12$$
$$y = \frac{3}{4}x - 3$$

The slope is $\frac{3}{4}$ and the y-intercept is (0, −3).

SECOND EQUATION

$$6x - 8y = 24$$
$$-8y = -6x + 24$$
$$y = \frac{3}{4}x - 3$$

The slope is $\frac{3}{4}$ and the y-intercept is (0, −3).

Rewrite each equation in slope-intercept form. Note that both equations have the same slope-intercept form, revealing that they are equations of the same line.

The two equations produce the same line.

NUMERICALLY	GRAPHICALLY
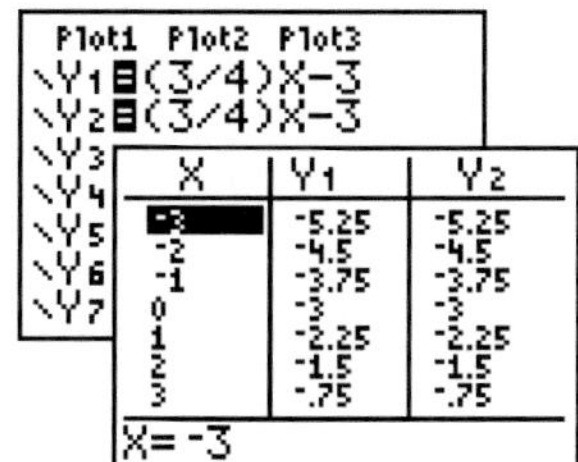	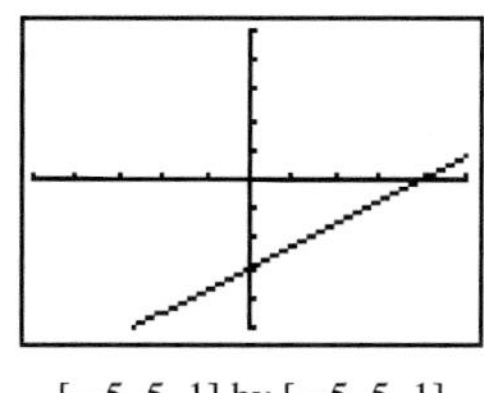 $[-5, 5, 1]$ by $[-5, 5, 1]$
For each x-value the Y_1 and Y_2 values are the same.	The equations produce the same line.

Answer: There are an infinite number of solutions; the equations are dependent.

Although the original equations in Example 5 look different, the lines coincide. Thus every point satisfying one equation will also satisfy the other. However, only the points on this line are solutions of the linear system of equations. All other points are not solutions of the system.

It is now time to reflect on the three classifications of systems of linear equations. Compare the slope-intercept forms of the equations in the following three classifications of linear systems. Can you predict the classification by using the slope-intercept form?

Classification of Systems of Linear Equations

The linear system $\begin{cases} y = m_1x + b_1 \\ y = m_2x + b_2 \end{cases}$ is:

VERBALLY	ALGEBRAICALLY	NUMERICAL EXAMPLE	GRAPHICAL EXAMPLE
1. A **consistent system of independent linear equations** having exactly one solution	$m_1 \neq m_2$ *Example:* $\begin{cases} y = 2x - 1 \\ y = -3x - 6 \end{cases}$	Only one x-value has matching y-values. X: -5, -4, -3, -2, -1, 0, 1 Y1: -11, -9, -7, -5, -3, -1, 1 Y2: 9, 6, 3, 0, -3, -6, -9 X=-1	One distinct point of intersection Intersection X=-1 Y=-3 $[-10, 10, 1]$ by $[-10, 10, 1]$
2. An **inconsistent system of linear equations** having no solution	$m_1 = m_2$ and $b_1 \neq b_2$ *Example:* $\begin{cases} y = \frac{1}{2}x + 1 \\ y = \frac{1}{2}x - 3 \end{cases}$	For each x-value, Y_2 is 4 units less than Y_1. X: -3, -2, -1, 0, 1, 2, 3 Y1: -.5, 0, .5, 1, 1.5, 2, 2.5 Y2: -4.5, -4, -3.5, -3, -2.5, -2, -1.5 X=-3	Parallel lines with no point of intersection $[-10, 10, 1]$ by $[-10, 10, 1]$
3. A **consistent system of dependent linear equations** having an infinite number of solutions	$m_1 = m_2$ and $b_1 = b_2$ *Example:* $\begin{cases} y = \frac{3}{2}x + \frac{1}{2} \\ y = \frac{3}{2}x + \frac{1}{2} \end{cases}$	For each x-value the Y_1 and Y_2 values are the same. X: -3, -2, -1, 0, 1, 2, 3 Y1: -4, -2.5, -1, .5, 2, 3.5, 5 Y2: -4, -2.5, -1, .5, 2, 3.5, 5 X=-3	Infinite number of common points on these coincident lines $[-10, 10, 1]$ by $[-10, 10, 1]$

Example 6 uses the slope-intercept form and the information from the previous box to determine the number of solutions of each of the following systems. Note that this work does not tell us which points satisfy both equations, only how many points will be solutions.

EXAMPLE 6 Determining the Number of Solutions of a Linear System

Determine the number of solutions of each of the following systems.

SOLUTION

(a) $\begin{Bmatrix} y = \frac{2}{3}x - 5 \\ y = \frac{3}{4}x + 7 \end{Bmatrix}$

FIRST EQUATION	SECOND EQUATION	
$y = \frac{2}{3}x - 5$	$y = \frac{3}{4}x + 7$	Determine the slope from the slope-intercept form $y = mx + b$.
$m_1 = \frac{2}{3}$	$m_2 = \frac{3}{4}$	Because the slopes are different the lines will intersect at exactly one point.

Answer: Because $m_1 \neq m_2$, the system has exactly one solution.

(b) $\begin{Bmatrix} 3x - y = 5 \\ 2y = 6x - 10 \end{Bmatrix}$

FIRST EQUATION	SECOND EQUATION	
$3x - y = 5$ $-y = -3x + 5$ $y = 3x - 5$ $m_1 = 3, b_1 = -5$	$2y = 6x - 10$ $y = 3x - 5$ $m_2 = 3, b_2 = -5$	First write each equation in slope-intercept form. Then use the slopes and y-intercepts to determine the number of solutions.

Answer: Because $m_1 = m_2$ and $b_1 = b_2$, the equations are dependent and the system of equations has an infinite number of solutions.

Both equations represent the same line. Only the points on this line are solutions of the system of linear equations.

(c) $\begin{Bmatrix} 2x - 3y = 15 \\ 2x - 3y = -6 \end{Bmatrix}$

FIRST EQUATION	SECOND EQUATION	
$2x - 3y = 15$ $-3y = -2x + 15$ $y = \frac{2}{3}x - 5$ $m_1 = \frac{2}{3}, b_1 = -5$	$2x - 3y = -6$ $-3y = -2x - 6$ $y = \frac{2}{3}x + 2$ $m_2 = \frac{2}{3}, b_2 = 2$	Because the slopes are equal the lines are parallel. Because the y-intercepts are different the lines do not coincide.

Answer: Because $m_1 = m_2$ but $b_1 \neq b_2$, the system is inconsistent and has no solution.

SELF-CHECK 3.3.5

Determine the number of solutions to each of the following systems of linear equations.

1. $\begin{Bmatrix} y_1 = 2x - 7 \\ y_2 = 2x + 7 \end{Bmatrix}$
2. $\begin{Bmatrix} y_1 = 2x - 7 \\ y_2 = 7x - 2 \end{Bmatrix}$
3. $\begin{Bmatrix} y_1 = 2x - 7 \\ y_2 = 2x - 7 \end{Bmatrix}$

Example 7 is included to reveal some of the weaknesses of the graphical method and an overreliance on graphing calculators. Before you read through the text explanation, try to solve the system $\begin{Bmatrix} 2x - 3y = 10 \\ 3x - 5y = 0 \end{Bmatrix}$ by using a graphing calculator. Please note the solution you obtain and any difficulties you encounter.

We will use Calculator Perspective 2.3.3 to change the calculator window to the one shown in the adjusted window.

EXAMPLE 7 Using a Graphing Calculator to Solve a System of Linear Equations

Solve the linear system $\begin{Bmatrix} 2x - 3y = 10 \\ 3x - 5y = 0 \end{Bmatrix}$ by using a graphing calculator.

SOLUTION

FIRST EQUATION

$$2x - 3y = 10$$
$$-3y = -2x + 10$$
$$y = \frac{2}{3}x - \frac{10}{3}$$

SECOND EQUATION

$$3x - 5y = 0$$
$$-5y = -3x$$
$$y = \frac{3}{5}x$$

Write each equation in slope-intercept form to prepare the equations for entry into a graphing calculator.

FIRST VIEW WINDOW

Plot1 Plot2 Plot3
\Y1=(2/3)X-10/3
\Y2=(3/5)X
\Y3=
\Y4=
\Y5=
\Y6=
\Y7=

[−10, 10, 1] by [−10, 10, 1]

You may find it easier to determine this point of intersection by using the algebraic methods in Sections 3.4 and 3.5.

Although the lines may appear to be parallel on this window, we know the lines are not parallel because they do not have the same slope. Thus the lines must have a point of intersection that is not revealed on this window.

ADJUSTED WINDOW

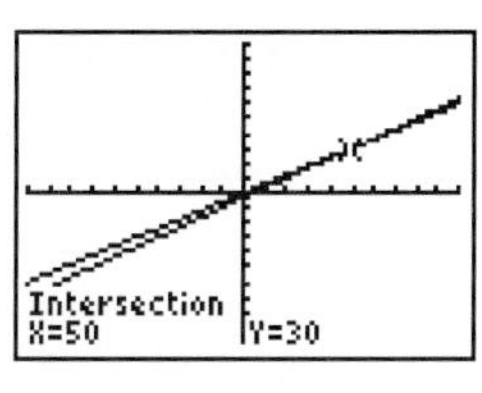

[−100, 100, 10] by [−100, 100, 10]

The new viewing rectangle contains the point of intersection which is located by using the **Intersect** feature on a graphing calculator.

The point of intersection is (50, 30).

Check:

FIRST EQUATION

$$2x - 3y = 10$$
$$2(50) - 3(30) \stackrel{?}{=} 10$$
$$100 - 90 \stackrel{?}{=} 10$$
$$10 \stackrel{?}{=} 10 \text{ checks.}$$

SECOND EQUATION

$$3x - 5y = 0$$
$$3(50) - 5(30) \stackrel{?}{=} 0$$
$$150 - 150 \stackrel{?}{=} 0$$
$$0 \stackrel{?}{=} 0 \text{ checks.}$$

Answer: (50, 30).

SELF-CHECK 3.3.6

1. Solve the linear system $\begin{Bmatrix} 2x + y = 10 \\ 3x + y = 0 \end{Bmatrix}$ by using a graphing calculator window of [−20, 5, 5] by [−5, 40, 5].

SELF-CHECK ANSWERS

3.3.1

1. $(-4.2, -0.6)$

3.3.2

1. 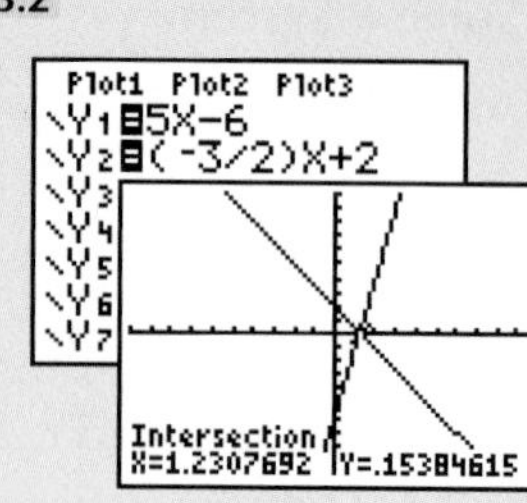

X▸Frac 16/13
Y▸Frac 2/13

The solution is the ordered pair $\left(\frac{16}{13}, \frac{2}{13}\right)$.

3.3.3

1. The smaller angle is 30°, and the larger angle is 60°.

3.3.4

1. The total daily costs are the same for the two machines when 80 units are produced.

3.3.5

1. No solution.
2. Exactly one solution.
3. An infinite number of solutions.

3.3.6

1. $(-10, 30)$

USING THE LANGUAGE AND SYMBOLISM OF MATHEMATICS 3.3

1. An ordered pair that satisfies each equation in a system of linear equations is called a __________ of the system of linear equations.
2. Graphically, the solution to a system of two linear equations is represented by their point of __________.
3. An __________ system of two linear equations has no solution. Graphically, these lines are __________ with no point in common.
4. A consistent system of two independent linear equations has exactly __________ solution. Graphically, these lines intersect at exactly __________ point.
5. A consistent system of two dependent linear equations has an __________ number of solutions. Graphically, these lines __________.
6. Use the linear system $\begin{Bmatrix} y = m_1x + b_1 \\ y = m_2x + b_2 \end{Bmatrix}$ to select the correct choice that describes the relationship between the two lines given.
   a. $m_1 = m_2$ and $b_1 = b_2$ The lines intersect at **zero/one/infinitely many** point(s).
   b. $m_1 = m_2$ and $b_1 \neq b_2$ The lines **are/are not** parallel. The lines intersect at **zero/one/infinitely many** point(s).
   c. $m_1 \neq m_2$ The lines **are/are not** parallel. The lines intersect at **zero/one/infinitely many** point(s).

EXERCISES 3.3

In Exercises 1–4, determine by inspection the point of intersection of the two lines.

1. 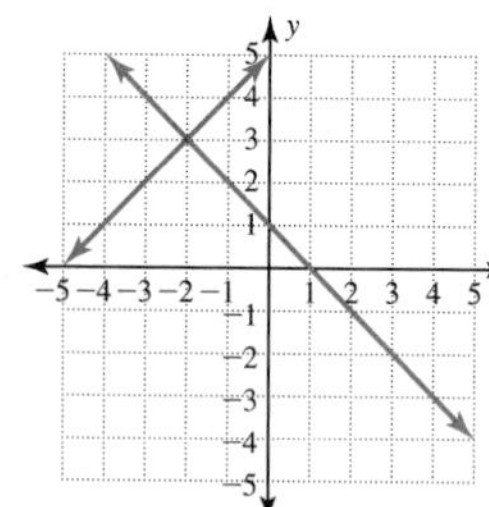

2. 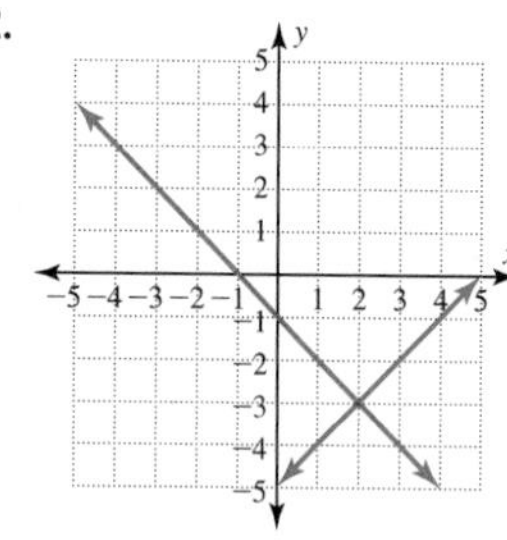
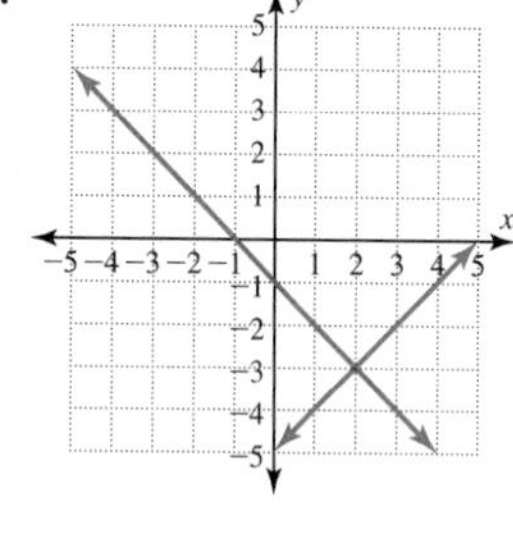

3. 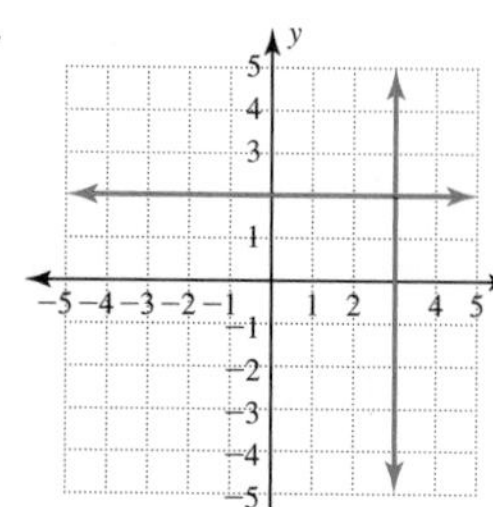

4. 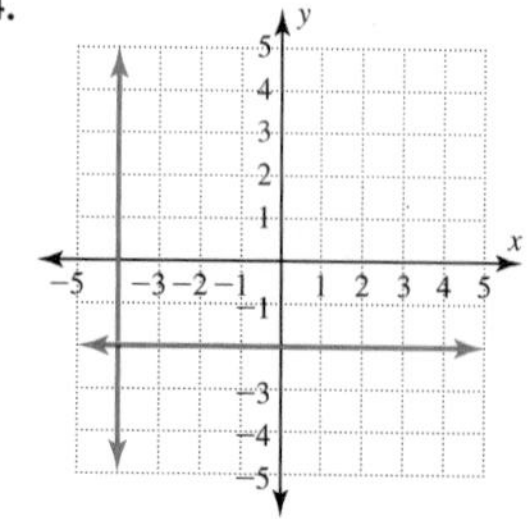

In Exercises 5–12, determine whether the given ordered pair is a solution of the system of linear equations.

5. $(-1, -2)$
   $7x - 3y + 1 = 0$
   $5x - 4y - 3 = 0$
6. $(-4, 2)$
   $4x + 5y + 6 = 0$
   $6x + 5y + 14 = 0$
7. $\left(\frac{1}{2}, \frac{1}{3}\right)$
   $4x - 3y = 1$
   $6x - 6y = 1$
8. $\left(\frac{1}{4}, -\frac{2}{3}\right)$
   $8x + 3y = 0$
   $4x - 3y = 3$
9. $(0.1, -0.2)$
   $4x - 3y = 1$
   $2x + y = 0$
10. $(-0.3, 0.5)$
    $5x + 3y = 0$
    $6x - y = -2.3$
11. $(0, 6)$
    $2x - y = -6$
    $3x + y = 3$
12. $(9, 0)$
    $3x - 8y = -24$
    $2x + 9y = 18$

In Exercises 13–16, determine the point of intersection. Then check this ordered pair in both equations to verify that it is a solution of the system of linear equations.

13.

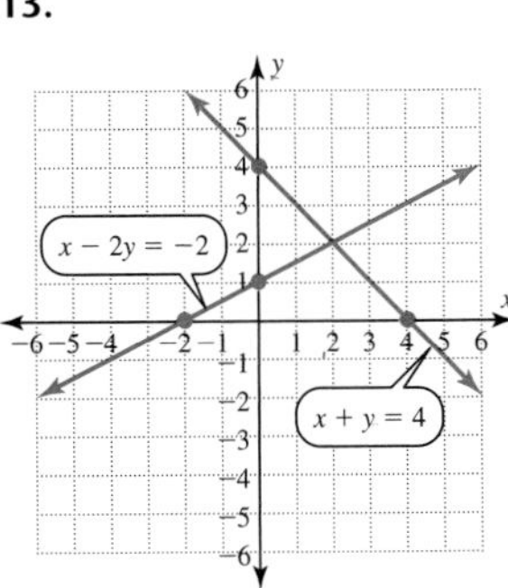

14.

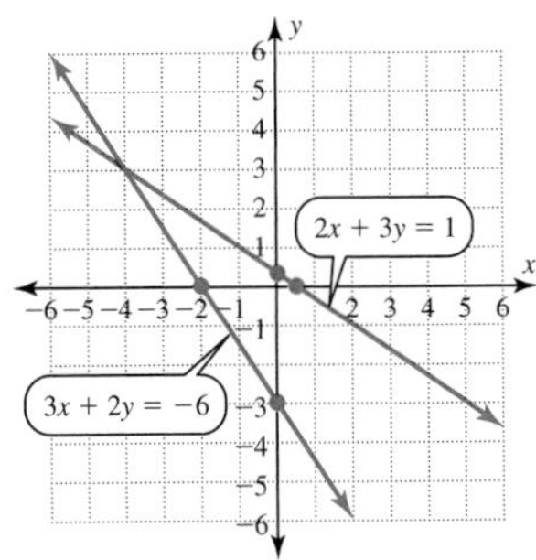

15.

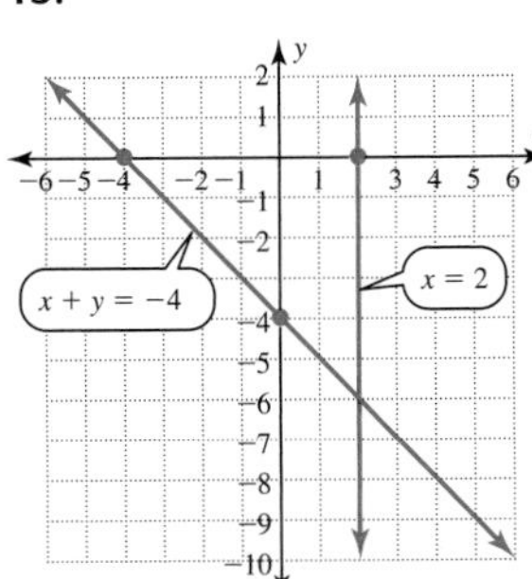

16.

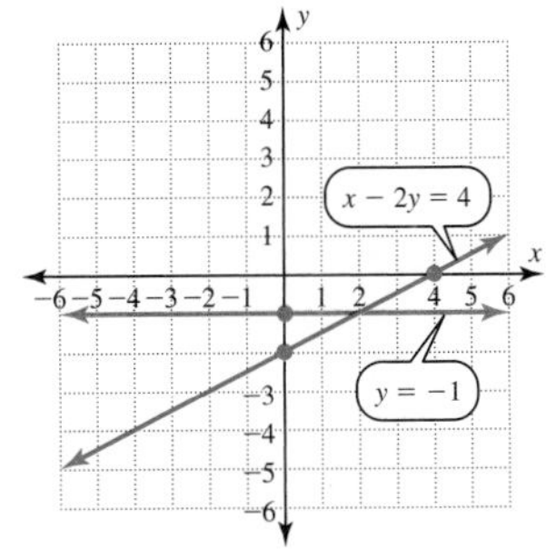

In Exercises 17–20, use the tables shown on the graphing calculator display to solve each system of linear equations.

17.

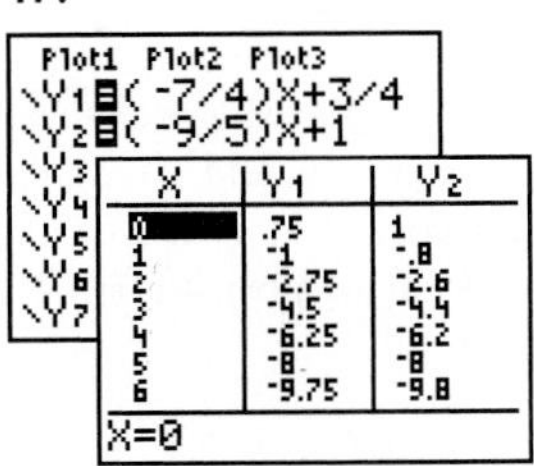
Y1=(-7/4)X+3/4
Y2=(-9/5)X+1

X	Y1	Y2
0	.75	1
1	-1	-.8
2	-2.75	-2.6
3	-4.5	-4.4
4	-6.25	-6.2
5	-8	-8
6	-9.75	-9.8

X=0

18.

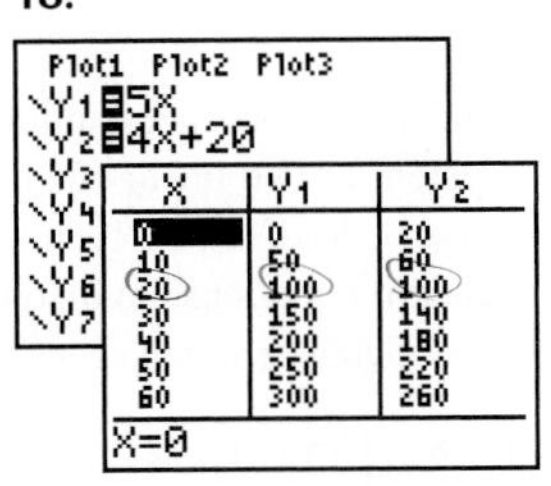
Y1=5X
Y2=4X+20

X	Y1	Y2
0	0	20
10	50	60
20	100	100
30	150	140
40	200	180
50	250	220
60	300	260

X=0

19.

Y1=(-20/3)X+2/3
Y2=(-15/2)X+1/4

X	Y1	Y2
-1	7.3333	7.75
-.9	6.6667	7
-.8	6	6.25
-.7	5.3333	5.5
-.6	4.6667	4.75
-.5	4	4
-.4	3.3333	3.25

X=-1

20.

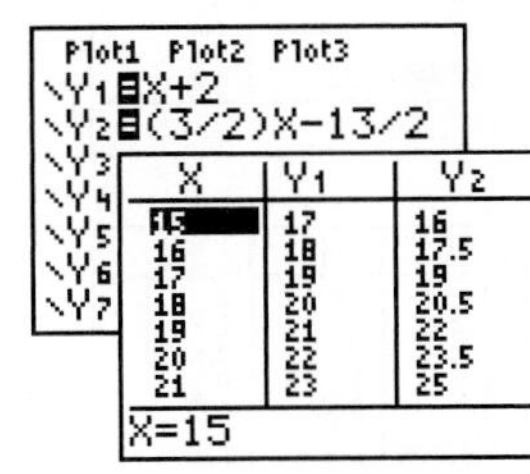
Y1=X+2
Y2=(3/2)X-13/2

X	Y1	Y2
15	17	16
16	18	17.5
17	19	19
18	20	20.5
19	21	22
20	22	23.5
21	23	25

X=15

21. **Rental Truck Costs** The following graph compares the cost of a one-day rental for a moving truck at two different rental companies. The cost is based upon the number of miles driven. Give the solution to the corresponding system of equations. Then interpret the meaning of the x- and y-coordinates of this solution.

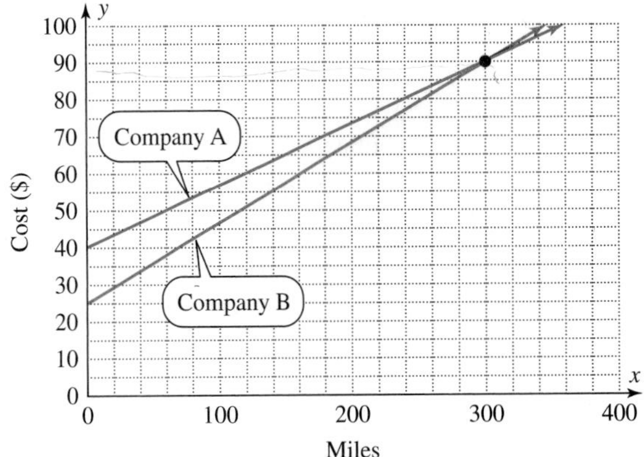

22. **Digital Phone Plans** The following graph compares the monthly cost of two different digital phone plans based on the number of minutes of use. Give the solution to the corresponding system of equations. Then interpret the meaning of the x- and y-coordinates of this solution.

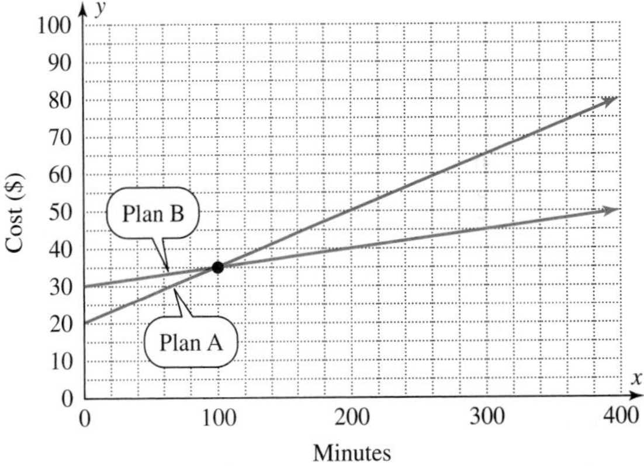

In Exercises 23–36, solve each system of linear equations by graphing either on paper or on a graphing calculator.

23. $y = -\frac{x}{3} + 2$
$y = -\frac{x}{2} + 1$

24. $y = -\frac{x}{2} + 2$
$y = \frac{x}{2}$

25. $x + 3 = 0$
$y - 2 = 0$

26. $y = 2$
$x - 5 = 0$

27. $y = -2x + 10$
$x = 6$

28. $y = 2x - 6$
$y = -4$

29. $y = 2x - 7$
$y = -\frac{3}{2}x + 7$

30. $y = -3x + 7$
$y = -\frac{2}{5}x + \frac{9}{5}$

31. $2x - 3y - 6 = 0$
$4x - 3y + 3 = 0$

32. $5x - 4y + 16 = 0$
$3x - 2y - 6 = 0$

33. $y = 3x$
$y = 3x - 6$

34. $y = 2x + 1$
$y = 2x - 1$

35. $x - 2y - 2 = 0$
$2x - 4y - 4 = 0$

36. $2x - 3y - 3 = 0$
$6x - 9y - 9 = 0$

Calculator Exercises

In Exercises 37–40, solve each system of linear equations by using a calculator to graph each equation on the coordinate system shown; then determine the point of intersection. Check the coordinates of this point in both of the linear equations. (*Hint:* Use the **Zoom 4** option and see Calculator Perspective 2.3.2.)

[−4.7, 4.7, 1] by [−3.1, 3.1, 1]

37. $y = 2x - 5$
$y = -2x - 1$

38. $y = x - 5$
$y = -x + 3$

39. $y = 2x - 1$
$y = \frac{x}{2} + 2$

40. $y = -x + 5$
$y = \frac{x}{3} + 1$

In Exercises 41–44, solve each system of linear equations and express both coordinates of the answer in fractional form. (*Hint:* See Calculator Perspective 3.3.1.)

41. $14x - 7y = -5$
$7x + 21y = 29$

42. $13x + 7y = 8$
$13x - 14y = -7$

43. $6x - 11y = 7$
$3x + 22y = -1$

44. $6x + 12y = -7$
$6x - 8y = 5$

In Exercises 45–48, solve each system of linear equations by using the given restriction on the variable x and the **ZoomFit** option and the **Intersect** feature on a calculator. (*Hint:* See Calculator Perspective 3.3.2.)

45. $\begin{Bmatrix} x + y = 1{,}000 \\ 2x + 3y = 2{,}700 \end{Bmatrix}$
The variable x is restricted to the interval [0, 1,000].

46. $\begin{Bmatrix} x + y = 8{,}000 \\ 0.06x + 0.03y = 435 \end{Bmatrix}$
The variable x is restricted to the interval [0, 8,000].

47. $\begin{Bmatrix} x + y = 1 \\ 8x + 11y = 10.1 \end{Bmatrix}$
The variable x is restricted to the interval [0, 1].

48. $\begin{Bmatrix} x + y = 1 \\ 4x - 5y = 2.65 \end{Bmatrix}$
The variable x is restricted to the interval [0, 1].

In Exercises 49–52, solve each system of linear equations by using a calculator to complete the following table of values.

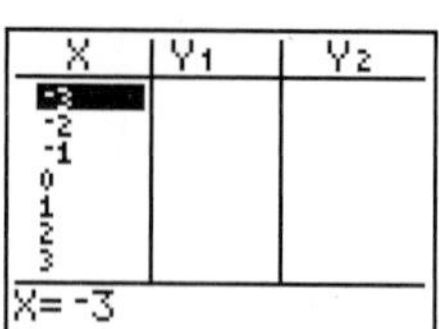

X	Y1	Y2
-3		
-2		
-1		
0		
1		
2		
3		

X=-3

49. $y = 2x - 5$
$y = -2x - 1$

50. $y = 2x - 5$
$y = -2x - 5$

51. $y = 5x + 3$
$y = -5x + 3$

52. $y = 5x + 3$
$y = -5x - 7$

In Exercises 53 and 54, use a calculator to complete each table and to solve the given system of linear equations.

53. $4x - y = 80$
$3x + y = 130$

54. $5x + 2y = 5$
$7x - 2y = 175$

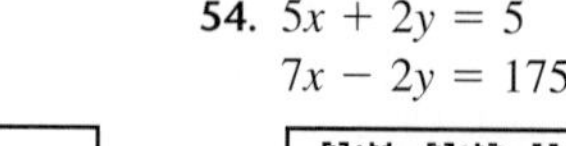

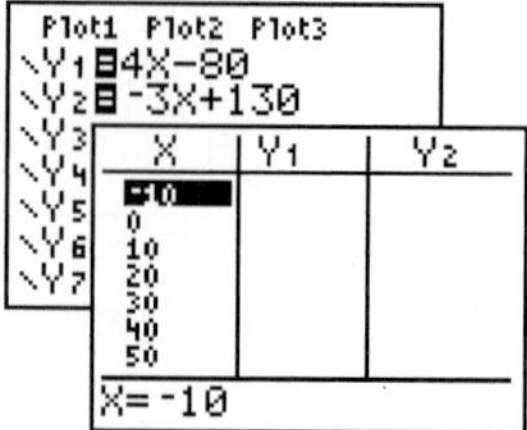

Plot1 Plot2 Plot3
\Y1■4X−80
\Y2■−3X+130

X	Y1	Y2
-10		
0		
10		
20		
30		
40		
50		

X=-10

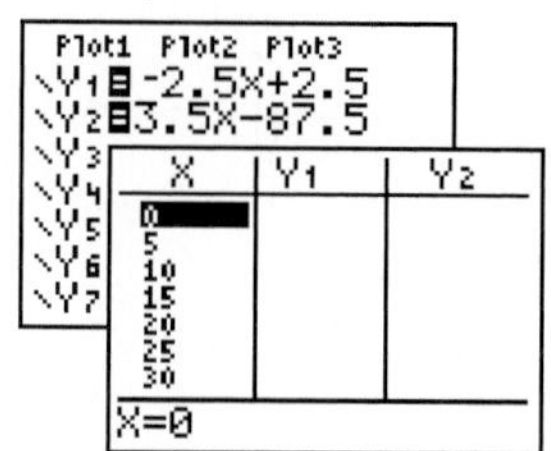

Plot1 Plot2 Plot3
\Y1■−2.5X+2.5
\Y2■3.5X−87.5

X	Y1	Y2
0		
5		
10		
15		
20		
25		
30		

X=0

In Exercises 55–62, use the slope-intercept form of each line to determine the number of solutions of the system; then classify each system as a consistent system of independent equations, an inconsistent system, or a consistent system of dependent equations.

55. $y = \frac{3}{7}x - 17$
$y = -\frac{7}{8}x + 8$

56. $y = \frac{5}{9}x + 4$
$y = \frac{9}{5}x - 4$

57. $y = \frac{3}{8}x + 11$
$y = \frac{3}{8}x - 5$

58. $y = -\frac{9}{5}x + 7$
$y = -\frac{9}{5}x + 4$

59. $6x + 2y = 4$
$9x = 6 - 3y$

60. $8x = 12 - 20y$
$6x + 15y = 9$

61. $2x + 3y = 3$
$4x - 3y = -3$

62. $\frac{x}{6} + \frac{y}{4} = \frac{1}{3}$
$\frac{x}{3} + \frac{y}{2} = \frac{1}{4}$

In Exercises 63–66, the graphing calculator display shows a table for a system of two linear equations $y_1 = m_1x + b_1$ and $y_2 = m_2x + b_2$. Classify each system as a consistent system of independent equations, an inconsistent system, or a consistent system of dependent equations.

63.

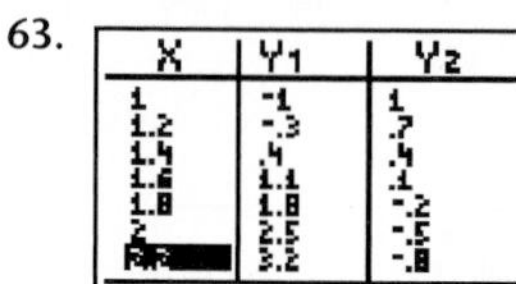

X	Y1	Y2
1	-1	1
1.2	-.3	.7
1.4	.4	.4
1.6	1.1	.1
1.8	1.8	-.2
2	2.5	-.5
2.2	3.2	-.8

X=2.2

64.

X	Y1	Y2
1	-1	-1
1.2	-.3	-.3
1.4	.4	.4
1.6	1.1	1.1
1.8	1.8	1.8
2	2.5	2.5
2.2	3.2	3.2

X=2.2

65.

X	Y1	Y2
1	-1	1
1.2	-.3	1.7
1.4	.4	2.4
1.6	1.1	3.1
1.8	1.8	3.8
2	2.5	4.5
2.2	3.2	5.2

X=2.2

66.

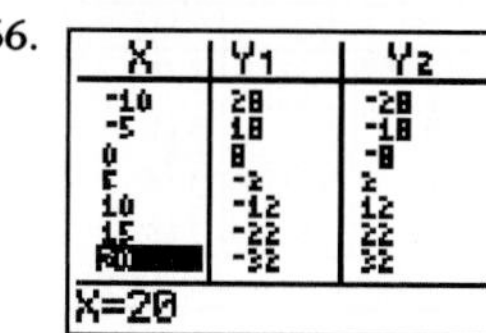

X	Y1	Y2
-10	28	-28
-5	18	-18
0	8	-8
5	-2	2
10	-12	12
15	-22	22
20	-32	32

X=20

Equivalent Costs

In Exercises 67 and 68, determine when both options will have the same cost and what this cost will be.

67. The input variable x represents the number of units produced, and the output variable $f(x)$ represents the cost of this production.

Option A: $f(x) = x + 7$
Option B: $f(x) = 2x + 3$

68. The input variable x represents the number of units produced, and the output variable $f(x)$ represents the cost of this production.

Option A: $f(x) = 2x + 9$

Option B: $f(x) = 3x + 6$

69. A Table of Flight Distances Two planes are leaving Chicago's O'Hare Airport. A small single-engine plane that travels 150 mi/h leaves at noon. Two hours later a larger commercial jet that travels 450 mi/h takes off and flies in the same direction. The tables display the distance in miles flown by each plane, where x is the number of hours past 2 p.m. Give the solution to the corresponding system of equations. Then interpret the meaning of the x- and y-coordinates of this solution.

Small Plane

HOURS x	DISTANCE y
0	300
0.5	375
1.0	450
1.5	525
2.0	600
2.5	675

Large Jet

HOURS x	DISTANCE y
0	0
0.5	225
1.0	450
1.5	675
2.0	900
2.5	1,125

70. A Table of Bus Fares When planning a student trip, the student council has to choose between two busing services. Service A charges $4.50 per person, while service B charges a fee of $200 plus $0.50 per person. The tables display the charges in dollars for each service based upon the number of students on the trip. Give the solution to the corresponding system of equations. Then interpret the meaning of the x- and y-coordinates of this solution.

Service A

STUDENTS x	COST y
10	45
20	90
30	135
40	180
50	225
60	270

Service B

STUDENTS x	COST y
10	205
20	210
30	215
40	220
50	225
60	230

71. Angles in Two Parts The two angles shown represent parts that should be assembled so that the angles are supplementary and the larger angle is 5 more than 3 times the smaller angle. Determine the number of degrees in each angle.

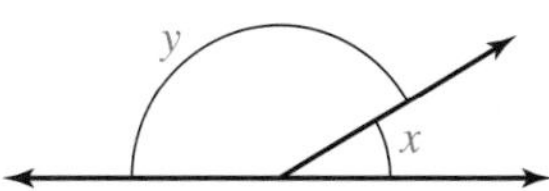

72. Angles in Two Parts The two angles shown in Exercise 71 represent parts that should be assembled so that the angles are supplementary and the larger angle is 6 less than twice the smaller angle. Determine the number of degrees in each angle.

73. Fixed and Variable Costs One machine has a fixed daily cost of $75 and a variable cost of $3 per item produced, whereas a second machine has a fixed daily cost of $60 and a variable cost of $4.50 per item produced. Using y to represent the total daily costs of these items, determine the number of items x for which the total daily costs will be the same.

74. Fixed and Variable Costs A small manufacturing plant makes custom replacement parts for plastic molding equipment. The maximum production is limited to 100 units per day. The company has two machines available to make this part. One machine has a fixed daily cost of $450 and a variable cost of $12.50 per unit of this part produced. A second machine has a fixed daily cost of $200 and a variable cost of $17.50 per unit of this part produced. Using y to represent the total daily cost of producing x units of this part, determine the number of units for which the total daily cost of producing this part will be the same for both machines.

Multiple Representations

In Exercises 75 and 76, complete the numerical and graphical descriptions and then write a full sentence that answers the question.

75. Verbally

One number is seven more than twice another number. The sum of these numbers is thirteen. Find these numbers.

Algebraically

$y = 2x + 7$
$x + y = 13$

Numerically

x	$y_1 = 2x + 7$	$y_2 = 13 - x$
−3	1	16
−2		
−1		
0		
1		
2		
3	13	10

Graphically

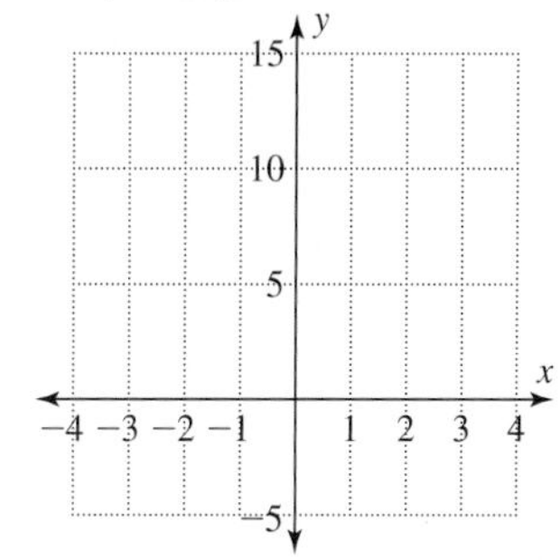

76.

Verbally

One number is five less than three times another number. The sum of these numbers is eleven. Find these numbers.

Algebraically

$y = 3x - 5$
$x + y = 11$

Numerically

x	$y_1 = 3x - 5$	$y_2 = 11 - x$
−1	−8	12
−0		
1		
2		
3		
4		
5	10	6

Graphically

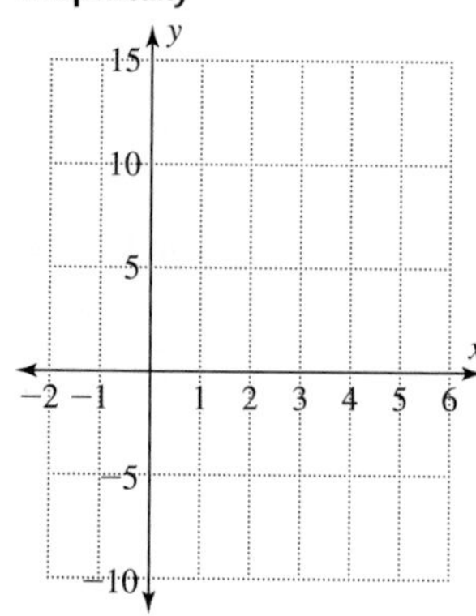

77. Write a system of two linear equations with a solution of $(3, 5)$ so that one line is horizontal and the other is vertical.

78. Write a system of two linear equations with a solution of $(-4, -2)$ so that one line is horizontal and the other is vertical.

79. Write a system of two linear equations with a solution of $(0, 3)$ so that one line has a slope of $\frac{1}{2}$ and the other line has a slope of -4.

80. Write a system of two linear equations with a solution of $(0, -2)$ so that one line has a slope of $\frac{1}{3}$ and the other line has a slope of -1.

Group Discussion Questions

81. Discovery Question

a. Consider a system of two linear equations of the form $Ax + By = C$. Can the system have zero solutions? If so, sketch an example.

b. Consider a system of two linear equations of the form $Ax + By = C$. Can the system have exactly one solution? If so, sketch an example.

c. Consider a system of two linear equations of the form $Ax + By = C$. Can the system have exactly two solutions? If so, sketch an example.

d. Consider a system of two linear equations of the form $Ax + By = C$. Can the system have exactly three solutions? If so, sketch an example.

e. Consider a system of two linear equations of the form $Ax + By = C$. Can the system have more than three solutions? If so, sketch an example.

82. Discovery Question

a. Consider a system of three linear equations of the form $Ax + By = C$. Can the system have zero solutions? If so, sketch an example.

b. Consider a system of three linear equations of the form $Ax + By = C$. Can the system have exactly one solution? If so, sketch an example.

c. Consider a system of three linear equations of the form $Ax + By = C$. Can the system have exactly two solutions? If so, sketch an example.

d. Consider a system of three linear equations of the form $Ax + By = C$. Can the system have exactly three solutions? If so, sketch an example.

e. Consider a system of three linear equations of the form $Ax + By = C$. Can the system have more than three solutions? If so, sketch an example.

83. Error Analysis A student graphed the system shown below and concluded the lines were parallel and the system was inconsistent with no solution. Describe the error the student has made, and then solve this system of equations.

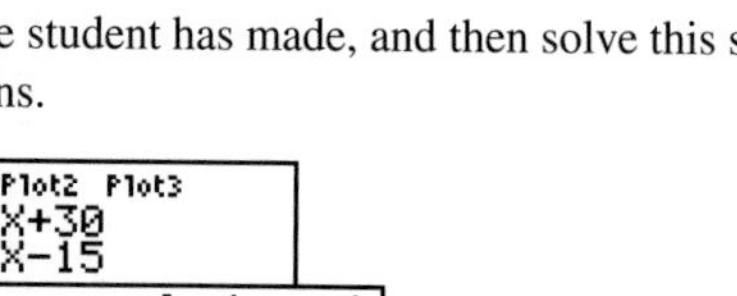

[−10, 10, 2] by [−50, 50, 10]

84. Error Analysis A student graphed the following system and concluded that the system had no solution. Describe the error the student has made, and then solve this system of equations.

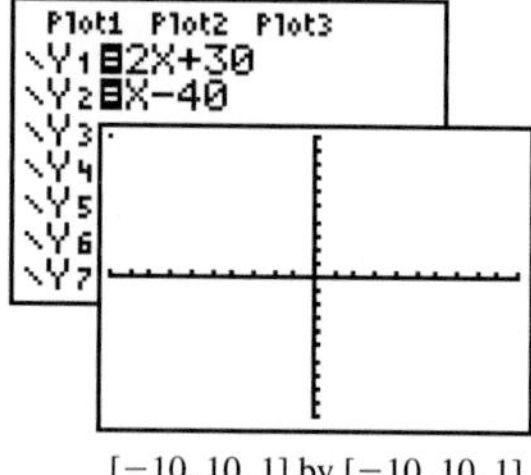

[−10, 10, 1] by [−10, 10, 1]

85. Challenge Question Use the graph below to determine the solution of the system of linear equations, and write the equation of both lines in general form.

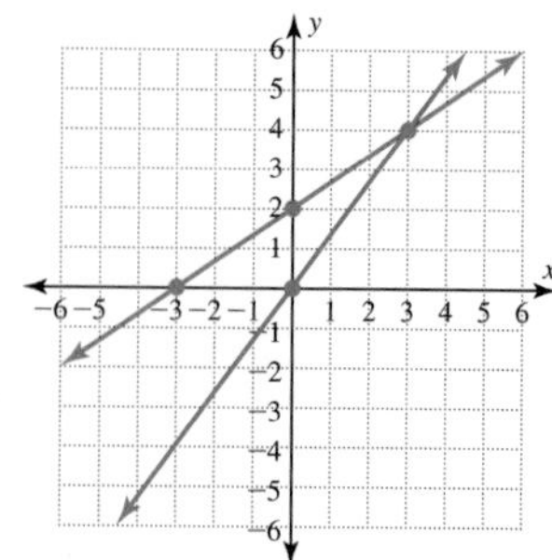

Section 3.4 Solving Systems of Linear Equations in Two Variables by the Substitution Method

Objective: 9. Solve a system of linear equations by the substitution method.

A Mathematical Note

Grace Hopper (1906–1992) was one of the foremost pioneers in the field of computer programming. She was one of the developers of the COBOL programming language. She retired from the U.S. Navy in 1986 as a rear admiral 43 years after joining as a lieutenant. In 1977 the author James Hall had the pleasure of hearing her humorously describe what she called the first computer bug—a moth that in 1945 flew into a relay of the early Mark II computer and caused it to malfunction. For a photo of this moth see the website listed on the Additional Resources link on the MathZone website at **www.mhhe.com/hallmercer.**

Although the graphical method is an excellent means of visualizing the solution of a system of equations, it has some limitations. Even with a computer or graphing calculator, this method can be time-consuming. This method also can produce some error because of limitations in estimating the point of intersection. Thus algebraic methods that are quicker and yield exact solutions often are preferred. The first algebraic method presented in this chapter is based on the **substitution principle,** which states that a quantity may be substituted for its equal. The substitution method is particularly appropriate when it is easy to solve one equation for either x or y.

Substitution Method

STEP 1. Solve one of the equations for one variable in terms of the other variable.

STEP 2. Substitute the expression obtained in step 1 into the other equation (eliminating one of the variables), and solve the resulting equation.

STEP 3. Substitute the value obtained in step 2 into the equation obtained in step 1 (back-substitution) to find the value of the other variable.

The ordered pair obtained in steps 2 and 3 is the solution.

The substitution method, which is shown here for systems of linear equations, can also be used to solve some nonlinear systems. The addition method for solving linear systems is covered in Section 3.5.

EXAMPLE 1 Solving a Linear System by the Substitution Method

Solve $\begin{Bmatrix} 3x + 4y = 8 \\ x + 2y = 10 \end{Bmatrix}$ by the substitution method.

SOLUTION

1 $x + 2y = 10$

$x = 10 - 2y$

Solve the second equation for x. (Note that this is relatively easy because x has a coefficient of 1 in this equation.)

2 $3x + 4y = 8$

$3(10 - 2y) + 4y = 8$

$30 - 6y + 4y = 8$

$30 - 2y = 8$

$-2y = 8 - 30$

$-2y = -22$

$y = 11$

Substitute for x in the first equation. (Note that x has been eliminated.) Then solve this equation for y.

This is the y-coordinate of the solution.

3 $x = 10 - 2y$

$x = 10 - 2(11)$

$x = 10 - 22$

$x = -12$

Back-substitute 11 for y into the equation that was solved for x in step 1.

This is the x-coordinate of the solution.

Check:

FIRST EQUATION	SECOND EQUATION
$3x + 4y = 8$	$x + 2y = 10$
$3(-12) + 4(11) \stackrel{?}{=} 8$	$-12 + 2(11) \stackrel{?}{=} 10$
$-36 + 44 \stackrel{?}{=} 8$	$-12 + 22 \stackrel{?}{=} 10$
$8 \stackrel{?}{=} 8$ checks	$10 \stackrel{?}{=} 10$ checks

Answer: $(-12, 11)$

The answer to a consistent system of independent linear equations should be written in ordered-pair notation. The notation $(-12, 11)$ emphasizes that the system in Example 1 has only one solution and that the x- and y-coordinates must be used in the correct order.

Example 2 revisits Example 2 from Section 3.3. Solving even a relatively simple system of linear equations by using a graphing calculator can take some time as we decide what input values will generate a useful table or graph. Fortunately, the substitution method provides an algebraic solution that is quick and exact.

EXAMPLE 2 Determining the Number of Degrees in Two Supplementary Angles

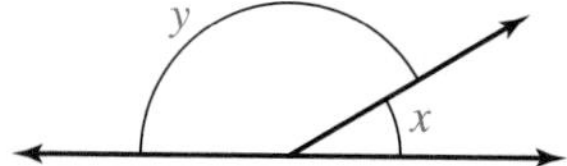

A metal fabrication shop must assemble a part so that the angles shown in the figure are supplementary. For it to function properly, the specifications for the part require that the larger angle be twice the smaller angle. Determine the number of degrees in each angle.

SOLUTION

Let x = number of degrees in smaller angle
y = number of degrees in larger angle

Select a variable to represent each unknown.

VERBALLY	ALGEBRAICALLY	
1. The angles are supplementary.	$x + y = 180$	Angles are supplementary if the total of their measures is 180°.
2. The larger angle is twice the smaller angle.	$y = 2x$	
	$x + 2x = 180$	Solve the system by using the substitution method. Substitute $2x$ for y in the first equation and then solve for x.
	$3x = 180$	
	$x = 60$	
	$y = 2x$	
	$y = 2(60)$	Back-substitute 60 for x in the second equation.
	$y = 120$	

Answer: The smaller angle is 60° and the larger angle is 120°.

Does this answer check?

SELF-CHECK 3.4.1

1. Rework Example 2 with the same conditions, except that the angles must be complementary (sum is 90°).
2. Solve $\begin{cases} 3x + y = 6 \\ 2x + y = 2 \end{cases}$ by the substitution method.

A system of equations that contains fractional coefficients may be easier to solve if we first convert the coefficients to integers by multiplying through by the LCD. For example, multiplying both sides of $\frac{x}{4} + \frac{y}{3} = 1$ by 12 yields $3x + 4y = 12$, which has integer coefficients.

EXAMPLE 3 Solving a Linear System with Fractional Coefficients

Solve $\begin{cases} \frac{x}{2} + \frac{y}{6} = \frac{2}{3} \\ \frac{x}{4} - \frac{y}{5} = \frac{7}{4} \end{cases}$

SOLUTION

FIRST EQUATION

$$\frac{x}{2} + \frac{y}{6} = \frac{2}{3}$$
$$6\left(\frac{x}{2} + \frac{y}{6}\right) = 6\left(\frac{2}{3}\right)$$
$$6\left(\frac{x}{2}\right) + 6\left(\frac{y}{6}\right) = 6\left(\frac{2}{3}\right)$$

(1) $3x + y = 4$

SECOND EQUATION

$$\frac{x}{4} - \frac{y}{5} = \frac{7}{4}$$
$$20\left(\frac{x}{4} - \frac{y}{5}\right) = 20\left(\frac{7}{4}\right)$$
$$20\left(\frac{x}{4}\right) - 20\left(\frac{y}{5}\right) = 20\left(\frac{7}{4}\right)$$

(2) $5x - 4y = 35$

We start by producing a simpler system of equations with integer coefficients.

Multiply both sides of the first equation by the LCD of 6 and both sides of the second equation by the LCD of 20.

(1), **(2)** $\begin{cases} 3x + y = 4 \\ 5x - 4y = 35 \end{cases}$

This system with integer coefficients has the same solution as the original system of equations.

(1)
$$3x + y = 4$$
$$y = 4 - 3x$$

Solve Equation (1) for y since y has a coefficient of 1.

(2)
$$5x - 4y = 35$$
$$5x - 4(4 - 3x) = 35$$
$$5x - 16 + 12x = 35$$
$$17x - 16 = 35$$
$$17x = 51$$
$$x = 3$$

Substitute for y in Equation (2). Then solve this equation for x.

This is the x-coordinate of the solution.

(3)
$$y = 4 - 3x$$
$$y = 4 - 3(3)$$
$$y = 4 - 9$$
$$y = -5$$

Back-substitute 3 for x in the equation that was solved for y in step 1.

This is the y-coordinate of the solution.

Answer: $(3, -5)$

Does this solution check in the original equations?

SELF-CHECK 3.4.2

1. Solve $\begin{cases} \frac{x}{2} + \frac{y}{3} = -4 \\ \frac{x}{6} - \frac{y}{6} = -3 \end{cases}$.

Examples 4 and 5 illustrate what will happen when the substitution method is used to solve an inconsistent system of linear equations or a consistent system of dependent linear equations.

EXAMPLE 4 Solving an Inconsistent System

Solve $\begin{Bmatrix} y = 3x - 6 \\ 6x - 2y = 9 \end{Bmatrix}$ by the substitution method.

SOLUTION

$y = 3x - 6$	The first equation is already solved for y.
$6x - 2y = 9$	
$6x - 2(3x - 6) = 9$	Substitute for y in the second equation.
$6x - 6x + 12 = 9$	Note that the simplification of this equation eliminates all the variables and produces a contradiction.
$12 = 9$ (a contradiction)	The conditions given by the two equations are contradictory. Thus this system has no solution; it is an inconsistent system.

Answer: There is no solution.

EXAMPLE 5 Solving a Consistent System of Dependent Equations

Solve $\begin{Bmatrix} 2x = 4y + 6 \\ 3x - 6y = 9 \end{Bmatrix}$ by the substitution method.

SOLUTION

$2x = 4y + 6$	Solve the first equation for x.
$x = 2y + 3$	
$3x - 6y = 9$	
$3(2y + 3) - 6y = 9$	Then substitute this value for x into the second equation.
$6y + 9 - 6y = 9$	Note that the simplification of this equation eliminates all variables and produces an identity.
$9 = 9$ (an identity)	This identity means that the system contains dependent equations and has an infinite number of solutions. Points $(3, 0)$, $\left(0, -\frac{3}{2}\right)$ and $(-1, -2)$ are three of the points that lie on the common line and satisfy both equations. Points $(0, 0)$, $(1, 1)$, and $(2, 3)$ are three points that do not lie on the common line and do not satisfy either equation.

Answer: There are an infinite number of solutions.

SELF-CHECK 3.4.3

1. Solve $\begin{Bmatrix} 5y = 4x - 10 \\ 12x - 15y = 30 \end{Bmatrix}$.

The following box summarizes what happens when each of the three types of linear systems is solved by the substitution method.

Algebraic Solution of the Three Types of Linear Systems

1. **Consistent system of independent equations:** The solution process will produce unique x- and y-values.
2. **Inconsistent system:** The solution process will produce a contradiction.
3. **Consistent system of dependent equations:** The solution process will produce an identity.

Example 6 compares the algebraic, numerical, and graphical solutions of a consistent system of dependent equations.

EXAMPLE 6 Using Multiple Perspectives to Examine a System of Dependent Equations

Solve $\left\{\begin{aligned} y &= 2x - 3 \\ 6x - 3y &= 9 \end{aligned}\right\}$ algebraically, numerically, and graphically. Then describe the answer verbally.

SOLUTION

ALGEBRAICALLY

Substitution Method

$$\begin{aligned} (1) \quad y &= 2x - 3 \\ (2) \quad 6x - 3y &= 9 \\ 6x - 3(2x - 3) &= 9 \\ 6x - 6x + 9 &= 9 \\ 9 &= 9 \end{aligned}$$

This identity signals a system of dependent equations.

Writing Both Equations in Slope-Intercept Form

$$\begin{aligned} (1) \quad y &= 2x - 3 \\ (2) \quad 6x - 3y &= 9 \\ -3y &= -6x + 9 \\ y &= 2x - 3 \end{aligned}$$

Equation (2) simplified to exactly the same form as Equation (1). Thus this is a system of dependent equations.

NUMERICALLY

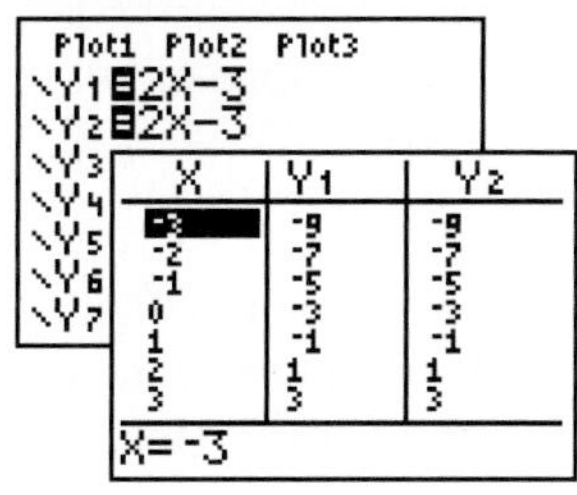

GRAPHICALLY

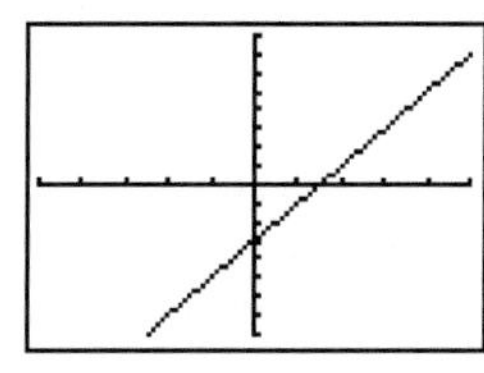

[−5, 5, 1] by [−8, 8, 1]

VERBALLY

Answer: There are an infinite number of solutions that satisfy both equations. Because both equations are just different forms of the same equation, they produce exactly the same output values, the same table of values, and the same graph.

The mean of a set of test scores is one type of average for these scores. The range is the difference between the high and low scores. Example 7 involves both the mean and the range of two test scores.

EXAMPLE 7 Determining Test Scores with a Given Mean and Range

A student has two test scores in an algebra class. The mean of these test scores is 82, and their range is 12. Use this information to determine both test scores.

SOLUTION

Let x = higher test score
y = lower test score

VERBALLY	ALGEBRAICALLY
(1) The mean of the two test scores is 82.	$\frac{x+y}{2} = 82$
(2) The range of the two test scores is 12.	$x - y = 12$

1 $x - y = 12$ Solve the second equation for x.

$x = y + 12$

2 $\frac{x+y}{2} = 82$ Simplify the first equation by multiplying both sides by the LCD of 2.

$x + y = 164$

$(y + 12) + y = 164$ Then substitute for x in Equation (2).

$2y + 12 = 164$ Simplify this equation and solve for y.

$2y = 152$

$y = 76$ This is the lower of the two test scores.

3 $x = y + 12$

$x = 76 + 12$ Back-substitute 76 for y into the equation that was solved for x in step 1.

$x = 88$ This is the higher of the two test scores.

Answer: The two scores are 88 and 76.

The answer to a word problem should be written as a full sentence. The question in Example 7 did not mention either x or y, so the answer should not use these symbols since they were introduced in the solution process.

SELF-CHECK 3.4.4

1. Two test scores have a mean of 82 and a range of 20. Determine both test scores.

Example 8 is solved by using the substitution method. The solution process includes the use of fractions, which many students would prefer to avoid. This is one reason that we will examine the addition method for solving this system in Section 3.5.

EXAMPLE 8 Solving a Linear System by the Substitution Method

Solve $\begin{Bmatrix} 3x - 2y = 5 \\ 9x + 4y = -10 \end{Bmatrix}$ by the substitution method.

SOLUTION

1 $3x - 2y = 5$ The coefficient of y is the smallest in magnitude, so we elect to solve for y.

$-2y = -3x + 5$

$y = \frac{3}{2}x - \frac{5}{2}$

2 $9x + 4y = -10$ Substitute this expression for y in the second equation. Then solve this equation for x.

$9x + 4\left(\frac{3}{2}x - \frac{5}{2}\right) = -10$

$9x + 6x - 10 = -10$

$15x = 0$

$x = \frac{0}{15}$

$x = 0$ This is the x-coordinate of the solution.

3 $y = \frac{3}{2}x - \frac{5}{2}$

$y = \frac{3}{2}(0) - \frac{5}{2}$ Back-substitute 0 for x in the equation that was solved for y in step 1.

$y = 0 - \frac{5}{2}$

$y = -\frac{5}{2}$ This is the y-coordinate of the solution.

Answer: $\left(0, -\frac{5}{2}\right)$ Does this answer check?

SELF-CHECK ANSWERS

3.4.1
1. The smaller angle is 30° and the larger angle is 60°.
2. $(4, -6)$

3.4.2
1. $(-12, 6)$

3.4.3
1. An infinite number of solutions; three of these solutions are $(2.5, 0)$, $(0, -2)$, and $(1.25, -1)$.

3.4.4
1. The two scores are 92 and 72.

USING THE LANGUAGE AND SYMBOLISM OF MATHEMATICS 3.4

1. The substitution principle states that a quantity may be substituted for its __________.
2. The solution for a system of two linear equations in two variables should be written in __________ __________ notation.
3. A contradiction is an equation that is always __________.
4. An identity is an equation that is always __________.
5. The solution of a consistent system of independent linear equations by the substitution method will produce **a conditional equation/a contradiction/an identity.** (Select the correct choice.)
6. The solution of a consistent system of dependent linear equations by the substitution method will produce **a conditional equation/a contradiction/an identity.** (Select the correct choice.)
7. The solution of an inconsistent system of independent linear equations by the substitution method will produce **a conditional equation/a contradiction/an identity.** (Select the correct choice.)
8. Using the substitution method produced the equations $x = 3$ and $y = 4$. Thus the system of two linear equations is a __________ system of __________ linear equations with one solution that is the ordered pair __________.
9. Using the substitution method produced the equation $3 = 4$. Thus the system of two linear equations is an __________ system of __________ linear equations with __________ solution(s).
10. Using the substitution method produced the equation $3 = 3$. Thus the system of two linear equations is a __________ system of __________ linear equations with an __________ number of solutions.

EXERCISES 3.4

In Exercises 1–42, solve each system of linear equations by the substitution method.

1. $y = 2x - 3$
   $x + y = 9$
2. $y = 3x - 4$
   $x + 2y = 6$
3. $y = 5x - 10$
   $2x + 3y = 4$
4. $y = 9 - 3x$
   $x + 2y = 8$
5. $x = 6 - 5y$
   $2x + 9y = 4$
6. $x = 4y - 9$
   $2x + 3y = 15$
7. $3x - y = 1$
   $x + 2y = 2$
8. $x - 3y = -14$
   $2x + y = 7$
9. $2x - y - 6 = 0$
   $x - y - 6 = 0$
10. $2x + y - 11 = 0$
    $x + 2y - 13 = 0$
11. $2x + 3y = -7$
    $x + 4y = -6$
12. $3x + 4y = -2$
    $x - 4y = -6$
13. $2x - 5y = 9$
    $x - 3 = 0$
14. $4x - 7y = -1$
    $x + 2 = 0$
15. $3x - 2y = 1$
    $y + 7 = 0$
16. $5x - 4y = 3$
    $y - 4 = 0$
17. $y = -2x$
    $5x - y = -7$
18. $y = 6x$
    $8x - 5y = -11$
19. $3x - 2y = 0$
    $11x - 9y = 5$
20. $5x + 4y = 0$
    $6x + 5y = 2$
21. $x + y = 21$
    $x - y = 3$
22. $2x + y = 3$
    $x + 2y = 9$

23. $2x - 7y = 42$
$x = 0$

24. $5x - 9y = 35$
$y = 0$

25. $5x - 2y = 11$
$3x + 3y = 15$

26. $3x + 4y + 2 = 0$
$5x - 20y + 30 = 0$

27. $5x - 4y + 12 = 0$
$2x - 3y + 2 = 0$

28. $3x + 4y + 5 = 0$
$-2x + 5y - 11 = 0$

29. $y = x + 3$
$\frac{x}{2} - \frac{y}{5} = 3$

30. $y = x + 4$
$\frac{x}{4} - \frac{y}{2} = 0$

31. $x = 2y + 4$
$3x - 6y = 12$

32. $y = 4 - 2x$
$6x + 3y = 12$

33. $y = 5x + 3$
$10x - 2y = 6$

34. $x = 5y - 3$
$4x - 20y = -1$

35. $4x + 7y = 0$
$7x - 4y = 0$

36. $9x - 5y = 0$
$5x + 9y = 0$

37. $x = 2y - 5$
$\frac{x}{6} + \frac{y}{8} = 1$

38. $\frac{x}{6} + \frac{y}{12} = \frac{1}{6}$
$\frac{x}{8} - \frac{y}{4} = -2$

39. $\frac{x}{2} + \frac{y}{5} = \frac{4}{5}$
$\frac{x}{6} - \frac{y}{2} = \frac{5}{6}$

40. $\frac{x}{8} - \frac{y}{5} = \frac{1}{10}$
$y = \frac{5}{8}x - \frac{1}{4}$

41. $\frac{x}{4} - \frac{y}{2} = \frac{7}{24}$
$\frac{x}{3} + \frac{y}{2} = 0$

42. $\frac{x}{2} - \frac{y}{3} = -\frac{7}{12}$
$\frac{x}{8} + \frac{y}{9} = 0$

In Exercises 43–46, select the system of linear equations (choices A, B, C, or D) that represents the word problem. Then solve this system by the substitution method and answer the problem, using a full sentence.

Verbally	**Algebraically**
43. Find two numbers whose sum is 120 and whose difference is 30.	**A.** $x + y = 20$, $x - y = 30$
44. Find two numbers whose sum is 30 and the first number is 120 more than the second number.	**B.** $x + y = 120$, $x - y = 30$
45. Find two numbers whose sum is 30 and whose difference is 30.	**C.** $x + y = 30$, $x - y = 30$
46. Find two numbers whose sum is 20 and whose difference is 30.	**D.** $x + y = 30$, $x - y = 120$

Multiple Representations

In Exercises 47 and 48, complete the numerical and graphical descriptions for each problem and then write a full sentence that answers the question.

47. Verbally
The sum of two numbers is 8 and their difference is 2. Find these numbers.

Algebraically
$x + y = 8$
$x - y = 2$

Numerically

x	$y_1 = 8 - x$	$y_2 = x - 2$
2	6	0
3		
4		
5		
6		
7		
8	0	6

Graphically

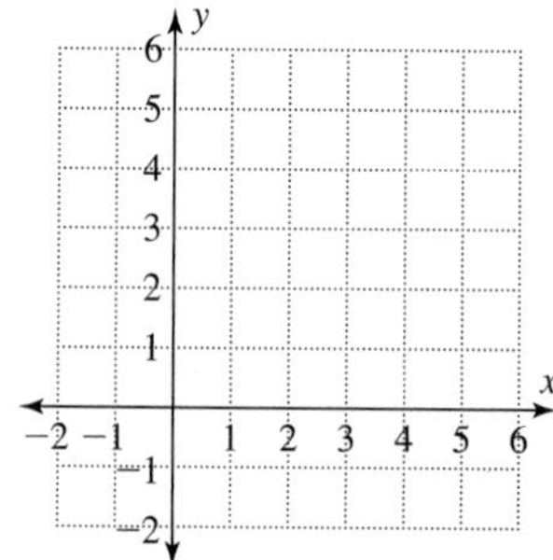

48. Verbally
The sum of two numbers is 2 and their difference is 6. Find these numbers.

Algebraically
$x + y = 2$
$x - y = 6$

Numerically

x	$y_1 = 2 - x$	$y_2 = x - 6$
0	2	−6
1		
2		
3		
4		
5		
6	−4	0

Graphically

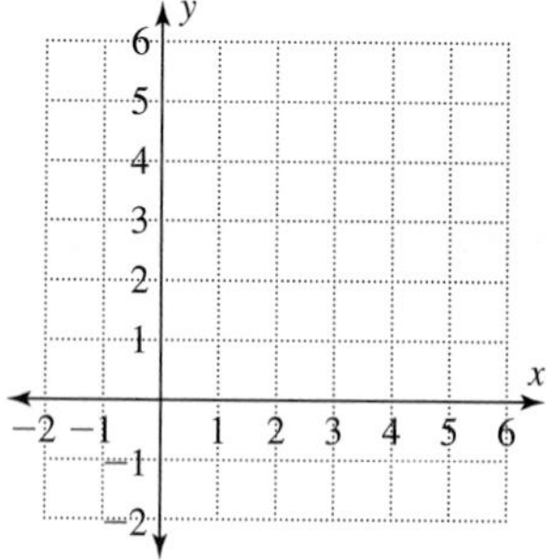

49. **Angles in Two Parts** To function properly, two parts must be assembled so that the two angles are complementary and one angle is 18° larger than the other angle. Determine the number of degrees in each angle. (See Exercise 50.)

50. **Angles in Two Parts** To function properly, two parts must be assembled so that the two angles are complementary, as shown in the figure, and one angle is 12° more than twice the other angle. Determine the number of degrees in each angle.

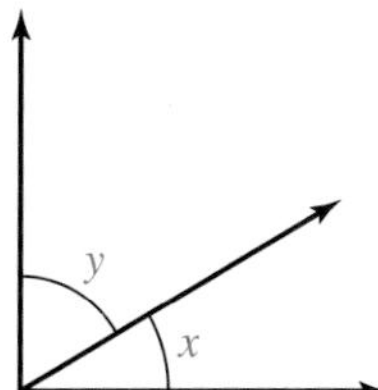

Mean and Range

In Exercises 51–54, write a system of linear equations using the variables x and y, and use this system of equations to solve the problem.

51. A student has two test scores in a psychology class. The mean of these scores is 76 and their range is 28. Use this information to determine both test scores.

52. A student has two test scores in a history class. The mean of these scores is 90 and their range is 12. Use this information to determine both test scores.

53. Tiger Woods had identical scores for three of the first four rounds at a PGA golf tournament. After he shot a tournament record in the final round, his mean score was 66 and the range of his scores was 8. Use this information to determine his score in each of the four rounds.

54. An amateur playing in a pro-am golf tournament in Scottsdale, Arizona, had identical scores for three of the first four rounds to enter the final day of play in first place. However, her score on the final day was her highest of the four rounds, and she finished fifth for the tournament. Her mean score was 73 and the range of her scores was 8. Use this information to determine her score in each of the four rounds.

55. **Dosages for Children** When a doctor knows the adult dosage of a medication but not a child's dosage, there are several formulas available for calculating an appropriate dosage for the child. Two formulas for calculating a child's dosage are Fried's rule and Cowling's rule. We use x to represent the age of the child and y to represent the child's dosage in milliliters. If the adult dosage of a medication is 48 mL, then these two formulas simplify to

Fried's rule: $y = 3.84x$

Cowling's rule: $y = 2x + 2$

a. Will these two formulas ever give the same dosage?

b. If so, at what age will the dosages be equal?

56. **Tuition at Two Colleges** The college tuition at one state university is \$95 per credit-hour plus a student activity fee of \$200. The tuition at the archrival across the state is \$100 per credit-hour plus a student activity fee of \$125. Use x to represent the number of credit-hours and y to represent the total cost of tuition and fees.

a. Write a linear equation to model the cost of a semester of tuition and fees at the state university.

b. Write a linear equation to model the cost of a semester of tuition and fees at the rival university.

c. Use this system of equations to determine the conditions for which these costs will be the same.

d. What is this cost?

57. **Taxi Fares** The following graph compares the cost of a taxi ride for two different taxicab companies. The cost is based upon the number of miles driven.

a. Use the y-intercept and an additional point to determine the equation of the line that models the cost for company A.

b. Use the y-intercept and an additional point to determine the equation of the line that models the cost for company B.

c. Solve this system of equations.

d. Does this algebraic solution seem consistent with the point of intersection shown on the graph?

e. Interpret the meaning of the x- and y-coordinates of this solution.

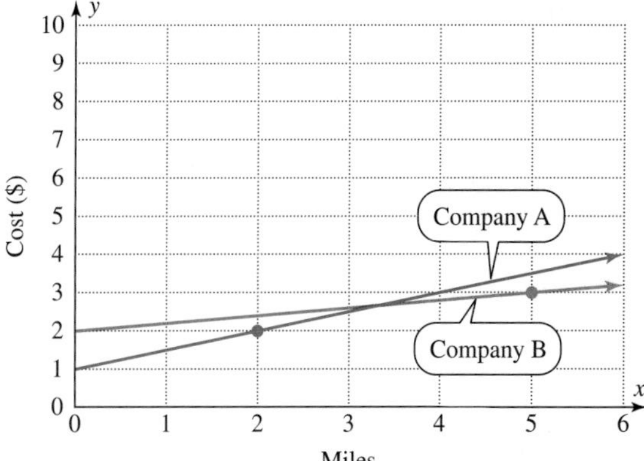

58. **Population Growth** Plainsville and Springfield were both founded in 1900. The following graph compares the populations of each city based upon the number of years since 1900.

a. Use the y-intercept and an additional point to determine the equation of the line that models the population for Plainsville.

b. Use the y-intercept and an additional point to determine the equation of the line that models the population for Springfield.

c. Solve the system of equations from parts **a** and **b.**

d. Does this algebraic solution seem consistent with the point of intersection shown on the graph?

e. Interpret the meaning of the x- and y-coordinates of this solution.

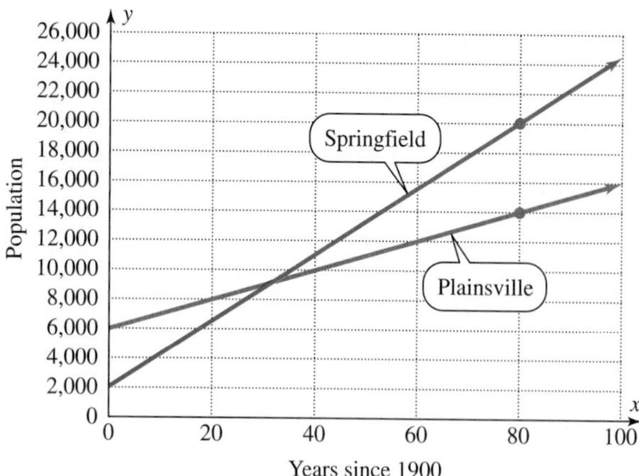

59. Investment Growth The following tables display the values of two different investments made in 1975. Investment A was an initial deposit of \$1,500 and has earned a simple interest rate of 12%. Investment B was an initial deposit of \$2,000 and has earned a simple interest rate of 7%. The variable x represents the number of years since 1975.

a. Use the y-intercept and an additional point to determine the equation of the line that models the value of investment A.

b. Use the y-intercept and an additional point to determine the equation of the line that models the value of investment B.

c. Solve the system of equations from parts **a** and **b.**

d. Does this algebraic solution seem consistent with the values shown in the tables?

e. Interpret the meaning of the x- and y-coordinates of this solution.

Investment A

YEAR x	VALUE y
0	1,500
5	2,400
10	3,300
15	4,200
20	5,100
25	6,000

Investment B

YEAR x	VALUE y
0	2,000
5	2,700
10	3,400
15	4,100
20	4,800
25	5,500

60. Sales Growth The following tables display the sales in millions of dollars of two companies founded in 1980. The variable x represents the number of years since 1980.

a. Use the y-intercept and an additional point to determine the equation of the line that models the sales for company A.

b. Use the y-intercept and an additional point to determine the equation of the line that models the sales for company B.

c. Solve the system of equations from parts **a** and **b.**

d. Does this algebraic solution seem consistent with the values shown in the tables?

e. Interpret the meaning of the x- and y-coordinates of this solution.

Company A

YEAR x	SALES y (\$ in millions)
0	5
5	22.5
10	40
15	57.5
20	75

Company B

YEAR x	SALES y (\$ in millions)
0	17
5	27
10	37
15	47
20	57

61. The table illustrates that the x-coordinate of the solution of the system $\begin{Bmatrix} x + 2y = 9 \\ x - y = -2 \end{Bmatrix}$ is between the integers __________ and __________ because Y_2 switches from less than Y_1 to more than Y_1 in this x-interval. This indicates the lines cross in this x-interval and signals a solution of the system of equations. Use the substitution method to confirm this fact and to find the exact coordinates of this solution.

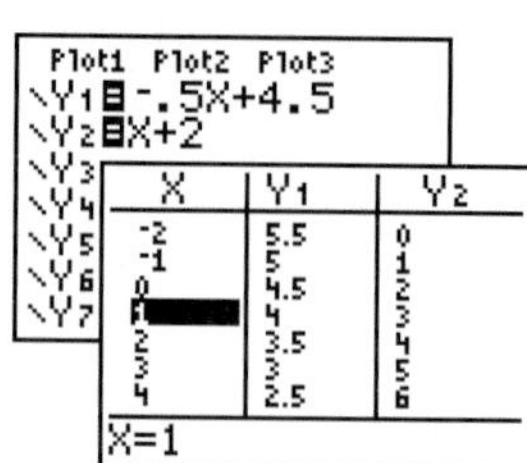

62. The table illustrates that the x-coordinate of the solution of the system $\begin{Bmatrix} 7x + y = 1 \\ 7x - y = -9 \end{Bmatrix}$ is between the integers __________ and __________ because Y_2 switches from less than Y_1 to more than Y_1 in this x-interval. This indicates the lines cross in this x-interval and signals a solution of the system of equations. Use the substitution method to confirm this fact and to find the exact coordinates of this solution.

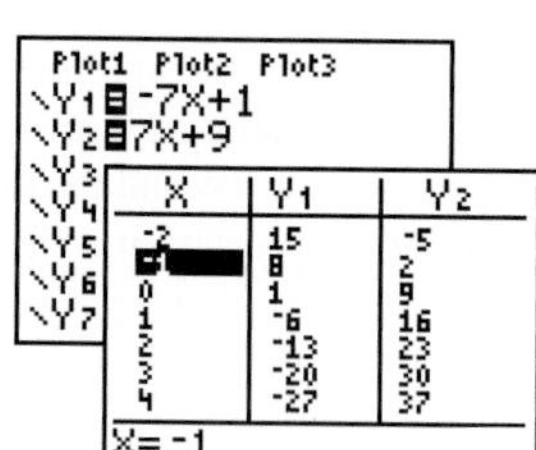

Inconsistent Systems

In Exercises 63 and 64, determine the value of B so that each system of equations will be an inconsistent system.

63. $4x - 8y = 12$
$3x + By = 10$

64. $4x + 2y = 10$
$10x + By = 20$

Systems of Dependent Equations

In Exercises 65 and 66, determine the value of C so that each system of equations will have dependent equations.

65. $6x - 15y = 9$
$4x - 10y = C$

66. $5x - 15y = 10$
$4x - 12y = C$

Calculator Usage

In Exercises 67–70, use the substitution method and a calculator to solve each system of linear equations.

67. $y = 1.05x - 2.1325$
$7.14x - 8.37y = 10.1835$

68. $y = 4.55x - 10.696$
$2.08x - 1.17y = 1.0972$

69. $x = 4.91y - 5.899$
$2.1x - 9.9y = -12.84$

70. $2.409x + 2.409y = 4.818$
$3.56x - 3.56y = 14.24$

Group Discussion Questions

71. **Discovery Question** Given the system $\begin{Bmatrix} x - 5y = -1 \\ 3x - y = 11 \end{Bmatrix}$:
   a. Solve this system by solving the first equation for x and then substituting for x in the second equation.
   b. Solve this system by solving the second equation for y and then substituting for y in the first equation.
   c. Compare these solutions and generalize about implementing the substitution method for other systems of linear equations.

72. **Discovery Question** Given the system $\begin{Bmatrix} x + 4y = -5 \\ 2x + 7y = 10 \end{Bmatrix}$, have part of your group solve this system by using a graphical method and have the rest of the group solve this system by using the substitution method. Time each group to determine which method seems more efficient for this problem. Did both groups get the same answer?

73. **Challenge Question** Solve each of these systems for (x, y) in terms of a and b for $a \neq 0$.
   a. $x - ay = b$
      $x + ay = 2b$
   b. $3ax + y = b$
      $2ax - y = 4b$

Section 3.5 Solving Systems of Linear Equations in Two Variables by the Addition Method

Objective: 10. Solve a system of linear equations by the addition method.

As noted in Section 3.4, the substitution method is well suited to systems that contain at least one variable with a coefficient of 1 or -1. For other systems it may be easier to use the addition method, which is described in the following box. This method is based on the **addition-subtraction principle of equality,** which states that equal values can be added to or subtracted from both sides of an equation to produce an equivalent equation. This method is also called the **elimination method** because the strategy is to eliminate a variable in one of the equations.

Addition Method

STEP 1. Write both equations in the general form $Ax + By = C$.

STEP 2. If necessary, multiply each equation by a constant so that the equations have one variable for which the coefficients are additive inverses.

STEP 3. Add the new equations to eliminate a variable and then solve the resulting equation.

STEP 4. Substitute this value into one of the original equations (back-substitution), and solve for the other variable.

The ordered pair obtained in steps 3 and 4 is the solution that should check in both equations.

EXAMPLE 1 Solving a Linear System by the Addition Method

Since $3x - y = 5$, $3x - y$ can be added to one side and 5 can be added to the other side of a balanced scale and maintain the balance.

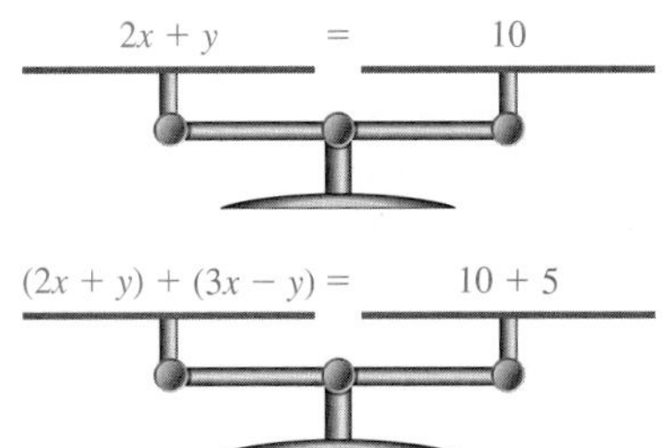

Solve $\begin{Bmatrix} 2x + y = 10 \\ 3x - y = 5 \end{Bmatrix}$ by the addition method.

SOLUTION

$$\begin{aligned} 2x + y &= 10 \\ 3x - y &= 5 \\ \hline 5x \quad &= 15 \\ x &= 3 \end{aligned}$$

The equations are already in the proper form, and the coefficients of y are additive inverses. Add these equations to eliminate y. Solve this equation for x.

This is the x-coordinate of the solution.

$$\begin{aligned} 2x + y &= 10 \\ 2(3) + y &= 10 \\ 6 + y &= 10 \\ y &= 4 \end{aligned}$$

Back-substitute 3 for x in the first equation. Then solve for y.

This is the y-coordinate of the solution.

Check:

FIRST EQUATION	SECOND EQUATION
$2x + y = 10$	$3x - y = 5$
$2(3) + 4 \stackrel{?}{=} 10$	$3(3) - 4 \stackrel{?}{=} 5$
$6 + 4 \stackrel{?}{=} 10$	$9 - 4 \stackrel{?}{=} 5$
$10 \stackrel{?}{=} 10$ checks.	$5 \stackrel{?}{=} 5$ checks.

Answer: (3, 4) — Remember to write the solution by using ordered-pair notation.

Example 2 revisits Example 8 from Section 3.4. When you compare these two solutions, you will likely prefer the addition method shown next.

EXAMPLE 2 Solving a Linear System by the Addition Method

Solve $\begin{cases} 3x - 2y = 5 \\ 9x + 4y = -10 \end{cases}$.

SOLUTION

$$3x - 2y = 5 \qquad\qquad 6x - 4y = 10$$
$$9x + 4y = -10 \qquad\qquad 9x + 4y = -10$$
$$15x = 0$$
$$x = 0$$

Multiply both sides of the first equation by 2 to obtain coefficients of y that are additive inverses. Add these equations to eliminate y.

Divide both sides of the equation by 15 to solve for x. This is the x-coordinate of the solution.

$$3x - 2y = 5$$
$$3(0) - 2y = 5$$
$$-2y = 5$$

Back-substitute 0 for x in the first equation of the original system. Then solve for y.

$$y = -\frac{5}{2}$$

This is the y-coordinate of the solution.

Answer: $\left(0, -\frac{5}{2}\right)$

This is the same solution we obtained in Example 8 in Section 3.4.

In Example 3 both equations are multiplied by a constant so that the coefficients of y will be additive inverses. The choice to eliminate y is arbitrary since it would be just as easy to eliminate x. Although any common multiple of the coefficients of y can be used to eliminate the y-variable, we encourage you to use the least common multiple of the coefficients.

EXAMPLE 3 Solving a Linear System by the Addition Method

Solve $\begin{cases} 5x + 3y = -6 \\ 3x + 2y = -5 \end{cases}$ by the addition method.

SOLUTION

$$5x + 3y = -6 \qquad\qquad 10x + 6y = -12$$

Multiply both sides by 2.

$$3x + 2y = -5 \qquad\qquad -9x - 6y = 15$$

Multiply both sides by -3.

$$x = 3$$

Add these equations to eliminate y.

$$\begin{aligned} 5x + 3y &= -6 \\ 5(3) + 3y &= -6 \\ 15 + 3y &= -6 \\ 3y &= -21 \\ y &= -7 \end{aligned}$$

Back-substitute 3 for x in the first equation of the original system.

Then solve for y.

Answer: $(3, -7)$

Does this solution check?

In Example 4 the first step is the decision to eliminate the variable y. The coefficients of y are -3 and 4. Use this information to guide the steps taken to produce the least common multiple of 12.

EXAMPLE 4 Solving a Linear System by the Addition Method

Solve $\begin{cases} 7x = 3y + 8 \\ 4y = -5(x + 5) \end{cases}$ by the addition method.

SOLUTION

$$\begin{aligned} 7x &= 3y + 8 \\ 4y &= -5(x + 5) \end{aligned} \qquad \begin{aligned} 7x - 3y &= 8 \\ 5x + 4y &= -25 \end{aligned} \qquad \begin{aligned} 28x - 12y &= 32 \\ 15x + 12y &= -75 \\ \hline 43x \quad &= -43 \\ x &= -1 \end{aligned}$$

First write both equations in the form $Ax + By = C$.

To produce the coefficients of -12 and 12 for y, multiply the first equation by 4 and the second equation by 3. Add these equations to eliminate y.

Solve for x.

$$\begin{aligned} 4y &= -5(x + 5) \\ 4y &= -5(-1 + 5) \\ 4y &= -5(4) \\ 4y &= -20 \\ y &= -5 \end{aligned}$$

Back-substitute -1 for x in the second equation of the original system.

Then solve for y.

Answer: $(-1, -5)$

Does this solution check?

SELF-CHECK 3.5.1

1. Solve $\begin{cases} 4x + 3y = 7 \\ -2x - 3y = -11 \end{cases}$.
2. Solve $\begin{cases} 2x - 7y = 1 \\ x + 5y = 9 \end{cases}$.
3. Solve $\begin{cases} 8x - 3y = 5 \\ -5x + 7y = 43 \end{cases}$.

EXAMPLE 5 Factory Defects

A quality control worker in a factory weighed eight soda cans. The weights of the first seven were all equal, but the eighth can had a defect and was too light. The mean of these weights was 15.65 oz, and the range was 2.8 oz. Use this information to determine each weight.

SOLUTION

DEFINITION OF VARIABLES

Let x = weight in ounces of each of first seven cans
y = weight in ounces of eighth can

Select a variable to represent each unknown.

WORD EQUATIONS

(1) The mean weight is 15.65 oz.
(2) The range of the weights is 2.8 oz.

First form word equations, and then translate them into algebraic equations.

ALGEBRAIC EQUATIONS

(1) $\frac{7x + y}{8} = 15.65$

The mean of a set of measurements is found by dividing their sum by the number of measurements.

(2) $x - y = 2.8$

The range of a set of measurements is found by subtracting the smallest from the largest.

SOLVING ALGEBRAICALLY BY THE ADDITION METHOD

$$\begin{aligned}
(1)\quad 7x + y &= 125.2 \\
(2)\quad x - y &= 2.8 \\
\hline
8x &= 128 \\
x &= 16
\end{aligned}$$

Multiply both sides of Equation (1) above by 8. Then add the two equations to eliminate y.
Solve for x.

$$\begin{aligned}
x - y &= 2.8 \\
16 - y &= 2.8 \\
-y &= -13.2 \\
y &= 13.2
\end{aligned}$$

Back-substitute 16 for x in the second equation and then solve for y.

NUMERICAL CHECK

$$\frac{7(16) + 13.2}{8} = \frac{125.2}{8} = 15.65$$

$$16 - 13.2 = 2.8$$

These values produce a mean of 15.65 and a range of 2.8.

Answer: The first seven cans each weigh 16 oz, and the eighth can weighs 13.2 oz.

Example 6 involves fractional coefficients. Thus in the first step we multiply through by the LCD to produce equivalent equations with integer coefficients.

EXAMPLE 6 Solving a Linear System with Fractional Coefficients

Solve $\left\{\begin{aligned} \frac{x}{6} + \frac{y}{9} &= 2 \\ \frac{x}{8} - \frac{y}{3} &= 9 \end{aligned}\right\}$ by the addition method.

SOLUTION

(1) $\frac{x}{6} + \frac{y}{9} = 2$ $\qquad 18\left(\frac{x}{6} + \frac{y}{9}\right) = 18(2)$

Multiply both sides by 18, the LCD. Distribute the factor of 18 and simplify this equation.

(2) $\frac{x}{8} - \frac{y}{3} = 9$ $\qquad -24\left(\frac{x}{8} - \frac{y}{3}\right) = -24(9)$

Multiply both sides by -24, the opposite of the LCD. Distribute the factor of -24 and simplify this equation.

$$\begin{array}{lrcr} (1) & 3x + 2y &=& 36 \\ (2) & -3x + 8y &=& -216 \\ \hline & 10y &=& -180 \\ & y &=& -18 \end{array}$$

Add these equations to eliminate x.

Solve for y.

$$\begin{aligned} 3x + 2y &= 36 \\ 3x + 2(-18) &= 36 \\ 3x - 36 &= 36 \\ 3x &= 72 \\ x &= 24 \end{aligned}$$

Back-substitute -18 for y in the first equation with integer coefficients.

Then solve for x.

Answer: $(24, -18)$

Does this answer check in the original system?

Remember that the algebraic solution of an inconsistent system of linear equations will produce a contradiction. The algebraic solution of a consistent system of dependent linear equations will produce an identity. In Section 3.4 we examined these two possibilities by using the substitution method. Now we will see what happens when the addition method is used to solve an inconsistent system of linear equations or a consistent system of dependent linear equations.

EXAMPLE 7 Solving an Inconsistent System

Solve $\left\{\begin{array}{r} 5x + 10y = 11 \\ x + 2y = 3 \end{array}\right\}$ by the addition method.

SOLUTION

$$\begin{aligned} 5x + 10y &= 11 \\ x + 2y &= 3 \end{aligned} \qquad \begin{array}{rcr} 5x + 10y &=& 11 \\ -5x - 10y &=& -15 \\ \hline 0 &=& -4, \end{array} \text{ a contradiction}$$

Multiply both sides of the second equation by -5. Adding these equations eliminates both variables and produces a contradiction.

Answer: There is no solution.

Since the result is a contradiction, the original system is inconsistent and has no solution.

EXAMPLE 8 Solving a Consistent System of Dependent Equations

Solve $\left\{\begin{array}{r} 6x - 15y = 3 \\ -4x + 10y = -2 \end{array}\right\}$ by the addition method.

SOLUTION

$$\begin{aligned} 6x - 15y &= 3 \\ -4x + 10y &= -2 \end{aligned} \qquad \begin{array}{rcr} 12x - 30y &=& 6 \\ -12x + 30y &=& -6 \\ \hline 0 &=& 0, \end{array} \text{ an identity}$$

Multiply both sides of the first equation by 2, and then multiply both sides of the second equation by 3.

Adding these equations eliminates both variables and produces an identity.

Answer: There are an infinite number of solutions.

Since the result is an identity, the original equations are dependent, and the system has an infinite number of solutions.

SELF-CHECK 3.5.2

1. Solve $\begin{cases} \dfrac{x}{2} - \dfrac{y}{8} = 3 \\ \dfrac{x}{4} + \dfrac{y}{2} = -3 \end{cases}$.

2. Solve $\begin{cases} 4x - 8y = 20 \\ 3x - 6y = 15 \end{cases}$.

Example 9 involves two unknown numbers. Variables x and y are used to represent these numbers so that a system of two linear equations can be formed to describe this problem.

EXAMPLE 9 Solving a Numeric Word Problem

Find two numbers whose sum is 60 and whose difference is 14.

SOLUTION

Let $x =$ larger number
$y =$ smaller number

VERBALLY	ALGEBRAICALLY	
(1) Two numbers whose sum is sixty	$x + y = 60$	
(2) Two numbers whose difference is fourteen	$x - y = 14$	
	$2x = 74$	Add the two equations to eliminate y.
	$x = 37$	Divide both sides of the equation by 2 to solve for x.

$$x + y = 60$$
$$37 + y = 60$$
$$y = 23$$

Back-substitute 37 for x in the first equation and solve for y.

Answer: The numbers are 37 and 23.

Write the answer as a full sentence. Does this answer check?

SELF-CHECK ANSWERS

3.5.1
1. $(-2, 5)$
2. $(4, 1)$
3. $(4, 9)$

3.5.2
1. $(4, -8)$
2. An infinite number of solutions

USING THE LANGUAGE AND SYMBOLISM OF MATHEMATICS 3.5

1. The addition-subtraction principle of equality states that __________ values can be added to or subtracted from both sides of an equation to produce an equivalent equation.
2. The addition method for solving a system of linear equations is also sometimes referred to as the __________ method because the strategy is to __________ one of the variables when the two equations are added.
3. To eliminate the y-variable when adding two linear equations, we first must make the coefficients of y in the two equations __________ of each other.
4. Using the addition method to solve a system of two linear equations produced the equation $0 = 5$. Thus the system of equations is an __________ system with __________ solution(s).
5. Using the addition method to solve a system of two linear equations produced the equation $0 = 0$. Thus the system is a __________ system of __________ equations with an __________ number of solutions.

EXERCISES 3.5

In Exercises 1–30, solve each system of linear equations by the addition method.

1. $x + 2y = 6$
   $-x + 3y = 4$
2. $5x - y = 10$
   $2x + y = 4$
3. $5x + 2y = -26$
   $3x - 2y = -38$
4. $-5x + 3y = 27$
   $5x + 2y = -7$
5. $6x + 2y = -1$
   $12x - y = 3$
6. $5x - y = -1$
   $15x + 2y = 7$
7. $x + 2y = 1$
   $3x + 4y = 0$
8. $x + 6y = 0$
   $3x + 8y = 5$
9. $2x + 3y = -9$
   $-4x + 5y = -37$
10. $3x + 7y = 25$
    $-6x + 3y = 18$
11. $5x - 3y = 5$
    $4x + 6y = 46$
12. $2x + 15y = 3$
    $3x - 5y = -1$
13. $2x + 5y = -3$
    $3x + 8y = -5$
14. $2x - 3y = 4$
    $11x - 5y = -1$
15. $2x + 3y = 15$
    $5x + 4y = -1$
16. $4x + 3y = 11$
    $5x + 2y = -9$
17. $2x - 13y = 38$
    $5x + 27y = 95$
18. $23x - 2y = -30$
    $47x + 3y = 45$
19. $2x = 11y$
    $5x = 19y$
20. $3x = 17y$
    $4x = 9y$
21. $2x + 6 = 0$
    $3x + 2y = 1$
22. $2y + 14 = 0$
    $2x + 3y = -13$
23. $\frac{x}{2} + \frac{y}{3} = 5$
    $\frac{x}{3} - \frac{y}{2} = -1$
24. $\frac{x}{8} + \frac{y}{8} = 0$
    $\frac{x}{2} - \frac{y}{4} = -3$
25. $6x - 8y = 10$
    $-15x + 20y = -20$
26. $2x - 6y = 8$
    $-3x + 9y = -12$
27. $6x = 9 - 3y$
    $4y = 12 - 8x$
28. $3x = 6 - 12y$
    $7 - 4x = 16y$
29. $\frac{x}{2} - \frac{y}{2} = \frac{3}{4}$
    $\frac{x}{2} + \frac{y}{4} = \frac{5}{8}$
30. $\frac{x}{3} + \frac{y}{3} = 0$
    $10x + 5y = -1$

In Exercises 31–40, solve each system of linear equations by either the substitution method or the addition method.

31. $y = 2x - 1$
    $7x - 4y = -1$
32. $x = 3y - 5$
    $-5x + 6y = -11$
33. $3x + 7y = 20$
    $5x - 7y = -4$
34. $4x + 3y = 26$
    $5x - 2y = -25$
35. $y = -x$
    $5x - 7y = 6$
36. $y = \frac{5}{3}x$
    $x + y = 2$
37. $\frac{x}{4} - \frac{y}{3} = \frac{5}{12}$
    $\frac{x}{2} - \frac{2y}{3} = 1$
38. $\frac{x}{4} - \frac{y}{6} = 6$
    $\frac{x}{6} + \frac{y}{3} = -4$
39. $2x + 3y = 4x + 5y - 6$
    $x - 5y = 7x + 2y - 21$
40. $3x + 4(2y - 1) = x + 8y - 2$
    $5x + 7(y + 5) = 4(x + y + 5) + 1$

In Exercises 41–44, select the system of linear equations (choice A, B, C, or D) that represents the word problem. Then solve this system by the addition method and answer the problem, using a full sentence.

Verbally

41. The sum of three times one number and a second number is one. The difference of the first number minus the second number is seven. Find these numbers.
42. The sum of twice one number plus a second number is fourteen. The first number plus three times the second number is twenty-two. Find these numbers.
43. The sum of five times one number plus twice a second number is two. The sum of twice the first number plus five times the second number is twenty-six. Find the numbers.
44. The sum of twice one number plus a second number is fourteen. The difference of the first number minus the second number is four. Find the numbers.

Algebraically

A. $2x + y = 14$
   $x - y = 4$

B. $2x + y = 14$
   $x + 3y = 22$

C. $3x + y = 1$
   $x - y = 7$

D. $5x + 2y = 2$
   $2x + 5y = 26$

In Exercises 45–48, write a system of linear equations using the variables x and y and use this system to solve the problem.

45. Find two numbers whose sum is 88 and whose difference is 28.
46. Find two numbers whose sum is 133 and whose difference is 11.
47. Find two numbers whose sum is 102 if one number is twice the other number.
48. Find two numbers whose sum is 213 if one number is twice the other number.

Mean and Range

In Exercises 49 and 50, write a system of linear equations using the variables x and y and use this system of equations to solve the problem.

49. **Basketball Scoring** A basketball player had four games with identical scores and a fifth game that was his high for the whole season. His mean score for the five games was 25 and the range was 15. Use this information to determine his score in each of the five games.
50. **Factory Defects** A quality control worker in a factory measured 10 wheel rims. The diameters of the first nine were all equal, but the tenth had a defect and was too small. The mean of these measurements was 24.9 cm and the range was 1 cm. Use this information to determine each measurement.
51. **College Enrollment Patterns** In 1985, 38% of all graduates at Mason High School attended a 4-year college and 18% of all graduates attended a community college. In 2001, of all graduates 30% attended a 4-year college and 42% attended a community college.
    a. Use the two points (1985, 38) and (2001, 30) to write an equation for the line showing the percent of graduates attending a 4-year college.

b. Use the two points (1985, 18) and (2001, 42) to write an equation for the line showing the percent of graduates attending a community college.
c. Solve the system of equations from parts **a** and **b.**
d. Interpret the meaning of the x- and y-coordinates of the solution from part **c.**

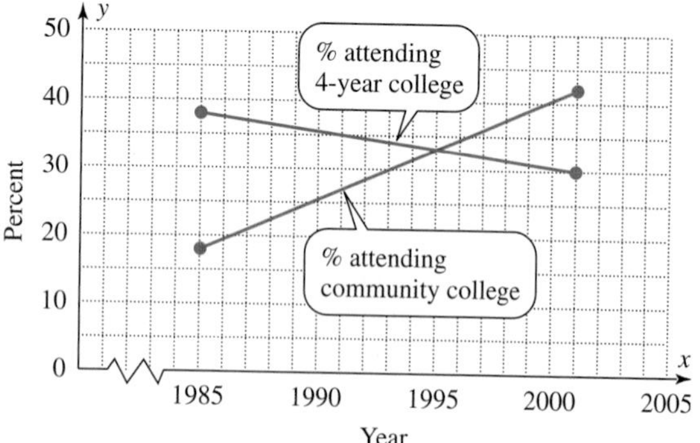

52. Generation of Electricity from the Wind In 1990, only 3% of all electricity needed in one community was generated by wind, and 39% was generated by coal. By 2000, of the electricity needed 20% was generated by wind and 16% was generated by coal.

a. Use the two points (1990, 3) and (2000, 20) to write an equation for the line showing the percent of electricity generated by wind.
b. Use the two points (1990, 39) and (2000, 16) to write an equation for the line showing the percent of electricity generated by coal.
c. Solve the system of equations from parts **a** and **b.**
d. Interpret the meaning of the x- and y-coordinates of the solution from part **c.**

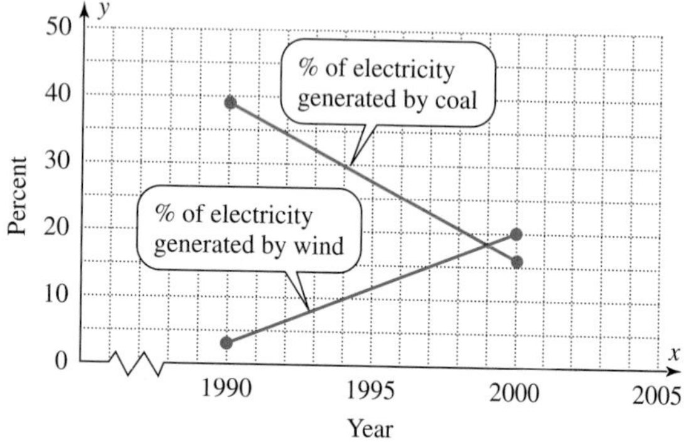

53. The table illustrates that the x-coordinate of the solution of the system $\left\{\begin{array}{l} 11x - 2y = 1 \\ 11x + 3y = 11 \end{array}\right\}$ is between the integers ________ and ________ because Y_2 switches from more than Y_1 to less than Y_1 in this x-interval. This indicates the lines cross in this x-interval and signals a solution of the system of equations. Use the addition method to confirm this fact and to find the exact coordinates of this solution.

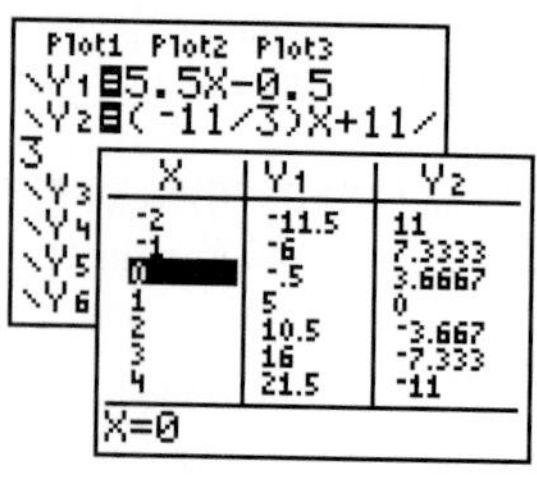

54. The table illustrates that the x-coordinate of the solution of the system $\left\{\begin{array}{l} 3x + 4y = 8 \\ -3x + 2y = -20 \end{array}\right\}$ is between the integers ________ and ________ because Y_2 switches from less than Y_1 to more than Y_1 in this x-interval. This indicates the lines cross in this x-interval and signals a solution of the system of equations. Use the addition method to confirm this fact and to find the exact coordinates of this solution.

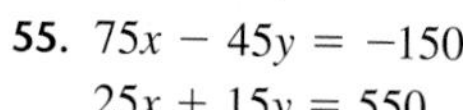

Calculator Usage

In Exercises 55–58, use a calculator to assist you in solving each system of linear equations.

55. $75x - 45y = -150$
$25x + 15y = 550$

56. $22x + 19y = 29$
$33x + 14y = -29$

57. $8.93x - 7.21y = -0.363$
$6.04x + 7.21y = 66.231$

58. $2.14x - 3.15y = -4.131$
$3.07x + 4.21y = 0.184$

Group Discussion Questions

59. Using x to represent the number of years after 2000, an equation that models the percent of students enrolling by phone is $y_1 = -2.054x + 18.804$. An equation that models the percent of students enrolling by the Web is $y_2 = 7.839x - 0.732$.

YEAR	x	PHONE y_1	WEB y_2
2000	0	18.0	0.0
2001	1	18.5	4.0
2002	2	14.5	13.5
2003	3	11.0	25.0
2004	4	11.0	35.0
2005	5	9.5	40.0
2006	6	6.0	42.0

a. Determine the point of intersection of these two lines.
b. Interpret the meaning of the x- and y-coordinates of this point of intersection.
c. Which of these functions is an increasing function?
d. Which of these functions is a decreasing function?
e. The phone system and the Web-based system each cost the college over $100,000 to operate each year. What would you advise the college to do? Why?

60. Communicating Mathematically

a. Two linear equations produce the same line when they are graphed. Write a sentence describing what would happen if you used the addition method to solve this system.
b. Two linear equations produce distinct parallel lines when they are graphed. Write a sentence describing what would happen if you used the addition method to solve this system.

61. Communicating Mathematically Write a numeric word problem for each system of linear equations.

a. $3x - 4y = 5$
$4x + y = 13$

b. $6x + 2y = -16$
$5x + y = -18$

62. **Challenge Question** Solve this system for (x, y) in terms of a, b, and c. Assume a, b, and c are all nonzero.

$ax + by = c$

$2ax - by = 2c$

63. **Discovery Question** Given the system $\begin{cases} 7x - 11y = -2 \\ -7x + 22y = 5 \end{cases}$, have part of your group solve this system by using a graphical method and the rest of the group solve this system by using the addition method. Time each group to determine which method seems more efficient for this problem. Did both groups get exactly the same answer? If not, which group was more accurate?

Section 3.6 More Applications of Linear Systems

Objective: 11. Use systems of linear equations to solve word problems.

A Mathematical Note

The problem-solving strategy given here and used throughout this book is not new or unique. George Polya (1887–1985) is well known for teaching problem-solving techniques. His best-selling book *How to Solve It* includes four steps to problem solving:

Step 1 Understand the problem.
Step 2 Devise a plan.
Step 3 Carry out the plan.
Step 4 Check back.

Compare these steps to those given in the box.

As we noted in Section 2.8, the mathematics that most students will encounter after college will be stated in words. To help prepare you for this, we use this section to further develop the word problem strategy which is restated below for your reference.

Strategy for Solving Word Problems

STEP 1. Read the problem carefully to determine what you are being asked to find.

STEP 2. Select a variable to represent each unknown quantity. Specify precisely what each variable represents, and note any restrictions on each variable.

STEP 3. If necessary, make a sketch and translate the problem into a word equation or a system of word equations. Then translate each word equation into an algebraic equation.

STEP 4. Solve the equation or system of equations, and answer the question asked by the problem.

STEP 5. Check the reasonableness of your answer.

As a rule of thumb, we usually use as many variables as we have unknowns. For a problem with two unknown quantities, it is often easier to solve the problem by using two variables than it is by using just one variable. Make and label sketches to create a visual representation of the problem.

We suggest that you continue to use the steps in this strategy as a checklist when you work word problems. Writing word equations is a key step that many students prefer to skip. However, students who practice this step often make major gains in their problem-solving skills. We strongly urge you to write word equations for the problems in this section.

The major new feature of the word problems in this section is that they contain two unknowns rather than just one unknown. Thus we will use two variables and a system of two equations to solve these problems.

If a word problem involves any shape or design, it is wise to make a sketch of this shape. Observing the visual relationship of the variables can assist you in writing equations for the problem. Example 1 illustrates the steps in the strategy for solving word problems with a problem involving two supplementary angles.

EXAMPLE 1 Modeling the Number of Degrees in Two Supplementary Angles

A metal brace on the platform on the side of a crane is designed so that the two angles shown in the figure are supplementary. The specifications require that the larger angle be 40° more than the smaller angle. Determine the number of degrees in each angle.

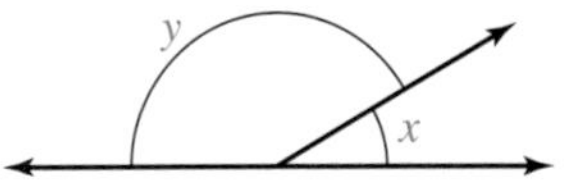

SOLUTION

DEFINITION OF VARIABLES

Let x = number of degrees in smaller angle
y = number of degrees in larger angle

Select a variable to represent each unknown. (The restrictions on the variables are considered when the system is solved graphically.)

WORD EQUATIONS

(1) The total of the measures of the two angles is 180°.
(2) The larger angle is 40° more than the smaller angle.

Angles are supplementary if the total of their measures is 180°. Specifications require that the larger angle be 40° more than the smaller angle.

ALGEBRAIC EQUATIONS

(1) $x + y = 180$
(2) $y = x + 40$

Translate each word equation into an algebraic equation.

ALGEBRAICALLY

(1)
$$\begin{aligned} x + (x + 40) &= 180 \\ 2x + 40 &= 180 \\ 2x &= 140 \\ x &= 70 \end{aligned}$$

(2)
$$\begin{aligned} y &= x + 40 \\ y &= 70 + 40 \\ y &= 110 \end{aligned}$$

To solve this system algebraically, use the substitution method, and substitute $x + 40$ for y in Equation (1). Then solve this equation for x. Back-substitute 70 for x in Equation (2).

GRAPHICALLY

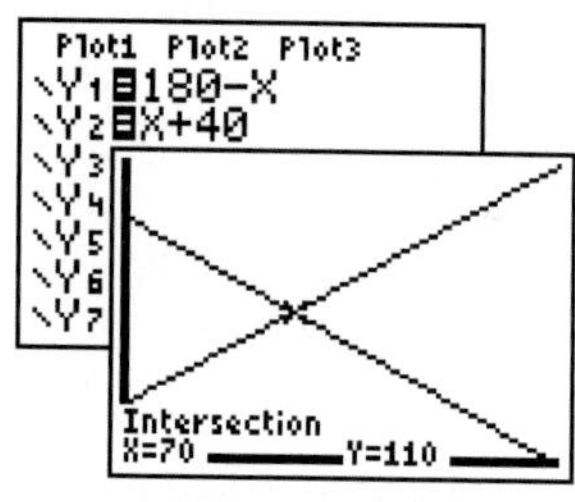

[0, 180, 1] by [0, 220, 1]

To solve this system by using a graphing calculator, solve each equation for the y-variable.
(1) $y = 180 - x$
(2) $y = x + 40$

Knowing the restriction on the variables can help us select an appropriate window for graphing these equations.
Because the angles are supplementary (total 180°), select the x-values [0, 180, 1] for the viewing window. The **ZoomFit** option selects the y-values and graphs these equations on this window.
Use the **Intersect** feature to find that these lines intersect where $x = 70$ and $y = 110$.

NUMERICAL CHECK

X	Y1	Y2
30	150	70
40	140	80
50	130	90
60	120	100
70	110	110
80	100	120
90	90	130

X=70

The table verifies that the x-value of 70 produces y-values of 110. Also note that other values of x do not check.

Answer: The smaller angle is 70°, and the larger angle is 110°.

SELF-CHECK 3.6.1

1. Two parts that brace part of a playground slide are joined at a common point. The two angles that they form at this common vertex must be complementary (sum is 90°). If the larger angle is 20° more than the smaller angle, determine the number of degrees in each angle.

One important measure of your academic success is your **grade-point average (GPA).** Your GPA is calculated by dividing your grade points by the number of academic hours you have completed. Your grade points are calculated by multiplying the number of academic hours of a course by the numeric grade (e.g., 4 for an A) for the course. We will use Example 2 to assist one student with some academic planning.

EXAMPLE 2 Computing a GPA

A student who has already completed 30 semester-hours with a GPA of 2.50 at Lamar State College in Orange, Texas, needs to raise her GPA to 3.00 to qualify for a scholarship. How many semester-hours with a grade of A would be required to meet her goal? How many grade points will she have if she meets her goal? (Assume an A = 4 points.)

SOLUTION

DEFINITION OF VARIABLES

Let x = number of semester-hours that student will need to take with straight A's
y = grade points earned

WORD EQUATIONS

(1) Actual grade points earned
= Current grade points + New grade points

(2) Desired grade points = Grade points with a 3.00 GPA

If we have two variables, then we will need to write two word equations.

ALGEBRAIC EQUATIONS

(1) $y = 2.50(30) + 4x$

To translate these word equations into algebraic form, multiply the current GPA by 30 semester-hours to obtain the current grade points. Multiply a numerical grade of 4 by the number of hours x to obtain the new grade points.

(2) $y = 3.00(30 + x)$

The desired GPA is 3.00 when $30 + x$ semester-hours have been completed.

$$\begin{aligned} 3.00(30 + x) &= 2.50(30) + 4x \\ 90.00 + 3.00x &= 75.00 + 4x \\ 15 &= x \end{aligned}$$

To determine when the desired grade points will equal the actual grade points, solve this system of equations. Substitute $3.00(30 + x)$ for y in the first equation.

(1) $y = 2.50(30) + 4x$
$y = 75 + 4(15)$
$y = 75 + 60$
$y = 135$

Back-substitute 15 for x in Equation (1). Then solve this equation for y.

Answer: She would need to earn all A's in 15 semester-hours to raise the cumulative GPA to 3.00. She would then have 135 grade points.

Does this seem reasonable? Does the answer check?

SELF-CHECK 3.6.2

The first semester a student enrolled in college he earned the following grades: C in a 3-hour English course, A in a 4-hour mathematics class, B in a 3-hour speech class, and C in a 4-hour psychology class.

1. Determine his grade points for this semester.
2. Determine his GPA for this semester.
3. The student planned to take a course during an intersession to try to raise his GPA to 3.00. How many semester-hours with a grade of A would be required to meet this goal? (Assume an A = 4 points.)

In Section 2.8 we examined a variety of applications of the mixture principle which is restated below.

Mixture principle: Amount in first + Amount in second = Amount in mixture

Another general principle used to model word problems is the rate principle, which is given in the following box.

Rate Principle

Amount = Rate × Base $\quad A = R \cdot B$

Applications of the Rate Principle

1. Variable cost = Cost per item × Number of items
2. Interest = Principal invested × Rate × Time
3. Distance = Rate × Time
4. Amount of active ingredient = Rate of concentration × Amount of mixture
5. Work = Rate × Time

Example 3 uses both the mixture principle and the rate principle. The mixture principle is used to express the total cost as the sum of the variable cost plus the fixed cost. Then the rate principle is used to determine the variable cost. The rate in this problem is the variable cost per pizza, and the base for this problem is the number of pizzas made.

EXAMPLE 3 Modeling the Fixed and Variable Costs for a Pizza Business

A family pizza business opened a new store. The daily cost for making pizzas includes a fixed daily cost and a variable cost per pizza. The total cost for making 200 pizzas on Monday was \$750. The total cost for making 250 pizzas on Tuesday was \$900. Determine the fixed daily cost and variable cost per pizza for this business.

SOLUTION

DEFINITION OF VARIABLES

Let x = variable cost per pizza

y = fixed dollar cost per day

Select a variable to represent each unknown.

WORD EQUATIONS

(1) **Monday:** Variable cost + Fixed cost = Total cost

(2) **Tuesday:** Variable cost + Fixed cost = Total cost

Note that each of these word equations is based on the mixture principle. Using the rate principle, the variable cost for each day is the number of pizzas times the cost per pizza.

ALGEBRAIC EQUATIONS

(1) $200x + y = 750$

(2) $250x + y = 900$

ALGEBRAICALLY

$$\begin{aligned}
(1)\quad -200x - y &= -750\\
(2)\quad 250x + y &= 900\\
\hline
50x &= 150\\
x &= 3
\end{aligned}$$

$$\begin{aligned}
(2)\quad 250x + y &= 900\\
250(3) + y &= 900\\
750 + y &= 900\\
y &= 150
\end{aligned}$$

To solve this system algebraically, multiply each side of Equation (1) by -1 and then use the addition method to eliminate y. Solve this equation for x. Then back-substitute 3 for x in Equation (2) and solve for y.

GRAPHICALLY

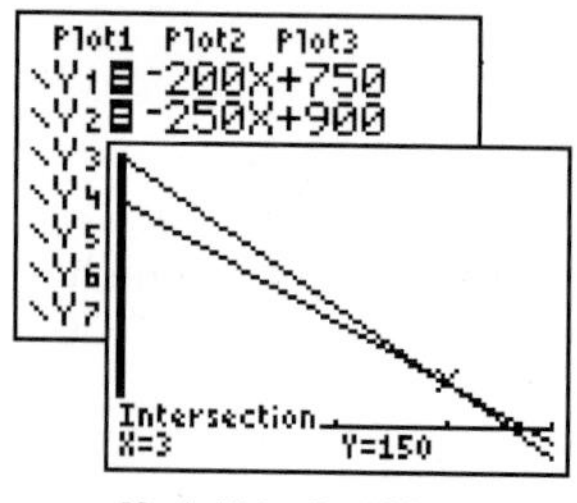

[0, 4, 1] by [−100, 900, 1]

To solve this system by using a graphing calculator, solve each equation for the y-variable.

(1) $y = -200x + 750$

(2) $y = -250x + 900$

A rough estimate of the restrictions on x would be to divide the total cost of about \$800 (\$750 rounded for a quick mental estimate) by the number of pizzas on a given day (\$800/200 pizzas = \$4/pizza). Thus, select the x-values [0, 4, 1] for the viewing window. The **ZoomFit** option selects the y-values and graphs these equations on this window. Use the **Intersect** feature to find that these lines intersect where $x = 3$ and $y = 150$.

NUMERICAL CHECK

X	Y1	Y2
0	750	900
1	550	650
2	350	400
3	150	150
4	-50	-100
5	-250	-350
6	-450	-600

X=3

The table verifies that the x-value of 3 produces y-values of 150. Also note that other values of x do not check.

Answer: The fixed cost per day is \$150, and \$3 is the variable cost per pizza.

SELF-CHECK 3.6.3

1. An umbrella manufacturer has daily costs that are both fixed and variable. The company produced 500 umbrellas on Monday at a cost of \$1,450 and 600 umbrellas Tuesday at a cost of \$1,700. Determine the fixed daily cost and the variable cost per umbrella.

Example 4 uses the mixture principle to form both equations. The total principal is composed of two separate investments, and the total interest is the combined interest from the two investments. The rate principle is also used in Example 4 to calculate the interest earned on each investment. The rate in this application is the interest rate, and the formula is $I = PRT$.

EXAMPLE 4 Modeling the Income from Two Investments

A student saving for college was given \$6,000 by her grandparents. She invested part of this money in a savings account that earned interest at the rate of 3% per year. The rest was invested in a bond that paid interest at the rate of 5.5% per year. If the combined interest at the end of 1 year was \$305, how much was invested at each rate?

SOLUTION

DEFINITION OF VARIABLES

Let x = principal invested in savings account
y = principal invested in bond

Select a variable to represent each unknown. (The restrictions on the variables are considered when the system is solved graphically.)

WORD EQUATIONS

Both word equations are based on the mixture principle.

(1) Principal in savings account + Principal in bond = Total principal
(2) Interest on savings account + Interest on bond = Total interest

ALGEBRAIC EQUATIONS

(1) $x + y = 6{,}000$
(2) $0.03x + 0.055y = 305$

By using $I = PRT$ and 1 year for time, the interest on the savings account is $I = (x)(0.03)(1) = 0.03x$, and the interest on the bond is $I = (y)(0.055)(1) = 0.055y$.

ALGEBRAICALLY

$$\begin{aligned}
(1)\quad -0.03x - 0.030y &= -180\\
(2)\quad 0.03x + 0.055y &= 305\\
0.025y &= 125\\
y &= \frac{125}{0.025}\\
y &= 5{,}000
\end{aligned}$$

$$\begin{aligned}
(1)\quad x + y &= 6{,}000\\
x + 5{,}000 &= 6{,}000\\
x &= 1{,}000
\end{aligned}$$

To solve this system algebraically, multiply each side of Equation (1) by -0.03 and then use the addition method to solve this system. Divide both sides by 0.025 to solve for y. Back-substitute 5,000 for y in Equation (1) and solve for x.

GRAPHICALLY

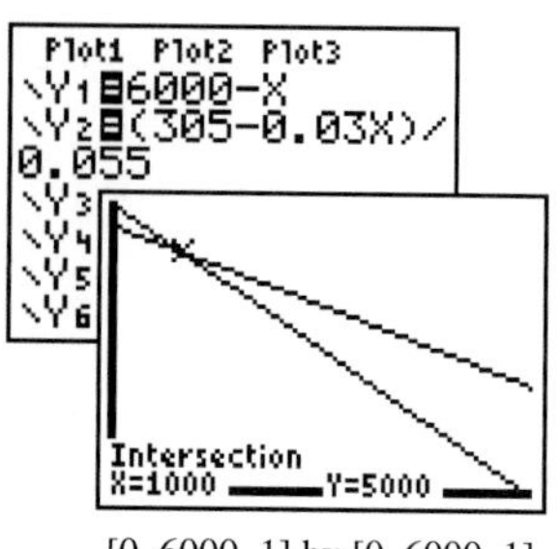

[0, 6000, 1] by [0, 6000, 1]

To solve this system by using a graphing calculator, solve each equation for the y-variable.

(1) $y = 6{,}000 - x$

(2) $y = \dfrac{305 - 0.03x}{0.055}$

It is not necessary to simplify an equation to enter it into a calculator. See the form of Equation (2) entered into the calculator as Y_2.

The amount invested in the savings account cannot be less than \$0 or more than the \$6,000 she has. Thus, select the x-values [0, 6,000, 1] for the viewing window. The **ZoomFit** option selects the y-values and graphs these equations on this window. Use the **Intersect** feature to find that these lines intersect where $x = 1{,}000$ and $y = 5{,}000$.

NUMERICAL CHECK

X	Y1	Y2
0	6000	5545.5
500	5500	5272.7
1000	5000	5000
1500	4500	4727.3
2000	4000	4454.5
2500	3500	4181.8
3000	3000	3909.1

X=1000

The table verifies that the x-value of 1,000 produces y-values of 5,000. Also note that other values of x do not check.

Answer: She invested \$1,000 in the savings account and \$5,000 in the bond.

SELF-CHECK 3.6.4

1. A broker made two separate investments on behalf of a client. The first investment earned 8% and the second earned 5% for a total gain of \$5,600. If these investments had earned 5% and 6%, respectively, then the gain would have been \$5,800. How much was actually invested at each rate?

EXAMPLE 5 Determining the Speed Flown by Two Airplanes

Two airplanes depart from an airport simultaneously, one flying 100 km/h faster than the other. These planes travel in opposite directions, and after 1.5 hours they are 1,275 km apart. Determine the speed of each plane. (*Hint:* $D = RT$; distance equals rate times the time.)

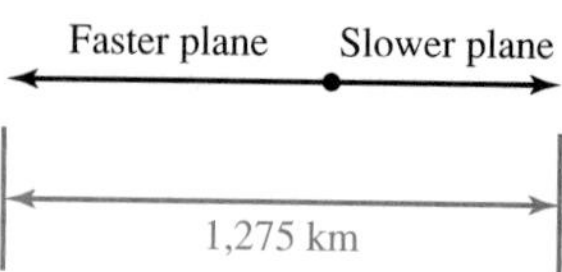

SOLUTION

DEFINITION OF VARIABLES

Let r_1 = rate of slower plane, km/h
r_2 = rate of faster plane, km/h

Select a variable to represent each unknown. (The restrictions on the variables are considered when the system is solved graphically.)

WORD EQUATIONS

(1) Rate of faster plane = Rate of slower plane + 100 km/h

(2) Distance by slower plane + Distance by faster plane = Total distance between planes

Note the second word equation is based on the mixture principle.

ALGEBRAIC EQUATIONS

(1) $r_2 = r_1 + 100$
(2) $1.5r_1 + 1.5r_2 = 1{,}275$

By using $D = RT$ and a time of 1.5 hours, the distance is $1.5r_1$ for the slower plane and $1.5r_2$ for the faster plane.

ALGEBRAICALLY

$$\begin{aligned}(1)\ r_1 - r_2 &= -100\\ (2)\ r_1 + r_2 &= 850\\ \hline 2r_1 &= 750\\ r_1 &= 375\end{aligned}$$

$$\begin{aligned}(1)\ r_2 &= r_1 + 100\\ r_2 &= 375 + 100\\ r_2 &= 475\end{aligned}$$

To solve this system algebraically, rearrange the terms in Equation (1) and simplify Equation (2) by dividing both sides of the equation by 1.5.
Then use the addition method to solve this system. Back-substitute 375 for r_1 in Equation (1).

GRAPHICALLY

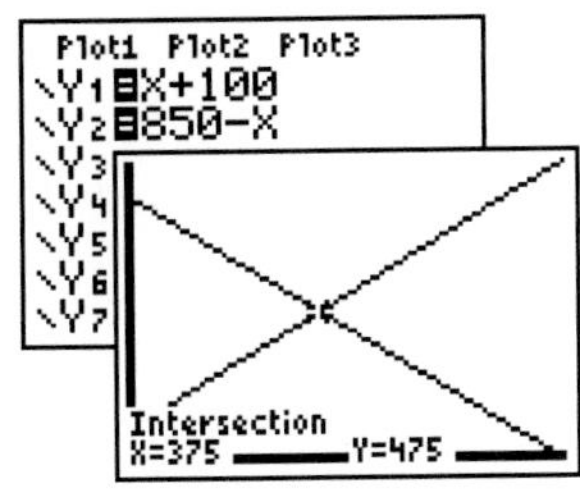

[0, 850, 1] by [0, 950, 1]

To solve this system by using a graphing calculator, use x to represent r_1 and y to represent r_2, and solve each equation for the y-variable.
(1) $y = x + 100$
(2) $y = 850 - x$

To establish restrictions on x, consider what r_1 would have to be for the first plane to cover the entire 1,275 km by itself in 1.5 hours (1,275 km/1.5 h = 850 km/h). Therefore this rate must be at least 0 km/h and at most 850 km/h. Thus, select the x-values [0, 850, 1] for the viewing window. The **ZoomFit** option selects the y-values and graphs these equations on this window. Use the **Intersect** feature to find that these lines intersect where $x = 375$ and $y = 475$.

NUMERICAL CHECK

X	Y1	Y2
300	400	550
325	425	525
350	450	500
375	475	475
400	500	450
425	525	425
450	550	400

X=375

The table verifies that the x-value of 375 produces y-values of 475. Also note that other values of x do not check.

Answer: The planes traveled at 375 and 475 km/h.

SELF-CHECK 3.6.5

1. Two trains depart from a station, traveling in opposite directions. The train that departs at 7:00 a.m. travels 20 km/h faster than the train that departs at 6 a.m. At 7:30 a.m. they are 170 km apart. Determine the rate of each train.

When two liquids of different concentrations are mixed, the total volume is found by adding the volumes of the liquids that are mixed. The total of the chemicals found in the mixture is calculated by adding the chemicals found in each of the individual solutions. This is illustrated in Example 6 where we also use the rate principle to determine the amount of chlorine in each solution.

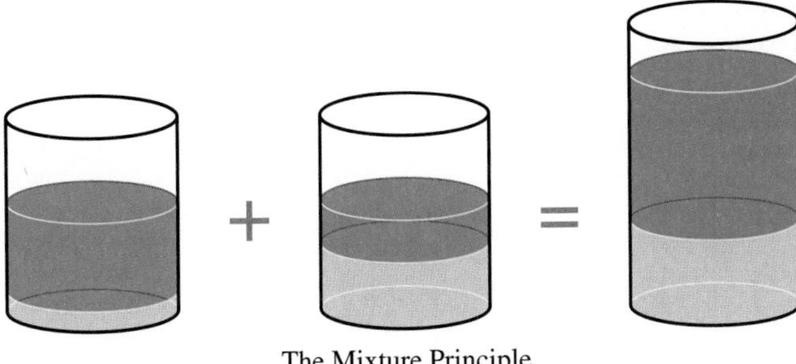

The Mixture Principle

EXAMPLE 6 Calculating the Amount of Two Solutions to Mix

A water treatment plant uses batch deliveries of two levels of chlorine concentrations to mix to create the level of concentration needed on any given day. Determine how many liters of a 3% chlorine solution and how many liters of a 5% chlorine solution should be mixed to produce 500 L of a 4.4% chlorine solution.

SOLUTION

DEFINITION OF VARIABLES

Let x = number of liters of 3% chlorine solution
y = number of liters of 5% chlorine solution

Select a variable to represent each unknown. (The restrictions on the variables are considered when the system is solved graphically.)

WORD EQUATIONS

(1) Volume of first solution + Volume of second solution = Total volume of mixture

(2) Volume of chlorine in first solution + Volume of chlorine in second solution = Total chlorine in mixture

Note both word equations are based on the mixture principle.

The volume of chlorine in each solution is calculated by using the rate principle.

ALGEBRAIC EQUATIONS

(1) $x + y = 500$
(2) $0.03x + 0.05y = 0.044(500)$

The volume of chlorine in the x L of 3% solution is $0.03x$. The volume of chlorine in the y L of the 5% solution is $0.05y$. The volume of chlorine in 500 L of 4.4% solution is $0.044(500)$.

ALGEBRAICALLY

$$\begin{aligned} (1)\quad -0.03x - 0.03y &= -15 \\ (2)\quad 0.03x + 0.05y &= 22 \\ \hline 0.02y &= 7 \\ \frac{0.02y}{0.02} &= \frac{7}{0.02} \\ y &= 350 \end{aligned}$$

$$\begin{aligned} (1)\quad x + y &= 500 \\ x + 350 &= 500 \\ x &= 150 \end{aligned}$$

To solve this system algebraically, multiply both sides of Equation (1) by -0.03 and then use the addition method to solve this system.
Divide both sides of this equation by 0.02. Back-substitute 350 for y in Equation (1) and solve for x.

GRAPHICALLY

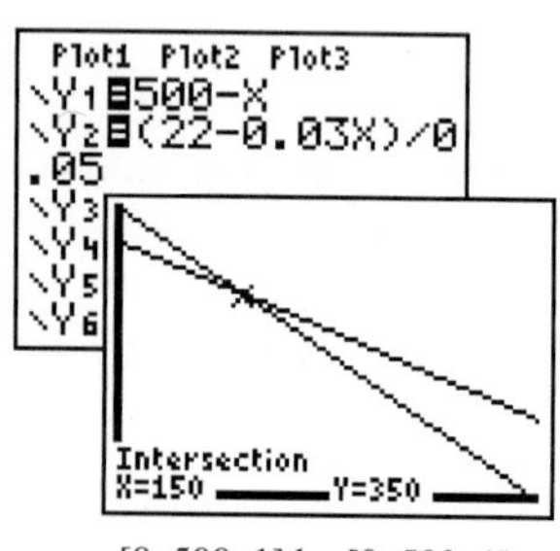

[0, 500, 1] by [0, 500, 1]

To solve this system by using a graphing calculator, solve each equation for the y-variable.
(1) $y = 500 - x$
(2) $y = \frac{22 - 0.03x}{0.05}$

The number of liters of the 3% chlorine solution cannot be less than 0 or more than the total of 500 L. Thus select the x-values [0, 500, 1] for the viewing window. The **ZoomFit** option selects the y-values and graphs these equations on this window. Use the **Intersect** feature to find that these lines intersect where $x = 150$ and $y = 350$.

NUMERICAL CHECK

X	Y1	Y2
0	500	440
50	450	410
100	400	380
150	350	350
200	300	320
250	250	290
300	200	260

X=150

The table verifies that the x-value of 150 produces y-values of 350. Also note that other values of x do not check.

Answer: Use 150 L of the 3% solution and 350 L of the 5% solution to produce 500 L of a 4.4% chlorine solution.

SELF-CHECK 3.6.6

1. A water treatment plant uses batch deliveries of two levels of chlorine concentrations to mix to create the level of concentration needed on any given day. Determine how many liters of a 3% chlorine solution and how many liters of a 5% chlorine solution should be mixed to produce 500 L of a 4.5% chlorine solution.

Example 7 illustrates another application of the rate principle. The amount of work done is equal to the rate of work times the time worked, or $W = RT$.

EXAMPLE 7 Determining the Rates of Two Workers

A small building contractor plans to add a bricklayer to his full-time crew. He has two bricklayers on a current job that he is considering for this position. On Monday he observed that these two bricklayers each worked 7 hours and laid a total of 3,150 bricks. On Tuesday the older bricklayer worked 6 hours and the younger bricklayer worked 5 hours, and they laid a total of 2,500 bricks. Determine for the contractor the rate of work for each bricklayer, assuming that both bricklayers work at a fairly consistent rate of work.

SOLUTION

DEFINITION OF VARIABLES

Let r_1 = rate of work for younger bricklayer, bricks/h
r_2 = rate of work for older bricklayer, bricks/h

Select a variable to represent each unknown. (The restrictions on the variables are considered when the system is solved graphically.)

WORD EQUATIONS

(1) Work by younger worker + Work by older worker = Total work

(2) Work by younger worker + Work by older worker = Total work

The mixture principle is used to form the word equation for Monday and another word equation for Tuesday.

ALGEBRAIC EQUATIONS

(1) $7r_1 + 7r_2 = 3{,}150$
(2) $5r_1 + 6r_2 = 2{,}500$

The amount of work done by each bricklayer is determined by using the rate principle in the form $W = RT$.

ALGEBRAICALLY

(1) $r_1 + r_2 = 450$
(2) $5r_1 + 6r_2 = 2{,}500$

(1) $-5r_1 - 5r_2 = -2{,}250$
(2) $5r_1 + 6r_2 = 2{,}500$
$r_2 = 250$

(1) $r_1 + r_2 = 450$
$r_1 + 250 = 450$
$r_1 = 200$

To solve this system algebraically, divide both sides of Equation (1) by 7. Then multiply both sides of this equation by -5 and use the addition method to solve this system.
Back-substitute 250 for r_2 in Equation (1) and solve for r_1.

GRAPHICALLY

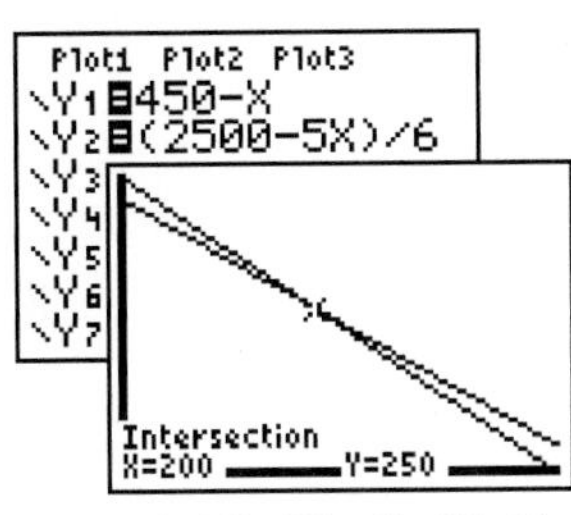

[0, 450, 1] by [0, 450, 1]

On a calculator use x to represent r_1 and use y to represent r_2.

To solve this system by using a graphing calculator, solve each equation for the y-variable.

(1) $y = 450 - x$

(2) $y = \dfrac{2{,}500 - 5x}{6}$

To establish restrictions on x, consider what r_1 would be if the younger bricklayer laid all 3,150 bricks in 7 hours (3,150 bricks/7 h = 450 bricks/h). Therefore this rate must be at least 0 bricks/h and at most 450 bricks/h. Thus, we select the x-values [0, 450, 1] for the viewing window. The **ZoomFit** option selects the y-values and graphs these equations on this window. Use the **Intersect** feature to find that these lines intersect where $x = 200$ and $y = 250$.

NUMERICAL CHECK

X	Y1	Y2
0	450	416.67
50	400	375
100	350	333.33
150	300	291.67
200	250	250
250	200	208.33
300	150	166.67

X=200

The table verifies that the x-value of 200 produces y-values of 250. Also note that other values of x do not check.

Answer: The younger worker lays 200 bricks/h, and the older worker lays 250 bricks/h.

SELF-CHECK 3.6.7

1. Two machines are used to produce pencils. During a 2-hour period in the morning, the two machines produced a total of 1,560 pencils. In the afternoon, machine 1 operated for 4 hours and machine 2 operated for 3 hours, and they produced 2,640 pencils. Determine the rate of production for each machine.

SELF-CHECK ANSWERS

3.6.1

1. The smaller angle is 35° and the larger angle is 55°.

3.6.2

1. He would have 39 grade points.
2. His GPA would be approximately 2.79.
3. He would need to earn an A in 3 semester-hours.

3.6.3

1. The fixed cost per day is $200, and the variable cost per umbrella is $2.50.

3.6.4

1. The first investment was for $20,000, and the second investment was for $80,000.

3.6.5

1. The first train travels at 80 km/h and the second at 100 km/h.

3.6.6

1. Use 125 L of the 3% solution and 375 L of the 5% solution to produce 500 L of a 4.5% chlorine solution.

3.6.7

1. Machine 1 produces 300 pencils/h, and machine 2 produces 480 pencils/h.

USING THE LANGUAGE AND SYMBOLISM OF MATHEMATICS 3.6

1. The first step in the word problem strategy given in this book is to read the problem carefully to determine what you are being asked to __________.
2. The second step in the word problem strategy given in this book is to select a __________ to represent each unknown quantity.
3. The third step in the word problem strategy given in this book is to model the problem verbally with word equations and then to translate these word equations into __________ equations.
4. The fourth step in the word problem strategy given in this book is to __________ the equation or system of equations and to answer the question asked by the problem.
5. The fifth step in the word problem strategy given in this book is to check your answer to make sure the answer is __________.
6. The __________ principle states that the amount obtained by combining two parts is equal to the amount obtained from the first part plus the amount obtained from the second part.
7. The __________ principle states that the amount obtained is equal to the rate times the base to which this rate is applied.
8. In the formula $D = RT$, D represents distance, R represents __________, and T represents __________.
9. In the formula $I = PRT$, I represents interest, P represents __________, R represents __________, and T represents __________.
10. In the formula $W = RT$, W represents the amount of __________, R represents the rate of work, and T represents the __________ worked.

EXERCISES 3.6

In Exercises 1 and 2, solve each problem by completing each of the given steps.

1. **Dimensions of a Rectangular Poster** The perimeter of a rectangular poster is 204 cm. Find the dimensions of the poster if the length is 6 cm more than the width.
   a. Select a variable to represent each of the unknown quantities, and identify each variable, including the units of measurement.
   b. Write a pair of word equations that model this problem.
   c. Translate the word equations from part **b** into algebraic equations.
   d. Solve the system of equations from part **c.**
   e. Is the solution from part **d** within the restrictions on the variables, and does it seem reasonable?
   f. Write a sentence that answers the problem.
2. **Dimensions of a Rectangular Pane of Glass** The perimeter of a rectangular pane of glass is 68 cm. Find the dimensions of the pane of glass if the length is 2 cm less than 3 times the width.
   a. Select a variable to represent each of the unknown quantities, and identify each variable, including the units of measurement.
   b. Write a pair of word equations that model this problem.
   c. Translate the word equations from part **b** into algebraic equations.
   d. Solve the system of equations from part **c.**
   e. Is the solution from part **d** within the restrictions on the variables, and does it seem reasonable?
   f. Write a sentence that answers the problem.

In Exercises 3–40, solve each problem by using a system of equations and the word problem strategy developed in this section.

3. **Numeric Word Problem** Find two numbers whose sum is 100 if one number is 16 more than twice the smaller number.
4. **Numeric Word Problem** Find two numbers whose sum is 100 if one number is 16 more than three times the smaller number.
5. **Dimensions of an Isosceles Triangle** An isosceles triangle has two equal sides. The triangle shown here has a perimeter of 72 cm. The base is 12 cm longer than the two equal sides. Find the dimensions of this triangle.

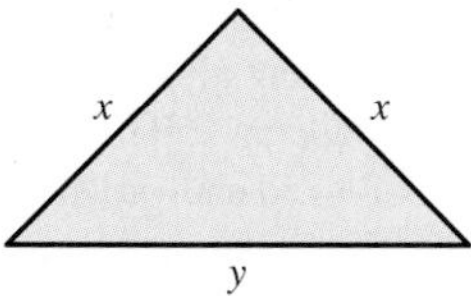

6. **Angles of an Isosceles Triangle** The sum of the interior angles of a triangle is 180°. An isosceles triangle has two equal angles. Find the angles of the isosceles triangle shown here if angle y is 33° larger than angle x.

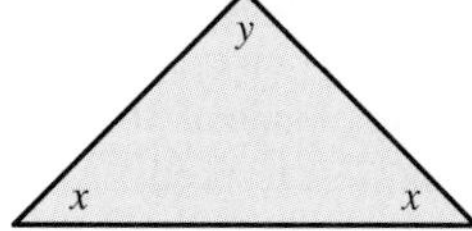

7. **Coded Messages** To receive a message, a spy first must send a pair of authorization numbers that satisfies the conditions of the two independent supervisors on duty that day. Supervisor A must receive the code and confirm that the sum of seven times the first number plus four times the second number is two hundred ninety-four. Supervisor B must receive the code and confirm that six times the first number minus the second number is ninety-seven. What two numbers should be sent when supervisors A and B are on duty?
8. **Coded Messages** To receive a message, an engineer traveling abroad first must send a pair of authorization numbers to his home office that satisfies the conditions of the two independent supervisors on duty that day. Supervisor A must receive the code and confirm that the sum of five times the first number plus three times the second number is eighty-four. Supervisor B must receive the code and confirm that four times the first number minus the second number is twenty-three. What two numbers should be sent when supervisors A and B are on duty?
9. **Mean and Range** A golfer recorded five scores for five rounds of golf one week. Four of these rounds were the same, and one was 10 strokes higher than the others. If the average score for the weeks was 83, what was the score for each of the five rounds?
10. **Mean and Range** A company has six employees. Five of these employees make exactly the same amount each month, which is less than their supervisor makes. The mean (average) hourly salary of all employees is \$26/h, and the range of these salaries is \$12/h. Determine the hourly salary of each employee.
11. **GPA** A student enrolled in college for the first semester in the spring of 2007. She earned the following grades: C in a 3-hour English course, B in a 4-hour mathematics class, C in a 3-hour speech class, and C in a 3-hour Spanish class.
    a. Determine her grade points for this semester.
    b. Determine her GPA for this semester.
    c. The student planned to take courses during the summer to try to raise her GPA to 3.00. How many semester-hours with a grade of A would be required to meet this goal? (Assume an A = 4 points.)
12. **GPA** A student enrolled in college for the first semester in the spring of 2007. He earned the following grades: B in a 3-hour English course, C in a 4-hour business class, C in a 3-hour speech class, and B in a 3-hour French class.
    a. Determine his grade points for this semester.
    b. Determine his GPA for this semester.
    c. The student planned to take courses during the summer to try to raise his GPA to 3.00. How many semester-hours with a grade of A would be required to meet this goal? (Assume an A = 4 points.)
13. **GPA** A student who has already completed 30 semester-hours with a GPA of 2.50 at Harper College needs to raise her GPA to 2.75 to meet transfer requirements into one college program. How many semester-hours with a grade of A would be required to meet this goal? How many grade points will the student have if this goal is met? (Assume an A = 4 points.)

14. **GPA** A student who has already completed 36 semester-hours with a GPA of 2.75 at Parkland College needs to raise his GPA to 3.00 to meet transfer requirements into one college program. How many semester-hours with a grade of A would be required to meet this goal? How many grade points will the student have if this goal is met? (Assume an A = 4 points.)
15. **Cable TV Bills** A cable TV bill consists of a fixed monthly charge and a variable charge, which depends on the number of pay-per-view movies ordered. In January the \$44 bill included four pay-per-view movies. In February the \$54.50 bill included seven pay-per-view movies. Determine the fixed monthly charge and the charge for each pay-per-view movie.
16. **Rental Truck Costs** The cost of a rental truck used to move furniture includes a fixed daily cost and a variable cost based on miles driven. For a customer driving 180 mi, the cost for 1 day will be \$75. For a customer driving 150 mi, the cost for 1 day will be \$67.50. Determine the fixed daily cost and the cost per mile for this truck.
17. **Costs for Union Employees** A union contract requires at least one union steward for every 20 union employees or fraction thereof. Regular union employees earn \$120/day, and union stewards earn \$144/day. A company has a need for 50 total employees and has a daily payroll budget of \$6,072. How many of each type of employee will meet the company's needs and consume all the budget?
18. **Equipment Costs** A bulldozer can move 25 tons/h of material, and an end loader can move 18 tons/h. The bulldozer costs \$75/h to operate, and the end loader costs \$50/h to operate. During 1 hour a construction company has moved 204 tons at a cost of \$600. How many bulldozers and how many end loaders were operating during this hour?

19. **Number of Basketball Tickets** The total receipts for a basketball game are \$1,400 for 788 tickets sold. Adults paid \$2.50 for admission and students paid \$1.25. The ticket takers counted only the number of tickets, not the type of tickets. Determine the number of each type of ticket sold.
20. **Loads of Topsoil** Topsoil sells for \$55 per truckload, and fill dirt sells for \$40 per truckload. A landscape architect estimates that 20 truckloads will be needed for a certain job. The estimated cost is \$920. How many loads of topsoil and how many loads of fill dirt are planned for this job?
21. **Complementary Angles** The angles shown in Exercise 22 are complementary, and one angle is 28° larger than the other angle. Determine the number of degrees in each angle.
22. **Complementary Angles** The angles shown in the figure are complementary, and one angle is 15° more than twice the other angle. Determine the number of degrees in each angle.

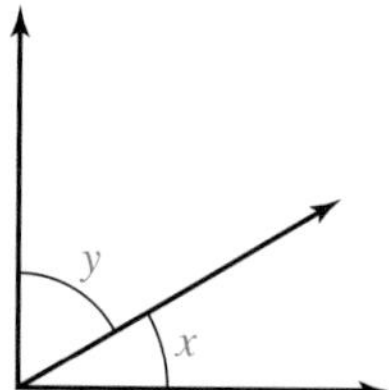

23. **Supplementary Angles** The two angles shown in Exercise 24 represent parts that should be assembled so that the angles are supplementary and the larger angle is 5° more than 4 times the smaller angle. Determine the number of degrees in each angle.
24. **Supplementary Angles** The two angles shown here represent parts that should be assembled so that the angles are supplementary and the larger angle is 9° less than twice the smaller angle. Determine the number of degrees in each angle.

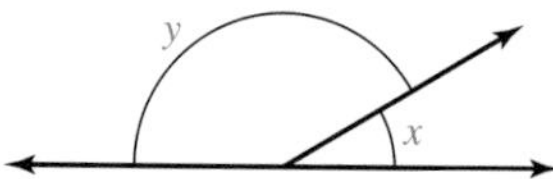

25. **Order for Left- and Right-Handed Desks** When ordering new desks for classrooms, the business manager for a college uses the fact that there are approximately 9 times as many right-handed people as left-handed people. How many desks of each type must be ordered if 600 new desks are needed?
26. **Length of a Board** A board 12 ft long is to be cut into two pieces so that one piece is 3 ft longer than the other piece. Determine the length of each piece.

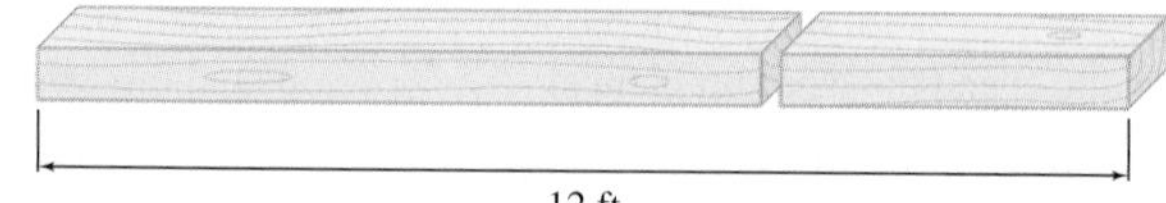

27. **Costs for Lightbulbs** An incandescent 60-W lightbulb costs \$0.50 to buy and \$0.12 per day to use (24 hours of usage). A compact fluorescent lamp (CFL) that produces the equivalent amount of light costs \$2.50 to buy and \$0.02 per day to use. Determine the number of days of usage when these costs are equivalent. What is this equivalent cost? (Source: Salt River Project Power Utility.)
28. **Fixed and Variable Costs** A family business makes custom rocking chairs, which it markets as retirement gifts. One month the total fixed and variable costs for producing 250 rockers were \$32,250. The next month the total fixed and variable costs for producing 220 rockers were \$28,500. Determine the fixed cost and the variable cost per rocking chair.
29. **Interest on Two Investments** Part of a \$12,000 investment earned interest at a rate of 7%, and the rest earned interest at a rate of 9%. The combined interest earned at the end of 1 year was \$890. How much was invested at each rate?
30. **Interest on Two Investments** A broker made two separate investments on behalf of a client. The first investment earned 8% and the second earned 5%, for a total gain of \$600. If these investments had earned 4% and 7%, respectively, then the gain would have been \$480. How much was actually invested at each rate?

31. **Interest on Two Investments** An investment of $10,000 earned a net income of $345 in 1 year. Part of the investment was in bonds and earned income at a rate of 8%. The rest of the investment was in stocks and lost money at the rate of 5%. How much was invested in bonds, and how much was invested in stocks?

32. **Interest on Two Investments** A broker made two separate investments on behalf of a client. The first investment earned 12% and the second investment lost 5%, for a total gain of $1,040. If the amounts in these two investments had been switched, then the result would have been a gain of $360. Determine how much was actually invested at each rate.

33. **Rates of Two Trains** Two trains depart simultaneously from a station, traveling in opposite directions. One averages 10 km/h more than the other. After $\frac{1}{2}$ hour they are 89 km apart. Determine the speed of each train.

34. **Rates of Two Airplanes** A jetplane and a tanker that are 525 mi apart head toward each other so that the jet can refuel. The jet flies 200 mi/h faster than the tanker. Determine the speed of each aircraft if they meet in 45 minutes. (*Hint:* Use consistent units of measurement.)

35. **Rates of Two Trains** Two trains depart from a station, traveling in opposite directions. The train that departs at 6:30 a.m. travels 15 km/h faster than the train that departs at 6 a.m. At 7 a.m. they are 135 km apart. Determine the rate of each train.

36. **Rates of Two Buses** Two buses that are 90 mi apart travel toward each other on an interstate highway. The slower bus, which departs at 9:30 a.m., travels 5 mi/h slower than the bus that departs at 9:45 a.m. At 10:30 a.m. the buses pass each other. Determine the rate of each bus.

37. **Rate of a River Current** When a boat travels downstream with a river current, the rate of the boat in still water and the rate of the current are added. When a boat travels upstream, the rate of the current is subtracted from the rate of the boat.

A paddlewheel riverboat takes 1 hour to go 24 km downstream and another 4 hours to return upstream. Determine the rate of the boat and the rate of the current.

38. **Rate of a River Current** A small boat can go 40 km downstream in 1 hour but only 10 km upstream in 1 hour. Determine the rate of the boat and the rate of the current. (*Hint:* See Exercise 37.)

39. **Mixture of Disinfectant** A hospital needs 80 L of a 12% solution of disinfectant. How many liters of a 33% solution and a 5% solution should be mixed to obtain this 12% solution?

40. **Mixture of Medicine** The dosage of a medicine ordered by a doctor is 40 mL of a 16% solution. A nurse has available both a 20% solution and a 4% solution of this medicine. How many milliliters of each could be mixed to prepare this 40-mL dosage?

41. **Gold Alloy** A goldsmith has 80 g of an alloy that is 50% pure gold. How many grams of an alloy that is 80% pure gold must be combined with the 80 g of alloy that is 50% pure gold to form an alloy that will be 72% pure gold?

42. **Saline Mixture** A 30% salt solution is prepared by mixing a 20% salt solution and a 45% salt solution. How many liters of each must be used to produce 60 L of the 30% salt solution?

43. **Mixture of Fruit Drinks** A fruit drink concentrate is 15% water. How many liters of pure water should be added to 12 L of concentrate to produce a mixture that is 50% water?

44. **Insecticide Mixture** A nurseryman is preparing an insecticide by mixing a 95% solution with water (0% solution). How much solution and how much water are needed to fill a 500-gal tank with 3.8% solution?

45. **Rates of Two Workers** A small building contractor plans to add a bricklayer to his full-time crew. He has two bricklayers on a current job that he is considering for this position. On Monday he observed that these two bricklayers each worked 8 hours and laid a total of 4,000 bricks. On Tuesday the older bricklayer worked 6 hours, the younger bricklayer worked 7 hours, and they laid a total of 3,220 bricks. Determine for the contractor the rate of work for each bricklayer, assuming that both bricklayers work at a fairly consistent rate.

46. **Rates of Two Machines** A small machine shop is considering buying one of two metal presses at an auction. The owner visited the company selling these presses on two different days. On Monday she observed that these two presses were used 8 hours each and they produced a total of 88 stamped parts. On Tuesday the older machine worked 4 hours and the newer machine worked 6 hours, and they produced a total of 56 stamped parts. Determine for the owner of the machine shop the rate of work for each metal press, assuming that both presses work at a fairly consistent rate.

47. **Scientific Articles** An important indicator of the innovation and the health of a country's economy is the publication of articles in scholarly journals. These articles are often followed by patents and new products. This table gives the publications in thousands in the United States and in Europe. (Source: National Science Foundation.)

YEAR	U.S. PUBLICATIONS	WESTERN EUROPEAN PUBLICATIONS
1990	190	160
1992	200	175
1994	205	190
1996	195	205
1998	195	220
2000	200	225

By using x to represent the number of years since 1990, the following equations can be used to model these data.

(1) United States $y = 200$
(2) Europe $y = 6.8x + 160$

a. Solve this system of equations.
b. Interpret the meaning of this solution.

48. **International Travel** This table gives the number in millions of U.S. travelers to other countries and the number of foreign visitors to the United States.

YEAR	U.S. TRAVELERS (MILLIONS)	FOREIGN VISITORS TO UNITED STATES
1985	30	40
1990	40	42
1995	50	45
2000	60	50
2005	65	52

By using x to represent the number of years since 1985, the following equations can be used to model these data.

(1) U.S. travelers $y = 1.8x + 30$
(2) Foreign visitors to United States $y = 0.64x + 39.4$

a. Solve this system of equations.
b. Interpret the meaning of this solution.

49. **Computer Lease Options** The graph compares the costs of two options for leasing the computer equipment for a small engineering firm. The cost is based upon the number of months the computer equipment is used.

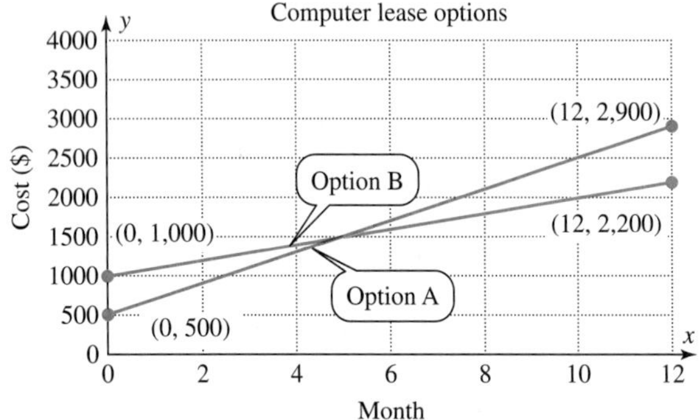

a. Use the y-intercept and an additional point to determine the equation of the line that models the cost for option A.
b. Use the y-intercept and an additional point to determine the equation of the line that models the cost for option B.
c. Solve the system of equations from parts **a** and **b.**
d. Does the algebraic solution from part **c** seem consistent with the point of intersection shown on the graph?
e. Interpret the meaning of the x- and y-coordinates of the solution from part **d.**

50. **Fish Populations** The graph compares the populations of bass and catfish in Lake Shelbyville after 1990.

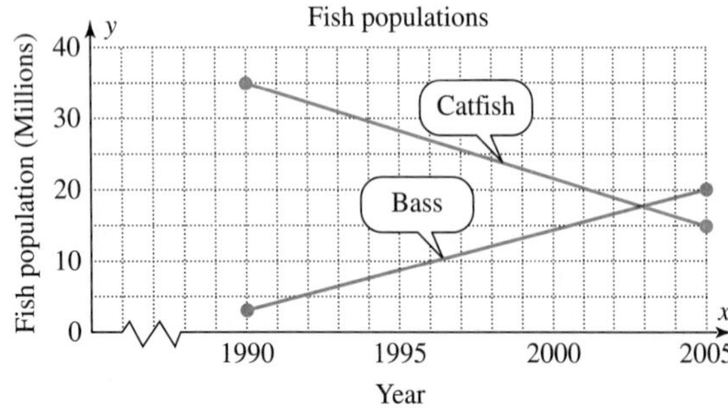

a. Use the two points (1990, 3) and (2005, 20) to write an equation for the line showing the bass population.
b. Use the two points (1990, 35) and (2005, 15) to write an equation for the line showing the catfish population.
c. Solve the system of equations from parts **a** and **b.**
d. Interpret the meaning of the x- and y-coordinates of the solution from part **c.**

In Exercises 51 and 52, use the given information to solve each part of the problem.

51. **Fixed and Variable Costs** A tennis shoe manufacturer has daily costs that are both fixed and variable. The variable costs depend upon the number of shoes produced that day. Determine the fixed daily cost and the variable cost per pair of shoes for each of the following situations. Check to confirm each situation is possible and that the proposed answer meets the restrictions on the variables.
a. The company produced 400 pairs of shoes on Monday at a cost of $5,700 and 525 pairs on Tuesday at a cost of $6,700.
b. The company produced 400 pairs of shoes on Monday at a cost of $9,100 and 525 pairs on Tuesday at a cost of $12,100.

52. **Investment Choices** A broker invested a total of $25,000 in two separate investments on behalf of a client. For the first year the rate of return was 5% on the first investment and 7% on the second investment. Determine how much was invested in each of these investments for each of the following situations. Check to confirm each situation is possible and that the proposed answer meets the restrictions on the variables.
a. The total income from the two investments was $1,450.
b. The total income from the two investments was $1,770.

53. **Point of No Return** A rescue helicopter is flying from an aircraft carrier 350 mi to Pearl Harbor hospital. The point of no return is the point where it would take as much time to fly back to the aircraft carrier as it would to fly on to Pearl Harbor. The airspeed of the helicopter is 175 mi/h, and it is flying into a 25-mi/h headwind.
a. Calculate the total time for the helicopter to fly to Pearl Harbor.
b. Once the helicopter is at the point of no return, determine the time it will take to complete the trip.
c. Use the results from parts **a** and **b** to determine the time for the helicopter to reach the point of no return.

54. **Point of No Return** Rework Exercise 53, assuming the helicopter is flying with a 25-mi/h tailwind to Pearl Harbor.

Group Discussion Questions

55. **Communicating Mathematically** Write a word problem that can be solved by using the system of equations $\begin{cases} x + y = 500 \\ 5x + 8y = 3{,}475 \end{cases}$ and that models
a. A numeric problem
b. A problem involving ticket sales (see Exercise 19)
c. A rate-of-work problem

56. **Communicating Mathematically** Write a word problem that can be solved by using the system of equations $\begin{cases} x + y = 500 \\ 0.05x + 0.08y = 34.75 \end{cases}$ and that models
a. A combined investment problem (see Exercise 29)
b. A mixture of two piles of copper ore (see Exercise 41)

57. **Discovery Question**
a. Compare the two systems of equations in Exercises 55 and 56. Are these systems equivalent; that is, can you convert one system to the other? If so, does this mean that this one system can represent all five of the word problems created in Exercises 55 and 56?

b. Discuss the use of one general mathematical principle, such as the mixture principle, to solve many seemingly unrelated problems.
c. Although pure numeric-based word problems have limited direct application, they are widely used in textbooks. Discuss a rationale of why this is helpful for students.

58. **Spreadsheet Exploration** Use Spreadsheet Exploration 3.6 from the website to explore the relationship among the speed of the current, the speed of a boat going downstream, and the speed of a boat going upstream. Use the sliders to observe the patterns in the table of values for the speed and distance traveled by each boat over a 1-hour time period. (For more information on this topic go to the Additional Resources link on the MathZone website at **www.mhhe.com/hallmercer.**)

KEY CONCEPTS FOR CHAPTER 3

1. **Slope of a Line**
   - The slope m of a line through (x_1, y_1) and (x_2, y_2) with $x_1 \neq x_2$ is $m = \frac{\text{change in } y}{\text{change in } x} = \frac{y_2 - y_1}{x_2 - x_1} = \frac{\text{rise}}{\text{run}}$.

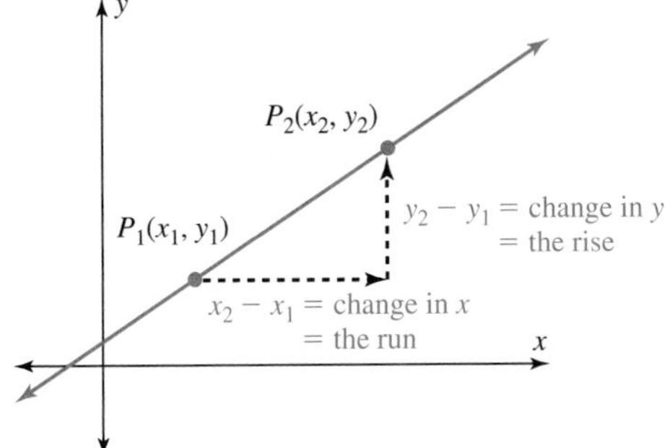

   - Δx, read "delta x," denotes a change in x; and Δy, read "delta y," denotes a change in y; $m = \frac{\Delta y}{\Delta x}$.
   - The slope of a line gives the change in y for each 1-unit change in x.
   - A line with positive slope goes upward to the right.
   - A line with negative slope goes downward to the right.
   - The slope of a horizontal line is 0.
   - The slope of a vertical line is undefined.
   - The slopes of parallel lines are the same.
   - The slopes of perpendicular lines are opposite reciprocals.
   - The product of the slopes of perpendicular lines is -1.
   - The slope of $y = mx + b$ is m.
   - The slope of $y - y_1 = m(x - x_1)$ is m.
   - The common difference of an arithmetic sequence is the same as the slope of the line through the points formed by this sequence.
   - The grade of a highway refers to the slope of the highway.
2. **Forms of Linear Equations:** A linear equation in x and y is first-degree in both x and y.
   - Slope-intercept form: $y = mx + b$ or $f(x) = mx + b$ with slope m and y-intercept $(0, b)$
   - Vertical line: $x = a$ for a real constant a
   - Horizontal line: $y = b$ for a real constant b
   - General form: $Ax + By = C$
   - Point-slope form: $y - y_1 = m(x - x_1)$ through (x_1, y_1) with slope m
3. **System of Equations:** When we consider two or more equations at the same time, we refer to this as a system of equations.
4. **Solution of a 2 × 2 System of Linear Equations**
   - A solution is an ordered pair that satisfies each equation in the system.
   - A solution of a system of linear equations is represented graphically by a point that is on both lines in the system.
   - A solution is represented in a table of values as an x-value for which both y-values are the same.
5. **Terminology Describing Systems of Linear Equations**
   - Consistent system: The system of equations has at least one solution.
   - Inconsistent system: The system of equations has no solution.
   - Independent equations: The equations cannot be reduced and rewritten as the same equation.
   - Dependent equations: The equations can both be written in exactly the same form.
6. **Classifications of Systems of Linear Equations**
   - Exactly one solution: A consistent system of independent equations
   - No solution: An inconsistent system of independent equations
   - An infinite number of solutions: A consistent system of dependent equations
7. **Using the Slope-Intercept Form to Classify Systems of Linear Equations:** For the linear system
   $\begin{Bmatrix} y = m_1x + b_1 \\ y = m_2x + b_2 \end{Bmatrix}$ with
   - $m_1 \neq m_2$
     The system is consistent with independent equations and exactly one solution.
   - $m_1 = m_2$ and $b_1 \neq b_2$
     The system is inconsistent with independent equations and no solution.
   - $m_1 = m_2$ and $b_1 = b_2$
     The system is consistent with dependent equations and an infinite number of solutions.
8. **Substitution Principle:** A quantity may be substituted for its equal.
9. **Addition-Subtraction Principle of Equality:** Equal values can be added to or subtracted from both sides of an equation to produce an equivalent equation.
10. **Methods for Solving Systems of Linear Equations in Two Variables**
    - Graphical method; the point(s) on both lines
    - Numerical method; for the same x-value the two y-values are equal.
    - Algebraically by the substitution method
    - Algebraically by the addition method

11. **Algebraic Solution of the Three Types of 2 × 2 Systems of Linear Equations**
    - Consistent system of independent equations: The solution process will produce unique x- and y-coordinates.
    - Inconsistent system: The solution process will produce a contradiction.
    - Consistent system of dependent equations: The solution process will produce an identity.
12. **Solving a System of Linear Equations Containing Fractions:** It may be easier to start by multiplying both sides of the equation by the least common denominator (LCD) of all the fractions in the equation.
13. **Strategy for Solving Word Problems**
    Step 1 Read the problem carefully to determine what you are being asked to find.
    Step 2 Select a variable to represent each unknown quantity. Specify precisely what each variable represents, and note any restrictions on each variable.
    Step 3 If necessary, make a sketch and translate the problem into a word equation or a system of word equations. Then translate each word equation into an algebraic equation.
    Step 4 Solve the equation or system of equations, and answer the question asked by the problem.
    Step 5 Check the reasonableness of your answer.
14. **Mixture Principle**

$$\begin{matrix}\text{Amount} \\ \text{in first}\end{matrix} + \begin{matrix}\text{Amount} \\ \text{in second}\end{matrix} = \begin{matrix}\text{Amount} \\ \text{in mixture}\end{matrix}$$

15. **Rate Principle:** The amount obtained is equal to the rate times the base to which this rate is applied.

$$\text{Amount} = \text{Rate} \times \text{Base} \qquad A = RB$$

REVIEW EXERCISES FOR CHAPTER 3

In Exercises 1–10, calculate the slope of each line.

1. The line through $(1, -3)$ and $(4, 3)$
2. The line through $(1, 4)$ and $(6, 1)$
3. $f(x) = \frac{4}{7}x + 3$
4. $y - 2 = -(x - 8)$
5. $y = -7$
6. $2x - 5y = 10$
7.

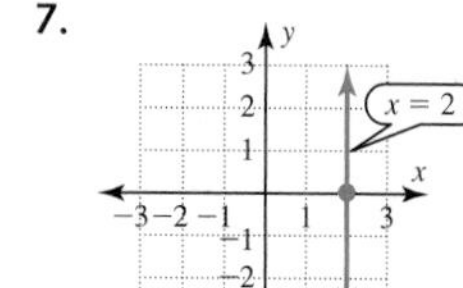

8. The line that contains the points in this table

x	y
−3	4
−2	4
−1	4
0	4
1	4
2	4
3	4

9. The line that contains the points in this table

x	y
4	−3
4	−2
4	−1
4	0
4	1
4	2
4	3

10. The line that contains the points in this table

X	Y1
-3	-.5
-2	0
-1	.5
0	1
1	1.5
2	2
3	2.5

X=3

11. Complete the following table involving the change in x, the change in y, and the slope of the line $f(x) = mx + b$.

	CHANGE IN x	CHANGE IN y	SLOPE
a.	5		$\frac{4}{5}$
b.		4	$\frac{4}{5}$
c.	−5		$\frac{4}{5}$
d.	1		$\frac{4}{5}$
e.	−1		$\frac{4}{5}$

12. Complete each table (without using a calculator) so that the points all lie on a line with a slope of $m = \frac{3}{5}$.

a.

x	y
0	6
5	
10	
15	
20	

b.

x	y
0	6
10	
20	
30	
40	

c.

x	y
0	6
−5	
−10	
−15	
−20	

13. **A Sequence of House Payments** The equation $a_n = 800n + 3{,}000$ gives the total dollars paid for a house for n months.
    a. Determine the sequence for the first 4 months.
    b. Graph this sequence.
    c. Determine the slope of the line through these points. Interpret the meaning of the slope in this application.
    d. Calculate a_0, and interpret the meaning of a_0.

14. Slope of a Wheelchair Ramp Determine the slope of the wheelchair ramp shown in the figure.

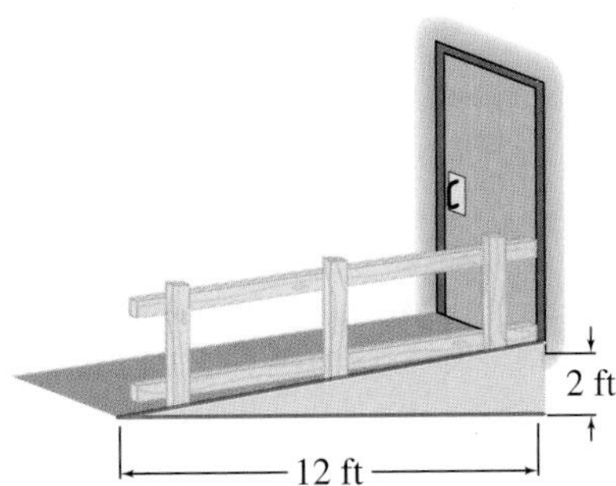

15. Determine whether the first line l_1 is parallel to, perpendicular to, or neither parallel nor perpendicular to the second line l_2.

a. Line l_1 passes through $(4, -2)$ and $(6, 2)$.
Line l_2 passes through $(5, 3)$ and $(9, 11)$.

b. Line l_1 defined by $y = 2$
Line l_2 defined by $x = -2$

c. Line l_1 defined by $y = 5x - 4$
Line l_2 defined by $y - 4 = -5(x + 1)$

In Exercises 16–21, write the equation of each line in slope-intercept form.

16. A line with y-intercept $(0, 6)$ and a slope of $-\frac{1}{2}$

17. A line with x-intercept $(3, 0)$ and y-intercept $(0, -2)$

18. A line that passes through $(2, 5)$ and $(4, -1)$

19. A line that passes through $(0, -4)$ and goes up 3 units for every 2-unit increase in x

20. A line that passes through the origin and is parallel to $y = 5x - 9$

21. A line that passes through $(2, 5)$ and is perpendicular to $y = -2x + 7$

22. Write the equation of a horizontal line that passes through $(4, 9)$.

23. Write the equation of a vertical line that passes through $(4, 9)$.

24. Without using a calculator, graph the line that has a slope of $m = -\frac{3}{4}$ and a y-intercept of $(0, 1)$.

25. Without using a calculator, graph the line defined by $y = \frac{2}{5}x - 4$.

In Exercises 26 and 27, mentally estimate the slope of each line and then use a calculator to determine the slope.

26.

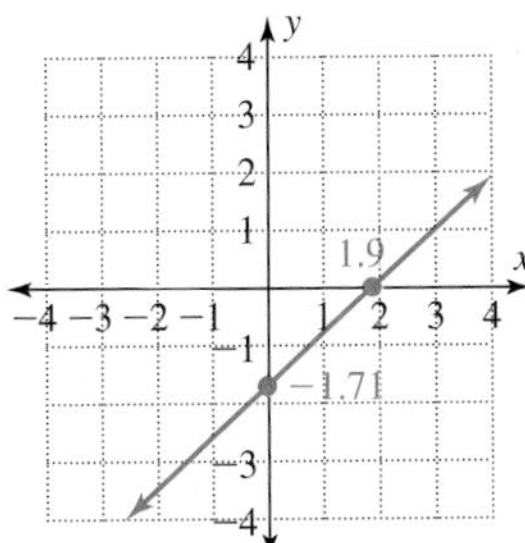

27.

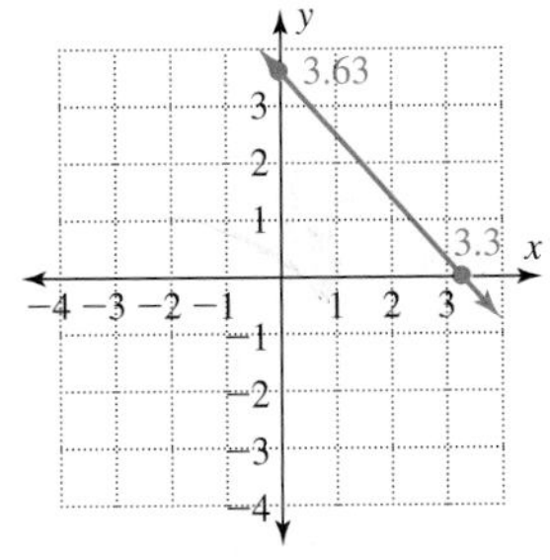

28. The following graphing calculator display shows a table of values for a linear function. Use this table to determine the following:

a. y-intercept of the line

b. Value of Δx shown in the table

c. Value of Δy shown in the table

d. Slope of the line

e. Equation of the line in slope-intercept form

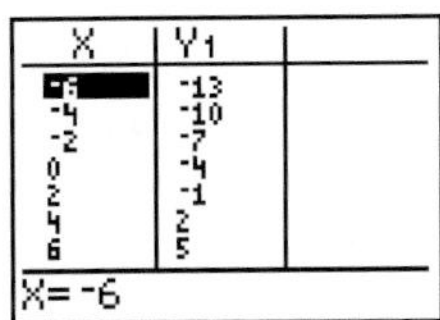

X	Y1
-6	-13
-4	-10
-2	-7
0	-4
2	-1
4	2
6	5

X=-6

29. Determine which points satisfy the equation $2x - 5y = 10$.

a. $(5, 0)$ **b.** $(10, 2)$ **c.** $(2, 10)$ **d.** $(0, -2)$

30. Determine which points satisfy the system of equations $\begin{Bmatrix} 3x + y = -2 \\ x + 2y = 6 \end{Bmatrix}$.

a. $(1, -5)$ **b.** $(2, 2)$ **c.** $(0, 1)$ **d.** $(-2, 4)$

In Exercises 31 and 32, use the graphs to solve the system of linear equations represented by each graph.

31.

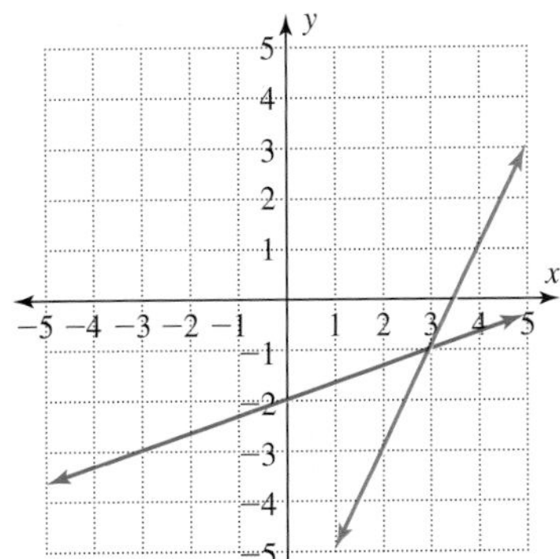

32.

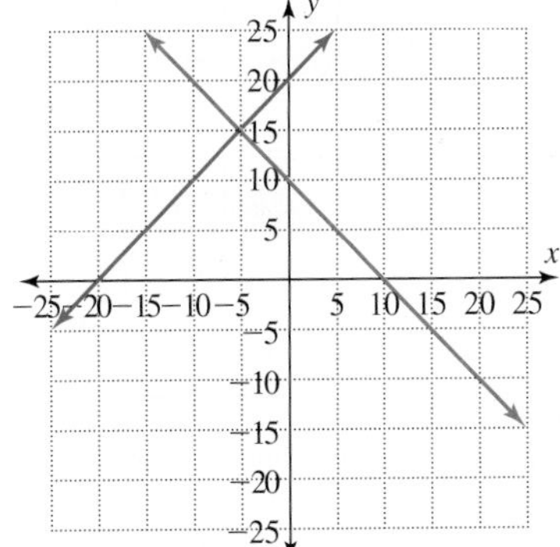

In Exercises 33 and 34, use the graphs to identify each system of linear equations as either an inconsistent system or a consistent system of dependent equations. Assume that each graph shows enough information to make this decision.

33.

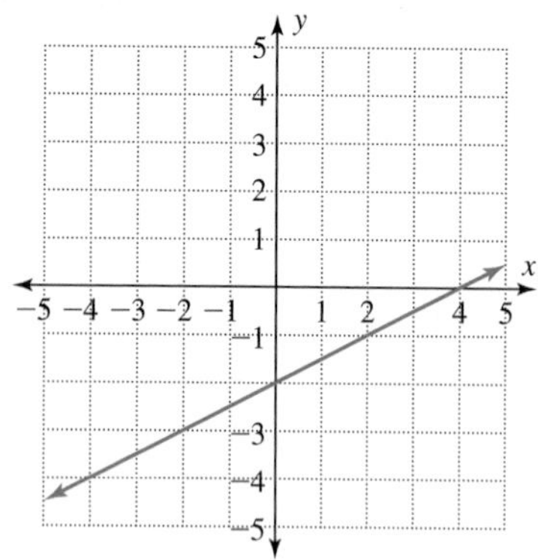

34.

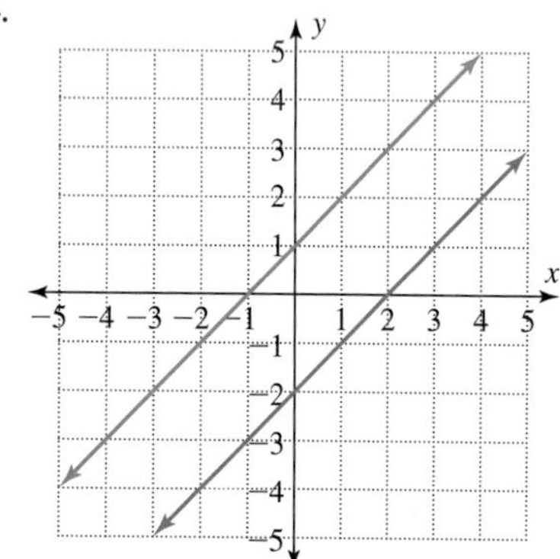

Exercises 35–38, give tables of values for each linear equation in a system of linear equations. Use these tables to solve each system.

35.

X	Y_1	Y_2
0	1	-4
1	1.6667	-2.5
2	2.3333	-1
3	3	.5
4	3.6667	2
5	4.3333	3.5
6	5	5

X=0

36.

X	Y_1	Y_2
0	-11	-11
1	-4	-4
2	3	3
3	10	10
4	17	17
5	24	24
6	31	31

X=0

37.

X	Y_1	Y_2
0	-11	-9
1	-4	-2
2	3	5
3	10	12
4	17	19
5	24	26
6	31	33

X=0

38.

X	Y_1	Y_2
-2	-3.5	-4.25
-1.5	-2.25	-2.75
-1	-1	-1.25
-.5	.25	.25
0	1.5	1.75
.5	2.75	3.25
1	4	4.75

X= -2

In Exercises 39–42, solve each system of linear equations by using the substitution method.

39. $2x + y = 15$
$3x + 5y = -2$

40. $5x + 15y = 11$
$x + 3y = 11$

41. $x = 3y - 4$
$0.8x + 2.4y = 3.2$

42. $x = 4y + 3$
$0.8x + 3.2y = -4.0$

In Exercises 43–48, solve each system of linear equations by using the addition method.

43. $2x - y = -1$
$3x - 2y = -7$

44. $3x + 2y = 5$
$5x - 4y = 8$

45. $\frac{x}{2} + \frac{y}{3} = 1$
$\frac{x}{4} + \frac{y}{5} = 8$

46. $1.2x + 2.3y = 168$
$2.5x - 3.1y = -123.5$

47. $4x - 7y = 3$
$4x - 7y = 8$

48. $7x + 10y = 4$
$14x - 5y = -7$

In Exercises 49–52, use the slope-intercept form of each line to determine the number of solutions of the system. Then classify each system as a consistent system of independent equations, an inconsistent system, or a consistent system of dependent equations.

49. $y = \frac{2}{9}x - 7$
$y = -\frac{2}{9}x + 7$

50. $y = \frac{2}{9}x - 7$
$y = \frac{2}{9}x + 7$

51. $6x + 2y = 4$
$15x = 10 - 5y$

52. $\frac{x}{6} + \frac{y}{4} = \frac{1}{3}$
$\frac{x}{3} + \frac{y}{2} = \frac{1}{4}$

In Exercises 53–56, select the system of linear equations (choice A, B, C or D) that represents the word problem. Then solve this system by the substitution method and answer the problem, using a full sentence.

Verbally	Algebraically
53. Find two numbers whose sum is 100 and whose difference is 20.	**A.** $x + y = 120$ $x - y = 20$
54. Find two numbers whose sum is 20 if the first number is 100 more than the second number.	**B.** $x + y = 100$ $x - y = 20$
55. Find two numbers whose sum is 20 and whose difference is 20.	**C.** $x + y = 20$ $x - y = 20$
56. Find two numbers whose sum is 120 and whose difference is 20.	**D.** $x + y = 20$ $x - y = 100$

57. A student who has already completed 45 credit-hours with a GPA of 2.80 at Chemeketa Community College needs to raise her GPA to 3.00 to qualify for a scholarship. How many credit-hours with a grade of A would be required to meet her goal? How many grade points will she have if she meets her goal? (Assume an A = 4 points.)

In Exercises 58–62, solve each problem by using a system of equations and each step of the word problem strategy outlined in this chapter.

58. Fixed and Variable Printing Costs A mathematics department is preparing a color supplement for its beginning algebra class. Two options are presented to the department chair. Option 1 is to pay 35 cents per page to a local printer. Option 2 is to buy a color printer for \$2,200 and print the supplement in the department for a variable cost of 13 cents per page. What total page count would produce the same cost for both options?

59. Supplementary Angles The angles labeled x and y in the figure are supplementary. (Their sum is 180°.) The larger angle is 42° more than twice the smaller angle. Determine the number of degrees in each angle.

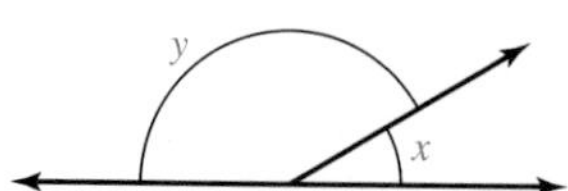

60. Mean and Range of Factory Measurements A quality-control worker in a factory measured 10 wheels for shopping carts. The diameters of the first nine were all equal, but the tenth had a defect and was too large. The mean of these measurements was 25.05 cm and the range was 0.75 cm. Use this information to determine each measurement.

61. Interest on Two Loans A student borrowed a total of \$9,000 on two loans—an 8% car loan and a 5% educational loan. The first month her total monthly interest on these two loans was \$45. Determine the amount of each loan.

62. Dimensions of an Isosceles Triangle The perimeter of an isosceles triangle (two equal sides) is 37 cm. The base is 5 cm shorter than each of the other two sides. Find the length of the base. (See the figure.)

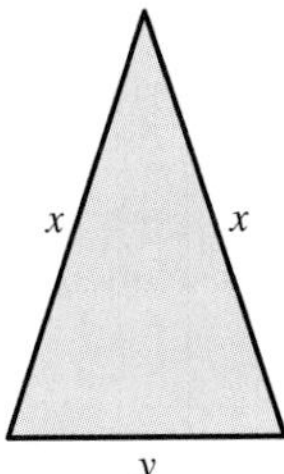

63. Rates of Two Trains Two trains depart simultaneously from a station, traveling in opposite directions. One averages 8 km/h more than the other, and after $\frac{1}{2}$ hour they are 100 km apart. Determine the speed of each train.

64. Rates of Two Workers A union steward observed two bricklayers working on a current job. On Monday she observed that these two bricklayers each worked 8 hours and laid a total of 384 bricks. On Tuesday the older bricklayer worked 7 hours and the younger bricklayer worked 5 hours, and they laid a total of 292 bricks. Determine for the union steward the rate of work for each bricklayer, assuming that both bricklayers work at a fairly consistent rate.

65. Mixture of Chemicals Petrochemicals worth \$6/L are mixed with other chemicals worth \$9/L to produce 100 L of chemicals worth \$7.65/L. How many liters of each are used?

66. Costs of Two Music Clubs Two popular music clubs have an introductory offer. Club A charges a \$6 initiation fee plus \$0.95 per CD. Club B does not charge any initiation fee but charges \$1.95 per CD. The following graph compares the cost of membership in each club based on the number of CDs purchased.

a. Give the equation in slope-intercept form for the line representing the costs for club A.

b. Interpret the meaning of the slope and the y-intercept of the line for club A.

c. Give the equation in slope-intercept form for the line representing the costs for club B.

d. Interpret the meaning of the slope and the y-intercept of the line for club B.

e. Determine the exact solution to the corresponding system of equations.

f. Interpret the meaning of the x- and y-coordinates of this solution.

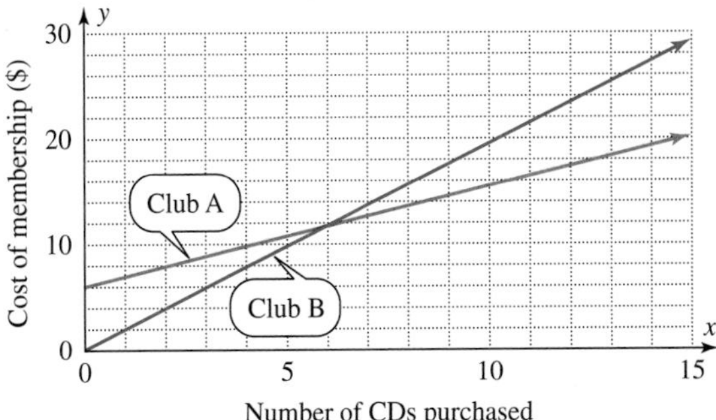

67. Automotive Service Charges The following tables display the charges by two automotive repair shops based on the number of hours required for a repair.

a. Give the equation in slope-intercept form for the line representing the costs for shop A.

b. Interpret the meaning of the slope and the y-intercept of the line for shop A.

c. Give the equation in slope-intercept form for the line representing the costs for shop B.

d. Interpret the meaning of the slope and the y-intercept of the line for shop B.

e. Give the solution to the corresponding system of equations.

f. Interpret the meaning of the x- and y-coordinates of this solution.

Shop A

HOURS x	COST y ($)
1	50
2	80
3	110
4	140

Shop B

HOURS x	COST y ($)
1	56
2	82
3	108
4	134

MASTERY TEST FOR CHAPTER 3

[3.1] **1.** Calculate the slope of the line through the given points.

a. (3, −1) and (5, 3)

b. (3, −1) and (1, 5)

c. Determine the slope of the line graphed.

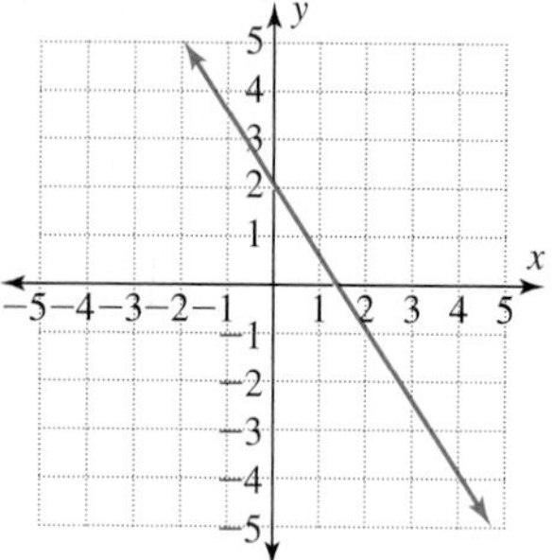

d. Determine the slope of the line containing all the points in the table.

x	y
−5	1
5	5
10	7
20	11
35	17
50	23

e. Determine the slope of the line defined by $y = -3x + 6$.

f. Determine the slope of the line defined by $y - 4 = 7(x + 1)$.

[3.1] **2. Rate of Ascent** The altitude of a model airplane is recorded every 5 seconds after liftoff. These data are displayed in the given table.

	A	B
1	Time (s)	Altitude (ft)
2	0	0
3	5	20
4	10	40
5	15	60
6		

a. What is the value of Δx shown in column A of the table?

b. What is the value of Δy shown in column B of the table?

c. Determine the slope of the line containing these points.

d. Interpret the meaning of this rate of change.

e. At this rate, how long after the plane begins its ascent will it reach an altitude of 1,200 ft?

[3.1] **3.** Determine whether the line defined by the first equation is parallel to, perpendicular to, or neither parallel nor perpendicular to the line defined by the second equation.

a. $y = \frac{1}{2}x - 3$
$y = -2x + 3$

b. $y = \frac{1}{2}x - 3$
$y = \frac{1}{2}x + 3$

c. $y = 3$
$x = 4$

d. $y = 2x + 3$
$y = -2x + 3$

[3.2] **4.** Write in slope-intercept form the equation of a line satisfying the given conditions.

a. Through (1, 4) with slope −2

b. Through (−1, 3) with slope 4

c. y-intercept (0, 5) and slope $\frac{2}{3}$

d. y-intercept (0, −2) and slope $-\frac{5}{3}$

[3.2] **5. a.** Write the equation of a horizontal line through (−4, 3).

b. Write the equation of a vertical line through (−4, 3).

c. Write the equation of a line through (−4, 3) and parallel to $x = 5$.

d. Write the equation of a line through (−4, 3) and perpendicular to $x = 5$.

[3.2] **6.** Graph a line satisfying the given conditions.

a. Through (3, 1) with slope −1

b. Through (0, 0) with slope $\frac{1}{3}$

c. y-intercept (0, 2) and slope $-\frac{3}{4}$

d. y-intercept (0, −4) and slope $\frac{5}{3}$

[3.3] **7.** Graphically determine the simultaneous solution for each system of linear equations.

a.

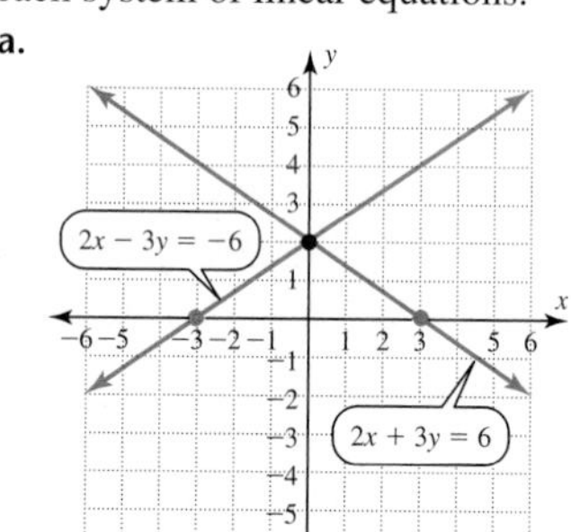

b. $f(x) = 2x + 1$
$f(x) = x + 3$

Use a numerical table to determine the simultaneous solution for each system of linear equations.

c.

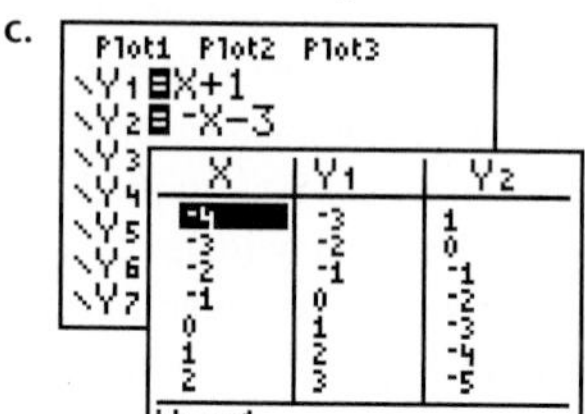

d. $y = 2x - 19$
$y = 3x - 31$

[3.3] **8.** Select the choice A, B, or C that best describes each system of linear equations.

a. $f(x) = 2x - 3$
$f(x) = 3x + 7$

b. $f(x) = 2x - 3$
$f(x) = 2x + 7$

c. $f(x) = 2x - 3$
$f(x) = 2x - 3$

A. A consistent system of dependent equations

B. A consistent system of independent equations

C. An inconsistent system of independent equations

[3.4] **9.** Solve each system of linear equations by the substitution method.

a. $y = -4x - 1$
$2x + y = 3$

b. $y = 4x - 1$
$2x + 3y = -2$

c. $x + 3y = 1$
$2x - 3y = 5$

d. $x - 2y = 5$
$5x + 10y = 12$

[3.5] **10.** Solve each system of linear equations by the addition method.

a. $3x - 7y = -13$
$2x + 7y = 3$

b. $3x + 19y = 16$
$6x - 11y = -17$

c. $3x - 2y = 6$
$-6x + 4y = -12$

d. $y = \frac{3}{5}x + 2$
$9x - 15y = 20$

[3.6] **11.** Use systems of linear equations to solve each word problem.

a. Complementary angles: Two angles are complementary. If the larger angle is 12° more than 5 times the smaller, determine the number of degrees in each angle.

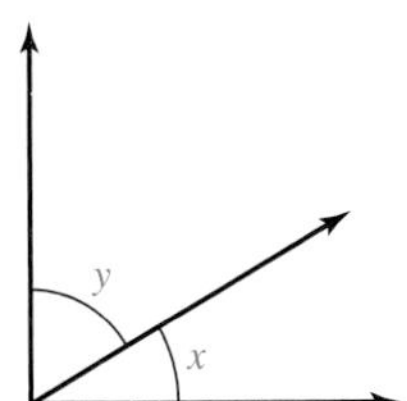

b. Value of two investments: An investment of \$10,000 earned a net income of \$625 in 1 year. Part of the investment was in bonds and earned income at a rate of 7%. The rest of the investment was in a savings account that earned income at a rate of 4%. How much was invested in bonds and how much in the savings account?

c. Mixture of two medicines: The dosage of medicine ordered by a doctor is 25 mL of a 50% solution. A nurse has available a 30% solution and an 80% solution of this medicine. How many milliliters of each should be mixed to produce this 25-mL dosage?

d. Rates of two planes: A jetplane and a tanker plane are 350 mi apart. They head toward each other so the jet can refuel. The jet flies 250 mi/h faster than the tanker. Determine the speed of each aircraft if they meet in 30 minutes.

REALITY CHECK FOR CHAPTER 3

One disadvantage of wind power has been that electricity generated from other sources costs less. The table gives the cost per kilowatt-hour for wind-generated electricity for several years. By using x to represent the number of years since 1950, the equation $y = -1.5x + 86$ can be used to model these data.

Cost for Wind-Generated Electricity

YEAR	CENTS/kWh
1980	40
1984	37
1988	31
1992	25
1996	20
2000	10

1. Determine the slope of this line and interpret the meaning of this rate of change.
2. Use this equation to estimate the cost of wind-generated electricity in 1970.
3. Use this equation to estimate the cost of wind-generated electricity in 2004.

GROUP PROJECT FOR CHAPTER 3

OSHA* Standards for Stair-Step Rise and Run

This project requires a tape measure, a ruler, or a yardstick.

The OSHA standards for stairs for single-family residences vary from the standards for other buildings. The standards for single-family residences set a maximum on the rise of 8 in and a minimum on the run of 9 in.

OSHA Standard 1910.24

Definitions:

Rise: *The vertical distance from the top of a tread to the top of the next-higher tread.*

Run: *The horizontal distance from the leading edge of a tread to the leading edge of an adjacent tread.*

*Occupational Safety and Health Act.

1. There are 10 steps to the first landing in the stairs of the home of the author James Hall. Each of these steps has a rise of 8 in and a run of $9\frac{3}{4}$ in.
   a. Do these stairs meet the stated standard for single-family residences?
   b. What is the total rise in feet and inches of these 10 steps?
   c. What is the total run in feet and inches of these 10 steps?
   d. Calculate to the nearest hundredth the slope of the line from the leading edge of one step to the leading edge of the next step.
2. a. Using the OSHA definitions, measure the rise and run of a set of stairs on your campus.
   b. How many steps are on this set of stairs?
   c. What is the total rise of these steps?
   d. What is the total run of these steps?
   e. Calculate the slope of the line from the leading edge of one step to the leading edge of the next step.
   f. Which stairs are steeper, those given for the Hall house or those in your school?

CUMULATIVE REVIEW FOR CHAPTERS 1 THROUGH 3

The answer to each of these questions follows this diagnostic review. Each answer is keyed to an example in this book. You can refer to these examples to find explanations and additional exercises for practice.

Arithmetic Review

In Exercises 1–9, calculate the value of each expression without using a calculator.

1. a. $12 + (-4)$ b. $12 - (-4)$
   c. $12(-4)$ d. $12 \div (-4)$
2. a. $-12 + (-6)$ b. $-12 - (-6)$
   c. $-12(-6)$ d. $-12 \div (-6)$
3. a. $-12 + 0$ b. $-12 - 0$
   c. $-12(0)$ d. $-12 \div 0$
4. a. $\frac{4}{5} + \frac{3}{10}$ b. $\frac{4}{5} - \frac{3}{10}$
   c. $\frac{4}{5}\left(\frac{3}{10}\right)$ d. $\frac{4}{5} \div \frac{3}{10}$
5. a. 2^3 b. 3^2
   c. $(-1)^6$ d. $\left(\frac{2}{3}\right)^2$
6. a. $|27|$ b. $|-27|$
   c. $-|27|$ d. $|27 - 27|$
7. a. $\sqrt{49}$ b. $\sqrt{\frac{4}{9}}$
   c. $\sqrt{9} + \sqrt{16}$ d. $\sqrt{9 + 16}$
8. a. $\sqrt{2}\sqrt{8}$ b. $\sqrt{2}\sqrt{32}$
   c. $\frac{\sqrt{63}}{\sqrt{7}}$ d. $\frac{\sqrt{48}}{\sqrt{3}}$
9. a. $-0.9 + 0.03$ b. $-0.9 - 0.03$
   c. $-0.9(0.03)$ d. $-0.9 \div 0.03$

Order of Operations

In Exercises 10–14, calculate the value of each expression without using a calculator.

10. a. $-13 + 14 - 7 + 8$ b. $-13 + 14 - (7 + 8)$
    c. $-(13 + 14) - (7 + 8)$ d. $-(13 + 14 - 7) + 8$
11. a. $6 - 4(11 - 8)$ b. $(6 - 4)11 - 8$
    c. $(6 - 4)(11 - 8)$ d. $6 - 4(11) - 8$
12. a. $\frac{2 + 3}{6 + 9}$ b. $2 + \frac{3}{6 + 9}$
    c. $\frac{2}{6} + \frac{3}{9}$ d. $2 + \frac{3}{6} + 9$
13. a. $-5^2 + 3^2 + 4^2$ b. $(-5)^2 + 3^2 + 4^2$
    c. $(-5 + 3 + 4)^2$ d. $-(5 + 3 + 4)^2$
14. a. $4 - 6 \div 2 + 5^2$ b. $(4 - 6) \div 2 + 5^2$
    c. $4 - (6 \div 2 + 5)^2$ d. $(4 - 6 \div 2 + 5)^2$

Evaluating Algebraic Expressions

In Exercises 15–18, evaluate each expression for $x = -4$, $y = -9$, and $z = -16$ without using a calculator.

15. a. $x + y$ b. $x - y$
    c. xy d. $\frac{z}{x}$
16. a. $x + y - z$ b. $x - (y + z)$
    c. $-(x + y + z)$ d. $|x| + |y| + |z|$
17. a. $2x - y$ b. $x^2 - y$
    c. $x^2 - y^2$ d. $(x - y)^2$
18. a. $-2y + z$ b. $-2y - z$
    c. $x^2 - 3x - 5$ d. $\frac{y + z}{x + 1}$

Properties and Subsets of the Real Numbers

Identify all the numbers from the set

$\left\{-7, -4.73, -\pi, -\sqrt{4}, -\frac{3}{7}, 0, \sqrt{7}, 17\right\}$ that are

19. a. Rational numbers b. Irrational numbers
    c. Integers d. Natural numbers

Name the property that justifies each statement.

20. a. $5(x + z) = 5x + 5z$ b. $5(x + z) = 5(z + x)$
    c. $5(x + z) = (x + z)(5)$ d. $5 + (x + z) = (5 + x) + z$
21. a. The additive identity is ________.
    b. The additive inverse of 7 is ________.
22. a. The multiplicative inverse of 7 is ________.
    b. The multiplicative inverse of $-\frac{3}{5}$ is ________.

Interval Notation

Write the interval notation for each of these inequalities.

23. a. $2 \le x < 4$ b. $-2 < x \le 5$
    c. $-1 \le x \le 0$ d. $x \ge 3$

Write the interval notation for each of these intervals.

24. a. (number line) −4 −3 −2 −1 0 1 2
    b. (number line) −1 0 1 2 3 4 5
    c. $x < -1$ d. x is greater than or equal to 5

Sequences

25. Calculate the first five terms of each sequence starting with $n = 1$.
    a. $a_n = n + 3$ b. $a_n = 3n$
    c. $a_n = n^2 - 1$ d. $a_n = 8 - n$
26. Determine whether each sequence is an arithmetic sequence. If the sequence is arithmetic, write the common difference.
    a. 0, 3, 6, 9, 12 b. 8, 6, 4, 2, 0
    c. 1, 3, 9, 27, 81 d. 3, 3, 3, 3, 3

Functions and Function Notation

27. Evaluate each expression given $f(x) = 5x - 3$.
    a. $f(0)$ b. $f(1)$
    c. $f(-1)$ d. $f(10)$
28. Use this table to evaluate each expression.

x	$f(x)$
−2	1
0	3
2	6
3	−2

    a. $f(-2)$ b. $f(3)$
    c. $f(x) = 3; x = ?$ d. $f(x) = -2; x = ?$

Function Models

29. A 16-ft board has two pieces of length x cut off.

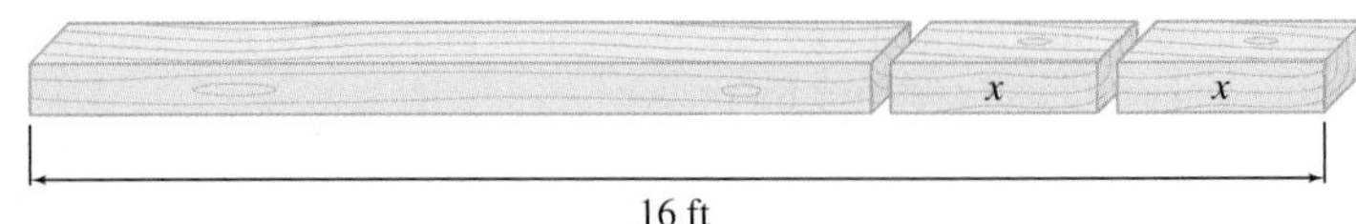

a. Write a function for the length of the remaining piece in terms of x.
b. What restrictions should be placed on the variable x?
c. Evaluate and interpret $f(3)$.

30. A college student started a tuition savings account with an initial \$1,000 deposit on January 1, 2006, and then made contributions of \$150 on the last Friday of each month.
a. Write an equation for the nth term of a sequence that gives the total contributions made to the account by the student by the end of the nth month.
b. Use this equation to complete this table.

n	3	6	9	12
a_n				

Solving Equations

31. Check $x = 5$ to determine whether it is a solution of each equation.
a. $2x - 4 = x + 1$ **b.** $2(x - 4) = x - 7$

In Exercises 32 and 33 solve each linear equation.

32. a. $2x + 4 = x + 1$ **b.** $2(x - 4) = x - 7$
c. $2(x - 5) = 2x + 9$ **d.** $2(x - 5) = 2x - 10$

33. a. $-x = 23$ **b.** $3(x - 4) = x - 2$
c. $5(x - 4) = 2(x - 1)$ **d.** $\frac{x}{2} = \frac{x}{3} - 2$

34. Solve each equation for y.
a. $4x - 2y = 6$ **b.** $4(x - 2y) = 3(2x - y) + 5$

35. Use the following table to solve $1.15x + 2.5 = 3.65x - 1.5$.

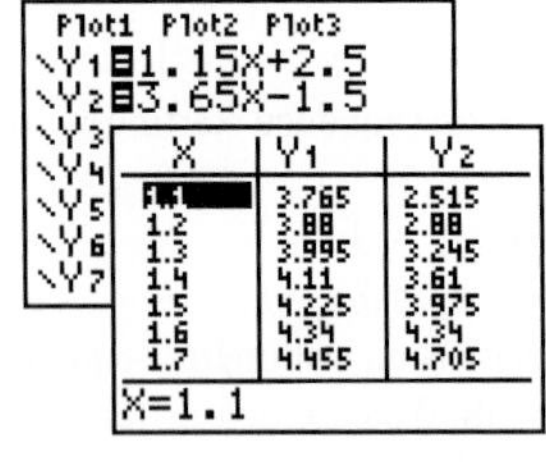

36. Use the following graph to solve $1.5x + 5.75 = -0.5x - 0.25$.

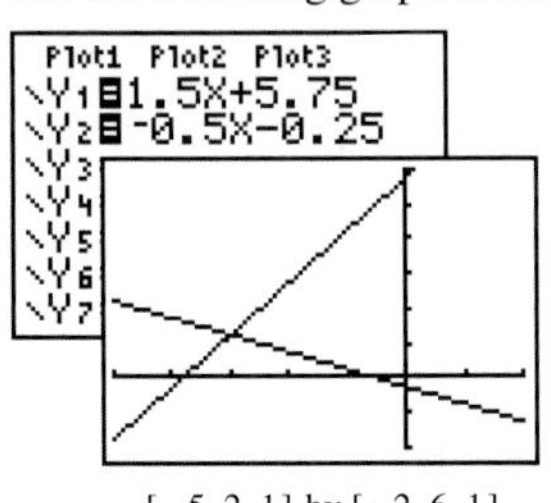

[−5, 2, 1] by [−2, 6, 1]

In Exercises 37 and 38 simplify each expression in part **a** and solve each equation in part **b**.

	Simplify	Solve
37.	**a.** $2(x - 3) + 4(x - 3)$	**b.** $2(x - 3) + 4(x - 3) = 0$
38.	**a.** $5(x + 4) + 3(x + 4)$	**b.** $5(x + 4) + 3(x + 4) = 0$

39. Determine whether the ordered pair $(-1, 2)$ is a solution of each linear equation.
a. $3x + y = -1$ **b.** $y = -2x + 4$
c. $x = -1$ **d.** $y = 2$

Graphs of Linear Equations

40. Determine the x- and y-intercepts of each line.

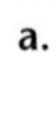

a.

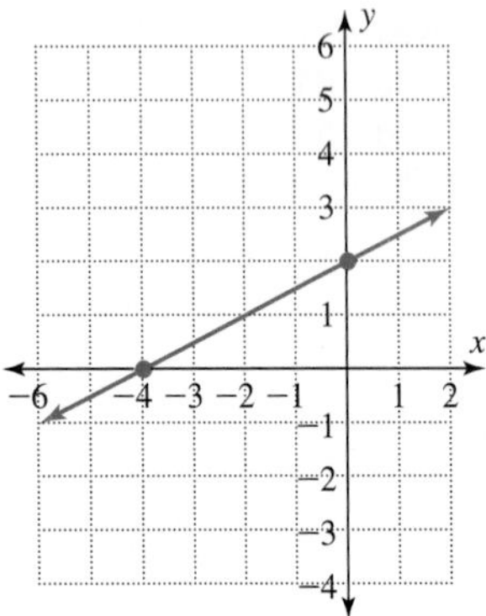

b.

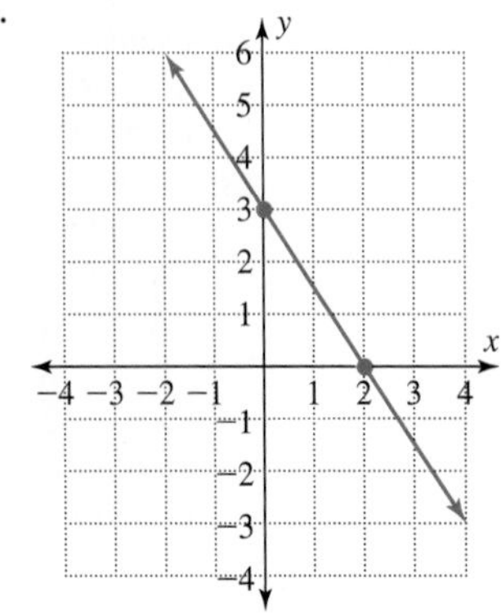

c. $y = 5x + 3$ **d.** $3x - 4y = 12$

41. Determine the point of intersection of these lines.

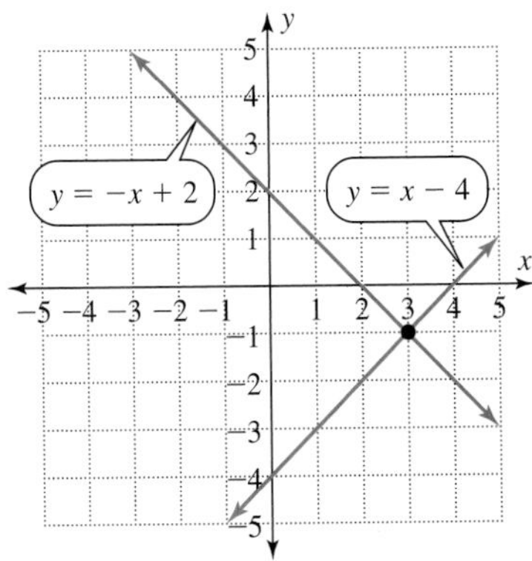

42. The table below gives values for two linear equations with $Y_1 = 5x + 3$ and $Y_2 = 4x + 2$. Determine the point that satisfies both equations.

Plot1 Plot2 Plot3
\Y1 = 5X+3
\Y2 = 4X+2

X	Y1	Y2
-3	-12	-10
-2	-7	-6
-1	-2	-2
0	3	2
1	8	6
2	13	10
3	18	14

X=-3

43. Graph each line.
   a. The line through (2, 4) and (−1, −3)
   b. The line with slope $m = \frac{3}{4}$ and with a y-intercept of (0, 4)
   c. The line with slope $m = -\frac{3}{2}$ and through (1, 2)
   d. The line defined by $3x - y = 6$

44. Match each equation with the description that best describes the graph of this linear equation.
   a. $y = \frac{1}{2}x + 5$
   b. $y = -2x + 3$
   c. $y = -2$
   d. $x = -3$

   A. A horizontal line
   B. A vertical line
   C. A line parallel to $y = -2x + 5$
   D. A line perpendicular to $y = -2x + 5$

45. Determine the slope of each line.
   a. The line through (2, 3) and (4, 1)
   b. $y = \frac{3}{5}x - 4$
   c. $y = 4$
   d. $x = 4$

46. Write the equation of a line with a slope $m = 3$ and a y-intercept $(0, -4)$.

47. Write the equation of a line with a slope $m = -2$ through (2, 6).

48. Write the equation of a line through (1, 1) and (2, 3).

49. Write the equation of a horizontal line through (2, 3).

50. Write the equation of a vertical line through (2, 3).

51. Write the equation of a line through (0, 4) and parallel to $y = -3x + 7$.

Systems of Linear Equations

52. Select the choice that best describes each system of linear equations.
   a. $y = 4x - 5$
      $y = 5x + 4$
   b. $y = 4x - 5$
      $y = 4x + 5$
   c. $y = 4x - 5$
      $y = 4x - 5$

   A. A consistent system of independent equations
   B. A consistent system of dependent equations
   C. An inconsistent system of independent equations

53. Solve this system of linear equations by the substitution method.

$$y = 2x + 3$$
$$3x + 2y = 20$$

54. Solve this system of linear equations by the addition method.

$$3x - 5y = 7$$
$$3x + 5y = -13$$

55. Solve this system of linear equations by the addition method.

$$-3x + 5y = 8$$
$$6x + 7y = 1$$

Applications of Equations

56. **Direct Variation** If y varies directly as x and y is 8 when x is $\frac{1}{4}$, find y when x is 2.

57. **Complementary Angles** Two angles are complementary. If the larger angle is 6° more than 6 times the smaller, determine the number of degrees in each angle.

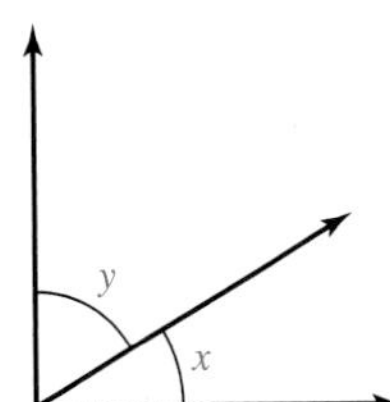

58. **Amount of Two Investments** A high school student invested $2,000 of summer income to save for college. A portion was deposited in a savings account that earned only 4%. The rest was invested in a local electrical utility and earned 8%. The total income for 1 year was $136. How much did the student put in the savings account?

59. **Rates of Two Airplanes** A jetplane and a tanker plane that are 450 mi apart head toward each other so that the jet can refuel. The jet flies 100 mi/h faster than the tanker. Determine the speed of each aircraft if they meet in 45 minutes. (*Hint:* Use consistent units of measurement.)

60. **Mixture of Medicine** The dosage of a medicine ordered by a doctor is 40 mL of a 22% solution. A nurse has available a 10% solution and a 25% solution of this medicine. How many milliliters of each should be mixed to prepare this 40-mL dosage?

61. **Tree Shadow** A man 2 m tall casts a 6-m shadow. Nearby is a tree that simultaneously casts a 39-m shadow. How tall is the tree?

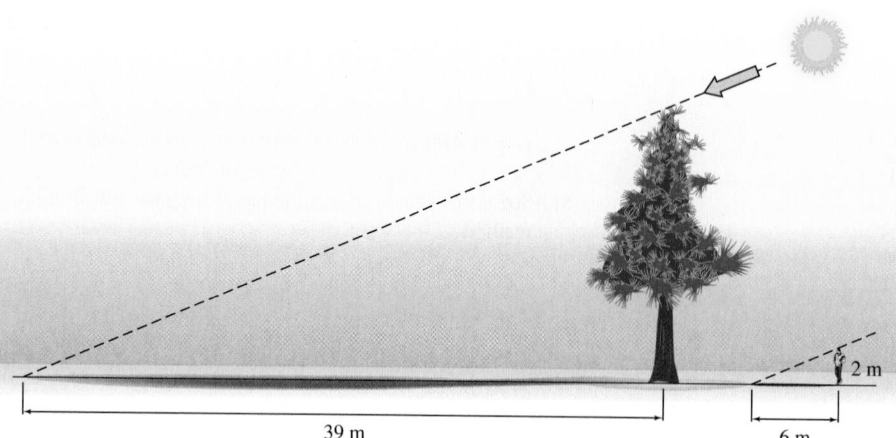

ANSWERS for the Cumulative Review for Chapters 1–3

Question	Answer	Reference Example
1. a	8	[1.3–2]
1. b	16	[1.4–2]
1. c	−48	[1.5–2]
1. d	−3	[1.6–1]
2. a	−18	[1.3–2]
2. b	−6	[1.4–2]
2. c	72	[1.5–2]
2. d	2	[1.6–1]
3. a	−12	[1.2–2]
3. b	−12	[1.4–3]
3. c	0	[1.5–2]
3. d	Undefined	[1.6–3]
4. a	$\frac{11}{10}$	[1.3–4]
4. b	$\frac{1}{2}$	[1.4–4]
4. c	$\frac{6}{25}$	[1.5–2]
4. d	$\frac{8}{3}$	[1.6–1]
5. a	8	[1.5–7]
5. b	9	[1.5–7]
5. c	1	[1.5–7]
5. d	$\frac{4}{9}$	[1.5–7]
6. a	27	[1.2–3]
6. b	27	[1.2–3]
6. c	−27	[1.2–3]
6. d	0	[1.2–3]
7. a	7	[1.2–6]
7. b	$\frac{2}{3}$	[1.2–7]
7. c	7	[1.2–6]
7. d	5	[1.2–6]
8. a	4	[1.5–4]
8. b	8	[1.5–4]
8. c	3	[1.6–4]
8. d	4	[1.6–4]
9. a	−0.87	[1.3–3]
9. b	−0.93	[1.4–4]
9. c	−0.027	[1.5–2]
9. d	−30	[1.6–1]
10. a	2	[1.7–2]
10. b	−14	[1.7–2]
10. c	−42	[1.7–2]
10. d	−12	[1.7–2]
11. a	−6	[1.7–2]
11. b	14	[1.7–2]
11. c	6	[1.7–2]
11. d	−46	[1.7–2]
12. a	$\frac{1}{3}$	[1.7–1]
12. b	$\frac{11}{5}$	[1.7–1]
12. c	$\frac{2}{3}$	[1.7–1]

Question	Answer	Reference Example
12. d	$\frac{23}{2}$	[1.7–1]
13. a	0	[1.7–3]
13. b	50	[1.7–3]
13. c	4	[1.7–3]
13. d	−144	[1.7–3]
14. a	26	[1.7–2]
14. b	24	[1.7–2]
14. c	−60	[1.7–2]
14. d	36	[1.7–2]
15. a	−13	[1.3–8]
15. b	5	[1.3–8]
15. c	36	[1.5–4]
15. d	4	[1.6–4]
16. a	3	[1.3–8]
16. b	21	[1.4–7]
16. c	29	[1.4–7]
16. d	29	[1.3–8]
17. a	1	[1.7–1]
17. b	25	[1.7–1]
17. c	−65	[1.7–1]
17. d	25	[1.7–1]
18. a	2	[1.7–1]
18. b	34	[1.7–1]
18. c	23	[1.7–1]
18. d	$\frac{25}{3}$	[1.7–1]
19. a	$-7, -4.73, -\sqrt{4}, -\frac{3}{7}, 0, 17$	[1.2–8]
19. b	$-\pi, \sqrt{7}$	[1.2–8]
19. c	$-7, -\sqrt{4}, 0, 17$	[1.2–8]
19. d	17	[1.2–8]
20. a	Distributive property of multiplication over addition	[1.7–5]
20. b	Commutative property of addition	[1.3–6]
20. c	Commutative property of multiplication	[1.5–1]
20. d	Associative property of addition	[1.3–6]
21. a	0	[1.2–2]
21. b	−7	[1.2–1]
22. a	$\frac{1}{7}$	[1.5–5]
22. b	$-\frac{5}{3}$	[1.5–5]
23. a	[2, 4)	[1.2–5]
23. b	(−2, 5]	[1.2–5]
23. c	[−1, 0]	[1.2–5]
23. d	[3, ∞)	[1.2–5]
24. a	(−3, 1]	[1.2–5]
24. b	(−∞, 2)	[1.2–5]
24. c	(−∞, −1)	[1.2–5]

continued

Question	Answer	Reference Example
24. d	$[5, \infty)$	[1.2–5]
25. a	4, 5, 6, 7, 8	[2.1–7]
25. b	3, 6, 9, 12, 15	[2.1–7]
25. c	0, 3, 8, 15, 24	[2.1–7]
25. d	7, 6, 5, 4, 3	[2.1–7]
26. a	Arithmetic, $d = 3$	[2.1–6]
26. b	Arithmetic, $d = -2$	[2.1–6]
26. c	Not arithmetic	[2.1–6]
26. d	Arithmetic, $d = 0$	[2.1–6]
27. a	-3	[2.2–1]
27. b	2	[2.2–1]
27. c	-8	[2.2–1]
27. d	47	[2.2–1]
28. a	1	[2.2–4]
28. b	-2	[2.2–4]
28. c	0	[2.2–4]
28. d	3	[2.2–4]
29. a	$f(x) = 16 - 2x$	[2.2–6]
29. b	The variable x must be restricted to values in the interval $[0, 8]$.	[2.8–2]
29. c	$f(3) = 10$; cutting off two 3-ft pieces from the 16-ft board will leave a 10-ft piece remaining.	[2.2–6]
30. a	$a_n = 150n + 1{,}000$	[2.1–7]
30. b	n: 3, 6, 9, 12; a_n: 1,450, 1,900, 2,350, 2,800	[2.1–7]
31. a	Solution	[1.4–8]
31. b	Not a solution	[1.4–8]
32. a	$x = -3$	[2.5–6]
32. b	$x = 1$	[2.5–6]
32. c	No solution	[2.5–6]
32. d	Every real number is a solution.	[2.5–6]
33. a	$x = -23$	[2.5–6]
33. b	$x = 5$	[2.5–6]
33. c	$x = 6$	[2.5–6]
33. d	$x = -12$	[2.5–8]
34. a	$y = 2x - 3$	[2.6–1]
34. b	$y = -\frac{2}{5}x - 1$	[2.6–2]
35.	$x = 1.6$	[2.4–8]
36.	$x = -3$	[2.4–8]
37. a	$6x - 18$	[1.7–7]
37. b	$x = 3$	[2.5–6]
38. a	$8x + 32$	[1.7–7]
38. b	$x = -4$	[2.5–6]
39. a	Solution	[2.3–1]
39. b	Not a solution	[2.3–1]
39. c	Solution	[2.3–1]
39. d	Solution	[2.3–1]
40. a	$(-4, 0), (0, 2)$	[2.3–4]
40. b	$(2, 0), (0, 3)$	[2.3–4]
40. c	$\left(-\frac{3}{5}, 0\right), (0, 3)$	[2.6–3]
40. d	$(4, 0), (0, -3)$	[2.6–3]
41.	$(3, -1)$	[2.3–6]
42.	$(-1, -2)$	[2.3–8]

Question	Answer	Reference Example
43. a	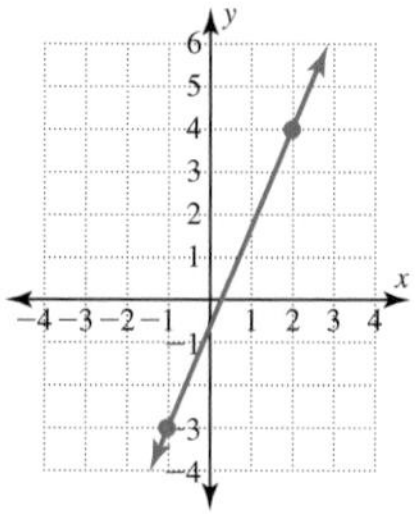 [graph]	[3.1–4]
43. b	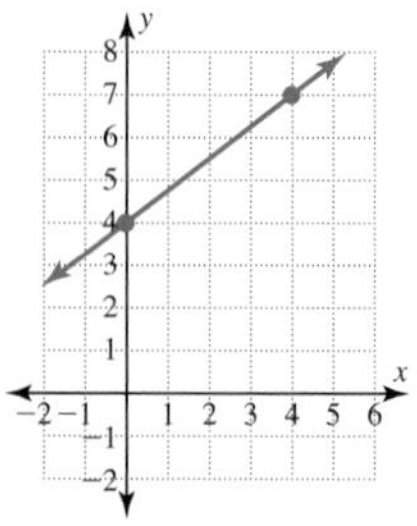 [graph]	[3.1–4]
43. c	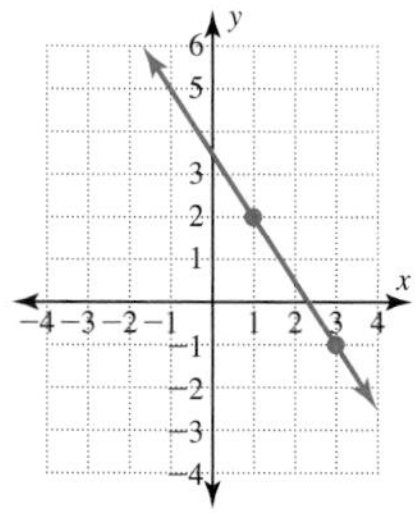 [graph]	[3.1–4]
43. d	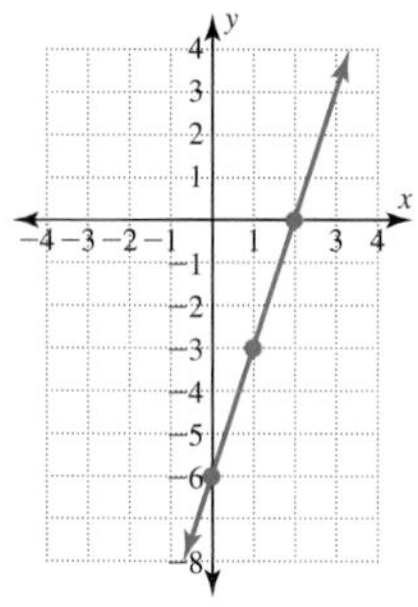 [graph]	[3.1–6]
44. a	D	[3.2–9]
44. b	C	[3.2–9]
44. c	A	[3.6–6]
44. d	B	[3.2–6]
45. a	-1	[3.1–1]
45. b	$\frac{3}{5}$	[3.2–2]
45. c	0	[3.2–6]
45. d	Undefined	[3.2–6]
46.	$y = 3x - 4$	[3.2–1]
47.	$y = -2x + 10$	[3.2–8]
48.	$y = 2x - 1$	[3.2–10]
49.	$y = 3$	[3.2–6]
50.	$x = 2$	[3.2–6]
51.	$y = -3x + 4$	[3.2–1]
52. a	A	[3.3–6]
52. b	C	[3.3–6]
52. c	B	[3.3–7]

Question	Answer	Reference Example
53.	(2, 7)	[3.4–1]
54.	$(-1, -2)$	[3.5–1]
55.	$(-1, 1)$	[3.5–2]
56.	$y = 64$	[2.7–6]
57.	12°, 78°	[3.6–1]
58.	$600 was invested in the savings account.	[3.6–4]
59.	The tanker flew 250 mi/h, and the jet flew 350 mi/h.	[3.6–5]
60.	Use 8 mL of 10% solution and 32 mL of 25% solution.	[3.6–6]
61.	The tree is 13 m tall.	[2.7–7]

chapter 4

Linear Inequalities and Systems of Linear Inequalities

CHAPTER OUTLINE

Golden Gate Bridge:
Width: 90 ft
Total length: 8,981 ft (1.7 mi)
Weight of bridge: 894,500 tons
Supporting cables: 2
Wires in each cable: 27,572
Diameter of each cable: 36.375 in

We rely on the strength of cables more than we may realize. Cranes, elevators, ski lifts, and many theme park rides use cables to move cargo or occupants safely from place to place. Mathematics plays a key role in designing cables for required strengths.

Gigantic cables support many of the most magnificent bridges in the world. Anyone driving across the graceful span of the Golden Gate Bridge places a vital trust in the two huge cables that suspend the bridge high over San Francisco Bay. These cables had to be designed and built to hold the enormous weight of the bridge as well as the traffic moving across it. The graph gives the minimum breaking strength for different size strands of cable.

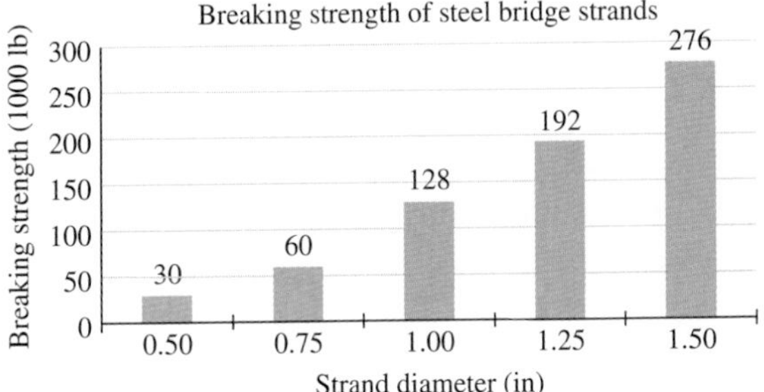

For more information on this topic, go to the Additional Resources link on the MathZone website at **www.mhhe.com/hallmercer.**

REALITY CHECK

Some bridge cables are composed of several strands with each strand containing several wires. The strength of a standard galvanized steel bridge strand depends on the number and size of the individual wires in the strand. Use the graph to determine the strength of a single strand of diameter 0.50 in as shown in the figure.

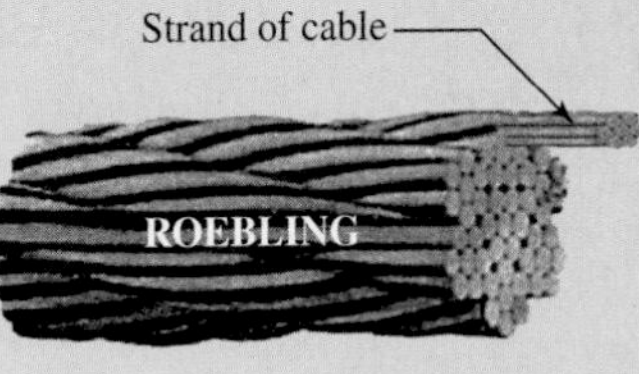

Section 4.1 Solving Linear Inequalities by Using the Addition-Subtraction Principle

Objectives:

1. Check a possible solution of an inequality.
2. Use tables and graphs to solve linear inequalities in one variable.
3. Solve linear inequalities in one variable by using the addition-subtraction principle for inequalities.

A Mathematical Note

The symbols > and < for greater than or less than are due to Thomas Harriot (1631). These symbols were not immediately accepted, as many mathematicians preferred the symbols ⌐ and ¬.

In Exercise 14 at the end of this section, we examine the fiber cable strength needed for a sailboat. The manufacturer specifies that each cable must support a minimum of 200 lb. We will use linear inequalities to examine this problem.

A linear equation in one variable can be written in the form $Ax + B = C$. If we replace this equality symbol with an inequality symbol, the result is a **linear inequality in one variable** of the form $Ax + B < C$, $Ax + B \le C$, $Ax + B > C$, or $Ax + B \ge C$. In this section we present three methods for solving linear inequalities—a graphical method, a numerical method, and an algebraic method. Although the steps used to solve linear inequalities are very similar to the steps used to solve linear equations, there are some important distinctions. We will build on these similarities and carefully point out these distinctions.

Linear Inequalities

VERBALLY	ALGEBRAICALLY	ALGEBRAIC EXAMPLES	GRAPHICALLY
A linear inequality in one variable is an inequality that is first-degree in that variable.	For real constants A, B, and C, with $A \neq 0$.		
	$Ax + B > C$	$x > 2$	(at 2, shaded right; −1 0 1 2 3 4 5
	$Ax + B \ge C$	$x \ge 2$	[at 2, shaded right; −1 0 1 2 3 4 5
	$Ax + B < C$	$x < 2$	) at 2, shaded left; −1 0 1 2 3 4 5
	$Ax + B \le C$	$x \le 2$	] at 2, shaded left; −1 0 1 2 3 4 5

In the previous box, remember that a parenthesis indicates that an endpoint is not included in the interval. A bracket indicates that the endpoint is included in the interval.

A **conditional inequality** contains a variable and is true for some, but not all, real values of the variable. A **solution of an inequality** is a value that makes the inequality a true statement. To solve an inequality is to find all the solutions of the inequality.

EXAMPLE 1 Checking Possible Solutions of an Inequality

Determine whether the given values are solutions of $6x - 2 < 5x - 4$.

SOLUTION

(a) $x = -3$

$$6x - 2 < 5x - 4$$
$$6(-3) - 2 \stackrel{?}{<} 5(-3) - 4$$
$$-18 - 2 \stackrel{?}{<} -15 - 4$$
$$-20 \stackrel{?}{<} -19 \text{ is true}$$

Substitute the given value for x in the inequality and determine if this makes the inequality a true statement.

Answer: -3 is a solution.

(b) $x = -2$

$$6x - 2 < 5x - 4$$
$$6(-2) - 2 \stackrel{?}{<} 5(-2) - 4$$
$$-12 - 2 \stackrel{?}{<} -10 - 4$$
$$-14 \stackrel{?}{<} -14 \text{ is false}$$

If the inequality were $6x - 2 \leq 5x - 4$, then $x = -2$ would be a solution.

Answer: -2 is not a solution.

SELF-CHECK 4.1.1

1. Determine which of the values -5, -4, 4, and 5 are solutions of the inequality in Example 1.

Since inequalities often have an infinite interval of solutions, it is common to represent these solutions with a graph or by using interval notation (see Section 1.2).

EXAMPLE 2 Identifying Solutions of an Inequality from a Graph

Determine which of the values -3, -2, -1, 0, 1, 2, and 3 are solutions of the graphed inequality.

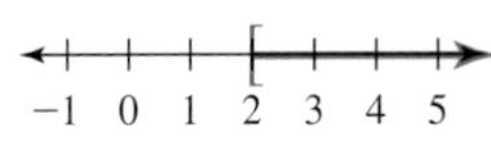

SOLUTION

2 and 3 are solutions

This graph represents the inequality $x \geq 2$. The graph includes 2 and all real numbers to the right of 2. The values -3, -2, -1, 0, and 1 are not shaded and are not solutions of the inequality.

The solution of a linear inequality in one variable can be determined graphically by letting y_1 represent the left side of the inequality and y_2 represent the right side of the inequality. In Example 3 we will solve $2x - 5 \leq 1$ by letting y_1 represent $2x - 5$ and y_2 represent 1 and then using a graph to determine where $y_1 \leq y_2$.

EXAMPLE 3 Using a Graph to Solve a Linear Inequality

Solve $2x - 5 \leq 1$ by graphing $y_1 = 2x - 5$ and $y_2 = 1$.

SOLUTION

The key point is to determine where the graph of y_1 is below or intersects the graph of y_2. Then write the x-values that make this true.

First graph both $y_1 = 2x - 5$ and $y_2 = 1$. Then determine the point of intersection of these two lines; in this case the point (3, 1). Observe where $y_1 \leq y_2$. This is where the graph of y_1 is below y_2 or intersects y_2; in this case for $x \leq 3$.

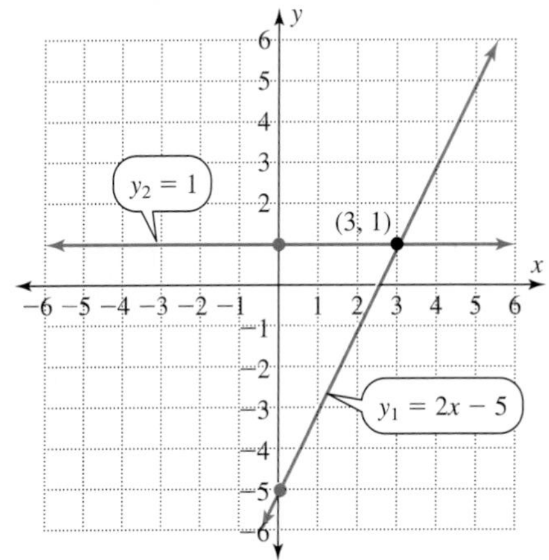

Answer: $(-\infty, 3]$

Note that the solution consists of the interval of values (number line with endpoint 3, shaded to the left). This is the portion of the x-axis for which y_1 is below y_2; $y_1 \leq y_2$. The endpoint of this interval is the x-value where y_1 and y_2 are equal.

SELF-CHECK 4.1.2

1. Solve $1.5x + 1 \geq -2$ by graphing $y_1 = 1.5x + 1$ and $y_2 = -2$.
2. Determine which of the values $-3, -2, 0, 2$, and 3 are solutions of this inequality.

Tables can also be used to solve linear inequalities. In the table shown we use y_1 to represent the left side of $2x + 1 > -x + 4$ and y_2 to represent the right side. By filling in the third column of this table with $<$, $=$, or $>$, we can determine the order relation between y_1 and y_2 and determine the x-values that satisfy $2x + 1 > -x + 4$.

Original Table

x	y_1	$<, =,$ OR $>$	y_2
−3	−5		7
−2	−3		6
−1	−1		5
0	1		4
1	3		3
2	5		2
3	7		1

Table with Inequality Symbols

	x	y_1	$<, =,$ OR $>$	y_2
For these x-values $y_1 < y_2$.	−3	−5	$<$	7
	−2	−3	$<$	6
	−1	−1	$<$	5
	0	1	$<$	4
For this x-value $y_1 = y_2$.	1	3	$=$	3
For these x-values $y_1 > y_2$.	2	5	$>$	2
	3	7	$>$	1

Note from the table that $y_1 = y_2$ at $x = 1$ and that $y_1 > y_2$ for $x > 1$. Thus the solution of $2x + 1 > -x + 4$ is the interval of x-values $(1, \infty)$.

Caution: The solution of inequalities by tables is limited to an approximation if the point where $y_1 = y_2$ is not displayed in the table.

If you use a calculator to generate a table similar to the one just illustrated, you may still want to write in the inequality symbols between the y_1 and y_2 columns to assist you in determining the solution of the inequality. If the table does not reveal the x-value where $y_1 = y_2$, then you cannot determine the exact solution from the table. This is especially true if $y_1 = y_2$ at a value of x such as $x = \frac{3}{7}$.

Example 4 examines a linear inequality graphically and numerically. Note the connections between these perspectives. Then we begin to discuss the use of algebraic methods to solve inequalities.

EXAMPLE 4 Solving a Linear Inequality Graphically and Numerically

Use a graphing calculator to solve $x - 2 > -x - 6$ both graphically and numerically.

SOLUTION

Let $Y_1 = x - 2$
$Y_2 = -x - 6$

Let Y_1 represent the left side of the inequality and Y_2 represent the right side of the inequality.

GRAPHICALLY

Then select an appropriate table and window for your calculator.

The graph shows that $Y_1 = Y_2$ at $x = -2$ and Y_1 is above Y_2 for $x > -2$.

This is confirmed by using an algebraic method to solve the equation that gives the x-coordinate of the point of intersection.

$$x - 2 = -x - 6$$
$$x - 2 + x = -x - 6 + x$$
$$2x - 2 = -6$$
$$2x - 2 + 2 = -6 + 2$$
$$2x = -4$$
$$x = -2$$

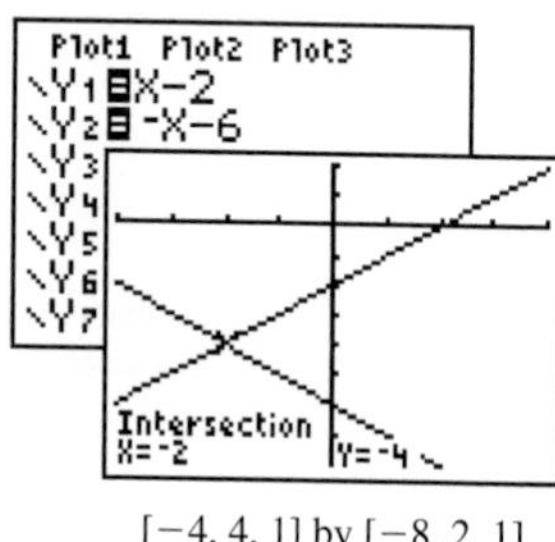

[−4, 4, 1] by [−8, 2, 1]

NUMERICALLY

X	Y1	Y2
-3	-5	-3
-2	-4	-4
-1	-3	-5
0	-2	-6
1	-1	-7
2	0	-8
3	1	-9

X=-3

Table with Inequality Symbols

x	Y_1	<, =, OR >	Y_2
-3	-5	$<$	-3
-2	-4	$=$	-4
-1	-3	$>$	-5
0	-2	$>$	-6
1	-1	$>$	-7
2	0	$>$	-8
3	1	$>$	-9

The table also shows that $Y_1 = Y_2$ at $x = -2$ and $Y_1 > Y_2$ for $x > -2$.

VERBALLY

The graph of Y_1 is above the graph of Y_2 for $x > -2$. In the table $Y_1 = Y_2$ at $x = -2$ and $Y_1 > Y_2$ for $x > -2$.

Answer: $(-2, \infty)$ whose graph is

-2

SELF-CHECK 4.1.3

The following table was created on a TI-84 Plus calculator.

1. Place $<$, $=$, or $>$ between the Y_1 and Y_2 columns.
2. Determine the x-values that satisfy $Y_1 < Y_2$.

X	Y1	Y2
1	7	8
1.1	7.3	8.1
1.2	7.6	8.2
1.3	7.9	8.3
1.4	8.2	8.4
1.5	8.5	8.5
1.6	8.8	8.6

X=1

The procedure for solving a linear inequality algebraically is very similar to the procedure we already used to solve linear equations. The basic idea is to isolate the variable terms on one side of the inequality and the constant terms on the other side. Inequalities that have the same solution are called **equivalent inequalities.** We use the addition-subtraction principle to produce equivalent inequalities that are simpler than the original inequality.

Addition-Subtraction Principle for Inequalities

VERBALLY	ALGEBRAICALLY*	NUMERICAL EXAMPLE
If the same number is added to or subtracted from both sides of an inequality, the result is an equivalent inequality.	If a, b, and c are real numbers, then $a < b$ is equivalent to $a + c < b + c$ and to $a - c < b - c$.	$x - 2 < 5$ is equivalent to $x - 2 + 2 < 5 + 2$ and to $x < 7$.

*Similar statements can be made for the inequalities $\leq$, $>$, and $\geq$.

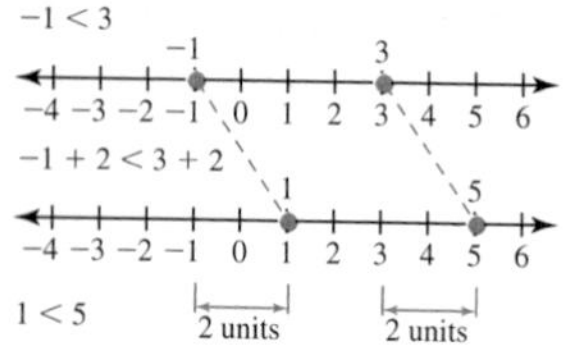

Figure 4.1.1

The logic behind the addition-subtraction principle can be explained by examining the two graphs in Figure 4.1.1. Adding 2 to both sides of an inequality shifts both points to the right 2 units. Thus their position relative to each other is preserved. Likewise, subtracting 3 from both sides of an inequality shifts both points to the left 3 units. Thus their position relative to each other is also preserved.

In Example 5 we first subtract $5v$ from both sides of the inequality and then we add 2 to both sides of the inequality.

EXAMPLE 5 Solving a Linear Inequality with Variables on Both Sides

Solve $6v - 2 < 5v - 3$, and graph the solution set.

SOLUTION

$$6v - 2 < 5v - 3$$

$$6v - 5v - 2 < 5v - 5v - 3 \quad \text{Subtract } 5v \text{ from both sides of the inequality.}$$

$$v - 2 < -3$$

$$v - 2 + 2 < -3 + 2 \quad \text{Add 2 to both sides of the inequality.}$$

$$v < -1$$

Answer: $(-\infty, -1)$ whose graph is [number line: −4 −3 −2 −1 0 1 2, parenthesis at −1, shaded left]

The strategy for solving linear inequalities is much the same as the one used to solve linear equations. At each step, try to produce simpler expressions than on the previous step. In Example 6 we start by using the distributive property to remove parentheses. The solution $2 \geq x$ is then rewritten in the equivalent form $x \leq 2$. It is common to write answers so that x is the subject of the sentence, as in "x is less than or equal to 2."

EXAMPLE 6 Solving a Linear Inequality Containing Parentheses

Solve $5(x + 2) \geq 3(2x + 3) - 1$ and graph the solution set.

SOLUTION

$$5(x + 2) \geq 3(2x + 3) - 1$$

$$5x + 10 \geq 6x + 9 - 1$$

$$5x + 10 \geq 6x + 8$$

Use the distributive property to remove the parentheses, and then combine like terms.

$$5x + 10 - 5x \geq 6x + 8 - 5x$$

Subtract $5x$ from both sides of the inequality.

$$10 \geq x + 8$$

$$10 - 8 \geq x + 8 - 8$$

Subtract 8 from both sides of the inequality.

$$2 \geq x \quad \text{or} \quad x \leq 2$$

The statements $a \geq b$ and $b \leq a$ have the same meaning.

Answer: $(-\infty, 2]$ whose graph is [number line: −1 0 1 2 3 4 5, bracket at 2, shaded left]

SELF-CHECK 4.1.4

Solve each inequality and graph the solution set.

1. $2x + 4 > x$ **2.** $3(x - 1.5) - 2 < 2(x - 2)$

All the exercises in this section can be solved by using the addition-subtraction principle to produce inequalities whose variables have a coefficient of 1. We will work with other coefficients, including negative ones, in Section 4.2. In Example 7 the same result is obtained algebraically, graphically, and numerically.

EXAMPLE 7 Using Multiple Perspectives to Solve a Linear Inequality

Solve $\frac{x}{2} - 1 \geq 1 - \frac{x}{2}$ algebraically, graphically, and numerically and describe the solution verbally.

SOLUTION

ALGEBRAICALLY

$$\frac{x}{2} - 1 \geq 1 - \frac{x}{2}$$
$$\frac{x}{2} - 1 + \frac{x}{2} \geq 1 - \frac{x}{2} + \frac{x}{2}$$
$$x - 1 \geq 1$$
$$x - 1 + 1 \geq 1 + 1$$
$$x \geq 2$$

GRAPHICALLY

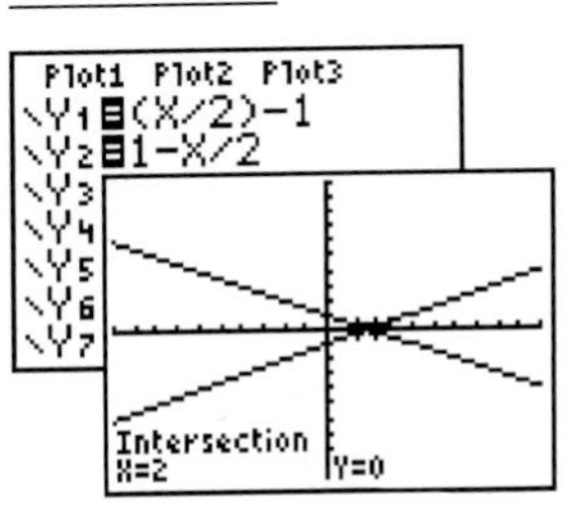

[−10, 10, 1] by [−10, 10, 1]

NUMERICALLY

X	Y1	Y2
0	-1	1
1	-.5	.5
2	0	0
3	.5	-.5
4	1	-1
5	1.5	-1.5
6	2	-2

X=2

VERBALLY

All values of x greater than or equal to 2 satisfy this inequality. From the graph, Y_1 is on or above Y_2 for $x \geq 2$. In the table $Y_1 \geq Y_2$ for $x \geq 2$.

Answer: $[2, \infty)$ whose graph is [number line from −1 to 5, shaded from 2 to the right with a bracket at 2]

−1 0 1 2 3 4 5

SELF-CHECK 4.1.5

1. Use a graphing calculator to complete the given table and graph.

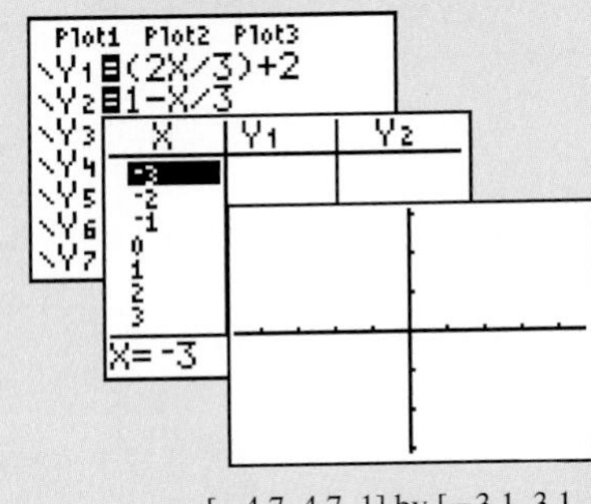

[−4.7, 4.7, 1] by [−3.1, 3.1, 1]

2. Use this graph and table to solve $\frac{2x}{3} + 2 \leq 1 - \frac{x}{3}$.

Some of the more common phrases used to indicate inequalities are given in the following box.

Phrases Used to Indicate Inequalities

PHRASE	VERBAL MEANING	INEQUALITY NOTATION	INTERVAL NOTATION	GRAPHICAL NOTATION
■ "x is at least a" ■ "x is a minimum of a"	x is greater than or equal to a	$x \geq a$	$[a, \infty)$	a
■ "x is at most a" ■ "x is a maximum of a" ■ "x never exceeds a"	x is less than or equal to a	$x \leq a$	$(-\infty, a]$	a
■ "x exceeds a"	x is greater than a	$x > a$	(a, ∞)	a
■ "x is smaller than a"	x is less than a	$x < a$	$(-\infty, a)$	a

Remember that correct usage of interval notation requires the smaller value on the left and the larger value on the right. For example, the notation $(2, 7]$ is correct while $[7, 2)$ is incorrect.

EXAMPLE 8 Translating Verbal Statements into Algebraic Inequalities

Write each of these statements in both inequality notation and interval notation, and then graph the solution of this inequality.

SOLUTION

	INEQUALITY	INTERVAL	GRAPH
(a) x is at least 4	$x \geq 4$	$[4, \infty)$	1 2 3 4 5 6 7
(b) x is at most 3	$x \leq 3$	$(-\infty, 3]$	0 1 2 3 4 5 6
(c) x exceeds 2	$x > 2$	$(2, \infty)$	−1 0 1 2 3 4 5
(d) x never exceeds 2	$x \leq 2$	$(-\infty, 2]$	−1 0 1 2 3 4 5
(e) x is a minimum of 5	$x \geq 5$	$[5, \infty)$	2 3 4 5 6 7 8

SELF-CHECK 4.1.6

Match each phrase with the corresponding algebraic inequality. (Some choices can be used more than once, and others may not be used.)

1. x exceeds 8 **A.** $x < 8$
2. x is at most 8 **B.** $x \leq 8$
3. x is at least 8 **C.** $x > 8$
4. x is a maximum of 8 **D.** $x \geq 8$

In Example 9 first we translate the verbally stated inequality to algebraic form. Then we solve this inequality. Note that the answer is written as a full sentence.

EXAMPLE 9 Solving a Verbally Stated Inequality

Determine all values of x so that four times the quantity x plus three exceeds eleven more than three times x.

SOLUTION

4 times the quantity x plus 3	exceeds	11 more than 3 times x
$4(x + 3)$	$>$	$3x + 11$

Write a verbal inequality and then translate this verbal inequality into an algebraic inequality.

$$4x + 12 > 3x + 11$$

Use the distributive property to remove the parentheses on the left side of the inequality.

$$4x + 12 - 3x > 3x + 11 - 3x$$
$$x + 12 > 11$$

Subtract $3x$ from both sides of the inequality.

$$x + 12 - 12 > 11 - 12$$
$$x > -1$$

Subtract 12 from both sides of the inequality.

Answer: The values of x are greater than or equal to -1.

It might be wise to check a couple of values. For example, -2 doesn't check and 2 does check.

SELF-CHECK ANSWERS

4.1.1

1. -5 and -4 are solutions.

4.1.2

1. $[-2, \infty)$
2. $-2, 0, 2,$ and 3 are solutions.

4.1.3

1. **Table with Inequality Symbols**

x	Y_1	<, =, or >	Y_2
1	7	<	8
1.1	7.3	<	8.1
1.2	7.6	<	8.2
1.3	7.9	<	8.3
1.4	8.2	<	8.4
1.5	8.5	=	8.5
1.6	8.8	>	8.6

2. In the table $Y_1 = Y_2$ at $x = 1.5$ and $Y_1 < Y_2$ for $x < 1.5$; this is the interval $(-\infty, 1.5)$ whose graph is

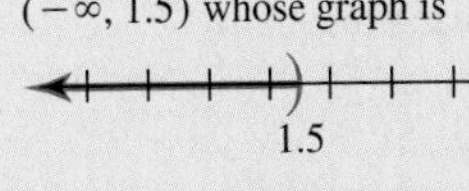

4.1.4

1. $(-4, \infty)$ whose graph is

−7 −6 −5 −4 −3 −2 −1

2. $(-\infty, 2.5)$ whose graph is

2.5

4.1.5

1. 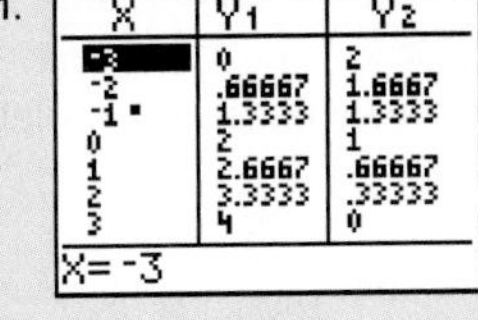

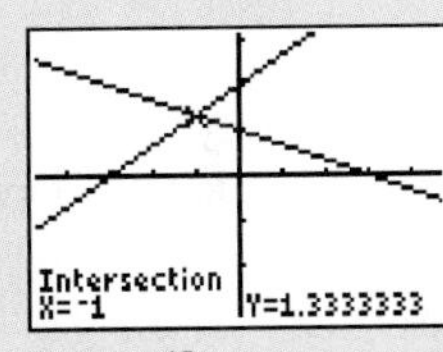

2. $(-\infty, -1]$

4.1.6

1. C 2. B 3. D 4. B

USING THE LANGUAGE AND SYMBOLISM OF MATHEMATICS 4.1

1. An inequality that is first degree in x is called a ________ inequality in x.
2. An inequality is first degree in x if the exponent on x is ________.
3. Inequalities with the same solution are called ________ inequalities.
4. A ________ inequality contains a variable and is true for some, but not all, real values of the variable.
5. A value that makes an inequality a true statement is a ________ of the inequality.
6. By the ________-subtraction principle, $a < b$ is equivalent to $a + c < b + c$.
7. By the addition-________ principle, $a < b$ is equivalent to $a - c < b - c$.
8. **a.** If $x > y$, then $x + 2$ ________ $y + 2$.
   b. If $x > y$, then $x - 2$ ________ $y - 2$.
   c. If $x < y$, then y ________ x.
9. The ________ property is often used to help us remove parentheses from expressions in an inequality.

10. The statement "x is at least five" is represented by the inequality x __________ 5.
11. The statement "x is at most five" is represented by the inequality x __________ 5.
12. The statement "x exceeds five" is represented by the inequality x __________ 5.
13. The statement "x is a maximum of five" is represented by the inequality x __________ 5.
14. The statement "x is a minimum of five" is represented by the inequality x __________ 5.

EXERCISES 4.1

In Exercises 1 and 2, determine which of the following are linear inequalities in one variable.

1. a. $6x - 3$ b. $6x - 3 \le 4x$
   c. $y = 6x - 3$ d. $x^2 \ge 36$
2. a. $4x + 1$ b. $x^2 > 4$
   c. $4x + 1 > 9$ d. $y \le 4x + 1$

In Exercises 3 and 4, express each interval by using interval notation.

3. a. $x \le -4$ b. $x > 5$
   c. (number line with a right parenthesis at 3, shaded to the left) d. x is at least -1
4. a. $x \ge 1.5$ b. $x < -3$
   c. (number line with a left parenthesis at -2, shaded to the right) d. x is at most 5

5. Determine if $x = 3$ is a solution of each inequality.
   a. $x < 3$ b. $x \le 3$
   c. $x > 3$ d. $x \ge 3$
6. Determine if $x = 2$ is a solution of each inequality.
   a. $x \le 2$ b. $x < 2$
   c. $x \ge 2$ d. $x > 2$
7. Determine if $x = 3$ is a solution of each inequality.
   a. $x > -3$ b. $x > 5$
   c. $3x - 5 \ge 2(x - 1)$ d. $5x - 3 \le 2x + 3$
8. Determine if $x = 2$ is a solution of each inequality.
   a. $x > 3$ b. $x \ge -1$
   c. $3x - 5 \ge 2(x - 1)$ d. $5x - 3 \le 2x + 3$

In Exercises 9–12, use the graph to determine the x-values that satisfy each equation and inequality.

9. a. $y_1 = y_2$
   b. $y_1 < y_2$
   c. $y_1 > y_2$

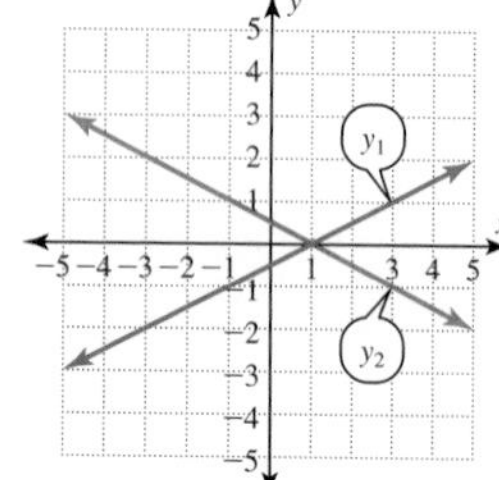

10. a. $y_1 = y_2$
    b. $y_1 < y_2$
    c. $y_1 > y_2$

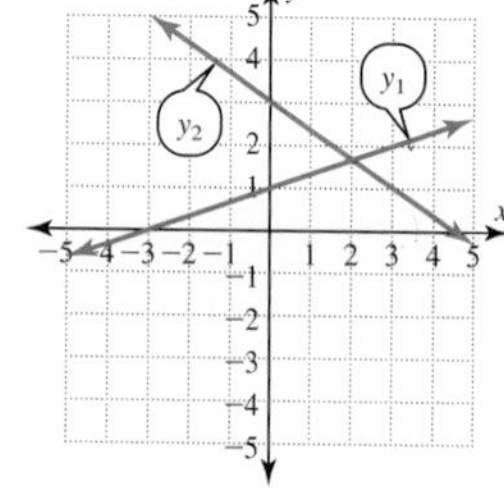

11. a. $y_1 = y_2$
    b. $y_1 \le y_2$
    c. $y_1 \ge y_2$

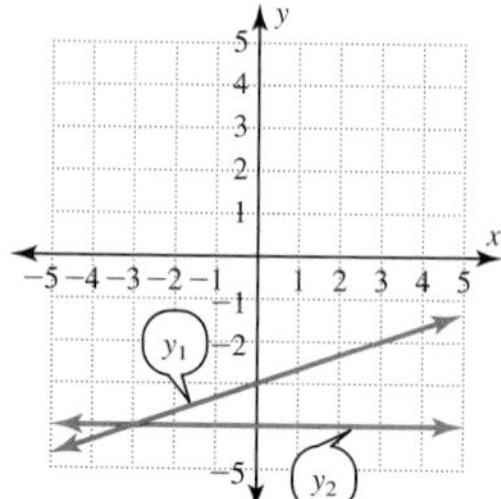

12. a. $y_1 = y_2$
    b. $y_1 \le y_2$
    c. $y_1 \ge y_2$

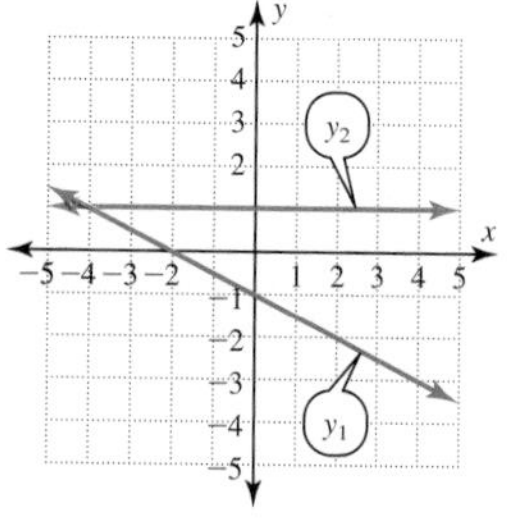

13. **Using a Graph That Models Overtime Spending Limits** Robinson Construction Company has budgeted a spending limit of $15,000 per year on overtime for its foremen. The graph of $y_1 = 15{,}000$ displays the spending limit, and the graph of $y_2 = f(x)$ displays the cost for overtime based upon the number of overtime hours worked by the foremen. The maximum possible number of hours of overtime by all the foremen is 800 hours. Use the graphs below to solve:
    a. $y_1 = y_2$ b. $y_1 < y_2$ c. $y_1 > y_2$
    d. Interpret the meaning of each of these solutions.

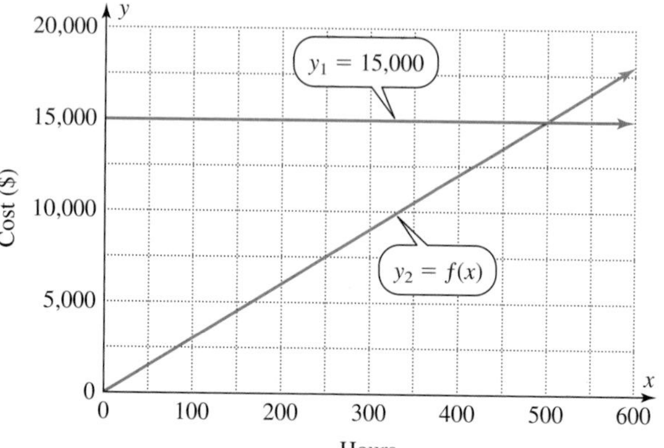

14. **Using a Graph That Models Strength Limits for a Cable** The contract for a composite fiber cable ordered for a sailboat from a cable manufacturer specifies that each cable must support a minimum load of 200 lb. The graph of $y_1 = 200$ displays the strength limit, and the graph of $y_2 = f(x)$

displays the strength of a cable based upon the number of separate fiber strands braided into the cable. Use the graphs below to solve:

a. $y_1 = y_2$ **b.** $y_1 < y_2$ **c.** $y_1 > y_2$

d. Interpret the meaning of each of these solutions.

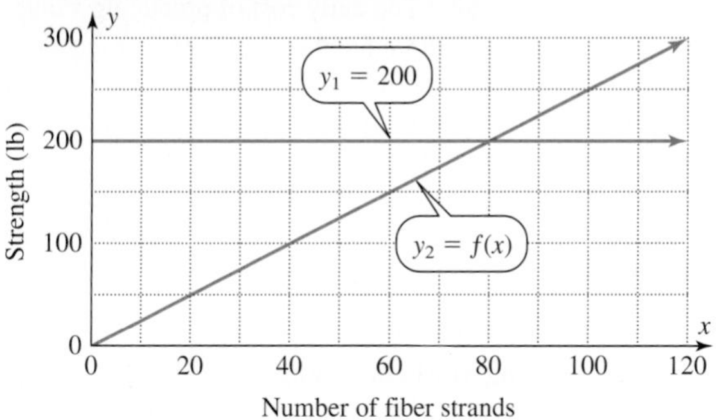

In Exercises 15–18, use the table of values to determine the x-values that satisfy each equation and inequality.

15. **a.** $Y_1 = Y_2$
b. $Y_1 \le Y_2$
c. $Y_1 \ge Y_2$

X	Y1	Y2
0	2	12
1	7	15
2	12	18
3	17	21
4	22	24
5	27	27
6	32	30
X=0		

16. **a.** $Y_1 = Y_2$
b. $Y_1 \le Y_2$
c. $Y_1 \ge Y_2$

X	Y1	Y2
0	-5	-9
1	-1	-3
2	3	3
3	7	9
4	11	15
5	15	21
6	19	27
X=0		

17. **a.** $Y_1 = Y_2$
b. $Y_1 < Y_2$
c. $Y_1 > Y_2$

X	Y1	Y2
-20	-67	-97
-15	-52	-72
-10	-37	-47
-5	-22	-22
0	-7	3
5	8	28
10	23	53
X= -20		

18. **a.** $Y_1 = Y_2$
b. $Y_1 < Y_2$
c. $Y_1 > Y_2$

X	Y1	Y2
1	-3	0
1.2	-2.4	-.4
1.4	-1.8	-.8
1.6	-1.2	-1.2
1.8	-.6	-1.6
2	0	-2
2.2	.6	-2.4
X=1		

19. Using Tables That Model Supply Costs The following two tables give the costs, including delivery, for ordering copier paper from two different office supply firms. Use these tables to solve:

a. $y_1 = y_2$ **b.** $y_1 < y_2$ **c.** $y_1 > y_2$

d. Interpret the meaning of each of the solutions from parts **a** to **c.**

Ace Office Supply

BOXES OF PAPER x	COST y_1 ($)
4	112
8	224
12	336
16	448
20	560
24	672

Max Office Supply

BOXES OF PAPER x	COST y_2 ($)
4	136
8	224
12	312
16	400
20	488
24	576

20. Using Tables That Model Bottle Production The following two tables give the number of glass bottles produced after a number of minutes by two different machines. For each machine there is an initial start-up time before any bottles will come out of either machine. Use these tables to solve:

a. $y_1 = y_2$ **b.** $y_1 < y_2$ **c.** $y_1 > y_2$

d. Interpret the meaning of each of the solutions from parts **a** to **c.**

Machine 1

MINUTES AFTER START-UP x	BOTTLES PRODUCED y_1
4	40
6	80
8	120
10	160
12	200
14	240
16	280

Machine 2

MINUTES AFTER START-UP x	BOTTLES PRODUCED y_2
4	20
6	70
8	120
10	170
12	220
14	270
16	320

In Exercises 21–24, algebraically solve each inequality and show each step of your solution process.

21. $x + 7 > 11$
22. $x - 7 < -11$
23. $2x + 3 \le x + 4$
24. $5x - 3 \ge 4x - 9$

Multiple Representations

In Exercises 25–28, solve each inequality algebraically, graphically, and numerically and describe the solution verbally. (*Hint:* See Example 7.)

25. $2x - 1 \ge x + 2$
26. $0.5x + 2 > -0.5x - 2$
27. $-x + 3 < -2x + 5$
28. $3(x - 1) \le 2(x - 2)$

In Exercises 29–44, solve each inequality algebraically.

29. $x - 7 \ge -3$
30. $x - 3 \le -7$
31. $5x - 8 < 4x - 8$
32. $11x + 12 > 10x + 12$
33. $7 \le x$
34. $-4 \ge x$
35. $x - 4 > 2x - 3$
36. $4x + 9 < 5x + 3$
37. $5 - 9x < 6 - 10x$
38. $12 - 11x < 10 - 12x$
39. $5(x - 2) \ge 4(x - 2)$
40. $7(x - 3) \ge 6(x - 4)$

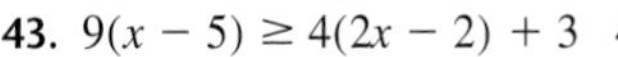

41. $5(2 - x) > 4(3 - x)$
42. $2(3 - 2x) > 3(2 - x)$
43. $9(x - 5) \ge 4(2x - 2) + 3$
44. $11(x + 4) \ge 5(2x + 3) - 7$

Estimation and Calculator Skills

In Exercises 45–48, mentally estimate the solution of each inequality to the nearest integer, and then use a calculator to calculate the solution.

PROBLEM	MENTAL ESTIMATE	CALCULATOR SOLUTION
45. $x - 0.727 > 0.385$		
46. $x + 29.783 \le 70.098$		
47. $2x + 25.87 < x + 3.94$		
48. $5x - 1.61 \ge 4x + 5.42$		

In Exercises 49–60, solve each inequality by the method of your choice.

49. $5(w - 2) + 14 > 4(w + 3)$
50. $7(x - 3) + 16 \ge 6(x - 2)$
51. $3(5n - 2) \le 7(2n - 1)$
52. $13(2n - 1) < 5(5n - 3)$
53. $4(6m + 7) - 2m \ge 7(3m + 1) + 1$
54. $5(2m + 8) + 3m > 3(4m + 5) - 4$
55. $-3(2 - m) < -4(3 - m) + 4.5$
56. $-11(2 - w) - 10.5 \le -12(3 - w)$
57. $4y - 4(5 + 2y) > y - 6(3 + y)$
58. $19 - 3(4 - y) \ge 11 - 4(1 - y)$
59. $\frac{1}{2}x - \frac{1}{3} \ge \frac{5}{3} - \frac{1}{2}x$
60. $\frac{2}{3}x - \frac{4}{7} < \frac{3}{7} - \frac{1}{3}x$

Multiple Representations

In Exercises 61–66, first write an algebraic inequality for each verbal statement, and then solve this inequality.

61. Five minus two times x is less than or equal to seven minus three times x.
62. Six plus four times x is greater than or equal to three times x plus five.
63. Twice w plus three results in a minimum of eleven more than w.
64. Five times w never exceeds three more than four times w.
65. Three times the quantity m plus five exceeds twelve more than twice m.
66. Six times the quantity m minus eleven is a maximum of ninety-six more than five times m.
67. Total Employees Because of cutbacks in the state budget, a driver's examination office is limited to employing at most 12 people. The office currently has 1 supervisor, 1 receptionist, and 2 clerks. The rest of the employees are driver's license examiners. How many driver's license examiners E can this office have?
68. Total Grade Points To earn a grade of B, a student needs her exam points to total at least 320 points. Her first three test grades are 71, 79, and 78. What score must she earn on the fourth test to obtain a B?
69. Perimeter of a Rectangular Pen A farmer has 84 ft of woven wire fence available to enclose three sides of the rectangular pen shown below. What are the possible values for the length x?

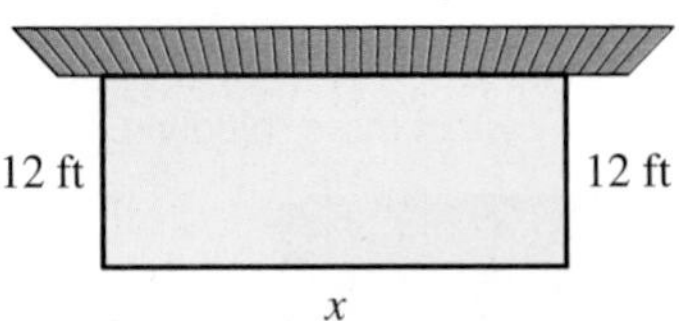

70. Perimeter of a Triangular Pen A buried pet security system is placed in the triangular corner of a yard. The pet owner can afford only 54 m of the buried wire. What are the possible values for the length x?

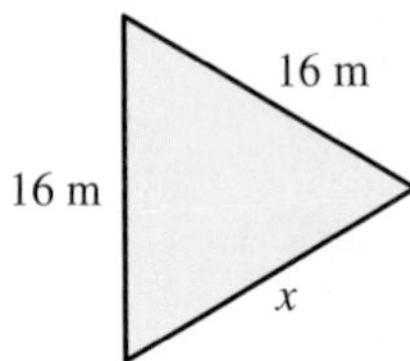

71. Costs and Revenue The daily cost of producing x units of computer mice includes a fixed cost of \$450 per day and a variable cost of \$4 per unit. The income produced by selling x units is \$5 per unit. Letting y_1 represent the income and y_2 represent the cost, graph y_1 and y_2. Determine the values of x for which $y_1 < y_2$, the loss interval for this company. Also determine the values of x for which $y_1 > y_2$, the profit interval for this company.
72. Costs and Revenue The daily cost of producing x units of cellular phones includes a fixed cost of \$600 per day and a variable cost of \$14 per unit. The income produced by selling x units is \$15 per unit. Letting y_1 represent the income and y_2 represent the cost, graph y_1 and y_2. Determine the values of x for which $y_1 < y_2$, the loss interval for this company. Also determine the values of x for which $y_1 > y_2$, the profit interval for this company.

Group Discussion Questions

73. Communicating Mathematically The sum of the lengths of any two sides of a triangle is greater than the length of the third side. Write three different inequalities expressing this concept for the triangle given below.

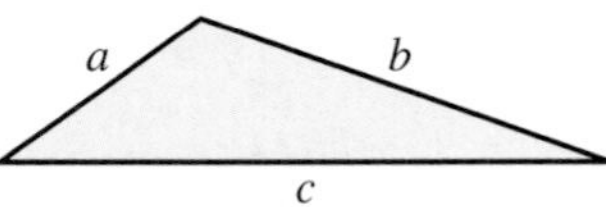

74. Communicating Mathematically Write an algebraic expression for each verbal expression. Then explain the different meanings of these expressions.
a. 3 more than x
b. 3 is more than x
c. 2 less than x
d. 2 is less than x
75. Challenge Question You have solved an inequality and tested one value from your solution. This value checked, yet your solution is incorrect. Give two distinct examples that illustrate how this is possible.

Section 4.2 Solving Linear Inequalities by Using the Multiplication-Division Principle

Objective: 4. Solve linear inequalities by using the multiplication-division principle for inequalities.

The specifications for many products place limits on the length, volume, weight, or strength of the product. For a tent it is important that each supporting cable meet minimum strength requirements. Engineers frequently use inequality notation to indicate these strength requirements. In Example 8 we use inequality notation to determine the number of synthetic fibers to put into a tent cable to meet the minimum strength requirements.

In this section we will expand our ability to solve linear inequalities in one variable. We use the addition-subtraction principle to isolate the variable terms on one side of an inequality and the constant terms on the other side. Then we use the multiplication-division principle to obtain a coefficient of 1 for the variable.

Multiplication-Division Principle for Inequalities

VERBALLY	ALGEBRAICALLY*	NUMERICAL EXAMPLES
Order-Preserving: If both sides of an inequality are multiplied or divided by a *positive number,* the result is an inequality that has the same solution as the original inequality.	If a, b, and c are real numbers and $c > 0$, then $a > b$ is equivalent to $ac > bc$.	$\frac{x}{2} > 3$ is equivalent to $2\left(\frac{x}{2}\right) > 2(3)$ and to $x > 6$
Order-Reversing: If both sides of an inequality are multiplied or divided by a *negative number and the order of inequality is reversed,* the result is an inequality that has the same solution as the original inequality.	If a, b, and c are real numbers and $c < 0$, then $a > b$ is equivalent to $ac < bc$.	$-\frac{x}{3} > 5$ is equivalent to $(-3)\left(-\frac{x}{3}\right) < (-3)(5)$ and to $x < -15$

*Similar statements can be made for division and for the inequalities $<$, $\leq$, and $\geq$.

The logic behind the multiplication-division principle can be understood by examining the two graphs in Figures 4.2.1 and 4.2.2.

An Order-Preserving Multiplication

$1 < 3$

$1(2) < 3(2)$

$2 < 6$

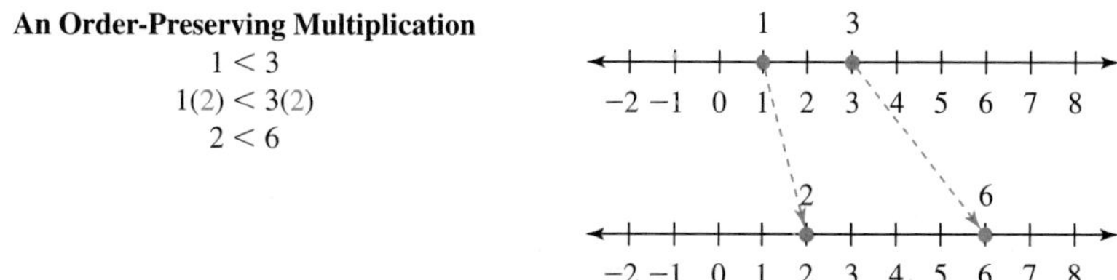

Figure 4.2.1 Multiplication-division principle with positive numbers.

Multiplying both sides of the inequality by +2 appears to stretch out the distance between the points, while their position relative to each other is preserved. Likewise,

One way to visually illustrate the order-reversing case is with a meter stick. Label the meter stick with $-$, 0, and $+$ and points a and b with a to the left of b. Multiplying by a -1 changes the $-$ side to $+$, and the $+$ side to $-$. Illustrate this by rotating the meter stick 180° about the 0 point. This reverses the order that points a and b now occur on the meter stick.

dividing by $+2$ would preserve the order relation, although the distance between the points would appear to shrink.

An Order-Reversing Multiplication

$$-1 < 3$$
$$-1(-2) > 3(-2)$$
$$2 > -6$$

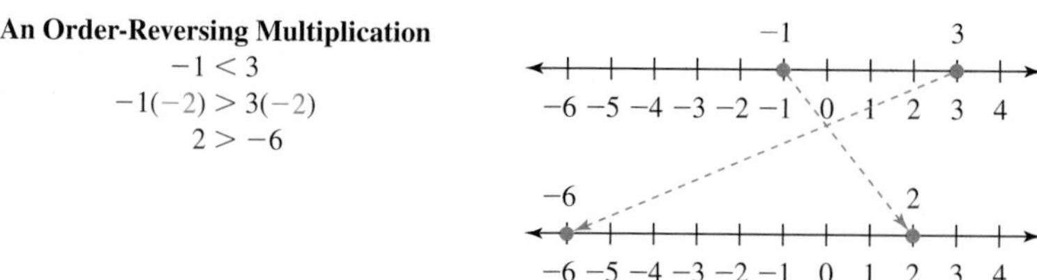

Figure 4.2.2 Multiplication-division principle with negative numbers.

Multiplying both sides of the inequality by -2 changes the sign of each side of the inequality and thus reverses the order of their positions relative to each other. Likewise, dividing by -2 would reverse the order relation.

EXAMPLE 1 Solving a Linear Inequality with a Positive Coefficient

Solve $3x \le 6$ and graph the solution set.

SOLUTION

$$3x \le 6$$
$$\frac{3x}{3} \le \frac{6}{3}$$ Dividing both sides of the inequality by $+3$ preserves the order.
$$x \le 2$$

Answer: $(-\infty, 2]$ whose graph is 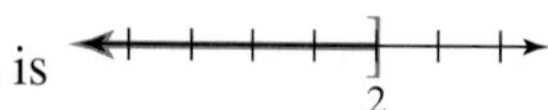

To solve an inequality for the variable w, we must isolate w on one side of the inequality, not $-w$. In Example 2 we solve for w by multiplying both sides of the inequality by -1. Note that this step reverses the order of the inequality.

EXAMPLE 2 Solving a Linear Inequality with a Coefficient of −1

Solve $-w < 15$, and graph the solution set.

SOLUTION

$$-w < 15$$
$$(-1)(-w) > (-1)(15)$$
$$w > -15$$

Multiplying both sides of the inequality by -1 reverses the order of the inequality. You can obtain the same result by dividing both sides of the inequality by -1.

Answer: $(-15, \infty)$ whose graph is (graph with parenthesis at -15)

−15

SELF-CHECK 4.2.1

Solve each inequality and graph the solution.

1. $6w \ge -12$ **2.** $-x < -13$ **3.** $-\frac{1}{2}y < 2$

If an inequality contains variables on both sides of the inequality, first we want to isolate the variable terms on one side of the inequality and the constant terms on the other side of the inequality. This is illustrated in Example 3.

EXAMPLE 3 Solving a Linear Inequality with Variables on Both Sides

Solve $5m - 7 < 3m - 13$.

SOLUTION

$$5m - 7 < 3m - 13$$

$$5m - 7 - 3m < 3m - 13 - 3m \quad \text{Subtracting } 3m \text{ preserves the order.}$$

$$2m - 7 < -13$$

$$2m - 7 + 7 < -13 + 7 \quad \text{Adding 7 preserves the order.}$$

$$2m < -6$$

$$\frac{2m}{2} < \frac{-6}{2} \quad \text{Dividing by } +2 \text{ preserves the order.}$$

$$m < -3$$

Answer: $(-\infty, -3)$ whose graph is [number line: −6 −5 −4 −3 −2 −1 0, shaded left of −3 with parenthesis at −3]

With practice some of the steps in Example 4 can be combined. This is illustrated by the Alternative Solution.

EXAMPLE 4 Solving a Linear Inequality with Variables on Both Sides

Solve $6x - 2 \geq 8x + 12$.

SOLUTION

$$6x - 2 \geq 8x + 12$$

$$6x - 2 - 8x \geq 8x + 12 - 8x \quad \text{Subtracting } 8x \text{ preserves the order.}$$

$$-2x - 2 \geq 12$$

$$-2x - 2 + 2 \geq 12 + 2 \quad \text{Adding 2 preserves the order.}$$

$$-2x \geq 14$$

$$\frac{-2x}{-2} \leq \frac{14}{-2} \quad \text{Dividing by } -2 \text{ reverses the order.}$$

$$x \leq -7$$

ALTERNATIVE SOLUTION

A shortened version of the solution process follows.

$$6x - 2 \geq 8x + 12$$

$$-2x \geq 14 \quad \text{Subtracting } 8x \text{ and adding 2 preserve the order.}$$

$$x \leq -7 \quad \text{Dividing by } -2 \text{ reverses the order.}$$

Answer: $(-\infty, -7]$ whose graph is [number line: −10 −9 −8 −7 −6 −5 −4, shaded left of −7 with bracket at −7]

SELF-CHECK 4.2.2

1. Solve $4y - 20 \leq 9y + 15$ and graph the solution.

The strategy for solving linear inequalities is much the same as that used to solve linear equations. At each step we try to produce simpler expressions than on the previous step. This is illustrated in Example 5, which begins by using the distributive property to remove parentheses.

EXAMPLE 5 Solving a Linear Inequality Containing Parentheses

Solve $6(3 - 4n) < 5(2 - 3n) - 28$.

A good rule of thumb to use when solving a linear inequality is to try to produce simpler expressions with each step.

SOLUTION

$6(3 - 4n) < 5(2 - 3n) - 28$

$18 - 24n < 10 - 15n - 28$ — Use the distributive property to remove the parentheses, and then combine like terms.

$18 - 24n < -15n - 18$ — Subtracting 18 preserves the order.

$-24n < -15n - 36$ — Adding $15n$ preserves the order.

$-9n < -36$ — Dividing by -9 reverses the order.

$n > 4$ — The solution is all values of n greater than 4.

Answer: $(4, \infty)$ whose graph is [number line with open parenthesis at 4, shaded to the right]

Linear inequalities with fractions, similar to linear equations with fractions, can be simplified by converting the inequality to an equivalent inequality that does not involve fractions. This can be accomplished by multiplying both sides of the inequality by the LCD (least common denominator) of all the terms.

EXAMPLE 6 Solving a Linear Inequality Containing Fractions

Solve $\frac{x}{15} \le \frac{x}{10} - \frac{5}{6}$.

SOLUTION

$\frac{x}{15} \le \frac{x}{10} - \frac{5}{6}$ — $15 = 3 \cdot 5$, $10 = 2 \cdot 5$, $6 = 2 \cdot 3$

$30\left(\frac{x}{15}\right) \le 30\left(\frac{x}{10}\right) - 30\left(\frac{5}{6}\right)$ — LCD $= 2 \cdot 3 \cdot 5 = 30$. Multiplying by 30, the LCD, preserves the order.

$2x \le 3x - 25$ — Subtracting $3x$ from both sides of the inequality preserves the order.

$-x \le -25$

$(-1)(-x) \ge (-1)(-25)$ — Multiplying by -1 reverses the order.

$x \ge 25$

Answer: $[25, \infty)$ whose graph is [number line with bracket at 25, shaded to the right]

SELF-CHECK 4.2.3

Solve each inequality and give the answer by using interval notation.

1. $3(2m - 5) \ge 4(1 - 2m) - 5$
2. $-\frac{14}{15} > -\frac{2}{3}x$

The ability to solve inequalities graphically and numerically is even more important when the inequalities become more complicated. Note that in Example 7 all perspectives yield the same result.

EXAMPLE 7 Using Multiple Perspectives to Solve a Linear Inequality

Solve $\frac{x}{2} + \frac{1}{6} \le \frac{x}{6} + \frac{1}{2}$ algebraically, graphically, and numerically and give the answer by using interval notation.

SOLUTION

ALGEBRAICALLY

$$\frac{x}{2} + \frac{1}{6} \le \frac{x}{6} + \frac{1}{2}$$
$$6\left(\frac{x}{2}\right) + 6\left(\frac{1}{6}\right) \le 6\left(\frac{x}{6}\right) + 6\left(\frac{1}{2}\right)$$
$$3x + 1 \le x + 3$$
$$3x \le x + 2$$
$$2x \le 2$$
$$x \le 1$$

GRAPHICALLY

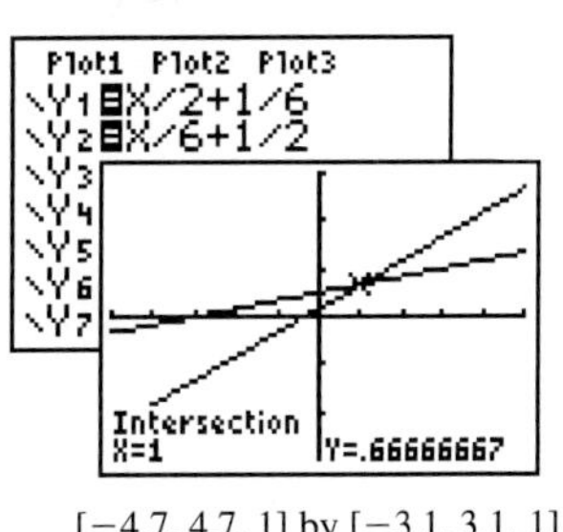

[−4.7, 4.7, 1] by [−3.1, 3.1, 1]

NUMERICALLY

X	Y1	Y2
-3	-1.333	0
-2	-.8333	.16667
-1	-.3333	.33333
0	.16667	.5
1	.66667	.66667
2	1.1667	.83333
3	1.6667	1

X= -3

VERBALLY

All values of x less than or equal to 1 satisfy this inequality. On the graph Y_1 is below or on Y_2 for $x \le 1$. Thus $Y_1 \le Y_2$ for $x \le 1$. From the table Y_1 is less than or equal to Y_2 for $x \le 1$.

Answer: $(-\infty, 1]$ whose graph is [number line shaded to the left of 1, with bracket at 1]

SELF-CHECK 4.2.4

1. Use a graphing calculator to complete this table.

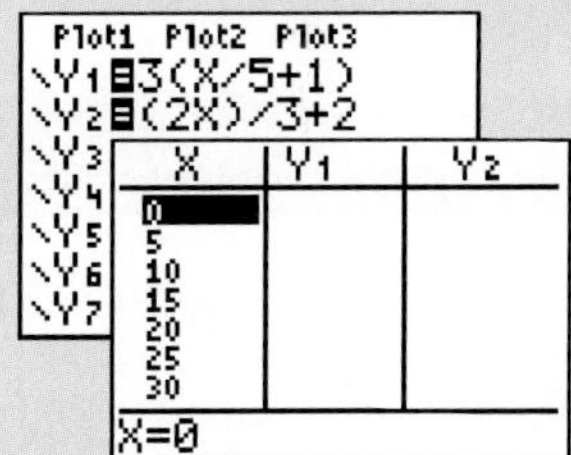

2. Use this table to solve $3\left(\frac{x}{5} + 1\right) \le \frac{2x}{3} + 2$ and give the answer by using interval notation.

Example 8 examines the limits placed on an order for a new cable.

EXAMPLE 8 Strength Limits on a Cable

The contract for a cable used to support a large tent requires that the cable be able to support a minimum load of 750 lb. The composite cable consists of a synthetic fiber wound around a hemp core that can support a load of 20 lb. Each synthetic fiber added to the cable adds 25 lb to the load that the cable can support. Determine the number of synthetic fibers to put in this cable.

SOLUTION

Let n = number of synthetic fibers to put in this cable.

The restrictions on n, the number of fibers in the cable, are $\{0, 1, 2, \ldots\}$. The number of fibers must be an integer and cannot be less than 0 fibers.

VERBAL INEQUALITY

Load that synthetic fibers can support	+	Load that hemp core can support	≥	750

The total load that the cable can support must be 750 lb or more.

ALGEBRAIC INEQUALITY

$$25n + 20 \geq 750$$
$$25n \geq 730$$
$$\frac{25n}{25} \geq \frac{730}{25}$$
$$n \geq 29.2$$

First subtract 20 from both sides of the inequality. Then divide both sides of the inequality by 25.

Because the number of fibers must be an integer greater than 29, the number of fibers must be 30 or greater.

Answer: The cable must contain at least 30 synthetic fibers.

SELF-CHECK ANSWERS

4.2.1

1. $[-2, \infty)$ 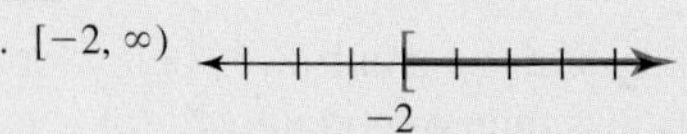
-2

2. $(13, \infty)$
13

3. $(-4, \infty)$
-4

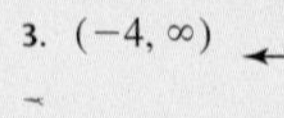

4.2.2

1. $[-7, \infty)$
−10 −9 −8 −7 −6 −5 −4

4.2.3

1. $[1, \infty)$ 2. $\left(\frac{7}{5}, \infty\right)$

4.2.4

1. 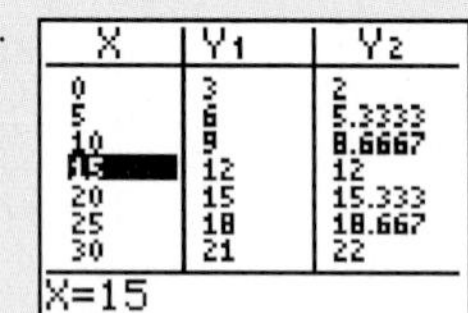

2. $[15, \infty)$

USING THE LANGUAGE AND SYMBOLISM OF MATHEMATICS 4.2

1. By the __________-division principle for inequalities, $x > y$ is equivalent to $2x > 2y$.
2. By the __________-division principle for inequalities, $x > y$ is equivalent to $-2x < -2y$.
3. If a, x, and y are real numbers and $a > 0$, then $x > y$ is equivalent to ax __________ ay.
4. If a, x, and y are real numbers and $a < 0$, then $x > y$ is equivalent to ax __________ ay.
5. To clear an inequality of fractions, we multiply both sides of the inequality by the __________ __________ __________ of all the terms in the inequality.
6. The __________ property is often used to help us remove parentheses from expressions in an inequality.

EXERCISES 4.2

In Exercises 1 and 2, fill in each blank with the correct inequality symbol.

1. a. If $x > y$, then $x + 2$ __________ $y + 2$.
   b. If $x > y$, then $x - 2$ __________ $y - 2$.
   c. If $x > y$, then $\frac{x}{2}$ __________ $\frac{y}{2}$.
   d. If $x > y$, then $-2x$ __________ $-2y$.
2. a. If $x < y$, then $2x$ __________ $2y$.
   b. If $x < y$, then $\frac{x}{-2}$ __________ $\frac{y}{-2}$.
   c. If $x < y$, then y __________ x.
   d. If $x < y$, then $-x$ __________ $-y$.

In Exercises 3–6, use the graph to solve each equation or inequality.

3. **a.** $y_1 = y_2$
b. $y_1 < y_2$
c. $y_1 > y_2$

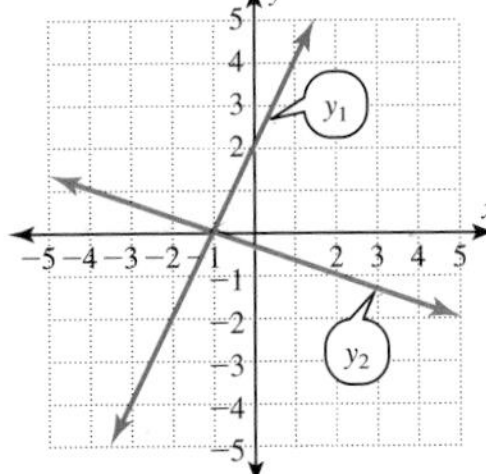

4. **a.** $y_1 = y_2$
b. $y_1 < y_2$
c. $y_1 > y_2$

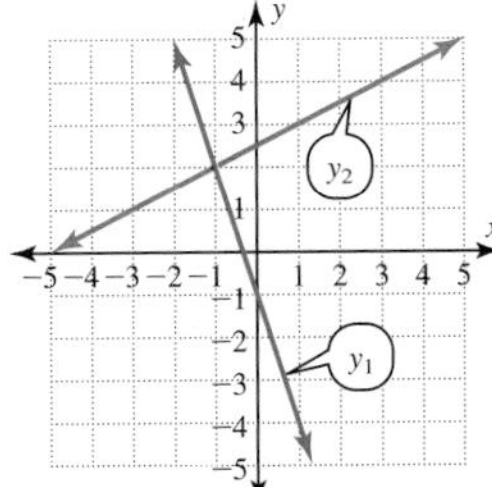

5. **a.** $y_1 = y_2$
b. $y_1 \le y_2$
c. $y_1 \ge y_2$

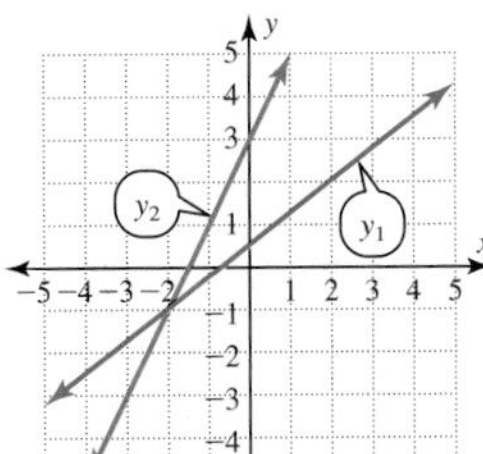

6. **a.** $y_1 = y_2$
b. $y_1 \le y_2$
c. $y_1 \ge y_2$

7. Using a Graph That Models Rental Costs The costs for renting a rug-shampooing machine from two different rental companies are given by the graphs below. The graph of $y_1 = f_1(x)$ gives the cost by Dependable Rental Company based upon the number of hours of use. The graph of $y_2 = f_2(x)$ gives the cost by Anytime Rental Company based upon the number of hours of use. Use these graphs to solve

a. $y_1 = y_2$
b. $y_1 < y_2$
c. $y_1 > y_2$
d. Interpret the meaning of each of these solutions.

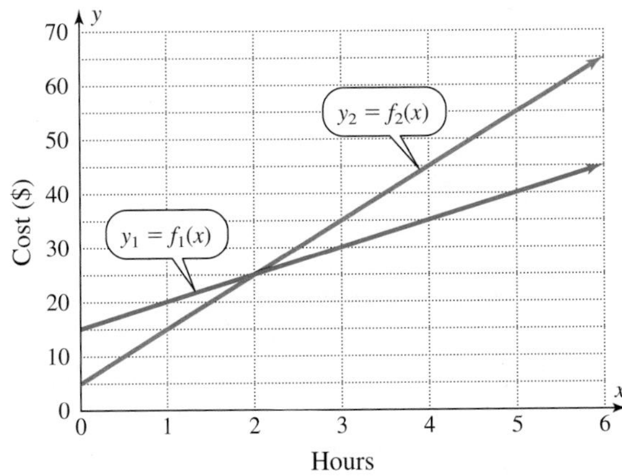

8. Using a Graph That Models the Strength of Two Cables A cable manufacturer makes cables that are wound around two different cores and braided from different fibers. Part of the tensile strength comes from the core, and the rest of the strength comes from the fiber strands. The graph of $y_1 = f_1(x)$ gives the strength of the first cable based upon the number of separate fiber strands braided into this cable. The graph of $y_2 = f_2(x)$ gives the strength of the second cable based upon the number of separate fiber strands braided into this cable. Use these graphs to solve

a. $y_1 = y_2$
b. $y_1 < y_2$
c. $y_1 > y_2$
d. Interpret the meaning of each of these solutions.

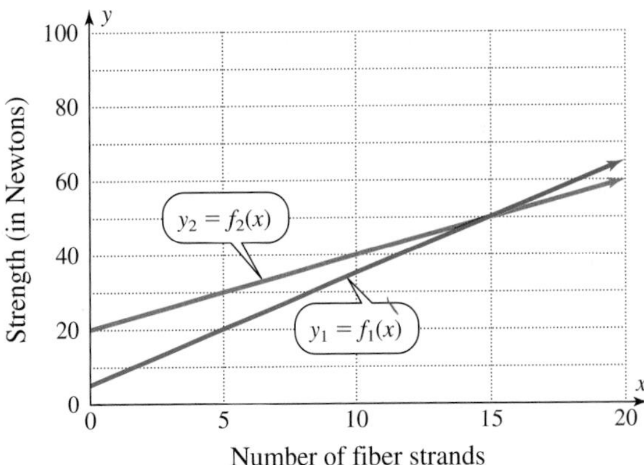

In Exercises 9–12, use the table of values to solve each equation or inequality.

9. **a.** $Y_1 = Y_2$
b. $Y_1 < Y_2$
c. $Y_1 > Y_2$

X	Y1	Y2
0	-3	3
1	2	6
2	7	9
3	12	12
4	17	15
5	22	18
6	27	21
X=0		

10. **a.** $Y_1 = Y_2$
b. $Y_1 < Y_2$
c. $Y_1 > Y_2$

X	Y1	Y2
-3	-11	-14
-2	-7	-7
-1	-3	0
0	1	7
1	5	14
2	9	21
3	13	28
X= -3		

11. **a.** $Y_1 = Y_2$
b. $Y_1 \le Y_2$
c. $Y_1 \ge Y_2$

X	Y1	Y2
-20	-67	-97
-15	-52	-72
-10	-37	-47
-5	-22	-22
0	-7	3
5	8	28
10	23	53
X= -20		

12. **a.** $Y_1 = Y_2$
b. $Y_1 \le Y_2$
c. $Y_1 \ge Y_2$

X	Y1	Y2
-2	7	-23
-1	5	-19
0	3	-15
1	1	-11
2	-1	-7
3	-3	-3
4	-5	1
X= -2		

13. Sweatshirt Costs The following two tables give the costs for ordering college logo sweatshirts from two sporting goods stores. In each case there is a setup charge for designing the logo as well as a cost per sweatshirt. Use these tables to solve

a. $y_1 = y_2$
b. $y_1 < y_2$
c. $y_1 > y_2$
d. Interpret the meaning of each of these solutions.

MEL'S SPORTS

NUMBER OF SWEATSHIRTS x	COST y_1 ($)
10	310
20	560
30	810
40	1,060
50	1,310
60	1,560

MICHAEL'S SPORTING GOODS

NUMBER OF SWEATSHIRTS x	COST y_2 ($)
10	320
20	565
30	810
40	1,055
50	1,300
60	1,545

14. Using Tables That Model Bottle Production The following two tables give the number of glass bottles produced after a number of minutes by two different machines. For each machine there is an initial start-up time before any bottles will come out of either machine. Use these tables to solve

a. $y_1 = y_2$
b. $y_1 < y_2$
c. $y_1 > y_2$
d. Interpret the meaning of each of these solutions.

MACHINE 1

MINUTES AFTER START-UP x	BOTTLES PRODUCED y_1
4	40
6	80
8	120
10	160
12	200
14	240
16	280

MACHINE 2

MINUTES AFTER START-UP x	BOTTLES PRODUCED y_2
4	20
6	70
8	120
10	170
12	220
14	270
16	320

In Exercises 15–18, algebraically solve each inequality and show each step of your solution process.

15. $-2x < 4$ **16.** $-3x \ge -12$
17. $-3x + 4 \ge 7$ **18.** $-5x - 7 \le 32$

Multiple Representations

In Exercises 19–22, solve each inequality algebraically, graphically, and numerically, and describe the solution verbally. (*Hint:* See Example 7.)

19. $2x < 5x + 6$ **20.** $3x + 1 < -x + 5$
21. $3(x - 5) \le 1 - x$ **22.** $2(x - 1) \ge 4 - x$

In Exercises 23–56, solve each inequality algebraically and give the answer by using interval notation.

23. $2x \ge 4$ **24.** $4x < 2$
25. $\frac{1}{2}v < -4$ **26.** $\frac{1}{3}v > -\frac{2}{3}$
27. $-\frac{3}{4}w < -12$ **28.** $-\frac{4}{3}w \le -12$
29. $-x \ge 0$ **30.** $-x \le 5$
31. $1.1m < 2.53$ **32.** $2.1m \le 1.68$
33. $-\frac{n}{4} > \frac{1}{2}$ **34.** $-\frac{n}{6} \ge \frac{2}{3}$
35. $-8z \ge 40$ **36.** $-9z < -36$
37. $-\frac{t}{3} \ge 0$ **38.** $-\frac{2t}{5} > 0$
39. $-10 > -5x$ **40.** $-21 > -7x$
41. $5t \le 21 - 2t$ **42.** $8t < 3t + 35$
43. $9x + 7 > 5x - 13$ **44.** $23x - 14 > 18x + 56$
45. $11y - 8 > 14y + 7$ **46.** $8y + 19 \ge 2y - 11$
47. $5 - 9y \le 19 - 2y$ **48.** $13 - 7y \le 1 - y$
49. $2(3t - 4) > 2(t - 2)$ **50.** $7(2t - 4) \ge 3(3t + 2)$
51. $4(3 - x) \ge 7(2 - x)$ **52.** $5(1 - 2x) < 9(3 - x)$
53. $3(2y - 1) \le 5(3y + 4) + 4$
54. $6(3y + 2) \le 7(2y - 3) + 1$
55. $-2(3w + 7) < 3(5 - w) + 2$
56. $-5(4w - 2) > 3(3w + 5) - 5$

In Exercises 57–60, solve each inequality by the method of your choice. Give each answer by using interval notation.

57. $-3(5 - 2x) > -2(5x + 1) + 3$
58. $-4(2 - 3x) \ge -7(3 - x) - 2$
59. $6(7x - 8) - 4(3x + 2) \le 8(5x + 3)$
60. $-5y - 7(2y - 3) < 3y - 2(5 - y) - 5$

Estimation and Calculator Skills

In Exercises 61–64, mentally estimate the solution of each inequality, and then use a calculator to calculate the solution.

PROBLEM	MENTAL ESTIMATE	CALCULATOR SOLUTION
61. $5.02x \le 35.642$		
62. $-4.95x \ge 31.185$		
63. $-0.47x > -4.183$		
64. $-0.99x < 118.8$		

Multiple Representations

In Exercises 65–68, first write an algebraic inequality for each verbal statement, and then solve this inequality.

65. Twice the quantity x plus five is less than six.
66. Three times the quantity x minus four is greater than fifteen.
67. Four times the quantity y minus seven is greater than or equal to six times the quantity y plus three.
68. Five times the quantity y plus nine is less than or equal to eight times the quantity y minus three.
69. Temperature Limit The temperature in a computer room must be at least 41° Fahrenheit (41°F). Find the allowable temperature in degrees Celsius. The relationship between Fahrenheit and Celsius temperatures is $F = \frac{9}{5}C + 32$.

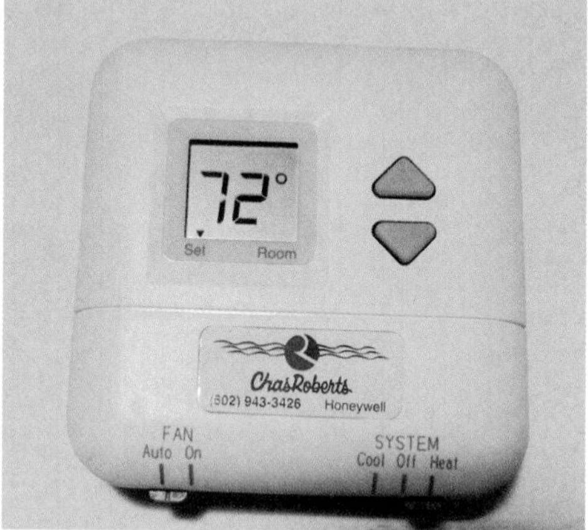

70. Budgetary Limit A company has set a limit of at most \$450 to be spent on its monthly electric bill. If electricity costs \$0.12/kW, find the number of kilowatts the company can use each month.
71. Interest Payment The most a family can budget each month for an interest payment is \$450. Their interest payment for one month will be approximately 0.75% of the amount they can borrow. How much can the family borrow?

72. **Taxi Fare** The minimum charge for a taxi ride is \$3.50, plus an additional \$0.25 per 0.1 mi. How far can you travel if you have at most \$20.50 to spend on taxi fare?

73. **Modeling Costs and Revenue** The cost of producing an order of bricks includes a fixed cost of \$250 and a variable cost of \$0.50 per brick. All bricks are custom-ordered, and the charge for manufacturing these bricks includes a setup fee of \$150 plus a charge of \$1 per brick. Let y_1 represent the income received from an order for x bricks and y_2 represent the cost of producing x bricks.
   a. Determine the values of x for which $y_1 < y_2$, the loss interval for this order.
   b. Determine the values of x for which $y_1 > y_2$, the profit interval for this order.

74. **Modeling Costs and Revenue** The cost of producing x flooring tiles includes a fixed cost of \$400 and a variable cost of \$1.25 per tile. All tiles are custom-ordered, and the charge for manufacturing these tiles includes a setup fee of \$300 plus a charge of \$2 per tile. Let y_1 represent the income received from an order for x tiles and y_2 represent the cost of producing x tiles.
   a. Determine the values of x for which $y_1 < y_2$, the loss interval for this order.
   b. Determine the values of x for which $y_1 > y_2$, the profit interval for this order.

Group Discussion Questions

75. **Error Analysis** Examine the following argument and explain the error in reasoning.

$x < 0$	
$x + x < x$	Add x to both sides.
$2x < x$	
$2 < 1$	Divide both side by x.

76. **Challenge Question** If x and y are real numbers and $x > y$, is there a real number a such that $ax > ay$ is false and $ax < ay$ is also false? If this is possible, list all values of a that make this possible.

77. **Challenge Question** Solve each inequality for x.
   a. $2x + 3y < 5y - 2x$
   b. $8x + 5y > 4x - 3y$
   c. $2(x + 3y) \le -3(x + 3y)$
   d. $5(x - 2y) > 2(7x - 5y)$

Section 4.3 Solving Compound Inequalities

Objectives:
5. Identify an inequality that is a contradiction or an unconditional inequality.
6. Solve compound inequalities involving intersection and union.

Unconditional Inequalities and Contradictions

Most of the inequalities that we consider are conditional inequalities. **Conditional inequalities** contain a variable and are true for some, but not all, real values of the variable. An inequality that is always true is called an **unconditional inequality,** whereas an inequality that is always false is called a **contradiction.**

EXAMPLE 1 Classifying Inequalities

Identify each inequality as a conditional inequality, an unconditional inequality, or a contradiction.

SOLUTION

(a) $x < x + 1$ — This inequality is true for every value of x and therefore is an unconditional inequality.

Every real number is less than 1 more than that number. Thus the solution set is the set of all real numbers $\mathbb{R}$. The graph of the solution set is the entire real number line.

(b) $y < y - 1$ — This inequality is false for every value of y and therefore is a contradiction.

No real number is less than the number 1 unit less than itself. The solution set is the empty set, a set with no elements. There are no solutions to plot on the number line.

(c) $z > 4$ — This inequality is a conditional inequality since it is true for real numbers greater than 4 but is false if 4 or any value less than 4 is substituted for z.

The graph of the solution set is

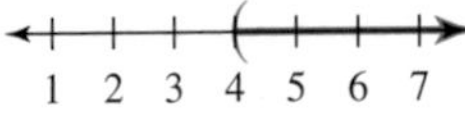

If simplifying an inequality results in an inequality with only constants and no variable, then the inequality is either a contradiction or an unconditional inequality. This is illustrated in Example 2.

EXAMPLE 2 Solving an Inequality When the Variable Is Eliminated

Solve each inequality.

SOLUTION

(a) $3x + 4 < 3x + 1$

$$3x + 4 < 3x + 1$$
$$3x + 4 - 3x < 3x + 1 - 3x$$
$$4 < 1 \text{ is a false statement.}$$

Subtracting $3x$ preserves the order. Because the last inequality is a contradiction, the original inequality is also a contradiction.

Answer: There is no solution. Since the inequality is a contradiction, there are no points to graph.

(b) $4v + 3 \geq 4v + 1$

$$4v + 3 \geq 4v + 1$$
$$4v + 3 - 4v \geq 4v + 1 - 4v$$
$$3 \geq 1 \text{ is a true statement.}$$

Subtracting $4v$ preserves the order. Since the last inequality is an unconditional inequality, the original inequality is also an unconditional inequality.

Answer: $\mathbb{R}$ (the set of all real numbers). Because the inequality is an unconditional inequality, the graph of the solution set is the set of all real numbers.

SELF-CHECK 4.3.1

Solve each inequality.

1. $2x + 5 \leq 2x - 1$ **2.** $2x + 5 \leq 3x - 1$ **3.** $2x + 5 \geq 2x - 1$

The slope-intercept form $y = mx + b$ makes it easy to determine the y-intercept of a line. In Calculator Perspective 4.3.1 we use the y-intercept of each line to determine that the graph of Y_1 is above the graph of Y_2.

CALCULATOR PERSPECTIVE 4.3.1 **Using a Table and a Graph to Solve Contradictions and Unconditional Inequalities**

To solve the inequality $3x + 4 < 3x + 1$ from Example 2(a) on a TI-84 Plus calculator, enter the following keystrokes:

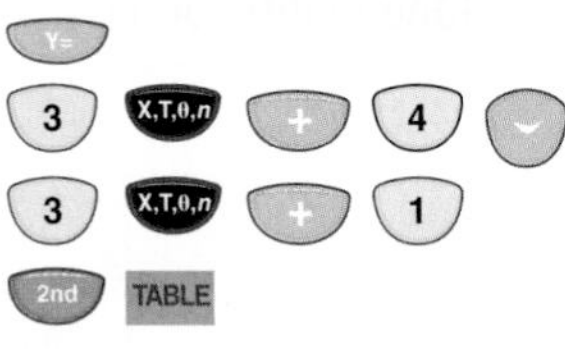

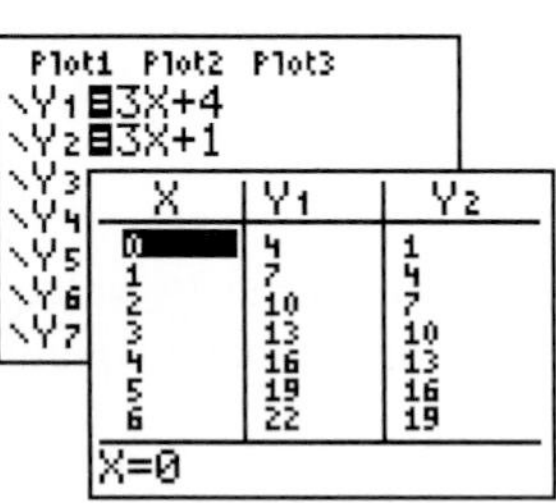

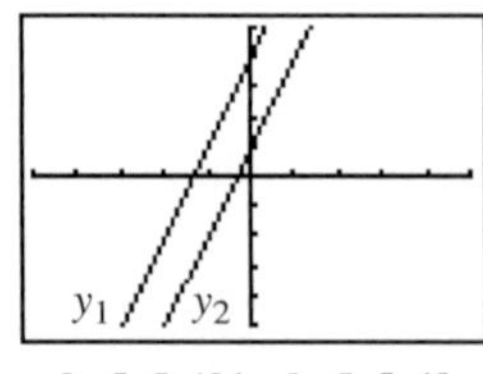

[−5, 5, 1] by [−5, 5, 1]

Note: The table shows that for every value of x, Y_1 is larger than Y_2 by 3. The graph of Y_1 is also above the graph of Y_2 for each x-value. There are no values of x for which $Y_1 < Y_2$. The inequality $3x + 4 < 3x + 1$ is a contradiction.

Answer: No solution.

To solve the inequality $4v + 3 \geq 4v + 1$ from Example 2(b) on a TI-84 Plus calculator, enter the following keystrokes:

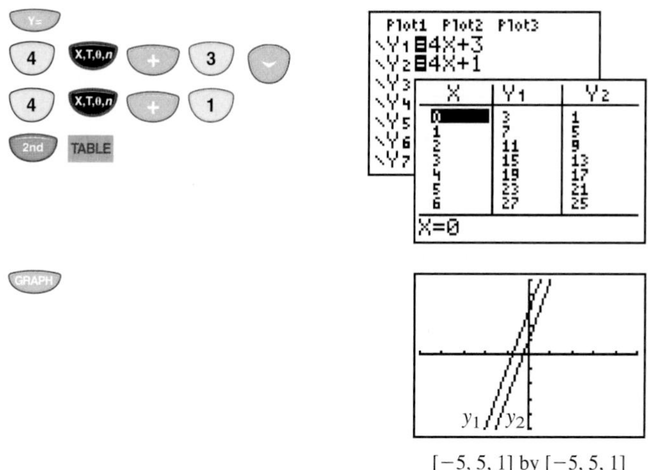

[−5, 5, 1] by [−5, 5, 1]

Note: The table shows that for every value of x, Y_1 is larger than Y_2 by 2. The graph of Y_1 is also above the graph of Y_2 for each x-value. For each value of x, $Y_1 \geq Y_2$. The inequality $4v + 3 \geq 4v + 1$ is an unconditional inequality.

Answer: $\mathbb{R}$ (the set of all real numbers)

Compound Inequalities

A compound inequality is formed when we place more than one restriction on a variable—for example, on the golf hole shown in the following figure, a golfer must hit a drive more than 153 yd but less than 195 yd to clear the water and to land safely on this island green.

The compound inequality $153 < x < 195$ can be read, "x is between 153 and 195," or it can be read "x is greater than 153 *and* x is less than 195." The key word connecting these two inequalities is the word *and.* The word *and* indicates the intersection of these two sets.

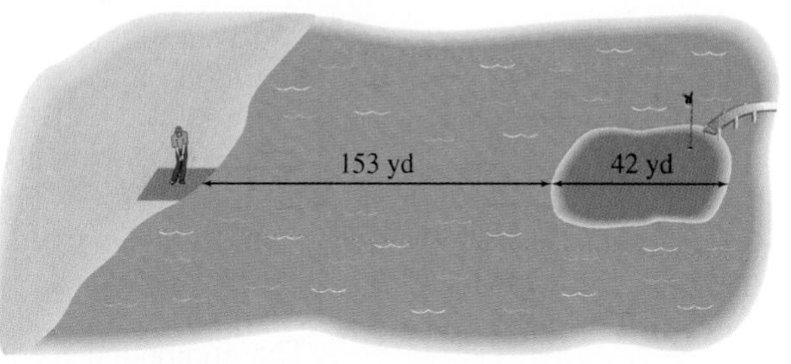

The intersection of two highways consists of the road common to both highways. The intersection of two sets consists of the elements common to both sets.

Intersection of Two Sets

ALGEBRAIC NOTATION	VERBALLY	NUMERICAL EXAMPLE	GRAPHICAL EXAMPLE
$A \cap B$	The intersection of A and B is the set that contains the elements that are in *both A and B*.	$(-3, 4) \cap [0, 6]$ $= [0, 4)$	The intersection of (−3 to 4) and [0 to 6] is [0 to 4) (number lines labeled −3, 4; 0, 6; −3, 0, 4, 6)

In Example 3 remember the convention: A parenthesis indicates that an endpoint is not included in the interval; a bracket indicates that an endpoint is included in the interval.

EXAMPLE 3 Interpreting Compound Inequalities

Graph each compound inequality, give a verbal statement for the inequality, and write the inequality by using interval notation.

SOLUTION

	GRAPH	VERBALLY	INTERVAL NOTATION
(a) $(2, 7] \cap [3, 9)$	[number line 2–8: bracket at 3, bracket at 7]	This intersection is the set of all x-values that are both greater than or equal to 3 and less or equal to 7.	$(2, 7] \cap [3, 9) = [3, 7]$
(b) $x \geq 3$ and $x < 8$	[number line 2–9: bracket at 3, parenthesis at 8]	This is the intersection of all x-values that are greater than or equal to 3 and less than 8.	$[3, \infty) \cap (-\infty, 8) = [3, 8)$
(c) $-1 \leq x < 4$	[number line −2–5: bracket at −1, parenthesis at 4]	x is greater than or equal to -1 and less than 4.	$[-1, 4)$
(d) $0 < x \leq 5$	[number line −1–6: parenthesis at 0, bracket at 5]	x is greater than 0 and less than or equal to 5.	$(0, 5]$

SELF-CHECK 4.3.2

1. Write $w > -3$ and $w \leq 4$ as a single compound inequality.
2. Graph $-2 \leq x < 1$.
3. Write the inequality corresponding to the interval $[-5, 1]$.
4. Write the interval notation for the statement "x is between -2 and 7."

The compound inequality $a \leq x \leq b$ is equivalent to $x \geq a$ *and* $x \leq b$. Compound inequalities can be solved by considering each of the component inequalities by using the graphical, numerical, or algebraic method. This is illustrated in Example 4.

EXAMPLE 4 Solving a Compound Inequality

Solve $-3 < 2x - 1 \leq 5$ algebraically, numerically, and graphically.

SOLUTION

ALGEBRAICALLY

$$-3 < 2x - 1 \leq 5$$
$$-3 < 2x - 1 \quad \text{and} \quad 2x - 1 \leq 5$$
$$-2 < 2x \quad \text{and} \quad 2x \leq 6$$
$$-1 < x \quad \text{and} \quad x \leq 3$$
$$-1 < x \leq 3$$

$-3 < 2x - 1 \leq 5$ is equivalent to $-3 < 2x - 1$ and $2x - 1 \leq 5$. Solve each of these inequalities separately and then form their intersection.

NUMERICALLY

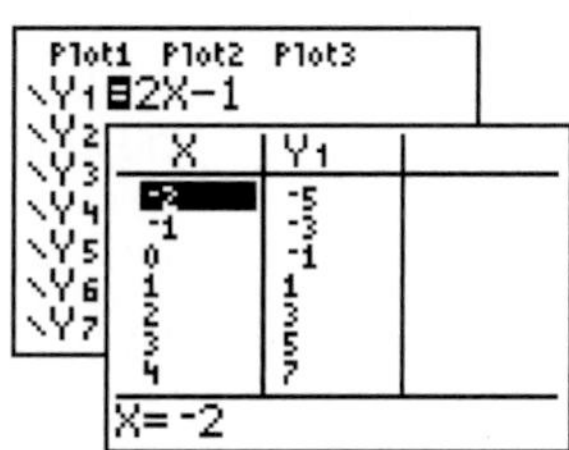

At $x = -1$, $Y_1 = -3$ and at $x = 3$, $Y_1 = 5$
For values of x between -1 and 3, Y_1 is between -3 and 5.

Hint: You may find it useful to highlight the x-values in the table that have y-values within the desired limits or to highlight the portion of the graph between the desired y limits.

GRAPHICALLY

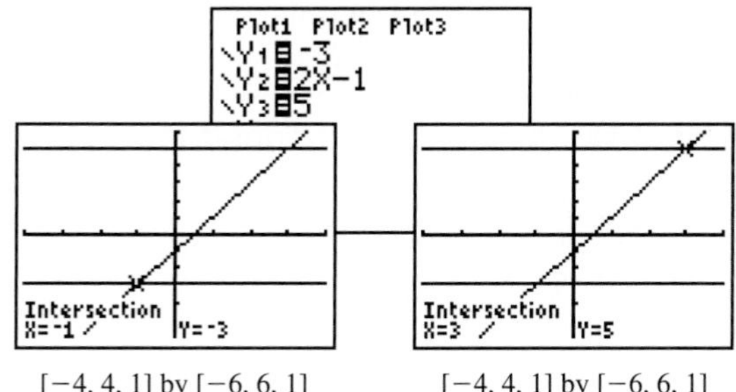

$[-4, 4, 1]$ by $[-6, 6, 1]$ $[-4, 4, 1]$ by $[-6, 6, 1]$

The graph of Y_2 is between the graphs of $Y_1 = -3$ and $Y_3 = 5$ for $-1 < x \leq 3$.

Answer: $(-1, 3]$ whose graph is
−2 −1 0 1 2 3 4

SELF-CHECK 4.3.3

1. Use this table to solve the inequality for the x-values for which $-5 \leq Y_1 < 4$.

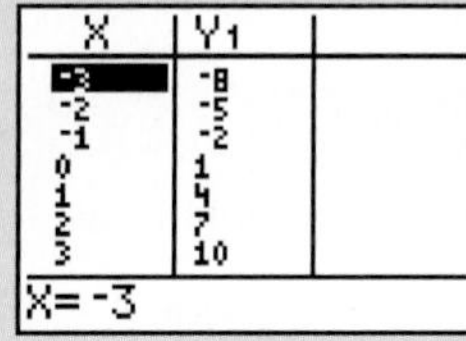

X	Y1
-3	-8
-2	-5
-1	-2
0	1
1	4
2	7
3	10

X=-3

In Example 4 the solution of $-3 < 2x - 1 \leq 5$ was obtained by rewriting this inequality as the intersection of two separate inequalities. Example 5 illustrates how this procedure can sometimes be condensed by performing the same operation on each member of the compound inequality.

EXAMPLE 5 Solving a Compound Inequality

Solve each inequality.

SOLUTION

(a) $-4 < w - 2 \leq 1$

$$-4 < w - 2 \leq 1$$
$$-4 + 2 < w - 2 + 2 \leq 1 + 2$$
$$-2 < w \leq 3$$

Adding 2 to each member of the inequality preserves the order.

Answer: $(-2, 3]$ whose graph is
−3 −2 −1 0 1 2 3

(b) $-4 < -2x \leq 2$

$$-4 < -2x \leq 2$$
$$\frac{-4}{-2} > \frac{-2x}{-2} \geq \frac{2}{-2}$$ Dividing by -2 reverses the order.
$$2 > x \geq -1$$
$$-1 \leq x < 2$$ The inequalities $b > x > a$ and $a < x < b$ have the same meaning. Both of these inequalities state that x is between a and b.

Answer: $[-1, 2)$ whose graph is [number line from −2 to 4 with bracket at −1 and parenthesis at 2]

It is customary to write the smaller number in the left member of an inequality and the larger value in the right member—in the same order as these numbers occur on the number line. In Example 6 the value of -3 is written on the left and -1 is written on the right.

EXAMPLE 6 Solving a Compound Inequality with Two Middle Terms

Solve $-5 \leq 3v + 4 \leq 1$.

SOLUTION

$$-5 \leq 3v + 4 \leq 1$$
$$-5 - 4 \leq 3v + 4 - 4 \leq 1 - 4$$ First subtract 4 from each member of the inequality to isolate the variable in the middle term. Subtracting preserves the order.
$$-9 \leq 3v \leq -3$$
$$\frac{-9}{3} \leq \frac{3v}{3} \leq \frac{-3}{3}$$ Dividing by $+3$ preserves the order.
$$-3 \leq v \leq -1$$

Answer: $[-3, -1]$ whose graph is [number line from −5 to 1 with brackets at −3 and −1]

SELF-CHECK 4.3.4

Solve these inequalities.

1. $0 \leq -z < 2$ **2.** $-1 < 2m + 1 \leq 3$

For some compound inequalities the variable cannot be isolated between the inequalities as in Example 5. In such cases the compound inequality must be split into two simple inequalities that can be solved individually. The final solution is then formed by the intersection of these individual sets.

EXAMPLE 7 Solving an Inequality Whose Solution Is the Intersection of Two Sets

Solve $5v - 8 \leq 2v + 4 < 4v - 2$.

SOLUTION

This compound inequality is equivalent to

$5v - 8 \leq 2v + 4$	and	$2v + 4 < 4v - 2$	Solve each of the inequalities individually.
$3v - 8 \leq 4$		$-2v + 4 < -2$	
$3v \leq 12$		$-2v < -6$	
$v \leq 4$		$v > 3$	

The intersection of $v \leq 4$ and $v > 3$ is $3 < v \leq 4$.

Answer: $(3, 4]$ whose graph is [number line from 0 to 6 with parenthesis at 3 and bracket at 4]

In many problems we are asked to form the *union* of the solution sets of two inequalities. The union of two sets is indicated often by the connecting word *or* since the union of set A and set B consists of the elements in set A *or* in set B.

The United States is a union of states, the set containing all these states.
Elements are listed only once in the union, even if they occur in both sets.

Union of Two Sets

ALGEBRAIC NOTATION	VERBALLY	NUMERICAL EXAMPLE	GRAPHICAL EXAMPLE
$A \cup B$	The union of A and B is the set that contains the elements in either *A or B or both.*	$(-3, 4) \cup [0, 6]$ $= (-3, 6]$	The union of (number line: −3, 4) and (number line: 0, 6) is (number line: −3, 0, 4, 6)

A Mathematical Note

The set symbols $\cap$ and $\cup$ were introduced by Giuseppe Peano in 1888.

The word *or* can be misunderstood if you think of this word only in the exclusive sense (as in "black *or* white") rather than in the inclusive sense used in the definition of the union of two sets.

EXAMPLE 8 Solving an Inequality Whose Solution Is the Union of Two Sets

Solve $2(w - 5) \geq 3w - 8$ or $3(w + 2) > w + 8$.

SOLUTION

$2(w - 5) \geq 3w - 8$	or	$3(w + 2) > w + 8$	Solve each of these inequalities individually. Then graph all the points that are in one set or the other set.
$2w - 10 \geq 3w - 8$		$3w + 6 > w + 8$	
$2w \geq 3w + 2$		$3w > w + 2$	
$-w \geq 2$		$2w > 2$	
$w \leq -2$		$w > 1$	

Answer: $(-\infty, -2] \cup (1, \infty)$ whose graph is (number line: −3 −2 −1 0 1 2 3)

The union symbol $\cup$ is used to indicate that the numbers can be in either the first interval or the second interval.

SELF-CHECK 4.3.5

Determine these sets.

1. $[2, 6) \cap (4, 7]$
2. $[2, 6) \cup (4, 7]$

SELF-CHECK ANSWERS

4.3.1

1. No solution.
2. $[6, \infty)$
3. All real numbers $\mathbb{R}$

4.3.2

1. $-3 < w \leq 4$
2. (number line: −3 −2 −1 0 1 2 3)
3. $-5 \leq x \leq 1$
4. $(-2, 7)$

4.3.3

1. $[-2, 1)$ whose graph is (number line: −3 −2 −1 0 1 2 3)

4.3.4

1. $(-2, 0]$ whose graph is (number line: −4 −3 −2 −1 0 1 2)
2. $(-1, 1]$ whose graph is (number line: −3 −2 −1 0 1 2 3)

4.3.5

1. $(4, 6)$ whose graph is (number line: 1 2 3 4 5 6 7)
2. $[2, 7]$ whose graph is (number line: 1 2 3 4 5 6 7)

USING THE LANGUAGE AND SYMBOLISM OF MATHEMATICS 4.3

1. A __________ inequality is true for some values of the variable in the inequality but false for other values.
2. An inequality that is always true is an __________ inequality.
3. An inequality that is always false is a __________.
4. The __________ inequality $a \le x \le b$ is equivalent to $x \ge a$ __________ $x \le b$.
5. The __________ of sets A and B is the set of elements that are in both A and B.
6. The __________ of sets A and B is the set of elements that are either in set A or in set B.

EXERCISES 4.3

In Exercises 1–3, match each inequality with the choice which best describes this inequality.

1. $x > x + 5$
2. $x > 5$
3. $x < x + 5$

A. A conditional inequality
B. An unconditional inequality
C. A contradiction

In Exercises 4–9, write an inequality for each statement.

4. a. x is greater than or equal to 1 and less than 3.
   b. x is less than 1 or greater than 3.
5. a. x is greater than -5 and less than or equal to 1.
   b. x is less than -5 or greater than 1.
6. a. x is greater than or equal to -4 and less than or equal to -2.
   b. x is between -4 and 9.
7. a. x is greater than 0 and less than 6.
   b. x is between 6 and 7.
8. a. x is at least 30 and at most 45.
   b. x is greater than 4 and less than or equal to 11.
9. a. x is at least 150 and at most 450.
   b. x is greater than or equal to 5 and less than 9.

In Exercises 10–13, rewrite each inequality as two separate inequalities, using the word *and* to connect the inequalities.

10. a. $-7 < x < 13$ b. $-3 \le x < 17$
11. a. $2 < x \le 8$ b. $-5 \le x \le 22$
12. a. $-1 < 2x + 3 \le 11$ b. $2 \le 3x - 1 \le 14$
13. a. $-13 < 5x + 2 < 7$ b. $-6 \le 5x - 1 < 14$

In Exercises 14–16, represent each union of intervals by two separate inequalities, using the word *or* to connect the inequalities.

14. a. $(-\infty, 5] \cup (7, \infty)$ b. $(-2, 5] \cup (7, 10)$
15. a. $(-\infty, -2) \cup (3, \infty)$ b. $[-4, 2] \cup (5, 9]$
16. a. $(-\infty, -8) \cup [0, \infty)$ b. $[-8, -5) \cup [-3, 1]$

In Exercises 17–22, determine $A \cap B$ and $A \cup B$ for the given intervals A and B.

17. $A = (-4, 7], B = [5, 11)$
18. $A = (-3, 6), B = [2, 9]$
19. $A = (-4, \infty), B = [5, 11)$
20. $A = (-3, \infty), B = [2, 9]$
21. $A = (-\infty, -2], B = [-4, \infty)$
22. $A = (-\infty, 5), B = (1, \infty)$

In Exercises 23 and 24, complete the following table:

INEQUALITY NOTATION	INTERVAL NOTATION	GRAPH	VERBALLY
23. $-1 \le x < 6$			
24. $-4 < x \le 2$			

In Exercises 25 and 26, write each inequality expression as a single compound inequality.

25. a. $x > -2$ and $x \le 5$ b. $x \ge 0$ and $x \le 4$
26. a. $x \ge -1$ and $x < \pi$ b. $x > -\pi$ and $x < 11$

In Exercises 27–30, use the graph to solve each compound inequality.

27. $2 \le x + 3 < 5$

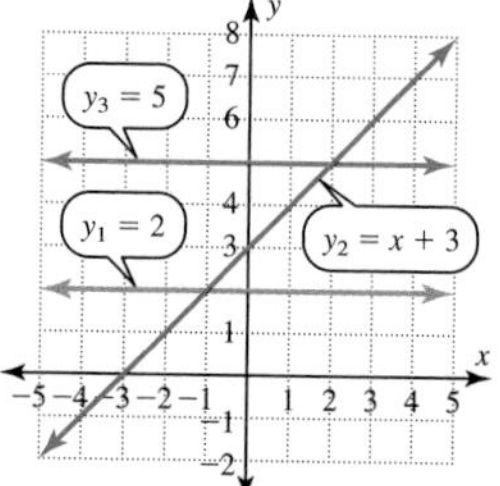

28. $-4 \le x - 2 \le 2$

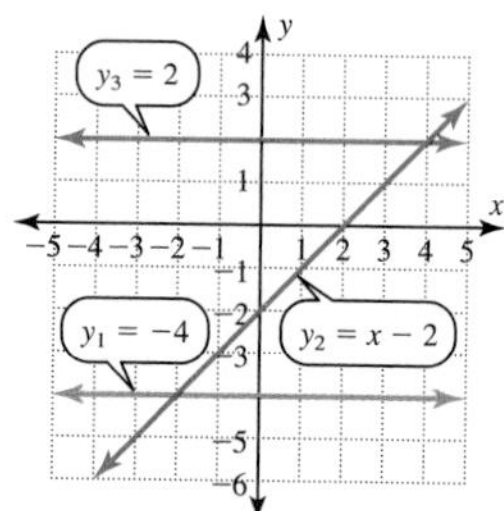

29. $x - 2 < 2x - 1 < x + 3$

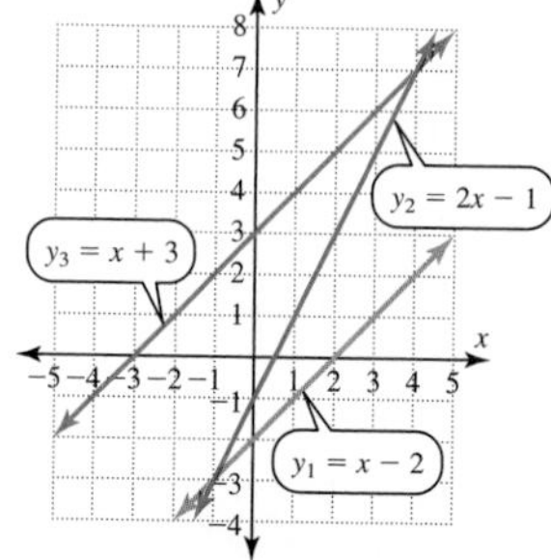

30. $-\frac{x}{2} - 3 < \frac{x}{2} + 2 < -\frac{x}{2} + 4$

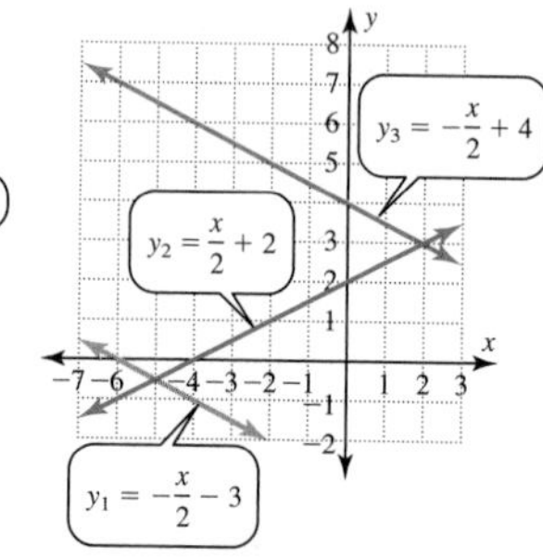

31. **Length of a Golf Shot** Write a compound inequality that expresses the length of the tee shot that a golfer must make to clear the water and to land safely on the green of the golf hole shown here.

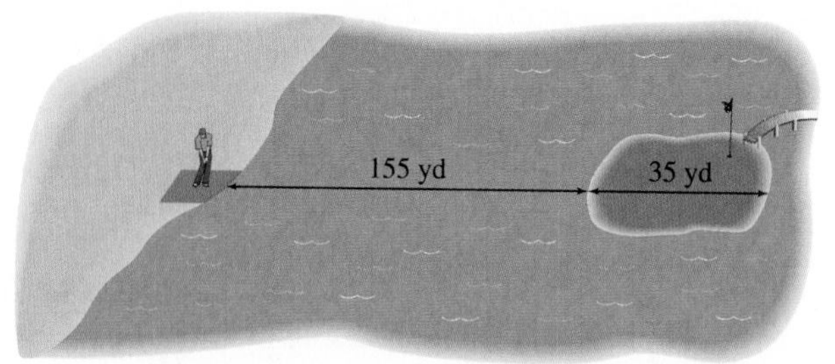

32. **Length of a Golf Shot** A golfer must hit a shot at least 170 yd to clear a lake in front of a green. The depth of the green is only 20 yd. Write a compound inequality that expresses the length of the tee shot that a golfer must make to clear the water and to land safely on the green of this golf hole. (*Hint:* See the figure in Exercise 31.)

33. **Length of a Golf Shot** A golfer is faced with a decision on the 15th fairway of the Lake of the Woods golf course. A safe shot of at most 150 yd will land short of a lake surrounding the green. A more aggressive shot must carry at least 220 yd and at most 270 yd to clear the lake and to stay on the green. Write a union of two intervals that gives the lengths of shots that will not land in the lake.

34. **Length of a Golf Shot** A golfer is faced with a decision on the third fairway of the Springfield golf course. A safe shot of at most 100 yd will land short of a lake surrounding the green. A more aggressive shot must carry at least 195 yd and at most 220 yd to clear the lake and to stay on the green. Write a union of two intervals that gives the lengths of shots that will not land in the lake.

35. **Elevator Cable Strength** The engineering specifications for many products require a safety factor that exceeds the expected value ever needed for the product to perform safely. The strength needed for one of the cables used on an elevator is 2,000 lb. Using a safety factor of 2 would require the cable to have a strength of 4,000 lb, and a safety factor of 3 would require the cable to have a strength of 6,000 lb. Each steel wire added to the cable will increase the strength of the cable by approximately 250 lb.
   a. Using x to represent the number of wires, write a single compound inequality that represents the number of wires that should be put in the cable to produce a safety factor from 2 to 3.
   b. Solve this inequality for x and interpret this solution.

36. **Shoe Sizes** A shoe store that caters to long-distance runners imports shoes from Europe for its customers. The store plans to stock men's U.S. sizes 7 through 13. If the European size is represented by x, then the U.S. size is given by $0.8x - 24.5$.
   a. Solve the inequality $7 \le 0.8x - 24.5 \le 13$ to determine all European shoe sizes the store must stock in order to fit U.S. sizes from 7 to 13.
   b. Shaquille O'Neal's European shoe size is 58. Does the store plan to stock his size?

37. **Strength of a Belt** The drive belt for a wood chipper needs a minimum strength of 200 lb. However, there is also a required maximum strength of 400 lb as the equipment is designed for the belt to shear to prevent major damage to other components if the machine encounters a foreign object such as a steel bar. The graphs of $y_1 = 200$ and $y_2 = 400$ display the strength limits, and the graph of $y_3 = f(x)$ displays the strength of the belt based on the number of separate fiber strands embedded into the belt. Use the following graphs to solve $200 \le y_3 \le 400$ and to interpret the meaning of this solution.

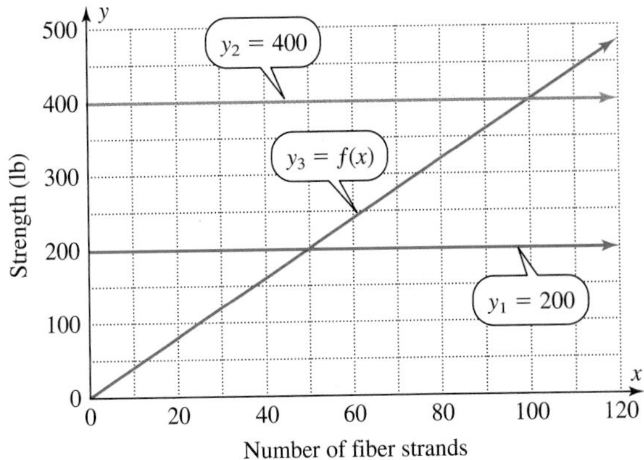

38. **Company Bonuses** A local automobile dealership calculated a holiday bonus for each salesperson based on the number of automobiles that the person sold during the year. The bonus records show that each salesperson received at least $400 and that the most anyone received was $1,500. The graphs of $y_1 = 400$ and $y_2 = 1{,}500$ display the minimum and maximum bonus amounts, and the graph of $y_3 = f(x)$ displays the bonus paid based on the number of cars sold. Use the following graph to solve $400 \le y_3 \le 1{,}500$ and to interpret the meaning of this solution.

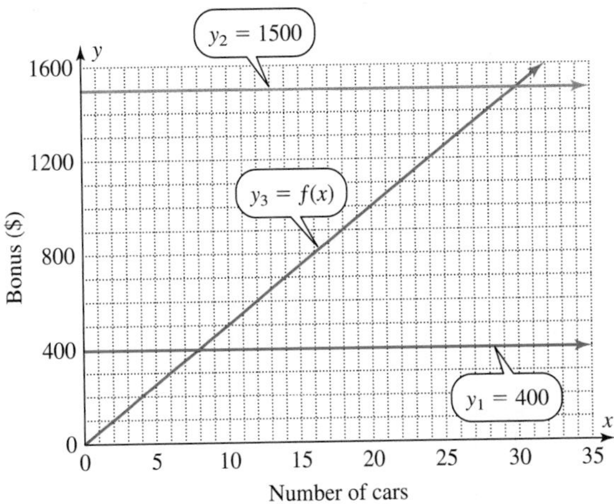

In Exercises 39–42, use the table to solve each inequality.

39. $-2 < 4 - x < 3$

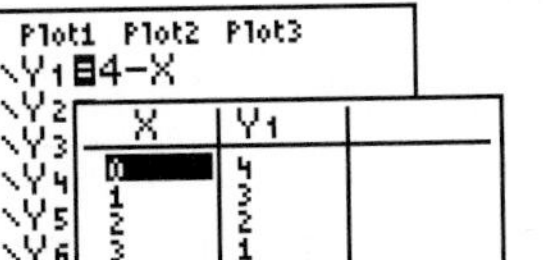

40. $0 < 3 - x < 4$

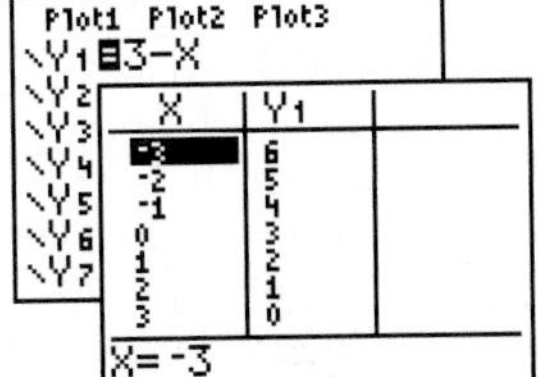

41. $-3 < 2x + 1 \le 7$

42. $0 \le \frac{x}{2} + 1 \le 2$

Plot1 Plot2 Plot3
\Y1 = 2X+1

X	Y1
-3	-5
-2	-3
-1	-1
0	1
1	3
2	5
3	7

X=-3

Plot1 Plot2 Plot3
\Y1 = X/2+1

X	Y1
-3	-.5
-2	0
-1	.5
0	1
1	1.5
2	2
3	2.5

X=-3

In Exercises 43–48, each inequality is a conditional inequality, an unconditional inequality, or a contradiction. Identify the type of each inequality and solve it.

43. $2x + 3 < 2(x + 1) + 1$

44. $2(x + 3) \ge 3(x - 2)$

45. $3x + 2 \le 3(x + 1)$

46. $2(x + 3) < 2(x - 2)$

47. $4x + 1 \le 3(x + 1)$

48. $2(x + 3) < 2(x + 4)$

Multiple Representations

In Exercises 49–52, solve each inequality algebraically, graphically, and numerically and describe the solution verbally.

49. $0 \le x + 5 < 6$

50. $-1 < x - 3 \le 1$

51. $-7 \le x - 8 \le -6$

52. $6 < x + 7 < 9$

In Exercises 53–74, solve each inequality.

53. $-30 < 5x < -20$

54. $21 \le 7x \le 42$

55. $-4 < -4x \le 8$

56. $-15 \le -5x < 5$

57. $-14 \le -\frac{7v}{2} \le 0$

58. $-15 \le -\frac{5v}{3} \le 0$

59. $3 \le 2w + 3 < 11$

60. $1 < 3w - 2 \le 13$

61. $5 < 4x - 3 \le 29$

62. $9 \le 7x + 2 \le 37$

63. $-4 < 6 - 5t < 11$

64. $-1 \le 5 - 6t < 17$

65. $-2 \le \frac{m}{2} - 3 \le -1$

66. $-3 < \frac{m}{2} - 2 < -1$

67. $-2 < \frac{m - 3}{2} \le -1$

68. $-3 \le \frac{m - 2}{3} < -1$

69. $\frac{x - 3}{10} \ge \frac{x - 2}{3}$

70. $\frac{x - 3}{3} < \frac{x + 1}{9}$

71. $x - 1 < 2x < x + 2$

72. $2x - 1 \le 3x \le 2x + 5$

73. $3x - 4 \le 4x + 1 \le 3x + 2$

74. $x - 1 \le 3x + 1 \le x + 5$

Estimation and Calculator Skills

In Exercises 75–78, mentally estimate the solution of each inequality and then use a calculator to determine the solution.

PROBLEM	MENTAL ESTIMATE	CALCULATOR SOLUTION
75. $5.719 < 3.01x < 9.331$		
76. $2.01 \le x + 0.99 \le 5.08$		
77. $-4.179 \le -1.99x < 5.771$		
78. $7.81 < -7.1x \le 13.49$		

In Exercises 79–84, solve each inequality.

79. $7x - 1 < 5x + 7 < 6x + 3$

80. $x - 1 < 2$ and $2 > x + 1$

81. $5(x + 4) < 28$ and $3(x - 4) > -11$

82. $x - 1 < \frac{2x + 1}{3}$ or $\frac{4x - 1}{3} \ge x - 1$

83. $2x - 1 < 3x + 1$ or $5x - 6 < 3x - 12$

84. $7(4x - 5) - 16 < 5(8x - 3)$ and
$11 - 4(2x - 3) \ge -3(2 - 7x)$

In Exercises 85–88, translate each problem into a linear inequality and then solve this inequality.

85. **Temperature Limits** The warranty for a computer specifies that the temperature of its environment must be maintained between 41 and 95°F. Find the temperatures that are within this range, expressed in degrees Celsius.
Solve $41 < \frac{9}{5}C + 32 < 95$ to find the acceptable range of Celsius temperatures.

86. **Glass Blowing Temperatures** The range of acceptable Celsius temperatures for working one type of glass by a glass blower is from 650 to 700°C. Solve
$650 \le \frac{5}{9}(F - 32) \le 700$ to find the acceptable range of Fahrenheit temperatures.

87. **Perimeter of a Triangle** Two sides of a triangle must be 8 and 12 m. The perimeter must be greater than 24 m and less than 40 m. Determine the length that can be used for the third side.

88. **Perimeter of a Rectangle** The width of a rectangle must be 9 m. If the perimeter must be between 44 and 64 m, determine the length that can be used for the rectangle.

Group Discussion Questions

89. **Communicating Mathematically** If possible, list all values of x for which $0.9 < x < 1.0$. If this is not possible, explain why.

90. **Challenge Question** Can you determine, without direct calculations, which is larger, $25(\pi - 2)$ or $25(4 - \pi)$?

91. **Challenge Question** Can you determine, without direct calculation, which is greater, the height or the circumference of the can of three tennis balls shown? (Assume that one ball is touching the top of the can and one ball is touching the bottom with no extra space between the balls.)

Section 4.4 Solving Absolute Value Equations and Inequalities

Objectives:

7. Solve absolute value equations and inequalities algebraically.
8. Solve absolute value equations and inequalities using tables and graphs.

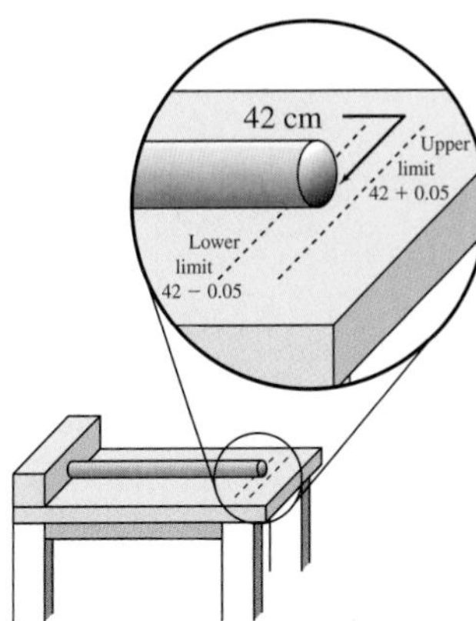

In industrial applications there is generally an allowance for a small variation, or leeway, between the standard size for a part or component and the actual size. This acceptable variation is called the *tolerance*. For a 42-cm steel rod, for example, the tolerance may be 0.05 cm. A worker on an assembly line might merely lay the rod on a table that is marked to indicate the upper and lower limits of tolerance, as shown in the figure. However, an engineer doing calculations would need to describe this tolerance algebraically. One way to write this tolerance algebraically is to use absolute value notation.

The absolute value of x, denoted by $|x|$, is the distance on the number line between 0 and x. Thus the absolute value inequality $|x| < 4$ represents the interval of points for which the distance from the origin is less than 4 units. Likewise, the absolute value inequality $|x| > 4$ represents the two intervals containing points for which the distance from the origin is more than 4 units. The following box illustrates these concepts.

Absolute Value Expressions

For any real number x and any nonnegative real number d:

ALGEBRAICALLY	VERBALLY	ALGEBRAIC EXAMPLE	GRAPHICAL EXAMPLE
$\|x\| = d$	The distance from 0 to x is d units.	$\|x\| = 4$ if $x = -4$ or $x = 4$	4 units left of 0; 4 units right of 0; −4, 0, 4
$\|x\| < d$	The distance from 0 to x is less than d units.	$\|x\| < 4$ if $-4 < x < 4$; $x > -4$ and $x < 4$	Points less than 4 units from 0; −4, 0, 4
$\|x\| > d$	The distance from 0 to x is more than d units.	$\|x\| > 4$ if $x < -4$ or $x > 4$	Points more than 4 units left of 0; Points more than 4 units right of 0; −4, 0, 4

SELF-CHECK 4.4.1

Use interval notation to represent the real numbers that are solutions of these inequalities.

1. $|x| < 5$ 2. $|x| \geq 5$

The distance between two real numbers on the number line can be determined by subtraction.

Questions: Can you determine the distance between these real numbers on the number line?

A: The distance from 2 to 5.
B: The distance from -5 to -2.
C: The distance from -2 to -5.

	Graphically	Absolute Value of Difference
Answers: **A:**	3 units (number line -2 to 6, points at 2 and 5)	$\|5 - 2\| = \|3\| = 3$
B:	3 units (number line -6 to 2, points at -5 and -2)	$\|-2 - (-5)\| = \|3\| = 3$
C:	3 units (number line -6 to 2, points at -5 and -2)	$\|-5 - (-2)\| = \|-3\| = 3$

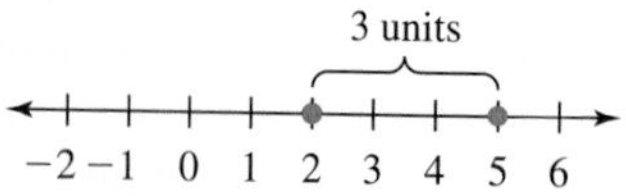

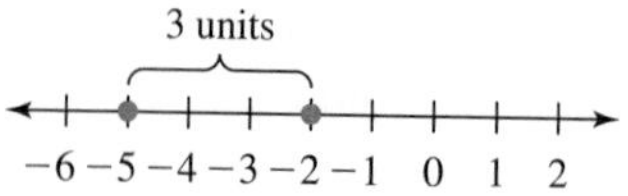

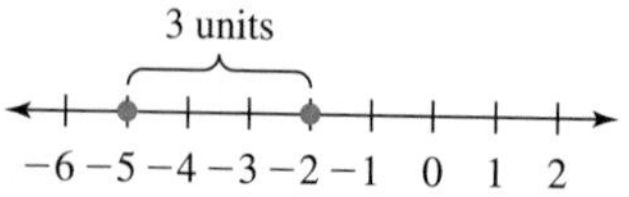

The distance between two points is always positive. The distance between two points is the same in either direction.

If a is larger than b, then the distance from a to b is given by the difference $a - b$. To indicate that the distance between a and b is always nonnegative, we can denote this distance by $|a - b|$. In particular, $|x - 0|$, or $|x|$, can be interpreted as the distance between x and the origin. Likewise, $|a + b| = |a - (-b)|$ equals the distance between a and $-b$. This concept of distance is examined in Examples 1 and 2.

EXAMPLE 1 Representing Distance Using Absolute Value Notation

Use absolute value notation to represent the distance between each pair of points.

SOLUTION

(a) -5 and 1 $\quad |1 - (-5)| = |6| = 6$ or $|-5 - 1| = |-6| = 6$

6 units (number line with points at -5 and 1)

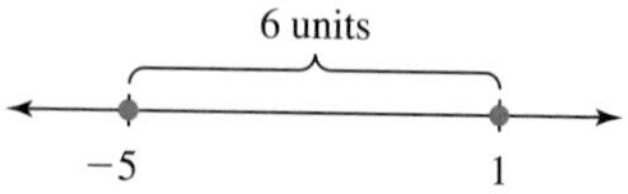

(b) v and w $\quad |v - w|$ or $|w - v|$ $\quad$ Note that $|v - w| = |w - v|$.

SELF-CHECK 4.4.2

Use absolute value notation to represent the distance between these real numbers.

1. 8 and x **2.** x and y **3.** v and $-w$

EXAMPLE 2 Using Distance to Interpret an Absolute Value Expression

Using the geometric concept of distance, interpret **(a)** $|x - 3| = 4$, **(b)** $|x - 3| < 4$, and **(c)** $|x - 3| > 4$.

SOLUTION

When you see a simple absolute value equation or inequality, try to think of this expression in terms of distance.

(a) $|x - 3| = 4$ indicates that the distance between x and 3 is 4 units. This means that x is either 4 units to the left of 3 or 4 units to the right of 3.

ALGEBRAICALLY

Left 4 Units	*or*	*Right 4 Units*
$x - 3 = -4$	or	$x - 3 = +4$
$x = -1$	or	$x = 7$

GRAPHICALLY

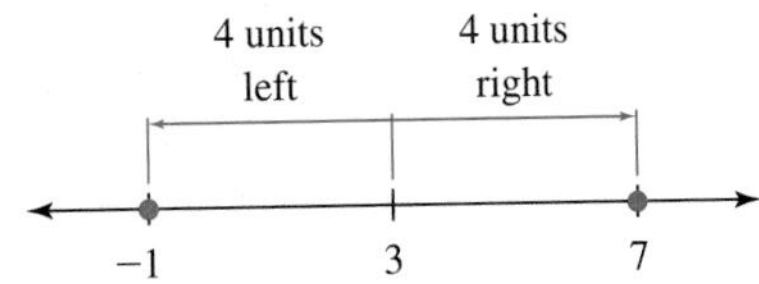

Answer: $x = -1$ or $x = 7$

Do these values check?

(b) $|x - 3| < 4$ indicates that the distance between x and 3 is less than 4 units. This means that x is less than 4 units to the left of 3 and less than 4 units to the right of 3.

ALGEBRAICALLY

Left Less than 4 Units *and* *Right Less than 4 Units*

$$-4 < x - 3 < 4$$
$$-1 < x < 7$$

GRAPHICALLY

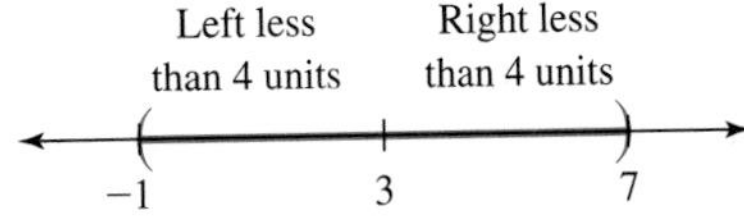

Answer: $(-1, 7)$

Can you check this interval?

(c) $|x - 3| > 4$ indicates that the distance between x and 3 is more than 4 units. This means that x is either more than 4 units to the left of 3 or more than 4 units to the right of 3.

ALGEBRAICALLY

Left More than 4 Units	*or*	*Right More than 4 Units*
$x - 3 < -4$	or	$x - 3 > 4$
$x < -1$	or	$x > 7$

GRAPHICALLY

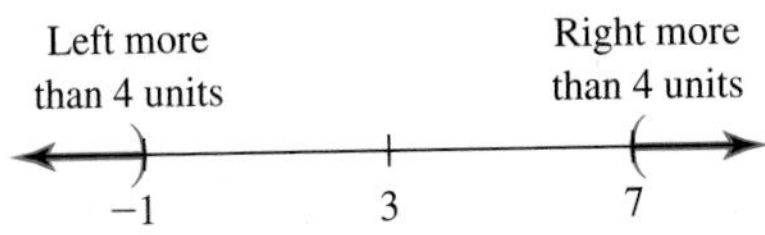

Answer: $(-\infty, -1) \cup (7, \infty)$

Can you check the union of these two intervals?

The results observed in Example 2 are summarized in the following box. Note that $|x - a| > d$ corresponds to the union of two separate intervals. This requires the word *or* to describe $x - a < -d$ or $x - a > d$. Again the visual aid of a graph and a description in terms of distance should help you to use the correct notation.

Solving Absolute Value Equations and Inequalities*

For any real numbers x and a and positive real number d:

ABSOLUTE VALUE EXPRESSION	VERBALLY	GRAPHICALLY	EQUIVALENT EXPRESSION
$\|x - a\| = d$	x is d units either left or right of a.	$a - d$, a, $a + d$	$x - a = -d$ or $x - a = +d$
$\|x - a\| < d$	x is less than d units from a.	$a - d$, a, $a + d$	$x - a > -d$ and $x - a < d$, $-d < x - a < +d$
$\|x - a\| > d$	x is more than d units from a.	$a - d$, a, $a + d$	$x - a < -d$ or $x - a > +d$

*Similar statements also can be made about the order relations less than or equal to (≤) and greater than or equal to (≥). Expressions with d negative are examined in the group exercises at the end of this section.

The results in this box can be applied to solve more complicated expressions within absolute value symbols, as illustrated in Example 3.

EXAMPLE 3 Solving an Absolute Value Equation

Solve $|2x - 3| = 31$.

SOLUTION

Left 31 Units	*or*	*Right 31 Units*
$2x - 3 = -31$	or	$2x - 3 = +31$
$2x = -28$		$2x = 34$
$x = -14$		$x = 17$

To obtain all solutions for this equation, we must consider both of these equations.

Answer: $x = -14$ or $x = 17$

Do these values check?

SELF-CHECK 4.4.3

1. Solve $|4x - 6| = 10$.

We can use two-dimensional graphs to solve absolute value equations and inequalities in one variable. This is the same procedure used in Section 2.4 to solve linear equations in one variable. Calculator Perspective 4.4.1 uses Y_1 to represent the absolute value expression.

CALCULATOR PERSPECTIVE 4.4.1 Using a Graph and a Table to Solve an Absolute Value Equation or Inequality

To solve $|x - 3| = 4$, $|x - 3| < 4$, and $|x - 3| > 4$ from Example 2 by using a TI-84 Plus calculator, enter the following keystrokes:

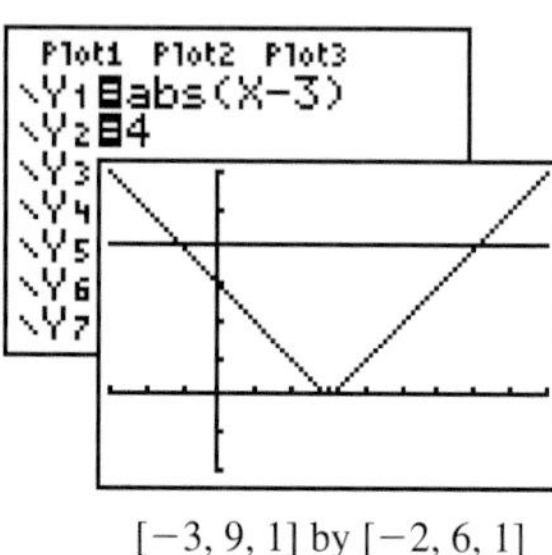

[−3, 9, 1] by [−2, 6, 1]

Press the down arrow key repeatedly to obtain the next screen.

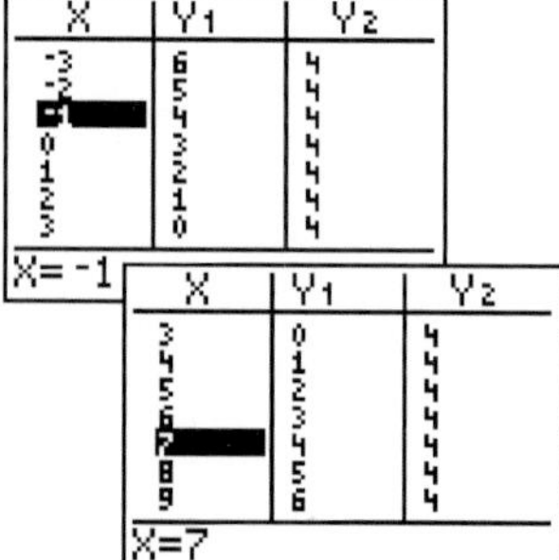

X	Y1	Y2
-3	6	4
-2	5	4
-1	4	4
0	3	4
1	2	4
2	1	4
3	0	4

X= -1

X	Y1	Y2
3	0	4
4	1	4
5	2	4
6	3	4
7	4	4
8	5	4
9	6	4

X=7

Note: Both in the table and on the graph $|x - 3| = 4$ for $x = -1$ or $x = 7$. Also $|x - 3| < 4$ for the x-values between -1 and 7, the interval $(-1, 7)$; and $|x - 3| > 4$ for $x < -1$ or for $x > 7$, the set $(-\infty, -1) \cup (7, \infty)$.

EXAMPLE 4 Using Multiple Perspectives to Solve an Absolute Value Inequality

Solve $|4x - 6| < 10$ algebraically, numerically, and graphically.

SOLUTION

ALGEBRAICALLY

Left Less than 10 Units and *Right Less than 10 Units*

$$-10 < 4x - 6 < 10$$
$$-4 < 4x < 16$$
$$-1 < x < 4$$

$|x - a| < d$ is equivalent to $-d < x - a < d$. Adding 6 to each member of an inequality preserves the order of the inequality. Dividing each member of an inequality by 4 preserves the order of the inequality.

NUMERICALLY

```
Plot1 Plot2 Plot3
\Y1=abs(4X-6)
\Y2=10
\Y3
\Y4
\Y5
\Y6
\Y7
```

X	Y1	Y2
-2	14	10
-1	10	10
0	6	10
1	2	10
2	2	10
3	6	10
4	10	10

X= -2

Table with Inequality Symbols

x	Y_1	<, =, OR >	Y_2
−2	14	>	10
−1	10	=	10
0	6	<	10
1	2	<	10
2	2	<	10
3	6	<	10
4	10	=	10

In the table $Y_1 < Y_2$ for $-1 < x < 4$.

GRAPHICALLY

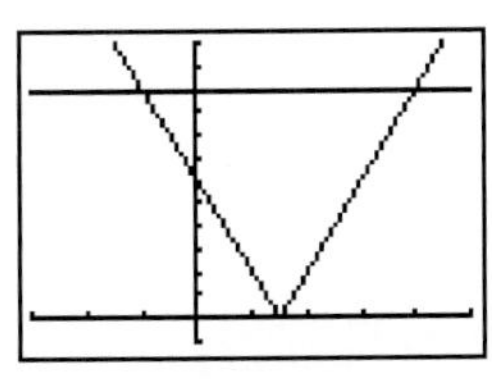

$[-3, 5, 1]$ by $[-1, 12, 1]$

The graph of Y_1 is below Y_2 for $-1 < x < 4$

Answer: $(-1, 4)$

To solve an absolute value equation or inequality, first isolate the absolute value expression on the left side of the expression, as illustrated in Example 5.

EXAMPLE 5 Using Multiple Perspectives to Solve an Absolute Value Inequality

Solve $\left|\frac{x+2}{2}\right| + 5 > 6$.

SOLUTION

ALGEBRAICALLY

$$\left|\frac{x+2}{2}\right| + 5 > 6$$
$$\left|\frac{x+2}{2}\right| + 5 - 5 > 6 - 5$$
$$\left|\frac{x+2}{2}\right| > 1$$

Subtract 5 from both sides of the inequality to isolate the absolute value expression on the left side of the inequality. Subtracting 5 preserves the order of the inequality.

Left More than 1 Unit	*or*	*Right More than 1 Unit*
$\frac{x+2}{2} < -1$	or	$\frac{x+2}{2} > 1$
$x + 2 < -2$	or	$x + 2 > 2$
$x < -4$	or	$x > 0$

Multiplying both sides of an inequality by 2 preserves the order of the inequality. Subtracting 2 from both sides of an inequality preserves the order of the inequality.

NUMERICALLY

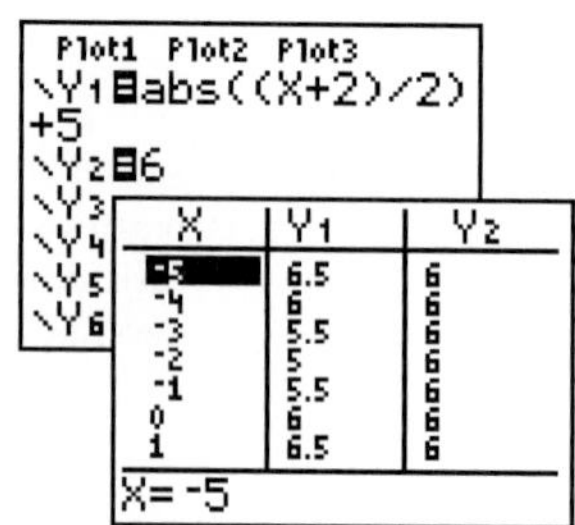

Table with Inequality Symbols

x	Y_1	<, =, OR >	Y_2
−5	6.5	>	6.0
−4	6.0	=	6.0
−3	5.5	<	6.0
−2	5.0	<	6.0
−1	5.5	<	6.0
0	6.0	=	6.0
1	6.5	>	6.0

In the table $Y_1 > Y_2$ for $x < -4$ or for $x > 0$.

GRAPHICALLY

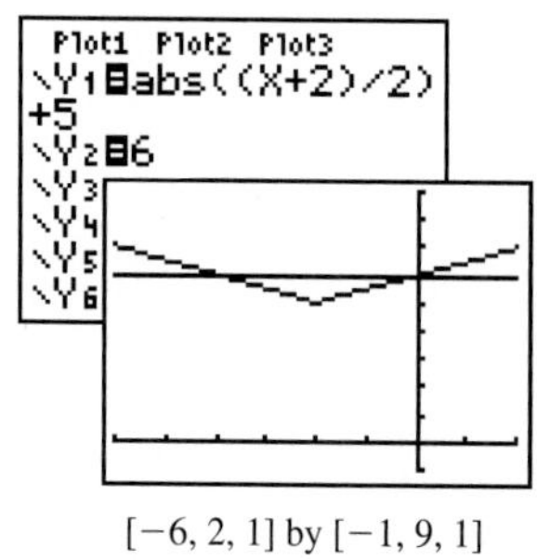

[−6, 2, 1] by [−1, 9, 1]

The graph of Y_1 is above Y_2 for $x < -4$ or for $x > 0$.

Answer: $(-\infty, -4) \cup (0, \infty)$

SELF-CHECK 4.4.4

Use the tables and graphs in Examples 4 and 5 to solve the following.

1. $|4x - 6| \geq 10$ **2.** $\left|\frac{x+2}{2}\right| + 5 \leq 6$

If $|a| = |b|$, then a and b are equal in magnitude but their signs can either agree or disagree. Thus $|a| = |b|$ is equivalent to $a = b$ or $a = -b$.

EXAMPLE 6 Using Multiple Perspectives to Solve an Equation Involving Two Absolute Value Expressions

Solve $|3x - 5| = |5x - 7|$.

SOLUTION

ALGEBRAICALLY

$$3x - 5 = 5x - 7 \quad \text{or} \quad 3x - 5 = -(5x - 7)$$
$$-2x = -2 \qquad\qquad 3x - 5 = -5x + 7$$
$$x = 1 \qquad\qquad 8x = 12$$
$$x = \frac{3}{2}$$

NUMERICALLY

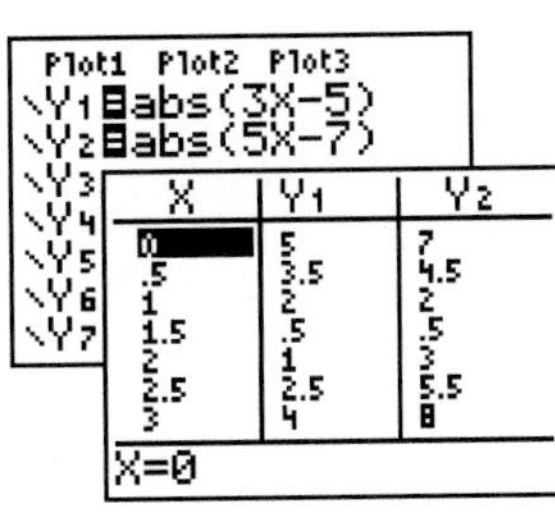

GRAPHICALLY

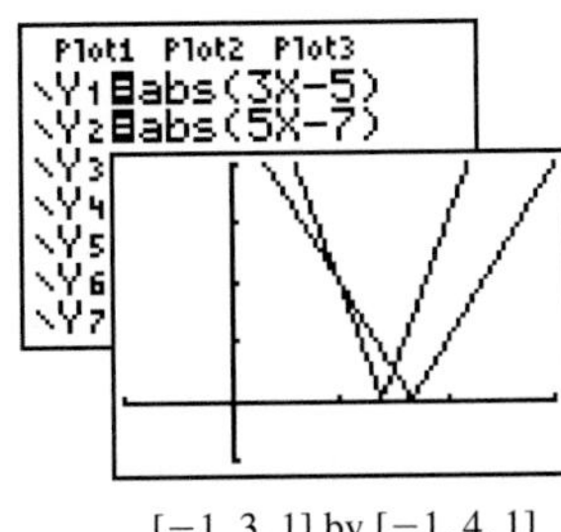

[−1, 3, 1] by [−1, 4, 1]

VERBALLY

From the table $Y_1 = Y_2$ for $x = 1$ and $x = 1.5$ and the graphs of Y_1 and Y_2 intersect at $x = 1$ and $x = 1.5$.

Answer: $x = 1$ or $x = \frac{3}{2}$

Some of the earlier examples in this section start with an absolute value inequality and then produce the interval of values that satisfy this inequality. Example 7 goes in the other direction. It starts with an interval of values and then produces an absolute value inequality that represents this interval.

EXAMPLE 7 Representing an Interval by Using Absolute Value Notation

Write an absolute value inequality to represent the interval $(-3, 7)$.

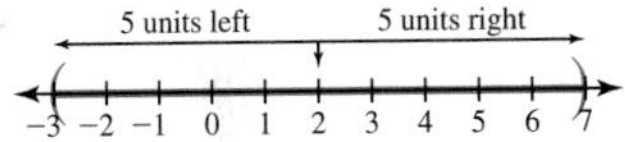

SOLUTION

The distance from one end of this interval to the other is $|7 - (-3)| = |10| = 10$. Thus the midpoint is 5 units to the right of -3 and 5 units to the left of 7. This midpoint is 2 since $-3 + 5 = 2$ and $7 - 5 = 2$. The interval $(-3, 7)$ consists of points less than 5 units from 2. In absolute value notation this is represented by $|x - 2| < 5$.

Answer: $|x - 2| < 5$ represents the interval $(-3, 7)$.

SELF-CHECK 4.4.5

1. Solve $|x - 3| = |2x|$.
2. Use Example 7 to write an absolute value inequality to represent $(-\infty, -3] \cup [7, \infty)$.

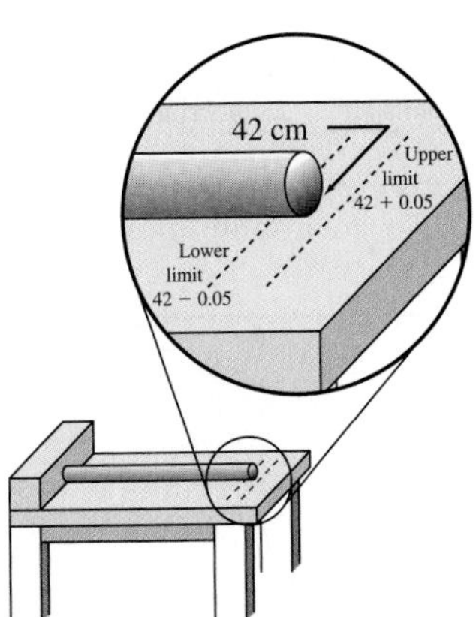

At the beginning of this section we introduced **tolerance,** the acceptable variation between the standard size for a part or component and the actual size. For example, the tolerance may be 0.05 cm for a 42-cm steel rod. An engineer can describe this tolerance algebraically as $|r - 42| \le 0.05$. This is an algebraic statement that the length of the rod r and the desired length of 42 cm can differ by at most 0.05 cm. In Example 8 we examine the tolerance for a drug dosage.

EXAMPLE 8 Representing a Tolerance Interval by Using Absolute Value Notation

The amount of a drug that is placed in an IV bag for a patient should be 20 mL, with a tolerance of ± 0.5 mL. Express this interval as a compound linear inequality and as an absolute value inequality, using d to represent the volume of the drug.

SOLUTION

ALGEBRAICALLY	GRAPHICALLY	VERBALLY
$-0.5 \le d - 20 \le 0.5$ $\lvert d - 20 \rvert \le 0.5$	$20 - 0.5$ (—— 20 ——) $20 + 0.5$	The drug can be from 0.5 mL below 20 mL to 0.5 mL above 20 mL.

SELF-CHECK ANSWERS

4.4.1
1. $(-5, 5)$
2. $(-\infty, -5] \cup [5, \infty)$

4.4.2
1. $|x - 8|$
2. $|x - y|$
3. $|v + w|$

4.4.3
1. $x = -1$ or $x = 4$

4.4.4
1. $(-\infty, -1] \cup [4, \infty)$
2. $[-4, 0]$

4.4.5
1. $x = -3$ or $x = 1$
2. $|x - 2| \ge 5$

USING THE LANGUAGE AND SYMBOLISM OF MATHEMATICS 4.4

1. The absolute value of x is the ________ on the number line between 0 and x.
2. Fill in these blanks with $=$, $<$, or $>$.
   a. The distance from 0 to x is less than d units. $|x|$ ________ d
   b. The distance from 0 to x is d units. $|x|$ ________ d
   c. The distance from 0 to x is greater than d units. $|x|$ ________ d
3. Write an absolute value equation that indicates x is d units left or right of a. ________
4. The acceptable variation between the standard size for a part and the actual size of the part is called the ________.
5. To solve an absolute value equation or inequality, first ________ the absolute value expression on the left side of the expression.

EXERCISES 4.4

In Exercises 1–6, write an absolute value equation or inequality to represent each set of points.

1. a.
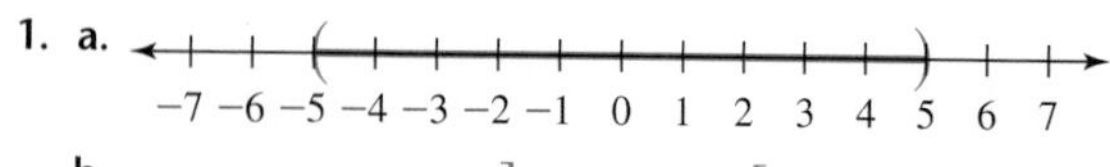

b.
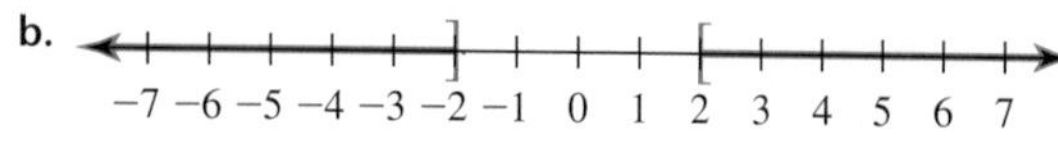

c.
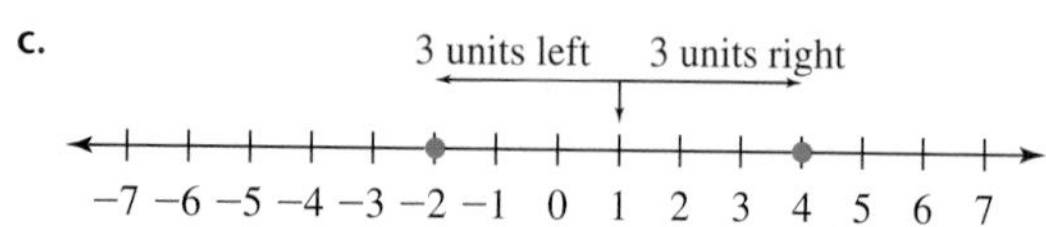

d. The points between −4 and 4.

2. a. −7 −6 −5 −4 −3 −2 −1 0 1 2 3 4 5 6 7

b.
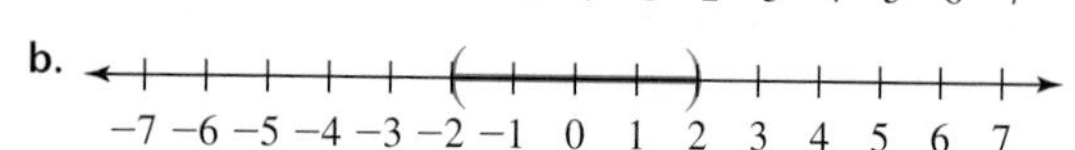

c.
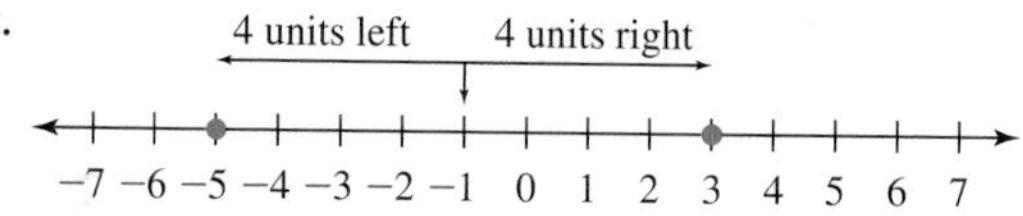

d. The numbers that are at least −3 and at most 3.

3. a.
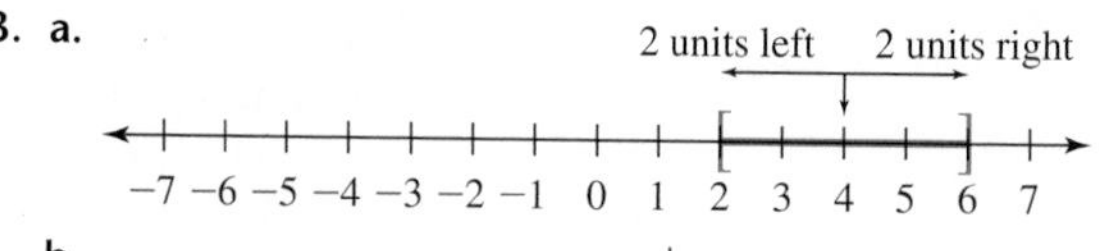

b.
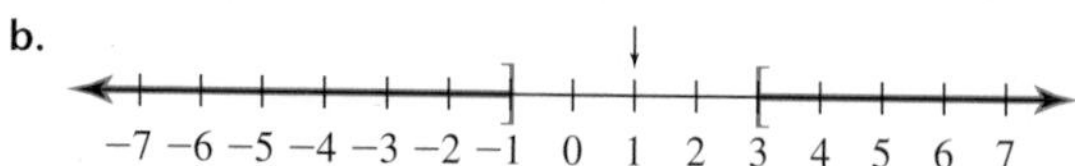

c.
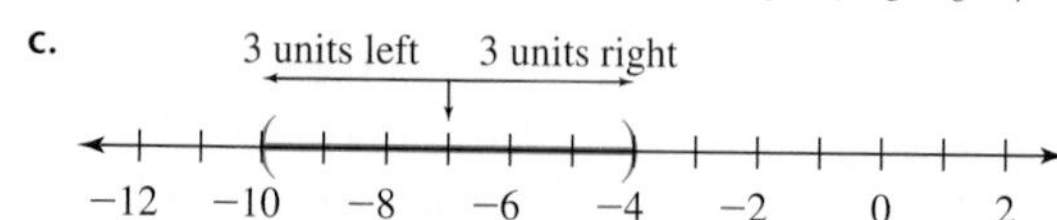

d. The numbers that are at least −5 and at most 5.

4. a.

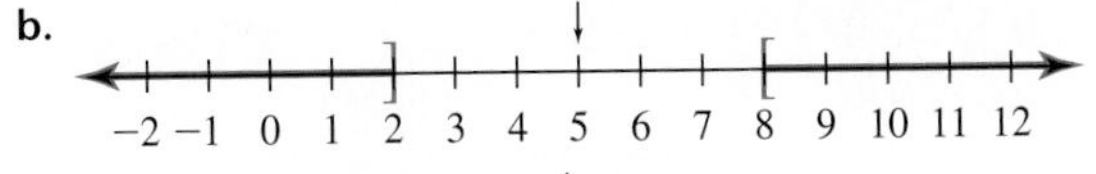

b.

−2 −1 0 1 2 3 4 5 6 7 8 9 10 11 12

c.

−6 −4 −2 0 2 4 6 8 10 12 14 16 18 20 22

d. The points between −8 and 8.

5. a. $[-3, 3]$ **b.** $(-\infty, -3) \cup (3, \infty)$ **c.** $(-7, 7)$

6. a. $(-9, 9)$ **b.** $(-\infty, -9] \cup [9, \infty)$ **c.** $[-\pi, \pi]$

In Exercises 7 and 8, solve each equation.

7. a. $|x| = 6$ **b.** $|x| = 0$
c. $|x - 2| = 5$ **d.** $|x + 2| = 5$

8. a. $|x| = 13$ **b.** $|x - 7| = 8$
c. $|2x| = 0$ **d.** $|x + 7| = 8$

In Exercises 9–12, use the graph to solve each equation and inequality.

9. a. $|x - 2| = 4$
b. $|x - 2| < 4$
c. $|x - 2| > 4$

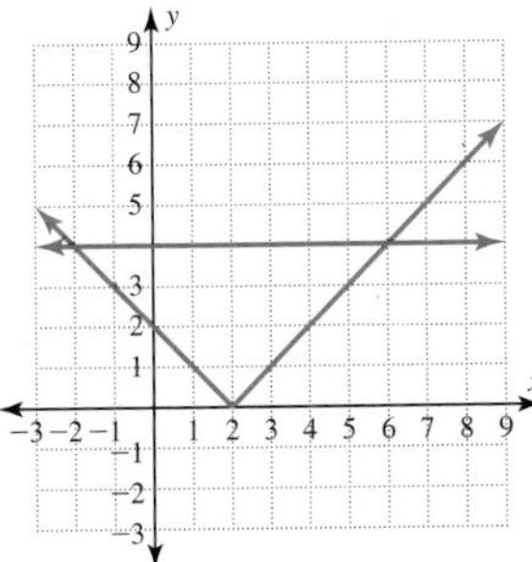

10. a. $|x + 3| = 2$
b. $|x + 3| \le 2$
c. $|x + 3| > 2$

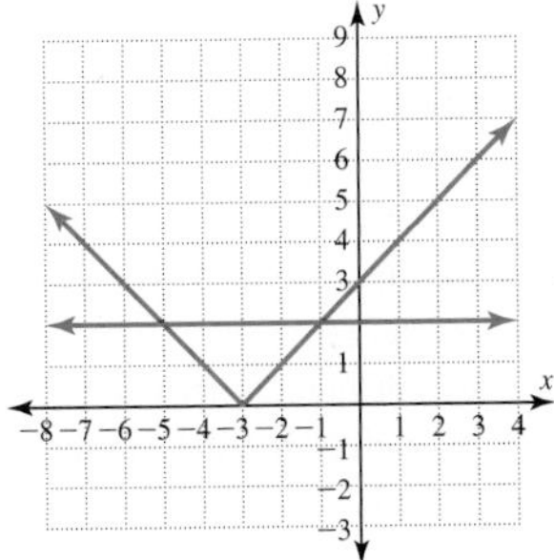

11. a. $\left|\dfrac{x - 4}{2}\right| = 1$
b. $\left|\dfrac{x - 4}{2}\right| < 1$
c. $\left|\dfrac{x - 4}{2}\right| \ge 1$

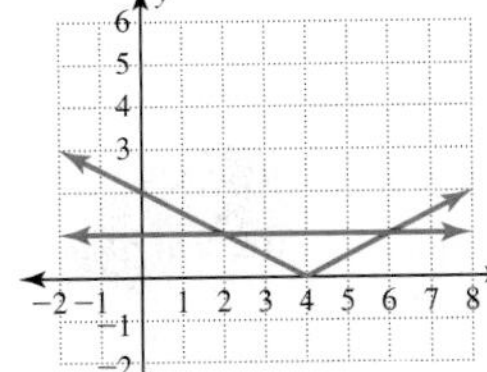

12. a. $|3x - 6| = 3$
b. $|3x - 6| \le 3$
c. $|3x - 6| \ge 3$

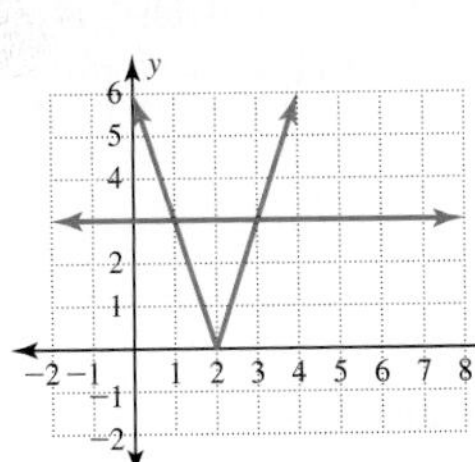

In Exercises 13 and 14, complete the table by placing <, =, or > in the third column. Then determine the x-values that satisfy each equation and inequality. In each case y_1 represents the left side of the expression and y_2 represents the right side of the expression.

13. a. $|x + 2| = 2$
b. $|x + 2| \le 2$
c. $|x + 2| \ge 2$

x	y_1	<, =, or >	y_2
−5	3		2
−4	2		2
−3	1		2
−2	0		2
−1	1		2
0	2		2
1	3		2

14. a. $|2x + 1| = 3$
b. $|2x + 1| < 3$
c. $|2x + 1| > 3$

x	y_1	<, =, or >	y_2
−3	5		3
−2	3		3
−1	1		3
0	1		3
1	3		3
2	5		3
3	7		3

In Exercises 15 and 16, use the table of values to determine the x-values that satisfy each equation and inequality.

15. a. $|x - 5| = 10$
b. $|x - 5| < 10$
c. $|x - 5| \ge 10$

Plot1 Plot2 Plot3
\Y1 = abs(X−5)

X	Y1
-10	15
-5	10
0	5
5	0
10	5
15	10
20	15

X=-10

16. a. $|x + 25| = 50$
b. $|x + 25| \le 50$
c. $|x + 25| > 50$

Plot1 Plot2 Plot3
\Y1 = abs(X+25)

X	Y1
-100	75
-75	50
-50	25
-25	0
0	25
25	50
50	75

X=-100

In Exercises 17–20, solve each inequality algebraically, graphically, and numerically, and describe the solution verbally.

17. $|x + 1| < 1$ **18.** $|x + 2| < 1$
19. $|2x - 1| < 3$ **20.** $|2x + 3| < 5$

In Exercises 21–40, solve each equation and inequality.

21. $|3x + 4| \ge 2$ **22.** $|3x - 5| \ge 1$

23. $\left|\dfrac{-7x}{2}\right| \le 14$ **24.** $\left|\dfrac{-5y}{11}\right| \le 55$

25. $|2x + 3| - 4 < 1$ **26.** $|2x - 5| - 2 < 3$

27. $|2(x + 3) - 4| < 1$ **28.** $|2(x - 5) - 2| < 3$

29. $|2x + 1| + 4 > 11$ **30.** $|2x - 1| + 6 > 9$

31. $\left|\dfrac{x - 1}{7}\right| \ge 14$ **32.** $\left|\dfrac{2x - 1}{3}\right| \ge 6$

33. $|2(2x - 1) - (x - 3)| < 11$

34. $|3(2x + 1) - 4(3x - 2)| < 5$

35. $|2x| = |x - 1|$ **36.** $|3x| = |4 - x|$

37. $|x - 1| = \left|\dfrac{x + 1}{2}\right|$ **38.** $|x + 3| = \left|\dfrac{x - 1}{2}\right|$

39. $|2x + 3| = |3x - 1|$ **40.** $|4x + 2| = |5x - 1|$

Estimation and Calculator Skills

In Exercises 41–44, mentally estimate the solution of each equation to the nearest integer, and then use a calculator to calculate the solution.

PROBLEM	MENTAL ESTIMATE	CALCULATOR SOLUTION
41. $\|2.01x\| = 9.849$		
42. $\|3.9x\| = 39.39$		
43. $\|x + 7.93\| = 15.03$		
44. $\|x - 9.01\| = 29.13$		

In Exercises 45–50, write an absolute value inequality to represent these intervals.

45. $(-12, 6)$ **46.** $(-3, 11)$

47. $[-4, 26]$ **48.** $[12, 32]$

49. $(-\infty, -2) \cup (6, \infty)$ **50.** $(-\infty, -6] \cup [4, \infty)$

51. Length of a Truck Spring The engineering specifications for a truck specify that one of the springs in the suspension system should compress to absorb shocks and extend when the wheels go into holes. The distance x in inches between the bottom of the spring and the ground is given by $|x - 22| < 6$. Write an interval that gives all possible distances satisfying the engineering specifications.

52. Length of a Transmission Cable The engineering specifications for an electrical transmission cable specify that the cable that spans the distance between two towers and crosses an interstate highway must be able to contract during cold weather and expand during warm weather. The distance x in feet between the bottom of the cable over the interstate and the highway below is given by $|x - 35| < 3$. Write an interval that gives all possible distances satisfying the engineering specifications.

Using Absolute Value Inequalities to Model Tolerance Intervals

In Exercises 53–58, express the tolerance interval as an absolute value inequality, and determine the lower and upper limits of the interval.

53. Tolerance of a Piston Rod The length of a piston rod in an automobile engine is specified by contract to the manufacturer to be 15 cm with a tolerance of ±0.001 cm.

54. Tolerance of a False Tooth A dental clinic has received an order for a dental implant that is 1.3 cm long with a tolerance of ±0.05 cm.

55. Tolerance of a Soft Drink Bottle The specifications for a bottle-filling machine in a soft drink plant call for the machine to dispense 16 oz into a bottle with a tolerance of ±0.25 oz.

56. Tolerance of a Fuel Tank A military contract for an aircraft fuel tank specifies a capacity of 420 L with a tolerance of ±1 L.

57. Tolerance of a Temperature Control A temperature control at a pharmaceutical manufacturing plant is set to keep the temperature at 25°C with a tolerance of ±2°C.

58. Tolerance of a Voltage Control The electrical supply to an important Internet provider is monitored to keep the line feed at 120 V with a tolerance of ±1 V.

59. Strength Limits on a Drive Belt The drive belt for a mulching machine must be able to handle a load of at least 3,000 lb. However, the design for the machine is for the belt to break on any load of 9,000 lb or greater. This is done so the belt is sacrificed to prevent damage to the rest of the machine if foreign material is input into the mulching machine. The belt consists of synthetic strands attached to a nylon core that can support a load of 250 lb. Each synthetic strand added to the cable adds 500 lb to the load the belt can handle.

a. Write a compound inequality to model this problem.

b. Determine the number of synthetic strands to put in this belt.

60. Strength Limits on a Drive Belt Rework Exercise 59, assuming that the drive belt for a mulching machine must be able to handle a load of at least 4,000 lb and that the design for the machine is for the belt to break on any load of 10,000 lb or greater.

a. Write a compound inequality to model this problem.

b. Determine the number of synthetic strands to put in this belt.

Group Discussion Questions

61. Discovery Question Solve these special-case absolute value equations and inequalities.

a. $|x| = -1$ **b.** $|x| < -1$ **c.** $|x| > -1$

Then generalize these results to give the solution of each of these inequalities for real numbers x and a and any negative real number d.

d. $|x - a| = d$ **e.** $|x - a| < d$ **f.** $|x - a| > d$

62. Error Analysis A student solved $|3x - 5| = 26$ by using the following steps:

$$\begin{aligned} |3x - 5| &= 26 \\ 3x + 5 &= 26 \\ 3x &= 21 \\ x &= 7 \end{aligned}$$

Identify any steps where this student made an error, and then determine the correct solution to this equation.

Section 4.5 Graphing Systems of Linear Inequalities in Two Variables

Objectives:

9. Graph a linear inequality in two variables.
10. Graph a system of linear inequalities.

A Mathematical Note

Winifred Edgerton Merrill (1862–1951) was the first American woman to receive a Ph.D. in mathematics. She studied mathematical astronomy and was later instrumental in the foundation of Barnard College for Women.

In this section we examine linear inequalities in two variables and systems of these inequalities. This topic is a natural extension of our work with systems of linear equations in Chapter 3 and our work in this chapter with inequalities involving only one variable. In Exercise 62, we will examine how inequalities can be used to describe possible options for businesses.

A **solution to a linear inequality** in two variables is an ordered pair of values that, when substituted into the inequality, makes a true statement.

EXAMPLE 1 Checking Possible Solutions of an Inequality

Determine whether each ordered pair is a solution of $2x + 3y \le 6$.

SOLUTION

(a) (5, 0)

$$2x + 3y \le 6$$
$$2(5) + 3(0) \stackrel{?}{\le} 6$$
$$10 \stackrel{?}{\le} 6 \text{ is false.}$$

Substitute the given coordinates into the inequality and determine whether a true statement results.

Answer: (5, 0) is not a solution.

(b) (0, 2)

$$2x + 3y \le 6$$
$$2(0) + 3(2) \stackrel{?}{\le} 6$$
$$6 \stackrel{?}{\le} 6$$

The inequality $\le$, less than or equal to, includes the equality, so the point (0, 2) is a solution.

Answer: (0, 2) is a solution.

(c) (0, 0)

$$2x + 3y \le 6$$
$$2(0) + 3(0) \stackrel{?}{\le} 6$$
$$0 \stackrel{?}{\le} 6$$

Answer: (0, 0) is a solution.

SELF-CHECK 4.5.1

Determine whether these points are solutions of $5x + y > 10$.

1. (2, 1) **2.** (2, 0) **3.** (0, 0)

The line $Ax + By = C$ separates the plane into two regions called half-planes. (See the following figure.) One half-plane will satisfy $Ax + By < C$, and the other half-plane will satisfy $Ax + By > C$. If the inequality is $\le$ or $\ge$, the line will be part of the solution. If the inequality is $<$ or $>$, the line will be the boundary separating the half-planes, but it will not be part of the solution.

Standard convention is to use a solid line for $\leq$ or $\geq$ when the line is part of the solution and to use a dashed line for $<$ or $>$ when the line is not part of the solution.

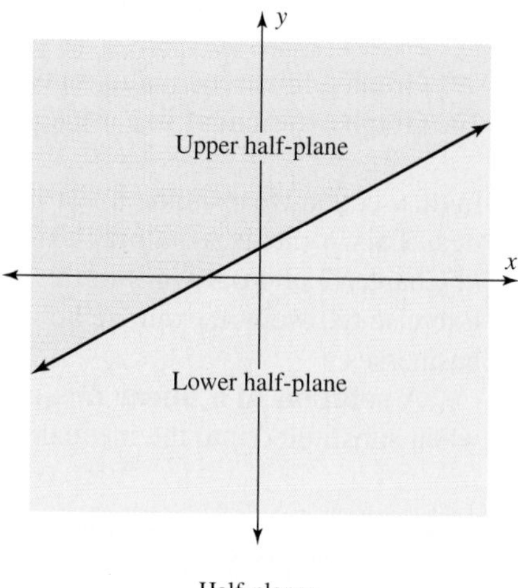

Half-planes

To graph a linear inequality, first we graph the line that forms the boundary for the half-planes. Then we use a test point to determine which half-plane satisfies the inequality.

EXAMPLE 2 Checking Possible Solutions of an Inequality

Determine which of the points A through F shown in the graph below satisfy the inequality $y > \frac{x}{2} - 1$.

SOLUTION

Equality: The dashed line represents all ordered pairs whose y-coordinate equals $\frac{x}{2} - 1$.

Upper half-plane: The upper half-plane represents all ordered pairs where the y-coordinate is greater than (above) $\frac{x}{2} - 1$. Thus A, B, and C represent solutions of $y > \frac{x}{2} - 1$.

Lower half-plane: The lower half-plane represents all ordered pairs where the y-coordinate is less than (below) $\frac{x}{2} - 1$. Thus D, E, and F are *not* solutions of $y > \frac{x}{2} - 1$.

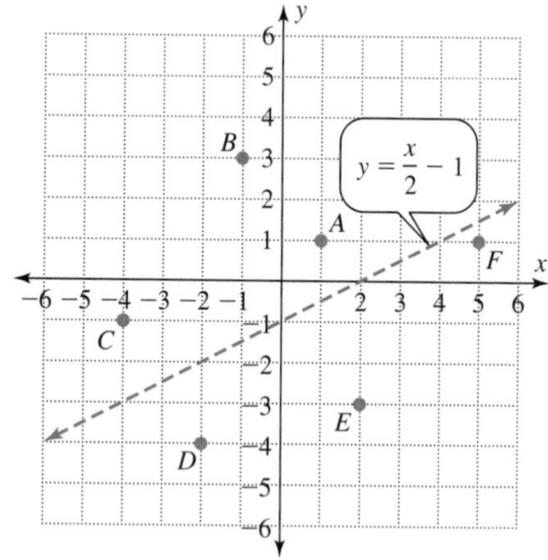

You also can check these points by substituting their coordinates into the given inequality.

In Example 2 you could never list all the solutions of the inequality $y > \frac{x}{2} - 1$. Thus we must represent these points algebraically or graphically.

Graphing a Linear Inequality

STEP 1 Graph the equality $Ax + By = C$, by using
- **a.** A solid line for $\leq$ or $\geq$
- **b.** A dashed line for $<$ or $>$

STEP 2 Choose an arbitrary test point not on the line; (0, 0) is often convenient. Substitute this test point into the inequality.

STEP 3
- **a.** If the test point satisfies the inequality, shade the half-plane containing this point.
- **b.** If the test point does not satisfy the inequality, shade the other half-plane.

Examples 3 and 4 illustrate each step of this procedure.

EXAMPLE 3 Graphing a Linear Inequality

Graph the solution of $3x - 5y < -15$.

SOLUTION

[1] Draw a dashed line for $3x - 5y = -15$ because the equality is not part of the solution. The line passes through the intercepts $(-5, 0)$ and $(0, 3)$.

[2] Test the origin:

$$3x - 5y < -15$$
$$3(0) - 5(0) \overset{?}{<} -15$$
$$0 \overset{?}{<} -15 \text{ is false.}$$

[3] Shade the half-plane that does *not* include the test point (0, 0).

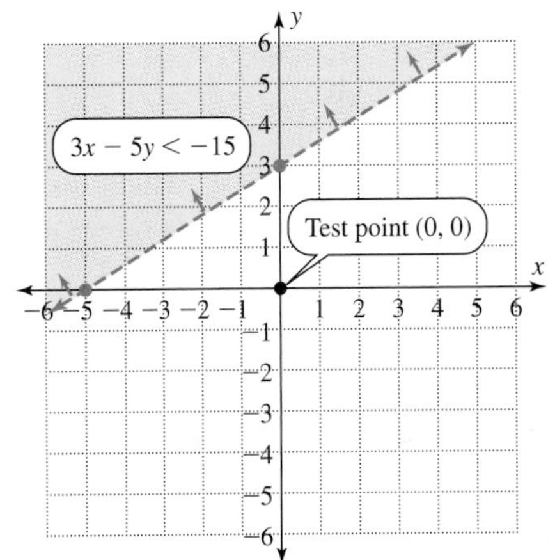

EXAMPLE 4 Graphing a Linear Inequality

Graph the solution of $5x \geq 3y$.

SOLUTION

[1] Draw a solid line for $5x = 3y$ since the original statement includes the equality. The line passes through (0, 0) and (3, 5).

[2] Test the point (1, 0). [Do not test (0, 0) since the origin lies on the line.]

$$5x \geq 3y$$
$$5(1) \overset{?}{\geq} 3(0)$$
$$5 \overset{?}{\geq} 0 \text{ is true.}$$

[3] Shade the half-plane that includes the test point (1, 0).

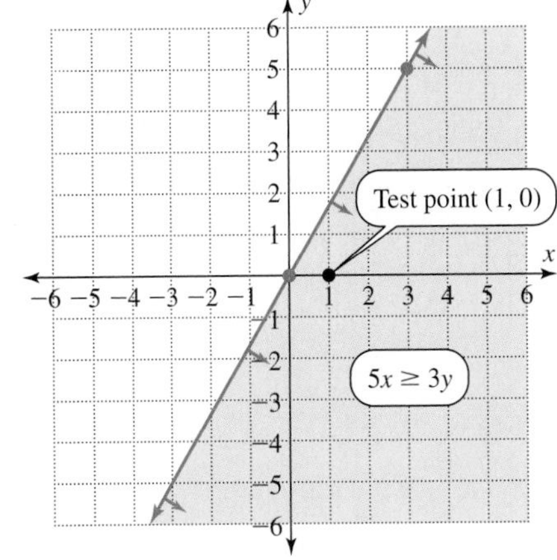

Observe that we used (0, 0) as a test point in Example 3 but not in Example 4. We always select a test point that is not on the line separating the two half-planes.

To graph a linear inequality on a TI-84 Plus calculator, first solve the inequality for y. Then enter the right side of the inequality next to Y_1, and select the shading that represents the appropriate half-plane. This is illustrated in Calculator Perspective 4.5.1

CALCULATOR PERSPECTIVE 4.5.1 **Graphing a Linear Inequality**

To graph the solution of $3x - 5y < -15$ from Example 3 on a TI-84 Plus calculator, first rewrite the inequality as $y > \frac{3}{5}x + 3$ and then enter the following keystrokes:

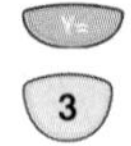

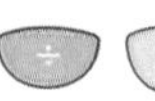

Note that on a TI-84 Plus calculator the symbol ◥ denotes that the upper half-plane should be shaded and the symbol ◣ denotes that the lower half-plane should be shaded.

Use the arrow keys to move the cursor to the left of Y_1 and press ENTER until the marker indicates the graph will be shaded above the line.

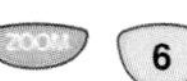

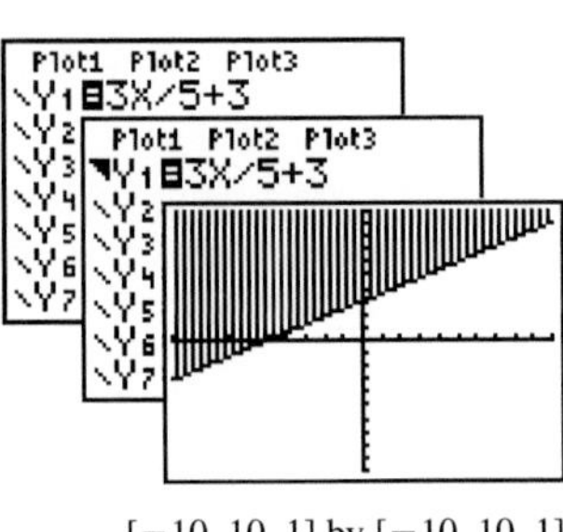

[−10, 10, 1] by [−10, 10, 1]

Note: For ≥ and > the graph will be shaded above the line, and for ≤ and < the graph will be shaded below the line. The calculator does not distinguish between ≥ and >, nor does it distinguish between ≤ and <.

EXAMPLE 5 Using a Graphing Calculator to Solve Inequalities

Use a graphing calculator to solve **(a)** $2x + 3y \leq 6$ and **(b)** $2x + 3y \geq 6$.

SOLUTION

(a) $2x + 3y \leq 6$

$$3y \leq -2x + 6$$

$$y \leq -\frac{2}{3}x + 2$$

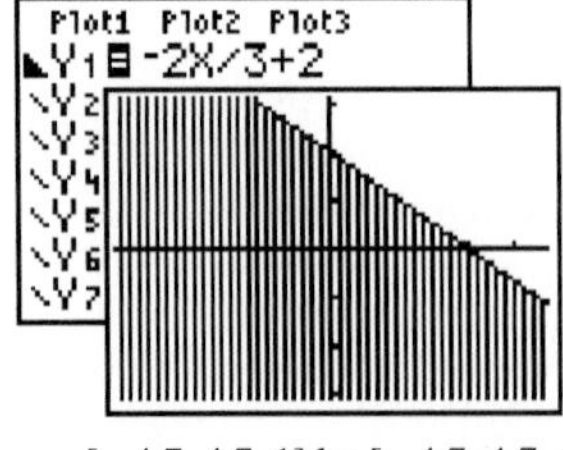

[−4.7, 4.7, 1] by [−4.7, 4.7, 1]

(b) $2x + 3y \geq 6$

$$3y \geq -2x + 6$$

$$y \geq -\frac{2}{3}x + 2$$

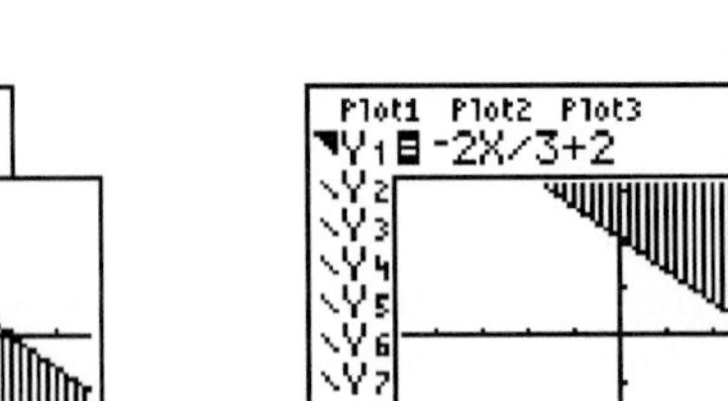

[−4.7, 4.7, 1] by [−4.7, 4.7, 1]

To graph these inequalities on a calculator, first solve for y. Then enter the equation into $\mathbf{Y_1}$ and select the appropriate shading for each inequality. The solution of the first inequality is the lower half-plane, and the solution of the second inequality is the upper half-plane.

SELF-CHECK 4.5.2

1. Graph $2x + y \leq 10$.
2. Use a graphing calculator to solve $4x - 5y \leq 20$.

To graph a system of linear inequalities, first we graph each inequality on the same coordinate system. The solution of the system is the intersection of these two individual regions. To clarify which points satisfy each inequality, the TI-84 Plus calculator uses vertical and horizontal lines to indicate the solutions of the individual inequalities. The solution of the system is the crosshatched region where these lines intersect. When we graph these inequalities by using pencil and paper, we use small arrows to indicate each individual half-plane. Then we shade the intersection of these half-planes as illustrated in Example 6.

EXAMPLE 6 Graphing a System of Linear Inequalities

Graph the solution of $\left\{\begin{matrix} x \geq -3 \\ x \leq 2 \end{matrix}\right\}$.

SOLUTION

Inspection reveals that both $x = -3$ and $x = 2$ represent vertical lines. The region containing (0, 0) satisfies both inequalities. Thus the solution set is the violet strip between these solid lines.

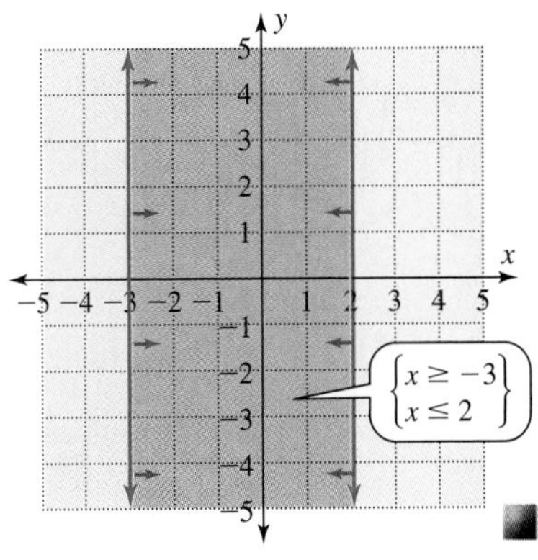

The solution to Example 7 proceeds step by step. First one inequality is graphed, and then the second inequality is graphed. Finally the intersection of these regions is shaded as the solution for the system.

EXAMPLE 7 Graphing a System of Linear Inequalities

Graph the solution of $\left\{\begin{matrix} x + y \geq 2 \\ 3x - 2y < 6 \end{matrix}\right\}$.

SOLUTION

(a) 1 To graph $x + y \geq 2$, draw a solid line for $x + y = 2$. Use the intercepts (2, 0) and (0, 2) to draw the line.

2 Test the point (0, 0) in $x + y \geq 2$

$0 + 0 \not\geq 2$ is false.

3 Shade the half-plane that does *not* contain (0, 0) (shown here in light red)

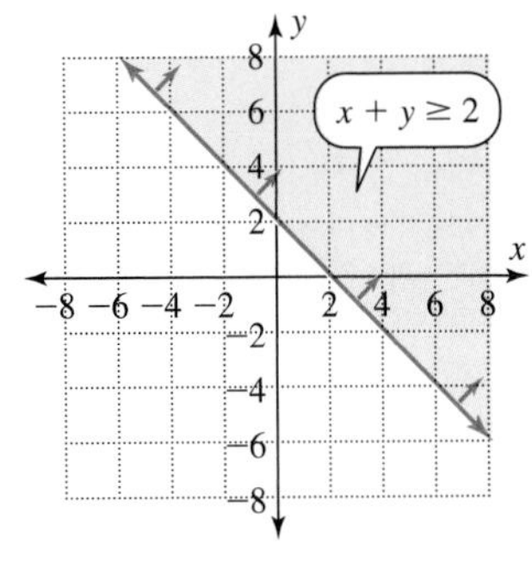

(b) 1 To graph $3x - 2y < 6$, draw a dashed line for $3x - 2y = 6$. Use the intercepts (2, 0) and (0, −3) to draw the line.

2 Test the point (0, 0) in

$$3x - 2y < 6$$
$$3(0) - 2(0) \overset{?}{<} 6$$
$$0 \overset{?}{<} 6 \text{ is true.}$$

3 Shade the half-plane that contains (0, 0) (shown here in light blue).

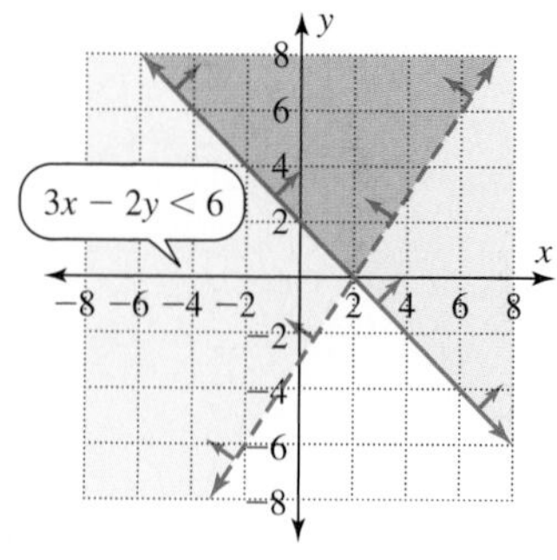

(c) The intersection of these two regions is the set of all the points satisfying this system of inequalities (shown here in violet).

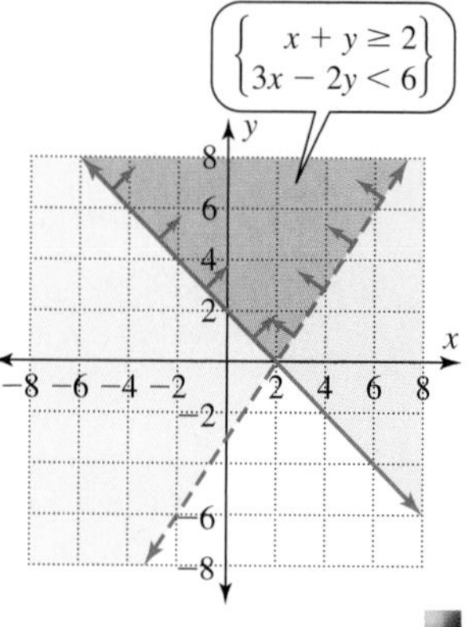

EXAMPLE 8 Using a Graphing Calculator to Solve a System of Inequalities

Use a graphing calculator to graph the solution of $\begin{Bmatrix} x + y \geq 2 \\ 3x - 2y < 6 \end{Bmatrix}$. This is the same system as in Example 7.

SOLUTION

$$x + y \geq 2$$
$$y \geq 2 - x$$

$$3x - 2y < 6$$
$$-2y < -3x + 6$$
$$y > \frac{3}{2}x - 3$$

To graph these inequalities on a calculator, first solve each inequality for y.

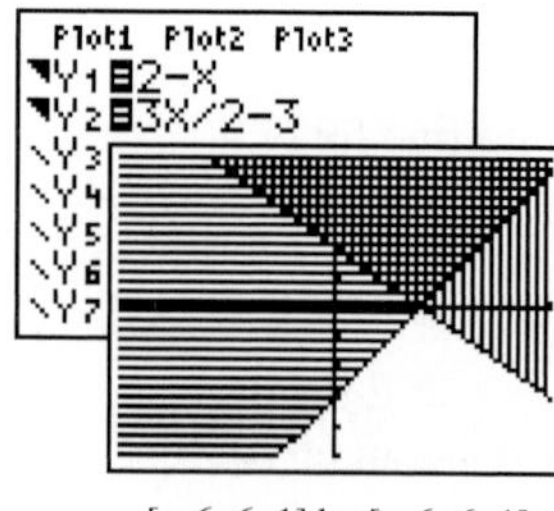

[−6, 6, 1] by [−6, 6, 1]

A graphing calculator displays the solution of a system of inequalities as the intersection of the different shading patterns.

Graphing calculators are powerful tools, but you must interpret properly the information they display. For inequalities, be careful to determine from the given inequalities whether the boundary lines on the display should be included in the solution set.

Answer: The solution set is represented by the crosshatched area. To interpret this solution properly, note that the line representing $y = \frac{3}{2}x - 3$ is not part of the solution.

SELF-CHECK 4.5.3

1. Graph the solution of $\begin{Bmatrix} y \le 4 \\ y > -3 \end{Bmatrix}$.

2. Use a graphing calculator to graph the solution of $\begin{Bmatrix} x + \ \ y \le 3 \\ 2x - 3y \le 6 \end{Bmatrix}$.

Example 9 presents an application that is modeled by a system of three linear inequalities. The solution for this system is graphed by using pencil-and-paper methods.

EXAMPLE 9 Length of Two Pieces of Optic Fiber

An installer for a telecommunications company used two pieces of optic fiber from a spool containing 100 m. Lengths x and y must both be positive, and their total is at most 100 m. Write a system of linear inequalities to model this problem, and then graph the solution of this system.

SOLUTION

VERBALLY	ALGEBRAICALLY	
x is positive.	$x > 0$	Translate each verbal statement into an algebraic inequality.
y is positive.	$y > 0$	
The total of x and y is at most 100.	$x + y \le 100$	

GRAPHICALLY

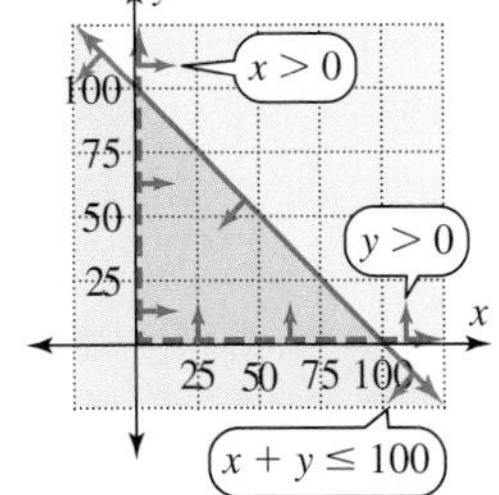

First graph each inequality on the same coordinate system. Then test the point (25, 25) in each inequality.

$x > 0$ $\qquad$ $y > 0$ $\qquad$ $x + y \le 100$

$25 \stackrel{?}{>} 0$ is true. $\qquad$ $25 \stackrel{?}{>} 0$ is true. $\qquad$ $25 + 25 \stackrel{?}{\le} 100$

$50 \stackrel{?}{\le} 100$ is true.

Use arrows to indicate each individual region formed by the linear boundaries, and then use violet shading to indicate the triangular region that is the solution of the system.

All x- and y-values within the violet shaded region are solutions of this system of inequalities.

SELF-CHECK ANSWERS

4.5.1

1. Solution
2. Not a solution
3. Not a solution

4.5.2

1.

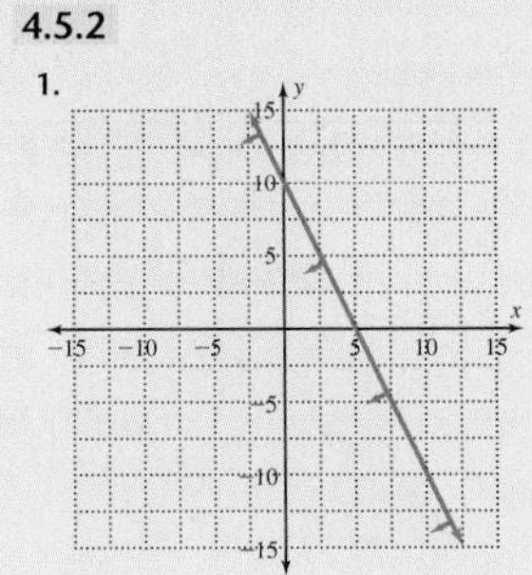

2.

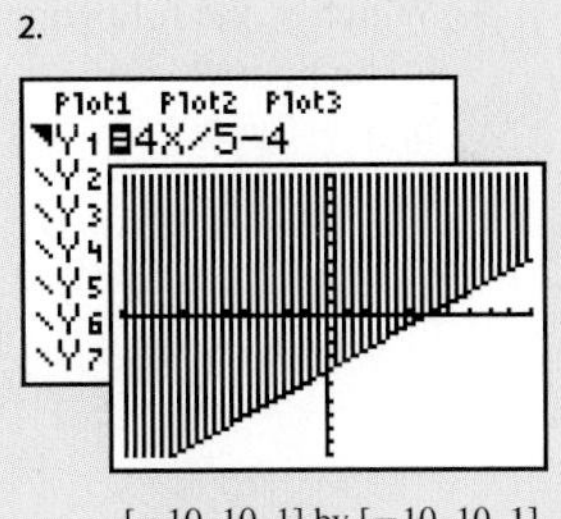

[−10, 10, 1] by [−10, 10, 1]

4.5.3

1.

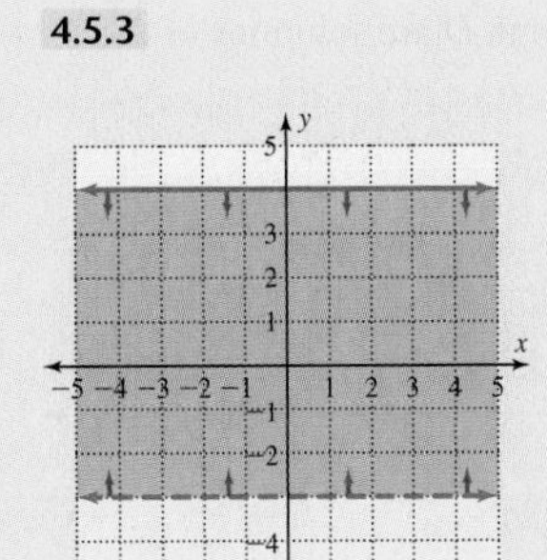

2.

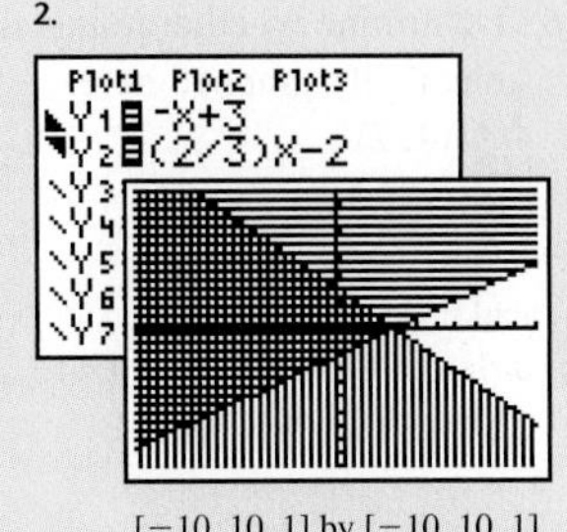

[−10, 10, 1] by [−10, 10, 1]

USING THE LANGUAGE AND SYMBOLISM OF MATHEMATICS 4.5

1. A solution to a linear inequality in two variables is an __________ __________ of values that, when substituted into the inequality, makes a __________ statement.
2. The line $Ax + By = C$ separates the plane into two regions called __________-__________.
3. For the linear inequalities $\leq$ and $\geq$ we graph the boundary equation by using a __________ line.
4. For the linear inequalities $<$ and $>$ we graph the boundary equations by using a __________ line.
5. If a test point satisfies an inequality, shade the half-plane that __________ this point.
6. If a test point does not satisfy an inequality, shade the half-plane that does __________ __________ this point.
7. If ◥ $y_1 = 3x - 7$ is used to graph an inequality on a graphing calculator, then the calculator will shade the __________ half-plane.
8. If ◣ $y_1 = 3x - 7$ is used to graph an inequality on a graphing calculator, then the calculator will shade the __________ half-plane.

EXERCISES 4.5

In Exercises 1–4, determine whether the given point is a solution of the inequality.

1. Check (0, 0) in these inequalities.
   a. $2x + 3y < 1$
   b. $2x + 3y \leq 1$
   c. $2x + 3y > 1$
   d. $2x + 3y \geq 1$
2. Check (1, 2) in these inequalities.
   a. $2x + 3y < 8$
   b. $2x + 3y \leq 8$
   c. $2x + 3y > 8$
   d. $2x + 3y \geq 8$
3. Check $(2, -3)$ in these inequalities.
   a. $3x - y < 9$
   b. $3x - y \leq 9$
   c. $3x - y > 9$
   d. $3x - y \geq 9$
4. Check $(-4, 1)$ in these inequalities.
   a. $x + 5y < 2$
   b. $x + 5y \leq 2$
   c. $x + 5y > 2$
   d. $x + 5y \geq 2$
5. Determine whether points A through D are solutions of the inequality graphed here.
   A (0, 0)
   B $(4, -2)$
   C (0, 2)
   D $(-5, 0)$

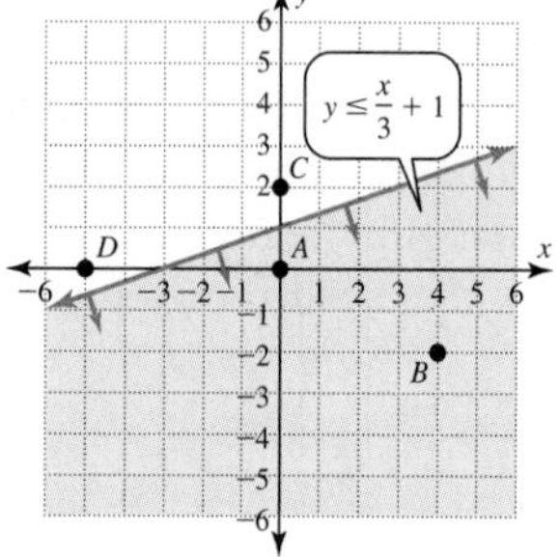

6. Determine whether points A through D are solutions of the inequality graphed here.
   A (4, 2)
   B (0, 4)
   C $(-2, 0)$
   D $(-4, -4)$

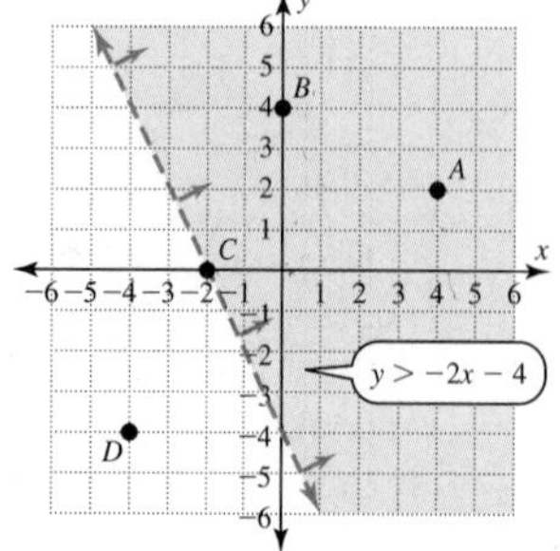

7. Determine whether points A through D are solutions of the inequality graphed here.

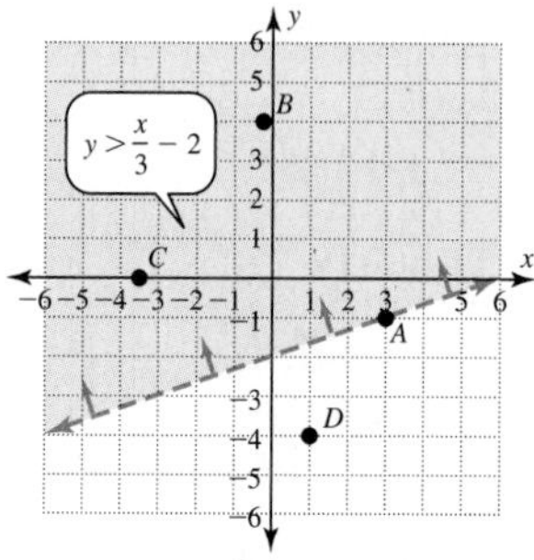

8. Determine whether points A through D are solutions of the inequality graphed here.

9. Which of the following inequalities corresponds to the graph in the figure?
   a. $x - 3y \leq 3$
   b. $x - 3y < 3$
   c. $x - 3y \geq 3$
   d. $x - 3y > 3$

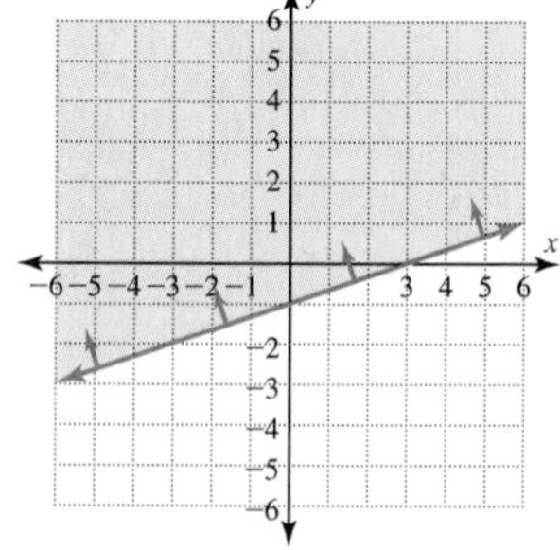

10. Which of the following inequalities corresponds to the graph in the figure?
a. $x + 2y \le 4$
b. $x + 2y < 4$
c. $x + 2y \ge 4$
d. $x + 2y > 4$

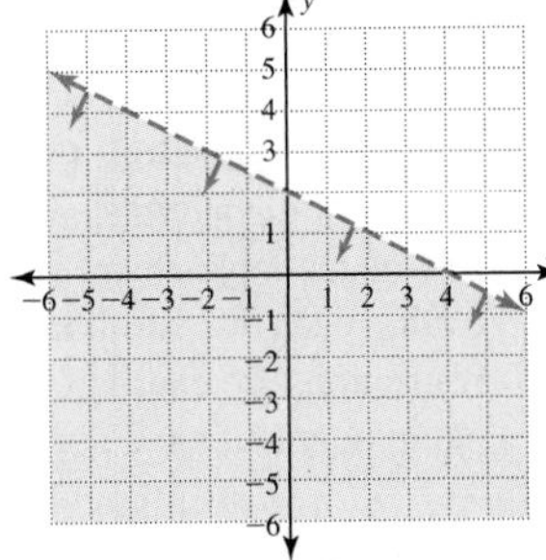

11. Determine which of points A through D satisfy the system of inequalities $\begin{Bmatrix} -x + 2y > 4 \\ 2x - y \ge -2 \end{Bmatrix}$.
A (4, 5) B (0, 2)
C (2, 6) D (−3, −2)

12. Determine which of points A through D satisfy the system of inequalities $\begin{Bmatrix} x - y \ge -4 \\ x + y < 4 \end{Bmatrix}$.
A (0, 4) B (−4, 0)
C (0, −4) D (0, 0)

In Exercises 13–16, shade the regions on the graphs of $y_1 = m_1x + b_1$ and $y_2 = m_2x + b_2$ that satisfy each system of inequalities. In Exercises 15 and 16, $m_1 = m_2$.

a. $\begin{Bmatrix} y_1 \ge m_1x + b_1 \\ y_2 \ge m_2x + b_2 \end{Bmatrix}$ b. $\begin{Bmatrix} y_1 \ge m_1x + b_1 \\ y_2 \le m_2x + b_2 \end{Bmatrix}$

c. $\begin{Bmatrix} y_1 \le m_1x + b_1 \\ y_2 \le m_2x + b_2 \end{Bmatrix}$ d. $\begin{Bmatrix} y_1 \le m_1x + b_1 \\ y_2 \ge m_2x + b_2 \end{Bmatrix}$

13.

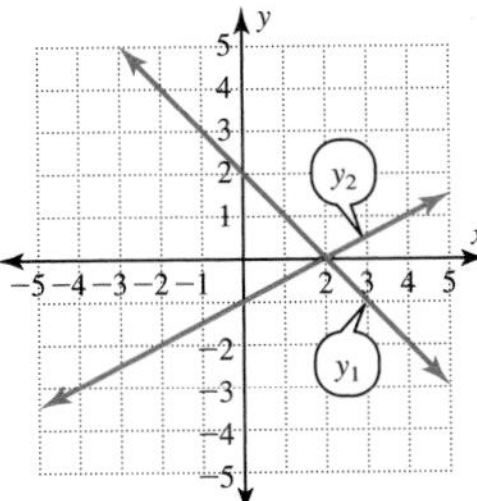

14.

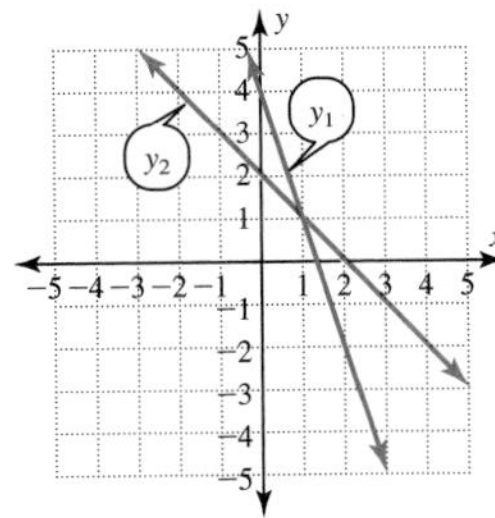

15.

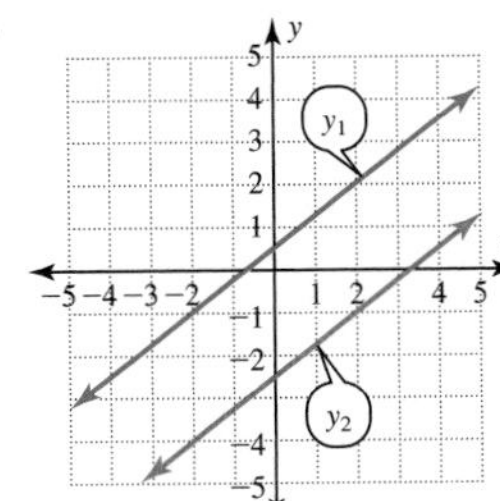

16.

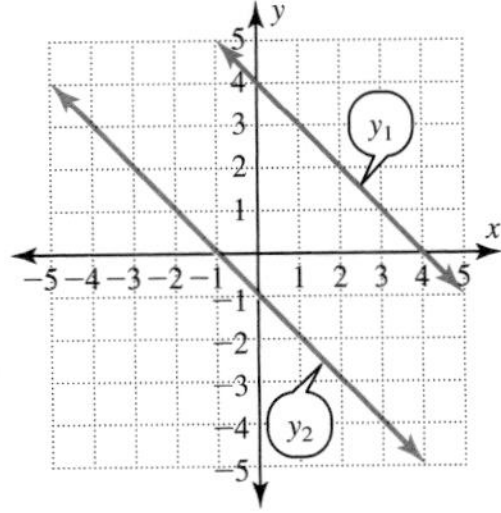

In Exercises 17–24, graph each linear inequality.

17. $x - y \ge 5$
18. $2x + y \le 6$
19. $3x - 2y - 12 < 0$
20. $-5x + 2y + 10 < 0$
21. $x > 3y$
22. $5x < -y$
23. $\frac{1}{2}x + \frac{1}{3}y \le 1$
24. $\frac{1}{4}x - \frac{1}{5}y \ge 1$

In Exercises 25–46, graph the solution of each system of linear inequalities.

25. $x - y \ge 4$
$2x - y < 6$

26. $x + 2y \le 2$
$2x - y \ge 4$

27. $2x - 5y - 10 < 0$
$2x - y - 7 \ge 0$

28. $3x + 2y - 12 > 0$
$2x + 5y - 10 < 0$

29. $\frac{x}{2} - \frac{y}{2} < 1$
$\frac{x}{2} + \frac{y}{2} > -1$

30. $\frac{x}{3} - \frac{y}{3} \ge -1$
$\frac{x}{4} + \frac{y}{4} \ge -1$

31. $x \ge 1$
$x < 4$

32. $x > -5$
$x \le -2$

33. $y < -2$
$y \ge -6$

34. $y > 3$
$y < 4$

35. $x > -2$
$x \le 3$

36. $y < 5$
$y > 3$

37. $y \ge -2$
$y \le -4$

38. $x \le -4$
$x \ge -1$

39. $x \ge 0$
$y \ge 0$
$2x + 3y < 6$

40. $x \ge 0$
$y \ge 0$
$5x + 2y < 10$

41. $x + y > 0$
$x - y < 0$
$y < 4$

42. $2x + 3y \le 0$
$3x - 2y \ge 0$
$y \ge -4$

43. $x \ge -2$
$x \le 2$
$y \ge -3$
$y \le 3$

44. $x > 1$
$x < 3$
$y > -5$
$y < -2$

45. $3x + 3y - 15 \le 0$
$6x + 2y - 18 \le 0$
$x \ge 0$
$y \ge 0$

46. $x - y - 2 \le 0$
$2x + 2y - 8 \le 0$
$x \ge 0$
$y \ge 0$

In Exercises 47 and 48, write the system of inequalities that is graphed on the given graphing calculator display. (Assume that the inequalities are either $\le$ or $\ge$ and are not $<$ or $>$.)

47.

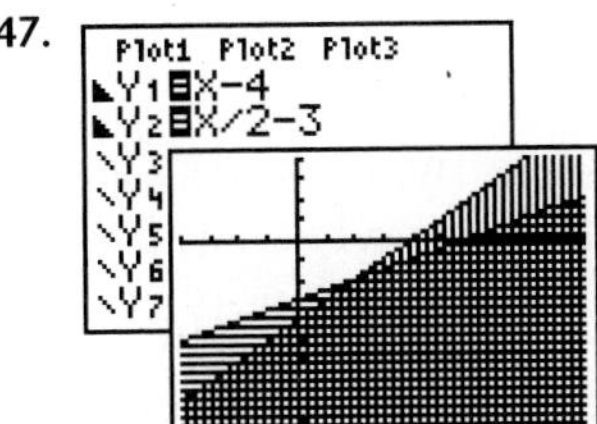

[−4, 10, 1] by [−10, 4, 1]

48.

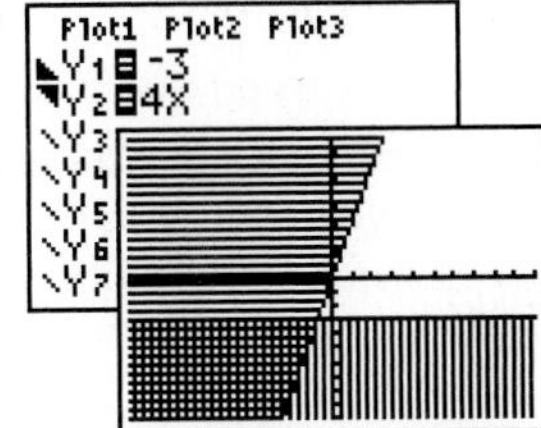

[−10, 10, 1] by [−10, 10, 1]

Multiple Representations

In Exercises 49–52, write an algebraic inequality for each verbal statement.

49. The x-coordinate is at least two more than the y-coordinate.
50. The y-coordinate is at least three more than the x-coordinate.
51. The y-coordinate is at most four more than the x-coordinate.
52. The x-coordinate is at most one more than the y-coordinate.

In Exercises 53–56, write a system of algebraic inequalities for these verbal statements.

53. Both the x- and y-coordinates are positive, and the sum of the two coordinates does not exceed ten.
54. Both the x- and y-coordinates are positive, and the sum of the two coordinates does not exceed eight.
55. Both the x- and y-coordinates are nonnegative, and the sum of x and twice y is at most five.
56. Both the x- and y-coordinates are nonnegative, and the sum of twice x and three times y is at most six.

In Exercises 57–60, write a system of inequalities that represents each situation and then graph this system of inequalities.

57. **Length of Two Pieces of Rope** A store sold two pieces of rope from a spool containing 150 m. Lengths x and y must both be positive, and their total is at most 150 m.

58. **Theater Tickets** A theater can seat at most 2,500 customers. Neither the number of adult tickets x nor the number of child tickets y can be negative, and their total is at most 2,500.

59. **Production Units** A company makes a profit of \$40 for each of the x stereos it ships and \$50 for each of the y televisions it ships. The number of units of each item shipped is nonnegative, and the profit per day for the factory has never exceeded \$4,400.

60. **Factory Production** A printer makes a profit of \$2 for each of the x books produced and \$0.25 for each of the y magazines produced. The number produced of each is nonnegative, and the profit per day for the printer has never exceeded \$15,000.

Group Discussion Questions

61. **Communicating Mathematically** Write two different word problems that describe a situation that can be modeled by this system of inequalities.

$$\begin{cases} x \geq 0 \\ y \geq 0 \\ x + y \leq 100 \end{cases}$$

62. **Challenge Question: Production Choices** The following system of inequalities models the restrictions on the number x of wooden table chairs and the number y of wooden rocking chairs made at a furniture factory each week. These restrictions are caused by limitations on the factory's equipment and labor supply.

$$\begin{cases} x \geq 0 \\ y \geq 0 \\ x + y \leq 50 \\ 3x + y \leq 90 \end{cases}$$

a. Determine each corner point of the region formed by graphing this system of inequalities.
b. The factory makes \$20 for each table chair and \$25 for each rocking chair. Write an expression for the profit involving x and y.
c. Evaluate the profit expression from part **b** at each of the corner points from part **a.** Which of these points produces the greatest profit for the factory?

63. **Challenge Question** Write a system of linear inequalities that is represented by the following graph.

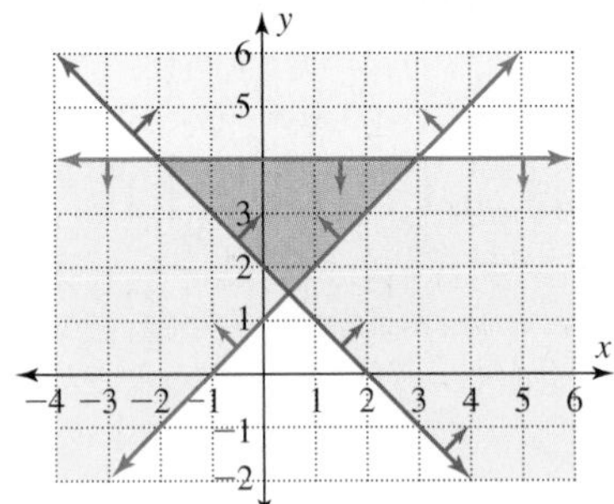

KEY CONCEPTS FOR CHAPTER 4

1. **Solution of an Inequality**
   - A solution of an inequality in one variable is a value of the variable that makes the inequality a true statement.
   - A solution of an inequality in two variables is an ordered pair that makes the inequality a true statement.
2. **Types of Inequalities**
   - Conditional inequality: An inequality that contains a variable and is true for some, but not all, real values of the variable(s).
   - Unconditional inequality: An inequality that is always true.
   - Contradiction: An inequality that is always false.
   - Equivalent inequalities: Inequalities that have the same solution are called equivalent inequalities.
   - Linear inequality: An inequality that is the first degree in each variable.
3. **Principles Used to Solve Inequalities***
   - **Addition-Subtraction Principle:** If a, b, and c are real numbers, then $a < b$ is equivalent to $a + c < b + c$ and to $a - c < b - c$.
   - **Multiplication-Division Principle:** If a, b, and c are real numbers and $c > 0$, then $a < b$ is equivalent to $ac < bc$. If $c < 0$, then $a < b$ is equivalent to $ac > bc$.†
4. **Intersection and Union of Two Sets**
   - Intersection: $A \cap B$ is the set of points in both A *and* B.
   - Union: $A \cup B$ is the set of points in either set A *or* B (or both).
5. **Compound Inequality:** $a \leq x \leq b$ is equivalent to $x \geq a$ and $x \leq b$.
6. **Absolute Value Equations and Inequalities:** For any real numbers x and a and positive real number d:
   - $|x - a| = d$ is equivalent to $x - a = -d$ or $x - a = d$.
   - $|x - a| < d$ is equivalent to $-d < x - a < d$.

*Similar statements can be made for the inequalities $\leq$, $>$, and $\geq$.
†Similar statements can be made for division.

- $|x - a| > d$ is equivalent to $x - a < -d$ or $x - a > d$.
- $|x - a| = -d$ is a contradiction and has no solution.
- $|x - a| < -d$ is a contradiction and has no solution.
- $|x - a| > -d$ is an unconditional inequality, and the solution set is the set of all real numbers.

7. **Solving an Absolute Value Equation of Inequality:** To solve an absolute value equation or inequality such as $|ax - b| + c < d$, first isolate the absolute value expression on the left side of the expression.
8. **Half-Planes:** The line $Ax + By = C$ separates the plane into two regions called half-planes. One half-plane satisfies $Ax + By < C$, and the other half-plane satisfies $Ax + By > C$.

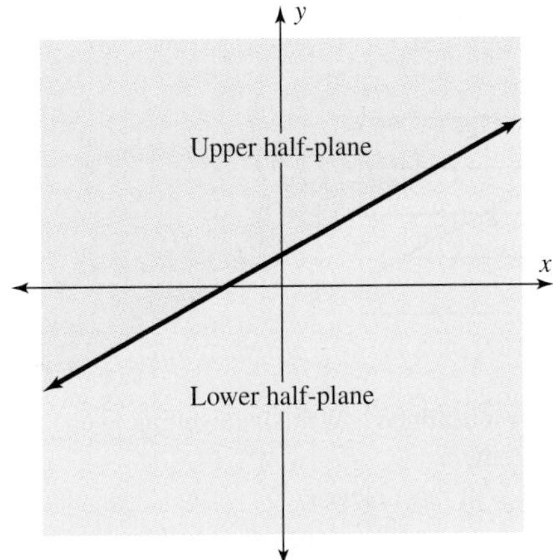

9. **Graphing a Linear Inequality**
   Step 1. Graph the equality $Ax + By = C$ by using
   a. A solid line for $\leq$ or $\geq$.
   b. A dashed line for $<$ or $>$.
   Step 2. Choose an arbitrary test point not on the line; (0, 0) is often convenient. Substitute this test point into the inequality.
   Step 3. a. If the test point satisfies the inequality, shade the half-plane containing this point.
   b. If the test point does not satisfy the inequality, shade the other half-plane.
10. **Graphing a System of Linear Inequalities:** Graph each inequality in the system on the same coordinate system. Use arrows to indicate each individual region formed by these lines. Then use shading to indicate the intersection of these regions. The solution of the system is represented by this intersection.

REVIEW EXERCISES FOR CHAPTER 4

1. Determine if $x = -4$ is a solution of each of these inequalities.
   a. $2x > -8$
   b. $x + 3 \leq -1$
   c. $5x + 3 < 2x - 3$
   d. $5(x - 2) \geq 3(2x - 4) + 7$
2. Determine which of points A through D satisfy the inequality $y < \frac{x}{3} - 2$.

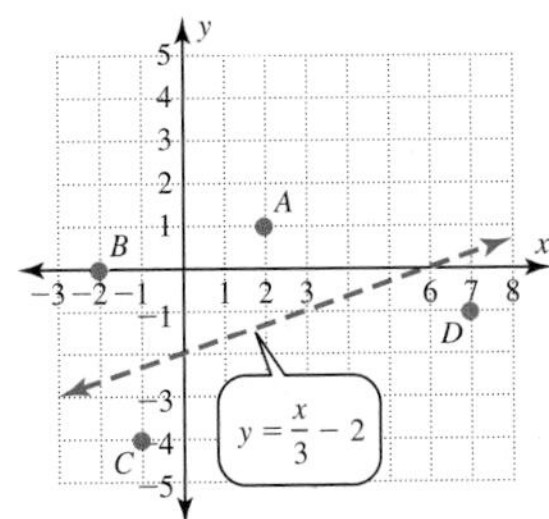

3. Determine if the point $(-2, 5)$ is a solution of each of these inequalities.
   a. $y \leq 2x - 12$
   b. $y \geq 3x - 12$
   c. $3x + 5y < 15$
   d. $-2x + 4y > 22$
4. Use the graphs of $Y_1 = 0.2x + 0.4$ and $Y_2 = 1$ to solve each equation and inequality.
   a. $0.2x + 0.4 = 1$
   b. $0.2x + 0.4 > 1$
   c. $0.2x + 0.4 < 1$

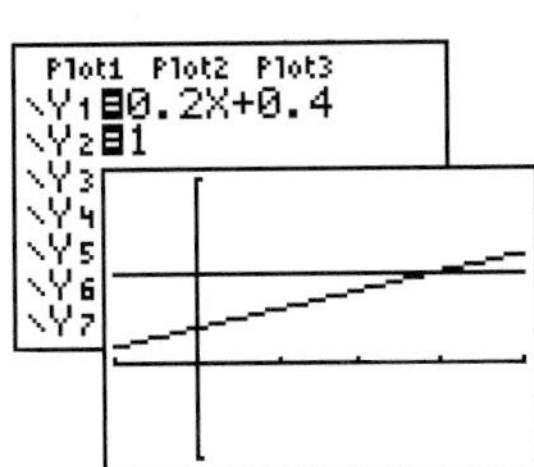

[−1, 4, 1] by [−1, 2, 1]

5. Complete the table for $y_1 = 5(x + 2)$ and $y_2 = 2x + 7$ by placing $<$, $=$, or $>$ in the third column. Then determine the x-values that satisfy each equation and inequality.
   a. $5(x + 2) = 2x + 7$
   b. $5(x + 2) < 2x + 7$
   c. $5(x + 2) > 2x + 7$

x	y_1	<, =, OR >	y_2
−3	−5		1
−2	0		3
−1	5		5
0	10		7
1	15		9
2	20		11
3	25		13

6. **Modeling Electricity Production from Wind and Coal** In 1990, only 3% of all electricity needed in one community was generated by wind and 43% was generated by coal. By 2000, 30% of the electricity needed was generated by wind and 20% was generated by coal. The graph of y_1 represents the percent of electricity generated by wind. The graph of y_2

represents the percent of electricity generated by coal. Use the graphs that follow to solve the following.

a. $y_1 = y_2$
b. $y_1 < y_2$
c. $y_1 > y_2$
d. Interpret the meaning of each of these solutions.

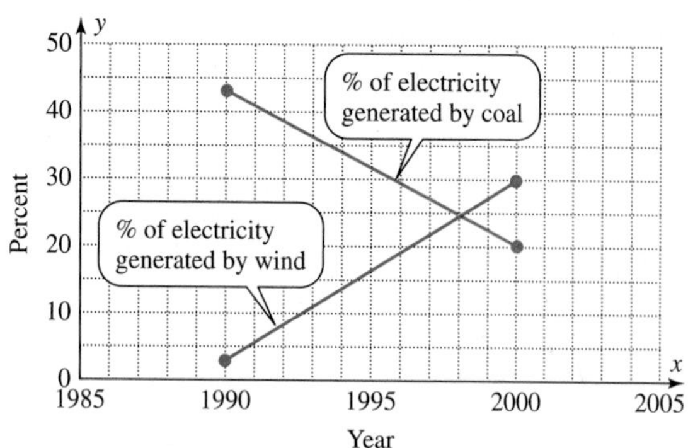

7. **Using Tables That Model Repair Shop Charges** The following tables display the charges by two TV repair shops based on the number of hours required for a repair. Use these tables to solve the following.

a. $y_1 = y_2$
b. $y_1 < y_2$
c. $y_1 > y_2$
d. Interpret the meaning of each of these solutions.

Shop A

HOURS x	COST y_1 ($)
1	50
2	85
3	120
4	155

Shop B

HOURS x	COST y_2 ($)
1	55
2	85
3	115
4	145

8. **Multiple Perspectives** Complete the following table.

VERBALLY	INEQUALITY NOTATION	INTERVAL NOTATION	GRAPH
a. x exceeds 3			
b.	$x \le 4$		
c.		$[2, +\infty)$	
d.			2 3 4 5 6 7 8

9. **Multiple Perspectives** Complete the following table.

VERBALLY	INEQUALITY NOTATION	INTERVAL NOTATION	GRAPH
a. x is greater than 2 and less than 6			
b.	$-3 < x \le 4$		
c.		$[-2, 0)$	
d.			−2 −1 0 1 2 3 4

10. Use the following table for $Y_1 = \dfrac{x-1}{2}$ to solve $-1 \le \dfrac{x-1}{2} < 1$.

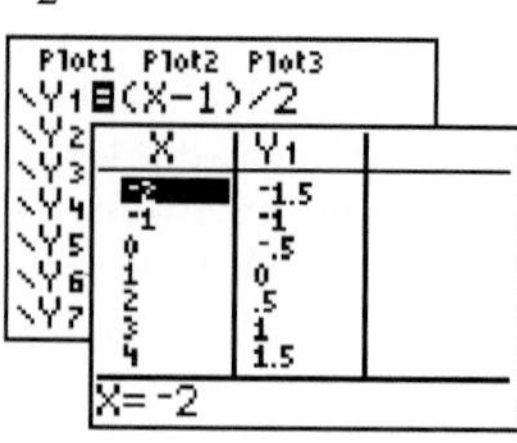

11. Use the given graph to solve $-1 \le \dfrac{2-x}{3} \le 1$.

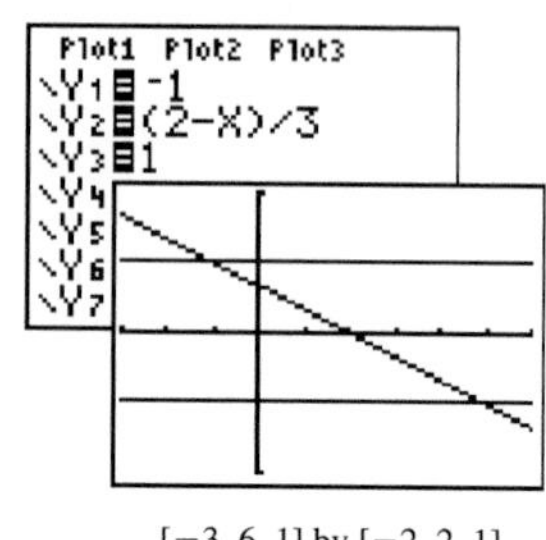

$[-3, 6, 1]$ by $[-2, 2, 1]$

12. **Multiple Representations** Write an algebraic inequality for each verbal statement.

a. x plus seven is at most eleven.
b. Twice the quantity x minus one exceeds thirteen.
c. Three x minus five is at least seven.
d. Four x plus nine never exceeds twenty-one.

Multiple Perspectives

In Exercises 13–18, solve each inequality algebraically, graphically, and numerically and describe the solution verbally.

13. $x + 3 > 1$
14. $\dfrac{x}{2} - 1 \le 1$
15. $\dfrac{x}{2} + 4 \ge 1 - \dfrac{x}{2}$
16. $2x - 1 < x + 2$
17. $2(x + 2) \le 3(x + 1)$
18. $|x - 2| < 2$

In Exercises 19–42, solve each inequality.

19. $x + 7 < 5$
20. $x - 11 > -10$
21. $-10 \ge 2x$
22. $-3x \le 12$
23. $3x + 7 > 5x + 13$
24. $7 \ge 7 - 9y$
25. $-\dfrac{x}{2} < \dfrac{1}{4}$
26. $\dfrac{3x}{7} + \dfrac{4}{5} > \dfrac{3x}{5} + \dfrac{2}{7}$
27. $7y + 14 \ge 2(3y + 8)$
28. $-5(2y - 2) \le 7 - 11y$
29. $2(3t - 4) > 3(t - 6) + 1$
30. $2(11t - 3) < 5(3t + 2) - 20$
31. $5(x - 4) + 6 < (x + 4) - 30$
32. $7(x - 3) + 6 \ge 12(x + 4) + 2$
33. $4(y - 1) - 7(y + 1) \le -3(y + 2) - 5$
34. $1 - 4(y - 2) \ge 3(y - 7) - 5(4 - y)$
35. $\dfrac{v}{2} - \dfrac{3v + 4}{4} > \dfrac{v + 40}{4}$
36. $\dfrac{v}{8} + 6 > \dfrac{v}{12} + 5$
37. $9 < x + 7 \le 13$
38. $0 \le \dfrac{4}{5}x < 20$
39. $42 \le \dfrac{-3m}{7} < 60$
40. $-5 \le 9 - x \le 5$
41. $-7 \le 5x + 3 \le 13$
42. $-10 < 5 - 3x < 8$

In Exercises 43–46, solve each absolute value equation or inequality.

43. $|x - 3| = 4$ **44.** $|2x - 5| + 6 = 9$

45. $|5x - 2| > 8$ **46.** $|x - 5| < 3$

In Exercises 47–50, each inequality is a conditional inequality, an unconditional inequality, or a contradiction. Identify the type of each inequality and solve it.

47. $3v + 3v > 5v$ **48.** $3v + 3v \le 5v$

49. $3(v + 1) < 3v + 4$ **50.** $5(3v - 1) > 3(5v - 1)$

In Exercises 51–54, determine $A \cap B$ and $A \cup B$ for the given intervals A and B.

51. $A = (1, 4), B = (2, 7)$

52. $A = [2, 5), B = [3, 6]$

53. $A = (-\infty, 2], B = [-3, \infty)$

54. $A = (-\infty, 5), B = [-2, 6)$

In Exercises 55 and 56, write each inequality expression as a compound inequality.

55. $x > -3$ and $x \le 4$

56. $x \ge 0$ and $x < \pi$

In Exercises 57 and 58, solve each inequality.

57. $x - 2 < 3$ and $3 > x + 2$

58. $x - 2 < 4x + 2$ or $6x - 6 < 4x - 12$

In Exercises 59–64, graph the solution of each system of inequalities.

59. $x \ge -1$
$x \le 3$

60. $y \le 4$
$y \ge 1$

61. $2x - 3y < 6$
$2x + 5y > 10$

62. $2x + y < 4$
$2x + y > 1$

63. $3x + 4y \le 12$
$3x - 4y \le 12$

64. $2x + y \le 4$
$x \ge 0$
$y \ge 0$

In Exercises 65–68, write an absolute value inequality to represent each interval.

65. $(-5, 5)$ **66.** $[-4, 4]$

67. $(-2, 10)$ **68.** $[-4, 20]$

Multiple Representations

In Exercises 69–71, write an algebraic inequality for each verbal statement, and then solve this inequality.

69. Four times the quantity x plus three is greater than eight. Solve for x.

70. Three times the quantity x minus seventeen is less than or equal to twice the quantity x plus eleven. Solve for x.

71. Three x minus two is greater than or equal to seven and is less than nineteen. Solve for x.

In Exercises 72–77, solve each word problem.

72. Basketball Average A basketball player scored 17 points, 27 points, and 18 points in the first three games of the season. How many points will she have to score in the fourth game to average at least 20 points for the first four games?

73. Average Salary When she was hired, an employee was guaranteed that she would average at least \$500 per week. The first three weeks she made \$560, \$450, and \$480. How much money must she make the fourth week to meet the guarantee?

74. Perimeter of a Rectangle The width of a rectangle must be exactly 12 cm. If the perimeter must be between 44 and 64 cm, determine the length that can be used for the rectangle.

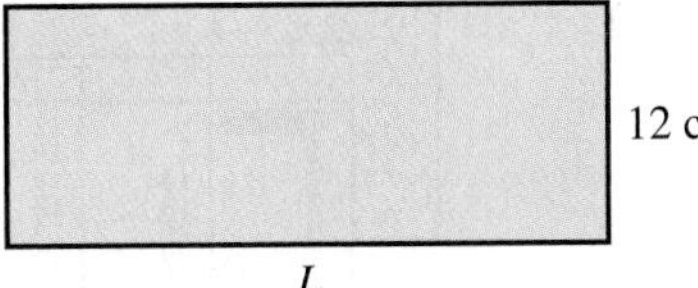

75. The Length of a Golf Shot Write a compound inequality that expresses the length of the tee shot that a golfer must make for the ball to clear the water and to land safely on the green of the golf hole shown here.

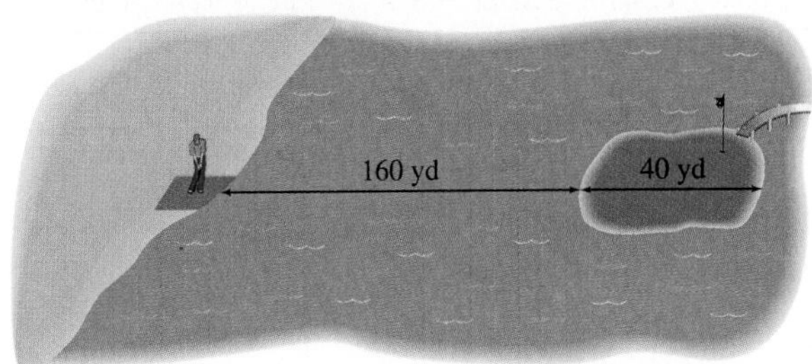

76. **Tolerance of Rocket Fuel** To adjust the flight of a Mars probe, a rocket is allowed to burn 7.2 g of fuel with a tolerance of 0.25 g. Express this tolerance interval as an absolute value inequality, and determine the lower and upper limits of the interval.

77. **Costs and Revenue** The cost of printing x advertising posters includes a fixed cost of \$75 and a variable cost of \$0.50 per poster. All posters are custom-ordered, and the charge for printing these posters includes a setup fee of \$50 plus a charge of \$1 per poster. Let y_1 represent the income received from an order for x posters and y_2 represent the cost of printing x posters. Determine the values of x for which $y_1 < y_2$, the loss interval for this order. Also determine the values of x for which $y_1 > y_2$, the profit interval for this order.

In Exercises 78–80, mentally estimate the solution of each inequality and then use a calculator to calculate the solution.

PROBLEM	MENTAL ESTIMATE	CALCULATOR SOLUTION
78. $x - 2.53 < 7.52$		
79. $-2.53x \geq 5.0853$		
80. $10.89 < 9.9x \leq 29.601$		

MASTERY TEST FOR CHAPTER 4

[4.1] **1.** Determine whether $x = 10$ is a solution of each inequality.

a. $4x - 1 \geq x + 29$

b. $2(x + 15) > -2(x + 5)$

c. $5(x + 2) < 2(x + 20)$

d. $5(x - 2) \leq 3(x - 10)$

[4.1] **2.** In parts **a** and **b** use the graph to solve each inequality.

a. $\dfrac{x}{3} - 2 < -\dfrac{1}{2}(x - 1)$

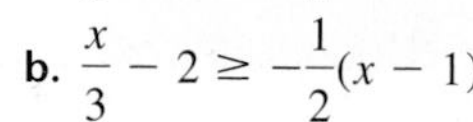

b. $\dfrac{x}{3} - 2 \geq -\dfrac{1}{2}(x - 1)$

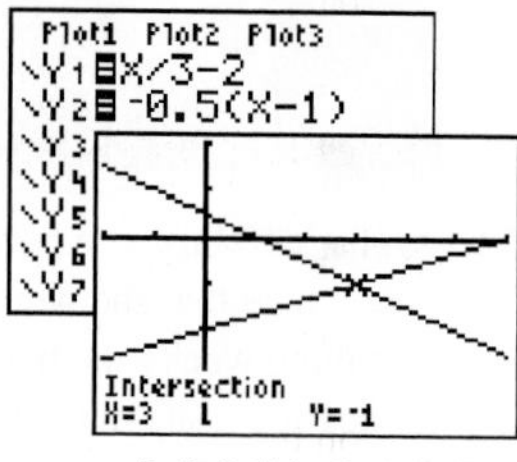

[−2, 6, 1] by [−4, 2, 1]

In parts **c** and **d** use the table to solve each inequality.

c. $4(x + 1) \leq 2(x + 6)$

d. $4(x + 1) > 2(x + 6)$

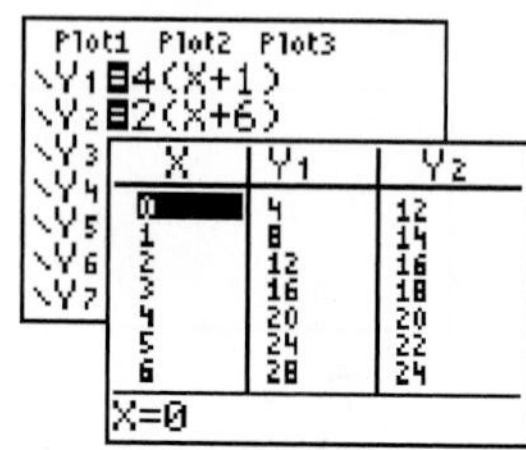

[4.1] **3.** Algebraically solve each of these inequalities.

a. $x + 12 \leq 17$

b. $x - 9 < -11$

c. $4x + 3 > 3x - 8$

d. $-2x - 5 \geq -3x - 1$

[4.2] **4.** Algebraically solve each of these inequalities.

a. $-11x \leq 165$

b. $\dfrac{3x}{7} > 105$

c. $8(x + 3) < 4(x + 4) - (10 - 2x)$

d. $-5 \leq \dfrac{4x - 9}{3}$

[4.3] **5.** Identify each of these inequalities as a conditional inequality, an unconditional inequality, or a contradiction, and then solve the inequality.

a. $x + x < x$

b. $2(3x + 5) \geq 3(2x - 6)$

c. $3(8x + 1) > 4(6x + 5)$

d. $2x + 3 \leq 3x + 3$

[4.3] **6.** Solve each of these inequalities.

a. $-30 < -6x \leq 48$

b. $-30 \leq x - 6 < 48$

c. $-1 \leq 4x + 3 \leq 19$

d. $2x + 1 \leq -7$ or $3x - 2 \geq 7$

[4.4] **7.** Algebraically solve each equation and inequality.

a. $|x| = 8$

b. $|2x + 5| = 49$

c. $|x - 5| > 4$

d. $|5x - 3| + 2 \leq 10$

[4.4] **8.** In parts **a** and **b**, use the graph to solve each inequality.

a. $|2x - 3| < 3$

b. $|2x - 3| > 3$

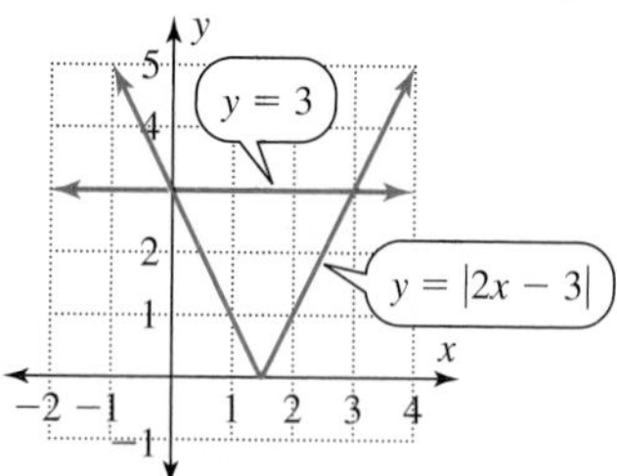

In parts **c** and **d** use the table to solve each inequality.

c. $\left|\dfrac{x}{2} + 1\right| \leq 1$

d. $\left|\dfrac{x}{2} + 1\right| \geq 1$

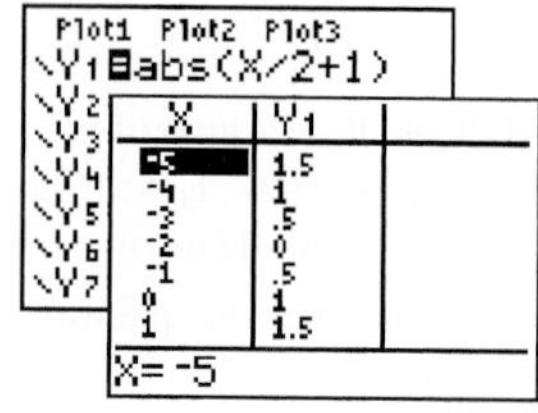

[4.5] **9.** Graph each of these linear inequalities.

a. $2x - 3y < 6$

b. $y \geq \dfrac{x}{2}$

c. $x \leq -2$

d. $y > -2$

[4.5] **10.** Graph the solution to each system of inequalities.

a. $\begin{cases} x \geq 2 \\ x \leq 4 \end{cases}$

b. $\begin{cases} y > -2 \\ y < 3 \end{cases}$

c. $\begin{cases} x + 2y \geq 4 \\ -2x + y < 4 \end{cases}$

d. $\begin{cases} x \geq 0 \\ y \geq 0 \\ 2x + y \leq 4 \end{cases}$

REALITY CHECK FOR CHAPTER 4

The design specifications for a bridge call for using a cable that can handle a load of 50,000 lb. Using a safety factor of 2 would require the cable to have a strength of 100,000 lb, and a safety factor of 5 would require the cable to have a strength of 250,000 lb. For one type of cable in this load range, the equation $y = 250x - 150$ where x is the strand diameter in inches, can be used to approximate the load strength y in thousands of pounds.

a. Write a compound linear inequality to represent strand diameters that would satisfy safety factors between 2 and 5 for this bridge cable.

b. Solve this inequality and interpret your answer.

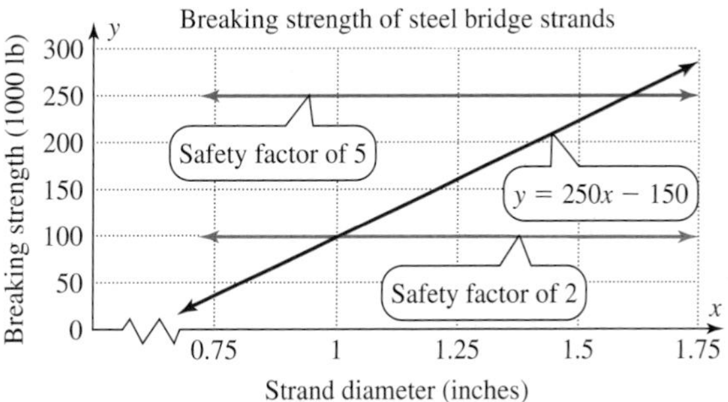

GROUP PROJECT FOR CHAPTER 4

Tolerance Intervals and Business Decisions

An engineer is responsible for the installation on the machinery that produces rollers for laser jet printers. Part of his job is to ensure that the rollers produced meet the specifications in a contract with one of the largest customers for these rollers. These rollers are supposed to be 2.50 cm in diameter, and 98.5% or more of the delivered rollers must satisfy the tolerance interval $|x - 2.50| < 0.015$ where the units are centimeters.

1. **a.** Express this tolerance interval in inequality notation. ______________
   b. What is the lower limit of this tolerance interval? ______________
   c. What is the upper limit of this tolerance interval? ______________
2. As part of the daily quality control check, workers randomly select 20 rollers each hour to carefully measure. The table shown here gives the measurements taken between 8:00 and 9:00 on a Monday morning.

a. Complete the table by marking each measurement as acceptable or unacceptable.

ROLLER NO.	MEASUREMENT (cm)	ACCEPTABLE (✓)	UNACCEPTABLE (✓)
1	2.51		
2	2.48		
3	2.48		
4	2.49		
5	2.50		
6	2.50		
7	2.49		
8	2.50		
9	2.51		
10	2.52		
11	2.50		
12	2.49		
13	2.48		
14	2.50		
15	2.50		
16	2.49		
17	2.47		
18	2.50		
19	2.51		
20	2.49		

b. How many of these 20 rollers were acceptable? ______________________

c. How many of these 20 rollers were unacceptable? ______________________

d. What percent of these 20 rollers was acceptable? ______________________

3. Based on data collected over several days, the engineer in charge of this machinery actually expects 99.5% of the rollers produced to be within the tolerance interval $|x - 2.50| < 0.015$. Suppose that data collected from 100 rollers during the first 5 hours of operation on Monday suggest trouble with the machinery. The samples yield 95 rollers that are acceptable and 5 rollers that are unacceptable. The next routine downtime for maintenance is not scheduled until midnight on Friday night.

The current plant schedule is to produce 20,000 rollers on the Monday through Friday shifts. All this production will be shipped the next day to the plant's primary customer for use in its production on the following Monday.

The engineer who has received this information must recommend a course of action for the plant manager. There are three options:

Option A: *Continue to operate the plant as it is now and ship the next 20,000 rollers to the customer, requiring that 98.5% or more of the delivered rollers must satisfy the tolerance interval* $|x - 2.50| < 0.015$ *where the units are centimeters. If these rollers fail this customer's inspections, the order may be rejected and a \$13,000,000 contract can be canceled.*

Option B: *Continue production but use a backup plan of putting quality control inspectors on the packaging line to examine the product. Past data show that 99% of the rollers screened by inspectors that are marked as acceptable are actually acceptable. (Even the inspectors make some errors.) Hiring a full crew of inspectors will cost the company \$2,500 to pay the inspectors and an additional \$3,500 in lost materials because of the discarded rejects that are produced. The cost of discarded rollers could increase if the problem with the machinery gets worse instead of staying the same.*

Option C: *Stop production immediately and readjust the machinery to bring it back to specifications. Estimated time to readjust the equipment is 4 hours. This will mean one shift is sent home 3 hours early and another shift will be paid for 1 hour of work before production can resume. This will also result in another 4 hours of overtime to make up for lost production. The cost for the added maintenance work is estimated to be \$1,250, and the cost for the extra production labor is \$5,000.*

Which option would you recommend? ______________________________________

Describe why you would select this option rather than one of the other two. ______________

____

____

____

chapter 5

Exponents and Operations with Polynomials

CHAPTER OUTLINE

- 5.1 **Product and Power Rules for Exponents**
- 5.2 **Quotient Rule and Zero Exponents**
- 5.3 **Negative Exponents and Scientific Notation**
- 5.4 **Adding and Subtracting Polynomials**
- 5.5 **Multiplying Polynomials**
- 5.6 **Special Products of Binomials**
- 5.7 **Dividing Polynomials**

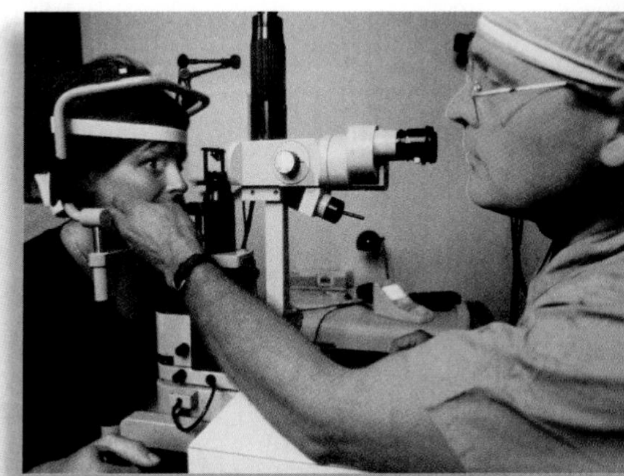

The curvature of the cornea and lens in a person's eye can be modeled by polynomial equations. In a person over the age of 40, the curvature of the corneas often becomes flatter; that is, the cornea equation changes. The resulting shape can cause problems with close-up vision. One solution to this is the use of reading glasses or contact lenses. Another solution is to use surgery to reshape the corneas.

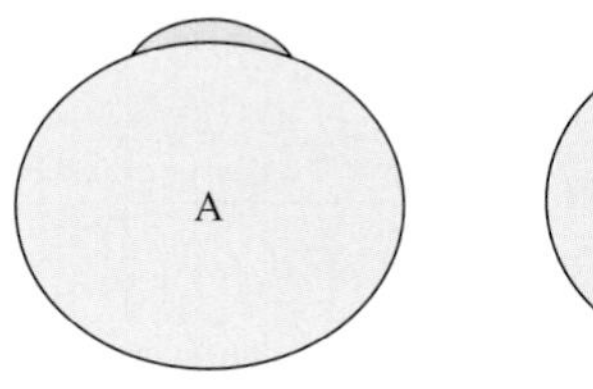

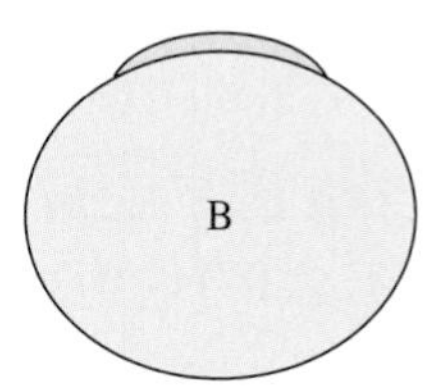

REALITY CHECK

Which of the two eyes pictured has a flatter curvature of the cornea and is from a patient with close-up vision problems?

Section 5.1 Product and Power Rules for Exponents

Objectives:

1. Use the product rule for exponents.
2. Use the power rule for exponents.

A Mathematical Note

In his authoritative *History of Mathematics,* David Eugene Smith attributes our present use of integral exponents to Renè Descartes (1596–1650).

One of the themes of this book is that algebra presents the language and symbolism of mathematics—a language and symbolism invented by humans to make things easier to represent and examine. Obviously, a student of algebra must study this notation in order to use it effectively—but please keep in mind that without algebra, mathematical concepts will be very cumbersome and lengthy to describe. Exponential notation, first examined in Section 1.5, is one example of a compact algebraic notation. This notation is reviewed in the box below.

Exponential Notation

ALGEBRAICALLY	VERBALLY	NUMERICAL EXAMPLES
For any natural number n. $b^n = \underbrace{b \cdot b \cdot \cdots \cdot b}_{n \text{ factors of } b}$ with base b and exponent n.	For any natural number n. b^n is the product of b used as a factor n times. The expression b^n is read as "b to the nth power."	$7^4 = 7 \cdot 7 \cdot 7 \cdot 7$ $(-3)^2 = (-3)(-3)$

Question: To determine the value of each \$1 invested at a 7% annual rate of return for a period of 10 years, we can use 1.07 as a factor 10 times (1.07 is 100% plus 7%). How would you represent this computation algebraically? Then use a calculator to approximate the value of this expression.

Answer: We can represent this product in either the expanded form or the exponential form.

EXPANDED FORM	EXPONENTIAL FORM
(1.07) (1.07) (1.07) (1.07) (1.07) (1.07) (1.07) (1.07) (1.07) (1.07)	$(1.07)^{10}$

The calculator display confirms that both results are identical. In either case, the value of \$1 nearly doubles in 10 years, producing a new value of about \$1.97.

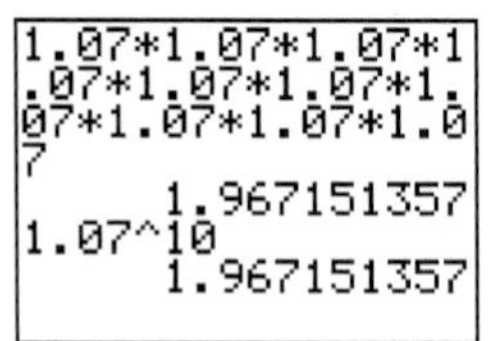

Try this yourself. Which of these two algebraic representations do you prefer?

In this section we examine the product and power rules, which enable us to quickly multiply exponential expressions with the same base. Before examining these rules, we will practice interpreting exponential notation.

The most common error made in evaluating exponential expressions is the use of the wrong base. If there are no symbols of grouping, only the constant or variable immediately to the left of the exponent is the base. Recall from Section 1.5 that $(-3)^2 = (-3)(-3) = 9$, whereas $-3^2 = -(3 \cdot 3) = -9$. The subtle but significant differences in the expressions in each part of Example 1 are very important for you to understand.

EXAMPLE 1 Identifying the Base of an Exponential Expression

Write each exponential expression in expanded form.

SOLUTION

(a) x^4	$x \cdot x \cdot x \cdot x$	The base is x, and the exponent is 4.
(b) $-x^4$	$-x \cdot x \cdot x \cdot x$	The base is x, not $-x$, and the exponent is 4.
(c) $(-x)^4$	$(-x)(-x)(-x)(-x)$	The base is $-x$, and the exponent is 4.
(d) $(xy)^3$	$(xy)(xy)(xy)$	The base is xy, and the exponent is 3.
(e) xy^3	$x \cdot y \cdot y \cdot y$	The base of the exponent 3 is y, not xy.

To understand the properties of exponents, it is important that you note the base of the exponent in each part of Examples 1 and 2. You can also practice this skill by completing Self-Check 5.1.1.

EXAMPLE 2 Identifying Bases and Exponents

Write each expression in exponential form.

SOLUTION

(a) $x \cdot x \cdot y \cdot y$	x^2y^2
(b) $x \cdot x + y \cdot y$	$x^2 + y^2$
(c) $(x + y)(x + y)$	$(x + y)^2$
(d) $-3 \cdot a \cdot a \cdot a \cdot a$	$-3a^4$
(e) $(-3a)(-3a)(-3a)(-3a)$	$(-3a)^4$

SELF-CHECK 5.1.1

Write each exponential expression in expanded form.

1. $-5a^3$ **2.** $(-5a)^3$ **3.** a^3b^2

Write each expression in exponential form.

4. $-x \cdot x$ **5.** $(-x)(-x)$ **6.** $x \cdot y \cdot y \cdot y$

Product Rule

The product rule and the quotient rule are shortcuts that will greatly simplify your work with exponential expressions. These properties of exponents follow directly from the definition of natural number exponents. We start by examining the logic behind the product rule for exponents. The product $x^3 \cdot x^4$ is written here in expanded form.

Three factors of x plus four more factors of x results in seven factors of x.

$$x^3x^4 = \underbrace{x \cdot x \cdot x}_{\text{Three factors of } x} \cdot \underbrace{x \cdot x \cdot x \cdot x}_{\text{Four factors of } x} = \underbrace{x \cdot x \cdot x \cdot x \cdot x \cdot x \cdot x}_{\text{Seven factors of } x} = x^7$$

Thus, $x^3 \cdot x^4 = x^{3+4} = x^7$. The fact that x^3x^4 and x^7 are equivalent expressions means they have the same value for all values of x. This is confirmed by comparing the values of Y_1 and Y_2 in the following table.

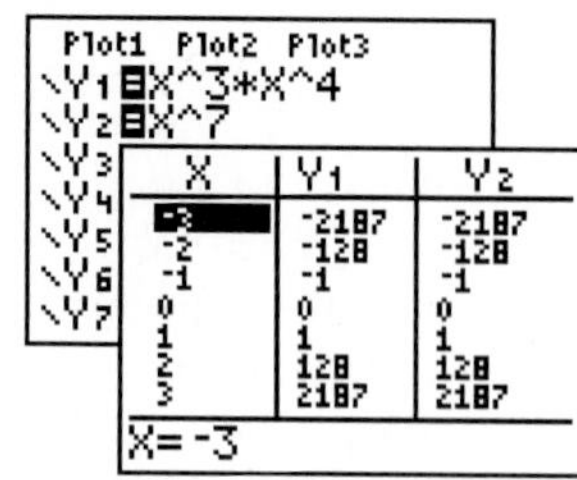

In general, the product $x^mx^n = x^{m+n}$ has the base x used as a factor a total of $m + n$ times.

m factors of *x* plus *n* more factors of *x* results in $m + n$ factors of *x*.

$$x^mx^n = \overbrace{(x \cdot x \cdots x)}^{m \text{ factors of } x} \cdot \overbrace{(x \cdot x \cdots x)}^{n \text{ factors of } x} = \overbrace{x \cdot x \cdots x}^{m+n \text{ factors of } x} = x^{m+n}$$

Although it is useful to memorize "add the exponents" for multiplying two factors, it is also important to remember that the bases must be the same.

Product Rule for Exponents

For any real number x and natural numbers m and n:

ALGEBRAICALLY	VERBALLY	ALGEBRAIC EXAMPLE
$x^m \cdot x^n = x^{m+n}$	To multiply two factors with the same base, use the common base and add the exponents.	$x^3 \cdot x^4 = x^7$

EXAMPLE 3 Using the Product Rule

Use the product rule to simplify each expression.

SOLUTION

(a) z^3z^{10}

$z^3z^{10} = z^{3+10}$
$= z^{13}$

Add the exponents with the common base z.

(b) $v^2v^4v^5$

$v^2v^4v^5 = v^{2+4+5}$
$= v^{11}$

Add the exponents with the common base v. The product rule can be extended to any number of factors that have the same base.

(c) $-x^7x^8$

$-x^7x^8 = -x^{7+8}$
$= -x^{15}$

$-x^7x^8 = -(x^7x^8)$; the base for each exponent is x, not $-x$.

(d) $(3y^4)(5y^5)$

$(3y^4)(5y^5) = (3)(5)y^{4+5}$
$= 15y^9$

Add the exponents with the common base y.

(e) y^3x^2

$y^3x^2 = x^2y^3$

Since these factors do not have the same base, this expression cannot be simplified. However, factors are usually written in alphabetical order.

SELF-CHECK 5.1.2

Simplify each expression.

1. $x^2 \cdot x^3$
2. $y^{15}y^{16}$
3. m^2n^4
4. $-z^2z^9$

Power Rule

The next important property of exponents that we examine is the power rule for exponents. We examine first the logic for raising a power to a power. The expression $(x^2)^4$ is expanded here.

$$(x^2)^4 = \overbrace{x^2 \cdot x^2 \cdot x^2 \cdot x^2}^{\text{Four factors of } x^2}$$

$$= \overbrace{(x \cdot x)(x \cdot x)(x \cdot x)(x \cdot x)}^{\text{Four groups of two factors of } x}$$

Four groups of two factors of x means that there are $2 \cdot 4$ factors of x.

$$= \overbrace{x \cdot x \cdot x \cdot x \cdot x \cdot x \cdot x \cdot x}^{2 \cdot 4 \text{ factors of } x}$$

$$= x^8$$

Thus, $(x^2)^4 = x^8$. This fact is confirmed by comparing the values of Y_1 and Y_2 in the following table:

```
Plot1 Plot2 Plot3
\Y1=(X^2)^4
\Y2=X^8
\Y3
\Y4
\Y5
\Y6
\Y7
```

X	Y1	Y2
-3	6561	6561
-2	256	256
-1	1	1
0	0	0
1	1	1
2	256	256
3	6561	6561

X=-3

In general,

$$(x^m)^n = \overbrace{x^m \cdot x^m \cdots x^m}^{n \text{ factors of } x^m}$$

$$= \overbrace{(x \cdot x \cdots x)(x \cdot x \cdots x) \cdots (x \cdot x \cdots x)}^{n \text{ groups of } m \text{ factors of } x}$$

$$= \overbrace{x \cdot x \cdot x \cdots x \cdot x \cdot x}^{m \cdot n \text{ factors of } x}$$

$$= x^{mn}$$

Power Rule for Exponents

For any real number x and natural numbers m and n:

ALGEBRAICALLY	VERBALLY	ALGEBRAIC EXAMPLE
$(x^m)^n = x^{mn}$	To raise a power to a power, multiply the exponents.	$(x^2)^4 = x^8$

EXAMPLE 4 Using the Power Rule

Simplify each expression.

SOLUTION

(a) $(b^3)^4$ $\qquad (b^3)^4 = b^{3\cdot 4}$ $\qquad$ To raise a power to a power, multiply the exponents.

$$= b^{12}$$

(b) $-(x^3)^2$ $\qquad -(x^3)^2 = -x^{3\cdot 2}$ $\qquad$ Multiply the exponents. The base is x, not $-x$.

$$= -x^6$$

SELF-CHECK 5.1.3

Simplify each expression.

1. $(x^2)^5$
2. $(x^5)^2$
3. $(-x^5)^2$
4. $(-x^2)^5$
5. $-(x^5)^2$

The power rule can be applied to products and quotients as described in the following box. We develop the logic for these rules immediately following this box.

Raising Products and Quotients to a Power

For any real numbers x and y and natural number m:

ALGEBRAICALLY	VERBALLY	ALGEBRAIC EXAMPLE
$(xy)^m = x^m y^m$	To raise a product to a power, raise each factor to this power.	$(xy)^3 = x^3y^3$
$\left(\frac{x}{y}\right)^m = \frac{x^m}{y^m}$ for $y \neq 0$	To raise a quotient to a power, raise both the numerator and the denominator to this power.	$\left(\frac{x}{y}\right)^3 = \frac{x^3}{y^3}$

A Product to a Power: As an example of raising a product to a power, we use the expression $(xy)^3$. It can be expanded as

$$(xy)^3 = \overbrace{(xy)(xy)(xy)}^{\text{Three factors of } xy}$$

$$= \overbrace{(x \cdot x \cdot x)}^{\text{Three factors of } x} \cdot \overbrace{(y \cdot y \cdot y)}^{\text{Three factors of } y}$$

$$= x^3y^3$$

Thus, $(xy)^3 = x^3y^3$. In general,

$$(xy)^m = \overbrace{(xy)(xy)\cdots(xy)}^{m \text{ factors of } xy}$$

$$= \overbrace{(x \cdot x \cdots x)}^{m \text{ factors of } x}\overbrace{(y \cdot y \cdots y)}^{m \text{ factors of } y}$$

$$= x^m y^m$$

A Quotient to a Power: As an example of raising a quotient to a power, we use the expression $\left(\frac{x}{y}\right)^3$. It can be expanded as

$$\left(\frac{x}{y}\right)^3 = \overbrace{\left(\frac{x}{y}\right)\left(\frac{x}{y}\right)\left(\frac{x}{y}\right)}^{\text{Three factors of } \frac{x}{y}}$$

$$= \frac{\overbrace{x \cdot x \cdot x}^{\text{Three factors of } x}}{\underbrace{y \cdot y \cdot y}_{\text{Three factors of } y}}$$

$$= \frac{x^3}{y^3}$$

Thus, $\left(\frac{x}{y}\right)^3 = \frac{x^3}{y^3}$.

In general,

$$\left(\frac{x}{y}\right)^m = \overbrace{\left(\frac{x}{y}\right)\left(\frac{x}{y}\right)\cdots\left(\frac{x}{y}\right)}^{m \text{ factors of } \frac{x}{y}}$$

$$= \frac{\overbrace{x \cdot x \cdots x}^{m \text{ factors of } x}}{\underbrace{y \cdot y \cdots y}_{m \text{ factors of } y}}$$

$$= \frac{x^m}{y^m}$$

In Example 5 we use the rule for raising a product to a power. Then in Example 6 we use the rule for raising a quotient to power.

EXAMPLE 5 Simplifying a Product to a Power

Simplify each expression.

SOLUTION

(a) $(ab)^5$	$(ab)^5 = a^5b^5$	For a product to a power, $(xy)^m = x^my^m$.
(b) $(a^2b)^5$	$(a^2b)^5 = (a^2)^5b^5$	For a product to a power, $(xy)^m = x^my^m$.
	$= a^{10}b^5$	Use the power rule $(x^m)^n = x^{m \cdot n}$.
(c) $(2t)^3$	$(2t)^3 = 2^3t^3$	For a product to a power, $(xy)^m = x^my^m$.
	$= 8t^3$	
(d) $(-3t)^4$	$(-3t)^4 = (-3)^4t^4$	Notice the subtle distinctions between
	$= 81t^4$	parts (d), (e), and (f). In part (d) the base is $-3t$.
(e) $-(3t)^4$	$-(3t)^4 = -(3^4 \cdot t^4)$	In part (e) the base is $3t$.
	$= -(81t^4)$	
	$= -81t^4$	
(f) $-3t^4$	$-3t^4 = -3t^4$	In part (f) the exponent applies to only the base of t; thus this expression is already in simplified form.

EXAMPLE 6 Simplifying a Quotient to a Power

Simplify each expression, assuming each variable is not zero.

SOLUTION

(a) $\left(\frac{a}{b}\right)^5$ $\qquad \left(\frac{a}{b}\right)^5 = \frac{a^5}{b^5}$ $\qquad$ Use the quotient to a power rule $\left(\frac{x}{y}\right)^m = \frac{x^m}{y^m}$.

(b) $\left(\frac{v^2}{w}\right)^3$ $\qquad \left(\frac{v^2}{w}\right)^3 = \frac{(v^2)^3}{w^3}$ $\qquad$ Use the quotient to a power rule $\left(\frac{x}{y}\right)^m = \frac{x^m}{y^m}$.

$= \frac{v^6}{w^3}$ $\qquad$ Use the product to a power rule $(x^m)^n = x^{m \cdot n}$.

(c) $\left(\frac{3}{v}\right)^4$ $\qquad \left(\frac{3}{v}\right)^4 = \frac{3^4}{v^4}$ $\qquad$ Use the quotient to a power rule $\left(\frac{x}{y}\right)^m = \frac{x^m}{y^m}$.

$= \frac{81}{v^4}$

(d) $\left(-\frac{5}{y}\right)^2$ $\qquad \left(-\frac{5}{y}\right)^2 = \left(\frac{-5}{y}\right)^2$ $\qquad$ The fraction $-\frac{a}{b}$ equals $\frac{-a}{b}$.

$= \frac{(-5)^2}{y^2}$ $\qquad$ Use the quotient to a power rule $\left(\frac{x}{y}\right)^m = \frac{x^m}{y^m}$.

$= \frac{25}{y^2}$

(e) $-\left(\frac{5}{y}\right)^2$ $\qquad -\left(\frac{5}{y}\right)^2 = -\left(\frac{5^2}{y^2}\right)$ $\qquad$ Use the quotient to a power rule $\left(\frac{x}{y}\right)^m = \frac{x^m}{y^m}$.

$= -\frac{25}{y^2}$ $\qquad$ The exponent applies to only the base $\frac{5}{y}$, not $-\frac{5}{y}$.

SELF-CHECK 5.1.4

Simplify each expression.

1. $(xy)^5$ **2.** $(x^3y)^5$ **3.** $\left(\frac{x}{y}\right)^5$ **4.** $\left(\frac{x}{y^4}\right)^5$

Example 7 reviews some problems that were covered when the order of operations was discussed in Section 1.7. Although these problems are not difficult, they easily can be misinterpreted. It is important to pay careful attention to the small changes in notation that dictate the correct order of operations.

EXAMPLE 7 Evaluating Algebraic Expressions by Using the Correct Order of Operations

Evaluate each expression for $x = -3$ and $y = -5$.

SOLUTION

(a) $(-x)^2$ $\qquad (-x)^2 = [-(-3)]^2$

$= (3)^2$

$= 9$

Note in parts (a) and (b) that the difference in notation leads to different answers. In part (a) the base for the exponent is $-x$, but in part (b) the base is just x.

(b) $-x^2$

$$-x^2 = -(x^2) = -[(-3)^2] = -(9) = -9$$

(c) $x^2 + y^2$

$$x^2 + y^2 = (-3)^2 + (-5)^2 = 9 + 25 = 34$$

(d) $(x + y)^2$

$$(x + y)^2 = [-3 + (-5)]^2 = (-8)^2 = 64$$

In parts (c) and (d) the different notations indicate distinctive orders of operation. This produces different answers.

SELF-CHECK 5.1.5

Evaluate each expression for $x = -3$ and $y = -5$.

1. xy^2 **2.** $(xy)^2$ **3.** $x + y^2$ **4.** $(x - y)^2$

In Section 1.7 we used the distributive property to add like terms. In Example 8, we review this topic and compare it to multiplying factors by using the product rule.

EXAMPLE 8 Comparing Addition and Multiplication

Add the like terms and simplify the products.

SOLUTION

	ADD	MULTIPLY	SUM	PRODUCT
(a)	$x + x$	$x \cdot x$	$x + x = 2x$	$x \cdot x = x^2$
(b)	$v^3 + v^3$	$v^3 \cdot v^3$	$v^3 + v^3 = 2v^3$	$v^3 \cdot v^3 = v^6$
(c)	$3x^2 + 5x^2$	$(3x^2)(5x^2)$	$3x^2 + 5x^2 = 8x^2$	$(3x^2)(5x^2) = (3)(5)(x^2)(x^2) = 15x^4$
(d)	$x^2 + x^4$	$x^2 \cdot x^4$	$x^2 + x^4$ cannot be simplified further as these terms are unlike.	$x^2 \cdot x^4 = x^6$

SELF-CHECK ANSWERS

5.1.1
1. $-5 \cdot a \cdot a \cdot a$
2. $(-5a)(-5a)(-5a)$
3. $a \cdot a \cdot a \cdot b \cdot b$
4. $-x^2$
5. $(-x)^2$
6. xy^3

5.1.2
1. x^5 2. y^{31}
3. m^2n^4 4. $-z^{11}$

5.1.3
1. x^{10} 2. x^{10}
3. x^{10} 4. $-x^{10}$
5. $-x^{10}$

5.1.4
1. x^5y^5 2. $x^{15}y^5$
3. $\frac{x^5}{y^5}$ 4. $\frac{x^5}{y^{20}}$

5.1.5
1. -75 2. 225
3. 22 4. 4

USING THE LANGUAGE AND SYMBOLISM OF MATHEMATICS 5.1

1. In the exponential expression x^n, x is the ________ and n is the ________.
2. In the product $x \cdot x$, both x and x are called ________.
3. In the sum $x + x$, both x and x are called ________ or ________.
4. In the exponential expression $-x^4$ the base is ________.

5. In the exponential expression $(-x)^4$ the base is ________.
6. In the exponential expression $(xy)^4$ the base is ________.
7. In the exponential expression xy^4 the base of exponent 4 is ________.
8. In the expression $x^5 = x \cdot x \cdot x \cdot x \cdot x$, x^5 is referred to as the ________ form and $x \cdot x \cdot x \cdot x \cdot x$ is referred to as the ________ form.
9. The product rule for exponents states that $x^m \cdot x^n =$ ________ for any real number x and natural numbers m and n.
10. The power rule for exponents states that $(x^m)^n =$ ________ for any real number x and natural numbers m and n.
11. For any real numbers x and y and any natural number m, $(xy)^m =$ ________.
12. For any real numbers x and y, $y \neq 0$, and any natural number m, $\left(\frac{x}{y}\right)^m =$ ________.

EXERCISES 5.1

Multiple Representations

In Exercises 1–4, write each expression in exponential form.

1. a. $x \cdot x \cdot x \cdot x \cdot x$ b. $x \cdot y \cdot y \cdot y \cdot y$ c. $(x \cdot y)(x \cdot y)(x \cdot y)$
2. a. $-3 \cdot a \cdot a \cdot a \cdot a$ b. $(-3a)(-3a)(-3a)(-3a)$ c. $-(3a)(3a)(3a)(3a)$
3. a. $(a + b)(a + b)(a + b)$ b. $a \cdot a \cdot a + b \cdot b \cdot b$ c. $a + b \cdot b \cdot b$
4. a. $a \cdot a \cdot a + b$ b. $(a + b)(a + b)$ c. $(ab)(ab)(ab)(ab)$

In Exercises 5–10, write each exponential expression in expanded form.

5. a. m^6 b. $(-m)^6$ c. $-m^6$
6. a. mn^6 b. $(mn)^6$ c. $6m^2$
7. a. $(m + n)^2$ b. $m^2 + n^2$ c. $m + n^2$
8. a. $m^2 + n^3$ b. m^2n^3 c. $(m + n)^3$
9. a. $\frac{a^2}{b^3}$ b. $\left(\frac{a}{b}\right)^4$ c. $\frac{a}{b^4}$
10. a. $\frac{x^3}{y^2}$ b. $\left(\frac{x}{y}\right)^3$ c. $\frac{x^3}{y}$

In Exercises 11–16, simplify each expression.

11. a. 6^2 b. $(-6)^2$ c. -6^2 d. -1^4
12. a. 4^2 b. $(-4)^2$ c. -4^2 d. $(-1)^5$
13. a. -1^{10} b. $(-1)^{10}$ c. $(-1)^{11}$ d. 10^3
14. a. -1^{12} b. $(-1)^{12}$ c. $(-1)^{13}$ d. 10^4
15. a. $5^2 + 12^2$ b. $(5 + 12)^2$
16. a. $11^2 - 8^2$ b. $(11 - 8)^2$

In Exercises 17–20, use a calculator to complete each table. Then state the equation that is confirmed by each table.

17.
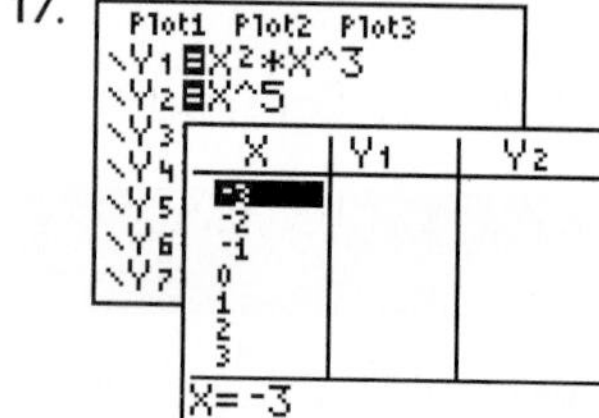

18.
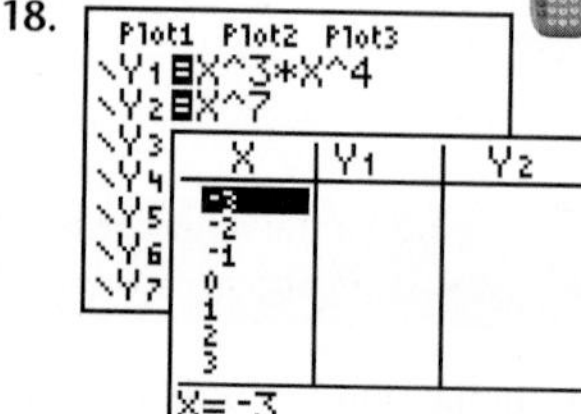

19.
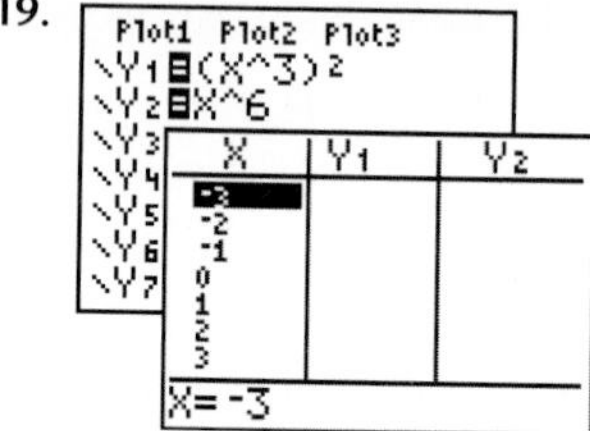

20.
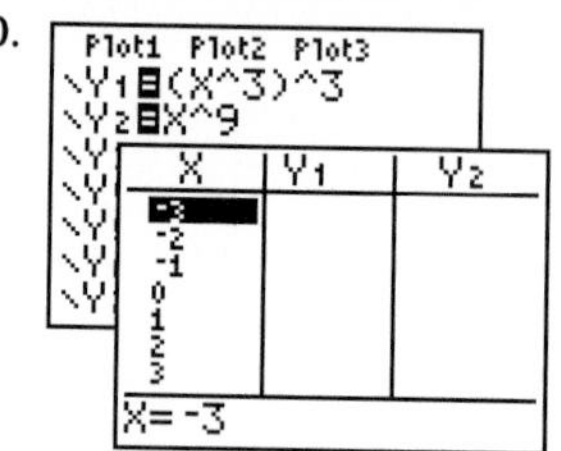

In Exercises 21–42, simplify each expression. Assume that variables are restricted to values that prevent division by zero.

21. a. x^9x^{11} b. $(x^9)^{11}$
22. a. y^7y^3 b. $(y^7)^3$
23. a. $(3m^4)(5m^6)$ b. $(6m^5)(4m^3)$
24. a. $(-7v^4)(-6v^5)$ b. $(-4v^7)(-5v^6)$
25. a. $-x(2x^2)(3x^3)$ b. $-3x^3(2x^2)(x)$
26. a. $-y(3y^3)(5y^5)$ b. $-5y^5(3y^3)(y)$
27. a. $(3xy)^2$ b. $(2xy)^3$
28. a. $(5xy)^2$ b. $(2xy)^5$
29. a. $(-n^2)^6$ b. $-(n^2)^6$
30. a. $(-3m)^4$ b. $-(3m)^4$
31. a. $5(abc)^2$ b. $(5abc)^2$
32. a. $-5(xyz)^2$ b. $(-5xyz)^2$
33. a. $(xy^2)^3$ b. $(xy^3)^2$
34. a. $(m^2n^3)^4$ b. $(m^4n^3)^2$
35. a. $\left(\frac{3}{4}\right)^2$ b. $\left(\frac{2}{3}\right)^4$
36. a. $\left(\frac{4}{5}\right)^2$ b. $\left(\frac{2}{5}\right)^4$
37. a. $\left(\frac{4}{w}\right)^3$ b. $\left(\frac{3}{w}\right)^4$
38. a. $-\left(\frac{3}{y^3}\right)^2$ b. $\left(-\frac{2}{y^2}\right)^3$
39. a. $\left(\frac{x^2}{y^3}\right)^{11}$ b. $\left(\frac{x^2}{y^{11}}\right)^3$
40. a. $\left(\frac{m}{n^5}\right)^7$ b. $\left(\frac{m}{n^7}\right)^5$
41. a. $\left(-\frac{2x^2}{3y^3}\right)^4$ b. $\left(-\frac{2x^2}{4y^4}\right)^3$
42. a. $\left(-\frac{5a^5}{3b^3}\right)^2$ b. $\left(-\frac{5a^5}{2b^2}\right)^3$

In Exercises 43–54, evaluate each expression for $x = 2$ and $y = 3$.

43. a. $-x^2$ b. $(-x)^2$
44. a. $-y^2$ b. $(-y)^2$

45. a. $(xy)^2$ b. xy^2
46. a. $(xy)^3$ b. xy^3
47. a. $x^3 + y^3$ b. $(x + y)^3$
48. a. $x^2 + y^2$ b. $(x + y)^2$
49. a. $5x^2y^2$ b. $(5xy)^2$
50. a. $(3xy)^2$ b. $3x^2y^2$
51. a. $x^3 - y^3$ b. $(x - y)^3$
52. a. x^y b. y^x
53. a. $6y^x$ b. $5x^y$
54. a. $-\left(\frac{x}{y}\right)^2$ b. $\left(-\frac{x}{y}\right)^2$

In Exercises 55–58, add the like terms and simplify the products.

Add	Multiply
55. a. $v + v + v$	b. $v \cdot v \cdot v$
56. a. $w^2 + w^2$	b. $w^2 \cdot w^2$
57. a. $4m^3 + 6m^3$	b. $(4m^3)(6m^3)$
58. a. $2x + 3x + 10x$	b. $(2x)(3x)(10x)$

Estimation and Calculator Skills

In Exercises 59–64, mentally estimate the value of each expression and then use a calculator to calculate the value.

PROBLEM	MENTAL ESTIMATE	CALCULATOR VALUE
59. -6.98^2		
60. $(-9.9)^4$		
61. $(1.01 + 1.01)^2$		
62. $1.01^2 + 1.01^2$		
63. $7.99^2 - 5.9^2$		
64. $(7.99 - 5.9)^2$		

Compound Interest

In Exercises 65 and 66, use the formula $A = P(1 + r)^t$ to determine the value of a principal P that is compounded for t years at an interest rate r.

65. What is the value of a principal P of \$1 compounded for 10 years at a rate of 8%?
66. What is the value of a principal P of \$1 compounded for 12 years at a rate of 7%?

Modeling the Generation of Electricity by a Windmill

Use the following information to work Exercises 67–70.

There are several environmental reasons to promote greater generation of electricity by windmills. There are also several concerns with this generation method: noise, visual pollution, reliability, efficiency, cost, and the variability of the wind from hour to hour. To make sound business decisions, a company needs the facts related to all these concerns. One manufacturer of windmills has prepared the following table that gives the number of kilowatt-hours generated from a given speed of wind. The output of the windmill actually varies directly as the cube of the speed of the wind. Thus the windmill is rather inefficient at low wind speeds and very efficient at higher wind speeds that are within its safe operating range. The function the company used to create the table below is $f(x) = 0.01x^3$.

WIND SPEED x (mi/h)	$f(x)$ (kW OF ELECTRICITY)
0	0.00
5	1.25
10	10.00
15	33.75
20	80.00
25	156.25
30	270.00

67. a. Evaluate and interpret $f(20)$.
    b. Evaluate and interpret $f(27)$.
68. What value of x will produce $f(x) = 270$? Interpret this result.
69. Compare the results of a company installing one of these windmills in an area with a wind speed of 10 mi/h with a company that installs the same type of windmill in an area with a wind speed of 20 mi/h. How many times greater will be the power generated by the higher 20-mi/h wind?
70. Suppose the company made an error and the function it should have used to produce this table is $f(x) = 0.01x^2$. Use this formula to produce a new table of values for the same input values of x.

In Exercises 71–78, simplify each expression. Assume that variables are restricted to values that prevent division by zero.

71. $(-5x^2y^4)^3$
72. $(-5x^2y^7)^2$
73. $(-2xy^2)(-3x^2y^4)$
74. $(-6x^2y^3)(7x^4y^5)$
75. $(5x^2)^2(2x^3)^3$
76. $(2x^5)^2(3x^4)^3$
77. $\left(\frac{3x}{2y^2}\right)^3$
78. $\left(\frac{-2x^3}{3y}\right)^4$
79. Use the figure below to determine the area of each region.

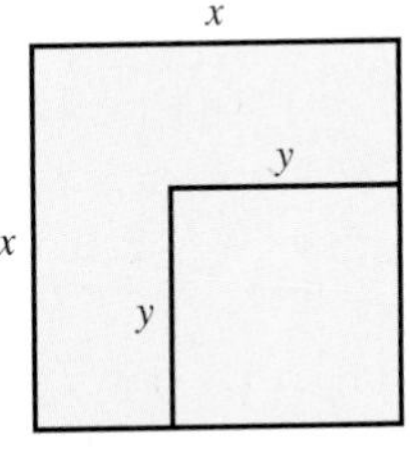

a. The area of the larger square.
b. The area of the smaller square.
c. The area inside the larger square and outside the smaller square.

80. Each side of the larger square is double that of the smaller square. Compare their areas.

81. Each side of the larger cube is double that of the smaller cube. Compare their volumes.

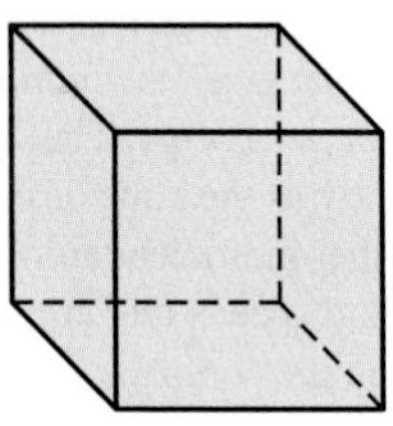
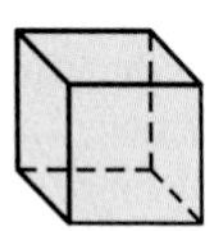

Group Discussion Questions

82. **Challenge Question** Use the product and power rules for exponents to simplify each expression, assuming m is a natural number.
 a. $x^m x^2$ **b.** $(x^m)^2$ **c.** $x^{m+1}x^{m+2}$ **d.** $(x^{m+1})^2$

83. **Challenge Question** One person becomes infected with a new strain of bacteria. Over the course of the next 24 hours the victim will infect 4 more persons. On the following day the original victim and the new victims will each infect 4 more people. Assume this infection continues in this manner for 8 days. Determine the total number of people infected on each of the 8 days in this table. Can you determine a formula for the number infected on the nth day?

DAY n	POPULATION INFECTED P
0	1
1	$1 + 4 = 5$
2	
3	
4	
5	
6	
7	
8	

84. **Discovery Question** Let Y_1 represent each expression, and use a graphing calculator to complete this table for each expression. Based upon your observations, make a conjecture about how you can simplify $\frac{x^m}{x^n}$.

X	Y_1
-3	
-2	
-1	
0	
1	
2	
3	

X= -3

 a. x^2 **b.** x^3 **c.** x^4
 d. $\frac{x^6}{x^2}$ **e.** $\frac{x^6}{x^3}$ **f.** $\frac{x^8}{x^4}$

Section 5.2 Quotient Rule and Zero Exponents

Objectives:
3. Use the quotient rule for exponents.
4. Simplify expressions with zero exponents.
5. Combine the properties of exponents to simplify expressions.

Quotient Rule

Another property of exponents that allows us to simplify some computations involving exponential expressions is the quotient rule. The logic behind the quotient rule is now examined. The quotient of $\frac{x^5}{x^3}$ is written below in expanded form.

For $x \neq 0$, the quotient is

$$\frac{x^5}{x^3} = \frac{\overbrace{x \cdot x \cdot x \cdot x \cdot x}^{\text{Five factors of } x}}{\underbrace{x \cdot x \cdot x}_{\text{Three factors of } x}}$$

The fraction is reduced by dividing both the numerator and the denominator by three factors of x.

$$= \overbrace{x \cdot x}^{\text{Two factors of } x}$$
$$= x^2$$

Thus $\frac{x^5}{x^3} = x^2$. The fact that these two expressions are equivalent for $x \neq 0$ is confirmed by comparing the values of Y_1 and Y_2 in the following table.

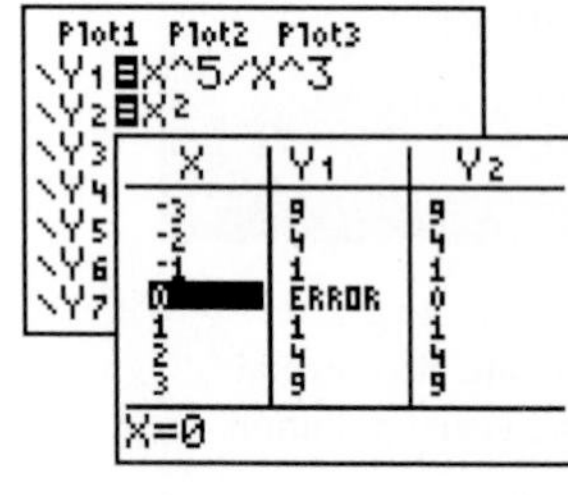

Note that $\frac{x^5}{x^3}$ is undefined for $x = 0$; for all other values $\frac{x^5}{x^3} = x^2$.

In general, if m and n are natural numbers and $m > n$, then the quotient $\frac{x^m}{x^n}$ can be reduced by dividing out each of the n common factors of x. This will leave $m - n$ factors of x in the numerator; that is, $\frac{x^m}{x^n} = x^{m-n}$ for $x \neq 0$ and $m > n$. As we note in the following box, $\frac{x^m}{x^n}$ is undefined for $x = 0$.

Quotient Rule for Exponents

For any real number x and natural numbers m and n with $m > n$:*

ALGEBRAICALLY	VERBALLY	ALGEBRAIC EXAMPLE
$\frac{x^m}{x^n} = x^{m-n}$ for $x \neq 0$ $\frac{x^m}{x^n}$ is undefined for $x = 0$	To divide two expressions with the same base, use the common base and subtract the exponents.	$\frac{x^5}{x^3} = x^2$

*We will soon note that the restriction $m > n$, is unnecessary when we extend the quotient rule.

EXAMPLE 1 Using the Quotient Rule

Use the quotient rule to simplify each expression. Assume the denominators are not zero.

SOLUTION

(a) $\frac{z^{10}}{z^3}$

$$\frac{z^{10}}{z^3} = z^{10-3}$$
$$= z^7$$

Subtract the exponents with the same base.

(b) $\frac{m^4n^7}{m^2n^6}$

$$\frac{m^4n^7}{m^2n^6} = m^{4-2}n^{7-6}$$
$$= m^2n^1$$
$$= m^2n$$

Subtract the exponents on the common bases m and n.

(c) $-\frac{7^8}{7^6}$

$$-\frac{7^8}{7^6} = -(7^{8-6})$$
$$= -(7^2)$$
$$= -49$$

Note how much easier it is to apply the quotient rule than it is to evaluate 7^8 and 7^6 and then divide.

(d) $\frac{x^5}{y^3}$

$$\frac{x^5}{y^3} = \frac{x^5}{y^3}$$

Because the numerator and the denominator do not have the same base, this expression cannot be simplified.

SELF-CHECK 5.2.1

Simplify each expression.

1. $\frac{m^{14}}{m^9}$
2. $\frac{x^5y^5}{x^3y^2}$
3. $\frac{5^8}{5^6}$
4. $\frac{111^{113}}{111^{112}}$

The quotient rule, as developed so far, allows us to simplify $\frac{x^m}{x^n}$ when $m > n$. The expression $\frac{x^m}{x^n}$ can also be simplified when $m < n$. The reasoning is similar to that used when $m > n$. If $x \neq 0$ and m and n are natural numbers, then $\frac{x^m}{x^n} = \frac{1}{x^{n-m}}$ for $m < n$. The example $\frac{x^4}{x^7}$ is examined below. This example is closely related to negative exponents, which are covered in the next section.

$$\frac{x^4}{x^7} = \frac{\overset{1}{\cancel{x \cdot x \cdot x \cdot x}}}{\underset{1}{\cancel{x \cdot x \cdot x \cdot x}} \cdot x \cdot x \cdot x}$$

$$= \frac{1}{x \cdot x \cdot x}$$

$$= \frac{1}{x^3}$$

Thus $\frac{x^4}{x^7} = \frac{1}{x^3}$. The fact that these two expressions are equivalent for $x \neq 0$ is confirmed by comparing the values of Y_1 and Y_2 in the following table.

Plot1 Plot2 Plot3
\Y1 ■ X^4/X^7
\Y2 ■ 1/X^3
\Y3 =
\Y4 =
\Y5 =
\Y6 =
\Y7 =

X	Y1	Y2
-3	-.037	-.037
-2	-.125	-.125
-1	-1	-1
0	ERR:	ERR:
1	1	1
2	.125	.125
3	.03704	.03704

X=0

Both expressions are undefined for $x = 0$.

The properties of exponents are so important that many exercises are given on each individual property. These properties will be revisited extensively throughout the book. In Example 2 we use the quotient rule to simplify $\frac{x^m}{x^n}$. In each case this is accomplished by subtracting the smaller exponent from the larger exponent.

EXAMPLE 2 Simplifying Exponential Expressions

Simplify each of the following expressions.

SOLUTION

(a) $\frac{x^3}{x^5}$

$$\frac{x^3}{x^5} = \frac{1}{x^{5-3}}$$

$$= \frac{1}{x^2}$$

Reduce the fraction by dividing both the numerator and the denominator by the three common factors of x. This is accomplished by subtracting the smaller exponent from the larger exponent.

(b) $\frac{x^4y^4}{x^6y}$

$$\frac{x^4y^4}{x^6y} = \frac{y^{4-1}}{x^{6-4}}$$

$$= \frac{y^3}{x^2}$$

For each base, subtract the smaller exponent from the larger exponent.

(c) $\frac{x^8}{y^9}$ $\frac{x^8}{y^9} = \frac{x^8}{y^9}$ The bases are not the same, so this expression cannot be simplified.

(d) $\frac{11^8}{11^{10}}$ $\frac{11^8}{11^{10}} = \frac{1}{11^{10-8}}$ Note that it is easier to first apply the properties of exponents than to start by evaluating 11^8.

$= \frac{1}{11^2}$

$= \frac{1}{121}$

SELF-CHECK 5.2.2

Simplify each expression.

1. $\frac{m^3}{m^{11}}$ **2.** $\frac{x^3y^5}{x^4y^4}$ **3.** $\frac{10^9}{10^{12}}$ **4.** $\frac{47^{48}}{47^{49}}$

Zero Exponent

To develop a definition of zero exponents, consider two approaches to simplifying $\frac{x^m}{x^m}$ for $x \neq 0$.

Option 1:

$$\frac{x^m}{x^m} = 1$$

In option 1 we reduced the fraction by dividing both the numerator and the denominator by the common factor of x^m.

Option 2:

$$\frac{x^m}{x^m} = x^{m-m}$$
$$= x^0$$

In option 2 we examine what will happen when we use the quotient rule to subtract the exponents.

For these two options to obtain the same result, we must define $x^0 = 1$ for $x \neq 0$. This is further confirmed by examining Y_1 in the following table.

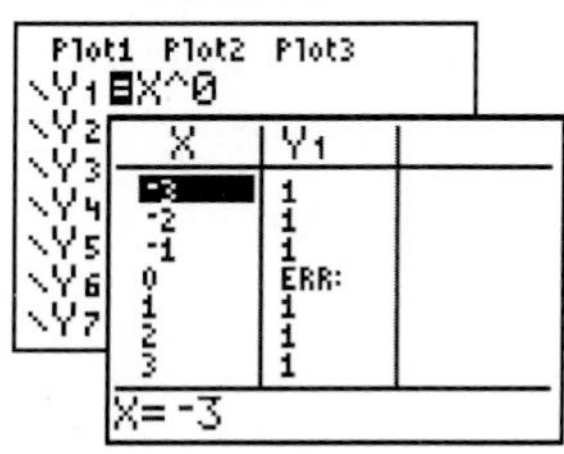

In the screen shot above note that 0^0 is undefined. This is also emphasized in the following box.

Definition of Zero Exponent

For any real number x:

ALGEBRAICALLY	VERBALLY	ALGEBRAIC EXAMPLE
$x^0 = 1$ for $x \neq 0$ 0^0 is undefined	Any nonzero real number raised to the 0 power is 1.	$17^0 = 1$

Example 3 illustrates why we must be very careful to correctly identify the base when we use exponential expressions.

EXAMPLE 3 Using Zero Exponents

Simplify each expression, assuming all bases are nonzero.

SOLUTION

(a) 3^0	$3^0 = 1$	The base is 3.
(b) $(-3)^0$	$(-3)^0 = 1$	The base is -3.
(c) -3^0	$-3^0 = -(3)^0 = -1$	The base is 3 (not -3).
(d) $3x^0$	$3x^0 = 3(1) = 3$	The base is x.
(e) $(3x)^0$	$(3x)^0 = 1$	The base is $3x$.
(f) $(x + y)^0$	$(x + y)^0 = 1$	The base is $x + y$.
(g) $x^0 + y^0$	$x^0 + y^0 = 1 + 1 = 2$	The different results in parts (f) and (g) are a result of the different order of operations.

SELF-CHECK 5.2.3

Simplify each expression, assuming all bases are nonzero.

1. $(5v)^0$ **2.** $5v^0$ **3.** $(5v + w)^0$ **4.** $5v^0 + w^0$

The rules for exponents are summarized in the following box. In Section 5.3 these rules will be applied to all integral exponents: positive, zero, and negative. So far we have developed these rules only for whole-number exponents. Remember the whole numbers include 0, 1, 2, 3, 4, ...

Summary of the Properties of Exponents

For any nonzero real numbers x and y and whole-number exponents m and n,

Product rule: $x^m \cdot x^n = x^{m+n}$

Power rules: $(x^m)^n = x^{mn}$

$(xy)^m = x^m y^m$

$\left(\frac{x}{y}\right)^m = \frac{x^m}{y^m}$

Quotient rule: $\frac{x^m}{x^n} = x^{m-n}$ for $m > n$

$\frac{x^m}{x^n} = \frac{1}{x^{n-m}}$ for $m < n$

Zero exponent: $x^0 = 1$ for $x \neq 0$

0^0 is undefined

Problems that involve several of these properties can sometimes be simplified correctly by using more than one sequence of steps. It is often wise to simplify the expressions inside grouping symbols first. Study the examples in the text and those given by your instructor. Try to use a sequence of steps that will minimize your effort as well as produce a correct result.

EXAMPLE 4 Simplifying Exponential Expressions

Simplify each expression, assuming $x \neq 0$ and $y \neq 0$.

SOLUTION

(a) $\left(\frac{x^{12}}{x^4}\right)^2$

$\left(\frac{x^{12}}{x^4}\right)^2 = (x^{12-4})^2$ Use the quotient rule; subtract the exponents.

$= (x^8)^2$ Use the power rule; multiply the exponents.

$= x^{16}$

(b) $[(2a^7)(5a^4)]^3$

$[(2a^7)(5a^4)]^3 = [(2)(5)a^{7+4}]^3$ Use the product rule; add the exponents.

$= (10a^{11})^3$

$= 10^3a^{(3)(11)}$ Use the power rule; multiply the exponents.

$= 1{,}000a^{33}$

(c) $\frac{12x^7y^7}{18x^5y^8}$

$\frac{12x^7y^7}{18x^5y^8} = \frac{2x^{7-5}}{3y^{8-7}}$ Divide both the numerator and the denominator by 6. Also use the quotient rule and subtract the exponents on the common bases.

$= \frac{2x^2}{3y}$

SELF-CHECK 5.2.4

Simplify each expression for $x \neq 0$.

1. $\left(\frac{x^7}{x^5}\right)^3$
2. $[(2x^3)(3x^2)]^2$

The ability to estimate calculations is useful in many contexts. One important use is to check calculator results to see if they are reasonable. Many calculator keystroke errors, especially misplacing a decimal point or keying in an incorrect operation, can produce significant errors. On the other hand, good mental approximations should produce relatively small errors. (Relative error was first covered in Section 1.6.) Thus mental estimations can help you spot your own calculator errors.

EXAMPLE 5 Estimating a Quotient and Calculating the Relative Error

Estimate $\frac{5.01^6}{4.98^4}$ and then determine the relative error of this estimate.

SOLUTION

ESTIMATED VALUE

$5.01 \approx 5$ $\qquad \frac{5^6}{5^4} = 5^2$

$4.98 \approx 5$ $\qquad = 25$

CALCULATOR VALUE

```
(5.01^6)/(4.98^4
)
          25.71040892
```

MENTAL ESTIMATE	CALCULATOR APPROXIMATION
$\frac{5.01^6}{4.98^4} \approx 25$	$\frac{5.01^6}{4.98^4} \approx 25.71$

Error of estimate	=	Estimated value	−	Actual value	The estimated value is smaller than the actual value, so the error of the estimate is negative.
	≈	25	−	25.71	
	≈	−0.71			
Relative error	=	Error of estimate	÷	Actual value	This relative error, whose magnitude is nearly 3%, is likely much smaller than an error produced by an order-of-operations error or an incorrect calculator keystroke. For error checking, this mental estimate is very useful.
	≈	(−0.71)	÷	25.71	
	≈	−0.0276			
	≈	−2.8%			

Answer: The estimate of 25 has a relative error of about −2.8% compared to the actual value of approximately 25.71.

Example 6 compares the subtraction of like terms to the division of expressions with a common base. Please note the distinction between these operations.

EXAMPLE 6 Comparing Subtraction and Division

Subtract the like terms and simplify the quotients. Assume $x \neq 0$.

SOLUTION

	SUBTRACT	DIVIDE	DIFFERENCE	QUOTIENT
(a)	$10x^2 - 2x^2$	$\frac{10x^2}{2x^2}$	$10x^2 - 2x^2 = 8x^2$	$\frac{10x^2}{2x^2} = 5$
(b)	$8x^2 - 4x$	$\frac{8x^2}{4x}$	$8x^2 - 4x$ cannot be simplified further as these terms are unlike.	$\frac{8x^2}{4x} = 2x$

The compound interest formula $A = P(1 + r)^t$ can be used to find the value A of an initial investment of principal P at an annual interest rate r compounded for t years. In Example 7 we use this formula and a calculator to compute the value of an investment over a 6-year period. Note that the expression $P(1 + r)^t$ involves an exponent that is a variable. Example 7 illustrates one way to evaluate this expression for different values of t. We will examine expressions with variable exponents further in Chapter 10.

EXAMPLE 7 Calculating the Value of an Investment

Use the formula $A = P(1 + r)^t$ to find the value of a $7,500 investment compounded at 7% at the end of each year over a 6-year period.

SOLUTION

$A = P(1 + r)^t$
$A = 7{,}500(1 + 0.07)^t$
$A = 7{,}500(1.07)^t$

Using the compound interest formula, substitute $7,500 for the principal P and 0.07 for the interest rate of 7%.

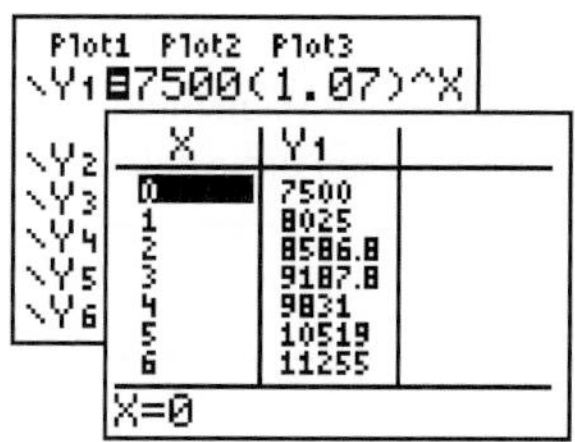

Enter this formula into a calculator, using x to represent the number of years t from the formula and Y_1 to represent A.

Note for a zero exponent $(1.07)^0 = 1$ and the value of $P = \$7{,}500$. This is the value at $t = 0$, the initial time of the investment.

The last row shows that the value of the investment after 6 years is approximately \$11,255. You can obtain additional significant digits by moving the cursor over this value and obtaining 11,255.4776389.

SELF-CHECK 5.2.5

1. Use the formula $A = P(1 + r)^t$ to find the value of a \$7,500 investment compounded at 8% at the end of each year over a 6-year period.

SELF-CHECK ANSWERS

5.2.1

1. m^5 2. x^2y^3
3. 25 4. 111

5.2.2

1. $\frac{1}{m^8}$ 2. $\frac{y}{x}$
3. $\frac{1}{1{,}000}$ 4. $\frac{1}{47}$

5.2.3

1. 1 2. 5
3. 1 4. 6

5.2.4

1. x^6 2. $36x^{10}$

5.2.5

1.

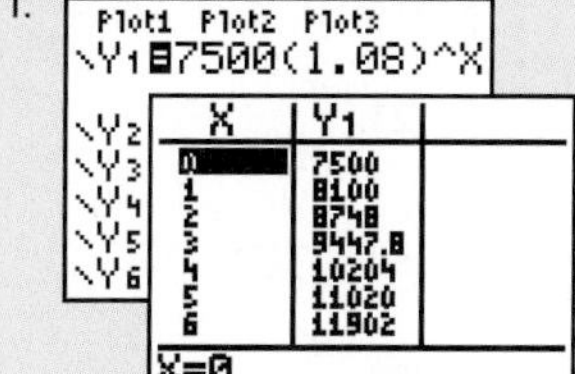

The last row shows that the value of the investment after 6 years is approximately \$11,902.

USING THE LANGUAGE AND SYMBOLISM OF MATHEMATICS 5.2

1. Two numbers that are added are called __________ or __________.
2. Two numbers that are multiplied are called __________.
3. The answer to a division of two numbers is called the __________.
4. The quotient rule for exponents states that $\frac{x^m}{x^n} =$ __________ for any real number x for $x \neq$ __________ and for natural numbers m and n with $m > n$.
5. If $m < n$, then we can use the quotient rule to write $\frac{x^m}{x^n} =$ __________.
6. To divide two expressions with the same base, use the common base and __________ the exponents.
7. The value of x^0 is __________ for $x \neq$ __________.
8. 0^0 is __________.

EXERCISES 5.2

In Exercises 1–14, simplify each expression.

1. a. $0 - 5$ b. $0 \cdot 5$ c. 5^0
   d. 0^5 e. $5 \div 0$
2. a. $0 + 6$ b. $6 \cdot 0$ c. 6^0
   d. 0^6 e. $6 \div 0$
3. a. $\frac{10^8}{10^5}$ b. $\frac{10^5}{10^5}$ c. $\frac{10^5}{10^8}$
4. a. $\frac{5^6}{5^4}$ b. $\frac{5^6}{5^6}$ c. $\frac{5^4}{5^6}$
5. a. 7^0 b. -7^0 c. $(-7)^0$
6. a. 10^0 b. -10^0 c. $(-10)^0$
7. a. $2^0 + 8^0$ b. $(2 + 8)^0$ c. $2^0 - 8^0$
8. a. $4^0 + 6^0$ b. $(4 + 6)^0$ c. $4^0 - 6^0$
9. a. 3^0 b. 0^3 c. -3^0
10. a. 5^0 b. 0^5 c. -5^0
11. a. x^0 for $x \neq 0$ b. x^0 for $x = 0$
12. a. $5y^0$ for $y \neq 0$ b. $(5y)^0$ for $y = 0$
13. a. $-3x^0$ for $x \neq 0$ b. $(-3x)^0$ for $x = 0$
14. a. $-9y^0$ for $y \neq 0$ b. $(-9y)^0$ for $y = 0$

In Exercises 15–18, use a calculator to complete each table. Then state the equation that is confirmed by each table.

15. **16.** **17.** **18.**

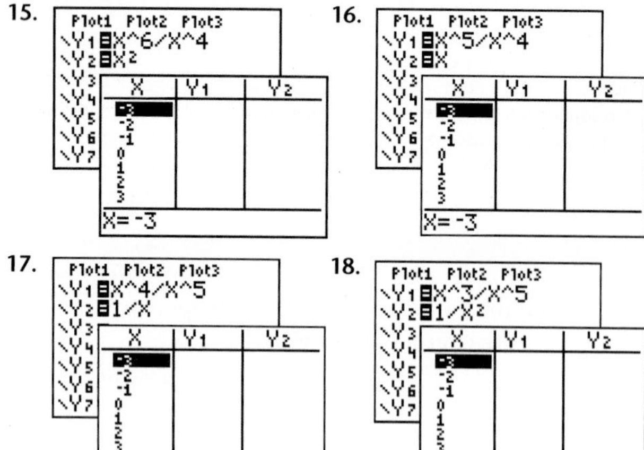

In Exercises 19–50, simplify each expression. Assume all bases are nonzero.

19. $\frac{x^{15}}{x^5}$ **20.** $\frac{y^{18}}{y^6}$ **21.** $\frac{m^{42}}{m^{18}}$

22. $\frac{m^{83}}{m^{72}}$ **23.** $\frac{y^{48}}{y^{48}}$ **24.** $\frac{v^{113}}{v^{113}}$

25. $\frac{-x^{11}}{x^7}$ **26.** $\frac{-y^{23}}{y^{17}}$ **27.** $\frac{t^8}{t^{12}}$

28. $\frac{w^5}{w^{11}}$ **29.** $\frac{v^{10}}{v^{15}}$ **30.** $\frac{n^{15}}{n^{18}}$

31. $-\frac{a^6}{a^9}$ **32.** $\frac{-b^6}{b^8}$ **33.** $\frac{a^3}{b^4}$

34. $\frac{a^4}{b^3}$ **35.** $\frac{14m^5}{21m^2}$ **36.** $\frac{35m^8}{42m^3}$

37. $\frac{36v^5}{66v^8}$ **38.** $\frac{20v^6}{12v^2}$ **39.** $\frac{15a^2b^3}{20a^2b^2}$

40. $\frac{33a^5b^5}{22a^3b^8}$ **41.** $\frac{8m^7n^7}{6m^9n^2}$ **42.** $\frac{6x^6y^6}{3x^3y^{10}}$

43. $-\frac{27v^4w^9}{45v^6w^6}$ **44.** $-\frac{35a^9b^7}{28a^8b^9}$ **45.** $\left(\frac{x^5}{x^2}\right)^2$

46. $\left(\frac{a^7}{a^4}\right)^2$ **47.** $\left(\frac{m^3}{m^9}\right)^2$ **48.** $\left(\frac{b^2}{b^6}\right)^3$

49. $[(2x^5)(3x^4)]^2$ **50.** $[(5x^2)(2x^4)]^3$

In Exercises 51–54, subtract the like terms and simplify the quotients. Assume the variables are nonzero.

	Subtract	Divide
51.	**a.** $38x - 2x$	**b.** $\frac{38x}{2x}$
52.	**a.** $x^3 - x^3$	**b.** $\frac{x^3}{x^3}$
53.	**a.** $3x^2 - 3x$	**b.** $\frac{3x^2}{3x}$
54.	**a.** $8x^3 - 2x^2$	**b.** $\frac{8x^3}{2x^2}$

Estimation and Calculator Skills

In Exercises 55–58, mentally estimate the value of each expression. Then use a calculator to calculate the value.

PROBLEM	MENTAL ESTIMATE	CALCULATOR VALUE
55. $\frac{8.014^2}{7.99}$		
56. $\frac{1.99^2}{2.01^4}$		
57. $(1.99)^3(5.02)^2$		
58. $[(1.99)(5.02)]^4$		

In Exercises 59–80, simplify each expression. Assume the variables are nonzero.

59. a. $(4x)^0$ **b.** $4x^0$ **c.** $4 + x^0$

60. a. $(-5y)^0$ **b.** $-5y^0$ **c.** $-5 + y^0$

61. a. $(4x - 3y)^0$ **b.** $(4x)^0 - (3y)^0$ **c.** $4x^0 - 3y^0$

62. a. $(-5y + 2z)^0$ **b.** $(-5y)^0 + (2z)^0$ **c.** $-5y^0 + 2z^0$

63. a. $\frac{2^{45}}{2^{43}}$ **b.** $\frac{2^{43}}{2^{45}}$ **c.** $\frac{2^{43}}{2^{43}}$

64. a. $\frac{5^{87}}{5^{84}}$ **b.** $\frac{5^{84}}{5^{87}}$ **c.** $\frac{5^{87}}{5^{87}}$

65. $(5a + 3b)^0 + (5a)^0 + (3b)^0 + 5a^0 + 3b^0$

66. $(7v - 8w)^0 + (7v)^0 + (-8w)^0 + 7v^0 - 8w^0$

67. $[(3x^2)(2x^3)]^2$ **68.** $[(2b^4)(5b^3)]^3$

69. $\left(\frac{10x^7}{5x^4}\right)^5$ **70.** $\left(\frac{21v^9}{7v^4}\right)^3$

71. $[(6m^4)(2m^5)]^2$ **72.** $[(5m^7)(3m^5)]^2$

73. $(5x^3)^2(2x^5)^3$ **74.** $(10v^4)^3(6v^5)^2$

75. $\left(\frac{12x^5}{6x^3}\right)\left(\frac{15x^7}{5x^3}\right)$ **76.** $\left(\frac{24a^9}{8a^3}\right)\left(\frac{42a^7}{14a^6}\right)$

77. $\frac{36a^7b^8}{12a^3b^3}$ **78.** $\frac{64m^9n^9}{16mn^7}$

79. $\frac{(4x^2y)^3}{(8xy^2)^2}$ **80.** $\frac{(2x^3y^2)^5}{(4x^2y^3)^3}$

Compound Interest

In Exercises 81 and 82, use the given table to determine the value of a \$5,000 investment for $x = 5$ years. In Exercise 81 the interest rate is 8.5%, and in Exercise 82 the interest rate is 6%.

81.

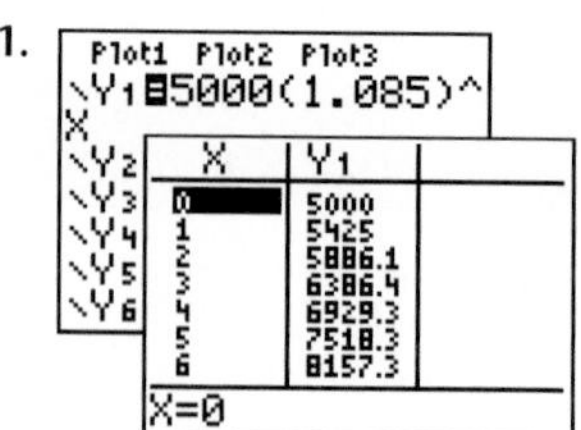

82.

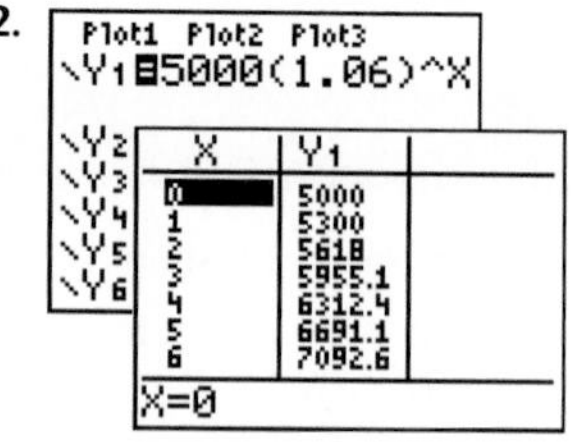

Compound Interest

In Exercises 83 and 84, use the formula $A = P(1 + r)^t$ to determine the value of each investment for $t = 5$ years. Variable A represents the value after t years of an original amount P invested at an interest rate r.

83. \$4,000 at 5% **84.** \$4,000 at 8%

Group Discussion Questions

85. **Challenge Question** Use the properties of exponents to simplify each expression, assuming m is a natural number and x is a nonzero real number.
    a. $\dfrac{x^{m+3}}{x^{m+1}}$ b. $\dfrac{x^{3m}}{x^m}$ c. $\left(\dfrac{x^{2m+1}}{x^m}\right)^3$
    d. $[(x^{m+1})(x^{m+2})]^4$ e. $\dfrac{(-1)^{m+2}}{(-1)^m}$
86. **Challenge Question** Insert parentheses in the expression $3 + 4 \cdot 5^2$ to create an order of operations that produces a result of
    a. 175 b. 403 c. 529 d. 1,225
87. **Discovery Question** Use a calculator to complete the following table for x^{-1}; then discuss what you think this notation represents.

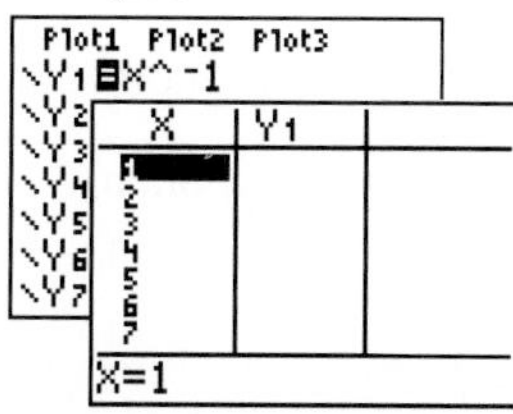

88. **Discovery Question**
    a. Evaluate 6^0, 5^0, 4^0, 3^0, 2^0, and 1^0. If you were defining 0^0 based on these observations, what value would you give for 0^0?
    b. Evaluate 0^6, 0^5, 0^4, 0^3, 0^2, and 0^1. If you were defining 0^0 based on these observations, what value would you give for 0^0?
    c. Use a calculator to evaluate 0^0. What result did you get?
    d. 0^0 is undefined. Based on your observations in parts **a** and **b**, explain why you think 0^0 is undefined.

Section 5.3 Negative Exponents and Scientific Notation

Objectives:
6. Simplify expressions with negative exponents.
7. Use scientific notation.

Planning for space flights to Mars and other planets must include many factors and solve many technical problems. If a flight crew has an emergency and needs to communicate with their ground crew on Earth, how long will it take for an emergency message to reach Earth from Mars?

Due to the vast distance and the speed of the signal, we will use scientific notation to examine this problem. In Example 8 we will represent the distance the signal must travel as 3.78×10^{11} m.

To use scientific notation we must first examine negative exponents.

Negative Exponents

To develop a definition of negative exponents, consider two approaches to simplifying $\dfrac{x^4}{x^5}$ for $x \neq 0$.

Option 1:

$$\frac{x^4}{x^5} = \frac{x \cdot x \cdot x \cdot x}{x \cdot x \cdot x \cdot x \cdot x} = \frac{1}{x}$$

In option 1 we reduce the fraction by dividing both the numerator and the denominator by the four common factors of x.

Option 2:

$$\frac{x^4}{x^5} = \frac{x^{4-5}}{1} = x^{-1}$$

In option 2 we examine what will happen when we use the quotient rule to subtract the exponents.

For these two options to obtain the same result, we must define $x^{-1} = \frac{1}{x}$ for $x \neq 0$. This is further confirmed by examining Y_1 and Y_2 in the following table:

X	Y_1	Y_2
0	ERR:	ERR:
1	1	1
2	.5	.5
3	.33333	.33333
4	.25	.25
5	.2	.2
6	.16667	.16667

X=0

(Screen: Plot1 Plot2 Plot3; $Y_1 = X^{-1}$; $Y_2 = 1/X$)

x^{-1} and $\frac{1}{x}$ are both undefined for $x = 0$.

A Mathematical Note

John Wallis, in *Arithmetica Infinitorum* (1655), was the first writer to explain the use of zero and negative exponents.

Similarly, $\frac{x^3}{x^5} = \frac{1}{x^2}$ or $\frac{x^3}{x^5} = x^{-2}$. Thus, we define $x^{-2} = \frac{1}{x^2}$ for $x \neq 0$. This suggests the following general definition of negative exponents.

Definition of Negative Exponents

For any real number x and natural number n:

ALGEBRAICALLY		VERBALLY	ALGEBRAIC EXAMPLE
$x^{-n} = \frac{1}{x^n}$	for $x \neq 0$	A nonzero base with a negative exponent can be rewritten by using the reciprocal of the base and the corresponding positive exponent.	$x^{-3} = \frac{1}{x^3}$
x^{-n} is undefined	for $x = 0$		

EXAMPLE 1 Simplifying Expressions with Negative Exponents

Simplify each of the following expressions.

SOLUTION

(a) 3^{-4}

$$3^{-4} = \frac{1}{3^4} = \frac{1}{81}$$

Take the reciprocal of the base. Note that the answer is positive.

(b) $(-3)^{-4}$

$$(-3)^{-4} = \frac{1}{(-3)^4} = \frac{1}{81}$$

The base is -3. Take the reciprocal of the base.

(c) -3^{-4}

$$-3^{-4} = -(3^{-4}) = -\frac{1}{3^4} = -\frac{1}{81}$$

The base is 3 (not -3). Take the reciprocal of the base. The negative coefficient makes the answer negative.

(d) $2^{-1} + 3^{-1}$

$$2^{-1} + 3^{-1} = \frac{1}{2} + \frac{1}{3} = \frac{3}{6} + \frac{2}{6} = \frac{5}{6}$$

Simplify each term, convert to a common denominator, and then add these fractions.

(e) $(2 + 3)^{-1}$

$$(2 + 3)^{-1} = (5)^{-1} = \frac{1}{5}$$

First simplify inside the parentheses and then take the reciprocal of the base.

SELF-CHECK 5.3.1

Simplify each expression.

1. 2^{-3} **2.** -2^3 **3.** $3^{-1} + 4^{-1}$ **4.** $(3 + 4)^{-1}$

A special case of the definition of negative exponents is useful for working with fractions. The fact that

$\left(\frac{x}{y}\right)^{-1} = \frac{y}{x}$ for $x \neq 0$ and $y \neq 0$ is shown as follows:

$$\left(\frac{x}{y}\right)^{-1} = \frac{1}{\frac{x}{y}} = 1 \div \frac{x}{y} = \frac{1}{1} \cdot \frac{y}{x} = \frac{y}{x}$$

The general case is given in the following box.

Fraction to a Negative Power

For any real numbers x and y and natural number n:

ALGEBRAICALLY	VERBALLY	NUMERICAL EXAMPLE
$\left(\frac{x}{y}\right)^{-n} = \left(\frac{y}{x}\right)^{n}$ for $x \neq 0$ and $y \neq 0$	A nonzero fraction to a negative exponent can be rewritten by taking the reciprocal of the fraction and using the corresponding positive exponent.	$\left(\frac{3}{5}\right)^{-2} = \left(\frac{5}{3}\right)^{2} = \frac{25}{9}$

EXAMPLE 2 Simplifying Expressions with Negative Exponents

Simplify the following expressions for $x \neq 0$ and $y \neq 0$.

SOLUTION

(a) $\left(\frac{2}{3}\right)^{-1}$

$$\left(\frac{2}{3}\right)^{-1} = \frac{3}{2}$$

Take the reciprocal of the base.

(b) $\left(\frac{5}{3}\right)^{-2}$

$$\left(\frac{5}{3}\right)^{-2} = \left(\frac{3}{5}\right)^{2} = \left(\frac{3}{5}\right)\left(\frac{3}{5}\right) = \frac{9}{25}$$

Take the reciprocal of the base, and then square this fraction.

We encourage you to rewrite $\left(\frac{5}{3}\right)^{-2}$ as $\left(\frac{3}{5}\right)^{2}$ rather than as $\frac{5^{-2}}{3^{-2}}$. Both are correct, but you should find the first simplification easier to use.

(c) $\left(\frac{x}{y}\right)^{-3}$ $\quad \left(\frac{x}{y}\right)^{-3} = \left(\frac{y}{x}\right)^{3}$ Take the reciprocal of the base and then apply the power rule.

$= \frac{y^3}{x^3}$

(d) $\frac{x^{-3}}{y}$ $\quad \frac{x^{-3}}{y} = \frac{x^{-3}}{1}\frac{1}{y}$ The base for the exponent -3 is just x.

$= \left(\frac{1}{x^3}\right)\left(\frac{1}{y}\right)$ Take the reciprocal of this base but not of the y.

$= \frac{1}{x^3y}$

Be careful to distinguish between negative exponents and negative coefficients.

One of the most common errors made is to confuse the meaning of negative coefficients with the meaning of negative exponents. Negative coefficients designate negative numbers. Negative exponents designate reciprocals, *not* negative numbers.

EXAMPLE 3 Evaluating Algebraic Expressions

Evaluate each expression for $x = 3$ and $y = 5$.

SOLUTION

(a) $-x^4$ $\quad -x^4 = -(3)^4 = -81$

(b) x^{-4} $\quad x^{-4} = 3^{-4} = \frac{1}{3^4} = \frac{1}{81}$

Compare parts (a) and (b) to note the distinction between a negative coefficient and a negative exponent.

(c) $(x + y)^{-1}$ $\quad (x + y)^{-1} = (3 + 5)^{-1} = 8^{-1} = \frac{1}{8}$

Using the correct order of operations, first simplify inside the parentheses and then take the reciprocal of the base of 8.

(d) $x^{-1} + y^{-1}$ $\quad x^{-1} + y^{-1} = 3^{-1} + 5^{-1} = \frac{1}{3} + \frac{1}{5} = \frac{5}{15} + \frac{3}{15} = \frac{8}{15}$

Simplify each term and then add these fractions, using the LCD of 15.

SELF-CHECK 5.3.2

Simplify each expression.

1. $\left(\frac{10}{3}\right)^{-1}$ 2. $\left(\frac{2}{7}\right)^{-2}$ 3. $\frac{5^{-2}}{4}$

Evaluate each expression for $x = 2$ and $y = 4$.

4. $-x^5$ 5. x^{-5} 6. $x^{-2} + y^{-2}$
7. $-(x + y)^2$ 8. $(x + y)^{-2}$

Rules for Negative Exponents

All the rules for exponents given earlier apply to all real exponents including zero and the negative integers. (Fractional exponents will be examined in Chapter 9.) The product, power, and quotient rules for exponents are restated below in algebraic form.

Properties of Integer Exponents

For any nonzero real number x and integers m and n:

Product rule: $x^m \cdot x^n = x^{m+n}$

Power rule: $(x^m)^n = x^{mn}$

Quotient rule: $\dfrac{x^m}{x^n} = x^{m-n}$

Use the properties of exponents whenever possible instead of applying the definitions. The properties of exponents are shortcuts that can save you time. When more than one property of exponents is involved, there may be several ways to simplify the expression. As you practice, try to gain efficiency by studying the examples given by your instructor and presented in this book.

EXAMPLE 4 Using the Properties of Exponents

Simplify each of the following expressions. Assuming each variable is not 0.

SOLUTION

(a) $\dfrac{x^4}{x^{-7}}$

$$\frac{x^4}{x^{-7}} = x^{4-(-7)}$$
$$= x^{11}$$

Use the quotient rule. Subtract the smaller exponent from the larger exponent to obtain an expression with a positive exponent.

(b) $(x^{-4})^{-2}$

$$(x^{-4})^{-2} = x^{(-4)(-2)}$$
$$= x^8$$

Use the power rule. Multiply the exponents.

(c) $(2a^{-2}b^4)^{-3}$

$$(2a^{-2}b^4)^{-3} = 2^{-3}(a^{-2})^{-3}(b^4)^{-3}$$

Product to a power rule.

$$= 2^{-3}a^6b^{-12}$$

Use the power rule: Multiply the exponents.

$$= \frac{a^6}{2^3b^{12}}$$

Express in terms of positive exponents by taking the reciprocal of the bases.

$$= \frac{a^6}{8b^{12}}$$

(d) $\left(\dfrac{12x^{-2}y^4}{15x^5y^{-6}}\right)^{-2}$

$$\left(\frac{12x^{-2}y^4}{15x^5y^{-6}}\right)^{-2} = \left(\frac{4y^{10}}{5x^7}\right)^{-2}$$
$$= \left(\frac{5x^7}{4y^{10}}\right)^2$$

First simplify the expression inside the parentheses, observing carefully the order of operations. Use the quotient rule on each base by subtracting the smaller exponent from the larger exponent. Then take the reciprocal of the base to remove the negative exponent.

$$= \frac{5^2(x^7)^2}{4^2(y^{10})^2}$$

Raise each factor to the second power.

$$= \frac{25x^{14}}{16y^{20}}$$

SELF-CHECK 5.3.3

Simplify each expression to a form using only positive exponents. Assume $x \neq 0$ and $y \neq 0$.

1. x^4x^{-7} **2.** $\frac{x^2y^{-3}}{x^{-1}y^4}$ **3.** $\left(\frac{14x^{-3}y^2}{35x^2y^{-4}}\right)^{-2}$

A Mathematical Note

Extremely small numbers and extremely large numbers are difficult to comprehend. For example, try to grasp the value of each of the following amounts: a million dollars, a billion dollars, and a trillion dollars. Spent at the rate of a dollar a second, a million dollars would last almost 12 days, a billion dollars would last almost 32 years, and a trillion dollars would last almost 32,000 years.

Scientific Notation

One very useful application of exponents is scientific notation. Scientific notation is used extensively in the sciences to represent numbers of either macroscopic or microscopic proportions. For example, the mean distance between Earth and the planet Uranus is 2,870,000,000,000 m, whereas the width of one line etched on an Intel computer chip is 0.000000032 m.

Our decimal system of representing numbers is based on powers of 10. Scientific notation uses exponents to represent powers of 10 and thus provides an alternative format for representing numbers. Each of the powers of 10 from 10,000 to 0.001 is written here in exponential notation:

$$\begin{aligned} 10{,}000 &= 10^4 \\ 1{,}000 &= 10^3 \\ 100 &= 10^2 \\ 10 &= 10^1 \\ 1 &= 10^0 \\ 0.1 &= 10^{-1} \\ 0.01 &= 10^{-2} \\ 0.001 &= 10^{-3} \end{aligned}$$

As these numbers illustrate, the exponent on 10 determines the position of the decimal point. **Scientific notation** uses this fact to express any decimal number as a product of a number between 1 and 10 (or between -1 and -10 if the number is negative) and an appropriate power of 10. Table 5.3.1 shows some representative numbers written in scientific notation.

Scientific Notation Format

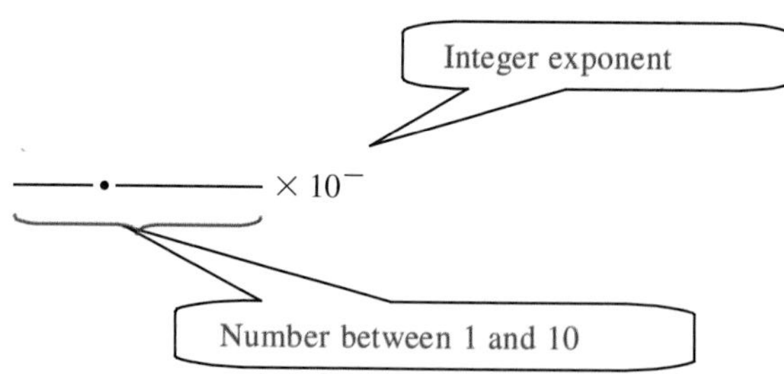

Table 5.3.1

NUMBER	SCIENTIFIC NOTATION
$12{,}345 = 1.2345 \times 10{,}000$	1.2345×10^4
$123.45 = 1.2345 \times 100$	1.2345×10^2
$12.345 = 1.2345 \times 10$	1.2345×10^1
$0.12345 = 1.2345 \times 0.1$	1.2345×10^{-1}
$0.0012345 = 1.2345 \times 0.001$	1.2345×10^{-3}
$-1{,}234.5 = -1.2345 \times 1{,}000$	-1.2345×10^3

If a number is written in scientific notation, then it can be rewritten in standard decimal notation by multiplying by the power of 10. This multiplication is easily accomplished by shifting the decimal point the number of places indicated by the exponent on the 10. Positive exponents indicate larger magnitudes and shift the decimal point to the right. Negative exponents indicate smaller magnitudes and shift the decimal point to the left.

Writing a Number in Standard Decimal Notation

VERBALLY	NUMERICAL EXAMPLES
Multiply out the two factors by using the given power of 10.	
1. If the exponent on 10 is positive, move the decimal point to the right.	**1.** $3.456 \times 10^2 = 3.456 \times 100 = 345.6$ The decimal point is moved two places to the right.
2. If the exponent on 10 is zero, do not move the decimal point.	**2.** $3.456 \times 10^0 = 3.456 \times 1 = 3.456$ The decimal point is not moved.
3. If the exponent on 10 is negative, move the decimal point to the left.	**3.** $3.456 \times 10^{-2} = 3.456 \times 0.01 = 0.03456$ The decimal point is moved two places to the left.

EXAMPLE 5 Writing Numbers in Standard Decimal Notation

Write each of the following numbers in standard decimal notation.

SOLUTION

(a) 5.789×10^2

$5.789 \times 10^2 = 578.9$

Two places right

Move the decimal point two places to the right.

(b) 8.6×10^4

$8.6 \times 10^4 = 86{,}000$

Four places right

Fill in zeros so that the decimal point can be moved four places to the right.

(c) 4.61×10^{-4}

$4.61 \times 10^{-4} = 0.000461$

Four places left

Fill in zeros so that the decimal point can be moved four places to the left.

(d) -1.4×10^7

$-1.4 \times 10^7 = -14{,}000{,}000$

Seven places right

Move the decimal point seven places to the right.

SELF-CHECK 5.3.4

Write each of these numbers in standard decimal format.

1. 7.23×10^5 **2.** 7.23×10^{-5} **3.** 1.4×10^{-7}

The steps for writing a number in scientific notation are given in the following box. Remember that positive exponents indicate larger magnitudes and negative exponents indicate smaller magnitudes.

Numbers with larger magnitudes have positive exponents on 10. Numbers whose magnitudes are between 0 and 1 have negative exponents on 10.

Writing a Number in Scientific Notation

VERBALLY	NUMERICAL EXAMPLES
1. Move the decimal point immediately to the right of the first nonzero digit of the number.	
2. Multiply by a power of 10 determined by counting the number of places the decimal point has been moved.	
a. The exponent on 10 is zero or positive if the magnitude of the original number is 1 or greater.	$3.456 = 3.456 \times 10^0$ $345.6 = 3.456 \times 10^2$
b. The exponent on 10 is negative if the magnitude of the original number is between 0 and 1.	$0.03456 = 3.456 \times 10^{-2}$

EXAMPLE 6 Writing Numbers in Scientific Notation

Write each of the following numbers in scientific notation.

SOLUTION

(a) 9,876

$9{,}876. = 9.876 \times 10^3$

Three places left

The decimal point is moved three places to position it behind the first nonzero digit, 9. The exponent 3 is positive, since the original number is greater than 1.

(b) 325.49

$325.49 = 3.2549 \times 10^2$

Two places left

The exponent on 10 is positive 2, since the decimal point was moved two places and the original number is greater than 1.

(c) 0.0009876

$0.0009876 = 9.876 \times 10^{-4}$

Four places right

The exponent on 10 is −4, since the decimal point was moved four places and the original number is less than 1.

Calculators often use a special syntax to represent scientific notation. Many calculators use E to precede the power of 10 that would be used to represent a number in scientific notation. This is illustrated in the following table.

STANDARD DECIMAL NOTATION	SCIENTIFIC NOTATION	CALCULATOR SYNTAX
12,345	1.2345×10^4	1.2345E4
1.2345	1.2345×10^0	1.2345E0
0.00567	5.67×10^{-3}	5.67E−3

When a value becomes sufficiently large or small, many graphing calculators will automatically display values in scientific notation. We can also use the EE feature to enter values into a calculator by using scientific notation.

CALCULATOR PERSPECTIVE 5.3.1 Using Scientific Notation

To evaluate the products 99×125 and $990{,}000 \times 125{,}000$ on a TI-84 Plus calculator, enter the following keystrokes:

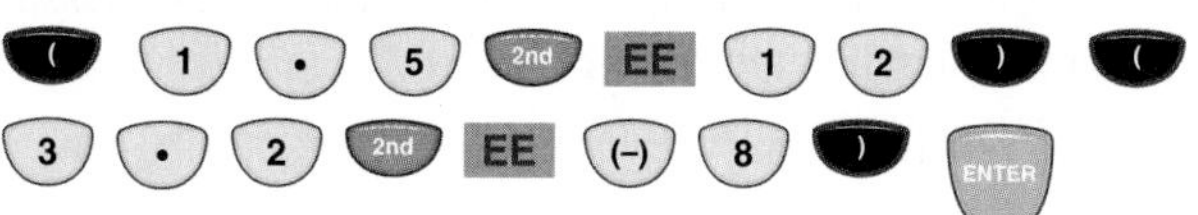

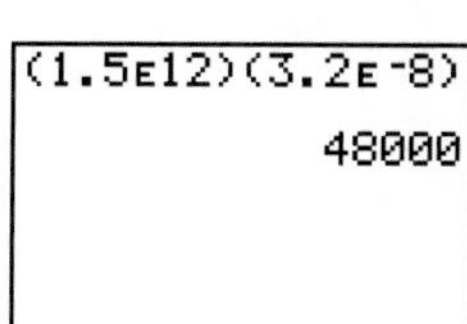

Note: Scientific notation is not needed to display 12,375 but is automatically used by the calculator to display the product of 123,750,000,000.

The EE feature is the secondary function of the , key.

To evaluate the product $(1.5 \times 10^{12})(3.2 \times 10^{-8})$ on a TI-84 Plus calculator, enter the following keystrokes:

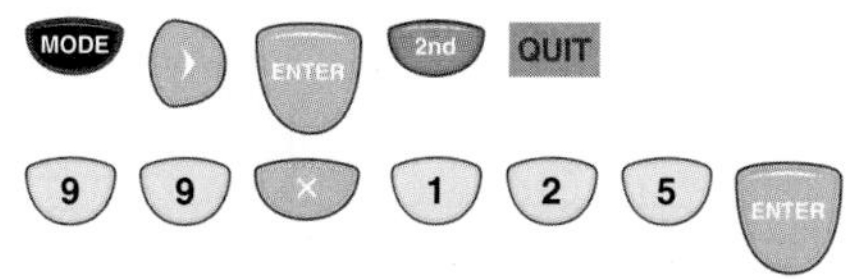

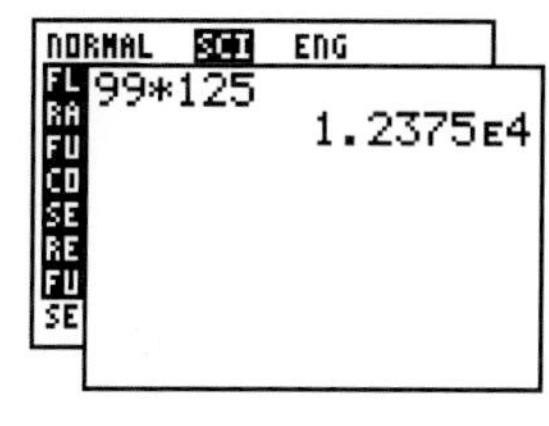

Note: Although the values entered were given in scientific notation, the magnitude of the result did not require scientific notation on the display. The result is shown in standard decimal notation.

To set a TI-84 Plus calculator to display all results in scientific notation and to evaluate the product 99×125, enter the following keystrokes:

MODE ▸ ENTER 2nd QUIT

9 9 × 1 2 5 ENTER

NORMAL SCI ENG
99*125
1.2375E4

Note: In **Normal** mode, as shown previously, this result would be in standard decimal form. To display all results in scientific notation, set the calculator to **Sci** mode.

Scientific notation can also be used to perform pencil-and-paper calculations or to perform quick estimates of otherwise lengthy calculations.

EXAMPLE 7 Using Scientific Notation to Estimate a Product

Use scientific notation to perform a pencil-and-paper estimate of (149,736)(0.003999).

SOLUTION

	ESTIMATED VALUE	CALCULATOR VALUE
$149{,}736 \approx 150{,}000$	$(1.5 \times 10^5)(4.0 \times 10^{-3})$	
$= 1.5 \times 10^5$	$= (1.5)(4)(10^5)(10^{-3})$	
$0.003999 \approx 0.004$	$= 6 \times 10^2$	
$= 4.0 \times 10^{-3}$	$= 600$	

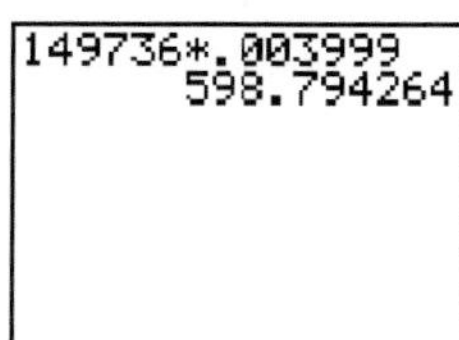

Answer: The calculator product of 598.794264 seems reasonable based upon our estimate of 600.

SELF-CHECK 5.3.5

Write each of these numbers in scientific notation.

1. 80,000 2. 0.008 3. 0.723 4. 72.3
5. Use scientific notation to estimate (4,997,483)(0.030147).

The results of calculations involving very small or very large numbers often need to be rounded to avoid reporting answers with an unrealistic number of significant digits. (Optional material on significant digits and precision is given in Appendix A.) This is illustrated in Example 8.

EXAMPLE 8 Calculating Time for an Emergency Message from Mars

If a flight crew orbiting Mars has an emergency and needs to communicate with the ground crew on Earth, how long will it take for an emergency message to reach Earth from Mars? Assume that the distance the signal must travel is 3.78×10^{11} m and the speed of the signal is approximately 2.998×10^8 m/s.

SOLUTION

Let t = time in seconds for the message to reach Earth.

Time = Distance ÷ Rate

Rearrange the formula $D = RT$ to form the word equation.

$$t = \frac{3.78 \times 10^{11}}{2.998 \times 10^8}$$

Then substitute in the given values, and use a calculator to approximate t.

```
(3.78E11)/(2.998
E8)
          1260.84056
```

$t \approx 1{,}260$ seconds

The answer in seconds has been rounded to three significant digits.

$$t \approx \frac{1{,}260}{60} \text{ minutes}$$

$t \approx 21$ minutes

Answer: It will take an emergency message approximately 21 minutes for the signal to travel from Mars to Earth.

SELF-CHECK ANSWERS

5.3.1

1. $\frac{1}{8}$ 2. -8 3. $\frac{7}{12}$ 4. $\frac{1}{7}$ 5. $\frac{1}{32}$ 6. $\frac{5}{16}$ 7. -36 8. $\frac{1}{36}$ (4. -32)

5.3.2

1. $\frac{3}{10}$ 2. $\frac{49}{4}$ 3. $\frac{1}{100}$

5.3.3

1. $\frac{1}{x^3}$ 2. $\frac{x^3}{y^7}$ 3. $\frac{25x^{10}}{4y^{12}}$

5.3.4

1. 723,000 2. 0.0000723 3. 0.00000014

5.3.5

1. 8.0×10^4
2. 8.0×10^{-3}
3. 7.23×10^{-1}
4. 7.23×10^1
5. 150,000

USING THE LANGUAGE AND SYMBOLISM OF MATHEMATICS 5.3

1. For any real number x, $x \neq$ __________, and natural number n, $x^{-n} =$ __________.
2. For any nonzero real number x and integers m and n, $x^m \cdot x^n =$ __________.
3. For any nonzero real number x and integers m and n, $(x^m)^n =$ __________.
4. For any nonzero real number x and integers m and n, $\frac{x^m}{x^n} =$ __________.

5. Scientific notation is used to express a decimal number as a product of a number between 1 and 10 (or between -1 and -10 if the number is negative) and an appropriate power of ________.

6. In scientific notation the exponent on ________ determines the position of the decimal point.

EXERCISES 5.3

In Exercises 1–4, use a calculator to complete each table. Then state the equation that is confirmed by each table.

1. 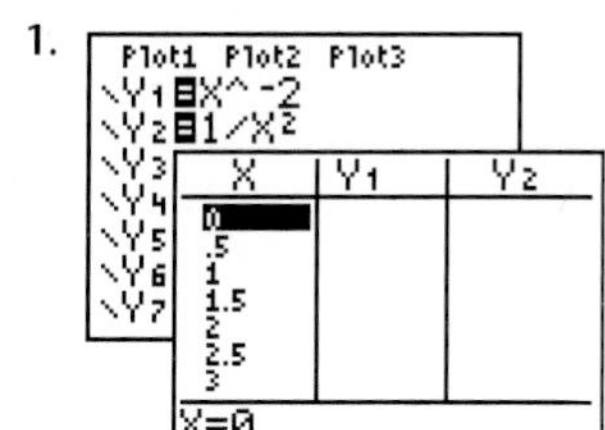

2. 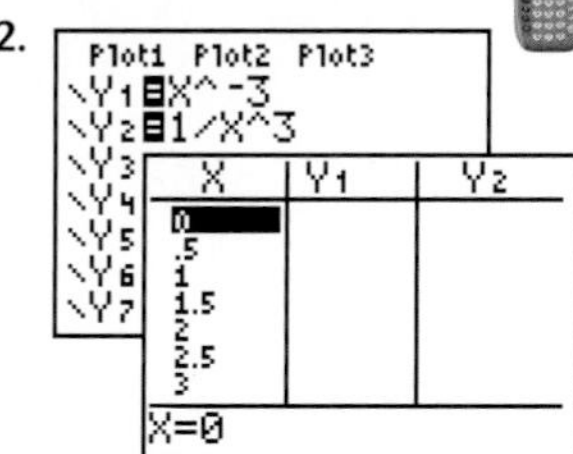

3. 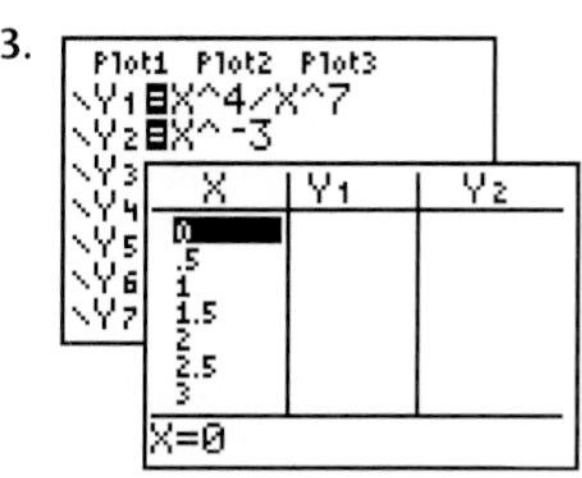

4. 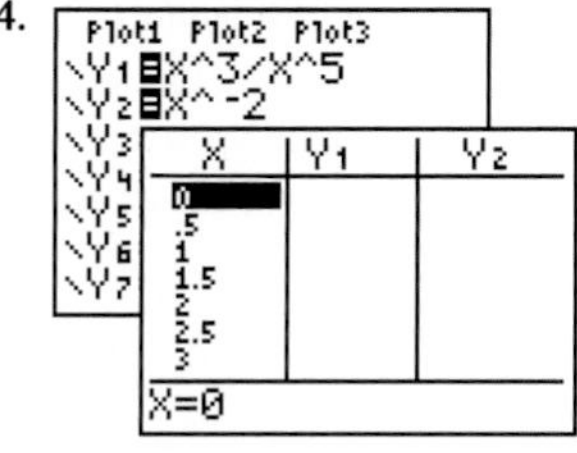

In Exercises 5–20, simplify each expression by writing it in a form free of negative exponents. Assume that all bases are nonzero real numbers.

5. a. 2^{-5} b. -2^5 c. -5^2
6. a. 4^{-3} b. 3^{-4} c. -4^3
7. a. $\left(\frac{6}{5}\right)^{-1}$ b. $\left(\frac{6}{5}\right)^{-2}$ c. $\left(\frac{6}{5}\right)^{0}$
8. a. $\left(\frac{3}{4}\right)^{-1}$ b. $\left(\frac{3}{4}\right)^{-2}$ c. $\left(\frac{3}{4}\right)^{0}$
9. a. 10^{-2} b. 10^{-3} c. -10^{-2}
10. a. 2^{-4} b. -2^{-4} c. -4^2
11. a. $(2 + 5)^{-1}$ b. $2^{-1} + 5^{-1}$ c. $-2 + 5^{-1}$
12. a. $(5 + 10)^{-1}$ b. $5^{-1} + 10^{-1}$ c. $-5 + 10^{-1}$
13. a. $\left(\frac{1}{2} + \frac{1}{5}\right)^{-1}$ b. $\left(\frac{1}{2}\right)^{-1} + \left(\frac{1}{5}\right)^{-1}$ c. $\left(\frac{1}{2}\right)^{0} + \left(\frac{1}{5}\right)^{0}$
14. a. $\left(\frac{1}{3} + \frac{1}{4}\right)^{-1}$ b. $\left(\frac{1}{3}\right)^{-1} + \left(\frac{1}{4}\right)^{-1}$ c. $\left(\frac{1}{3} + \frac{1}{4}\right)^{0}$
15. a. $\left(\frac{x}{y}\right)^{-1}$ b. $\left(\frac{x}{y}\right)^{-2}$ c. $\frac{x^{-2}}{y}$
16. a. $\left(\frac{v}{w}\right)^{-1}$ b. $\left(\frac{v}{w}\right)^{-2}$ c. $\frac{v}{w^{-2}}$
17. a. $(3x)^{-2}$ b. $3x^{-2}$ c. $(-3x)^2$
18. a. $(2x)^{-3}$ b. $2x^{-3}$ c. $(-2x)^3$
19. a. $(m + n)^{-1}$ b. $m^{-1} + n^{-1}$ c. $m + n^{-1}$
20. a. $(2x + y)^{-1}$ b. $2x + y^{-1}$ c. $2x^{-1} + y^{-1}$

In Exercises 21–24, write each number in standard decimal notation.

21. a. 4.58×10^4 b. 4.58×10^{-4}
c. 4.58×10^6 d. 4.58×10^{-6}
22. a. 1.7×10^1 b. 1.7×10^{-1}
c. 1.7×10^5 d. 1.7×10^{-5}
23. a. 8.1×10^3 b. 8.1×10^{-3}
c. -8.1×10^3 d. -8.1×10^{-7}
24. a. 3.6×10^7 b. 3.6×10^{-7}
c. -3.6×10^7 d. -3.6×10^{-7}

In Exercises 25–28, write each number in scientific notation.

25. a. 9,700 b. 97,000,000
c. 0.97 d. 0.00097
26. a. 18,900 b. 189
c. 0.0000189 d. 0.00189
27. a. 35,000,000,000 b. 0.0000000035
c. $-3{,}500$ d. -0.035
28. a. 470,000,000,000 b. 0.000000047
c. $-47{,}000$ d. -0.0047

Applications of Scientific Notation

In Exercises 29–32, write the numbers in each statement in standard decimal notation.

29. **Distance to Neptune** The mean distance from our sun to the planet Neptune, which was examined by the *Voyager* 2 spacecraft in 1989, is 4.493×10^9 km.

30. **Operations by a Computer** A 1990 advertisement for one computer claimed that the computer could perform 4.5×10^8 floating-point operations per second.

31. **Computer Switching Speed** The switching speeds of some of the most sophisticated computers are measured in picoseconds. A picosecond is one-trillionth of a second, or 1.0×10^{-12} second.

32. **Wavelength of Red Light** The wavelength of red light is 7,000 angstroms, which is 7.0×10^{-7} m.

In Exercises 33–36, write the numbers in each statement in scientific notation.

33. **Temperature of the Sun** Temperatures inside the sun are estimated to be 14,000,000°C.

34. **Speed of Light** The speed of light is approximately 299,790,000 m/s.

35. **Pain Signals** A burn to a human finger will trigger a pain message to the brain. This pain signal will travel approximately 1 ft in 0.0000568 second.

36. **Tidal Friction** The friction between the ocean and the ocean floor due to tides is causing the earth's rotation to slow down by about 0.00000002 second/day.

37. **Moore's Law** Gordon Moore gave a talk in 1965, four years after the first integrated circuit was developed, which predicted that the number of transistors per integrated circuit would double every 18 months. Remarkably, this doubling prediction has proved remarkably accurate for 40 plus years, yielding fantastic advances in computers. This prediction has become known as Moore's law. Use the following table to answer each part of this question. (Source: For more

information on this topic go to the Additional Resources link on the MathZone website at **www.mhhe.com/hallmercer.**)

YEAR	NUMBER OF TRANSISTORS
1971	2.3×10^3
1978	2.9×10^4
1985	2.8×10^5
1993	3.1×10^6
2000	4.2×10^7
2003	2.2×10^8
2006	1.1×10^9

a. Write in standard decimal notation the number of transistors on an integrated circuit in 1971.
b. Write in standard decimal notation the number of transistors on an integrated circuit in 2003.
c. Newspaper articles have claimed that there were 100,000 times as many transistors on an integrated circuit in 2003 as in 1971. Use the data in this table to determine whether this is an accurate claim.

38. **Modeling Time with a Cesium Clock** The NIST (National Institute of Standards & Technology) F-1 Cesium fountain clock built in 1999 is accurate to within 1 second in 20 million years. Accuracy of this level is important for many scientific processes such as global positioning systems on airplanes and the design of microelectronic devices. (Source: For more information on this topic go to the Additional Resources link on the MathZone website at **www.mhhe.com/hallmercer.**)

How many seconds does the clock vary from the actual time in a

a. Century b. Year c. Day

39. **Time for an Emergency Message** If a flight crew orbiting Mars has an emergency and needs to communicate with the ground crew on Earth, how long will it take for an emergency message to reach Earth from Mars? Assume that the distance the signal must travel is 3.78×10^{11} m and the speed of the signal is approximately 2.998×10^8 m/s.

40. **Storage Capacity of a Personal Music Player** Each song on a personal music player requires approximately 4 megabytes (MB) (4.0×10^6 bytes) of memory. If you purchase a personal music player with 60 gigabytes (GB) (6.0×10^{10} bytes) of memory, approximately how many songs can you store on this player?

Estimation and Calculator Skills

In Exercises 41–46, use scientific notation and pencil and paper to estimate the value of each expression. Then use a calculator to approximate the value of each expression.

PROBLEM	PENCIL AND PAPER ESTIMATE	CALCULATOR APPROXIMATION
41. (10,013)(0.00007943)		
42. (9,853,493)(0.00061)		
43. $\dfrac{0.000005034}{0.00009893}$		
44. $\dfrac{901{,}053{,}792}{0.02987}$		
45. $(0.02973)^3$		
46. $(0.0005012)^2$		

In Exercises 47 and 48, write each result on the calculator screen in standard decimal notation.

47.
```
2.8E6*6.3E-11
        1.764E-4
```

48. 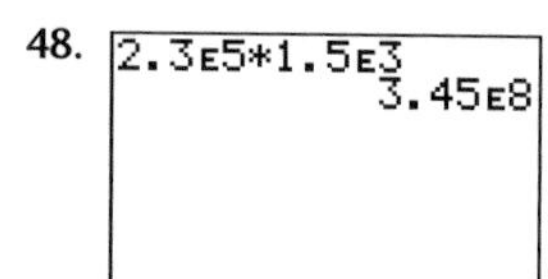

Calculator Skills

In Exercises 49 and 50, use a calculator to approximate each expression accurate to three significant digits. Write the answer in standard decimal notation. (*Hint:* See Appendix A on significant digits.)

49. a. $(4.32 \times 10^{15})(8.49 \times 10^{-17})$ b. $(7.16 \times 10^3)^2$

50. a. $\dfrac{9.71 \times 10^{13}}{6.53 \times 10^{11}}$ b. $[(6.7)(4.98 \times 10^2)]^2$

In Exercises 51–58, simplify each expression, writing each in a form free of negative exponents. Assume all variables are nonzero.

51. a. $v^{-3}v^{12}$ b. $\dfrac{v^{12}}{v^{-3}}$ c. $\dfrac{v^{-12}}{v^3}$

52. a. $w^{21}w^{-7}$ b. $\dfrac{w^{21}}{w^{-7}}$ c. $\dfrac{w^{-21}}{w^7}$

53. a. $x^3x^0x^{-7}$ b. $(x^3x^{-7})^0$ c. $x^0x^{-3}x^{-7}$

54. a. y^0yy^{-9} b. $(y^9y^{-1})^0$ c. $y^0y^{-1}y^{-9}$

55. a. $\dfrac{6x^{-4}}{2x^{-3}}$ b. $(2x^{-3})(6x^4)$ c. $(2x^{-3})(6x^{-4})$

56. a. $\dfrac{15x^7}{5x^{-2}}$ b. $(15x^7)(5x^{-2})$ c. $(15x^{-7})(5x^{-2})$

57. a. $\left(\dfrac{3x}{5y}\right)^2$ b. $\left(\dfrac{3x}{5y}\right)^{-2}$ c. $[(3x)(5y)]^2$

58. a. $\left(\dfrac{5v}{7w}\right)^2$ b. $\left(\dfrac{5v}{7w}\right)^{-2}$ c. $[(5v)(7w)]^2$

In Exercises 59–72, simplify each expression, assuming all variables are nonzero.

59. $[(5x^3)(-4x^{-2})]^{-1}$

60. $[(6x^{-4})(5x^7)]^{-1}$

61. $(2x^{-3}y^4)^3(3x^4y^{-2})^2$

62. $(5x^3y^{-4})(2x^{-5}y^3)^4$

63. $\dfrac{-6x^6}{4x^{-4}}$

64. $\dfrac{-12x^{-3}}{-3x^{-12}}$

65. $\dfrac{24x^3y^{-8}}{-6x^{-12}y^{24}}$

66. $\dfrac{-16x^{-8}y^{-5}}{8x^4y^{-10}}$

67. $\left(\dfrac{12x^3}{6x^{-2}}\right)^2$

68. $\left(\dfrac{24x^5}{8x^{-3}}\right)^3$

69. $\left(\dfrac{m^3n^{-7}}{m^7n^{-11}}\right)^{-3}$

70. $\left(\dfrac{v^{-6}w^4}{v^{-9}w^{-7}}\right)^{-5}$

71. $\dfrac{(2x^{-1}y^2)(4x^2y^{-3})^{-2}}{(12x^{-2}y^{-2})^{-1}}$

72. $\left[\dfrac{(5x^{-3}y^4)^{-2}(6x^2y^{-5})}{15x^2y^{-4}}\right]^{-2}$

In Exercises 73–76, use the given values of x and y to evaluate each expression.

	$x^2 + y^2$	$(x + y)^2$	$x^{-1} + y^{-1}$	$(x + y)^{-1}$
73. $x = 2, y = 3$				
74. $x = 3, y = 4$				
75. $x = -3, y = -4$				
76. $x = -2, y = -4$				

Group Discussion Questions

77. **Risks and Choices**
 a. If you live to be 1 billion seconds old, how many years will you have lived?
 b. If you live to be 80 years old, how many seconds will you have lived?
 c. If smoking cigarettes decreases one individual's lifetime from 80 years to 65, how many seconds has this individual lost of the potential lifetime?
 d. Assume in part **c** that this person smoked $20 \times 365 \times 50$ cigarettes. How many seconds of life has each cigarette cost this person?

78. **Multiple Representations** Engineering notation is similar to scientific notation. Engineering notation expresses a decimal number as a product of a number between 1 and 1,000 (or -1 and $-1{,}000$ if the number is negative) and a power of 10 which is a multiple of 3, as in 12.4×10^3 or 453.7×10^6. Set a calculator first to scientific notation **(SCI)**, then to engineering notation **(ENG)**, and complete this table.

STANDARD DECIMAL NOTATION	SCIENTIFIC NOTATION	ENGINEERING NOTATION
a. 45		
b. 45,678		
c. 456,789		
d. 4,567,890		
e. 0.045		
f. 0.00045		

79. **Discovery Question** If x is negative and m is an integer:
 a. Can x^m be positive? If so, give the values of m for which x^m is positive.
 b. Can x^m be negative? If so, give the values of m for which x^m is negative.
 c. Can x^{-2} represent a negative value? If so, give an example.
 d. Can x^{-3} represent a negative value? If so, give an example.

Section 5.4 Adding and Subtracting Polynomials

Objectives:
8. Use the terminology associated with polynomials.
9. Add and subtract polynomials.

This section examines important algebraic expressions called polynomials. We have already used polynomials such as $2x + 3$, $\frac{5}{9}(F - 32)$, $2\pi r$, and πr^2 throughout the previous sections of this book. We now extend our earlier work on operations with polynomials and define some key terminology.

Monomials and Polynomials

Factors are multiplied; terms are added or subtracted.

Recall that factors are constants or variables that are multiplied together to form a product. A single number or an indicated product of factors is called a **term.** The terms in an algebraic expression are separated from one another by plus or minus symbols.

The expression $7x^2 + 9xy - 3y^2$, which also can be written as $7x^2 + 9xy + (-3y^2)$, has three terms: $7x^2$, $9xy$, and $-3y^2$. The second term, $9xy$, has three factors: 9, x, and y.

A **monomial** is a real number, a variable, or a product of real numbers and variables. Because a variable can be used repeatedly as a factor, whole-number exponents can occur in monomials. A **polynomial** is a monomial or a sum of a finite number of monomials.

Monomials and Polynomials

	VERBALLY	ALGEBRAIC EXAMPLES
Monomials	A monomial is a real number, a variable, or a product of real numbers and variables with whole-number exponents.	-5, π, x, A, $5x$, $7xy$, and πr^2 are monomials.
Polynomials	A polynomial is a monomial or a sum of a finite number of monomials.	-5, $7xy$, and $7x^2 + 9xy - 3y^2$ are polynomials.

A monomial in the variable x has the form ax^n, where a is the constant coefficient. Although the word *coefficient* is not usually applied to a constant term, it can be. In Example 1 we note that the coefficient of 5 is 5. The exponent n in the monomial ax^n can be any whole number 0, 1, 2, 3, ... Because the exponent n cannot be negative, a monomial cannot have a variable in the denominator.

EXAMPLE 1 Identifying Monomials

Determine which of the following expressions are monomials.

SOLUTION

(a) $-7x^4$	Monomial	The coefficient of x^4 is -7.
(b) $\frac{4}{7}x^3y^2$	Monomial	The numerical coefficient is the constant $\frac{4}{7}$. The exponent on x is $+3$ and on y is $+2$.
(c) $\frac{4x^3}{7y^2}$ or $\frac{4}{7}x^3y^{-2}$	Not a monomial	This expression contains a variable in the denominator; in the optional form the exponent on y is not a whole number.
(d) 5	Monomial	All constants are monomials. Also 5 can be written as $5x^0$. So the numerical coefficient is 5.
(e) $5x^{1/2}$	Not a monomial	$5x^{1/2}$ (or $5\sqrt{x}$)* has an exponent that is not a whole number.

*The relationship between exponential and radical notation will be examined in Section 9.5.

Polynomials containing one, two, and three terms are called **monomials, binomials,** and **trinomials,** respectively. Although the prefix *poly* means many, polynomials can have a single term. Example 2 illustrates these classifications.

EXAMPLE 2 Identifying Polynomials

Determine whether each expression is a polynomial. Classify each polynomial according to the number of terms it contains.

SOLUTION

(a) $5x^2 + 3x - 7$	Trinomial	The three terms are $5x^2$, $3x$, and -7.
(b) $x^5y - 4x^3y^2$	Binomial	The two terms are x^3y and $-4x^3y^2$.
(c) $\frac{x + 5}{x - 5}$	Not a polynomial	A polynomial cannot contain a variable in the denominator.

(d) $\frac{7xy}{9}$	Monomial	The polynomial has only one term with a coefficient of $\frac{7}{9}$.
(e) 18.93	Monomial	All constants are monomials.
(f) $3x^4 - 9x^3 + 7x^2 + 8x - 1$	Polynomial with five terms	Polynomials with more than three terms are not assigned special names.

The **degree of a monomial** is the sum of the exponents for all the variables in this term. A nonzero constant is understood to have degree zero ($4 = 4x^0$ with exponent 0), but no degree is assigned to the monomial 0. One place the degree of a polynomial plays a key role is determining the number of possible solutions to an nth-degree polynomial equation. For example, a first-degree linear equation has one solution and a second-degree equation has two solutions.

EXAMPLE 3 Determining the Degree of a Monomial

Determine the coefficient and the degree of each monomial.

SOLUTION

	COEFFICIENT	DEGREE	
(a) $-5x^3$	-5	3	The numerical coefficient is -5, not 5.
(b) $5x^3y^7$	5	10	The sum of the exponents is $3 + 7 = 10$.
(c) 5	5	0	$5 = 5x^0$ with exponent 0.
(d) $-x$	-1	1	$-x = -1x^1$ with coefficient of -1 and exponent of 1 (usually not written).
(e) mn^3	1	4	The coefficient is understood to be 1. The sum of the exponents is $1 + 3 = 4$.

SELF-CHECK 5.4.1

Determine whether each expression is a polynomial. Classify each polynomial according to the number of terms it contains.

1. $7x + 3$
2. $\frac{7x}{3}$
3. $\frac{3}{7x}$
4. $7x^3 + x - 9$

Determine the coefficient and the degree of each monomial.

5. $-3x^4$
6. xyz
7. π

The **degree of a polynomial** is the same as the degree of the term with the highest degree. To find this highest degree, examine each term individually—do ***not*** sum the degrees of the terms.

EXAMPLE 4 Determining the Degree of a Polynomial

Determine the degree of each of the following polynomials.

SOLUTION

(a) $5x^3 + 7x^2$	3	The degrees of the individual terms are 3 and 2.
(b) $-11x^2 + 7x + 8$	2	The degrees of the individual terms are 2, 1, and 0.

(c) $4x^6y - 3x^3y^5$	8	The degrees of the individual terms are 7 and 8.
(d) $a^3 + 5a^2b - 3ab^2 + b^3$	3	Each of these terms is of degree 3.

A polynomial in x is in **descending order** if the exponents on x decrease from left to right. A polynomial in x is in **ascending order** if the exponents on x increase from left to right. For example, $3x^2 - 7x^3 + 4 - 9x$ can be written in descending order as $-7x^3 + 3x^2 - 9x + 4$ or in ascending order as $4 - 9x + 3x^2 - 7x^3$. For some algebraic work, polynomials are easier to compare and manipulate if they are written in standard form. A polynomial is in **standard form** if (1) the variables in each term are written in alphabetical order and (2) the terms are arranged in descending powers of the first variable. The **leading term** is the first term of a polynomial in standard form and the term of highest degree. In Section 5.6 it is definitely best to write polynomials in standard form before performing long division of polynomials.

EXAMPLE 5 Writing Polynomials in Standard Form

Write each of the following polynomials in standard form.

SOLUTION

(a) $8y^3z^2x$	$8xy^3z^2$	Write the factors in alphabetical order.
(b) $2x^3 + 7 + x^4 - 5x^2$	$x^4 + 2x^3 - 5x^2 + 7$	Arrange the terms in descending order. The leading term is x^4.
(c) $y^2 + 4yx + x^2$	$x^2 + 4xy + y^2$	Write each term in alphabetical order, and then arrange the terms in decreasing powers of x.

SELF-CHECK 5.4.2

Write each polynomial in standard form, and give the degree of each.

1. $-9 + 3x^2 + 8x$
2. $-9y^2zx^5$
3. $-4v + 9v^5 + 3v^2 - v^3 + 1$

A polynomial whose only variable is x is called a **polynomial in *x*.** The standard form of an nth-degree polynomial in x is $a_nx^n + a_{n-1}x^{n-1} + \cdots + a_1x + a_0$. If each of the coefficients of this polynomial is a real number, then this polynomial is called a **real polynomial.** A polynomial function in x can be represented by the function notation $f(x)$. For example, $f(x) = 5x^2 - 13$ represents a polynomial function. Sometimes we use letters other than f to represent functions. For example, we can let $P(x) = 2x^2 - 4x + 11$. To evaluate this polynomial for $x = 3$, we then substitute 3 for x in the polynomial: $P(3) = 2(3)^2 - 4(3) + 11 = 17$.

EXAMPLE 6 Evaluating a Profit Polynomial

The profit in dollars made by producing and selling x units is given by the polynomial $P(x) = -x^2 + 14x - 33$. Evaluate and interpret each expression.

SOLUTION

(a) $P(0)$

$$P(0) = -(0)^2 + 14(0) - 33$$
$$= -33$$

Selling 0 units results in a loss of \$33.

(b) $P(3)$ $\quad P(3) = -(3)^2 + 14(3) - 33$
$= -9 + 42 - 33$
$= 0$

The seller breaks even if 3 units are sold.

(c) $P(10)$ $\quad P(10) = -(10)^2 + 14(10) - 33$
$= -100 + 140 - 33$
$= 7$

The seller makes \$7 if 10 units are sold.

Adding and Subtracting Polynomials

The distributive property justifies adding the coefficients of like terms because $ax + bx = (a + b)x$.

We add or subtract polynomials by combining like terms. **Like terms** or **similar terms** have exactly the same variable factors. For example, $5x^2y$ and $-11x^2y$ have the same variable factors and are like terms whereas $5x^2y$ and $11xy$ have different variable factors and are unlike terms. We first used the distributive property in Section 1.7 to add like terms and to remove parentheses from a group of terms. We will use the distributive property extensively to add and subtract polynomials.

Some individuals find a vertical format helpful to organize the work for adding or subtracting two polynomials. If you use a vertical format, align only like terms in the same column. To do this, you may need to leave blanks in some rows (or insert zero coefficients if you write terms in these locations). Example 7 illustrates both the horizontal and the vertical format.

EXAMPLE 7 Adding Polynomials

Add $5x^2 - 7x + 9$ and $3x^2 + 6x - 8$.

SOLUTION

HORIZONTAL FORMAT

$(5x^2 - 7x + 9) + (3x^2 + 6x - 8)$ — Use a plus symbol to indicate the addition.
$= 5x^2 - 7x + 9 + 3x^2 + 6x - 8$ — Remove the parentheses.
$= 5x^2 + 3x^2 - 7x + 6x + 9 - 8$ — Use the commutative and associative properties of addition to reorder the terms.
$= (5 + 3)x^2 + (-7 + 6)x + (9 - 8)$ — Then use the distributive property to combine like terms.
$= 8x^2 - x + 1$ — The answer is written in standard form.

VERTICAL FORMAT

$$\begin{array}{r} 5x^2 - 7x + 9 \\ \underline{3x^2 + 6x - 8} \\ 8x^2 - x + 1 \end{array}$$

Align only like terms in the same column.

Although use of the distributive property is not as apparent in this format, it is still the key justification for combining the like terms. The answer is written in standard form.

Answer: $(5x^2 - 7x + 9) + (3x^2 - 6x - 8) = 8x^2 - x + 1$

SELF-CHECK 5.4.3

1. Use the profit polynomial $P(x) = -x^2 + 17x - 30$ and a calculator to evaluate $P(5)$.
2. Add $5x - 9$ and $8x + 7$.
3. Add $-9x^4 + 8x^3 - 9$ and $6x^4 + 8x^2 - 3x + 1$.

The opposite of $7x^2 - 5x + 4$ also can be written as $-1(7x^2 - 5x + 4)$. Distributing the factor of -1, we obtain $-1(7x^2) + (-1)(-5x) + (-1)(4) = -7x^2 + 5x - 4$.

To subtract polynomials, we also combine like terms. Subtraction can be defined as the addition of an additive inverse or opposite. The opposite of a polynomial is formed by taking the opposite of each term of the polynomial. For example, the opposite of

$7x^2 - 5x + 4$ is denoted by $-(7x^2 - 5x + 4)$, and $-(7x^2 - 5x + 4) = -7x^2 + 5x - 4$. Example 8 illustrates how to subtract the polynomial $7x^2 - 5x + 4$ by adding its opposite.

EXAMPLE 8 Subtracting Polynomials

Subtract $7x^2 - 5x + 4$ from $3x^2 + 6x - 8$.

SOLUTION

HORIZONTAL FORMAT

$(3x^2 + 6x - 8) - (7x^2 - 5x + 4)$	Use a minus symbol to indicate the subtraction.
$= 3x^2 + 6x - 8 - 7x^2 + 5x - 4$	Remove the parentheses by using the distributive property.
$= 3x^2 - 7x^2 + 6x + 5x - 8 - 4$ $= (3 - 7)x^2 + (6 + 5)x + (-8 - 4)$	Reorder the terms. Then use the distributive property to combine like terms.
$= -4x^2 + 11x - 12$	The answer is written in standard form.

VERTICAL FORMAT

$$\begin{array}{r} 3x^2 + 6x - 8 \\ + \ -7x^2 + 5x - 4 \\ \hline -4x^2 + 11x - 12 \end{array}$$

To subtract by using the vertical format, align like terms. To subtract, add the opposite of $7x^2 - 5x + 4$. Note that this is accomplished by changing the sign of each term of $7x^2 - 5x + 4$.

Answer: $(3x^2 + 6x - 8) - (7x^2 - 5x + 4) = -4x^2 + 11x - 12$

SELF-CHECK 5.4.4

1. Subtract $2x^2 - 4$ from $5x^2 + 7x$.
2. Subtract $-3x^4 - 4x^3 + 2x^2 - 4$ from $x^3 - 5x^2 + 7x$.

If two polynomials are equal, then a table of values will show identical values for the two polynomials and their graphs will be identical. In Example 9 we use tables and graphs as a check on the sum of two polynomials. We will examine the graphs of polynomials further in Chapter 6.

EXAMPLE 9 Using Tables and Graphs to Compare Two Polynomials

Add $2(3x + 5)$ to $4(2x - 3)$ and check the sum by using a table of values.

SOLUTION

ALGEBRAICALLY

$2(3x + 5) + 4(2x - 3)$ $= 6x + 10 + 8x - 12$	Indicate the addition and then use the distributive property to remove the parentheses.
$= 6x + 8x + 10 - 12$	Reorder the terms.
$= (6 + 8)x - 2$	Then use the distributive property to combine like terms.
$= 14x - 2$	

NUMERICAL CHECK

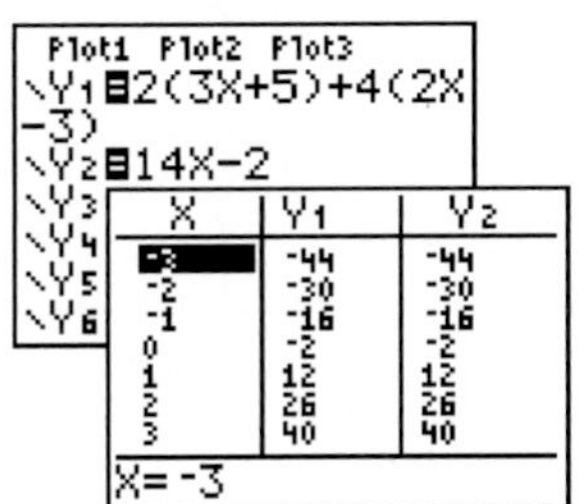

Let Y_1 represent the indicated sum of the two terms and Y_2 represent the calculated sum. Since the table values of Y_1 and Y_2 are identical, this indicates that $2(3x + 5) + 4(2x - 3)$ is equal to $14x - 2$.

GRAPHICAL CHECK

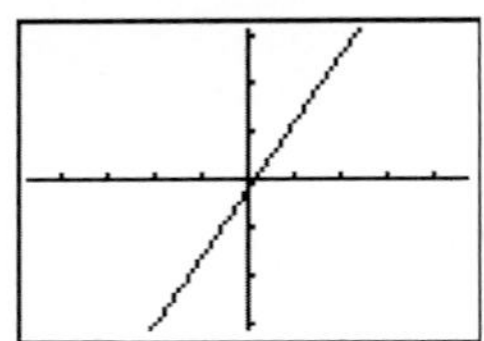

[−4.7, 4.7, 1] by [−31, 31, 10]

Since the graphs of Y_1 and Y_2 are identical, this indicates that $2(3x + 5) + 4(2x - 3)$ is equal to $14x - 2$.

Answer: $2(3x + 5) + 4(2x - 3) = 14x - 2$.

We must use caution when using tables or graphs to compare two polynomials. Two polynomials can share several points without being the same polynomial. In general, two nth-degree polynomials can share up to n points and still be distinct polynomials. If these two nth-degree polynomials share $n + 1$ points, then the polynomials are equal. In Self-Check 5.4.5, these two distinct second-degree polynomials share exactly two common points—they have different values for all other values of x.

SELF-CHECK 5.4.5

1. A student subtracted $4x^2 - 5x + 2$ from $3x^2 - 2x + 6$ and obtained $x^2 - 3x - 4$. Check this result by using a table of values.

SELF-CHECK ANSWERS

5.4.1

1. Binomial
2. Monomial
3. Not a polynomial
4. Trinomial
5. -3; degree 4
6. 1; degree 3
7. π; degree 0

5.4.2

1. $3x^2 + 8x - 9$; degree 2
2. $-9x^5y^2z$; degree 8
3. $9v^5 - v^3 + 3v^2 - 4v + 1$; degree 5

5.4.3

1. $P(5) = 30$
2. $13x - 2$
3. $-3x^4 + 8x^3 + 8x^2 - 3x - 8$

5.4.4

1. $3x^2 + 7x + 4$
2. $3x^4 + 5x^3 - 7x^2 + 7x + 4$

5.4.5

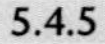

1. 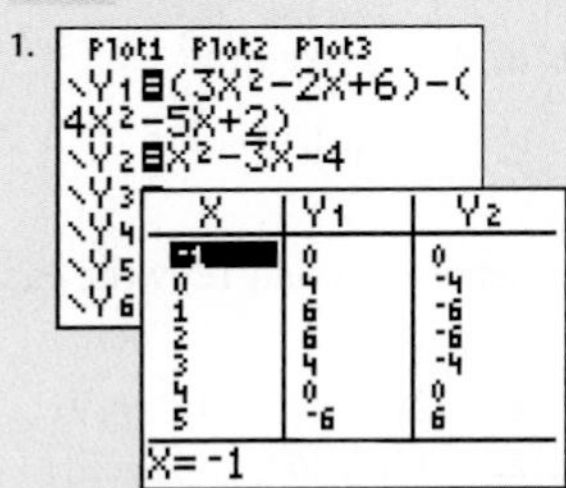

Because $Y_2 = -Y_1$, this indicated the student subtracted the polynomials in the wrong order.

USING THE LANGUAGE AND SYMBOLISM OF MATHEMATICS 5.4

1. ___________ are constants or variables that are multiplied together to form a product.
2. Terms in an algebraic expression are separated from each other by ___________ or ___________ symbols.
3. The whole numbers are ___________ .
4. A ___________ is a real number, a variable, or a product of real numbers and variables with whole-number exponents.
5. A ___________ is a monomial or a sum of monomials.

6. In the monomial ax^n, a is a __________ of x^n.
7. A polynomial containing exactly one term is a __________.
8. A polynomial containing exactly two terms is a __________.
9. A polynomial containing exactly three terms is a __________.
10. The degree of a monomial is the __________ of the exponents for all the variables in the monomial.
11. The degree of a polynomial is the same as the degree of the term with the __________ degree.
12. The degree of a nonzero constant is __________.
13. No degree is assigned to the constant __________.
14. A polynomial is in standard form if (1) the variables in each term are written in __________ order and (2) the terms are arranged in __________ powers of the first variable.
15. The coefficient of x is __________.
16. The coefficient of $-x$ is __________.
17. The degree of $-x$ is __________.
18. If two polynomials are equal, then a table of values will show __________ values for the two polynomials and their graphs will be __________.

EXERCISES 5.4

In Exercises 1–6, determine whether each expression is a monomial. If the expression is a monomial, give its coefficient and its degree.

1. a. -12 b. $-12x$ c. $-x^{12}$ d. x^{-12}
2. a. 3 b. $3x$ c. x^3 d. x^{-3}
3. a. $\frac{2}{5}x^4$ b. $\frac{5x^4}{2}$ c. $-\frac{5x^4}{2}$ d. $\frac{2}{5x^4}$
4. a. $\sqrt{2}$ b. $\sqrt{2}x$ c. $2\sqrt{x}$ d. $\frac{x}{2}$
5. a. $5x^2y^2$ b. $-5x^2y^2$ c. $\frac{5x^2}{y^2}$ d. $5x^2 + y^2$
6. a. $9a^2b^2c^2$ b. $\frac{a^2b^2c^2}{9}$ c. $\frac{9a^2b^2}{c^2}$ d. $9a^2 + b^2c^2$

In Exercises 7–10, determine whether each expression is a polynomial. Classify each polynomial according to the number of terms it contains, and give its degree.

7. a. $3x^2 - 7x$ b. $x^3 - 8x^2 + 9$
   c. $-9x^4$ d. -9
8. a. $\frac{-4x^5}{7}$ b. $x^4 - 5$
   c. $x^2 - 5x + 7$ d. 0
9. a. $11xy$ b. $11x + y$
   c. $\frac{11x}{y}$ d. $x^3 + y^2 - 11$
10. a. $4x^2 - 3xy + y^2$ b. $-3xy$
    c. $-3x + y$ d. $\frac{3x}{y}$

In Exercises 11 and 12, write each polynomial in standard form.

11. a. $8yx^2z^3$ b. $-9 + x^2 - 5x$
    c. $-2x^2 + 7 - 4x^3 + 9x$
12. a. $-11c^3a^4b$ b. $-7x + 8 - 3x^2$
    c. $13x + 7x^2 - 9x^3 - 11$

In Exercises 13–16, determine if the polynomials defined by Y_1 and Y_2 are equal. Each polynomial is fifth degree or less.

13.

X	Y_1	Y_2
-3	18	18
-2	8	8
-1	0	0
0	-6	-6
1	-10	-10
2	-12	-12
3	-12	-12

X= -3

14.

X	Y_1	Y_2
-3	80	80
-2	88	-8
-1	86	20
0	80	80
1	76	112
2	80	80
3	98	-28

X= -3

15.

X	Y_1	Y_2
-2	2	2
-1	2	2
0	2	2
1	2	2
2	2	2
3	122	-238
4	722	-718

X= -2

16.

X	Y_1	Y_2
-3	36	36
-2	24	24
-1	14	14
0	6	6
1	0	0
2	-4	-4
3	-6	-6

X= -3

In Exercises 17 and 18, identify each pair of terms as like or unlike.

17. a. $7xy$ and $-xy$
    b. $-4x^3y^2$ and $6x^2y^3$
    c. $5xy^2z^3$ and $5xy^2z^2$
18. a. $5v^4w^3$ and $8v^3w^4$
    b. $3vw^2$ and πvw^2
    c. $\frac{3}{5}xy^2z^3$ and $-0.9xy^2z^3$

In Exercises 19 and 20, write the opposite or additive inverse of each polynomial.

19. a. $7xy$
    b. $7x - y$
    c. $-2x^2 + 3xy - 7y^2$
20. a. $-5vw$
    b. $-5v + w$
    c. $5v^2 - 6vw - w^2$

In Exercises 21–26, determine each sum and difference.

21. a. $7x + (5x + 9x)$ b. $7x - (5x + 9x)$
22. a. $8y + (3y + 12y)$ b. $8y - (3y + 12y)$
23. a. $(3x^2 - 7x) + (4x^2 - 5x)$ b. $(3x^2 - 7x) - (4x^2 - 5x)$
24. a. $(-5x + 3) + (8x - 11)$ b. $(-5x + 3) - (8x - 11)$
25. a. $(5x^2 - 7x + 9) + (4x^2 + 6x - 3)$
    b. $(5x^2 - 7x + 9) - (4x^2 + 6x - 3)$
26. a. $(7x^2 + 6x - 13) + (3x^2 - 9x - 4)$
    b. $(7x^2 + 6x - 13) - (3x^2 - 9x - 4)$

In Exercises 27–58, determine each sum or difference and write the result in standard form.

27. $(6x + 5) + (8x - 7)$
28. $(-5x + 3) + (4x - 11)$
29. $(-7v + 3) - (2v - 11)$
30. $(-4v - 13) - (-7v + 11)$
31. $(5a^2 - 7a + 9) + (4a^2 + 6a - 3)$
32. $(7a^2 + 6a - 13) + (3a^2 - 9a - 4)$
33. $(8x^2 - 3x - 4) - (2x^2 - 2x - 9)$
34. $(11x^2 + 7x - 1) - (3x^2 - 12x + 8)$
35. $(-w^2 + 4) - (-4w^2 + 7w - 6)$
36. $(17w - 13) - (3w^2 - 4w + 6)$
37. $(-2m^4 + 3m^3 + m^2 - 1) + (m^4 - m^2 - 2m + 5)$
38. $(4m^3 + 7m^2 - 9m - 1) + (2m^3 - m^2 + 7)$
39. $(-3n^4 + 7n^3 + n - 8) - (n^4 - 2n^3 - n^2 + 4)$
40. $(-2n^4 - n^2 + 3n - 5) - (5n^4 + 7n^3 - 2n + 5)$

41. $(-7y^4 + 3y^6 - y + 4y^5 - 11 + 2y^2) - (3y^5 - 5y + 7y^6 - 9y^2 + 7 - y^4)$
42. $(8y^3 + 3 - 5y^2 + y^6 - 4y^4 + 9y) - (5 + 3y^2 - 2y^6 + y - 2y^4 - y^5)$
43. $2(x^2 - 3x - 5) - 7(x^2 + 2x - 1)$
44. $8(x^2 + 5x - 2) - 5(2x^2 - 3x - 4)$
45. $3(2a^2 - 5a + 1) - 2(3a^2 - a - 5)$
46. $(7a^2 - a + 3) - 5(5a^2 + a - 2)$
47. $(5x + 7) + (4x - 8) + (3x - 11)$
48. $(6x - 4) + (2x - 3) + (5x + 9)$
49. $(13x - 8) - (7x - 9)$
50. $(17x + 5) - (14x - 7)$
51. $(5x^2 + 7xy - 9y^2) + (6x^2 - 3xy + y^2)$
52. $(11x^2 - 9xy - 4y^2) + (8x^2 + xy - 5y^2)$
53. $(-7x^2 + 6xy + 8y^2) - (13x^2 - xy - 3y^2)$
54. $(4x^2 - 5xy + 12y^2) - (7x^2 - 8xy - 11y^2)$
55. $(5x^3 - 7x + 9 + 5x^2) + (2x^2 + 13 + x^3 - x) - (4x - 3x^3 + x^2 - 8)$
56. $(6x - 11x^3 + 7 - x^2) - (14 - 2x^2 - x^3 - 5x) + (9x - 12 - x^3 + 8x^2)$
57. $(x^4 - 2x^3y + x^2y^2 + xy^3 - 3y^4) - (2x^4 + x^3y - 5x^2y^2 - 7y^4) - (7x^4 + 3x^2y^2 + 2y^4)$
58. $(3x^4 - 9x^3y + 7x^2y^2 - y^4) - (7x^4 + 4x^2y^2 - 11y^4) - (-11x^4 - 9x^3y + y^4)$

In Exercises 59 and 60, evaluate each expression for $P(x) = 5x^2 + 3x - 2$.

59. a. $P(0)$ b. $P(2)$ c. $P(10)$ d. $P(8)$
60. a. $P(1)$ b. $P(5)$ c. $P(20)$ d. $P(100)$

Profit Polynomial

In Exercises 61–66, evaluate and interpret each expression. The profit in dollars made by producing and selling x units is given by $P(x) = -x^2 + 22x - 40$.

61. $P(0)$ 62. $P(1)$ 63. $P(2)$
64. $P(10)$ 65. $P(20)$ 66. $P(22)$

In Exercises 67 and 68, use a calculator to evaluate the given polynomial for each x value.

67. 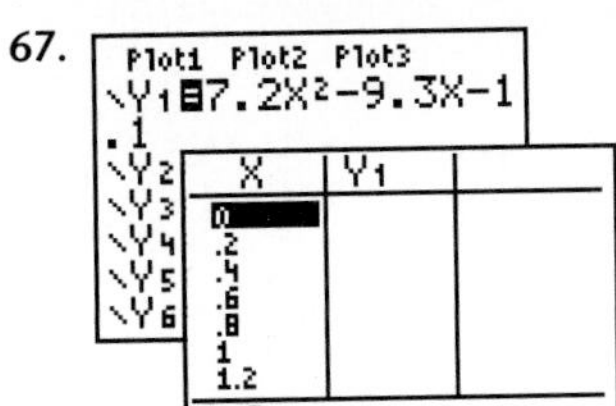

68. 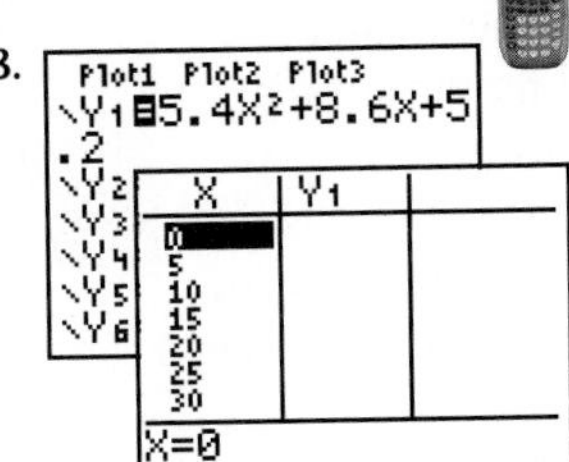

In Exercises 69 and 70, calculate each sum. Use Y_1 to represent the given problem and Y_2 to represent the calculated sum. Then check your result by using a calculator to create a table of values from $x = -3$ to 3.

69. $(3.49x^2 + 7.81x) + (5.53x^2 - 3.79x)$
70. $(5.76x^2 + 4.3x - 1.9) + (2.04x^2 + 5.7x - 2.1)$

In Exercises 71 and 72, calculate each sum. Use Y_1 to represent the given problem and Y_2 to represent the calculated sum. Then check your result by using a calculator to graph Y_1 and Y_2. (*Hint:* Use a window of $[-2, 2, 1]$ by $[-10, 15, 5]$.)

71. $(6.11x^2 - 5.98x - 3.51) + (5.07x^2 + 4.01x - 2.47)$
72. $(4.5x - 9.1) + (3.7x + 6.7) + (6.3x - 4.6)$

Perimeter of a Polygon

In Exercises 73–76, write a polynomial for the perimeter of each polygon.

73. Quadrilateral

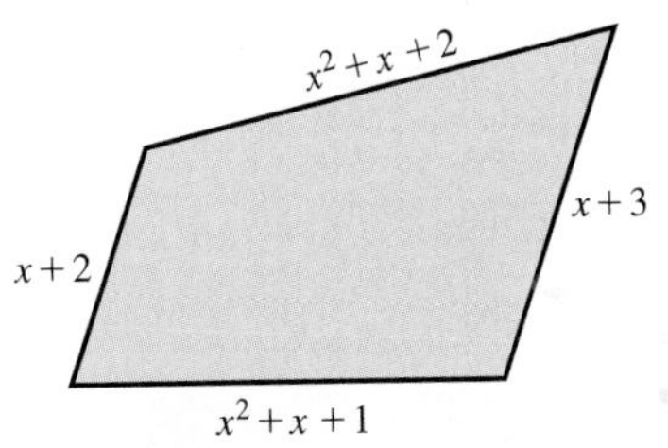

74. Trapezoid

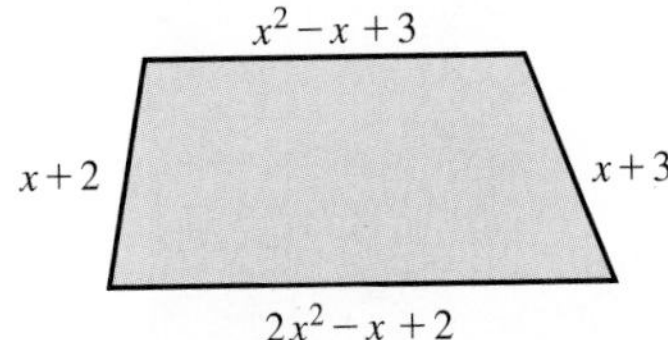

75. Pentagon

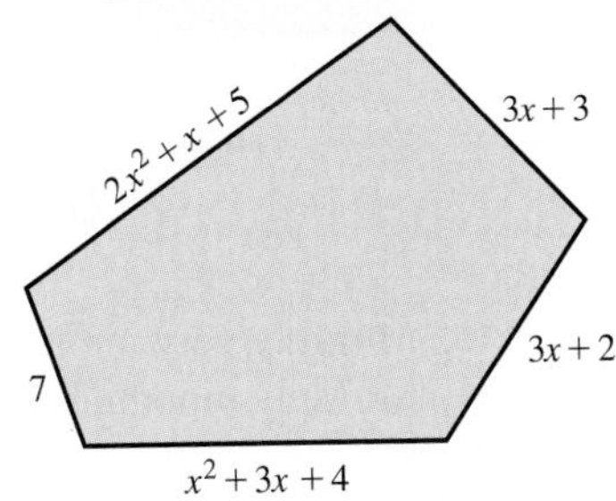

76. Hexagon

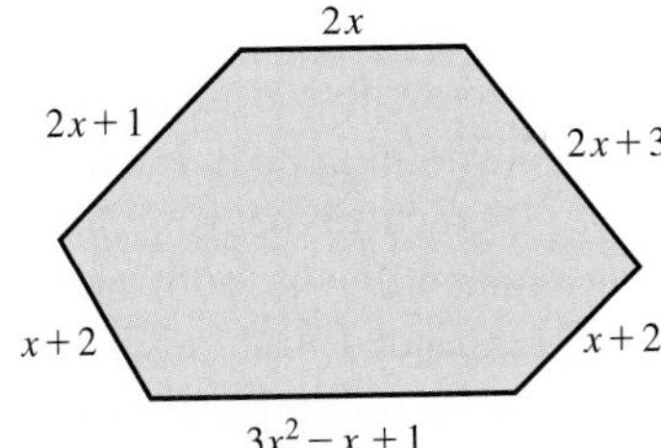

Multiple Representations

In Exercises 77–84, write a polynomial for each verbal expression.

77. Two times x cubed minus five x plus eleven.
78. Nine times y squared plus seven y minus nine.
79. Three times the sixth power of w plus five times the fourth power of w.
80. Eight times the fifth power of v minus v cubed plus eleven v.
81. x squared minus y squared.
82. a cubed plus b cubed.
83. Seven more than twice w.
84. Eight less than the opposite of four y.

World Production of Chlorofluorocarbons

The 1987 Montreal Protocol on Substances That Deplete the Ozone Layer restricted the use of CFCs which damage the ozone layer above the Earth. The table on the next page gives the world production of CFCs for selected years after 1950 as measured in thousands of tons. (Source: For more information, go to the Additional Resources link on the MathZone website at **www.mhhe.com/hallmercer.**)

YEARS x AFTER 1950	$C(x)$ (THOUSANDS OF TONS OF CFCs)
0	60
10	100
20	700
30	900
40	1,250
50	800

85. These data can be modeled by the function $C(x) = -0.05x^3 + 3.0x^2 - 13.5x + 47.7$. Use this function to approximate the tons of CFC produced in 1965.

86. Use the function in Exercise 85 to predict the tons of CFC produced in 2005.

Group Discussion Questions

87. Challenge Question Give an example that satisfies the given conditions.

a. A fifth-degree monomial in x with a coefficient of 2
b. A monomial of degree 0
c. A first-degree binomial in x with a constant term of -3
d. A second-degree binomial in x with a constant term of -3
e. Two first-degree monomials in x whose sum is a constant
f. Two first-degree monomials in x whose difference is a constant
g. Two fourth-degree binomials in x whose sum is a third-degree monomial.

88. Error Analysis A classmate subtracted $3x^2 - x - 8$ from $7x^2 + 10x + 9$ by writing the expression $7x^2 + 10x + 9 - 3x^2 - x - 8$ to obtain $4x^2 + 9x + 1$. Explain the *error* in this work, and make a suggestion as to how the classmate could organize the work to avoid this error.

Section 5.5 Multiplying Polynomials

Objectives:

10. Multiply polynomials.
11. Multiply binomials by inspection.

Polynomials are used to describe many applications, especially those involving geometric shapes. Exercise 84 examines a polynomial expression that represents the area of concrete around a rectangular swimming pool. Each factor of this expression has a practical meaning that may be important to those planning the pool area. On the other hand, the expanded form might be preferred by those calculating the concrete that will be needed.

The product of two monomials was considered when the product rule for exponents was discussed. Recall that the product rule for exponents states that $x^m x^n = x^{m+n}$.

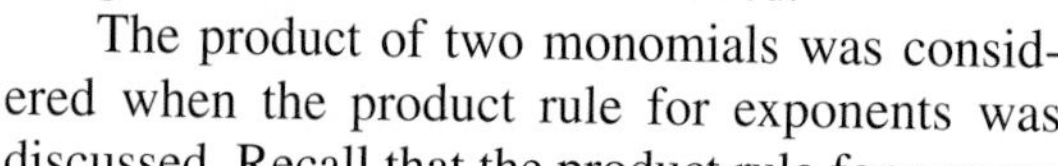

EXAMPLE 1 Multiplying Monomials

Simplify $(7x^3y^2)(-8x^4y^6)$.

SOLUTION

$$(7x^3y^2)(-8x^4y^6) = (7)(-8)(x^3x^4)(y^2y^6)$$

Regroup and reorder the factors, using the associative and commutative properties of multiplication.

$$= -56x^{3+4}y^{2+6}$$
$$= -56x^7y^8$$

Using the product rule for exponents, add the exponents on the common bases.

Note that the coefficients are multiplied and the exponents are added.

The distributive property is used to find the product of a monomial and a polynomial. This use of the distributive property is illustrated in Example 2 by the product of a monomial and a binomial.

EXAMPLE 2 Multiplying a Monomial by a Polynomial

Multiply these polynomials.

SOLUTION

(a) $5x^3(7x - 9)$

$5x^3(7x - 9) = 5x^3(7x) - 5x^3(9)$
$= (5)(7)(x^3x) - (5)(9)x^3$
$= 35x^4 - 45x^3$

Distribute the multiplication of $5x^3$ to each term of the binomial. Then simplify each term, using the product rule for exponents.

(b) $(4x^2 - 5xy + 11y^2)(2xy)$

$(4x^2 - 5xy + 11y^2)(2xy)$
$= 4x^2(2xy) - 5xy(2xy) + 11y^2(2xy)$
$= 8x^3y - 10x^2y^2 + 22xy^3$

Distribute the multiplication of $2xy$ to each term of the trinomial. Then simplify, using the product rule for exponents.

The procedure for multiplying a monomial by a polynomial is summarized in the following box.

Multiplying a Monomial Times a Polynomial

VERBALLY	ALGEBRAIC EXAMPLE
To multiply a monomial by a polynomial, use the distributive property to multiply the monomial by each term of the polynomial.	$5x(2x^2 - 3x + 8) = 5x(2x^2) - 5x(3x) + 5x(8)$ $= 10x^3 - 15x^2 + 40x$

SELF-CHECK 5.5.1

Multiply the polynomial factors.

1. $(-9a^2bc)(-4ab^2c^3)$
2. $9x^3(8x - 5)$
3. $-3x^2(7x^2 - 2xy - 3y^2)$
4. $(7x^2 - 2xy - 3y^2)(4y)$

Multiplying two polynomials with more than one term each requires the repeated use of the distributive property. An illustration is provided in Example 3.

EXAMPLE 3 Multiplying by a Binomial

Multiply the following factors.

SOLUTION

(a) $(7x + 3)(2x + 5)$

$(7x + 3)(2x + 5)$
$= 7x(2x + 5) + 3(2x + 5)$
$= 7x(2x) + 7x(5) + 3(2x) + 3(5)$
$= 14x^2 + 35x + 6x + 15$
$= 14x^2 + 41x + 15$

First distribute the multiplication of $2x + 5$ to each term of $7x + 3$. Then distribute $7x$ times each term of $2x + 5$; also distribute 3 times each term of $2x + 5$. Finally, simplify by combining like terms.

(b) $(3x - 2)(4x^2 - 5x + 6)$

$$\begin{aligned}(3x - 2)(4x^2 - 5x + 6) &= 3x(4x^2 - 5x + 6) - 2(4x^2 - 5x + 6)\\ &= 3x(4x^2) - 3x(5x) + 3x(6) - 2(4x^2) - (-2)(5x) - 2(6)\\ &= 12x^3 - 15x^2 + 18x - 8x^2 + 10x - 12\\ &= 12x^3 - 23x^2 + 28x - 12\end{aligned}$$

The distributive property justifies forming the product of each term of $4x^2 - 5x + 6$ by each term of $3x - 2$.

Simplify by combining like terms.

Fortunately, all the steps shown in Example 3 to illustrate the use of the distributive property can be shortened by using the procedure in the following box.

Multiplying Two Polynomials

To multiply one polynomial by another, multiply each term of the first polynomial by each term of the second polynomial and then combine the like terms.

We have examined the multiplication of polynomials by using algebraic properties. Now we give a geometric viewpoint.

A Mathematical Note

Plato (c. 427–348 B.C.) was a pupil of Socrates. In 388 B.C. he formed a school called the Academy in Athens. Inscribed above the gate to this academy was the following motto: "Let no man ignorant of geometry enter." Many of our algebraic formulas are based upon early geometric discoveries.

Multiplying Binomials: A Geometric Viewpoint

A geometric viewpoint of the multiplication of two binomials is shown next. The area of the rectangle is considered in two different ways. First the area of the whole rectangle is calculated, and then the area is computed by adding the areas of the four parts. Since the area is the same either way, these two polynomials must be equal.

Area of Whole Rectangle

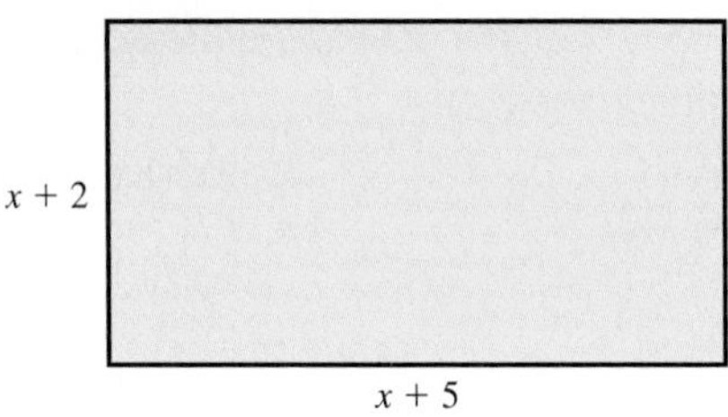

Areas of Four Parts

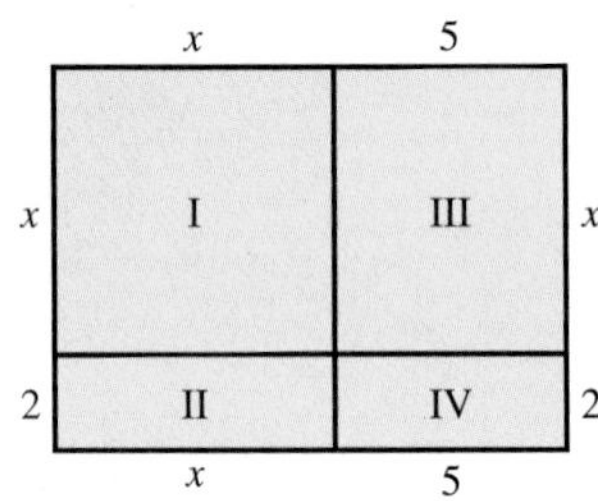

Note that each term of $x + 5$ is multiplied by each term of $x + 2$.

Area = Length × Width
$A = (x + 5)(x + 2)$

Total area = Area I + Area II + Area III + Area IV
$A = x^2 + 2x + 5x + 2(5)$
$A = x^2 + 7x + 10$

Thus $(x + 5)(x + 2) = x^2 + 7x + 10$.

It is very useful to be able to multiply two binomials as efficiently as possible. With practice, the procedure of multiplying each term of the first polynomial by each term of the second polynomial can be used to multiply two binomials by inspection. One way to remember the steps for performing this mental multiplication is to use the **FOIL** method. FOIL is an acronym for **F**irst, **O**uter, **I**nner, and **L**ast. The logic of the FOIL method is illustrated below for the product $(2x + 5)(x - 6)$.

$$(2x + 5)(x - 6) = 2x(x - 6) + 5(x - 6)$$
$$= 2x^2 - 12x + 5x - 30$$

Using the distributive property results in multiplying each term of $2x + 5$ times each term of $x - 6$.

Think:

F irst terms, $2x(x)$

O uter terms, $2x(-6)$

I nner terms, $5(x)$

L ast terms, $5(-6)$

When the middle two terms are like terms, we combine these like terms. In the example above, the sum of the middle terms is $-12x + 5x = -7x$. Thus $(2x + 5)(x - 6) = 2x^2 - 7x - 30$.

An extra step is shown in parts (a) and (b) of Example 4 to illustrate the FOIL method. Usually the inner and outer products are added mentally and the product is written in its final form. Once you feel comfortable omitting this extra step, you should do so.

EXAMPLE 4 Multiplying Binomials by Inspection

Use the FOIL method to multiply each pair of binomials.

SOLUTION

(a) $(x + 9)(x - 5)$

$$(x + 9)(x - 5) = x^2 - 5x + 9x - 45$$
$$= x^2 + 4x - 45$$

F $(x)(x)$
O $(x)(-5)$
I $(9)(x)$
L $(9)(-5)$

The step of forming the product of the outer and inner terms is shown in this example. The middle term is the sum of $-5x$ and $9x$.

(b) $(7v - 3)(4v + 1)$

$$(7v - 3)(4v + 1) = 28v^2 + 7v - 12v - 3$$
$$= 28v^2 - 5v - 3$$

F $(7v)(4v)$
O $(7v)(1)$
I $(-3)(4v)$
L $(-3)(1)$

The step of forming the product of the outer and inner terms is shown in this example. The middle term is the sum of $7v$ and $-12v$.

(c) $(5w + 4)(w + 2)$

$$(5w + 4)(w + 2) = 5w^2 + 14w + 8$$

The middle term is the sum of $10w$ and $4w$.

(d) $(3v - 8w)(2v - 5w)$

$$(3v - 8w)(2v - 5w) = 6v^2 - 31vw + 40w^2$$

The middle term is the sum of $-15vw$ and $-16vw$.

SELF-CHECK 5.5.2

Multiply these binomials by inspection.

1. $(x + 4)(x + 7)$
2. $(3v + 5)(4v - 1)$
3. $(2v - 5w)(3v - 4w)$

The ability to use inspection to multiply binomials not only facilitates your work with multiplication, but also gives you a strong preparation to factor trinomials in Chapter 6.

Next we examine the product of two trinomials. Again we start by multiplying each term of the first trinomial by each term of the second trinomial. It is generally easier to keep our steps organized and to obtain accurate results if we write each polynomial factor in standard form. Example 5 illustrates an optional vertical format for organizing the steps used to multiply these polynomials.

EXAMPLE 5 Multiplying Two Trinomials

Multiply $(x + 1 + x^2)$ by $(4 + x^2 - 2x)$.

SOLUTION

ALGEBRAICALLY

$(x + 1 + x^2)(4 + x^2 - 2x) = (x^2 + x + 1)(x^2 - 2x + 4)$ — First write each factor in standard form.

$$\begin{array}{rrrrr} & & x^2 & - 2x & + 4 \\ & & x^2 & + x & + 1 \\ \hline x^4 & - 2x^3 & + 4x^2 & & \\ & x^3 & - 2x^2 & + 4x & \\ & & x^2 & - 2x & + 4 \\ \hline x^4 & - x^3 & + 3x^2 & + 2x & + 4 \end{array}$$

Then write one factor on the first row and the second factor on the next row.

Multiply each term on the second row by every term on the first row. First multiply by x^2, then by x, and then by 1.

Note that only similar terms are aligned in the same column. Add the similar terms in each column, and write the product in standard form.

Answer: $(x + 1 + x^2)(4 + x^2 - 2x) = x^4 - x^3 + 3x^2 + 2x + 4$

SELF-CHECK 5.5.3

Multiply the polynomial factors.

1. $(2x + 3y)(2x - 3y)$
2. $(x - 2)(x^2 + 2x + 4)$
3. Multiply $3y + y^2 - 1$ by $2 + y^2 - 2y$

Example 6 illustrates the multiplication of three binomial factors. We start by multiplying the first two factors, and then we multiply this product by the third factor.

EXAMPLE 6 Finding the Product of Three Binomials

Multiply $(x - 3)(x + 2)(x - 5)$.

SOLUTION

ALGEBRAICALLY

$$\begin{aligned} (x - 3)(x + 2)(x - 5) &= [(x - 3)(x + 2)](x - 5) \\ &= [x^2 + 2x - 3x - 6](x - 5) \\ &= (x^2 - x - 6)(x - 5) \\ &= x^2(x - 5) - x(x - 5) - 6(x - 5) \\ &= x^3 - 5x^2 - x^2 + 5x - 6x + 30 \\ &= x^3 - 6x^2 - x + 30 \end{aligned}$$

First find the product of $(x - 3)(x + 2)$ by using FOIL.

Then distribute $x - 5$ to each term of $x^2 - x - 6$.

Answer: $(x - 3)(x + 2)(x - 5) = x^3 - 6x^2 - x + 30$

The number of units that can be sold for some products depends on the price of the item. For example, lowering the price of new cars will generally generate more sales. Lowering interest rates (the cost of borrowing money) is a standard tool of the Federal Reserve to promote more borrowing in the economy. The revenue generated by selling a number of units is found by multiplying the price per unit by the number of units sold. Thus some companies can exert control over their revenue by controlling their prices. Example 7 expresses the revenue for automobile sales as a function of the price of the car.

EXAMPLE 7 Modeling the Revenue for Automobile Sales

For a price of x dollars per car (x from \$20,000 to \$30,000), an automobile manufacturer estimates that it can sell $10{,}000 - 0.2x$ cars. Write a revenue polynomial $R(x)$ based on this estimate, and evaluate $R(20{,}000)$ and $R(30{,}000)$. Then use a graphing calculator to check these values.

SOLUTION

Let x = price per car in dollars
$10{,}000 - 0.2x$ = number of cars that will sell

Revenue = Number of items sold · Price per item — Write a word equation as the basis of an algebraic equation.

$$R(x) = (10{,}000 - 0.2x)(x)$$

This is the revenue polynomial.

$$\begin{aligned} R(20{,}000) &= [10{,}000 - 0.2(20{,}000)](20{,}000) \\ &= [10{,}000 - 4{,}000](20{,}000) \\ &= (6{,}000)(20{,}000) \\ &= 120{,}000{,}000 \end{aligned}$$

Evaluate $R(x)$ when the price per car is \$20,000.

The revenue produced by selling cars at \$20,000 each is estimated to be \$120,000,000.

$$\begin{aligned} R(30{,}000) &= [10{,}000 - 0.2(30{,}000)](30{,}000) \\ &= (10{,}000 - 6{,}000)(30{,}000) \\ &= (4{,}000)(30{,}000) \\ &= 120{,}000{,}000 \end{aligned}$$

Evaluate $R(x)$ when the price per car is \$30,000.

The revenue produced by selling cars at \$30,000 each is also estimated to be \$120,000,000.

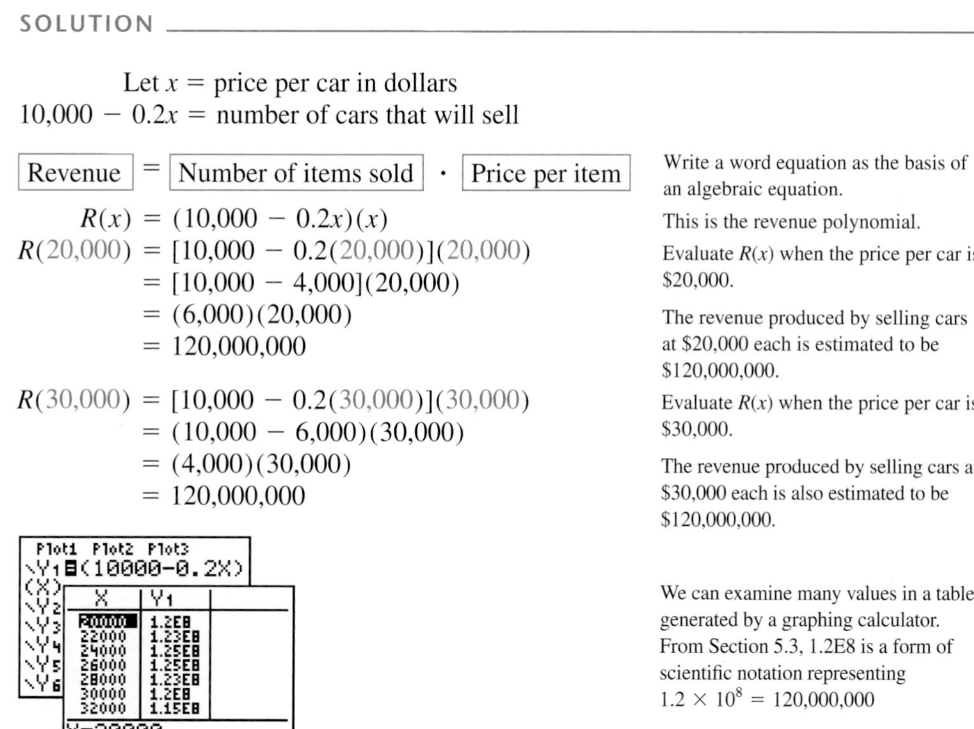

We can examine many values in a table generated by a graphing calculator. From Section 5.3, 1.2E8 is a form of scientific notation representing $1.2 \times 10^8 = 120{,}000{,}000$

Answer: The revenue polynomial $R(x) = (10{,}000 - 0.2x)(x)$ yields a revenue of \$120,000,000 for both a price of \$20,000 and a price of \$30,000. However, the revenue is not constant for all input values. Note that a price of \$26,000 will generate a revenue of \$125,000,000.

SELF-CHECK 5.5.4

For a price of x dollars per truck (x from \$25,000 to \$40,000) an automobile manufacturer estimates that it can sell $9{,}000 - 0.15x$ trucks.

1. Write a revenue polynomial $R(x)$ based on this estimate.
2. Use this revenue polynomial to evaluate $R(30{,}000)$.

Equal polynomials will have identical output values for all input values. In Example 8 we use tables and graphs to compare two polynomials as a check that these polynomials are equal. Working a problem by one method and then using another method to check your work is a very useful test strategy.

EXAMPLE 8 Using Tables and Graphs to Compare Two Polynomials

Multiply $(2x + 3)(2x - 3)$ and check this product by using a table of values and a graph.

SOLUTION

ALGEBRAICALLY

$(2x + 3)(2x - 3) = 4x^2 - 9$

Multiplying these binomials, by inspection we note that the middle term is the sum of $-6x$ and $6x$. Thus the product is $4x^2 + 0x - 9 = 4x^2 - 9$.

NUMERICAL CHECK

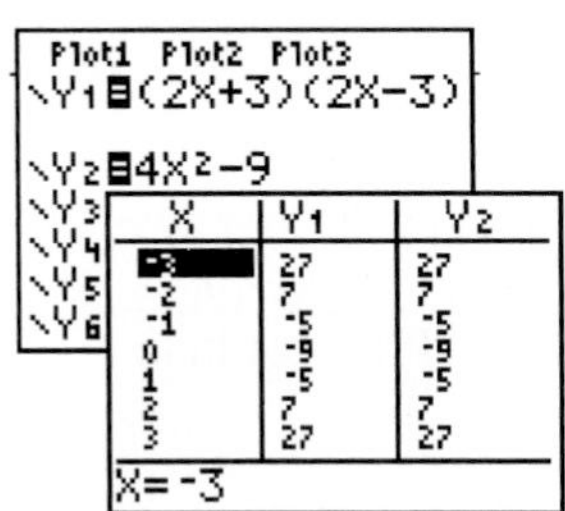

Let Y_1 represent the indicated product of the two factors and Y_2 represent the calculated product. Because the values of Y_1 and Y_2 in the table are identical, this indicates that the polynomials $(2x + 3)(2x - 3)$ and $4x^2 - 9$ are equal.

GRAPHICAL CHECK

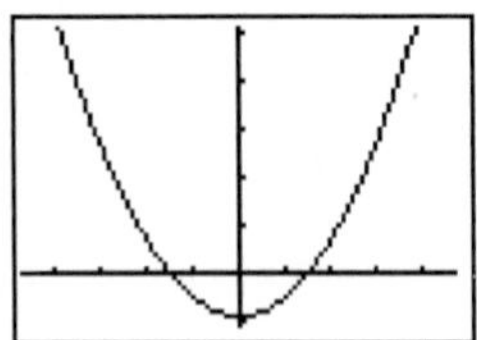

$[-4.7, 4.7, 1]$ by $[-11, 51, 10]$

Since the graphs of Y_1 and Y_2 are identical, this indicates that the polynomials $(2x + 3)(2x - 3)$ and $4x^2 - 9$ are equal. We will examine these graphs in greater detail in Chapter 7.

Answer: $(2x + 3)(2x - 3) = 4x^2 - 9$

The distributive property is one of the most used properties of algebra, a property used both to multiply and expand expressions and to factor expressions. In Example 6 of Section 1.7 we used this property to expand the expression $9(x + 4) = 9x + 36$ and to factor the expression $5x - 5y = 5(x - y)$. Both parts of Example 9 use the distributive property, first to expand $2x(5x - 4) + 3(5x - 4)$ and then to factor this expression. In this section we have concentrated on multiplying polynomials and will examine factoring in greater detail in Chapter 6.

EXAMPLE 9 Comparison of Expanding and Factoring a Polynomial

Use the distributive property first to expand $2x(5x - 4) + 3(5x - 4)$ and then to factor this expression.

SOLUTION

EXPAND

$2x(5x - 4) + 3(5x - 4)$
$= 2x(5x) - 2x(4) + 3(5x) - 3(4)$
$= 10x^2 - 8x + 15x - 12$
$= 10x^2 + 7x - 12$

FACTOR

$2x(5x - 4) + 3(5x - 4)$
$= (2x + 3)(5x - 4)$

To expand, distribute the factor $2x$ to each term of $5x - 4$ and the factor 3 to each term of $5x - 4$. Use the product rule for exponents and then add like terms.
To factor, use the distributive property to factor out the common factor of $5x - 4$.

SELF-CHECK 5.5.5

Use the distributive property to factor these polynomials.

1. $5x^2 - 10x$ 2. $7x(2x - 5) - 3(2x - 5)$

SELF-CHECK ANSWERS

5.5.1
1. $36a^3b^3c^4$
2. $72x^4 - 45x^3$
3. $-21x^4 + 6x^3y + 9x^2y^2$
4. $28x^2y - 8xy^2 - 12y^3$

5.5.2
1. $x^2 + 11x + 28$
2. $12v^2 + 17v - 5$
3. $6v^2 - 23vw + 20w^2$

5.5.3
1. $4x^2 - 9y^2$
2. $x^3 - 8$
3. $y^4 + y^3 - 5y^2 + 8y - 2$

5.5.4
1. $R(x) = (9{,}000 - 0.15x)(x)$
2. $R(30{,}000) = 135{,}000{,}000$; a price of \$30,000 will generate a revenue of \$135,000,000.

5.5.5
1. $5x(x - 2)$
2. $(7x - 3)(2x - 5)$

USING THE LANGUAGE AND SYMBOLISM OF MATHEMATICS 5.5

1. The degree of $5x^2$ is __________, and its coefficient is __________.
2. The product rule for exponents says $x^m x^n =$ __________.
3. In the product $(5x^3)(8x^4) = 40x^7$, the coefficients are **added/multiplied** and the exponents are **added/multiplied.** (Select the correct choices.)
4. To multiply a monomial by a polynomial, use the __________ property to multiply the monomial by each term of the polynomial.
5. To multiply one polynomial by another, multiply each term of the first polynomial by __________ term of the second polynomial and then combine like terms.
6. A polynomial with two terms is called a __________.
7. A polynomial with three terms is called a __________.
8. The acronym FOIL represents __________, __________, __________, __________.
9. a. The polynomial $4x - 5$ has ______ terms.
   b. The polynomial $3x + 2$ has ______ terms.
   c. The product of $4x - 5$ and $3x + 2$ is the polynomial $12x^2 - 7x - 10$ which has __________ terms.
10. a. The binomial $4x - 5$ is of degree __________.
    b. The binomial $3x + 2$ is of degree __________.
    c. The product of $4x - 5$ and $3x + 2$ is the trinomial $12x^2 - 7x - 10$ of degree __________.
11. The polynomial $P(x) = x^2 - x + 3$ is read "P of __________ equals x squared minus x plus 3."

EXERCISES 5.5

In Exercises 1–38, multiply the polynomial factors.

1. $-9v^2(-7v^3)$
2. $11w^3(-w^5)$
3. $(a^3b^4)(4a^6b)$
4. $(-11x^2y^3)(-8xy^6)$
5. $4x^2(6x - 5)$
6. $7x^3(9x + 4)$
7. $-4v^2(8v + 3)$
8. $-9v^4(2v - 11)$
9. $2x(7x^2 - 9x - 4)$
10. $3x(8x^2 + 4x - 5)$
11. $-11x^2(4x^2 - 5x - 8)$
12. $-14x^2(2x^2 - 3x + 5)$
13. $(2x^2 - xy - y^2)(5xy)$
14. $(3x^2 + xy - y^2)(2xy)$
15. $3a^2b(2a^2 + 4ab - b^2)$
16. $5v^2w^2(8v^2 - vw + 3w^2)$
17. $(y + 2)(y + 3)$
18. $(y - 4)(y - 6)$
19. $(2v + 5)(3v - 4)$
20. $(2v - 3)(5v + 2)$
21. $(6x + 5)(9x + 3)$
22. $(7x + 11)(4x + 6)$
23. $(x + y)(2x - y)$
24. $(x - y)(3x + y)$
25. $(2x + 4y)(5x - 7y)$
26. $(2x - 5y)(3x + 4y)$
27. $(x + 2)(x^2 - 3x - 4)$
28. $(2x - 1)(x^2 + x - 3)$
29. $\begin{array}{r} m^2 + 3m + 2 \\ \times \quad 2m - 5 \\ \hline \end{array}$
30. $\begin{array}{r} m^2 - 2m - 7 \\ \times \quad 3m + 4 \\ \hline \end{array}$
31. $\begin{array}{r} x^2 - xy + y^2 \\ \times \quad x + y \\ \hline \end{array}$
32. $\begin{array}{r} x^2 + xy + y^2 \\ \times \quad x - y \\ \hline \end{array}$
33. $\begin{array}{r} 2x^2 - 5x + 11 \\ \times\ 3x^2 + 4x - 9 \\ \hline \end{array}$
34. $\begin{array}{r} -6x^2 + 3x - 8 \\ \times\ 5x^2 - 2x + 7 \\ \hline \end{array}$
35. $5x(x + 3)(x - 2)$
36. $7x(x - 4)(x + 2)$
37. $(2x - 1)(x + 3)(x + 4)$
38. $(3x + 2)(x - 1)(x + 5)$

In Exercises 39–50, multiply the binomials by inspection. (*Hint:* See Example 4 which illustrates the FOIL method.)

39. $(m + 3)(m + 4)$
40. $(m + 7)(m + 11)$
41. $(n - 5)(n + 8)$
42. $(n + 9)(n - 4)$
43. $(5x + 7)(4x - 3)$
44. $(4x - 5)(3x + 8)$
45. $(9y - 1)(y - 7)$
46. $(8y - 3)(y - 4)$
47. $(3a - 2b)(4a + 7b)$
48. $(5a + 9b)(2a - b)$
49. $(9x - 7y)(10x + 3y)$
50. $(5x + 7y)(7x - 5y)$

Area of a Region

In Exercises 51–56, write a polynomial for the area of each figure. See the formulas on the inside back cover of this book.

51.

52.

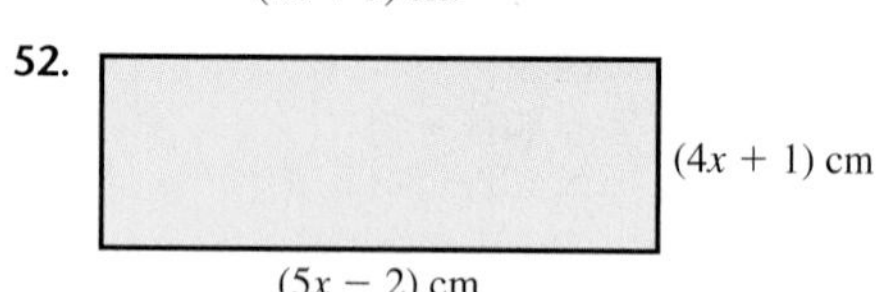

53.

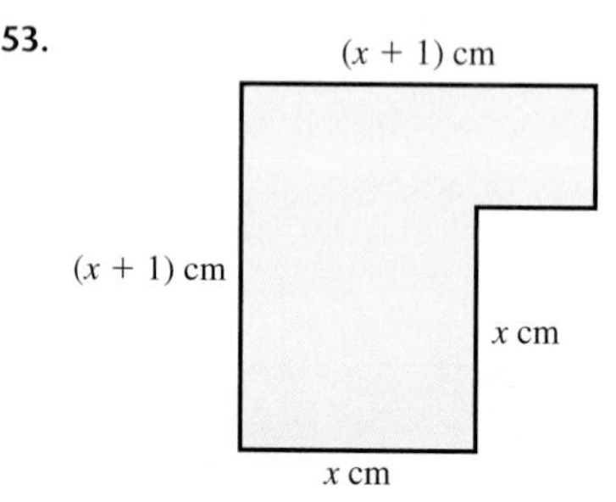

54.

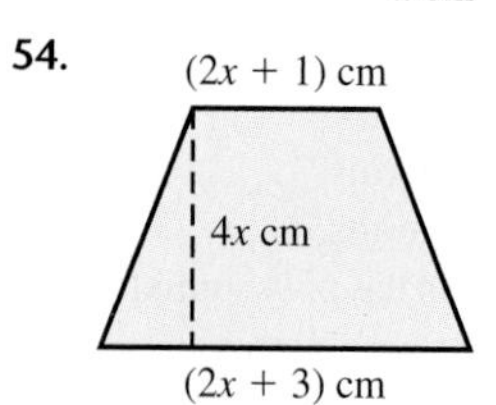

55.

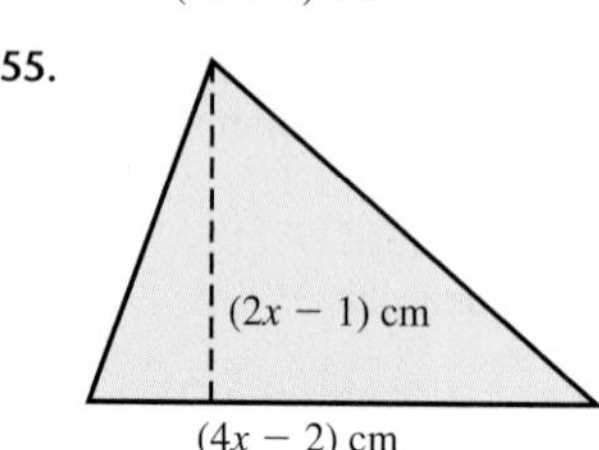

56.

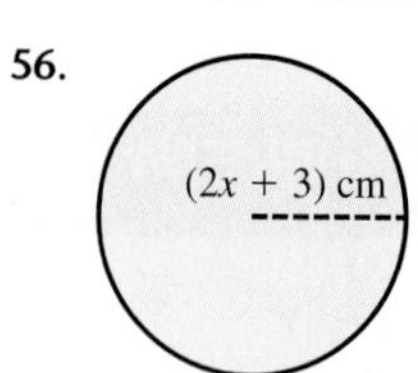

In Exercises 57–59, write a polynomial for the area of the shaded region. Assume the curved portions are semicircles of the same size.

57.

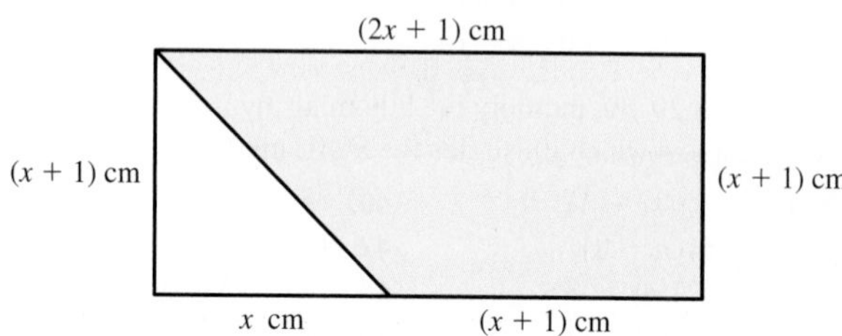

58.

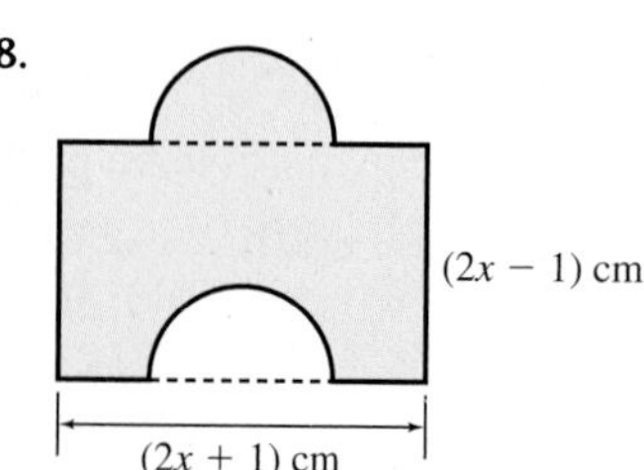

59.

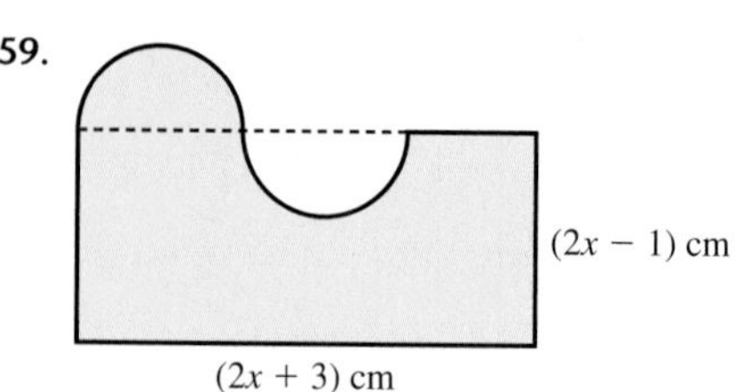

Volume of a Solid

In Exercises 60 and 61, write a polynomial for the volume of each solid.

60.

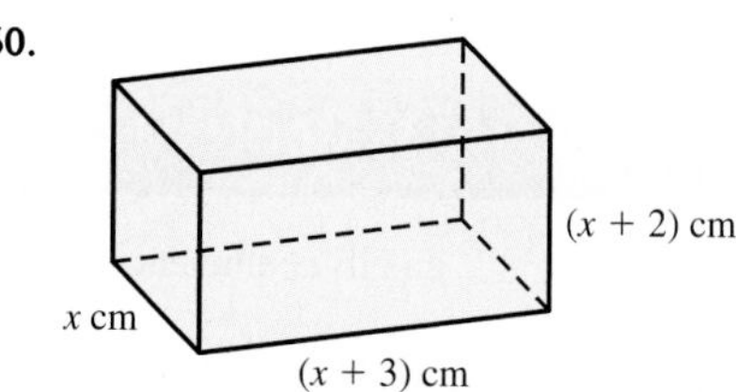

61.

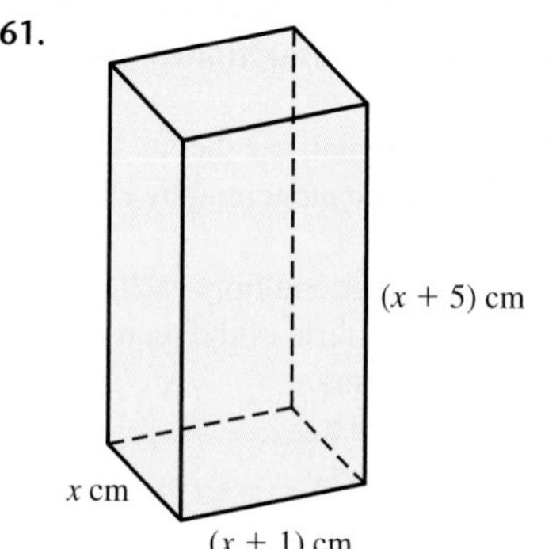

In Exercises 62 and 63, use a graphing calculator to fill in the table of values for Y_1 and Y_2. Do the tables indicate that these polynomials are equal?

62. **63.**

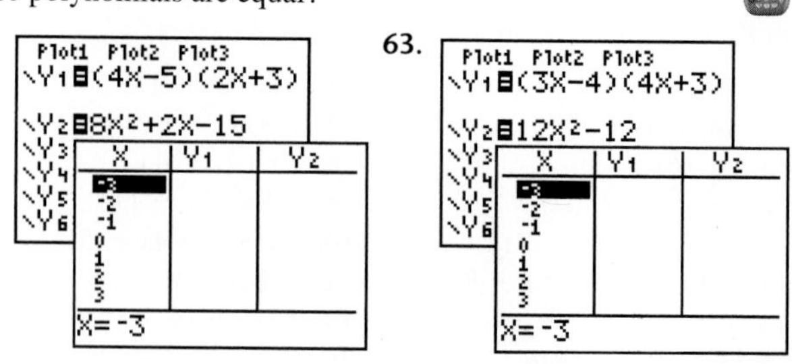

In Exercises 64 and 65, use a graphing calculator to graph Y_1 and Y_2 on the given window. Do the graphs indicate that these polynomials are equal?

64. 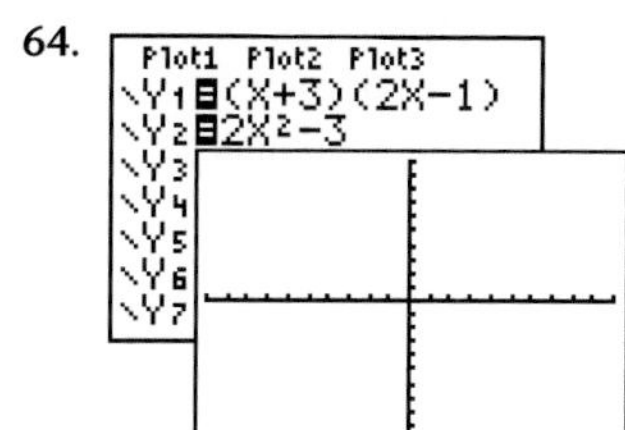

[−10, 10, 1] by [−10, 10, 1]

65. 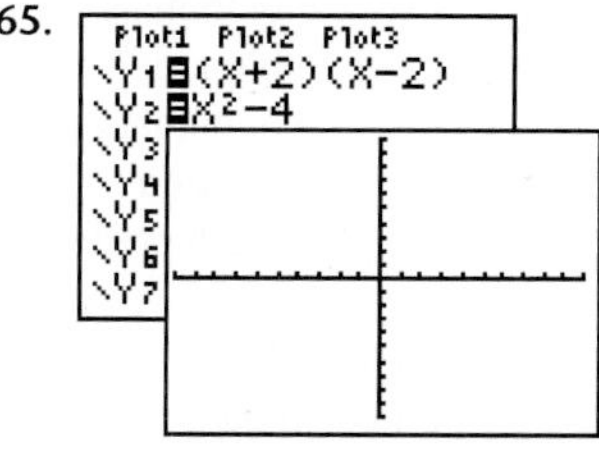

[−10, 10, 1] by [−10, 10, 1]

In Exercises 66–74, use the distributive property to expand the expression in the first column and to factor the greatest common factor out of the polynomial in the second column.

	Expand	Factor
66.	a. $4x(9x + 2)$	b. $36x^2 + 8x$
67.	a. $5v(8v - 7)$	b. $40v^2 - 35v$
68.	a. $8xy^2(3x^2 - 2y)$	b. $24x^3y^2 - 16xy^3$
69.	a. $2mn(m^2 - mn + 5n^2)$	b. $2m^3n - 2m^2n^2 + 10mn^3$
70.	a. $3ab(a^2 + 2ab - 6b^2)$	b. $3a^3b + 6a^2b^2 - 18ab^3$
71.	a. $x(x - 3) + 2(x - 3)$	b. $x(x - 3) + 2(x - 3)$
72.	a. $x(x + 4) - 5(x + 4)$	b. $x(x + 4) - 5(x + 4)$
73.	a. $3x(5x - 2) - 4(5x - 2)$	b. $3x(5x - 2) - 4(5x - 2)$

74. a. $(2x + 7)(5x) + (2x + 7)(3)$
b. $(2x + 7)(5x) + (2x + 7)(3)$

In Exercises 75–82, multiply the polynomial factors.

75. $(5 - 2x^2)(3x^2 + 4)$
76. $(2 - x)(x - x^2 + 3)$
77. $(x + 4 - x^2)(5 - x + x^2)$
78. $(3 + 2x - x^2)(4x - x^2 + 1)$
79. $(v + 5)^2$
80. $(v - 6)^2$
81. $(x + 2)^3$
82. $(x - 5)^3$

83. **Volume of a Tray** A 24-in by 12-in metal sheet has squares x on a side cut from each corner. Then the sides are bent upward and perpendicular to the base to form a metal tray. By using the formula $V = LWH$, the volume of the metal tray will be $V = (24 - 2x)(12 - 2x)(x)$. Write an expanded form of this polynomial.

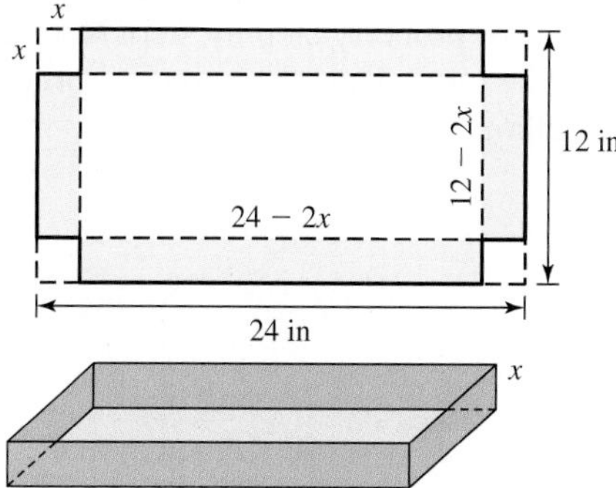

84. **Area of a Concrete Pad** Concrete of width x ft is poured around all four sides of a 20-ft by 40-ft pool.
a. One way to calculate the area covered by the concrete is to take the total area occupied by the pool and concrete and subtract the area of the pool. A polynomial representing this area is $A = (40 + 2x)(20 + 2x) - (40)(20)$. Expand and simplify this polynomial.
b. Another way to calculate the area covered by the concrete is to add the areas covered by concrete on each end and on the two sides. A polynomial representing this area is $A = 2[x(20 + 2x) + 40x]$. Expand and simplify this polynomial.
c. Compare the polynomials in parts **a** and **b.**

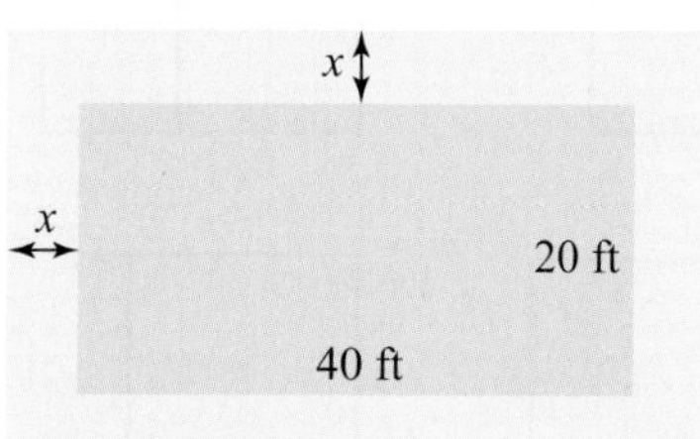

85. **Area of a Concrete Pad** A concrete sidewalk of width x ft is poured around a circular pool of radius 15 ft. From the formula $A = \pi r^2$, the area covered by the concrete will be $A = \pi(x + 15)^2 - \pi(15)^2$. Expand and simplify this polynomial.

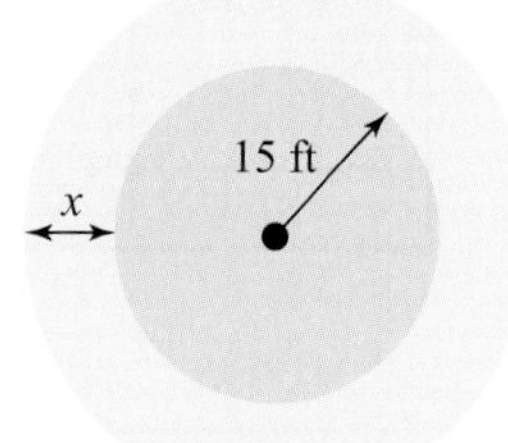

86. **Area of a Concrete Pad** A concrete sidewalk of width 4 ft is poured around a circular pool of radius x ft. From the formula $A = \pi r^2$, the area covered by the concrete will be $A = \pi(x + 4)^2 - \pi x^2$. Expand and simplify this polynomial.

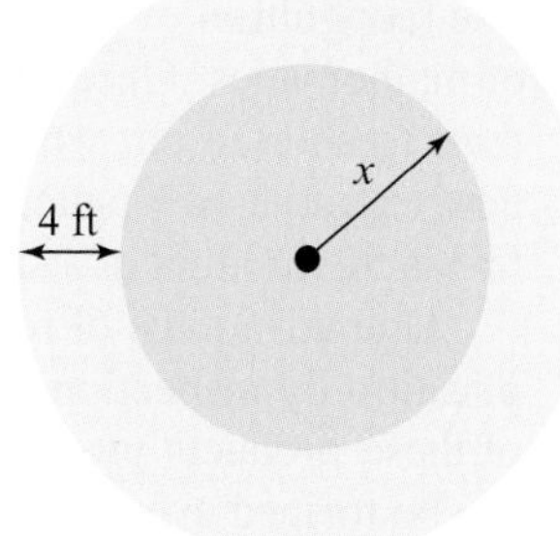

Revenue

87. The price per unit is x, and the number of units that will sell at the price is $280 - 2x$.
a. Write a revenue polynomial $R(x)$ for this product.
b. Evaluate $R(50)$.

88. The price per unit is x, and the number of units that will sell at that price is $540 - 3x$.
a. Write a revenue polynomial $R(x)$ for this product.
b. Evaluate $R(50)$.

Group Discussion Questions

89. **Challenge Question** For each rectangular region, first write a polynomial that represents the area of the whole rectangle.

Then write a polynomial that represents the sum of the areas of all the parts. Compare these polynomials. Are they equal?

Area of Whole Rectangle | **Area of Component Parts**

a.

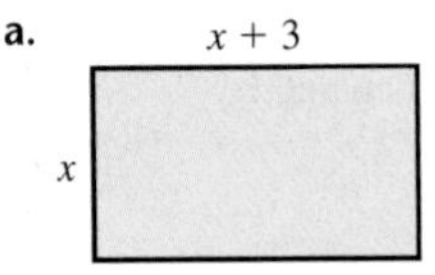

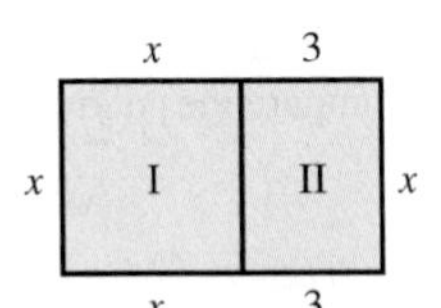

b.

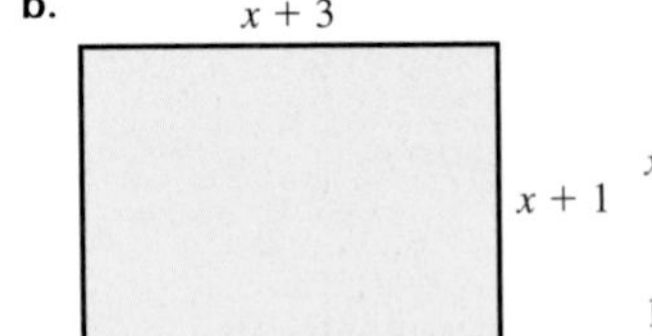

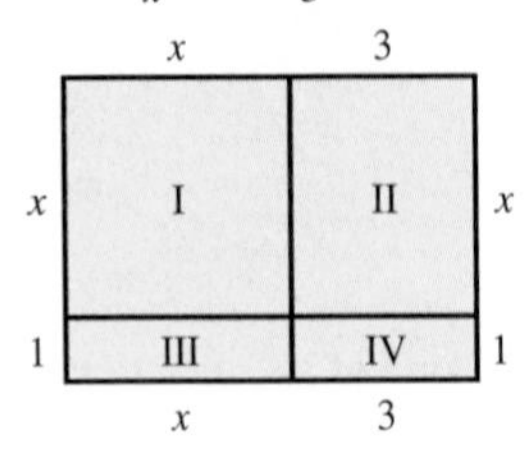

c.

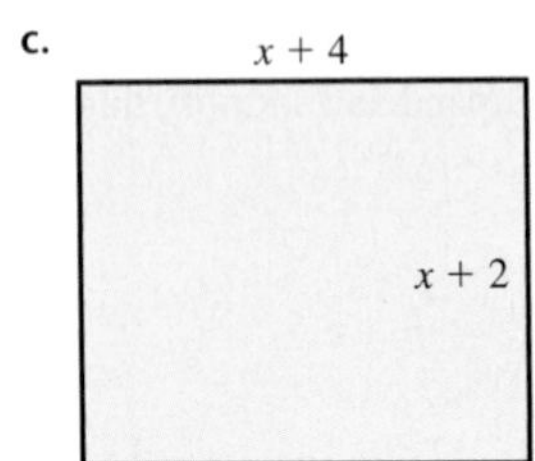

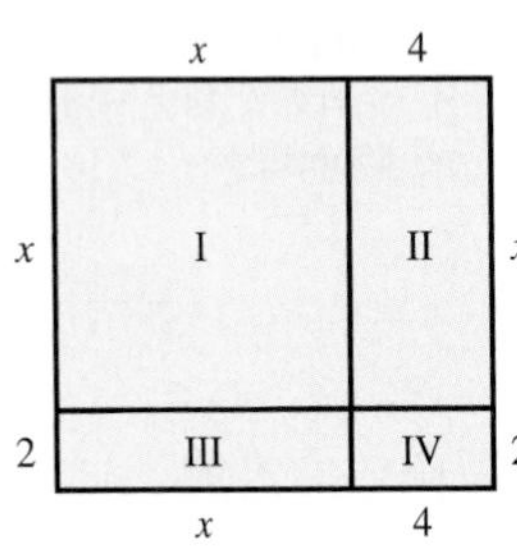

90. **Challenge Question**
    a. Give two binomials whose sum is a monomial.
    b. Give two binomials whose sum is a binomial.
    c. Give two binomials whose sum is a trinomial.
    d. Give two binomials whose product is a trinomial.
    e. Give two binomials whose product is a binomial.
91. **Discovery Question**
    a. Complete each product.
       i. $(x + 1)(x - 1) =$ ________
       ii. $(x + 2)(x - 2) =$ ________
       iii. $(x + 3)(x - 3) =$ ________
       iv. $(x + 4)(x - 4) =$ ________
       v. $(x + 5)(x - 5) =$ ________
    b. Make a general conjecture based on this work. State this conjecture in algebraic form and be prepared to describe this conjecture verbally to your teacher.
    c. Test this conjecture on $(x + 12)(x - 12)$.
    d. Test this conjecture on $(2v + 3)(2v - 3)$.
92. **Challenge Question** Expand the product of the factors $(x - a)(x - b)(x - c) \cdots (x - y)(x - z)$.

Section 5.6 Special Products of Binomials

Objectives:

12. Multiply by inspection a sum by a difference.
13. Multiply by inspection the square of a binomial.

The recognition of patterns algebraically, numerically, or graphically is an important part of mathematics. Once patterns are recognized, we can often take advantage of these patterns to shorten our work or to gain insights on the problems we are examining. We will now examine two patterns that occur frequently in the multiplication of binomials. The first is the product of a sum and a difference.

One advantage of recognizing these special factors is that we can perform the multiplication by inspection without applying the distributive property step by step. Note each of these products produces a binomial with no middle term. Thus these special products can be formed even quicker than using the FOIL method.

Product of a Sum and a Difference

Factors	Products
$(x + 2)(x - 2)$	$= x^2 - 4$
$(x + 3)(x - 3)$	$= x^2 - 9$
$(x + y)(x - y)$	$= x^2 - y^2$

Product of a Sum and a Difference

ALGEBRAICALLY	VERBALLY	ALGEBRAIC EXAMPLE
$(x + y)(x - y) = x^2 - y^2$	This product of a sum and a difference is the difference of their squares.	$(x + 11)(x - 11) = x^2 - 121$

This relationship is illustrated geometrically by the areas of the figures shown next.

A Sum Times a Difference: A Geometric Viewpoint

The difference between the areas of two squares with sides of x and y is $x^2 - y^2$. This is illustrated by the L-shaped region at the end of the first row of figures. This region consists of two rectangular regions, both of width $x - y$. One has length x, and the other has length y. These two rectangles can be combined to form a rectangle with width $x - y$ and length $x + y$. The area of this rectangle is $(x - y)(x + y)$.

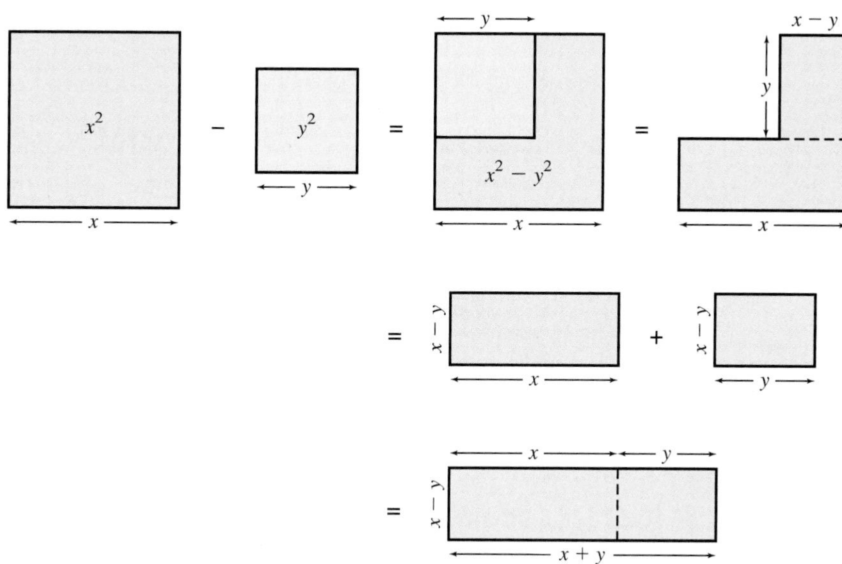

Thus $x^2 - y^2 = (x + y)(x - y)$.

It is not necessary for you to write the extra step shown in Example 1. Try to think this step, and write only the final product.

EXAMPLE 1 Multiplying a Sum by a Difference

Multiply these binomial factors by inspection.

SOLUTION

(a) $(x + 9)(x - 9)$

$(x + 9)(x - 9) = (x)^2 - (9)^2$
$= x^2 - 81$

The product is the difference of the squares of x and 9.

(b) $(3v - 8)(3v + 8)$

$(3v - 8)(3v + 8) = (3v)^2 - 8^2$
$= 9v^2 - 64$

Since multiplication is commutative, this product equals $(3v + 8)(3v - 8)$. The product is the difference of the squares.

(c) $(5x + 7y)(5x - 7y)$

$(5x + 7y)(5x - 7y) = (5x)^2 - (7y)^2$
$= 25x^2 - 49y^2$

SELF-CHECK 5.6.1

Multiply these binomial factors.

1. $(w + 12)(w - 12)$ 2. $(8v - 1)(8v + 1)$ 3. $(2a + 3b)(2a - 3b)$

As a look ahead to Chapter 6 we will emphasize the strong connection between multiplying and factoring polynomials. Expanding and factoring are reverse processes. These two processes are two different uses of the same algebraic information. Note that the steps used in Example 2 to factor $25v^2 - 1$ are the same steps in reverse order that we use to expand $(5v + 1)(5v - 1)$.

EXAMPLE 2 Comparing Expanding and Factoring

Use the special product $(x + y)(x - y) = x^2 - y^2$ to expand $(5v + 1)(5v - 1)$ and to factor $25v^2 - 1$.

SOLUTION

EXPAND

$$(5v + 1)(5v - 1) = (5v)^2 - (1)^2$$
$$= 25v^2 - 1$$

FACTOR

$$25v^2 - 1 = (5v)^2 - (1)^2$$
$$= (5v + 1)(5v - 1)$$

SELF-CHECK 5.6.2

1. Expand $(x + 11)(x - 11)$.
2. Factor $x^2 - 121$.
3. Expand $(2a + 7b)(2a - 7b)$.
4. Factor $x^2 - 36y^2$.

Square of a Binomial

The second special pattern that we examine is the square of a binomial, the product of a binomial multiplied by itself. Each of the products listed below is the square of a binomial. Note that these products all form a pattern. By taking advantage of this pattern we can square binomials by inspection.

Factors	Products
$(x + 2)^2$	$= x^2 + 4x + 4$
$(x + 5)^2$	$= x^2 + 10x + 25$
$(x + y)^2$	$= x^2 + 2xy + y^2$
$(x - 2)^2$	$= x^2 - 4x + 4$
$(x - 5)^2$	$= x^2 - 10x + 25$
$(x - y)^2$	$= x^2 - 2xy + y^2$

These patterns are summarized in the following box.

Square of a Binomial

ALGEBRAICALLY	VERBALLY	ALGEBRAIC EXAMPLE
Square of a sum: $(x + y)^2 = x^2 + 2xy + y^2$	The square of a binomial is a trinomial that has	
	1. A first term that is a square of the first term of the binomial.	$(x + 1)^2 = x^2 + 2x + 1$
Square of a difference: $(x - y)^2 = x^2 - 2xy + y^2$	2. A middle term that is twice the product of the two terms of the binomial.	$(x - 1)^2 = x^2 - 2x + 1$
	3. A last term that is a square of the last term of the binomial.	

Square of a Binomial: A Geometric Viewpoint

The formula for the square of a sum is illustrated geometrically by the areas of the following figures. The total area inside the square is equal to the sum of the areas of the four parts shown.

Area of a Whole Square

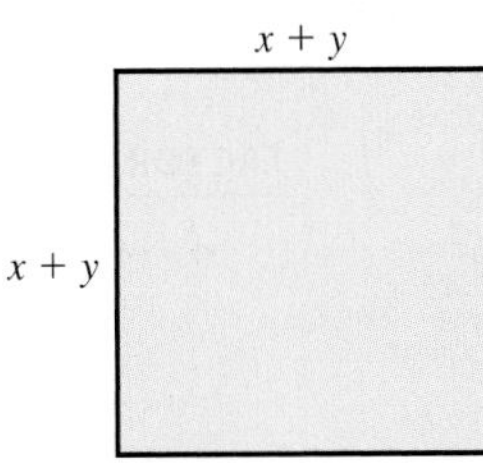

Areas of Four Parts

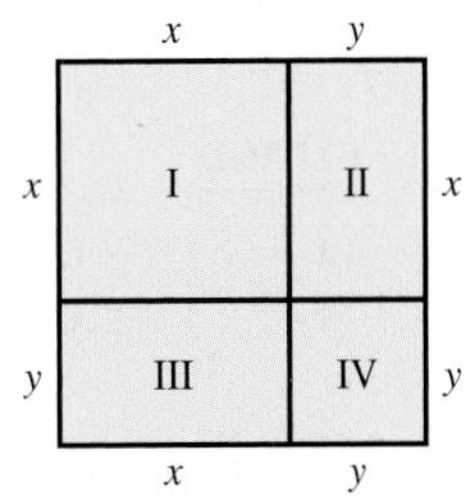

Area $= (x + y)(x + y)$
$A = (x + y)^2$

Total area = Area I + Area II + Area III + Area IV
$A = x^2 + xy + xy + y^2$
$A = x^2 + 2xy + y^2$

Thus $(x + y)^2 = x^2 + 2xy + y^2$.

EXAMPLE 3 Squaring Binomials

Square these binomials.

SOLUTION

(a) $(x + 7)^2$

$(x + 7)^2 = x^2 + 2(7)(x) + (7)^2$

The square of x
Twice the product of x and 7
The square of 7

$= x^2 + 14x + 49$

Use the form $(x + y)^2 = x^2 + 2xy + y^2$ to square this binomial.

(b) $(v - 4)^2$

$(v - 4)^2 = v^2 - 2(4)(v) + (4)^2$

The square of v
The opposite of twice the product of v and 4
The square of 4

$= v^2 - 8v + 16$

Use the form $(x - y)^2 = x^2 - 2xy + y^2$ to square this binomial.

(c) $(2x + 3y)^2$

$(2x + 3y)^2 = (2x)^2 + 2(2x)(3y) + (3y)^2$
$= 4x^2 + 12xy + 9y^2$

Use the form $(x + y)^2 = x^2 + 2xy + y^2$ to square this binomial. The first term is the square of $2x$. The second term is twice the product of $2x$ and $3y$. The third term is the square of $3y$.

SELF-CHECK 5.6.3

Square these binomials.

1. $(x - 3)^2$ **2.** $(7m + n)^2$ **3.** $(3a - 4b)^2$

We will examine the factoring of polynomials in detail in Chapter 6. The main purpose of Example 4 is to continue our emphasis that expanding and factoring are reverse processes that both rely heavily on the distributive property.

EXAMPLE 4 Comparison of Expanding and Factoring

Use the special product $(x + y)^2 = x^2 + 2xy + y^2$ to expand $(x + 8)^2$ and to factor $x^2 + 16x + 64$.

SOLUTION

EXPAND	FACTOR
$(x + 8)^2 = (x)^2 + 2(x)(8) + (8)^2$	$x^2 + 16x + 64 = (x)^2 + 2(x)(8) + (8)^2$
$= x^2 + 16x + 64$	$= (x + 8)^2$

SELF-CHECK 5.6.4

1. Expand $(x - 11)^2$.
2. Factor $x^2 - 22x + 121$.
3. Expand $(2a + 7b)^2$.
4. Factor $x^2 - 12xy + 36y^2$.

Problems involving operations with polynomials are often written using parentheses. Thus it is extremely important to follow the order of operations first given in Section 1.7 and restated below.

Order of Operations

STEP 1 Start with the expression within the innermost pair of grouping symbols.

STEP 2 Perform all exponentiations.

STEP 3 Perform all multiplications and divisions as they appear from left to right.

STEP 4 Perform all additions and subtractions as they appear from left to right.

In Example 5 note the distinct meanings of $(x + y)^2$ and $x^2 + y^2$.

EXAMPLE 5 Using the Correct Order of Operations to Simplify Polynomials

Simplify $(x + y)^2 + (x^2 + y^2)$.

SOLUTION

$$\begin{aligned}(x + y)^2 + (x^2 + y^2) &= (x^2 + 2xy + y^2) + (x^2 + y^2)\\ &= 2x^2 + 2xy + 2y^2\end{aligned}$$

First square the binomial to obtain $x^2 + 2xy + y^2$. Then add $x^2 + y^2$.

Example 6 presents two polynomial expressions that share a similar appearance but indicate a distinct order of operations.

EXAMPLE 6 Using the Correct Order of Operations to Simplify Polynomials

Simplify each expression.

SOLUTION

(a) $3x - 4(3x + 4)$ $\quad 3x - 4(3x + 4) = 3x - 12x - 16$
$= -9x - 16$

Multiplication by -4 is distributed to each term of $3x + 4$. Then like terms are added.

(b) $(3x - 4)(3x + 4)$ $\quad (3x - 4)(3x + 4) = (3x)^2 - (4)^2$
$= 9x^2 - 16$

The parentheses indicate the product of two binomials. This product is a special form that can be multiplied by inspection.

SELF-CHECK 5.6.5

Simplify each expression.

1. $(3x - 5)^2 - [(3x)^2 - (5)^2]$
2. $6x + 5(6x - 5)$
3. $(6x + 5)(6x - 5)$

To avoid careless order-of-operation errors, it is important to write your intermediate steps clearly and carefully. On problems as involved as Example 7 it is unwise to try to do too many steps mentally.

EXAMPLE 7 Using the Correct Order of Operations to Simplify Polynomials

Simplify $3(4x - 5y)^2 - 2(4x + 5y)^2$.

SOLUTION

$3(4x - 5y)^2 - 2(4x + 5y)^2$
$= 3(16x^2 - 40xy + 25y^2) - 2(16x^2 + 40xy + 25y^2)$ First square each of the binomials by inspection.

$= 48x^2 - 120xy + 75y^2 - 32x^2 - 80xy - 50y^2$ Next distribute the factor of 3 and the factor of -2.

$= 16x^2 - 200xy + 25y^2$ Then combine like terms.

SELF-CHECK ANSWERS

5.6.1
1. $w^2 - 144$
2. $64v^2 - 1$
3. $4a^2 - 9b^2$

5.6.2
1. $x^2 - 121$
2. $(x + 11)(x - 11)$
3. $4a^2 - 49b^2$
4. $(x + 6y)(x - 6y)$

5.6.3
1. $x^2 - 6x + 9$
2. $49m^2 + 14mn + n^2$
3. $9a^2 - 24ab + 16b^2$

5.6.4
1. $x^2 - 22x + 121$
2. $(x - 11)^2$
3. $4a^2 + 28ab + 49b^2$
4. $(x - 6y)^2$

5.6.5
1. $-30x + 50$
2. $36x - 25$
3. $36x^2 - 25$

USING THE LANGUAGE AND SYMBOLISM OF MATHEMATICS 5.6

1. Match each of these polynomials with the description that best fits each polynomial,
   a. $a^2 - b^2$ A. The square of a sum
   b. $(a - b)^2$ B. The sum of two squares
   c. $(a + b)^2$ C. The difference of two squares
   d. $a^2 + b^2$ D. The square of a difference
2. A binomial is a polynomial with __________ terms.
3. A trinomial is a polynomial with __________ terms.
4. Expanding and __________ are reverse processes.
5. The property that allows us to rewrite $2x(3x + 5)$ as $2x(3x) + 2x(5)$ is the __________ property.
6. Expanding $(x + 2)(x - 2)$ produces __________ .
7. Factoring $x^2 - 4$ produces __________ .

EXERCISES 5.6

In Exercises 1–12, multiply each sum by a difference by inspection.

1. $(7a + 1)(7a - 1)$
2. $(v + 9)(v - 9)$
3. $(z - 10)(z + 10)$
4. $(w - 13)(w + 13)$
5. $(2w - 3)(2w + 3)$
6. $(3w + 5)(3w - 5)$
7. $(9a + 4b)(9a - 4b)$
8. $(7a + 3b)(7a - 3b)$
9. $(4x - 11y)(4x + 11y)$
10. $(5x - 12y)(5x + 12y)$
11. $(x^2 - 2)(x^2 + 2)$
12. $(x^2 + 3)(x^2 - 3)$

In Exercises 13–30, square each binomial by inspection.

13. $(4m + 1)^2$
14. $(7m + 1)^2$
15. $(n - 9)^2$
16. $(n - 10)^2$
17. $(3t + 2)^2$
18. $(8t + 3)^2$
19. $(5v - 8)^2$
20. $(7v - 3)^2$
21. $(4x + 5y)^2$
22. $(7x + 2y)^2$
23. $(6a - 11b)^2$
24. $(8a - 7b)^2$
25. $(a + bc)^2$
26. $(a - bc)^2$
27. $(x^2 - 3)^2$
28. $(x^2 + 5)^2$
29. $(x^2 + y)^2$
30. $(x^2 - 3y)^2$

In Exercises 31–44, use the order of operations to simplify each expression.

31. a. $5x - 6(5x + 6)$ b. $(5x - 6)(5x + 6)$
32. a. $4x + 7(4x - 7)$ b. $(4x + 7)(4x - 7)$
33. a. $(4x - 9)^2$ b. $(4x)^2 - (9)^2$
34. a. $(7x + 10)^2$ b. $(7x)^2 + (10)^2$
35. $(v + 7)^2 - (v^2 - 7^2)$
36. $(w - 5)^2 - (w^2 - 5^2)$
37. $(2x + y)^2 - [(2x)^2 + y^2]$
38. $(x + y)^2 - (x - y)^2$
39. $(v + w)(v - w) - (v - w)^2$
40. $(3v - w)(3v + w) - (3v - w)^2$
41. $(a - b)^2 - (b - a)^2$
42. $(a - b)^2 + (b - a)^2$
43. $2(3x - 2y)^2 - (3x + 2y)^2$
44. $5(2x + 7y)^2 - 3(2x - 7y)^2$

Multiple Representations

In Exercises 45–50, write each verbal statement in algebraic form and then perform the indicated operation.

45. Square the quantity $3x + 7y$.
46. Square the quantity $8x - 9y$.
47. Subtract the square of $x - 6$ from the square of $x + 8$.
48. Subtract the square of $x + 7$ from the square of $x - 9$.
49. Multiply $a^2 + 1$ by the product of $a + 1$ and $a - 1$.
50. Multiply $a^2 + 4$ by the product of $a - 2$ and $a + 2$.

In Exercises 51 and 52, use a graphing calculator to fill in the table of values for Y_1 and Y_2. Do these results indicate that Y_2 is the expanded form of Y_1?

51.

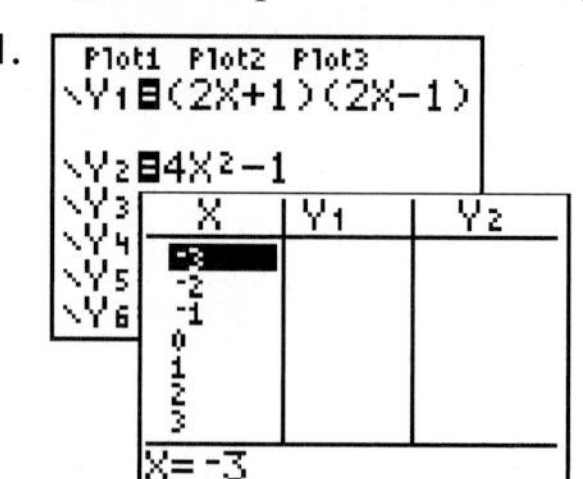

52.

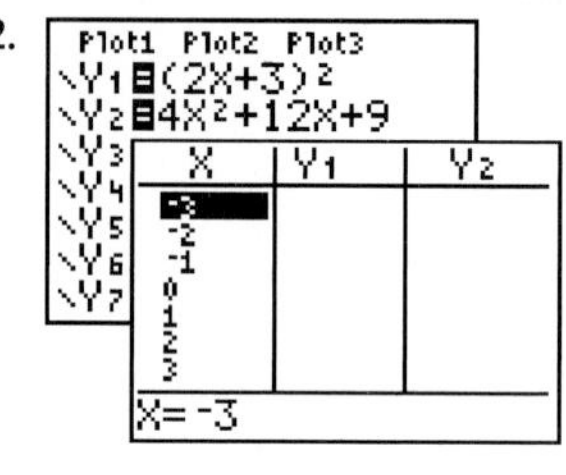

In Exercises 53 and 54, use a graphing calculator to complete each graph for Y_1 and Y_2. Do these results indicate that Y_2 is the factored form of Y_1?

53. 54.

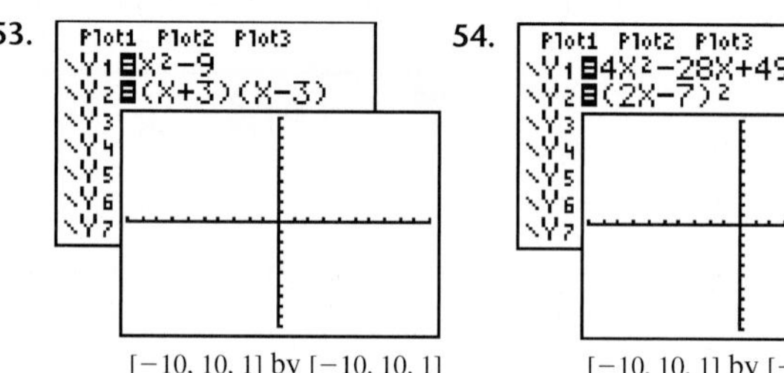

[−10, 10, 1] by [−10, 10, 1] [−10, 10, 1] by [−10, 10, 1]

In Exercises 55–58, expand each expression in the first column and factor each expression in the second column.

	Expand	Factor
55.	$(9x - 1)(9x + 1)$	$81x^2 - 1$
56.	$(7x - 11)(7x + 11)$	$49x^2 - 121$
57.	$(10w - 3x)^2$	$100w^2 - 60wx + 9x^2$
58.	$(8m + 5n)^2$	$64m^2 + 80mn + 25n^2$

Group Discussion Questions

59. **Mental Arithmetic** Sometimes impressive feats of mental arithmetic have a very simple algebraic basis.
   a. Use the fact that $(103)(97) = (100 + 3)(100 - 3)$ to compute this product mentally.
   b. Use the fact that $(96)(104) = (100 - 4)(100 + 4)$ to compute this product mentally.
   c. Use the fact that $(99)(99) = (100 - 1)^2$ to compute this product mentally.
   d. Use the fact that $(101)(101) = (100 + 1)^2$ to compute this product mentally.
   e. Make up two problems of your own that you can compute mentally.

60. **Discovery Question** Substitute $-y$ for y in the equation $(x + y)^2 = x^2 + 2xy + y^2$ and simplify the result. What do you observe?

61. **Discovery Question**

   a. Let Y_1 represent each of these polynomials, and then graph Y_1, using the viewing window $[-6, 6, 1]$ by $[-20, 10, 5]$. What relationship do you observe between the factors of Y_1 and attributes of the graph?

      i. $(x + 1)(x - 1)$

      ii. $(x + 2)(x - 2)$

      iii. $(x + 2)(x - 3)$

      iv. $(x + 4)(x - 1)$

   b. Given the graph below of $y = P(x)$, can you predict the factors of $P(x)$? Test your prediction with a graphing calculator.

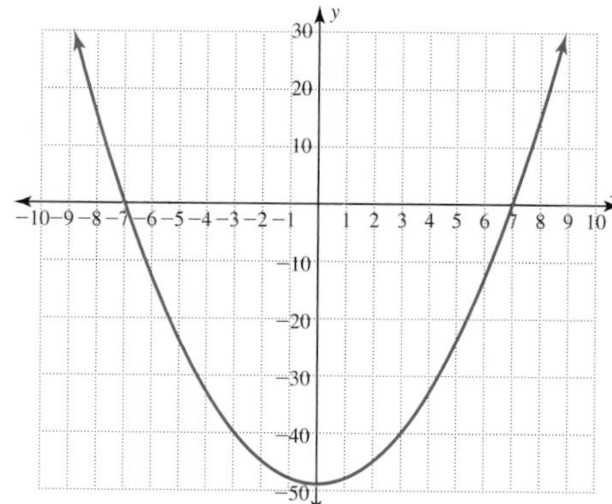

62. **Error Analysis** A student multiplied $2x + 3$ by $2x - 3$ and obtained $4x^2 + 9$. The student then checked this work on a graphing calculator by comparing the graphs of Y_1 and Y_2 as shown below. Seeing only one parabola on the display, the student concluded this meant that $Y_1 = Y_2$ and that the answer checked. Describe the calculator error the student has made, and then work this multiplication problem correctly.

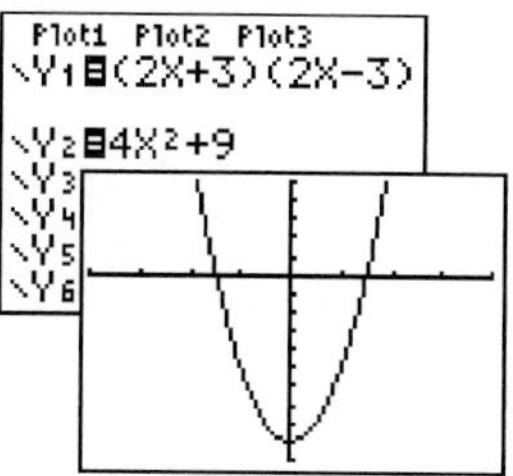

$[-4, 4, 1]$ by $[-10, 5, 1]$

Section 5.7 Dividing Polynomials

Objective: 14. Divide polynomials.

A Mathematical Note

George Boole (1815–1864), an English mathematician, created a whole new area of algebra now called Boolean algebra. Boolean algebra plays a key role in the logic and design of computer circuitry.

This section examines the division of polynomials. We build on our familiarity with the division of real number and monomials. Recall that $\frac{15}{3} = 5$ because 15 equals the product of the factors 3 and 5. Thus we can check the division of polynomials by multiplying two factors (the divisor by the quotient). This also helps us prepare for factoring polynomials in Chapter 6.

We will start by reviewing the quotient of two monomials which was first considered in Section 5.2. Recall that the quotient rule for exponents states that $\frac{x^m}{x^n} = x^{m-n}$ for $x \neq 0$. We assume throughout this section that the variables are restricted to values that will avoid division by zero.

EXAMPLE 1 Dividing a Monomial by a Monomial

Simplify $\frac{8x^2y^5}{4xy^3}$ and then check your answer by multiplying the divisor and the quotient. Assume that $x \neq 0$ and $y \neq 0$ to avoid division by 0.

SOLUTION

$$\frac{8x^2y^5}{4xy^3} = \frac{2x^{2-1}y^{5-3}}{1}$$

Divide both the numerator and the denominator by 4. Using the quotient rule for exponents, subtract the exponents on the common bases. Note that the coefficients are divided and the exponents are subtracted.

$$= 2xy^2$$

This is the quotient.

Check: $(4xy^3)(2xy^2) = 8x^2y^5$

The answer checks because $8x^2y^5$ equals the product of two factors, the divisor and the quotient.

The quotient of a polynomial divided by a monomial is often written in fractional form. Since $\frac{a + b}{c} = \frac{a}{c} + \frac{b}{c}$ $(c \neq 0)$, each term of the polynomial in the numerator is divided by the monomial in the denominator. Then each fraction is written in reduced form.

Dividing a Polynomial by a Monomial

VERBALLY	ALGEBRAIC EXAMPLE
To divide a polynomial by a monomial, divide each term of the polynomial by the monomial.	$\frac{6x^2 + 4x}{2x} = \frac{6x^2}{2x} + \frac{4x}{2x}$ $= 3x + 2$

EXAMPLE 2 Dividing a Binomial by a Monomial

Divide $15x^3 + 12x^2$ by $3x^2$. Then check your answer by multiplying the divisor by the quotient. Assume that $x \neq 0$ to avoid division by 0.

SOLUTION

$$\frac{15x^3 + 12x^2}{3x^2} = \frac{15x^3}{3x^2} + \frac{12x^2}{3x^2}$$

Divide each term of the numerator by $3x^2$, and then simplify.

$$= 5x + 4$$

This is the quotient.

Check: $3x^2(5x + 4) = (3x^2)(5x) + (3x^2)(4)$
$= 15x^3 + 12x^2$

This quotient checks.

SELF-CHECK 5.7.1

1. Reduce $\frac{18x^3y^2}{45xy^5}$ to lowest terms.
2. Divide $21x^3y^2$ by $3x^2y^2$ and use multiplication to check your answer.
3. Simplify $\frac{10x^3y + 4x^2y^2}{2x^2y}$ and use multiplication to check your answer.

Some divisions do not have a zero remainder. In fact, the quotient $-6y^2 + 9y - \frac{4}{y}$ in Example 3 is not even a polynomial $\left(-\frac{4}{y} \text{ has a variable in the denominator}\right)$. Nonetheless, the answers to all division problems can be checked by multiplication.

EXAMPLE 3 Dividing a Trinomial by a Monomial

Find $(20y + 30y^4 - 45y^3) \div (-5y^2)$ and check your result. Assume that $y \neq 0$ to avoid division by 0.

SOLUTION

$$\frac{30y^4 - 45y^3 + 20y}{-5y^2} = \frac{30y^4}{-5y^2} - \frac{45y^3}{-5y^2} + \frac{20y}{-5y^2}$$

First write the division in fractional form, expressing the numerator in descending order.

$$= -6y^2 + 9y - \frac{4}{y}$$

Divide each term of the numerator by $-5y^2$, and then simplify.

Check: $(-5y^2)\left(-6y^2 + 9y - \frac{4}{y}\right)$

The answer to a division problem can be checked by multiplication.

$$= (-5y^2)(-6y^2) + (-5y^2)(9y) - (-5y^2)\left(\frac{4}{y}\right)$$

Distribute the multiplication of $-5y^2$ to each term, and then simplify.

$$= 30y^4 - 45y^3 + 20y$$

This answer checks.

The procedure for dividing one polynomial by another polynomial with two or more terms is very similar to the long-division procedure used to divide integers. We refer to this as long division of polynomials. To use this procedure, be sure to first write both polynomials in standard form.

Long Division of Polynomials

STEP 1. Write the polynomials in long-division format, expressing each in standard form.

STEP 2. Divide the first term of the divisor into the first term of the dividend. The result is the first term of the quotient.

STEP 3. Multiply the first term of the quotient by every term in the divisor, and write this product under the dividend, aligning like terms.

STEP 4. Subtract this product from the dividend, and bring down the next term.

STEP 5. Use the result of step 4 as a new dividend, and repeat steps 2 through 4 until either the remainder is 0 or the degree of the remainder is less than the degree of the divisor.

To illustrate the close relationship between the long division of integers and the long division of polynomials, we compare these procedures side by side. Before you examine this comparison, note that if 10 is substituted for x in $(6x^2 + 7x + 2) \div (2x + 1)$, we obtain $[6(10)^2 + 7(10) + 2] \div [2(10) + 1]$, or $(600 + 70 + 2) \div (20 + 1) = 672 \div 21$. Thus for this value of x both problems denote the same thing.

Long Division of Integers	**Long Division of Polynomials**	
Problem: Divide 672 by 21.	**Problem:** Divide $6x^2 + 7x + 2$ by $2x + 1$.	
Step 1. Write the division in the long-division format.		
$21\overline{)672}$	$2x + 1\overline{)6x^2 + 7x + 2}$	
Step 2. Divide to obtain the first term in the quotient.		
$\begin{array}{r} 3 \\ 21\overline{)672} \end{array}$	$\begin{array}{r} 3x \\ 2x + 11\overline{)6x^2 + 7x + 2} \end{array}$	$6x^2$ divided by $2x$ is $3x$.
Step 3. Multiply the first term in the quotient by the divisor.		
$\begin{array}{r} 3 \\ 21\overline{)672} \\ \underline{63} \end{array}$	$\begin{array}{r} 3x \\ 2x + 1\overline{)6x^2 + 7x + 2} \\ \underline{6x^2 + 3x} \end{array}$	Note the alignment of similar terms.
Step 4. Subtract this product from the dividend, and bring down the next term.		
$\begin{array}{r} 3 \\ 21\overline{)672} \\ \underline{63} \\ 42 \end{array}$	$\begin{array}{r} 3x \\ 2x + 1\overline{)6x^2 + 7x + 2} \\ \underline{6x^2 + 3x} \\ 4x + 2 \end{array}$	
Step 5. Repeat Steps 2–4 to obtain the next term in the quotient.		
$\begin{array}{r} 32 \\ 21\overline{)672} \\ \underline{63} \\ 42 \\ \underline{42} \\ 0 \end{array}$	$\begin{array}{r} 3x + 2 \\ 2x + 1\overline{)6x^2 + 7x + 2} \\ \underline{6x^2 + 3x} \\ 4x + 2 \\ \underline{4x + 2} \\ 0 \end{array}$	Because the remainder is 0, the division is finished.
Answer: $672 \div 21 = 32$	$(6x^2 + 7x + 2) \div (2x + 1) = 3x + 2$	

Example 4 illustrates the use of long division of polynomials to divide a trinomial by a binomial.

EXAMPLE 4 Using Long Division to Divide a Trinomial by a Binomial

Divide $2x^2 + x - 15$ by $x + 3$. Assume $x \neq -3$ so that the divisor $x + 3$ is not 0.

SOLUTION

Step 1. $x + 3\overline{)2x^2 + x - 15}$

Set up the format for long division, writing both polynomials in standard form.

Step 2. $\begin{array}{r} 2x \\ x + 3\overline{)2x^2 + x - 15} \end{array}$

Divide $2x^2$ (the first term of the dividend) by x (the first term of the divisor) to obtain $2x$ (the first term of the quotient). Align similar terms.

Step 3. $\begin{array}{r} 2x \\ x + 3\overline{)2x^2 + x - 15} \\ \underline{2x^2 + 6x} \end{array}$

Multiply $2x$ by every term in the divisor, aligning under similar terms in the dividend.

Step 4.

$$x + 3\overline{)2x^2 + x - 15}$$ quotient $2x$; $2x^2 + 6x$; $-5x - 15$

Subtract $2x^2 + 6x$ from the dividend by changing the sign of each term and adding.

Step 5.

$$\begin{array}{r} 2x - 5 \\ x + 3\overline{)2x^2 + x - 15} \\ \underline{2x^2 + 6x} \\ -5x - 15 \\ \underline{-5x - 15} \\ 0 \end{array}$$

Divide $-5x$ (the first term in the last row) by x (the first term of the divisor) to obtain -5 (the next term in the quotient). Then multiply each term of the divisor by -5, aligning under similar terms in the dividend. Subtract to obtain the remainder of 0.

Answer: $\underbrace{(2x^2 + x - 15)}_{\text{Dividend}} \div \underbrace{(x + 3)}_{\text{Divisor}} = \underbrace{2x - 5}_{\text{Quotient}}$

To check your answer, multiply $(x + 3)(2x - 5)$.

The first step in the long-division procedure is to write the polynomials in standard form. In Example 5 we write $7v + 6v^3 + 2 - 19v^2$ in standard form before setting up the long-division format.

EXAMPLE 5 Using Long Division of Polynomials

Divide $(7v + 6v^3 + 2 - 19v^2)$ by $3v - 2$. Assume $v \neq \frac{2}{3}$ so that the divisor $3v - 2$ is not 0.

SOLUTION

$$\begin{array}{r} 2v^2 \\ 3v - 2\overline{)6v^3 - 19v^2 + 7v + 2} \\ \underline{6v^3 - 4v^2} \\ -15v^2 + 7v \end{array}$$

Write both polynomials in standard form. Divide to obtain the first term in the quotient, $\frac{6v^3}{3v} = 2v^2$. Multiply $2v^2$ by $3v - 2$ to obtain $6v^3 - 4v^2$, and then subtract.

$$\begin{array}{r} 2v^2 - 5v \\ 3v - 2\overline{)6v^3 - 19v^2 + 7v + 2} \\ \underline{6v^3 - 4v^2} \\ -15v^2 + 7v \\ \underline{-15v^2 + 10v} \\ -3v + 2 \end{array}$$

Divide to obtain the second term in the quotient, $-\frac{15v^2}{3v} = -5v$.

Multiply $-5v$ by $3v - 2$ to obtain $-15v^2 + 10v$, and then subtract.

$$\begin{array}{r} 2v^2 - 5v - 1 \\ 3v - 2\overline{)6v^3 - 19v^2 + 7v + 2} \\ \underline{6v^3 - 4v^2} \\ -15v^2 + 7v \\ \underline{-15v^2 + 10v} \\ -3v + 2 \\ \underline{-3v + 2} \\ 0 \end{array}$$

Divide to obtain the third term in the quotient, $\frac{-3v}{3v} = -1$.

Multiply -1 by $3v - 2$ to obtain $-3v + 2$, and then subtract.

Answer: $\frac{6v^3 - 19v^2 + 7v + 2}{3v - 2} = 2v^2 - 5v - 1$

To check your answer, multiply $(3v - 2)(2v^2 - 5v - 1)$.

If we divide 22 by 5, we get 4 with a remainder of 2. Thus,

$$\frac{22}{5} = 4 + \frac{2}{5}$$

Because 5 does not divide 22 evenly, we say that 5 is not a factor of 22. In Example 6 we divide two polynomials that produce a nonzero remainder.

EXAMPLE 6 Performing a Division That Has a Nonzero Remainder

Divide $x^2 + 2x - 13$ by $x + 5$. Assume $x \neq -5$ so that the divisor $x + 5$ is not 0.

SOLUTION

$$\begin{array}{r} x - 3 \\ x + 5\overline{)x^2 + 2x - 13} \\ \underline{x^2 + 5x} \quad\quad \\ -3x - 13 \\ \underline{-3x - 15} \\ 2 \end{array}$$

Divide x^2 by x to obtain x.

Multiply x by $x + 5$ to obtain $x^2 + 5x$ and then subtract.

Divide $-3x$ by x to obtain -3

Multiply -3 by $x + 5$ to obtain $-3x - 15$, and then subtract to obtain the remainder of 2.

Answer: $\dfrac{x^2 + 2x - 13}{x + 5} = x - 3 + \dfrac{2}{x + 5}$

The quotient is expressed as $x - 3$ plus the remainder 2, divided by $x + 5$.

Check: $(x + 5)\left(x - 3 + \dfrac{2}{x + 5}\right)$

$= (x + 5)(x - 3) + (x + 5)\left(\dfrac{2}{x + 5}\right)$

$= x^2 - 3x + 5x - 15 + 2$

$= x^2 + 2x - 13$

To check, we multiply the divisor $x + 5$ by the quotient $x - 3 + \dfrac{2}{x + 5}$.

The answer checks.

Remember that tables and graphs can also be used to check the answers to these division problems. In Self-Check 5.7.2 you can enter the given problem as Y_1 and the quotient that you obtain as Y_2. You can check your answer by comparing either tables or graphs of Y_1 and Y_2.

SELF-CHECK 5.7.2

1. Divide $12x^2 + x - 35$ by $3x - 5$.
2. Divide $15x - 31x^2 + 12x^3 - 2$ by $4x - 1$.
3. a. Divide $x^2 + x - 12$ by $x + 4$.
   b. Check your answer by using graphs.
   c. Check your answer by using tables.

The first step in the long-division procedure is to write the polynomials in standard form. Some polynomials have missing terms. In Example 7 the polynomial $x^3 - 5x - 14$ is missing the x^2 term. This polynomial can be rewritten as $x^3 + 0x^2 - 5x - 14$. This format can help prevent careless errors in the long-division procedure.

EXAMPLE 7 Dividing a Polynomial with Missing Terms

Divide $x^3 - 5x - 14$ by $x - 3$. Assume $x \neq 3$ so that the divisor $x - 3$ is not 0.

SOLUTION

$$\begin{array}{r|l} x - 3 \overline{)x^3 + 0x^2 - 5x - 14} & \text{quotient: } x^2 + 3x + 4 \\ \underline{x^3 - 3x^2} & \\ 3x^2 - 5x & \\ \underline{3x^2 - 9x} & \\ 4x - 14 & \\ \underline{4x - 12} & \\ -2 & \end{array}$$

Use 0 as the coefficient of the missing x^2 term.
Divide x^3 by x to obtain x^2.
Multiply x^2 by $x - 3$ to obtain $x^3 - 3x^2$, and then subtract.
Divide $3x^2$ by x to obtain $3x$. Multiply $3x$ by $x - 3$ to obtain $3x^2 - 9x$, and then subtract.
Divide $4x$ by x to obtain 4. Multiply 4 by $x - 3$ to obtain $4x - 12$. Subtract to obtain the remainder of -2.

Answer: $\dfrac{x^3 - 5x - 14}{x - 3} = x^2 + 3x + 4 - \dfrac{2}{x - 3}$

Because the remainder is not zero, $x - 3$ is not a factor of $x^3 - 5x - 14$.

Example 8 examines a problem with a second-degree trinomial as a divisor.

EXAMPLE 8 Dividing by a Second-Degree Trinomial

Divide $x^4 - x^3 + 6x^2 - 7x + 15$ by $x^2 - 2x + 3$.

SOLUTION

$$\begin{array}{r|l} x^2 - 2x + 3 \overline{)x^4 - x^3 + 6x^2 - 7x + 15} & \text{quotient: } x^2 + x + 5 \\ \underline{x^4 - 2x^3 + 3x^2} & \\ x^3 + 3x^2 - 7x & \\ \underline{x^3 - 2x^2 + 3x} & \\ 5x^2 - 10x + 15 & \\ \underline{5x^2 - 10x + 15} & \\ 0 & \end{array}$$

Divide x^4 by x^2 to obtain the first term of x^2 in the quotient.
Multiply x^2 by $x^2 - 2x + 3$ to obtain $x^4 - 2x^3 + 3x^2$, and then subtract.
Divide x^3 by x^2 to obtain x.
Multiply x by $x^2 - 2x + 3$ to obtain $x^3 - 2x^2 + 3x$, and then subtract.
Divide $5x^2$ by x^2 to obtain 5.
Multiply 5 by $x^2 - 2x + 3$ to obtain $5x^2 - 10x + 15$, and then subtract to produce the remainder of 0.

Answer: $\dfrac{x^4 - x^3 + 6x^2 - 7x + 15}{x^2 - 2x + 3} = x^2 + x + 5$

SELF-CHECK 5.7.3

1. Divide $8x^3 - 27$ by $2x - 3$.
2. Divide $2x^4 + 6x^3 + 7x^2 - 3x - 4$ by $x^2 + 3x + 4$.

The **average cost per unit** for units produced in a factory is the total cost of the units divided by the number of units. In Example 9 we determine the average cost per unit for the production of portable music players.

EXAMPLE 9 Determining the Average Cost per Unit

A company recorded the number of portable music players produced and the cost for producing these units for different periods of time. Using x to represent the number of units produced, the company determined that the cost function for this music player is $C(x) = 35x + 12{,}000$.

(a) Write an algebraic expression $A(x)$ for the average cost of producing x units.
(b) Evaluate and interpret $C(1{,}000)$.
(c) Evaluate and interpret $A(1{,}000)$.

SOLUTION

(a) Average cost = Total cost ÷ Number of units

$$A(x) = (35x + 12{,}000) \div x$$
$$A(x) = \frac{35x + 12{,}000}{x}$$
$$A(x) = 35 + \frac{12{,}000}{x}$$

Translate this word equation into an algebraic representation, using the given information.

(b)
$$C(x) = 35x + 12{,}000$$
$$C(1{,}000) = 35(1{,}000) + 12{,}000$$
$$C(1{,}000) = 35{,}000 + 12{,}000$$
$$C(1{,}000) = 47{,}000$$

Substitute 1,000 into the cost function to determine the cost of 1,000 units.

The cost of producing 1,000 portable music players is $47,000.

(c)
$$A(x) = 35 + \frac{12{,}000}{x}$$
$$A(1{,}000) = 35 + \frac{12{,}000}{1{,}000}$$
$$A(1{,}000) = 35 + 12$$
$$A(1{,}000) = 47$$

Substitute 1,000 into the average cost function to determine the average cost of 1,000 units.

The average cost of producing 1,000 portable music players is $47 per unit. ■

SELF-CHECK ANSWERS

5.7.1

1. $\frac{2x^2}{5y^3}$
2. $7x$; $(3x^2y^2)(7x) = 21x^3y^2$
3. $5x + 2y$; $(2x^2y)(5x + 2y) = 10x^3y + 4x^2y^2$

5.7.2

1. $4x + 7$
2. $3x^2 - 7x + 2$

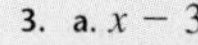

3. a. $x - 3$
b. The graphs are identical for $x \neq -4$.

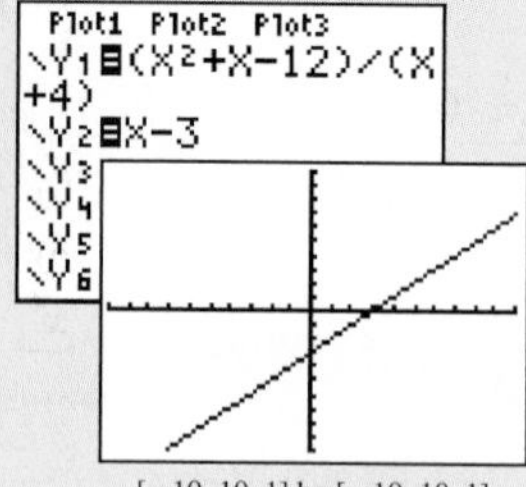

[−10, 10, 1] by [−10, 10, 1]

c. The tables are identical for $x \neq -4$.

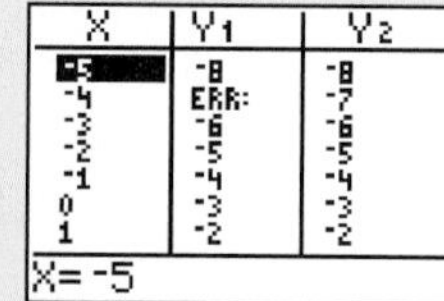

X	Y1	Y2
-5	-8	-8
-4	ERR:	-7
-3	-6	-6
-2	-5	-5
-1	-4	-4
0	-3	-3
1	-2	-2

X= -5

5.7.3

1. $4x^2 + 6x + 9$
2. $2x^2 - 1$

USING THE LANGUAGE AND SYMBOLISM OF MATHEMATICS 5.7

1. In the division $\frac{15}{3} = 5$, the dividend is 15, the divisor is 3, and the __________ is 5.
2. In the division $\frac{x^2 - x - 6}{x - 3} = x + 2$, the dividend is $x^2 - x - 6$, the divisor is__________, and the quotient is $x + 2$.
3. The division $\frac{35}{5} = 7$ can be checked by writing $5 \cdot 7 = 35$. In the equation $5 \cdot 7 = 35$, both 5 and 7 are factors and 35 is their __________.
4. The division $\frac{x^2 - x - 6}{x - 3} = x + 2$ can be checked by writing $(x - 3)(x + 2) = x^2 - x - 6$. In the equation $(x - 3)(x + 2) = x^2 - x - 6$, the factors are __________ and__________, and the product is __________.
5. The first step in the long-division procedure for dividing polynomials is to write both the divisor and the dividend in __________ form.
6. After stating the long-division procedure for dividing polynomials, we continue the steps until either the remainder is __________ or the degree of the remainder is less than the degree of the __________.
7. The polynomial $3x^4 + 5x^2 - 7x + 2$ is said to have a missing x^3 term. If this polynomial is rewritten with an x^3 term, the coefficient of x^3 will be __________.
8. The __________ cost per unit is the total cost of the units divided by the number of units.

EXERCISES 5.7

All exercises are assumed to restrict the variables to values that avoid division by zero.

In Exercises 1–18, find each quotient.

1. $\frac{18x^3}{6x}$
2. $\frac{24x^5}{3x^2}$
3. $\frac{35a^3b^2}{-7ab}$
4. $\frac{48a^4b^4}{-8ab^2}$
5. $\frac{15a^2 - 20a}{5a}$
6. $\frac{24a^5 - 16a^3}{4a}$
7. $\frac{9a^3 - 15a^2 + 6a}{3a^2}$
8. $\frac{24a^5 - 6a^4 - 12a^2}{6a^3}$
9. $\frac{16m^4 - 8m^3 + 10m^2 + 6m}{2m}$
10. $\frac{24m^4 - 18m^3 + 36m^2 - 6m}{3m}$
11. $\frac{45x^5y^2 - 54x^4y^5 + 99x^2y^4}{9x^2y^2}$
12. $\frac{-77x^7y^2 + 55x^5y^4 - 33x^3y^6}{11x^3y^2}$
13. $\frac{54v^2 - 36v^4 + 12v - 18v^3}{-6v}$
14. $\frac{15v^3 - 6v + 9v^4 + 27v^2}{-3v}$
15. $(35v^3w - 7v^2w - 28vw^2) \div 7vw$
16. $(54m^3n - 18m^2n + 27mn^2) \div 9mn$
17. $(100x^{20} - 50x^{12} + 30x^9) \div 10x^9$
18. $(66x^{21} - 48x^{14} - 18x^7) \div 6x^7$

In Exercises 19–22, find each quotient, and then use multiplication to check your answer.

19. $x + 2\overline{)x^2 + 9x + 14}$
20. $x + 5\overline{)x^2 + 8x + 15}$
21. $\frac{v^2 + 2v - 24}{v - 4}$
22. $\frac{v^2 - 2v - 35}{v - 7}$

In Exercises 23 and 24, perform the indicated division and then use multiplication to check your answer. (*Hint:* See Example 46.)

23. $\frac{4m^3 - 9m^2 + 10m + 7}{m - 3}$
24. $\frac{6m^3 - 19m^2 - 23m + 15}{m - 4}$

In Exercises 25–44, find each quotient.

25. $\frac{6w^2 + w - 12}{2w + 3}$
26. $\frac{12w^2 + 14w - 10}{3w + 5}$
27. $\frac{20m^2 - 43m + 14}{5m - 2}$
28. $\frac{28m^2 - 27m + 5}{4m - 1}$
29. $\frac{63y^2 - 130y + 63}{7y - 9}$
30. $\frac{32y^2 + 36y - 35}{8y - 5}$
31. $\frac{30x^2 - 45x^3 + 10x + 35x^4}{5x}$
32. $\frac{14x^2 - 35x^3 + 28x^4 - 42x}{7x}$
33. $(x^2 - 12x + 35) \div (x - 5)$
34. $(x^2 - 16x + 60) \div (x - 6)$
35. $(6a^2 - 11a - 10) \div (2a - 5)$
36. $(15a^2 - 19a - 10) \div (5a + 2)$
37. $(3v^3 - 4v^2 - 8) \div (v - 2)$
38. $(4v^3 - 6v^2 - 54) \div (v - 3)$
39. $(x^3 - 8) \div (x - 2)$
40. $(x^3 + 125) \div (x + 5)$
41. $(20y^3 + 16y^2 - 3y - 54) \div (5y - 6)$
42. $(18y^3 - 63y^2 + 106y - 55) \div (6y - 5)$
43. $\frac{21x^4 - 62x^2 - 7x^3 + 9x + 45}{3x^2 - x - 5}$
44. $\frac{7x^3 - 19x - 20 + 6x^4 - x^2}{3x + 4 + 2x^2}$
45. **Area of a Rectangle** The area of the rectangle shown is $(4x^2 + 4x - 3)$ cm^2. Find the width of this rectangle.

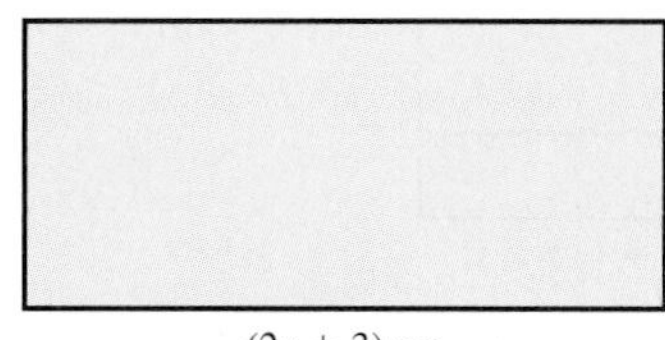

46. Area of a Rectangle The area of the rectangle shown is $(9x^2 - 1)$ cm^2. Find the width of this rectangle.

$(3x + 1)$ cm

47. Area of a Triangle The area of the triangle shown is $(12x^2 - 17x - 7)$ cm^2. Find the altitude of this triangle as shown by the dashed line.

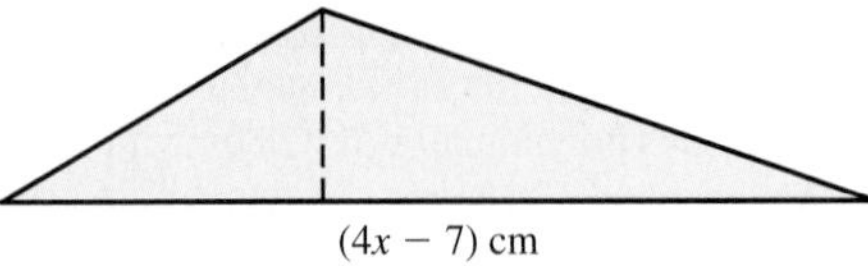

$(4x - 7)$ cm

48. Area of a Triangle The area of the triangle shown is $(5x^2 + 13x + 6)$ cm^2. Find the altitude of this triangle as shown by the dashed line.

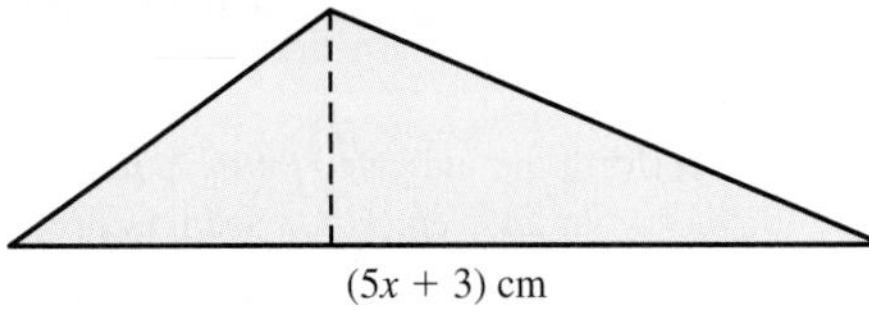

$(5x + 3)$ cm

49. Volume of a Box The volume of the solid shown below is $(8x^3 + 14x^2 + 3x)$ cm^3. Find the height of this solid.

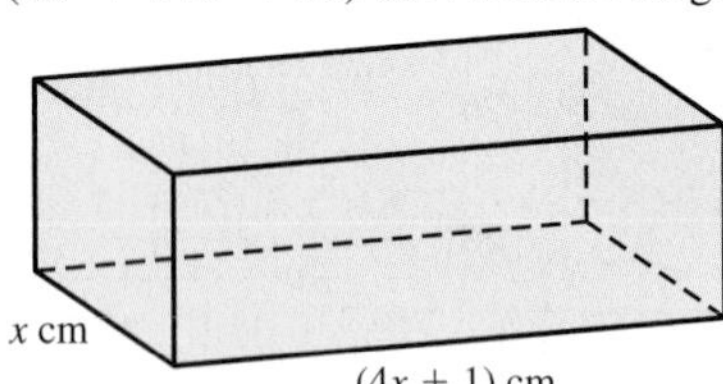

x cm

$(4x + 1)$ cm

50. Volume of a Box The volume of the solid shown below is $(18x^3 + 33x^2 + 5x)$ cm^3. Find the height of this solid.

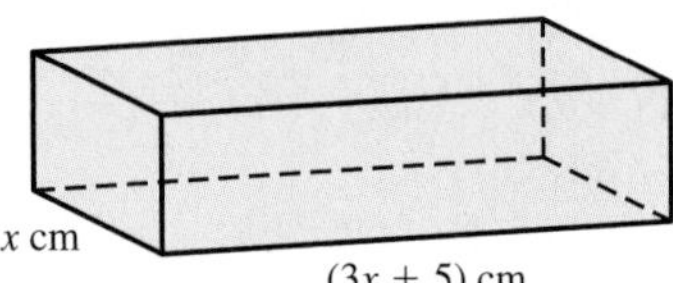

x cm

$(3x + 5)$ cm

In Exercises 51 and 52, use a graphing calculator to fill in the table of values and to graph $Y_1 = f_1(x)$ and $Y_2 = f_2(x)$, using the indicated window. Do these results indicate that $f_1(x) = f_2(x)$?

51.

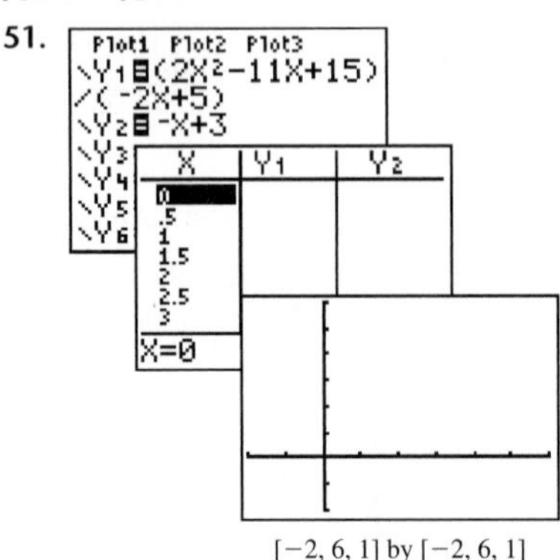

[−2, 6, 1] by [−2, 6, 1]

52.

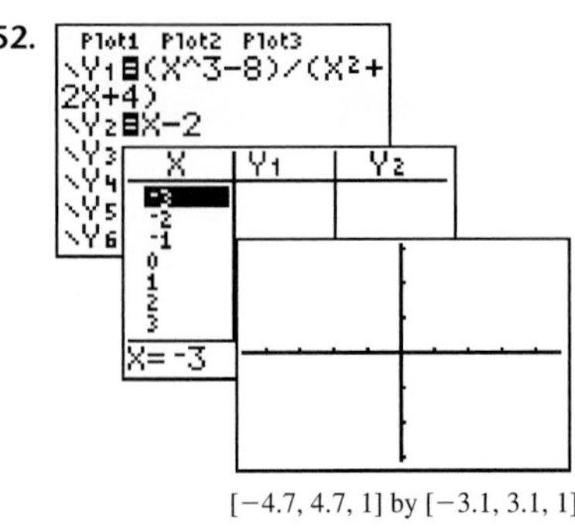

[−4.7, 4.7, 1] by [−3.1, 3.1, 1]

53. Entries in a Spreadsheet Table A two-dimensional table in an Excel spreadsheet has $30k^2 + 39k + 12$ entries. If there are $6k + 3$ rows in the table, how many columns are in the table?

54. Entries in a Spreadsheet Table A two-dimensional table in an Excel spreadsheet has $4k^2 + 81k + 20$ entries. If there are $k + 20$ columns in the table, how many rows are in the table?

55. What polynomial, when divided by $x + 3$, yields a quotient of $x^2 - 3x - 5$?

56. What polynomial, when divided by $x - 7$, yields a quotient of $2x^2 + x - 6$?

57. What polynomial, when divided into $6x^2 - 2x - 20$, produces a quotient of $2x - 4$?

58. What polynomial, when divided into $20x^2 + 7x - 6$, produces a quotient of $5x - 2$?

59. a. If a fourth-degree trinomial in x is added to a second-degree binomial in x, the degree of the sum is __________.

b. If a fourth-degree trinomial in x is multiplied by a second-degree binomial in x, the degree of the product is __________.

c. If a fourth-degree trinomial in x is divided by a second-degree binomial in x, the degree of the quotient is __________.

60. a. If a third-degree trinomial in x is added to a second-degree binomial in x, the degree of the sum is __________.

b. If a third-degree trinomial in x is multiplied by a second-degree binomial in x, the degree of the product is __________.

c. If a third-degree trinomial in x is divided by a second-degree binomial in x, the degree of the quotient is __________.

61. When a polynomial is divided by $2x + 3$, the quotient is $x + 2 + \dfrac{4}{2x + 3}$. Determine this polynomial.

62. When a polynomial is divided by $3x - 2$, the quotient is $2x + 1 + \dfrac{2}{3x - 2}$. Determine this polynomial.

63. One factor of $28x^2 + 11x - 30$ is $7x - 6$. Determine the other factor.

64. One factor of $40x^2 + 63x + 18$ is $8x + 3$. Determine the other factor.

65. a. One factor of $6x^3 - 29x^2 - 17x + 60$ is $2x + 3$. Determine the other factor.

b. Another factor of $6x^3 - 29x^2 - 17x + 60$ is $3x - 4$. Complete this equation:
$6x^3 - 29x^2 - 17x + 60 = (2x + 3)(3x - 4)(?)$.

66. a. One factor of $20x^3 + 8x^2 - 125x - 50$ is $5x + 2$. Determine the other factor.

b. Another factor of $20x^3 + 8x^2 - 125x - 50$ is $2x - 5$. Complete this equation:
$20x^3 + 8x^2 - 125x - 50 = (5x + 2)(2x - 5)(?)$.

Average Cost per Unit

67. Using x to represent the number of units produced, a factory determined that the cost in dollars to produce these units was $C(x) = 125x + 960$.
    a. Write an algebraic expression for $A(x)$, the average cost of producing x units.
    b. Evaluate and interpret $A(50)$.
68. Using x to represent the number of units produced, a factory determined that the cost in dollars to produce these units was $C(x) = 60x + 720$.
    a. Write an algebraic expression for $A(x)$, the average cost of producing x units.
    b. Evaluate and interpret $A(90)$.
69. A company manufactures lawn chairs. The company determined that the cost in dollars to produce x chairs is $C(x) = 8x + 1{,}200$.

    a. Write an algebraic expression for $A(x)$, the average cost of producing x chairs.
    b. Evaluate and interpret $C(200)$.
    c. Evaluate and interpret $A(200)$.
    d. Evaluate and interpret $C(250)$.
    e. Evaluate and interpret $A(250)$.
    f. What is the change in cost when production is increased from 200 to 250 chairs?
    g. What is the change in the average cost per unit when production is increased from 200 to 250 chairs?
70. A company manufactures baseball gloves. The company determined that the cost in dollars to produce x baseball gloves is $C(x) = 11x + 800$.
    a. Write an algebraic expression for $A(x)$, the average cost of producing x chairs.
    b. Evaluate and interpret $C(100)$.
    c. Evaluate and interpret $A(100)$.
    d. Evaluate and interpret $C(200)$.
    e. Evaluate and interpret $A(200)$.
    f. What is the change in cost when production is increased from 100 to 200 gloves?
    g. What is the change in the average cost per unit when production is increased from 100 to 200 gloves?

Group Discussion Questions

71. **Challenge Question**
    a. Can you give two binomials whose quotient is a monomial?
    b. Can you give two binomials whose quotient is a binomial?
    c. Can you give two binomials whose quotient is a trinomial?
72. **Discovery Question** Complete each of these divisions.
    a. $(x^2 - 1) \div (x - 1)$
    b. $(x^3 - 1) \div (x - 1)$
    c. $(x^4 - 1) \div (x - 1)$
    d. Based upon parts **a** through **c,** predict the quotient when $x^5 - 1$ is divided by $x - 1$. Then perform this division to check your prediction.
    e. Use inspection and this pattern to divide $x^8 - 1$ by $x - 1$.
73. **Discovery Question**
    a. Use a graphing calculator to complete the table of values shown to the right.
    b. Then use the **Trace** feature on a graphing calculator to examine this graph on the window indicated.

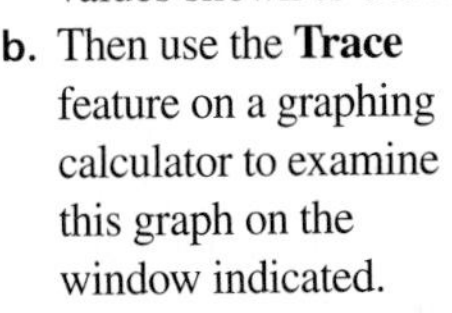

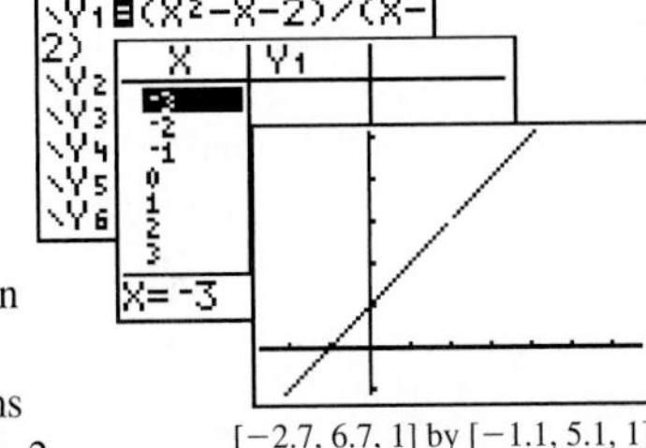

[−2.7, 6.7, 1] by [−1.1, 5.1, 1]

    c. From the table, what is the value of Y_1 when $x = 2$?
    d. Describe what happens to the graph when $x = 2$.
    e. This graph is not a straight line. Explain why it is not a straight line.
    f. Compare $Y_1 = \dfrac{x^2 - x - 2}{x - 2}$ to $Y_2 = x + 1$.
    How are they the same? How are they different?

KEY CONCEPTS FOR CHAPTER 5

1. **Exponential Notation:** For any real base x and natural number n,
    a. $x^n = x \cdot x \cdot \cdots \cdot x$ (n factors of x)
    b. $x^0 = 1$ for $x \neq 0$; 0^0 is undefined.
    c. $x^{-n} = \dfrac{1}{x^n}$ for $x \neq 0$.
2. **Base of an Exponent:** If there are no symbols of grouping, only the constant or variable immediately to the left of an exponent is the base. Thus $-x^2$ and $(-x)^2$ have distinct meanings.
3. **Summary of the Properties of Exponents:** For any nonzero real numbers x and y and integral exponents m and n,

Product rule:	$x^m \cdot x^n = x^{m+n}$
Power rule:	$(x^m)^n = x^{mn}$
Product to a power:	$(xy)^m = x^m y^m$
Quotient to a power:	$\left(\dfrac{x}{y}\right)^m = \dfrac{x^m}{y^m}$
Quotient rule:	$\dfrac{x^m}{x^n} = x^{m-n}$
Negative power:	$\left(\dfrac{x}{y}\right)^{-n} = \left(\dfrac{y}{x}\right)^n$

4. **Scientific Notation**
   a. A number is in scientific notation when it is expressed as the product of a number between 1 and 10 (or between -1 and -10 if negative) and an appropriate power of 10.
   b. On many calculators the EE feature is used to enter the power of 10.
   c. If a number such as 6.89E5 appears on a calculator display, this represents 6.89×10^5 in scientific notation.
5. **To Write a Number in Standard Decimal Notation:** Multiply the two factors by using the given power of 10.
   a. If the exponent on 10 is positive, move the decimal point to the right.
   b. If the exponent on 10 is zero, do not move the decimal point.
   c. If the exponent on 10 is negative, move the decimal point to the left.
6. **To Write a Number in Scientific Notation**
   a. Move the decimal point immediately to the right of the first nonzero digit of the number.
   b. Multiply by a power of 10 determined by counting the number of places the decimal point has been moved.
      i. The exponent on 10 is 0 or positive if the magnitude of the original number is 1 or greater.
      ii. The exponent on 10 is negative if the magnitude of the original number is between 0 and 1.
7. **Monomial:** A monomial is a real number, a variable, or a product of real numbers and variables.
8. **Polynomial:** A polynomial is a monomial or a sum of a finite number of monomials.
9. **Classification of Polynomials**
   a. Monomials contain one term.
   b. Binomials contain two terms.
   c. Trinomials contain three terms.
10. **Degree of a Monomial:** The degree of a monomial is the sum of the exponents on all the variables in the term.
11. **Degree of a Polynomial:** The degree of a polynomial is the same as the degree of the term of highest degree.
12. **A Polynomial Is in Standard Form If**
   a. The variables in each term are written in alphabetical order.
   b. The terms are arranged in descending powers of the first variable.
13. **Real Polynomial in x:** An nth-degree real polynomial in x can be written in the standard form $a_nx^n + a_{n-1}x^{n-1} + \cdots + a_1x + a_0$, where each of the coefficients of this polynomial is a real number.
14. **Equal Polynomials:** Equal polynomials in x have exactly the same values for each value of x. The tables and graphs for two equal polynomials are identical.
15. **Adding and Subtracting Polynomials:** To add or subtract polynomials, combine like terms.
16. **Product of Polynomials**
   a. To multiply a monomial by a polynomial, use the distributive property to multiply the monomial by each term of the polynomial.
   b. To multiply two polynomials, use the distributive property to multiply each term of the first polynomial by each term of the second polynomial, and then combine like terms.
   c. To multiply binomials by inspection, you can use FOIL to keep track of the steps. FOIL is an acronym for **F**irst, **O**uter, **I**nner, and **L**ast.
17. **Special Products**
   a. Product of a sum and a difference $\quad (x + y)(x - y) = x^2 - y^2$
   b. The square of a sum $\quad (x + y)^2 = x^2 + 2xy + y^2$
   c. The square of a difference $\quad (x - y)^2 = x^2 - 2xy + y^2$
18. **Quotient of Polynomials**
   a. To divide a polynomial by a monomial, divide each term of the polynomial by the monomial.
   b. To divide a polynomial by a polynomial of more than one term, use long division of polynomials.
19. **Average Cost per Unit:** The average cost per unit for units that are manufactured is the total cost of the units divided by the number of units.

REVIEW EXERCISES FOR CHAPTER 5

All exercises are assumed to restrict the variables to values that avoid division by 0 and avoid 0^0.

1. Write each expression in exponential form.
   a. $xyyy$ b. $(xy)(xy)(xy)$
   c. $xx + yy$ d. $(x + y)(x + y)$
2. Write each exponential expression in expanded form.
   a. $x^2 + y^2$ b. $(x + y)^2$
   c. $-x^4$ d. $(-x)^4$

In Exercises 3–12, simplify each expression.

3. a. 3^2 b. 3^{-2}
   c. -3^2 d. $(-3)^{-2}$
4. a. 3^0 b. 0^3
   c. $3^0 + 4^0$ d. $(3 + 4)^0$
5. a. $\left(\frac{1}{2}\right)^{-1} + \left(\frac{1}{3}\right)^{-1}$ b. $\left(\frac{1}{2} + \frac{1}{3}\right)^{-1}$
   c. $\left(\frac{2}{3}\right)^{-1}$ d. $\frac{2^{-1}}{3}$
6. a. -1^6 b. $(-1)^6$
   c. 6^{-1} d. -6^{-1}
7. a. $x^3 + x^3$ b. x^3x^3
   c. $\frac{x^3}{x^3}$ d. $x^3 - x^3$
8. a. 10^2 b. $(-10)^2$
   c. 10^{-2} d. -10^{-2}
9. a. $5^0 + 6^0 + 7^0$ b. $(5 + 6)^0 + 7^0$
   c. $(5 + 6 + 7)^0$ d. $0^5 + 0^6 + 0^7$
10. a. $2x^0 + 3y^0 + 4z^0$ b. $(2x + 3y + 4z)^0$
   c. $(2x)^0 + (3y)^0 + (4z)^0$ d. $2x^0 + (3y + 4z)^0$
11. a. x^2x^5 b. $\frac{x^5}{x^2}$
   c. $(x^2)^5$ d. $\left(\frac{x^5}{x}\right)^2$
12. a. $(4x^3y^2)(12x^5y^4)$ b. $\frac{12x^5y^4}{4x^3y^2}$
   c. $(4x^3y^2)^2$ d. $\left(\frac{12x^5y^4}{4x^3y^2}\right)^3$

In Exercises 13–18, evaluate each expression for $x = 3$ and $y = 4$.

13. a. $-x^2$ b. $(-x)^2$
14. a. xy^2 b. $(xy)^2$
15. a. $x^2 + y^2$ b. $(x + y)^2$
16. a. $x^{-1} + y^{-1}$ b. $(x + y)^{-1}$
17. a. $x^0 + y^0$ b. $(x + y)^0$
18. a. xy^{-2} b. $(xy)^{-2}$

In Exercises 19–32, simplify each expression.

19. $(5x^2y^3z^4)(6xy^4z^7)$
20. $(5x^2y^3z^4)^2$
21. $\dfrac{24a^2b^4c^5}{40ab^7c^3}$
22. $\left(\dfrac{12m^3}{4n^2}\right)^{-2}$
23. $(3x^2)(4x^3)(5x^{11})$
24. $\dfrac{(3x^2)^4}{9x^5}$
25. $(-2x^5)^3(-5x^3)^2$
26. $(7y^8)(8y^{-6})$
27. $\dfrac{16w^{-4}}{8w^{-6}}$
28. $(5m^{-6})^{-4}$
29. $(6x)^{-3}\left(\dfrac{x}{6}\right)^{-4}$
30. $\left(\dfrac{x^{188}}{x^{-439}}\right)^0$
31. $\dfrac{(2a^3b^3)(8a^2b^5)}{(4ab^2)^2}$
32. $(x^2y^{-3}z^{-4})^{-2}(x^4y^{-1}z^{-2})^{-1}$

33. **Strength of Satellite Signals** The strength of the signals from *Voyager 2* to Earth on its 1989 flyby of Neptune was only 1.0×10^{-13} W. Write this strength in standard decimal notation.
34. **Speed of Satellite Signals** The speed of the signals sent by *Voyager 2* was about the speed of light, 2.99×10^8 m/s. Write this speed in standard decimal notation.
35. Express the number shown on the following calculator display in standard decimal notation.

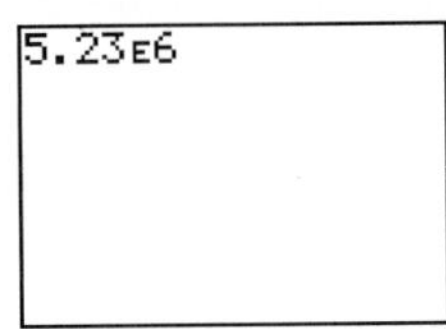

36. Express the number shown on the following calculator display in standard decimal notation.

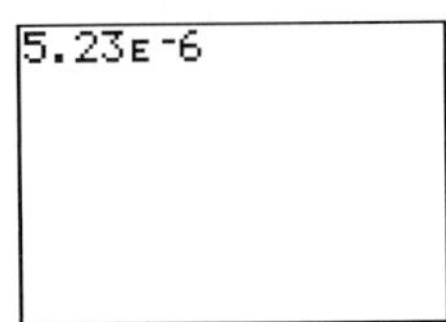

37. **Time for a Satellite Signal** If the distance from *Voyager 2* to Earth on its 1989 flyby of Neptune was 4.5×10^{12} m, determine the time for a *Voyager 2* signal traveling at 2.99×10^8 m/s to reach Earth.
38. **Estimation Skills** Use scientific notation to mentally select the best approximation of (41,010)(0.0000000001989).
    a. 8.0×10^{-8} b. 8.0×10^{-6}
    c. 8.0×10^{-4} d. 6.0×10^{-5}
    e. 6.0×10^5

In Exercises 39–42, classify each polynomial according to the number of terms it contains, and give its degree.

39. π
40. $3x^6 - 17x^2$
41. $-9x^3 + 7x^2 + 8x - 11$
42. $x^5y - 7x^4y^2 + 23x^3y^3$
43. Write $11x^4 - 3x^2 + 9x^3 + 7x^5 + 4 - 8x$ in standard form.
44. Write a fifth-degree monomial in x with a coefficient of negative seven.
45. Write a second-degree binomial in x whose leading coefficient is one and with a constant term of negative three.

In Exercises 46–64, perform the indicated operations.

46. $(7x^2 - 9x + 13) + (4x^2 + 6x - 11)$
47. $(9x^3 - 5x^2 - 7) - (4x^3 + 8x - 11)$
48. $(3x^4 - 8x^3 + 7x^2 + 9x - 4) + (2x^4 + 6x^2 + 9)$
49. $7x^5 + 9x^3 + 6x - 3 - (4x^5 - 3x^4 - 7x^3 - x^2 - x + 8)$
50. $(x^2 - 8x + 7) - (2x^2 + 7x + 11) + (3x^2 + 4x - 8)$
51. $5x^2(7x^3 - 9x^2 + 3x + 1)$
52. $(5v + 1)(7v - 1)$
53. $(5y - 7)^2$
54. $(9y + 5)^2$
55. $(3a + 5b)(3a - 5b)$
56. $(3m + 5)(2m^2 - 6m + 7)$
57. $\dfrac{-36a^3b^7}{9a^4b^4}$
58. $\dfrac{15m^5n^4 - 21m^4n^5 - 3m^3n^6}{3m^2n^3}$
59. $\dfrac{v^2 - 6v + 8}{v - 4}$
60. $\dfrac{21w^2 - 40w - 21}{3w - 7}$
61. $\dfrac{(a + b)^7}{(a + b)^6}$
62. $(3x + 4y)^2 - (3x + 4y)(3x - 4y)$
63. $(2x + 3y)^2 - (2x - 3y)^2$
64. $\dfrac{6x^2 + x - 2}{2x - 1} - \dfrac{8x^2 + 18x - 35}{4x - 5}$
65. The answer to $\dfrac{x^3 + 64}{x + 4}$ is $x^2 - 4x + 16$. Check this answer.
66. The answer to $\dfrac{x^2 + 4x - 5}{x - 2}$ is $x + 6 + \dfrac{7}{x - 2}$. Check this answer.

In Exercises 67–70, use the distributive property and the special products to expand the first expression and to factor the second expression.

	Expand	Factor
67.	a. $3x(x - 4)$	b. $3x^2 - 12x$
68.	a. $5x(x + 3) - 7(x + 3)$	b. $5x(x + 3) - 7(x + 3)$
69.	a. $(4x + 3)(4x - 3)$	b. $16x^2 - 9$
70.	a. $(6x - 7)^2$	b. $36x^2 - 84x + 49$

In Exercises 71–76, multiply these binomial factors by inspection.

71. $(9x - 2)(5x + 3)$
72. $(4x - 3y)(5x + 6y)$
73. $(5x - 6y)(5x + 6y)$
74. $(5x - 6y)^2$
75. $(7x + 2y)^2$
76. $(x + 3)(x - 3)(x^2 + 9)$

In Exercises 77–80, use the order of operations to simplify each expression.

77. a. $3x - 7(4x + 1)$ b. $(3x - 7)(4x + 1)$
78. a. $(8x)^2 - (5y)^2$ b. $(8x - 5y)^2$
79. $(7a - 2b)^2 + (7a + 2b)^2$
80. $2(a - 3b)^2 + 3(a + 3b)^2 + 4a^2 + 5(3b)^2$
81. Divide $8x^4 - 2x^3 + 12x^2 - 31x + 7$ by $2x^3 + 3x - 7$.
82. Divide $81x^4 - 16$ by $3x - 2$.
83. **Area of a Rectangle** Determine the area of the rectangle shown below.

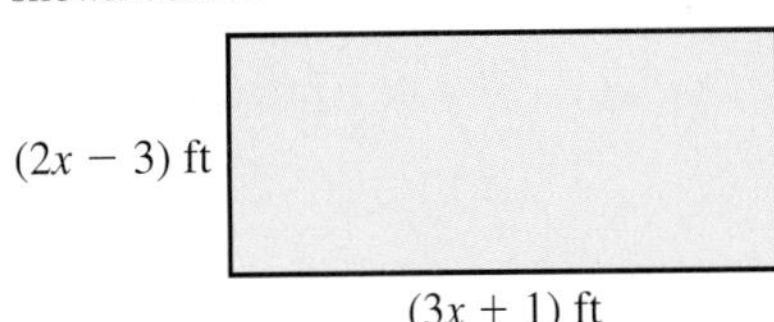

84. Height of a Box The volume of the box shown below is $(2x^3 + 2x^2)$ cm^3. Determine the height of this box.

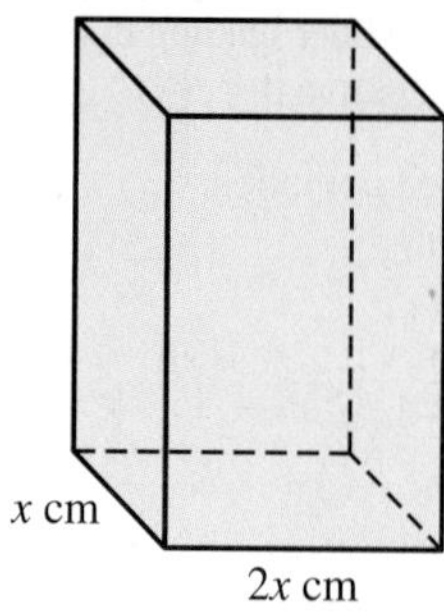

85. Revenue The price per unit of an item is x dollars, and the number of units that will sell at this price is estimated to be $350 - 4x$.

a. Write $R(x)$, a revenue polynomial for this item.

b. Evaluate and interpret $R(0)$.

c. Evaluate and interpret $R(40)$.

d. Evaluate and interpret $R(80)$.

86. Value of an Investment Use the formula $A = P(1 + r)^t$ to find the value of a \$10,000 investment compounded at 7.5% for 8 years.

87. Average Cost A company manufactures range finders for golfers. The company determined that the cost in dollars to produce x range finders is $C(x) = 15x + 1{,}000$.

a. Write an algebraic expression for $A(x)$, the average cost of producing x range finders.

b. Evaluate and interpret $C(100)$.

c. Evaluate and interpret $A(100)$.

d. Evaluate and interpret $C(200)$.

e. Evaluate and interpret $A(200)$.

f. What is the change in cost when production is increased from 100 to 200 range finders?

g. What is the change in the average cost per unit when production is increased from 100 to 200 range finders?

88. Use a graphing calculator to complete the following four tables and to determine which, if any, of the polynomials are equal.

a.

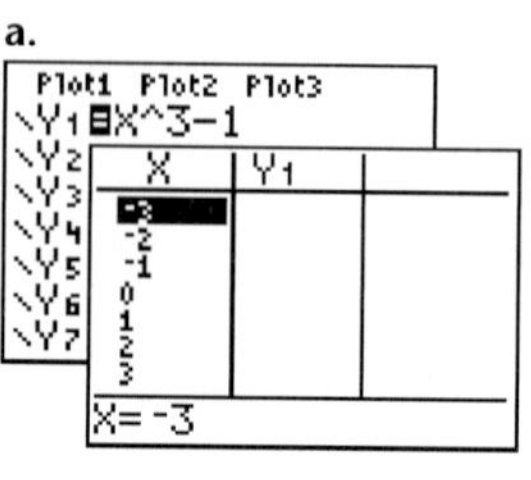

b.

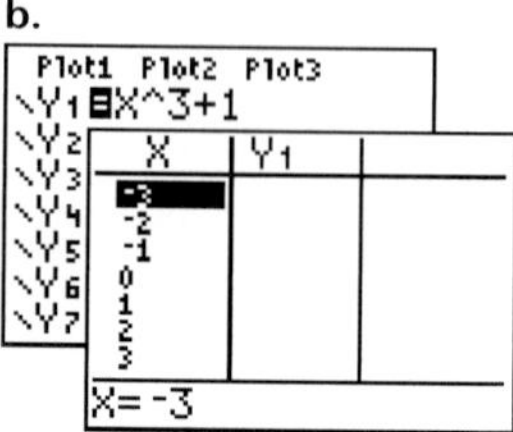

c.

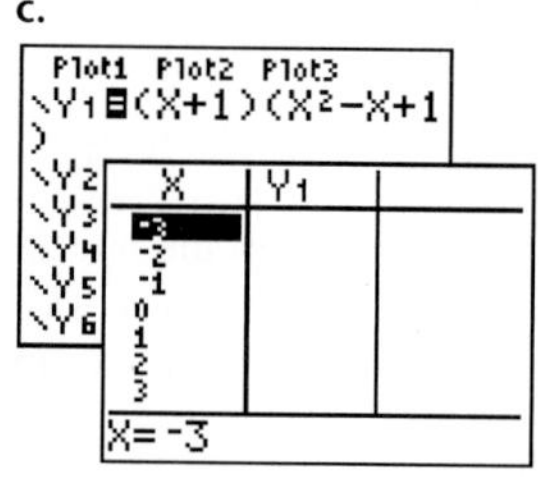

d.

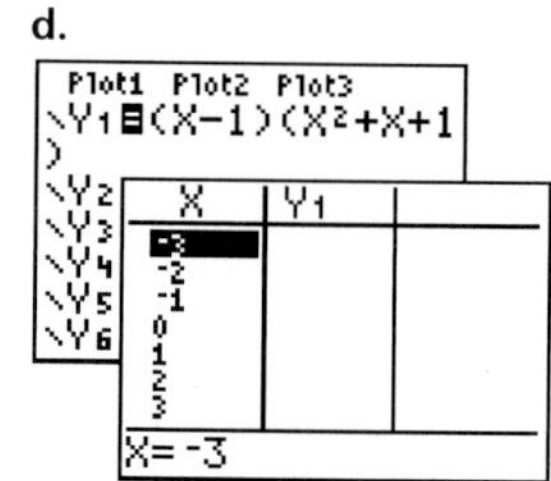

MASTERY TEST FOR CHAPTER 5

All exercises are assumed to restrict the variables to values that avoid division by 0 and avoid 0^0.

[5.1] **1.** Simplify each of these expressions.

a. x^5x^8 b. $(5y^4)(8y^3)$

c. $-v^4v^5v^6$ d. $(mn^2)(m^3n)$

[5.1] **2.** Simplify each of these expressions.

a. $(w^5)^8$ b. $(2y^4)^5$

c. $(-v^2w^3)^4$ d. $\left(\frac{-3m}{2n}\right)^3$

[5.2] **3.** Simplify each of these expressions.

a. $\frac{z^8}{z^2}$ b. $\frac{4x^9}{2x^3}$

c. $-\frac{v^{23}}{v^{23}}$ d. $\frac{-6a^4b^6}{3a^2b^2}$

[5.2] **4.** Simplify each of these expressions.

a. 1^0 b. $\left(\frac{2}{5}\right)^0$

c. $(3x + 5y)^0$ d. $3x^0 + 5y^0$

[5.2] **5.** Simplify each of these expressions.

a. $(x^3y^5)(x^2y^6)$ b. $\left(\frac{x^2y^3}{xy}\right)^3$

c. $\frac{(6x^2)^2}{(3x^3)^3}$ d. $[(2x^2)(3x^4)]^2$

[5.3] **6.** Simplify each of these expressions.

a. 3^{-1} b. 7^{-2}

c. $\left(\frac{2}{3}\right)^{-2}$ d. $\left(\frac{1}{3}\right)^{-1} + \left(\frac{1}{6}\right)^{-1}$

[5.3] **7.** Write these numbers in standard decimal notation.

a. 3.57×10^5 b. 7.35×10^{-5}

Write these numbers in scientific notation.

c. 0.000509 d. 93,050,000

[5.4] **8.** Classify each of these polynomials according to the number of terms it contains, and give its degree.

a. $-5y^3 - 13y$ b. 273

c. $2x^2 - 7x + 1$ d. $17x^5 - 4x^3 + 9x + 8$

[5.4] **9.** Simplify each of these expressions.

a. $(4x - 9y) + (3x + 8y)$

b. $(2x^2 - 3x + 7) - (5x^2 - 9x - 11)$

c. $(5x^4 - 9x^3 + 7x^2 + 13) + (4x^4 + 12x^2 + 9x - 5)$

d. $(3x^3 - 7x + 1) - (12x^2 + 9x - 5)$

[5.5] **10.** Find each of the indicated products.

a. $(5x^2y^3)(11x^4y^7)$

b. $-3x^2(5x^3 - 2x^2 + 7x - 9)$

c. $(x + 5)(x^2 + 3x + 1)$

d. $(x - 3y)(x^2 + xy + y^2)$

[5.5] **11.** Multiply these binomials by inspection.

a. $(x + 8)(x + 11)$ b. $(x - 9)(x - 5)$

c. $(2x + 5y)(3x - 4y)$ d. $(5x - 3y)(2x + 7y)$

[5.6] **12.** Multiply these factors by inspection.

a. $(6x + 1)(6x - 1)$ **b.** $(x - 9y)(x + 9y)$
c. $(5x + 11)(5x - 11)$ **d.** $(3x - 8y)(3x + 8y)$

[5.6] **13.** Square these binomials by inspection.

a. $(6x + 1)^2$ **b.** $(x - 9y)^2$
c. $(5x - 11)^2$ **d.** $(3x + 8y)^2$

[5.7] **14.** Find each of the indicated quotients.

a. $\dfrac{15a^4b^2}{5a^2b^2}$ **b.** $\dfrac{36m^4n^3 - 48m^2n^5}{12m^2n^2}$

c. $x - 2\overline{)x^2 + 9x - 22}$ **d.** $\dfrac{6x^3 + 10x^2 - 32}{3x - 4}$

REALITY CHECK FOR CHAPTER 5

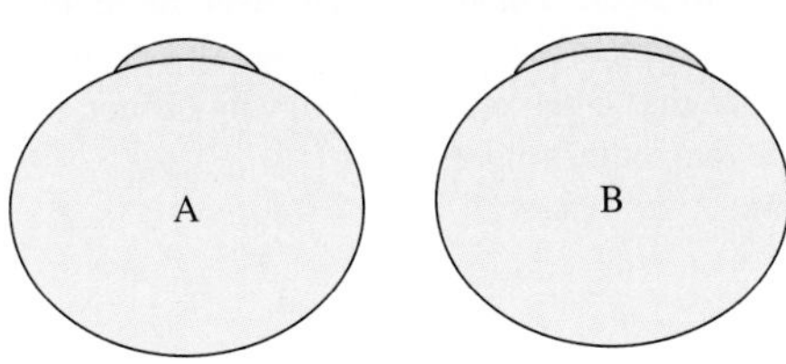

The equations $y = -0.0657x^2 + 0.394x$ and $y = -0.0867x^2 + 0.52x - 0.1$ model the cross sections of the two eyes shown (units for x and y are millimeters). One of these is from a patient with normal vision, and the other is from a patient with close-up vision problems due to a flatter cornea. Use a graphing calculator to graph each equation, using the viewing window [0, 6, 1] by [0, 2, 1]. Match each eye with the equation that models its shape.

GROUP PROJECT FOR CHAPTER 5

A Polynomial Model for the Construction of a Box

Supplies Needed for Each Group

- Heavy-weight paper (120-lb weight is suggested) for constructing a box
- Metric rulers with centimeters marked for measuring and to use as an aid in folding the paper
- Scissors for cutting out the corners of the paper
- Masking tape to use to tape up the sides of the box at the corners
- Rice to place into the box and to be measured for volume
- Graduated cylinders marked in milliliters ($1 \text{ mL} = 1 \text{ cm}^3$)
- Funnels to be used to pour rice into the graduated cylinders

I. LAB
Constructing a Box and Measuring the Volume of Rice That the Box Contains

1. Measure the width of your sheet of paper to the nearest tenth of a centimeter. ______
2. Measure the length of your sheet of paper to the nearest tenth of a centimeter. ______
3. Assign different groups different-sized squares to cut out (e.g., 4 cm, 4.5 cm, 5 cm). Mark off equal squares to be cut from each corner, and record to the nearest tenth of a centimeter the width of each square. ______
4. Carefully cut the squares from each corner of your sheet of paper as illustrated.
5. Using a ruler to maintain a straight fold, fold the paper along lines that you have drawn between the corners.
6. Using the masking tape, tape up the corners as shown in the illustration. Take care to form vertical sides.

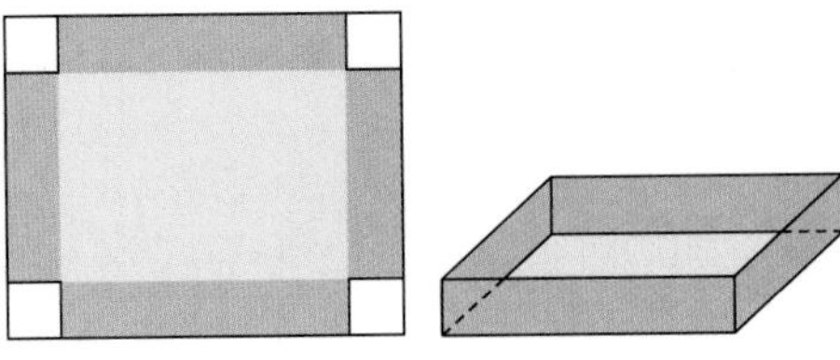

7. Using graduated cylinders, measure the rice in milliliters, and carefully pour the rice into your box until it is full (level the top of the rice with your ruler as needed). Record the volume of rice your box contains in cubic centimeters. ______
8. Calculate the volume of the box, using the formula $V = l \cdot w \cdot h$ and the measurements of the length, width, and height recorded above. ______
9. Compare the volume of rice measured to the calculated volume of the box. What is the relative error of the rice measured to that predicted by the formula? ______ Is your relative error acceptable? Be prepared to defend your answer or to explain any excessive error.*

*Relative error is the error of the measurement divided by the value predicted by the formula. See Section 1.6 for more information on relative error.

10. Create a table of values using the results from the box of each group. Use x to represent the width of the square cut from each corner of the sheet of paper. Use y to represent the volume of rice contained in each box. (Discard any values with excessive error.)

II. Creating a Mathematical Model for the Volume of a Box

1. Use x to represent the height of each box. Write an interval to express the restrictions on x for the sheet of paper used by your group. __________
2. Write an equation that gives the width of the box in terms of the height x. $W =$ __________
3. Write an equation that gives the length of the box in terms of the height x. $L =$ __________
4. Write an equation that gives the volume of the box in terms of the height x. $V =$ __________
5. Use the volume equation to create a table of values over the set of x-values listed in part **1** above.
6. Use the volume equation to create a graph over the set of x-values listed in part 1 above.
7. Do all values of x produce the same volume? If so, why? If not, is there any value that seems to produce greater volume than other values of x?

DIAGNOSTIC REVIEW OF BEGINNING ALGEBRA

The purpose of this diagnostic review is to help you gauge your mastery of beginning algebra, material that is needed for the intermediate algebra portion of this book in Chapters 6 through 11. This review is intended to give you a realistic assessment of your areas of strength and weakness. A few questions in this review may cover topics not discussed at your school. You may wish to ask your instructor for topics that your school stresses. Some of these questions require you to interpret graphs or calculator screens.

The left column contains a list of selected key concepts, and in the right column there are review exercises. The answer to each question follows this diagnostic review. Each answer is keyed to an example in this book. You can refer to these examples to find explanations and additional exercises for practice.

Order of Operations

Step 1 Start with the expression within the innermost pair of grouping symbols.
Step 2 Perform all exponentiations.
Step 3 Perform all multiplications and divisions as they appear from left to right.
Step 4 Perform all additions and subtractions as they appear from left to right.

Operations with Fractions

Addition: $\frac{a}{b} + \frac{c}{b} = \frac{a + c}{b}$ for $b \neq 0$

Subtraction: $\frac{a}{b} - \frac{c}{b} = \frac{a - c}{b}$ for $b \neq 0$

Multiplication: $\frac{a}{b} \cdot \frac{c}{d} = \frac{a \cdot c}{b \cdot d}$ for $b \neq 0$ and $d \neq 0$

Division: $\frac{a}{b} \div \frac{c}{d} = \frac{a}{b} \cdot \frac{d}{c} = \frac{a \cdot d}{b \cdot c}$ for $b \neq 0$, $c \neq 0$, and $d \neq 0$

Reducing fractions: $\frac{ac}{bc} = \frac{a}{b}$ for $b \neq 0$ and $c \neq 0$

Subsets of the Real Numbers

Real Numbers: The real numbers consist of the rational numbers and the irrational numbers.
Rational Numbers: Numbers that can be written as a ratio of two integers are rational numbers. Expressed decimally, rational numbers are either terminating decimals or repeating decimals.
Irrational Numbers: Real numbers, such as $\sqrt{2}$ or π, that cannot be written as a ratio of two integers are irrational numbers. Expressed decimally, irrational numbers are infinite nonrepeating decimals.
Subsets of the Rational Numbers:

Natural numbers: $1, 2, 3, 4, 5, 6, \ldots$
Whole numbers: $0, 1, 2, 3, 4, 5, 6, \ldots$
Integers: $\ldots, -3, -2, -1, 0, 1, 2, 3, \ldots$

Inequality and Interval Notation

INEQUALITY NOTATION	MEANING	INTERVAL NOTATION
$x > a$	x is greater than a.	(a, ∞)
$x \geq a$	x is greater than or equal to a.	$[a, \infty)$
$x < a$	x is less than a.	$(-\infty, a)$
$x \leq a$	x is less than or equal to a.	$(-\infty, a]$
$a < x < b$	x is greater than a and less than b.	(a, b)
$a < x \leq b$	x is greater than a and less than or equal to b.	$(a, b]$
$a \leq x < b$	x is greater than or equal to a and less than b.	$[a, b)$
$a \leq x \leq b$	x is greater than or equal to a and less than or equal to b.	$[a, b]$

Arithmetic and Order of Operations

In Exercises 1–6, calculate the value of each expression without using a calculator.

1. a. $15 + (-3)$ b. $15 - (-3)$ c. $15(-3)$ d. $15 \div (-3)$
2. a. $-0.6 + 0.02$ b. $-0.6 - 0.02$ c. $-0.6(0.02)$ d. $-0.6 \div 0.02$
3. a. $\frac{1}{6} + \frac{3}{4}$ b. $\frac{1}{6} - \frac{3}{4}$ c. $\frac{1}{6}\left(\frac{3}{4}\right)$ d. $\frac{1}{6} \div \frac{3}{4}$
4. a. $3 + 0$ b. $3 - 0$ c. $3(0)$ d. $3 \div 0$
5. a. $6 - 12 \div 3 + 4^2$ b. $(6 - 12) \div 3 + 4^2$
6. a. $\frac{2(4) \div 4(-2)}{8 \div 2(-4)}$ b. $\frac{5 - 2(3 - 7)}{(10 + 3)2}$

Properties and Subsets of the Real Numbers

7. Identify all the numbers from the set $\left\{-5, -3.28, -\sqrt{3}, -\frac{1}{3}, 0, \sqrt{9}, 11\right\}$ that are
   a. Rational numbers b. Irrational numbers
   c. Integers d. Natural numbers
8. Name the property that justifies each statement.
   a. $2(x + y) = 2x + 2y$
   b. $2(x + y) = 2(y + x)$
   c. $2(x + y) = (x + y)(2)$
   d. $2 + (x + y) = (2 + x) + y$
9. a. The additive identity is ______. b. The additive inverse of 5 is ______.
10. a. The multiplicative identity is ______. b. The multiplicative inverse of 5 is ______.

Interval Notation

11. Write the interval notation for each inequality.
    a. $x \leq 5$ b. $x > -2$
    c. $-3 \leq x < 0$ d. $3 < x \leq 9$

Strategy for Solving Linear Equations Algebraically

Step 1. Simplify each side of the equation.

a. If the equation contains fractions, simplify by multiplying both sides of the equation by the least common denominator (LCD) of all the fractions.

b. If the equation contains grouping symbols, simplify by using the distributive property to remove the grouping symbols and then combine like terms.

Step 2. Using the addition-subtraction principle of equality, isolate the variable terms on one side of the equation and the constant terms on the other side.

Step 3. Using the multiplication-division principle of equality, solve the equation produced in step 2.

Solving Equations

In Exercises 12 and 13, simplify each expression in part **a** and solve each equation in part **b.**

	Simplify	**Solve**
12.	**a.** $3(x - 2) - 5(x - 4)$	**b.** $3(x - 2) - 5(x - 4) = 0$
13.	**a.** $2(x + 5) - (x + 3)$	**b.** $2(x + 5) - (x + 3) = 0$

14. Check $x = 5$ to determine whether it is a solution of each equation.

a. $2x - 3 = x + 1$ **b.** $2(x - 4) = x - 3$

Solve each linear equation.

15. a. $5x + 6 = 2(x - 9)$ **b.** $5(x - 3) = 3(x + 1)$

16. a. $\frac{x}{4} - 1 = \frac{2x}{5}$ **b.** $2x - \frac{3x + 1}{2} = \frac{1}{3}$

17. Solve each equation for y.

a. $2x - 3y = 12$ **b.** $\frac{x}{3} - \frac{y}{5} = 1$

18. Use the following table to solve $0.712x - 0.328 = -0.671x + 5.204$.

Plot1 Plot2 Plot3
\Y1■0.712X−0.328
\Y2■−0.671X+5.204
\Y3
\Y4
\Y5

X	Y1	Y2
0	−.328	5.204
1	.384	4.533
2	1.096	3.862
3	1.808	3.191
4	2.52	2.52
5	3.232	1.849
6	3.944	1.178

X=0

19. Use the following graph to solve $\frac{2x}{3} + 3 = -x - 7$.

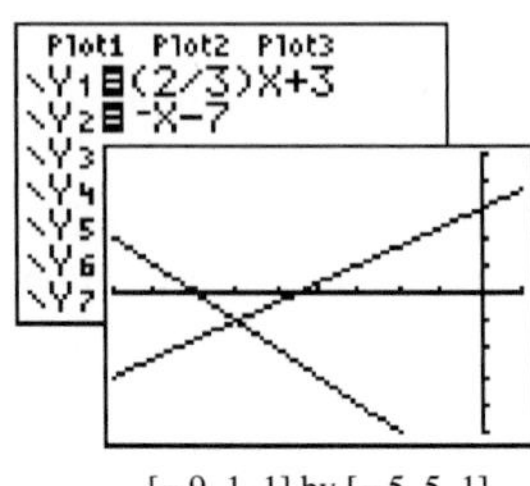

[−9, 1, 1] by [−5, 5, 1]

Intercepts of a Line

- The x-intercept, often denoted by $(a, 0)$, is the point where the line crosses the x-axis.
- The y-intercept, often denoted by $(0, b)$, is the point where the line crosses the y-axis.

Intercepts

20. Determine the x and y-intercepts of each line.

a.

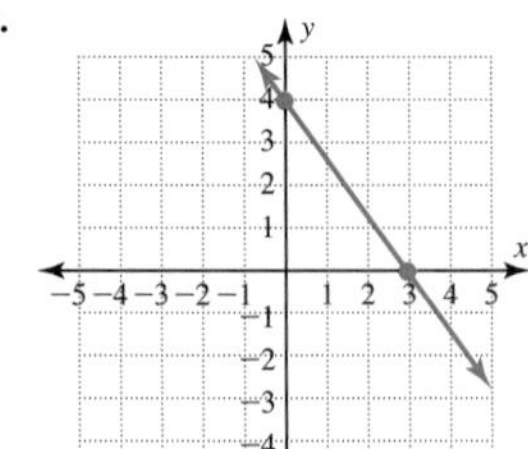

b. $3x - 2y = 18$

Proportions

- A proportion is an equation that states two ratios are equal.
- In the proportion $\frac{a}{b} = \frac{c}{d}$, the terms a and d are called the *extremes* and the terms b and c are called the *means*.

Direct Variation

If x and y are variables and k is a nonzero constant, then stating "y varies directly as x with constant of variation k" means $y = kx$.

Proportions and Direct Variation

21. The amount of concession sales at a new ballpark varies directly as the number of fans in attendance. Concession sales are \$1,000,000 at a game where 40,000 people attend.

a. Write an equation relating the concession sales at a game to attendance at the game.

b. What does the constant of variation mean?

c. Use this equation to complete the table of values.

ATTENDANCE	CONCESSION SALES (\$)
34,000	
38,000	
42,000	

22. Use similar triangles and proportional sides to find the value of x.

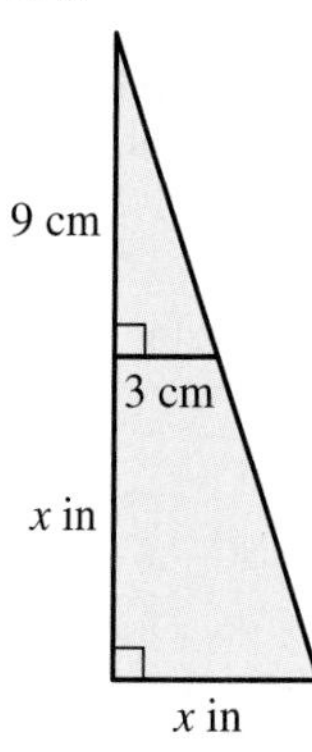

Sequence

- A sequence is an ordered set of numbers with a first term, a second term, a third term, etc.
- An arithmetic sequence is a sequence with a constant change from term to term.
- Subscript notation can be used to represent a sequence as $a_1, a_2, \ldots, a_n$.

23. Sequences Calculate the first five terms of each sequence, starting with $n = 1$.

a. $a_n = n - 2$ **b.** $a_n = -4n$

c. $a_n = 3n - 5$ **d.** $a_n = 2 - 5n$

24. Determine whether each sequence is an arithmetic sequence. If the sequence is arithmetic, write the common difference.

a. 0, 4, 8, 12, 16 **b.** 12, 9, 6, 3, 0

c. 1, 2, 4, 8, 16 **d.** 5, 5, 5, 5, 5

Function Notation

- The notation $f(x)$ is read as "f of x" or "the value of f at x."
- The letter f names the function.
- The variable x represents an input value from the domain.
- The notation $f(x)$ represents a unique output value corresponding to x.

Function Notation

25. Evaluate each expression, given $f(x) = 5x - 2$.

a. $f(0)$ **b.** $f(-3)$

c. $f(4)$ **d.** $f(10)$

26. Use the given graph to determine the missing input and output values.

a. $f(-4) =$ ______

b. $f(0) =$ ______

c. $f(x) = -1; x =$ ______

d. $f(x) = 0; x =$ ______

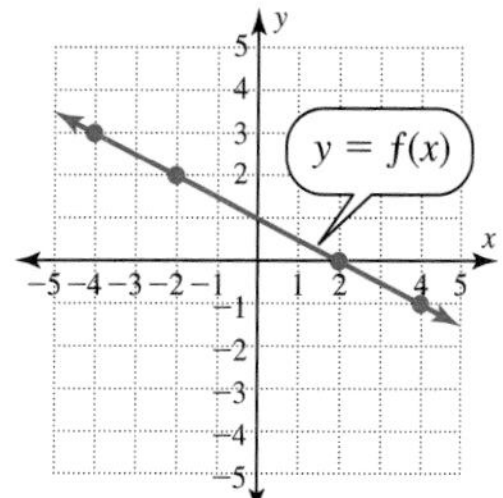

27. **Price Discount** A discount store had a holiday special with a 20% discount on the price of all sporting goods.
    a. Write a function for the amount of discount on an item with an original price of x dollars.
    b. Write a function for the new price of an item with an original price of x dollars.
    c. Complete the following table for the new price of each item whose original price is given.

x	$f(x)$
20	16
48	
75	
90	
110	

Solution of a Linear Equation

A solution of a linear equation of the form $y = mx + b$ is an ordered pair (x, y) that makes the equation a true statement.

Slope of a Line

The slope m of a line through (x_1, y_1) and (x_2, y_2) with

$$x_1 \neq x_2 \text{ is } m = \frac{\text{change in } y}{\text{change in } x} = \frac{y_2 - y_1}{x_2 - x_1}$$

$$m = \frac{\Delta y}{\Delta x} = \frac{\text{rise}}{\text{run}}.$$

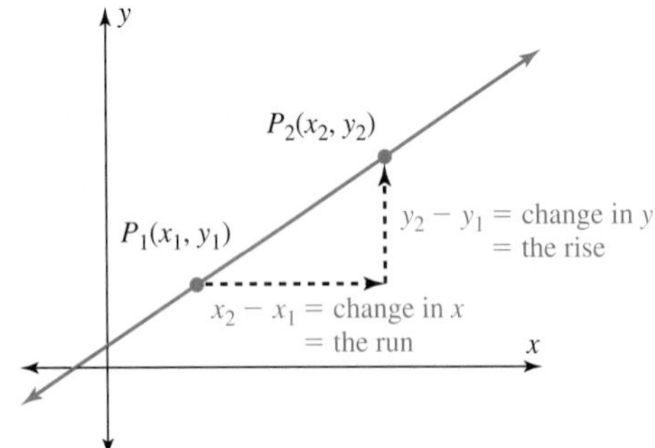

Forms of Linear Equations

A linear equation in x and y is first degree in both x and y.

- Slope-intercept form: $y = mx + b$ or $f(x) = mx + b$ with slope m and y-intercept $(0, b)$.
- Vertical line: $x = a$ for a real constant a.
- Horizontal line: $y = b$ for a real constant b.
- General form: $Ax + By = C$.
- Point-slope form: $y - y_1 = m(x - x_1)$ through (x_1, y_1) with slope m.

Graphs of Linear Equations

28. Determine whether the ordered pair$(1, -3)$ is a solution of each linear equation.
    a. $5x + 2y = -1$
    b. $y = -2x + 4$
    c. $x = 1$
    d. $y = 2$
29. Graph each of these lines.
    a. The line through $(3, 1)$and $(-2, -4)$
    b. The line with slope $m = \frac{2}{3}$ and with a y-intercept $(0, -1)$
    c. The line with slope $m = -\frac{3}{4}$ and through $(-1, 2)$
    d. The line defined by $2x - y = 8$
    e. The line defined by $y - 2 = \frac{1}{2}(x + 3)$
30. Match each equation with the best description of the graph of this linear equation.
    a. $y = -\frac{3}{2}x + 4$ A. A horizontal line
    b. $y = \frac{2}{3}x + 7$ B. A vertical line
    c. $y = 4$ C. A line parallel to $y = -\frac{3}{2}x + 1$
    d. $x = -1$ D. A line perpendicular to $y = -\frac{3}{2}x + 1$
31. Determine the slope of each line.
    a. The line through $(3, 7)$ and $(-2, 1)$
    b. $y = \frac{3}{4}x - 2$
    c. $y = 2$
    d. $x = 2$
32. Write the equation in slope-intercept form of a line with a slope $m = 3$ and a y-intercept $(0, -2)$.
33. Write the equation in slope-intercept form of a line with a slope $m = -2$ through $(3, 1)$.
34. Write the equation in slope-intercept form of a line through $(-1, 1)$ and $(2, 3)$.
35. Write the equation of a horizontal line through $(3, 4)$.
36. Write the equation of a vertical line through $(3, 4)$.
37. Write the equation in slope-intercept form of a line through $(0, 2)$ and parallel to $y = -\frac{2}{5}x + 7$.

Classifications of Systems of Linear Equations

- Exactly one solution: A consistent system of independent equations
- No solution: An inconsistent system of independent equations
- An infinite number of solutions: A consistent system of dependent equations

Strategy for Solving Word Problems

Step 1 Read the problem carefully to determine what you are being asked to find.

Step 2 Select a variable to represent each unknown quantity. Specify precisely what each variable represents and note any restrictions on each variable.

Step 3 If necessary, make a sketch and translate the problem into a word equation or a system of word equations. Then translate each word equation into an algebraic equation.

Step 4 Solve the equation or system of equations, and answer the question asked by the problem.

Step 5 Check the reasonableness of your answer.

Mixture Principle

$$\begin{matrix}\text{Amount} \\ \text{in first}\end{matrix} + \begin{matrix}\text{Amount} \\ \text{in second}\end{matrix} = \begin{matrix}\text{Amount} \\ \text{in mixture}\end{matrix}$$

Rate Principle

The amount obtained is equal to the rate times the base to which this rate is applied.

$$\text{Amount} = \text{Rate} \times \text{Base}$$
$$A = RB$$

Principles Used to Solve Inequalities*

- **Addition-Subtraction Principle:** If a, b, and c are real numbers, then $a < b$ is equivalent to $a + c < b + c$ and to $a - c < b - c$.
- **Multiplication-Division Principle:** If a, b, and c are real numbers and $c > 0$, then $a < b$ is equivalent to $ac < bc$. If $c < 0$, then $a < b$ is equivalent to $ac > bc$.

*Similar statements can be made for the inequalities $\leq$, $>$, and $\geq$.

Systems of Linear Equations

38. Select the choice that best describes each system of linear equations.

a. $y = 3x - 5$
$y = 2x + 1$

b. $y = 3x - 2$
$y = 3x + 2$

c. $y = 3x - 2$
$y = 3x - 2$

A. A consistent system of independent equations

B. A consistent system of dependent equations

C. An inconsistent system of independent equations

39. Solve this system of linear equations by the substitution method.

$$y = 3x + 5$$
$$3x + 2y = 28$$

40. Solve this system of linear equations by the addition method.

$$4x - 7y = 40$$
$$4x + 7y = -16$$

41. Solve this system of linear equations by the addition method.

$$-3x + 4y = 19$$
$$5x + 7y = -18$$

Applications of Linear Equations and Systems of Linear Equations

42. **Travel Time** Two boats leave a dock at the same time. One travels downstream at a rate of 15 mi/h, and the other travels upstream 6 mi/h slower than the first boat. Determine the number of hours it will take the boats to be 72 mi apart.

43. **Rates of Two Trains** Two trains depart simultaneously from a station, traveling in opposite directions. One averages 8 km/h less than the other. After $\frac{1}{2}$ hour they are 99 km apart. Determine the speed of each train.

44. **Amount of Acid in a Solution** A chemist needs to create a 20% acid solution. She has 5 L of a 25% acid solution in a 10-L container. How much water should she mix with this solution to decrease the concentration to the desired 20%?

45. **Amount of Acid in a Solution** A chemist needs 6 L of a 20% acid solution. He has available a 25% acid solution and a 10% acid solution. How much of each solution should he mix to create this 20% solution?

Linear Inequalities

46. Use the table below to solve each inequality.

a. $4x - 1 \geq 3x + 1$

b. $4x - 1 \leq 3x + 1$

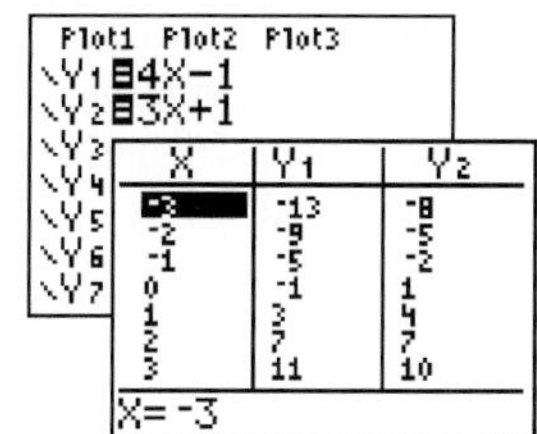

Types of Inequalities

- Conditional inequality: A conditional inequality contains a variable and is true for some, but not all, real values of the variable(s).
- Unconditional inequality: An inequality that is always true.
- Contradiction: An inequality that is always false.
- Equivalent inequalities: Inequalities that have the same solution are called equivalent inequalities.
- Linear inequality: An inequality that is the first degree in each variable.

Absolute Value Equations and Inequalities

For any real numbers x and a and positive real number d:

- $|x - a| = d$ is equivalent to $x - a = -d$ or $x - a = d$.
- $|x - a| < d$ is equivalent to $-d < x - a < d$.
- $|x - a| > d$ is equivalent to $x - a < -d$ or $x - a > d$.
- $|x - a| = -d$ is a contradiction and has no solution.
- $|x - a| < -d$ is a contradiction and has no solution.
- $|x - a| > -d$ is an unconditional inequality, and the solution set is the set of all real numbers.

Intersection and Union of Two Sets

- **Intersection:** $A \cap B$ is the set of points in both A *and* B.
- **Union:** $A \cup B$ is the set of points in either A *or* B (or both).

Graphing a Linear Inequality

Step 1. Graph the equality $Ax + By = C$, using
  a. A solid line for $\leq$ or $\geq$.
  b. A dashed line for $<$ or $>$.

Step 2. Choose an arbitrary test point not on the line; (0, 0) is often convenient. Substitute this test point into the inequality.

Step 3. **a.** If the test point satisfies the inequality, shade the half-plane containing this point.
  b. If the test point does not satisfy the inequality, shade the other half-plane.

Graphing a System of Linear Inequalities

Graph each inequality in the system on the same coordinate system. Use arrows to indicate each individual region formed by these lines. Then use shading to indicate the intersection of these regions. The solution of the system is represented by this intersection.

47. Use the graph below to solve each inequality.
  a. $2x + 1 \geq -x - 2$
  b. $2x + 1 \leq -x - 2$

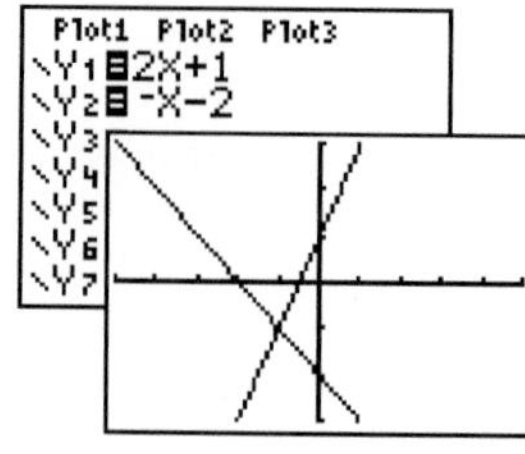

[−5, 5, 1] by [−3, 3, 1]

48. **a.** Solve $5(x - 3) \geq 2x + 9$.
  b. Solve $2(x + 3) < 4(x - 1) - 3(x + 2)$.

49. Identify each of these inequalities as a conditional inequality, an unconditional inequality, or a contradiction. Then solve each inequality.
  a. $x + x < 0$
  b. $x + x < 2x$
  c. $x + x \leq 2x$

50. Solve each equation and inequality.
  a. $|2x - 1| = 7$
  b. $|2x + 1| < 5$
  c. $|x - 3| \geq 5$

51. Solve $2 < 3x - 4 \leq 5$.

52. Solve $2x + 1 \leq -5$ or $3x - 2 \geq 4$.

53. Graph the solution of $3x - 4y \leq 12$.

54. Graph the solution of this system of inequalities.

$$x + 2y \geq 2$$
$$-x + 2y < 4$$

55. Graph the solution of this system of inequalities.

$$x \geq 0$$
$$y \geq 0$$
$$3x + y \leq 3$$

Zero and Negative Exponents

For any real base x and exponent n:

1. $x^0 = 1$ for $x \neq 0$.
2. 0^0 is undefined.
3. $x^{-n} = \dfrac{1}{x^n}$ for $x \neq 0$.
4. $\left(\dfrac{x}{y}\right)^{-n} = \left(\dfrac{y}{x}\right)^n$ for $x \neq 0$ and $y \neq 0$.

Summary of the Properties of Exponents

For any nonzero real numbers x and y and real exponents m and n:

Product rule: $x^m \cdot x^n = x^{m+n}$

Power rule: $(x^m)^n = x^{mn}$

Product to a power: $(xy)^m = x^m y^m$

Quotient to a power: $\left(\dfrac{x}{y}\right)^m = \dfrac{x^m}{y^m}$

Quotient rule: $\dfrac{x^m}{x^n} = x^{m-n}$

Scientific Notation

1. A number is in scientific notation when it is expressed as the product of a number between 1 and 10 (or between -1 and -10 if negative) and an appropriate power of 10.
2. On many calculators the **EE** feature is used to enter the power of 10.
3. If a number such as 6.89E5 appears on a calculator display, this represents 6.89×10^5 in scientific notation.

To Write a Number in Standard Decimal Notation

Multiply out the two factors by using the given power of 10.

1. If the exponent on 10 is positive, move the decimal point to the right.
2. If the exponent on 10 is zero, do not move the decimal point.
3. If the exponent on 10 is negative, move the decimal point to the left.

To Write a Number in Scientific Notation

1. Move the decimal point immediately to the right of the first nonzero digit of the number.
2. Multiply by a power of 10 determined by counting the number of places the decimal point has been moved.
   a. The exponent on 10 is zero or positive if the magnitude of the original number is 1 or greater.
   b. The exponent on 10 is negative if the magnitude of the original number is between 0 and 1.

Adding and Subtracting Polynomials

To add or subtract polynomials, combine like terms.

Product of Polynomials

1. To multiply a monomial times a polynomial, use the distributive property to multiply the monomial times each term of the polynomial.
2. To multiply two polynomials, use the distributive property to multiply each term of the first polynomial times each term of the second polynomial, and then combine like terms.

Using Properties of Exponents

In Exercises 56–64, simplify each expression.

56. a. $-5^0 + 3^0 + 4^0$ b. $(-5)^0 + 3^0 + 4^0$
c. $(-5 + 3 + 4)^0$ d. $-(5 + 3 + 4)^0$

57. a. $\left(\dfrac{1+1}{2+3}\right)^{-1}$ b. $\left(\dfrac{1}{2+3}\right)^{-1}$
c. $\left(\dfrac{1}{2} + \dfrac{1}{3}\right)^{-1}$ d. $\left(\dfrac{1}{2}\right)^{-1} + \left(\dfrac{1}{3}\right)^{-1}$

58. a. 5^0 b. 0^5
c. 6^{-1} d. $(-1)^6$

59. a. $\left(\dfrac{2}{3}\right)^0$ b. $\left(\dfrac{2}{3}\right)^{-2}$
c. $\left(\dfrac{2}{3}\right)^{-1}$ d. $\left(\dfrac{2}{3}\right)^2$

60. a. 10^2 b. 10^4
c. 10^0 d. 10^{-1}

61. a. $(3x + 7y)^0$ b. $(3x)^0 + (7y)^0$
c. $3(x + 7y)^0$ d. $3x^0 + 7y^0$

62. a. $x^3 + x^3$ b. $x^2 \cdot x^6$
c. $\dfrac{x^6}{x^2}$ d. $(x^2)^6$

63. a. $7x^2y - 5x^2y$ b. $(8x^2y^3)(4xy)$
c. $\dfrac{8x^2y^3}{4xy}$ d. $(8x^2y^3)^2$

64. a. $x^{-2}x^5$ b. $\left(\dfrac{x^{-3}}{x^{-4}}\right)^{-2}$
c. $(3x^{-3}y^4)^{-2}$ d. $\dfrac{-6x^{-6}y^2}{12x^{12}y^{-8}}$

Write each number in standard decimal notation.

65. a. 1.23×10^{-2} b. 1.23×10^4

Write each number in scientific notation.

66. a. 0.0000435 b. 1,870,000

Operations with Polynomials

In Exercises 67–70, simplify each expression.

67. a. $(3x + 7y) + (2x - 4y)$
b. $(3x + 7y) - (2x - 4y)$
c. $(3x + 7y)(2x - 4y)$

68. a. $(3x + 7y) - 2(x - 4y)$
b. $(4x^2 - 5x + 3) + (3x^2 + 4x - 9)$
c. $(4x^2 - 5x + 3) - (3x^2 + 4x - 9)$

69. a. $(x + 4)(x^2 - 2x - 3)$
b. $(4x + 7)(4x - 7)$
c. $(x - 3)(x + 2)(x - 4)$

70. a. $\dfrac{6x^3 + 4x^2}{2x}$ b. $\dfrac{x^2 - x - 12}{x + 3}$ c. $\dfrac{2x^4 - 4x^3 + 2x^2 - 8}{2x - 4}$

Question	Answer	Reference Example
1. a	12	[1.3–2]
1. b	18	[1.4–2]
1. c	−45	[1.5–2]
1. d	−5	[1.6–1]
2. a	−0.58	[1.3–3]
2. b	−0.62	[1.4–4]
2. c	−0.012	[1.5–2]
2. d	−30	[1.6–1]
3. a	$\frac{11}{12}$	[1.3–4]
3. b	$-\frac{7}{12}$	[1.4–4]
3. c	$\frac{1}{8}$	[1.5–2]
3. d	$\frac{2}{9}$	[1.6–1]
4. a	3	[1.2–2]
4. b	3	[1.4–3]
4. c	0	[1.5–2]
4. d	Undefined	[1.6–3]
5. a	18	[1.7–2]
5. b	14	[1.7–2]
6. a	$\frac{1}{4}$	[1.7–2]
6. b	$\frac{1}{2}$	[1.7–2]
7. a	$-5, -3.28, -\frac{1}{3}, 0, \sqrt{9}, 11$	[1.2–8]
7. b	$-\sqrt{3}$	[1.2–8]
7. c	$-5, 0, \sqrt{9}, 11$	[1.2–8]
7. d	$\sqrt{9}, 11$	[1.2–8]
8. a	Distributive property of multiplication over addition	[1.7–5]
8. b	Commutative property of addition	[1.3–6]
8. c	Commutative property of multiplication	[1.5–1]
8. d	Associative property of addition	[1.3–6]
9. a	0	[1.2–2]
9. b	−5	[1.2–1]
10. a	1	[1.5–5]
10. b	$\frac{1}{5}$	[1.5–5]
11. a	$(-\infty, 5]$	[1.2–5]
11. b	$(-2, \infty)$	[1.2–5]
11. c	$[-3, 0)$	[1.2–5]
11. d	$(3, 9]$	[1.2–5]
12. a	$-2x + 14$	[1.7–7]
12. b	$x = 7$	[2.5–6]
13. a	$x + 7$	[1.7–7]
13. b	$x = -7$	[2.5–6]
14. a	No	[1.4–8]
14. b	Yes	[1.4–8]
15. a	$x = -8$	[2.5–6]
15. b	$x = 9$	[2.5–6]
16. a	$x = -\frac{20}{3}$	[2.5–8]
16. b	$x = \frac{5}{3}$	[2.5–8]

Question	Answer	Reference Example
17. a	$y = \frac{2}{3}x - 4$	[2.6–1]
17. b	$y = \frac{5}{3}x - 5$	[2.6–2]
18.	$x = 4$	[2.4–8]
19.	$x = -6$	[2.4–8]
20. a	x-intercept: (3, 0) y-intercept: (0, 4)	[2.3–4]
20. b	x-intercept: (6, 0) y-intercept: (0, −9)	[2.6–3]
21. a	$y = 25x$	[2.7–5]
21. b	Concession sales average $25 per person.	[2.7–5]
21. c	(see table below)	[2.7–5]

Attendance	Concession Sales
34,000	850,000
38,000	950,000
42,000	1,050,000

Question	Answer	Reference Example
22.	$x = \frac{9}{2}$	[2.7–7]
23. a	−1, 0, 1, 2, 3	[2.1–7]
23. b	−4, −8, −12, −16, −20	[2.1–7]
23. c	−2, 1, 4, 7, 10	[2.1–7]
23. d	−3, −8, −13, −18, −23	[2.1–7]
24. a	Yes, $d = 4$	[2.1–6]
24. b	Yes, $d = -3$	[2.1–6]
24. c	No	[2.1–6]
24. d	Yes, $d = 0$	[2.1–6]
25. a	$f(0) = -2$	[2.2–1]
25. b	$f(-3) = -17$	[2.2–1]
25. c	$f(4) = 18$	[2.2–1]
25. d	$f(10) = 48$	[2.2–1]
26. a	$f(-4) = 3$	[2.2–4]
26. b	$f(0) = 1$	[2.2–4]
26. c	$x = 4$	[2.2–4]
26. d	$x = 2$	[2.2–4]
27. a	$f(x) = 0.20x$	[2.2–6]
27. b	$f(x) = 0.80x$	[2.2–6]
27. c	(see table below)	[2.2–6]

x	$f(x)$
20	16
48	38.4
75	60
90	72
110	88

Question	Answer	Reference Example
28. a	Yes	[2.3–1]
28. b	No	[2.3–1]
28. c	Yes	[2.3–1]
28. d	No	[2.3–1]
29. a	(graph)	[3.1–4]

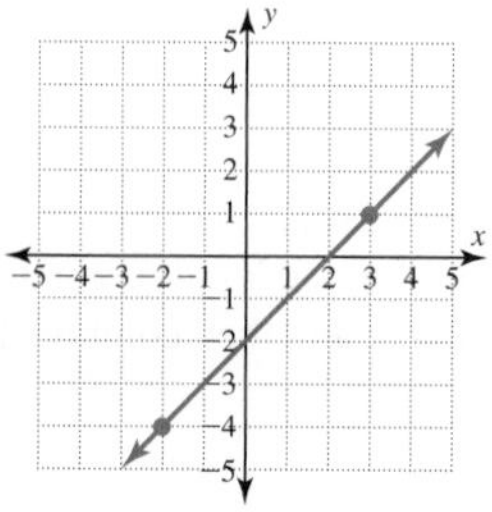

Question	Answer	Reference Example
29. b	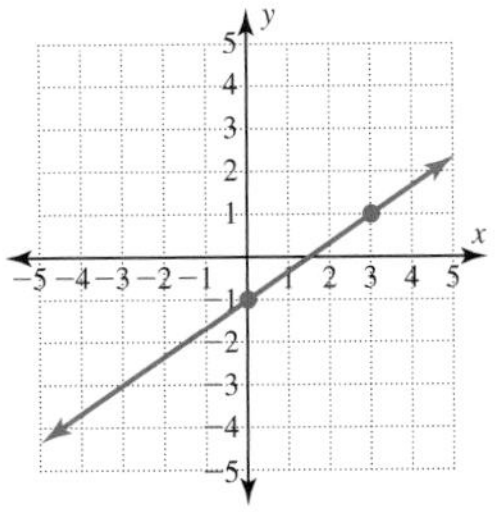	[3.1–4]
29. c	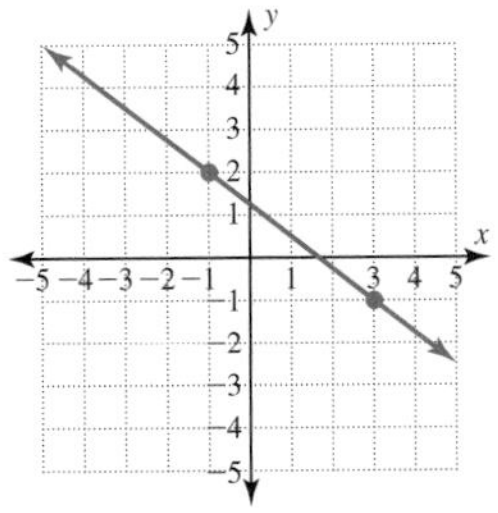	[3.1–4]
29. d	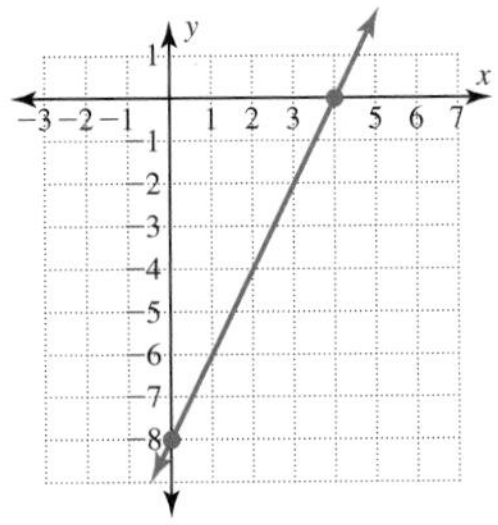	[3.1–6]
29. e	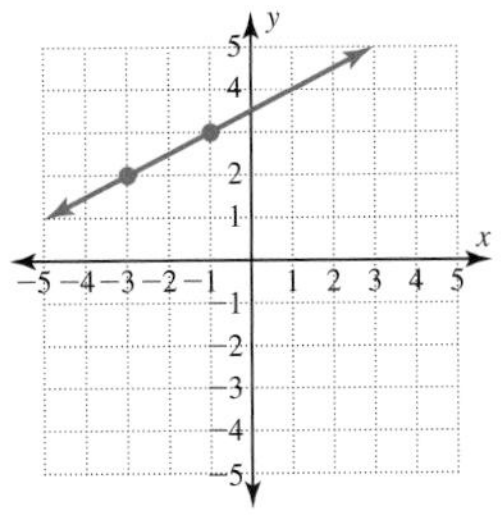	[3.2–8]
30. a	C	[3.2–9]
30. b	D	[3.2–9]
30. c	A	[3.2–6]
30. d	B	[3.2–6]
31. a	$m = \frac{6}{5}$	[3.1–1]
31. b	$m = \frac{3}{4}$	[3.2–2]
31. c	$m = 0$	[3.2–6]
31. d	Undefined	[3.2–6]
32.	$y = 3x - 2$	[3.2–1]
33.	$y = -2x + 7$	[3.2–8]
34.	$y = \frac{2}{3}x + \frac{5}{3}$	[3.2–10]
35.	$y = 4$	[3.2–6]
36.	$x = 3$	[3.2–6]
37.	$y = -\frac{2}{5}x + 2$	[3.2–1]
38. a	A	[3.3–6]
38. b	C	[3.3–6]
38. c	B	[3.3–7]

Question	Answer	Reference Example
39.	(2,11)	[3.4–1]
40.	(3,−4)	[3.5–1]
41.	(−5,1)	[3.5–2]
42.	After 3 hours the boats will be 72 mi apart.	[2.8–4]
43.	The speed of the faster train is 103 km/h, and the speed of the slower train is 95 km/h.	[3.6–5]
44.	Use 1.25 L of water.	[2.8–5]
45.	Mix 4 L of 25% solution and 2 L of 10% solution.	[3.6–6]
46. a	$[2, \infty)$	[4.1–4]
46. b	$(-\infty, 2]$	[4.1–4]
47. a	$[-1, \infty)$	[4.1–4]
47. b	$(-\infty, -1]$	[4.1–4]
48. a	$[8, \infty)$	[4.2–3]
48. b	$(-\infty, -16)$	[4.2–5]
49. a	Conditional inequality; $(-\infty, 0)$	[4.3–1]
49. b	Contradiction; no solution	[4.3–1]
49. c	Unconditional inequality; $\mathbb{R}$	[4.3–1]
50. a	$x = -3, x = 4$	[4.4–3]
50. b	$(-3, 2)$	[4.4–4]
50. c	$(-\infty, -2] \cup [8, \infty)$	[4.4–5]
51.	$(2, 3]$	[4.3–4]
52.	$(-\infty, -3] \cup [2, \infty)$	[4.3–8]
53.	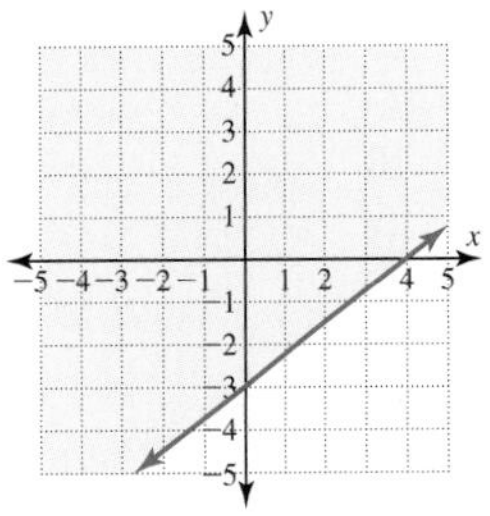	[4.5–3]
54.	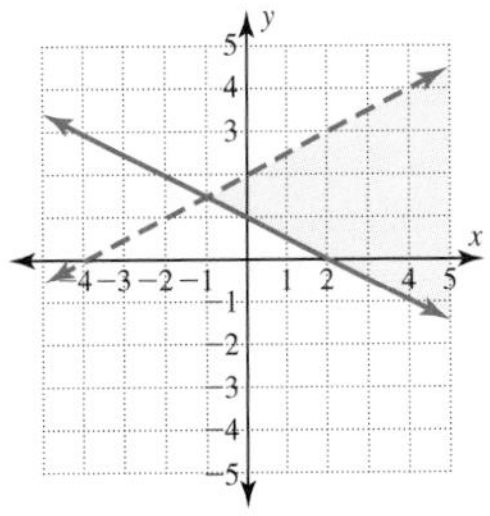	[4.5–7]
55.	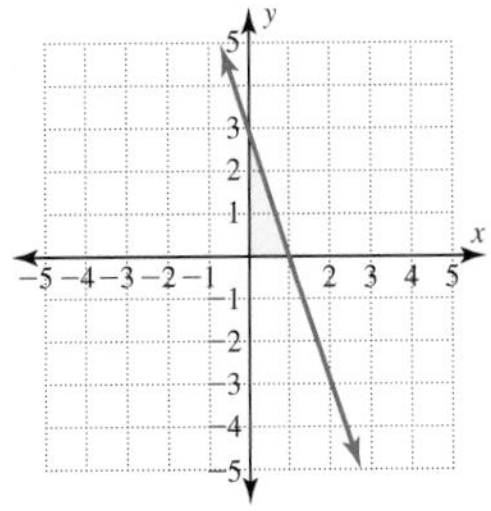	[4.5–9]
56. a	1	[5.2–3]
56. b	3	[5.2–3]
56. c	1	[5.2–3]
56. d	−1	[5.2–3]

continued

Question	Answer	Reference Example
57. a	$\frac{5}{2}$	[5.3–2]
57. b	5	[5.3–2]
57. c	$\frac{6}{5}$	[5.3–2]
57. d	5	[5.3–2]
58. a	1	[5.2–3]
58. b	0	[1.5–7]
58. c	$\frac{1}{6}$	[5.3–1]
58. d	1	[1.5–7]
59. a	1	[5.2–3]
59. b	$\frac{9}{4}$	[5.3–2]
59. c	$\frac{3}{2}$	[5.3–2]
59. d	$\frac{4}{9}$	[1.5–7]
60. a	100	[1.5–7]
60. b	10,000	[1.5–7]
60. c	1	[5.2–3]
60. d	$\frac{1}{10}$	[5.3–1]
61. a	1	[5.2–3]
61. b	2	[5.2–3]
61. c	3	[5.2–3]
61. d	10	[5.2–3]
62. a	$2x^3$	[5.1–8]
62. b	x^8	[5.1–8]
62. c	x^4	[5.2–1]
62. d	x^{12}	[5.1–4]
63. a	$2x^2y$	[5.4–7]
63. b	$32x^3y^4$	[5.1–3]
63. c	$2xy^2$	[5.2–4]
63. d	$64x^4y^6$	[5.2–4]
64. a	x^3	[5.3–4]
64. b	$\frac{1}{x^2}$	[5.3–4]
64. c	$\frac{x^6}{9y^8}$	[5.3–4]
64. d	$-\frac{y^{10}}{2x^{18}}$	[5.3–4]
65. a	0.0123	[5.3–5]
65. b	12,300	[5.3–5]
66. a	4.35×10^{-5}	[5.3–6]
66. b	1.87×10^6	[5.3–6]
67. a	$5x + 3y$	[5.4–7]
67. b	$x + 11y$	[5.4–8]
67. c	$6x^2 + 2xy - 28y^2$	[5.5–3]
68. a	$x + 15y$	[5.4–8]
68. b	$7x^2 - x - 6$	[5.4–7]
68. c	$x^2 - 9x + 12$	[5.4–8]
69. a	$x^3 + 2x^2 - 11x - 12$	[5.5–3]
69. b	$16x^2 - 49$	[5.5–3]
69. c	$x^3 - 5x^2 - 2x + 24$	[5.5–6]
70. a	$3x^2 + 2x$	[5.7–2]
70. b	$x - 4$	[5.7–4]
70. c	$x^3 + x + 2$	[5.7–7]

chapter 6

Factoring Polynomials

CHAPTER OUTLINE

Polynomials are used to describe many applications, especially those involving geometric shapes. The volume of dirt that must be moved to create the base of a new roadway is very important to contractors who bid on a construction project. The cross section of this roadway can be approximated by a trapezoid. Using the desired dimensions of the roadway and design limitations which include safety considerations, engineers can form each factor in the formula for the volume of dirt that must be moved.

REALITY CHECK

Why do you think engineers use a trapezoidal cross section for a roadway instead of a rectangular cross section?

Section 6.1 An Introduction to Factoring Polynomials

Objectives:

1. Factor out the GCF of a polynomial.
2. Factor by grouping.
3. Use the zeros of a polynomial $P(x)$ and the x-intercepts of the graph of $y = P(x)$ to factor the polynomial.

A Mathematical Note

The term *factor* comes from the Latin root *facere* meaning "to do." Things that are necessary in order to do something are known as its factors. Factors of polynomials are polynomials that are put together to make a product. Similarly we use the word *factory* as a place where a business puts things together to make a final product.

The ability to factor polynomials allows us to examine these polynomials piece by piece. Complicated problems can be broken down into simpler components. Developing the skills to factor some polynomials algebraically will deepen your mathematical ability. Although some calculators can now perform most of the factorizations shown in this chapter, the ability to perform mental factorizations will give you greater insight and will speed your work in other mathematical topics. An important part of this chapter is the building of the connections that factoring has to other topics.

The purpose of this section is to examine the factoring of polynomials from algebraic, numerical, and graphical viewpoints. We use the distributive property to factor some polynomials by grouping. This work lays the foundation for factoring trinomials in Sections 6.2 and 6.3.

We start by factoring the **greatest common factor (GCF)** out of each term of a polynomial. Loosely speaking, the GCF of a polynomial is the "largest" factor common to each term. To be more precise, the **GCF of a polynomial** is the common factor that contains both of the following:

1. Largest possible numerical coefficient
2. Largest possible exponent on each variable factor

The best way to start factoring most polynomials is to determine the GCF and then factor out this GCF. For many problems you will be able to determine the GCF by inspection. Example 1 shows the GCF as a factor of each term and then uses the distributive property to factor the polynomial. The distributive property $a(b + c) = ab + ac$ not only justifies multiplying $a(b + c)$ to obtain $ab + ac$, but also justifies factoring $ab + ac$ as $a(b + c)$.

EXAMPLE 1 Factoring Out the Greatest Common Factor

Factor the GCF out of each polynomial.

SOLUTION

(a) $20x^2y - 30xy^2$

$$20x^2y - 30xy^2 = 10xy(2x) + 10xy(-3y) = 10xy(2x - 3y)$$

The GCF is $10xy$. Use the distributive property $ab + ac = a(b + c)$ to factor the GCF out of this binomial.

(b) $12a^3b - 18a^2b^2 + 30ab^3$

$$12a^3b - 18a^2b^2 + 30ab^3 = 6ab(2a^2) + 6ab(-3ab) + 6ab(5b^2) = 6ab(2a^2 - 3ab + 5b^2)$$

The GCF is $6ab$. Use the distributive property to factor this GCF out of this trinomial.

(c) $x(x - 3) + 2(x - 3)$

$$x(x - 3) + 2(x - 3) = (x + 2)(x - 3)$$

The GCF is $x - 3$. Use the distributive property to factor the GCF out of this binomial. This expression can be thought of as $xz + 2z = (x + 2)z$, where $z = (x - 3)$.

SELF-CHECK 6.1.1

Factor the GCF out of each polynomial.

1. $10x^3 - 15x^2$
2. $6x^3 - 9x^2 + 12x$
3. $x(x + 5) - 7(x + 5)$

The GCF in Example 1(*c*) is the binomial factor $x - 3$. Some polynomials with four or more terms can be factored by strategically grouping the terms to produce a common binomial factor. This method is called **factoring by grouping.** To factor by grouping, carefully use parentheses to group terms, making sure the grouping symbols *do not* change the order of operations.

To illustrate why care must be taken with grouping symbols, note that $10 - 3 + 5$ and $10 - (3 + 5)$ do not have the same value.

$$10 - 3 + 5 = 7 + 5 \qquad 10 - (3 + 5) = 10 - 8$$
$$= 12 \qquad\qquad\qquad = 2$$

Also note that

$$x^3 + 4x^2 - 5x - 20 \neq (x^3 + 4x^2) - (5x - 20)$$

Instead,

$$x^3 + 4x^2 - 5x - 20 = (x^3 + 4x^2) - (5x + 20)$$

To factor a polynomial by grouping pairs of terms, we group terms that we already know how to factor with the hope of producing a common factor of each group. This is described in the following box. Section 6.5 examines other grouping possibilities.

Factoring a Four-Term Polynomial by Grouping Pairs of Terms

1. Be sure you have factored out the GCF if it is not 1.
2. Use grouping symbols to pair the terms so that each pair has a common factor other than 1.
3. Factor the GCF out of each pair of terms.
4. If there is a common binomial factor of these two groups, factor out this GCF. If there is no common binomial factor, try to use step 2 again with a different pairing of terms. If all possible pairs fail to produce a common binomial factor, the polynomial will not factor by this method.

EXAMPLE 2 Factoring by Grouping

Factor each polynomial by using the grouping method.

SOLUTION

(a) $ax + bx + 3a + 3b$

$$\begin{aligned} ax + bx + 3a + 3b &= (ax + bx) + (3a + 3b) \\ &= x(a + b) + 3(a + b) \\ &= (x + 3)(a + b) \end{aligned}$$

Note the original expression has four terms. Group the first two terms, and group the second two terms. Factor an x out of the first pair of terms and a 3 out of the second pair of terms. Then factor the common factor $(a + b)$ out of each group.

(b) $x^3 + 3x^2 - 7x - 21$

$$\begin{aligned} x^3 + 3x^2 - 7x - 21 &= (x^3 + 3x^2) - (7x + 21) \\ &= x^2(x + 3) - 7(x + 3) \\ &= (x^2 - 7)(x + 3) \end{aligned}$$

Group the first two terms. Note that grouping the second pair of terms requires factoring out a -1 from the third and fourth terms. Factor an x^2 out of the first group and a 7 out of the second group. Then factor the common factor $(x + 3)$ out of each group.

(c) $x^2 + 4x - 6x - 24$

$$\begin{aligned} x^2 + 4x - 6x - 24 &= (x^2 + 4x) - (6x + 24) \\ &= x(x + 4) - 6(x + 4) \\ &= (x - 6)(x + 4) \end{aligned}$$

Group the first two terms. Note that grouping the second pair of terms requires factoring out a -1 from the third and fourth terms. Factor an x out of the first group and a 6 out of the second group. Once you gain confidence with this method, you will factor the -1 and the 6 out of the third and fourth terms in one step. Then factor the common factor $(x + 4)$ out of each group.

(d) $5x^2 - 20x + 2x - 8$

$$\begin{aligned} 5x^2 - 20x + 2x - 8 &= (5x^2 - 20x) + (2x - 8) \\ &= 5x(x - 4) + 2(x - 4) \\ &= (5x + 2)(x - 4) \end{aligned}$$

Group the first two terms, and group the second two terms. Factor a $5x$ out of the first two terms and a 2 out of the second pair of terms. Then factor the common factor $(x - 4)$ out of each group. ■

SELF-CHECK 6.1.2

Factor each polynomial by using the grouping method.

1. $ax - ay + bx - by$ 2. $x^2 + 3x - 7x - 21$

It is important to read directions before "working" a math problem. You could be given the expression $x(x - 3) + 2(x - 3)$ and asked to do two different tasks: to expand or to factor. Both of these options are displayed here.

EXPAND	FACTOR
$x(x - 3) + 2(x - 3) = x^2 - 3x + 2x - 6$ $= x^2 - x - 6$	$x(x - 3) + 2(x - 3) = (x + 2)(x - 3)$

Note that all three forms shown above — $x^2 - x - 6$, $x(x - 3) + 2(x - 3)$, and $(x + 2)(x - 3)$ — are different representations of the same polynomial. Equal polynomials have identical graphs and identical tables of values. Example 3 uses graphs and tables to compare $x^2 - x - 6$ to its factored form.

EXAMPLE 3 Using Graphs and Tables to Compare Two Polynomials

Use a graph and a table to compare $Y_1 = x^2 - x - 6$ to $Y_2 = (x + 2)(x - 3)$.

SOLUTION

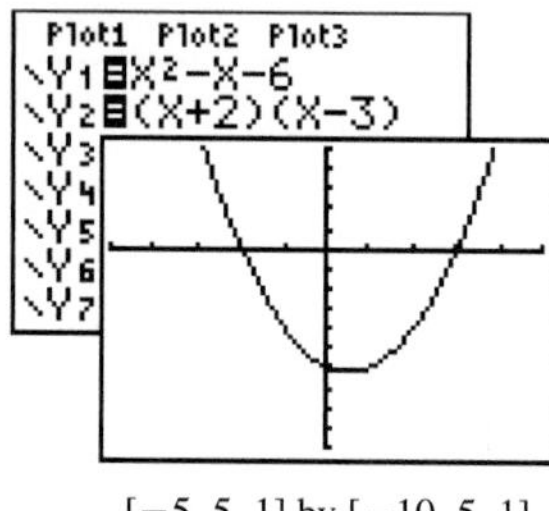

[−5, 5, 1] by [−10, 5, 1]

The graphs of Y_1 and Y_2 are identical. We will examine these graphs in detail in Section 7.2.

X	Y1	Y2
-2	0	0
-1	-4	-4
0	-6	-6
1	-6	-6
2	-4	-4
3	0	0
4	6	6

X=-2

The table of values confirms our observation because both Y_1 and Y_2 are identical for each of the x-values shown.

Answer: $x^2 - x - 6 = (x + 2)(x - 3)$ ■

An examination of the table in Example 3 reveals that both Y_1 and Y_2 are 0 for $x = -2$ and $x = 3$. If $P(x)$ represents a polynomial, we call an x-value that produces $P(x) = 0$ a **zero of the polynomial.** For the second-degree polynomial $x^2 - x - 6$ there are two zeros. Both -2 and 3 are zeros of this polynomial because $P(-2) = 0$ and $P(3) = 0$. Also note that the graph in Example 3 has x-intercepts of $(-2, 0)$ and $(3, 0)$.

Each of these zeros and intercepts corresponds to a factor of $x^2 - x - 6$. A summary of this information for the polynomial $P(x) = x^2 - x - 6$ is given here.

FACTORS OF $P(x) = x^2 - x - 6$	ZEROS OF $P(x) = x^2 - x - 6$	INTERCEPTS OF THE GRAPH OF $Y_1 = x^2 - x - 6$
$x + 2$ is a factor of $x^2 - x - 6$	-2 is a zero of $P(x) = x^2 - x - 6$	$(-2, 0)$ is an x-intercept of the graph of $Y_1 = x^2 - x - 6$
$x - 3$ is a factor of $x^2 - x - 6$	3 is a zero of $P(x) = x^2 - x - 6$	$(3, 0)$ is an x-intercept of the graph of $Y_1 = x^2 - x - 6$

A generalization of this relationship is given in the following table. The fact that these statements are equivalent means that any time we know any one of these three pieces of information, we also know the other two pieces of information. A polynomial that has real numbers as the coefficients for each term is a real polynomial. Each real polynomial of degree n has exactly n zeros. What can vary from polynomial to polynomial is how many of these zeros are real numbers. The zeros that are not real numbers will be imaginary numbers. In Chapter 7 we examine complex numbers and imaginary numbers.

Equivalent Statements About Linear Factors of a Polynomial

For a real constant c and a real polynomial $P(x)$, the following statements are equivalent.

ALGEBRAICALLY	NUMERICALLY	GRAPHICALLY
$x - c$ is a factor of $P(x)$	$P(c) = 0$, that is, c is a zero of $P(x)$	$(c, 0)$ is an x-intercept of the graph of $y = P(x)$

Note that if $x - c$ is evaluated for $x = c$, the result is $c - c = 0$. This means that when $x - c$ is a factor of $P(x)$, than c must be a zero of $P(x)$. In Example 4 we use the zeros of a polynomial to factor this polynomial. Note that this second-degree polynomial has two zeros.

EXAMPLE 4 Using Zeros to Factor a Polynomial

Use a graphing calculator to find the zeros of $x^2 - 7x + 12$. Then use these zeros to factor $x^2 - 7x + 12$.

SOLUTION

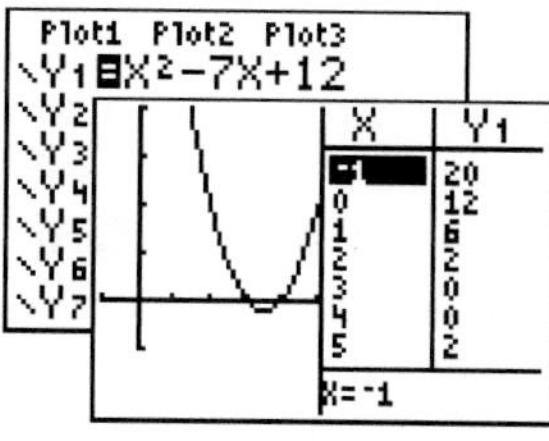

$[-1, 5, 1]$ by $[-1, 5, 1]$

Enter $x^2 - 7x + 12$ as Y_1 and use the **Graph-Table** feature (see Calculator Perspective 2.2.2) to create a graph and a table of values for Y_1.

By inspection it appears that the graph has x-intercepts of $(3, 0)$ and $(4, 0)$. This is confirmed by the table which shows zeros for $x = 3$ and $x = 4$.

The zeros of $x^2 - 7x + 12$ are $x = 3$ and $x = 4$.
Thus the factors of $x^2 - 7x + 12$ are $x - 3$ and $x - 4$.

We could also use the x-intercepts to factor $x^2 - 7x + 12$.

Answer: $x^2 - 7x + 12 = (x - 3)(x - 4)$

You can check this factorization by multiplying $x - 3$ by $x - 4$.

As a follow-up to Example 4, evaluate the polynomial expression $x^2 - 7x + 12$ for $x = 3$. We can do this by substituting 3 for x either into the original expression $x^2 - 7x + 12$ or into the factored form $(x - 3)(x - 4)$.

USING THE FORM $x^2 - 7x - 12$	USING THE FORM $(x - 3)(x - 4)$
$(3)^2 - 7(3) + 12 = 9 - 21 + 12$ $= 0$	$(3 - 3)(3 - 4) = (0)(-1)$ $= 0$

Note that we could evaluate the factored form by inspection. The factored form makes it much easier to recognize the zeros of a polynomial by inspection.

Example 5 presents a third-degree polynomial with three real zeros. These zeros are used to factor this polynomial.

EXAMPLE 5 Using Zeros to Factor a Polynomial

Use a graphing calculator to find the zeros of $x^3 - 5x^2 - x + 5$. Then factor this polynomial.

SOLUTION

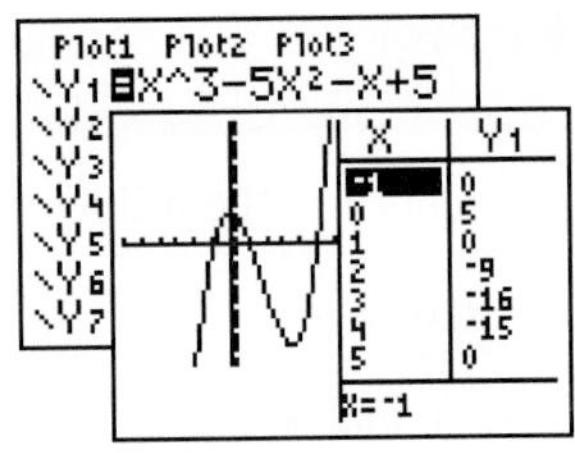

$[-6, 6, 1]$ by $[-20, 20, 1]$

Enter $x^3 - 5x^2 - x + 5$ as Y_1, and create a graph. The x-intercepts appear to occur at $(-1, 0)$, $(1, 0)$, and $(5, 0)$. This suggests the x-values to examine in the table are -1, 1, and 5.

The table confirms that the polynomial is 0 at $x = -1$, $x = 1$, and $x = 5$.

The zeros of $x^3 - 5x^2 - x + 5$ are -1, 1, and 5. Thus the factors are $x + 1$, $x - 1$, and $x - 5$.

The zeros are -1, 1, and 5. The corresponding factors are $x + 1$, $x - 1$, and $x - 5$.

Answer: $x^3 - 5x^2 - x + 5 = (x + 1)(x - 1)(x - 5)$

You can check this factorization by multiplying these three factors. ■

SELF-CHECK 6.1.3

1. Use a graphing calculator to determine the zeros of $x^2 + x - 2$.
2. Use a graphing calculator to determine the x-intercepts of the graph of $Y_1 = x^2 + x - 2$.
3. Use this information to factor the second-degree trinomial $x^2 + x - 2$.
4. Use a graphing calculator to determine the zeros of $x^3 + 3x^2 - 4x$.
5. Then factor this polynomial.

In Examples 4 and 5 we used a graph and a table to determine the zeros of a polynomial. Determining the x-intercepts by inspection is very restrictive because the x-intercepts may not always occur at nice integer values. Using the **Zero** feature on a graphing calculator, we can have a calculator assist us in determining the zeros and the x-intercepts. In Calculator Perspective 6.1.1 we use this feature to determine the zeros of $x^2 - 4x - 32$. Before you read this calculator perspective, enter $Y_1 = x^2 - 4x - 32$ and set the viewing window as $[-10, 10, 1]$ by $[-40, 20, 5]$.

CALCULATOR PERSPECTIVE 6.1.1 Calculating the Zeros of a Polynomial

To calculate one of the zeros of $Y_1 = x^2 - 4x - 32$ on a TI-84 Plus calculator, enter the following keystrokes:

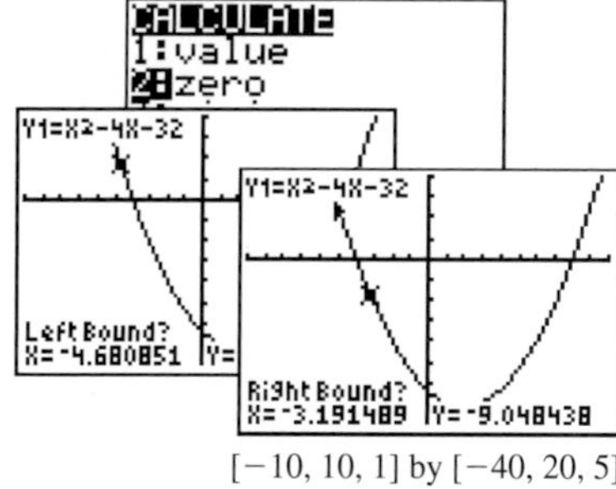

When the screen requests a left bound, use the arrow key ◀ or ▶ to move the cursor until it is located slightly to the left side of the zero. Then press ENTER.

When the screen requests a right bound, use the right arrow key to move the cursor until it is located slightly on the right side of the zero. Then press ENTER.

[−10, 10, 1] by [−40, 20, 5]

When the screen requests a guess, use the left arrow key to move the cursor until it is close to the zero. Then press ENTER.

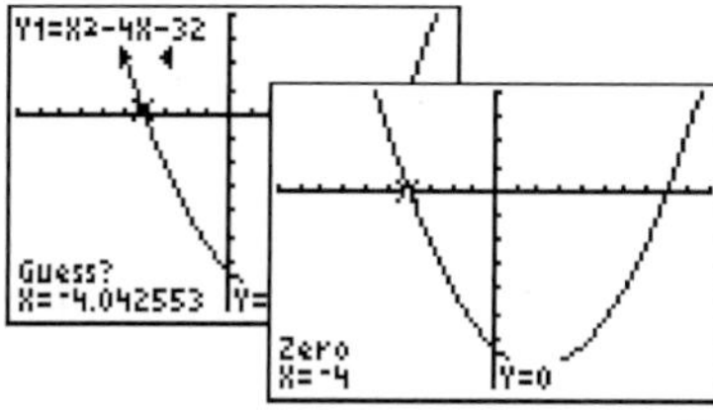

Thus −4 is a zero of $x^2 - 4x - 32$.

Note: We could use the same procedure to determine that the other zero is 7.

In Example 6 we use a graphing calculator to assist us in factoring $x^2 + 4x - 221$. Rather than go through tables of values by trial and error to find the exact zeros of $x^2 + 4x - 221$, we use Calculator Perspective 6.1.1 to calculate the zeros.

EXAMPLE 6 Using a Calculator to Calculate Zeros and to Factor a Polynomial

Use a graphing calculator to graph $Y_1 = x^2 + 4x - 221$ and to determine the zeros of $x^2 + 4x - 221$ and the x-intercepts of this graph. Then factor $x^2 + 4x - 221$.

SOLUTION

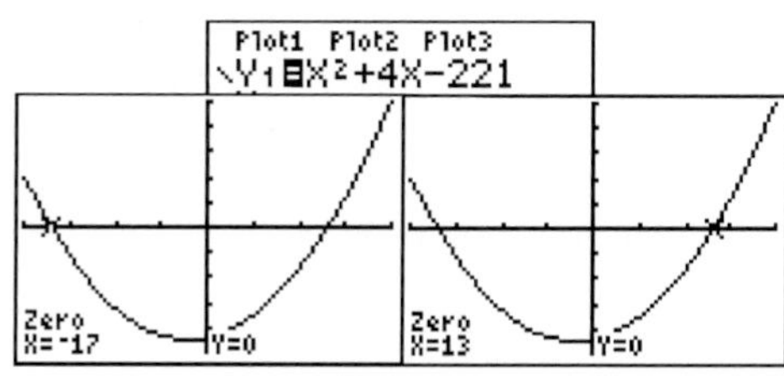

[−20, 20, 5] by [−250, 250, 50]

Enter $x^2 + 4x - 221$ as Y_1 and create a graph of $Y_1 = x^2 + 4x - 221$.

Use a graphing calculator to calculate the zeros and thus the x-intercepts.

The zeros are −17 and 13.
The x-intercepts of $Y_1 = x^2 + 4x - 221$ are $(-17, 0)$ and $(13, 0)$. Thus the factors of $x^2 + 4x - 221$ are $x + 17$ and $x - 13$.

Answer: $x^2 + 4x - 221 = (x + 17)(x - 13)$

You can check this factorization by multiplying $x + 17$ by $x - 13$.

Example 7 examines a fourth-degree polynomial whose graph has four real x-intercepts. The polynomial has four corresponding linear factors.

EXAMPLE 7 Using x-Intercepts to Factor a Polynomial

Use a graphing calculator to graph $y = x^4 - 20x^2 + 64$ and to determine the x-intercepts of this graph.

Then factor $x^4 - 20x^2 + 64$.

SOLUTION

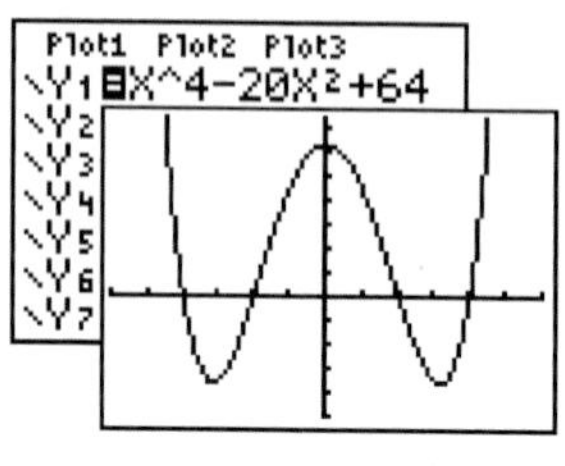

[−6, 6, 1] by [−50, 75, 10]

Enter $Y_1 = x^4 - 20x^2 + 64$ as Y_1 and create a graph of $Y_1 = x^4 - 20x^2 + 64$.

Use a graphing calculator to calculate the zeros, which are $x = -4$, $x = -2$, $x = 2$, and $x = 4$. These zeros correspond to the x-intercepts of the graph.

The x-intercepts of $y = x^4 - 20x^2 + 64$ are $(-4, 0)$, $(-2, 0)$, $(2, 0)$, and $(4, 0)$.

Answer: $x^4 - 20x^2 + 64 = (x + 4)(x + 2)(x - 2)(x - 4)$

You can check this factorization by multiplying these factors.

SELF-CHECK 6.1.4

1. Use a graphing calculator to graph $Y_1 = x^2 - 2x - 288$ and to determine the x-intercepts of this graph.
2. Factor $x^2 - 2x - 288$.
3. Use a graphing calculator to graph $Y_1 = x^3 - 6x^2 - x + 30$ and to determine the x-intercepts of this graph.
4. Factor $x^3 - 6x^2 - x + 30$.

Observe the pattern illustrated by the factorizations given below. We can use this pattern to factor polynomials that contain more than one variable.

FACTORIZATIONS OF POLYNOMIALS IN x	FACTORIZATIONS OF POLYNOMIALS IN x AND y
$x^2 - 4 = (x + 2)(x - 2)$	$x^2 - 4y^2 = (x + 2y)(x - 2y)$
$x^2 - 3x - 18 = (x + 3)(x - 6)$	$x^2 - 3xy - 18y^2 = (x + 3y)(x - 6y)$
$x^2 - 7x + 12 = (x - 3)(x - 4)$	$x^2 - 7xy + 12y^2 = (x - 3y)(x - 4y)$

In Example 8 we use the factorization of $x^2 - 5x - 24$ to factor $x^2 - 5xy - 24y^2$.

EXAMPLE 8 Factoring a Polynomial with Two Variables

Use a graphing calculator to assist you in factoring $x^2 - xy - 56y^2$.

SOLUTION

Use a graphing calculator to factor $x^2 - x - 56$; then use this pattern to factor $x^2 - xy - 56y^2$.

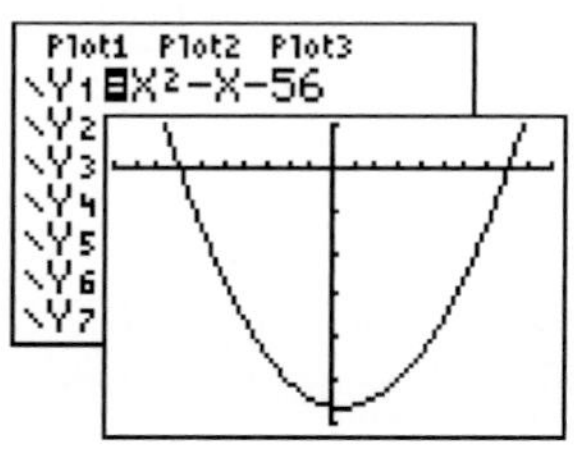

$[-10, 10, 1]$ by $[-60, 10, 10]$

Enter $x^2 - x - 56$ as Y_1 and create a graph of $Y_1 = x^2 - x - 56$.

Use a graphing calculator to determine the x-intercepts $(-7, 0)$ and $(8, 0)$.

The x-intercepts of $Y_1 = x^2 - x - 56$ are $(-7, 0)$ and $(8, 0)$.
Thus $x^2 - x - 56 = (x + 7)(x - 8)$.

You can check this factorization by multiplying these factors.

Answer: $x^2 - xy - 56y^2 = (x + 7y)(x - 8y)$

Then use the factorization of $x^2 - x - 56$ to factor $x^2 - xy - 56y^2$.

SELF-CHECK 6.1.5

1. Factor $x^2 + 6xy - 55y^2$.

SELF-CHECK ANSWERS

6.1.1

1. $5x^2(2x - 3)$
2. $3x(2x^2 - 3x + 4)$
3. $(x - 7)(x + 5)$

6.1.2

1. $(a + b)(x - y)$
2. $(x - 7)(x + 3)$

6.1.3

1. $-2, 1$
2. $(-2, 0), (1, 0)$
3. $x^2 + x - 2 = (x + 2)(x - 1)$
4. $-4, 0$, and 1
5. $x^3 + 3x^2 - 4x = x(x + 4)(x - 1)$

6.1.4

1. $(-16, 0), (18, 0)$
2. $x^2 - 2x - 288 = (x + 16)(x - 18)$
3. $(-2, 0), (3, 0), (5, 0)$
4. $x^3 - 6x^2 - x + 30 = (x + 2)(x - 3)(x - 5)$

6.1.5

1. $(x + 11y)(x - 5y)$

USING THE LANGUAGE AND SYMBOLISM OF MATHEMATICS 6.1

1. The GCF of a polynomial is the __________ __________ factor of the polynomial.
2. The GCF of a polynomial is a common factor of each term of the polynomial that has the largest possible numerical __________ and the largest possible __________ on each variable factor.
3. When we factor out a common factor from each term of a polynomial, we are using the __________ property.
4. Some polynomials with four or more terms can be factored by strategically __________ the terms to produce a common binomial factor.
5. An x-intercept of a graph is a point on the graph whose y-coordinate is __________.
6. If $P(x)$ is a polynomial and $P(5) = 0$, then 5 is called a __________ of the polynomial.
7. If $x + 4$ is a factor of the polynomial $P(x)$, then __________ is an x-intercept of the graph of $y = P(x)$.
8. If $P(x)$ is a polynomial and $(-2, 0)$ is an x-intercept of the graph of $y = P(x)$, then __________ is a zero of $P(x)$.
9. If $P(x)$ is a polynomial and $(8, 0)$ is an x-intercept of the graph of $y = P(x)$, then __________ is a factor of $P(x)$.

10. Each coefficient of a real polynomial $P(x)$ is a __________ number.
11. Each real polynomial function of degree n has exactly __________ zeros. Some (perhaps all) of these zeros can be real numbers. If some of the zeros are not real numbers, then they will be imaginary numbers. (We will examine imaginary numbers and complex numbers in Section 7.7.)

QUICK REVIEW 6.1

1. Simplify $3x^3y^4(6x^2y^5)$.
2. Simplify $\frac{12x^6y^3}{2x^3y^3}$.
3. Expand $-4x^2y^3(3x^2 - xy + 5y^2)$.
4. Expand $3xy^2(5x^2 - 2xy + 4y^2)$.
5. Simplify $\frac{5x^3y^2 + 15x^2y^3 - 10xy^4}{2xy^2}$.

EXERCISES 6.1

In Exercises 1–10, complete each factorization by factoring out the GCF.

1. $12x^3 - 18x^2 = 6x^2(\ )$
2. $15x^4 + 20x^3 = 5x^3(\ )$
3. $10a^2b + 6ab^2 = 2ab(\ )$
4. $21a^3b - 15a^2b^2 = 3a^2b(\ )$
5. $10x^3 - 15x^2y + 20xy^2 = 5x(\ )$
6. $12x^3 + 16x^2y + 20xy^2 = 4x(\ )$
7. $2x(x - 2y) + y(x - 2y) = (\ )(x - 2y)$
8. $5x(x + 3y) + 4y(x + 3y) = (\ )(x + 3y)$
9. $7x(2x - 5) - 9(2x - 5) = (\ )(2x - 5)$
10. $(3x + 4)(5v) - (3x + 4)(4w) = (3x + 4)(\ )$

In Exercises 11–20, factor the GCF out of each polynomial.

11. $14x^2 + 77x$
12. $22x^3 - 33x^2$
13. $8x^2y - 20xy^2$
14. $18x^2y^2 - 30xy^2$
15. $6x^3 - 8x^2 + 10x$
16. $6x^3 - 9x^2 - 21x$
17. $(5x - 2)(x) + (5x - 2)(3)$
18. $(3x + 4)(7x) - (3x + 4)(5)$
19. $(x + 2y)(6x) - (x + 2y)(7y)$
20. $(5x - 2y)(3x) + (5x - 2y)(4y)$

In Exercises 21–28, factor each polynomial by using the grouping method.

21. $(5ax - 2a) + (15x - 6)$
22. $(3ax + a) + (6x + 2)$
23. $(2ax - 7a) - (2bx - 7b)$
24. $(12ax - 8a) - (21bx - 14b)$
25. $3x^2 - 5x + 12x - 20$
26. $2x^2 - 8x + 7x - 28$
27. $12x^2 + 4x - 15x - 5$
28. $10x^2 - 15x - 14x + 21$

In Exercises 29–32, factor the expression in the left column. Then expand and simplify the expression in the right column.

FACTOR	EXPAND
29. $x(x + 4) + 2(x + 4)$	$x(x + 4) + 2(x + 4)$
30. $x(x - 5) + 3(x - 5)$	$x(x - 5) + 3(x - 5)$
31. $2x(5x - 2) - 7(5x - 2)$	$2x(5x - 2) - 7(5x - 2)$
32. $3x(2x + 5) - 8(2x + 5)$	$3x(2x + 5) - 8(2x + 5)$

In Exercises 33–36, use the polynomial $P(x)$ and the graph and table for $Y_1 = P(x)$.

a. List the x-intercepts of the graph of $y = P(x)$.
b. List the zeros of $P(x)$.
c. Use these zeros to factor each polynomial.

33.

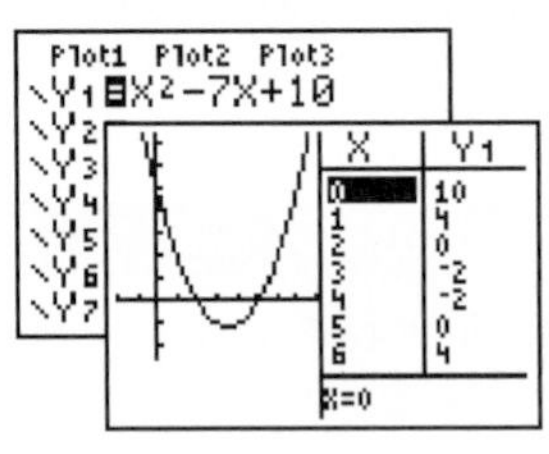
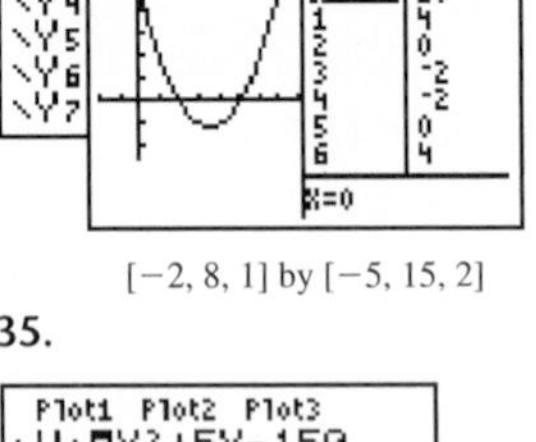

[−2, 8, 1] by [−5, 15, 2]

34.

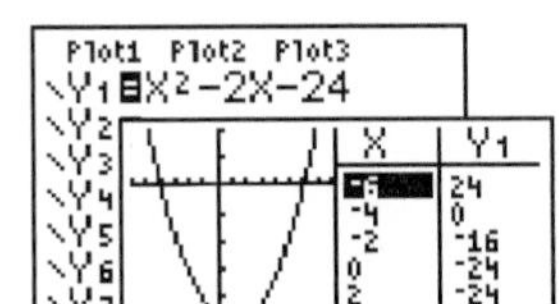

[−6, 8, 1] by [−30, 10, 5]

35.

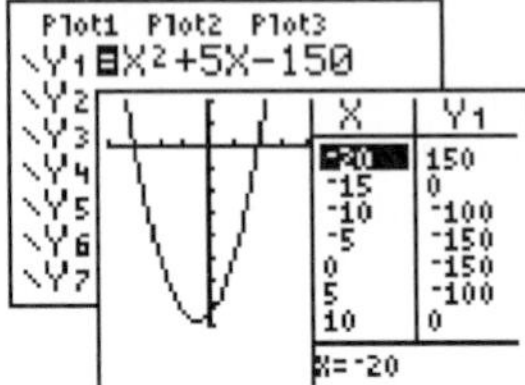

[−20, 20, 5] by [−160, 40, 20]

36.

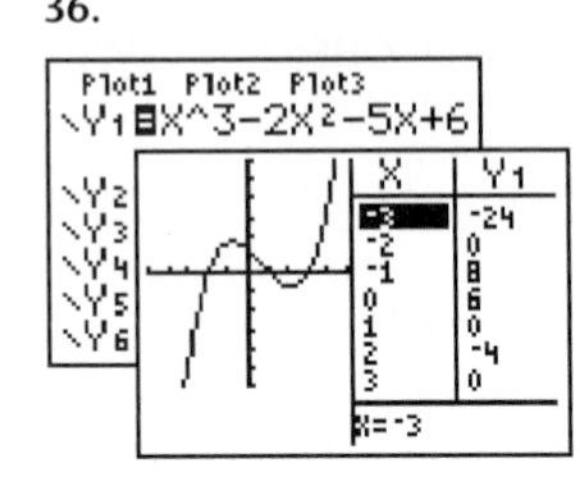

[−5, 5, 1] by [−30, 30, 5]

In Exercises 37–40, use the polynomial $P(x)$ and the table for $Y_1 = P(x)$.

a. List the zeros of $P(x)$.
b. List the x-intercepts of the graph of $Y_1 = P(x)$.
c. Use these zeros to factor each polynomial.

37.

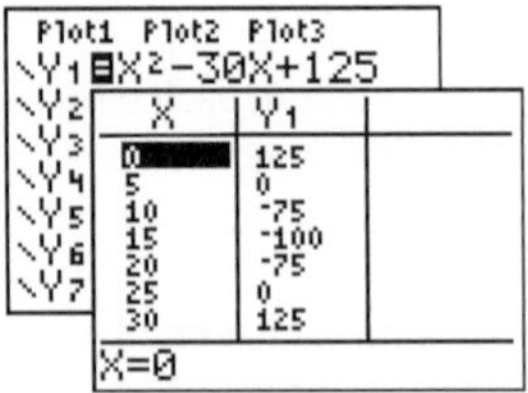

38.

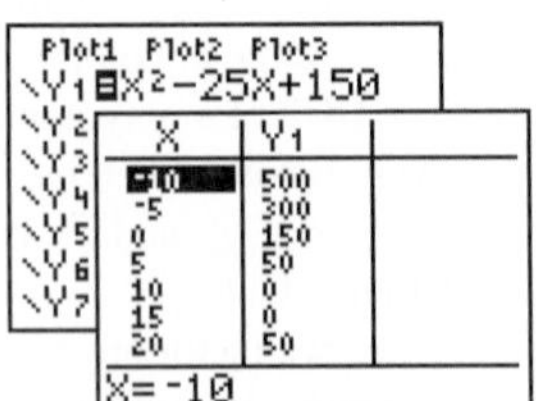

39.

Plot1 Plot2 Plot3
Y1 = X^3−2X²−16X+32

X	Y1
-4	0
-2	48
0	32
2	0
4	0
6	80
8	288

X= -4

40.

Plot1 Plot2 Plot3
Y1 = X^3+X²−4X−4

X	Y1
-3	-10
-2	0
-1	0
0	-4
1	-6
2	0
3	20

X= -3

In Exercises 41–44, use the graph of $y = P(x)$.

a. Determine the x-intercepts of the graph.

b. Factor the polynomial $P(x)$.

41. $P(x) = x^2 - 7x + 10$

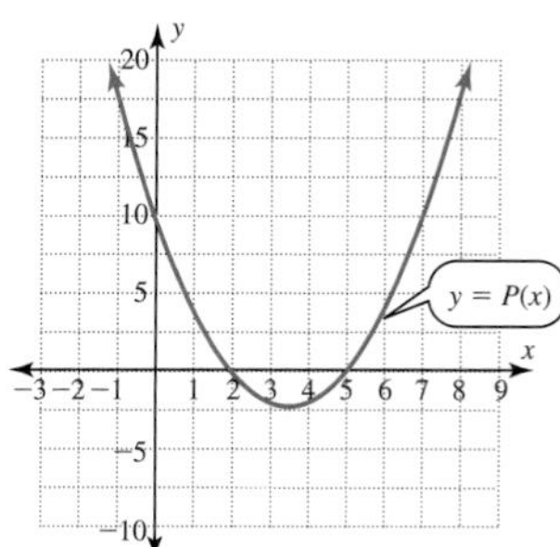

42. $P(x) = x^2 - 5x - 24$

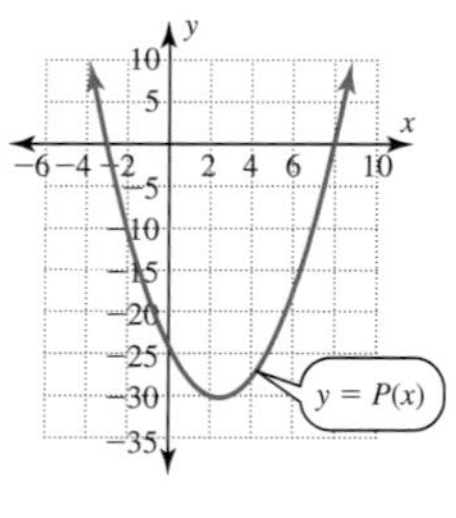

43. $P(x) = x^3 - x^2 - 6x$

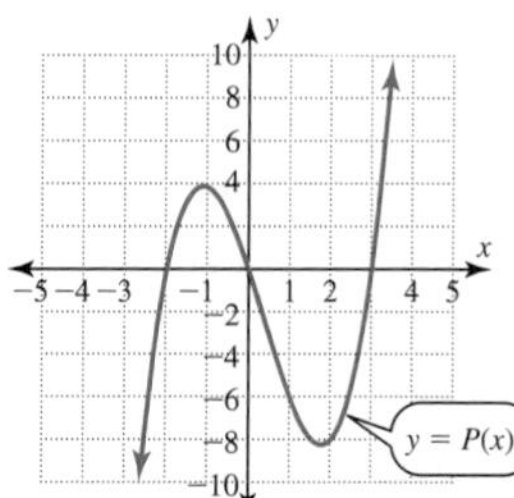

44. $P(x) = x^3 - x^2 - 12x$

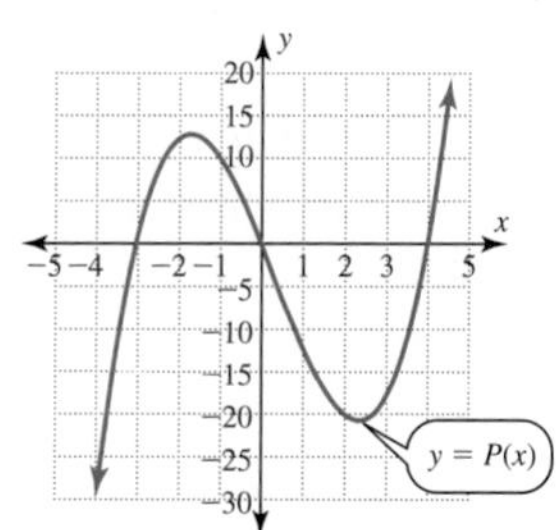

In Exercises 45–50, use the factored form of each polynomial $P(x)$.

a. List the zeros of $P(x)$.

b. List the x-intercepts of the graph of $y = P(x)$.

45. $P(x) = (x + 5)(x - 9)$

46. $P(x) = (x + 7)(x - 11)$

47. $P(x) = x(x + 3) - 17(x + 3)$ (*Hint:* Complete the factorization first.)

48. $P(x) = x(x + 4) - 15(x + 4)$ (*Hint:* Complete the factorization first.)

49. $P(x) = (x + 3)(x + 1)(x - 8)$

50. $P(x) = (x + 5)(x - 6)(x - 12)$

In Exercises 51–54, use the given polynomial and a graphing calculator.

a. Complete each table.

b. Use these results to factor the polynomial.

51. $P(x) = x^2 - 35x + 300$

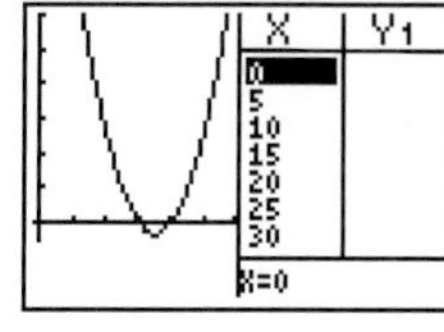
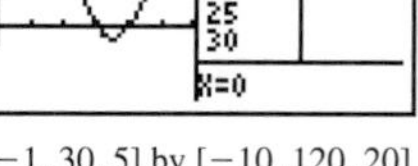

[−1, 30, 5] by [−10, 120, 20]

52. $P(x) = x^2 + 26x + 168$

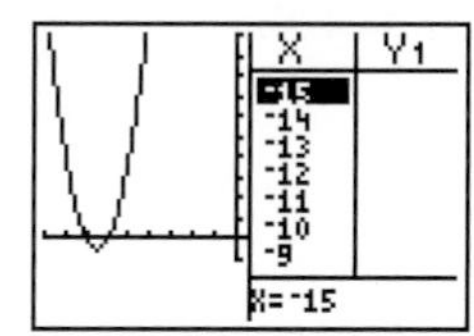

[−18, 1, 2] by [−2, 20, 2]

53. $P(x) = x^3 - 10x^2 + 24x$

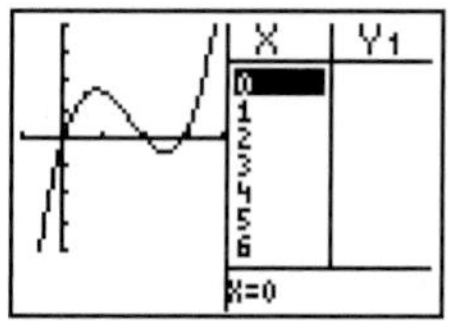

[−2, 8, 2] by [−40, 40, 10]

54. $P(x) = x^3 - 35x^2 + 350x - 1{,}000$

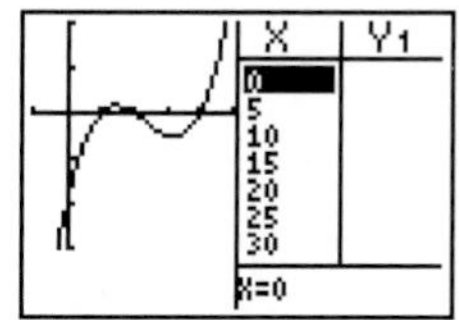

[−5, 25, 5] by [−1500, 1000, 500]

Exercises 55–58, depict some polynomials that arise from applications in Chapter 7. These polynomials involve larger coefficients that make using a calculator an attractive option. Use your calculator and the window on the graph shown for each exercise to determine the zeros of each polynomial, and then factor this polynomial. (*Hint:* See Calculator Perspective 6.1.1.)

55.

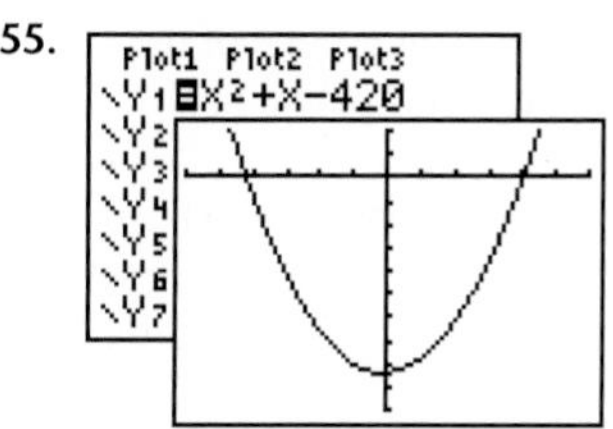

[−30, 30, 5] by [−500, 100, 50]

56.

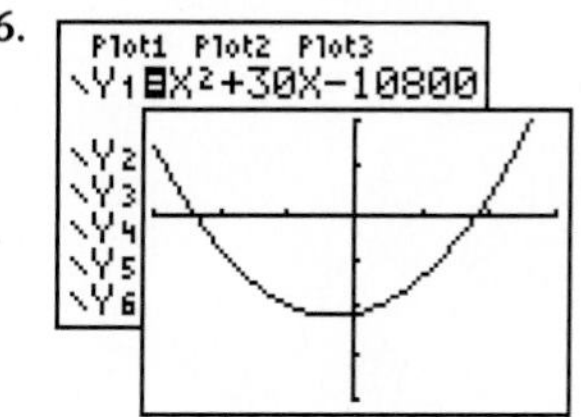

[−150, 150, 50] by [−20000, 10000, 5000]

57.

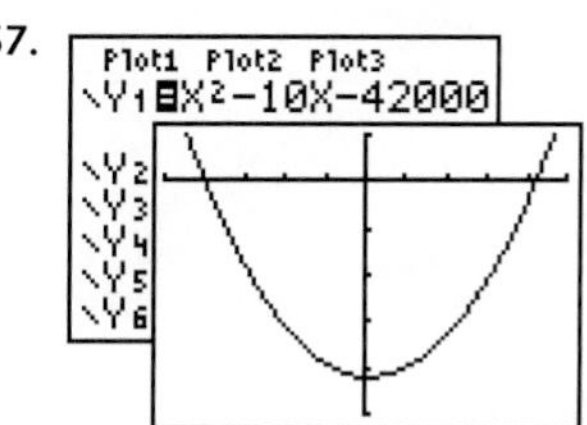

[−250, 250, 50] by [−50000, 10000, 10000]

58.

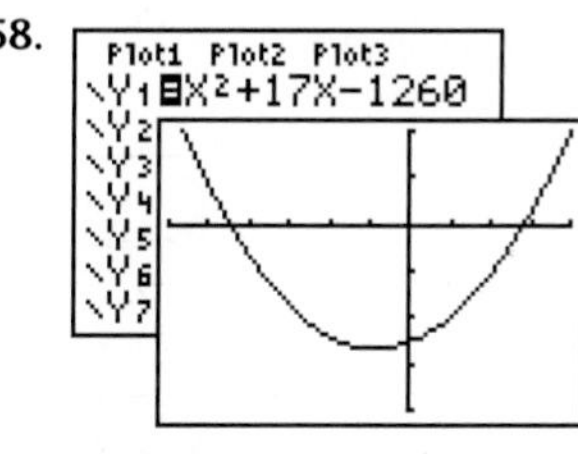

[−60, 40, 10] by [−2000, 1000, 500]

In Exercises 59–64, use the given polynomial $P(x)$, a graphing calculator, and the graph of $y = P(x)$ to factor $P(x)$.

59. $P(x) = x^2 + 20x - 125$

60. $P(x) = x^2 - 2x - 48$

61. $P(x) = x^3 + 3x^2 - 49x + 45$

62. $P(x) = x^3 + 8x^2 - 4x - 32$

63. $P(x) = x^4 - 5x^2 + 4$

64. $P(x) = x^4 - 10x^2 + 9$

In Exercises 65–68, complete the following table for each polynomial.

POLYNOMIAL $P(x)$	FACTORED FORM OF $P(x)$	ZEROS OF $P(x)$	x-INTERCEPTS OF THE GRAPH OF $y = P(x)$
65. $x^2 + 2x - 63$	$(x - 7)(x + 9)$		
66. $x^2 - 17x + 66$		6 and 11	
67. $x^2 - 2x - 80$			$(-8, 0)$, $(10, 0)$
68. $x^3 - x^2 - 30x$		-5, 0, and 6	

69. Factor $x^2 - xy - 72y^2$ given that $x^2 - x - 72 = (x + 8)(x - 9)$.

70. Factor $x^2 + 20xy + 96y^2$ given that $x^2 + 20x + 96 = (x + 12)(x + 8)$.

71. Factor $x^2 - 24xy + 143y^2$ given that $x^2 - 24x + 143 = (x - 11)(x - 13)$.

72. Factor $x^2 - 25xy - 150y^2$ given that $x^2 - 25x - 150 = (x + 5)(x - 30)$.

73. Investment Value If a principal of P dollars earns interest at an annual interest rate r, then the amount of dollars that represents the value of this investment at the end of the year is given by $A = P + Pr$. Factor this polynomial expression.

74. Number of Games A Little League consists of n teams. The league director plans to create a schedule so that each team will play each of the other teams 4 times during the season. The number of games required to accomplish this is given by $2n^2 - 2n$.

a. Factor this polynomial expression.
b. If there are only two teams in the league, evaluate $2n^2 - 2n$.
c. If there are six teams in the league, evaluate $2n^2 - 2n$.

75. Surface Area The surface area of a cylindrical soda can is given by $A = 2\pi r^2 + 2\pi rh$ for a can with radius r and height h. Factor this polynomial expression.

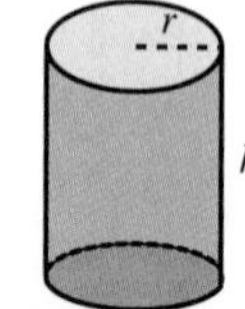

76. Area A concrete pad near a small storage building is in the shape of a trapezoid. The surface area of this pad is $x^2 + 3x$, where the area of the square portion of the figure is x^2 and the area of the triangular portion is $3x$. Factor this polynomial expression.

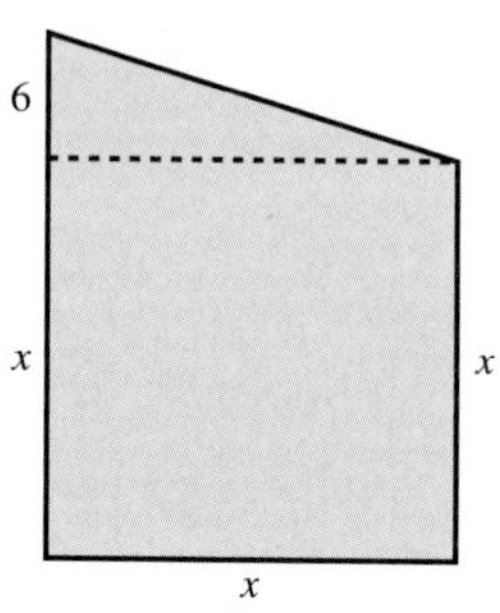

Group Discussion Questions

77. Discovery Question Two forms of a polynomial $P(x)$ are given here.

Factored Form	Expanded Form
$(x + 5)(x - 7)$	$x^2 - 2x - 35$

a. What is the degree of this polynomial?
b. Which of these forms did you use to determine the answer in part **a**?
c. What are the x-intercepts of the graph of $y = x^2 - 2x - 35$? (Be sure to write your answers as ordered pairs.)
d. Which of these forms did you use to determine the answer in part **c**?
e. What is the y-intercept of the graph of $y = x^2 - 2x - 35$? (Be sure to write your answer as an ordered pair.)
f. Which of these forms did you use to determine the answer in part **e**?
g. What are the zeros of $x^2 - 2x - 35$?
h. Which of these forms did you use to determine the answer in part **g**?

78. Error Analysis A student graphed the equation as displayed and concluded that the zeros of $(x + 4)(x - 6)$ are -2 and 3. Describe the error the student has made, and give the correct zeros.

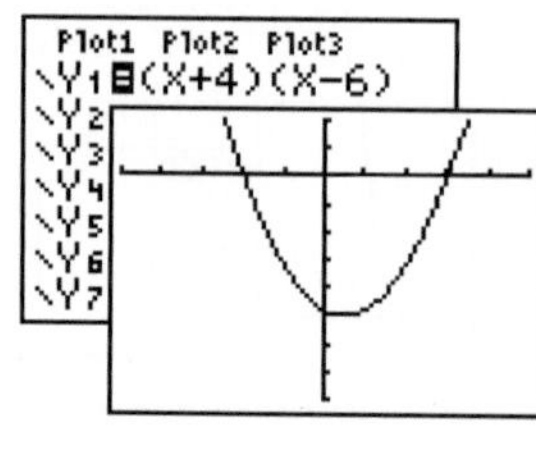

$[-10, 10, 2]$ by $[-40, 10, 5]$

79. Error Analysis A student graphed the equation as displayed and concluded that the zeros of $(11x + 12)(11x - 32)$ are -1 and 3. Describe the error the student has made, and give the correct zeros.

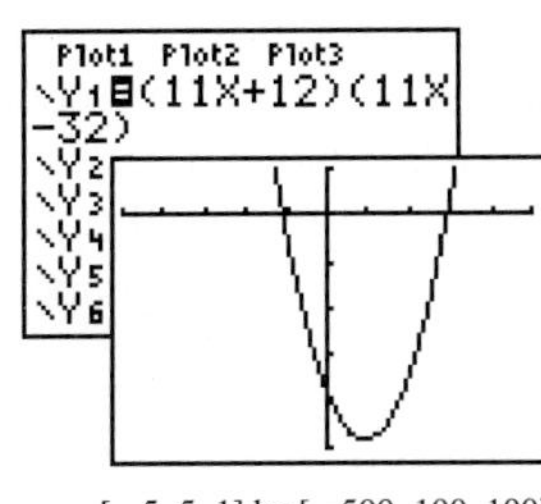

$[-5, 5, 1]$ by $[-500, 100, 100]$

80. Discovery Question Determine the zeros for each of these polynomials.

a. $(2x + 1)(x - 4)$
b. $(2x - 1)(x + 3)$
c. $(4x - 1)(x + 2)$
d. $(4x + 1)(x - 1)$

Use your observations from parts **a–d** to write the factors of a polynomial with these zeros.

e. $\frac{1}{3}$ and 5 f. $-\frac{1}{3}$ and -5 g. $\frac{1}{6}$ and -2 h. $\frac{2}{3}$ and $\frac{1}{3}$

Explain what you have observed to your instructor.

81. Discovery Question

a. Graph $y = x^2 - x - 42$, and use this graph to factor $x^2 - x - 42$.
b. Graph $y = -x^2 + x + 42$, and use this graph to factor $-x^2 + x + 42$.
c. Graph $y = x^2 + 5x - 24$, and use this graph to factor $x^2 + 5x - 24$.
d. Graph $y = -x^2 - 5x + 24$, and use this graph to factor $-x^2 - 5x + 24$.
e. What is the relationship between the graph of $y = P(x)$ and the graph of $y = -P(x)$?
f. What is the relationship between the zeros of $P(x)$ and of $-P(x)$?
g. What is the relationship between the factored form of $P(x)$ and of $-P(x)$?

Section 6.2 Factoring Trinomials of the Form $x^2 + bxy + cy^2$

Objectives:

4. Factor trinomials of the form $x^2 + bx + c$.
5. Factor trinomials of the form $x^2 + bxy + cy^2$.

A Mathematical Note

Prime numbers play a key role in cryptography. In the frequently used RSA algorithm, prime numbers are used both for encryption to input a message and for decryption to decode this message. The security of this procedure comes from the computational difficulty of factoring extremely large numbers. To be secure, the prime numbers must have at least 100 digits.

In Section 6.1 we factored some trinomials of the form $x^2 + bx + c$ by using tables to display the zeros of the polynomial or by using graphs to display the x-intercepts of the graph of $y = x^2 + bx + c$. We now use the grouping method from Section 6.1 as the basis for a procedure that can be used to factor any trinomial of the form $x^2 + bx + c$. This procedure enables us to either factor the trinomial over the integers or determine if the trinomial is prime. A polynomial is **prime over the integers** if its only factorization must involve either 1 or -1 as one of the factors.

The trinomial $ax^2 + bx + c$ is referred to as a **quadratic trinomial.** The ax^2 term is called the **second-degree term,** or the **quadratic term;** bx is the **first-degree term** or the **linear term:** and c is the **constant term.** For example, $6x^2 - 7x - 5$ is a quadratic trinomial with $a = 6$, $b = -7$, and $c = -5$. The quadratic term is $6x^2$, the linear term is $-7x$, and the constant term is -5.

Different methods for factoring trinomials are presented in some textbooks. It is likely that your class will have students who have studied different methods. We urge you to follow the advice of your instructor. Work toward understanding the method you select as you practice it to gain speed and proficiency. As more factoring is done with calculators, it is this understanding of factoring and its connections with other topics that has lasting importance.

Because of the importance of connecting concepts, we present a method for factoring trinomials that builds on our earlier work, using the distributive property to factor out the greatest common factor (GCF) of a polynomial.

Factoring $x^2 + bx + c$

To develop the logic we will use to factor trinomials of the form $x^2 + bx + c$, first we examine the following three products.

$$\begin{aligned}(x + 24)(x + 1) &= x(x + 1) + 24(x + 1)\\ &= x^2 + x + 24x + 24\\ &= x^2 + 25x + 24\end{aligned}$$

$$\begin{aligned}(x + 2)(x + 12) &= x(x + 12) + 2(x + 12)\\ &= x^2 + 12x + 2x + 24\\ &= x^2 + 14x + 24\end{aligned}$$

$$\begin{aligned}(x + 3)(x + 8) &= x(x + 8) + 3(x + 8)\\ &= x^2 + 8x + 3x + 24\\ &= x^2 + 11x + 24\end{aligned}$$

Note in each of these three products that the coefficients of the two middle terms have a product of 24. In the first product $1 \cdot 24 = 24$, in the second product $12 \cdot 2 = 24$, and in the third product $8 \cdot 3 = 24$. In general $(x + c_1)(x + c_2) = x^2 + (c_1 + c_2)x + c_1c_2$. Thus a factorization of $x^2 + bx + c$ must have $c_1c_2 = c$ and $c_1 + c_2 = b$. The method of factoring $x^2 + bx + c$ that we now illustrate examines the factors of c in search of a pair of factors whose sum is b.

Factorable Trinomials

ALGEBRAICALLY	VERBALLY	EXAMPLE
$x^2 + bx + c = (x + c_1)(x + c_2)$ where $c_1c_2 = c$ and $c_1 + c_2 = b$.	A trinomial $x^2 + bx + c$ is factorable into a pair of binomial factors with integer coefficients if and only if there are two integers whose product is c and whose sum is b. Otherwise, the trinomial is prime over the integers.	$x^2 + 14x + 24 = (x + 2)(x + 12)$ where $2 \cdot 12 = 24$ and $2 + 12 = 14$.

Factoring reverses the steps used in multiplication. Both use the distributive property.

Thus to factor $x^2 + 10x + 24$ in Example 1, we can examine all possible pairs of factors of 24 in search of a pair of factors whose sum is 10. Note that the steps used in this example to factor the trinomial into two binomials are identical to the steps used to multiply the two binomials—but in the reverse order.

EXAMPLE 1 Factoring a Quadratic Trinomial

Factor $x^2 + 10x + 24$.

SOLUTION

Examine the factors of 24 for a pair of factors whose sum is 10.

FACTORS OF 24		SUM OF FACTORS
1	24	25
2	12	14
3	8	11
4	6	10

The only factors of 24 with a sum of 10 are 4 and 6. Thus we rewrite the middle term of $10x$ as $6x + 4x$.

$$\begin{aligned} x^2 + 10x + 24 &= x^2 + 6x + 4x + 24 \\ &= x(x + 6) + 4(x + 6) \\ &= (x + 4)(x + 6) \end{aligned}$$

After rewriting the linear term $10x$ as $6x + 4x$, use the distributive property to factor the GCF of x out of the first two terms and the GCF of 4 out of the last two terms. Then use the distributive property to factor the GCF of $x + 6$ out of both terms.

Answer: $x^2 + 10x + 24 = (x + 4)(x + 6)$

Does this factorization check?

The procedure used in Example 1 is summarized in the following box. A key step in this procedure is to use the sign pattern of the terms to determine the sign pattern for each possible pair of factors of the constant term.

Factoring Trinomials of the Form $x^2 + bx + c$ by Tables and Grouping

PROCEDURE	EXAMPLE
STEP 1. Factor out the GCF if it is not 1. (The GCF is 1 for $x^2 + bx + c$.)	Factor $x^2 + 5x - 24$.
STEP 2. Find a pair of factors of c whose sum is b. If there is not a pair of factors whose sum is b, the trinomial is prime over the integers. • If the constant c is positive, the factors of c must have the same sign. These factors will share the same sign as the linear coefficient b. • If the constant c is negative, the factors of c must be opposite in sign. The sign of b will determine the sign of the factor with the larger absolute value.	See the table of factors of −24 below.
STEP 3. Rewrite the linear term of $x^2 + bx + c$ so that b is the sum of the pair of factors from step 2.	$x^2 + 5x - 24 = x^2 + 8x - 3x - 24$
STEP 4. Factor the polynomial from step 3 by grouping the terms and factoring the GCF out of each pair of terms.	$= x(x + 8) - 3(x + 8)$ $= (x - 3)(x + 8)$

FACTORS OF −24		SUM OF FACTORS
−1	24	23
−2	12	10
−3	8	5
−4	6	2

Each step of the factoring procedure is shown in the box above. If you can mentally select the correct pair of factors without using all these steps, that is great. However, the harder the problem, the more useful this procedure will be. This procedure does not involve any guessing; it is a systematic procedure that always works. It is particularly useful when the trinomial is prime.

EXAMPLE 2 Factoring a Quadratic Trinomial

Factor $x^2 + 7x - 60$.

SOLUTION

Examine the factors of −60 for a pair of factors whose sum is 7.

FACTORS OF −60		SUM OF FACTORS
−1	60	59
−2	30	28
−3	20	17
−4	15	11
−5	12	7
−6	10	4

The constant term −60 is negative, so the factors of −60 are of opposite sign. Because 7, the coefficient of the linear term, is positive, the positive factor of −60 must be larger in absolute value than the negative factor.

$$\begin{aligned} x^2 + 7x - 60 &= x^2 - 5x + 12x - 60 \\ &= x(x - 5) + 12(x - 5) \\ &= (x + 12)(x - 5) \end{aligned}$$

Use this table to rewrite the linear term $7x$ as $-5x + 12x$. Factor the GCF of x from the first two terms and factor the GCF of 12 from the last two terms. Then use the distributive property again to factor $x - 5$ out of both terms.

Answer: $x^2 + 7x - 60 = (x + 12)(x - 5)$

Does this factorization check?

SELF-CHECK 6.2.1

Factor each trinomial.

1. $v^2 + 8v + 12$ **2.** $m^2 - 7m + 12$ **3.** $n^2 - 11n - 12$

Example 3 provides further practice factoring a trinomial of the form $x^2 + bx + c$. In this example both b and c are negative.

EXAMPLE 3 Factoring a Quadratic Trinomial

Factor $x^2 - 2x - 48$.

SOLUTION

Examine the factors of -48 for a pair of factors whose sum is -2.

FACTORS OF −48		SUM OF FACTORS
1	−48	−47
2	−24	−22
3	−16	−13
4	−12	−8
6	−8	−2

The constant term -48 is negative, so the factors of -48 are of opposite sign. Because -2, the coefficient of the linear term, is negative, the negative factor of -48 must be larger in absolute value than the positive factor.

$$\begin{aligned} x^2 - 2x - 48 &= x^2 + 6x - 8x - 48 \\ &= x(x + 6) - 8(x + 6) \\ &= (x - 8)(x + 6) \end{aligned}$$

Rewrite the linear term $-2x$ as $6x - 8x$. Factor the GCF of x from the first two terms and the GCF of -8 from the last two terms. Then use the distributive property again to factor $x + 6$ out of both terms.

Answer: $x^2 - 2x - 48 = (x - 8)(x + 6)$

Does this factorization check?

It is more efficient to memorize frequently used phone numbers than it is to look them up each time we want to use them. Similarly, it is easier to mentally multiply 3×12 to get 36 than it is to get out a calculator or to use a pencil and paper to perform this multiplication. As you use the method of tables and grouping to factor trinomials like those in Examples 1 and 2, also try to observe patterns that will allow you to factor these same trinomials by inspection. Not only is this more efficient, but also it will help you develop algebraic skills for other problems.

Options for Factoring Trinomials of the Form $x^2 + bx + c$

By inspection: We encourage you to practice until you can factor many of the trinomials in this section by inspection.

By tables and grouping: If you cannot factor the trinomial by inspection, then this systematic procedure can be used to factor each trinomial in this section or to identify the trinomial as prime.

Example 4 compares factoring by using tables and grouping to factoring by inspection. In this example the coefficients b and c of $x^2 + bx + c$ are both negative. This is key information in determining the sign pattern for the binomial factors.

EXAMPLE 4 Factoring a Quadratic Trinomial

Factor $x^2 - 4x - 45$ (**a**) by using tables and grouping and (**b**) by inspection.

SOLUTION

(a) Examine the factors of -45 for a pair of factors whose sum is -4.

FACTORS OF −45		SUM OF FACTORS
1	−45	−44
3	−15	−12
5	−9	−4

The constant term -45 is negative, so the factors of -45 are of opposite sign. Because -4, the coefficient of the linear term, is negative, the negative factor of -45 must be larger in absolute value than the positive factor. Although you could list other factors, there is no need. Once the correct pair of factors is found, you can stop the list and use these factors.

$$\begin{aligned} x^2 - 4x - 45 &= x^2 + 5x - 9x - 45 \\ &= x(x + 5) - 9(x + 5) \\ &= (x - 9)(x + 5) \end{aligned}$$

Rewrite the linear term $-4x$ as $5x - 9x$.

Factor the GCF of x from the first two terms and the GCF of -9 from the last two terms. Then use the distributive property again to factor $x + 5$ out of both terms.

(b)
$$\begin{aligned} x^2 - 4x - 45 &= (x - ?)(x + ?) \\ &= (x - 9)(x + 5) \end{aligned}$$

The constant term -45 is negative, so the factors of -45 must be of opposite signs. The sign pattern is $(x - ?)(x + ?)$. The middle term has a coefficient of -4, so the negative factor of -45 will be larger in absolute value than the positive factor. By inspection you can determine that the factors of -45 that total -4 are 5 and -9.

Check:
$$\begin{aligned} (x - 9)(x + 5) &= x(x + 5) - 9(x + 5) \\ &= x^2 + 5x - 9x - 45 \\ &= x^2 - 4x - 45 \end{aligned}$$

This factorization checks.

Example 5 illustrates how to factor a trinomial by inspection. The sign pattern information in step 2 of the grouping method plays a key role in this mental process.

EXAMPLE 5 Factoring Trinomials by Inspection

Factor the trinomial $x^2 + 11x + 24$ by inspection.

SOLUTION

$$\begin{aligned} x^2 + 11x + 24 &= (x + ?)(x + ?) \\ &= (x + 3)(x + 8) \end{aligned}$$

The constant term 24 is positive, so the factors of 24 must be of the same sign. The middle term has a coefficient of $+11$, so both factors of 24 are positive and the sign pattern is $(x + ?)(x + ?)$. By inspection you can determine that the factors of 24 that total 11 are 3 and 8.

Multiplying to check your factors is a wise step when you are factoring by inspection.

Check:
$$\begin{aligned} (x + 3)(x + 8) &= x(x + 8) + 3(x + 8) \\ &= x^2 + 8x + 3x + 24 \\ &= x^2 + 11x + 24 \end{aligned}$$

This factorization checks.

On an exam or in your future work using factoring, you now have two methods to factor trinomials of the form $x^2 + bx + c$. If you can factor them quickly by inspection, this is advantageous. If not, then you can rely on using tables and grouping to factor all trinomials of this form.

SELF-CHECK 6.2.2

Factor each trinomial by inspection.

1. $v^2 + 7v + 12$ **2.** $m^2 - 13m + 12$ **3.** $n^2 - n - 12$

One thing that can be difficult to do by inspection or by examining a graph is to determine whether the polynomial is prime over the integers. This can be accomplished systematically by using the method of tables and grouping. This is illustrated in Example 6. In this example, the list of all possible factors does not produce the correct middle term of $9v$. Therefore $v^2 + 9v + 12$ must be prime.

EXAMPLE 6 Identifying Prime Trinomials

Factor $v^2 + 9v + 12$.

SOLUTION

Examine the factors of 12 for a pair of factors whose sum is 9.

FACTORS OF 12		SUM OF FACTORS
1	12	13
2	6	8
3	4	7

Both factors must be positive because both 12 and 9 are positive. None of the pairs of factors of 12 have a sum of 9. Because all integral factor pairs of 12 have been systematically examined and eliminated, this polynomial is prime over the integers.

Answer: $v^2 + 9v + 12$ is prime over the integers.

SELF-CHECK 6.2.3

Factor each trinomial.

1. $x^2 + 18x + 48$ **2.** $x^2 + 14x + 48$

As we noted in Section 6.1, we can use the procedure for factoring $x^2 + bx + c$ to factor trinomials of the form $x^2 + bxy + cy^2$. This is illustrated next and in Examples 7 and 8.

FACTORIZATIONS OF POLYNOMIALS IN x	FACTORIZATIONS OF POLYNOMIALS IN x AND y
$x^2 + 10x + 24 = (x + 6)(x + 4)$	$x^2 + 10xy + 24y^2 = (x + 6y)(x + 4y)$
$x^2 + 7x - 60 = (x + 12)(x - 5)$	$x^2 + 7xy - 60y^2 = (x + 12y)(x - 5y)$
$v^2 + 9v + 12$ is prime.	$v^2 + 9vw + 12w^2$ is prime.

EXAMPLE 7 Factoring a Trinomial with Two Variables

Factor $x^2 + 19xy + 48y^2$.

SOLUTION

$$x^2 + 19xy + 48y^2 = (x + ?y)(x + ?y)$$

To factor this trinomial by inspection, first determine the correct sign pattern.

$$= (x + 3y)(x + 16y)$$

Then select factors of 48 with a sum of 19.

Answer: $x^2 + 19xy + 48y^2 = (x + 3y)(x + 16y)$

Does this answer check?

EXAMPLE 8 Factoring a Trinomial with Two Variables

Factor $m^2 - 4mn - 32n^2$.

SOLUTION

$$m^2 - 4mn - 32n^2 = (m + ?n)(m - ?n)$$

To factor this trinomial by inspection, first determine the correct sign pattern.

$$= (m + 4n)(m - 8n)$$

Then select factors of -32 with a sum of -4. Because -4, the coefficient of the middle term, is negative, the negative factor of -32 must be larger in absolute value than the positive factor.

Answer: $m^2 - 4mn - 32n^2 = (m + 4n)(m - 8n)$

Does this answer check?

SELF-CHECK 6.2.4

Factor each trinomial.

1. $x^2 + 8xy + 15y^2$ **2.** $a^2 - 2ab - 48b^2$

If each term of a trinomial has a GCF other than 1, then the first step in factoring this trinomial is to factor out the GCF. In Example 9 the GCF of the polynomial is $2x$.

EXAMPLE 9 Completely Factoring a Trinomial

Factor $2x^3 + 6x^2 - 80x$.

SOLUTION

$$\begin{aligned} 2x^3 + 6x^2 - 80x &= 2x(x^2 + 3x - 40) \\ &= 2x(x + \ ?)(x - \ ?) \\ &= 2x(x + 8)(x - 5) \end{aligned}$$

First factor out $2x$, the GCF of the trinomial. Then set up the correct sign pattern to factor $x^2 + 3x - 40$ by inspection. Because 3, the coefficient of the middle term, is positive, the positive factor of -40 must be larger in absolute value than the negative factor.

Answer: $2x^3 + 6x^2 - 80x = 2x(x + 8)(x - 5)$

Does this factorization check?

Summary

Sections 6.1 and 6.2 have covered a variety of methods for factoring trinomials of the form $x^2 + bx + c$. The box on the following page lists some of the advantages and disadvantages of each method.

METHODS FOR FACTORING $x^2 + bx + c$	ADVANTAGES	DISADVANTAGES
Graphs	This method visually displays the x-intercepts that correspond to the factors of the trinomial. If there are no x-intercepts, this indicates the polynomial is prime over the integers.	It can be time-consuming to select the appropriate viewing window and to approximate the x-intercepts. If the x-intercepts are near integer values but not exactly on these values, it is easy to be misled by the graph and to select incorrect factors.
Tables	Users can see the zeros of the trinomial and can observe numerical patterns that are important in many applications. Spreadsheets allow us to use the power of computers to exploit this method.	Often it requires insight or some trial and error to select the most appropriate table.
Tables and grouping	This is a precise step-by-step process that can factor any trinomial of the form $x^2 + bx + c$ or can identify the trinomial as prime. This method has the same steps used to multiply binomial factors, except that the steps are reversed.	Many trinomials with small integer coefficients can be factored by inspection, and it is not necessary to write the table and all the steps of this method.
Inspection	This method takes advantage of patterns and insights to quickly factor trinomials with small integer coefficients. Observing the mathematical patterns in these trinomials can improve your foundation for other topics.	This can become a trial-and-error process that requires considerable insight. For novices this process can be very challenging. It is important to examine all possibilities before you decide a trinomial is prime.

SELF-CHECK ANSWERS

6.2.1
1. $(v + 6)(v + 2)$
2. $(m - 3)(m - 4)$
3. $(n + 1)(n - 12)$

6.2.2
1. $(v + 3)(v + 4)$
2. $(m - 1)(m - 12)$
3. $(n + 3)(n - 4)$

6.2.3
1. $x^2 + 18x + 48$ is prime
2. $(x + 6)(x + 8)$

6.2.4
1. $(x + 3y)(x + 5y)$
2. $(a + 6b)(a - 8b)$

USING THE LANGUAGE AND SYMBOLISM OF MATHEMATICS 6.2

1. A polynomial is __________ over the integers if its only factorization must involve either 1 or -1 as one of the factors.
2. A trinomial of the form $ax^2 + bx + c$ is called a __________ trinomial.
3. In the trinomial $ax^2 + bx + c$:
   a. The second-degree term ax^2 is called the __________ term.
   b. The first-degree term bx is called the __________ term.
   c. The term c is called the __________ term.
4. Factoring reverses the steps used in multiplication. Both use the __________ property.
5. A trinomial $x^2 + bx + c$ is factorable over the integers if there are two integers whose __________ is c and whose __________ is b.
6. Factoring $x^2 + bx + c$ by the method of tables and grouping creates a table of the factors of __________ in search of a pair of factors whose sum is __________.
7. If the constant c is positive in the trinomial $x^2 + bx + c$, the factors of c must have the __________ sign.
8. If the constant c is negative in the trinomial $x^2 + bx + c$, the factors of c must have __________ signs.
9. Factoring $x^2 + bx + c$ directly without writing any steps to determine the factors of c whose sum is b is called factoring by __________.

QUICK REVIEW 6.2

1. Multiply and simplify $(x + 3)(x - 5)$.
2. Multiply and simplify $(x + 4y)(x + 7y)$.
3. Multiply and simplify $x(x - 5)(x - 2)$.
4. Multiply and simplify $-2xy^2(x + y)(x - 5y)$.
5. Multiply and simplify $(x - 1)(x - 3)(x + 5)$.

EXERCISES 6.2

In Exercises 1–4, match each trinomial with the appropriate sign pattern for the factors of this trinomial.

1. $x^2 - 10x - 11$
2. $x^2 - 18x + 17$
3. $x^2 + 6x + 5$
4. $x^2 + 3x - 4$

A. $(x + ?)(x + ?)$
B. $(x + ?)(x - ?)$
C. $(x - ?)(x - ?)$

In Exercises 5–8, use inspection and the given sign pattern for the binomial factors to complete the factorization of each trinomial.

5. $x^2 - 6x + 8 = (x - ?)(x - ?)$
6. $x^2 + 9x + 8 = (x + ?)(x + ?)$
7. $x^2 - 11x - 12 = (x + ?)(x - ?)$
8. $x^2 + 4x - 12 = (x + ?)(x - ?)$

In Exercises 9–20, factor each trinomial by inspection.

9. $x^2 + 3x + 2$
10. $x^2 + 4x + 3$
11. $x^2 - x - 2$
12. $x^2 - 2x - 3$
13. $x^2 + x - 20$
14. $x^2 + x - 6$
15. $x^2 - 9x + 20$
16. $x^2 - 5x + 6$
17. $x^2 - 7x - 18$
18. $x^2 - 3x - 18$
19. $x^2 + 2x - 24$
20. $x^2 + 10x - 24$

In Exercises 21–26, fill in the blanks to complete the factorization of each trinomial.

21. $x^2 - 12x + 35 = x^2 - 5x - 7x + 35$
$= x(\quad) - 7(\quad)$
$= (x - 7)(\quad)$

22. $x^2 - 6x - 72 = x^2 + 6x - 12x - 72$
$= x(\quad) - 12(\quad)$
$= (x - 12)(\quad)$

23. $x^2 + 2x - 80 = x^2 - 8x +$ ____ $- 80$
$= x(\quad) + 10(\quad)$
$= (x + 10)(\quad)$

24. $x^2 + 12x + 32 = x^2 + 4x +$ ____ $+ 32$
$= x(\quad) + 8(\quad)$
$= (x + 8)(\quad)$

25. $x^2 - 5xy - 36y^2 = x^2 + 4xy - 9xy - 36y^2$
$= x(\quad) - 9y(\quad)$
$= (x - 9y)(\quad)$

26. $x^2 - 12xy + 35y^2 = x^2 - 7xy - 5xy + 35y^2$
$= x(\quad) - 5y(\quad)$
$= (x - 5y)(\quad)$

In Exercises 27–32, use the given table to factor each polynomial. (*Hint:* Some of these trinomials are prime.)

27. a. $x^2 + 7x + 12$ b. $x^2 + 8x + 12$
c. $x^2 + 10x + 12$ d. $x^2 + 13x + 12$

FACTORS OF 12		SUM OF FACTORS
1	12	13
2	6	8
3	4	7

28. a. $x^2 + x - 12$ b. $x^2 + 2x - 12$
c. $x^2 + 4x - 12$ d. $x^2 + 11x - 12$

FACTORS OF −12		SUM OF FACTORS
−1	12	11
−2	6	4
−3	4	1

29. a. $x^2 - 3x - 28$ b. $x^2 - 12x - 28$
c. $x^2 - 15x - 28$ d. $x^2 - 27x - 28$

FACTORS OF −28		SUM OF FACTORS
1	−28	−27
2	−14	−12
4	−7	−3

30. a. $x^2 - 10x + 28$ b. $x^2 - 11x + 28$
c. $x^2 - 16x + 28$ d. $x^2 - 29x + 28$

FACTORS OF 28		SUM OF FACTORS
−1	−28	−29
−2	−14	−16
−4	−7	−11

31. a. $x^2 - 36$ b. $x^2 + 9x - 36$
c. $x^2 + 16x - 36$ d. $x^2 + 18x - 36$

FACTORS OF −36		SUM OF FACTORS
−1	36	35
−2	18	16
−3	12	9
−4	9	5
−6	6	0

32. a. $x^2 + 36$ b. $x^2 + 12x + 36$
c. $x^2 + 13x + 36$ d. $x^2 + 20x + 36$

FACTORS OF 36		SUM OF FACTORS
1	36	37
2	18	20
3	12	15
4	9	13
6	6	12

In Exercises 33–54, factor each trinomial.

33. $a^2 - 10a - 11$
34. $b^2 + 6b + 5$
35. $y^2 - 18y + 17$
36. $x^2 + 3x - 4$
37. $x^2 - 8x + 7$
38. $x^2 + 14x + 13$
39. $y^2 - 2y - 99$
40. $y^2 + 30y - 99$
41. $p^2 - 29p + 100$
42. $p^2 + 20p + 100$
43. $t^2 + 14t + 48$
44. $t^2 - 26t + 48$
45. $n^2 + 5n - 36$
46. $n^2 - 9n - 36$
47. $x^2 + 5xy + 6y^2$
48. $x^2 - 7xy + 6y^2$
49. $x^2 + xy - 6y^2$
50. $x^2 - 3xy - 10y^2$
51. $x^2 - 10xy + 25y^2$
52. $x^2 - 14xy + 49y^2$
53. $a^2 - 16ab - 36b^2$
54. $a^2 + 5ab - 36b^2$

In Exercises 55–64, some of the polynomials are factorable over the integers and others are prime over the integers. Factor those that are factorable, and label the others as prime.

55. $x^2 + 33x + 90$
56. $x^2 - 13x - 90$
57. $x^2 - 22x + 90$
58. $x^2 + 10x - 90$
59. $m^2 - 21mn + 98n^2$
60. $m^2 + 20mn + 98n^2$
61. $m^2 - 7mn - 9n^2$
62. $m^2 - 12mn - 98n^2$
63. $v^2 + 25$
64. $v^2 - 25$

In Exercises 65–74, first factor out the GCF and then complete the factorization.

65. $5x^3 + 15x^2 + 10x$
66. $7y^3 - 14y^2 - 21y$
67. $4ax^3 - 20ax^2 + 24ax$
68. $7bx^3 - 56bx^2 - 63bx$
69. $10az^2 + 290az + 1{,}000a$
70. $4xy^2 + 8xy - 192x$
71. $2abx^2 + 4abx - 96ab$
72. $33v^2w - 99vw + 66w$
73. $x^2(a + b) - 6x(a + b) + 8(a + b)$
74. $(2a - 3)y^2 + 9(2a - 3)y - 22(2a - 3)$

In Exercises 75–78, make a table of the factors of c and use this table to factor each trinomial of the form $x^2 + bx + c$.

75. $x^2 - 24x + 80$
76. $x^2 - 18x + 80$
77. $x^2 - 18x - 144$
78. $x^2 + 10x - 144$

In Exercises 79–84, first factor -1 out of each polynomial and then complete the factorization.

79. $-x^2 + 2x + 35$
80. $-x^2 + 34x + 35$
81. $-m^2 - 3m + 18$
82. $-m^2 + m + 12$
83. $-x^2 + 12xy + 45y^2$
84. $-x^2 - xy + 20y^2$

In Exercises 85 and 86, use the given graph and table to factor each polynomial entered into the calculator as Y_1.

85.

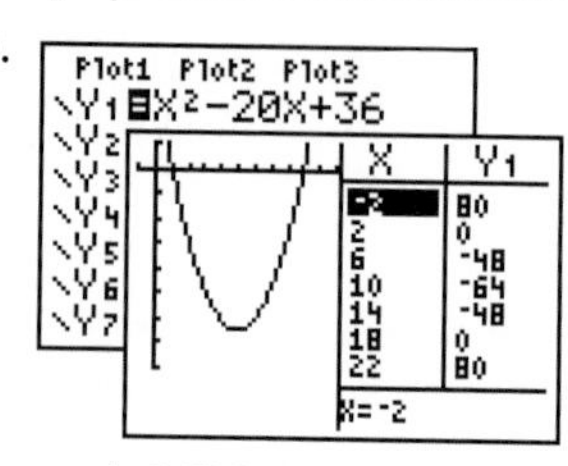

[−2, 22, 2] by [−80, 10, 10]

86.

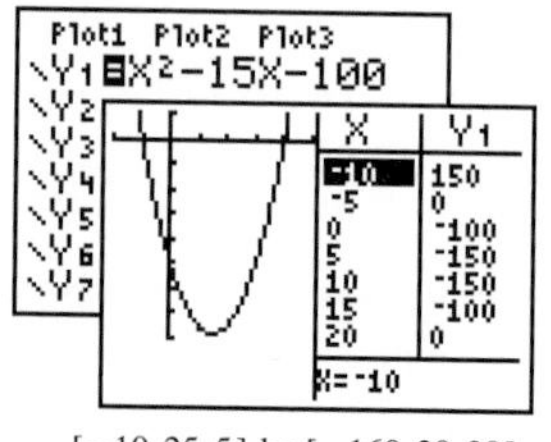

[−10, 25, 5] by [−160, 20, 20]

In Exercises 87 and 88, expand and simplify the expression in the left column and factor the expression in the right column.

	EXPAND	FACTOR
87.	$x(x - 3) - 2(x - 3)$	$x(x - 3) - 2(x - 3)$
88.	$x(x + 1) - 8(x + 1)$	$x(x + 1) - 8(x + 1)$

In Exercises 89 and 90, complete the following table for each polynomial.

POLYNOMIAL $P(x)$	FACTORED FORM	ZEROS OF $P(x)$	x-INTERCEPTS OF THE GRAPH OF $y = P(x)$
89. $x^2 + 2x - 15$			
90. $x^2 - 17x + 60$			

Group Discussion Questions

91. **Discovery Question** Determine the zeros of each polynomial function.
    a. $y = (2x + 1)(3x - 1)$
    b. $y = (4x + 1)(3x - 2)$
    c. $y = (4x - 3)(5x + 2)$

    Use your observations from parts **a** to **c** to write in factored form a polynomial function with the given zeros.

    d. $\frac{2}{3}$ and $\frac{1}{5}$
    e. $-\frac{1}{2}$ and $-\frac{3}{5}$
    f. $\frac{2}{5}$ and $-\frac{1}{7}$
    g. Explain what you observed in parts **a** to **c** to your instructor.
    h. Use the given graph and table to factor $8x^2 + 2x - 3$. (*Hint:* First write each zero as a fraction.)

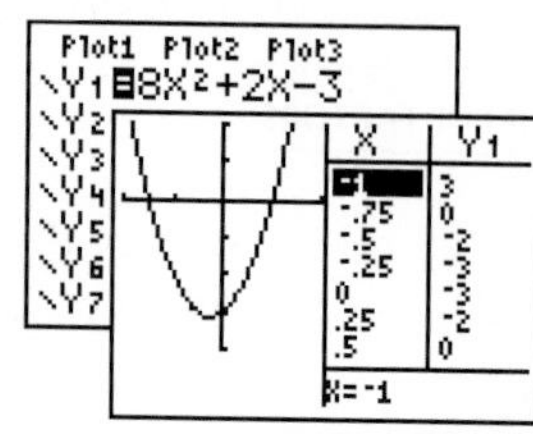

[−1, 1, 0.5] by [−4, 2, 1]

92. **Calculator Discovery** To factor $x^2 + 21x - 72$, we need to find a pair of factors of -72 whose sum is 21. Let x and Y_1 represent this pair of factors. Thus $xY_1 = -72$ or $Y_1 = -72/x$. Let $Y_2 = x + Y_1$ and use a table to find where Y_2 is 21.

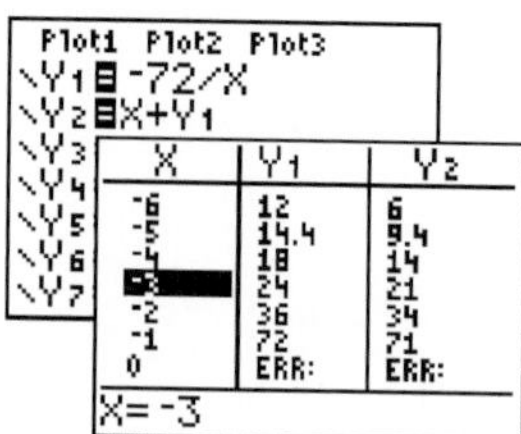

We see that $Y_2 = 21$ when $x = -3$ and $Y_1 = 24$. So,

$$\begin{aligned} x^2 + 21x - 72 &= x^2 - 3x + 24x - 72 \\ &= x(x - 3) + 24(x - 3) \\ &= (x + 24)(x - 3) \end{aligned}$$

Try this method to factor $x^2 + 10x - 96$.

Section 6.3 Factoring Trinomials of the Form $ax^2 + bxy + cy^2$

Objectives:
6. Factor trinomials of the form $ax^2 + bx + c$.
7. Factor trinomials of the form $ax^2 + bxy + cy^2$.

This section builds on our work in Section 6.2 and extends the method of factoring trinomials by grouping. We will now factor trinomials in which the leading coefficient is not 1. To develop the logic to factor trinomials of the form $ax^2 + bx + c$, we start by examining the pattern shown in the following three products.

$$\begin{aligned}(2x+5)(3x+1) &= 2x(3x+1) + 5(3x+1)\\ &= 6x^2 + 2x + 15x + 5\\ &= 6x^2 + 17x + 5\end{aligned}$$

$$\begin{aligned}(6x+1)(x+5) &= 6x(x+5) + 1(x+5)\\ &= 6x^2 + 30x + x + 5\\ &= 6x^2 + 31x + 5\end{aligned}$$

$$\begin{aligned}(6x+5)(x+1) &= 6x(x+1) + 5(x+1)\\ &= 6x^2 + 6x + 5x + 5\\ &= 6x^2 + 11x + 5\end{aligned}$$

A Mathematical Note

Who won the last Nobel Prize in mathematics? This is a trick question for your next trivia party. There is no Nobel Prize in mathematics. Alfred Nobel (1833–1896), the inventor of dynamite, established the Nobel Prizes in his will because he was concerned about creating a product that had caused so many injuries and deaths. These prizes are given in chemistry, physics, physiology or medicine, literature, world peace, and economics. There is a Fields Medal that is considered to be the equivalent of a Nobel Prize for mathematics. It was established by the Canadian mathematician John Charles Fields (1863–1932).

Note in each of these three products that the coefficients of the two middle terms have a product of 30. In the first product $2 \cdot 15 = 30$, in the second product $30 \cdot 1 = 30$, and in the third product $6 \cdot 5 = 30$. In each product the two middle terms consist of a pair of factors of ac. In each of these three examples, $ac = 30$.

Thus to factor $6x^2 + 13x + 5$, we can examine all possible pairs of factors of 30 in search of a pair of factors whose sum is 13. This is illustrated in Example 1. Note that the steps used in Example 1 to factor the trinomial into two binomials are identical to the steps used to multiply the two binomials—but in the reverse order.

EXAMPLE 1 Factoring a Quadratic Trinomial

Factor $6x^2 + 13x + 5$.

SOLUTION

Examine the factors of 30 for a pair of factors whose sum is 13.

FACTORS OF 30		SUM OF FACTORS
1	30	31
2	15	17
3	10	13
5	6	11

First note that $6 \cdot 5 = 30$. Then examine the positive factors of 30.

$$\begin{aligned}6x^2 + 13x + 5 &= 6x^2 + 3x + 10x + 5\\ &= 3x(2x+1) + 5(2x+1)\\ &= (3x+5)(2x+1)\end{aligned}$$

The only factors of 30 with a sum of 13 are 3 and 10.
Thus, rewrite the middle term of $13x$ as $3x + 10x$. Use the distributive property to factor the GCF of $3x$ out of the first two terms and the GCF of 5 out of the last two terms. Then use the distributive property to factor the GCF of $2x + 1$ out of both terms.

Answer: $6x^2 + 13x + 5 = (3x + 5)(2x + 1)$

Does this factorization check?

The procedure used to factor the trinomial in Example 1 is a generalization of the procedure used to factor trinomials of the form $ax^2 + bx + c$ with $a = 1$. This procedure is summarized in the following box.

Factoring Trinomials of the Form $ax^2 + bx + c$ by Tables and Grouping

PROCEDURE	EXAMPLE
STEP 1. Factor out the GCF if it is not 1.	Factor $6x^2 - x - 12$.
STEP 2. Find a pair of factors of ac whose sum is b. If there is not a pair of factors whose sum is b, the trinomial is prime over the integers. • If the constant c is positive, the factors of ac must have the same sign. These factors will share the same sign as the linear coefficient b. • If the constant c is negative, the factors of ac must be opposite in sign. The sign of b will determine the sign of the factor with the larger absolute value.	see table below
STEP 3. Rewrite the linear term of $ax^2 + bx + c$ so that b is the sum of the pair of factors from step 2.	$6x^2 - x - 12 = 6x^2 + 8x - 9x - 12$ $= 2x(3x + 4) - 3(3x + 4)$ $= (2x - 3)(3x + 4)$
STEP 4. Factor the polynomial from step 3 by grouping the terms and factoring the GCF out of each pair of terms.	

FACTORS OF −72		SUM OF FACTORS
1	−72	−71
2	−36	−34
3	−24	−21
4	−18	−14
6	−12	−6
8	−9	−1

Example 2 provides further practice in factoring a trinomial of the form $ax^2 + bx + c$. In $15x^2 - 41x + 14$, b is negative and ac is positive, so both factors of ac must be negative.

EXAMPLE 2 Factoring a Quadratic Trinomial

Factor $15x^2 - 41x + 14$.

SOLUTION

The product of $15 \cdot 14 = 210$. Examine the factors of 210 for a pair of factors whose sum is −41.

FACTORS OF 210		SUM OF FACTORS
−1	−210	−211
−2	−105	−107
−3	−70	−73
−5	−42	−47
−6	−35	−41
−7	−30	−37
−10	−21	−31
−14	−15	−29

The last term, 14, is positive, so the factors of 210 have the same sign. Because the linear coefficient −41 is negative, both factors of 210 must be negative.

Obviously when you determine the correct factors, you can stop your list. The whole list is shown here so that you have it for reference.

$$15x^2 - 41x + 14 = 15x^2 - 6x - 35x + 14$$
$$= 3x(5x - 2) - 7(5x - 2)$$
$$= (3x - 7)(5x - 2)$$

Rewrite the linear term $-41x$ as $-6x - 35x$. Then complete the factorization.

Answer: $15x^2 - 41x + 14 = (3x - 7)(5x - 2)$

Does this factorization check?

SELF-CHECK 6.3.1

Factor each trinomial.

1. $5x^2 + 12x + 7$
2. $15x^2 - 37x + 14$
3. $15x^2 - 107x + 14$

One thing that can be difficult to do by just examining a table of values or the graph of a polynomial function is to determine if the polynomial is prime over the integers. This can be accomplished systematically by using the method of tables and grouping. This is illustrated in Example 3. In this example, the list of all possible factors does not produce the correct middle term of $15x$. Therefore $4x^2 + 15x + 6$ must be prime.

EXAMPLE 3 Identifying a Prime Trinomial

Factor $4x^2 + 15x + 6$.

SOLUTION

Examine the factors of $(4)(6) = 24$ for a pair of factors whose sum is 15.

FACTORS OF 24		SUM OF FACTORS
1	24	25
2	12	14
3	8	11
4	6	10

None of the factor pairs of 24 have a sum of 15. Because all integral factor pairs of 24 have been systematically examined and eliminated, this polynomial is prime over the integers.

Answer: $4x^2 + 15x + 6$ is prime over the integers.

SELF-CHECK 6.3.2

Factor each trinomial.

1. $4x^2 + 9x + 5$
2. $4x^2 + 10x + 5$

Factoring $ax^2 + bx + c$ by inspection is challenging unless a and c are relatively small. When the leading coefficient a is prime, the number of options that we must consider is more manageable. This is illustrated in Example 4.

EXAMPLE 4 Factoring a Trinomial by Inspection

Factor $3x^2 + 17x + 10$ by inspection.

SOLUTION

$$3x^2 + 17x + 10 = (3x + ?)(x + ?)$$
$$= (3x + 2)(x + 5)$$

Because 3 is prime, the only factors of $3x^2$ must be $3x$ and x. The constant term 10 is positive, and the middle coefficient 17 is positive, so the sign pattern must be $(3x + ?)(x + ?)$. After mentally examining the factors of 10, select the factors of 2 and 5 as the factors of the constant term 10.

Check: $(3x + 2)(x + 5) = 3x(x + 5) + 2(x + 5)$
$$= 3x^2 + 15x + 2x + 10$$
$$= 3x^2 + 17x + 10$$

This factorization checks.

In Example 5 first we factor out a factor of -1, and then we complete the factorization of the polynomial. This is a customary step when the leading coefficient is negative.

EXAMPLE 5 Factoring a Trinomial with a Lead Coefficient of −5

Factor $-5a^2 + 3a + 2$.

SOLUTION

$$\begin{aligned} -5a^2 + 3a + 2 &= -1(5a^2 - 3a - 2) \\ &= -1(?a + ?)(?a - ?) \\ &= -1(5a + 2)(a - 1) \end{aligned}$$

First factor out a −1 so that the leading coefficient of the trinomial will be positive. To factor $5a^2 - 3a - 2$ by inspection, note that because −2 is negative the factors of −2 must have opposite signs. Also because 5 is prime, the only factors of $5a^2$ must be $5a$ and a. After mentally examining the possibilities, place the factors of −2 as shown here.

Check: $-1(5a + 2)(a - 1) = -1(5a^2 - 3a - 2)$
$= -5a^2 + 3a + 2$

This factorization checks.

In Example 6, we factor a trinomial with two variables by inspection.

EXAMPLE 6 Factoring a Trinomial with Two Variables

Factor $6v^2 - 19vw - 7w^2$.

SOLUTION

$$\begin{aligned} 6v^2 - 19vw - 7w^2 &= (?v + ?w)(?v - ?w) \\ &= (3v + w)(2v - 7w) \end{aligned}$$

To factor $6v^2 - 19vw - 7w^2$ by inspection, note that because −7 is negative the factors of −7 must have opposite signs. Also because 7 is prime, the only factors of $-7w^2$ must be either $7w$ and $-w$ or $-7w$ and w. Possible factors of the coefficient of 6 are either 1 and 6 or 2 and 3. After mentally examining the possibilities, select the factors as shown here.

Check: $(3v + w)(2v - 7w) = 6v^2 - 19vw - 7w^2$

This factorization checks.

SELF-CHECK 6.3.3

Factor each trinomial.

1. $-5a^2 + 8a - 3$
2. $11v^2 - 20vw - 4w^2$

If each term of a trinomial has a GCF other than 1, then the first step in factoring this trinomial is to factor out the GCF.

EXAMPLE 7 Completely Factoring a Trinomial

Factor $22ax^2 - 28axy + 6ay^2$.

SOLUTION

$$\begin{aligned} 22ax^2 - 28axy + 6ay^2 &= 2a(11x^2 - 14xy + 3y^2) \\ &= 2a(?x - ?y)(?x - ?y) \\ &= 2a(11x - 3y)(x - y) \end{aligned}$$

First factor out $2a$, the GCF of the trinomial. Then set up the sign pattern to factor $11x^2 - 14xy + 3y^2$ by inspection. The only factors of 11 are 1 and 11, and the only factors of 3 are 1 and 3. Place these factors as shown here.

Check: $2a(11x - 3y)(x - y) = 2a(11x^2 - 14xy + 3y^2)$
$= 22ax^2 - 28axy + 6ay^2$

This factorization checks.

With sophisticated mathematical software programs available for computers and high-end calculators, there is less emphasis on factoring trinomials with larger coefficients. For your reference Example 8 illustrates factoring $12x^2 - 29x + 15$ both by grouping and tables and by inspection. If you do this one by inspection, it is more likely to be trial and error of the possibilities than doing the whole problem mentally.

EXAMPLE 8 Factoring a Trinomial

Factor $12x^2 - 29x + 15$.

SOLUTION

BY GROUPING AND TABLES

Examine the factors of 180 for a pair of factors whose sum is -29.

FACTORS OF 180		SUM OF FACTORS
−1	−180	−181
−2	−90	−92
−3	−60	−63
−4	−45	−49
−5	−36	−41
−6	−30	−36
−9	−20	−29

First note that $12 \cdot 15 = 180$. Then examine all the negative factors of 180 in search of a pair of factors with a sum of -29.

$$\begin{aligned} 12x^2 - 29x + 15 &= 12x^2 - 9x - 20x + 15 \\ &= 3x(4x - 3) - 5(4x - 3) \\ &= (3x - 5)(4x - 3) \end{aligned}$$

After rewriting the linear term $-29x$ as $-9x - 20x$, use the distributive property to factor the GCF of $3x$ out of the first two terms and the GCF of -5 out of the last two terms. Then use the distributive property to factor the GCF of $4x - 3$ out of both terms.

BY INSPECTION

$$12x^2 - 29x + 15 = (?x - ?)(?x - ?)$$

Set up the sign pattern to factor $12x^2 - 29x + 15$ by inspection. There are several possibilities to consider that come from the factors of 12 and 15.

Factors of 12		*Factors of 15*	
1	12	1	15
2	6	3	5
3	4		

Possible Factors	*Resulting Linear Term*
$(x - 1)(12x - 15)$	$-27x$
$(x - 15)(12x - 1)$	$-181x$
$(x - 3)(12x - 5)$	$-41x$
$(x - 5)(12x - 3)$	$-63x$
$(2x - 1)(6x - 15)$	$-36x$
$(2x - 15)(6x - 1)$	$-92x$
$(2x - 3)(6x - 5)$	$-28x$
$(2x - 5)(6x - 3)$	$-36x$
$(3x - 1)(4x - 15)$	$-49x$
$(3x - 15)(4x - 1)$	$-63x$
$(3x - 3)(4x - 5)$	$-27x$
$(3x - 5)(4x - 3)$	$-29x$

Then start examining the possibilities by trial and error. Only write as many of the steps as you need. However, you do need to be systematic in eliminating possibilities so you do not overlook the correct factors.

Answer: $12x^2 - 29x + 15 = (3x - 5)(4x - 3)$

Does this factorization check?

SELF-CHECK 6.3.4

Factor each trinomial.

1. $3ax^2 - 24axy + 21ay^2$ **2.** $12a^2 - 7ab - 12b^2$

SELF-CHECK ANSWERS

6.3.1
1. $(5x + 7)(x + 1)$
2. $(15x - 7)(x - 2)$
3. $(15x - 2)(x - 7)$

6.3.2
1. $(4x + 5)(x + 1)$
2. $4x^2 + 10x + 5$ is prime

6.3.3
1. $-(5a - 3)(a - 1)$
2. $(11v + 2w)(v - 2w)$

6.3.4
1. $3a(x - 7y)(x - y)$
2. $(3a - 4b)(4a + 3b)$

USING THE LANGUAGE AND SYMBOLISM OF MATHEMATICS 6.3

1. The first step in factoring $ax^2 + bx + c$ is to factor out the __________ if it is not 1.
2. When factoring the trinomial $ax^2 + bx + c$ by grouping, we use a pair of factors of ac whose sum is __________ .
3. When we factor the trinomial $ax^2 + bx + c$, the factors of ac are of the __________ sign when $ac > 0$.
4. When we factor the trinomial $ax^2 + bx + c$, the factors of ac are of __________ sign when $ac < 0$.
5. A polynomial is prime over the integers if its only factorization must involve either __________ or __________ as one of the factors.

QUICK REVIEW 6.3

1. Multiply and simplify $(2x + 7)(3x - 1)$.
2. Multiply and simplify $(5x + 4y)(2x + 7y)$.
3. Multiply and simplify $(2x + 5)(3x - 4)$.
4. Multiply and simplify $-3xy(3x + y)(6x - 5y)$.
5. Multiply and simplify $(3x - 2)(2x + 1)(x + 5)$.

EXERCISES 6.3

In Exercises 1–4, match each trinomial with the appropriate sign pattern for the factors of this trinomial.

1. $16x^2 - 50x - 21$
2. $16x^2 + 10x - 21$
3. $16x^2 + 38x + 21$
4. $16x^2 - 50x + 21$

A. $(?x + ?)(?x + ?)$
B. $(?x + ?)(?x - ?)$
C. $(?x - ?)(?x - ?)$

In Exercises 5–8, fill in the missing + and − symbols to complete the factorization of each trinomial.

5. a. $3x^2 + 4x + 1 = (3x\_1)(x\_1)$
   b. $3x^2 - 2x - 1 = (3x\_1)(x\_1)$
   c. $3x^2 - 4x + 1 = (3x\_1)(x\_1)$
   d. $3x^2 + 2x - 1 = (3x\_1)(x\_1)$
6. a. $3x^2 + 7x + 2 = (3x\_1)(x\_2)$
   b. $3x^2 + 5x + 2 = (3x\_2)(x\_1)$
   c. $3x^2 + 5x - 2 = (3x\_1)(x\_2)$
   d. $3x^2 - x - 2 = (3x\_2)(x\_1)$
7. a. $2x^2 + x - 1 = (2x\_1)(x\_1)$
   b. $2x^2 - 3x + 1 = (2x\_1)(x\_1)$
   c. $2x^2 - x - 1 = (2x\_1)(x\_1)$
   d. $2x^2 + 3x + 1 = (2x\_1)(x\_1)$
8. a. $3x^2 - 5x + 2 = (3x\_2)(x\_1)$
   b. $3x^2 + x - 2 = (3x\_2)(x\_1)$
   c. $3x^2 - 5x - 2 = (3x\_1)(x\_2)$
   d. $3x^2 - 7x + 2 = (3x\_1)(x\_2)$

In Exercises 9–16, fill in the missing information to complete the factorization of each trinomial by inspection.

9. $5x^2 + 9x - 2 = (5x - ?)(x + ?)$
10. $5x^2 - 3x - 2 = (5x + ?)(x - ?)$
11. $5x^2 - 7x + 2 = (5x - ?)(x - ?)$
12. $5x^2 - 11x + 2 = (5x - ?)(x - ?)$
13. $6v^2 - 5v - 6 = (3v + 2)(\ ?\ )$
14. $6v^2 + 5v - 6 = (3v - 2)(\ ?\ )$
15. $6v^2 - 37v + 6 = (6v - 1)(\ ?\ )$
16. $6v^2 - 35v - 6 = (6v + 1)(\ ?\ )$

In Exercises 17–26, factor each trinomial by inspection.

17. $5x^2 + 6x + 1$
18. $11x^2 + 12x + 1$
19. $7x^2 - 8x + 1$
20. $2x^2 - 3x + 1$
21. $7x^2 - 6x - 1$
22. $5x^2 - 4x - 1$
23. $11x^2 + 10x - 1$
24. $7x^2 + 6x - 1$
25. $2x^2 + 5x + 3$
26. $2x^2 + 7x + 3$

In Exercises 27–30, fill in the blanks to complete the factorization of each trinomial.

27. $3a^2 + 17a + 10 = 3a^2 + 15a + 2a + 10$
    $= 3a(\ ) + 2(\ )$
    $= (3a + 2)(\ )$
28. $3a^2 - 11a + 10 = 3a^2 - 6a - 5a + 10$
    $= 3a(\ ) - 5(\ )$
    $= (3a - 5)(\ )$
29. $3m^2 - 10m - 8 = 3m^2 - 12m + \_\_\_\_\_ - 8$
    $= 3m(\ ) + 2(\ )$
    $= (3m + 2)(\ )$
30. $3m^2 + 23m - 8 = 3m^2 - m + \_\_\_\_\_ - 8$
    $= m(\ ) + 8(\ )$
    $= (m + 8)(\ )$

In Exercises 31–36, use the given table to factor each polynomial. (*Hint:* Some of these trinomials are prime.)

31. a. $3m^2 + 10m + 8$ **b.** $3m^2 + 11m + 8$
c. $3m^2 + 14m + 8$ **d.** $3m^2 + 25m + 8$

FACTORS OF 24		SUM OF FACTORS
1	24	25
2	12	14
3	8	11
4	6	10

32. a. $8v^2 - 14v + 3$ **b.** $6v^2 - 11v + 4$
c. $12v^2 - 25v + 2$ **d.** $6v^2 - 25v + 4$

FACTORS OF 24		SUM OF FACTORS
−1	−24	−25
−2	−12	−14
−3	−8	−11
−4	−6	−10

33. a. $4n^2 + 39n - 10$ **b.** $2n^2 + 3n - 20$
c. $5n^2 + 6n - 8$ **d.** $40n^2 + 8n - 1$

FACTORS OF −40		SUM OF FACTORS
−1	40	39
−2	20	18
−4	10	6
−5	8	3

34. a. $4n^2 - 3n - 10$ **b.** $4n^2 - 39n - 10$
c. $5n^2 - 18n - 8$ **d.** $5n^2 - 12n - 8$

FACTORS OF −40		SUM OF FACTORS
1	−40	−39
2	−20	−18
4	−10	−6
5	−8	−3

35. a. $3x^2 - 25xy + 12y^2$ **b.** $4x^2 - 15xy + 9y^2$
c. $2x^2 - 13xy + 18y^2$ **d.** $6x^2 - 37xy + 6y^2$

FACTORS OF 36		SUM OF FACTORS
−1	−36	−37
−2	−18	−20
−3	−12	−15
−4	−9	−13
−6	−6	−12

36. a. $3x^2 + 16xy - 12y^2$ **b.** $18x^2 + 9xy - 2y^2$
c. $6x^2 + 11xy - 6y^2$ **d.** $9x^2 + 35xy - 4y^2$

FACTORS OF −36		SUM OF FACTORS
−1	36	35
−2	18	16
−3	12	9
−4	9	5
−6	6	0

In Exercises 37–58, factor each polynomial.

37. $2x^2 + 15x - 7$
38. $11m^2 + 100m + 9$
39. $4w^2 - 9w + 5$
40. $9w^2 + 9w - 10$
41. $9z^2 - 9z - 10$
42. $3z^2 - 35z + 50$
43. $10b^2 + 29b + 10$
44. $15b^2 - 22b + 8$
45. $12m^2 - mn - 35n^2$
46. $12m^2 + 13mn - 35n^2$
47. $18y^2 + 55y - 28$
48. $18y^2 + 3y - 28$
49. $55y^2 - 29y - 12$
50. $24v^2 - 43v + 18$
51. $20x^2 + 37xy + 15y^2$
52. $15x^2 - 37xy + 20y^2$
53. $14x^2 - 39xy + 10y^2$
54. $15x^2 + 61xy + 22y^2$
55. $33m^2 + 13mn - 6n^2$
56. $35m^2 + 4mn - 4n^2$
57. $6v^2 + 13vw + 6w^2$
58. $10v^2 - 29vw + 10w^2$

In Exercises 59–64, first factor −1 out of each polynomial and then complete the factorization.

59. $-12x^2 + 20x - 3$
60. $-12x^2 - 20x - 3$
61. $-35y^2 + 4y + 4$
62. $-35y^2 + 24y - 4$
63. $-12z^2 - 47z - 40$
64. $-12z^2 + 17z + 40$

In Exercises 65–74, first factor out the GCF and then completely factor the polynomial.

65. $-38ax^2 + 32ax + 6a$
66. $-51ax^2 - 99ax + 6a$
67. $60x^5 - 145x^4 + 75x^3$
68. $48x^4 + 112x^3 + 60x^2$
69. $15m^3n - 6m^2n^2 - 21mn^3$
70. $44m^3n - 8m^2n^2 - 52mn^3$
71. $210av^3 + 77av^2w - 210avw^2$
72. $350av^2w + 240avw^2 - 350aw^3$
73. $35x^2(a + b) - 29x(a + b) + 6(a + b)$
74. $21x^2(a - b) + 41x(a - b) + 10(a - b)$

Group Discussion Questions

75. Discovery Question Write each of these polynomials as a trinomial in the form $ax^2 + 0x + c$ and then factor this trinomial by inspection.

a. $x^2 - 25$ **b.** $x^2 - 64$
c. $4x^2 - 1$ **d.** $4x^2 - 25$

Use your observations from these problems to write the factors of the following polynomials by inspection.

e. $x^2 - 49$ **f.** $9x^2 - 1$
g. $9x^2 - 16$

Explain what you have observed to your instructor.

76. Discovery Question

a. Use $2x^2 + 3x + 8x + 12$ to factor $2x^2 + 11x + 12$.
b. Use $2x^2 + 8x + 3x + 12$ to factor $2x^2 + 11x + 12$ and compare the results to part **a**.
c. Use $-2x^2 + 3x - 4x + 6$ to factor $-2x^2 - x + 6$.
d. Use $-2x^2 - 4x + 3x + 6$ to factor $-2x^2 - x + 6$ and compare the results to part **c**.
e. What conclusion can you reach from this work?

Explain your conclusion to your instructor.

77. Error Analysis A student graphed the equation as displayed and concluded that $100x^3 - 301x + 198$ has two zeros which are −2 and 1. How many zeros does this polynomial really have? Find these zeros.

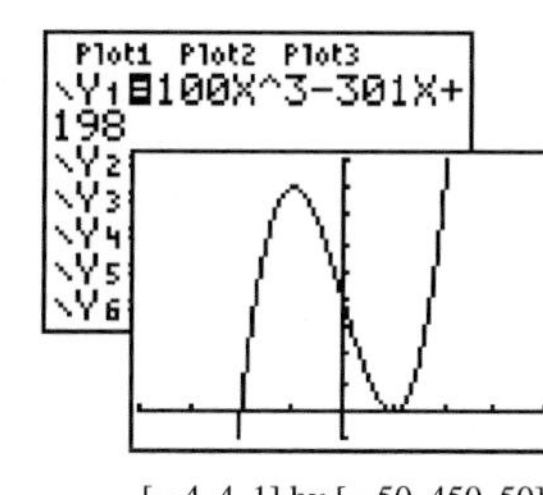

[−4, 4, 1] by [−50, 450, 50]

78. Challenge Question Part of the polynomial that defines Y_1 is blocked out on the screen shown to the right. Determine the polynomial that defines Y_1.

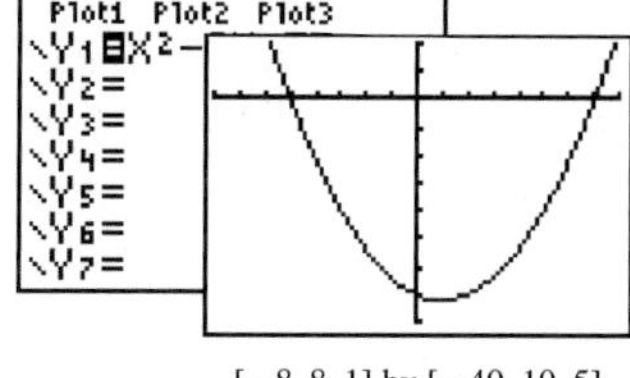

[−8, 8, 1] by [−40, 10, 5]

Section 6.4 Factoring Special Forms

Objectives:

8. Factor perfect square trinomials.
9. Factor the difference of two squares.
10. Factor the sum or difference of two cubes. (Optional)

The connection of factoring to other topics in mathematics is one of the primary reasons we study factoring. The special forms examined in this section occur frequently in many different contexts throughout mathematics. You can save considerable time and effort by memorizing these special forms. Although you can continue to use tables and grouping to factor quadratic trinomials, you can gain insight and efficiency by recognizing these patterns.

Perfect Square Trinomials

We start by reexamining the special forms first introduced in Section 5.6. A perfect square trinomial factors as the square of a binomial.

Perfect Square Trinomials

ALGEBRAICALLY	VERBALLY	ALGEBRAIC EXAMPLE
Square of a sum $x^2 + 2xy + y^2 = (x + y)^2$ **Square of a difference** $x^2 - 2xy + y^2 = (x - y)^2$	A trinomial that is the square of a binomial has 1. A first term that is a square of the first term of the binomial. 2. A middle term that is twice the product of the two terms of the binomial. 3. A last term that is a square of the last term of the binomial.	$x^2 + 12x + 36 = (x + 6)^2$ $x^2 - 12x + 36 = (x - 6)^2$

It is very useful to be able to determine whether a trinomial is a perfect square trinomial. If it is, then it can be factored according to one of the special patterns in the preceding box; and if it isn't, then it can be factored by using either inspection or tables and grouping. Example 1 illustrates how to identify perfect square trinomials.

EXAMPLE 1 Factoring Perfect Square Trinomials

Determine which of these polynomials are perfect square trinomials, and factor those that fit this form.

SOLUTION

(a) $25v^2 + 10v + 1$

$$25v^2 + 10v + 1 = (5v)^2 + 10v + (1)^2$$
$$= (5v)^2 + 2(5v)(1) + (1)^2$$
$$= (5v + 1)^2$$

Write the first and last terms as perfect squares. Then check to see if the middle term fits the form $x^2 + 2xy + y^2$.
Use the form $x^2 + 2xy + y^2 = (x + y)^2$ to write the factored form.

(b) $m^2 - 12m + 36$

$$m^2 - 12m + 36 = m^2 - 12m + (6)^2$$
$$= m^2 - 2(m)(6) + (6)^2$$
$$= (m - 6)^2$$

This trinomial fits the form $x^2 - 2xy + y^2 = (x - y)^2$ with $x = m$ and $y = 6$.

(c) $x^2 - 13x + 36$

This trinomial is not a perfect square.

Although the last term is 36, a perfect square, this trinomial does not fit the form $x^2 - 2xy + y^2$. This trinomial does factor as $x^2 - 13x + 36 = (x - 4)(x - 9)$.

(d) $x^2 - 10x - 25$ This trinomial is not a perfect square. Because the last term has a negative coefficient it cannot be a perfect square of an integer.

(e) $x^2 + 8x + 10$ This trinomial is not a perfect square. The last term 10 is not a perfect square of an integer.

SELF-CHECK 6.4.1

Factor each of these trinomials.

1. $x^2 - 18x + 81$ 2. $121w^2 + 22w + 1$ 3. $25v^2 + 60v + 36$

Difference of Two Squares

Another special form that is encountered frequently in subsequent mathematics courses is the difference of two squares. This form was first examinal in Section 5.6.

Difference of Two Squares

ALGEBRAICALLY	VERBALLY	ALGEBRAIC EXAMPLE
$x^2 - y^2 = (x + y)(x - y)$	The difference of the squares of two terms factors as the sum of these terms times their difference.	$x^2 - 49 = (x + 7)(x - 7)$

In Example 2 we practice identifying the difference of two squares.

EXAMPLE 2 Factoring the Difference of Two Squares

Determine which of these polynomials are the difference of two squares, and factor those that fit this form.

SOLUTION

(a) $100a^2 - 9$

$100a^2 - 9 = (10a)^2 - (3)^2$
$= (10a + 3)(10a - 3)$

First write this binomial in the form $x^2 - y^2$, and then factor as a sum times a difference.

(b) $9v^2 - 16w^2$

$9v^2 - 16w^2 = (3v)^2 - (4w)^2$
$= (3v + 4w)(3v - 4w)$

(c) $9v^2 + w^2$ This binomial is not the difference of two squares.

This binomial is the sum of two perfect squares, not their difference.

(d) $x^2 - 5$ This binomial is not the difference of two squares.

5 is not a perfect square integer. However, you could factor this binomial by using irrational factors as $x^2 - 5 = x^2 - (\sqrt{5})^2 = (x + \sqrt{5})(x - \sqrt{5})$

Sum of Two Squares

The sum of two squares is another special form that can be advantageous to recognize. Consider the binomial $x^2 + y^2$ in the trinomial form $x^2 + 0xy + y^2$. An attempt to factor this by the method of tables and grouping cannot produce a middle term of 0. In the form $ax^2 + bxy + cy^2$, the factors of ac must be of the same sign. Thus their sum cannot be 0. Therefore, the sum of two squares is prime over the integers.

A Mathematical Note

Pierre deFermat's last theorem is one of the most famous theorems in mathematics. It states that there do not exist positive integers x, y, z, and n such that $x^n + y^n = z^n$ for $n > 2$. Two of the features that make this theorem so interesting are as follows: (1) It is easily understood—it even resembles the Pythagorean theorem $x^2 + y^2 = z^2$ and thus shows the Pythagorean theorem does not generalize to a higher-degree equation. (2) It resisted the best attempts of mathematicians to prove until 1993 when Andrew Wiles, a British-American mathematician, proved this result.

Sum of Two Squares

ALGEBRAICALLY	VERBALLY	ALGEBRAIC EXAMPLE
$x^2 + y^2$ is prime*	The sum of two squares is prime.	$16x^2 + 9$ is prime.

* For binomials of degree greater than two this statement is not always true.

Each one of the polynomials in Example 3 could be factored by inspection or by tables and grouping. (We can consider $4a^2 - 25b^2$ as $4a^2 + 0ab - 25b^2$.) However, note how much more efficient it is to use the special forms to quickly select the correct factors.

EXAMPLE 3 Factoring Special Forms

Factor each of these polynomials.

SOLUTION

(a) $4a^2 - 20ab + 25b^2$

$$\begin{aligned}&4a^2 - 20ab + 25b^2\\ &= (2a)^2 - 2(2a)(5b) + (5b)^2\\ &= (2a - 5b)^2\end{aligned}$$

First write this trinomial in the form $x^2 - 2xy + y^2$ and then factor this perfect square trinomial.

(b) $4a^2 - 25b^2$

$$\begin{aligned}&4a^2 - 25b^2\\ &= (2a)^2 - (5b)^2\\ &= (2a + 5b)(2a - 5b)\end{aligned}$$

First write this binomial in the form $x^2 - y^2$ and then factor this difference of two squares.

(c) $4a^2 + 25b^2$

$4a^2 + 25b^2$ is prime over the integers.

This binomial can be written as $(2a)^2 + (5b)^2$. As the sum of two squares, it must be prime over the integers.

SELF-CHECK 6.4.2

Factor each of these binomials.

1. $36y^2 - 25$ 2. $100a^2 - 121b^2$ 3. $16a^2b^2 - 1$

Factor each of these polynomials.

4. $169v^2 - 100w^2$ 5. $169v^2 + 100w^2$ 6. $169v^2 + 260vw + 100w^2$

Sum or Difference of Two Cubes (Optional)

Two other special forms for binomials are $x^3 + y^3$ and $x^3 - y^3$. These forms are known as the sum of two cubes and the difference of two cubes, respectively. We start by examining $x^3 - 8$, which factors as $x^3 - 8 = (x - 2)(x^2 + 2x + 4)$. This factorization of $x^3 - 8$ can be checked by multiplication.

$$\begin{aligned}(x - 2)(x^2 + 2x + 4) &= x(x^2 + 2x + 4) - 2(x^2 + 2x + 4)\\ &= x^3 + 2x^2 + 4x - 2x^2 - 4x - 8\\ &= x^3 - 8\end{aligned}$$

Note in the graph of $y = x^3 - 8$ that (2, 0) is an x-intercept of the graph. This is visual confirmation that $x - 2$ is a linear factor of $x^3 - 8$. The graph also suggests that there are no other x-intercepts corresponding to the factor $x^2 + 2x + 4$. Is $x^2 + 2x + 4$ factorable, or is it prime? We use the method of tables and grouping to determine this.

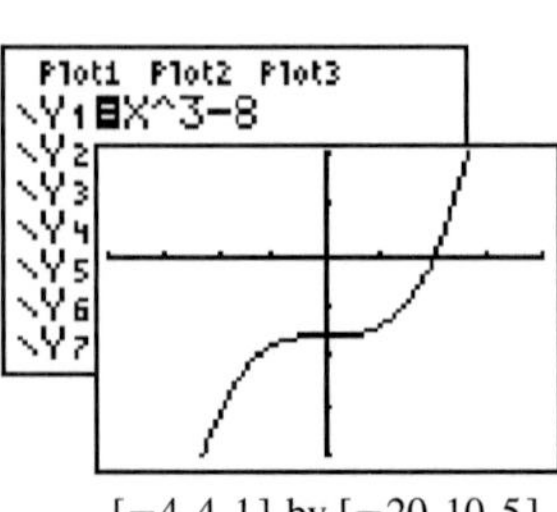

[−4, 4, 1] by [−20, 10, 5]

Does $x^2 + 2x + 4$ factor?

FACTORS OF 4		SUM OF FACTORS
1	4	5
2	2	4

To attempt to factor $x^2 + 2x + 4$, examine the factors of 4 for a pair of factors whose sum is 2. None of the pairs of factors of 4 have a sum of 2. Therefore $x^2 + 2x + 4$ is prime over the integers.

Since $x^2 + 2x + 4$ is prime over the integers, the complete factorization of $x^3 - 8$ is $x^3 - 8 = (x - 2)(x^2 + 2x + 4)$.

We now list some other products and factors that fit the form of the sum or difference of two cubes. You may wish to multiply these factors to verify these products.

PRODUCTS	FACTORS
$(x + 1)(x^2 - x + 1) = x^3 + 1$	$x^3 + 1 = (x + 1)(x^2 - x + 1)$
$(x - 1)(x^2 + x + 1) = x^3 - 1$	$x^3 - 1 = (x - 1)(x^2 + x + 1)$
$(x + 2)(x^2 - 2x + 4) = x^3 + 8$	$x^3 + 8 = (x + 2)(x^2 - 2x + 4)$
$(x - 2)(x^2 + 2x + 4) = x^3 - 8$	$x^3 - 8 = (x - 2)(x^2 + 2x + 4)$

The pattern exhibited by these polynomials is summarized in the following box.

Sum or Difference of Two Cubes

ALGEBRAICALLY	VERBALLY	ALGEBRAIC EXAMPLES
$x^3 + y^3 = (x + y)(x^2 - xy + y^2)$ $x^3 - y^3 = (x - y)(x^2 + xy + y^2)$	Write the binomial factor as the sum (or difference) of the two cube roots. Then use the binomial factor to obtain each term of the trinomial: **1.** The square of the first term of the binomial is the first term of the trinomial factor. **2.** The opposite of the product of the two terms of the binomial is the second term of the trinomial factor. **3.** The square of the last term of the binomial is the third term of the trinomial factor.	$x^3 + 8 = (x + 2)(x^2 - 2x + 4)$ $x^3 - 8 = (x - 2)(x^2 + 2x + 4)$

It is helpful to be able to recognize the perfect squares 4, 9, 16, 25, and so on, when you are examining a binomial to determine whether it is a sum or a difference of two squares. Likewise, it is helpful to be able to recognize the perfect cubes 8, 27, 64, 125, and so on, when you are examining a binomial to determine whether it is a sum or a difference of two cubes. We practice this skill in Example 4.

EXAMPLE 4 Identifying the Sum or Difference of Two Cubes

Examine each binomial to determine whether it is a sum or difference of two cubes or neither.

SOLUTION

(a) $x^3 + 64$ This binomial is the sum of two cubes. In the form $x^3 + y^3$, $x^3 + 64 = x^3 + 4^3$.

(b) $x^3 + 16$ This binomial is not the sum of two cubes. The last term, 16, is not a perfect cube.

(c) $27x^3 - 1$ This binomial is the difference of two cubes. $27x^3 - 1 = (3x)^3 - 1^3$

(d) $8x^3 - 125$ This binomial is the difference of two cubes. $8x^3 - 125 = (2x)^3 - 5^3$ ■

In Examples 5 and 6 note that we first write the binomial factor and then use this binomial to obtain each term of the trinomial factor.

EXAMPLE 5 Factoring the Sum of Two Cubes

Factor $x^3 + 64$.

SOLUTION

$$\begin{aligned} x^3 + 64 &= x^3 + 4^3 \\ &= (x + 4)[(\ \ )^2 - (\ \ )(\ \ ) + (\ \ )^2] \\ &= (x + 4)[x^2 - (x)(4) + 4^2] \\ &= (x + 4)(x^2 - 4x + 16) \end{aligned}$$

Square of the first term, x.
Opposite of the product of the two terms.
Square of the last term, 4.

First express this binomial in the form $x^3 + y^3$. Write the binomial factor and then use this factor and the form $x^3 + y^3 = (x + y)(x^2 - xy + y^2)$ to complete the factorization.
The trinomial $x^2 - xy + y^2$ is prime over the integers. ■

EXAMPLE 6 Factoring the Difference of Two Cubes

Factor $27x^3 - 1$.

SOLUTION

$$\begin{aligned} 27x^3 - 1 &= (3x)^3 - (1)^3 \\ &= (3x - 1)[(\ \ )^2 + (\ \ )(\ \ ) + (\ \ )^2] \\ &= (3x - 1)[(3x)^2 + (3x)(1) + (1)^2] \\ &= (3x - 1)(9x^2 + 3x + 1) \end{aligned}$$

Square of the first term, $3x$.
Opposite of the product of the two terms.
Square of the last term, 1.

First express this binomial in the form $x^3 - y^3$. Write the binomial factor and then use this factor and the form $x^3 - y^3 = (x - y)(x^2 + xy + y^2)$ to complete the factorization.
The trinomial $9x^2 + 3x + 1$ is prime over the integers. ■

Example 7 contains two variables and fits the form of the difference of two cubes.

EXAMPLE 7 Factoring the Difference of Two Cubes

Factor $125a^3 - 64b^3$.

SOLUTION

$$\begin{aligned} 125a^3 - 64b^3 &= (5a)^3 - (4b)^3 \\ &= (5a - 4b)[(\ \ )^2 + (\ \ )(\ \ ) + (\ \ )^2] \\ &= (5a - 4b)[(5a)^2 + (5a)(4b) + (4b)^2] \\ &= (5a - 4b)(25a^2 + 20ab + 16b^2) \end{aligned}$$

Square of the first term, $5a$.
Opposite of the product of the two terms.
Square of the last term, $4b$.

First express this binomial in the form $x^3 - y^3$.
Write the binomial factor and then use this factor and the form $x^3 - y^3 = (x - y)(x^2 + xy + y^2)$ to complete the factorization. ■

SELF-CHECK 6.4.3

Factor these binomials.

1. $x^3 + 125$ 2. $64x^3 - 1$ 3. $8x^3 - 125$ 4. $1{,}000x^3 + 27y^3$

Some polynomials with degree greater than 2 can also be factored as special forms. When you factor polynomials of higher degree, be sure to continue your work until the polynomial is factored completely. A polynomial is factored completely if each factor other than a constant factor is prime over the integers. This is illustrated in Example 8 with a fourth-degree binomial.

EXAMPLE 8 Factoring a Fourth-Degree Binomial

Factor $16y^4 - 1$.

SOLUTION

$$16y^4 - 1 = (4y^2)^2 - 1^2$$

First write this binomial in the form $x^2 - y^2$ and then factor as a sum times a difference.

$$= (4y^2 + 1)(4y^2 - 1)$$

$4y^2 + 1$ is the sum of two squares and is prime.

$$= (4y^2 + 1)[(2y)^2 - 1^2]$$

$4y^2 - 1$ is the difference of two squares and factors as a sum times a difference.

$$= (4y^2 + 1)(2y + 1)(2y - 1)$$

SELF-CHECK ANSWERS

6.4.1
1. $(x - 9)^2$
2. $(11w + 1)^2$
3. $(5v + 6)^2$

6.4.2
1. $(6y + 5)(6y - 5)$
2. $(10a + 11b)(10a - 11b)$
3. $(4ab + 1)(4ab - 1)$
4. $(13v + 10w)(13v - 10w)$
5. $169v^2 + 100w^2$ is prime.
6. $(13v + 10w)^2$

6.4.3
1. $(x + 5)(x^2 - 5x + 25)$
2. $(4x - 1)(16x^2 + 4x + 1)$
3. $(2x - 5)(4x^2 + 10x + 25)$
4. $(10x + 3y)(100x^2 - 30xy + 9y^2)$

USING THE LANGUAGE AND SYMBOLISM OF MATHEMATICS 6.4

Match each of these forms with the most appropriate description.

1. $x^2 + 2xy + y^2 = (x + y)^2$
2. $x^2 - 2xy + y^2 = (x - y)^2$
3. $x^2 - y^2 = (x + y)(x - y)$
4. $x^3 + y^3 = (x + y)(x^2 - xy + y^2)$
5. $x^3 - y^3 = (x - y)(x^2 + xy + y^2)$

A. Difference of two squares
B. Difference of two cubes
C. Sum of two cubes
D. Square of a sum
E. Square of a difference

QUICK REVIEW 6.4

1. Multiply and simplify $(x + 7)(x - 7)$.
2. Multiply and simplify $(5x + 2y)(5x + 2y)$.
3. Multiply and simplify $(6x - 5)(6x - 5)$.
4. Multiply and simplify $(3x - 2)(9x^2 + 6x + 4)$.
5. Perform the indicated operations and simplify $(3x + 7)^2 - (3x + 7)(3x - 7)$.

EXERCISES 6.4

In Exercises 1–4, fill in the blanks to complete the factorization of each perfect square trinomial.

1. $a^2 + 6a + 9 = a^2 + 2(\ )(a) + (\ )^2$
   $= (a + \ )^2$
2. $a^2 - 10a + 25 = a^2 - 2(\ )(a) + (\ )^2$
   $= (a - \ )^2$
3. $4w^2 - 12w + 9 = (\ )^2 - 2(\ )(3) + 3^2$
   $= (\ -3)^2$
4. $9w^2 + 30w + 25 = (\ )^2 + 2(\ )(5) + 5^2$
   $= (\ + 5)^2$

In Exercises 5–18, completely factor each trinomial, using the forms for perfect square trinomials. If necessary, first factor out the GCF.

5. $m^2 - 2m + 1$
6. $m^2 + 2m + 1$
7. $36v^2 + 12v + 1$
8. $100v^2 - 20v + 1$
9. $25x^2 - 20xy + 4y^2$
10. $81x^2 + 180xy + 100y^2$
11. $121m^2 + 88mn + 16n^2$
12. $169m^2 - 78mn + 9n^2$
13. $-x^2 + 16xy - 64y^2$
14. $-25x^2 + 60xy - 36y^2$
15. $9 + 16x^2 - 24x$
16. $121 + x^2 - 22x$
17. $2x^3 + 16x^2 + 32x$
18. $3x^3 + 42x^2 + 147x$

In Exercises 19–32, completely factor each binomial, using the form for the difference of two squares. If necessary, first factor out the GCF.

19. $w^2 - 49$
20. $w^2 - 36$
21. $9v^2 - 1$
22. $16v^2 - 1$
23. $81m^2 - 25$
24. $36m^2 - 49$
25. $4a^2 - 9b^2$
26. $100a^2 - 49b^2$
27. $16v^2 - 121w^2$
28. $25v^2 - 144w^2$
29. $36 - m^2$
30. $81 - m^2$
31. $20x^2 - 45y^2$
32. $75x^2 - 48y^2$

In Exercises 33–40, some of the polynomials are factorable over the integers, and others are prime over the integers. Factor those that are factorable, and label the others as prime. If necessary, first factor out the GCF.

33. a. $x^2 - 6x + 9$
    b. $x^2 + 9$
    c. $x^2 - 10x + 9$
34. a. $x^2 + 4x + 4$
    b. $x^2 + 5x + 4$
    c. $x^2 + 4$
35. a. $x^2 - 12xy + 36y^2$
    b. $x^2 - 36y^2$
    c. $x^2 - 13xy + 36y^2$
36. a. $64x^2 - y^2$
    b. $64x^2 + 20xy + y^2$
    c. $64x^2 + 16xy + y^2$
37. a. $4x^2 + 9y^2$
    b. $4x^2 - 12xy + 9y^2$
    c. $4x^2 + 15xy + 9y^2$
38. a. $16x^2 + 56xy + 49y^2$
    b. $16x^2 - 70xy + 49y^2$
    c. $16x^2 + 49y^2$
39. a. $25x^2 - 64y^2$
    b. $25x^2 - 80xy + 64y^2$
    c. $25x^2 - 100xy + 64y^2$
40. a. $100x^2 - 81y^2$
    b. $100x^2 + 225xy + 81y^2$
    c. $100x^2 + 180xy + 81y^2$

In Exercises 41–44, determine the missing term so each given trinomial is a perfect square trinomial.

41. a. $x^2 +$ ______ $+ 36$
    b. $x^2 + 12x +$ ______
42. a. $x^2 -$ ______ $+ 25$
    b. $x^2 - 10x +$ ______
43. a. $x^2 -$ ______ $+ 100$
    b. $x^2 - 20x +$ ______
44. a. $x^2 +$ ______ $+ 49$
    b. $x^2 + 14x +$ ______

45. Use the graph and table.
    a. Determine the x-intercepts of this graph.
    b. Determine the zeros of the polynomial.
    c. Factor this polynomial.

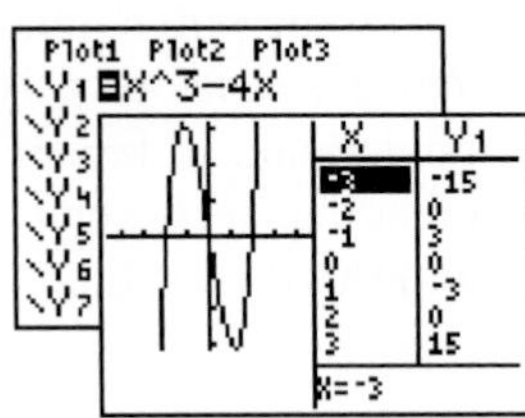

$[-4.7, 4.7, 1]$ by $[-3.1, 3.1, 1]$

46. Use the graph and table.
    a. Determine the x-intercepts of this graph.
    b. Determine the zeros of the polynomial.
    c. Factor this polynomial.

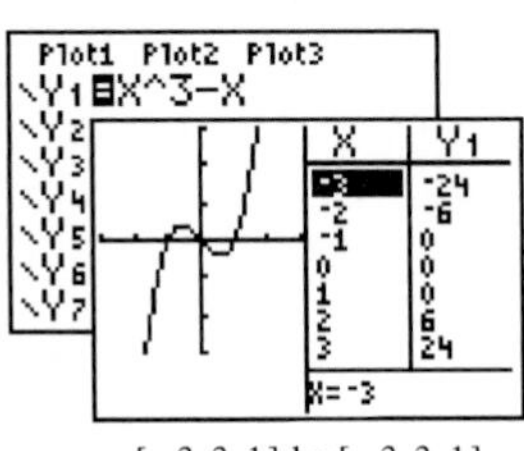

$[-3, 3, 1]$ by $[-3, 3, 1]$

In Exercises 47–54, completely factor each binomial, using the form for the sum or difference of two cubes. (Optional)

47. $x^3 + 27$
48. $x^3 - 27$
49. $m^3 - 125$
50. $m^3 + 125$
51. $64a^3 - b^3$
52. $64a^3 + b^3$
53. $125x^3 + 8y^3$
54. $64a^3 - 125b^3$

In Exercises 55–72, completely factor each polynomial.

55. $x_1^2 - x_2^2$ (the area between two squares)

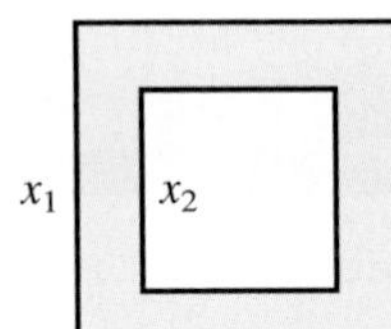

56. $\pi r_1^2 - \pi r_2^2$ (the area between two concentric circles)

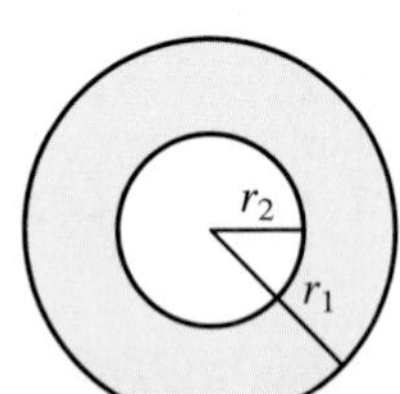

57. $\pi r_1^2 h - \pi r_2^2 h$ (the volume of steel in a steel pipe)

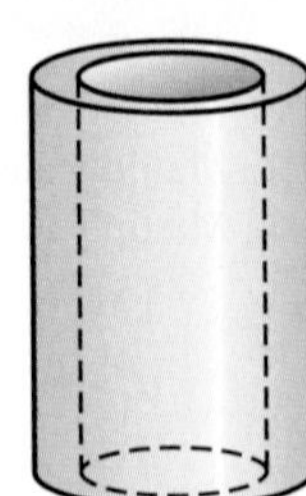

A General Strategy for Factoring Polynomials

58. The area of a window is given by the function $A(x) = 7x - x^2$, where x represents the width of the window.
   a. Write this function in factored form.
   b. Explain the meaning of each factor. (*Hint:* $A = l \cdot w$.)
   c. Is this function more meaningful in expanded form or in factored form? Explain.

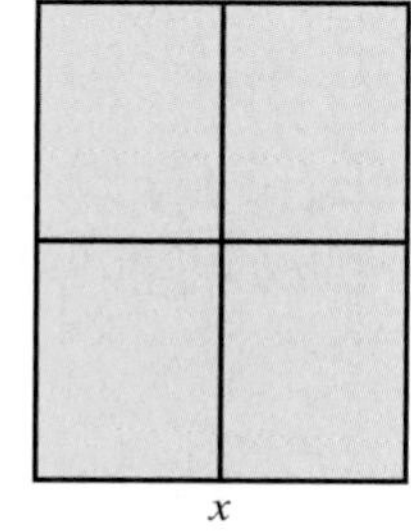

59. A steel sheet 7 in wide by 9 in long has squares x cm on a side cut from each corner, and then the sides are turned up to form the steel tray shown. The volume of this tray is given by the function $V(x) = 63x - 32x^2 + 4x^3$.
   a. Write this function in factored form. (*Hint:* Do not reorder the terms.)
   b. Explain the meaning of each factor. (*Hint:* $V = l \cdot w \cdot h$.)
   c. Is this function more meaningful in expanded form or in factored form? Explain.

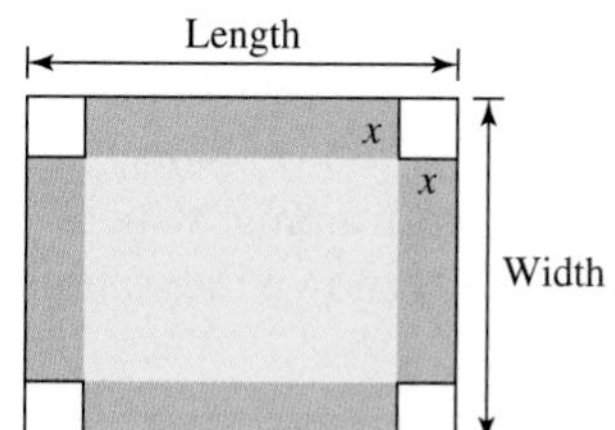

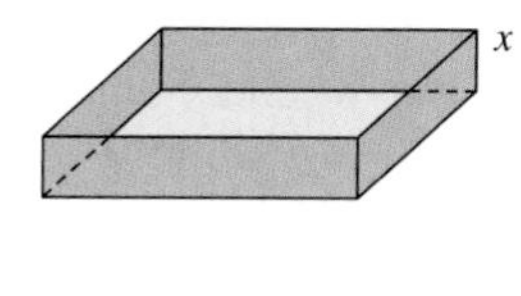

In Exercises 60–72, completely factor each polynomial.

60. $-16t^2 + 96t - 144$
61. $x^4 - 9$
62. $x^6 - 36$
63. $25y^4 - 1$
64. $x^4 - 16$
65. $81y^4 - 1$
66. $m^2n^2 - 25$
67. $(a + 2b)^2 - (2a + b)^2$
68. $(a - 5b)x^2 - (a - 5b)y^2$
69. $(a + b)x^2 + 6(a + b)x + 9(a + b)$
70. $(a - b)x^2 - 10(a - b)x + 25(a - b)$
71. $(x^2 - y^2)a^2 - (x^2 - y^2)b^2$
72. $(x^2 + 20xy + 100y^2)a^2 - 16(x^2 + 20xy + 100y^2)$

In Exercises 73–76, factor each binomial. (Optional)

73. $z^3 - 1{,}000$
74. $8ax^3 - 216ay^3$
75. $343v^3 + w^3$
76. $32a^3 + 4$

Calculator Usage

In Exercises 77–80, use your calculator to factor out the GCF and then complete each factorization.

77. $7.46x^2 - 89.52x + 268.56 = 7.46(\quad) = 7.46(\quad)^2$
78. $5.83x^2 + 34.98xy + 52.47y^2 = 5.83(\quad) = 5.83(\quad)^2$
79. $3.64v^2 - 91 = 3.64(\quad) = 3.64(\quad)(\quad)$
80. $-9.21a^2 + 147.36 = -9.21(\quad) = -9.21(\quad)(\quad)$

Group Discussion Questions

81. **Challenge Question** Factor each polynomial completely, assuming that m and n are natural numbers.
   a. $81y^{2n} - 16$
   b. $x^{2m} - y^{2n}$
   c. $4x^{2m} + 20x^my^n + 25y^{2n}$
   d. $10x^{2m} + 4x^my^n - 6y^{2n}$
82. **Challenge Question** The binomial $a^6 + b^6$ can be written as the sum of two squares $(a^3)^2 + (b^3)^2$. Yet $a^6 + b^6$ is not prime. Factor $a^6 + b^6$.

Section 6.5 Factoring by Grouping and a General Strategy for Factoring Polynomials

Objectives:

11. Factor polynomials by the method of grouping.
12. Determine the most appropriate technique for factoring a polynomial.

The factoring techniques developed in previous sections of this chapter are used primarily with binomials and trinomials. For polynomials with four or more terms a technique called **factoring by grouping** is often used. Instead of examining the entire polynomial at one time, we first group terms that we already know how to factor. Thus factoring by grouping relies heavily on the ability to spot common factors and special forms. This grouping method is an extension of the method used to factor trinomials.

EXAMPLE 1 Factoring by Grouping

Factor $3ac - bc + 6a - 2b$.

SOLUTION

$$\begin{aligned} 3ac - bc + 6a - 2b &= (3ac - bc) + (6a - 2b) \\ &= c(3a - b) + 2(3a - b) \\ &= (c + 2)(3a - b) \end{aligned}$$

There is no common factor other than 1 for all four terms. Thus we group the first two terms, which have a GCF of c. Also we group the last two terms, which have a GCF of 2. Then we factor $3a - b$ out of both terms.

To determine an advantageous grouping for a polynomial, examine the polynomial for groups that are special forms. This is illustrated in Example 2, where the first two terms form the difference of two squares.

EXAMPLE 2 Factoring by Grouping

Factor $x^2 - y^2 + x - y$.

SOLUTION

$$\begin{aligned} x^2 - y^2 + x - y &= (x^2 - y^2) + (x - y) \\ &= (x - y)(x + y) + (x - y)(1) \\ &= (x - y)(x + y + 1) \end{aligned}$$

Note that the first two terms form the difference of two squares. Group them and factor $x^2 - y^2$. Then factor $x - y$ out of both terms.

Caution: This term of 1 must be included to account for the term $(x - y)(1)$ in the previous step.

The terms of a polynomial can be grouped in several ways. It is possible that one grouping will lead to a factorization, whereas other groupings will prove useless. For example, it is tempting to group $z^2 - a^2 + 4ab - 4b^2$ as $(z^2 - a^2) + (4ab - 4b^2)$, which would result in $(z + a)(z - a) + 4b(a - b)$. However, this grouping, which has a GCF of 1, is not useful. A more useful grouping of this polynomial is shown in Example 3.

EXAMPLE 3 Factoring by Grouping

Factor $z^2 - a^2 + 4ab - 4b^2$.

SOLUTION

$$\begin{aligned} z^2 - a^2 + 4ab - 4b^2 &= z^2 - (a^2 - 4ab + 4b^2) \\ &= z^2 - (a - 2b)^2 \\ &= [z + (a - 2b)][z - (a - 2b)] \\ &= (z + a - 2b)(z - a + 2b) \end{aligned}$$

Group the last three terms, noting the sign change of each term within the parentheses. Then factor the perfect square trinomial within the parentheses.
This polynomial fits the form of the difference of two squares. Factor this special form as a sum times a difference and then simplify.

Sometimes it is advantageous to reorder the terms of a polynomial to facilitate a useful grouping. This is illustrated in Example 4.

EXAMPLE 4 Factoring by Grouping

Factor $y^3 + x + x^3 + y$.

SOLUTION

$$\begin{aligned} y^3 + x + x^3 + y &= (x^3 + y^3) + (x + y) \\ &= (x + y)(x^2 - xy + y^2) + (x + y)(1) \\ &= (x + y)[(x^2 - xy + y^2) + 1] \\ &= (x + y)(x^2 - xy + y^2 + 1) \end{aligned}$$

Reorder the terms and group the sum of the two perfect cubes. Then factor this special form. Now factor out the GCF of $x + y$.
Be sure to include the last term of 1.

SELF-CHECK 6.5.1

Factor each polynomial.

1. $rt - st + rv - sv$ **2.** $x^2 - 10x + 25 - 36y^2$

It is important to practice factoring a variety of polynomials so that you can quickly select the appropriate technique. The exercises at the end of this section contain a mix of problems intended to promote the development of a general factoring strategy. The following box outlines the methods of factoring covered earlier in this chapter. A key step in determining how to factor a polynomial is to note the number of terms in the polynomial.

Strategy for Factoring a Polynomial Over the Integers

After factoring out the GCF (greatest common factor), proceed as follows.

Binomials

Factor special forms:

$x^2 - y^2 = (x + y)(x - y)$	Difference of two squares
$x^3 - y^3 = (x - y)(x^2 + xy + y^2)$	Difference of two cubes
$x^3 + y^3 = (x + y)(x^2 - xy + y^2)$	Sum of two cubes
$x^2 + y^2$ is prime	The sum of two squares is prime if x^2 and y^2 are only second-degree terms and have no common factor other than 1.

Trinomials

Factor the forms that are perfect squares:

$x^2 + 2xy + y^2 = (x + y)^2$	Perfect square trinomial; a square of a sum
$x^2 - 2xy + y^2 = (x - y)^2$	Perfect square trinomial; a square of a difference

Factor trinomials that are not perfect squares by inspection if possible; otherwise, use the method of tables and grouping.

Polynomials of Four or More Terms

Factor by grouping.

The first step in Example 5 is to factor out the GCF of $5xy$ from each term.

EXAMPLE 5 Completely Factoring a Binomial

Factor $5x^3y - 5xy^3$.

SOLUTION

$5x^3y - 5xy^3 = 5xy(x^2 - y^2)$	Factor out the GCF of $5xy$.
$= 5xy(x + y)(x - y)$	Noting that $x^2 - y^2$ is the difference of two perfect squares, factor this special form.

The first step in Example 6 is to factor out the GCF of $6ab^2$ from each term.

EXAMPLE 6 Completely Factoring a Trinomial

Factor $6ab^2x^2 + 30ab^2xy + 36ab^2y^2$.

SOLUTION

$$\begin{aligned} 6ab^2x^2 + 30ab^2xy + 36ab^2y^2 &= 6ab^2(x^2 + 5xy + 6y^2) \\ &= 6ab^2(x + 2y)(x + 3y) \end{aligned}$$

Factor out the GCF of $6ab^2$. Next factor the trinomial by inspection. Two factors of 6 with a sum of 5 are 2 and 3.

SELF-CHECK 6.5.2

Completely factor each polynomial.

1. $12ax^3 - 75axy^2$ 2. $44x^3y - 220x^2y^2 + 275xy^3$ 3. $8x^4y - 12x^3y^2 - 20x^2y^3$

The polynomial in Example 7 contains four terms. Factor out the GCF, and then use the grouping method to factor the polynomial.

EXAMPLE 7 Completely Factoring a Polynomial

Factor $3x^3y^2 - 3xy^4 - 3x^2y^2 + 3xy^3$.

SOLUTION

$3x^3y^2 - 3xy^4 - 3x^2y^2 + 3xy^3$

$= 3xy^2(x^2 - y^2 - x + y)$ Factor out the GCF of $3xy^2$.

$= 3xy^2[(x^2 - y^2) - 1(x - y)]$ Group the first two terms that fit the form of a difference of two squares. Also factor -1 out of the last two terms $-x + y$ to rewrite them as $-1(x - y)$.

$= 3xy^2[(x + y)(x - y) - 1(x - y)]$

$= 3xy^2(x + y - 1)(x - y)$

Factor the difference of two squares as a sum times a difference. Then factor out the common factor of $x - y$. Be sure to include the last term of -1 in the last factor.

The substitution shown in Example 8 is not necessary and may be skipped. However, the substitution is shown to emphasize that this polynomial can be considered a trinomial of the form $a^2 - 8ab + 15b^2$. To apply the special forms and the methods of factoring presented here, it is important to recognize the variety of expressions that fit these forms.

EXAMPLE 8 Using Substitution to Factor a Polynomial

Factor $(x + 3y)^2 - 8b(x + 3y) + 15b^2$.

SOLUTION

The polynomial $(x + 3y)^2 - 8b(x + 3y) + 15b^2$ is of the form $a^2 - 8ab + 15b^2$.

Substitute a for the quantity $x + 3y$.

To factor $a^2 - 8ab + 15b^2$, notice that

$$\begin{aligned} a^2 - 8ab + 15b^2 &= (a - ?b)(a - ?b) \\ &= (a - 5b)(a - 3b) \end{aligned}$$

The constant term 15 is positive, so the factors of 15 must be of the same sign. The middle term has a coefficient of -8, so both factors of 15 are negative and the sign pattern is $(a - ?b)(a - ?b)$. By inspection you can determine that the factors of 15 with a sum of -8 are -3 and -5.

$$a^2 - 8ab + 15b^2 = (a - 5b)(a - 3b)$$
$$(x + 3y)^2 - 8(x + 3y)b + 15b^2 = [(x + 3y) - 5b][(x + 3y) - 3b]$$
$$= (x + 3y - 5b)(x + 3y - 3b)$$

Substitute $x + 3y$ back for a. Then simplify each factor.

Answer: $(x + 3y)^2 - 8b(x + 3y) + 15b^2 = (x + 3y - 5b)(x + 3y - 3b)$

SELF-CHECK 6.5.3

Completely factor each polynomial.

1. $7m^3 - 35m^2 - 42m$
2. $(a + 3b)^2 - 6(a + 3b) + 8$

A polynomial is prime over the integers if its only factorization must involve 1 or -1 as one of the factors. You should continue all factorizations until each factor, other than a monomial factor, is prime. The polynomials presented in the examples and exercises in this section either are prime or can be factored by the strategy given in this section. Other methods of factoring are usually considered in college algebra courses.

SELF-CHECK ANSWERS

6.5.1
1. $(r - s)(t + v)$
2. $(x + 6y - 5)(x - 6y - 5)$

6.5.2
1. $3ax(2x + 5y)(2x - 5y)$
2. $11xy(2x - 5y)^2$
3. $4x^2y(2x - 5y)(x + y)$

6.5.3
1. $7m(m + 1)(m - 6)$
2. $(a + 3b - 2)(a + 3b - 4)$

USING THE LANGUAGE AND SYMBOLISM OF MATHEMATICS 6.5

1. The method often used to factor polynomials with four or more terms is called factoring by __________.
2. GCF stands for __________ __________ __________.
3. The binomial $x^2 + y^2$ is a __________ polynomial.
4. The perfect square trinomial $x^2 + 2xy + y^2$ is the square of a __________.
5. The perfect square trinomial $x^2 - 2xy + y^2$ is the square of a __________.
6. The form $x^3 + y^3$ is the sum of two __________.
7. The form $x^3 - y^3$ is the __________ of two cubes.

QUICK REVIEW 6.5

1. Factor out the GCF. $a(x - y) - 3a$
2. Factor out the GCF. $2bx - 6b(y - 1)$
3. Factor out the GCF. $(3a - b)x^2 - 3(3a - b)x + 9(3a - b)$
4. Factor the following difference of two squares. $(a + 2b)^2 - 9y^2$
5. Factor the following difference of two squares. $25x^2 - (a - 3b)^2$

EXERCISES 6.5

In Exercises 1–4, fill in the blanks to complete the factorization of each polynomial by the method of grouping.

1. $x^2 - xy + 5x - 5y = x(\ ) + 5(\ )$
   $= (x + 5)(\ )$
2. $xy + xz - 2y - 2z = x(\ ) - 2(\ )$
   $= (x - 2)(\ )$
3. $x^2 - y^2 + 2y - 1 = x^2 - (\quad)$
   $= x^2 - (\ )^2$
   $= [x + (\ )][x - (\ )]$
   $= (x + y - 1)(\quad)$
4. $x^2 - 5xy - 6y^2 + x - 6y = (x^2 - 5xy - 6y^2) + (x - 6y)$
   $= (x - 6y)(\ ) + (x - 6y)(1)$
   $= (x - 6y)(\quad)$

In Exercises 5–22, factor each polynomial completely using the technique of grouping.

5. $ac + bc + ad + bd$
6. $xy + xz + 2y + 2z$
7. $3a - 6b + 5ac - 10bc$
8. $6a^2 + 3ab + 2a + b$
9. $ab + bc - ad - cd$
10. $ac + bc + a + b$
11. $v^2 - vw - 7v + 7w$
12. $mn - 7m - n + 7$
13. $4a^2 + 12a + 9 - 16b^2$
14. $16x^2 - a^2 - 2a - 1$
15. $3mn + 15m - kn - 5k$
16. $az^2 + bz^2 + aw^2 + bw^2$
17. $az^3 + bz^3 + aw^2 + bw^2$
18. $s^3 + 11s^2 + s + 11$
19. $9b^2 - 24b + 16 - a^2$
20. $ax - ay - az + bx - by - bz$
21. $ay^2 + 2ay - y + a - 1$
22. $x^2 - 14x + 49 - 16y^2$

In Exercises 23–72, completely factor each polynomial by using the strategy outlined in this section.

23. $64y^2 - 9z^2$
24. $25a^2 - 144b^2$
25. $16x^2 + 49y^2$
26. $3ax^2 + 33ax + 72a$
27. $12x^2 - 27x + 15$
28. $121m^2 + 49n^2$
29. $49a^2 - 28a + 4$
30. $25a^2 - 10a + 1$
31. $x(a - b) + y(a - b)$
32. $a(x - y) - b(x - y)$
33. $10w^2 - 6w - 21$
34. $3s^2 + 3s + 3t - 3t^2$
35. $25v^2 - vw + 36w^2$
36. $32x^3 + 12x^2 - 20x$
37. $4x^{10} + 12x^5y^3 + 9y^6$
38. $9s^2 - 63$
39. $12x^3y - 12xy^3$
40. $25y^2 - 30yz + 9z^2$
41. $cx + cy + dx + dy$
42. $5a^2bc - 5b^3c$
43. $ax^2 + ax + bxy + by$
44. $20x^3y - 245xy^3$
45. $x^6 + 4x^3y + 4y^2$
46. $3ax^2 + 3ay^2$
47. $64ax^2 - 104ax + 40a$
48. $100s^4 + 120s^3t + 36s^2t^2$
49. $5x^2 - 55$
50. $200x^2 + 2$
51. $63a^3b - 175ab$
52. $49b^2 + 126bc + 81c^2$
53. $-12x^2 + 12xy - 3y^2$
54. $-71ax^4 + 71a$
55. $144x^2 + 81$
56. $x^2 + 2xy + y^2 - 16z^2$
57. $9x^2 - 6x + 1 - 25y^2$
58. $2x^2 + 2x + 2y - 2y^2$
59. $12ax^2 - 10axy - 12ay^2$
60. $8ax^2 - 648ay^4$
61. $7s^5t - 7st^5$
62. $-6ax^3 + 24ax$
63. $3ax^2 - 3ay^2 + 6ay - 3a$
64. $x^3 + 4x^2y + 4xy^2$
65. $2ax^3 + 10ax^2 - 28ax$
66. $30ax^3 + 5ax^2 - 10ax$
67. $a^2 + 2a + 1 + ab + b$
68. $30x^3 - 57x^2 + 18x$
69. $(a - 2b)x^2 - 2(a - 2b)x - 24(a - 2b)$
70. $36(a - b)x^2 - 6(a - b)xy - 20(a - b)y^2$
71. $(4x^2 - 12xy + 9y^2) + (72ay - 48ax) - 25a^2$
72. $(25x^2 - 10xy + y^2) + (10xz - 2yz) - 24z^2$

In Exercises 73 and 74, use a graphing calculator to graph $y = P(x)$. Use the graph to determine the x-intercepts of this graph, the zeros of $P(x)$, and the factors of this polynomial.

POLYNOMIAL $P(x)$	x-INTERCEPTS OF THE GRAPH OF $y = P(x)$	ZEROS OF $P(x)$	FACTORED FORM
73. $x^3 - 3x^2 - 10x + 24$			
74. $(x^3 - 4x) + (5x^2 - 20)$			

In Exercises 75–78, completely factor each polynomial by using the strategy outlined in this section. (Optional)

75. $5ab^3 - 5a$
76. $bw^3 + b$
77. $x^3 + y^3 + x^2 - y^2$
78. $x^3 + y^3 - 3x - 3y$

Group Discussion Questions

79. **Challenge Question** Factor each polynomial completely, assuming that m and n are natural numbers.
   a. $x^{m+2} + 4x^m$
   b. $x^{3m} + 6x^{2m} + 9x^m$

80. **Challenge Question** Use the distributive property to expand the first expression and to factor the second expression.

Expand	Factor
a. $x^{-4}(x^2 - 3)$	$x^{-2} - 3x^{-4} = x^{-4}(\quad)$
b. $2x^{-5}(x^2 - 2)$	$2x^{-3} - 4x^{-5} = 2x^{-5}(\quad)$
c. $x^{-4}(2x - 1)(x - 3)$	$2x^{-2} - 7x^{-3} + 3x^{-4} = x^{-4}(\quad)$
	$= x^{-4}(\quad)(\quad)$

81. **Challenge Question** First factor $2{,}025x^4 + 2{,}700x^2y + 900y^2$ by factoring this perfect square trinomial as the square of a sum. Then start over and factor the GCF from this trinomial. Complete each of these factorizations and compare the results, stating which factorization you find easier.

Section 6.6 Solving Equations by Factoring

Objectives:

13. Use factoring to solve selected second- and third-degree equations.
14. Solve a quadratic inequality by using the x-intercepts of the corresponding graph.

A Mathematical Note

The importance of quadratic equations has been noted by many ancient civilizations. A surviving document from Egypt (c. 2000 B.C.) contains quadratic equations. The Hindu mathematician Brahmagupta (c. 628) included a version of the quadratic formula for solving quadratic equations in his works.

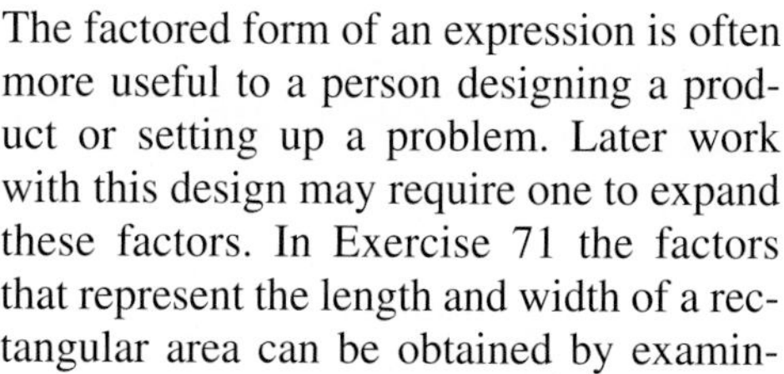

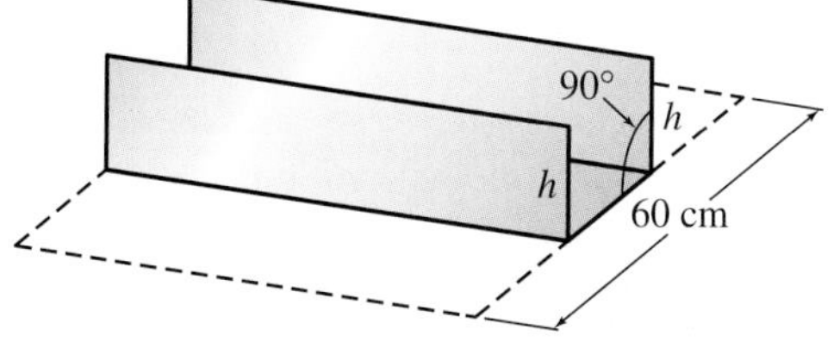

The factored form of an expression is often more useful to a person designing a product or setting up a problem. Later work with this design may require one to expand these factors. In Exercise 71 the factors that represent the length and width of a rectangular area can be obtained by examining the figure. After writing the expression for the area in factored form, you will need to expand this expression to solve a second-degree equation for a given area.

A quadratic equation in x is a second-degree equation that can be written in the standard form $ax^2 + bx + c = 0$, where a, b, and c represent real constants and $a \neq 0$. (If we allowed a to equal zero, the equation would not be quadratic; instead it would be the linear equation $bx + c = 0$.) In the quadratic equation $ax^2 + bx + c = 0$, ax^2 is called the **quadratic term,** bx is called the **linear term,** and c is called the **constant term.**

Quadratic Equation

ALGEBRAICALLY	VERBALLY	ALGEBRAIC EXAMPLE
If a, b, and c are real constants and $a \neq 0$, then $ax^2 + bx + c = 0$ is the standard form of a quadratic equation in x.	A quadratic equation in x is a second-degree equation in x.	$3x^2 - 4x + 1 = 0$ is a quadratic equation with Quadratic term: $3x^2$ Linear term: $-4x$ Constant term: 1

EXAMPLE 1 Identifying Quadratic Equations

Determine which of these equations are quadratic equations in one variable. Write each quadratic equation in standard form.

SOLUTION

(a) $5y - 6 = -y^2$

This equation is a quadratic equation in y and can be written in standard form as $y^2 + 5y - 6 = 0$.

In this standard form $a = 1$, $b = 5$, and $c = -6$. It is customary to write the equation so that a is positive.

(b) $5w = 12$

This equation is a linear equation in w; it is not a quadratic equation.

Quadratic equations are second-degree. Linear equations are first-degree.

(c) $5v^2 = 3v - 4$

This equation is a quadratic equation in v and can be written in standard form as $5v^2 - 3v + 4 = 0$.

The quadratic term is $5v^2$, the linear term is $-3v$, and the constant term is 4.

(d) $x^2 + 7x = x^3 - 5$

This equation is a third-degree equation; it is not a quadratic equation.

(e) $x^2 + 3x + 7$

This is not an equation, just a polynomial expression.

(f) $x^2 = 6x$

This equation is a quadratic equation and can be written in standard form as $x^2 - 6x = 0$.

In standard form, $a = 1$, $b = -6$, and c is understood to be 0.

Caution: Some quadratic equations cannot be solved by factoring over the integers.

The easiest method for solving some quadratic equations is to use factoring and the zero-factor principle. Because some quadratic equations are not factorable over the integers, some quadratic equations cannot be solved by factoring over the integers. We will cover methods for solving those quadratic equations in Section 7.4, but first we state the zero-factor principle.

Zero-Factor Principle

For real numbers a, b, and x:

ALGEBRAICALLY	VERBALLY	ALGEBRAIC EXAMPLE
If $a = 0$ or $b = 0$, then $ab = 0$.	The product of zero and any other factor is zero.	$0 \cdot x = 0$
If $ab = 0$, then $a = 0$ or $b = 0$.	If the product of two or more factors is zero, at least one of the factors must be zero.	If $(x - 3)(x + 4) = 0$, then either $x - 3 = 0$ or $x + 4 = 0$.

The word *or* is used in the inclusive sense to mean either $a = 0$ or $b = 0$ (or both a and b are 0).

EXAMPLE 2 Solving a Quadratic Equation in Factored Form

Solve $(v - 2)(v + 3) = 0$.

SOLUTION

$$(v - 2)(v + 3) = 0$$

$$v - 2 = 0 \quad \text{or} \quad v + 3 = 0$$

By the zero-factor principle, either the first factor or the second factor is zero.

$$v = 2 \qquad\qquad v = -3$$

Notice that there are two solutions.

Answer: $v = 2$, $v = -3$

Do both of these values check?

The quadratic equation in Example 2 was carefully selected so we could easily apply the zero-factor principle. We now examine a wider variety of quadratic equations by using the strategy described in the following box. The steps in this procedure are illustrated in Example 3.

Solving Quadratic Equations by Factoring

VERBALLY	ALGEBRAIC EXAMPLE
1. Write the equation in standard form, with the right side zero.	$x^2 + x = 6$ $x^2 + x - 6 = 0$
2. Factor the left side of the equation.	$(x - 2)(x + 3) = 0$
3. Set each factor equal to zero.	$x - 2 = 0$ or $x + 3 = 0$
4. Solve the resulting first-degree equations.	$x = 2$ $\quad$ $x = -3$

EXAMPLE 3 Solving a Quadratic Equation by Factoring

Solve $6x^2 + 7x = 5$.

SOLUTION

$6x^2 + 7x = 5$	Write the equation in standard form, with the right side 0.
$6x^2 + 7x - 5 = 0$	Factor the left side of the equation.
$(?x + ?)(?x - ?) = 0$ $(3x + 5)(2x - 1) = 0$	To factor $6x^2 + 7x - 5$ by inspection, note that the factors of -5 must have opposite signs. Also because 5 is prime, the only factors of 5 must be 1 and 5. The possible factor pairs of 6 are 1 and 6 or 2 and 3.
$3x + 5 = 0$ or $2x - 1 = 0$	Set each factor equal to 0.
$3x = -5$ $\quad 2x = 1$	Solve the resulting first-degree equations.
$x = -\frac{5}{3}$ $\quad x = \frac{1}{2}$	Notice that there are two solutions.
Answer: $x = -\frac{5}{3}, x = \frac{1}{2}$	Do both of these values check?

If a quadratic equation contains fractions, then we can simplify it by converting it to an equivalent equation that does not involve fractions. To do this, multiply both sides of the equation by the LCD of all the terms. The LCD in Example 4 is 8.

EXAMPLE 4 Solving a Quadratic Equation Containing Fractions

Solve $\frac{1}{8}x^2 - \frac{1}{4}x = 1$.

SOLUTION

A common theme in algebra is to take a given problem and convert it to an equivalent but easier problem. Multiplying both sides of the equation by the LCD produces an equivalent equation that does not contain fractions.

$\frac{1}{8}x^2 - \frac{1}{4}x = 1$	
$\frac{1}{8}x^2 - \frac{1}{4}x - 1 = 0$	First write the equation in standard form.
$8\left(\frac{1}{8}x^2 - \frac{1}{4}x - 1\right) = 8(0)$	Multiply both sides by the LCD of 8 to obtain integer coefficients.
$x^2 - 2x - 8 = 0$ $(x - 4)(x + 2) = 0$	Use the sign pattern $(x - ?)(x + ?)$ to factor $x^2 - 2x - 8$ by inspection.
$x - 4 = 0$ or $x + 2 = 0$	Set each factor equal to 0.
$x = 4$ $\quad x = -2$	Solve each of these linear equations.
Answer: $x = 4, x = -2$	Do both these values check?

SELF-CHECK 6.6.1

Solve each quadratic equation.

1. $(2x - 9)(x + 5) = 0$ **2.** $6x^2 - 37x + 22 = 0$ **3.** $x^2 - \frac{5}{2}x - \frac{3}{2} = 0$

It is important that one side of a quadratic equation be 0 before you try to apply the zero-factor principle. Failure to observe this is the source of many student errors.

EXAMPLE 5 Solving a Quadratic Equation Not in Standard Form

Caution: The zero-factor principle works only if the product is 0.

Solve $(6x + 1)(x - 2) = 8$.

SOLUTION

$$(6x + 1)(x - 2) = 8$$
$$6x^2 - 11x - 2 = 8$$
$$6x^2 - 11x - 10 = 0$$

The first step is to multiply the factors on the left side of the equation and write the equation in standard form with the right side equal to 0. Then factor the left side of the equation.

$$(?x - ?)(?x + ?) = 0$$
$$(2x - 5)(3x + 2) = 0$$

To factor $6x^2 - 11x - 10$ by inspection, note that the factors of -10 must have opposite signs. The possible factor pairs of 6 are 1 and 6 or 2 and 3. The possible factor pairs of 10 are 1 and 10 or 2 and 5.

$$2x - 5 = 0 \quad \text{or} \quad 3x + 2 = 0$$
$$2x = 5 \qquad\qquad 3x = -2$$
$$x = \frac{5}{2} \qquad\qquad x = -\frac{2}{3}$$

Set each factor equal to 0. Then solve each of the linear equations for x.

Answer: $x = \frac{5}{2}, x = -\frac{2}{3}$

Do both these solutions check?

Recall that a solution to an equation is an x-value that causes the left side of the equation to be equal to the right side of the equation. Calculator Perspective 6.6.1 illustrates another way to take advantage of the **Table** features of a graphing calculator.

CALCULATOR PERSPECTIVE 6.6.1 Using the Ask Mode to Check Solutions of an Equation

To use a TI-84 Plus calculator to check the solutions $x = \frac{5}{2}$ and $x = -\frac{2}{3}$ in the equation $(6x + 1)(x - 2) = 8$ from Example 5, enter the following keystrokes:

(Set the independent variable x to **Ask** mode.)

[Enter $Y_1 = (6x + 1)(x - 2)$ and $Y_2 = 8$ on the Y= screen.]

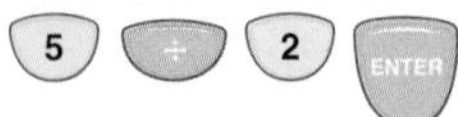

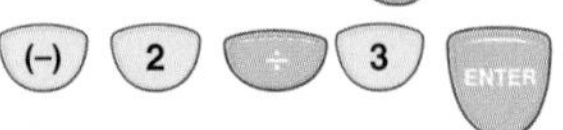

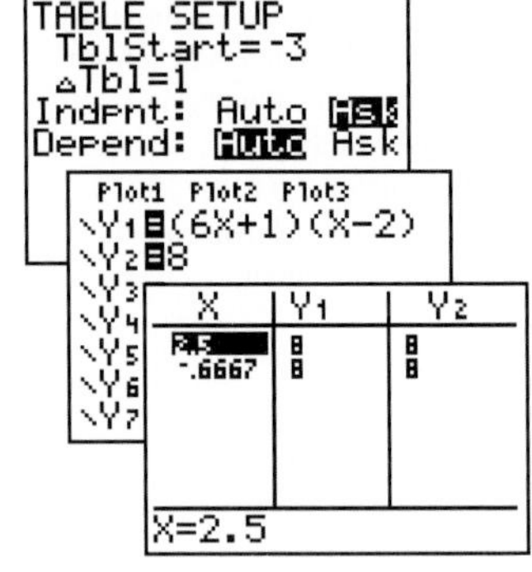

When $x = \frac{5}{2}$ and $x = -\frac{2}{3}$, the value of the left side of the equation is equal to the value of the right side of the equation ($Y_1 = Y_2$). Thus, $x = \frac{5}{2}$ and $x = -\frac{2}{3}$ both check as solutions of $(6x + 1)(x - 2) = 8$.

Note: Entering one side of the equation as Y_1 and the other side as Y_2 is essential if there are variables on both sides of the equation.

The zero-factor principle can also be extended to solve equations of higher degree. This is illustrated in Example 6 which compares the algebraic, numerical, and graphical methods for solving a cubic equation.

EXAMPLE 6 Using Multiple Perspectives to Solve a Cubic Equation

Solve $x^3 - 4x^2 - 4x + 16 = 0$ algebraically, numerically, and graphically.

SOLUTION

ALGEBRAICALLY

$$x^3 - 4x^2 - 4x + 16 = 0$$
$$x^2(x - 4) - 4(x - 4) = 0$$

First use the grouping method to factor the left side of the equation.

$$(x^2 - 4)(x - 4) = 0$$
$$(x + 2)(x - 2)(x - 4) = 0$$

Then note that $x^2 - 4$ is a difference of two squares and factors as a sum times a difference.

$$x + 2 = 0, x - 2 = 0, \text{ or } x - 4 = 0$$
$$x = -2, \quad x = 2, \text{ or } \quad x = 4$$

Set each factor equal to zero, and solve the resulting linear equations.

NUMERICALLY

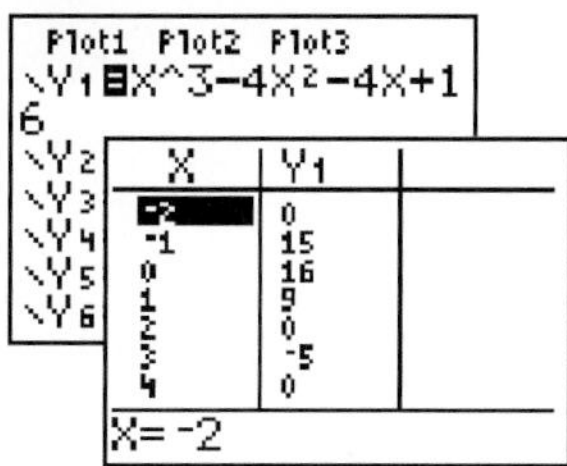

Enter the cubic polynomial into a graphing calculator as Y_1. Then use the **TABLE** feature to find the zeros of this polynomial.
Note that this polynomial is 0 for $x = -2$, $x = 2$, and $x = 4$.

The zeros of $y = x^3 - 4x^2 - 4x + 16$ are -2, 2, and 4.

GRAPHICALLY

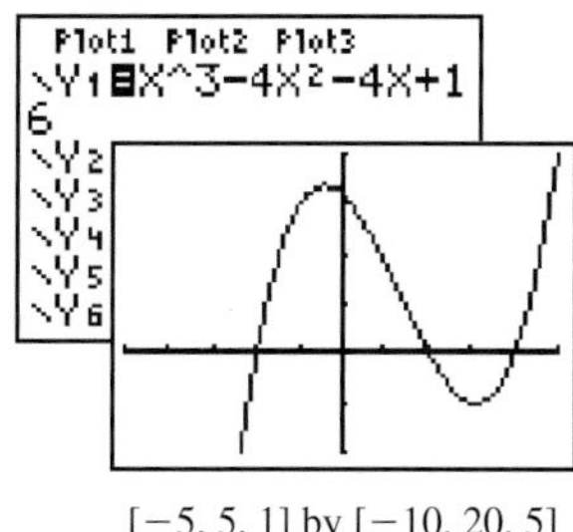

$[-5, 5, 1]$ by $[-10, 20, 5]$

Use a graphing calculator to graph $Y_1 = x^3 - 4x^2 - 4x + 16$. Then examine this graph to determine its x-intercepts.
This graph has x-intercepts of $(-2, 0)$, $(2, 0)$, and $(4, 0)$.
Algebraically we used the factors of $x^3 - 4x^2 - 4x + 16$ to solve $x^3 - 4x^2 - 4x + 16 = 0$. Numerically we found the zeros of $y = x^3 - 4x^2 - 4x + 16$ to solve $x^3 - 4x^2 - 4x + 16 = 0$.
Graphically, we determined the x-intercepts of the graph of $y = x^3 - 4x^2 - 4x + 16$ to solve $x^3 - 4x^2 - 4x + 16 = 0$.

The x-intercepts of the graph of $y = x^3 - 4x^2 - 4x + 16$ are $(-2, 0)$, $(2, 0)$, and $(4, 0)$.

Answer: The solutions of $x^3 - 4x^2 - 4x + 16 = 0$ are $x = -2$, $x = 2$, and $x = 4$.

SELF-CHECK 6.6.2

1. Solve $(x - 4)(x - 3) = 2$.
2. Use the **Ask** mode on a graphing calculator to check your solutions.
3. Factor $x^3 + 5x^2 - x - 5$.
4. Solve $x^3 + 5x^2 - x - 5 = 0$.

In Example 6 we factored $x^3 - 4x^2 - 4x + 16$ as $(x + 2)(x - 2)(x - 4)$. We can also make the following observations about $x^3 - 4x^2 - 4x + 16$.

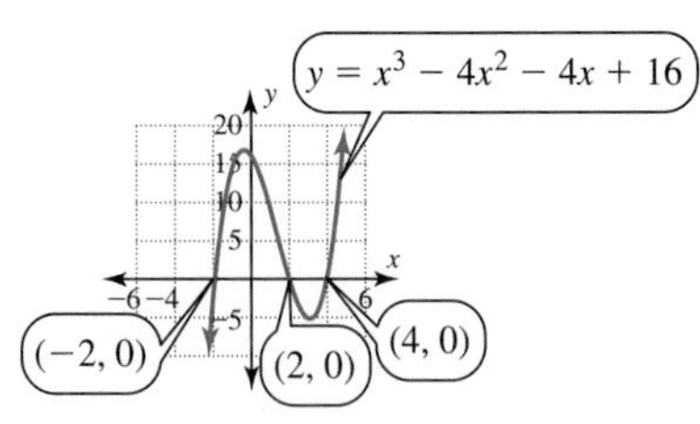

FACTORS OF $x^3 - 4x^2 - 4x + 16$	SOLUTIONS OF $x^3 - 4x^2 - 4x + 16 = 0$	ZEROS OF $x^3 - 4x^2 - 4x + 16$	x-INTERCEPTS OF THE GRAPH OF $y = x^3 - 4x^2 - 4x + 16$
$x + 2$	$x = -2$	-2	$(-2, 0)$
$x - 2$	$x = 2$	2	$(2, 0)$
$x - 4$	$x = 4$	4	$(4, 0)$

A generalization of this relationship is given in the following table. We will examine this relationship further in Section 7.3.

Equivalent Statements About Linear Factors of a Polynomial

For a real constant c and a real polynomial $P(x)$, the following statements are equivalent.

ALGEBRAICALLY		NUMERICALLY	GRAPHICALLY
$x - c$ is a factor of $P(x)$.	$x = c$ is a solution of $P(x) = 0$.	$P(c) = 0$, that is, c is a zero of $P(x)$.	$(c, 0)$ is an x-intercept of the graph of $y = P(x)$.

In Example 7 we use the graph of $y = x^2 + 2x - 35$ to solve the quadratic equation $x^2 + 2x - 35 = 0$. On the graph the x-intercepts occur where $y = 0$. Thus the x-intercepts identify the solutions of $x^2 + 2x - 35 = 0$.

EXAMPLE 7 Using x-Intercepts to Solve a Quadratic Equation

Use the graph and table for $y = x^2 + 2x - 35$ to solve $x^2 + 2x - 35 = 0$.

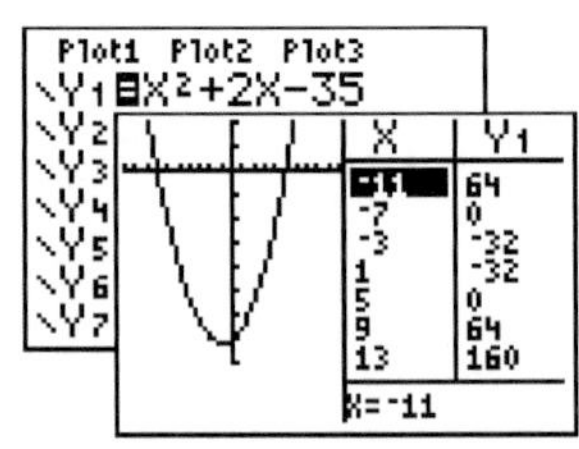

$[-10, 10, 1]$ by $[-40, 10, 5]$

SOLUTION

The x-intercepts are $(-7, 0)$ and $(5, 0)$.

You can approximate these x-intercepts by inspection and can confirm them by noting the zeros of the polynomial from the table. The zeros of this polynomial are -7 and 5.

Answer: $x = -7$ and $x = 5$

Each x-intercept $(c, 0)$ corresponds to a solution $x = c$.

The graph of $y = x^2 + 2x - 35$ in Example 7 actually reveals much more than we used to solve $x^2 + 2x - 35 = 0$. By examining where the y-coordinate is positive we can solve $x^2 + 2x - 35 > 0$. Likewise by examining where the y-coordinate is negative we can solve $x^2 + 2x - 35 < 0$. This information is summarized in the following box.

Solutions of Equations and Inequalities

If $P(x)$ is a real polynomial and c is a real number, then:

VERBALLY	ALGEBRAICALLY	GRAPHICALLY
c is a solution of $P(x) = 0$	$P(c) = 0$	$(c, 0)$ is an x-intercept of the graph of $y = P(x)$.
c is a solution of $P(x) < 0$	$P(c) < 0$	At $x = c$, the graph is below the x-axis.
c is a solution of $P(x) > 0$	$P(c) > 0$	At $x = c$, the graph is above the x-axis.

In Example 8 we use the graph of $y = x^2 + 2x - 35$ to solve two quadratic inequalities.

EXAMPLE 8 Using x-Intercepts to Solve Quadratic Inequalities

Use the graph and table for $y = x^2 + 2x - 35$ to solve

(a) $x^2 + 2x - 35 < 0$

(b) $x^2 + 2x - 35 > 0$

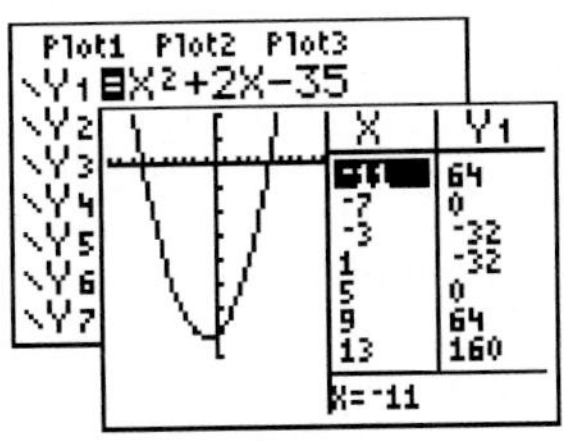

$[-10, 10, 1]$ by $[-40, 10, 5]$

SOLUTION

(a) $(-7, 5)$

x-values for which $y < 0$

−7 5

The graph of $Y_1 = x^2 + 2x - 35$ is below the x-axis for the x-values given by $(-7, 5)$. In the table, the value of Y_1 is less than or equal to 0 for the x-values between -7 and 5.

(b) $(-\infty, -7) \cup (5, \infty)$

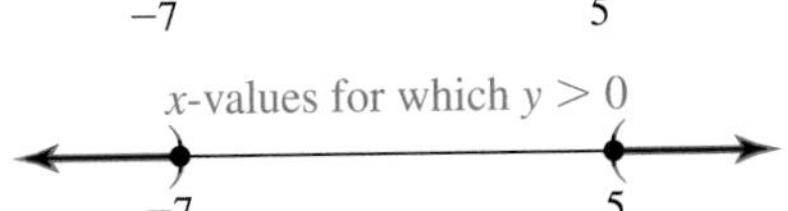

The graph of $Y_1 = x^2 + 2x - 35$ is above the x-axis for the x-values given by $(-\infty, -7) \cup (5, \infty)$.

In the table, the value of Y_1 is greater than 0 for the x-values less than -7 or greater than 5. ■

SELF-CHECK 6.6.3

1. Use a graphing calculator to graph $y = 2x^2 - 11x - 6$.
2. Determine the x-intercepts of this graph.
3. Determine the zeros of $2x^2 - 11x - 6$.
4. Solve $2x^2 - 11x - 6 = 0$.
5. Factor $2x^2 - 11x - 6$.
6. Solve $2x^2 - 11x - 6 \geq 0$.
7. Solve $2x^2 - 11x - 6 < 0$.

Many applications involve quadratic equations. This is especially true for applications involving area. This is illustrated by Example 9.

EXAMPLE 9 Determining the Dimensions of a Rectangle

Forty meters of rope are used to enclose a rectangular area of 91 square meters (m^2). Find the dimensions of this rectangular area.

w | 91 m^2

SOLUTION

Let w = width of rectangle
$20 - w$ = length of rectangle

The dimensions are the width and length. Because the perimeter is 40 meters, $2w + 2l = 40$ and $l = 20 - w$.

VERBALLY

Area equals 91 square meters.
(Width)(Length) = 91

The area of a rectangle is given by $A = w \cdot l$.

ALGEBRAICALLY

$$w(20 - w) = 91$$
$$20w - w^2 = 91$$
$$0 = w^2 - 20w + 91$$

Substitute the expressions identified above for the width and length.

$$w^2 - 20w + 91 = 0$$

Write this quadratic equation in standard form.

$$(w - 7)(w - 13) = 0$$

Solve this quadratic equation by factoring.

$$w - 7 = 0 \quad \text{or} \quad w - 13 = 0$$
$$w = 7 \quad \text{or} \quad w = 13$$
$$20 - w = 13 \quad \text{or} \quad 20 - w = 7$$

Either the width is 7 m and the length is 13 m or the width is 13 m and the length is 7 m.

Answer: The dimensions of this rectangular region are 7 m by 13 m.

Do these dimensions check both for the perimeter and for the area?

In Chapter 7 we examine other algebraic methods for solving quadratic equations. The factoring method used in this chapter has been used to find real solutions that are rational numbers. In Chapter 7 we examine quadratic equations with irrational solutions and quadratic equations with imaginary solutions, and we also solve quadratic inequalities.

SELF-CHECK ANSWERS

6.6.1

1. $x = \frac{9}{2}, x = -5$
2. $x = \frac{2}{3}, x = \frac{11}{2}$
3. $x = -\frac{1}{2}, x = 3$

6.6.2

1. $x = 2, x = 5$
2. Both solutions check.

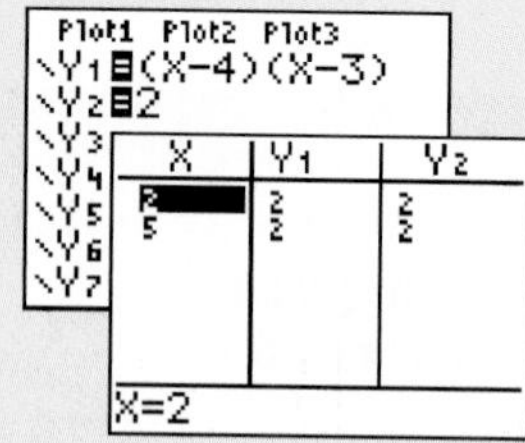

3. $(x - 1)(x + 1)(x + 5)$
4. $x = 1, x = -1, x = -5$

6.6.3

1.

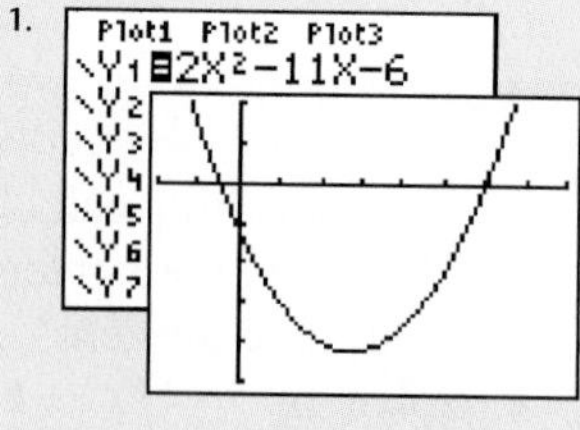

$[-2, 8, 1]$ by $[-25, 10, 5]$

2. $\left(-\frac{1}{2}, 0\right)$, $(6, 0)$
3. $-\frac{1}{2}$ and 6
4. $x = -\frac{1}{2}$ and $x = 6$
5. $(2x + 1)(x - 6)$
6. $\left(-\infty, -\frac{1}{2}\right] \cup [6, \infty)$
7. $\left(-\frac{1}{2}, 6\right)$

USING THE LANGUAGE AND SYMBOLISM OF MATHEMATICS 6.6

1. A second-degree equation in x that can be written as $ax^2 + bx + c = 0$ is called a __________ equation.
2. In the equation $ax^2 + bx + c = 0$, ax^2 is called the __________ term, bx is called the __________ term, and c is called the __________ term.
3. By the zero-factor principle, if $(x - 3)(x + 6) = 0$, then __________ = 0 or __________ = 0.
4. To clear an equation of fractions, multiply each term by the __________ of all the terms.

In Exercises 5–9, assume $P(x)$ represents a real polynomial in x.

5. If c is a real solution of $P(x) = 0$, then __________ is an x-intercept of the graph of $y = P(x)$.
6. If c is a real solution of $P(x) = 0$, then __________ is a factor of $P(x)$.
7. If the graph of $y = P(x)$ crosses the x-axis at $x = c$, then $P(c)$ ______ 0.
8. If the graph of $y = P(x)$ is above the x-axis at $x = c$, then $P(c)$ ______ 0.
9. If the graph of $y = P(x)$ is below the x-axis at $x = c$, then $P(c)$ ______ 0.

QUICK REVIEW 6.6

1. Solve $3x - 45 = 0$.
2. Solve $2x + 14 = 0$.
3. Solve $5x - 7 = 0$.
4. Determine whether $x = \frac{2}{3}$ is a solution of $(3x - 2)(x + 5) = 0$.
5. Determine whether $x = 3$ is a solution of $(x - 3)(x + 5) = 2$.

EXERCISES 6.6

Exercises 1 and 2, match each algebraic expression with the most appropriate description.

1. a. $2x - 3 = 0$
   b. $2x - 3 = 2x + 3$
   c. $(2x - 3)(x + 4) = 0$
   d. $2x^2 + 5x - 12$

   A. A quadratic equation with two distinct real solutions
   B. A quadratic polynomial but not an equation
   C. A linear equation with no real solution
   D. A linear equation with one real solution

2. a. $x^2 - 25 = 0$
   b. $-10x + 25 = 0$
   c. $x^2 - 10x + 25$
   d. $x^3 - 25x = 0$

   A. A linear equation with one real solution
   B. A quadratic equation with two distinct real solutions
   C. A cubic equation with three solutions
   D. A quadratic polynomial but not an equation

In Exercises 3 and 4, write each quadratic equation in the standard form $ax^2 + bx + c = 0$ and identify a, b, and c.

3. a. $2x^2 - 7x + 3 = 0$ b. $8x^2 = 3x$
   c. $7x^2 = -5$ d. $(x - 1)(2x + 1) = 3$
4. a. $3x^2 = 5x - 2$ b. $3x^2 = 17$
   c. $2x^2 = 17x$ d. $(3x + 1)(x - 2) = 3$

In Exercises 5–50, solve each equation.

5. $(m - 8)(m + 17) = 0$
6. $(m + 15)(m - 3) = 0$
7. $(2n - 5)(3n + 1) = 0$
8. $(5n + 2)(n - 6) = 0$
9. $(z - 1)(z + 2)(2z - 7) = 0$
10. $z(z + 6)(6z + 1) = 0$
11. $v^2 - 121 = 0$
12. $v^2 - 169 = 0$
13. $x^2 + 3x + 2 = 0$
14. $x^2 + 5x + 4 = 0$
15. $y^2 - 3y = 18$
16. $t^2 + 4t = 12$
17. $3v^2 = -v$
18. $5v^2 = v$
19. $2w^2 = 7w + 15$
20. $3w^2 = 17w + 6$
21. $6x^2 + 19x + 10 = 0$
22. $10x^2 - 17x + 3 = 0$
23. $x^2 = 11x - 24$
24. $x^2 = 2x + 48$
25. $70z^2 = 5z + 15$
26. $8v^2 + 24v = 0$
27. $9x^2 = 25$
28. $4x^2 = 9$
29. $r(r + 3) = 10$
30. $(r + 6)(r - 1) = -10$
31. $\frac{m^2}{18} - \frac{m}{6} - 1 = 0$
32. $\frac{m^2}{20} - \frac{m}{4} + \frac{1}{5} = 0$
33. $\frac{x^2}{12} - \frac{x}{3} - 1 = 0$
34. $\frac{x^2}{18} + \frac{x}{6} - 1 = 0$
35. $(v - 12)(v + 1) = -40$
36. $(v + 5)(v + 3) = 5v + 25$
37. $(2w - 3)^2 = 25$
38. $(3w + 2)^2 = 16$
39. $(3x - 8)(x + 1) = (x + 1)(x - 3)$
40. $(4x + 5)(x + 5) = 45x$
41. $(3x + 2)(3x - 4) = -3(4x + 3)$
42. $(t + 1)(t - 1)(3t + 1) = 0$
43. $v(v^2 - 5v - 24) = 0$
44. $v(v^2 + 14v + 24) = 0$
45. $w(6w^2 + 5w - 6) = 0$
46. $w(5w^2 + w - 4) = 0$
47. $14y^3 = 3y - 19y^2$
48. $10y^3 = 29y^2 - 10y$
49. $9m^2 = 42m - 49$
50. $25m^2 = 90m - 81$
51. Use the graph of $y = x^2 + 6x - 91$.
    a. Determine the x-intercepts of this graph.
    b. Factor $x^2 + 6x - 91$.
    c. Determine the zeros of $x^2 + 6x - 91$.
    d. Solve $x^2 + 6x - 91 = 0$.

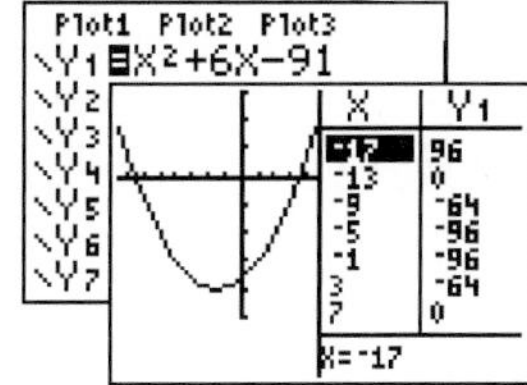

[−15, 9, 2] by [−125, 75, 25]

52. Use the graph of $y = x^2 - 6x - 247$.
- **a.** Determine the x-intercepts of this graph.
- **b.** Factor $x^2 - 6x - 247$.
- **c.** Determine the zeros of $x^2 - 6x - 247$.
- **d.** Solve $x^2 - 6x - 247 = 0$.

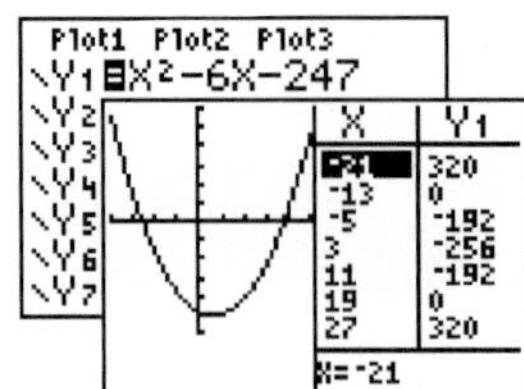

[−20, 25, 5] by [−300, 300, 50]

53. Use the graph of $y = x^3 - x^2 - 2x$.
- **a.** Determine the x-intercepts of this graph.
- **b.** Factor $x^3 - x^2 - 2x$.
- **c.** Determine the zeros of $x^3 - x^2 - 2x$.
- **d.** Solve $x^3 - x^2 - 2x = 0$.

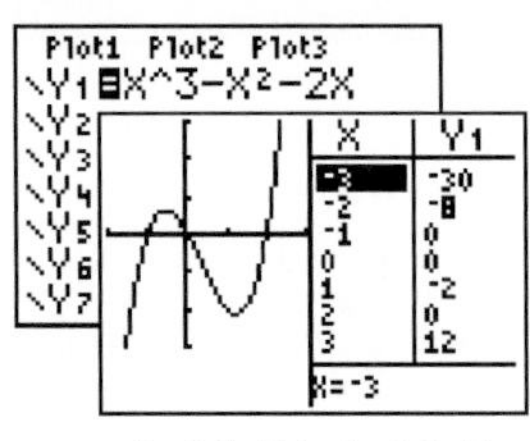

[−2, 3, 1] by [−3, 3, 1]

54. Use the graph of $y = x^3 - 3x^2 + 2x$.
- **a.** Determine the x-intercepts of this graph.
- **b.** Factor $x^3 - 3x^2 + 2x$.
- **c.** Determine the zeros of $x^3 - 3x^2 + 2x$.
- **d.** Solve $x^3 - 3x^2 + 2x = 0$.

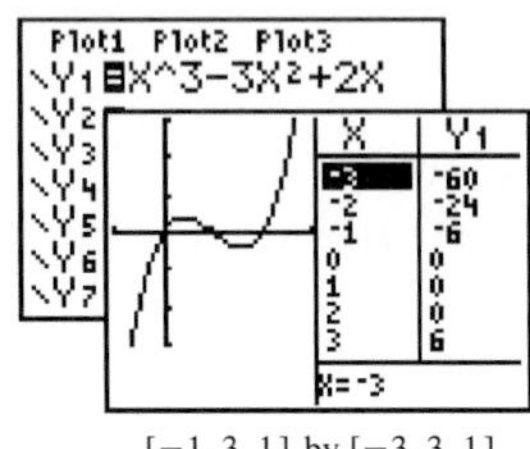

[−1, 3, 1] by [−3, 3, 1]

Exercises 55–58, have two parts. In part **a** solve the equation, and in part **b** perform the indicated operations and simplify the result.

	Solve	Simplify
55.	**a.** $(5m - 3)(m - 2) = 0$	**b.** $(5m - 3)(m - 2)$
56.	**a.** $(4x - 3)(5x + 2) = 0$	**b.** $(4x - 3)(5x + 2)$
57.	**a.** $4(x - 3)(5x + 2) = 0$	**b.** $4(x - 3)(5x + 2)$
58.	**a.** $x(x - 3)(5x + 2) = 0$	**b.** $x(x - 3)(5x + 2)$

In Exercises 59–62, each graph was created by letting $Y_1 = P(x)$. Determine the x-values that satisfy each equation and inequality.

59. $P(x) = x^2 - 2x - 15$
- **a.** $x^2 - 2x - 15 = 0$
- **b.** $x^2 - 2x - 15 < 0$
- **c.** $x^2 - 2x - 15 > 0$

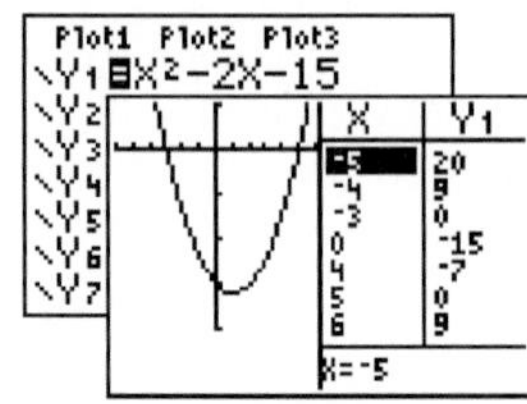

[−6, 6, 1] by [−20, 5, 5]

60. $P(x) = x^2 + 2x - 8$
- **a.** $x^2 + 2x - 8 = 0$
- **b.** $x^2 + 2x - 8 < 0$
- **c.** $x^2 + 2x - 8 > 0$

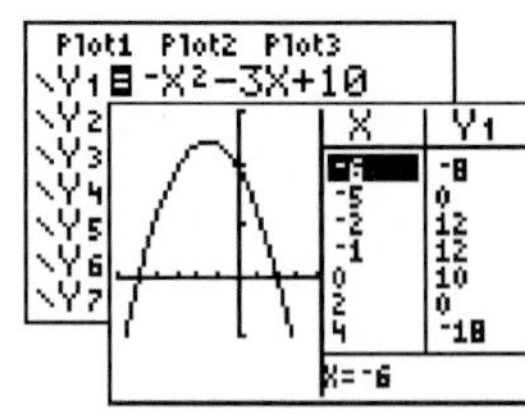

[−6, 4, 1] by [−10, 5, 5]

61. $P(x) = -x^2 - 3x + 10$
- **a.** $-x^2 - 3x + 10 = 0$
- **b.** $-x^2 - 3x + 10 \le 0$
- **c.** $-x^2 - 3x + 10 \ge 0$

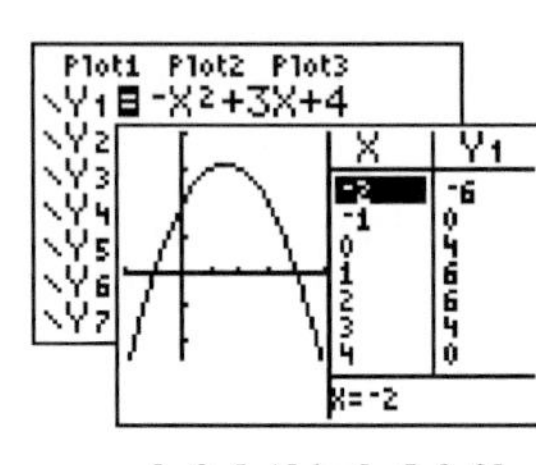

[−6, 4, 1] by [−5, 15, 5]

62. $P(x) = -x^2 + 3x + 4$
- **a.** $-x^2 + 3x + 4 = 0$
- **b.** $-x^2 + 3x + 4 \le 0$
- **c.** $-x^2 + 3x + 4 \ge 0$

[−2, 5, 1] by [−5, 8, 2]

In Exercises 63–66, use a graphing calculator to graph $y = P(x)$ and complete this table.

POLYNOMIAL $P(x)$	SOLUTIONS OF $P(x) = 0$	SOLUTIONS OF $P(x) < 0$	SOLUTIONS OF $P(x) > 0$
63. $(x - 2)(x + 7)$			
64. $-2(x + 4)(x - 2)$			
65. $-3x^2 - 19x + 14$			
66. $x^2 + 7x - 8$			

67. The length of the rectangle shown in the figure is 2 cm more than 3 times the width. Find the dimensions of this rectangle if the area is 33 square centimeters (cm^2).

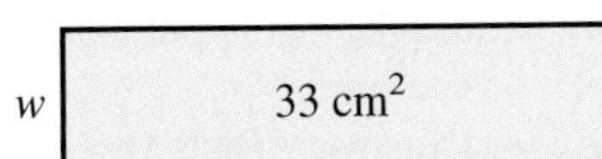

68. The length of the rectangle shown in the figure is 5 cm more than twice the width. Find the dimensions of this rectangle if the area is 52 cm^2.

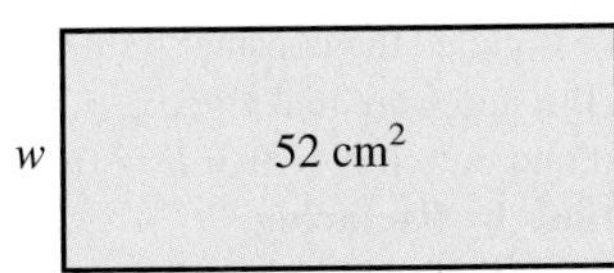

69. The base of a triangle in the figure is 2 m longer than the height. Find the base if the area of this triangle is 24 m^2.

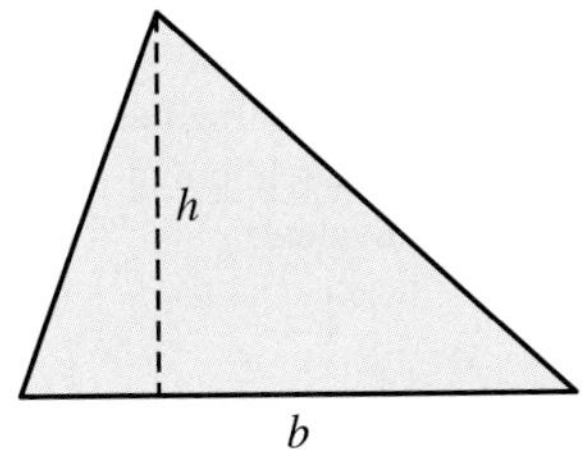

70. The base of the triangle shown in the figure is 10 cm longer than the height. Determine the height of the triangle if its area is 48 cm^2.

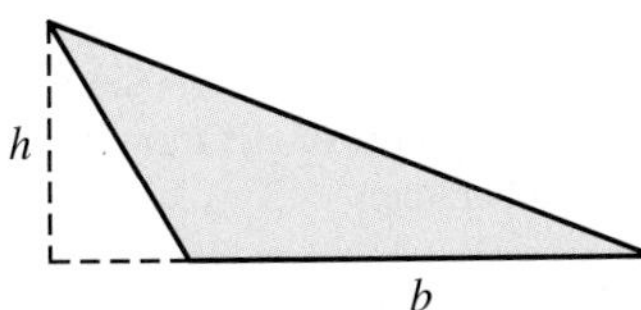

71. A metal sheet 60 cm wide is used to form a trough by bending up each side as illustrated in the figure. Determine the height of each side if the cross-sectional area is 450 cm^2.

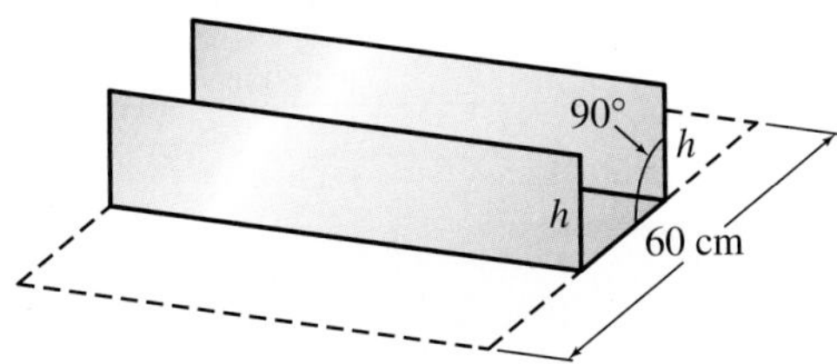

72. A rectangular pad of concrete is outlined by a 52-m stripe of paint. Find the dimensions of this pad if the surface area is 168 m^2.

Group Discussion Questions

73. **Discovery Question** Each of these polynomials is prime over the integers, but can be factored over the irrational numbers. Factor these polynomials and solve each equation.

$P(x)$	FACTORS OF $P(x)$	SOLUTIONS OF $P(x) = 0$
Example $x^2 - 3$	$(x + \sqrt{3})(x - \sqrt{3})$	$x = -\sqrt{3}, x = \sqrt{3}$
a. $x^2 - 5$	$(x + \sqrt{5})(x - \sqrt{5})$	
b. $x^2 - 7$		$x = -\sqrt{7}, x = \sqrt{7}$
c. $x^2 - 11$		

74. **Error Analysis** A student solved the equation $(x - 3)(x + 5) = 9$ as follows:

$$(x - 3)(x + 5) = 9$$
$$x - 3 = 9 \qquad x + 5 = 9$$
$$x = 12 \qquad x = 4$$

Find the error in this solution and determine the correct solution.

75. **Challenge Question** Solve each equation for x.
   a. $(x + y)(x - 3y) = 0$
   b. $x^2 - 5xy - 6y^2 = 0$
   c. $6x^2 - xy = 2y^2$
   d. $x^2 + 2xy = 15y^2$

76. **Discovery Question**
   a. Graph each of these equations and then describe the relationship of the graph to the x-axis: $y = x^2 - 2x + 1$, $y = x^2 + 2x + 1$, $y = x^2 - 4x + 4$, and $y = x^2 + 4x + 4$.
   b. Factor each of these polynomials and then describe any relationship that you observe between the factors and the relationship of the corresponding graph to the x-axis: $x^2 - 2x + 1$, $x^2 + 2x + 1$, $x^2 - 4x + 4$, and $x^2 + 4x + 4$.
   c. Factor $x^2 + 18x + 81$ and predict the relationship of the graph of $y = x^2 + 18x + 81$ to the x-axis.

KEY CONCEPTS FOR CHAPTER 6

1. **Quadratic Trinomial:** $ax^2 + bx + c$ is a quadratic trinomial; ax^2 is called the second-degree term or the quadratic term, bx is the first-degree term or the linear term, and c is the constant term.
2. **Factors and Terms:** Factors are constants or variables that are multiplied together to form a product. A single number or an indicated product of factors is called a term. The terms in an algebraic expression are separated from each other by plus or minus symbols.
3. **Factored Form of a Polynomial:** The factored form of a polynomial $P(x)$ is useful for quickly determining the solutions of $P(x) = 0$, the zeros of $P(x)$, and the x-intercepts of the graph of $y = P(x)$.
4. **Prime Polynomial:** A polynomial is prime over the integers if its only factorization with integer coefficients must involve 1 or -1 as one of the factors.
5. **Completely Factoring a Polynomial:** A polynomial is factored completely if each factor other than a common monomial factor is prime.

6. **Factorable Trinomials:** A trinomial $ax^2 + bx + c$ is factorable into a pair of binomial factors with integer coefficients if and only if there are two integers whose product is ac and whose sum is b. Otherwise the trinomial is prime over the integers.
7. **GCF of a Polynomial:** The greatest common factor of a polynomial is the factor of each term that contains both of the following:
   a. Largest possible numerical coefficient
   b. Largest possible exponent on each variable factor
8. **Factoring $ax^2 + bx + c$ by Tables and Grouping**
   Step 1 Factor out the GCF if it is not 1.
   Step 2 Find two factors of ac whose sum is b.
   Step 3 Rewrite the linear term of $ax^2 + bx + c$ so that b is the sum of the factors from step 2.
   Step 4 Factor the polynomial from step 3 by grouping the terms and factoring the GCF out of each pair of terms.
9. **Factoring $ax^2 + bx + c$ by Inspection:** A suggested first step in factoring a trinomial by inspection is to determine the sign pattern for each of the possible binomial factors.
10. **Factoring by Grouping:** To factor polynomials of four or more terms, we can first group terms together that we already know how to factor. This method relies on the ability to spot common factors and special forms.
11. **Strategy for Factoring a Polynomial over the Integers:** After factoring out the GCF (greatest common factor), proceed as follows:
    a. **Binomials:** Factor special forms:
    - Difference of two squares
      $x^2 - y^2 = (x + y)(x - y)$
    - Difference of two cubes
      $x^3 - y^3 = (x - y)(x^2 + xy + y^2)$
    - Sum of two cubes
      $x^3 + y^3 = (x + y)(x^2 - xy + y^2)$
    - $x^2 + y^2$ is prime
      The sum of two squares is prime if x^2 and y^2 are only second-degree terms and have no common factor other than 1.

    b. **Trinomials:** Factor the forms that are perfect squares:
    - Perfect square trinomial; a square of a sum
      $x^2 + 2xy + y^2 = (x + y)^2$
    - Perfect square trinomial; a square of a difference
      $x^2 - 2xy + y^2 = (x - y)^2$
    - Factor trinomials that are not perfect squares by inspection if possible; otherwise, use the method of tables and grouping.

    c. **Polynomials of Four or More Terms:** Factor by grouping.
12. **Quadratic Equation:** If a, b, and c are real constants and $a \neq 0$, then $ax^2 + bx + c = 0$ is the standard form of a quadratic equation in x; ax^2 is called the quadratic term, bx is called the linear term, and c is called the constant term.
13. **Zero-Factor Principle:** If a and b are real algebraic expressions, then $ab = 0$ if and only if $a = 0$ or $b = 0$.
14. **Solving Quadratic Equations by Factoring**
    Step 1 Write the equation in standard form, with the right side 0.
    Step 2 Factor the left side of the equation.
    Step 3 Set each factor equal to 0.
    Step 4 Solve the resulting first-degree equations.
15. **Equivalent Statements About Linear Factors of a Polynomial:** For a real constant c and a real polynomial $P(x)$, the following statements are equivalent:
    - $x - c$ is a factor of $P(x)$.
    - $x = c$ is a solution of $P(x) = 0$.
    - $P(c) = 0$; that is, c is a zero of $P(x)$.
    - $(c, 0)$ is an x-intercept of the graph of $y = P(x)$.
16. **Solutions of a Quadratic Inequality**
    - The solution of $ax^2 + bx + c > 0$ is the set of x-values for which the graph of $y = ax^2 + bx + c$ is above the x-axis.
    - The solution of $ax^2 + bx + c < 0$ is the set of x-values for which the graph of $y = ax^2 + bx + c$ is below the x-axis.
17. **Solutions of Equations and Inequalities:** If $P(x)$ is a real polynomial and c is a real number, then

VERBALLY	ALGEBRAICALLY	GRAPHICALLY
c is a solution of $P(x) = 0$.	$P(c) = 0$	$(c, 0)$ is an x-intercept of the graph of $y = P(x)$.
c is a solution of $P(x) < 0$.	$P(c) < 0$	At $x = c$, the graph is below the x-axis.
c is a solution of $P(x) > 0$.	$P(c) > 0$	At $x = c$, the graph is above the x-axis.

REVIEW EXERCISES FOR CHAPTER 6

In Exercises 1–4, factor the GCF out of each polynomial.

1. $4x - 36$
2. $2x^2 - 10x$
3. $12ax^2 - 24ax$
4. $5x^3 - 15x$

In Exercises 5–8, use grouping to factor each polynomial.

5. $ax + bx + ay + by$
6. $ax + 2ay - 3x - 6y$
7. $6x^2 + 21x - 10x - 35$
8. $24x^2 - 20x + 18x - 15$

In Exercises 9–40, factor each polynomial completely over the integers.

9. $x^2 - 4$
10. $4x^2 - 1$
11. $7x^2 - 28$
12. $2m^2 - 128$
13. $11x^2 - 44y^2$
14. $4ax^2 - 9ay^2$
15. $x^2 - 11x + 18$
16. $m^2 + 10m + 21$
17. $x^2 + 4x + 4$
18. $m^2 + 10m + 25$
19. $x^2 - xy - 42y^2$
20. $x^2 + 26xy + 25y^2$
21. $4x^2 - 12xy + 9y^2$
22. $9y^2 - 30y + 25$
23. $ax + 2ay - 7x - 14y$
24. $x^3y - 5x^2y + 4xy$
25. $x^2(a + b) - 25(a + b)$
26. $11x^2 - 11y^2 + 33x + 33y$
27. $v^2 + 2v + 1 - w^2$
28. $5a(2x - 3y) + 3b(3y - 2x)$
29. $v^2 + 9w^2$
30. $v^2 - 9w^2$
31. $100x^2 - 49y^2$
32. $100x^2 + 49y^2$
33. $4x^2 + 2xy - 30y^2$
34. $6x^2 - 7xy - 20y^2$
35. $20x^2 + 50x - 30$
36. $6x^2 + 61xy + 10y^2$
37. $v^4 - 1$
38. $81w^4 - 16$
39. $a^2(x + 5y) + 2ab(x + 5y) + b^2(x + 5y)$
40. $9x^2 - y^2 + 3x - y$

In Exercises 41–44, use the given table to factor each polynomial completely over the integers.

41. a. $x^2 - 48x - 100$ **b.** $x^2 - 15x - 100$
c. $x^2 - 100$ **d.** $x^2 - 20x - 100$

FACTORS OF −100		SUM OF FACTORS
1	−100	−99
2	−50	−48
4	−25	−21
5	−20	−15
10	−10	0

42. a. $x^2 + 22x - 48$ **b.** $x^2 + 13x - 48$
c. $x^2 + 2x - 48$ **d.** $x^2 + x - 48$

FACTORS OF −48		SUM OF FACTORS
−1	48	47
−2	24	22
−3	16	13
−4	12	8
−6	8	2

43. a. $100x^2 + 52x + 1$ **b.** $4x^2 + 52x + 25$
c. $5x^2 + 29xy + 20y^2$ **d.** $10x^2 + 29xy + 10y^2$

FACTORS OF 100		SUM OF FACTORS
1	100	101
2	50	52
4	25	29
5	20	25
10	10	20

44. a. $48x^2 - 19x + 1$ **b.** $3x^2 - 26x + 16$
c. $2x^2 - 49xy + 24y^2$ **d.** $16x^2 - 16xy + 3y^2$

FACTORS OF 48		SUM OF FACTORS
−1	−48	−49
−2	−24	−26
−3	−16	−19
−4	−12	−16
−6	−8	−14

In Exercises 45 and 46, use the graph and table for the polynomial to factor the polynomial.

45.

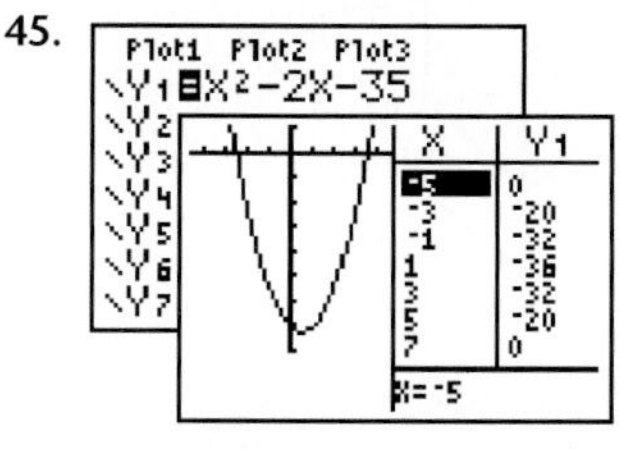

[−9, 9, 2] by [−40, 5, 5]

46.

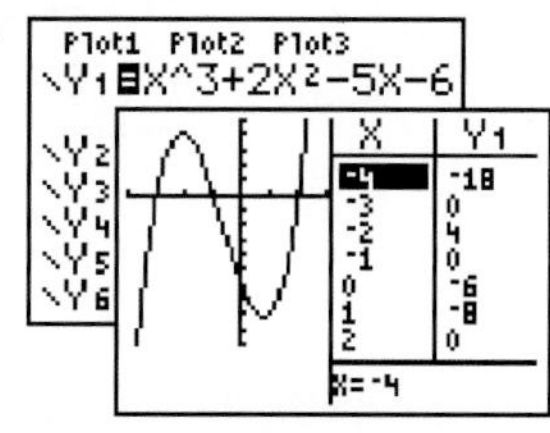

[−4, 3, 1] by [−10, 5, 1]

In Exercises 47–50, factor each polynomial completely over the integers. (Optional)

47. $8x^3 - y^3$ **48.** $x^3 + 64$
49. $2x^4 - 16x$ **50.** $v^4 - v^3 - v + 1$

In Exercises 51–64, solve each equation.

51. $(x - 5)(x + 1) = 0$ **52.** $(2x - 3)(3x + 2) = 0$
53. $x(x - 7)(7x - 2) = 0$ **54.** $v^2 - 4v - 21 = 0$
55. $10y^2 + 13y - 3 = 0$ **56.** $6w^2 = 11w + 21$
57. $\frac{v^2}{30} = \frac{v}{15} + \frac{1}{2}$ **58.** $(x + 6)(x - 2) = 9$
59. $x(x^2 - 36) = 0$ **60.** $2x(x^2 - 10x + 25) = 0$
61. $\frac{z^2}{12} = \frac{z + 3}{3}$ **62.** $\frac{x^2}{9} + \frac{9}{4} = -x$
63. $(3w + 2)(w + 1) = (2w + 3)(w - 2)$
64. $(x - 3)(x + 2) = (x + 9)(2x - 9)$

Exercises 65 and 66 have two parts. In part **a** solve the equation, and in part **b** perform the indicated operations and simplify the result.

	Solve	Simplify
65.	**a.** $(x - 3)(x + 12) = 0$	**b.** $(x - 3)(x + 12)$
66.	**a.** $(3x - 1)(2x + 5) = 0$	**b.** $(3x - 1)(2x + 5)$

67. Use the given graph and table.
a. Factor $x^2 + 2x - 24$.
b. Determine the zeros of $x^2 + 2x - 24$.
c. Solve $x^2 + 2x - 24 = 0$.
d. Solve $x^2 + 2x - 24 < 0$.
e. Solve $x^2 + 2x - 24 > 0$.

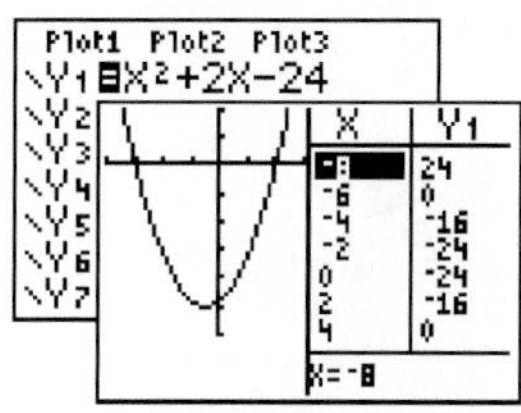

[−8, 6, 2] by [−30, 10, 5]

68. Use the given graph and table.
a. Factor $x^2 - 9x - 36$.
b. Determine the zeros of $x^2 - 9x - 36$.
c. Solve $x^2 - 9x - 36 = 0$.
d. Solve $x^2 - 9x - 36 \le 0$.
e. Solve $x^2 - 9x - 36 \ge 0$.

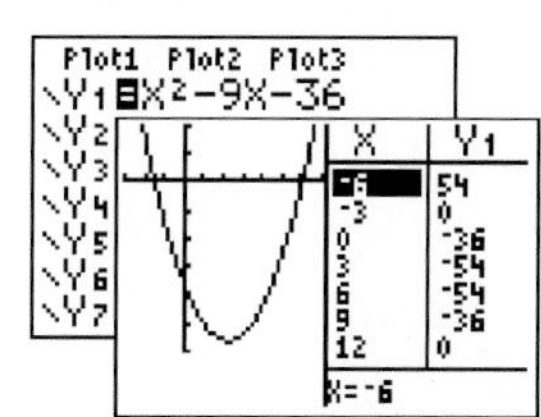

[−6, 14, 2] by [−60, 20, 10]

69. Use the given graph and table.
a. Factor $x^3 + 3x^2 - x - 3$.
b. Determine the zeros of $x^3 + 3x^2 - x - 3$.
c. Solve $x^3 + 3x^2 - x - 3 = 0$.
d. Solve $x^3 + 3x^2 - x - 3 \le 0$.
e. Solve $x^3 + 3x^2 - x - 3 \ge 0$.

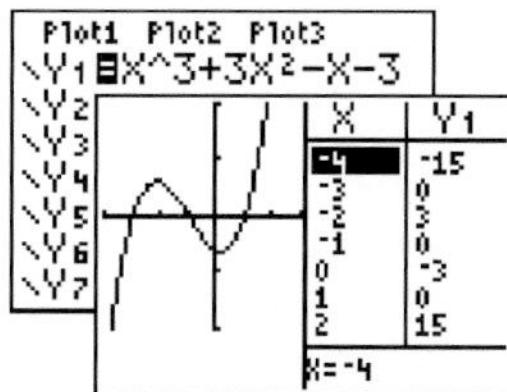

[−4, 3, 1] by [−10, 10, 5]

70. Use the given graph and table.
   a. Factor $x^3 + 8x^2 + 4x - 48$.
   b. Determine the zeros of $x^3 + 8x^2 + 4x - 48$.
   c. Solve $x^3 + 8x^2 + 4x - 48 = 0$.
   d. Solve $x^3 + 8x^2 + 4x - 48 < 0$.
   e. Solve $x^3 + 8x^2 + 4x - 48 > 0$.

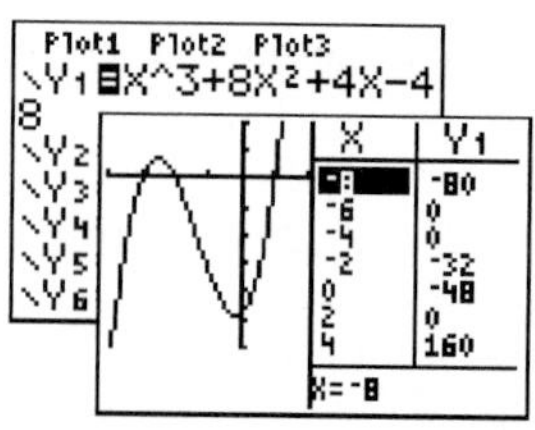

[−8, 4, 2] by [−60, 20, 10]

In Exercises 71 and 72, use a graphing calculator to graph $y = P(x)$ and then use this graph to assist you in completing this table.

POLYNOMIAL $P(x)$	x-INTERCEPTS OF THE GRAPH	FACTORED FORM	ZEROS OF $P(x)$	SOLUTIONS OF $P(x) = 0$
71. $x^2 + 4x - 77$				
72. $x^2 - 20x - 525$				

73. The difference in the distance traveled in 1 revolution by two bicycle tires of radii r_1 and r_2 is $2\pi r_1 - 2\pi r_2$. Factor this polynomial.

74. The lengths of two booms on circular irrigation systems are L_1 and L_2. The difference in the areas covered by these irrigation systems is $\pi L_1^2 - \pi L_2^2$. Factor this polynomial.

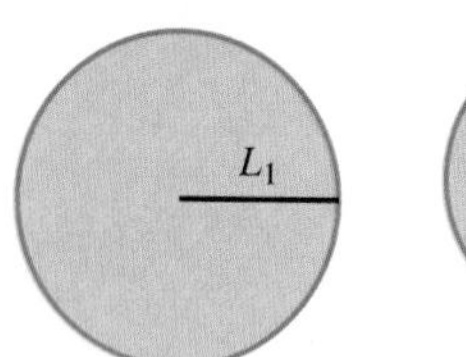

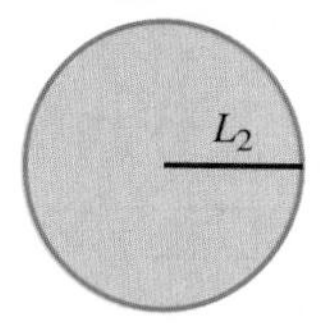

75. The difference in the volumes of two spherical oxygen tanks is $\frac{4}{3}\pi r_1^3 - \frac{4}{3}\pi r_2^3$. Factor this polynomial.

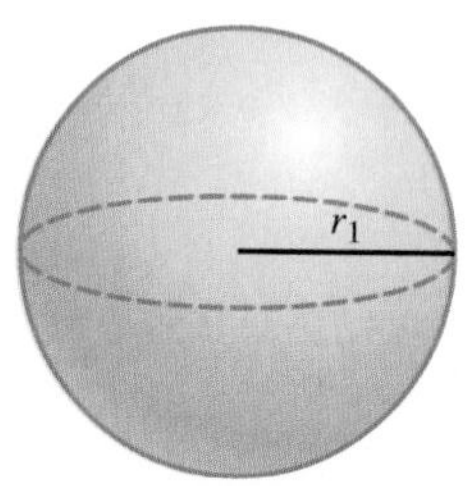

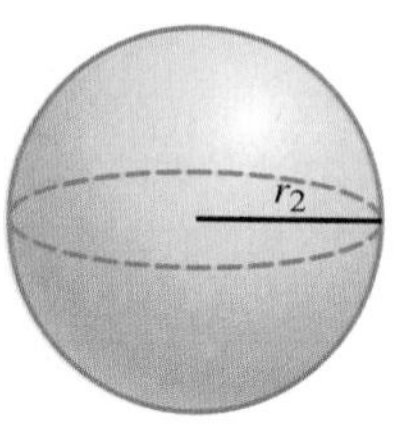

76. The base of the triangle shown in the figure below is 3 cm longer than the height. Determine the height of the triangle if its area is 14 cm^2.

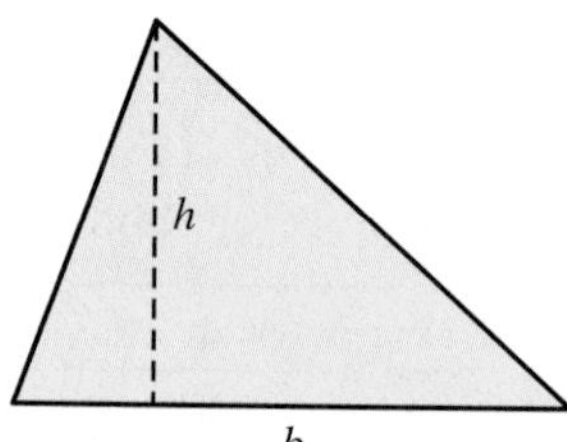

77. A metal sheet 60 cm wide is used to form a trough by bending up each side, as illustrated in the figure below. Determine the height of each side if the cross-sectional area is 400 cm^2.

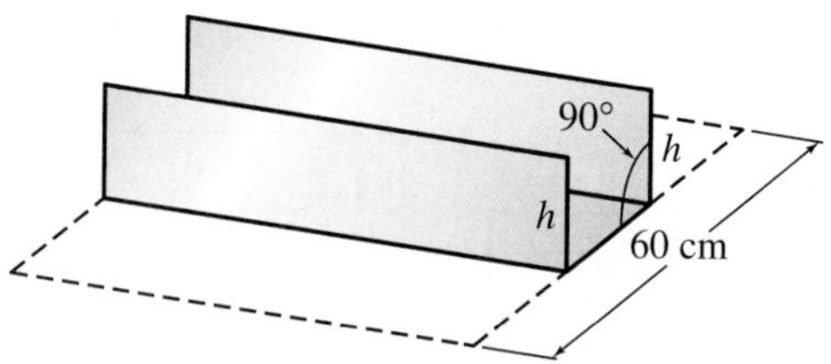

MASTERY TEST FOR CHAPTER 6

[6.1] 1. Factor out the GCF from each polynomial.
   a. $18x^3 - 99x^2$
   b. $4x^3y^2 - 6x^2y^3 + 10xy^4$
   c. $(7x - 2)(3x) + (7x - 2)(4)$
   d. $(2x + 3y)(x) - (2x + 3y)(5y)$

[6.1] 2. Factor each polynomial by using the grouping method.
   a. $(7ax + a) + (21x + 3)$
   b. $(12ax - 8a) - (15bx - 10b)$
   c. $2x^2 - 6x + 4x - 12$
   d. $10x^2 - 15x - 12x + 18$

[6.1] **3.** Use the graph and table for the polynomial function $Y_1 = P(x)$.
a. List the x-intercepts of the graph of $Y_1 = P(x)$.
b. List the zeros of $P(x)$.
c. Determine the solutions of $P(x) = 0$.
d. Factor $P(x)$.
$P(x) = x^3 + 5x^2 - 34x - 80$

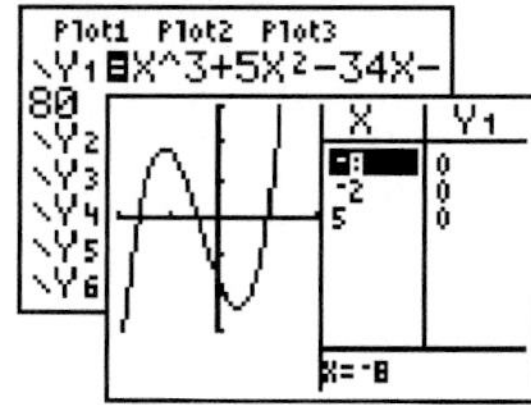

[−10, 10, 5] by [−150, 150, 50]

[6.2] **4.** Completely factor each trinomial.
a. $w^2 - 4w - 45$
b. $w^2 + 14w + 45$
c. $v^2 - 10v + 24$
d. $v^2 + 5v - 36$

[6.2] **5.** Completely factor each trinomial.
a. $x^2 - xy - 12y^2$
b. $x^2 - 13xy + 12y^2$
c. $a^2 + 30ab + 144b^2$
d. $a^2 + 2ab - 48b^2$

[6.3] **6.** Completely factor each trinomial.
a. $6x^2 - 17x - 14$
b. $6x^2 - 31x + 14$
c. $6x^2 + 85x + 14$
d. $10x^2 + 17x - 6$

[6.3] **7.** Completely factor each trinomial.
a. $12x^2 + 25xy + 12y^2$
b. $12x^2 - 7xy - 12y^2$
c. $12x^2 - 145xy + 12y^2$
d. $40x^2 + 7xy - 3y^2$

[6.4] **8.** Factor each perfect square trinomial.
a. $x^2 + 14x + 49$
b. $x^2 - 16x + 64$
c. $9x^2 + 60x + 100$
d. $25x^2 - 110x + 121$

[6.4] **9.** Completely factor each difference of two squares.
a. $x^2 - 4y^2$
b. $400a^2 - b^2$
c. $16v^2 - 49w^2$
d. $36x^2 - 25y^2$

[6.4] **10.** Completely factor each sum or difference of two cubes. (Optional)
a. $64v^3 - 1$
b. $v^3 + 125$
c. $8x^3 + 125y^3$
d. $27a^3 - 1{,}000b^3$

[6.5] **11.** Use grouping to factor each polynomial.
a. $2ax + 3a + 2bx + 3b$
b. $14ax - 6bx - 35ay + 15by$
c. $a^2 - 4b^2 + a - 2b$
d. $x^2 + 10xy + 25y^2 - 4$

[6.5] **12.** Completely factor each polynomial.
a. $5x^2 - 245$
b. $2ax^2 + 20ax + 50a$
c. $3x^3 - 15x^2 - 42x$
d. $5a^2x + 5a^2y - 15abx - 15aby$

[6.6] **13.** Solve each equation.
a. $6x^2 + x - 12 = 0$
b. $4x^2 + 25 = 20x$
c. $(x + 6)(x - 7) = -22$
d. $2x^3 + 5x^2 - 3x = 0$

[6.6] **14.** Use the graph of $y = x^2 - 5x - 6$.
a. Determine the x-intercepts of this graph.
b. Solve $x^2 - 5x - 6 = 0$.
c. Solve $x^2 - 5x - 6 < 0$.
d. Solve $x^2 - 5x - 6 > 0$.

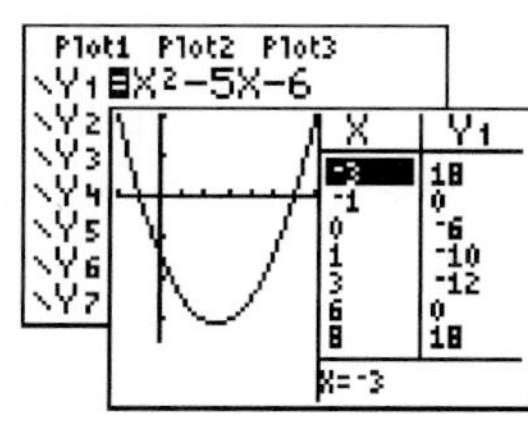

[−2, 7, 1] by [−14, 8, 4]

REALITY CHECK FOR CHAPTER 6

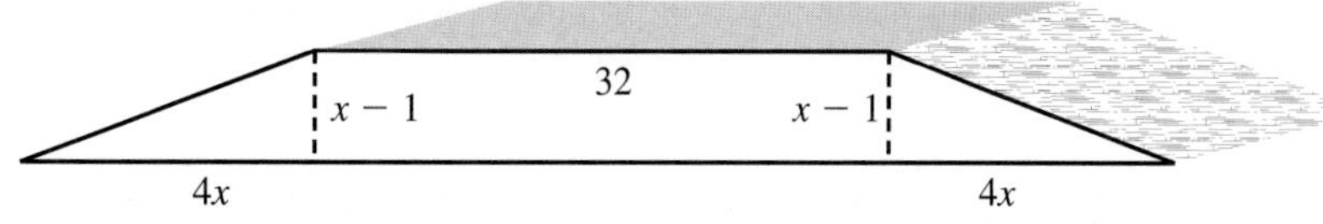

For a roadway where the concrete and rock base are 1 ft thick, the depth of the trapezoidal cross section will be 1 ft less than the planned height of the road. This is shown in the figure for a roadway with height x. Use the formula for the area of a trapezoid $A = \frac{1}{2}h(a + b)$ to write an equation for the cross-sectional area of a section of roadway in terms of x. Use $x = 5$ to determine the area of this section. Then determine the slope of the embankment on the right side of the roadway.

GROUP PROJECT FOR CHAPTER 6

Risks and Choices: Creating a Mathematical Model and Using This Model to Find an Optimal Solution

A fencing contractor in Sacramento, California, was contracted to install 150 ft of chain-link fencing to surround a playground area next to an existing grade school building. This is illustrated in the figure to the right. For security reasons, the specifications indicate that the only entrances to this area come through the school. The length of the wall along the school where no fencing is needed is 180 ft. The question that we will explore is: What dimensions should the contractor use to enclose the maximum possible rectangular area inside this fence?

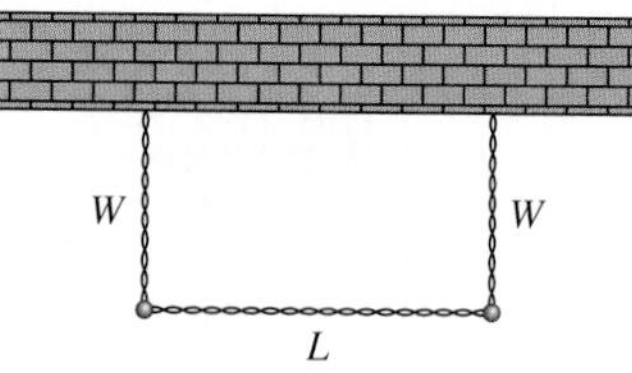

Exploration: Examining the relationship between the width of the playground and the area of the playground.

I. **Creating Numerical, Graphical, and Algebraic Models for This Problem**

1. Complete the table shown to the right to form a numerical model for this problem.

WIDTH W (ft)	LENGTH L (ft)	AREA A (ft^2)
5.0	140	700
10.0		
15.0		
20.0		
25.0		
30.0		
35.0		
37.5		
40.0		
45.0		
50.0		
55.0		
60.0		
65.0		
70.0		

2. a. Using the first two columns of this table, plot these data on a scatter diagram, using an appropriate scale for each axis. This will produce a graphical model for the length as a function of the width.
   b. Now write an equation that gives the length L of the rectangular playground area in terms of the width W. This will produce an algebraic model for the length in terms of the width. (*Suggestion:* Use your algebraic model to double-check your table and your graph.)
   c. What are the restrictions on the variable W based on your equation in part **2(b)**? (*Hint:* There is only so much fencing.)
   d. What are the restrictions on the variable L based on the input values listed in part **2(c)**?
3. a. Using the first and third columns of this table, plot these data on a scatter diagram, using an appropriate scale for each axis. This will produce a graphical model for the area in terms of the width.
   b. Now write an equation for A that gives the area of the rectangular playground in terms of the width W. This will produce an algebraic model for the area in terms of the width. (*Suggestion:* Use your algebraic model to double-check your table and your graph.)
   c. What are the restrictions on the variable W based on your equation in part **3(b)**?
   d. What is the interval of values of A that corresponds to the values of W listed in part **3(c)**?

II. **Using the Models to Answer Questions About This Problem**

1. a. Determine the length of the playground when the width is 32 ft.
   b. Determine the area of the playground when the width is 32 ft.
2. a. Determine the length of the playground when the width is 0 ft.
   b. Determine the area of the playground when the width is 0 ft.
3. a. Determine the length of the playground when the width is 75 ft.
   b. Determine the area of the playground when the width is 75 ft.
4. Determine the exact value of W for which the length is 105 ft.
5. Determine the exact value of W for which the area is 2,772 ft^2.
6. Determine the width and the length of the fence that will enclose the maximum possible playground area. What is the maximum possible area?
7. Explain how you determined the dimensions that yield the maximum area. Which of the three types of mathematical models (numerical, graphical, or algebraic) did you use to answer this question? Why did you use this model?

chapter 7

Functions: Linear, Absolute Value, and Quadratic

CHAPTER OUTLINE

7.1 **Functions and Representations of Functions**
7.2 **Linear and Absolute Value Functions**
7.3 **Linear and Quadratic Functions and Curve Fitting**
7.4 **Using the Quadratic Formula to Find Real Solutions**
7.5 **The Vertex of a Parabola and Max-Min Applications**
7.6 **More Applications of Quadratic Equations**
7.7 **Complex Numbers and Solving Quadratic Equations with Complex Solutions**

The path followed by water as it leaves a fire hose is that of a parabola. There are two factors that affect the distance the water travels. One is the pressure, and the other is the direction in which the nozzle is aimed. Once the water has been turned on, the firefighters adjust the aim of the nozzle to ensure the water hits the desired location.

The two parabolic water paths shown are based on the same water pressure with different nozzle directions.

REALITY CHECK

Which parabolic path do you think would be more appropriate for putting out a fire inside the building?

A.

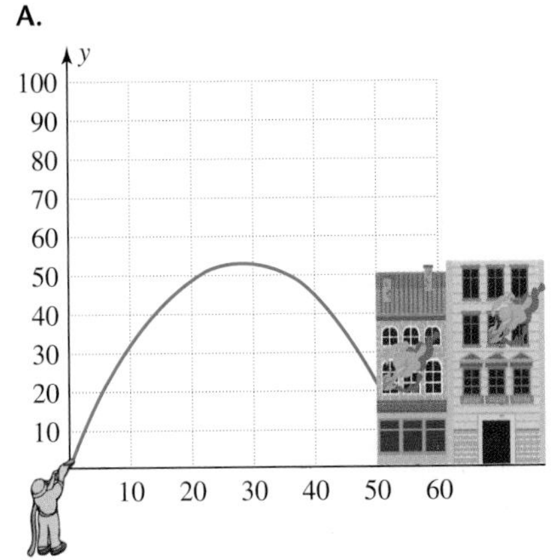

B.

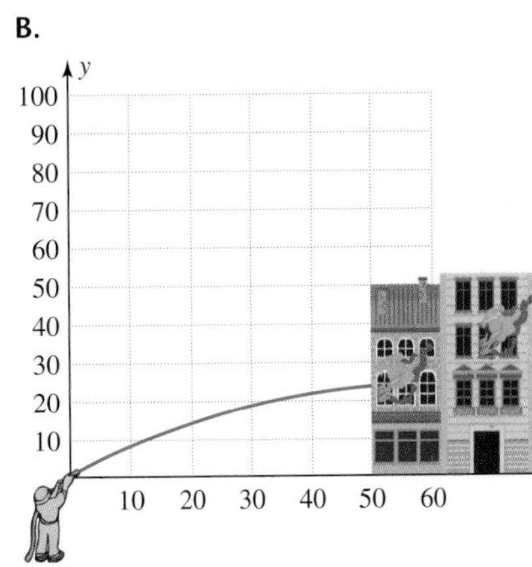

Section 7.1 Functions and Representations of Functions

Objectives:

1. Determine whether a relation is a function.
2. Determine the domain and range of a function.
3. Use function notation.

A Mathematical Note

The word *function* was introduced by the German mathematician Gottfried Wilhelm Freiherr von Leibniz (c. 1682). Leibniz is credited as a developer of calculus, together with the English mathematician Sir Isaac Newton (c. 1680). Leibniz contributed both original terminology and notation as well as his results in the areas of algebra and calculus.

Function notation and linear functions were first examined in Section 2.2. Many mathematicians view the concept of a function as the single most important concept in mathematics.* Because of the importance of this function concept, we will define a function, reexamine function notation, and examine functions from multiple perspectives:

- Mapping representation of a function
- Ordered-pair representation of a function
- Table representation of a function
- Graphical representation of a function
- Function notation representation of a function

Definition of a Function

In mathematics we use the word *function* to designate a correspondence that has very specific properties. A function must match each input value with exactly one value of an output variable.

Function

A **function** is a correspondence that matches each input value with exactly one value of the output variable. The set of all possible input values is called the **domain** of the function or the set of **independent** values. The set of all output values is called the **range** of the function or the set of **dependent** values.

When you first begin studying the function concept, you may want to think about three different interrelated parts of a function:

A key point is to understand the importance of the words *exactly one.* Each input value is paired with one and only one output value.

1. The domain of the function, the set of all possible input values
2. The range of the function, the set of all possible output values
3. The correspondence (rule or formula), which pairs each domain element with exactly one range element

Mapping Representation of a Function

Functions can be denoted in a variety of ways. Example 1 uses mapping notation. The mapping or arrow notation clarifies the pairing that a function creates between the domain elements and the range elements. Compare parts (b) and (c) of Example 1. It is important to understand the distinction between these examples.

*See Ed Dubinsky and Gvershon Harel, "The Concept of Function: Aspects of Epistemology and Pedagogy," *MAA Notes*, vol. 25, 1992.

EXAMPLE 1 Identifying Functions by Using Mapping Notation

Classify each correspondence as either a function or a correspondence that is not a function. For each function identify the domain D and range R.

SOLUTION

(a)

D		R
3	→	15
9	→	39
13	→	55

This correspondence is a function.
Domain $D = \{3, 9, 13\}$
Range $R = \{15, 39, 55\}$

Each domain element is paired with exactly one output value in the range.

(b)

D		R
16	→	4
	↘	−4
0	→	0

This correspondence is not a function.

The domain element 16 is not paired with exactly one element in the range; it is paired with both 4 and −4.

(c)

D		R
−4	→	16
4	↗	
0	→	0

This correspondence is a function.
Domain $D = \{-4, 0, 4\}$
Range $R = \{0, 16\}$

−4 is paired only with 16; 4 is paired only with 16; and 0 is paired only with 0. Thus each element in D is paired with exactly one element in R.

SELF-CHECK 7.1.1

Classify each correspondence as either a function or a correspondence that is not a function. Give the domain and range of each function.

1.

D		R
0	→	0
3	→	3
−3	↗	

2.

D		R
0	→	0
3	→	3
	↘	−3

Ordered-Pair Representation of a Function

The mapping notation helps emphasize that a function can be viewed as a process that takes input values and produces a unique output.

The arrows in the mapping notation for functions have the advantage of clearly stressing the active nature of a function in pairing input values with output values. Conceptually this is a great notation. Practically, mapping notation has some limitations. If there are many domain values, it can take a long time to draw all the arrows for the function. Arrows are also more awkward to type than other notations. The ordered-pair notation shown in Example 2 can convey the same information more concisely. The first coordinate of an ordered pair is the input value, the second coordinate is the output value, and the parentheses establish how these elements are paired.

A **relation** is defined as any set of ordered pairs. Thus some relations are functions and others aren't. To be a function, a relation must pair each input value with exactly one output value.

EXAMPLE 2 Comparing Mapping Notation and Ordered-Pair Notation

(a) Convert this function, given in mapping notation, to ordered-pair notation.

SOLUTION

Mapping Notation	*Ordered-Pair Notation*
D → R	
$-2 \rightarrow -1$	$(-2, -1)$
$-1 \rightarrow 3$	$(-1, 3)$
$0 \rightarrow 5$	$(0, 5)$
$1 \rightarrow 7$	$(1, 7)$
$2 \rightarrow 9$	$(2, 9)$

Both notations indicate a function whose domain is $D = \{-2, -1, 0, 1, 2\}$ and whose range is $R = \{-1, 3, 5, 7, 9\}$. The two notations also pair the elements exactly the same way.

This function can be written as the set of ordered pairs $\{(-2, -1), (-1, 3), (0, 5), (1, 7), (2, 9)\}$.

(b) Convert this function, given as a set of ordered pairs, to mapping notation.

SOLUTION

In an ordered pair (x, y) it is crucial to give the coordinates in the correct order. A set is just a collection of elements—elements that we can list in any order. Thus $\{-5, 5, -7, 7\} = \{-7, -5, 5, 7\}$.

Ordered-Pair Notation

$\{(-5, 5), (5, 5), (-7, 7), (7, 7)\}$

Mapping Notation

D		R
-5	→	5
5	↗	
-7	→	7
7	↗	

Both notations indicate a function whose domain is $D = \{-7, -5, 5, 7\}$ and whose range is $R = \{5, 7\}$.

SELF-CHECK 7.1.2

Classify each relation as either a function or a correspondence that is not a function. Give the domain and range of each function.

1. $\{(1, 6), (3, 8), (\pi, 10), (9, 12)\}$
2. $\{(1, 6), (1, 8), (\pi, 10), (9, 12)\}$

Graphical Representation of a Function

The Cartesian coordinate system provides a pictorial means of presenting the relationship between two variables. Since each point in a plane can be uniquely identified by an ordered pair (x, y), graphs can be used to represent mathematical relations. The x-coordinate of the ordered pair (x, y) is called the **independent variable,** and the y-coordinate is called the **dependent variable.** In an experiment, the experimenter has the freedom to select different input values. To reflect this freedom, we refer to the variable representing these input values as the independent variable. When the experiment is run, the corresponding output values are determined by the input values that were used. Thus we refer to the variable representing the output values as the dependent variable.

EXAMPLE 3 Graphing Relations

Graph each of these relations and determine whether the relation is a function.

SOLUTION

(a) $\{(2, -2), (2, -1), (2, 0), (2, 1), (2, 2)\}$

Not a function.

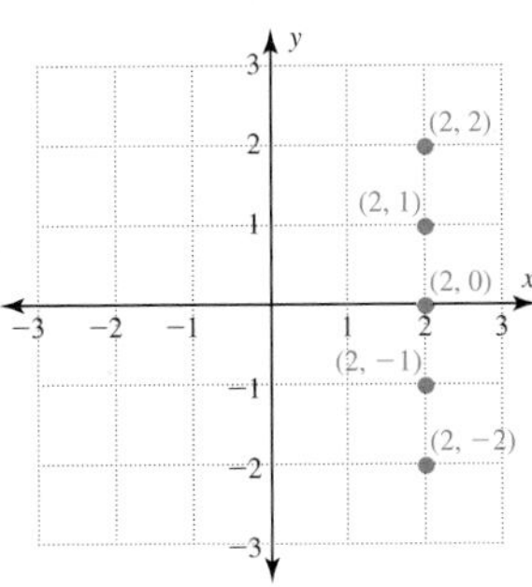

This relation is not a function. The input of 2 has more than one output paired with it.

(b)

D		*R*
−2	→	−3
−1	→	−1
0	→	1
1	→	3
2	→	5

Function.

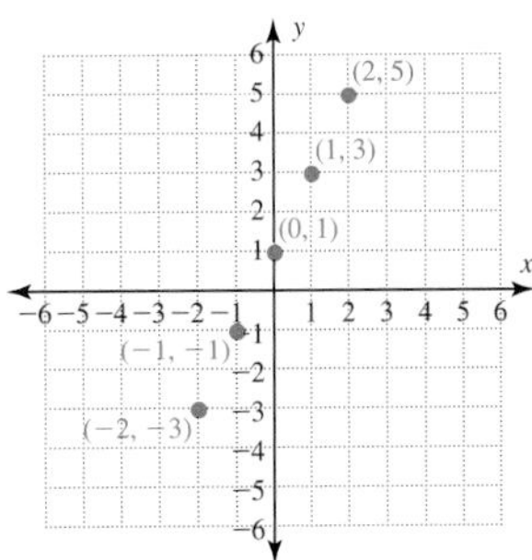

This relation is a function with domain $D = \{-2, -1, 0, 1, 2\}$ and range $R = \{-3, -1, 1, 3, 5\}$.

(c)

x	*y*
−2	2
−1	2
0	2
1	2
2	2

Function.

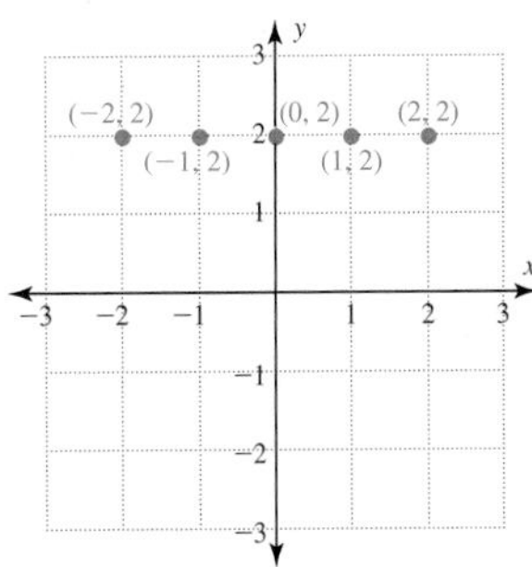

This relation is a function with domain $D = \{-2, -1, 0, 1, 2\}$ and range $R = \{2\}$. Each input value has exactly one output paired with it. Each input is paired with 2.

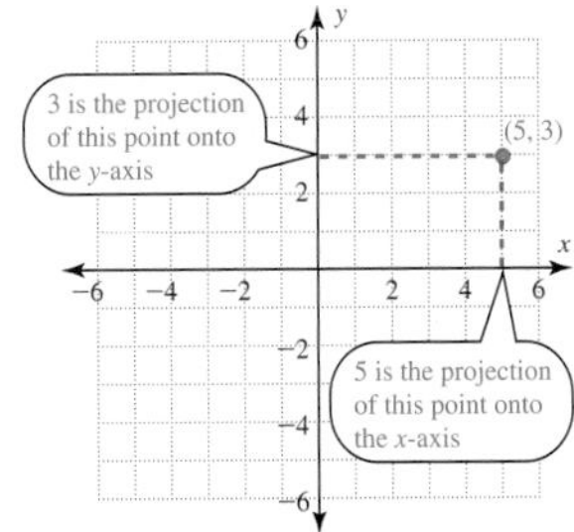

If a function is defined by a graph, then the ordered pairs can be determined by examining the points that form the graph. The domain can be found by projecting these points onto the x-axis, and the range can be found by projecting these points onto the y-axis. This is illustrated by the point (5, 3) shown in the graph.

Domain and Range from the Graph of a Function

Domain: The domain of a function is the projection of its graph onto the x-axis.
Range: The range of a function is the projection of its graph onto the y-axis.

EXAMPLE 4 Determining the Domain and Range from the Graph of a Function

Write the domain and the range of the functions defined by these graphs.

SOLUTION

	DOMAIN: PROJECTION ONTO x-AXIS	RANGE: PROJECTION ONTO y-AXIS
(a)	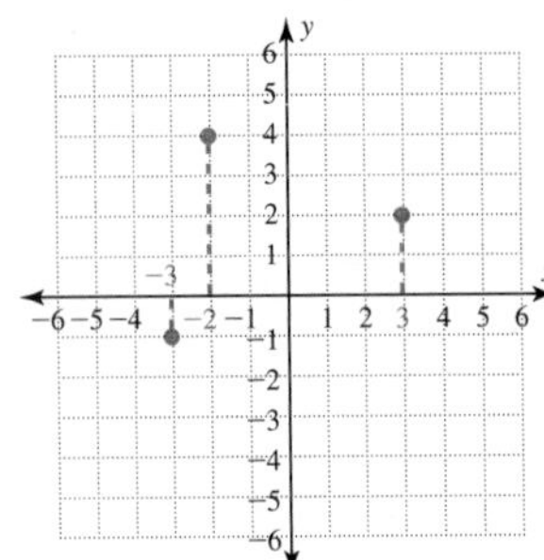 Domain $= \{-3, -2, 3\}$ This function consists of the ordered pairs $\{(-3, -1), (-2, 4), (3, 2)\}$.	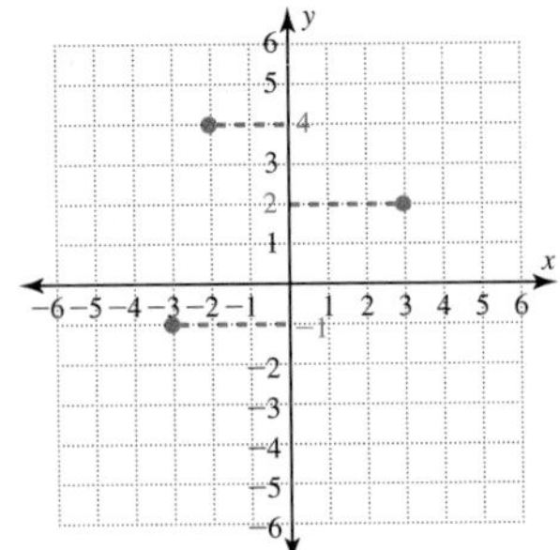 Range $= \{-1, 2, 4\}$
(b) 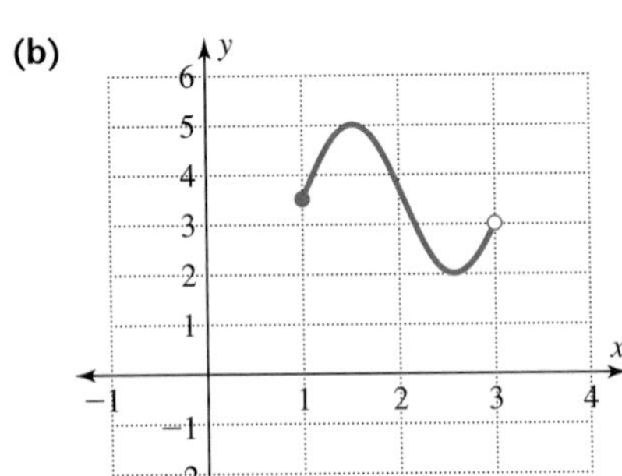	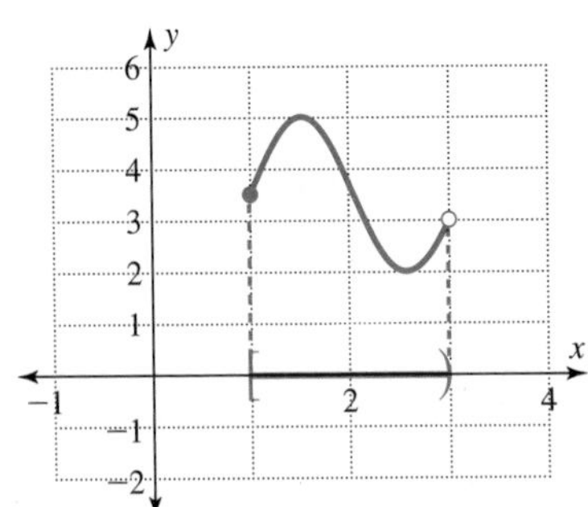 Domain $= [1, 3)$ The domain includes the endpoint 1 but does not include the endpoint 3.	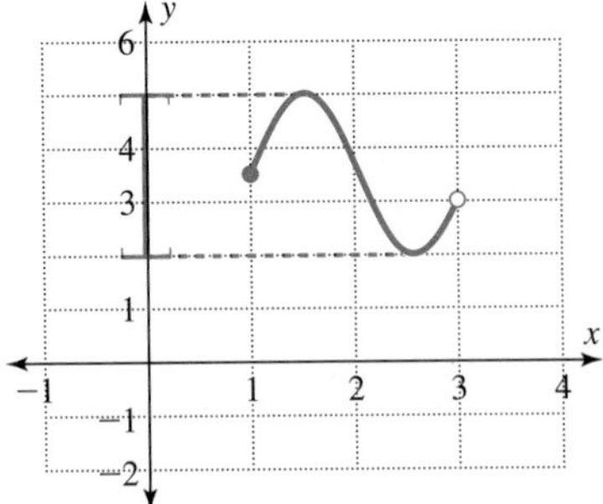 Range $= [2, 5]$
(c) 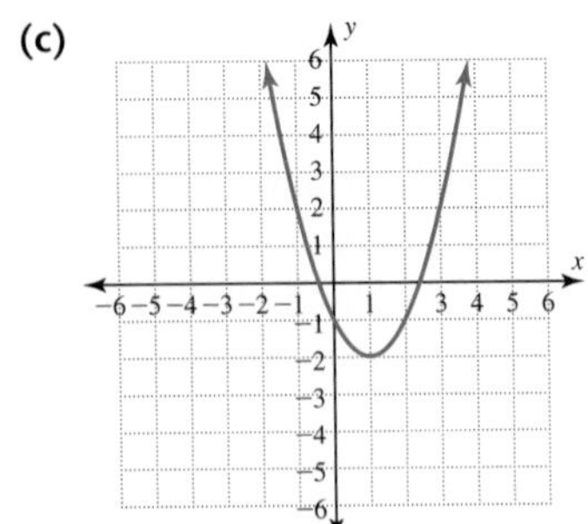	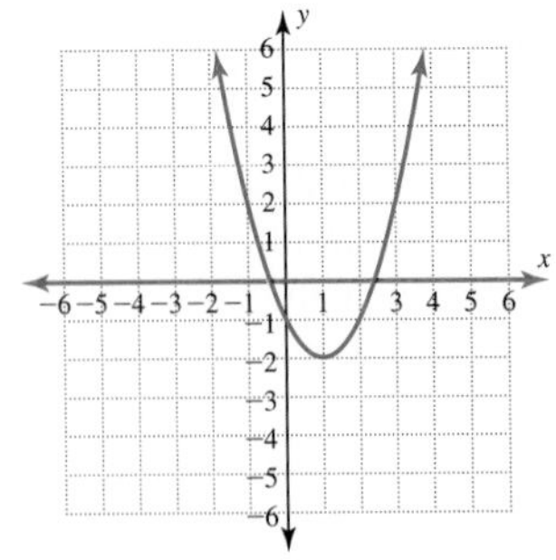 Domain $= \mathbb{R}$, the set of all real numbers.	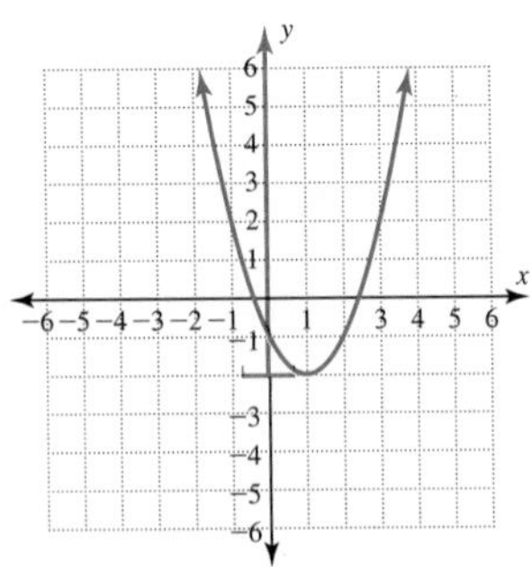 Range $= [-2, \infty)$

SELF-CHECK 7.1.3

Write the domain and the range of the functions defined by these graphs.

1.

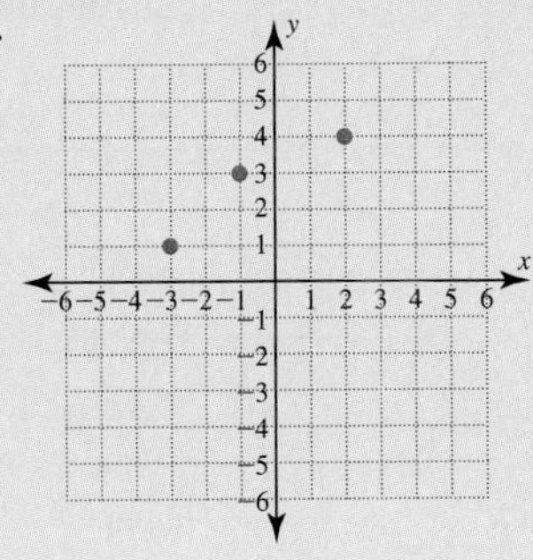

2.

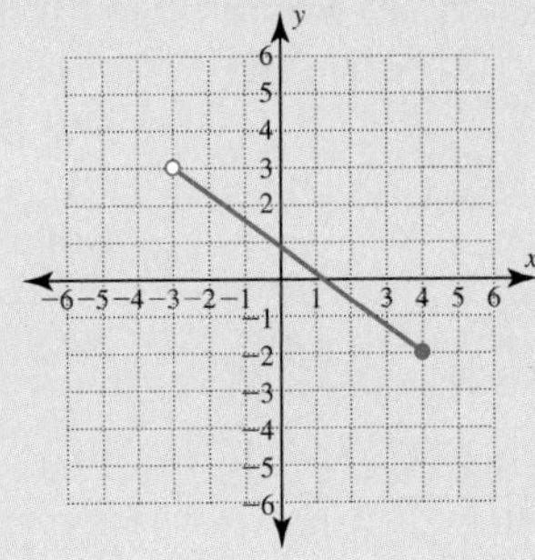

The Vertical Line Test

An easy way to determine if a graph represents a function is to imagine a vertical line sweeping across the graph. A vertical line hitting a graph in exactly one point pairs that input value of x with exactly one output value of y. A vertical line hitting a graph in two points pairs that input value of x with two output values of y—thus the graph does not represent a function.

A Graph of a Function

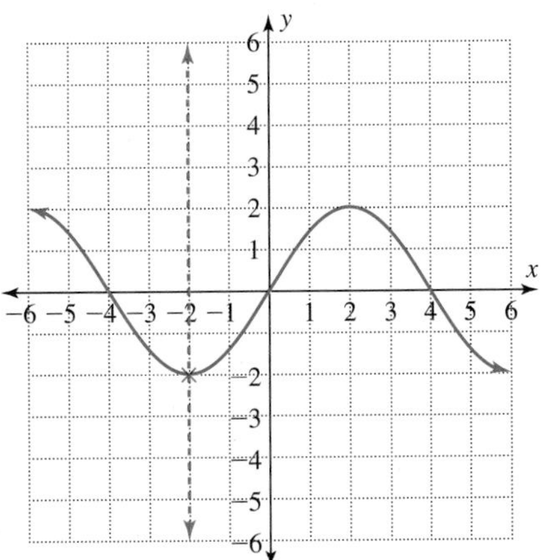

A Graph That Is Not a Function

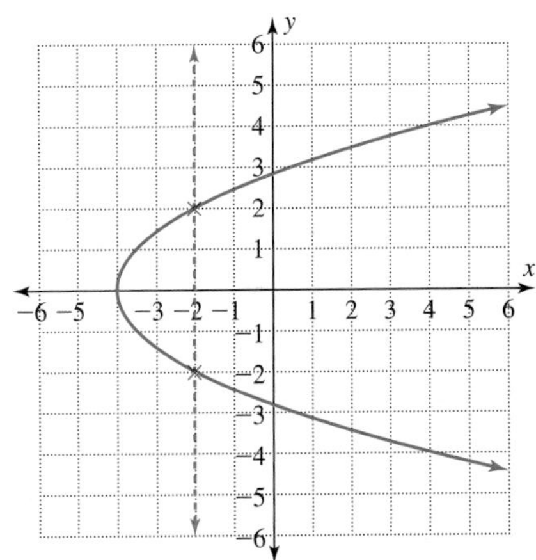

The Vertical Line Test

A graph represents a function if it is impossible to have any vertical line intersect the graph at more than one point.

EXAMPLE 5 Using the Vertical Line Test

Use the vertical line test to determine if the graphs of these relations represent functions.

SOLUTION

(a)

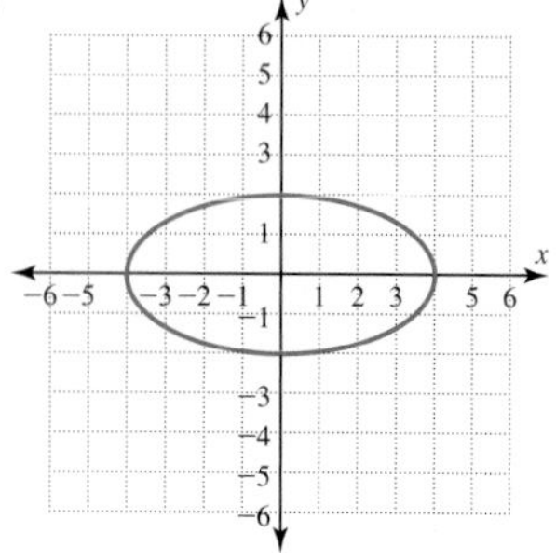

Not a function.

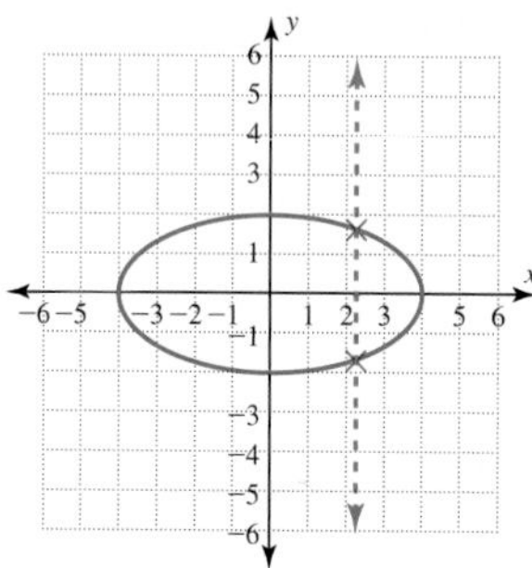

It is possible to draw a vertical line, as the one shown here, that intersects the graph in two points.

(b)

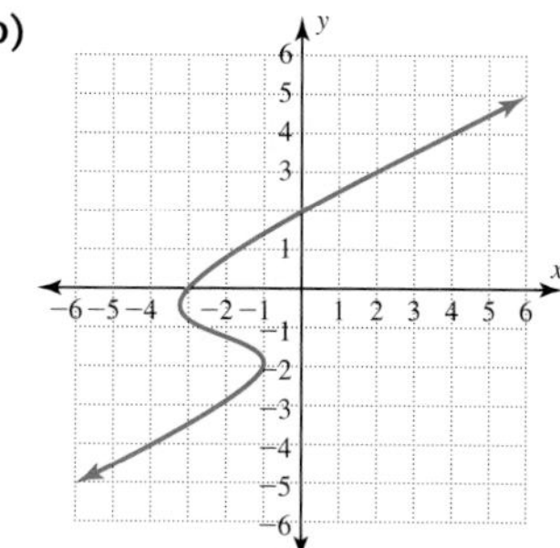

Not a function.

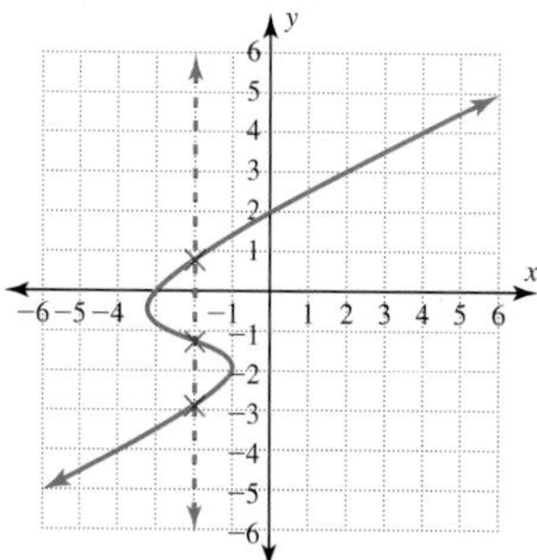

The vertical line shown intersects the graph at three points.

(c)

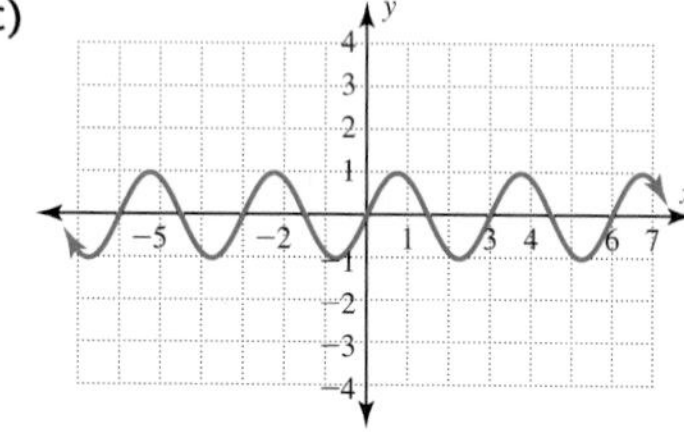

A function.

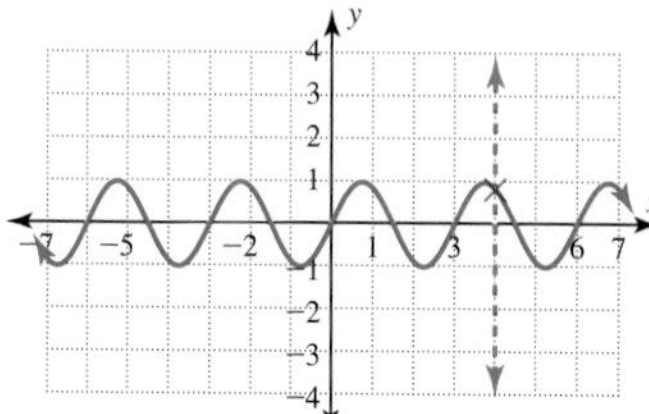

Every vertical line will intersect this graph in exactly one point. This function pairs each input value with exactly one output value.

SELF-CHECK 7.1.4

Use the vertical line test to determine whether each relation is a function.

1.

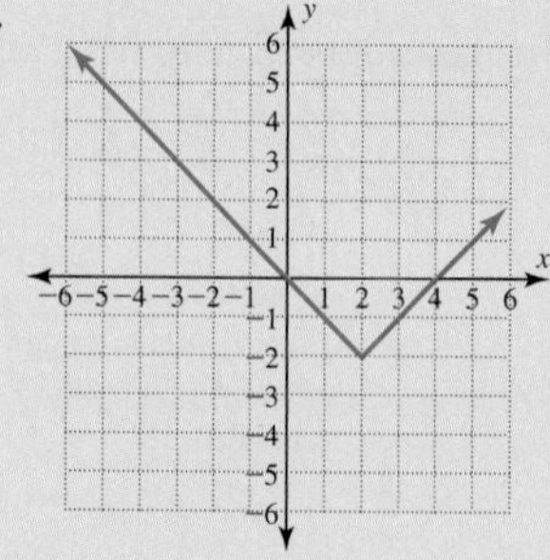

2.

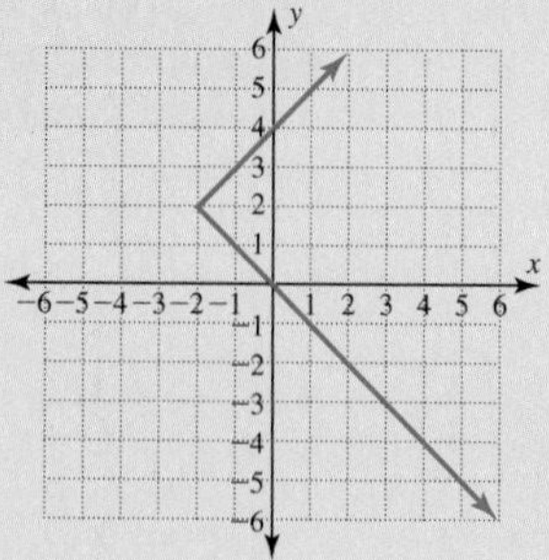

Function Notation Representation of a Function

A table of values or a graph of a few points can each give us valuable representations for showing the relationship between a set of x-y pairs. For many x-y pairs it may be inconvenient or impossible to put all these points in a table. Thus it is useful to be able to express the relationship between x and y by using an equation. The notation $f(x)$ is referred to as **function notation** and is read **"f of x"** or **"$f(x)$ is the output value for an input value of x."** Function notation, first introduced in Section 2.2, is used to give an equation that describes a unique output that we can calculate for each input of x. To evaluate $f(x)$ for a specific value of x, replace x on both sides of the equation by this specific value.

A Mathematical Note

The publication *Beyond Crossroads* by the American Mathematical Association of Two-Year Colleges (AMATC) strongly recommends that functions be examined verbally, numerically, graphically, and symbolically.

Function Notation

The expression $f(x)$ is read as "f of x" or "the value of f at x." The letter f names the function, and the variable x represents an input value from the domain. The notation $f(x)$ represents a specific output value corresponding to x.

Caution: $f(x)$ does not mean f times x in this context. It represents an output value for a specific input value of x.

It is common to use the letter f to represent the function, although any letter can be used. Example 6 illustrates how function notation is used to represent a linear function.

This book has been written to follow this AMATYC guideline and to use technology to enhance your learning. Nonetheless, it is important that you understand how to produce these tables and sketch these graphs by hand. We arbitrarily selected our input values of x. Because the input values can be any real number, the graph is one continuous line.

EXAMPLE 6 Evaluating Function Notation

Evaluate $f(x) = 2x - 4$ for each input value, and then graph this linear function.

SOLUTION

(a) $x = 0$

$f(x) = 2x - 4$

$f(0) = 2(0) - 4$ — Substitute each input value of x into the formula $f(x) = 2x - 4$.

$f(0) = 0 - 4$

$f(0) = -4$ — $(0, -4)$ is a point on the graph.

(b) $x = 2$

$f(x) = 2x - 4$

$f(2) = 2(2) - 4$

$f(2) = 4 - 4$

$f(2) = 0$ — $(2, 0)$ is a point on the graph.

(c) $x = 3$

$f(x) = 2x - 4$

$f(3) = 2(3) - 4$

$f(3) = 6 - 4$

$f(3) = 2$ — $(3, 2)$ is a point on the graph.

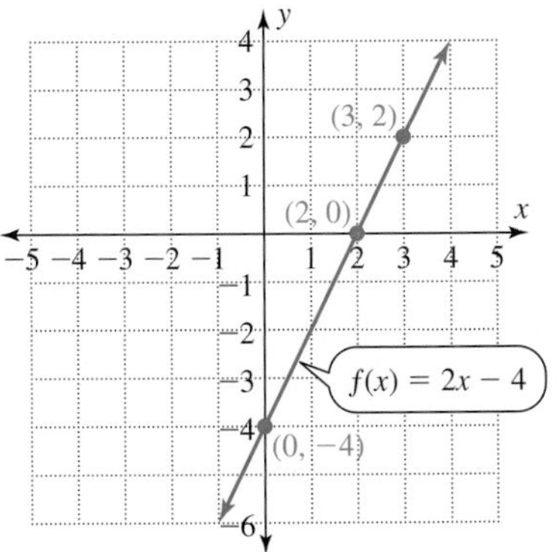

The equation $f(x) = 2x - 4$ is in the slope-intercept form ($y = mx + b$) of a line with slope $m = 2$ and y-intercept $(0, -4)$.

The graph of $f(x) = 2x - 4$ is one continuous line, not just the three distinct points calculated here. These points allow us to represent the graph of all the points on this line.

The letter x represents the input value. The output value is represented either by y or by $f(x)$.

In Example 7 the equation that produces this graph is given by $y = f(x)$. For each point (x, y) on the graph, x represents an input value and y represents $f(x)$; $f(x)$ is the output of the function f for the input x.

EXAMPLE 7 Using a Graph to Evaluate Input and Output Values

Use the given graph to complete the following.

(a) Evaluate $f(-3)$.
(b) Evaluate $f(1)$.
(c) Determine the value of x for which $f(x) = 1$.
(d) Determine the value of x for which $f(x) = 3$.

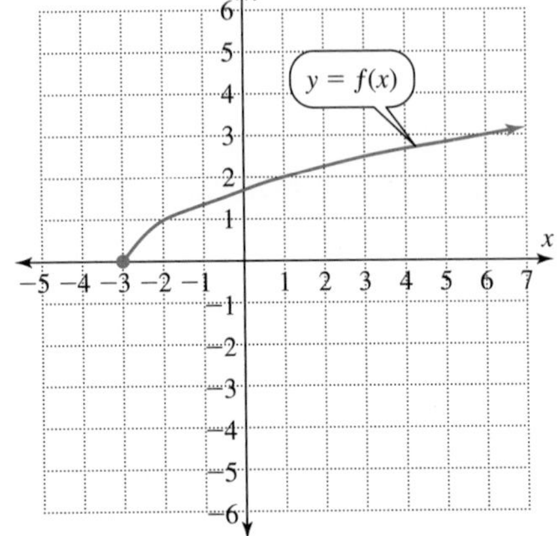

SOLUTION

(a) $f(-3) = 0$ — $(-3, 0)$ is a point on the graph. Thus $f(-3) = 0$.
(b) $f(1) = 2$ — $(1, 2)$ is a point on the graph. Thus $f(1) = 2$.
(c) $x = -2$ — $(-2, 1)$ is a point on the graph. Thus $f(-2) = 1$.
(d) $x = 6$ — $(6, 3)$ is a point on the graph. Thus $f(6) = 3$.

SELF-CHECK 7.1.5

Use $f(x) = -3x + 7$ to evaluate each expression.

1. $f(-2)$ **2.** $f(0)$ **3.** $f(4)$

Use the given graph to complete the following.

4. Evaluate $f(-3)$.
5. Evaluate $f(2)$.
6. Determine the value(s) of x for which $f(x) = -3$.
7. Determine the value(s) of x for which $f(x) = 2$.

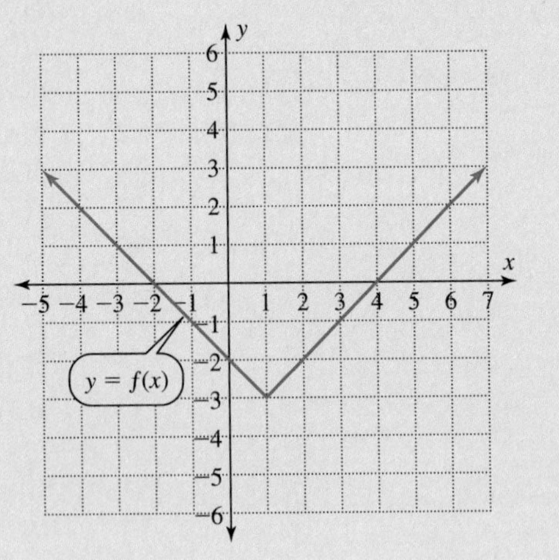

The equation that defines a function can be used to produce a graph of the function, and then this graph can be used to determine the domain and range of the function. This can be facilitated by the use of a graphing calculator.

EXAMPLE 8 Using a Graphing Calculator to Determine the Domain and Range of a Function

Use a graphing calculator to determine the domain and range of $f(x) = x^2 - 2$.

SOLUTION

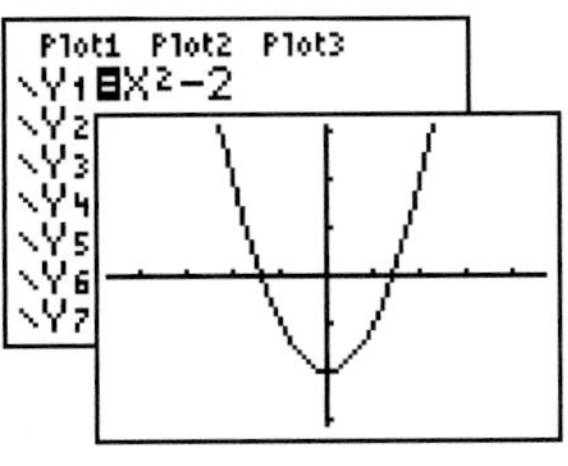

$[-4.7, 4.7, 1]$ by $[-3.1, 3.1, 1]$

Use $y = x^2 - 2$ to represent the equation. [Again either y or $f(x)$ can represent the output variable.]

The graph is continuing to spread to both the left and right, so the projection of the graph onto the x-axis will include the whole x-axis. This means the domain is the set of all real numbers.

The range, the projection of the graph onto the y-axis, is the interval $[-2, \infty)$. This notation indicates that -2 is included in the range.

Answer: Domain $= \mathbb{R}$
Range $= [-2, \infty)$

SELF-CHECK 7.1.6

Use the graph in Example 7 to answer these questions.

1. Determine the domain of this function.
2. Determine the range of this function.

Multiple Representations of a Function

As we noted at the beginning of this section, there are many ways to represent a function. Example 9 compares the representations given in this section.

EXAMPLE 9 Using Multiple Representations for a Function

A Mathematical Note

AMATYC's publication *Beyond Crossroads* recommends that topics on discrete mathematics be integrated throughout introductory mathematics.

An investment in a bond tripled in value over a 25-year period. Four individuals invested respectively \$50, \$100, \$200, and \$500. Using these four values as the input values, represent this function by using function notation, mapping notation, a table, ordered pairs, and a graph.

SOLUTION

VERBAL REPRESENTATION

Triple each input value to obtain the output value.

FUNCTION NOTATION

$f(x) = 3x$ for each x in the domain
$D = \{50, 100, 200, 500\}$

MAPPING NOTATION

x		$f(x)$
50	$\rightarrow$	150
100	$\rightarrow$	300
200	$\rightarrow$	600
500	$\rightarrow$	1,500

This example contains only selected discrete points. Thus the graph contains only these discrete points. The dashed line is not part of the graph but does help show the linear pattern.

TABLE

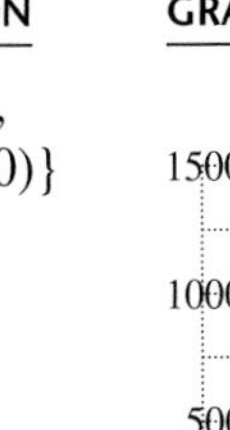

x	$f(x)$
50	150
100	300
200	600
500	1,500

ORDERED-PAIR NOTATION

$\{(50, 150), (100, 300), (200, 600), (500, 1,500)\}$

GRAPH

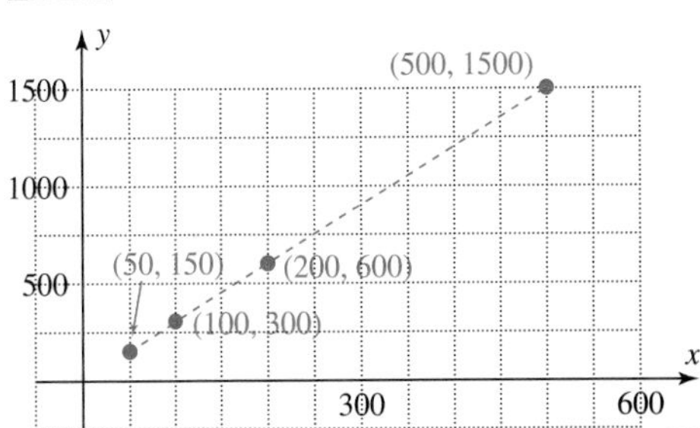

SELF-CHECK 7.1.7

Use the domain $D = \{-2, 0, 1, 3\}$ and $f(x) = -2x + 5$.

1. Give a table to represent this function.
2. Represent this function as a set of ordered-pairs.

SELF-CHECK ANSWERS

7.1.1

1. Function with $D = \{-3, 0, 3\}$ and $R = \{0, 3\}$
2. Not a function

7.1.2

1. Function with $D = \{1, 3, \pi, 9\}$ and $R = \{6, 8, 10, 12\}$
2. This relation is not a function; 1 is paired with two outputs, 6 and 8.

7.1.3

1. $D = \{-3, -1, 2\}$, $R = \{1, 3, 4\}$
2. $D = (-3, 4]$, $R = [-2, 3)$

7.1.4

1. A function
2. A relation but not a function

7.1.5

1. $f(-2) = 13$
2. $f(0) = 7$
3. $f(4) = -5$
4. $f(-3) = 1$
5. $f(2) = -2$
6. $x = 1$
7. $x = -4$ and $x = 6$

7.1.6

1. $D = [-3, \infty)$
2. $R = [0, \infty)$

7.1.7

1.

x	$f(x)$
-2	9
0	5
1	3
3	-1

2. $\{(-2, 9), (0, 5), (1, 3), (3, -1)\}$

USING THE LANGUAGE AND SYMBOLISM OF MATHEMATICS 7.1

1. A function is a correspondence that matches each input value with exactly __________ value of the __________ variable.
2. The set of all input values of a function is called the __________ of the function.
3. The set of all output values of a function is called the __________ of the function.
4. The independent values are represented by the __________ variable.
5. The dependent values are represented by the __________ variable.
6. In ordered-pair notation, the domain of a function is represented by the set of __________-coordinates.
7. In ordered-pair notation, the range of a function is represented by the set of __________-coordinates.
8. If a function is defined by a graph, the domain of the function is represented by the projection of the graph onto the __________-axis.
9. If a function is defined by a graph, the range of the function is represented by the projection of the graph onto the __________-axis.
10. The __________ line test can be used to visually inspect a graph to determine if it represents a function.
11. The notation $f(x)$ is called __________ notation.
12. The notation $f(x) = 3x^2$ is read "__________ of __________ equals three x squared." In this notation the input variable is represented by __________, and $f(x)$ represents the __________ variable. The letter __________ names the function.

QUICK REVIEW 7.1

The purpose of this quick review is to help you recall skills needed in this section.

1. Use the function $f(x) = 3x^2 - 5x - 1$ to evaluate each of these expressions.
   a. $f(0)$ b. $f(10)$ c. $f(-10)$
2. Graph the ordered pair $(-1, 3)$.
3. Graph the interval $(-1, 3)$.
4. Represent by using interval notation.

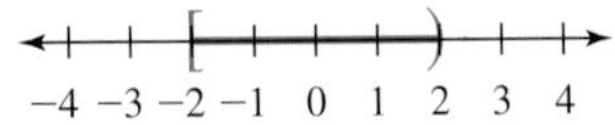

5. Represent by using interval notation.

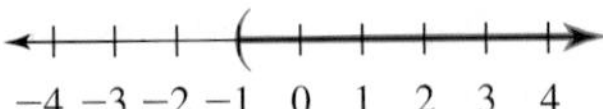

EXERCISES 7.1

In Exercises 1–4, determine whether each relation is a function. For each function identify the domain and range.

1. a.

D		*R*
1	→	−1
1	→	1
2	→	2
3	→	3

b.

D		*R*
−1	→	1
1	→	1
2	→	2
−3	→	3

c.

D		*R*
3	→	−1
2	→	1
1	→	3
0	→	0

2. a.

D		*R*
3	→	−5
3	→	0
3	→	4

b.

D		*R*
−5	→	3
0	→	3
4	→	3

c.

D		*R*
4	→	2
1	→	1
0	→	0
−1	→	0

3. a. $\{(1, 1), (-1, 1), (2, 0)\}$

b. $\{(1, 1), (1, -1), (0, 2)\}$

c.

x	*y*
−3	−1
−2	0
−1	1
0	2
1	3

4. a. $\{(2, 3), (2, -3), (0, 13), (13, 0)\}$

b. $\{(3, 2), (-3, 2), (0, 13), (13, 0)\}$

c.

x	*y*
−2	−8
−1	−1
0	0
1	1
2	8

In Exercises 5–10, use the vertical line test to determine whether each graph represents a function.

5. a.

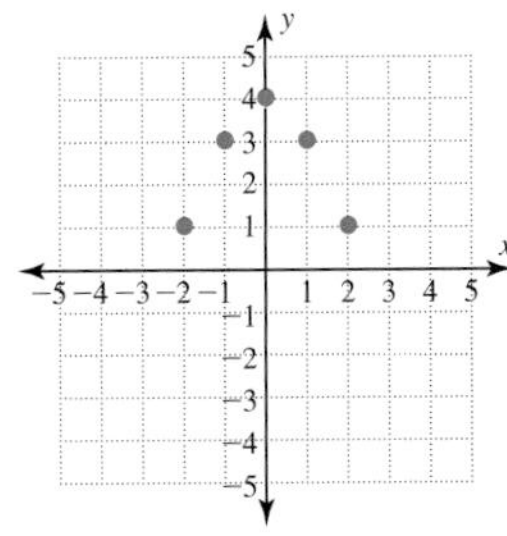

b.

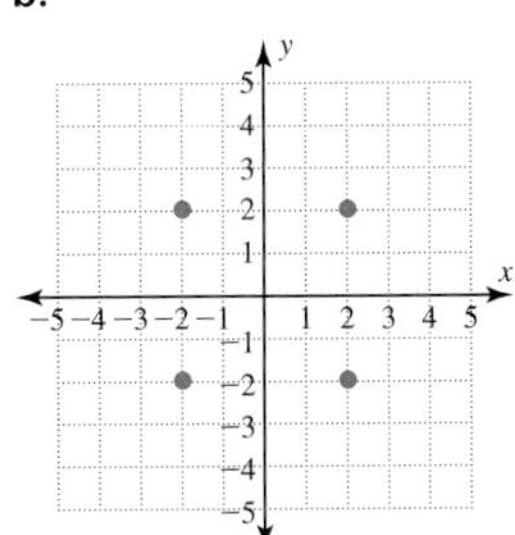

c.

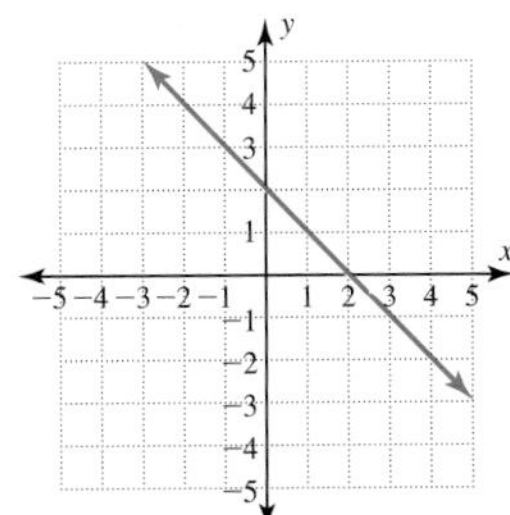

6. a.

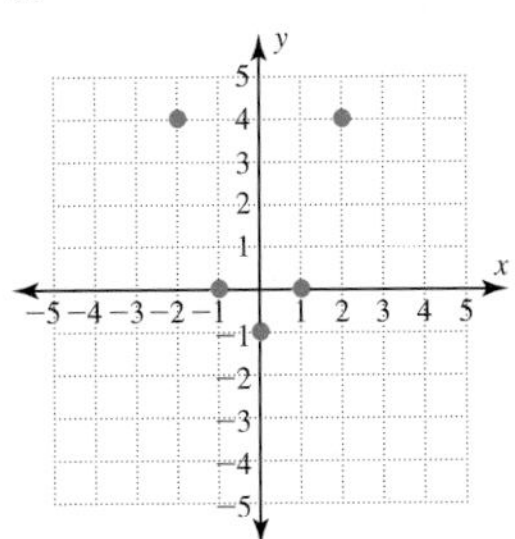

b.

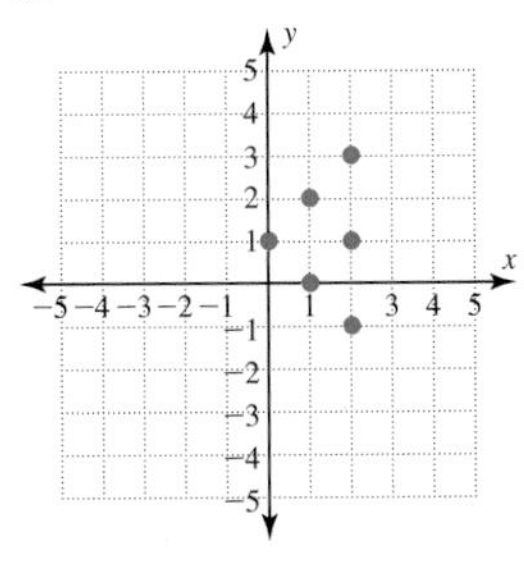

c.

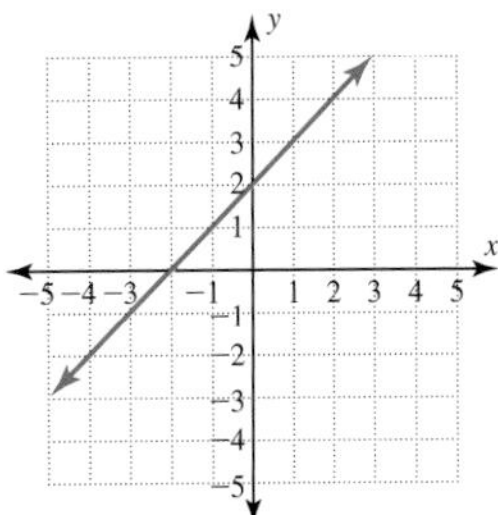

7. a.

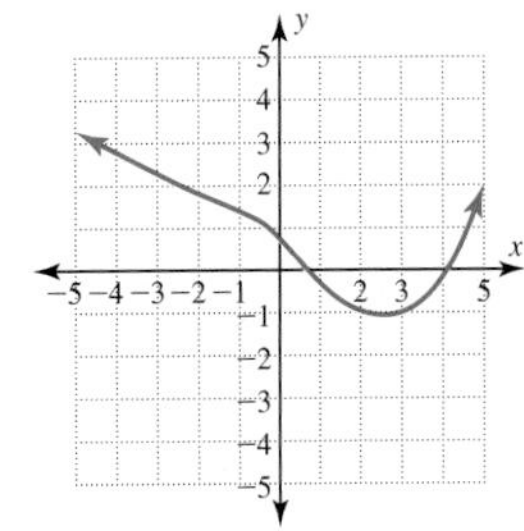

b.

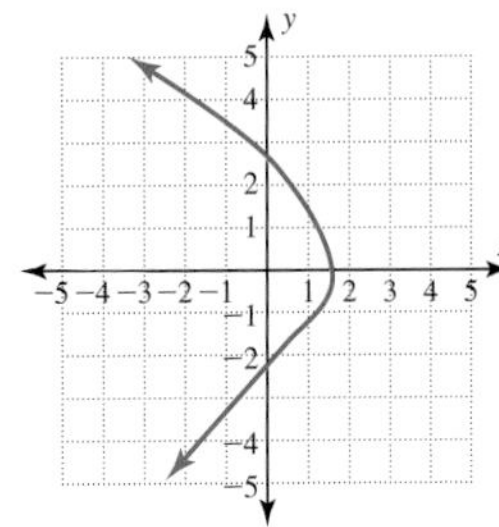

c.

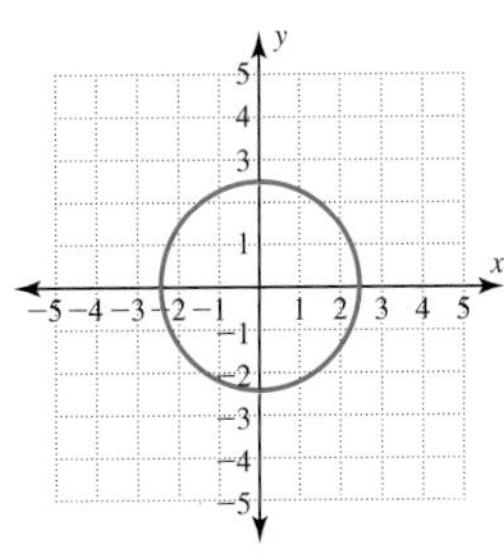

8. a.

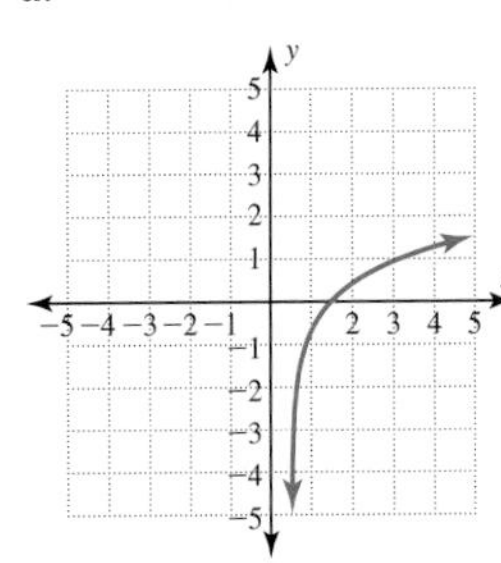

b.

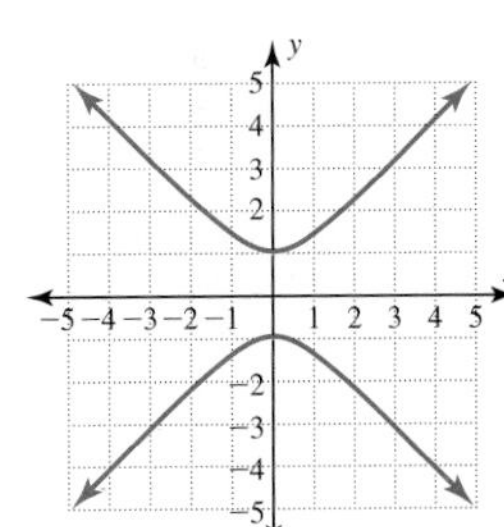

c.

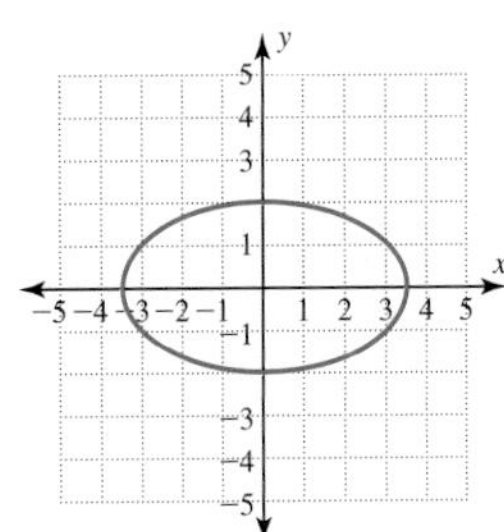

9. a.

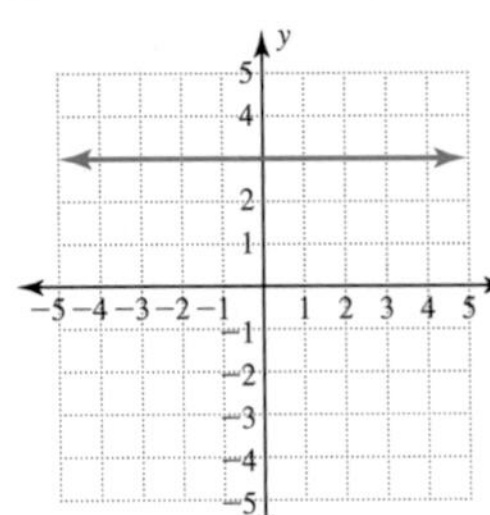

b.

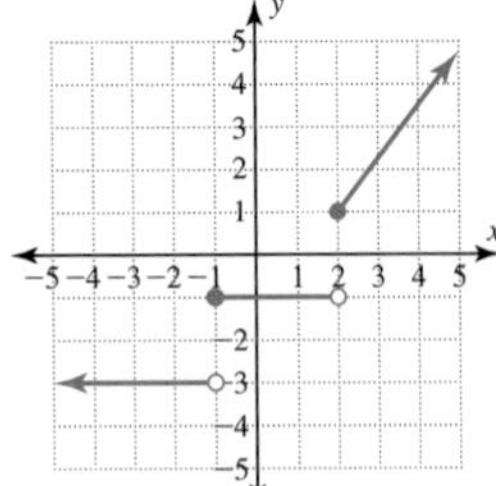

c.

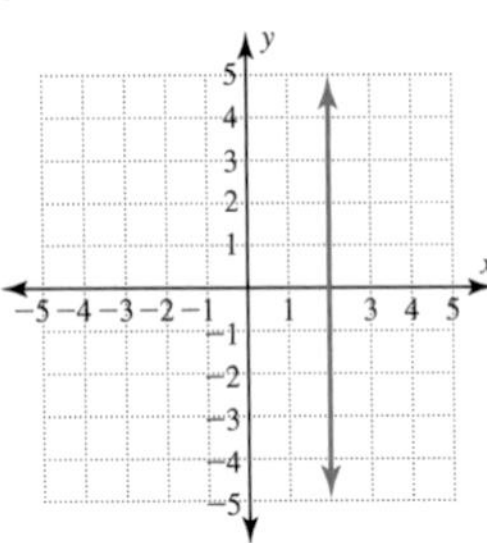

10. a.

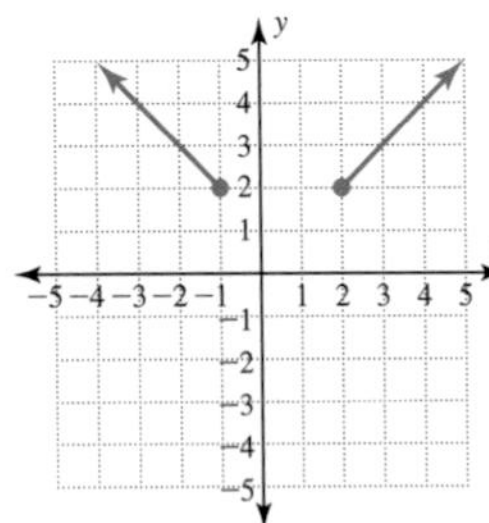

b.

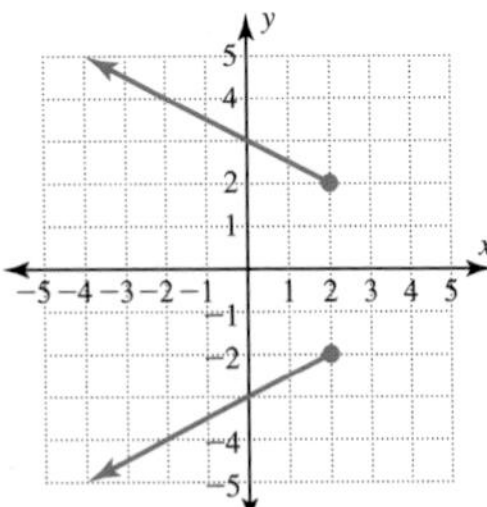

c.

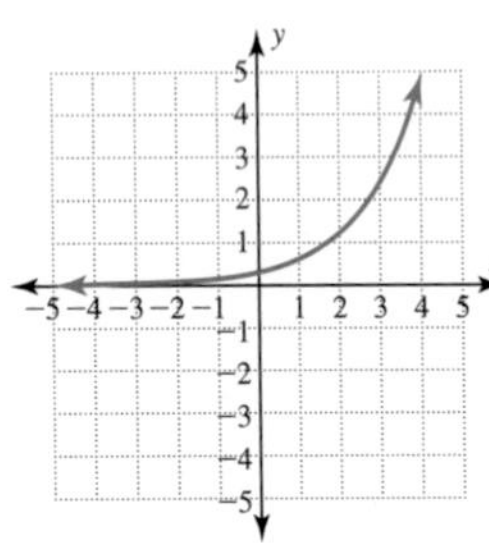

Given $f(x) = 2x - 7$, evaluate each expression in Exercises 11 and 12.

11. a. $f(-10)$ b. $f(0)$ c. $f(5)$

12. a. $f(-4)$ b. $f(-1)$ c. $f(10)$

Given $f(x) = \sqrt{x}$, evaluate each expression in Exercises 13 and 14.

13. a. $f(0)$ b. $f(9)$ c. $f\left(\frac{1}{4}\right)$

14. a. $f(16)$ b. $f(100)$ c. $f\left(\frac{9}{25}\right)$

15. Use the table below to answer each question.

x	y
−1	1
2	3
5	5
8	7

a. Is 3 an input value or an output value?
b. Is −1 an input value or an output value?
c. If 2 is the input value, what is the output value?
d. If the output value is 7, what was the input value?

16. Use the set of ordered pairs below to answer each question.
$\{(0, 4), (3, 5), (5, 1), (2, 0)\}$
a. Is 3 an input value or an output value?
b. Is 4 an input value or an output value?
c. If 5 is the input value, what is the output value?
d. If the output value is 5, what was the input value?

17. Use the graph below to answer each question.

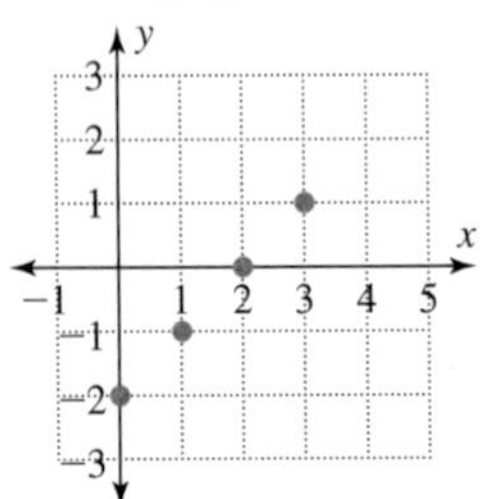

a. Is 3 an input value or an output value?
b. Is −1 an input value or an output value?
c. If 2 is the input value, what is the output value?
d. If the output value is 1, what was the input value?

18. Use the graph below to answer each question.

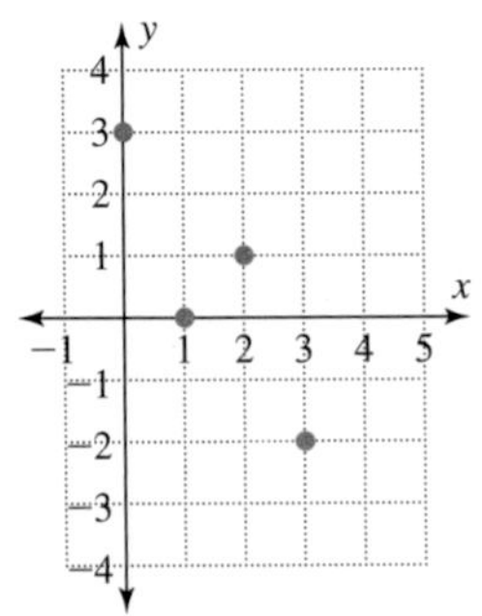

a. Is −2 an input value or an output value?
b. Is 2 an input value or an output value?
c. If 2 is the input value, what is the output value?
d. If the output value is 0, what was the input value?

In Exercises 19 and 20, use the given table to evaluate each value of $f(x)$.

19. a. $f(-3)$ b. $f(1)$ c. $f(2)$

20. a. $f(-2)$ b. $f(0)$ c. $f(3)$

In Exercises 21 and 22, use the given table to determine the value of x that produces the given value of $f(x)$.

21. a. $f(x) = -1$ b. $f(x) = 0$ c. $f(x) = 8$

22. a. $f(x) = 3$ b. $f(x) = -2$ c. $f(x) = -10$

x	f(x)
−3	8
−2	3
−1	0
0	−1
1	−2
2	−5
3	−10

Table for Exercises 19–22

In Exercises 23 and 24, use the given graph to evaluate each value of $f(x)$.

23. a. $f(-2)$ b. $f(0)$ c. $f(3)$

24. a. $f(-1)$ b. $f(1)$ c. $f(2)$

In Exercises 25 and 26, use the given graph to determine the value of x that produces the given value of $f(x)$.

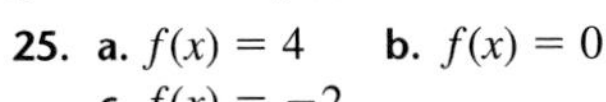

25. a. $f(x) = 4$ b. $f(x) = 0$
c. $f(x) = -2$

26. a. $f(x) = 2$ b. $f(x) = -4$
c. $f(x) = -3$

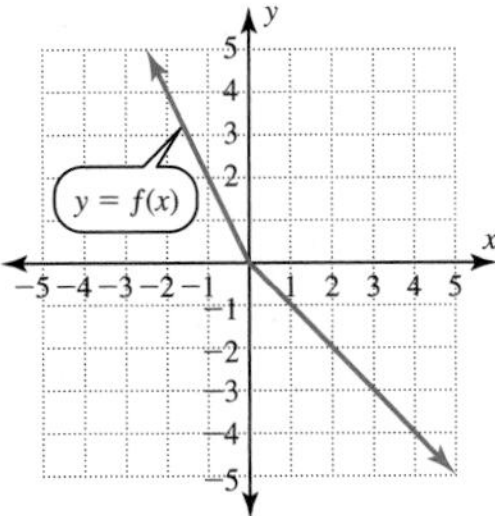

Graph for Exercises 23–26

In Exercises 27–30, use the given graph to determine the domain and range of each function.

27. a.

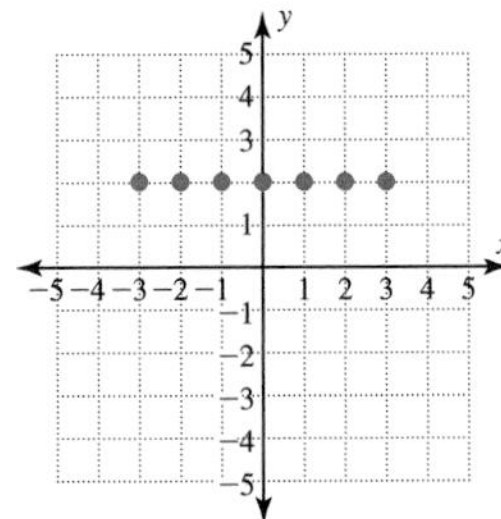

b.

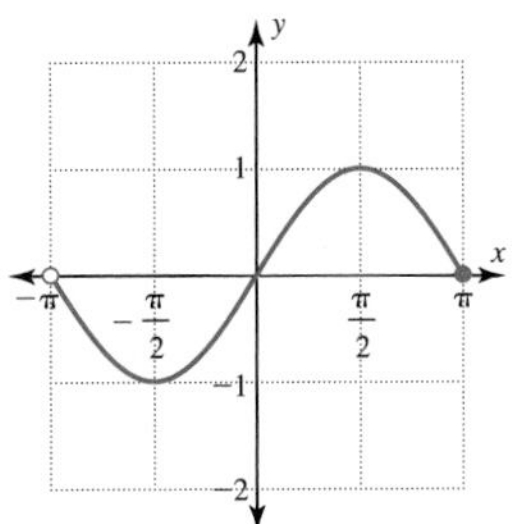

c.

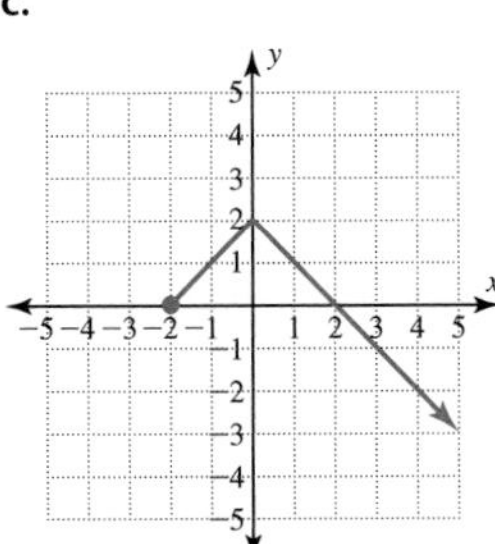

28. a.

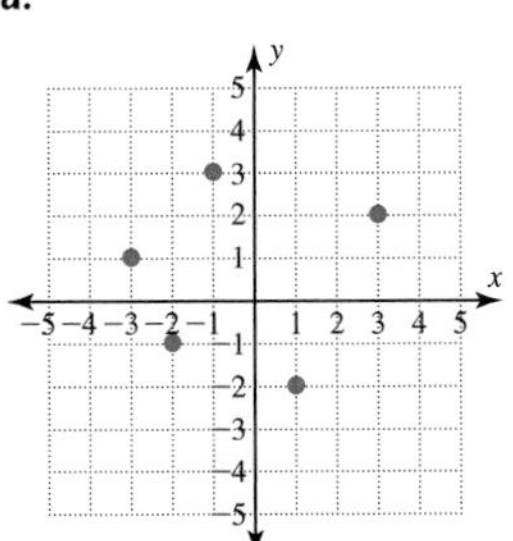

b.

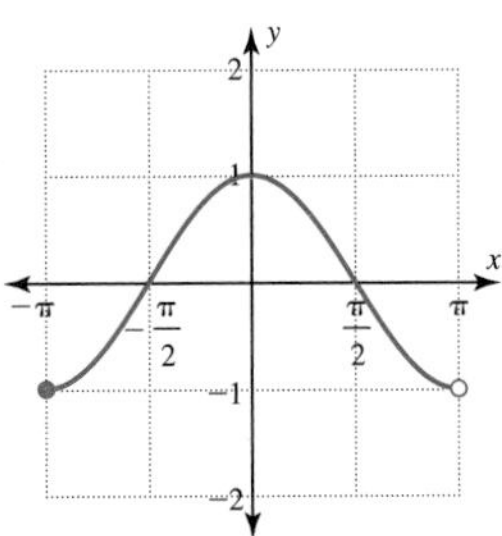

c.

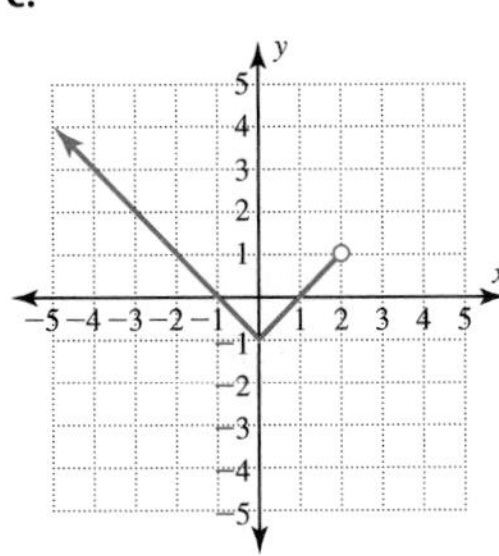

29. a.

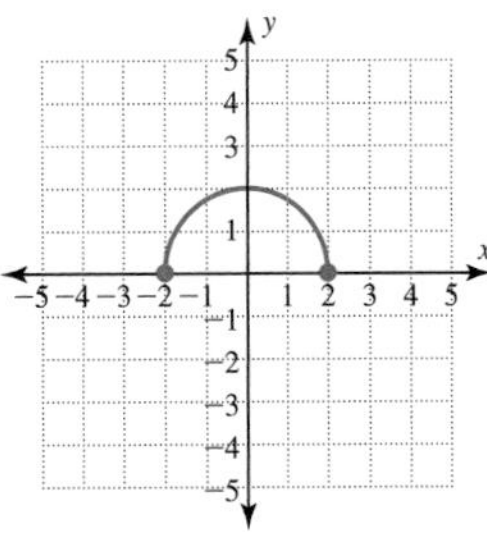

b.

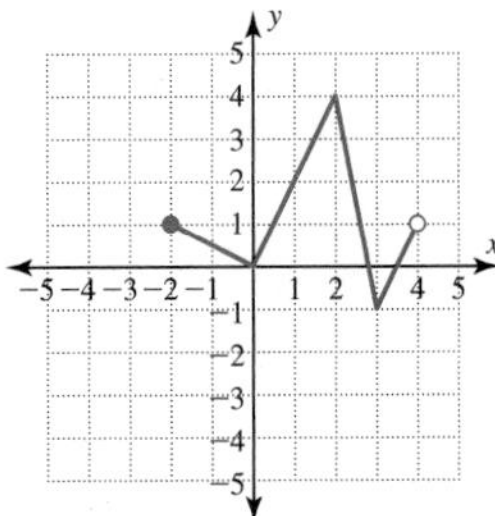

c.

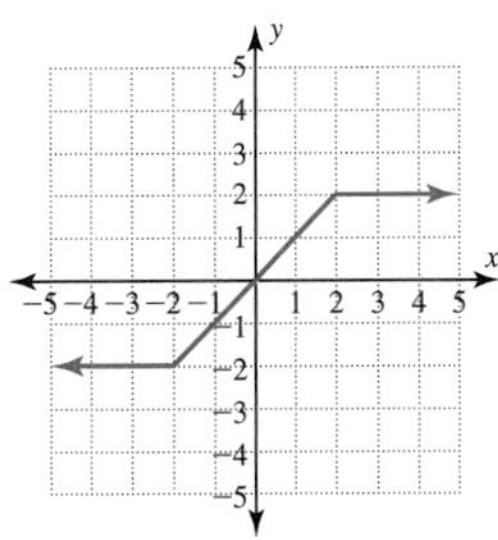

30. a.

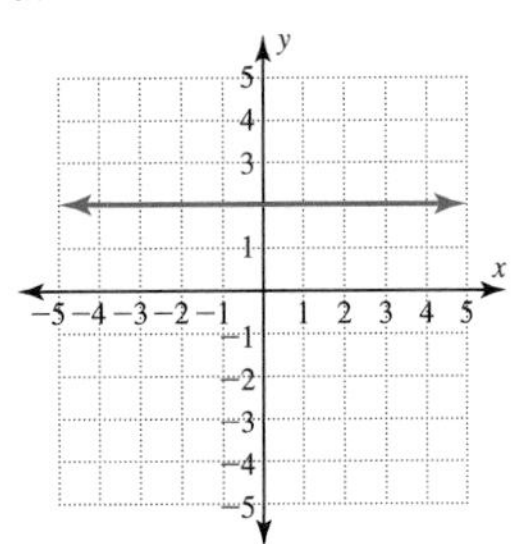

b.

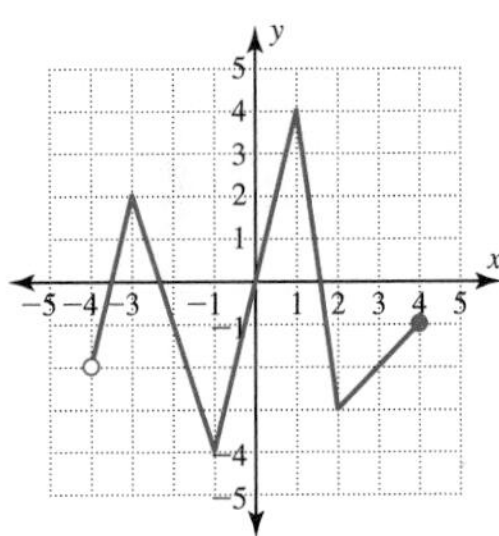

c.

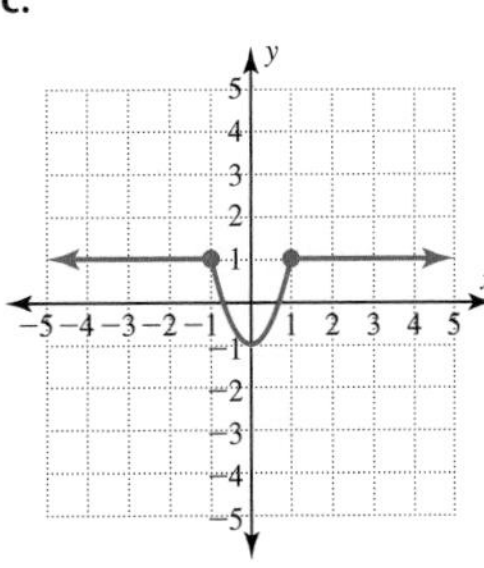

Multiple Representations

In Exercises 31–38, represent each function in multiple ways as described in each exercise.

31. Use the function defined by the table below to complete each part of this exercise.

x	−5	−3	−2	0	1	4
y	4	2	0	−2	−3	−4

a. Express this function by using mapping notation.
b. Express this function by using ordered-pair notation.
c. Graph this function.

32. Use the function defined by the given table to complete each part of this exercise.

x	-5	-3	-2	0	1	4
y	3	3	3	3	3	3

a. Express this function by using mapping notation.
b. Express this function by using ordered-pair notation.
c. Graph this function.

33. Use the function defined by the graph below to complete each part of this exercise.

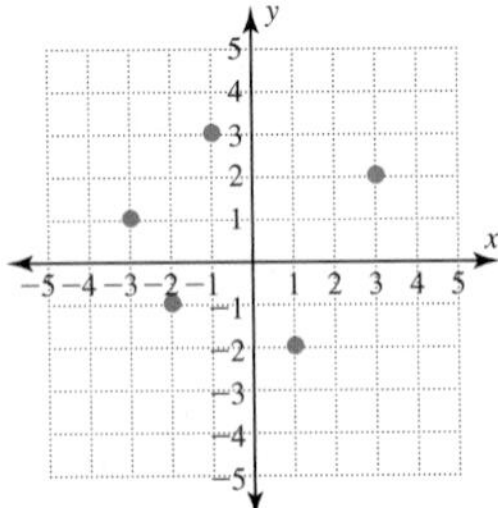

a. Express this function by using mapping notation.
b. Express this function by using ordered-pair notation.
c. Express this function by using a table format.

34. Use the function defined by the graph below to complete each part of this exercise.

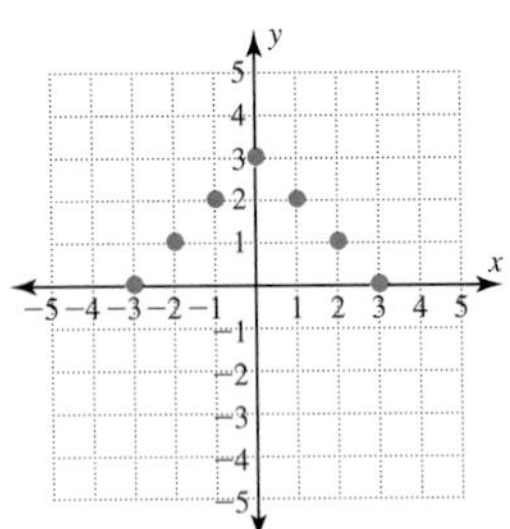

a. Express this function by using mapping notation.
b. Express this function by using ordered-pair notation.
c. Express this function by using a table format.

35. Use the function defined by the given mapping notation to complete each part of this exercise.

D		*R*
-1	$\rightarrow$	1
1	$\rightarrow$	3
2	$\rightarrow$	-1
4	$\rightarrow$	-2

a. Express this function by using a table format.
b. Express this function by using ordered-pair notation.
c. Graph this function.

36. Use the function defined by the given mapping notation to complete each part of this exercise.

D		*R*
-4	$\rightarrow$	-4
-2	$\rightarrow$	-2
0	$\rightarrow$	0
1	$\rightarrow$	1
3	$\rightarrow$	3

a. Express this function by using a table format.
b. Express this function by using ordered-pair notation.
c. Graph this function.

37. Use the function $\{(-5, 4), (-3, 4), (1, 4), (2, 4), (3, 4)\}$ to complete each part of this exercise.
a. Express this function by using a table format.
b. Express this function by using mapping notation.
c. Graph this function.

38. Use the function $\{(-4, -2), (-2, 4), (4, -3), (3, 1)\}$ to complete each part of this exercise.
a. Express this function by using a table format.
b. Express this function by using mapping notation.
c. Graph this function.

39. **Value of an Investment** An investor purchased four properties for \$25,000, \$40,000, \$50,000, and \$65,000. Over a 7-year period the value of each property doubled. Use these four values as the input values for a function to calculate the current value of each property.
a. Represent this function by using function notation.
b. Represent this function by using mapping notation.
c. Represent this function by using a table.
d. Represent this function by using ordered pairs.

40. **Diet Plan Predictions** A diet plan promoted by a nationwide business promised a 5% weight loss for its customers during the first 2 weeks. One local franchise of this business had five new customers sign contracts in 1 week. The sign-in weights in pounds of these customers were 140, 160, 190, 230, and 280.
a. Using function notation, represent the promised new weight of each customer after 2 weeks on this plan.
b. Represent this function by using mapping notation.
c. Represent this function by using a table.
d. Represent this function by using ordered pairs.

41. **Pressure on a Gas (Boyle's Law)** Boyle's law gives the relationship between the pressure and the volume of a gas at a constant temperature. The given histogram shows the pressure and volume of a gas. Use the function defined by this histogram to complete each part of this exercise.

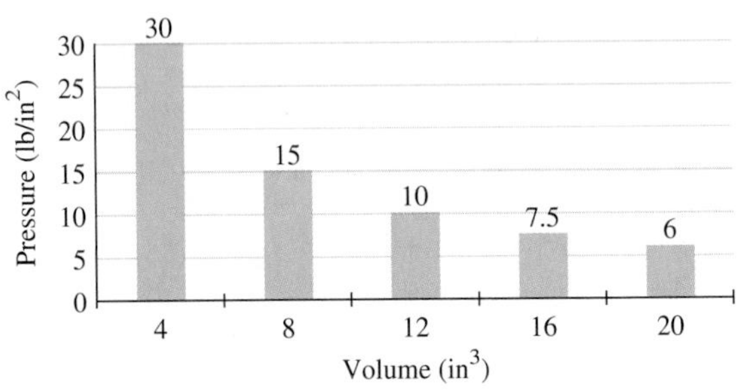

a. If the input x represents the volume of gas, the output y represents __________.
b. What is the volume of gas if the pressure is 7.5 lb/in^3?
c. What is the pressure of a gas that has a volume of 8 in^3?

42. **World Series Scores** The Florida Marlins defeated the New York Yankees in six games to win the 2003 World Series. The given histogram shows the number of runs scored by the Marlins in each of the six games. Use the function defined by this histogram to complete each part of this exercise.

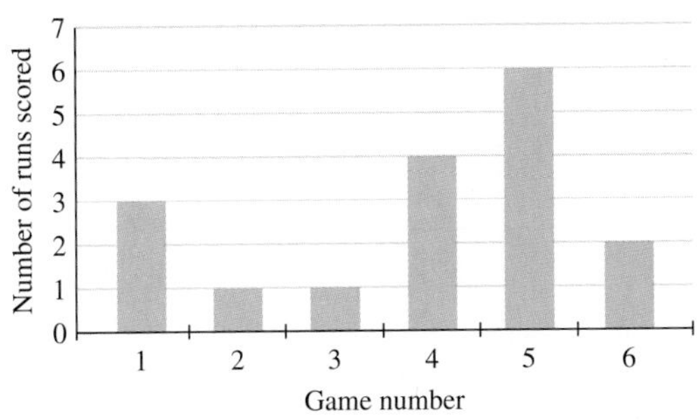

a. If the input x represents the game number, the output y represents __________.
b. How many runs did the Marlins score in game 4?
c. In which game(s) did the Marlins score only 1 run?

43. **Smoking Trends** The following table shows the percent of U.S. adults who smoke.

YEAR x	PERCENT OF U.S. ADULTS WHO SMOKE y
1965	41.9
1974	37.0
1985	29.9
1990	25.3
1995	24.6
2000	24.1

Source: Center for Disease Control. (For more information on this topic, go to the Additional Resources link on the MathZone website at **www.mhhe.com/hallmercer.**)

a. Graph these points.
b. What percent of U.S. adults smoked in 1985?
c. In what year did 37% of U.S. adults smoke?

44. **Salaries by Gender** The following table shows women's annual earnings as a percent of men's annual earnings in the United States.

YEAR x	ANNUAL EARNINGS FOR WOMEN AS A PERCENT OF MEN'S y
1979	59.7
1984	63.7
1989	68.7
1994	72.0
1999	72.2
2002	76.0

Source: U.S. Department of Labor. (For more information on this topic, go to the Additional Resources link on the MathZone website at **www.mhhe.com/hallmercer.**)

a. Graph these points.
b. What were the earnings for women as a percent of men's in 1989?
c. In what year did women earn 76% of what men earned in the United States?

45. **Sales Tax** The sales tax in Houston, Texas, is 8.25% (6.25% state tax, 1% city tax, and 1% for the Houston MTA). (For more information on this topic, go to the Additional Resources link on the MathZone website at **www.mhhe.com/hallmercer.**)
a. If x represents the amount of a purchase at a Houston clothing store, write a function T so that $T(x)$ represents the tax on this purchase.
b. Use the function from part **a** to complete the table shown.

X	Y1
0	
100	
200	
300	
400	
500	
600	

X=0

46. **College Tuition and Fees** The cost of tuition and fees at 2-year colleges in the United States can be approximated by the function $C(x) = 200x + 1{,}000$, where x represents the number of years since 1970. (*Source:* U.S. Department of Education.)
a. Use this function to calculate the tuition for years 1970, 1980, 1990, and 2000. Then graph these four points.
b. Estimate the cost of tuition and fees in 2010.
c. Determine the approximate year that the cost of tuition and fees was $5,800.

47. **Cost of a Car Lease** The total cost to lease a car for x months is given by the function $C(x) = 300x + 1{,}500$.
a. Use this function to calculate the total cost to lease a car for each of the first 6 months. Then graph these six points.
b. Give the cost to lease the car for 24 months.
c. The lessee has already spent $4,800 on this car lease. How many months of the lease has this person paid?

48. **Jet Fuel Use** According to Captain Michael Ragsdale, the number of pounds of jet fuel used in an hour by an MD88 passenger jet at cruising altitude is 9,000 lb.
a. If x represents the number of hours the plane has been at cruising altitude, write a function F so that $F(x)$ represents the number of pounds of jet fuel that the plane will use during this portion of its flight.
b. Use the function from part **a** to complete the table shown.

X	Y1
0	
.25	
.5	
.75	
1	
1.25	
1.5	

X=0

In Exercises 49–54, use a graphing calculator to graph each function and to determine the domain and range of each function.

49. $f(x) = \sqrt{x}$
50. $f(x) = -\sqrt{x}$
51. $f(x) = -x^2 + 1$
52. $f(x) = (x - 1)^2$
53. $f(x) = \sqrt{-x}$
54. $f(x) = \sqrt{2 - x}$

In Exercises 55–58, match each graph with its domain and range.

55.

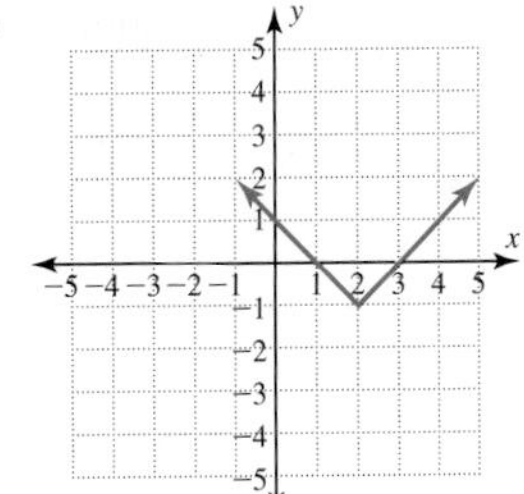

56.

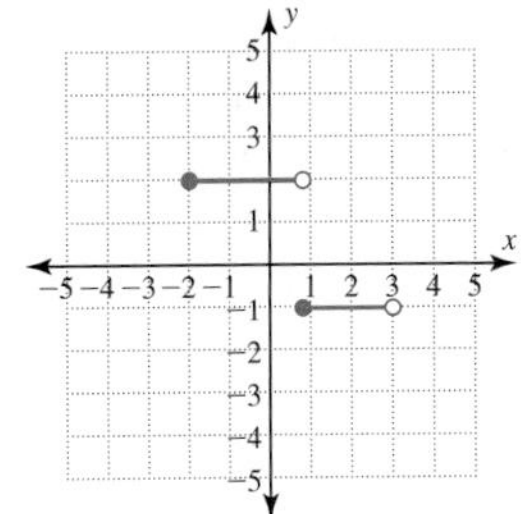

57. 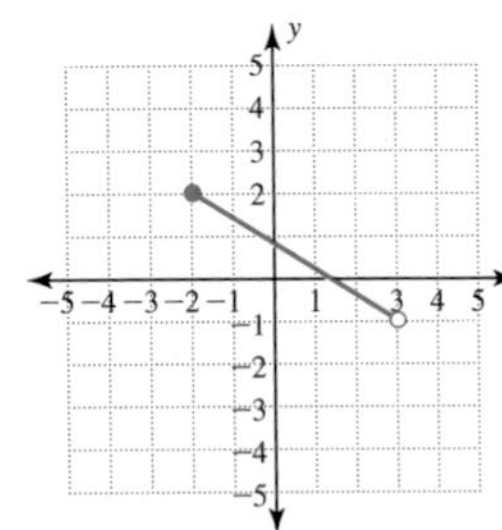

58. 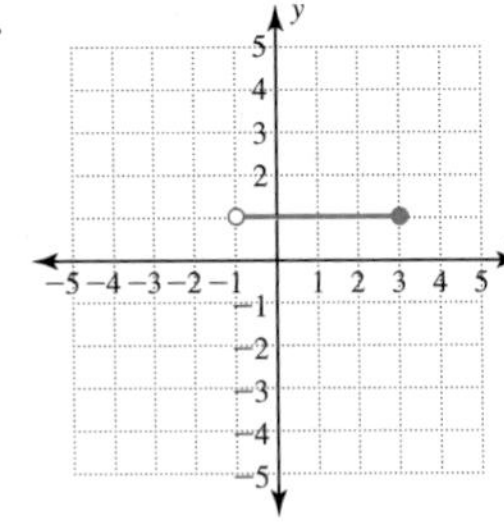

A. $D = (-1, 3], R = \{1\}$
B. $D = [-2, 3), R = (-1, 2]$
C. $D = [-2, 3), R = \{-1, 2\}$
D. $D = \mathbb{R}$ (all real numbers), $R = [-1, \infty)$

In Exercises 59–68, graph a function with the given domain D and range R. For some exercises there may be several correct graphs. (*Hint:* For Exercises 63 and 64, see Exercise 56.)

59. $D = \{2\}, R = \{3\}$
60. $D = \{-3\}, R = \{2\}$
61. $D = (-3, 2], R = \{-2\}$
62. $D = [-4, 4), R = \{3\}$
63. $D = (-3, 3), R = \{-1, 2\}$
64. $D = [-2, 2], R = \{-4, 4\}$
65. $D = [-4, 3), R = (-3, 4]$
66. $D = (-3, 4], R = [-2, 2)$
67. $D = \mathbb{R}$ (all real numbers), $R = [-2, \infty)$
68. $D = \mathbb{R}$ (all real numbers), $R = (-\infty, 2]$
69. a. Draw a semicircle that represents a function.
    b. Draw a semicircle that does not represent a function.
70. Can a circle ever be a function? Explain your answer.

Group Discussion Questions

71. **Calculator Discovery** Each of these graphs is produced by an equation of the form $y = |x| + c$ for a specific value of c. Write the equation for each graph, and use a graphing calculator to test each function.

a. 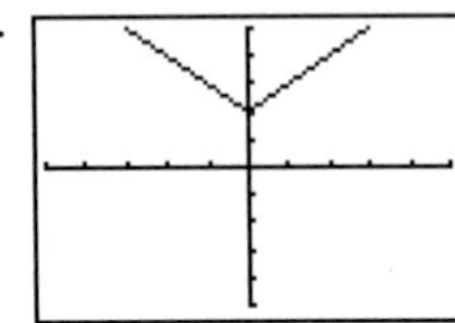

[−5, 5, 1] by [−5, 5, 1]

b. 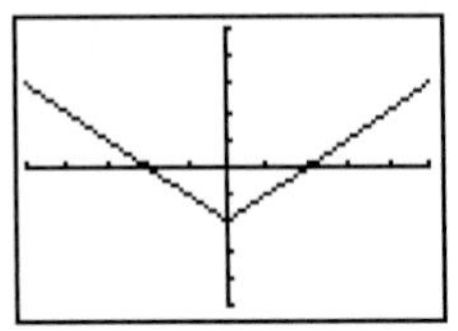

[−5, 5, 1] by [−5, 5, 1]

c. 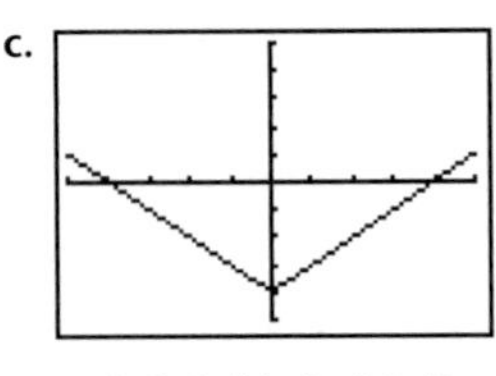

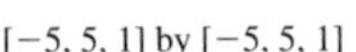
[−5, 5, 1] by [−5, 5, 1]

d. 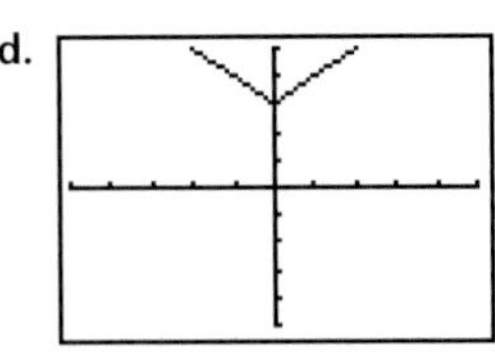

[−5, 5, 1] by [−5, 5, 1]

72. **Calculator Discovery** Use each set of data to write a function $y = f(x)$ that will produce the given table of values. Use a graphing calculator to test each function.

a. 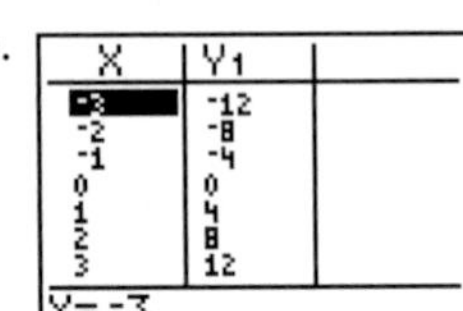

b. 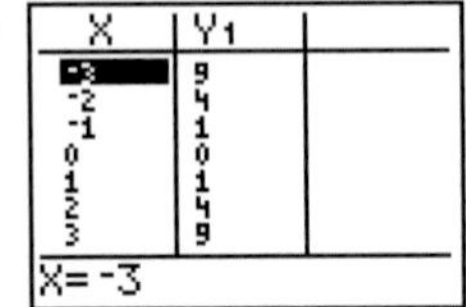

c. 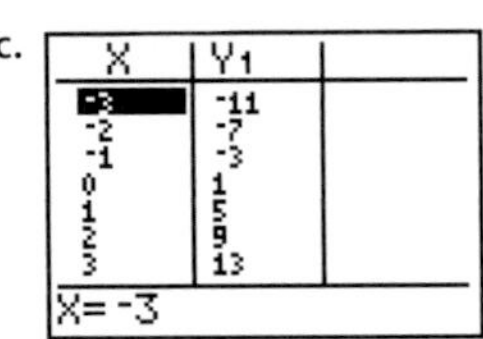

d. 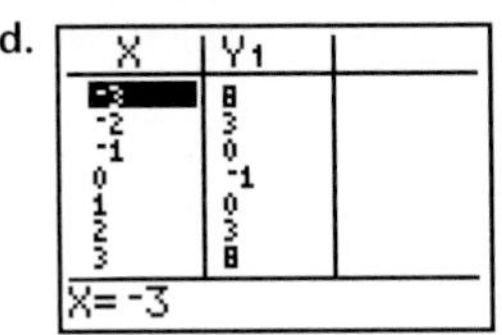

73. **Error Analysis** A student examined the graph below and concluded that it passed the vertical line test and therefore the graph represents a function. Correct this answer and explain the error in this reasoning.

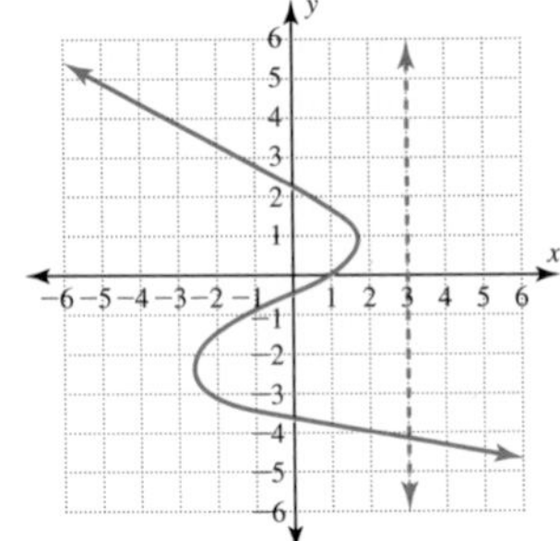

Section 7.2 Linear and Absolute Value Functions

Objectives:
4. Use the slope-intercept form of a linear equation.
5. Identify a function as an increasing or a decreasing function.
6. Determine where a function is positive or negative.
7. Analyze the graph of an absolute value function.

Although the variety of functions that you can encounter in the workplace is considerable, it is remarkable that most of these functions will belong to a few classic families of functions. Each member of a family will share many important characteristics with other members of that family. Thus by examining only a few families of functions you can be well prepared for many contingencies that you may encounter. We will examine some of the key characteristics of these functions from algebraic, graphical, and numerical perspectives.

In this section we will review linear functions (see Sections 2.2, 3.1, and 3.2) and examine absolute value functions.

Linear Function, Shape of a Line, and First-Degree Equation

Linear functions are not only among the most common and useful functions, but also the easiest to study. The graph of a **linear function** is an easily drawn straight line, and algebraically each of these functions can be written in the slope-intercept form $f(x) = mx + b$ or $y = mx + b$. In this form, m represents the slope of the line and b represents the y-coordinate of the y-intercept $(0, b)$.

Linear Function

ALGEBRAICALLY	NUMERICAL EXAMPLE	GRAPHICAL EXAMPLE	VERBALLY
A function of the form $f(x) = mx + b$ is called a linear function.	$f(x) = 2x - 5$	$f(x) = 2x - 5$	In the table for $y = 2x - 5$, each 1-unit increase in x produces a 2-unit increase in y. The graph of $y = 2x - 5$ is a straight line. Each point on this line satisfies this equation. This line rises 2 units for each 1-unit move to the right.

x	$y = f(x)$
-1	-7
0	-5
1	-3
2	-1
3	1
4	3
5	5

Example 1 illustrates how to use the slope-intercept form to graph a line.

EXAMPLE 1 Graphing a Line by Using the Slope-Intercept Form

Use the slope and y-intercept to graph the line defined by $f(x) = \frac{2}{3}x - 1$.

SOLUTION

$f(x) = \frac{2}{3}x - 1$

Use the slope-intercept form $f(x) = mx + b$ to determine the slope $m = \frac{2}{3}$ and the y-intercept is $(0, -1)$. The slope $m = \frac{\Delta y}{\Delta x}$ represents the change in y for a given change in x.

Slope: $m = \frac{2}{3}$

y-Intercept: $(0, -1)$

Second point: $(0 + 3, -1 + 2) = (3, 1)$

Plot the y-intercept $(0, -1)$ and then use the slope to plot a second point. Do this by increasing y by 2 units for a 3-unit increase in x.

Sketch the line through these two points.

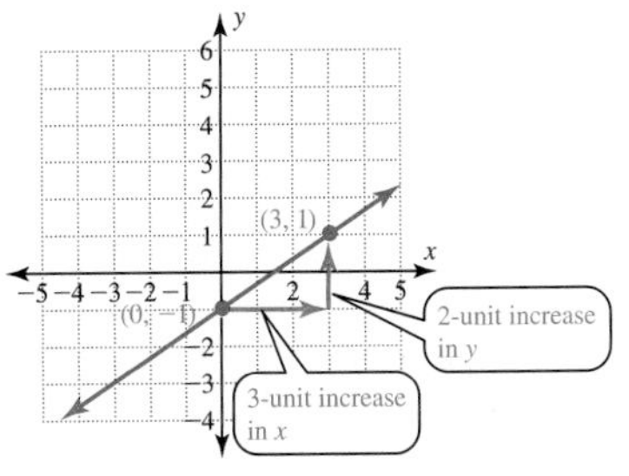

SELF-CHECK 7.2.1

Use the linear function defined by $f(x) = \frac{5}{2}x - 3$ to complete each of these exercises.

1. Determine the slope of the line defined by this function.
2. Determine the y-intercept of the line defined by this function.
3. Use the slope and y-intercept to graph this line.

One of the things that we can observe quickly from the slope of a line is whether the line is sloping upward to the right or downward to the right. A line with positive slope goes upward as it moves from left to right. A line with negative slope goes downward as it moves from left to right. A line with zero slope is a horizontal line that does not move up or down.

We can generalize this concept to analyze the behavior of other functions. We say that a function is **increasing** if its graph rises as it moves from left to right. As the x-values increase, the y-values also increase. We say that a function is **decreasing** if its graph drops as it moves from left to right. As the x-values increase, the y-values decrease. A function is called an **increasing function** if it increases over its entire domain, and a function is called a **decreasing function** if it decreases over its entire domain.

For an increasing function, a positive value of Δx produces a positive value of Δy.

For a decreasing function, a positive value of Δx produces a negative value of Δy.

Positive Slope: An Increasing Linear Function

Negative Slope: A Decreasing Linear Function

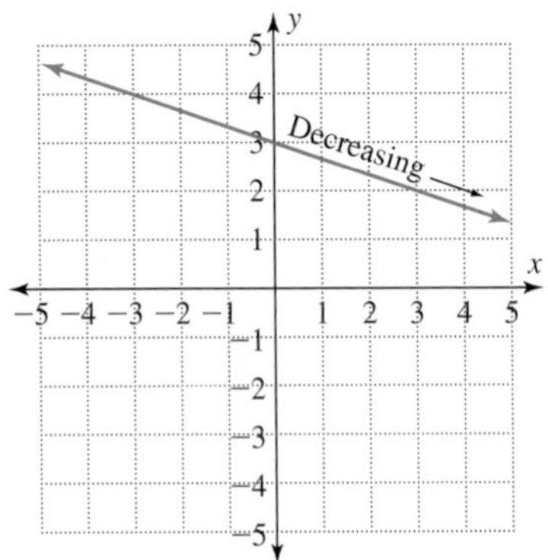

Increasing and Decreasing Functions

	GRAPHICAL EXAMPLE	VERBALLY
Increasing function		A function is increasing over its entire domain if the graph rises as it moves from left to right. For all x-values, as the x-values increase, the y-values also increase.
Decreasing function		A function is decreasing over its entire domain if the graph drops as it moves from left to right. For all x-values, as the x-values increase, the y-values decrease.

Example 2 illustrates how to identify increasing and decreasing functions.

EXAMPLE 2 Identifying Increasing and Decreasing Functions

Identify each function as an increasing function or as a decreasing function.

SOLUTION

(a)

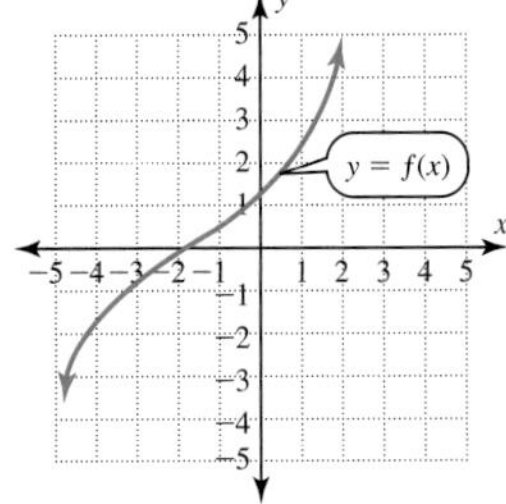

Increasing function

The graph of this function rises as the graph moves from left to right. Thus this function increases over its entire domain.

(b)

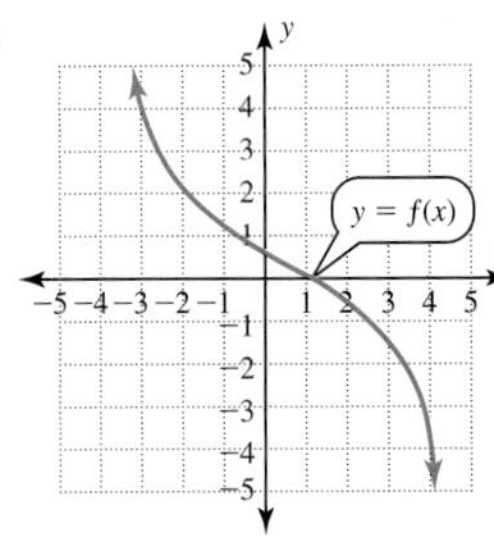

Decreasing function

The graph of this function drops as the graph moves from left to right. Thus this function decreases over its entire domain.

(c) $f(x) = 3x - 17$

Increasing function

This linear function is given in the slope-intercept form $f(x) = mx + b$ with a slope of 3. This is an increasing function because the slope is positive.

(d) $f(x) = 7$

Neither increasing nor decreasing

This is a horizontal line with a slope of 0. The line is neither rising nor dropping.

(e)

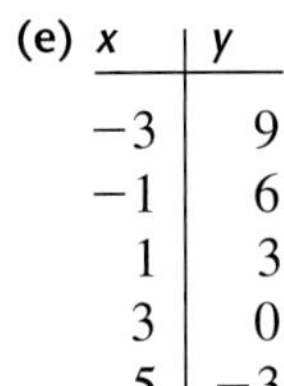

x	y
-3	9
-1	6
1	3
3	0
5	-3

Decreasing function

As the x-values increase, the y-values decrease.

SELF-CHECK 7.2.2

Identify each function defined by these graphs and equations as either an increasing function or a decreasing function.

1.

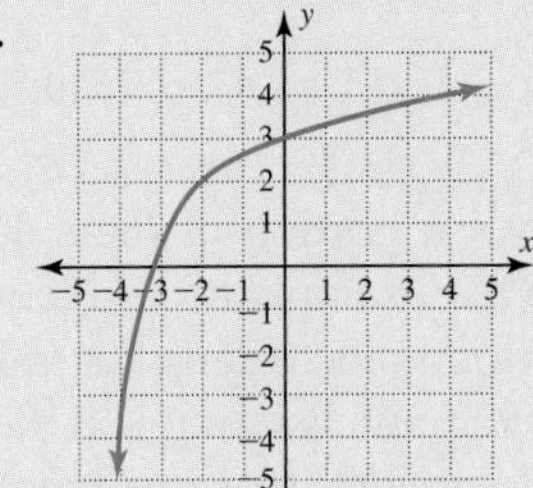

2.

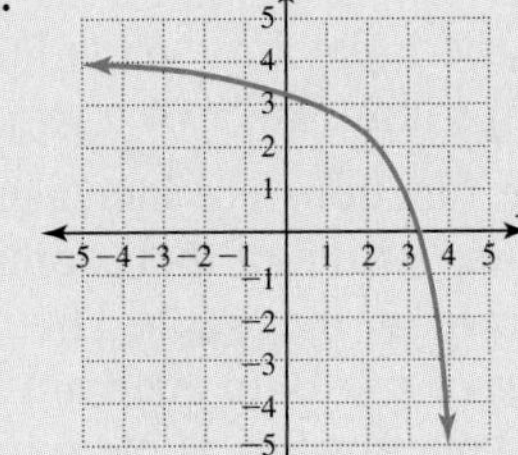

3. $f(x) = -4x - 1$

4. $f(x) = 4x - 19$

Example 3 examines engine performance. This is an example of a decreasing function. This application of linear functions uses data on engine performance taken from the September 2001 *Popular Science* magazine.

EXAMPLE 3 Writing a Linear Equation to Model Engine Performance

The horsepower of a 100-hp engine falls by about 3 hp for every 1,000 ft above sea level that you travel.

(a) Write a linear equation to describe the horsepower of an automobile engine as a function of elevation.

(b) Use this function to calculate the horsepower of an engine at the top of the road to Pikes Peak, elevation 14,000 ft.

(c) Use this function to prepare a horsepower performance table for this engine for elevations from sea level to 6,000 ft in increments of 1,000 ft.

Turbocharging the air pressure back to that at sea level can compensate for this loss of power. Aircraft engines often have turbochargers for this purpose. (*Source:* Page 76 of September 2001 *Popular Science* magazine.)

SOLUTION

(a) Let $x =$ elevation of engine, ft
$y =$ horsepower of engine, hp

y-Intercept: $(0, 100)$

Slope: $m = \dfrac{-3}{1{,}000}$

$m = -0.003$

Equation: $P(x) = -0.003x + 100$

$y = -0.003x + 100$

We are using elevation as the input variable and horsepower as the output variable.

Sea level is at an elevation of 0 ft. The horsepower at that level is 100 hp.

There is a 3-hp decrease for each 1,000-ft increase in elevation.

Write the equation in the form $P(x) = mx + b$ or $y = mx + b$ to represent the horsepower at x ft. Use the y-intercept of (0, 100) and the slope of -0.003.

(b) $P(x) = -0.003x + 100$

$P(14{,}000) = -0.003(14{,}000) + 100$

$P(14{,}000) = 58$

The engine has 58 hp at 14,000 ft.

To determine the horsepower at 14,000 ft, substitute 14,000 into the function $P(x) = -0.003x + 100$.

The horsepower of this engine has decreased to 58 hp by the time it has climbed to 14,000 ft.

(c)

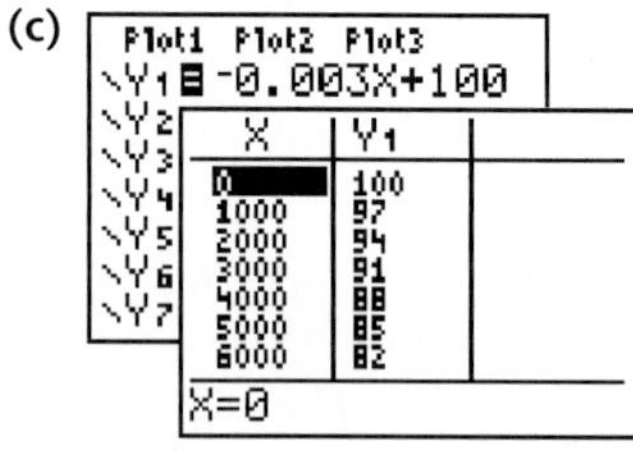

V-Shaped Absolute Value Function and Equation That Includes the Absolute Value of a Variable

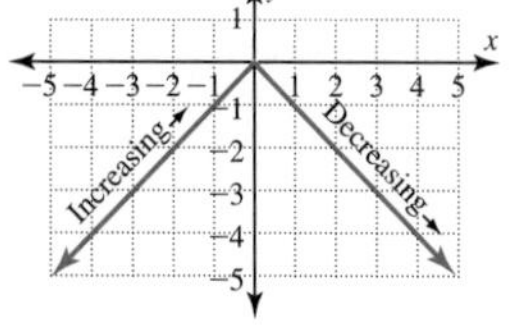

Some functions are increasing over a portion of their domain and then decreasing over another portion of their domain. One example of a function that has this type of behavior is shown on the graph. This function is increasing for values of $x < 0$ and decreasing for values of $x > 0$. In quadrant III this graph is rising and the function is increasing. Thus for $x < 0$ the function is increasing. In quadrant IV this graph is dropping and the function is decreasing. Thus for $x > 0$ the function is decreasing.

Note that the V-shaped graph opens downward with a highest point occurring at the origin (0, 0)—the point at the tip of the V shape. If a graph has a highest point, the y-value of this point is the **maximum y-value,** which is as large as or larger than any other y-value for the function. If a graph has a lowest point, the y-value of this point is the **minimum y-value,** which is as small as or smaller than any other y-value for the function.

We now examine the family of absolute value functions. These functions are named for the algebraic equation that defines the function. Every absolute value function has a characteristic V shape. The **vertex** of the graph of an absolute value function is the point at the tip of the V shape. If the V shape opens upward, the vertex will be the lowest point on the graph and the y-value there will be the minimum y-value. If the V shape opens downward, the vertex will be the highest point on the graph and the y-value there will be the maximum y-value.

Absolute Value Function

ALGEBRAICALLY

$f(x) = |x|$

NUMERICALLY

x	$f(x) = \|x\|$
−3	3
−2	2
−1	1
0	0
1	1
2	2
3	3

GRAPHICALLY

VERBALLY

This V-shaped graph opens upward with the vertex of the V shape at (0, 0). The minimum y-value is 0. The domain, the projection of this graph onto the x-axis, is $D = \mathbb{R}$. The range, the projection of this graph onto the y-axis, is $R = [0, \infty)$. The function is decreasing for $x < 0$ and increasing for $x > 0$.

We can analyze the behavior of the graph of an absolute value function by first creating a table of values for the function and then graphing the function by plotting these points. This is illustrated in Example 4.

EXAMPLE 4 Analyzing the Graph of an Absolute Value Function

Use the function $f(x) = -|x|$ to complete each part of this example.

(a) Create a table of values for this function.
(b) Plot these points, and sketch a graph of $y = f(x)$.
(c) Does this graph open upward or downward?
(d) Determine the vertex.
(e) Determine the minimum or maximum y-value.
(f) Determine the domain of this function.
(g) Determine the range of this function.
(h) Determine the x-values for which the function is increasing.
(i) Determine the x-values for which the function is decreasing.

SOLUTION

ALGEBRAICALLY

$f(x) = -|x|$

NUMERICALLY

(a)

x	$f(x) = -\|x\|$
−3	−3
−2	−2
−1	−1
0	0
1	−1
2	−2
3	−3

GRAPHICALLY

(b)

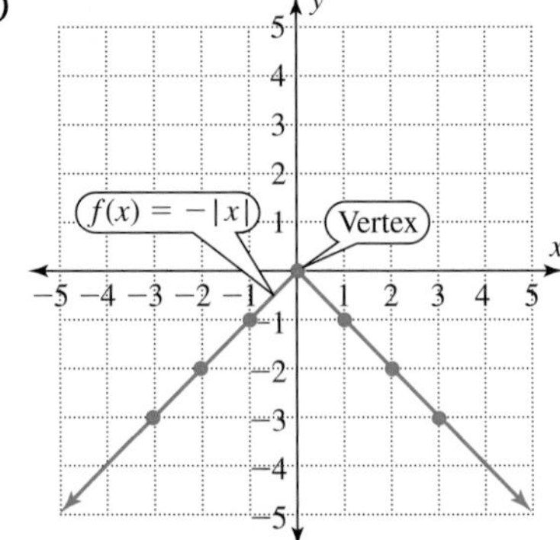

VERBALLY

(c) This V-shaped graph opens downward.
(d) The vertex of the V shape is at (0, 0).
(e) The maximum y-value is 0. This value is located at the vertex.
(f) The domain, the projection of this graph onto the x-axis, is $D = \mathbb{R}$.
(g) The range, the projection of this graph onto the y-axis, is $R = (-\infty, 0]$.
(h) The function is increasing for $x < 0$; that is for the x-values $(-\infty, 0)$.
(i) The function is decreasing for $x > 0$; that is for the x-values $(0, \infty)$.

Recall from Section 1.2 that in interval notation the smaller value is always written first. $(-\infty, 0]$ is the only correct interval notation for the range.

Example 5 is very similar to Example 4. Although the function in Example 5 is slightly more complicated, this function is analyzed in the same way.

EXAMPLE 5 Analyzing the Graph of an Absolute Value Function

Use the function $f(x) = |x + 1| - 2$ to complete each part of this example.

(a) Create a table of values for this function.
(b) Plot these points, and sketch a graph of $y = f(x)$.
(c) Does this graph open upward or downward?
(d) Determine the vertex.
(e) Determine the minimum or maximum y-value.
(f) Determine the domain of this function.
(g) Determine the range of this function.
(h) Determine the x-values for which the function is increasing.
(i) Determine the x-values for which the function is decreasing.

SOLUTION

ALGEBRAICALLY

$f(x) = |x + 1| - 2$

NUMERICALLY

(a)

x	$f(x) = \|x + 1\| - 2$
−3	0
−2	−1
−1	−2
0	−1
1	0
2	1
3	2

GRAPHICALLY

(b)

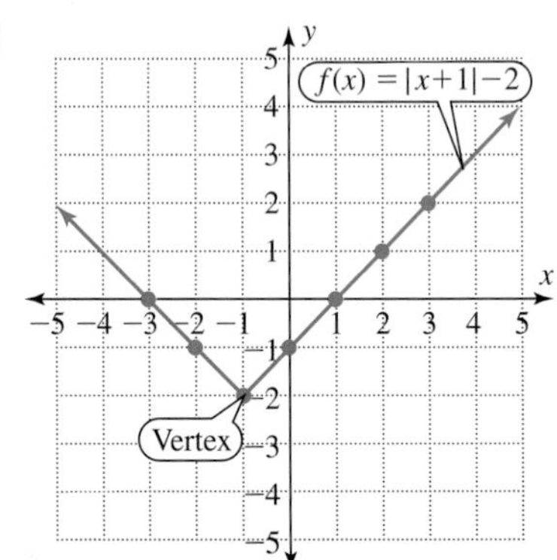

VERBALLY

(c) This V-shaped graph opens upward.
(d) The vertex of the V shape is at $(-1, -2)$.
(e) The minimum y-value is -2. This value is located at the vertex.
(f) The domain, the projection of this graph onto the x-axis, is $D = \mathbb{R}$.
(g) The range, the projection of this graph onto the y-axis, is $R = [-2, \infty)$.
(h) The function is increasing for $x > -1$; that is for the x-values $(-1, \infty)$.
(i) The function is decreasing for $x < -1$; that is for the x-values $(-\infty, -1)$.

SELF-CHECK 7.2.3

Use the function $f(x) = |x - 2|$ to complete each of these exercises.

1. Sketch a graph of $y = f(x)$.
2. Does this graph open upward or downward?
3. Determine the vertex.
4. Determine the minimum or maximum y-value.
5. Determine the domain of this function.
6. Determine the range of this function.
7. Determine the x-values for which the function is increasing.
8. Determine the x-values for which the function is decreasing.

A **function** $y = f(x)$ **is positive** when the output value $f(x)$ is positive. On the graph of $y = f(x)$ this occurs at points above the x-axis—positive y-values. A **function** $y = f(x)$ **is negative** when the output value $f(x)$ is negative. On the graph of $y = f(x)$ this occurs at points below the x-axis—negative y-values. Example 6 illustrates this.

EXAMPLE 6 Determining When a Function Is Positive and When a Function Is Negative

Determine the x-values for which each function is positive and the x-values for which each function is negative.

SOLUTION

(a)

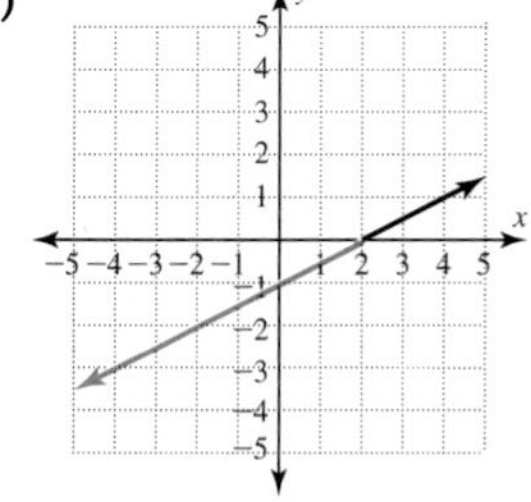

This function is positive (in the black) for x-values in the interval $(2, \infty)$.

The function is above the x-axis to the right of the x-intercept $(2, 0)$. Thus the y-values are positive for $x > 2$.

This function is negative (in the red) for x-values in the interval $(-\infty, 2)$.

The function is below the x-axis to the left of the x-intercept $(2, 0)$. Thus the y-values are negative for $x < 2$.

(b)

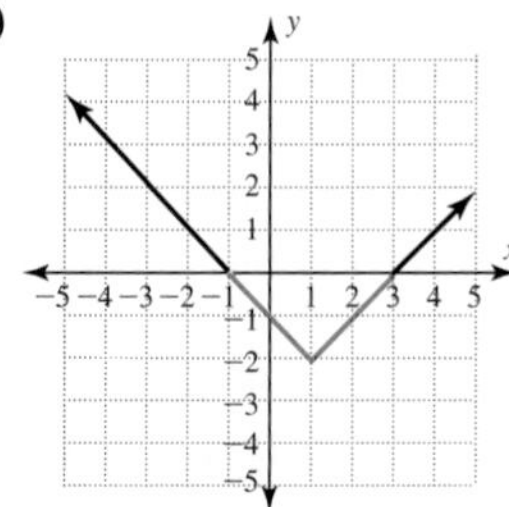

This function is positive (in the black) for x-values in the interval $(-\infty, -1) \cup (3, \infty)$.

The function is above the x-axis to the left of the x-intercept $(-1, 0)$ or to the right of the x-intercept $(3, 0)$. Thus the y-values are positive for $x < -1$ or $x > 3$.

This function is negative (in the red) for x-values in the interval $(-1, 3)$.

The function is below the x-axis between the x-intercepts $(-1, 0)$ and $(3, 0)$. Thus the y-values are negative for $-1 < x < 3$.

SELF-CHECK 7.2.4

Determine the x-values for which each function is positive and those for which each function is negative.

1.

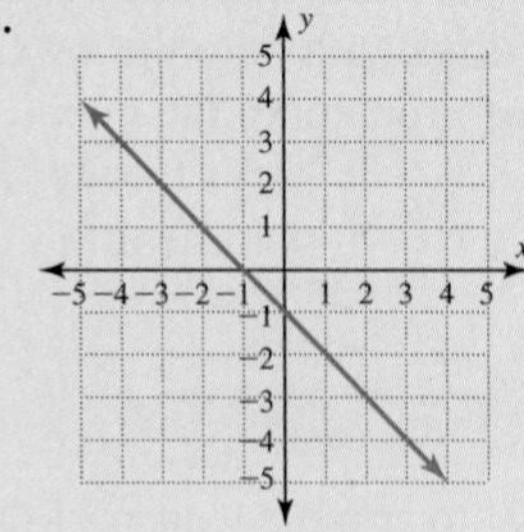

2.

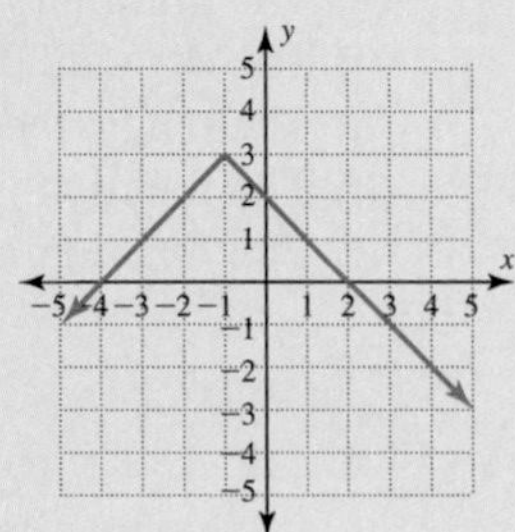

In Section 2.3, we looked at the significance of intercepts in an application involving the overhead costs and the break-even value for production using a machine. The overhead costs are the costs involved if no units are produced. The break-even value is the number of units a machine must make before a profit is realized. That is, the break-even value will result in a profit of $0, not a loss or a gain. Producing less than this break-even number of units will result in a loss. The set of input values that will produce a loss is called the **loss interval.** Producing more than this break-even number of units will result in a profit. The set of input values that will produce a profit is called the **profit interval.** The profit and loss intervals are illustrated in Example 7.

EXAMPLE 7 Determining Profit and Loss Intervals for Machine Production

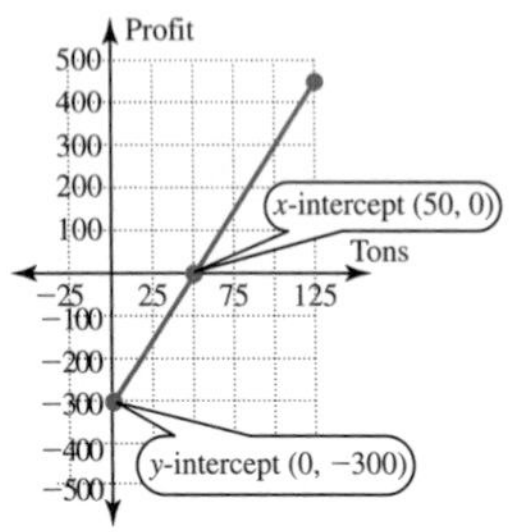

In the given graph, the input value of x is the number of tons of refined ore that a machine produces in 1 day. The output y is the profit generated by the sale of this ore when it is produced. The machine is capable of producing a maximum of 125 tons in 1 day, but it is not always able to produce this maximum number. The break-even value for this machine is 50 tons produced in 1 day. Determine **(a)** the loss interval and **(b)** the profit interval for this machine.

SOLUTION

(a) The loss interval is $[0, 50)$.

(b) The profit interval is $(50, 125]$

The machine could produce 0 tons. The bracket is used to indicate that 0 is included in the loss interval. Producing up to but not including 50 tons will result in a loss. Producing more than 50 tons will result in a profit. The maximum this machine can produce is 125 tons. The bracket indicates that 125 is included in the profit interval.

Some production or manufacturing processes result in output that can be counted—-output that corresponds to the whole numbers 0, 1, 2, 3, ... For example, a company manufacturing windows can produce 0, 1, 2, 3, . . . windows. Other companies produce or manufacture a product that can be measured—-output that can correspond to an interval of positive real numbers. In Example 7 ore production in tons per day was the interval $[0, 125]$. Depending on the product involved, the volume of production can be represented by either just whole numbers or an interval of real numbers. Nonetheless, it is customary in business and economics to refer to profit and loss intervals in both situations.

We first solved absolute value equations and inequalities in Section 4.4. In Example 8 we review this topic by solving an absolute value inequality algebraically, graphically, and numerically. This example illustrates the use of a graphing calculator to graph the absolute value function. Recall from Calculator Perspective 4.4.1 that we can access the absolute value function as one of the **NUM** options under the **MATH** menu on a TI-84 Plus calculator. Notice that the absolute value function is denoted algebraically by $f(x) = |x|$ and on a graphing calculator by $Y_1 = \text{abs}(x)$.

EXAMPLE 8 Using Multiple Perspectives to Solve an Absolute Value Inequality

Solve $|3x - 3| > 6$.

SOLUTION

ALGEBRAICALLY

Left More Than 6 Units	**or**	**Right More Than 6 Units**
$3x - 3 < -6$	or	$3x - 3 > 6$
$x - 1 < -2$	or	$x - 1 > 2$
$x < -1$	or	$x > 3$

GRAPHICALLY AND NUMERICALLY

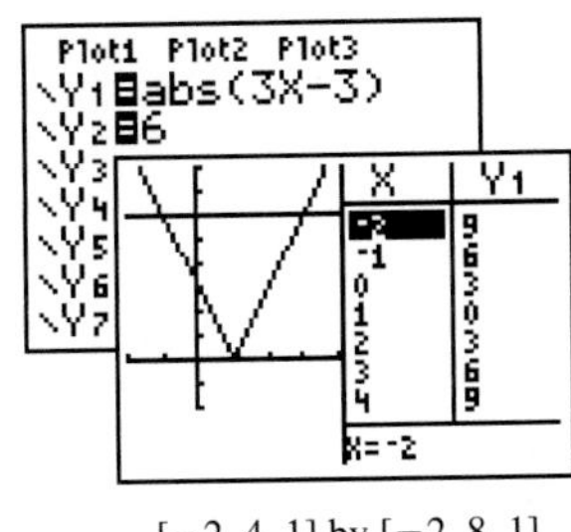

[−2, 4, 1] by [−2, 8, 1]

From the graph Y_1 is above Y_2 for $x < -1$ or for $x > 3$. This is confirmed by the values shown in the table.

Answer: $(-\infty, -1) \cup (3, \infty)$

SELF-CHECK 7.2.5

1. Solve $|2x - 1| \le 5$.

SELF-CHECK ANSWERS

7.2.1

1. $m = \frac{5}{2}$
2. $(0, -3)$
3.

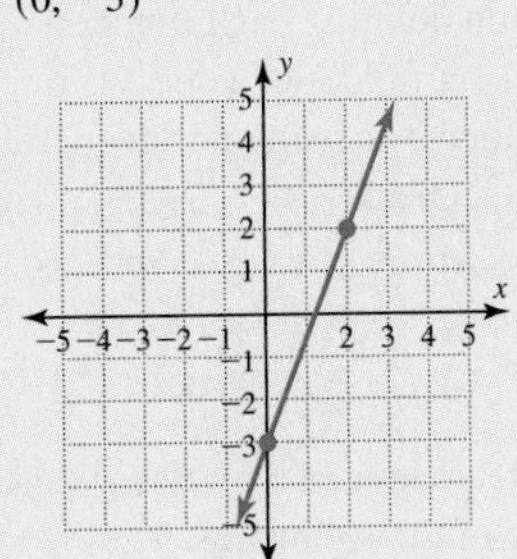

7.2.2

1. Increasing function
2. Decreasing function
3. Decreasing function
4. Increasing function

7.2.3

1.

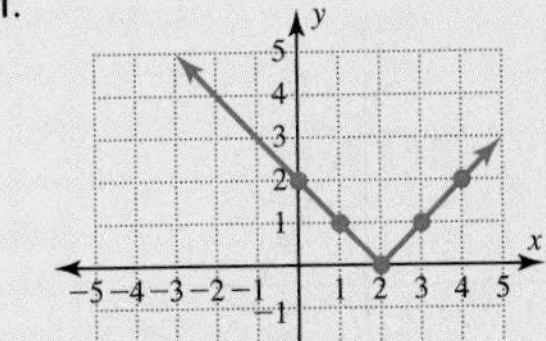

2. Opens upward
3. $(2, 0)$
4. Minimum y-value is 0,
5. $\mathbb{R}$
6. $[0, \infty)$
7. Increasing for $x > 2$
8. Decreasing for $x < 2$

7.2.4

1. Positive: $(-\infty, -1)$
   Negative: $(-1, \infty)$
2. Positive: $(-4, 2)$
   Negative: $(-\infty, -4) \cup (2, \infty)$

7.2.5

1. $[-2, 3]$

USING THE LANGUAGE AND SYMBOLISM OF MATHEMATICS 7.2

1. A function of the form $f(x) = mx + b$ is called a __________ function.
2. The graph of $f(x) = 2x + 5$ has a slope of __________ and a y-intercept of __________.
3. We say that a function is __________ if its graph rises as it moves from left to right.
4. We say that a function is __________ if its graph drops as it moves from left to right.

5. A function is called an **increasing function** if it __________ over its entire domain.
6. A function is called a **decreasing function** if it __________ over its entire domain.
7. A line with __________ slope is an increasing function.
8. A line with __________ slope is a decreasing function.
9. The __________ point on a graph has a **maximum y-value** that is as large as or larger than any other y-value for the function.
10. The __________ point on a graph has a **minimum y-value** that is as small as or smaller than any other y-value for the function.
11. The graph of an absolute value function has a __________ shape.
12. The __________ of the graph of an absolute value function is the point at the tip of the V shape.
13. A function is positive for points __________ the x-axis—positive y-values.
14. A function is negative for points __________ the x-axis—negative y-values.
15. The set of input values that will produce a loss is called the __________ __________.
16. The set of input values that will produce a profit is called the __________ __________.

QUICK REVIEW 7.2

The purpose of this quick review is to help you recall skills needed in this section.

1. The slope of a line is defined by the rise over the __________.
2. Calculate the slope of a line through $(-2, 4)$ and $(3, 6)$.
3. Complete this table so that a line through these points will have a slope of $\frac{3}{7}$.

x	y
-14	5
-7	
0	
7	
14	

4. Evaluate $|-17 + 8| + |-17| + |8|$.
5. Use interval notation to represent the real numbers that satisfy $x < -4$.

EXERCISES 7.2

In Exercises 1 and 2, use the slope and y-intercept to graph each line.

1. $f(x) = -\frac{3}{7}x + 2$
2. $f(x) = \frac{2}{5}x - 3$

In Exercises 3 and 4, use the given graph to determine

a. The y-intercept of the line
b. The slope of the line
c. The equation of the line in slope-intercept form

3.
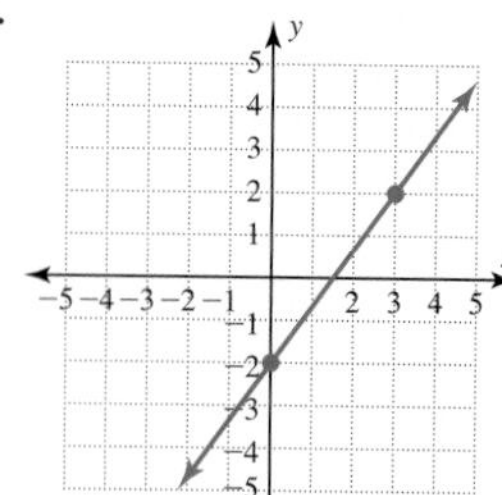

4.
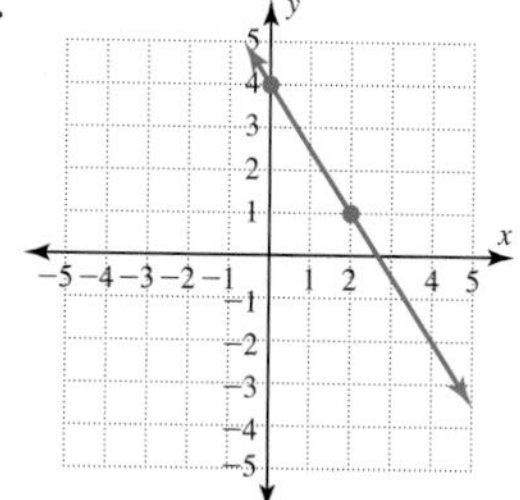

In Exercises 5–8, use the graph of each function to identify the function as an increasing function or a decreasing function.

5.
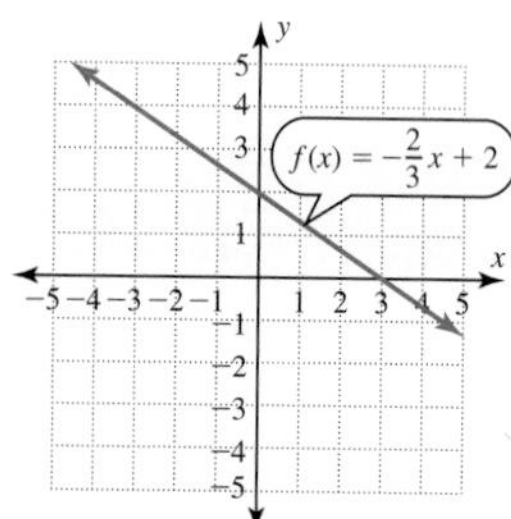

6.
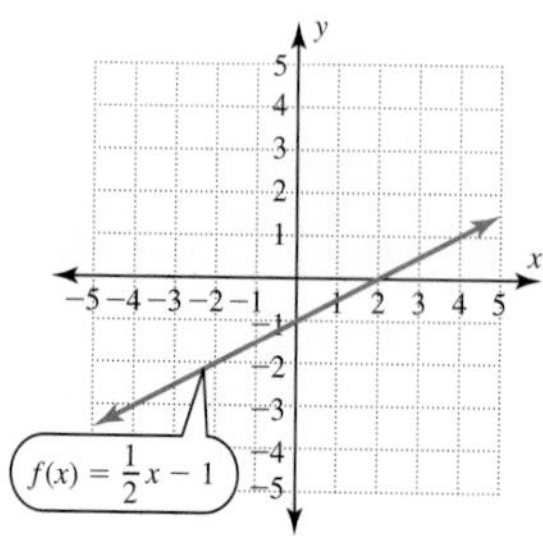

7.
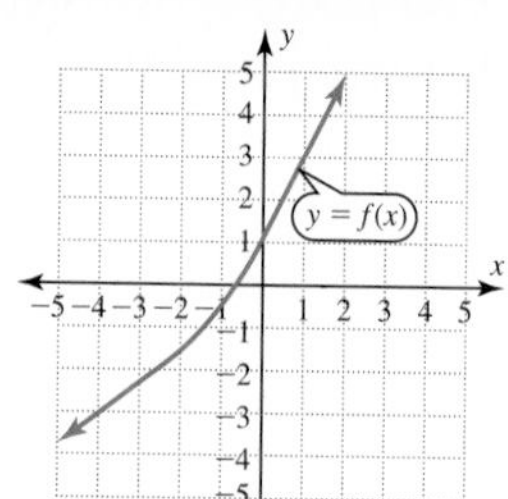

8.
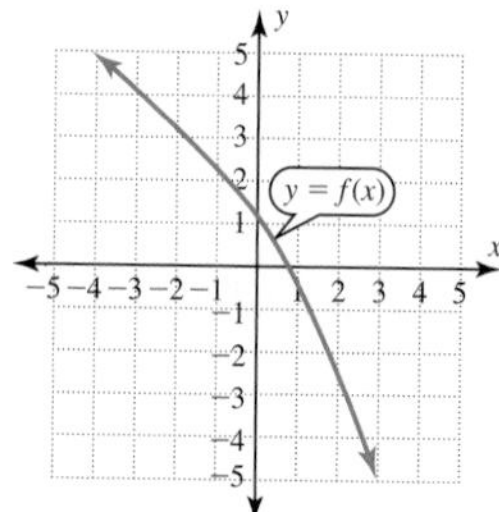

In Exercises 9–14, use the equation defining each function to identify the function as an increasing function, a decreasing function, or a function that is neither increasing nor decreasing. Use a graphing calculator only as a check.

9. $f(x) = -2x + 5$
10. $f(x) = 4x - 2$
11. $f(x) = 2x - 5$
12. $f(x) = -4x + 2$
13. $f(x) = -3$
14. $f(x) = 3$

In Exercises 15–18, use each table generated by a linear function $f(x) = mx + b$ to complete each of the following.

a. Determine the slope of each line.
b. Identify the function as an increasing function or a decreasing function.
c. Determine the equation in slope-intercept form of the line through these points.

15.

X	Y1
0	3
1	3.5
2	4
3	4.5
4	5
5	5.5
6	6

X=0

16.

X	Y1
-6	13
-4	10
-2	7
0	4
2	1
4	-2
6	-5

X= -6

17. 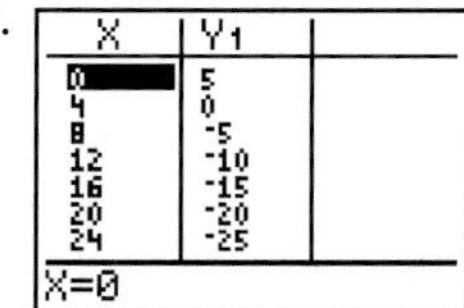

X	Y1
0	5
4	0
8	-5
12	-10
16	-15
20	-20
24	-25

X=0

18. 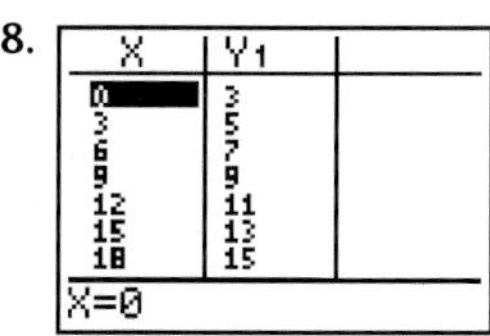

X	Y1
0	3
3	5
6	7
9	9
12	11
15	13
18	15

X=0

In Exercises 19–22, complete the given table of values, and then use these points to sketch the graph of each absolute value function. Determine the domain and the range of the function and the vertex of each graph.

x	-3	-2	-1	0	1	2	3
y							

19. $f(x) = |x + 2|$ **20.** $f(x) = |x| + 2$

21. $f(x) = |x| - 3$ **22.** $f(x) = |x - 2|$

In Exercises 23–26, use a graphing calculator to assist you in matching each function with its graph.

23. $f(x) = |x + 1| - 2$ **24.** $f(x) = |x - 2| + 3$

25. $f(x) = |x - 1| + 2$ **26.** $f(x) = |x + 2| - 3$

A.

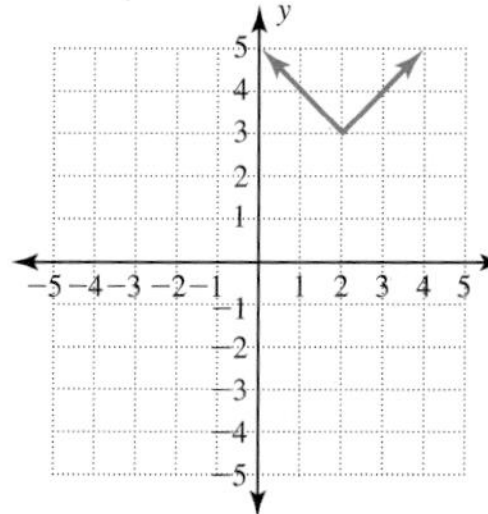

B.

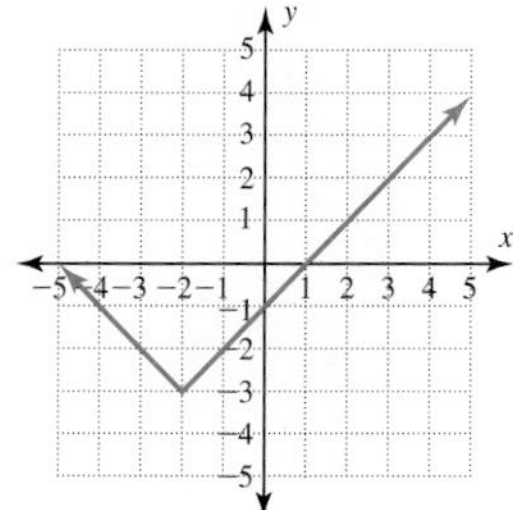

C.

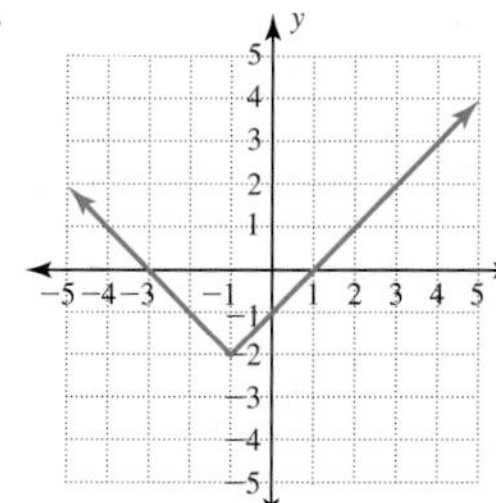

D.

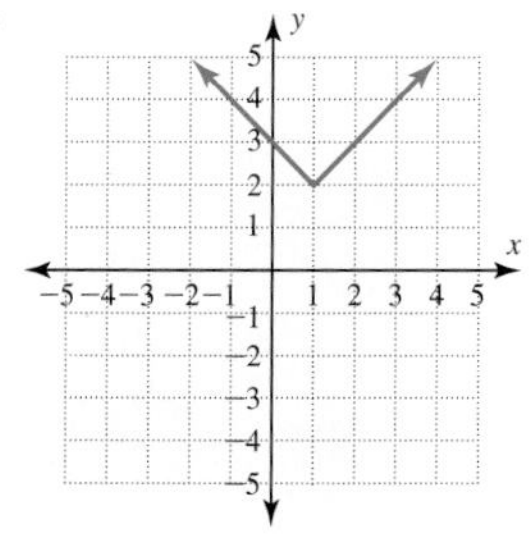

In Exercises 27–30, use a graphing calculator to graph each linear function, and then determine the domain and range of the function.

27. $f(x) = 2x + 3$ **28.** $f(x) = -2x - 3$

29. $f(x) = 3$ **30.** $f(x) = -3$

In Exercises 31–34, use a graphing calculator to graph each absolute value function, and then determine the domain and range of the function and the vertex of the graph.

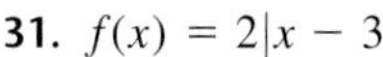

31. $f(x) = 2|x - 3|$ **32.** $f(x) = \frac{1}{2}|x + 3|$

33. $f(x) = -|x - 1| + 2$ **34.** $f(x) = -|x + 2| - 1$

In Exercises 35–38, use the graph of each absolute value function to determine the interval of x-values for which the function is increasing and the interval of x-values for which the function is decreasing.

35.

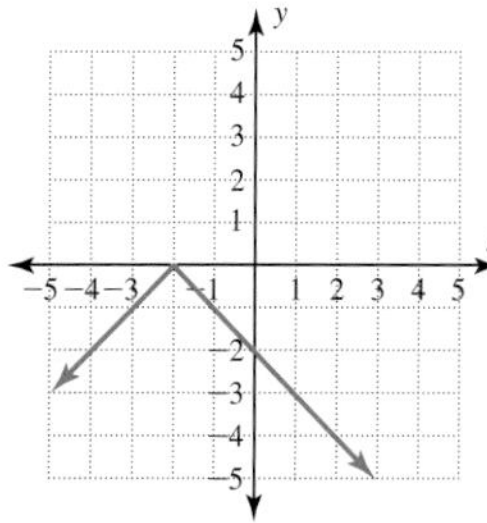

36.

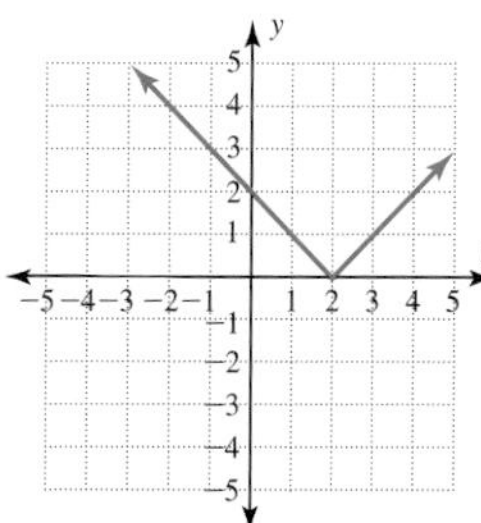

37.

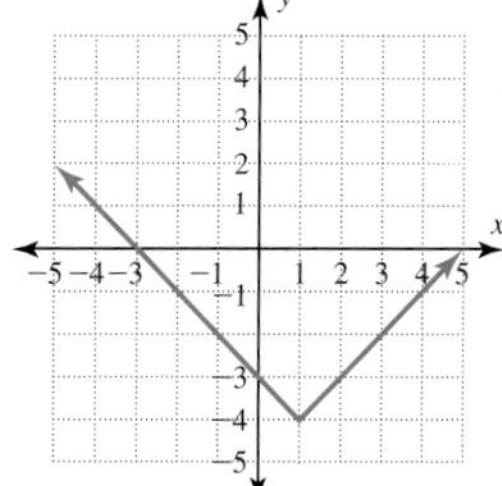

38.

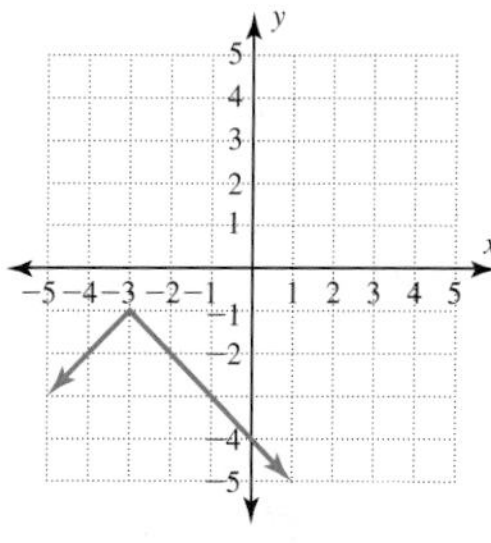

In Exercises 39–42, use the graph of each function to determine the interval of x-values for which the function is positive and the interval of x-values for which the function is negative.

39.

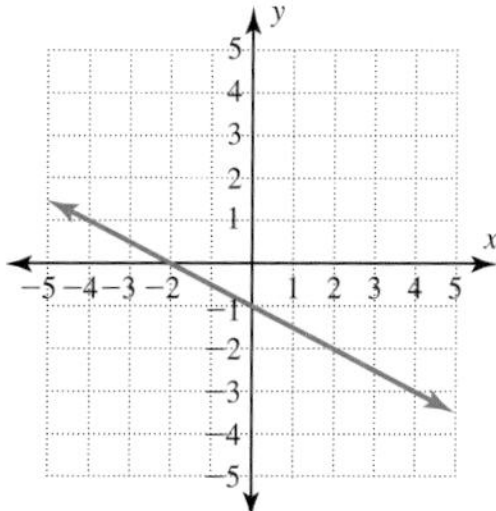

40.

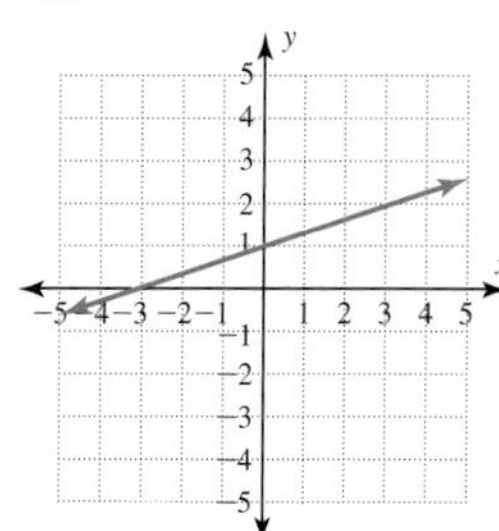

41.

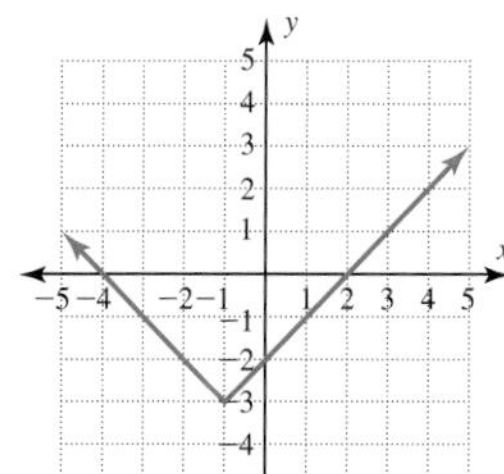

42.

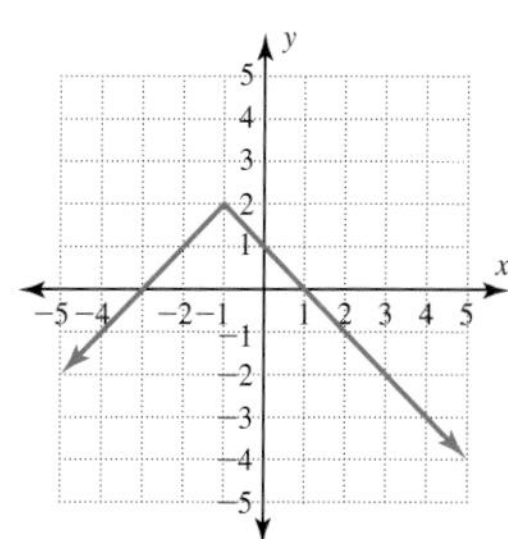

In Exercises 43 and 44, use the graph of each absolute value function to determine the following.

a. Vertex of the graph

b. Maximum y-value of the function

c. The x-value at which the maximum y-value occurs

43.

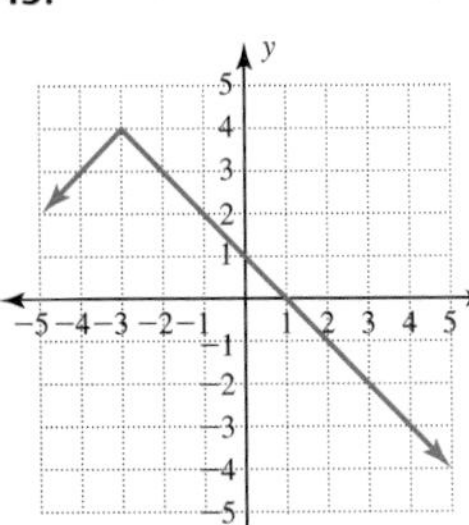

44.

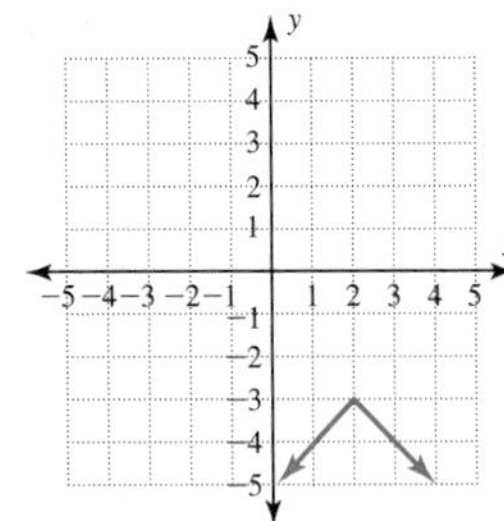

In Exercises 45 and 46, use the given absolute value function and a graphing calculator to complete each of the following parts.

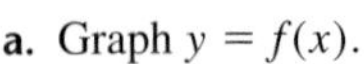

a. Graph $y = f(x)$.
b. Determine the vertex of this graph.
c. Determine the minimum or maximum y-value.
d. Determine the domain of this function.
e. Determine the range of this function.
f. Determine the x-values for which the function is decreasing.
g. Determine the x-values for which the function is increasing.
h. Determine the x-intercepts.
i. Determine the y-intercept.
j. Determine the x-values for which the function is positive.
k. Determine the x-values for which the function is negative.

45. $f(x) = -|x - 1| + 3$ 46. $f(x) = |x + 2| - 4$

Profit and Loss Intervals

Determine the profit and loss intervals for each of the profit functions in Exercises 47 and 48. The x-variable represents the number of units of production, and the y-variable represents the profit in dollars generated by the sale of this production.

47.

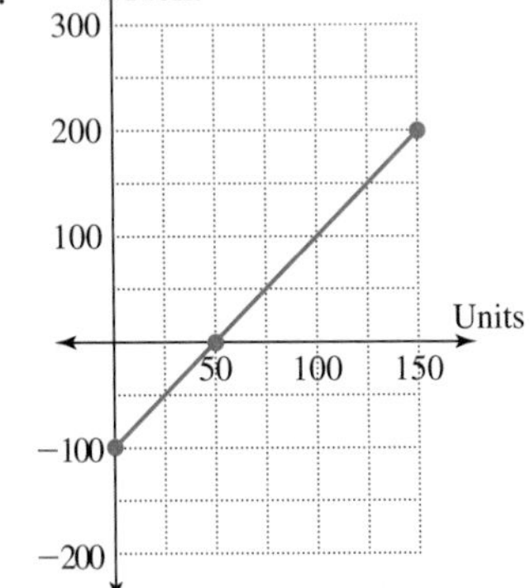

48.

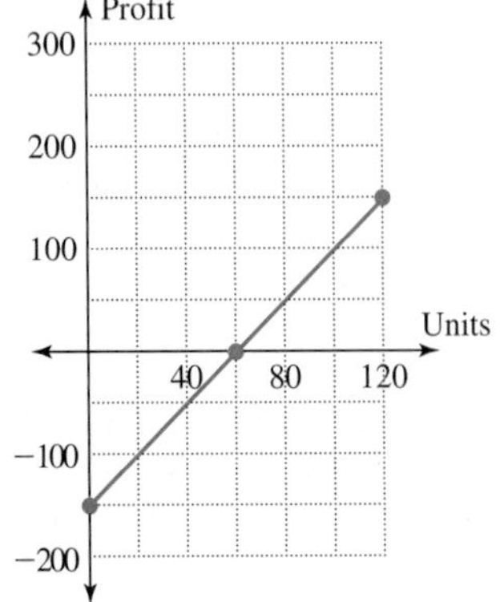

In Exercises 49 and 50, solve each equation and inequality. (*Hint:* See Example 8.)

49. a. $|3x - 1| = 5$ b. $|3x - 1| < 5$ c. $|3x - 1| > 5$
50. a. $|2x + 3| = 7$ b. $|2x + 3| \leq 7$ c. $|2x + 3| \geq 7$
51. **Windchill Temperature** Temperature and wind combine to cause heat loss from body surfaces. With a 10-mi/h wind, the linear function $T(x) = 1.2x - 21.6$ can be used to approximate the windchill temperature, where x is the actual Fahrenheit temperature and $T(x)$ is the temperature that your body is sensing due to the wind. *Source:* National Weather Service. (For more information on this topic go to the Additional Resources link on the MathZone website at **www.mhhe.com/hallmercer.**) Evaluate and interpret each of the expressions in parts **a** to **c**.
a. $T(0)$ b. $T(5)$ c. $T(20)$
d. Is this an increasing or a decreasing function?
e. As the actual temperature x increases, the temperature $T(x)$ that your body senses is __________.

52. **Smoking Trends** The percent of U.S. adults who smoke can be approximated by the function $P(n) = -0.61n + 42.04$, where n is the number of years past 1965. *Source:* Centers for Disease Control. (For more information on this topic go to the Additional Resources link on the MathZone website at **www.mhhe.com/hallmercer.**) Evaluate and interpret each of the expressions in parts **a** to **c**.
a. $P(0)$ b. $P(15)$ c. $P(35)$
d. Is this an increasing or a decreasing function?
e. As x, the number of years past 1965, increases, the percent of U.S. adults who smoke is ____________.
53. **Rolling a Pair of Dice** When a pair of dice is rolled, the probability that the sum of the numbers on those dice is x is

given by $P(x) = \dfrac{6 - |7 - x|}{36}$, where x can be any integer from 2 to 12. Evaluate and interpret each of the expressions in parts **a** to **c.**

a. $P(2)$ b. $P(7)$ c. $P(12)$

d. List each value in the domain of this function.

e. Which has a greater probability, a sum of 2 or a sum of 7?

54. **Electrical Voltage** The voltage in an electric circuit designed by a beginning electronics class is a function of time. This function is given by $V(t) = -120|2t - 1| + 120$, where t is the time in seconds and $V(t)$ is the voltage. Evaluate and interpret each of these expressions:

a. $V(0)$ b. $V(0.5)$ c. $V(1.0)$

55. **Health Care Costs** A patient has an insurance policy that requires a 20% copayment by the patient for all medical expenses. Let x represent the dollar cost of a medical bill.

a. Write a function that models the patient's responsibility for this bill.

b. Use this function and a calculator to complete this table.

X	Y1	
0		
500		
1000		
1500		
2000		
2500		
3000		

X=0

c. Determine the patient's responsibility if the medical bill is $1,000.

d. Determine the cost of a medical bill if the cost to the patient is $425.

56. **Health Care Costs** A patient has an insurance policy that requires a 20% copayment by the patient for all medical expenses. Let x represent the dollar cost of a medical bill.

a. Write a function that models the responsibility of the insurance company for this bill.

b. Use this function and a calculator to complete this table.

X	Y1	
0		
500		
1000		
1500		
2000		
2500		
3000		

X=0

c. Determine the responsibility of the insurance company if the medical bill is $1,000.

d. Determine the cost of a medical bill if the insurance company's responsibility is $2,250.

57. **Depreciation of Carpentry Tools** The value of a $500 set of tools depreciates by $50 per year.

a. Write a linear function V so that $V(x)$ gives the value of these tools after x years.

b. Use this function to calculate the value of these tools after 7 years.

c. Determine the number of years before the value of the tools becomes $50.

d. Use this function to prepare a table of values for this function from year 0 to year 6.

e. Is V an increasing or a decreasing function?

58. **Length of Hair** The length of a human hair grows by approximately 1.3 cm per month. A young girl has a ponytail 20 cm long and wants a ponytail much longer.

a. Write a linear function L so that $L(x)$ gives the length of her ponytail after x months.

b. Determine the length of her ponytail after 6 months.

c. Use this function to determine how many months it will take her to have a ponytail that is 50 cm long.

d. Use this function to prepare a table of values for this function from 0 to 3 years in increments of 6 months.

e. Is L an increasing or a decreasing function?

59. **Cost of a Collect Call** The given table displays the dollar cost of a collect telephone call, based upon a call of x minutes.

a. Use this table to determine the linear equation $C(x) = mx + b$ for these data points.

b. Interpret the meaning of m in this problem.

c. Is this an increasing or a decreasing function?

d. Interpret the meaning of b in this problem.

MINUTES x	COST y ($)
0	2.00
3	3.80
7	6.20
10	8.00

60. **Printing Costs** The given table displays the total dollar cost of a science manual, based upon the number of manuals printed for the order.

a. Use this table to determine the linear equation $C(x) = mx + b$ for these data points.

b. Interpret the meaning of m in this problem.

c. Is this an increasing or a decreasing function?

d. Interpret the meaning of b in this problem.

MANUALS x	COST y ($)
0	125
25	325
50	525
75	725

Group Discussion Questions

61. **Communicating Mathematically** The monthly interest charge of 1.5% of the unpaid balance on a credit card is modeled by the function $f(x) = 0.015x$.

a. What does the input value x represent?

b. What does the output value $f(x)$ represent?

c. Describe the meaning of both x and $f(x)$ when $x = 100$.

d. Describe the meaning of both x and $f(x)$ when $f(x) = 100$.

62. **Communicating Mathematically**

a. The function $f(t) = 100t + 300$ models the cost to rent and operate a bulldozer for t hours in a day. As you read this function, the story that it is telling you is that it costs _________ dollars to rent this bulldozer even if it is not used this day. When it is used, each hour of usage costs an additional _________ dollars.

b. The function $P(x) = 25x - 200$ models the profit made by producing x glass tabletops per day in a factory. As you read this function, the story that it is telling you is that _________.

63. **Calculator Discovery** Use a graphing calculator to graph each of the functions $f(x) = |x|$, $f(x) = |x - 1|$, $f(x) = |x - 2|$, $f(x) = |x - 3|$, and $f(x) = |x - 4|$. Describe the relation between the graphs of $f(x) = |x|$ and $f(x) = |x - c|$.

64. **Error Analysis** A student produced the following calculator window for the function shown and concluded that the function is a linear function and that the range of the function is the set of all real numbers. Determine the correct analysis and the source of the error in this answer.

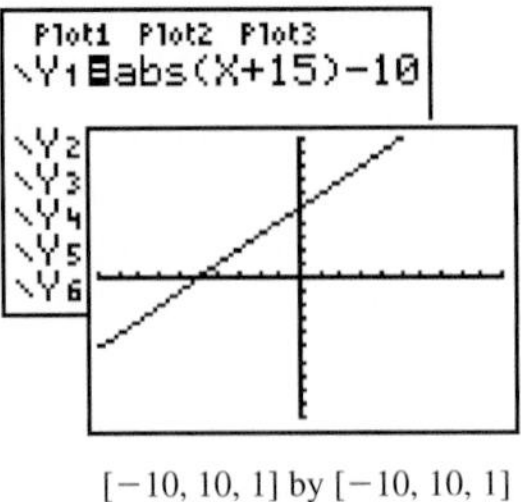

$[-10, 10, 1]$ by $[-10, 10, 1]$

65. **Error Analysis** A student graphed $y = \frac{3}{5}x - 2$ and $y = -\frac{5}{3}x + 3$ on the following calculator window and concluded that the lines are not perpendicular. Determine the correct analysis and the source of the error in this answer.

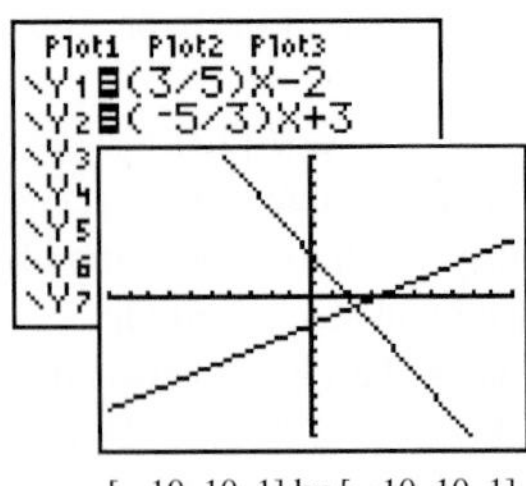

$[-10, 10, 1]$ by $[-10, 10, 1]$

66. **Spreadsheet Exploration** Use spreadsheet Exploration 7.2 from the website **www.mhhe.com/hallmercer** to explore changes to the form $y = |x - h| + k$. Use the sliders to observe the patterns in the table of values for h and k and the graph of the absolute value function defined by $y = |x - h| + k$.

Section 7.3 Linear and Quadratic Functions and Curve Fitting

Objectives:

8. Graph a quadratic function and identify key features of the resulting parabola.
9. Select a linear function that best fits a set of data points.
10. Select a quadratic function that best fits a set of data points.

Parabolic shapes are all around us in our daily lives. The photo illustrates the parabolic path of a stream of water. The path of a thrown ball will be parabolic (assuming that air resistance is negligible). To consider the design and behavior of ejection seats, NASA examined parabolic paths for the seats after they blast away from an aircraft. Many human-made objects display a parabolic shape. Some cables supporting bridges hang in a parabolic shape, and parabolic reflectors are used for spotlights, on microphones at sporting events, and in satellite receiver dishes.

Parabolic Satellite Dish

All the functions graphed here are quadratic functions, and the graph of each is a parabola.

EXAMPLES OF QUADRATIC FUNCTIONS

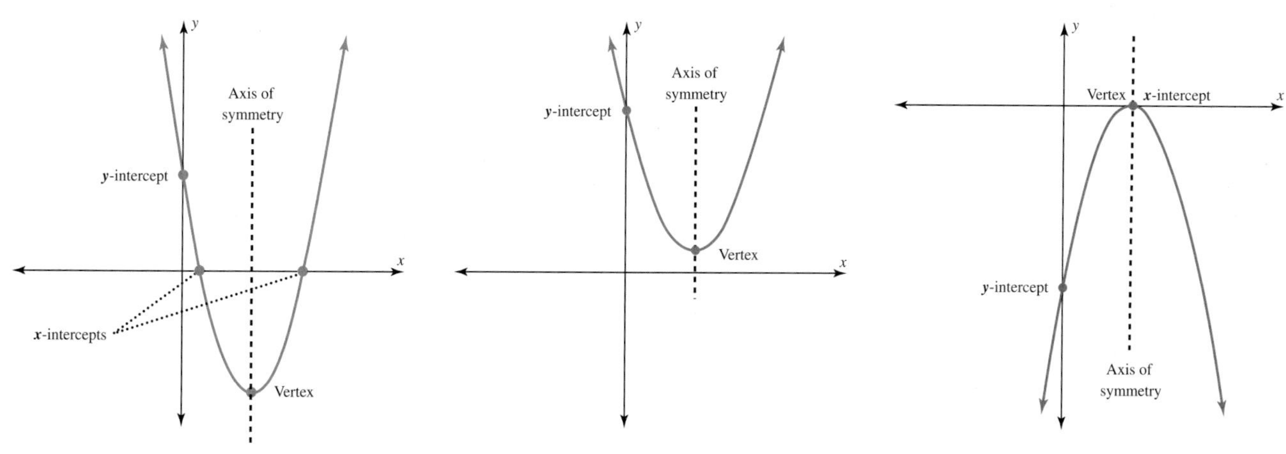

Quadratic Function, Shape of a Parabola, and Second-Degree Equation

A second-degree polynomial function can be written in the form $f(x) = ax^2 + bx + c$ and is called a **quadratic function** if $a \neq 0$. The term ax^2 is called the **quadratic** or the **second-degree term,** bx is called the **linear term,** and c is called the **constant term.** The graph of a quadratic function of this form is always a **parabola** that is either opening upward or opening downward. Parabolas opening upward are called *concave up,* and those opening downward are called *concave down.* The **vertex** of a parabola opening upward is the lowest point on the parabola. The vertex of a parabola opening downward is the highest point on the parabola. The **axis of symmetry** of a parabola passes through the vertex. The portion of the parabola to the left of the axis of symmetry is a mirror image of the portion to the right of this axis of symmetry. Quadratic functions have applications in many areas including the business application given in Example 4.

Each family of functions has its own distinctive graph, its own numerical pattern, and its own algebraic definition. Key characteristics of quadratic functions are summarized in the following box.

Second-Degree Functions—Quadratic Functions

If x is a real variable and a, b, and c are real constants with $c \neq 0$, then:

ALGEBRAICALLY	GRAPHICALLY	NUMERICALLY	EXAMPLE
$y = ax^2 + bx + c$	A parabola	The y-values form a symmetric pattern about the vertex.	Plot1 Plot2 Plot3 \Y1=X²−4X+4 X: −1, 0, 1, 2, 3, 4, 5; Y1: 9, 4, 1, 0, 1, 4, 9 [−1, 5, 1] by [−1, 9, 1]

The parabola defined by $f(x) = ax^2 + bx + c$ opens upward if $a > 0$ and downward if $a < 0$.

Example 1 examines two quadratic functions. The function $f(x) = x^2$ is of the form $f(x) = ax^2 + bx + c$ with $a = 1, b = 0,$ and $c = 0$. Note that because $a > 0$ the parabola defined by $f(x) = x^2$ opens upward. The function defined by $f(x) = -x^2$ is of the form $f(x) = ax^2 + bx + c$ with $a = -1, b = 0$, and $c = 0$, and its graph opens downward because $a < 0$. We can use the defining equation to create a table of values, plot these points, and then smoothly connect them with a parabolic shape. This process is illustrated in Example 1.

EXAMPLE 1 Graphing a Quadratic Function

Create a table of values for the quadratic functions $f(x) = x^2$ and $f(x) = -x^2$, plot these points, and sketch the graph of the parabola through these points. Determine the domain and the range of each function.

SOLUTION

ALGEBRAICALLY / **NUMERICALLY** / **GRAPHICALLY** / **VERBALLY**

(a) $f(x) = x^2$

x	$f(x) = x^2$
−3	9
−2	4
−1	1
0	0
1	1
2	4
3	9

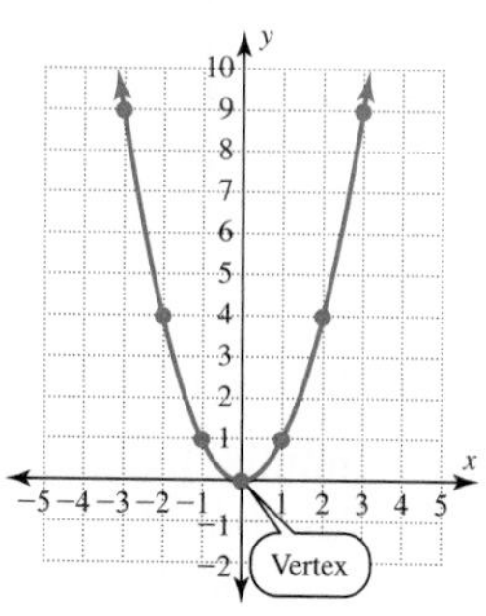

This parabola opens upward with a vertex of (0, 0). This function is defined for all real numbers. The domain $D = \mathbb{R}$. The range, the projection of the graph onto the y-axis, is $R = [0, \infty)$.

(b) $f(x) = -x^2$

x	$f(x) = -x^2$
−3	−9
−2	−4
−1	−1
0	0
1	−1
2	−4
3	−9

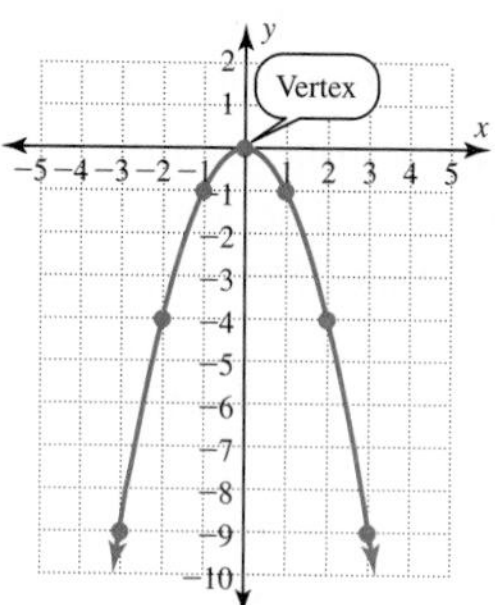

This parabola opens downward with a vertex of (0, 0). This function is defined for all real numbers. The domain $D = \mathbb{R}$. The range, the projection of the graph onto the y-axis, is $R = (-\infty, 0]$.

A Mathematical Note

Ian Stewart writes in *Nature's Numbers* (a 1995 book published by Basic Books) that his "book ought to have been called *Nature's Numbers and Shapes.*" Even though shapes can always be reduced to numbers, "it is often better to think of shapes that make use of our powerful visual capabilities."

It is important to be able to predict the shape of a graph from the defining equation. Knowing the shape of a graph can help us to determine an appropriate viewing window to use on a graphing calculator. Without this knowledge we may select an inappropriate window that fails to show key features of the graph. When we are working with real applications, knowing the shape of the graph can help us select a model that fits the behavior that we are trying to describe and explain. Example 2 illustrates how to use the defining equation to describe the graph of each linear or quadratic function.

EXAMPLE 2 Identifying Lines and Parabolas

Identify the graph of each function as a line or a parabola. If the graph is a line, determine whether the slope is positive or negative. If the graph is a parabola, determine whether the parabola opens upward or downward.

SOLUTION

(a) $f(x) = 4x - 11$ Line with positive slope

This first-degree function is a linear function in the slope-intercept form $f(x) = mx + b$. The slope $m = 4$ is positive.

(b) $f(x) = -5x + 6$	Line with negative slope	This first-degree function is a linear function in the slope-intercept form $f(x) = mx + b$. The slope $m = -5$ is negative.
(c) $f(x) = -2x^2 + x - 9$	Parabola that opens downward	This second-degree function is a quadratic function in the form $f(x) = ax^2 + bx + c$. Because $a = -2$ this parabola opens downward.
(d) $f(x) = 3x^2 - 5x - 1$	Parabola that opens upward	This second-degree function is a quadratic function in the form $f(x) = ax^2 + bx + c$. Because $a = 3$ this parabola opens upward.
(e) $f(x) = 0x^2 + 0x + 7$	Line with slope of 0	This is not a quadratic function in the form $f(x) = ax^2 + bx + c$ because $a = 0$. It is a linear function in the slope-intercept form $f(x) = mx + b$ with $m = 0$. The graph of this constant function will be the horizontal line with a y-intercept of (0, 7). Thus the slope is neither positive nor negative; it is 0.

SELF-CHECK 7.3.1

Identify the graph of each function as a line or a parabola. If the graph is a line, determine whether the slope is positive or negative. If the graph is a parabola, determine whether the parabola opens upward or downward.

1. $f(x) = 2x^2 + 0x - 1$
2. $f(x) = 0x^2 + 2x - 1$
3. $f(x) = 0x^2 - x + 2$
4. $f(x) = -2x^2 + x + 0$

The vertex of a parabola is a key point on the parabola.

Example 3 uses the graph of $f(x) = x^2 + 2x - 8$ to describe the key features of a parabola. The most important point on the parabola is the vertex; the coordinates of this point are approximated from the graph. The intercepts of the parabola also often play key roles, which we examine in Example 4.

The symmetry exemplified by a quadratic function is readily observable from the parabolic shape of its graph. This symmetry makes it easy to visually identify the vertex. The numerical symmetry of the y-values on both sides of the vertex can also be used to identify the location of the vertex if we have an appropriate table of values. This is explained further in Section 7.5.

EXAMPLE 3 Determining Key Features of a Parabola

Use the graph of $f(x) = x^2 + 2x - 8$ shown to determine the following information about this graph.

(a) Does the parabola open upward or downward?
(b) Vertex
(c) The y-intercept
(d) The x-intercepts
(e) Domain of f
(f) Range of f
(g) Interval where f is decreasing
(h) Interval where f is increasing

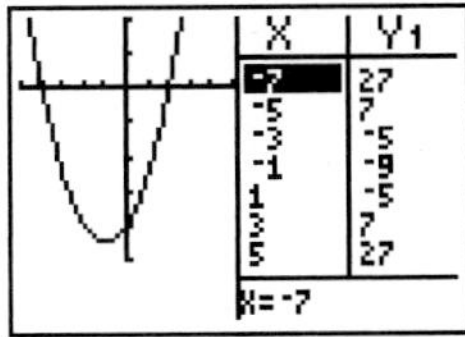

[−5, 5, 1] by [−10, 4, 2]

SOLUTION

(a) The parabola opens upward.

In the function $f(x) = x^2 + 2x - 8$ the coefficient of x^2 is 1. Because this coefficient is positive the parabola opens upward.

(b) Vertex: $(-1, -9)$

On this graph the vertex is the lowest point on the parabola which by inspection is approximately $(-1, 9)$. Calculator Perspective 7.5.1 will illustrate how to use a calculator to approximate the coordinates of the vertex.

(c) The y-intercept: $(0, -8)$ — The y-intercept is the point where the graph crosses the y-axis. The x-coordinate of the y-intercept is 0.

(d) The x-intercepts: $(-4, 0)$ and $(2, 0)$ — Approximate the x-intercepts directly from the graph or algebraically by determining the x-values where the y-coordinate is 0. Solve $x^2 + 2x - 8 = 0$ by factoring (see Section 6.6). Factoring and solving $(x + 4)(x - 2) = 0$ makes it possible to determine the x-intercepts are at $x = -4$ and $x = 2$.

(e) Domain $= \mathbb{R}$ — The projection of this graph onto the x-axis is the set of all real numbers.

(f) Range $= [-9, \infty)$ — The projection of this graph onto the y-axis is the set of y-values given by the interval $[-9, \infty)$.

(g) Decreasing on the interval $(-\infty, -1)$ — The graph is decreasing for all values to the left of -1.*

(h) Increasing on the interval $(-1, \infty)$ — The graph is increasing for all values to the right of -1.*

SELF-CHECK 7.3.2

Use the graph of $f(x) = -x^2 + 2x + 3$ shown to determine the following information about this graph.

1. Does the parabola open upward or downward?
2. Vertex
3. The y-intercept
4. The x-intercepts
5. Domain of f
6. Range of f
7. Interval where f is decreasing
8. Interval where f is increasing

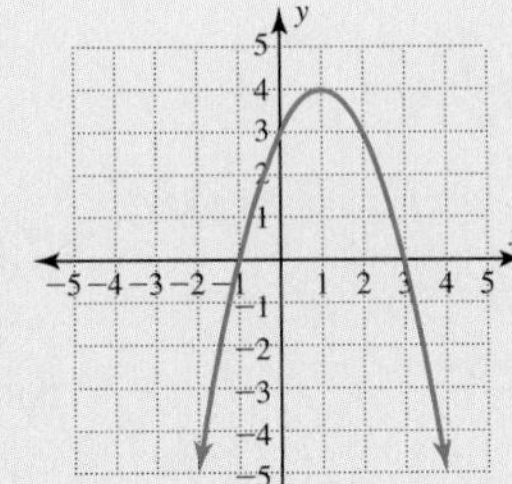

EXAMPLE 4 Analyzing a Quadratic Model for Profits

As part of his diversified production, a farmer can harvest and sell up to 5,000 lb of catfish per week. Each week he has overhead costs, and when he increases his harvest, he must hire extra help and use additional equipment. All these factors are accounted for in the following function that gives his weekly profit in dollars as a function of the number of pounds x that he sells. His weekly profit is modeled by the function $P(x) = -0.0001x^2 + 0.6x - 500$.

(a) Determine $P(0)$ and interpret this value.
(b) Use a calculator to determine the x-intercepts of the graph of this function, and interpret these values.
(c) Determine the profit interval for this function.
(d) Approximate the vertex of this graph by inspection, and interpret the meaning of both coordinates.

SOLUTION

(a)
$$P(x) = -0.0001x^2 + 0.6x - 500$$
$$P(0) = -0.0001(0)^2 + 0.6(0) - 500$$
$$P(0) = -500$$
If he sells 0 lb, he will lose \$500.

Evaluate $P(0)$ by substituting 0 for x. Because of the overhead costs, he will lose money if he does not sell any catfish.
The y-intercept of the graph is $(0, -500)$.

*The definition of increasing and decreasing in most calculus books would include the endpoints of these intervals.

(b) Plot1 Plot2 Plot3
\Y1 = -0.0001X²+0.6X-500

Zero X=1000 Y=0 | Zero X=5000 Y=0

[0, 6000, 1000] by [−500, 500, 100]

Graph this function on a graphing calculator, and then use the **Zero** feature on the **CALC** menu to calculate the x-value where y is zero in order to determine the x-intercepts of the function.

The x-intercepts are (1,000, 0) and (5,000, 0). If the farmer sells 1,000 or 5,000 lb, he will break even.

(c) The farmer will make a profit with sales from 1,000 to 5,000 lb.

The graph is above the x-axis, and the profit is positive for the x-values in the interval (1,000, 5,000).

(d)

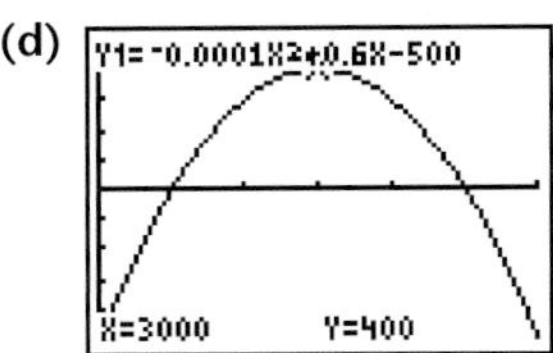

[0, 6000, 1000] by [−500, 500, 100]

The vertex is (3,000, 400).

If the farmer sells 3,000 lb of catfish, he will make $400. This is the maximum amount he can make in a week from his catfish sales.

Note that the x-coordinate of the vertex is midway between the x-intercepts. We will examine this in greater detail in Section 7.5.

SELF-CHECK 7.3.3

Use the function $f(x) = x^2 - 4x + 3$ and a graphing calculator.

1. Determine the x-intercepts for the graph of this function.
2. Determine by inspection the vertex of the graph of this function.

Curve Fitting

If we have the equation defining a function, then we can use this equation to create a table of values or to graph the function. What we will do now is much more sophisticated. We will take a table of values and then determine the equation of the line or parabola that best fits the points—even when some of the points do not lie on this curve. The topic of selecting a curve that best fits a set of data points is called **curve fitting.**

We start with scatter diagrams, which we first examined in Section 2.1. The scatter diagram for a set of points often reveals a visual pattern, which can help us select the type of function that best describes the data. We can then use a graphing calculator to calculate the equation for the function that best fits the data. The data for a real application often do not lie exactly on a curve due to experimental error and other factors. However, these data points may all lie very close to a curve. The scatter diagrams that follow illustrate three possibilities. The first set of data displays a linear pattern; the second set of data displays a parabolic pattern typical of a second-degree function; and the third set of data displays the pattern of exponential growth. We will examine linear patterns and parabolic patterns in this section and exponential patterns in Chapter 10.

Linear Pattern

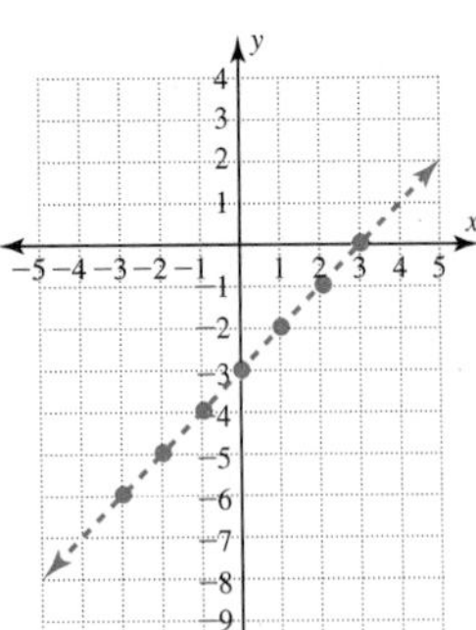

Parabolic Pattern

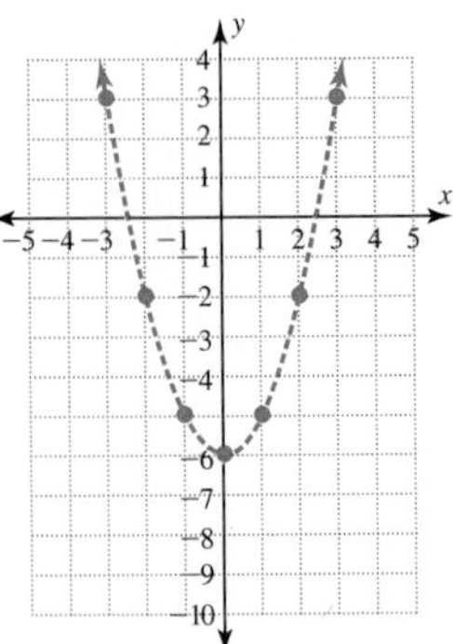

Exponential Pattern

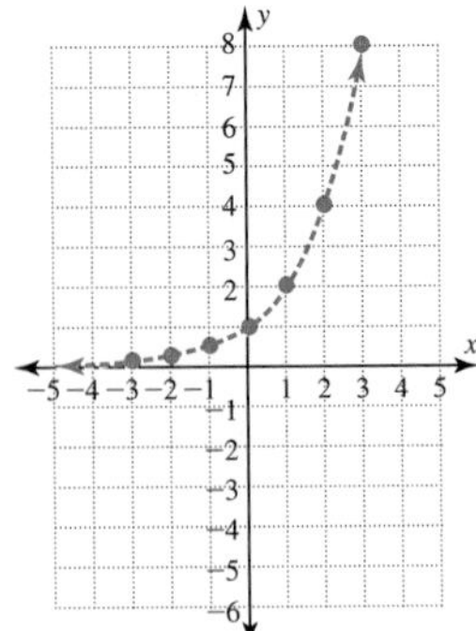

Calculator Perspectives 7.3.1 and 7.3.2 illustrate how to use a graphing calculator to draw a scatter diagram for the data points in the following table.

x	-3	-2	-1	0	1	2	3
y	-9	-7	-5	-3	-1	1	3

CALCULATOR PERSPECTIVE 7.3.1 Using Lists to Store a Set of Data Points

To enter the data points from this table on a TI-84 Plus calculator, enter the following keystrokes:

(to **Edit** the lists L1 and L2)

Input the x-values under L1.

. . .

Press ▶ to move to the L2 column and then input the y-values.

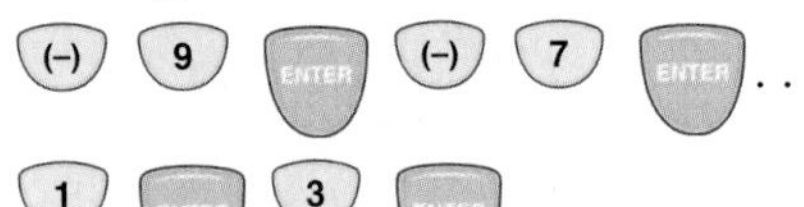

Once data points are entered into a graphing calculator, we can use the calculator to draw the scatter diagram. This is illustrated by Calculator Perspective 7.3.2. The STAT PLOT feature is the secondary function of the Y= key. It is a good idea to remember to turn off **Plot1** when you are working problems that do not require the STAT PLOT feature. By selecting **ZoomStat** the calculator will automatically select an appropriate window to display the data points.

CALCULATOR PERSPECTIVE 7.3.2 Drawing a Scatter Diagram

To draw the scatter diagram for the data points from this table on a TI-84 Plus calculator, enter the following keystrokes:

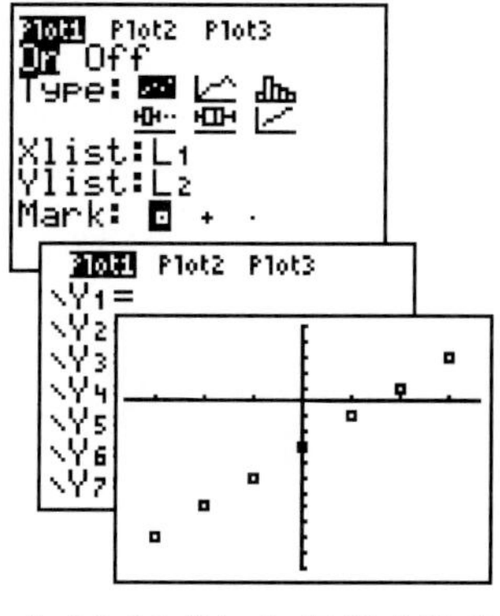

Then press ENTER to turn **Plot1** on and confirm that the menu is set up as shown.

Y= (Check to see that the Y= screen is blank; otherwise clear all entries.)

Select **ZoomStat** by pressing ZOOM 9.

[−3.6, 3.6, 1] by [−11.04, 5.04, 1]

Example 5 is used to examine a scatter diagram that illustrates the relationship between the speed in miles per hour of an automobile and its rate of fuel consumption, measured in miles per gallon. At higher speeds there is greater wind resistance and less engine efficiency. Thus greater speeds generally mean less fuel efficiency, as illustrated by the data in Example 5.

EXAMPLE 5 Using a Graphing Calculator to Draw a Scatter Diagram

Use a graphing calculator to draw a scatter diagram for the data points in the given table. Then describe the visual pattern formed by these points.

x ml/h	y mi/gal
20	23
25	22
30	21
35	20
40	19
45	18
50	17

SOLUTION

First enter the data points, using the STAT-EDIT feature.

L1	L2	L3 1
20	23	------
25	22	
30	21	
35	20	
40	19	
45	18	
50	17	

L1(1)=20

Then use the STAT-PLOT feature to draw a scatter diagram of the data points listed in **L1** and **L2.**

Answer: These points appear to form a linear pattern with a negative slope. The meaning of the negative slope is that fuel efficiency is decreasing as speed increases.

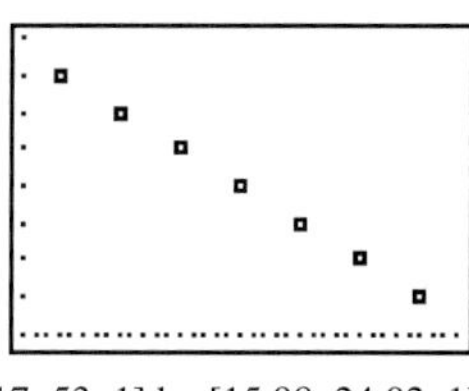

[17, 53, 1] by [15.98, 24.02, 1]

SELF-CHECK 7.3.4

1. Use a graphing calculator to draw a scatter diagram for the data points in this table.

x	y
-3	-3
-2	-1
-1	1
0	5
1	7
2	9
3	12

A Mathematical Note

Sir Francis Galton introduced the concept of linear regression in 1877. He was studying data which showed that tall parents have children whose heights tend to "regress" or revert to the mean height of the population.

We have now shown how to store data points in a graphing calculator, and we have examined how to use a calculator to draw a scatter diagram of these points. We now add the last piece to our curve-fitting skills. We examine how to select a line that best fits a set of data points. The topic of fitting an equation to a set of data points is known in mathematics as **regression analysis.** We limit our discussion of the details of this topic and take advantage of the capabilities of a graphing calculator to perform this work for us.

Calculator Perspective 7.3.3 illustrates how to use a calculator to determine the line of best fit for the data in Example 5 and to then enter this equation into the Y= screen, using the slope and y-intercept just calculated. (Although there are shortcuts for transferring this equation directly to the Y= screen, we show the entry here to emphasize that the calculator has found the slope and y-intercept for the line best fitting these data.) Before you read Calculator Perspective 7.3.3, enter the data from Example 5 in a graphing calculator and form the scatter diagram.

CALCULATOR PERSPECTIVE 7.3.3 Calculating and Graphing the Line of Best Fit

To calculate the line of best fit for the data in Example 5 on a TI-84 Plus calculator and then to graph this line, enter the following keystrokes:

 [Execute the **LinReg(ax + b)** command from the **STAT CALC** menu.]

 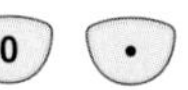   

Equation: $y = -0.2x + 27$

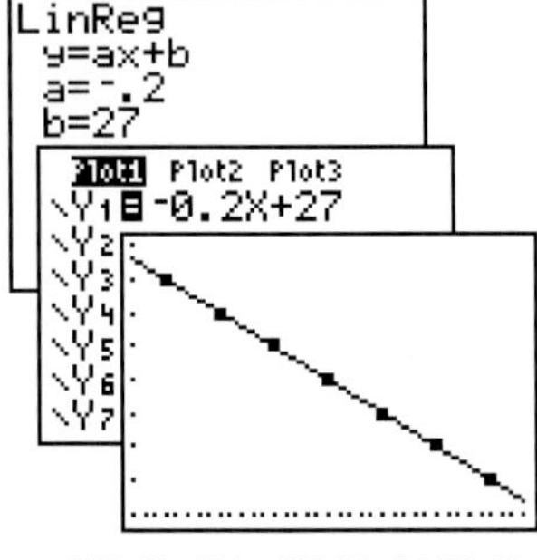

[17, 53, 1] by [15.98, 24.02, 1]

Note that the equation of the line of best fit can be inserted into the Y= screen from the memory of the calculator by pressing. VARS 5 > > ENTER.

Note: The line of best fit and the data points are both displayed on the graph. For each 1 mi/h the speed increases the fuel economy decreases by 0.2 mi/gal. ■

The criteria for deciding which curve best fits a set of data is covered in most introductory statistics courses. We use calculators to help us focus on the key concepts here, rather than on the theoretical background.

Real Data for Real People: The data in Example 5 are phony in the sense that the authors made them up to exactly fit a line—the data were manipulated to meet an ideal model. Textbooks are full of idealized data. It is easier to teach concepts piece by piece with nice clean data. Students do not get lost in the messy details and can concentrate on the concepts. However, students often sense that problems contain phony data—even though there are very good reasons for these problems. It is important to realize we can use the same procedures with real data—it just may be a little messier. Example 6 addresses working with real data. Real data rarely lie perfectly on an idealized curve. Thus we try to select the curve that best fits the data, even when the curve will miss some of the data points.

The three scatter diagrams in the following box each contain the same data points. There is no line that will pass through all these points. Look at the three options. One of these lines is the line of best fit. Which of the three lines do you think best fits the data?

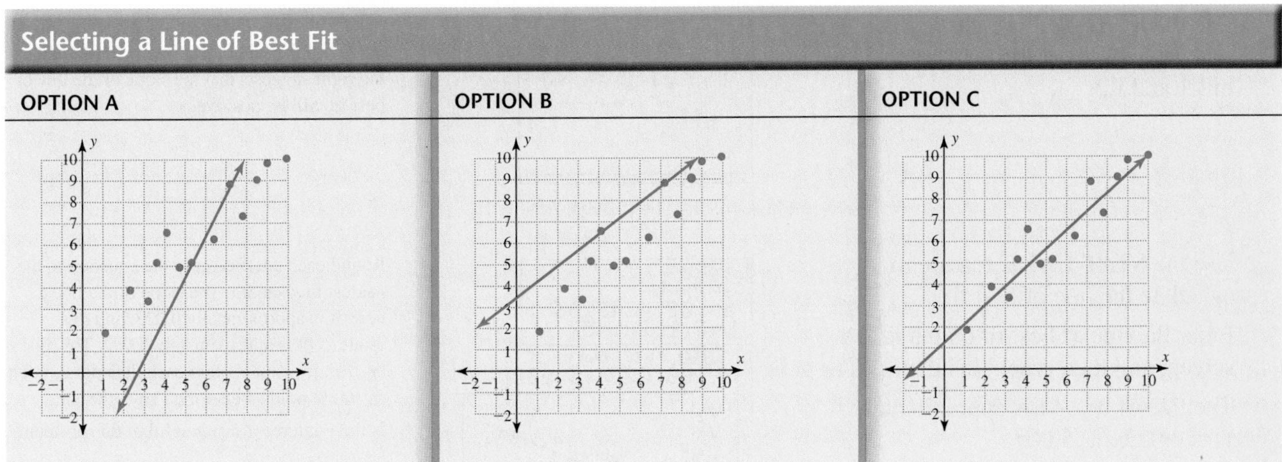

Most people select option C because it visually appears closer to more points than the other options. This intuitive understanding of the line of best fit is actually consistent with what is happening mathematically when a calculator is determining the line of best fit. The calculator has been programmed to calculate the one line that overall best fits the data points.

Some of the reasons we use curve fitting to analyze data are to better describe the relationship between the independent and dependent variables and to either understand the past or to predict the future. Example 6 examines the women's long-jump records over several years and then illustrates how to use the line of best fit for these data to make a prediction for the record for a future time.

EXAMPLE 6 Modeling Athletic Records by Using a Line of Best Fit

The given table displays the winning lengths for the Olympic women's long jump from 1948 through 2000. Each distance is given in meters. The actual dates are not used in the calculations. The x-values represent the number of years after 1900. These smaller x-values allow us to round the coefficients of the equation of best fit without introducing an unacceptable amount of error.

(a) Draw a scatter diagram for these data and then determine the line of best fit.
(b) Interpret the meaning of the slope of this line.
(c) Use the equation of the line of best fit to estimate the missing entry in the table from 1972. How does this estimation compare to the actual 1972 record of 6.78 m?
(d) Use the equation of the line of best fit to predict the women's long-jump record for the year 2004. How does this prediction compare with the actual record of 7.07 m?

YEAR	x	y m
1948	48	5.69
1952	52	6.24
1956	56	6.35
1960	60	6.37
1964	64	6.76
1968	68	6.82
1976	76	6.72
1980	80	7.06
1984	84	6.96
1988	88	7.40
1992	92	7.14
1996	96	7.12
2000	100	6.99

SOLUTION

(a) Enter the data points by using the **STAT-EDIT** feature.

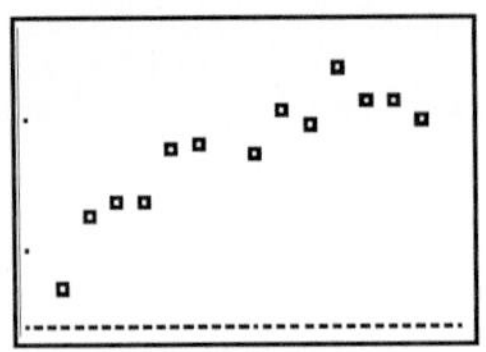

Be sure to enter all the data points since these points will fill more than one display screen.

Use the **STAT-PLOT** feature to draw a scatter diagram of the data points listed in L1 and L2.

[42.8, 105.2, 1] by [5.3993, 7.6907, 1]

Note that the points in this scatter diagram do not lie on the same line. The upward trend to the right suggests that the slope of the line of best fit will be positive.

Use the **STAT-CALC** feature to calculate the line of best fit.

Enter the line of best fit on the Y= screen, and then graph this line.

The line of best fit is approximately $y = 0.023x + 5.002$.

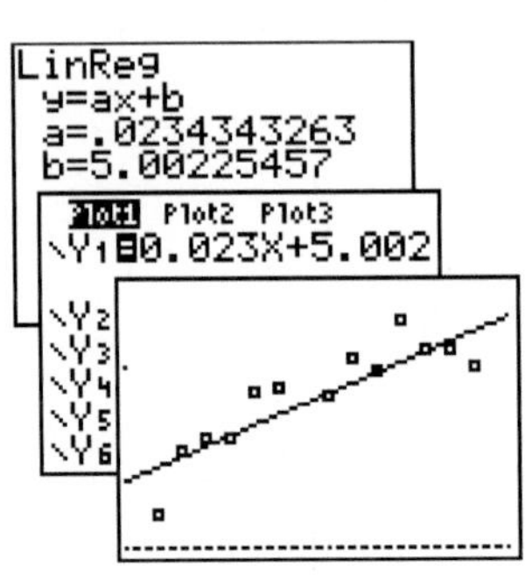

[42.8, 105.2, 1] by [5.3993, 7.6907, 1]

Round each coefficient in the equation to the nearest thousandth. The slope of the line of best fit is approximately 0.023, and the y-intercept is approximately 5.002. This line of best fit does not contain each data point, but it fits these points better than any other line.

Is the y-intercept meaningful in this problem?

(b) The slope of the line of best fit is positive which indicates that this linear function is increasing. The trend is for the record to increase by about 2.3 cm/yr.

Caution: Although the long-term trend is for the record to increase, this does not mean that the record does increase at each Olympics. In fact, for some Olympics the long-jump record is less than that at the previous Olympics.

0.023 m = 2.3 cm

(c) For 1972 the line of best fit yields an estimation of 6.66 m. The actual 1972 record of 6.78 m is 12 cm longer than that given by this linear model.

(d) For 2004 the line of best fit yields an estimation of 7.39 m. The actual 2004 record of 7.07 m is 32 cm less than that predicted by this linear model.

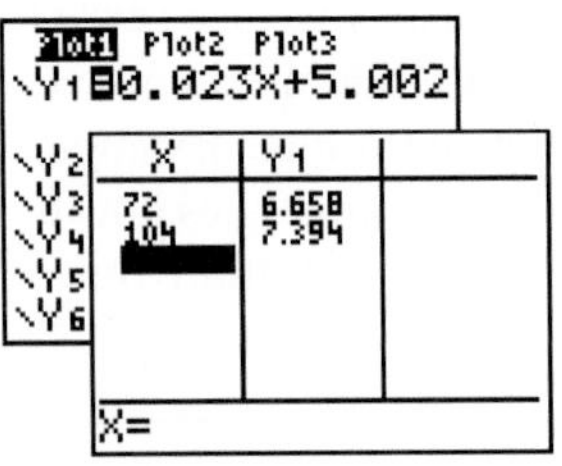

To estimate the long-jump record for 1972, substitute 72 for x in the equation for the line of best fit. To estimate the long-jump record for 2004, substitute 104 for x in the equation for the line of best fit.

The estimation error of 12 cm is about the same as the width of about six adult fingers.

Both calculations are made by using the **Ask** mode in a table for this linear function.

Many factors could affect this prediction. Do you think a linear model is reasonable for this problem?

SELF-CHECK 7.3.5

1. Use a graphing calculator to determine the line of best fit for these data points.

x	-3	-2	-1	0	1	2	3
y	-6	-3	-2	2	2	6	6

Write your answer in slope-intercept form with m and b rounded to the nearest thousandth.

2. Use this linear equation to estimate y when $x = -1.3$.

Caution: Rounding can sometimes create significant differences when we use the quadratic function of best fit.

If the scatter diagram for a set of data points displays an overall linear pattern, then we will likely examine linear curve fitting. If the scatter diagram for a set of data points displays an overall parabolic pattern, then we will likely examine quadratic curve fitting. The data points given in Example 7 form a parabolic pattern. Therefore a graphing calculator is used to determine the quadratic function of best fit. The keystrokes for calculating the parabola of best fit are identical to those for calculating the line of best fit except that we select option **5** instead of **4** under the **STAT-CALC** menu. Option **5** is the **QuadReg** option. The coefficients in this function will be rounded to the nearest thousandth.

EXAMPLE 7 Using a Graphing Calculator to Determine a Parabola of Best Fit

Enter these points into a graphing calculator, draw the scatter diagram, and calculate the parabola of best fit. Use this quadratic equation to approximate y when $x = 6$.

x	y
0.1	−7.7
0.9	−2
1.9	0.4
3.3	2.1
5.2	4.5
6.7	2.9
8.9	−4.6
10.0	−8.2

SOLUTION

1. Enter the data points by using the **STAT-EDIT** feature.

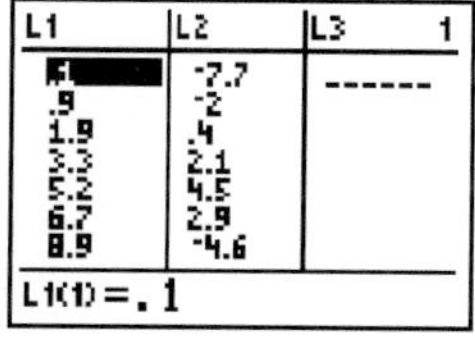

Be sure to enter all points even though they cannot all be displayed on one screen.

2. Turn **Plot 1** on to draw a scatter diagram of the data points listed in L1 and L2. Use the visual pattern to select the most appropriate family of curves. In this case, the visual pattern is parabolic.

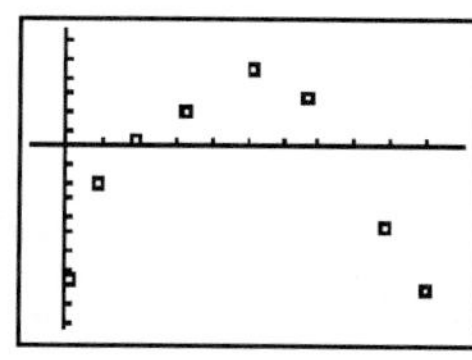

[−0.89, 10.99, 1] by [−10.359, 6.659, 1]

Since the points appear to form a parabola opening downward, try to find a quadratic function that best fits these data.

3. Use the **STAT-CALC** feature to calculate the parabola of best fit.

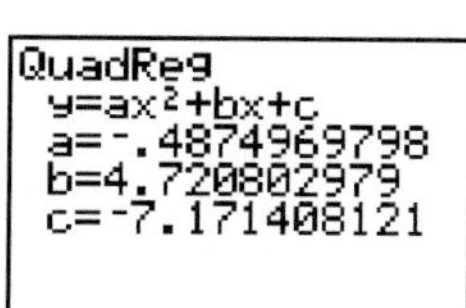

To calculate the parabola of best fit, select option 5 under the STAT **CALC** menu.

4. Enter the parabola of best fit on the Y= screen, and then graph this curve. Round each coefficient to the nearest thousandth.

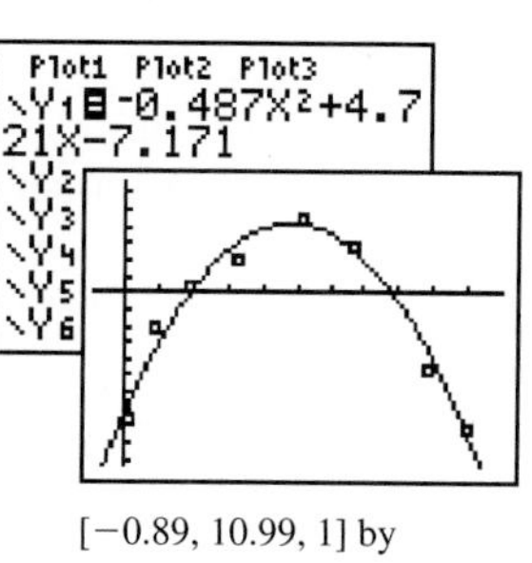

[−0.89, 10.99, 1] by [−10.359, 6.659, 1]

This parabola of best fit does *not* contain each data point, but it does fit these points better than any other parabola.

Note that the equation of the parabola of best fit also can be inserted into the Y= screen from the memory of the calculator by pressing VARS 5 › › ENTER.

5. 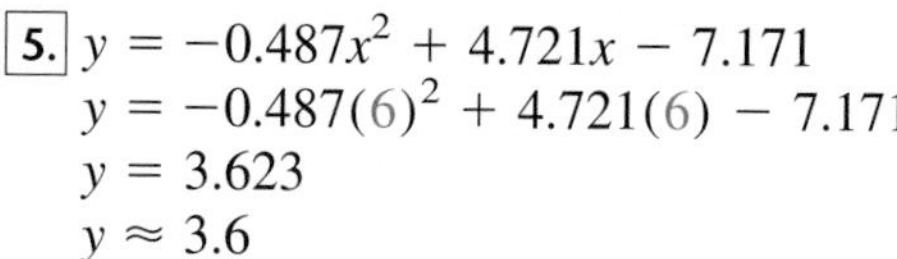

$$y = -0.487x^2 + 4.721x - 7.171$$
$$y = -0.487(6)^2 + 4.721(6) - 7.171$$
$$y = 3.623$$
$$y \approx 3.6$$

To approximate y for $x = 6$, substitute 6 into the equation of the parabola of best fit. Since the original data were accurate only to the tenths place, the estimate is rounded to the nearest tenth.

Answer: $y = -0.487x^2 + 4.721x - 7.171$ is approximately the parabola of best fit. For $x = 6$, $y \approx 3.6$.

The equation of the parabola of best fit is written in $y = ax^2 + bx + c$ form with each coefficient rounded to the nearest thousandth.

SELF-CHECK 7.3.6

1. Use the **QuadReg** feature on a graphing calculator to determine the equation of the parabola of the form $y = ax^2 + bx + c$ that contains the points (1, 24), (2, 16), and (3, −24).

SELF-CHECK ANSWERS

7.3.1

1. Parabola that opens upward
2. Line with positive slope
3. Line with negative slope
4. Parabola that opens downward

7.3.2

1. The parabola opens downward.
2. Vertex: (1, 4)
3. y-intercept: (0, 3)
4. x-intercepts: (−1, 0) and (3, 0)
5. Domain = ℝ
6. Range = $(-\infty, 4]$
7. Decreasing on the interval $(1, \infty)$
8. Increasing on the interval $(-\infty, 1)$

7.3.3

1. (1, 0), (3, 0)
2. (2, −1)

7.3.4

1.

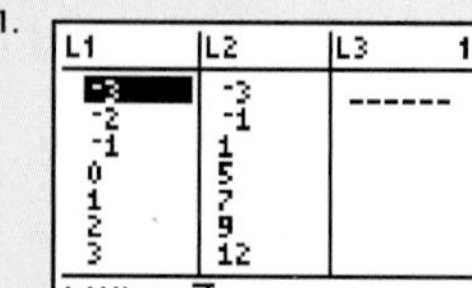

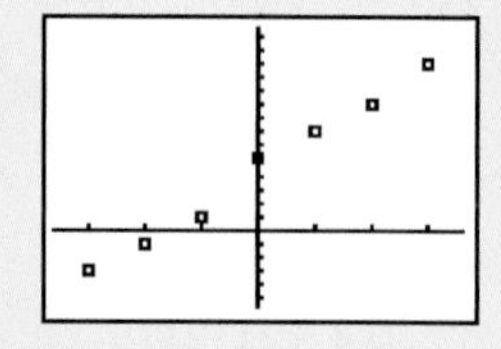

[−3.6, 3.6, 1] by [−5.55, 14.55, 1]

7.3.5

1. $y = 2.071x + 0.714$
2. For $x = -1.3$, $y \approx -2$.

7.3.6

1. $y = -16x^2 + 40x$

USING THE LANGUAGE AND SYMBOLISM OF MATHEMATICS 7.3

1. A function of the form $f(x) = ax^2 + bx + c$ with $a \neq 0$ has a graph whose shape is a __________.
2. A function of the form $f(x) = ax^2 + bx + c$ with $a \neq 0$ is called a __________ function.
3. The highest point on a parabola that opens downward is called the __________ of the parabola.
4. The vertex is the lowest point on a parabola that opens __________.
5. A parabola defined by $f(x) = ax^2 + bx + c$ is symmetric about a vertical line passing through its vertex. This vertical line is called the __________ of __________.
6. In the function $f(x) = ax^2 + bx + c$, ax^2 is called the __________ term, bx is called the __________ term, and c is called the __________ term.
7. A function that is defined by a first-degree polynomial has a graph that is __________.
8. A parabola is defined by a function that is a __________-degree polynomial.
9. If a line is defined by $y = mx + b$ and $m > 0$, then the line slopes __________ to the right.
10. If a parabola is defined by $f(x) = ax^2 + bx + c$ and $a > 0$, then the parabola opens __________.
11. If a parabola is defined by $f(x) = ax^2 + bx + c$ and $a < 0$, then the parabola opens __________.
12. The topic of selecting a curve that best fits a set of data points is called __________ fitting.
13. A __________ diagram for a set of data points is formed by graphing these points on a coordinate system.
14. If the pattern formed by the points of a scatter diagram is approximately __________, then we will select an equation of the form $y = mx + b$.
15. If the pattern formed by the points of a scatter diagram is approximately a __________, then we will select an equation of the form $y = ax^2 + bx + c$.

QUICK REVIEW 7.3

The purpose of this quick review is to help you recall skills needed in this section.

1. The standard form of an nth-degree __________ in x is $a_nx^n + a_{n-1}x^{n-1} + \cdots + a_1x + a_0$.
2. The equation $y - y_1 = m(x - x_1)$ is known as the __________-__________ form of a line.
3. The fixed costs or the __________ costs for a company can include insurance, rent, electricity, and other fixed expenses when 0 units are produced.
4. If (x, y) lies in quadrant II, then x is __________ and y is positive.
5. If x and y have the same sign, then the point (x, y) lies in quadrant I or quadrant __________.

EXERCISES 7.3

In Exercises 1–4, use the shape of each graph to match it with the function that defines this graph.

1. 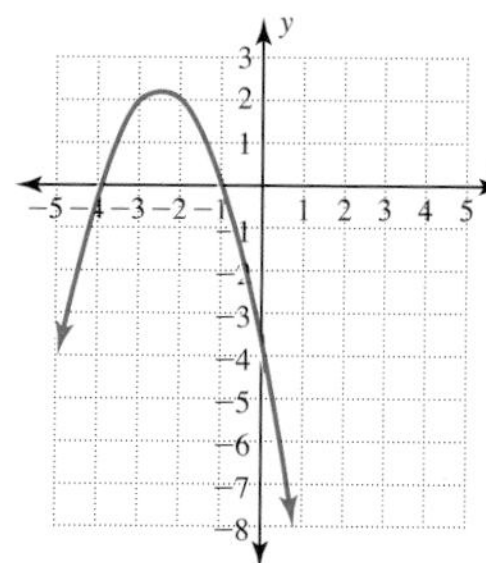

A. $f(x) = 3x - 4$

2. 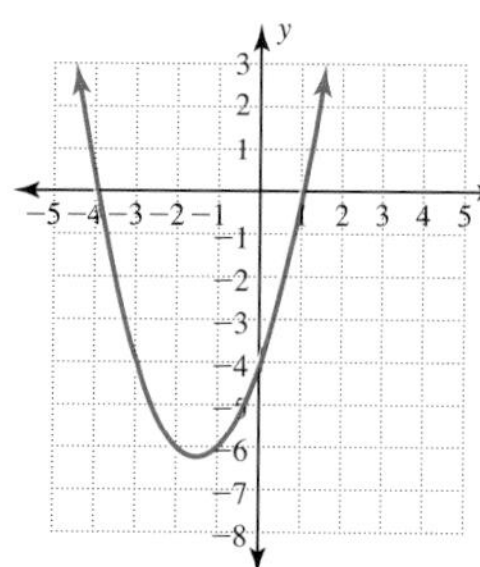

B. $f(x) = -x - 4$

3. 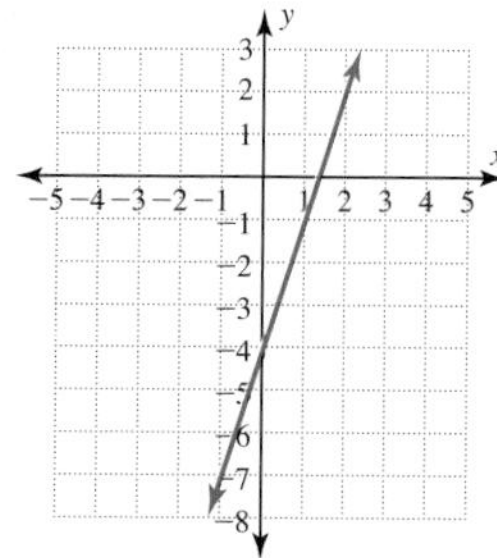

C. $f(x) = x^2 + 3x - 4$

4. 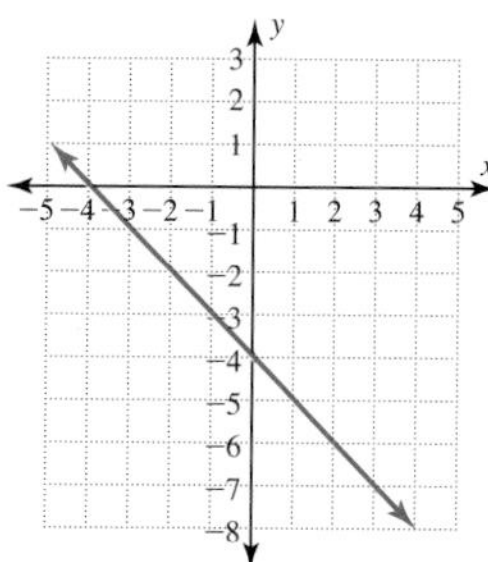

D. $f(x) = -x^2 - 5x - 4$

In Exercises 5–14, identify the graph of each function as a line or a parabola. If the graph is a line, determine whether the slope is negative, zero, or positive. If the graph is a parabola, determine whether the parabola opens upward or downward.

5. $y = -3x + 4$
6. $y = -2x - 1$
7. $y = 4x - 3$
8. $y = 2x - 1$
9. $y = 8x^2$
10. $y = -4x^2$
11. $y = -3x^2 + 8x + 9$
12. $y = 2x^2 + 4x + 5$
13. $f(x) = 7$
14. $f(x) = -7$

In Exercises 15–20, use the given quadratic function.

a. Complete the table.
b. Plot these points on a graph.
c. Connect these points with a smooth parabolic curve.

x	y
−3	
−2	
−1	
0	
1	
2	
3	

15. $y = x^2 - 2$
16. $y = x^2 - 4$
17. $y = -x^2 + 3$
18. $y = -x^2 + 2$
19. $y = x^2 + x - 6$
20. $y = -x^2 + x + 5$

In Exercises 21–26, use the given graphs to determine the following key features of each parabola.

a. Is the parabola opening upward or downward?
b. Vertex
c. The y-intercept
d. The x-intercepts
e. Domain of f
f. Range of f
g. Interval where f is decreasing
h. Interval where f is increasing

21. 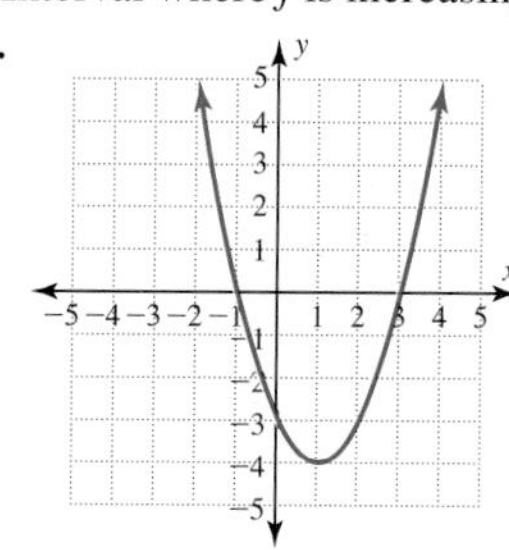

22. 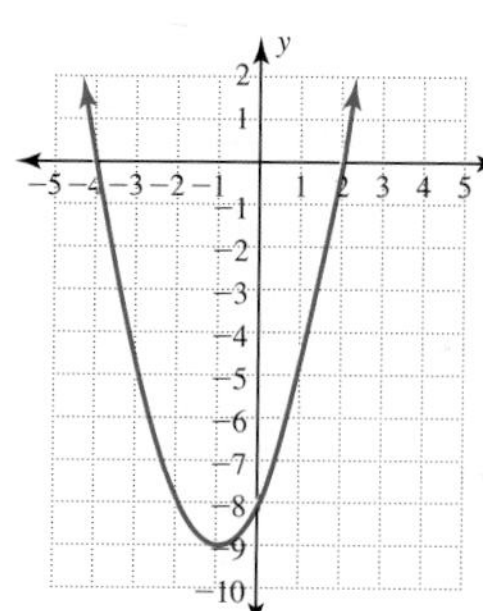

23. 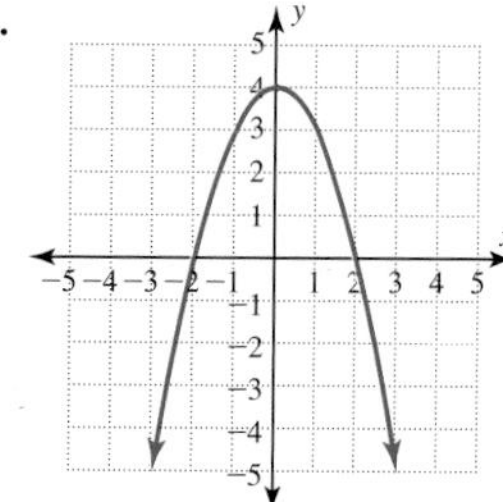

24. 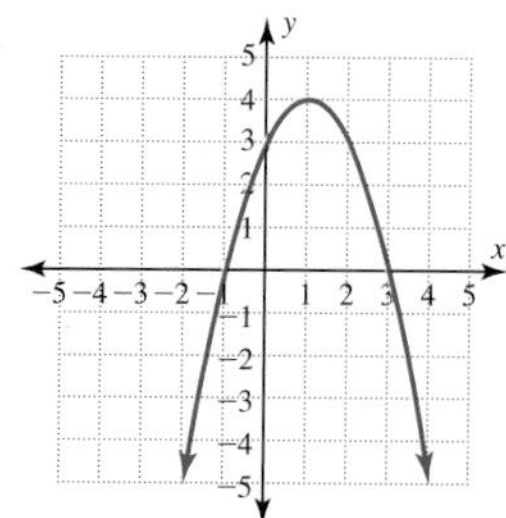

25. 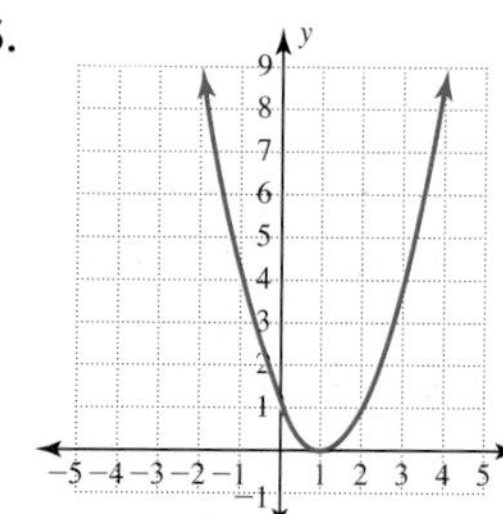

26. 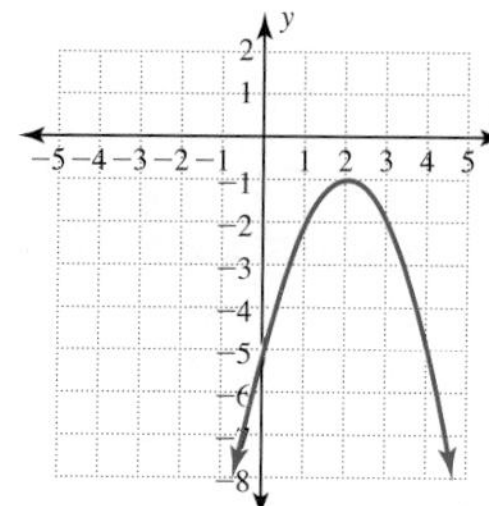

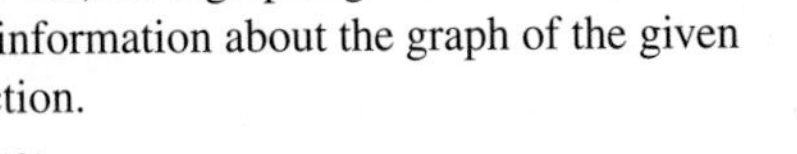

In Exercises 27–30, use a graphing calculator to determine the following information about the graph of the given quadratic function.

a. Evaluate $f(0)$.
b. Write the y-intercept of the parabola.
c. Determine the x-intercepts of the parabola.
d. Write the solutions of the equation formed by setting $f(x) = 0$.

27. $f(x) = x^2 - 16$ **28.** $f(x) = -x^2 + 25$
29. $f(x) = -x^2 - 6x - 5$ **30.** $f(x) = x^2 + 5x + 4$

In Exercises 31–34, match the given graph and table for a quadratic function with the corresponding information about the parabola defined by this function.

31. 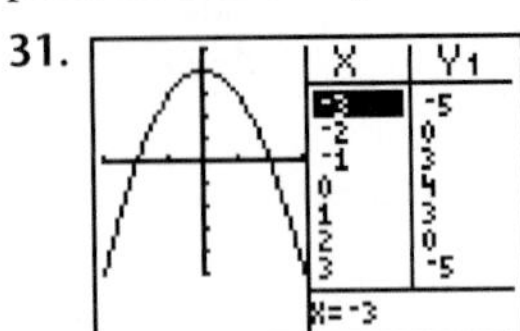

[−3, 3, 1] by [−5, 5, 1]

32. 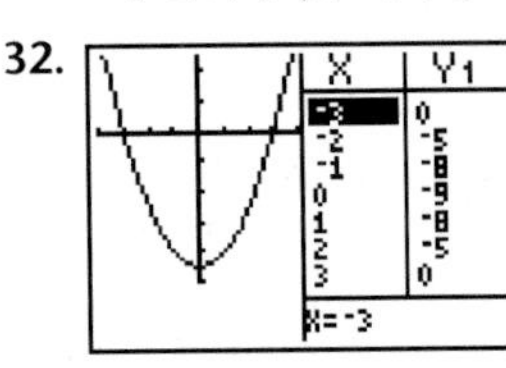

[−4, 4, 1] by [−10, 5, 2]

33. 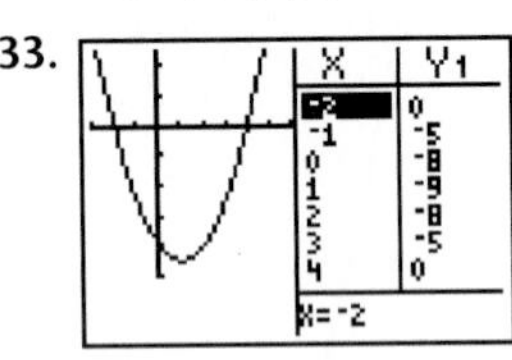

[−3, 6, 1] by [−10, 5, 2]

34. 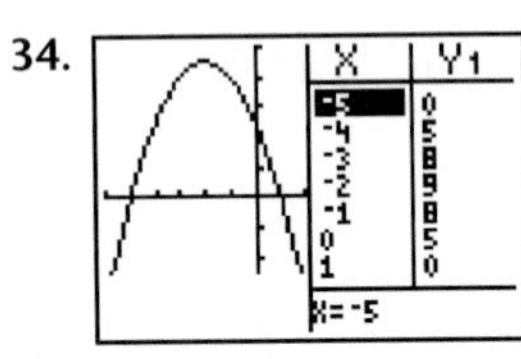

[−6, 2, 1] by [−5, 10, 2]

A. Vertex: $(0, -9)$
y-intercept: $(0, -9)$
x-intercepts: $(-3, 0)$, $(3, 0)$

B. Vertex: $(-2, 9)$
y-intercept: $(0, 5)$
x-intercepts: $(-5, 0)$, $(1, 0)$

C. Vertex: $(0, 4)$
y-intercept: $(0, 4)$
x-intercepts: $(-2, 0)$, $(2, 0)$

D. Vertex: $(1, -9)$
y-intercept: $(0, -8)$
x-intercepts: $(-2, 0)$, $(4, 0)$

In Exercises 35–38, the given graph and table for a quadratic function contain key points on the graph of the parabola. For each parabola determine the following.

a. The vertex
b. The y-intercept
c. The x-intercepts

35. 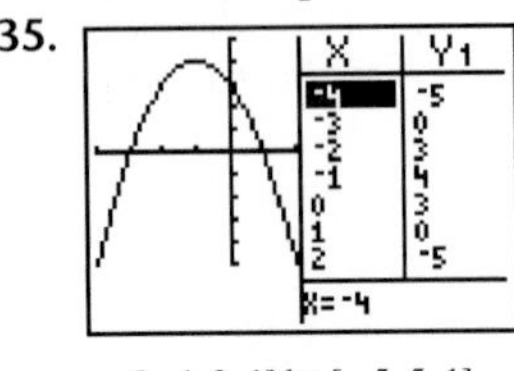

[−4, 2, 1] by [−5, 5, 1]

36. 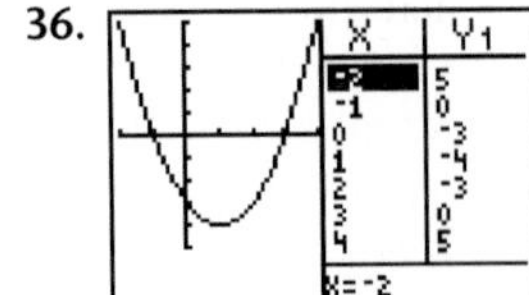

[−2, 4, 1] by [−5, 5, 1]

37. 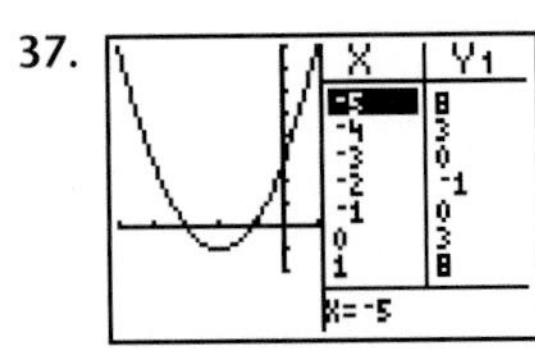

[−5, 1, 1] by [−2, 8, 1]

38. 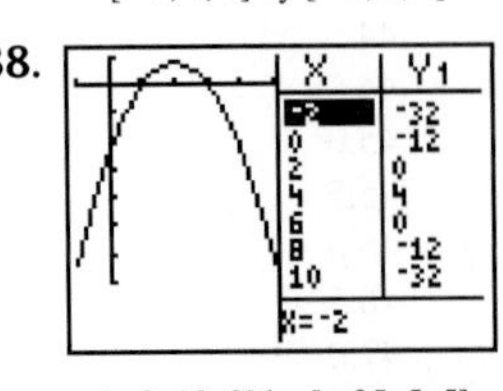

[−2, 10, 2] by [−35, 5, 5]

Break-even Values

In Exercises 39–42, $P(x)$ gives the profit in dollars when x units are produced and sold. Use the graph of the profit function to estimate the following.

a. Overhead costs [*Hint:* Evaluate $P(0)$.]
b. Break-even values [*Hint:* When does $P(x) = 0$?]
c. Maximum profit that can be made and the number of units to sell to create this profit

39. 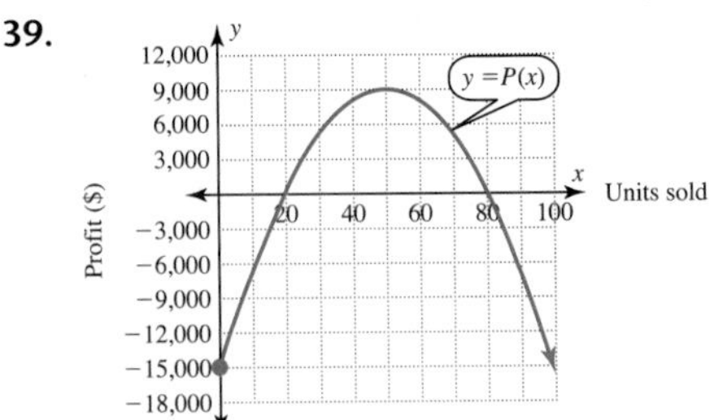

40. 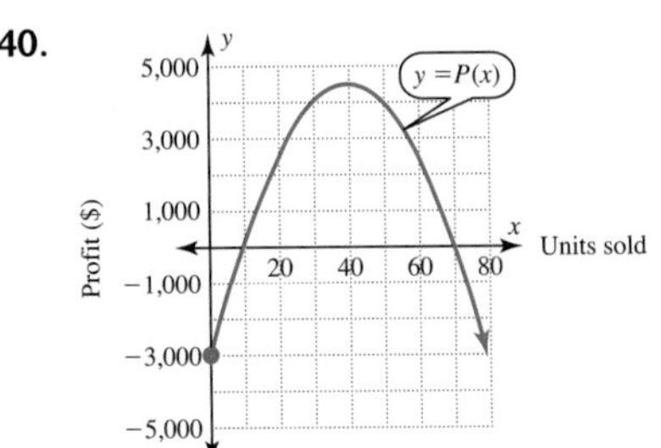

41. 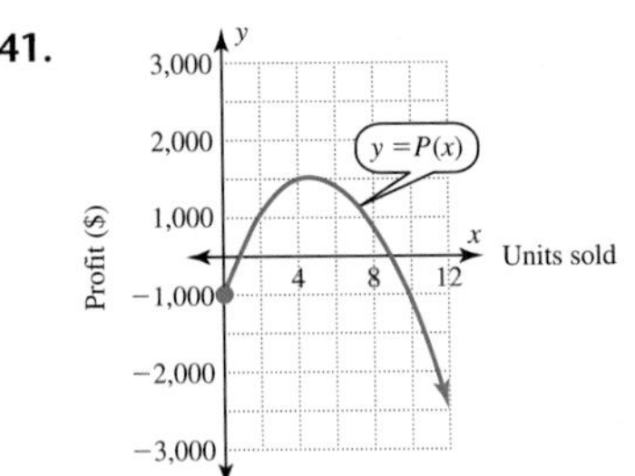

42. 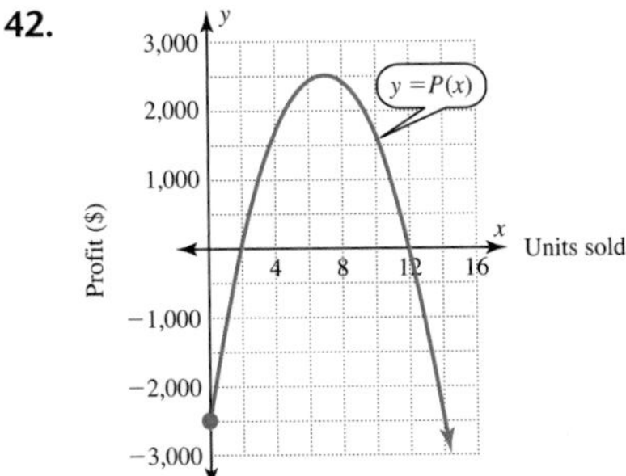

In Exercises 43–46, match each description of the graph of a function with the scatter diagram of the corresponding description. All graphs are displayed with the window [0, 10, 1] by [0, 10, 1].

43. A line with positive slope
44. A line with negative slope
45. A parabola opening upward
46. A parabola opening downward

A. 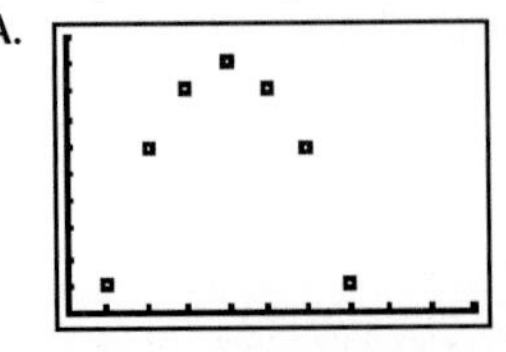

B. 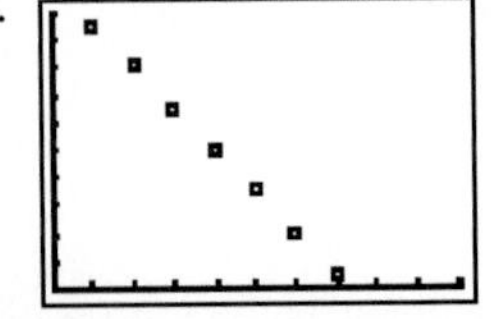

C.

D. 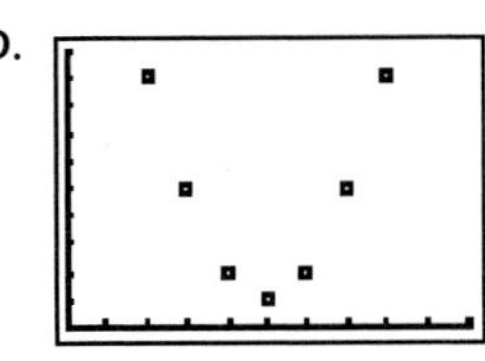

In Exercises 47–50, use a graphing calculator to draw a scatter diagram for each set of data points and then predict whether the line of best fit will have a positive slope, indicating an increasing function, or a negative slope, indicating a decreasing function.

47.

x	y
−3	4
−2	4
−1	3
0	3
1	2
2	2
3	1

48.

x	y
−3	16
−2	8
−1	0
0	6
1	−8
2	−8
3	−10

49.

x	y
1942	12.8
1950	13.9
1960	15.2
1970	15.7
1980	16.9
1990	16.5
2000	17.8

50.

x	y
53	41
65	40
72	45
78	51
83	50
94	58
98	59

In Exercises 51–54, write in slope-intercept form the equation of a line that satisfies the given conditions.

51. The line with slope $m = \frac{4}{7}$ and y-intercept (0, 3)

52. The line with slope $m = -\frac{3}{2}$ and y-intercept (0, 5)

53. The line that goes down 3 units for every one unit increase in x and passes through (0, −6)

54. The line that goes up 2 units for every one unit increase in x and passes through (0, −4)

In Exercises 55–58, use the point-slope form $y - y_1 = m(x - x_1)$ and the given information to write the equation of the line. Write each answer in slope-intercept form. (*Hint:* See Example 10 in Section 3.2.)

55. A line through (2, −4) with slope $-\frac{2}{3}$

56. A line through (−3, 5) with slope $\frac{4}{5}$

57. A line through (2, −3) and (−5, 1)

58. A line through (−1, 4) and (2, 2)

In Exercises 59–62, use a graphing calculator and the points given in the table.

a. Draw a scatter diagram of these points.

b. Determine the linear function that best fits these points. (Round the coefficients to the nearest thousandth.)

59.

x	y
1	3
2	5
3	8
4	9
5	11
6	12

c. Use this linear equation to estimate y when x is 3.5.

60.

x	y
10	18
15	16
19	12
26	9
29	6
34	2

c. Use this linear equation to estimate y when x is 22.

61.

x	y
−10	5
−7	1
−5	−3
−2	−7
0	−10
3	−14

c. Use this linear equation to estimate y when x is 1.

62.

x	y
1	4
4	7
9	8
12	11
15	15
19	19

c. Use this linear equation to estimate y when x is 10.

In Exercises 63–66, use a graphing calculator and the points given in the table.

a. Draw a scatter diagram of these points.

b. Determine the quadratic function that best fits these points. (Round the coefficients to three significant digits.)

63.

x	y
0	11
1	2
2	−4
3	−5
4	−3
5	1

c. Use this quadratic equation to estimate y when x is 2.5.

64.

x	y
2	5.3
3	3.1
5	1.0
6	1.4
7	2.9
8	5.5

c. Use this quadratic equation to estimate y when x is 4.

65.

x	y
−5	−18
−4	−12
−3	−9
−2	−8
−1	−9
0	−13

c. Use this quadratic equation to estimate y when x is 1.

66.

x	y
−5	−5.1
−3	−2.8
−1	−1.9
1	−2.6
3	−5.2
5	−9.1

c. Use this quadratic equation to estimate y when x is 6.

67. Effect of Smoking on Birth Weights This table compares the number of cigarettes smoked per day by eight expectant mothers to the birth weight of their children.

NUMBER OF CIGARETTES x	BIRTH WEIGHT y (kg)
5	3.20
10	3.15
15	3.10
20	3.06
25	3.09
30	3.05
35	3.01
40	3.00

a. Draw a scatter diagram for these data points, and determine the line of best fit. (Round the coefficients to the nearest thousandth.)

b. Interpret the meaning of the slope of the line of best fit.

c. Use this line of best fit to estimate the birth weight of a child whose mother smoked 18 cigarettes per day.

d. Use this line of best fit to estimate the birth weight of a child whose mother smoked 45 cigarettes per day.

68. Body Height Based on Femur Length Forensic scientists can estimate the height of a person based upon the length of different bones of that person. This table of values compares the length of a man's femur (the thigh bone) to the height of that man.

LENGTH OF FEMUR x (cm)	HEIGHT y (cm)
43.0	164
43.5	166
44.0	166
44.5	168
45.0	170
45.5	169
46.0	172
46.5	173
47.0	175
47.5	175

a. Draw a scatter diagram for these data points and determine the line of best fit. (Round the coefficients to the nearest thousandth.)
b. Interpret the meaning of the slope of the line of best fit.
c. Use the line of best fit from part **a** to estimate the height of a man whose femur length is 44.7 cm.
d. Use the line of best fit from part **a** to estimate the height of a man whose femur length is 48.5 cm.

69. **Olympic Men's Long-Jump Records** The given table displays the winning lengths for the Olympic men's long jump for selected years from 1900 through 2000. Each distance is given in meters. (Do not use the actual dates in your calculations. Use the x-values so that you can approximate the coefficients of the equation of best fit without introducing an unacceptable amount of error.)
a. Draw a scatter diagram for these data points and then determine the line of best fit. (Round the coefficients to the nearest thousandth.)
b. Interpret the meaning of the slope of the line of best fit.
c. Use the equation of the line of best fit to estimate the missing entry in the table from 1984. How does this estimation compare to the actual 1984 record of 8.54 m?
d. Use the equation of the line of best fit to predict the year in which the men's long-jump record will become 9.00 m. (Remember the Olympics are only held every 4 years.)

YEAR	x	y (m)
1900	0	7.18
1912	12	7.60
1920	20	7.15
1932	32	7.64
1948	48	7.82
1960	60	8.12
1972	72	8.24
1980	80	8.54
1996	96	8.50
2000	100	8.55

70. **Olympic Men's Marathon Records** The given table displays the winning times for the Olympic men's marathon for selected years from 1900 through 2000. Because each time is given as 2 hours and some number of minutes and seconds, the y-values in the table include only the time in seconds in excess of 2 hours. (Do not use the actual dates in your calculations. Use the x-values so that you can approximate the coefficients of the equation of best fit without introducing an unacceptable amount of error.)

YEAR	x	y SECONDS AFTER 2 HOURS
1900	0	3,585
1912	12	2,214
1920	20	1,956
1932	32	1,896
1948	48	2,091
1960	60	916
1972	72	740
1980	80	663
1996	96	756
2000	100	611

a. Draw a scatter diagram for these data points and then determine the line of best fit. (Round the coefficients to the nearest thousandth.)
b. Interpret the meaning of the slope of the line of best fit.
c. Use the equation of the line of best fit to estimate the missing entry in the table from 1984. How does this estimation compare to the actual 1984 record of 2 hours, 9 minutes, 21 seconds?
d. Use the equation of the line of best fit to predict the year in which the men's marathon record will become 2 hours. (Remember the Olympics are only held every 4 years.)

71. **Number of Women in the U.S. House** The given table displays the number of women in the U.S. House of Representatives for selected years from 1980 through 2000. (Do not use the actual dates in your calculations. Use the x-values so that you can approximate the coefficients of the equation of best fit without introducing an unacceptable amount of error.)

YEAR	x	WOMEN IN HOUSE y
1980	80	19
1982	82	21
1984	84	22
1986	86	23
1988	88	25
1990	90	28
1992	92	47
1994	94	47
1996	96	51
1998	98	56
2000	100	59
2002	102	59

a. Draw a scatter diagram for these data and then determine the line of best fit. (Round the coefficients to the nearest thousandth.)
b. Interpret the meaning of the slope of the line of best fit from part **a.**
c. Use the equation of the line of best fit to estimate the number of women Representatives in 1976. There were 18 women Representatives in 1976. How close was your estimate?
d. Use the equation of the line of best fit to predict the number of women Representatives in the year 2008.

72. **Profit Based on Sales** A new company recorded its sales units versus its profit (or loss) for eight consecutive quarters.

UNITS x SOLD IN HUNDREDS	PROFIT y IN THOUSANDS OF DOLLARS
0	−30
25	50
40	70
50	70
60	75
75	40
80	30
100	−60

a. Use a calculator to determine the parabola of best fit for these data.

b. Use this equation to estimate the profit in dollars generated by the sale of 7,000 units in a quarter. (*Hint:* Be careful with units.)

73. **Load Strength of a Beam** The following experimental data relate the maximum load in kilograms that a beam can support to its depth in centimeters.

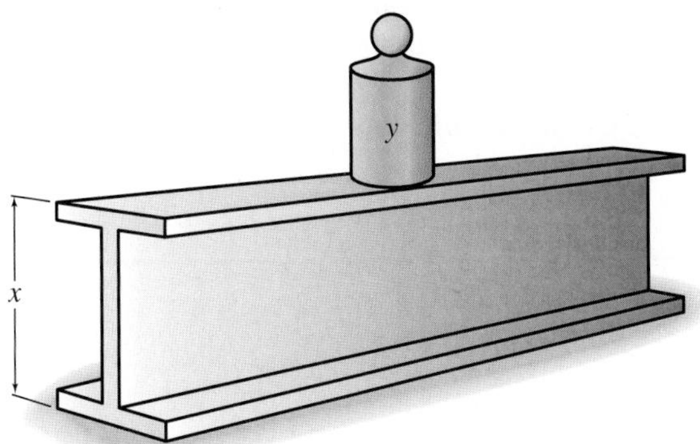

a. Use a calculator to determine the parabola of best fit for these data.

b. Use this quadratic equation to estimate the load that the beam can support if its depth is 15 cm.

DEPTH x (cm)	LOAD y (kg)
10	4,900
12	7,200
14	9,900
16	12,500
18	16,000
20	20,000
22	24,000
24	28,500

Group Discussion Questions

74. **Discovery Question**

a. Graph $y = x^2 + 1$, $y = x^2 + 3$, and $y = x^2 + 4$, and verbally compare these graphs to the graph of $y = x^2$.

b. On the basis of these observations, what would you predict about the graph of $y = x^2 + 6$ as compared to the graph of $y = x^2$?

75. **Discovery Question**

a. Graph $y = (x + 1)^2$, $y = (x + 3)^2$, and $y = (x + 4)^2$, and verbally compare these graphs to the graph of $y = x^2$.

b. On the basis of these observations, what would you predict about the graph of $y = (x + 6)^2$ as compared to the graph of $y = x^2$?

76. **Discovery Question** Compare the change in linear functions to the change in quadratic functions.

a. For a linear function such as $y = 2x - 3$, the change in y for each 1-unit change in x is constant. In this example this constant change in y is __________. This constant change represents the __________ of this line.

x	$y = 2x - 3$	Change in y
0	−3	—
1	−1	2
2	1	2
3	3	2
4	5	2
5	7	2
6	9	2

b. Before you complete this table, predict what each change in y will be for a 1-unit change in x. Then complete a table similar to the one in part **a** for $y = -3x + 1$. The constant change in y is __________.

x	$y = -3x + 1$	Change in y
0		—
1		
2		
3		
4		
5		
6		

c. For a quadratic function such as $y = (x - 4)^2$, the changes in y vary for each 1-unit change in x. The graph is not a line with a constant slope, but a parabola. Use the rightmost column in the table to determine the change in the changes for each 1-unit change in x.

x	$y = (x - 4)^2$	Change in y	Change in the Changes
0	16	—	—
1	9	−7	—
2	4	−5	2
3	1	−3	2
4	0	−1	2
5	1	1	2
6	4	3	2

d. Complete a table similar to the one in part **c** for $y = -x^2 + 5$. Use the rightmost column in the table to determine the change in the changes for each 1-unit change in x.

x	$y = -x^2 + 5$	Change in y	Change in the Changes
−3		—	—
−2			—
−1			
0			
1			
2			
3			

77. **Challenge Question** Collect a pair of data items from approximately 15 people of the same gender and display these data in a table. For example, measure on the same individual both the distance from the tip of an elbow to the end of an index finger and the length of the individual's right foot.

a. Before you collect these data, predict whether or not a line of best fit will be a good model.

b. Do you expect this line of best fit to have a positive or negative slope? Why?

c. Draw the scatter diagram, and calculate the line of best fit.

d. Did your predictions match the results you found?

Section 7.4 Using the Quadratic Formula to Find Real Solutions

Objectives:

11. Use the quadratic formula to solve quadratic equations with real solutions.
12. Use the discriminant to determine the nature of the solutions of a quadratic equation.

The x-intercepts of the graph of $y = ax^2 + bx + c$ can be determined by solving $ax^2 + bx + c = 0$. In applications, these x-intercepts often have an important meaning. For example, in the profit polynomial $P(x) = -50(x - 10)(x - 100)$, the x-intercepts $(10, 0)$ and $(100, 0)$ identify the break-even values for the number of units sold. Some simple quadratic equations with integer coefficients can be solved by factoring; but in most applications the equations are not this simple, and it is useful to have a formula to solve these equations. We now develop the quadratic formula for this purpose. Although we can use the quadratic formula to solve any quadratic equation, we limit our discussion in this section to equations with solutions that are real numbers. We will consider complex solutions with imaginary parts in Section 7.7.

Extraction of Roots

The notation $x = \pm\sqrt{k}$ is read "x equals plus or minus the square root of k." The notation $\sqrt{k}$ represents the positive square root of k (called the principal root), and $-\sqrt{k}$ represents the negative square root of k.

As a first step in developing the quadratic formula, we examine quadratic equations of the form $x^2 = k$. The solutions of $x^2 = k$ are $x = \sqrt{k}$ and $x = -\sqrt{k}$. To denote both possible square roots of k, we write $x = \pm\sqrt{k}$. The notation $\pm$ is read "plus or minus." Solving $x^2 = k$ as $x = \pm\sqrt{k}$ is called solving by **extraction of roots.** Example 1 uses this method to solve four different quadratic equations.

EXAMPLE 1 Solving Quadratic Equations by Extraction of Roots

Solve each of these quadratic equations by extraction of roots.

SOLUTION

(a) $x^2 = 9$

$$x^2 = 9$$
$$x = \pm\sqrt{9}$$
$$x = -3 \quad \text{or} \quad x = 3$$

Solve for both solutions of the quadratic equation by extracting both the positive and the negative square roots of 9. The two solutions are -3 and 3.

(b) $x^2 = 2$

$$x^2 = 2$$
$$x = \pm\sqrt{2}$$
$$x = -\sqrt{2} \quad \text{or} \quad x = \sqrt{2}$$

Both real solutions of this quadratic equation, $-\sqrt{2}$ and $\sqrt{2}$, are irrational numbers. These two solutions are exact. Approximate these solutions with a calculator if approximations are needed.

(c) $(x - 1)^2 = 3$

$$(x - 1)^2 = 3$$
$$x - 1 = \pm\sqrt{3}$$
$$x = 1 \pm \sqrt{3}$$
$$x = 1 - \sqrt{3} \quad \text{or} \quad x = 1 + \sqrt{3}$$

Start by extracting both square roots of 3. Then add 1 to both sides of the equation. These two irrational solutions are the exact solutions.

(d) $(x - 3)^2 = 12$

$$(x - 3)^2 = 12$$
$$x - 3 = \pm\sqrt{12}$$
$$x = 3 - \sqrt{12} \quad \text{or} \quad x = 3 + \sqrt{12}$$
$$x = 3 - 2\sqrt{3} \qquad\quad x = 3 + 2\sqrt{3}$$

Start by extracting both square roots of 12. Then add 3 to both sides of the equation.
Simplify $\sqrt{12}$: $\sqrt{12} = \sqrt{4 \cdot 3} = \sqrt{4}\,\sqrt{3} = 2\sqrt{3}$. (For more on this, see the following paragraph.) These two irrational solutions are the exact solutions. ■

Note in Example 1(d) that the radical $\sqrt{12}$ can be simplified to $2\sqrt{3}$. This simplification uses the fact from Section 1.5 that for $a > 0$ and $b > 0$, $\sqrt{a \cdot b} = \sqrt{a}\sqrt{b}$. Thus $\sqrt{12} = \sqrt{4 \cdot 3} = \sqrt{4}\sqrt{3} = 2\sqrt{3}$. Before the advent of calculators, algebra texts spent considerable time in simplifying radical expressions. Although this material still has value in simplifying radical expressions in calculus and other areas, we limit our discussion of simplifying radicals in this chapter. Instead we assume that we can approximate any exact solutions satisfactorily with a graphing calculator.

SELF-CHECK 7.4.1

Solve each of these quadratic equations by extraction of roots.

1. $x^2 = 16$ 2. $x^2 = 17$ 3. $(x + 2)^2 = 5$

To use extraction of roots to solve a quadratic equation $ax^2 + bx + c = 0$, we need to have the left side of the equation written as a perfect square. Using the process of completing the square, we can always rewrite the left side of $ax^2 + bx + c = 0$ as a perfect square.

A Mathematical Note

Many methods that we now use quickly and efficiently in algebraic form have their roots in ancient geometric methods. In particular, the method of completing the square was used by both Greek and Arab mathematicians in the geometric form shown in the following geometric viewpoint. Al-Khowârizmi (c. 825) was a noted Arab mathematician, astronomer, and author who illustrated this method in his writings.

Completing the Square: A Geometric Viewpoint

The left side of a quadratic equation can always be written as a perfect square. The process of writing the left side of the equation in this form is called **completing the square.** The term *completing the square* has a very natural geometric basis that is illustrated in the following geometric viewpoint. In this box, we examine $x^2 + 6x$. This expression will be the left side of the equation that we examine in Example 2(a).

COMPLETING THE SQUARE FOR $x^2 + 6x$

ALGEBRAICALLY	GEOMETRICALLY
1. Start with: $x^2 + 6x$	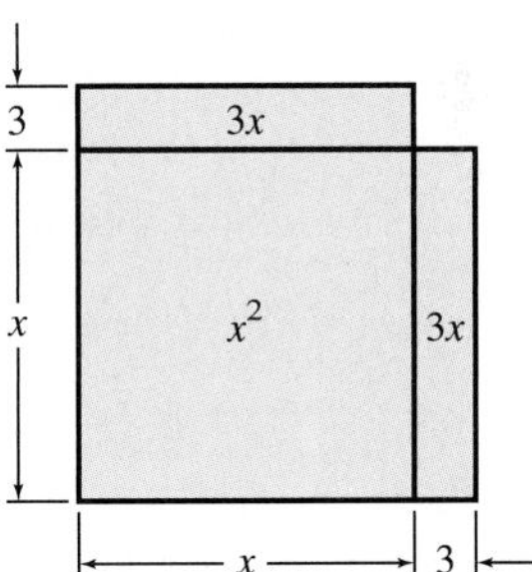
2. Add 9: $x^2 + 6x + 9$	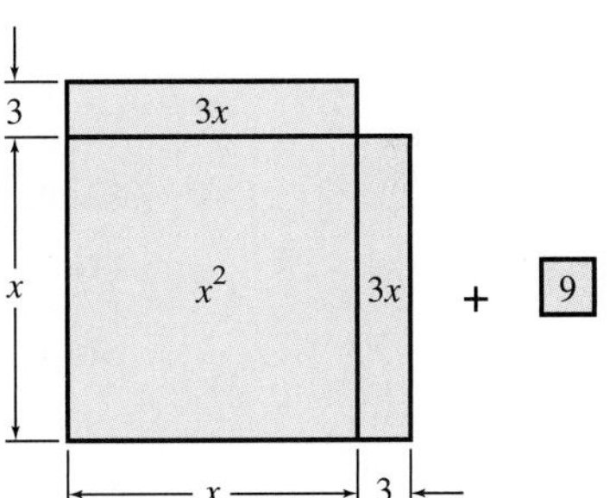
3. Write as a perfect square: $x^2 + 6x + 9 = (x + 3)^2$	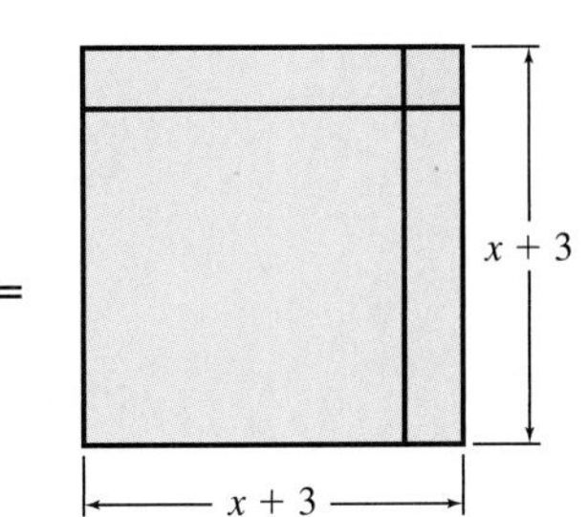

The expanded form of the square $(x + a)^2$ is $x^2 + 2ax + a^2$. To complete the square of the expression $x^2 + 2ax$, we must add a^2. In effect we must add the square of one-half the coefficient of x. This is illustrated in each part of Example 2. We are using the method of completing the square to put the left side of a quadratic equation in a form that allows us to use the method of extraction of roots.

EXAMPLE 2 Solving Quadratic Equations by Completing the Square

Rewrite each equation so that the left side of the equation is a perfect square. Then solve this equation by extraction of roots.

SOLUTION

(a) $x^2 + 6x = -5$

$$\begin{aligned} x^2 + 6x &= -5 \\ x^2 + 6x + 9 &= -5 + 9 \\ (x+3)^2 &= 4 \\ x + 3 &= \pm 2 \\ x &= -3 \pm 2 \end{aligned}$$

$x = -3 - 2$ or $x = -3 + 2$

$x = -5$ $\quad$ $x = -1$

The constant we add to both sides of the equation is the square of one-half the coefficient of x: $\left(\frac{6}{2}\right)^2 = 3^2 = 9$. After you write the left side of the equation as a perfect square, solve the equation by extraction of roots. Do these solutions check?

(b) $x^2 - 8x = -11$

$$\begin{aligned} x^2 - 8x &= -11 \\ x^2 - 8x + 16 &= -11 + 16 \\ (x-4)^2 &= 5 \\ x - 4 &= \pm\sqrt{5} \\ x &= 4 \pm \sqrt{5} \end{aligned}$$

$x = 4 - \sqrt{5}$ or $x = 4 + \sqrt{5}$

$x \approx 1.76$ $\quad$ $x \approx 6.24$

The constant we add to both sides of the equation is the square of one-half the coefficient of x: $\left(\frac{-8}{2}\right)^2 = (-4)^2 = 16$. After you write the left side of the equation as a perfect square, solve the equation by extraction of roots. The exact solutions have then been approximated to the nearest hundredth with a calculator.

SELF-CHECK 7.4.2

Rewrite each equation so that the left side is written as a perfect square.

1. $x^2 + 2x = 4$ **2.** $x^2 - 10x = -17$ **3.** $x^2 + 4x = -3$ **4.** $x^2 + 5x = -5$

The process of completing the square can be used to solve any quadratic equation. The key steps in this process are shown in the following box.

Solving Quadratic Equations by Completing the Square

VERBALLY	ALGEBRAIC EXAMPLE
Step 1 Write the equation with the constant term on the right side.	$x^2 + 3x - 4 = 0$ $x^2 + 3x = 4$
Step 2 Divide both sides of the equation by the coefficient of x^2 to obtain a coefficient of 1 for x^2.	$x^2 + 3x + \left(\frac{3}{2}\right)^2 = 4 + \left(\frac{3}{2}\right)^2$
Step 3 Take one-half of the coefficient of x, square this number, and add the result to both sides of the equation.	$\left(x + \frac{3}{2}\right)^2 = \frac{16}{4} + \frac{9}{4}$
Step 4 Write the left side of the equation as a perfect square.	$\left(x + \frac{3}{2}\right)^2 = \frac{25}{4}$ $x + \frac{3}{2} = \pm\frac{5}{2}$
Step 5 Solve this equation by extraction of roots.	$x = -\frac{3}{2} \pm \frac{5}{2}$ $x = -\frac{3}{2} - \frac{5}{2}$ or $x = -\frac{3}{2} + \frac{5}{2}$ $x = -4$ $\quad$ $x = 1$

The steps outlined in this box will be used to solve the equation in Example 3. Then we use this procedure to create a formula that can be used to solve any quadratic equation.

EXAMPLE 3 Using the Method of Completing the Square to Determine Two Irrational Solutions

Solve $4x^2 - 20x + 7 = 0$ by completing the square.

SOLUTION

$$4x^2 - 20x + 7 = 0$$

$$4x^2 - 20x = -7$$ Shift the constant to the right side of the equation.

$$x^2 - 5x = -\frac{7}{4}$$ Divide both sides of the equation by the coefficient of x^2, which is 4.

$$x^2 - 5x + \left(-\frac{5}{2}\right)^2 = -\frac{7}{4} + \left(-\frac{5}{2}\right)^2$$ Take one-half of the coefficient of x: $\frac{1}{2}(-5) = -\frac{5}{2}$.

$$\left(x - \frac{5}{2}\right)^2 = -\frac{7}{4} + \frac{25}{4}$$ Square this number: $\left(-\frac{5}{2}\right)^2 = \frac{25}{4}$. Then add this result to both sides of the equation. Write the left side as a perfect square.

$$\left(x - \frac{5}{2}\right)^2 = \frac{9}{2}$$ Extract the roots.

$$x - \frac{5}{2} = \pm\sqrt{\frac{9}{2}}$$ Add $\frac{5}{2}$ to both sides of the equation.

$$x = \frac{5}{2} \pm \frac{3\sqrt{2}}{2}$$ Simplify, noting that $\sqrt{\frac{9}{2}} = \frac{3}{\sqrt{2}} \cdot \frac{\sqrt{2}}{\sqrt{2}} = \frac{3\sqrt{2}}{2}$.

Exact answers: $x = \frac{5 - 3\sqrt{2}}{2}, x = \frac{5 + 3\sqrt{2}}{2}$ The exact answers can be approximated with a calculator to the desired accuracy.

Approximate answers: $x \approx 0.38, x \approx 4.62$ These answers have been approximated to the nearest hundredth.

SELF-CHECK 7.4.3

1. Rewrite $x^2 + 2x = 8$ so that the left side of the equation is a perfect square.
2. Solve this equation by extraction of roots.

Now we use the method of completing the square to develop the quadratic formula.

The Quadratic Formula

We could continue to solve each individual quadratic equation by completing the square. Rather than repeat these steps for each problem, we can solve the general quadratic equation $ax^2 + bx + c = 0$ by completing the square and then use this general solution as a formula that can be applied to any quadratic equation.

$$ax^2 + bx + c = 0$$ Start with a quadratic equation in standard form with a, b, and c real numbers and $a > 0$.

$$ax^2 + bx = -c$$ Shift the constant to the right side of the equation.

$$x^2 + \frac{b}{a}x = -\frac{c}{a}$$ Divide both sides by the coefficient of x^2, which is a. To divide by a, we assume $a \neq 0$.

$$x^2 + \frac{b}{a}x + \left(\frac{b}{2a}\right)^2 = -\frac{c}{a} + \left(\frac{b}{2a}\right)^2$$ Add the square of one-half the coefficient of x to both sides of the equation.

$$\left(x + \frac{b}{2a}\right)^2 = -\frac{4ac}{4a^2} + \frac{b^2}{4a^2}$$ Write the left side as a perfect square, and the right side in terms of a common denominator.

$$\left(x + \frac{b}{2a}\right)^2 = \frac{b^2 - 4ac}{4a^2}$$ Simplify by combining the terms on the right side of the equation.

$$x + \frac{b}{2a} = \pm\sqrt{\frac{b^2 - 4ac}{4a^2}}$$ Extract the roots.

$$x = -\frac{b}{2a} \pm \frac{\sqrt{b^2 - 4ac}}{2a}$$ Simplify the radical.

$$x = \frac{-b \pm \sqrt{b^2 - 4ac}}{2a}$$ This is the quadratic formula.

The quadratic formula can be rewritten as $x = \dfrac{-b + \sqrt{b^2 - 4ac}}{2a}$ or $x = \dfrac{-b - \sqrt{b^2 - 4ac}}{2a}$.

The quadratic formula is given in the following box. You should memorize this formula. Note that the symbol $\pm$ is used to express the two solutions concisely.

If $a = 0$, the equation $ax^2 + bx + c = 0$ becomes the linear equation $bx + c = 0$.

Quadratic Formula

The solutions of the quadratic equation $ax^2 + bx + c = 0$ with real coefficients a, b, and c, when $a \neq 0$, are

$$x = \frac{-b \pm \sqrt{b^2 - 4ac}}{2a}$$

EXAMPLE 4 Using the Quadratic Formula to Determine Two Rational Solutions

Use the quadratic formula to solve $x^2 - 5x - 6 = 0$ and to determine the x-intercepts of the graph of $y = x^2 - 5x - 6$.

SOLUTION

ALGEBRAICALLY

$$x^2 - 5x - 6 = 0$$

$$x = \frac{-(-5) \pm \sqrt{(-5)^2 - 4(1)(-6)}}{2(1)}$$ Substitute $a = 1$, $b = -5$, and $c = -6$ into $x = \dfrac{-b \pm \sqrt{b^2 - 4ac}}{2a}$.

$$x = \frac{5 \pm \sqrt{25 + 24}}{2}$$ Simplify, and find both solutions.

$$x = \frac{5 \pm \sqrt{49}}{2}$$

$$x = \frac{5 \pm 7}{2}$$

$$x = \frac{5 - 7}{2} \quad \text{or} \quad x = \frac{5 + 7}{2}$$

$$x = -1 \qquad\qquad x = 6$$

GRAPHICALLY AND NUMERICALLY

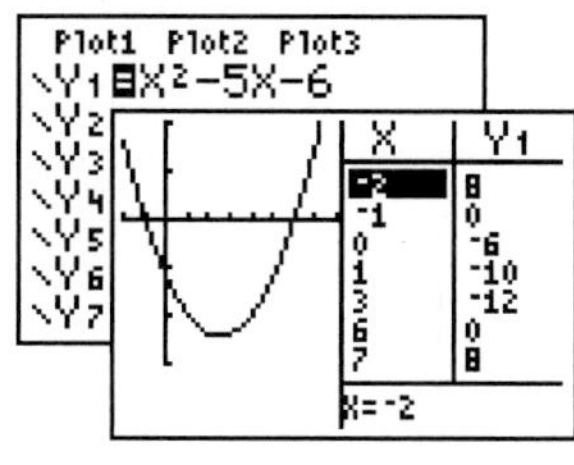

$[-2, 8, 1]$ by $[-15, 10, 5]$

The x-intercepts of the parabola defined by $y = x^2 - 5x - 6$ correspond to the solutions of $x^2 - 5x - 6 = 0$. The graph and the table show that the x-intercepts are $(-1, 0)$ and $(6, 0)$. These x-intercepts confirm that the solutions of $x^2 - 5x - 6 = 0$ are $x = -1$ and $x = 6$.

Answer: The solutions are $x = -1$ and $x = 6$. The x-intercepts of the graph are $(-1, 0)$ and $(6, 0)$.

Note that we also could have solved the equation in Example 4 by factoring it as $(x + 1)(x - 6) = 0$. The primary advantage of the quadratic formula is that it can be used to solve problems that cannot be solved by factoring over the integers. This is illustrated in Example 5.

EXAMPLE 5 Using the Quadratic Formula to Determine Two Irrational Solutions

Before you use the quadratic formula, be sure to write the equation in the standard form $ax^2 + bx + c = 0$. This will help prevent making an error in the sign of a, b, or c.

Use the quadratic formula to determine the exact solutions of $4x^2 - 4x = 1$, and approximate these solutions to the nearest hundredth. Then solve this equation graphically.

SOLUTION

ALGEBRAICALLY

$$4x^2 - 4x = 1$$
$$4x^2 - 4x - 1 = 0$$

First write the equation in standard form.

$$x = \frac{-(-4) \pm \sqrt{(-4)^2 - 4(4)(-1)}}{2(4)}$$

Substitute $a = 4$, $b = -4$, and $c = -1$ into the quadratic formula.

$$x = \frac{4 \pm \sqrt{16 + 16}}{8}$$
$$x = \frac{4 \pm \sqrt{32}}{8}$$
$$x = \frac{4 \pm 4\sqrt{2}}{8}$$

Note that $\sqrt{32} = \sqrt{16} \cdot \sqrt{2} = 4\sqrt{2}$.

$$x = \frac{4(1 \pm \sqrt{2})}{8}$$
$$x = \frac{1 \pm \sqrt{2}}{2}$$

Simplify, and write both solutions separately.

Exact solutions: $x = \frac{1 - \sqrt{2}}{2}, x = \frac{1 + \sqrt{2}}{2}$

Approximate solutions: $x \approx -0.21, x \approx 1.21$

The exact solutions are approximated to the nearest hundredth with a calculator.

GRAPHICALLY

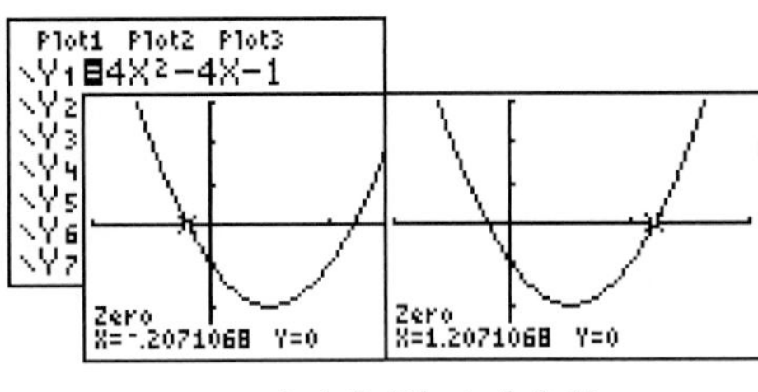

$[-1, 2, 1]$ by $[-3, 3, 1]$

At the x-intercepts of $y = 4x^2 - 4x - 1$, $y = 0$. Examine the x-intercepts of this graph and solve $4x^2 - 4x - 1 = 0$.
To determine the x-intercepts, use the **Zero** feature in the **CALC** menu.
These x-intercepts confirm that the solutions of $4x^2 - 4x = 1$ are $x \approx -0.21$ and $x \approx 1.21$.

Approximate solutions: $x \approx -0.21$, $x \approx 1.21$

SELF-CHECK 7.4.4

1. Use the quadratic formula to determine the exact solutions of $x^2 - 2x = 5$.
2. Approximate these solutions to the nearest hundredth.

The x-intercepts of a graph have a y-coordinate of 0. Example 5 illustrated how we can examine the x-intercepts to confirm the solutions of $4x^2 - 4x - 1 = 0$. We can reverse the roles of these two related pieces of information and use the solutions of $ax^2 + bx + c = 0$ to determine the exact x-intercepts of the graph of $y = ax^2 + bx + c$. We start by setting $y = 0$ and then solving $ax^2 + bx + c = 0$. This is illustrated in Example 6.

EXAMPLE 6 Determining the Exact Intercepts of a Parabola

Algebraically determine the exact y-intercept and x-intercepts of the parabola defined by $y = x^2 - 4x + 1$. Then examine these intercepts graphically and numerically.

SOLUTION

ALGEBRAICALLY

To find the y-intercept, set $x = 0$ and determine y.

$$y = x^2 - 4x + 1$$
$$y = (0)^2 - 4(0) + 1$$
$$y = 1$$

The graph crosses the y-axis if the x-coordinate is 0.

y-intercept: $(0, 1)$

To find the x-intercepts, set $y = 0$ and then solve for x.

The graph crosses the x-axis if the y-coordinate is zero.

$0 = x^2 - 4x + 1$ or $x^2 - 4x + 1 = 0$

$$x = \frac{-(-4) \pm \sqrt{(-4)^2 - 4(1)(1)}}{2(1)}$$
$$x = \frac{4 \pm \sqrt{16 - 4}}{2}$$
$$x = \frac{4 \pm \sqrt{12}}{2}$$
$$x = \frac{4 \pm 2\sqrt{3}}{2}$$
$$x = 2 \pm \sqrt{3}$$

Use the quadratic formula $x = \frac{-b \pm \sqrt{b^2 - 4ac}}{2a}$ to solve $x^2 - 4x + 1 = 0$. In this equation $a = 1, b = -4$, and $c = 1$.

$\sqrt{12} = \sqrt{4}\sqrt{3} = 2\sqrt{3}$

Exact x-intercepts: $(2 - \sqrt{3}, 0)$ and $(2 + \sqrt{3}, 0)$
Approximate x-intercepts: $(0.3, 0)$ and $(3.7, 0)$

The exact values are approximated to the nearest tenth with a calculator.

GRAPHICALLY

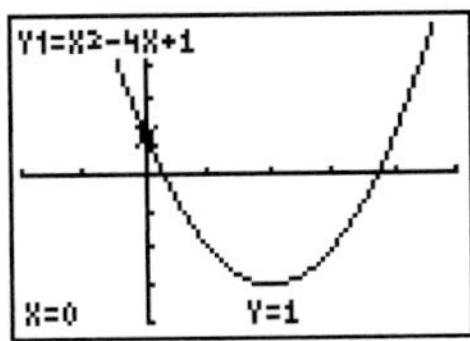

$[-2, 5, 1]$ by $[-4, 4, 1]$

Start by graphing $y = x^2 - 4x + 1$ on an appropriate viewing window. Each of the screen shots uses the window $[-2, 5, 1]$ by $[-4, 4, 1]$.

To determine the y-intercept, use the **TRACE** feature and enter 0 for x.

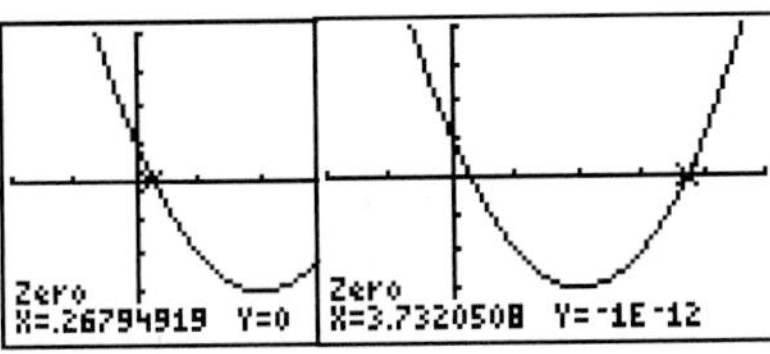

To determine the x-intercepts use the **zero** feature in the **CALC** menu to calculate the x-values for which y is zero.

Note that $-1\text{E} - 12$ represents -1×10^{-12}. This is a number in scientific notation that is very close to zero.

y-intercept: $(0, 1)$
Approximate x-intercepts: $(0.3, 0)$ and $(3.7, 0)$

NUMERICALLY

X	Y1
0	1
1	-2
2	-3
3	-2
4	1
5	6
6	13

X=0

y-intercept: $(0, 1)$
Approximate x-intercepts: $(0.3, 0)$ and $(3.7, 0)$

The y-intercept is found by looking for the point with an x-coordinate of 0.

The x-intercepts have a y-coordinate of 0. Although these values are not displayed in this table, note that the y-coordinate changes sign between 0 and 1. This is consistent with the x-intercept of $(0.3, 0)$. Also note that the y-coordinate changes sign between 3 and 4. This is consistent with the x-intercept of $(3.7, 0)$.

Examine tables with more values between 0 and 1 or 3 and 4 to further refine estimates of the x-intercepts. ■

SELF-CHECK 7.4.5

1. Determine the exact intercepts of the parabola defined by $y = 2x^2 - x - 2$.
2. Approximate the x-intercepts to the nearest tenth.

A Mathematical Note

James Joseph Sylvester (1814–1897) was born in England as James Joseph. He changed his last name to Sylvester when he moved to the United States. At Johns Hopkins University he led efforts to establish graduate work in mathematics in the United States. He also founded the *American Journal of Mathematics.* Among his lasting contributions to mathematics are the many new terms he introduced, including the term *discriminant.*

The Discriminant

Every quadratic equation has either two distinct solutions or a double real solution. The two distinct solutions may be either real numbers or complex numbers with imaginary parts (see Section 7.7). We can determine the nature of the solutions algebraically by examining the radicand $b^2 - 4ac$ of the quadratic formula

$$x = \frac{-b \pm \sqrt{b^2 - 4ac}}{2a}$$

Because $b^2 - 4ac$ can be used to discriminate between real solutions and complex solutions with imaginary parts, it is called the **discriminant.** Notice that we also can use the parabola defined by $y = ax^2 + bx + c$ to determine the nature of the solutions of $ax^2 + bx + c = 0$.

The Nature of the Solutions of a Quadratic Equation

There are three possibilities for the solutions of $ax^2 + bx + c = 0$.

VALUE OF THE DISCRIMINANT	SOLUTIONS OF $ax^2 + bx + c = 0$	THE PARABOLA $y = ax^2 + bx + c$	GRAPHICAL EXAMPLE
1. $b^2 - 4ac > 0$	Two distinct real solutions	Two x-intercepts	$y = x^2 - x - 2$
2. $b^2 - 4ac = 0$	A double real solution	One x-intercept with the vertex on the x-axis	$y = x^2 - 4x + 4$
3. $b^2 - 4ac < 0$	Neither solution is real; both solutions are complex numbers with imaginary parts. These solutions will be complex conjugates. (Complex numbers are covered in Section 7.7.)	No x-intercepts	$y = x^2 + 1$

EXAMPLE 7 Determining the Nature of the Solutions of a Quadratic Equation

Use the discriminant to determine the nature of the solutions of each of these quadratic equations.

SOLUTION

(a) $x^2 - 6x + 8 = 0$

$$\begin{aligned} b^2 - 4ac &= (-6)^2 - 4(1)(8) \\ &= 36 - 32 \\ &= 4 \end{aligned}$$

There are two distinct real solutions.

Substitute $a = 1$, $b = -6$, and $c = 8$ into the discriminant. The discriminant is greater than 0, so there are two real solutions as revealed by the two x-intercepts shown in the following graph.

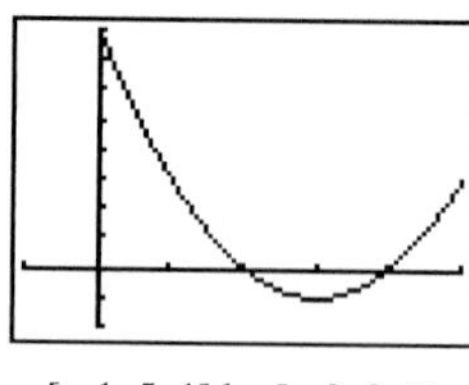

[−1, 5, 1] by [−2, 8, 1]

(b) $x^2 - 6x + 9 = 0$

$$\begin{aligned} b^2 - 4ac &= (-6)^2 - 4(1)(9) \\ &= 36 - 36 \\ &= 0 \end{aligned}$$

There is a double real solution.

Substitute $a = 1$, $b = -6$, and $c = 9$ into the discriminant. The discriminant is 0, so there is a double real solution as revealed in the following graph by the point of tangency on the x-axis.

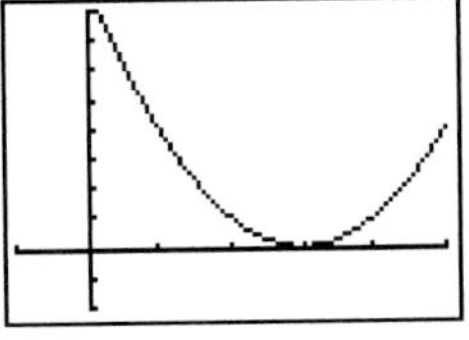

[−1, 5, 1] by [−2, 8, 1]

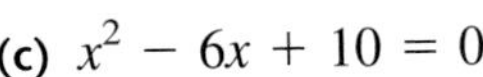

(c) $x^2 - 6x + 10 = 0$

$$\begin{aligned} b^2 - 4ac &= (-6)^2 - 4(1)(10) \\ &= 36 - 40 \\ &= -4 \end{aligned}$$

The solutions are complex numbers with imaginary parts.

Substitute $a = 1$, $b = -6$, and $c = 10$ into the discriminant. The discriminant is negative, so the two solutions are complex numbers with imaginary parts. The following graph has no x-intercepts and reveals that there are no real values for which $x^2 - 6x + 10 = 0$.

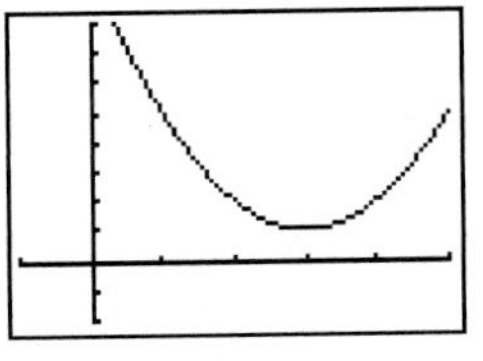

[−1, 5, 1] by [−2, 8, 1]

SELF-CHECK 7.4.6

Use the discriminant to determine the nature of the solutions of each equation.

1. $4x^2 = 8x - 4$ **2.** $4x^2 = 8x - 5$ **3.** $x^2 + x = 1$

We now revisit a problem similar to Example 9 from Section 6.6. This time we use the quadratic formula to determine the dimensions of the rectangle.

EXAMPLE 8 Determining the Dimensions of a Rectangle

An archeologist used 30 m of rope to enclose a rectangular area of 55 m². Find the dimensions of this rectangular area.

w 55 m^2

SOLUTION

Let $w =$ width of rectangle in meters
$15 - w =$ length of rectangle in meters

The dimensions are the width and length. Because the perimeter is 30 m, $2w + 2l = 30$ and $l = 15 - w$.

VERBALLY

Area equals fifty-five square meters.
(Width)(length) $= 55$

The area of a rectangle is given by $A = w \cdot l$.

ALGEBRAICALLY

$$\begin{aligned} w(15 - w) &= 55 \\ 15w - w^2 &= 55 \end{aligned}$$

Substitute the expressions identified above for the width and length.

$$0 = w^2 - 15w + 55$$
$$w^2 - 15w + 55 = 0$$

Write this quadratic equation in standard form.

$$w = \frac{-(-15) \pm \sqrt{(-15)^2 - 4(1)(55)}}{2(1)}$$

Solve this quadratic equation by using the quadratic formula.

$$w = \frac{15 \pm \sqrt{5}}{2}$$

$$w = \frac{15 - \sqrt{5}}{2} \quad \text{or} \quad w = \frac{15 + \sqrt{5}}{2}$$

$$w \approx 6.38 \qquad w \approx 8.62$$

$$15 - w \approx 8.62 \qquad 15 - w \approx 6.38$$

Either the width is 6.38 m and the length is 8.62 m, or the width is 8.62 m and the length is 6.38 m.

Answer: The approximate dimensions of this rectangular region are 6.38 m by 8.62 m.

Do these dimensions check both for the perimeter and for the area?

Summary

We have now covered a variety of methods for solving quadratic equations. The following box lists some of the advantages and disadvantages of each method. Graphical methods have the advantage of helping us "see" the solutions but often lack some of the precision and speed of algebraic methods. Other than for very simple quadratic equations, the algebraic approach that we recommend is the quadratic formula.

Most business applications and many in the social sciences produce quadratic equations with only real solutions. Applications with solutions that are complex numbers with imaginary parts are more likely to be found in the sciences. The quadratic formula will also be used to solve these equations in Section 7.7.

METHODS FOR SOLVING $ax^2 + bx + c = 0$	ADVANTAGES	DISADVANTAGES
Graphs	Visually displays the x-intercepts that correspond to real solutions. If there are no x-intercepts, this indicates the solutions are imaginary.	It can be time-consuming to select the appropriate viewing window and to approximate the real solutions. This method cannot determine complex solutions with imaginary parts.
Tables	Tables allow us to observe numerical patterns that are important in many applications. Spreadsheets allow us to use the power of computers to exploit this method.	Often it requires insight or some trial and error to select the most appropriate table.
Factoring	Factoring is easiest to use when the coefficients of a, b, and c are small integers. Factoring also makes it easy to identify the x-intercepts of the corresponding parabola.	Most quadratic equations that arise from applications do not have small integers for the coefficients a, b, and c. This method works best only for real solutions that are rational.
Extraction of roots	This method is very easy to use if the left side of the quadratic equation is written as a perfect square.	If the left side of the quadratic equation is not written as a perfect square, then one must first complete the square.
Completing the square	This method can be used to solve any quadratic equation. This method is used to develop the quadratic formula and is the basis for work done in later mathematics courses.	It requires one to repeat all the steps used to develop the quadratic formula.
Quadratic formula	This method can be used to solve any quadratic equation. It is routine to substitute a, b, and c into the quadratic formula. This formula can easily be programmed into calculators and computers.	For some easy equations factoring may be faster.

SELF-CHECK ANSWERS

7.4.1

1. $x = -4, x = 4$
2. $x = -\sqrt{17}, x = \sqrt{17}$
3. $x = -2 - \sqrt{5}$, $x = -2 + \sqrt{5}$

7.4.2

1. $(x + 1)^2 = 5$
2. $(x - 5)^2 = 8$
3. $(x + 2)^2 = 1$
4. $\left(x + \frac{5}{2}\right)^2 = \frac{5}{4}$

7.4.3

1. $(x + 1)^2 = 9$
2. $x = -4, x = 2$

7.4.4

1. $x = 1 - \sqrt{6}, x = 1 + \sqrt{6}$
2. $x \approx -1.45, x \approx 3.45$

7.4.5

1. y-intercept: $(0, -2)$; x-intercepts: $\left(\frac{1 - \sqrt{17}}{4}, 0\right)$ and $\left(\frac{1 + \sqrt{17}}{4}, 0\right)$
2. $(-0.8, 0)$ and $(1.3, 0)$

7.4.6

1. A double real solution
2. Complex numbers with imaginary parts
3. Two distinct real solutions

USING THE LANGUAGE AND SYMBOLISM OF MATHEMATICS 7.4

1. The symbol $\pm$ is read __________ __________ __________.
2. There are two square roots of 3, denoted by $-\sqrt{3}$ and $\sqrt{3}$. The positive square root of 3 is denoted by $\sqrt{3}$; this square root is called the __________ square root of 3.
3. Solving $x^2 = k$ as $x = \pm\sqrt{k}$ is called solving by __________ __________ __________.
4. The act of rewriting an expression or equation so that it contains a perfect square trinomial is called __________ __________ __________.
5. One method to solve the equation $ax^2 + bx + c = 0$ is to use the formula $x = \frac{-b \pm \sqrt{b^2 - 4ac}}{2a}$, which is called the __________ __________.
6. Before you use the quadratic formula to solve a quadratic equation, it is wise to first write the quadratic equation in __________ form.
7. The expanded from of $(x + a)^2$ is __________.
8. The standard form of the quadratic equation $5x = 7 - 3x^2$ is __________.
9. From the quadratic formula, $b^2 - 4ac$ is called the __________.

QUICK REVIEW 7.4

The purpose of this quick review is to help you recall skills needed in this section.

1. A rational number is a real number x that can be written in the form $x = \frac{a}{b}$ where a and b are ________ and $b \neq 0$.
2. A real number that is not rational is called __________.
3. In decimal form, a rational number is either a _________ decimal or an infinite repeating decimal.
4. In decimal form, an irrational number is an infinite __________ decimal.
5. Multiply $\sqrt{3}\sqrt{12}$ and simplify the result.
6. Factor $x^2 + 10x + 25$.
7. Factor $x^2 - 16x + 64$.

EXERCISES 7.4

In Exercises 1–12, use extraction of roots to determine the exact solutions of each quadratic equation.

1. $v^2 = 81$
2. $v^2 = 169$
3. $x^2 = 7$
4. $x^2 = 6$
5. $z^2 = 18$
6. $z^2 = 50$
7. $(x - 1)^2 = 4$
8. $(x + 3)^2 = 25$
9. $(2x + 1)^2 = 9$
10. $(2x - 1)^2 = 36$
11. $(3x - 2)^2 = 3$
12. $(3x + 2)^2 = 5$

In Exercises 13–20, use the quadratic formula to determine the exact solutions of each quadratic equation.

13. $8t^2 - 2t - 1 = 0$
14. $2t^2 + 5t + 3 = 0$
15. $-5m^2 = -6m$
16. $-6m^2 = 5m$
17. $w^2 - 9 = 0$
18. $5w^2 - 4 = 0$
19. $9v^2 + 12v + 4 = 0$
20. $25v^2 - 30v + 9 = 0$

In Exercises 21–24, use the quadratic formula to determine the exact solutions of each quadratic equation. Then approximate each solution to the nearest hundredth.

21. $y^2 - 5 = 0$
22. $4y^2 - 7 = 0$
23. $5w^2 = 2w + 1$
24. $3(w^2 + 1) = -7w$

In Exercises 25–28, determine the exact y-intercept and x-intercepts of each parabola. Then approximate these intercepts to the nearest hundredth.

25.

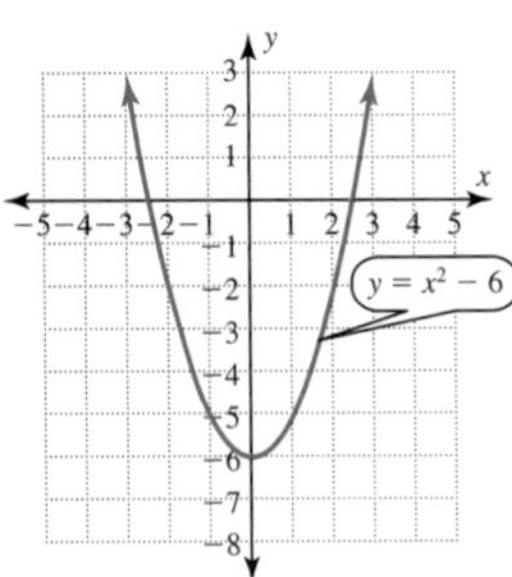

26.

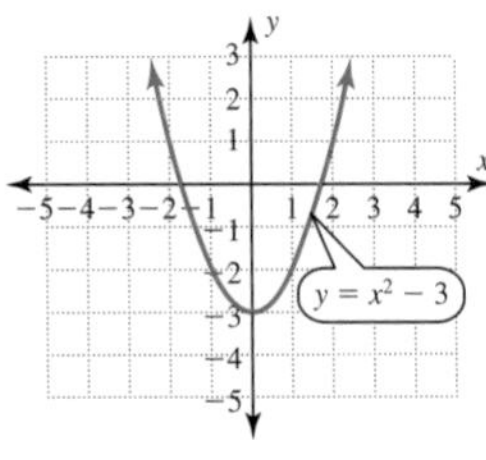

27.

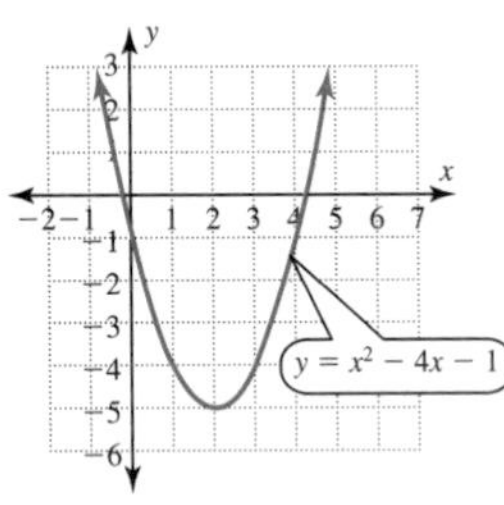

28.

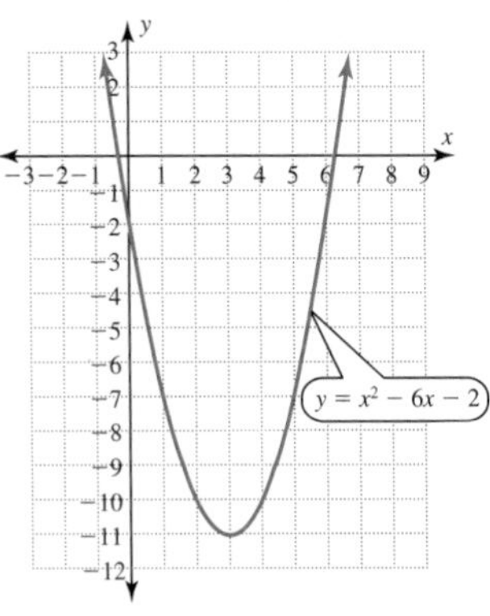

29. One number is 6 more than another number. Find these numbers if their product is 8.

30. The product of a number and twice this number is 98. Find this number.

31. The rectangle shown in the figure is 4 cm longer than it is wide. Find the dimensions if the area is 8 cm^2.

32. The perimeter of the square shown in the figure is numerically 2 more than the area. Find the length in meters of one side of the square.

33. Break-even Values Given the profit function $P(x) = -80x^2 + 5{,}280x - 61{,}200$ where x represents the number of units produced and $P(x)$ represents the profit in dollars, determine the following.

a. Overhead costs [evaluate $P(0)$]

b. Break-even values [determine to the nearest unit when $P(x) = 0$]

34. Break-even Values Given the profit function $P(x) = -75x^2 + 5{,}925x - 47{,}250$ where x represents the number of units produced and $P(x)$ represents the profit in dollars, determine the following.

a. Overhead costs [evaluate $P(0)$]

b. Break-even values [determine to the nearest unit when $P(x) = 0$]

35. Path of a Model Rocket The path of a model rocket is given by $f(x) = -16x^2 + 96x - 20$, where y represents the height of the rocket in feet and x is the number of seconds that have elapsed since the rocket was launched from a pit 20 ft below ground level.

a. Determine the times at which the rocket is at ground level.

b. Determine the interval of times for which the rocket is above ground level.

36. Path of a Model Rocket The path of a model rocket is given by $f(x) = -16x^2 + 48x - 18$, where y represents the height of the rocket in feet and x is the number of seconds that have elapsed since the rocket was launched from a pit 18 ft below ground level.

a. Determine the times at which the rocket is at ground level.

b. Determine the interval of times for which the rocket is above ground level.

In Exercises 37–39, match the graph of each quadratic function with the corresponding quadratic equation. All graphs are displayed with the window $[-5, 5, 1]$ by $[-5, 5, 1]$.

37. A quadratic equation whose discriminant is zero

38. A quadratic equation whose discriminant is negative

39. A quadratic equation whose discriminant is positive

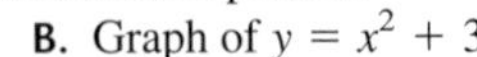

A. Graph of $y = x^2 - 5x + 4$

B. Graph of $y = x^2 + 3$

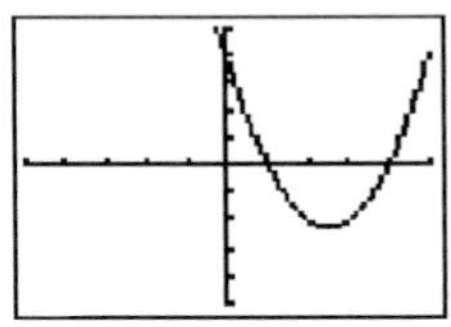

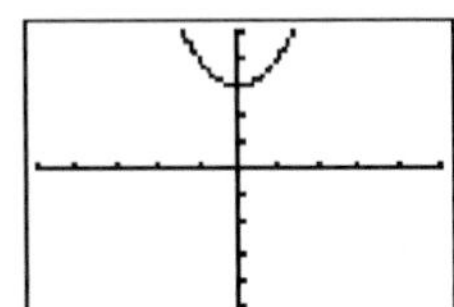

C. Graph of $y = x^2 + 4x + 4$

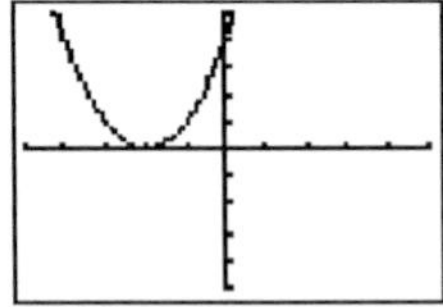

In Exercises 40–42, calculate the discriminant of each quadratic equation, and determine the nature of the solutions of each equation.

40. $x^2 - 6x + 9 = 0$ **41.** $x^2 - 6 = 0$ **42.** $x^2 + 6 = 0$

In Exercises 43–48, rewrite each equation so that the left side of the equation is expressed as a perfect square.

43. $x^2 + 10x = -22$ **44.** $x^2 - 10x = -22$

45. $x^2 - 14x = -48$ **46.** $x^2 + 14x = -24$

47. $x^2 - 18x = 9$ **48.** $x^2 + 20x = 1$

In Exercises 49 and 50, replace the question mark in each step. The solution of the given equation is outlined by using the method of completing the square.

49. $x^2 + 2x - 4 = 0$
a. $x^2 + 2x = ?$
b. $x^2 + 2x + ? = 5$
c. $(x + ?)^2 = 5$
d. $x + ? = \pm\sqrt{5}$
e. $x = ?$

50. $x^2 + 6x - 5 = 0$
a. $x^2 + 6x = ?$
b. $x^2 + 6x + ? = 14$
c. $(x + ?)^2 = 14$
d. $x + ? = \pm\sqrt{14}$
e. $x = ?$

In Exercises 51–56, solve each quadratic equation by completing the square.

51. $z^2 - 4z = 0$
52. $z^2 - 8z = 0$
53. $x^2 + 4x - 5 = 0$
54. $x^2 + 2x = 4$
55. $z^2 + 4z + 2 = 0$
56. $z^2 + 6z + 6 = 0$

57. Modeling Profits A business collected the data in the given table.

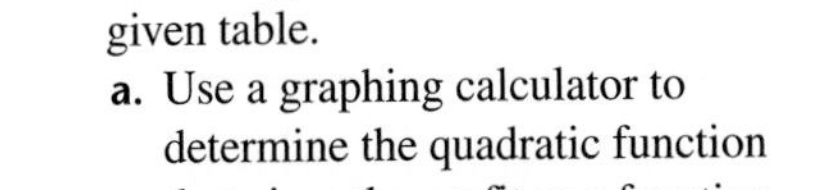

a. Use a graphing calculator to determine the quadratic function that gives the profit as a function of the number of units produced per day. (*Hint:* See Example 7 in Section 7.3.) Round the coefficients to three significant digits.

b. Use this function to approximate the break-even values for this business.

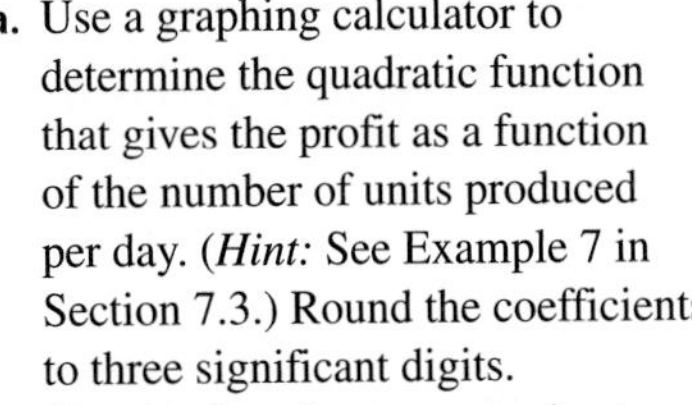

x UNITS	y PROFIT ($)
0	−350
10	−80
20	100
30	300
40	400
50	400
60	350
70	280

58. Modeling Profits A business collected the data in the given table.

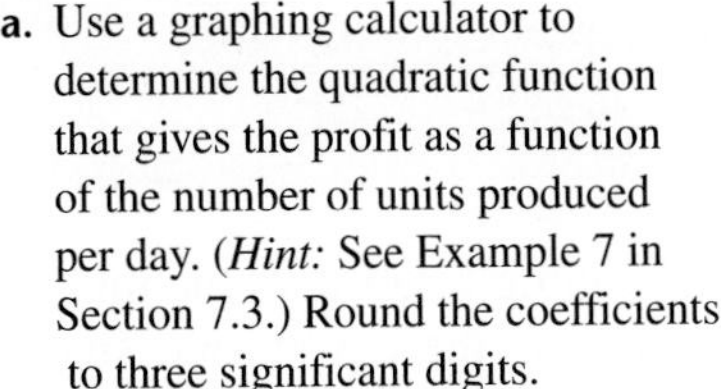

a. Use a graphing calculator to determine the quadratic function that gives the profit as a function of the number of units produced per day. (*Hint:* See Example 7 in Section 7.3.) Round the coefficients to three significant digits.

b. Use this function to approximate the break-even values for this business.

x UNITS	y PROFIT ($)
0	−70
10	70
20	140
30	210
40	230
50	170
60	100
70	50

In Exercises 59–64, solve each cubic equation.

59. $x(x - 4)(x + 7) = 0$
60. $x(x + 5)(x - 9) = 0$
61. $x(x^2 - 5x + 4) = 0$
62. $x(x^2 - 5x - 6) = 0$
63. $x(x^2 - 10) = 0$
64. $x(x^2 - 10x + 23) = 0$

In Exercises 65–68, use the given parabola to solve each equation and inequality.

65. a. $x^2 - 4x - 5 = 0$
b. $x^2 - 4x - 5 < 0$
c. $x^2 - 4x - 5 > 0$

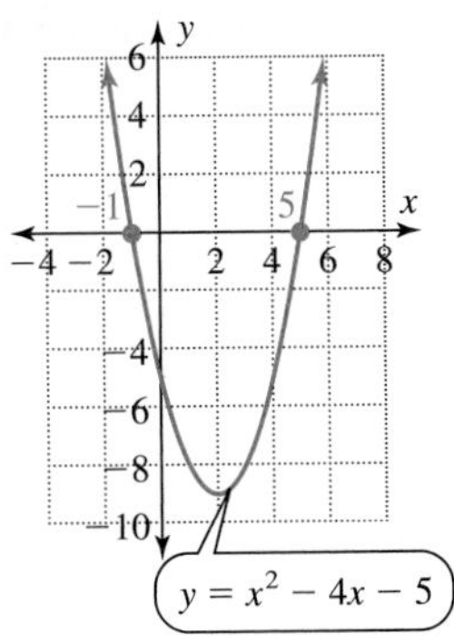

66. a. $-x^2 - 2x + 8 = 0$
b. $-x^2 - 2x + 8 < 0$
c. $-x^2 - 2x + 8 > 0$

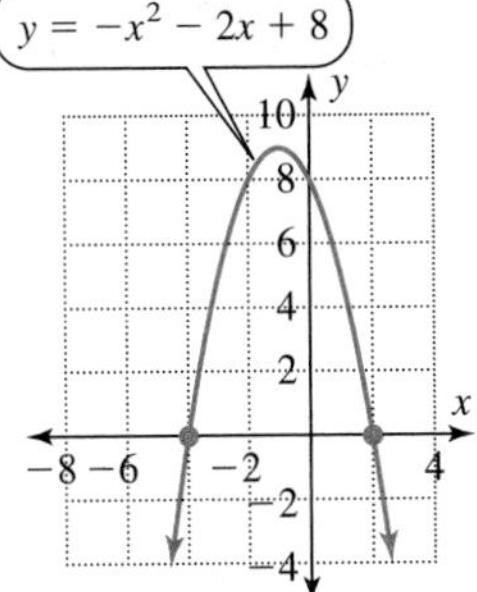

67. a. $-x^2 + 2x + 6 = 0$
b. $-x^2 + 2x + 6 \le 0$
c. $-x^2 + 2x + 6 \ge 0$

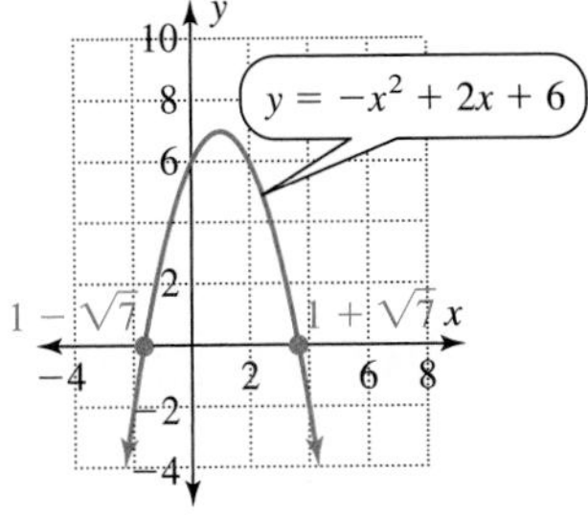

68. a. $4x^2 - 11 = 0$
b. $4x^2 - 11 \le 0$
c. $4x^2 - 11 \ge 0$

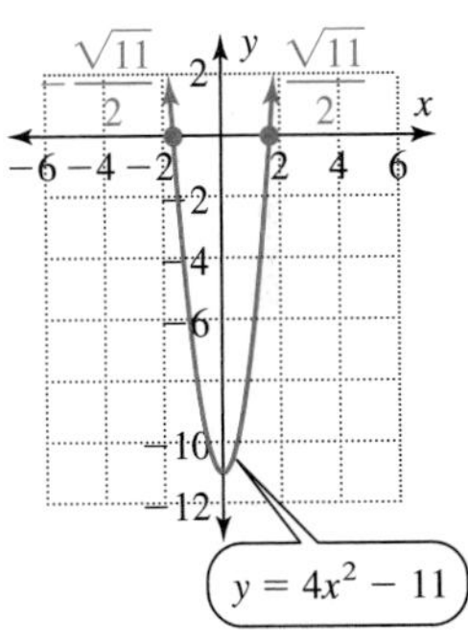

In Exercises 69–72, determine the exact solution of each equation, and then use a graphing calculator to help you solve each inequality.

69. a. $2x^2 + 9x - 5 = 0$
b. $2x^2 + 9x - 5 < 0$
c. $2x^2 + 9x - 5 > 0$

70. a. $2x^2 - 11x - 6 = 0$
b. $2x^2 - 11x - 6 < 0$
c. $2x^2 - 11x - 6 > 0$

71. a. $x^2 - 7 = 0$
b. $x^2 - 7 < 0$
c. $x^2 - 7 > 0$

72. a. $x^2 - 11 = 0$
b. $x^2 - 11 < 0$
c. $x^2 - 11 > 0$

Group Discussion Questions

73. Discovery Question Solve the equation $x^2 - 36 = 0$ by
a. Factoring
b. Extraction of roots
c. The quadratic formula
d. Graphing $y = x^2 - 36$ with a graphics calculator
e. Which method did you prefer? Explain why you prefer this method.

74. **Discovery Question** Complete the square of $x^2 + 10x$ both algebraically and geometrically.

ALGEBRAICALLY	GEOMETRICALLY
$x^2 + 10x$	

75. **Challenge Question** Solve each equation for x.
   a. $x^2 - 4y^2 = 0$
   b. $x^2 - 3y^2 = 0$
   c. $x^2 + 4xy + 3y^2 = 0$
   d. $x^2 + 5xy + 3y^2 = 0$

76. **Discovery Question** The quadratic formula $x = \dfrac{-b \pm \sqrt{b^2 - 4ac}}{2a}$ can also be written as $x = \dfrac{-b}{2a} \pm \dfrac{\sqrt{b^2 - 4ac}}{2a}$. Examine the graph of each of the quadratic functions given here. Then describe any connections that you note among $\dfrac{-b}{2a}$, $\dfrac{\sqrt{b^2 - 4ac}}{2a}$, and these graphs.
   a. $y = x^2 + 2x - 8$
   b. $y = x^2 - 4x - 5$
   c. $y = x^2 - 2x - 15$
   d. $y = x^2 + 2x - 24$

Section 7.5 The Vertex of a Parabola and Max-Min Applications

Objectives:
13. Calculate the vertex of a parabola.
14. Solve problems involving a maximum or minimum value.
15. Construct an equation with given solutions.

The amount of power an engine produces depends on the speed at which it is running. The graph relating speed to power is called a power curve. For some engines this power curve is approximately parabolic. The point of maximum power is important because engines usually perform better by running faster than the maximum power point than they do by running slower than the maximum power point. Suppose the engine on a boat is temporarily put under extra load due to a wave or debris on the propeller. For an engine running faster than the maximum power level, this will briefly slow down the engine, increase the power, and smoothly overcome the extra load. In Example 5 we use data collected on an engine to determine its power curve and its maximum power level.

If a problem is modeled by quadratic function, then this function will have a maximum or minimum value that occurs at the vertex of the parabola defined by this function. The ability to determine relative minima and maxima is one of the most powerful applications of algebraic models of problems. This ability allows us to pick among many possible options and select the one that has optimal results. These optimal results could be minimum cost, minimum production of pollutants, maximum size, or maximum profit. We will illustrate this in Examples 3 and 4.

Maximum Value

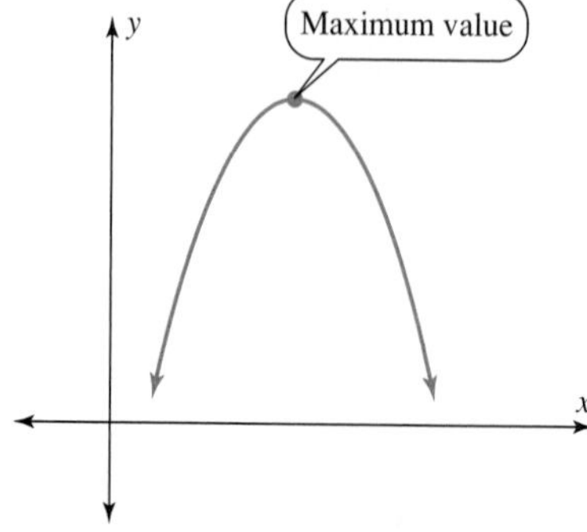

Minimum Value

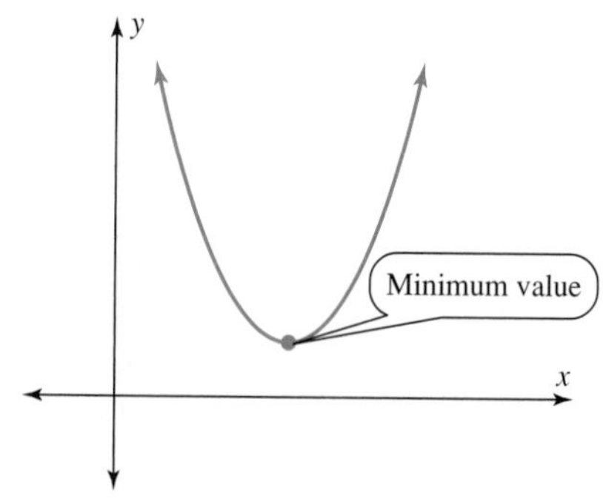

The symmetry of a parabola can be used to develop an algebraic method for determining its vertex. To start, note that the vertex lies on the axis of symmetry and that the x-intercepts are equally spaced on either side of this axis of symmetry.

In the example shown, the x-intercepts $(1, 0)$ and $(5, 0)$ are both 2 units from the axis of symmetry. The axis of symmetry is the vertical line $x = 3$. The vertex $(3, -4)$ lies on the axis of symmetry.

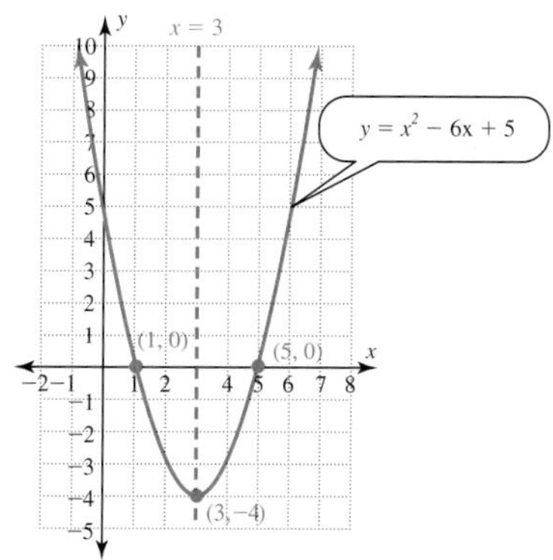

This information is used in Example 1 to determine the vertex of a parabola. We do so in three steps:

1. Determine the x-intercepts.
2. Determine the x-coordinate of the vertex.
3. Calculate the y-coordinate of the vertex.

EXAMPLE 1 Determining the Vertex of a Parabola

Determine the vertex of the parabola defined by $f(x) = -x^2 + 7x + 8$.

SOLUTION

(a) Determine the x-intercepts.

$$f(x) = -x^2 + 7x + 8$$
$$0 = -x^2 + 7x + 8$$
$$x^2 - 7x - 8 = 0$$
$$(x + 1)(x - 8) = 0$$
$$x + 1 = 0 \quad \text{or} \quad x - 8 = 0$$
$$x = -1 \quad \text{or} \quad x = 8$$

The x-intercepts are $(-1, 0)$ and $(8, 0)$.

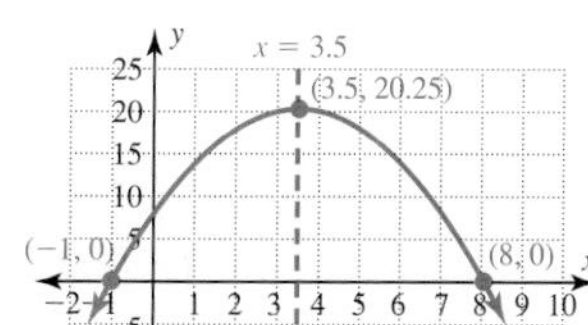

To determine the x-intercepts, set the y-coordinate equal to 0.

Write this equation in standard form and then solve it by factoring.

(b) Determine the x-coordinate of the vertex.

$$x = \frac{-1 + 8}{2} = \frac{7}{2} = 3.5$$

The x-coordinate of the vertex is midway between the x-intercepts. This value is found by adding the two x-intercepts and dividing by 2.

(c) Calculate the y-coordinate of the vertex.

$$f(x) = -x^2 + 7x + 8$$
$$f(3.5) = -(3.5)^2 + 7(3.5) + 8$$
$$f(3.5) = 20.25$$

Substitute the x-coordinate 3.5 into the function $f(x) = -x^2 + 7x + 8$.

Answer: The vertex is $(3.5, 20.25)$.

The x-coordinate of the vertex is 3.5, and the y-coordinate is 20.25.

SELF-CHECK 7.5.1

1. Determine the x-intercepts of the graph of $f(x) = x^2 - 5x - 36$.
2. Determine the vertex of this parabola.

The procedure used to find the vertex in Example 1 can be generalized to give the vertex of any parabola including those that do not have any x-intercepts. From the quadratic formula, a parabola with two x-intercepts will have the x-coordinates of these x-intercepts located at $\left(-\frac{b}{2a} - \frac{\sqrt{b^2 - 4ac}}{2a}, 0\right)$ and $\left(-\frac{b}{2a} + \frac{\sqrt{b^2 - 4ac}}{2a}, 0\right)$. The x-coordinate of the vertex is located midway between these x-intercepts. Adding $-\frac{b}{2a} - \frac{\sqrt{b^2 - 4ac}}{2a}$ and $-\frac{b}{2a} + \frac{\sqrt{b^2 - 4ac}}{2a}$ and dividing by 2, we obtain $-\frac{b}{2a}$. To determine the y-coordinate of the vertex, evaluate $f\left(-\frac{b}{2a}\right)$.

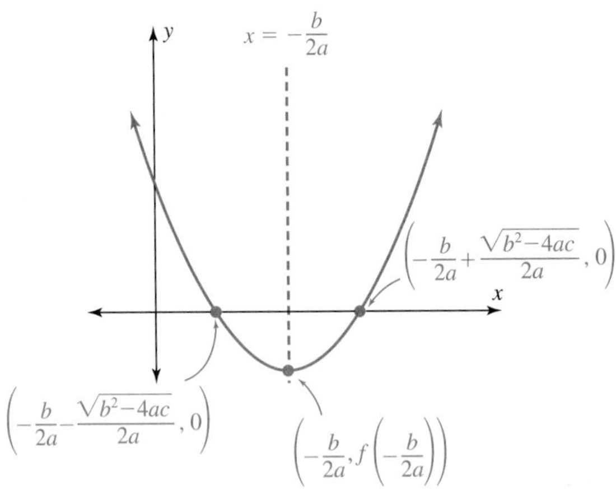

Vertex of the Parabola Defined by $f(x) = ax^2 + bx + c$

ALGEBRAICALLY	VERBALLY	NUMERICALLY	EXAMPLE
$\left(-\frac{b}{2a}, f\left(-\frac{b}{2a}\right)\right)$	The vertex is either the highest or the lowest point on the parabola.	The y-values form a symmetric pattern about the vertex.	$Y_1 = -X^2+7X+8$ [−2, 10, 1] by [−10, 25, 5] Vertex: (3.5, 20.25)

EXAMPLE 2 Determining the Vertex of a Parabola

Determine the vertex of the parabola defined by $f(x) = 2x^2 + 5x - 12$.

SOLUTION

Calculate the x-coordinate of the vertex.

$$-\frac{b}{2a} = -\frac{5}{2(2)}$$
$$= -1.25$$

The x-coordinate of the vertex is $-\frac{b}{2a}$. For the function $f(x) = 2x^2 + 5x - 12$, $a = 2$ and $b = 5$.

Calculate the y-coordinate of the vertex.

$$f(x) = 2x^2 + 5x - 12$$
$$f\left(-\frac{b}{2a}\right) = f(-1.25)$$
$$= 2(-1.25)^2 + 5(-1.25) - 12$$
$$= -15.125$$

The y-coordinate of the vertex is $f\left(-\frac{b}{2a}\right)$.

Answer: Vertex is $(-1.25, -15.125)$.

You can use a graphing calculator to see if this answer seems reasonable.

SELF-CHECK 7.5.2

Determine the vertex of the parabola defined by each function.

1. $f(x) = x^2 + 2x - 120$ **2.** $f(x) = -5x^2 + 7x + 6$

EXAMPLE 3 Maximum Height of a Ball

As part of a dynamic exhibit at a children's museum, hard plastic balls are released down a curved ramp. After the balls leave the end of the ramp, their height is a function of time. The height of each ball in meters is given by $h(t) = -4.9t^2 + 10t + 4$, where t is given in seconds. Determine the maximum height of each ball.

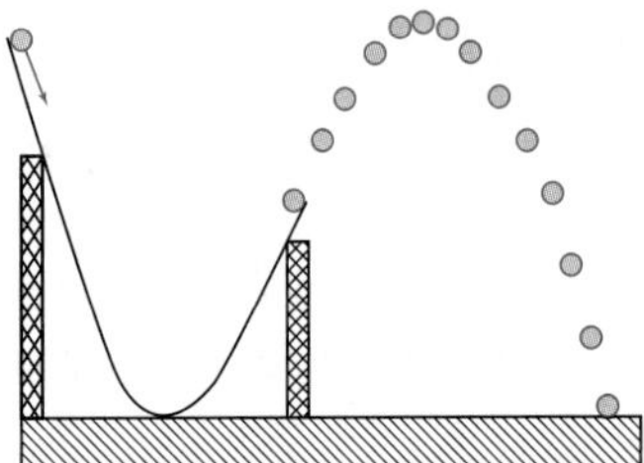

SOLUTION

Calculate the t-coordinate of the vertex.

$$-\frac{b}{2a} = -\frac{10}{2(-4.9)}$$
$$\approx 1.02$$

Because the leading coefficient of $h(t) = -4.9t^2 + 10t + 4$ is negative, the parabola is concave downward. The maximum height will occur at the vertex of this parabola $\left(-\frac{b}{2a}, h\left(-\frac{b}{2a}\right)\right)$.

For this function $a = -4.9$ and $b = 10$.

Calculate the y-coordinate of the vertex.

$$h(t) = -4.9t^2 + 10t + 4$$
$$h\left(-\frac{b}{2a}\right) \approx h(1.02)$$
$$\approx -4.9(1.02)^2 + 10(1.02) + 4$$
$$\approx 9.10$$

The vertex is approximately (1.02, 9.10).

Answer: The maximum height of approximately 9.1 m is reached after 1.02 seconds.

You can use a graphing calculator to confirm that this answer seems reasonable.

SELF-CHECK 7.5.3

The path of a golf ball is given by $h(t) = -16t^2 + 70t$, where $h(t)$ represents the height of the ball in feet and t is the number of seconds that have elapsed since the ball was hit.

1. Determine the highest point the ball reaches.
2. Determine how many seconds into the flight the maximum height is reached.

EXAMPLE 4 Maximum Grapefruit Production

Research at several test plots has revealed that 50 grapefruit trees per plot will average 200 grapefruit per tree. Increasing the number of trees on the plot increases competition among the trees for resources and diminishes the number of grapefruit per tree by 3 grapefruit for each tree added. Determine the number of trees that should be planted on a plot this size to maximize the number of grapefruit the plot will yield.

SOLUTION

Let t = number of extra (in excess of 50) trees to plant on plot
$50 + t$ = total number of trees on plot
$200 - 3t$ = number of grapefruit per tree
N = total number of grapefruit yielded by plot

The yield of 200 grapefruit per tree is diminished by 3 grapefruit per tree added. When the number of trees becomes $50 + t$, the number per tree will become $200 - 3t$.

$$\boxed{\text{Total number of grapefruit}} = \boxed{\text{Number per tree}} \cdot \boxed{\text{Number of trees}}$$

$N(t) = (200 - 3t)(50 + t)$
$N(t) = -3t^2 + 50t + 10{,}000$

Calculate the t-coordinate of the vertex.

$$-\frac{b}{2a} = -\frac{50}{2(-3)}$$
$$\approx 8 \quad \text{(rounded to the nearest whole number)}$$

If 8 trees are added to the plot, then the total number of trees on the plot will be 58 trees.

Calculate the y-coordinate of the vertex.

$N(t) = -3t^2 + 50t + 10{,}000$
$N(8) = -3(8)^2 + 50(8) + 10{,}000$
$N(8) = 10{,}208$

The vertex is approximately (8, 10,208).

Answer: The maximum number of grapefruit that is projected for the plot is 10,208. This number can be obtained by planting 58 trees.

You can use a graphing calculator to confirm that this answer seems reasonable.

We can also use a graphing calculator to approximate the coordinates of a vertex. In Calculator Perspective 7.5.1 we illustrate how to approximate the coordinates of a point at a relative maximum. The steps for approximating a relative minimum are nearly identical. Before you read the next calculator perspective, enter into your calculator $y = -2x^2 + 5x + 25$ and select $[-4, 6, 1]$ by $[-20, 40, 10]$ for the viewing window.

CALCULATOR PERSPECTIVE 7.5.1 Calculating the Coordinates of a Relative Maximum

To approximate to the nearest thousandth the coordinates of the vertex of $y = -2x^2 + 5x + 25$ on a TI-84 Plus calculator, enter the following keystrokes:

  4

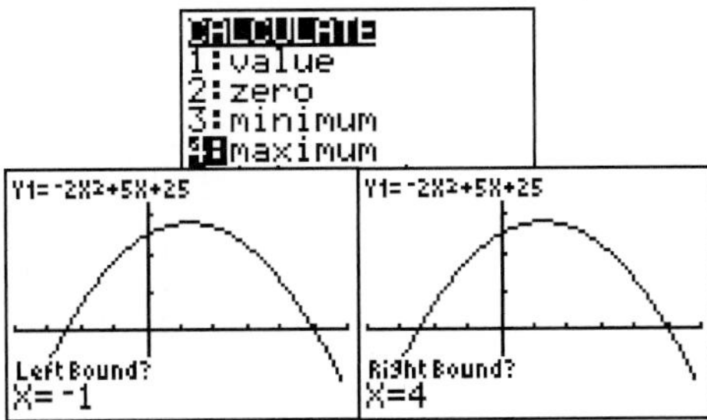

By inspection, the location of the maximum appears to be between $x = -1$ and $x = 4$. The **maximum** feature under the **CALC** menu first asks for the left bound. Input -1. We are then asked for the right bound. Input 4. Finally, we are asked for a guess. Input 1.

Note: The x-inputs given above must be between -4 and 6 because of the window selected.

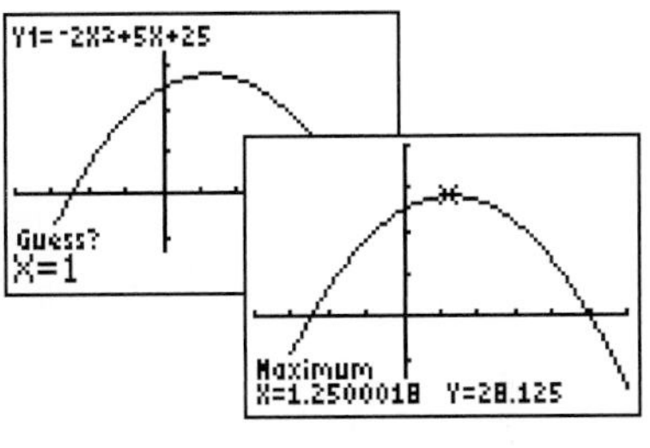

Answer: (1.250, 28.125)

$[-4, 6, 1]$ by $[-20, 40, 10]$

The coordinates of the vertex have been rounded to the nearest thousandth.

SELF-CHECK 7.5.4

1. Use a graphing calculator to determine to the nearest hundredth the coordinates of the vertex of the parabola defined by $y = -1.28x^2 + 3.54x + 7.29$.

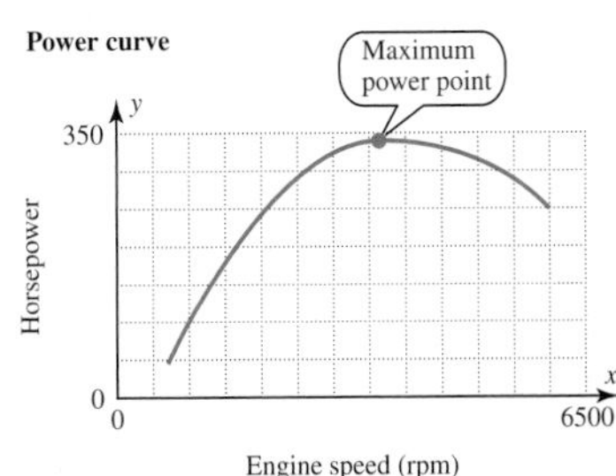

The amount of power an engine produces depends on the speed at which it is running. The graph relating speed to power is called a power curve. For some engines this power curve is approximately parabolic. The point of maximum power is important because engines usually perform better when running faster than the maximum power point than they do when running slower than the maximum power point. Suppose the engine on a boat is temporarily put under extra load due to a wave or debris on the propeller. For an engine running faster than the maximum power level, this will briefly slow down the engine, increase the power, and smoothly overcome the extra load. We can use data collected on an engine to determine its power curve and its maximum power level. This is illustrated in Example 5.

SPEED x (rpm)	HORSEPOWER $y = f(x)$(hp)
0	0
500	130
1,000	240
2,000	340
3,000	280
3,500	180
4,000	50

EXAMPLE 5 Modeling the Power Curve for the Engine on a Boat

(a) The data points shown in the table give the horsepower of one engine running at some tested speed, in revolutions per minute (rpm). Create a power curve that is a quadratic function that best fits these data. (Round the coefficients of this quadratic function to three significant digits.)

(b) Use this power curve to determine the point of maximum power for this engine.
(c) Interpret both coordinates of this maximum power point.

SOLUTION

(a)

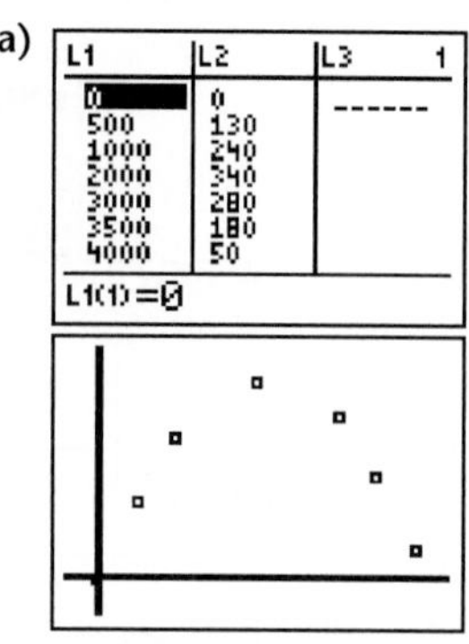

L1	L2	L3
0	0	------
500	130	
1000	240	
2000	340	
3000	280	
3500	180	
4000	50	

L1(1)=0

[−400, 4400, 1] by [−57.8, 397.8, 1]

Use a graphing calculator to plot the given points on a scatter diagram, and then use the **QuadReg** feature under the **CALC** submenu of the **STAT** menu to calculate the equation of the parabola of best fit.

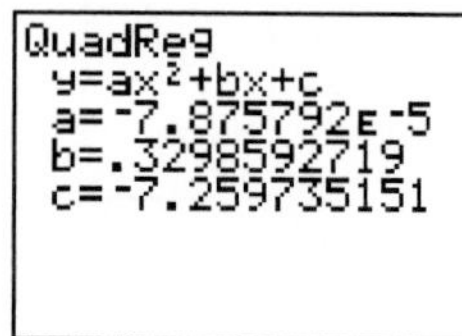

The equation of the parabola of best fit (with coefficients rounded to three significant digits) is $f(x) = -0.0000788x^2 + 0.330x - 7.26$.

Power curve:
$f(x) = -0.0000788x^2 + 0.330x - 7.26$

(b)

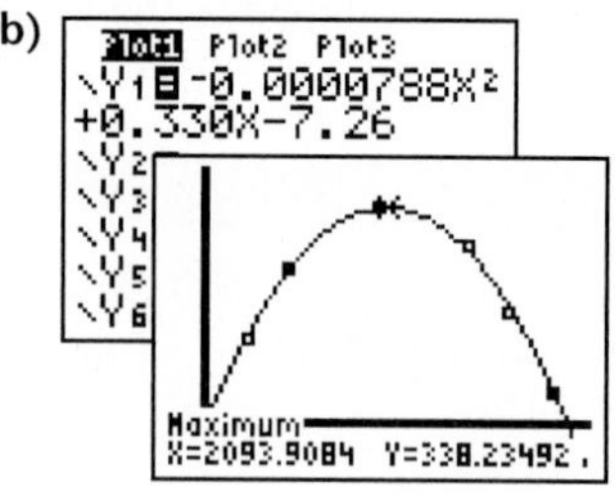

[−400, 4400, 1] by [−57.8, 397.8, 1]

Graph this parabola, and then use the **maximum** feature in the **CALC** menu to determine the vertex of the parabola.

Maximum power point: The vertex of this parabola of best fit is approximately at (2,100, 340).

The coordinates of the vertex have been rounded to two significant digits.

(c) To achieve the maximum possible horsepower of approximately 340 hp, run the engine at approximately 2,100 rev/min.

From the preceding paragraph it might be wise to actually run the engine at about 2,400 rev/min. The horsepower at this level is about 330 hp; and when an extra load is placed on the engine at this speed, the engine will slow down and respond with more power.

$$f(x) = -0.0000788x^2 + 0.330x - 7.26$$
$$f(2{,}400) = -0.0000788(2{,}400)^2 + 0.330(2{,}400) - 7.26$$
$$f(2{,}400) \approx 330$$

We now examine another connection between the solutions of a quadratic equation and the factors of the corresponding quadratic polynomial. If we can factor a quadratic equation in standard form, then we can find its solution. If we reverse this procedure, we can create a quadratic equation whose solutions are given.

SOLVING A QUADRATIC EQUATION			CREATING A QUADRATIC EQUATION		
$3x^2 - 10x - 8 = 0$			$x = -\frac{2}{3}$	or	$x = 4$
$(3x + 2)(x - 4) = 0$			$3x = -2$		$x = 4$
$3x + 2 = 0$	or	$x - 4 = 0$	$3x + 2 = 0$	or	$x - 4 = 0$
$3x = -2$		$x = 4$	$(3x + 2)(x - 4) = 0$		
$x = -\frac{2}{3}$	or	$x = 4$	$3x^2 - 10x - 8 = 0$		

Engineers can create products with desired behavior characteristics by creating equations to model the products they are designing.

This procedure is used routinely by teachers and authors to create equations with known answers and a known level of difficulty. This procedure is also understood by engineers who wish to design a product with known behavior characteristics. Engineers start with the behaviors they want and create equations that model products which exhibit the desired behaviors.

When you finish reading Example 6, look at the steps in reverse to confirm that these are exactly the steps that we would use to solve $35x^2 + 6x - 8 = 0$.

EXAMPLE 6 Constructing a Quadratic Equation with Given Solutions

Construct a quadratic equation in x with solutions of $x = -\frac{4}{7}$ and $x = \frac{2}{5}$.

SOLUTION

$$x = -\frac{4}{7} \quad \text{or} \quad x = \frac{2}{5}$$

The solutions are $-\frac{4}{7}$ and $\frac{2}{5}$.

$$7x = -4 \qquad 5x = 2$$

$$7x + 4 = 0 \qquad 5x - 2 = 0$$

Rewrite each equation so that the right side is zero.

$$(7x + 4)(5x - 2) = 0$$

These factors equal zero, so their product is zero.

$$35x^2 + 6x - 8 = 0$$

Answer: $35x^2 + 6x - 8 = 0$ is a quadratic equation with solutions of $-\frac{4}{7}$ and $\frac{2}{5}$.

You can check these solutions in the equation to confirm this equation works. *Caution:* As noted in the following paragraph, this equation is not unique.

The answer in Example 6 is not unique because any nonzero multiple of $35x^2 + 6x - 8 = 0$ would have the same solutions. For example, $350x^2 + 60x - 80 = 0$ is obtained by multiplying both sides of this equation by 10; it has the same solutions.

We can extend this procedure to construct polynomial equations of higher degree. In Example 7 we construct a cubic equation that has three given solutions.

EXAMPLE 7 Constructing a Cubic Equation with Given Solutions

Construct a cubic equation in x with solutions of $x = -2$, $x = 2$, and $x = 5$.

SOLUTION

$$x = -2 \quad \text{or} \quad x = 2 \quad \text{or} \quad x = 5$$

The solutions are -2, 2, and 5.

$$x + 2 = 0 \qquad x - 2 = 0 \qquad x - 5 = 0$$

Rewrite each equation so that the right side is zero.

$$(x + 2)(x - 2)(x - 5) = 0$$

These factors equal zero, so their product is zero.

$$(x^2 - 4)(x - 5) = 0$$

$$x^3 - 5x^2 - 4x + 20 = 0$$

Answer: $x^3 - 5x^2 - 4x + 20 = 0$ is a cubic equation with solutions of $x = -2$, $x = 2$, and $x = 5$.

You can check these solutions in the equation to confirm this equation works. *Caution:* As noted in the paragraph above, this equation is not unique.

SELF-CHECK 7.5.5

1. Write a quadratic equation with solutions of $x = 4$ and $x = \frac{3}{7}$.
2. Write a cubic equation with solutions of $x = -3$, $x = 4$, and $x = 10$.

EXAMPLE 8 Constructing a Quadratic Equation with Given Irrational Solutions

Construct a quadratic equation in x with solutions of $x = 1 - \sqrt{2}$ and $x = 1 + \sqrt{2}$.

SOLUTION

$$x = 1 - \sqrt{2} \quad \text{or} \quad x = 1 + \sqrt{2}$$

The solutions are $1 - \sqrt{2}$ and $1 + \sqrt{2}$.

$$x - 1 + \sqrt{2} = 0 \qquad x - 1 - \sqrt{2} = 0$$

Rewrite each equation so that the right side is zero.

$$(x - 1 + \sqrt{2})(x - 1 - \sqrt{2}) = 0$$

These factors equal zero, so their product is zero.

$$[(x - 1) + \sqrt{2}][(x - 1) - \sqrt{2}] = 0$$

Note that the multiplication as shown takes advantage of the special product of a sum and a difference.

$$(x - 1)^2 - 2 = 0$$
$$x^2 - 2x + 1 - 2 = 0$$
$$x^2 - 2x - 1 = 0$$

Answer: $x^2 - 2x - 1 = 0$ is a quadratic equation with solutions of $1 - \sqrt{2}$ and $1 + \sqrt{2}$.

You can check these solutions in the equation to confirm this equation works. *Caution:* As noted earlier, this equation is not unique.

Suppose that you see a parabola on a test or in the workplace and would like to explore this graph in greater detail. Can you quickly write the equation of this quadratic function without using curve fitting? If you know the x-intercepts, you may be able to do this quickly. This process is illustrated in Example 9.

EXAMPLE 9 Writing the Quadratic Function for a Given Graph

Write the equation of a quadratic function for the parabola shown.

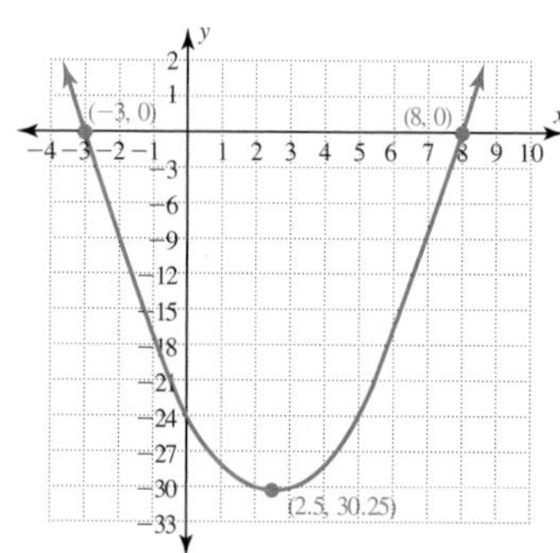

SOLUTION

The x-intercepts are $(-3, 0)$ and $(8, 0)$.

$$x = -3 \quad \text{or} \quad x = 8$$
$$x + 3 = 0 \qquad x - 8 = 0$$

First identify the x-intercepts from the graph. For each x-intercept there is a solution for the corresponding quadratic equation.

$$(x + 3)(x - 8) = 0$$
$$x^2 - 5x - 24 = 0$$

Rewrite each equation so that the right side is zero. These factors equal zero, so their product is zero.

$f(x) = x^2 - 5x - 24$
$f(2.5) = (2.5)^2 - 5(2.5) - 24$
$f(2.5) = 6.25 - 12.5 - 24$
$f(2.5) = -30.25$

Since all nonzero multiples of this function also have these x-intercepts, we check to confirm that this function also produces the vertex shown on this graph.

Answer: $f(x) = x^2 - 5x - 24$

This point checks, so this is the equation of this parabola.

SELF-CHECK 7.5.6

1. Determine the x-intercepts of this parabola.
2. Because this parabola is concave downward, its leading coefficient is __________.
3. Write the equation of a quadratic function for this parabola.

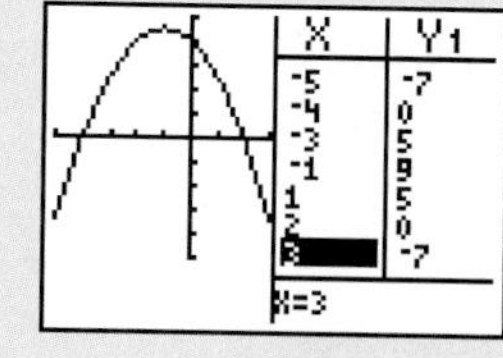

[−5, 3, 1] by [−10, 10, 2]

SELF-CHECK ANSWERS

7.5.1
1. (−4, 0), (9, 0)
2. (2.5, −42.25)

7.5.2
1. (−1, −121)
2. (0.7, 8.45)

7.5.3
1. The highest point is 76.6 ft.
2. The highest point is reached after 2.19 seconds.

7.5.4
1. (1.38, 9.74)

7.5.5
1. $7x^2 - 31x + 12 = 0$
2. $x^3 - 11x^2 - 2x + 120 = 0$

7.5.6
1. (−4, 0), (2, 0)
2. Negative
3. $y = -x^2 - 2x + 8$

USING THE LANGUAGE AND SYMBOLISM OF MATHEMATICS 7.5

1. The __________ of a parabola that is concave downward is the highest point on the parabola.
2. The vertex of a parabola lies on its axis of __________.
3. The x-coordinate of the vertex of a parabola defined by $f(x) = ax^2 + bx + c$ is given by $x =$ __________.
4. The y-coordinate of the vertex of a parabola defined by $f(x) = ax^2 + bx + c$ is given by $y =$ __________.
5. A quadratic equation in x is a __________-degree equation in x.
6. A cubic equation in x is a __________-degree equation in x.

QUICK REVIEW 7.5

The purpose of this quick review is to help you recall skills needed in this section.

1. Round 123.4567 to the nearest hundredth.
2. Round 123.4567 to the nearest hundred.
3. Round 123.4567 to the two significant digits. (*Hint:* See Appendix A.)
4. The ____-factor principle states that if $ab = 0$, then $a = 0$ or $b = 0$.
5. If $a = 0$ or $b = 0$, then $ab =$ ______.

EXERCISES 7.5

In Exercises 1–4, use the given graph and table to determine the following by inspection.

a. The y-intercept of the parabola
b. The x-intercepts (if any) of the parabola
c. The vertex of the parabola

1.

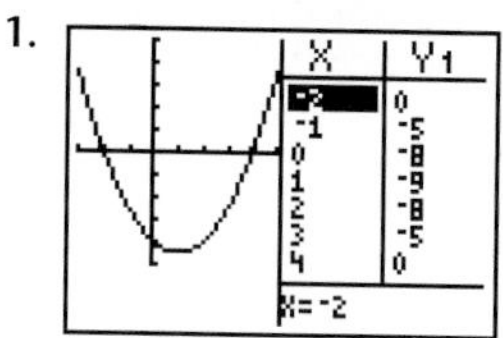

[−3, 5, 1] by [−10, 10, 2]

2.

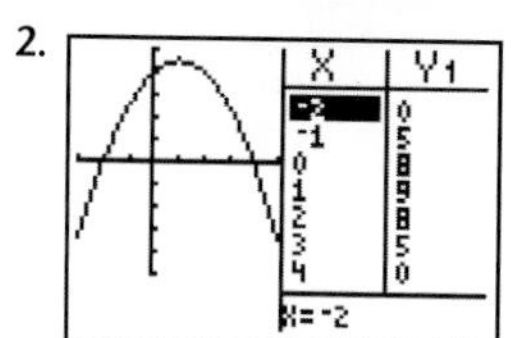

[−3, 5, 1] by [−10, 10, 2]

3.

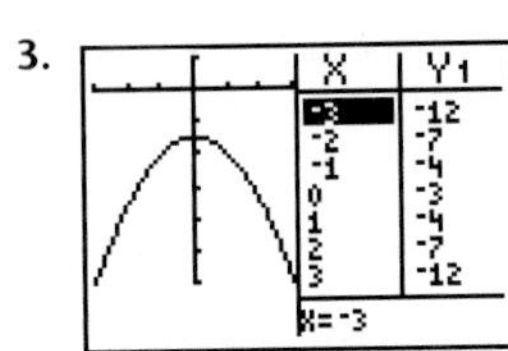

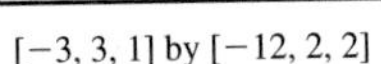
[−3, 3, 1] by [−12, 2, 2]

4.

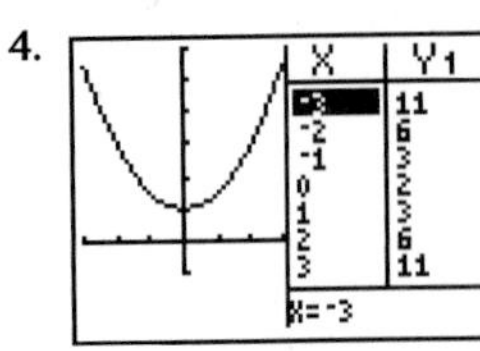

[−3, 3, 1] by [−2, 12, 2]

5. The x-intercepts of a parabola are $(6, 0)$ and $(-2, 0)$. Determine the x-coordinate of the vertex.
6. The x-intercepts of a parabola are $(-5, 0)$ and $(3, 0)$. Determine the x-coordinate of the vertex.

In Exercises 7–10, calculate the vertex of the parabola defined by each function.

7. $f(x) = x^2 - 6x + 11$
8. $f(x) = x^2 + 4x - 9$
9. $f(x) = -5x^2 + 10x + 6$
10. $f(x) = -3x^2 + 6x + 7$

In Exercises 11–14, use the given equation to calculate the following.

a. The y-intercept of the parabola
b. The x-intercepts of the parabola
c. The vertex of the parabola

11. $y = 2x^2 - 3x - 2$
12. $y = 5x^2 - x - 4$
13. $y = (2x - 3)(x + 4)$
14. $y = (5x - 2)(x + 3)$

In Exercises 15–26, construct a quadratic equation with the given solutions.

15. $x = -3$ and $x = 3$
16. $x = -2$ and $x = 5$
17. $x = 0$ and $x = 6$
18. $x = -4$ and $x = 0$
19. $x = -\frac{7}{5}$ and $x = 2$
20. $x = -2$ and $x = \frac{3}{7}$
21. $x = -\frac{2}{5}$ and $x = \frac{5}{2}$
22. $x = \frac{2}{3}$ and $x = \frac{5}{4}$
23. $x = -\sqrt{5}$ and $x = \sqrt{5}$
24. $x = -\sqrt{7}$ and $x = \sqrt{7}$
25. $x = 2 - \sqrt{5}$ and $x = 2 + \sqrt{5}$
26. $x = 1 - \sqrt{7}$ and $x = 1 + \sqrt{7}$

In Exercises 27–30, construct a cubic equation with the given solutions.

27. $x = 0, x = 2,$ and $x = 5$
28. $x = 0, x = -3,$ and $x = 4$
29. $x = -1, x = 1,$ and $x = 6$
30. $x = -2, x = 2,$ and $x = 3$

In Exercises 31–34, write the equation of a quadratic function for the parabola shown.

31.

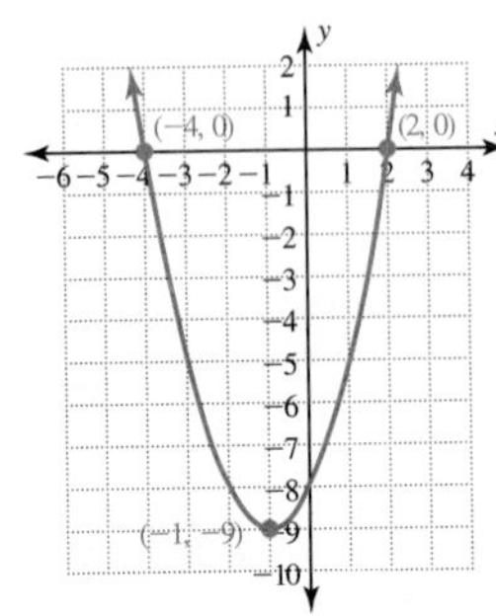

32.

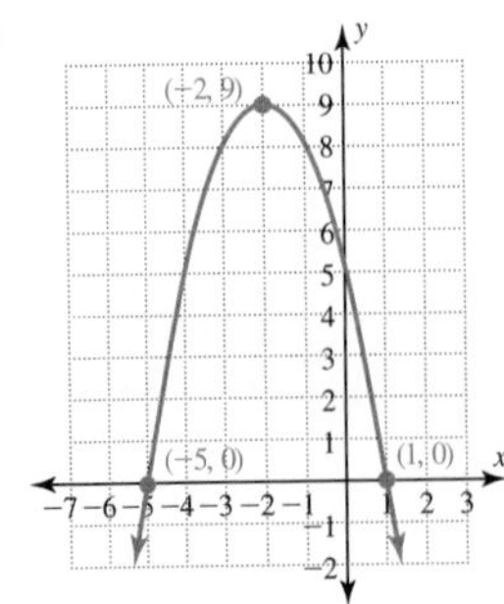

33.

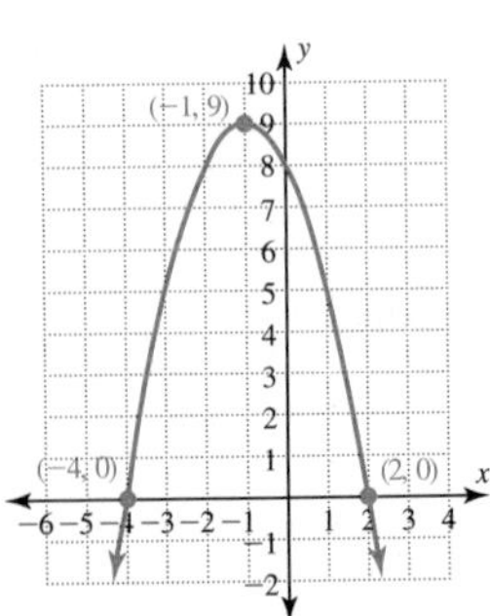

34.

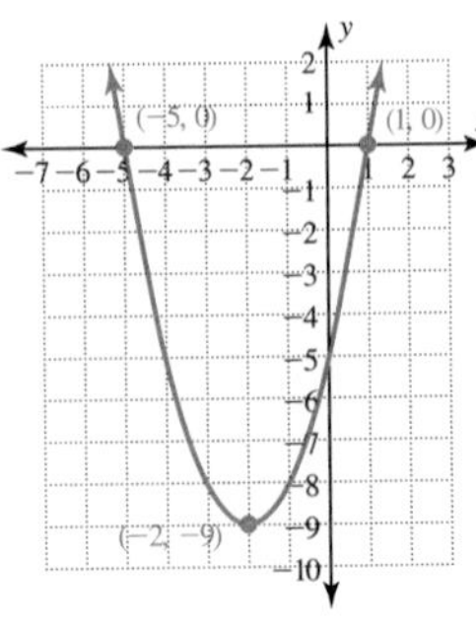

Break-even Values

In Exercises 35–38, $P(x)$ gives the profit in dollars when x tons are produced and sold. Determine the following.

a. Overhead costs [*Hint:* Evaluate $P(0)$.]
b. Break-even values [*Hint:* When does $P(x) = 0$?]
c. Maximum profit that can be made and the number of tons to sell to create this profit

35. $P(x) = -100x^2 + 14{,}000x - 240{,}000$
36. $P(x) = -50x^2 + 7{,}000x - 65{,}000$
37. $P(x) = -140x^2 + 7{,}000x - 28{,}665$
38. $P(x) = -160x^2 + 8{,}000x - 39{,}160$
39. **Path of a Golf Ball** The path of a golf ball is given by $y = -16x^2 + 96x$, where y represents the height of the ball in feet and x is the number of seconds that have elapsed since the ball was hit. Determine the highest point that the ball reaches. Determine how many seconds into the flight the maximum height is reached.
40. **Path of a Softball** The path of a softball is given by $y = -16x^2 + 48x + 8$, where y represents the height of the ball in feet and x is the number of seconds that have elapsed since the ball was released. Determine the highest point that the ball reaches. Determine how many seconds into the flight the maximum height is reached.
41. **Maximum Area** Determine the maximum rectangular area that can be enclosed by 40 m of fencing.
42. **Maximum Area** A lumberyard plans to enclose a rectangular storage area for storing treated lumber. One side of the storage area will be formed by the wall of an existing building, and the other three sides will be formed by using up to 60 m of fencing. Write a function that models the area as a function of the width w. Then find the maximum area that can be enclosed with this fencing.

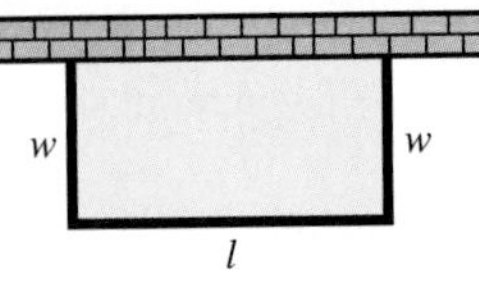

43. **Engine Power Curve** A power boat engine has a power curve approximated by $y = -\frac{x^2}{5{,}000} + \frac{21x}{25} - 32$, where x is the number of revolutions per minute and y is the horsepower generated. At what number of revolutions per minute is the engine putting out maximum horsepower? What is the maximum horsepower?

44. **Engine Power Curve** An engine in a stunt plane has a power curve approximated by $y = -\frac{x^2}{15{,}000} + \frac{2x}{5} - 12$, where x is the number of revolutions per minute and y is the horsepower generated. At what number of revolutions per minute is the engine putting out maximum horsepower? What is the maximum horsepower?

45. **Engine Power Curve**
   a. The data points shown in the table give the horsepower of one engine running at some tested value of revolutions per minute. Create a power curve that is a quadratic function that best fits these data. (Round the coefficients of this quadratic function to three significant digits.)

x (REV/MIN)	HORSEPOWER $y = f(x)$ (hp)
0	0
500	120
1,000	220
2,000	310
3,000	250
3,500	160
4,000	45

   b. Use this power curve to determine the point of maximum power for this engine.
   c. Interpret both coordinates of this maximum power point.

46. **Profit as a Function of Production** A business collected the data in the given table.
   a. Use a graphing calculator to determine the quadratic function that gives the profit in dollars as a function of the number of units produced per day. Round the coefficients to three significant digits.
   b. Use this function to approximate the break-even values for this business.
   c. Use this function to approximate the maximum profit the business can make per day and the number of units it should produce to generate this profit.

x UNITS	PROFIT y ($)
0	−350
10	−80
20	100
30	300
40	400
50	400
60	350
70	280

47. **Profit as a Function of Production** A business collected the data in the given table.
   a. Use a graphing calculator to determine the quadratic function that gives the profit in dollars as a function of the number of units produced per day. Round the coefficients to three significant digits.
   b. Use this function to approximate the break-even values for this business.
   c. Use this function to approximate the maximum profit the business can make per day and the number of units it should produce to generate this profit.

x UNITS	PROFIT y ($)
0	−70
10	70
20	140
30	210
40	230
50	170
60	100
70	50

Home Run Distance

The technology of tracking home run distances in baseball has improved greatly over the years. When a home run lands 40 ft above the ground in the bleachers 350 ft from home plate, it can be very difficult to determine exactly how far it would have traveled. Today special cameras are placed near home plate and at each of the foul poles to record data on the exact location of the ball. Then computers analyze the data to determine how far the ball would have traveled. In Exercises 48 and 49, x represents the horizontal distance the ball travels from home plate, and y represents the height of the ball. Use a graphing calculator and the points given in the table to do the following.

a. Draw a scatter diagram of these points.
b. Determine the quadratic function that best fits these points. (Round the coefficients to three significant digits.)
c. Use this function f to approximate $f(325)$.
d. Interpret the meaning of the value found in part **c**.
e. Determine the x-intercepts of the graph of this function.
f. Interpret the meaning of the positive x-intercept found in part **e**. (The negative x-intercept has no practical meaning.)
g. Determine the vertex of this parabola.
h. Interpret the meaning of the vertex found in part **g**.

48.

x ft	y ft
0	4
100	75
200	90
300	49

49.

x ft	y ft
0	3
100	110
200	160
300	153

Group Discussion Questions

50. **Challenge Question**
   a. Sketch a parabola that is defined by $y = ax^2 + bx + c$ with a vertex in quadrant II and that has a positive discriminant.
   b. Sketch a parabola that is defined by $y = ax^2 + bx + c$ with a vertex in quadrant II and that has a negative discriminant.
   c. Sketch a parabola that is defined by $y = ax^2 + bx + c$ and does *not* have any points in quadrant II and that has a positive discriminant.

d. Sketch a parabola that is defined by $y = ax^2 + bx + c$ and does *not* have any points in quadrant II and that has a negative discriminant.
e. Sketch a parabola that is defined by $y = ax^2 + bx + c$ and does *not* have any points in quadrant II and that has a zero discriminant.

51. **Discovery Question**
a. Graph each of these functions and identify the vertex of each parabola. $f(x) = x^2$, $f(x) = x^2 + 1$, $f(x) = x^2 + 2$, $f(x) = x^2 - 3$, and $f(x) = x^2 - 4$
b. Use this information to predict the vertex of $f(x) = x^2 + c$ for a constant c.

52. **Discovery Question**
a. Graph each of these functions and identify the vertex of each parabola. $f(x) = x^2$, $f(x) = (x + 1)^2$, $f(x) = (x + 2)^2$, $f(x) = (x - 3)^2$, and $f(x) = (x - 4)^2$
b. Use this information to predict the vertex of $f(x) = (x + c)^2$ for a constant c.

53. **Spreadsheet Exploration** A spreadsheet cell can contain a formula that calculates the value displayed in the cell. Each spreadsheet formula starts with an = symbol. For example, the formula `= A5*B5`, when entered in cell C5, actually means C5=A5*B5. In algebra we must write both sides of an equation, but in a spreadsheet the left side of the equation is implied by the cell in which the formula is stored.

The function $f(x) = 120 - 2x$ is used to represent the length of a rectangular lot shown here where x is the width of the lot and 120 ft of fencing is available to enclose three sides of this lot. This spreadsheet examines the area enclosed by this fencing.

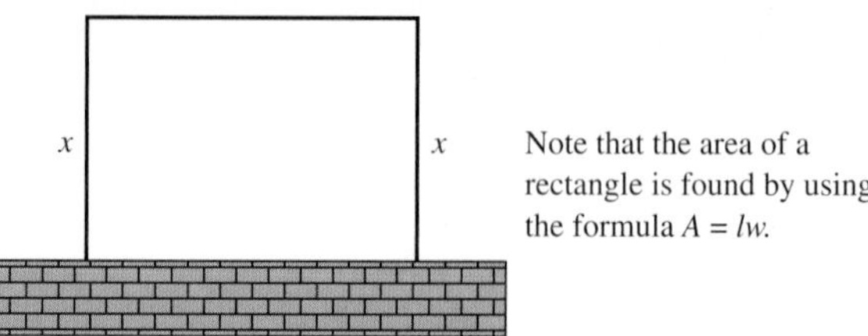

B4 | f_x =120-2*A4

	A	B	C
1	Fencing used for the width	Fencing remaining for the length	Area enclosed by the fence
2	0	120	0
3	10	100	1000
4	20		1600
5	30	60	
6	40	40	1600
7	50	20	1000
8	60	0	0
9			

Cell B4 = **120 – 2*A4**

Selecting cell B4 reveals the formula shown above.

Cell C5 = **A5*B5**

a. The formula in cell B4 is `= 120-2*A4`. Determine the numerical value that is displayed in this cell.
b. Write the formula that goes in cell B5.
c. Write the formula that goes in cell B6.
d. The formula in cell C5 is C5 is `= A5*B5`. Determine the numerical value that is displayed in this cell.
e. Write the formula that goes in cell C7.
f. What width and length in the table yield the largest area?

Section 7.6 More Applications of Quadratic Equations

Objective: 16. Use quadratic equations to solve word problems.

Most college graduates encounter mathematics in the context of verbally stated problems—problems that need to be translated into mathematics to be solved. It is rare that anyone will ask you to solve a quadratic equation. More likely someone will ask, "How do you . . ." or "What is the cheapest way to . . . ." These are the real world problems that will challenge your problem-solving skills.

To benefit from your algebra skills, you must be able to work with word problems. Each sport or profession requires practice time on the basics in order to execute efficiently in a real performance. Even Tiger Woods practices chipping and putting to hone these skills for tournaments. Likewise, we will include some basic drill exercises here as well as some applications that you may find more challenging. The basic strategy used to solve these word problems is summarized below. It is the same strategy we used in Section 2.8 to solve word problems by using linear equations. This time the equations we use will be quadratic equations.

Strategy for Solving Word Problems

STEP 1. Read the problem carefully to determine what you are being asked to find.

STEP 2. Select a variable to represent each unknown quantity. Specify precisely what each variable represents.

STEP 3. If necessary, translate the problem into word equations. Then translate the word equations into algebraic equations.

STEP 4. Solve the equation(s), and answer the question asked by the problem.

STEP 5. Check the reasonableness of your answer.

In Section 2.8 we noted that it is important to pay attention to the restrictions on the variable—the values of the variable that are permissible in a given application. Another terminology that is used in the context of writing a function to model an application is the **practical domain** of the function. After solving the equations that you form to model a problem, always inspect the solutions to see if they are appropriate for the original problem. For example, check for negative lengths, negative interest rates, negative rates of speed, more than 24 hours in 1 day, or more than 31 days in 1 month. That is, make sure that the values given as answers are in the practical domain of the function.

In some problems there are well-known formulas that can be used with established variables to represent the unknown quantities. For example, we often use A for area, V for volume, and r for an interest rate. Example 1 uses the well-known formula $A = P\left(1 + \frac{r}{2}\right)^2$ and presents a possible solution for this equation that is not appropriate for the original problem.

EXAMPLE 1 Calculating an Interest Rate

The formula for computing the amount A of an investment of principal P invested at interest rate r for 1 year and compounded semiannually is $A = P\left(1 + \frac{r}{2}\right)^2$. Approximately what interest rate is necessary for \$800 to grow to \$848.72 in 1 year if interest is compounded semiannually?

SOLUTION

$$A = P\left(1 + \frac{r}{2}\right)^2$$

The variables and equation are given in the statement of the problem.

$$848.72 = 800\left(1 + \frac{r}{2}\right)^2$$

Substitute the given values into the formula for compound interest. The resulting amount of the investment is $A = 848.72$, and the original principal is $P = 800$.

$$1.0609 = \left(1 + \frac{r}{2}\right)^2$$

Divide both sides of this quadratic equation by 800.

$$\pm 1.03 = 1 + \frac{r}{2}$$

Solve this quadratic equation by extraction of roots. Note that this problem is very well suited to using extraction of roots.

$$\pm 2.06 = 2 + r$$

Multiply both sides of the equation by 2.

$$r = -2 \pm 2.06$$

$$r = 0.06 \quad \text{or} \quad r = -4.06$$

$$r = 6\%$$

A negative interest rate is not meaningful in this problem. Only 6% is a meaningful answer.

Answer: At a rate of 6% compounded semiannually, a principal of \$800 will grow to \$848.72 in 1 year.

Does this rate check, and is it a reasonable answer?

SELF-CHECK 7.6.1

1. Approximately what interest rate is necessary for \$800 to grow to \$856.98 in 1 year if the interest is compounded semiannually?

In this section we use one variable to solve the applications even though most of the problems involve two unknowns. It is perfectly permissible to use two variables and to use substitution to reduce the problem to a one-variable problem. In Example 2, it would be permissible to let y be the length in meters of each side of the larger square. Then note that $y = x + 1$ and substitute $x + 1$ for y throughout the rest of the problem.

EXAMPLE 2 Finding the Dimensions of Two Tarps

A square tarp is used to shield a sports area from the rain. A new design for the tarp is 1 m longer on each side. The combined coverage area of the two tarps is 61 m^2. Find the dimensions of each tarp.

SOLUTION

Let x = length of each side of smaller tarp, m
$x + 1$ = length of each side of larger tarp, m

The new tarp is 1 m longer on each side.

VERBALLY

The sum of their areas is 61 m^2.

ALGEBRAICALLY

$$x^2 + (x + 1)^2 = 61$$
$$x^2 + x^2 + 2x + 1 = 61$$
$$2x^2 + 2x - 60 = 0$$
$$x^2 + x - 30 = 0$$
$$(x - 5)(x + 6) = 0$$
$$x - 5 = 0 \quad \text{or} \quad x + 6 = 0$$
$$x = 5 \qquad\qquad x = -6$$
$$x + 1 = 6 \qquad\qquad x + 1 = -5$$

Form a word equation. Then use the formula $A = s^2$, the formula for the area of a square, to translate this statement into an algebraic equation.

Write the quadratic equation in standard form, and simplify by dividing both sides of the equation by 2. Then solve this equation by factoring. The form $(x - ?)(x + ?)$ is used to factor the left side of the equation by inspection.

Be sure to find $x + 1$ as well as x.

The negative values -6 and -5 are not meaningful dimensions.

NUMERICALLY

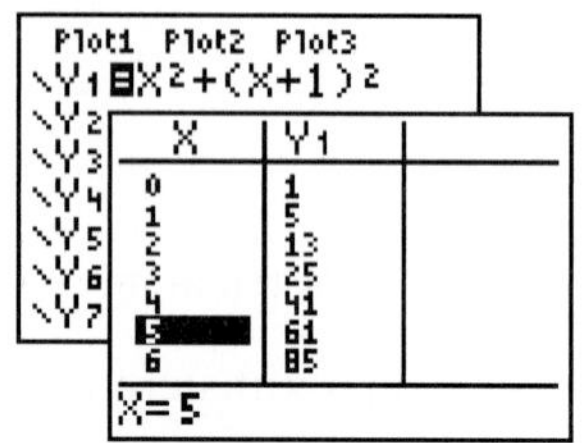

The table reveals the value of 5 for x. Then $x + 1$ must be 6.

Answer: The dimensions of the tarps are, respectively, 5 and 6 m on each side.

Check:

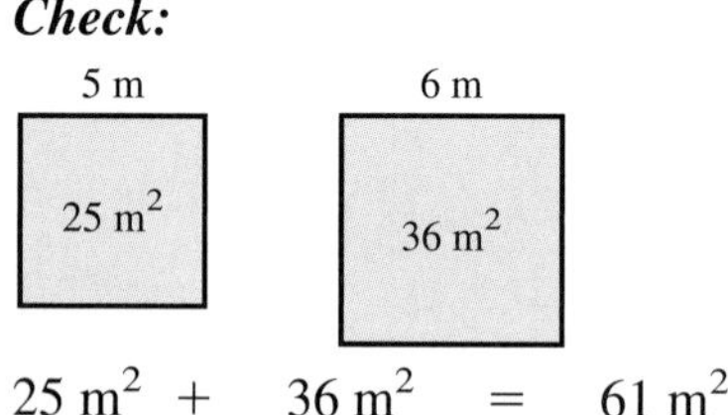

$$25 \text{ m}^2 + 36 \text{ m}^2 = 61 \text{ m}^2$$

The danger in just guessing the solution to a word problem, even if you are correct and the answer checks, is that you can overlook other solutions. The solutions you overlook can actually be preferable if they are cheaper or better in some other way.

Sometimes it is easy to determine an answer to a word problem without taking time to write the equation. These answers may even check. What you risk by not using algebra to form an equation is to overlook other possible solutions. In a practical problem, this may mean an engineer or a businessperson has overlooked another solution that may be cheaper or has preferable attributes over the obvious solution. Example 3 has two pairs of solutions. It is common for some students to overlook one of these pairs.

EXAMPLE 3 Finding Consecutive Even Integers with a Known Product

Find two consecutive even integers whose product is 48.

SOLUTION

Let n = smaller of two even integers
$n + 2$ = larger of two even integers

Consecutive even integers (such as 2, 4, 6, and 8) differ by 2 units.

VERBALLY

Smaller number times larger number equals 48.

Form a word equation, and translate it into an algebraic equation.

ALGEBRAICALLY

$$n(n + 2) = 48$$
$$n^2 + 2n = 48$$
$$n^2 + 2n - 48 = 0$$
$$(n - 6)(n + 8) = 0$$

Write the quadratic equation in standard form, and then solve it by factoring. The form $(n - ?)(n + ?)$ is used to factor the left side of the equation by inspection.

$n - 6 = 0$ or $n + 8 = 0$
$n = 6$ $\quad n = -8$
$n + 2 = 8$ $\quad n + 2 = -6$

Be sure to find $n + 2$ as well as n.

Answer: The consecutive even integers are either -8 and -6 or 6 and 8.

Both pairs of even integers check because their product is 48. Be careful if you guess or use a table to check values. It is easy to overlook one pair of answers.

SELF-CHECK 7.6.2

1. One number is 3 more than another, and their product is 40. Find both numbers.

EXAMPLE 4 Determining the Height of Fireworks

A faulty fireworks rocket is launched vertically with an initial velocity of 96 ft/s and then falls to the earth unexploded. Its height h in feet after t seconds is given by $h(t) = -16t^2 + 96t$. During what time interval after launch will the height of the rocket exceed 90 ft? The given figure shows the relationship of the height to the time. It does not show the path of the rocket, which goes straight up and then down.

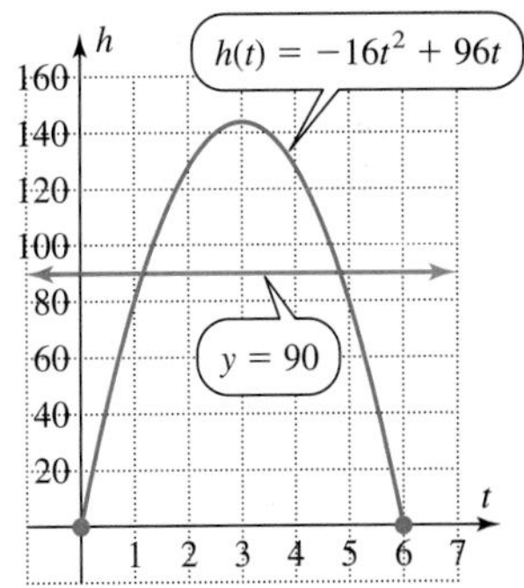

SOLUTION

$$h(t) = -16t^2 + 96t$$
$$90 = -16t^2 + 96t$$

First determine the times when the height is 90 ft.

$$16t^2 - 96t + 90 = 0$$
$$8t^2 - 48t + 45 = 0$$

Write the quadratic equation in standard form, and then divide both sides of the equation by 2.

$$t = \frac{48 \pm \sqrt{(-48)^2 - 4(8)(45)}}{2(8)}$$

Solve this equation by using the quadratic formula.

$$t = \frac{48 \pm \sqrt{864}}{16}$$

$t = \dfrac{48 - \sqrt{864}}{16}$ or $t = \dfrac{48 + \sqrt{864}}{16}$

These are the exact values of t.

$t \approx 1.2$ $\quad t \approx 4.8$

Use a calculator to approximate these values to the nearest tenth of a second. At these times the height of the rocket is 90 ft.

Answer: The height of the rocket will exceed 90 ft between 1.2 and 4.8 seconds after launch.

Using the graph above, note the parabola is above $y = 90$ for t values between 1.2 and 4.8. The height will be over 90 ft from 1.2 to 4.8 seconds after launch.

A Mathematical Note

The Greek mathematician Pythagoras (c. 500 B.C.) taught orally and required secrecy of his initiates. Thus records of the society formed by Pythagoras are anecdotal. His society produced a theory of numbers that was part science and part numerology. It assigned numbers to many abstract concepts, for example, 1 for reason, 2 for opinion, 3 for harmony, and 4 for justice. The star pentagon was the secret symbol of the Pythagoreans.

SELF-CHECK 7.6.3

1. Rework Example 4 to determine the time interval for which the rocket is above 80 ft.

The triangle is one of the most basic shapes in geometry and in the world around us. A **right triangle** is a triangle containing a 90° angle; that is, two sides of the triangle are perpendicular. In a right triangle, the two shorter sides are called the **legs** and the longer side is called the **hypotenuse.** The lengths of the three sides can be denoted, respectively, by a, b, and c. (See the figure.)

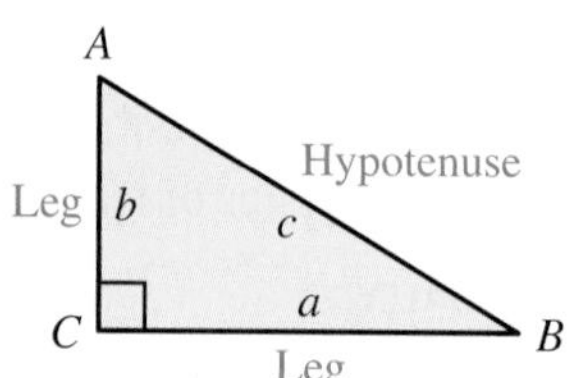

Right triangles and the Pythagorean theorem play a key role in solving many applied problems. The Pythagorean theorem is given in the following box.

Pythagorean Theorem

If triangle ABC is a right triangle, then:

ALGEBRAICALLY	GEOMETRIC EXAMPLE	VERBALLY
$a^2 + b^2 = c^2$ (The converse of this theorem is also true. If $a^2 + b^2 = c^2$, then triangle ABC is a right triangle.)	3 5 4 $3^2 + 4^2 = 5^2$ $9 + 16 = 25$	The sum of the areas of the squares formed on the legs of a right triangle is equal to the area of the square formed on the hypotenuse.

Examples 5, 6, and 7 all involve the Pythagorean theorem and quadratic equations.

EXAMPLE 5 Using the Pythagorean Theorem to Solve an Application

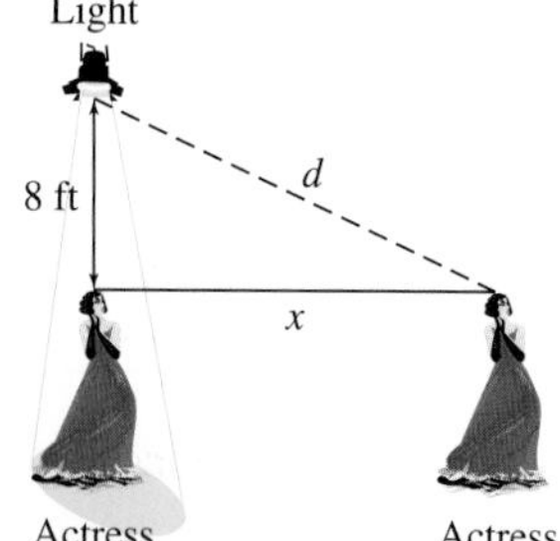

An actress has a spotlight located 8 ft above her head. If the limit of the light's effective range from her head is 17 ft, how far can she move from underneath this light?

SOLUTION

Let x = number of feet she can move across the floor
d = number of feet from spotlight to top of her head

$d^2 = x^2 + 8^2$ $d^2 = x^2 + 64$	Use the Pythagorean theorem and the values identified in the given sketch.
$17^2 = x^2 + 64$	Substitute 17 for d.
$289 = x^2 + 64$ $225 = x^2$ $x^2 = 225$ $x = \pm\sqrt{225}$	Simplify this quadratic equation, and use extraction of roots to solve this equation.
$x = 15$ or $x = -15$ (-15 is not a meaningful answer)	Since the distance moved is interpreted to be positive, use only the answer of $+15$ ft.

Answer: She can move 15 ft in any direction and stay within the effective range of the light.

Example 6 is based on the analysis of structural components for a building. The shape that results may not be exactly a right triangle. However, this mathematical model is close enough to the actual result to yield useful information.

EXAMPLE 6 Determining the Deflection of a Beam

A steel beam that is 20 ft long is placed in a hydraulic press for testing. Under extreme pressure the beam deflects upward so that the distance between the hydraulic rams decreases by 2 in. Assume the deflected shape is approximated by the figure shown. Determine the height of the bulge in the middle of the beam.

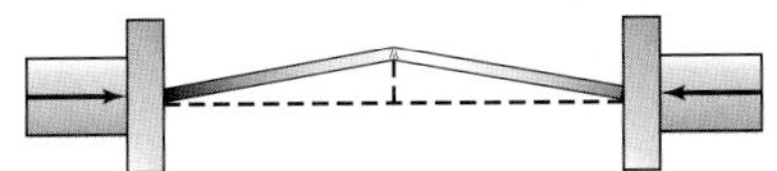

SOLUTION

Let x = height of bulge, in

First identify the unknown value with a variable. Then note that the shape in the sketch is a right triangle, so you can apply the Pythagorean theorem.

VERBALLY

$$\left(\begin{array}{c}\text{Length to}\\ \text{center point}\end{array}\right)^2 + \left(\begin{array}{c}\text{Height of}\\ \text{bulge}\end{array}\right)^2 = \left(\begin{array}{c}\text{Length of left}\\ \text{half of beam}\end{array}\right)^2$$

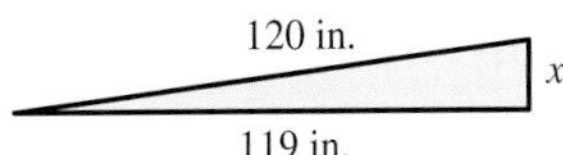

ALGEBRAICALLY

$$119^2 + x^2 = 120^2$$
$$14{,}161 + x^2 = 14{,}400$$
$$x^2 = 239$$
$$x = \pm\sqrt{239}$$

One-half the length of the beam is 10 ft or 120 in. One-half the distance between the presses is 119 in, which is 1 in less than 120 in. Substitute these values into the verbal equation.

$x = \sqrt{239}$ or $x = -\sqrt{239}$
$x \approx 15.5$ $\quad x \approx -15.5$

Solve this quadratic equation by extraction of roots. Then approximate these solutions.

The beam could deflect either up or down 15.46 in.

Answer: The bulge would be approximately 15.5 in.

This large a bulge is not tolerable in most building projects. Therefore engineers plan for expansion and contraction spaces in their designs.

SELF-CHECK 7.6.4

1. Two legs of a right triangle are 5 and 12 cm. Determine the length of the hypotenuse.
2. The hypotenuse of a right triangle is 25 cm, and one leg is 7 cm. How long is the other leg?
3. In Example 6, determine the height of the bulge if the rams decrease the distance from 20 ft to 19 ft 11 in.

Example 7 uses both the Pythagorean theorem and the formula $D = R \cdot T$ for calculating the distance for a given rate and time.

EXAMPLE 7 Applying the Pythagorean Theorem

Two airplanes depart simultaneously from an airport. One flies due south; the other flies due east at a rate 50 mi/h faster than that of the first airplane. After 2 hours, radar indicates that the airplanes are 500 mi apart. What is the ground speed of each airplane?

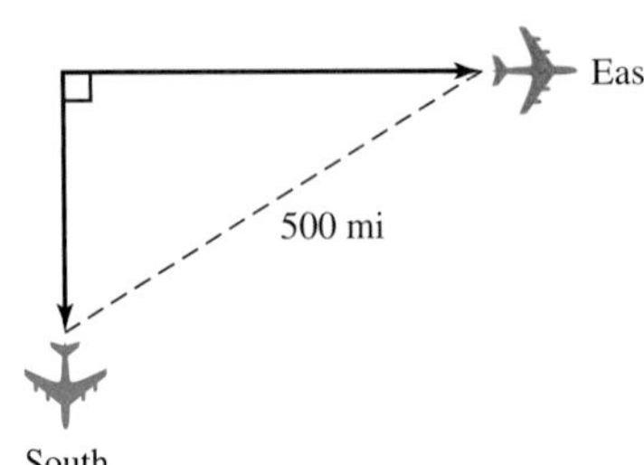

Make a sketch that models the problem.

SOLUTION

Let r = ground speed of first plane, mi/h
$r + 50$ = ground speed of second plane, mi/h

Represent each unknown quantity by using a variable. To be consistent with the formula $D = R \cdot T$, we use the variable r to represent speed.

VERBALLY

$$\begin{pmatrix}\text{Distance first}\\ \text{plane travels}\end{pmatrix}^2 + \begin{pmatrix}\text{Distance second}\\ \text{plane travels}\end{pmatrix}^2 = \begin{pmatrix}\text{Distance between}\\ \text{two planes}\end{pmatrix}^2$$

The word equation is based on the Pythagorean theorem.

ALGEBRAICALLY

Using $D = R \cdot T$ and a time of 2 hours, we find the first plane travels $2r$ mi and the second travels $2(r + 50)$ mi.

$$(2r)^2 + [2(r + 50)]^2 = (500)^2$$
$$(2r)^2 + (2r + 100)^2 = (500)^2$$

Substitute the distances into the word equation and simplify.

$$4r^2 + (4r^2 + 400r + 10{,}000) = 250{,}000$$
$$8r^2 + 400r - 240{,}000 = 0$$

Write the quadratic equation in standard form.

$$r^2 + 50r - 30{,}000 = 0$$

Divide both sides of the equation by 8.

$$(r - 150)(r + 200) = 0$$

Solve this quadratic equation by factoring.

$r - 150 = 0$ or $r + 200 = 0$

A negative rate is not meaningful for this problem.

$r = 150$ $\quad$ $r = -200$

$r + 50 = 200$ $\quad$ (-200 is not meaningful)

Find both r and $r + 50$.

Answer: The first plane is flying south at 150 mi/h, and the second is flying east at 200 mi/h.

Do these rates check, and are they reasonable?

SELF-CHECK 7.6.5

1. Two military airplanes depart simultaneously from an aircraft carrier. One flies due south; the other flies due east at a rate 100 mi/h faster than that of the first airplane. After $\frac{1}{2}$ hour, radar indicates that the airplanes are 250 mi apart. What is the ground speed of each airplane?

The most common description for the size of a computer monitor is the diagonal measurement of the screen. Based on that specification, two different monitors could have the same diagonal dimension yet have different viewable areas. Another technical specification typically listed for a computer monitor is the aspect ratio. The aspect ratio is the ratio of the width to the height of the screen. In Example 8 the listed specifications for a monitor are used to calculate the dimensions of the screen.

EXAMPLE 8 Using the Aspect Ratio and Screen Size to Determine Screen Dimensions

A computer monitor is advertised as a 21-inch monitor with an 8:5 aspect ratio. Determine the actual dimensions and viewable area of this monitor.

SOLUTION

Let x = height of screen

$\frac{8x}{5}$ = width of screen

An aspect ratio of 8:5 indicates the width is $\frac{8}{5}$ as large as the height.

VERBALLY

$$\begin{pmatrix}\text{Screen}\\\text{width}\end{pmatrix}^2 + \begin{pmatrix}\text{Screen}\\\text{height}\end{pmatrix}^2 = \begin{pmatrix}\text{Screen}\\\text{diagonal}\end{pmatrix}^2$$

The word equation is based on the Pythagorean theorem.

ALGEBRAICALLY

$$\left(\frac{8x}{5}\right)^2 + x^2 = 21^2$$

$$\frac{64x^2}{25} + x^2 = 441$$

Substitute the dimensions into the word equation and simplify.

$$25\left(\frac{64x^2}{25} + x^2\right) = 25(441)$$

$$64x^2 + 25x^2 = 11{,}025$$

Multiply both sides of the equation by 25 to clear the equation of fractions.

$$89x^2 = 11{,}025$$

$$x^2 = \frac{11{,}025}{89}$$

$$x = \pm\sqrt{\frac{11{,}025}{89}}$$

Combine like terms and use extraction of roots to solve for x.

$$x = \sqrt{\frac{11{,}025}{89}} \quad \text{or} \quad x = -\sqrt{\frac{11{,}025}{89}}$$

$$x \approx 11.1 \qquad\qquad x \approx -11.1$$

$$\frac{8x}{5} \approx 17.8 \qquad\qquad (-11.1 \text{ is not meaningful})$$

A negative value for the screen height is not meaningful. Find both x and $\frac{8x}{5}$.

Answer: The height of the screen is 11.1 in, and the width of the screen is 17.8 in. The viewable area of this monitor is 197.6 in^2.

The viewable area is found by multiplying the screen width by the screen height.

SELF-CHECK 7.6.6

1. A computer monitor is advertised as a 21-inch monitor with a 4:3 aspect ratio. Determine the actual dimensions and viewable area of this monitor.

SELF-CHECK ANSWERS

7.6.1

1. At a rate of 7% compounded semiannually, a principal of \$800 will grow to \$856.98 in 1 year.

7.6.2

1. The numbers are either -8 and -5 or 5 and 8.

7.6.3

1. The height of the fireworks rocket will exceed 80 ft between 1 and 5 seconds after launch.

7.6.4

1. 13 cm 2. 24 cm
3. The bulge would be approximately 10.9 in.

7.6.5

1. The first plane is flying south at 300 mi/h, and the second is flying east at 400 mi/h.

7.6.6

1. The height of the screen is 12.6 in, and the width of the screen is 16.8 in. The viewable area is 211.7 in^2.

USING THE LANGUAGE AND SYMBOLISM OF MATHEMATICS 7.6

1. The first step in the word problem strategy given in this book is to read the problem carefully to determine what you are being asked to __________.
2. The second step in the word problem strategy given in this book is to select a __________ to represent each unknown quantity.
3. The third step in the word problem strategy given in this book is to model the problem verbally with word equations and then to translate these word equations into __________ equations.
4. The fourth step in the word problem strategy given in this book is to __________ the equation or system of equations and answer the question asked by the problem.
5. The fifth step in the word problem strategy given in this book is to check your answer to make sure the answer is __________.
6. In the formula $D = RT$, D represents distance, R represents __________, and T represents __________.
7. In a right triangle, the two shorter sides are called the __________ and the longer side is called the __________.
8. The Pythagorean theorem states that for a right triangle with hypotenuse c and sides of length a and b, __________.
9. The __________ domain of a function is the set of values of the input variable that are permissible in an application.

QUICK REVIEW 7.6

The purpose of this quick review is to help you recall skills needed in this section.

1. If n represents an even integer, represent the next two integers in terms of n.
2. If n represents an even integer, represent the next two even integers in terms of n.
3. Solve $x^2 = 64$ by extraction of roots.
4. The function $L(x) = 100 - 5x$ models the length of wire remaining on a 100-ft roll after an electrician has cut off five equal pieces of length x. Identify the practical domain for this function.
5. The function $L(x) = 100 - 5x$ models the length of wire remaining on a 100-ft roll after an electrician has cut off x pieces of length 5 ft. Identify the practical domain for this function.

EXERCISES 7.6

In Exercises 1–46, solve each problem.

Numeric Word Problems (Exercises 1–8)

1. Find two consecutive integers whose product is 156.
2. Find two consecutive integers whose product is 240.
3. Find two consecutive even integers whose product is 288.
4. Find two consecutive odd integers whose product is 99.
5. The sum of the squares of two consecutive integers is 113. Find these integers.
6. The sum of the squares of three consecutive integers is 50. Find these integers.
7. One number is 4 more than three times another number. Find these numbers if their product is 175.
8. One number is 3 less than four times another number. Find these numbers if their product is 10.

Dimensions of a Rectangle

9. The length of a rectangle (see the figure) is 2 cm more than three times the width. Find the dimensions of this rectangle if the area is 21 cm^2.

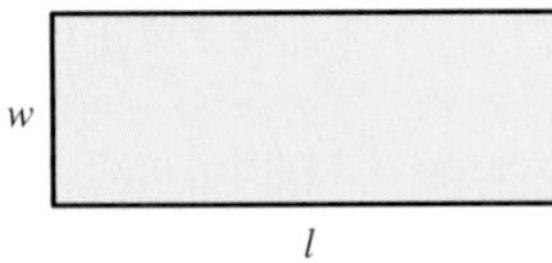

10. The length of a rectangle (see the figure) is 10 m less than twice the width. Find the dimensions of this rectangle if the area is 72 m^2.

Dimensions of a Triangle

11. The base of a triangle (see the figure) is 3 m longer than the height. Find the base if the area of this triangle is 77 m^2.

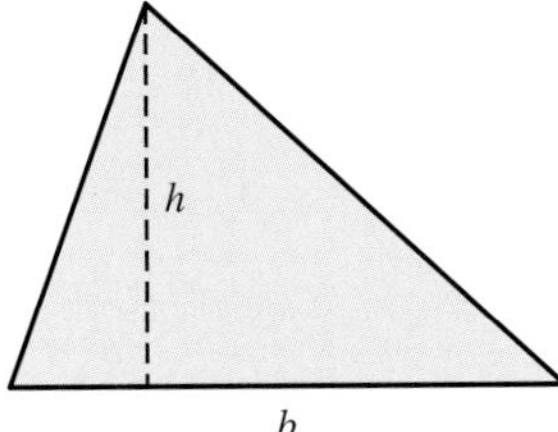

12. The base of the triangle shown in the figure is 5 cm longer than the height. Determine the height of the triangle if its area is 12 cm^2.

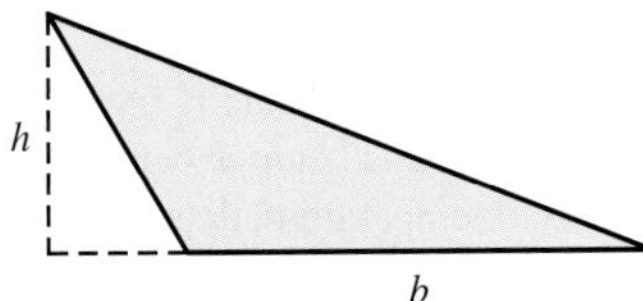

Dimensions of a Square

13. If each side of a square is increased by 5 cm, the total area of both the new square and the original square will be 200 cm^2. Approximate to the nearest tenth of a centimeter the length of each side of the original square.

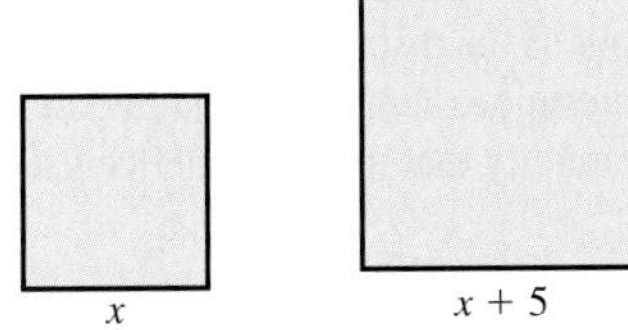

14. If each side of a square is increased by 2 cm, the total area of both the new square and the original square will be 400 cm^2. Approximate to the nearest tenth of a centimeter the length of each side of the original square.

15. **Trajectory of a Projectile** The height in meters of a ball released from a ramp is given by the function $h(t) = -4.9t^2 + 29.4t + 34.3$, where t represents the time in seconds since the ball was released from the end of the ramp. Determine the time interval that the ball is above 50 m.

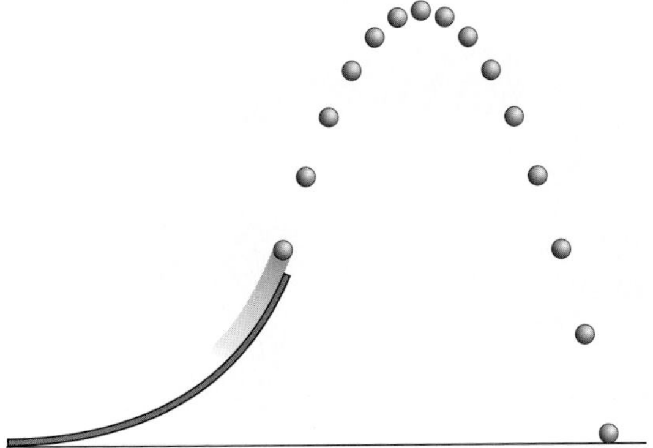

16. **Height of a Golf Ball** The height in feet of a golf ball hit from an elevated tee box is given by $h(t) = -16t^2 + 60t + 25$, where t represents the time in seconds the ball is in flight.
    a. Determine the number of seconds before the golf ball hits the ground at a height level of 0 ft.
    b. Determine the time interval for which the ball is above 25 ft.

17. **Profit Polynomial** The business manager of a small windmill manufacturer projects that the profit in dollars from making x windmills per week will be $P(x) = -x^2 + 70x - 600$. How many windmills should be produced each week in order to generate a profit? (*Hint:* Only an integral number of units can be produced.)

18. **Average Cost** The president of a company producing water pumps has gathered data suggesting that the average cost in dollars of producing x units per hour of a new style of pump will be $C(x) = x^2 - 22x + 166$. Determine the number of pumps per hour to produce to keep the average cost below \$75 per pump.

19. **Radius of a Circle** The area enclosed between two concentric circles is 56π cm^2. The radius of the larger circle is 1 cm less than twice the radius of the smaller circle. Determine the length of the shorter radius. (See the figure.)

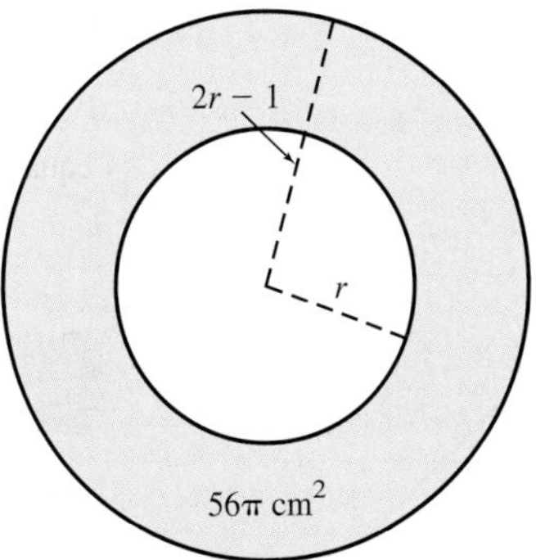

20. **Dimensions of a Mat** A square mat used for athletic exercises has a uniform red border on all four sides. The rest of the mat is blue. The width of the blue square is three-fourths the width of the entire square. If the area colored red is 28 m^2, determine the length of each side of the mat.

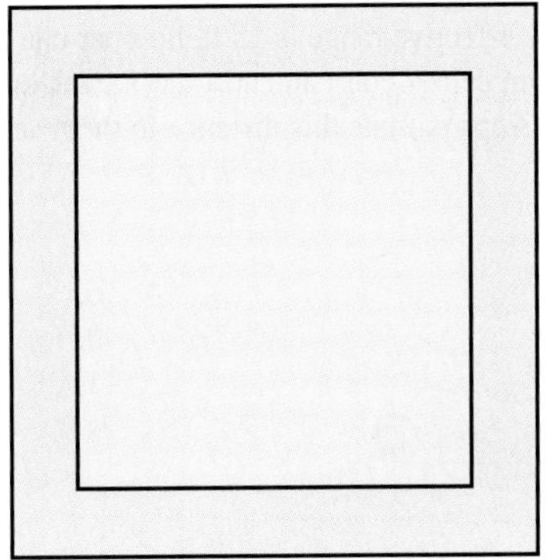

21. If the bottom of a 17-ft ladder is 8 ft from the base of the chimney, how far is it from the bottom of the chimney to the top of the ladder?

22. **Length of a Guy Wire** A television tower has a guy wire attached 40 ft above the base of the tower. If the anchor point of the guy wire is 42 ft from the base of the tower, how long is the guy wire?

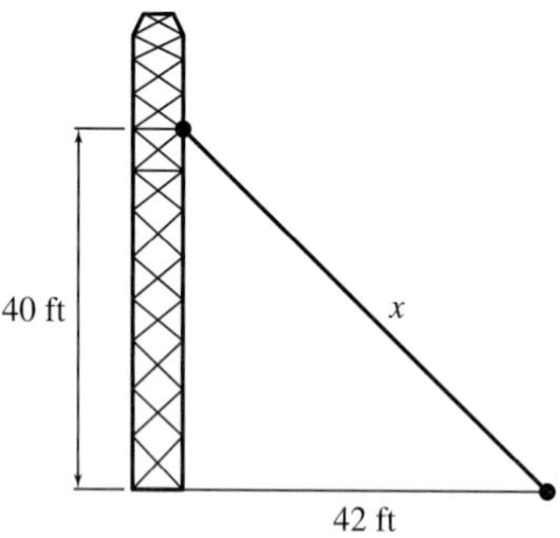

23. **Positioning Effective Lighting** The lighting system planned for a basketball court has bulbs that should be placed at least 30 ft above the floor for safety reasons. One bulb is located directly above the center of the court at a height of 30 ft. A ball is located at a position x ft from center court. If the limit of the light's effective range is 35 ft, how far can the ball be positioned from center court and still receive adequate light from this bulb? Approximate this distance to the nearest tenth of a foot.

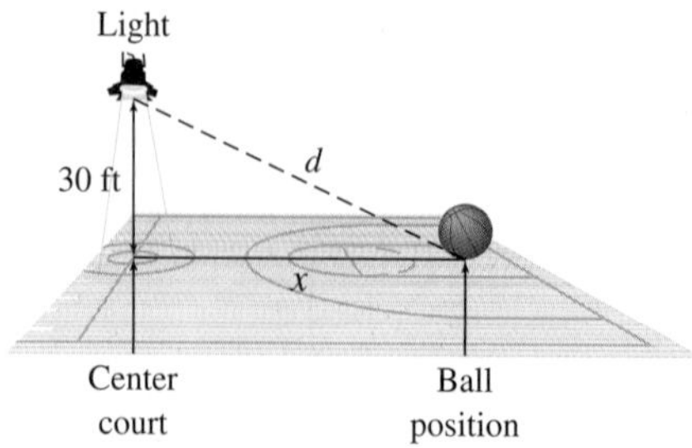

24. **Positioning an Effective Sound System** The sound system planned for an airport concourse has speakers that should be placed at least 18 ft above the floor and thus approximately 12 ft above the heads of most customers. A customer is standing at a position x ft from a reference point directly under one speaker that is 18 ft above the floor. If the limit of the speaker's effective range is 20 ft, approximate to the nearest foot the distance a customer can walk from the reference point and stay within the effective range of the speaker.

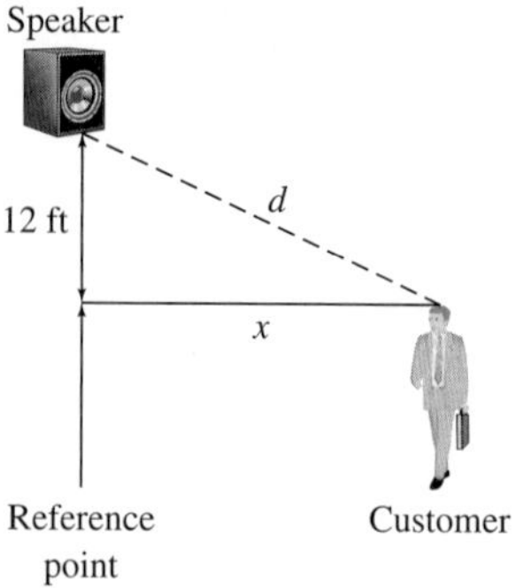

25. **Deflection of a Beam** A 40-ft concrete beam is fixed between two rigid anchor points. If this beam expands by 0.5 inch due to a 50° increase in temperature, determine the bulge in the middle of the beam. (Assume for simplicity that the beam deflects as shown in this diagram.)

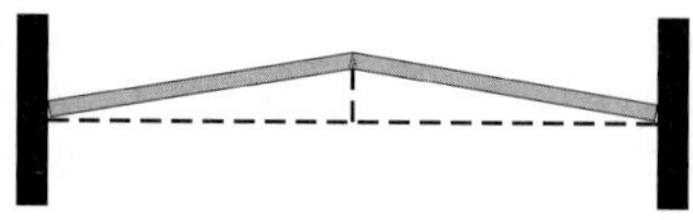

26. **Deflection of a Beam** A 28-ft aluminum beam is tested between two hydraulic rams. If the distance between the rams is decreased by 0.25 in, determine the bulge in the middle of the beam. (Assume for simplicity that the beam deflects as shown in this diagram.)

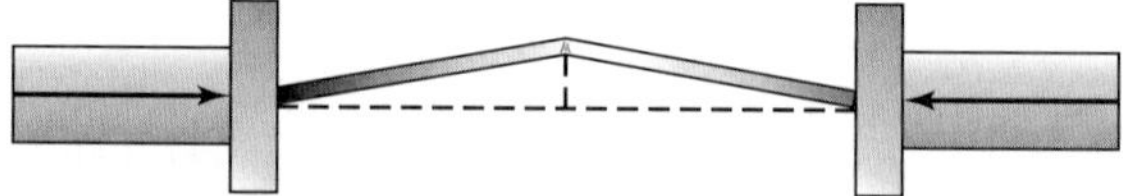

27. **Screen Size** The size of a computer monitor is usually given as the length of a diagonal of the screen. A new computer comes with a 17-in monitor. Give the height of the screen if the width is 13 in.

28. **Screen Size** A handheld organizer offers a 3.5-in screen (diagonal length). Approximate its width to the nearest hundredth if its height is 3.0 in.

29. **Screen Size** A computer monitor is advertised as a 17-in monitor with a 4:3 aspect ratio. Determine the actual dimensions and viewable area of this monitor.

30. **Screen Size** A computer monitor is advertised as a 19-in monitor with a 5:4 aspect ratio. Determine the actual dimensions and viewable area of this monitor.
31. **Pizza Size** A couple plans to order a 10-in pizza when another couple they know joins them. Approximately what size pizza should they order to double the amount in a 10-in pizza?
32. **Spaghetti Recipe** A recipe for spaghetti for four people suggests using enough spaghetti to fill a 1-in-diameter opening on a spaghetti measuring device. What size hole would be recommended for two people?
33. **Distance by Plane** Upon leaving an airport, an airplane flew due south and then due east. After it had flown 17 mi farther east than it had flown south, it was 25 mi from the airport. How far south had it flown?

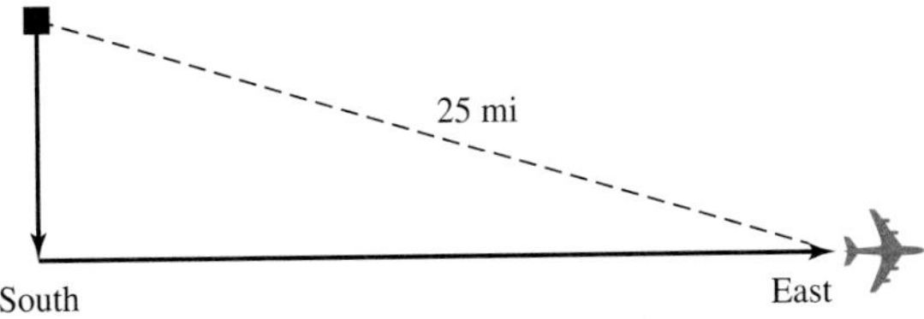

34. **Distance by Plane** Upon leaving an airport, an airplane flew due west and then due north. After it had flown 1 mi farther north than it had flown west, it was 29 mi from the airport. How far west had it flown?

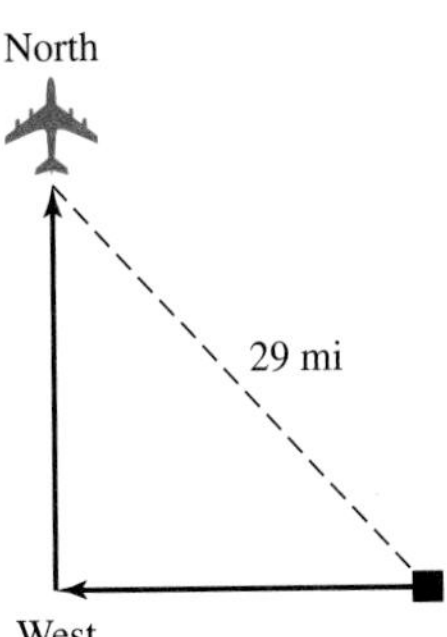

35. **Speed of an Airplane** Two airplanes depart simultaneously from an airport. One flies due south; the other flies due east at a rate 30 mi/h faster than that of the first airplane. After 3 hours, radar indicates that the airplanes are 450 mi apart. What is the ground speed of each airplane?

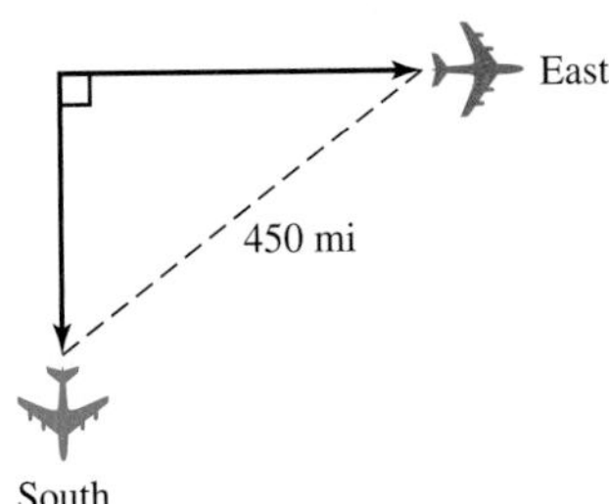

36. **Speed of an Airplane** Two airplanes depart simultaneously from an airport. One flies due south; the other flies due east at a rate 10 mi/h faster than that of the first airplane. After 1 hour, radar indicates that the airplanes are 290 mi apart. What is the ground speed of each airplane?

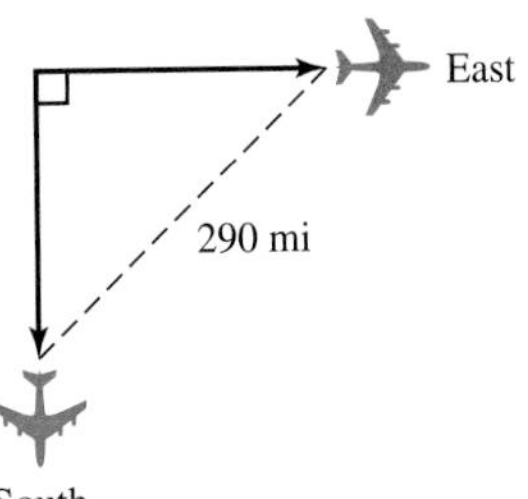

37. **Room Width** Examining the blueprints for a rectangular room that is 17 ft longer than it is wide, an electrician determines that a wire run diagonally across this room will be 53 ft long. What is the width of the room?

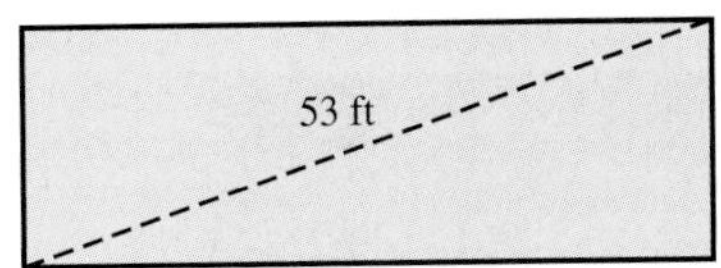

38. **Diameter of a Storage Bin** The length of the diagonal brace in the cylindrical storage bin shown in the figure is 17 m. If the height of the cylindrical portion of the bin is 7 m more than the diameter, determine the diameter.

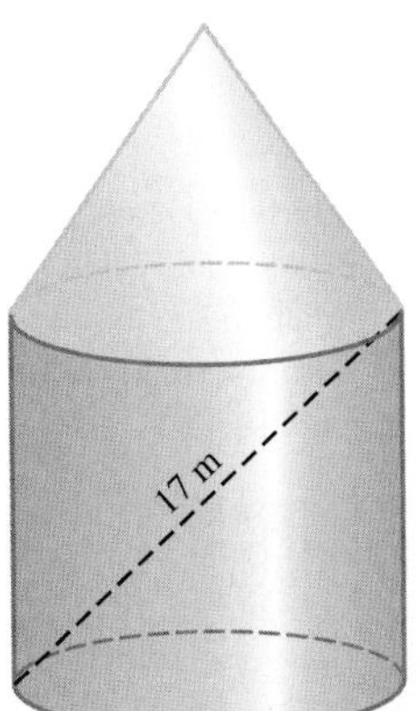

39. **Rope Length** The length of one piece of rope is 8 m less than twice the length of another piece of rope. Each rope is used to enclose a square region. The area of the region enclosed by the longer rope is 279 m^2 more than the area of the region enclosed by the shorter rope. Determine the length of the shorter rope.

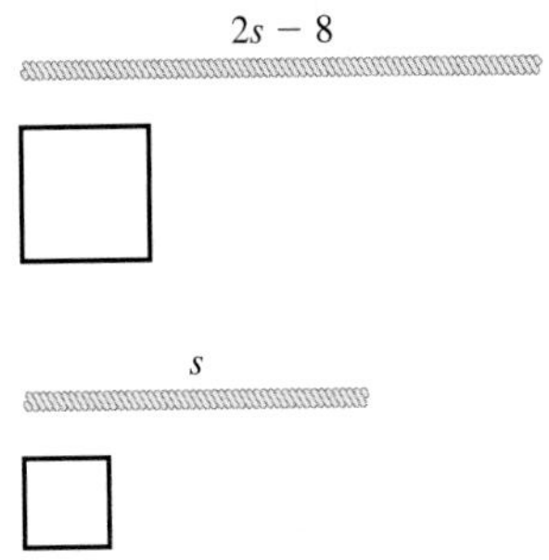

40. **Dimensions of Poster Board** A rectangular piece of poster board is 6.0 cm longer than it is wide. A 1-cm strip is cut off each side. The area of the remaining poster board is 616 cm^2. Find the original dimensions.

41. **Distances on a Baseball Diamond** The bases on a baseball diamond are placed at the corners of a square whose sides are 90 ft long. How much farther does a catcher have to throw the ball to get it from home plate to second base than from home plate to third base?

42. **Speed of a Baseball** If a catcher on a baseball team throws a baseball at 120 ft/s (over 80 mi/h), approximately how long will it take his throw to go from home plate to second base? Approximately how long will it take his throw to go from home plate to third base? (*Hint:* See Exercise 41.)

43. **Interest Rate** The formula for computing the amount A of an investment of principal P invested at interest rate r for 1 year and compounded semiannually is $A = P\left(1 + \frac{r}{2}\right)^2$. Approximately what interest rate is necessary for $1,000 to grow to $1,095 in 1 year if the interest is compounded semiannually?

44. **Interest Rate** The formula for computing the amount A of an investment of principal P invested at interest rate r for 1 year and compounded semiannually is $A = P\left(1 + \frac{r}{2}\right)^2$. Approximately what interest rate is necessary for $1,000 to grow to $1,075 in 1 year if the interest is compounded semiannually?

45. **Distance to the Horizon** The radius of the Earth is approximately 4,000 mi. Approximate to within 10 mi the distance from the horizon to a plane flying at an altitude of 4 mi.

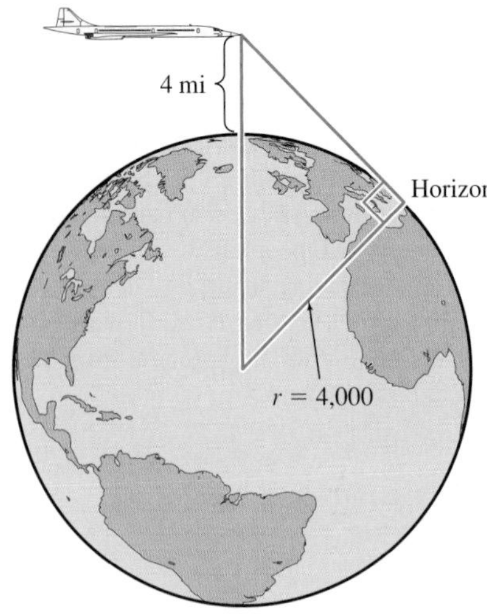

46. **Distance to the Horizon** The radius of the Earth is approximately 4,000 mi. Approximate to within 10 mi the distance from the horizon to a plane flying at an altitude of 5 mi. (See the figure for Exercise 45.)

47. **Metal Machining** A round stock of metal that is $18\sqrt{2}$ cm in diameter is milled into a square piece of stock. How long are the sides of the largest square that can be milled?

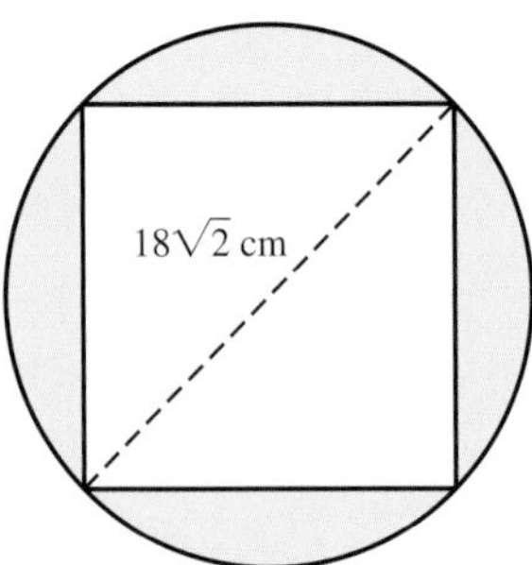

48. **Metal Machining** A round stock of metal that is 30 cm in diameter is milled into a square piece of stock. How long are the sides of the largest square that can be milled? (See the figure in Exercise 47.)

49. **Dimensions of a Trough** A metal sheet 60 cm wide is used to form a trough by bending up each side as illustrated in the given figure.
   a. Write a function $A(h)$ that gives the cross-sectional area for a height h.
   b. Determine the practical domain for this function.
   c. Evaluate and interpret $A(5)$.
   d. Determine the height of each side if the cross-sectional area is 400 cm^2.
   e. Determine the maximum cross-sectional area and the height that produces this area.

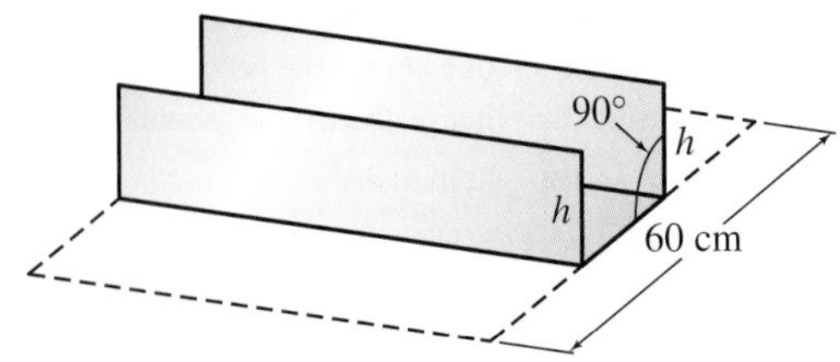

50. **Dimensions of a Trough** A metal sheet 80 cm wide is used to form a trough by bending up each side as illustrated in the figure for Exercise 49.
    a. Write a function $A(h)$ that gives the cross-sectional area for a height h.
    b. Determine the practical domain for this function.
    c. Evaluate and interpret $A(5)$.
    d. Determine the height of each side if the cross-sectional area is 750 cm^2.
    e. Determine the maximum cross-sectional area and the height that produces this area.

Golf Ball Distance

To make more realistic games, the computer gaming industry uses data to generate equations to model the flight of a golf ball. This is done by using very sophisticated cameras and sensors. In Exercises 51 and 52, x represents the horizontal distance the golf ball travels away from the golfer, and y represents the height to which the golf ball rises. Use a graphing calculator and the points given in the table.

a. Draw a scatter diagram of these points.
b. Determine the quadratic function that best fits these points. (Round the coefficients to three significant digits.)
c. Use this function f to approximate $f(50)$.
d. Interpret the meaning of the value found in part **c**.
e. Determine the x-intercepts of the graph of this function.
f. Interpret the meaning of these x-intercepts found in part **e**.
g. Determine the vertex of this parabola.
h. Interpret the meaning of the vertex found in part **g**.

51.

x yd	y yd
0	0
40	12
80	18
120	18

52.

x yd	y yd
0	0
40	22
80	24
120	6

The graphs in Exercises 53 and 54 show the height of a ball in feet based on time.

53. Use the graph to estimate the following.
    a. The initial height of the ball
    b. The time when the ball will hit the ground
    c. The time interval when the height of the ball is increasing
    d. The time interval when the height of the ball is decreasing
    e. The time when the ball will reach its maximum height
    f. The maximum height of the ball

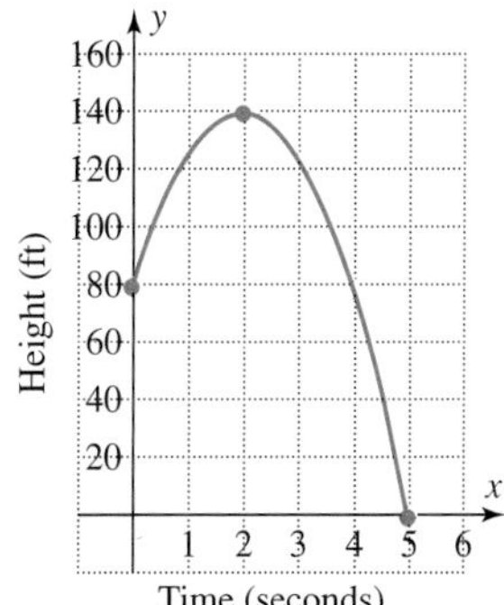

54. Use the graph to estimate the following.
    a. The initial height of the ball
    b. The time when the ball will hit the ground
    c. The time interval when the height of the ball is increasing
    d. The time interval when the height of the ball is decreasing
    e. The time when the ball will reach its maximum height
    f. The maximum height of the ball

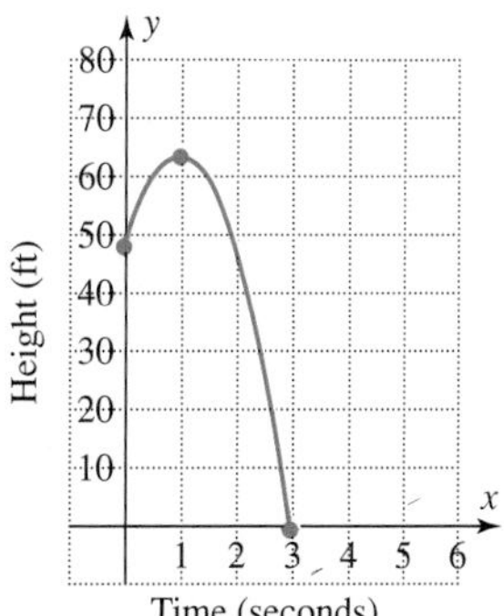

55. **Height of a Ball** The height of a ball in feet after t seconds is given by $h(t) = -16t^2 + 90t + 8$.
    a. When will the ball hit the ground?
    b. When will the ball be above 100 ft?
    c. When will the ball be above 120 ft?
    d. When will the ball be above 130 ft?
    e. What is the maximum height reached by this ball?
    f. When will the ball reach its maximum height?

Group Discussion Questions

56. **Challenge Question** Given the equation $x^2 + (x + 2)^2 = 244$:
    a. Write a numeric word problem that is modeled by this equation.
    b. Write a word problem that models the areas of two squares. (See Exercise 13.)
    c. Write a word problem that models the distance between two airplanes. (See Exercise 33.)
57. **Discovery Question**

    a. Draw a quadrilateral (4 sides). Then determine the number of diagonals a quadrilateral has.
    b. Draw a pentagon (5 sides). Then determine the number of diagonals a pentagon has.
    c. Draw a hexagon (6 sides). Then determine the number of diagonals a hexagon has.

d. Use these data and a graphing calculator with **QuadReg** to calculate the equation of the parabola of best fit for these data.
e. Use this formula to complete the table for polygons with 7, 8, 9, and 10 sides. Do these data check?
f. Use this formula to determine how many sides a polygon has if it has 65 diagonals.
g. Assume that each vertex represents a city. Discuss why a package delivery company or a communications company might be interested in the number of sides and diagonals a polygon has.

NUMBER OF SIDES x	NUMBER OF DIAGONALS y
4	
5	
6	
7	
8	
9	
10	
	65

Section 7.7 Complex Numbers and Solving Quadratic Equations with Complex Solutions

Objectives:

17. Express complex numbers in standard form.
18. Add, subtract, multiply, and divide complex numbers.
19. Solve a quadratic equation with imaginary solutions.

A Mathematical Note

Jean-Victor Poncelet (1788–1867) was a lieutenant in Napoleon's army. After being left for dead on the Russian front, he was captured and marched for months across frozen plains in subzero weather. To survive the boredom of captivity, he decided to reproduce as much mathematics as he could. When he returned to France after the war, he had developed projective geometry and a geometric interpretation of imaginary numbers.

Complex Numbers

Some quadratic equations do not have real number solutions. For example, $x^2 = -1$ has no real solutions. The square of a negative number is positive, the square of 0 is 0, and the square of a positive number is positive. Thus there is no real number that satisfies $x^2 = -1$. In the 17th century, mathematicians first defined the number i so that $i^2 = -1$. We use i to represent $\sqrt{-1}$ and $-i$ to represent $-\sqrt{-1}$. Both $i^2 = -1$ and $(-i)^2 = -1$.

When these numbers were first developed, they were called imaginary numbers because mathematicians were not familiar with them and did not know concrete applications for them. One of the first concrete applications of imaginary numbers was developed in 1892, when Charles P. Steinmetz used them in his theory of alternating currents. We often use imaginary numbers in the computation of problems whose final answers are real numbers, just as we use fractions in the computation of problems whose answers are natural numbers.

The square root of any negative number is imaginary and can be expressed in terms of the imaginary unit i. For $x > 0$, we define $\sqrt{-x}$ to be $i\sqrt{x}$.

The Imaginary Number i

$i = \sqrt{-1}$ so $i^2 = -1$
$-i = -\sqrt{-1}$ so $(-i)^2 = i^2 = -1$
For any positive real number x, $\sqrt{-x} = i\sqrt{x}$.

$\sqrt{-1}$ has two square roots: the principal square root is i and the other square root is $-i$.

EXAMPLE 1 Writing Imaginary Numbers by Using the i Notation

Write each imaginary number in terms of i.

SOLUTION

(a) $\sqrt{-25}$

$$\sqrt{-25} = i\sqrt{25} \quad \sqrt{-x} = i\sqrt{x}$$
$$= i(5)$$
$$= 5i$$

(b) $\sqrt{-3}$

$$\sqrt{-3} = i\sqrt{3}$$

We usually write $5i$, not $i5$, just as we usually write $7x$, not $x7$. We usually write $i\sqrt{3}$ instead of $\sqrt{3}i$ so we do not accidentally interpret this expression as $\sqrt{3i}$.

Imaginary numbers and complex numbers really do exist. To the layperson who does not need to use irrational numbers such as $\sqrt{2}$ or an imaginary number i, these numbers may seem mysterious. One goal of this section is to remove some of this mystery.

Using the real numbers, the imaginary number i, and the operations of addition, subtraction, multiplication, and division, we obtain numbers that can be written in the form $a + bi$, where a and b are real numbers.

Any number that can be written in the **standard form** $a + bi$ is called a **complex number.** If $b = 0$, then $a + bi$ is just the real number a. If $b \neq 0$, then $a + bi$ is called **imaginary.** If $a = 0$ and $b \neq 0$, then bi is called **pure imaginary.** Thus the complex numbers include both the real numbers and the pure imaginary numbers.

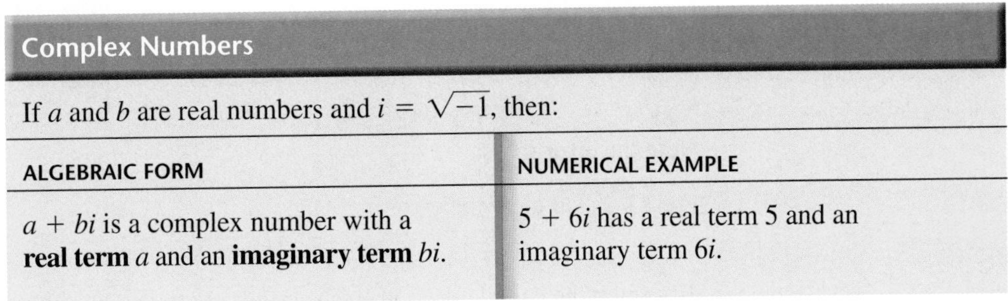

Complex Numbers

If a and b are real numbers and $i = \sqrt{-1}$, then:

ALGEBRAIC FORM	NUMERICAL EXAMPLE
$a + bi$ is a complex number with a **real term** a and an **imaginary term** bi.	$5 + 6i$ has a real term 5 and an imaginary term $6i$.

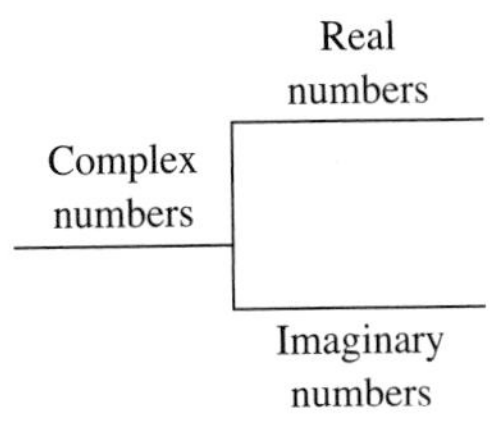

Figure 7.7.1

Every real number is a complex number, and every imaginary number is a complex number. The real numbers and the imaginary numbers are distinct subsets of the complex numbers with no elements in common. The relationships among the real, complex, and imaginary numbers are shown in Figure 7.7.1 and illustrated by the examples in Table 7.7.1.

Every number in this table is a complex number.

A common misunderstanding is that the imaginary numbers contain only the pure imaginary numbers. As these examples of complex numbers illustrate, $3 - 4i$ is an imaginary number. [Robert James and Glenn James, *Mathematics Dictionary*, 4th ed. (multilingual edition), New York: Van Nostrand Reinhold, 1976.]

Table 7.7.1 Classifying Complex Numbers

COMPLEX NUMBER	STANDARD FORM	REAL TERM	COEFFICIENT OF THE IMAGINARY TERM	CLASSIFICATION
6	$6 + 0i$	6	0	Real
$-7i$	$0 - 7i$	0	-7	Pure imaginary
$3 - 4i$	$3 - 4i$	3	-4	Imaginary
0	$0 + 0i$	0	0	Real
$-\sqrt{25}$	$-5 + 0i$	-5	0	Real
$\sqrt{-25}$	$0 + 5i$	0	5	Pure imaginary

Many graphing calculators have the ability to work with complex numbers. The TI-84 Plus calculator has a **REAL** mode and an $\boldsymbol{a + bi}$ (complex) mode. Attempting to evaluate $\sqrt{-25}$ in **REAL** mode will generate an error because $\sqrt{-25}$ is not a real number. When the calculator is set to complex mode, the calculator correctly evaluates $\sqrt{-25}$ as $5i$. This is illustrated in Calculator Perspective 7.7.1.

CALCULATOR PERSPECTIVE 7.7.1 Switching Between Real Mode and Complex Mode

To evaluate $\sqrt{-25}$ and set a TI-84 Plus calculator to complex mode, enter the following keystrokes.

MODE 2nd QUIT (First check that the calculator is in **REAL** mode.)

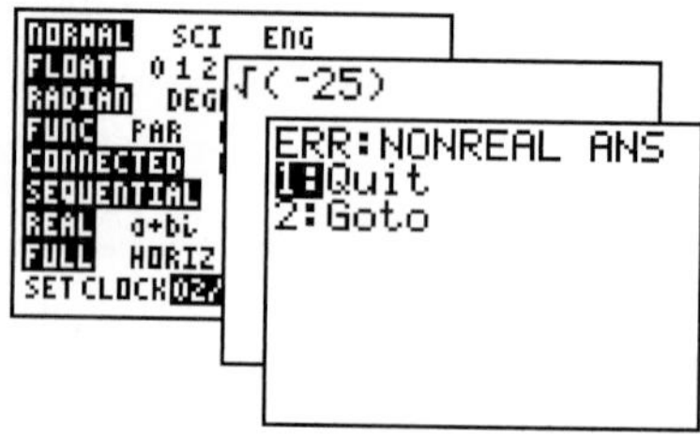

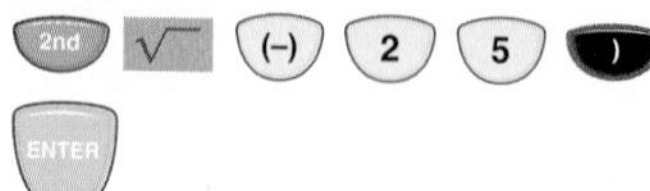

Note the error message.

Press ENTER to clear the error message.

To switch the calculator to ***a + bi*** (complex) mode, enter the following keystrokes:

(Use the arrow keys and the ENTER key to change modes as shown.) 2nd QUIT

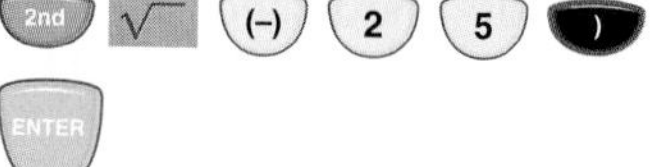

Note the value of $\sqrt{-25}$ is displayed as $5i$.

Two complex numbers are equal if and only if both their real terms and their imaginary terms are equal. That is, $a + bi = c + di$ if and only if $a = c$ and $b = d$. For example, $a + 9i = -5 + bi$ if and only if $a = -5$ and $b = 9$.

SELF-CHECK 7.7.1

Determine a and b such that the following complex numbers are equal.

1. $3 + bi = a - 17i$
2. $-\sqrt{36} + \sqrt{-49} = a + bi$

Addition of Complex Numbers

Since complex numbers consist of two terms—a real term and an imaginary term—the arithmetic of complex numbers is very similar to the arithmetic of binomials. The similarity is illustrated in the following examples by using the operation of addition.

The arithmetic of complex numbers is very similar to the arithmetic of binomials.

ADDITION OF BINOMIALS	ADDITION OF COMPLEX NUMBERS
$(2x + 3y) + (4x + 7y)$	$(2 + 3i) + (4 + 7i)$
$= (2x + 4x) + (3y + 7y)$	$= (2 + 4) + (3i + 7i)$
$= 6x + 10y$	$= 6 + 10i$

Add complex numbers by adding the real terms and the imaginary terms separately. That is,

$$(a + bi) + (c + di) = (a + c) + (b + d)i$$

Subtraction is performed similarly.

EXAMPLE 2 Adding and Subtracting Complex Numbers

Perform the indicated operations.

SOLUTION

(a) $(2 + 5i) + (8 + 4i)$

$$(2 + 5i) + (8 + 4i) = 2 + 5i + 8 + 4i$$
$$= (2 + 8) + (5 + 4)i$$
$$= 10 + 9i$$

Group like terms together, and then add like terms. Add the real terms, and then add the imaginary terms.

(b) $6 + (11 - 4i)$

$$6 + (11 - 4i) = (6 + 11) + (0 - 4)i$$
$$= 17 - 4i$$

Also 6 can be written in the form $6 + 0i$.

(c) $(8 - 5i) - 6i$

$$(8 - 5i) - 6i = (8 + 0) + (-5 - 6)i$$
$$= 8 - 11i$$

Also $-6i$ can be written in the form $0 - 6i$.

(d) $(-7 + 6i) - (4 + 10i)$

$$(-7 + 6i) - (4 + 10i) = -7 + 6i - 4 - 10i$$
$$= (-7 - 4) + (6 - 10)i$$
$$= -11 - 4i$$

Use the distributive property to remove the second pair of parentheses. Then add like terms.

Since $i^2 = -1$, higher powers of i can always be simplified to $i, -1, -i$, or 1. The first four powers of i are keys to simplifying higher powers to standard form and therefore should be memorized.

FIRST FOUR POWERS OF i

$i^1 = i$
$i^2 = i \cdot i = -1$
$i^3 = i^2 \cdot i = (-1)i = -i$
$i^4 = i^2 \cdot i^2 = (-1)(-1) = 1$

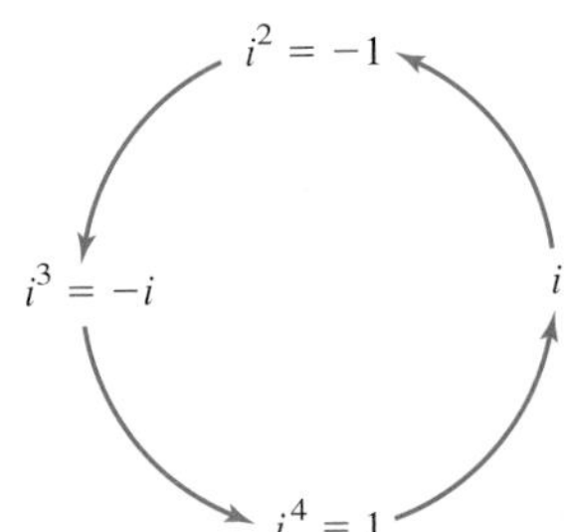

The powers of i repeat in cycles of four: i^4, i^8, i^{12}, and so on; all are equal to 1. We will use this fact to simplify i^n, where n is any integer exponent, to $i, -1, -i$, or 1.

EXAMPLE 3 Powers of i

Simplify each power of i.

SOLUTION

(a) i^7

$$i^7 = i^4 \cdot i^3$$
$$= (1)(-i)$$
$$= -i$$

First extract the largest multiple of 4 from each exponent. Then simplify, replacing i^4, i^8, i^{12}, etc., by 1.

(b) i^{13}

$$i^{13} = i^{12}i^1$$
$$= (1)(i)$$
$$= i$$

(c) i^{406}

$$i^{406} = i^{404} \cdot i^2$$
$$= (i^4)^{101} \cdot (-1)$$
$$= (1)^{101}(-1)$$
$$= (1)(-1)$$
$$= -1$$

$$\begin{array}{r} 101 \\ 4\overline{)406} \\ \underline{404} \\ 2 \end{array}$$

SELF-CHECK 7.7.2

Simplify each of these expressions.

1. $(21 + 4i) - (9 - 7i)$ **2.** i^{33}

We will examine properties of radicals in greater detail in Chapter 9. For now, please note that for positive values of x and y, $\sqrt{x}\sqrt{y} = \sqrt{xy}$; but for negative values of x and y, $\sqrt{x}\sqrt{y} \neq \sqrt{xy}$. For example,

$$\begin{aligned}\sqrt{-2}\sqrt{-3} &= (i\sqrt{2})(i\sqrt{3}) \\ &= i^2\sqrt{6} \\ &= -\sqrt{6}\end{aligned} \qquad \text{whereas} \qquad \begin{aligned}\sqrt{(-2)(-3)} &= \sqrt{+6} \\ &= \sqrt{6}\end{aligned}$$

Note that, in the equation above, we wrote $i\sqrt{2}$ rather than $\sqrt{2}i$. The latter form could easily be confused with $\sqrt{2i}$, which has i under the radical symbol. To make it clear that the factor i is not in the radicand, it is best to put the i in front of the radical.

Another example illustrating $\sqrt{x}\sqrt{y} \neq \sqrt{xy}$ for negative values of x and y is given here.

$$\begin{aligned}(\sqrt{-5})^2 &= (i\sqrt{5})^2 \\ &= 5i^2 \\ &= -5\end{aligned} \qquad \text{whereas} \qquad \begin{aligned}\sqrt{(-5)^2} &= \sqrt{25} \\ &= 5\end{aligned}$$

To avoid the potential errors noted earlier, we recommend that you write every complex number in standard form before proceeding with any operations. We can multiply complex numbers as if they were binomials with a real term and an imaginary term. We use the distributive property extensively to form these products. This is illustrated in Example 4.

EXAMPLE 4 Multiplying Complex Numbers

Calculate each product.

SOLUTION

(a) $3(4 - 5i)$

$$\begin{aligned}3(4 - 5i) &= 3(4) - 3(5i) \\ &= 12 - 15i\end{aligned}$$

Distribute the factor 3.

(b) $\sqrt{-2}(\sqrt{3} - \sqrt{-2})$

$$\begin{aligned}\sqrt{-2}(\sqrt{3} - \sqrt{-2}) &= i\sqrt{2}(\sqrt{3} - i\sqrt{2}) \\ &= i\sqrt{2}(\sqrt{3}) - i\sqrt{2}(i\sqrt{2}) \\ &= i\sqrt{6} - 2i^2 \\ &= i\sqrt{6} - 2(-1) \\ &= 2 + i\sqrt{6}\end{aligned}$$

First write each factor in the standard $a + bi$ form.
Then distribute the factor $i\sqrt{2}$.
Simplify each term, noting that $\sqrt{2}\sqrt{2} = 2$.
Replace i^2 by -1.
Reorder the terms to write the answer in the standard form.

(c) $(5 - 6i)(5 + 6i)$

$$\begin{aligned}(5 - 6i)(5 + 6i) &= (5)^2 - (6i)^2 \\ &= 25 - 36i^2 \\ &= 25 - 36(-1) \\ &= 61\end{aligned}$$

Note that this expression is of the form $(x - y)(x + y) = x^2 - y^2$.
Replace i^2 by -1.

(d) $(2 - 7i)(5 + 3i)$

$$\begin{aligned}(2 - 7i)(5 + 3i) &= 2(5 + 3i) - 7i(5 + 3i) \\ &= 10 + 6i - 35i - 21i^2 \\ &= 10 - 29i - 21(-1) \\ &= 10 - 29i + 21 \\ &= 31 - 29i\end{aligned}$$

Distribute the factor of $5 + 3i$. Then distribute the factors of 2 and of $-7i$.
Combine like terms and replace i^2 by -1.

(e) $(a + bi)(a - bi)$

$$\begin{aligned}(a + bi)(a - bi) &= a^2 - (bi)^2 \\ &= a^2 - b^2i^2 \\ &= a^2 + b^2\end{aligned}$$

Note that this expression is of the form $(x + y)(x - y) = x^2 - y^2$.
Replace i^2 by -1.

SELF-CHECK 7.7.3

1. Calculate the product $(5 - 3i)(6 + 2i)$.

A Mathematical Note

The term *conjugates* for $a + bi$ and $a - bi$ was suggested by Augustin-Louis Cauchy (1789–1857). Cauchy was a prolific writer in mathematics, perhaps second only to Euler.

Complex Conjugates

The **conjugate** of any complex number $a + bi$ is $a - bi$. This definition is illustrated in the following table.

COMPLEX NUMBER	COMPLEX NUMBER IN STANDARD FORM	COMPLEX CONJUGATE IN STANDARD FORM	COMPLEX CONJUGATE
$5 - 6i$	$5 - 6i$	$5 + 6i$	$5 + 6i$
$\sqrt{2} + i$	$\sqrt{2} + i$	$\sqrt{2} - i$	$\sqrt{2} - i$
$7i$	$0 + 7i$	$0 - 7i$	$-7i$
8	$8 + 0i$	$8 - 0i$	8

The product of a complex number and its conjugate is always a real number.

The product of a complex number and its conjugate uses the special product of a sum and a difference to yield the difference of two squares. This product is always a real number, as we illustrated in Example 4(e). We will now look at some specific examples of the product of conjugates.

EXAMPLE 5 Multiplying Complex Conjugates

Multiply each complex number by its conjugate and simplify the result.

SOLUTION

(a) $3 + 4i$

$$\begin{aligned}(3 + 4i)(3 - 4i) &= 9 - 16i^2 \\ &= 9 + 16 \\ &= 25\end{aligned}$$

The conjugate of $3 + 4i$ is $3 - 4i$. Multiply, using the fact that this expression is of the form $(x + y)(x - y) = x^2 - y^2$.

(b) $5 - 2i$

$$\begin{aligned}(5 - 2i)(5 + 2i) &= 25 - 4i^2 \\ &= 25 + 4 \\ &= 29\end{aligned}$$

The conjugate of $5 - 2i$ is $5 + 2i$.

(c) $7i$

$$\begin{aligned}(7i)(-7i) &= -49i^2 \\ &= 49\end{aligned}$$

The conjugate of $7i$ is $-7i$.

(d) 8

$$8 \cdot 8 = 64$$

The conjugate of 8 is 8.

Dividing Complex Numbers

The fact that the product of a complex number and its conjugate is always a real number plays a key role in the division of complex numbers, as outlined in the following box.

Division of Complex Numbers

VERBALLY	NUMERICAL EXAMPLE
Step 1 Write the division problem as a fraction.	$20 \div (1 + 3i) = \dfrac{20}{1 + 3i}$
Step 2 Multiply both the numerator and the denominator by the conjugate of the denominator.	$= \dfrac{20}{1 + 3i} \cdot \dfrac{1 - 3i}{1 - 3i}$ $= \dfrac{20(1 - 3i)}{1 + 9}$
Step 3 Simplify the result, and express it in standard $a + bi$ form.	$= 2(1 - 3i)$ $= 2 - 6i$

EXAMPLE 6 Dividing Complex Numbers

Simplify $(8 - i) \div (1 - 2i)$.

SOLUTION

$(8 - i) \div (1 - 2i) = \dfrac{8 - i}{1 - 2i}$ Write the division problem as a fraction.

$= \dfrac{8 - i}{1 - 2i} \cdot \dfrac{1 + 2i}{1 + 2i}$ Multiply the numerator and the denominator by $1 + 2i$, the conjugate of the denominator.

$= \dfrac{8 + 15i - 2i^2}{1 - 4i^2}$ Multiply the numerators and multiply the conjugates in the denominator.

$= \dfrac{10 + 15i}{5}$ Simplify, replacing i^2 by -1.

$= \dfrac{10}{5} + \dfrac{15}{5}i$ Write the result in standard $a + bi$ form.

$= 2 + 3i$ You can check this answer by multiplying $2 + 3i$ by $1 - 2i$. Is the product $8 - i$?

SELF-CHECK 7.7.4

1. Multiply $11 + 10i$ by its conjugate.
2. Calculate the quotient $\dfrac{3 + 2i}{2 - 3i}$.

Calculator Perspective 7.7.2 illustrates how to evaluate expressions with complex numbers. Before you use this Calculator Perspective, use Calculator Perspective 7.7.2 to switch the mode to $\boldsymbol{a + bi}$. The i key is the secondary function of the $\cdot$ key.

CALCULATOR PERSPECTIVE 7.7.2 Performing Operations with Complex Numbers

To evaluate the expressions with complex numbers from examples in this section on a TI-84 Plus calculator, enter the following keystrokes:

Example 2(a): $(2 + 5i) + (8 + 4i)$

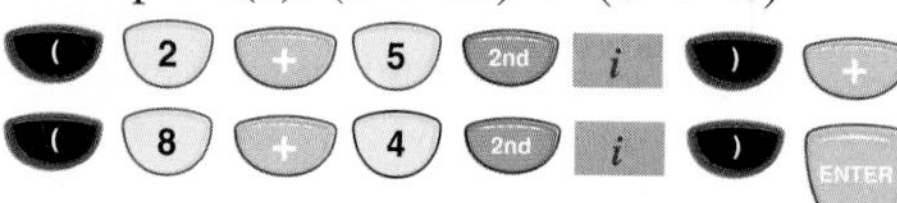

```
(2+5i)+(8+4i)
            10+9i
```

Example 4(d): $(2 - 7i)(5 + 3i)$

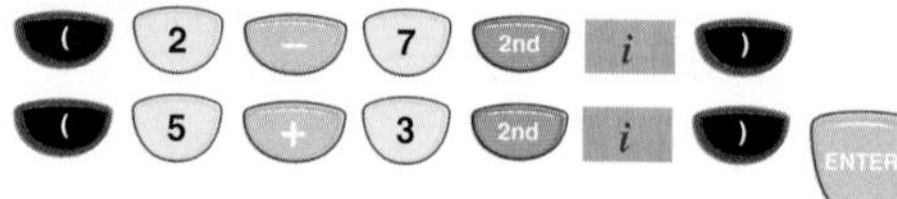

```
(2-7i)(5+3i)
           31-29i
```

Example 6: $(8 - i) \div (1 - 2i)$

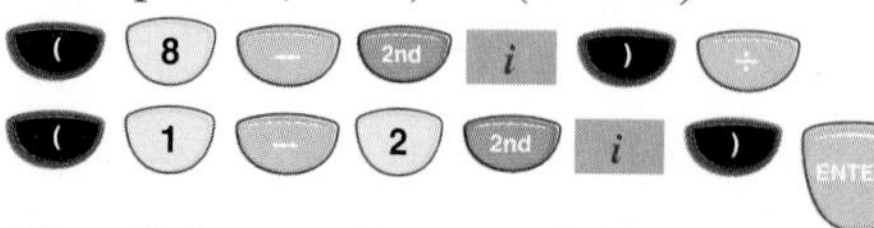

```
(8-i)/(1-2i)
             2+3i
```

Note: It is sometimes useful to press

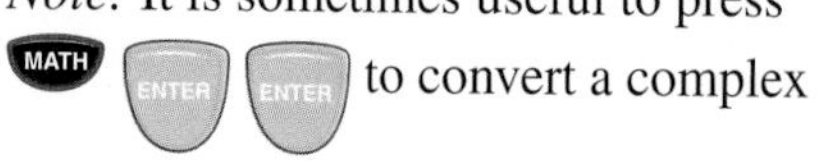

to convert a complex number in decimal form to fractional form.

Calculators are powerful tools that allow us to leverage our mathematical power. It is important for us to know how to use our calculator and to know some of its limitations. One example of this is illustrated in Example 7.

EXAMPLE 7 Evaluating Powers of *i* and Interpreting Calculator Results

The TI-84 Plus calculator screen shown displays an evaluation of i^{27}. Evaluate i^{27} by hand, and interpret this calculator screen.

i^27
-3E-13-i

SOLUTION

$i^{27} = i^{24} \cdot i^3$ — First extract the largest multiple of 4 from the exponent of 27.

$= (1)(-i)$ — Replace i^{24} by 1 and i^3 by $-i$.

$= -i$

The calculator display in $a + bi$ form is $(-3.0 \times 10^{-13}) - i \approx 0 - i$ — The calculator display gives the result in $a + bi$ form with the real part a given in scientific notation. $-3.0 \times 10^{-13} = -0.0000000000003 \approx 0$

Therefore the calculator has introduced a small error because the real part is exactly zero; $i^{27} = 0 - i$.

Answer: $i^{27} = -i$

SELF-CHECK 7.7.5

1. Evaluate i^{44} by hand and interpret this calculator screen.

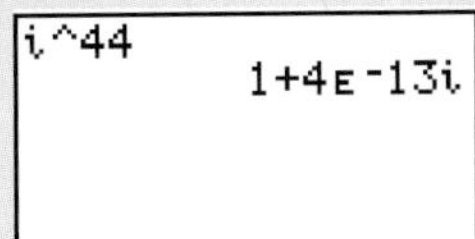

One of the uses you are most likely to have for complex numbers is as solutions of quadratic equations. Every quadratic equation has two complex solutions (a double real solution is counted as two solutions). Some of these complex solutions are real numbers and others are imaginary numbers. We start by using extraction of roots to solve the equations given in Example 8. Problems with imaginary solutions occur more in engineering and the sciences than they do in business or the social sciences.

EXAMPLE 8 Solving Quadratic Equations by Extraction of Roots

Solve each of these quadratic equations by extraction of roots.

SOLUTION

(a) $x^2 = -9$

$x^2 = -9$

$x = \pm\sqrt{-9}$

$x = \pm 3i$

$x = -3i$ or $x = 3i$

Solve the equation by extracting both square roots of -9. These two imaginary solutions are complex conjugates.

(b) $x^2 = -3$

$$x^2 = -3$$
$$x = \pm\sqrt{-3}$$
$$x = \pm i\sqrt{3}$$
$$x = -i\sqrt{3} \quad \text{or} \quad x = i\sqrt{3}$$

Solve the equation by extracting both square roots of -3. These two imaginary solutions are complex conjugates.

(c) $(2x - 3)^2 = -4$

$$(2x - 3)^2 = -4$$
$$2x - 3 = \pm\sqrt{-4}$$
$$2x = 3 \pm 2i$$
$$x = 1.5 - i \quad \text{or} \quad x = 1.5 + i$$

Solve the equation by extracting both square roots of -4. Then add 3 to both sides of the equation. Divide by 2 to solve for x. These two imaginary solutions are complex conjugates.

Example 9 gives a quadratic equation with imaginary solutions. It is not possible to obtain these solutions by factoring over the integers or by graphing the corresponding parabola on the real coordinate plane. This quadratic equation illustrates the importance of the quadratic formula as a method for solving any quadratic equation.

EXAMPLE 9 Using the Quadratic Formula to Determine Two Imaginary Solutions

Use the quadratic formula to solve $x^2 - 4x + 5 = 0$.

SOLUTION

$$x^2 - 4x + 5 = 0$$
$$x = \frac{-(-4) \pm \sqrt{(-4)^2 - 4(1)(5)}}{2(1)}$$

The equation is in the standard form $ax^2 + bx + c = 0$ with $a = 1, b = -4$, and $c = 5$. Substitute these values into the quadratic formula:

$$x = \frac{-b \pm \sqrt{b^2 - 4ac}}{2a}$$

$$x = \frac{4 \pm \sqrt{16 - 20}}{2}$$
$$x = \frac{4 \pm \sqrt{-4}}{2}$$
$$x = \frac{4 \pm 2i}{2}$$

Because the discriminant $b^2 - 4ac$ is negative, these solutions will be imaginary. Simplify and replace $\sqrt{-4}$ by $2i$.

$$x = \frac{4}{2} \pm \frac{2i}{2}$$
$$x = 2 \pm i$$

Write each of these complex numbers in standard $a + bi$ form. Note that these imaginary solutions are complex conjugates. Do these values check?

Answer: $x = 2 - i$ or $x = 2 + i$

Because of the $\pm$ symbol in the quadratic formula, imaginary solutions will always appear as complex conjugates.

SELF-CHECK 7.7.6

Solve each of these quadratic equations by extraction of roots.

1. $x^2 = 49$ **2.** $x^2 = 5$ **3.** $x^2 = -4$ **4.** $x^2 = -7$

5. Use the quadratic formula to solve $x^2 - 6x + 34 = 0$.

SELF-CHECK ANSWERS

7.7.1

1. $a = 3, b = -17$
2. $a = -6, b = 7$

7.7.2

1. $12 + 11i$ 2. i

7.7.3

1. $36 - 8i$

7.7.4

1. 221 2. i

7.7.5

1. $1; 1 + (4 \times 10^{-13})i \approx 1 + 0i = 1$

7.7.6

1. $x = -7$ or $x = 7$
2. $x = -\sqrt{5}$ or $x = \sqrt{5}$
3. $x = -2i$ or $x = 2i$
4. $x = -i\sqrt{7}$ or $x = i\sqrt{7}$
5. $x = 3 - 5i$ or $x = 3 + 5i$

USING THE LANGUAGE AND SYMBOLISM OF MATHEMATICS 7.7

1. The imaginary number i is used to represent ________.
2. The complex number $a + bi$ has a real term ________, and an imaginary term ________.
3. A complex number is **always/sometimes/never** a real number.
4. A real number is **always/sometimes/never** a complex number.
5. An imaginary number is **always/sometimes/never** a complex number.
6. A complex number is **always/sometimes/never** an imaginary number.
7. A real number is **always/sometimes/never** an imaginary number.
8. The standard form of a complex number is ________.
9. Powers of i repeat in cycles of ________.
10. The complex ________ of $3 + 5i$ is $3 - 5i$.
11. Every quadratic equation has ________ complex solutions (a double real solution is counted as two solutions).
12. If $7 + 2i$ is a solution of $ax^2 + bx + c = 0$ and a, b, and c are real numbers, then ________ is also a solution of this equation.

QUICK REVIEW 7.7

The purpose of this quick review is to help you recall skills needed in this section.

1. Write 37,469.21 by using scientific notation.
2. Write 4.06×10^{-5} in standard decimal notation.
3. Use the distributive property to expand and simplify $5x(2x - 3) + 4(2x - 3)$.
4. Use the distributive property to factor $5x(2x - 3) + 4(2x - 3)$.
5. Multiply $(5x + 3y)(5x - 3y)$.

EXERCISES 7.7

In Exercises 1–10, simplify each expression and write the result in the standard $a + bi$ form.

1. a. $-\sqrt{36}$ b. $\sqrt{-36}$
   c. $-\sqrt{-36}$ d. $\sqrt{36}$
2. a. $-\sqrt{49}$ b. $\sqrt{-49}$
   c. $-\sqrt{-49}$ d. $\sqrt{49}$
3. a. $\sqrt{-9} + \sqrt{16}$ b. $\sqrt{9} + \sqrt{-16}$
   c. $-\sqrt{9} - \sqrt{-16}$ d. $\sqrt{-9} - \sqrt{16}$
4. a. $\sqrt{4} + \sqrt{-25}$ b. $\sqrt{-4} - \sqrt{25}$
   c. $-\sqrt{4} - \sqrt{-25}$ d. $-\sqrt{-4} + \sqrt{-25}$
5. a. $\sqrt{-9 - 16}$ b. $-\sqrt{-9 - 16}$
   c. $\sqrt{-6}\sqrt{-6}$ d. $\sqrt{(-6)(-6)}$
6. a. $\sqrt{-25 - 144}$ b. $\sqrt{-25} + \sqrt{-144}$
   c. $\sqrt{-7}\sqrt{-7}$ d. $\sqrt{(-7)(-7)}$
7. a. $\sqrt{-4}\sqrt{-25}$ b. $\sqrt{(-4)(-25)}$
   c. $-\sqrt{4}\sqrt{-25}$ d. $(-\sqrt{4})(-\sqrt{25})$
8. a. $\sqrt{-9}\sqrt{-100}$ b. $\sqrt{(-9)(-100)}$
   c. $-\sqrt{9}\sqrt{-100}$ d. $(-\sqrt{9})(-\sqrt{100})$
9. a. $\sqrt{-\frac{25}{9}}$ b. $\frac{\sqrt{-25}}{\sqrt{9}}$
   c. $\frac{\sqrt{25}}{\sqrt{-9}}$ d. $\sqrt{\frac{-25}{-9}}$
10. a. $\sqrt{\frac{-36}{-49}}$ b. $\frac{\sqrt{36}}{\sqrt{-49}}$
    c. $\sqrt{-\frac{36}{49}}$ d. $\frac{\sqrt{-36}}{\sqrt{49}}$

In Exercises 11–36, perform the indicated operations and express the result in the standard $a + bi$ form.

11. $(1 + 2i) + (8 - 3i)$
12. $(6 - 7i) + (13 - 4i)$
13. $(5 + 3i) - (2 + 2i)$
14. $(7 + i) - (5 - 2i)$
15. $6(10 + 4i)$
16. $-9(3 - 12i)$
17. $(3 + 3i) + \frac{1}{2}(8 - 6i)$
18. $(3 + 3i) - \frac{1}{3}(6 - 9i)$
19. $(2i)(3i)$
20. $(5i)(7i)$
21. $2i(3 - 5i)$
22. $-3i(4 - 2i)$
23. $i(3 + 2i) + 2(5 - 3i)$
24. $4(3 - 2i) - i(5 + 3i)$
25. $(2 - 7i)(2 + 7i)$
26. $(6 + 9i)(6 - 9i)$
27. $(5 - 2i)(4 + 7i)$
28. $(6 + i)(3 - 5i)$
29. $(5 + i)^2$
30. $(6 - i)^2$
31. $(4 - 3i)^2$
32. $(3 + 5i)^2$
33. $(3 + 2i)^2 - [3^2 + (2i)^2]$
34. $(5 - 4i)^2 - [5^2 + (-4i)^2]$
35. $\sqrt{4} + \sqrt{-9} - \sqrt{9} - \sqrt{-25}$
36. $\sqrt{64} + \sqrt{-36} - \sqrt{9} - \sqrt{-1}$

In Exercises 37–40, give the conjugate of each number and then multiply each complex number by its conjugate.

37. $2 + 5i$
38. $3 - 8i$
39. $13i$
40. 13

In Exercises 41–46, perform the indicated operations and express the result in standard $a + bi$ form.

41. $\frac{4}{1 + i}$
42. $\frac{6}{1 - i}$
43. $\frac{4 - i}{4 + i}$
44. $\frac{5 + i}{5 - i}$
45. $85 \div (7 - 6i)$
46. $185 \div (11 + 8i)$

In Exercises 47–50, simplify each power of i.

47. i^9
48. i^{11}
49. i^{58}
50. i^{81}

In Exercises 51–56, calculate the discriminant of each quadratic equation and determine the nature of the solutions of this equation.

51. $x^2 - 10x + 25 = 0$
52. $x^2 - 10x + 24 = 0$
53. $x^2 - 10x + 26 = 0$
54. $x^2 - 2 = 0$
55. $x^2 + 2 = 0$
56. $x^2 + 2x = 0$

Calculator Usage

In Exercises 57–61, use a calculator to evaluate each expression.

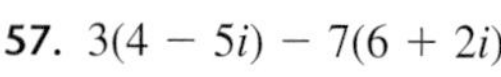

57. $3(4 - 5i) - 7(6 + 2i)$
58. $\dfrac{19 + 8i}{6 + 7i}$
59. $(1 - i)^4$
60. $\left(-\dfrac{1}{2} - \dfrac{\sqrt{3}}{2}i\right)^3$
61. $\dfrac{58}{\sqrt{4} - \sqrt{-25}}$

In Exercises 62 and 63, interpret each calculator display.

62.
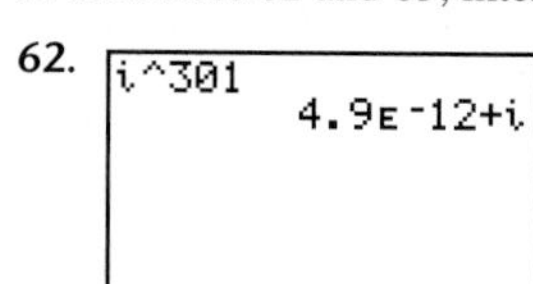

63.
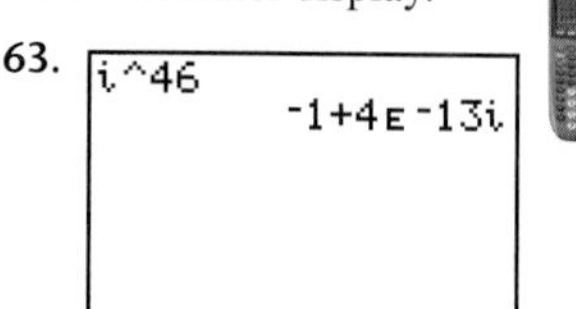

64. Given $z = 3 - i$:
   a. What is the conjugate of z?
   b. What is the additive inverse of z?
   c. What is the multiplicative inverse of z?

In Exercises 65–68, solve each quadratic equation by extraction of roots.

65. $z^2 = -100$
66. $v^2 = -36$
67. $(2x + 3)^2 = -9$
68. $(2x - 3)^2 = -25$

In Exercises 69–76, use the quadratic formula to solve each quadratic equation.

69. $2v^2 - 19v - 33 = 0$
70. $5v^2 + 41v - 36 = 0$
71. $x^2 - 4x + 29 = 0$
72. $x^2 - 8x + 25 = 0$
73. $x^2 = 6x - 13$
74. $x^2 = 4x - 6$
75. $-2w(w - 3) = 5$
76. $3w(w + 1) = -2 - w$

In Exercises 77 and 78, use the quadratic formula to solve each equation.

77. a. $x^2 + 2x - 3 = 0$
   b. $x^2 + 2x + 1 = 0$
   c. $x^2 + 2x + 2 = 0$

78. a. $x^2 + 6x + 5 = 0$
   b. $x^2 + 6x + 9 = 0$
   c. $x^2 + 6x + 10 = 0$

In Exercises 79–82, construct a quadratic equation in x that has the given solutions.

79. $-4i$ and $4i$
80. $-7i$ and $7i$
81. $3 - 2i$ and $3 + 2i$
82. $2 - 5i$ and $2 + 5i$

In Exercises 83 and 84, find all solutions of each cubic equation.

83. $(x - 2)(x^2 + x + 1) = 0$
84. $(x + 2)(x^2 - x + 1) = 0$

Group Discussion Question

85. **Discussion Question**
   a. Is $\sqrt{3}$ a real number?
   b. Is $\sqrt{3}$ an imaginary number?
   c. Is $\sqrt{3}$ a complex number?
   d. Is $\sqrt{-3}$ a real number?
   e. Is $\sqrt{-3}$ an imaginary number?
   f. Is $\sqrt{-3}$ a complex number?

86. **Discussion Question**
   a. What can you say about a complex number that is equal to its conjugate?
   b. Give an example of two imaginary numbers whose sum is a real number.
   c. Give an example of two imaginary numbers whose product is a real number.
   d. Name a complex number that cannot be used as a divisor.

87. **Discovery Question** Complete the following diagram for i to i^{12} to illustrate that powers of i repeat in cycles of four.

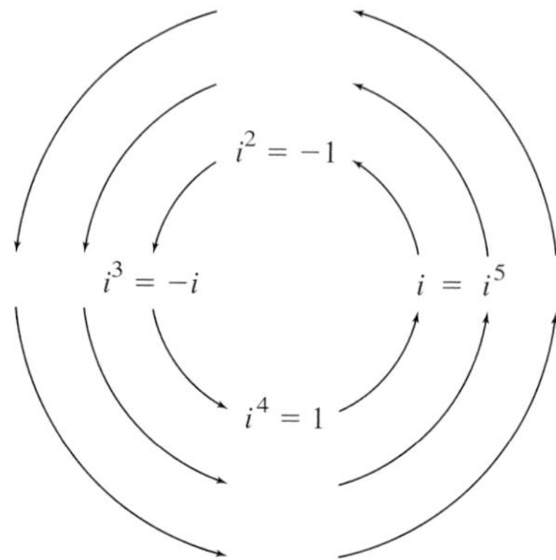

88. **Challenge Question** Use a calculator to answer these questions.
   a. Is 2 a solution of $x^3 - 8 = 0$?
   b. Is $-1 + i\sqrt{3}$ a solution of $x^3 - 8 = 0$?
   c. Is $-1 - i\sqrt{3}$ a solution of $x^3 - 8 = 0$?
   d. How many cube roots does 8 have? What are these cube roots, and which is the principal cube root of 8?

89. **Challenge Question**
   a. Simplify $i + i^2 + i^3 + i^4 + \cdots + i^{99} + i^{100}$.
   b. Explain how this sum can be computed mentally.

90. **Challenge Question** There are four fourth roots of 16. That is, the equation $x^4 = 16$ has four solutions. Solve $x^4 - 16 = 0$ to find these four fourth roots. (*Hint:* Start by factoring the left side of this equation.)

KEY CONCEPTS FOR CHAPTER 7

1. **Functions**
   - A function is a correspondence that matches each input value with exactly one value of the output variable.
   - The input variable is also called the independent variable.
   - The output variable is also called the dependent variable.
   - The domain of a function is the set of all input values.
   - The range of a function is the set of all output values.

2. **Notations for Representing Functions**
   - Mapping notation
   - Ordered-pair notation
   - Table of values
   - Graphs
   - Function notation
   - Verbally

3. **Function Notation**
   - The notation $f(x)$ is read as "*f* of *x*" or "the value of *f* at *x*."
   - The letter *f* names the function.
   - The variable x represents an input value from the domain.
   - The notation $f(x)$ represents a unique output value corresponding to x.
4. **Domain and Range from the Graph of a Function**
   - The domain of a function is the projection of its graph onto the x-axis.
   - The range of a function is the projection of its graph onto the y-axis.
5. **Vertical Line Test** A graph represents a function if it is impossible to have any vertical line intersect the graph at more than one point.
6. **Increasing and Decreasing Functions**
   - A function is increasing on a portion of its graph if as the x-values increase, the y-values also increase.
   - A function is increasing on a portion of its graph if the graph rises as it moves from left to right.
   - A function is decreasing on a portion of its graph if as the x-values increase, the y-values decrease.
   - A function is decreasing on a portion of its graph if the graph drops as it moves from left to right.
   - A line with positive slope is an increasing function.
   - A line with negative slope is a decreasing function.
7. **Positive and Negative Functions**
   - A function is positive when $f(x)$ is positive. On the graph of $y = f(x)$ this occurs at points above the x-axis.
   - A function is negative when $f(x)$ is negative. On the graph of $y = f(x)$ this occurs at points below the x-axis.
8. **Maximum and Minimum Values**
   - A maximum y-value of a function is at the highest point on the graph of this function.
   - A minimum y-value of a function is at the lowest point on the graph of this function.
9. **Linear Functions**
   - A function of the form $f(x) = mx + b$ is a linear function.
   - The shape of a linear function is a line.
10. **Absolute Value Functions**
   - The shape of the graph of an absolute value function is V-shaped.
   - The vertex of an absolute value function is at the tip of the V shape.
   - If the graph of an absolute value function opens upward, the minimum y-value will be at the vertex.
   - If the graph of an absolute value function opens downward, the maximum y-value will be at the vertex.
11. **Quadratic Functions**
   - A function of the form $f(x) = ax^2 + bx + c$ is a quadratic function.
   - The shape of a quadratic function is a parabola.
   - ax^2 is called the quadratic term or the second-degree term.
   - bx is called the linear term or the first-degree term.
   - c is called the constant term.
   - For $a > 0$ the parabola opens upward.
   - For $a < 0$ the parabola opens downward.
   - The vertex is the lowest point on a parabola that opens upward.
   - The vertex is the highest point on a parabola that opens downward.
   - The axis of symmetry is a vertical line passing through the vertex.
   - The vertex is located at $\left(-\frac{b}{2a}, f\left(-\frac{b}{2a}\right)\right)$.
12. **Curve Fitting**
   - A scatter diagram for a set of data points is a graph of these points.
   - Selecting a curve that best fits a set of data is called curve fitting.
   - The topic of fitting an equation to a set of data points is regression analysis.
   - On a graphing calculator use linear regression, **LinReg,** to calculate a line of best fit for a set of data. Use **QuadReg** to calculate a parabola of best fit for a set of data.
13. **Standard Form of a Quadratic Equation** If x is a real variable and a, b, and c are real constants with $a \neq 0$, the standard form of a quadratic equation in x is $ax^2 + bx + c = 0$.
14. **Methods of Solving Quadratic Equations**
   - Graphically
   - Numerically
   - Factoring
   - Extraction of roots
   - Completing the square
   - The quadratic formula: $x = \frac{-b \pm \sqrt{b^2 - 4ac}}{2a}$
15. **Discriminant** In the quadratic formula $x = \frac{-b \pm \sqrt{b^2 - 4ac}}{2a}$ the expression $b^2 - 4ac$ is called the discriminant. The discriminant, which is the expression under the radical symbol, can be used to discriminate between the real solutions and imaginary solutions of a quadratic equation.
16. **Solution of Multiplicity 2** If both solutions of a quadratic equation are the same, we call the solution a double solution or a solution of multiplicity 2.

17. Nature of the Solutions of a Quadratic Equation There are three possibilities for the solutions of $ax^2 + bx + c = 0$.

VALUE OF THE DISCRIMINANT	SOLUTIONS OF $ax^2 + bx + c = 0$	THE PARABOLA $y = ax^2 + bx + c$	GRAPHICAL EXAMPLE
$b^2 - 4ac > 0$	Two distinct real solutions	Two x-intercepts	$y = x^2 - x - 2$
$b^2 - 4ac = 0$	A double real solution	One x-intercept with the vertex on the x-axis	$y = x^2 - 4x + 4$
$b^2 - 4ac < 0$	Neither solution is real; both solutions are complex numbers with imaginary parts. These solutions will be complex conjugates.	No x-intercepts	$y = x^2 + 1$

18. Solutions of a Quadratic Inequality
- The solution of $ax^2 + bx + c > 0$ is the set of x-values for which the graph of $y = ax^2 + bx + c$ is above the x-axis.
- The solution of $ax^2 + bx + c < 0$ is the set of x-values for which the graph of $y = ax^2 + bx + c$ is below the x-axis.

19. Constructing a Quadratic Equation with Given Solutions If $x = r_1$ and $x = r_2$ are two solutions of a quadratic equation, then $(x - r_1)(x - r_2) = 0$ is a factored form of this quadratic equation.

20. Pythagorean Theorem and Converse
- A right triangle is a triangle containing a 90° angle.
- Triangle ABC is a right triangle if, and only if, $a^2 + b^2 = c^2$.

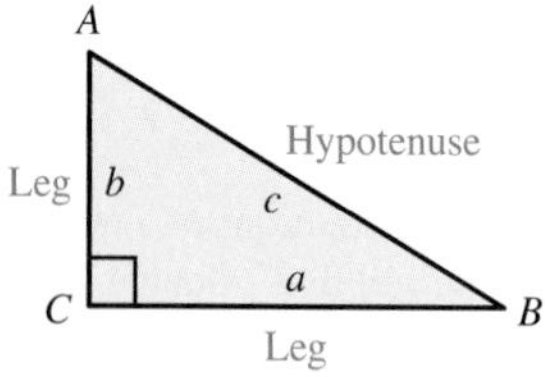

21. The Imaginary Number i
- $i = \sqrt{-1}$ so $i^2 = -1$
  $-i = -\sqrt{-1}$ so $(-i)^2 = i^2 = -1$
- For any positive real number x, $\sqrt{-x} = i\sqrt{x}$.
- The first four powers of i are
  $i^1 = i, i^2 = -1, i^3 = -i, i^4 = 1$.

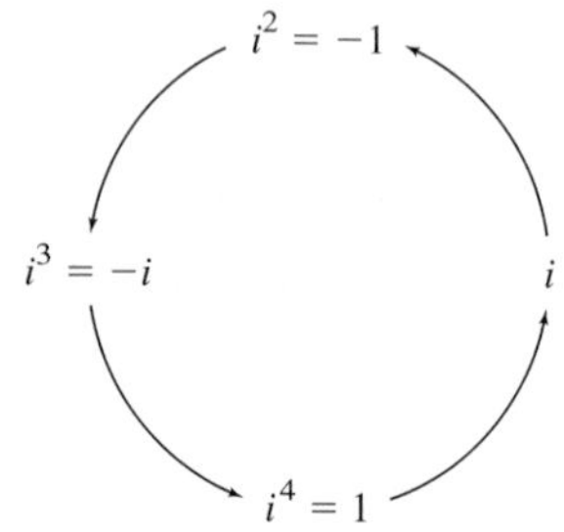

22. Complex Numbers
- If a and b are real numbers and $i = \sqrt{-1}$, then $a + bi$ is a complex number with real term a and imaginary term bi.
- $a + bi$ and $a - bi$ are complex conjugates.

23. Operations with Complex Numbers The addition, subtraction, and multiplication of complex numbers are similar to the corresponding operations with binomials. Division by a complex number is accomplished by multiplying both the numerator and the denominator by the conjugate of the denominator and then simplifying the result.

24. Subsets of the Complex Numbers The relationships of important subsets of the set of complex numbers are summarized in the tree diagram, where a and b are real numbers and $i = \sqrt{-1}$.

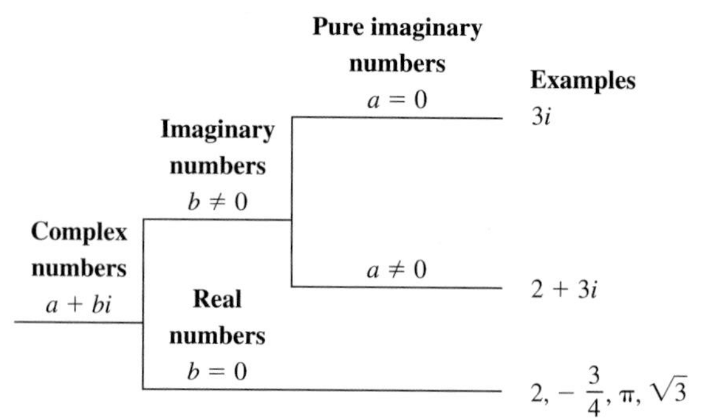

REVIEW EXERCISES FOR CHAPTER 7

In Exercises 1 and 2, determine whether each relation is a function.

1. a.

x	y
9	1
8	2
6	4
3	5

b. x → y: 9 → 1, 8 → 1, 3 → 1

c. $\{(1, 3), (1, 8), (1, 9)\}$

d. $\{(1, \pi), (2, 4\pi), (3, 9\pi)\}$

2. a.

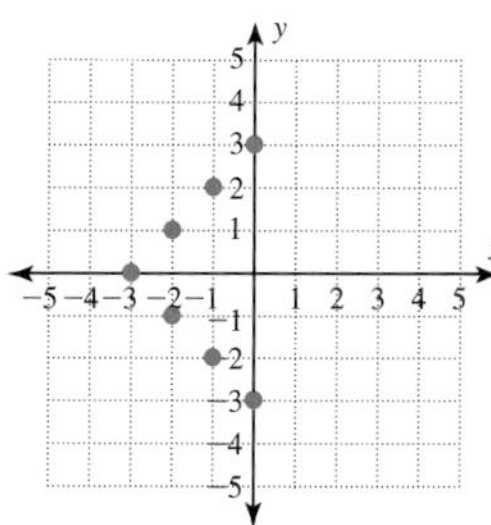

b.

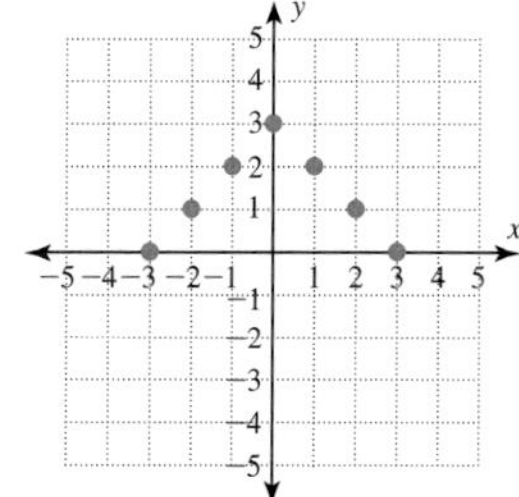

c.

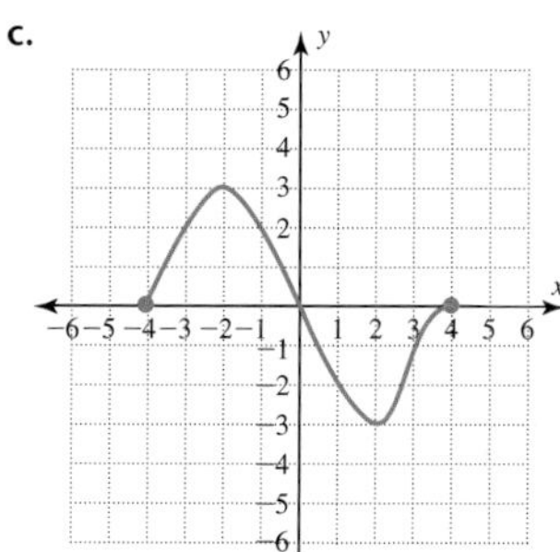

d.

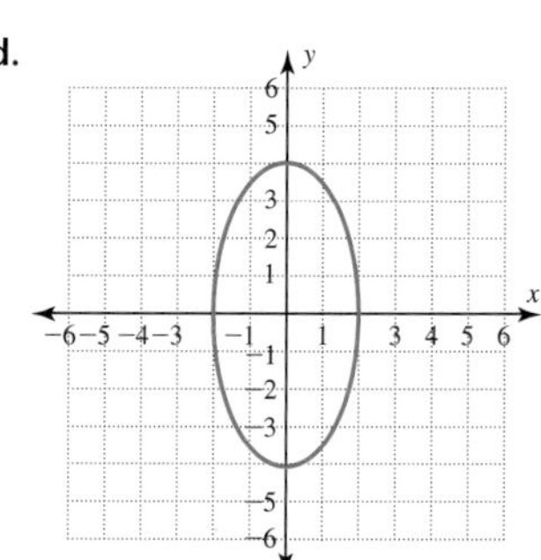

3. Determine the domain and range of each function.

a. $\{(5, 8), (6, 8), (7, 8)\}$

b.

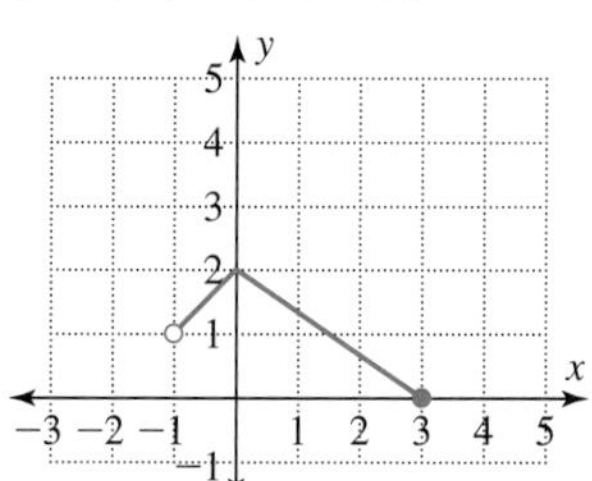

c.

x	−3	−2	−1	0	1
y	11	9	4	1	6

d. $f(x) = |x + 3| + 2$

4. Evaluate each expression for $f(x) = (x + 2)^2 + 3$.

a. $f(-2)$ b. $f(0)$

c. $f(1)$ d. $f(8)$

5. Evaluate each expression in parts **a** and **b**, given the table of values for $f(x)$.

x	y
−3	5
−2	4
−1	3
0	2
1	6
2	8
3	10

a. $f(-2)$ b. $f(0)$

c. Determine the input value of x for which $f(x) = 8$.

d. Determine the input value of x for which $f(x) = 3$.

6. Evaluate each expression in parts **a** and **b**, given the graph of $y = f(x)$.

a. $f(-4)$ b. $f(3)$

c. Determine the input value of x for which $f(x) = 0$.

d. Determine the input value of x for which $f(x) = -1$.

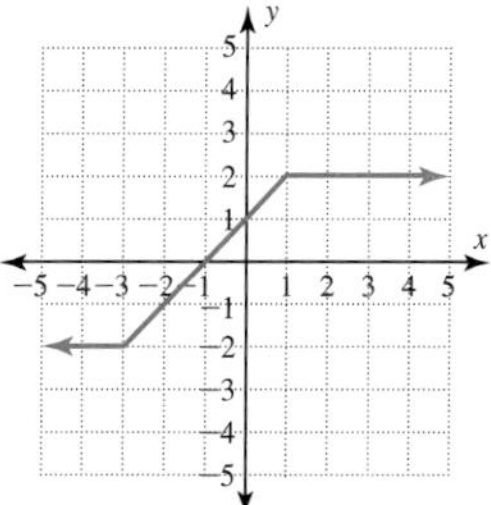

7. Use the function defined by the given table to complete each part of this exercise.

x	y
−10	4
−5	3
0	2
5	1
10	0
15	−1
20	−2

a. Express this function by using mapping notation.

b. Express this function by using ordered-pair notation.

c. Graph this function by making a scatter diagram of these data.

d. Write an equation for the line containing these points.

8. Use the function graphed in the given scatter diagram to complete each part of this exercise.

a. Express this function by using ordered-pair notation.

b. Express this function by using a table format.

c. Express this function by using mapping notation.

d. Write the equation of the line containing these points.

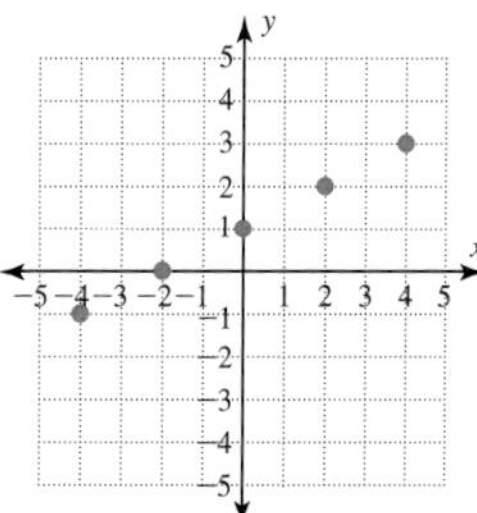

In Exercises 9–11, write the equation of each line in slope-intercept form.

9. A line with y-intercept $(0, 4)$ and slope of $-\frac{1}{2}$.

10. A line passes through the point $(0, -4)$ and goes up 2 units for every 3-unit increase in x.

11. Write the equation of a line through the origin and parallel to $y = 2x - 3$.

12. Without using a calculator, graph the line that has a slope of $m = -\frac{4}{3}$ and a y-intercept of $(0, 1)$.

13. Without using a calculator, graph the line defined by $y = \frac{3}{5}x - 2$.

14. The following graphing calculator display shows a table of values for a linear function. Use this table to determine the following.

a. The y-intercept of the line
b. Value of Δx shown in the table
c. Value of Δy shown in the table
d. Slope of the line
e. Equation of the line in slope-intercept form

X	Y1	
-10	-14	
-5	-10	
0	-6	
5	-2	
10	2	
15	6	
20	10	

X=-10

In Exercises 15–18, identify each function as an increasing function or a decreasing function.

15.

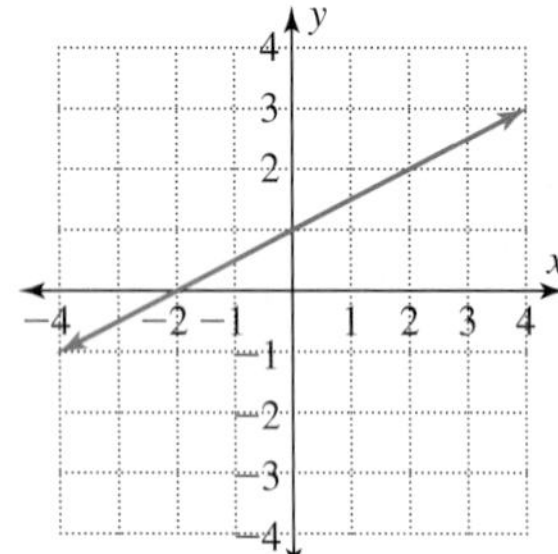

16.

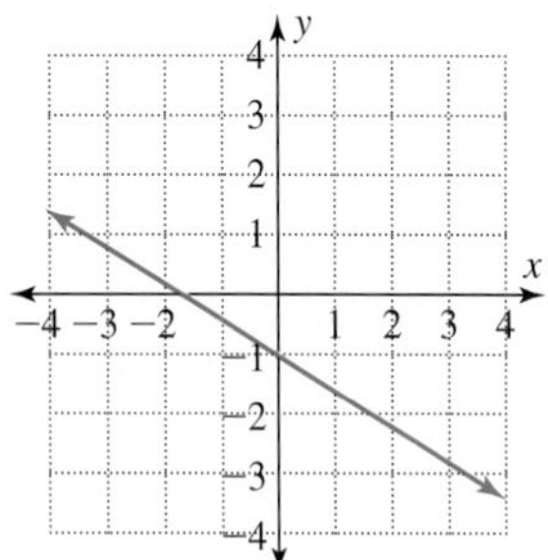

17. $f(x) = -4.1x + 6.3$

18. $f(x) = 4.1x - 6.3$

In Exercises 19 and 20, determine the intervals for which the function is increasing and the intervals for which the function is decreasing.

19.

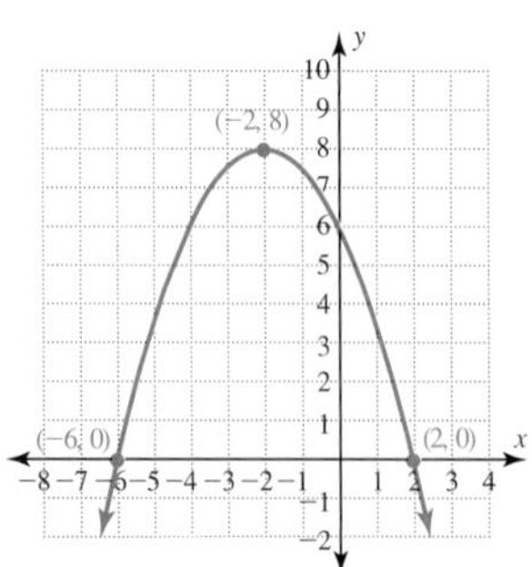

20.

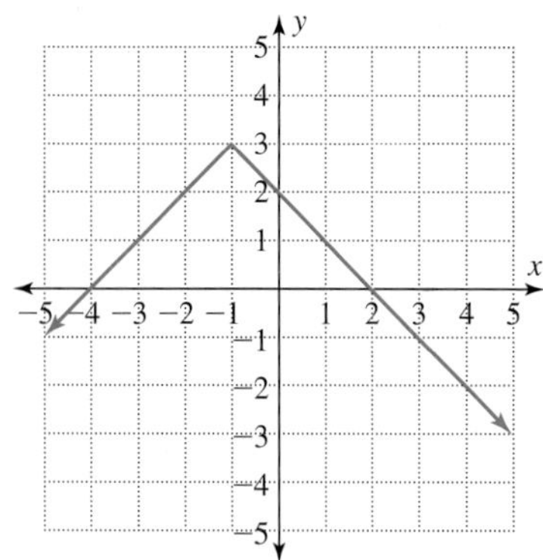

In Exercises 21 and 22, determine the x-values for which each function is positive and those for which the function is negative.

21.

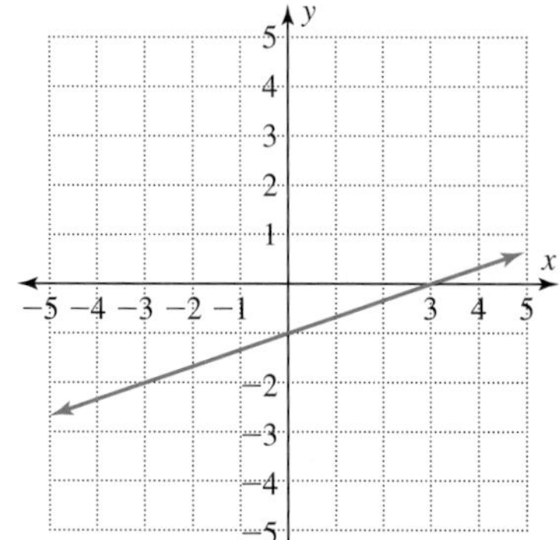

22.

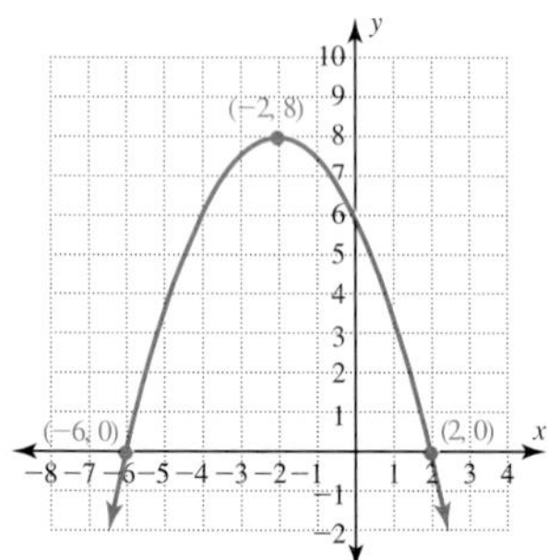

In Exercises 23–34, use the absolute value function $f(x) = -|x + 3| + 4$ to answer each question.

23. Complete the following table of values.

x	$f(x) = -\lvert x + 3\rvert + 4$
−8	
−7	
−3	
−1	
0	
1	
3	

24. Graph this function.
25. What are the x-intercepts of the graph in Exercise 24?
26. What is the y-intercept of the graph in Exercise 24?
27. What point is the vertex of the graph in Exercise 24?
28. What y-value is the maximum output value for this function?
29. What is the domain of this function?
30. What is the range of this function?
31. For what interval of input values of x is the function increasing?
32. For what interval of input values of x is the function decreasing?
33. For what interval of input values of x is the function positive?
34. For what interval of input values of x is the function negative?

In Exercises 35 and 36, use a graphing calculator to draw a scatter diagram for the data points and to determine the line of best fit. Express your answer in the form $y = mx + b$ with m and b rounded to the nearest hundredth.

35.

x	y
−2.5	−2.5
−2.0	−1.0
−1.5	0.0
−1.0	1.5
−0.5	2.5
0.0	4.0
0.5	5.0

36.

x	y
−1.4	9.0
−0.9	7.4
−0.4	5.8
0.1	4.0
0.6	2.5
1.1	1.0
1.6	−0.5

37. **Stretch of a Steel Spring** The given table displays data points taken from an experiment in which a mass was hung from a spring. The distance the spring stretched was measured for each mass.
    a. Draw a scatter diagram for these data, and then determine the line of best fit. (Round the coefficients to the nearest thousandth.)
    b. Interpret the meaning of the slope of the line of best fit.
    c. Use the equation of the line of best fit to estimate the stretch for a mass of 12 kg.
    d. Use the equation of the line of best fit to predict the number of kilograms of mass hung on the spring when the stretch is 2.4 cm.

MASS ON SPRING x (kg)	STRETCH BY SPRING y (cm)
5	0.3
10	0.9
15	1.4
20	1.6
25	2.2
30	2.5
35	2.7

In Exercises 38–40, use the given parabola to estimate the

a. Vertex b. y-intercept c. x-intercepts

38.

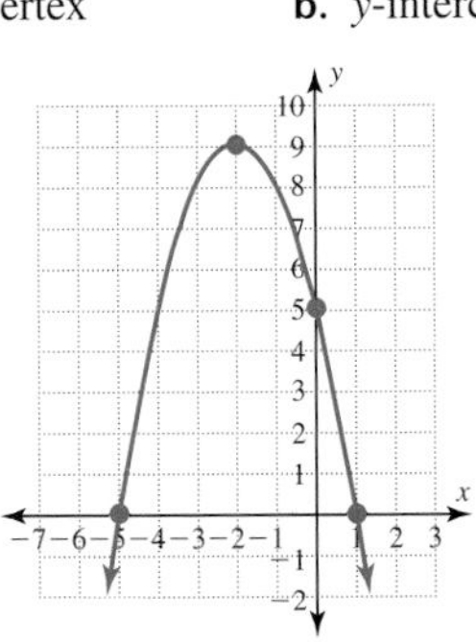

39.

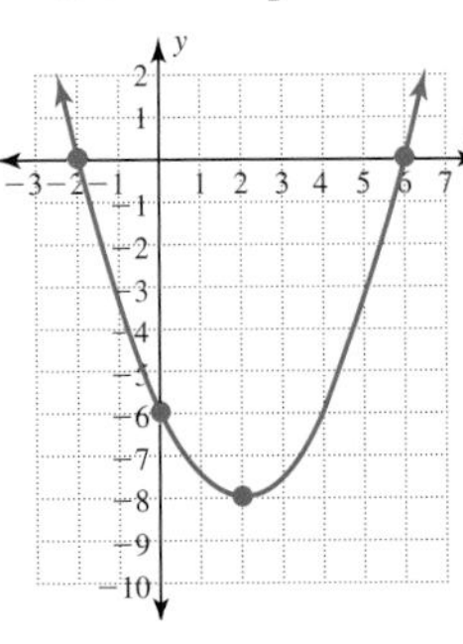

40.

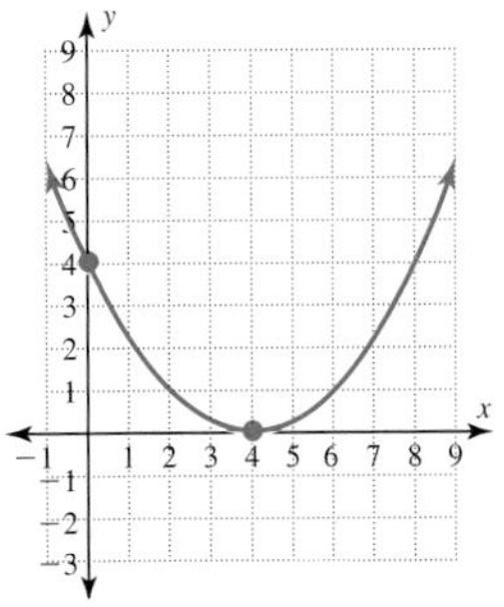

In Exercises 41–44, match each quadratic function with the appropriate parabola.

41. $y = (x + 2)(x - 4)$
42. $y = -(x + 2)(x - 4)$
43. $y = -(x - 1)(x + 7)$
44. $y = (x - 1)(x + 7)$

A.

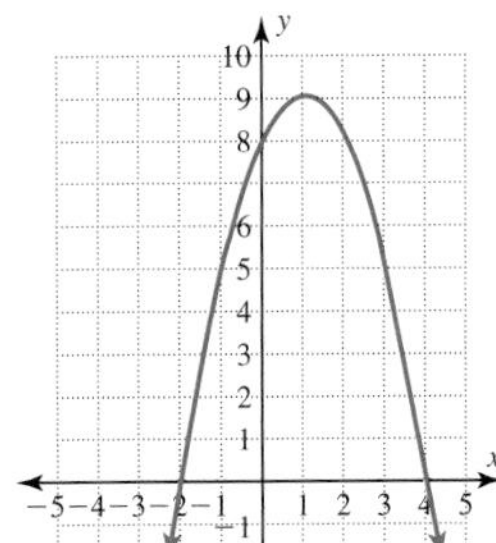

B.

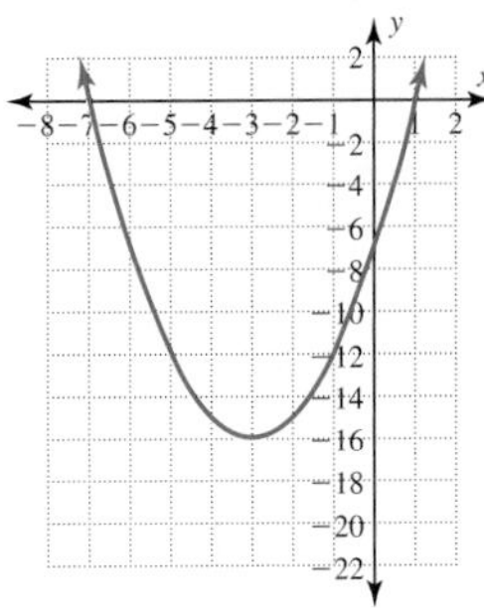

C.

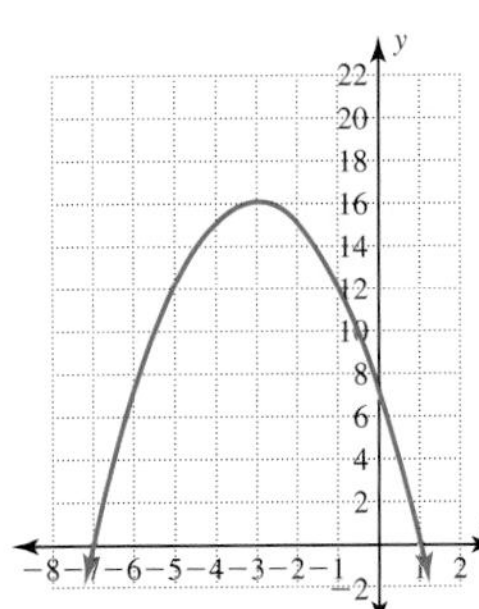

D.

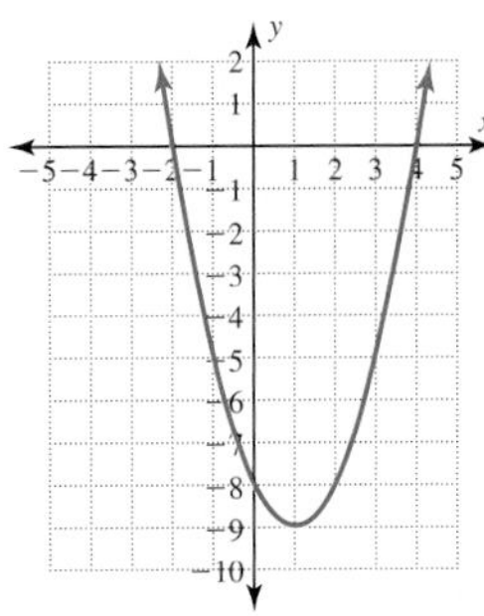

In Exercises 45 and 46, use a graphing calculator to graph the parabola. Then determine the following.

a. Does the parabola open upward or downward?
b. Vertex
c. y-intercept
d. x-intercepts
e. Find the interval where the function is increasing.
f. Find the interval where the function is decreasing.

45. $y = -0.5x^2 + 2x + 6$
46. $y = 0.25x^2 - 2x - 2.25$

In Exercises 47 and 48, solve each quadratic equation by factoring.

47. $x^2 - 7x + 12 = 0$
48. $(2x - 1)(4x - 5) = 20$

In Exercises 49 and 50, solve each quadratic equation by extraction of roots.

49. $x^2 = 64$
50. $(2y - 3)^2 = 1$

In Exercises 51–54, solve each quadratic equation by using the quadratic formula.

51. $v^2 - 9v + 20 = 0$
52. $w^2 = 4w - 2$

53. $10(m^2 - 1) = 21m$ **54.** $\frac{2}{3}x^2 = 2x + 3$

55. Use the given graph to solve $x^2 - 5x - 150 = 0$.

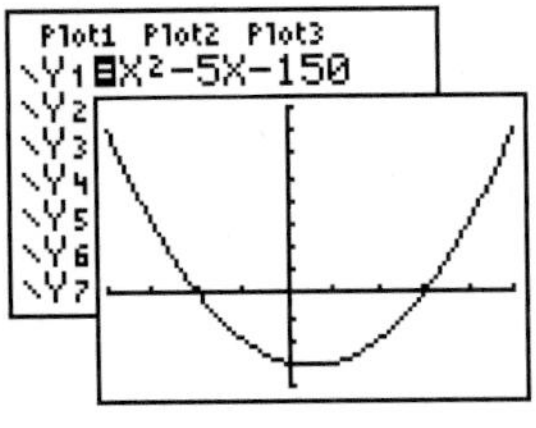

$[-20, 25, 5]$ by $[-200, 400, 50]$

56. Use the given table to solve $x^2 + 18x + 77 = 0$.

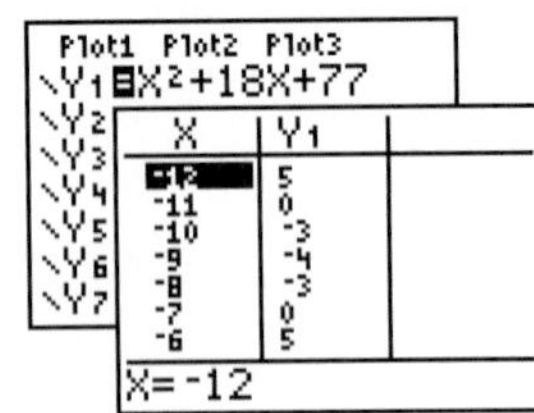

57. Use the given graph to solve each equation and inequality.

a. $x^2 + 3x - 10 = 0$
b. $x^2 + 3x - 10 > 0$
c. $x^2 + 3x - 10 < 0$

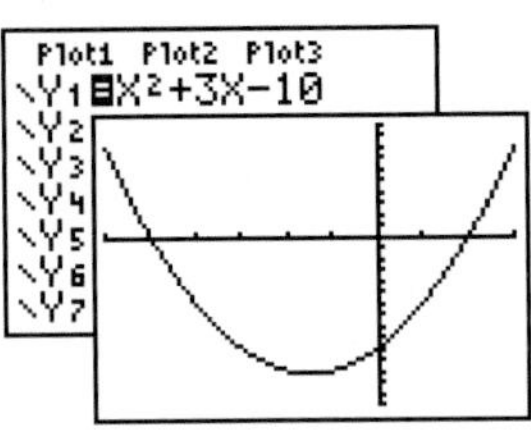

$[-6, 3, 1]$ by $[-15, 10, 1]$

In Exercises 58 and 59, use a graphing calculator to graph $y = P(x)$, to determine the x-intercepts of this graph, and to complete this table.

POLYNOMIAL $P(x)$	x-INTERCEPTS OF THE GRAPH OF $y = P(x)$	FACTORED FORM OF $P(x)$	ZEROS OF $P(x)$	SOLUTIONS OF $P(x) = 0$
58. $-3x^2 + 8x + 3$				
59. $x^3 + x^2 - 2x$				

60. Dimensions of a Right Triangle The length of the hypotenuse of the right triangle shown in the figure is 2 cm more than the length of the longer leg. If the longer leg is 7 cm longer than the short leg, determine the length of each side.

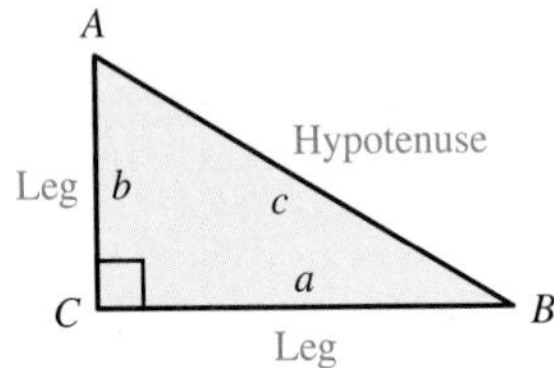

61. Deflection of a Beam A 40-ft concrete beam is fixed between two rigid anchor points. If this beam expands by 0.4 inch due to a 45° increase in temperature, determine the bulge in the middle of the beam. (Assume for simplicity the beam deflects as shown in this diagram.)

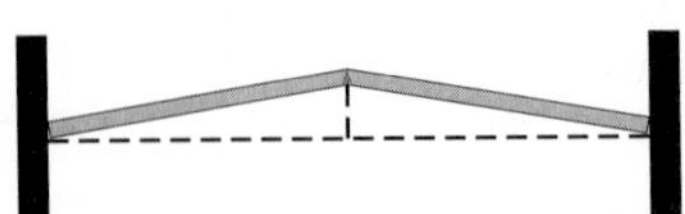

62. Positioning an Effective Sound System The sound system planned for an airport concourse has speakers that should be placed at least 14 ft above the floor and thus approximately 8 ft above the heads of most customers. A customer is standing at a position x ft from a reference point directly under one speaker that is 14 ft above the floor.

a. If a customer is standing 16 ft from the reference point, how far is it from the speaker to the customer's ears?

b. If the limit of the speaker's effective range is 20 ft, approximate to the nearest foot the distance a customer can walk from the reference point and stay within the effective range of the speaker.

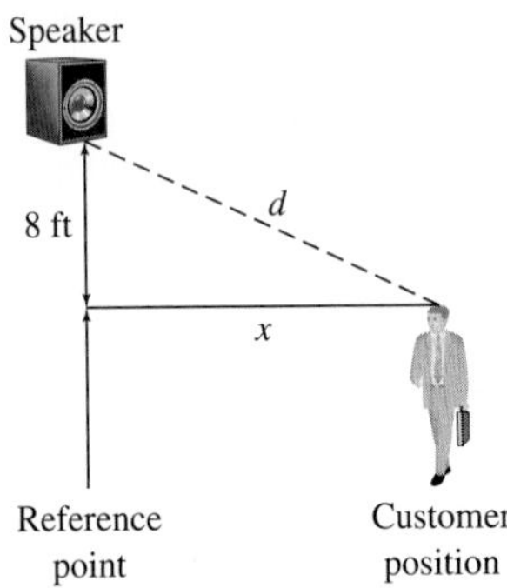

63. Speed of an Airplane Two airplanes depart simultaneously from an airport. One flies due south; the other flies due east at a rate 20 mi/h faster than that of the first airplane. After 1 hour, radar indicates that the airplanes are approximately 450 mi apart. What is the ground speed of each airplane?

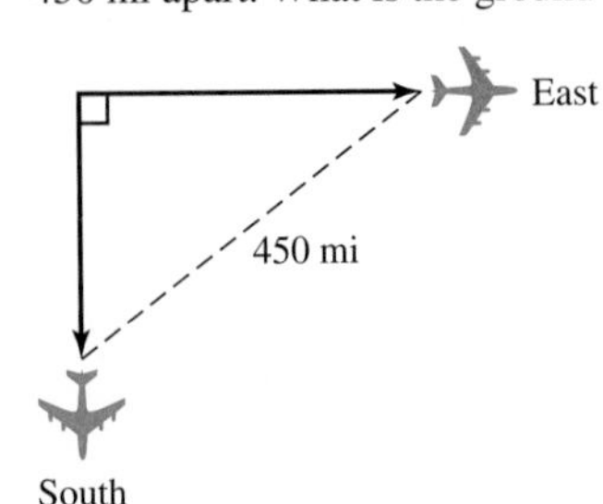

64. **Pasta Recipe** A recipe for pasta for eight people suggests using enough angel hair pasta to fill a 6-cm-diameter opening on a pasta-measuring device. Approximately what size hole would be recommended for four people?

65. **Profit** The net income in dollars produced by selling x units of a product is given by $P(x) = -x^2 + 45x - 200$. If $P(x) > 0$, there is a profit.
  a. Determine the profit interval, the values of x that will generate a profit.
  b. What is the maximum profit that can be produced?

66. **Height of a Baseball** The height h in feet of a baseball t seconds after being hit by a batter is given by $h(t) = -16t^2 + 80t + 3$.

  a. Determine to the nearest hundredth of a second when the ball will hit the ground.
  b. What is the maximum height that the ball reaches?
  c. When does it reach this height?
  d. During what time interval after the baseball is hit will its height exceed 67 ft?

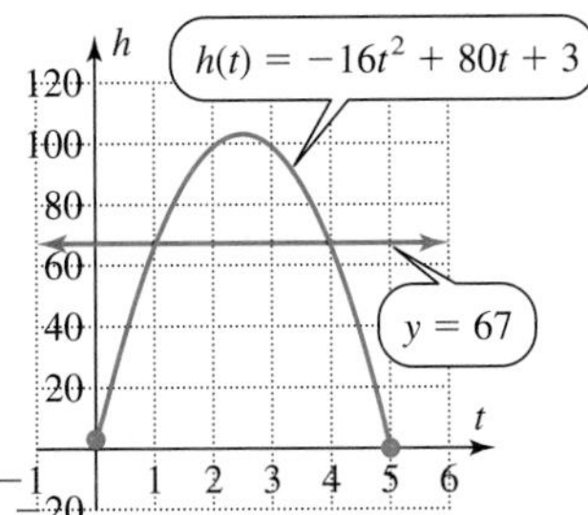

67. **Compound Interest** The formula for computing the amount A of an investment of principal P at interest rate r for 1 year compounded semiannually is $A = P\left(1 + \frac{r}{2}\right)^2$. Approximately what interest rate is necessary for \$1,000 to grow to \$1,050 in 1 year if the interest is compounded semiannually?

68. **Curve Fitting** Use a graphing calculator to draw a scatter diagram for these data points and to determine the parabola of best fit. Express your answer in the form $y = ax^2 + bx + c$ with a, b, and c rounded to the nearest thousandth.

x	y
0	4.5
1	8
2	14
3	22
4	32
5	45
6	60

69. **Modeling Length of a Pendulum** The following experimental data were collected which relate the length of a pendulum to its period.
  a. Use a calculator to determine the parabola of best fit for these data. Round the coefficients to three significant digits.

TIME x (sec)	LENGTH y (cm)
0.8	15
1.1	30
1.4	45
1.5	60
1.7	75
1.9	90
2.2	105
2.3	120

  b. Use this quadratic equation to estimate the length of a pendulum needed for a clock to produce a period of 2 seconds.

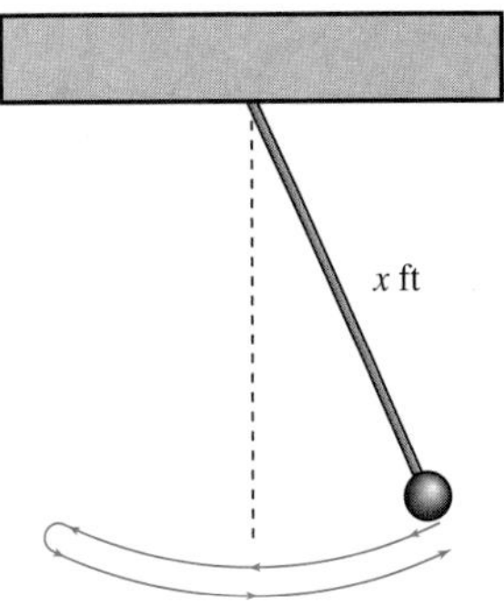

70. **Modeling Profit Based on Sales** A new company recorded its sales units versus its profit (or loss) for eight consecutive quarters.

UNITS SOLD x	PROFIT y (\$ hundreds)
0	−50
25	1,000
50	1,800
75	2,400
100	3,000
125	2,900
150	2,900
175	2,000

  a. Use a calculator to determine the parabola of best fit for these data. Round the coefficients to three significant digits.
  b. Use the quadratic equation to estimate the profit in dollars generated by the sale of 80 units in a quarter.
  c. Determine the break-even values.
  d. Determine the maximum profit that can be made and the number of units sold that generates this profit.

In Exercises 71–73, write a quadratic equation in standard form with the given solutions.

71. -1 and $\frac{3}{5}$
72. $-\sqrt{3}$ and $\sqrt{3}$
73. A double solution of $\frac{5}{7}$

In Exercises 74–86, simplify each expression and write the result in standard $a + bi$ form.

74. $\sqrt{-100}$
75. $\sqrt{64} - \sqrt{-64}$
76. $(5 - 6i) - (3 - 2i)$
77. $2(4 - 3i) - 5(2 + 6i)$
78. $2i(5 - 6i)$
79. $(5 - 7i)(6 + 3i)$
80. $(7 - 3i)^2$
81. $(5 - 2i)(5 + 2i)$
82. i^5
83. $i^6 + i^7$
84. $\frac{1 + i}{1 - i}$
85. $\frac{58}{2 + 5i}$
86. $\frac{3 + 2i}{i}$

87. Use the calculator display shown to the right to evaluate each expression.
  a. i^{25} b. i^{-11} c. i^{32}

```
i^25
          -5E-13+i
i^(-11)
           1E-13+i
i^32
          1+2E-13i
```

88. Use the parabola defined by $y = ax^2 + bx + c$ to determine whether the discriminant for $ax^2 + bx + c = 0$ is negative, zero, or positive. Then identify the nature of the solutions of the quadratic equation as distinct real solutions, a double real solution, or complex solutions with imaginary parts.

a.

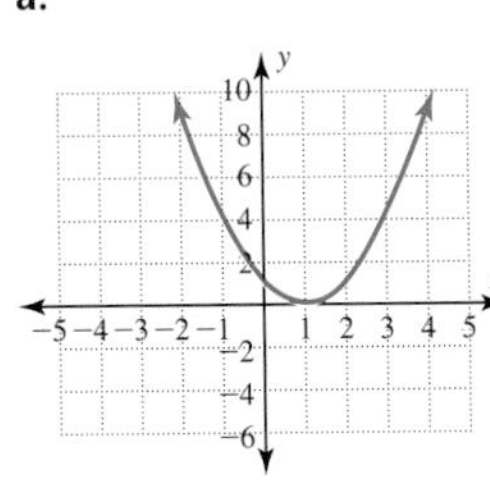

b.

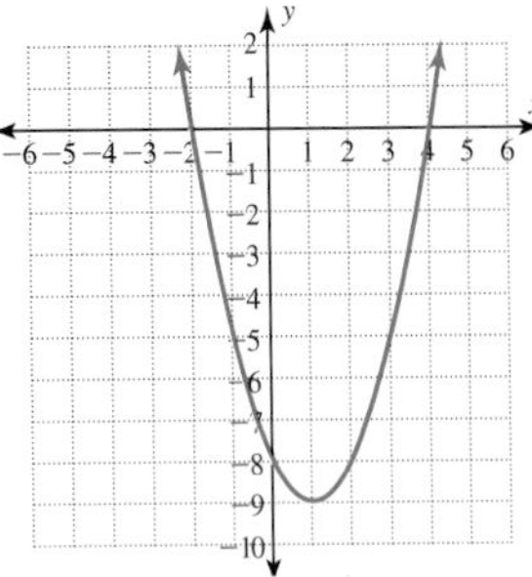

c.

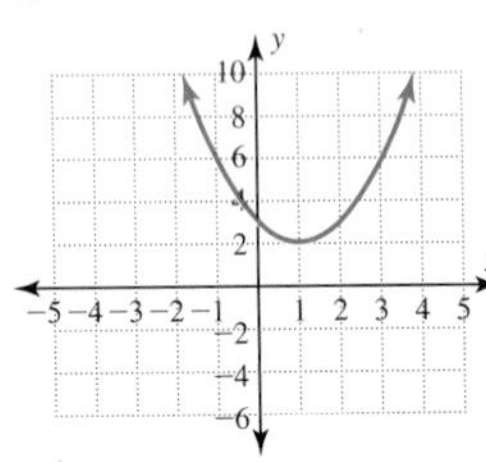

In Exercises 89–91, use the discriminant to determine the nature of the solutions of each quadratic equation.

89. $-3x^2 = 75 - 30x$

90. $4v^2 + 2v + 1 = 0$

91. $x^2 + \sqrt{11}x = 2$

In Exercises 92 and 93, solve each quadratic equation by extraction of roots.

92. $z^2 = -49$

93. $(2z + 1)^2 = -9$

In Exercises 94 and 95, solve each quadratic equation by using the quadratic formula.

94. $x^2 = 6x - 10$

95. $(x - 1)(x - 5) = -6$

In Exercises 96 and 97, write a quadratic equation with the given solutions.

96. $-11i$ and $11i$

97. $5 - i\sqrt{3}$ and $5 + i\sqrt{3}$

98. Solve $x^2 + 6x = -5$ by completing the square. (Optional)

MASTERY TEST FOR CHAPTER 7

[7.1] **1.** Determine whether each relation is a function.

a.

D	R
−3 →	9
3 ↗	
−4 →	16
4 ↗	

b.

x	y
9	−3
9	3
16	−4
16	4

c.

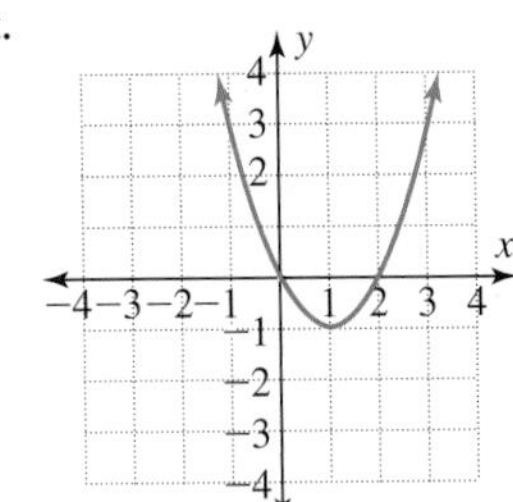

d.

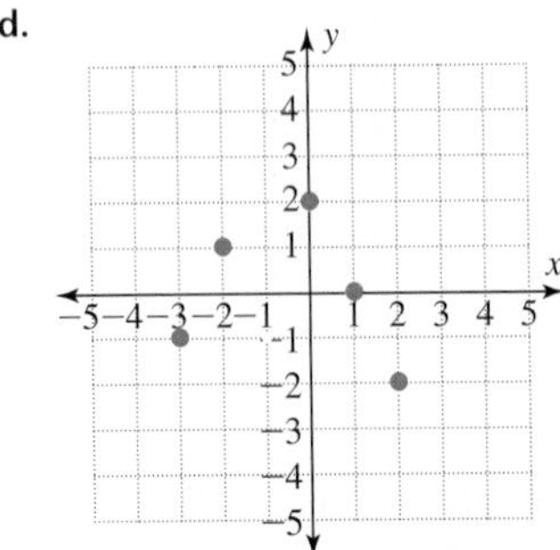

e.

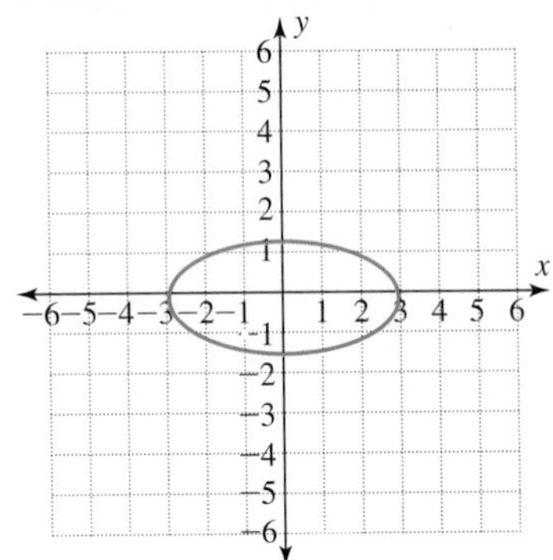

f. $\{(6, 6), (5, 6), (3, 6)\}$

[7.1] **2.** Determine the domain and range of each of these functions.

a. $\{(1, 8), (2, 4), (7, 11), (8, 13)\}$

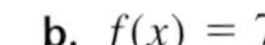

b. $f(x) = 7$

c.

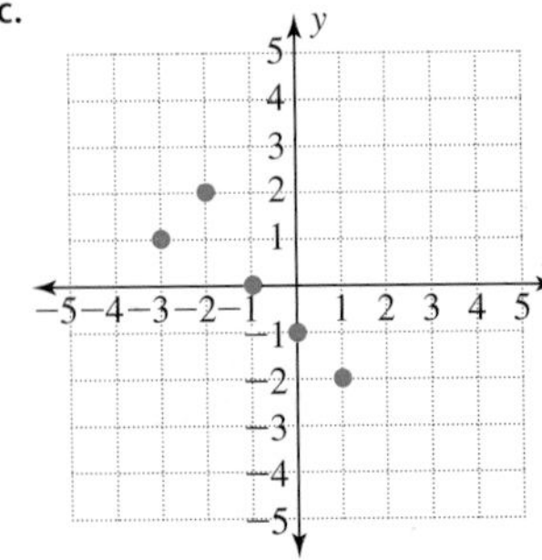

d.

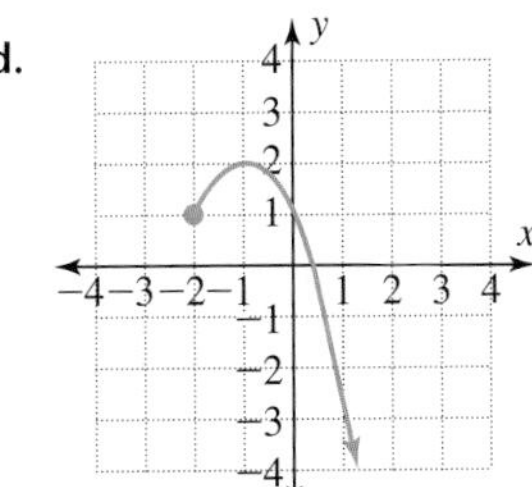

e.

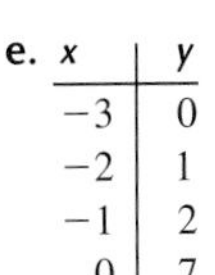

x	y
−3	0
−2	1
−1	2
0	7

f. $f(x) = \sqrt{x - 3}$
(*Hint:* Use a calculator to graph this function.)

[7.1] 3. Evaluate each of the following expressions for $f(x) = 3x^2 + x - 12$.

a. $f(2)$ b. $f(-2)$

Use the given graph to determine the missing input and output values.

c. $f(0) =$ ______ d. $f(x) = 0$; $x =$ ______

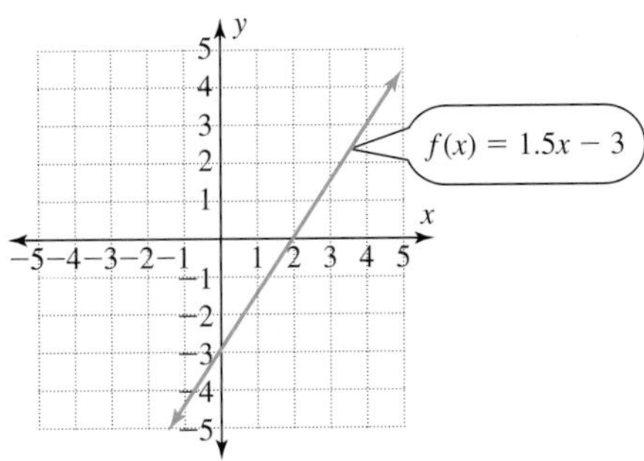

Use the given table of values to determine the missing input and output values.

e. $f(-10) =$ ______

f. $f(x) = 12$; $x =$ ______

x	f(x)
−10	−18
−5	−10.5
0	−3
5	4.5
10	12
15	19.5
20	27

[7.2] 4. a. Graph a line with a y-intercept of (0, 2) and a slope of $-\frac{3}{4}$.

b. Graph the line defined by $y = \frac{x}{2} - 3$.

[7.2] 5. Identify each function as an increasing or a decreasing function.

a.

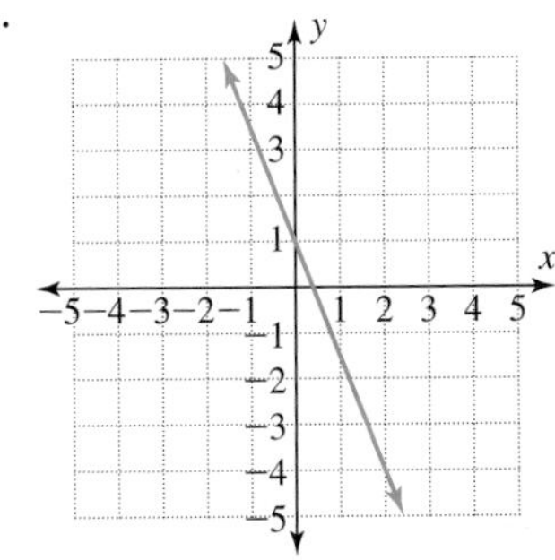

b.

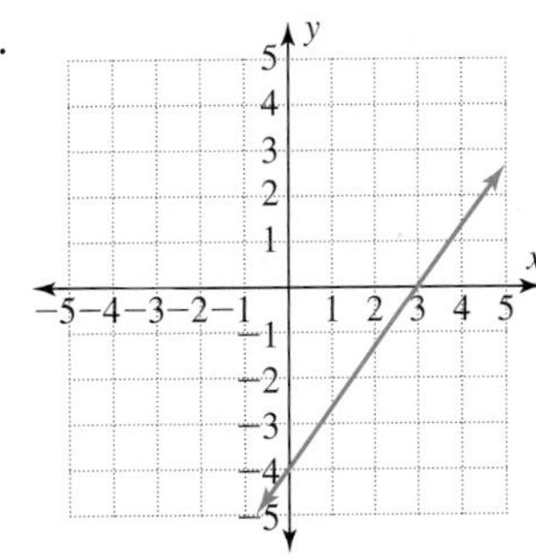

c. $f(x) = 3.7x - 4.9$ d. $f(x) = -4.9x + 3.7$

[7.2] 6. Determine the x-values for which each function is positive and those for which each function is negative.

a.

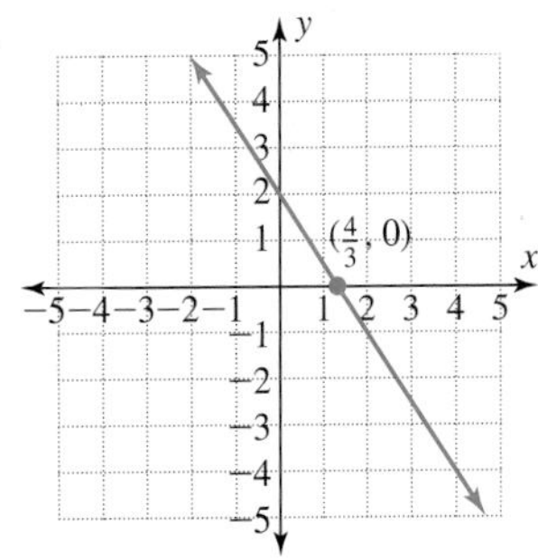

b.

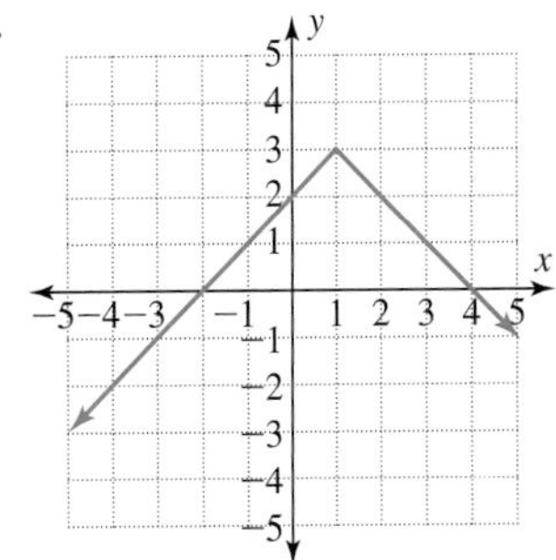

[7.2] 7. Use the absolute value function $f(x) = |x + 1| - 3$ to complete each part of this exercise.

a. Complete the following table of values.

x	f(x) = |x + 1| − 3
−5	
−3	
−1	
1	
3	
5	
7	

b. Graph this function.

c. What are the x-intercepts of the graph in part **b**?

d. What is the y-intercept of the graph in part **b**?

e. What point is the vertex of the graph in part **b**?

f. What y-value is the minimum output value for this function?

g. What is the domain of this function?

h. What is the range of this function?

i. For what interval of input values of x is the function decreasing?

j. For what interval of input values of x is the function increasing?

k. For what interval of input values of x is the function positive?

l. For what interval of input values of x is the function negative?

[7.3] 8. For the parabolas shown in parts **a** and **b,** estimate the vertex, the y-intercept, and the x-intercepts.

a.

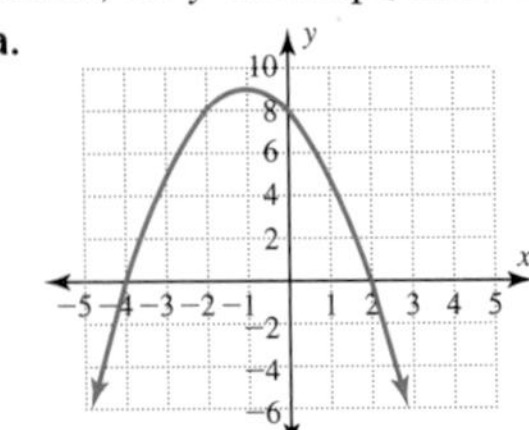

b.

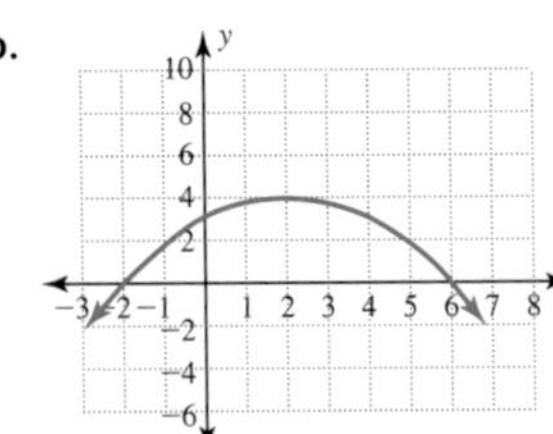

In parts **c** and **d** use the equation to predict whether the parabola defined by this function opens upward or downward. Then use a graphing calculator to graph the parabola and determine its vertex, y-intercept, and x-intercepts.

c. $f(x) = x^2 + 4x - 21$

d. $f(x) = -x^2 + 2x + 15$

[7.3] 9. **Olympic Women's High-Jump Records** The given table displays the winning heights for the Olympic women's high jump from 1948 through 2000. Each distance is given in meters. (The actual dates are not used in the calculations. The x-values represent the number of years after 1900. These smaller x-values allow us to round the coefficients of the equation of best fit without introducing an unacceptable amount of error.)

a. Draw a scatter diagram for these data and then determine the line of best fit. (Round the coefficients to the nearest thousandth.)

b. Interpret the meaning of the slope of this line.

c. Use the equation of the line of best fit to estimate the missing entry in the table from 1972. How does this estimation compare to the actual 1972 record of 1.92 m?

d. Use the equation of the line of best fit to predict the women's high-jump record for the year 2004. How does this compare to the actual 2004 record?

YEAR	x	y (m)
1948	48	1.68
1952	52	1.67
1956	56	1.76
1960	60	1.85
1964	64	1.90
1968	68	1.82
1976	76	1.93
1980	80	1.97
1984	84	2.02
1988	88	2.03
1992	92	2.02
1996	96	2.05
2000	100	2.01

[7.3] 10. Draw a scatter diagram of the points in the following table. Then determine the quadratic function that best fits these points. (Round the coefficients to three significant digits.) Use this function f to approximate $f(7)$.

x	−6	−4	−2	0	2	4	6	8
y	7	2	−2	−3	−4	−2	1	6

[7.4] 11. Use the quadratic formula to solve each quadratic equation.

a. $6x^2 - 19x + 10 = 0$ b. $w^2 = 2 - 2w$

c. $4v^2 + 49 = 28v$ d. $(x - 1)^2 = 2(x + 1)$

[7.4] 12. Use the discriminant to identify the nature of the solutions of each quadratic equation as distinct real solutions, a double real solution, or complex solutions with imaginary parts.

a. $5x^2 + 5x + 1 = 0$ b. $7y^2 = 84y - 252$

c. $3w^2 + 2w - 1 = 0$ d. $(x + 1)(x - 2) = -3$

[7.5] 13. Determine the vertex of the parabola defined by each of these equations.

a. $y = x^2 - 4x - 21$ b. $y = 2x^2 + x - 15$

c. $y = x^2 - 4x + 8$ d. $y = (5x - 4)(x + 6)$

[7.5] 14. **Height of a Softball**

The path of a softball is given by $y = -16x^2 + 48x + 4$, where y represents the height of the ball in feet and x is the number of seconds that have elapsed since the ball was hit. Determine the highest point that the ball reaches. Determine at how many seconds into the flight the maximum height is reached.

[7.5] 15. Construct a quadratic equation with the given solutions.

a. $x = 7$ and $x = -\dfrac{2}{9}$

b. $x = -\sqrt{7}$ and $x = \sqrt{7}$

c. $x = 2 - 3i$ and $x = 2 + 3i$

d. Construct a cubic equation with solutions $x = 2$, $x = 3i$, and $x = -3i$.

[7.6] 16. a. **Pizza Size** Two couples planned to order a 14-in pizza, but one couple had to unexpectedly leave. What size pizza should the remaining couple order and still have approximately the same amount of pizza per person?

b. **Length of a Side of a Square** The side of one square is 5 cm longer than the side of another square. Their total area is 193 cm^2. Determine the length of a side of each square.

c. **Profit Interval** $P(x) = -x^2 + 190x - 925$ is a profit polynomial that gives the profit in dollars made by selling x units of a product. Determine the profit interval and the maximum profit that can be made.

d. **Airplane Speeds** Two airplanes depart simultaneously from an airport. One flies due south; the other flies due east at a rate 100 mi/h faster than that of the first airplane. After 2 hours, radar indicates that the airplanes are 600 mi apart. What is the ground speed of each airplane?

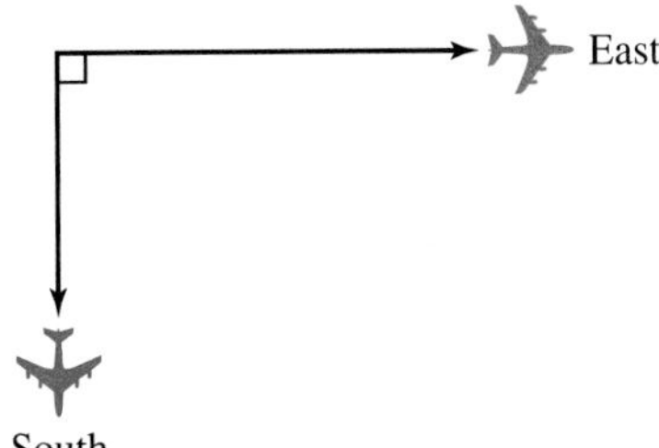

[7.7] **17.** Write each complex number in standard $a + bi$ form.

a. $\sqrt{-81}$

b. $\sqrt{16 - 25}$

c. $\sqrt{-16} - \sqrt{25}$

d. $i^2 + i^3$

[7.7] **18.** Perform each operation, and write the result in standard $a + bi$ form.

a. $2(4 - 5i) - 3(3 - 4i)$

b. $(4 - 5i)(2 - 4i)$

c. $\dfrac{4 - 5i}{2 - 4i}$

d. $(3 - i)^2$

[7.7] **19.** Solve these quadratic equations by extraction of roots.

a. $(y - 1)^2 = 36$

b. $(y - 1)^2 = -36$

Solve these quadratic equations by using the quadratic formula.

c. $x^2 - 10x + 29 = 0$

d. $(x - 2)(x - 4) = -26$

REALITY CHECK FOR CHAPTER 7

The path followed by a stream of water as it leaves a fire hose is parabolic. The following three graphs illustrate three possible paths using the same hose and the same water pressure. Discuss the reason(s) why you think this occurs. Which path will allow the water to travel the greatest horizontal distance? Use the Spreadsheet Exploration from the website www.mhhe.com/hallmercer for Chapter 7 on water paths to test your ideas.

A.

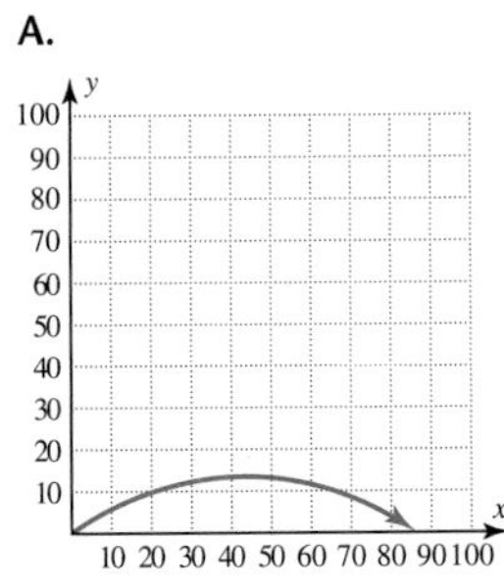

B.

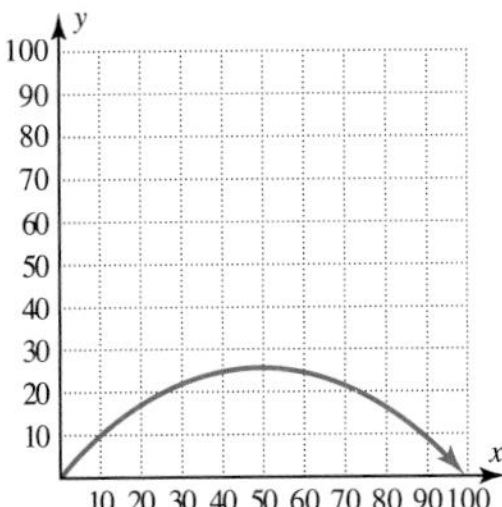

C.

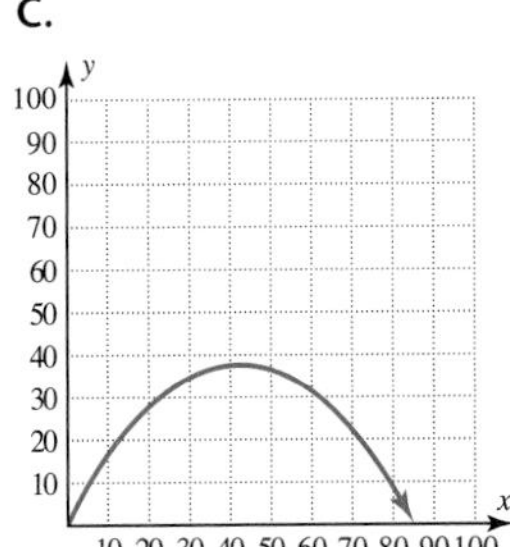

GROUP PROJECT FOR CHAPTER 7

An Algebraic Model for Real Data

Supplies needed for the lab:

I. A clear cylindrical container such as a 2-L soft drink container with a hole drilled near the bottom of the straight-walled portion and with the height labeled in centimeters on an attached paper strip or on the plastic with a permanent marker. Label the numerical scale from top to bottom, so that 0 is on the hole. *Suggestion:* Experiment with the hole size—approximately $\frac{1}{4}$ in—until the total drain time is less than 3 minutes.

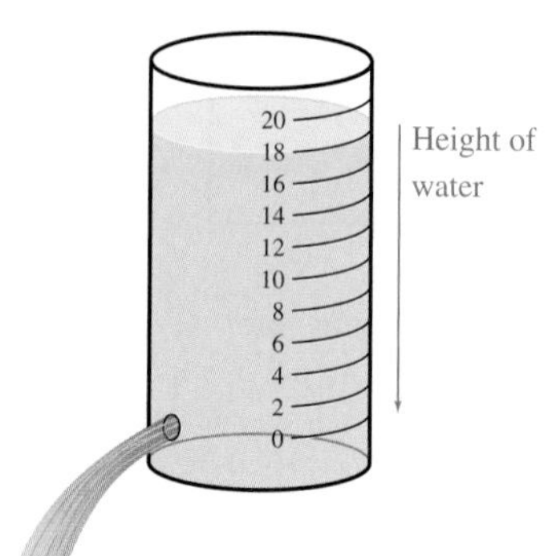

II. Water and a bucket to drain the water into.

III. A watch that displays seconds.

IV. A graphing calculator with a linear and a quadratic regression feature.

Lab: Recording the Time to Drain a Cylindrical Container

Place water into a cylindrical container up to the fill line of the cylindrical portion of this container while keeping the drain hole closed. Set your time to 0 seconds, and record this time and the initial height of the water. Then open the drain hole and remove the cap from the top of the container. As the water drains from the container, record the time for each height in the table. (You may decide to repeat this experiment by recording the height every 10 or more seconds.)

TIME x (seconds)	HEIGHT y (cm)
0	20
	18
	16
	14
	12
	10
	8
	6
	4
	2
	0

1. Record your data in a table similar to the one shown.
2. Plot these data on a scatter diagram, using an appropriate scale for each axis.

Creating an Algebraic Model for this Data

3. a. Which of these models, a linear model or a quadratic model, do you think is the better model for this experiment?
   b. What information in the table or in the graph supports your choice in part **a**? Why do you believe this model is better?
   c. Without referring to any of the data that you collected, describe any observations that you made during the course of the experiment that you think support your choice in part **a**.
4. a. Using your graphing calculator, calculate either the line of the best fit or the parabola of the best fit (depending on which is the better choice). (*Hint:* Watch out for scientific notation.)
   b. Graph this curve on the scatter diagram.
5. Let $y = H(x)$ represent the curve of best fit from **4a**:
   a. Evaluate and interpret $H(50)$.
   b. Evaluate and interpret $H(70)$.
   c. Determine the value of x for which $H(x) = 11$. Interpret the meaning of this value.

chapter 8

Rational Functions

CHAPTER OUTLINE

The weight of an object depends upon the mass of the object, the distance from the object to the center of the nearest planet or moon, and the mass of this planet or moon. This is known as the universal law of gravitation. As the distance from the object to the nearest planet or moon increases, the weight of the object decreases.

Since the mass of the Earth is much larger than that of the Moon, the same object will weigh more on the Earth than on the Moon. The graph of the first function gives the weight in newtons (N) of an astronaut based upon a distance in kilometers above the surface of the Earth. The *y*-intercept of this graph indicates the weight of the astronaut on the surface of the Earth. The graph of the second function gives the weight of this astronaut based upon a distance above the surface of the Moon. The *y*-intercept of this graph indicates the weight of the astronaut on the surface of the Moon.

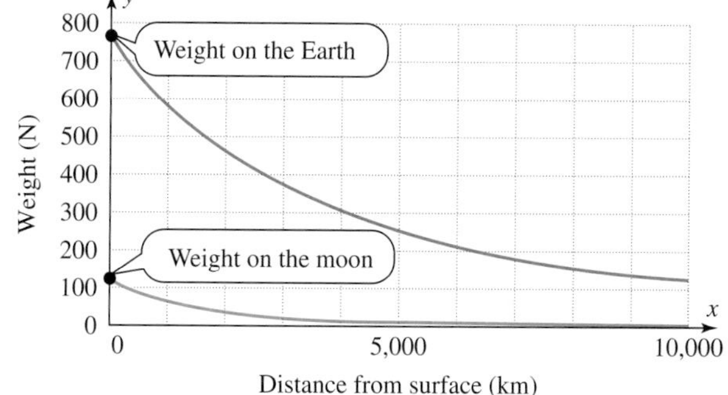

REALITY CHECK

Some of the proposals for space exploration of Mars and other planets suggest establishing a launch site on the Moon to produce materials and fuel for these launches. What major advantage does a launch from the Moon have over a launch from the Earth?

Section 8.1 Graphs of Rational Functions and Reducing Rational Expressions

Objectives:

1. Determine the domain of a rational function.
2. Identify the vertical asymptotes of the graph of a rational function.
3. Reduce a rational expression to lowest terms.

Each family of functions has its own distinctive numerical pattern, its own graph, and its own algebraic definition. Note the distinctive numerical patterns exhibited in the following two tables. The first table gives the distance that a spring stretches when a mass is attached to it, and the second table gives the time to travel a distance of 100 mi based upon the rate of travel.

As x increases, y increases.

Distance a Spring Stretches

MASS x (kg)	DISTANCE y (cm)
0	0
10	1.25
20	2.50
30	3.75
40	5.00
50	6.25
60	7.50
70	8.75
80	10.00

Time to Travel 100 mi

SPEED x (mi/h)	TIME y (h)
5	20.00
10	10.00
20	5.00
25	4.00
50	2.00
100	1.00
200	0.50
400	0.25
500	0.20

As x increases, y decreases.

In the first table, the distance that the spring stretches increases as the mass attached to the spring increases. In the second table, the time to cover 100 mi decreases as the speed increases. These two numerical patterns are vastly different and produce completely different shapes and algebraic equations. This is illustrated by the following two graphs.

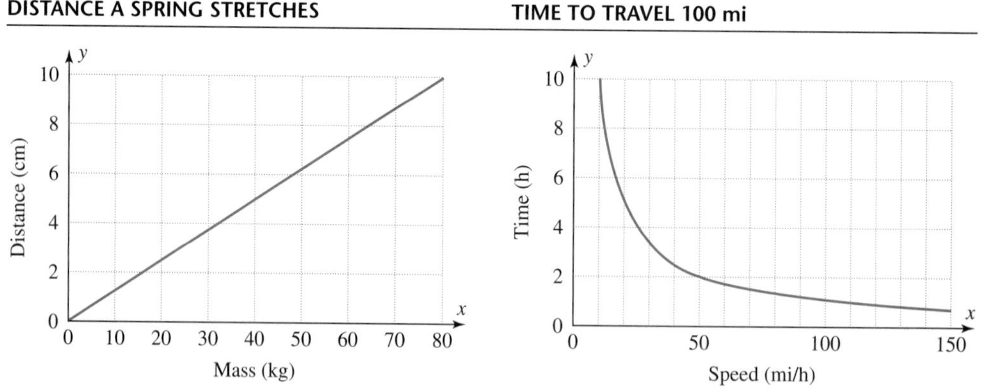

The graph on the left exhibits a linear pattern while the graph on the right does not have the characteristics of a polynomial function. Before we analyze the behavior of this graph, we first summarize some of the characteristics of polynomial functions. These are characteristics exemplified both by first-degree functions of the form $y = mx + b$ whose graphs have a characteristic linear shape and by second-degree functions of the form $y = ax^2 + bx + c$ whose graphs have a characteristic parabolic shape.

Characteristics of the Graphs of Polynomial Functions

- Each polynomial function is continuous (there are no breaks or gaps between points on the graph).
- Each polynomial function is smooth (there are no sharp points or corners on the graph).
- The domain of each polynomial function is the set of all real numbers (the graph extends without limit both to the left and to the right).
- Some polynomial functions have relative (localized) maximum or minimum points (there are high or low points on the graph)—there are more possibilities for these points for higher-degree polynomials.

Now we examine rational functions. These functions can have breaks in the graph, and their domain may not be all real numbers.

Rational Functions

A fraction that is the ratio of two polynomials is called a **rational expression.** We can use rational expressions to define rational functions. The function given for the time to travel a distance of 100 mi based upon the rate of travel is an example of a rational function. These functions have graphs that are considerably more complicated than those of polynomial functions, and thus the algebraic expressions that define these functions are also more involved.

Rational Function

ALGEBRAICALLY	VERBALLY	ALGEBRAIC EXAMPLE	GRAPHICAL EXAMPLE
$f(x) = \dfrac{P(x)}{Q(x)}$ is a rational function if $P(x)$ and $Q(x)$ are polynomials and $Q(x) \neq 0$.	A rational function is defined by the ratio of two polynomials.	$f(x) = \dfrac{1}{x}$	Graph of $f(x) = \dfrac{1}{x}$

EXAMPLE 1 Identifying Rational Functions

Determine whether each algebraic expression defines a rational function.

SOLUTION

(a) $y = \dfrac{x - 1}{x + 3}$ This is a rational function.

Both the numerator $x - 1$ and denominator $x + 3$ are polynomials. Thus this is a rational function.

(b) $y = 7x^2 - 5x + 12$ This is a rational function.

This expression can be written in the form $y = \dfrac{7x^2 - 5x + 12}{1}$. Both the numerator $7x^2 - 5x + 12$ and the denominator 1 are polynomials. Thus this is a rational function.

As part (b) illustrates, all polynomial functions are also rational functions. (They can be rewritten with a denominator of 1.)

(c) $y = \dfrac{2x + 3}{\sqrt{x - 5}}$ This is not a rational function.

The denominator $\sqrt{x - 5}$ is not a polynomial. Thus this is not a rational function.

SELF-CHECK 8.1.1

Determine whether each algebraic expression defines a rational function.

1. $f(x) = 2x^2 - x + 1$ 2. $f(x) = \dfrac{3x - 1}{x + 2}$ 3. $f(x) = \dfrac{\sqrt{x}}{7}$ 4. $f(x) = \dfrac{5}{x}$

The graph of $f(x) = \dfrac{1}{x}$ illustrates some of the essential characteristics that are possible with rational functions. One of the classic behaviors possible with rational functions is that their graphs can approach and get very near to vertical and horizontal lines. A vertical line that a graph approaches but does not touch is called a **vertical asymptote.** A horizontal line that a graph approaches is called a **horizontal asymptote.***

Characteristics of the Graph of $f(x) = \dfrac{1}{x}$

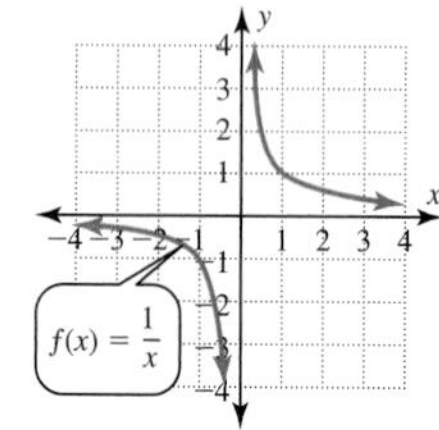

- The domain of this function does not include 0.
- The function is discontinuous; there is a break in the graph at $x = 0$. The graph consists of two unconnected branches.
- The graph approaches the vertical line $x = 0$ asymptotically.
- The graph approaches the horizontal line $y = 0$ asymptotically.

The domain of a rational function must exclude values that would cause division by 0.

The notation $\mathbb{R} \sim \{2\}$ is read "the set of all real numbers except for 2."

We can determine the domain of a rational function algebraically by solving for the values that make the denominator 0 and then excluding these values from the domain. In a numerical table there will be error messages at the excluded values due to division by 0. Graphically there will be a break in the graph of a rational function at an excluded value of the domain, and the function can have a vertical asymptote at this x-value. To denote that the number 2 is excluded from the domain of a function, we can write the domain by using interval notation as $(-\infty, 2) \cup (2, \infty)$. Another notation to represent this domain is $\mathbb{R} \sim \{2\}$, which is read "the set of all real numbers except for 2."

Domain of a Rational Function

The domain of a rational function must exclude values that would cause division by zero.

ALGEBRAIC EXAMPLE	NUMERICAL EXAMPLE	GRAPHICAL EXAMPLE	VERBAL EXAMPLE
The domain of $f(x) = \dfrac{1}{x}$ is $\mathbb{R} \sim \{0\}$.	Plot1 Plot2 Plot3 \Y1=1/X X: -3, -2, -1, 0, 1, 2, 3 Y1: -.3333, -.5, -1, ERR:, 1, .5, .33333 X= -3	[−4.7, 4.7, 1] by [−3.1, 3.1, 1]	Only zero is excluded from the domain to prevent division by zero. Note that zero causes an error message to appear in the table. Also note that there is a break in the graph at the vertical asymptote $x = 0$.

*We limit our analysis of vertical and horizontal asymptotes to an intermediate algebra–level discussion. Rational functions are examined in greater depth in college algebra and calculus. For example, the graph of a rational function can have a hole at $x = c$ instead of a vertical asymptote (see Exercise 90 at the end of this section) if each factor of $x - c$ in the denominator is also in the numerator. Although the graph of a rational function can never touch a vertical asymptote, the behavior with respect to a horizontal asymptote is more complex—the graph can cross a horizontal asymptote before eventually approaching and staying very near to this horizontal line. The graphs of rational functions can also approach lines that are neither horizontal nor vertical.

The values that cause division by 0 in a rational function (see the previous footnote for information on holes) also can indicate where the graph of the function has vertical asymptotes. This is illustrated in Example 2.

EXAMPLE 2 Using Multiple Perspectives to Determine the Domain of a Rational Function

The key to determining the domain of a rational function is to first determine the values that are not allowed because they would cause division by 0.

Determine the domain of $f(x) = \frac{x}{x^2 - 4}$ algebraically, numerically, and graphically. Also determine the vertical asymptotes of the graph of this function.

SOLUTION

ALGEBRAICALLY

$$x^2 - 4 = 0$$
$$(x + 2)(x - 2) = 0$$
$$x + 2 = 0 \quad \text{or} \quad x - 2 = 0$$
$$x = -2 \qquad\qquad x = 2$$

To determine the domain, first calculate the values that would cause division by 0. Do this by setting the denominator equal to 0. Solve this quadratic equation by factoring.

The function is undefined at $x = -2$ and $x = 2$.

The domain includes all real numbers except -2 and 2. We can denote this by $\mathbb{R} \sim \{-2, 2\}$.

NUMERICALLY

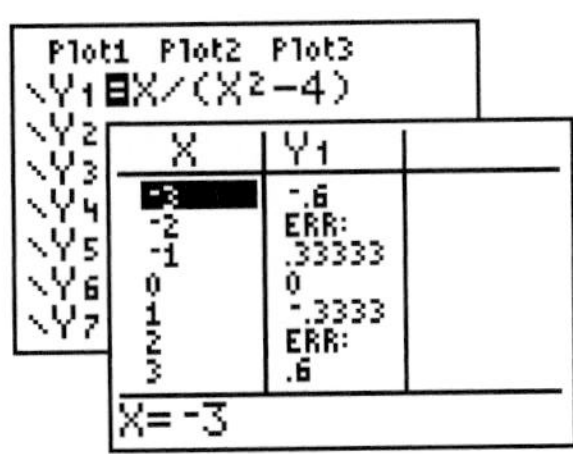

The error messages in the table occur because of division by 0. This indicates the function is undefined at $x = -2$ and $x = 2$.

Since the denominator is second degree, these are the only two values that make the denominator 0.

The error messages mean the function is undefined at $x = -2$ and $x = 2$.

The domain is the set of all real numbers except -2 and 2.

GRAPHICALLY

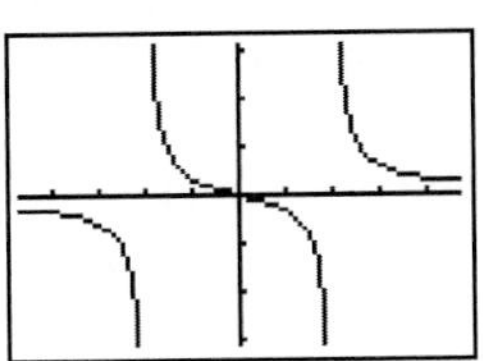

[−4.7, 4.7, 1] by [−3.1, 3.1, 1]

The graph of $y = \frac{x}{x^2 - 4}$ consists of three unconnected branches. The graph approaches the vertical lines $x = -2$ and $x = 2$ asymptotically.

There are no points on the graph for the input values of -2 and 2.

The graph has vertical asymptotes of $x = -2$ and $x = 2$.

The domain contains all real numbers except -2 and 2.

Answer: Domain: $\mathbb{R} \sim \{-2, 2\}$; the vertical asymptotes are $x = -2$ and $x = 2$.

Note that the vertical lines that the graph approaches asymptotically are not part of the graph. You can sketch these asymptotes as a guideline for sketching or analyzing the function. ■

Problems that involve ratios, rates, and averages can all produce rational expressions. Example 3 revisits the problem from the start of this section that involved the time to travel a distance of 100 mi based upon the rate of travel.

EXAMPLE 3 Time to Travel 100 Miles

Write a function that gives the time it takes to travel 100 mi as a function of the rate of travel. Then analyze the asymptotic behavior of this function.

SOLUTION

$D = R \cdot T$ — Distance = Rate · Time

$100 = R \cdot T$

$\frac{100}{R} = T$ or $T = \frac{100}{R}$ — Substitute in the given distance of 100 mi, and then solve for T.

The domain must exclude 0 to avoid division by zero. The practical domain will consider only rates that are positive.

Although there is some upper bound on the rate of travel, this bound is unknown; therefore, an upper limit on the values in the domain is not used.

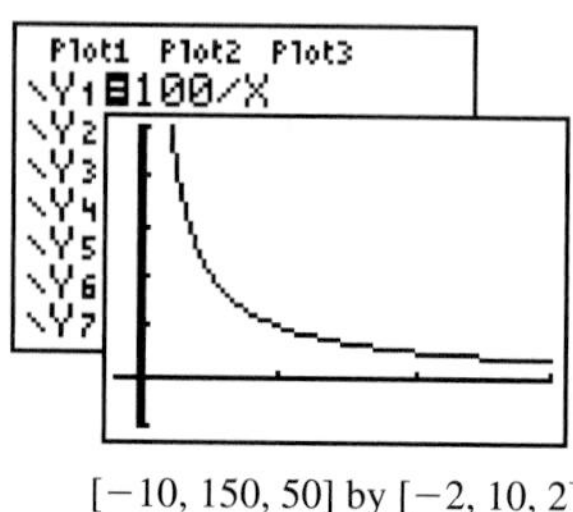

[−10, 150, 50] by [−2, 10, 2]

To use a calculator, rewrite the function as $y = \frac{100}{x}$.

This graph has a vertical asymptote of $x = 0$ and a horizontal asymptote of $y = 0$.

The asymptotic behavior near the vertical asymptote means that as the rate gets very slow, the time to travel 100 mi becomes very large.

The asymptotic behavior near the horizontal asymptote means that as the rate gets very fast, the time to travel 100 mi becomes very small.

If the graph of a rational function has a vertical asymptote, it is important that the graph accurately reflect the behavior of the function close to this asymptote. If you graph a rational function using pencil and paper, you may want to sketch each asymptote using a dashed line. An asymptote can be an aid to sketching the graph as it approaches the asymptote. Whether you graph a rational function using pencil and paper or a graphing calculator, it is important to understand that the vertical asymptote is *not* part of the graph of the function.

SELF-CHECK 8.1.2

Use the function $f(x) = \frac{x^2 - 2x - 15}{x + 4}$ to complete these questions.

1. Evaluate $f(0)$. 2. Evaluate $f(-3)$. 3. Evaluate $f(-4)$.
4. Determine the domain of this function.
5. Determine the vertical asymptote of the graph of this function.

Use the rational function $f(x) = \frac{x + 3}{x - 2}$ to answer each of these questions.

6. Determine the domain of this function.
7. Determine the vertical asymptote of the graph of this function.
8. Use a graphing calculator to graph this function on a window of $[-7.4, 11.4, 4]$ by $[-10.4, 14.4, 4]$.

To analyze rational functions, we need to be able to reduce rational expressions to lowest terms and to perform operations with rational expressions. We now examine the procedure for simplifying rational expressions. This procedure is identical to that used to simplify rational numbers in Chapter 1. A rational expression is in **lowest terms** when the numerator and the denominator have no common factor other than -1 or 1. The basic principle for reducing rational expressions is given in the following box.

To reduce a rational expression, the numerator and denominator must have a common factor.

Note that $\frac{C}{C} = 1$, the multiplicative identity for $C \neq 0$.

Reducing a Rational Expression to Lowest Terms

ALGEBRAICALLY	VERBALLY	ALGEBRAIC EXAMPLE
If A, B, and C are polynomials and $B \neq 0$ and $C \neq 0$, then $\frac{AC}{BC} = \frac{A}{B}$.	1. Factor both the numerator and the denominator of the rational expression. 2. Divide the numerator and the denominator by any common nonzero factors.	$\frac{3x - 6}{x^2 - 2x} = \frac{3(x - 2)}{x(x - 2)}$ $= \frac{3}{x}$ for $x \neq 2$

To divide the numerator and denominator by a common factor, it is helpful to first write each as a product. Example 4 illustrates this method and reviews the rules for exponents from Chapter 5.

EXAMPLE 4 Reducing a Rational Expression with a Monomial Denominator

Reduce $\frac{15x^3y^2}{25x^4y^2}$ to lowest terms. Assume that x and y are not zero.

SOLUTION

$$\frac{15x^3y^2}{25x^4y^2} = \frac{5x^3y^2(3)}{5x^3y^2(5x)}$$

Factor the GCF out of the numerator and the denominator. Note that x and y must both be nonzero to avoid division by zero.

$$= \frac{3}{5x}$$

This is the same answer you would get by using the properties of exponents given in Chapter 5.

Remember that equal algebraic expressions have the same outputs for all input values of the variables. Looking at a limited set of values in a table does not confirm that the expressions are equal for all values, but it does serve as a good check that is likely to catch potential errors from an algebraic computation. This is illustrated in Example 5.

EXAMPLE 5 Reducing a Rational Expression and Using a Table as a Check

Reduce $\frac{x^2 + 3x}{x^2 + x - 6}$ and compare this result to the original expression using a table of values.

SOLUTION

ALGEBRAICALLY

$$\frac{x^2 + 3x}{x^2 + x - 6} = \frac{x(x+3)}{(x-2)(x+3)}$$

Factor both the numerator and the denominator. Note that the excluded values are -3 and 2.

$$= \frac{x}{x-2} \quad \text{for } x \neq 2, x \neq -3$$

Divide both the numerator and the denominator by the common factor $x + 3$. For $x \neq -3$, $x + 3$ is not zero.

NUMERICAL CHECK

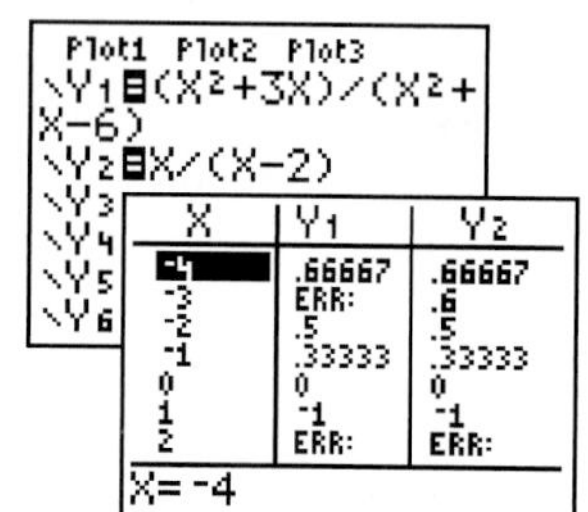

Let Y_1 represent the original expression and Y_2 represent the reduced expression.

The expressions are identical except for $x = -3$. The original expression has excluded values of -3 and 2. The reduced expression was obtained under these restrictions.

Answer: $\frac{x^2 + 3x}{x^2 + x - 6} = \frac{x}{x-2}$ for $x \neq 2, x \neq -3$

It is understood that this equality means that these expressions are identical for all values for which they are both defined. For $x \neq -3$ and $x \neq 2$ these expressions are identical.

We will assume that the simplified form of all rational expressions excludes values that cause division by zero. Then you will not need to write these restrictions with each of your answers.

Two polynomials are **opposites** if every term of the first polynomial is the opposite of the corresponding term in the second polynomial. The ratio of opposites is -1.

EXAMPLE 6 Reducing a Rational Expression When the Denominator Is the Opposite of the Numerator

If two polynomials are opposites, their ratio is -1.

Reduce each of these rational expressions to lowest terms.

SOLUTION

(a) $\frac{2x - 3y}{3y - 2x}$

$$\frac{2x - 3y}{3y - 2x} = \frac{2x - 3y}{-(2x - 3y)}$$
$$= -1$$

The denominator is the opposite of the numerator. The ratio of these opposites is -1.

(b) $\frac{-4a + 3b + c}{4a - 3b - c}$

$$\frac{-4a + 3b + c}{4a - 3b - c} = \frac{-(4a - 3b - c)}{4a - 3b - c}$$
$$= -1$$

This expression equals -1, because the numerator and the denominator are opposites.

SELF-CHECK 8.1.3

Reduce each rational expression to lowest terms. Assume that variables are restricted to values that prevent division by zero.

1. $\dfrac{22a^2b^3}{33a^3b^2}$
2. $\dfrac{x^2 - 9}{4x - 12}$

Reduce each rational expression to lowest terms.

3. $\dfrac{5r - 10t}{r^2 - 4t^2}$
4. $\dfrac{5r - 15t}{3t - r}$

Warning: Reduce only by dividing by common factors—do *not* cancel terms.

To reduce a rational expression to lowest terms, we divide both the numerator and the denominator by a common nonzero factor. Sometimes students use the word *cancel* to describe this step. This can be dangerous because students often cancel where it is not applicable: you cannot cancel terms. A classic error in Example 7 is to cancel the $3x$ terms in the numerator and the denominator.

EXAMPLE 7 Recognizing a Rational Expression Already in Reduced Form

Do *not* cancel the $3x$ terms in the numerator and the denominator.

Reduce $\dfrac{3x - 7}{3x + 11}$ to lowest terms.

SOLUTION

Answer: $\dfrac{3x - 7}{3x + 11}$ — This expression is already in reduced form, because the greatest common factor of the numerator and the denominator is 1.

The fundamental principle of fractions, the property that $\dfrac{ac}{bc} = \dfrac{a}{b}$ for $b \neq 0$ and $c \neq 0$, is used both to reduce fractions to lowest terms and to express fractions so that they have a common denominator. In Example 8 both the numerator and the denominator are multiplied by the same nonzero value, called the **building factor.** This step is important in adding rational expressions.

EXAMPLE 8 Converting a Rational Expression to a Given Denominator

Convert $\dfrac{x + 3}{x - 2}$ to an equivalent expression with a denominator of $x^2 - 3x + 2$.

SOLUTION

$$\frac{x + 3}{x - 2} = \frac{?}{x^2 - 3x + 2}$$

$$\frac{x + 3}{x - 2} = \frac{?}{(x - 2)(x - 1)}$$

To determine the building factor, first factor $x^2 - 3x + 2$. The other factor needed to build the denominator that we want is $x - 1$.

$$\frac{x + 3}{x - 2} = \frac{x + 3}{x - 2} \cdot \frac{x - 1}{x - 1}$$

Multiply $\dfrac{x + 3}{x - 2}$ by the multiplicative identity 1 in the form $\dfrac{x - 1}{x - 1}$.

$$\frac{x + 3}{x - 2} = \frac{x^2 + 2x - 3}{x^2 - 3x + 2}$$

This expression is equivalent to $\dfrac{x + 3}{x - 2}$ and has a denominator of $x^2 - 3x + 2$.

SELF-CHECK 8.1.4

1. Determine the missing numerator in $\dfrac{3x-2}{2x-3} = \dfrac{?}{10x^2 - 7x - 12}$.

SELF-CHECK ANSWERS

8.1.1

1. A polynomial function that is also a rational function
2. A rational function
3. Not a rational function
4. A rational function
5. $x = -4$
6. $\mathbb{R} \sim \{2\}$
7. $x = 2$
8.

```
Plot1 Plot2 Plot3
\Y1■(X+3)/(X-2)
\Y2
\Y3
\Y4
\Y5
\Y6
\Y7
```

[−7.4, 11.4, 4] by [−10.4, 14.4, 4]

8.1.2

1. $-\dfrac{15}{4}$
2. 0
3. Undefined
4. $\mathbb{R} \sim \{-4\}$

8.1.3

1. $\dfrac{2b}{3a}$
2. $\dfrac{x+3}{4}$
3. $\dfrac{5}{r+2t}$
4. -5

8.1.4

1. $\dfrac{3x-2}{2x-3} = \dfrac{15x^2 + 2x - 8}{10x^2 - 7x - 12}$

USING THE LANGUAGE AND SYMBOLISM OF MATHEMATICS 8.1

1. A rational expression is defined as the __________ of two polynomials.
2. Division by 0 is __________.
3. A function defined by $f(x) = \dfrac{P(x)}{Q(x)}$, where $P(x)$ and $Q(x)$ are polynomials and $Q(x) \neq 0$, is called a __________ function.
4. To determine the values excluded from the domain of a rational function, we look at the values for which the __________ is equal to __________.
5. A vertical line that a graph approaches but does not touch is called a vertical __________.
6. A horizontal line that a graph approaches is called a horizontal __________.
7. The graph of the rational function $f(x) = \dfrac{5}{x-4}$ will have the line $x = 4$ as a vertical __________.
8. A rational expression is in __________ __________ when the numerator and denominator have no common factor other than -1 or 1.
9. If A, B, and C are polynomials and $B \neq 0$ and $C \neq 0$, then $\dfrac{AC}{BC} =$ __________.
10. If two polynomials are opposites, their ratio is __________.

QUICK REVIEW 8.1

The purpose of this quick review is to help you recall skills needed in this section.

1. Factor 56 as the product of prime integers.
2. Reduce $\dfrac{36}{90}$ to lowest terms.

Factor each expression.

3. $-3x^3 + 4x^2$
4. $x^3 - 6x^2 - 40x$
5. $6x^2 + xy - 35y^2$

EXERCISES 8.1

1. If $f(x) = \dfrac{2x+1}{2x-1}$, evaluate each expression.

   a. $f(0)$ b. $f(1)$ c. $f\left(\dfrac{1}{2}\right)$

2. If $f(x) = \dfrac{5x-11}{7x-14}$, evaluate each expression.

   a. $f(0)$ b. $f(1)$ c. $f(2)$

3. Given the rational function $f(x) = \dfrac{5}{x-4}$:

   a. What value is excluded from the domain of this function?
   b. What is the domain of this function?
   c. What is the equation of the vertical asymptote of the graph of this function?

4. Given the rational function $f(x) = \dfrac{x+1}{x+6}$:
   a. What value is excluded from the domain of this function?
   b. What is the domain of this function?
   c. What is the equation of the vertical asymptote of the graph of this function?

In Exercises 5–8, match the values excluded from the domain of the rational function $y = f(x)$ with the domain of this function.

Excluded Values	**Domain**
5. -3	A. $\mathbb{R} \sim \{0\}$ or equivalently $(-\infty, 0)\cup(0, \infty)$
6. 0	B. $\mathbb{R} \sim \{0, 3\}$ or equivalently $(-\infty, 0)\cup(0, 3)\cup(3, \infty)$
7. -3 and 0	C. $\mathbb{R} \sim \{-3\}$ or equivalently $(-\infty, -3)\cup(-3, \infty)$
8. 0 and 3	D. $\mathbb{R} \sim \{-3, 0\}$ or equivalently $(-\infty, -3)\cup(-3, 0)\cup(0, \infty)$

In Exercises 9–12, match the excluded values with the corresponding rational expression.

9. $f(x) = \dfrac{2x-1}{x+1}$ — A. Excluded value 0.5

10. $f(x) = \dfrac{x+1}{2x-1}$ — B. Excluded values 0.5 and 1

11. $f(x) = \dfrac{2x-1}{(x+1)(2x+1)}$ — C. Excluded values -0.5 and -1

12. $f(x) = \dfrac{2x+1}{(x-1)(2x-1)}$ — D. Excluded value -1

In Exercises 13 and 14, determine the domain of each rational function.

13. a. $f(x) = \dfrac{2x+1}{2x-1}$ b. $f(x) = \dfrac{3x-7}{x^2-81}$

14. a. $f(x) = \dfrac{5x-11}{7x-14}$ b. $f(x) = \dfrac{x^2-9}{x^2-x-42}$

In Exercises 15–18, use the given table to determine the domain of each rational function.

15. $f(x) = \dfrac{x+4}{2x-3}$

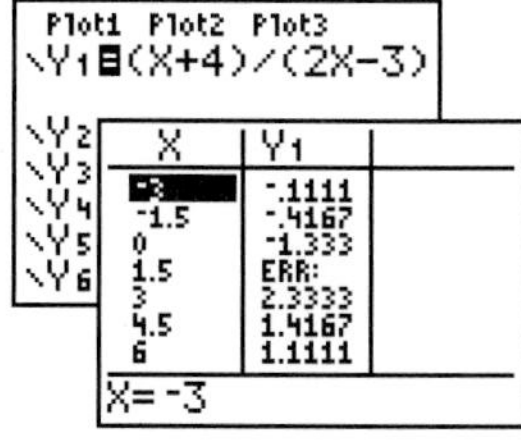

16. $f(x) = \dfrac{2x-3}{x+4}$

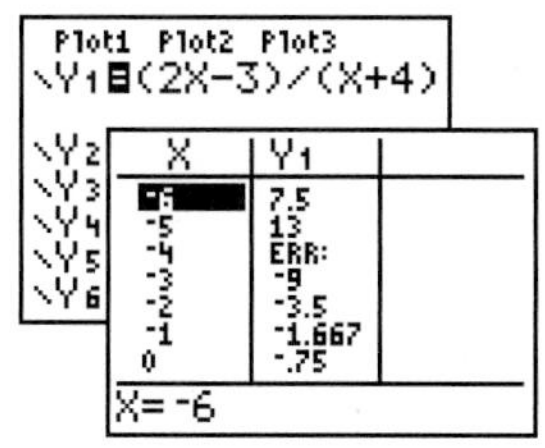

17. $f(x) = \dfrac{x^2+4}{x^2+x-6}$

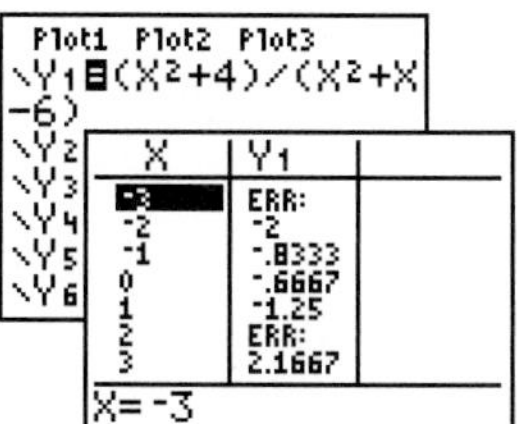

18. $f(x) = \dfrac{2x^2+3}{x^2-3x-4}$

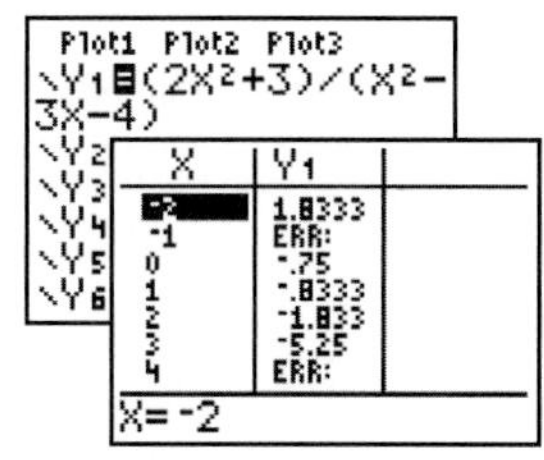

In Exercises 19–22, match each function with the graph of the function.

19. $f(x) = \dfrac{1}{x+2}$

20. $f(x) = \dfrac{1}{x-2}$

21. $f(x) = \dfrac{1}{x-4}$

22. $f(x) = \dfrac{1}{x+4}$

A.

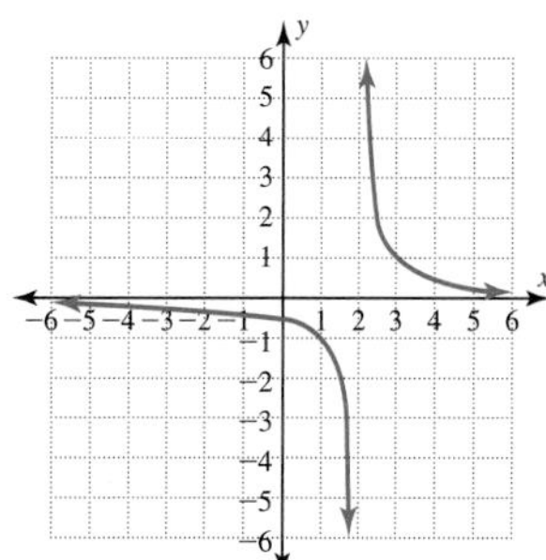

B.

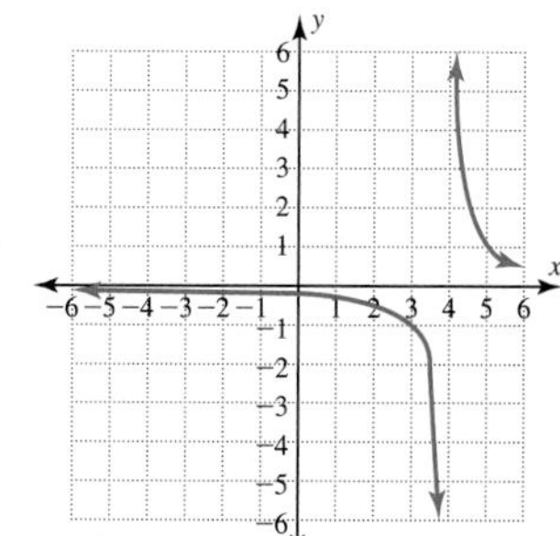

C.

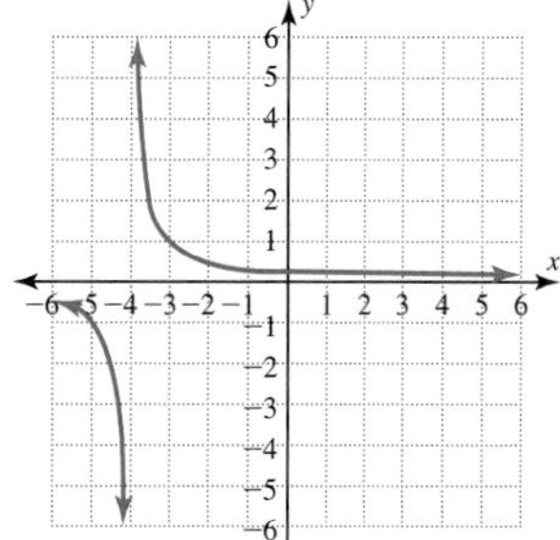

D.

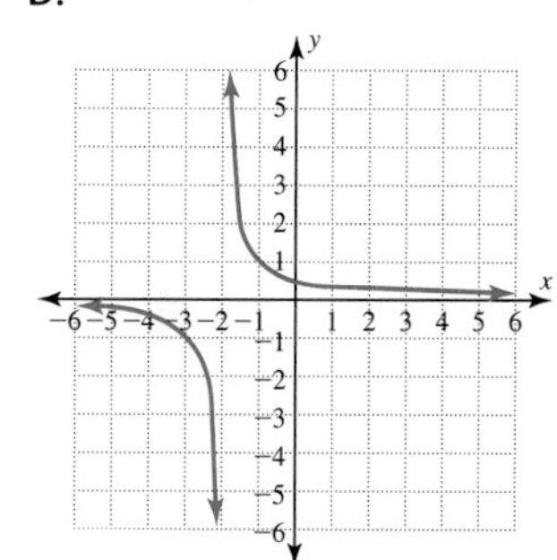

In Exercises 23–26, match each graph with the vertical asymptotes for this function.

23. $x = -1$

24. $x = -3$

25. $x = -1$ and $x = 1$

26. $x = -3$ and $x = 2$

A.

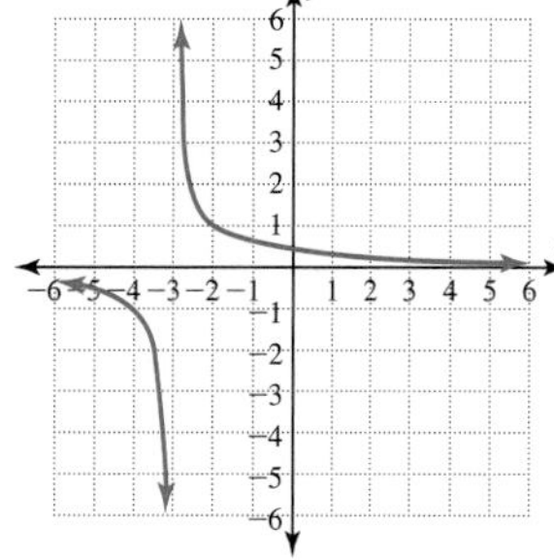

B.

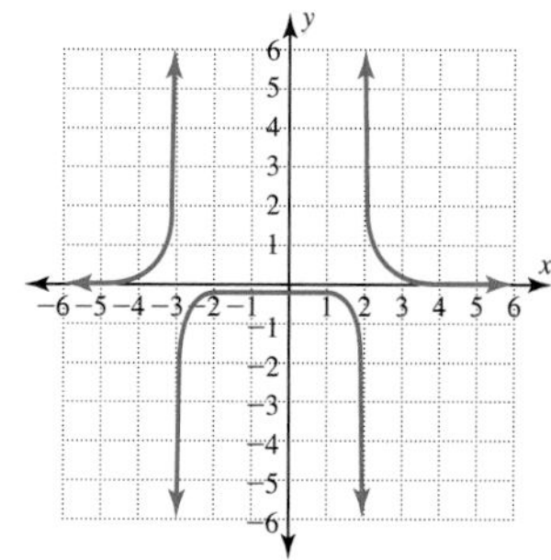

C.

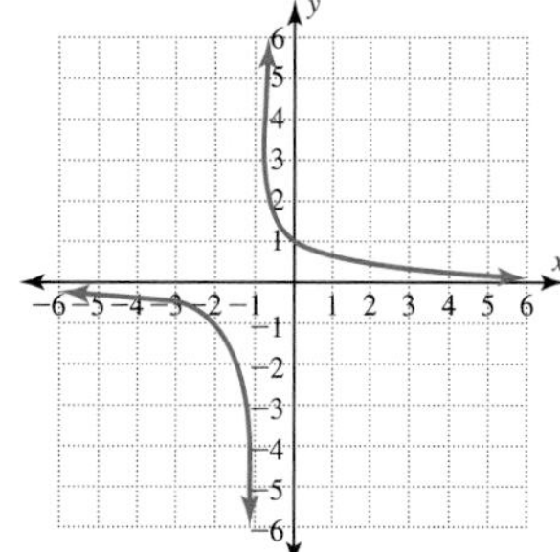

D.

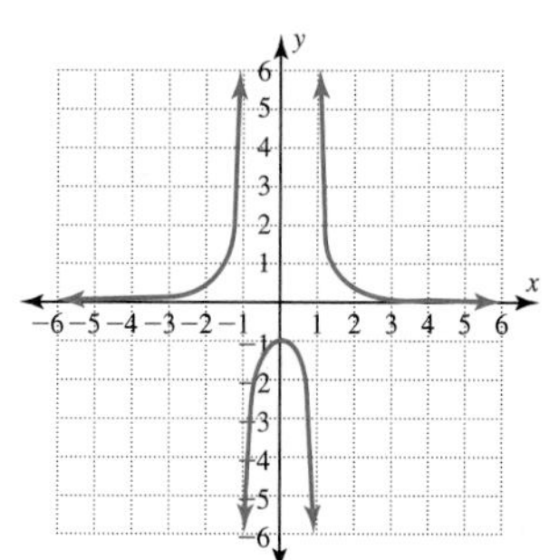

In Exercises 27–70, reduce each rational expression to lowest terms. Assume that the variables are restricted to values that prevent division by 0.

27. $\frac{22a^2b^3}{33a^3b}$

28. $\frac{21x^7y^5}{35x^4y^8}$

29. $\frac{30x^2 - 45x}{5x}$

30. $\frac{46x^4 - 69x^3}{23x^2}$

31. $\frac{7x}{14x^2 - 21x}$

32. $\frac{6m^2}{9m^4 - 15m^3}$

33. $\frac{a^2b(2x - 3y)}{-ab^2(2x - 3y)}$

34. $\frac{ab(2x - 3y)^2}{ab(2x - 3y)^3}$

35. $\frac{7x - 8y}{8y - 7x}$

36. $\frac{3a - 5b}{5b - 3a}$

37. $\frac{ax - ay}{by - bx}$

38. $\frac{5x - 10}{8 - 4x}$

39. $\frac{(x - 2y)(x + 5y)}{(x + 2y)(x + 5y)}$

40. $\frac{(x - 4y)(x + 4y)}{(x - 4y)(x + 7y)}$

41. $\frac{x^2 - y^2}{3x + 3y}$

42. $\frac{v^2 - w^2}{7v - 7w}$

43. $\frac{25x^2 - 4}{14 - 35x}$

44. $\frac{4x^2 - 9}{9 - 6x}$

45. $\frac{3x^2 - 2xy}{3x^2 - 5xy + 2y^2}$

46. $\frac{x^2 - 2xy + y^2}{x^2 - y^2}$

47. $\frac{4x^2 + 12xy + 9y^2}{14x + 21y}$

48. $\frac{x^2 + 9x + 14}{x^2 + 5x - 14}$

49. $\frac{14ax + 21a}{35ay + 42a}$

50. $\frac{3ax - 3ay}{3ax + 3ay}$

51. $\frac{ax - y - z}{y + z - ax}$

52. $\frac{x^2 - x - 3}{x + 3 - x^2}$

53. $\frac{2a^2 - ab - b^2}{a^2 - b^2}$

54. $\frac{ax + bx - ay - by}{5x - 5y}$

55. $\frac{vx + vy - wx - wy}{v^2 - w^2}$

56. $\frac{2x^2 - 9x - 5}{4x^2 - 1}$

57. $\frac{x^2 - 25}{3x^2 + 14x - 5}$

58. $\frac{2x^2 + xy - 3y^2}{3x^2 - 5xy + 2y^2}$

59. $\frac{5a^2 + 4ab - b^2}{5a^2 - 6ab + b^2}$

60. $\frac{4a^2 - 4ab + b^2}{2a^2 + ab - b^2}$

61. $\frac{(b^2 - 2b + 1) - a^2}{5ab - 5a + 5a^2}$

62. $\frac{(a^2 - 4ab + 4b^2) - z^2}{5a - 10b + 5z}$

63. $\frac{4z^2 - 5w}{5w - 4z^2}$

64. $\frac{x^4 - y^3}{y^3 - x^4}$

65. $\frac{12x^2 + 24xy + 12y^2}{16x^2 - 16y^2}$

66. $\frac{6y^2 + 11yz - 7z^2}{9y^2 + 42yz + 49z^2}$

67. $\frac{14x^2 - 9xy + y^2}{y^2 - 7xy}$

68. $\frac{x^2 - 25y^2}{x^2 - 5xy}$

69. $\frac{b^4 - 1}{5b + 5}$

70. $\frac{w^4 - 1}{w^3 + w}$

71. **Canoeing Time Upstream** The time it takes a camper to paddle 12 mi upstream in a river that flows 2 mi/h is given by $T(x) = \frac{12}{x - 2}$, where x is the speed the camper can paddle in still water. Evaluate and interpret each of these expressions.
    a. $T(6)$ b. $T(8)$ c. $T(2)$

72. **Canoeing Time Downstream** The time it takes a camper to paddle 12 mi downstream in a river that flows 3 mi/h is given by $T(x) = \frac{12}{x + 3}$, where x is the speed the camper can paddle in still water. Evaluate and interpret each of these expressions.
    a. $T(0)$ b. $T(1)$ c. $T(3)$

73. **Average Cost** The average cost of producing x m of copper cable is displayed in the following graph.
    a. Describe what happens to the average cost when the number of meters of cable produced approaches 0 m.
    b. Describe what happens to the average cost when the number of meters of cable produced becomes very large.

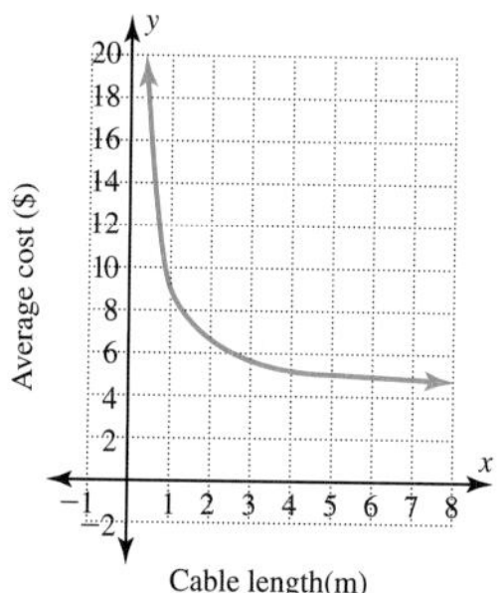

74. **Average Cost** The average cost of producing x kg of detergent is displayed in the following graph.
    a. Describe what happens to the average cost when the number of kilograms of detergent produced approaches 0 kg.
    b. Describe what happens to the average cost when the number of kilograms of detergent produced becomes very large.

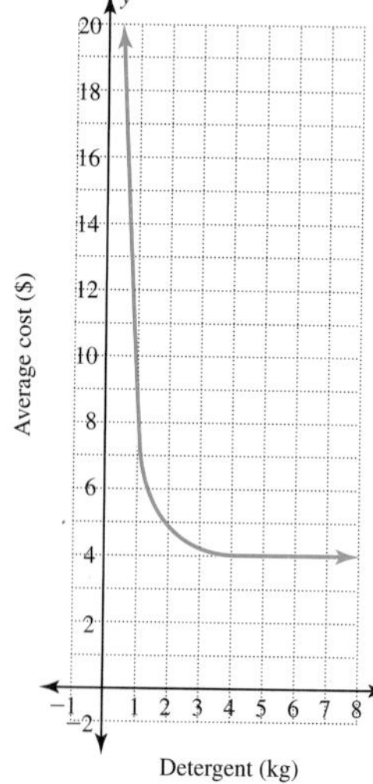

In Exercises 75–86, fill in the missing numerator or denominator.

75. $\frac{7}{12} = \frac{?}{24}$

76. $\frac{3}{5} = \frac{?}{20}$

77. $\frac{2x}{x - 5} = \frac{?}{7(x - 5)}$

78. $\frac{3x}{x + 4} = \frac{?}{5(x + 4)}$

79. $\dfrac{7x - 8y}{3a - 5b} = \dfrac{?}{10b - 6a}$

80. $\dfrac{3x - 2y}{b - 3a} = \dfrac{?}{9a - 3b}$

81. $\dfrac{10}{a + b} = \dfrac{?}{a^2 - b^2}$

82. $\dfrac{6}{x - y} = \dfrac{?}{x^2 - y^2}$

83. $\dfrac{5}{x - 6} = \dfrac{?}{x^2 - 7x + 6}$

84. $\dfrac{2x - y}{x + 3y} = \dfrac{?}{x^2 + 5xy + 6y^2}$

85. $\dfrac{2x - y}{x + 3y} = \dfrac{2x^2 + xy - y^2}{?}$

86. $\dfrac{x - 2y}{x + 3y} = \dfrac{x^2 - 4y^2}{?}$

Group Discussion Questions

87. **Error Analysis** Examine each student's work, and correct any mistakes you find.

a. Student A

$$\frac{x - 3}{x + 5} = \frac{-3}{5}$$

b. Student B

$$\frac{x^2 - 9}{x^2 - 4x + 3} = \frac{(x - 3)(x + 3)}{(x - 3)(x - 1)} = \frac{x + 3}{x - 1} = \frac{3}{-1} = -3$$

88. **Challenge Question** Reduce each rational expression to lowest terms. Assume that m is a natural number and that all values of the variables that cause division by zero are excluded.

a. $\dfrac{x^{2m} - 25}{7x^m + 35}$ b. $\dfrac{x^{m+1} - 4x}{x^{2m} - 16}$

89. **Challenge Question**

a. Factor x^{-2} out of both the numerator and the denominator of $\dfrac{2 + x^{-1}y - 15x^{-2}y^2}{1 + 6x^{-1}y + 9x^{-2}y^2}$, and then reduce the expression to lowest terms.

b. Factor $x^{-1}y^{-1}$ out of both the numerator and the denominator of $\dfrac{4xy^{-1} + 12 + 9x^{-1}y}{6xy^{-1} + 5 - 6x^{-1}y}$, and then reduce the expression to lowest terms.

90. **Discovery Question**

a. Reduce $\dfrac{x^2 - 5x + 6}{x - 2}$ to lowest terms.

b. Determine the value excluded from the domain of $f(x) = \dfrac{x^2 - 5x + 6}{x - 2}$.

c. Graph $f(x) = \dfrac{x^2 - 5x + 6}{x - 2}$.

d. Graph $g(x) = x - 3$.

e. What is domain of $f(x)$?

f. What is domain of $g(x)$?

g. The graphs of $y = f(x)$ and $y = g(x)$ are different (although you may not see the difference on a graphing calculator). Describe the difference between these graphs.

91. **Challenge Question**

a. Have each person in your group create a different rational function whose domain is all real numbers except for 2 and 5.

b. Compare these functions algebraically.

c. Use a graphing calculator to compare the graphs of these functions.

92. **Challenge Question**

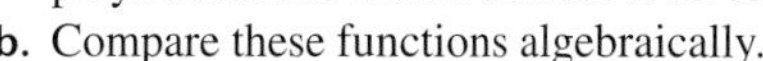

a. Have each person in your group create a rational function whose denominator is a second-degree polynomial and whose domain is all real numbers.

b. Compare these functions algebraically.

c. Use a graphing calculator to compare the graphs of these functions.

93. Use your graphing calculator to complete the following table.

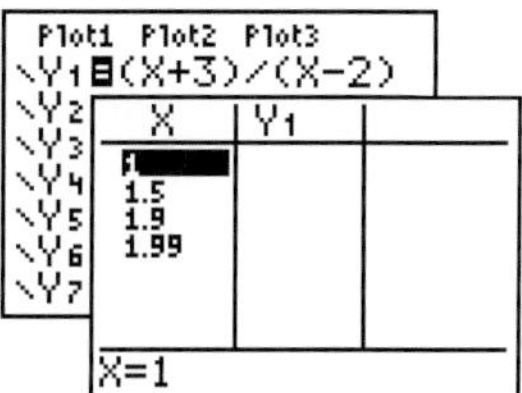

a. Describe the trend you observe in the x-values in the table.

b. Describe the trend you observe in the y-values after you complete the table.

Use your graphing calculator to complete the following table.

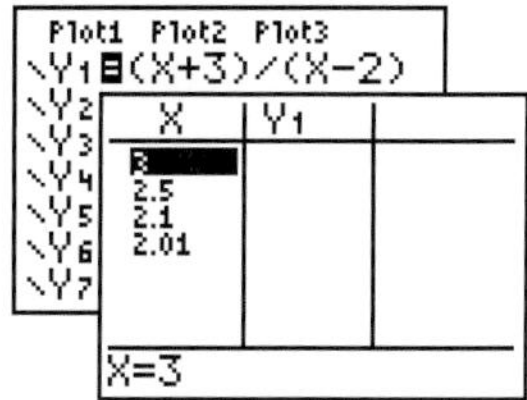

c. Describe the trend you observe in the x-values in the table.

d. Describe the trend you observe in the y-values after you complete the table.

e. What is the vertical asymptote of this function?

94. Use your graphing calculator to complete the following table.

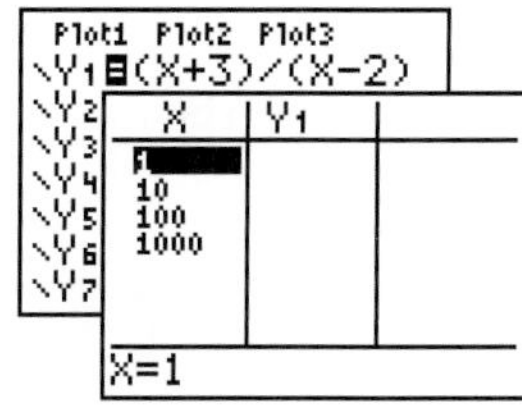

a. Describe the trend you observe in the x-values in the table.

b. Describe the trend you observe in the y-values after you complete the table.

Use your graphing calculator to complete the following table.

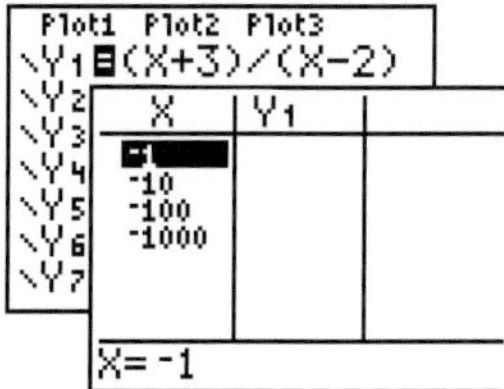

c. Describe the trend you observe in the x-values in the table.

d. Describe the trend you observe in the y-values after you complete the table.

e. What is the horizontal asymptote of this function?

Section 8.2 Multiplying and Dividing Rational Expressions

Objective: 4. Multiply and divide rational expressions.

Since the variables in a rational expression represent real numbers, the rules and procedures for performing multiplication and division with rational expressions are the same as those used in Chapter 1 for performing these operations with arithmetic fractions.

When you are multiplying arithmetic fractions, there are two methods that will produce the correct result. Method 1 involves simply multiplying the numerators and multiplying the denominators and then reducing the resulting fraction. Method 2 involves first factoring each numerator and denominator. The product is still formed by multiplying the numerators and multiplying the denominators, but leaving these products in factored form and reducing the fraction. Both methods are illustrated below to multiply $\frac{15}{8} \cdot \frac{12}{25}$.

METHOD 1

$$\frac{15}{8} \cdot \frac{12}{25} = \frac{180}{200} = \frac{9 \cdot \cancel{20}}{10 \cdot \cancel{20}} = \frac{9}{10}$$

METHOD 2

$$\frac{15}{8} \cdot \frac{12}{25} = \frac{3 \cdot 5}{2 \cdot 4} \cdot \frac{3 \cdot 4}{5 \cdot 5} = \frac{3 \cdot \cancel{5} \cdot 3 \cdot \cancel{4}}{2 \cdot \cancel{4} \cdot 5 \cdot \cancel{5}} = \frac{9}{10}$$

Analyzing these two methods, you will notice that the numbers you are required to deal with in Method 2 are smaller. When dealing with rational expressions, you will notice that Method 2 is much more efficient because actually multiplying the polynomial expressions in the numerators and multiplying the polynomial expressions in the denominators can produce very complicated expressions that may be very difficult to reduce.

Multiplication and division will be covered before addition and subtraction so that you can develop some of the skills needed to add rational expressions with different denominators. The rule for multiplying rational expressions is given in the following box.

Multiplying Rational Expressions

ALGEBRAICALLY	VERBALLY	ALGEBRAIC EXAMPLE
If A, B, C, and D are real polynomials and $B \neq 0$ and $D \neq 0$, then $\frac{A}{B} \cdot \frac{C}{D} = \frac{AC}{BD}$.	1. Factor the numerators and the denominators. Then write the product as a single fraction, indicating the product of the numerators and the product of the denominators. 2. Reduce this fraction by dividing the numerator and the denominator by any common nonzero factors.	$\frac{x^2 - xy}{4a - 12b} \cdot \frac{3a - 9b}{5x - 5y}$ $= \frac{x(x-y)}{4(a-3b)} \cdot \frac{3(a-3b)}{5(x-y)}$ $= \frac{3x\overset{1}{\cancel{(x-y)}}\overset{1}{\cancel{(a-3b)}}}{20\underset{1}{\cancel{(x-y)}}\underset{1}{\cancel{(a-3b)}}}$ $= \frac{3x}{20}$

A Mathematical Note

The Babylonians left considerable evidence of their mathematical abilities. Early in the 19th century approximately 500,000 clay tablets were excavated. All these tablets were written in *cuneiform* script. About 400 of these tablets contain mathematical problems and tables. Tablet number 322 of the G. A. Plimpton collection at Columbia University is famous because it contains a list of Pythagorean triples for the lengths of the three sides of a right triangle.

EXAMPLE 1 Multiplying Rational Expressions

Find the product of $\frac{2x - 2y}{35}$ and $\frac{15}{3x - 3y}$. Assume the variables are restricted to values that prevent division by 0.

SOLUTION

$$\frac{2x - 2y}{35} \cdot \frac{15}{3x - 3y} = \frac{2(x - y)}{35} \cdot \frac{15}{3(x - y)}$$

Factor the numerators and the denominators. (Usually you do not factor the constant coefficients.)

$$= \frac{2\overset{1}{\cancel{(x - y)}}\overset{1}{\cancel{(3)}}\overset{1}{\cancel{(5)}}}{\underset{1}{\cancel{5}}(7)\underset{1}{\cancel{(3)}}\underset{1}{\cancel{(x - y)}}}$$

Then write the product as a single fraction, indicating the product of these factors.

Reduce by dividing out the common factors.

$$= \frac{2}{7}$$

As Example 2 illustrates, once the polynomials in rational expressions are in factored form, the simplification of a product is easy.

EXAMPLE 2 Multiplying Rational Expressions

Multiply $\frac{(x - 2)(x - 3)}{(x + 1)(x - 2)} \cdot \frac{(x + 1)(x - 4)}{(x - 3)(x + 4)}$. Assume that the variables are restricted to values that prevent division by 0.

SOLUTION

$$\frac{(x - 2)(x - 3)}{(x + 1)(x - 2)} \cdot \frac{(x + 1)(x - 4)}{(x - 3)(x + 4)} = \frac{\overset{1}{\cancel{(x - 2)}}\overset{1}{\cancel{(x - 3)}}\overset{1}{\cancel{(x + 1)}}(x - 4)}{\underset{1}{\cancel{(x + 1)}}\underset{1}{\cancel{(x - 2)}}\underset{1}{\cancel{(x - 3)}}(x + 4)}$$

$$= \frac{x - 4}{x + 4}$$

The numerators and denominators are already in factored form. Thus write the product as a single fraction and then reduce by dividing out the common factors.

Example 2 illustrates how easy it is to multiply rational expressions once all the polynomials in the numerator and the denominator are factored. Thus the majority of work in multiplying rational expressions is in factoring the numerators and the denominators.

In Example 3 note that each factor in the numerator also occurs in the denominator. Thus the reduced form has only a factor of 1 remaining in the numerator.

EXAMPLE 3 Multiplying Rational Expressions

Multiply $\frac{2x - y}{4x^2 - 4y^2} \cdot \frac{x^2 + 2xy + y^2}{2x^2 + xy - y^2}$. Assume the variables are restricted to values that prevent division by 0.

Reduce before multiplying out: Multiplication and division problems with rational expressions will be much easier if you divide both the numerator and the denominator by any common factors before multiplying any of the factors.

SOLUTION

$$\frac{2x - y}{4x^2 - 4y^2} \cdot \frac{x^2 + 2xy + y^2}{2x^2 + xy - y^2} = \frac{2x - y}{4(x^2 - y^2)} \cdot \frac{(x + y)(x + y)}{(2x - y)(x + y)}$$

Factor the numerators and the denominators. Then write the product as a single fraction.

$$= \frac{(2x - y)(x + y)(x + y)}{4(x - y)(x + y)(2x - y)(x + y)}$$

Reduce this fraction by dividing out the common factors.

$$= \frac{1}{4(x - y)}$$

SELF-CHECK 8.2.1

Find each product. Assume the variables are restricted to values that prevent division by 0.

1. $\frac{2x - 5y}{7xy} \cdot \frac{35x^2}{25y - 10x}$
2. $\frac{10x^2 - 6x}{7xy - 7y} \cdot \frac{5x^2 - 2x - 3}{25x^2 - 9}$

If two rational expressions are equal, then tables for these two expressions will be equal for all values for which both expressions are defined. Looking at a limited set of values in a table does not confirm that the expressions are equal for all values, but it does serve as a good check that is likely to catch potential errors from an algebraic computation. This is illustrated in Example 4.

EXAMPLE 4 Using Tables to Compare Two Rational Expressions

Writing an equal symbol between two rational expressions means that these expressions are identical for all values for which both expressions are defined.

Multiply $\frac{x^3 + x}{x^3 - 3x^2} \cdot \frac{x^2 - 4x + 3}{5x^2 + 5}$ and compare the product to the original expression by using a table of values. Assume the variables are restricted to values that prevent division by 0.

SOLUTION

ALGEBRAICALLY

$$\frac{x^3 + x}{x^3 - 3x^2} \cdot \frac{x^2 - 4x + 3}{5x^2 + 5} = \frac{x(x^2 + 1)}{x^2(x - 3)} \cdot \frac{(x - 1)(x - 3)}{5(x^2 + 1)}$$

Factor the numerators and the denominators.

$$= \frac{x(x^2 + 1)(x - 1)(x - 3)}{x^2(x - 3)(5)(x^2 + 1)}$$

Then write the product as a single fraction. Reduce this fraction by dividing out the common factors.

$$= \frac{x - 1}{5x}$$

NUMERICAL CHECK

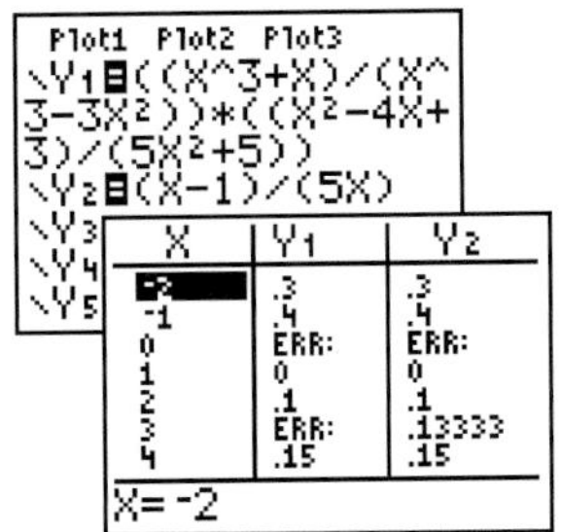

Let Y_1 represent the original expression and Y_2 represent the simplified product.

Both expressions are undefined for $x = 0$.

The expressions are identical except for $x = 3$.

The original expression has excluded values of 0 and 3. The reduced expression has an excluded value of 0.

Answer: $\dfrac{x^3 + x}{x^3 - 3x^2} \cdot \dfrac{x^2 - 4x + 3}{5x^2 + 5} = \dfrac{x - 1}{5x}$

It is understood that this equality means that these expressions are identical for all values for which they are both defined. For $x \neq 0$ and $x \neq 3$ these expressions are identical.

We now examine the division of rational expressions. The rule for dividing rational expressions is given in the following box.

Dividing Rational Expressions

ALGEBRAICALLY	VERBALLY	ALGEBRAIC EXAMPLE
If A, B, C, and D are real polynomials and $B \neq 0$, $C \neq 0$, and $D \neq 0$, then $\dfrac{A}{B} \div \dfrac{C}{D} = \dfrac{A}{B} \cdot \dfrac{D}{C}$ $= \dfrac{AD}{BC}$	1. Rewrite the division problem as the product of the dividend and the reciprocal of the divisor. 2. Perform the multiplication by using the rule for multiplying rational expressions.	$\dfrac{12x}{10x - 35} \div \dfrac{15x}{4x - 14}$ $= \dfrac{12x}{5(2x - 7)} \cdot \dfrac{2(2x - 7)}{15x}$ $= \dfrac{\overset{4}{\cancel{(12x)}}(2)\overset{1}{\cancel{(2x - 7)}}}{5\underset{1}{\cancel{(2x - 7)}}\underset{5}{\cancel{(15x)}}}$ $= \dfrac{8}{25}$

EXAMPLE 5 Dividing Rational Expressions

Divide $\dfrac{10ab}{3y - x} \div \dfrac{5a^3}{7x - 21y}$. Assume the variables are restricted to values that prevent division by 0.

SOLUTION

$$\frac{10ab}{3y - x} \div \frac{5a^3}{7x - 21y} = \frac{10ab}{3y - x} \cdot \frac{7x - 21y}{5a^3}$$

Rewrite the division problem as the product of the dividend and the reciprocal of the divisor.

$$= \frac{10ab}{(-1)(x - 3y)} \cdot \frac{7(x - 3y)}{5a^3}$$

Factor the binomials in the numerator and in the denominator. Note that -1 is factored out of $3y - x$.

$$= \frac{\overset{2b}{\cancel{(10ab)}}(7)\overset{1}{\cancel{(x - 3y)}}}{(-1)\underset{1}{\cancel{(x - 3y)}}\underset{a^2}{\cancel{(5a^3)}}} = \frac{(2b)(7)}{(-1)a^2} = -\frac{14b}{a^2}$$

Write this product as a single fraction. Then reduce by dividing out the common factors.

SELF-CHECK 8.2.2

1. Divide $\frac{10x - 15}{x^2 - 5x} \div \frac{14x - 21}{10 - 2x}$. Assume that the variables are restricted to values that prevent division by 0.

EXAMPLE 6 Dividing Rational Expressions

Divide $\frac{x^2 - 9}{x^2 - 25}$ by $\frac{x - 3}{x + 5}$. Assume the variables are restricted to values that prevent division by 0.

SOLUTION

With practice you may be able to combine one or two of the steps illustrated in Example 6. Be sure your proficiency is well developed before you try to perform too many steps mentally. The steps that are skipped are the source of many student errors—especially errors in sign.

$$\frac{x^2 - 9}{x^2 - 25} \div \frac{x - 3}{x + 5} = \frac{x^2 - 9}{x^2 - 25} \cdot \frac{x + 5}{x - 3}$$

Rewrite the division problem as the product of the dividend and the reciprocal of the divisor.

$$= \frac{(x + 3)(x - 3)}{(x + 5)(x - 5)} \cdot \frac{x + 5}{x - 3}$$

Factor the numerator and denominator of the first fraction.

$$= \frac{(x + 3)\overset{1}{\cancel{(x - 3)}}\overset{1}{\cancel{(x + 5)}}}{\underset{1}{\cancel{(x + 5)}}(x - 5)\underset{1}{\cancel{(x - 3)}}}$$

Then reduce by dividing out the common factors.

$$= \frac{x + 3}{x - 5}$$

Examples 7 and 8 involve both multiplication and division. Compare these two examples to note the correct order of operations in each example.

EXAMPLE 7 Simplifying a Rational Expression Involving Both Multiplication and Division

Simplify $\frac{12x}{x + y} \div \left(\frac{5x - 5}{x^2 - y^2} \cdot \frac{3xy}{xy - y}\right)$. Assume the variables are restricted to values that prevent division by 0.

SOLUTION

$$\frac{12x}{x + y} \div \left(\frac{5x - 5}{x^2 - y^2} \cdot \frac{3xy}{xy - y}\right) = \frac{12x}{x + y} \div \left[\frac{(5x - 5)(3xy)}{(x^2 - y^2)(xy - y)}\right]$$

The grouping symbols indicate that the multiplication inside the brackets has priority over the division. Multiply the numerators and the denominators within the brackets first.

$$= \frac{12x}{x + y} \cdot \frac{(x^2 - y^2)(xy - y)}{(5x - 5)(3xy)}$$

Then invert the divisor within the brackets and multiply.

$$= \frac{12x}{x + y} \cdot \frac{(x - y)(x + y)(y)(x - 1)}{(5)(x - 1)(3xy)}$$

Factor the numerator and the denominator.

$$= \frac{\overset{4}{\cancel{(12x)}}(x - y)\overset{1}{\cancel{(x + y)}}\overset{1}{\cancel{(y)}}\overset{1}{\cancel{(x - 1)}}}{\underset{1}{\cancel{(x + y)}}(5)\underset{1}{\cancel{(x - 1)}}\underset{\underset{1}{\cancel{y}}}{\cancel{(3xy)}}} = \frac{4(x - y)}{5}$$

Then reduce by dividing out the common factors.

EXAMPLE 8 Simplifying a Rational Expression Involving Both Multiplication and Division

Simplify $\frac{x^2 - 25}{x^2 - x - 12} \div \frac{x^2 - x - 20}{3x - 3} \cdot \frac{x^2 - 16}{x^2 + 4x - 5}$. Assume the variables are restricted to values that prevent division by 0.

SOLUTION

$$\frac{x^2 - 25}{x^2 - x - 12} \div \frac{x^2 - x - 20}{3x - 3} \cdot \frac{x^2 - 16}{x^2 + 4x - 5}$$

Following the correct order of operations from left to right, invert only the fraction in the middle.

$$= \frac{x^2 - 25}{x^2 - x - 12} \cdot \frac{3x - 3}{x^2 - x - 20} \cdot \frac{x^2 - 16}{x^2 + 4x - 5}$$

$$= \frac{(x+5)(x-5)}{(x+3)(x-4)} \cdot \frac{3(x-1)}{(x-5)(x+4)} \cdot \frac{(x+4)(x-4)}{(x+5)(x-1)}$$

Factor each numerator and denominator and then write this product as a single fraction.

$$= \frac{\overset{1}{\cancel{(x+5)}}\overset{1}{\cancel{(x-5)}}(3)\overset{1}{\cancel{(x-1)}}\overset{1}{\cancel{(x+4)}}\overset{1}{\cancel{(x-4)}}}{(x+3)\underset{1}{\cancel{(x-4)}}\underset{1}{\cancel{(x-5)}}\underset{1}{\cancel{(x+4)}}\underset{1}{\cancel{(x+5)}}\underset{1}{\cancel{(x-1)}}}$$

Reduce by dividing out the common factors.

$$= \frac{3}{x+3}$$

SELF-CHECK 8.2.3

1. Simplify $\frac{5x^2}{x^2 - 3x + 2} \cdot \frac{x^2 - 4x + 4}{x^2 - x} \div \frac{x^2 - 2x}{x^2 - 2x + 1}$. Assume the variables are restricted to values that prevent division by 0.

SELF-CHECK ANSWERS

8.2.1

1. $-\frac{x}{y}$ 2. $\frac{2x}{7y}$

8.2.2

1. $-\frac{10}{7x}$

8.2.3

1. 5

USING THE LANGUAGE AND SYMBOLISM OF MATHEMATICS 8.2

1. If A, B, C, and D are real polynomials and $B \neq 0$ and $D \neq 0$, then $\frac{A}{B} \cdot \frac{C}{D} =$ ________.
2. To multiply two rational expressions:
   a. Factor the numerators and the denominators. Then write the product as a single ________, indicating the product of the numerators and the product of the denominators.
   b. Reduce this fraction by dividing the numerator and the denominator by any common ________ factors.
3. If A, B, C, and D are real polynomials and $B \neq 0$, $C \neq 0$, and $D \neq 0$, then $\frac{A}{B} \div \frac{C}{D} =$ ________.
4. To divide two rational expressions:
   a. Rewrite the division problem as the product of the dividend and the ________ of the divisor.
   b. Perform the multiplication by using the rule for multiplying rational expressions.
5. Writing an equal symbol between two rational expressions means that these expressions are identical for all values for which ________ expressions are defined.

QUICK REVIEW 8.2

The purpose of this quick review is to help you recall skills needed in this section.

Perform the indicated operations without using a calculator.

1. $\frac{5}{3} \cdot \frac{2}{5} \cdot \frac{8}{5}$

2. $\frac{5}{3} \div \frac{2}{5} \cdot \frac{8}{5}$

3. $\frac{5}{3} \div \left(\frac{2}{5} \cdot \frac{8}{5}\right)$

Determine the missing fraction without using a calculator.

4. $\frac{3}{10} \cdot ? = \frac{6}{20}$

5. $\frac{5}{8} \cdot ? = \frac{35}{56}$

EXERCISES 8.2

In Exercises 1–46, perform the indicated operations and reduce all results to lowest terms. Assume the variables are restricted to values that prevent division by 0.

1. $\frac{14}{15} \cdot \frac{55}{42}$

2. $\frac{33}{10} \cdot \frac{28}{77}$

3. $\frac{9}{35} \div \frac{27}{55}$

4. $\frac{36}{49} \div \frac{9}{4}$

5. $\frac{7m}{9} \cdot \frac{6}{14m^2}$

6. $\frac{5m^3}{21} \cdot \frac{14}{15m^9}$

7. $\frac{66x^2}{30y} \div \frac{77x^8}{45y^3}$

8. $\frac{36x^5}{143} \div \frac{45x^7}{22}$

9. $\frac{x - 2y}{6} \cdot \frac{3}{2y - x}$

10. $\frac{15}{3a - 2b} \cdot \frac{2b - 3a}{12}$

11. $\frac{2x(x - 1)}{6(x + 3)} \cdot \frac{3(x + 3)}{x^2(x + 1)}$

12. $\frac{x^2(x + 5)}{12(x - 7)} \cdot \frac{36(x + 7)}{5x(x + 5)}$

13. $\frac{(x - 2)(x - 3)}{10(x - 2)} \cdot \frac{5(x - 3)}{(x + 3)(x - 3)}$

14. $\frac{3(x^2 - 4)}{14(x - 2)} \cdot \frac{7x}{11(x + 2)}$

15. $\frac{x^2 - y^2}{x^2 - 2xy + y^2} \div \frac{3x + 3y}{7x - 21}$

16. $\frac{4x^2 + 12x + 9}{5x^3 - 2x^2} \div \frac{14x^2 + 21x}{10x^2y^2}$

17. $\frac{x^2 - 3x + 2}{x^2 - 4x + 4} \div \frac{x^2 - 2x + 1}{3x^2 - 12}$

18. $\frac{a^2 - 9b^2}{4a^2 - b^2} \cdot \frac{4a^2 - 4ab + b^2}{2a^2 - 7ab + 3b^2}$

19. $\frac{7(c - y) - a(c - y)}{2c^2 - 7cy + 5y^2} \cdot \frac{2c - 5y}{c - y}$

20. $\frac{5(a + b) - x(a + b)}{2a^2 - 2b^2} \cdot \frac{6ax}{15 - 3x}$

21. $\frac{10y - 14x}{x - 1} \cdot \frac{x^2 - 2x + 1}{21x - 15y}$

22. $\frac{3w - 5z}{6w^2z} \cdot \frac{18w^2z + 30wz^2}{9w^2 - 25z^2}$

23. $\frac{9x^2 - 9xy + 9y^2}{5x^2y + 5xy^2} \cdot \frac{2x^3y - 2xy^3}{3x^2 - 3xy + 3y^2}$

24. $\frac{4x^2 - 1}{18xy} \div \frac{6x - 3}{16x^2 + 8x}$

25. $\frac{4x^2 - 4}{18xy} \cdot \frac{30x^2y}{5 - 5x^2}$

26. $\frac{2x^2 - 4xy + 2y^2}{6xy} \cdot \frac{25xy^2}{2xy - x^2 - y^2}$

27. $(3x^2 - 14x - 5) \cdot \frac{x^2 - 2x - 35}{3x^2 - 20x - 7}$

28. $(6x^2 - 19x + 10) \div \frac{6x^2 - 11x - 10}{9x^2 + 12x + 4}$

29. $\frac{20x^2 + 3x - 9}{21x - 35x^2} \div (12x^2 - 11x - 15)$

30. $\frac{4a^2 - ab - 5b^2}{ax + by + ay + bx} \div (8a - 10b)$

31. $\frac{-4}{7} \cdot \left(\frac{15}{6} \div \frac{-10}{14}\right)$

32. $-\frac{2}{9} \div \left(\frac{22}{12} \cdot \frac{-20}{33}\right)$

33. $\frac{18x^2}{5a} \cdot \frac{15ax}{81a^2} \cdot \frac{44ax}{24x^3}$

34. $\frac{14ab^2}{18x} \cdot \frac{27x^3}{15a^3b^3} \div \frac{7a^2x}{25ab}$

35. $\frac{x^2 + x - y^2 - y}{3x^2 - 3y^2} \div \frac{5x + 5y + 5}{7x^2y + 7xy^2}$

36. $\frac{a(2x - y) - b(2x - y)}{a^2 - b^2} \div \frac{6x - 3y}{4a + 4b}$

37. $\frac{2x - 5}{2x^2 - x - 15} \cdot \frac{-2x^2 - x + 10}{3x + 4} \cdot \frac{-15x - 20}{2x^2 - 9x + 10}$

38. $\frac{x^4 - y^4}{-2x - 3y} \cdot \frac{2x + 3y}{7x^2 + 7y^2} \cdot \frac{12xy}{8x^2 + 8xy}$

39. $\frac{5a - b}{a^2 - 5ab + 4b^2} \div \left(\frac{6ab}{3a - 12b} \cdot \frac{b^2 - 5ab}{4a - 4b}\right)$

40. $\frac{x - y}{x^2 - y^2} \div \frac{6x}{x + y} \cdot \frac{15x^2}{4}$

41. $\frac{x^2 - xy}{5x^2y^2} \div \frac{4x - 4y}{3xy} \cdot \frac{20y}{11}$

42. $\frac{4v^2 + 11v + 6}{2v^2 - v - 10} \div \left(\frac{4v^2 - 21v - 18}{2v^2 - 3v - 5} \cdot \frac{4v^2 - 7v + 3}{4v^2 - 27v + 18}\right)$

43. $\frac{x(a - b + 2c) - 2y(a - b + 2c)}{x^2 - 4xy + 4y^2} \div \frac{2a^2 - 2ab + 4ac}{7xy^2 - 14y^3}$

44. $\dfrac{a(2x - y) - b(2x - y)}{a^2 - b^2} \div \dfrac{6x - 3y}{4a + 4b}$

45. $\dfrac{5x^2 - 5xy}{6x^2 - 6y^2} \div \dfrac{10xy}{x^2 + 2xy + y^2}$

46. $\dfrac{45x^2 - 45xy}{18x^2y^3} \cdot \dfrac{x^2 + 2xy + y^2}{x^2 - y^2} \div \dfrac{9x^2y + 9xy^2}{36y^4}$

Estimation Skills and Calculator Skills

In Exercises 47–50, mentally estimate the value of each expression for $x = 10.01$, and then use a calculator to approximate each value to the nearest hundredth.

PROBLEM	MENTAL ESTIMATE	CALCULATOR APPROXIMATION
47. $\dfrac{(x + 6)(x - 7)}{(x - 7)(x - 6)}$		
48. $\dfrac{(x - 2)(x + 5)}{(x - 2)(x - 5)}$		
49. $\dfrac{x^2 + 50}{x^2 - 50}$		
50. $\dfrac{x^2 - 80}{x - 6}$		

In Exercises 51–54, use the tables of values for each function to match each rational expression with its reduced form.

51. $Y_1 = \dfrac{3x^2 + 6x + 3}{(x + 1)^2}$

X	Y1
-3	3
-2	3
-1	ERR:
0	3
1	3
2	3
3	3

X= -3

52. $Y_1 = \dfrac{-x^3 + 4x^2 - 4x}{(x - 2)^2}$

X	Y1
-3	3
-2	2
-1	1
0	0
1	-1
2	ERR:
3	-3

X= -3

53. $Y_1 = \dfrac{x^3 + 4x^2 + 4x}{(x + 2)^2}$

X	Y1
-3	-3
-2	ERR:
-1	-1
0	0
1	1
2	2
3	3

X= -3

54. $Y_1 = \dfrac{6x^2 - 6}{2 - 2x^2}$

X	Y1
-3	-3
-2	-3
-1	ERR:
0	-3
1	ERR:
2	-3
3	-3

X= -3

A. $Y_2 = -3$

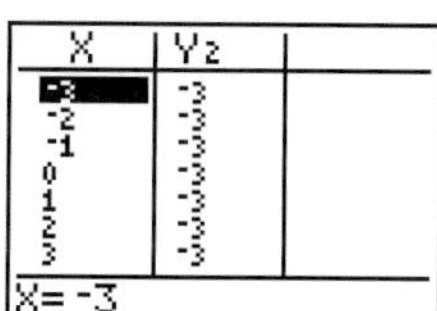

X	Y2
-3	-3
-2	-3
-1	-3
0	-3
1	-3
2	-3
3	-3

X= -3

B. $Y_2 = x$

X	Y2
-3	-3
-2	-2
-1	-1
0	0
1	1
2	2
3	3

X= -3

C. $Y_2 = -x$

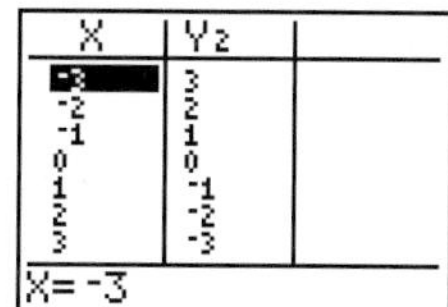

X	Y2
-3	3
-2	2
-1	1
0	0
1	-1
2	-2
3	-3

X= -3

D. $Y_2 = 3$

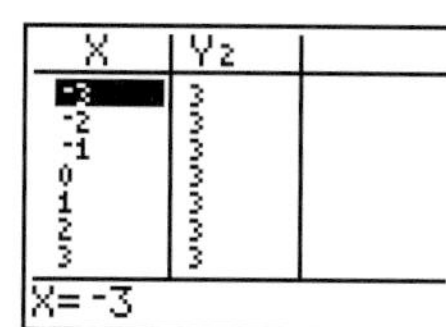

X	Y2
-3	3
-2	3
-1	3
0	3
1	3
2	3
3	3

X= -3

In Exercises 55–64, determine the unknown expression in each equation.

55. $12 \cdot (?) = 60$

56. $12 \div (?) = 4$

57. $(x^2y^3) \div (?) = xy$

58. $(x^2y^3) \cdot (?) = x^4y^4$

59. $(x^2 - 7x - 8) \div (?) = x + 1$

60. $(2x - 3) \cdot (?) = 4x^2 - 9$

61. $\left(\dfrac{x + 2}{x - 3}\right) \cdot (?) = \dfrac{x^2 - x - 6}{x^2 - 6x + 9}$

62. $\left(\dfrac{x + 5}{x - 5}\right) \cdot (?) = \dfrac{x^2 + 7x + 10}{x^2 - 3x - 10}$

63. $\left(\dfrac{2x^2 - 5x + 2}{3x^2 - 4x + 1}\right) \div (?) = \dfrac{2x - 1}{3x - 1}$

64. $\dfrac{x^2 - 2x - 15}{x^2 + x - 12} \div (?) = \dfrac{x + 3}{x - 3}$

65. **Average Cost of Lamps** The number of lamps that a factory can assemble in a day is approximated by the function $N(t) = 2t^2 - 2t$, where t is the number of hours the factory operates. The total cost of operating the plant for t hours is given by $C(t) = 200t + 200$. If the plant is shut down overnight, then it takes a period of time to resume production.

a. The average cost per lamp produced is given by $A(t) = C(t) \div N(t)$. Write a formula for the function $A(t)$.

b. Use this function to complete the following table.

t	2	6	10	14	18	22
$A(t)$						

c. Describe what happens to the average cost when the number of hours the factory operates is reduced to almost 1 hour.

d. Describe what happens to the average cost when the number of hours the factory operates is increased to almost 24 hours.

66. **Average Cost of Lamps** The number of lamps that a factory can assemble in a day is approximated by the function $N(t) = 2.5t^2 - 2.5t$, where t is the number of hours the factory operates. The total cost of operating the plant for t hours is given by $C(t) = 750t + 750$. If the plant is shut down overnight, then it takes a period of time to resume production.

a. The average cost per lamp produced is given by $A(t) = C(t) \div N(t)$. Write a formula for the function $A(t)$.

b. Use this function to complete the following table.

t	2	6	10	14	18	22
$A(t)$						

c. Describe what happens to the average cost when the number of hours the factory operates is reduced to almost 1 hour.

d. Describe what happens to the average cost when the number of hours the factory operates is increased to almost 24 hours.

Group Discussion Questions

67. Challenge Question Perform the indicated operations, and reduce all results to lowest terms. Assume that m is a natural number and that all values of the variables that cause division by 0 are excluded.

a. $\dfrac{x^{2m+1}}{x^m - 1} \cdot \dfrac{3x^m - 3}{x}$ **b.** $\dfrac{x^{2m} - 1}{x^{2m} + x^m} \div \dfrac{x^m - 1}{x^{2m}}$

c. $\dfrac{x^{2m} - 1}{x^m - 1} \cdot \dfrac{3x^m + 3}{(x^m + 1)^2}$ **d.** $\dfrac{x^{3m} - x^m}{x^{2m} + x^m} \div \dfrac{(x^m + 1)^3}{x^{2m} + 2x^m + 1}$

68. Challenge Question

Reduce $\dfrac{x}{x - 1} \cdot \dfrac{x - 1}{x - 2} \cdot \dfrac{x - 2}{x - 3} \cdots \dfrac{x - 8}{x - 9} \cdot \dfrac{x - 9}{x - 10}$, and list all the excluded values.

69. Error Analysis A student graphed $f(x) = \dfrac{x^2 - x}{x - 1}$ on a TI-84 Plus calculator and obtained the graph shown below.

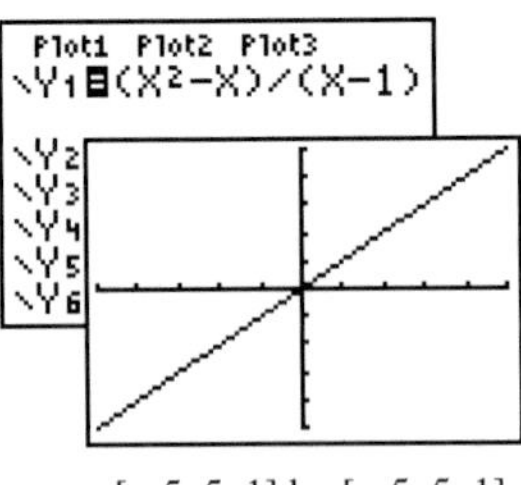

$[-5, 5, 1]$ by $[-5, 5, 1]$

Because no breaks appear in the graph of this function, the student concluded the domain of $f(x) = \dfrac{x^2 - x}{x - 1}$ is $\mathbb{R}$.

This is incorrect.

a. What is the actual domain?

b. Sketch this graph by hand to properly show the correct domain.

c. Graph this function on a TI-84 Plus calculator, using the ZOOM 4 option. Then carefully examine this graph. Using the **TRACE** feature, what do you observe at the excluded value of this function?

d. Why do you think the excluded value is not revealed by a break in the graph on the window $[-5, 5, 1]$ by $[-5, 5, 1]$?

Section 8.3 Adding and Subtracting Rational Expressions

Objective: 5. Add and subtract rational expressions.

The rules for adding and subtracting rational expressions are the same as the rules given in Chapter 1 for adding and subtracting rational numbers. The rules for adding and subtracting rational expressions with the same denominator are given in the following box. The result should be reduced to lowest terms.

Adding and Subtracting Rational Expressions with the Same Denominator

ALGEBRAICALLY	VERBALLY	ALGEBRAIC EXAMPLE
If A, B, and C are real polynomials and $C \neq 0$, then $\dfrac{A}{C} + \dfrac{B}{C} = \dfrac{A + B}{C}$ and	To add two rational expressions with the same denominator, add the numerators and use this common denominator.	$\dfrac{7}{24x} + \dfrac{5}{24x} = \dfrac{7 + 5}{24x}$ $= \dfrac{12}{24x}$ $= \dfrac{1}{2x}$
$\dfrac{A}{C} - \dfrac{B}{C} = \dfrac{A - B}{C}$	To subtract two rational expressions with the same denominator, subtract the numerators and use this common denominator.	$\dfrac{7}{24x} - \dfrac{5}{24x} = \dfrac{7 - 5}{24x}$ $= \dfrac{2}{24x}$ $= \dfrac{1}{12x}$

It is easy to make errors in the order of operations if you are not careful when adding or subtracting rational expressions. Example 1 illustrates how parentheses can be used to avoid these errors.

EXAMPLE 1 Subtracting Rational Expressions with the Same Denominator

Simplify $\frac{5x+7}{2x+2} - \frac{3x-1}{2x+2}$. Assume the variables are restricted to values that prevent division by 0.

SOLUTION

$$\frac{5x+7}{2x+2} - \frac{3x-1}{2x+2} = \frac{(5x+7)-(3x-1)}{2x+2}$$

Note: A common error is to write the numerator as $5x + 7 - 3x - 1$. To avoid this error, enclose the terms of the numerator in parentheses before combining like terms.

$$= \frac{5x+7-3x+1}{2x+2}$$

$$= \frac{2x+8}{2x+2}$$

$$= \frac{\overset{1}{\cancel{2}}(x+4)}{\underset{1}{\cancel{2}}(x+1)}$$

Factor, and then reduce to lowest terms.

$$= \frac{x+4}{x+1}$$

This sum is defined for all real numbers other than -1. So $x = -1$ is the only excluded value.

EXAMPLE 2 Subtracting Rational Expressions with the Same Denominator

Simplify $\frac{x^2}{x^2-1} - \frac{2x-1}{x^2-1}$. Assume the variables are restricted to values that prevent division by 0.

SOLUTION

$$\frac{x^2}{x^2-1} - \frac{2x-1}{x^2-1} = \frac{x^2-(2x-1)}{x^2-1}$$

Use parentheses in the numerator to avoid an error in sign.

$$= \frac{x^2-2x+1}{x^2-1}$$

Simplify the numerator.

$$= \frac{\overset{1}{\cancel{(x-1)}}(x-1)}{\underset{1}{\cancel{(x-1)}}(x+1)}$$

Factor, and then reduce to lowest terms. The excluded values are 1 and -1.

$$= \frac{x-1}{x+1}$$

This result is defined for all real numbers other than 1 and -1. The only excluded values are $x = -1$ and $x = 1$.

Remember that $-\frac{a}{b} = \frac{-a}{b} = \frac{a}{-b}$. These alternative forms are often useful when you are working with two fractions whose denominators are opposites of each other.

EXAMPLE 3 Adding Rational Expressions Whose Denominators Are Opposites

Simplify $\frac{5x}{3x-6} + \frac{-4x}{6-3x}$. Assume the variables are restricted to values that prevent division by 0.

SOLUTION

$$\frac{5x}{3x-6} + \frac{-4x}{6-3x} = \frac{5x}{3x-6} + \frac{-4x}{6-3x} \cdot \frac{-1}{-1}$$

Note that the denominators $3x - 6$ and $6 - 3x$ are opposites. Convert $\frac{-4x}{6-3x}$ to the equivalent form $\left(\frac{-4x}{6-3x}\right) \cdot \left(\frac{-1}{-1}\right) = \frac{4x}{3x-6}$.

$$= \frac{5x}{3x-6} + \frac{4x}{3x-6}$$

$$= \frac{5x+4x}{3x-6}$$

Now that the denominators are the same, you can add the numerators of these like terms.

$$= \frac{\overset{3}{\cancel{9}}x}{\underset{1}{\cancel{3}}(x-2)}$$

Then factor the denominator, and reduce this result by dividing both the numerator and the denominator by 3.

$$= \frac{3x}{x-2}$$

This sum is defined for all real numbers other than 2.

SELF-CHECK 8.3.1

Simplify each expression. Assume the variables are restricted to values that prevent division by 0.

1. $\frac{2x-1}{x^2} + \frac{4x+1}{x^2}$
2. $\frac{9x-2}{2x+1} - \frac{5x-4}{2x+1}$
3. $\frac{2y-1}{7y-3} - \frac{5y-2}{3-7y}$

Fractions can be added or subtracted only if they are expressed in terms of a common denominator. Although any common denominator can be used, we can simplify our work considerably by using the least common denominator (LCD). Sometimes the LCD can be determined by inspection. In other cases it may be necessary to use the procedure given in the following box.

Finding the LCD of Two or More Rational Expressions

VERBALLY	ALGEBRAIC EXAMPLE
	Determine the LCD of $\frac{7}{6xy^2}$ and $\frac{2}{15x^2y}$.
1. Factor each denominator completely, including constant factors. Express repeated factors in exponential form.	$6xy^2 = 2 \cdot 3 \cdot x \cdot y^2$
2. List each factor to the highest power to which it occurs in any single factorization.	$15x^2y = 3 \cdot 5 \cdot x^2 \cdot y$
3. Form the LCD by multiplying the factors listed in step 2.	$\text{LCD} = 2 \cdot 3 \cdot 5 \cdot x^2 \cdot y^2$ $\text{LCD} = 30x^2y^2$

The procedure for adding or subtracting rational expressions with different denominators is given in the following box. The common denominator in the example below was calculated in the previous box.

Adding (Subtracting) Rational Expressions

VERBALLY	ALGEBRAIC EXAMPLE
1. Determine the LCD of all the terms. 2. Convert each term to an equivalent rational expression whose denominator is the LCD. 3. Retain the LCD as the denominator and add (subtract) the numerators to form the sum (difference). 4. Reduce the expression to lowest terms.	$\frac{7}{6xy^2} + \frac{2}{15x^2y} = \frac{?}{30x^2y^2} + \frac{?}{30x^2y^2}$ $= \frac{7}{6xy^2} \cdot \frac{5x}{5x} + \frac{2}{15x^2y} \cdot \frac{2y}{2y}$ $= \frac{35x}{30x^2y^2} + \frac{4y}{30x^2y^2}$ $= \frac{35x + 4y}{30x^2y^2}$

EXAMPLE 4 Subtracting Two Rational Expressions with Different Denominators

Simplify $\frac{5}{x + 4} - \frac{4}{x - 6}$ and compare the difference to the original expression by using a table of values. Assume the variables are restricted to values that prevent division by 0.

SOLUTION

ALGEBRAICALLY

$$\frac{5}{x+4} - \frac{4}{x-6} = \frac{5}{x+4} \cdot \frac{x-6}{x-6} - \frac{4}{x-6} \cdot \frac{x+4}{x+4}$$

The LCD is $(x + 4)(x - 6)$. Convert each fraction so that it has this LCD as its denominator.

$$= \frac{5(x-6)}{(x+4)(x-6)} - \frac{4(x+4)}{(x+4)(x-6)}$$

$$= \frac{5(x-6) - 4(x+4)}{(x+4)(x-6)}$$

$$= \frac{5x - 30 - 4x - 16}{(x+4)(x-6)}$$

Add like terms and simplify the numerator.

$$= \frac{x - 46}{(x+4)(x-6)}$$

For many applications it is more useful to leave the denominator in factored form.

NUMERICAL CHECK

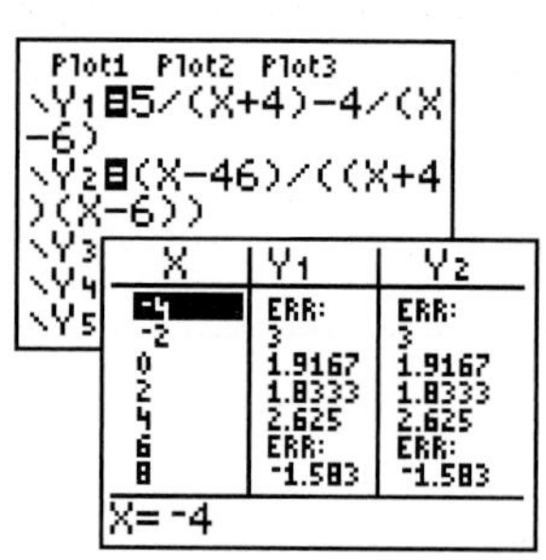

The denominator of a sum or difference is usually left in factored form because this form reveals the most information about the rational expression.

Let Y_1 represent the original expression and Y_2 represent the difference.

These tables are identical for all values of x. Both are undefined at -4 and 6.

Answer: $\frac{5}{x+4} - \frac{4}{x-6} = \frac{x-46}{(x+4)(x-6)}$

SELF-CHECK 8.3.2

1. Simplify $\frac{x}{x-5} + \frac{3}{x+2}$. Assume the variables are restricted to values that prevent division by 0.

First factor the quadratic trinomials in Example 5 to determine the LCD of these two rational expressions.

EXAMPLE 5 Adding Rational Expressions with Different Denominators

Add $\frac{x}{6x^2 - 30x + 36} + \frac{7}{9x^2 + 9x - 108}$. Assume the variables are restricted to values that prevent division by 0.

SOLUTION

$$\frac{x}{6x^2 - 30x + 36} + \frac{7}{9x^2 + 9x - 108}$$

$$= \frac{x}{6(x-2)(x-3)} + \frac{7}{9(x-3)(x+4)}$$

$$= \frac{x}{6(x-2)(x-3)} \cdot \frac{3(x+4)}{3(x+4)} + \frac{7}{9(x-3)(x+4)} \cdot \frac{2(x-2)}{2(x-2)}$$

$$= \frac{3x(x+4)}{18(x-2)(x-3)(x+4)} + \frac{14(x-2)}{18(x-2)(x-3)(x+4)}$$

First factor each denominator. Then determine the LCD of these two fractions:
$2 \cdot 3(x-2)(x-3)$
$3^2(x-3)(x+4)$
$\text{LCD} = 2 \cdot 3^2(x-2)(x-3)(x+4)$
$\text{LCD} = 18(x-2)(x-3)(x+4)$
Convert each term to a fraction with this LCD as its denominator.

$$= \frac{3x(x+4) + 14(x-2)}{18(x-2)(x-3)(x+4)}$$

$$= \frac{3x^2 + 12x + 14x - 28}{18(x-2)(x-3)(x+4)}$$

Leave the denominator in factored form, and simplify the numerator by multiplying and then combining like terms.

$$= \frac{3x^2 + 26x - 28}{18(x-2)(x-3)(x+4)}$$

The numerator is prime, so this expression is in reduced form.

Some of the steps illustrated in Example 5 can be combined to decrease the amount of writing necessary in these problems. Although most of the steps are still shown in the following examples, some have been combined. Be sure your proficiency is well developed before you try to perform too many steps mentally. The steps that are skipped are the source of many student errors—especially errors in sign.

EXAMPLE 6 Combining Three Rational Expressions with Different Denominators

Simplify $\frac{2v+11}{v^2+v-6} - \frac{2}{v+3} + \frac{3}{2-v}$. Assume the variables are restricted to values that prevent division by 0.

SOLUTION

$$\frac{2v+11}{v^2+v-6} - \frac{2}{v+3} + \frac{3}{2-v}$$

$$= \frac{2v+11}{(v+3)(v-2)} - \frac{2}{v+3} + \frac{-3}{v-2}$$

$$= \frac{2v+11}{(v+3)(v-2)} - \frac{2}{v+3}\cdot\frac{v-2}{v-2} + \frac{-3}{v-2}\cdot\frac{v+3}{v+3}$$

Factor each denominator. Noting that the factor $v-2$ in the first denominator is the opposite of the denominator $2-v$ in the last term, change $\frac{3}{2-v}$ to the equivalent form $\left(\frac{3}{2-v}\right)\left(\frac{-1}{-1}\right) = \frac{-3}{v-2}$.

Now convert each term so that it has the LCD of $(v+3)(v-2)$ as its denominator.

$$= \frac{2v+11}{(v+3)(v-2)} - \frac{2(v-2)}{(v+3)(v-2)} + \frac{-3(v+3)}{(v+3)(v-2)}$$

$$= \frac{(2v+11) - 2(v-2) - 3(v+3)}{(v+3)(v-2)}$$

$$= \frac{2v+11-2v+4-3v-9}{(v+3)(v-2)}$$

Simplify the numerator by multiplying and then combining like terms.

$$= \frac{-3v+6}{(v+3)(v-2)}$$

$$= \frac{-3\cancel{(v-2)}^{1}}{(v+3)\cancel{(v-2)}_{1}}$$

Factor the numerator, and reduce by dividing out the common factor $v-2$.

$$= \frac{-3}{v+3}$$

$$= -\frac{3}{v+3}$$

To multiply or divide rational expressions: first reduce as much as possible and then perform the multiplication or division.

Note that when we multiply or divide rational expressions, we first reduce as much as possible and then perform the multiplication or division. On the other hand, when we add or subtract rational expressions, we add or subtract first, using a common denominator, and then reduce afterward.

To add or subtract rational expressions: first add or subtract the terms, using a common denominator, and then reduce afterward.

SELF-CHECK 8.3.3

1. Simplify $\frac{3v-1}{v^2-1} + \frac{3}{1-v} - \frac{2}{v+1}$. Assume the variables are restricted to values that prevent division by 0.

SELF-CHECK ANSWERS

8.3.1

1. $\frac{6}{x}$ 2. 2 3. 1

8.3.2

1. $\frac{x^2 + 5x - 15}{(x - 5)(x + 2)}$

8.3.3

1. $-\frac{2}{v - 1}$

USING THE LANGUAGE AND SYMBOLISM OF MATHEMATICS 8.3

1. If A, B, and C are real polynomials and $C \neq 0$, then $\frac{A}{C} + \frac{B}{C} =$ __________ and $\frac{A}{C} - \frac{B}{C} =$ __________.
2. To add two rational expressions with the same denominator, __________ the numerators and use this common denominator.
3. To subtract two rational expressions with the same denominator, __________ the numerators and use this common denominator.
4. To find the least common denominator of two or more rational expressions:
   a. Factor each __________ completely, including constant factors. Express repeated factors in __________ form.
   b. List each factor to the __________ power to which it occurs in any single factorization.
   c. Form the LCD by __________ the factors listed in step **b.**
5. To add or subtract two rational expressions:
   a. Determine the __________ of all the terms.
   b. Convert each term to an equivalent rational expression whose __________ is the LCD.
   c. Retaining the LCD as the denominator, add (subtract) the __________ to form the sum (difference).
   d. Reduce the expression to __________ __________.

QUICK REVIEW 8.3

The purpose of this quick review is to help you recall skills needed in this section.

Perform the indicated operations without using a calculator.

1. $\frac{7}{12} + \frac{3}{12}$
2. $\frac{17}{24} - \frac{5}{24}$
3. $\frac{2}{3} - \frac{1}{12}$
4. $\frac{3}{24} + \frac{5}{16}$
5. $\frac{1}{6} + \frac{2}{3} - \frac{1}{4}$

EXERCISES 8.3

In Exercises 1–8, perform the indicated operations, and reduce the results to lowest terms. Assume the variables are restricted to values that prevent division by 0.

1. $\frac{5b + 13}{3b^2} + \frac{b - 4}{3b^2}$
2. $\frac{3x^2 + 1}{8x^3} - \frac{1 - 3x^2}{8x^3}$
3. $\frac{7}{x - 7} - \frac{x}{x - 7}$
4. $\frac{a - 1}{a + 7} + \frac{8}{a + 7}$
5. $\frac{3s + 7}{s^2 - 9} + \frac{s + 5}{s^2 - 9}$
6. $\frac{5t - 22}{t^2 - 5t + 6} + \frac{4t - 5}{t^2 - 5t + 6}$
7. $\frac{x + a}{x(a + b) + y(a + b)} - \frac{x - b}{x(a + b) + y(a + b)}$
8. $\frac{2y^2 + 1}{2y^2 - 5y - 12} - \frac{4 - y}{2y^2 - 5y - 12}$

In Exercises 9–14, fill in the blanks to complete each addition or subtraction.

9. $$\begin{aligned}\frac{1}{15xy^2} + \frac{2}{27x^2y} &= \frac{1}{15xy^2}\left(\frac{?}{9x}\right) + \frac{2}{27x^2y}\left(\frac{?}{5y}\right)\\ &= \frac{9x}{135x^2y^2} + \frac{?}{135x^2y^2}\\ &= \frac{?}{135x^2y^2}\end{aligned}$$

10. $$\begin{aligned}\frac{3}{35x^3y} + \frac{2}{55x^2y^3} &= \frac{3}{35x^3y}\left(\frac{?}{11y^2}\right) + \frac{2}{55x^2y^3}\left(\frac{?}{7x}\right)\\ &= \frac{?}{385x^3y^3} + \frac{14x}{385x^3y^3}\\ &= \frac{?}{385x^3y^3}\end{aligned}$$

11. $\dfrac{a+b}{21(3a-b)} - \dfrac{a-b}{14(3a-b)} = \dfrac{a+b}{21(3a-b)}\left(\dfrac{?}{?}\right) - \dfrac{a-b}{14(3a-b)}\left(\dfrac{?}{?}\right)$
$= \dfrac{?}{42(3a-b)} - \dfrac{?}{42(3a-b)}$
$= \dfrac{?}{42(3a-b)}$

12. $\dfrac{2x-y}{12(x+y)} - \dfrac{x-y}{15(x+y)} = \dfrac{2x-y}{12(x+y)}\left(\dfrac{?}{?}\right) - \dfrac{x-y}{15(x+y)}\left(\dfrac{?}{?}\right)$
$= \dfrac{?}{60(x+y)} - \dfrac{?}{60(x+y)}$
$= \dfrac{?}{60(x+y)}$

13. $\dfrac{2}{(x-3)(x-15)} + \dfrac{1}{(x-3)(x+3)} = \dfrac{2}{(x-3)(x-15)}\left(\dfrac{?}{x+3}\right) + \dfrac{1}{(x-3)(x+3)}\left(\dfrac{?}{x-15}\right)$
$= \dfrac{?}{(x-3)(x-15)(x+3)} + \dfrac{?}{(x-3)(x-15)(x+3)}$
$= \dfrac{?}{(x-3)(x-15)(x+3)}$
$= \dfrac{?}{(x-15)(x+3)}$

14. $\dfrac{3}{(2x-1)(x+4)} - \dfrac{1}{(2x-1)(x+1)} = \dfrac{3}{(2x-1)(x+4)}\left(\dfrac{?}{x+1}\right) - \dfrac{1}{(2x-1)(x+1)}\left(\dfrac{?}{x+4}\right)$
$= \dfrac{?}{(2x-1)(x+4)(x+1)} - \dfrac{?}{(2x-1)(x+4)(x+1)}$
$= \dfrac{?}{(2x-1)(x+4)(x+1)}$
$= \dfrac{?}{(x+4)(x+1)}$

In Exercises 15–40, perform the indicated operations, and reduce the results to lowest terms. Assume the variables are restricted to values that prevent division by 0.

15. $\dfrac{4}{9w} - \dfrac{7}{6w}$

16. $\dfrac{4}{5z} - \dfrac{1}{2z}$

17. $\dfrac{3v-1}{7v} - \dfrac{v-2}{14v}$

18. $\dfrac{12a-7}{35a} - \dfrac{a-1}{5a}$

19. $\dfrac{4}{b} + \dfrac{b}{b+4}$

20. $\dfrac{5}{c} - \dfrac{c}{c-7}$

21. $5 - \dfrac{1}{x}$

22. $7 + \dfrac{3}{y}$

23. $\dfrac{1}{x} - \dfrac{2}{x^2} + \dfrac{3}{x^3}$

24. $\dfrac{3}{y} + \dfrac{5}{y^2} - \dfrac{7}{y^3}$

25. $\dfrac{3}{x-2} - \dfrac{2}{x+3}$

26. $\dfrac{2}{x-5} - \dfrac{3}{x+6}$

27. $\dfrac{x}{x-1} - \dfrac{x}{x+1}$

28. $\dfrac{x}{x+2} - \dfrac{x}{x-2}$

29. $\dfrac{x+5}{x-4} - \dfrac{x+3}{x+2}$

30. $\dfrac{x-3}{x-2} - \dfrac{x-2}{x-3}$

31. $\dfrac{x-1}{x+2} + \dfrac{x+2}{x-1}$

32. $\dfrac{x+1}{x-3} - \dfrac{x-3}{x+1}$

33. $\dfrac{x}{77x-121y} - \dfrac{y}{49x-77y}$

34. $\dfrac{a}{6a-9b} - \dfrac{b}{4a-6b}$

35. $\dfrac{2}{(m+1)(m-2)} + \dfrac{3}{(m-2)(m+3)}$

36. $\dfrac{5}{(m-1)(m+2)} + \dfrac{2}{(m+2)(m-3)}$

37. $\dfrac{m+2}{m^2-6m+8} - \dfrac{8-3m}{m^2-5m+6}$

38. $\dfrac{3n+8}{n^2+6n+8} - \dfrac{4n+2}{n^2+n-12}$

39. $\dfrac{1}{a-3b} + \dfrac{b}{a^2-7ab+12b^2} + \dfrac{1}{a-4b}$

40. $\dfrac{5x}{5x-y} + \dfrac{2y^2}{25x^2-y^2} - \dfrac{5x}{5x+y}$

41. **Total Area of Two Regions** Find the sum of the areas of the triangle and rectangle shown.

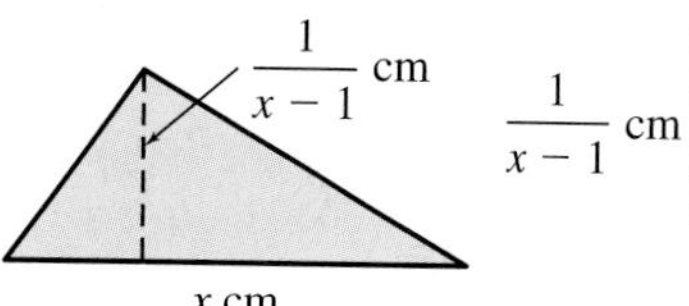

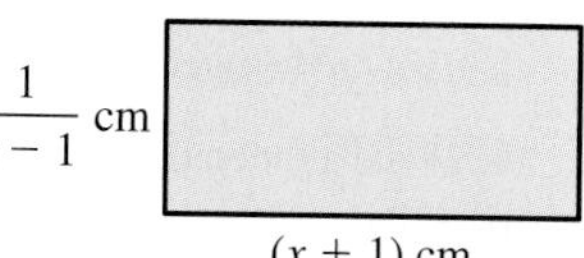

42. Total Area of Two Regions Find the sum of the areas of the parallelogram and rectangle.

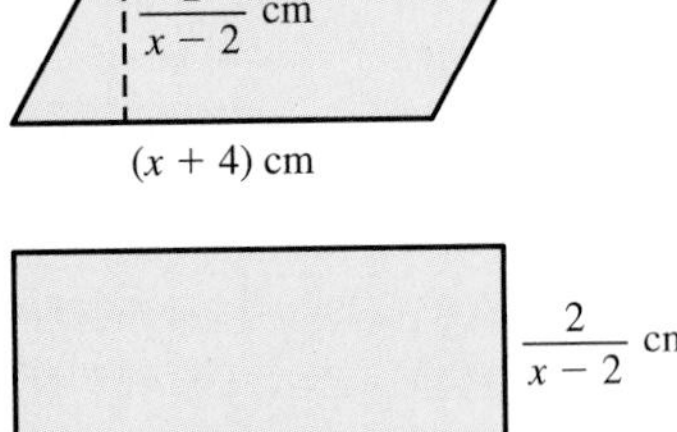

$\frac{2}{x-2}$ cm

(x + 5) cm

43. Total Work by Two Painters One painter can paint a house in t hours, while a second painter would take $(t + 5)$ hours to do the same house. In 8 hours the fractional portion of the job done by each is $\frac{8}{t}$ and $\frac{8}{t+5}$, respectively.

a. Write a rational function that gives the total amount of work completed by the two painters working together for 8 hours.

b. Use the graph of this function to describe what happens to the total amount of work done as t becomes very large.

44. Total Work by Two Pipes One pipe can fill a cooling tank in t hours. A second pipe would take $(t + 3)$ hours to fill the same tank. If both pipes are used, the fractional portion of the tank that can be filled by each pipe in 10 hours is $\frac{10}{t}$ and $\frac{10}{t+3}$, respectively.

a. Write a rational function that gives the total amount of the cooling tank filled by the two pipes working together for 10 hours.

b. Use the graph of this function to describe what happens to the total amount of the cooling tank filled by the two pipes as t becomes very large.

45. Total Time of an Automobile Trip An automobile traveled 90 mi at a rate r on a gravel road in a time of $\frac{90}{r}$ hours. Then the automobile traveled 165 mi on paved roads in a time of $\frac{165}{r+10}$ hours.

a. Write a rational function for the total time of this trip.

b. What is the total time of the 255-mi trip if the average speed of the car on the gravel road is 45 mi/h?

46. Total Time of a Delivery Truck The route for a delivery truck includes both interstate highway and local two-way highways. The truck can travel 120 mi on the two-way highway at a rate r in a time of $\frac{120}{r}$ hours. It can also travel 180 mi on the interstate in a time of $\frac{180}{r+15}$ hours.

a. Write a rational function for the total time of this trip.

b. Approximate the total time of the 300-mi trip if the average speed of the truck on the two-way highway is 50 mi/h.

In Exercises 47–50, use the table of values for each function to match each expression with its sum.

47. $Y_1 = \frac{5}{x-1} + \frac{3}{x+2}$

X	Y_1
-3	-4.25
-2	ERR:
-1	.5
0	-3.5
1	ERR:
2	5.75
3	3.1

X= -3

48. $Y_1 = \frac{5}{x-1} - \frac{3}{x+2}$

X	Y_1
-3	1.75
-2	ERR:
-1	-5.5
0	-6.5
1	ERR:
2	4.25
3	1.9

X= -3

49. $Y_1 = \frac{2x}{x-1} - \frac{2x}{x+2}$

X	Y_1
-3	-4.5
-2	ERR:
-1	3
0	0
1	ERR:
2	3
3	1.8

X= -3

50. $Y_1 = \frac{x}{x-1} + \frac{x}{x+2}$

X	Y_1
-3	3.75
-2	ERR:
-1	-.5
0	0
1	ERR:
2	2.5
3	2.1

X= -3

A. $Y_2 = \frac{8x + 7}{(x-1)(x+2)}$

X	Y_1
-3	-4.25
-2	ERR:
-1	.5
0	-3.5
1	ERR:
2	5.75
3	3.1

X= -3

B. $Y_2 = \frac{x(2x+1)}{(x-1)(x+2)}$

X	Y_1
-3	3.75
-2	ERR:
-1	-.5
0	0
1	ERR:
2	2.5
3	2.1

X= -3

C. $Y_2 = \frac{2x + 13}{(x-1)(x+2)}$

X	Y_1
-3	1.75
-2	ERR:
-1	-5.5
0	-6.5
1	ERR:
2	4.25
3	1.9

X= -3

D. $Y_2 = \frac{6x}{(x-1)(x+2)}$

X	Y_1
-3	-4.5
-2	ERR:
-1	3
0	0
1	ERR:
2	3
3	1.8

X= -3

In Exercises 51–68, perform the indicated operations, and reduce the results to lowest terms. Assume the variables are restricted to values that prevent division by 0.

51. $\frac{x^2 + 5}{x^2 - 3x + 4} - 2$

52. $\frac{2x^2 - 5x + 3}{3x^2 + 7x + 9} + 2$

53. $\frac{5}{2x+2} + \frac{x+5}{2x^2-2} - \frac{3}{x-1}$

54. $\frac{p+6}{p^2-4} - \frac{4}{p+2} - \frac{2}{p-2}$

55. $\frac{1}{x-5} + \frac{1}{x+5} - \frac{10}{x^2-25}$

56. $\frac{42}{x^2-49} + \frac{3}{x+7} + \frac{3}{x-7}$

57. $\frac{3}{s-5t} + \frac{7}{s-2t} - \frac{9t}{s^2-7st+10t^2}$

58. $\frac{y+8}{y^2-2y-8} + \frac{1}{y+2} - \frac{4}{4-y}$

59. $\frac{2z + 11}{z^2 + z - 6} + \frac{2}{z + 3} - \frac{3}{z - 2}$

60. $\frac{3}{2w - 1} + \frac{4}{2w + 1} - \frac{10w - 3}{4w^2 - 1}$

61. $\frac{v + w}{vw} + \frac{w}{v^2 - vw} - \frac{1}{w}$

62. $\frac{5v}{v^3 - 5v^2 + 6v} - \frac{4}{v - 2} - \frac{3v}{3 - v}$

63. $\frac{9w + 2}{3w^2 - 2w - 8} - \frac{7}{4 - w - 3w^2}$

64. $\frac{2x + y}{(x - y)(x - 2y)} - \frac{x + 4y}{(x - 3y)(x - y)} - \frac{x - 7y}{(x - 3y)(x - 2y)}$

65. $\frac{2w - 7}{w^2 - 5w + 6} - \frac{2 - 4w}{w^2 - w - 6} + \frac{5w + 2}{4 - w^2}$

66. $\frac{6m}{m + 2} + \frac{2}{m - 2} - \frac{8m}{m^2 - 4}$

67. $\frac{4m}{1 - m^2} + \frac{2}{m + 1} - 2$

68. $\frac{y - 5}{(x^2 + 5x) + (xy + 5y)} + \frac{1}{x + y} - \frac{2}{x + 5}$

69. The sum of a rational expression and $\frac{1}{x + 5}$ is $\frac{4x}{2x^2 + 5x - 25}$. Determine this rational expression.

70. The difference formed by a rational expression minus $\frac{1}{x + 5}$ is $\frac{27}{4x^2 + 13x - 35}$. Determine this rational expression.

Group Discussion Questions

71. **Error Analysis** Discuss the error made by the student whose work is shown below.

$$\frac{x}{x + 1} - \frac{1}{x + 1} \cdot \frac{2x}{x - 1} = \frac{x - 1}{x + 1} \cdot \frac{2x}{x - 1}$$
$$= \frac{2x}{x + 1}$$

72. **Challenge Question** Simplify each expression. Assume that m is a natural number and that all values of the variables that cause division by 0 are excluded.

a. $\frac{x^m}{x^{2m} - 4} - \frac{2}{x^{2m} - 4}$

b. $\frac{2x^m}{x^{2m} - 6x^m + 9} - \frac{6}{x^{2m} - 6x^m + 9}$

c. $\frac{3x^m}{x^{2m} - x^m - 20} - \frac{15}{x^{2m} - x^m - 20}$

73. **Challenge Question** Assume that m is an even integer. Simplify $\frac{1}{m + 1} + \frac{2}{m + 1} + \frac{3}{m + 1} + \cdots + \frac{m}{m + 1}$. Then test your result by evaluating both the original expression and your result for $m = 4, 6,$ and 8.

Section 8.4 Combining Operations and Simplifying Complex Rational Expressions

Objectives:

6. Simplify rational expressions in which the order of operations must be determined.
7. Simplify complex fractions.

This section shows how to combine the operations already covered in the earlier sections of this chapter and provides further exercises involving rational expressions. The correct order of operations, from Section 1.7, is as follows:

A Mathematical Note

Evelyn Boyd Granville completed her Ph.D. in functional analysis in 1949 and became one of the first African-American women to earn a doctorate in mathematics. She later worked on both the Mercury and Apollo space programs.

Order of Operations

STEP 1. Start with the expression within the innermost pair of grouping symbols.

STEP 2. Perform all exponentiations.

STEP 3. Perform all multiplications and divisions as they appear from left to right.

STEP 4. Perform all additions and subtractions as they appear from left to right.

EXAMPLE 1 Using the Correct Order of Operations When Simplifying Rational Expressions

Simplify $\frac{2x}{5} + \frac{x^3}{15} \cdot \frac{9}{x^2}$. Assume the variables are restricted to values that prevent division by 0.

SOLUTION

$$\frac{2x}{5} + \frac{x^3}{15} \cdot \frac{9}{x^2} = \frac{2x}{5} + \frac{\overset{3x}{\cancel{9x^3}}}{\underset{5}{\cancel{15x^2}}}$$

Multiplication has higher priority than addition, so perform the indicated multiplication first. Reduce the product to lowest terms by dividing both the numerator and the denominator by $3x^2$.

$$= \frac{2x}{5} + \frac{3x}{5}$$

$$= \frac{2x + 3x}{5}$$

Then perform the indicated addition, and reduce this sum to lowest terms.

$$= \frac{\overset{1}{\cancel{5}}x}{\underset{1}{\cancel{5}}}$$

$$= x$$

The expressions in Examples 1 and 2 look very much alike. However, the pair of parentheses in Example 2 completely changes the order of operations used in Example 1.

EXAMPLE 2 Using the Correct Order of Operations When Simplifying Rational Expressions

Simplify $\left(\frac{2x}{5} + \frac{x^3}{15}\right) \cdot \frac{9}{x^2}$. Assume the variables are restricted to values that prevent division by 0.

SOLUTION

$$\left(\frac{2x}{5} + \frac{x^3}{15}\right) \cdot \frac{9}{x^2} = \left(\frac{2x}{5} \cdot \frac{3}{3} + \frac{x^3}{15}\right) \cdot \frac{9}{x^2}$$

First add the terms inside the parentheses, using a common denominator of 15.

$$= \left(\frac{6x}{15} + \frac{x^3}{15}\right) \cdot \frac{9}{x^2}$$

$$= \left(\frac{6x + x^3}{15}\right) \cdot \frac{9}{x^2}$$

Then multiply these factors. Reduce the product to lowest terms.

$$= \frac{\overset{1}{\cancel{x}}(x^2 + 6)}{\underset{5}{\cancel{15}}} \cdot \frac{\overset{3}{\cancel{9}}}{\underset{x}{\cancel{x^2}}}$$

$$= \frac{3(x^2 + 6)}{5x}$$

It is customary to leave the simplified form with both the numerator and the denominator in factored form.

SELF-CHECK 8.4.1

Simplify each expression. Assume that x is restricted to values that prevent division by 0.

1. $\left(\frac{x}{3} - \frac{2}{x^2}\right) \cdot \frac{x^3}{4}$

2. $\frac{x}{3} - \frac{2}{x^2} \cdot \frac{x^3}{4}$

Example 3, which involves division of one rational expression by another, is used as a lead-in to the topic of complex rational expressions.

EXAMPLE 3 Using the Correct Order of Operations When Simplifying Rational Expressions

Simplify $\left(\frac{2}{x} - \frac{1}{3}\right) \div \left(\frac{2}{x} + \frac{1}{3}\right)$ and compare the result to the original expression by using a table of values. Assume the variables are restricted to values that prevent division by 0.

SOLUTION

ALGEBRAICALLY

$$\left(\frac{2}{x} - \frac{1}{3}\right) \div \left(\frac{2}{x} + \frac{1}{3}\right) = \left[\frac{2(3) - 1(x)}{3x}\right] \div \left[\frac{2(3) + 1(x)}{3x}\right]$$

First simplify the terms inside each set of parentheses. The LCD inside each set of parentheses is $3x$.

$$= \left(\frac{6 - x}{3x}\right) \div \left(\frac{6 + x}{3x}\right)$$

Simplify each numerator. Note that 0 is an excluded value to prevent division by 0.

$$= \frac{6 - x}{\cancel{3x}_1} \cdot \frac{\cancel{3x}^1}{6 + x}$$

To divide, invert the divisor and then multiply. Note that -6 is also an excluded value.

Reduce by dividing out the common factor of $3x$.

$$= \frac{6 - x}{6 + x}$$

$$= \frac{-(x - 6)}{x + 6}$$

To rewrite the numerator so that x has a positive coefficient, factor a -1 out of the numerator.

$$= -\frac{x - 6}{x + 6}$$

NUMERICAL CHECK

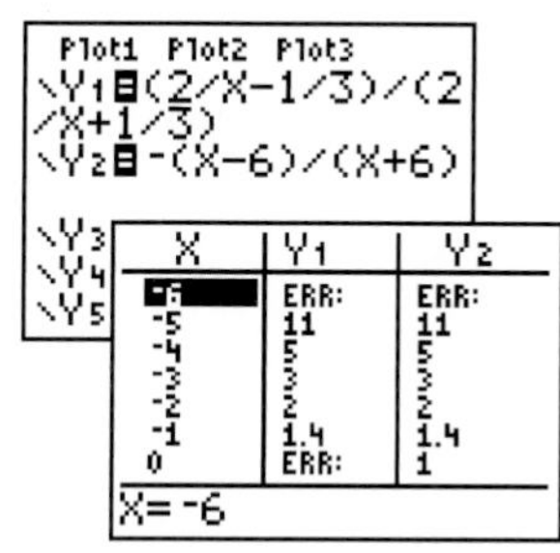

Let Y_1 represent the original expression and Y_2 represent the result.

One or both of the functions generate an error message at the excluded values of -6 and 0. The table indicates that these expressions are identical for all other real numbers. This supports the fact that these two expressions are equal.

Answer: $\left(\frac{2}{x} - \frac{1}{3}\right) \div \left(\frac{2}{x} + \frac{1}{3}\right) = -\frac{x - 6}{x + 6}$

SELF-CHECK 8.4.2

1. Simplify $\left(\frac{2}{x} - \frac{1}{y}\right) \div \left(\frac{6}{x} - \frac{3}{y}\right)$. Assume the variables are restricted to values that prevent division by 0.

Complex Rational Expressions

Make sure the main fraction bar is longer than the other fraction bars to clearly denote the meaning of the expression.

The original expression in Example 3 can be written as either

$$\left(\frac{2}{x} - \frac{1}{3}\right) \div \left(\frac{2}{x} + \frac{1}{3}\right) \qquad \text{or} \qquad \frac{\frac{2}{x} - \frac{1}{3}}{\frac{2}{x} + \frac{1}{3}}$$

The second form above is called a complex rational expression. A **complex rational expression** is a rational expression whose numerator or denominator, or both, is also a rational expression. Each of the following fractions contains more than one fraction bar and is therefore a complex fraction:

$$\frac{7 + \frac{1}{x}}{y} \qquad \frac{8}{19 + \frac{1}{2}} \qquad \frac{\frac{m - n}{2m + 3}}{\frac{m^2 - 5n}{m^2 - 2mn + n^2}}$$

EXAMPLE 4 Simplifying Two Complex Fractions

Simplify each complex fraction.

SOLUTION

(a) $\dfrac{2}{\frac{3}{4}}$

$$\frac{2}{\frac{3}{4}} = 2 \div \frac{3}{4}$$

In part (a) the main fraction bar indicates the numerator is 2 and the denominator is $\frac{3}{4}$.

$$= \frac{2}{1} \cdot \frac{4}{3}$$

To divide by $\frac{3}{4}$, invert the divisor and multiply by $\frac{4}{3}$.

$$= \frac{8}{3}$$

(b) $\dfrac{\frac{2}{3}}{4}$

$$\frac{\frac{2}{3}}{4} = \frac{2}{3} \div 4$$

In part (b) the main fraction bar indicates the numerator is $\frac{2}{3}$ and the denominator is 4.

$$= \frac{\overset{1}{\cancel{2}}}{3} \cdot \frac{1}{\underset{2}{\cancel{4}}}$$

To divide by 4, invert the divisor and multiply by $\frac{1}{4}$.

$$= \frac{1}{6}$$

Note that the expressions in parts (a) and (b) have different values.

SELF-CHECK 8.4.3

Simplify each complex fraction.

1. $\dfrac{\frac{3}{4}}{\frac{5}{6}}$ 2. $\dfrac{\frac{3}{4}}{5}$ 3. $\dfrac{3}{\frac{4}{5}}$

Rewriting Complex Fractions by Using the Division Symbol

There are two useful methods of simplifying complex fractions. Both are worth learning, since some problems can be worked more easily by one method than by the other. Because the fraction bar represents division we can rewrite a complex fraction by using the division symbol. This is the first method we use to simplify complex fractions.

EXAMPLE 5 Simplifying a Complex Rational Expression

Simplify $\dfrac{\frac{a^2 - b^2}{5a + 5b}}{15a}$ by rewriting the expression as a division problem. Assume the variables are restricted to values that prevent division by 0.

SOLUTION

$$\frac{\frac{a^2 - b^2}{5a + 5b}}{15a} = \left(\frac{a^2 - b^2}{1}\right) \div \left(\frac{5a + 5b}{15a}\right)$$

Rewrite this complex rational expression as a division problem. A denominator of 1 is understood for $a^2 - b^2$.

$$= \left(\frac{a^2 - b^2}{1}\right) \cdot \left(\frac{15a}{5a + 5b}\right)$$

Then rewrite the division problem as the product of the dividend and the reciprocal of the divisor.

$$= \frac{\overset{1}{\cancel{(a + b)}}(a - b)}{1} \cdot \frac{\overset{3}{\cancel{15}}a}{\underset{1}{\cancel{5}}\underset{1}{\cancel{(a + b)}}}$$

Factor and reduce by dividing out the common factors.

$$= \frac{3a(a - b)}{1}$$

$$= 3a^2 - 3ab$$

The use of the equal symbol in these examples means that the expressions are equal for all values, except the excluded values.

In Example 6 we first simplify the denominator of the main fraction and then replace the main fraction bar by a division symbol.

EXAMPLE 6 Simplifying a Complex Rational Expression

Simplify $\dfrac{\dfrac{x^2+b^2}{x^2-b^2}}{\dfrac{x-b}{x+b}+\dfrac{x+b}{x-b}}$ by rewriting the expression as a division problem. Assume the variables are restricted to values that prevent division by 0.

SOLUTION

$$\frac{\dfrac{x^2+b^2}{x^2-b^2}}{\left(\dfrac{x-b}{x+b}\right)+\left(\dfrac{x+b}{x-b}\right)}=\frac{\dfrac{x^2+b^2}{x^2-b^2}}{\dfrac{(x-b)^2+(x+b)^2}{(x+b)(x-b)}}$$

First add the two terms in the denominator of the main fraction.

$$=\frac{x^2+b^2}{x^2-b^2}\div\frac{(x^2-2bx+b^2)+(x^2+2bx+b^2)}{(x+b)(x-b)}$$

Then rewrite this complex rational expression as a division problem.

$$=\frac{x^2+b^2}{x^2-b^2}\div\frac{2x^2+2b^2}{(x+b)(x-b)}$$

Simplify the numerator of the divisor.

$$=\frac{\overset{1}{\cancel{x^2+b^2}}}{\underset{1}{\cancel{(x+b)}}\underset{1}{\cancel{(x-b)}}}\cdot\frac{\overset{1}{\cancel{(x+b)}}\overset{1}{\cancel{(x-b)}}}{2\underset{1}{\cancel{(x^2+b^2)}}}$$

Rewrite the division problem by inverting the divisor and multiplying.

$$=\frac{1}{2}$$

Factor and reduce by dividing out the common factors.

SELF-CHECK 8.4.4

Simplify each expression. Assume the variables are restricted to values that prevent division by 0.

1. $\dfrac{\dfrac{x^2-y^2}{15xy^2}}{\dfrac{x-y}{6x^2y}}$

2. $\dfrac{\dfrac{3}{m}-1}{m^2-9}$

Simplifying Complex Fractions by Multiplying by the LCD

Another method for simplifying complex rational expressions is illustrated in Example 7. To use this method, multiply both the numerator and the denominator of the complex fraction by the LCD of all the fractions that occur in both the numerator and the denominator of the complex fraction. The effect of this is to remove all the individual fractions that occur in the numerator and the denominator.

EXAMPLE 7 Simplifying a Complex Rational Expression

Multiplying both the numerator and the denominator by the same value is equivalent to multiplying the fraction by the multiplicative identity 1.

Simplify $\dfrac{1 + \dfrac{2}{x} + \dfrac{1}{x^2}}{1 - \dfrac{1}{x^2}}$ by multiplying both the numerator and the denominator by the LCD of all terms. Assume the variables are restricted to values that prevent division by 0.

SOLUTION

$$\frac{1 + \frac{2}{x} + \frac{1}{x^2}}{1 - \frac{1}{x^2}} = \frac{1 + \frac{2}{x} + \frac{1}{x^2}}{1 - \frac{1}{x^2}} \cdot \frac{x^2}{x^2}$$

By inspection, the LCD of all the fractions in the numerator and the denominator is x^2. Multiply both the numerator and the denominator by this LCD.

$$= \frac{\left(1 + \frac{2}{x} + \frac{1}{x^2}\right)x^2}{\left(1 - \frac{1}{x^2}\right)x^2}$$

$$= \frac{(1)x^2 + \left(\frac{2}{x}\right)x^2 + \left(\frac{1}{x^2}\right)x^2}{(1)x^2 - \left(\frac{1}{x^2}\right)x^2}$$

Using the distributive property, multiply each term by x^2.

$$= \frac{x^2 + 2x + 1}{x^2 - 1}$$

Simplify the numerator and the denominator.

$$= \frac{\cancel{(x + 1)}(x + 1)}{\cancel{(x + 1)}(x - 1)}$$

Factor the numerator and denominator.

$$= \frac{x + 1}{x - 1}$$

Then reduce the fraction to lowest terms.

SELF-CHECK 8.4.5

Simplify each complex fraction. Assume that x is restricted to values that prevent division by 0.

1. $\dfrac{\dfrac{5}{4} - \dfrac{2}{3}}{\dfrac{1}{2} + 6}$

2. $\dfrac{1 - \dfrac{4}{x} + \dfrac{3}{x^2}}{1 - \dfrac{1}{x} - \dfrac{6}{x^2}}$

Because $a^{-n} = \dfrac{1}{a^n}$, fractions that involve negative exponents can be expressed as complex fractions. One way to simplify some expressions involving negative exponents is to first rewrite these expressions in terms of positive exponents. Then you can apply the methods just described for complex fractions.

EXAMPLE 8 Simplifying a Fraction with Negative Exponents

Simplify $\frac{x^{-1} - x^{-2}}{x^{-1} + x^{-2}}$. Assume the variables are restricted to values that prevent division by 0.

SOLUTION

$$\frac{x^{-1} - x^{-2}}{x^{-1} + x^{-2}} = \frac{\frac{1}{x} - \frac{1}{x^2}}{\frac{1}{x} + \frac{1}{x^2}}$$

Rewrite this expression as a complex fraction, using the definition of negative exponents.

$$= \frac{\frac{1}{x} - \frac{1}{x^2}}{\frac{1}{x} + \frac{1}{x^2}} \cdot \frac{x^2}{x^2}$$

Multiply both the numerator and the denominator by x^2, the LCD of all fractions that occur in the numerator and the denominator of the complex fraction.

$$= \frac{\left(\frac{1}{x}\right)x^2 - \left(\frac{1}{x^2}\right)x^2}{\left(\frac{1}{x}\right)x^2 + \left(\frac{1}{x^2}\right)x^2}$$

Use the distributive property to multiply each term by x^2.

$$= \frac{x - 1}{x + 1}$$

Simplify the numerator and the denominator.

In Example 8 the LCD of all the fractions within the numerator and the denominator of the complex fraction is x^2. Thus to simplify this complex fraction, we multiplied both the numerator and the denominator by x^2. This is exactly what we do in Example 9, but without going through the work of writing the expression as a complex fraction. The alternative strategy illustrated in Example 9 is particularly appropriate when negative exponents are applied only to monomials. The key is to select the appropriate expression to multiply by both the numerator and the denominator of the given fraction. The appropriate LCD to use as a factor can be determined by inspecting all the negative exponents in the original expression.

EXAMPLE 9 Using an Alternative Method for Simplifying a Fraction with Negative Exponents

From definitions in Chapter 1 we know that for any nonzero real number x and exponent n, $x^0 = 1$ and $x^{-n} = \frac{1}{x^n}$.

Simplify $\frac{x^{-1} - x^{-2}}{x^{-1} + x^{-2}}$. Assume the variables are restricted to values that prevent division by 0.

SOLUTION

$$\frac{x^{-1} - x^{-2}}{x^{-1} + x^{-2}} = \frac{x^{-1} - x^{-2}}{x^{-1} + x^{-2}} \cdot \frac{x^2}{x^2}$$

Multiply both the numerator and the denominator by x^2. Note that this is the lowest power of x that will eliminate all the negative exponents on x in both the numerator and the denominator.

$$= \frac{(x^{-1})x^2 - (x^{-2})x^2}{(x^{-1})x^2 + (x^{-2})x^2}$$

Use the distributive property to multiply each term by x^2.

$$= \frac{x - x^0}{x + x^0}$$

$$= \frac{x - 1}{x + 1}$$

Then simplify by replacing x^0 with 1.
Note that this result is the same as the result in Example 8.

SELF-CHECK 8.4.6

Assume that x is restricted to values that prevent division by 0.

1. Determine the lowest power of x to multiply by both the numerator and the denominator to eliminate all the negative exponents in $\frac{x^{-1} - 4x^{-2} - 21x^{-3}}{x^{-1} + 6x^{-2} + 9x^{-3}}$.
2. Reduce the expression in part 1 to lowest terms.

SELF-CHECK ANSWERS

8.4.1 1. $\frac{x^4 - 6x}{12}$ 2. $-\frac{x}{6}$

8.4.2 1. $\frac{1}{3}$

8.4.3 1. $\frac{9}{10}$ 2. $\frac{3}{20}$ 3. $\frac{15}{4}$

8.4.4 1. $\frac{2x^2 + 2xy}{5y}$ 2. $-\frac{1}{m(m + 3)}$

8.4.5 1. $\frac{7}{78}$ 2. $\frac{x - 1}{x + 2}$

8.4.6 1. x^3 2. $\frac{x - 7}{x + 3}$

USING THE LANGUAGE AND SYMBOLISM OF MATHEMATICS 8.4

1. When one is simplifying a rational expression, the correct order of operations is essential. This order is as follows:
   a. Start with the expression within the innermost pair of ________ ________.
   b. Perform all ________.
   c. Perform all ________ and ________ as they appear from left to right.
   d. Perform all ________ and ________ as they appear from left to right.
2. A complex rational expression is a rational expression whose ________ or ________, or both, is also a rational expression.
3. Dividing by a fraction is equivalent to multiplying by the ________ of that fraction.
4. Two rational expressions must have the same denominator when combined by using the operations of ________ or ________.

QUICK REVIEW 8.4

The purpose of this quick review is to help you recall skills needed in this section.

Determine the LCD of each pair of rational expressions.

1. $\frac{3}{20}$ and $\frac{17}{45}$
2. $\frac{x}{x - y}$ and $\frac{y}{x + y}$

Simplify each expression without using a calculator.

3. $-5^2 + 2(11 - 4)$
4. $(-5)^2 + 2(11) - 4$
5. Rewrite $x^{-1} + 3x^{-2}$ by using positive exponents.

EXERCISES 8.4

In Exercises 1–40, simplify each expression. Assume the variables are restricted to values that prevent division by 0.

1. a. $\frac{2x}{3} + \frac{x^2}{15} \cdot \frac{5}{x}$ b. $\left(\frac{2x}{3} + \frac{x^2}{15}\right) \cdot \frac{5}{x}$
2. a. $\frac{3x}{4} - \frac{x^2}{12} \cdot \frac{3}{x}$ b. $\left(\frac{3x}{4} - \frac{x^2}{12}\right) \cdot \frac{3}{x}$
3. a. $-\frac{2y}{3} + \frac{y^2}{6} \div \frac{y}{9}$ b. $\left(-\frac{2y}{3} + \frac{y^2}{6}\right) \div \frac{y}{9}$
4. a. $\frac{3y}{5} - \frac{2y^2}{3} \div \frac{y}{10}$ b. $\left(\frac{3y}{5} - \frac{2y^2}{3}\right) \div \frac{y}{10}$
5. a. $\frac{v}{v - 6} - \frac{3}{v + 2} \cdot \frac{2v + 4}{v^2 - 6v}$ b. $\left(\frac{v}{v - 6} - \frac{3}{v + 2}\right) \cdot \frac{2v + 4}{v^2 - 6v}$
6. a. $\frac{v}{v + 10} + \frac{5v}{v - 3} \cdot \frac{2v - 6}{v^2 + 10v}$ b. $\left(\frac{v}{v + 10} + \frac{5v}{v - 3}\right) \cdot \frac{2v - 6}{v^2 + 10v}$
7. a. $\dfrac{\frac{4}{6}}{\frac{5}{8}}$ b. $\dfrac{4}{\frac{6}{5}}$

8. a. $\dfrac{\frac{3}{4}}{\frac{5}{6}}$ b. $\dfrac{\frac{4}{6}}{5}$

9. $\dfrac{1+\frac{1}{5}}{1-\frac{1}{5}}$

10. $\dfrac{\frac{1}{2}+\frac{2}{3}}{\frac{3}{2}+2}$

11. $\dfrac{\frac{3}{5}-\frac{5}{3}}{\frac{1}{3}+\frac{1}{5}}$

12. $\dfrac{\frac{7}{10}-\frac{3}{4}}{\frac{3}{8}+\frac{1}{3}}$

13. $\dfrac{2-3x}{2x-3}+\dfrac{8-2x}{3x-6}\div\dfrac{4x-6}{3x-6}$

14. $\dfrac{6z+3}{3z-1}-\dfrac{5}{9z^2-1}\div\dfrac{1}{3z+1}$

15. $\left(\dfrac{2}{v+3}+v\right)\left(v-\dfrac{3}{v+2}\right)$

16. $\left(v-\dfrac{6v+35}{v+4}\right)\left(v-\dfrac{44}{v-7}\right)$

17. $\left(x-2-\dfrac{3}{x}\right)\div\left(1+\dfrac{1}{x}\right)$

18. $\left(\dfrac{w+3}{2w+1}-2w\right)\div\left(\dfrac{w-1}{2w+1}+2w\right)$

19. $1-\dfrac{4}{x}-\left(1-\dfrac{2}{x}\right)^2$

20. $4-\dfrac{12}{x}-\left(2-\dfrac{3}{x}\right)^2$

21. $\left(1+\dfrac{3}{5x}\right)^2-\left(1-\dfrac{3}{5x}\right)^2$

22. $\left(1+\dfrac{2}{3x}\right)^2-\left(1-\dfrac{2}{3x}\right)^2$

23. $\dfrac{\frac{12x^2}{5y}}{\frac{16x^2}{15y^2}}$

24. $\dfrac{\frac{18a^3}{25x}}{\frac{24a^3}{35x^2}}$

25. $\dfrac{\frac{w-z}{x^2y^2}}{\frac{w^2-z^2}{2xy}}$

26. $\dfrac{\frac{a^2-b^2}{6a^2b^3}}{\frac{a+b}{9a^3b}}$

27. $\dfrac{\frac{x^2-9x+14}{34x^4}}{\frac{5x^2-20}{17x^5}}$

28. $\dfrac{\frac{5x^2+10xy+5y^2}{15x^2-15y^2}}{\frac{9x^2-9y^2}{25x^2-50xy+25y^2}}$

29. $\dfrac{2-\frac{1}{x}}{4-\frac{1}{x^2}}$

30. $\dfrac{\frac{1}{x^2}-49}{7-\frac{1}{x}}$

31. $\dfrac{vw}{\frac{1}{v}+\frac{1}{w}}$

32. $\dfrac{\frac{1}{v}-\frac{1}{w}}{vw}$

33. $\dfrac{\frac{3}{x^2}-\frac{6}{x}+3}{18-\frac{18}{x^2}}$

34. $\dfrac{\frac{1}{x}-\frac{8}{x^2}+\frac{15}{x^3}}{1-\frac{5}{x}}$

35. $\dfrac{3+\frac{9}{x}}{\frac{15}{x^3}+\frac{8}{x^2}+\frac{1}{x}}$

36. $\dfrac{x+2-\frac{6}{2x+3}}{x+\frac{8x}{2x-1}}$

37. $\dfrac{\frac{w-a}{w+a}-\frac{w+a}{w-a}}{\frac{w^2+a^2}{w^2-a^2}}$

38. $\dfrac{w-3+\frac{2}{w-6}}{w-2-\frac{22}{w+7}}$

39. $\dfrac{\frac{12}{v^2}+\frac{1}{v}-1}{\frac{24}{v^2}-\frac{2}{v}-1}$

40. $\dfrac{\frac{12}{v^2}-\frac{1}{v}-1}{\frac{8}{v^2}+\frac{6}{v}+1}$

Estimation Skills and Calculator Skills

In Exercises 41–44, mentally estimate the value of each expression, and then use a calculator to approximate each value to the nearest thousandth.

PROBLEM	MENTAL ESTIMATE	CALCULATOR APPROXIMATION
41. $\dfrac{\frac{211}{429}}{\frac{107}{325}}$		
42. $\dfrac{\frac{211}{107}}{\frac{899}{298}}$		
43. $\dfrac{\frac{401}{99}}{\frac{499}{101}}$		
44. $\dfrac{\frac{1{,}022}{350}}{\frac{1{,}017}{204}}$		

In Exercises 45–60, simplify each expression. Assume the variables are restricted to values that prevent division by 0.

45. $\dfrac{a-b}{ab}+\dfrac{(a+b)^2}{a^3b^3}\cdot\dfrac{a^2b^2}{4a^2+8ab+4b^2}$

46. $\dfrac{a+2}{a+3}-\dfrac{(a-2)^3}{a^2-9}\div\dfrac{a^2-4a+4}{a-3}$

47. $\dfrac{x^{-2}}{x^{-2}+y^{-2}}$

48. $\dfrac{x^{-2}-y^{-2}}{y^{-2}}$

49. $\dfrac{v^{-1}+w^{-1}}{v^{-1}-w^{-1}}$

50. $\dfrac{m^2-n^2}{m^{-1}+n^{-1}}$

51. $\dfrac{m^2-n^2}{m^{-1}-n^{-1}}$

52. $\dfrac{(x+y)^2}{x^{-1}+y^{-1}}$

53. $\dfrac{a^{-2}-b^{-2}}{a^{-1}-b^{-1}}$

54. $\dfrac{b-a^{-1}}{a-b^{-1}}$

55. $\dfrac{x^{-1}y^{-2}+x^{-2}y^{-1}}{y^{-2}-x^{-2}}$

56. $\dfrac{x^{-1}y^{-2}+x^{-2}y^{-1}}{x^{-1}+y^{-1}}$

57. $\dfrac{\dfrac{\dfrac{1}{a} - \dfrac{1}{b}}{ab}}{\dfrac{1 - \dfrac{a}{b}}{1 + \dfrac{a}{b}}}$

58. $\dfrac{\dfrac{\dfrac{1}{x} + \dfrac{1}{y}}{\dfrac{1}{x}}}{\dfrac{y}{x} + 1}$

59. $\dfrac{\dfrac{1 + \dfrac{x}{y}}{1 - \dfrac{x}{y}}}{\dfrac{x + y}{x - \dfrac{y^2}{x}}}$

60. $\dfrac{\dfrac{1 + \dfrac{x}{y}}{1 - \dfrac{x}{y}}}{\dfrac{x - \dfrac{y^2}{x}}{x + y}}$

In Exercises 61–64, use the table of values for each expression to match each expression with its simplified form.

61. $\dfrac{3}{x} + \dfrac{1}{4} \div \dfrac{3}{x} - \dfrac{1}{4}$

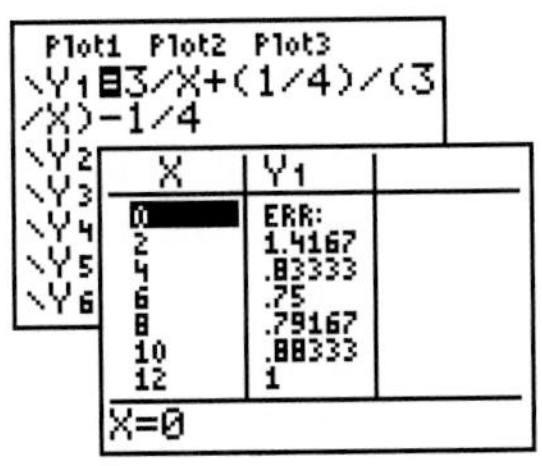

62. $\left(\dfrac{3}{x} + \dfrac{1}{4}\right) \div \left(\dfrac{3}{x} - \dfrac{1}{4}\right)$

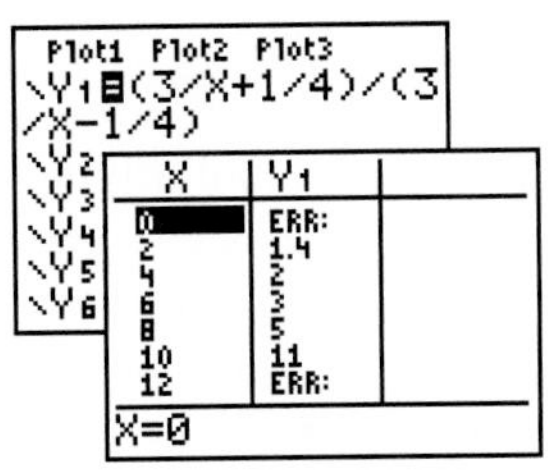

63. $\dfrac{3}{x} + \dfrac{1}{4} \div \left(\dfrac{3}{x} - \dfrac{1}{4}\right)$

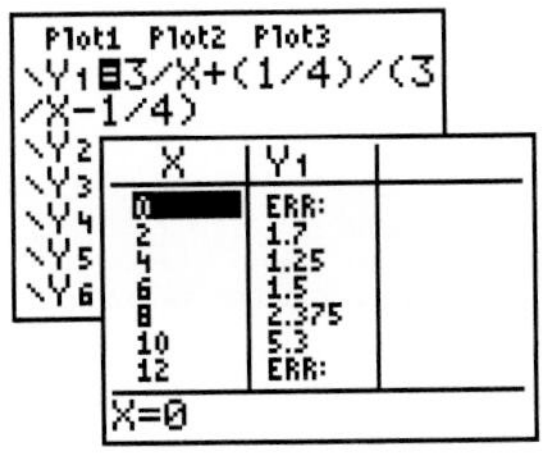

64. 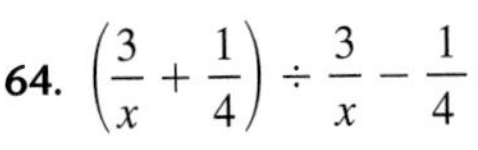
$\left(\dfrac{3}{x} + \dfrac{1}{4}\right) \div \dfrac{3}{x} - \dfrac{1}{4}$

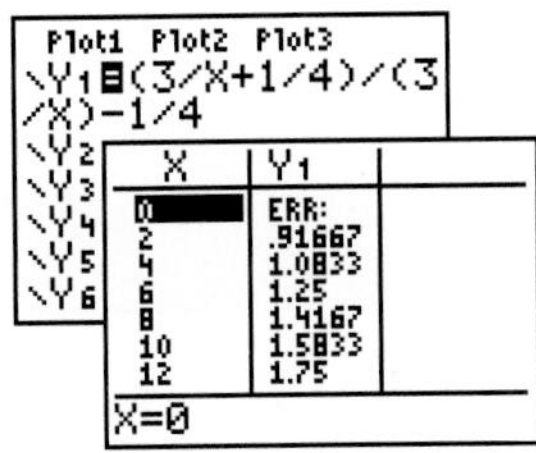

A. $\dfrac{x + 9}{12}$

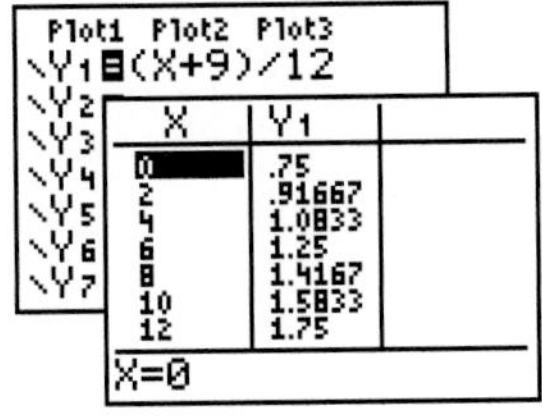

B. 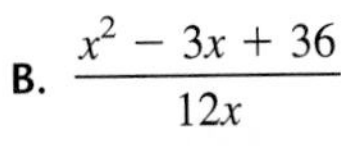
$\dfrac{x^2 - 3x + 36}{12x}$

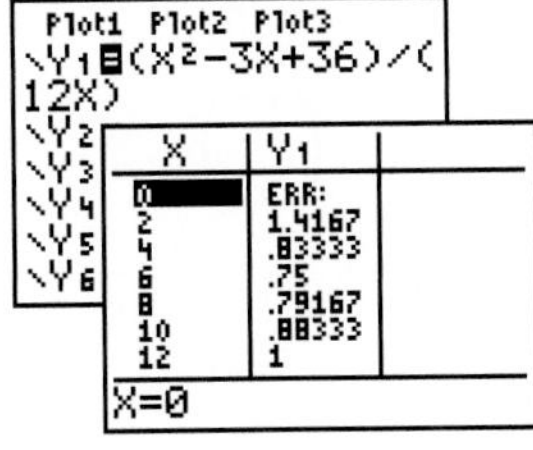

C. $-\dfrac{x + 12}{x - 12}$

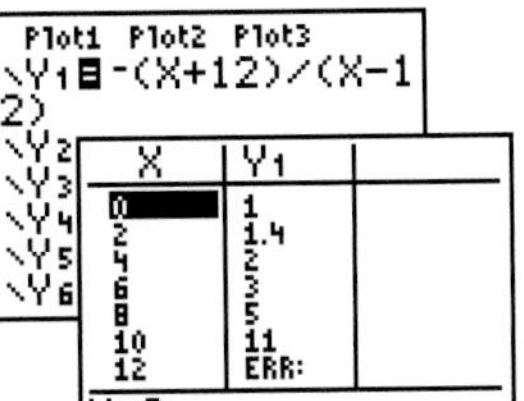

D. 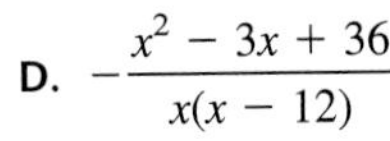
$-\dfrac{x^2 - 3x + 36}{x(x - 12)}$

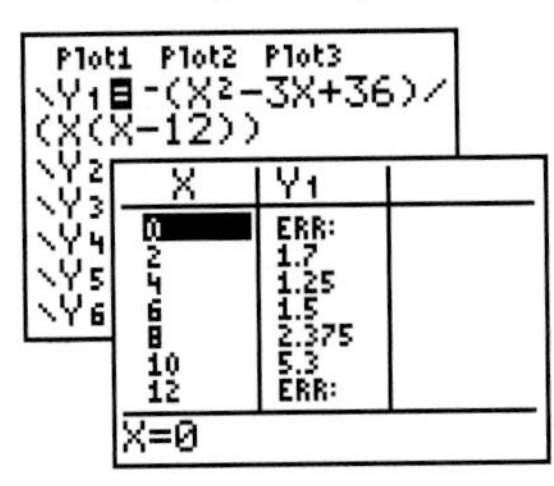

65. **Area of a Trapezoid** The formula for the area of a trapezoid is $A = \dfrac{h}{2}(a + b)$. Use this formula to find an expression that represents the area of the trapezoid to the right.

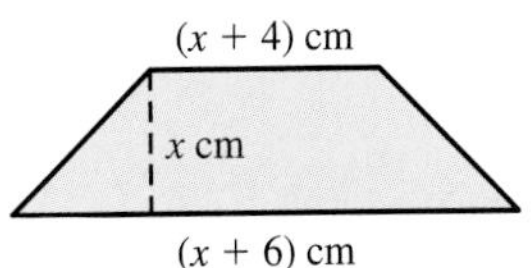

66. **Total Area of Two Regions** Write an expression that represents the total area of the rectangle and triangle shown.

$\dfrac{x}{x^2 - 1}$ cm

x cm

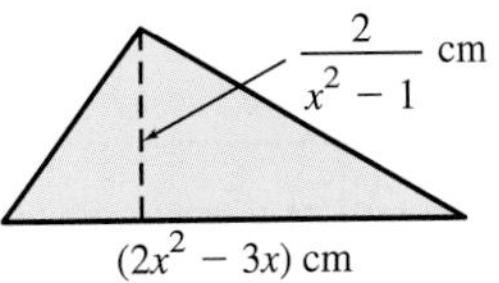

67. **Average Cost** The number of units of patio chairs that a factory can make by operating t h/wk is given by $N(t) = 5t - 10$. The cost of operating the factory t h/wk is given by $C(t) = 8{,}400 - \dfrac{16{,}800}{t}$.

a. Write a rational function $A(t)$ that gives the average cost per unit when the plant operates t h/wk.

b. Determine the practical domain for the values of t.

c. Use the graph of this function to describe what happens to the average cost as t becomes larger.

68. **Average Cost** The number of units of patio tables a factory can make by operating t h/wk is given by $N(t) = 2t - 4$. The cost of operating the factory t h/wk is given by $C(t) = 5{,}040 - \dfrac{10{,}080}{t}$.

a. Write a rational function $A(t)$ that gives the average cost per unit when the plant operates t h/wk.

b. Determine the practical domain for the values of t.

c. Use the graph of this function to describe what happens to the average cost as t becomes larger.

69. **Lens Formula** The relationship between the focal length f of a lens, the distance d_o of an object from the lens, and the distance d_i of an image from the lens is $\dfrac{1}{f} = \dfrac{1}{d_o} + \dfrac{1}{d_i}$. So $f = \dfrac{1}{\dfrac{1}{d_o} + \dfrac{1}{d_i}}$.

a. Give a simplified rational expression for the focal length f.

b. Determine f when the object distance is 20 ft and the image distance is 0.5 ft. Round to the nearest thousandth.

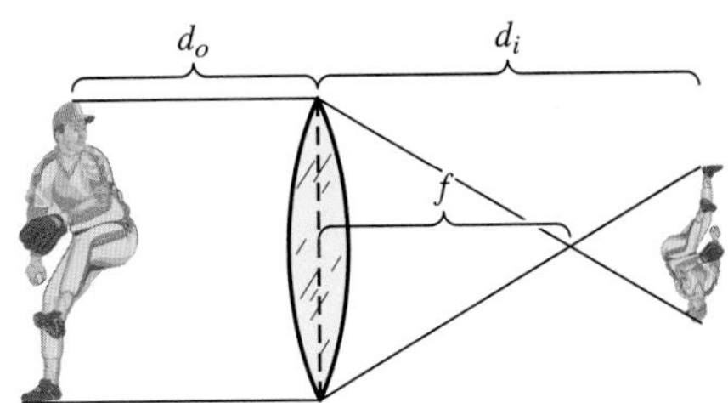

70. **Electrical Resistance** The total resistance R in a parallel circuit with two individual resistors r_1 and r_2 can be calculated by using the formula $\frac{1}{R} = \frac{1}{r_1} + \frac{1}{r_2}$. So

$$R = \frac{1}{\frac{1}{r_1} + \frac{1}{r_2}}.$$

a. Give a simplified rational expression for the total resistance R.

b. Determine R when the $r_1 = 8$ ohms (Ω) and $r_2 = 56\ \Omega$.

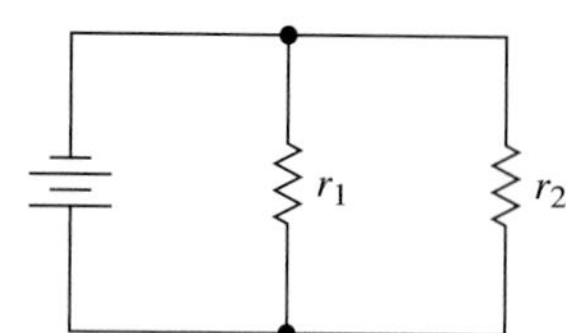

Group Discussion Questions

71. **Challenge Question** Simplify each expression. Assume that m and n are natural numbers and that x and y are restricted to values that prevent division by 0.

a. $\dfrac{x^{-m} + y^{-n}}{x^{-m}y^{-n}}$ b. $\dfrac{x^{-m} - y^{-n}}{x^m - y^n}$

c. $\dfrac{x^{-m} - x^{-2m}}{x^{-m} + x^{-2m}}$ d. $\dfrac{x^{-m} + 3x^{-2m}}{x^{-m} - 2x^{-2m}}$

72. **Error Analysis** A student simplified the expression $\dfrac{x^{-1} - y^{-1}}{x^{-2} - y^{-2}}$ as follows:

$$\frac{x^{-1} - y^{-1}}{x^{-2} - y^{-2}} = \frac{x^2 - y^2}{x - y} = \frac{(x + y)(x - y)}{x - y} = x + y$$

Correct this error and then finish simplifying the expression.

Section 8.5 Solving Equations Containing Rational Expressions

Objective: 8. Solve equations containing rational expressions.

Equivalent equations are equations with exactly the same set of solutions. Multiplying both sides of an equation by any nonzero number produces an equivalent equation. Multiplying both sides of an equation by 0 produces $0 = 0$.

In Section 2.5 we solved fractional equations by multiplying both sides of the equation by the least common denominator. This method produces an equation equivalent to the original one as long as we do not multiply by 0. Multiplying by an expression that is equal to 0 can produce an equation that is *not* equivalent to the original equation and thus can produce an extraneous value. An **extraneous value** is a value obtained in the solution procedure that is not a solution of the original equation.

When we solve an equation with a variable in the denominator, we must check each possible solution. We do this to verify that our answer does not include a value that would cause division by 0.

Solving an Equation Containing Rational Expressions

VERBALLY	ALGEBRAIC EXAMPLE	
1. Multiply both sides of the equation by the LCD.	Solve: $\frac{21}{x} - \frac{15}{x} = 2$	The only excluded value is 0.
2. Solve the resulting equation.	$x\left(\frac{21}{x} - \frac{15}{x}\right) = x(2)$	Multiply both sides of the equation by the LCD, x.
3. Check the solution to determine whether it is an excluded value and therefore extraneous.	$x\left(\frac{21}{x}\right) - x\left(\frac{15}{x}\right) = 2x$	
	$21 - 15 = 2x$	
	$6 = 2x$	
	$3 = x$	Since 3 is not an excluded value, it will check.
	Check: $\frac{21}{3} - \frac{15}{3} \stackrel{?}{=} 2$	
	$7 - 5 \stackrel{?}{=} 2$	
	$2 \stackrel{?}{=} 2$ checks	
	Answer: $x = 3$	

EXAMPLE 1 Solving an Equation Containing Rational Expressions

The solution of an equation cannot include any excluded values.

Solve $\dfrac{3}{x-7} + 5 = \dfrac{8}{x-7}$ algebraically and numerically by using a table of values.

SOLUTION

ALGEBRAICALLY

$$\frac{3}{x-7} + 5 = \frac{8}{x-7}$$

Note that $x = 7$ is the only excluded value.

$$(x-7)\left(\frac{3}{x-7} + 5\right) = (x-7)\left(\frac{8}{x-7}\right)$$

Multiply both sides of the equation by the LCD, $x - 7$. The LCD is nonzero since 7 is an excluded value.

$$\cancel{(x-7)}\left(\frac{3}{\cancel{x-7}}\right) + (x-7)(5) = \cancel{(x-7)}\left(\frac{8}{\cancel{x-7}}\right)$$

Use the distributive property to multiply each term by $x - 7$. Then combine like terms.

$$3 + 5x - 35 = 8$$
$$5x = 40$$
$$x = 8$$

Since $x = 8$ is not an excluded value, this solution will check.

NUMERICALLY

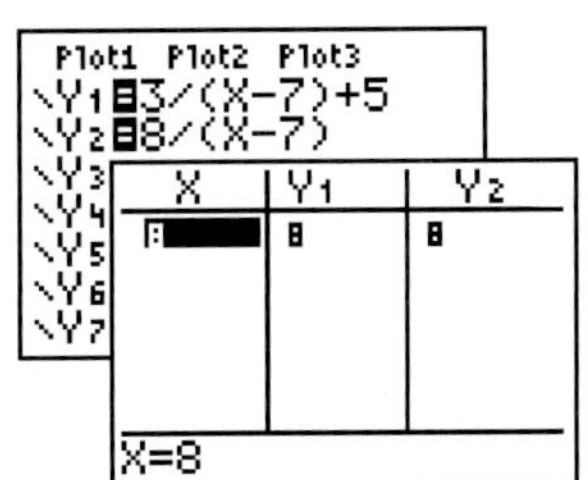

Let Y_1 represent the left side of the equation and Y_2 represent the right side of the equation.

Use the **ASK** mode to confirm that for $x = 8$, Y_1 and Y_2 are equal. This confirms that the algebraic solution above checks. *Caution:* Unless you already know how many solutions this equation has, this table will not confirm that $x = 8$ is the only solution.

Caution: A table can locate or confirm a solution of an equation. However, unless we already know how many solutions the equation has, we will not know if the solution shown in the table is the only solution.

Answer: $x = 8$

The solution process in Example 2 produces an excluded value for the variable. This value cannot be a solution; it is extraneous.

EXAMPLE 2 Solving an Equation with an Extraneous Value

Solve $\dfrac{v}{v-3} = 4 - \dfrac{3}{3-v}$.

SOLUTION

$$\frac{v}{v-3} = 4 - \frac{3}{3-v}$$

$$\frac{v}{v-3} = 4 + \frac{3}{v-3}$$

Note that the only excluded value is $v = 3$. Noting that the denominators are opposites, change $-\dfrac{3}{3-v}$ to $\dfrac{3}{v-3}$.

$$(v-3)\left(\frac{v}{v-3}\right) = (v-3)\left(4 + \frac{3}{v-3}\right)$$

Multiply both sides of the equation by the nonzero LCD, $v - 3$. The LCD is nonzero since 3 is an excluded value.

$$\cancel{(v-3)}\left(\frac{v}{\cancel{v-3}}\right) = (v-3)(4) + \cancel{(v-3)}\left(\frac{3}{\cancel{v-3}}\right)$$

Use the distributive property to multiply each term by $v - 3$.

$$v = 4v - 12 + 3$$
$$-3v = -9$$
$$v = 3$$

Solve the resulting equation for v.

Since $v = 3$ is the excluded value noted above, this value is extraneous.

This value causes division by 0 in the original equation so there is no solution.

Answer: There is no solution.

SELF-CHECK 8.5.1

Solve each equation.

1. $\frac{x-12}{x-4} = \frac{2x}{4-x} + 2$ **2.** $\frac{2y}{y+2} = \frac{y}{y+2} - 3$

The solution process in Example 3 produces two values. One of these is an excluded value and cannot be a solution while the other solution will check.

EXAMPLE 3 Solving an Equation with an Extraneous Value

Solve $\frac{y+1}{y+2} + \frac{y}{y-2} + 1 = \frac{8}{y^2-4}$.

SOLUTION

$$\frac{y+1}{y+2} + \frac{y}{y-2} + 1 = \frac{8}{y^2-4}$$

$$\frac{y+1}{y+2} + \frac{y}{y-2} + 1 = \frac{8}{(y+2)(y-2)}$$

Factor the denominator of the right side of the equation to determine the LCD, $(y+2)(y-2)$, and the excluded values -2 and 2.

$$(y+2)(y-2)\left[\frac{y+1}{y+2} + \frac{y}{y-2} + 1\right] = (y+2)(y-2)\left[\frac{8}{(y+2)(y-2)}\right]$$

Multiply both sides of the equation by the LCD, $(y+2)(y-2)$. The LCD is nonzero since -2 and 2 are excluded values.

$$(y+2)(y-2)\left(\frac{y+1}{y+2}\right) + (y+2)(y-2)\left(\frac{y}{y-2}\right) + (y+2)(y-2)(1) = 8$$

Use the distributive property to multiply each term by $(y+2)(y-2)$.

$$(y-2)(y+1) + (y+2)(y) + (y+2)(y-2)(1) = 8$$

$$y^2 - y - 2 + y^2 + 2y + y^2 - 4 = 8$$

$$3y^2 + y - 14 = 0$$

$$(3y+7)(y-2) = 0$$

$$3y + 7 = 0 \quad \text{or} \quad y - 2 = 0$$

$$y = -\frac{7}{3} \qquad y = 2$$

Write the resulting quadratic equation in standard form, and then solve it by factoring.

The value $y = 2$ causes division by 0 in the original equation and is therefore an extraneous value.

Because $y = 2$ is an excluded value, it cannot be a solution.

Answer: $y = -\frac{7}{3}$

The value $y = -\frac{7}{3}$ is not an excluded value. Therefore it should check.

Does $x = -\frac{7}{3}$ check?

If the steps in the solution of an equation produce a contradiction, then the original equation has no solution. Example 4 shows how to interpret the answer when the steps in the solution of an equation produce an identity.

EXAMPLE 4 Solving an Equation That Simplifies to an Identity

Solve $\frac{3}{x-4} - \frac{2}{x-2} = \frac{x+2}{x^2-6x+8}$.

SOLUTION

$$\frac{3}{x-4} - \frac{2}{x-2} = \frac{x+2}{x^2-6x+8}$$

Factor the denominator of the right side of the equation to determine the LCD, $(x-4)(x-2)$, and the excluded values 2 and 4.

$$\frac{3}{x-4} - \frac{2}{x-2} = \frac{x+2}{(x-4)(x-2)}$$

$$(x-4)(x-2)\left[\frac{3}{x-4} - \frac{2}{x-2}\right] = \cancel{(x-4)(x-2)}\left[\frac{x+2}{\cancel{(x-4)(x-2)}}\right]$$

$$\cancel{(x-4)}(x-2)\left(\frac{3}{\cancel{x-4}}\right) - (x-4)\cancel{(x-2)}\left(\frac{2}{\cancel{x-2}}\right) = x+2$$

Multiply both sides of the equation by the LCD, $(x-4)(x-2)$. The LCD is nonzero since 2 and 4 are excluded values. Use the distributive property to multiply each term by $(x-4)(x-2)$.

$$3(x-2) - 2(x-4) = x+2$$
$$3x - 6 - 2x + 8 = x+2$$
$$x + 2 = x + 2$$

Simplify, and then solve the equation.

$$0 = 0 \quad \text{(an identity)}$$

This simplified equation is an identity. Thus all real numbers are solutions of the original equation except for the values excluded to avoid division by 0. The solution is all real numbers except 2 and 4. In interval notation the answer is $(-\infty, 2) \cup (2, 4) \cup (4, \infty)$.

Answer: $\mathbb{R} \sim \{2, 4\}$

SELF-CHECK 8.5.2

Solve each equation.

1. $\frac{2}{x} + \frac{1}{x-1} = \frac{3x-2}{x^2-x}$

2. $\frac{2}{x} + \frac{1}{x-1} = \frac{3x+2}{x^2-x}$

3. $\frac{2}{x} + \frac{1}{x-1} = \frac{4}{x^2-x}$

The resistance R in a parallel circuit with two individual resistors r_1 and r_2 can be calculated by using the formula $\frac{1}{R} = \frac{1}{r_1} + \frac{1}{r_2}$. We use this formula to determine appropriate values for r_1 and r_2.

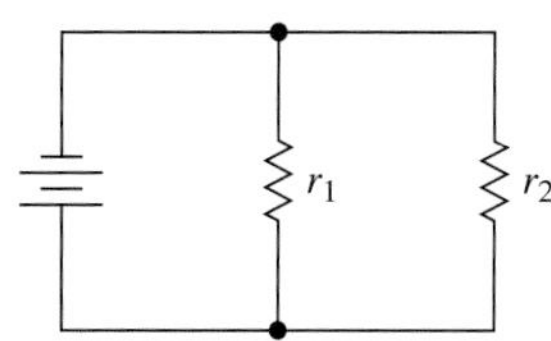

EXAMPLE 5 Calculating Resistances in a Parallel Circuit

The resistance planned by an electrical technician for a parallel electric circuit with two resistors is 4 Ω. The resistance of r_2 must be 15 Ω greater than that of r_1. Determine the resistance of each resistor.

SOLUTION

Let $r =$ resistance of first resistor, Ω
$r + 15 =$ resistance of second resistor, Ω

The resistance of the second resistor is 15 Ω greater than that of the first resistor. Note that we use one variable r rather than two variables r_1 and r_2. We use r in place of r_1 in the formula and $r + 15$ in place of r_2.

$$\frac{1}{r_1} + \frac{1}{r_2} = \frac{1}{R}$$

$$\frac{1}{r} + \frac{1}{r+15} = \frac{1}{4}$$

Substitute the total resistance of 4 Ω and the identified variables into the resistance formula.

$$4r(r+15)\left(\frac{1}{r} + \frac{1}{r+15}\right) = 4r(r+15)\left(\frac{1}{4}\right)$$

Multiply both sides of the equation by the LCD, $4r(r + 15)$.

$$4r(r+15)\left(\frac{1}{r}\right) + 4r(r+15)\left(\frac{1}{r+15}\right) = r(r+15)$$

Use the distributive property to multiply each term by $4r(r + 15)$.

$$4(r+15) + 4r = r^2 + 15r$$

$$4r + 60 + 4r = r^2 + 15r$$

$$0 = r^2 + 7r - 60 \quad \text{or} \quad r^2 + 7r - 60 = 0$$

$$(r-5)(r+12) = 0$$

Simplify and write this quadratic equation in standard form. Then solve this equation by factoring.

$$r - 5 = 0 \quad \text{or} \quad r + 12 = 0$$

$$r = 5 \quad \text{or} \quad r = -12$$

$$r + 15 = 20$$

The value of 5 Ω seems reasonable for r (we will check below). However the resistance of −12 Ω is not reasonable because resistance cannot be negative.

Check: $\frac{1}{5} + \frac{1}{20} \stackrel{?}{=} \frac{1}{4}$

$0.25 \stackrel{?}{=} 0.25$ checks

Answer: The resistance of the first resistor is 5 Ω, and the resistance of the second resistor is 20 Ω.

In Section 2.6 we practiced rewriting formulas to solve for a specified variable. Example 6 involves rewriting a formula that includes a rational expression. To solve for b, we must isolate this variable in the numerator on one side of the equation.

EXAMPLE 6 Solving for a Specified Variable

Solve each equation for b.

(a) $\frac{2}{a} = \frac{3}{b} - \frac{1}{ab}$ **(b)** $\frac{1}{a} + \frac{1}{c} = \frac{1}{b}$

SOLUTION

(a)

$$\frac{2}{a} = \frac{3}{b} - \frac{1}{ab}$$

$$ab\left(\frac{2}{a}\right) = ab\left(\frac{3}{b} - \frac{1}{ab}\right)$$

To get b in the numerator, multiply both sides of the equation by the LCD, ab.

$$ab\left(\frac{2}{a}\right) = ab\left(\frac{3}{b}\right) - ab\left(\frac{1}{ab}\right)$$

Use the distributive property to multiply each term by ab.

$$2b = 3a - 1$$

$$b = \frac{3a-1}{2}$$

Simplify, and then divide both sides of the equation by 2 to isolate b on the left side with a coefficient of 1.

(b)

$$\frac{1}{a} + \frac{1}{c} = \frac{1}{b}$$

$$abc\left(\frac{1}{a} + \frac{1}{c}\right) = abc\left(\frac{1}{b}\right)$$ To get b in the numerator, multiply both sides of the equation by the LCD, abc.

$$abc\left(\frac{1}{a}\right) + abc\left(\frac{1}{c}\right) = abc\left(\frac{1}{b}\right)$$ Use the distributive property to multiply each term by abc.

$$bc + ab = ac$$
$$b(c + a) = ac$$ Simplify and then use the distributive property to factor b out of each term on the left side of the equation.

$$\frac{b(c + a)}{c + a} = \frac{ac}{c + a}$$ Divide both sides of the equation by $c + a$ to isolate b on the left side of the equation.

$$b = \frac{ac}{a + c}$$ Simplify the left side of the equation, and on the right side write the terms in the denominator in alphabetical order.

SELF-CHECK 8.5.3

Solve each equation for x.

1. $z = \dfrac{vw}{x}$
2. $\dfrac{w}{x} = \dfrac{v}{y}$
3. $\dfrac{2v}{x + 3y} = z$
4. $\dfrac{1}{x} + \dfrac{1}{w} = \dfrac{1}{y}$

SELF-CHECK ANSWERS

8.5.1
1. There is no solution.
2. $y = -\dfrac{3}{2}$

8.5.2
1. $\mathbb{R} \sim \{0, 1\}$
2. There is no solution.
3. $x = 2$

8.5.3
1. $x = \dfrac{vw}{z}$
2. $x = \dfrac{wy}{v}$
3. $x = \dfrac{2v - 3yz}{z}$
4. $x = \dfrac{wy}{w - y}$

USING THE LANGUAGE AND SYMBOLISM OF MATHEMATICS 8.5

1. Equivalent equations have exactly the __________ set of solutions.
2. To produce equivalent equations, we can multiply both sides of the equation by the same number as long as we do not multiply by __________.
3. To solve an equation containing rational expressions:
   a. Multiply both sides of the equation by the __________.
   b. __________ the resulting equation.
   c. Check the solution to determine whether it is an excluded value and therefore __________.
4. An excluded value for an equation containing rational expressions is one that causes division by __________.

QUICK REVIEW 8.5

The purpose of this quick review is to help you recall skills needed in this section.

Solve each linear equation.

1. $\dfrac{x}{2} - \dfrac{x}{3} = 1$
2. $x = \dfrac{36}{15} + \dfrac{x}{5}$
3. $\dfrac{x - 1}{3} - \dfrac{x + 3}{4} = 2$
4. $\dfrac{2x + 3}{5} - 1 = \dfrac{x - 2}{3}$
5. $\dfrac{x}{2} + 5 = \dfrac{x}{2}$

EXERCISES 8.5

In Exercises 1 and 2, determine the values excluded from the domain of the variable because they would cause division by 0.

1. a. $\dfrac{3}{m-3} + 5 = \dfrac{2}{m-2}$

   b. $\dfrac{3y}{(2y+3)(3y-2)} - \dfrac{7}{2y+3} = \dfrac{y-1}{3y-2}$

   c. $\dfrac{4y-5}{2y^2+5y-3} = \dfrac{5y-4}{6y^2-y-1}$

2. a. $\dfrac{2}{n+2} + 3 = \dfrac{2}{n-5}$

   b. $\dfrac{y+2}{4y^2-13y+3} = \dfrac{y}{4y-1} - \dfrac{5}{y-3}$

   c. $\dfrac{7y}{6y^2-5y-6} = \dfrac{9y-1}{6y^2-11y-10}$

In Exercises 3–6, each equation has exactly one solution. Use the given table to determine this solution.

3. $\dfrac{x}{x+3} = \dfrac{1}{x+3} + \dfrac{1}{3}$

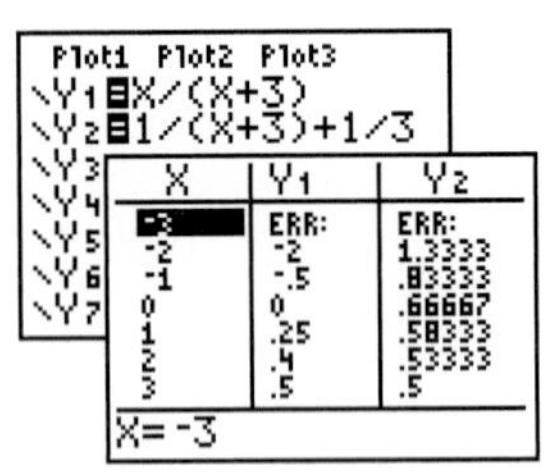

4. $\dfrac{3}{x-1} = \dfrac{1}{x-1} - 1$

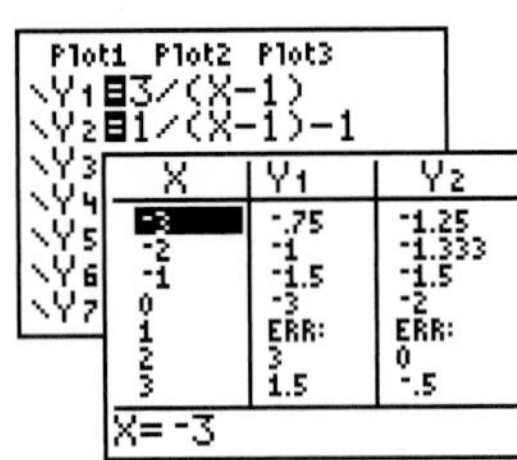

5. $\dfrac{x}{x-3} + \dfrac{x-1}{x+3} = \dfrac{6x}{x^2-9}$

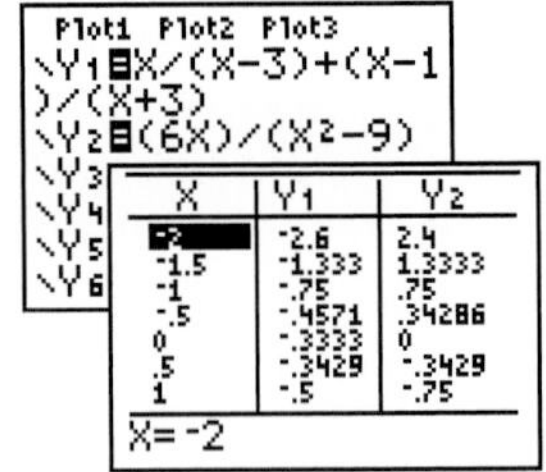

6. $\dfrac{x-20}{x^2-5x-6} + \dfrac{12}{2x^2-13x+6} = \dfrac{7x-2}{2x^2+x-1}$

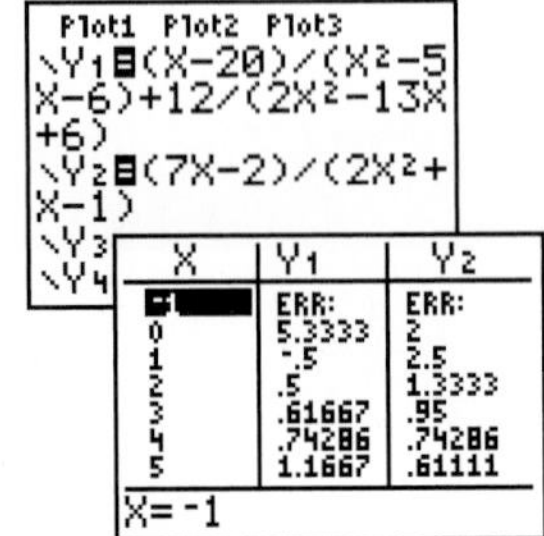

In Exercises 7–40, solve each equation.

7. $\dfrac{3}{z-1} + 2 = \dfrac{5}{z-1}$

8. $\dfrac{5}{z+3} - 2 = \dfrac{4}{z+3}$

9. $\dfrac{6w-1}{2w-1} - 5 = \dfrac{2w-3}{1-2w}$

10. $\dfrac{m}{m-2} - 5 = \dfrac{2}{m-2}$

11. $\dfrac{-3}{p+2} = \dfrac{-8}{p-3}$

12. $\dfrac{7}{p-4} = \dfrac{2}{p+1}$

13. $\dfrac{7}{3n-1} = \dfrac{2}{n+2}$

14. $\dfrac{10}{r-3} = \dfrac{34}{2r+1}$

15. $\dfrac{4}{k+2} = \dfrac{1}{3k+6} + \dfrac{11}{9}$

16. $\dfrac{5}{4k+1} = \dfrac{3}{8k+2} + 1$

17. $\dfrac{3y}{(y+4)(y-2)} = \dfrac{5}{y-2} + \dfrac{2}{y+4}$

18. $\dfrac{y^2+18}{(2y+3)(y-3)} = \dfrac{5}{y-3} - \dfrac{1}{2y+3}$

19. $1 - \dfrac{14}{y^2+4y+4} = \dfrac{7y}{y^2+4y+4}$

20. $1 + \dfrac{6w}{w^2-6w+9} = \dfrac{18}{w^2-6w+9}$

21. $\dfrac{4}{x-5} + \dfrac{5}{x-2} = \dfrac{x+6}{3x-6}$

22. $\dfrac{w+1}{w} + \dfrac{14}{w-7} = \dfrac{3w-7}{w^2-7w}$

23. $\dfrac{1}{(t-1)^2} - 3 = \dfrac{2}{1-t}$

24. $\dfrac{8}{(t+4)^2} + \dfrac{2}{t+4} = 3$

25. $\dfrac{z}{(z-2)(z+1)} - \dfrac{z}{(z+1)(z+3)} = \dfrac{3z}{(z+3)(z-2)}$

26. $\dfrac{4}{(z+1)(z-1)} = \dfrac{6-z}{(z+1)(z-2)} - \dfrac{8}{(z-2)(z-1)}$

27. $\dfrac{2v-5}{3v^2-v-14} + \dfrac{7}{3v-7} = \dfrac{8}{v+2}$

28. $\dfrac{v^2-2v+2}{6v^2+23v-4} + \dfrac{2}{v+4} = \dfrac{v}{6v-1}$

29. $\dfrac{m+4}{6m^2+5m-6} = \dfrac{m}{3m-2} - \dfrac{m}{2m+3}$

30. $\dfrac{2m+17}{2m^2+11m+14} + \dfrac{m-2}{m+2} = \dfrac{m-3}{2m+7}$

31. $\dfrac{x+1}{3x^2-4x+1} - \dfrac{x+1}{2x^2+x-3} = \dfrac{2}{6x^2+7x-3}$

32. $\dfrac{4y}{6y^2-7y-3} + \dfrac{2}{3y^2-2y-1} = \dfrac{y+2}{2y^2-5y+3}$

33. $\dfrac{2}{m+2} - \dfrac{1}{m+1} = \dfrac{1}{m}$

34. $\dfrac{m^2-1}{2m+1} = \dfrac{1-m}{3}$

35. $\dfrac{z-2}{2z^2-5z+3} + \dfrac{3}{3z^2-2z-1} = \dfrac{3z}{6z^2-7z-3}$

36. $\dfrac{n-3}{n^2+5n+4} + \dfrac{n-2}{n^2+3n+2} = \dfrac{n^2-12}{(n+1)(n+2)(n+4)}$

37. $\dfrac{1}{n^2-5n+6} - \dfrac{1}{n^2-n-2} + \dfrac{3}{n^2-2n-3} = 0$

38. $\dfrac{12}{2w^2-13w+6} - \dfrac{7w-2}{2w^2+w-1} + \dfrac{w-20}{w^2-5w-6} = 0$

39. $\dfrac{x^2}{x^2 - x - 2} = \dfrac{2x}{x^2 + x - 6}$

40. $\dfrac{6v + 6}{2v^2 + 7v - 4} = \dfrac{3v}{v^2 + 2v - 8} - \dfrac{5v - 7}{2v^2 - 5v + 2}$

In Exercises 41–46, simplify the expression in part (a) and solve the equation in part (b) for all real solutions.

	Simplify	Solve
41.	a. $\dfrac{1}{p-1} - \dfrac{3}{p+1}$	b. $\dfrac{1}{p-1} = \dfrac{3}{p+1}$
42.	a. $\dfrac{m-1}{m+1} - \dfrac{m-3}{m-2}$	b. $\dfrac{m-1}{m+1} = \dfrac{m-3}{m-2}$
43.	a. $\dfrac{x-1}{x+1} - 1 - \dfrac{x-6}{x-2}$	b. $\dfrac{x-1}{x+1} - 1 = \dfrac{x-6}{x-2}$
44.	a. $\dfrac{x^2}{x+2} + \dfrac{x-1}{x-3}$	b. $\dfrac{x^2}{x+2} + \dfrac{x-1}{x-3} = 0$

45. a. $\dfrac{2x - 8}{6x^2 + x - 2} - \dfrac{4}{3x + 2} + \dfrac{2}{2x - 1}$

b. $\dfrac{2x - 8}{6x^2 + x - 2} = \dfrac{4}{3x + 2} - \dfrac{2}{2x - 1}$

46. a. $\dfrac{w^2 - w - 3}{2w^2 - 9w + 9} + \dfrac{1}{3 - w} - \dfrac{w}{2w - 3}$

b. $\dfrac{w^2 - w - 3}{2w^2 - 9w + 9} + \dfrac{1}{3 - w} = \dfrac{w}{2w - 3}$

47. **Radius of a Window** A window in an office is in the shape of a semicircle. One foot from the center of the semicircle a 4-ft metal security bar is placed in the window. Determine the radius of this window to the nearest hundredth of a foot. $\left(\textit{Hint: } \dfrac{a}{b} = \dfrac{b}{c}\right)$

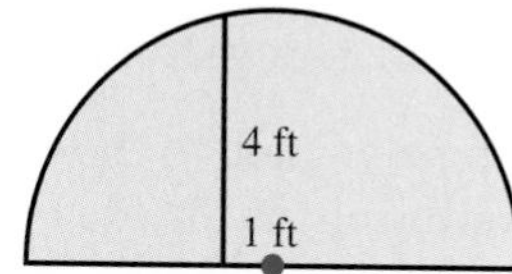

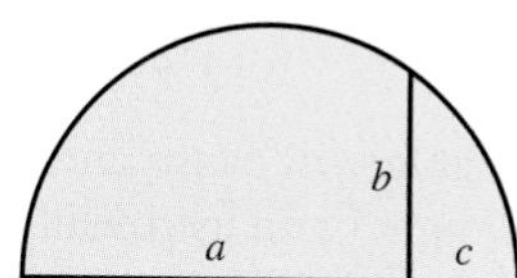

48. **Length of a Security Bar** A window in an office is in the shape of a semicircle with a 5-ft radius. Two feet from the center of the semicircle a metal security bar is placed in the window. Determine the length of this security bar to the nearest hundredth of a foot.

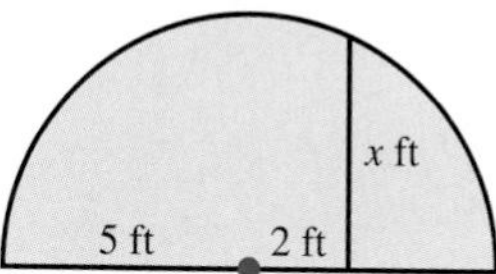

49. **Resistance in a Parallel Circuit** The resistance planned by an electrical technician for a parallel electric circuit with two resistors is 5 Ω. The resistance of r_2 must be 5 times the resistance of r_1. Determine the resistance of each resistor. The resistance R in a parallel circuit with two individual resistors r_1 and r_2 can be calculated by using the formula $\dfrac{1}{R} = \dfrac{1}{r_1} + \dfrac{1}{r_2}$.

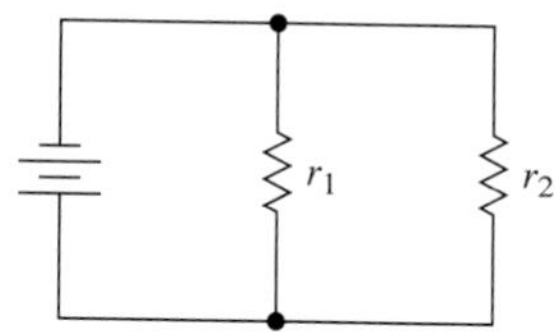

50. **Resistance in a Parallel Circuit** The resistance planned by an electrical technician for a parallel electric circuit with two resistors is 3 Ω. The resistance of r_2 must be 8 Ω greater than that of r_1. Determine the resistance of each resistor. (See Exercise 49.)

In Exercises 51–62, solve each equation for the variable specified.

51. $\dfrac{a}{b} = \dfrac{c}{d}$ for b

52. $a = \dfrac{bc}{d}$ for d

53. $\dfrac{a}{b-1} = \dfrac{c}{d+1}$ for b

54. $\dfrac{a}{b-1} = \dfrac{c}{d+1}$ for d

55. $I = \dfrac{k}{d}$ for d

56. $F = \dfrac{k}{r}$ for r

57. $\dfrac{1}{R} = \dfrac{1}{r_1} + \dfrac{1}{r_2}$ for R

58. $\dfrac{1}{R} = \dfrac{1}{r_1} + \dfrac{1}{r_2}$ for r_1

59. $h = \dfrac{2A}{B + b}$ for B

60. $h = \dfrac{2A}{B + b}$ for b

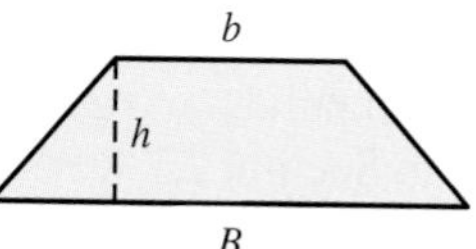

61. $\dfrac{1}{x} = \dfrac{1}{y} - \dfrac{1}{z}$ for z

62. $I = \dfrac{E}{r_1 + r_2}$ for r_1

In Exercises 63–70, solve each equation.

63. a. $\dfrac{5v + 4}{2v^2 + v - 15} + \dfrac{3}{5 - 2v} - \dfrac{1}{v + 3} = 0$

b. $\dfrac{5v + 3}{2v^2 + v - 15} + \dfrac{3}{5 - 2v} - \dfrac{1}{v + 3} = 0$

64. a. $\dfrac{w}{3w^2 - 11w + 10} + \dfrac{5}{3w - 5} = \dfrac{2}{w - 2}$

b. $\dfrac{w + 1}{3w^2 - 11w + 10} + \dfrac{5}{3w - 5} = \dfrac{2}{w - 2}$

65. $\dfrac{x^2}{2x^2 + 9x - 5} + \dfrac{2x}{x^2 + 2x - 15} = \dfrac{4x}{(2x - 1)(x + 5)(x - 3)}$

66. $\dfrac{x}{x - 3} + \dfrac{1}{x - 2} - \dfrac{1}{x + 2} = \dfrac{x - 12}{x^3 - 3x^2 - 4x + 12}$

67. $\dfrac{z - 2}{4z^2 - 29z + 30} - \dfrac{z + 2}{5z^2 - 27z - 18} = \dfrac{z + 1}{20z^2 - 13z - 15}$

68. $\dfrac{3n - 7}{n^2 - 5n + 6} + \dfrac{2n + 8}{9 - n^2} - \dfrac{n + 2}{n^2 + n - 6} = 0$

69. $\dfrac{z - 1}{z^2 - 2z - 3} + \dfrac{z + 1}{z^2 - 4z + 3} = \dfrac{z + 8}{z^2 - 1} + \dfrac{20}{(z - 1)(z + 1)(z - 3)}$

70. $\frac{3w-8}{w^2-5w+6}+\frac{w+2}{w^2-6w+8}=\frac{5-2w}{w^2-7w+12}+\frac{12}{(w-4)(w-3)(w-2)}$

Group Discussion Questions

71. **Error Analysis** Determine the error in the following argument:

Let $x = 1$
Then $3x = 2x + 1$
$3x - 3 = 2x - 2$
$3(x-1) = 2(x-1)$
$\frac{3(x-1)}{x-1}=\frac{2(x-1)}{x-1}$
Thus $3 = 2$

72. **Challenge Question**
   a. Complete this equation so that the solution is all real numbers except for -1 and 2.
   $\frac{5}{x-2}+\frac{2}{x+1}=\frac{?}{x^2-x-2}$
   b. Complete this equation so that the solution is all real numbers except $-\frac{2}{3}$ and $\frac{3}{2}$.
   $\frac{?}{6x^2-5x-6}=\frac{4}{3x+2}-\frac{1}{2x-3}$

73. **Challenge Question** Complete each equation so that the equation is a contradiction with no solution.
   a. $\frac{5}{x-2}+\frac{2}{x+1}=\frac{7x+?}{x^2-x-2}$
   b. $\frac{5x+?}{6x^2-5x-6}=\frac{4}{3x+2}-\frac{1}{2x-3}$

Section 8.6 Inverse and Joint Variation and Other Applications Yielding Equations with Fractions

Objectives:

9. Solve problems involving inverse and joint variation.
10. Solve applied problems that yield equations with fractions.

The problem-solving skills developed in earlier chapters are now used to solve applications that yield equations containing rational expressions. The problems selected here illustrate some of the variety that you may encounter outside the classroom. Some of these problems are skill builders to gradually develop your expertise for other applications.

Numbers with a large magnitude or absolute value are located relatively far from the origin, while numbers with a small magnitude or absolute value are located relatively close to the origin.

In Section 2.7 we examined problems that involved direct variation. If y varies directly as x with a constant of variation $k \neq 0$, then $y = kx$. When two variables vary directly, an increase in magnitude of one variable will cause an increase in magnitude of the other variable. We now examine problems involving inverse variation. If **y varies inversely as x** with a constant of variation $k \neq 0$, then $y = \frac{k}{x}$ for $x \neq 0$. When two variables vary inversely, an increase in magnitude of one variable will cause a decrease in magnitude of the other variable. Exercise 7 states that the weight of an astronaut varies inversely as the square of the distance from the center of the Earth. As the distance of the astronaut from the center of the Earth increases, the weight of the astronaut will decrease. Direct and inverse variation have quite contrasting behaviors. This is described in the following table.

Comparison of Direct and Inverse Variation

If x and y are real variables, k is a real constant, and $k \neq 0$, then:

VERBALLY	ALGEBRAICALLY	NUMERICAL EXAMPLE	GRAPHICAL EXAMPLE
y varies **directly** as x. As the magnitude of x increases, the magnitude of y increases linearly.	$y = kx$ *Example:* $y = 2x$	x | $y = 2x$ -3 | -6 -1 | -2 0 | 0 1 | 2 3 | 6	$y = 2x$
y varies **inversely** as x. As the magnitude of x increases, the magnitude of y decreases.	$y = \frac{k}{x}$ for $x \neq 0$ *Example:* $y = \frac{12}{x}$	x | $y = \frac{12}{x}$ 1 | 12 2 | 6 3 | 4 4 | 3 5 | 2.4	$y = \frac{12}{x}$

EXAMPLE 1 Translating Statements of Variation

Translate each of these equations into a verbal statement of variation.

SOLUTION

(a) $V = \frac{k}{P}$ The volume V of a fixed amount of gas (at a given temperature) varies inversely with the pressure P exerted on the gas.

This is a statement of Boyle's law from chemistry.

(b) $A(t) = \frac{C}{N(t)}$ The average overhead cost per item $A(t)$ of a production varies inversely with the number of items produced $N(t)$.

C is the fixed overhead cost. $N(t)$ is the number of units produced. $A(t)$ is the average overhead cost per unit.

Sometimes proportions (see Section 2.7) are also used to describe the relation between variables that vary directly or inversely. If x and y are **directly proportional,** then y varies directly as x. If x and y are **inversely proportional,** then y varies inversely as x.

Example 2 works through a numeric word problem involving direct and inverse variation. This allows us to compare these two types of variation before the more involved application shown in Example 3.

EXAMPLE 2 Solving Problems with Direct and Inverse Variation

The variable y equals 48 when x equals 4. Find y when $x = 6$ if

(a) y varies directly as x.

(b) y varies inversely as x.

SOLUTION

(a) **VERBALLY** y varies directly as x

ALGEBRAICALLY

$$y = kx$$
$$48 = k(4)$$
$$12 = k$$

Translate the word equation into algebraic form, using k for the constant of variation. Substitute in the given values of 48 for y and 4 for x. Then solve for the constant k.

$$y = 12x$$
$$y = 12(6)$$
$$y = 72$$

Use this constant to write the equation of variation. Then substitute in the value of 6 for x and solve for y.

(b) **VERBALLY** y varies inversely as x

ALGEBRAICALLY

$$y = \frac{k}{x}$$

Translate the word equation into algebraic form, using k for the constant of variation.

$$48 = \frac{k}{4}$$

Substitute in the given values of 48 for y and 4 for x.

$$192 = k$$

Then solve for the constant k.

$$y = \frac{192}{x}$$

Use this constant to write the equation of variation.

$$y = \frac{192}{6}$$
$$y = 32$$

Then substitute in the value of 6 for x and solve for y.

SELF-CHECK 8.6.1

1. y varies directly as x; and $y = 6$ when $x = 3$. Find y when $x = 2$.
2. y varies inversely as x; and $y = 6$ when $x = 3$. Find y when $x = 2$.

Sphere: $S = 4\pi r^2$

The energy sent out from a single point source often radiates in a spherical pattern. This is true whether the energy is heat, light, sound, or a radio signal. The surface area of the sphere receiving this energy at any instant is given by $S = 4\pi r^2$. As the distance r from the energy source increases, the area sharing this fixed amount of energy increases as the square of r. Thus the intensity of the energy received at any place on this sphere varies inversely as the square of the distance from the energy source. Example 3 illustrates this concept by examining the illumination given by a car's headlight.

EXAMPLE 3 Using Inverse Variation

The illumination provided by a car's headlight varies inversely as the square of the distance from the headlight. A headlight produces 15 footcandles (fc) at a distance of 20 ft.

(a) Determine the constant of variation for this headlight.
(b) Write an equation relating the illumination provided by this headlight to the distance from the headlight.
(c) Use this equation to create a table of values displaying the illumination in 5-ft increments starting at 10 ft.
(d) What will the illumination be at 40 ft?

SOLUTION

Let I = illumination of headlight, fc
d = distance of an object from the headlight, ft
k = constant of variation for this headlight

Identify each unknown with a variable. Be sure to include units in this identification.

VERBALLY The illumination varies inversely as the square of the distance.

Precisely state an equation that can be translated into algebraic form.

ALGEBRAICALLY

$$I = \frac{k}{d^2}$$

Translate the word equation into algebraic form by using the variables identified above.

$$15 = \frac{k}{20^2}$$

Substitute in the given values for I and d, and solve for the constant k.

$$15 = \frac{k}{400}$$

$$15(400) = k$$

$$k = 6{,}000$$

(a) The constant of variation is 6,000.

(b) $I = \dfrac{6{,}000}{d^2}$

(c)

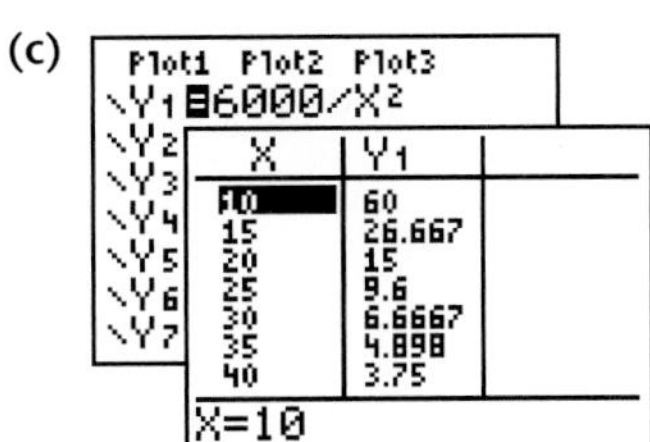

To generate a table of values, enter the equation $I = \dfrac{6{,}000}{d^2}$ as $Y_1 = \dfrac{6{,}000}{x^2}$. Then use 10 for **TblStart** and 5 for **ΔTbl**.

(d) $I = \dfrac{6{,}000}{40^2}$

$I = 3.75$

The illumination will be 3.75 fc.

Using the constant of variation for this headlight, find I when the distance is 40 ft by substituting $d = 40$ into the equation $I = \dfrac{6{,}000}{d^2}$. This value is also confirmed by the table.

It is important to keep the contrasting behaviors of direct and inverse variation in mind as you analyze real-world variables. If y varies directly as x, then the magnitude of y increases linearly as the magnitude of x increases. If y varies inversely as x, then the magnitude of y decreases as the magnitude of x increases. This is illustrated in Example 4.

EXAMPLE 4 Selecting Direct or Inverse Variation

Determine for each relationship whether the variation between the two variables is more likely to be direct variation or inverse variation.

SOLUTION

(a) The average weight of a group of people of the same age varies __________ as the height of this group of people.

Directly

Taller people are overall bigger people and also weigh more. Thus, these variables are more likely to vary directly.

(b) The average vertical jumping height of a group of adults of approximately the same size varies __________ as the age of this group of people.

Inversely

After maturity and past age 25, the athletic ability of people declines as they age. Their jumping height will decrease as their age increases. Thus, these variables are more likely to vary inversely.

The language of variation is used to describe relationships between real-world variables. In many cases there are more than two variables involved. In these cases we often will see some variables related directly and others inversely. We can also see examples of joint variation. To say that the variable z **varies jointly** as x and y means that z varies directly as the product of x and y.

Joint Variation

If x, y, and z are real variables, k is a real constant, and $k \neq 0$, then:

VERBALLY	ALGEBRAICALLY	EXAMPLE
z varies jointly as x and y.	$z = kxy$	$z = 5xy$

EXAMPLE 5 Translating Statements of Variation

Translate each statement of variation into an equation.

SOLUTION

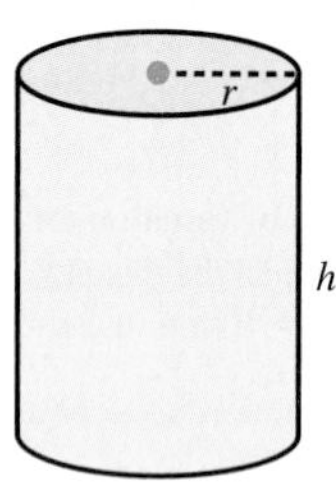

(a) The time T for a trip varies directly as the distance D and inversely as the rate of travel R.

$T = k\frac{D}{R}$ or $T = \frac{D}{R}$

If the units are selected appropriately, then the constant of variation is 1. This equation can also be written as $D = RT$.

(b) The volume of a cylinder varies jointly as the square of its radius and its height.

$V = kr^2h$ or $V = \pi r^2 h$

If the units are selected appropriately, then the constant of variation is π.

SELF-CHECK 8.6.2

Translate each statement of variation into an equation.

1. The current I in an electric circuit varies inversely as the resistance R.
2. The interest I on an investment varies jointly as the rate R and the time T.

The formula $D = RT$ states the distance traveled equals the product of the rate of travel and the time traveled.

Many problems involving rates and times involve inverse variation. For example, the formula for distance $D = RT$ can be rewritten as $R = \frac{D}{T}$. This form states that the rate of travel varies directly as the distance traveled and inversely as the time traveled. As the time you have available for a trip increases, you can travel at a slower rate.

The formula $W = RT$ states the work done is the product of the rate of work and the time worked.

Likewise, the formula for work $W = RT$ can be rewritten as $R = \frac{W}{T}$. For a fixed amount of work the rate of work varies inversely as the time required to do this work. For less time you must work faster—for more time you can work slower.

The basic strategy used to solve these word problems is summarized in the following box. It is the same strategy we employed earlier to solve word problems by using linear equations and quadratic equations.

Strategy for Solving Word Problems

STEP 1. Read the problem carefully to determine what you are being asked to find.

STEP 2. Select a variable to represent each unknown quantity. Specify precisely what each variable represents, and note any restrictions on each variable.

STEP 3. If necessary, make a sketch and translate the problem into a word equation or a system of word equations. Then translate each word equation into an algebraic equation.

STEP 4. Solve the equation or system of equations, and answer the question asked by the problem.

STEP 5. Check the reasonableness of your answer.

After solving the equations that you form, always inspect the solutions to see if they are appropriate for the original problem. Check not only for extraneous values but also for answers that may not be meaningful in the application, such as negative amounts of time.

Example 6 involves the shared work of two people working together to complete a job. This problem assumes that when the painters work together, there is no gain or loss of efficiency of either painter. Both painters will continue to work at the same rate as when they were working alone. Another key point in this example is that information about the time to do a job is also information about the rate of work for this job. This is true because of the inverse relationship between the time and the rate of work.

EXAMPLE 6 Determining the Time to Paint a Sign

Working alone, painter A can paint a sign in 6 hours less time than it would take painter B working alone. When painters A and B work together, the job takes only 4 hours. How many hours would it take each painter working alone to paint the sign?

This problem assumes that when the painters work together, there is no gain or loss of efficiency for either painter.

SOLUTION

Let t = time for painter B to paint sign working alone, hours

$t - 6$ = time for painter A to paint sign working alone, hours

$\frac{1}{t-6}$ = rate of work for painter A

$\frac{1}{t}$ = rate of work for painter B

$\frac{1}{4}$ = rate of work when both painters work together

First identify the time for each painter to paint the sign while working alone.

The rate of work varies inversely as the time; that is, $R = \frac{W}{T}$. The work equals 1 sign painted; thus this equation becomes $R = \frac{1}{T}$.

VERBALLY

[A's rate of work] + [B's rate of work] = [Total rate of work when working together]

The word equation is based on the mixture principle. When the two painters share the work, the total rate of work is found by adding their individual rates of work.

ALGEBRAICALLY

$$\frac{1}{t-6}+\frac{1}{t}=\frac{1}{4}$$

Substitute the rates of each painter into the word equation.

$$4t(t-6)\frac{1}{t-6}+4t(t-6)\frac{1}{t}=4t(t-6)\left(\frac{1}{4}\right)$$

Multiply both sides of the equation by the LCD, $4t(t-6)$.

$$4t+4(t-6)=t(t-6)$$
$$4t+4t-24=t^2-6t$$
$$8t-24=t^2-6t$$
$$0=t^2-14t+24$$
$$0=(t-12)(t-2)$$

Simplify this quadratic equation, and write it in standard form. Then solve this equation by factoring.

$t-12=0$ or $t-2=0$

$t=12$ $\quad$ $t=2$

$t-6=6$ $\quad$ $t-6=-4$

This value is not appropriate.

While 2 hours may seem meaningful for painter B, the value of −4 hours is not meaningful for painter A.

Check: The rates for each painter are $\frac{1}{6}$ sign/hour for A and $\frac{1}{12}$ sign/hour for B. In 4 hours they paint a total of

$$\frac{1}{6}(4)+\frac{1}{12}(4)=\frac{4}{6}+\frac{4}{12}$$
$$=\frac{2}{3}+\frac{1}{3}$$
$$=1 \text{ sign}$$

Answer: Working alone, painter A could paint the sign in 6 hours and painter B could paint the sign in 12 hours.

SELF-CHECK 8.6.3

1. Working alone, a welder can weld a metal framework 5 hours faster than an apprentice can. How long will it take the apprentice to do the job alone if the welder and the apprentice can do it in 6 hours when they share the work?

It is wise to include the units of measurement when you identify your variables. This is especially true if there are different units within the same problem (e.g., both minutes and hours).

Example 7 has many similarities to the $W = RT$ problem in Example 6. However, in this example the formula is $D = RT$, and the rate involved is the rate of travel instead of the rate of work.

EXAMPLE 7 Determining the Speed of a Boat in a Still Water

Two tugboats that travel at the same speed in still water travel in opposite directions in a river with a constant current of 8 mi/h. The tugboats depart at the same time from a refueling station; and after a period of time one has traveled 30 mi downstream, and the other has traveled 6 mi upstream. Determine the rate of each boat in still water.

SOLUTION

Let r = rate of each boat in still water, mi/h

$r+8$ = rate of boat going downstream, mi/h

$r-8$ = rate of boat going upstream, mi/h

$\frac{30}{r+8}$ = time of boat going downstream, hours

$\frac{6}{r-8}$ = time of boat going upstream, hours

First identify the quantities being sought with a variable—the rates of the two boats.

The formula $D = RT$ can also be written as $T=\frac{D}{R}$. Substitute the given distances and the rates $r+8$ and $r-8$ into this formula to also identify the time for each boat.

VERBALLY

Time of boat going downstream	=	Time of boat going upstream

Although the times are unknown, we do know they are the same because the boats left at the same time.

ALGEBRAICALLY

$$\frac{30}{r+8} = \frac{6}{r-8}$$

Substitute the times as identified above into the word equation.

$$(r+8)(r-8)\frac{30}{r+8} = (r+8)(r-8)\frac{6}{r-8}$$

Multiply both sides of the equation by the LCD, $(r + 8)(r - 8)$.

$$30(r-8) = 6(r+8)$$
$$30r - 240 = 6r + 48$$
$$24r = 288$$
$$r = 12$$

Then simplify this equation and solve for r.

Check: $r + 8 = 20$
$r - 8 = 4$

Going downstream at 20 mi/h, the tugboat will take $\frac{30}{20} = 1.5$ hours to go 30 mi. Going upstream at 4 mi/h, the tugboat will take $\frac{6}{4} = 1.5$ hours to go 6 mi. These times are equal.

Answer: The rate of each tugboat in still water is 12 mi/h.

SELF-CHECK 8.6.4

1. Two tugboats that travel at the same speed in still water travel in opposite directions in a river with a constant current of 5 mi/h. The tugboats depart at the same time from a dock; and after a period of time one has traveled 30 mi downstream, and the other has traveled 6 mi upstream. Determine the rate of travel of each boat in still water.

Example 8 also involves a rate problem. For this application the rate is the interest rate charged for borrowed money. The formula $I = PRT$ can be rewritten as $P = \frac{I}{RT}$. For a fixed time of 1 year, this formula can be simplified to $P = \frac{I}{R}$, as is done in Example 8. Recall that P represents the principal, R represents the interest rate, T represents the time, and I represents the interest.

EXAMPLE 8 Determining Two Interest Rates

A local bank pays interest on both checking accounts and savings accounts. The interest rate for a savings account is 1% higher than that for a checking account. A customer calculates that a deposit in a checking account would earn yearly interest of \$80, whereas this same deposit would earn yearly interest of \$100 in a savings account. What is the interest rate on each account?

SOLUTION

Let r = rate of interest on checking account
$r + 0.01$ = rate of interest on savings account
$\frac{80}{r}$ = principal in checking account
$\frac{100}{r + 0.01}$ = principal in savings account

First identify the quantities being sought with a variable—the interest rates of the two accounts.

$I = PRT$; for $T = 1$ we can write $P = \frac{I}{R}$. Substitute the given interest for each account and the rates identified to label the principal in each account.

One question to ask when you are trying to form the word equation for a problem is, "What things are equal or the same?"

VERBALLY

Principal in a checking account	=	Principal in a savings account

The deposit is the same for both options.

ALGEBRAICALLY

$$\frac{80}{r} = \frac{100}{r + 0.01}$$

Substitute the values identified above into the word equation.

$$r(r + 0.01)\left(\frac{80}{r}\right) = r(r + 0.01)\left(\frac{100}{r + 0.01}\right)$$

Multiply both sides of the equation by the LCD, $r(r + 0.01)$.

$$\begin{aligned} 80(r + 0.01) &= 100r \\ 80r + 0.8 &= 100r \\ 0.8 &= 20r \\ 0.04 &= r \\ r &= 0.04 \\ r + 0.01 &= 0.05 \end{aligned}$$

Simplify, and then solve for r.

The checking account rate is 4%.

The savings account rate is 5%.

Answer: The interest rate on the checking account is 4% and on the savings account is 5%.

Do these values check?

SELF-CHECK 8.6.5

1. A local bank pays interest on both checking accounts and savings accounts. The interest rate for a savings account is 0.5% higher than that for a checking account. A customer calculates that a deposit in a checking account would earn yearly interest of \$80, whereas this same deposit would earn yearly interest of \$100 in a savings account. What is the interest rate on each account?

The last problem we examine is a numeric word problem involving reciprocals. These problems do not have the context of a real-world problem, but exercises like this can help us to develop our problem-solving skills.

EXAMPLE 9 Solving a Numeric Word Problem

The sum of the reciprocals of two consecutive even integers is $\frac{13}{84}$. Find these integers.

SOLUTION

Let n = smaller integer

$\frac{1}{n}$ = reciprocal of smaller integer

$n + 2$ = larger integer

$\frac{1}{n + 2}$ = reciprocal of larger integer

Identify each number and its reciprocal.

Consecutive even integers differ by 2.

VERBALLY

Reciprocal of first integer	+	Reciprocal of second integer	=	$\frac{13}{84}$

Write the word equation.

ALGEBRAICALLY

$$\frac{1}{n} + \frac{1}{n+2} = \frac{13}{84}$$

Substitute the values identified above into the word equation.

$$84n(n+2)\left(\frac{1}{n} + \frac{1}{n+2}\right) = 84n(n+2)\left(\frac{13}{84}\right)$$

Multiply both sides of the equation by the LCD, $84n(n + 2)$.

$$84(n+2) + 84n = 13n(n+2)$$
$$84n + 168 + 84n = 13n^2 + 26n$$
$$13n^2 - 142n - 168 = 0$$

Simplify, and write the quadratic equation in standard form.

$$(n - 12)(13n + 14) = 0$$

Factor the left side of the equation.

$$n - 12 = 0 \quad \text{or} \quad 13n + 14 = 0$$

Set each factor equal to 0.

$$n = 12 \qquad n = -\frac{14}{13}$$
$$n + 2 = 14$$

Then solve for the smaller integer n and the larger integer $n + 2$.

This value is not an integer.

Check: $\frac{1}{12} + \frac{1}{14} = \frac{7}{84} + \frac{6}{84} = \frac{13}{84}$.

Answer: The integers are 12 and 14.

SELF-CHECK ANSWERS

8.6.1

1. $y = 4$
2. $y = 9$

8.6.2

1. $I = \frac{k}{R}$
2. $I = kRT$ or $I = PRT$ with the principal P being the constant of variation.

8.6.3

1. The apprentice can do the job alone in 15 hours. (No other values check in this problem.)

8.6.4

1. The rate of each tugboat in still water is 7.5 mi/h.

8.6.5

1. The interest rate on a checking account is 2% and on a savings account is 2.5%.

USING THE LANGUAGE AND SYMBOLISM OF MATHEMATICS 8.6

1. If a varies directly as b, then $a =$ ___________.
2. If a varies inversely as b, then $a =$ ___________.
3. If a varies jointly as b and c, then $a =$ ___________.
4. If a varies directly as b and b increases in magnitude, then a will ___________ in magnitude.
5. If a varies directly as b and b decreases in magnitude, then a will ___________ in magnitude.
6. If a varies inversely as b and b increases in magnitude, then a will ___________ in magnitude.
7. If a varies inversely as b and b decreases in magnitude, then a will ___________ in magnitude.
8. In the formula $D = RT$, D represents distance, R represents ___________, and T represents ___________.
9. In the formula $I = PRT$, I represents interest, P represents ___________, R represents ___________, and T represents ___________.
10. In the formula $W = RT$, W represents work, R represents ___________, and T represents ___________.

QUICK REVIEW 8.6

The purpose of this quick review is to help you recall skills needed in this section.

1. Determine the time it will take to drive 240 mi while averaging 60 mi/h.
2. Determine the time it will take to drive 240 mi while averaging 50 mi/h.
3. Determine the rate of travel of a vehicle that travels 240 mi in 30 minutes.
4. Determine the interest on a \$5,000 investment for 1 year at 6% annual interest.
5. A \$12,000 investment earns \$80 interest the first month. What is the annual rate of interest?

EXERCISES 8.6

In Exercises 1–12, write an equation for each statement of variation. Use k as the constant of variation.

1. a. m varies directly as n.
   b. m varies inversely as p.
   c. m varies directly as n and inversely as p.
   d. m varies jointly as n and p.
2. a. w varies directly as z.
   b. w varies inversely as v.
   c. w varies directly as z and inversely as v.
   d. w varies jointly as z and v.
3. a. v varies directly as the square root of w.
   b. v varies inversely as the square of x.
   c. v varies directly as the square root of w and inversely as the square of x.
   d. v varies jointly as the square root of w and the square of x.
4. a. y varies directly as x cubed.
   b. y varies inversely as the cube root of z.
   c. y varies directly as x cubed and inversely as the cube root of z.
   d. y varies jointly as x cubed and the cube root of z.
5. **Ohm's Law** The electric current I varies directly as the voltage V.
6. **Electricity from a Windmill** The number of kilowatts W of electricity that can be produced by a windmill varies directly as the cube of the speed v of the wind.
7. **Weight of an Astronaut** The weight w of an astronaut varies inversely as the square of the distance d from the center of the Earth.

8. **Travel Time** The time t required to drive between two towns varies inversely as the rate r.
9. **Wind Resistance** The force of the wind resistance R on a moving automobile varies directly as the square of the velocity v of the automobile.
10. **Period of a Pendulum** The period T (time for one complete swing) of a pendulum varies directly as the square root of the length of the pendulum L.
11. **Volume of a Cone** The volume V of a cone varies jointly with the square of the radius r and the height h. (What is the constant of variation for this problem?)
12. **Universal Law of Gravitation** Every object in the universe attracts every other object with a force F that varies jointly as their masses m_1 and m_2 and inversely as the square of the distance d between the objects.

In Exercises 13–18, use the given statement of variation to solve each problem.

13. a. y varies directly as x, and $y = 24$ when $x = 8$. Find y when $x = 10$.
    b. y varies inversely as x, and $y = 24$ when $x = 8$. Find y when $x = 10$.
14. a. a varies directly as b, and $a = 12$ when $b = 3$. Find a when $b = 2$.
    b. a varies inversely as b, and $a = 12$ when $b = 3$. Find a when $b = 2$.
15. a varies directly as b and inversely as c, and $a = 3$ when $b = 9$ and $c = 12$. Find a when $b = 15$ and $c = 6$.
16. a varies directly as b and inversely as c, and $a = 6$ when $b = 2$ and $c = 24$. Find a when $b = 5$ and $c = 60$.
17. a varies jointly as b and c, and $a = 27$ when $b = 9$ and $c = 12$. Find a when $b = 15$ and $c = 6$.
18. a varies jointly as b and c, and $a = 8$ when $b = 2$ and $c = 24$. Find a when $b = 5$ and $c = 60$.

In Exercises 19–22, match each table with the corresponding statement of variation.

19. y varies directly as x with a positive constant of variation.
20. y varies directly as x with a negative constant of variation.
21. y varies inversely as x with a positive constant of variation.
22. y varies inversely as x with a negative constant of variation.

A.

X	Y1
1	12
2	6
3	4
4	3
5	2.4
6	2
7	1.7143

X=1

B.

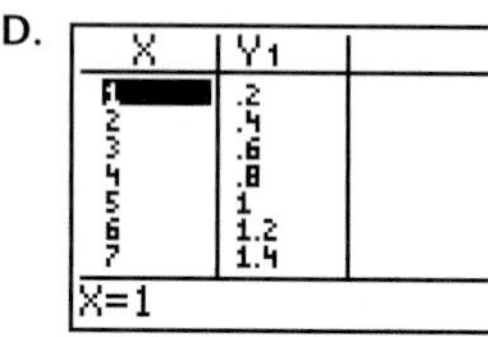

X	Y1
1	-12
2	-6
3	-4
4	-3
5	-2.4
6	-2
7	-1.714

X=1

C.

X	Y1
1	-2
2	-4
3	-6
4	-8
5	-10
6	-12
7	-14

X=1

D.

X	Y1
1	.2
2	.4
3	.6
4	.8
5	1
6	1.2
7	1.4

X=1

In Exercises 23–28, match each graph with the corresponding statement of variation. All graphs are displayed with the window $[-10, 10, 1]$ by $[-10, 10, 1]$.

23. y varies directly as x with a constant of variation of $\frac{1}{3}$.
24. y varies directly as x with a constant of variation of $-\frac{1}{3}$.
25. y varies directly as x with a constant of variation of 3.
26. y varies directly as x with a constant of variation of -3.
27. y varies inversely as x with a constant of variation of 10.
28. y varies inversely as x with a constant of variation of -10.

A.

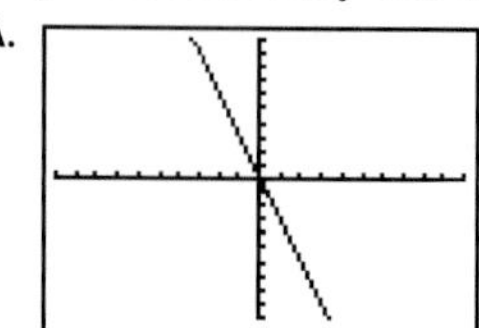

B.

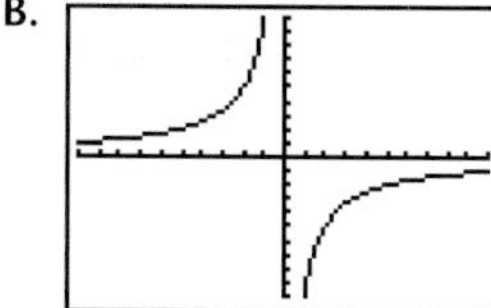

C.

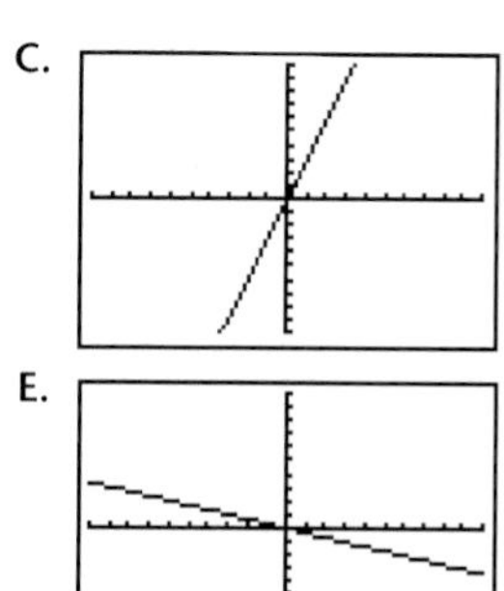

D.

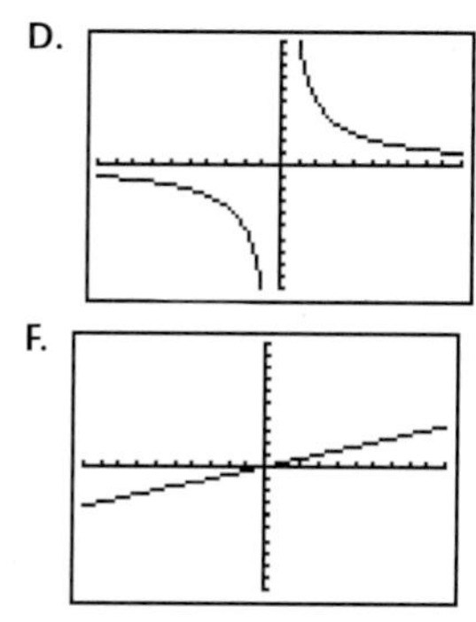

E.

F.

29. **Use of Newsprint** The use of newsprint varies directly as the number of people using the newsprint. The newsprint used to supply the needs of 1,000 people is approximately 34,800 kg. (*Source:* North American Newsprint Producers Association.)
    a. Determine the constant of variation.
    b. Write an equation relating the kilograms of newsprint to the number of people using the newsprint.
    c. Use this equation to create a table of values displaying the kilograms of newsprint required to supply the needs of various populations starting with 700,000 people and increasing in increments of 10,000 people.
    d. How many kilograms would be needed to supply Columbus, Ohio, which has a population of approximately 720,000?
30. **Stopping Distance of a Car** The distance required for an emergency stop for a car varies directly as the square of the car's speed. A car traveling at 50 mi/h requires 140 ft to stop.

    a. Determine the constant of variation.
    b. Write an equation relating the stopping distance to the speed of this car.
    c. Use this equation to create a table of values displaying the stopping distances required for this car traveling at various speeds starting at 45 mi/h and increasing in increments of 5 mi/h.
    d. How many feet will be required for this car to make an emergency stop if it was traveling at 70 mi/h?
31. **Time to Distribute Fliers** The time required to distribute a batch of political fliers in a neighborhood varies inversely as the number of people who help pass out these fliers. It will take 8 hours for 5 workers to distribute a batch of fliers.
    a. Determine the constant of variation.
    b. Write an equation relating the time required to distribute fliers to the number of people who help pass out these fliers.
    c. Use this equation to create a table of values displaying the times required for distributing fliers in this neighborhood, starting with 5 people and increasing in increments of 5 people.
    d. How long will it take to distribute these fliers in this neighborhood if 20 people are helping?
32. **Force to Prevent Skidding** The force required to keep a car from skidding on a curve varies inversely as the radius of the curve. A force of 7,500 lb is needed to keep a car from skidding on a curve of radius 800 ft.
    a. Determine the constant of variation.
    b. Write an equation relating the force required to keep a car from skidding to the radius of the curve.
    c. Use this equation to create a table of values displaying the force required to keep a car from skidding on various curves, starting at a radius of 500 ft and increasing in increments of 100 ft.
    d. What force will be required to keep a car from skidding on a radius of 500 ft?
33. **Time to Get a Sunburn** The time it takes to get a sunburn varies inversely as the UV (ultraviolet light index) rating. At a UV rating of 6 the time it takes to obtain a sunburn can be as little as 15 minutes. Use this information to determine the time it will take to get a sunburn when the UV rating is 10. (*Source:* CBS News online Consumer Tips.)
34. **Volume of a Gas** The volume of a gas varies inversely as the pressure applied to the gas. Under a pressure of 38 kg/cm^2 the volume of a gas is 240 cm^3. Approximate the volume when the pressure is increased to 42 kg/cm^2 (Boyle's law).
35. **Resistance of a Wire** The electrical resistance of a wire varies inversely as the square of the diameter of the wire. The resistance of a wire with a diameter of 4 mm is 4.5 Ω. Find the resistance of another wire of the same length and made from similar materials with a diameter of 6 mm.
36. **Illumination** The illumination I received at an object from a light source varies inversely as the square of the distance from the source. If the illumination 5 m from a light source is 4 lumens (lm), find the illumination 10 m from this same light source.
37. **Load on a Beam** The load that a rectangular beam can support varies jointly as its width and the square of its height and inversely as its length. A 2×6 (assume actual measurements are really 2 in and 6 in) pine beam that is 120 in long can carry a load of 800 lb when installed properly with the width of 2 in. Approximate the load this beam can support if improperly installed with a width of 6 in and a height of 2 in.
38. **Volume of a Pyramid** The volume of a right pyramid varies jointly as the height and area of the base of the pyramid. The Great Pyramid of Giza has a height of 145 m and a base area of 52,500 m^2. The volume of a smaller pyramid with a height of 55 m and an area of 7,200 m^2 is 132,000 m^3. Determine the constant of variation and the volume of the Great Pyramid of Giza.

In Exercises 39–72, solve each problem.

39. The sum of the reciprocals of two consecutive integers is $\frac{11}{30}$. Find these integers.

40. The sum of the reciprocals of two consecutive even integers is $\frac{7}{24}$. Find these integers.

41. The sum of the reciprocals of two consecutive odd integers is 16 times the reciprocal of their product. Find these integers.

42. The sum of the reciprocals of two consecutive even integers is 10 times the reciprocal of their product. Find these integers.

43. The denominator of a fraction is an integer that is 4 more than the square of the numerator. If the fraction is reduced, it equals $\frac{3}{20}$. Find this numerator.

44. The denominator of a fraction is an integer that is 3 more than the square of the numerator. If the fraction is reduced, it equals $\frac{1}{4}$. Find this numerator.

45. The sum of a number and its reciprocal is $\frac{13}{6}$. Find this number.

46. The difference of a number and its reciprocal is $\frac{9}{20}$. Find this number.

47. **Ratio of Gauge Readings** The ratio of two readings from a gauge is $\frac{4}{5}$. The first reading is 3 units above normal, and the second reading is 5 units above normal. What is the normal reading?

48. **Ratio of Temperatures** The ratio of two temperature readings is $\frac{7}{8}$. The first reading is 7° below normal, whereas the second reading is 2° above normal. What is the normal reading?

49. **Wire Length** A wire 16 m long is cut into two pieces whose lengths have a ratio of 3 to 1. Find the length of each piece.

50. **Rope Length** A rope 20 m long is cut into two pieces whose lengths have a ratio of 4 to 1. Find the length of each piece.

51. **Electrical Resistance** The resistance of the parallel circuit with two resistors shown in the figure is 40 Ω. If one resistor has twice the resistance of the other, what is the resistance of each?

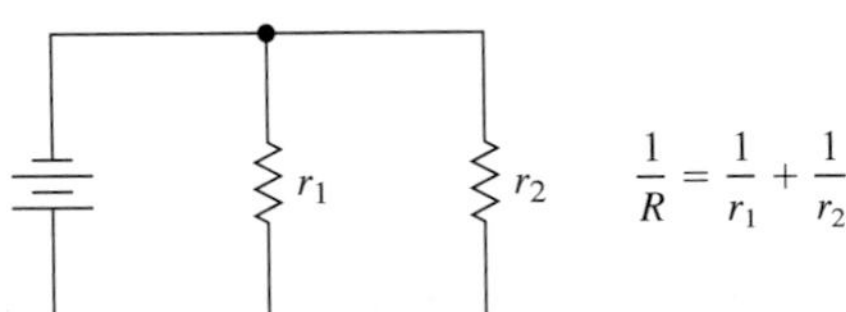

52. **Electrical Resistance** The resistance of one resistor in a parallel circuit is 3 Ω greater than that of the second resistor in the circuit. Find the resistance of each resistor if the resistance of the circuit is $5\frac{1}{7}$ Ω.

53. **Complementary Angles** The ratio of the measures of two complementary angles is 3 to 2. Find the measure of each angle.

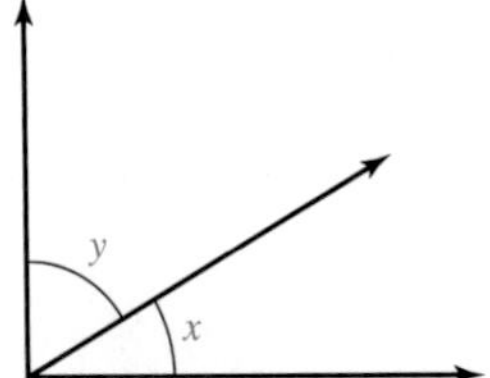

54. **Supplementary Angles** The ratio of the measures of two supplementary angles is 4 to 1. Find the measure of each angle.

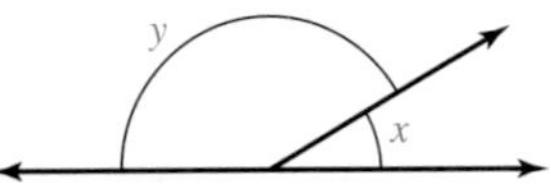

55. **Investment Ratio** An investor invests $12,000 in a combination of secure funds and high-risk funds. The ratio of dollars invested in secure funds to dollars invested in high-risk funds is 7 to 3. Find the amount invested in secure funds.

56. **Investment Ratio** By charter a mutual fund must invest its funds in bonds and stocks in a 5 to 3 ratio, respectively. If the mutual fund has $16,000,000, how much is invested in bonds?

57. **Ratio of Assets to Liabilities** The assets of a small automobile dealership are approximated by $500x + 10{,}000$, and its liabilities are approximated by $100x + 14{,}000$, where x represents the number of vehicles sold. One measure of the strength of this business is the ratio of its assets to its liabilities. For the month of January the ratio was 5 to 4. How many vehicles were sold in January?

58. **Ratio of Assets to Liabilities** The assets of a heating and air-conditioning business are approximated by $300x + 5{,}000$, and its liabilities are approximated by $200x + 6{,}000$, where x represents the number of furnaces sold. One measure of the strength of this business is the ratio of its assets to its liabilities. For the month of February the ratio was $\frac{7}{6}$. How many furnaces were sold in February?

59. **Boats in a Flowing River** Two boats that travel the same speed in still water depart simultaneously from a dock, traveling in opposite directions on a river that has a current of 6 km/h. After a period of time one boat is 51 km downstream, and the other boat is 15 km upstream. What is the speed of each boat in still water?

60. **Boats in a Flowing River** Two boats that travel the same speed in still water depart simultaneously from a dock, traveling in opposite directions on a river that has a current of 7 km/h. After a period of time one boat is 30 km downstream and the other boat is 9 km upstream. What is the speed of each boat in still water?

61. **Airplanes Traveling in an Airstream** Two planes departed from the same airport at the same time, flying in opposite directions. After a period of time the slower plane has traveled 1,080 mi and the faster plane has traveled 1,170 mi. The faster plane is traveling 40 mi/h faster than the slower plane. Determine the rate of each plane. (*Hint:* What is the same for each plane?)

62. **Airplanes Traveling in an Airstream** Two planes traveling at the same airspeed depart simultaneously from an airport, flying in opposite directions. One plane flies directly into a 40-mi/h wind, and the other flies in the same direction as this wind. After a period of time one plane has traveled 540 mi and the other has traveled 660 mi. What is the airspeed of each plane?

63. **Time for Pipes to Fill a Tank** One pipe can fill a tank in 15 hours, and a larger pipe can fill the same tank in 10 hours. If both pipes are used simultaneously, how many hours will it take to fill this tank?

64. **Time for Assembly Line to Fill an Order** Working alone, one assembly line can complete an order in 36 hours. A newer type of assembly line in this same factory can complete the same order in 18 hours. If both assembly lines are used

simultaneously, how many hours will it take to complete this order?

65. **Time for Pipes to Fill a Tank** Working together, two pipes can fill one tank in 4 hours. Working alone, the smaller pipe would take 6 hours longer than the larger pipe to fill the tank. How many hours would it take the larger pipe to fill the tank, working alone?

66. **Time for Assembly Line to Fill an Order** Working together, two assembly lines can complete an order in 18 hours. Working alone, the older assembly line would take 15 hours longer than the newer assembly line to complete the order. How many hours would it take the newer assembly line to complete the order, working alone?

67. **Search Time for Planes** Search plane A can search an area for a crash victim in 50 hours. Planes A and B can jointly search the area in 30 hours. How many hours would it take plane B to search the area alone?

68. **Generating a Mailing List** Printer A takes twice as long as printer B to generate a mailing list. If a particular list can be generated in 9 hours when both printers are used, how many hours would it take each printer to generate the list, working separately?

69. **Interest Rates on Two Accounts** A 1-year investment will earn \$120 in interest in a savings account or \$200 in interest in a money market account that pays 2% more than the savings account. Determine each interest rate.

70. **Interest Rates on Two Bonds** Two bonds pay yearly interest. The longer-term bond pays 1.5% more than the shorter-term bond. The yearly interest on an investment will be \$275 if invested in the shorter-term bond or \$350 if invested in the longer-term bond. Determine each interest rate.

71. **Comparison of Mortgage Rates** A prospective homeowner is looking for a mortgage for a new house. For a 30-year mortgage the interest for the first month will be \$850. For a 15-year mortgage the interest for the first month will be \$700. The rate is 1.5% less for the 15-year mortgage than for the 30-year mortgage. Determine each interest rate.

72. **Comparison of Mortgage Rates** A prospective homeowner is looking for a mortgage for a new house. For a 30-year mortgage the interest for the first month will be \$720. For a 15-year mortgage the interest for the first month will be \$630. The rate is 1% less for the 15-year mortgage than for the 30-year mortgage. Determine each interest rate.

Group Discussion Questions

73. **Communicating Mathematically** For each of the following pairs of variables determine whether it is more reasonable that these variables vary directly or vary inversely.
    a. The volume of paint to be painted on a wall and the time to brush this paint on the wall
    b. The area covered by a bucket of paint and the thickness with which the paint is applied
    c. The resistance of a pipe to the flow of water and the square of the radius of the pipe
    d. The radius of a household water pipe and the volume of water flowing through this pipe

74. **Challenge Question**
    a. Write a numerical word problem that is modeled by $\frac{1}{n} + \frac{1}{n+2} = \frac{12}{35}$.
    b. Write a word problem involving the work done by two farm tractors that is modeled by the equation $\frac{4}{t} + \frac{4}{t+6} = 1$.
    c. Write a word problem involving two investments that is modeled by the equation $\frac{600}{r} = \frac{400}{r - 0.02}$.
    d. Write a word problem involving the distance traveled by two planes that is modeled by the equation $\frac{600}{r + 50} = \frac{400}{r - 50}$.

75. **Discovery Question**
    a. Describe two variables that you believe should vary directly.
    b. Collect 10 pairs of data involving these variables and construct a scatter diagram for these points.
    c. Do these data points support your claim?

76. **Discovery Question**
    a. Describe two variables that you believe should vary inversely.
    b. Collect 10 pairs of data involving these variables and construct a scatter diagram for these points.
    c. Do these data points support your claim?

KEY CONCEPTS FOR CHAPTER 8

1. **Rational Expression** A rational expression is the ratio of two polynomials. Values of the variables that would cause division by 0 must be excluded.

2. **Reducing Rational Expressions** If A, B, and C are real algebraic expressions and $B \neq 0$ and $C \neq 0$, then $\frac{AC}{BC} = \frac{A}{B}$. Note $\frac{C}{C} = 1$, the multiplicative identity. A rational expression is in lowest terms when the numerator and the denominator have no common factor other than -1 or 1.

3. **Least Common Denominator (LCD)** The least common denominator of two or more fractions is the product formed by using each factor the greatest number of times it occurs in any of the denominators.

4. **Operations with Rational Expressions** If A, B, C, and D are real polynomials, then:
    - $\frac{A}{C} + \frac{B}{C} = \frac{A+B}{C}$ for $C \neq 0$
    - $\frac{A}{C} - \frac{B}{C} = \frac{A-B}{C}$ for $C \neq 0$
    - $\frac{A}{B} \cdot \frac{C}{D} = \frac{AC}{BD}$ for $B \neq 0$ and $D \neq 0$
    - $\frac{A}{B} \div \frac{C}{D} = \frac{A}{B} \cdot \frac{D}{C} = \frac{AD}{BC}$ for $B \neq 0$, $C \neq 0$, and $D \neq 0$

5. **Signs of Rational Expressions** If A and B are rational expressions, then:
   - $-\frac{A}{B} = \frac{-A}{B} = \frac{A}{-B}$ for $B \neq 0$
   - $\frac{1}{B - A} = -\frac{1}{A - B}$ for $A \neq B$
   - $\frac{A - B}{B - A} = -1$ for $A \neq B$
6. **Order of Operations**
   Step 1. Start with the expression within the innermost pair of grouping symbols.
   Step 2. Perform all exponentiations.
   Step 3. Perform all multiplications and divisions as they appear from left to right.
   Step 4. Perform all additions and subtractions as they appear from left to right.
7. **Complex Rational Expression** A complex rational expression is a rational expression whose numerator or denominator (or both) is also a rational expression.
8. **Simplifying Complex Fractions**
   - Complex fractions can be simplified by rewriting the complex fraction as a division problem. Divide the main numerator by the main denominator.
   - Complex fractions can be simplified by multiplying both the numerator and the denominator by the LCD of all the fractions that occur in the numerator and denominator of the complex fraction.
9. **Rational Function** A rational function f is defined by the ratio of polynomials: $f(x) = \frac{P(x)}{Q(x)}$, where $P(x)$ and $Q(x)$ are polynomials and $Q(x) \neq 0$.
10. **Domain of a Rational Function** The domain of a rational function consists of all real numbers except those that are excluded to prevent division by 0.
11. **Notation for Excluded Values** If 2 is the only value excluded from the domain of a function, we can write the domain by using interval notation as $(-\infty, 2)\cup(2, \infty)$. Another notation to represent this domain is $\mathbb{R} \sim \{2\}$ which is read "the set of all real numbers except for 2."
12. **Linear Asymptotes**
   - **Vertical asymptote** A vertical line that a graph approaches but does not touch is called a vertical asymptote.
   - **Horizontal asymptote** A horizontal line that a graph approaches is called a horizontal asymptote.
13. **Vertical Asymptote of a Rational Function** The graph of a rational function will have a vertical asymptote of $x = c$ if $x - c$ is a factor of the denominator of the reduced form of this function.
14. **Variation**
   - **Direct variation** If x and y are variables and $k \neq 0$ is a constant, then stating "y varies directly as x with constant of variation k" means $y = kx$.
   - **Inverse variation** If x and y are variables and $k \neq 0$ is a constant, then stating "y varies inversely as x with constant of variation k" means $y = \frac{k}{x}$ for $x \neq 0$.
   - **Joint variation** If x, y, and z are variables and $k \neq 0$ is a constant, then stating "z varies jointly as x and y with constant of variation k" means $z = kxy$.
15. **Extraneous Value** An extraneous value is a value produced in the solution process that does not check in the original equation.
16. **Solving an Equation Containing Rational Expressions**
   Step 1. Multiply both sides of the equation by the LCD.
   Step 2. Solve the resulting equation.
   Step 3. Check the solution to determine whether it is an excluded value and therefore extraneous.
17. **Strategy for Solving Word Problems**
   Step 1. Read the problem carefully to determine what you are being asked to find.
   Step 2. Select a variable to represent each unknown quantity. Specify precisely what each variable represents and note any restrictions on each variable.
   Step 3. If necessary, make a sketch and translate the problem into a word equation or a system of word equations. Then translate each word equation into an algebraic equation.
   Step 4. Solve the equation or system of equations, and answer the question asked by the problem.
   Step 5. Check the reasonableness of your answer.
18. **Rate of Work** If one job can be done in a time t, then the rate of work is $\frac{1}{t}$ of the job per unit of time.

REVIEW EXERCISES FOR CHAPTER 8

In Exercises 1–4, evaluate each expression for $f(x) = \frac{2x - 9}{x^2 - 1}$.

1. $f(0)$
2. $f(1)$
3. $f(4)$
4. $f(10)$
5. Use the table below to determine the domain of $f(x) = \frac{3x + 1}{x^2 - x - 6}$.

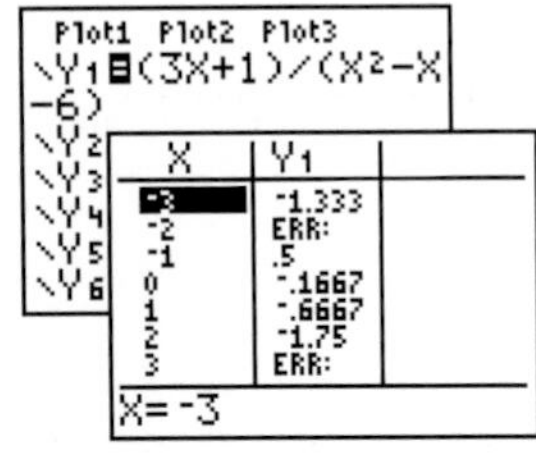

6. Use the graph below to determine the domain of $f(x) = \frac{x + 3}{x^2 + 2x - 15}$.

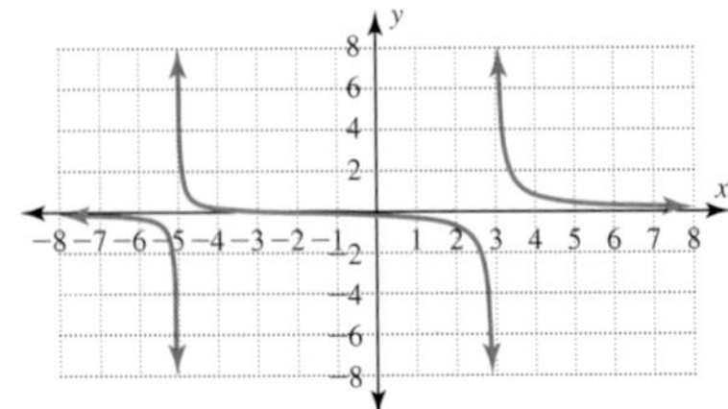

In Exercises 7 and 8, determine the domain of each function.

7. $f(x) = \dfrac{(5x + 1)(x - 3)}{(x + 5)(3x - 1)}$

8. $f(x) = \dfrac{x^2 - 11}{x^2 + 13}$

In Exercises 9–11, determine the vertical asymptotes of the graph of each rational function.

9. $f(x) = \dfrac{7x - 9}{4x - 2}$

10. $f(x) = \dfrac{5x^2 - 7x + 11}{x^2 - 36}$

11. $f(x) = \dfrac{6x^2 - 23x + 21}{x^2 - 9x}$

In Exercises 12–20, reduce each expression to lowest terms.

12. $\dfrac{36x^2y}{12xy^3}$

13. $\dfrac{6a - 18b}{12b - 4a}$

14. $\dfrac{15x^2 - 15}{25x + 25}$

15. $\dfrac{x^2 + x - 30}{2x^2 + 11x - 6}$

16. $\dfrac{cx - cy}{ax - ay + bx - by}$

17. $\dfrac{3 - 7m}{14m^2 - 6m}$

18. $\dfrac{10x^2 + 29xy + 10y^2}{6x^2 + 13xy - 5y^2}$

19. $\dfrac{9x^2 - 24xy + 16y^2}{12x^2 - 25xy + 12y^2}$

20. $\dfrac{x^2 - 4y^2}{x^2 - 4xy + 4y^2}$

In Exercises 21 and 22, find the missing numerators.

21. $\dfrac{2x + 1}{x - 3} = \dfrac{?}{x^2 - 6x + 9}$

22. $\dfrac{y - 3}{y + 4} = \dfrac{?}{2y^2 + 9y + 4}$

In Exercises 23–26, use the table of values for each expression to match each expression with its result.

23. $\dfrac{2x - 1}{x - 2} + \dfrac{x - 2}{2x - 1}$

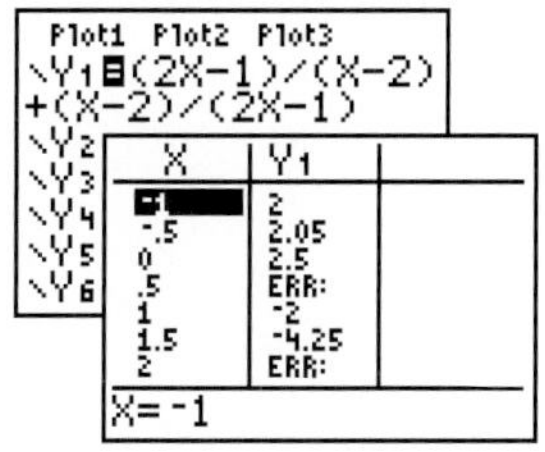

24. $\dfrac{2x - 1}{x - 2} - \dfrac{x - 2}{2x - 1}$

Plot1 Plot2 Plot3
\Y1■(2X-1)/(X-2)-(X-2)/(2X-1)
X Y1
-1 0
-.5 -.45
0 -1.5
.5 ERR:
1 0
1.5 -3.75
2 ERR:
X=-1

25. $\dfrac{2x - 1}{x - 2} \cdot \dfrac{x - 2}{2x - 1}$

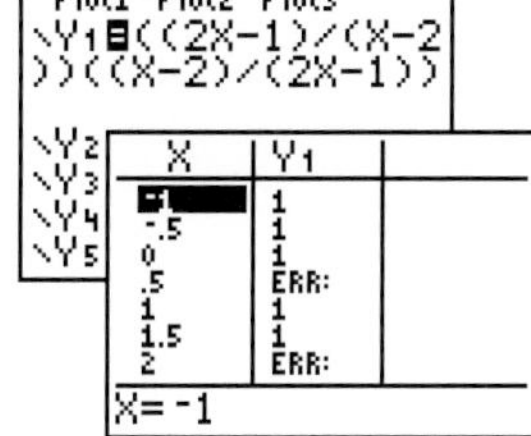

26. $\dfrac{2x - 1}{x - 2} \div \dfrac{x - 2}{2x - 1}$

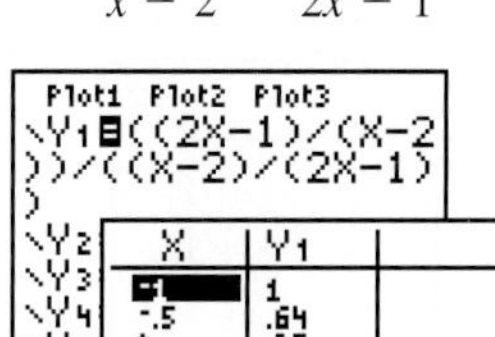

A. 1

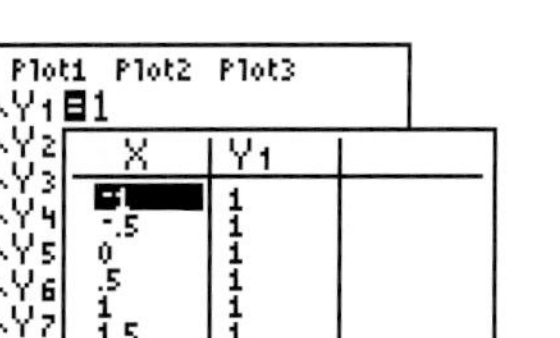

B. $\dfrac{4x^2 - 4x + 1}{(x - 2)^2}$

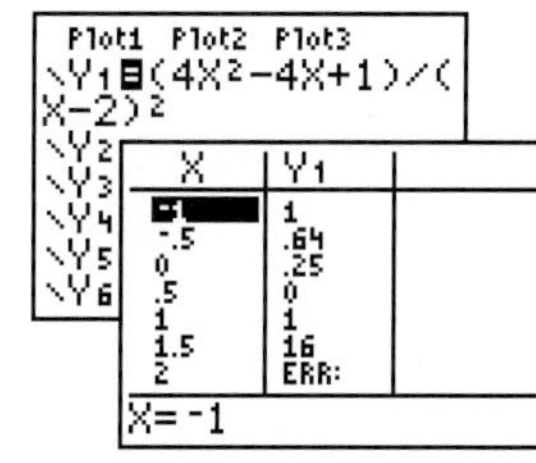

C. $\dfrac{3x^2 - 3}{(x - 2)(2x - 1)}$

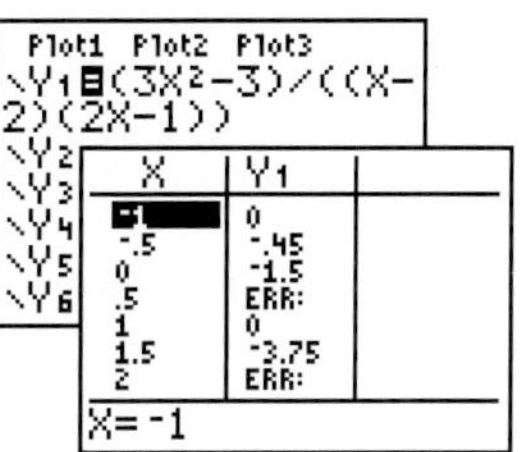

D. $\dfrac{5x^2 - 8x + 5}{(x - 2)(2x - 1)}$

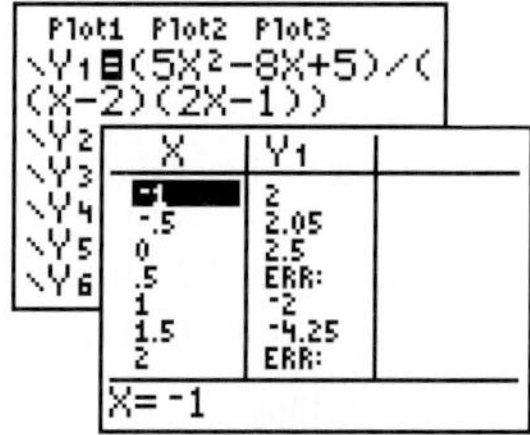

In Exercises 27–48, simplify each expression as completely as possible.

27. $\dfrac{36x^2 - 24x}{6x}$

28. $\dfrac{6xy}{2x - y} \cdot \dfrac{4x^2 - y^2}{3x^2}$

29. $\dfrac{9t^2 - 4}{16st} \div \dfrac{3t + 2}{16st^2 + 8st}$

30. $\dfrac{v^2 + 9vw + 8w^2}{v^2 - w^2} \div \dfrac{v^2 + 7vw - 8w^2}{v^2 + 5vw - 6w^2}$

31. $\dfrac{3x}{6x^2 + x - 1} - \dfrac{1}{6x^2 + x - 1}$

32. $\dfrac{1}{w + 1} - \dfrac{w}{w - 2} + \dfrac{w^2 + 2}{w^2 - w - 2}$

33. $\dfrac{3v}{3v^2 - 5v + 2} - \dfrac{2v}{2v^2 - v - 1}$

34. $\dfrac{6x}{2x - 3} - \dfrac{9x + 18}{4x^2 - 9} \cdot \dfrac{2x^2 - x - 6}{x^2 - 4}$

35. $\dfrac{2w + 4}{w^2 + 4w - 12} + \dfrac{w^2 - 169}{2w^2 - 13w + 21} \div \dfrac{w^2 - 15w + 26}{2w - 7}$

36. $\dfrac{6v^2 - 25v + 4}{6v^2 + 5v - 6} \div \dfrac{2v^2 - 3v - 20}{4v^2 + 16v + 15}$

37. $\left(\dfrac{y}{4} - \dfrac{4}{y}\right)\left(y - \dfrac{y^2}{y + 4}\right)$

38. $\dfrac{5z + 5}{6z^2 + 13z + 6} - \dfrac{1 - 4z}{3z^2 - 7z - 6} - \dfrac{3z}{2z^2 - 3z - 9}$

39. $\dfrac{\dfrac{1}{a} + \dfrac{1}{a + 1}}{\dfrac{1}{a} + \dfrac{1}{a^2}}$

40. $\dfrac{x + \dfrac{44}{x + 5} - 10}{x + \dfrac{33}{x + 5} - 9}$

41. $\left(\frac{3v^2 + 11v + 6}{3v^2 + 5v + 2}\right)^2 \div \frac{v^2 - 9}{v^2 + 2v + 1}$

42. $\frac{x^{-2} - x^{-1}}{x^{-2} + x^{-1}}$

43. $\frac{x + y^{-1}}{x - y^{-1}}$

44. $\frac{36w^{-2} + 23w^{-1} - 8}{24 - 5w^{-1} - 36w^{-2}}$

45. $\left(m - \frac{15}{m + 2}\right) \div \left(m - 1 - \frac{10}{m + 2}\right)$

46. $\frac{6y}{3y + 2} + \frac{20y + 4}{15y^2 + 7y - 2} \div \frac{5y^2 - 24y - 5}{5y^2 - 26y + 5}$

47. $\frac{3y}{25y^2 - 1} + \frac{2y - 1}{25y^2 - 1}$

48. $\frac{z^2 - 2z + 1}{z^5 - z^4} \cdot \frac{2z^4}{z^2 - 1} + \frac{2z^2 + 2z}{z^2 + 2z + 1}$

In Exercises 49–52, simplify the expression in the left column, and solve the equation in the right column.

	Simplify	Solve
49.	a. $\frac{1}{m + 2} - \frac{3}{m + 4}$	b. $\frac{1}{m + 2} = \frac{3}{m + 4}$
50.	a. $\frac{x - 2}{x + 1} - \frac{x - 4}{x - 6}$	b. $\frac{x - 2}{x + 1} = \frac{x - 4}{x - 6}$
51.	a. $\frac{x}{x - 3} + \frac{x + 4}{x - 2}$	b. $\frac{x}{x - 3} = \frac{x + 4}{x - 2}$
52.	a. $\frac{2x}{x^2 - 1} + \frac{2}{x^2 - 1}$	b. $\frac{2x}{x^2 - 1} = \frac{2}{x^2 - 1}$

In Exercises 53–60, solve each equation.

53. $\frac{7}{2x} = \frac{2}{x} - \frac{3}{2}$

54. $\frac{15}{w^2 + 5w} + \frac{w + 4}{w + 5} = \frac{w + 3}{w}$

55. $\frac{1}{y^2 + 5y + 6} - \frac{2}{y + 3} = \frac{7}{y + 2}$

56. $1 - \frac{14}{y^2 + 4y + 4} = \frac{7y}{y^2 + 4y + 4}$

57. $\frac{w}{w^2 - 9w + 20} + \frac{14}{w^2 - 3w - 4} = \frac{18}{w^2 - 4w - 5}$

58. $\frac{4}{v^2 + 7v + 10} - \frac{3}{v + 2} + \frac{8}{v + 5} = 0$

59. $\frac{5z + 11}{2z^2 + 7z - 4} = \frac{1}{z + 4} - \frac{3}{1 - 2z}$

60. $\frac{1}{2z^2 - 9z - 5} - \frac{1}{2z^2 - 6z - 20} = \frac{3}{(2z + 1)(2z + 4)(z - 5)}$

61. Solve $\frac{a}{b + 1} = \frac{a + 1}{b}$ for b.

62. Solve $I = \frac{E}{r_1 + r_2}$ for r_2.

Estimation Skills and Calculator Skills

In Exercises 63 and 64, mentally estimate the value of each expression for $f(x) = \frac{(x + 5)(x - 8)}{x^2 - 4}$ and then use a calculator to approximate each value to the nearest hundredth.

PROBLEM	MENTAL ESTIMATE	CALCULATOR APPROXIMATION
63. $f(1.01)$		
64. $f(-4.99)$		

65. **Perimeter and Area of a Trapezoid**
   a. Determine the perimeter of the trapezoid shown in the figure below.
   b. Determine the area of this trapezoid by using the formula $A = \frac{h}{2}(a + b)$.

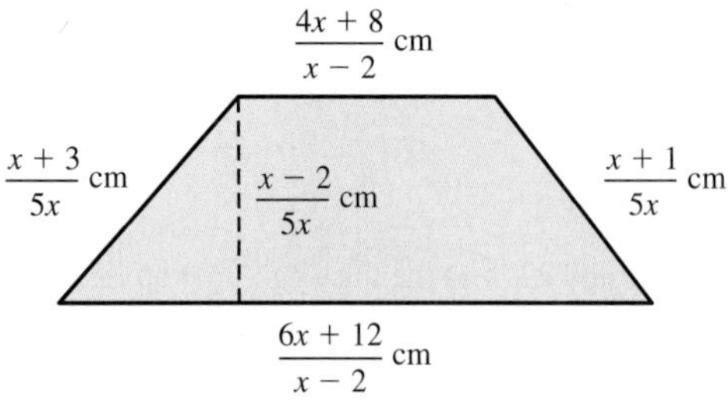

66. **Area of a Region** Determine the area of the region shown below.

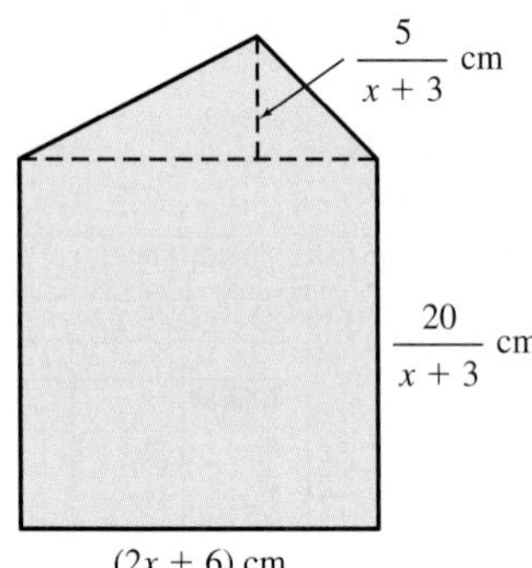

67. **Volume of a Box** Determine the volume of the box shown below.

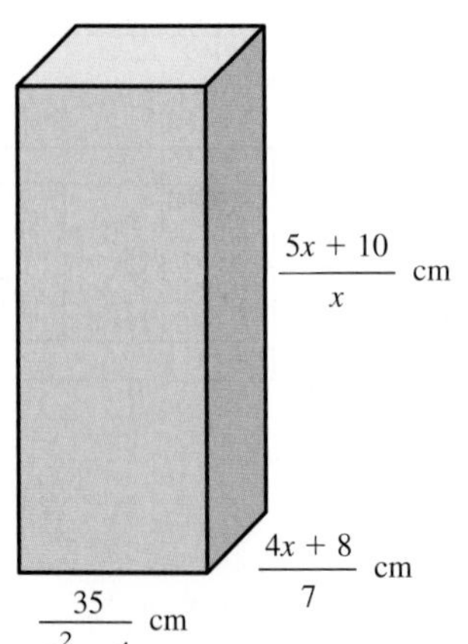

68. **Average Cost** The number of units that a plant can produce by operating t h/day is given by $N(t) = 0.5t^2 + 10t$. The cost in dollars of operating the plant for t hours is given by $C(t) = 100t + 500$.
   a. What is the practical domain for the values of t?
   b. Determine $A(t)$, the average cost per unit when the plant is operated for t h/day.
   c. Evaluate and interpret $N(10)$.
   d. Evaluate and interpret $C(10)$.
   e. Evaluate and interpret $A(10)$.
   f. Describe what happens to the average cost when the number of hours the plant operates approaches 0 hours.
   g. Describe what happens to the average cost when the plant approaches operating round the clock.
69. **Travel Time** The amount of time required for a bus to travel from St. Louis, Missouri, to Chicago, Illinois, varies inversely with the average speed of the bus. The trip takes 5 hours if the bus averages 60 mi/h. How long will the trip take at an average speed of 50 mi/h? What is the significance of the constant of variation k?
70. **Revolutions per Minute of Two Pulleys** The number of revolutions per minute (rev/min) of two pulleys connected by a belt varies inversely as their diameters. A pump is belt-driven by an electric motor that runs at 2,000 rev/min. If the pump has a pulley of 6-in diameter and must operate at 1,500 rev/min, what size pulley is required on the motor?
71. **Force of the Wind** The force of the wind blowing into the front of a car varies jointly as the effective area of the front of the car and the square of the speed of the wind. A car with an effective area of 25 ft^2 has a force of 80 lb exerted when the car is driving 40 mi/h in still air. Find the force exerted if the car goes 60 mi/h.
72. The denominator of a fraction is 6 more than the numerator, and the fraction equals -1. Find the numerator.
73. The ratio of one number to another is $\frac{5}{2}$. If the first number is 5 less than 3 times the second, find both numbers.
74. The sum of the reciprocals of two consecutive odd integers is 12 times the reciprocal of their product. Find these integers.
75. **Electrical Resistance** The resistance of the parallel circuit with two resistors shown in the figure is 3 Ω. If one resistor has 3 times the resistance of the other, what is the resistance of each?

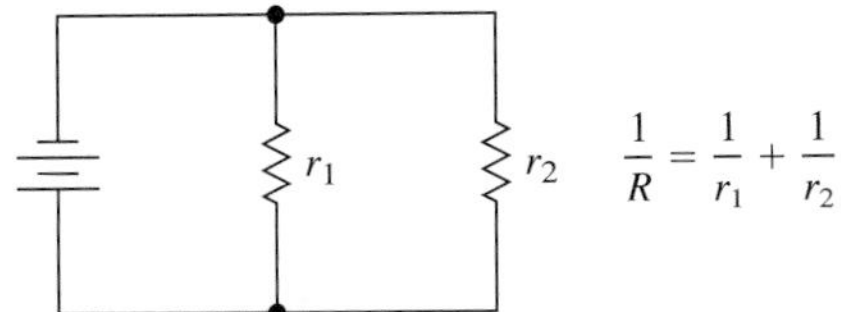

76. **Filling a Bathtub** A hot water faucet takes 7 minutes longer than the cold water faucet to fill a bathtub. If both faucets were turned on, the tub would be half full in 6 minutes. How many minutes would it take to fill an empty tub with cold water if the hot water faucet were turned off?
77. **Average Cost** A furniture company can produce x units of a desk for \$54 per desk, plus an initial start-up investment of \$20,000. To compete in the market, it must be able to achieve an average cost of only \$74 per unit. How many units must the company produce to reach an average cost of \$74 per unit?
78. **Dimensions of a Metal Sheet** A rectangular piece of metal is 6 cm longer than it is wide. A 0.5-cm strip is cut off of each side, leaving an area thirteen-sixteenths of the original area. Find the original dimensions.

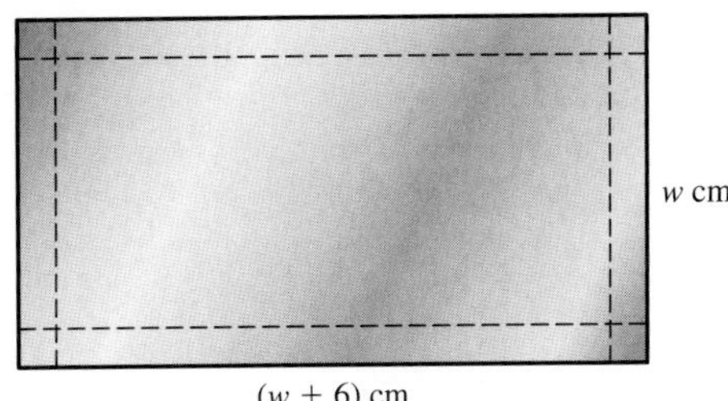

79. **Speed of Boats** Two boats that travel the same speed in still water depart from a dock at the same time. The boats travel in opposite directions in a river that has a current of 6 mi/h. After a period of time one boat is 54 mi downstream, and the other boat is 30 mi upstream. What is the speed of each boat in still water?
80. **Selecting the Appropriate Type of Variation** For each of the following pairs of variables, determine whether it is more reasonable that these variables vary directly or inversely.
   a. The distance a car must travel between two cities and the distance between these two cities on a map
   b. The time interval on a train schedule and the rate of travel of the train between two stations
   c. A weather balloon filled with helium is pulled under water by a small rope. The number of feet the balloon is submerged and the volume of the balloon
   d. A weather balloon filled with helium is pulled under water by a small rope. The number of feet the balloon is submerged and the pressure on the helium in the balloon

MASTERY TEST FOR CHAPTER 8

[8.1] 1. Determine the domain of each rational function.

a. $f(x) = \dfrac{x - 5}{2x - 12}$ b. $f(x) = \dfrac{x + 5}{(x - 1)(x + 6)}$

c. $f(x) = \dfrac{x + 7}{x^2 - 64}$ d. $f(x) = \dfrac{2x + 19}{x^2 + 1}$

[8.1] 2. Determine the vertical asymptotes of the graphs of each of these rational functions.

a. $f(x) = \dfrac{x - 5}{2x - 12}$ b. $f(x) = \dfrac{x + 5}{(x - 1)(x + 6)}$

c.

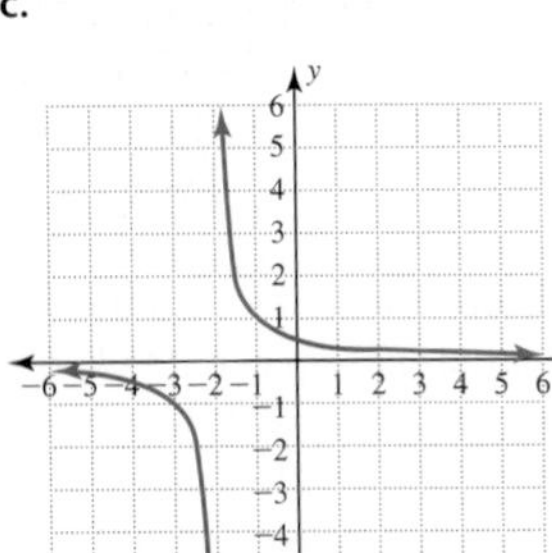

d.

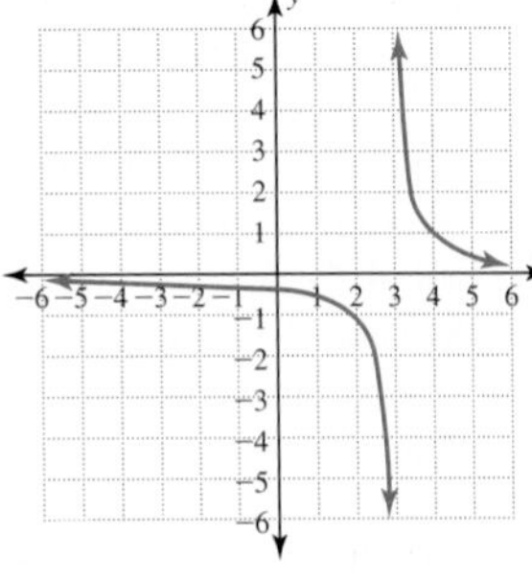

[8.1] **3.** Reduce each rational expression to lowest terms.

a. $\dfrac{6x}{12x^2 - 18x}$ b. $\dfrac{2x - 6}{x^2 - 9}$

c. $\dfrac{2x^2 - 7x - 15}{x^2 - 25}$ d. $\dfrac{4x^2 - 12xy + 9y^2}{3ay - 6by - 2ax + 4bx}$

[8.2] **4.** Perform each multiplication or division, and reduce the result to lowest terms.

a. $\dfrac{ax - bx}{x^2} \cdot \dfrac{5x}{4a - 4b}$

b. $\dfrac{x^2 - 1}{x + 1} \div \dfrac{x^2 - 3x + 2}{x - 2}$

c. $\dfrac{v^2 - v - 20}{v^2 - 16} \div \dfrac{3v - 15}{2v - 8}$

d. $\dfrac{x^2 - 4}{x^2 - 4x - 21} \cdot \dfrac{x^2 - 2x - 35}{x^2 - 7x + 10}$

[8.3] **5.** Perform each addition or subtraction, and reduce the result to lowest terms.

a. $\dfrac{3x - 4}{2x - 3} + \dfrac{x - 2}{2x - 3}$

b. $\dfrac{3x + 7}{x^2 + x - 12} - \dfrac{2x + 3}{x^2 + x - 12}$

c. $\dfrac{7}{w^2 + w - 12} + \dfrac{2}{w^2 - 8w + 15}$

d. $\dfrac{1}{x + y} + \dfrac{3}{x - y} + \dfrac{2y}{x^2 - y^2}$

[8.4] **6.** Simplify each expression.

a. $\dfrac{4x}{5} + \dfrac{x^2}{15} \cdot \dfrac{3}{x}$ b. $\left(\dfrac{4x}{5} + \dfrac{x^2}{15}\right) \cdot \dfrac{3}{x}$

c. $\dfrac{x - y}{x + y} + \dfrac{3x - 21y}{x^2 - y^2} \div \dfrac{15x^2 - 105xy}{10x^2y - 10xy^2}$

d. $\left(\dfrac{w}{w - 2} + \dfrac{1}{w - 3}\right) \div \dfrac{w^2 - 2w - 2}{w^2 - w - 2}$

[8.4] **7.** Simplify each expression.

a. $\dfrac{\frac{4}{5}}{6}$ b. $\dfrac{\frac{x}{3} - 2 + \frac{3}{x}}{1 - \frac{3}{x}}$

c. $\dfrac{36x^{-2} - 3x^{-1} - 18}{12x^{-2} - 25x^{-1} + 12}$ d. $\dfrac{\frac{v}{v - 4} - \frac{2v}{v + 3}}{\frac{v^2 - 4v}{2v + 6}}$

[8.5] **8.** Solve each of the following equations.

a. $\dfrac{z - 4}{z - 2} = \dfrac{1}{z - 2}$

b. $\dfrac{z - 2}{z - 1} + \dfrac{z - 3}{2z - 5} = 1$

c. $\dfrac{z + 11}{z^2 - 5z + 4} = \dfrac{5}{z - 4} + \dfrac{3}{1 - z}$

d. Solve $\dfrac{x + 4}{x - 5} = y$ for x.

[8.6] **9.** **a.** If y varies inversely as x, and y is 12 when x is 6, find y when x is 4.

b. The volume of a gas varies inversely with the pressure when temperature is held constant. If the pressure of 4 L of a gas is 6 newtons per square centimeter (N/cm^2), what will the pressure be if the volume is 3 L?

c. z varies jointly as x and y, and z is 12 when x is 6 and y is 4. Find z when x is 8 and y is 10.

d. The force of a wind blowing directly into a wall varies jointly as the area of the wall and the square of the speed of the wind. A wall that measures 8 ft by 10 ft has a 20-mi/h wind exert a force of 240 lb. Find the force exerted on the same wall by a 45-mi/h wind.

[8.6] **10.** Solve each of the following problems.

a. Ratio of Economic Indicators The ratio of two monthly economic indicators is $\frac{2}{3}$. The first monthly indicator is 2 units below normal, and the second monthly indicator is 2 units above normal. What is the normal number of units for this indicator?

b. Work by Two Machines When members of a construction crew use two end loaders at once, they can move a pile of sand in 6 hours. If they use only the larger end loader, they can do the job in 5 hours less time than it would take if they used the smaller machine. How many hours would it take to do the job while using the larger machine?

c. Interest Rates on Two Bonds Two bonds pay yearly interest. The longer-term bond pays 2.5% more than the shorter-term bond. The yearly interest on an investment would be \$680 if invested in the longer-term bond or \$480 if invested in the shorter-term bond. Determine each interest rate.

d. Airplanes Traveling in an Airstream Two planes depart from the same airport at the same time, flying in opposite directions. After a period of time, the slower plane has traveled 950 mi and the faster plane has traveled 1,050 mi. The faster plane is traveling 50 mi/h faster than the slower plane. Determine the rate of each plane. (*Hint:* What is the same for each plane?)

REALITY CHECK FOR CHAPTER 8

The weight and mass of an object are two commonly confused ideas. The weight of an object is an indication of the gravitational force between it and a planet or moon. An astronaut's weight will decrease as the astronaut moves farther away from a planet. The mass of an astronaut will be the same in space as it is on the surface of the Earth. The function $E(x) = \dfrac{3.189 \times 10^{16}}{(x + 6{,}378{,}100)^2}$ gives the weight of an 80-kg astronaut as a function of the distance x of the astronaut above the surface of the Earth. The function $M(x) = \dfrac{3.929 \times 10^{14}}{(x + 1{,}738{,}000)^2}$ gives the weight of the same astronaut as a function of the distance x of the astronaut above the surface of the Moon. Determine the weight of this astronaut on the surface of the Earth and on the surface of the Moon. The units for weight will be newtons.

GROUP PROJECT FOR CHAPTER 8

An Algebraic Model for Average Cost

A factory that manufactures custom steering wheels collected data over several weeks. For selected values, the table gives t, the number of hours the factory operated on a given day. The table also displays $N(t)$, the number of steering wheels produced on this day, and $C(t)$, the cost in dollars of operating the factory that day.

t	$N(t)$	$C(t)$
2	20	3,000
4	60	5,000
6	120	8,000
8	180	10,000
12	300	15,000
16	440	20,000
20	600	25,000
24	700	30,000

1. Use the **QuadReg** feature on a graphing calculator to determine the equation of a quadratic function for $N(t)$. Round the coefficients to three significant digits.
2. Use the **LinReg** feature on a graphing calculator to determine the equation of a linear function for $C(t)$. Round the coefficients to three significant digits.
3. Use the functions from parts 1 and 2 to write an equation for the average cost per steering wheel as a function of the number of hours per day the factory operates: $A(t) = \dfrac{C(t)}{N(t)}$.
4. What is the practical domain for this function?
5. Use this function to approximate and interpret $A(10)$.
6. Describe what happens to the average cost when the number of hours the plant operates per day approaches approximately 2 hours.
7. Describe what happens to the average cost when the plant approaches operating round the clock.

CUMULATIVE REVIEW FOR CHAPTERS 1 THROUGH 8

The answer to each of these questions follows this diagnostic review. Each answer is keyed to an example in this book. You can refer to these examples to find explanations and additional exercises for practice.

Additive Inverse, Absolute Value, and Square Roots

1. Write the additive inverse of each of the following real numbers.
   a. -5 b. $\frac{1}{2}$
2. Evaluate each of these absolute value expressions.
   a. $|7|$ b. $|-7|$
3. Evaluate each expression for $x = 3$, $y = -9$, and $z = 7$.
   a. $|-x|$ b. $-|-x|$
   c. $\sqrt{-y}$ d. $\sqrt{z - y}$
4. Mentally estimate to the nearest integer the value of each of the following expressions, and then use your calculator to approximate each value to the nearest hundredth.
   a. $\sqrt{63}$ b. $\sqrt{50}$
   c. $\sqrt{10 + 26}$ d. $\sqrt{10} + \sqrt{26}$

Properties and Subsets of the Real Numbers

5. Match each number with the most appropriate description.
   a. $-\frac{5}{3}$ A. A whole number that is not a natural number
   b. -2 B. A natural number
   c. 0 C. A rational number that is not an integer
   d. $\sqrt{7}$ D. A positive number that is not rational
   e. $\sqrt{16}$ E. An integer that is not a whole number
6. a. The property that states that $5 + (4 + x) = (5 + 4) + x$ is the __________ property of __________.
   b. The property that states that $3(2 + x) = 3(x + 2)$ is the __________ property of __________.
   c. The property that states that $2(5 \cdot 3) = (2 \cdot 5) \cdot 3$ is the __________ property of __________.
   d. The property that states that $5 \cdot (x - 4) = (x - 4) \cdot 5$ is the __________ property of __________.
7. Use the distributive property of multiplication over addition to complete these exercises.
   a. Expand $5(2x - 3)$. b. Factor $25x + 30$.
   c. Expand $-3(4x - 7)$. d. Factor $18x - 27$.

Arithmetic and Order of Operations

Calculate the value of each expression without using a calculator.

8. a. $-\frac{5}{12} + \frac{1}{12}$ b. $-\frac{3}{4} - \left(-\frac{1}{9}\right)$
   c. $-\frac{3}{8} \div \left(-\frac{3}{4}\right)$ d. $-2(-5)(-3)$
9. a. $(-4)^2$ b. -4^2
   c. 5^0 d. $(-1)^{36}$
10. a. $12 - 2(5 - 2)$ b. $\frac{12 \div 2 \cdot 4}{12 - 2 \cdot 4}$
    c. $(3 + 4)^2$ d. $4 - 2[7 - 3(5 - 8)]$

Using the Properties of Exponents

Simplify each expression, assuming that all variables are restricted to values that avoid division by 0 and avoid 0^0.

11. a. $\frac{-6a^4b^{-6}}{3a^{-2}b^2}$ b. $(-2v^2w^3)^4$
    c. $\left(\frac{x^2y^3}{xy}\right)^3$ d. $[(2x^2)(3x^4)]^2$
12. a. $\frac{5}{2^0}$ b. $\left(\frac{2}{3}\right)^{-2}$
    c. $\left(\frac{1}{3}\right)^{-1} + \left(\frac{1}{6}\right)^{-1}$ d. $\left(\frac{1}{3} + \frac{1}{6}\right)^{-1}$
13. Write these numbers in standard decimal notation.
    a. 1.23×10^4 b. 3.89×10^{-6}
    Write these numbers in scientific notation.
    c. 0.0000157 d. 2,450,000,000

Linear Equations in One Variable and Direct Variation

14. Solve each linear equation.
    a. $-2x = 3x - 15$ b. $\frac{6}{7} = \frac{3x}{14}$
    c. $4(3x + 2) = 5(x + 3) + 14$
    d. $\frac{x + 1}{5} = \frac{x - 7}{3}$
15. To solve the equation $2.5x - 1.6 = 3.4x - 6.1$, the following table was formed on a graphing calculator by letting Y_1 represent the left side of the equation and Y_2 represent the right side of this equation. Use the table to solve this equation for x.

X	Y_1	Y_2
1	.9	-2.7
2	3.4	.7
3	5.9	4.1
4	8.4	7.5
5	10.9	10.9
6	13.4	14.3
7	15.9	17.7
X=1		

16. Each of these equations is a conditional equation, an identity, or a contradiction. Identify the type of each equation and then solve it.
    a. $3(x + 5) = 3x + 5$ b. $3(x + 5) = 3x + 15$
    c. $3(x + 5) = 5(x + 3)$
17. Solve the proportions in parts **a** and **b**.
    a. $\frac{x - 5}{x + 2} = \frac{3}{4}$ b. $\frac{x - 3}{8} = \frac{x + 3}{5}$
    c. The recipe for one batch of cookies calls for $\frac{3}{4}$ cup of sugar. How many batches of cookies can be made by using 3 cups of sugar?
18. a. If y varies directly as x, and y is 24 when x is 18, find y when x is 6.
    b. The number of Mexican pesos varies directly as the number of U.S. dollars exchanged. In 2005, one could exchange 20 U.S. dollars for 219.64 pesos. What was the constant of variation (the exchange rate) at that time?

19. Solve each of these equations for a.

a. $P = a + 2b + c$ b. $S = 2ab + 2ac + 2bc$

20. Solve the linear equation $2x + y - 7 = 4(2 + 3x - y)$ for y, and then use a graphing calculator to complete this table.

X	Y1
-3	
-2	
-1	
0	
1	
2	
3	

X=-3

The Rectangular Coordinate System and Arithmetic Sequences

21. Identify the coordinates of the points A through D in the figure and give the quadrant in which each point is located.

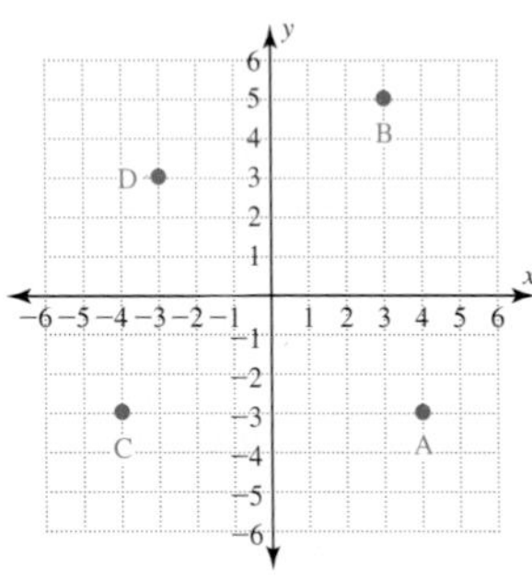

22. Draw a scatter diagram for the data points in the table.

x	y
−1	5
0	−1
1	0
2	2
3	−3

23. Determine whether each sequence is an arithmetic sequence. If the sequence is arithmetic, write the common difference d.

a. 1, 4, 7, 10, 13 b. 2, 4, 8, 16, 32

Functions and Representations of Functions

24. Determine whether each relation is a function.

a. *D* → *R*

2 → 7
2 → 9
5 → 13
5 → 12

b.

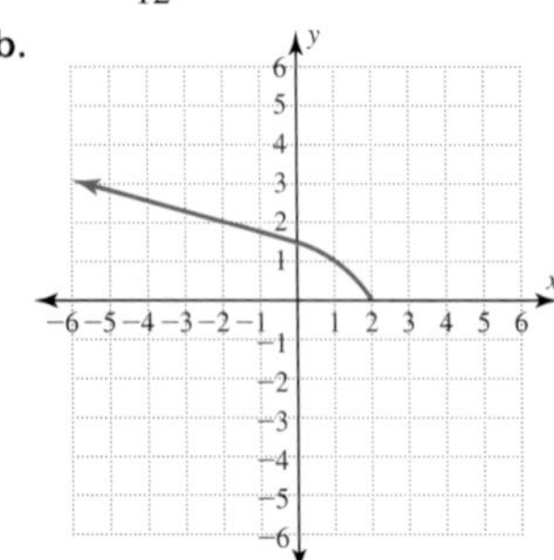

c. $\{(1, 5), (2, 5), (3, 5)\}$

25. Determine the domain and range of each of these functions.

a. $\{(5, 3), (1, 4), (6, 10), (7, 11)\}$

b.

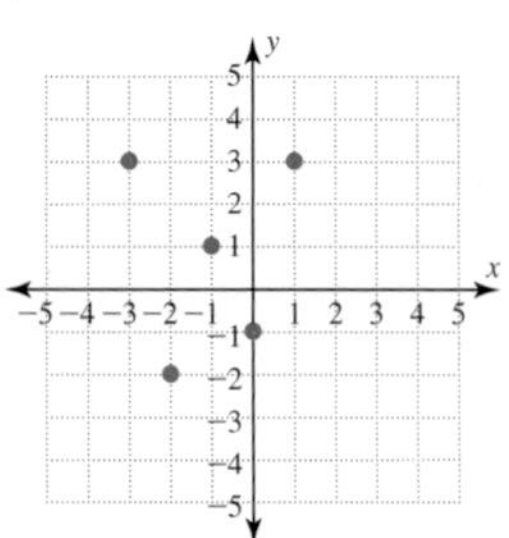

c. $f(x) = \sqrt{2 - x}$ (*Hint:* Use a calculator to graph this function.)

26. Evaluate each of the following expressions for $f(x) = 2x^2 + 3x - 5$.

a. $f(-3)$ b. $f(4)$

27. Use the given graph to determine the missing input and output values.

a. $f(0) =$ ________

b. $f(x) = 0$; $x =$ ________

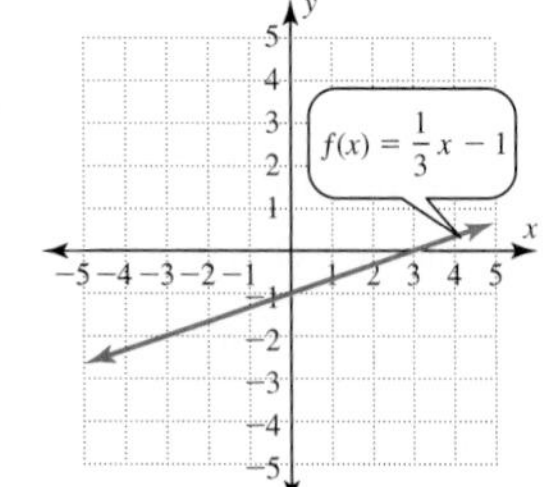

28. Use the given table of values to determine the missing input and output values.

a. $f(-3) =$ ________

b. $f(x) = 3$; $x =$ ________

x	$f(x)$
−6	−3
−3	−2
0	−1
3	0
6	1
9	2
12	3

Linear Functions

29. Determine whether $(-1, 4)$ is a solution of each equation.

a. $y = 2x + 7$ b. $y = -3x + 1$

30. Use each equation to complete a table with input values of 1, 2, 3, 4, and 5, and then graph these points and the line through these points.

a. $y = x - 1$ b. $y = -x + 3$

31. a. Determine the x- and y-intercepts of this line.

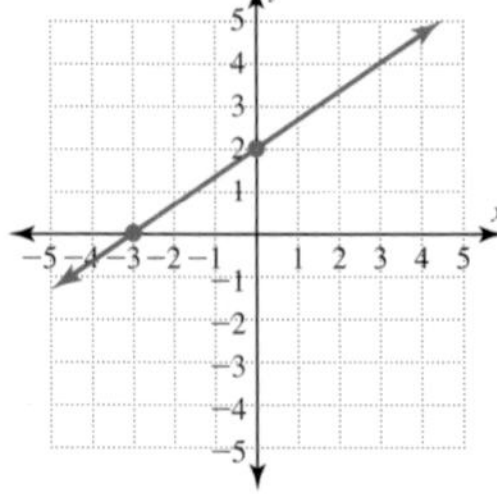

b. Determine the x- and y-intercepts of the line containing the points in the table.

x	$f(x)$
−8	−7
−4	0
0	7
4	14
8	21
12	28
16	35

c. Determine the x- and y-intercepts of the line defined by this equation: $y = -2x + 8$.

32. Total Paid on a Car Loan A new car loan requires a \$500 down payment and a monthly payment of \$225.

a. Write a function that gives the total paid by the end of the xth month.

b. Use the function obtained in part **a** to complete this table.

x	0	6	12	18	24	30
$f(x)$						

c. Evaluate and interpret $f(20)$.

d. Determine the value of x for which $f(x) = 9{,}500$. Interpret this value.

33. a. Calculate the slope of the line through the points $(5, -2)$ and $(3, 4)$.

b. Determine the slope of the line graphed.

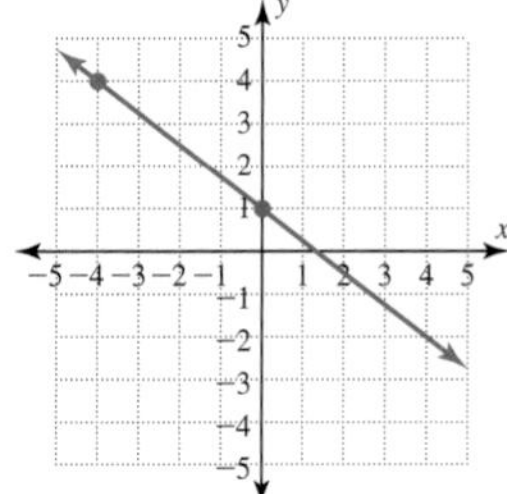

c. Determine the slope of the line defined by $y = -2x + 5$.

34. Determine whether the line defined by the first equation is parallel to, perpendicular to, or neither parallel nor perpendicular to the line defined by the second equation.

a. $y = 4x + 1$
$y = -4x + 1$

b. $y = 2$
$x = -1$

c. $y = \frac{1}{3}x - 2$
$y = \frac{1}{3}x + 2$

d. $y = \frac{1}{3}x - 2$
$y = -3x + 2$

35. Rate of Descent The altitude of a hot-air balloon is recorded every 5 seconds after it starts its descent. These data are displayed in the table.

	A	B
1	x, Time (sec)	y, Altitude (ft)
2	0	400
3	5	380
4	10	360
5	15	340
6	20	320
7		

a. Determine the value of Δx shown in the table.

b. Determine the value of Δy shown in the table.

c. Calculate the slope of the line containing these points.

d. Interpret the meaning of the rate of change obtained in part **c.**

36. Write in slope-intercept form the equation of a line satisfying the given conditions.

a. Through $(1, 4)$ with slope -2

b. y-intercept $(0, 5)$ and slope $\frac{2}{3}$

c. Through $(-1, 3)$ and parallel to $y = \frac{2}{3}x - 4$

37. a. Write the equation of a horizontal line through $(3, 5)$.

b. Write the equation of a vertical line through $(3, 5)$.

38. Graph a line satisfying the given conditions.

a. Through $(2, 3)$ with slope -2

b. y-intercept $(0, 2)$ and slope $\frac{3}{5}$

Properties of the Graphs of Linear and Absolute Value Functions

39. Identify each function as an increasing or a decreasing function.

a.

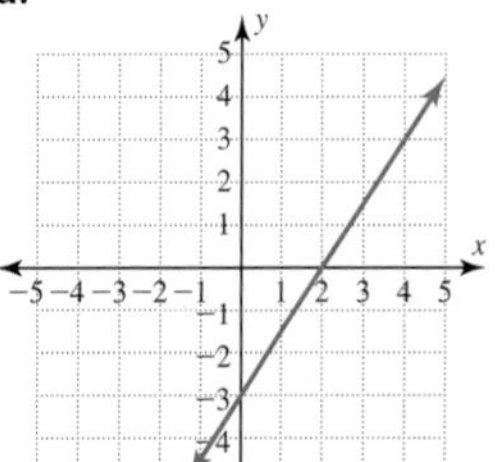

b. $f(x) = -5x + 4$

40. Determine the x-values for which each function is positive and those for which each function is negative.

a.

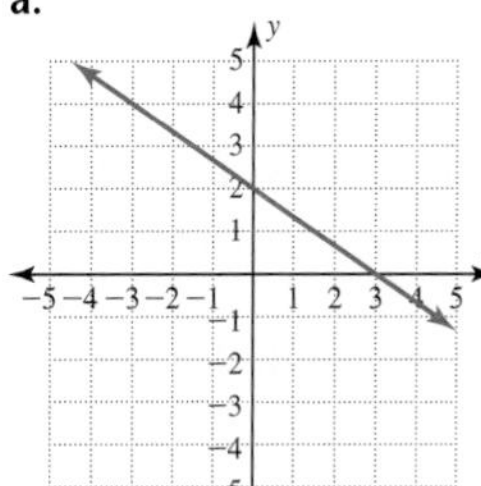

b.

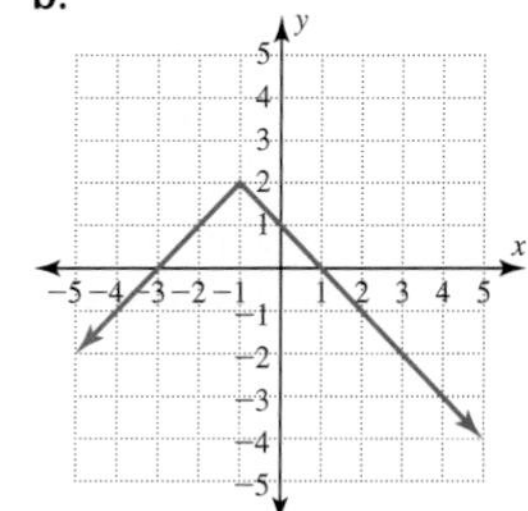

41. Use the absolute value function $f(x) = |x - 1| - 2$ to answer the following.

a. Complete the following table of values.

x	$f(x) = \|x - 1\| - 2$
−2	
−1	
0	
1	
2	
3	
4	

b. Graph this function.

c. What are the x-intercepts of this graph?

d. What is the y-intercept of this graph?

e. What point is the vertex of this graph?
f. What y-value is the minimum output value for this function?
g. What is the domain of this function?
h. What is the range of this function?
i. For what interval of input values of x is the function decreasing?
j. For what interval of input values of x is the function increasing?
k. For what interval of input values of x is the function positive?
l. For what interval of input values of x is the function negative?

Curve Fitting

42. **Copy Machine Repair Costs** To track the durability of a copy machine, the manufacturer collects data on repair costs based on the age of the machine. This table compares the years of usage on a copy machine with the yearly repair costs on that type of machine.

Years of Usage x	Repair Costs y
1	365
3	675
6	1,110
7	1,275
10	1,750

a. Draw a scatter diagram for these data points and determine the line of best fit. (Round the coefficients to three significant digits.)
b. Interpret the meaning of the slope of the line of best fit.
c. Use this line of best fit to estimate the repair costs of a copy machine that has been used for 4 years.
d. Use this line of best fit to estimate the years of usage on a copy machine that has \$1,425 in repair costs this year.

43. a. Draw a scatter diagram of the points in the given table. Then determine the quadratic function that best fits these points. (Round the coefficients to three significant digits.)
b. Use this function f to approximate $f(5)$.

x	−6	−4	−2	0	2	4	6	8
y	8	3	−1	−2	−3	−2	−1	3

Restrictions on a Variable

44. Determine the restrictions on a variable for each problem.
a. The function $C(x) = 2.20x$ models the cost of filling a 14-gal gas tank with x gal of gasoline.
b. The function $f(x) = 180 - 6x$ models the length of aluminum remaining on a roll after x pieces, each 6 ft long, have been removed from this spool.

Linear Systems

45. Graphically determine the simultaneous solution for each system of linear equations.

a.

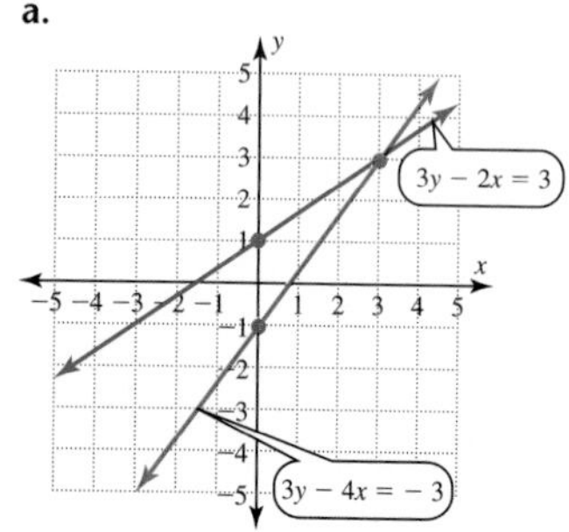

b. $f(x) = 3x - 4$
$f(x) = -2x + 11$

46. Use a numerical table to determine the simultaneous solution for each system of linear equations.

a.

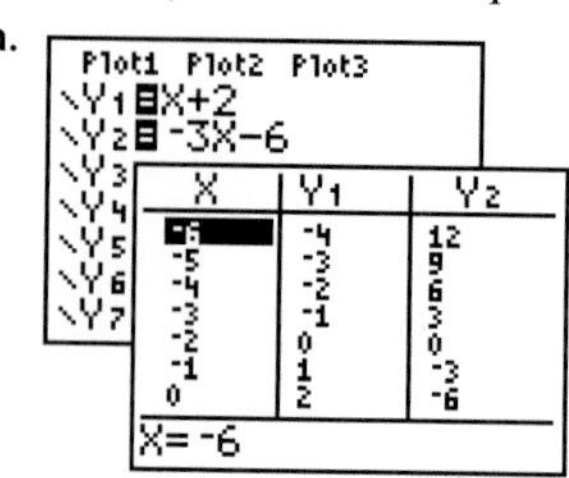

b. $y = 2x + 4$
$y = -x - 8$

47. Select the choice A, B, or C that best describes each system of linear equations.

a. $y = 2x - 3$
$y = 2x + 7$
b. $y = 3x - 4$
$y = 3x - 4$
c. $y = 3x + 1$
$y = 4x - 2$

A. A consistent system of dependent equations
B. A consistent system of independent equations
C. An inconsistent system of equations

48. Solve each system of linear equations by the substitution method.
a. $y = -2x + 1$
$3x + y = 3$
b. $x + 4y = 2$
$2x - 4y = 7$

49. Solve each system of linear equations by the addition method.
a. $x - 5y = -15$
$3x + 5y = -5$
b. $2x - 5y = -3$
$-6x + 15y = 10$

50. Use systems of linear equations to solve each word problem.
a. **Value of Two Investments** An investment of \$12,000 earned a net income of \$600 in 1 year. Part of the investment was in bonds and earned income at a rate of 6%. The rest of the investment was in a savings account that earned income at a rate of 3%. How much was invested in bonds, and how much was invested in the savings account?
b. **Mixture of Two Acids** A lab technician wants 30 mL of a 64% acid solution. She has available a 40% acid solution and an 80% acid solution. How many milliliters of each should be mixed to produce this 30-mL solution?

Interval Notation

Write the interval notation for each of these inequalities.

51. a. $-6 \le x < 1$ b. $-1 < x \le 5$
c. $x < 2$ d. $x \ge -2$

Write the interval notation for each of these intervals.

52. a. (number line: bracket at −2, parenthesis at 3) b. (number line: bracket at 4, shaded to the left)
−2 3 4

c. $[4, 9) \cup (6, 12]$ d. $[4, 9) \cap (6, 12]$

Linear Inequalities

53. Use the table to solve each inequality.
a. $2x + 1 \ge 3x - 4$
b. $2x + 1 < 3x - 4$

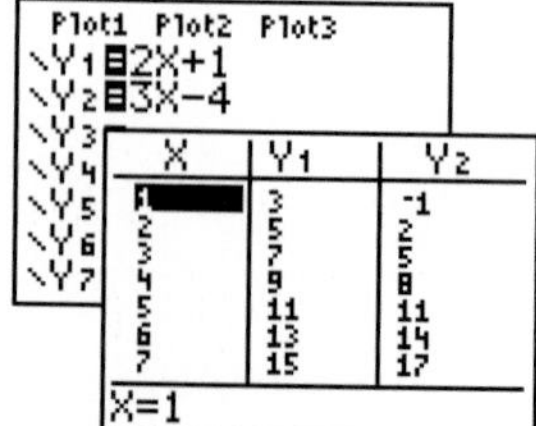

54. Use the graph to solve each inequality.
a. $x + 4 > -x - 2$
b. $x + 4 \leq -x - 2$

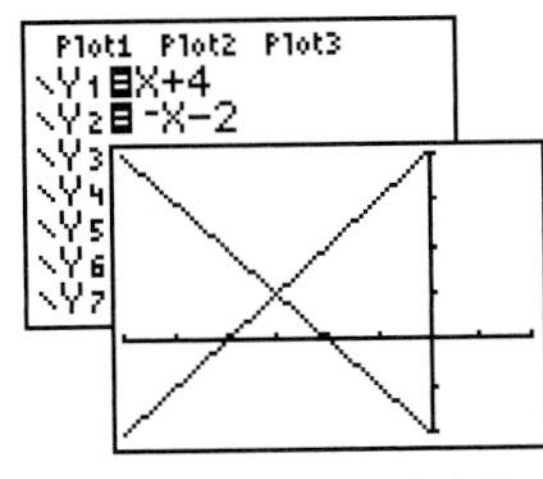

[−6, 2, 1] by [−2, 4, 1]

55. Algebraically solve $3(x + 2) < 3 + 4(x - 5)$.
56. Identify each of these inequalities as a conditional inequality, an unconditional inequality, or a contradiction. Then solve each inequality.
a. $2x + 1 < x - 1$
b. $2x + 1 < 2(x + 1)$
c. $2x + 1 \leq 2(x - 1)$
57. Solve each equation and inequality.
a. $|2x - 1| = 3$
b. $|2x - 1| < 3$
c. $|2x - 1| > 3$
58. Solve $-3 < 2x - 1 \leq 5$.
59. Solve $3x + 1 \leq -5$ or $2x - 3 \geq 5$.
60. Graph the solution of $3x + 6y \leq 12$.
61. Graph the solution of this system of inequalities.
$2x + y \geq 4$
$-x + y < 1$
62. Graph the solution of this system of inequalities.
$x \geq 0$
$y \geq 0$
$2x + y \leq 6$

Polynomials

63. Classify each polynomial according to the number of terms it contains and give its degree.
a. $3x^3 + 7x^2$ b. $5x^2$
c. $5x^4 - 2x^2 + 3$
d. $12x^4 - 7x^3 + 5x^2 + 8x + 2$
64. Simplify each sum or difference, and write the result in standard form.
a. $(2x - 7) + (5x + 8)$
b. $(5x^2 - 2x + 1) - (3x^2 - 10x - 12)$
65. Simplify each of these products.
a. $-5x^2(4x^3 + 3x^2 - 6x - 8)$
b. $(x - 3)(x^2 + 4x + 2)$
66. Multiply these factors by inspection.
a. $(x + 11)(x - 11)$ b. $(3x - 7)(3x + 7)$
c. $(x + 11)^2$ d. $(x - 5y)^2$

Factoring

67. Use the graph and table for the polynomial function $Y_1 = P(x)$.
a. List the x-intercepts of the graph of $y = P(x)$.
b. List the zeros of $P(x)$.
c. Factor $P(x)$.

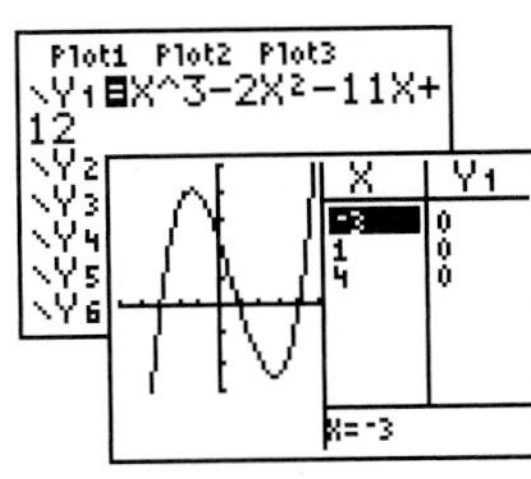

[−5, 5, 1] by [−15, 25, 5]

68. a. Use the graph of $y = P(x)$ to factor $P(x) = x^2 - 3x - 18$.

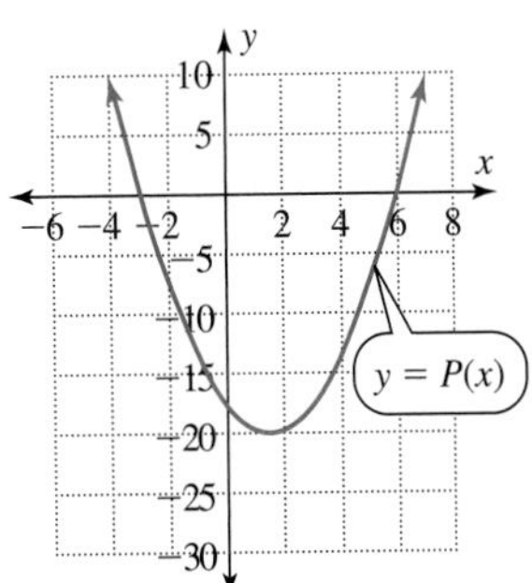

b. Use the table of values to factor the polynomial entered into the calculator as Y_1.

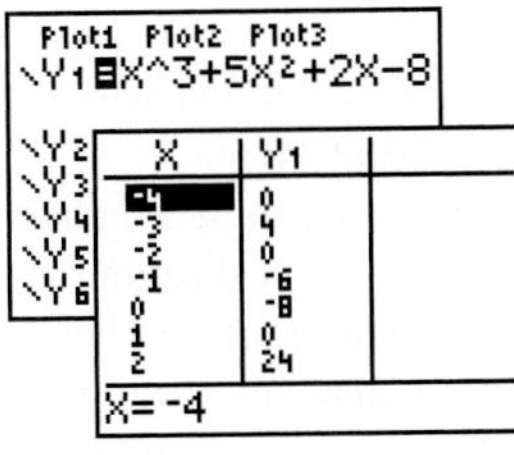

69. Factor out the GCF from each polynomial.
a. $12x^4 - 30x^2$ b. $2x(3x - 5) - 7(3x - 5)$
70. Factor each polynomial by using the grouping method.
a. $(3ax + 2a) + (15x + 10)$
b. $8x^2 - 10x - 12x + 15$
71. Completely factor each trinomial.
a. $w^2 - 12w - 45$ b. $w^2 + 13w + 35$
c. $v^2 - 11v + 24$ d. $v^2 + 9v - 36$
72. Completely factor each trinomial.
a. $8x^2 - 11x - 10$ b. $8x^2 - 21x + 10$
c. $8x^2 + 81x + 10$ d. $8x^2 + 31x + 10$
73. Factor each perfect square trinomial.
a. $x^2 + 18x + 81$ b. $x^2 - 10x + 25$
c. $25x^2 + 70x + 49$ d. $4x^2 - 44x + 121$
74. Factor each binomial.
a. $x^2 - 9y^2$ b. $49x^2 - 81y^2$
c. $x^3 - 125$ d. $27x^3 + 125y^3$
75. Use grouping to factor each polynomial.
a. $3ax + 2a + 6bx + 4b$
b. $15ax - 12bx - 20ay + 16by$
c. $a^2 - 9b^2 + a - 3b$
d. $x^2 + 12xy + 36y^2 - 25$
76. Completely factor each polynomial.
a. $5x^2 - 20$ b. $2bx^2 + 40bx + 72b$
c. $4x^3 - 20x^2 - 24x$ d. $a^3b + 2a^3c - 4ab - 8ac$
77. Solve each equation.
a. $(3x - 1)(x + 4) = 0$
b. $(x + 2)(x - 1)(3x - 5) = 0$
c. $x(x + 2) - 3(x + 2) = 0$
d. $x^2 + 3x - 8x - 24 = 0$
78. Solve each equation.
a. $3x^2 + x - 24 = 0$ b. $x^2 + 49 = 14x$
c. $(x - 5)(x + 4) = 22$ d. $2x^3 + 3x^2 - 5x = 0$

Quadratic Functions

79. a. Give the vertex, the y-intercept, and the x-intercepts.

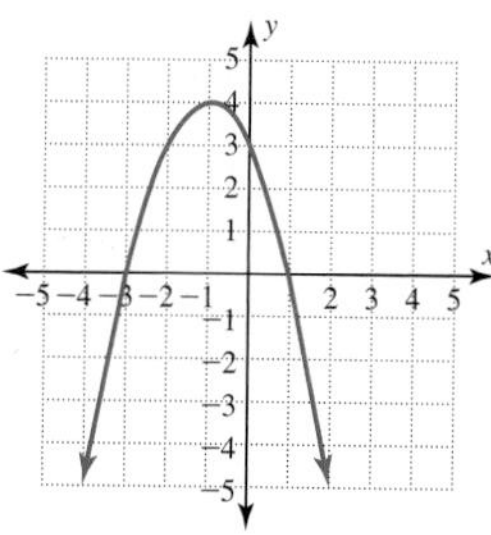

b. Use the equation $f(x) = x^2 + 6x - 7$ to predict whether the parabola defined by this function opens upward or downward. Then use a graphing calculator to graph the parabola and determine its vertex, y-intercept, and x-intercepts.

80. Use the parabola defined by $y = ax^2 + bx + c$ to determine whether the discriminant for $ax^2 + bx + c = 0$ is negative, zero, or positive. Then identify the nature of the solutions of the quadratic equation as distinct real solutions, a double real solution, or complex solutions with imaginary parts.

a.

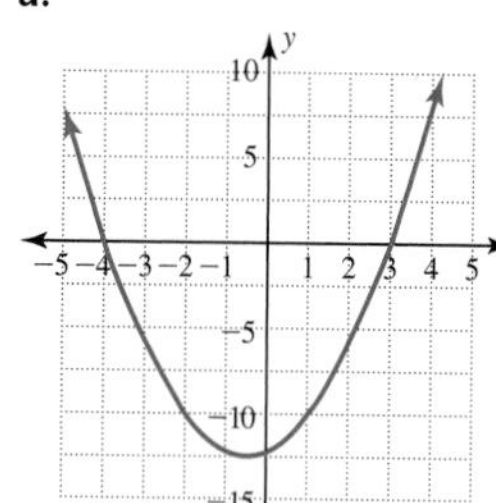

b.

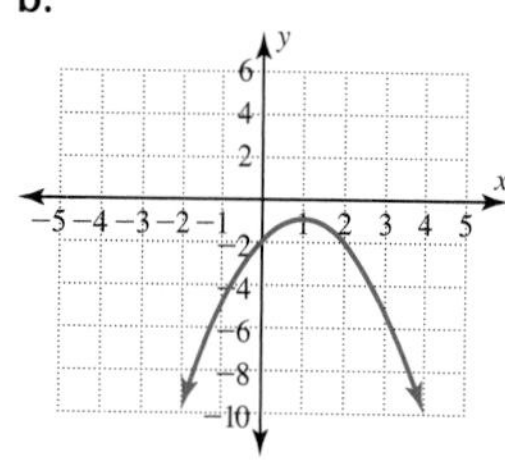

c.

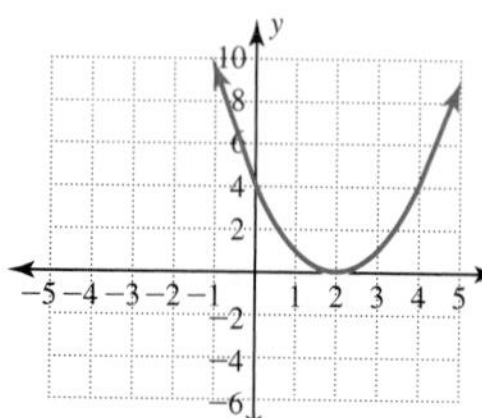

d.

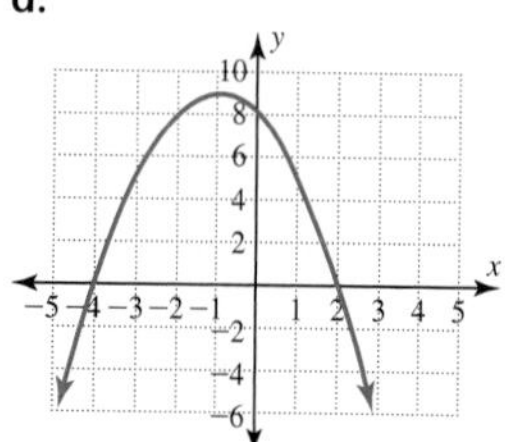

81. Determine the vertex of the parabola defined by each of these equations.

a. $y = x^2 - 6x - 4$ b. $y = x^2 + 2x - 7$

82. **Height of a Softball** The path of a softball is given by $y = -16x^2 + 48x + 3$, where y represents the height of the ball in feet and x is the number of seconds that have elapsed since the ball was hit. Determine the highest point that the ball reaches. Determine how many seconds into the flight the maximum height is reached.

83. Construct a quadratic equation with the given solutions.

a. $x = 3$ and $x = -\dfrac{1}{5}$

b. $x = -\sqrt{6}$ and $x = \sqrt{6}$

c. $x = 1 - 2i$ and $x = 1 + 2i$

84. Use factoring to find the zeros of $x^2 - 3x - 70$ and then solve each equation and inequality.

a. $x^2 - 3x - 70 = 0$ b. $x^2 - 3x - 70 < 0$

c. $x^2 - 3x - 70 \geq 0$

85. Solve these quadratic equations by extraction of roots.

a. $(x - 3)^2 = 16$ b. $(x - 3)^2 = -16$

86. Use the quadratic formula to solve each quadratic equation.

a. $3x^2 - 8x + 5 = 0$ b. $4x^2 - 19x = x - 25$

c. $(x - 1)(x + 5) = -25$

87. a. **Length of a Side of a Square** The side of one square is 3 cm longer than the side of another square. The total area of the two squares is 149 cm^2. Determine the length of a side of each square.

x cm

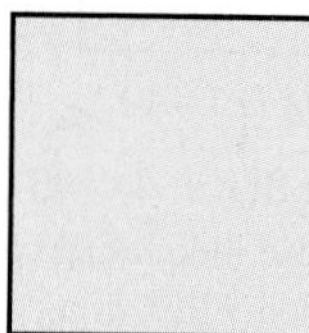

$(x + 3)$ cm

b. **Profit Interval** $P(x) = -x^2 + 190x - 2{,}625$ is a profit polynomial that gives the profit in dollars made by selling x units of a product. Determine the profit interval and the maximum profit that can be made.

88. Write each complex number in standard $a + bi$ form.

a. $\sqrt{-49}$ b. $\sqrt{36 - 100}$

c. $\sqrt{-36} - \sqrt{100}$ d. $i^3 + i^4$

89. Perform each operation, and write the result in standard $a + bi$ form.

a. $3(2 - 7i) - 2(5 - i)$ b. $(5 - 3i)(2 + 7i)$

c. $\dfrac{3 - 5i}{1 - 3i}$ d. $(2 - 5i)^2$

90. Find the domain of each function.

a. $f(x) = \dfrac{x + 2}{2x - 6}$ b. $f(x) = \dfrac{2x + 16}{x^2 + 4}$

91. Reduce each rational expression to lowest terms.

a. $\dfrac{7x}{14x^2 - 21x}$ b. $\dfrac{2x^2 - 5x - 42}{2x^2 - 72}$

92. Perform each multiplication or division and reduce the result to lowest terms.

a. $\dfrac{2x^2 - 2x - 12}{x^2 - 4} \cdot \dfrac{5x^2 - 10x}{4x - 12}$

b. $\dfrac{3x^2 - 27}{x + 3} \div \dfrac{6x^2 - 12x - 18}{x + 1}$

93. Perform each addition or subtraction and reduce the result to lowest terms.

a. $\dfrac{x}{x^2 - 1} - \dfrac{4x + 3}{x^2 - 1}$ b. $\dfrac{2x}{x + 3} + \dfrac{3}{x - 3} + \dfrac{8x - 12}{x^2 - 9}$

94. Simplify each expression.

a. $\left(1 + \dfrac{x + 6}{x}\right) \div \left(\dfrac{x}{9} - \dfrac{1}{x}\right)$

b. $\left(\dfrac{x}{x - 3} - \dfrac{1}{x + 2}\right) \cdot \dfrac{x^2 + x - 2}{x^2 + x + 3}$

95. Simplify each expression.

a. $\dfrac{28x^{-2} + 10x^{-1} - 2}{21x^{-2} + 11x^{-1} - 2}$ b. $\dfrac{\dfrac{2x}{x + 3} - \dfrac{x}{x - 4}}{\dfrac{x^2 - 4x}{x^2 - x - 12}}$

96. Solve each of the following equations.

a. $\dfrac{x+2}{x-2} = \dfrac{8}{x^2-2x} + \dfrac{1}{x}$

b. $\dfrac{x+1}{x^2-3x+2} = \dfrac{3}{x-2} + \dfrac{1}{1-x}$

97. The volume of a gas varies inversely with the pressure when temperature is held constant. If the pressure of 4 L of a gas is 12 N/cm^2, what will the pressure be if the volume is 3 L?

98. Solve each of the following problems.

a. **Ratio of Economic Indicators** The ratio of two monthly economic indicators is $\dfrac{3}{4}$. The first monthly indicator is 1 unit below normal, and the second monthly indicator is 1 unit above normal. What is the normal number of units for this indicator?

b. **Work by Two Machines** When members of a construction crew use two end loaders at once, they can move a pile of sand in 4 hours. If they use only the larger end loader, they can do the job in 6 hours less time than it would take if they used only the smaller machine. How many hours would it take to do the job while using the larger machine?

ANSWERS for the Cumulative Review for Chapters 1–8

Question	Answer	Reference Example
1. a	5	[1.2–1]
1. b	$-\frac{1}{2}$	[1.2–1]
2. a	7	[1.2–3]
2. b	7	[1.2–3]
3. a	3	[1.3–8]
3. b	-3	[1.3–8]
3. c	3	[1.3–8]
3. d	4	[1.3–8]
4. a	8; 7.94	[1.2–6]
4. b	7; 7.07	[1.2–6]
4. c	6; 6.00	[1.2–6]
4. d	8; 8.26	[1.2–6]
5. a	C	[1.2–8]
5. b	E	[1.2–8]
5. c	A	[1.2–8]
5. d	D	[1.2–8]
5. e	B	[1.2–8]
6. a	Associative, addition	[1.3–6]
6. b	Commutative, addition	[1.3–6]
6. c	Associative, multiplication	[1.5–1]
6. d	Commutative, multiplication	[1.5–1]
7. a	$10x - 15$	[1.7–5]
7. b	$5(5x + 6)$	[1.7–6]
7. c	$-12x + 21$	[1.7–5]
7. d	$9(2x - 3)$	[1.7–6]
8. a	$-\frac{1}{3}$	[1.3–4]
8. b	$-\frac{23}{36}$	[1.4–4]
8. c	$\frac{1}{2}$	[1.6–1]
8. d	-30	[1.5–2]
9. a	16	[1.5–8]
9. b	-16	[1.5–8]
9. c	1	[5.2–3]
9. d	1	[1.5–8]
10. a	6	[1.7–1]
10. b	6	[1.7–1]
10. c	49	[1.7–1]
10. d	-28	[1.7–1]
11. a	$\frac{-2a^6}{b^8}$	[5.3–4]
11. b	$16v^8w^{12}$	[5.2–4]
11. c	x^3y^6	[5.2–4]
11. d	$36x^{12}$	[5.2–4]
12. a	5	[5.2–3]
12. b	$\frac{9}{4}$	[5.3–2]
12. c	9	[5.3–2]
12. d	2	[5.3–2]
13. a	12,300	[5.3–5]
13. b	0.00000389	[5.3–5]
13. c	1.57×10^{-5}	[5.3–6]
13. d	2.45×10^9	[5.3–6]
14. a	$x = 3$	[2.5–5]

Question	Answer	Reference Example
14. b	$x = 4$	[2.7–2]
14. c	$x = 3$	[2.5–6]
14. d	$x = 19$	[2.7–2]
15.	$x = 5$	[2.4–8]
16. a	Contradiction; no solution	[2.4–7]
16. b	Identity; all real numbers	[2.4–6]
16. c	Conditional equation; $x = 0$	[2.4–5]
17. a	$x = 26$	[2.7–1]
17. b	$x = -13$	[2.7–1]
17. c	4 batches	[2.7–3]
18. a	$y = 8$	[2.7–6]
18. b	10.982 Mexican pesos for $1 U.S.	[2.7–5]
19. a	$a = P - 2b - c$	[2.6–5]
19. b	$a = \frac{S - 2bc}{2b + 2c}$	[2.6–5]
20.	$y = 2x + 3$	[2.6–2]
21.	A (4, −3); IV B (3, 5); I C (−4, −3); III D (−3, 3); II	[2.1–2]
22.	(graph)	[2.1–3]
23. a	Yes; 3	[2.1–6]
23. b	No	[2.1–6]
24. a	No	[7.1–1]
24. b	Yes	[7.1–5]
24. c	Yes	[7.1–3]
25. a	$D = \{1, 5, 6, 7\}$, $R = \{3, 4, 10, 11\}$	[7.1–2]
25. b	$D = \{-3, -2, -1, 0, 1\}$, $R = \{-2, -1, 1, 3\}$	[7.1–4]
25. c	$D = (-\infty, 2]$, $R = [0, \infty)$	[7.1–8]
26. a	4	[7.1–6]
26. b	39	[7.1–6]
27. a	-1	[7.1–7]
27. b	3	[7.1–7]
28. a	-2	[7.1–7]
28. b	12	[7.1–7]
29. a	No	[2.3–1]
29. b	Yes	[2.3–1]

Calculator table for 20:

X	Y1
-3	-3
-2	-1
-1	1
0	3
1	5
2	7
3	9

X= -3

Question	Answer	Reference Example
30. a	$\begin{array}{c\|ccccc} x & 1 & 2 & 3 & 4 & 5 \\ \hline y & 0 & 1 & 2 & 3 & 4 \end{array}$	[2.2–2]
30. b	$\begin{array}{c\|ccccc} x & 1 & 2 & 3 & 4 & 5 \\ \hline y & 2 & 1 & 0 & -1 & -2 \end{array}$	[2.2–2]
31. a	x-intercept: $(-3, 0)$; y-intercept: $(0, 2)$	[2.3–4]
31. b	x-intercept: $(-4, 0)$; y-intercept: $(0, 7)$	[2.3–4]
31. c	x-intercept: $(4, 0)$; y-intercept: $(0, 8)$	[2.6–3]
32. a	$f(x) = 225x + 500$	[2.2–6]
32. b	$\begin{array}{c\|cccccc} x & 0 & 6 & 12 & 18 & 24 & 30 \\ \hline f(x) & 500 & 1{,}850 & 3{,}200 & 4{,}550 & 5{,}900 & 7{,}250 \end{array}$	[2.2–6]
32. c	$f(20) = 5{,}000$; at the end of the 20th month, the total amount paid on the loan will be \$5,000.	[2.2–6]
32. d	$x = 40$; at the end of the 40th month, the total amount paid on the loan will be \$9,500.	[2.2–6]
33. a	$m = -3$	[3.1–1]
33. b	$m = -\frac{3}{4}$	[3.1–2]
33. c	$m = -2$	[3.1–6]
34. a	Neither	[3.2–5]
34. b	Perpendicular	[3.2–5]
34. c	Parallel	[3.2–5]
34. d	Perpendicular	[3.2–5]
35. a	5	[3.1–3]
35. b	−20	[3.1–3]
35. c	−4	[3.1–3]
35. d	The hot-air balloon is descending at a rate of 4 ft/s.	[3.1–7]
36. a	$y = -2x + 6$	[3.2–10]
36. b	$y = \frac{2}{3}x + 5$	[3.2–1]
36. c	$y = \frac{2}{3}x + \frac{11}{3}$	[3.2–9]

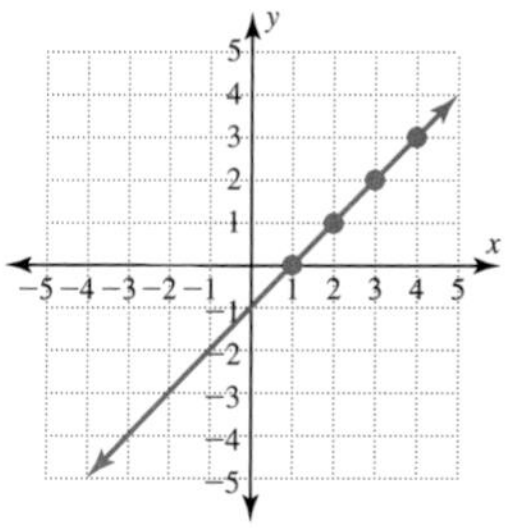

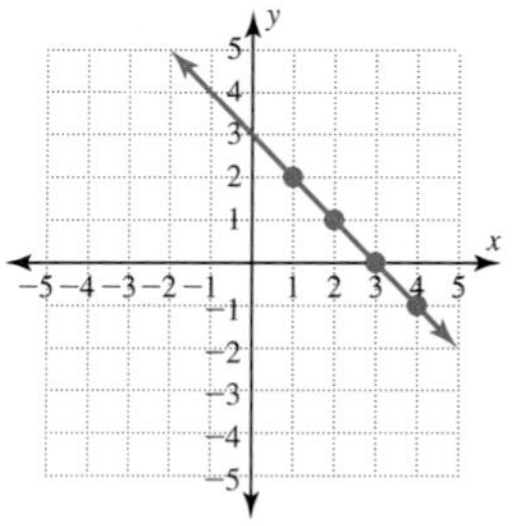

Question	Answer	Reference Example
37. a	$y = 5$	[3.2–6]
37. b	$x = 3$	[3.2–6]
38. a		[3.2–10]
38. b		[3.1–4]
39. a	Increasing	[7.2–2]
39. b	Decreasing	[7.2–2]
40. a	$f(x)$ is positive for the x-interval $(-\infty, 3)$; $f(x)$ is negative for the x-interval $(3, \infty)$.	[7.2–6]
40. b	$f(x)$ is positive for the x-interval $(-3, 1)$; $f(x)$ is negative for the x-interval $(-\infty, -3) \cup (1, \infty)$.	[7.2–6]
41. a	$\begin{array}{c\|c} x & f(x) = \|x-1\| - 2 \\ \hline -2 & 1 \\ -1 & 0 \\ 0 & -1 \\ 1 & -2 \\ 2 & -1 \\ 3 & 0 \\ 4 & 1 \end{array}$	[7.2–5]
41.b		[7.2–5]
41. c	$(-1, 0), (3, 0)$	[7.2–5]
41. d	$(0, -1)$	[7.2–5]
41. e	$(1, -2)$	[7.2–5]
41. f	−2	[7.2–5]
41. g	$D = \mathbb{R}$	[7.2–5]
41. h	$R = [-2, \infty)$	[7.2–5]
41. i	$(-\infty, 1)$	[7.2–5]
41. j	$(1, \infty)$	[7.2–5]
41. k	$(-\infty, -1) \cup (3, \infty)$	[7.2–6]
41. l	$(-1, 3)$	[7.2–6]

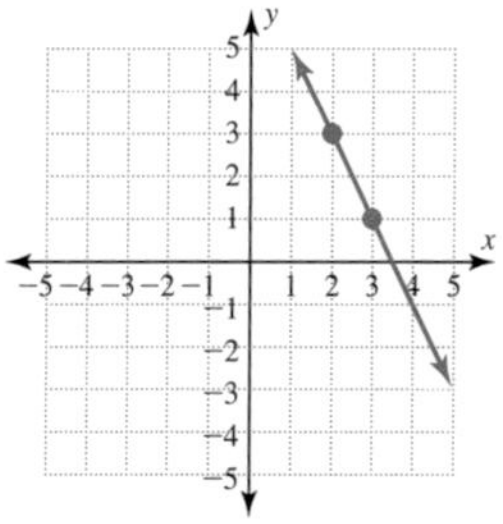

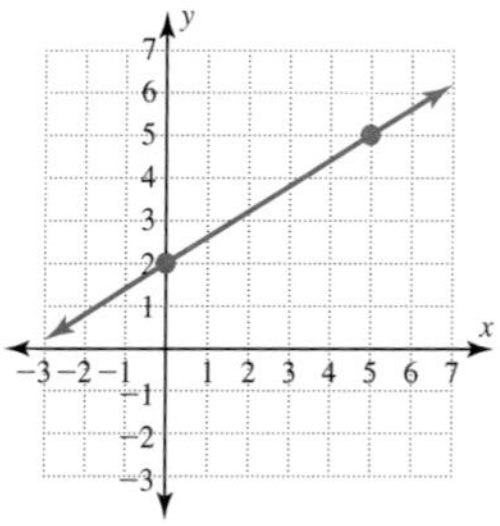

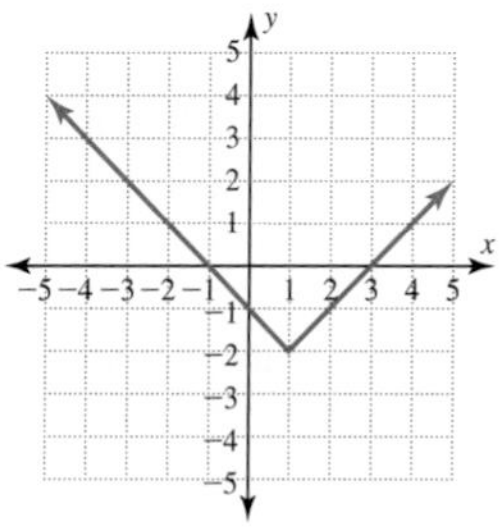

continued

Question	Answer	Reference Example
42. a	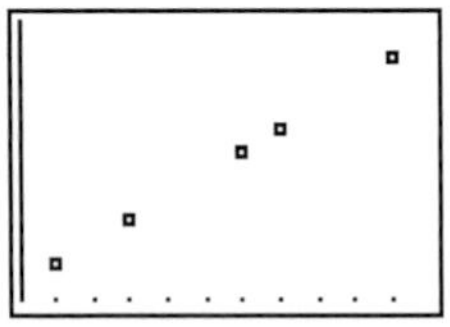 [0.1, 10.9, 1] by [129.55, 1985.45, 1] $y = 153x + 209$	[7.3–6]
42. b	The estimated repair costs increase by about \$153 per year.	[7.3–6]
42. c	The estimated repair cost of a machine that has been used for 4 years is \$821.	[7.3–6]
42. d	Using this equation, a copy machine that has \$1,425 in repair costs for the year has been used for 8 years.	[7.3–6]
43. a	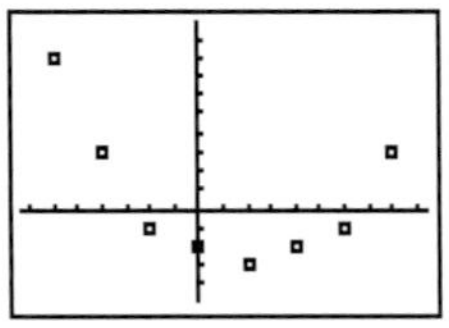 [−7.4, 9.4, 1] by [−4.87, 9.87, 1] $f(x) = 0.168x^2 - 0.688x - 2.39$	[7.3–7]
43. b	$f(5) = -1.63$	[7.3–7]
44. a	$[0, 14]$	[2.8–2]
44. b	$\{0, 1, 2, \ldots, 30\}$	[2.8–2]
45. a	$(3, 3)$	[3.3–1]
45. b	$(3, 5)$	[3.3–1]
46. a	$(-2, 0)$	[3.3–1]
46. b	$(-4, -4)$	[3.3–1]
47. a	C	[3.4–4]
47. b	A	[3.4–5]
47. c	B	[3.4–1]
48. a	$(2, -3)$	[3.4–1]
48. b	$\left(3, -\frac{1}{4}\right)$	[3.4–1]
49. a	$(-5, 2)$	[3.5–2]
49. b	Inconsistent system; no solution	[3.5–7]
50. a	\$8,000 was invested in bonds, and \$4,000 was invested in the savings account.	[3.6–4]
50. b	12 mL of the 40% acid solution and 18 mL of the 80% acid solution should be mixed.	[3.6–6]
51. a	$[-6, 1)$	[1.2–5]
51. b	$(-1, 5]$	[1.2–5]
51. c	$(-\infty, 2)$	[1.2–5]
51. d	$[-2, \infty)$	[1.2–5]
52. a	$[-2, 3)$	[1.2–5]
52. b	$(-\infty, 4]$	[1.2–5]
52. c	$[4, 12]$	[4.3–8]
52. d	$(6, 9)$	[4.3–3]
53. a	$(-\infty, 5]$	[4.1–4]
53. b	$(5, \infty)$	[4.1–4]
54. a	$(-3, \infty)$	[4.1–4]
54. b	$(-\infty, -3]$	[4.1–4]
55.	$(23, \infty)$	[4.1–6]
56. a	Conditional; $(-\infty, -2)$	[4.3–1]
56. b	Unconditional; $\mathbb{R}$	[4.3–1]
56. c	Contradiction; no solution	[4.3–1]
57. a	$x = -1, x = 2$	[4.4–2]
57. b	$(-1, 2)$	[4.4–2]
57. c	$(-\infty, -1) \cup (2, \infty)$	[4.4–2]
58.	$(-1, 3]$	[4.3–4]
59.	$(-\infty, -2] \cup [4, \infty)$	[4.3–8]
60.	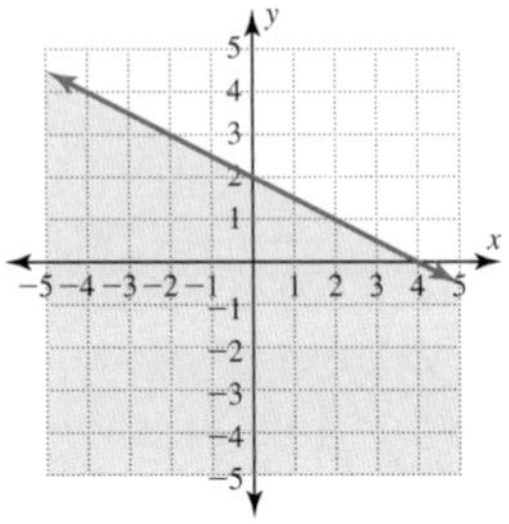	[4.5–3]
61.	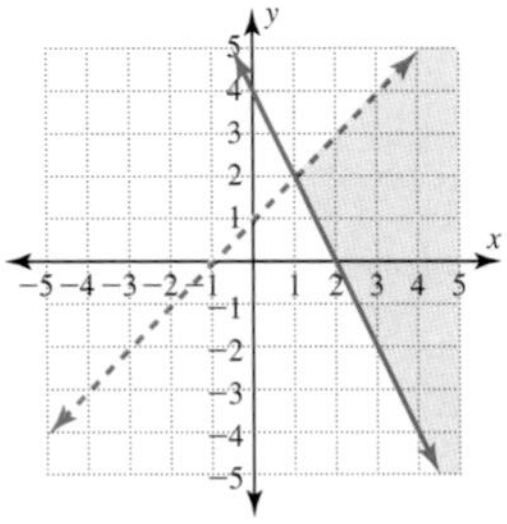	[4.5–7]
62.	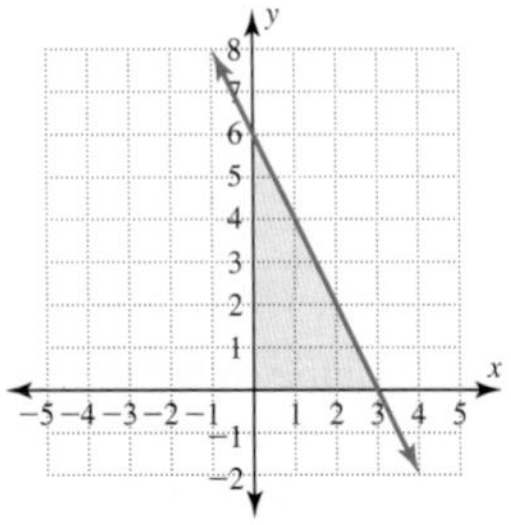	[4.5–9]
63. a	Binomial; degree = 3	[5.4–2]
63. b	Monomial; degree = 2	[5.4–2]
63. c	Trinomial; degree = 4	[5.4–2]
63. d	Polynomial; degree = 4	[5.4–2]
64. a	$7x + 1$	[5.4–7]
64. b	$2x^2 + 8x + 13$	[5.4–8]
65. a	$-20x^5 - 15x^4 + 30x^3 + 40x^2$	[5.5–2]
65. b	$x^3 + x^2 - 10x - 6$	[5.5–3]
66. a	$x^2 - 121$	[5.6–1]
66. b	$9x^2 - 49$	[5.6–1]
66. c	$x^2 + 22x + 121$	[5.6–3]
66. d	$x^2 - 10xy + 25y^2$	[5.6–3]
67. a	$(-3, 0), (1, 0), (4, 0)$	[6.1–4]
67. b	$-3, 1, 4$	[6.1–4]
67. c	$(x + 3)(x - 1)(x - 4)$	[6.1–4]
68. a	$(x + 3)(x - 6)$	[6.1–4]
68. b	$(x + 4)(x + 2)(x - 1)$	[6.1–4]
69. a	$6x^2(2x^2 - 5)$	[6.1–1]
69. b	$(2x - 7)(3x - 5)$	[6.1–1]
70. a	$(a + 5)(3x + 2)$	[6.5–1]
70. b	$(2x - 3)(4x - 5)$	[6.2–4]

Question	Answer	Reference Example
71. a	$(w - 15)(w + 3)$	[6.2–4]
71. b	Prime	[6.2–6]
71. c	$(v - 3)(v - 8)$	[6.2–4]
71. d	$(v + 12)(v - 3)$	[6.2–4]
72. a	$(8x + 5)(x - 2)$	[6.3–2]
72. b	$(8x - 5)(x - 2)$	[6.3–2]
72. c	$(8x + 1)(x + 10)$	[6.3–2]
72. d	Prime	[6.3–3]
73. a	$(x + 9)^2$	[6.4–1]
73. b	$(x - 5)^2$	[6.4–1]
73. c	$(5x + 7)^2$	[6.4–1]
73. d	$(2x - 11)^2$	[6.4–1]
74. a	$(x - 3y)(x + 3y)$	[6.4–2]
74. b	$(7x - 9y)(7x + 9y)$	[6.4–2]
74. c	$(x - 5)(x^2 + 5x + 25)$	[6.4–6]
74. d	$(3x + 5y)(9x^2 - 15xy + 25y^2)$	[6.4–5]
75. a	$(a + 2b)(3x + 2)$	[6.5–1]
75. b	$(3x - 4y)(5a - 4b)$	[6.5–1]
75. c	$(a - 3b)(a + 3b + 1)$	[6.5–2]
75. d	$(x + 6y + 5)(x + 6y - 5)$	[6.5–3]
76. a	$5(x + 2)(x - 2)$	[6.5–5]
76. b	$2b(x + 2)(x + 18)$	[6.5–6]
76. c	$4x(x - 6)(x + 1)$	[6.5–6]
76. d	$a(a + 2)(a - 2)(b + 2c)$	[6.5–6]
77. a	$x = \frac{1}{3}, x = -4$	[6.6–2]
77. b	$x = -2, x = 1, x = \frac{5}{3}$	[6.6–2]
77. c	$x = -2, x = 3$	[6.6–2]
77. d	$x = -3, x = 8$	[6.6–2]
78. a	$x = -3, x = \frac{8}{3}$	[6.6–3]
78. b	$x = 7$	[6.6–3]
78. c	$x = -6, x = 7$	[6.6–5]
78. d	$x = -\frac{5}{2}, x = 0, x = 1$	[6.6–6]
79. a	Vertex: $(-1, 4)$; y-intercept: $(0, 3)$; x-intercepts: $(-3, 0)$, $(1, 0)$	[7.3–3]
79. b	Opens upward because $a > 0$; vertex: $(-3, -16)$; y-intercept: $(0, -7)$; x-intercepts: $(-7, 0)$, $(1, 0)$	[7.3–3]
80. a	Positive; two distinct real solutions	[7.4–7]
80. b	Negative; complex solutions with imaginary parts	[7.4–7]
80. c	Zero; a double real solution	[7.4–7]
80. d	Positive; two distinct real solutions	[7.4–7]
81. a	$(3, -13)$	[7.5–2]
81. b	$(-1, -8)$	[7.5–2]
82.	The highest point is 39 ft. This occurs 1.5 seconds after the ball was hit.	[7.5–3]
83. a	$5x^2 - 14x - 3 = 0$	[7.5–6]
83. b	$x^2 - 6 = 0$	[7.5–8]
83. c	$x^2 - 2x + 5 = 0$	[7.7–8]
84. a	$x = -7, x = 10$	[6.6–8]
84. b	$(-7, 10)$	[6.6–8]
84. c	$(-\infty, -7] \cup [10, \infty)$	[6.6–8]
85. a	$x = -1, x = 7$	[7.4–1]
85. b	$x = 3 \pm 4i$	[7.7–8]
86. a	$x = 1, x = \frac{5}{3}$	[7.4–4]
86. b	$x = \frac{5}{2}$	[7.4–4]
86. c	$x = -2 \pm 4i$	[7.7–9]
87. a	7 cm and 10 cm	[7.6–2]
87. b	Between 15 units and 175 units; \$6,400	[7.6–4]
88. a	$7i$	[7.7–1]
88. b	$8i$	[7.7–1]
88. c	$-10 + 6i$	[7.7–1]
88. d	$1 - i$	[7.7–3]
89. a	$-4 - 19i$	[7.7–2]
89. b	$31 + 29i$	[7.7–4]
89. c	$\frac{9}{5} + \frac{2}{5}i$	[7.7–6]
89. d	$-21 - 20i$	[7.7–4]
90. a	$\mathbb{R} \sim \{3\}$	[8.1–2]
90. b	$\mathbb{R}$	[8.1–2]
91. a	$\frac{1}{2x - 3}$	[8.1–5]
91. b	$\frac{2x + 7}{2(x + 6)}$	[8.1–5]
92. a	$\frac{5x}{2}$	[8.2–3]
92. b	$\frac{1}{2}$	[8.2–6]
93. a	$-\frac{3}{x - 1}$	[8.3–1]
93. b	$\frac{2x - 1}{x - 3}$	[8.3–6]
94. a	$\frac{18}{x - 3}$	[8.4–2]
94. b	$\frac{x - 1}{x - 3}$	[8.4–3]
95. a	$\frac{2(x + 2)}{2x + 3}$	[8.4–8]
95. b	$\frac{x - 11}{x - 4}$	[8.4–6]
96. a	$x = -3$	[8.5–3]
96. b	There is no solution.	[8.5–4]
97.	The pressure will be 16 N/cm^2.	[8.6–3]
98. a	The normal value of this indicator is 7 units.	[8.6–8]
98. b	It would take 6 h by using only the larger machine.	[8.6–6]

chapter 9

Square Root and Cube Root Functions and Rational Exponents

CHAPTER OUTLINE

A geosynchronous satellite completes one orbit around the Earth each day. Thus these satellites remain above the same point on the equator of the Earth. Special care must be taken when a geosynchronous satellite is placed into orbit. The speed and altitude must be just right to ensure that the satellite will remain in the proper position. In the Reality Check at the end of this chapter we use radical equations to calculate the radius needed for a geosynchronous orbit of the Earth.

REALITY CHECK

Why is a geosynchronous orbit a desirable orbit for a telecommunications satellite?

Section 9.1 Evaluating Radical Expressions and Graphs of Square Root and Cube Root Functions

Objectives:

1. Graph square root and cube root functions.
2. Determine the domain of a square root function.
3. Evaluate radical expressions.

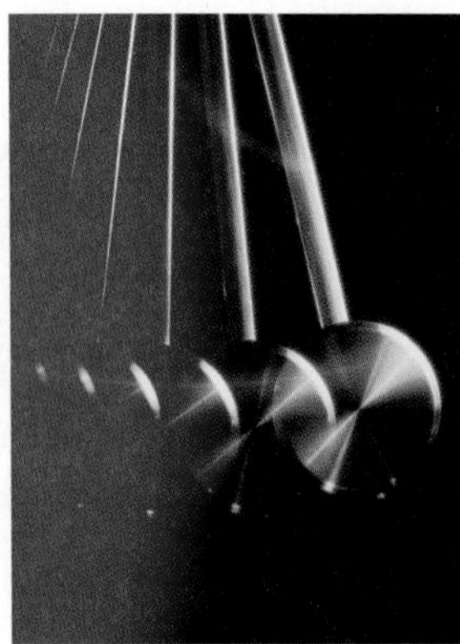

The use of square roots and cube roots is common in many applications in business, engineering, and the sciences. In Exercises 37 and 38 the period of a pendulum is modeled by a square root function.

Each family of functions has its own distinctive graph, its own numerical pattern, and its own algebraic definition. In this section we examine the basic shape of square root and cube root functions. To be able to use these functions, we use this chapter to develop the properties of radicals and operations with radical expressions.

EXAMPLES OF SQUARE ROOT AND CUBE ROOT FUNCTIONS

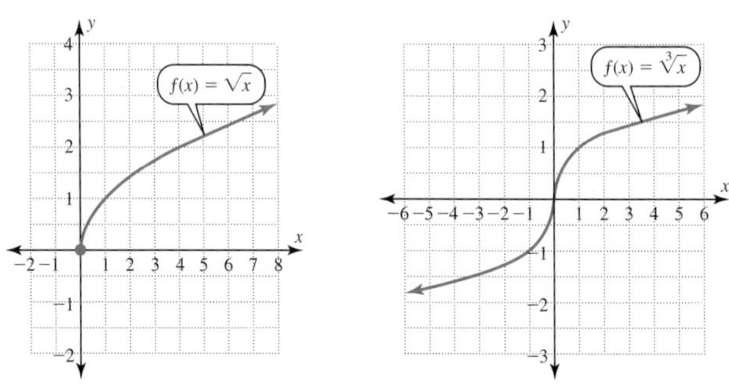

Note that these two graphs have at least one major difference. The square root function is defined only for nonnegative values, while the cube root function is defined for all real numbers. We now examine why these domains are different.

In this book we examine only functions that have real values in both the domain and the range. These functions are called **real-valued functions.** Thus we restrict the input values of square root functions to nonnegative values. This will ensure that we obtain real values as output values in the range. If we did allow negative numbers in the domain of $f(x) = \sqrt{x}$, then the output values would be imaginary numbers. By restricting the domain of $f(x) = \sqrt{x}$ to $[0, \infty)$, we can have this function be a real-valued function.

Both -2 and 2 are square roots of 4. However, only the positive value of 2 is called the principal square root of 4. Recall that in Section 1.2 we defined $\sqrt{x}$ to denote the **principal square root of *x*** for $x \geq 0$. That is, $\sqrt{x}$ denotes the nonnegative real number r so that $r^2 = x$. For example, $\sqrt{4}$ denotes the principal square root of 4, that is, $\sqrt{4} = 2$. The other square root of 4 is denoted by $-\sqrt{4}$, that is, $-\sqrt{4} = -2$. The following table examines the definition of $\sqrt{x}$ algebraically, verbally, numerically, and graphically.

Principal Square Root

For any real number $x \geq 0$:

ALGEBRAICALLY	VERBALLY	NUMERICALLY	GRAPHICALLY
$y = \sqrt{x}$ if $y^2 = x$ and $y \geq 0$.	$\sqrt{x}$ denotes the principal square root of x.	Plot1 Plot2 Plot3 \Y1 = √(X) X: 0, 1, 2, 3, 4, 5, 6 Y1: 0, 1, 1.4142, 1.7321, 2, 2.2361, 2.4495 X=0	Plot1 Plot2 Plot3 \Y1 = √(X) $[-1, 10, 1]$ by $[-1, 5, 1]$

Example 1 reviews the evaluation of some square roots. Notice the subtle differences in notation—subtleties that are the source of many common errors.

EXAMPLE 1 Evaluating Square Roots

Evaluate each of the following expressions.

SOLUTION

(a) $\sqrt{9}$ — $\sqrt{9} = 3$ — $3^2 = 9$; 3 is the positive square root of 9.

(b) $-\sqrt{9}$ — $-\sqrt{9} = -3$ — $(-3)^2 = 9$; -3 is the negative square root of 9.

(c) $\sqrt{-9}$ — $\sqrt{-9} = i\sqrt{9} = 3i$ — This is an imaginary number, not a real number. Imaginary numbers were covered in Section 7.7.

The notation $\pm\sqrt{x}$ is read "plus or minus the (principal) square root of x."

(d) $\pm\sqrt{9}$ — $\pm\sqrt{9} = \pm 3$ (3 or -3) — The notation $\pm\sqrt{x}$ is read "plus or minus the square root of x" and is used to represent both square roots of x.

(e) $\sqrt{9 + 16}$ — $\sqrt{9 + 16} = \sqrt{25} = 5$ — Following the correct order of operations, we get different results for parts (e) and (f).

(f) $\sqrt{9} + \sqrt{16}$ — $\sqrt{9} + \sqrt{16} = 3 + 4 = 7$ — In general $\sqrt{x + y} \neq \sqrt{x} + \sqrt{y}$.

(g) $\sqrt{4^2}$ — $\sqrt{4^2} = \sqrt{16} = 4$ — $\sqrt{16}$ represents the principal square root of 16, which is 4.

(h) $\sqrt{(-4)^2}$ — $\sqrt{(-4)^2} = \sqrt{16} = 4$ — Note from parts (g) and (h) that $\sqrt{4^2}$ and $\sqrt{(-4)^2}$ both equal 4. In general $\sqrt{x^2} = |x|$.

SELF-CHECK 9.1.1

Evaluate each of the following expressions.

1. $\sqrt{36} + \sqrt{64}$
2. $\sqrt{36 + 64}$
3. $\sqrt{(-5)^2}$
4. $-\sqrt{5^2}$

In Example 2 note the distinctive shape of this square root function. The shape of a square root function $f(x) = \sqrt{ax + b} + c$ can be altered by changing the constants in this function, but the basic shape will stay the same. Remember the domain of a function is the projection of its graph onto the x-axis.

EXAMPLE 2 Determining the Domain of a Square Root Function

Determine the domain of $f(x) = \sqrt{2x - 6}$.

SOLUTION

ALGEBRAICALLY

Solve $2x - 6 \geq 0$

$2x \geq 6$

$x \geq 3$

The radicand of a square root function must be greater than or equal to 0 to produce real values as outputs. Thus to obtain the domain of this real-valued function, set $2x - 6 \geq 0$. Then solve for the x-values for which this is true. Thus the domain is the interval $[3, \infty)$.

GRAPHICALLY AND NUMERICALLY

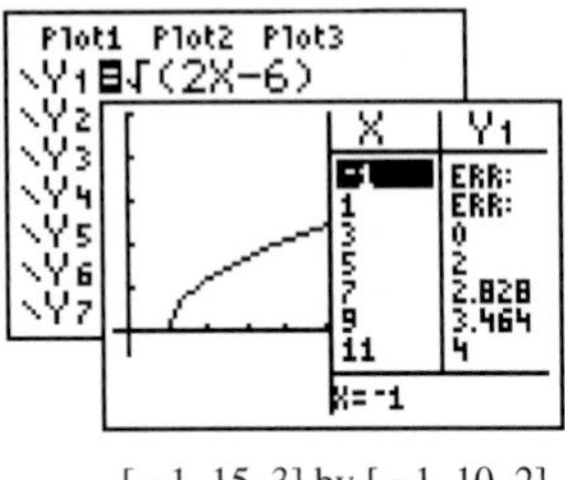

$[-1, 15, 3]$ by $[-1, 10, 2]$

The projection of this graph onto the x-axis is the interval $[3, \infty)$. Thus the domain $D = [3, \infty)$. Note that the x-values in the table that are less than 3 are undefined for this function. Also note that the range, the projection of the graph onto the y-axis, is $R = [0, \infty)$.

Answer: $D = [3, \infty)$

SELF-CHECK 9.1.2

1. Use a graphing calculator to graph $f(x) = \sqrt{4 - x}$.
2. Determine the domain and range of this function.

A Mathematical Note

The radical symbol $\sqrt{}$ is composed of two parts: $\sqrt{}$ and ‾‾‾‾. The symbol $\sqrt{}$ comes from the letter r, the first letter of the Latin word *radix,* which means root. Thus the $\sqrt{}$ indicates that a root is to be taken of the quantity underneath the bar (also called the *vinculum*).

On many calculators we can access the symbol $\sqrt{}$ but not the bar. To indicate the square root of a quantity, we must use parentheses instead of the bar. For example, $\sqrt{2x + 1}$ can be represented on calculators by $\sqrt{}\ (2x + 1)$.

Although square roots are the most common roots used in applications, you should also be familiar with higher-order roots. The number r is an nth root of x if $r^n = x$. The **principal *n*th root of *x*** is denoted by $\sqrt[n]{x}$. For example, 2 is a cube root (third root) of 8 because $2^3 = 8$. In the following table we describe radical notation for nth roots of any order.

Radical Notation

The number r is an nth root of x if $r^n = x$.

$\sqrt[n]{x}$ denotes the principal nth root of x.

- x is the **radicand.**
- $\sqrt{}$ is the **radical symbol,** or the radical.
- The natural number n is the **index** or the **order of the radical.**

Note in the following table the importance of distinguishing between even- and odd-order radicals when we examine principal nth roots.

If no index is written, as in $\sqrt{x}$, then the index is understood to be 2 and this denotes the principal square root of x.

Principal *n*th Root

The principal nth root of the real number x is denoted by $\sqrt[n]{x}$.

VERBALLY		NUMERICAL EXAMPLES	
For $x > 0$:	The principal nth root is positive for all natural numbers n.	$\sqrt{9} = 3$ $\sqrt[3]{8} = 2$	Check: $3^2 = 9$ Check: $2^3 = 8$
For $x = 0$:	The principal nth root of 0 is 0.	$\sqrt{0} = 0$ $\sqrt[3]{0} = 0$	Check: $0^2 = 0$ Check: $0^3 = 0$
For $x < 0$:	If n is odd, the principal nth root is negative.	$\sqrt[3]{-8} = -2$ $\sqrt[5]{-1} = -1$	Check: $(-2)^3 = -8$ Check: $(-1)^5 = -1$
	If n is even, there is no real nth root. The nth roots will be imaginary numbers, as covered in Section 7.7.	$\sqrt{-49} = 7i$	Check: $(7i)^2 = 49i^2$ $= 49(-1)$ $= -49$

When working with square roots, we should be thinking of perfect squares, such as 1, 4, 9, 16, and 25. When working with cube roots, we should be thinking of perfect cubes, such as 1, 8, 27, 64, and 125.

EXAMPLE 3 Interpreting and Using Radical Notation

Interpret and evaluate each expression.

SOLUTION

	RADICAL NOTATION	VERBAL INTERPRETATION	
(a) $\sqrt{25}$	$\sqrt{25} = 5$	The principal square root of 25 is 5.	Check: $5 \cdot 5 = 25$
(b) $\sqrt[3]{27}$	$\sqrt[3]{27} = 3$	The principal cube root of 27 is 3.	Check: $3 \cdot 3 \cdot 3 = 27$
(c) $\sqrt[3]{-27}$	$\sqrt[3]{-27} = -3$	The principal cube root of -27 is -3.	Check: $(-3)(-3)(-3) = -27$
(d) $\sqrt[4]{16}$	$\sqrt[4]{16} = 2$	The principal fourth root of 16 is 2.	Check: $(2)(2)(2)(2) = 16$
(e) $\sqrt[5]{1}$	$\sqrt[5]{1} = 1$	The principal fifth root of 1 is 1.	Check: $(1)(1)(1)(1)(1) = 1$
(f) $\sqrt{-25}$	$\sqrt{-25} = i\sqrt{25}$ $= 5i$	The principal square root of -25 is $5i$.	Check: $(5i)(5i) = 25i^2$ $= 25(-1)$ $= -25$

SELF-CHECK 9.1.3

Evaluate each of the following expressions.

1. $\sqrt{81}$ 2. $\sqrt[4]{81}$ 3. $\sqrt[5]{-32}$ 4. $-\sqrt[3]{-8}$

The expressions in Example 3 were carefully selected to illustrate the notation but also so that they could be evaluated by using pencil and paper. Calculator Perspective 9.1.1 illustrates how to use a calculator to approximate cube roots and fourth roots.

CALCULATOR PERSPECTIVE 9.1.1 Evaluating Radical Expressions

To evaluate $\sqrt[3]{27}$, $\sqrt[3]{-27}$, and $\sqrt[4]{16}$ from Example 3 on a TI-84 Plus calculator, enter the following keystrokes:

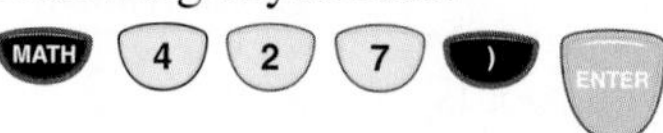

(The cube root is the fourth option on this menu.)

MATH 4 (–) 2 7) ENTER

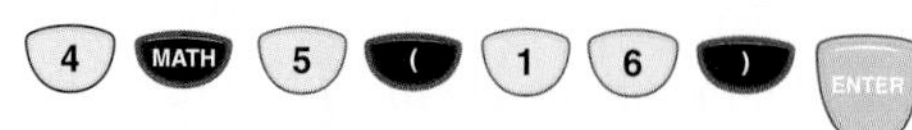

(First enter the index 4, and then the nth root symbol is the fifth option on this menu.)

```
³√(27)
                3
³√(-27)
               -3
4 ˣ√(16)
                2
```

We also can use a calculator to approximate the value of a radical expression to the nearest thousandth. This is illustrated in Example 4 with noninteger answers that are rounded.

EXAMPLE 4 Approximating Radical Expressions with a Calculator

Use a graphing calculator to approximate each expression to the nearest thousandth.

(a) $\sqrt[3]{40}$ **(b)** $\sqrt[4]{50}$ **(c)** $\sqrt[5]{-45}$

SOLUTION

```
³√(40)
      3.419951893
4 ˣ√(50)
      2.659147948
5 ˣ√(-45)
     -2.141127368
```

Use a graphing calculator to approximate each value, and then round these results to the nearest thousandth.

Answers: **(a)** $\sqrt[3]{40} \approx 3.420$
(b) $\sqrt[4]{50} \approx 2.659$
(c) $\sqrt[5]{-45} \approx -2.141$

SELF-CHECK 9.1.4

Use a calculator to approximate each expression to the nearest hundredth.

1. $\sqrt[7]{28}$ **2.** $\sqrt[7]{-28}$ **3.** $\sqrt[4]{28}$ **4.** $\sqrt[4]{0.0028}$

Sphere: $V = \frac{4}{3}\pi r^3$

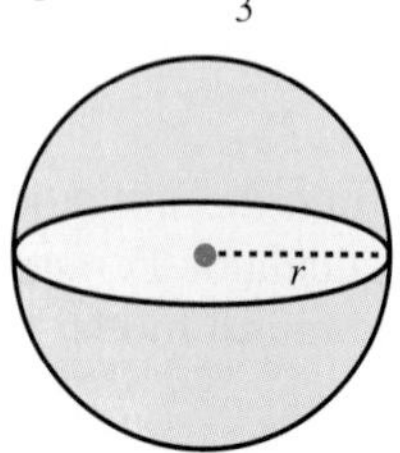

Example 5 illustrates how to evaluate a radical expression containing variables. This equation is another form of $V = \frac{4}{3}\pi r^3$ which gives the volume of a sphere with a radius r.

EXAMPLE 5 Determining the Radius of a Spherical Container

The radius of a sphere with a given volume can be determined by using the formula $r = \sqrt[3]{\frac{3V}{4\pi}}$. Use a calculator to approximate to the nearest tenth of a centimeter the radius of a spherical container needed to hold 6,880 cm^3 of water.

SOLUTION

$$r = \sqrt[3]{\frac{3V}{4\pi}}$$

$$= \sqrt[3]{\frac{3(6{,}880)}{4\pi}}$$

$$\approx 11.8$$

Use the given equation and substitute 6,880 for V. Then use a calculator to approximate r.

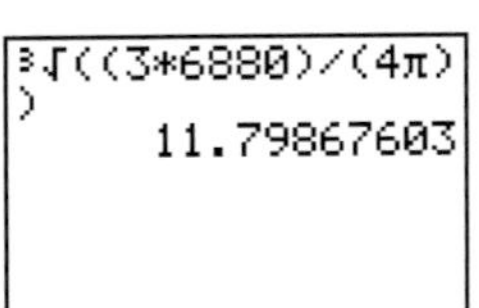

Answer: The radius of the sphere should be approximately 11.8 cm.

SELF-CHECK 9.1.5

1. The time in seconds for one complete period of a pendulum of length x cm is modeled by the function $T(x) = 0.064\sqrt{x}$. Evaluate and interpret $T(81)$.

Because $(-4)(-4)(-4) = -64$, the principal cube root of -64 is -4. That is, $\sqrt[3]{-64} = -4$. However $\sqrt{-64}$ is not a real number; $\sqrt{-64}$ equals the imaginary number $8i$. Therefore the **cube root function** $f(x) = \sqrt[3]{x}$ is defined for all real numbers while the square root function $f(x) = \sqrt{x}$ is restricted to values that make the radicand nonnegative. The following table examines the definition of $\sqrt[3]{x}$ algebraically, verbally, numerically, and graphically.

Principal Cube Root

For any real number x:

ALGEBRAICALLY	VERBALLY	NUMERICALLY	GRAPHICALLY
$y = \sqrt[3]{x}$ if $y^3 = x$ and y is a real number.	$\sqrt[3]{x}$ denotes the principal cube root of x.	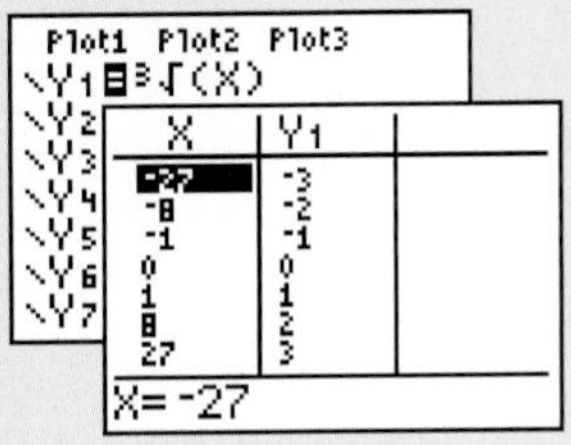	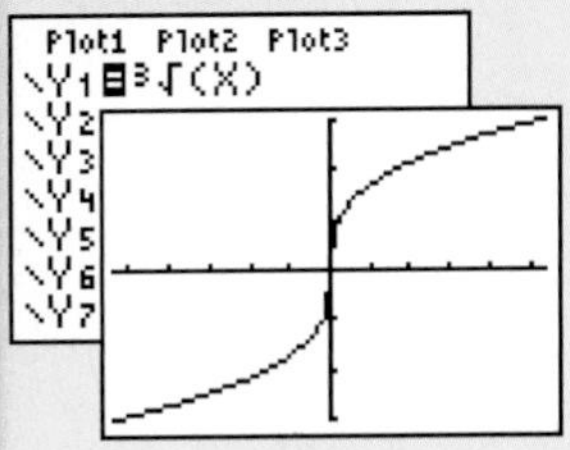 [−27, 27, 5] by [−3, 3, 1]

The domain of $f(x) = \sqrt{x}$ is $D = [0, \infty)$, and the domain of $f(x) = \sqrt[3]{x}$ is $D = \mathbb{R}$.

From the graphs of $y = \sqrt{x}$ and $y = \sqrt[3]{x}$ it is clear that these functions have different domains. We now examine the domains of real-valued functions from an algebraic perspective.

Domain of $y = f(x)$

The domain of a real-valued function $y = f(x)$ is the set of all real numbers for which $f(x)$ is also a real number. This means the domain excludes all values that

1. Cause division by 0
2. Cause a negative number under a square root symbol

Negative numbers are excluded from the radicand of any even-order radical including square roots, fourth roots, etc.

In Example 6 the domain of each function is determined algebraically. We urge you to compare the domain that is determined algebraically to the domain that can be determined from the graph of the function. Remember, the domain of a graph is its projection onto the x-axis.

EXAMPLE 6 Determining the Domain of a Function

Determine the domain of each of these functions.

SOLUTION

	ALGEBRAICALLY	GRAPHICALLY
(a) $f(x) = \dfrac{x+1}{x-3}$	To avoid division by 0, we must determine when the denominator is 0. $x - 3 = 0$ $x = 3$ The domain is the set of all real numbers except for 3. $D = \mathbb{R} \sim \{3\}$	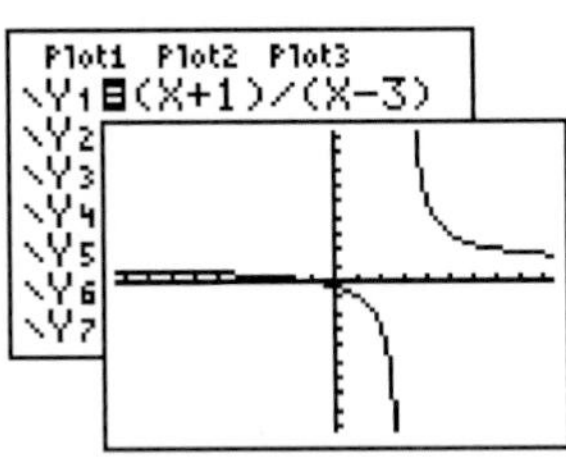 $[-9.4, 9.4, 1]$ by $[-9.3, 9.3, 1]$ The projection of the graph onto the x-axis is the set of all real numbers except for 3. $D = \mathbb{R} \sim \{3\}$
(b) $f(x) = \sqrt{5x + 30}$	The expression under the radical of a square root must be nonnegative. $5x + 30 \geq 0$ $5x \geq -30$ $x \geq -6$ $D = [-6, \infty)$	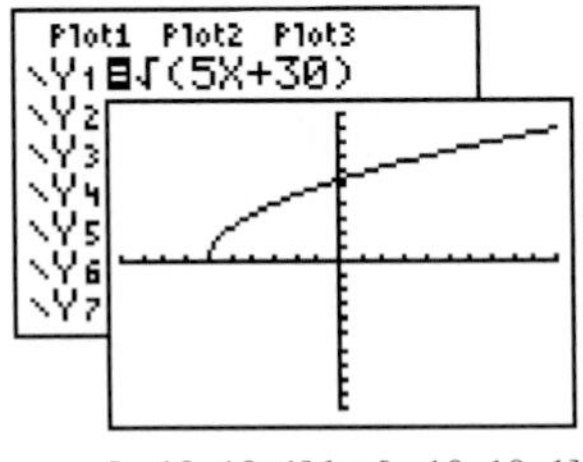 $[-10, 10, 1]$ by $[-10, 10, 1]$ The projection of the graph onto the x-axis is the interval starting at -6 and extending to the right. $D = [-6, \infty)$

(c) $f(x) = \sqrt[3]{5x + 30}$

The expression under the radical of a cube root can be any real number.

$D = \mathbb{R}$

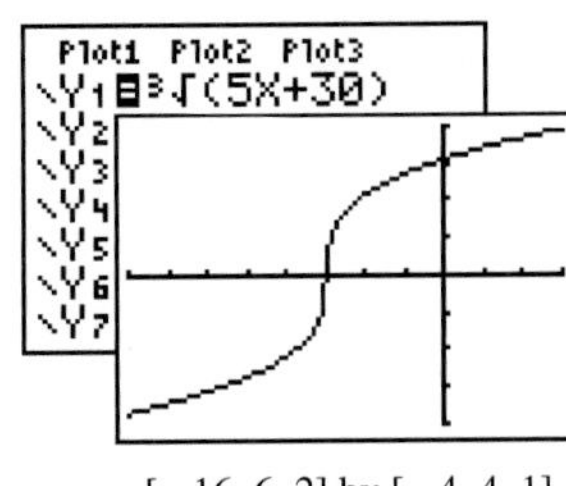

[−16, 6, 2] by [−4, 4, 1]

The projection of the graph onto the x-axis is the set of all real numbers.

$D = \mathbb{R}$

SELF-CHECK 9.1.6

Determine the domain of each of these functions.

1. $f(x) = 2x - 8$
2. $f(x) = \dfrac{1}{2x - 8}$
3. $f(x) = \sqrt{2x - 8}$
4. $f(x) = \sqrt[3]{2x - 8}$

In Example 1 we noted that $\sqrt{4^2} = \sqrt{16} = 4$ and that $\sqrt{(-4)^2} = \sqrt{16} = 4$. In general $\sqrt{x^2} = |x|$. This is further illustrated by the graphs of $y = \sqrt{x^2}$ and $y = |x|$ shown in Example 7.

EXAMPLE 7 Comparing the Functions $y = \sqrt{x^2}$ and $y = |x|$

Compare the tables and graphs for $Y_1 = \sqrt{x^2}$ and $Y_2 = |x|$.

SOLUTION

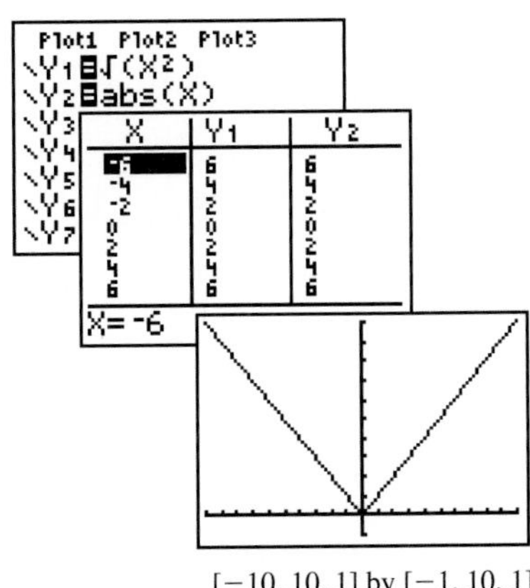

[−10, 10, 1] by [−1, 10, 1]

The values in this table for $Y_1 = \sqrt{x^2}$ and $Y_2 = |x|$ are identical. This indicates (but does not prove) that these functions are equal.

Note that the graph of $Y_1 = \sqrt{x^2}$ has the classic V shape of an absolute value function. This graph appears identical to the graph of $Y_2 = |x|$.

Answer: Both the graph and the table support the conclusion that $\sqrt{x^2} = |x|$ for all real values of x.

We now summarize some of the different families of functions that we have examined so far in this book. Remember that each family of functions has its own characteristic shape.

NAME OF FUNCTION	ALGEBRAIC EXAMPLE	GRAPHIC EXAMPLE	DOMAIN OF EXAMPLE
Linear function	$f(x) = x$	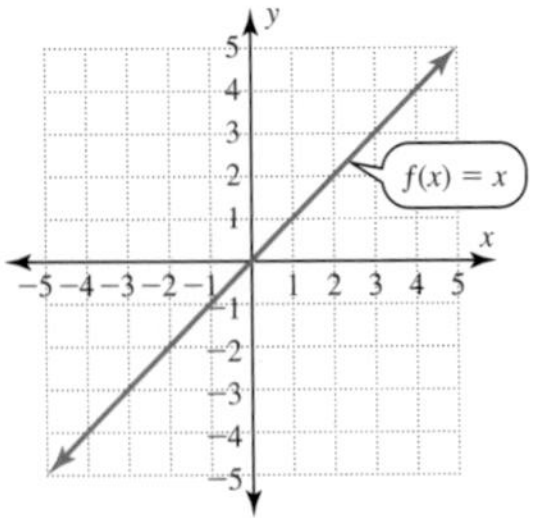	$D = \mathbb{R}$
Absolute value function	$f(x) = \lvert x \rvert$	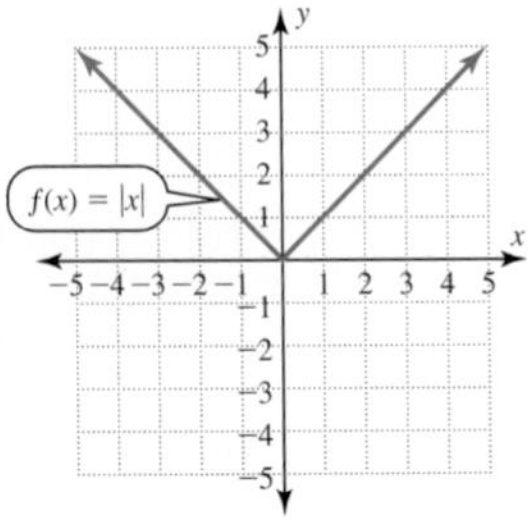	$D = \mathbb{R}$
Quadratic function	$f(x) = x^2$	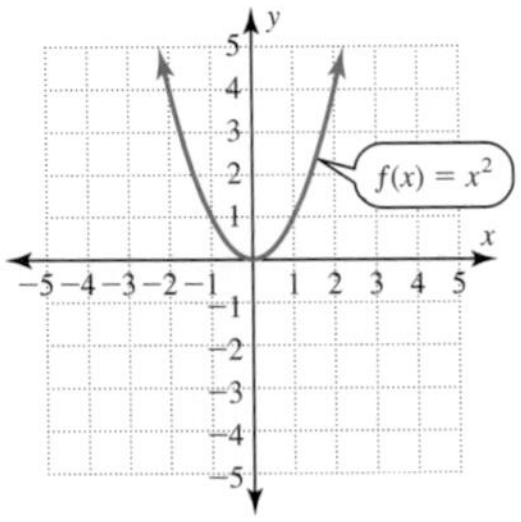	$D = \mathbb{R}$
Rational function	$f(x) = \dfrac{1}{x}$	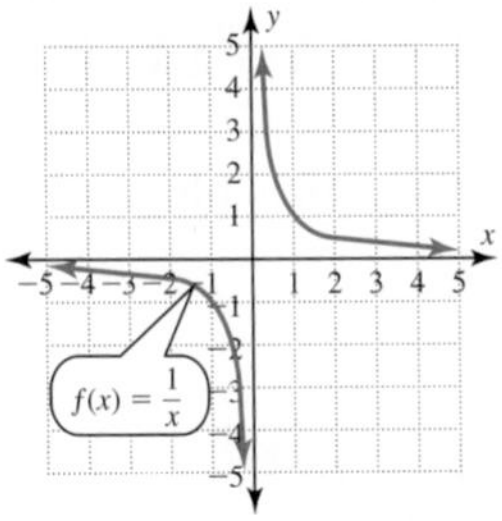	$D = \mathbb{R} \sim \{0\}$
Square root function	$f(x) = \sqrt{x}$	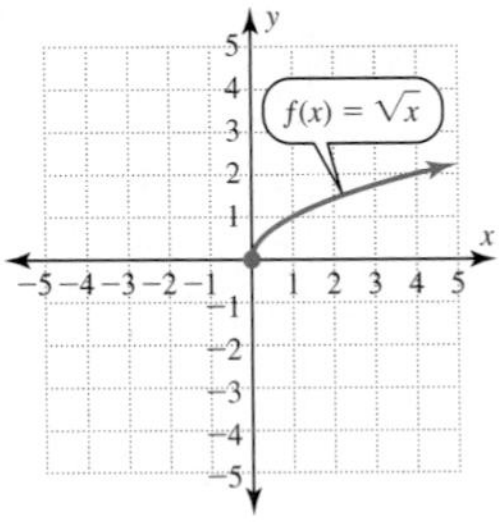	$D = [0, \infty)$
Cube root function	$f(x) = \sqrt[3]{x}$	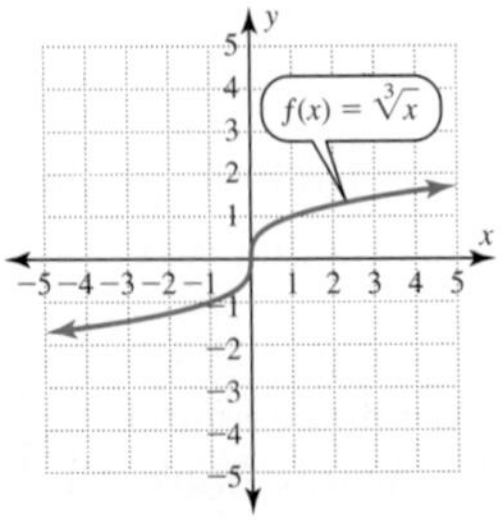	$D = \mathbb{R}$

SELF-CHECK ANSWERS

9.1.1

1. 14
2. 10
3. 5
4. -5

9.1.2

1.

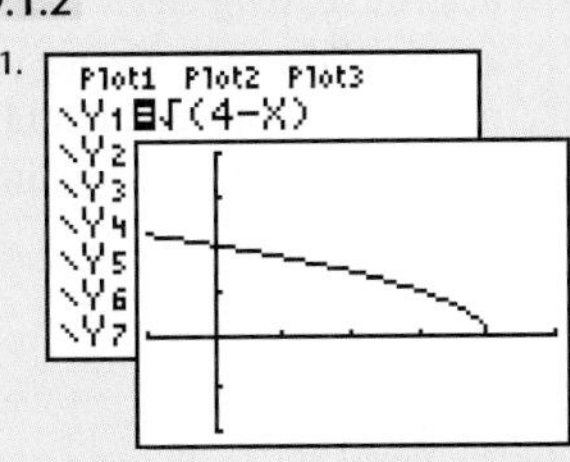

$[-1, 5, 1]$ by $[-2, 4, 1]$

2. $D = (-\infty, 4]$, $R = [0, \infty)$

9.1.3

1. 9 2. 3 3. -2 4. 2

9.1.4

1. 1.61
2. -1.61
3. 2.30
4. 0.23

9.1.5

1. A pendulum of length 81 cm will have a period of 0.576 second.

9.1.6

1. $D = \mathbb{R}$
2. $D = \mathbb{R} \sim \{4\}$
3. $D = [4, \infty)$
4. $D = \mathbb{R}$

USING THE LANGUAGE AND SYMBOLISM OF MATHEMATICS 9.1

1. In the radical notation $\sqrt[n]{x}$, x is called the __________.
2. In the radical notation $\sqrt[n]{x}$, $\sqrt{}$ is called the __________ symbol.
3. In the radical notation $\sqrt[n]{x}$, n is called the __________ or the __________ of the radical.
4. The principal nth root of x is denoted by __________.
5. $\sqrt[3]{-125} = -5$ is read "the (principal) __________ root of -125 equals -5."
6. $\sqrt[4]{16} = 2$ is read "the (principal) __________ root of 16 equals 2."
7. The principal nth root of x is not a real number if n is __________ and x is __________.
8. The notation $\pm\sqrt{x}$ is read "__________ or __________ the (principal) square root of x."
9. A real-valued function $y = f(x)$ has real values of x in the domain and is restricted to __________ numbers in the range.
10. The domain of a real-valued function excludes all values that cause division by __________.
11. The domain of a real-valued function excludes all values that cause the radicand of a square root to be __________.

QUICK REVIEW 9.1

The purpose of this quick review is to help you recall skills needed in this section.

1. Simplify $4^2 + (-4)^2$.
2. Simplify $-3^2 + (3i)^2$.
3. Determine the domain of $f(x) = \dfrac{x}{2x - 6}$.
4. Determine the domain of the function with the given graph.
5. Determine the range of the function with the given graph.

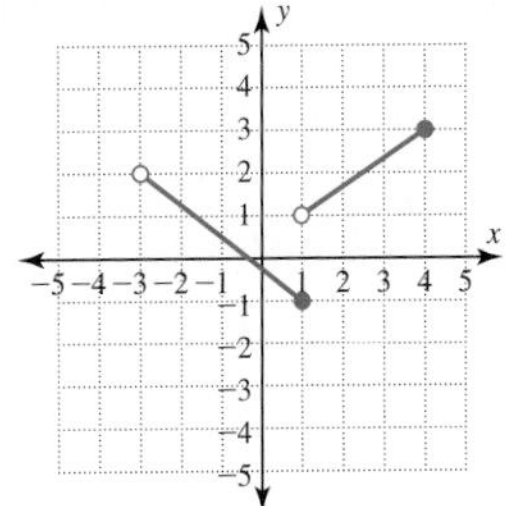

EXERCISES 9.1

Exercises 1 and 2, represent each verbal expression by using radical notation. Assume that all variables are nonnegative.

1. a. The principal cube root of w
   b. The principal fourth root of x
   c. The principal seventh root of v
2. a. The principal sixth root of x
   b. The principal ninth root of v
   c. The principal twelfth root of y

In Exercises 3 and 4, represent each verbal expression by using radical notation and then evaluate this expression.

3. a. The principal square root of 4
   b. The principal cube root of 1,000
   c. The principal fifth root of 32
4. a. The principal square root of 100
   b. The principal fourth root of 81
   c. The principal sixth root of 1

In Exercises 5–20, mentally evaluate each radical expression. Use a calculator only to check your answer.

5. a. $\sqrt{16}$ b. $\sqrt[4]{16}$ c. $\sqrt[3]{0}$
6. a. $\sqrt{64}$ b. $\sqrt[3]{64}$ c. $\sqrt[6]{0}$
7. a. $\sqrt{36}$ b. $-\sqrt{36}$ c. $\sqrt[4]{1}$
8. a. $\sqrt{100}$ b. $-\sqrt{100}$ c. $\sqrt[5]{1}$
9. a. $\sqrt[3]{8}$ b. $\sqrt[3]{-8}$ c. $\sqrt{0.64}$
10. a. $\sqrt[3]{125}$ b. $\sqrt[3]{-125}$ c. $\sqrt[3]{0.125}$
11. a. $\sqrt{100}$ b. $\sqrt{10{,}000}$ c. $\sqrt[4]{10{,}000}$
12. a. $\sqrt[3]{1{,}000}$ b. $\sqrt[3]{1{,}000{,}000}$ c. $\sqrt[6]{1{,}000{,}000}$

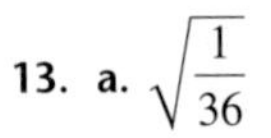

13. a. $\sqrt{\frac{1}{36}}$ b. $\sqrt{\frac{16}{49}}$ c. $\sqrt[3]{\frac{-1}{8}}$
14. a. $\sqrt{\frac{1}{49}}$ b. $\sqrt{\frac{25}{121}}$ c. $\sqrt[3]{\frac{-27}{1{,}000}}$
15. a. $\sqrt{25} + \sqrt{144}$ b. $\sqrt{25 + 144}$
    c. $\sqrt{25} - \sqrt{16}$
16. a. $\sqrt{64} + \sqrt{81} + \sqrt{144}$ b. $\sqrt{64 + 81 + 144}$
    c. $\sqrt{25 - 16}$
17. a. $\sqrt{9} + \sqrt{16} + \sqrt{144}$ b. $\sqrt{9 + 16 + 144}$
    c. $\sqrt{1 + 1 + 1 + 1}$
18. a. $\sqrt{16}$ b. $\sqrt{16^2}$ c. $\sqrt{(-16)^2}$
19. a. $\sqrt{25}$ b. $-\sqrt{25^2}$ c. $-\sqrt{(-25)^2}$
20. a. $\sqrt{4^3}$ b. $\sqrt[3]{8^2}$ c. $\sqrt[15]{-1}$

Estimation and Calculator Skills

In Exercises 21–26, mentally estimate the value of each radical expression, and then use a calculator to approximate each value to the nearest thousandth.

PROBLEM	MENTAL ESTIMATE	CALCULATOR APPROXIMATION
21. $\sqrt{66}$		
22. $\sqrt[3]{66}$		
23. $7 - 2\sqrt{15}$		
24. $15 + 2\sqrt{50}$		
25. $11 + \sqrt[3]{9}$		
26. $25 - \sqrt[3]{990}$		

27. Given $f(x) = \sqrt{4 - x}$, evaluate each expression.
    a. $f(-21)$ b. $f(-5)$ c. $f(0)$ d. $f(4)$
28. Given $f(x) = \sqrt{8 - 2x}$, evaluate each expression.
    a. $f(-4)$ b. $f\left(-\frac{1}{2}\right)$ c. $f(2)$ d. $f(3.5)$
29. Given $f(x) = \sqrt[3]{x + 3}$, evaluate each expression.
    a. $f(5)$ b. $f(-2)$ c. $f(-11)$ d. $f(24)$
30. Given $f(x) = \sqrt[3]{3x - 2}$, evaluate each expression.
    a. $f(1)$ b. $f\left(\frac{1}{3}\right)$ c. $f(22)$ d. $f(-41)$

In Exercises 31 and 32, evaluate the quadratic formula $x = \dfrac{-b \pm \sqrt{b^2 - 4ac}}{2a}$ for each quadratic equation of the form $ax^2 + bx + c = 0$.

31. $x^2 + 2x - 35 = 0$ 32. $x^2 - 7x - 44 = 0$

In Exercises 33 and 34, use the formula $c = \sqrt{a^2 + b^2}$ (from the Pythagorean theorem) to find the exact length of the hypotenuse c for each of these right triangles.

33.

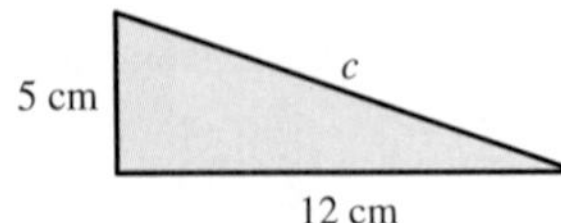

34.

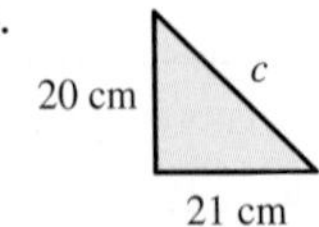

In Exercises 35 and 36, use the formula $s = \sqrt[3]{V}$ to determine to the nearest tenth of a centimeter the dimensions of a cubical container with the given volume.

35. $V = 216 \text{ cm}^3$
36. $V = 2{,}744 \text{ cm}^3$

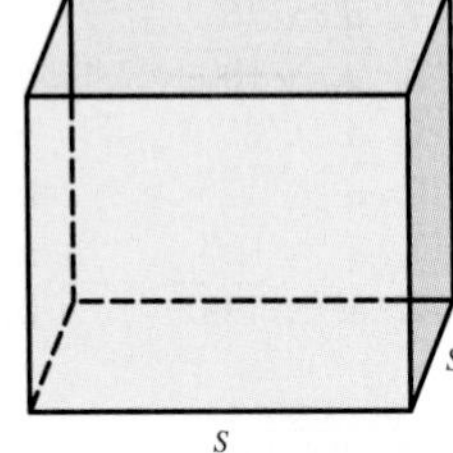

In Exercises 37 and 38, use the formula $T(x) = 0.064\sqrt{x}$ to determine to the nearest tenth of a second the time for one complete period of a pendulum with the given length x in centimeters.

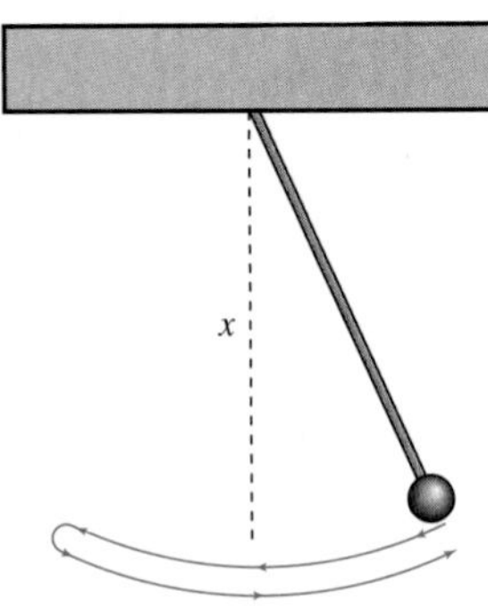

37. $x = 60$ cm
38. $x = 110$ cm

In Exercises 39 and 40, use the formula $r = \sqrt[3]{\frac{3V}{4\pi}}$ to determine to the nearest tenth of a centimeter the radius of a spherical container with the given volume.

39. $V = 216 \text{ cm}^3$
40. $V = 2{,}744 \text{ cm}^3$

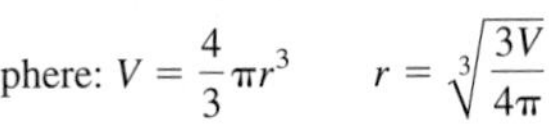
Sphere: $V = \frac{4}{3}\pi r^3$ $r = \sqrt[3]{\frac{3V}{4\pi}}$

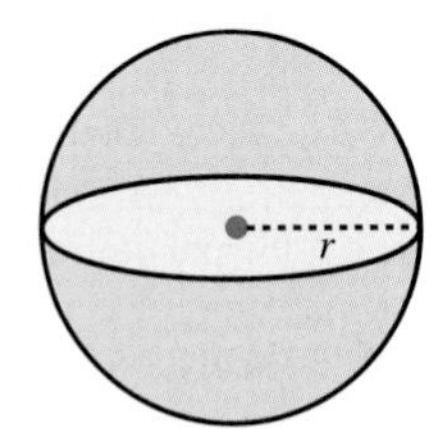

In Exercises 41–48, match each description of a function with its graph.

41. A linear function whose graph has a positive slope
42. A linear function whose graph has a negative slope
43. An absolute value function whose graph opens upward
44. An absolute value function whose graph opens downward
45. A quadratic function whose graph opens upward
46. A quadratic function whose graph opens downward
47. A square root function
48. A cube root function

A.

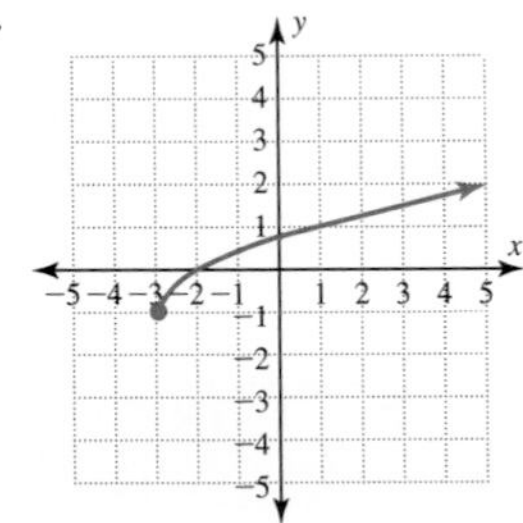

B.

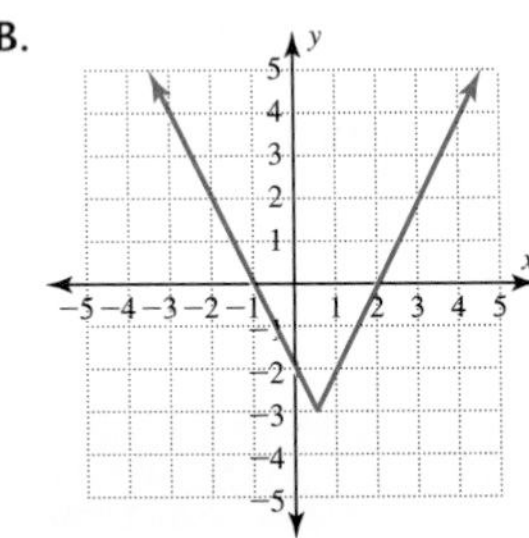

C.

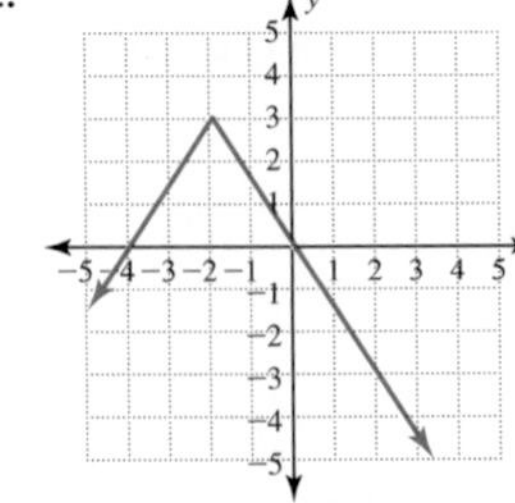

D.

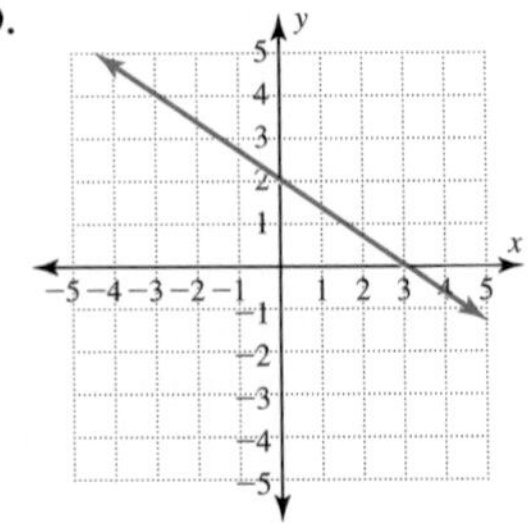

E.

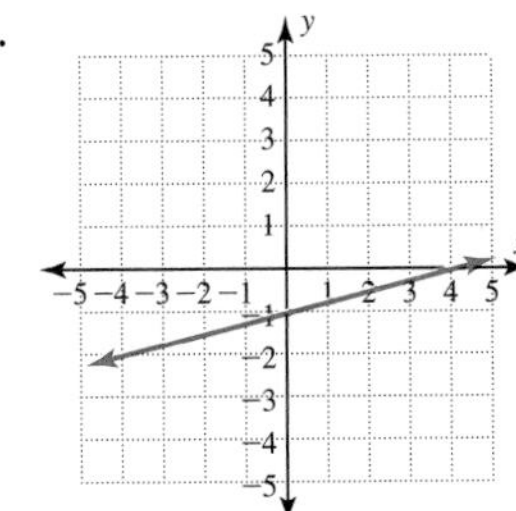

F.

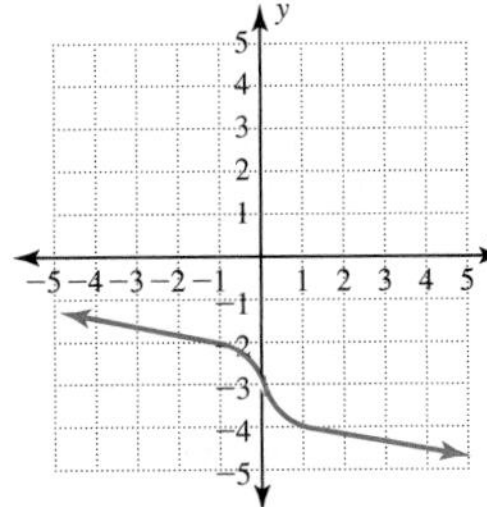

G.

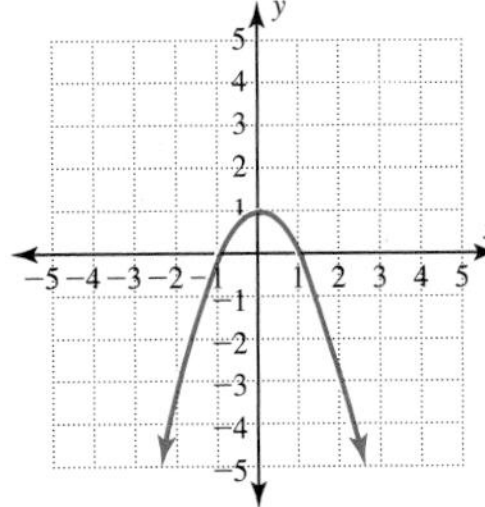

H.

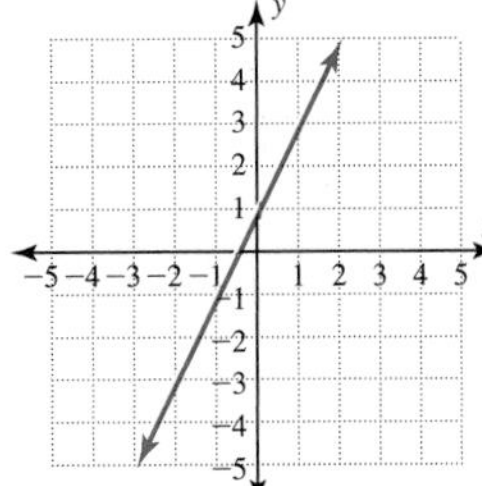

G.

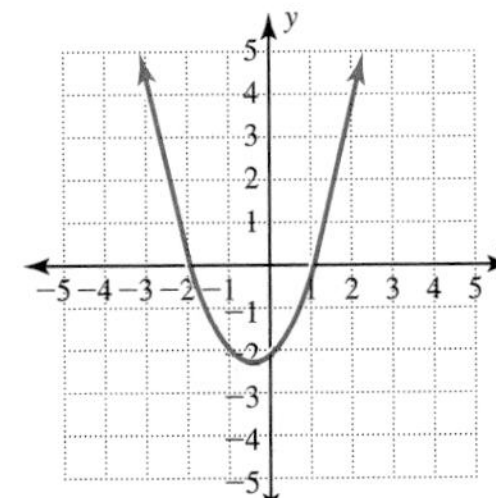

H.

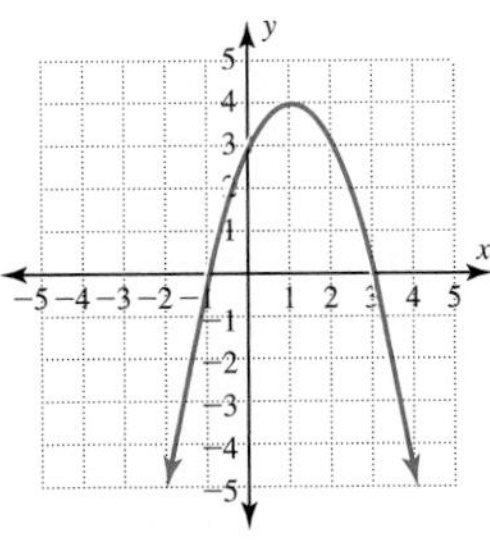

I.

In Exercises 49–57, match each function with its graph and determine its domain. (*Hint:* Work each problem by recognizing the shape of the graph from the defining equation.) Use a graphing calculator only as a check on your work.

49. $f(x) = 2x + 1$

50. $f(x) = -2x + 1$

51. $f(x) = \sqrt{2 - x}$

52. $f(x) = \sqrt{x - 2}$

53. $f(x) = |x - 2| + 1$

54. $f(x) = -x^2 + 1$

55. $f(x) = -|x + 1| + 2$

56. $f(x) = \sqrt[3]{x} + 2$

57. $f(x) = -\sqrt[3]{x} + 2$

A.

B.

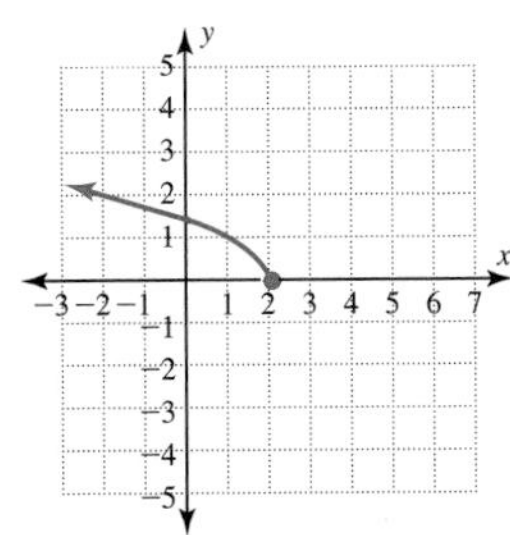

C.

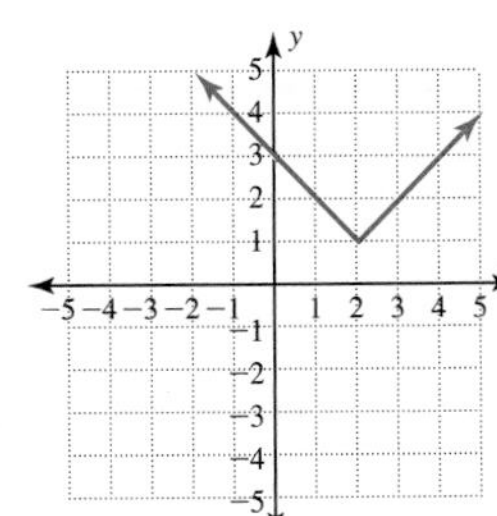

D.

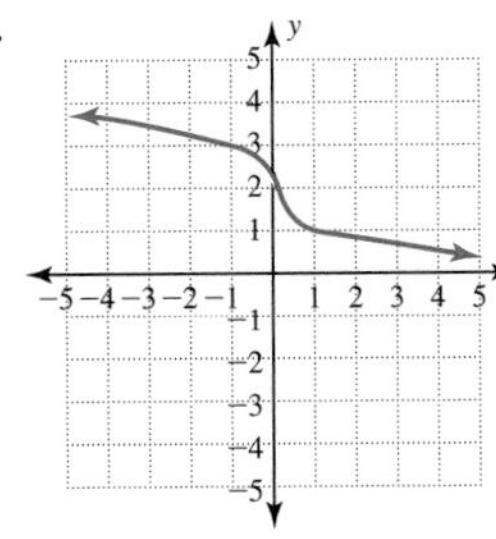

E.

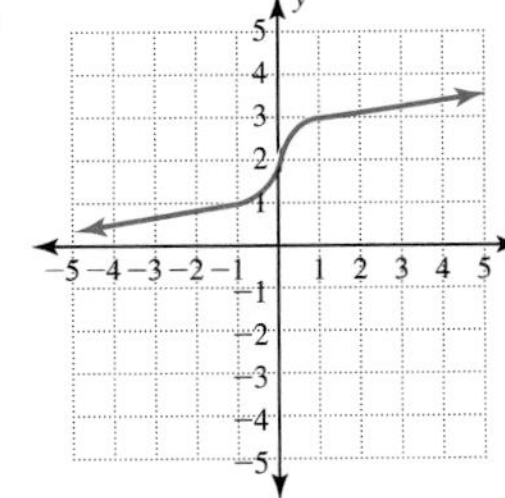

F.

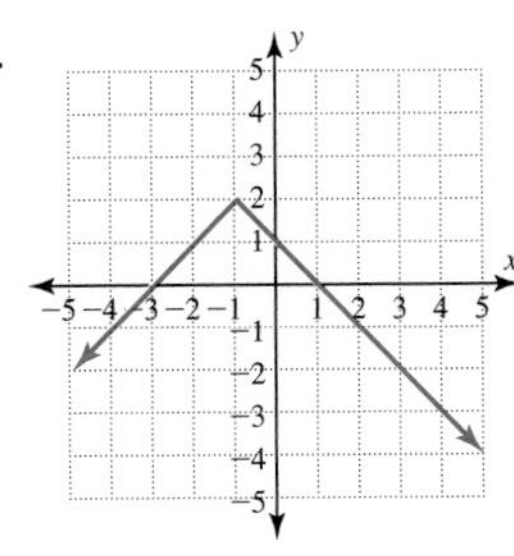

In Exercises 58–64, determine the domain of each function by inspecting the defining equation. Then graph the equation by using a graphing calculator, and use this graph to check your work.

58. $f(x) = x^2 - 5$

59. $f(x) = 4 - x^2$

60. $f(x) = -|x + 2| + 1$

61. $f(x) = |x - 3| + 2$

62. $f(x) = -\sqrt{x + 3}$

63. $f(x) = \sqrt{x} - 3$

64. $f(x) = \sqrt[3]{x} - 3$

65. Cost to Run a Power Line The cost in dollars to run a power line from a power station on one side of a river to a factory on the other side of the river and x mi downstream is given by $C(x) = 1{,}200\sqrt{x^2 + 1}$. Evaluate and interpret each of these expressions.

a. $C(0)$ b. $C(2)$ c. $C(5)$

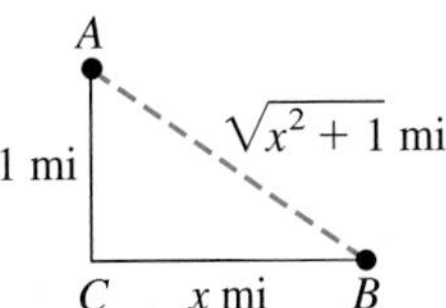

Group Discussion Questions

66. Calculator Discovery Use a graphing calculator to graph each of the following functions: $f(x) = \sqrt{x}$, $f(x) = \sqrt{x + 1}$, $f(x) = \sqrt{x + 2}$, $f(x) = \sqrt{x + 3}$, and $f(x) = \sqrt{x + 4}$. Describe the relationship between the graphs of $f(x) = \sqrt{x}$ and $f(x) = \sqrt{x + c}$ for $c > 0$.

67. Error Analysis A student graphing $f(x) = \sqrt[3]{x} + 2$ obtained the following display. Explain how you can tell by inspection that an error has been made.

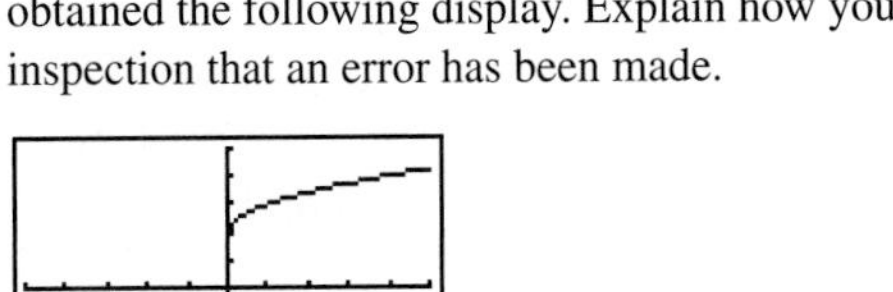

[−5, 5, 1] by [−5, 5, 1]

68. Calculator Discovery Complete the following table, and then verbally compare $Y_1 = \sqrt[4]{x}$ and $Y_2 = \sqrt{\sqrt{x}}$. (*Hint:* Use the **Ask** mode in the **TBLSET** menu.)

X	Y1	Y2
1		
16		
81		
256		
625		
1296		

X=

69. Calculator Discovery

a. Complete the following tables, and then compare the functions $Y_1 = \sqrt{x}$ and $Y_2 = x^2$. How do the shapes of these graphs compare?

For $Y_1 = \sqrt{x}$,

X	Y1
0	
1	
4	
9	
16	
25	
36	

X=0

For $Y_2 = x^2$,

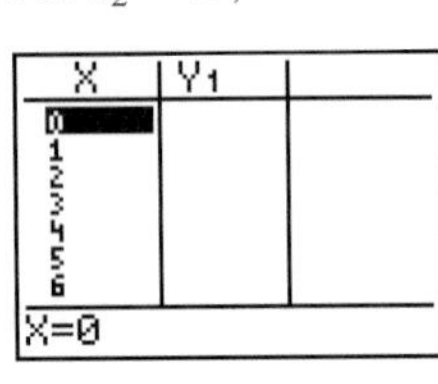

X	Y1
0	
1	
2	
3	
4	
5	
6	

X=0

b. Complete the following tables, and then compare the functions $Y_1 = \sqrt[3]{x}$ and $Y_2 = x^3$. How do the shapes of these graphs compare?

For $Y_1 = \sqrt[3]{x}$,

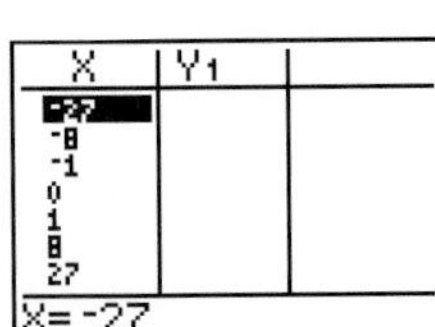

X	Y1
-27	
-8	
-1	
0	
1	
8	
27	

X=-27

For $Y_2 = x^3$,

X	Y1
-3	
-2	
-1	
0	
1	
2	
3	

X=-3

70. Challenge Question

a. Give an example of x so that $\sqrt{x} = -\sqrt{x}$.

b. Give an example of x so that $\sqrt{x}$ and $\sqrt[3]{x}$ are real numbers but $\sqrt{-x}$ is not a real number.

c. Give an example of x so that $\sqrt{-x}$ and $\sqrt[3]{x}$ are real numbers but $\sqrt{x}$ is not a real number.

d. Give an example of x so that $\sqrt{x^2} \neq x$.

71. Challenge Question Match each radical expression with the appropriate simplified form. (*Hint:* If you are not sure, test some different values for x.)

a. $\sqrt{4x^2}$ for $x \geq 0$ A. $-4x$

b. $\sqrt{4x^2}$ for $x < 0$ B. $4x$

c. $\sqrt[3]{64x^3}$ for $x < 0$ C. $-2x$

d. $\sqrt[3]{-64x^3}$ for $x \geq 0$ D. $2x$

Section 9.2 Adding and Subtracting Radical Expressions

Objectives:

4. Add and subtract radical expressions.
5. Simplify radical expressions.

Like Radicals

We add like radicals in exactly the same way as we add like terms of a polynomial. **Like radicals** are radical expressions that have the same index and the same radicand; only the coefficients of the like radicals can differ.

EXAMPLE 1 Classifying Radicals as Like or Unlike

Classify these radicals as like or unlike.

SOLUTION

(a) $-3\sqrt{2}$, $5\sqrt{2}$, and $\frac{\sqrt{2}}{3}$	Like radicals	The radicand of each of these square roots is 2.
(b) $-\sqrt{xy^2}$ and $4\sqrt{xy^2}$	Like radicals	The radicand of each of these square roots is xy^2. The coefficients of the radicals are -1 and 4.
(c) $7\sqrt{5}$ and $7\sqrt[3]{5}$	Unlike radicals	These radicals are unlike because they do not have the same index. One is a square root, and the other is a cube root.
(d) $11\sqrt{7}$ and $11\sqrt{8}$	Unlike radicals	These square roots are unlike because the radicands 7 and 8 are different.
(e) $\sqrt[3]{xy^2}$ and $\sqrt[3]{x^2y}$	Unlike radicals	These radicals are unlike because the radicands are not equal ($xy^2 \neq x^2y$).

Addition and Subtraction of Radicals

To add like radicals, we add their coefficients.

Using the distributive property, we can rewrite $3\sqrt{2} + 4\sqrt{2}$ as $(3 + 4)\sqrt{2}$, which equals $7\sqrt{2}$. Note that we add like radicals by adding their coefficients, just as we add like terms of a polynomial. This similarity is illustrated by the examples in the following box.

Comparison of the Addition of Polynomials and the Addition of Radical Terms

ADDITION OF POLYNOMIALS	ADDITION OF RADICALS
$3x + 4x = 7x$	$3\sqrt{2} + 4\sqrt{2} = 7\sqrt{2}$
$9x + 5y - 6x = 3x + 5y$	$9\sqrt{7} + 5\sqrt{11} - 6\sqrt{7} = 3\sqrt{7} + 5\sqrt{11}$

EXAMPLE 2 Adding and Subtracting Radicals

Perform the following additions and subtractions.

SOLUTION

(a) $\sqrt{3} + 5\sqrt{3}$

$\sqrt{3} + 5\sqrt{3} = (1 + 5)\sqrt{3}$
$= 6\sqrt{3}$

Use the distributive property to add the coefficients of these like terms.

(b) $\sqrt[3]{7} - 5\sqrt[3]{7} + 12\sqrt[3]{7}$

$\sqrt[3]{7} - 5\sqrt[3]{7} + 12\sqrt[3]{7} = (1 - 5 + 12)\sqrt[3]{7}$
$= 8\sqrt[3]{7}$

Use the distributive property to combine like terms.

(c) $4\sqrt{5} + 7\sqrt[3]{5}$

$4\sqrt{5} + 7\sqrt[3]{5}$

This expression cannot be simplified since the terms are unlike.

(d) $3\sqrt[3]{x^2y} - 4\sqrt[3]{x^2y}$

$3\sqrt[3]{x^2y} - 4\sqrt[3]{x^2y} = (3 - 4)\sqrt[3]{x^2y}$
$= -\sqrt[3]{x^2y}$

Subtract the coefficients of these like terms.

(e) $(5\sqrt{11} + 6\sqrt[3]{11}) - (2\sqrt[3]{11} - 7\sqrt{11})$

$(5\sqrt{11} + 6\sqrt[3]{11}) - (2\sqrt[3]{11} - 7\sqrt{11})$
$= 5\sqrt{11} + 6\sqrt[3]{11} - 2\sqrt[3]{11} + 7\sqrt{11}$
$= (5\sqrt{11} + 7\sqrt{11}) + (6\sqrt[3]{11} - 2\sqrt[3]{11})$
$= (5 + 7)\sqrt{11} + (6 - 2)\sqrt[3]{11}$
$= 12\sqrt{11} + 4\sqrt[3]{11}$

First remove the parentheses, and then group like terms together. Then add the coefficients of the like terms.

SELF-CHECK 9.2.1

Classify each pair of radicals as either like or unlike.

1. $-\dfrac{\sqrt{7}}{2}$ and $\dfrac{\sqrt{7}}{3}$
2. $\sqrt[3]{7}$ and $-\sqrt{7}$
3. $\sqrt[7]{2}$ and $\sqrt[7]{3}$

Perform the indicated operations.

4. $4\sqrt{6} - 9\sqrt{6}$
5. $8\sqrt[3]{xy} - 4\sqrt[3]{xy} + \sqrt[3]{xy}$

Simplifying Radical Expressions

One way to simplify a radical expression is to use the properties in the following table to make the radicand as small as possible. *Warning:* If $\sqrt[n]{x}$ is not a real number, then the properties in this table are not true.

Properties of Radicals

If $\sqrt[n]{x}$ and $\sqrt[n]{y}$ are both real numbers,* then:

RADICAL NOTATION	VERBALLY	RADICAL EXAMPLE
$\sqrt[n]{xy} = \sqrt[n]{x}\sqrt[n]{y}$	The nth root of a product equals the product of the nth roots.	$\sqrt{50} = \sqrt{25}\sqrt{2}$
$\sqrt[n]{\dfrac{x}{y}} = \dfrac{\sqrt[n]{x}}{\sqrt[n]{y}}$ for $y \neq 0$	The nth root of a quotient equals the quotient of the nth roots.	$\sqrt{\dfrac{4}{9}} = \dfrac{\sqrt{4}}{\sqrt{9}}$

*If either $\sqrt[n]{x}$ or $\sqrt[n]{y}$ is not a real number, these properties do not hold true.

The properties in the previous box are used to simplify the expressions in Example 3. The key to using these properties is to recognize perfect nth powers. When you are working with square roots, look for perfect square factors, such as 4, 9, 16 and 25. Similarly, when you are working with cube roots, look for perfect cube factors, such as 8, 27, 64, and 125.

EXAMPLE 3 Simplifying Radicals

Simplify each of these radical expressions.

SOLUTION

Caution: Leaving off the index of a radical symbol completely changes the value of the expression. For example, $2\sqrt[3]{5} \neq 2\sqrt{5}$.

(a) $\sqrt{50}$

$$\sqrt{50} = \sqrt{25 \cdot 2}$$ 25 is a perfect square factor of 50.
$$= \sqrt{25}\sqrt{2}$$ $\sqrt{xy} = \sqrt{x}\sqrt{y}$
$$= 5\sqrt{2}$$ $5\sqrt{2}$ is considered a simpler form than $\sqrt{50}$ because it has a smaller radicand.

(b) $\sqrt{45}$

$$\sqrt{45} = \sqrt{9 \cdot 5}$$ 9 is a perfect square factor of 45.
$$= \sqrt{9}\sqrt{5}$$ $\sqrt{xy} = \sqrt{x}\sqrt{y}$
$$= 3\sqrt{5}$$

(c) $\sqrt[3]{40}$

$$\sqrt[3]{40} = \sqrt[3]{8 \cdot 5}$$ 8 is a perfect cube factor of 40.
$$= \sqrt[3]{8}\sqrt[3]{5}$$ $\sqrt[3]{xy} = \sqrt[3]{x}\sqrt[3]{y}$
$$= 2\sqrt[3]{5}$$ *Caution:* Do not make the classic error of writing the answer incorrectly as $2\sqrt{5}$.

In Example 3(a), the expressions $\sqrt{50}$ and $5\sqrt{2}$ are equal. Any calculator approximation of these irrational numbers will not be exact but can be used to check if the expressions appear to be equal. We illustrate this in Self-Check 9.2.2.

SELF-CHECK 9.2.2

1. Use a calculator to compare the values of $\sqrt{50}$ and $5\sqrt{2}$ from Example 3(a).

A second way to simplify a radical expression is to use the property for the radical of a quotient to remove radicals from the denominator. Sometimes this property can also be used to remove fractions from the radicand.

EXAMPLE 4 Simplifying Radicals

Simplify each of these radical expressions.

SOLUTION

(a) $\frac{\sqrt{75}}{\sqrt{3}}$ $\quad \frac{\sqrt{75}}{\sqrt{3}} = \sqrt{\frac{75}{3}}$ $\quad \frac{\sqrt{x}}{\sqrt{y}} = \sqrt{\frac{x}{y}}$

$= \sqrt{25}$ Reduce the fraction in the radicand, and then take the square root.

$= 5$

(b) $\sqrt{\frac{4}{9}}$ $\quad \sqrt{\frac{4}{9}} = \frac{\sqrt{4}}{\sqrt{9}}$ $\quad \sqrt{\frac{x}{y}} = \frac{\sqrt{x}}{\sqrt{y}}$

$= \frac{2}{3}$

(c) $\sqrt[3]{\frac{25}{8}}$ $\quad \sqrt[3]{\frac{25}{8}} = \frac{\sqrt[3]{25}}{\sqrt[3]{8}}$ $\quad \sqrt[3]{\frac{x}{y}} = \frac{\sqrt[3]{x}}{\sqrt[3]{y}}$ The numerator $\sqrt[3]{25}$ cannot be reduced further since it has no perfect cube factor other than 1.

$= \frac{\sqrt[3]{25}}{2}$ $\quad \frac{\sqrt[3]{25}}{2}$ is considered a simpler form than $\sqrt[3]{\frac{25}{8}}$ because it does not contain a radical in the denominator.

SELF-CHECK 9.2.3

Simplify each of these radical expressions.

1. $\sqrt{27}$ 2. $\sqrt[3]{16}$ 3. $\sqrt{\frac{36}{169}}$ 4. $\frac{\sqrt[3]{16}}{\sqrt[3]{2}}$

One reason for always writing radicals in simplified form is to identify like terms. Radicals that appear to be unlike can sometimes be simplified so that it is clear that the terms are like terms. Remember to look for perfect square factors when you are working with square roots and for perfect cube factors when working with cube roots.

EXAMPLE 5 Simplifying Radicals and Then Combining Like Radicals

First simplify these radicals, and then combine like radicals.

SOLUTION

(a) $7\sqrt{20} - 2\sqrt{45}$

$$\begin{aligned} 7\sqrt{20} - 2\sqrt{45} &= 7\sqrt{4 \cdot 5} - 2\sqrt{9 \cdot 5} \\ &= 7\sqrt{4}\sqrt{5} - 2\sqrt{9}\sqrt{5} \\ &= 7(2\sqrt{5}) - 2(3\sqrt{5}) \\ &= 14\sqrt{5} - 6\sqrt{5} \\ &= 8\sqrt{5} \end{aligned}$$

Note that 4 is a perfect square factor of 20 and that 9 is a perfect square factor of 45. Simplify by using the property that $\sqrt{xy} = \sqrt{x}\sqrt{y}$. Then combine like terms.

(b) $11\sqrt[3]{16} - 2\sqrt[3]{54}$

$$\begin{aligned} 11\sqrt[3]{16} - 2\sqrt[3]{54} &= 11\sqrt[3]{8 \cdot 2} - 2\sqrt[3]{27 \cdot 2} \\ &= 11\sqrt[3]{8}\sqrt[3]{2} - 2\sqrt[3]{27}\sqrt[3]{2} \\ &= 11(2\sqrt[3]{2}) - 2(3\sqrt[3]{2}) \\ &= 22\sqrt[3]{2} - 6\sqrt[3]{2} \\ &= 16\sqrt[3]{2} \end{aligned}$$

Note that 8 is a perfect cube factor of 16 and that 27 is a perfect cube factor of 54. Simplify these radicals. Then combine like terms.

(c) $\sqrt{18} + \sqrt{12}$

$$\sqrt{18} + \sqrt{12} = \sqrt{9 \cdot 2} + \sqrt{4 \cdot 3}$$
$$= \sqrt{9}\sqrt{2} + \sqrt{4}\sqrt{3}$$
$$= 3\sqrt{2} + 2\sqrt{3}$$

9 is a perfect square factor of 18, and 4 is a perfect square factor of 12.

These radicals are unlike, so the expression cannot be simplified further.

SELF-CHECK 9.2.4

Simplify each expression.

1. $2\sqrt{50} - 3\sqrt{200}$
2. $\sqrt[3]{24} - \sqrt[3]{3,000}$

We must exercise care when taking even roots of variables because in some problems the variables can assume negative values. Expressions such as $\sqrt{-1}$, $\sqrt[4]{-1}$, and other even roots of negative values are imaginary numbers rather than real numbers. As we illustrate next, some of the properties that hold when $\sqrt[n]{x}$ is a real number will fail when $\sqrt[n]{x}$ is an imaginary number. The following situation illustrates a case where $\sqrt[n]{x^n} \neq x$:

$$\sqrt{(-5)^2} = \sqrt{25} = 5$$

Thus $\sqrt{(-5)^2} \neq -5$.

If $x < 0$ and n is an even number, then $\sqrt[n]{x^n} \neq x$. Since x^2 is positive both when x is positive and when x is negative, the principal square root of x^2 will be a positive real number. We can use absolute value notation to correctly denote the result in all cases:

$$\sqrt{x^2} = |x|$$

The definition of $|x|$ given in Section 1.2 is

$$|x| = \begin{cases} x & \text{if } x \text{ is nonnegative} \\ -x & \text{if } x \text{ is negative} \end{cases}$$

In Section 9.1 we used Example 7 to compare $Y_1 = \sqrt{x^2}$ with $Y_2 = |x|$. We noted that these functions are equal. We now generalize this property to radicals of higher order. To handle the general nth root of both positive and negative values of x, we can define $\sqrt[n]{x^n} = |x|$ when n is even. This is highlighted in the following box.

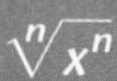

For any real number x and natural number n:

EXAMPLES

ALGEBRAICALLY	NUMERICALLY	GRAPHICALLY	VERBALLY
If n is even, $\sqrt[n]{x^n} = \lvert x \rvert$. *Example* $\sqrt{x^2} = \lvert x \rvert$	Plot1 Plot2 Plot3 \Y1=√(X²) \Y2=abs(X) X: -3, -2, -1, 0, 1, 2, 3 Y1: 3, 2, 1, 0, 1, 2, 3 Y2: 3, 2, 1, 0, 1, 2, 3 X=-3	[−10, 10, 1] by [−1, 10, 1]	These functions are identical for all real numbers x, $\sqrt{x^2} = \lvert x \rvert$. The tables are the same for all values of x, and the graphs are identical.
If n is odd, $\sqrt[n]{x^n} = x$. *Example* $\sqrt[3]{x^3} = x$	Plot1 Plot2 Plot3 \Y1=³√(X^3) \Y2=X X: -3, -2, -1, 0, 1, 2, 3 Y1: -3, -2, -1, 0, 1, 2, 3 Y2: -3, -2, -1, 0, 1, 2, 3 X=-3	[−10, 10, 1] by [−10, 10, 1]	These functions are identical for all real numbers x, $\sqrt[3]{x^3} = x$. The tables are the same for all values of x, and the graphs are identical.

There are two correct ways to handle $\sqrt[n]{x^n}$. The first option is the simpler one: we merely avoid the difficulty by restricting x to positive values. However, since odd roots pose no difficulty, the second option is to allow x to be negative and use absolute value notation for the restricted case $\sqrt[n]{x^n}$ when n is even.

EXAMPLE 6 Evaluating Radical Expressions

Simplify each of these radical expressions. Use absolute value notation wherever necessary.

SOLUTION

(a) $\sqrt{(-7)^2}$

$\sqrt{(-7)^2} = |-7|$
$= 7$

$\sqrt[n]{x^n} = |x|$ if n is even: $\sqrt{(-7)^2} = \sqrt{49} = 7$

(b) $\sqrt[3]{(-2)^3}$

$\sqrt[3]{(-2)^3} = -2$

$\sqrt[n]{x^n} = x$ if n is odd: $\sqrt[3]{(-2)^3} = \sqrt[3]{-8} = -2$

(c) $\sqrt{64x^6}$ for $x > 0$

$\sqrt{64x^6} = \sqrt{(8x^3)^2}$
$= 8x^3$

First write $64x^6$ as the perfect square of $8x^3$. Absolute value notation is not needed because $x > 0$. If we know that $x > 0$, then $|8x^3| = 8x^3$.

(d) $\sqrt{64x^6}$ if x is any real number

$\sqrt{64x^6} = |8x^3|$
$= 8|x|^3$

Because $\sqrt[n]{x^n} = |x|$ if n is even, we must use absolute value notation in this problem.

(e) $\sqrt[3]{64x^6}$

$\sqrt[3]{64x^6} = \sqrt[3]{(4x^2)^3}$
$= 4x^2$

First write $64x^6$ as the perfect cube of $4x^2$.
$\sqrt[n]{x^n} = x$ if n is odd.

SELF-CHECK 9.2.5

Simplify each of these radical expressions. Use absolute value notation wherever necessary.

1. $\sqrt[3]{y^6}$
2. $\sqrt{y^6}$
3. $\sqrt[5]{-32y^5}$

If we restrict the variables to values that keep the radicand positive, then we do not need to use absolute value notation. This is the situation in Example 7.

EXAMPLE 7 Simplifying Radical Expressions

Simplify $\sqrt{x^2 + 2xy + y^2} + \sqrt{x^2 - 2xy + y^2}$, assuming that $x > y > 0$.

SOLUTION

$$\begin{aligned}\sqrt{x^2 + 2xy + y^2} + \sqrt{x^2 - 2xy + y^2} &= \sqrt{(x + y)^2} + \sqrt{(x - y)^2} \\ &= (x + y) + (x - y) \\ &= 2x\end{aligned}$$

Express the radicands as perfect squares. If $x > y > 0$, then $x + y > 0$ and $x - y > 0$; thus absolute value notation is not needed for these square roots. Then combine the like terms.

Carpenters, cooks, and many other people who make things often perform mental computations either to check their calculations or to make quick decisions about the task in front of them. Example 8 involves estimating the length of a side of a square.

The length of a side of the square must be positive. Therefore we use only the principal square root of A for the value of s in Example 8.

EXAMPLE 8 Estimating the Dimensions of a Square

Estimate the length of each side of a square piece of plywood needed to cover a damaged window with an area of 85 cm^2. Then use a calculator to approximate this length.

SOLUTION

$A = s^2$ Formula for the area of a square.

$s = \sqrt{A}$ Formula for the length of a side of a square.

ESTIMATED ANSWER

$85 \approx 81$

$\sqrt{85} \approx \sqrt{81} = 9$

$s \approx 9$ cm

CALCULATOR APPROXIMATION

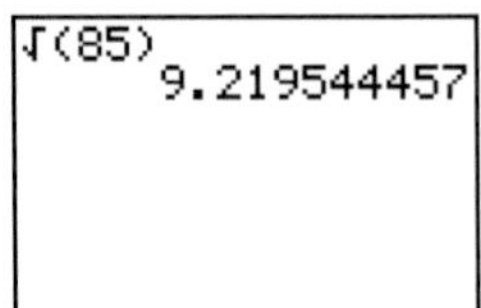

Answer: The calculator approximation of 9.219544457 seems reasonable based upon the mental estimate of 9. The length of each side of the square piece of plywood is approximately 9.2 cm.

Because $81 < 85$, the estimate of 9 is less than the actual value of $\sqrt{85}$.

SELF-CHECK ANSWERS

9.2.1

1. Like
2. Unlike
3. Unlike
4. $-5\sqrt{6}$
5. $5\sqrt[3]{xy}$

9.2.2

1. $\sqrt{50} = 5\sqrt{2}$

```
√(50)
          7.071067812
5√(2)
          7.071067812
```

9.2.3

1. $3\sqrt{3}$
2. $2\sqrt[3]{2}$
3. $\frac{6}{13}$
4. 2

9.2.4

1. $-20\sqrt{2}$
2. $-8\sqrt[3]{3}$

9.2.5

1. y^2
2. $|y^3|$ or $|y|^3$
3. $-2y$

USING THE LANGUAGE AND SYMBOLISM OF MATHEMATICS 9.2

1. Like radicals must have the same __________ and the same __________.
2. The statement that "the nth root of a product equals the product of the nth roots" states that $\sqrt[n]{xy} =$ __________.
3. The statement that "the nth root of a quotient equals the quotient of the nth roots" states that $\sqrt[n]{\frac{x}{y}} =$ __________.
4. If n is even, $\sqrt[n]{x^n} =$ __________.
5. If n is odd, $\sqrt[n]{x^n} =$ __________.

QUICK REVIEW 9.2

The purpose of this quick review is to help you recall skills needed in this section.

Simplify each expression.

1. $(5x + 3y) + (4x - 9y)$
2. $3(2x - y) - (4x - 9y)$
3. $|17|$
4. $|-22|$
5. $|17 - 22|$

EXERCISES 9.2

In Exercises 1 and 2, determine which radical is a like radical to the given radical.

1. $7\sqrt{5}$ a. $2\sqrt[3]{5}$ b. $3\sqrt{7}$ c. $7\sqrt{3}$ d. $-2\sqrt{5}$
2. $4\sqrt[3]{7}$ a. $4\sqrt{7}$ b. $11\sqrt[3]{7}$ c. $5\sqrt[3]{4}$ d. $-5\sqrt[3]{4}$

In Exercises 3–20, find the sum or difference. Assume that all variables represent positive real numbers so that absolute value notation is not necessary.

3. $\sqrt{25} + \sqrt{49}$
4. $\sqrt{36} + \sqrt{64}$
5. $2\sqrt{9} - 3\sqrt{16}$
6. $5\sqrt{100} - 3\sqrt{4}$
7. $19\sqrt{2} + 11\sqrt{2}$
8. $7\sqrt{3} - 8\sqrt{3}$
9. $9\sqrt{7} - 13\sqrt{7}$
10. $6\sqrt{5} + 8\sqrt{5}$
11. $4\sqrt[3]{6} + 2\sqrt[3]{6} - 11\sqrt[3]{6}$
12. $5\sqrt[3]{11} - 2\sqrt[3]{11} + 12\sqrt[3]{11}$
13. $5\sqrt[4]{13} + 6\sqrt[4]{13} - \sqrt[4]{13}$
14. $\sqrt[5]{9} - 8\sqrt[5]{9} - 7\sqrt[5]{9}$
15. $7\sqrt{5} - 5\sqrt{7} - \sqrt{5} + \sqrt{7}$
16. $7\sqrt{2} - \sqrt{3} + 3\sqrt{2} + 2\sqrt{3}$
17. $6\sqrt{7x} - 9\sqrt{7x}$
18. $4\sqrt{11y} - 7\sqrt{11y}$
19. $7\sqrt[4]{17w} - 8\sqrt[4]{17w} + \sqrt[4]{17w}$
20. $6\sqrt[7]{5w} - 11\sqrt[7]{5w} + 5\sqrt[7]{5w}$

In Exercises 21–28, simplify each radical expression.

21. $\sqrt{75}$
22. $\sqrt{28}$
23. $\sqrt{63}$
24. $\sqrt{72}$
25. $\sqrt[3]{24}$
26. $\sqrt[3]{40}$
27. $\sqrt[4]{48}$
28. $\sqrt[5]{96}$

In Exercises 29–50, find the sum or difference. Assume that all variables represent positive real numbers so that absolute value notation is not necessary.

29. $\sqrt{28} + \sqrt{63}$
30. $\sqrt{12} - \sqrt{27}$
31. $\sqrt{75} - \sqrt{48}$
32. $\sqrt{44} + \sqrt{99}$
33. $3\sqrt{50v} - 7\sqrt{32v}$
34. $6\sqrt{45v} - 7\sqrt{320v}$
35. $5\sqrt{28w} - 4\sqrt{63w}$
36. $3\sqrt{27b} - \sqrt{75b}$
37. $\sqrt[3]{24} - \sqrt[3]{375}$
38. $\sqrt[3]{54} - \sqrt[3]{128}$
39. $20\sqrt[3]{-81t^3} - 7\sqrt[3]{24t^3}$
40. $11\sqrt[3]{-16t^3} - 5\sqrt[3]{54t^3}$
41. $9\sqrt[3]{40z^2} - 2\sqrt[3]{5{,}000z^2}$
42. $7\sqrt{1{,}210w^3} - 5\sqrt{1{,}440w^3}$
43. $\sqrt{\frac{5}{4}} + \sqrt{\frac{5}{9}}$
44. $\sqrt{\frac{7}{9}} - \sqrt{\frac{7}{25}}$
45. $2\sqrt{\frac{11x}{25}} - \sqrt{\frac{11x}{49}}$
46. $3\sqrt{\frac{13x}{49}} - 5\sqrt{\frac{13x}{4}}$
47. $7\sqrt{0.98} + 2\sqrt{0.75} - \sqrt{0.12} + 5\sqrt{0.72}$
48. $4\sqrt{0.20} - 3\sqrt{0.90} - 7\sqrt{0.80} + \sqrt{1.60}$
49. $\sqrt{x^2 + 2xy + y^2} - \sqrt{x^2} - \sqrt{y^2}$
50. $\sqrt{x^2} + \sqrt{y^2} - \sqrt{x^2 - 2xy + y^2}$ for $x > y > 0$

In Exercises 51–58, simplify each expression. Use absolute value notation only when it is needed.

	FOR $x \ge 0$	FOR ANY REAL NUMBER x
51. $\sqrt{25x^2}$	a.	b.
52. $\sqrt{36x^2}$	a.	b.
53. $\sqrt[3]{8x^3}$	a.	b.
54. $\sqrt[3]{-125x^3}$	a.	b.
55. $\sqrt{1{,}000{,}000x^6}$	a.	b.
56. $\sqrt[3]{1{,}000{,}000x^6}$	a.	b.
57. $\sqrt[5]{x^{30}}$	a.	b.
58. $\sqrt[6]{x^{30}}$	a.	b.

In Exercises 59–62, match each function with its graph. All graphs are displayed with the window $[-5, 5, 1]$ by $[-5, 5, 1]$.

59. $f(x) = \sqrt{(x + 2)^2}$
60. $f(x) = |x| + 2$
61. $f(x) = \sqrt[3]{(x - 3)^3}$
62. $f(x) = x + 3$

A. B.

C. D.

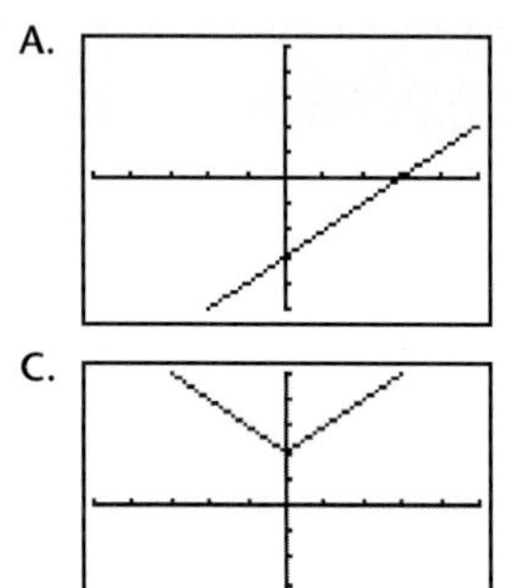

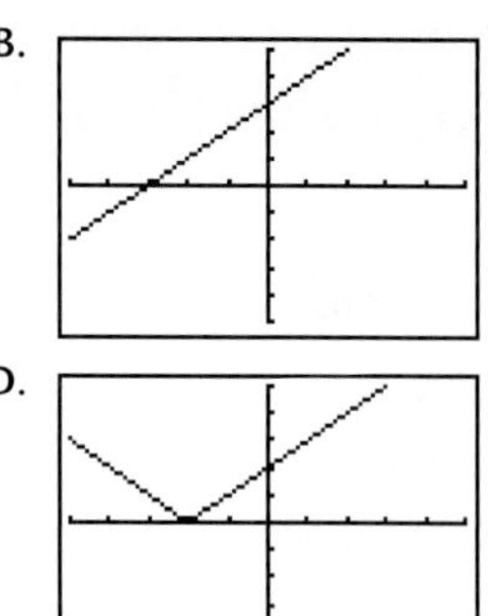

Calculator Exercises

In Exercises 63–68, complete the following table.

a. Estimate each radical expression to the nearest integer.

b. Determine whether this integer estimate is less than or greater than the actual value.

c. Use a calculator to approximate each expression to the nearest thousandth.

	INTEGER ESTIMATE	INEQUALITY	APPROXIMATION
Example $\sqrt[4]{79}$	3	$3 > \sqrt[4]{79}$	2.981
63. $\sqrt[3]{63}$			
64. $\sqrt[4]{82}$			
65. $\frac{1 + \sqrt{9.05}}{2}$			
66. $\frac{1 - \sqrt{9.05}}{2}$			
67. $\frac{-4 - \sqrt{35.97}}{2}$			
68. $\frac{-4 + \sqrt{35.97}}{2}$			

69. Select the best mental estimate of the length of each side of a square of area 26 cm^2. (See the figure.)

a. 3 cm b. 4 cm c. 5 cm
d. 12 cm e. 13 cm

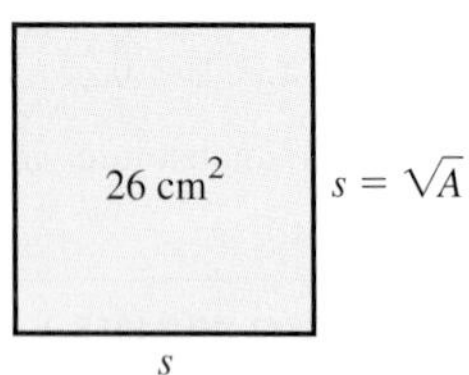

70. Select the best mental estimate of the length of each side of a cubic box whose volume is 26 cm^3. (See the figure.)

a. 3 cm b. 4 cm c. 5 cm
d. 12 cm e. 13 cm

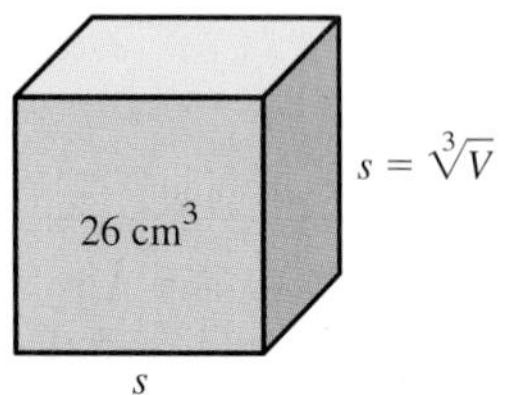

71. Determine to the nearest tenth of a centimeter the length c of the hypotenuse of the right triangle shown in the figure below. The length c can be calculated by $c = \sqrt{a^2 + b^2}$.

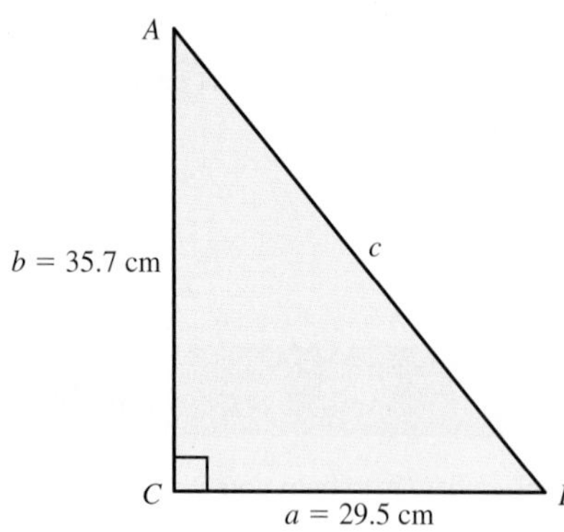

72. Length of a Brace The length d of a diagonal brace from the lower corner of a rectangular storage box to the opposing upper corner is given by

$$L = \sqrt{w^2 + l^2 + h^2}$$

where w, l, and h are, respectively, the width, length, and height of the box. Determine to the nearest tenth of a meter the length of a diagonal brace of a box with width 5.2 m, length 9.4 m, and height 6.5 m.

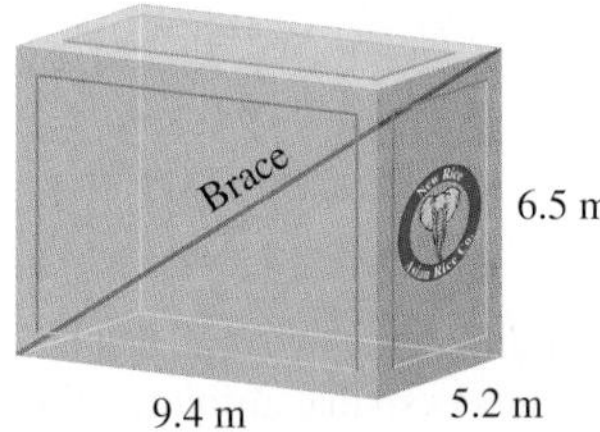

Group Discussion Questions

73. Discovery Question

a. Use a graphing calculator to compare tables of values for $f(x) = \sqrt[4]{x^2}$ and $f(x) = \sqrt{x}$.

b. Use a graphing calculator to compare tables of values for $f(x) = \sqrt[9]{x^3}$ and $f(x) = \sqrt[3]{x}$.

c. Using your observations from parts **a** and **b,** predict how you can simplify $\sqrt[6]{x^2}$ to a radical of lower index than 6. Then use a graphing calculator to compare $f(x) = \sqrt[6]{x^2}$ to the radical function defined by the expression you predicted.

74. Discovery Question

a. Does $\sqrt{\sqrt{256}} = \sqrt[4]{256}$? Explain how you determined your answer.

b. Does $\sqrt{\sqrt{\sqrt{256}}}$ equal either $\sqrt[6]{256}$ or $\sqrt[8]{256}$? Explain how you determined your answer.

c. What is another way of writing $\sqrt{\sqrt{\sqrt{\sqrt{x}}}}$? Explain your reasoning to your teacher.

75. Challenge Question

a. Using a calculator, evaluate $\sqrt{10}$. Then take the square root of this result, then the square root of this result, and so on, until you have taken 10 square roots.

b. Using a calculator, evaluate $\sqrt{0.1}$. Then take the square root of this result, then the square root of this result, and so on, until you have taken 10 square roots.

c. Make a conjecture based on your observations in parts **a** and **b.**

Section 9.3 Multiplying and Dividing Radical Expressions

Objectives:
6. Multiply radical expressions.
7. Divide and simplify radical expressions.

To multiply and divide some radical expressions, we use the following properties:

$$\sqrt[n]{x}\sqrt[n]{y} = \sqrt[n]{xy} \quad \text{for } x \geq 0 \text{ and } y \geq 0$$

$$\frac{\sqrt[n]{x}}{\sqrt[n]{y}} = \sqrt[n]{\frac{x}{y}} \quad \text{for } x \geq 0 \text{ and } y > 0$$

The restriction that $y \neq 0$ is needed in the expression $\frac{\sqrt[n]{x}}{\sqrt[n]{y}} = \sqrt[n]{\frac{x}{y}}$ to prevent division by 0.

When you are multiplying or dividing radicals using these properties, it is important to remember two things:

1. The radicals must be of the same order before these properties can be applied.
2. Do not apply these properties if one or both of the radicands are negative and n is even.

EXAMPLE 1 Multiplying Radicals

Perform each indicated multiplication, and then simplify the product. Assume $x \geq 0$.

SOLUTION

(a) $\sqrt{6}\sqrt{15}$

$$\begin{aligned}\sqrt{6}\sqrt{15} &= \sqrt{90} && \sqrt{x}\sqrt{y} = \sqrt{xy}\\ &= \sqrt{9(10)} && \text{9 is a perfect square factor of 90.}\\ &= \sqrt{9}\sqrt{10} && \sqrt{xy} = \sqrt{x}\sqrt{y}\\ &= 3\sqrt{10}\end{aligned}$$

(b) $\sqrt{9x}\sqrt{x}$

$$\begin{aligned}\sqrt{9x}\sqrt{x} &= \sqrt{9x^2} && \sqrt{x}\sqrt{y} = \sqrt{xy}\\ &= 3x && \text{Absolute value notation is not needed for the answer since } x \geq 0.\end{aligned}$$

The property that $\sqrt[n]{x}\sqrt[n]{y} = \sqrt[n]{xy}$ for $x \geq 0$ and $y \geq 0$ can also be used to separate a single radical expression into two factors. In Example 2 we use this property to simplify the radical expressions.

EXAMPLE 2 Simplifying Radicals

Simplify each radical. Assume all variables are greater than or equal to 0.

SOLUTION

(a) $\sqrt{w^3}$

$$\begin{aligned}\sqrt{w^3} &= \sqrt{w^2}\sqrt{w}\\ &= w\sqrt{w}\end{aligned}$$

Use the property that $\sqrt{x}\sqrt{y} = \sqrt{xy}$ for $x \geq 0$ and $y \geq 0$.

(b) $\sqrt[4]{x^9y^3}$

$$\begin{aligned}\sqrt[4]{x^9y^3} &= \sqrt[4]{x^8}\sqrt[4]{xy^3}\\ &= x^2\sqrt[4]{xy^3}\end{aligned}$$

Factor the given radical into two factors, one of which is a perfect fourth power.

(c) $(\sqrt[3]{20y^2})(\sqrt[3]{50y^2})$

$$\begin{aligned}(\sqrt[3]{20y^2})(\sqrt[3]{50y^2}) &= \sqrt[3]{1{,}000y^4}\\ &= \sqrt[3]{(1{,}000y^3)(y)}\\ &= \sqrt[3]{1{,}000y^3}\sqrt[3]{y}\\ &= 10y\sqrt[3]{y}\end{aligned}$$

$\sqrt[3]{x}\sqrt[3]{y} = \sqrt[3]{xy}$
$1{,}000y^3$ is a perfect cube factor of $1{,}000y^4$. Because this root is odd, this simplification would also work if $y \leq 0$.

The multiplication of some radical expressions is similar to the multiplication of polynomials. Watch for special forms that can be multiplied by inspection, and use the distributive property when it is appropriate.

EXAMPLE 3 Multiplying Radicals with More Than One Term

Perform each indicated multiplication, and then simplify the product. Assume that the variables represent positive real numbers so that absolute value notation is not necessary.

SOLUTION

(a) $2\sqrt{3}(3\sqrt{3} - 5)$

$$\begin{aligned} 2\sqrt{3}(3\sqrt{3} - 5) &= (2\sqrt{3})(3\sqrt{3}) - (2\sqrt{3})(5) \\ &= 6(\sqrt{3})^2 - 10\sqrt{3} \\ &= 6(3) - 10\sqrt{3} \\ &= 18 - 10\sqrt{3} \end{aligned}$$

Distribute the multiplication of $2\sqrt{3}$, and then simplify.

(b) $(4\sqrt{11} - 2)(3\sqrt{11} + 7)$

$$\begin{aligned} &(4\sqrt{11} - 2)(3\sqrt{11} + 7) \\ &= 4\sqrt{11}(3\sqrt{11}) + 4\sqrt{11}(7) - 2(3\sqrt{11}) - 2(7) \\ &= 12(\sqrt{11})^2 + 28\sqrt{11} - 6\sqrt{11} - 14 \\ &= 12(11) + 22\sqrt{11} - 14 \\ &= 132 + 22\sqrt{11} - 14 \\ &= 118 + 22\sqrt{11} \end{aligned}$$

Use the distributive property to multiply each term in the first factor by each term in the second factor.

Simplify and then add the like terms.

(c) $(\sqrt{2a} + \sqrt{3b})(\sqrt{2a} - \sqrt{3b})$

$$\begin{aligned} (\sqrt{2a} + \sqrt{3b})(\sqrt{2a} - \sqrt{3b}) &= (\sqrt{2a})^2 - (\sqrt{3b})^2 \\ &= 2a - 3b \end{aligned}$$

This is a special product of the form $(x + y)(x - y) = x^2 - y^2$ with $x = \sqrt{2a}$ and $y = \sqrt{3b}$.

(d) $(\sqrt[3]{a} - 1)(\sqrt[3]{a^2} + \sqrt[3]{a} + 1)$

$$\begin{aligned} (\sqrt[3]{a} - 1)(\sqrt[3]{a^2} + \sqrt[3]{a} + 1) &= (\sqrt[3]{a})^3 - (1)^3 \\ &= a - 1 \end{aligned}$$

This is a special product of the form $(x - y)(x^2 + xy + y^2) = x^3 - y^3$ with $x = \sqrt[3]{a}$ and $y = 1$.

SELF-CHECK 9.3.1

Simplify each product. Assume $x \geq 0$.

1. $\sqrt{6x}\sqrt{10x}$
2. $\sqrt[3]{4x^2}\sqrt[3]{2x^2}$
3. $(2\sqrt{3} - \sqrt{5})^2$
4. $(5\sqrt{x} + 4)(5\sqrt{x} - 4)$

Conjugate Radicals

The multiplication of conjugate radicals produces a product that does not contain a radical expression.

The special products play an important role in simplifying many algebraic expressions. Once again we will use the product $(x + y)(x - y) = x^2 - y^2$. The radical expressions $x + \sqrt{y}$ and $x - \sqrt{y}$ are called **conjugates** of each other. Knowing that $(x + \sqrt{y})(x - \sqrt{y}) = x^2 - y$ is very useful because the product is an expression free of radicals.

RADICAL EXPRESSION	CONJUGATE RADICAL EXPRESSION
$2 + \sqrt{3}$	$2 - \sqrt{3}$
$2 - \sqrt{3}$	$2 + \sqrt{3}$
$-3 + 4\sqrt{11}$	$-3 - 4\sqrt{11}$
$1 + 6\sqrt{3x}$	$1 - 6\sqrt{3x}$

EXAMPLE 4 Multiplying Conjugates

Find the product of $7 + \sqrt{5}$ and its conjugate.

SOLUTION

$$(7 + \sqrt{5})(7 - \sqrt{5}) = (7)^2 - (\sqrt{5})^2$$
$$= 49 - 5$$
$$= 44$$

The conjugate of $7 + \sqrt{5}$ is $7 - \sqrt{5}$. Note that both $7 + \sqrt{5}$ and $7 - \sqrt{5}$ are irrational numbers, but the product of these conjugates is the natural number 44.

SELF-CHECK 9.3.2

1. Write the conjugate of $5 + \sqrt{2a}$.
2. Write the conjugate of $7 - \sqrt{y}$.
3. Multiply $-5 + \sqrt{3}$ by its conjugate.
4. Check the product in part 3 on a calculator.

Dividing Radicals

In Section 9.2 we simplified the radical expression $\sqrt{\frac{4}{9}}$ as follows:

$$\sqrt{\frac{4}{9}} = \frac{\sqrt{4}}{\sqrt{9}} = \frac{2}{3}$$

Rationalizing the denominator of an expression produces an expression whose denominator does not contain a radical expression.

However, we do not always obtain a perfect square when we divide. Sometimes we wish to remove the radical from the denominator when obtaining something other than a perfect square in the denominator. The process of doing this is called **rationalizing the denominator.** Because $\sqrt{x^2} = x$ for $x \geq 0$, we can multiply a square root by itself to obtain a perfect square and thus remove the radical in the denominator. This is illustrated in Example 5.

EXAMPLE 5 Dividing Radicals Involving Square Roots

Perform the indicated divisions and express the quotients in rationalized form. Assume $a \geq 0$ and $b > 0$.

SOLUTION

(a) $\sqrt{7} \div \sqrt{11}$

$$\sqrt{7} \div \sqrt{11} = \frac{\sqrt{7}}{\sqrt{11}}$$
$$= \frac{\sqrt{7}}{\sqrt{11}} \cdot \frac{\sqrt{11}}{\sqrt{11}}$$
$$= \frac{\sqrt{77}}{11}$$

Write the quotient in fractional form. Multiply both the numerator and the denominator by $\sqrt{11}$ to make the radicand in the denominator a perfect square. Multiplying by the multiplicative identity in the form $\frac{\sqrt{11}}{\sqrt{11}}$ does not change the value of this fraction. Note $\sqrt{11}\sqrt{11} = 11$ and simplify the denominator. The denominator is rationalized because the denominator is now the rational number 11 instead of the irrational number $\sqrt{11}$.

(b) $\frac{\sqrt{2a}}{\sqrt{3b}}$

$$\frac{\sqrt{2a}}{\sqrt{3b}} = \frac{\sqrt{2a}}{\sqrt{3b}} \cdot \frac{\sqrt{3b}}{\sqrt{3b}}$$
$$= \frac{\sqrt{6ab}}{3b}$$

Multiply both the numerator and the denominator by $\sqrt{3b}$ to make the radicand in the denominator a perfect square. Multiplying by the multiplicative identity in the form $\frac{\sqrt{3b}}{\sqrt{3b}}$ does not change the value of this fraction.

For a radicand to be a perfect cube, every factor in the radicand must be a perfect cube. Example 6 illustrates how to rationalize an expression involving cube roots.

EXAMPLE 6 Dividing Radicals Involving Cube Roots

Divide $\frac{\sqrt[3]{v}}{\sqrt[3]{wx^2}}$ and express the quotient in rationalized form. Assume $w \neq 0$ and $x \neq 0$.

SOLUTION

$$\frac{\sqrt[3]{v}}{\sqrt[3]{wx^2}} = \frac{\sqrt[3]{v}}{\sqrt[3]{wx^2}} \cdot \frac{\sqrt[3]{w^2x}}{\sqrt[3]{w^2x}}$$

Multiply both the numerator and the denominator by the factors needed to produce the perfect cube w^3x^3.

$$= \frac{\sqrt[3]{vw^2x}}{\sqrt[3]{w^3x^3}}$$

Simplify the denominator.

$$= \frac{\sqrt[3]{vw^2x}}{wx}$$

This form is rationalized because the denominator is written free of radicals.

SELF-CHECK 9.3.3

1. Use a calculator to compare, from Example 5(a), the values of $\frac{\sqrt{7}}{\sqrt{11}}$ and $\frac{\sqrt{77}}{11}$.
2. Determine an exact rationalized form for $\sqrt[3]{\frac{2}{25}}$ and then check this value on your calculator.

We now examine division by radical expressions with two terms. The key to simplifying these expressions is to use conjugate pairs because the product of the conjugate pair $\sqrt{x} + \sqrt{y}$ and $\sqrt{x} - \sqrt{y}$ contains no radical. For $x \geq 0$ and $y \geq 0$, $(\sqrt{x} + \sqrt{y})(\sqrt{x} - \sqrt{y}) = x - y$.

EXAMPLE 7 Rationalizing a Denominator with Two Terms

Simplify $\frac{6}{\sqrt{7} - \sqrt{5}}$ by rationalizing the denominator.

SOLUTION

$$\frac{6}{\sqrt{7} - \sqrt{5}} = \frac{6}{\sqrt{7} - \sqrt{5}} \cdot \frac{\sqrt{7} + \sqrt{5}}{\sqrt{7} + \sqrt{5}}$$

$$= \frac{6(\sqrt{7} + \sqrt{5})}{7 - 5}$$

Multiply the numerator and the denominator by $\sqrt{7} + \sqrt{5}$. This factor is used because $\sqrt{7} - \sqrt{5}$ and $\sqrt{7} + \sqrt{5}$ are a conjugate pair. Multiplying by the multiplicative identity 1 in the form $\frac{\sqrt{7} + \sqrt{5}}{\sqrt{7} + \sqrt{5}}$ does not change the value of this fraction. $(\sqrt{7} - \sqrt{5})(\sqrt{7} + \sqrt{5}) = 7 - 5$

$$= \frac{6(\sqrt{7} + \sqrt{5})}{2}$$

$$= 3(\sqrt{7} + \sqrt{5})$$

$$= 3\sqrt{7} + 3\sqrt{5}$$

Divide the numerator and the denominator by their common factor 2.

Example 8 involves a denominator with variables in the radicands. The restriction that $x \neq y$ is made to prevent division by 0.

EXAMPLE 8 Rationalizing a Denominator with Variables

Simplify $\frac{\sqrt{x}}{\sqrt{x} + \sqrt{y}}$ by rationalizing the denominator. Assume that x and y represent positive real numbers and that $x \neq y$.

SOLUTION

$$\frac{\sqrt{x}}{\sqrt{x} + \sqrt{y}} = \frac{\sqrt{x}}{\sqrt{x} + \sqrt{y}} \cdot \frac{\sqrt{x} - \sqrt{y}}{\sqrt{x} - \sqrt{y}}$$

Multiply both the numerator and the denominator by $\sqrt{x} - \sqrt{y}$. The product of the conjugate pair $\sqrt{x} + \sqrt{y}$ and $\sqrt{x} - \sqrt{y}$ yields an expression free of radicals.

$$= \frac{\sqrt{x}(\sqrt{x} - \sqrt{y})}{(\sqrt{x} + \sqrt{y})(\sqrt{x} - \sqrt{y})}$$

$$= \frac{x - \sqrt{xy}}{x - y}$$

Distribute the factor of $\sqrt{x}$ in the numerator, and multiply the conjugates in the denominator.

One context in which we frequently encounter radical expressions is as solutions of quadratic equations. Example 9 checks a radical expression to determine whether it is a solution of the given quadratic equation.

EXAMPLE 9 Checking a Solution of a Quadratic Equation

Determine whether $1 + \sqrt{3}$ is a solution of $x^2 - 2x - 2 = 0$.

SOLUTION

$$x^2 - 2x - 2 = 0$$

$$(1 + \sqrt{3})^2 - 2(1 + \sqrt{3}) - 2 \stackrel{?}{=} 0$$

$$1 + 2\sqrt{3} + 3 - 2 - 2\sqrt{3} - 2 \stackrel{?}{=} 0$$

Substitute the given value of x into the equation to determine if it checks.

$$0 \stackrel{?}{=} 0$$

This is a true statement.

Because $1 + \sqrt{3}$ checks, it is a solution of the equation.

SELF-CHECK 9.3.4

Simplify each of these expressions.

1. $\frac{12}{\sqrt{6}}$
2. $\frac{12}{\sqrt[3]{6}}$
3. $\frac{12}{\sqrt{6} - 2}$
4. Determine whether the conjugate of $1 + \sqrt{3}$ is a solution of $x^2 - 2x - 2 = 0$. (Note Example 9.)

SELF-CHECK ANSWERS

9.3.1

1. $2x\sqrt{15}$
2. $2x\sqrt[3]{x}$
3. $17 - 4\sqrt{15}$
4. $25x - 16$

9.3.2

1. $5 - \sqrt{2a}$
2. $7 + \sqrt{y}$
3. $(-5 + \sqrt{3})(-5 - \sqrt{3}) = 22$
4.

```
(-5+√(3))(-5-√(3
))
                22
```

9.3.3

1. $\dfrac{\sqrt{7}}{\sqrt{11}} = \dfrac{\sqrt{77}}{11}$

```
√(7)/√(11)
       .7977240352
√(77)/11
       .7977240352
```

2. $\sqrt[3]{\dfrac{2}{25}} = \dfrac{\sqrt[3]{10}}{5}$

```
3√(2/25)
        .430886938
3√(10)/5
        .430886938
```

9.3.4

1. $2\sqrt{6}$
2. $2\sqrt[3]{36}$
3. $6\sqrt{6} + 12$
4. $1 - \sqrt{3}$ is also a solution of $x^2 - 2x - 2 = 0$.

USING THE LANGUAGE AND SYMBOLISM OF MATHEMATICS 9.3

1. Two properties that are sometimes used to multiply and divide radicals:
   a. $\sqrt[n]{x}\sqrt[n]{y} =$ __________ for $x \geq 0$ and $y \geq 0$
   b. $\dfrac{\sqrt[n]{x}}{\sqrt[n]{y}} =$ __________ for $x \geq 0$ and $y > 0$
2. Before the properties mentioned in part 1 can be applied, the radicals must be of the same __________.
3. The properties mentioned in part 1 cannot be applied if one or both of the radicands are negative and n is __________.
4. The radical expressions $x + \sqrt{y}$ and $x - \sqrt{y}$ are called __________ of each other.
5. The radical expressions $\sqrt{x} + \sqrt{y}$ and $\sqrt{x} - \sqrt{y}$ are a conjugate __________.
6. The process of removing the radical from the denominator of an expression is called __________ the denominator.

QUICK REVIEW 9.3

The purpose of this quick review is to help you recall skills needed in this section.

In Exercises 1–4, multiply these polynomials and simplify the results.

1. $5xy(3x - 2y)$
2. $(2x + 5y)(7x - 3y)$
3. $(4x + 5y)(4x - 5y)$
4. $(4x + 5y)^2$
5. Write the conjugate of $5 - 7i$

EXERCISES 9.3

In Exercises 1–4, perform each indicated multiplication and simplify the product. Then use a calculator to evaluate both the original expression and your answer as a check on your answer.

1. $\sqrt{2}\sqrt{6}$
2. $\sqrt{3}\sqrt{6}$
3. $\sqrt{2}(5\sqrt{2} - 1)$
4. $\sqrt{7}(2\sqrt{7} + 1)$

In Exercises 5–20, perform the indicated multiplication and simplify the product. Assume that the variables represent nonnegative real numbers, so that absolute value notation is not necessary.

5. $(2\sqrt{3})(4\sqrt{5})$
6. $(3\sqrt{7})(4\sqrt{2})$
7. $(7\sqrt{15})(4\sqrt{21})$
8. $(4\sqrt{10})(9\sqrt{15})$
9. $\sqrt{3}(\sqrt{2} + \sqrt{5})$
10. $\sqrt{5}(\sqrt{3} + \sqrt{7})$
11. $3\sqrt{5}(2\sqrt{15} - 7\sqrt{35})$
12. $2\sqrt{7}(5\sqrt{14} - \sqrt{21})$
13. $\sqrt{2}(5\sqrt{6})(8\sqrt{3})$
14. $-\sqrt{5}(3\sqrt{15})(\sqrt{3})$
15. $\sqrt{8w}\sqrt{2w}$
16. $\sqrt{6v}\sqrt{21v}$
17. $(2\sqrt{6z})(4\sqrt{3z})$
18. $(5\sqrt{14z})(11\sqrt{7z})$
19. $3\sqrt{x}(2\sqrt{x} - 5)$
20. $4\sqrt{x}(3\sqrt{x} - 11)$

In Exercises 21–24, simplify each expression. Assume that the variables represent nonnegative real numbers so that absolute value notation is not necessary.

21. $\sqrt{12x^7}$
22. $\sqrt{18x^9}$
23. $\sqrt[3]{x^7y^5}$
24. $\sqrt[3]{x^4y^{11}}$

In Exercises 25–42, perform the indicated multiplication and simplify the product. Assume that the variables represent nonnegative real numbers so that absolute value notation is not necessary.

25. $(\sqrt{3x} - \sqrt{y})(\sqrt{3x} + 4\sqrt{y})$
26. $(\sqrt{2x} + \sqrt{y})(\sqrt{2x} - 5\sqrt{y})$
27. $(2\sqrt{3} - \sqrt{2})^2$
28. $(3\sqrt{2} - \sqrt{3})^2$
29. $(\sqrt{a} + 5\sqrt{3b})^2$
30. $(\sqrt{3a} + 3\sqrt{5b})^2$
31. $(\sqrt{v-2} + 3)(\sqrt{v-2} - 3)$
32. $(\sqrt{2v+3} - 5)(\sqrt{2v+3} + 5)$
33. $\sqrt[3]{7}\sqrt[3]{49}$
34. $\sqrt[3]{36}\sqrt[3]{6}$
35. $\sqrt[3]{9v}\sqrt[3]{-3v^2}$
36. $\sqrt[3]{-4v}\sqrt[3]{2v^2}$
37. $-\sqrt[3]{4}(2\sqrt[3]{2} + \sqrt[3]{5})$
38. $-\sqrt[3]{6}(3\sqrt[3]{4} - 2\sqrt[3]{9})$
39. $(\sqrt[3]{xy^2})^2$
40. $(\sqrt[3]{4ab^2})^2$
41. $-\sqrt{2}(3\sqrt{2} - 5\sqrt{6} - 7)$
42. $3(5\sqrt{2} - \sqrt{6} + 8)$

In Exercises 43–48, multiply each radical expression by its conjugate and simplify the result. Assume that the radicands represent nonnegative real numbers, so that absolute value notation is unnecessary.

43. $2 - \sqrt{2}$
44. $7 + \sqrt{6}$
45. $x + \sqrt{3y}$
46. $2x - \sqrt{y}$
47. $v + \sqrt{3v - 1}$
48. $v - \sqrt{4v - 3}$

In Exercises 49–56, simplify each indicated division. Rationalize the denominator only if this step is necessary. Then use a calculator to evaluate both the original expression and your answer as a check on your answer.

49. $\sqrt{\dfrac{3}{4}}$
50. $\sqrt{\dfrac{7}{36}}$
51. $\dfrac{2}{\sqrt{6}}$
52. $\dfrac{5}{\sqrt{15}}$
53. $\dfrac{\sqrt{5}}{\sqrt{8}}$
54. $\dfrac{\sqrt{11}}{\sqrt{13}}$
55. $\dfrac{\sqrt{10}}{\sqrt{2}}$
56. $\dfrac{\sqrt{33}}{\sqrt{3}}$

In Exercises 57–78, perform each indicated division by rationalizing the denominator and then simplifying. Assume that all variables represent positive real numbers.

57. $18 \div \sqrt{6}$
58. $25 \div \sqrt{5}$
59. $\dfrac{26}{\sqrt{10}}$
60. $\dfrac{9}{\sqrt{15}}$
61. $\dfrac{15}{\sqrt{3x}}$
62. $\dfrac{14}{\sqrt{7y}}$
63. $\dfrac{3}{1 + \sqrt{7}}$
64. $\dfrac{12}{\sqrt{5} - 3}$
65. $\dfrac{36}{\sqrt{13} - 5}$
66. $\dfrac{2}{1 + \sqrt{3}}$
67. $\dfrac{-15}{\sqrt{7} - \sqrt{2}}$
68. $\dfrac{-6}{\sqrt{5} - \sqrt{3}}$
69. $\dfrac{\sqrt{a}}{\sqrt{a} - \sqrt{b}}$
70. $\dfrac{\sqrt{b}}{\sqrt{a} + \sqrt{b}}$
71. $\dfrac{12}{\sqrt[3]{3}}$
72. $\dfrac{4}{\sqrt[3]{2}}$
73. $\sqrt[3]{\dfrac{3}{4}}$
74. $\sqrt[3]{\dfrac{2}{25}}$
75. $\sqrt[3]{\dfrac{v}{9w^2}}$
76. $\sqrt[3]{\dfrac{2v^2}{3w}}$
77. $\dfrac{5}{\sqrt{3x} + \sqrt{2y}}$
78. $\dfrac{7}{\sqrt{5x} - \sqrt{3y}}$

79. Determine whether $1 + \sqrt{2}$ is a solution of $x^2 - 2x - 1 = 0$.
80. Determine whether $1 - \sqrt{2}$ is a solution of $x^2 - 2x - 1 = 0$.

In Exercises 81 and 82, perform the indicated operations and simplify the result.

81. Write out each step to verify that $-\dfrac{1}{2} + \dfrac{\sqrt{3}}{2}i$ is a solution of $x^3 = 1$.
82. Write out each step to verify that $-\dfrac{1}{2} - \dfrac{\sqrt{3}}{2}i$ is a solution of $x^3 = 1$.

Calculator Exercises

In Exercises 83–86, complete the following table.

a. Estimate each radical expression to the nearest integer.
b. Determine whether this integer estimate is less than or greater than the actual value.
c. Use a calculator to approximate each expression to the nearest thousandth.

	INTEGER ESTIMATE	INEQUALITY	APPROXIMATION
Example $\dfrac{6}{\sqrt{4.1}}$	3	$3 > \dfrac{6}{\sqrt{4.1}}$	2.963
83. $\dfrac{100}{\sqrt{24}}$			
84. $\dfrac{144}{\sqrt{65}}$			
85. $\dfrac{50}{\sqrt{37} - \sqrt{0.98}}$			
86. $\dfrac{10}{\sqrt{50} - \sqrt{24}}$			

Group Discussion Questions

87. **Discovery Question** **Approximations of π** Use a calculator to determine which of these approximations is closest to π.
a. $\dfrac{22}{7}$ **b.** $\dfrac{355}{113}$ **c.** $\sqrt{\sqrt{\dfrac{2{,}143}{22}}}$

88. **Challenge Question**
a. A circle and a square have the same area. Determine the ratio of the perimeter of the circle to the perimeter of the square.

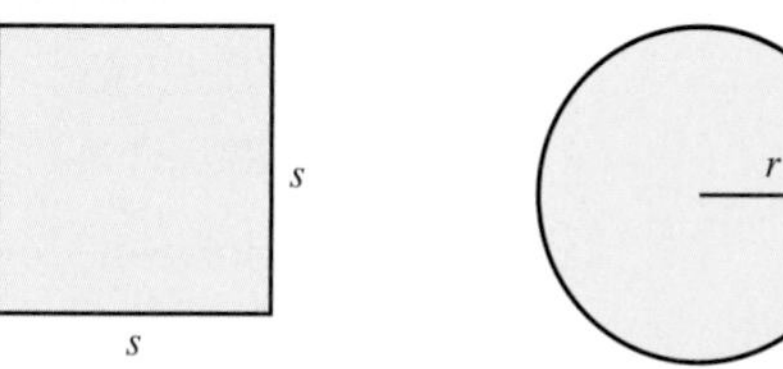

b. A circle and a square have the same perimeter. Determine the ratio of the area of the circle to the area of the square.

89. **Challenge Question**
a. Write each step to verify that $x_1 = \dfrac{-b + \sqrt{b^2 - 4ac}}{2a}$ is a solution of the quadratic equation $ax^2 + bx + c = 0$.
b. x_2 is the conjugate of x_1 from part **a.** Write each step to verify that x_2 is a solution of $ax^2 + bx + c = 0$.

Section 9.4 Solving Equations Containing Radical Expressions

Objectives:

8. Solve equations containing radical expressions.
9. Calculate the distance between two points.

A Mathematical Note

Niels Abel (1802–1829) was a Norwegian mathematician who was raised in extreme poverty and died of tuberculosis at the age of 26. Nonetheless, he established several new mathematical concepts. Among his accomplishments was his proof, using radicals, that the general fifth-degree polynomial equation is impossible to solve exactly.

Equations that contain variables in a radicand are called **radical equations.** Since these equations occur frequently in various disciplines, it is important to know how to solve them. In this section we examine radical equations that result in either linear equations or quadratic equations. An example of such a radical equation is $\sqrt{x} = 3$. It is easy to verify that $x = 9$ is a solution of this equation because $\sqrt{9} = 3$. The key to solving radical equations is to raise both sides of the equation to the same power.

Power Theorem

For any real numbers x and y and natural number n:

ALGEBRAICALLY	VERBALLY	EXAMPLE
If $x = y$, then $x^n = y^n$.	If two expressions are equal, then their nth powers are equal.	If $\sqrt{x} = 3$, then $(\sqrt{x})^2 = (3)^2$ $x = 9$
Caution: The equations $x = y$ and $x^n = y^n$ are not always equivalent. The equation $x^n = y^n$ can have a solution that is not a solution of $x = y$.		$(-3)^2 = 3^2$ but $-3 \neq 3$

Because $x^n = y^n$ can be true when $x \neq y$, we must check all possible solutions in the original equation. Any value that occurs as a solution of the last equation of a solution process but is not a solution of the original equation is called an **extraneous value.**

Solving an equation with radicals is a two-stage process. Stage 1 is to solve for possible solutions. Stage 2 is to check each of these possibilities to eliminate any extraneous values.

Solving Radical Equations Containing a Single Radical

VERBALLY	ALGEBRAIC EXAMPLE
Step 1. Isolate a radical term on one side of the equation.	Solve $\sqrt{x-1} + 2 = 6$ $\sqrt{x-1} = 4$
Step 2. Raise both sides to the nth power.	$(\sqrt{x-1})^2 = 4^2$
Step 3. Solve the resulting equation. (If this equation contains a radical, repeat steps 1 and 2.)	$x - 1 = 16$ $x = 17$
Step 4. Check the possible solutions in the original equation to determine whether they are really solutions or are extraneous.	Check: $\sqrt{17-1} + 2 \stackrel{?}{=} 6$ $\sqrt{16} + 2 \stackrel{?}{=} 6$ $4 + 2 \stackrel{?}{=} 6$ $6 \stackrel{?}{=} 6$ is true Answer: $x = 17$

EXAMPLE 1 Using Multiple Perspectives to Solve a Radical Equation

Solve $\sqrt{7 - x} = 4$ algebraically, numerically, and graphically.

SOLUTION

ALGEBRAICALLY

$$\sqrt{7 - x} = 4$$

$$(\sqrt{7 - x})^2 = 4^2$$ Square both sides of the equation.

$$7 - x = 16$$ Simplify, and then solve the equation for x.

$$-x = 9$$

$$x = -9$$

Check: $\sqrt{7 - (-9)} \stackrel{?}{=} 4$ Substitute -9 back into the original equation to check this value.

$$\sqrt{16} \stackrel{?}{=} 4$$

$$4 \stackrel{?}{=} 4 \text{ checks}$$

$x = -9$ checks

NUMERICALLY

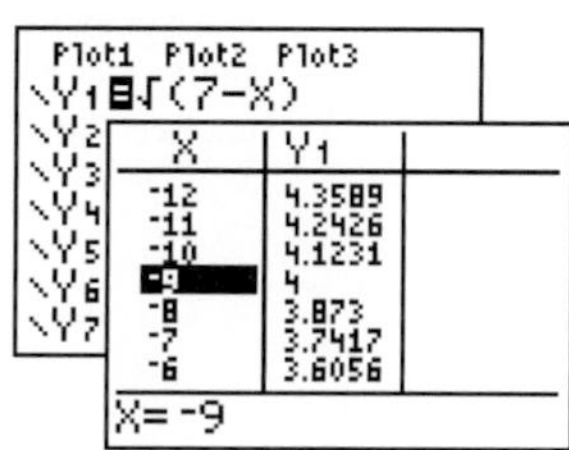

Enter $\sqrt{7 - x}$ for Y_1 and examine the table for a value of x that results in a value of 4 for Y_1.
This function equals 4 for an x-value of -9.
What this table does not answer is whether there are other solutions. The algebraic method shows that there is only one solution.

For $x = -9$, $Y_1 = 4$.

GRAPHICALLY

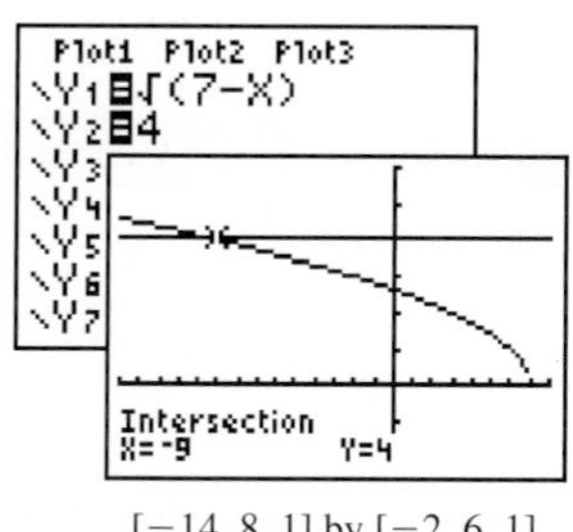

$[-14, 8, 1]$ by $[-2, 6, 1]$

Use a graphing calculator to graph $Y_1 = \sqrt{7 - x}$ and $Y_2 = 4$. Then determine the x-value of the point of intersection of these two graphs.
These graphs intersect for an x-value of -9.

The x-coordinate of the point of intersection is $x = -9$.

Answer: $x = -9$

Example 2 illustrates a problem that yields an extraneous value.

EXAMPLE 2 Solving a Radical Equation with No Solution

Solve $\sqrt{2x + 3} = -5$ algebraically and graphically.

SOLUTION

ALGEBRAICALLY

$$\sqrt{2x + 3} = -5$$
$$(\sqrt{2x + 3})^2 = (-5)^2$$
$$2x + 3 = 25$$
$$2x = 22$$
$$x = 11$$

This equation cannot have a solution since a principal root is always nonnegative. However, continue the solution process to see what happens.

Square both sides of the equation.

Simplify, and then solve for x.

The step of checking a possible solution is not a luxury; it is a necessity to avoid extraneous values when you solve a radical equation.

Check: $\sqrt{2(11) + 3} \stackrel{?}{=} -5$

$\sqrt{25} \stackrel{?}{=} -5$

$5 \stackrel{?}{=} -5$ does not check

$x = 11$ does not check.

Substitute 11 back into the original equation to check this value. The principal square root of 25 is $+5$, not -5. Thus 11 is an extraneous value of the original equation.

GRAPHICALLY

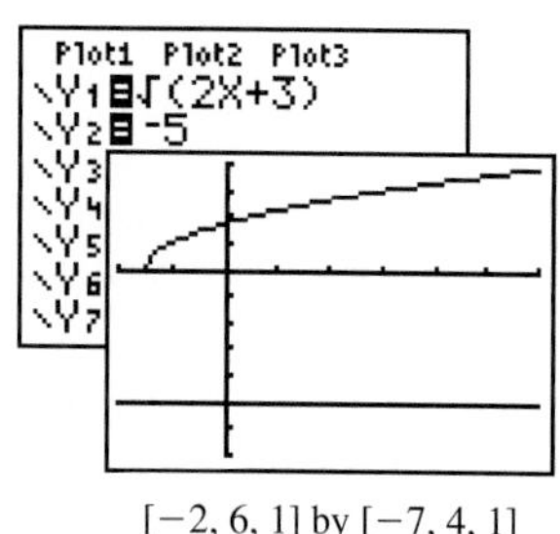

[−2, 6, 1] by [−7, 4, 1]

Use a graphing calculator to graph $Y_1 = \sqrt{2x + 3}$ and $Y_2 = -5$. Note that these two graphs do not intersect. Thus there is no solution for $\sqrt{2x + 3} = -5$.

The graphs of the two functions do not intersect.

Answer: No solution

SELF-CHECK 9.4.1

Solve each equation.

1. $\sqrt{t - 3} = 5$
2. $\sqrt{1 - p} = -1$

Example 3 involves a cube root. The basic procedure for solving this equation is the same as that used for square roots.

EXAMPLE 3 Solving an Equation with One Radical Term

Solve $\sqrt[3]{a - 1} = 4$.

SOLUTION

$$\sqrt[3]{a - 1} = 4$$
$$(\sqrt[3]{a - 1})^3 = 4^3$$
$$a - 1 = 64$$
$$a = 65$$

Cube both sides of the equation.

Simplify, and then solve the linear equation.

Check: $\sqrt[3]{65 - 1} \stackrel{?}{=} 4$

$\sqrt[3]{64} \stackrel{?}{=} 4$

Answer: $a = 65$

The solution process for solving the radical equation in Example 4 produces a quadratic equation. Although the quadratic equation has two solutions, only one of these values checks in the original equation. The other value is extraneous.

EXAMPLE 4 Solving an Equation with One Radical Term

Solve $x = \sqrt{x + 6}$.

SOLUTION

$x = \sqrt{x+6}$	Square both sides of the equation.
$x^2 = x + 6$	Write the quadratic equation in standard form. Factor the left side of the equation.
$x^2 - x - 6 = 0$	
$(x - 3)(x + 2) = 0$	
$x - 3 = 0$ or $x + 2 = 0$	Set each factor equal to 0.
$x = 3$ $\quad$ $x = -2$	Solve for x.

Check:

$x = 3$: $3 \stackrel{?}{=} \sqrt{3 + 6}$ $\quad$ $x = -2$: $-2 \stackrel{?}{=} \sqrt{-2 + 6}$

$3 \stackrel{?}{=} \sqrt{9}$ $\quad$ $-2 \stackrel{?}{=} \sqrt{4}$

$3 \stackrel{?}{=} 3$ is true $\quad$ $-2 \stackrel{?}{=} 2$ is false $\quad$ -2 is an extraneous value.

Answer: $x = 3$

SELF-CHECK 9.4.2

Solve each equation.

1. $\sqrt[3]{x - 5} = -2$ $\qquad$ **2.** $x = \sqrt{3x + 4}$

Whenever a radical equation contains another term on the same side of the equation as the radical, we begin by isolating the radical term on one side of the equation. Then we simplify both sides by squaring if the radicals are square roots, cubing if the radicals are cube roots, etc.

EXAMPLE 5 Solving an Equation with One Radical Term

Solve $\sqrt{2w - 3} + 9 = w$.

SOLUTION

$\sqrt{2w - 3} + 9 = w$	First isolate the radical term on the left side of the equation by subtracting 9 from both sides.
$\sqrt{2w - 3} = w - 9$	
$2w - 3 = w^2 - 18w + 81$	Square both sides of the equation. (Be careful not to omit the middle term when you square the binomial on the right side.)
$w^2 - 20w + 84 = 0$	Write the quadratic equation in standard form.
$(w - 14)(w - 6) = 0$	
$w - 14 = 0$ or $w - 6 = 0$	Factor, and solve for w.
$w = 14$ $\quad$ $w = 6$	

Check: $w = 14$: $\sqrt{2(14) - 3} + 9 \stackrel{?}{=} 14$
$\sqrt{25} + 9 \stackrel{?}{=} 14$
$5 + 9 \stackrel{?}{=} 14$
$14 \stackrel{?}{=} 14$ is true

$w = 6$: $\sqrt{2(6) - 3} + 9 \stackrel{?}{=} 6$
$\sqrt{9} + 9 \stackrel{?}{=} 6$
$3 + 9 \stackrel{?}{=} 6$
$12 \stackrel{?}{=} 6$ is false — 6 is an extraneous value.

Answer: $w = 14$

SELF-CHECK 9.4.3

1. Solve $\sqrt{4x + 15} - 2 = x$

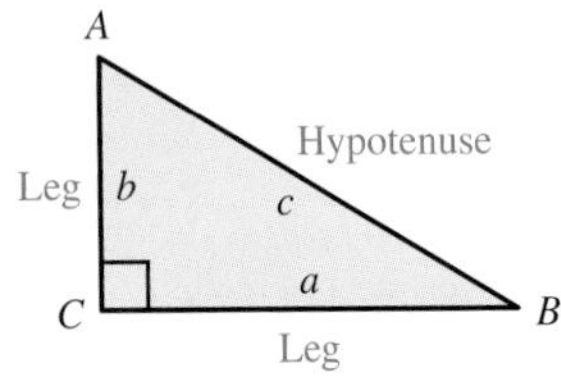

The Pythagorean theorem states, for a right triangle with sides a, b, and c, that $a^2 + b^2 = c^2$. (See the figure.)

An important formula that can be developed by using the Pythagorean theorem is the formula for the distance between two points in a plane. We develop this formula by first considering the horizontal and vertical changes between the two points. Then we compute the direct distance between them by using the Pythagorean theorem.

DISTANCE BETWEEN TWO POINTS

SPECIAL CASE	GENERAL CASE
Calculate the distance between $P(2, 2)$ and $Q(5, 6)$.	Calculate the distance between $P(x_1, y_1)$ and $Q(x_2, y_2)$.

Step 1 Find the horizontal change from P to Q.

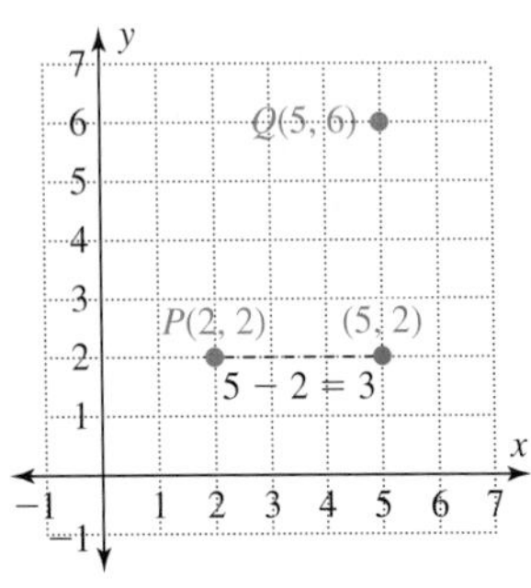

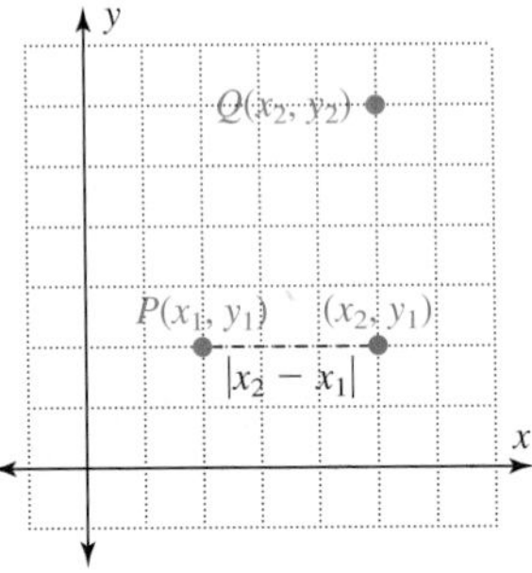

Horizontal distance $= 5 - 2 = 3$

Horizontal distance $= |x_2 - x_1|$

Step 2 Find the vertical change from P to Q.

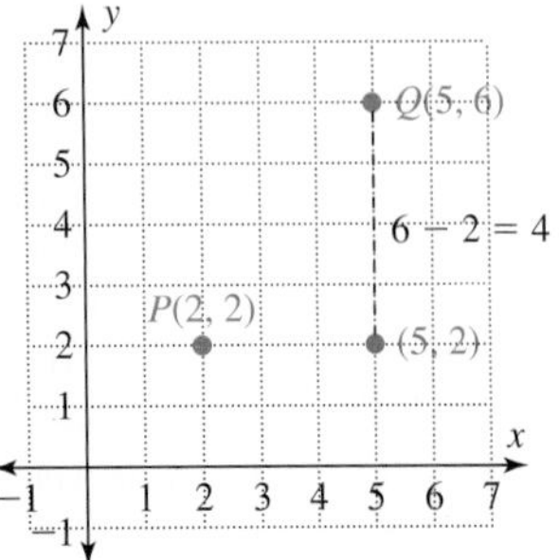

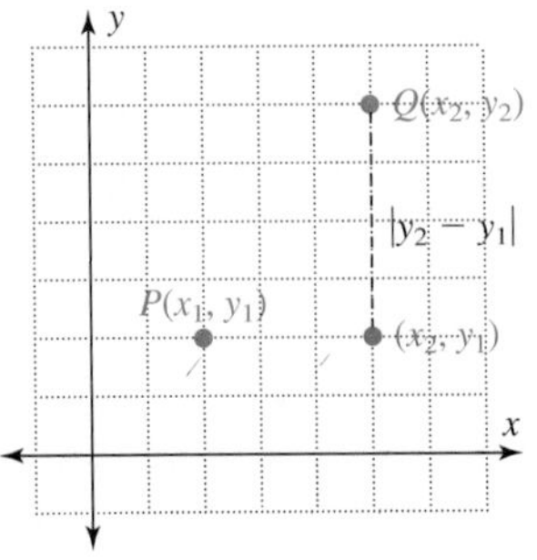

Vertical distance $= 6 - 2 = 4$

Vertical distance $= |y_2 - y_1|$

Step 3 Use the Pythagorean theorem to find the length of hypotenuse PQ.

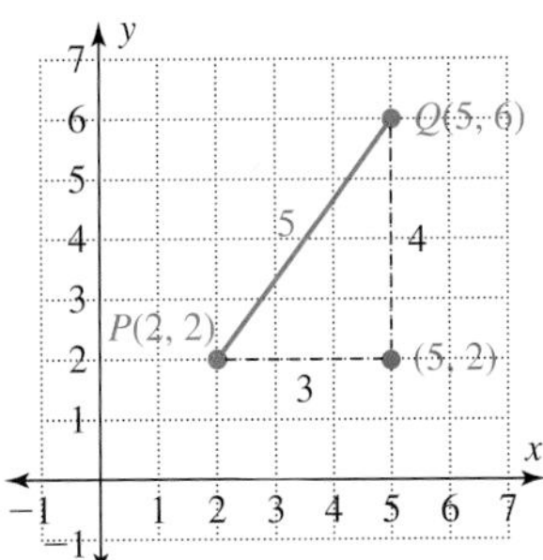

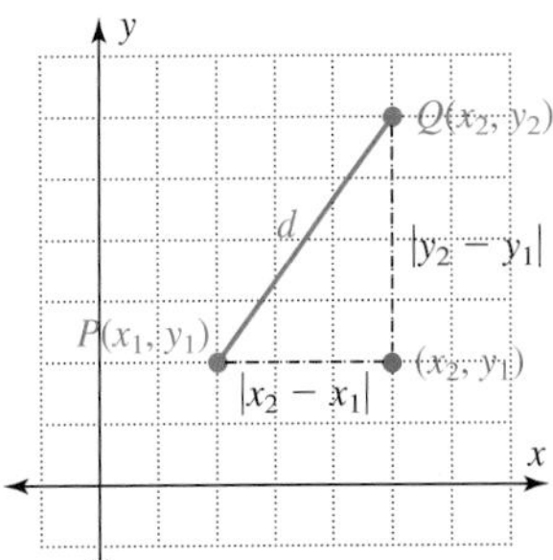

$$\begin{aligned} d^2 &= 3^2 + 4^2 \\ d &= \sqrt{9 + 16} \\ &= \sqrt{25} \\ &= 5 \end{aligned}$$

$$\begin{aligned} d^2 &= |x_2 - x_1|^2 + |y_2 - y_1|^2 \\ d &= \sqrt{|x_2 - x_1|^2 + |y_2 - y_1|^2} \\ d &= \sqrt{(x_2 - x_1)^2 + (y_2 - y_1)^2} \end{aligned}$$

Absolute value notation is not needed in the distance formula because the squares in this formula must be nonnegative. The formula is applicable in all cases, even if P and Q are on the same horizontal or vertical line.

We refer to the distance *between* (x_1, y_1) and (x_2, y_2) rather than the distance *from* (x_1, y_1) to (x_2, y_2) because the distance between these points is the same in either direction. In the formula, note that $(x_2 - x_1)^2 = (x_1 - x_2)^2$ and $(y_2 - y_1)^2 = (y_1 - y_2)^2$.

Distance Formula

The distance d between (x_1, y_1) and (x_2, y_2) is given by $d = \sqrt{(x_2 - x_1)^2 + (y_2 - y_1)^2}$.

EXAMPLE 6 Calculating the Distance Between Two Points

Calculate the distance between $(-3, 1)$ and $(5, -1)$.

SOLUTION

$$\begin{aligned} d &= \sqrt{(x_2 - x_1)^2 + (y_2 - y_1)^2} \\ d &= \sqrt{[5 - (-3)]^2 + (-1 - 1)^2} \\ d &= \sqrt{8^2 + (-2)^2} \\ d &= \sqrt{64 + 4} \\ d &= \sqrt{68} \\ d &= \sqrt{4}\sqrt{17} \\ d &= 2\sqrt{17} \end{aligned}$$

Substitute the given values into the distance formula.

This exact distance is illustrated in the figure to the right. A calculator approximation of this distance is $2\sqrt{17} \approx 8.25$.

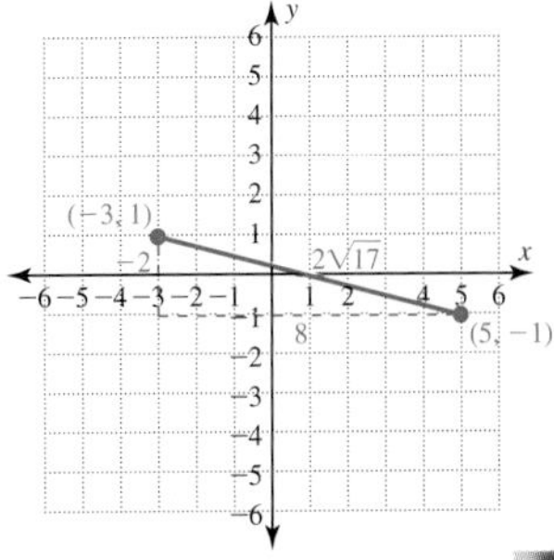

SELF-CHECK 9.4.4

1. Calculate the distance between $(2, -3)$ and $(-13, 5)$.

Engineering, architecture, and construction often involve placing new structures in a specific location relative to other building and utility lines. Example 7 is a skill builder for problems of this type.

EXAMPLE 7 Applying the Distance Formula

Find all points with an x-coordinate of 6 that are 5 units from (3, 3).

SOLUTION

Identify the desired point(s) by $(6, y)$.

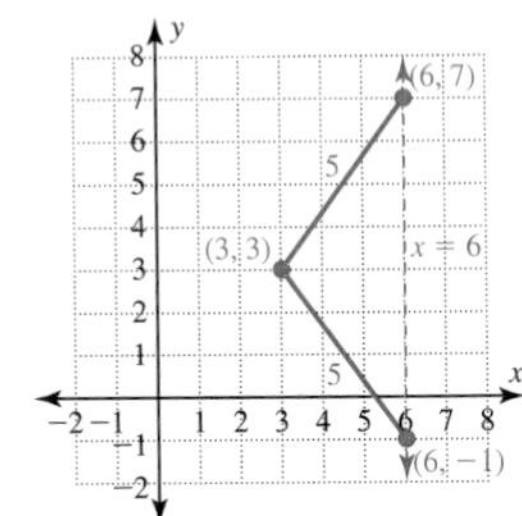

$d = \sqrt{(x_2 - x_1)^2 + (y_2 - y_1)^2}$ All points with an x-coordinate of 6 can be written as $(6, y)$.

$5 = \sqrt{(6 - 3)^2 + (y - 3)^2}$ Substitute the given values into the distance formula.

$5 = \sqrt{9 + y^2 - 6y + 9}$ Then simplify the radicand.

$5 = \sqrt{y^2 - 6y + 18}$ Solve the radical equation for y.

$25 = y^2 - 6y + 18$ Square both sides of the equation.

$0 = y^2 - 6y - 7$ Subtract 25 from both sides of the equation.

$(y + 1)(y - 7) = 0$ Factor the trinomial.

$y + 1 = 0$ or $y - 7 = 0$ Set each factor equal to 0.

$y = -1$ $\quad y = 7$ Both values check.

Answer: The points $(6, -1)$ and $(6, 7)$ are both 5 units from (3, 3).

The strategy that we used to develop the distance formula can also be used to solve some problems directly, as illustrated in Example 8.

EXAMPLE 8 Length of an Exercise Rope

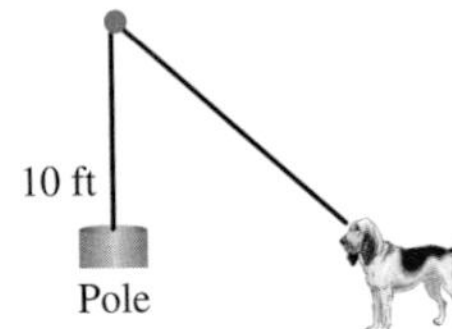

One end of a retractable exercise rope for a dog is fastened to the swivel at the top of a 10-ft pole. The other end is fastened to the collar of this dog.

(a) Write a function $L(x)$ that gives the length of the rope as a function of x, where x is the distance in feet of the dog from the base of the pole. (Assume the base of the pole is at the same height as the dog's head.)

(b) Evaluate and interpret $L(20)$.

(c) What is the area that the dog can occupy if the rope is 25 ft?

SOLUTION

(a) $L(x) = \sqrt{x^2 + 100}$

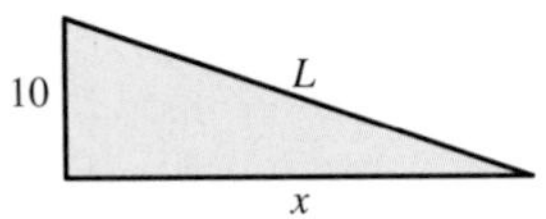

From the Pythagorean theorem,

$$L^2 = x^2 + 10^2$$
$$L = \sqrt{x^2 + 100}$$

(b)
$$\begin{aligned} L(x) &= \sqrt{x^2 + 100} \\ L(20) &= \sqrt{20^2 + 100} \\ &= \sqrt{400 + 100} \\ &= \sqrt{500} \\ &\approx 22.4 \end{aligned}$$

Substitute 20 in for x and then approximate this value to the nearest tenth of a foot.

When the dog is 20 ft from the base of the pole, the length of the rope is 22.4 ft.

(c) $L = 25$

$25^2 = x^2 + 10^2$

$625 = x^2 + 100$

$525 = x^2$

$x = \sqrt{525}$

For a rope of length 25 ft, determine the maximum distance the dog can move its head from the pole. This will be the radius of the circular area that the dog can occupy.

$A = \pi r^2$

$A = \pi(\sqrt{525})^2$

$= 525\pi$

$\approx 1{,}650$

Substitute the radius of $\sqrt{525}$ into the formula for the area of a circle.

When the rope is 25 ft long, the dog can occupy a circular area of approximately 1,650 ft^2.

SELF-CHECK ANSWERS

9.4.1	9.4.2	9.4.3	9.4.4
1. $t = 28$	1. $x = -3$	1. $x = \sqrt{11}$	1. 17
2. No solution	2. $x = 4$		

USING THE LANGUAGE AND SYMBOLISM OF MATHEMATICS 9.4

1. A radical equation is an equation that contains variables in a __________.
2. The power theorem states that if $x = y$, then __________, provided that x and y are real numbers and n is a natural number.
3. The reason that we must check all possible solutions of a radical equation in the original equation is that when we raise both sides to the nth power, we can introduce __________ values.
4. In a right triangle, the two shorter sides are called the __________ and the longer side is called the __________.
5. The Pythagorean theorem states that for a right triangle with hypotenuse c and legs a and b, __________.
6. The distance between points (x_1, y_1) and (x_2, y_2) is given by $d =$ __________.

QUICK REVIEW 9.4

The purpose of this quick review is to help you recall skills needed in this section.

1. Solve $5x - 1 = 64$.
2. Solve $(5x - 1)^2 = 64$.
3. Is a triangle with sides of 5, 12, and 13 a right triangle?
4. Is 1 a solution of $\sqrt{2x - 1} + 2 = x$?
5. Is 5 a solution of $\sqrt{2x - 1} + 2 = x$?

EXERCISES 9.4

In Exercises 1–4, use the given graphs to solve each radical equation.

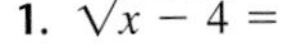

1. $\sqrt{x - 4} = 1$

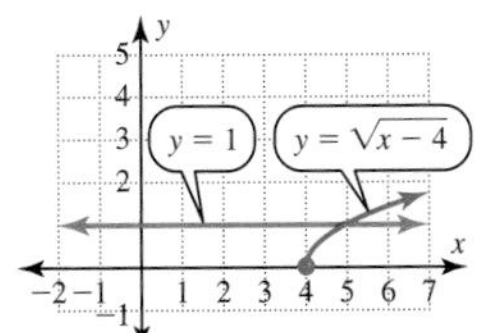

2. $\sqrt{5x - 1} = 3$

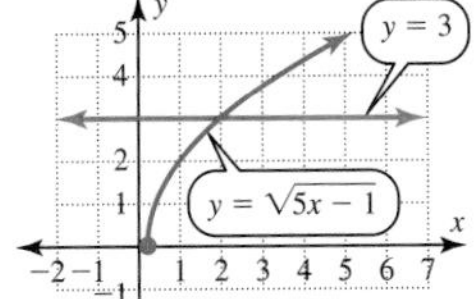

3. $\sqrt[3]{3x - 5} + 2 = 0$

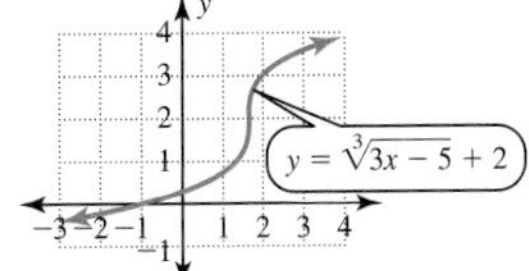

4. $\sqrt[3]{3x - 4} - 2 = 0$

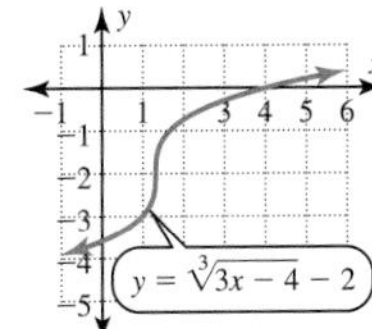

In Exercises 5–8, use the given table to solve each radical equation.

5. $\sqrt{2x - 11} = 3$

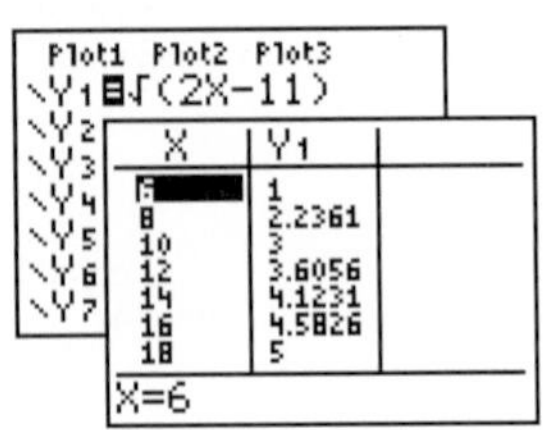

6. $\sqrt{5x + 19} = 7$

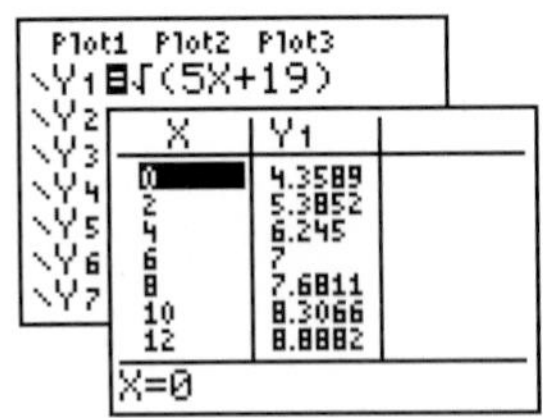

7. $\sqrt{4x + 1} = x - 1$

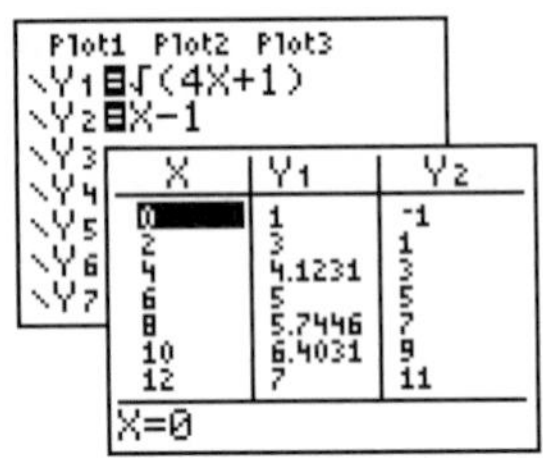

8. $\sqrt{2x + 1} = x - 1$

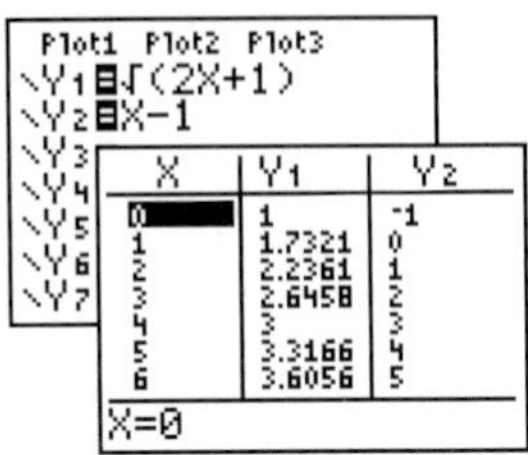

In Exercises 9–30, solve each equation.

9. $\sqrt{t - 4} = 3$
10. $\sqrt{t - 2} = 12$
11. $\sqrt{c + 7} + 23 = 43$
12. $\sqrt{c + 5} - 3 = 6$
13. $\sqrt{3x - 21} + 7 = 2$
14. $\sqrt{25 + 2x} + 11 = 5$
15. $\sqrt[3]{2w + 1} = -2$
16. $\sqrt[3]{5w + 2} = -3$
17. $\sqrt[4]{6v - 2} = 2$
18. $\sqrt[4]{7v - 3} = 3$
19. $\sqrt[5]{w^2 - 4w} = 2$
20. $\sqrt[5]{2w^2 + w - 16} = -1$
21. $\sqrt{y^2 - y + 13} - y = 1$
22. $\sqrt{y^2 + y + 11} - y = 1$
23. $\sqrt{7t + 2} = 2t$
24. $\sqrt{-9t - 2} = 2t$
25. $\sqrt{w^2 - 2w + 1} = 2w$
26. $\sqrt{2w + 1} = w + 1$
27. $\sqrt{2x + 1} + 5 = 2x$
28. $\sqrt{7x - 3} + 1 = 3x$
29. $\sqrt{6u + 7} = \sqrt{11u + 7}$
30. $\sqrt{14u - 12} = \sqrt{10u - 8}$

In Exercises 31–34, determine the exact x- and y-intercepts of the graph of each function.

31. $y = \sqrt{x + 9} - 2$
32. $y = \sqrt{x + 4} - 5$
33. $y = \sqrt[3]{x + 1} + 2$
34. $y = \sqrt[3]{x - 8} - 1$

In Exercises 35–40, use the Pythagorean theorem to find the length of the side that is not given.

35.

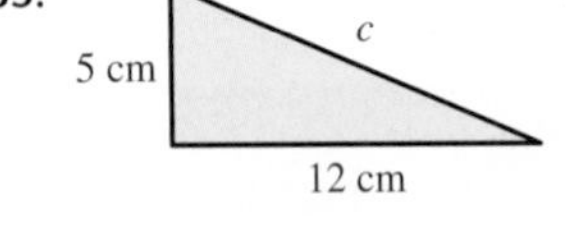

36.

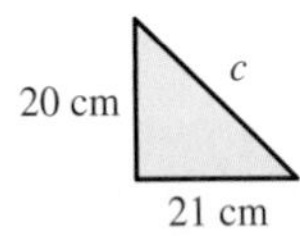

37.

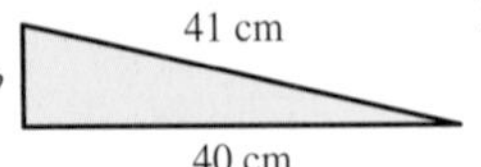

38.

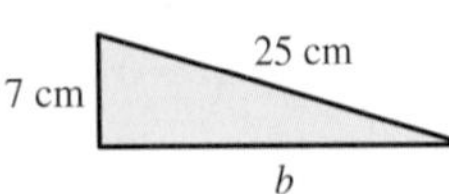

39.

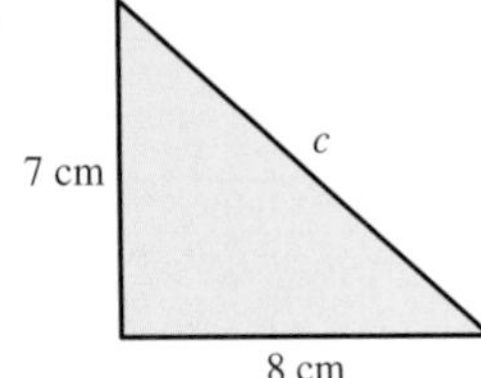

40.

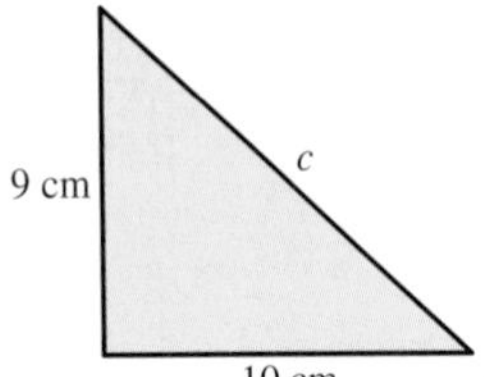

In Exercises 41–48, use the distance formula to calculate the distance between each pair of points.

41. $(-2, 6)$ and $(3, 6)$
42. $(-3, 2)$ and $(1, -1)$
43. $(2, -7)$ and $(-6, 8)$
44. $(0, -20)$ and $(-9, 20)$
45. $\left(-\frac{1}{2}, \frac{2}{3}\right)$ and $\left(\frac{1}{2}, -\frac{1}{3}\right)$
46. $\left(\frac{4}{5}, \frac{3}{2}\right)$ and $\left(-\frac{1}{5}, \frac{1}{2}\right)$
47. $(0, 0)$ and $\left(-\sqrt{2}, \sqrt{7}\right)$
48. $\left(\sqrt{11}, -\sqrt{14}\right)$ and $(0, 0)$

In Exercises 49–52, use the distance formula to determine the perimeter of the triangle formed by these points. Then use the Pythagorean theorem to determine if the triangle formed is a right triangle.

49. $(-3, -1)$, $(4, -4)$, and $(-1, 1)$
50. $(-6, -1)$, $(2, -1)$, and $(0, 1)$
51. $(-5, -1)$, $(2, -2)$, and $(4, 4)$
52. $(3, -3)$, $(-2, -1)$, and $(5, 2)$

53. Measuring a Buried Pipe A pipe cleaning firm contracted to clean a pipe buried in a lake. Access points to the pipe are at points A and B on the edge of the lake, as shown in the figure. The contractor placed a stake as a reference point and then measured the coordinates in meters from this reference point to A and B. The coordinates of A and B are, respectively, (3.0, 5.2) and (37.8, 29.6). Approximate to the nearest tenth of a meter the distance between A and B.

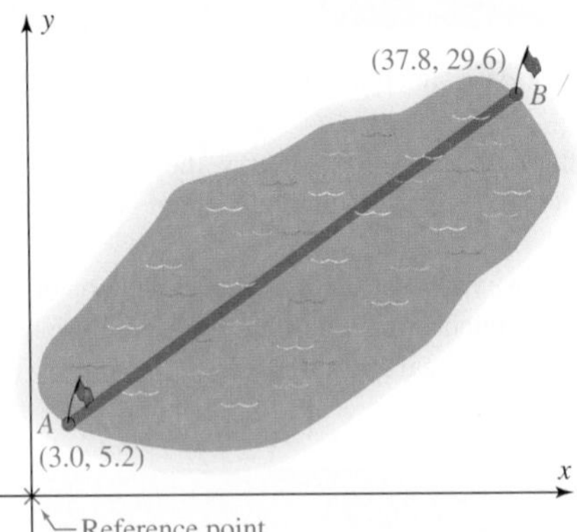

54. Measuring an Engine Block A machinist is measuring a metal block that is part of an automobile engine. To find the distance from the center of hole A to the center of hole B, the machinist determines the coordinates shown on the drawing with respect to a reference point at the lower left corner of the metal block. Approximate the distance from the center of A to the center of B. The dimensions are given in centimeters.

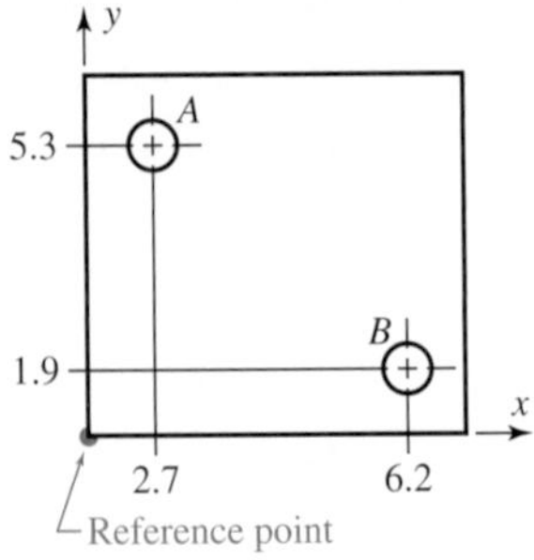

55. Find all points with an x-coordinate of 5 that are 10 units from the point (13, 2).

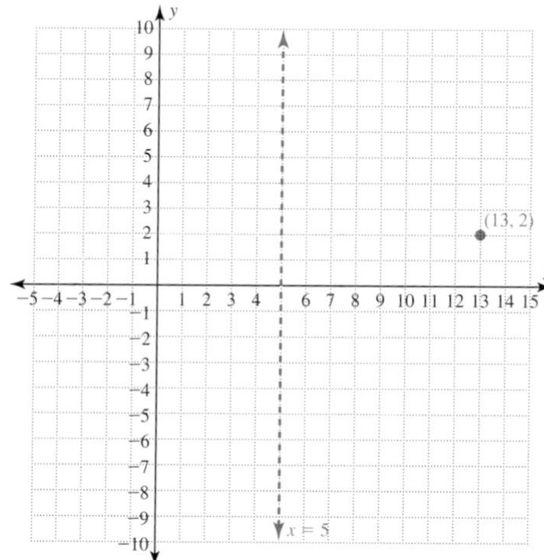

56. Find all points with a y-coordinate of 2 that are 13 units from the point $(-2, 7)$.

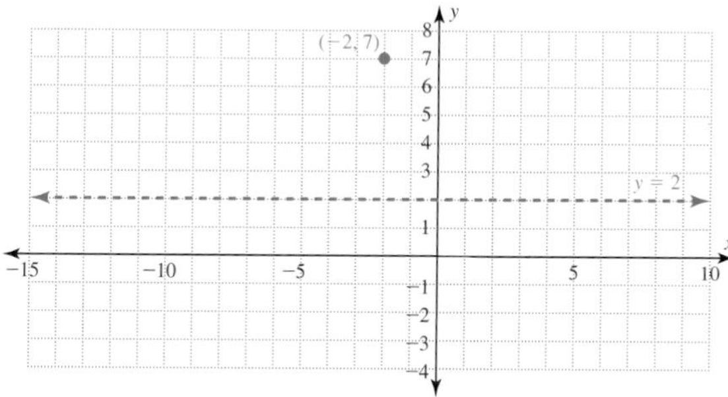

In Exercises 57–60, solve each problem for x.

57. The square root of the quantity x plus five is equal to three.
58. The cube root of the quantity x plus two is equal to minus two.
59. Four times the square root of x equals three times the quantity x minus five.
60. If the square root of the quantity x plus four is added to two, the result equals x.

In Exercises 61–68, solve each equation.

61. $\sqrt{2x + 6} = x$
62. $\sqrt{x + 5} = x$
63. $\sqrt{2x^2 + 3x - 1} = x$
64. $\sqrt{2x^2 + 6x + 3} - 2 = x$
65. $\sqrt[3]{x^3 - 6x^2 + 12x} = x$
66. $\sqrt[3]{x^3 + 9x^2 + 27x} = x$
67. $\sqrt{4w^2 + 12w + 6} = 2w$
68. $\sqrt{9w^2 - 12w + 8} = 3w$

Production of Solar Cells

In Exercises 69–72, use the fact that the dollar cost C of producing n solar cells per shift is given by the formula $C = 18\sqrt[3]{n^2} + 450$.

69. Find the overhead cost for one shift.
70. Find the cost of producing 27 solar cells.
71. Find the number of solar cells produced when the cost is $738.
72. Find the number of solar cells produced when the cost is $1,098.

Strength of a Box Beam

In Exercises 73–76, use the fact that the load strength of a square box beam is related to its volume by the function $S(V) = 750\sqrt[3]{V^2}$. In this formula V is given in cubic centimeters and S is given in newtons.

73. Evaluate and interpret $S(0)$.
74. Evaluate and interpret $S(1{,}000)$.
75. If $S(V) = 750$, calculate and interpret V.
76. If $S(V) = 300{,}000$, calculate and interpret V.
77. **Height of a Wall** A 17-ft ladder is leaning against a chimney.
    a. Express the distance from the base of the chimney to the top of the ladder as a function of x, the distance from the base of the ladder to the base of the chimney.
    b. If the bottom of the ladder is 8 ft from the base of the chimney, how far is it from the bottom of the chimney to the top of the ladder?

78. **Length of a Guy Wire** A television tower has a guy wire attached 40 ft above the base of the tower.
    a. Express the length of the guy wire as a function of x, the distance from the base of the tower to the anchor point of the guy wire.
    b. If the anchor point of the guy wire is 42 ft from the base of the tower, how long is the guy wire?

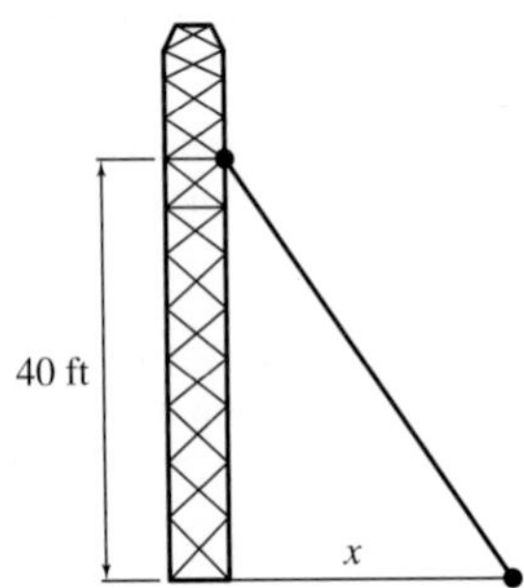

79. **Length of an Exercise Rope** One end of a retractable exercise rope for a quarter horse is fastened to the swivel at the top of a 16-ft pole. The other end is fastened to the neck of the horse at a height of 5 ft.
   a. Write a function $L(x)$ that gives the length of the rope as a function of x, where x is the distance in feet from the pole to the horse.
   b. Evaluate and interpret $L(30)$.
   c. Approximate the area that the horse can occupy if the rope is 35 ft.

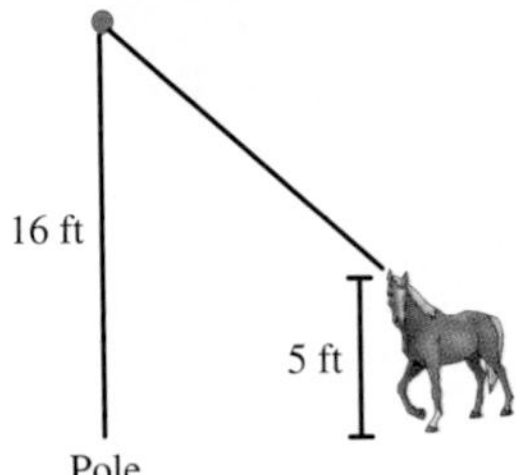

80. **Length of an Extension Cord** One end of a retractable extension cord is fastened to a swivel 12 ft above the floor on the ceiling of an automotive repair shop. The other end of the cord is an outlet that can be positioned around the floor of the shop.
   a. Write a function $L(x)$ that gives the length of the cord that is extended as a function of x, where x is the distance in feet from a point directly under the swivel to the end of the cord on the shop floor.
   b. Evaluate and interpret $L(15)$.
   c. Solve $L(x) = 16$ for x, and interpret this result.

81. **Path of a Power Cable** The path of a power cable between booster stations located at A and C is shown below. The maximum length of cable between booster stations is 12 mi. Determine the distance x so that the path from A to C through B will be 12 mi.

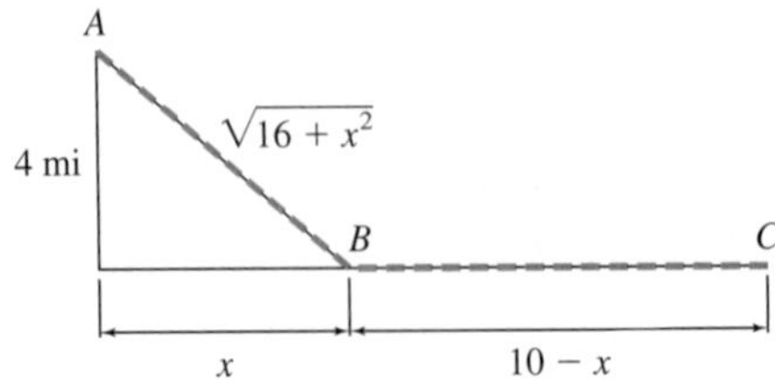

82. **Path of a Fiber Optics Cable** The path of an underground fiber-optic cable between buildings located at A and C in a business district is shown below. The maximum length of cable allowed for connections of this type is 17 km. Determine the distance x so that the path from A to C through B will be 17 km.

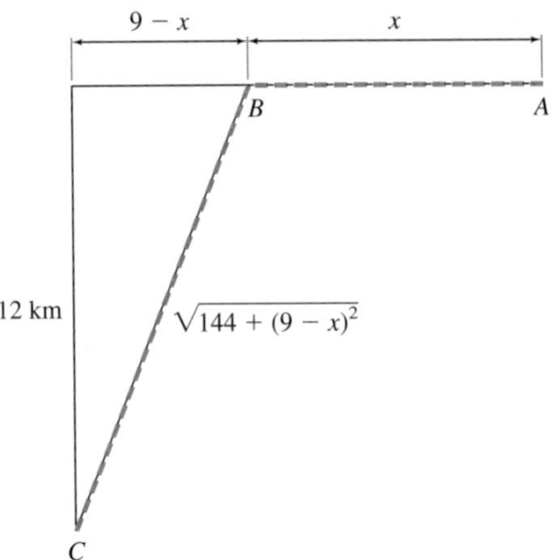

Group Discussion Questions

83. **Discovery Question**
   a. Write the equation of all points 3 units from the origin. Describe the shape formed by these points.
   b. Write the equation of all points 2 units from the point (3, 5). Describe the shape formed by these points.
   c. Write the equation of all points r units from the point (h, k). Describe the shape formed by these points.

84. **Challenge Question** Solve each equation for x.
   a. $y = \dfrac{\sqrt[3]{2x - z}}{3}$
   b. $y = \sqrt{v^2 - w^2 + x}$
   c. $5 = \sqrt{\dfrac{v + x}{v - x}}$

85. **Challenge Question** The square shown in the figure was etched in stone, along with the single word *BEHOLD*, by a Hindu mathematician, who meant that this figure offers visual proof of the Pythagorean theorem. Describe how this square can be used to prove the Pythagorean theorem. (*Hint:* Add the areas of the four triangles and the inner square.)

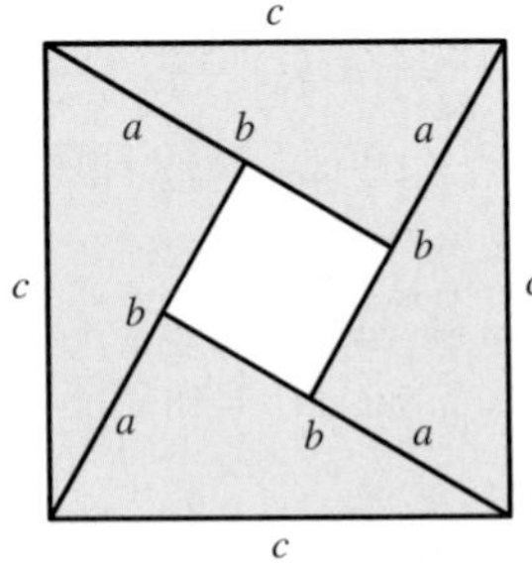

Section 9.5 Rational Exponents and Radicals

Objective: 10. Interpret and use rational exponents.

A Mathematical Note

In his authoritative *History of Mathematics,* David Eugene Smith attributes our present use of exponents to René Descartes (1596–1650).

One classic engineering problem that involves fractional exponents is the relationship between the volume of a box beam and the strength of the beam. Bridges built to scale based upon a scaled-down model may not be a good idea because the strength of the beam and weight of the beam do not grow at the same rate. We examine this problem in Example 8 at the end of this section.

Fractional Exponents

To develop a definition of fractional exponents, we start by examining $x^{1/2}$. We want $x^{1/2}$ to obey the same rules for exponents as those given in Chapter 5 for integer exponents. Thus we want to be able to use the product rule to simplify $x^{1/2} \cdot x^{1/2}$.

$$x^{1/2} \cdot x^{1/2} = x^{1/2+1/2}$$ Use the product rule $x^m \cdot x^n = x^{m+n}$

$$= x^1$$ to add the exponents.

A number multiplied by itself that yields x as a product is known as the square root of x. Thus it is reasonable to define $x^{1/2}$ as $\sqrt{x}$, the principal square root of x.

The following table examines the definition of $x^{1/2}$ algebraically, numerically, graphically, and verbally.

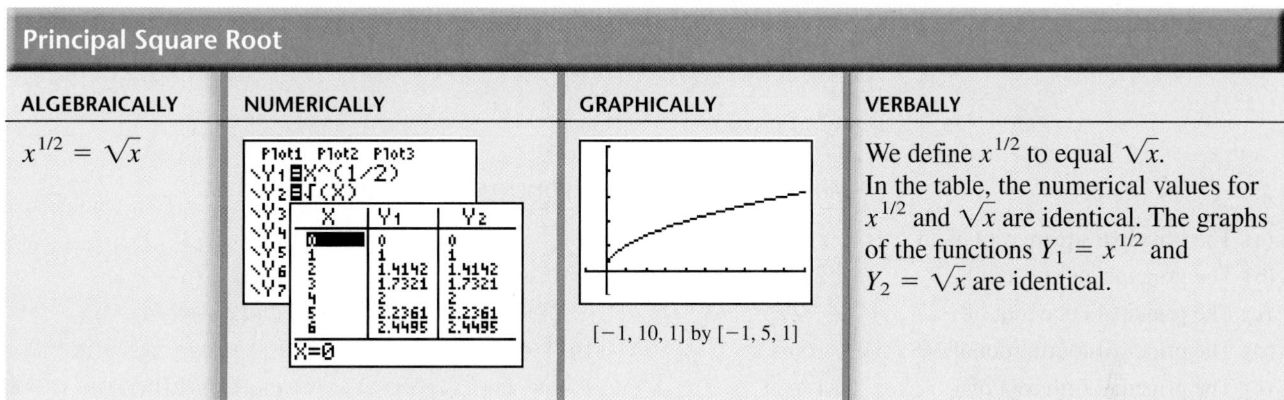

Principal Square Root

ALGEBRAICALLY	NUMERICALLY	GRAPHICALLY	VERBALLY
$x^{1/2} = \sqrt{x}$		[−1, 10, 1] by [−1, 5, 1]	We define $x^{1/2}$ to equal $\sqrt{x}$. In the table, the numerical values for $x^{1/2}$ and $\sqrt{x}$ are identical. The graphs of the functions $Y_1 = x^{1/2}$ and $Y_2 = \sqrt{x}$ are identical.

We now generalize the meaning of fractional exponents in order to define $x^{1/n}$ for any natural number n and fraction $\frac{1}{n}$. We want to be able to use the power rule for exponents with $x^{1/n}$ as shown below.

$$(x^{1/n})^n = x^{n/n}$$ Use the power rule $(x^m)^n = x^{mn}$ to

$$= x^1$$ multiply the exponents.

Because $x^{1/n}$ is used as a factor n times in the expression $(x^{1/n})^n$, it is reasonable to define $x^{1/n}$ as $\sqrt[n]{x}$, **the principal nth root of x.**

Principal nth Root

The principal nth root of the real number x is denoted by either $x^{1/n}$ or $\sqrt[n]{x}$.

VERBALLY		EXAMPLES: RADICAL NOTATION	EXAMPLES: EXPONENTIAL NOTATION
For $x > 0$:	The principal nth root is positive for all natural numbers n.	$\sqrt{9} = 3$ $\sqrt[3]{8} = 2$	$9^{1/2} = 3$ $8^{1/3} = 2$
For $x = 0$:	The principal nth root of 0 is 0.	$\sqrt{0} = 0$ $\sqrt[3]{0} = 0$	$0^{1/2} = 0$ $0^{1/3} = 0$
For $x < 0$:	If n is odd, the principal nth root is negative.	$\sqrt[3]{-8} = -2$ $\sqrt[5]{-1} = -1$	$(-8)^{1/3} = -2$ $(-1)^{1/5} = -1$
	If n is even, there is no real nth root. (The nth roots are imaginary.)	$\sqrt{-1}$ is not a real number.	$(-1)^{1/2}$ is not a real number.

EXAMPLE 1 Representing and Evaluating Principal nth Roots

Represent each of these principal nth roots by using both radical notation and exponential notation, and evaluate each expression.

SOLUTION

	RADICAL NOTATION	EXPONENTIAL NOTATION	
(a) The principal square root of 25	$\sqrt{25} = 5$	$25^{1/2} = 5$	Check: $5 \cdot 5 = 25$
(b) The principal cube root of 27	$\sqrt[3]{27} = 3$	$27^{1/3} = 3$	Check: $3 \cdot 3 \cdot 3 = 27$
(c) The principal cube root of -27	$\sqrt[3]{-27} = -3$	$(-27)^{1/3} = -3$	Check: $(-3)(-3)(-3) = -27$
(d) The principal fourth root of 16	$\sqrt[4]{16} = 2$	$16^{1/4} = 2$	Check: $(2)(2)(2)(2) = 16$
(e) The principal fifth root of 1	$\sqrt[5]{1} = 1$	$1^{1/5} = 1$	Check: $(1)(1)(1)(1)(1) = 1$

SELF-CHECK 9.5.1

Write each of these expressions in radical notation and evaluate each expression.

1. $64^{1/2}$ 2. $64^{1/3}$ 3. $64^{1/6}$ 4. $-64^{1/2}$ 5. $(-64)^{1/2}$ 6. $(-64)^{1/3}$

To extend the definition of exponents to include any rational number $\frac{m}{n}$, we start with $x^{1/n} = \sqrt[n]{x}$ and then use the power rule for exponents.

$$x^{1/n} = \sqrt[n]{x}$$

$$(x^{1/n})^m = (\sqrt[n]{x})^m \quad \text{Raise both sides of the equation to the } m\text{th power.}$$

$$x^{m/n} = (\sqrt[n]{x})^m \quad \text{Then use the power rule for exponents to multiply the exponents.}$$

In keeping with our earlier work with negative exponents, we also interpret $x^{-m/n}$ as $\frac{1}{x^{m/n}}$. These new definitions are summarized in the following box.

Rational Exponents

For a real number x and natural numbers m and n:

ALGEBRAICALLY	EXAMPLES: RADICAL NOTATION	EXAMPLES: EXPONENTIAL NOTATION
If $x^{1/n}$ is a real number*, then $x^{m/n} = (x^{1/n})^m = (\sqrt[n]{x})^m$ or	$(-8)^{2/3} = (\sqrt[3]{-8})^2$ $= (-2)^2$ $= 4$	$(-8)^{2/3} = [(-8)^{1/3}]^2$ $= (-2)^2$ $= 4$
$x^{m/n} = (x^m)^{1/n} = \sqrt[n]{x^m}$	$(y^3)^{2/7} = \sqrt[7]{(y^3)^2}$ $= \sqrt[7]{y^6}$	$(y^3)^{2/7} = y^{3/1 \cdot 2/7}$ $= y^{6/7}$
$x^{-m/n} = \frac{1}{x^{m/n}} = \frac{1}{\sqrt[n]{x^m}}$ $x \neq 0$	$16^{-3/4} = \frac{1}{(\sqrt[4]{16})^3}$ $= \frac{1}{2^3}$ $= \frac{1}{8}$	$16^{-3/4} = \frac{1}{16^{3/4}}$ $= \frac{1}{(16^{1/4})^3}$ $= \frac{1}{2^3}$ $= \frac{1}{8}$

*If $x < 0$ and n is even, then $x^{1/n}$ is an imaginary number.

These definitions are illustrated in Example 2.

EXAMPLE 2 Evaluating Expressions with Fractional Exponents

Represent each of these expressions in radical notation and exponential notation, and evaluate each expression.

When you use negative exponents, think "take the reciprocal."

SOLUTION

	RADICAL NOTATION	EXPONENTIAL NOTATION
(a) $27^{2/3}$	$(\sqrt[3]{27})^2 = 3^2 = 9$	$27^{2/3} = (27^{1/3})^2 = 3^2 = 9$
(b) $32^{3/5}$	$(\sqrt[5]{32})^3 = 2^3 = 8$	$32^{3/5} = (32^{1/5})^3 = 2^3 = 8$
(c) $9^{-1/2}$	$\frac{1}{\sqrt{9}} = \frac{1}{3}$	$9^{-1/2} = \frac{1}{9^{1/2}} = \frac{1}{3}$
(d) $27^{-2/3}$	$\frac{1}{(\sqrt[3]{27})^2} = \frac{1}{3^2} = \frac{1}{9}$	$27^{-2/3} = \frac{1}{27^{2/3}} = \frac{1}{(27^{1/3})^2} = \frac{1}{3^2} = \frac{1}{9}$

The expressions in Example 2 were carefully selected to illustrate the notation but also so that they could be evaluated by using pencil and paper. Calculator Perspective 9.5.1 and

Example 3 illustrates how to use exponential notation to evaluate higher-order radical expressions.

CALCULATOR PERSPECTIVE 9.5.1 Evaluating Expressions with Rational Exponents

To evaluate $27^{2/3}$ and $9^{-1/2}$ from Example 2 on a TI-84 Plus calculator, enter the following keystrokes:

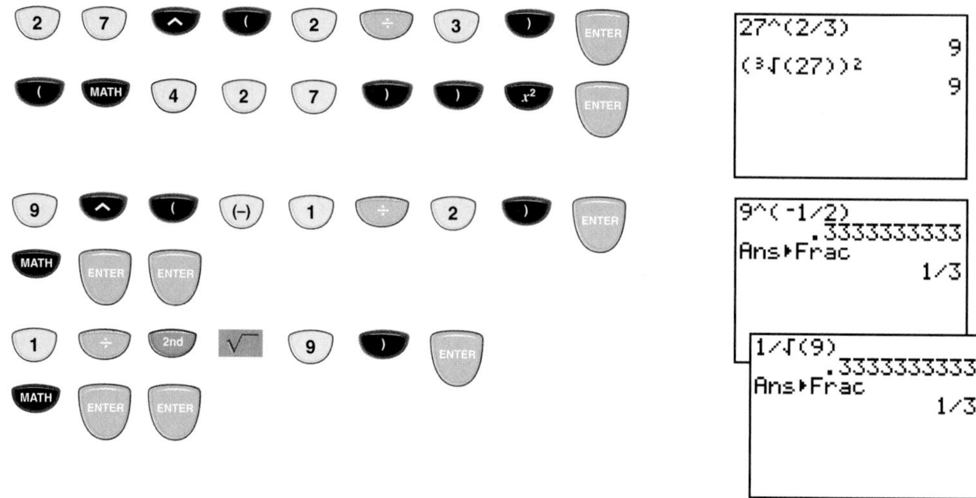

Many students find it easier to use fractional exponents than to use the radical form when evaluating higher-order radicals with a calculator. This is illustrated in Example 3.

EXAMPLE 3 Approximating Radical Expressions with a Calculator

Write each of these radical expressions in exponential form, and then use a calculator to approximate each expression to the nearest thousandth.

(a) $\sqrt[3]{40}$ (b) $\sqrt[4]{50}$ (c) $\sqrt[5]{-45}$

SOLUTION

(a) $\sqrt[3]{40} = 40^{1/3} \approx 3.420$
(b) $\sqrt[4]{50} = 50^{1/4} \approx 2.659$
(c) $\sqrt[5]{-45} = (-45)^{1/5} \approx -2.141$

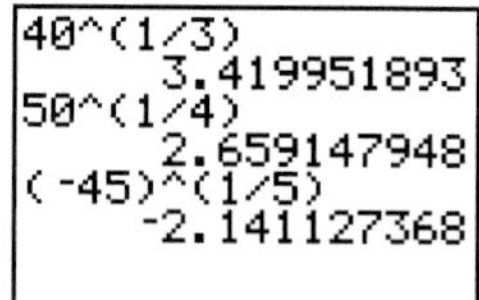

SELF-CHECK 9.5.2

Write each of these expressions in radical notation and evaluate each expression.

1. $64^{3/2}$ 2. $64^{2/3}$ 3. $(-64)^{2/3}$ 4. $64^{-2/3}$

Use a calculator to approximate each expression to the nearest thousandth.

5. $\sqrt[6]{100}$ 6. $\sqrt[9]{-99}$ 7. $41^{4/7}$ 8. $41^{7/4}$

We have defined fractional exponents in such a way that all the properties of integral exponents now apply to fractional exponents. We have not established it in this book, but these properties apply to all real number exponents. These properties are summarized in the following box.

One reason it is useful to examine radical expressions in exponential form is that we can apply the properties of exponents to simplify these expressions.

Properties of Exponents

Let m, and n be real numbers and x, x^m, x^n, y, y^m, and y^n be nonzero real numbers.

Product rule: $x^m \cdot x^n = x^{m+n}$

Power rule: $(x^m)^n = x^{mn}$

Product to a power: $(xy)^m = x^m y^m$

Quotient to a power: $\left(\frac{x}{y}\right)^m = \frac{x^m}{y^m}$

Quotient rule: $\frac{x^m}{x^n} = x^{m-n}$

Negative power: $\left(\frac{x}{y}\right)^{-n} = \left(\frac{y}{x}\right)^n$

To avoid the special care that must be exercised when the base is negative and we are evaluating an even number root, we restrict the variables to positive values in Examples 4 to 8.

EXAMPLE 4 Using the Properties of Exponents

Simplify each of the following expressions. Express all answers in terms of positive exponents. Assume that z represents a positive real number.

SOLUTION

(a) $9^{3/8} \cdot 9^{1/8}$

$$9^{3/8} \cdot 9^{1/8} = 9^{(3/8)+(1/8)}$$
$$= 9^{1/2}$$
$$= 3$$

Use the product rule to add the exponents: $\frac{3}{8} + \frac{1}{8} = \frac{4}{8} = \frac{1}{2}$

The principal square root of 9 is 3.

(b) $8^{1/5} \cdot 4^{1/5}$

$$8^{1/5} \cdot 4^{1/5} = (8 \cdot 4)^{1/5}$$
$$= 32^{1/5}$$
$$= 2$$

Product to a power: $x^m y^m = (xy)^m$.

The principal fifth root of 32 is 2.

(c) $\frac{625^{7/8}}{625^{1/8}}$

$$\frac{625^{7/8}}{625^{1/8}} = 625^{7/8-1/8}$$
$$= 625^{3/4}$$
$$= (625^{1/4})^3$$
$$= 5^3$$
$$= 125$$

Use the quotient rule to subtract the exponents: $\frac{7}{8} - \frac{1}{8} = \frac{6}{8} = \frac{3}{4}$

The principal fourth root of 625 is 5.

(d) $(z^{-2/3})^{-3}$

$$(z^{-2/3})^{-3} = z^{(-2/3)(-3)}$$
$$= z^2$$

Use the power rule to multiply the exponents: $\left(-\frac{2}{3}\right)(-3) = 2$.

EXAMPLE 5 Using the Properties of Exponents

Simplify each expression. Express each answer in terms of positive exponents. Assume that the variables represent only positive real numbers.

SOLUTION

(a) $(a^{-8}b^{12})^{-1/4}$

$$(a^{-8}b^{12})^{-1/4} = (a^{-8})^{-1/4}(b^{12})^{-1/4}$$

Use the product to a power rule for exponents $(xy)^m = x^m y^m$ to take the $-\frac{1}{4}$ power of each factor.

$$= a^2b^{-3}$$

Use the power rule to multiply the exponents: $-8\left(-\frac{1}{4}\right) = 2$ and $12\left(-\frac{1}{4}\right) = -3$.

$$= \frac{a^2}{b^3}$$

Rewrite the expression without negative exponents: $b^{-3} = \frac{1}{b^3}$.

(b) $\left(\frac{16x^4}{25}\right)^{-3/2}$

$$\left(\frac{16x^4}{25}\right)^{-3/2} = \left[\left(\frac{16x^4}{25}\right)^{1/2}\right]^{-3}$$

$x^{m/n} = (x^{1/n})^m$

$$= \left[\frac{16^{1/2}x^{4/2}}{25^{1/2}}\right]^{-3}$$

$$= \left[\frac{4x^2}{5}\right]^{-3}$$

Take the $\frac{1}{2}$ power of each factor in the numerator and the denominator, and then simplify. Note that $16^{1/2} = 4$ and $25^{1/2} = 5$.

$$= \left[\frac{5}{4x^2}\right]^{3}$$

Rewrite the expression without negative exponents.

$$= \frac{5^3}{4^3(x^2)^3}$$

$$= \frac{125}{64x^6}$$

Raise each factor to the third power. To raise a power to a power, multiply exponents.

SELF-CHECK 9.5.3

Simplify each expression. Assume that x is a positive real number.

1. $(8^{-4/5})^{-5/2}$
2. $25^{5/16} \cdot 25^{3/16}$
3. $25^{1/3} \cdot 5^{1/3}$
4. $\left(\frac{8x^6}{125}\right)^{2/3}$

Example 6 uses the distributive property to simplify an expression involving both fractional and negative exponents.

EXAMPLE 6 Multiplying by Using the Distributive Property

Simplify $2x^{2/3}(3x^{1/3} - 5x^{-2/3})$. Express the answer in terms of positive exponents. Assume that all variables represent positive real numbers.

SOLUTION

$$2x^{2/3}(3x^{1/3} - 5x^{-2/3}) = (2x^{2/3})(3x^{1/3}) - (2x^{2/3})(5x^{-2/3})$$

First use the distributive property.

$$= 6x^{3/3} - 10x^0$$

$$= 6x - 10$$

Then simplify each term, adding the exponents of the factors with the same base. Note that $x^0 = 1$.

The product of exponential expressions with two or more terms can be found by using some of the skills developed for multiplying polynomials. In particular, we will use special forms, such as the product of a sum and a difference: $(x + y)(x - y) = x^2 - y^2$.

EXAMPLE 7 Multiplying a Sum by a Difference

Simplify $(a^{1/2} + b^{1/2})(a^{1/2} - b^{1/2})$. Assume that all variables represent positive real numbers.

SOLUTION

$$(a^{1/2} + b^{1/2})(a^{1/2} - b^{1/2}) = (a^{1/2})^2 - (b^{1/2})^2$$

By inspection, observe that this expression fits the special form $(x + y)(x - y) = x^2 - y^2$, with $x = a^{1/2}$ and $y = b^{1/2}$.

$$= a - b$$

To raise a power to a power, multiply exponents.

SELF-CHECK 9.5.4

Simplify each expression.

1. $4x^{5/8}(3x^{11/8} - 5x^{3/8} - 7x^{-5/8})$
2. $(z^{1/2} - 5)(z^{1/2} + 5)$

The history of science and engineering has many examples of "great" ideas that did not "scale up," that is, ideas that seemed to work great with a model in the lab but did not work in real-world applications. One of the problems that engineers must face is that scale models may not behave as the real product does if not all the variables involved increase at the same rate. Example 8 investigates the strength of a box beam to illustrate this problem.

EXAMPLE 8 Strength of Box Beam

The strength of a box beam varies directly as the cross-sectional area. However, part of the stress on the beam comes from its own weight, which is directly related to the volume of the beam. Thus, the stress that a beam puts on itself increases at a greater rate than its strength. For the sample beam shown here, the following functions give the strength S and the volume V of the beam: $S = 180{,}000x^2$ and $V = 27{,}000x^3$. In this example, x is given in centimeters, V in cubic centimeters, and S in newtons.

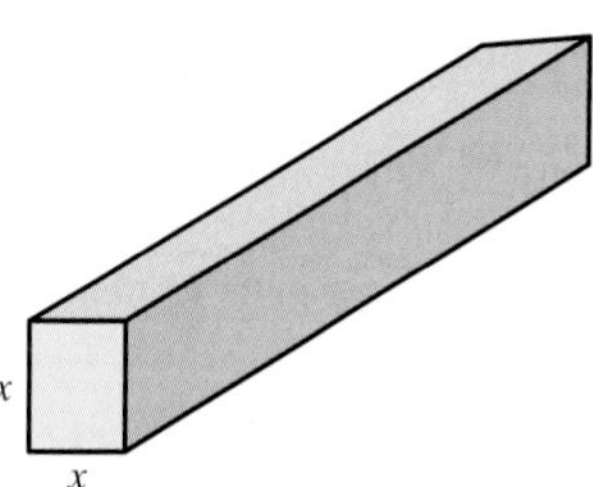

(a) Write S as a function of V.

(b) Using the function from part (a), evaluate and interpret $S(1{,}000)$.

SOLUTION

(a) $V = 27{,}000x^3$

$\dfrac{V}{27{,}000} = x^3$

$x = \left(\dfrac{V}{27{,}000}\right)^{1/3}$

$x = \dfrac{V^{1/3}}{30}$

$S = 180{,}000x^2$

$S = 180{,}000\left(\dfrac{V^{1/3}}{30}\right)^2$

$S = 200V^{2/3}$

$S(V) = 200V^{2/3}$

S is currently written as a function of x. Solve the formula $V = 27{,}000x^3$ for x.

First divide both sides of the equation by 27,000 and then raise both sides to the $\frac{1}{3}$ power.

Note that $27{,}000^{1/3} = 30$.

To write S as a function of V substitute $\frac{V^{1/3}}{30}$ for x.

(b) $S(V) = 200V^{2/3}$

$S(1{,}000) = 200(1{,}000)^{2/3}$

$= 200(10)^2$

$= 200(100)$

$= 20{,}000$

Substitute 1,000 for V and then simplify this expression.

When the volume is 1,000 cm^3, the strength will be 20,000 N.

SELF-CHECK ANSWERS

9.5.1

1. $\sqrt{64} = 8$
2. $\sqrt[3]{64} = 4$
3. $\sqrt[6]{64} = 2$
4. $-\sqrt{64} = -8$
5. $\sqrt{-64}$ is an imaginary number; $\sqrt{-64} = 8i$
6. $\sqrt[3]{-64} = -4$

9.5.2

1. $64^{3/2} = (\sqrt{64})^3 = 8^3 = 512$
2. $64^{2/3} = (\sqrt[3]{64})^2 = 4^2 = 16$
3. $(-64)^{2/3} = (\sqrt[3]{-64})^2 = (-4)^2 = 16$
4. $64^{-2/3} = \dfrac{1}{(\sqrt[3]{64})^2} = \dfrac{1}{4^2} = \dfrac{1}{16}$
5. 2.154
6. −1.666
7. 8.348
8. 664.311

9.5.3

1. 64
2. 5
3. 5
4. $\dfrac{4x^4}{25}$

9.5.4

1. $12x^2 - 20x - 28$
2. $z - 25$

USING THE LANGUAGE AND SYMBOLISM OF MATHEMATICS 9.5

1. In radical notation, the expression $x^{1/5}$ can be written as ________.
2. In exponential notation, the expression $\sqrt[7]{x}$ can be written as ________.
3. In radical notation, the expression $x^{2/5}$ can be written as ________.
4. In exponential notation, the expression $\sqrt[4]{x^3}$ can be written as ________.
5. The principal nth root of x is denoted by either $\sqrt[n]{x}$ or ________.
6. The principal nth root of x is not a real number if n is even and x is ________.
7. In the radical notation $\sqrt[n]{x}$, n is called the index or the order of the radical and x is called the ________.

QUICK REVIEW 9.5

The purpose of this quick review is to help you recall skills needed in this section.

Simplify each expression. Assume $x \neq 0$.

1. $\left(\dfrac{4}{5}\right)^{-2}$
2. x^7x^{11}
3. $(x^{11})^7$
4. $\dfrac{x^{11}}{x^7}$
5. $\left(\dfrac{x^5}{x^{-3}}\right)^2$

EXERCISES 9.5

In Exercises 1 and 2, represent each expression by using exponential notation. Assume that x and y are positive real numbers.

1. a. The principal cube root of w
   b. The principal fourth root of x
   c. The principal seventh root of v
2. a. The principal sixth root of x
   b. The principal ninth root of v
   c. The principal twelfth root of y

In Exercises 3 and 4, represent each expression by using both radical notation and exponential notation, and evaluate each expression.

3. a. The principal square root of 100
   b. The principal cube root of 125
   c. The principal fifth root of −32
4. a. The principal square root of 81
   b. The principal fourth root of 81
   c. The principal fifth root of −100,000

In Exercises 5–8, represent each expression by using exponential notation, and evaluate each expression.

5. a. $\sqrt{16}$ b. $\sqrt[4]{16}$ c. $\sqrt[5]{0}$
6. a. $\sqrt{49}$ b. $\sqrt[3]{125}$ c. $\sqrt[7]{1}$
7. a. $\sqrt[3]{-8}$ b. $\sqrt[5]{\dfrac{-1}{32}}$ c. $\sqrt{0.01}$
8. a. $\sqrt[3]{-0.001}$ b. $\sqrt{\dfrac{4}{9}}$ c. $\sqrt[5]{-32}$

In Exercises 9–26, represent each expression by using radical notation, and evaluate each expression.

9. a. $36^{1/2}$ b. $36^{-1/2}$ c. $-36^{1/2}$
10. a. $100^{1/2}$ b. $100^{-1/2}$ c. $-100^{1/2}$
11. a. $27^{1/3}$ b. $(-27)^{1/3}$ c. $27^{-1/3}$
12. a. $16^{1/4}$ b. $16^{-1/4}$ c. $-16^{1/4}$
13. a. $0.09^{1/2}$ b. $-0.09^{1/2}$ c. $400^{-1/2}$
14. a. $0.49^{1/2}$ b. $-0.49^{1/2}$ c. $900^{-1/2}$
15. a. $0.008^{1/3}$ b. $(-0.008)^{1/3}$ c. $(0.008)^{-1/3}$
16. a. $0.027^{1/3}$ b. $(-0.027)^{1/3}$ c. $125^{-1/3}$
17. a. $64^{1/3}$ b. $64^{-1/3}$ c. $-64^{1/3}$
18. a. $81^{1/4}$ b. $81^{-1/4}$ c. $-81^{1/2}$
19. a. $\left(\dfrac{8}{125}\right)^{2/3}$ b. $\left(\dfrac{9}{4}\right)^{3/2}$ c. $\left(\dfrac{8}{125}\right)^{-2/3}$
20. a. $\left(\dfrac{16}{81}\right)^{3/4}$ b. $\left(\dfrac{49}{25}\right)^{3/2}$ c. $\left(\dfrac{16}{81}\right)^{-3/4}$
21. a. $25^{1/2} + 144^{1/2}$ b. $(25 + 144)^{1/2}$ c. $(25 - 16)^{1/2}$
22. a. $9^{1/2} + 16^{1/2}$ b. $(9 + 16)^{1/2}$ c. $25^{1/2} - 16^{1/2}$
23. a. $8^{1/3} + 1^{1/3}$ b. $(26 + 1)^{1/3}$ c. $(1{,}001 - 1)^{1/3}$
24. a. $27^{1/3} - 1^{1/3}$ b. $(65 - 1)^{1/3}$ c. $(7 + 1)^{1/3}$
25. a. $0.000001^{1/2}$ b. $0.000001^{1/3}$ c. $0.000001^{1/6}$
26. a. $0.0001^{1/2}$ b. $0.0001^{1/4}$ c. $(-0.00001)^{1/5}$

In Exercises 27–30, use a calculator to approximate each expression to the nearest thousandth.

27. a. $75^{1/4}$ b. $63^{2/7}$
28. a. $85^{1/6}$ b. $29^{4/9}$
29. a. $\sqrt[6]{361}$ b. $\sqrt[9]{31^4}$
30. a. $\sqrt[11]{691}$ b. $\sqrt[7]{22^3}$

Estimation and Calculator Skills

In Exercises 31 and 32, mentally estimate the value of each expression to the nearest integer, and then use a calculator to approximate each value to the nearest thousandth.

PROBLEM	MENTAL ESTIMATE	CALCULATOR APPROXIMATION
31. $30^{1/3}$		
32. $30^{1/5}$		

In Exercises 33–56, simplify each expression. Express all answers in terms of positive exponents. Assume that all variables represent positive real numbers.

33. $5^{1/2} \cdot 5^{3/2}$
34. $7^{5/3} \cdot 7^{1/3}$
35. $(8^{5/3})^{2/5}$
36. $(32^{7/10})^{2/7}$
37. $\dfrac{11^{4/3}}{11^{1/3}}$
38. $\dfrac{9^{5/4}}{9^{3/4}}$
39. $(27^{1/12} \cdot 27^{-5/12})^{-2}$
40. $(4^{1/5} \cdot 4^{2/5})^{5/2}$
41. $x^{1/3} \cdot x^{1/2}$
42. $y^{1/4} \cdot y^{1/5}$
43. $\dfrac{x^{1/2}}{x^{1/3}}$
44. $\dfrac{y^{1/4}}{y^{1/5}}$
45. $(z^{3/4})^{2/7}$
46. $(z^{5/12})^{4/15}$
47. $\dfrac{w^{-2/3}}{w^{-5/3}}$
48. $\dfrac{w^{3/4}}{w^{-3/8}}$
49. $(v^{-10}w^{-15})^{-1/5}$
50. $(v^{-26}w^{-39})^{-1/13}$
51. $(16v^{-2/5})^{3/2}$
52. $(25v^{-4/9})^{3/2}$
53. $\left(\dfrac{16n^{2/3}}{81n^{-2/3}}\right)^{-3/4}$
54. $\left(\dfrac{27n^{3/5}}{n^{-3/5}}\right)^{-2/3}$
55. $\dfrac{(27x^2y)^{1/2}(3xy)^{1/2}}{5x^{1/2}y^2}$
56. $\dfrac{(25x^2y^3z)^{1/3}(5xy^2z^2)^{1/3}}{3x^2y^{-1/3}z^{-2}}$

In Exercises 57–70, perform the indicated multiplications, and express the answers in terms of positive exponents. Assume that all variables represent positive real numbers.

57. $x^{3/5}(x^{2/5} - x^{-3/5})$
58. $x^{2/3}(x^{4/3} + x^{-2/3})$
59. $y^{-7/4}(2y^{11/4} - 3y^{7/4})$
60. $y^{-9/7}(5y^{23/7} - 6y^{16/7})$
61. $3w^{5/11}(2w^{17/11} - 5w^{6/11} - 9w^{-5/11})$
62. $4w^{-5/3}(3w^{11/3} - 7w^{8/3} + 2w^{5/3})$
63. $(a^{1/2} + 3)(a^{1/2} - 3)$
64. $(2a^{1/2} - 3b^{1/2})(2a^{1/2} + 3b^{1/2})$
65. $(b^{3/5} - c^{5/3})(b^{3/5} + c^{5/3})$
66. $(b^{3/5} + c^{5/3})^2$
67. $(b^{3/5} - c^{5/3})^2$
68. $(x^{1/2} - x^{-1/2})^2$
69. $(x^{-1/2} + x^{1/2})^2$
70. $(x^{2/3} + x)(x^{2/3} - x)$

Strength of a Box Beam

In Exercises 71–74, use the fact that the load strength of a square box beam is related to its volume by the function $S(V) = 750V^{2/3}$. In this formula V is given in cubic centimeters and S is given in newtons.

71. Evaluate and interpret $S(1{,}000)$.
72. If $S(V) = 750$, calculate and interpret V.
73. If the volume in Exercise 71 is increased by a factor of 8 to 8,000 cm^3, determine the factor by which the strength has increased.
74. If the strength in Exercise 72 is increased by a factor of 16 to 12,000 N, determine the factor by which the volume has increased.

Group Discussion Questions

75. **Challenge Question**
 a. Give an example of a real number for which $(x^2)^{1/2} \neq x$.
 b. Give an example of a real number for which $x^{1/2}$ is defined but $x^{-1/2}$ is undefined.
76. **Challenge Question** Use the properties of exponents to simplify each expression, assuming m is a natural number and that x is a positive real number.
 a. $x^{m/3}x^{m/2}$ b. $\dfrac{16x^{m/2}}{2x^{m/3}}$
 c. $(x^{m/3}x^{m/2})^6$
77. **Challenge Question** Use the distributive property to expand the first expression and to factor the second expression.

Expand	Factor
a. $x^{1/2}(x^2 + 3x - 4)$	$x^{5/2} + 3x^{3/2} - 4x^{1/2} = x^{1/2}(\quad)$ $= x^{1/2}(\quad)(\quad)$
b. $x^{-1/2}(x^2 - 25)$	$x^{3/2} - 25x^{-1/2} = x^{-1/2}(\quad)$ $= x^{-1/2}(\quad)(\quad)$

78. **Challenge Question** The results shown below were obtained on a TI-84 Plus calculator. Explain these results. (*Hint:* Write each expression in radical notation.)
 a. (-70) ^ $.2 \approx -2.34$
 b. (-70) ^ $.3$ is undefined
 c. (-70) ^ $.4 \approx 5.47$
 d. (-70) ^ $.5$ is undefined
 e. (-70) ^ $.6 \approx -12.80$
79. **Challenge Question** Use inspection to match each function with its graph.
 a. $f(x) = x^0$
 b. $f(x) = x^{1/4}$
 c. $f(x) = x^{1/5}$

A.
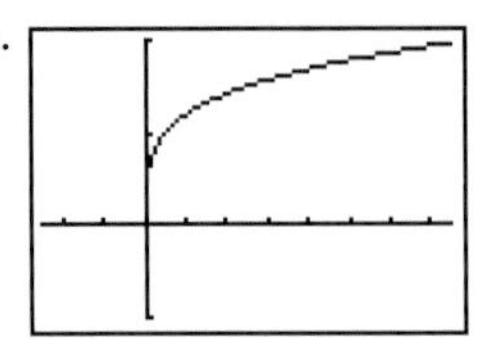
[−5, 15, 2] by [−1, 2, 1]

B.
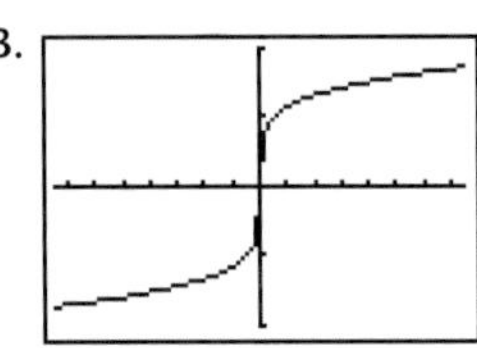
[−15, 15, 2] by [−2, 2, 1]

C.
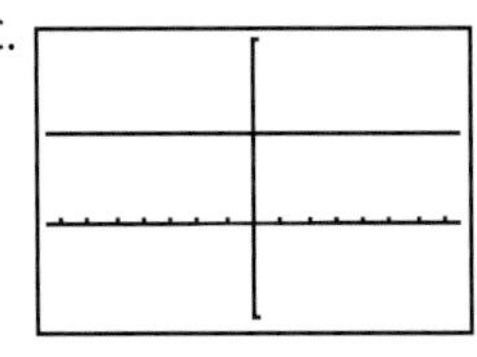
[−15, 15, 2] by [−1, 2, 1]

80. **Challenge Question** Perform the indicated multiplications by taking advantage of the factored forms for the sum and difference of two cubes.
 a. $(y^{1/3} + 2)(y^{2/3} - 2y^{1/3} + 4)$
 b. $(3y^{1/3} - 5)(9y^{2/3} + 15y^{1/3} + 25)$

KEY CONCEPTS FOR CHAPTER 9

1. **Radical Notation** $\sqrt[n]{x}$ is read "the principal nth root of x." x is called the radicand, $\sqrt{\ }$ is called the radical symbol, and n is called the index or the order of the radical. If $\sqrt[n]{x}$ is used as a factor n times, the product is x.
2. **Principal nth Root** The principal nth root of the real number x is denoted by either $x^{1/n}$ or $\sqrt[n]{x}$.
 a. For $x > 0$: The principal root is positive for all natural numbers n.
 b. For $x < 0$: If n is odd, the principal root is negative. If n is even, there is no principal real nth root; it is imaginary.
 c. For $x = 0$: The principal root is 0.
3. **Rational Exponents $x^{m/n}$** For a real number x and natural numbers m and n,
 $x^{m/n} = (x^{1/n})^m = (x^m)^{1/n}$ if $x^{1/n}$ is a real number
 $x^{-m/n} = \dfrac{1}{x^{m/n}}$ if $x \neq 0$ and $x^{1/n}$ is a real number
 If $x < 0$ and n is even, $x^{1/n}$ is not a real number and the equalities listed above are not true. For example, $\sqrt{-5}$ is not a real number, and $\sqrt{(-5)^2} \neq (\sqrt{-5})^2$.

4. $\sqrt[n]{x^n}$ For any real number x and natural number n,

$\sqrt[n]{x^n} = |x|$ if n is even

$\sqrt[n]{x^n} = x$ if n is odd

5. **Properties of Radicals** If $\sqrt[n]{x}$ and $\sqrt[n]{y}$ are both real numbers, then

$$\sqrt[n]{xy} = \sqrt[n]{x}\sqrt[n]{y} \quad \text{and} \quad \sqrt[n]{\frac{x}{y}} = \frac{\sqrt[n]{x}}{\sqrt[n]{y}} \quad \text{for } y \neq 0$$

6. **Like Radicals** Like radicals have the same index and the same radicand; only the coefficients of like radicals can differ.
7. **Conjugates** The radical expressions $x + \sqrt{y}$ and $x - \sqrt{y}$ are called conjugates of each other.
8. **Operations with Radical Expressions** The addition, subtraction, and multiplication of radical expressions are similar to the corresponding operations with polynomials. Division by a radical expression is accomplished by rationalizing the denominator.
9. **Rationalizing the Denominator** If an expression contains a radical in the denominator, the process of rewriting the expression so that there is no radical in the denominator is called rationalizing the denominator.
10. **Power Theorem** For any real numbers x and y and natural number n:
    - If $x = y$, then $x^n = y^n$.
    - *Caution:* $x^n = y^n$ can be true when $x \neq y$. For example, $(-3)^2 = 3^2$ but $-3 \neq 3$.
11. **Extraneous Value** Any value that occurs as a solution of the last equation of a solution process but is not a solution of the original equation is called an extraneous value.
12. **Solving Radical Equations**
    a. Isolate a radical term on one side of the equation.
    b. Raise both sides to the nth power.
    c. Solve the resulting equation. (If this equation contains a radical, repeat steps a and b.)
    d. Check each possible solution in the original equation to determine whether it is a solution or an extraneous value.
13. **Pythagorean Theorem and Converse** ABC is a right triangle if, and only if, $a^2 + b^2 = c^2$.

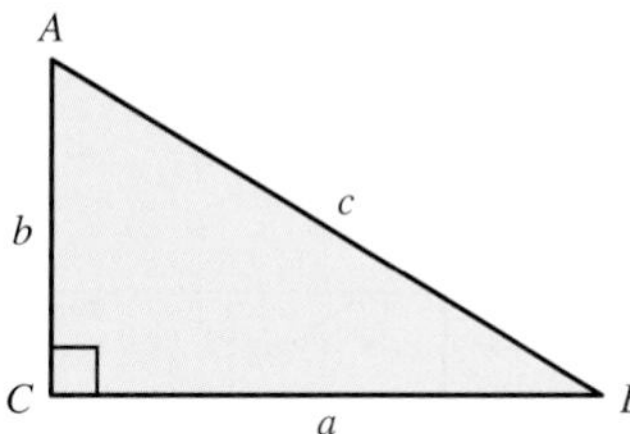

14. **Formula for the Distance Between Two Points in a Plane** The distance between (x_1, y_1) and (x_2, y_2) is given by $d = \sqrt{(x_2 - x_1)^2 + (y_2 - y_1)^2}$.
15. **Real-Valued Function** A function that produces only real output values for real input values is called a real-valued function.
16. **Domain of a Real-Valued Function** The domain of a real-valued function $f(x)$ is the set of all real numbers x for which $f(x)$ is also a real number. This means the domain excludes all values that cause either of the following:
    - Division by 0
    - A negative number under a square root symbol or any other even-order radical
17. **Basic Functions Examined in This Chapter** Each type of function has an equation of a standard form and a characteristic shape.
    - Square root functions
    - Cube root functions

REVIEW EXERCISES FOR CHAPTER 9

In Exercises 1–3, simplify each expression without using a calculator. Use a calculator only to check your answer.

1. a. $\sqrt{625 - 576}$ b. $\sqrt{625} - \sqrt{576}$ c. $\sqrt{\dfrac{576}{625}}$
2. a. $\sqrt[5]{-32}$ b. $\sqrt[7]{-1}$ c. $\sqrt[8]{0}$
3. a. $\sqrt{(-7)^2}$ b. $\sqrt[3]{(-7)^3}$ c. $\sqrt[4]{(-5)^4}$
4. Plot $\sqrt{27}$, $\sqrt{-27}$, $\sqrt[3]{27}$, $\sqrt[3]{-27}$ on a real number line. If the number is not located on the real number line, explain why not.

   −30 −25 −20 −15 −10 −5 0 5 10 15 20 25 30

5. Use a calculator to approximate each expression to the nearest hundredth.
   a. $\sqrt{70}$ b. $\sqrt[3]{70}$ c. $\sqrt[5]{-70}$

In Exercises 6–32, simplify each expression without using a calculator. Assume that all variables represent positive real numbers.

6. $12\sqrt{3} + 8\sqrt{3}$
7. $12\sqrt{3} - 8\sqrt{3}$
8. $(5\sqrt{2} - 7\sqrt{3}) - (\sqrt{2} - 4\sqrt{3})$
9. $6\sqrt[3]{5} - 2\sqrt[3]{5}$
10. $\dfrac{6\sqrt[3]{5}}{2\sqrt[3]{5}}$
11. $(6\sqrt[3]{5})(2\sqrt[3]{5})$
12. $\sqrt{20}$
13. $\sqrt[3]{24}$
14. $3\sqrt{72} - 2\sqrt{98}$
15. $5\sqrt{3x} + 9\sqrt{3x}$
16. $2\sqrt{50v} - 3\sqrt{8v}$
17. $2\sqrt[3]{8v} - \sqrt[3]{125v}$
18. $\dfrac{\sqrt{175}}{\sqrt{7}}$
19. $\sqrt[3]{\dfrac{27x^3}{y^6}}$
20. $\sqrt[3]{\dfrac{25}{27}}$
21. $3\sqrt{8} - 7\sqrt{50}$
22. $3\sqrt{3}(2\sqrt{12} - 9\sqrt{75})$
23. $(2\sqrt{2} - 5\sqrt{3})(2\sqrt{2} + 5\sqrt{3})$
24. $(3\sqrt{14})(15\sqrt{2})$
25. $\dfrac{3\sqrt{14}}{15\sqrt{2}}$
26. $\dfrac{15\sqrt{2}}{3\sqrt{14}}$
27. $(\sqrt{2} - \sqrt{3})^2$
28. $\dfrac{\sqrt{2} - \sqrt{3}}{\sqrt{2} + \sqrt{3}}$
29. $\dfrac{12}{\sqrt{7} - \sqrt{3}}$
30. $(2\sqrt{3})(3\sqrt{2})(5\sqrt{6})$
31. $\dfrac{(15\sqrt{2})(2\sqrt{3})}{10\sqrt{6}}$
32. $(2\sqrt{3})(3\sqrt{2}) - 5\sqrt{6}$

In Exercises 33–35, simplify each expression. Assume that the variables represent nonnegative real numbers so that absolute value notation is not necessary.

33. $\sqrt{32x^7}$ **34.** $\sqrt[3]{32x^7}$ **35.** $\sqrt{12x^3y^{11}}$

36. Match each function with its graph. All graphs are displayed with the window $[-10, 10, 1]$ by $[-10, 10, 1]$.

a. $f(x) = x^2 - 4$ **b.** $f(x) = |x - 4|$
c. $f(x) = \sqrt{x - 4}$ **d.** $f(x) = \sqrt{4 - x}$
e. $f(x) = 4 - x$ **f.** $f(x) = x - 4$
g. $f(x) = \sqrt[3]{x - 4}$

A.
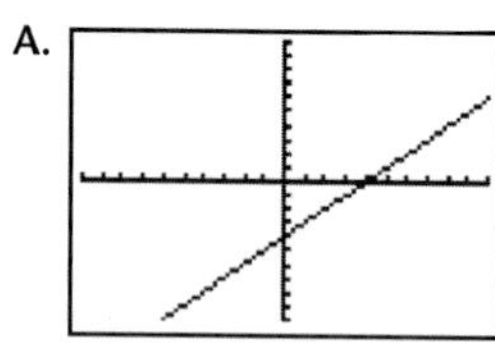

B.
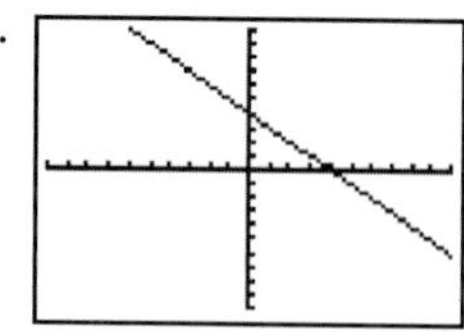

C.
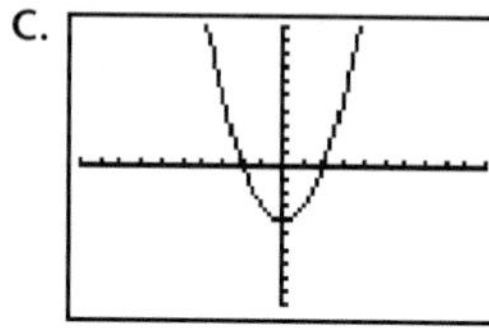

D.
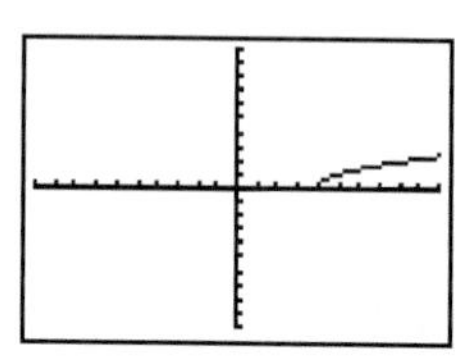

E.
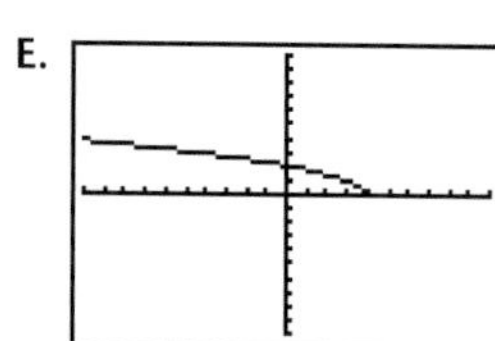

F.
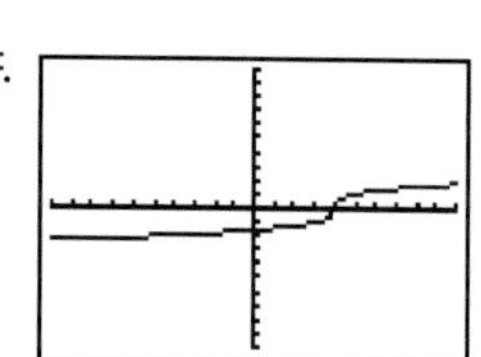

G.
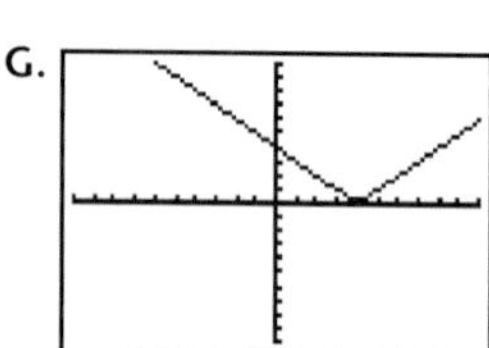

37. Match each function with a table for this function.

a. $f(x) = 2x$

A.
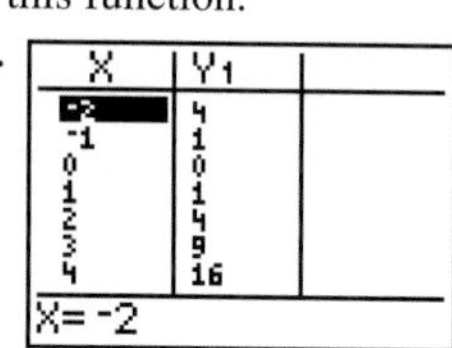

X	Y1
-2	4
-1	1
0	0
1	1
2	4
3	9
4	16

X= -2

b. $f(x) = -2x$

B.
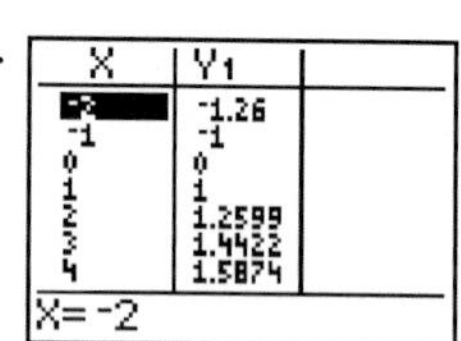

X	Y1
-2	-1.26
-1	-1
0	0
1	1
2	1.2599
3	1.4422
4	1.5874

X= -2

c. $f(x) = x^2$

C.
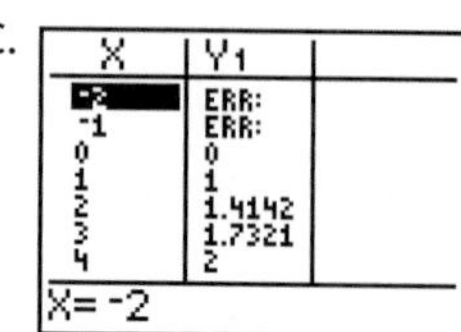

X	Y1
-2	ERR:
-1	ERR:
0	0
1	1
2	1.4142
3	1.7321
4	2

X= -2

d. $f(x) = \sqrt{x}$

D.

X	Y1
-2	4
-1	2
0	0
1	-2
2	-4
3	-6
4	-8

X= -2

e. $f(x) = \sqrt[3]{x}$

E.

X	Y1
-2	-4
-1	-2
0	0
1	2
2	4
3	6
4	8

X= -2

In Exercises 38–43, use a graphing calculator to graph each function and to determine the domain and range of each function.

38. $f(x) = -x^2 + 3$ **39.** $f(x) = x^3 + 3$
40. $f(x) = \sqrt{4 - x}$ **41.** $f(x) = \sqrt[3]{4 - x}$
42. $f(x) = |x| - 3$ **43.** $f(x) = 1$

In Exercises 44–51, determine algebraically the domain of each function. Use a graphing calculator only to check your answer.

44. $f(x) = x - 5$ **45.** $f(x) = x^2 - 5$
46. $f(x) = x^3 - 5$ **47.** $f(x) = \sqrt[3]{x - 5}$
48. $f(x) = \sqrt{x - 5}$ **49.** $f(x) = |x - 5|$
50. $f(x) = \dfrac{1}{x - 5}$ **52.** $f(x) = \dfrac{x - 5}{x^2 - 9}$

In Exercises 52–55, solve each equation.

52. $\sqrt{x - 5} = 4$ **53.** $\sqrt{x + 12} - x = 0$
54. $\sqrt{2v - 1} + 2 = v$ **55.** $\sqrt[3]{4w + 5} = -3$

56. Determine the exact x- and y-intercepts of $y = \sqrt{x + 25} - 7$.

57. Calculate the distance between $(-4, 8)$ and $(1, -4)$.

58. The points $(-2, 4)$, $(4, 6)$, and $(2, 2)$ form a triangle. First calculate the perimeter of this triangle, and then determine if it is a right triangle.

59. Dimensions of a Right Triangle The length of the hypotenuse of the right triangle shown in the figure is 2 cm more than the length of the longer leg. If the longer leg is 7 cm longer than the shorter leg, determine the length of each side.

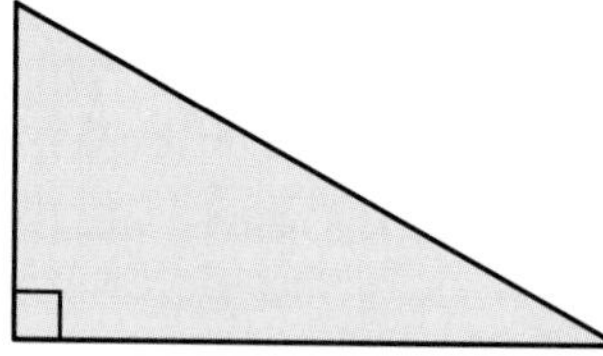

60. Find all the points with an x-coordinate of 8 that are 5 units from the point $(4, 4)$.

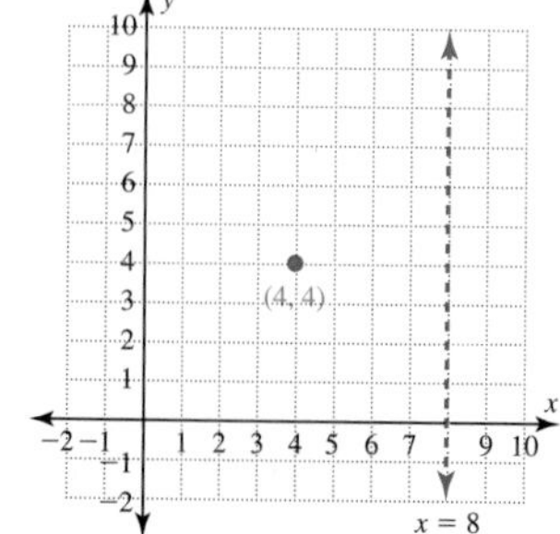

61. Write each of these radical expressions in exponential form.

a. $\sqrt{x}$ b. $\sqrt[3]{x^2}$ c. $\dfrac{1}{\sqrt[3]{x}}$

62. Write each of these exponential expressions in radical form.

a. $x^{1/4}$ b. $x^{-1/4}$ c. $x^{3/4}$

63. Represent each expression by using both radical notation and exponential notation.

a. Twice the principal square root of w

b. The principal square root of two w

c. The principal cube root of the quantity x plus four

In Exercises 64–72, simplify each expression.

64. $49^{1/2}$ 65. $-49^{1/2}$

66. $49^{-1/2}$ 67. $\left(\dfrac{27}{125}\right)^{2/3}$

68. $\left(\dfrac{-27}{125}\right)^{2/3}$ 69. $\left(\dfrac{27}{125}\right)^{-2/3}$

70. $1{,}000{,}000^{1/2}$ 71. $1{,}000{,}000^{1/3}$

72. $1{,}000{,}000^{1/6}$

In Exercises 73–81, simplify each expression. Assume all variables are positive real numbers.

73. $16^{5/8}16^{3/8}$ 74. $\dfrac{16^{5/8}}{16^{1/8}}$

75. $(16^{3/8})^{2/3}$ 76. $(8x^{-6}y^9)^{2/3}$

77. $(-32x^{5/3}y^{3/2})^{2/5}$ 78. $\dfrac{(x^2y^2z^2)^{1/3}}{(xyz)^{-4/3}}$

79. $2x^{1/2}(3x^{1/2} - 5x^{-1/2})$ 80. $(3x^{1/2} + 5)(3x^{1/2} - 5)$

81. $(x^2 + 2xy + y^2)^{1/2}$

MASTERY TEST FOR CHAPTER 9

[9.1] 1. Graph each function, and use the graph to determine the domain of the function.

a. $f(x) = \sqrt{3x - 6}$ b. $f(x) = \sqrt{6 - 2x}$

c. $f(x) = \sqrt[3]{4x - 12}$

Match each function with its graph. All graphs are displayed with the window $[-2, 8, 1]$ by $[-5, 5, 1]$.

d. $f(x) = (x - 3)^2 - 1$ e. $f(x) = -\dfrac{x}{3} + 1$

f. $f(x) = \sqrt{3 - x}$ g. $f(x) = \sqrt[3]{x - 3}$

h. $f(x) = |x - 3| + 1$

A.

B.

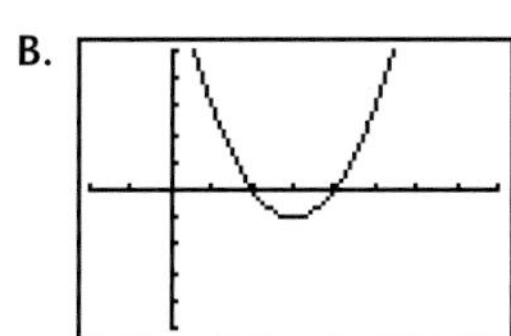

C.

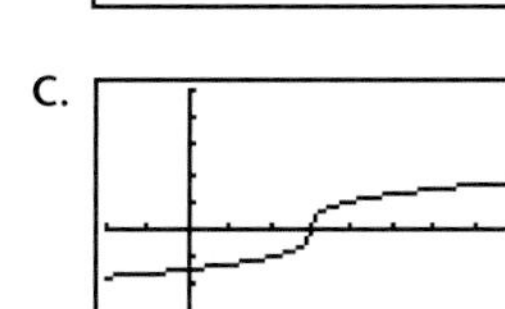

D.

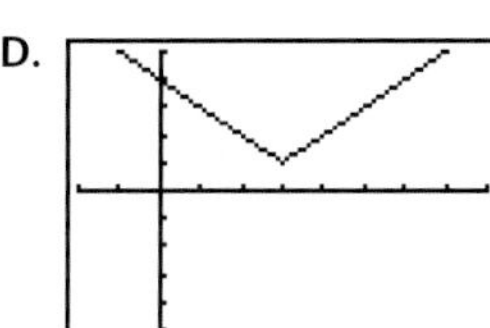

E.

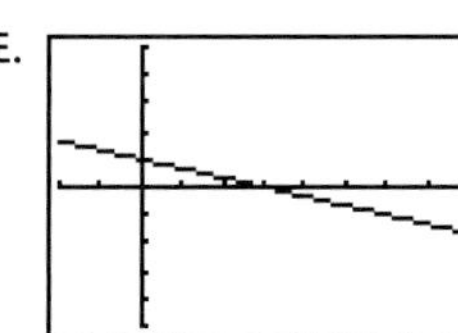

[9.1] 2. Algebraically determine the domain of each function.

a. $f(x) = 2x - 12$ b. $f(x) = \dfrac{7}{2x - 12}$

c. $f(x) = \sqrt[3]{2x - 12}$ d. $f(x) = \sqrt{2x - 12}$

[9.1] 3. Evaluate each expression without using a calculator.

a. $\sqrt{81}$ b. $\sqrt[4]{81}$

c. $\sqrt[3]{-0.008}$ d. $\sqrt[5]{-32}$

e. $\sqrt{25} + \sqrt{144}$ f. $\sqrt{25 + 144}$

Mentally estimate each expression to the nearest integer, and then use a calculator to approximate each value to the nearest hundredth.

g. $\sqrt{33}$ h. $\sqrt[5]{33}$

i. $\sqrt{17}$ j. $\sqrt[4]{17}$

[9.2] 4. Simplify each sum or difference without using a calculator. Assume $x > 0$.

a. $8\sqrt{7} - 3\sqrt{7}$

b. $(5\sqrt{2} - 3\sqrt{5}) - (2\sqrt{2} - 7\sqrt{5})$

c. $4\sqrt[3]{7} - 11\sqrt[3]{7} + 6\sqrt[3]{7}$

d. $13\sqrt{2x} - 5\sqrt{2x}$

[9.2] 5. Simplify each expression without using a calculator.

a. $\sqrt{40}$ b. $\sqrt[3]{24}$

c. $3\sqrt{28} - 5\sqrt{63}$ d. $\dfrac{\sqrt{200}}{\sqrt{18}}$

[9.3] 6. Perform each multiplication without using a calculator, and then simplify the product. Assume $x > 0$.

a. $(3\sqrt{2})(5\sqrt{2})$

b. $2\sqrt{5}(3\sqrt{5} - 2\sqrt{10})$

c. $\sqrt[3]{16x^5}\sqrt[3]{-4x}$

d. $(2\sqrt{3x} - \sqrt{5})(2\sqrt{3x} + \sqrt{5})$

[9.3] 7. Perform each division by rationalizing the denominator, and then simplify each quotient without using a calculator.

a. $\dfrac{\sqrt{18}}{\sqrt{2}}$ b. $\dfrac{18}{\sqrt{6}}$

c. $\dfrac{40}{\sqrt{7} - \sqrt{2}}$ d. $\dfrac{12}{2 - \sqrt{7}}$

[9.4] 8. Solve each equation.

a. $\sqrt{x - 3} = 11$ b. $\sqrt[3]{2x - 17} = -3$

c. $\sqrt{x + 4} = x + 11$ d. $\sqrt{3x + 1} + 3 = x$

[9.4] 9. Calculate the distance between each pair of points.

a. $(3, 9)$ and $(-5, 9)$

b. $(-4, 7)$ and $(-4, -5)$

c. $(1, -2)$ and $(7, 6)$

d. $(1, 1)$ and $(-4, 3)$

[9.5] 10. Simplify each expression. Assume $x > 0$.

a. $81^{1/2}$ b. $(-1{,}000)^{1/3}$

c. $81^{3/4}$ d. $\left(\dfrac{x^{5/12}}{x^{1/12}}\right)^{3/7}$

e. $100^{1/2} - 64^{1/2}$ f. $(100 - 64)^{1/2}$

REALITY CHECK FOR CHAPTER 9

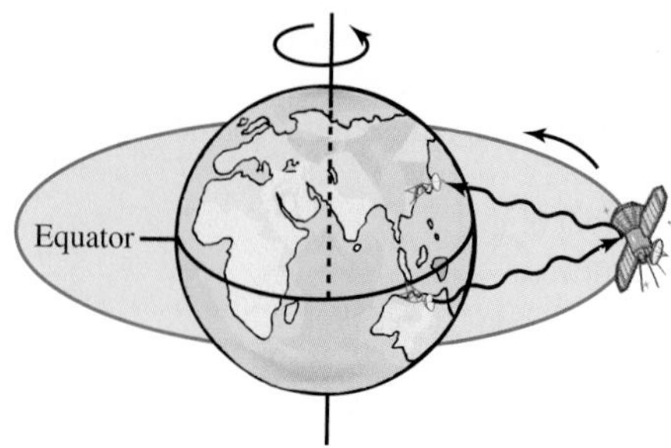

The formula $T = (3.1469 \times 10^{-7})r^{3/2}$ relates the period T of the orbit of a satellite in seconds and the radius r of the circular orbit around the center of the Earth in meters. A satellite placed in a geosynchronous orbit of the Earth will make 1 complete orbital revolution each day (once every 86,400 seconds). Use $T = 86{,}400$ to determine the radius of the geosynchronous orbit around the center of the Earth. Given that the radius of the Earth is 6,378,100 m, determine the distance above the Earth's surface at which a geosynchronous satellite must be placed in orbit.

GROUP PROJECT FOR CHAPTER 9

An Algebraic Model for Real Data

Supplies needed for the lab:

- A strong string at least 2 m long; a connector on the top to fasten to a ceiling, doorway frame, or stepladder; and a connector to attach a weight to the bottom of the string.
- A weight that is small but is relatively heavy, for example, a large steel nut.
- A metric ruler to measure the length of the string.
- Scissors to shorten the length of the string.
- A stopwatch.
- A graphing calculator with a Power Regression feature **(PwrReg).**

Lab: Recording the Time for the Period of a Pendulum

Suspend the top of the string from an elevated position such as a ceiling or the top of a doorframe. (Your Physics lab may have a setup for this type of experiment.) Measure the length of string and attach a weight that is much heavier than the string. Using clock positions as a reference system, start with the top of the string in the 12:00 position and the bottom of the string in the 6:00 position. Then move the bottom of the string and the weight to approximately the 4:30 position. Simultaneously release the weight and start the stopwatch. Stop the stopwatch after this pendulum has completed 10 full periods. Divide this time by 10 to determine the time for one period. Then record the length of the string in centimeters and the time in seconds for one period. Shorten the string by about 15 cm and repeat this experiment about 11 times.

LENGTH x (cm)	TIME y (seconds)
200	
185	
170	
155	
140	
125	
110	
95	
80	
65	
50	

1. Record your data in a table similar to the one shown to the right.
2. Plot these data on a scatter diagram, using an appropriate scale for each axis.

Creating an Algebraic Model for These Data

1. Using your graphing calculator, calculate the power function of best fit. This function will be of the form $T(x) = ax^b$.
2. Let $y = T(x)$ represent the curve of best fit that you just calculated.
   a. Evaluate and interpret $T(165)$.
   b. Evaluate and interpret $T(70)$.
   c. Determine the value of x for which $T(x) = 4$. Interpret the meaning of this value.

chapter 10

Exponential and Logarithmic Functions

CHAPTER OUTLINE

Financial advisers agree that investing for the future, including retirement, should be an essential item in the budget of every worker. The graph illustrates why starting to invest early in your career is so important. It shows the exponential growth of the value of a $500 investment earning 7% interest compounded continuously over a 40-year time period.

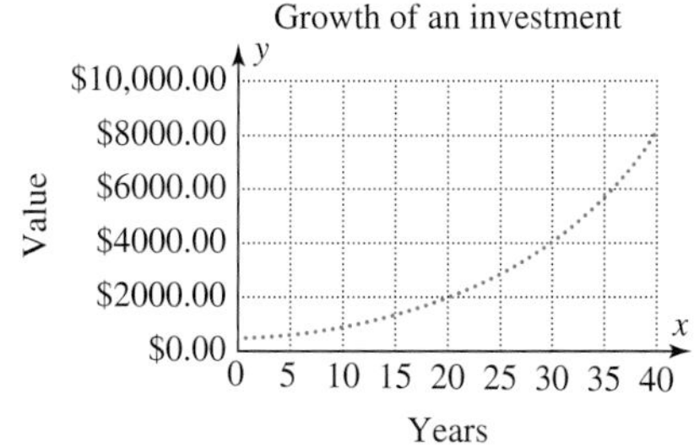

REALITY CHECK

Approximately how many years would it take this investment to be worth $4,000? Approximately how many years would it take this investment to be worth $8,000? How many years would it take this investment to double in value?

Section 10.1 Geometric Sequences and Graphs of Exponential Functions

Objectives:
1. Identify a geometric sequence.
2. Graph and use exponential functions.

Exponential and logarithmic functions are used to solve many types of growth and decay problems. These functions are used by bankers to compute compound interest, by sociologists to predict population growth, and by archaeologists to compute the age of ancient wood samples through carbon-14 dating. One specific example that uses exponential growth is the computation of the safe shelf life of products such as milk. By calculating how long it will take bacteria to reach an unacceptable level, we can determine how many days this perishable product is safe.

This chapter examines applications of both exponential and logarithmic functions. Because exponential and logarithmic functions are inverses of each other, we also examine inverse functions in this chapter. We start by examining exponential functions and geometric sequences. We then end the chapter with a Group Project that involves collecting data on the height of a bouncing ball. An exponential function can be used to model these data. We use a graphing calculator to determine this exponential model.

The following tables illustrate two distinctive types of sequences. The first sequence gives the yearly salary amounts for an employee over a 10-year period, assuming a starting salary of $40,000 and annual raises of $2,400. The second sequence gives the yearly salary amounts for an employee over a 7-year period, assuming a starting salary of $40,000 and annual raises of 6%.

ANNUAL SALARY INCREASE OF $2,400

NUMBER OF PRIOR YEARS x	ANNUAL SALARY y ($)
0	40,000
1	42,400
2	44,800
3	47,200
4	49,600
5	52,000
6	54,400
7	56,800
8	59,200
9	61,600

Salary increased by **adding** a constant value

ANNUAL SALARY INCREASE OF 6%

NUMBER OF PRIOR YEARS x	ANNUAL SALARY y ($)
0	40,000
1	42,400
2	44,944
3	47,641
4	50,499
5	53,529
6	56,741
7	60,145
8	63,754
9	67,579

Salary increased by **multiplying** by a constant value

An arithmetic sequence has a constant change from term to term (a common difference). This common difference is often denoted by d.

The first sequence is an arithmetic sequence with a common difference $d = 2{,}400$. As we noted in Section 2.1, arithmetic sequences have a constant change from term to term and produce a linear pattern when graphed. The second sequence is an example of a geometric sequence. Geometric sequences have a constant factor from term to term. A geometric sequence differs significantly from an arithmetic sequence in both its numerical pattern and its graphical shape. We now examine geometric sequences.

A geometirc sequence has a constant ratio from term to term (a common ratio). This common ratio is often denoted by r.

Geometric Sequences

The sequence 1, 2, 4, 8, 16, 32 has a constant ratio from term to term. Each term is twice the preceding term. This is an example of a geometric sequence. A **geometric sequence** is a sequence with a constant ratio from term to term. That is, the ratio of any two consecutive terms is constant. This constant is called the **common ratio** and is usually denoted by r. The common ratio for the sequence 1, 2, 4, 8, 16, 32 is 2. The ratio of consecutive terms $\frac{2}{1} = \frac{4}{2} = \frac{8}{4} = \frac{16}{8} = \frac{32}{16} = 2$.

EXAMPLE 1 Writing the Terms of Arithmetic and Geometric Sequences

Write the first five terms of each sequence.

SOLUTION

(a) An arithmetic sequence with $a_1 = 5$ and a common difference of 4.

$a_1 = 5$
$a_2 = 5 + 4 = 9$
$a_3 = 9 + 4 = 13$
$a_4 = 13 + 4 = 17$
$a_5 = 17 + 4 = 21$

Each term of this arithmetic sequence is obtained by adding the common difference of 4 to the previous term.

(b) A geometric sequence with $a_1 = 5$ and a common ratio of 4.

$a_1 = 5$
$a_2 = 5(4) = 20$
$a_3 = 20(4) = 80$
$a_4 = 80(4) = 320$
$a_5 = 320(4) = 1280$

Each term of this geometric sequence is obtained by multiplying the common ratio of 4 by the previous terms.

In Example 2 we examine each sequence to determine if the sequence is geometric. If a sequence is geometric, the ratio of consecutive terms will be constant.

EXAMPLE 2 Identifying Geometric Sequences

Determine which of these sequences are geometric sequences. For those that are geometric, determine the common ratio r.

(a) 2, 3, 4.5, 6.75, 10.125, 15.1875
(b) 32, 16, 8, 4, 2, 1
(c) 2, 3.5, 5, 6.5, 8, 9.5

SOLUTION

The graph of each sequence contains only discrete points. Each dashed curve is shown to emphasize the pattern formed by these points.

(a) NUMERICALLY: RATIO OF CONSECUTIVE TERMS

$$\frac{3}{2} = 1.5$$
$$\frac{4.5}{3} = 1.5$$
$$\frac{6.75}{4.5} = 1.5$$
$$\frac{10.125}{6.75} = 1.5$$
$$\frac{15.1875}{10.125} = 1.5$$
$$r = 1.5$$

VERBALLY

Because there is a common ratio $r = 1.5$, this is a geometric sequence.

GRAPHICALLY

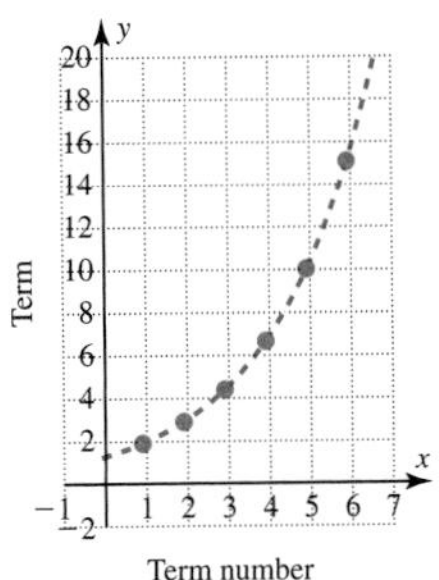

The graph of these points exhibits the shape that we will define as geometric growth.

(b) NUMERICALLY: RATIO OF CONSECUTIVE TERMS

$\frac{16}{32} = 0.5$

$\frac{8}{16} = 0.5$

$\frac{4}{8} = 0.5$

$\frac{2}{4} = 0.5$

$\frac{1}{2} = 0.5$

$r = 0.5$

VERBALLY

Because there is a common ratio $r = 0.5$, this is a geometric sequence.

GRAPHICALLY

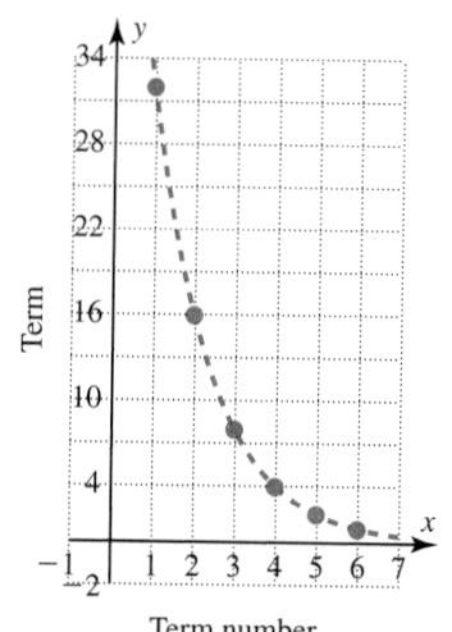

The graph of these points exhibits the shape that we will define as geometric decay.

(c) NUMERICALLY: RATIO OF CONSECUTIVE TERMS

$\frac{3.5}{2} = 1.75$

$\frac{5}{3.5} \approx 1.43$

$1.75 \neq 1.43$

VERBALLY

This is not a geometric sequence because the ratio from term to term is not constant. In fact, this is an arithmetic sequence with a common difference $d = 1.5$.

GRAPHICALLY

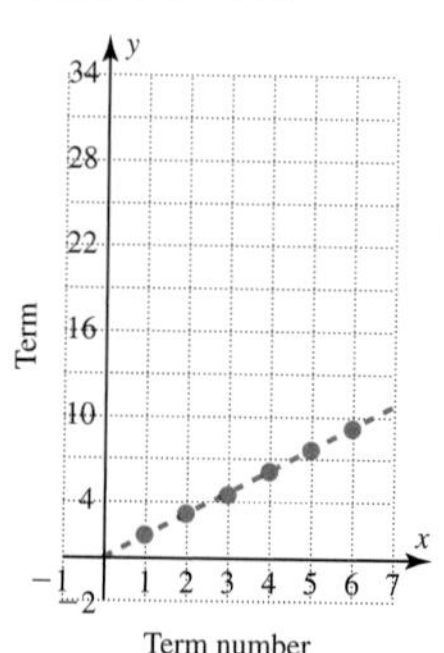

Note these points exhibit a linear pattern.

SELF-CHECK 10.1.1

1. Write an arithmetic sequence of seven terms with $a_1 = 2$ and a common difference $d = 3$.
2. Write a geometric sequence of seven terms with $a_1 = 2$ and a common ratio $r = 3$.
3. Write a geometric sequence of seven terms with $a_1 = 2$ and a common ratio $r = -3$.

If the common ratio r of a geometric sequence is greater than 1, then we refer to the sequence as **geometric growth.** Each successive term will grow by a factor of r. If the common ratio r of a geometric sequence is positive and less than 1, then we refer to the sequence as **geometric decay.** Each successive term will decrease or decay by a factor of r.

Geometric Growth and Decay

GROWTH	DECAY
If $r > 1$, then a geometric sequence exhibits geometric growth.	If $0 < r < 1$, then a geometric sequence exhibits geometric decay.
$r = 2$	$r = 0.5$
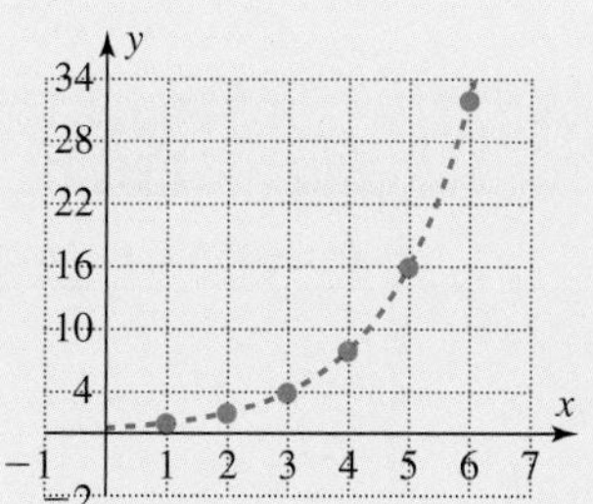	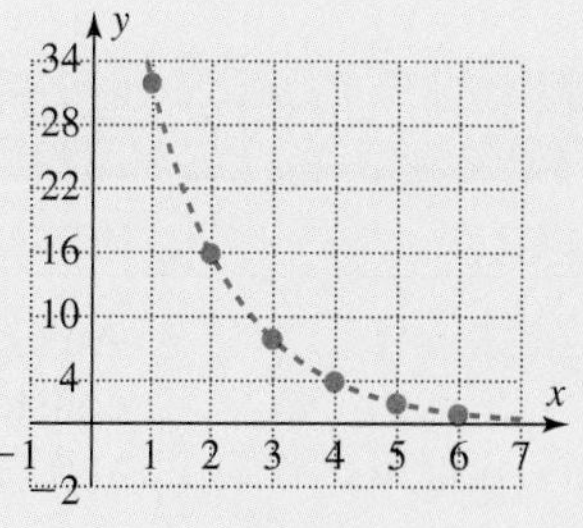

Populations of people, bacteria, and other organisms often exhibit geometric growth. Example 3 examines an application that exhibits geometric decay.

EXAMPLE 3 Determining a Sequence that Models the Height of a Bouncing Ball

A golf ball dropped onto a concrete cart path from a height of 10 ft. Each bounce is 0.6 as high as the height of the previous bounce. Determine the height of the first four bounces.

SOLUTION

$a_0 = 10$
$a_1 = 0.6(10) = 6$
$a_2 = 0.6(6) = 3.6$
$a_3 = 0.6(3.6) = 2.16$
$a_4 = 0.6(2.16) = 1.296$

The sequence exhibits geometric decay. The common ratio $r = 0.6 < 1$. The height of each bounce is decreasing geometrically by a factor of 0.6, and a_0 denotes the initial position of 10 ft.

Answer: The heights of the first four bounces are 6 ft, 3.6 ft, 2.16 ft, and 1.296 ft.

SELF-CHECK 10.1.2

One share of a particular stock on the NASDAQ is currently priced at $128.

1. Determine the price of the share at the end of each of the next 4 years if the price exhibits geometric growth with $r = 1.5$.
2. Determine the price of the share at the end of each of the next 4 years if the price exhibits geometric decay with $r = 0.5$.

Exponential Functions

The terms of a geometric sequence can be obtained by repeated multiplication by the common ratio r. If a_1 is the first term, then we can write the first six terms as

$$a_1, a_1r, a_1r^2, a_1r^3, a_1r^4, a_1r^5$$

The graph of a geometric sequence consists of individual discrete points whereas the graph of an exponential function is continuous.

Since exponentiation was originally developed to represent repeated multiplication like that just presented, we now examine exponential functions. In the work that follows, the exponent can be any real number, including an irrational number. Thus the graphs of exponential functions are connected and continuous over the whole domain of all real numbers, while the graph of a geometric sequence is a set of disconnected discrete points. Some applications such as mortgage payments occur at discrete intervals, while the growth of a tree is a continuous process.

Exponential Function: $f(x) = b^x$

ALGEBRAICALLY	VERBALLY	ALGEBRAIC EXAMPLE	GRAPHICAL EXAMPLE
If $b > 0$ and $b \neq 1$, then $f(x) = b^x$ is an exponential function with base b.	An exponential function has a constant for the base, and the variable is in the exponent.	$y = 2^x$	y, x, 65, 60, 55, 50, 45, 40, 35, 30, 25, 20, 15, 10, 5, −5, −1, 1, 2, 3, 4, 5, 6, 7, $y = 2^x$

EXAMPLE 4 Identifying Exponential Functions

Determine which of these equations define exponential functions.

SOLUTION

(a) $f(x) = 3^x$ — An exponential function — This function has a base of 3.

(b) $f(x) = x^3$ — Not an exponential function — The exponent is the constant 3, not a variable. This is called a cubic polynomial function.

(c) $f(x) = \left(\frac{4}{5}\right)^x$ — An exponential function — This function has a base of $\frac{4}{5}$.

(d) $f(x) = (-5)^x$ — Not an exponential function — The base of this exponential expression is -5. Since the base is negative, this equation does not represent an exponential function.

(e) $y = 1^x$ — Not an exponential function — $y = 1^x$ simplifies to $y = 1$, since 1 to any power equals 1. Thus it is a constant function. ■

The domain of an exponential function $f(x) = b^x$ is understood to be the set of all real numbers. You may be able to evaluate some selected values such as 2^4 by hand, but you will need a calculator to approximate values such as 2^{π}. This is illustrated in Example 5.

EXAMPLE 5 Evaluating an Exponential Function

Given $f(x) = 2^x$, evaluate each expression.

SOLUTION

(a) $f(4)$
(b) $f(-2)$
(c) $f(\pi)$

```
2^4
                16
2^(-2)
                .25
2^π
       8.824977827
```

Answers: (a) $f(4) = 2^4 = 16$

(b) $f(-2) = 2^{-2} = \frac{1}{2^2} = \frac{1}{4} = 0.25$

(c) $f(\pi) = 2^{\pi} \approx 8.825$

SELF-CHECK 10.1.3

Given $f(x) = 3^x$, evaluate each expression.

1. $f(2)$ 2. $f(-1)$ 3. $f(\sqrt{2})$

The exponential decay function $f(x) = \left(\frac{1}{2}\right)^x$ can also be written as $f(x) = 2^{-x}$.

If the base b of an exponential function is greater than 1, as in $f(x) = 2^x$, then the graph of the function rises rapidly to the right as the base is used repeatedly as a factor. Thus an exponential function $f(x) = b^x$ with $b > 1$ is an increasing function, called an **exponential growth function.** If $0 < b < 1$, then the graph of $f(x) = b^x$ declines rapidly to the right and is called an **exponential decay function.** The exponential growth function in the following box is $f(x) = 2^x$, and the exponential decay example is $f(x) = \left(\frac{1}{2}\right)^x$ or $f(x) = 2^{-x}$.

A function whose graph rises to the right is called an **increasing function.** Similarly, a function whose graph falls to the right is called a **decreasing function.** A line with positive slope represents an increasing function, and a line with negative slope represents a decreasing function.

Recall that the domain of a function is represented by the projection of its graph onto the x-axis and the range is represented by the projection of its graph onto the y-axis. Note that this is illustrated by the growth and decay functions given in the box.

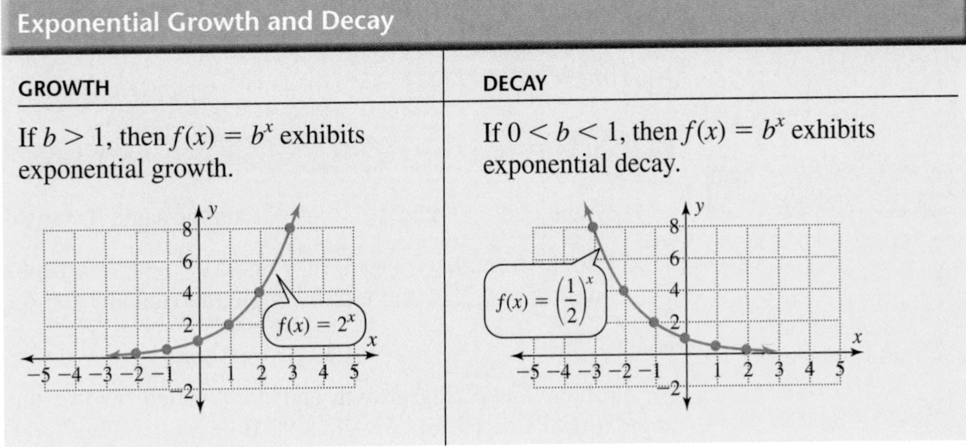

The graphs of $f(x) = 2^x$ and $f(x) = \left(\frac{1}{2}\right)^x$ just given reveal several properties and features of exponential functions. We will list some of these properties below.

Properties of the Graphs of Exponential Functions

For the exponential function $f(x) = b^x$ with $b > 0$ and $b \neq 1$:

1. The domain of f is the set of all real numbers $\mathbb{R}$.
2. The range of f is the set of all positive real numbers.
3. There is no x-intercept.
4. The y-intercept is $(0, 1)$.
5. a. The growth function is asymptotic to the negative portion of the x-axis (it approaches but does not touch the x-axis).
   b. The decay function is asymptotic to the positive portion of the x-axis.
6. a. The growth function rises (or grows) from left to right.
   b. The decay function falls (or decays) from left to right.

Newspapers and magazines often use the term *exponential growth* incorrectly. They sometimes use the term simply to indicate rapid growth. If this term is used correctly, the data would need to fit an exponential function.

Some financial advisers and insurance salespeople promoting annuities and other investments refer to the "magic of compound interest." The "magic" they are referring to is exponential growth—growth in which the growth is always a factor of the previous base. Thus as the terms grow, so does the change from term to term. Example 6 involves exponential growth with a base $b = 1.0925$.

EXAMPLE 6 Modeling Compound Interest

The formula $A = P(1 + r)^t$ can be used to compute the total amount of money A that accumulates when a principal P is invested at a yearly interest rate r and left to compound annually for t years. Use this growth formula to do the following.

(a) Write a function for A for an investment of \$5,000 at 9.25% for t years.
(b) Evaluate this function for $t = 8$ years.

SOLUTION

(a)
$$A = P(1 + r)^t$$
$$A(t) = 5{,}000(1 + 0.0925)^t$$
$$A(t) = 5{,}000(1.0925)^t$$

Substitute the given values into the compound interest formula: $P = 5{,}000$ and $r = 0.0925$. This is an exponential function of the form $f(x) = ab^x$ with a base of $b = 1.0925$.

(b)
$$A(8) = 5{,}000(1.0925)^8$$
$$A(8) \approx 5{,}000(2.029418267)$$
$$A(8) \approx 10{,}147$$

To determine the value of this investment after 8 years, evaluate $A(t)$ for $t = 8$.

The value of this investment after 8 years is approximately \$10,147.

This calculator approximation has been rounded to the nearest dollar.

SELF-CHECK 10.1.4

1. Graph $f(x) = 3^x$ and $f(x) = \left(\frac{1}{3}\right)^x$ on the same rectangular coordinate system.
2. Use the formula $A = P(1 + r)^t$ and a calculator to determine the value of \$450 invested at 8.75% and left to compound annually for 7 years.

A Mathematical Note

Swiss-born Leonard Euler (1701–1783) was hired by Catherine the Great of Russia to write the elementary mathematics textbooks for Russian schools. He wrote prolifically on various mathematical topics. From Euler's textbook *Introducio* came many symbols, such as i for $\sqrt{-1}$, π for the ratio of the circumference of a circle to its diameter, and e for the base of natural logarithms.

Problems involving growth and decay often involve the irrational number e that is approximately equal to 2.718281828. Both π and e are fundamental constants that show up over and over in our study of the world and universe around us.

```
π
          3.141592654
e
          2.718281828
```

These calculator approximations are just that—approximations. Because π and e are irrational their decimal representations never repeat or terminate.

CALCULATOR PERSPECTIVE 10.1.1 Evaluating Expressions Involving e^x

To evaluate the expression $1{,}000e^{-0.01245}$ that occurs in Example 7 on a TI-84 Plus calculator, enter the following keystrokes:

1 0 0 0 2nd e^x

(–) 0 . 0 1 2 4 5)

ENTER

```
1000e^(-0.01245)
         987.6271806
```

The e^x feature is the secondary function of the LN key.

Example 7 involves e and a function that models radioactive decay.

EXAMPLE 7 Determining the Amount of Radioactive Decay

The amount of a 1,000-g sample of radioactive carbon-14 remaining after t years is given by the function $A(t) = 1{,}000e^{-0.0001245t}$. How much of a 1,000-g sample will remain a century later?

SOLUTION

$$A(t) = 1{,}000e^{-0.0001245t}$$
$$A(100) = 1{,}000e^{-0.0001245(100)}$$ A century is 100 years. Substitute this value of t into the given function.
$$A(100) = 1{,}000e^{-0.01245}$$
$$A(100) \approx 987.6$$ Approximate this value by using a calculator.

Answer: After 100 years, approximately 988 g of carbon-14 remains.

SELF-CHECK 10.1.5

1. Approximate e^2 to the nearest thousandth.
2. Approximate e^{-2} to the nearest thousandth.
3. Use the formula in Example 7 to determine the amount of carbon-14 remaining after 1,000 years.

The equations in Example 8 can all be solved easily without the use of a calculator because they involve familiar powers of 2, 3, 4, or other small integers. The properties we will use are given in the following box.

Properties of Exponential Functions

ALGEBRAICALLY	VERBALLY
For real exponents x and y and bases $a > 0, b > 0$:	
1. For $b \neq 1$, $b^x = b^y$ if and only if $x = y$.	1. The exponents must be equal because the expressions are equal and have the same base.
2. For $x \neq 0$, $a^x = b^x$ if and only if $a = b$.	2. The bases must be equal because the expressions are equal and have the same exponents.

EXAMPLE 8 Solving Exponential Equations

Solve each equation.

SOLUTION

(a) $3^{x-4} = 9$

$$3^{x-4} = 9$$

$$3^{x-4} = 3^2$$ Substitute 3^2 for 9 in order to express both members in terms of the common base 3.

$$x - 4 = 2$$ The exponents are equal since the bases are the same.

Answer: $x = 6$

(b) $4^x = \frac{1}{8}$

$$4^x = \frac{1}{8}$$ Express both 4 and $\frac{1}{8}$ in terms of the common base 2.

$$(2^2)^x = 2^{-3}$$

$$2^{2x} = 2^{-3}$$ Use the power rule for exponents to simplify the left side of this equation.

$$2x = -3$$ The exponents are equal since the bases are the same.

Answer: $x = -\frac{3}{2}$

(c) $(b + 2)^3 = 125$

$$(b + 2)^3 = 125$$

$$(b + 2)^3 = 5^3$$ Substitute 5^3 for 125.

$$b + 2 = 5$$ The bases are equal since the exponents are the same.

Answer: $b = 3$

SELF-CHECK 10.1.6

Solve each equation.

1. $2^{x+7} = 16$
2. $27^x = 9$
3. $b^{-3} = \frac{1}{64}$

We will solve more complicated exponential equations in Section 10.6 after we develop some skills with logarithms.

SELF-CHECK ANSWERS

10.1.1

1. 2, 5, 8, 11, 14, 17, 20
2. 2, 6, 18, 54, 162, 486, 1,458
3. 2, −6, 18, −54, 162, −486, 1,458

10.1.2

1. \$192, \$288, \$432, \$648
2. \$64, \$32, \$16, \$8

10.1.3

1. $f(2) = 3^2 = 9$
2. $f(-1) = 3^{-1} = \frac{1}{3} \approx 0.333$
3. $f(\sqrt{2}) = 3^{\sqrt{2}} \approx 4.729$

10.1.4

1.

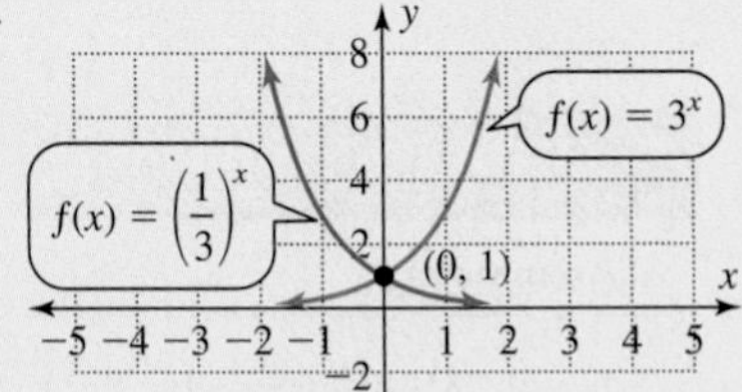

2. $A(7) \approx 809.50$; after 7 years the investment is worth \$809.50.

10.1.5

1. $e^2 \approx 7.389$
2. $e^{-2} \approx 0.135$
3. After 1,000 years, approximately 883 g of carbon-14 remains.

10.1.6

1. $x = -3$
2. $x = \frac{2}{3}$
3. $b = 4$

USING THE LANGUAGE AND SYMBOLISM OF MATHEMATICS 10.1

1. An arithmetic sequence is a sequence with a constant __________ from term to term.
2. A geometric sequence is a sequence with a constant __________ from term to term.
3. If the common ratio r of a geometric sequence is greater than 1, then we say the sequence exhibits geometric __________.
4. If the common ratio r of a geometric sequence is between 0 and 1, then we say the sequence exhibits geometric __________.
5. An exponential function is of the form $f(x) =$ __________ with a constant for the __________ and a variable in the __________.
6. An exponential function $f(x) = b^x$ with $b > 1$ is called an exponential __________ function.
7. An exponential function $f(x) = b^x$ with $0 < b < 1$ is called an exponential __________ function.
8. The graph of an exponential function $f(x) = b^x$ is asymptotic to a portion of the __________-axis.
9. The irrational number $e \approx$ __________ (to the nearest thousandth).
10. For real exponents x and y and base $b > 0$ and $b \neq 1$, $b^x = b^y$ if and only if __________ = __________.
11. For a real exponent x and bases $a > 0$ and $b > 0$ and $x \neq 0$, $a^x = b^x$ if and only if __________ = __________.

QUICK REVIEW 10.1

The purpose of this quick review is to help you recall skills needed in this section.

1. The subscript notation a_n is read a __________ n.

Use the function shown to answer Exercises 2–5.

2. Determine the interval where this function is increasing.
3. Determine the interval where this function is decreasing.
4. Determine the domain of this function.
5. Determine the range of this function.

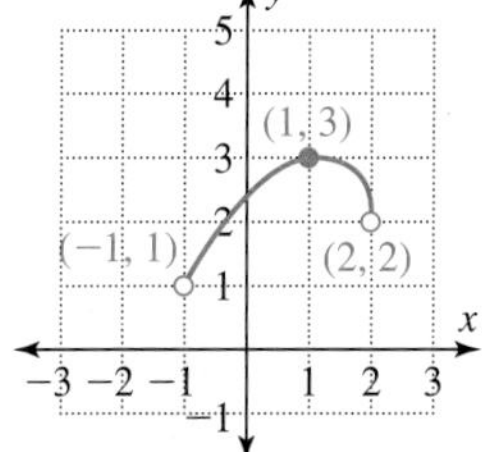

EXERCISES 10.1

In Exercises 1–4, determine whether each sequence is geometric. If the sequence is geometric determine the common ratio r.

1. a. 1, 5, 25, 125, 625, 3125
   b. 3125, 625, 125, 25, 5, 1
   c. 1, −5, 25, −125, 625, −3125
   d. 3125, −625, 125, −25, 5, −1
2. a. 4, 12, 36, 108, 324, 972
   b. 972, 324, 108, 36, 12, 4
   c. −4, 12, −36, 108, −324, 972
   d. 972, −324, 108, −36, 12, −4
3. a. 36, 24, 16, $\frac{32}{3}$, $\frac{64}{9}$
   b. 36, 24, 12, 0, −12
   c. 4, 8, 12, 16, 20
   d. 4, 8, 16, 32, 64
4. a. 2, 10, 40, 200, 1000
   b. 2, $2\sqrt{2}$, 4, $4\sqrt{2}$, 8
   c. 1, 10, 100, 1000, 10000
   d. 100, 90, 80, 70, 60
5. a. Write an arithmetic sequence of five terms with a first term $a_1 = 10$ and a common difference $d = 2$.
   b. Write an arithmetic sequence of five terms with a first term $a_1 = 10$ and a common difference $d = -2$.
   c. Write a geometric sequence of five terms with a first term $a_1 = 10$ and a common ratio $r = 2$.
   d. Write a geometric sequence of five terms with a first term $a_1 = 10$ and a common ratio $r = -2$.
6. a. Write an arithmetic sequence of five terms with a first term of $a_1 = 12$ and a common difference $d = \frac{1}{2}$.
   b. Write an arithmetic sequence of five terms with a first term of $a_1 = 12$ and a common difference $d = -\frac{1}{2}$.
   c. Write a geometric sequence of five terms with a first term of $a_1 = 12$ and a common ratio $r = \frac{1}{2}$.
   d. Write a geometric sequence of five terms with a first term of $a_1 = 12$ and a common ratio $r = -\frac{1}{2}$.

In Exercises 7–12, match each graph with the corresponding description.

7. A linear growth function (slope positive)
8. A geometric growth sequence
9. An exponential growth function
10. A linear decay function (slope negative)
11. A geometric decay sequence
12. An exponential decay function

A.
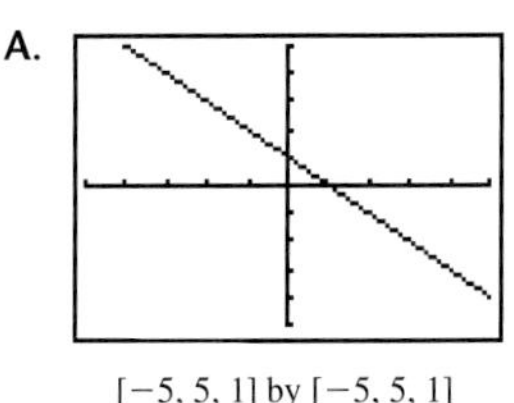
[−5, 5, 1] by [−5, 5, 1]

B.
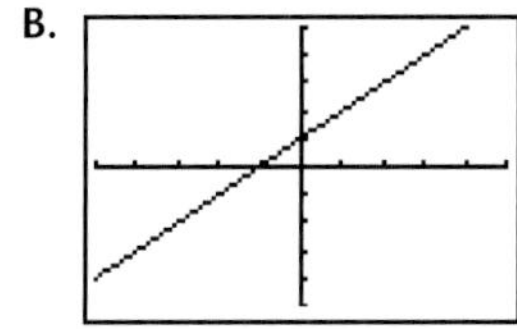
[−5, 5, 1] by [−5, 5, 1]

C.
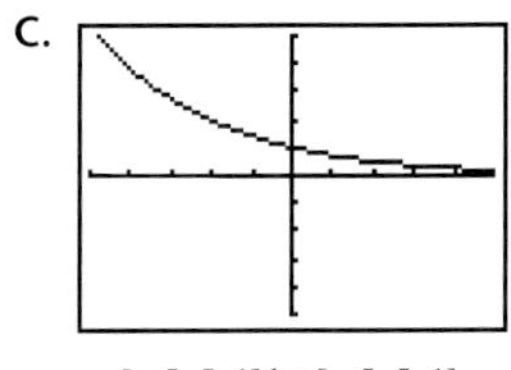
[−5, 5, 1] by [−5, 5, 1]

D.
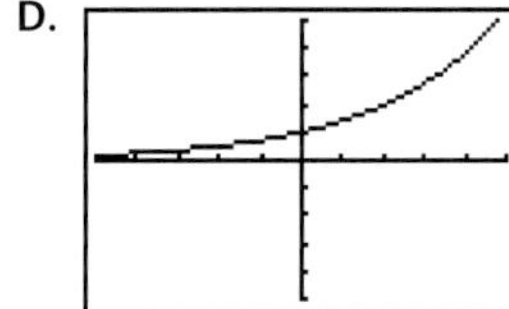
[−5, 5, 1] by [−5, 5, 1]

E. 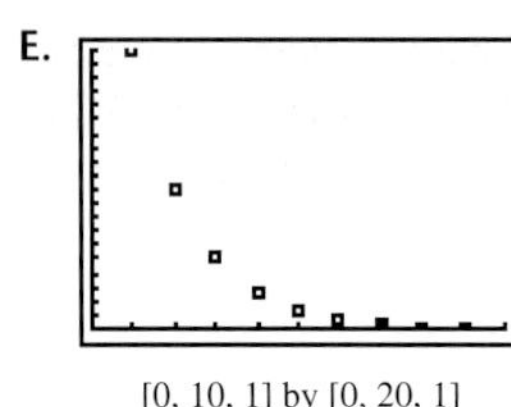
[0, 10, 1] by [0, 20, 1]

F. 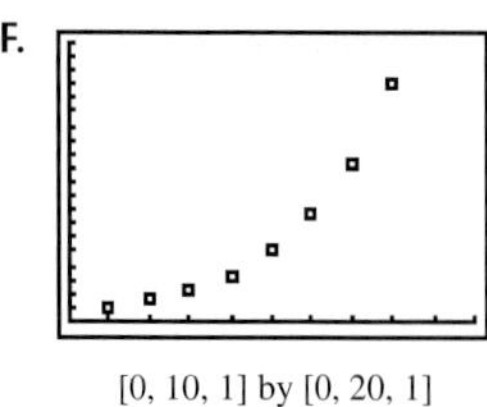
[0, 10, 1] by [0, 20, 1]

In Exercises 13–16, complete the next three terms of each sequence

a. So that the sequence will be arithmetic
b. So that the sequence will be geometric

13. 3, 6, __, __, __
14. 5, 15, __, __, __
15. 2, 2, __, __, __
16. −4, −4, __, __, __

17. Given $f(x) = 4^x$, mentally evaluate each expression.
a. $f(1)$ b. $f(3)$ c. $f(0)$
d. $f(-2)$ e. $f\left(\frac{1}{2}\right)$

18. Given $f(x) = 8^x$, mentally evaluate each expression.
a. $f(1)$ b. $f(2)$ c. $f(0)$
d. $f(-2)$ e. $f\left(\frac{1}{3}\right)$

Estimation Skills and Calculator Skills

In Exercises 19–22, mentally estimate the value of each expression and then use a calculator to approximate each value to the nearest hundredth. Use the function $f(x) = (3.9)^x$.

PROBLEM	MENTAL ESTIMATE	CALCULATOR APPROXIMATION
19. $f(2)$		
20. $f(0.99)$		
21. $f(-1)$		
22. $f(0.5)$		

23. Given $f(x) = 2.7^x$, use a calculator to approximate each expression to the nearest thousandth.
a. $f(8)$ b. $f(-2.3)$ c. $f(\pi)$

24. Given $f(x) = 0.83^x$, use a calculator to approximate each expression to the nearest thousandth.
a. $f(3)$ b. $f(-3)$ c. $f(\pi)$

25. Given $f(x) = e^x$, use a calculator to approximate each expression to the nearest thousandth.
a. $f(\sqrt{8})$ b. $f(-1.6)$ c. $f(\pi)$

26. Given $f(x) = e^{-0.2x}$, use a calculator to approximate each expression to the nearest thousandth.
a. $f(\sqrt{7})$ b. $f(-2.4)$ c. $f(\pi)$

In Exercises 27–30, use the given function to calculate by hand a table of values. Use this table of values to sketch the function. You may then check your work by using a calculator.

27. $f(x) = 4^x$
28. $f(x) = \left(\frac{1}{4}\right)^x$
29. $f(x) = 5^x$
30. $f(x) = \left(\frac{1}{5}\right)^x$

In Exercises 31–64, solve each equation without using a calculator.

31. a. $2x = 64$ b. $x^2 = 64$ c. $2^x = 64$
32. a. $3x = 1$ b. $3^x = \frac{1}{3}$ c. $3^x = 1$

33. $4^x = 16$
34. $2^y = 32$
35. $5^{v+1} = 125$
36. $7^{v+5} = 49$
37. $\left(\frac{2}{3}\right)^w = \frac{8}{27}$
38. $\left(\frac{3}{4}\right)^w = \frac{9}{16}$
39. $2^m = \frac{1}{2}$
40. $5^m = \frac{1}{5}$
41. $5^n = 1$
42. $2^n = 1$
43. $25^x = 5$
44. $27^{2y} = 3$
45. $49^x = 7$
46. $8^{5y} = 2$
47. $7^x = \frac{1}{49}$
48. $5^x = \frac{1}{25}$
49. $\left(\frac{2}{5}\right)^w = \frac{125}{8}$
50. $\left(\frac{2}{7}\right)^w = \frac{49}{4}$
51. $32^{3n-5} = 2$
52. $64^{2n+9} = 2$
53. $3^v = \sqrt{3}$
54. $7^v = \sqrt[3]{7}$
55. $10^x = 10{,}000$
56. $10^x = 100{,}000$
57. $10^y = 0.001$
58. $10^y = 0.00001$
59. $5^{x+7} = 25$
60. $7^{x-3} = 49$
61. $7^{3x} = \sqrt{7}$
62. $11^{2x-1} = \sqrt[3]{11}$
63. $\left(\frac{3}{7}\right)^x = \frac{49}{9}$
64. $\left(\frac{5}{4}\right)^{-x} = \frac{64}{125}$

65. **Stock Prices** A share of stock on the NASDAQ is priced at \$100. Determine the price per share at the end of each of the next 4 years under each of the following assumptions.
a. Arithmetic growth of \$10 per year
b. Arithmetic decay of \$10 per year
c. Geometric growth of 10% per year
d. Geometric decay of 10% per year

66. **Stock Prices** A share of stock on the NASDAQ is priced at \$100. Determine the price per share at the end of each of the next 4 years under each of the following assumptions.
a. Arithmetic growth of \$15 per year
b. Arithmetic decay of \$15 per year
c. Geometric growth of 15% per year
d. Geometric decay of 15% per year

67. **Height of a Bouncing Ball** A golf ball is dropped from 36 m onto a synthetic surface that causes it to rebound to one-half its previous height on each bounce. Determine the heights of the first four bounces.

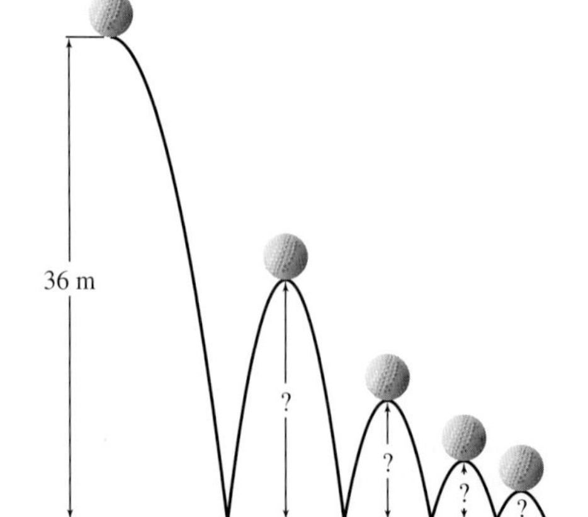

68. **Radioactive Decay** A nuclear chemist starts an experiment with 100 g of a radioactive material. At the end of each time

period only one-half of the amount present at the start of the period is left. Determine the amount of this material left at the end of each of the first five time periods.

In Exercises 69 and 70, use the compound interest formula $A = P(1 + r)^t$ with A representing the total amount of money that accumulates when a principal P is invested at an annual interest rate r and left to compound for a time of t years.

69. **Compound Interest** Write A as a function of t when \$1,000 is invested at 6% for t years. Then evaluate this function for $t = 7$ years.
70. **Compound Interest** Write A as a function of t when \$4,000 is invested at 5.5% for t years. Then evaluate this function for $t = 10$ years.
71. **Radioactive Decay** The strontium-90 in a nuclear reactor decays continuously. If 100 mg is present initially, the amount present after t years is given by $A(t) = 100e^{-0.0248t}$. Approximate to the nearest tenth of a milligram the amount left after 10 years.
72. **Radioactive Decay** The strontium-90 in a nuclear reactor decays continuously. If 100 mg is present initially, the amount present after t years is given by $A(t) = 100e^{-0.0248t}$. Approximate to the nearest tenth of a milligram the amount left after 50 years.

Group Discussion Questions

73. **Challenge Question** $f(x) = b^x$ is an exponential function if $b > 0$ and $b \neq 1$.
    a. Discuss the nature of the function $f(x) = b^x$ if $b = 0$.
    b. Discuss the nature of the function $f(x) = b^x$ if $b = 1$.
74. **Challenge Question** A question showing the growth potential of geometric growth is given by the classic problem of a blacksmith shoeing a horse. A blacksmith attaches each horseshoe with eight nails. The blacksmith offers to charge by the nail for shoeing all four hooves. The cost for the first nail would be 1 cent; the second, 2 cents; the third, 4 cents; etc. At this rate, complete the following table.

Nail	1	2	4	8	10	16	20	30	32
Cost of nail, \$	0.01	0.02							

75. **Discovery Question** An employee is offered two salary options. Option 1 starts at \$20,000 per year with a \$3,000 raise after each of the first 9 years. Option 2 starts at \$20,000 per year with a 10% raise after each of the first 9 years. Complete the two tables below.

OPTION 1

YEAR	SALARY (\$)	RAISE (\$)	RAISE (%)
1	20,000		
2	23,000	3,000	15
3	26,000	3,000	13
4	29,000		
5			
6			
7			
8			
9			
10			

OPTION 2

YEAR	SALARY (\$)	RAISE (\$)	RAISE (%)
1	20,000		
2	22,000	2,000	10
3	24,200	2,200	10
4	26,620		
5			
6			
7			
8			
9			
10			

Discuss the advantages and disadvantages of each option, using the terms *arithmetic growth* and *geometric growth* where these terms are appropriate. Which option is better for a person who plans to quit this job after 5 years, after 10 years, after 20 years?

Section 10.2 Inverse Functions

Objectives:

3. Write the inverse of a function.
4. Graph the inverse of a function.

A function pairs each input value with exactly one output value.

Loan amount	Function $\longrightarrow$	Monthly payment
Monthly payment	Inverse $\longrightarrow$	Loan amount

A function (see Section 7.1) is a correspondence between two sets of values; a function matches each input value with exactly one output value. Because a function clearly pairs two sets of values, the information available from this pairing can often be used both ways—from x to y and from y to x.

For example, we can write a formula for a function that for a fixed time and interest rate will pair each car loan amount with the required monthly payment. By reversing this function we can use the monthly payment we can afford to determine the loan amount that is affordable. This reversing of a function produces what we will call the inverse of a function.

Inverse of a Function

VERBALLY	ALGEBRAICALLY	NUMERICAL EXAMPLE	GRAPHICAL EXAMPLE
If f is a function that matches each input value x with an output value y, then the inverse of f reverses this correspondence to match this y-value with the x-value.	For each point (x, y) of the function f, (y, x) is a point of the inverse of f.	Function: x | y −2 | −5 −1 | −3 0 | −1 1 | 1 2 | 3 Inverse: x | y −5 | −2 −3 | −1 −1 | 0 1 | 1 3 | 2	Function: Inverse:

In Section 7.1 we examined several representations for functions including mapping notation, ordered pairs, tables, graphs, and function notation. We now examine the inverse of a function by using each of these representations. Note in each case that the inverse undoes what f does.

EXAMPLE 1 Determining the Inverse of a Function

Determine the inverse of each function.

SOLUTION

Function f / *The inverse of f*

(a)

x		y
7	⟶	1
8	⟶	3
9	⟶	5

x		y
1	⟶	7
3	⟶	8
5	⟶	9

In the mapping notation, the inverse of f reverses the arrows and the order of the input-output pairs.

(b)

x	y
4	−1
7	2
−π	5
0	6

x	y
−1	4
2	7
5	−π
6	0

In the table, the inverse of f reverses the x-y pairing.

(c) $\{(0, 1), (2, 7), (-5, 4)\}$

$\{(1, 0), (7, 2), (4, -5)\}$

The inverse of f reverses the order of each (x, y) pair.

(d)

LOAN AMOUNT ($)	MONTHLY PAYMENT ($)
5,000	122.06
10,000	244.13
15,000	366.19
20,000	488.26

MONTHLY PAYMENT ($)	LOAN AMOUNT ($)
122.06	5,000
244.13	10,000
366.19	15,000
488.26	20,000

The given function pairs each loan amount with the payment required to pay off an 8% loan in 4 years.

The inverse pairs a payment with the amount of loan this payment will pay off.

Note that all the samples in Example 1 illustrate that the inverse of f reverses the roles of the domain and range of the function f. In Example 1(c) the domain of $f = \{(0, 1), (2, 7), (-5, 4)\}$ is the set $\{0, 2, -5\}$, and the range is the set $\{1, 7, 4\}$. The domain of the inverse of f, $\{(1, 0), (7, 2), (4, -5)\}$, is the set $\{1, 7, 4\}$, and the range is the set $\{0, 2, -5\}$.

SELF-CHECK 10.2.1

For the function f:

x		y
8	$\longrightarrow$	6
4	$\longrightarrow$	11
−2	$\longrightarrow$	0

1. Give the domain and range of f.
2. Give f^{-1}.
3. Give the domain and range of f^{-1}.

We now investigate the conditions under which the inverse of a function f is also a function. Remember, a function must pair each input value with exactly one output value.

The inverse of the function $f = \{(3, 4), (5, 4)\}$ is given by $\{(4, 3), (4, 5)\}$. In this case the inverse of f does not match each input with exactly one output—4 is paired with both 3 and with 5. Examining $f = \{(3, 4), (5, 4)\}$, we note that f has two input values with an output of 4. This causes the inverse of f to have two outputs for the input of 4. Thus the inverse of f is not a function.

However, if different inputs of f have different outputs, then the inverse of f also will be a function. We call functions with this property **one-to-one functions.**

One-to-One Function

VERBALLY

A function is one-to-one if no two x-values correspond to the same y-value.

That is, different input values produce different output values.

ALGEBRAICALLY

The only way that $f(x_1) = f(x_2)$ is if $x_1 = x_2$.

That is, if $x_1 \neq x_2$, then $f(x_1) \neq f(x_2)$.

NUMERICAL EXAMPLE

Function: $y = x - 1$

x	y
−2	−3
−1	−2
0	−1
1	0
2	1

No two x-values correspond to the same y-value.

GRAPHICAL EXAMPLE

Function: $y = x - 1$

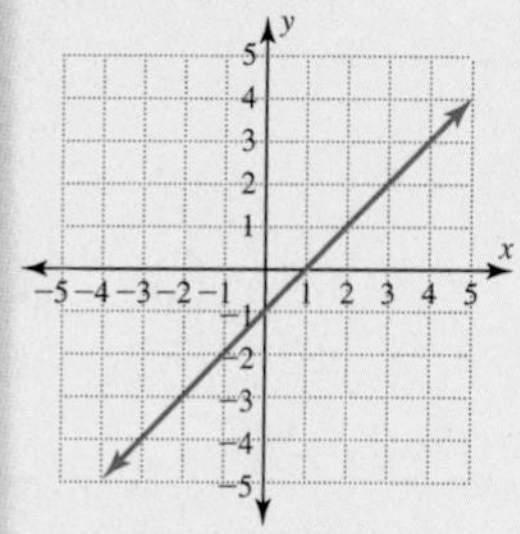

Since this graph is increasing, different input values will produce different output values.

If a function f is one-to-one, then the inverse of f will also be a function. In this case we denote the inverse of f by f^{-1} and call f^{-1} an **inverse function.**

The inverse of a function f is an inverse function if f is one-to-one. We denote this inverse by f^{-1}.

EXAMPLE 2 Identifying One-to-One Functions

Determine whether the function f is one-to-one and whether the inverse of f is an inverse function.

SOLUTION

Caution: Treat f^{-1} as a single symbol that undoes what the function f does. This is *not* exponential notation and does not represent $\frac{1}{f}$. The context of the notation should make the intent of the notation clear.

(a) $f = \{(-5, 6), (2, 7), (8, 9), (9, 3)\}$ This function is one-to-one, and thus $f^{-1} = \{(6, -5), (7, 2), (9, 8), (3, 9)\}$ is an inverse function.

(b) $f = \{(-3, 9), (0, 0), (3, 9)\}$ This function is not one-to-one because both -3 and 3 are paired with 9. Thus the inverse of f, $\{(9, -3), (0, 0), (9, 3)\}$, is not a function.

SELF-CHECK 10.2.2

Determine whether f is one-to-one and whether the inverse of f is an inverse function.

1. $f = \{(3, \pi), (e, 4), (8, -1)\}$ **2.** $f = \{(5, 9), (7, 6), (-3, 9)\}$

The vertical line test covered in Section 7.1 is a quick method for determining whether a graph represents a function. This test allows us to determine visually whether each x input value is paired with exactly one y output value. Similarly, the horizontal line test is a quick method for determining whether the graph of a function represents a one-to-one function.

The Horizontal Line Test

A graph of a function represents a one-to-one function if it is impossible to have any horizontal line intersect the graph at more than one point.

EXAMPLE 3 Using the Horizontal Line Test

Use the horizontal line test to determine whether each graph represents a one-to-one function.

SOLUTION

Use the vertical line test to determine whether a graph represents a function. Use the horizontal line test to determine whether the graph of a function represents a one-to-one function.

(a)

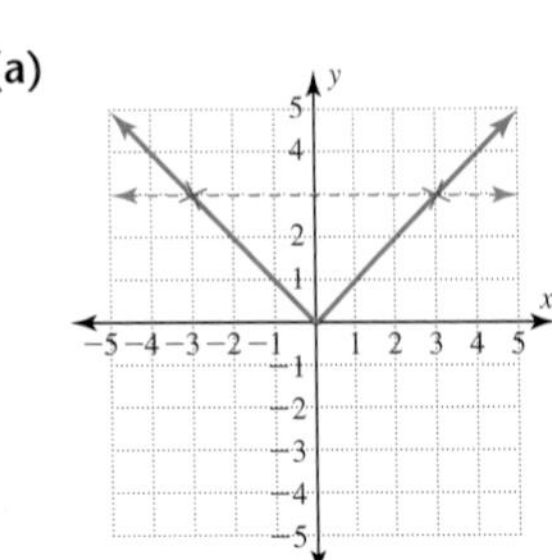

Not a one-to-one function

The horizontal line passes through more than one point of the graph of this function.

(b) A one-to-one function

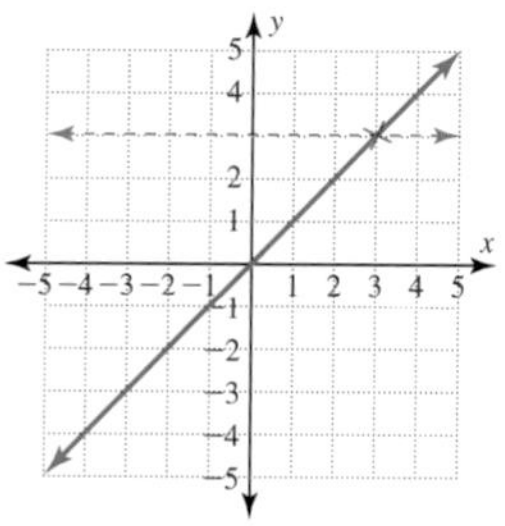

Any horizontal line will pass through exactly one point of the graph of this function.

A key concept to understand about f^{-1} is that f^{-1} undoes what the function f does. This is illustrated in Example 4.

EXAMPLE 4 Describing Inverse Functions

Describe each function and its inverse verbally and algebraically. Give a numerical example from the domain and range of each function.

SOLUTION

	VERBALLY	ALGEBRAICALLY	NUMERICAL EXAMPLE
(a) $f(x) = 2x$	f doubles x	$f(x) = 2x$	$8 \xrightarrow{f} 16$
	f^{-1} takes one-half of x	$f^{-1}(x) = \frac{x}{2}$	$16 \xrightarrow{f^{-1}} 8$
(b) $f(x) = \sqrt[3]{x}$	f takes the cube root of x	$f(x) = \sqrt[3]{x}$	$8 \xrightarrow{f} 2$
	f^{-1} cubes x	$f^{-1}(x) = x^3$	$2 \xrightarrow{f^{-1}} 8$

SELF-CHECK 10.2.3

Use a graphing calculator to graph each function and to determine whether the function is one-to-one.

1. $f(x) = \sqrt{4x - 1}$
2. $f(x) = |x - 1| - 2$

Describe f^{-1} verbally and write both f and f^{-1} by using function notation.

3. f triples x.
4. f increases x by 5.

$f^{-1}(x)$ undoes what f does.

If f is a one-to-one function defined by an equation, then the inverse function f^{-1} is also indirectly given by this equation. This is illustrated in Example 4 where simple functions are used to describe how f^{-1} undoes what f does. Another example of this is as follows:

ALGEBRAICALLY	VERBALLY
$f(x) = 2x + 1$	$f: 5 \longrightarrow$ [First double] $\xrightarrow{10}$ [Then add 1] $\longrightarrow 11$
$f^{-1}(x) = \frac{x-1}{2}$	$f^{-1}: 11 \longrightarrow$ [First subtract 1] $\xrightarrow{10}$ [Then take one-half of this value] $\longrightarrow 5$

It is customary to use x to represent the input value for both f and f^{-1}. Thus the formulas for both f and f^{-1} are often expressed by using the variable x. Because the inverse function f^{-1} reverses the order of each ordered pair (x, y), we interchange the roles of x and y in the formula for f in order to find a formula for f^{-1} in terms of x. The following box describes this procedure and reexamines the function $f(x) = 2x + 1$ given in the display above.

Finding an Equation for an Inverse Function

To find the inverse of a one-to-one function $y = f(x)$:

VERBALLY	ALGEBRAIC EXAMPLE	
Step 1. Replace $f(x)$ by y. Write this as a function of y in terms of x.	Function:	$f(x) = 2x + 1$ $y = 2x + 1$
Step 2. To form the inverse, replace each x with y and each y with x.	Inverse function:	$x = 2y + 1$
Step 3. If possible, solve the resulting equation for y.		$x - 1 = 2y$ $\frac{x-1}{2} = y$
Step 4. The inverse can then be written in function notation by replacing y with $f^{-1}(x)$.		$f^{-1}(x) = \frac{x-1}{2}$

Example 5 illustrates this procedure.

EXAMPLE 5 Determining an Inverse Function

Determine the inverse of $f(x) = 3x - 5$ algebraically and compare f and f^{-1} both verbally and graphically.

SOLUTION

ALGEBRAICALLY

Function: $f(x) = 3x - 5$ — First replace $f(x)$ by y to express y as a function of x.

$y = 3x - 5$

Inverse function: $x = 3y - 5$ — Then interchange x and y in the equation to form the inverse.

$x + 5 = 3y$ — Solve for y by adding 5 to both sides of the equation and then dividing both sides by 3.

$y = \frac{x + 5}{3}$

$f^{-1}(x) = \frac{x + 5}{3}$ — Rewrite the inverse by using functional notation.

VERBALLY

$f(x) = 3x - 5 \quad f: x \rightarrow$ [First triple x] $\xrightarrow{3x}$ [Then subtract 5] $\rightarrow 3x - 5$

$f^{-1}(x) = \frac{x+5}{3} \quad f^{-1}: x \rightarrow$ [First add 5] $\xrightarrow{x+5}$ [Then divide by 3] $\rightarrow \frac{x+5}{3}$

NUMERICAL EXAMPLE

$f(2) = 1$: $2 \rightarrow$ [First triple 2] $\xrightarrow{6}$ [Then subtract 5] $\rightarrow 1$

$f^{-1}(1) = 2$: $1 \rightarrow$ [First add 5] $\xrightarrow{6}$ [Then divide by 3] $\rightarrow 2$

GRAPHICALLY

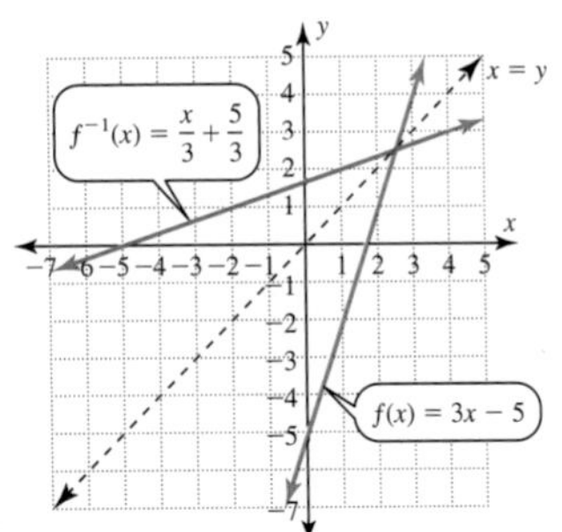

The graphs of both f and f^{-1} are straight lines. Inspection of the slope-intercept form reveals that f has slope 3 and y-intercept -5 and f^{-1} has slope $\frac{1}{3}$ and y-intercept $\frac{5}{3}$. Note that these graphs are symmetric about the line $x = y$.

SELF-CHECK 10.2.4

1. Determine the inverse of $f(x) = \dfrac{2x - 1}{3}$.

Symmetry of (a, b) and (b, a) about $x = y$

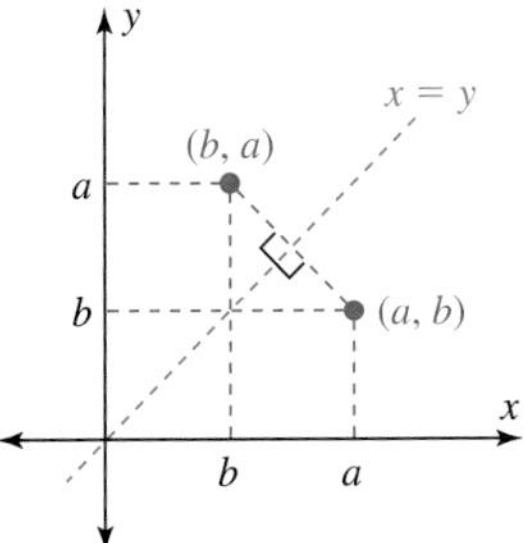

As illustrated by Example 5, the graphs of f and f^{-1} are symmetric about the line $x = y$. In the figure to the left note that the points (a, b) and (b, a) are mirror images about the graph of $x = y$. Thus f and f^{-1} will always be symmetric about the line $x = y$. Example 6 further illustrates this symmetry.

EXAMPLE 6 Graphing the Inverse of a Function

The graph of the function $f = \{(2, 1), (4, 2), (6, 3)\}$ is shown to the right. Graph the inverse of this function.

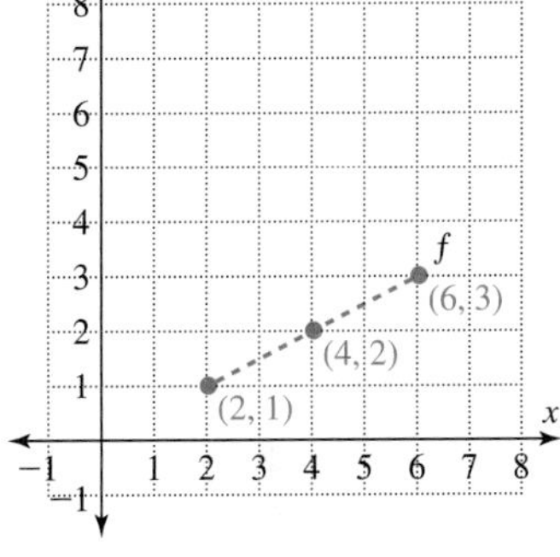

SOLUTION

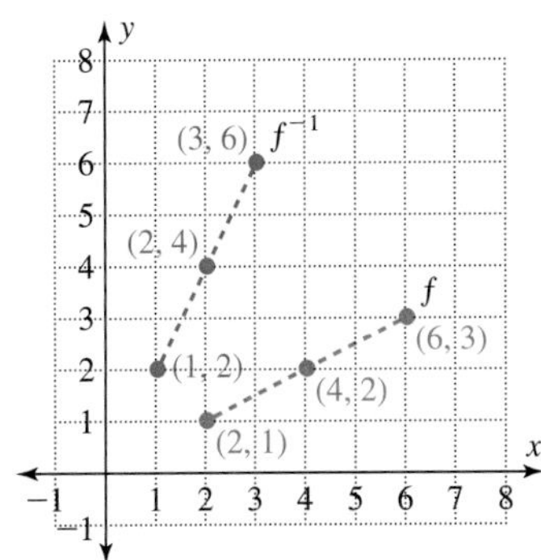

The inverse of $f = \{(2, 1), (4, 2), (6, 3)\}$ is $f^{-1} = \{(1, 2), (2, 4), (3, 6)\}$. In fact, the inverse of all the points on the dashed line segment through the points of f is the set of all the points on the dashed line segment through the points of f^{-1}. You can check some of these points yourself.

We can take advantage of the symmetry of f and f^{-1} about the line $x = y$. Using this symmetry, we can graph f^{-1} directly from the graph of f without knowing the equation for f^{-1}. This is illustrated in Example 7.

EXAMPLE 7 Using Symmetry to Graph f^{-1}

Graph $f(x) = 2^x$ and use symmetry about the line $x = y$ to graph $y = f^{-1}(x)$.

SOLUTION

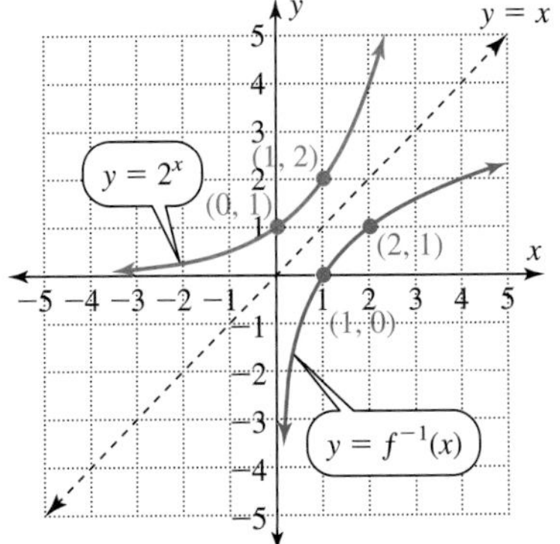

First graph the exponential function $f(x) = 2^x$ and sketch in the dashed line $x = y$. Then use symmetry to create a symmetric image for $f(x) = 2^x$ about $x = y$. This symmetric image is the graph of $y = f^{-1}(x)$.

Note that the symmetric image of $(0, 1)$ is $(1, 0)$ and of $(1, 2)$ is $(2, 1)$.

Some graphing calculators have the ability to draw the inverse of a function. This is illustrated in the Calculator Perspective 10.2.1.

CALCULATOR PERSPECTIVE 10.2.1 Drawing the Inverse of a Function

To draw the inverse of the function $f(x) = 3x - 5$ from Example 5 on a TI-84 Plus calculator, enter the following keystrokes:

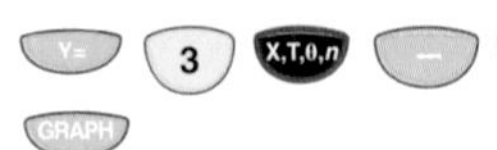

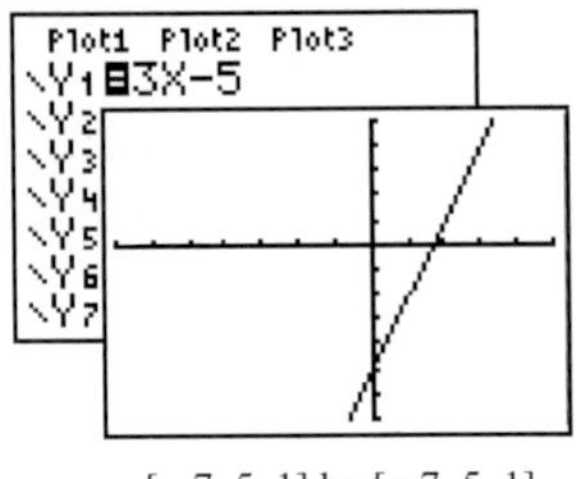

[−7, 5, 1] by [−7, 5, 1]

The DRAW feature is the secondary function of the PRGM key.

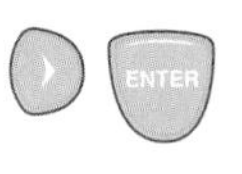

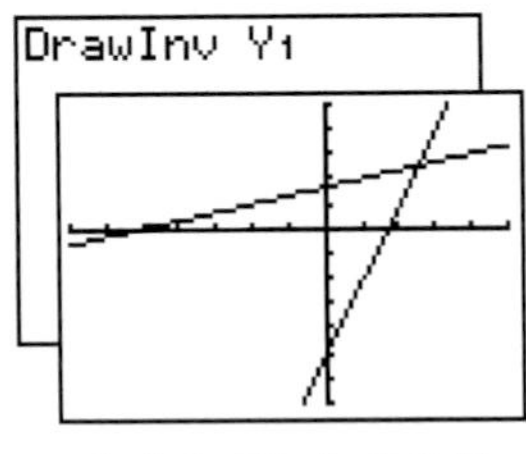

[−7, 5, 1] by [−7, 5, 1]

SELF-CHECK 10.2.5

1. Use a graphing calculator to check Example 7. First graph $f(x) = 2^x$. Then use **DrawInv** to graph $y = f^{-1}(x)$.

SELF-CHECK ANSWERS

10.2.1

1. $D = \{8, 4, -2\}$, $R = \{6, 11, 0\}$
2.

x		y
6	⟶	8
11	⟶	4
0	⟶	−2

3. $D = \{6, 11, 0\}$, $R = \{8, 4, -2\}$

10.2.2

1. f is one-to-one; the inverse of f, $\{(\pi, 3), (4, e), (-1, 8)\}$, is an inverse function.
2. f is not one-to-one; the inverse of f, $\{(9, 5), (6, 7), (9, -3)\}$, is the inverse of a function, but it is not an inverse function.

10.2.3

1. A one-to-one function
2. Not a one-to-one function
3. f^{-1} takes one-third of x; $f(x) = 3x$; $f^{-1}(x) = \frac{x}{3}$
4. f^{-1} decreases x by 5; $f(x) = x + 5$; $f^{-1}(x) = x - 5$

10.2.4

1. $f^{-1}(x) = \frac{3x + 1}{2}$

10.2.5

1.

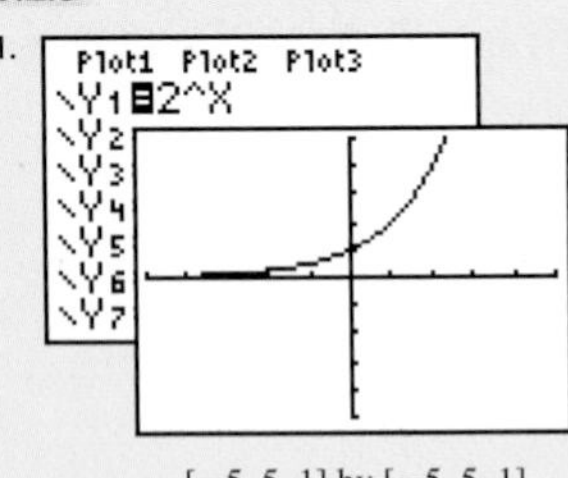

[−5, 5, 1] by [−5, 5, 1]

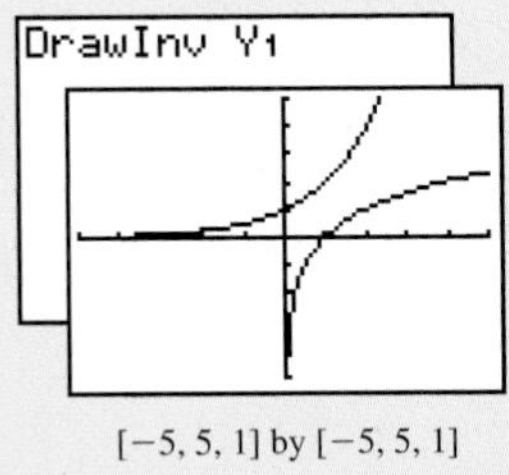

[−5, 5, 1] by [−5, 5, 1]

USING THE LANGUAGE AND SYMBOLISM OF MATHEMATICS 10.2

1. If f is a function that matches an input value x with an output value y, then the inverse of f reverses this correspondence to match this ___________-value with the ___________-value.
2. A function is one-to-one if different input values produce ___________ output values.
3. A graph of a function represents a ___________-___________-___________ function if it is impossible to have any horizontal line intersect the graph at more than one point.
4. If f is a one-to-one function, then the inverse of f is a function denoted by ___________.
5. The graphs of f and f^{-1} are symmetric about the line ___________.

QUICK REVIEW 10.2

The purpose of this quick review is to help you recall skills needed in this section.

In Exercises 1–3, determine whether the graph of each relation represents a function.

1.

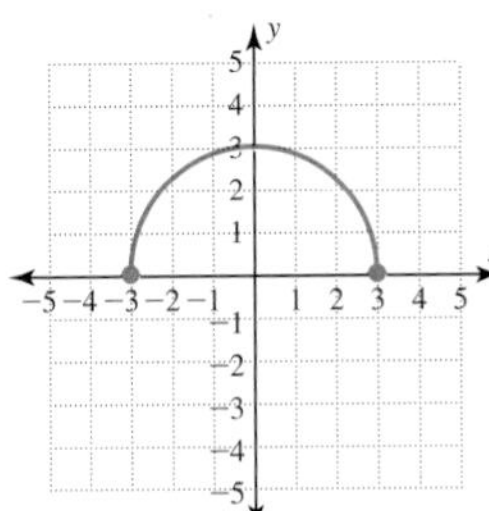

2.

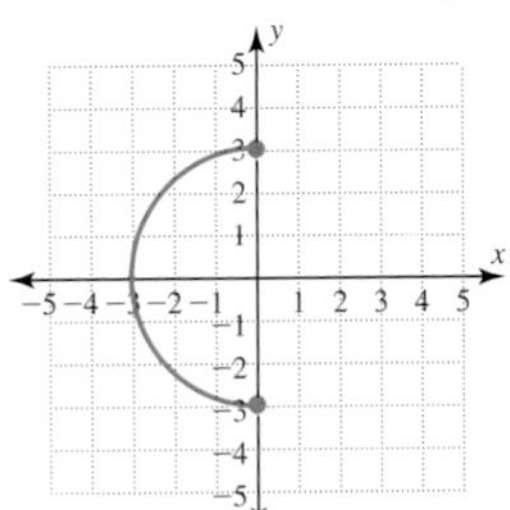

3.

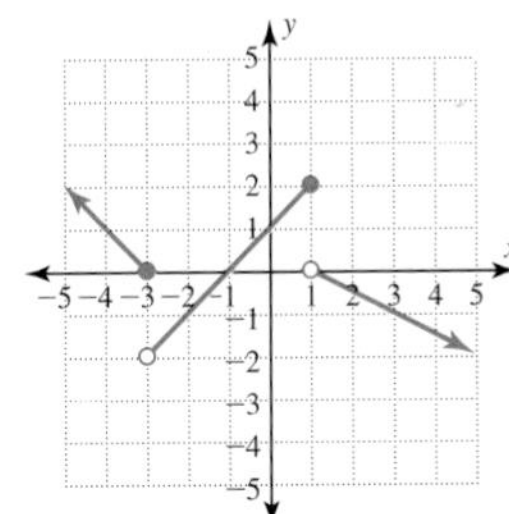

In Exercises 4 and 5, use symmetry to complete the graph of each parabola.

4.

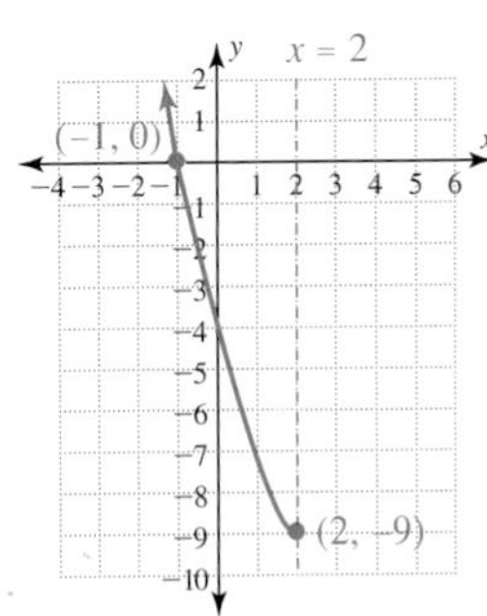

5.

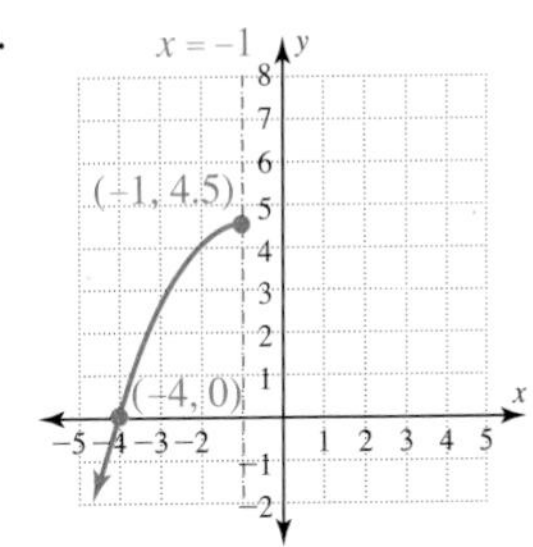

EXERCISES 10.2

In Exercises 1–10, write the inverse of each function by using ordered-pair notation.

1. $\{(1, 4), (3, 11), (8, 2)\}$
2. $\{(2, -1), (0, 3), (-4, 6)\}$
3. $\{(-3, 2), (-1, 2), (0, 2), (2, 2)\}$
4. $\{(-0.7, \pi), (-0.5, \pi), (1.3, \pi)\}$
5. $\{(a, b), (c, d)\}$
6. $\{(w, x), (y, z)\}$
7.

x		y
4	→	7
e	→	9
−3	→	−2

8.

x		y
−8	→	1
−9	→	3
6	→	−4

9.

x	y
9	0
4	3
6	−2
−4	5

10.

x	y
−2	3
−1	5
0	7
1	9

In Exercises 11 and 12, first express the function f whose graph is given in ordered-pair notation. Then write f^{-1} by using ordered-pair notation.

11.

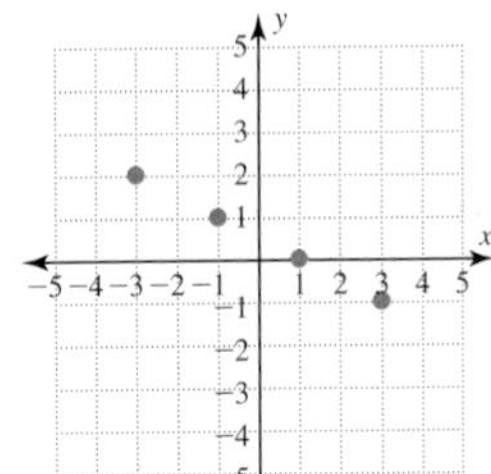

12.

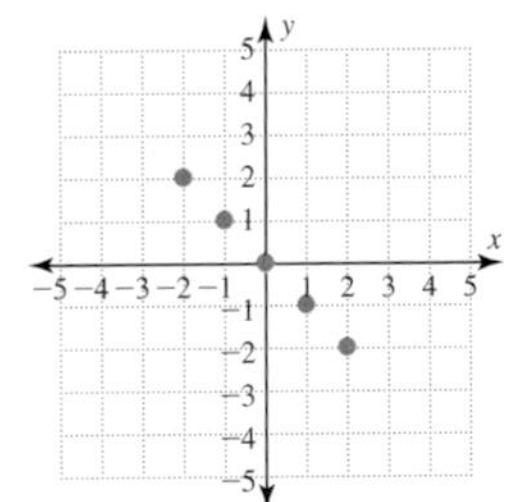

In Exercises 13–16, state the domain and range of the given function f and its inverse f^{-1}.

13.

x		y
0	$\to$	−3
1	$\to$	−2
2	$\to$	−1
3	$\to$	0
4	$\to$	1

14.

x		y
−1	$\to$	−6
0	$\to$	−4
1	$\to$	−2
2	$\to$	0
3	$\to$	2

15.

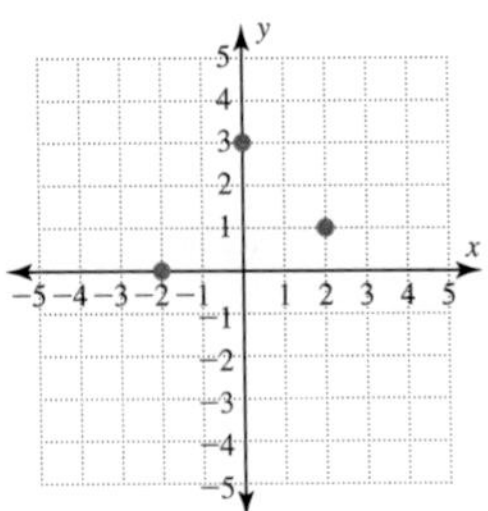

16.

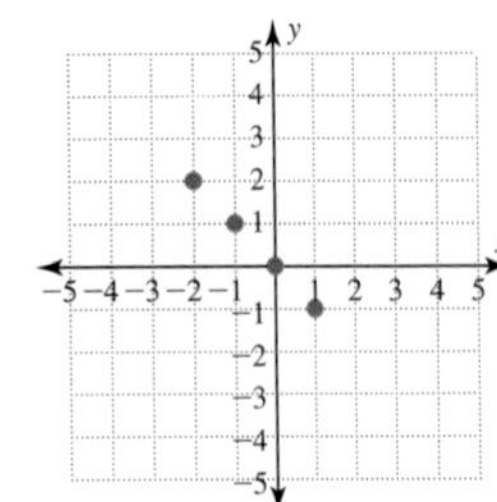

In Exercises 17–20, determine whether each function is one-to-one.

17. $\{(-\pi, 4), (3, \pi), (7, -9)\}$

18. $\{(-8, 9), (7, 13), (8, 9)\}$

19. $\{(7.5, 7), (7.8, 7), (8.3, 8)\}$

20. $\{(-5, 2), (-4, -1), (3, 7), (8, 8)\}$

In Exercises 21–30, use the horizontal line test to determine whether each graph represents a one-to-one function.

21.

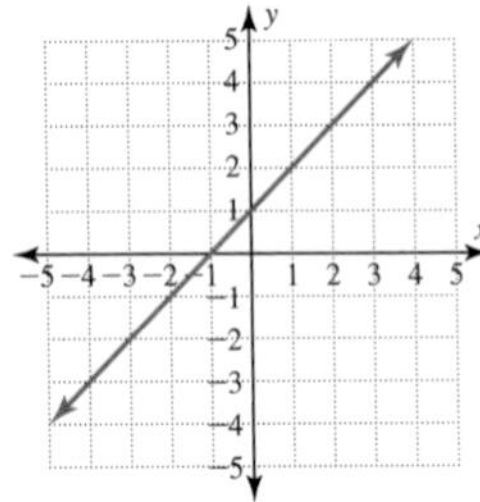

22.

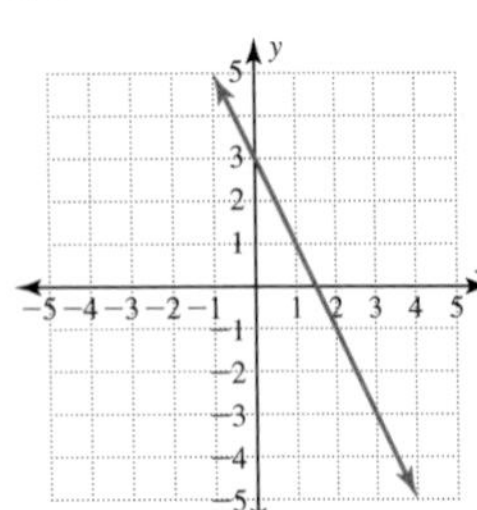

23.

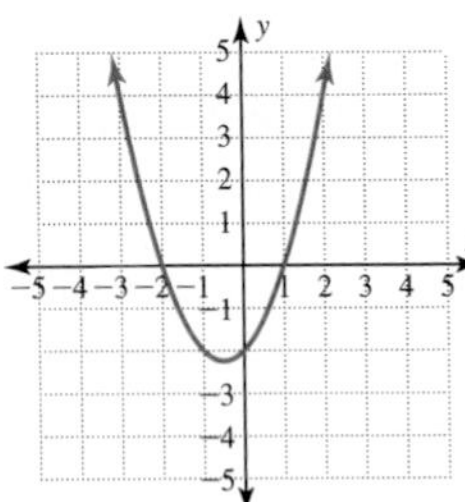

24.

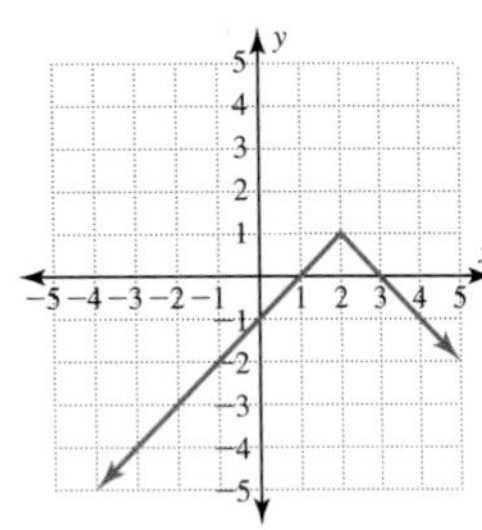

25.

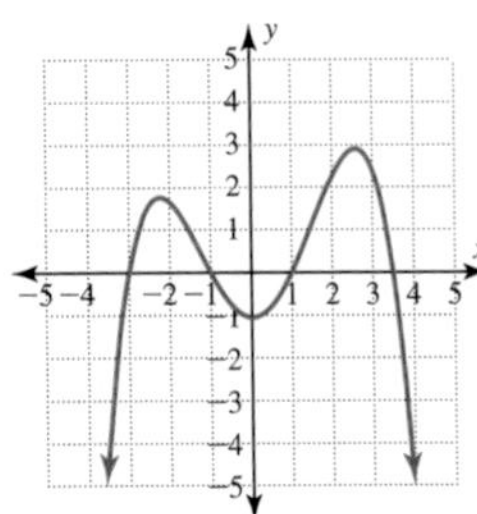

26.

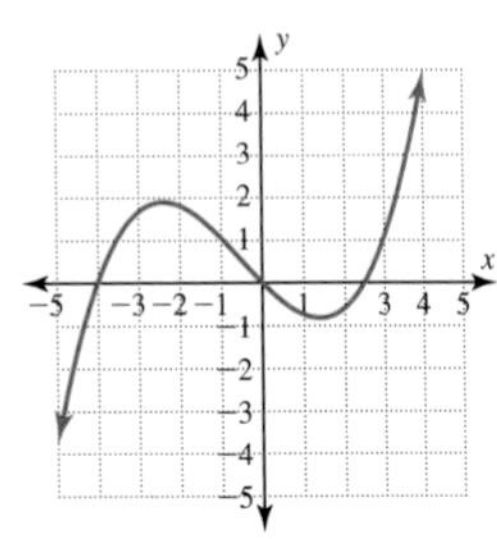

27.

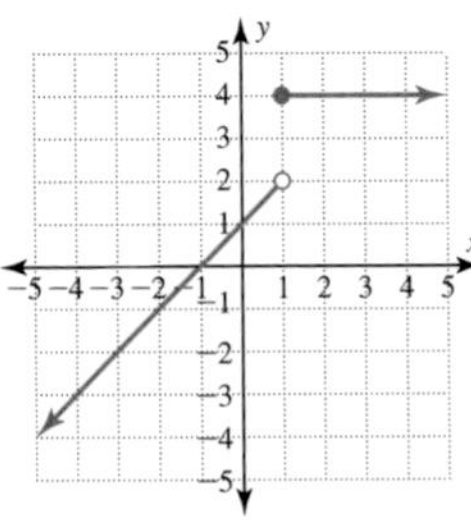

28.

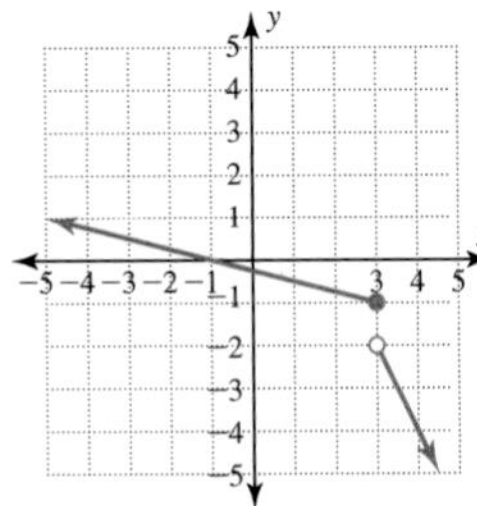

29.

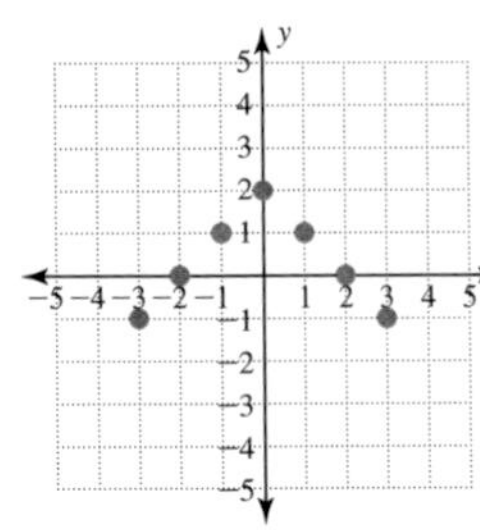

30.

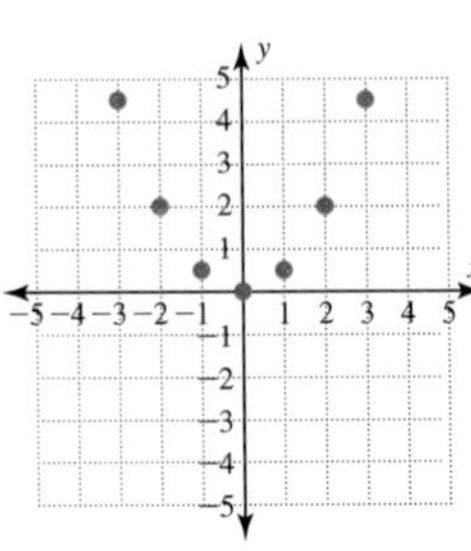

In Exercises 31–40, sketch the inverse of each function by using the symmetry of f and f^{-1} about the line $x = y$. (*Hint:* For Exercises 33–40 pick a few key points on f and f^{-1} and then sketch the rest of f^{-1}.)

31.

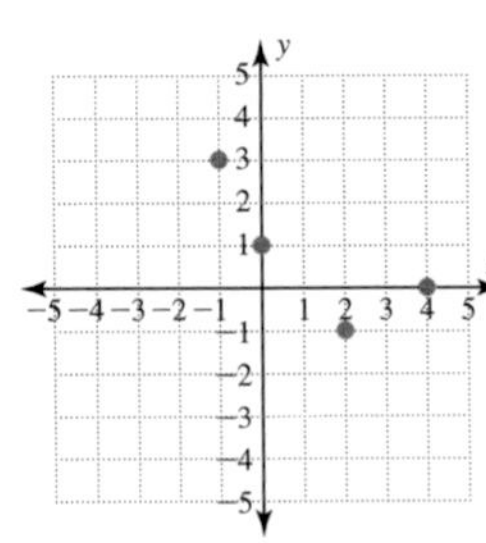

32.

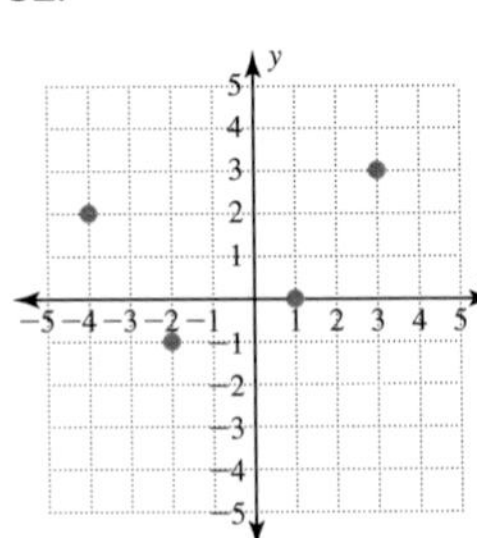

33.

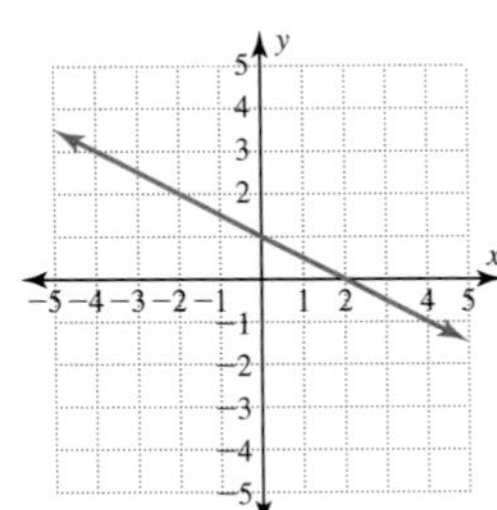

34.

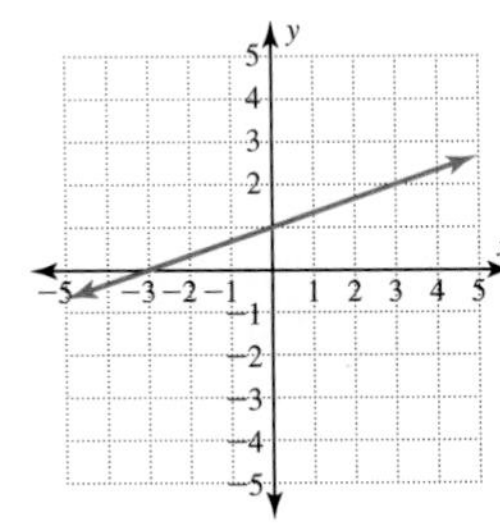

35.

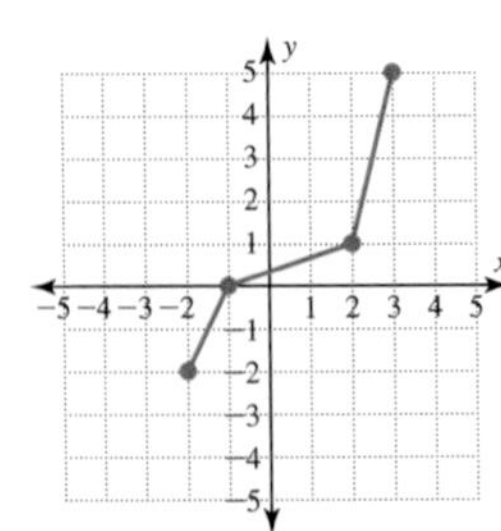

36.

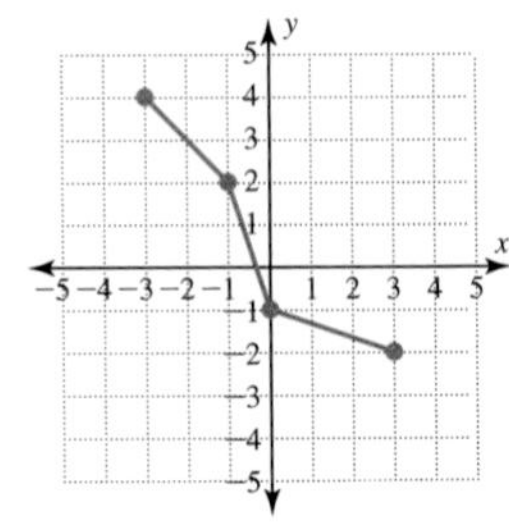

37.

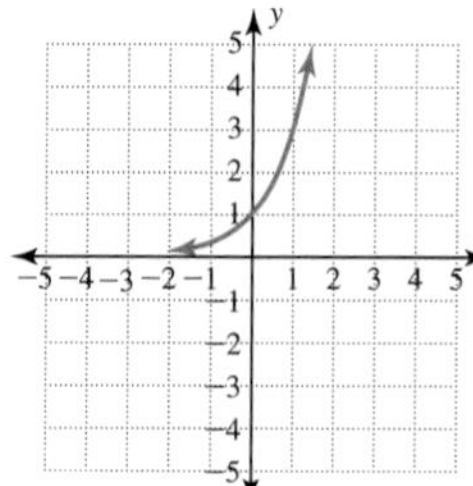

38.

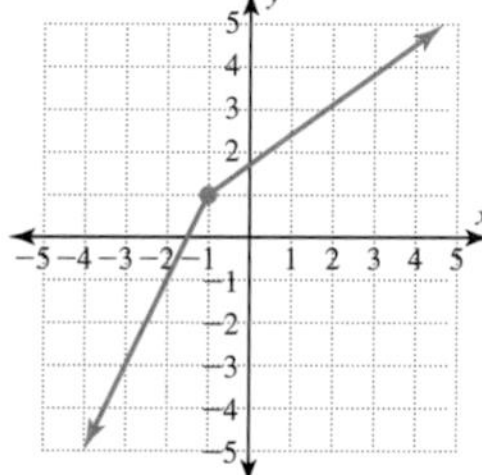

39.

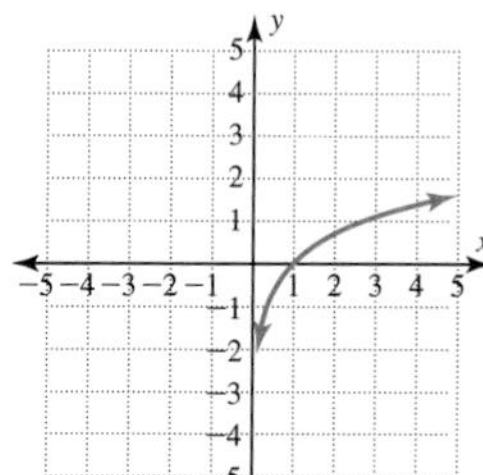

40.

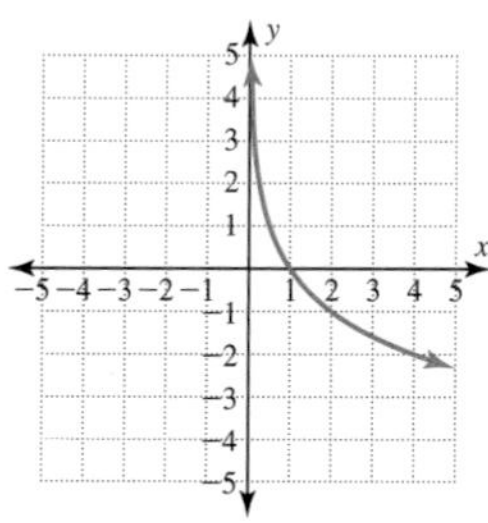

In Exercises 41–44, use a graphing calculator with the **DrawInv** feature to graph $y = f(x)$ and to draw $f^{-1}(x)$.

41. $y = \frac{1}{3}x + 1$ **42.** $y = 3x - 3$

43. $y = \left(\frac{1}{2}\right)^x$ **44.** $y = 3^x$

In Exercises 45–50, the function f is described verbally and algebraically. Describe f^{-1} both verbally and algebraically.

VERBALLY	ALGEBRAICALLY
45. f decreases x by 2	$f(x) = x - 2$
46. f increases x by 7	$f(x) = x + 7$
47. f quadruples x	$f(x) = 4x$
48. f takes one-fifth of x	$f(x) = \frac{x}{5}$
49. f doubles x and then subtracts 3	$f(x) = 2x - 3$
50. f takes one-half of x and then adds 5	$f(x) = \frac{1}{2}x + 5$

In Exercises 51–58, write the inverse of each function by using function notation.

51. $f(x) = 5x + 2$ **52.** $f(x) = 2x - 5$

53. $f(x) = \frac{1}{3}x - 7$ **54.** $f(x) = -\frac{1}{4}x + 3$

55. $f(x) = \sqrt[3]{x} + 2$ **56.** $f(x) = \sqrt[3]{x + 2}$

57. $f(x) = -x$ **58.** $f(x) = \frac{1}{x}$

In Exercises 59–64, use $f(x) = 4x - 3$ and $f^{-1}(x) = \frac{x + 3}{4}$ to evaluate each expression.

59. a. $f(5)$ **b.** $f^{-1}(17)$

60. a. $f(0)$ **b.** $f^{-1}(-3)$

61. a. $f(2)$ **b.** $f^{-1}(5)$

62. a. $f\left(\frac{3}{4}\right)$ **b.** $f^{-1}(0)$

63. a. $f(1)$ **b.** $f^{-1}(1)$

64. a. $f(10)$ **b.** $f^{-1}(37)$

In Exercises 65–70, each table defines a function. Form a table that defines the inverse of each function. State conditions for which you think the inverse function might be more useful than the original function.

65. Shoe Sizes

U.S. SIZE	EUROPEAN SIZE
7	39
8	40.5
9	42
10	43
11	44

66. Environmental Application

YEAR	U.S. LAND USED FOR RECREATION (MILLIONS OF ACRES)
1977	66
1982	71
1987	84
1992	87
1997	96

Source: U.S. Department of Agriculture. For more information on this topic, go to the Additional Resources link on the MathZone website at **www.mhhe.com/hallmercer.**

67. Car Loan at 7.5% for 4 Years

LOAN AMOUNT ($)	MONTHLY PAYMENT ($)
5,000	120.89
10,000	241.79
15,000	362.68
20,000	483.58
25,000	604.47

68. Car Loan at 5% for 4 Years

LOAN AMOUNT ($)	MONTHLY PAYMENT ($)
5,000	115.15
10,000	230.29
15,000	345.44
20,000	460.59
25,000	575.73

69. Temperature Conversions

TEMPERATURE °F	TEMPERATURE °C
0	−17.8
20	−6.7
40	4.4
60	15.6
80	26.7
100	37.8

70. Tuition at Gulf Coast Community College

SEMESTER HOURS	TUITION ($)
1	50
2	100
3	150
4	200
5	250
10	500
15	800

71. **Production Cost** The cost of producing x units of a product is given by the function $C(x) = 12x + 350$.
    a. What does the variable x represent in $C(x)$?
    b. What does the output of $C(x)$ represent?
    c. Determine the formula for $C^{-1}(x)$.
    d. What does the variable x represent in $C^{-1}(x)$?
    e. What does the output of $C^{-1}(x)$ represent?
    f. Determine the cost of producing 100 units.
    g. Determine the number of units that can be produced for \$1,934.

72. **Sales Commission** A car dealership hires new salespeople and calculates the monthly salary for each person for each of the first 6 months by using the formula $S(x) = 500x + 1{,}000$, where x represents the number of cars this person sold for the month.
    a. Evaluate and interpret $S(0)$.
    b. Evaluate and interpret $S(8)$.
    c. Determine the formula for $S^{-1}(x)$.
    d. What does the variable x represent in $S^{-1}(x)$?
    e. Evaluate and interpret $S^{-1}(2{,}500)$.
    f. Evaluate and interpret $S^{-1}(6{,}000)$.

Group Discussion Questions

73. **Discovery Question**
    a. Make two identical copies of the graph of a function $y = f(x)$ on two sheets of clear plastic. Also, dash in the line $x = y$ on each sheet of plastic.
    b. Place one sheet of plastic on an overhead projector.
    c. Then flip the second sheet of plastic over and turn the sheet 90°, so that the upside-down sheet has its positive y-axis align with the positive x-axis on the other sheet.
    d. Try this for three or four functions, and compare these results to using the **DrawInv** feature on a graphics calculator.
    e. Describe to your teacher what you have observed and why you think this is so.

74. **Challenge Question**
    a. Is $f(x) = x^2$ a one-to-one function?
    b. If we restrict $f(x) = x^2$ to the domain $D = [0, \infty)$, is this restricted function one-to-one?
    c. Graph the restricted function in part **b** and its inverse.
    d. Write the equation of the inverse of the restricted function in part **b.**
    e. If we restrict $f(x) = x^2$ to the domain $D = (-\infty, 0]$, is this restricted function one-to-one?
    f. Graph the restricted function in part **e** and its inverse.
    g. Write the equation of the inverse of the restricted function in part **e.**

75. **Discovery Question**
    a. Use a graphing calculator to graph $Y_1 = e^x$.
    b. Use a graphing calculator to graph $Y_2 = \ln x$. (*Hint:* Use the LN key.)
    c. Compare these two graphs and make a conjecture on the relationship between the functions $Y_1 = e^x$ and $Y_2 = \ln x$.

Section 10.3 Logarithmic Functions

Objectives:

5. Interpret and use logarithmic notation.
6. Simplify logarithmic expressions by using the basic identities.

The following graphs of exponential functions illustrate that an exponential function $f(x) = b^x$ is a one-to-one function for both exponential growth and exponential decay. Thus each exponential function has an inverse function.

Also note that the projection of the graph of an exponential function onto the x-axis is the entire real axis. Thus the domain of $f(x) = b^x$ is $\mathbb{R}$, the set of all real numbers. The projection of each graph onto the y-axis is the interval $(0, \infty)$. Thus the range of $f(x) = b^x$ is $(0, \infty)$.

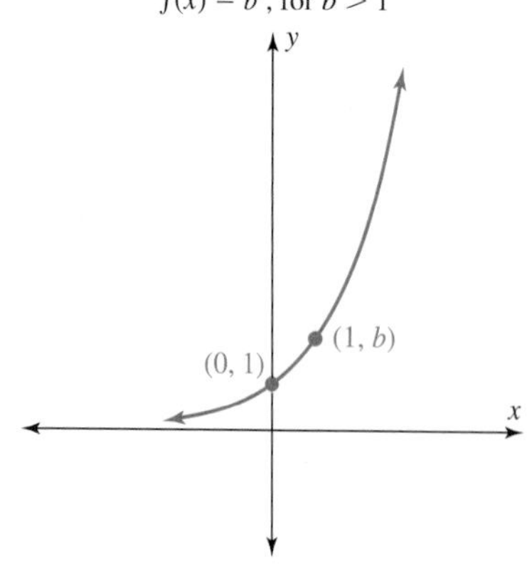

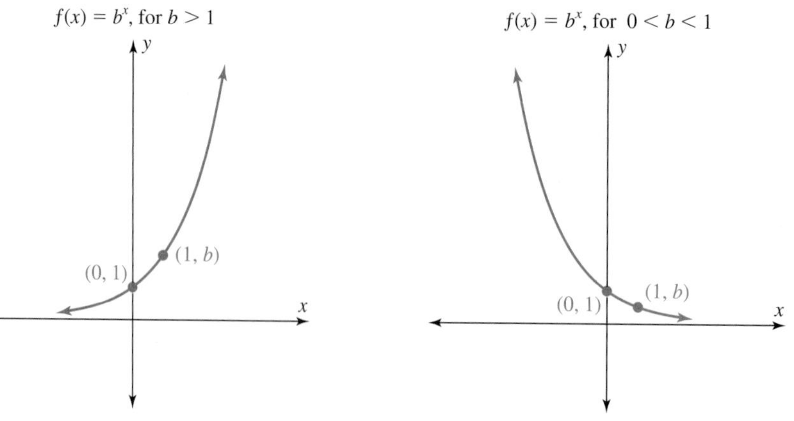

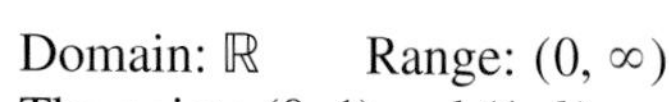

The domain of an exponential function $f(x) = b^x$ is $\mathbb{R}$, and the range is $(0, \infty)$.

Domain: $\mathbb{R}$ Range: $(0, \infty)$

The points $(0, 1)$ and $(1, b)$ are on the graph of $f(x) = b^x$.

The symmetry of f and f^{-1} about the line $x = y$ is used to sketch the inverse of $f(x) = b^x$ in the following graphs. Note that the domain of each inverse function is $(0, \infty)$ and the range is $\mathbb{R}$.

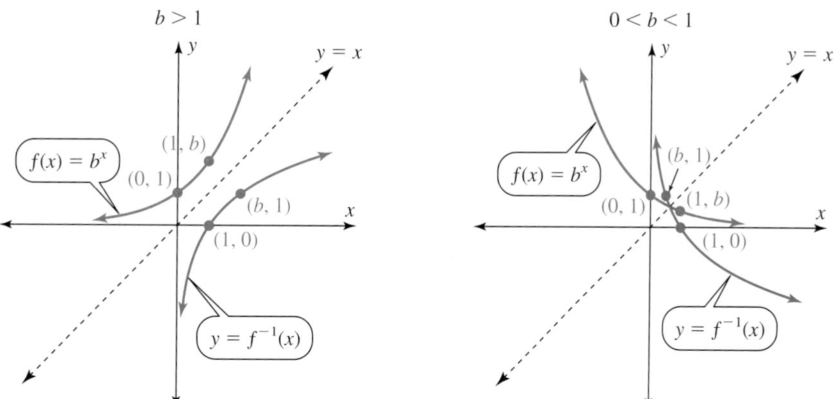

The domain of the inverse of $f(x) = b^x$ is $(0, \infty)$, and the range is $\mathbb{R}$.

Domain of $y = f^{-1}(x)$: $(0, \infty)$ Range: $\mathbb{R}$
The points $(1, 0)$ and $(b, 1)$ are on the graph of $y = f^{-1}(x)$.

The inverse of $f(x) = b^x$ cannot be rewritten by using the basic operations to solve $x = b^y$ for y. Thus we introduce a new notation to denote the inverse of $f(x) = b^x$. This inverse is denoted by $f^{-1}(x) = \log_b x$, which is read "the log base b of x." Logarithmic functions are defined in the following box.

Logarithmic Function $f(x) = \log_b x$

ALGEBRAICALLY	VERBALLY	ALGEBRAIC EXAMPLE	GRAPHICAL EXAMPLE
For $x > 0$, $b > 0$, and $b \neq 1$, $y = \log_b x$ if and only if $b^y = x$.	$y = \log_b x$ is read "y equals the log base b of x."	$y = \log_2 x$	$y = \log_2 x$

A function and its inverse both contain the same information but from different perspectives—the ordered pairs in the original function are reversed to obtain the ordered pairs in the inverse. As you work with $y = b^x$ and $y = \log_b x$, remember that these functions are inverses of each other and contain the same information.

**Think:* Logarithms are exponents.

Exponential Form **Logarithmic Form**

$b^y = x$ $y = \log_b x$

Exponent or logarithm*

Base

The relationship between exponential and logarithmic forms is examined further in Examples 1 and 2.

A Mathematical Note

Logarithms were invented by John Napier (1550–1617), a Scottish mathematician. His first publication on logarithms was published in 1614.

Napier's other inventions include "Napier's Bones," a mechanical calculating device that is an ancestor of the slide rule, and a hydraulic screw and revolving axle for controlling the water level in coal pits. He also worked on plans to use mirrors to burn and destroy enemy ships.

EXAMPLE 1 Translating Logarithmic Equations to Exponential Form

Write each logarithmic equation in verbal form, and translate this equation to the equivalent exponential form.

SOLUTION

LOGARITHMIC FORM	VERBALLY	EXPONENTIAL FORM
(a) $\log_5 25 = 2$	The log base 5 of 25 is 2.	$5^2 = 25$
(b) $\log_2 8 = 3$	The log base 2 of 8 is 3.	$2^3 = 8$
(c) $\log_3\left(\frac{1}{3}\right) = -1$	The log base 3 of $\frac{1}{3}$ is -1.	$3^{-1} = \frac{1}{3}$
(d) $\log_7 \sqrt{7} = \frac{1}{2}$	The log base 7 of $\sqrt{7}$ is $\frac{1}{2}$.	$7^{1/2} = \sqrt{7}$

EXAMPLE 2 Translating Exponential Equations to Logarithmic Form

Translate each exponential equation to logarithmic form.

SOLUTION

EXPONENTIAL FORM	LOGARITHMIC FORM
(a) $10^{-3} = 0.001$	$\log_{10} 0.001 = -3$
(b) $e^2 \approx 7.389$	$\log_e 7.389 \approx 2$
(c) $A = (1 + r)^x$	$\log_{(1+r)} A = x$

SELF-CHECK 10.3.1

Write each logarithmic equation in exponential form and each exponential equation in logarithmic form.

1. $\log_7 49 = 2$
2. $\log_6 \sqrt[5]{6} = \frac{1}{5}$
3. $3^{-4} = \frac{1}{81}$
4. $6^0 = 1$

To evaluate $\log_b x$, think, "What exponent on b is needed to obtain x?" Problems involving familiar powers of small integers can often be solved easily without the use of a calculator. For problems involving more complicated values, a calculator is generally used.

EXAMPLE 3 Evaluating Logarithms by Inspection

Determine the value of each logarithm by inspection.

SOLUTION

(a) $\log_5 125$	$\log_5 125 = 3$	In exponential form, $5^3 = 125$.
(b) $\log_2 \frac{1}{8}$	$\log_2 \frac{1}{8} = -3$	In exponential form, $2^{-3} = \frac{1}{8}$.
(c) $\log_{49} 7$	$\log_{49} 7 = \frac{1}{2}$	In exponential form, $49^{1/2} = 7$.

(d) $\log_5 0$	$\log_5 0$ is undefined.	The input of a logarithmic function cannot be 0.
(e) $\log_5 (-25)$	$\log_5 (-25)$ is undefined.	The input of a logarithmic function cannot be negative.

Four properties that result from the definition of a logarithm are given in the following box. These important properties are used often in solving the logarithmic equations that arise in the solution of some word problems.

Properties of Logarithmic Functions

For $b > 0$ and $b \neq 1$:

LOGARITHMIC FORM	EXPONENTIAL FORM	NUMERICAL EXAMPLE	
1. $\log_b 1 = 0$	$b^0 = 1$	$\log_5 1 = 0$	$5^0 = 1$
2. $\log_b b = 1$	$b^1 = b$	$\log_5 5 = 1$	$5^1 = 5$
3. $\log_b \frac{1}{b} = -1$	$b^{-1} = \frac{1}{b}$	$\log_5 \frac{1}{5} = -1$	$5^{-1} = \frac{1}{5}$
4. $\log_b b^x = x$	$b^x = b^x$	$\log_5 5^4 = 4$	$5^4 = 5^4$

Example 4 uses these properties to evaluate logarithms. Other properties of logarithms are examined in the following sections of this chapter.

EXAMPLE 4 Using Properties to Evaluate Logarithms

Determine the value of each logarithm by inspection.

SOLUTION

(a) $\log_{19} 1$	$\log_{19} 1 = 0$	In exponential form $19^0 = 1$.
(b) $\log_7 7$	$\log_7 7 = 1$	In exponential form $7^1 = 7$.
(c) $\log_8 \frac{1}{8}$	$\log_8 \frac{1}{8} = -1$	In exponential form $8^{-1} = \frac{1}{8}$.
(d) $\log_8 8^{2y}$	$\log_8 8^{2y} = 2y$	In exponential form $8^{2y} = 8^{2y}$.

SELF-CHECK 10.3.2

Determine the value of each logarithm.

1. $\log_2 64$
2. $\log_4 64$
3. $\log_8 64$
4. $\log_7 \frac{1}{7}$
5. $\log_e 1$
6. $\log_e e$
7. $\log_e \frac{1}{e}$
8. $\log_e e^8$

The equations in Example 5 can be solved without use of a calculator because the numbers involve well-known powers. We examine equations of this type in greater detail in Section 10.6.

EXAMPLE 5 Solving Equations

Solve each equation for x.

SOLUTION

(a) $\log_3(x - 1) = 2$

$\log_3(x - 1) = 2$

$x - 1 = 3^2$ First rewrite this equation in exponential form.

$x - 1 = 9$ Substitute 9 for 3^2, and then solve for x.

$x = 10$

(b) $\log_x 9 = -2$

$\log_x 9 = -2$

$x^{-2} = 9$ First rewrite this equation in exponential form.

$x^{-2} = 3^2$ Now work toward expressing each side in terms of the same exponent.

$x^{-2} = \left(\frac{1}{3}\right)^{-2}$

$x = \frac{1}{3}$ The bases are equal since the exponents are equal.

(c) $\log_4 8 = x$

$\log_4 8 = x$

$4^x = 8$ First rewrite this equation in exponential form.

$(2^2)^x = 2^3$ Then express each side in terms of the common base of 2. Use the power rule for exponents to rewrite the left side of the equation.

$2^{2x} = 2^3$

$2x = 3$ Then equate the exponents since the bases are the same.

$x = \frac{3}{2}$

SELF-CHECK 10.3.3

Solve each equation for x.

1. $\log_2 x = 3$
2. $\log_{11} \sqrt{11} = x$
3. $\log_x 125 = 3$

Linear functions with a positive slope exhibit a constant rate of growth. By contrast, exponential growth functions exhibit a rate of growth that eventually becomes extremely rapid. On the other extreme, logarithmic growth functions eventually exhibit extremely slow growth. Example 6 uses the fact that $f(x) = \log_e x$ is the inverse of $f(x) = e^x$ and illustrates how to use a graphing calculator and the **DrawInv** feature to graph both of these growth functions. In Section 10.4 we illustrate how to access logarithmic functions directly on a graphing calculator by using the LN and LOG keys.

EXAMPLE 6 Graphing a Logarithmic Growth Function

To graph $f(x) = \log_e x$, first use a graphing calculator to graph the corresponding exponential function $y = e^x$. Then use the **DrawInv** feature to sketch the graph of $f(x) = \log_e x$.

SOLUTION

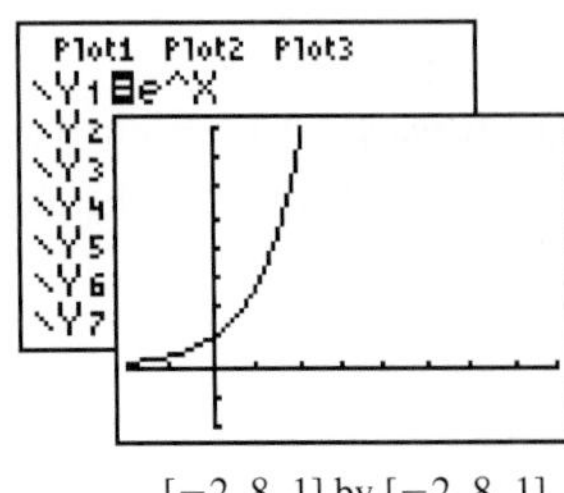

[−2, 8, 1] by [−2, 8, 1]

Because $f(x) = \log_e x$ and $Y_1 = e^x$ are inverses of each other, you can use the graph of Y_1 and **DrawInv** to sketch the graph of $f(x) = \log_e x$.

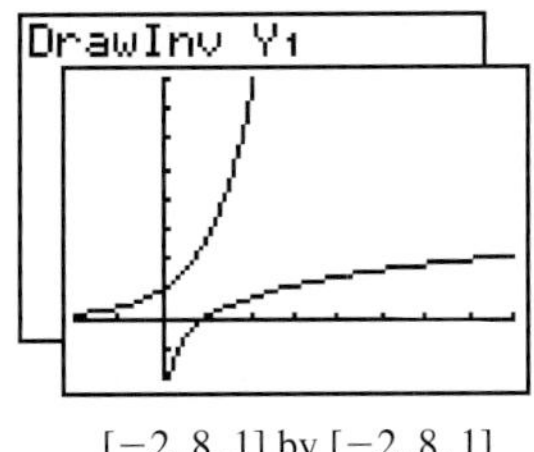

[−2, 8, 1] by [−2, 8, 1]

For larger values of x, note the extremely slow growth rate for $f(x) = \log_e x$.

The following graphs compare exponential growth to logarithmic growth. We examine each of the patterns again with the applications in Section 10.7.

Exponential growth functions eventually grow very rapidly, while logarithmic growth functions eventually grow very slowly.

Exponential Growth

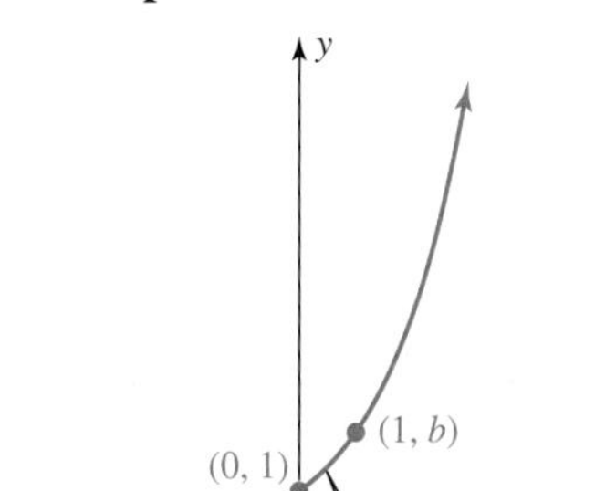

Logarithmic Growth

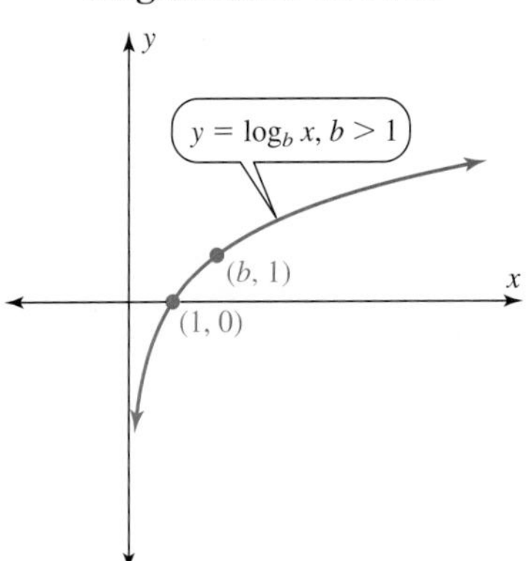

Features:

- For large values of x, extremely rapid growth
- x-intercept: None
- y-intercept: $(0, 1)$
- Key point: $(1, b)$
- Asymptotic to negative x-axis

Features:

- For large values of x, extremely slow growth
- x-intercept: $(1, 0)$
- y-intercept: None
- Key point: $(b, 1)$
- Asymptotic to negative y-axis

It is important to keep this distinction between the rapid growth of exponential functions and the slow growth of logarithmic functions in mind as you analyze real-world problems. This is illustrated by Example 7.

EXAMPLE 7 Selecting an Exponential or Logarithmic Growth Model

Determine for each relationship whether the growth over time is more likely to be an exponential growth function or a logarithmic growth function.

SOLUTION

(a) A frog embryo grows __________ over a period of time.

Exponentially

At the beginning the first cell doubles to form two cells; then these cells grow and split to form four cells. This growing and splitting continues to form millions of cells and a live creature within a few weeks.

(b) The productivity gain of an employee on an assembly line grows __________.

Logarithmically

Employees normally make their greatest gains of efficiency from their novice state to becoming seasoned employees. After they have experienced all the unique features of this job, there is little new to learn and any new gains of productivity are minimal.

SELF-CHECK ANSWERS

10.3.1

1. $7^2 = 49$
2. $6^{1/5} = \sqrt[5]{6}$
3. $\log_3\left(\frac{1}{81}\right) = -4$
4. $\log_6 1 = 0$

10.3.2

1. 6
2. 3
3. 2
4. -1
5. 0
6. 1
7. -1
8. 8

10.3.3

1. $x = 8$
2. $x = \frac{1}{2}$
3. $x = 5$

USING THE LANGUAGE AND SYMBOLISM OF MATHEMATICS 10.3

1. For $x > 0$, $b > 0$, and $b \neq 1$, the inverse of the function $f(x) = b^x$ is written as $f^{-1}(x) =$ __________.
2. $\log_b x = y$ is read "the log base __________ of __________ equals __________."
3. For $b > 0$ and $b \neq 1$, $\log_b 0$ is __________.
4. For $b > 0$ and $b \neq 1$, $\log_b 1 =$ __________.
5. For $b > 0$ and $b \neq 1$, $\log_b b =$ __________.
6. For $b > 0$ and $b \neq 1$, $\log_b \frac{1}{b} =$ __________.
7. For $b > 0$ and $b \neq 1$, $\log_b b^x =$ __________.
8. For $b > 1$ the graph of $f(x) = \log_b x$ is asymptotic to the negative portion of the __________-axis.

QUICK REVIEW 10.3

The purpose of this quick review is to help you recall skills needed in this section.

In Exercises 1–3, simplify each expression.

1. x^0 for $x \neq 0$
2. x^0 for $x = 0$
3. $\left(\frac{2}{3}\right)^{-1}$

In Exercises 4 and 5, solve each equation.

4. $2x - 1 = 8$
5. $x^2 - 1 = 8$

EXERCISES 10.3

In Exercises 1–16, use the given information to complete each entry in the table, as illustrated by the example in the first row.

LOGARITHMIC FORM	VERBAL DESCRIPTION	EXPONENTIAL FORM
Example: $\log_7 49 = 2$	The log base 7 of 49 is 2.	$7^2 = 49$
1. $\log_5 125 = 3$		
2. $\log_2 16 = 4$		
3. $\log_3 \sqrt{3} = \frac{1}{2}$		
4. $\log_5 \sqrt[3]{5} = \frac{1}{3}$		
5. $\log_5 \frac{1}{5} = -1$		
6. $\log_2 \frac{1}{4} = -2$		
7.	The log base 16 of 4 is $\frac{1}{2}$.	
8.	The log base 16 of $\frac{1}{16}$ is -1.	
9.		$8^{2/3} = 4$
10.		$4^{3/2} = 8$
11.		$3^{-2} = \frac{1}{9}$
12.		$\left(\frac{1}{3}\right)^{-4} = 81$
13.		$m^p = n$
14. $\log_x y = z$		
15.	The log base n of m is k.	
16. $\log_{10} 0.0001 = -4$		

In Exercises 17–46, determine the value of each logarithm by inspection.

17. $\log_{10} 100$ **18.** $\log_{10} 1{,}000$ **19.** $\log_{10} 0.001$
20. $\log_{10} 0.01$ **21.** $\log_{10} \sqrt[3]{10}$ **22.** $\log_{10} \sqrt{10}$
23. $\log_2 32$ **24.** $\log_{11} 121$ **25.** $\log_3 \frac{1}{3}$
26. $\log_7 \frac{1}{7}$ **27.** $\log_{18} 1$ **28.** $\log_{25} 1$
29. $\log_{3/4} \frac{4}{3}$ **30.** $\log_{2/7} \frac{7}{2}$ **31.** $\log_3 81$
32. $\log_9 81$ **33.** $\log_2 64$ **34.** $\log_{64} 64$
35. $\log_3 27$ **36.** $\log_{27} 3$ **37.** $\log_{16} 2$
38. $\log_2 16$ **39.** $\log_{25} 125$ **40.** $\log_9 27$
41. $\log_{19} 19$ **42.** $\log_{\pi} \pi$ **43.** $\log_5 5^{2.7}$
44. $\log_3 3^{4.1}$ **45.** $\log_{3/4} \frac{16}{9}$ **46.** $\log_{5/7} \frac{49}{25}$

In Exercises 47–50, determine which expressions are defined and which are not defined. Do not evaluate these expressions.

47. a. $\log_7 10$ **b.** $\log_7 0$
48. a. $\log_0 7$ **b.** $\log_7 7$
49. a. $\log_5 (-5)$ **b.** $\log_5 1$
50. a. $\log_{-3} 3$ **b.** $\log_{0.3} 3$

In Exercises 51–68, solve each equation without using a calculator.

51. $\log_6 36 = x$ **52.** $\log_{11} 121 = x$ **53.** $\log_6 x = 1$
54. $\log_{11} x = 1$ **55.** $\log_6 x = -1$ **56.** $\log_{11} x = -1$
57. $\log_6 x = \frac{1}{2}$ **58.** $\log_{11} x = \frac{1}{3}$ **59.** $\log_x 4 = 2$
60. $\log_x 4 = \frac{1}{2}$ **61.** $\log_x 125 = 3$ **62.** $\log_x 8 = 3$
63. $\log_b \sqrt[5]{b} = x$ **64.** $\log_b \sqrt[5]{b^2} = x$
65. $\log_3 1 = 2x + 1$ **66.** $\log_3 3 = 4x - 1$
67. $\log_2 (3x - 4) = 3$ **68.** $\log_5 (3x + 1) = 2$

In Exercises 69–72, match each graph (A through D) with the function that defines this graph. All graphs are displayed with the window [−5, 5, 1] by [−5, 5, 1].

69. $f(x) = \left(\frac{1}{3}\right)^x$ **70.** $f(x) = 2^x$
71. $f(x) = \log_3 x$ **72.** $f(x) = \log_{1/3} x$

A.
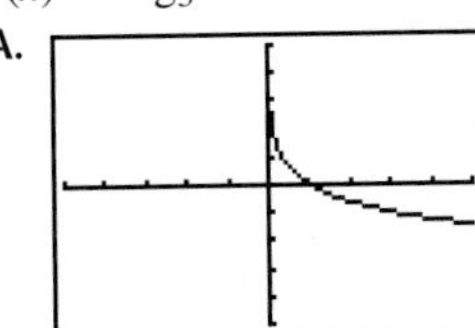

B.
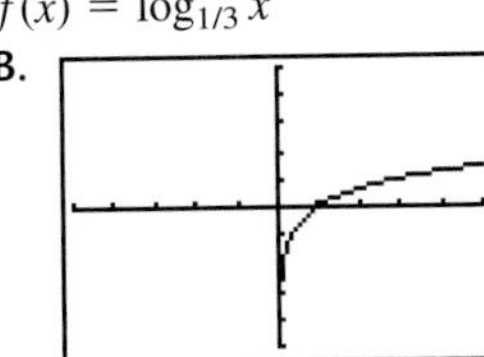

C.

D.
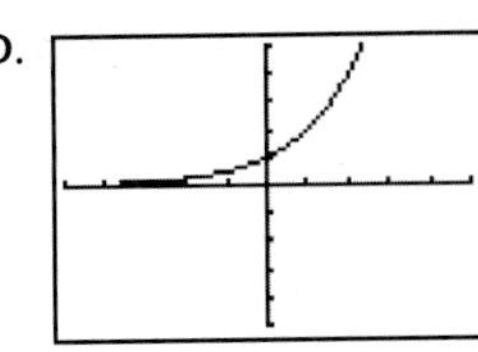

73. Use the graph of $f(x) = \log_2 x$ shown in the figure to determine the domain and range of this function.

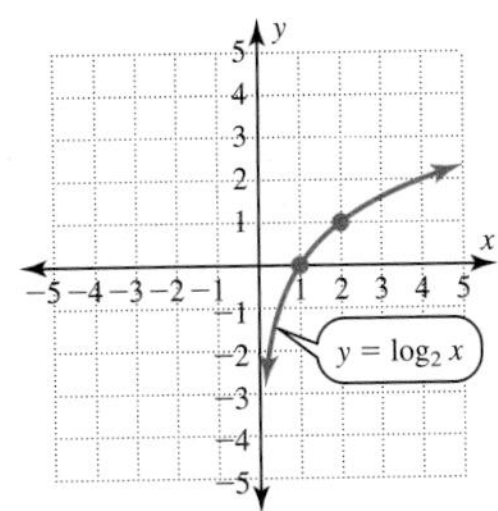

74. Use the graph of $f(x) = \log_4 x$ shown in the figure to determine the domain and range of this function.

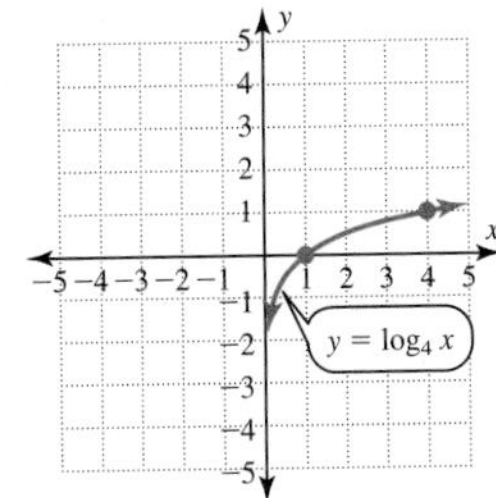

In Exercises 75 and 76, first use a graphing calculator to graph the corresponding exponential function. Then use the **DrawInv** feature to sketch the graph of the logarithmic function.

75. $f(x) = \log_{1.5} x$ **76.** $f(x) = \log_{2.5} x$

Group Discussion Questions

77. Challenge Question Determine for each relationship whether the growth over time is more likely to be an exponential growth function or a logarithmic growth function.
a. The value of an investment in a U.S. Treasury note over a 30-year period
b. The amount of learning by an individual over his or her 85-year lifetime
c. The population of locusts from the start of a growing season until the end of the growing season
d. The heights obtained by a pole-vaulter over a 20-year career

Section 10.4 Evaluating Logarithms

Objective: **7.** Evaluate common and natural logarithms with a calculator.

Although $\log_b x$ is defined for any base $b > 0$ and $b \neq 1$, the only two bases commonly used are base 10 and base e. Since our number system is based upon powers of 10, the most convenient base for many computations is base 10. Logarithms to the base 10 are called **common logarithms.** The abbreviated form $\log x$ is used to denote $\log_{10} x$.

Mathematicians and scientists use base e extensively because many formulas are stated most easily by using this base. The number e is irrational:

Remember $e \approx 2.718$.

$e \approx 2.718\,281\,828\,459\,045 \cdots$

You should remember that e is approximately equal to 2.718. Logarithms to the base e are called **natural logarithms.** The abbreviated form $\ln x$ is used to denote $\log_e x$.

A Mathematical Note

The earliest use of natural logarithms was in 1618. Natural logarithms are used frequently because the base of e occurs in many natural applications from astronomy to zoology.

Common and Natural Logarithms

COMMON LOGARITHMS	VERBALLY	NUMERICAL EXAMPLE
$\log x$ means $\log_{10} x$	$\log x$ is read "the (common) log of x" or "the log base 10 of x."	$\log 100 = 2$ because $10^2 = 100$
NATURAL LOGARITHMS		
$\ln x$ means $\log_e x$	$\ln x$ is read "the natural log of x" or "the log base e of x."	$\ln 100 \approx 4.605$ because $e^{4.605} \approx 100$

EXAMPLE 1 Evaluating Logarithms by Inspection

Determine the value of each expression by inspection.

SOLUTION

(a) $\log 10{,}000$	$\log 10{,}000 = 4$	In exponential form, $10^4 = 10{,}000$.
(b) $\log 0.1$	$\log 0.1 = -1$	In exponential form, $10^{-1} = \frac{1}{10} = 0.1$.
(c) $\ln e^5$	$\ln e^5 = 5$	In exponential form, $e^5 = e^5$.
(d) $\ln e^{-4}$	$\ln e^{-4} = -4$	In exponential form, $e^{-4} = e^{-4}$.

Only relatively simple logarithms, such as $\log 100 = 2$, can be determined by inspection. Thus common and natural logarithms have historically been determined through use of tables, slide rules, and other devices. Today calculators are usually used to determine these values. The examples in this section assume calculator usage. Calculator Perspective 10.4.1 illustrates how to approximate common and natural logarithms.

CALCULATOR PERSPECTIVE 10.4.1 Evaluating Common and Natural Logarithms

To evaluate the expressions from Example 1 on a TI-84 Plus calculator, enter the following keystrokes:

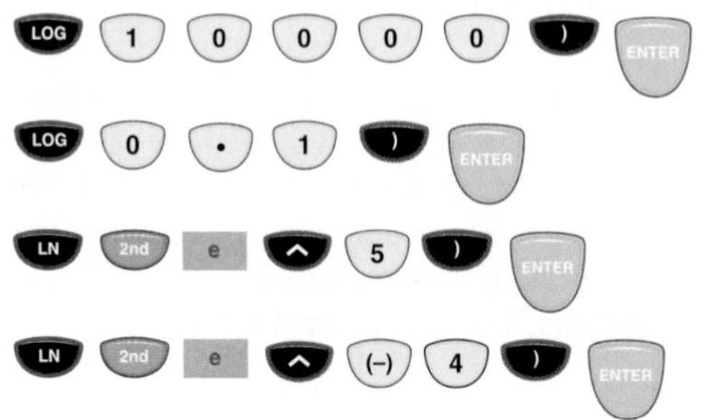

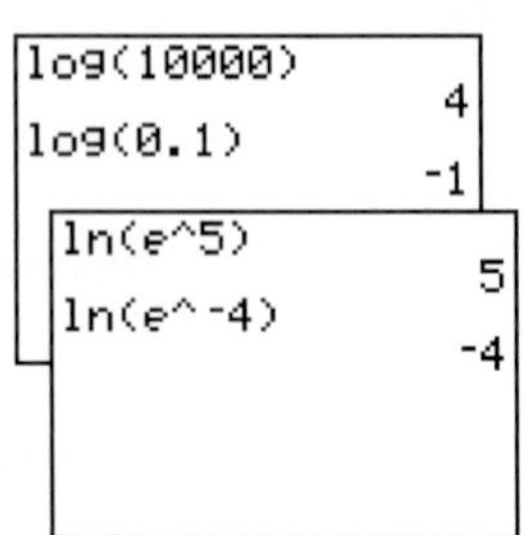

Example 2 illustrates two expressions in which it is a clear advantage to be able to use a calculator. Although the two expressions in Example 2 look very similar, the values for common and natural logarithms are quite distinct.

EXAMPLE 2 Evaluating Common and Natural Logarithms

Use a calculator to approximate each expression to the nearest thousandth. Check each answer by using the equivalent exponential form.

SOLUTION

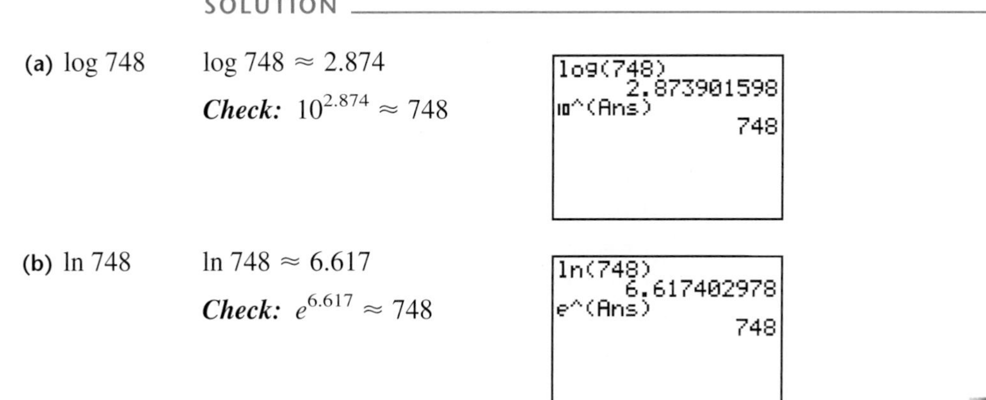

The 10^x key is the secondary function of the LOG key.

The e^x key is the secondary function of the LN key.

(a) log 748 $\quad \log 748 \approx 2.874$

Check: $10^{2.874} \approx 748$

(b) ln 748 $\quad \ln 748 \approx 6.617$

Check: $e^{6.617} \approx 748$

Logarithmic functions are also available in many computer languages; however, these functions may not mean what you expect. Note the following warning.

Warning for Computer Users

Some computer languages provide only the natural logarithmic function, in which case this function may be called LOG instead of ln, even though the base is e. If you are unsure of the meaning of LOG on a particular computer, test a known value such as LOG 10 to determine which base is being used. We know $\log 10 = 1$ but $\ln 10 \approx 2.30$.

SELF-CHECK 10.4.1

Use a calculator to approximate each expression to the nearest ten-thousandth.

1. log 0.47912 **2.** ln 0.47912 **3.** log 512.8 **4.** ln 512.8

Example 3 illustrates how to use a calculator to complete a table of logarithmic values.

EXAMPLE 3 Completing a Table of Logarithms

Use a calculator to complete this table of logarithmic values.

x	$Y_1 = \log x$	$Y_2 = \ln x$
10		
20		
30		
40		
50		
60		
70		

SOLUTION

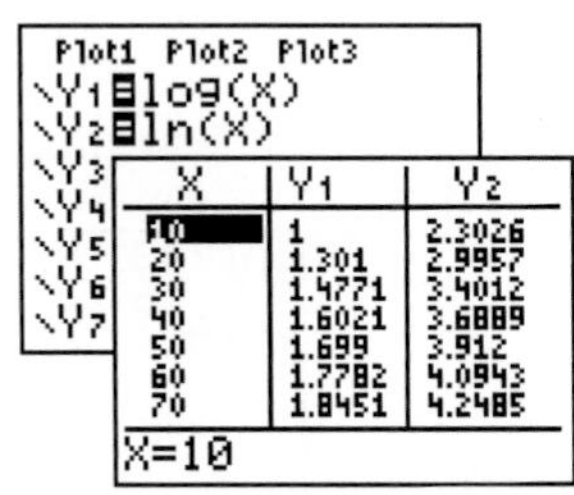

We often use scientific notation (see Calculator Perspective 5.3.1 in Section 5.3) to write numbers whose magnitude is either very large or very small. This is illustrated in Example 4.

EXAMPLE 4 Using Scientific Notation with Logarithmic Expressions

Use scientific notation to enter the argument of each logarithm. Use a calculator to approximate each expression to the nearest ten-thousandth.

SOLUTION

(a) ln 894,000,000,000

$\ln 894{,}000{,}000{,}000$
$= \ln(8.94 \times 10^{11})$
≈ 27.5190

Check: $e^{27.5190} \approx 8.94 \times 10^{11}$

```
ln(8.94E11)
         27.51897161
e^(Ans)
             8.94E11
```

(b) log 0.000 000 001 234

$\log 0.000\ 000\ 001\ 234$
$= \log(1.234 \times 10^{-9})$
≈ -8.9087

Check: $10^{-8.9087} \approx 1.234 \times 10^{-9}$

```
log(1.234E-9)
         -8.90868484
10^(Ans)
            1.234E-9
```

SELF-CHECK 10.4.2

Use a calculator to approximate each expression to the nearest thousandth.

1. $\log(5.81 \times 10^4)$
2. $\ln(3.96 \times 10^{-5})$

To avoid order-of-operations errors, it is wise to use sufficient parentheses to clearly denote the correct interpretation of logarithmic expressions. Note the similar but distinct meanings of the expressions in Example 5. It is better to use extra parentheses than not to use enough to denote the meaning that you intend.

EXAMPLE 5 Using a Calculator to Evaluate Logarithmic Expressions

Rewrite each expression by using parentheses, and then use a calculator to approximate each expression to the nearest ten-thousandth.

SOLUTION

(a) $\dfrac{\log 12}{5}$ $\qquad \dfrac{\log 12}{5} = (\log(12)) \div 5 \approx 0.2158$

(b) $\log \dfrac{12}{5}$ $\qquad \log \dfrac{12}{5} = \log\left(\dfrac{12}{5}\right) \approx 0.3802$

(c) $\dfrac{\log 12}{\log 5}$ $\qquad \dfrac{\log 12}{\log 5} = \dfrac{\log(12)}{\log(5)} \approx 1.5440$

```
(log(12))/5
              .2158362492
log(12/5)
              .3802112417
log(12)/log(5)
              1.543959311
```

We will use the fact that logarithmic functions and exponential functions are inverses of each other to solve for x in Example 6.

EXAMPLE 6 Solving Logarithmic Equations

Use a calculator to approximate the value of x to the nearest thousandth.

SOLUTION

(a) $\log x = 0.69897$

$$\log x = 0.69897$$
$$x = 10^{0.69897}$$
$$x \approx 5.000$$

First rewrite the equation in exponential form with a base of 10. Then approximate x by using a calculator.

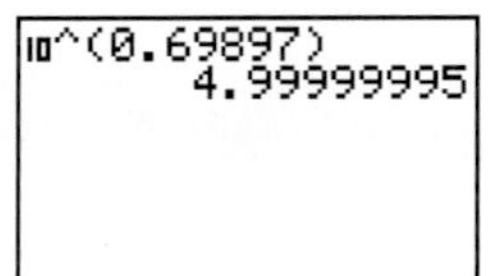

(b) $\ln x = -1.045$

$$\ln x = -1.045$$
$$x = e^{-1.045}$$
$$x \approx 0.352$$

First rewrite the equation $\log_e x = -1.045$ in exponential form with a base of e. Then approximate x by using a calculator.

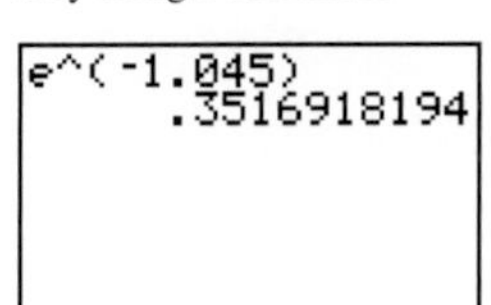

SELF-CHECK 10.4.3

Use a calculator to approximate the value of each expression to the nearest hundredth.

1. $\dfrac{5}{\log 7}$ 2. $\log \dfrac{5}{7}$ 3. $\dfrac{\log 5}{\log 7}$

Use a calculator to approximate x to the nearest thousandth.

4. $x = 10^{1.4583}$ 5. $x = e^{1.4583}$ 6. $\log x = 1.4583$ 7. $\ln x = 1.4583$

It is important to know that the values that we calculate are reasonable and not in error because of some careless keystroke. Example 7 illustrates two estimation problems. We examine a formula that allows us to calculate logarithms to any base in Section 10.5.

EXAMPLE 7 Estimation Skills

Mentally estimate the value of each of these expressions.

SOLUTION

(a) $2^{3.98765}$

$2^{3.98765} \approx 2^4$
$2^{3.98765} \approx 16$

Since 3.98765 is only slightly less than 4, $2^{3.98765}$ is slightly less than 2^4, which is 16.

(b) $\log_7 50.0017$

$\log_7 50.0017 \approx \log_7 49$
$\log_7 50.0017 \approx 2$

Since 50.0017 is slightly larger than 49, $\log_7 50.0017$ is slightly larger than $\log_7 49$. And $\log_7 49 = 2$ because $7^2 = 49$.

SELF-CHECK ANSWERS

10.4.1
1. −0.3196 2. −0.7358
3. 2.7099 4. 6.2399

10.4.2
1. 4.764 2. −10.137

10.4.3
1. 5.92 2. −0.15
3. 0.83 4. 28.728
5. 4.299 6. 28.728
7. 4.299

USING THE LANGUAGE AND SYMBOLISM OF MATHEMATICS 10.4

1. The irrational number $e \approx$ ________ (to the nearest thousandth).
2. The common logarithm of x, denoted $\log x$, means ________.
3. The natural logarithm of x, denoted $\ln x$, means ________.

QUICK REVIEW 10.4

The purpose of this quick review is to help you recall skills needed in this section.

In Exercises 1–3, simplify each expression.

1. $\dfrac{0}{10}$ 2. $\dfrac{10}{0}$ 3. 10^{-4}

In Exercises 4 and 5, write each number in scientific notation.

4. 75,345,247 5. 0.000087

EXERCISES 10.4

In Exercises 1–8, determine the value of the expression in part **a** by inspection. Then use a calculator to approximate the value of the expressions in parts **b** and **c** to the nearest thousandth.

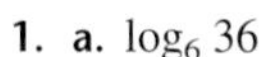

1. a. $\log_6 36$ b. $\log 36$ c. $\ln 36$
2. a. $\log_3 27$ b. $\log 27$ c. $\ln 27$
3. a. $\log_3 \frac{1}{9}$ b. $\log \frac{1}{9}$ c. $\ln \frac{1}{9}$
4. a. $\log_5 \frac{1}{125}$ b. $\log \frac{1}{125}$ c. $\ln \frac{1}{125}$
5. a. $\log_8 1$ b. $\log 1$ c. $\ln 1$
6. a. $\log 1{,}000$ b. $\log 900$ c. $\ln 900$
7. a. $\ln e^3$ b. $\log e^3$ c. $\ln 1000$
8. a. $\ln e^{-3}$ b. $\log e^{-3}$ c. $\ln 10^{-3}$

In Exercises 9–12, determine the value of each expression by inspection.

9. a. $\log 100{,}000$ b. $\log 0.0001$ c. $\log 10^9$
10. a. $\log 1{,}000{,}000$ b. $\log 0.01$ c. $\log 10^{-6}$
11. a. $\ln e^5$ b. $\ln e^{-5}$ c. $\ln \sqrt{e}$
12. a. $\ln e^7$ b. $\ln e^{-7}$ c. $\ln \sqrt[3]{e}$

In Exercises 13–42, use a calculator to approximate the value of each expression to the nearest thousandth.

13. a. $\log 47$ b. $10^{1.672}$
14. a. $\log 2.47$ b. $10^{0.393}$
15. a. $\ln 47$ b. $e^{3.850}$
16. a. $\ln 2.47$ b. $e^{0.904}$
17. a. $\log 10$ b. $\ln 10$
18. a. $\log 100$ b. $\ln 100$
19. a. $\log e$ b. $\ln e$
20. a. $\log e^2$ b. $\ln e^2$
21. a. $\log 113$ b. $10^{2.053}$
22. a. $\ln 113$ b. $e^{4.727}$
23. a. $\ln 0.00621$ b. $e^{-5.082}$
24. a. $\log 0.00621$ b. $10^{-2.207}$
25. a. $\log \pi$ b. $\ln \pi$
26. a. $\log \sqrt{7}$ b. $\ln \sqrt{7}$
27. a. $\log 0.00009$ b. $\ln 0.00009$
28. a. $\log (3.87 \times 10^{-7})$ b. $\ln (3.87 \times 10^{-7})$
29. a. $\log (5.93 \times 10^{-12})$ b. $\ln (5.93 \times 10^{-12})$
30. a. $\log (4.07 \times 10^5)$ b. $\ln (4.07 \times 10^5)$
31. a. $\log (6.93 \times 10^{11})$ b. $\ln (6.93 \times 10^{11})$
32. a. $\log (\ln 9)$ b. $\ln (\log 9)$
33. a. $\log (\ln 13)$ b. $\ln (\log 13)$
34. a. $(\log 7.52)^2$ b. $\log (7.52^2)$
35. a. $(\ln 0.84)^2$ b. $\ln (0.84^2)$
36. a. $\ln \frac{11}{4}$ b. $\frac{\ln 11}{\ln 4}$
37. a. $\log \frac{17}{5}$ b. $\frac{\log 17}{\log 5}$
38. a. $(\log 11)(\log 4)$ b. $\log (11 \cdot 4)$
39. a. $(\ln 5)(\ln 17)$ b. $\ln (5 \cdot 17)$
40. a. $\ln 5 + \ln 17$ b. $\ln (5 + 17)$
41. a. $\log 11 - \log 4$ b. $\log (11 - 4)$
42. a. $\ln 13 - \ln 5$ b. $\ln (13 - 5)$

In Exercises 43–48, identify whether each expression is defined. Then use a calculator to approximate to the nearest thousandth the value of each expression that is defined.

43. a. $-\ln 8$ b. $\ln (-8)$
44. a. $-\log 11$ b. $\log (-11)$
45. a. $0 \cdot \log 23$ b. $\log (23 \cdot 0)$
46. a. $\ln 17 - \ln 17$ b. $\ln (17 - 17)$
47. a. $\ln 17 - \ln 19$ b. $\ln (17 - 19)$
48. a. $\frac{\log 41}{\log 1}$ b. $\frac{\log 41}{\log 10}$

In Exercises 49–56, use a calculator to approximate the value of x to the nearest thousandth.

49. $\log x = 2.477$
50. $\log x = 2.778$
51. $\log x = -0.301$
52. $\log x = -0.699$
53. $\ln x = 2.079$
54. $\ln x = 2.773$
55. $\ln x = -2.996$
56. $\ln x = -4.711$

Estimation Skills

In Exercises 57–60, mentally estimate the value of each expression to the nearest integer.

PROBLEM	MENTAL ESTIMATE
57. $\log_8 63$	
58. $\log_2 17$	
59. $2.99^{4.01}$	
60. $5.01^{1.99}$	

Group Discussion Questions

61. **Discovery Question**
   a. Evaluate each of these expressions: $10^{\log 33}$, $e^{\ln 19}$, $2^{\log_2 8}$, $3^{\log_3 81}$
   b. Based on your observations in part **a**, make a conjecture stated as an algebraic equation.
   c. Test your conjecture in part **b**, and describe it to your instructor.
62. **Discovery Question**
   a. Evaluate each of these expressions: $\log (5 \cdot 6)$, $\log 5 + \log 6$, $\ln (7 \cdot 11)$, $\ln 7 + \ln 11$
   b. Based on your observations in part **a**, make a conjecture stated as an algebraic equation.
   c. Test your conjecture in part **b**, and describe it to your instructor.
63. **Discovery Question**
   a. Evaluate each of these expressions: $\log \frac{15}{5}$, $\log 15 - \log 5$, $\ln \frac{24}{6}$, $\ln 24 - \ln 6$
   b. Based on your observations in part **a**, make a conjecture stated as an algebraic equation.
   c. Test your conjecture in part **b**, and describe it to your instructor.
64. **Discovery Question**
   a. Evaluate each of these expressions: $\log 5^7$, $7 \log 5$, $\log \pi^3$, $3 \log \pi$, $\ln 10^9$, $9 \ln 10$
   b. Based on your observations in part **a**, make a conjecture stated as an algebraic equation.
   c. Test your conjecture in part **b**, and describe it to your instructor.

Section 10.5 Properties of Logarithms

Objectives:
8. Use the product rule, the quotient rule, and the power rule for logarithms.
9. Use the change-of-base formula to evaluate logarithms.

The properties of logarithms follow directly from the definition of a logarithmic function as the inverse of an exponential function. Thus for every exponential property there is a corresponding logarithmic property. The logarithmic properties provide a means of simplifying some algebraic expressions and make it easier to solve exponential equations.

The logarithmic and exponential properties are stated here side by side so that you can compare them. (The capital letters X and Y are used in the logarithmic form to show that these variables are not the same as x and y in the exponential form.)

In exponential form, we add exponents; in logarithmic form, we add logarithms.
In exponential form, we subtract exponents; in logarithmic form, we subtract logarithms.
In exponential form, we multiply exponents; in logarithmic form, we multiply the power p by the log of X.

	EXPONENTIAL FORM	LOGARITHMIC FORM
Product rule:	$b^x b^y = b^{x+y}$	$\log_b XY = \log_b X + \log_b Y$
Quotient rule:	$\frac{b^x}{b^y} = b^{x-y}$	$\log_b \frac{X}{Y} = \log_b X - \log_b Y$
Power rule:	$(b^x)^p = b^{xp}$	$\log_b X^p = p \log_b X$

The proof of the product rule is presented here. The quotient rule and the power rule can be proved in a similar fashion; their proofs are left to you in Exercises 82 and 83 at the end of this section.

Proof of the Product Rule for Logarithms Let $X = b^m$ and $Y = b^n$.

$\log_b X = m$ and $\log_b Y = n$ — Rewrite these statements in logarithmic form.

Now examine XY.

$XY = b^m b^n$ — Substitute b^m for X and b^n for Y.
$XY = b^{m+n}$ — Use the product rule for exponents.
$\log_b XY = m + n$ — Rewrite this statement in logarithmic form.
$\log_b XY = \log_b X + \log_b Y$ — Substitute $\log_b X$ for m and $\log_b Y$ for n.

Properties of Logarithms

For $x, y > 0$, $b > 0$, and $b \neq 1$:

	ALGEBRAICALLY	VERBALLY	NUMERICAL EXAMPLE
Product rule:	$\log_b xy = \log_b x + \log_b y$	The log of a product is the sum of the logs.	$\log_2 32 = \log_2(4 \cdot 8) = \log_2 4 + \log_2 8$ $5 = 2 + 3$
Quotient rule:	$\log_b \frac{x}{y} = \log_b x - \log_b y$	The log of a quotient is the difference of the logs.	$\log_2 4 = \log_2 \frac{32}{8} = \log_2 32 - \log_2 8$ $2 = 5 - 3$
Power rule:	$\log_b x^p = p \log_b x$	The log of the pth power of x is p times the log of x.	$\log_2 64 = \log_2 4^3 = 3 \log_2 4$ $6 = 3 \cdot 2$

EXAMPLE 1 Using a Calculator to Examine the Properties of Logarithms

Use the properties of logarithms to rewrite each expression. Then use a calculator to verify each equation.

SOLUTION

(a) $\ln 17^5$

$\ln 17^5 = 5 \ln 17$
$14.166 \approx 14.166$

Use the power rule:
$\log_b x^p = p \log_b x$

```
ln(17^5)
         14.16606672
5ln(17)
         14.16606672
```

(b) $\ln (15 \cdot 91)$

$\ln (15 \cdot 91) = \ln 15 + \ln 91$
$7.219 \approx 7.219$

Use the product rule:
$\log_b xy = \log_b x + \log_b y$

```
ln(15*91)
         7.218909708
ln(15)+ln(91)
         7.218909708
```

(c) $\ln \dfrac{86}{37}$

$\ln \dfrac{86}{37} = \ln 86 - \ln 37$
$0.843 \approx 0.843$

Use the quotient rule:
$\log_b \dfrac{x}{y} = \log_b x - \log_b y$

```
ln(86/37)
         .8434293836
ln(86)-ln(37)
         .8434293836
```

The properties of logarithms are used both to rewrite logarithmic expressions in a simpler form and to combine logarithms. These properties also are used to solve logarithmic equations and to simplify expressions in calculus.

The input value of $y = f(x)$ also is called the **argument** of the function. In the logarithmic function $y = \log(x + 2)$, we refer to $x + 2$ as the argument of the logarithmic function. Likewise, the argument of $y = \ln \dfrac{x}{x - 1}$ is $\dfrac{x}{x - 1}$. This terminology is often used for clarification when we are taking a logarithm of an expression other than just x.

EXAMPLE 2 Using the Properties of Logarithms to Form Simpler Expressions

Use the properties of logarithms to write these expressions in terms of logarithms of simpler expressions. Assume that the argument of each logarithm is a positive real number.

SOLUTION

(a) $\log xyz$ $\quad$ $\log xyz = \log x + \log y + \log z$ $\quad$ Product rule (generalized to three factors): Add the logarithms.

(b) $\ln x^7$ $\quad$ $\ln x^7 = 7 \ln x$ $\quad$ Power rule: Multiply the power 7 by the logarithm.

(c) $\ln x^4y^5$ $\quad \ln x^4y^5 = \ln x^4 + \ln y^5$ — Product rule: Add the logarithms.

$= 4 \ln x + 5 \ln y$ — Power rule: Multiply each power by the logarithm.

(d) $\log_5 \frac{x+3}{x-2}$ $\quad \log_5 \frac{x+3}{x-2} = \log_5(x+3) - \log_5(x-2)$ — Quotient rule: Subtract the logarithms.

(e) $\ln\sqrt{\frac{x}{y}}$ $\quad \ln\sqrt{\frac{x}{y}} = \ln\left(\frac{x}{y}\right)^{1/2}$

$= \frac{1}{2}\ln\frac{x}{y}$ — Power rule: Multiply the power $\frac{1}{2}$ by the logarithm.

$= \frac{1}{2}(\ln x - \ln y)$ — Quotient rule: Subtract the logarithms.

Example 3 is essentially the reverse of Example 2. The properties of logarithms are used to combine logarithmic terms.

EXAMPLE 3 Using the Properties of Logarithms to Combine Expressions

Combine the logarithmic terms in each expression into a single logarithmic expression with a coefficient of 1. Assume that the argument of each logarithm is a positive real number.

SOLUTION

(a) $2 \ln x + 3 \ln y - 4 \ln z$ $\quad 2 \ln x + 3 \ln y - 4 \ln z = \ln x^2 + \ln y^3 - \ln z^4$ — Power rule

$= \ln x^2y^3 - \ln z^4$ — Product rule

$= \ln \frac{x^2y^3}{z^4}$ — Quotient rule

(b) $5 \log x + \frac{1}{3}\log y$ $\quad 5 \log x + \frac{1}{3}\log y = \log x^5 + \log y^{1/3}$ — Power rule

$= \log x^5 + \log \sqrt[3]{y}$

$= \log x^5\sqrt[3]{y}$ — Product rule

SELF-CHECK 10.5.1

Express each logarithm in terms of logarithms of simpler expressions. Assume that the argument of each logarithm is a positive real number.

1. $\log_{11} x^3y^2$
2. $\log_6 \frac{(x+1)^5}{(y-2)^3}$

Combine the logarithmic terms in each expression into a single logarithmic expression with a coefficient of 1. Assume that the argument of each logarithm is a positive real number.

3. $\log_{12}(x+3) + \log_{12}(2x-9)$
4. $3 \log_b x + 2 \log_b y - \frac{1}{2}\log_b z$

Example 4 can serve as a good source of practice in observing patterns and using the properties of logarithms. These values are recalculated in Example 5 by using the change-of-base formula for logarithms. However, the purpose of Example 4 is to give practice using the product, quotient, and power rules for logarithms.

EXAMPLE 4 Logarithmic Properties

Use $\log_5 2 \approx 0.4307$, $\log_5 3 \approx 0.6826$, and the properties of logarithms to approximate to the nearest thousandth the value of each of these logarithmic expressions.

SOLUTION

(a) $\log_5 6$

$$\begin{aligned}\log_5 6 &= \log_5(2 \cdot 3) && \text{Product rule}\\ &= \log_5 2 + \log_5 3 && \text{Substitute the given values for } \log_5 2 \text{ and } \log_5 3.\\ &\approx 0.4307 + 0.6826\\ &\approx 1.113\end{aligned}$$

(b) $\log_5 9$

$$\begin{aligned}\log_5 9 &= \log_5 3^2 && \text{Power rule}\\ &= 2\log_5 3 && \text{Substitute the given value of } \log_5 3.\\ &\approx 2(0.6826)\\ &\approx 1.365\end{aligned}$$

(c) $\log_5 0.4$

$$\begin{aligned}\log_5 0.4 &= \log_5 \frac{2}{5} && \text{Quotient rule}\\ &= \log_5 2 - \log_5 5 && \text{The value of } \log_5 2 \text{ is given, and } \log_5 5 \text{ can be determined by inspection.}\\ &\approx 0.4307 - 1\\ &\approx -0.569\end{aligned}$$

SELF-CHECK 10.5.2

Use the values given in Example 4 to approximate to the nearest thousandth the value of each expression.

1. $\log_5 10$ **2.** $\log_5 8$ **3.** $\log_5 1.5$

Calculators have special keys for both $\log x$ and $\ln x$. However, no other logarithmic keys are needed. The change-of-base formula for logarithms can be used to convert any logarithmic expression to either base 10 or base e.

Although any base b can be used in the formula $\log_a x = \dfrac{\log_b x}{\log_b a}$, we usually use either $\dfrac{\log x}{\log a}$ or $\dfrac{\ln x}{\ln a}$. One reason we use these bases is that they are available on most graphing calculators.

Change-of-Base Formula

For $a, b > 0$, $a \neq 1$, and $b \neq 1$:

FOR LOGARITHMS		EXAMPLES		
$\log_a x = \dfrac{\log_b x}{\log_b a}$	for $x > 0$	$\log_2 x = \dfrac{\log x}{\log 2}$	or	$\log_2 x = \dfrac{\ln x}{\ln 2}$
FOR EXPONENTS				
$a^x = b^{x \log_b a}$		$2^x = e^{(\ln 2)x}$		

Proof of the Change-of-Base Formula for Logarithms

Let $\log_a x = y$. Then

$x = a^y$	Rewrite this equation in exponential form.
$\log_b x = \log_b a^y$	Take the log to base b of both sides of this equation.
$\log_b x = y \log_b a$	Use the power rule for logarithms.
$y = \dfrac{\log_b x}{\log_b a}$	Solve for y.
$\log_a x = \dfrac{\log_b x}{\log_b a}$	Substitute $\log_a x$ for y.

The most useful forms of this identity are

$$\log_a x = \frac{\log x}{\log a} \quad \text{and} \quad \log_a x = \frac{\ln x}{\ln a}$$

The proof of the change-of-base formula for exponents is similar to the proof given for logarithms.

EXAMPLE 5 Using the Change-of-Base Formula for Logarithms

Use the change-of-base formula for logarithms and a calculator to approximate the value of each expression to the nearest thousandth.

SOLUTION

(a) $\log_5 6$ $\quad \log_5 6 = \dfrac{\ln 6}{\ln 5} \approx 1.113$

(b) $\log_5 9$ $\quad \log_5 9 = \dfrac{\ln 9}{\ln 5} \approx 1.365$

(c) $\log_5 0.4$ $\quad \log_5 0.4 = \dfrac{\ln 0.4}{\ln 5} \approx -0.569$

Use the change-of-base formula $\log_a x = \dfrac{\ln x}{\ln a}$ to convert to natural logs. (Converting to common logs would produce the same results.)

```
ln(6)/ln(5)
          1.113282753
ln(9)/ln(5)
          1.365212389
ln(0.4)/ln(5)
         -.5693234419
```

Note that these results confirm the results calculated in Example 4.

SELF-CHECK 10.5.3

1. Estimate $\log_3 80$ to the nearest integer.
2. Use the change-of-base formula for logarithms and a calculator to approximate $\log_3 80$ to the nearest thousandth.

The change-of-base formula for exponents can be used to convert any exponential growth or decay function to base e. In scientific applications base e is used almost exclusively.

EXAMPLE 6 Using the Change-of-Base Formula for Exponents

Use the change-of-base formula for exponents to write $f(x) = 3^x$ as an exponential function with base e.

SOLUTION

ALGEBRAICALLY

$y = 3^x$

$y = e^{(\ln 3)x}$ — Use the change-of-base formula $a^x = e^{(\ln a)x}$ to convert to base e.

$y \approx e^{1.099x}$ — Then approximate ln 3 with a calculator.

NUMERICAL CHECK

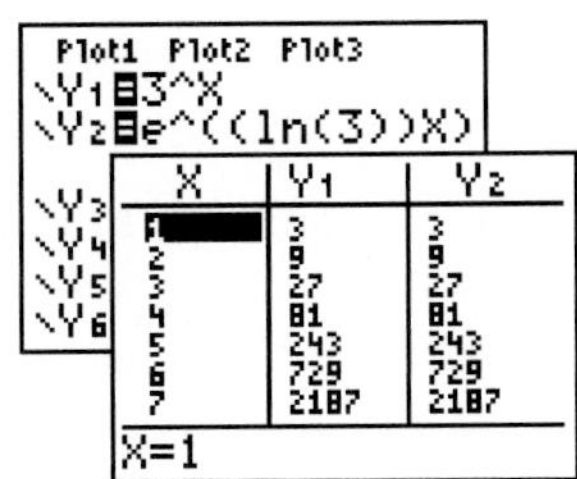

Let Y_1 represent the original expression 3^x. Let Y_2 represent the new expression $e^{(\ln 3)x}$.

Because Y_1 and Y_2 have identical values, these two functions are equal. A comparison of their graphs would reveal that their graphs are identical.

Example 7 proves an identity involving logarithms and exponents.

EXAMPLE 7 Verifying an Identity Involving Logarithms

Verify that $b^{\log_b x} = x$ for any positive real number x.

SOLUTION

Let $y = \log_b x$. Then

$b^y = x$ and — Rewrite the logarithmic expression in exponential form.

$b^{\log_b x} = x$ — Substitute $\log_b x$ for y.

Another Property of Logarithms

For $x > 0$, $b > 0$, and $b \neq 1$:

ALGEBRAICALLY	NUMERICAL EXAMPLE
$b^{\log_b x} = x$	$23^{\log_{23} 19} = 19$

EXAMPLE 8 Using the Properties of Logarithms

Determine the value of $e^{\ln 22}$, and check this result with a calculator.

SOLUTION

$e^{\ln 22} = 22$

This value can be determined by inspection by using the formula $b^{\log_b x} = x$; $e^{\ln 22} = e^{\log_e 22} = 22$. The calculator check confirms that the value of this expression is 22.

CALCULATOR CHECK

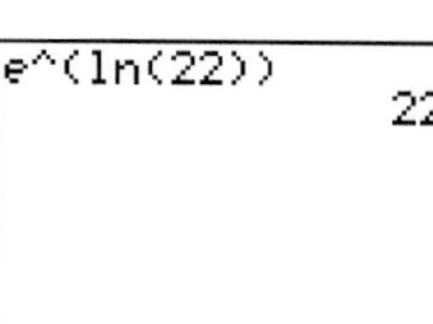

This property of logarithms is used in Example 9 to evaluate some expressions by inspection.

EXAMPLE 9 Using the Properties of Logarithms

Determine the value of each expression by inspection.

SOLUTION

(a) $7^{\log_7 8}$ $\quad 7^{\log_7 8} = 8$

(b) $3^{\log_3 0.179}$ $\quad 3^{\log_3 0.179} = 0.179$

(c) $a^{\log_a 231}$ $\quad a^{\log_a 231} = 231 \quad$ for $a > 0, a \neq 1$

Each of these expressions can be determined by inspection, by using the identity $b^{\log_b x} = x$.

SELF-CHECK 10.5.4

Determine the value of each expression by inspection.

1. $e^{\ln 31}$
2. $e^{\ln 0.07}$
3. $10^{\log 86}$
4. $10^{\log 0.041}$

SELF-CHECK ANSWERS

10.5.1

1. $3 \log_{11} x + 2 \log_{11} y$
2. $5 \log_6(x+1) - 3 \log_6(y-2)$
3. $\log_{12}(x+3)(2x-9)$
4. $\log_b \dfrac{x^3y^2}{\sqrt{z}}$

10.5.2

1. $\log_5 10 = \log_5(2 \cdot 5) = \log_5 2 + \log_5 5 \approx 1.431$
2. $\log_5 8 = \log_5 2^3 = 3 \log_5 2 \approx 1.292$
3. $\log_5 1.5 = \log_5 \dfrac{3}{2} = \log_5 3 - \log_5 2 \approx 0.252$

10.5.3

1. $\log_3 80 \approx \log_3 81 = 4$
2. $\log_3 80 = \dfrac{\ln 80}{\ln 3} \approx 3.989$

10.5.4

1. 31
2. 0.07
3. 86
4. 0.041

USING THE LANGUAGE AND SYMBOLISM OF MATHEMATICS 10.5

1. By the product rule for exponents, $b^xb^y =$ ________.
2. By the product rule for logarithms, $\log_b xy =$ ________.
3. By the quotient rule for exponents, $\dfrac{b^x}{b^y} =$ ________.
4. By the quotient rule for logarithms, $\log_b \dfrac{x}{y} =$ ________.

5. By the power rule for exponents, $(b^x)^p =$ __________.
6. By the power rule for logarithms, $\log_b x^p =$ __________.
7. The change-of-base formula for logarithms is $\log_a x =$ __________.
8. The change-of-base formula for exponents is $a^x =$ __________.
9. By the change-of-base formula for exponents, $e^{(\ln 7)x} =$ __________.
10. If $x > 0$, $b > 0$, and $b \neq 1$, then $b^{\log_b x} =$ __________.
11. In the logarithmic function $y = \log_2 (5x - 1)$ the base is __________ and the argument of the logarithm is __________.

QUICK REVIEW 10.5

The purpose of this quick review is to help you recall skills needed in this section.

In Exercises 1–3, use the properties of exponents to simplify each expression.

1. $3x^4y^2(5x^3y^7)$
2. $\dfrac{15x^4y^2}{5x^3y^7}$
3. $(5x^3y^7)^2$

In Exercises 4 and 5, write each expression by using exponential notation.

4. $x\sqrt{y}$
5. $\sqrt[3]{xy^2}$

EXERCISES 10.5

In Exercises 1–20, use the properties of logarithms to express each logarithm in terms of logarithms of simpler expressions. Assume that the argument of each logarithm is a positive real number.

1. $\log abc$
2. $\log xyz$
3. $\ln \dfrac{x}{11}$
4. $\ln \dfrac{y}{7}$
5. $\log xy^5$
6. $\log x^3y^4$
7. $\ln x^2y^3z^4$
8. $\ln (2x + 3)(x + 7)$
9. $\ln \dfrac{2x + 3}{x + 7}$
10. $\ln \dfrac{5x + 8}{3x + 4}$
11. $\log\sqrt{4x + 7}$
12. $\log\sqrt[3]{6x + 1}$
13. $\ln \dfrac{\sqrt{x + 4}}{(y + 5)^2}$
14. $\ln \dfrac{(x + 9)^3}{\sqrt{y + 1}}$
15. $\log\sqrt{\dfrac{xy}{z - 8}}$
16. $\log\sqrt[4]{\dfrac{x^2y}{z^3}}$
17. $\log \dfrac{x^2(2y + 3)^3}{z^4}$
18. $\log \dfrac{(x + 1)^3(y - 2)^2}{(z + 4)^5}$
19. $\ln \left(\dfrac{xy^2}{z^3}\right)$
20. $\ln \left(\dfrac{x^2y^3}{z^5}\right)^3$

In Exercises 21–30, combine the logarithmic terms into a single logarithmic expression with a coefficient of 1. Assume that the argument of each logarithm is a positive real number.

21. $2 \log x + 5 \log y$
22. $7 \ln x + 3 \ln y$
23. $3 \ln x + 7 \ln y - \ln z$
24. $4 \log x + 9 \log y - 5 \log z$
25. $\dfrac{1}{2} \log (x + 1) - \log (2x + 3)$
26. $\ln (3x + 8) - \dfrac{1}{2} \ln (5x + 1)$
27. $\dfrac{1}{3}[\ln (2x + 7) + \ln (7x + 1)]$
28. $\dfrac{1}{4}[\log (x + 3) - \log (2x + 9)]$
29. $2 \log_5 x + \dfrac{2}{3} \log_5 y$
30. $\dfrac{3}{5} \log_7 x - 2 \log_7 y$

In Exercises 31–40, use the properties of logarithms to determine the value of each expression. Assume that x and y are positive.

31. $17^{\log_{17} 34}$
32. $83^{\log_{83} 51}$
33. $1.93^{\log_{1.93} 0.53}$
34. $4.6^{\log_{4.6} 0.068}$
35. $10^{\log 37}$
36. $10^{\log 0.0093}$
37. $e^{\ln 0.045}$
38. $e^{\ln 453}$
39. $e^{\ln x}$
40. $10^{\log y}$

In Exercises 41–52, use $\log_b 2 \approx 0.3562$, $\log_b 5 \approx 0.8271$, and the properties of logarithms to approximate the value of each logarithmic expression.

41. $\log_b 10$
42. $\log_b 4$
43. $\log_b 25$
44. $\log_b\sqrt{5}$
45. $\log_b\sqrt[3]{2}$
46. $\log_b 5b$
47. $\log_b 2b$
48. $\log_b 2.5$
49. $\log_b 0.4$
50. $\log_b \dfrac{b}{5}$
51. $\log_b \dfrac{b}{2}$
52. $\log_b\sqrt{10b}$

In Exercises 53–62, match each expression with an equal expression from choices A through F. Note choices A and F can be the correct response to more than one exercise. Assume that the argument of each logarithm is a positive real number.

53. $\log \dfrac{x}{y}$
54. $\dfrac{\log x}{\log y}$
55. $\log xy$
56. $(\log x)(\log y)$
57. $\log (x - y)$
58. $\log (x + y)$
59. $\log x^y$
60. $\log y^x$
61. $10^{\log x}$
62. $e^{\ln x}$

A. x
B. $\log x + \log y$
C. $\log x - \log y$
D. $x \log y$
E. $y \log x$
F. This expression cannot be simplified.

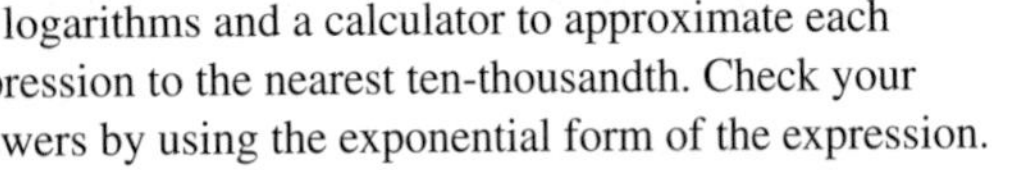

In Exercises 63–70, use the change-of-base formula for logarithms and a calculator to approximate each expression to the nearest ten-thousandth. Check your answers by using the exponential form of the expression.

63. $\log_5 37.1$ **64.** $\log_7 81.8$

65. $\log_{13} 7.08$ **66.** $\log_{11} 4.31$

67. $\log_{6.8} 0.085$ **68.** $\log_{7.2} 0.0032$

69. $\log_{0.49} 3.86$ **70.** $\log_{0.61} 18.4$

Estimation Skills and Calculator Skills

In Exercises 71–74, mentally estimate the value of each expression to the nearest integer, and then use a calculator to approximate each value to the nearest thousandth.

PROBLEM	MENTAL ESTIMATE	CALCULATOR APPROXIMATION
71. $\log_4 63$		
72. $\log_2 33$		
73. $1.99^{5.02}$		
74. $10.01^{1.999}$		

In Exercises 75 and 76, use the change-of-base formula for exponents to rewrite each exponential function as a function with base e.

75. **a.** $f(x) = 5^x$ **b.** $f(x) = \left(\frac{1}{5}\right)^x$

76. **a.** $f(x) = 10^x$ **b.** $f(x) = (0.1)^x$

In Exercises 77–79, the first column contains a student's answer to a calculus problem. The second column contains the book's answer. Use the properties of logarithms to show that these expressions are equal.

STUDENT'S ANSWER	BOOK'S ANSWER
77. $\ln\left(\frac{1}{2}\right)$	$-\ln 2$
78. $\log 24^x - \log 60^x - \log 0.2^{2x}$	x
79. $\ln y = kt + c_1$ (*Hint:* c_1 and c_2 are different constants.)	$y = c_2e^{kt}$

Group Discussion Questions

80. Challenge Question Every exponential growth or decay function can be expressed in terms of base e.

a. Select an exponential growth function, and write it in terms of base e.

b. Select an exponential decay function, and write it in terms of base e.

c. Describe how one can tell by inspection whether an exponential function base e represents growth or decay.

d. Prepare a justification for your answer in part **c** to present to your instructor.

81. Challenge Question Write $\ln x + t = 0$ as an exponential equation that does not contain logarithms.

82. Challenge Question Prove the quotient rule for logarithms. (*Hint:* See the book's proof at the beginning of this section for the product rule.)

83. Challenge Question Prove the power rule for logarithms.

84. Prove that $\log_b a = \dfrac{1}{\log_a b}$ for $a > 0$, $b > 0$, and $a \neq 1$ and $b \neq 1$.

Section 10.6 Solving Exponential and Logarithmic Equations

Objective: 10. Solve exponential and logarithmic equations.

An **exponential equation** contains a variable in the exponent. A **logarithmic equation** contains a variable in the argument of a logarithm. We solve both types of equations in this section.

Only relatively simple exponential equations, such as the one in Example 1, result in nice integer solutions. For other equations we often use a calculator to approximate the solutions.

Later in this section we also examine some identities involving exponents and logarithms.

EXAMPLE 1 Solving an Exponential Equation

Solve $2^{3x+2} = 32$.

SOLUTION

$2^{3x+2} = 32$

$2^{3x+2} = 2^5$ Express both sides of the equation in terms of base 2.

$3x + 2 = 5$ The exponents are equal since the bases are the same.

$3x = 3$

$x = 1$

Answer: $x = 1$

If both sides of an exponential equation cannot be expressed easily in terms of a common base, then we solve the equation by first taking logarithms of both sides of the equation. If two expressions are equal, then their logarithms are equal.

EXAMPLE 2 Solving an Exponential Equation

Algebraically determine the exact solution of $5^x = 30$. Also approximate this solution to the nearest ten-thousandth and determine this solution graphically.

SOLUTION

ALGEBRAICALLY

$5^x = 30$	
$\log 5^x = \log 30$	Take the common log of both sides of the equation.
$x \log 5 = \log 30$	Use the power rule for logarithms. Then divide both sides of the equation by log 5. This is the exact solution. $\left(\textit{Caution: } \frac{\log 30}{\log 5} \text{ is not the same as } \log \frac{30}{5}.\right)$ Then approximate this value with a calculator.
$x = \frac{\log 30}{\log 5}$	
$x \approx 2.1133$	

GRAPHICALLY

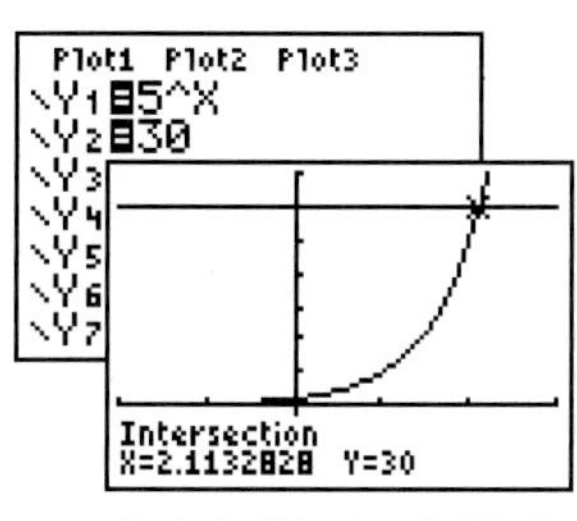

$[-2, 3, 1]$ by $[-10, 35, 5]$

Let $Y_1 = 5x$ and $Y_2 = 30$, and use the **intersect** feature on a calculator to determine the point of intersection.

The approximate x-coordinate of the point of intersection is $x \approx 2.1133$.

Answer: $x = \frac{\log 30}{\log 5} \approx 2.1133$ ***Check:*** $5^{2.1133} \approx 30.000833 \approx 30$

SELF-CHECK 10.6.1

1. Graphically approximate the solution of $2^x = 7$ to the nearest thousandth.

The procedure used in Examples 1 and 2 is summarized in the following box.

Solving Exponential Equations Algebraically

VERBALLY	ALGEBRAIC EXAMPLE
Step 1. a. If it is obvious that both sides of the equation are powers of the same base, express each side of the equation as a power of this base and then equate the exponents.	a. $2^x = 8$ $2^x = 2^3$ $x = 3$
b. Otherwise, take the logarithm of both sides of the equation and use the power rule to form an equation that does not contain variable exponents.	b. $2^x = 7$ $\log 2^x = \log 7$ $x \log 2 = \log 7$
Step 2. Solve the equation formed in step 1.	$x = \frac{\log 7}{\log 2}$

When we take the logarithm of both sides of an equation, we usually use either common logs or natural logs. We do this because these functions are available on calculators. Example 3 uses natural logs.

EXAMPLE 3 Solving an Exponential Equation

Solve $6^{z+3} = 8^{2z-1}$ and approximate this solution to the nearest ten-thousandth.

SOLUTION

$$6^{z+3} = 8^{2z-1}$$

$$\ln 6^{z+3} = \ln 8^{2z-1}$$

Because 6 and 8 are not powers of the same base, take the natural log of both sides of the equation.

$$(z + 3)\ln 6 = (2z - 1)\ln 8$$

Use the power rule for logarithms.

$$z \ln 6 + 3 \ln 6 = 2z \ln 8 - \ln 8$$

$$z \ln 6 - 2z \ln 8 = -\ln 8 - 3 \ln 6$$

Multiply by using the distributive property, and then rewrite the equation so that the variable terms are on the left side of the equation.

$$z(\ln 6 - 2 \ln 8) = -(\ln 8 + 3 \ln 6)$$

$$z = -\frac{\ln 8 + 3 \ln 6}{\ln 6 - 2 \ln 8}$$

Factor out z, and then divide both sides of the equation by the coefficient of z. This is the exact solution.

$$z \approx 3.1492736$$

Use a calculator to determine an approximation of the exact answer obtained in the previous step.

Answer: $z \approx 3.1493$ Does this value check?

SELF-CHECK 10.6.2

1. Solve $3^{2z+1} = 5^z$ to the nearest thousandth.

As Examples 2 and 3 illustrate, either common logs or natural logs can be used to solve exponential equations. If the exponential equation involves base e, then we usually use natural logarithms.

EXAMPLE 4 Solving an Exponential Equation

Solve $3e^{x^2+2} = 49.287$, and then approximate this solution to the nearest ten-thousandth.

SOLUTION

$$3e^{x^2+2} = 49.287$$

$$e^{x^2+2} = 16.429$$

Divide both sides of the equation by 3.

$$x^2 + 2 = \ln 16.429$$

Rewrite this exponential equation in logarithmic form with base e.

$$x^2 = \ln 16.429 - 2$$

Subtract 2 from both sides of the equation.

$$x = \pm\sqrt{\ln 16.429 - 2}$$

Find the exact roots of this quadratic equation by extraction of roots.

$$x \approx \pm 0.8938948853$$

Use a calculator to approximate this value.

Answer: $x \approx -0.8939$ or $x \approx 0.8939$ Do these values check?

Since exponential and logarithmic functions are inverses of each other, it is not surprising that we can use logarithms to solve some exponential equations. Likewise, we can solve some logarithmic equations by first rewriting them in exponential form.

EXAMPLE 5 Solving a Logarithmic Equation with One Logarithmic Term

Solve $\log(3x + 7) = 2$ algebraically, numerically, and graphically.

SOLUTION

ALGEBRAICALLY

$$\begin{aligned} \log(3x+7) &= 2 \\ 3x + 7 &= 10^2 \\ 3x + 7 &= 100 \\ 3x &= 93 \\ x &= 31 \end{aligned}$$

Rewrite this logarithmic equation in exponential form with base 10.

Then solve for x.

NUMERICALLY

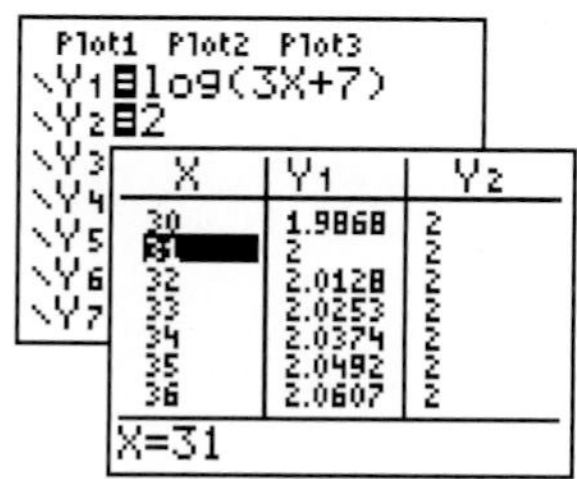

Let $Y_1 = \log(3x + 7)$ and $Y_2 = 2$, and then examine these functions both by tables and by graphs.

For $x = 31$, $Y_1 = Y_2$.

GRAPHICALLY

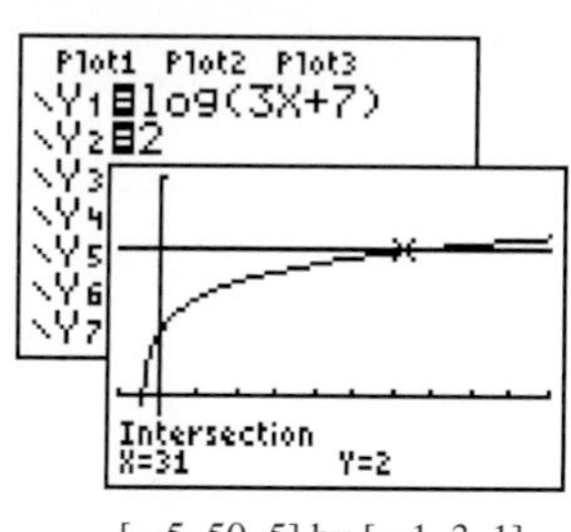

[−5, 50, 5] by [−1, 3, 1]

The x-coordinate of the point of intersection is $x = 31$.

Answer: $x = 31$

Check: $\log[3(31) + 7] \stackrel{?}{=} 2$
$\log 100 \stackrel{?}{=} 2$
$2 \stackrel{?}{=} 2$ checks.

Logarithmic functions are one-to-one functions.

Logarithmic functions are one-to-one functions. (The horizontal line test can be applied to the graph of $y = \log_b x$ to show that this function is one-to-one.) Thus if $x, y > 0$ and $\log_b x = \log_b y$, then $x = y$. The fact that the arguments are equal when logarithms are equal allows us to replace a logarithmic equation with one that does not involve logarithms. Because real logarithms are defined only for positive arguments, each possible solution must be checked to make sure that it is not an extraneous value.

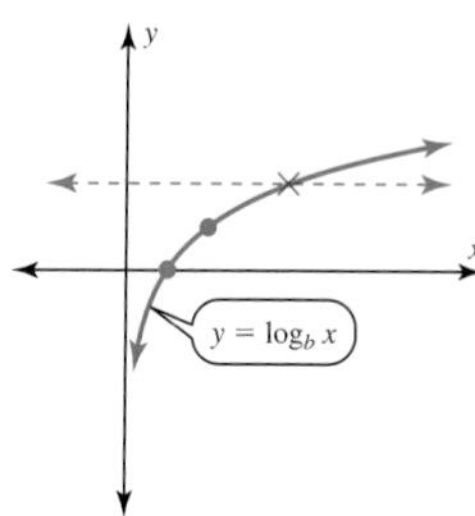

EXAMPLE 6 Solving a Logarithmic Equation with Two Logarithmic Terms

Solve $\ln y = \ln(4y + 6)$.

SOLUTION

$\ln y = \ln(4y + 6)$	
$y = 4y + 6$	Since the logarithms are equal, the arguments are equal.
$-3y = 6$	
$y = -2$	Now solve for y.
Check: $\ln y = \ln(4y + 6)$	
$\ln(-2) \stackrel{?}{=} \ln[4(-2) + 6]$	Logarithms of negative argument values are not real numbers.
$\ln(-2)$ is not a real number.	
Answer: There is no solution.	-2 is an extraneous value.

SELF-CHECK 10.6.3

Solve these logarithmic equations.

1. $\log(t - 1) = 1$
2. $\log(v - 5) = \log(1 - v)$

Solving Logarithmic Equations Algebraically

VERBALLY	ALGEBRAIC EXAMPLE
Step 1. a. If possible, rewrite the logarithmic equation in exponential form. b. Otherwise, use the properties of logarithms to write each side of the equation as a single logarithmic term with the same base. Then form a new equation by equating the arguments of these logarithms. **Step 2.** Solve the equation formed in step 1. **Step 3.** Check all possible solutions for extraneous values; logarithms of zero or negative arguments are undefined.	**a.** $\log(x + 1) = 2$ $x + 1 = 10^2$ $x = 99$ *Check:* $\log(99 + 1) \stackrel{?}{=} 2$ $\log(100) \stackrel{?}{=} 2$ $2 \stackrel{?}{=} 2$ checks. **b.** $\ln 2x = \ln(x + 1)$ $2x = x + 1$ $x = 1$ *Check:* $\ln 2(1) \stackrel{?}{=} \ln(1 + 1)$ $\ln 2 \stackrel{?}{=} \ln 2$ checks.

If a logarithmic equation has more than one logarithmic term on one side of the equation, then we may need to use the properties of logarithms to rewrite these terms as a single logarithm.

EXAMPLE 7 Solving a Logarithmic Equation with Three Logarithmic Terms

Solve $\ln(3 - w) + \ln(1 - w) = \ln(11 - 6w)$.

SOLUTION

$$\ln(3 - w) + \ln(1 - w) = \ln(11 - 6w)$$
$$\ln[(3 - w)(1 - w)] = \ln(11 - 6w)$$
$$(3 - w)(1 - w) = 11 - 6w$$
$$3 - 4w + w^2 = 11 - 6w$$

Express the left side of the equation as a single logarithm, by using the product rule for logarithms. Since the natural logarithms are equal, the arguments are equal.

$$w^2 + 2w - 8 = 0$$
$$(w + 4)(w - 2) = 0$$
$$w + 4 = 0 \quad \text{or} \quad w - 2 = 0$$
$$w = -4 \qquad\qquad w = 2$$

Combine like terms and write the quadratic equation in standard form. Then factor and solve for w.

Be sure to check possible solutions to a logarithmic equation to make sure the values are not extraneous.

Check: For $w = -4$,

$$\ln[3 - (-4)] + \ln[1 - (-4)] \stackrel{?}{=} \ln[11 - 6(-4)]$$
$$\ln 7 + \ln 5 \stackrel{?}{=} \ln(11 + 24)$$
$$\ln 35 \stackrel{?}{=} \ln 35 \text{ checks.}$$

Check: For $w = 2$,

$$\ln(3 - 2) + \ln(1 - 2) \stackrel{?}{=} \ln[11 - 6(2)]$$
$$\ln 1 + \ln(-1) \stackrel{?}{=} \ln(-1)$$
$$\ln(-1) \text{ is undefined.}$$

Thus $w = 2$ is an extraneous value.

Answer: $w = -4$

SELF-CHECK 10.6.4

1. Solve for z.
   $\ln(2z + 5) + \ln z = \ln 3$.

EXAMPLE 8 Solving a Logarithmic Equation with Two Logarithmic Terms

Solve $\log(2v + 1) - \log(v - 4) = 1$.

SOLUTION

$$\log(2v + 1) - \log(v - 4) = 1$$
$$\log\left(\frac{2v + 1}{v - 4}\right) = 1$$

Express the left side of the equation as a single logarithm, by using the quotient rule for logarithms.

$$\frac{2v + 1}{v - 4} = 10^1$$

Rewrite this equation in exponential form, by using a base of 10.

Then solve for v.

$$2v + 1 = 10(v - 4)$$
$$2v + 1 = 10v - 40$$
$$-8v = -41$$
$$v = \frac{41}{8}$$

$$\textit{Check: } \log\left[2\left(\frac{41}{8}\right)+1\right]-\log\left(\frac{41}{8}-4\right)\stackrel{?}{=}1$$
$$\log\left(\frac{45}{4}\right)-\log\left(\frac{9}{8}\right)\stackrel{?}{=}1$$
$$\log\left(\frac{45}{4}\div\frac{9}{8}\right)\stackrel{?}{=}1$$
$$\log 10\stackrel{?}{=}1$$
$$1\stackrel{?}{=}1 \text{ checks.}$$

Answer: $v=\dfrac{41}{8}$

The exponential and logarithmic equations that we have examined in this section have been either contradictions with no solution or conditional equations with one or two solutions. We now examine some exponential and logarithmic equations that are identities.

EXAMPLE 9 Verifying Logarithmic and Exponential Identities

Verify the following identities.

SOLUTION

(a) $\log_7 14^x - \log_7 2^x = x$

$\log_7 14^x - \log_7 2^x = x\log_7 14 - x\log_7 2$ — Rewrite the left side of the equation by using the power rule for logarithms.

$\log_7 14^x - \log_7 2^x = x(\log_7 14 - \log_7 2)$ — Then factor out x.

$\log_7 14^x - \log_7 2^x = x\log_7 \dfrac{14}{2}$ — Simplify this expression by using the quotient rule for logarithms.

$\log_7 14^x - \log_7 2^x = x\log_7 7$

$\log_7 14^x - \log_7 2^x = x(1)$ — Thus you have verified that the left side of the equation does equal the right side of the equation.

$\log_7 14^x - \log_7 2^x = x$

(b) $e^{-\ln x}=\dfrac{1}{x}$ for $x>0$

$e^{-\ln x}=e^{\ln(x^{-1})}$ — Rewrite the left side of the equation by using the power rule for logarithms: $p\ln x=\ln x^p$ with $p=-1$.

$e^{-\ln x}=x^{-1}$ — Simplify by using the identity $b^{\log_b y}=y$, with $b=e$ and $y=x^{-1}$. Replace x^{-1} with $\dfrac{1}{x}$ to obtain the right side of this identity.

$e^{-\ln x}=\dfrac{1}{x}$

SELF-CHECK ANSWERS

10.6.1
1. $x\approx 2.807$

10.6.2
1. $z\approx -1.869$

10.6.3
1. $t=11$
2. No solution; $v=3$ is extraneous.

10.6.4
1. $z=\dfrac{1}{2}$

USING THE LANGUAGE AND SYMBOLISM OF MATHEMATICS 10.6

1. An __________ equation contains a variable in the exponent.
2. A __________ equation contains a variable in the argument of a logarithm.
3. If $x>0$, $y>0$, and $\log_b x=\log_b y$, then __________ = __________.
4. An __________ value is a value produced in the solution process that does not check in the original equation.

QUICK REVIEW 10.6

The purpose of this quick review is to help you recall skills needed in this section.

In Exercises 1–3, use a calculator to approximate to the nearest thousandth the value of each expression.

1. $\log\left(\frac{48}{5}\right)$ **2.** $\frac{\log 48}{5}$ **3.** $\frac{48}{\ln 5}$

4. Use the tables shown to solve this system of linear equations.
$y = 2.38x - 13.9$
$y = -4.71x + 21.55$

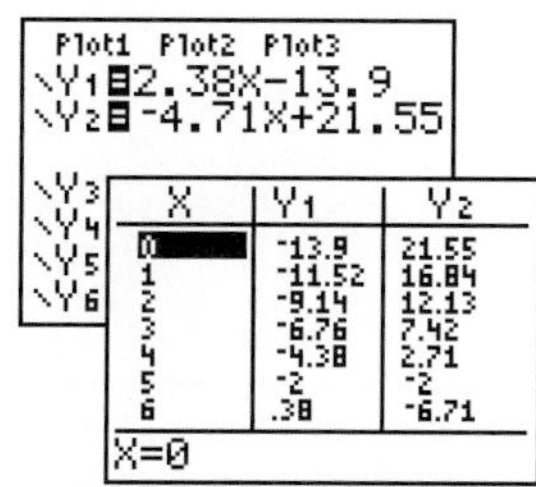

5. Use the graph shown to solve this system of linear equations.
$y = 2.4x + 0.1$
$y = -3.2x + 5.7$

Plot1 Plot2 Plot3
\Y1■2.4X+0.1
\Y2■-3.2X+5.7
Intersection
X=1 Y=2.5

[−2, 5, 1] by [−1, 5, 1]

EXERCISES 10.6

In Exercises 1–24, solve each equation without using a calculator.

1. $3^{w-5} = 27$ **2.** $7^{2w-1} = \sqrt{7}$

3. $\left(\frac{2}{3}\right)^{x^2} = \frac{16}{81}$ **4.** $5^{x^2-6} = \frac{1}{25}$

5. $\log(y - 5) = 1$ **6.** $\log(2y + 1) = 1$

7. $\log(3n - 5) = 2$ **8.** $\log(5n - 4) = 0$

9. $\ln(3m - 7) = \ln(2m + 9)$

10. $\log(5y + 6) = \log(2y - 9)$

11. $\log(4w + 3) = \log(8w + 5)$

12. $\ln(5y - 7) = \ln(2y + 1)$

13. $\ln(3 - x) = \ln(1 - 2x)$

14. $\log(7 - 5x) = \log(4 - 8x)$

15. $\log(t + 3) + \log(t - 1) = \log 5$

16. $\ln(7t + 3) - \ln(t + 1) = \ln(6t + 2)$

17. $\ln(v^2 - 9) - \ln(v + 3) = \ln 7$

18. $\ln(2v + 6) - \ln(v + 1) = \ln(v + 3)$

19. $\ln(5x - 7) - \ln(2x + 3) = \ln 3$

20. $\log(7x + 13) - \log(4x + 13) = \log 5$

21. $\log(1 - y) + \log(4 - y) = \log(18 - 10y)$

22. $\ln(2 - y) + \ln(1 - y) = \ln(32 - 4y)$

23. $\log(w - 3) - \log(w^2 + 9w - 32) = -1$

24. $\log(w^2 + 1) - \log(w - 2) = 1$

In Exercises 25 and 26, use the given tables to solve each equation.

25. $4^x = 32$ **26.** $32^x = 64$

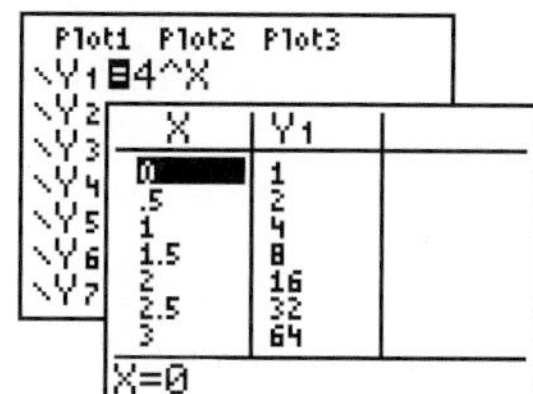

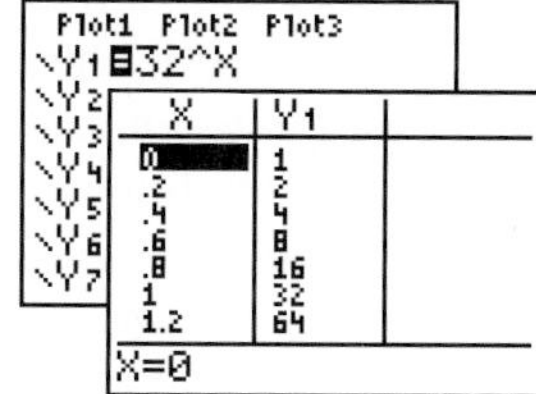

In Exercises 27 and 28, use the given graphs to solve each equation.

27. $8^{(x+2)/3} = 16$ **28.** $27^{(2x-4)/3} = 9$

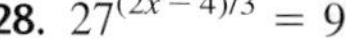

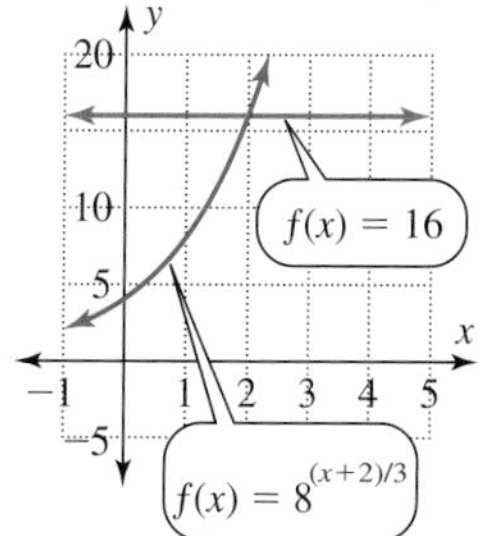

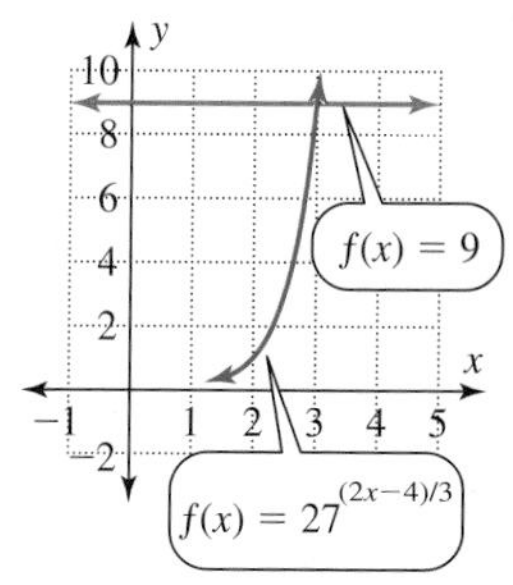

In Exercises 29–44, solve each equation and then use a calculator to approximate the solution to the nearest thousandth.

29. $4^x = 15$ **30.** $5^x = 18$

31. $3^{-w+7} = 22$ **32.** $6^{-w+3} = 81$

33. $9.2^{2t+1} = 11.3^t$ **34.** $8.7^{3t-1} = 10.8^{2t}$

35. $7.6^{-2z} = 5.3^{2z-1}$ **36.** $8.1^{-3z} = 6.5^{1-2z}$

37. $e^{3x} = 78.9$ **38.** $e^{3x+5} = 15.9$

39. $10^{2y+1} = 51.3$ **40.** $10^{3y-4} = 73.8$

41. $0.83^{v^2} = 0.68$ **42.** $0.045^{v^2} = 0.0039$

43. $3.7e^{x^2+1} = 689.7$ **44.** $2.5e^{x^2+4} = 193.2$

In Exercises 45 and 46, use the given tables to solve each equation.

45. $\log(5x - 9.99) = -2$ **46.** $\log(90x + 100) = 3$

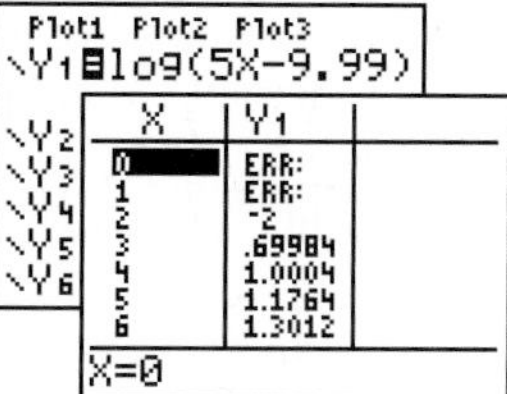

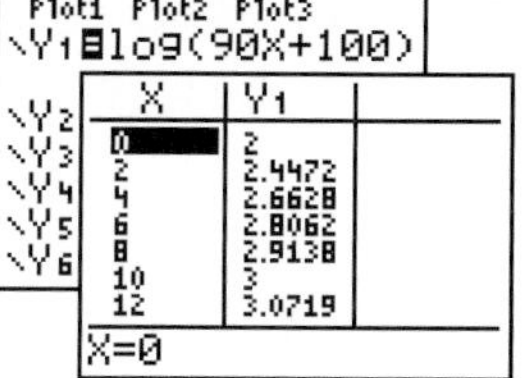

In Exercises 47 and 48, use the given graphs to solve each equation.

47. $\log(100x^2) = 2$

48. $\log(100x + 1{,}000) = 2$

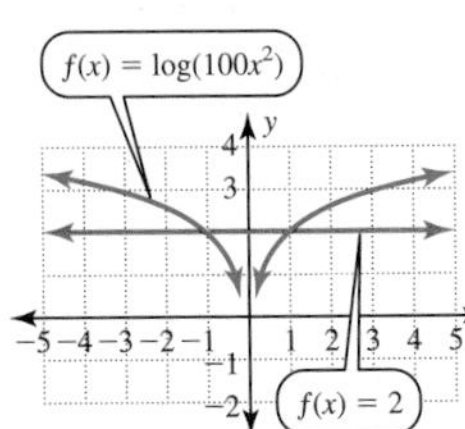

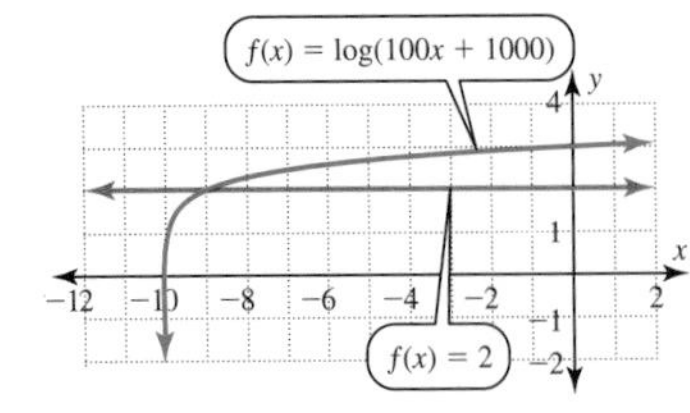

In Exercises 49–58, solve each equation and then use a calculator to approximate each solution to the nearest thousandth.

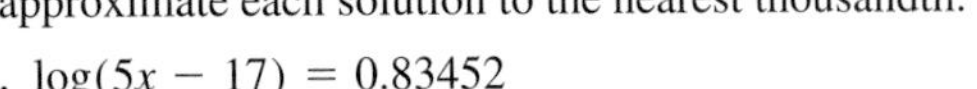

49. $\log(5x - 17) = 0.83452$
50. $\ln(11x - 3) = 1.44567$
51. $\ln(\ln y) = 1$
52. $\log(\log y) = 0.48913$
53. $\ln(v - 4) + \ln(v - 3) = \ln(5 - v)$
54. $\log(3x + 1) + \log(x + 2) = \log x$
55. $\ln(11 - 5x) - \ln(x - 2) = \ln(x - 6)$
56. $\ln(3 - x) - \ln(x - 2) = \ln(x + 1)$
57. $(\ln x)^2 = \ln x^2$
58. $(\log x)^2 = \log x^2$

In Exercises 59–62, use a graphing calculator to approximate the solution(s) of each equation to the nearest thousandth.

59. $\log(398x + 5) = x + 0.5$
60. $\ln(18x + 3) = 3x - 5$
61. $0.1e^x = 8 - 2x$
62. $e^{0.5x} = 4 - x^2$

In Exercises 63–70, verify each identity, assuming that x is a positive real number.

63. $10^{-\log x} = \dfrac{1}{x}$
64. $100^{\log x} = x^2$
65. $e^{-x \ln 3} = \left(\dfrac{1}{3}\right)^x$
66. $e^{(\ln x)/2} = \sqrt{x}$
67. $\log 60^x - \log 6^x = x$
68. $\log 5^x + \log 2^x = x$
69. $\ln\left(\dfrac{4}{5}\right)^x + \ln\left(\dfrac{5}{3}\right)^x + \ln\left(\dfrac{3}{4}\right)^x = 0$
70. $\ln\left(\dfrac{2}{3}\right)^x + \ln\left(\dfrac{5}{2}\right)^x - \ln\left(\dfrac{5}{3}\right)^x = 0$

71. **Depreciation** The value V of an industrial lathe after t years of depreciation is given by the formula $V = 35{,}000e^{-0.2t} + 1{,}000$. Approximately how many years will it take for the value to depreciate to \$10,000?

72. **Depreciation** The value V of an irrigation system after t years of depreciation is given by the formula $V = 59{,}000e^{-0.2t} + 3{,}000$. Approximately how many years will it take for the value to depreciate to \$7,000?

Group Discussion Questions

73. **Challenge Question** Which is larger, 2001^{2002} or 2002^{2001}?

74. **Error Analysis** The steps used to solve a certain logarithmic equation produced one possible solution that was a negative number. One student claimed this number could not check in the equation, since it was negative. Discuss the logic of this student's claim.

75. **Discovery Question**
   a. Create an exponential equation whose only solution is 5.
   b. Create an exponential equation whose only solutions are -2 and 5.

76. **Discovery Question**
   a. Create a logarithmic equation with $x = -2$ as an extraneous value and that has no solution.
   b. Create a logarithmic equation with $x = -2$ as an extraneous value and with a solution of 5.
   c. Create a logarithmic equation with solutions of -2 and 5.

Section 10.7 Exponential Curve Fitting and Other Applications of Exponential and Logarithmic Equations

Objective: 11. Use exponential and logarithmic equations to solve applied problems.

Exponential and logarithmic equations can be used to describe an increase or a decrease in an amount of money, the growth or decline of populations, and the decay of radioactive elements. Logarithmic scales are often used to measure natural phenomena. For example, decibels measure the intensity of sounds, and the Richter scale measures the intensity of earthquakes. In this section we examine all these applications.

Two important formulas that we apply in the following examples are the formulas for periodic growth and continuous growth. Growth that occurs at discrete intervals (such as yearly, monthly, or weekly) is called **periodic growth;** growth that occurs continually at each instant is called **continuous growth.** The formula for continuous growth is a good model for periodic growth when the number of growth periods is relatively large. For example, the population of rabbits over a period of years can be accurately estimated by using the continuous-growth formula, even though female rabbits do not have offspring continuously.

Growth and Decay Formulas

	VERBALLY	ALGEBRAIC EXAMPLE
Periodic growth formula: $A = P\left(1 + \frac{r}{n}\right)^{nt}$	The amount A that is produced by an original amount P growing at an annual rate r with periodic compounding n times a year for t years.	$A = 1{,}000\left(1 + \frac{0.07}{4}\right)^{(4)(10)}$ This is the amount resulting from investing \$1,000 at 7% compounded quarterly for 10 years.
Continuous growth and decay formula: $A = Pe^{rt}$	The amount A that is produced by an original amount P growing (decaying) continuously at a rate r for a time t. If $r > 0$, there is growth. If $r < 0$, there is decay.	$A = 1{,}000e^{(0.07)(10)}$ This is the amount resulting from investing \$1,000 at 7% compounded continuously for 10 years.

It is customary that most statements of interest rates are given as annual rates. In Example 1 the rate of 11% means an annual rate of 11%.

EXAMPLE 1 Periodic Growth of an Investment

How many years will it take an investment of \$1,000 to double in value if interest on the investment is compounded semiannually at a rate of 11%?

SOLUTION

$$A = P\left(1 + \frac{r}{n}\right)^{nt}$$

$$2{,}000 = 1{,}000\left(1 + \frac{0.11}{2}\right)^{2t}$$ Substitute $A = 2{,}000$, $P = 1{,}000$, $r = 0.11$, and $n = 2$ into the periodic-growth formula and then simplify.

$$2 = (1.055)^{2t}$$

$$\ln 2 = \ln(1.055)^{2t}$$ Take the natural log of both sides of the equation.

$$\ln 2 = 2t \ln 1.055$$ Use the power rule for logarithms.

$$t = \frac{\ln 2}{2 \ln 1.055}$$ Solve for t by dividing both sides of the equation by $2 \ln 1.055$.

$$t \approx 6.4730785$$ Use a calculator to approximate this value.

Answer: The investment will double in value in approximately 6.5 years.

SELF-CHECK 10.7.1

1. Determine the value of \$1,000 invested at 7% interest compounded quarterly for 10 years.
2. Compute the number of years it will take an investment of \$1 to triple in value if the interest is compounded annually at 8%.

A comparison of periodic growth to continuous growth is given in Table 10.7.1. This table shows various compounding periods and the amount to which \$1 accumulates in each situation after 1 year. The interest formula, with $P = 1$ and $r = 1.00$, is $A = 1\left(1 + \frac{1}{n}\right)^{n(1)} = \left(1 + \frac{1}{n}\right)^n$.

Note that compounding monthly produces a big gain over simple interest. However, further gains are relatively small, and as n increases, the expression $\left(1 + \frac{1}{n}\right)^n$ approaches the irrational number e.

Table 10.7.1 Periodic Compounding and Continuous Compounding for 1 Year

PERIODIC COMPOUNDING	n	$A = \left(1 + \frac{1}{n}\right)^n$ (TO SIX PLACES)	A (TO THE NEAREST PENNY)
Annually	1	2.000000	\$2.00
Semiannually	2	2.250000	2.25
Quarterly	4	2.441406	2.44
Monthly	12	2.613035	2.61
Weekly	52	2.692597	2.69
Daily	365	2.714567	2.71
Hourly	8,760	2.718127	2.72
Every minute	525,600	2.718279	2.72
Every second	31,536,000	2.718282	2.72
⋮	⋮	⋮	
Continuously	Infinite	e	
CONTINUOUS COMPOUNDING	$A = Pe^{rt} = e$	with $P = 1$, $r = 1$, and $t = 1$.	

The formula for continuous compounding provides a reasonable approximation of periodic compounding when there are many growth periods.

A key point to note is that continuous compounding provides a reasonable approximation for some periodic events; this is especially true when there are many growth periods. Because the formula $A = Pe^{rt}$ is easy to use and can provide a good approximation of periodic growth, it is used widely in growth and decay problems. Examples 2 through 4 illustrate some of the possibilities.

EXAMPLE 2 Comparing Periodic Growth to Continuous Growth

Compute the value of \$1,000 invested at 6% for 5 years under two conditions.

(a) The interest is compounded daily.

(b) The interest is compounded continuously.

SOLUTION

(a) $A = P\left(1 + \frac{r}{n}\right)^{nt}$ — Use the formula for periodic growth. Substitute in the given values. Use 365 days for each year (assume that there is no change due to leap year).

$A = 1{,}000\left(1 + \frac{0.06}{365}\right)^{(365)(5)}$

$A \approx 1{,}000(1.000164384)^{1825}$ — Use a calculator to approximate this value.

$A \approx \$1{,}349.83$

(b) $A = Pe^{rt}$ — Use the formula for continuous growth. Substitute in the given values.

$A = 1{,}000e^{(0.06)(5)}$

$A = 1{,}000e^{0.30}$

$A \approx \$1{,}349.86$ — Use a calculator to approximate this value.

Note that the results in Example 2 differ by only 3 cents over a 5-year period. Thus continuous growth provided a good model for growth compounded daily.

> **SELF-CHECK 10.7.2**
>
> Compute the value of \$5,000 invested at 7% for 8 years under two conditions.
>
> 1. The interest is compounded monthly.
> 2. The interest is compounded continuously.

EXAMPLE 3 Continuous Growth of Bacteria

A new culture of bacteria grows continuously at the rate of 20% per day. If a culture of 10,000 bacteria isolated in a laboratory is allowed to multiply, how many bacteria will there be at the end of 1 week?

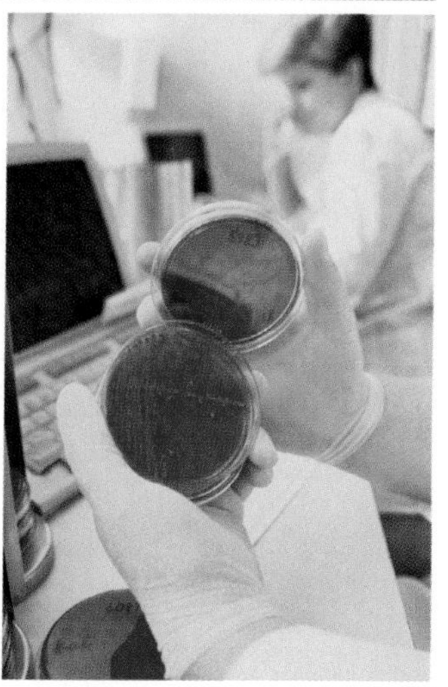

SOLUTION

$A = Pe^{rt}$	Substitute $P = 10{,}000$, $r = 0.20$, and $t = 7$ (1 week $= 7$ days) into the continuous-growth formula, and then simplify.
$A = 10{,}000e^{0.20(7)}$	
$A = 10{,}000e^{1.4}$	
$A \approx 40{,}552$	Use a calculator to approximate this value.

Answer: At the end of the week there will be approximately 41,000 bacteria.

For physical phenomena such as radioactive decay, exponential decay models have been experimentally tested and provide an excellent mathematical model for predicting the future or describing the past. This is illustrated in Example 4.

EXAMPLE 4 Continuous Decay of Carbon-14

Carbon-14 decays continuously at the rate of 0.01245% per year. When living tissue dies, it no longer absorbs carbon-14; any carbon-14 present decays and is not replaced. An archaeologist has determined that only 20% of the carbon-14 originally in a plant specimen remains. Estimate the age of this specimen.

SOLUTION

$A = Pe^{rt}$	Substitute the given values into the continuous-decay formula. The 0.01245% decay rate means that $r = -0.0001245$, and 20% of the original amount P left means that $A = 0.20P$.
$0.20P = Pe^{-0.0001245t}$	
$0.20 = e^{-0.0001245t}$	Divide both sides of the equation by P.
$\ln 0.20 = -0.0001245t$	Rewrite this statement in logarithmic form.
$t = \dfrac{\ln 0.20}{-0.0001245}$	Solve this linear equation for t.
$t \approx 12{,}927.21215$	Use a calculator to approximate this value.

Answer: This specimen is approximately 13,000 years old.

SELF-CHECK 10.7.3

1. The population of an island is growing continuously at the rate of 5% per year. Estimate to the nearest thousand the population of the island in 10 years, given that the population is now 100,000.
2. The population of a species of whales is estimated to be decreasing continuously at a rate of 5% per year ($r = -0.05$). If this rate of decrease continues, in how many years will the population have declined from its current level of 20,000 to 8,000?

Logarithmic growth functions grow very slowly. For example, log 100,000 = 5, while log 1,000,000 = 6. The increase from 100,000 to 1,000,000 causes an increase in the logarithm of only 1 unit from 5 to 6. Scientists take advantage of this property to describe phenomena whose numerical measure covers a wide range of scores. By converting these measures to a logarithmic scale, the numbers become smaller and more comprehensible. Two examples of this are the Richter scale and the decibel scale. These scales are illustrated in Examples 5 and 6.

Seismologists use the Richter scale to measure the magnitude of earthquakes. The equation $R = \log \frac{A}{a}$ compares the amplitude A of the shock wave of an earthquake to the amplitude a of a reference shock wave of minimal intensity.

EXAMPLE 5 Using the Richter Scale

In the Denver, Colorado, area in the 1960s an earthquake with an amplitude 40,000 times the reference amplitude occurred after a liquid was injected under pressure into a well more than 2 mi deep. Calculate the magnitude of this earthquake on the Richter scale.

SOLUTION

$R = \log\left(\frac{A}{a}\right)$

$R = \log\left(\frac{40{,}000a}{a}\right)$ Substitute the given amplitude into the Richter scale equation.

$R = \log 40{,}000$ Simplify the fraction.

$R \approx 4.6020600$ Use a calculator to approximate this value.

Answer: This earthquake measured approximately 4.6 on the Richter scale.

The human ear can hear a vast range of sound intensities, so it is more practical to use a logarithmic scale than to use an absolute scale to measure the intensity of sound. The unit of measurement on this scale is the **decibel.** The number of decibels D of a sound is given by the formula $D = 10 \log \frac{I}{I_o}$, which compares the intensity I of the sound to the reference intensity I_o, which is at the threshold of hearing ($I_o \approx 10^{-16}$ W/cm^2).

A Mathematical Note

A decibel is not a unit of loudness but rather a measure of the energy of the sound. A decibel (abbreviatal dB) is one-tenth of a bel—a unit of measure named after Alexander Graham Bell, the inventor of the telephone.

EXAMPLE 6 Determining the Decibel Level of Music

If a store in a mall plays background music at an intensity of 10^{-14} W/cm^2, determine the number of decibels (dB) of the music you hear.

SOLUTION

$D = 10 \log \frac{I}{I_o}$

$D = 10 \log \frac{10^{-14}}{10^{-16}}$ Substitute 10^{-14} for I and 10^{-16} for I_o into the decibel equation.

$D = 10 \log 10^2$ Simplify and then solve for D by inspection.

$D = 10(2)$

$D = 20$

Answer: 20 dB

SELF-CHECK 10.7.4

1. Compute the number of decibels of a child's cry if its intensity is 10^{-7} W/cm^2.

Chemists use pH to measure the hydrogen ion concentration in a solution. Distilled water has a pH of 7, acids have a pH of less than 7, and bases have a pH of more than 7. The formula for the pH of a solution is $\text{pH} = -\log \text{H}^+$, where H^+ measures the concentration of hydrogen ions in moles per liter.

EXAMPLE 7 Determining the pH of a Beer

Determine the pH of a light beer if its H^+ is measured at 6.3×10^{-5} mol/L.

SOLUTION

$\text{pH} = -\log \text{H}^+$

$\text{pH} = -\log(6.3 \times 10^{-5})$ Substitute $\text{H}^+ = 6.3 \times 10^{-5}$ into the pH equation.

$\text{pH} \approx 4.2007$ Approximate this value with a calculator.

Answer: $\text{pH} \approx 4.2$

A Mathematical Note

In a November 2004 article in *Technology Review,* Rodney Brooks states that a stable system can become unstable when even one component experiences exponential growth. Digital memory is growing exponentially and dramatically changing the marketplace. In 2004, an iPod could store 20,000 books—more than most people will ever read in a lifetime. At this rate a teenager in 2025 will likely be able to own a \$400 device that can hold every movie ever made.

Example 8 examines data that are best modeled by exponential growth and illustrates how to use a graphing calculator to approximate the exponential curve of best fit. The exponential regression feature **ExpReg,** which is in the **CALC** submenu of the **STAT** menu on a TI-84 Plus calculator, can be used by following almost exactly the same steps given in Calculator Perspective 7.3.3.

EXAMPLE 8 Using a Graphing Calculator to Determine an Exponential Curve of Best Fit

Enter these points into a graphing calculator, draw the scatter diagram, and calculate the exponential curve of best fit. Each x-value represents a year after 1970, and each y-value represents the number of transistors on an integrated circuit. Use this exponential equation to approximate the number of transistors that were on an integrated circuit in 1985.

YEAR	NUMBER OF YEARS AFTER 1970 x	NUMBER OF TRANSISTORS y
1971	1	2,250
1972	2	2,500
1974	4	5,000
1978	8	29,000
1982	12	120,000

SOLUTION

1 Enter the data points by using the **STAT-EDIT** feature.

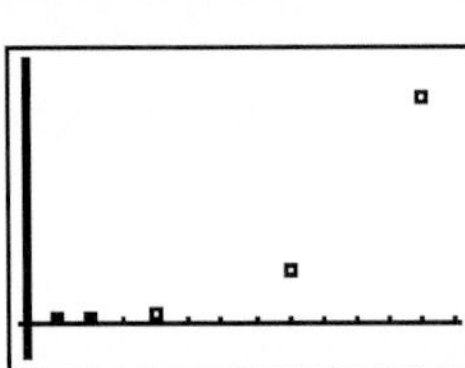

Enter all the (x, y) points.

2 Turn on **Plot 1** to draw a scatter diagram of the data points listed in L1 and L2. Use the visual pattern to select the most appropriate family of curves. In this case, the visual pattern is exponential growth.

[−0.1, 13.1, 1] by
[−17767, 140017, 1]

Since the points appear to form an exponential growth curve, try to find an exponential function that best fits these data.

3 Use the **STAT-CALC** feature to calculate the exponential curve of best fit.

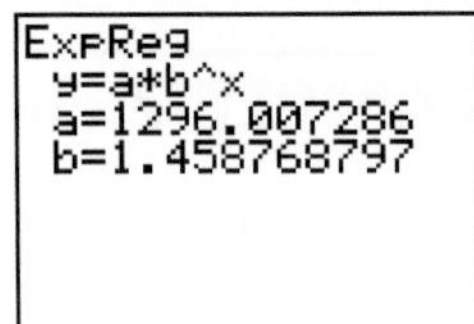

To calculate the exponential curve of best fit, select option 0 under the STAT **CALC** menu. This exponential curve of best fit does *not* contain each data point, but it does fit these points better than any other exponential curve. Note that the equation of the curve of best fit can be inserted into the Y= screen from the memory of the calculator by pressing

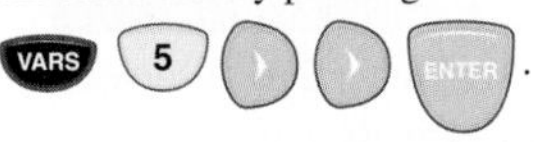

4 Enter the exponential curve of best fit on the Y= screen, and then graph this curve. Round each coefficient to three significant digits.

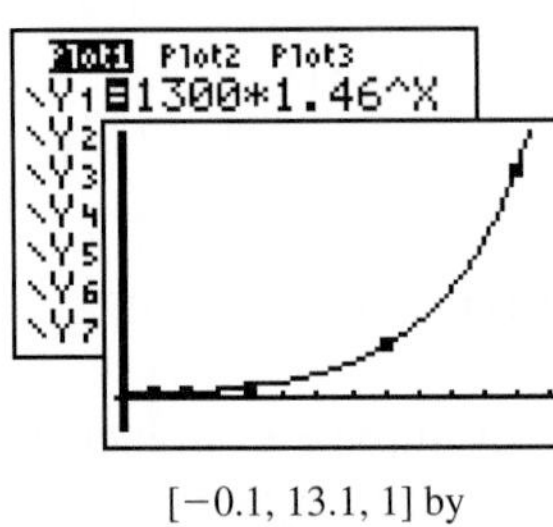

[−0.1, 13.1, 1] by
[−17767, 140017, 1]

The equation of the exponential growth function of best fit is written in $y = a \cdot b^x$ form with each coefficient rounded to three significant digits.

5 $y = 1{,}300(1.46^x)$
$y = 1{,}300(1.46^{15})$
$y \approx 380{,}000$

To approximate y for 1985, substitute 15 into the equation of the exponential function. The estimate is rounded to three significant digits.

Answer: $y = 1{,}300(1.46^x)$ is approximately the exponential growth function of best fit. The formula predicts that in 1985 there would be 380,000 transistors on an integrated circuit. The actual number given on Intel's website is 375,000.

SELF-CHECK ANSWERS

10.7.1
1. \$2,001.60
2. Approximately 14.3 years

10.7.2
1. \$8,739.13
2. \$8,753.36

10.7.3
1. 165,000
2. 18 years (to the nearest year)

10.7.4
1. 90 decibels

USING THE LANGUAGE AND SYMBOLISM OF MATHEMATICS 10.7

1. Growth that occurs at discrete intervals is called __________ growth.
2. Growth that occurs continually at each instant is called __________ growth.
3. The periodic growth formula is $A =$ __________.
4. The continuous-growth formula is $A =$ __________.
5. Seismologists use the Richter scale to measure the magnitude of __________.
6. The decibel is a unit of measure for the intensity of __________.

QUICK REVIEW 10.7

The purpose of this quick review is to help you recall skills needed in this section.

1. Write $10^4 = 10{,}000$ in logarithmic form.
2. Write $\log 0.001 = -3$ in exponential form.
3. Write $\ln 78 \approx 4.357$ in exponential form.
4. Use a calculator to approximate $\log(2.58 \times 10^4)$ to the nearest hundredth.
5. Use a calculator to approximate $\log(2.58 \times 10^{-4})$ to the nearest hundredth.

EXERCISES 10.7

Compound Interest In Exercises 1–8, use the formula for periodic growth to solve each problem. All interest rates are stated as annual rates of interest.

1. Find the value of \$150 invested at 9% with interest compounded quarterly for 5 years.
2. Find the value of \$210 invested at 7% with interest compounded monthly for 9 years.
3. How many years will it take an investment to double in value if interest is compounded annually at 8%?
4. How many years will it take an investment to double in value if interest is compounded semiannually at 10%?
5. How many years will it take a savings account to triple in value if interest is compounded monthly at 6%?
6. How many years will it take a zero-coupon bond to triple in value if interest is compounded monthly at 7.5%?
7. If an investment on which interest is compounded monthly doubles in value in 8 years, what is the annual rate of interest?
8. If an investment on which interest is compounded quarterly doubles in value in 9 years, what is the annual rate of interest?

In Exercises 9–22, use the formula for continuous growth and decay to solve each problem.

9. **Rate of Inflation** If prices will double in 10 years at the current rate of inflation, what is the current rate of inflation? Assume that the effect of inflation is continuous.
10. **Rate of Inflation** If prices will double in 9 years at the current rate of inflation, what is the current rate of inflation? Assume that the effect of inflation is continuous.
11. **Continuous Compound Interest** How many years will it take an investment to double in value if interest is compounded continuously at 7%?
12. **Continuous Compound Interest** How many years will it take an investment to double in value if interest is compounded continuously at 9%?
13. **Carbon-14 Dating** Carbon-14 decays continuously at the rate of 0.01245% per year. An archaeologist has determined that only 5% of the original carbon-14 from a plant specimen remains. Estimate the age of this specimen.
14. **Carbon-14 Dating** Carbon-14 decays continuously at the rate of 0.01245% per year. An archaeologist has determined that only 10% of the original carbon-14 from a plant specimen remains. Estimate the age of this specimen.
15. **Modeling the Safe Storage Time for Milk** The safe level of psychrotrophic bacteria in a gallon of milk is 100 units. In a refrigerator set at 38°F the number of units of these bacteria in a gallon of skim milk is approximated by the exponential function $B(t) = 4.0e^{0.24t}$, where t is the time in days. Evaluate and interpret each expression.
    a. $B(0)$ b. $B(5)$ c. $B(10)$ d. $B(t) = 100, t = ?$

16. **Modeling the Safe Storage Time for Milk** The safe level of psychrotrophic bacteria in a gallon of milk is 100 units. In a refrigerator set at 38°F the number of units of these bacteria in a gallon of whole milk is approximated by the exponential function $B(t) = 4.0e^{0.26t}$, where t is the time in days. Evaluate and interpret each expression.
    a. $B(0)$ b. $B(5)$ c. $B(10)$ d. $B(t) = 100, t = ?$
17. **Radioactive Decay** The radioactive material used to power a satellite decays at a rate that decreases the available power by 0.05% per day. When the power supply reaches $\frac{1}{100}$ of its original level, the satellite is no longer functional. Approximately how many days should the power supply last?
18. **Radioactive Decay** The radioactive material used to power a satellite decays at a rate that decreases the available power by 0.03% per day. When the power supply reaches $\frac{1}{100}$ of its original level, the satellite is no longer functional. Approximately how many days should the power supply last?
19. **Population Decline** The population of a species of whales is estimated to be decreasing at a rate of 4% per year ($r = -0.04$). The current population is approximately 15,000. First estimate the population 10 years from now, and then determine how many years from now the population will have declined to 5,000.
20. **Population Decline** If the number of white owls in Illinois has decreased from 750 to 500 in 10 years, approximate the annual rate of decrease.

21. **Population Growth** If the population of a town has grown from 1,200 to 1,800 in 3 years, approximate the annual rate of increase.
22. **Population Growth** The human population in a remote area has doubled in the last 18 years. Approximate the annual rate of increase.

Magnitude of an Earthquake In Exercises 23–26, use the formula $R = \log \frac{A}{a}$ to solve each problem.

23. The amplitude of the September 19, 1985, earthquake in Mexico City was 63,100,000 times the reference amplitude. Calculate the magnitude of this earthquake on the Richter scale.
24. The amplitude of the September 20, 1965, earthquake in Mexico City was 20,000,000 times the reference amplitude. Calculate the magnitude of this earthquake on the Richter scale.
25. The damage done by an earthquake is related to its magnitude, the population of the area affected, and the quality of the buildings in this area. The August 17, 1999, earthquake that hit Izmit, Turkey, is rated by some (The Learning Channel, "Ultimate Natural Disasters") as the number 7 worst natural disaster to ever occur. At 7.4 on the Richter scale it is the greatest earthquake to hit a major city since the 1906 San Francisco earthquake, which was estimated to be 8.25. Calculate how many times more intense the 1906 San Francisco earthquake was than the 1999 Izmit earthquake.
26. Calculate how many times more intense an earthquake with a Richter scale reading of 8.6 is than an earthquake with a Richter scale reading of 8.3.

Noise Levels In Exercises 27–30, use the formula $D = 10 \log \frac{I}{I_o}$ to solve each problem ($I_o = 10^{-16}$ W/cm^2).

27. Find the number of decibels of a whisper if its intensity is 3×10^{-14} W/cm^2.
28. Find the number of decibels of city traffic if its intensity is 8.9×10^{-7} W/cm^2.
29. The noise level in a bar measures 85 dB. Calculate the intensity of this noise.
30. The decibel reading near a jet aircraft is 105 dB. Calculate the intensity of this noise.

pH Measurements In Exercises 31–34, use the formula $\text{pH} = -\log \text{H}^+$ to solve each problem.

31. Determine the pH of grape juice that has an H^+ concentration of 0.000109 mol/L.
32. Determine the pH of saccharin, a sugar substitute, which has an H^+ concentration of 4.58×10^{-7} mol/L.
33. A leading shampoo has a pH of 9.13. What is the H^+ concentration in moles per liter?
34. Blood is buffered (kept constant) at a pH of 7.35. What is the H^+ concentration in moles per liter?

The monthly payment P required to pay off a loan of amount A at an annual interest rate R in n years is given by the formula

$$P = \frac{A\left(\frac{R}{12}\right)}{1 - \left(1 + \frac{R}{12}\right)^{-12n}}$$

In Exercises 35 and 36, use this formula to calculate the monthly payment.

35. Loan Payments Determine the monthly payment necessary to pay off a \$47,400 home loan that is financed for 30 years at 9.875%.

36. Loan Payments Determine the monthly payment necessary to pay off a \$47,400 home loan that is financed for 30 years at 10.5%.

The number of monthly payments of amount P required to completely pay off a loan of amount A borrowed at interest rate R is given by the formula

$$n = -\frac{\log\left(1 - \frac{AR}{12P}\right)}{\log\left(1 + \frac{R}{12}\right)}$$

In Exercises 37 and 38, use this formula to calculate the number of monthly payments.

37. Car Payments Determine the number of monthly car payments of \$253.59 required to pay off a \$7,668.00 car loan when the interest rate is 11.7%.

38. Car Payments Determine the number of monthly car payments of \$200.80 required to pay off a \$7,668.00 car loan when the interest rate is 11.7%.

In Exercises 39 and 40, use a graphing calculator and the points given in the table.

a. Draw a scatter diagram of these points.
b. Determine the exponential function that best fits these points. (Round the coefficients to three significant digits.)
c. Does this function represent exponential growth or exponential decay?
d. Use this equation to estimate y when x is 2.5.

39.

x	y
1	2.5
2	2.0
3	1.5
4	1.2
5	1.0
6	0.8

40.

x	y
1	6.5
2	9.5
3	14.5
4	21.5
5	32.5
6	50.0

41. Data Storage Requirements The Sweet Tooth Cookie Company recorded its data storage requirements in gigabytes (GB) for 5 years.
a. Draw a scatter diagram for these data.
b. Determine the exponential function of best fit.
c. Use this function to estimate the storage requirements for 2006.

YEARS x AFTER 2000	STORAGE CAPACITY y (GB)
0	40
1	60
2	90
3	150
4	200
5	300

42. Revenue The revenue for wireless service is given for 6 years for one telecommunications company.
a. Draw a scatter diagram for these data.
b. Determine the exponential function of best fit.
c. Use this function to estimate the revenue for 2006.

YEARS x AFTER 2000	REVENUE y (\$1,000,000)
0	250
1	370
2	500
3	550
4	1,000
5	1,500

Group Discussion Questions

43. Communicating Mathematically Which type of function, exponential or logarithmic, would provide a better model for the number of skills a dog acquires during the first 10 years of its life? Explain your answer.

44. Challenge Question Suppose that a bacteria culture doubles in volume every day. After 30 days, the population of these bacteria has grown to fill a laboratory jar. On what day was the jar half full of these bacteria?

KEY CONCEPTS FOR CHAPTER 10

1. Geometric Sequence
- A geometric sequence is a sequence with a constant ratio for any two consecutive terms. This common ratio is often denoted by r.
- If the common ratio $r > 1$, the geometric sequence exhibits geometric growth.
- If the common ratio $0 < r < 1$, the geometric sequence exhibits geometric decay.
- The graph of the terms of a geometric sequence consists of discrete points that lie on the graph of an exponential function.

2. Exponential Function $f(x) = b^x$
- If $b > 0$ and $b \neq 1$, then $f(x) = b^x$ is an exponential function with base b.
- If $b > 1$, then $f(x) = b^x$ is called an exponential growth function.
- If $0 < b < 1$, then $f(x) = b^x$ is called an exponential decay function.
- For $b \neq 1$, $b^x = b^y$ if and only if $x = y$.
- For $a, b > 0$, and $x \neq 0$, $a^x = b^x$ if and only if $a = b$.

3. **Exponential Growth and Decay**
   - **Exponential Growth Function** If $b > 1$, then $f(x) = b^x$ exhibits exponential growth.

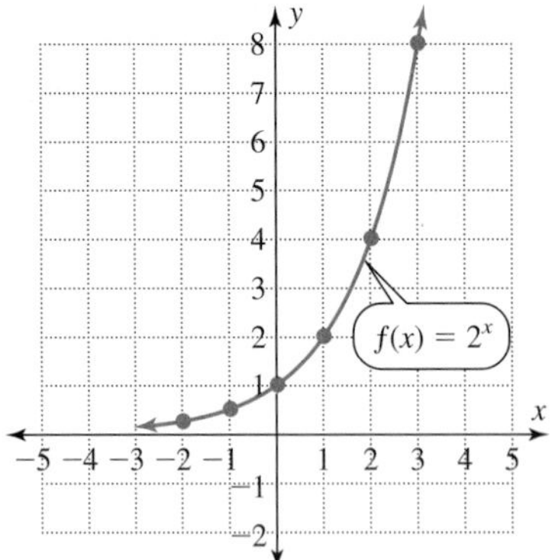

   - **Exponential Decay Function** If $0 < b < 1$, then $f(x) = b^x$ exhibits exponential decay.

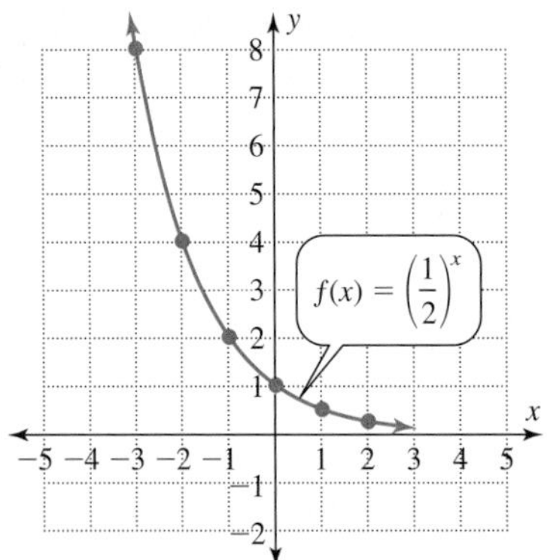

4. **Properties of the Graphs of Exponential Functions** For the exponential function $f(x) = b^x$ with $b > 0$ and $b \neq 1$:
   - The domain of f is the set of all real numbers $\mathbb{R}$.
   - The range of f is the set of all positive real numbers.
   - There is no x-intercept.
   - The y-intercept is $(0, 1)$.
   - Key point: $(1, b)$
   - The growth function is asymptotic to the negative portion of the x-axis (it approaches but does not touch the x-axis).
   - The decay function is asymptotic to the positive portion of the x-axis.
   - The growth function rises (or grows) from left to right.
   - The decay function falls (or decays) from left to right.
5. **The Irrational Constant *e***
   $e \approx 2.718\,281\,828 \cdots$
   Remember $e \approx 2.718$.
6. **A One-to-One Function**
   - A function is one-to-one if different input values produce different output values.
   - The vertical line test can be used to determine whether a graph represents a function.
   - The horizontal line test can be used to determine whether the graph of a function represents a one-to-one function.
7. **Horizontal Line Test** The graph of a function represents a one-to-one function if it is impossible to have any horizontal line intersect the graph at more than one point.
8. **Inverse of a Function**
   - If f is a function that matches each input value x with an output value y, then the inverse of f reverses this correspondence to match this y-value with the x-value.
   - The inverse of a one-to-one function f is also a function. In this case we call f^{-1} an inverse function.
   - The graphs of f and f^{-1} are symmetric about the line $x = y$.
   - The **DrawInv** feature on a graphing calculator can be used to sketch the inverse of $Y_1 = f(x)$.
9. **Finding an Equation for the Inverse of a Function**
   Step 1. Replace $f(x)$ by y. Write this as a function of y in terms of x.
   Step 2. To form the inverse, replace each x with y and each y with x.
   Step 3. If possible, solve the resulting equation for y.
   Step 4. The inverse then can be written in function notation by replacing y with $f^{-1}(x)$.
10. **Logarithmic and Exponential Functions**
    - Logarithmic and exponential functions are inverses of each other.
    - For the logarithmic function $f(x) = \log_b x$, the input value x is called the argument of the function.
11. **Logarithmic Function $f(x) = \log_b x$**
    - For $x > 0$, $b > 0$, and $b \neq 1$, $y = \log_b x$ if and only if $b^y = x$.
    - Common logarithms: $\log x$ means $\log_{10} x$.
    - Natural logarithms: $\ln x$ means $\log_e x$.
    - $y = \log x$ is equivalent to $x = 10^y$.
    - $y = \ln x$ is equivalent to $x = e^y$.
12. **Properties of Logarithms** For $x > 0$, $y > 0$, $b > 0$, and $b \neq 1$:
    - $\log_b 1 = 0$
    - $\log_b b = 1$
    - $\log_b \frac{1}{b} = -1$
    - $\log_b b^x = x$
    - $b^{\log_b x} = x$
    - Product rule: $\log_b xy = \log_b x + \log_b y$
    - Quotient rule: $\log_b \frac{x}{y} = \log_b x - \log_b y$
    - Power rule: $\log_b x^p = p \log_b x$
13. **Logarithmic Growth Function** If $x > 0$ and $b > 1$, then $f(x) = \log_b x$ exhibits logarithmic growth.

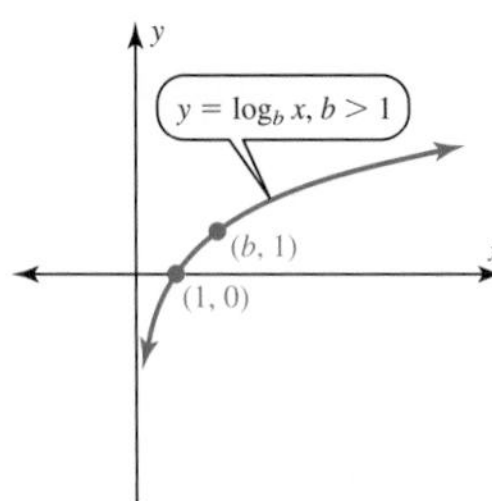

14. **Properties of the Graphs of Logarithmic Functions, $f(x) = \log_b x$** For the logarithmic function $f(x) = \log_b x$ with $x > 0$, $b > 0$, and $b \neq 1$:
    - The domain of f is the set of all positive real numbers $(0, \infty)$.
    - The range of f is the set of all real numbers $\mathbb{R}$.
    - The x-intercept is $(1, 0)$.
    - There is no y-intercept.
    - Key point: $(b, 1)$

- The growth function is asymptotic to the negative portion of the y-axis.
- The decay function is asymptotic to the positive portion of the y-axis.
- The growth function rises (or grows) from left to right, exhibiting extremely slow growth for large values of x.
- The decay function falls (or decays) from left to right, exhibiting extremely slow decay for large values of x.

15. Change-of-Base Formulas For $a, b > 0$ and $a, b \neq 1$:

- $\log_a x = \dfrac{\log_b x}{\log_b a}$ for $x > 0$
- $a^x = b^{x \log_b a}$

16. Extraneous Value An extraneous value is a value produced in the solution process that does not check in the original equation.

17. Solving Exponential Equations

a. If it is obvious that both sides of an equation are powers of the same base, express each side as a power of this base and then equate the exponents. Otherwise, take the logarithm of both sides of the equation and use the power rule to form an equation that does not contain variable exponents.

b. Then solve the equation formed.

18. Solving Logarithmic Equations

a. If possible, rewrite the logarithmic equation in exponential form. Otherwise, use the properties of logarithms to write the two sides of the equation as two single logarithms with the same base. Form a new equation by equating the arguments of these logarithms.

b. Then solve the equation formed.

c. Check all possible solutions for extraneous values, since logarithms of 0 and negative arguments are undefined.

19. Growth and Decay Formulas

- Periodic-growth formula: $A = P\left(1 + \dfrac{r}{n}\right)^{nt}$
- Continuous-growth formula: $A = Pe^{rt}$

REVIEW EXERCISES FOR CHAPTER 10

In Exercises 1 and 2, determine whether each sequence is arithmetic, geometric, both, or neither. If the sequence is arithmetic, write the common difference d. If the sequence is geometric, write the common ratio r.

1. a. 2, 4, 6, 8, 10, . . .
   b. 2, 4, 8, 16, 32, . . .
   c. 2, 4, 6, 8, 10, 16, 26, . . .
2. a. 5, 5, 5, 5, 5, . . .
   b. 7, 3, −1, −5, −9, . . .
   c. 5, −5, 5, −5, 5, . . .

3. **Dilution of a Mixture** A tank contains 100 gal of a cleaning solvent. Then 10 gal is drained and replaced with pure water. The contents are throughly mixed. Then the process is repeated by draining 10 gal and replacing this mixture with pure water. Determine the volume of solvent in the tank after each of the first of five such drainings.
   100, 90, 81, ________, ________, ________

4. Use $f(x) = 9^x$ to evaluate each expression.
   a. $f(0)$ b. $f(-1)$ c. $f(2)$ d. $f\left(\dfrac{1}{2}\right)$

5. Use $f(x) = \pi^x$ and a calculator to approximate each expression to the nearest thousandth.
   a. $f(-1)$ b. $f(2)$ c. $f(\pi)$ d. $f(e)$

In Exercises 6–8, match each graph with the most appropriate description.

6. Not a function of x
7. A function of x but not a one-to-one function
8. A one-to-one function of x

A.

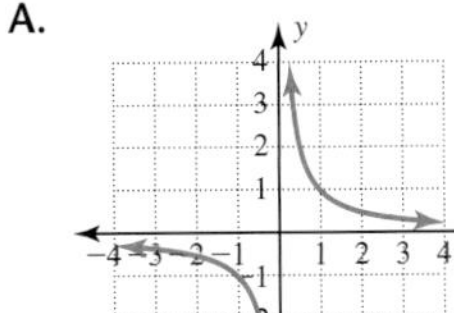

B.

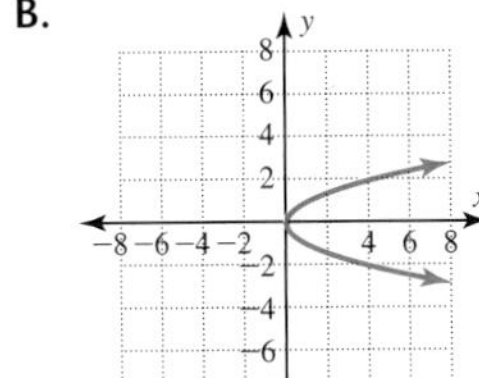

C.

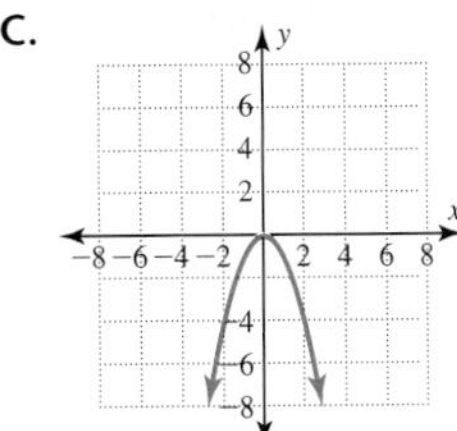

9. Graph each function and its inverse on the same coordinate system.
   a. $f(x) = \left(\dfrac{4}{3}\right)^x$ b. $f(x) = \left(\dfrac{3}{4}\right)^x$
   c. $f(x) = \ln x$

In Exercises 10 and 11, the function f is described verbally and algebraically. Describe f^{-1} both verbally and algebraically.

	Verbally	Algebraically
10.	$f(x)$ triples x and then subtracts 2.	$f(x) = 3x - 2$
11.	$f(x)$ takes one-fourth of x and then adds 6.	$f(x) = \dfrac{x}{4} + 6$

In Exercises 12 and 13 write the inverse of each function by using ordered-pair notation.

12. $f = \{(-4, -5), (-3, -3), (0, 3), (2, 7), (3, 9)\}$

13.

x	−2	−1	0	1	2	3
y	0	2	4	6	8	10

In Exercises 14 and 15, write the equation of the inverse of each function.

14. $f(x) = \dfrac{1}{3}x - 4$
15. $f(x) = 3^x$
16. **Product Costs** The cost of producing x pizzas at a small pizzeria is given by the function $C(x) = 2x + 250$.
   a. What does the variable x represent in $C(x)$?
   b. Determine the formula for $C^{-1}(x)$.

c. What does the variable x represent in $C^{-1}(x)$?
d. What does the output of $C^{-1}(x)$ represent?
e. Determine the cost of producing 100 pizzas.
f. Determine the number of pizzas that can be produced for \$398.

In Exercises 17 and 18, graph the inverse of the given function.

17.

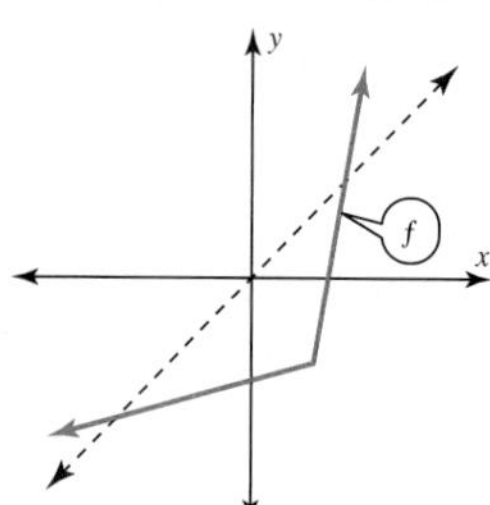

18.

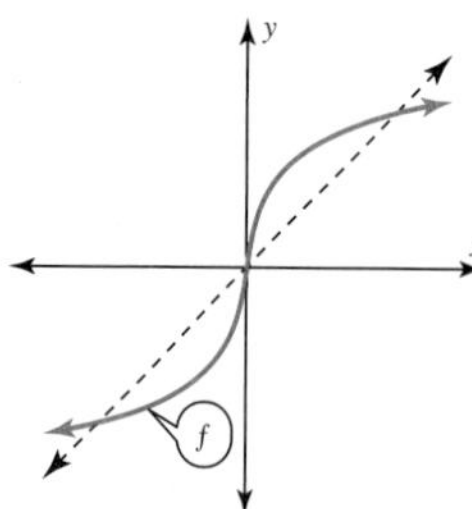

In Exercises 19 and 20, write each logarithmic equation in exponential form.

19. a. $\log_6 \sqrt{6} = \frac{1}{2}$ **b.** $\log_{17} 1 = 0$ **c.** $\log_8 \frac{1}{64} = -2$

20. a. $\log_b a = c$ **b.** $\ln a = c$ **c.** $\log c = d$

In Exercises 21 and 22, write each exponential equation in logarithmic form.

21. a. $7^3 = 343$ **b.** $19^{1/3} = \sqrt[3]{19}$ **c.** $\left(\frac{4}{7}\right)^{-2} = \frac{49}{16}$

22. a. $e^{-1} = \frac{1}{e}$ **b.** $10^{-4} = 0.0001$ **c.** $8^x = y$

In Exercises 23 and 24, evaluate each expression by inspection.

23. a. $\log_{12} 144$ **b.** $\log_{12} \sqrt{12}$ **c.** $\log_{12} \frac{1}{12}$

24. a. $\log_{12} 1$ **b.** $\log_{12} 12^{0.23}$ **c.** $\log_{12} \frac{1}{\sqrt[3]{12}}$

In Exercises 25–32, use a calculator to approximate each expression to the nearest thousandth.

25. a. $(\sqrt{7})^{\pi}$ **b.** $e^{7.83}$ **c.** $10^{-0.8107}$
26. a. $\pi^{\sqrt{7}}$ **b.** $e^{-2.4897}$ **c.** $10^{4.83}$
27. a. $\log 113.58$ **b.** $\ln 113.58$ **c.** $\log_5 113.58$
28. a. $\log(8.1 \times 10^{-4})$ **b.** $\ln(8.1 \times 10^{-4})$ **c.** $\log_5(8.1 \times 10^{-4})$
29. a. $\frac{\ln 23}{\ln 19}$ **b.** $\ln \frac{23}{19}$ **c.** $\frac{\ln 23}{19}$
30. a. $\ln(6 + 5)$ **b.** $\ln(6 \cdot 5)$ **c.** $\ln 6 + \ln 5$
31. a. $\log_7 50$ **b.** $\log_8 9$ **c.** $\log_2 9$
32. a. $\log_4 63$ **b.** $\log_{12} 150$ **c.** $\log_3 8$

In Exercises 33–66, solve each equation without using a calculator. Note that some expressions are undefined.

33. $11^x = \frac{1}{121}$ **34.** $125^x = 25$

35. $2^y = \sqrt[3]{4}$ **36.** $\left(\frac{4}{9}\right)^y = \frac{3}{2}$

37. $2^{4w-1} = 8$ **38.** $9^{2v+1} = 27$
39. $2^{x^2-1} = 8$ **40.** $9^{x^2} = 3^{x+1}$
41. $\log_8 64 = z$ **42.** $\log_{16} 64 = x$
43. $\log_2 x = 2$ **44.** $\log_{13} x = \frac{1}{2}$
45. $\log_3 w = -2$ **46.** $\log_3(-2) = w$
47. $\log_{-2} 3 = y$ **48.** $\log_t 169 = 2$
49. $\log_t 8 = -3$ **50.** $5^{\log_5 11} = x$
51. $\log_5 5^{17} = x$ **52.** $\log_7 x = 1$
53. $\ln 0 = x$
54. $\log(3n - 4) = \log(2n - 1)$
55. $\log_3 20 + \log_3 7 = \log_3 y$
56. $\log(3x + 1) = 2$
57. $\log_3(x^2 - 19) = 4$
58. $\ln(w + 2) + \ln w = \ln 3$
59. $\ln(1 - w) + \ln(1 - 2w) = \ln(7 - 4w)$
60. $\log(5 - 2v) - \log(1 - v) = \log(3 - 2v)$
61. $\ln(5v + 3) = \ln(3v + 9)$
62. $\log_5 27 - \log_5 2 = \log_5 y$
63. $\log(2 - 6v) - \log(2 - v) = \log(1 - v)$
64. $\ln(5 - x) + \ln(x + 1) = \ln(3x - 1)$
65. $\log x + \log(x - 9) = 1$
66. $\log_2 x + \log_2(x - 2) = 3$

In Exercises 67–70, express each logarithm in terms of logarithms of simpler expressions. Assume that the argument of each logarithm is a positive real number.

67. $\log x^3y^5$ **68.** $\ln \frac{7x - 9}{2x + 3}$

69. $\ln \frac{\sqrt{2x + 1}}{5x + 9}$ **70.** $\log \sqrt{\frac{x^2y^3}{z}}$

In Exercises 71–74, combine the logarithmic terms into a single logarithmic expression with a coefficient of 1. Assume that the argument of each logarithm is a positive real number.

71. $2 \ln x + 3 \ln y$ **72.** $5 \ln x - 4 \ln y$

73. $\ln(x^2 - 3x - 4) - \ln(x - 4)$ **74.** $\frac{1}{2}(\ln x - \ln y)$

In Exercises 75 and 76, approximate the solution to each equation to the nearest thousandth.

75. $\log(5x - 2) + \log(x - 1) = \log 10$
76. $\ln(2w + 3) + \ln(w + 1) = \ln(w + 2)$
77. Convert $y = 5^x$ to an exponential function with base e.
78. Verify that $\log 50^x + \log 6^x - \log 3^x = 2x$ is an identity.
79. Verify that $1{,}000^{\log x} = x^3$ is an identity.

Estimation and Calculator Skills

In Exercises 80–83, mentally estimate the value of each expression to the nearest integer, and then use a calculator to approximate the value to the nearest thousandth.

PROBLEM	MENTAL ESTIMATE	CALCULATOR APPROXIMATION
80. $\log 990$		
81. $\ln 3$		
82. $\log_3 30$		
83. $\log_5 120$		

84. Given $f(x) = 5x - 2$, evaluate each expression.

a. $f(3)$ **b.** $f^{-1}(13)$ **c.** $\frac{1}{f(3)}$

In Exercises 85–89, match each graph with the most appropriate description. All graphs are displayed by using the window [−5, 5, 1] by [−5, 5, 1].

85. A linear growth function

86. A logarithmic growth function

87. An exponential growth function

88. An exponential decay function

89. A constant function

A. 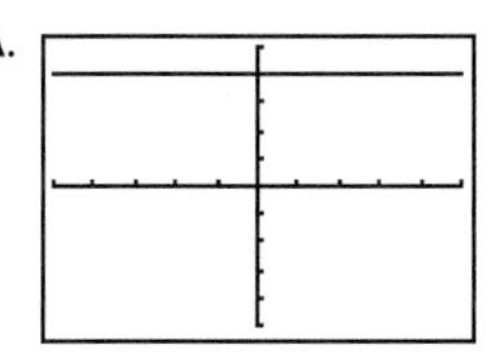

B. 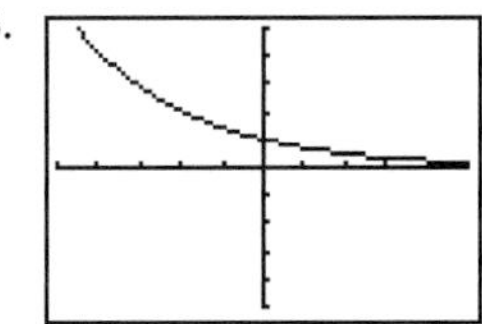

C. 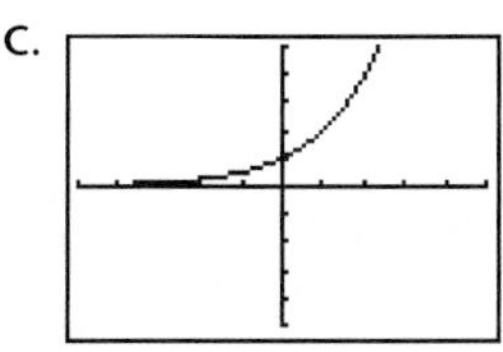

D. 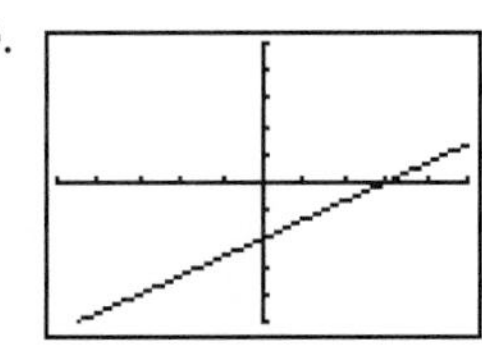

E. 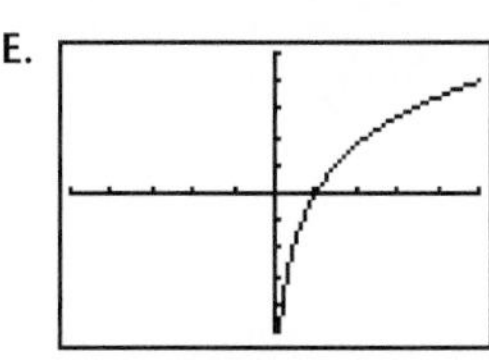

90. Compound Interest A $2,000 investment is invested at 8% for 5 years. Determine the value of the investment under each condition.

a. The interest is compounded yearly.

b. The interest is compounded monthly.

c. The interest is compounded continuously.

91. Radioactive Decay The radioactive material used to power a satellite decays at a rate that decreases the available power by 0.045% per day. When the power supply reaches $\frac{1}{100}$ of its original level, the satellite is no longer functional. Approximately how many days should the power supply last?

92. Doubling Period for an Investment How many years will it take a savings bond to double in value if interest is compounded annually at 7.5%?

93. Rate of Interest If an investment on which interest is compounded continuously doubles in value in 8 years, what is the rate of interest?

94. Richter Scale The 2001 Kodiak Island, Alaska, earthquake had an amplitude A that was 6,310,000 times the reference amplitude a. Use the formula $R = \log \frac{A}{a}$ to calculate the magnitude of this earthquake on the Richter scale. (For more information on this topic, go to the Additional Resources link on the MathZone website at **www.mhhe.com/hallmercer**.)

95. Modeling the Safe Storage Time for Milk The safe level of psychrotrophic bacteria in a gallon of milk is 100 units. In a refrigerator set at 40°F, the number of units of these bacteria in 1 gal of whole milk is approximated by the exponential function $B(t) = 3.0e^{0.29t}$, where t is the time in days. Evaluate and interpret each expression.

a. $B(0)$ b. $B(5)$

c. $B(10)$ d. $B(t) = 100, t = ?$

MASTERY TEST FOR CHAPTER 10

[10.1] **1.** Determine whether each sequence is a geometric sequence. If the sequence is geometric, write the common ratio r.

a. 1, 5, 25, 125, 625 b. 1, 5, 9, 13, 17

c. 48, 24, 12, 6, 3 d. 0.9, 0.09, 0.009, 0.0009, 0.00009

[10.1] **2.** Graph each function.

a. $f(x) = \left(\frac{5}{2}\right)^x$ b. $f(x) = \left(\frac{2}{5}\right)^x$

Evaluate each expression, given $f(x) = 16^x$.

c. $f(-1)$ d. $f\left(\frac{1}{2}\right)$

[10.2] **3.** Write the inverse of each function. For parts **a, c,** and **d** use ordered-pair notation.

a. $\{(-1, 4), (8, 9), (-7, 11)\}$

b. Write the equation for the inverse of $f(x) = 3x - 6$.

c. 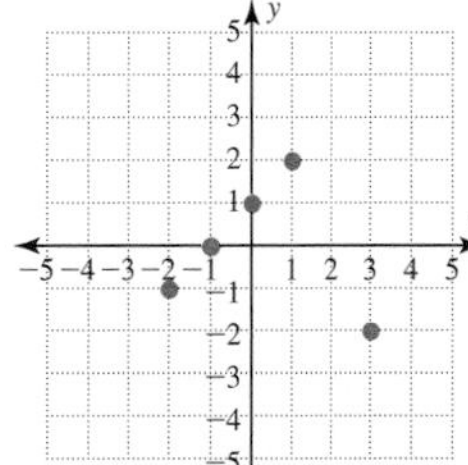

d.

x	y
$\frac{1}{2}$	2
$\frac{1}{3}$	3
6	$\frac{1}{6}$
1	1

[10.2] **4.** Graph each function and its inverse on the same coordinate system.

a. $f = \{(-3, 2), (-2, 1), (-1, 3), (2, 4)\}$

b. $f(x) = \frac{3}{2}x + 2$

c. $f(x) = \left(\frac{1}{2}\right)^x$

d. $f(x) = x^2$ for $x \geq 0$

[10.3] 5. Translate the following logarithmic equations to exponential form.

a. $\log_5 \sqrt[3]{25} = \frac{2}{3}$

b. $\log_5 \frac{1}{125} = -3$

c. $\log_b (y + 1) = x$

d. $\log_b x = y + 1$

[10.3] 6. Determine the value of each logarithm by inspection.

a. $\log_8 1$

b. $\log_8 8$

c. $\log_8 \frac{1}{8}$

d. $\log_8 8^{17}$

[10.4] 7. Use a calculator to approximate each expression to the nearest ten-thousandth.

a. $\log 19.1$

b. $\ln 19.1$

c. $\log(4.876 \times 10^{12})$

d. $\ln(3.04 \times 10^{-5})$

[10.5] 8. Use the properties of logarithms to express each logarithm in terms of logarithms of simpler expressions. Assume that x and y are positive real numbers.

a. $\log x^4y^5$

b. $\ln \frac{x^3}{\sqrt{y}}$

Use the properties of logarithms to combine these logarithmic terms into a single logarithmic expression with a coefficient of 1. Assume that the argument of each logarithm is a positive real number.

c. $2 \ln(7x + 9) - \ln x$

d. $\frac{1}{2} \log(x + 3) + \log x$

[10.5] 9. Use the change-of-base formula for logarithms and a calculator to approximate each expression to the nearest thousandth.

a. $\log_4 11$

b. $\log_\pi \sqrt{7}$

c. $\log_2 100$

d. $\log_\pi 5\pi$

[10.6] 10. Solve each equation without using a calculator.

a. $3^x = \frac{1}{81}$

b. $16^w = 64$

c. $\log_2 x = 4$

d. $\log_8 2 = y$

e. $\log(1 - 4t) - \log(5 + t) = \log 3$

f. $\ln(1 - z) + \ln(2 - z) = \ln(17 - z)$

Using a calculator, approximate the solution of each equation to the nearest thousandth.

g. $3^{4y+1} = 17.83$

h. $\ln(x + 1) + \ln(3x - 1) = \ln(6x)$

[10.7] 11. a. **Population Growth** Assume that the population of a new space colony is growing continuously at a rate of 5% per year. Approximately how many years will it take the population to grow from 500 to 3,000?

b. **Periodic Growth** How many years will it take an investment to double in value if interest is compounded monthly at 8.25%?

REALITY CHECK FOR CHAPTER 10

The graph at the beginning of the chapter shows the exponential growth of the value of a \$500 investment earning 7% interest compounded continuously over a 40-year time period. How much money would you have to invest to produce \$1,000,000 after 40 years if the investment earned 7% interest compounded continuously?

GROUP PROJECT FOR CHAPTER 10

An Algebraic Model for Real Data

Supplies Needed for the Lab:

- A golf ball or other hard ball
- A meterstick or a metric tape measure
- A chair or a short stepladder
- Masking tape and a marking pen
- Hard, smooth floor area next to a wall
- A graphing calculator with an exponential regression feature **(ExpReg)**

Lab: Recording the Height of a Bouncing Golf Ball

Move to a location with a relatively high ceiling and a smooth concrete or other hard floor. (Tile with cracks can cause problems with the bounce of the golf ball.) Ask the tallest member of the class to get on a chair or small stepladder. Run a strip of masking tape from the base of the floor to the top of the student's reach. Stick this masking tape to the wall in a straight line. (Starting from a height of 250 to 300 cm will make the data collection easier.) Using your meterstick and the pen, label the heights on the masking tape every 10 cm, starting with 0 cm at floor level.

Placing the bottom of the golf ball level with your starting point at the top of the tape (250 to 300 cm high) and about 5 cm away from the wall, have one student drop the golf ball and have a second student note the approximate point that marks the height of the golf ball at the top of its first bounce. It would be wise to have a third student catch the golf ball at the beginning of the second bounce. Repeat this experiment now that you know approximately where to look, and have the second student mark the height of this bounce by using his or her finger. Place a pen mark on the tape at this height. Using the reference marks on the tape and the meterstick, record the height of this bounce to the nearest centimeter.

Use the height just recorded as the new release point, and repeat the experiment. You may need to repeat drops from the same height before you are comfortable that you have recorded accurately the height of the first bounce. Repeat this experiment until you have recorded the results for 8 to 10 bounces.

1. Record your data in a table similar to the one shown.
2. Use these data to create a scatter diagram by using an appropriate scale for each axis.

BOUNCE NO. x	HEIGHT y (cm)
0	250
1	
2	
3	
4	
5	
6	
7	
8	
9	
10	

Creating an Algebraic Model for These Data

1. Using your graphing calculator, calculate the exponential function of best fit. This function will be of the form $f(x) = ab^x$.
2. Let $y = H(x)$ represent the curve of best fit that you just calculated.
   a. Evaluate and interpret $H(7)$. How does this value compare to the corresponding value in the table?
   b. Determine the value of x for which $H(x) = 100$. Interpret the meaning of this value.
3. In the function $f(x) = ab^x$ that models your data, what is the value of a? Interpret the meaning of this value in this application.
4. In the function $f(x) = ab^x$ that models your data, what is the value of b? Interpret the meaning of this value in this application.

CUMULATIVE REVIEW FOR CHAPTERS 1 THROUGH 10

The answer to each of these questions follows this diagnostic review. Each answer is keyed to an example in this book. You can refer to these examples to find explanations and additional exercises for practice.

Exponents and Order of Operations

Calculate the value of each expression without using a calculator.

1. a. $-\frac{5}{6}+\frac{1}{8}$ b. $\left(-\frac{3}{4}\right)\left(-\frac{1}{9}\right)\left(\frac{12}{13}\right)$
2. a. $(-7)^2$ b. -7^2
3. a. $\frac{|5-3(4)|}{2+8(3-5)}$ b. $\sqrt{64+36}-(\sqrt{64}+\sqrt{36})$
4. Simplify each expression, assuming that all variables are restricted to values that avoid division by 0 and avoid 0^0.
   a. $(x^3y^5)(x^2y^6)$ b. $\frac{(6x^2)^2}{(3x^3)^3}$

Linear Equations in One Variable and Direct Variation

5. Solve each linear equation.
   a. $7x+3=3x-21$ b. $\frac{x+5}{2}=\frac{3x-7}{4}$
6. a. If y varies directly as x, and y is 36 when x is 48, find y when x is 20.
   b. The number of euros varies directly as the number of U.S. dollars exchanged. In 2005, one could exchange 50 U.S. dollars for 41.28 euros. What was the constant of variation (the exchange rate) at that time?
7. Solve each equation for y, and then use a graphing calculator to complete this table.

X	Y1
-3	
-2	
-1	
0	
1	
2	
3	

X=-3

   a. $3y-6x=15$ b. $\frac{y-3}{x+2}=5$

Functions and Representations of Functions

8. Determine whether each relation is a function.
   a.

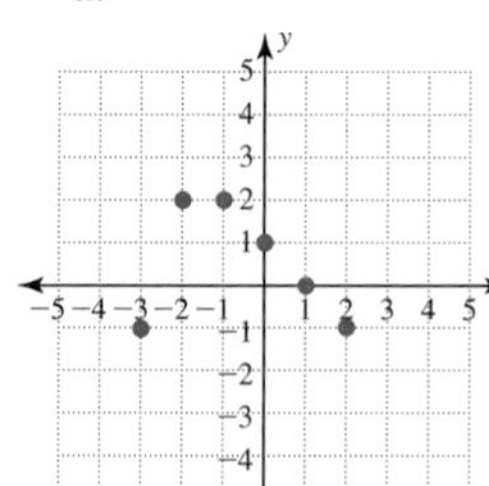

   b.

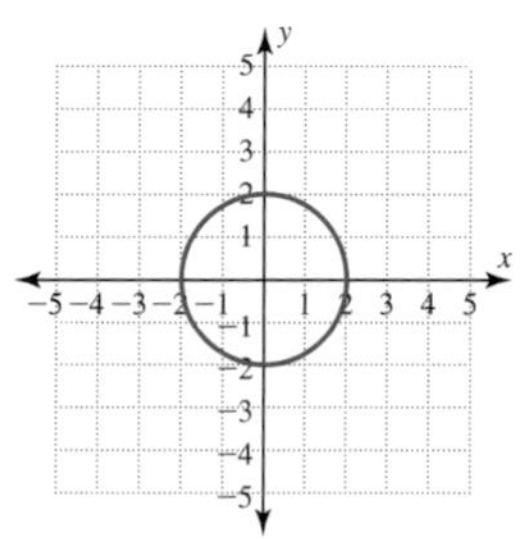

9. Determine the domain and range of each of these functions.
   a. $f(x)=-3$ b.

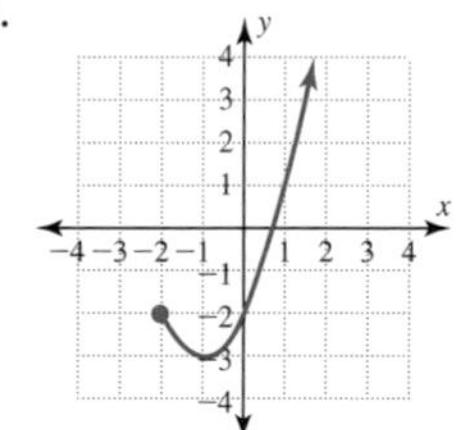

10. Evaluate each of the following expressions for $f(x)=3x^2+2x-4$.
    a. $f(-2)$ b. $f(5)$
11. Use the given graph to determine the missing input and output values.
    a. $f(0)=$ __________
    b. $f(x)=0$; $x=$ __________

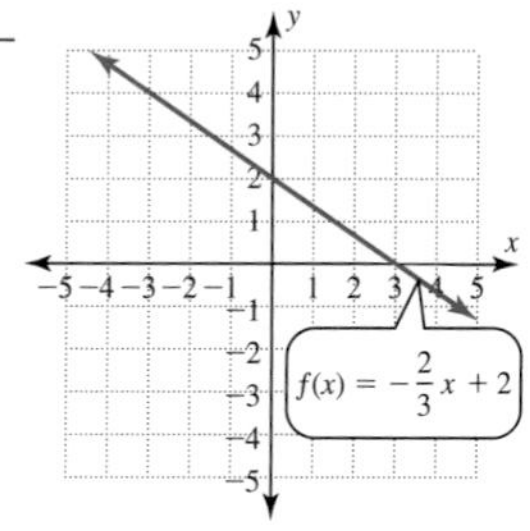

12. Use the given table of values to determine the missing input and output values.
    a. $f(2)=$ __________
    b. $f(x)=2$; $x=$ __________

x	$f(x)$
−2	6
−1	2
0	−2
1	−6
2	−10
3	−14
4	−18

Linear Functions

13. Use each equation to complete a table with input values of 1, 2, 3, 4, and 5, and then graph these points and the line through these points.
    a. $y=2x-5$ b. $y=-3x+4$
14. a. Determine the x- and y-intercepts of this line.

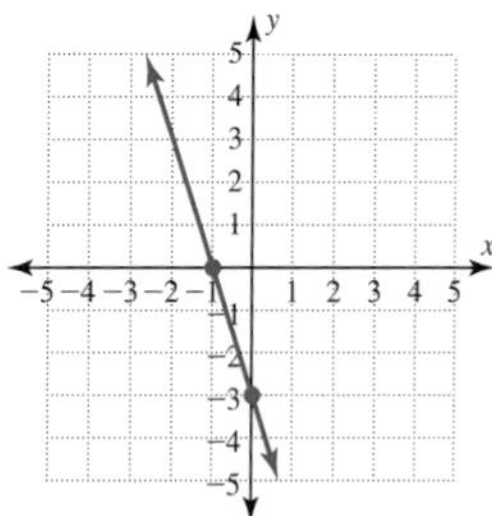

    b. Determine the x- and y-intercepts of the line containing the points in the table.

x	$f(x)$
−5	9
0	6
5	3
10	0
15	−3
20	−6
25	−9

c. Determine the x- and y-intercepts of the line defined by the equation $2x - 3y = 36$.

15. **Ice Cream Shop Expenses** The expenses required to run an ice cream shop include a fixed monthly cost of \$200 and a variable cost of \$0.25 per ice cream cone sold.

a. Write a function that gives the total cost of running this ice cream shop for a month when the shop sells x ice cream cones.

b. Use the function obtained in part **a** to complete this table.

x	0	50	100	150	200	250
$f(x)$						

c. Evaluate and interpret $f(75)$.

d. Determine the value of x for which $f(x) = 260$. Interpret this value.

16. a. Determine the slope of the line containing the points in the table.

x	y
−3	5
0	1
3	−3
6	−7
9	−11
12	−15

b. Determine the slope of the line defined by $y - 5 = \frac{2}{3}(x - 3)$.

17. Write in slope-intercept form the equation of a line satisfying the given conditions.

a. y-intercept $(0, -2)$ and slope $-\frac{5}{3}$

b. Through $(-2, 4)$ and perpendicular to $y = \frac{2}{3}x - 4$

18. Graph a line satisfying the given conditions.

a. Through $(0, 0)$ with slope $-\frac{1}{4}$

b. y-intercept $(0, -4)$ and slope $\frac{7}{4}$

Properties of the Graphs of Linear and Absolute Value Functions

19. Identify each function as an increasing or a decreasing function.

a.

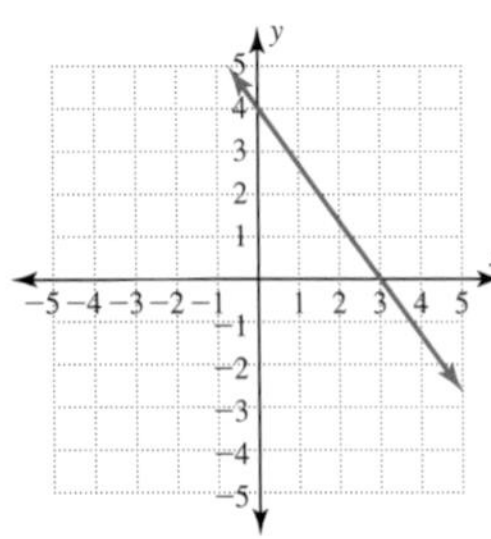

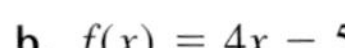
b. $f(x) = 4x - 5$

20. Use the absolute value function $f(x) = |x - 2| - 3$ to answer the following.

a. Complete the following table of values.

x	$f(x) = \|x - 2\| - 3$
−1	
0	
1	
2	
3	
4	
5	

b. Graph this function.

c. What are the x-intercepts of this graph?

d. What is the y-intercept of this graph?

e. What point is the vertex of this graph?

f. What y-value is the minimum output value for this function?

g. What is the domain of this function?

h. What is the range of this function?

i. For what interval of input values of x is the function decreasing?

j. For what interval of input values of x is the function increasing?

k. For what interval of input values of x is the function positive?

l. For what interval of input values of x is the function negative?

Linear Systems

21. Determine the simultaneous solution for each system of linear equations.

a.

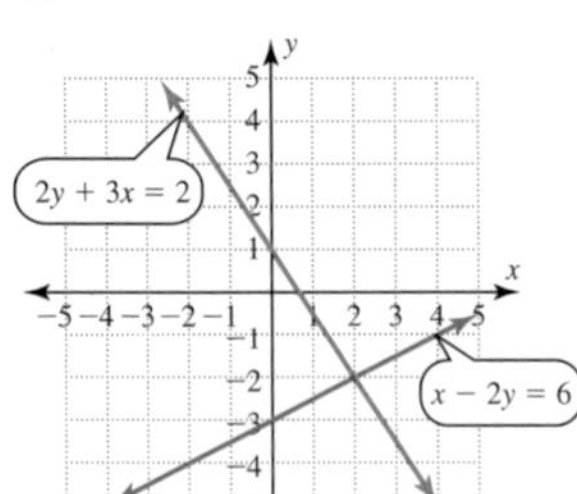

b.

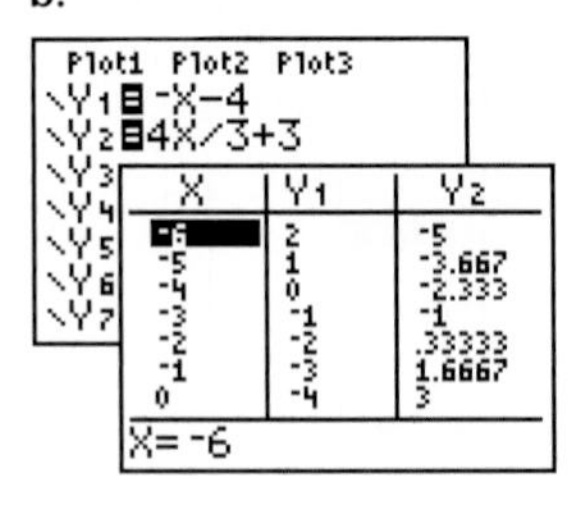

22. Solve each system of linear equations by the substitution method.

a. $y = 3x - 5$
$3x + 4y = 10$

b. $x + 4y = -3$
$3x + 12y = 2$

23. Solve each system of linear equations by the addition method.

a. $2x + 5y = 16$
$4x - 9y = -6$

b. $y = \frac{3}{5}x + 2$
$6x - 10y = -20$

24. Use systems of linear equations to solve each word problem.

a. **Rental Car Costs** The cost of a rental car includes a fixed daily cost and a variable cost based on miles driven. For a customer driving 180 mi the cost for one day will be \$52. For a customer driving 220 mi, the cost for one day will be \$58. Determine the fixed daily cost and the cost per mile for this car.

b. **Rate of a River Current** A boat can go 36 mi downstream in 2 hours but only 24 mi upstream in 2 hours. Determine the rate of the boat and the rate of the current.

Linear Inequalities

25. Use the table to solve each inequality.
   a. $x - 3 < -x - 5$
   b. $x - 3 \geq -x - 5$

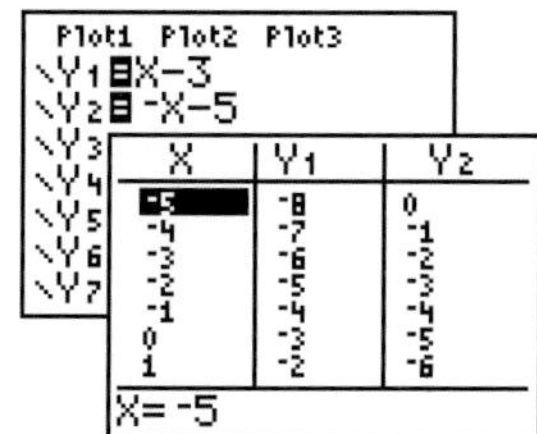

26. Use the graph to solve each inequality.
   a. $2x - 5 \leq -x + 7$
   b. $2x - 5 \geq -x + 7$

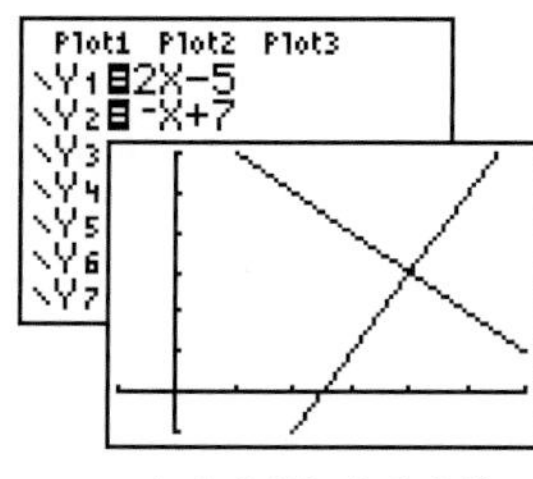

[−1, 6, 1] by [−1, 6, 1]

27. Algebraically solve $2(x + 7) < 12 - 4(x - 5)$.
28. Identify each of these inequalities as a conditional inequality, an unconditional inequality, or a contradiction. Then solve each inequality.
   a. $x - 2 > 8 - x$
   b. $3x - 1 > 3(x + 2)$
   c. $3x - 3 \geq 3(x - 1)$
29. Solve each equation and inequality.
   a. $|2x + 3| = 7$
   b. $|2x + 3| < 7$
   c. $|2x + 3| \geq 7$
30. Solve $-7 \leq 2x + 3 < 9$.
31. Solve $5x - 2 \leq -17$ or $7x - 4 \geq 3$.
32. Graph the solution of this system of inequalities.

$$x \geq 0$$
$$y \geq 0$$
$$3x + 2y \leq 12$$

Polynomials

33. Visually approximate the maximum value of y on the graph, and give the x-value at which this maximum occurs.

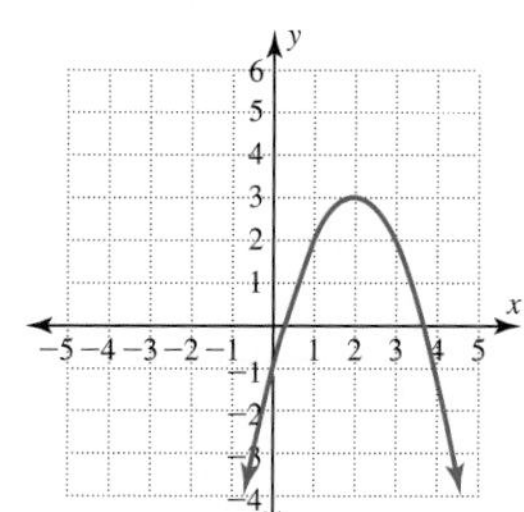

34. The height in feet of a baseball after t seconds is given by the polynomial function $h(t) = -16t^2 + 96t + 3$. Use a calculator to determine the maximum height that this baseball reaches.
35. Perform each operation, and simplify the result.
   a. $(x + 3y)(x - 3y) - (x - 3y)^2$
   b. $(x - 2)(x^2 - 3x + 5)$

Factoring

36. a. Use the graph of $y = P(x)$ to factor $P(x) = x^3 - 7x^2 + 7x + 15$.
   b. Use the table of values to factor the polynomial entered into the calculator as Y_1.

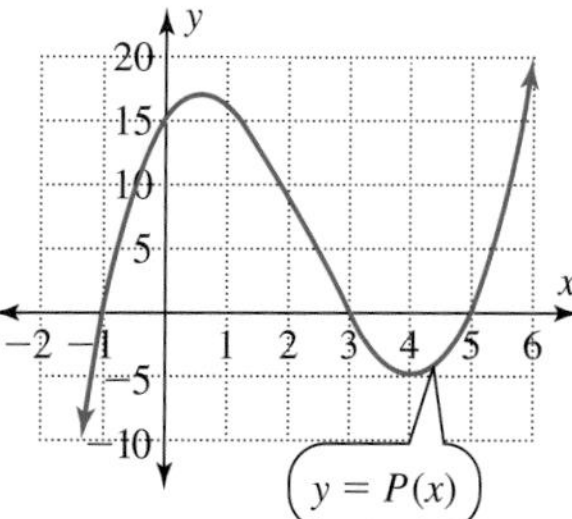

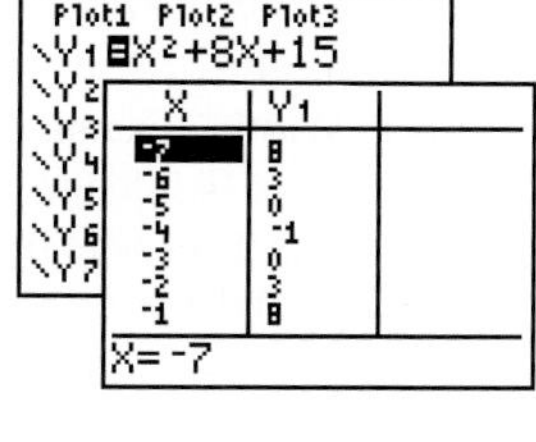

37. Factor out the GCF from each polynomial.
   a. $14x^3 - 35x^2$
   b. $5x(2x - 3) - 6(2x - 3)$
38. Factor each polynomial by using the grouping method.
   a. $2ax + 10bx + 3ay + 15by$
   b. $8x^2 - 24x - 5x + 15$
39. Completely factor each trinomial.
   a. $x^2 + 4x - 45$
   b. $x^2 - 10x - 24$
40. Completely factor each trinomial.
   a. $6x^2 - 17x - 10$
   b. $6x^2 - 13x - 10$
41. Factor each perfect square trinomial.
   a. $x^2 + 20x + 100$
   b. $x^2 - 14xy + 49y^2$
42. Factor each binomial.
   a. $x^2 - 64$
   b. $x^3 - 64$
43. Use grouping to factor each polynomial.
   a. $a^2 - b^2 + 5a - 5b$
   b. $a^2 - x^2 + 10x - 25$
44. Completely factor each polynomial.
   a. $5x^2 - 80$
   b. $-3ax^2 + 21ax - 18a$
45. Solve each equation.
   a. $2x^2 + 13x - 24 = 0$
   b. $(x + 3)(x - 7) = 24$

Quadratic Functions

46. a. Give the vertex, the y-intercept, and the x-intercepts.
   b. Use the equation $f(x) = -x^2 + 2x + 8$ to predict whether the parabola defined by this function opens upward or downward. Then use a graphing calculator to graph the parabola and determine its vertex, y-intercept, and x-intercepts.

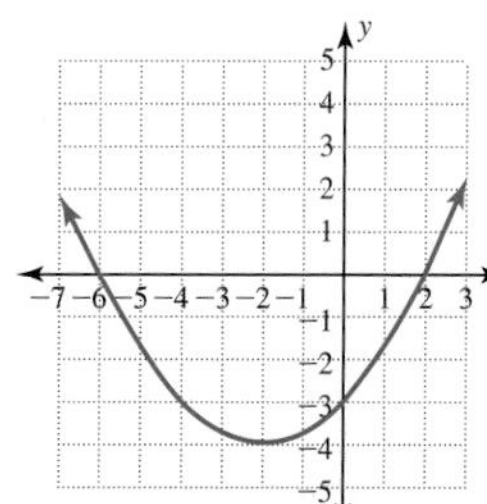

47. Use factoring to find the zeros of $3x^2 + 5x - 12$ and then solve.
   a. $3x^2 + 5x - 12 = 0$
   b. $3x^2 + 5x - 12 \leq 0$
   c. $3x^2 + 5x - 12 > 0$
48. Solve these quadratic equations by extraction of roots.
   a. $(2x - 3)^2 = 7$
   b. $(3x - 4)^2 = -9$

49. Use the discriminant to identify the nature of the solutions of each quadratic equation as distinct real solutions, a double real solution, or two imaginary solutions that are complex conjugates.
   a. $4x^2 + 3x - 2 = 0$
   b. $2x^2 - 6x - 1 = -8$
   c. $4x^2 - 12x = -9$

50. Use the quadratic formula to solve each quadratic equation.
   a. $4x^2 - 3x + 2 = 0$
   b. $4x^2 - 36x + 81 = 0$
   c. $3x^2 + 5x - 7 = 0$

51. **Room Width** Examining the blueprints for a rectangular room that is 14 ft longer than it is wide, an electrician determines that a wire run diagonally across this room will be 26 ft long. What is the width of the room? (See the figure.)

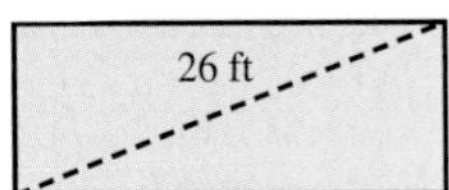

Complex Numbers

52. Write each complex number in standard $a + bi$ form.
   a. $\sqrt{-36} + \sqrt{-64}$ b. $\sqrt{-100}$

53. Perform each operation, and write the result in standard $a + bi$ form.
   a. $(2 - 3i)(4 + 7i)$
   b. $\dfrac{29 + 2i}{2 - 3i}$

Rational Functions

54. Determine the domain of each rational function.
   a. $f(x) = \dfrac{x - 2}{(x - 1)(x + 3)}$ b. $f(x) = \dfrac{3x + 1}{x^2 + 4}$

55. Determine the vertical asymptotes of the graphs of each of these rational functions.
   a. $f(x) = \dfrac{x + 5}{(x - 3)(x + 4)}$
   b.

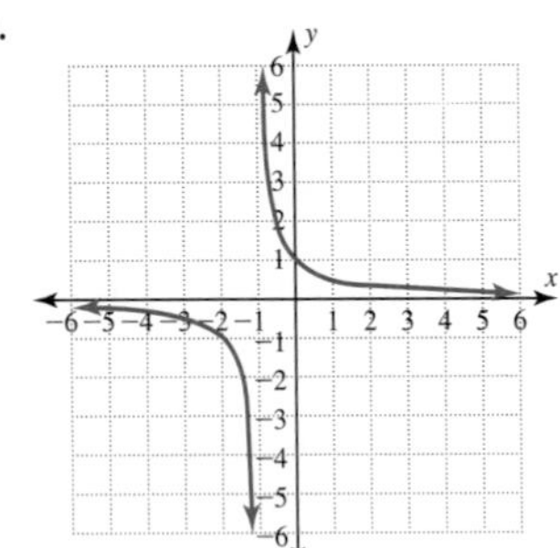

56. Reduce each rational expression to lowest terms.
   a. $\dfrac{4x}{12x^2 - 20x}$ b. $\dfrac{2x^2 - x - 15}{x^2 - 9}$

57. Perform each multiplication or division, and reduce the result to lowest terms.
   a. $\dfrac{2x^2 - 6x}{10x^2} \cdot \dfrac{5x^2}{x - 3}$
   b. $\dfrac{x^2 - 16}{x^2 - 2x - 24} \div \dfrac{x^2 - 7x + 12}{x^2 - 3x - 18}$

58. Perform each addition or subtraction, and reduce the result to lowest terms.
   a. $\dfrac{5x + 4}{3x - 5} + \dfrac{x - 14}{3x - 5}$ b. $\dfrac{3x}{3x - y} - \dfrac{3x}{3x + y} - \dfrac{2y^2}{9x^2 - y^2}$

59. Simplify each expression.
   a. $\dfrac{5x}{4} + \dfrac{x^3}{6} \cdot \dfrac{3}{2x^2}$
   b. $\left(\dfrac{4}{5x + 25} + \dfrac{1}{5x - 25}\right) \div \dfrac{x^2 - x - 6}{x + 5}$

60. Simplify each expression.
   a. $\dfrac{\dfrac{x}{4} - 2 + \dfrac{3}{x}}{1 - \dfrac{6}{x}}$ b. $\dfrac{\dfrac{3x}{x + 1} - \dfrac{2x}{x - 2}}{\dfrac{x^2 - 8x}{2x - 4}}$

61. Find each quotient.
   a. $\dfrac{x^2 - 13x - 48}{x + 3}$
   b. $(3x^3 + x^2 - 22x - 24) \div (3x + 4)$

62. Solve each of the following equations for x.
   a. $\dfrac{x^2 - 13x - 10}{x^2 + 2x - 15} = \dfrac{2}{x + 5} + \dfrac{5}{3 - x}$
   b. Solve $\dfrac{x - 3}{x + 2} = y$.

63. a. If y varies inversely as x, and y is 2 when x is 12, find y when x is 4.
   b. The volume of a gas varies inversely with the pressure when temperature is held constant. If the pressure of 6 L of a gas is 10 N/cm^2, what will the pressure be if the volume is 4 L?

64. Solve each of the following problems.
   a. **Work by Two Roofers** When two members of a roofing crew work together, they can complete a roof on a shed in 2 hours. Working alone, the inexperienced worker could complete the roof in 3 more hours than it would take the more experienced worker. How many hours would it take the more experienced worker to complete the roof?
   b. **Airplanes Traveling in an Airstream** Two planes departed from the same airport at the same time, flying in opposite directions. After a period of time, the slower plane has traveled 660 mi and the faster plane has traveled 780 mi. The faster plane is traveling 40 mi/h faster than the slower plane. Determine the rate of each plane. (*Hint:* What is the same for each plane?)

65. Match each function with its graph. All graphs are displayed with the window [−3, 7, 1] by [−5, 5, 1].
   a. $f(x) = (x - 2)^2 + 1$ b. $f(x) = \dfrac{x}{2} + 1$
   c. $f(x) = \sqrt{x - 2}$ d. $f(x) = \sqrt[3]{2 - x}$
   e. $f(x) = |x - 2| - 1$

A.

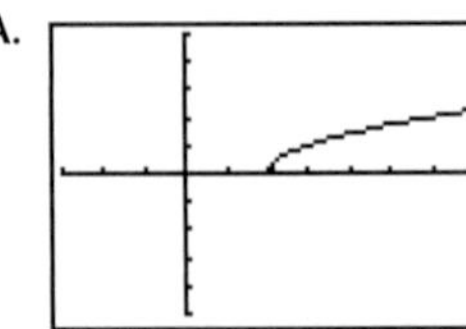

B.

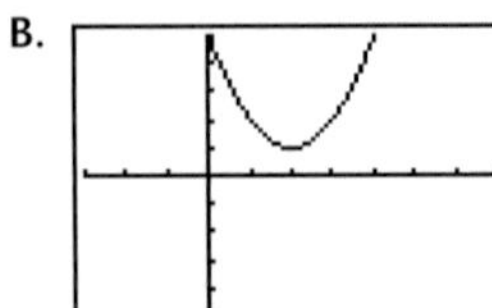

C.

D.

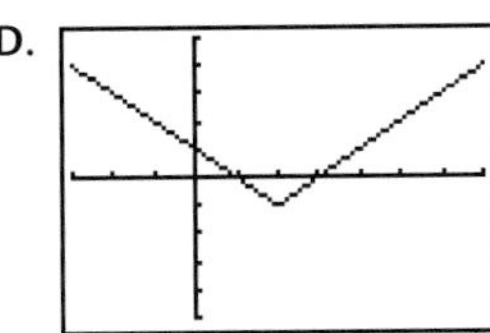

E.

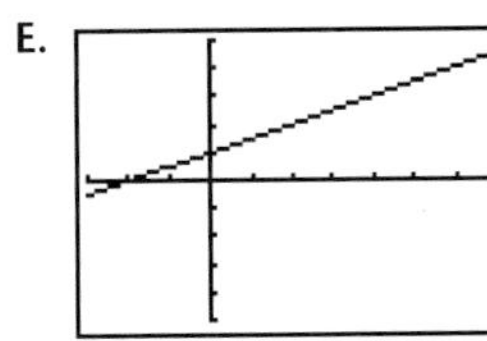

66. Algebraically determine the domain of each function.
a. $f(x) = 3x - 15$ b. $f(x) = \dfrac{2}{3x - 15}$
c. $f(x) = \sqrt[3]{3x - 15}$ d. $f(x) = \sqrt{3x - 15}$

Radicals

67. Evaluate each expression without using a calculator.
a. $\sqrt{64}$ b. $\sqrt[3]{64}$
c. $\sqrt[3]{-0.125}$ d. $\sqrt[4]{16}$

68. Mentally estimate each expression to the nearest integer, and then use a calculator to approximate each value to the nearest thousandth.
a. $\sqrt{26}$ b. $\sqrt[3]{26}$

69. Simplify each sum or difference without using a calculator. Assume $x > 0$.
a. $15\sqrt{3x} - 7\sqrt{3x}$ b. $5\sqrt{72} - 3\sqrt{8}$

70. Simplify each expression without using a calculator.
a. $\sqrt{48}$ b. $\sqrt[3]{48}$

71. Solve each equation.
a. $\sqrt{x + 8} = x - 4$ b. $\sqrt[3]{2x - 3} = 5$

72. Calculate the distance between each pair of points.
a. $(-2, 5)$ and $(3, -7)$ b. $(1, -3)$ and $(3, -7)$

73. Simplify each expression without using a calculator. Assume $x > 0$.
a. $-25^{1/2}$ b. $(x^{3/8}x^{1/8})^{2/3}$
c. $25^{3/2} - 9^{3/2}$ d. $(25 - 9)^{3/2}$

Exponential and Logarithmic Functions and Equations

74. Determine whether each sequence is a geometric sequence. If the sequence is geometric, write the common ratio r.
a. 2, 10, 50, 250, 1250 b. 2, 5, 8, 11, 14

75. Graph each function.
a. $f(x) = 7^x$ b. $f(x) = \left(\dfrac{1}{5}\right)^x$

76. Evaluate each expression, given $f(x) = 9^x$.
a. $f(-1)$ b. $f\left(\dfrac{1}{2}\right)$

77. Graph each function and its inverse on the same coordinate system.
a. $f(x) = 2x - 5$ b. $f(x) = 5^x$

78. Translate the following logarithmic equations to exponential form.
a. $\log_4 x = \dfrac{3}{2}$ b. $\log_3(y - 2) = x$

79. Determine the value of each logarithm by inspection.
a. $\log_4 1$ b. $\log_4 4$
c. $\log_4 \dfrac{1}{4}$ d. $\log_4 4^7$

80. Use a calculator to approximate each expression to the nearest thousandth.
a. $\log 25$ b. $\ln(2.89 \times 10^{-4})$

81. Use the properties of logarithms to express each logarithm in terms of logarithms of simpler expressions. Assume that x and y are positive real numbers.
a. $\log x^5y^3$ b. $\ln\dfrac{x^2}{\sqrt[3]{y}}$

82. Use the properties of logarithms to combine these logarithms into a single logarithmic expression with a coefficient of 1. Assume that the argument of each logarithm is a positive real number.
a. $2\ln x - \ln(x - 5)$ b. $\dfrac{1}{2}\log x + \log(x + 2)$

83. Use the change-of-base formula for logarithms and a calculator to approximate each expression to the nearest thousandth.
a. $\log_3 7$ b. $\log_5 26$

Solve each equation without using a calculator.

84. a. $2^x = \dfrac{1}{32}$ b. $9^x = 27$

85. a. $\log_3 x = 4$ b. $\log_{64} 4 = y$

86. a. $\log(x - 4) + \log(x + 3) = \log 8$
b. $\ln(x - 3) + \ln(x - 7) = \ln(3 - x)$

87. **Bacteria Growth**
Monitoring a culture of 10,000 bacteria that has been isolated and allowed to multiply, a lab technician recorded the number of bacteria each day for 7 days.
a. Draw a scatter diagram for these data points, and determine the exponential function of best fit. (Round the coefficients to three significant digits.)
b. Use this function to estimate the number of bacteria in the culture after 10 days.

x DAYS	y BACTERIA (THOUSANDS)
1	18
2	32
3	56
4	100
5	178
6	316
7	562

88. **Growth of an Investment** How many years will it take a $1,000 investment to increase in value to $2,500 if interest is compounded continuously at 8.25%?

ANSWERS for the Cumulative Review for Chapters 1–10

Question	Answer	Reference Example
1. a	$-\frac{17}{24}$	[1.3–4]
1. b	$\frac{1}{13}$	[1.5–2]
2. a	49	[1.5–8]
2. b	-49	[1.5–8]
3. a	$-\frac{1}{2}$	[1.7–1]
3. b	-4	[1.7–4]
4. a	x^5y^{11}	[5.2–4]
4. b	$\frac{4}{3x^5}$	[5.2–4]
5. a	$x = -6$	[2.5–5]
5. b	$x = 17$	[2.7–2]
6. a	$y = 15$	[2.7–6]
6. b	0.8256 euros for $1 U.S.	[2.7–5]
7. a	$y = 2x + 5$	[2.6–2]

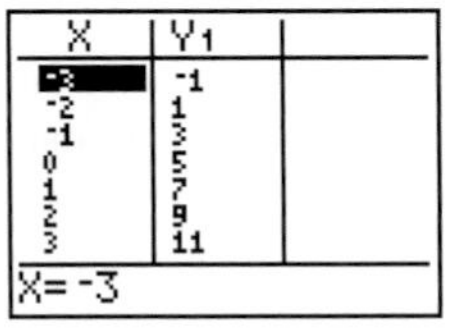

X	Y1
-3	-1
-2	1
-1	3
0	5
1	7
2	9
3	11

X=-3

Question	Answer	Reference Example
7. b	$y = 5x + 13$	[2.6–2]

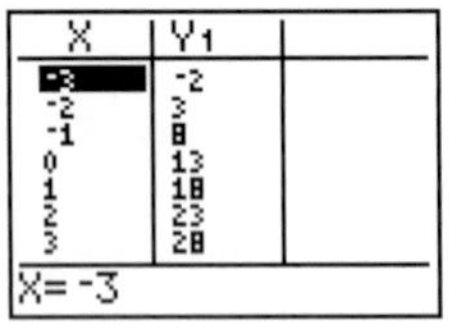

X	Y1
-3	-2
-2	3
-1	8
0	13
1	18
2	23
3	28

X=-3

Question	Answer	Reference Example
8. a	Yes	[7.1–5]
8. b	No	[7.1–5]
9. a	$D = \mathbb{R}$ $R = \{-3\}$	[7.1–8]
9. b	$D = [-2, \infty)$ $R = [-3, \infty)$	[7.1–4]
10. a	4	[7.1–6]
10. b	81	[7.1–6]
11. a	2	[7.1–7]
11. b	3	[7.1–7]
12. a	-10	[7.1–7]
12. b	-1	[7.1–7]

13. a [2.2–2]

x	1	2	3	4	5
y	-3	-1	1	3	5

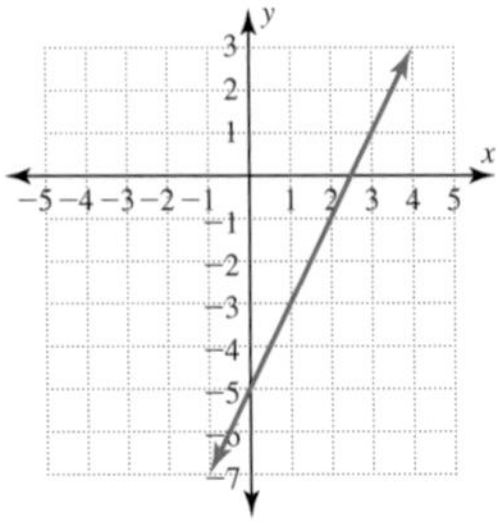

13. b [2.2–2]

x	1	2	3	4	5
y	1	-2	-5	-8	-11

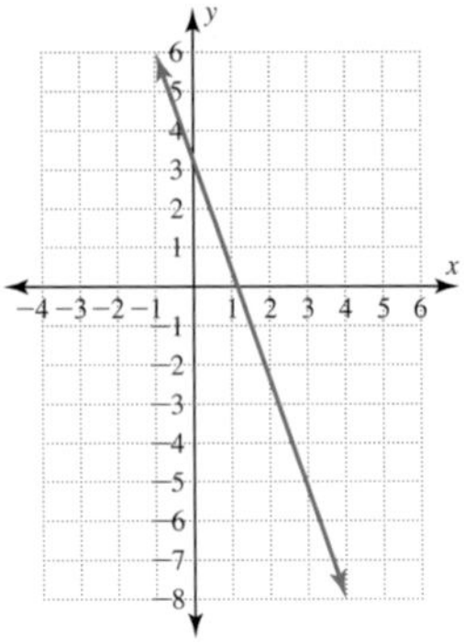

Question	Answer	Reference Example
14. a	x-intercept: $(-1, 0)$; y-intercept: $(0, -3)$	[2.3–4]
14. b	x-intercept: $(10, 0)$; y-intercept: $(0, 6)$	[2.3–4]
14. c	x-intercept: $(18, 0)$; y-intercept: $(0, -12)$	[2.6–3]
15. a	$f(x) = 0.25x + 200$	[2.2–6]

15. b [2.2–6]

x	0	50	100	150	200	250
$f(x)$	200	212.5	225	237.5	250	262.5

Question	Answer	Reference Example
15. c	$f(75) = 218.75$; if the ice cream shop sells 75 cones in the month, the total operating costs for the month will be $218.75.	[2.2–6]
15. d	$x = 240$; if the ice cream shop sells 240 cones in the month, the total operating costs for the month will be $260.00.	[2.2–6]
16. a	$m = -\frac{4}{3}$	[3.1–3]
16. b	$m = \frac{2}{3}$	[3.2–8]
17. a	$y = -\frac{5}{3}x - 2$	[3.2–1]
17. b	$y = -\frac{3}{2}x + 1$	[3.2–9]
18. a		[3.2–3]

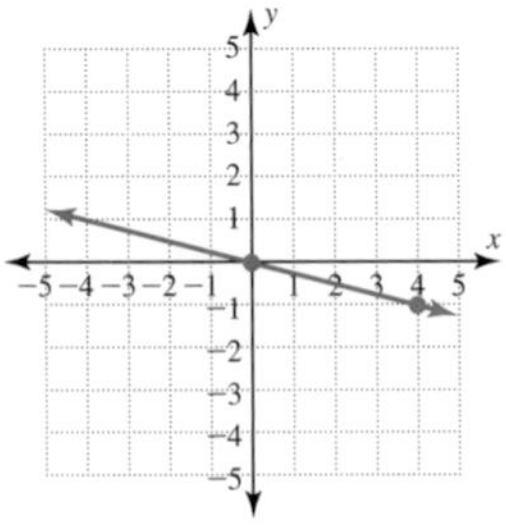

Question	Answer	Reference Example
18. b	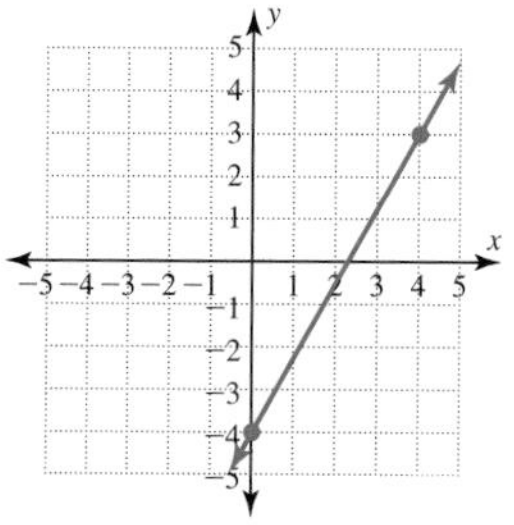 (graph)	[3.2–3]
19. a	Decreasing	[7.2–2]
19. b	Increasing	[7.2–2]
20. a	x | $f(x) = \|x - 2\| - 3$: -1 | 0; 0 | -1; 1 | -2; 2 | -3; 3 | -2; 4 | -1; 5 | 0	[7.2–5]
20. b	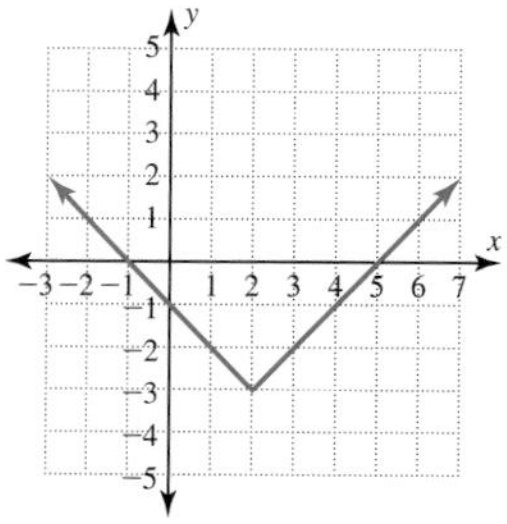 (graph)	[7.2–5]
20. c	$(-1, 0), (5, 0)$	[7.2–5]
20. d	$(0, -1)$	[7.2–5]
20. e	$(2, -3)$	[7.2–5]
20. f	-3	[7.2–5]
20. g	$D = \mathbb{R}$	[7.2–5]
20. h	$R = [-3, \infty)$	[7.2–5]
20. i	$(-\infty, 2)$	[7.2–5]
20. j	$(2, \infty)$	[7.2–5]
20. k	$(-\infty, -1)\cup(5, \infty)$	[7.2–6]
20. l	$(-1, 5)$	[7.2–6]
21. a	$(2, -2)$	[3.3–1]
21. b	$(-3, -1)$	[3.3–1]
22. a	$(2, 1)$	[3.4–1]
22. b	No solution	[3.4–4]
23. a	$(3, 2)$	[3.5–2]
23. b	Infinite number of solutions	[3.5–7]
24. a	The costs for renting the car are \$25 per day and \$0.15 per mile.	[3.6–3]
24. b	The rate of the boat is 15 mi/h, and the rate of the current is 3 mi/h.	[3.6–5]
25. a	$(-\infty, -1)$	[4.1–4]
25. b	$[-1, \infty)$	[4.1–4]
26. a	$(-\infty, 4]$	[4.1–4]
26. b	$[4, \infty)$	[4.1–4]
27.	$(-\infty, 3)$	[4.1–6]
28. a	Conditional; $(5, \infty)$	[4.3–1]
28. b	Contradiction; no solution	[4.3–1]
28. c	Unconditional; $\mathbb{R}$	[4.3–1]
29. a	$x = -5, x = 2$	[4.4–2]
29. b	$(-5, 2)$	[4.4–2]
29. c	$(-\infty, -5]\cup[2, \infty)$	[4.4–2]
30.	$[-5, 3)$	[4.3–4]
31.	$(-\infty, -3]\cup[1, \infty)$	[4.3–8]
32.	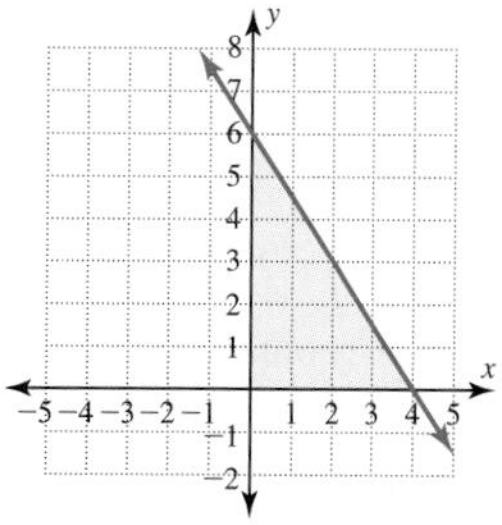 (graph)	[4.5–9]
33.	The maximum value of y is 3. This maximum occurs when x is 2.	[7.3–4]
34.	The maximum height is 147 ft.	[7.3–4]
35. a	$6xy - 18y^2$	[5.6–1]
35. b	$x^3 - 5x^2 + 11x - 10$	[5.5–3]
36. a	$(x + 1)(x - 3)(x - 5)$	[6.1–4]
36. b	$(x + 5)(x + 3)$	[6.1–4]
37. a	$7x^2(2x - 5)$	[6.1–1]
37. b	$(5x - 6)(2x - 3)$	[6.1–1]
38. a	$(2x + 3y)(a + 5b)$	[6.5–1]
38. b	$(x - 3)(8x - 5)$	[6.1–2]
39. a	$(x + 9)(x - 5)$	[6.2–4]
39. b	$(x - 12)(x + 2)$	[6.2–4]
40. a	$(3x - 10)(2x + 1)$	[6.3–2]
40. b	Prime	[6.3–3]
41. a	$(x + 10)^2$	[6.4–1]
41. b	$(x - 7y)^2$	[6.4–1]
42. a	$(x - 8)(x + 8)$	[6.4–2]
42. b	$(x - 4)(x^2 + 4x + 16)$	[6.4–6]
43. a	$(a - b)(a + b + 5)$	[6.5–2]
43. b	$(a + x - 5)(a - x + 5)$	[6.5–3]
44. a	$5(x + 4)(x - 4)$	[6.5–5]
44. b	$-3a(x - 6)(x - 1)$	[6.5–6]
45. a	$x = \frac{3}{2}, x = -8$	[6.6–3]
45. b	$x = 9, x = -5$	[6.6–5]
46. a	Vertex: $(-2, -4)$; y-intercept: $(0, -3)$; x-intercepts: $(-6, 0), (2, 0)$	[7.3–3]
46. b	Opens down because $a < 0$; vertex: $(1, 9)$; y-intercept: $(0, 8)$; x-intercepts: $(-2, 0), (4, 0)$	[7.3–3]
47. a	$x = -3, x = \frac{4}{3}$	[6.6–8]
47. b	$\left[-3, \frac{4}{3}\right]$	[6.6–8]
47. c	$(-\infty, -3)\cup\left(\frac{4}{3}, \infty\right)$	[6.6–8]
48. a	$x = \frac{3}{2} \pm \frac{\sqrt{7}}{2}$	[7.4–1]
48. b	$x = \frac{4}{3} \pm i$	[7.7–8]
49. a	$b^2 - 4ac = 41$; two distinct real solutions	[7.4–7]

continued

Question	Answer	Reference Example
49. b	$b^2 - 4ac = -20$; two complex solutions with imaginary parts	[7.4–7]
49. c	$b^2 - 4ac = 0$; double real solution	[7.4–7]
50. a	$x = \frac{3}{8} \pm \frac{i\sqrt{23}}{8}$	[7.7–9]
50. b	$x = \frac{9}{2}$	[7.4–4]
50. c	$x = -\frac{5}{6} \pm \frac{\sqrt{109}}{6}$	[7.4–4]
51.	The width of the room is 10 ft.	[7.6–5]
52. a	$14i$	[7.7–1]
52. b	$10i$	[7.7–1]
53. a	$29 + 2i$	[7.7–4]
53. b	$4 + 7i$	[7.7–6]
54. a	$D = \mathbb{R} \sim \{-3, 1\}$	[8.1–2]
54. b	$D = \mathbb{R}$	[8.1–2]
55. a	$x = -4, x = 3$	[8.1–2]
55. b	$x = -1$	[8.1–2]
56. a	$\frac{1}{3x - 5}$	[8.1–5]
56. b	$\frac{2x + 5}{x + 3}$	[8.1–5]
57. a	x	[8.2–3]
57. b	$\frac{x + 3}{x - 3}$	[8.2–5]
58. a	2	[8.3–1]
58. b	$\frac{2y}{3x + y}$	[8.3–6]
59. a	$\frac{3x}{2}$	[8.4–1]
59. b	$\frac{1}{(x - 5)(x + 2)}$	[8.4–2]
60. a	$\frac{x - 2}{4}$	[8.4–7]
60. b	$\frac{2}{x + 1}$	[8.4–6]
61. a	$x - 16$	[5.7–4]
61. b	$x^2 - x - 6$	[5.7–5]
62. a	$x = 7$	[8.5–3]
62. b	$x = \frac{2y + 3}{1 - y}$	[8.5–6]
63. a	$y = 6$	[8.6–2]
63. b	The pressure is 15 N/cm^2 if the volume is 4 L.	[8.6–3]
64. a	It will take the more experienced worker 3 hours to complete the roof.	[8.6–6]
64. b	The speed of the slower plane is 220 mi/h, and the speed of the faster plane is 260 mi/h.	[8.6–7]
65. a	B	[9.1–6]
65. b	E	[9.1–6]
65. c	A	[9.1–6]
65. d	C	[9.1–6]
65. e	D	[9.1–6]

Question	Answer	Reference Example
66. a	$D = \mathbb{R}$	[9.1–6]
66. b	$D = \mathbb{R} \sim \{5\}$	[9.1–6]
66. c	$D = \mathbb{R}$	[9.1–6]
66. d	$D = [5, \infty)$	[9.1–6]
67. a	8	[9.1–3]
67. b	4	[9.1–3]
67. c	-0.5	[9.1–3]
67. d	2	[9.1–3]
68. a	5; 5.099	[9.1–4]
68. b	3; 2.962	[9.1–4]
69. a	$8\sqrt{3x}$	[9.2–2]
69. b	$24\sqrt{2}$	[9.2–5]
70. a	$4\sqrt{3}$	[9.2–3]
70. b	$2\sqrt[3]{6}$	[9.2–3]
71. a	$x = 8$	[9.4–5]
71. b	$x = 64$	[9.4–3]
72. a	13	[9.4–6]
72. b	$2\sqrt{5}$	[9.4–6]
73. a	-5	[9.5–2]
73. b	$x^{1/3}$	[9.5–5]
73. c	98	[9.5–2]
73. d	64	[9.5–2]
74. a	Geometric; 5	[10.1–2]
74. b	Not geometric	[10.1–2]
75. a	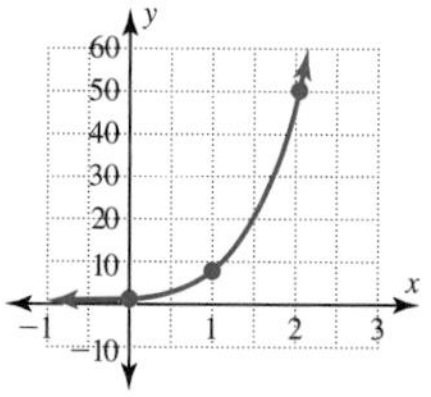	[10.1–5]
75. b	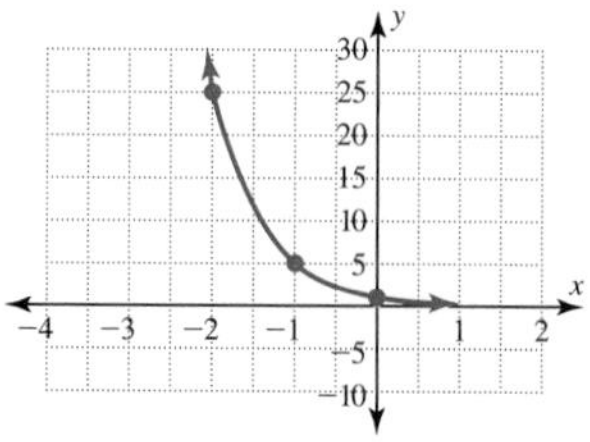	[10.1–5]
76. a	$f(-1) = \frac{1}{9}$	[10.1–5]
76. b	$f\left(\frac{1}{2}\right) = 3$	[10.1–5]
77. a	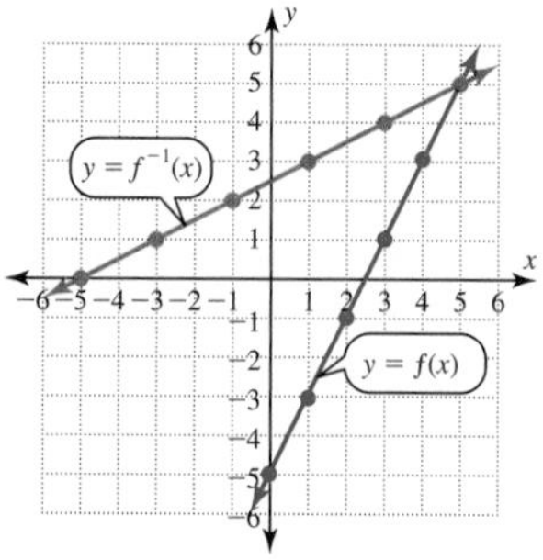 $y = f^{-1}(x)$; $y = f(x)$	[10.2–6]

Question	Answer	Reference Example
77. b	[graph of $y = f(x)$ and $y = f^{-1}(x)$]	[10.2–7]
78. a	$x = 4^{3/2}$	[10.3–1]
78. b	$y - 2 = 3^x$	[10.3–1]
79. a	0	[10.3–4]
79. b	1	[10.3–4]
79. c	-1	[10.3–4]
79. d	7	[10.3–4]
80. a	1.398	[10.4–2]
80. b	-8.149	[10.4–2]
81. a	$5 \log x + 3 \log y$	[10.5–2]
81. b	$2 \ln x - \frac{1}{3} \ln y$	[10.5–2]
82. a	$\ln\left(\frac{x^2}{x-5}\right)$	[10.5–3]

Question	Answer	Reference Example
82. b	$\log [\sqrt{x}(x + 2)]$	[10.5–3]
83. a	1.771	[10.5–5]
83. b	2.024	[10.5–5]
84. a	$x = -5$	[10.6–1]
84. b	$x = \frac{3}{2}$	[10.6–1]
85. a	$x = 81$	[10.6–5]
85. b	$y = \frac{1}{3}$	[10.6–5]
86. a	$x = 5$	[10.6–7]
86. b	No solution	[10.6–7]
87. a	[0.4, 7.6, 1] by [−74.48, 654.48, 1] $y = 10.1(1.77^x)$	[10.7–8]
87. b	There will be approximately 3,050,000 bacteria in the culture after 10 days.	[10.7–8]
88.	It will take approximately 11.1 years.	[10.7–3]

chapter 11

A Preview of College Algebra

CHAPTER OUTLINE

The axle of a train is solid so that the wheels must turn together. On the arc of a circular curve in the tracks, the wheels must travel different distances while turning the same number of times. This is accomplished by using a slightly slanted wheel. This allows the wheels to slide on the rails so that the point of contact with the rail varies the turning radius of each wheel.

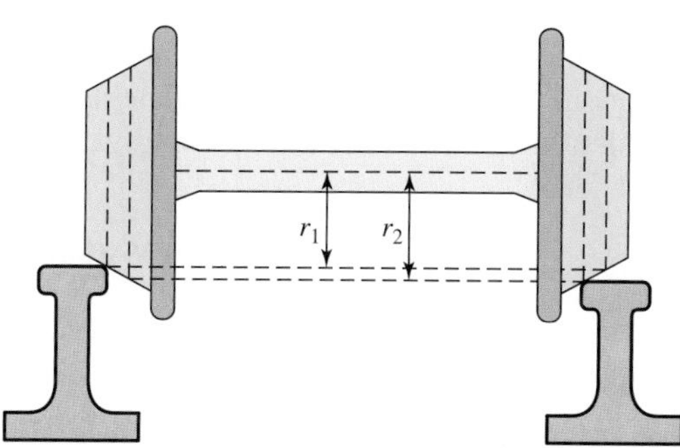

REALITY CHECK

In the figure, which wheel is the inside wheel on a curve and which is the outside wheel?

Section 11.1 Solving Systems of Linear Equations by Using Augmented Matrices

Objective: 1. Use augmented matrices to solve systems of two linear equations with two variables.

A Mathematical Note

James Joseph Sylvester (1814–1897) wrote a paper formulating many of the properties of matrices in 1850 and introduced the term *matrix*.

In chapter 3, we covered two algebraic methods for solving systems of linear equations, the substitution method and the addition method.

One question that you might have as we begin this section is, Why do we need another algebraic method for solving systems of linear equations? The answer to this question is that the substitution method and the addition method work fine for relatively simple systems, but these methods become increasingly complicated to execute for messy coefficients or for larger systems of equations. Graphs and tables are also limited to linear systems with only two variables.

Thus to handle larger systems of linear equations, we will develop over the next two sections the augmented matrix method. This method is very systematic and simplifies the solution of larger systems of linear equations. This method is also well suited to implementation by computer or calculator. After we have developed the augmented matrix method in this section, we illustrate in Section 11.2 how to use the TI-84 Plus calculator to implement this method.

The rows in a theater are arranged horizontally as are the rows of a matrix.
The columns on a building are vertical pillars. The columns of a matrix consist of entries arranged vertically.

We now develop the augmented matrix method, a method that has many similarities to the addition method. A **matrix** is a rectangular array of numbers arranged into rows and columns. Each **row** consists of entries arranged horizontally. Each **column** consists of entries arranged vertically. The **dimension** of a matrix is given by stating first the number of rows and then the number of columns. The dimension of the following matrix is 2×3, read "two by three," because it has two rows and three columns.

MATRIX	DIMENSION
$\begin{bmatrix} 5 & 6 & -1 \\ 4 & 3 & 1 \end{bmatrix}$	2×3 2 rows by 3 columns

The entries in an **augmented matrix** for a system of linear equations consist of the coefficients and constants in the equations. To form the augmented matrix for a system, first we align the similar variables on the left side of each equation and the constants on the right side. Then we form each row in the matrix from the coefficients and the constant of the corresponding equation. A 0 should be written in any position that corresponds to a missing variable in an equation.

SYSTEM OF LINEAR EQUATIONS	AUGMENTED MATRIX
$\begin{Bmatrix} 4x + y = 3 \\ 3x - 2y = 16 \end{Bmatrix}$	$\left[\begin{array}{cc\|c} 4 & 1 & 3 \\ 3 & -2 & 16 \end{array}\right]$

Coefficients of x
Coefficients of y
Brackets enclose the matrix.
Constants
Optional vertical bar to separate coefficients from the constant terms

EXAMPLE 1 Writing Augmented Matrices for Systems of Linear Equations

Write an augmented matrix for each system of linear equations.

SOLUTION

(a) $2x - 5y = 11$
$3x + 4y = 5$

$$\left[\begin{array}{cc|c} 2 & -5 & 11 \\ 3 & 4 & 5 \end{array}\right]$$

The x-coefficients 2 and 3 are in the first column. The y-coefficients -5 and 4 are in the second column. The constants 11 and 5 are in the third column.

(b) $\begin{aligned} 2x + 3y &= 29 \\ 5x - 4y + 8 &= 0 \end{aligned}$ $\qquad$ $\begin{aligned} 2x + 3y &= 29 \\ 5x - 4y &= -8 \end{aligned}$

$$\left[\begin{array}{cc|c} 2 & 3 & 29 \\ 5 & -4 & -8 \end{array}\right]$$

First write each equation with the constant on the right side of the equation. Then form the augmented matrix.

(c) $\begin{aligned} 2x - 4 &= 0 \\ 6x + y &= 4 \end{aligned}$ $\qquad$ $\begin{aligned} 2x + 0y &= 4 \\ 6x + y &= 4 \end{aligned}$

$$\left[\begin{array}{cc|c} 2 & 0 & 4 \\ 6 & 1 & 4 \end{array}\right]$$

Write each equation in the form $Ax + By = C$, using zero coefficients as needed. Then form the augmented matrix.

To solve systems of linear equations by using augmented matrices, we need to be able to represent these systems by using matrices and also be able to write the system of linear equations represented by a given augmented matrix.

EXAMPLE 2 Writing Systems of Linear Equations Represented by Augmented Matrices

Using the variables x and y, write a system of linear equations that is represented by each augmented matrix.

SOLUTION

Although any pair of variables can be used in a system of two linear equations in two variables, we use only the variables x and y in the examples in this section.

(a) $\left[\begin{array}{cc|c} 4 & 5 & 17 \\ 5 & -3 & 12 \end{array}\right]$ $\qquad$ $\begin{aligned} 4x + 5y &= 17 \\ 5x - 3y &= 12 \end{aligned}$

Each row of the matrix represents an equation with an x-coefficient, a y-coefficient, and a constant on the right side of the equation.

(b) $\left[\begin{array}{cc|c} 8 & 3 & -1 \\ 0 & 2 & 10 \end{array}\right]$ $\qquad$ $\begin{aligned} 8x + 3y &= -1 \\ 0x + 2y &= 10 \end{aligned}$

When the coefficient of a variable is 0, you do not have to write this variable in the equation. For example $0x + 2y = 10$ can be written as $2y = 10$.

(c) $\left[\begin{array}{cc|c} 1 & 0 & 7 \\ 0 & 1 & -8 \end{array}\right]$ $\qquad$ $\begin{aligned} x + 0y &= 7 \\ 0x + y &= -8 \end{aligned}$

The last pair of equations can be written as $x = 7$, $y = -8$. Thus the solution of this system is the ordered pair $(7, -8)$.

CALCULATOR PERSPECTIVE 11.1.1 Entering a Matrix into a Calculator

To enter the matrix from Example 2(a) into a TI-84 Plus calculator, enter the following keystrokes:

The MATRIX feature is the secondary function of the x^{-1} key.

    ENTER

```
NAMES MATH EDIT
1:[A] 2×3
2:[B]
3:[C]
4:[D]
5:[E]
6:[F]
7↓[G]

MATRIX[A] 2 ×3
[ 4   5   17 ]
[ 5  -3   12 ]
```

Enter the dimension of the matrix row by column and then enter each element of the matrix. Press ENTER after each value.

Press 2nd QUIT after you have finished entering the values.

```
[A]
      [[4  5  17]
       [5 -3  12]]
```

This will bring a matrix name into the home screen to use in other calculations. Pressing ENTER will display the actual matrix elements.

Note: This matrix has 2 rows and 3 columns so its dimension is 2×3.

The augmented matrix $\left[\begin{array}{cc|c} 1 & 0 & 7 \\ 0 & 1 & -8 \end{array}\right]$ in Example 2(c) is an excellent example of the type of matrices that we want to produce. It follows immediately from this matrix that the solution of the corresponding system of linear equations has $x = 7$ and $y = -8$. In ordered-pair notation the answer is $(7, -8)$. In general, the form $\left[\begin{array}{cc|c} 1 & 0 & k_1 \\ 0 & 1 & k_2 \end{array}\right]$ represents a system of linear equations whose solution is the ordered pair (k_1, k_2). The material that follows shows how to produce this form.

SELF-CHECK 11.1.1

1. Write the augmented matrix for the system $\begin{Bmatrix} 5x + y = -3 \\ 3x + 2y = 8 \end{Bmatrix}$.
2. Using the variables x and y write the system of linear equations represented by the augmented matrix
$\left[\begin{array}{cc|c} 4 & -3 & 1 \\ 2 & 9 & 4 \end{array}\right]$.
3. What is the solution of the system of linear equations represented by
$\left[\begin{array}{cc|c} 1 & 0 & 5 \\ 0 & 1 & 6 \end{array}\right]$?

Equivalent systems of equations have the same solution set. The strategy for solving a system of linear equations is to transform the system into an equivalent system composed of simpler equations. In Chapter 3 we used the properties of equality to justify the transformations we used to solve systems of linear equations. We review these transformations and give the corresponding elementary row operations on a matrix.

TRANSFORMATIONS RESULTING IN EQUIVALENT SYSTEMS	ELEMENTARY ROW OPERATIONS ON AUGMENTED MATRICES
1. Any two equations in a system may be interchanged.	1. Any two rows in the matrix may be interchanged.
2. Both sides of any equation in a system may be multiplied by a nonzero constant.	2. Any row in the matrix may be multiplied by a nonzero constant.
3. Any equation in a system may be replaced by the sum of itself and a constant multiple of another equation in the system.	3. Any row in the matrix may be replaced by the sum of itself and a constant multiple of another row.

The strategy that we use to solve a system of linear equations is to transform the system to an equivalent system composed of simpler equations.

We use the elementary row operations on an augmented matrix just as if the rows were the equations they represent. This is illustrated by the parallel development in Example 3.

EXAMPLE 3 Solving a System by Using the Addition Method and the Augmented Matrix Method

The goal we are working toward is to transform the augmented matrix to the form $\left[\begin{array}{cc|c} 1 & 0 & k_1 \\ 0 & 1 & k_2 \end{array}\right]$.

Solve the system $\begin{Bmatrix} 3x + 2y = -2 \\ x - 3y = 14 \end{Bmatrix}$.

SOLUTION

ADDITION METHOD	AUGMENTED MATRIX METHOD	
$\begin{cases} 3x + 2y = -2 \\ x - 3y = 14 \end{cases}$	$\left[\begin{array}{cc\|c} 3 & 2 & -2 \\ 1 & -3 & 14 \end{array}\right]$	
$\downarrow$	$\downarrow r_1 \leftrightarrow r_2$	This notation means that the first row is interchanged with the second row.
$\begin{cases} x - 3y = 14 \\ 3x + 2y = -2 \end{cases}$	$\left[\begin{array}{cc\|c} 1 & -3 & 14 \\ 3 & 2 & -2 \end{array}\right]$	
$\downarrow$	$\downarrow r_2' = r_2 - 3r_1$	Replace the second row with itself minus 3 times the first row. $\begin{array}{r} 3x + 2y = -2 \\ -3x + 9y = -42 \\ \hline 0x + 11y = -44 \end{array}$
$\begin{cases} x - 3y = 14 \\ 11y = -44 \end{cases}$	$\left[\begin{array}{cc\|c} 1 & -3 & 14 \\ 0 & 11 & -44 \end{array}\right]$	
$\downarrow$	$\downarrow r_2' = \frac{1}{11}r_2$	Multiply the second row by $\frac{1}{11}$.
$\begin{cases} x - 3y = 14 \\ y = -4 \end{cases}$	$\left[\begin{array}{cc\|c} 1 & -3 & 14 \\ 0 & 1 & -4 \end{array}\right]$	
$\downarrow$	$\downarrow r_1' = r_1 + 3r_2$	Replace the first row with itself plus 3 times the second row. $\begin{array}{r} x - 3y = 14 \\ 3y = -12 \\ \hline x + 0y = 2 \end{array}$
$\begin{cases} x = 2 \\ y = -4 \end{cases}$	$\left[\begin{array}{cc\|c} 1 & 0 & 2 \\ 0 & 1 & -4 \end{array}\right]$	This form gives the answer.
Answer: (2, −4)		Does this answer check?

Note that we use arrows, not equal symbols, to denote the flow from one matrix to the next. The matrices are not equal because entries in them have been changed. However, if we use the elementary row operations, they do represent equivalent systems of equations and will help us to determine the solution of the system of linear equations.

The recommended first step in transforming an augmented matrix to the reduced form $\left[\begin{array}{cc|c} 1 & 0 & k_1 \\ 0 & 1 & k_2 \end{array}\right]$ is to get a 1 to occur in the row 1, column 1 position.

Example 4 illustrates three ways to accomplish this. We suggest that you master the notation that is used to describe each step. This will help you better understand this topic and also prepare you to use calculator and computer commands to perform these elementary row operations.

EXAMPLE 4 Using the Notation for Elementary Row Operations

Use the elementary row operations to transform $\left[\begin{array}{cc|c} 3 & 1 & 2 \\ 1 & 2 & 9 \end{array}\right]$ into the form $\left[\begin{array}{cc|c} 1 & \_ & \_ \\ \_ & \_ & \_ \end{array}\right]$.

SOLUTION

(a) By interchanging rows 1 and 2.

$$\left[\begin{array}{cc|c} 3 & 1 & 2 \\ 1 & 2 & 9 \end{array}\right] \xrightarrow{r_1 \leftrightarrow r_2} \left[\begin{array}{cc|c} 1 & 2 & 9 \\ 3 & 1 & 2 \end{array}\right]$$

The notation $r_1 \leftrightarrow r_2$ denotes that rows 1 and 2 have been interchanged.

(b) By multiplying row 1 by $\frac{1}{3}$.

$$\left[\begin{array}{cc|c} 3 & 1 & 2 \\ 1 & 2 & 9 \end{array}\right] \xrightarrow{r_1' = \frac{r_1}{3}} \left[\begin{array}{cc|c} 1 & \frac{1}{3} & \frac{2}{3} \\ 1 & 2 & 9 \end{array}\right]$$

r_1' denotes the new row 1 obtained by multiplying r_1 by $\frac{1}{3}$.

(c) By replacing row 1 with the sum of row 1 and −2 times row 2.

$$\left[\begin{array}{cc|c} 3 & 1 & 2 \\ 1 & 2 & 9 \end{array}\right] \xrightarrow{r_1' = r_1 - 2r_2} \left[\begin{array}{cc|c} 1 & -3 & -16 \\ 1 & 2 & 9 \end{array}\right]$$

r_1' denotes the new row 1 obtained by adding row 1 and −2 times row 2.

The method used in Example 4(a) may be the easiest to apply, but it only works when there is a coefficient of 1 in another row to shift to this first row. The method used in part (c) is often used to avoid the fractions that can result from the method in part (b).

SELF-CHECK 11.1.2

Perform the indicated elementary row operations to produce new matrices.

1. $\left[\begin{array}{cc|c} 4 & 1 & 11 \\ 3 & 2 & 12 \end{array}\right] \xrightarrow{r_1' = r_1 - r_2} \left[\begin{array}{cc|c} \_ & \_ & \_ \\ 3 & 2 & 12 \end{array}\right]$

2. $\left[\begin{array}{cc|c} 1 & 1 & 9 \\ 2 & 3 & 23 \end{array}\right] \xrightarrow{r_2' = r_2 - 2r_1} \left[\begin{array}{cc|c} 1 & 1 & 9 \\ \_ & \_ & \_ \end{array}\right]$

3. $\left[\begin{array}{cc|c} 1 & 3 & 7 \\ 0 & 2 & 6 \end{array}\right] \xrightarrow{r_2' = \frac{r_2}{2}} \left[\begin{array}{cc|c} 1 & 3 & 7 \\ \_ & \_ & \_ \end{array}\right]$

Example 5 uses the elementary row operations to solve a system of linear equations.

EXAMPLE 5 Solving a System of Linear Equations by Using an Augmented Matrix

Use an augmented matrix and elementary row operations to solve $\begin{Bmatrix} 2x + 3y = 4 \\ x + 4y = -3 \end{Bmatrix}$.

SOLUTION

$\begin{Bmatrix} 2x + 3y = 4 \\ x + 4y = -3 \end{Bmatrix}$ $\left[\begin{array}{cc|c} 2 & 3 & 4 \\ 1 & 4 & -3 \end{array}\right]$ First form the augmented matrix.

$\xrightarrow{r_1 \leftrightarrow r_2} \left[\begin{array}{cc|c} 1 & 4 & -3 \\ 2 & 3 & 4 \end{array}\right]$ To work toward the reduced form $\left[\begin{array}{cc|c} 1 & 0 & k_1 \\ 0 & 1 & k_2 \end{array}\right]$, interchange rows 1 and 2 to place a 1 in row 1, column 1.

$\xrightarrow{r_2' = r_2 - 2r_1} \left[\begin{array}{cc|c} 1 & 4 & -3 \\ 0 & -5 & 10 \end{array}\right]$ Next add $-2r_1$ to r_2 to produce a 0 in row 2, column 1.

$\xrightarrow{r_2' = \frac{r_2}{-5}} \left[\begin{array}{cc|c} 1 & 4 & -3 \\ 0 & 1 & -2 \end{array}\right]$ Then divide row 2 by -5 to produce a 1 in row 2, column 2.

$\xrightarrow{r_1' = r_1 - 4r_2} \left[\begin{array}{cc|c} 1 & 0 & 5 \\ 0 & 1 & -2 \end{array}\right]$ To complete the transformation to the form $\left[\begin{array}{cc|c} 1 & 0 & k_1 \\ 0 & 1 & k_2 \end{array}\right]$, produce a 0 in row 1, column 2 by adding $-4r_2$ to r_1.

$x + 0y = 5$
$0x + y = -2$ Write the equivalent system of equations for the reduced form.

Answer: $(5, -2)$ Does this answer check?

Note that in Example 5 we first worked on column 1 and then on column 2. This column-by-column strategy is recommended for transforming all augmented matrices to reduced form. This strategy also is used in Example 6.

EXAMPLE 6 Solving a System of Linear Equations by Using an Augmented Matrix

Use an augmented matrix and elementary row operations to solve $\begin{cases} 4x + 3y = 3 \\ 2x + 9y = 4 \end{cases}$.

SOLUTION

$\begin{cases} 4x + 3y = 3 \\ 2x + 9y = 4 \end{cases}$ $\left[\begin{array}{cc|c} 4 & 3 & 3 \\ 2 & 9 & 4 \end{array}\right]$ First form the augmented matrix.

$\xrightarrow{r_1' = \frac{r_1}{4}} \left[\begin{array}{cc|c} 1 & \frac{3}{4} & \frac{3}{4} \\ 2 & 9 & 4 \end{array}\right]$ Then work on putting column 1 in the form $\left[\begin{array}{cc|c} 1 & - & - \\ 0 & - & - \end{array}\right]$. $\frac{r_1}{4}$ means to multiply row 1 by $\frac{1}{4}$.

$\xrightarrow{r_2' = r_2 - 2r_1} \left[\begin{array}{cc|c} 1 & \frac{3}{4} & \frac{3}{4} \\ 0 & \frac{15}{2} & \frac{5}{2} \end{array}\right]$ $r_2 - 2r_1$ means to subtract twice row 1 from row 2.

$\xrightarrow{r_2' = \frac{2}{15}r_2} \left[\begin{array}{cc|c} 1 & \frac{3}{4} & \frac{3}{4} \\ 0 & 1 & \frac{1}{3} \end{array}\right]$ Next work on column 2 to put the matrix in the reduced form $\left[\begin{array}{cc|c} 1 & 0 & k_1 \\ 0 & 1 & k_2 \end{array}\right]$.

$\frac{2}{15}r_2$ means to multiply row 2 by $\frac{2}{15}$.

$\xrightarrow{r_1' = r_1 - \frac{3}{4}r_2} \left[\begin{array}{cc|c} 1 & 0 & \frac{1}{2} \\ 0 & 1 & \frac{1}{3} \end{array}\right]$ $r_1 - \frac{3}{4}r_2$ means to subtract $\frac{3}{4}r_2$ from row 1.

$$x + 0y = \frac{1}{2}$$
$$0x + y = \frac{1}{3}$$

Write the equivalent system of equations for the reduced form.

Answer: $\left(\frac{1}{2}, \frac{1}{3}\right)$ Does this answer check?

SELF-CHECK 11.1.3

Write the solution of a system of linear equations represented by the following.

1. $\left[\begin{array}{cc|c} 1 & 0 & -4 \\ 0 & 1 & 8 \end{array}\right]$ **2.** $\left[\begin{array}{cc|c} 1 & 0 & -\frac{3}{5} \\ 0 & 1 & \frac{2}{3} \end{array}\right]$

Solve by using augmented matrices.

3. $\begin{cases} x + 5y = -3 \\ 2x + 9y = -7 \end{cases}$

The reduced form $\left[\begin{array}{cc|c} 1 & 0 & k_1 \\ 0 & 1 & k_2 \end{array}\right]$ always yields a unique solution for a consistent system of independent equations. For an inconsistent system with no solution or a consistent system of dependent equations with an infinite number of solutions, the reduced form is different. We now examine these two possibilities. Solving an inconsistent system by the addition method produces an equation that is a contradiction. Note the matrix equivalent of this in Example 7.

EXAMPLE 7 Solving an Inconsistent System by Using an Augmented Matrix

Use an augmented matrix and elementary row operations to solve $\begin{Bmatrix} x + 2y = 3 \\ 5x + 10y = 11 \end{Bmatrix}$.

SOLUTION

$\begin{Bmatrix} x + 2y = 3 \\ 5x + 10y = 11 \end{Bmatrix}$ $\left[\begin{array}{cc|c} 1 & 2 & 3 \\ 5 & 10 & 11 \end{array}\right]$ First form the augmented matrix.

$\xrightarrow{r_2' = r_2 - 5r_1} \left[\begin{array}{cc|c} 1 & 2 & 3 \\ 0 & 0 & -4 \end{array}\right]$ Then work on putting column 1 in the form $\left[\begin{array}{cc|c} 1 & - & - \\ 0 & - & - \end{array}\right]$.

$x + 2y = 3$
$0 + 0 = -4$

There is no need to proceed further because the last row in the matrix corresponds to an equation that is a contradiction. Thus the original system of equations is inconsistent and has no solution.

Answer: There is no solution.

We encourage you to compare this Example 7 to the Example 7 in Section 3.5 where the same problem was worked by the addition method. Example 8 examines a consistent system of dependent equations.

EXAMPLE 8 Solving a Consistent System of Dependent Equations by Using an Augmented Matrix

Use an augmented matrix and elementary row operations to solve $\begin{Bmatrix} -4x + 10y = -2 \\ 6x - 15y = 3 \end{Bmatrix}$.

SOLUTION

$\begin{Bmatrix} -4x + 10y = -2 \\ 6x - 15y = 3 \end{Bmatrix}$ $\left[\begin{array}{cc|c} -4 & 10 & -2 \\ 6 & -15 & 3 \end{array}\right]$ First form the augmented matrix.

$\xrightarrow{r_1' = -\frac{1}{4}r_1} \left[\begin{array}{cc|c} 1 & -\frac{5}{2} & \frac{1}{2} \\ 6 & -15 & 3 \end{array}\right]$ Then work on putting column 1 in the form $\left[\begin{array}{cc|c} 1 & - & - \\ 0 & - & - \end{array}\right]$.

$\xrightarrow{r_2' = r_2 - 6r_1} \left[\begin{array}{cc|c} 1 & -\frac{5}{2} & \frac{1}{2} \\ 0 & 0 & 0 \end{array}\right]$ The last row corresponds to an equation that is an identity. It cannot be used to simplify column 2 further.

$x - \frac{5}{2}y = \frac{1}{2}$
$0x + 0y = 0$

These two equations form a dependent system of equations with an infinite number of solutions.

$x = \frac{5}{2}y + \frac{1}{2}$

Because the coefficient of x in the first equation is 1, solve this equation for x in terms of y. You can also solve this equation for y in terms of x and express the answer in the form $\left(x, \frac{2}{5}x - \frac{1}{5}\right)$. You may wish to confirm that both of these represent the same set of points. All the solutions are points on the same line, points that can be written in this form.

Answer: There are an infinite number of solutions all having the form $\left(\frac{5}{2}y + \frac{1}{2}, y\right)$.

The **general solution** of a system of dependent linear equations describes all solutions of the system and is given by indicating the relationship between the coordinates of the solutions. The **particular solutions** obtained from the general solution contain only constant coordinates.

The general solution in Example 8 is $\left(\frac{5}{2}y + \frac{1}{2}, y\right)$. By arbitrarily selecting y-values, we can produce as many particular solutions as we wish. For $y = 0$ and $y = 1$, we obtain two particular solutions $\left(\frac{1}{2}, 0\right)$ and (3, 1).

SELF-CHECK 11.1.4

Write the solution of the system of linear equations represented by the following.

1. $\left[\begin{array}{cc|c} 1 & 0 & 7 \\ 0 & 0 & 6 \end{array}\right]$
2. $\left[\begin{array}{cc|c} 1 & -1 & 4 \\ 0 & 0 & 0 \end{array}\right]$

SELF-CHECK ANSWERS

11.1.1

1. $\left[\begin{array}{cc|c} 5 & 1 & -3 \\ 3 & 2 & 8 \end{array}\right]$
2. $\begin{cases} 4x - 3y = 1 \\ 2x + 9y = 4 \end{cases}$
3. (5, 6)

11.1.2

1. $\left[\begin{array}{cc|c} 1 & -1 & -1 \\ 3 & 2 & 12 \end{array}\right]$
2. $\left[\begin{array}{cc|c} 1 & 1 & 9 \\ 0 & 1 & 5 \end{array}\right]$
3. $\left[\begin{array}{cc|c} 1 & 3 & 7 \\ 0 & 1 & 3 \end{array}\right]$

11.1.3

1. $(-4, 8)$
2. $\left(-\frac{3}{5}, \frac{2}{3}\right)$
3. $(-8, 1)$

11.1.4

1. There is no solution; it is an inconsistent system.
2. A dependent system with a general solution $(y + 4, y)$; three particular solutions are (4, 0), (5, 1), and (6, 2).

USING THE LANGUAGE AND SYMBOLISM OF MATHEMATICS 11.1

1. A matrix is a __________ array of numbers.
2. The entries in an augmented matrix for a system of linear equations consist of the __________ and __________ in the equations.
3. __________ systems of equations have the same solution set.
4. The notation $r_1 \leftrightarrow r_2$ denotes that rows 1 and 2 for a matrix are __________.
5. The notation $r_1' = \frac{1}{2}r_1$ denotes that row __________ is being replaced by multiplying the current row __________ by __________.
6. The notation to represent that row 2 is being replaced by the sum of the current row 2 plus twice row 1 is __________.
7. The __________ solution of a system of dependent linear equations describes all solutions of the system and is given by indicating the relationship between the coordinates of the solutions.
8. The __________ solutions obtained from a general solution contain only constant coordinates.

QUICK REVIEW 11.1

The purpose of this quick review is to help you recall skills needed in this section.

1. Determine whether $(-2, 5)$ is a solution of the system $\begin{cases} 4x + 3y = 7 \\ 5x + 2y = 1 \end{cases}$.
2. Solve $\begin{cases} 3x - 4y = 8 \\ x - 2y = 2 \end{cases}$ by the substitution method.
3. Solve $\begin{cases} 2x + 5y = 1 \\ 3x - 4y = 13 \end{cases}$ by the addition method.

4. Use the graph to determine the solution of the system $\left\{\begin{array}{l} x + 3y = 5 \\ 3x - 2y = 4 \end{array}\right\}$.

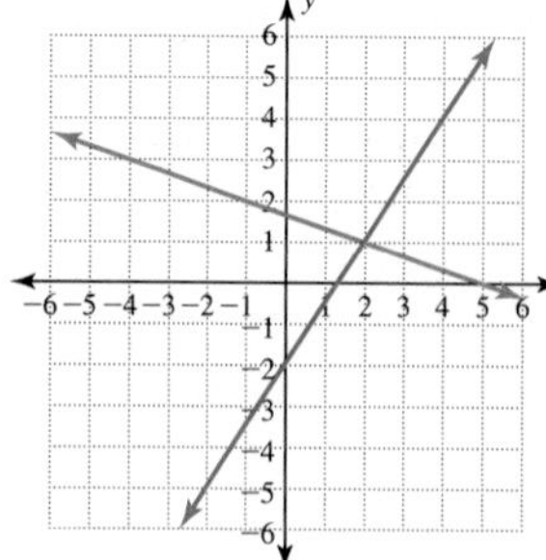

5. Use the table to determine the solution of the system $\left\{\begin{array}{l} y = 5x - 4 \\ y = 2x + 2 \end{array}\right\}$.

x	y_1	y_2
1	1	4
2	6	6
3	11	8
4	16	10
5	21	12

EXERCISES 11.1

In Exercises 1–6, write an augmented matrix for each system of linear equations.

1. $\left\{\begin{array}{l} 3x + y = 0 \\ 2x - y = -5 \end{array}\right\}$

2. $\left\{\begin{array}{l} 5x - y = 3 \\ 4x + 3y = 29 \end{array}\right\}$

3. $\left\{\begin{array}{r} 4x = 12 \\ 3x + 2y = 1 \end{array}\right\}$

4. $\left\{\begin{array}{r} 2x + 7y = 3 \\ -3y = 3 \end{array}\right\}$

5. $\left\{\begin{array}{l} x = 5 \\ y = -6 \end{array}\right\}$

6. $\left\{\begin{array}{l} x = -4 \\ y = 11 \end{array}\right\}$

In Exercises 7–12, write a system of linear equations in x and y that is represented by each augmented matrix.

7. $\left[\begin{array}{cc|c} 2 & 3 & 2 \\ 4 & -3 & 1 \end{array}\right]$

8. $\left[\begin{array}{cc|c} 5 & -1 & 0 \\ 3 & 2 & 13 \end{array}\right]$

9. $\left[\begin{array}{cc|c} 2 & 1 & 1 \\ 1 & 3 & 0 \end{array}\right]$

10. $\left[\begin{array}{cc|c} 2 & 3 & 5 \\ 6 & -4 & 2 \end{array}\right]$

11. $\left[\begin{array}{cc|c} 1 & 0 & 7 \\ 0 & 1 & -8 \end{array}\right]$

12. $\left[\begin{array}{cc|c} 1 & 0 & \frac{2}{3} \\ 0 & 1 & -\frac{4}{5} \end{array}\right]$

In Exercises 13–20, use the given elementary row operations to complete each matrix.

13. $\left[\begin{array}{cc|c} 2 & 1 & 1 \\ 1 & 3 & 0 \end{array}\right] \xrightarrow{r_1 \leftrightarrow r_2} \left[\begin{array}{cc|c} \_ & \_ & \_ \\ \_ & \_ & \_ \end{array}\right]$

14. $\left[\begin{array}{cc|c} -6 & 3 & 4 \\ 1 & 2 & 1 \end{array}\right] \xrightarrow{r_1 \leftrightarrow r_2} \left[\begin{array}{cc|c} \_ & \_ & \_ \\ \_ & \_ & \_ \end{array}\right]$

15. $\left[\begin{array}{cc|c} 1 & -3 & -2 \\ 2 & -5 & 4 \end{array}\right] \xrightarrow{r_2' = r_2 - 2r_1} \left[\begin{array}{cc|c} 1 & -3 & -2 \\ \_ & \_ & \_ \end{array}\right]$

16. $\left[\begin{array}{cc|c} 1 & -1 & 1 \\ 2 & -3 & -2 \end{array}\right] \xrightarrow{r_2' = r_2 - 2r_1} \left[\begin{array}{cc|c} 1 & -1 & 1 \\ \_ & \_ & \_ \end{array}\right]$

17. $\left[\begin{array}{cc|c} 3 & -1 & 3 \\ 6 & 4 & -6 \end{array}\right] \xrightarrow{r_1' = \frac{1}{3}r_1} \left[\begin{array}{cc|c} \_ & \_ & \_ \\ 6 & 4 & -6 \end{array}\right]$

18. $\left[\begin{array}{cc|c} 2 & 3 & 1 \\ 3 & -3 & -21 \end{array}\right] \xrightarrow{r_1' = \frac{1}{2}r_1} \left[\begin{array}{cc|c} \_ & \_ & \_ \\ 3 & -3 & -21 \end{array}\right]$

19. $\left[\begin{array}{cc|c} 1 & 5 & 16 \\ 0 & 1 & 3 \end{array}\right] \xrightarrow{r_1' = r_1 - 5r_2} \left[\begin{array}{cc|c} \_ & \_ & \_ \\ 0 & 1 & 3 \end{array}\right]$

20. $\left[\begin{array}{cc|c} 1 & -4 & 10 \\ 0 & 1 & -2 \end{array}\right] \xrightarrow{r_1' = r_1 + 4r_2} \left[\begin{array}{cc|c} \_ & \_ & \_ \\ 0 & 1 & -2 \end{array}\right]$

In Exercises 21–26, write the solution for the system of linear equations represented by each augmented matrix. If the matrix represents a consistent system of dependent equations, write the general solution and three particular solutions.

21. $\left[\begin{array}{cc|c} 1 & 0 & -5 \\ 0 & 1 & 9 \end{array}\right]$

22. $\left[\begin{array}{cc|c} 1 & 0 & 4 \\ 0 & 1 & 6 \end{array}\right]$

23. $\left[\begin{array}{cc|c} 1 & 4 & 3 \\ 0 & 0 & 7 \end{array}\right]$

24. $\left[\begin{array}{cc|c} 1 & -3 & 5 \\ 0 & 0 & -2 \end{array}\right]$

25. $\left[\begin{array}{cc|c} 1 & 3 & 5 \\ 0 & 0 & 0 \end{array}\right]$

26. $\left[\begin{array}{cc|c} 1 & -2 & 3 \\ 0 & 0 & 0 \end{array}\right]$

In Exercises 27–34, label the elementary row operation used to transform the first matrix to the second. Use the notation developed in this section.

27. $\left[\begin{array}{cc|c} 2 & 6 & 8 \\ 3 & 7 & 10 \end{array}\right] \xrightarrow{r_1' = ?} \left[\begin{array}{cc|c} 1 & 3 & 4 \\ 3 & 7 & 10 \end{array}\right]$

28. $\left[\begin{array}{cc|c} 2 & 5 & 9 \\ 3 & 2 & 8 \end{array}\right] \xrightarrow{?} \left[\begin{array}{cc|c} 1 & \frac{5}{2} & \frac{9}{2} \\ 3 & 2 & 8 \end{array}\right]$

29. $\left[\begin{array}{cc|c} 1 & 5 & 8 \\ 3 & 2 & -2 \end{array}\right] \xrightarrow{?} \left[\begin{array}{cc|c} 1 & 5 & 8 \\ 0 & -13 & -26 \end{array}\right]$

30. $\left[\begin{array}{cc|c} 1 & 3 & 8 \\ 2 & -3 & 7 \end{array}\right] \xrightarrow{?} \left[\begin{array}{cc|c} 1 & 3 & 8 \\ 0 & -9 & -9 \end{array}\right]$

31. $\left[\begin{array}{cc|c} 1 & -2 & 8 \\ 0 & 3 & -9 \end{array}\right] \xrightarrow{?} \left[\begin{array}{cc|c} 1 & -2 & 8 \\ 0 & 1 & -3 \end{array}\right]$

32. $\left[\begin{array}{cc|c} 1 & 6 & 2 \\ 0 & 3 & 1 \end{array}\right] \xrightarrow{?} \left[\begin{array}{cc|c} 1 & 6 & 2 \\ 0 & 1 & \frac{1}{3} \end{array}\right]$

33. $\left[\begin{array}{cc|c} 1 & 2 & 6 \\ 0 & 1 & 1 \end{array}\right] \xrightarrow{?} \left[\begin{array}{cc|c} 1 & 0 & 4 \\ 0 & 1 & 1 \end{array}\right]$

34. $\left[\begin{array}{cc|c} 1 & 3 & 2 \\ 0 & 1 & \frac{1}{3} \end{array}\right] \xrightarrow{?} \left[\begin{array}{cc|c} 1 & 0 & 1 \\ 0 & 1 & \frac{1}{3} \end{array}\right]$

In Exercises 35–50, use an augmented matrix and elementary row operations to solve each system of linear equations.

35. $x + 3y = 5$, $2x + y = -5$

36. $x - 2y = 9$, $3x + 4y = 7$

37. $x + 3y = 1$, $3x + 7y = 7$

38. $x + 2y = 7$, $4x + 3y = 3$

39. $2x + 5y = -4$, $4x + 3y = 6$

40. $2x - 3y = 17$, $4x + y = 13$

41. $4x - 9y = 5$, $3x + 12y = 10$

42. $8x + 3y = -39$, $7x - 2y = -11$

43. $3x - y = 2$, $2x + y = 6$

44. $3x + 4y = 0$, $2x - 3y = 16$

45. $6x + 4y = 11$, $10x + 6y = 17$

46. $5x - y = 7$, $2x + 4y = 5$

47. $3x + 4y = 7$
$6x + 8y = 10$

48. $4x + 3y = 2$
$16x + 12y = 7$

49. $2x - y = 5$
$4x - 2y = 10$

50. $3x - 6y = 12$
$4x - 8y = 16$

In Exercises 51–60, write a system of linear equations by using the variables x and y, and use this system to solve the problem.

51. Numeric Word Problem Find two numbers whose sum is 160 and whose difference is 4.

52. Numeric Word Problem Find two numbers whose sum is 260 if one number is 3 times the other number.

53. Complementary Angles The two angles shown are complementary, and one angle is 32° larger than the other. Determine the number of degrees in each angle.

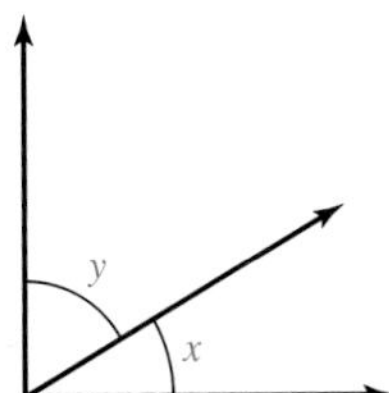

54. Supplementary Angles The two angles shown are supplementary, and one angle is 74° larger than the other. Determine the number of degrees in each angle.

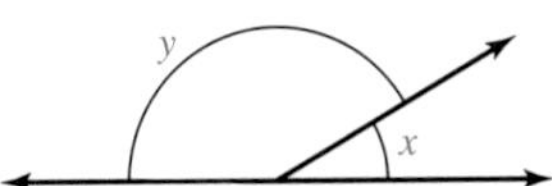

55. Fixed and Variable Costs A seamstress makes custom costumes for operas. One month the total of fixed and variable costs for making 20 costumes was $3,200. The next month the total of the fixed and variable costs for making 30 costumes was $4,300. Determine the fixed cost and the variable cost per costume.

56. Rates of Two Bicyclists Two bicyclists depart at the same time from a common location, traveling in opposite directions. One averages 5 km/h, more than the other. After 2 hours, they are 130 km apart. Determine the speed of each bicyclist.

57. Rate of a River Current A small boat can go 30 km downstream in 1 hour, but only 14 km upstream in 1 hour. Determine the rate of the boat and the rate of the current.

58. Mixture of Two Disinfectants A hospital needs 100 L of a 15% solution of disinfectant. How many liters of a 40% solution and a 5% solution should be mixed to obtain this 15% solution?

59. Mixture of a Fruit Drink A fruit concentrate is 15% water. How many liters of pure water and how many liters of concentrate should be mixed to produce 100 L of mixture that is 83% water?

60. Basketball Scores During one game for the Phoenix Suns, Steve Nash scored 39 points on 17 field goals. How many of these field goals were 2-pointers and how many were 3-pointers?

Group Discussion Questions

61. Challenge Question Solve $\begin{Bmatrix} a_1x + b_1y = c_1 \\ a_2x + b_2y = c_2 \end{Bmatrix}$ for (x, y) in terms of a_1, a_2, b_1, b_2, c_1, and c_2. Assume that $a_1b_2 - a_2b_1 \neq 0$.

62. Discovery Question

a. Extend the augmented matrix notation given for systems of two linear equations with two variables to write an augmented matrix for this system.

$$\begin{Bmatrix} x + 2y + 3z = 5 \\ 2x + y - z = -1 \\ 3x + y + z = 4 \end{Bmatrix}$$

b. Write a system of linear equations that is represented by this augmented matrix. Use the variables x, y, and z.

$$\left[\begin{array}{ccc|c} 2 & -3 & 3 & 2 \\ 4 & 0 & 2 & -1 \\ -2 & 4 & -3 & -2 \end{array}\right]$$

c. Write a system of linear equations that is represented by this augmented matrix. Use the variables w, x, y, and z.

$$\left[\begin{array}{cccc|c} 1 & -1 & 3 & -4 & 1 \\ -2 & 4 & -1 & 1 & -4 \\ 3 & 2 & -4 & 5 & 7 \\ 2 & -1 & -1 & 4 & 7 \end{array}\right]$$

Section 11.2 Systems of Linear Equations in Three Variables

Objective: 2. Solve a system of three linear equations in three variables.

A first-degree equation in two variables of the form $Ax + By = C$ is called a linear equation because its graph is a straight line if A and B are not both 0. Similarly, a first-degree equation in three variables of the form $Ax + By + Cz = D$ also is called a linear equation. However, this name is misleading because if A, B, and C are not all 0, the graph of $Ax + By + Cz = D$ is not a line but a plane in three-dimensional space.

The plane defined by $2x + 3y + 4z = 12$

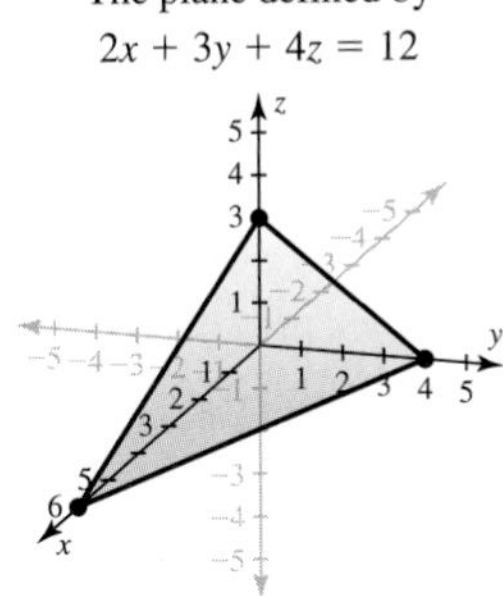

The graph of a three-dimensional space on two-dimensional paper is limited in its portrayal of the third dimension. Nonetheless, we can give the viewer a feeling for planes in a three-dimensional space by orienting the x-, y-, and z-axes as shown in the figure. This graph illustrates the plane $2x + 3y + 4z = 12$, whose x-intercept is $(6, 0, 0)$, whose y-intercept is $(0, 4, 0)$, and whose z-intercept is $(0, 0, 3)$. Drawing lines to connect these intercepts gives the view of the plane in the region where all coordinates are positive.

A system of three linear equations in three variables is referred to as a 3×3 (three-by-three) system. A **solution** of a system of equations with the three variables x, y, and z is an **ordered triple** (x, y, z) that is a solution of each equation in the system.

EXAMPLE 1 Determining Whether an Ordered Triple Is a Solution of a System of Linear Equations

Determine whether (2, −3, 5) is a solution of $\begin{Bmatrix} x + y - z = -6 \\ 2x - y + z = 12 \\ 3x - 2y - 2z = 3 \end{Bmatrix}$.

SOLUTION

FIRST EQUATION

$$x + y - z = -6$$
$$2 + (-3) - 5 \stackrel{?}{=} -6$$
$$-6 \stackrel{?}{=} -6 \text{ is true.}$$

SECOND EQUATION

$$2x - y + z = 12$$
$$2(2) - (-3) + 5 \stackrel{?}{=} 12$$
$$4 + 3 + 5 \stackrel{?}{=} 12$$
$$12 \stackrel{?}{=} 12 \text{ is true.}$$

THIRD EQUATION

$$3x - 2y - 2z = 3$$
$$3(2) - 2(-3) - 2(5) \stackrel{?}{=} 3$$
$$6 + 6 - 10 \stackrel{?}{=} 3$$
$$2 \stackrel{?}{=} 3 \text{ is false.}$$

Answer: (2, −3, 5) is not a solution of this system. To be a solution of this system, the point must satisfy all three equations.

SELF-CHECK 11.2.1

1. Determine whether (2, −3.25, 4.75) is a solution of the system in Example 1.

The graph of each linear equation $Ax + By + Cz = D$ is a plane in three-dimensional space unless A, B, and C are all 0.

A system of three linear equations in three variables can be viewed geometrically as the intersection of a set of three planes. These planes may intersect in one point, no points, or an infinite number of points. The illustrations in the following box show some of the ways we can obtain these solutions. Can you sketch other ways of obtaining these solution sets?

Types of Solution Sets for Linear Systems with Three Equations

The linear system $\begin{Bmatrix} A_1x + B_1y + C_1z = D_1 \\ A_2x + B_2y + C_2z = D_2 \\ A_3x + B_3y + C_3z = D_3 \end{Bmatrix}$ can have

ONE SOLUTION

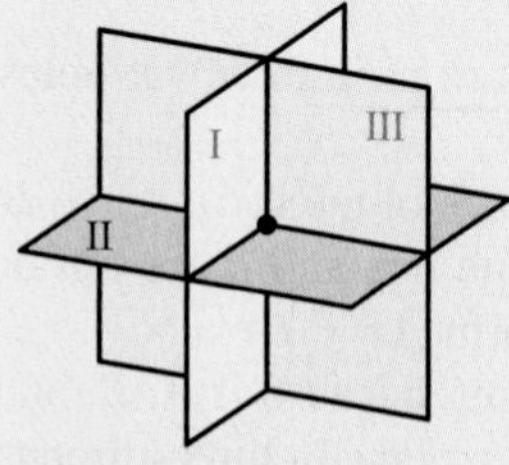

The planes intersect at a single point; the system is consistent and the equations are independent.

NO SOLUTION

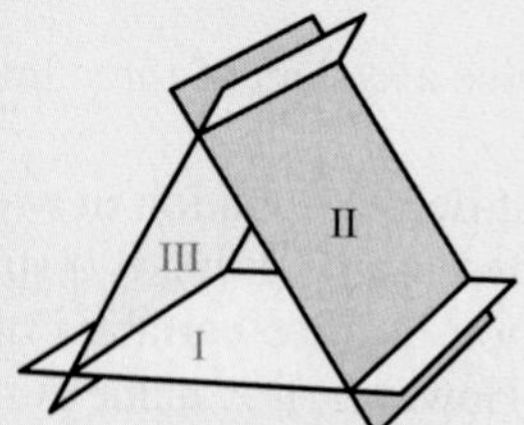

The planes have no point in common; the system is inconsistent.

AN INFINITE NUMBER OF SOLUTIONS

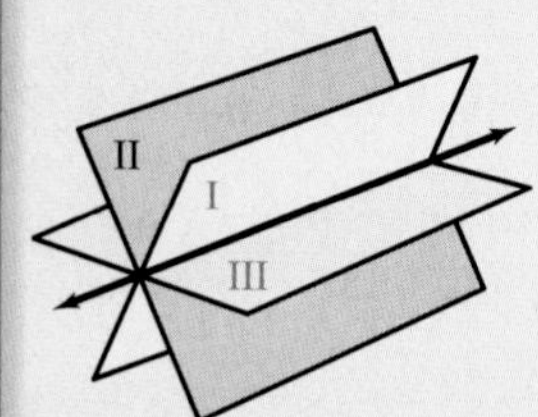

The planes intersect along a line and thus have an infinite number of common points; the system is consistent and the equations are dependent.

The figures in this box can give us an intuitive understanding of the possible solutions to these systems; however, it is not practical to actually solve these systems graphically. Thus we rely entirely on algebraic methods.

In this section we illustrate two methods for solving systems of three linear equations in three variables (3×3 systems). Example 2 extends the substitution and addition methods from Sections 3.4 and 3.5 to solve a 3×3 system. Later in the section we use augmented matrices, which were introduced in Section 11.1.

To solve a 3×3 system, first eliminate a variable to produce a 2×2 system. Then eliminate another variable from the 2×2 system to produce an equation that has only one variable.

Equivalent systems of equations have the same solution set. The general goal of each step of a solution process is to produce an equivalent system that is simpler than the previous step. By eliminating some of the variables we can reduce a 3×3 system to a system with only two variables, and then we can eliminate another variable to produce an equation with only one variable. We can then back-substitute to obtain the values of the other two variables. This strategy is outlined in the following box.

Strategy for Solving a 3 × 3 System of Linear Equations*

STEP 1. Write each equation in the general form $Ax + By + Cz = D$.

STEP 2. Select one pair of equations and use the substitution method or the addition method to eliminate one of the variables.

STEP 3. Repeat step 2 with another pair of equations. Be sure to eliminate the *same* variable as in step 2.

STEP 4. Eliminate another variable from the pair of equations produced in steps 2 and 3, and solve this 2×2 system of equations.

STEP 5. Back-substitute the values from step 4 into one of the original equations to solve for the third variable.

STEP 6. Does this solution check in all three of the original equations?

*If a contradiction is obtained in any of these steps, the system is inconsistent and has no solution. If an identity is obtained in any step, the system is either dependent with infinitely many solutions or inconsistent with no solution.

EXAMPLE 2 Solving a 3 × 3 System of Linear Equations

Solve the system
$$\begin{matrix}(1)\\(2)\\(3)\end{matrix}\begin{Bmatrix} x + y + z = 2 \\ -x + y - 2z = 1 \\ x + y - z = 0 \end{Bmatrix}.$$

SOLUTION

PRODUCE A 2 × 2 SYSTEM OF EQUATIONS

$$\begin{matrix}(1)\\(2)\\(3)\end{matrix}\begin{Bmatrix} x + y + z = 2 \\ -x + y - 2z = 1 \\ x + y - z = 0 \end{Bmatrix} \quad \begin{array}{lr}(1) & x + y + z = 2 \\ (2) & -x + y - 2z = 1 \\ \hline & 2y - z = 3\end{array} \quad \begin{array}{lr}(2) & -x + y - 2z = 1 \\ (3) & x + y - z = 0 \\ \hline & 2y - 3z = 1\end{array}$$

First decide which variable to eliminate. Here eliminate x by working first with Equations (1) and (2) and then with Equations (2) and (3).
Add Equations (1) and (2). Then add Equations (2) and (3).
Note that x has been eliminated from both of these equations.

The 2×2 system of equations is $\begin{Bmatrix} 2y - z = 3 \\ 2y - 3z = 1 \end{Bmatrix}$.

PRODUCE AN EQUATION WITH ONLY ONE VARIABLE

$$\begin{Bmatrix} 2y - z = 3 \\ 2y - 3z = 1 \end{Bmatrix} \quad \begin{array}{r} 2y - z = 3 \\ -2y + 3z = -1 \\ \hline 2z = 2 \\ z = 1 \end{array}$$

Subtract the second equation from the first equation to eliminate y.

Solve this equation for z.

BACK-SUBSTITUTE

$$\begin{aligned} 2y - z &= 3 \\ 2y - (1) &= 3 \\ 2y &= 4 \\ y &= 2 \end{aligned} \qquad (1) \quad \begin{aligned} x + y + z &= 2 \\ x + (2) + (1) &= 2 \\ x + 3 &= 2 \\ x &= -1 \end{aligned}$$

First substitute 1 for z in the equation $2y - z = 3$ and solve for y.

Then substitute 2 for y and 1 for z in the equation $x + y + z = 2$ and solve for x.

The answer as an ordered triple is $(-1, 2, 1)$.

Check: Check $(-1, 2, 1)$.

(1) $x + y + z = 2$
$-1 + 2 + 1 \stackrel{?}{=} 2$
$2 \stackrel{?}{=} 2$ is true.

(2) $-x + y - 2z = 1$
$-(-1) + 2 - 2(1) \stackrel{?}{=} 1$
$1 \stackrel{?}{=} 1$ is true.

(3) $x + y - z = 0$
$-1 + 2 - 1 \stackrel{?}{=} 0$
$0 \stackrel{?}{=} 0$ is true.

Answer: The solution $(-1, 2, 1)$ checks in all three equations.

We can also use augmented matrices to organize and expedite our work in solving 3×3 systems of linear equations. To prepare for this, we examine what matrices for 3×3 systems of linear equations look like.

In Example 3, the entries in the first column of the matrix are the coefficients of x, the entries in the second column are the coefficients of y, the entries in the third column are the coefficients of z, and the entries in the fourth column are the constants.

EXAMPLE 3 Writing Augmented Matrices for Systems of Linear Equations

Write an augmented matrix for each system of linear equations.

SOLUTION

(a) $\left\{\begin{aligned} x + y - z &= -6 \\ 2x - y + z &= 12 \\ 3x - 2y - 2z &= 3 \end{aligned}\right\}$ $\quad \left[\begin{array}{rrr|r} 1 & 1 & -1 & -6 \\ 2 & -1 & 1 & 12 \\ 3 & -2 & -2 & 3 \end{array}\right]$

The augmented matrix contains the coefficients and the constants from each equation.

(b) $\left\{\begin{aligned} 2x + y &= 7 \\ y - z &= 2 \\ x + z &= 2 \end{aligned}\right\}$ $\quad \left[\begin{array}{rrr|r} 2 & 1 & 0 & 7 \\ 0 & 1 & -1 & 2 \\ 1 & 0 & 1 & 2 \end{array}\right]$

Use zero coefficients as needed for any missing terms.

SELF-CHECK 11.2.2

1. Solve the system $\left\{\begin{aligned} x + y + z &= 6 \\ -x + y + 2z &= -1 \\ x + y - z &= 0 \end{aligned}\right\}$.
2. Write the augmented matrix for this system of linear equations.
$\left\{\begin{aligned} x + 2y + z &= -2 \\ 2x + 4y - 3z &= 0 \\ -x + 2y &= -3 \end{aligned}\right\}$
3. Write the system of linear equations in x, y, and z that is represented by the augmented matrix.
$\left[\begin{array}{rrr|r} 1 & -1 & 0 & 2 \\ 2 & 3 & -1 & 4 \\ 0 & 3 & -4 & 2 \end{array}\right]$

The reduced form for the augmented matrix associated with a consistent system of three independent linear equations with three variables is $\left[\begin{array}{ccc|c} 1 & 0 & 0 & k_1 \\ 0 & 1 & 0 & k_2 \\ 0 & 0 & 1 & k_3 \end{array}\right]$. The solution of this system is (k_1, k_2, k_3). Similar forms can be obtained for dependent or inconsistent systems. The reduced form defined here is also called **reduced row echelon form.**

Properties of the Reduced Row Echelon Form of a Matrix

1. The first nonzero entry in a row is a 1. All other entries in the column containing the leading 1 are 0s.
2. All nonzero rows are above any rows containing only 0s.
3. The first nonzero entry in a row is to the left of the first nonzero entry in the following row.

EXAMPLE 4 Using Elementary Row Operations

Use the elementary row operations to transform the matrix $\left[\begin{array}{ccc|c} 5 & 6 & -4 & -8 \\ 2 & 4 & -1 & 1 \\ 1 & 1 & -3 & 0 \end{array}\right]$ into the form $\left[\begin{array}{ccc|c} 1 & \_ & \_ & \_ \\ 0 & \_ & \_ & \_ \\ 0 & \_ & \_ & \_ \end{array}\right]$.

SOLUTION

$$\left[\begin{array}{ccc|c} 5 & 6 & -4 & -8 \\ 2 & 4 & -1 & 1 \\ 1 & 1 & -3 & 0 \end{array}\right] \xrightarrow{r_1 \leftrightarrow r_3} \left[\begin{array}{ccc|c} 1 & 1 & -3 & 0 \\ 2 & 4 & -1 & 1 \\ 5 & 6 & -4 & -8 \end{array}\right]$$

Place a 1 in the upper left position by interchanging the first and third rows.

$$\xrightarrow{r_2' = r_2 - 2r_1} \left[\begin{array}{ccc|c} 1 & 1 & -3 & 0 \\ 0 & 2 & 5 & 1 \\ 5 & 6 & -4 & -8 \end{array}\right]$$

Introduce 0s into column 1, rows 2 and 3.
To produce the new row 2, subtract twice row 1 from row 2.

$$\xrightarrow{r_3' = r_3 - 5r_1} \left[\begin{array}{ccc|c} 1 & 1 & -3 & 0 \\ 0 & 2 & 5 & 1 \\ 0 & 1 & 11 & -8 \end{array}\right]$$

To produce the new row 3, subtract 5 times row 1 from row 3.

The last two steps in Example 4 can be combined. This is illustrated in Example 5.

EXAMPLE 5 Solving a System of Linear Equations by Using an Augmented Matrix

Solve $\left\{\begin{array}{r} 2a + 3b - 2c = -8 \\ a - b + 2c = 25 \\ 4a + 6b - c = -7 \end{array}\right\}$.

SOLUTION

The first step is to form the augmented matrix.

$$\left[\begin{array}{rrr|r} 2 & 3 & -2 & -8 \\ 1 & -1 & 2 & 25 \\ 4 & 6 & -1 & -7 \end{array}\right] \xrightarrow{r_1 \leftrightarrow r_2} \left[\begin{array}{rrr|r} 1 & -1 & 2 & 25 \\ 2 & 3 & -2 & -8 \\ 4 & 6 & -1 & -7 \end{array}\right]$$

$$\xrightarrow[r_3' = r_3 - 4r_1]{r_2' = r_2 - 2r_1} \left[\begin{array}{rrr|r} 1 & -1 & 2 & 25 \\ 0 & 5 & -6 & -58 \\ 0 & 10 & -9 & -107 \end{array}\right]$$

Transform the first column into the form $\left[\begin{array}{ccc|c} 1 & \_ & \_ & \_ \\ 0 & \_ & \_ & \_ \\ 0 & \_ & \_ & \_ \end{array}\right]$.

$$\xrightarrow{r_2' = \frac{1}{5}r_2} \left[\begin{array}{rrr|r} 1 & -1 & 2 & 25 \\ 0 & 1 & -\frac{6}{5} & -\frac{58}{5} \\ 0 & 10 & -9 & -107 \end{array}\right]$$

$$\xrightarrow[r_3' = r_3 - 10r_2]{r_1' = r_1 + r_2} \left[\begin{array}{rrr|r} 1 & 0 & \frac{4}{5} & \frac{67}{5} \\ 0 & 1 & -\frac{6}{5} & -\frac{58}{5} \\ 0 & 0 & 3 & 9 \end{array}\right]$$

Transform the second column into the form $\left[\begin{array}{ccc|c} 1 & 0 & \_ & \_ \\ 0 & 1 & \_ & \_ \\ 0 & 0 & \_ & \_ \end{array}\right]$.

$$\xrightarrow{r_3' = \frac{1}{3}r_3} \left[\begin{array}{rrr|r} 1 & 0 & \frac{4}{5} & \frac{67}{5} \\ 0 & 1 & -\frac{6}{5} & -\frac{58}{5} \\ 0 & 0 & 1 & 3 \end{array}\right]$$

$$\xrightarrow[r_2' = r_2 + \frac{6}{5}r_3]{r_1' = r_1 - \frac{4}{5}r_3} \left[\begin{array}{rrr|r} 1 & 0 & 0 & 11 \\ 0 & 1 & 0 & -8 \\ 0 & 0 & 1 & 3 \end{array}\right]$$

Transform the third column into the form $\left[\begin{array}{ccc|c} 1 & 0 & 0 & \_ \\ 0 & 1 & 0 & \_ \\ 0 & 0 & 1 & \_ \end{array}\right]$.

The answer is displayed in the last column of the reduced form.

Answer: $(11, -8, 3)$

Does this answer check?

The overall strategy used in Example 5 can be summarized as "work from left to right and produce the leading 1s before you produce the 0s." This strategy is formalized in the following box.

Transforming an Augmented Matrix into Reduced Echelon Form

Work from left to right one column at a time.

STEP 1 $\begin{bmatrix} 1 & \cdots \\ 0 & \\ 0 & \\ \vdots & \vdots \\ 0 & \cdots \end{bmatrix}$ **Transform the first column** into this form by using the elementary row operations to

a. Produce a 1 in the top position.

b. Use the 1 in row 1 to produce 0s in the other positions of column 1.

Produce the leading 1s before you produce the 0s.

STEP 2 $\begin{bmatrix} 1 & 0 & \cdots \\ 0 & 1 & \\ 0 & 0 & \\ \vdots & \vdots & \vdots \\ 0 & 0 & \cdots \end{bmatrix}$ **Transform the next column,** if possible, into this form by using the elementary operations to

a. Produce a 1 in the next row.

b. Use the 1 in this row to produce 0s in the other positions of this column.

If it is not possible to produce a 1 in the next row, proceed to the next column.

STEP 3 Repeat step 2 column by column, always producing the 1 in the next row, until you arrive at the reduced form.

For emphasis, remember your goal is to produce the leading 1s before you produce the 0s. You may use shortcuts in this process whenever they are appropriate. This procedure is also appropriate for dependent and inconsistent systems.

SELF-CHECK 11.2.3

1. Produce a 1 in row 2, column 2 of the matrix $\left[\begin{array}{ccc|c} 1 & 1 & -3 & 0 \\ 0 & 2 & 5 & 1 \\ 0 & 1 & 11 & -8 \end{array}\right]$ by interchanging the second and third rows. Then introduce two 0s into the second column of this matrix.

EXAMPLE 6 Solving an Inconsistent System by Using an Augmented Matrix

Solve $\left\{\begin{aligned} r + 8s + 2t &= 20 \\ 11s + t &= 28 \\ -22s - 2t &= -55 \end{aligned}\right\}$.

SOLUTION

$$\left[\begin{array}{ccc|c} 1 & 8 & 2 & 20 \\ 0 & 11 & 1 & 28 \\ 0 & -22 & -2 & -55 \end{array}\right] \xrightarrow{r_3' = r_3 + 2r_2} \left[\begin{array}{ccc|c} 1 & 8 & 2 & 20 \\ 0 & 11 & 1 & 28 \\ 0 & 0 & 0 & 1 \end{array}\right]$$

Although this matrix is not in reduced form, the last row indicates that the system is inconsistent with no solution. The last row represents the equation $0r + 0s + 0t = 1$, which is a contradiction.

Answer: There is no solution.

Example 7 produces the general solution for a consistent system of dependent linear equations.

EXAMPLE 7 Solving a Consistent System of Dependent Equations by Using an Augmented Matrix

Solve $\left\{\begin{aligned} x_1 + 4x_2 + 5x_3 &= -2 \\ x_2 + 2x_3 &= -1 \\ -5x_2 - 10x_3 &= 5 \end{aligned}\right\}$.

SOLUTION

$$\left[\begin{array}{ccc|c} 1 & 4 & 5 & -2 \\ 0 & 1 & 2 & -1 \\ 0 & -5 & -10 & 5 \end{array}\right] \xrightarrow[r_3' = r_3 + 5r_2]{r_1' = r_1 - 4r_2} \left[\begin{array}{ccc|c} 1 & 0 & -3 & 2 \\ 0 & 1 & 2 & -1 \\ 0 & 0 & 0 & 0 \end{array}\right]$$

The new matrix is in reduced form. A row of 0s in an $n \times n$ consistent system indicates a dependent system with infinitely many solutions.

$$\longrightarrow \left\{\begin{aligned} x_1 + 0x_2 - 3x_3 &= 2 \\ 0x_1 + x_2 + 2x_3 &= -1 \\ 0x_1 + 0x_2 + 0x_3 &= 0 \end{aligned}\right\}$$

This is the system represented by the reduced matrix. The system is dependent, since the last equation is an identity satisfied by all values of x_1, x_2, and x_3.

Thus

$$x_1 = 3x_3 + 2$$
$$x_2 = -2x_3 - 1$$

The general solution is acquired by solving the first two equations for x_1 and x_2 in terms of x_3.

Answer: The general solution is $(3x_3 + 2, -2x_3 - 1, x_3)$; three particular solutions are $(8, -5, 2)$, $(2, -1, 0)$, and $(-7, 5, -3)$.

These particular solutions were found by letting x_3 be 2, 0, and -3, respectively.

Forms That Indicate a Dependent or Inconsistent $n \times n$ System of Equations

Let A be the augmented matrix of an $n \times n$ system of equations.

1. If the reduced form of A has a row of the form $[0\ 0\ 0 \ldots 0\ k]$, where $k \neq 0$, then the system is inconsistent and has no solution.
2. If the system is consistent and the reduced form of A has a row of the form $[0\ 0\ 0 \ldots 0\ 0]$ (all zeros), then the equations in the system are dependent and the system has infinitely many solutions.

SELF-CHECK 11.2.4

Write the solution for each system of linear equations represented by the following reduced augmented matrices.

1. $\left[\begin{array}{ccc|c} 1 & 0 & 0 & -1 \\ 0 & 1 & 0 & 1 \\ 0 & 0 & 1 & 3 \end{array}\right]$ **2.** $\left[\begin{array}{ccc|c} 1 & 0 & 1 & 5 \\ 0 & 1 & -1 & 2 \\ 0 & 0 & 0 & 0 \end{array}\right]$ **3.** $\left[\begin{array}{ccc|c} 1 & 0 & 0 & -2 \\ 0 & 1 & 2 & 1 \\ 0 & 0 & 0 & 1 \end{array}\right]$

The matrix method is also a convenient method for solving systems of linear equations that do not have the same number of variables as equations.

EXAMPLE 8 Solving a System of Equations with More Variables Than Equations

Find a general solution and three particular solutions for the 2×3 system $\begin{Bmatrix} 2x - 3y + 17z = 12 \\ 8x + 2y - 2z = 20 \end{Bmatrix}$.

SOLUTION

$$\left[\begin{array}{ccc|c} 2 & -3 & 17 & 12 \\ 8 & 2 & -2 & 20 \end{array}\right] \xrightarrow{r_1' = \frac{1}{2}r_1} \left[\begin{array}{ccc|c} 1 & -\frac{3}{2} & \frac{17}{2} & 6 \\ 8 & 2 & -2 & 20 \end{array}\right]$$

Use the elementary row operations to transform the matrix to reduced row echelon form.

$$\xrightarrow{r_2' = r_2 - 8r_1} \left[\begin{array}{ccc|c} 1 & -\frac{3}{2} & \frac{17}{2} & 6 \\ 0 & 14 & -70 & -28 \end{array}\right]$$

$$\xrightarrow{r_2' = \frac{1}{14}r_2} \left[\begin{array}{ccc|c} 1 & -\frac{3}{2} & \frac{17}{2} & 6 \\ 0 & 1 & -5 & -2 \end{array}\right]$$

$$\xrightarrow{r_1' = r_1 + \frac{3}{2}r_2} \left[\begin{array}{ccc|c} 1 & 0 & 1 & 3 \\ 0 & 1 & -5 & -2 \end{array}\right]$$

$$\begin{Bmatrix} x + z = 3 \\ y - 5z = -2 \end{Bmatrix}$$

This is the system represented by the reduced matrix.

$$x = 3 - z$$
$$y = 5z - 2$$

The general solution is obtained by solving these equations for x and y in terms of z.

Answer: The general solution is $(3 - z, 5z - 2, z)$; particular solutions are $(3, -2, 0)$, $(4, -7, -1)$, and $(1, 8, 2)$.

The particular solutions were found by letting z be 0, -1, and 2, respectively.

CALCULATOR PERSPECTIVE 11.2.1 Using rref (Reduced Row Echelon Form) to Solve a System of Linear Equations

To solve the system in Example 5 on a TI-84 Plus calculator, enter the following keystrokes:

Note: The fourth column of this matrix has been entered but is not visible on this display window. The arrow keys can be used to scroll through all the entries.

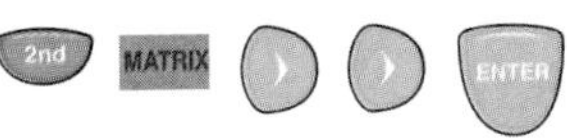

Then enter the dimension of the matrix and the elements of the matrix.

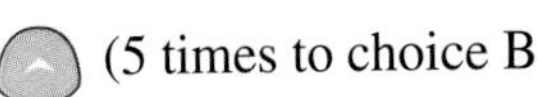

(5 times to choice B:)

Note: The **rref** feature skips all the intermediate steps and proceeds immediately to the reduced row echelon form. The solution to the system is $(11, -8, 3)$.

The graphical method and tables of values can be used to solve some systems of linear equations with two variables. Example 9 involves a 4×4 system with four equations and four variables. Graphs and tables are not appropriate tools for these larger systems because it is difficult to graphically depict more than three dimensions. For these larger systems, augmented matrices are often used with calculators or computers.

EXAMPLE 9 Solving a System of Four Linear Equations by Using an Augmented Matrix

Use an augmented matrix and a calculator to solve $\begin{cases} w + x + 2y - z = 2 \\ 2w + 4x - 6y + z = -4 \\ w - 4x + 5y - 3z = 11 \\ 3w + 3x - y - 2z = -5 \end{cases}$.

SOLUTION

$$\left[\begin{array}{cccc|c} 1 & 1 & 2 & -1 & 2 \\ 2 & 4 & -6 & 1 & -4 \\ 1 & -4 & 5 & -3 & 11 \\ 3 & 3 & -1 & -2 & -5 \end{array}\right]$$

```
[A]
[[1 1  2  -1 2...
 [2 4  -6 1  -4...
 [1 -4 5  -3 11...
 [3 3  -1 -2 -5...
```

First form the augmented matrix. Note that it is a 4×5 matrix. Enter this as matrix [A] on a calculator.

$$\longrightarrow \left[\begin{array}{cccc|c} 1 & 0 & 0 & 0 & 10 \\ 0 & 1 & 0 & 0 & -4 \\ 0 & 0 & 1 & 0 & 3 \\ 0 & 0 & 0 & 1 & 10 \end{array}\right]$$

```
rref([A])
  [[1 0 0 0 10]
   [0 1 0 0 -4]
   [0 0 1 0 3 ]
   [0 0 0 1 10]]
```

Then use the **rref** feature to transform this matrix to reduced row echelon form.

Obtain the answer from this reduced form.

Answer: $(10, -4, 3, 10)$

Does this answer check?

SELF-CHECK ANSWERS

11.2.1

1. (2, −3.25, 4.75) is a solution of the system.

11.2.2

1. (5, −2, 3)
2. $\left[\begin{array}{rrr|r} 1 & 2 & 1 & -2 \\ 2 & 4 & -3 & 0 \\ -1 & 2 & 0 & -3 \end{array}\right]$
3. $\left\{\begin{aligned} x - y &= 2 \\ 2x + 3y - z &= 4 \\ 3y - 4z &= 2 \end{aligned}\right\}$

11.2.3

1. $\left[\begin{array}{rrr|r} 1 & 1 & -3 & 0 \\ 0 & 2 & 5 & 1 \\ 0 & 1 & 11 & -8 \end{array}\right] \xrightarrow{r_2 \leftrightarrow r_3} \left[\begin{array}{rrr|r} 1 & 1 & -3 & 0 \\ 0 & 1 & 11 & -8 \\ 0 & 2 & 5 & 1 \end{array}\right] \xrightarrow[r_3' = r_3 - 2r_2]{r_1' = r_1 - r_2} \left[\begin{array}{rrr|r} 1 & 0 & -14 & 8 \\ 0 & 1 & 11 & -8 \\ 0 & 0 & -17 & 17 \end{array}\right]$

11.2.4

1. (−1, 1, 3)
2. General solution $(5 - z, z + 2, z)$
3. There is no solution.

USING THE LANGUAGE AND SYMBOLISM OF MATHEMATICS 11.2

1. The graph of $Ax + By + Cz = D$, if A, B, and C are not all 0, is a __________ in three-dimensional space.
2. A __________ of a system of linear equations with three variables is an ordered triple which satisfies each equation in the system.
3. A 3 × 3 system of linear equations contains __________ linear equations with __________ variables.
4. The reduced form for an augmented matrix is also called reduced row __________ form.
5. In reduced row echelon form, a row of the form $[0\ 0\ 0 \ldots 0\ k]$, where $k \neq 0$, represents an __________ system of linear equations.

QUICK REVIEW 11.2

The purpose of this quick review is to help you recall skills needed in this section.

1. Determine whether (4, −3) is a solution of the system $\left\{\begin{aligned} 2x + 3y &= -1 \\ 6x + 5y &= 9 \end{aligned}\right\}$.
2. Solve $\left\{\begin{aligned} 2x - y &= 5 \\ \frac{x}{2} - \frac{y}{3} &= 1 \end{aligned}\right\}$ by the substitution method.

Choose the letter that best describes each system of equations.

3. $\left\{\begin{aligned} y &= 2x - 1 \\ y &= 2x + 3 \end{aligned}\right\}$
4. $\left\{\begin{aligned} y &= 2x - 1 \\ y &= 2x - 1 \end{aligned}\right\}$
5. $\left\{\begin{aligned} y &= 2x - 1 \\ y &= x + 3 \end{aligned}\right\}$

A. A consistent system of independent equations
B. A consistent system of dependent equations
C. An inconsistent system

EXERCISES 11.2

In Exercises 1–14, solve each system of three linear equations in three variables.

1. $\begin{aligned} x + 2y + z &= 11 \\ -x - y + 2z &= 1 \\ 2x - y + z &= 4 \end{aligned}$
2. $\begin{aligned} x + 2y + 3z &= 7 \\ x - 3y - 2z &= -13 \\ 2x - y + 2z &= 5 \end{aligned}$
3. $\begin{aligned} x + y - z &= 1 \\ 2x + y + z &= 4 \\ x - y - 2z &= -2 \end{aligned}$
4. $\begin{aligned} 3x - y + 2z &= 4 \\ 2x + 2y - z &= 10 \\ x - y + 3z &= -4 \end{aligned}$
5. $\begin{aligned} 5x + y + 3z &= -1 \\ 2x - y + 4z &= -6 \\ 3x + y - 2z &= 7 \end{aligned}$
6. $\begin{aligned} 3x - y + z &= 8 \\ 2x + 3y - z &= 0 \\ 4x + 2y + z &= 7 \end{aligned}$
7. $\begin{aligned} x - 10y + 3z &= -5 \\ 2x - 15y + z &= 7 \\ 3x + 5y - 2z &= 8 \end{aligned}$
8. $\begin{aligned} x + 2y + 2z &= 4 \\ 2x + y + 2z &= 3 \\ 3x + y - 4z &= 2 \end{aligned}$
9. $\begin{aligned} x + y &= -2 \\ -y + z &= 2 \\ x - z &= -1 \end{aligned}$
10. $\begin{aligned} 2x + y &= 7 \\ y - z &= 2 \\ x + z &= 2 \end{aligned}$
11. $\begin{aligned} 2x + z &= 7 \\ y - 2z &= -5 \\ x + 2y &= 4 \end{aligned}$
12. $\begin{aligned} x + y &= 0 \\ x + 2z &= 5 \\ y + z &= 4 \end{aligned}$
13. $\begin{aligned} 2x - 4y + 2z &= 6 \\ 3x - 6y + 3z &= 10 \\ 4x - 8y + 4z &= 11 \end{aligned}$
14. $\begin{aligned} x + 2y + 2z &= 2 \\ 2x - y + z &= 1 \\ 4x + 3y + 5z &= 3 \end{aligned}$

In Exercises 15–18, write an augmented matrix for each system of linear equations.

15. $\left\{\begin{aligned} x + y + z &= 2 \\ -x + y - 2z &= 1 \\ x + y - z &= 0 \end{aligned}\right\}$
16. $\left\{\begin{aligned} x - 3y + 3z &= 10 \\ x + y - 4z &= -17 \\ -x + y + z &= 6 \end{aligned}\right\}$
17. $\left\{\begin{aligned} x + 2y - z &= 19 \\ 2x - y &= -1 \\ 3x - 2y + 4z &= -32 \end{aligned}\right\}$
18. $\left\{\begin{aligned} 6x - 3y &= 2 \\ 3y + 2z &= 1 \\ 2x - z &= 5 \end{aligned}\right\}$

In Exercises 19–24, write a system of linear equations in x, y, and z that is represented by each augmented matrix.

19. $\left[\begin{array}{ccc|c} 1 & -2 & 5 & 0 \\ 2 & 4 & -3 & 8 \\ 3 & 5 & 7 & 11 \end{array}\right]$ **20.** $\left[\begin{array}{ccc|c} 1 & 2 & 3 & 7 \\ 2 & -1 & 2 & 5 \\ 1 & -3 & -2 & -13 \end{array}\right]$

21. $\left[\begin{array}{ccc|c} 1 & 0 & 2 & 5 \\ 1 & 1 & 0 & 0 \\ 0 & 1 & 1 & 4 \end{array}\right]$ **22.** $\left[\begin{array}{ccc|c} 1 & 1 & 0 & 0 \\ 1 & 0 & 2 & 5 \\ 0 & 1 & 1 & 4 \end{array}\right]$

23. $\left[\begin{array}{ccc|c} 1 & 0 & 0 & 7 \\ 0 & 1 & 0 & -5 \\ 0 & 0 & 1 & 8 \end{array}\right]$ **24.** $\left[\begin{array}{ccc|c} 1 & 0 & 0 & 2 \\ 0 & 1 & 0 & 6 \\ 0 & 0 & 1 & -3 \end{array}\right]$

In Exercises 25–36, use the given elementary row operations to complete each matrix.

25. $\left[\begin{array}{ccc|c} 2 & 1 & -2 & -11 \\ 1 & 2 & 3 & 16 \\ 3 & 2 & 1 & 3 \end{array}\right] \xrightarrow{r_1 \leftrightarrow r_2} \left[\begin{array}{ccc|c} \underline{\quad} & \underline{\quad} & \underline{\quad} & \underline{\quad} \\ \underline{\quad} & \underline{\quad} & \underline{\quad} & \underline{\quad} \\ 3 & 2 & 1 & 3 \end{array}\right]$

26. $\left[\begin{array}{ccc|c} 2 & 3 & 4 & 5 \\ 1 & -1 & 3 & 6 \\ 3 & -2 & 2 & 10 \end{array}\right] \xrightarrow{r_1 \leftrightarrow r_2} \left[\begin{array}{ccc|c} \underline{\quad} & \underline{\quad} & \underline{\quad} & \underline{\quad} \\ \underline{\quad} & \underline{\quad} & \underline{\quad} & \underline{\quad} \\ 3 & -2 & 2 & 10 \end{array}\right]$

27. $\left[\begin{array}{ccc|c} 2 & 4 & 8 & 6 \\ 3 & 5 & 7 & 1 \\ 4 & 9 & 2 & 8 \end{array}\right] \xrightarrow{r_1' = \frac{1}{2} r_1} \left[\begin{array}{ccc|c} \underline{\quad} & \underline{\quad} & \underline{\quad} & \underline{\quad} \\ 3 & 5 & 7 & 1 \\ 4 & 9 & 2 & 8 \end{array}\right]$

28. $\left[\begin{array}{ccc|c} 1 & 2 & 5 & 1 \\ 0 & 5 & 6 & 4 \\ 0 & 8 & 2 & 5 \end{array}\right] \xrightarrow{r_2' = \frac{1}{5} r_2} \left[\begin{array}{ccc|c} 1 & 2 & 5 & 11 \\ \underline{\quad} & \underline{\quad} & \underline{\quad} & \underline{\quad} \\ 0 & 8 & 2 & 5 \end{array}\right]$

29. $\left[\begin{array}{ccc|c} 1 & 3 & 5 & 11 \\ 2 & 7 & 9 & 13 \\ 4 & 8 & 3 & 7 \end{array}\right] \xrightarrow{r_2' = r_2 - 2r_1} \left[\begin{array}{ccc|c} 1 & 3 & 5 & 11 \\ \underline{\quad} & \underline{\quad} & \underline{\quad} & \underline{\quad} \\ 4 & 8 & 3 & 7 \end{array}\right]$

30. $\left[\begin{array}{ccc|c} 1 & 3 & 5 & 11 \\ 2 & 7 & 9 & 13 \\ 4 & 8 & 3 & 7 \end{array}\right] \xrightarrow{r_3' = r_3 - 4r_1} \left[\begin{array}{ccc|c} 1 & 3 & 5 & 11 \\ 2 & 7 & 9 & 13 \\ \underline{\quad} & \underline{\quad} & \underline{\quad} & \underline{\quad} \end{array}\right]$

31. $\left[\begin{array}{ccc|c} 1 & 2 & 2 & 3 \\ 2 & 1 & 3 & -1 \\ 3 & -1 & 4 & -8 \end{array}\right] \xrightarrow[r_3' = r_3 - 3r_1]{r_2' = r_2 - 2r_1} \left[\begin{array}{ccc|c} 1 & 2 & 2 & 3 \\ 0 & \underline{\quad} & \underline{\quad} & \underline{\quad} \\ 0 & \underline{\quad} & \underline{\quad} & \underline{\quad} \end{array}\right]$

32. $\left[\begin{array}{ccc|c} 1 & 3 & 5 & 51 \\ -3 & 4 & 6 & 70 \\ 4 & -1 & 2 & 11 \end{array}\right] \xrightarrow[r_3' = r_3 - 4r_1]{r_2' = r_2 + 3r_1} \left[\begin{array}{ccc|c} 1 & 3 & 5 & 51 \\ 0 & \underline{\quad} & \underline{\quad} & \underline{\quad} \\ 0 & \underline{\quad} & \underline{\quad} & \underline{\quad} \end{array}\right]$

33. $\left[\begin{array}{ccc|c} 1 & 3 & -2 & 19 \\ 0 & 2 & 4 & -10 \\ 0 & 4 & 1 & 8 \end{array}\right] \xrightarrow{r_2' = \frac{1}{2} r_2} \left[\begin{array}{ccc|c} 1 & 3 & -2 & 19 \\ \underline{\quad} & \underline{\quad} & \underline{\quad} & \underline{\quad} \\ 0 & 4 & 1 & 8 \end{array}\right]$

34. $\left[\begin{array}{ccc|c} 1 & 0 & 2 & -1 \\ 0 & 1 & 3 & -8 \\ 0 & 0 & 4 & -12 \end{array}\right] \xrightarrow{r_3' = \frac{1}{4} r_3} \left[\begin{array}{ccc|c} 1 & 0 & 2 & -1 \\ 0 & 1 & 3 & -8 \\ \underline{\quad} & \underline{\quad} & \underline{\quad} & \underline{\quad} \end{array}\right]$

35. $\left[\begin{array}{ccc|c} 1 & 2 & 4 & 9 \\ 0 & 1 & 2 & 3 \\ 0 & -3 & 5 & 13 \end{array}\right] \xrightarrow[r_3' = r_3 + 3r_2]{r_1' = r_1 - 2r_2} \left[\begin{array}{ccc|c} 1 & 0 & \underline{\quad} & \underline{\quad} \\ 0 & 1 & 2 & 3 \\ 0 & 0 & \underline{\quad} & \underline{\quad} \end{array}\right]$

36. $\left[\begin{array}{ccc|c} 1 & 0 & -5 & -2 \\ 0 & 1 & 3 & 0 \\ 0 & 0 & 1 & 1 \end{array}\right] \xrightarrow[r_2' = r_2 - 3r_3]{r_1' = r_1 + 5r_3} \left[\begin{array}{ccc|c} 1 & 0 & \underline{\quad} & \underline{\quad} \\ 0 & 1 & \underline{\quad} & \underline{\quad} \\ 0 & 0 & 1 & 1 \end{array}\right]$

In Exercises 37–42, write the solution for the system of linear equations represented by each augmented matrix. If the matrix represents a dependent system, write the general solution and three particular solutions.

37. $\left[\begin{array}{ccc|c} 1 & 0 & 0 & -2 \\ 0 & 1 & 0 & 7 \\ 0 & 0 & 1 & 3 \end{array}\right]$ **38.** $\left[\begin{array}{ccc|c} 1 & 0 & 0 & 5 \\ 0 & 1 & 0 & -9 \\ 0 & 0 & 1 & 11 \end{array}\right]$

39. $\left[\begin{array}{ccc|c} 1 & 0 & 7 & 3 \\ 0 & 1 & 2 & -9 \\ 0 & 0 & 0 & -2 \end{array}\right]$ **40.** $\left[\begin{array}{ccc|c} 1 & 0 & 3 & 8 \\ 0 & 1 & -3 & 4 \\ 0 & 0 & 0 & 7 \end{array}\right]$

41. $\left[\begin{array}{ccc|c} 1 & 0 & 3 & 2 \\ 0 & 1 & 2 & -5 \\ 0 & 0 & 0 & 0 \end{array}\right]$ **42.** $\left[\begin{array}{ccc|c} 1 & 0 & -4 & 2 \\ 0 & 1 & 5 & 3 \\ 0 & 0 & 0 & 0 \end{array}\right]$

In Exercises 43–48, label the elementary row operation used to transform the first matrix to the second. Use the notation introduced in Section 11.1.

43. $\left[\begin{array}{ccc|c} 2 & 4 & 6 & -6 \\ 3 & 1 & -1 & 6 \\ 1 & 2 & -4 & 11 \end{array}\right] \xrightarrow{r_1' = ?} \left[\begin{array}{ccc|c} 1 & 2 & 3 & -3 \\ 3 & 1 & -1 & 6 \\ 1 & 2 & -4 & 11 \end{array}\right]$

44. $\left[\begin{array}{ccc|c} 1 & -2 & 5 & 25 \\ 0 & 3 & -6 & -18 \\ 0 & 4 & -3 & -9 \end{array}\right] \xrightarrow{r_2' = ?} \left[\begin{array}{ccc|c} 1 & -2 & 5 & 25 \\ 0 & 1 & -2 & -6 \\ 0 & 4 & -3 & -9 \end{array}\right]$

45. $\left[\begin{array}{ccc|c} 1 & 3 & -2 & -3 \\ -3 & 1 & 1 & -16 \\ 4 & 2 & -1 & 15 \end{array}\right] \longrightarrow \left[\begin{array}{ccc|c} 1 & 3 & -2 & -3 \\ 0 & 10 & -5 & -25 \\ 4 & 2 & -1 & 15 \end{array}\right]$

46. $\left[\begin{array}{ccc|c} 1 & 3 & -2 & -3 \\ 0 & 10 & -5 & -25 \\ 4 & 2 & -1 & 15 \end{array}\right] \longrightarrow \left[\begin{array}{ccc|c} 1 & 3 & -2 & -3 \\ 0 & 10 & -5 & -25 \\ 0 & -10 & 7 & 27 \end{array}\right]$

47. $\left[\begin{array}{ccc|c} 1 & -3 & 2 & 10 \\ 0 & 1 & 5 & -12 \\ 0 & -2 & -1 & 1 \end{array}\right] \longrightarrow \left[\begin{array}{ccc|c} 1 & 0 & 17 & -26 \\ 0 & 1 & 5 & -12 \\ 0 & -2 & -1 & 1 \end{array}\right]$

48. $\left[\begin{array}{ccc|c} 1 & 0 & 17 & -26 \\ 0 & 1 & 5 & -12 \\ 0 & -2 & -1 & 1 \end{array}\right] \longrightarrow \left[\begin{array}{ccc|c} 1 & 0 & 17 & -26 \\ 0 & 1 & 5 & -12 \\ 0 & 0 & 9 & -23 \end{array}\right]$

In Exercises 49–58, use an augmented matrix and elementary row operations to solve each system of linear equations.

49. $2x + y = 7$
$y - z = 2$
$x + z = 2$

50. $x + y = -2$
$-y + z = 2$
$x - z = -1$

51. $x + 2y + z = 11$
$-x - y + 2z = 1$
$2x - y + z = 4$

52. $3x - y + 2z = 4$
$2x + 2y - z = 10$
$x - y + 3z = -4$

53. $x_1 + 2x_2 - 2x_3 = -7$
$-2x_1 + 3x_2 - 17x_3 = -14$
$4x_1 + 2x_2 + 10x_3 = -4$

54. $x_1 - 2x_2 + 7x_3 = 3$
$2x_1 + 2x_2 - 3x_3 = -5$
$x_1 - 11x_2 + 24x_3 = 11$

55. $5a + b - 2c = -3$
$2a + 4b + c = -3$
$-3a + 5b - 6c = -21$

56. $6a - 3b + 3c = 3$
$3a + 3b - c = 5$
$5a + 2b - 2c = 4$

57. $r + 2s - 5t = 4$
$3r - s + 2t = 3$
$r + 9s - 22t = 10$

58. $r - 3s + 2t = 1$
$4r - 2s + t = 2$
$2r + 4s - 2t = 0$

In Exercises 59 and 60, find a general solution and two particular solutions for each system.

59. $2a - b - 3c = -5$
$3a + b - 2c = -10$

60. $3a - b + 5c = 4$
$2a + 2b - 2c = 0$

61. **Numeric Word Problem** The sum of three numbers is 108. The largest number is 16 less than the sum of the other two numbers. The sum of the largest and the smallest is twice the other number. Find the three numbers.

62. **Numeric Word Problem** The largest of three numbers is 7 times the second number. The second number is 7 times the smallest number. The sum of the numbers is 285. Find the three numbers.

63. **Dimensions of a Triangle** The perimeter of this triangle is 168 cm. The length of the longest side is twice that of the shortest side. The sum of the lengths of the shortest side and the longest side is 48 cm more than the length of side b. Find the length of each side. (See the figure.)

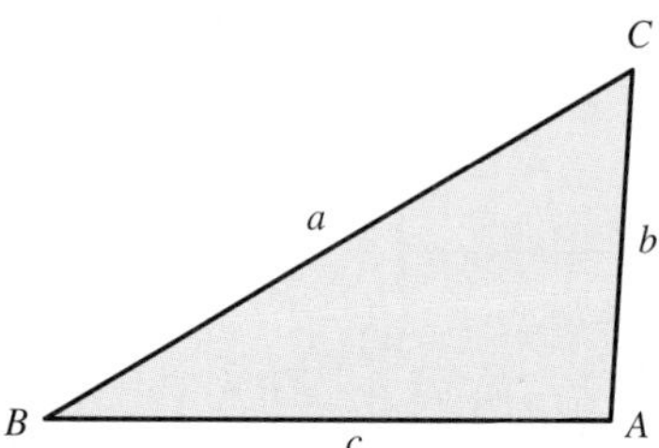

64. **Dimensions of a Triangle** Triangle ABC has sides a, b, and c with side a the longest side and side b the shortest side. The length of the longest side of a triangle is 12 cm less than the sum of the lengths of the other two sides. The length of the shortest side is 10 cm more than one-half the length of side c. Find the length of each side if the perimeter is 188 cm.

65. **Angles of a Triangle** Triangle ABC has angles A, B, and C with angle A the largest angle and angle B the smallest angle. Angle C is twice as large as the smallest angle. The largest angle is 9° larger than the sum of the other two angles. Find the number of degrees in each angle. (*Hint:* The sum of the angles of a triangle is 180°.)

66. **Angles of a Triangle** Triangle ABC has angles A, B, and C with angle A the largest angle and angle B the smallest angle. The smallest angle of this triangle is 78° less than the largest angle. Angle C is 3 times as large as the smallest angle. How many degrees are in each angle?

67. **Mixture of Foods** A zookeeper mixes three foods, the contents of which are described in the following table. How many grams of each food are needed to produce a mixture with 133 g of fat, 494 g of protein, and 1,700 g of carbohydrates?

	A	B	C
Fat (%)	6	4	5
Protein (%)	15	18	20
Carbohydrates (%)	45	65	70

68. **Use of Farmland** A farmer must decide how many acres of each of three crops to plant during this growing season. The farmer must pay a certain amount for seed and devote a certain amount of labor and water to each acre of crop planted, as shown in the following table.

	A	B	C
Seed cost ($)	120	85	80
Hours of labor	4	12	8
Gallons of water	500	900	700

The amount of money available to pay for seed is $26,350. The farmer's family can devote 2,520 hours to tending the crops, and the farmer has access to 210,000 gal of water for irrigation. How many acres of each crop would use *all* these resources?

In Exercises 69 and 70, match each graph with the linear equation that defines this plane.

69. $2x + y + 3z = 6$

70. $3x + 2y + 2z = 6$

A.

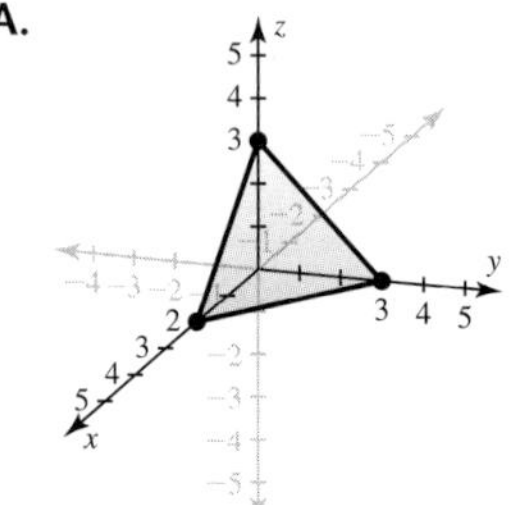

B.

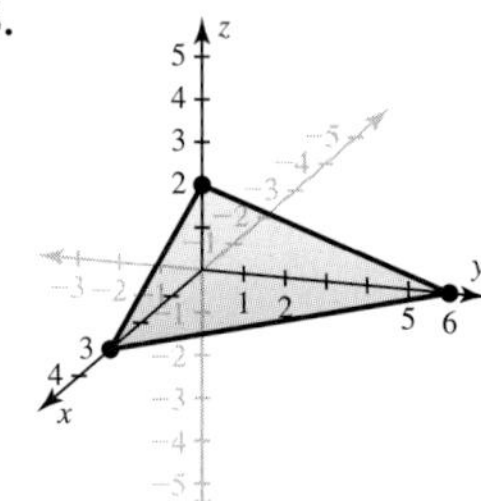

In Exercises 71 and 72, graph the plane defined by each linear equation.

71. $4x + 2y + 3z = 12$

72. $2x + 3y + 6z = 6$

Group Discussion Questions

73. **Discovery Question**

a. Solve $\left\{\begin{aligned} 3a - 5b &= 1 \\ a + 2b &= 4 \end{aligned}\right\}$.

b. Use the solution for part **a** to solve this nonlinear system: $\left\{\begin{aligned} \frac{3}{x} - \frac{5}{y} &= 1 \\ \frac{1}{x} + \frac{2}{y} &= 4 \end{aligned}\right\}$.

74. **Discovery Question**

a. Solve $\left\{\begin{aligned} a - 2b - c &= 5 \\ 2a - b + 3c &= 11 \\ -3a - 2b + 2c &= 7 \end{aligned}\right\}$.

b. Use the solution for part **a** to solve this nonlinear system: $\left\{\begin{aligned} \frac{1}{x} - \frac{2}{y} - \frac{1}{z} &= 5 \\ \frac{2}{x} - \frac{1}{y} + \frac{3}{z} &= 11 \\ -\frac{3}{x} - \frac{2}{y} + \frac{2}{z} &= 7 \end{aligned}\right\}$.

75. Challenge Question Find the constants a, b, and c such that $(1, -3, 5)$ is a solution of the linear system

$$\begin{Bmatrix} ax + by + cz = 5 \\ ax - by - cz = -1 \\ 2ax + 3by + 4cz = 13 \end{Bmatrix}$$

76. Challenge Question Solve the following system for (x, y, z) in terms of the nonzero constants a, b, and c.

$$\begin{Bmatrix} ax + by + cz = 0 \\ 2ax - by + cz = 14 \\ -ax + by + 2cz = -21 \end{Bmatrix}$$

Section 11.3 Horizontal and Vertical Translations of the Graphs of Functions

Objectives:

3. Predict the vertical shift in the graph of a function or the shift in the y-values in a table of values for a function.
4. Predict the horizontal shift in the graph of a function or the shift in the x-values in a table of values for a function.

Each function from a family of functions shares the same characteristic shape.

Each type of algebraic equation $y = f(x)$ generates a specific family of graphs. Some of the families that we have examined in earlier chapters are linear functions, absolute value functions, quadratic functions, rational functions, square root functions, exponential functions, and logarithmic functions. Each member of a family of functions shares a characteristic shape with all other members of this family. Now we will use one basic graph from a family to generate other graphs in this family.

A Mathematical Note

Funds donated to the American Mathematical Society in memory of Ruth Satter, a researcher for Bell Labs, were used to establish a prize for outstanding contributions in mathematics research by a woman. This prize has been awarded every 2 years, starting in 1991.

Vertical Shifts

We first examine shifting graphs up and down. We start by observing the patterns illustrated by the following functions. Observing patterns is an important part of mathematics. By recognizing patterns, we can gain insights that deepen our understanding and shorten our work. The functions compared next are $y_1 = |x|$ and $y_2 = |x| + 2$.

NUMERICALLY

x	$y_1 = \|x\|$	$y_2 = \|x\| + 2$
−3	3	5
−2	2	4
−1	1	3
0	0	2
1	1	3
2	2	4
3	3	5

GRAPHICALLY

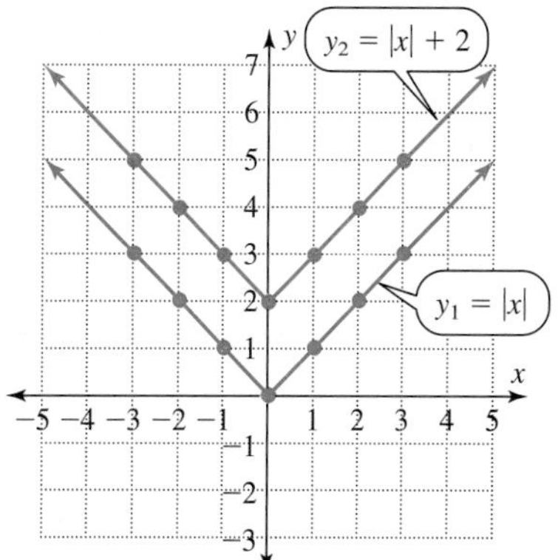

VERBALLY

Each y_2-value is 2 more than the corresponding y_1-value.

The graph of y_2 can be obtained by shifting the graph of y_1 up 2 units.

The graph of $y_2 = |x| + 2$ can be obtained by vertically shifting or translating the graph $y_1 = |x|$ up 2 units. **Vertical translations** or **vertical shifts** are described in the following box.

Vertical Shifts of $y = f(x)$

For a function $y_1 = f(x)$ and a positive real number c:

ALGEBRAICALLY	GRAPHICALLY	NUMERICALLY
$y_2 = f(x) + c$	To obtain the graph of $y_2 = f(x) + c$, shift the graph of $y_1 = f(x)$ up c units.	For the same x-value, $y_2 = y_1 + c$.
$y_3 = f(x) - c$	To obtain the graph of $y_3 = f(x) - c$, shift the graph of $y_1 = f(x)$ down c units.	For the same x-value, $y_3 = y_1 - c$.

EXAMPLE 1 Examining a Vertical Shift of a Quadratic Function

As noted in Section 7.3, each quadratic function in the form $f(x) = ax^2 + bx + c$ graphs as a parabola. The vertex of $y_1 = x^2$ is (0, 0). The vertex of $y_2 = x^2 - 4$ is (0, −4).

Compare the functions $y_1 = x^2$ and $y_2 = x^2 - 4$ numerically, graphically, and verbally.

SOLUTION

NUMERICALLY

x	$y_1 = x^2$	$y_2 = x^2 - 4$
−3	9	5
−2	4	0
−1	1	−3
0	0	−4
1	1	−3
2	4	0
3	9	5

GRAPHICALLY

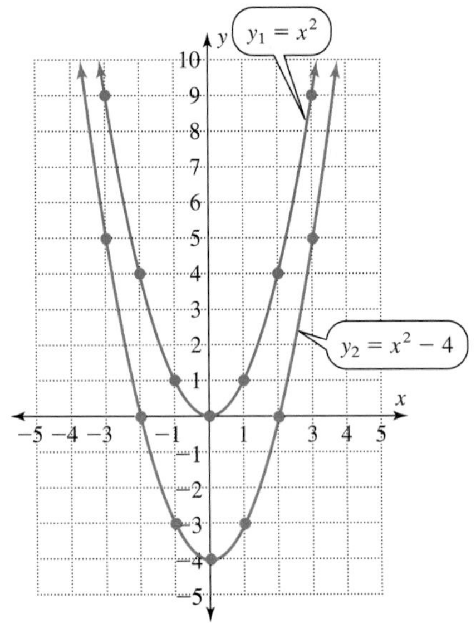

VERBALLY

For each value of x, y_2 is 4 units less than y_1.

The parabola defined by $y_2 = x^2 - 4$ can be obtained by shifting the parabola defined by $y_1 = x^2$ down 4 units. The range of y_1 is $[0, \infty)$. The range of y_2 is $[-4, \infty)$.

SELF-CHECK 11.3.1

1. Use a graphing calculator to compare $f(x) = x^2$ and $f(x) = x^2 + 3$, both numerically and graphically.

CALCULATOR PERSPECTIVE 11.3.1 Performing a Vertical Translation of a Graph

To perform the vertical translation of $Y_1 = x^2$ from Example 1 on a TI-84 Plus calculator, enter the following keystrokes:

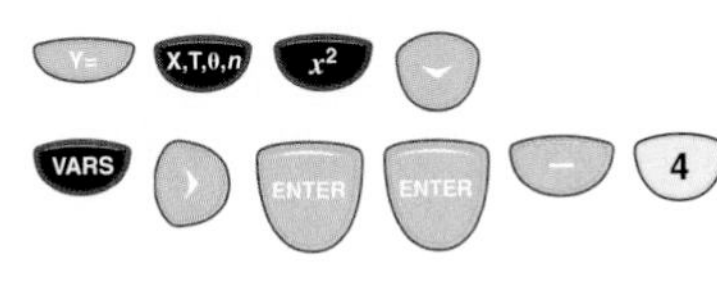

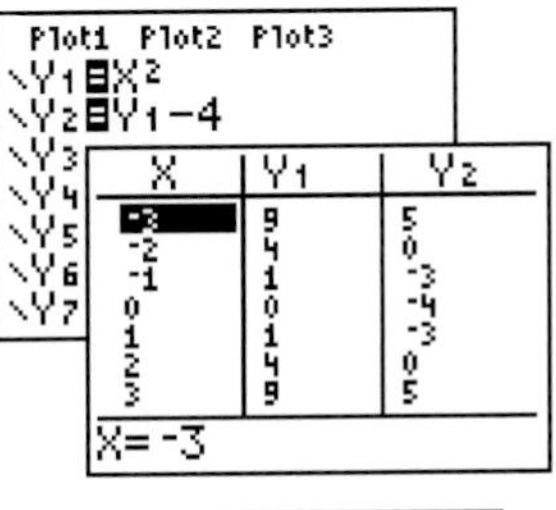

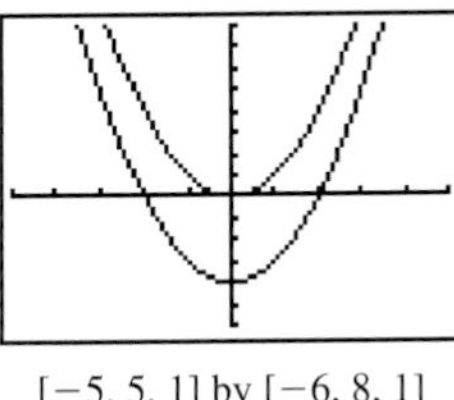

Note: The table illustrates that Y_2 is 4 units less than Y_1 at each x-value. The graphs show that the graph of Y_2 can be obtained by shifting the graph of Y_1 down 4 units.

[−5, 5, 1] by [−6, 8, 1]

Examples 2 and 3 illustrate how to recognize functions that are vertical shifts of each other. Example 2 examines two graphs, and Example 3 examines two tables of values.

EXAMPLE 2 Identifying a Vertical Shift

Use the graphs of $y_1 = f_1(x)$ and $y_2 = f_2(x)$ to write an equation for y_2 in terms of $f_1(x)$.

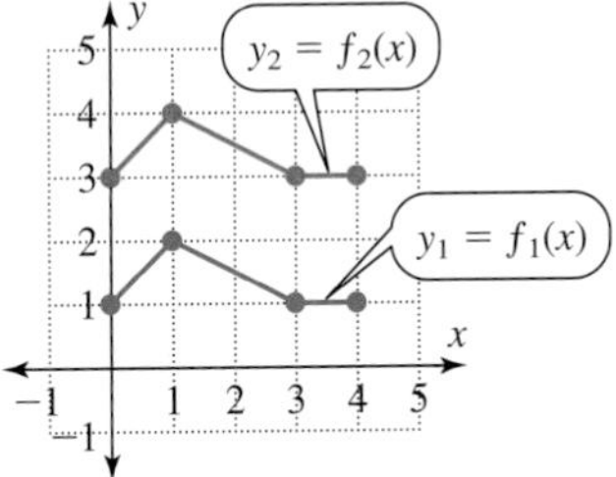

SOLUTION

The graph of $y_2 = f_2(x)$ can be obtained by shifting the graph of $y_1 = f_1(x)$ up 2 units. Thus $y_2 = f_1(x) + 2$.

Key points shifted are as follows:
(0, 1) is shifted up 2 units to (0, 3).
(1, 2) is shifted up 2 units to (1, 4).
(3, 1) is shifted up 2 units to (3, 3).
(4, 1) is shifted up 2 units to (4, 3).

Answer: $y_2 = f_1(x) + 2$

EXAMPLE 3 Identifying a Vertical Shift

Use the table of values for $y_1 = f_1(x)$ and $y_2 = f_2(x)$ to write an equation for y_2 in terms of $f_1(x)$.

x	$y_1 = f_1(x)$	$y_2 = f_2(x)$
−3	18	13
−2	6	1
−1	1	−4
0	0	−5
1	0	−5
2	−2	−7
3	−9	−14

SOLUTION

For each value of x, y_2 is 5 units less than y_1. Thus $y_2 = f_1(x) - 5$.

Answer: $y_2 = f_1(x) - 5$

SELF-CHECK 11.3.2

1. Use the graph of $y_1 = f(x)$ to graph $y_2 = f(x) + 2$.

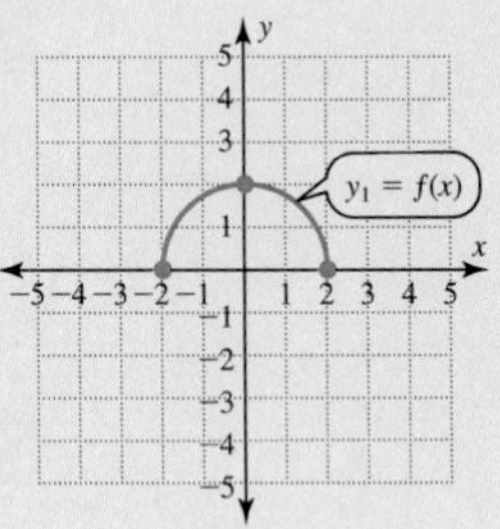

2. Use the table for $y_1 = f(x)$ to complete the table for $y_2 = f(x) + 1$.

x	y_1	y_2
−2	−8	
−1	−1	
0	0	
1	1	
2	8	

Horizontal Shifts

We have observed that vertical shifts affect the y-values of the points on a graph. Now we examine horizontal shifts that affect the x-values of the points on a graph. We start our inspection of horizontal shifts by examining the functions $y_1 = |x|$ and $y_2 = |x + 2|$.

NUMERICALLY

x	$y_1 = \|x\|$
−3	3
−2	2
−1	1
0	0
1	1
2	2
3	3

x	$y_2 = \|x + 2\|$
−5	3
−4	2
−3	1
−2	0
−1	1
0	2
1	3

GRAPHICALLY

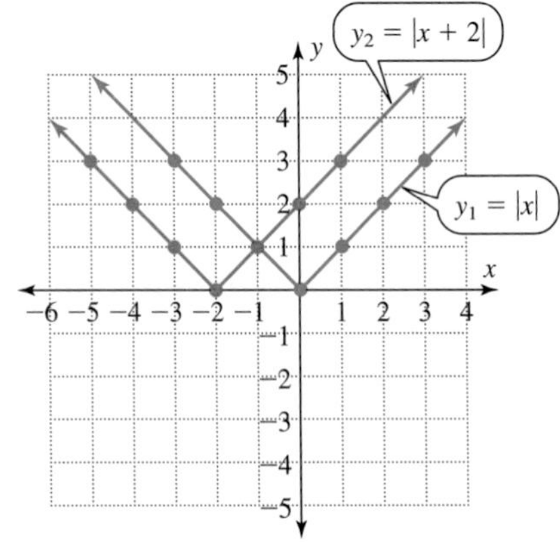

VERBALLY

For the same y-value in y_1 and y_2, the x-value is 2 units less for y_2 than for y_1.

The graph of y_2 can be obtained by shifting the graph of y_1 to the left 2 units.

Horizontal Shifts of $y = f(x)$

For a function $y_1 = f(x)$ and a positive real number c:

ALGEBRAICALLY	GRAPHICALLY	NUMERICALLY
$y_2 = f(x + c)$	To obtain the graph of $y_2 = f(x + c)$, shift the graph of $y_1 = f(x)$ to the left c units.	For the same y-value in y_1 and y_2, the x-value is c units less for y_2 than for y_1.
$y_3 = f(x - c)$	To obtain the graph of $y_3 = f(x - c)$, shift the graph of $y_1 = f(x)$ to the right c units.	For the same y-value in y_1 and y_2, the x-value is c units more for y_2 than for y_1.

Caution: Adding a positive c shifts the graph to the *left*. Reconsider $y = |x + 2|$. Note that output values of $y = |x + 2|$ occur 2 units sooner for x than they do for $y = |x|$. Two units sooner means 2 units to the left on the number line.

EXAMPLE 4 Examining a Horizontal Shift of a Quadratic Function

Compare the functions $f_1(x) = x^2$ and $f_2(x) = (x - 2)^2$ numerically, graphically, and verbally.

SOLUTION

NUMERICALLY

x	$f_1(x) = x^2$
−3	9
−2	4
−1	1
0	0
1	1
2	4
3	9

x	$f_2(x) = (x - 2)^2$
−1	9
0	4
1	1
2	0
3	1
4	4
5	9

GRAPHICALLY

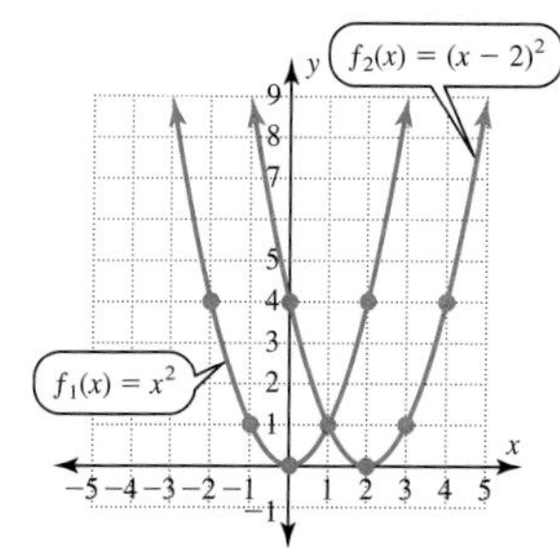

Both graphs are parabolas. The vertex of $f_1(x) = x^2$ is (0, 0). The vertex of $f_2(x) = (x - 2)^2$ is (2, 0).

VERBALLY

For the same y-value in f_1 and f_2, the x-value is 2 units more for f_2 than for f_1.

The graph of $f_2(x) = (x - 2)^2$ can be obtained by shifting the parabola defined by $f_1(x) = x^2$ to the right 2 units. The domain for both functions is $\mathbb{R}$, the set of all real numbers.

SELF-CHECK 11.3.3

1. Use a graphing calculator to compare the graph of $f(x) = x^2$ to the graph of $f(x) = (x + 1)^2$.

Example 5 examines two graphs that are horizontal translations or shifts of each other.

EXAMPLE 5 Identifying a Horizontal Shift

Use the graphs of $y_1 = f_1(x)$ and $y_2 = f_2(x)$ to write an equation for y_2 in terms of $f_1(x)$.

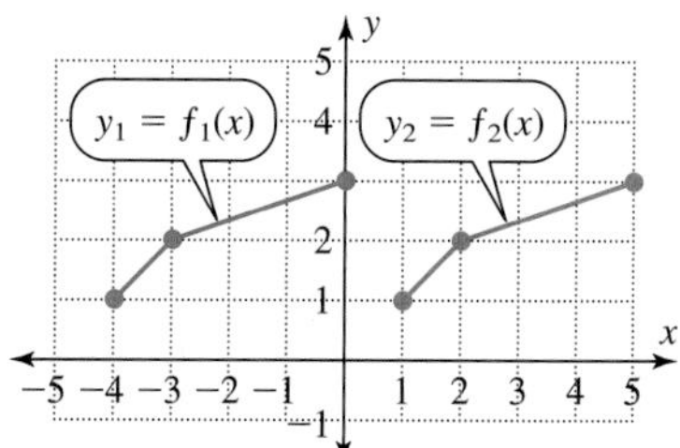

SOLUTION

Note that the graph of $y_2 = f_2(x)$ can be obtained by shifting the graph of $y_1 = f_1(x)$ to the right 5 units. Thus $y_2 = f_1(x - 5)$.

Key points shifted are as follows:
(−4, 1) is shifted right 5 units to (1, 1).
(−3, 2) is shifted right 5 units to (2, 2).
(0, 3) is shifted right 5 units to (5, 3).

Answer: $y_2 = f_1(x - 5)$

Example 6 examines horizontal shifts when two functions are defined by a table of values.

EXAMPLE 6 Using Translations to Report Sales Data

When working with data for which the input values are years, analysts often make use of a horizontal translation to work with smaller input values but with equivalent results. The following tables represent the sales figures for a company founded in 1997. Use the information in the tables to write an equation for $y_2 = f_2(x)$ as a translation of $y_1 = f_1(x)$.

YEAR x	SALES y_1 ($)
1997	100,000
1998	120,000
1999	150,000
2000	200,000
2001	270,000
2002	350,000

YEAR x	SALES y_2 ($)
0	100,000
1	120,000
2	150,000
3	200,000
4	270,000
5	350,000

SOLUTION

Note that 1997 corresponds to year 0, 1998 to year 1, and so on. We can obtain a table of values for y_2 based on the table for y_1 by shifting the input x-values to the left 1997 units. Thus $y_2 = f_1(x + 1997)$.

Answer: $y_2 = f_1(x + 1997)$ ■

SELF-CHECK 11.3.4

The following two tables show the number of students who completed a graphics design program that started in 1980 at a Midwestern community college.

YEAR x	STUDENTS y_1
1985	2
1990	6
1995	14
2000	20

YEAR x	STUDENTS y_2
5	2
10	6
15	14
20	20

Use the information in the tables to write an equation for $y_2 = f_2(x)$ as a translation of $y_1 = f_1(x)$.

Example 7 gives some additional practice recognizing horizontal and vertical shifts from the equation defining the function.

EXAMPLE 7 Identifying Horizontal and Vertical Shifts

Describe how to shift the graph of $y = f(x)$ to obtain the graph of each function.

SOLUTION

(a) $y = f(x) + 8$	Shift the graph of $y = f(x)$ up 8 units.
(b) $y = f(x) - 8$	Shift the graph of $y = f(x)$ down 8 units.
(c) $y = f(x + 8)$	Shift the graph of $y = f(x)$ left 8 units.
(d) $y = f(x - 8)$	Shift the graph of $y = f(x)$ right 8 units.

■

SELF-CHECK 11.3.5

1. Use the graph of $y_1 = f(x)$ to graph $y_2 = f(x - 3)$.

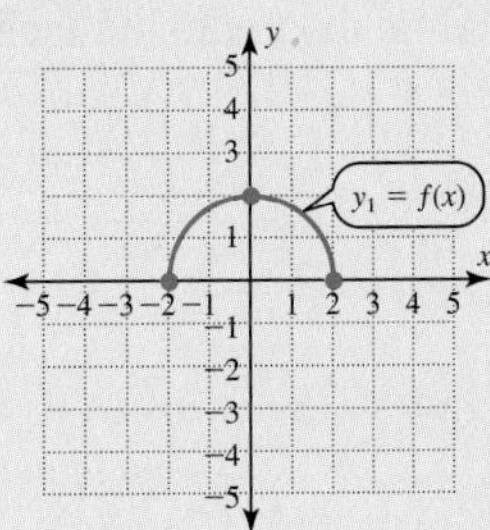

2. Use the table for $y_1 = f(x)$ to complete the table for $y_2 = f(x - 5)$.

x	f(x)
−3	18
−2	6
−1	1
0	0
1	0
2	−2
3	9

x	f(x − 5)
	18
	6
	1
	0
	0
	−2
	9

Combining Horizontal and Vertical Shifts

Example 8 illustrates how to combine horizontal and vertical shifts to produce a new graph.

EXAMPLE 8 Combining Horizontal and Vertical Shifts

Use the graphs of $y_1 = f_1(x)$ and $y_2 = f_2(x)$ to write an equation for y_2 in terms of $f_1(x)$. Give the domain and range of each function.

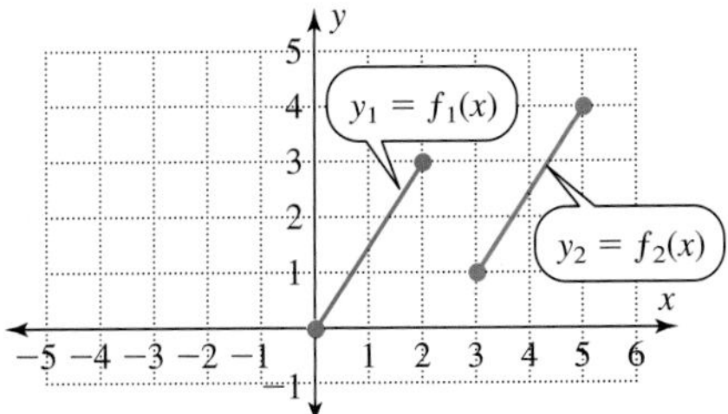

SOLUTION

The graph of $y_2 = f_2(x)$ can be obtained by shifting the graph of $y_1 = f_1(x)$ to the right 3 units and up 1 unit. Note how this shift affects the domain and range.

Key points shifted are as follows:
(0, 0) is shifted right 3 units and up 1 unit to (3, 1).
(2, 3) is shifted right 3 units and up 1 unit to (5, 4).

Answer: $y_2 = f_1(x - 3) + 1$

Domain of y_1: $D = [0, 2]$
Range of y_1: $R = [0, 3]$
Domain of y_2: $D = [3, 5]$
Range of y_2: $R = [1, 4]$

The parabola $f(x) = x^2$ has a vertex of (0, 0). This parabola can be shifted horizontally and vertically to produce many other parabolas. The vertex is a key point on each of these parabolas. Example 9 gives practice in identifying the vertex of a parabola from the defining equation.

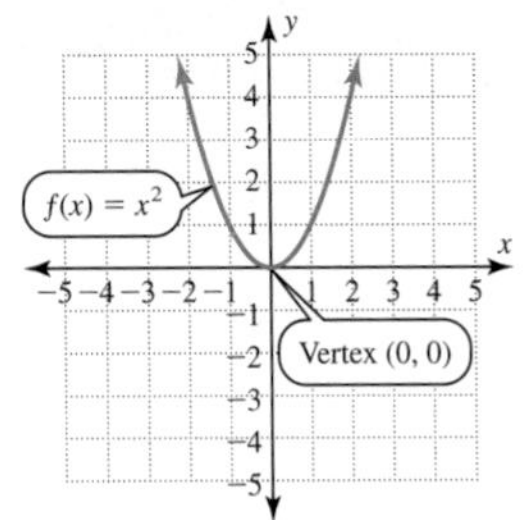

EXAMPLE 9 Determining the Vertex of a Parabola

Determine the vertex of each parabola by using the fact that the vertex of $f(x) = x^2$ is $(0, 0)$.

SOLUTION

(a) $f(x) = x^2 + 7$ Vertex of $(0, 7)$ This is a vertical shift 7 units up of every point on $f(x) = x^2$.

(b) $f(x) = (x + 7)^2$ Vertex of $(-7, 0)$ This is a horizontal shift 7 units left of every point on $f(x) = x^2$.

(c) $f(x) = (x - 7)^2 + 9$ Vertex of $(7, 9)$ This is a horizontal shift 7 units right and a vertical shift 9 units up of every point on $f(x) = x^2$.

(d) $f(x) = (x + 8)^2 - 6$ Vertex of $(-8, -6)$ This is a horizontal shift 8 units left and a vertical shift 6 units down of every point on $f(x) = x^2$.

In Example 9(c) you can think of the shifts following the same order as the order of operations in the expression. First we subtract inside the parentheses, producing a shift 7 units to the right. Later we add 9, producing a shift 9 units up.

SELF-CHECK 11.3.6

Write an equation to shift the graph of $y = \sqrt{x}$ as described in each problem.

1. Shift left 10 units.
2. Shift down 10 units.
3. Shift right 5 units and down 6 units.
4. Shift left 7 units and up 8 units.
5. Use a graphing calculator to graph the parabolas in Example 9 in order to check the vertex of each parabola.

SELF-CHECK ANSWERS

11.3.1

1.

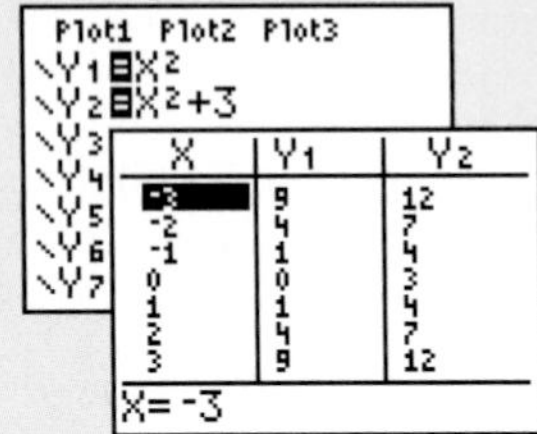

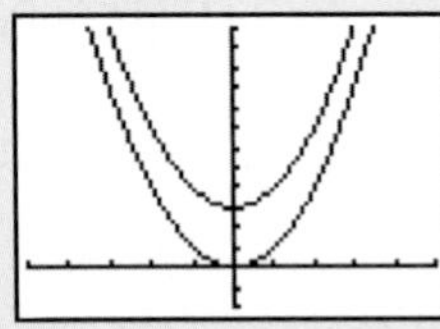

[−5, 5, 1] by [−2, 12, 1]

For each value of x, Y_2 is 3 units more than Y_1. The parabola defined by $f(x) = x^2 + 3$ can be obtained by shifting the parabola defined by $f(x) = x^2$ up 3 units.

11.3.2

1.

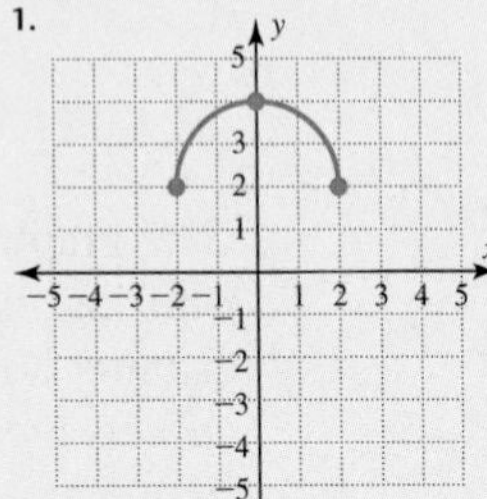

2.

x	y_1	y_2
−2	−8	−7
−1	−1	0
0	0	1
1	1	2
2	8	9

The parabola $f(x) = (x + 1)^2$ can be obtained by shifting the parabola defined by $f(x) = x^2$ to the left 1 unit.

11.3.3

1.

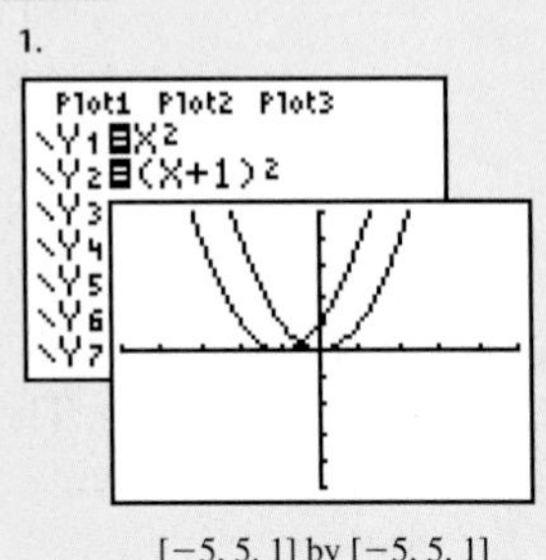

[−5, 5, 1] by [−5, 5, 1]

11.3.4

1. $y_2 = f_1(x + 1980)$

11.3.5

1.

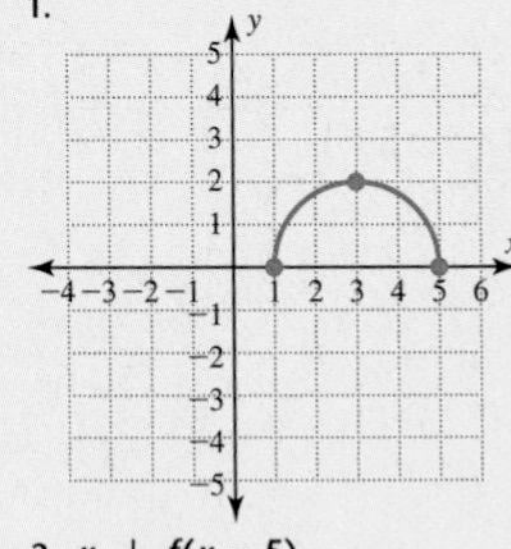

2.

x	$f(x - 5)$
2	18
3	6
4	1
5	0
6	0
7	−2
8	9

11.3.6

1. $y = \sqrt{x + 10}$
2. $y = \sqrt{x} - 10$
3. $y = \sqrt{x - 5} - 6$
4. $y = \sqrt{x + 7} + 8$

5a. Vertex of $(0, 7)$
5b. Vertex of $(-7, 0)$
5c. Vertex of $(7, 9)$
5d. Vertex of $(-8, -6)$

USING THE LANGUAGE AND SYMBOLISM OF MATHEMATICS 11.3

1. Each member of a family of functions has a graph with the same characteristic __________.
2. A shift of a graph up or down is called a __________ shift or a __________ translation.
3. A shift of a graph left or right is called a __________ shift or a __________ translation.
4. For a function $y = f(x)$ and a positive real number c:
   a. $y = f(x) + c$ will produce a __________ translation c units __________.
   b. $y = f(x) - c$ will produce a __________ translation c units __________.
   c. $y = f(x + c)$ will produce a __________ translation c units __________.
   d. $y = f(x - c)$ will produce a __________ translation c units __________.
5. For a function $y = f(x)$ and positive real numbers h and k:
   a. $y = f(x + h) + k$ will produce a __________ translation h units __________ and a __________ translation k units __________.
   b. $y = f(x - h) - k$ will produce a __________ translation h units __________ and a __________ translation k units __________.
6. The shape of the graph of $f(x) = x^2$ is called a __________, and the lowest point on this graph is called its __________.

QUICK REVIEW 11.3

The purpose of this quick review is to help you recall skills needed in this section.

Graph each function on a graphing calculator by using the viewing window $[-5, 5, 1]$ by $[-5, 5, 1]$. Then determine the domain and range of each function.

1. $f(x) = x^2$
2. $f(x) = x^2 - 2$
3. $f(x) = x^2 - 4$
4. $f(x) = \sqrt{x}$
5. $f(x) = \sqrt{x + 4}$

EXERCISES 11.3

Exercises 1–4 give functions that are translations of $f(x) = |x|$. Match each graph to the correct function. All graphs are displayed with the window $[-10, 10, 1]$ by $[-10, 10, 1]$.

1. $f(x) = |x| + 3$
2. $f(x) = |x + 3|$
3. $f(x) = |x| - 3$
4. $f(x) = |x - 3|$

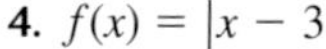

A.

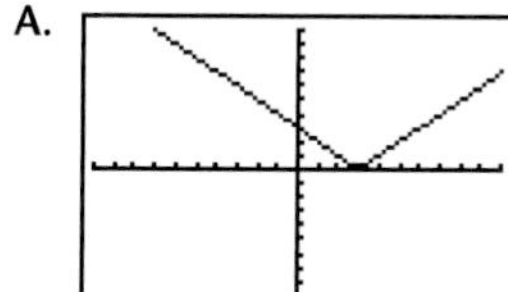

B.

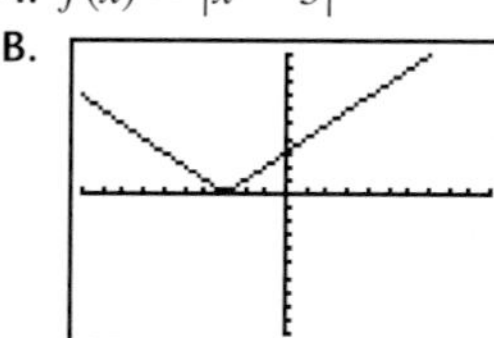

C.

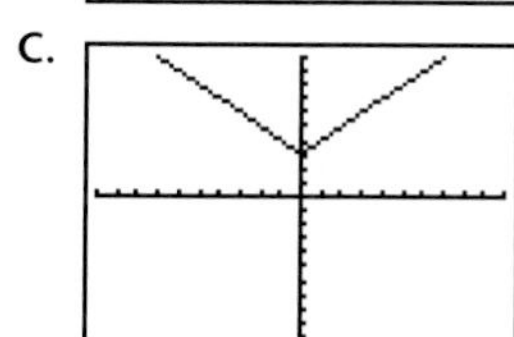

D.

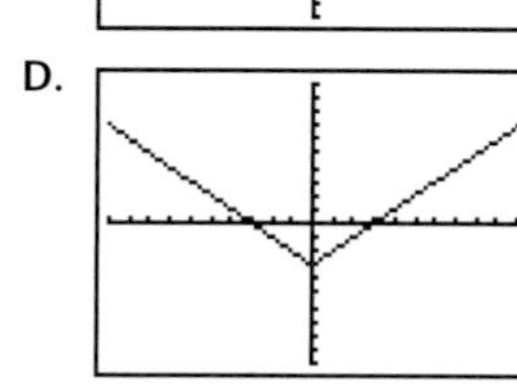

Exercises 5–8 give functions that are translations of $f(x) = x^2$. Match each graph to the correct function. All graphs are displayed with the window $[-10, 10, 1]$ by $[-10, 10, 1]$.

5. $f(x) = x^2 - 5$
6. $f(x) = (x - 5)^2$
7. $f(x) = (x + 5)^2$
8. $f(x) = x^2 + 5$

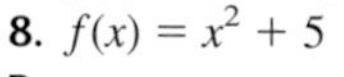

A.

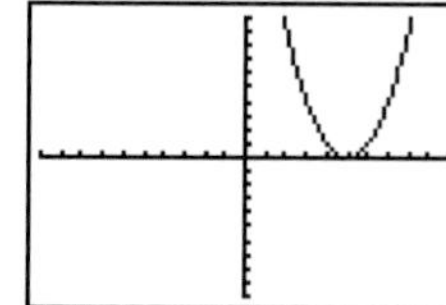

B.

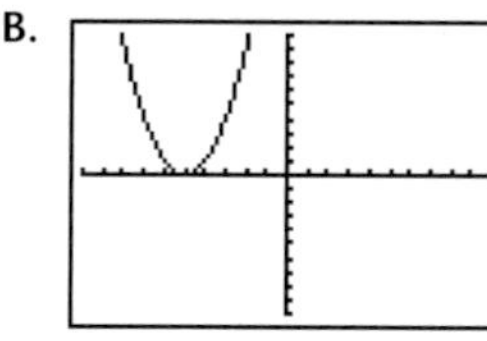

C.

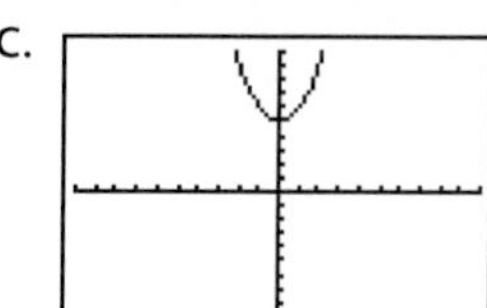

D.

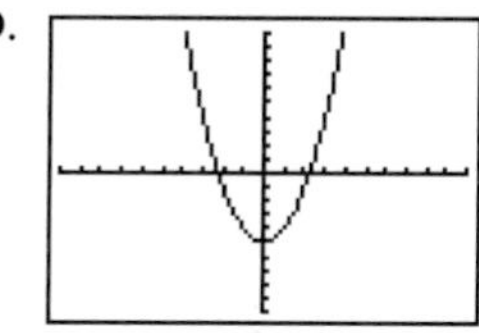

Exercises 9–12 give functions that are translations of $f(x) = \sqrt{x}$. Match each graph to the correct function. All graphs are displayed with the window $[-10, 10, 1]$ by $[-10, 10, 1]$.

9. $f(x) = \sqrt{x - 4}$
10. $f(x) = \sqrt{x + 4}$
11. $f(x) = \sqrt{x} + 4$
12. $f(x) = \sqrt{x} - 4$

A.

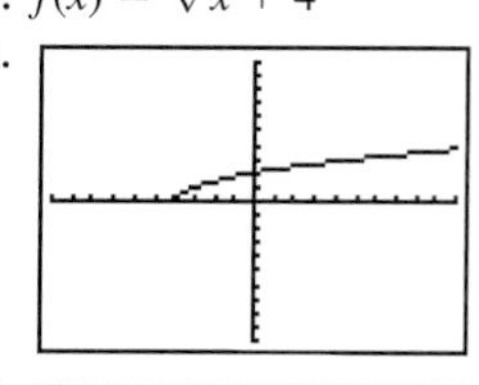

B.

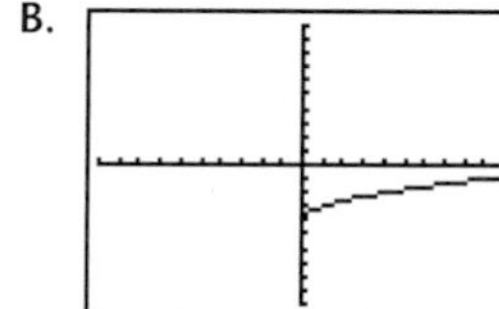

C.

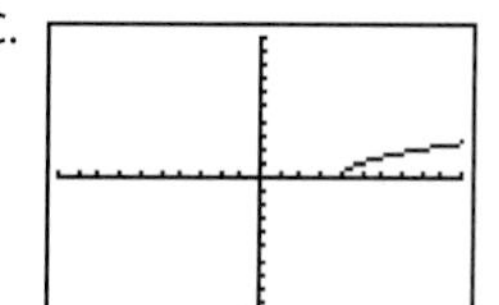

D.

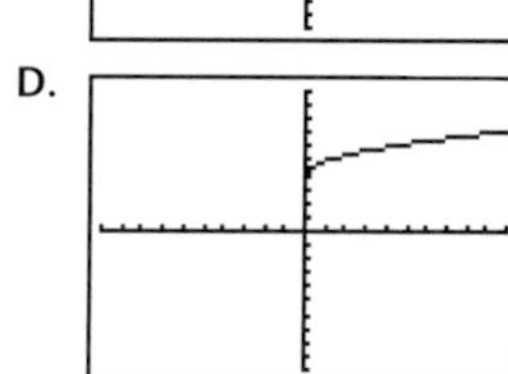

Exercises 13–16 give functions that are translations of $f(x) = |x|$. Match each graph to the correct function. All graphs are displayed with the window $[-10, 10, 1]$ by $[-10, 10, 1]$.

13. $f(x) = |x - 1| + 2$
14. $f(x) = |x + 1| - 2$
15. $f(x) = |x + 1| + 2$
16. $f(x) = |x - 1| - 2$

A.
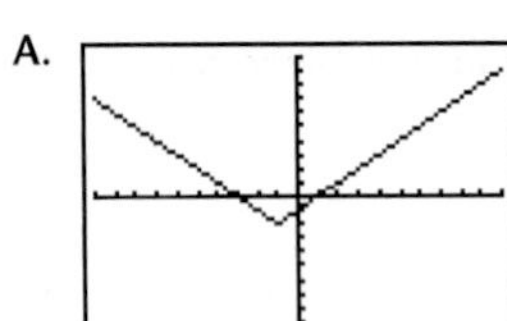

B.
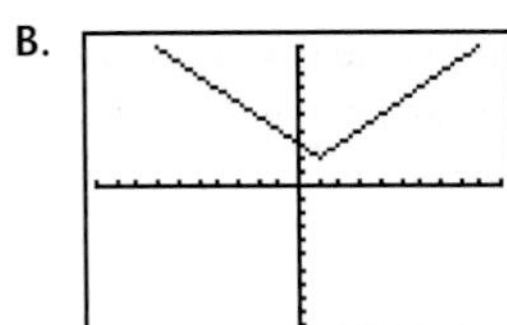

C.
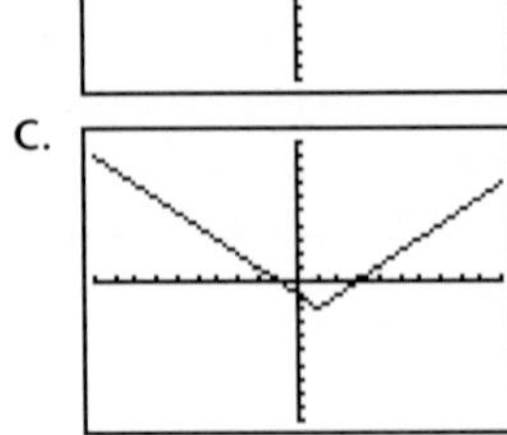

D.
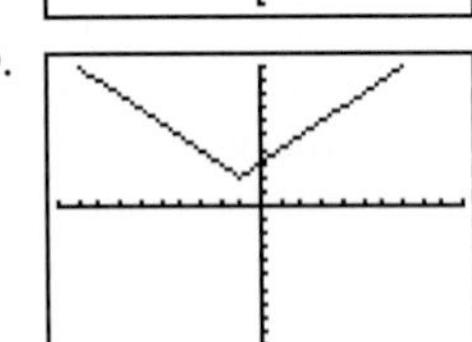

In Exercises 17–22, use the given graph of $y = f(x)$ and horizontal and vertical shifts to graph each function. (*Hint:* First translate the three points labeled on the graph.)

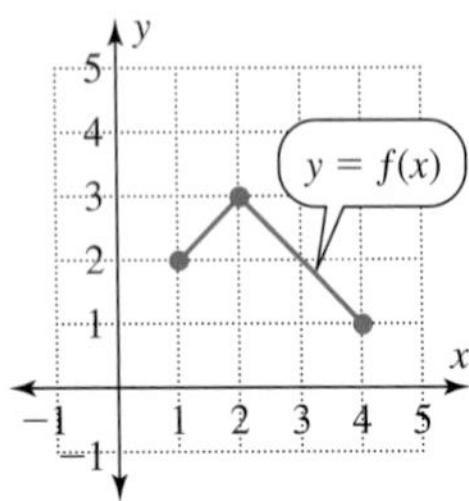

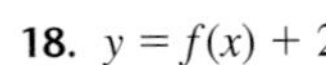

17. $y = f(x) - 3$

18. $y = f(x) + 2$

19. $y = f(x + 4)$

20. $y = f(x - 1)$

21. $y = f(x - 1) + 2$

22. $y = f(x + 1) - 2$

In Exercises 23 and 24, each graph is a translation of the graph of $y = \frac{x}{2}$. Write the equation of each graph. Identify the y-intercept of each graph.

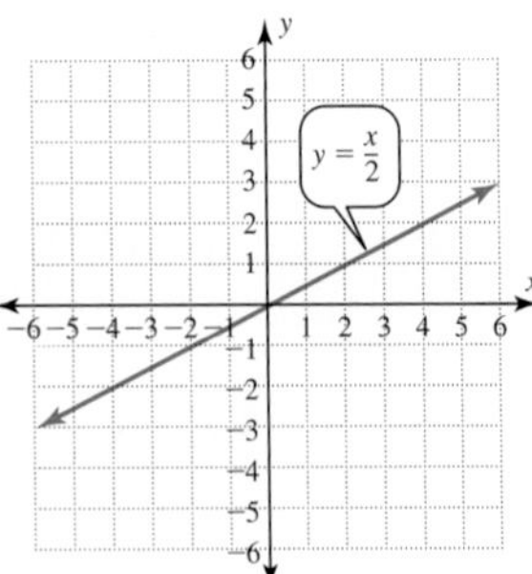

23.

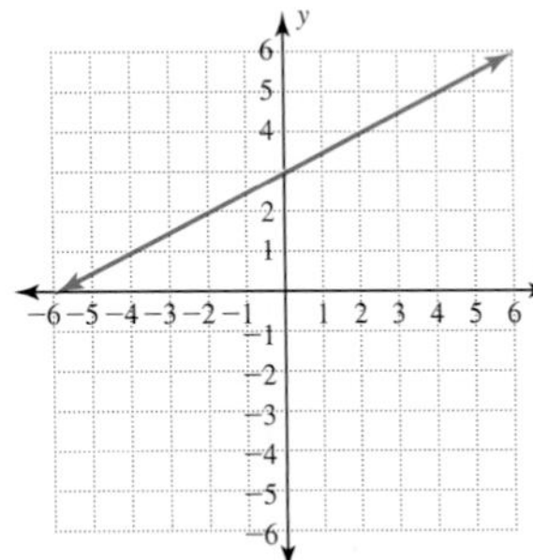

24.

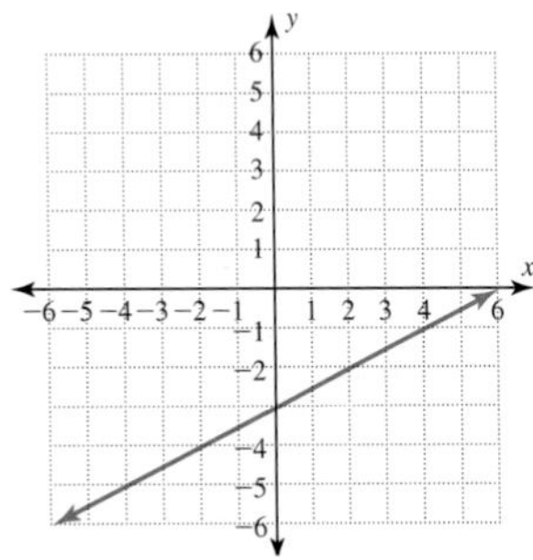

In Exercises 25 and 26, use the given table of values for $y = f(x)$ to complete each table.

x	y
-3	0
-2	8
-1	6
0	0
1	-4
2	0

25.

x	$y = f(x) + 10$
-3	
-2	
-1	
0	
1	
2	

26.

x	$y = f(x) - 10$
-3	
-2	
-1	
0	
1	
2	

In Exercises 27 and 28, use the given tables to write an equation for y_2 and y_3.

x	$y_1 = f(x)$
-3	4
-2	3
-1	8
0	7
1	0
2	-1

x	y_2
-3	-3
-2	-4
-1	1
0	0
1	-7
2	-8

x	y_3
-3	11
-2	10
-1	15
0	14
1	7
2	6

27. Write an equation for y_2 in terms of $f(x)$.

28. Write an equation for y_3 in terms of $f(x)$.

29. Use the given table for $f(x) = 2x - 3$ to complete the table for y_2.

x	$y_1 = f(x)$
0	-3
1	-1
2	1
3	3
4	5
5	7

x	$y_2 = f(x - 4)$
4	-3
	-1
	1
	3
	5
9	7

30. Use the given table for $f(x) = -|x|$ to complete the table for y_2.

x	$y_1 = f(x)$
-3	-3
-2	-2
-1	-1
0	0
1	-1
2	-2
3	-3

x	$y_2 = f(x + 1)$
-4	-3
	-2
	-1
	0
	-1
	-2
2	-3

In Exercises 31 and 32, use the given tables to write an equation for y_2 and y_3.

x	$y_1 = f(x)$
-3	3
-2	2
-1	1
0	0
1	1
2	2
3	3

x	y_2
0	3
1	2
2	1
3	0
4	1
5	2
6	3

x	y_3
-4	3
-3	2
-2	1
-1	0
0	1
1	2
2	3

31. Write an equation for y_2 in terms of $f(x)$.

32. Write an equation for y_3 in terms of $f(x)$.

In Exercises 33–38, determine the vertex of each parabola by using the fact that the vertex of $f(x) = x^2$ is at (0, 0). Use a graphing calculator to check your answers.

33. $f(x) = (x - 6)^2$

34. $f(x) = (x + 6)^2$

35. $f(x) = x^2 + 11$

36. $f(x) = x^2 - 11$

37. $f(x) = (x + 5)^2 - 8$

38. $f(x) = (x - 8)^2 + 5$

In Exercises 39–44, determine the vertex of the graph of each absolute value function by using the fact that the vertex of $f(x) = |x|$ is at (0, 0). Use a graphing calculator to check your answers.

39. $f(x) = |x| - 13$

40. $f(x) = |x| + 14$

41. $f(x) = |x + 15|$

42. $f(x) = |x - 10|$

43. $f(x) = |x - 7| - 6$

44. $f(x) = |x + 9| + 8$

Exercises 45–50 describe a translation of the graph $y = f(x)$. Match each description to the correct function.

45. A translation eleven units left
46. A translation eleven units right
47. A translation eleven units down
48. A translation eleven units up
49. A translation eleven units right and eleven units up
50. A translation eleven units left and eleven units down

A. $y = f(x - 11) + 11$
B. $y = f(x + 11) - 11$
C. $y = f(x - 11)$
D. $y = f(x) - 11$
E. $y = f(x + 11)$
F. $y = f(x) + 11$

In Exercises 51–56, determine the domain and range of each function f_2, given that the domain of a function f_1 is $D = [0, 5)$ and the range is $R = [2, 4)$.

51. The graph of f_2 is obtained from the graph of f_1 by shifting it 3 units right.
52. The graph of f_2 is obtained from the graph of f_1 by shifting it 5 units left.
53. The graph of f_2 is obtained from the graph of f_1 by shifting it 7 units down.
54. The graph of f_2 is obtained from the graph of f_1 by shifting it 8 units up.
55. The graph of f_2 is obtained from the graph of f_1 by shifting it 4 units left and 5 units up.
56. The graph of f_2 is obtained from the graph of f_1 by shifting it 6 units right and 4 units down.

In Exercises 57–64, match each function with its graph. Use the shape of the graph of each function and your knowledge of translations to make your choices. All graphs are displayed with the window $[-10, 10, 1]$ by $[-10, 10, 1]$.

57. $f(x) = x - 6$
58. $f(x) = x + 4$
59. $f(x) = |x + 5|$
60. $f(x) = |x| - 3$
61. $f(x) = \sqrt{x + 3}$
62. $f(x) = (x + 4)^3$
63. $f(x) = \sqrt[3]{x} - 5$
64. $f(x) = (x - 3)^2$

A.
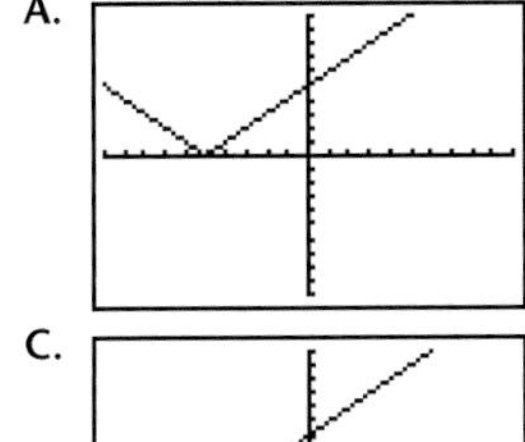

B.
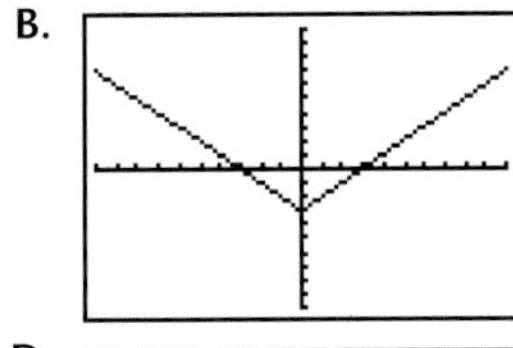

C.

D.

E.
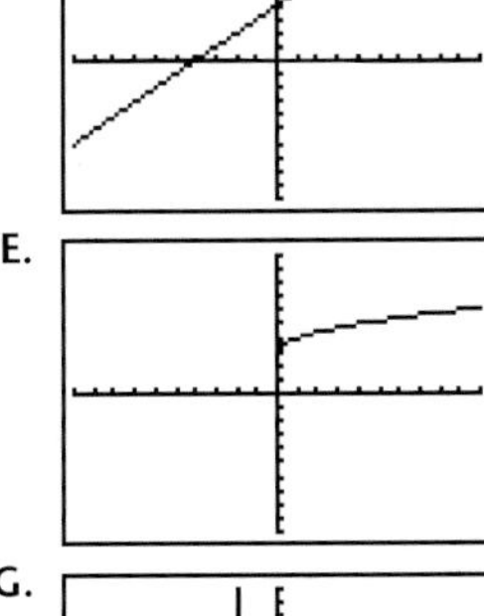

F.
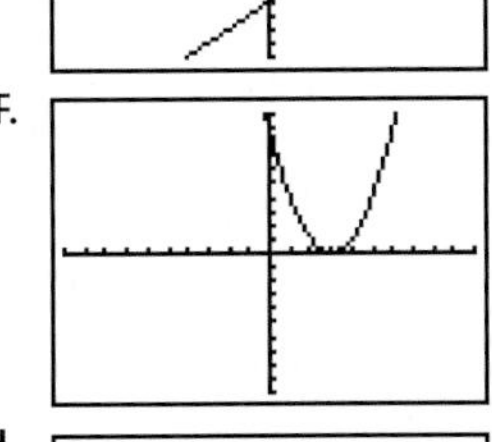

G.
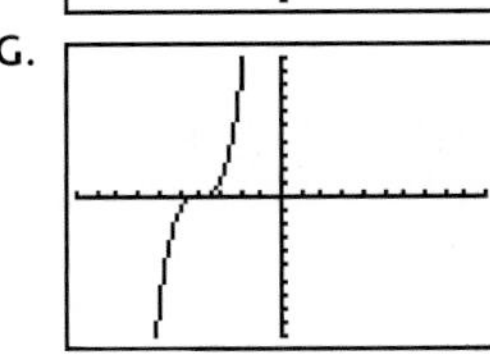

H.
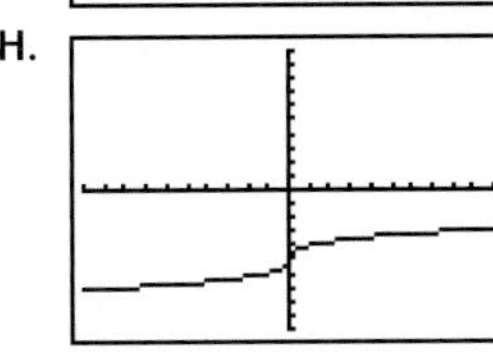

65. Write the first five terms of each sequence. (*Hint:* See Example 7 in Section 2.1.)
    a. $a_n = 2n$
    b. $a_n = 2n + 3$
    c. $a_n = 2(n + 3)$
    d. $a_n = -2n$

66. Write the first five terms of each sequence. (*Hint:* See Example 7 in Section 2.1.)
    a. $a_n = n^2$
    b. $a_n = n^2 - 9$
    c. $a_n = (n - 9)^2$
    d. $a_n = (n - 1)^2 + 3$

67. **Shifting the Temperature Scale** Many properties of nature are related to the temperature of the object being studied. The Celsius temperature scale (°C) is the same incrementally as the Kelvin temperature scale (K) except that it is 273° higher (water freezes at 0°C and 273°K). The equation $V(t) = 4t$ gives the volume in milliliters of a gas as a function of the temperature in degrees Celsius. Write an equation for the volume as a function of the temperature in kelvins and complete the following table.

t°C	$V(t) = 4t$
200	800
150	600
100	400
50	200
0	0

t°K	$V = ?$
	800
	600
	400
	200
273	0

68. **Shifting a Time Reference** A company has a function that approximates its revenue in terms of the year those revenues were generated. The company was founded in January 1995. Thus 1995 is taken as year 0 for the company, 1996 as year 1, and so on. A formula for its revenue per year in millions of dollars is given by $R(t) = t^2$, where t is the number of years the company has been in business. Convert this formula to give the revenue in terms of the calendar year.

YEAR t	REVENUE $R(t) = t^2$
0	0
1	1
2	4
3	9
4	16
5	25

69. **Shifting a Pricing Function** The price in dollars of five different types of networking cables is given in the table. Because of increased overhead cost, the price of each item is increased by $1. Form a table that gives the new price of each item.

ITEM NO. x	PRICE $P(x)$ ($)
1	24.95
2	33.79
3	12.98
4	26.78
5	19.95

70. **Shifting the Distance for Golf Clubs** Thanks to a new technology in making golf clubs, many golfers have been able to add 5 yd to the distance they can hit with each club. Form a table that gives the new distance for each new club.

IRON	DISTANCE FOR OLD CLUBS (yd)
2	200
3	190
4	180
5	165
6	150

Group Discussion Questions

71. **Calculator Discovery** Use a graphing calculator to graph each pair of functions on the same graphing screen. Then describe the relationship between each pair of functions.
    a. $Y_1 = x$, $Y_2 = -x$
    b. $Y_1 = |x|$, $Y_2 = -|x|$
    c. $Y_1 = x^2$, $Y_2 = -x^2$
    d. $Y_1 = \sqrt{x}$, $Y_2 = -\sqrt{x}$

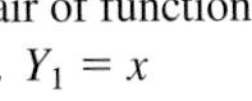

72. **Calculator Discovery** Use a graphing calculator to graph the functions: $f(x) = |x|$, $f(x) = 2|x|$, $f(x) = 3|x|$, and $f(x) = 4|x|$. Can you describe the relationship between the graph of $f(x) = |x|$ and $f(x) = c|x|$ for any positive constant c?

73. **Calculator Discovery** Use a graphing calculator to graph the functions: $f(x) = -x^2, f(x) = -2x^2$, $f(x) = -3x^2$, and $f(x) = -4x^2$. Can you describe the relationship between the graph of $f(x) = x^2$ and the graph of $f(x) = -cx^2$ for any positive constant c?

74. **Error Analysis** A student graphing $f(x) = |x + 2| + 3$ obtained the following display. Explain how you can tell by inspection that an error has been made.

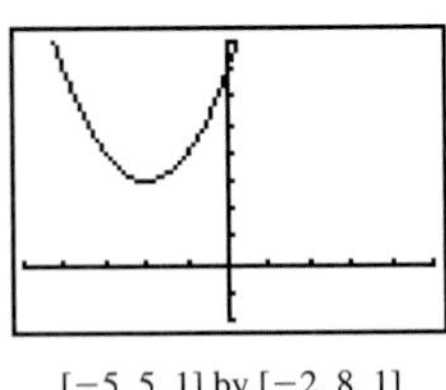

[−5, 5, 1] by [−2, 8, 1]

Section 11.4 Stretching, Shrinking, and Reflecting Graphs of Functions

Objectives:

5. Recognize the reflection of a graph.
6. Use stretching and shrinking factors to graph functions.

A key concept of algebra is that there are some basic families of graphs. In Section 11.3 we learned how to shift a graph to create an exact copy of this graph at another location in the plane. We now learn how to modify a given graph to create a stretching or a shrinking of this shape. We start by examining the reflection of a graph.

Reflecting a Graph Across the x-Axis

The reflection of a point (x,y) across the x-axis is a mirror image of the point on the opposite side of the x-axis. The reflection of the point (x, y) about the x-axis is the point $(x, -y)$.

We now examine the reflection of an entire graph by examining the functions $y_1 = |x|$ and $y_2 = -|x|$.

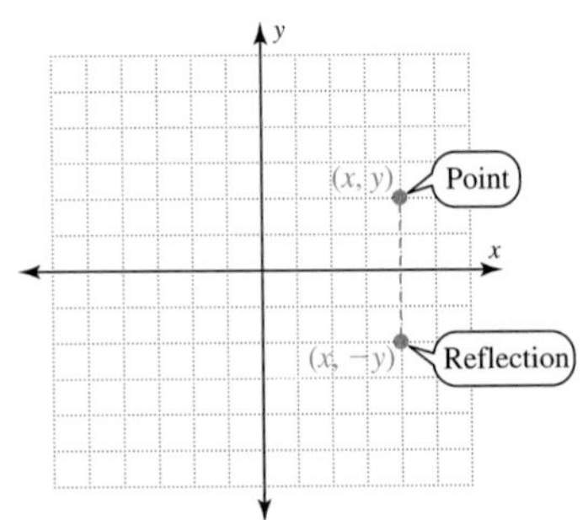

NUMERICALLY

x	$y_1 = \|x\|$	$y_2 = -\|x\|$
−3	3	−3
−2	2	−2
−1	1	−1
0	0	0
1	1	−1
2	2	−2
3	3	−3

GRAPHICALLY

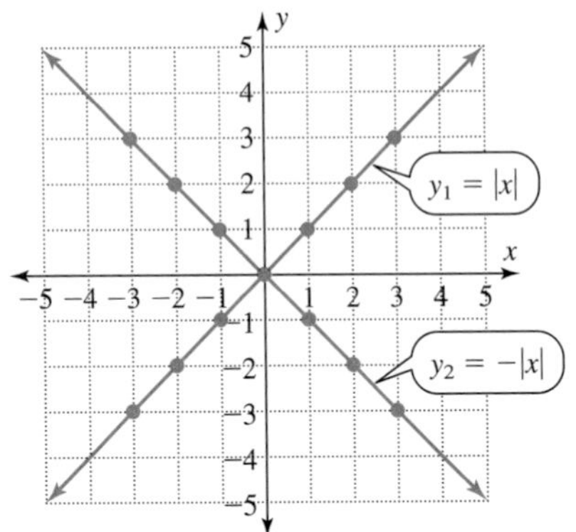

VERBALLY

Each y_2-value is the opposite of the corresponding y_1-value.

The graph of y_2 can be obtained by reflecting across the x-axis each point of the graph of y_1.

A reflection of a graph is simply a reflection of each point of the graph. The following box describes a reflection of a graph across the x-axis.

Reflection of $y = f(x)$ Across the x-Axis

GRAPHICALLY	NUMERICALLY	ALGEBRAICALLY
To obtain the graph of $y = -f(x)$, reflect the graph of $y = f(x)$ across the x-axis.	For each value of x, y_2 is the additive inverse of y_1. That is, $y_2 = -y_1$.	Original function: $y_1 = f(x)$ Reflection: $y_2 = -f(x)$

EXAMPLE 1 Examining a Reflection of a Quadratic Function

Compare the functions $y_1 = x^2$ and $y_2 = -x^2$ numerically, graphically, and verbally.

SOLUTION

NUMERICALLY

x	$y_1 = x^2$	$y_2 = -x^2$
−3	9	−9
−2	4	−4
−1	1	−1
0	0	0
1	1	−1
2	4	−4
3	9	−9

GRAPHICALLY

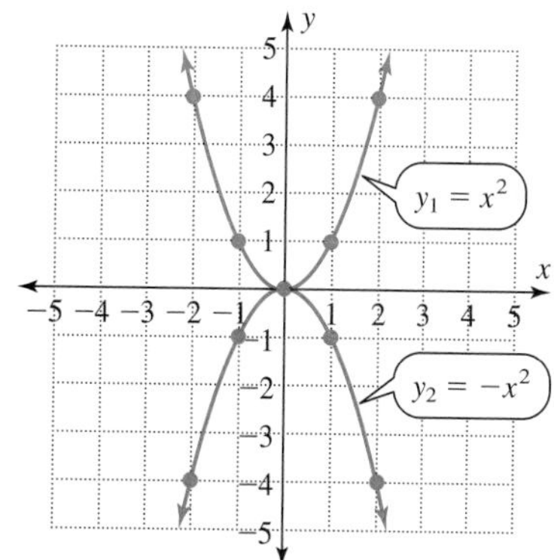

VERBALLY

For each value of x, $y_2 = -y_1$.

The parabola defined by $y_2 = -x^2$ can be obtained by reflecting the graph of $y_1 = x^2$ across the x-axis.

CALCULATOR PERSPECTIVE 11.4.1 Reflecting a Graph Across the x-Axis

To create the table and the graph of $Y_2 = -Y_1$ from Example 1 on a TI-84 Plus calculator, enter the following keystrokes:

2nd TABLE

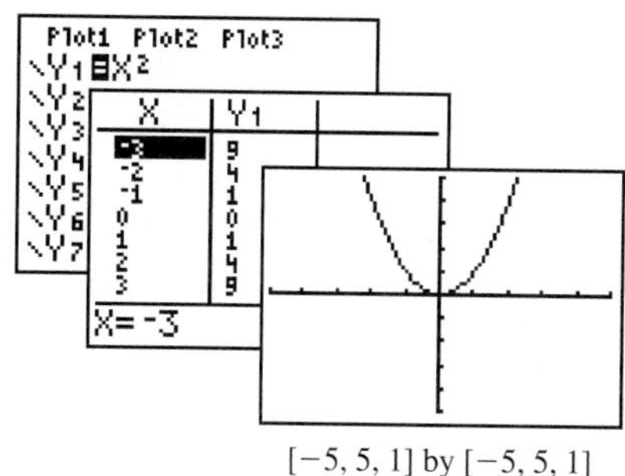

[−5, 5, 1] by [−5, 5, 1]

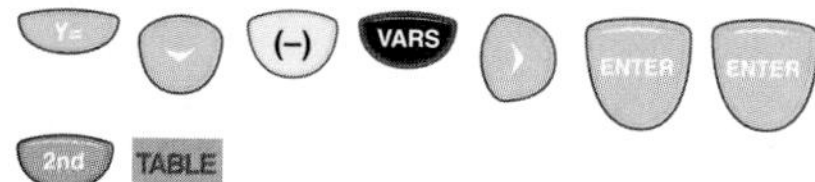

2nd TABLE

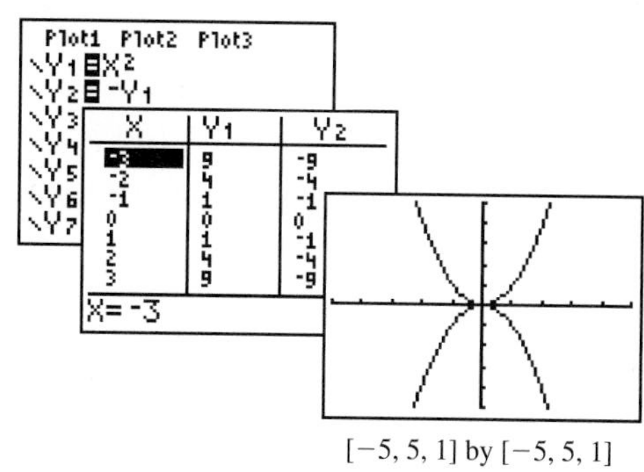

[−5, 5, 1] by [−5, 5, 1]

SELF-CHECK 11.4.1

1. Use a graphing calculator to graph $Y_1 = x^2 - 1$ and $Y_2 = -Y_1$ on the same display window.
2. Write an equation for Y_2 in terms of x.
3. Compare the graphs of Y_1 and Y_2.

Example 2 illustrates how to form the reflection of a graph when no formula is given for $y = f(x)$.

EXAMPLE 2 Forming the Reflection of a Graph

Use the graph of $y_1 = f(x)$ to graph $y_2 = -f(x)$.

SOLUTION

Example 2 clearly illustrates the reflection idea when we visually compare these two graphs. According to David Bock, Graphics Research Programmer for the National Center for Supercomputing Applications, "This concept is used extensively in three-dimensional computer graphics and animation to create and position symmetrical objects and models."

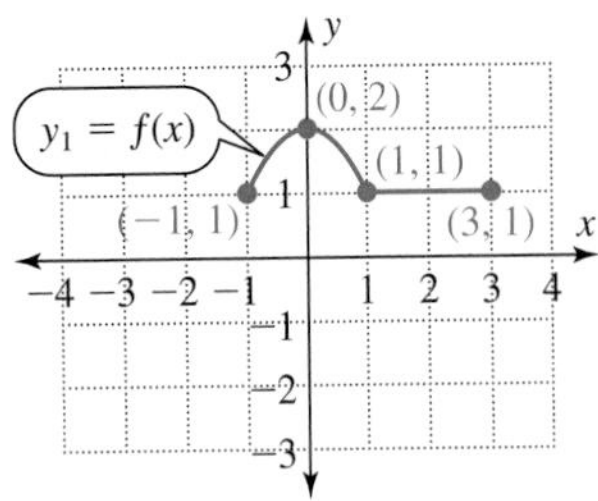

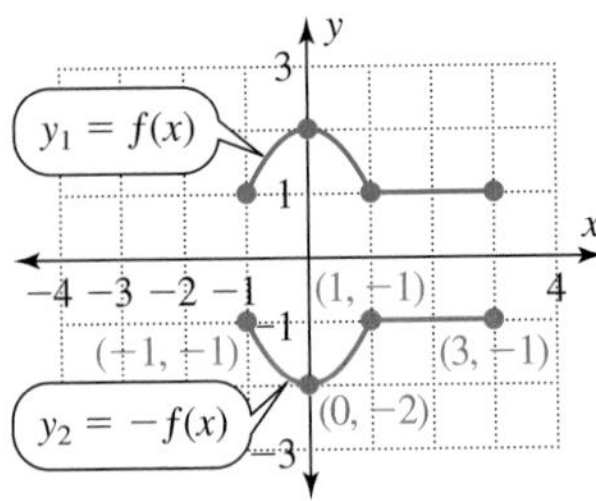

Start by reflecting the key points labeled on the graph of $y_1 = f(x)$. Then use the shape of $y_1 = f(x)$ to sketch its reflection across the x-axis.

Example 3 is used to examine reflections when two functions are defined by a table of values.

EXAMPLE 3 Identifying a Reflection Across the x-Axis

Use the table of values for $y_1 = x^3 - x^2$ and y_2 to write an equation for y_2.

x	$y_1 = x^3 - x^2$	y_2
−3	−36	36
−2	−12	12
−1	−2	2
0	0	0
1	0	0
2	4	−4
3	18	−18

SOLUTION

For each value of x, y_2 is the additive inverse of y_1.

Answer: $y_2 = -y_1$ or $y_2 = -x^3 + x^2$

SELF-CHECK 11.4.2

1. Use the graph of $y_1 = f(x)$ to graph $y_2 = -f(x)$.

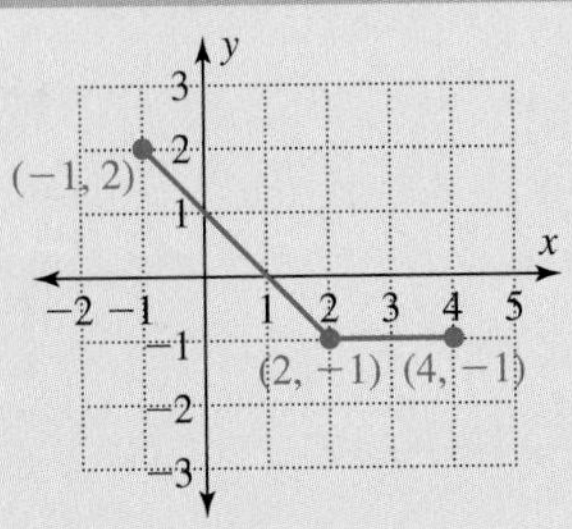

2. Use the table for $y_1 = f(x)$ to complete the table for $y_2 = -f(x)$.

x	y_1	y_2
−2	7	
−1	−4	
0	3	
1	9	
2	−6	

Stretching and Shrinking Graphs

We now examine how to change the size of the graph of a function by applying a stretching or shrinking factor. These stretching or shrinking factors affect the vertical scale of the graph. They do not affect the horizontal scale, and they are not rigid translations. We start our inspection of scaling factors by examining the functions $y_1 = |x|$ and $y_2 = 3|x|$.

NUMERICALLY

x	$y_1 = \|x\|$	$y_2 = 3\|x\|$
−3	3	9
−2	2	6
−1	1	3
0	0	0
1	1	3
2	2	6
3	3	9

GRAPHICALLY

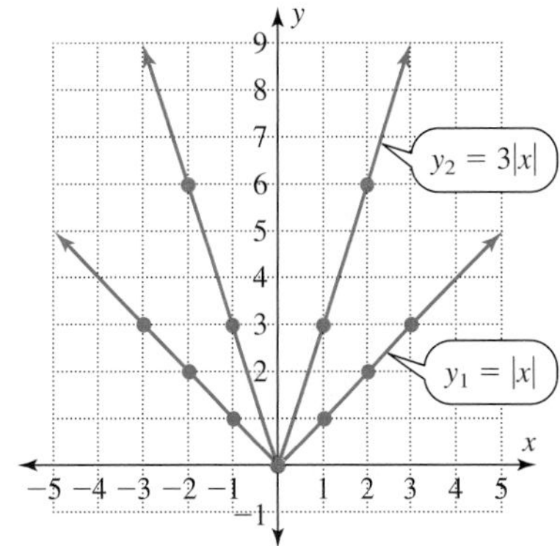

VERBALLY

Each value of y_2 is 3 times the corresponding y_1 value.

The graph of y_2 is the same basic V shape as y_1 but rises 3 times as rapidly.

The scaling factor c in $y = cf(x)$ affects the vertical scale of the graph—not the horizontal scale.

Comparing the graphs of $y_2 = 3|x|$ to $y_1 = |x|$, we call 3 a **stretching factor** because this factor vertically stretches the graph of $y_1 = |x|$. The following box describes **vertical scaling factors** that can either stretch or shrink a graph. Then Example 4 examines a scaling factor of $\frac{1}{4}$ that vertically shrinks the graph.

Vertical Stretching and Shrinking Factors of $y = f(x)$

For a function $y_1 = f(x)$ and a positive real number c:

ALGEBRAICALLY	GRAPHICALLY	NUMERICALLY
If $c > 1$:		
Original function: $y_1 = f(x)$ Scaled function: $y_2 = cf(x)$	To obtain the graph of $y_2 = cy_1$, vertically stretch the graph of $y_1 = f(x)$ by a factor of c.	For each value of x, $y_2 = cy_1$.
If $0 < c < 1$:		
Original function: $y_1 = f(x)$ Scaled function: $y_2 = cf(x)$	To obtain the graph of $y_2 = cy_1$, vertically shrink the graph of $y_1 = f(x)$ by a factor of c.	For each value of x, $y_2 = cy_1$.

EXAMPLE 4 Examining a Shrinking of a Quadratic Function

Compare the functions $y_1 = x^2$ and $y_2 = \frac{1}{4}x^2$ numerically, graphically, and verbally.

SOLUTION

NUMERICALLY

x	$y_1 = x^2$	$y_2 = \frac{1}{4}x^2$
-3	9	$\frac{9}{4}$
-2	4	1
-1	1	$\frac{1}{4}$
0	0	0
1	1	$\frac{1}{4}$
2	4	1
3	9	$\frac{9}{4}$

Note that points on the x-axis are not moved by scaling factors.

GRAPHICALLY

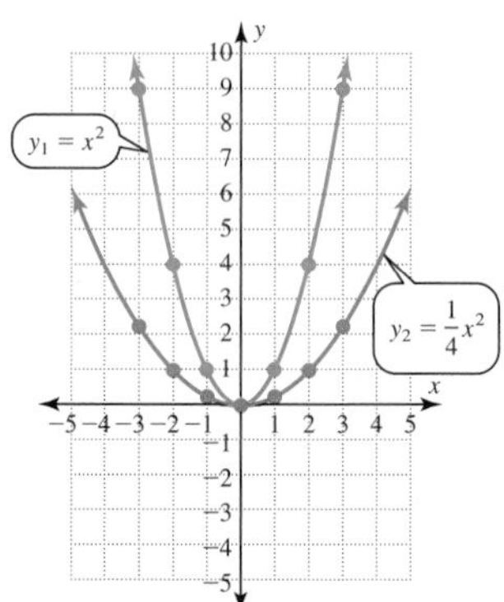

Both graphs are parabolas opening upward. At each x-value the height of y_2 is $\frac{1}{4}$ the height of y_1. The vertex of each parabola is $(0, 0)$.

VERBALLY

For each value of x, $y_2 = \frac{1}{4}y_1$.

The parabola defined by $y_2 = \frac{1}{4}x^2$ can be obtained by vertically shrinking the parabola defined by $y_1 = x^2$ by a factor of $\frac{1}{4}$. Note that $\frac{1}{4}(0) = 0$ and that $(0, 0)$ is the vertex of both parabolas.

EXAMPLE 5 Examining a Stretching of a Square Root Function

Use a graphing calculator to compare the functions $Y_1 = \sqrt{x}$ and $Y_2 = 2\sqrt{x}$ numerically, graphically, and verbally.

SOLUTION

NUMERICALLY

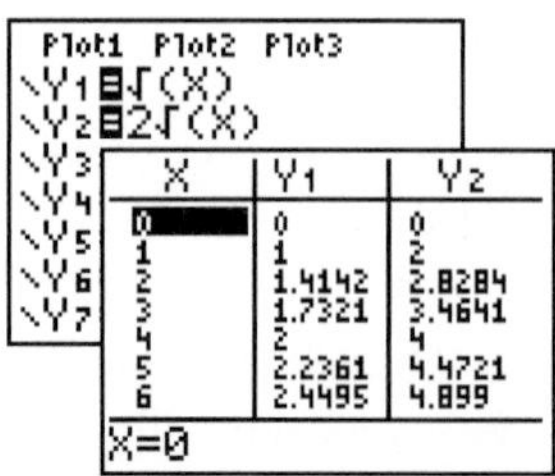

GRAPHICALLY

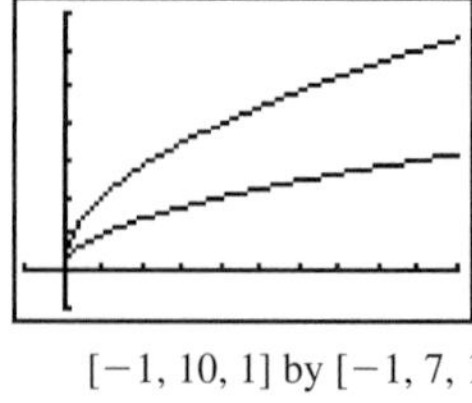

$[-1, 10, 1]$ by $[-1, 7, 1]$

Both graphs have the characteristic shape of the square root function. For each the domain $D = [0, \infty)$ and the range $R = [0, \infty)$.

VERBALLY

For each value of x, $Y_2 = 2Y_1$.

The graph of $Y_2 = 2\sqrt{x}$ can be obtained by vertically stretching by a factor of 2 the graph of $Y_1 = \sqrt{x}$.

Example 6 uses a table of values to identify a stretching factor.

EXAMPLE 6 Identifying a Stretching Factor

Use the table of values for $y_1 = f_1(x)$ and $y_2 = f_2(x)$ to write an equation for y_2 in terms of $f_1(x)$.

x	$y_1 = f_1(x)$	$y_2 = f_2(x)$
−2	5	15
−1	4	12
0	2	6
1	2	6
2	4	12

SOLUTION

For each value of x, $y_2 = 3y_1$. Thus $y_2 = 3f_1(x)$.

Answer: $y_2 = 3f_1(x)$

SELF-CHECK 11.4.3

1. Use a graphing calculator to compare $Y_1 = x^2$ and $Y_2 = 4x^2$ both numerically and graphically.
2. Use the graph of $y_1 = f(x)$ to graph $y_2 = 2f(x)$. *Hint:* Start by stretching the labeled points.

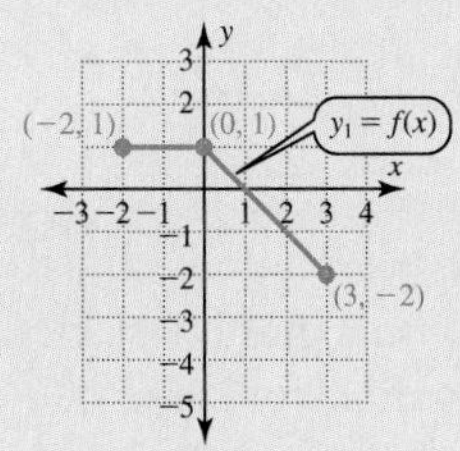

3. Use the table for $y_1 = f(x)$ to complete the table for $y_2 = 5f(x)$.

x	y_1	y_2
−2	4	
−1	3	
0	1	
1	−2	
2	−5	

Every graph in the family of absolute value functions can be obtained from the V-shaped graph of $y = |x|$. To obtain other members of this family, we can use horizontal and vertical translations, stretching and shrinking factors, and reflections. In Example 7, a function that both shrinks and reflects $y = |x|$ is examined.

EXAMPLE 7 Combining a Shrinking Factor and a Reflection

Use the graph of $Y_1 = |x|$ to graph $Y_2 = -\frac{1}{2}|x|$.

SOLUTION

NUMERICALLY

x	$Y_1 = \|x\|$	$Y_2 = -\frac{1}{2}\|x\|$
−4	4	−2
−2	2	−1
0	0	0
2	2	−1
4	4	−2

GRAPHICALLY

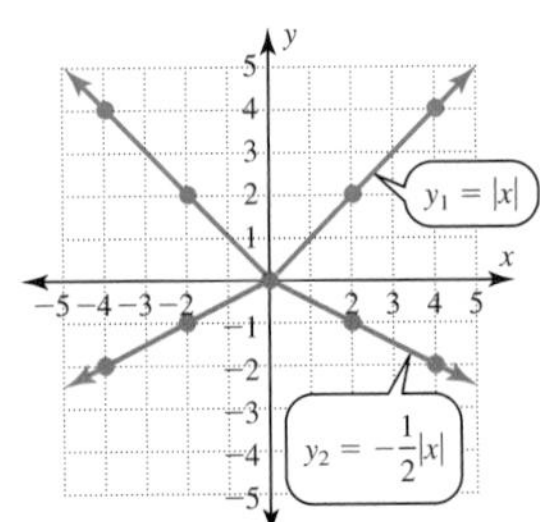

Both graphs have the characteristic V shape. The vertex of both V-shaped graphs is (0, 0), but they open in opposite directions.

VERBALLY

Every Y_2 value can be obtained by multiplying the corresponding Y_1 value by $-\frac{1}{2}$. $Y_2 = -\frac{1}{2}y_1$

The graph of $Y_2 = -\frac{1}{2}|x|$ can be obtained by vertically shrinking the graph of $Y_1 = |x|$ by a factor of $\frac{1}{2}$ and then reflecting this graph across the x-axis.

SELF-CHECK 11.4.4

1. Use a graphing calculator to graph $Y_1 = x^2$ and $Y_2 = -2x^2$ on the same calculator screen.
2. Compare the graphs of Y_1 and Y_2.

Example 8 illustrates how to analyze a graph that involves horizontal and vertical translations and a reflection of a parabola.

EXAMPLE 8 Combining Translations and a Reflection

Use the graph of $y_1 = x^2$ to write an equation for y_2. Give the domain and the range of each function.

We can think of the shifts and reflections following the same order as the order of operations in the expression. First we subtract 2 inside the parentheses to produce a shift 2 units to the right. Next we multiply by −1 to produce a reflection across the x-axis. Finally we add 1, to produce a shift 1 unit up.

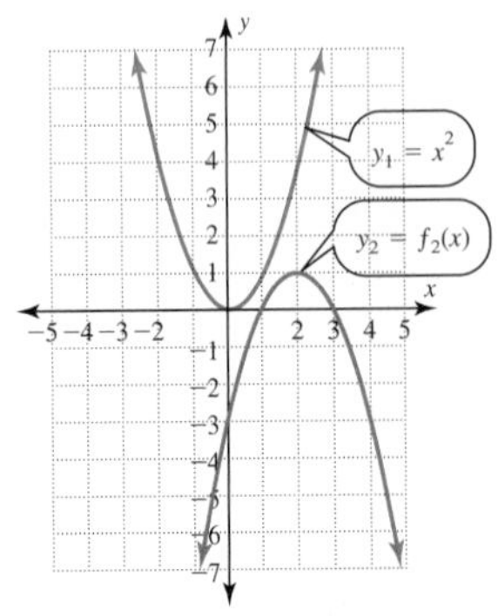

SOLUTION

The vertex of $y_1 = x^2$ is (0,0), and the vertex of $y_2 = f_2(x)$ is (2,1). The parabola defined by $y_2 = f_2(x)$ is the same size as the parabola defined by $y_1 = x^2$. There is no stretching or shrinking in this example. The graph of $y_2 = f_2(x)$ can be obtained from the graph of the parabola defined by $y_1 = x^2$ by first shifting this graph to the right 2 units, then reflecting this graph across the x-axis, and finally shifting this graph up 1 unit.

Answer: $y_2 = -(x - 2)^2 + 1$
For y_1 the domain $D = \mathbb{R}$ and the range $R = [0, \infty)$.
For y_2 the domain $D = \mathbb{R}$ and the range $R = (-\infty, 1]$.

The parabola $f(x) = x^2$ has a vertex of (0, 0). Example 9 provides practice identifying the vertex of a parabola from the defining equation.

EXAMPLE 9 Determining the Vertex of a Parabola

Determine the vertex of each parabola by using the fact that the vertex of $f(x) = x^2$ is at (0, 0).

You may wish to use a graphing calculator to visualize the results in Example 9.

SOLUTION

(a) $f(x) = 7x^2$ — Vertex of (0, 0) — The stretching factor of 7 does not move the vertex of $f(x) = x^2$. Points on the x-axis are not moved by scaling factors.

(b) $f(x) = (x + 9)^2$ — Vertex of $(-9, 0)$ — This is a horizontal shift 9 units left of every point on $f(x) = x^2$. The vertex is shifted from (0, 0) to $(-9, 0)$.

(c) $f(x) = -(x - 5)^2 + 8$ — Vertex of (5, 8) — The vertex (0, 0) is shifted to (5, 0) by the horizontal shift. The reflection leaves this vertex at (5, 0). The vertical translation shifts the vertex to (5, 8).

SELF-CHECK ANSWERS

11.4.1

1.

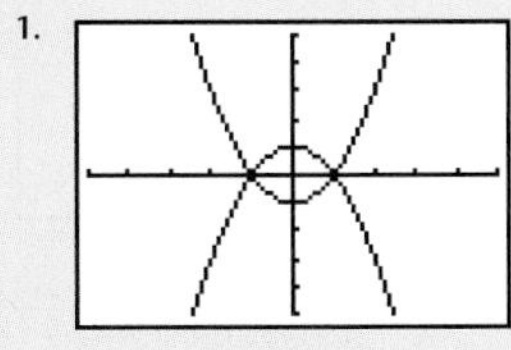

$[-5, 5, 1]$ by $[-5, 5, 1]$

2. $Y_2 = -x^2 + 1$
3. Y_2 is a reflection of Y_1 across the x-axis.

11.4.2

1.

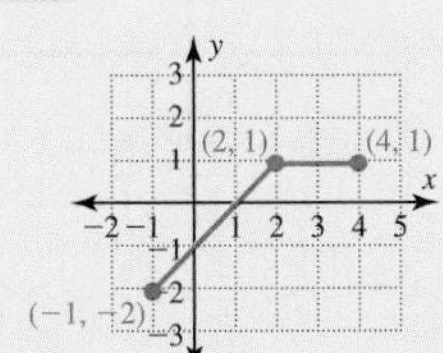

2.

x	y_1	y_2
−2	7	−7
−1	−4	4
0	3	−3
1	9	−9
2	−6	6

11.4.3

1.

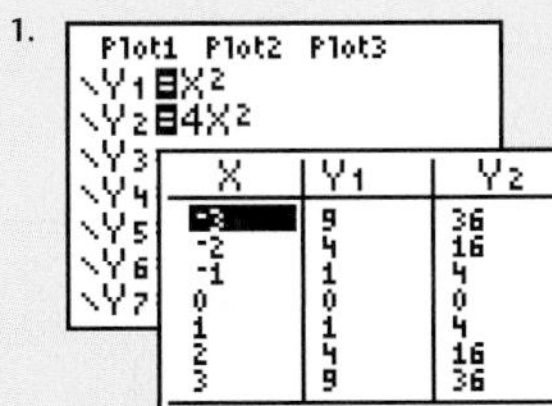

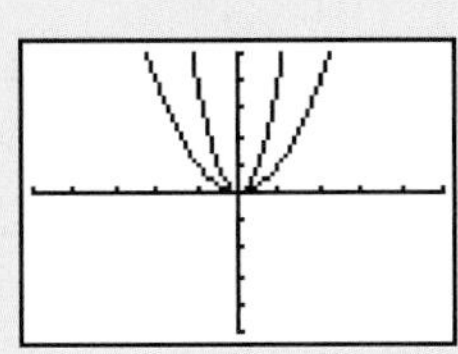

$[-5, 5, 1]$ by $[-5, 5, 1]$

For each value of x, $Y_2 = 4Y_1$. The parabola defined by $Y_2 = 4x^2$ can be obtained by vertically stretching by a factor of 4 the parabola defined by $Y_1 = x^2$.

2. (−2, 2) (0, 2) (3, −4)

3.

x	y_1	y_2
−2	4	20
−1	3	15
0	1	5
1	−2	−10
2	−5	−25

11.4.4

1.

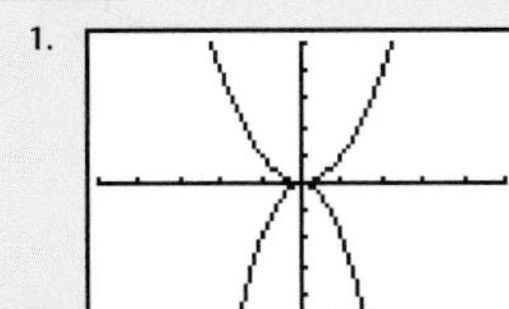

$[-5, 5, 1]$ by $[-5, 5, 1]$

2. The graph of $Y_2 = -2x^2$ can be obtained by vertically stretching the parabola defined by $Y_1 = x^2$ by a factor of 2 and then reflecting this graph across the x-axis.

USING THE LANGUAGE AND SYMBOLISM OF MATHEMATICS 11.4

1. Vertical and horizontal shifts of a graph create another graph whose shape is the __________ as that of the original graph and whose size is the __________ as that of the original graph.
2. The reflection of the point (x, y) across the x-axis is the point __________.
3. If $y = f(x)$ and $c > 1$, then we call c a __________ factor in the function $y = cf(x)$.
4. If $y = f(x)$ and $0 < c < 1$, then we call c a __________ factor in the function $y = cf(x)$.
5. If $y = f(x)$ and $c = -1$, then we call the graph of $y = cf(x)$ a __________ of the graph of $y = f(x)$ across the __________-axis.
6. If we examine the shapes of $y = f(x)$ and $y = 2f(x)$, then we will observe that the two graphs have the __________ basic shape but that $y = 2f(x)$ is obtained by vertically __________ $y = f(x)$ by a factor of __________.

QUICK REVIEW 11.4

The purpose of this quick review is to help you recall skills needed in this section.

1. Evaluate $f(2)$ for $f(x) = 3x^2 - 4$.
2. Evaluate $f(-2)$ for $f(x) = 3x^2 - 4$.

Graph each function on a graphing calculator by using the viewing window $[-5, 5, 1]$ by $[-5, 5, 1]$. Then determine the domain and range of each function.

3. $f(x) = |x|$
4. $f(x) = -|x|$
5. $f(x) = 0.5|x|$

EXERCISES 11.4

Exercises 1–8 give some basic functions and reflections of these functions. Match each graph to the correct function. All graphs are displayed with the window $[-5, 5, 1]$ by $[-5, 5, 1]$.

1. $f(x) = |x|$
2. $f(x) = -|x|$
3. $f(x) = x$
4. $f(x) = -x$
5. $f(x) = \sqrt{x}$
6. $f(x) = -\sqrt{x}$
7. $f(x) = x^3$
8. $f(x) = -x^3$

A.
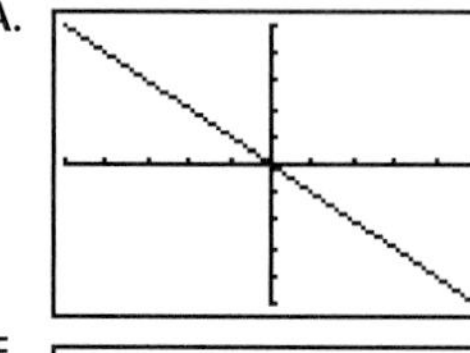

B.
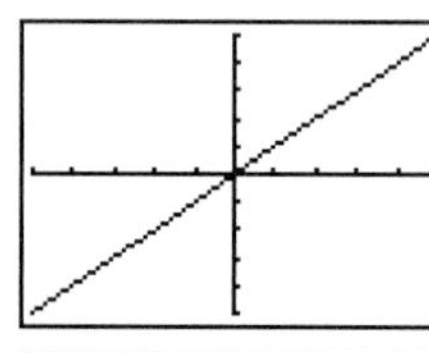

C.
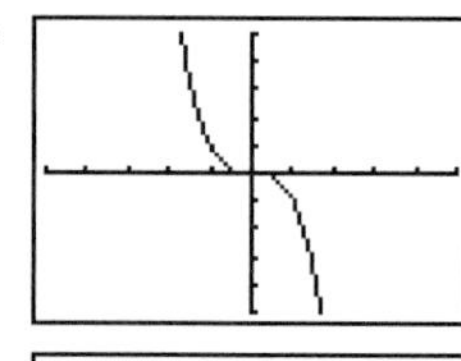

D.
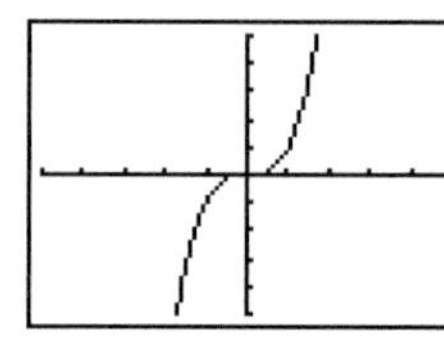

E.
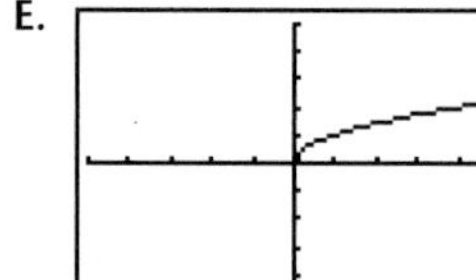

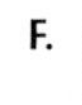
F.
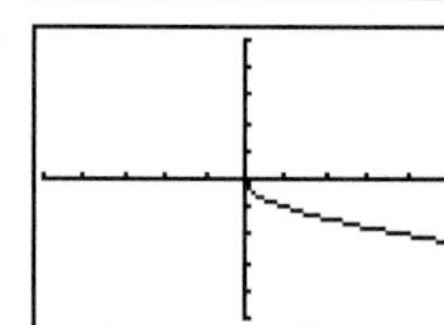

G.
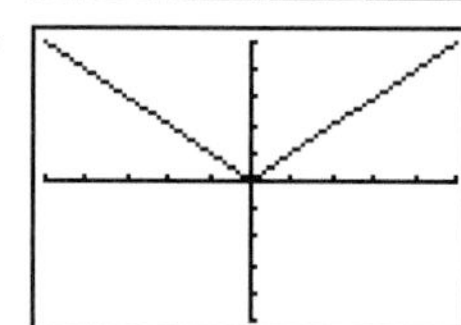

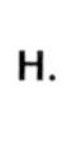
H.
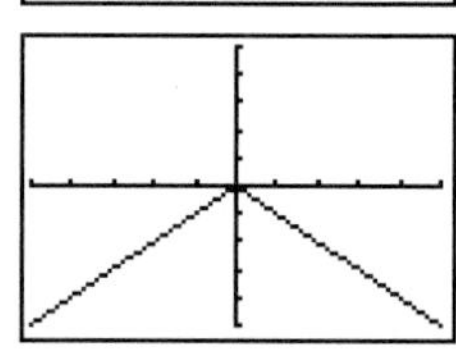

Exercises 9–12 give functions that are obtained by either stretching or shrinking the graph of $f(x) = x^2$. Match each graph to the correct function. All graphs are displayed with the window $[-5, 5, 1]$ by $[-5, 5, 1]$.

9. $f(x) = 2x^2$
10. $f(x) = 5x^2$
11. $f(x) = \frac{1}{2}x^2$
12. $f(x) = \frac{1}{5}x^2$

A.
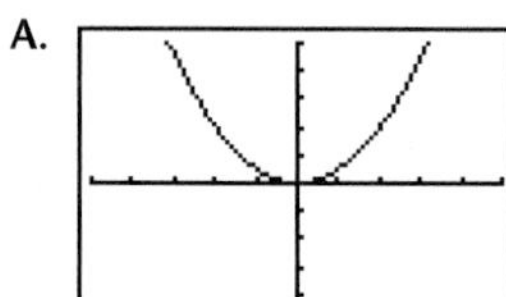

B.
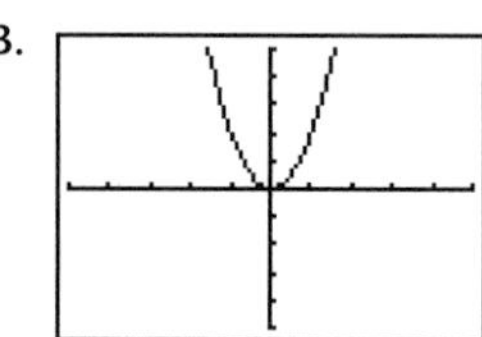

C.
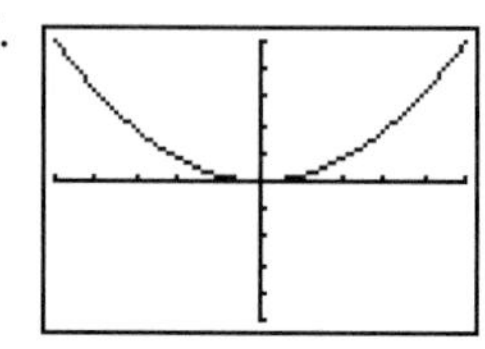

D.
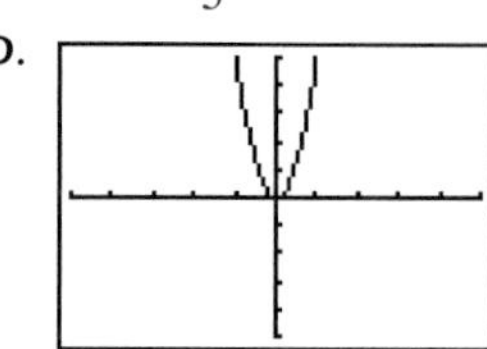

Exercises 13–16 give functions that are obtained by reflecting across the x-axis or stretching or shrinking the graph of $f(x) = x$. Match each graph to the correct function. All graphs are displayed with the window $[-5, 5, 1]$ by $[-5, 5, 1]$.

13. $f(x) = 5x$
14. $f(x) = \frac{1}{5}x$
15. $f(x) = -2x$
16. $f(x) = -\frac{1}{2}x$

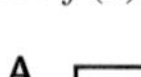
A.
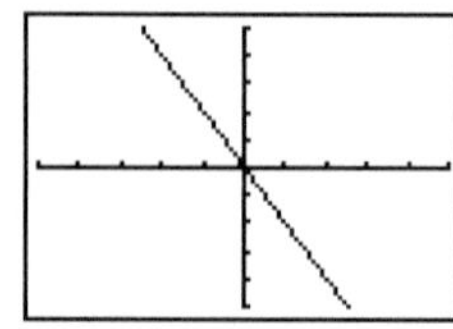

B.
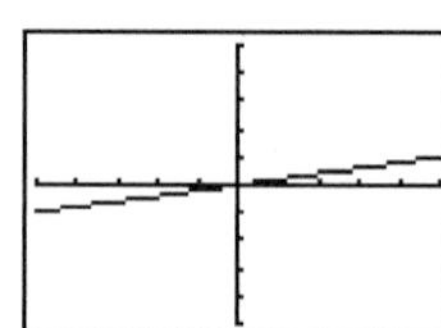

C.
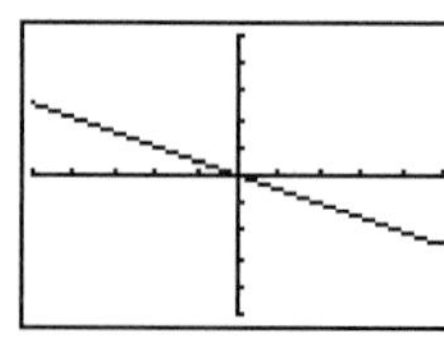

D.
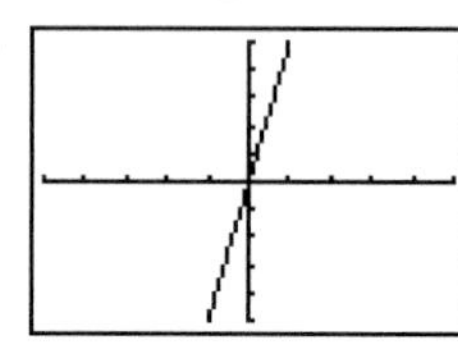

Exercises 17–20 give functions that are obtained by either stretching or shrinking the graph of $f(x) = |x|$. Match each graph to the correct function. All graphs are displayed with the window $[-5, 5, 1]$ by $[-5, 5, 1]$.

17. $f(x) = \frac{3}{4}|x|$
18. $f(x) = 3|x|$
19. $f(x) = \frac{3}{2}|x|$
20. $f(x) = \frac{1}{4}|x|$

A.
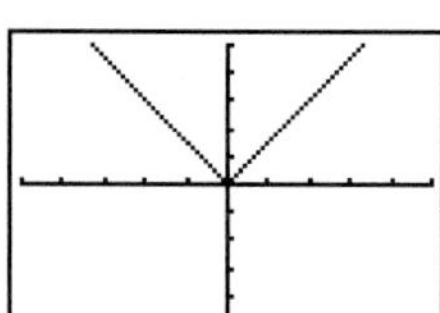

B.
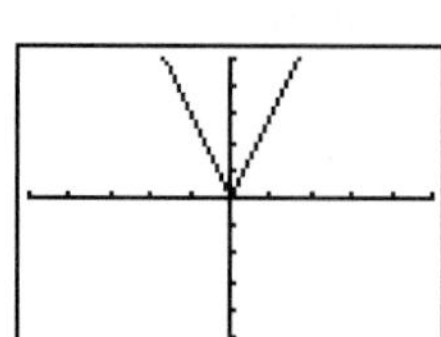

C.
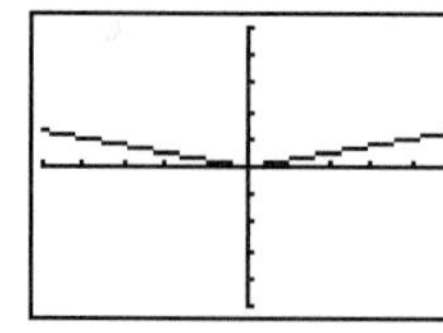

D.
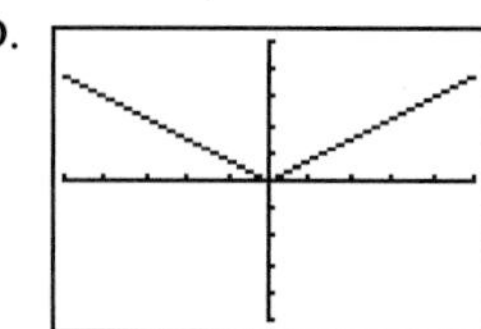

Exercises 21–24 give functions that are obtained by either stretching or shrinking the graph of $y = \sqrt{x}$. Match each graph to the correct function. All graphs are displayed with the window $[-1, 6, 1]$ by $[-1, 6, 1]$.

21. $f(x) = 4\sqrt{x}$

22. $f(x) = \frac{1}{4}\sqrt{x}$

23. $f(x) = \frac{3}{4}\sqrt{x}$

24. $f(x) = \frac{9}{4}\sqrt{x}$

A.

B.

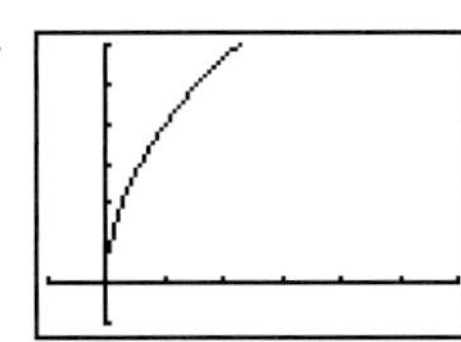

C.

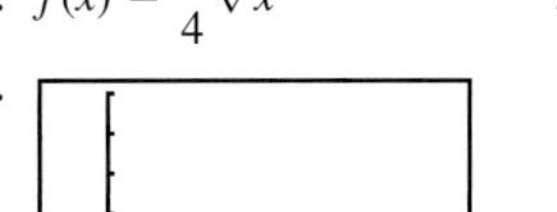

D.

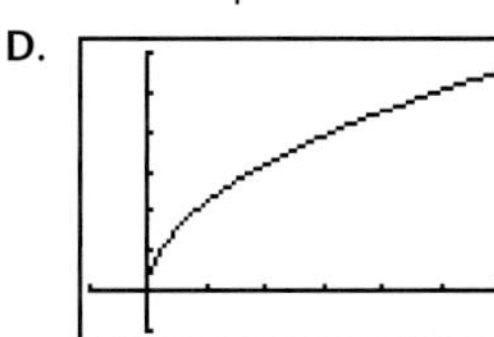

In Exercises 25–30, use the given graph of $y = f(x)$ to graph each function. (*Hint:* First graph the three key points on the new graph.)

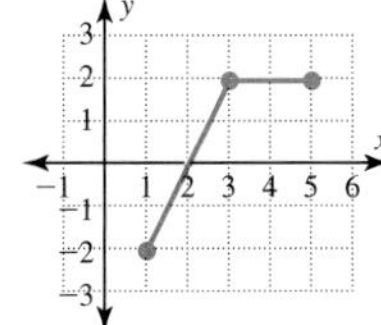

25. $y = -f(x)$

26. $y = \frac{1}{2}f(x)$

27. $y = 2f(x)$

28. $y = -3f(x)$

29. $y = f(x) + 2$

30. $y = f(x + 2)$

In Exercises 31–34, use the given table of values for $y = f(x)$ to complete each table.

x	$f(x)$
−2	18
−1	9
0	3
1	1
2	3

31.

x	$y = -f(x)$
−2	
−1	
0	
1	
2	

32.

x	$y = \frac{1}{3}f(x)$
−2	
−1	
0	
1	
2	

33.

x	$y = 2f(x)$
−2	
−1	
0	
1	
2	

34.

x	$y = -3f(x)$
−2	
−1	
0	
1	
2	

In Exercises 35–42, use the given table to write an equation for y_2 in terms of $f(x)$.

x	$y_1 = f(x)$
0	8
1	4
2	6
3	12
4	20
5	24

35.

x	y_2
0	4
1	2
2	3
3	6
4	10
5	12

36.

x	y_2
0	−8
1	−4
2	−6
3	−12
4	−20
5	−24

37.

x	y_2
0	−2
1	−1
2	−1.5
3	−3
4	−5
5	−6

38.

x	y_2
0	80
1	40
2	60
3	120
4	200
5	240

39.

x	y_2
0	0
1	−4
2	−2
3	4
4	12
5	16

40.

x	y_2
0	13
1	9
2	11
3	17
4	25
5	29

41.

x	y_2
−3	8
−2	4
−1	6
0	12
1	20
2	24

42.

x	y_2
2	8
3	4
4	6
5	12
6	20
7	24

In Exercises 43–50, determine the vertex of each parabola by using the fact that the vertex of $f(x) = x^2$ is at (0, 0). Use a graphing calculator to check your answer.

43. $f(x) = 8x^2$

44. $f(x) = \frac{1}{7}x^2$

45. $f(x) = 8x^2 - 7$

46. $f(x) = \frac{1}{7}(x - 7)^2$

47. $f(x) = (x - 8)^2 - 7$

48. $f(x) = (x + 6)^2 + 9$

49. $f(x) = -(x - 8)^2 + 7$

50. $f(x) = -(x + 6)^2 - 9$

51. Translate or reflect the graph of $f(x) = x^2$ as described by each equation.

a. Graph $f(x) = (x - 3)^2$, and describe how to obtain this graph from $f(x) = x^2$.

b. Graph $f(x) = -(x - 3)^2$, and describe how to obtain this graph from $f(x) = (x - 3)^2$.

c. Graph $f(x) = -(x - 3)^2 + 2$, and describe how to obtain this graph from $f(x) = -(x - 3)^2$.

52. Translate or reflect the graph of $f(x) = |x|$ as described by each equation.

a. Graph $f(x) = |x + 2|$, and describe how to obtain this graph from $f(x) = |x|$.

b. Graph $f(x) = -|x + 2|$, and describe how to obtain this graph from $f(x) = |x + 2|$.

c. Graph $f(x) = -|x + 2| - 3$, and describe how to obtain this graph from $f(x) = -|x + 2|$.

Exercises 53–58, describe a translation, a reflection, a stretching, or a shrinking of $y = f(x)$. Match each description to the correct function.

53. A vertical stretching of $y = f(x)$ by a factor of 7

54. A vertical shrinking of $y = f(x)$ by a factor of $\frac{1}{7}$

55. A reflection of $y = f(x)$ across the x-axis

56. A horizontal shift of $y = f(x)$ left 7 units

57. A horizontal shift of $y = f(x)$ right 7 units

58. A vertical shift of $y = f(x)$ up 7 units

A. $y = f(x + 7)$
B. $y = 7f(x)$
C. $y = \frac{1}{7}f(x)$
D. $y = -f(x)$
E. $y = f(x) + 7$
F. $y = f(x - 7)$

In Exercises 59–64, determine the range of each function, given that the range of $y = f(x)$ is [2, 6).

59. $y = f(x) + 2$ **60.** $y = f(x) - 2$

61. $y = 2f(x)$ **62.** $y = \frac{1}{2}f(x)$

63. $y = -f(x)$ **64.** $y = -3f(x)$

65. Write the first five terms of each sequence. (*Hint:* See Example 7 in Section 2.1.)
a. $a_n = n^3$ **b.** $a_n = -n^3$
c. $a_n = 2n^3$ **d.** $a_n = n^3 - 2$

66. Write the first five terms of each sequence. (*Hint:* See Example 7 in Section 2.1.)
a. $a_n = |n|$ **b.** $a_n = |n - 5|$
c. $a_n = -|n - 5|$ **d.** $a_n = 2|n - 5|$

67. Sequence of Retirement Bonuses As part of a bonus plan a company gives each secretary a number of shares in the company at retirement. The number of shares given equals the number of years the secretary has worked.
a. Write a formula for A_n, the sequence of the number of shares that would be given for a retirement after n years.
b. Write a formula for V_n, the value of the shares that would be given for a retirement after n years if the value of each share is \$50.

68. Comparing the Distance Traveled by Two Airplanes The formula $D = RT$ can be used to determine the distance D flown by an airplane flying at a rate R for time T. Complete the table if the rate of the second plane is double that of the first plane.

Plane 1

HOURS T	DISTANCE D
1	200
2	400
3	600
4	800

Plane 2

HOURS T	DISTANCE D
1	
2	
3	
4	

69. Comparing the Growth of Two Investments The formula $I = PRT$ can be used to determine the interest earned on an investment of P dollars for T years at simple interest rate R. A second investment of the same amount is invested at a rate that is three-fourths that of the first investment. Complete the table for the second investment.

Investment 1

YEARS T	INTEREST I
5	4,000
10	8,000
15	12,000
20	16,000

Investment 2

YEARS T	INTEREST I
5	
10	
15	
20	

70. Comparing the Production of Two Assembly Lines The following graph shows the number of lightbulbs produced by the workers at an assembly line during a very busy day shift. The night shift, which has fewer workers than the day shift, produces one-third as many bulbs as the day shift. Sketch the graph of the number of lightbulbs produced by the workers on the night shift.

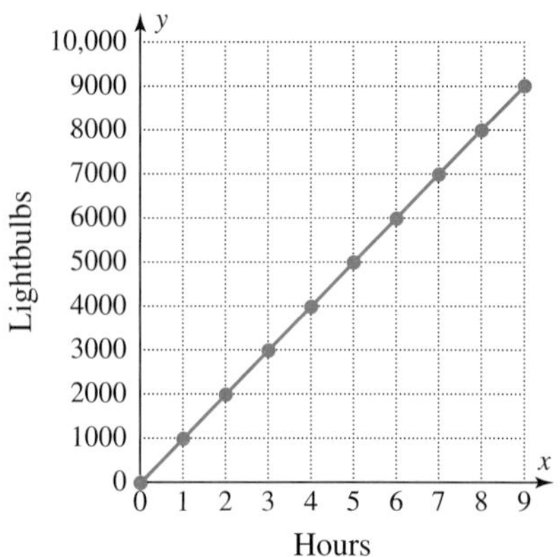

Group Discussion Questions

71. Discovery Question
a. Graph all the points in this table.
b. Sketch a parabola containing all these points.
c. Use your knowledge of translations, reflections, stretchings, and shrinkings to write the equation of this parabola.
d. Check your equation by using a graphing calculator.

x	y
−2	11
−1	4
0	−1
1	−4
2	−5
3	−4
4	−1

72. Discovery Question
a. Graph all the points in this table.
b. Sketch a parabola containing all these points.
c. Use your knowledge of translations, reflections, stretchings, and shrinkings to write the equation of this parabola.
d. Check your equation by using a graphing calculator.

x	y
−3	0.5
−2	−2
−1	−3.5
0	−4
1	−3.5
2	−2
3	0.5

73. Error Analysis A student graphing $y = 0.5|x|$ obtained the following display. Explain how you can tell by inspection that an error has been made.

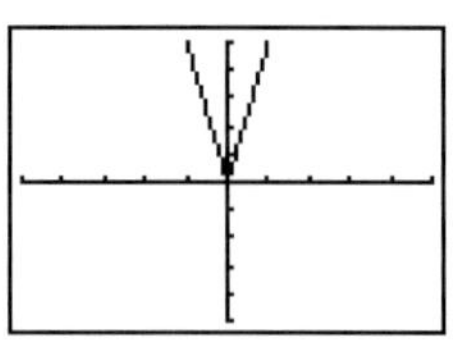

[−5, 5, 1] by [−5, 5, 1]

74. Error Analysis A student graphing $y = -3\sqrt{x}$ obtained the following display. Explain how you can tell by inspection that an error has been made.

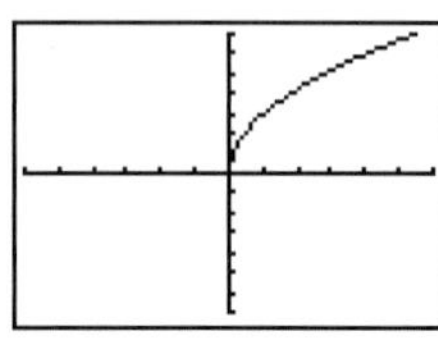

[−6, 6, 1] by [−7, 7, 1]

Section 11.5 Algebra of Functions

Objectives:

7. Add, subtract, multiply, and divide two functions.
8. Form the composition of two functions.

Problems, such as those in business, are often broken down into simpler components for analysis. For example, to determine the profit made by producing and selling an item, both the revenue and the cost must be known. Separate divisions of a business might be asked to find the revenue function and the cost function; the profit function would be found then by properly combining these two functions. We shall examine five ways to combine functions: sum, difference, product, quotient, and composition of functions.

We have added two real numbers to obtain another real number. We have added two polynomials to obtain another polynomial. Now we are adding two functions to obtain a new function.

The **sum of two functions** f and g, denoted by $f + g$, is defined as

$$(f + g)(x) = f(x) + g(x)$$

for all values of x that are in the domain of both f and g. Note that if either $f(x)$ or $g(x)$ is undefined, then $(f + g)(x)$ is also undefined.

EXAMPLE 1 Determining the Sum of Two Functions

Find the sum of $f(x) = x^2$ and $g(x) = 2$ algebraically, numerically, and graphically.

SOLUTION

ALGEBRAICALLY

$$(f + g)(x) = f(x) + g(x)$$
$$(f + g)(x) = x^2 + 2$$

The new function called $f + g$ is determined by adding $f(x)$ and $g(x)$. Substitute x^2 for $f(x)$ and 2 for $g(x)$.

NUMERICALLY

x	$f(x) = x^2$
−2	4
−1	1
0	0
1	1
2	4

x	$g(x) = 2$
−2	2
−1	2
0	2
1	2
2	2

x	$(f + g)(x) = x^2 + 2$
−2	6
−1	3
0	2
1	3
2	6

This table contains only a few values from the domain of input values from $\mathbb{R}$. These values illustrate that $(f + g)(x)$ can be determined by adding $f(x) + g(x)$ for each input x; $4 + 2 = 6$, $1 + 2 = 3$, $0 + 2 = 2$, $1 + 2 = 3$, and $4 + 2 = 6$.

GRAPHICALLY

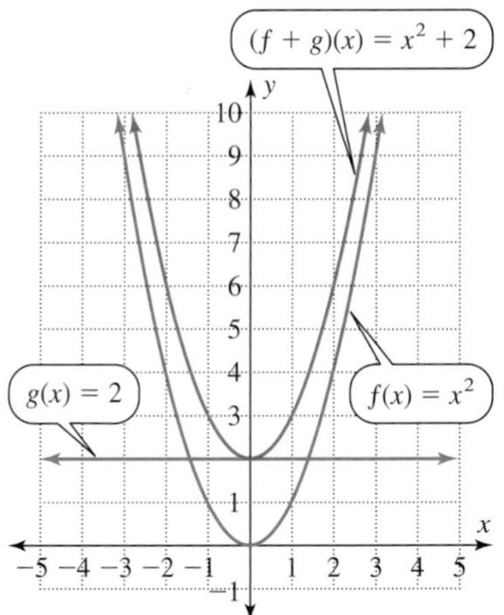

The graph of $f(x) = x^2$ is a parabola opening upward with its vertex at (0, 0).

The graph of $g(x) = 2$ is the horizontal line representing a constant output of 2.

The new function $f + g$ adds all output values. Since $g(x) = 2$, all output values of f are translated up 2 units when $g(x)$ is added to $f(x)$.

SELF-CHECK 11.5.1

1. Given $f(x) = x^2$ and $g(x) = -4x + 4$, determine $f + g$.

If two functions are equal, their input and output values are identical. If the domain of input values is the set of all real numbers, then we cannot list all the input-output pairs. However, a table of values can serve to check that two formulas yield the same values. Calculator Perspective 11.5.1 uses a table of values to check the result of $f + g$ in Example 1.

CALCULATOR PERSPECTIVE 11.5.1 Verifying the Sum of Two Functions

To verify the sum of the functions $f(x) = x^2$ and $g(x) = 2$ from Example 1 on a TI-84 Plus calculator, enter the following keystrokes:

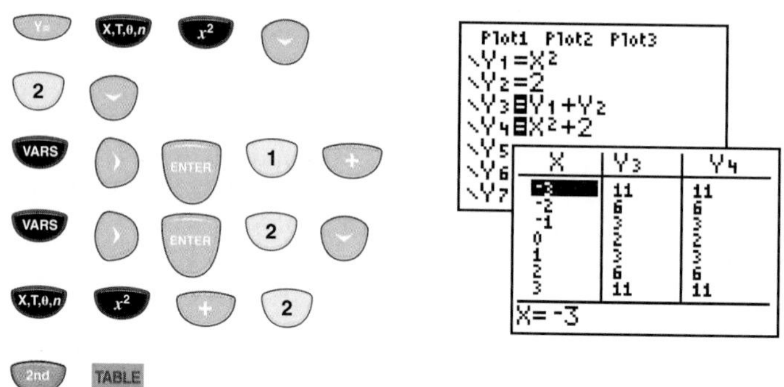

Note: Function Y_3 is formed by adding the functions Y_1 and Y_2. Function Y_4 is formed by adding x^2 and 2. Functions Y_3 and Y_4 are equal for each input value of x. Notice that Y_1 and Y_2 have been deselected on the Y= screen, so they do not appear in the table.

The **difference of two functions** f and g, denoted by $f - g$, is defined as

$$(f - g)(x) = f(x) - g(x)$$

for all values of x that are in the domain of both f and g.

EXAMPLE 2 Using the Difference of Two Functions to Model Profit

Suppose that the weekly revenue function for u units of a product sold is $R(u) = 5u^2 - 7u$ dollars, and the cost function for u units is $C(u) = 8u + 23$. Assume 0 is the fewest number of units that can be produced and 100 is the greatest number that can be marketed. Assuming profit can be determined by subtracting the cost from the revenue, find the profit function P and determine $P(4)$, the profit made by selling 4 units.

SOLUTION

VERBALLY

Profit = Revenue − Cost

First write a word equation to model this problem.

ALGEBRAICALLY

$P(u) = R(u) - C(u)$

$P(u) = (5u^2 - 7u) - (8u + 23)$ Substitute the given expressions for $R(u)$ and $C(u)$.

$P(u) = 5u^2 - 7u - 8u - 23$ The domain is $0 \le u \le 100$. (For some types of units, u may have to be an integer.)

$P(u) = 5u^2 - 15u - 23$ This is the profit function.

The profit on 4 units is

$P(4) = 5(4)^2 - 15(4) - 23$ Substitute 4 for x in the profit function P.

$P(4) = -3$

Thus \$3 will be lost if only 4 units are sold.

SELF-CHECK 11.5.2

Given $f(x) = 3x^2 - x + 5$ and $g(x) = 2x^2 + 4x + 7$, determine these two functions.

1. $(f + g)(x)$ **2.** $(f - g)(x)$

The operations of multiplication and division are defined similarly to addition except that the quotient $\frac{f}{g}$ is not defined if $g(x) = 0$.

Operations on Functions

	NOTATION	DEFINITION*	EXAMPLES: FOR $f(x) = x^2 - 4$ AND $g(x) = x - 2$:
Sum	$f + g$	$(f + g)(x) = f(x) + g(x)$	$(f + g)(x) = x^2 + x - 6$ for all real numbers
Difference	$f - g$	$(f - g)(x) = f(x) - g(x)$	$(f - g)(x) = x^2 - x - 2$ for all real numbers
Product	$f \cdot g$	$(f \cdot g)(x) = f(x)g(x)$	$(f \cdot g)(x) = x^3 - 2x^2 - 4x + 8$ for all real numbers
Quotient	$\frac{f}{g}$	$\left(\frac{f}{g}\right)(x) = \frac{f(x)}{g(x)}$	$\left(\frac{f}{g}\right)(x) = x + 2$ for all real numbers except 2

*The domain of all these functions except $\frac{f}{g}$ is the set of values in both the domain of f and the domain of g. For $\frac{f}{g}$, we also must have $g(x) \neq 0$.

EXAMPLE 3 Determining the Product and Quotient of Two Functions

Given $f(x) = x^2 + 5x$ and $g(x) = \frac{x + 5}{x}$, find the following.

SOLUTION

(a) $(f \cdot g)(x)$

$(f \cdot g)(x) = f(x) \cdot g(x)$

$(f \cdot g)(x) = (x^2 + 5x)\left(\frac{x + 5}{x}\right)$

$(f \cdot g)(x) = (x + 5)^2$

(b) The domain of $(f \cdot g)(x)$

f is defined for $\mathbb{R}$.

g is defined for $x \neq 0$.

$f \cdot g$ is defined for $x \neq 0$.

$(f \cdot g)(x)$ is defined only for values for which both f and g are defined.

(c) $\left(\frac{f}{g}\right)(x)$

$$\left(\frac{f}{g}\right)(x) = \frac{f(x)}{g(x)}$$

$$\left(\frac{f}{g}\right)(x) = \frac{\frac{x^2 + 5x}{x + 5}}{x}$$

$$\left(\frac{f}{g}\right)(x) = \frac{x(x + 5)}{1} \cdot \frac{x}{x + 5}$$

To simplify this complex rational expression, invert the divisor and multiply.

$$\left(\frac{f}{g}\right)(x) = x^2$$

(d) The domain of $\left(\frac{f}{g}\right)(x)$

f is defined for $\mathbb{R}$.

g is defined for $x \neq 0$.

$g(x) = 0$ for $x = -5$.

$\frac{f}{g}$ is defined for $x \neq 0$ and $x \neq -5$.

$\left(\frac{f}{g}\right)(x)$ is defined for values for which both f and g are defined and $g(x) \neq 0$.

SELF-CHECK 11.5.3

Given $f(x) = x^2 - 1$ and $g(x) = x^2 + 1$, determine these two functions.

1. $(f \cdot g)(x)$ **2.** $\left(\frac{f}{g}\right)(x)$

Functions are defined not only by their formulas, but also by the set of input values contained in the domains of the functions. We often allow the domain of a function to be implied by the formula. In this case, the domain is understood to be all real numbers for which the formula is defined and produces real number outputs. When we combine functions to produce new functions, we must take care that the formulas are used only for input values that are allowed for the new function. This is illustrated by Example 4. Two functions f and g are **equal** if the domain of f equals the domain of g and $f(x) = g(x)$ for each x in their common domain.

EXAMPLE 4 Comparing Functions to Determine If They Are Equal

Given $f(x) = x$, $g(x) = \frac{x^3 - 4x}{x^2 - 4}$, and $h(x) = \frac{x^3 + 4x}{x^2 + 4}$, determine whether the functions in (a) and (b) are equal.

SOLUTION

(a) $f = g$

$f(x) = x$	for all real x	
$g(x) = \frac{x^3 - 4x}{x^2 - 4}$	for $x \neq \pm 2$	Both -2 and 2 result in division by 0.
$g(x) = \frac{x(x^2 - 4)}{x^2 - 4}$	for $x \neq \pm 2$	
$g(x) = x$	for $x \neq \pm 2$	

$f \neq g$ because f is defined for $x = -2$ and $x = 2$ but g is not. $f(x)$ and $g(x)$ have the same values for all real numbers except $x = -2$ and $x = 2$.

(b) $f = h$

$f(x) = x$	for all real x	
$h(x) = \dfrac{x^3 + 4x}{x^2 + 4}$	for all real x	There are no real numbers for which the denominator of h is 0.
$h(x) = \dfrac{x(x^2 + 4)}{x^2 + 4}$	for all real x	
$h(x) = x$	for all real x	

$f = h$ because these functions produce the same output for each real number x.

As shown in the following figure, the graphs of the three functions f, g, and h in Example 4 are nearly identical. The only difference is that the graph of g has "holes" at $x = -2$ and $x = 2$ because these values are not in its domain.

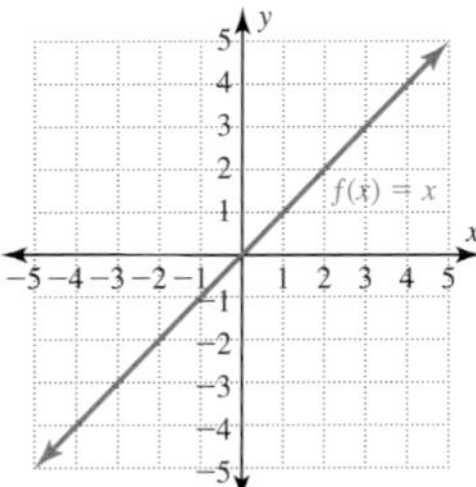

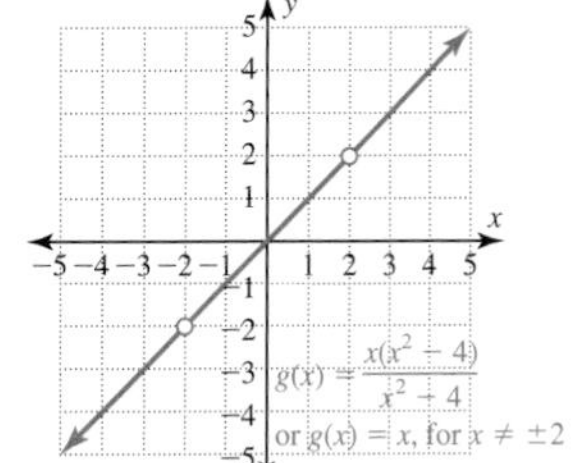

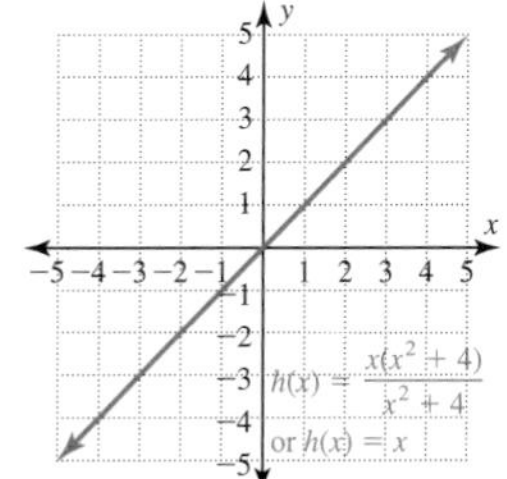

Functions, especially in formula form, are a powerful means of describing the relationship between two quantities. We can further amplify this power by "chaining" two functions together. This way of combining two functions is called **composition.**

Caution: The symbol $\circ$ denoting the composition of two functions, written $f \circ g$, should not be confused with the symbol for the multiplication of two functions $f \cdot g$.

Composite Function $f \circ g$

ALGEBRAICALLY	VERBALLY	ALGEBRAIC EXAMPLE
$(f \circ g)(x) = f[g(x)]$ The domain of $f \circ g$ is the set of x values from the domain of g for which $g(x)$ is in the domain of f.	$f \circ g$ denotes the composition of function f with function g. $f \circ g$ is read "f composed with g."	For $f(x) = x + 1$ and $g(x) = 2x$: $(f \circ g)(x) = f[g(x)]$ $= f(2x)$ $= 2x + 1$

Functions can be examined by using mapping notation, ordered-pair notation, tables, graphs, and function notation. In each case, to evaluate $(f \circ g)(x)$, we first evaluate $g(x)$ and then apply f to this result.

EXAMPLE 5 Determining the Composition of Two Functions

Find $f \circ g$ for the given functions f and g.

SOLUTION

g	f	VERBALLY	MAPPING NOTATION $f \circ g$	NUMERICAL EXAMPLE $f \circ g$
$6 \to 3$	$3 \to 9$	g maps 6 to 3; then f maps 3 to 9.	$6 \xrightarrow{g} 3 \xrightarrow{f} 9$	$6 \to 9$
$5 \to 4$	$4 \to 7$	g maps 5 to 4; then f maps 4 to 7.	$5 \xrightarrow{g} 4 \xrightarrow{f} 7$	$5 \to 7$
$-1 \to 0$	$0 \to 2$	g maps -1 to 0; then f maps 0 to 2.	$-1 \xrightarrow{g} 0 \xrightarrow{f} 2$	$-1 \to 2$

EXAMPLE 6 Determining the Composition of Two Functions

Given $f(x) = x^2$ and $g(x) = 3x$, evaluate these expressions.

SOLUTION

(a) $(f \circ g)(4)$

$$\begin{aligned}(f \circ g)(4) &= f[g(4)] = f[3(4)]\\ &= f(12)\\ &= (12)^2\\ &= 144\end{aligned}$$

First apply the formula for $g(x)$. Evaluate $g(4) = 3(4)$. Then apply the formula for $f(x)$ to 12. Evaluate $f(12) = 12^2$.

(b) $(g \circ f)(4)$

$$\begin{aligned}(g \circ f)(4) &= g[f(4)] = g(4^2)\\ &= g(16)\\ &= 3(16)\\ &= 48\end{aligned}$$

First apply the formula for $f(x)$. Evaluate $f(4) = 4^2$. Then apply the formula for $g(x)$ to 16. Evaluate $g(16) = 3(16)$.

(c) $(f \circ g)(x)$

$$\begin{aligned}(f \circ g)(x) &= f[g(x)] = f(3x)\\ &= (3x)^2\\ &= 9x^2\end{aligned}$$

First apply the formula for g to x. Then apply the formula for f to $3x$.

Note that in this example $f \circ g \neq g \circ f$.

(d) $(g \circ f)(x)$

$$\begin{aligned}(g \circ f)(x) &= g[f(x)] = g(x^2)\\ &= 3(x^2)\\ &= 3x^2\end{aligned}$$

First apply the formula for f to x. Then apply the formula for g to x^2.

In Example 6, $f \circ g \neq g \circ f$. Although $f \circ g$ can equal $g \circ f$ in special cases, in general the order in which we perform composition of functions is important.

SELF-CHECK 11.5.4

Given $f(x) = 3x - 1$ and $g(x) = \dfrac{x+1}{3}$, evaluate these expressions.

1. $(f + g)(2)$
2. $(f - g)(2)$
3. $(f \cdot g)(-1)$
4. $\left(\dfrac{f}{g}\right)(1)$
5. $(f \circ g)(2)$
6. $(g \circ f)(2)$

The functions $f(x) = 3x - 1$ and $g(x) = \dfrac{x+1}{3}$ from Self-Check 11.5.4 are inverses of each other. Note that

$(f \circ g)(2) = 2 \quad 2 \xrightarrow{g} 1 \xrightarrow{f} 2$ and

$(g \circ f)(2) = 2 \quad 2 \xrightarrow{f} 5 \xrightarrow{g} 2$

For any input value x,

$$(f \circ g)(x) = f[g(x)] = f\left(\frac{x+1}{3}\right) = 3\left(\frac{x+1}{3}\right) - 1 = (x+1) - 1 = x$$

The inverse of the function f reverses the order of each ordered pair (x, y) of f.

Likewise, $(g \circ f)(x) = x$. Thus the composition of functions gives us another way to characterize the inverse of a function.

Inverse of a Function

ALGEBRAICALLY	EXAMPLE
The functions f and f^{-1} are inverses of each other if and only if $(f \circ f^{-1})(x) = x$ for each input value of f^{-1}	$f(x) = 3x - 1$ and $f^{-1}(x) = \dfrac{x+1}{3}$ are inverses because $(f \circ f^{-1})(x) = f\left(\dfrac{x+1}{3}\right)$ $(f \circ f^{-1})(x) = 3\left(\dfrac{x+1}{3}\right) - 1$ $(f \circ f^{-1})(x) = x$
and $(f^{-1} \circ f)(x) = x$ for each input value of f	and $(f^{-1} \circ f)(x) = f^{-1}(3x - 1)$ $(f^{-1} \circ f)(x) = \dfrac{(3x-1)+1}{3}$ $(f^{-1} \circ f)(x) = x$

CALCULATOR PERSPECTIVE 11.5.2 Composing a Function with Its Inverse

To numerically check that $(f \circ g)(x) = x$ for the functions $f(x) = 3x - 4$ and $g(x) = \dfrac{x+4}{3}$ on a TI-84 Plus calculator, enter the following keystrokes:

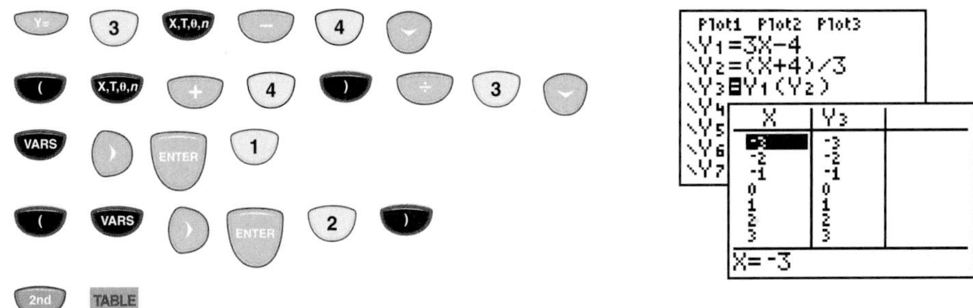

Note: $Y_1(Y_2)$ represents $f[g(x)]$. Observe that Y_3 has the same values as x, thus confirming that $(f \circ g)(x) = x$.

Example 7 illustrates how a problem can be broken down into pieces with individual functions as mathematical models. Then the overall relationship can be modeled by a composite function.

EXAMPLE 7 Composing Cost and Production Functions

The quantity of items a factory can produce weekly is a function of the number of hours it operates. For one company this is given by $q(t) = 40t$ for $0 \le t \le 168$. The dollar cost of manufacturing these items is a function of the quantity produced; in this case, $C(q) = q^2 - 40q + 750$ for $q \ge 0$. Evaluate and interpret the following expressions.

SOLUTION

(a) $q(8)$

$q(t) = 40t$

$q(8) = 40(8)$ — Substitute 8 into the formula for $q(t)$.

$q(8) = 320$

320 units can be produced in 8 hours.

(b) $C(320)$

$C(q) = q^2 - 40q + 750$
$C(320) = (320)^2 - 40(320) + 750$ Substitute 320 into the formula for $C(q)$.
$C(320) = 90{,}350$
\$90,350 is the cost of manufacturing 320 units.

(c) $(C \circ q)(8)$

$(C \circ q)(8) = C[q(8)]$ Substitute for $q(8)$ from part (a).
$(C \circ q)(8) = C(320)$ Substitute for $C(320)$ from part (b).
$(C \circ q)(8) = \$90{,}350$
\$90,350 is the cost of 8 hours of production.

(d) $(C \circ q)(t)$

$(C \circ q)(t) = C[q(t)]$
$(C \circ q)(t) = C[40t]$ Substitute $40t$ for $q(t)$. Then evaluate C for $40t$.
$(C \circ q)(t) = (40t)^2 - 40(40t) + 750$
$(C \circ q)(t) = 1{,}600t^2 - 1{,}600t + 750$
This is the cost of t hours of production.

Example 8 illustrates the analysis of a geometric problem by using a composite function.

EXAMPLE 8 Composing Area and Length Functions

A piece of wire 20 m long is cut into two pieces. The length of the shorter piece is s m, and the length of the longer piece is L m. The longer piece is then bent into the shape of a square of area A m^2.

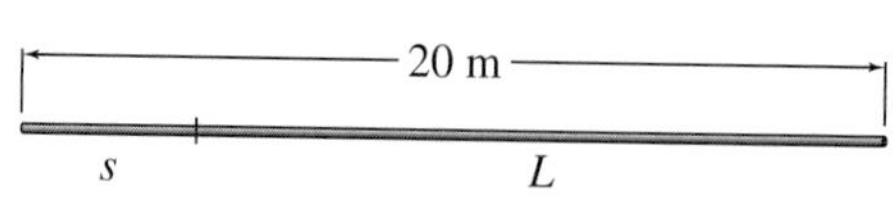

(a) Express the length L as a function of s.
(b) Express the area A as a function of L.
(c) Express the area A as a function of s.

SOLUTION

(a) $\begin{pmatrix}\text{Length of the}\\ \text{longer piece}\end{pmatrix} = \begin{pmatrix}\text{Total}\\ \text{length}\end{pmatrix} - \begin{pmatrix}\text{Length of the}\\ \text{shorter piece}\end{pmatrix}$ First write a word equation that describes the relationship between the length of the longer piece and the length of the shorter piece.

$L(s) = 20 - s$ This equation expresses the length as a function of s.

(b) Area = square of length of one side

$A(L) = \left(\dfrac{L}{4}\right)^2$ The length of one side of the square is one-fourth the perimeter.

(c) $(A \circ L)(s) = A[L(s)]$ The composition $A \circ L$ expresses A as a function of s.
$(A \circ L)(s) = A(20 - s)$ Substitute $20 - s$ for $L(s)$ from part (a).
$(A \circ L)(s) = \left(\dfrac{20 - s}{4}\right)^2$ Evaluate A by using the formula from part (b).

An important skill in calculus is the ability to take a given function and decompose it into simpler components.

EXAMPLE 9 Decomposing Functions into Simpler Components

Express each of these functions in terms of $f(x) = 2x - 3$ and $g(x) = \sqrt{x}$.

SOLUTION

(a) $h(x) = \sqrt{2x - 3}$

$h(x) = \sqrt{2x - 3} = \sqrt{f(x)}$ First substitute $f(x)$ for $2x - 3$. Then replace the square root function with the g function.
$= g[f(x)]$
$= (g \circ f)(x)$

(b) $h(x) = 2\sqrt{x} - 3$

$h(x) = 2\sqrt{x} - 3 = 2g(x) - 3$ First substitute $g(x)$ for $\sqrt{x}$. Then rewrite the expression by using the definition of the f function.
$= f[g(x)]$
$= (f \circ g)(x)$

SELF-CHECK 11.5.5

1. Express $h(x) = \dfrac{1}{x-6}$ in terms of $f(x) = x - 6$ and $g(x) = \dfrac{1}{x}$.

SELF-CHECK ANSWERS

11.5.1

1. $(f+g)(x) = x^2 - 4x + 4$

11.5.2

1. $(f+g)(x) = 5x^2 + 3x + 12$
2. $(f-g)(x) = x^2 - 5x - 2$

11.5.3

1. $(f \cdot g)(x) = x^4 - 1$
2. $\left(\dfrac{f}{g}\right)(x) = \dfrac{x^2-1}{x^2+1}$

11.5.4

1. 6
2. 4
3. 0
4. 3
5. 2
6. 2

11.5.5

1. $h(x) = (g \circ f)(x)$

USING THE LANGUAGE AND SYMBOLISM OF MATHEMATICS 11.5

1. The sum of two functions f and g, denoted by $f + g$, is defined as $(f+g)(x) =$ __________ for all values of x that are in the domain of both f and g.
2. The difference of two functions f and g, denoted by $f - g$, is defined as $(f-g)(x) =$ __________ for all values of x that are in the domain of both f and g.
3. The product of two functions f and g, denoted by $f \cdot g$, is defined as $(f \cdot g)(x) =$ __________ for all values of x that are in the domain of both f and g.
4. The quotient of two functions f and g, denoted by $\dfrac{f}{g}$, is defined as $\left(\dfrac{f}{g}\right)(x) =$ __________ for all values of x that are in the domain of both f and g, provided $g(x) \neq$ __________.
5. The composition of the function f with the function g, denoted by $f \circ g$, is defined by $(f \circ g)(x) =$ __________.
6. The domain of $f \circ g$ is the set of x values from the domain of g for which __________ is in the domain of __________.
7. The functions f and f^{-1} are inverses of each other if and only if $(f \circ f^{-1})(x) =$ __________ for each input value from the domain of __________ and $(f^{-1} \circ f)(x) =$ __________ for each input value from the domain of __________.

QUICK REVIEW 11.5

The purpose of this quick review is to help you recall skills needed in this section.

1. Add $(2x^2 - 7x - 15) + (2x + 3)$.
2. Subtract $(2x^2 - 7x - 15) - (2x + 3)$.
3. Multiply $(2x^2 - 7x - 15)(2x + 3)$.
4. Divide $(2x^2 - 7x - 15) \div (2x + 3)$.
5. Find the inverse of the function $f(x) = 2x + 3$.

EXERCISES 11.5

In Exercises 1–4, evaluate each expression, given $f(x) = x^2 - 1$ and $g(x) = 2x + 5$.

1. a. $(f+g)(2)$
   b. $(f-g)(2)$
   c. $(f \cdot g)(2)$
   d. $\left(\dfrac{f}{g}\right)(2)$
2. a. $(f+g)(-3)$
   b. $(f-g)(-3)$
   c. $(f \cdot g)(-3)$
   d. $\left(\dfrac{f}{g}\right)(-3)$
3. a. $(f \circ g)(2)$
   b. $(g \circ f)(2)$
   c. $(f \circ f)(2)$
   d. $(g \circ g)(2)$
4. a. $(f \circ g)(-3)$
   b. $(g \circ f)(-3)$
   c. $(f \circ f)(-3)$
   d. $(g \circ g)(-3)$

In Exercises 5–10, use the given tables for f and g to form a table of values for each function.

x	f(x)
0	−2
3	0
8	7

x	g(x)
0	−1
3	5
8	9

5. $f + g$
6. $f - g$
7. $g - f$
8. $f \cdot g$
9. $g \cdot f$
10. $\dfrac{f}{g}$

In Exercises 11–14, use $f = \{(-2, 3), (1, 5), (4, 7)\}$ and $g = \{(-2, 4), (1, -1), (4, 6)\}$ to form a set of ordered pairs for each function.

11. $f - g$ **12.** $f + g$ **13.** $\dfrac{f}{g}$ **14.** $f \cdot g$

In Exercises 15–18, use the given graphs for f and g to graph each function. (*Hint:* You first may want to write each function as a set of ordered pairs.)

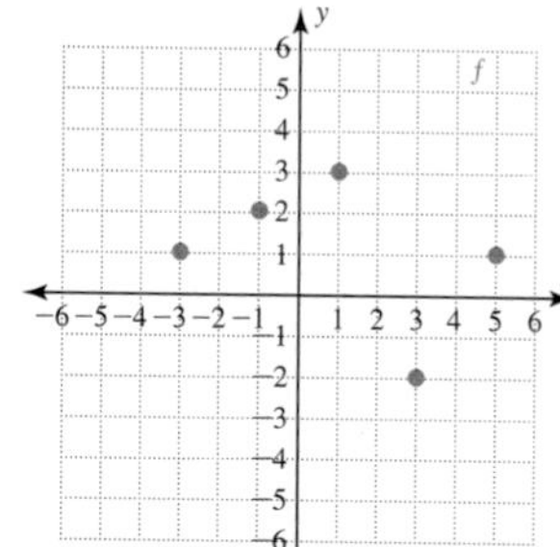
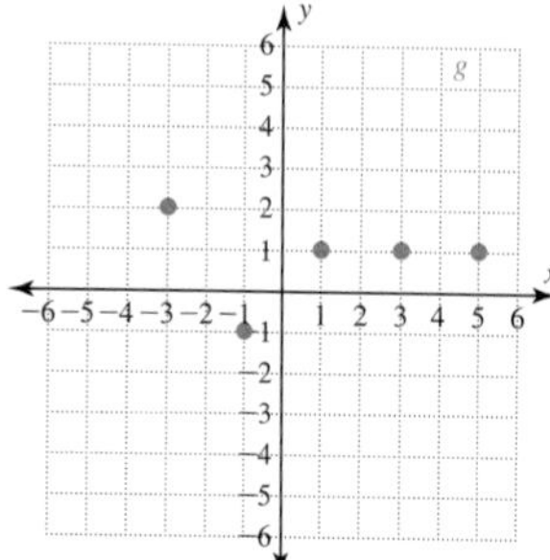

15. $f + g$ **16.** $f - g$ **17.** $f \cdot g$ **18.** $\dfrac{f}{g}$

In Exercises 19–22, determine $f + g, f - g, f \cdot g, \dfrac{f}{g}$, and $f \circ g$ for the given functions. State the domain of the resulting function.

19. $f(x) = 2x^2 - x - 3$
$g(x) = 2x - 3$

20. $f(x) = 6x^2 - x - 15$
$g(x) = 3x - 5$

21. $f(x) = \dfrac{x}{x + 1}$
$g(x) = \dfrac{1}{x}$

22. $f(x) = 5x - 7$
$g(x) = 3$

In Exercises 23–28, determine if the functions f and g are equal. If $f \neq g$, state the reason.

23. $f(x) = x - 3$
$g(x) = \dfrac{x^2 - 5x + 6}{x - 2}$

24. $f(x) = 1 - \dfrac{1}{x}$
$g(x) = \dfrac{x - 1}{x}$

25. $f = \{(-1, 1), (0, 0), (1, 1), (2, 4)\}$
$g(x) = x^2$

26. $f = \{(-7, 7), (0, 0), (8, 8)\}$
$g(x) = |x|$

27. $f(x) = x$
$g(x) = \dfrac{x^3 + x}{x^2 + 1}$

28. $f(x) = 5$
$g(x) = \dfrac{5x^2 + 15}{x^2 + 3}$

29. Use the given tables for f and g to complete the tables for $f \circ g$.

$x \to g(x)$	$x \to f(x)$	$x \longrightarrow g(x) \longrightarrow f[g(x)]$	$x \to f[g(x)]$
$3 \to 5$	$5 \to 9$	$3 \longrightarrow 5 \longrightarrow 9$	$3 \to 9$
$4 \to 2$	$2 \to 8$	$4 \longrightarrow 2 \longrightarrow ?$	$? \to ?$
$7 \to 1$	$1 \to 0$	$7 \longrightarrow ? \longrightarrow ?$	$? \to ?$
$8 \to 4$	$4 \to 6$	$8 \xrightarrow{g} ? \xrightarrow{f} ?$	$? \to ?$

30. Use $f = \{(1, 3), (4, 5), (6, 2)\}$ and $g = \{(3, 6), (5, 1), (2, 4)\}$ to form a set of ordered pairs for
a. $f \circ g$ **b.** $g \circ f$

31. Use the given graphs of f and g to graph $f \circ g$. (*Hint:* You first may want to write each function as a set of ordered pairs.)

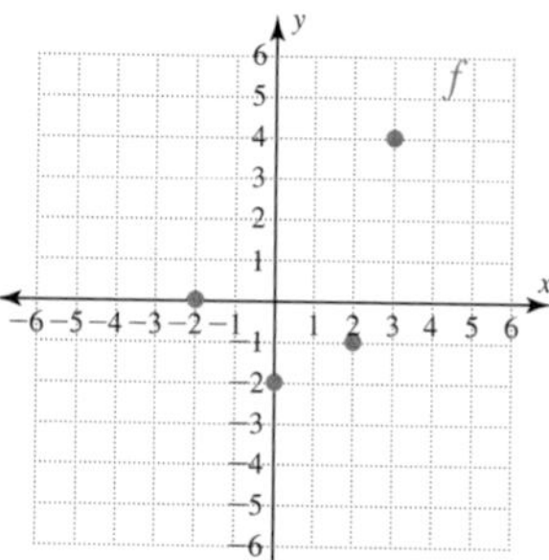
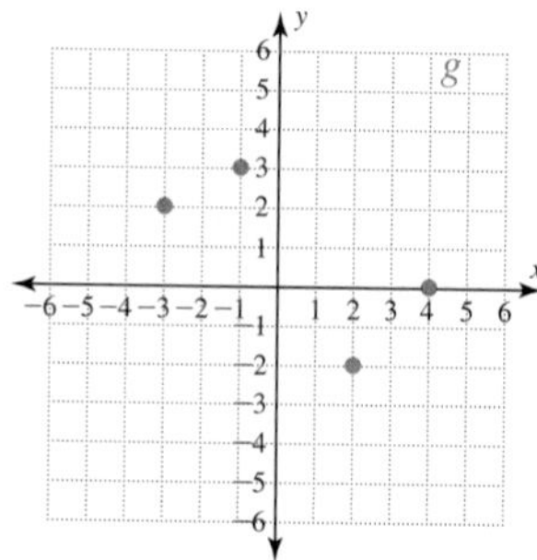

In Exercises 32–37, determine $f \circ g$ and $g \circ f$.

32. $f(x) = 3x - 4$
$g(x) = 4x + 3$

33. $f(x) = x^2 - 5x + 3$
$g(x) = 4x - 2$

34. $f(x) = |x|$
$g(x) = x - 8$

35. $f(x) = \sqrt{x}$
$g(x) = x + 5$

36. $f(x) = \dfrac{1}{x}$
$g(x) = x^3 + 1$

37. $f(x) = \dfrac{1}{x + 2}$
$g(x) = x^2 - 4$

In Exercises 38 and 39, use the graphs of f and g to graph $f + g$.

38. **39.**

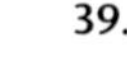

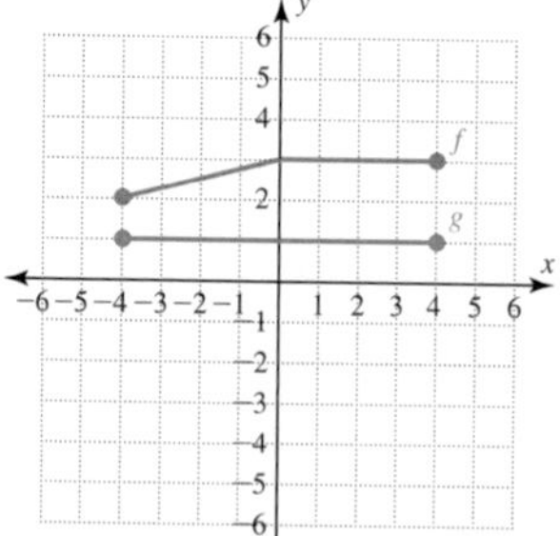
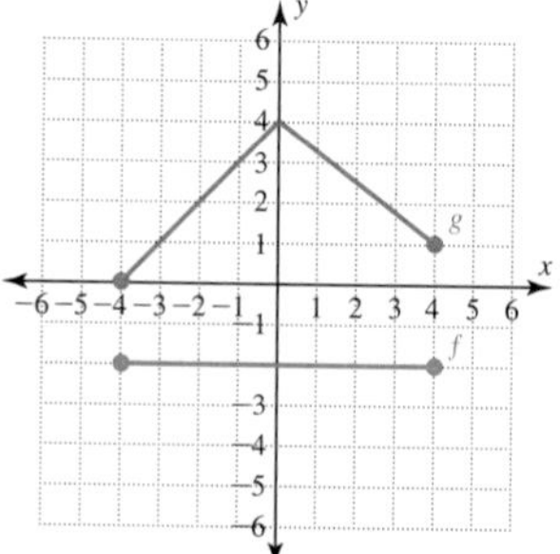

40. Profit as the Difference of Revenue Minus Cost The weekly revenue function for u units of a product sold is $R(u) = 4u^2 - 3u$ dollars, and the cost function for u units is $C(u) = 10u + 25$. Assume 0 is the least number of units that can be produced and 100 is the greatest number that can be marketed. Find the profit function P, and find $P(10)$, the profit made by selling 10 units.

41. Determining a Formula for Total Cost and Average Cost The fixed monthly cost F (rent, insurance, etc.) of a manufacturer is \$5,000. The variable cost (labor, materials, etc.) for producing u units is given by $V(u) = u^2 + 5u$ for $0 \leq u \leq 1{,}200$. The total cost of producing u units is $C(u) = F(u) + V(u)$. Determine

a. $F(500)$
b. $V(500)$
c. $C(500)$
d. The average cost of producing 500 units
e. A formula for $C(u)$
f. A formula for $A(u)$, the average cost of producing u units

42. Determining a Formula for Total Cost and Average Cost A manufacturer produces circuit boards for the electronics industry. The fixed cost F associated with this production is \$3,000 per week, and the variable cost V is

\$10 per board. The circuit boards produce revenue of \$12 each. Determine

a. $V(b)$, the variable cost of producing b boards per week
b. $F(b)$, the fixed cost of producing b boards per week
c. $C(b)$, the total cost of producing b boards per week
d. $A(b)$, the average cost of producing b boards per week
e. $R(b)$, the revenue from selling b boards per week
f. $P(b)$, the profit from selling b boards per week
g. $P(1{,}000)$
h. $P(1{,}500)$
i. $P(2{,}000)$
j. The break-even value (the value producing a profit of \$0)

43. **Determining the Product of a Price Function and a Demand Function** The number of items demanded by consumers is a function of the number of months that the product has been advertised. The price per item is varied each month as part of the marketing strategy. The number demanded during the mth month is $N(m) = 36m - m^2$, and the price per item during the mth month is $P(m) = 5m + 45$. The revenue for the mth month $R(m)$ is the product of the price per item and the number of items demanded. Determine
a. $N(7)$ b. $P(7)$
c. $R(7)$ d. $R(m)$

44. **Composing Cost and Production Functions** The number of sofas a factory can produce weekly is a function of the number of hours t it operates. This function is $S(t) = 5t$ for $0 \le t \le 168$. The cost of manufacturing s sofas is given by $C(s) = s^2 - 6s + 500$ for $s \ge 0$. Evaluate and interpret the following.
a. $S(10)$ b. $C(50)$
c. $(C \circ S)(10)$ d. $(C \circ S)(t)$
e. $(C \circ S)(40)$ f. $(C \circ S)(100)$

45. **Composing Cost and Markup Functions** The weekly cost C of making d doses of a vaccine is $C(d) = 0.30d + 400$. The company charges 150% of cost to its wholesaler for this drug; that is, $R(C) = 1.5C$. Evaluate and interpret the following.
a. $C(5{,}000)$ b. $R(1{,}900)$
c. $(R \circ C)(5{,}000)$ d. $(R \circ C)(d)$

46. **Composing an Area and a Radius Function** A circular concrete pad was poured to serve as the base for a grain bin. This pad was inscribed in a square plot as shown.

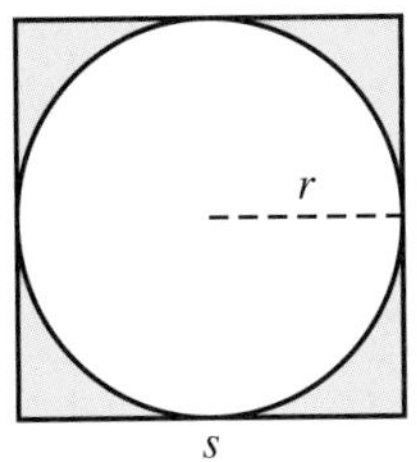

a. Express the radius r of this circle as a function of the length s of a side of the square.
b. Express the area A of this circle as a function of the radius r.
c. Determine $(A \circ r)(s)$ and interpret this result.

47. **Composing an Area and a Width Function** A border of uniform width x is trimmed from all sides of the square poster board as shown.

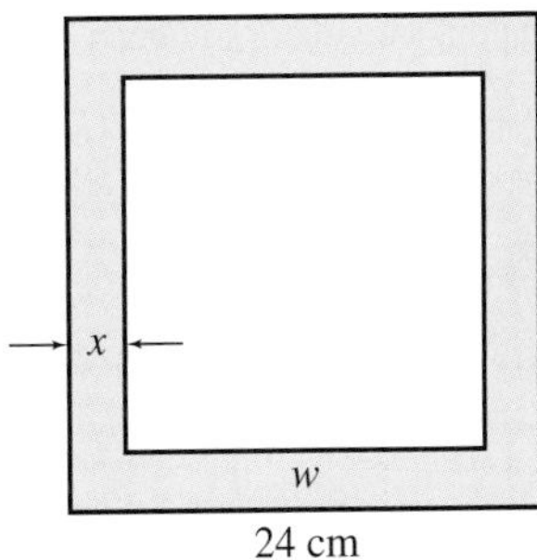

a. Express the area A of the border that is removed as a function of the remaining width w.
b. Express the remaining width w as a function of x.
c. Determine $(A \circ w)(x)$ and interpret this function.

48. **Composing an Area and a Width Function** A metal box with an open top can be formed by cutting squares of sides x cm from each corner of a square piece of sheet metal of width 44 cm, as shown.

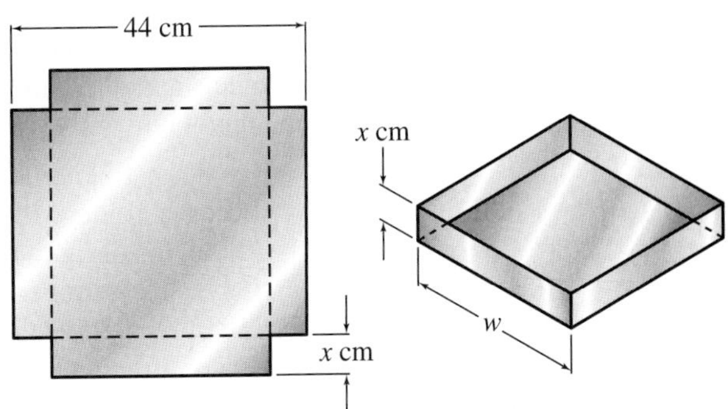

a. Express the area of the base of this box as a function of its width w.
b. Express w as a function of x.
c. Determine $(A \circ w)(x)$ and interpret this function.
d. Express the volume V of the box as a function of x.

In Exercises 49 and 50, determine both $(f \circ f^{-1})(x)$ and $(f^{-1} \circ f)(x)$.

49. $f(x) = 2x - 3$
$f^{-1}(x) = \dfrac{x + 3}{2}$

50. $f(x) = \dfrac{2x - 1}{3}$
$f^{-1}(x) = \dfrac{3x + 1}{2}$

51. Use $f(x) = 4x + 5$ to determine
a. $f^{-1}(x)$ b. $(f \circ f^{-1})(x)$
c. $(f^{-1} \circ f)(x)$

In Exercises 52–55, express each function in terms of $f(x) = x^2 + 1$ and $g(x) = \sqrt{x}$.

52. $h(x) = x^2 + \sqrt{x} + 1$
53. $h(x) = x + 1, x \ge 0$
54. $h(x) = \sqrt{x^2 + 1}$
55. $h(x) = \dfrac{\sqrt{x}}{x^2 + 1}$

In Exercises 56–59, decompose each function into functions f and g such that $h(x) = (f \circ g)(x)$. (Answers may vary.)

56. $h(x) = \sqrt[3]{x^2 + 4}$

57. $h(x) = \dfrac{1}{3x^2 - 7x + 9}$

58. $h(x) = (x + 2)^2 + 3(x + 2) + 5$

59. $h(x) = x^3 - 2 + \dfrac{1}{x^3 - 2}$

In Exercises 60 and 61, use a graphing calculator to obtain the graph of $y = (f \circ g)(x)$.

60. $f(x) = \sqrt{x}$, $g(x) = 2x + 3$

61. $f(x) = 3x - 1$, $g(x) = x^2$

Group Discussion Questions

62. Discovery Question For $f(x) = \log x$ and $g(x) = 10^x$, determine $(f \circ g)(x)$. What can you observe from this result?

63. Discovery Question

a. Write the inverse of the function $f = \{(1, 4), (2, 5), (3, 3), (4, 1), (5, 2)\}$. Does $f(x) = f^{-1}(x)$ for each input value of x?

b. Complete the function $f = \{(1, 7), (2, 6), (3, 5), (4, ?), (5, ?), (6, ?), (7, ?)\}$ so that $f(x) = f^{-1}(x)$.

c. Write the inverse of $f(x) = \dfrac{1}{x}$. Does $f(x) = f^{-1}(x)$ for each input value of x?

d. Have each member of your group write the equation of another function so that $f(x) = f^{-1}(x)$. Then examine the graphs of all these functions for similar characteristics.

64. Discovery Question

a. For $f(x) = \dfrac{3x - 2}{7x - 3}$, determine f^{-1}.

b. Compare f and f^{-1}.

c. What does this imply about the graph of $y = f(x)$?

Section 11.6 Sequences, Series, and Summation Notation

Objectives:

9. Calculate the terms of arithmetic and geometric sequences.

10. Use summation notation and evaluate the series associated with a finite sequence.

11. Evaluate an infinite geometric series.

Sequences

Objects and natural phenomena often form regular and interesting patterns. The study of these objects and phenomena generally involves data collected sequentially or in a systematic manner. Here are some examples of occurrences of sequences:

- Business: To enumerate the payments necessary to repay a loan
- Biology: To describe the growth pattern of a living organism
- Calculator design: To specify the sequence of terms used to calculate functions such as the exponential function e^x and the trigonometric function $\cos x$
- Calculus: To give the areas of a sequence of rectangles used to approximate the area of a region

We first examined arithmetic sequences in Section 2.1 and geometric sequences in Section 10.1. We now give a more formal definition of a sequence. A **sequence** is a function whose domain is a set of consecutive natural numbers. For example, the sequence 2, 5, 8, 11, 14, 17 can be viewed as the function $\{(1, 2), (2, 5), (3, 8), (4, 11), (5, 14), (6, 17)\}$ with input values $\{1, 2, 3, 4, 5, 6\}$ and output values $\{2, 5, 8, 11, 14, 17\}$.

ALTERNATIVE NOTATIONS FOR THE SEQUENCE 2, 5, 8, 11, 14, 17

MAPPING NOTATION	FUNCTION NOTATION	SUBSCRIPT NOTATION
$1 \rightarrow 2$	$f(1) = 2$	$a_1 = 2$
$2 \rightarrow 5$	$f(2) = 5$	$a_2 = 5$
$3 \rightarrow 8$	$f(3) = 8$	$a_3 = 8$
$4 \rightarrow 11$	$f(4) = 11$	$a_4 = 11$
$5 \rightarrow 14$	$f(5) = 14$	$a_5 = 14$
$6 \rightarrow 17$	$f(6) = 17$	$a_6 = 17$

The notation consisting of three dots, which is used to represent the sequence $a_1, a_2, \ldots, a_n$, is called ellipsis notation. This notation is used to indicate that terms in the sequence are missing in the listing but the pattern shown is continued.

A **finite sequence** has a last term. A sequence that continues without end is called an **infinite sequence.** A finite sequence with n terms can be denoted by $a_1, a_2, \ldots, a_n$, and an infinite sequence can be denoted by $a_1, a_2, \ldots, a_n, \ldots$.

Arithmetic and Geometric Sequences

We now review the definitions and descriptions of arithmetic and geometric sequences.

Arithmetic and Geometric Sequences

ALGEBRAICALLY	VERBALLY	NUMERICAL EXAMPLE	GRAPHICAL EXAMPLE
Arithmetic: $a_n - a_{n-1} = d$ or $a_n = a_{n-1} + d$	An arithmetic sequence has a constant difference d from term to term. The graph forms a set of discrete points lying on a straight line.	$-3, -1, 1, 3, 5, 7$	(1, −3), (2, −1), (3, 1), (4, 3), (5, 5), (6, 7)
Geometric: $\frac{a_n}{a_{n-1}} = r$ or $a_n = ra_{n-1}$	A geometric sequence has a constant ratio r from term to term. The graph forms a set of discrete points lying on an exponential curve.	$\frac{1}{4}, \frac{1}{2}, 1, 2, 4, 8$	$(1, \frac{1}{4})$, $(2, \frac{1}{2})$, (3, 1), (4, 2), (5, 4), (6, 8)

EXAMPLE 1 Identifying Arithmetic and Geometric Sequences

Determine whether each sequence is arithmetic, geometric, both, or neither. If the sequence is arithmetic, write the common difference d. If the sequence is geometric, write the common ratio r.

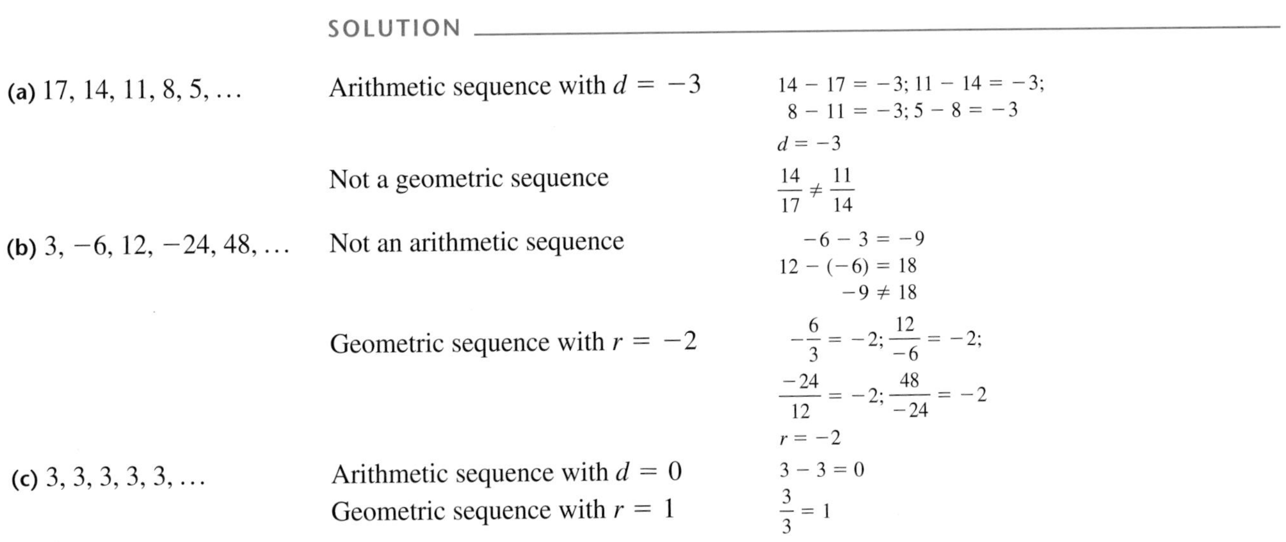

SOLUTION

(a) 17, 14, 11, 8, 5, ...	Arithmetic sequence with $d = -3$	$14 - 17 = -3; 11 - 14 = -3;$ $8 - 11 = -3; 5 - 8 = -3$ $d = -3$
	Not a geometric sequence	$\frac{14}{17} \neq \frac{11}{14}$
(b) $3, -6, 12, -24, 48, \ldots$	Not an arithmetic sequence	$-6 - 3 = -9$ $12 - (-6) = 18$ $-9 \neq 18$
	Geometric sequence with $r = -2$	$-\frac{6}{3} = -2; \frac{12}{-6} = -2;$ $\frac{-24}{12} = -2; \frac{48}{-24} = -2$ $r = -2$
(c) 3, 3, 3, 3, 3, ...	Arithmetic sequence with $d = 0$	$3 - 3 = 0$
	Geometric sequence with $r = 1$	$\frac{3}{3} = 1$

(d) 1, 1, 2, 3, 5, 8, ... Not an arithmetic sequence $1 - 1 = 0$, $2 - 1 = 1$, $0 \neq 1$

Not a geometric sequence $\frac{1}{1} = 1$, $\frac{2}{1} = 2$, $1 \neq 2$

Recursive Definitions

Recursively defined sequences are used extensively by computer programmers.

Since a_n can represent any term of the sequence, it is called the **general term.** The general term of a sequence sometimes is defined in terms of one or more of the preceding terms. A sequence defined in this manner is said to be **defined recursively.** Examples of recursive definitions for parts (a) and (b) of Example 1 are

(a) $a_n = a_{n-1} - 3$ with $a_1 = 17$ 17, 14, 11, 8, 5, ...
(b) $a_n = -2a_{n-1}$ with $a_1 = 3$ 3, −6, 12, −24, 48, ...

Because recursive definitions relate terms to preceding terms, there must be an initial term or condition given so that the sequence can get started. This initial or "seed" value is frequently used by computer programmers to construct looping structures within a program. Example 2 reexamines the sequence in Example 1(d). This recursive definition requires two initial conditions.

A Mathematical Note

The sequence 1, 1, 2, 3, 5, 8, 13, . . . is known as a *Fibonacci sequence,* in honor of the Italian mathematician Leonard Fibonacci (1170–1250). Fibonacci is known as the greatest mathematician of the 13th century. His sequence appears in many surprising ways in nature, including the arrangement of seeds in some flowers, the layout of leaves on the stems of some plants, and the spirals on some shells. There are so many applications that in 1963 the Fibonacci Association was founded and began to publish *The Fibonacci Quarterly.* In its first 3 years the association published nearly 1,000 pages of research.

EXAMPLE 2 Using a Recursive Definition

Write the first five terms of the sequence defined recursively by the formulas $a_1 = 1$, $a_2 = 1$, and $a_n = a_{n-2} + a_{n-1}$.

SOLUTION

$a_1 = 1$ The first two terms are given; for $n = 3$, a_n is a_3, a_{n-2} is a_1, and a_{n-1} is a_2.
$a_2 = 1$
$a_3 = a_1 + a_2$
$= 1 + 1$ Substitute 1 for a_1 and 1 for a_2.
$a_3 = 2$
$a_4 = a_2 + a_3$ For $n = 4$, a_n is a_4, a_{n-2} is a_2, and a_{n-1} is a_3.
$= 1 + 2$ Substitute 1 for a_2 and 2 for a_3.
$a_4 = 3$
$a_5 = a_3 + a_4$ For $n = 5$, a_n is a_5, a_{n-2} is a_3, and a_{n-1} is a_4.
$= 2 + 3$ Substitute 2 for a_3 and 3 for a_4.
$a_5 = 5$

Answer: 1, 1, 2, 3, 5

SELF-CHECK 11.6.1

Complete the finite sequence 3, 6, _____, _____, _____ so that the sequence will be

1. An arithmetic sequence
2. A geometric sequence
3. Use the recursive definition $a_1 = 1$, $a_2 = 1$, and $a_n = a_{n-2} + a_{n-1}$ to write the next four terms of 1, 1, 2, 3, 5, ____, ____, ____, ____.

CALCULATOR PERSPECTIVE 11.6.1 Using the seq Feature to Display the Terms of a Sequence

To display the first five terms of the sequence defined by $a_n = 2n + 3$ on a TI-84 Plus calculator, enter the following keystrokes:

The LIST feature is the secondary function of the STAT key.

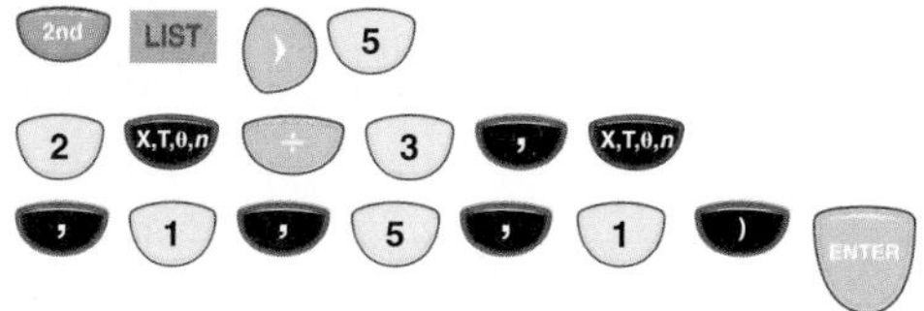

```
seq(2X+3,X,1,5,1
)
   {5 7 9 11 13}
```

Note: The **seq** feature has five parameters: seq(*expression, variable, first value, last value, increment*). In this case the variable used is x, the first value is 1, the last value is 5 (for the fifth term), and the increment is 1. If no increment is given, the default is to increase by 1.

Formulas for the *n*th Term of an Arithmetic and a Geometric Sequence

Recursive formulas are used extensively in computer science and in mathematics. However, these formulas do require us to compute a sequence term by term. For example, to compute the 100th term of $a_n = a_{n-1} - 3$, we first need the 99th term. It is also very useful to have a form that allows us to directly compute a_n from n. We now examine formulas for a_n for arithmetic and geometric sequences.

ARITHMETIC SEQUENCE	GEOMETRIC SEQUENCE	
$a_n = a_{n-1} + d$	$a_n = ra_{n-1}$	Start with the recursive formula for a_n and note the pattern that develops. Arithmetic sequences *add* a common difference for each new term, and geometric sequences *multiply* by a common ratio for each new term.
a_1	a_1	
$a_2 = a_1 + d$	$a_2 = a_1r$	
$a_3 = a_2 + d = (a_1 + d) + d = a_1 + 2d$	$a_3 = a_2r = (a_1r)r = a_1r^2$	
$a_4 = a_3 + d = (a_1 + 2d) + d = a_1 + 3d$	$a_4 = a_3r = (a_1r^2)r = a_1r^3$	
$\vdots$	$\vdots$	
$a_n = a_{n-1} + d = a_1 + (n - 1)d$	$a_n = a_{n-1}r = a_1r^{n-1}$	
$\mathbf{a_n = a_1 + (n - 1)d}$	$\mathbf{a_n = a_1r^{n-1}}$	

EXAMPLE 3 Calculating a_n for an Arithmetic and a Geometric Sequence

Use the formulas for a_n for an arithmetic sequence and a geometric sequence to calculate a_{10} for each sequence.

SOLUTION

(a) Determine a_{10} for an arithmetic sequence with $a_1 = 8$ and $d = 7$.

$a_n = a_1 + (n - 1)d$
$a_{10} = 8 + (10 - 1)(7)$
$a_{10} = 8 + 9(7)$
$a_{10} = 8 + 63$
$a_{10} = 71$

Substitute the given values into the formula for a_n for an arithmetic sequence.

(b) Determine a_{10} for a geometric sequence with $a_1 = 8$ and $r = 3$.

$a_n = a_1r^{n-1}$
$a_{10} = 8(3)^{10-1}$
$a_{10} = 8(3)^9$
$a_{10} = 8(19{,}683)$
$a_{10} = 157{,}464$

Substitute the given values into the formula for a_n for a geometric sequence.

Use a calculator to evaluate this expression.

(c) Determine a_{10} for a geometric sequence with $a_1 = 5$ and $r = -2$.

$a_n = a_1r^{n-1}$
$a_{10} = 5(-2)^{10-1}$
$a_{10} = 5(-2)^9$
$a_{10} = 5(-512)$
$a_{10} = -2{,}560$

Because r is negative, the terms in this geometric sequence will alternate in sign. All even-numbered terms will be negative.

The formula $a_n = a_1 + (n - 1)d$ relates a_n, a_1, n, and d. Therefore, we can find any of these four variables when the other three are known. A similar statement can be made for the formula $a_n = a_1 r^{n-1}$. This is illustrated in Example 4.

EXAMPLE 4 Determining the Common Difference and the Common Ratio

Use the formulas for a_n for an arithmetic sequence and a geometric sequence to calculate d and r.

SOLUTION

(a) Find d in an arithmetic sequence with $a_1 = -87$ and $a_{57} = 529$.

$$a_n = a_1 + (n - 1)d$$
$$529 = -87 + (57 - 1)d$$

Substitute the given values into the formula for a_n for an arithmetic sequence.

$$616 = 56d$$
$$d = 11$$

Then solve for d.

(b) Find r in a geometric sequence with $a_1 = 6$ and $a_3 = 24$.

$$a_n = a_1 r^{n-1}$$
$$24 = (6)(r)^{3-1}$$

Substitute the given values into the formula for a_n for a geometric sequence.

$$4 = r^2$$

Then solve this quadratic equation for both possible values of r.

$$r = 2 \quad \text{or} \quad r = -2$$

Check: The geometric sequences 6, 12, 24, 48, ... and 6, −12, 24, −48, ... both satisfy the given conditions.

Example 5 illustrates how to use the formulas for a_n to calculate the number of terms in an arithmetic sequence and a geometric sequence.

EXAMPLE 5 Determining the Number of Terms in a Sequence

Use the formulas for a_n for an arithmetic sequence and a geometric sequence to calculate the number of terms in each sequence.

SOLUTION

(a) −20, −13, −6, ..., 281

$$d = -13 - (-20) = 7$$
$$a_n = a_1 + (n - 1)d$$

First note that this is an arithmetic sequence with a common difference $d = 7$ and a first term of −20.

$$281 = -20 + (n - 1)7$$
$$301 = 7(n - 1)$$

Substitute $a_1 = -20$, $a_n = 281$, and $d = 7$ into the formula for a_n for an arithmetic sequence.

$$n - 1 = 43$$
$$n = 44$$

Then solve for n.

This arithmetic sequence has 44 terms.

(b) 3,125, 1,250, 500, ..., 32

$$r = \frac{1{,}250}{3{,}125} = \frac{500}{1{,}250} = 0.4$$

First note that this is a geometric sequence with a common ratio of $r = 0.4$.

$$a_n = a_1 r^{n-1}$$
$$32 = (3{,}125)(0.4)^{n-1}$$

Substitute $a_1 = 3{,}125$, $a_n = 32$, and $r = 0.4$ into the formula for a_n for a geometric sequence.

$$(0.4)^{n-1} = 0.01024$$

Divide both sides of the equation by 3,125.

$$\log 0.4^{n-1} = \log 0.01024$$

Take the common log of both members.

$$(n - 1)\log 0.4 = \log 0.01024$$

Simplify by using the power rule for logarithms.

$$n - 1 = \frac{\log 0.01024}{\log 0.4}$$

Divide both sides of the equation by log 0.4.

$$n - 1 = 5$$
$$n = 6$$

Evaluate the right side with a calculator.

This geometric sequence has six terms.

You can check this answer by writing the first six terms of this geometric sequence.

SELF-CHECK 11.6.2

Use the formulas $a_n = a_1 + (n - 1)d$ and $a_n = a_1r^{n-1}$ to determine

1. a_{21} for an arithmetic sequence with $a_1 = 4$ and $d = 5$.
2. d for an arithmetic sequence with $a_1 = 10$ and $a_{81} = 250$.
3. The number of terms in the arithmetic sequence with $a_1 = -12$, $a_n = 288$, and $d = 4$.
4. a_9 for a geometric sequence with $a_1 = 4$ and $r = 5$.
5. r for a geometric sequence with $a_1 = 5$ and $a_4 = 1{,}080$.
6. The number of terms in the geometric sequence with $a_1 = \frac{32}{3{,}125}$, $a_n = \frac{625}{16}$, and $r = \frac{5}{2}$.

Series

The sum of the terms of a sequence is called a **series.** If $a_1, a_2, \ldots, a_n$ is a finite sequence, then the indicated sum $a_1 + a_2 + \cdots + a_n$ is the series associated with this sequence. For arithmetic and geometric series there are formulas that serve as shortcuts to evaluating the series. Example 6 illustrates the meaning of a series. Then we examine the shortcuts.

EXAMPLE 6 Evaluating an Arithmetic Series and a Geometric Series

Find the value of the six-term series associated with these arithmetic and geometric sequences.

SOLUTION

(a) $a_n = 3n$

$$\begin{aligned} &a_1 + a_2 + a_3 + a_4 + a_5 + a_6 \\ &= 3(1) + 3(2) + 3(3) + 3(4) + 3(5) + 3(6) \\ &= 3 + 6 + 9 + 12 + 15 + 18 \\ &= 63 \end{aligned}$$

The series is the sum of the first six terms. Substitute the first six natural numbers into the formula $a_n = 3n$ to determine the first six terms, and then add the terms of this arithmetic series.

(b) $a_n = 3^n$

$$\begin{aligned} &a_1 + a_2 + a_3 + a_4 + a_5 + a_6 \\ &= 3^1 + 3^2 + 3^3 + 3^4 + 3^5 + 3^6 \\ &= 3 + 9 + 27 + 81 + 243 + 729 \\ &= 1{,}092 \end{aligned}$$

Calculate each of the six terms of this geometric sequence and then determine the sum.

Summation Notation

A convenient way of denoting a series is to use **summation notation,** in which the Greek letter Σ (*sigma,* which corresponds to the letter S for "sum") indicates the summation.

Summation Notation

ALGEBRAICALLY	VERBALLY	ALGEBRAIC EXAMPLE
$\sum_{i=1}^{n} a_i = a_1 + a_2 + \cdots + a_{n-1} + a_n$ Last value of the index: n; Index variable: i; Initial value of the index: 1; Formula for general term: a_i	The sum of a sub i from i equals 1 to i equals n.	$\sum_{i=1}^{5} a_i = a_1 + a_2 + a_3 + a_4 + a_5$

Generally the index variable is denoted by i, j, or k. The index variable is always replaced with successive integers from the initial value through the last value. Example 7 illustrates that the initial value can be a value other than 1. For example, in $\sum_{i=5}^{8} a_i$, i is replaced with 5, 6, 7, and then 8 to yield $\sum_{i=5}^{8} a_i = a_5 + a_6 + a_7 + a_8$.

EXAMPLE 7 Using Summation Notation to Evaluate an Arithmetic Series and a Geometric Series

Evaluate each series.

SOLUTION

(a) $\sum_{i=1}^{4} 5i$

$\sum_{i=1}^{4} 5i = 5(1) + 5(2) + 5(3) + 5(4)$ Replace i with 1, 2, 3, and then 4 and indicate the sum of these terms.

$= 5 + 10 + 15 + 20$ Evaluate each term of this arithmetic series and then add these terms.

$= 50$

(b) $\sum_{k=3}^{7} 2^k$

$\sum_{k=3}^{7} 2^k = 2^3 + 2^4 + 2^5 + 2^6 + 2^7$ Replace k with 3, 4, 5, 6, and then 7 and indicate the sum of these terms.

$= 8 + 16 + 32 + 64 + 128$ Evaluate each term of this geometric series and then add these terms.

$= 248$

SELF-CHECK 11.6.3

Evaluate each series.

1. $\sum_{i=1}^{5} 4i$ **2.** $\sum_{j=1}^{4} (j^2 - 2j)$ **3.** $\sum_{k=3}^{6} 5^k$

CALCULATOR PERSPECTIVE 11.6.2 Computing the Sum of a Sequence

To evaluate the series $\sum_{i=1}^{4} 5i = 5 + 10 + 15 + 20$ from Example 7(a) on a TI-84 Plus calculator, enter the following keystrokes:

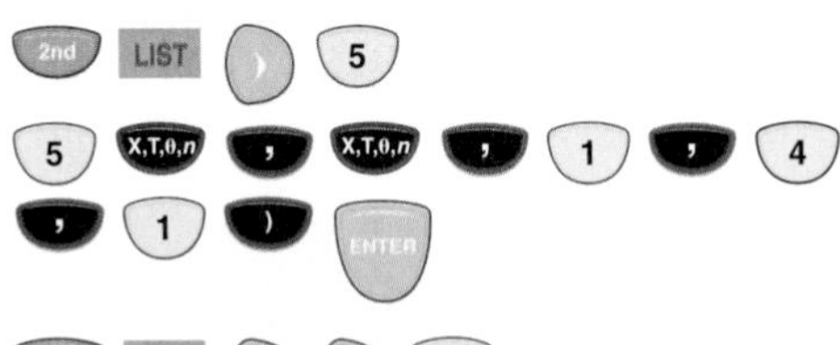

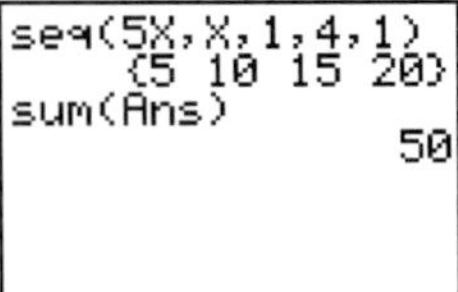

The ANS feature is the secondary function of the (−) key.

For a series with many terms, such as $\sum_{i=1}^{100} i$, it is useful to have a shortcut formula that allows us to calculate the sum without actually doing all the adding. We now develop formulas for an arithmetic series and a geometric series.

Development of a Formula for an Arithmetic Series

$$S_n = a_1 + (a_1 + d) + \cdots + [a_1 + (n - 2)d] + [a_1 + (n - 1)d]$$ Sum the terms from a_1 to a_n.

$$S_n = [a_1 + (n - 1)d] + [a_1 + (n - 2)d] + \cdots + (a_1 + d) + a_1$$ Sum the terms from a_n to a_1.

Add corresponding terms.

$$2S_n = [a_1 + a_1 + (n - 1)d] + [a_1 + a_1 + (n - 1)d] + \cdots + [a_1 + a_1 + (n - 1)d] + [a_1 + a_1 + (n - 1)d]$$

$$2S_n = (a_1 + a_n) + (a_1 + a_n) + \cdots + (a_1 + a_n) + (a_1 + a_n)$$ Substitute a_n for $a_1 + (n - 1)d$.

$$2S_n = n(a_1 + a_n)$$ Note that $(a_1 + a_n)$ is added n times.

$$S_n = \frac{n(a_1 + a_n)}{2}$$ Solve for S_n by dividing both sides by 2.

$$S_n = \frac{n[2a_1 + (n - 1)d]}{2}$$ Substitute $a_1 + (n - 1)d$ for a_n for an alternative form of this formula.

Development of a Formula for a Geometric Series

$$S_n = a_1 + a_1r + a_1r^2 + \cdots + a_1r^{n-2} + a_1r^{n-1}$$

$$rS_n = a_1r + a_1r^2 + a_1r^3 + \cdots + a_1r^{n-2} + a_1r^{n-1} + a_1r^n$$

$$S_n - rS_n = a_1 + 0 + 0 + \cdots + 0 + 0 - a_1r^n$$

To obtain the second equation, multiply both sides of the first equation by r and shift terms to the right to align similar terms. Subtract the second equation from the first equation.

$$S_n(1 - r) = a_1(1 - r^n)$$ Factor both sides.

$$S_n = \frac{a_1(1 - r^n)}{1 - r} \quad \text{for } r \neq 1$$ Divide both members by $1 - r$. If $r \neq 1$, $1 - r \neq 0$.

$$S_n = \frac{a_1 - a_1r^n}{1 - r}$$

$$S_n = \frac{a_1 - ra_n}{1 - r}$$ Substitute a_n for a_1r^{n-1} to obtain an alternative form of this formula.

Formulas for Arithmetic and Geometric Series $S_n = \sum_{i=1}^{n} a_i$

	ALGEBRAICALLY	ALGEBRAIC EXAMPLE
Arithmetic series:	$S_n = \frac{n}{2}(a_1 + a_n)$ or $S_n = \frac{n}{2}[2a_1 + (n - 1)d]$	$S_4 = \sum_{i=1}^{4} 5i = 5 + 10 + 15 + 20$ $S_4 = \frac{4}{2}(5 + 20)$ $S_4 = 2(25)$ $S_4 = 50$
Geometric series:	$S_n = \frac{a_1(1 - r^n)}{1 - r}$ or $S_n = \frac{a_1 - ra_n}{1 - r}$	$S_5 = \sum_{k=1}^{5} 2^k = 2 + 4 + 8 + 16 + 32$ $S_5 = \frac{2(1 - 2^5)}{1 - 2}$ $S_5 = \frac{2(-31)}{-1}$ $S_5 = 62$

Example 8 examines an application for both arithmetic and geometric series.

EXAMPLE 8 Using Arithmetic and Geometric Series to Model Applications

(a) Rolls of carpet are stacked in a warehouse, with 20 rolls on the first level, 19 on the second level, and so on. The top level has only 1 roll. How many rolls are in this stack? (See the figure.)

(b) If you could arrange to be paid \$1,000 at the end of January, \$2,000 at the end of February, \$4,000 at the end of March, and so on, what is the total amount you would be paid for the year?

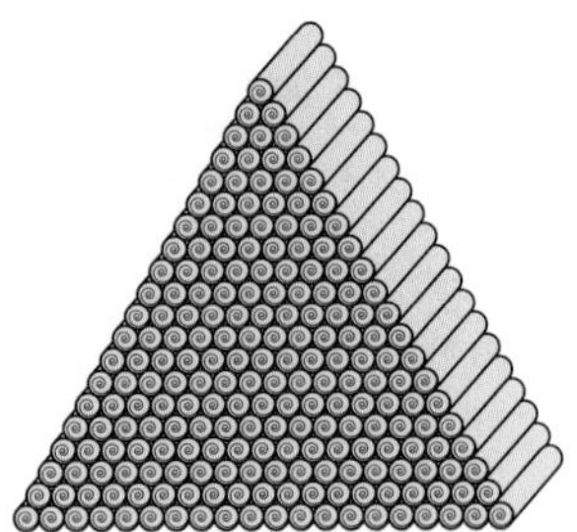

SOLUTION

(a) The number of rolls on the various levels form an arithmetic sequence with $a_1 = 20$, $n = 20$, and $a_{20} = 1$.

ALGEBRAICALLY

$$S_n = \frac{n(a_1 + a_n)}{2}$$
$$S_{20} = \frac{20(20 + 1)}{2}$$
$$S_{20} = 210$$

NUMERICALLY

$$S_n = a_1 + a_2 + \cdots + a_n$$
$$S_{20} = a_1 + a_2 + \cdots + a_{20}$$
$$S_{20} = 20 + 19 + 18 + \cdots + 3 + 2 + 1$$
$$S_{20} = 210$$

There is a constant difference of 1 roll between consecutive levels in this stack. Thus the numbers of rolls form an arithmetic sequence, and the total number of rolls is determined by adding the terms of this arithmetic series.

Answer: There are 210 rolls in this stack.

(b) The payments at the end of each month form a geometric sequence with $a_1 = 1{,}000$, $r = 2$, $n = 12$.

The total amount for the year is determined by adding the 12 terms of this geometric sequence.

ALGEBRAICALLY

$$S_n = \frac{a_1(1 - r^n)}{1 - r}$$
$$S_{12} = \frac{1{,}000(1 - 2^{12})}{1 - 2}$$
$$S_{12} = \frac{1{,}000(1 - 4{,}096)}{-1}$$
$$S_{12} = \frac{-4{,}095{,}000}{-1}$$
$$S_{12} = 4{,}095{,}000$$

NUMERICALLY

$$S_n = a_1 + a_2 + \cdots + a_n$$
$$S_{12} = a_1 + a_2 + \cdots + a_{12}$$
$$S_{12} = 1{,}000 + 2{,}000 + 4{,}000 + 8{,}000 + 16{,}000 + 32{,}000 + 64{,}000 + 128{,}000 + 256{,}000 + 512{,}000 + 1{,}024{,}000 + 2{,}048{,}000$$
$$S_{12} = 4{,}095{,}000$$

Simplify and calculate S_{12}.

Although this answer may seem unreasonable, it is the doubling pay scheme that makes the total unreasonable—not the arithmetic.

Answer: The total amount for the year would be \$4,095,000.

Compare the answers obtained algebraically in Example 8 with the answers obtained numerically by actually adding the terms in the series. Which method do you prefer? Which method would you prefer if there were 1,000 terms?

A Mathematical Note

At the age of 10, Carl Friedrich Gauss (1777–1855) mentally computed the sum $\sum_{i=1}^{100} i$. This was a problem that none of his fellow students were able to answer correctly by the end of the hour.

SELF-CHECK 11.6.4

Use the formulas for arithmetic and geometric series to determine these sums.

1. $\sum_{i=1}^{100} i$
2. $\sum_{i=1}^{20} 2^i$

If $|r| < 1$, then the absolute values of the terms of the geometric sequence are decreasing. For example, the geometric sequence $\frac{1}{2}, \frac{1}{4}, \frac{1}{8}, \frac{1}{16}, \ldots, \frac{1}{2^n}, \ldots$ is a decreasing, infinite geometric sequence with $r = \frac{1}{2}$. We already have a formula for finding the sum of a finite geometric sequence. Is the infinite sum $\sum_{i=1}^{\infty} \left(\frac{1}{2}\right)^i$ meaningful? Although we could never actually add an infinite number of terms, we can adopt a new meaning for this sum. If the sum $\sum_{i=1}^{n} a_i$ approaches some limiting value S as n becomes large, then we will call this value the **infinite sum.** Symbolically, $\sum_{i=1}^{\infty} a_i = S$ if $\sum_{i=1}^{n} a_i$ approaches S as n increases.

A general formula for the infinite sum can be obtained by examining the formula for S_n.

$$S_n = \frac{a_1(1 - r^n)}{1 - r}$$

If $|r| < 1$, then $|r|^n$ approaches 0 as n becomes larger. Thus $S_n = \frac{a_1(1 - r^n)}{1 - r}$ approaches $\frac{a_1(1 - 0)}{1 - r}$; that is, S_n approaches $\frac{a_1}{1 - r}$. The infinite sum S is the limiting value, so $S = \frac{a_1}{1 - r}$. If $|r| \geq 1$, the terms of a geometric series do not approach 0 and $\sum_{i=1}^{n} a_i$ does not approach any limit as n becomes large. In this case, we do not assign a value to $\sum_{i=1}^{\infty} a_i$.

Infinite Geometric Series $S = \sum_{i=1}^{\infty} a_i$

ALGEBRAICALLY	ALGEBRAIC EXAMPLE
If $\lvert r\rvert < 1$, $S = \frac{a_1}{1 - r}$. If $\lvert r\rvert \geq 1$, this sum does not exist.	$S = 0.333 \cdots$ $S = 0.3 + 0.03 + 0.003 + \cdots$ with $a_1 = 0.3$ and $r = 0.1$ $S = \frac{a_1}{1 - r}$ $S = \frac{0.3}{1 - 0.1}$ $S = \frac{0.3}{0.9}$ $S = \frac{1}{3}$

EXAMPLE 9 Writing a Repeating Decimal in Fractional Form

Write $0.272727\cdots$ as a fraction.

SOLUTION

$$0.272727\cdots = 0.27 + 0.0027 + 0.000027 + \cdots$$
$$= 27(0.01) + 27(0.01)^2 + 27(0.01)^3 + \cdots$$
$$= \sum_{i=1}^{\infty} 27(0.01)^i$$

This series is an infinite geometric series with $a_1 = 0.27$ and $r = 0.01$.

$$S = \frac{a_1}{1 - r}$$

This is the formula for the sum of an infinite geometric series.

$$S = \frac{0.27}{1 - 0.01}$$

Substitute 0.27 for a_1 and 0.01 for r.

$$S = \frac{0.27}{0.99}$$

Simplify and express S in fractional form.

$$S = \frac{3}{11}$$

Answer: $0.272727\cdots = \dfrac{3}{11}$

You can check this answer by dividing 3 by 11.

SELF-CHECK 11.6.5

1. Write $0.545454\cdots$ as a fraction.

SELF-CHECK ANSWERS

11.6.1

1. 3, 6, 9, 12, 15
2. 3, 6, 12, 24, 48
3. 1, 2, 3, 5, 8, 13, 21, 34

11.6.2

1. $a_{21} = 104$
2. $d = 3$
3. $n = 76$
4. $a_9 = 1{,}562{,}500$
5. $r = 6$
6. $n = 10$

11.6.3

1. 60
2. 10
3. 19,500

11.6.4

1. 5,050
2. 2,097,150

11.6.5

1. $0.545454\cdots = \dfrac{6}{11}$

USING THE LANGUAGE AND SYMBOLISM OF MATHEMATICS 11.6

1. A sequence is a function whose domain is a set of consecutive __________ numbers.
2. A sequence that has a last term is called a __________ sequence.
3. A sequence that continues without end is called an __________ sequence.
4. A sequence with $a_n = a_{n-1} + d$ is an __________ sequence.
5. A sequence with $a_n = ra_{n-1}$ is a __________ sequence.
6. The sequence 1, 1, 2, 3, 5, 8, 13, . . . is an example of a __________ sequence.
7. Since a_n can represent any term of a sequence, it is called the __________ term.
8. A sequence where the general term is defined in terms of one or more of the preceding terms is said to be defined __________.
9. The formula for a_n for an arithmetic sequence is $a_n =$ __________.
10. The formula for a_n for a geometric sequence is $a_n =$ __________.
11. In the __________ notation $\sum_{i=1}^{n} a_i$, the index variable is __________. The initial value of the index variable in this example is __________, and the last value of the index variable is __________.
12. The formula for an arithmetic series is $S_n =$ __________.
13. The formula for a geometric series is $S_n =$ __________.
14. The formula for an infinite geometric series is $S =$ __________ if $|r| < 1$.

QUICK REVIEW 11.6

The purpose of this quick review is to help you recall skills needed in this section.

1. Write an algebraic expression for "*a* sub 5 equals 23."
2. Write an algebraic expression for "*a* sub *n* equals four *n* plus eight."

Determine the first five terms of each sequence.

3. $a_n = 3n + 5$
4. $a_n = n^2 - 5n + 8$
5. $a_n = 5 \cdot 2^{n-1}$

EXERCISES 11.6

In Exercises 1–4, determine whether each sequence is arithmetic, geometric, both, or neither. If the sequence is arithmetic, write the common difference *d*. If the sequence is geometric, write the common ratio *r*.

1. **a.** 1, 6, 36, 216, 1,296, 7,776
   b. 1, 6, 11, 16, 21, 26
   c. 2, 3, 5, 8, 12, 13
   d. 6, 6, 6, 6, 6, 6
2. **a.** 144, 124, 104, 84, 64, 44
   b. 144, 72, 36, 18, 9, 4.5
   c. 9, −9, 9, −9, 9, −9
   d. 9, −8, 7, −6, 5, −4
3. **a.** $a_n = 3n - 5$
   b. $a_n = n^2$
   c. $a_n = 4^n$
   d. $a_n = 7$
4. **a.** $a_1 = 1$ and $a_n = 5a_{n-1}$ for $n > 1$
   b. $a_1 = 2$ and $a_n = (a_{n-1})^2$ for $n > 1$
   c. $a_1 = 3$ and $a_n = a_{n-1} + 5$ for $n > 1$
   d. $a_1 = 4$ and $a_n = 4$ for $n > 1$

In Exercises 5 and 6, write the first six terms of each sequence.

5. **a.** $a_n = 5n + 3$
   b. $a_n = n^2 - n + 1$
   c. $a_n = 16\left(\frac{1}{2}\right)^n$
   d. $a_n = 5(-2)^n$
6. **a.** $a_1 = 4$ and $a_n = a_{n-1} + 5$ for $n > 1$
   b. $a_1 = 4$ and $a_n = 5a_{n-1}$ for $n > 1$
   c. $a_1 = 4$ and $a_n = -a_{n-1}$ for $n > 1$
   d. $a_1 = 1$, $a_2 = 2$, and $a_n = 2a_{n-2} + a_{n-1}$ for $n > 1$

In Exercises 7 and 8, write the first six terms of each arithmetic sequence.

7. **a.** 18, 14, _____, _____, _____, _____
   b. 18, _____, 14, _____, _____, _____
   c. $a_1 = 7, d = -2$
   d. $a_1 = -9, d = 3$
8. **a.** $a_1 = -8, a_6 = 12$
   b. $a_n = 4n - 1$
   c. $a_n = 11 - 2n$
   d. $a_1 = 1.2$ and $a_n = a_{n-1} + 0.4$ for $n > 1$

In Exercises 9 and 10, write the first five terms of each geometric sequence.

9. **a.** 1, 4, _____, _____, _____
   b. 1, _____, 4, _____, _____
   c. $a_1 = 16, r = \frac{1}{2}$
   d. $a_1 = 5, r = -3$
10. **a.** $a_1 = 3, a_4 = 24$
    b. $a_n = 9\left(\frac{1}{3}\right)^n$
    c. $a_n = 96\left(-\frac{1}{2}\right)^n$
    d. $a_1 = 10$ and $a_n = 2a_{n-1}$ for $n > 1$
11. Use the given information to calculate the indicated term of the arithmetic sequence.
    a. $a_1 = 6, d = 5, a_{83} = ?$
    b. $a_1 = 17, a_2 = 15, a_{51} = ?$
    c. $a_{80} = 25, a_{81} = 33, a_{82} = ?$
    d. $a_1 = 2, a_3 = 12, a_{101} = ?$
12. Use the given information to calculate the indicated term of the geometric sequence.
    a. $a_1 = \frac{1}{256}, r = 2, a_{10} = ?$
    b. $a_1 = 4, a_2 = 12, a_8 = ?$
    c. $a_{50} = 8, a_{51} = 40, a_{52} = ?$
    d. $a_1 = 5, a_3 = 20, a_{10} = ?$

In Exercises 13–20, use the information given for the arithmetic sequence to find the quantities indicated.

13. 18, 14, 10, … , −62; $n = ?$
14. 48, 55, … , 496; $n = ?$
15. $a_{44} = 216, d = 12, a_1 = ?$
16. $a_{113} = -109, d = -2, a_1 = ?$
17. $a_{11} = 4, a_{31} = 14, d = ?$
18. $a_{47} = 23, a_{62} = 28, d = ?$
19. $a_1 = 5, a_n = 14, d = \frac{1}{5}, n = ?$
20. $a_1 = -12, a_n = 6, d = \frac{1}{2}, n = ?$

In Exercises 21–28, use the information given for the geometric sequences to find the quantities indicated.

21. 243, 81, … , $\frac{1}{3}$; $n = ?$
22. 1,024, 512, … , 1; $n = ?$
23. $a_5 = 24, r = 2, a_1 = ?$
24. $a_6 = 64, r = 4, a_1 = ?$
25. $a_9 = 32, a_{11} = 288, r = ?$
26. $a_{45} = 17, a_{47} = 425, r = ?$
27. $a_n = 5a_{n-1}, a_1 = \frac{1}{3{,}125}, a_9 = ?$
28. $a_n = 0.1a_{n-1}, a_1 = 7{,}000, a_8 = ?$

In Exercises 29 and 30, write the terms of each series and then add these terms.

29. a. $\sum_{i=1}^{6}(2i + 3)$ b. $\sum_{i=1}^{5}(i^2 - 1)$
c. $\sum_{j=1}^{6} 2^j$ d. $\sum_{k=1}^{10} 5$

30. a. $\sum_{k=1}^{5}\frac{k+3}{5}$ b. $\sum_{i=4}^{8}(i^2 + 1)$
c. $\sum_{j=4}^{7} 3^j$ d. $\sum_{k=10}^{16} 7$

In Exercises 31–40, use the information given to evaluate each arithmetic series.

31. $a_1 = 2, a_{40} = 80, S_{40} = ?$
32. $a_1 = 3, a_{51} = 153, S_{51} = ?$
33. $a_1 = \frac{1}{2}, a_{12} = \frac{1}{3}, S_{12} = ?$
34. $a_1 = 0.36, a_{18} = 0.64, S_{18} = ?$
35. $a_1 = 10, d = 4, S_{66} = ?$
36. $a_1 = 11, d = -3, S_{11} = ?$
37. $\sum_{i=1}^{61}(2i + 3)$
38. $\sum_{k=1}^{47}(3k - 2)$
39. $\sum_{k=1}^{24}\frac{k+3}{5}$
40. $\sum_{j=1}^{40}\frac{j-5}{3}$

In Exercises 41–52, use the information given to evaluate each geometric series.

41. $a_1 = 3, r = 2, S_7 = ?$
42. $a_1 = 2, r = 3, S_6 = ?$
43. $a_1 = 0.2, r = 0.1, S_5 = ?$
44. $a_1 = 0.5, r = 0.1, S_6 = ?$
45. $a_1 = 48, r = -\frac{1}{2}, S_8 = ?$
46. $a_1 = 729, r = -\frac{1}{3}, S_7 = ?$
47. $a_1 = 1, a_n = 3.71293, r = 1.3, S_n = ?$
48. $64 + 32 + \cdots + \frac{1}{8}$
49. $\sum_{i=1}^{7} 8(0.1)^i$
50. $\sum_{k=1}^{6} 7(0.1)^k$
51. $a_1 = 40, a_n = \frac{1}{5}a_{n-1}$ for $n > 1, S_6 = ?$
52. $a_1 = 81, a_n = \frac{1}{3}a_{n-1}$ for $n > 1, S_6 = ?$

In Exercises 53–64, use the information given to evaluate the infinite geometric series.

53. $a_1 = 5, r = \frac{2}{3}$
54. $a_1 = 16, r = \frac{1}{5}$
55. $a_1 = 14, r = -\frac{3}{4}$
56. $a_1 = 7, r = -\frac{2}{5}$
57. $a_1 = 0.12, r = \frac{1}{100}$
58. $a_1 = 0.9, r = \frac{1}{10}$
59. $\sum_{k=1}^{\infty}\left(\frac{4}{9}\right)^k$
60. $\sum_{j=1}^{\infty}\left(\frac{3}{7}\right)^j$
61. $6 - 4 + \frac{8}{3} - \frac{16}{9} + \cdots$
62. $16 - 12 + 9 - 6.75 + \cdots$
63. $a_1 = 24, a_n = \frac{3}{8}a_{n-1}$ for $n > 1$
64. $a_1 = -36, a_n = \left(-\frac{5}{9}\right)a_{n-1}$ for $n > 1$

In Exercises 65 and 66, write each repeating decimal as a fraction.

65. a. $0.444\cdots$ b. $0.212121\cdots$
c. $0.409409409\cdots$ d. $2.5555\cdots$
66. a. $0.555\cdots$ b. $0.363636\cdots$
c. $0.495495495\cdots$ d. $8.3333\cdots$

In Exercises 67–72, use the information given for the arithmetic sequences to find the quantities indicated.

67. $a_{77} = 19, d = -11, S_{77} = ?$
68. $S_n = 240, a_1 = 4, a_n = 16, n = ?$
69. $S_{30} = 1{,}560, a_{30} = 93, a_1 = ?$
70. $S_{17} = 527, a_1 = 15, a_{17} = ?$
71. $S_{40} = 680, a_1 = 11, d = ?$
72. $S_{25} = 60, a_1 = 17, d = ?$

In Exercises 73–78, use the information given for the geometric sequences to find the quantities indicated.

73. $r = 2, S_{10} = 6{,}138, a_1 = ?$
74. $a_1 = 12, r = 5, S_n = 46{,}872, n = ?$
75. $a_1 = 6, a_n = 24{,}576, S_n = 32{,}766, r = ?$
76. $\sum_{i=1}^{\infty} a_i = 27, a_1 = 12, r = ?$
77. $\sum_{k=1}^{\infty} a_k = 21, r = \frac{2}{9}, a_1 = ?$
78. $\sum_{k=1}^{\infty} a_i = 2, r = \frac{6}{7}, a_1 = ?$
79. **Stacks of Logs** Logs are stacked so that each layer after the first has 1 less log than the previous layer. If the bottom layer has 24 logs and the top layer has 8 logs, how many logs are in the stack? (See the figure.)

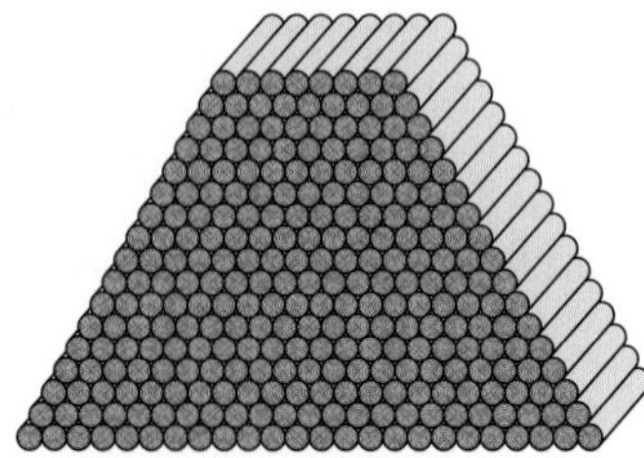

80. **Rolls of Insulation** Rolls of insulation are stacked so that each layer after the first has 8 fewer rolls than the previous layer. How many layers will a lumberyard need to use in order to stack 120 rolls if 40 rolls are placed on the bottom layer? (See the figure.)

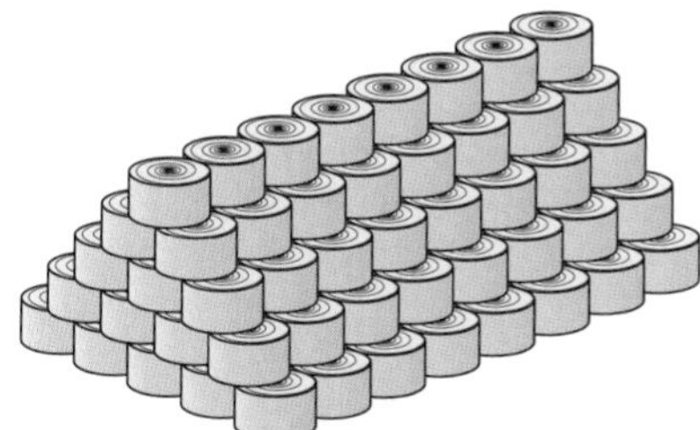

81. **Increased Productivity** The productivity gain from installing a new robot welder on a machinery assembly line is estimated to be \$4,000 the first month of operation, \$4,500 the second month, and \$5,000 the third month. If this trend continues, what will be the total productivity gain for the first 12 months of operation?
82. **Seats in a Theater** A theater has 20 rows of seats, with 100 seats in the back row. Each row has 2 fewer seats than the row immediately behind it. How many seats are in the theater?
83. **Chain Letter** A chain-letter scam requires that each participant persuade four other people to participate. If one person starts this venture as a first-generation participant, determine how many people will have been involved by the time the eighth generation has signed on but not yet contacted anyone.
84. **Bouncing Ball** A ball dropped from a height of 36 m rebounds to six-tenths its previous height on each bounce. How far has it traveled when it reaches the apex of its eighth bounce? Give your answer to the nearest tenth of a meter. (*Hint:* You may wish to consider the distances going up separately from the distances going down.)

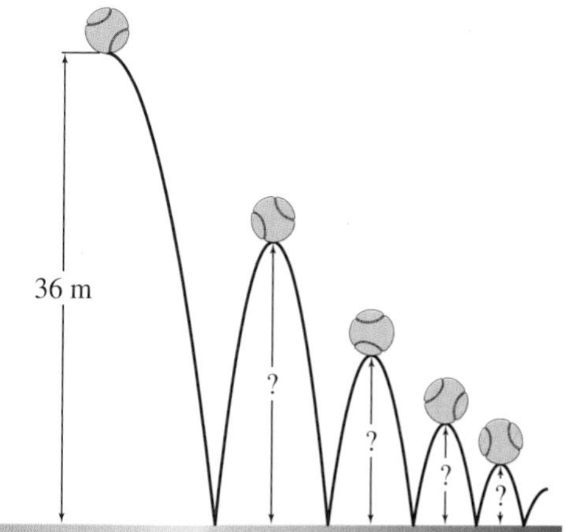

85. **Vacuum Pump** With each cycle a vacuum pump removes one-third of the air in a glass vessel. What percent of the air has been removed after eight cycles?
86. **Arc of a Swing** A child's swing moves through a 3-m arc. On each swing it travels only two-thirds the distance it traveled on the previous arc. How far does the swing travel before coming to rest?

87. **Multiplier Effect** City planners estimate that a new manufacturing plant located in their area will contribute \$600,000 in salaries to the local economy. They estimate that those who earn the salaries will spend three-fourths of this money within the community. The merchants, service providers, and others who receive this \$450,000 from the salary earners in turn will spend three-fourths of it in the community, and so on. Taking into account the multiplier effect, find the total amount of spending within the local economy that will be generated by the salaries from this new plant.

Group Discussion Questions

88. **Challenge Question** If 20 people in a room shake hands with each other exactly once, how many handshakes will take place?
89. **Discovery Question** The trichotomy property of real numbers states that exactly one of the following three statements must be true in each case. Determine which statement is true.

    a. i. $0.333\cdots < \frac{1}{3}$ ii. $0.333\cdots = \frac{1}{3}$ iii. $0.333\cdots > \frac{1}{3}$

    b. i. $0.666\cdots < \frac{2}{3}$ ii. $0.666\cdots = \frac{2}{3}$ iii. $0.666\cdots > \frac{2}{3}$

    c. i. $0.888\cdots < \frac{8}{9}$ ii. $0.888\cdots = \frac{8}{9}$ iii. $0.888\cdots > \frac{8}{9}$

    d. What is the meaning of the three dots in $0.999\cdots$?

    e. Which of the following is true?
    i. $0.999\cdots < 1$ ii. $0.999\cdots = 1$ iii. $0.999\cdots > 1$

    f. Are there any real numbers between $0.999\cdots$ and 1? If so, give an example.

    g. Multiply both sides of $\frac{1}{3} = 0.333\cdots$ by 3. What does this prove?

    h. Is the statement, "The number one-half on the number line can be represented by $\frac{1}{2}$, $\frac{2}{4}$, 50%, 0.5, and $0.499\cdots$" a true statement? If not, why is it false?
90. **Risks and Choices** One gimmick advertised in a newspaper was a "sure-fire" secret formula for becoming a millionaire. The secret discovered by those unwise enough to pay for it was as follows: "On the first day of the month save 1¢, on the second day save 2¢, on the third 4¢, etc." How many days would be required to save a total of at least \$1,000,000? At this rate, what amount would be saved on the last day?

Section 11.7 Conic Sections

Objectives:

12. Graph parabolas, circles, ellipses, and hyperbolas.
13. Write the equations of parabolas, circles, ellipses, and hyperbolas in standard form.

Parabolas, circles, ellipses, and hyperbolas can be formed by cutting a cone or a pair of cones with a plane. Therefore these figures collectively are referred to as conic sections. Conic sections originally were studied from a geometric viewpoint. (See Fig. 11.7.1.)

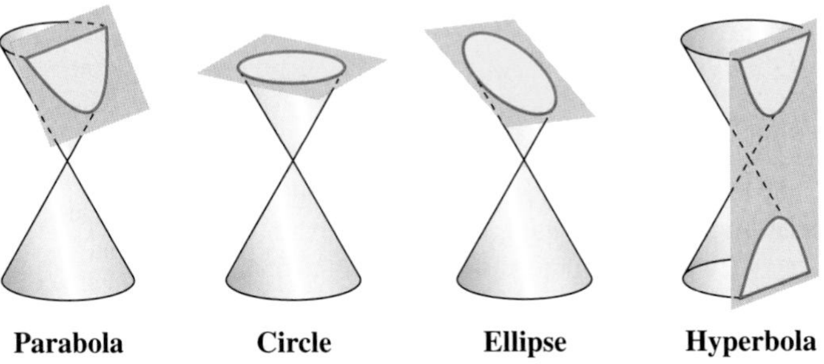

Figure 11.7.1 Conic sections.

Two formulas that are useful for examining relationships between points are the distance formula (Section 8.4) and the midpoint formula. Note that the midpoint is found by taking the average of the x-coordinates and the average of the y-coordinates.

Midpoint Formula:

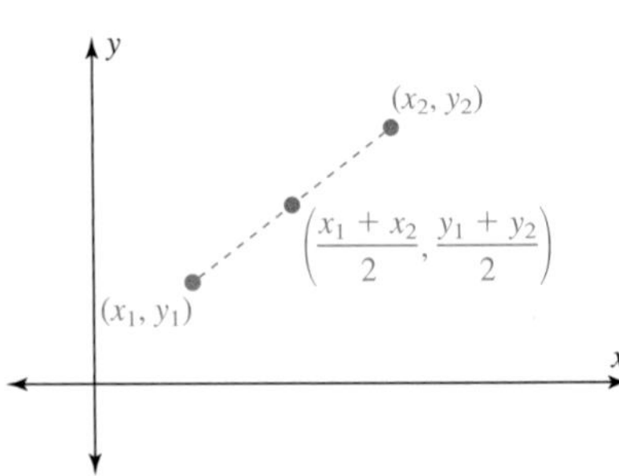

The distance and midpoint formulas are summarized in the following box.

Distance and Midpoint Formulas

If (x_1, y_1) and (x_2, y_2) are two points, then:

ALGEBRAICALLY	NUMERICAL EXAMPLE
Distance formula: $d = \sqrt{(x_2 - x_1)^2 + (y_2 - y_1)^2}$	For $(-2, 7)$ and $(4, -1)$: Distance between the points: $d = \sqrt{(4 - (-2))^2 + (-1 - 7)^2}$ $d = \sqrt{36 + 64} = \sqrt{100}$ $d = 10$
Midpoint formula: $(x, y) = \left(\frac{x_1 + x_2}{2}, \frac{y_1 + y_2}{2}\right)$	Midpoint between the points: $(x, y) = \left(\frac{-2 + 4}{2}, \frac{7 + (-1)}{2}\right) = \left(\frac{2}{2}, \frac{6}{2}\right)$ $(x, y) = (1, 3)$

Circle

Using the distance formula, we now develop the equation of a circle with center (h, k) and radius r. A **circle** is the set of all points in a plane that are a constant distance from a fixed point. (See Fig. 11.7.2). The fixed point is called the **center** of the circle, and the distance from the center to the points on the circle is called the length of a **radius.** A **diameter** is a line segment from one point on a circle through the center to another point on the circle. The length of a diameter is twice the length of a radius.

The distance r from any point (x, y) on the circle to the center (h, k) is given by

$$r = \sqrt{(x - h)^2 + (y - k)^2}$$

Squaring both sides of this equation gives an equation satisfied by the points on the circle

$$(x - h)^2 + (y - k)^2 = r^2$$

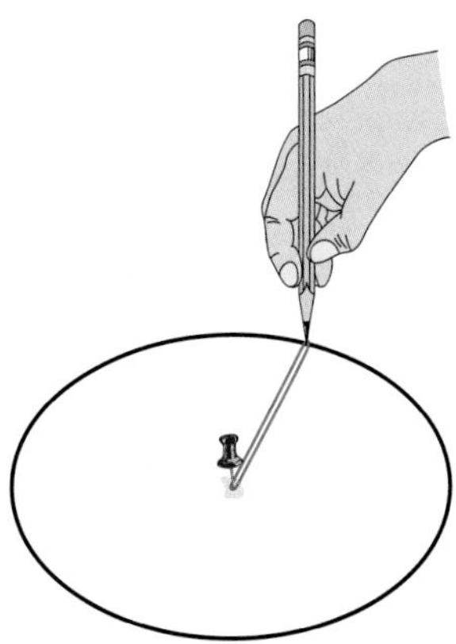

Figure 11.7.2 To draw a circle, fix a loop of string with a tack and draw the circle as illustrated.

Standard Form of the Equation of a Circle

ALGEBRAICALLY	GRAPHICALLY	ALGEBRAIC EXAMPLE
The equation of a circle with center (h, k) and radius r is $(x - h)^2 + (y - k)^2 = r^2$	y (x, y) r k (h, k) x h	The equation of a circle with center $(-1, 3)$ and radius 5 is $(x - (-1))^2 + (y - 3)^2 = 5^2$ $(x + 1)^2 + (y - 3)^2 = 25$

EXAMPLE 1 Determining the Equation of a Circle

Determine the equation of each graphed circle.

SOLUTION

(a)

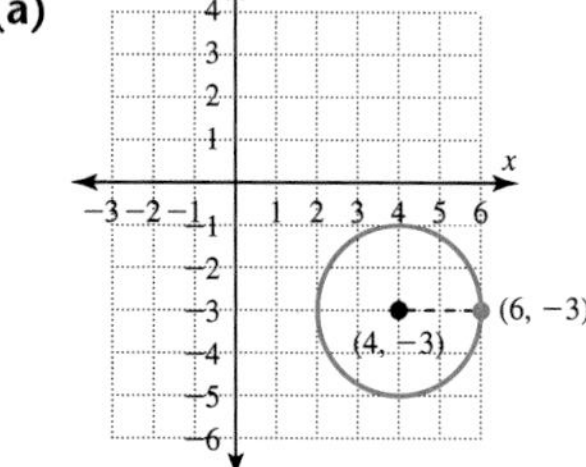

Center $(h, k) = (4, -3)$

Radius $r = \sqrt{(6-4)^2 + [-3-(-3)]^2}$

$r = \sqrt{2^2 + 0^2} = \sqrt{4}$

$r = 2$

Determine the center by inspection, and use the distance formula to calculate the length of the radius. The segment (4, −3) to (6, −3) is a radius of this circle.

Circle: $(x-h)^2 + (y-k)^2 = r^2$

$(x-4)^2 + [y-(-3)]^2 = 2^2$

$(x-4)^2 + (y+3)^2 = 4$

Substitute the center and radius into the standard form for the equation of a circle.

(b)

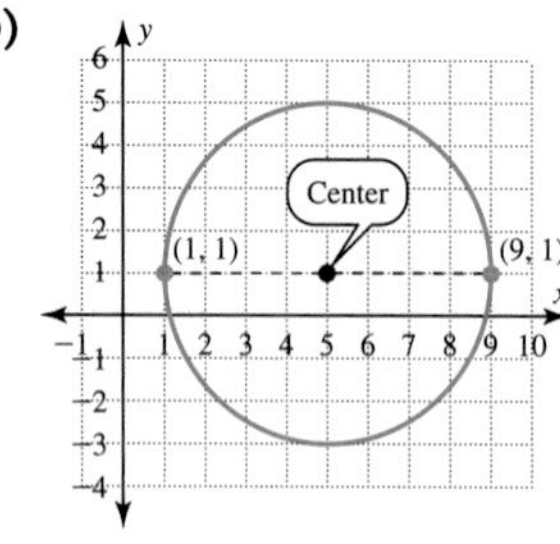

Center $(h, k) = \left(\frac{1+9}{2}, \frac{1+1}{2}\right)$

$(h, k) = (5, 1)$

Use the midpoint formula to calculate the center of the circle with a diameter from (1, 1) to (9, 1).

Radius $r = \sqrt{(9-5)^2 + (1-1)^2}$

$r = \sqrt{4^2 + 0^2} = \sqrt{16}$

$r = 4$

Then use the center, one of the points on the circle, and the distance formula to calculate the length of the radius.

Circle: $(x-h)^2 + (y-k)^2 = r^2$

$(x-5)^2 + (y-1)^2 = 4^2$

$(x-5)^2 + (y-1)^2 = 16$

Substitute the center and radius into the standard form for the equation of a circle.

SELF-CHECK 11.7.1

A circle has a diameter from (−3, 4) to (3, −4).

1. Determine the center of this circle.
2. Determine the length of this diameter.
3. Determine the length of a radius.
4. Determine the equation of this circle.

The standard form of the circle $(x-4)^2 + (y+3)^2 = 4$ can be expanded to give the general form $x^2 + y^2 - 8x + 6y + 21 = 0$. If we are given the equation of a circle in general form, we can use the process of completing the square to rewrite the equation in standard form so that the center and radius will be obvious.

EXAMPLE 2 Writing the Equation of a Circle in Standard Form

Determine the center and the radius of the circle defined by the equation $3x^2 + 3y^2 + 30x - 66y + 330 = 0$.

SOLUTION

$$3x^2 + 3y^2 + 30x - 66y + 330 = 0$$
$$x^2 + y^2 + 10x - 22y + 110 = 0 \quad \text{Divide both sides of the equation by 3.}$$
$$(x^2 + 10x) + (y^2 - 22y) = -110 \quad \text{Regroup terms.}$$
$$(x^2 + 10x + 25) + (y^2 - 22y + 121) = -110 + 25 + 121 \quad \text{Complete the square.}$$
$$(x+5)^2 + (y-11)^2 = 36 \quad \text{Write in standard form.}$$
$$[x-(-5)]^2 + (y-11)^2 = 6^2$$

Answer: The circle has center (−5, 11) and radius 6.

A Mathematical Note

The term *diameter* is formed by combining the Greek words *dia* meaning "across" and *metros* meaning to "measure." Euclid used diameter as an appropriate name for the chord that measures the distance across a circle through the circle's center.

Equations of the form $x^2 + y^2 = c$, for $c > 0$, define circles centered at the origin. The following box also describes the cases where $c = 0$ or $c < 0$.

Equations of the Form $x^2 + y^2 = c$

As shown in the figure,

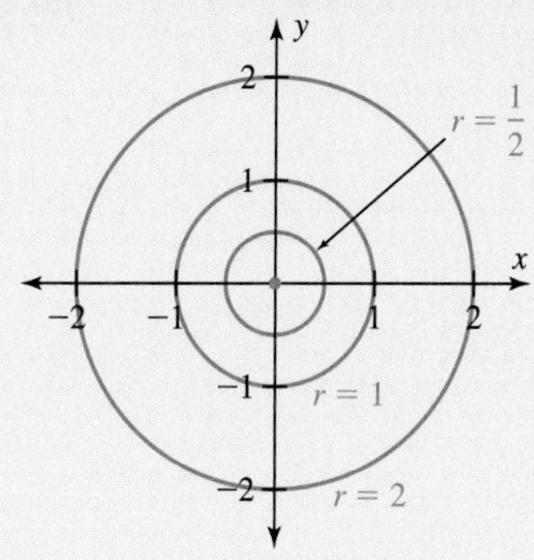

1. $x^2 + y^2 = 4$ is a circle with center (0, 0) and radius 2.
2. $x^2 + y^2 = 1$ is a circle with center (0, 0) and radius 1.
3. $x^2 + y^2 = \frac{1}{4}$ is a circle with center (0, 0) and radius $\frac{1}{2}$.
4. $x^2 + y^2 = 0$ is the degenerate case of a circle—a single point (0, 0).
5. $x^2 + y^2 = -1$ has no real solutions and thus no points to graph. Both x^2 and y^2 are nonnegative, so $x^2 + y^2$ also must be nonnegative.

SELF-CHECK 11.7.2

1. Write $4x^2 + 4y^2 - 24x + 32y + 99 = 0$ in standard form and identify the center and the radius.

By the vertical line test, circles are not functions. Thus graphing calculators do not graph circles with y as a function of x. Nonetheless, there are ways to get the graph of a circle to appear on a graphing calculator display. One way is to split a circle into two semicircles—an upper semicircle and a lower semicircle, both of which are functions. This is illustrated with the circle $x^2 + y^2 = 16$. Solving for y, we obtain $y = \pm\sqrt{16 - x^2}$. The upper semicircle is defined by $Y_1 = +\sqrt{16 - x^2}$, and the lower semicircle is defined by $Y_2 = -\sqrt{16 - x^2}$. These semicircles are both graphed in Calculator Perspective 11.7.1.

CALCULATOR PERSPECTIVE 11.7.1 Graphing a Circle by Using Two Functions

To graph the circle $x^2 + y^2 = 16$ on a TI-84 Plus calculator, graph the two semicircles $Y_1 = \sqrt{16 - x^2}$ and $Y_2 = -\sqrt{16 - x^2}$ by entering the following keystrokes:

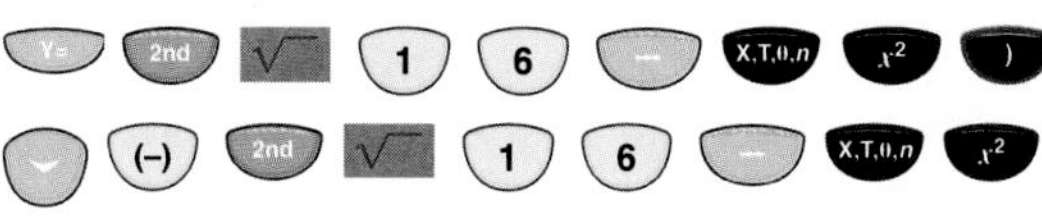

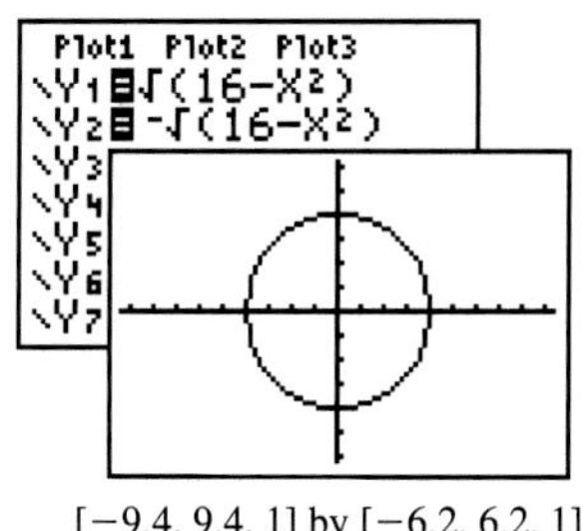

[−9.4, 9.4, 1] by [−6.2, 6.2, 1]

If a circle looks distorted on your calculator display, try using the **ZSquare** feature in the **ZOOM** menu to adjust for the fact that the display window is wider than it is tall.

Ellipse

An **ellipse** is the set of all points in a plane, the sum of whose distances from two fixed points is constant. (See Fig. 11.7.3.) The two fixed points $F_1(-c, 0)$ and $F_2(c, 0)$ are called **foci.** The **major axis** of the ellipse passes through the foci. The **minor axis** is shorter than the major axis and is perpendicular to it at the center. The ends of the major axis are called the **vertices,** and the ends of the minor axis are called the **covertices.** If the ellipse is centered at the origin, then the equation of the ellipse as shown in Fig. 11.7.4 is $\frac{x^2}{a^2} + \frac{y^2}{b^2} = 1$ where $a > b$. This formula is derived in the group exercises.

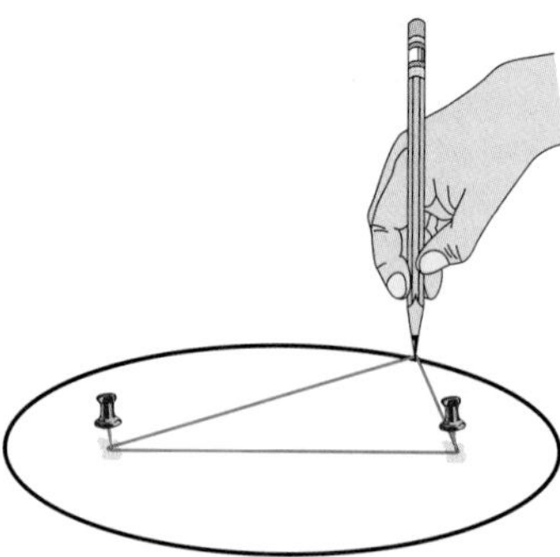

Figure 11.7.3 To draw an ellipse, loop a piece of string around two tacks and draw the ellipse as illustrated.

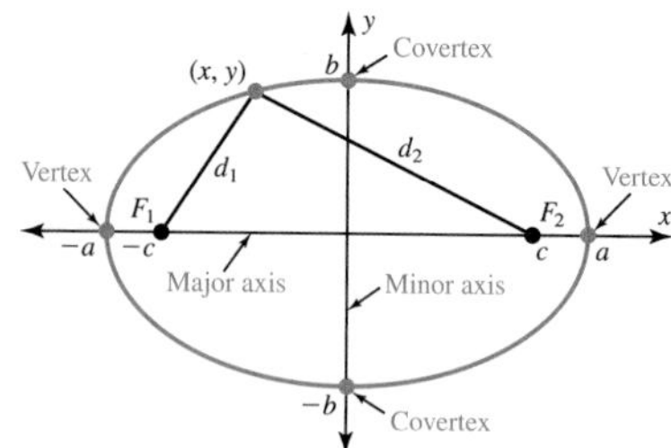

Figure 11.7.4 $d_1 + d_2 = 2a$.

EXAMPLE 3 Graphing an Ellipse Centered at the Origin

Graph $\frac{x^2}{49} + \frac{y^2}{16} = 1$.

SOLUTION

Plot the x-intercepts $(-7, 0)$ and $(7, 0)$.
Plot the y-intercepts $(0, -4)$ and $(0, 4)$.
Then use the known shape to complete the ellipse.

From the given equation $\frac{x^2}{49} + \frac{y^2}{16} = 1$ this is an ellipse centered at the origin with $a^2 = 49$ and $b^2 = 16$. Thus $a = 7$, $b = 4$, and the intercepts are $(-7, 0)$, $(7, 0)$, $(0, -4)$, and $(0, 4)$.

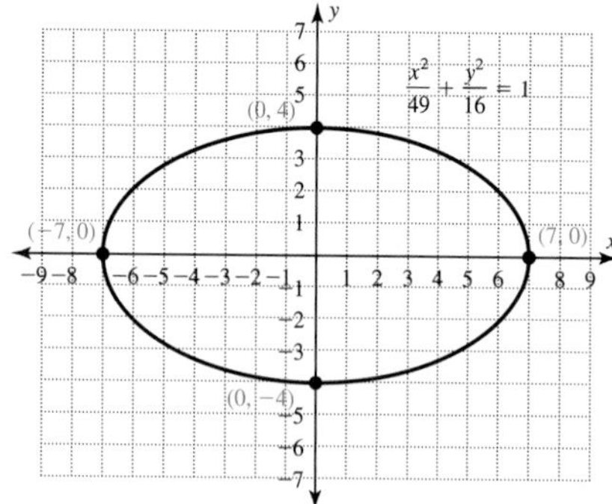

We now use translations (Section 11.3) to examine ellipses that are not centered at the origin. The graph of $\frac{(x-h)^2}{a^2} + \frac{(y-k)^2}{b^2} = 1$ is identical to the graph of $\frac{x^2}{a^2} + \frac{y^2}{b^2} = 1$, except that it has been translated so that the center is at (h, k) instead of $(0, 0)$.

Standard Form of the Equation of an Ellipse

The equation of an ellipse with center (h, k), major axis of length $2a$, and minor axis of length $2b$ is:

ALGEBRAICALLY	ALGEBRAIC EXAMPLE
Horizontal major axis: $\frac{(x-h)^2}{a^2} + \frac{(y-k)^2}{b^2} = 1$	$\frac{(x-2)^2}{49} + \frac{(y+3)^2}{16} = 1$
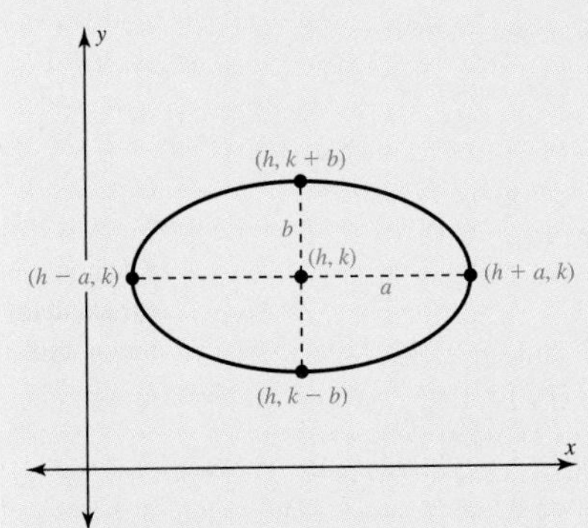	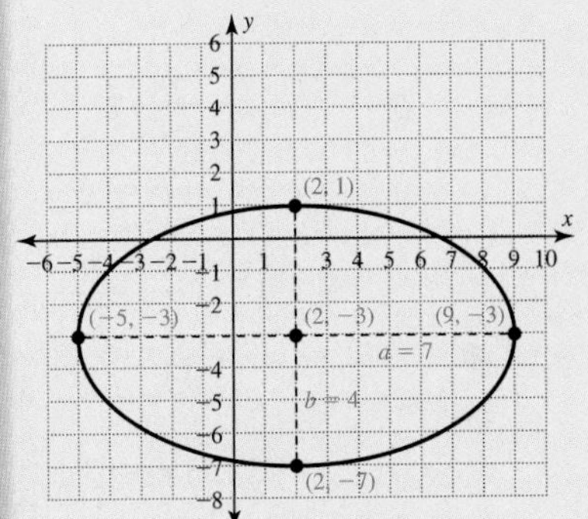
Vertical major axis: $\frac{(x-h)^2}{b^2} + \frac{(y-k)^2}{a^2} = 1$	$\frac{(x+4)^2}{16} + \frac{(y-5)^2}{36} = 1$
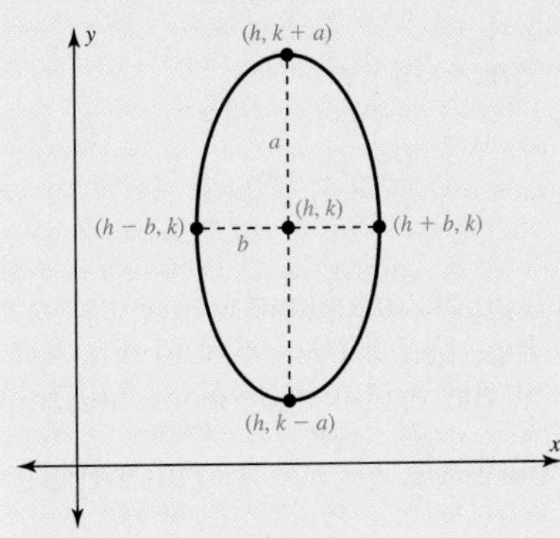	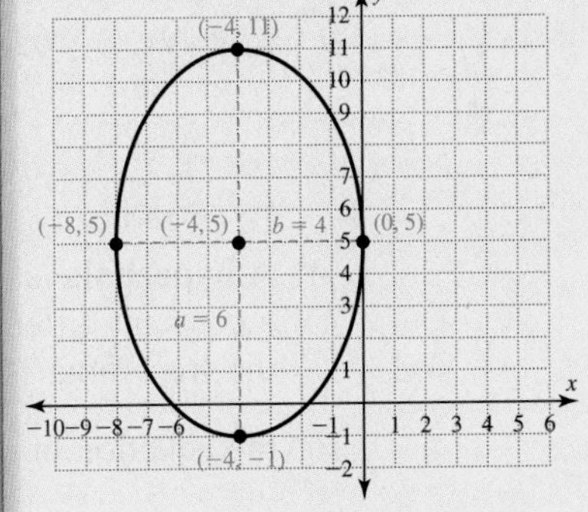

Note: $a > b > 0$.

EXAMPLE 4 Writing the Equation of an Ellipse in Standard Form

Determine the equation of the graphed ellipse.

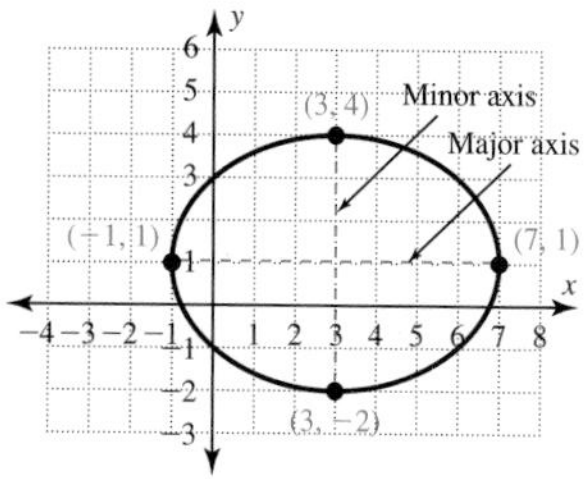

SOLUTION

$$(h, k) = \left(\frac{-1+7}{2}, \frac{1+1}{2}\right)$$
$$(h, k) = (3, 1)$$

The center of the ellipse is at the midpoint of the major axis.

$$2a = \sqrt{(7-(-1))^2 + (1-1)^2}$$
$$2a = \sqrt{8^2 + 0^2}$$
$$2a = \sqrt{64}$$
$$2a = 8$$
$$a = 4$$

The length of the major axis is $2a$, and the length of the minor axis is $2b$. Use the distance formula to determine these values.

$$2b = \sqrt{(3-3)^2 + (4-(-2))^2}$$
$$2b = \sqrt{0^2 + 6^2}$$
$$2b = \sqrt{36}$$
$$2b = 6$$
$$b = 3$$

$$\frac{(x-h)^2}{a^2} + \frac{(y-k)^2}{b^2} = 1$$

The horizontal axis is the major axis; thus we use the standard form of an ellipse with a horizontal major axis.

$$\frac{(x-3)^2}{4^2} + \frac{(y-1)^2}{3^2} = 1$$

Substitute a, b, h, and k into the standard form.

$$\frac{(x-3)^2}{16} + \frac{(y-1)^2}{9} = 1$$

SELF-CHECK 11.7.3

Write the equation of each ellipse in standard form and graph the ellipse.

1. $9x^2 + 36y^2 = 324$
2. $9x^2 + 16y^2 - 72x - 96y + 144 = 0$

Hyperbola

A **hyperbola** is the set of all points in a plane whose distances from two fixed points have a constant difference. (See Fig. 11.7.5.) The fixed points are the **foci** of the hyperbola. If the hyperbola is centered at the origin and opens horizontally as in Fig. 11.7.5, then the equation of the hyperbola is $\frac{x^2}{a^2} - \frac{y^2}{b^2} = 1$. This hyperbola is asymptotic to the lines $y = -\frac{b}{a}x$ and $y = \frac{b}{a}x$. As $|x|$ becomes larger, the hyperbola gets closer and closer to these lines. The asymptotes pass through the corners of the rectangle formed by (a, b), $(-a, b)$, $(-a, -b)$, and $(a, -b)$. This rectangle, which is shown in Fig. 11.7.6, is called the **fundamental rectangle** and is used to sketch quickly the linear asymptotes.

Figure 11.7.5 $|d_1 - d_2| = 2a$.

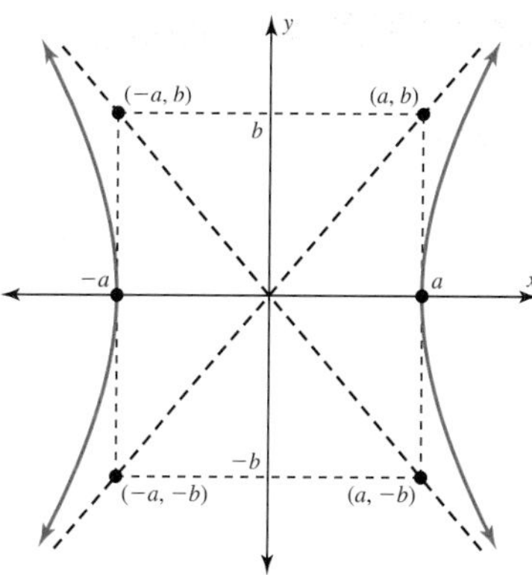

Figure 11.7.6 $\dfrac{x^2}{a^2} - \dfrac{y^2}{b^2} = 1$.

EXAMPLE 5 Graphing a Hyperbola Centered at the Origin

Graph $\dfrac{x^2}{16} - \dfrac{y^2}{9} = 1$.

SOLUTION

Plot the x-intercepts $(-4, 0)$ and $(4, 0)$. Sketch the fundamental rectangle with corners $(4, 3)$, $(-4, 3)$, $(-4, -3)$, and $(4, -3)$. Draw the asymptotes through the corners of this rectangle. Plot the x-intercepts and then sketch the hyperbola that opens horizontally, using the asymptotes as guidelines.

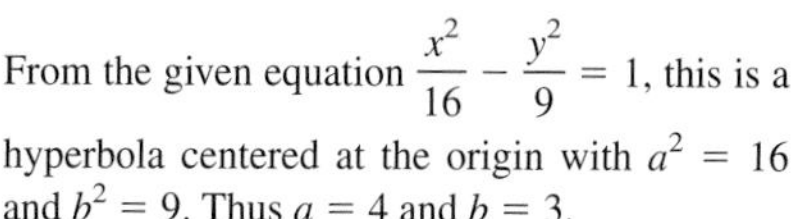

From the given equation $\dfrac{x^2}{16} - \dfrac{y^2}{9} = 1$, this is a hyperbola centered at the origin with $a^2 = 16$ and $b^2 = 9$. Thus $a = 4$ and $b = 3$.

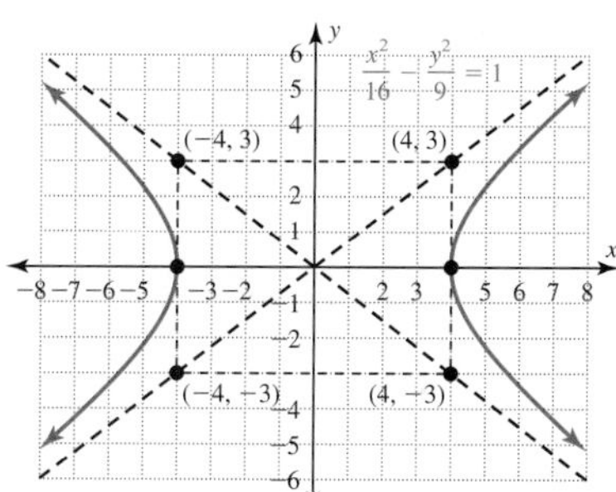

We now use translations to examine hyperbolas whose center is (h, k). The following box covers both hyperbolas that open horizontally and those that open vertically.

Standard Form of the Equation of a Hyperbola

The equation of a hyperbola with center (h, k) is

ALGEBRAICALLY	ALGEBRAIC EXAMPLE
Opening horizontally: $\dfrac{(x-h)^2}{a^2} - \dfrac{(y-k)^2}{b^2} = 1$	$\dfrac{(x+1)^2}{16} - \dfrac{(y-2)^2}{9} = 1$
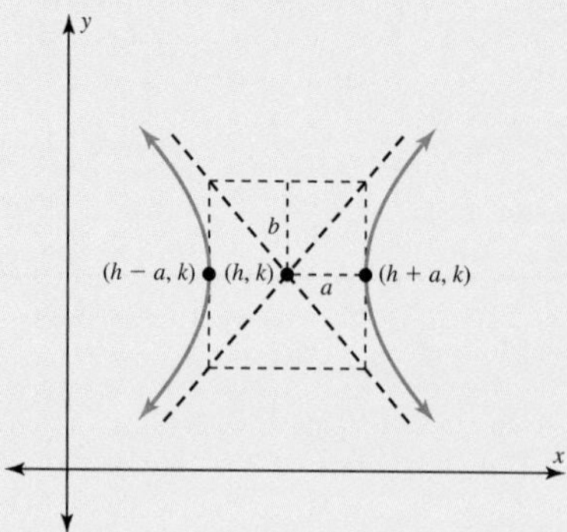	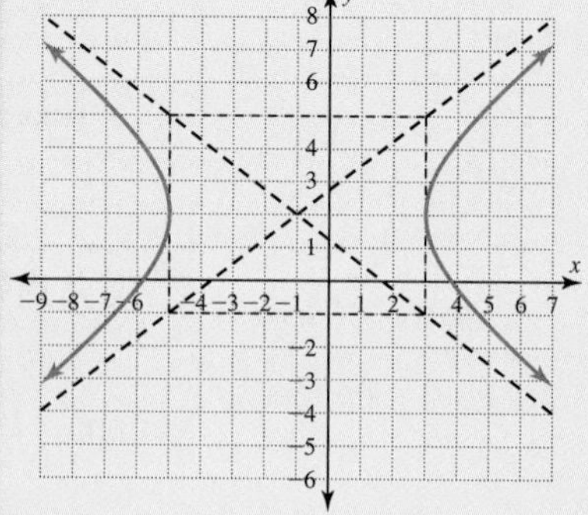
Vertices: $(h - a, k)$ and $(h + a, k)$. The fundamental rectangle has base $2a$ and height $2b$.	
Opening vertically: $\dfrac{(y-k)^2}{a^2} - \dfrac{(x-h)^2}{b^2} = 1$	$\dfrac{(y-2)^2}{9} - \dfrac{(x+1)^2}{16} = 1$
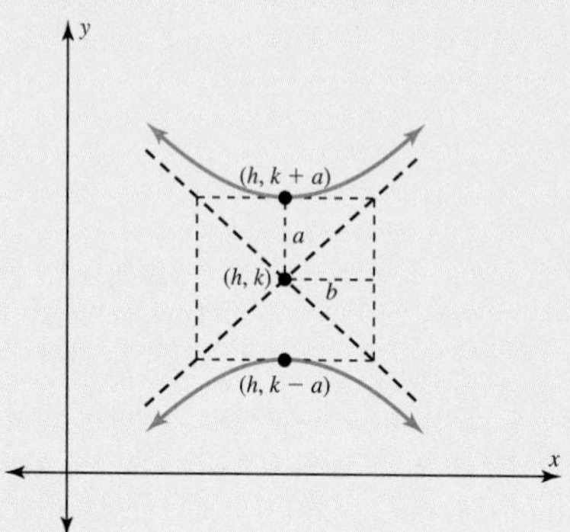	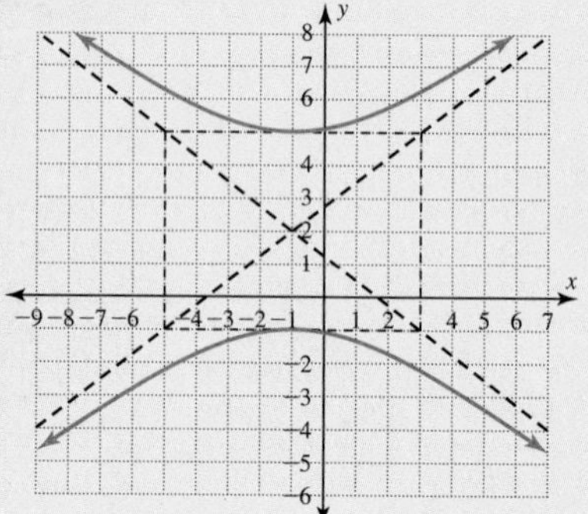
Vertices: $(h, k - a)$ and $(h, k + a)$. The fundamental rectangle has base $2b$ and height $2a$.	

EXAMPLE 6 Writing the Equation of a Hyperbola in Standard Form

Determine the equation of the hyperbola graphed in the figure.

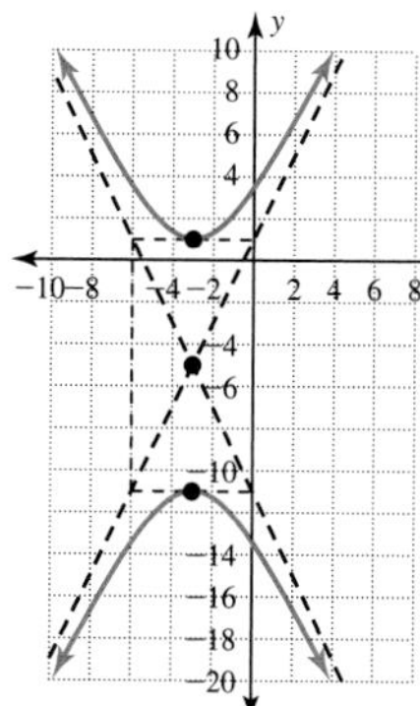

SOLUTION

$(h, k) = (-3, -5)$

$$\frac{(y-k)^2}{a^2} - \frac{(x-h)^2}{b^2} = 1$$

The center of the hyperbola was determined by inspection.
Select the form for a hyperbola that opens vertically.

$$2a = \sqrt{[-3-(-3)]^2 + [1-(-11)]^2}$$
$$2a = \sqrt{0 + 12^2} = \sqrt{144}$$
$$2a = 12$$
$$a = 6$$

Use points on the fundamental rectangle to compute the values of a and b.

$$2b = \sqrt{(-6-0)^2 + [-5-(-5)]^2}$$
$$2b = \sqrt{(-6)^2 + 0} = \sqrt{36}$$
$$2b = 6$$
$$b = 3$$

$$\frac{[y-(-5)]^2}{6^2} - \frac{[x-(-3)]^2}{3^2} = 1$$
$$\frac{(y+5)^2}{36} - \frac{(x+3)^2}{9} = 1$$

Substitute $(-3, -5)$ for (h, k) and 6 for a and 3 for b into the standard form for this hyperbola.

SELF-CHECK 11.7.4

Graph each hyperbola.

1. $\frac{x^2}{25} - \frac{y^2}{4} = 1$
2. $\frac{(y-2)^2}{4} - \frac{(x-3)^2}{25} = 1$

Parabolas

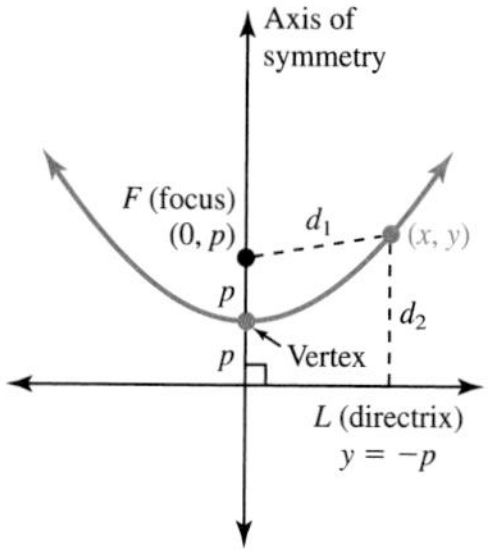

Figure 11.7.7 $d_1 = d_2$.

We already have graphed parabolas in several sections of this book (including Section 7.3). For completeness of this introduction to conic sections, we now note that a parabola can be formed by cutting a cone with a plane. A **parabola** is the set of all points in a plane the same distance from a fixed line L (the **directrix**) and from a fixed point F (the **focus**). See Fig. 11.7.7. Note that the axis of symmetry passes through the vertex, and it is perpendicular to the directrix. If the vertex of the parabola in Fig. 11.7.7 is at the origin, then the equation of this parabola can be written in the form $y = \frac{1}{4p}x^2$. If the vertex is translated to the point (h, k), then the equation becomes $y - k = \frac{1}{4p}(x-h)^2$.

The equation $y + 2 = \frac{1}{4\left(\frac{1}{4}\right)}(x-3)^2$ can also be rewritten in the form $y = (x-3)^2 - 2$. This form reveals the horizontal and vertical translations that we covered in Section 11.3.

EXAMPLE 7 Graphing a Parabola

Graph $y = (x - 3)^2 - 2$.

SOLUTION

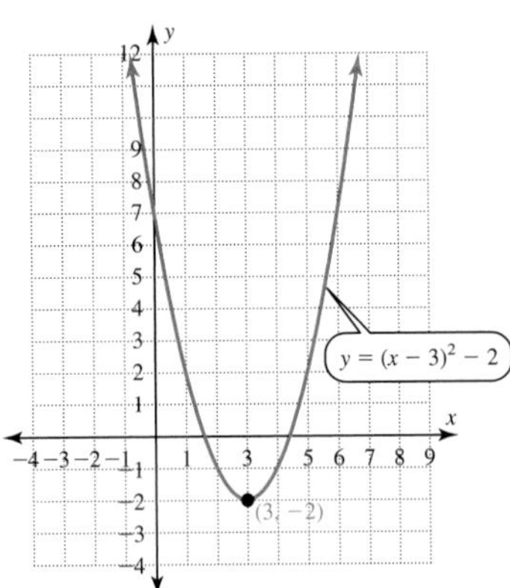

This parabola can be obtained by translating the graph of $y = x^2$ right 3 units and down 2 units. The vertex of this parabola is $(3, -2)$, and the y-intercept is $(0, 7)$.

SELF-CHECK 11.7.5

1. Graph $y = -(x + 2)^2 + 1$.

Example 8 involves a circle and a parabola and uses the graphs of these conic sections to solve the corresponding system of equations and inequalities.

EXAMPLE 8 Solving a Nonlinear System of Equations and a Nonlinear System of Inequalities

Use graphs to solve.

(a) $\begin{Bmatrix} x^2 + y^2 = 25 \\ y = x^2 - 5 \end{Bmatrix}$

(b) $\begin{Bmatrix} x^2 + y^2 \leq 25 \\ y \geq x^2 - 5 \end{Bmatrix}$

SOLUTION

(a)

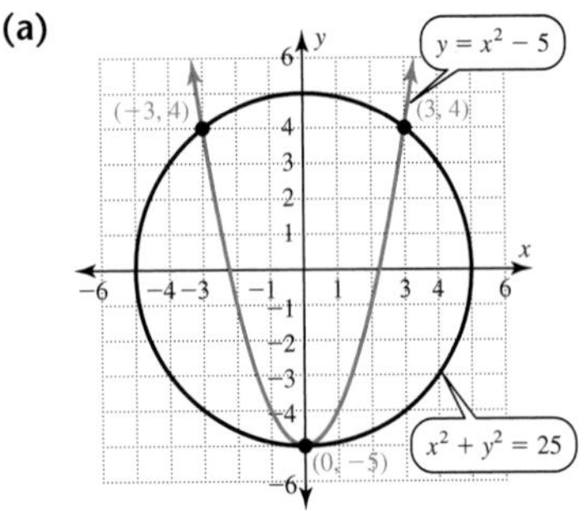

$x^2 + y^2 = 25$ is a circle centered at the origin with radius $r = 5$.
$y = x^2 - 5$ is a parabola opening upward with its vertex at $(0, -5)$.

$(-3, 4)$, $(0, -5)$, and $(3, 4)$ all satisfy both equations.

Do all these points check?

(b)

The points satisfying $x^2 + y^2 \leq 25$ lie on and inside the circle. Use (0, 0) as a test value. The points satisfying $y \geq x^2 - 5$ lie on and above the parabola. Use (0, 0) as a test value.

The shaded points satisfy both inequalities.

SELF-CHECK ANSWERS

11.7.1

1. (0, 0)
2. 10
3. 5
4. $x^2 + y^2 = 25$

11.7.2

1. $(x - 3)^2 + (y + 4)^2 = \frac{1}{4}$ with center $(3, -4)$ and radius $r = \frac{1}{2}$.

11.7.3

1. $\frac{x^2}{36} + \frac{y^2}{9} = 1$

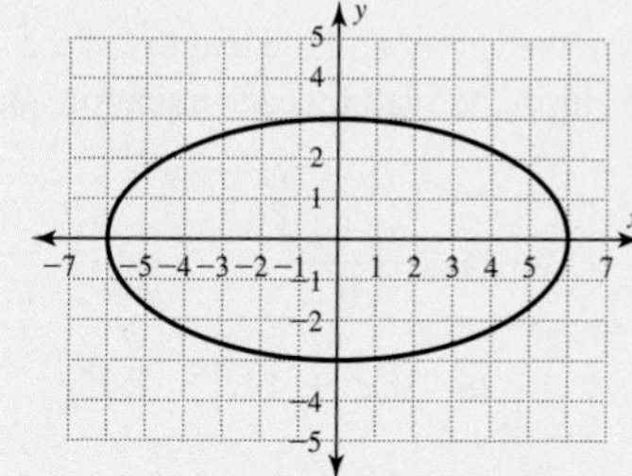

2. $\frac{(x - 4)^2}{16} + \frac{(y - 3)^2}{9} = 1$

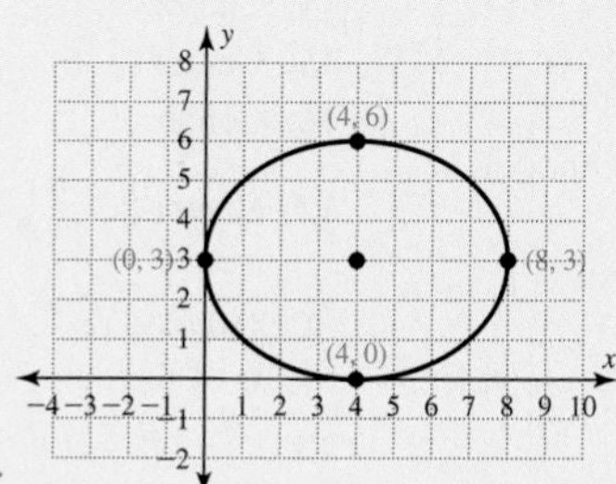

11.7.4

1.

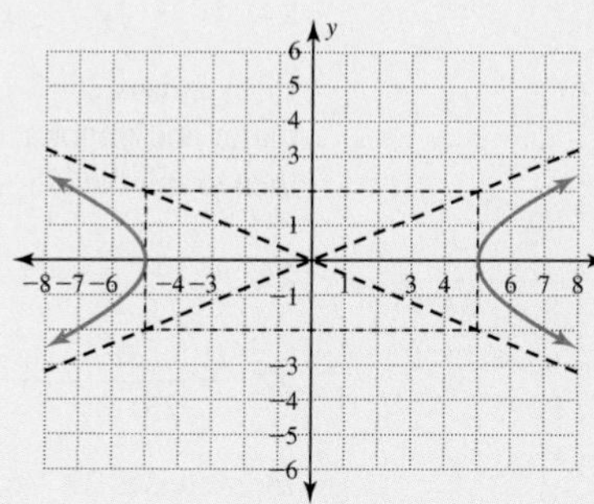

2.

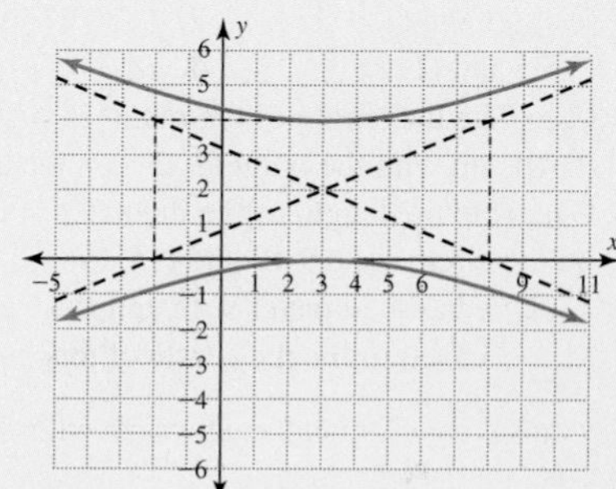

11.7.5

1.

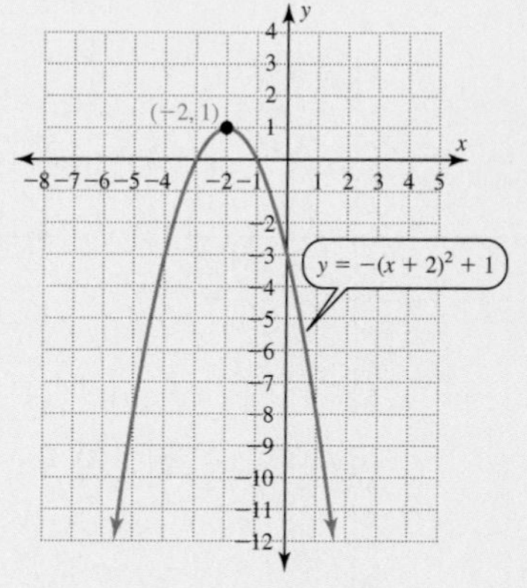

USING THE LANGUAGE AND SYMBOLISM OF MATHEMATICS 11.7

1. Parabolas, circles, ellipses, and hyperbolas are collectively referred to as __________ __________.
2. A __________ is the set of all points in a plane that are a constant distance from a fixed point.
3. An __________ is the set of all points in a plane the sum of whose distances from two fixed points is constant.
4. A __________ is the set of all points in a plane whose distances from two fixed points have a constant difference.
5. A __________ is the set of all points in a plane the same distance from a fixed line and a fixed point.
6. A line segment from the center of a circle to a point on the circle is a __________.
7. A line segment from one point on a circle through the center to another point on the circle is a __________.
8. The ends of the major axis of an ellipse are called __________.
9. The ends of the minor axis of an ellipse are called __________.
10. The hyperbola $\frac{x^2}{a^2} - \frac{y^2}{b^2} = 1$ is __________ to the lines $y = -\frac{b}{a}x$ and $y = \frac{b}{a}x$.

QUICK REVIEW 11.7

The purpose of this quick review is to help you recall skills needed in this section.

1. The vertex of the parabola defined by $ax^2 + bx + c = 0$ has an x-coordinate of __________ and a y-coordinate of $f\left(-\frac{b}{2a}\right)$.
2. The vertex of the parabola defined by $f(x) = 2x^2 + 4x + 3$ is __________.
3. Expand $9(x - 2)^2 + 4(y + 1)^2 - 36$ and simplify the result.
4. Solve $2x^2 + 5x + 7 = 0$ by the completing the square.
5. Complete the square to rewrite the function $f(x) = x^2 + 12x - 30$ in the form $f(x) = a(x - h)^2 + k$.

EXERCISES 11.7

In Exercises 1–4, calculate the distance between each pair of points and the midpoint between these points.

1. $(-3, 2)$ and $(1, -1)$
2. $(-6, 2)$ and $(6, 3)$
3. $(0, 1)$ and $(1, 2)$
4. $(a + 1, b)$ and $(a - 3, b + 3)$
5. Calculate the length of the radius of a circle with center at $(0, 0)$ and with the point $(7, 24)$ on the circle.
6. Calculate the length of the diameter of a circle with endpoints on the circle at $(-8, 0)$ and $(15, 0)$.
7. A circle has a diameter with endpoints at $(-1, 5)$ and $(5, -3)$. Determine the center of this circle.

In Exercises 8–12, match each graph with the corresponding equation.

8. $y = x^2 + 4$
9. $x^2 + y^2 = 4$
10. $\frac{x^2}{1} + \frac{y^2}{4} = 1$
11. $\frac{x^2}{1} - \frac{y^2}{4} = 1$
12. $y = x + 4$

A.

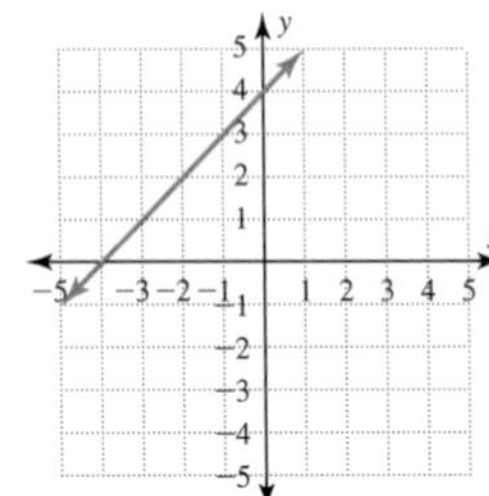

B.

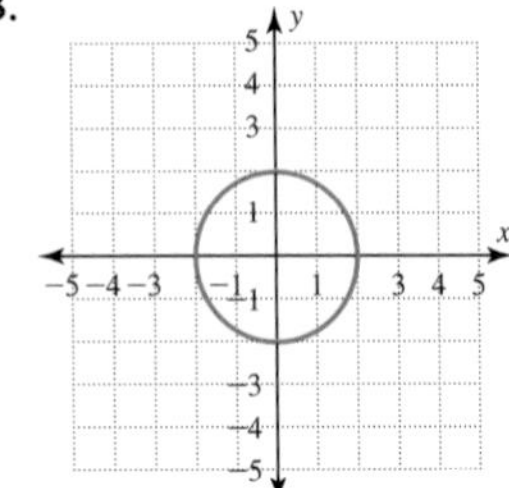

C.

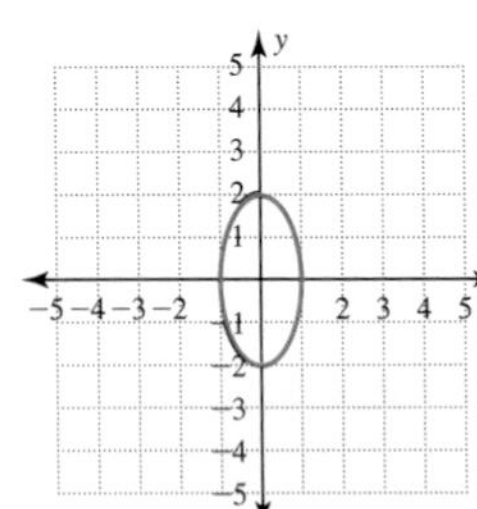

D.

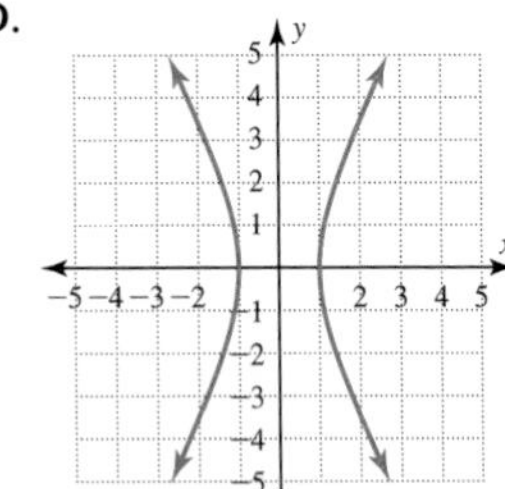

E.

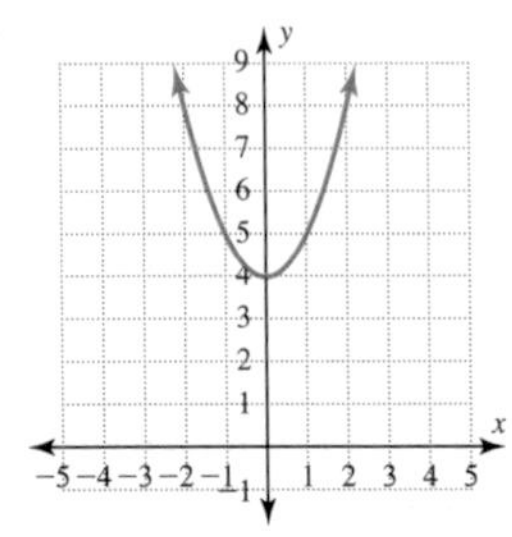

In Exercises 13–17, write in standard form the equation of the circle satisfying the given conditions.

13. Center $(0, 0)$, radius 10
14. Center $(5, -4)$, radius 4
15. Center $(2, 6)$, radius $\sqrt{2}$
16. Center $(1, 1)$, radius 0.1
17. Center $\left(0, \frac{1}{2}\right)$, radius $\frac{1}{2}$

In Exercises 18–24, determine the center and the length of the radius of the circle defined by each equation.

18. $x^2 + y^2 = 144$
19. $(x + 5)^2 + (y - 4)^2 = 64$
20. $(x - 4)^2 + (y + 5)^2 = 36$
21. $x^2 + y^2 - 6x = 0$
22. $x^2 + y^2 + 12y - 13 = 0$
23. $x^2 + y^2 - 2x + 10y + 22 = 0$
24. $4x^2 + 4y^2 - 8x - 40y + 103 = 0$

In Exercises 25–28, determine the center and the lengths of the major and minor axes of the ellipse defined by each equation, and then graph the ellipse.

25. $\frac{x^2}{36} + \frac{y^2}{16} = 1$
26. $\frac{(x - 1)^2}{9} + \frac{(y + 2)^2}{49} = 1$
27. $\frac{(y - 3)^2}{25} + \frac{(x + 4)^2}{16} = 1$
28. $36x^2 + 49y^2 = 1{,}764$

In Exercises 29–36, write in standard form the equation of the ellipse satisfying the given conditions.

29. Center $(0, 0)$, x-intercepts $(-9, 0)$ and $(9, 0)$, y-intercepts $(0, -5)$ and $(0, 5)$
30. Center $(0, 0)$, x-intercepts $(-5, 0)$ and $(5, 0)$, y-intercepts $(0, -3)$ and $(0, 3)$
31. Center $(3, 4)$, horizontal major axis of length 4, vertical minor axis of length 2
32. Center $(-5, 2)$, horizontal major axis of length 6, vertical minor axis of length 2
33. Center $(6, -2)$, vertical major axis of length 10, horizontal minor axis of length 6
34. Center $(-3, -4)$, vertical major axis of length 8, horizontal minor axis of length $\frac{1}{2}$
35. Center $(-3, -4)$, $a = 6$, $b = 2$, major axis vertical
36. Center $(2, 5)$, $a = 8$, $b = 4$, major axis horizontal

In Exercises 37–42, determine the center, the values of a and b, and the direction in which the hyperbola opens, and then sketch the hyperbola.

37. $\frac{x^2}{25} - \frac{y^2}{81} = 1$ **38.** $\frac{y^2}{36} - \frac{x^2}{16} = 1$

39. $25x^2 = 100y^2 + 100$ **40.** $\frac{(x-4)^2}{49} - \frac{(y-6)^2}{9} = 1$

41. $\frac{(y-6)^2}{9} - \frac{(x-4)^2}{49} = 1$

42. $16x^2 + 96x = 9y^2 - 126y + 153$

In Exercises 43–46, write the standard form of the equation of the hyperbola satisfying the given conditions.

43. The center is (0, 0), the hyperbola opens vertically, and the fundamental rectangle has height 10 and width 8.

44. The center is (0, 0), the hyperbola opens horizontally, and the fundamental rectangle has height 6 and width 12.

45. The hyperbola has vertices $(-3, 0)$ and $(3, 0)$, and the height of the fundamental rectangle is 14.

46. The hyperbola has vertices $(0, -5)$ and $(0, 5)$, and the width of the fundamental rectangle is 16.

In Exercises 47–50, graph each parabola.

47. **a.** $y = x^2$
b. $y = x^2 - 5$
c. $y = (x+2)^2 - 5$

48. **a.** $y = -x^2$
b. $y = -x^2 + 3$
c. $y = -(x-1)^2 + 3$

49. **a.** $y = x^2$
b. $y = \frac{1}{2}x^2$
c. $y = -2x^2$

50. **a.** $y = x^2$
b. $y = \frac{1}{3}x^2$
c. $y = -3x^2$

In Exercises 51–56, use a graphing calculator to graph each pair of functions.

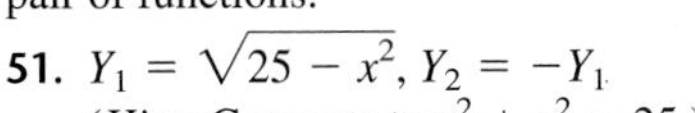

51. $Y_1 = \sqrt{25 - x^2}$, $Y_2 = -Y_1$
(*Hint:* Compare to $x^2 + y^2 = 25$.)

52. $Y_1 = \sqrt{16 - (x+2)^2}$, $Y_2 = -Y_1$
(*Hint:* Compare to $(x+2)^2 + y^2 = 16$.)

53. $Y_1 = \frac{1}{3}\sqrt{36 - 4x^2}$, $Y_2 = -Y_1$
$\left(\textit{Hint: }\text{Compare to } \frac{x^2}{9} + \frac{y^2}{4} = 1.\right)$

54. $Y_1 = \frac{1}{3}\sqrt{36 + 4x^2}$, $Y_2 = -Y_1$
$\left(\textit{Hint: }\text{Compare to } \frac{y^2}{4} - \frac{x^2}{9} = 1.\right)$

55. Split $x^2 + y^2 = 36$ into Y_1 and Y_2, which represent upper and lower semicircles. Then graph Y_1 and Y_2.

56. Split $\frac{x^2}{25} + \frac{y^2}{16} = 1$ into Y_1 and Y_2, which represent upper and lower semiellipses. Then graph Y_1 and Y_2.

In Exercises 57–60, use the given graphs to solve each system of equations and each system of inequalities.

57. **a.** $\begin{Bmatrix} y = x^2 - 2x - 1 \\ 2x + y = 3 \end{Bmatrix}$
b. $\begin{Bmatrix} y \geq x^2 - 2x - 1 \\ y \leq -2x + 3 \end{Bmatrix}$

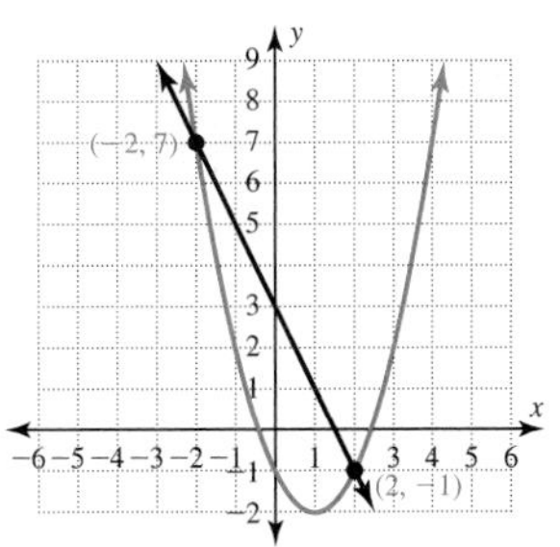

58. **a.** $\begin{Bmatrix} x + y = -1 \\ (x-4)^2 + (y+1)^2 = 40 \end{Bmatrix}$
b. $\begin{Bmatrix} y \leq -x - 1 \\ (x-4)^2 + (y+1)^2 \leq 40 \end{Bmatrix}$

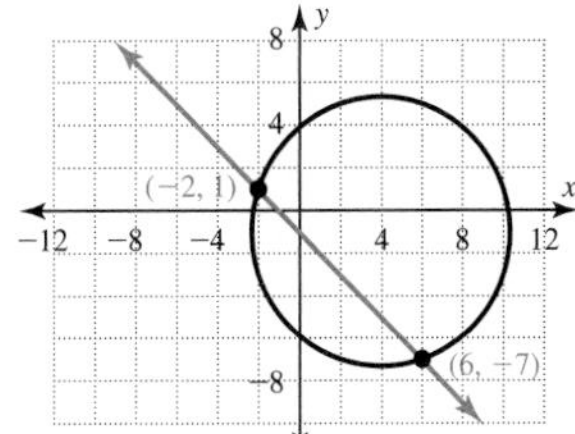

59. **a.** $\begin{Bmatrix} x^2 + y^2 = 8 \\ y = x \end{Bmatrix}$
b. $\begin{Bmatrix} x^2 + y^2 \leq 8 \\ y \geq x \end{Bmatrix}$

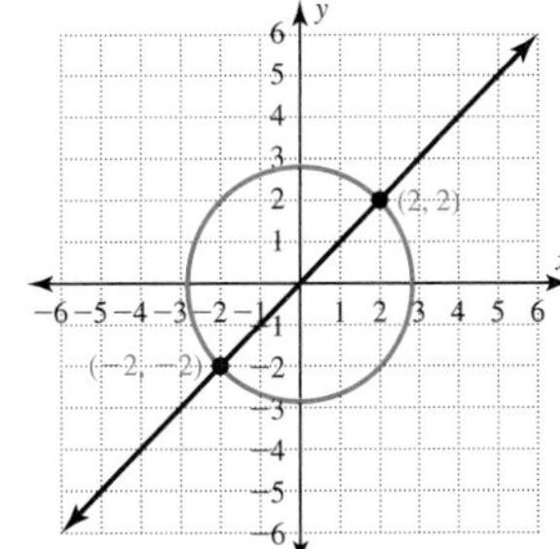

60. **a.** $\begin{Bmatrix} y = e^x \\ y = -x + 1 \end{Bmatrix}$
b. $\begin{Bmatrix} y \geq e^x \\ y \leq -x + 1 \end{Bmatrix}$

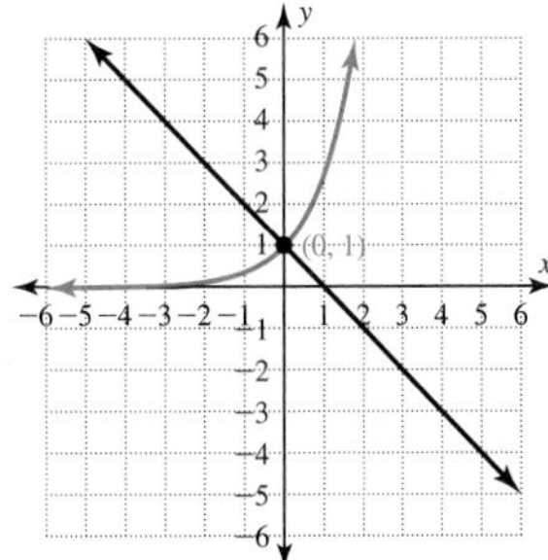

Group Discussion Questions

61. **Discovery Question** A line and a circle can intersect at 0, 1, or 2 points. Illustrate each possibility with a sketch.

62. **Discovery Question** A circle and an ellipse can intersect at 0, 1, 2, 3, or 4 points. Illustrate each possibility with a sketch.

63. **Discovery Question** A circle and a hyperbola can intersect at 0, 1, 2, 3, or 4 points. Illustrate each possibility with a sketch.

64. **Challenge Question** Use the following figure to complete each part of this question.

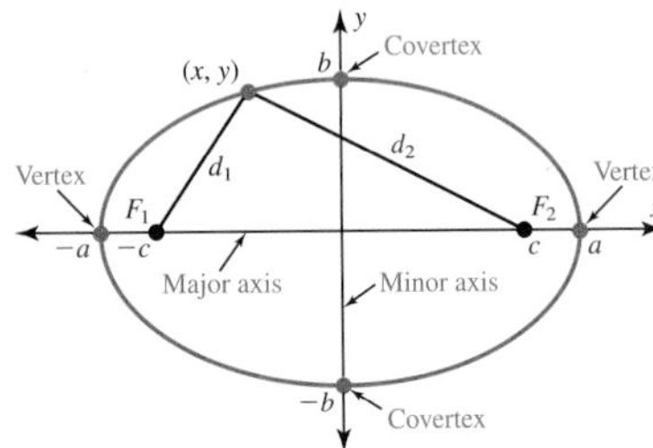

a. Place (x, y) at a convenient point on the figure, and present a justification for $d_1 + d_2 = 2a$.

b. Place (x, y) at a convenient point on the figure, and present a justification for $a^2 = b^2 + c^2$.

c. Use the distance formula to represent d_1 and d_2. Then substitute these values into $d_1 + d_2 = 2a$. Show that this equation can be simplified to $\frac{x^2}{a^2} + \frac{y^2}{b^2} = 1$.

KEY CONCEPTS FOR CHAPTER 11

1. **Methods for Solving 3 × 3 Systems of Linear Equations**
   - Use the addition method and the substitution method to eliminate variables.
   - Use the augmented matrix method.
2. **Augmented Matrix**
   - A matrix is a rectangular array of numbers consisting of rows and columns.
   - A row consists of entries arranged horizontally.
   - A column consists of entries arranged vertically.
   - The dimension of a matrix is given by first stating the number of rows and then the number of columns.
   - An augmented matrix for a system of linear equations consists of the coefficients and constants in the equation.
   - The augmented matrix for $\begin{cases} a_1x + b_1y + c_1z = d_1 \\ a_2x + b_2y + c_2z = d_2 \\ a_3x + b_3y + c_3z = d_3 \end{cases}$ is $\left[\begin{array}{ccc|c} a_1 & b_1 & c_1 & d_1 \\ a_2 & b_2 & c_2 & d_2 \\ a_3 & b_3 & c_3 & d_3 \end{array}\right]$.
3. **Elementary Row Operations on Augmented Matrices**
   - Any two rows in the matrix may be interchanged.
   - Any row in the matrix may be multiplied by a nonzero constant.
   - Any row in the matrix may be replaced by the sum of itself and a constant multiple of another row.
4. **Reduced Row Echelon Form of a Matrix**
   - The first nonzero entry in a row is 1. All other entries in the column containing the leading 1 are 0s.
   - All nonzero rows are above any rows containing only 0s.
   - The first nonzero entry in a row is to the left of the first nonzero entry in the following row.
5. **Equation of a Plane** The graph of a linear equation of the form $Ax + By + Cz = D$ is a plane in three-dimensional space.
6. **Basic Functions Examined in This Book** Each type of function has an equation of a standard form and a graph with a characteristic shape.
   - Linear functions: First-degree polynomial functions (The shape of the graph is a straight line.)
   - Quadratic functions: Second-degree polynomial functions (The shape of the graph is a parabola.)
   - Cubic functions: Third-degree polynomial functions
   - Absolute value functions (The graph is V-shaped.)
   - Square root functions
   - Cube root functions
   - Exponential functions
   - Logarithmic functions
   - Rational functions
7. **Vertical and Horizontal Translations of $y = f(x)$**
   If c is a positive real number,
   - $y = f(x) + c$ shifts the graph of $y = f(x)$ up c units.
   - $y = f(x) - c$ shifts the graph of $y = f(x)$ down c units.
   - $y = f(x + c)$ shifts the graph of $y = f(x)$ left c units.
   - $y = f(x - c)$ shifts the graph of $y = f(x)$ right c units.
8. **Stretching, Shrinking, and Reflecting $y = f(x)$**
   - $y = -f(x)$ reflects the graph of $y = f(x)$ across the x-axis.
   - $y = cf(x)$ stretches the graph of $y = f(x)$ vertically by a factor of c for $c > 1$.
   - $y = cf(x)$ shrinks the graph of $y = f(x)$ vertically by a factor of c for $0 < c < 1$.
9. **Operations on Functions** If f and g are functions, then for all input values in both the domain of f and g:
   - Sum $(f + g)(x) = f(x) + g(x)$
   - Difference $(f - g)(x) = f(x) - g(x)$
   - Product $(f \cdot g)(x) = f(x)g(x)$
   - Quotient $\left(\frac{f}{x}\right)(x) = \frac{f(x)}{g(x)}$ for $g(x) \neq 0$
10. **Composite Function $f \circ g$**
    - If f and g are functions, then $(f \circ g)(x) = f[g(x)]$ for all input values x in the domain of g for which $g(x)$ is an input value in the domain of f.
    - If f and f^{-1} are inverses of each other, then $(f \circ f^{-1})(x) = x$ for each input value of f^{-1} and $(f^{-1} \circ f)(x)$ for each input value of f.
11. **Sequences**
    - A sequence is a function whose domain is a set of consecutive natural numbers.
    - A finite sequence has a last term.
    - An infinite sequence continues without end.
    - A sequence is arithmetic if $a_n = a_{n-1} + d$; d is called the common difference.
    - The points of an arithmetic sequence lie on a line.
    - A sequence is geometric if $a_n = ra_{n-1}$; r is called the common ratio.
    - The points of a geometric sequence lie on an exponential curve.

12. **General Term of a Sequence**
- A formula for a_n is a formula for the general term of a sequence.
- A formula defining a_n in terms of one or more of the preceding terms is called a recursive definition for a_n.
- The Fibonacci sequence defined by $a_1 = 1$, $a_2 = 1$, and $a_n = a_{n-2} + a_{n-1}$ is an example of a recursive definition.
- A formula for the general term of an arithmetic sequence is $a_n = a_1 + (n - 1)d$.
- A formula for the general term of a geometric sequence is $a_n = a_1r^{n-1}$.

13. **Series**
- A series is a sum of the terms of a sequence.
- Summation notation is used to denote a series:
$$\sum_{i=1}^{n} a_i = a_1 + a_2 + \cdots + a_{n-1} + a_n$$
- The sum of the first n terms of an arithmetic sequence is
$$S_n = \frac{n}{2}(a_1 + a_n).$$
- The sum of the first n terms of a geometric sequence is
$$S_n = \frac{a_1(1 - r^n)}{1 - r}.$$
- If $|r| < 1$, the infinite geometric series $S = \sum_{i=1}^{\infty} a_i$ is $S = \frac{a_1}{1 - r}$. If $|r| \geq 1$, this sum does not exist.

14. **Distance and Midpoint Formulas** For two points (x_1, y_1) and (x_2, y_2):
- Distance: $d = \sqrt{(x_2 - x_1)^2 + (y_2 - y_1)^2}$
- Midpoint: $(x, y) = \left(\frac{x_1 + x_2}{2}, \frac{y_1 + y_2}{2}\right)$

15. **Standard Forms for Conic Sections**
- Circle: $(x - h)^2 + (y - k)^2 = r^2$
- Ellipse:
$\frac{(x - h)^2}{a^2} + \frac{(y - k)^2}{b^2} = 1$ with horizontal major axis
$\frac{(x - h)^2}{b^2} + \frac{(y - k)^2}{a^2} = 1$ with vertical major axis
- Hyperbola:
$\frac{(x - h)^2}{a^2} - \frac{(y - k)^2}{b^2} = 1$ opens horizontally
$\frac{(y - k)^2}{a^2} - \frac{(x - h)^2}{b^2} = 1$ opens vertically
- Parabola: $y - k = \frac{1}{4p}(x - h)^2$

REVIEW EXERCISES FOR CHAPTER 11

1. Write an augmented matrix for each system of linear equations.

a. $\begin{cases} 2x - 5y = 17 \\ 3x + 4y = 14 \end{cases}$ b. $\begin{cases} 3x - 4y + 2z = -11 \\ 2x + 2y + 3z = -3 \\ 4x - y + 5z = -13 \end{cases}$

2. Write the solution for the system of linear equations represented by each augmented matrix.

a. $\left[\begin{array}{cc|c} 1 & 0 & -2 \\ 0 & 1 & 6 \end{array}\right]$ b. $\left[\begin{array}{ccc|c} 1 & 0 & 0 & 4 \\ 0 & 1 & 0 & 5 \\ 0 & 0 & 1 & 8 \end{array}\right]$

3. Write the general solution and three particular solutions for the system of linear equations represented by this augmented matrix: $\left[\begin{array}{ccc|c} 1 & 0 & 3 & 5 \\ 0 & 1 & -2 & 4 \\ 0 & 0 & 0 & 0 \end{array}\right]$.

In Exercises 4–7 use the given elementary row operations to complete each matrix.

4. $\left[\begin{array}{cc|c} 3 & 5 & 13 \\ 1 & 4 & 2 \end{array}\right] \xrightarrow{r_1 \leftrightarrow r_2} \left[\begin{array}{cc|c} \_ & \_ & \_ \\ \_ & \_ & \_ \end{array}\right]$

5. $\left[\begin{array}{cc|c} 2 & 4 & -10 \\ 3 & 2 & -3 \end{array}\right] \xrightarrow{r_1' = \frac{1}{2}r_1} \left[\begin{array}{cc|c} \_ & \_ & \_ \\ 3 & 2 & -3 \end{array}\right]$

6. $\left[\begin{array}{cc|c} 1 & 2 & 14 \\ 3 & -1 & 13 \end{array}\right] \xrightarrow{r_2' = r_2 - 3r_1} \left[\begin{array}{cc|c} 1 & 2 & 14 \\ \_ & \_ & \_ \end{array}\right]$

7. $\left[\begin{array}{ccc|c} 1 & 3 & -2 & -6 \\ 0 & 1 & 2 & 3 \\ 0 & 2 & 3 & 4 \end{array}\right] \xrightarrow[r_3' = r_3 - 2r_2]{r_1' = r_1 - 3r_2} \left[\begin{array}{ccc|c} \_ & \_ & \_ & \_ \\ 0 & 1 & 2 & 3 \\ \_ & \_ & \_ & \_ \end{array}\right]$

In Exercises 8–14 solve each system of linear equations.

8. $x + 2y = 17$
$2x + 5y = 41$

9. $3x - 2y = -16$
$x + 3y = 13$

10. $3x + 4y = -1$
$2x - 3y = 5$

11. $x - 5y = 8$
$-3x + 15y = 10$

12. $x + 2y - z = -2$
$2x - y + z = 4$
$3x + 2y + 2z = 3$

13. $x - 3z = 10$
$2x + y = 6$
$2y + z = 5$

14. $2x - 3y + 4z = 11$
$3x + 2y - 4z = -4$
$4x + y - 3z = -1$

15. Graph the plane defined by the linear equation $5x + 2y + 2.5z = 10$.

16. **Weights of Pallets of Bricks and Blocks** A truck has an empty weight of 30,000 lb. On the first trip the truck delivered two pallets of bricks and three pallets of concrete blocks. The gross weight on this delivery was 48,000 lb. On the second trip the truck delivered four pallets of bricks and one pallet of concrete blocks. The gross weight on this

delivery was 49,600 lb. Use this information to determine the weight of a pallet of bricks and the weight of a pallet of concrete blocks.

17. Price of Calculators A department store chain with three store retails calculators of types A, B, and C. The table shows the number sold of each type of calculator and the total income from these sales at each store. Find the price of each type of calculator.

	A	B	C	TOTAL SALES ($)
Store 1	1	2	5	156
Store 2	2	4	8	294
Store 3	3	3	7	321

18. Match each function with its graph. All graphs are displayed with the window $[-6, 6, 1]$ by $[-6, 16, 1]$.

a. $f(x) = x^2$ **b.** $f(x) = x^2 - 5$
c. $f(x) = x^2 + 5$

A.
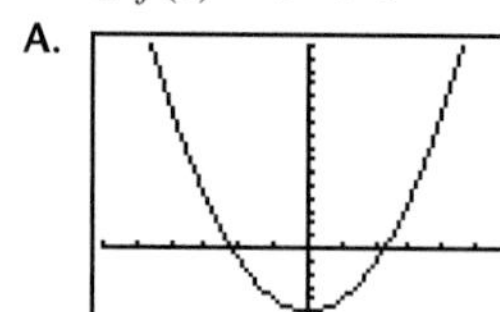

B.
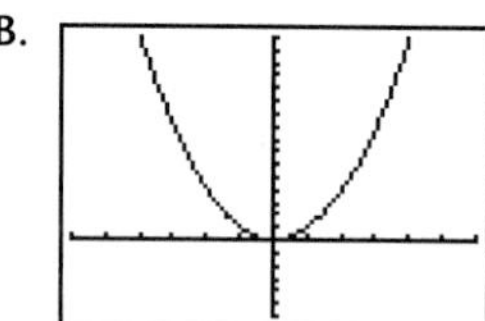

C.
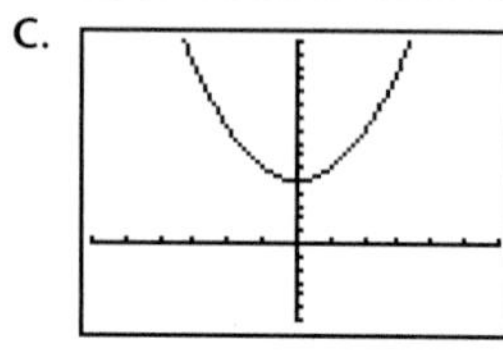

19. Match each function with its graph. All graphs are displayed with the window $[-5, 10, 1]$ by $[-5, 5, 1]$.

a. $f(x) = \sqrt{x}$ **b.** $f(x) = \sqrt{x - 4}$
c. $f(x) = \sqrt{x + 4}$

A.
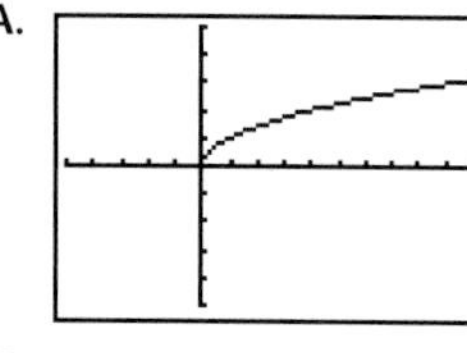

B.
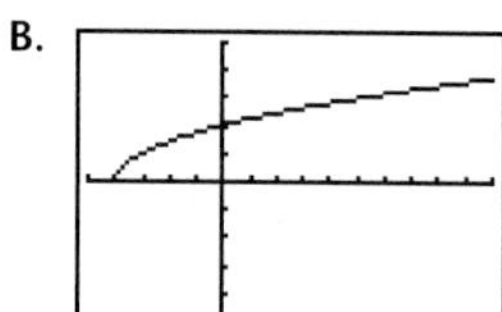

C.

20. Match each function with its graph. All graphs are displayed with the window $[-5, 5, 1]$ by $[-5, 5, 1]$.

a. $f(x) = |x - 1| + 2$ **b.** $f(x) = |x - 2| + 1$
c. $f(x) = |x + 2| - 1$

A.
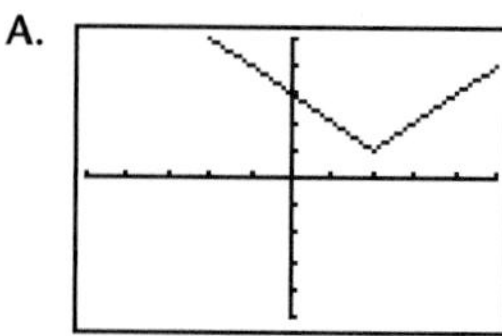

B.
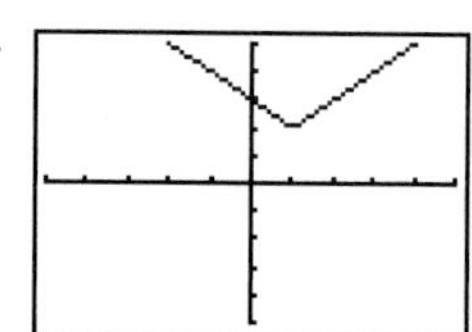

C.
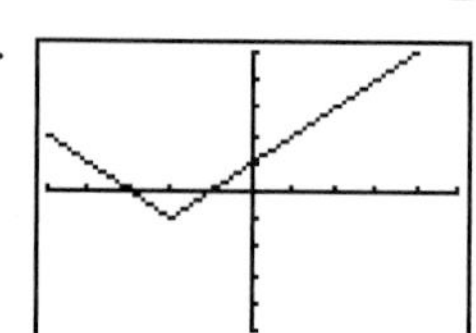

21. Use the given graph of $y = f(x)$ to graph each function.

a. $y = f(x - 3)$
b. $y = f(x) + 4$
c. $y = f(x - 3) + 4$

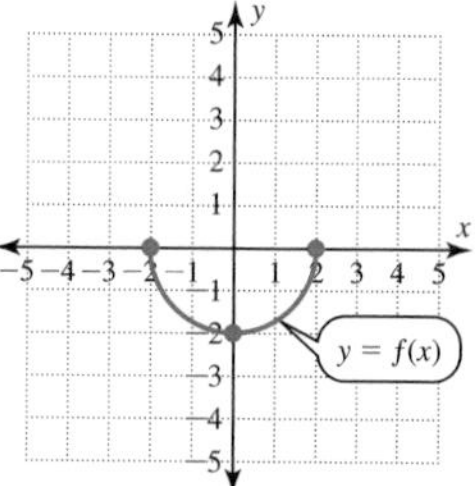

22. Determine the vertex of each parabola by using the fact that the vertex of $f(x) = x^2$ is at (0, 0).

a. $f(x) = x^2 + 7$ **b.** $f(x) = (x - 9)^2$
c. $f(x) = (x + 11)^2 - 12$

23. Use the given table of values to complete each table.

x	$y = f(x)$
−2	5
−1	9
0	3
1	−4
2	0

a.

x	$y = f(x) + 5$
−2	
−1	
0	
1	
2	

b.

x	$y = f(x + 5)$
	5
	9
	3
	−4
	0

24. The graph of each of these functions is a translation of the graph of $y = f(x)$. Match each function to the correct description.

a. $y = f(x) - 17$ **A.** A translation 17 units right
b. $y = f(x - 17)$ **B.** A translation 17 units up
c. $y = f(x) + 17$ **C.** A translation 17 units left
d. $y = f(x + 17)$ **D.** A translation 17 units down

25. Use the given table and match each function with its table.

x	$y = f(x)$
0	5
1	−3
2	0
3	2
4	−6

a. $y = f(x) + 2$

b. $y = f(x + 2)$

c. $y = f(x - 1) + 3$

d. $y = f(x + 3) - 1$

A.

x	y
−3	4
−2	−4
−1	−1
0	1
1	−7

B.

x	y
−2	5
−1	−3
0	0
1	2
2	−6

C.

x	y
1	8
2	0
3	3
4	5
5	−3

D.

x	y
0	7
1	−1
2	2
3	4
4	−4

26. Use the graph of $y = f(x)$ shown and match each function with its graph.

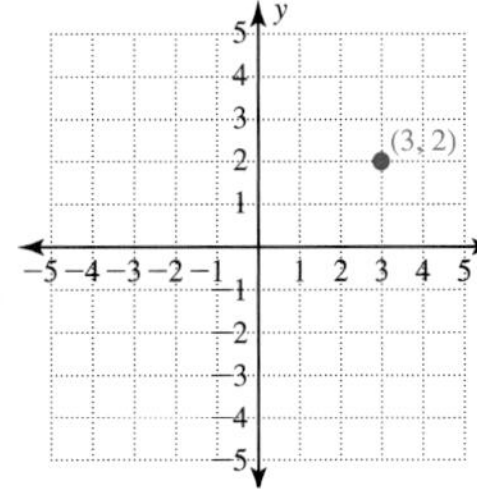

a. $y = -f(x)$

b. $y = 2f(x)$

c. $y = -2f(x)$

d. $y = \frac{1}{2}f(x)$

A.

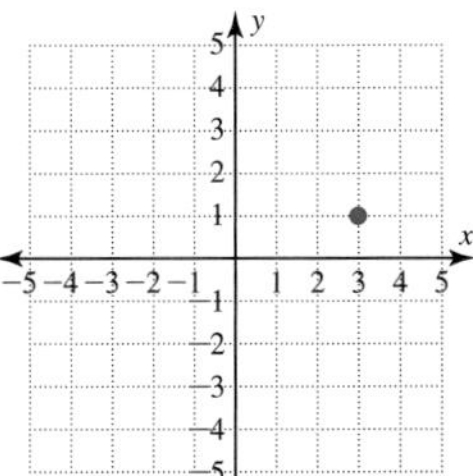

B.

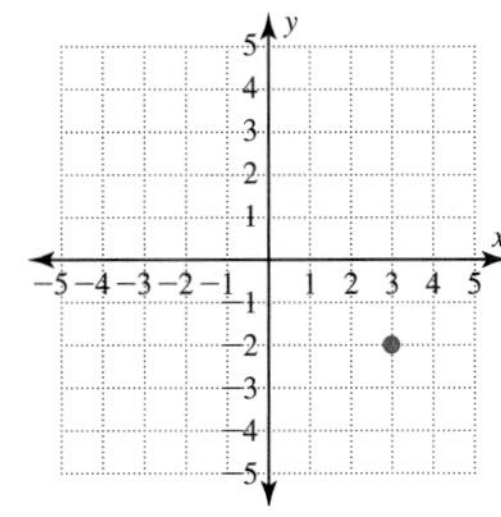

C.

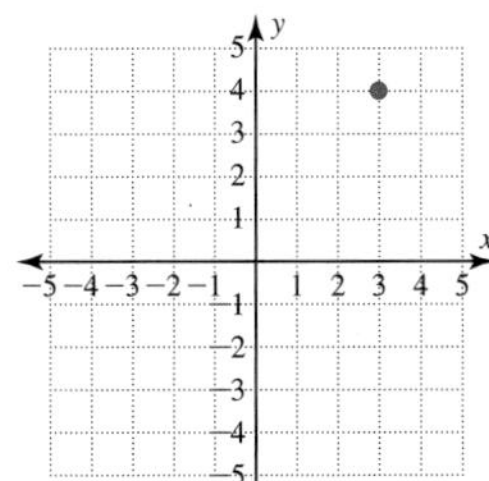

D.

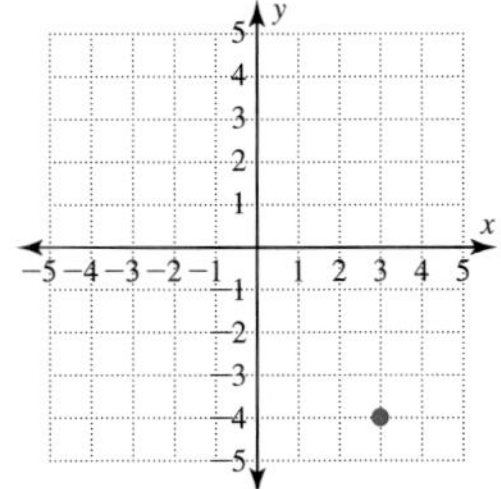

27. Use the graph of $y = f(x)$ shown and match each function with its graph.

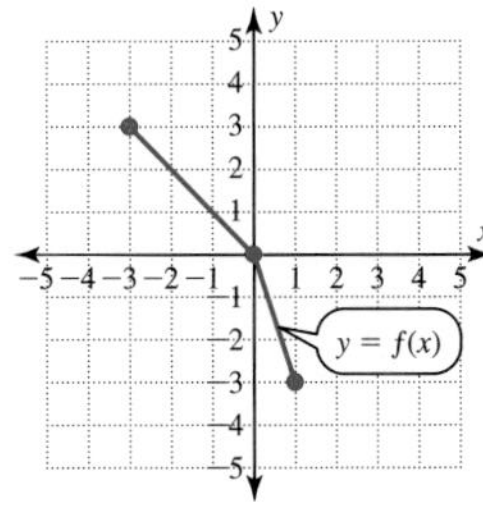

a. $y = -f(x)$

b. $y = -\frac{1}{3}f(x)$

c. $y = \frac{1}{3}f(x)$

d. $y = 2f(x)$

A.

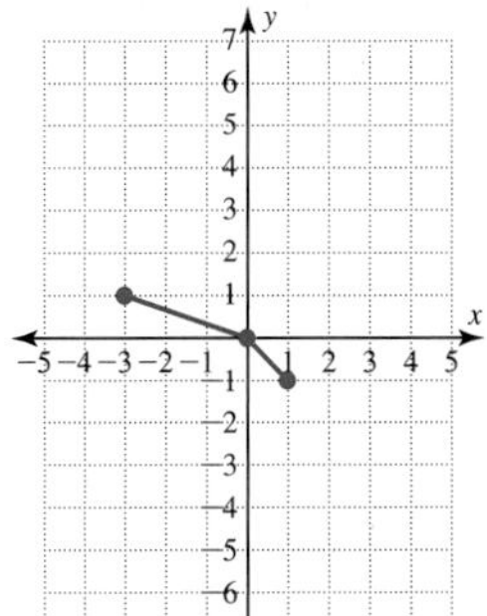

B.

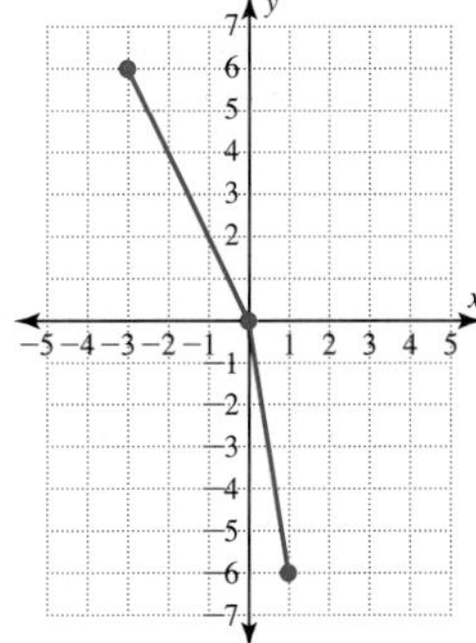

C.

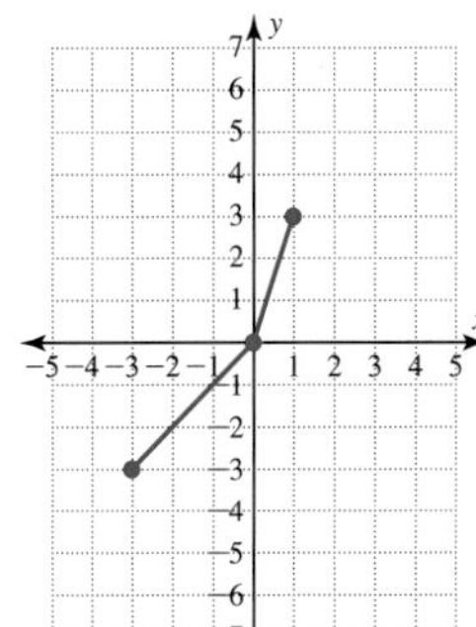

D.

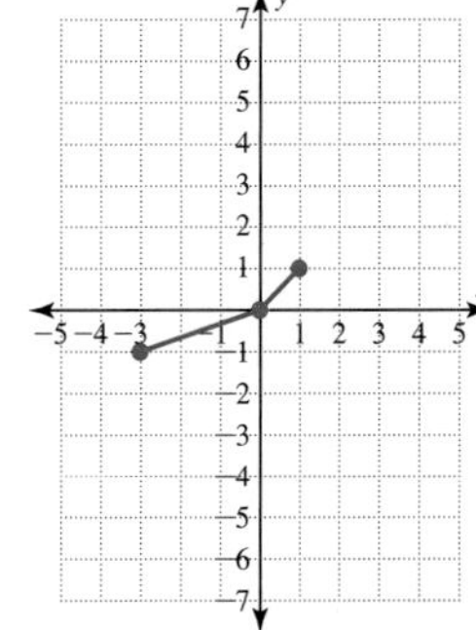

28. Use the given table of values for $y = f(x)$ to complete each table.

x	$y = f(x)$
0	0
1	8
2	4
3	12
4	24

a.

x	$y = -f(x)$
0	
1	
2	
3	
4	

b.

x	$y = \frac{1}{4}f(x)$
0	
1	
2	
3	
4	

c.

x	$y = 4f(x)$
0	
1	
2	
3	
4	

d.

x	$y = -3f(x)$
0	
1	
2	
3	
4	

29. Match each function with the description that compares its graph to the graph of $y = f(x)$.

a. $y = 2f(x)$

b. $y = -f(x)$

c. $y = f(x + 2)$

d. $y = f(x) + 2$

A. A reflection of $y = f(x)$ across the x-axis

B. A horizontal shift of $y = f(x)$ 2 units left

C. A vertical shift of $y = f(x)$ 2 units up

D. A vertical stretching of $y = f(x)$ by a factor 2

In Exercises 30–32, evaluate each expression, given $f(x) = 3x - 7$ and $g(x) = 2x^2 + 1$.

30. **a.** $f(10)$
b. $g(10)$
c. $(f + g)(10)$
d. $(f - g)(10)$

31. **a.** $(f \cdot g)(2)$
b. $\left(\frac{f}{g}\right)(2)$
c. $\left(\frac{g}{f}\right)(2)$
d. $(g - f)(2)$

32. **a.** $(f \circ g)(2)$
b. $(g \circ f)(2)$
c. $(f \circ f)(2)$
d. $(g \circ g)(2)$

33. Use the given tables for f and g to form a table of values for each function.

x	f(x)
0	−7
1	−4
2	−1
3	2

x	g(x)
0	1
1	3
2	9
3	19

a. $f + g$ **b.** $f - g$ **c.** $f \cdot g$ **d.** $\frac{f}{g}$

34. Use the given graphs for f and g to form a set of ordered pairs for each function.

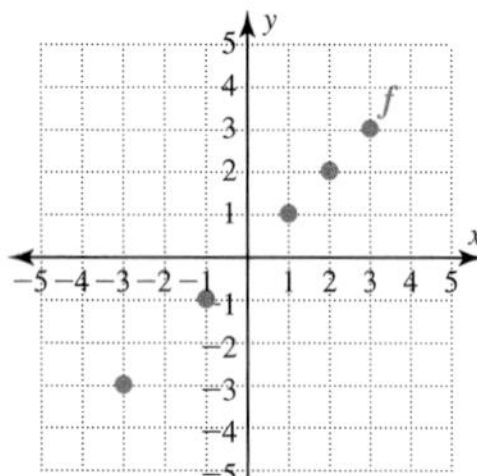

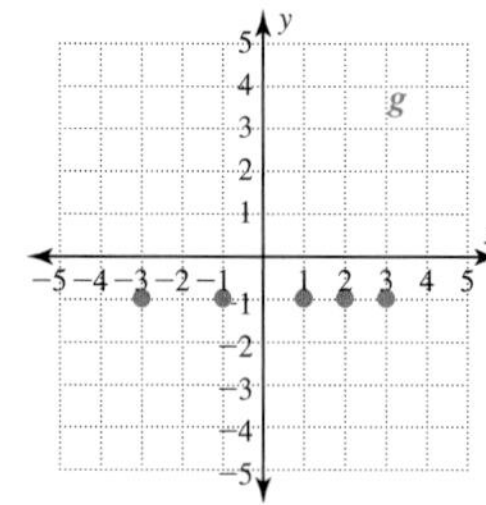

a. $f + g$ **b.** $f - g$ **c.** $f \cdot g$ **d.** $\frac{f}{g}$

35. Use the given tables for f and g to complete a table for $f \circ g$.

x	f(x)
0	9
2	7
4	5
6	11

x	g(x)
−1	4
2	6
3	2
5	0

36. Use the given graphs of f and g to graph $f + g$.

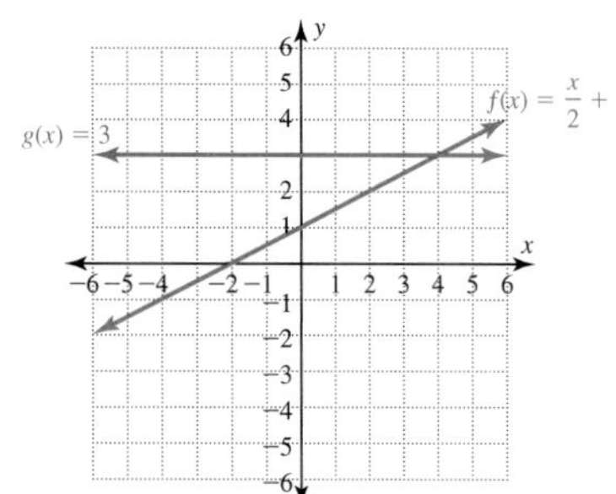

In Exercises 37 and 38, determine each function, given $f(x) = 4x - 2$ and $g(x) = 1 - 2x$. Give the domain for each function.

37. **a.** $f + g$ **b.** $f \cdot g$ **c.** $f \circ g$

38. **a.** $f - g$ **b.** $\frac{f}{g}$ **c.** $g \circ f$

39. Given $f(x) = 2x$ and $g(x) = \frac{2x^2 - 6x}{x - 3}$, state the reason that $f \neq g$.

40. Given $f(x) = 4x + 3$, determine f^{-1}. Then determine $(f \circ f^{-1})(x)$.

41. **Determining a Formula for Total Cost and Average Cost** The fixed weekly costs for a sandwich shop are \$400; $F(x) = 400$. The variable dollar cost for x sandwiches is $V(x) = 1.5x$. Determine
a. $F(100)$ **b.** $V(100)$
c. $C(x)$, the total cost of making x sandwiches in a week
d. $A(x)$, the average cost per sandwich when x sandwiches are made in a week

42. **Composing Cost and Production Functions** The number of desks a factory can produce weekly is a function of the number of hours t it operates. This function is $N(t) = 3t$ for $0 \le t \le 168$. The cost of manufacturing N desks is given by $C(N) = \left(150 - \frac{N}{4}\right)N$.

a. Evaluate and interpret $N(40)$.
b. Evaluate and interpret $C(120)$.
c. Evaluate and interpret $(C \circ N)(40)$.
d. Determine $(C \circ N)(t)$.
e. Explain the logic of the domain $0 \le t \le 168$.

43. Write the first six terms of an arithmetic sequence that satisfies the given conditions.
a. $a_1 = 5, d = 4$ **b.** $a_1 = 5, d = -4$
c. $a_1 = 5, a_2 = 8$ **d.** $a_1 = 5, a_6 = 20$

44. Write the first six terms of a geometric sequence that satisfies the given conditions.
a. $a_1 = 3, r = 2$ **b.** $a_1 = 3, r = -2$
c. $a_1 = 3, a_2 = 12$ **d.** $a_1 = 1, a_6 = 243$

45. Write the first five terms of each sequence.
a. $a_n = 2n + 7$ **b.** $a_n = n^2 + 3$
c. $a_n = 64\left(\frac{1}{2}\right)^n$
d. $a_1 = 3$ and $a_n = a_{n-1} + 7$ for $n > 1$

46. An arithmetic sequence has $a_1 = 11$ and $d = 3$. Find a_{101}.

47. A geometric sequence has $a_1 = \frac{1}{64}$ and $r = 2$. Find a_{11}.

48. An arithmetic sequence has $a_1 = -20$, $a_n = 300$, and $d = 8$. Find n.

49. A geometric sequence has $a_1 = \frac{1}{15{,}625}$, $a_n = 78{,}125$, and $r = 5$. Find n.

50. Write the terms of each series, and then add these terms.

a. $\sum_{i=1}^{6}(3i - 2)$ **b.** $\sum_{k=3}^{7}(k^2 - 2k)$

c. $\sum_{j=1}^{5} j^3$ **d.** $\sum_{i=1}^{6} 10$

51. An arithmetic sequence has $a_1 = 23$ and $a_{50} = 366$. Find S_{50}, the sum of the first 50 terms.

52. Write $0.121212\cdots$ as a fraction.

53. Evaluate $\sum_{k=1}^{\infty}\left(\frac{3}{5}\right)^k$.

54. Rungs of a Ladder The lengths of the rungs of a wooden ladder form an arithmetic sequence. There are 17 rungs ranging in length from 80 to 46 cm. Find the total length of these 17 rungs.

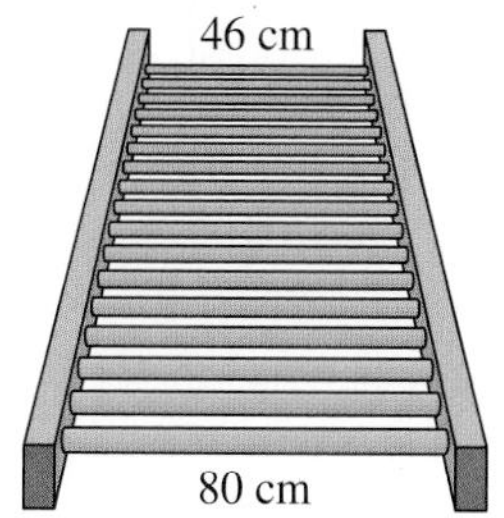

55. Perimeter of an Art Design An art design is formed by drawing a square with sides of 20 cm and then connecting the midpoints of the sides to form a second square. (See the following figure.) If this process is continued infinitely, determine the perimeter of the third square and the total perimeter of all the squares.

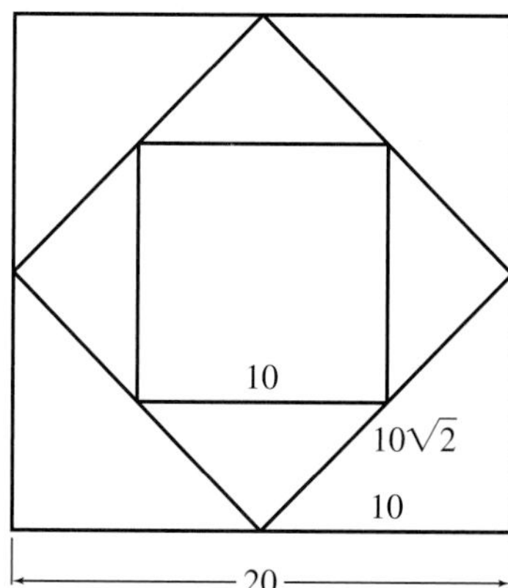

56. Determine the distance from $(-4, 7)$ to $(1, -5)$.

57. Determine the midpoint between $(-5, 2)$ and $(7, 12)$.

In Exercises 58–60, refer to the following figure.

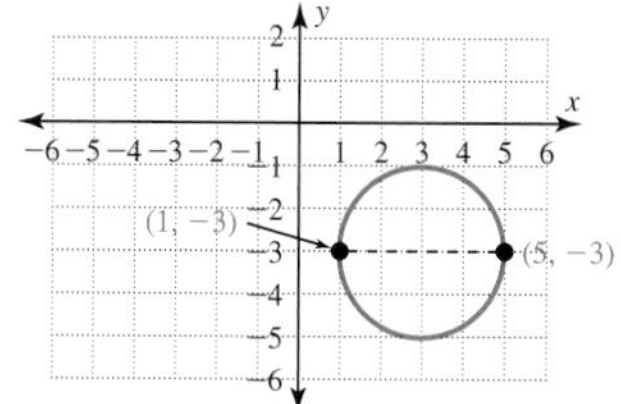

58. Determine the center of the graphed circle.

59. Determine the length of a radius of the graphed circle.

60. Determine the equation of the graphed circle.

In Exercises 61–66, match each graph with the corresponding equation.

61. $y = 2x - 4$

62. $y = 2(x - 1)^2$

63. $(x - 1)^2 + (y + 2)^2 = 4$

64. $\frac{(x - 1)^2}{9} + \frac{(y + 2)^2}{4} = 1$

65. $\frac{(x - 1)^2}{9} - \frac{(y + 2)^2}{4} = 1$

66. $y = \sqrt{4 - x^2}$

A.

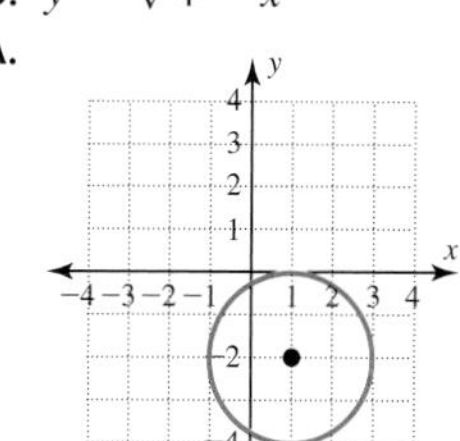

B.

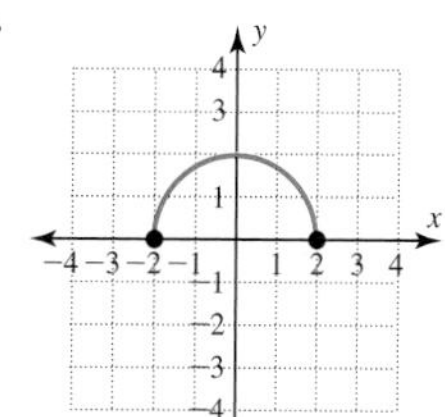

C.

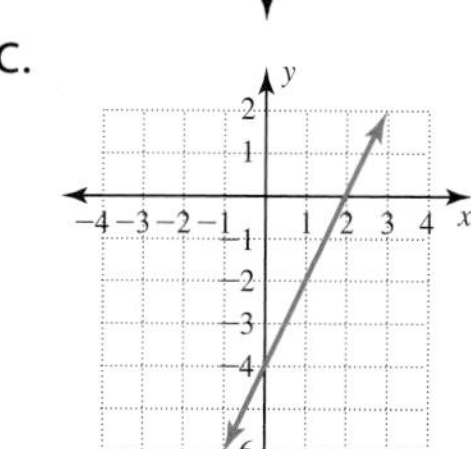

D.

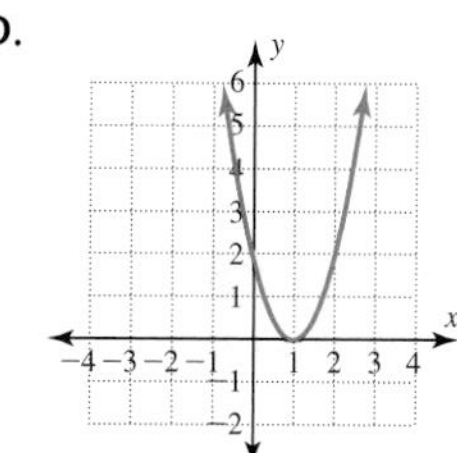

E.

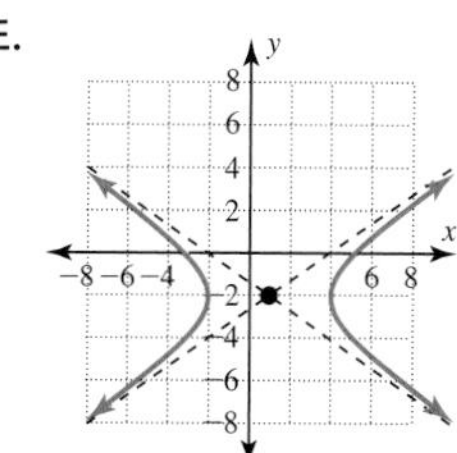

F.

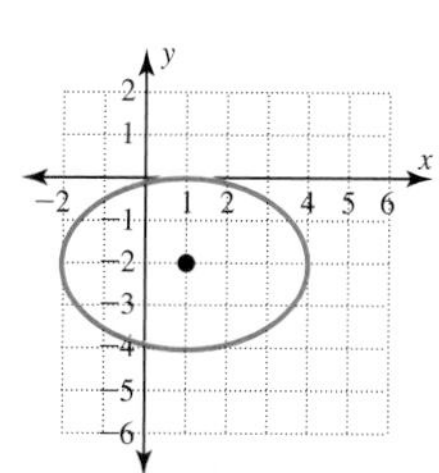

In Exercises 67–76, graph each equation.

67. $y = \frac{1}{2}x^2 - 3$

68. $y = \frac{1}{2}x - 3$

69. $y = -x^2 + 4$

70. $x^2 + y^2 = 36$

71. $(x + 3)^2 + (y - 2)^2 = 9$

72. $y = \sqrt{25 - x^2}$

73. $y = -\sqrt{25 - x^2}$

74. $\frac{x^2}{49} + \frac{y^2}{16} = 1$

75. $\frac{(x - 1)^2}{4} + \frac{(y + 3)^2}{9} = 1$

76. $\frac{x^2}{36} - \frac{y^2}{9} = 1$

In Exercises 77 and 78, refer to the following figure.

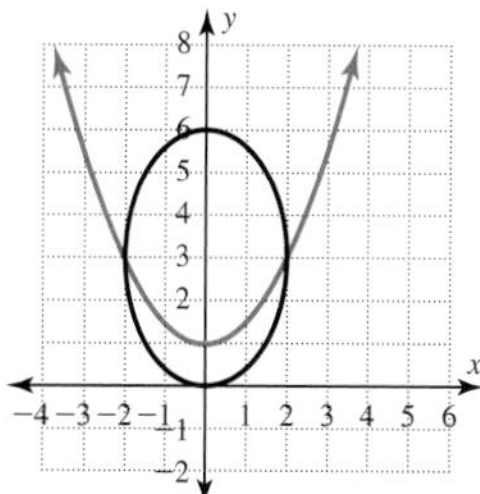

77. Use the graphs shown in the figure to solve this system of equations.

$$\begin{Bmatrix} y = \frac{1}{2}x^2 + 1 \\ \frac{x^2}{4} + \frac{(y-3)^2}{9} = 1 \end{Bmatrix}$$

78. Use the graphs shown in the figure to graph the solution of this system of inequalities.

$$\begin{Bmatrix} y \geq \frac{1}{2}x^2 + 1 \\ \frac{x^2}{4} + \frac{(y-3)^2}{9} \leq 1 \end{Bmatrix}$$

MASTERY TEST FOR CHAPTER 11

[11.1] **1. a.** Write the solution for the system of linear equations represented by $\left[\begin{array}{cc|c} 1 & 0 & -2 \\ 0 & 1 & 7 \end{array}\right]$.

b. Write the solution for the system of linear equations represented by $\left[\begin{array}{cc|c} 1 & 0 & -2 \\ 0 & 0 & 7 \end{array}\right]$.

c. Write the general solution and three particular solutions for the system of linear equations represented by $\left[\begin{array}{cc|c} 1 & 2 & 6 \\ 0 & 0 & 0 \end{array}\right]$.

d. Use augmented matrices to solve $\begin{Bmatrix} 2x + 4y = -6 \\ 3x - 5y = 68 \end{Bmatrix}$.

[11.2] **2.** Solve each system of linear equations.

a. $\begin{Bmatrix} x + 3y + 2z = 13 \\ 3x + 3y - 2z = 13 \\ 6x + 2y - 5z = 13 \end{Bmatrix}$

b. $\begin{Bmatrix} x + 3y - 2z = 2 \\ 2x - y + z = -1 \\ -5x + 6y - 5z = 5 \end{Bmatrix}$

c. $\begin{Bmatrix} x + y + z = 1 \\ -2x + y + z = -2 \\ 3x + 6y + 6z = 5 \end{Bmatrix}$

[11.3] **3.** For parts **a** and **b**, match each function with its graph. All graphs are displayed with the window $[-5, 5, 1]$ by $[-3, 7, 1]$.

a. $f(x) = |x| - 2$ **b.** $f(x) = |x| + 2$

A.

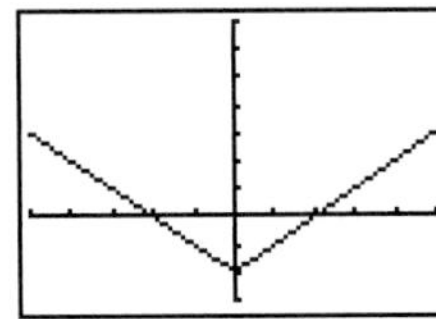

B.

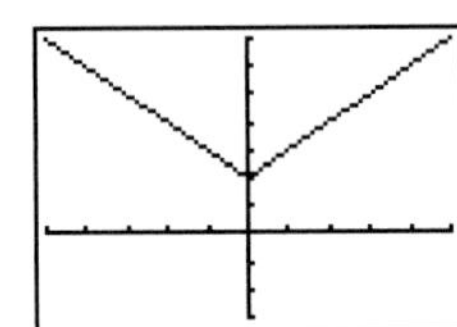

For parts **c** and **d**, use the given table to complete each table.

x	$f(x)$
0	−5
1	−4
2	3
3	22
4	59

c.

x	$f(x) - 2$
0	
1	
2	
3	
4	

d.

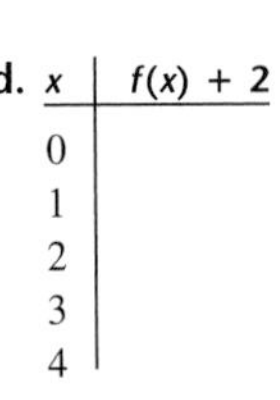

x	$f(x) + 2$
0	
1	
2	
3	
4	

[11.3] **4.** For parts **a** and **b**, match each function with its graph. All graphs are displayed with the window $[-7, 7, 1]$ by $[-3, 7, 1]$.

a. $f(x) = (x - 3)^2$ **b.** $f(x) = (x + 3)^2$

A.

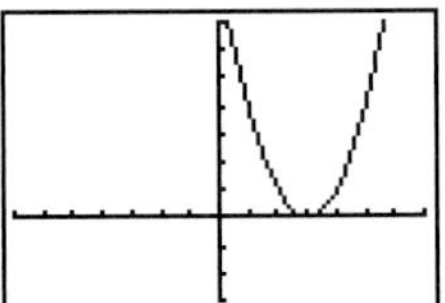

B.

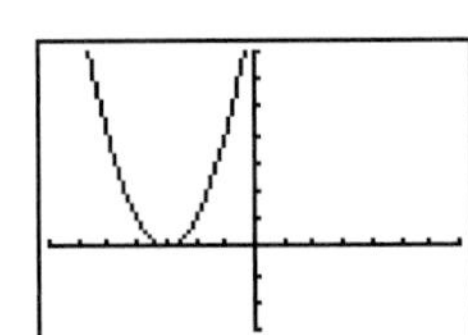

For parts **c** and **d**, use the given table to complete each table.

x	$f(x)$
0	−5
1	−4
2	3
3	22
4	59

c.

x	$f(x - 3)$
	−5
	−4
	3
	22
	59

d.

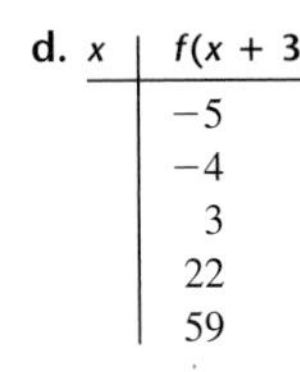

x	$f(x + 3)$
	−5
	−4
	3
	22
	59

[11.4] **5.** Determine the reflection across the x-axis of each graph.

a.

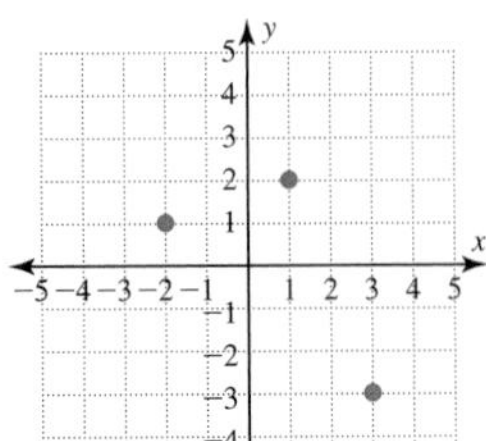

b.

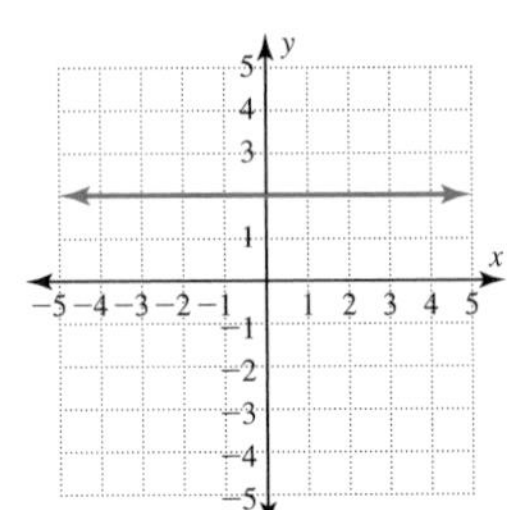

c. **d.**

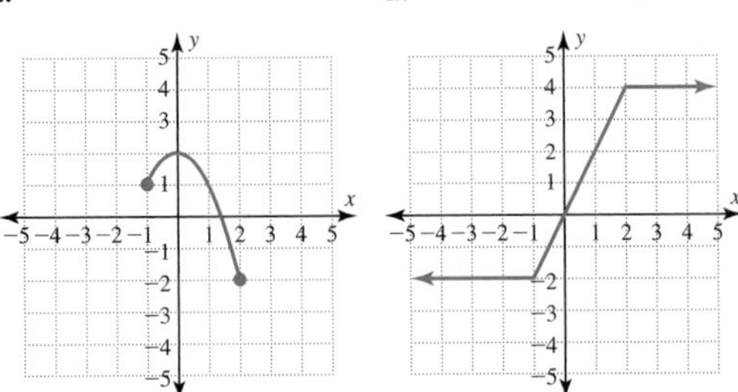

[11.4] **6.** Use the graph of $y = f(x)$ shown and match each function with its graph.

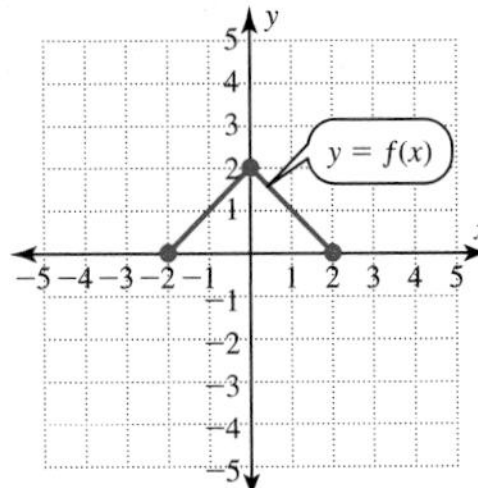

a. $y = 2f(x)$ **b.** $y = -2f(x)$

c. $y = \frac{1}{2}f(x)$ **d.** $y = -\frac{1}{2}f(x)$

A. **B.**

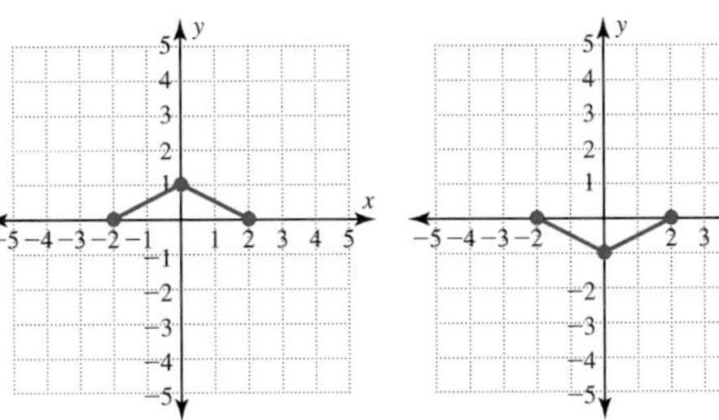

C. **D.**

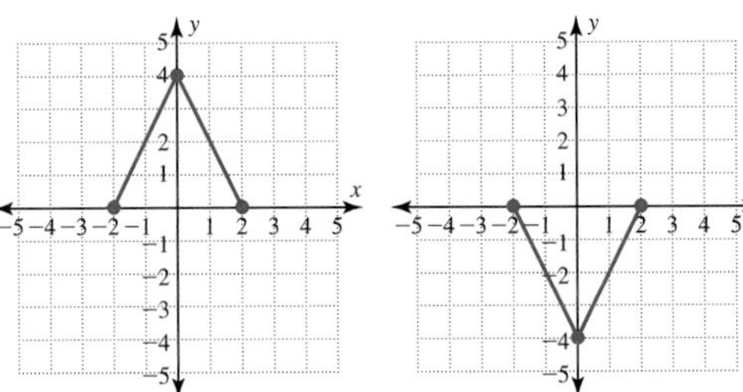

[11.5] **7.** Given $f(x) = 3x - 5$ and $g(x) = x + 2$, determine each function.

a. $f + g$ **b.** $f - g$

c. $f \cdot g$ **d.** $\frac{f}{g}$

[11.5] **8.** Given $f(x) = x^2 + 1$ and $g(x) = 2x - 1$, determine

a. $(f \circ g)(5)$ **b.** $(g \circ f)(5)$

c. $(f \circ g)(x)$ **d.** $(g \circ f)(x)$

[11.6] **9.** **a.** Write the first six terms of an arithmetic sequence with $a_1 = 4$ and $d = 3$.

b. Write the first six terms of a geometric sequence with $a_1 = 4$ and $r = 3$.

c. Write the first six terms of an arithmetic sequence with $a_1 = 2$ and $a_2 = 8$.

d. Write the first six terms of a geometric sequence with $a_1 = 2$ and $a_2 = 8$.

[11.6] **10.** Evaluate each series.

a. The sum of the first five terms of an arithmetic sequence with $a_1 = 2$ and $d = 5$

b. The sum of the first five terms of a geometric sequence with $a_1 = 2$ and $r = 5$

c. The sum of the first 100 terms of an arithmetic sequence with $a_1 = 10$ and $a_{100} = 21$

d. The sum of the first 21 terms of a geometric sequence with $a_1 = 6$ and $r = -2$

e. $\sum_{i=1}^{6} (3i^2 + i)$

[11.6] **11.** **a.** Evaluate $\sum_{i=1}^{10} 2\left(\frac{1}{2}\right)^i$.

b. Evaluate $\sum_{i=1}^{\infty} 2\left(\frac{1}{2}\right)^i$.

c. Write $0.888\cdots$ as a fraction.

d. Write $0.545454\cdots$ as a fraction.

[11.7] **12.** Graph each of these conic sections.

a. $y = (x + 3)^2 - 4$

b. $(x - 3)^2 + (y + 4)^2 = 16$

c. $\frac{(x - 3)^2}{25} + \frac{(y + 4)^2}{16} = 1$

d. $\frac{x^2}{25} - \frac{y^2}{16} = 1$

[11.7] **13.** Write the equation for each of these conic sections.

a.

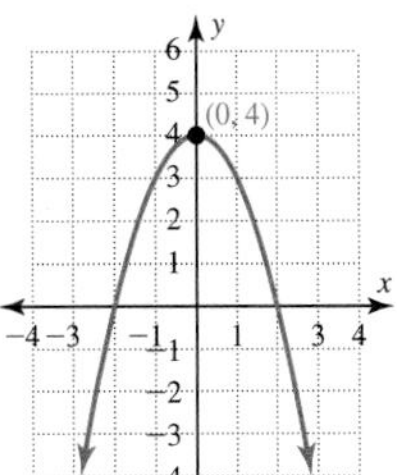

b.

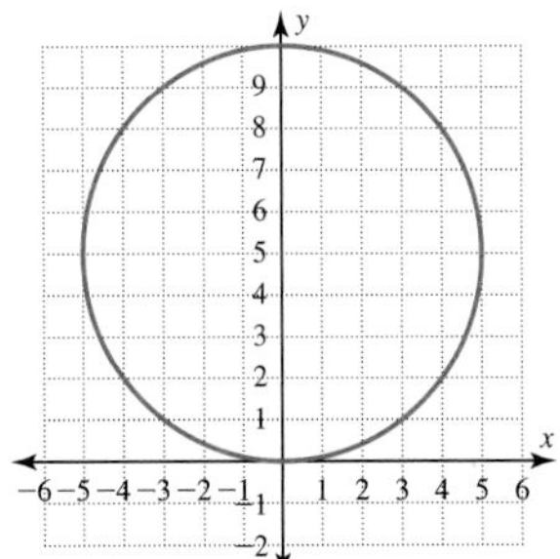

c.

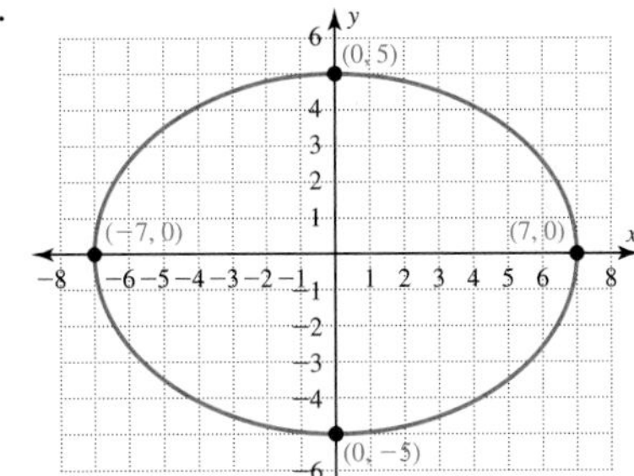

d.

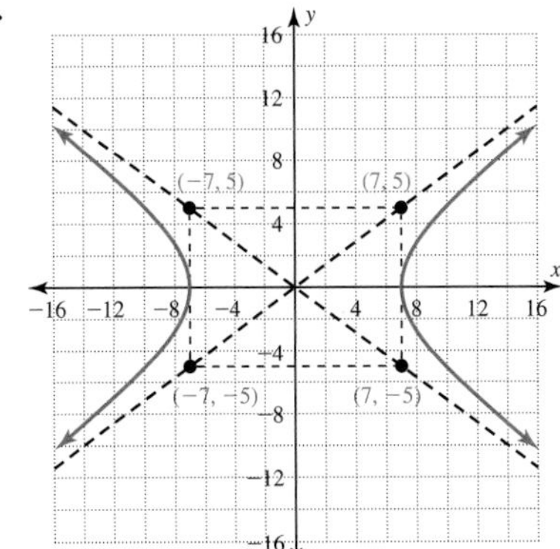

REALITY CHECK FOR CHAPTER 11

The axle of a train is solid so that the wheels must turn together. On the arc of a circular curve in the tracks, the wheels must travel different distances while turning the same number of times. If the difference in the effective radii of the two train wheels is $\frac{1}{32}$ in, find the difference in the distances covered with 1 rev of each wheel.

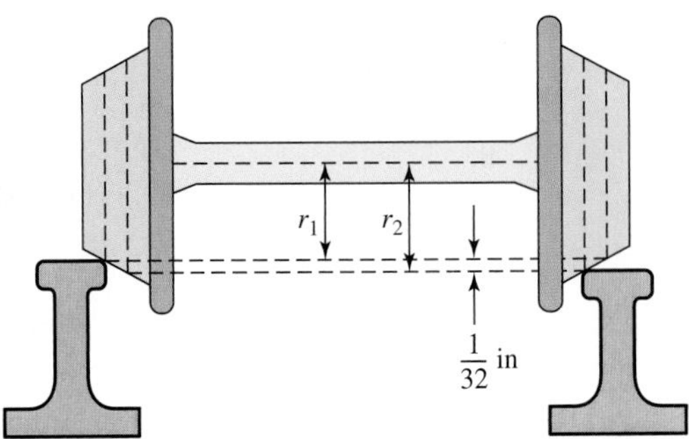

Appendix A

Accuracy, Precision, Error, and Unit Conversion

Accuracy and Precision

Calculators will typically display from 8 to 12 digits, and it is tempting to copy all these digits when you are giving answers. However, it is not appropriate to copy digits that are not significant. Roughly speaking, the answer to a problem cannot be assumed to be more accurate than the measurements that produced that answer.

The **accuracy** of a number refers to the number of **significant digits.** For numbers written in scientific notation, all digits written are considered significant. For numbers obtained by measurement, the digits considered reasonably trustworthy are those that are significant.

The **precision** of a measurement refers to the smallest unit used in the measuring device and thus to the position of the last significant digit. The accuracy and precision of some sample measurements are given in the box.

Accuracy and Precision

NUMBER OF METERS	SCIENTIFIC NOTATION	ACCURACY	PRECISION
0.045	4.5×10^{-2}	2 Significant digits	Thousandths of a meter
0.0450	4.50×10^{-2}	3 Significant digits	Ten-thousandths of a meter
45,000	4.5×10^{4}	2 Significant digits	Thousands of meters
123.45	1.2345×10^{2}	5 Significant digits	Hundredths of a meter
0.05	5×10^{-2}	1 Significant digit	Hundredths of a meter

There are many different rounding rules used for specific tasks and professions. All these rules are based on trying to report an answer that is reasonably trustworthy. Most of these rules are similar to the standardized rules given below.

Rounding Rules

- For calculations involving multiplication, division, or exponentiation, round the answer to the same number of significant digits as the measurement having the least number of significant digits.
- For calculations involving addition or subtraction, round the answer to the same precision as the least precise measurement.

Actual Error and Relative Error

Measurements, approximations, and estimations often introduce error into the reported results used by engineers, scientists, and statisticians. The **actual error** or **the error of the estimate** is the estimated value minus the actual value. **Relative error** is the actual error divided by the actual value. Many calculator errors are the result of keystrokes which produce incorrectly placed decimal points, missing digits, etc. These errors generate large relative errors and thus can often be detected by simply thinking about our results and asking ourselves if the results seem reasonable based upon our mental estimate.

EXAMPLE 1 Calculating the Relative Error

A contractor estimated the dimensions of the region shown to the right as 45 m by 60 m and calculated the approximate area. What was the relative error of this estimate?

44.1 m

58.7 m

SOLUTION

Actual area:

$A = l \cdot w$
$A = (44.1 \text{ m})(58.7 \text{ m})$
$A = 2{,}588.67 \text{ m}^2$

Use the formula for the area of a rectangle with the actual dimensions. The calculator value gives the area in square meters.

Estimated area:

$A = l \cdot w$
$A \approx (45 \text{ m})(60 \text{ m})$
$A \approx 2{,}700 \text{ m}^2$

Use the estimated values in the same formula.

Error of estimate = Estimated value − Actual value

$= 2{,}700 \text{ m}^2 - 2{,}588.67 \text{ m}^2$
$= +111.33 \text{ m}^2$

The estimated value is larger than the actual value.

$$\boxed{\text{Relative error}} = \boxed{\text{Error of estimate}} \div \boxed{\text{Actual value}}$$
$$= 111.33 \text{ m}^2 \div 2{,}588.67 \text{ m}^2$$
$$\approx 0.043$$
$$\approx 4.3\%$$

First, use a calculator to determine the relative error as a decimal and then convert to a percent.

Answer: The contractor's relative error was +4.3%. Many contractors prefer to be a little high on their estimate so that they do not run short on materials.

Unit Conversion

To convert from one unit to another, use the method of multiplying by a fraction equal to 1. This fraction is called a conversion factor. Since 12 in is equal to 1 ft, we can write 12 in = 1 ft. If we divide both sides of this equation by 1 ft, we get $\frac{12 \text{ in}}{1 \text{ ft}} = 1$. This means $\frac{12 \text{ in}}{1 \text{ ft}}$ is a conversion factor equal to 1. Also, we could have divided both sides of the equation 12 in = 1 ft by 12 in to obtain the conversion factor $\frac{1 \text{ ft}}{12 \text{ in}}$. Since multiplying by 1 does not change the value of an expression, we can use conversion factors to change the units and preserve the value of a measurement.

EXAMPLE 2 Using One Conversion Factor

Convert 3 ft to inches.

SOLUTION

$$3 \text{ ft} = 3 \cancel{\text{ft}} \cdot \frac{12 \text{ in}}{1 \cancel{\text{ft}}}$$

Multiply 3 ft by the conversion factor $\frac{12 \text{ in}}{1 \text{ ft}}$.

$$= 36 \text{ in}$$

Divide the common units of feet, leaving the desired units of inches.

The table shows unit conversions for some basic units of length and time. Conversions for many other units can be found in science texts or on the Web. A search for "unit conversion" will find many useful websites for converting units. It is important to point out that the rounding of numbers can cause slight differences in these conversions.

Length	Time
1 mi = 5,280 ft	1 h = 60 min
1 yd = 3 ft	1 min = 60 s
1 ft = 12 in	
1 in = 2.54 cm	

Sometimes it is necessary to use more than one step to make a conversion. This is known as a chain of conversion factors. Example 3 shows a conversion of yards to centimeters by converting yards to feet, then feet to inches, and then inches to centimeters.

EXAMPLE 3 Using a Chain of Conversion Factors

Convert 5 yd to centimeters.

SOLUTION

$$5 \text{ yd} = 5 \cancel{\text{yd}} \cdot \frac{3 \cancel{\text{ft}}}{1 \cancel{\text{yd}}} \cdot \frac{12 \cancel{\text{in}}}{1 \cancel{\text{ft}}} \cdot \frac{2.54 \text{ cm}}{1 \cancel{\text{in}}}$$
$$= 457.2 \text{ cm}$$

Multiply 5 yd by the conversion factor $\frac{3 \text{ ft}}{1 \text{ yd}}$.

Divide the common units of yards leaving units of feet. Then multiply by the conversion factor $\frac{12 \text{ in}}{1 \text{ ft}}$ and divide the common units of feet, leaving units of inches. Finally, multiply by the conversion factor $\frac{2.54 \text{ cm}}{1 \text{ in}}$ and divide the common units of inches, leaving the desires units of centimeters.

Example 4 illustrates how to convert units in the numerator and in the denominator.

EXAMPLE 4 Converting Units in the Numerator and Denominator

Convert 110 ft/s to miles per hour.

SOLUTION

$$\frac{110 \text{ ft}}{\text{s}} = \frac{110 \cancel{\text{ft}}}{1 \cancel{\text{s}}} \cdot \frac{1 \text{ mi}}{5{,}280 \cancel{\text{ft}}} \cdot \frac{60 \cancel{\text{s}}}{1 \cancel{\text{min}}} \cdot \frac{60 \cancel{\text{min}}}{1 \text{ h}}$$
$$= \frac{396{,}000 \text{ mi}}{5{,}280 \text{ h}}$$
$$= 75 \text{ mi/h}$$

Write the original problem as $\frac{110 \text{ ft}}{1 \text{ s}}$. To convert the units in the numerator to miles, multiply by the conversion factor $\frac{1 \text{ mi}}{5{,}280 \text{ ft}}$. Divide the common units of feet, leaving the desired units of miles in the numerator. To convert the units in the denominator to hours, we use a chain of conversion factors. Multiply by the conversion factor $\frac{60 \text{ s}}{1 \text{ min}}$ and divide the common units of seconds, leaving units of minutes in the denominator. Then, multiply by the conversion factor $\frac{60 \text{ min}}{1 \text{ h}}$ and divide the common units of minutes, leaving the desires units of hours in the denominator. Multiply the values in the numerators, multiply the values in the denominators, and simplify the result.

Student Answer Appendix

Answers to Selected Exercises

Chapter 1

Reality Check for Chapter 1 Opener

The monthly payment is slightly less on an 8% mortgage for 30 years.

Exercises 1.1

1–9. Answers can vary. **11.** Self **13.** Lecture **15.** MathZone **17.** Tutor **19.** Answers can vary.

Using the Language and Symbolism of Mathematics 1.2

1. Variable **2.** Additive inverse **3.** 0 **4.** a **5.** Seven; left **6.** Absolute value; distance **7.** Greater; equal **8.** Less; left **9.** Equal **10.** Approximately equal **11.** Square root **12.** Rational **13.** Rational **14.** Irrational **15.** Pi **16.** Infinity **17.** Is not **18.** Is **19.** $\mathbb{R}$ **20.** $[-3, 2]$ **21.** $(a, b]$

Exercises 1.2

1. a. -9 **b.** 13 **c.** $\frac{3}{5}$ **d.** -7.23 **e.** $\sqrt{13}$ **f.** 0 **3. a.** 7 **b.** -7 **c.** 7 **5. a.** 0 **b.** 0 **c.** 0 **d.** 0 **7. a.** 0 **b.** $6x$ **c.** 0 **d.** $-6x$ **9. a.** 29 **b.** 29 **c.** -29 **d.** -29 **11. a.** 5 **b.** -5 **c.** 0.5 **d.** 50 **13. b** **15. c** **17.** 7; $7 < \sqrt{50.6}$; 7.113 **19.** 1; $1 > \sqrt{0.976}$; 0.988 **21.** 4; $4 > \sqrt{15.1}$; 3.886 **23.** 7 **25.** 3 **27. a.** $>$ **b.** $<$ **29. a.** $<$ **b.** $>$ **31. a.** $<$ **b.** $>$ **33. a.** $=$ **b.** $=$ **35. a.** $>$ **b.** $>$ **37. a.** $>$ **b.** $<$ **39. a.** 15 **b.** 0, 15 **c.** $-11, -\sqrt{9}$, 0, 15 **d.** $-11, -4.8, -\sqrt{9}, 0, 1\frac{3}{5}$, 15 **e.** $\sqrt{5}$ **41.** $-3 < x \le 3$; x is greater than -3 and less than or equal to 3; $(-3, 3]$ **43.** x is greater than 1; [number line, 1]; $(1, \infty)$

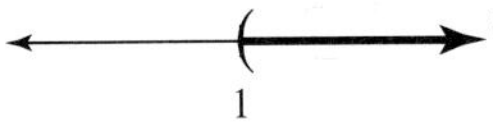

45. $x \ge -3$; [number line, -3]; $[-3, \infty)$

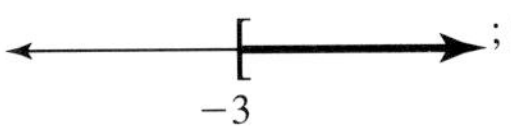

47. $x \le 4$; x is less than or equal to 4; [number line, 4]

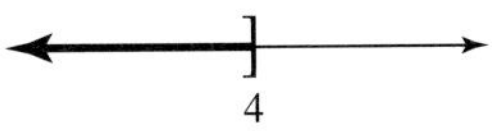

49. Integers, rational numbers **51.** Natural numbers, whole numbers, integers, rational numbers **53.** Irrational numbers **55.** Rational numbers

57. [number line: -5, -3.5, 0, $1\frac{3}{4}$, 4]

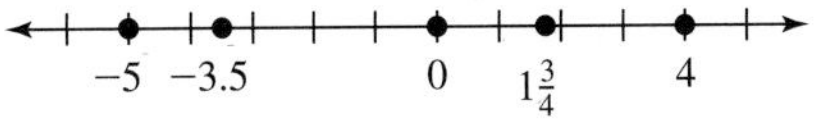

59. [number line: $-|4|$, -2.5, 0, $\frac{5}{4}$, $|-3|$]

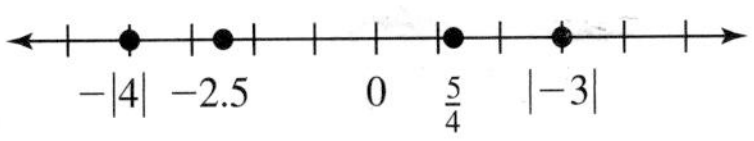

61. a. Rational **b.** Irrational **63. a.** Rational **b.** Irrational **65. a.** Rational **b.** Rational **67.** $|x| = y$ **69.** $3.14 < \pi < 3.15$ **71.** $-x \le -2$ **73.** Answers can vary. **a.** $x = 9$ **b.** $x = \frac{1}{4}$ **c.** $x = 0$ or $x = 1$ **75.** Answers can vary. **a.** 0 **b.** $\frac{3}{4}, \frac{20}{3}$ **c.** $\pi, \sqrt{5}$ **d.** $-2.5, -2.75$

77.

x	y_1
0	0
25	5
50	7.0711
75	8.6603
100	10
125	11.18
150	12.247

79. a. 41 **b.** 2001 **c.** 2004 **81.** Heating and cooling

Using the Language and Symbolism of Mathematics 1.3

1. Terms; addends; sum **2.** Commutative **3.** Associative **4.** Both **5.** Larger **6.** Commutative; associative **7.** $\frac{w + y}{x}$; 0 **8.** Evaluate **9.** $x + 5$ **10.** $x + y + z$

Exercises 1.3

1. a. 13 **b.** -3 **c.** 3 **d.** -13 **3. a.** 0 **b.** -16 **c.** 9 **d.** -43 **5. a.** -21 **b.** -19 **7. a.** -18 **b.** -17 **9.** -3 **11.** -1 **13. a.** 0 **b.** -0.3 **c.** 1.29 **15. a.** 34 **b.** 0 **17.** $\frac{4}{11}$ **19.** $\frac{7}{20}$ **21.** $-\frac{13}{12}$ **23.** Commutative **25.** $(11 + 12) + 13$ **27.** $(-1) + 5 = 4$; the sum of negative one and five is four. **29.** The sum of five and negative two is three;

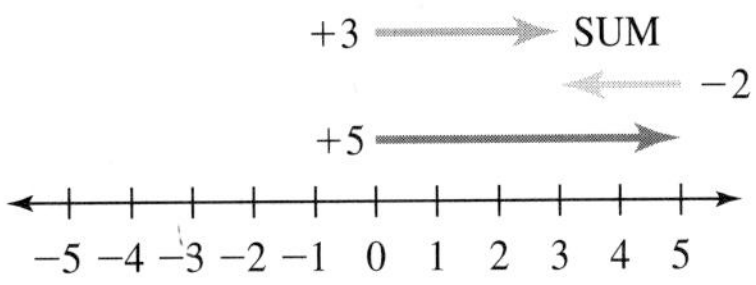

[arrows: $+3$ → SUM; -2; $+5$ → ; number line -5 -4 -3 -2 -1 0 1 2 3 4 5]

31. a. 3 **b.** -3 **c.** 3 **33. a.** -2 **b.** 2 **35. a.** -99.01 **b.** 12.4 **37. a.** 171.26 **b.** 54.75 **39.** 21.6 cm **41.** 70 cm **43.** 32 hours **45.** 4,540,000 units **47.** 48% **49.** 95% **51.** 20% **53.** 9°F **55.** 76.6 m **57.** 8 cm **59.** Q1: \$12,000; Q2: \$2,000; Q3: $-$\$2,000; Q4: $-$\$12,000 **61.** $+$; 5.8 **63.** $-$; $-\frac{47}{20}$ **65. b** **67. b** **69. b** **71. d** **73.** $\frac{2}{5}$ **75.** 0.375 **77. a.** $a + b = b + a$ **b.** $x + (-x) = 0$ **79. a.** $\pi \ne 3.14$ **b.** $\pi \approx 3.14$ **81.** $r + 50$ **83.** $n + 1$

Using the Language and Symbolism of Mathematics 1.4

1. Subtrahend; difference **2.** $x - 5$ **3.** $w - 17$ **4.** $\frac{w - y}{x}$; 0 **5.** $x_2 - x_1$ **6.** $y_2 - y_1$ **7.** a sub one; a sub n **8.** Sequence **9.** Solution **10.** Satisfy **11.** Checking **12.** Solution

Exercises 1.4

1. a. -6 **b.** 20 **c.** -20 **d.** 6 **3. a.** -7 **b.** 7 **c.** -37 **d.** 37 **5. a.** 0 **b.** $-\frac{7}{3}$ **7. a.** $\frac{43}{30}$ **b.** $\frac{7}{30}$ **9. a.** -57 **b.** 21 **11. a.** -34 **b.** 0 **13. a.** 0 **b.** -12, 348.56 **15. a.** 2 **b.** 164 **17. a.** -8 **b.** 5 **19. a.** 0 **b.** 10 **21.** -41.68 **23.** 55.71 **25.** 129.93 **27. a.** 4 is not a solution; 5 is a solution **b.** 4 is a solution; 5 is not a solution **29. a.** 4 is a solution; 5 is not a solution **b.** 4 is not a solution; 5 is a solution **31.** $-3, -2, -1, 0, 1$ **33.** $2, 1, 0, -1, -2$ **35.** 70 ft **37.** \$624 **39.** \$5,948,000,000 **41.** 41% **43.** -7°F **45. a.** $+6°$ **b.** $-6°$ **47. a.** $+9°$ **b.** $-9°$ **49. a.** $-4°$ **b.** $+4°$ **51.** 13.6%, -16.2% **53.** $+5.5$ yd **55.** \$97.79 **57.** 79.449 m **59.** $-$; -175.93 **61.** $+$; $\frac{47}{24}$ **63.** $+$; 4.303 **65. b** **67. b** **69. a** **71.** $x - 8 = z$ **73.** $x - y = 7$ **75.** $|x| - |-y| = 11$ **77.** $P - 2{,}000$ **79. a.** $n - 2$ **b.** $n + 2$ **81. a.** $r - 40$ **b.** $r + 40$

Using the Language and Symbolism of Mathematics 1.5

1. Factors; product **2.** PRT **3.** $0.14C$ **4.** Commutative **5.** Associative **6.** 0 **7.** Positive **8.** Negative **9.** Positive **10.** Negative **11.** 60 **12.** $\frac{wy}{xz}$; 0; 0 **13.** Reciprocals; inverses **14.** 1 **15.** 0 **16.** 1 **17.** $\sqrt{ab}$ **18.** w; 4 **19.** x^7 **20.** Fifth **21.** Factor **22.** $-x$ **23.** x

Exercises 1.5

1. a. −77 **b.** −77 **c.** 77 **d.** −77 **3. a.** −60 **b.** 60 **c.** −60 **d.** 60 **5. a.** 8 **b.** 8 **c.** −8 **d.** 0 **7. a.** −123.4 **b.** −123,400 **c.** −1.234 **d.** −1,234,000 **9. a.** $\frac{3}{10}$ **b.** $-\frac{3}{10}$ **11. a.** 1 **b.** $-\frac{1}{5}$ **13. a.** 6 **b.** 10 **15. a.** 5^4 **b.** $(-4)^3$ **c.** y^5 **d.** $(3z)^6$ **17. a.** $-3 \cdot 3$ **b.** $(-3)(-3)$ **c.** $3 \cdot x \cdot x$ **d.** $(3x)(3x)$ **19. a.** 8 **b.** 9 **c.** 9 **d.** −9 **21. a.** 0 **b.** 1 **c.** 1 **d.** −1 **23. a.** −100 **b.** 100 **c.** 0.001 **d.** −1,000 **25. a.** $\frac{1}{8}$ **b.** $-\frac{1}{8}$ **c.** $\frac{1}{9}$ **d.** $\frac{1}{9}$ **27. a.** −300 **b.** −300 **29. a.** 900 **b.** −1,200 **31. a.** −2 is a solution; 2 is not a solution **b.** −2 is not a solution; 2 is a solution **33. a.** 2, 4, 6, 8, 10 **b.** 1, 4, 9, 16, 25 **35.** 13; 40 **37.** −3; −40 **39.** π; 0 **41. a.** $8x$ **b.** x^8 **43. a.** $3P$ **b.** $P - 300$ **45. a.** Sum **b.** Difference **c.** Product **47. a.** Commutative **b.** Associative **49. a.** Multiplication **b.** Addition **51.** 180 cm² **53.** 3,360 in³ **55.** 14.82 cm **57. a.** \$280 **b.** \$290 **59.** 75 mL **61.** 0.675 L **63.** 1,260 mi **65.** 13 m **67.** 941.192 cm³ **69. a.** 339.8 cm² **b.** 589.0 cm³ **71.** \$31.25 **73.** \$304.56 **75.** \$64; \$32; \$75; \$150; total cost: \$321 **77.** \$16,800; \$19,200; \$18,200; \$16,000 **79.** +; 125.28 **81.** +; $\frac{28}{65}$ **83.** +; 2.449 **85.** a **87.** c

Using the Language and Symbolism of Mathematics 1.6

1. Divisor; Quotient **2.** $\frac{D}{R} = T$ **3.** $m{:}w = 3{:}5$ **4.** Positive **5.** Negative **6.** Multiplicative; reciprocal **7.** $\frac{wz}{xy}$; y; 0 **8.** Ratio **9.** Undefined **10.** $\sqrt{\frac{a}{b}}$ **11.** Range **12.** Mean **13.** Relative **14.** Units

Exercises 1.6

1. a. −8 **b.** 8 **c.** −8 **d.** 0 **3. a.** −96 **b.** −24 **c.** 144 **d.** 16 **5. a.** 1,230 **b.** −12,300 **c.** −0.123 **d.** 12.3 **7. a.** 4 **b.** −4 **c.** $\frac{1}{9}$ **d.** $-\frac{2}{3}$ **9. a.** 0 **b.** Undefined **c.** Undefined **d.** Undefined **11. a.** 3 **b.** 5 **13. a.** 3 **b.** −3 **c.** −3 **d.** 3 **15. a.** −6 is not a solution; 6 is a solution **b.** −6 is a solution; 6 is not a solution **17. a.** 120, 60, 40, 30, 24 **b.** $-\frac{1}{2}, -1, -\frac{3}{2}, -2, -\frac{5}{2}$ **19.** 18; 20 **21.** −9; −13 **23.** 6; 12 **25.** 1:4 **27.** 3:1 **29. a.** 1:5 **b.** 1:4 **31.** 8: 5 **33.** 64-oz bottle **35.** 82 **37. a.** 155.8 cm **b.** 160 cm **c.** 2.7% **39. a.** 50.9 cm **b.** 52 cm **c.** 2.2% **41. a.** \$108.31 **b.** \$156.77 **43.** 290 mi/h **45.** 24 cm/min **47.** 300 lightbulbs/h **49.** $\frac{1}{5}$ pool/h **51.** −; −3.28 **53.** +; $\frac{9}{20}$ **55.** −; −1.475 **57.** c **59.** c **61.** b **63.** \$103.13; \$62 **65.** Q1: \$0.12; Q2: \$0.10; Q3: \$0.10; Q4: \$.075 **67.** 14.5 cm **69. a.** $\frac{a}{5}$ **b.** $\frac{d}{t} = 7$ **71. a.** $\frac{x}{y} = 8$ **b.** $xy = 8$ **73. a.** $5 > \frac{x}{3}$ **b.** $-2 \le \frac{8}{w}$ **75.** −1

Using the Language and Symbolism of Mathematics 1.7

1. Multiplication **2.** Addition **3.** Exponentiation **4.** x; 5 **5.** $3x$; 5 **6.** $-x$; 4 **7.** x; 4 **8.** Distributive; multiplication; addition **9.** Coefficient **10.** Distributive **11.** Combining

Exercises 1.7

1. a. 21 **b.** 56 **3. a.** 2 **b.** 70 **5. a.** 100 **b.** 52 **7. a.** 25 **b.** 1 **9. a.** −19 **b.** 0 **11. a.** −45 **b.** 9 **13. a.** −27 **b.** −21 **15. a.** 4 **b.** 26 **17. a.** 0 **b.** 12 **19. a.** −117 **b.** −27 **21. a.** −1,000 **b.** 800,000 **23. a.** 17 **b.** $\frac{16}{17}$ **25. a.** $\frac{1}{16}$ **b.** $\frac{5}{12}$ **27. a.** 49 **b.** 25 **29. a.** 7 **b.** 5 **31. a.** −3 **b.** 3 **33. a.** 40 **b.** 6 **35.** 16 **37.** 106 **39.** 158 **41.** 56 **43.** $\frac{1}{2}$ **45. a.** $7x + 35$ **b.** $9(x + 4)$ **47. a.** $-6x + 14$ **b.** $-5(a + b)$ **49. a.** $10x - 15y$ **b.** $11(2x - 3y)$ **51. a.** $-b + 3$ **b.** $-1(b - 5)$ **53. a.** $7x$ **b.** $7\sqrt{3}$ **55. a.** $6w$ **b.** $9w - 3z$ **57. a.** $6a - 2b$ **b.** $-2a + 8b$ **59.** $6\sqrt{3} - 2\sqrt{5}$ **61.** −2 is a solution; 5 is not a solution **63.** −2 is not a solution; 5 is a solution **65.** −2 is not a solution; 5 is a solution **67.** > **69.** < **71.** 81; 80.407 **73.** 40; 41.954 **75.** 180,000 ft³ **77.** 21,205.75 ft³ **79. a.** $2x + 5y$ **b.** $2(x + 5y)$ **81. a.** $c = \sqrt{a + b}$ **b.** $c = \sqrt{a} + \sqrt{b}$ **83. a.** $\frac{y_2 - y_1}{x_2 - x_1}$ **b.** $m = \frac{y_2 - y_1}{x_2 - x_1}$ **85.** $r - 8$ **87.** $3(r - 8)$

Review Exercises for Chapter 1

1. a. −12 **b.** −20 **c.** −20 **d.** −12 **2. a.** −64 **b.** 64 **c.** −4 **d.** 4 **3. a.** −7 **b.** 0 **c.** 0 **d.** Undefined **4. a.** 18 **b.** 30 **c.** −144 **d.** −4 **5. a.** 9.01 **b.** 8.99 **c.** 0.09 **d.** 900 **6. a.** 995.5 **b.** −1,004.5 **c.** −4,500 **d.** −0.0045 **7. a.** −6 **b.** −24 **c.** 10 **d.** −8 **8. a.** −24 **b.** 6 **c.** −24 **d.** −10 **9. a.** −18 **b.** 27 **c.** −50 **d.** 70 **10. a.** −27 **b.** −3 **c.** −27 **d.** −27 **11. a.** 9 **b.** 5.25 **c.** 1.2 **d.** 18 **12. a.** 9 **b.** −9 **c.** −27 **d.** −27 **13. a.** 49 **b.** 25 **c.** 1 **d.** −7 **14. a.** 25 **b.** 32 **c.** 25 **d.** −32 **15. a.** 8 **b.** −8 **c.** −8 **d.** 14 **16. a.** 14 **b.** 10 **c.** 4 **d.** 2 **17. a.** 24 **b.** −120 **c.** 0 **d.** 720 **18. a.** 0 **b.** 0 **c.** −1 **d.** 1 **19. a.** $\frac{17}{12}$ **b.** $-\frac{1}{12}$ **c.** $\frac{1}{2}$ **d.** $\frac{8}{9}$ **20. a.** $-\frac{7}{75}$ **b.** $-\frac{133}{75}$ **c.** $-\frac{98}{125}$ **d.** $-\frac{10}{9}$ **21. a.** −44 **b.** −71 **c.** −38 **d.** −26 **22.** 22 **23.** 79 **24.** −2 **25.** −7 **26.** 6 **27.** 15 **28.** 10 **29.** B **30.** D **31.** A **32.** C **33.** −; −533 **34.** +; 31 **35.** +; 41,181 **36.** −; −315 **37.** −; −557 **38.** −; −5 **39.** 16; 16.049 **40.** 800; 799.4636 **41.** 3,000; 2,950.388 **42.** 6; 6.03 **43.** 125; 125.751501 **44.** 7; 7.1 **45.** $x > -2$; $(-2, \infty)$ **46.** $x \le 4$; [number line with] at 4] **47.** [number line with) at 3]; $(-\infty, 3)$ **48.** $-3 < x < 0$; [number line with (at −3 and) at 0] **49.** [number line with [at −7 and] at −1]; $[-7, -1]$ **50.** $4 \le x < 10$; [4, 10) **51.** $-x = 11$ **52.** $|x| \le 7$ **53.** $\sqrt{26} > 5$ **54.** $x + 5 = 4$ **55.** $x - 3 < y$ **56.** $-3y = -1$ **57.** $\frac{x}{3} = 12$ **58.** $\frac{x}{y} = \frac{3}{4}$ **59.** $5(3x - 4) = 13$ **60.** 7, 17, 27, 37, 47 **61.** −9, −6, −1, 6, 15 **62.** B **63.** A **64.** D **65.** E **66.** C **67.** 2 **68.** −4 **69.** 2 is not a solution; 3 is a solution **70.** 2 is a solution; 3 is not a solution **71. a.** 135 ft³ **b.** 159 ft² **72. a.** 54 m **b.** 162 m² **73. a.** 0.4 m **b.** 0.7% **74.** 20-oz bottle **75.** 17:18 **76.** \$1,039.20 **77. a.** 40 L **b.** 30 gal **78. a.** 200 golf balls/h **b.** $\frac{1}{4}$ vat/h **79. a.** $2r$ **b.** $r + 30$ **c.** $2(r + 30)$ **d.** $r - 30$ **e.** $2(r - 30)$ **80. a.** 369 **b.** 61.5 **c.** 30

Mastery Test for Chapter 1

1. Answers can vary. **2. a.** 2 **b.** 0 **c.** −1.5 **d.** −25 **3. a.** 23 **b.** 23 **c.** 0 **d.** 46 **4. a.** (−2,3] **b.** [−1,∞) **c.** (−∞,4) **d.** [5,9) **5. a.** 5; 5.10 **b.** 10; 9.95 **c.** 3; 2.83 **d.** 4; 4.00 **6. a.** B **b.** C **c.** A **d.** E **e.** D **7. a.** 6 **b.** −28 **c.** −6 **d.** $\frac{1}{2}$ **e.** $-\frac{1}{4}$ **f.** $-\frac{31}{40}$ **g.** 0 **h.** −12 **8. a.** Commutative; addition **b.** Associative; addition **c.** $x + (y + 5)$ **d.** $5(y + x)$ **9. a.** −6 **b.** −4 **c.** 7 **d.** 3 **10. a.** 7 **b.** −23 **c.** −7 **d.** 23 **e.** $\frac{2}{3}$ **f.** −1 **g.** $\frac{5}{6}$ **h.** $\frac{7}{30}$ **11. a.** 3, 4, 5, 6, 7 **b.** 11, 10, 9, 8, 7 **c.** 3, 5, 7, 9, 11 **d.** −4, −1, 4, 11, 20 **12. a.** Solution **b.** Not a solution **c.** Solution **d.** Not a solution **13. a.** −30 **b.** 30 **c.** −30 **d.** 0 **e.** $\frac{1}{15}$ **f.** $\frac{1}{3}$ **g.** $-\frac{3}{7}$ **h.** −24 **14. a.** Commutative; multiplication **b.** Associative; multiplication **c.** $5(xy)$ **d.** $(a + b)x$ **15. a.** 25 **b.** 32 **c.** 25 **d.** −25 **e.** 0 **f.** $\frac{9}{49}$ **g.** 219 **h.** −1 **16. a.** 60 m² **b.** 34 m **17. a.** −3 **b.** 3 **c.** −3 **d.** 0 **e.** Undefined **f.** $-\frac{3}{2}$ **g.** $\frac{5}{4}$ **h.** $-\frac{66}{245}$

18. a. $\frac{1}{5}$ **b.** 12-oz bottle **19. a.** 9 **b.** 27 **c.** 24 **d.** $-\frac{2}{33}$ **e.** 64 **f.** 34 **g.** -59 **h.** 2 **20. a.** Distributive; multiplication; addition **b.** $33x + 44$ **c.** $11(3x + 4)$ **d.** $-14x + 35$ **e.** $8(2x - 3)$ **21. a.** $12x$ **b.** $9x - 33$ **c.** $x - 7$ **d.** $4a - b + 3$

Reality Check for End of Chapter 1

1. $121,811.29 **2.** This option will result in substantial savings in interest paid and thus in the total cost of paying off this mortgage. **3.** This option has a lower monthly payment. It also extends by 15 years the tax break that homeowners get on their federal income tax.

Chapter 2

Reality Check for Chapter 2 Opener

$-22°$F

Using the Language and Symbolism of Mathematics 2.1

1. Cartesian **2.** x; y **3.** Origin; (0, 0) **4.** y **5.** x **6.** Counterclockwise **7.** I **8.** III **9.** Scatter **10.** Arithmetic **11.** Difference **12.** Linear

Exercises 2.1

1. A: $(-1, 5)$, II; B: $(5, -2)$, IV; C: (3, 2), I; D: $(-4, -1)$, III

3.

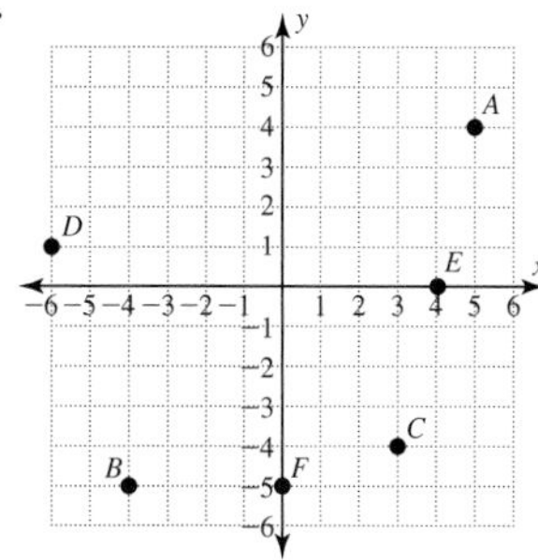

5. a. II **b.** III **c.** IV **d.** I **7. a.** y-axis **b.** x-axis **c.** x-axis **d.** y-axis

9. The points do not lie on one line, but they all lie close to a line.

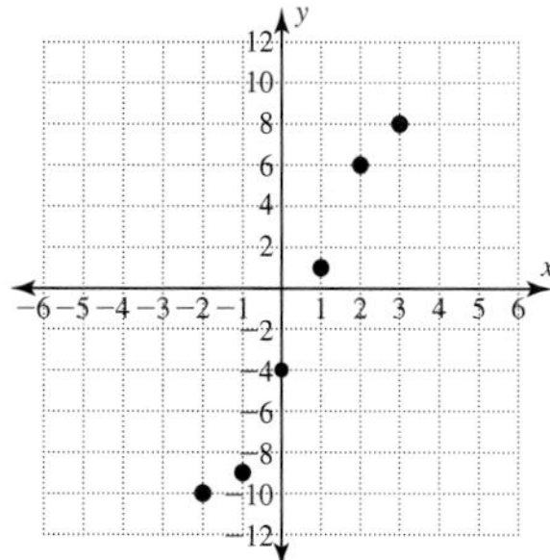

11. a. $a_1 = 5$
$a_2 = 13$
$a_3 = 19$
$a_4 = 8$
$a_5 = 6$

b.

x	y
1	5
2	13
3	19
4	8
5	6

13. a.

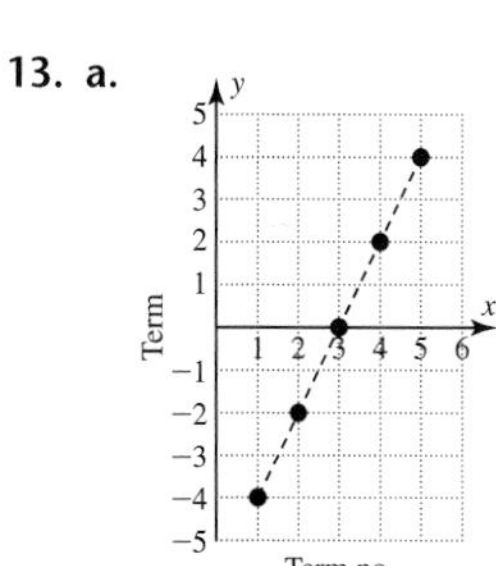

b. Yes **c.** Yes; 2

15. a.

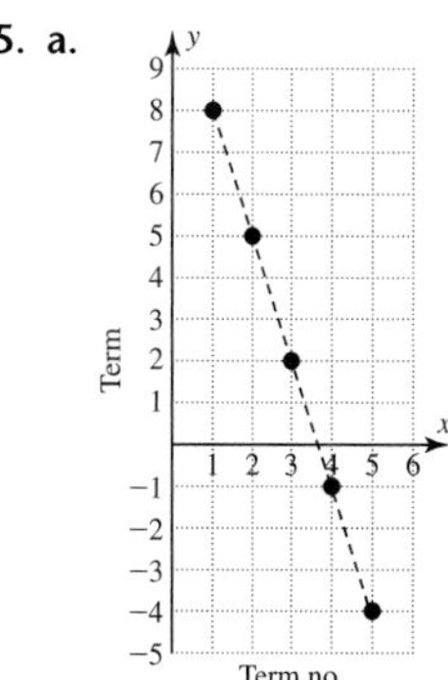

b. Yes **c.** Yes; -3

17. 4, 9, 14, 19, 24 **19.** $-4, -9, -14, -19, -24$ **21. a.** 7 **b.** 10 **c.** 19 **d.** 37 **23. a.** 4 **b.** -2 **c.** -20 **d.** -236 **25.** 1, 1.5, 2, 2.5, 3 **27.** 4, 1, $-2, -5, -8$ **29.** Arithmetic; $d = 4$ **31.** Arithmetic; $d = -10$ **33.** Not arithmetic **35. a.** Arithmetic; $d = 2$ **b.** Not arithmetic **37.** Arithmetic; $d = 2$ **39.** Not arithmetic **41.** Arithmetic; $d = -2$ **43.** Arithmetic; $d = 0$ **45.** $-4, -3, -2, -1, 0, 1, 2$; arithmetic **47.** 4, 4, 4, 4, 4; arithmetic **49.** $-9, -6, -1, 6$; not arithmetic **51. a.** 0; at the beginning no dollars have been paid. **b.** 2,550; after 6 months $2,550 has been paid. **c.** 5,100; after 12 months $5,100 has been paid. **53. a.** 2,850; after 1 month $2,850 has been paid. **b.** 10,500; after 18 months $10,500 has been paid. **c.** 13,200; after 24 months $13,200 has been paid. **55.** $>$; $>$ **57.** $<$; $<$ **59.** I; III **61.** Up **63.** Yes; $d = 2$ assuming these integers are listed in increasing order **65.** Area: 20 square units; perimeter: 18 units **67.** Yes; $d = 1$ **69.** Yes; $d = 150$ **71. a.** 240 lb **b.** 230 lb **c.** After 6 weeks **d.** Between weeks 2 and 4 **73. a.** B **b.** D **c.** A **d.** C

75. a. **b.** **c.** **d.**

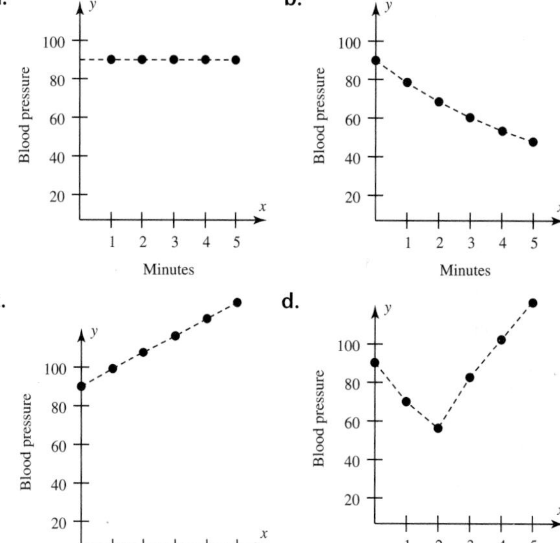

Using the Language and Symbolism of Mathematics 2.2

1. Function **2.** f; x **3.** x; output **4.** 5; 9 **5.** Line **6.** Linear **7.** Modeling

Exercises 2.2

1. 7 **3.** 4 **5.** 4 **7.** -56 **9.** 28 **11.** 3 **13.** 3 **15.** 6 **17.** 8, 6, 4, 2, 0 **19.** 1.5, 2, 2.5, 3 **21.**

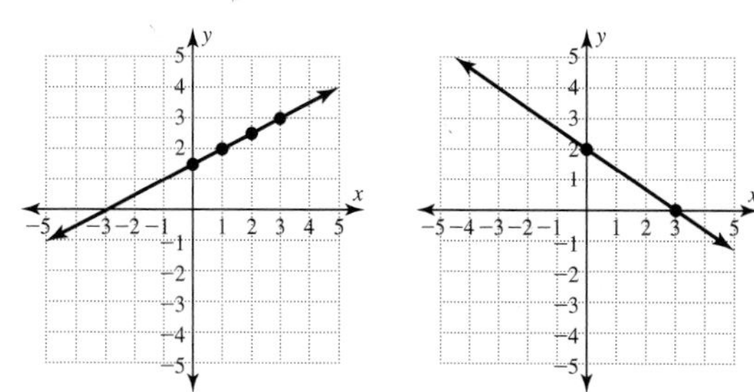

23. a. $-3, -1, 1, 3, 5, 7, 9$ **b.** $3, 23, 43, 63, 83, 103, 123$ **25. a.** $-4, -3, -2, -1, 0, 1, 2$ **b.** $-2, 3, 8, 13, 18, 23, 28$ **27.** $-3, -1, 1, 3, 5$ **29.** $-2, 0, 3, 5$ **31. a.** -8 **b.** 2 **c.** 8 **d.** 0 **33. a.** 3 **b.** 1 **c.** 0 **d.** 4

35. a.

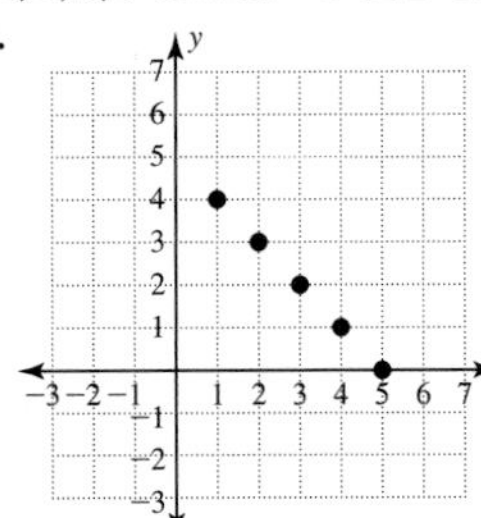

b.

c. The graph in part **a** consists of discrete points. The graph in part **b** is a connected line.

37. a. $3, 2, 1, 0, -1, -2, -3$

b.

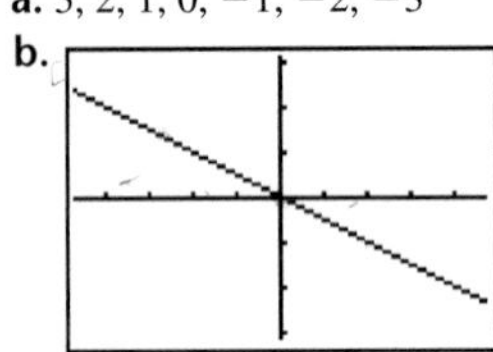

39.

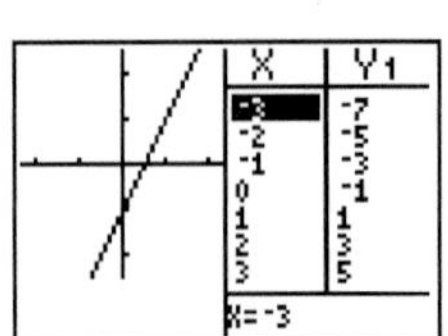

41. $f(x) = 18 - 3x$ **43.** $f(x) = 2x + 17$ **45. a.** $f(x) = 2x + 64$ **b.** $f(x) = 32x$ **47. a.** $f(x) = x + 30$ **b.** $f(x) = x - 30$ **c.** $f(x) = 2(x + 30)$ **49. a.** $f(x) = 0.02x$ **b.** $f(x) = 1.02x$ **c.** \$35.70, \$44.88, \$81.60, \$91.80 **51. a.** $f(x) = 100 - 5x$ **b.** 100, 75, 50, 25, 0, -25, -50 **c.** $f(20) = 0$; when 5 pieces of length 20 ft are cut from the roll, there is no wire left. **d.** $x = 16$; when 5 pieces of length 16 ft are cut from the roll, there will be 20 ft left. **e.** No, the maximum length that can be cut from the roll is 100 ft.

Using the Language and Symbolism of Mathematics 2.3

1. *x*-intercept **2.** *y*-intercept **3.** True **4.** Infinite **5.** Solution **6.** Intersection **7.** y **8.** x **9.** y **10.** x **11.** Overhead **12.** Break-even

Exercises 2.3

1. Solution **3.** Not a solution **5.** Solution **7.** Solution **9.** Not on the graph **11.** Not on the graph **13.** On the graph **15.** On the graph **17.** Points A and D are solutions. **19.** Point C is a solution.

21. *x*-intercept: $(3, 0)$
y-intercept: $(0, 3)$

23. *x*-intercept: $(2, 0)$
y-intercept: $(0, -4)$

25.

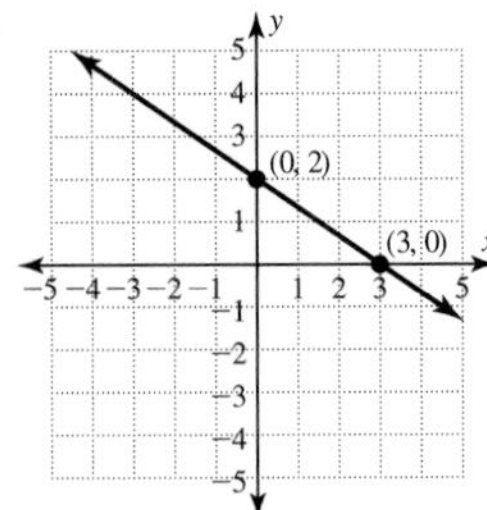

27.

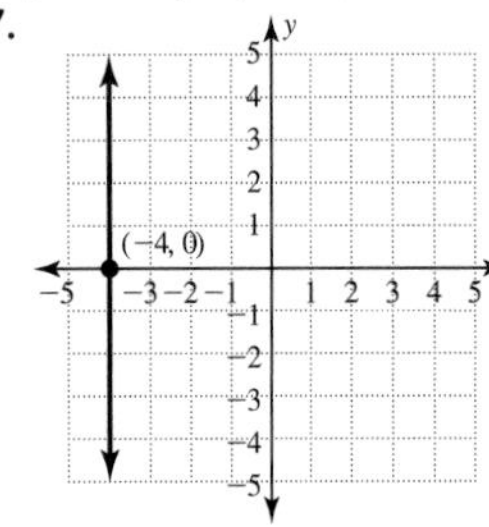

29.

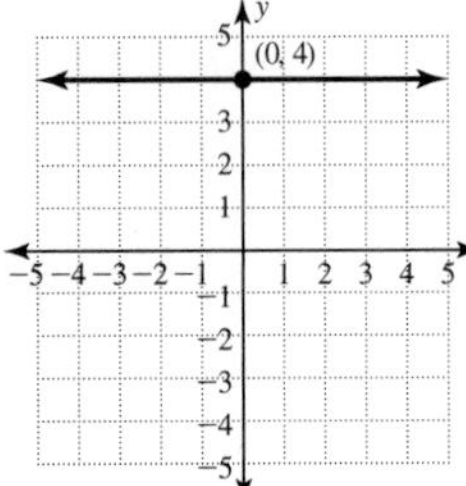

31.

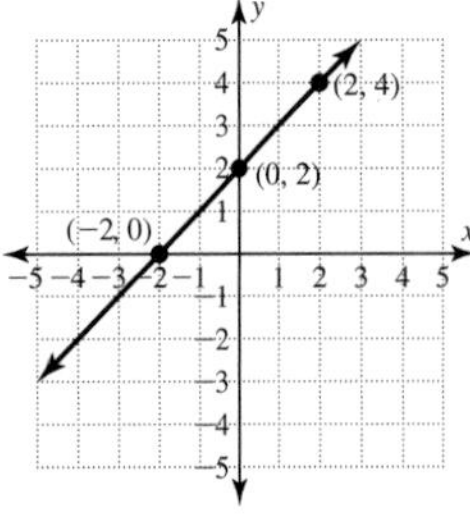

33.

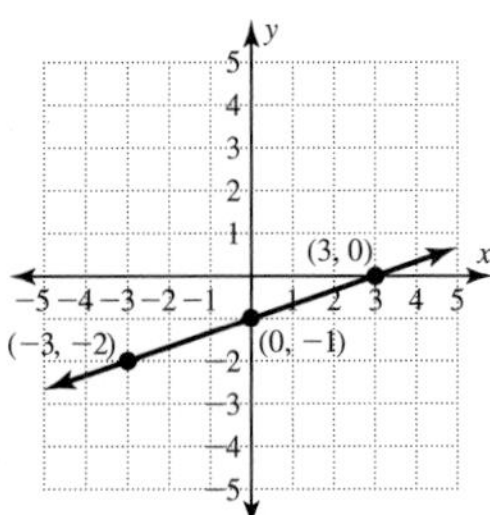

35.

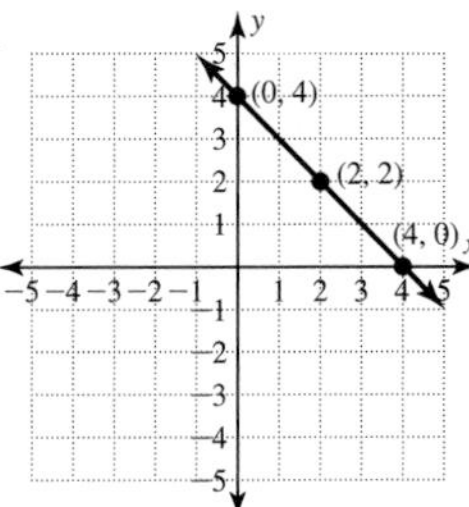

37.

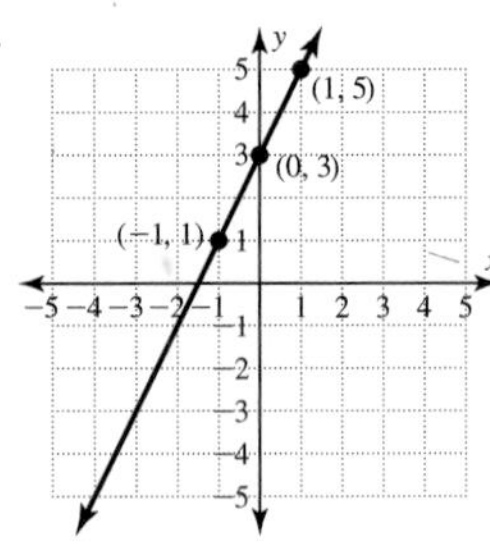

39.

x	y
-3	-18
-2	-15
-1	-12
0	-9
1	-6
2	-3
3	0

x-intercept: $(3, 0)$
y-intercept: $(0, -9)$

41. *x*-intercept: $(2, 0)$
y-intercept: $(0, 1)$

43. $(-2, 3)$ **45.** $(-1, -2)$ **47.** $(3, 2)$ **49.** $(2, 1)$ **51.** $(1, -3)$ **53.** $(2, 3)$ **55.** *x*-intercept: $(80, 0)$; *y*-intercept: $(0, -500)$: The *x*-intercept tells us that a profit of \$0.00 occurs when 80 units are produced; this is the break-even point. The *y*-intercept tells us that there is a loss of \$500.00 when no items are produced; \$500 is the overhead cost.

57.

x	$f(x)$
1	790.83
2	1,124.16
3	1,457.49
4	1,790.82
5	2,124.15
6	2,457.48
7	2,790.81
8	3,124.14
9	3,457.47
10	3,790.80
11	4,124.13
12	4,457.46

59. Units: 4
Cost: \$11.00

61. Months: 2
Cost: \$550.00

Using the Language and Symbolism of Mathematics 2.4

1. 1 **2.** Linear **3.** Linear **4.** Linear **5.** Equivalent **6.** Addition **7.** $b - c$ **8.** Conditional **9.** Identity **10.** Contradiction

Exercises 2.4

1. b **3.** $x = 24$ **5.** $v = -4$ **7.** $y = -1$ **9.** $z = -11$ **11.** $y = 4$ **13.** $x = 16$ **15.** $y = 63$ **17.** $v = 2$ **19.** $y = 0$ **21.** $m = 26$ **23.** $n = 99$ **25.** $t = 0$ **27.** $v = 0$ **29.** $x = 7$ **31.** $v = 25$ **33.** $w = 22$ **35.** $m = -5$ **37.** $n = 9$ **39.** $y = -12$ **41. a.** Conditional; $x = 0$ **b.** Contradiction; no solution **c.** Identity; every real number is a solution **43. a.** Contradiction; no solution **b.** Identity; every real number is a solution **c.** Conditional; $v = 4$ **45. a.** $9x - 5$ **b.** $x = -7$ **47. a.** $x - 7$ **b.** $x = 7$

49. a. $5.8x + 0.6$ **b.** $x = 4$ **51.** $x = 5$ **53.** $x = 2.3$ **55.** $x = -1$ **57.** $x = 2$ **59.** $x = 3$ **61.** $x = 2; x = 1.905$ **63.** $x = 500; x = 501.277$ **65.** $3m + 7 = 2m - 8; m = -15$ **67.** $12 - 9m = 2 - 10m; m = -10$ **69.** $2(3m - 9) = 5(m + 13); m = 83$ **71.** $a = 7$ **73.** $x = 253$

Using the Language and Symbolism of Mathematics 2.5

1. Equivalent **2.** Multiplication **3.** $\frac{b}{c}$ **4.** -9 **5.** Conditional **6.** Zero **7.** Distributive **8.** Least common denominator **9.** 4 **10.** -4

Exercises 2.5

1. a. $x = 6$ **b.** $x = 35$ **3. a.** $v = -8$ **b.** $v = 9$ **5. a.** $t = -48$ **b.** $t = -27$ **7.** $t = -4$ **9.** $x = -5$ **11.** $y = -5$ **13.** $y = 3$ **15.** $z = -3$ **17.** $m = -\frac{1}{9}$ **19.** $w = 0$ **21.** $v = -15$ **23.** $x = 0$ **25.** $y = -5$ **27.** $a = 110$ **29.** $x = -7$ **31.** $v = -16$ **33.** $x = 3$ **35.** $w = -10$ **37.** $x = 4$ **39.** $x = -\frac{3}{2}$ **41. a.** Contradiction; no solution **b.** Identity; every real number is a solution **c.** Conditional; $x = 0$ **43. a.** $x - 2$ **b.** $x = 2$ **45. a.** $21x - 19$ **b.** $x = \frac{19}{21}$ **47.** 6.25 mg **49.** $s \approx 25$ cm **51.** \$75.80 **53.** $x = -4; x = -4.2$ **55.** $x = 4; x = 4.1$ **57.** $x = 1.5$ **59.** $x = 0.7$ **61.** $5(x + 2) = 7(x - 3); x = \frac{31}{2}$ **63.** $2(3v - 2) = 4(v + 9); v = 20$ **65.** $2(3m - 5) = 3(m + 6) - 4; m = 8$ **67.** $\frac{1}{3}(2x + 5) = 4x + 2; x = -\frac{1}{10}$ **69.** $x = 6$ **71.** $a = \frac{1}{2}$ **73.** $x = 2$

Using the Language and Symbolism of Mathematics 2.6

1. Specified **2.** Distributive **3.** Break-even **4.** Overhead **5.** x **6.** y

Exercises 2.6

1. $y = -2x + 7$ **3.** $y = 3x - 2$ **5.** $y = 2x - 3$ **7.** $y = 2x + 4$ **9.** $y = -3x - 2$ **11.** $y = -4x + 4$ **13.** $y = 6x - 8$

15. $y = 2x - 3$

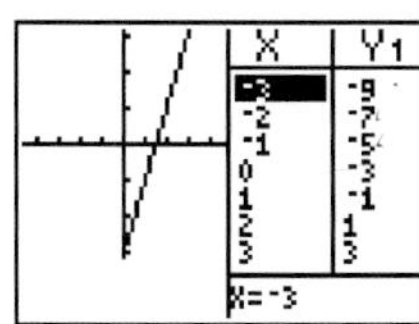

17. $y = -\frac{1}{2}x - 2$

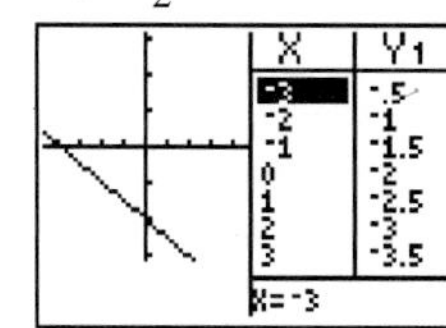

19. $l = \frac{A}{w}$ **21.** $r = \frac{C}{2\pi}$ **23.** $V_1 = \frac{V_2T_1}{T_2}$ **25.** $h = \frac{3V}{\pi r^2}$ **27.** $E = V + F - 2$ **29.** $b = \frac{2A}{h} - a$ **31.** $F = \frac{9}{5}C + 32$ **33.** $a = l - (n - 1)d$ **35.** $n = \frac{l - a}{d} + 1$ **37.** $a = S(1 - r)$ **39.** $m = \frac{y - b}{x}$ **41.** $w = \frac{P - 2l}{2}$

43. a. $P = \frac{T}{R}$ **b.** $P = \frac{T}{0.08}$

c.

T	P
100	1,250
200	2,500
300	3,750
400	5,000
500	6,250
600	7,500

45. a. $W = \frac{P - 2L}{2}$ **b.** $W = 400 - L$

L	W
100	300
120	280
140	260
160	240
160	240
180	220
200	200

47. a. $y = 0.75x - 120$ **b.** x-intercept: (160, 0); the ice cream shop will break even when 160 ice cream products are sold **c.** y-intercept: (0, -120); the overhead for the ice cream shop is \$120. (The shop will lose \$120 if no products are sold.)

49. $R = \frac{I}{PT}$

P	R
10,000	0.0600
12,000	0.0500
14,000	0.0429
16,000	0.0375
18,000	0.0333
20,000	0.0300
22,000	0.0273

Using the Language and Symbolism of Mathematics 2.7

1. Proportion **2.** Terms **3.** b; c **4.** a; d **5.** Directly **6.** Linear **7.** Increases **8.** Decreases **9.** Arithmetic

Exercises 2.7

1. $x = 12$ **3.** $x = 10$ **5.** $x = 3$ **7.** $y = 11$ **9.** $y = 16$ **11.** $a = 2$ **13.** $v = \frac{10}{3}$ **15.** $x = 3$ **17.** $z = -8$ **19.** $x = 9$ **21.** $v = 3$ **23.** $d = kt$ **25.** v varies directly as w **27.** $v = 10$ **29.** $g = 4$ **31.** $k = \frac{4}{5}$ **33. a.** 0.15 **b.** $C = 0.15S$ **c.** \$2,400, \$2,700, \$3,000, \$3,300, \$3,600, \$3,900, \$4,200

35. a. 0.82 **b.** $E = 0.82\ U$

c.

USD	EUR
50	41
100	82
150	123
200	164
250	205
300	246
350	287

37. a. $W = 76.38t$ **b.** Pounds per minute of fuel consumption

39. Algebraically: $y = \frac{1}{2}x$

Numerically:

x	y
1	0.5
2	1
3	1.5
4	2
5	2.5

Graphically:

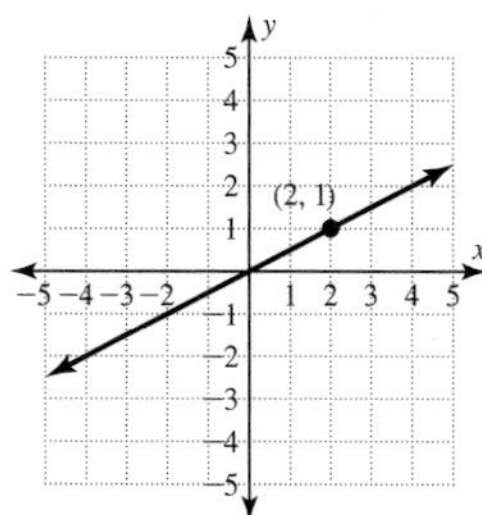

41. Numerically:

x	y
1	3
2	6
3	9
4	12
5	15

Graphically:

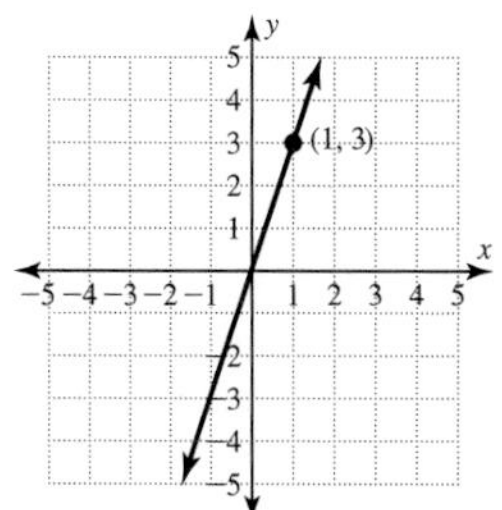

Verbally: y varies directly as x with a constant of variation 3

43. $4\frac{1}{2}$ cups **45.** $a = 18$ **47.** $x = 3$ **49.** 10 mm **51.** 50 bulbs **53.** 1,416 bricks **55.** 8.75m × 7.5m **57.** 13.5 lb **59.** 9 face cards **61.** 22.4 ft **63.** 17.5 cm **65.** 17.5 ft **67.** $53\frac{1}{3}$ ft **69.** 9 **71.** 24 doses **73. a.** 60° **b.** 60° **c.** 30° **d.** 24 **e.** $12\sqrt{3}$ **75. a.** $C = 0.80x$ **b.** \$400 **c.** \$625 **77.** 0.8

Using the Language and Symbolism of Mathematics 2.8

1. Restrictions **2.** Find **3.** Variable **4.** Algebraic **5.** Solve **6.** Reasonable **7.** Mixture **8.** Rate; time **9.** Principal; rate; time **10.** Fixed **11.** Variable

Exercises 2.8

1. B **3.** D **5.** $\{0, 1, 2, \ldots, 49, 50\}$ **7.** $[0, 225]$ **9. a.** Let x = number of months **b.** Current thickness minus wear equals 4.0 mm **c.** $11.2 - 0.2x = 4.0$ **d.** $x = 36$ **e.** Yes **f.** The brake pads have 36 months of wear left. **11.** 345 min **13.** 50 cushions **15.** 8% **17.** $57,500 **19.** 3 h 12 min **21.** 2 h 48 min **23.** 800 gal **25.** 150 gal **27.** 0.5 MW **29.** 22,500 lb **31. a.** 7,500 tiles **b.** The plant does not have the capacity to produce 10,000 tiles per day. **33. a.** 826 gal **b.** There is not enough capacity to add 1,267 gal. **35. a.** There is not enough capacity to add 400 gal. **b.** 150 gal **37. a.** $44.00 **b.** $1.10 **c.** 100 **d.** 120 **e.** 40

Review Exercises for Chapter 2

1. A: $(4, 0)$; B: $(2, 4)$; I; C: $(-6, 2)$, II; D: $(0, -2)$

2.

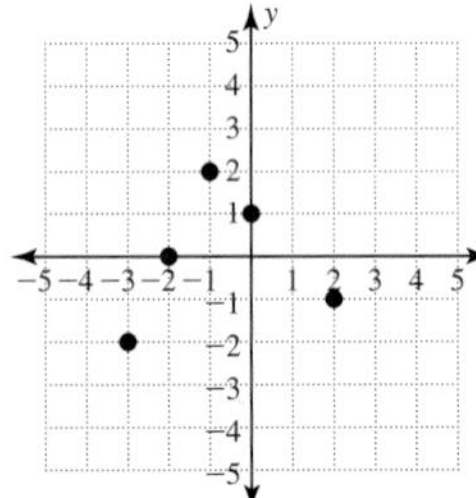

3.

n	a_n
1	1
2	−1
3	−3
4	−5
5	−7

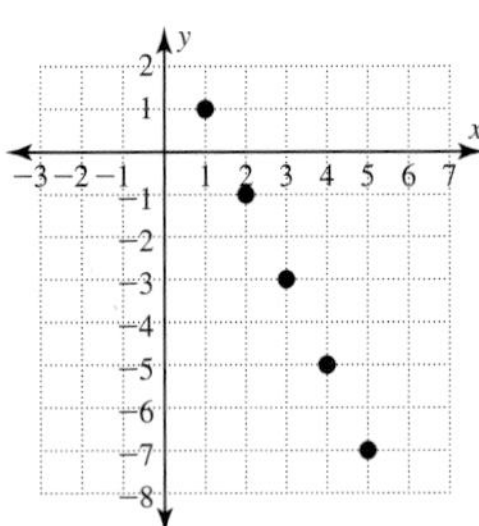

4.

n	a_n
1	2
2	2.5
3	3
4	3.5
5	4

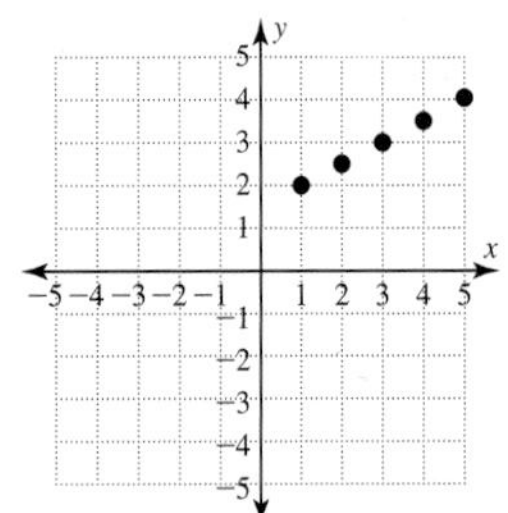

5. Arithmetic; $d = 3$. **6.** Not arithmetic **7.** Arithmetic; $d = -2$ **8.** Not arithmetic **9.** Not arithmetic **10.** Arithmetic; $d = 2$ **11.** Neither is a solution **12.** $(-3, 2)$ is a solution; $(3, -2)$ is not a solution **13.** $(-3, 2)$ is a solution; $(3, -2)$ is not a solution **14.** Neither is a solution **15.** x-intercept: $(3, 0)$; y-intercept: $(0, -2)$ **16.** x-intercept: $(-5, 0)$; y-intercept: $(0, 5)$ **17.** x-intercept: $(-2.75, 0)$; y-intercept: $(0, 11)$ **18.** B **19.** C **20.** D **21.** A **22.** $(1, 2)$ **23.** $(2, -3)$ **24.** $x = 2.5$ **25.** $x = 24$ **26.** $x = 2$ is not a solution. $x = 3$ is a solution. **27.** $x = 2$ is a solution. $x = 3$ is not a solution. **28.** $x = 15$ **29.** $v = 2$ **30.** $m = 0$ **31.** $x = -3$ **32.** $m = 2$ **33.** $n = -36$ **34.** $y = -98$ **35.** $w = -4$ **36.** $t = 11{,}700$ **37.** $b = 0$ **38.** $x = 7$ **39.** $z = \frac{3}{16}$ **40.** $r = 0$ **41.** $x = -4$ **42.** $x = 27$ **43.** $y = 3$ **44.** $y = 0$ **45.** $w = -31.6$ **46.** $t = -24$ **47.** $m = -9$ **48.** $n = -11$ **49.** $y = -1{,}295$ **50.** $x = -13.3198$ **51.** All real numbers **52.** $a = 0$ **53.** $x = 1.5$ **54.** $x = -2.4$ **55.** Contradiction; no solution **56.** Contradiction; no solution **57.** Identity; every real number is a solution **58.** Conditional; $v = 0$ **59. a.** -81 **b.** -11 **c.** 52 **d.** $7\pi - 11 \approx 10.991$ **60. a.** -74 **b.** -14 **c.** 106 **d.** 1,986 **61.** $x = v + w - y$ **62.** $x = \frac{vw}{y}$ **63.** $x = \frac{5y + 7z}{3}$ **64.** $x = -\frac{1}{8}y + \frac{19}{4}$ **65. a.** $8x + 10$ **b.** $x = -1$ **66. a.** $-2x - 2$ **b.** $x = \frac{1}{4}$ **67. a.** $2x - 1$ **b.** $x = \frac{1}{2}$ **68. a.** $-x - 7$ **b.** $x = -7$ **69.** $x = 10$; $x = 10.1$ **70.** $x = 20$; $x = 19.95$ **71.** $5m + 4 = 49$; $m = 9$ **72.** $\frac{1}{3}(m + 7) = 12$; $m = 29$ **73.** $2(m + 2) = -(m-3)$; $m = -\frac{1}{3}$ **74.** $m + (m + 1) + (m + 2) = 93$; $m = 30$ **75. a.** $P = 8w + 8$ **b.** $P = 104$ cm **c.** $8w + 8 = 48$ **d.** $w = 5$ **76. a.** $f(x) = 0.25x + 500$ **b.** 500, 750, 1,000, 1,250, 1,500, 1,750 **c.** $f(2{,}500) = 1{,}125$; it costs $1,125 to sell 2,500 snow cones. **77. a.** $C = 26x + 160$ **b.** $C = \$1{,}720$ **c.** $26x + 160 = 1{,}200$ **d.** $x = 40$ **78.** 210 mi **79.** $a = 4.5$ cm; $b = 13.5$ cm; $c = 7.5$ cm **80.** 4.4 m **81. a.** $v = 10$ **b.** $k = \frac{4}{5}$ **82.** 20 defective display panels **83.** 290 mi/h **84. a.** $y = 280x + 680$

b.

x	0	1	2	3	4	5	6
y	680	960	1,240	1,520	1,800	2,080	2,360

c.

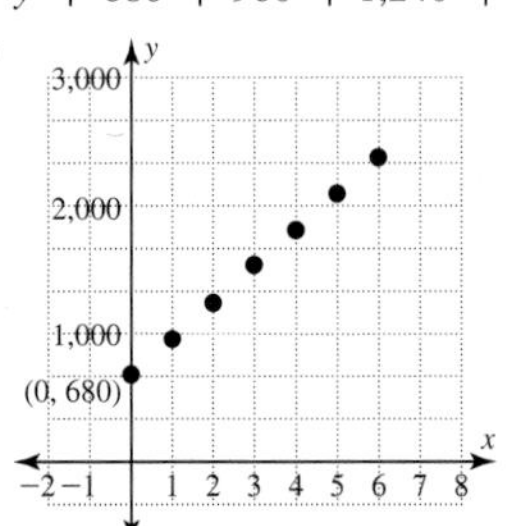

d. The sequence is arithmetic with $d = 280$ **85.** B **86.** C **87.** D **88.** A **89.** $3\frac{1}{2}$ hours **90.** Add 150 gal of water. There is enough room.

Mastery Test for Chapter 2

1. A: $(-3, 1)$ II; B: $(1, -3)$, IV; C: $(4, 2)$, I; D: $(-2, -2)$, III

2.

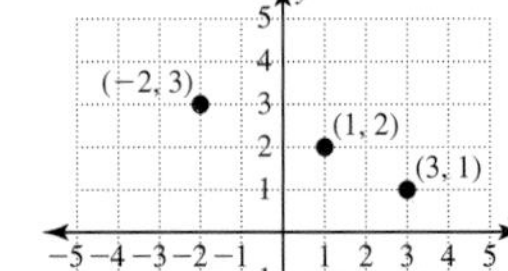
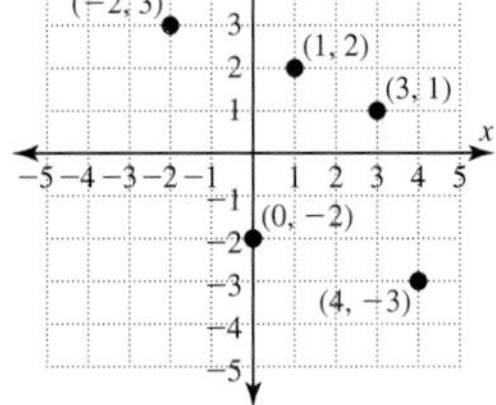

3. a. Arithmetic; $d = 2$ **b.** Not arithmetic **c.** Arithmetic; $d = 0$ **d.** Arithmetic; $d = -3$ **4. a.** -7 **b.** -18 **c.** 70 **d.** 103

5. a.

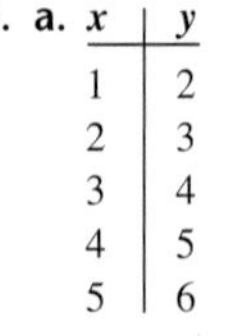

x	y
1	2
2	3
3	4
4	5
5	6

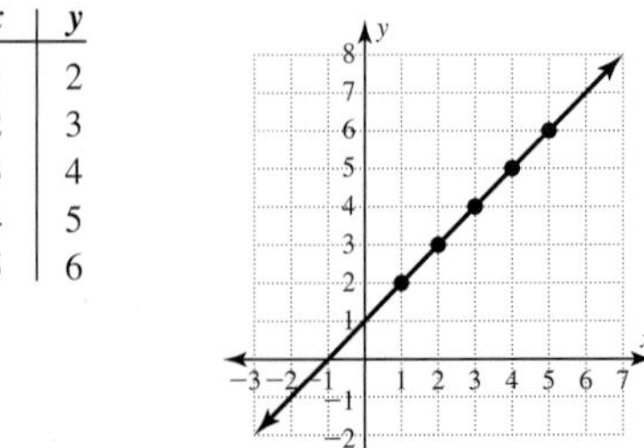

b.

x	y
1	1
2	0
3	−1
4	−2
5	−3

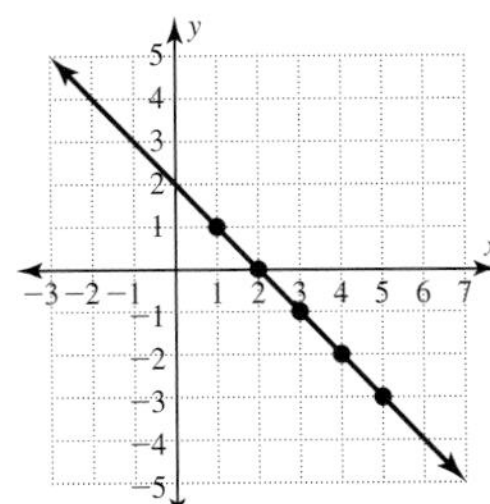

c.

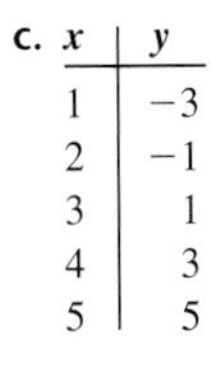

x	y
1	−3
2	−1
3	1
4	3
5	5

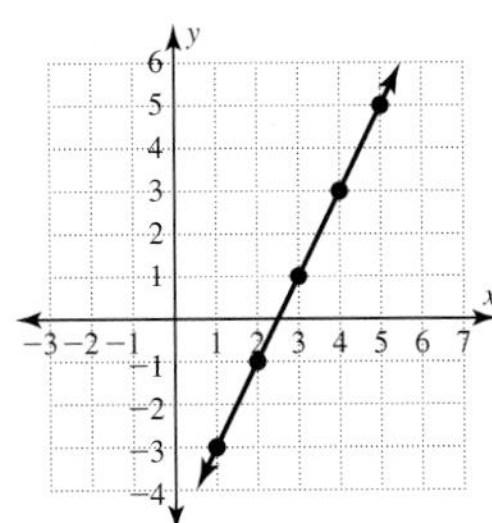

d.

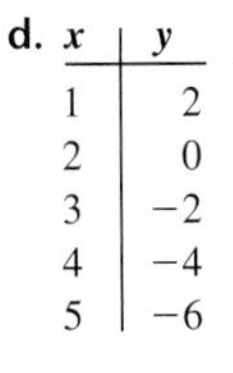

x	y
1	2
2	0
3	−2
4	−4
5	−6

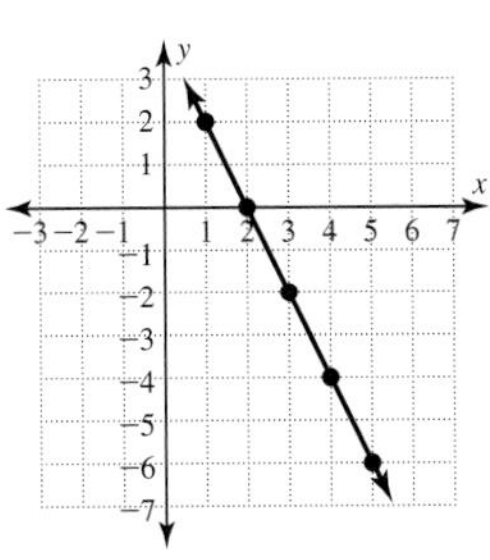

6. a. $f(x) = 375x + 800$ **b.**

x	0	6	12	18	24	30
$f(x)$	800	3,050	5,300	7,550	9,800	12,050

c. $f(25) = 10{,}175$; the total amount paid at the end of the 25th month is \$10,175 **7. a.** Solution **b.** Not a solution **c.** Solution **d.** Solution **8. a.** $(-2, 0)$; $(0, 3)$ **b.** $(-3, 0)$; $(0, -1)$ **c.** No x-intercept; $(0, 2)$ **d.** $(-1, 0)$; no y-intercept **9. a.** $(2, 1)$ **b.** $(-1, 3)$ **10. a.** $x = 4$ **b.** $x = 14$ **c.** $x = 0$ **d.** $x = 6$ **11. a.** Identity: every real number is a solution **b.** Contradiction; no solution **c.** Conditional equation; $x = 0$

12. a. $x = 2$ **b.** $x = -2$ **13. a.** $x = 3$ **b.** $y = -\frac{5}{9}$ **c.** $v = \frac{3}{2}$ **d.** $w = 13$

14. $y = 4x - 1$

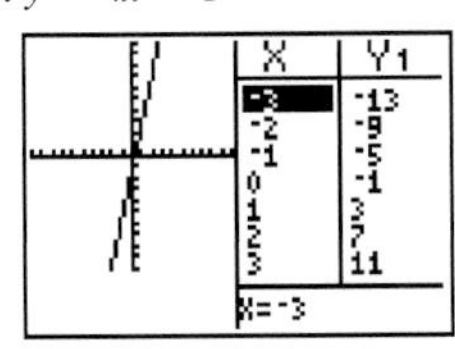

$[-10, 10, 1]$ by $[-10, 10, 1]$

15. a. $(3, 0)$; $(0, -6)$ **b.** No x-intercept; $(0, 4)$ **c.** $(5, 0)$; $(0, -2)$ **d.** $(-2, 0)$; $(0, 1)$ **16. a.** $m = -3$ **b.** $x = 11$ **c.** 187.5 km **17. a.** $y = 12$ **b.** 9.5 pesos for each U.S. dollar **18. a.** $[0, 30]$ **b.** $\{0, 1, 2, 3, \ldots, 29, 30\}$ **c.** $[0, 300]$ **d.** $\{0, 1, 2, 3, \ldots, 9, 10\}$ **19. a.** 20.8% **b.** 163 gal **c.** 278 gal

Reality Check for End of Chapter 2

The windchill at 5° with a 30-mi/h wind is −19° while only −11° at 0° with a 5-mi/h wind. Although both will feel very cold, most people would prefer 0° with a 5-mi/h wind.

Chapter 3

Reality Check for Chapter 3 Opener

1999

Using the Language and Symbolism of Mathematics 3.1

1. y **2.** x **3.** m **4.** $m = \frac{y_2 - y_1}{x_2 - x_1}$ **5.** 1 **6.** 0 **7.** Undefined **8.** Change **9.** Change **10.** Rate **11.** Increases **12.** Decreases **13.** Parallel **14.** Perpendicular **15.** Slope **16.** Grade

Exercises 3.1

1. a. 2 **b.** 2 **3. a.** $-\frac{11}{9}$ **b.** 0 **5. a.** $-\frac{3}{4}$ **b.** Undefined **7. a.** $\frac{2}{5}$ **b.** $\frac{3}{5}$ **9. a.** $\frac{1}{3}$ **b.** −2 **c.** Undefined **d.** 0 **11.** $\frac{3}{2}$ **13.** 8, 11, 14, 17 **15. a.** 3, 1, −1, −3 **b.** 1, −3, −7, −11 **17. a.** −2, 3, 8, 13, 18, 23, 28 **b.** 3 **c.** 5 **d.** $\frac{5}{3}$ **19. a.** $-\frac{3}{5}$ **b.** $\frac{5}{8}$ **c.** Undefined **d.** 0 **21.** $-\frac{8}{5}$ **23.** 2 **25.** $\frac{2}{3}$ **27.** 3 **29.** −9 **31.** 0

33. Answers can vary. Point (4, 5)

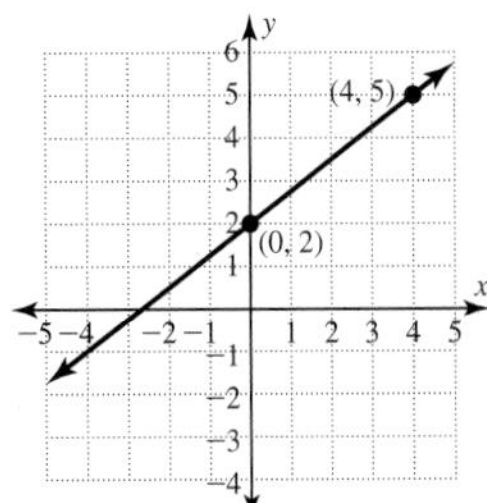

35. Answers can vary. Point (1, −5)

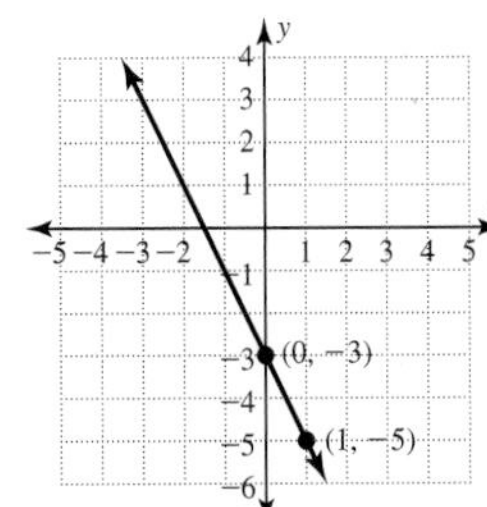

37.

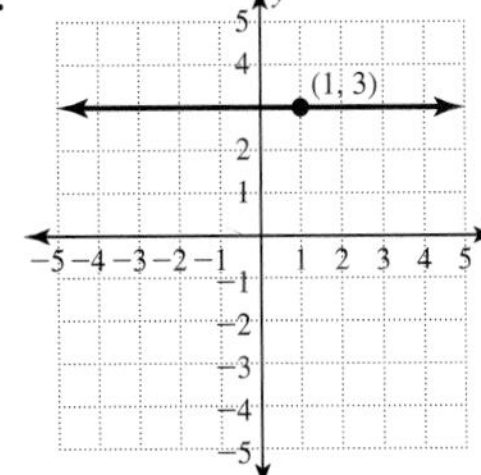

39.

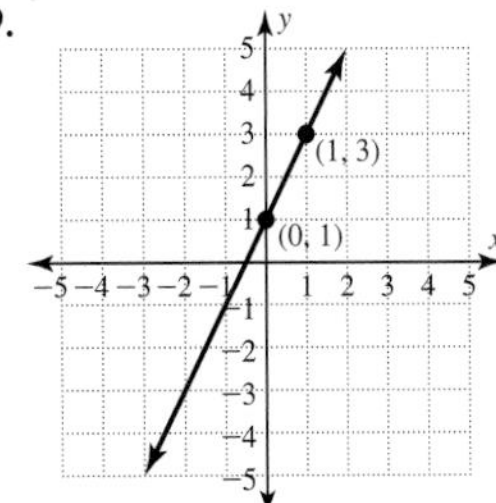

41.

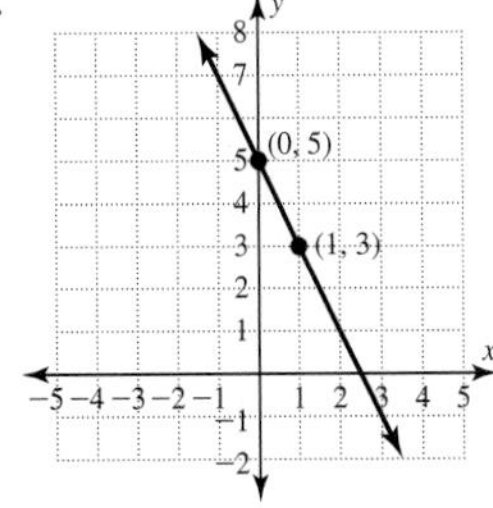

43. a. (2, 0) **b.** (0, 4) **c.** −2 **45.** $m = 65$; the speed of the car is 65 mi/h. **47.** $\frac{1}{2}$; 0.494 **49.** $-\frac{1}{3}$; −0.333 **51.** $m_2 = \frac{3}{7}$; $m_3 = -\frac{7}{3}$ **53.** $m_2 = 0$; m_3 is undefined **55.** $m_1 = \frac{8}{5}$; $m_3 = -\frac{5}{8}$ **57.** Parallel **59.** Perpendicular

61. Neither **63.** Perpendicular **65.** Parallel **67. a.** −3 gal/s **b.** Each second the volume of water decreases by 3 gal. **c.** 2 min 47 s

69. a. \$6,250, \$7,500, \$8,750, \$10,000 **b.**

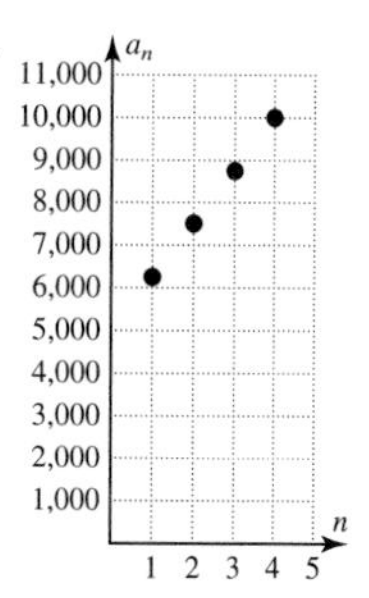

c. $m = 1{,}250$; monthly payment is \$1,250 **d.** $a_0 = 5{,}000$; down payment is \$5,000 **71.** $\frac{1}{7}$ **73.** 90 m **75.** 3.5 ft **77.** 0 **79.** Positive **81.** Negative **83.** Positive **85.** 2

Using the Language and Symbolism of Mathematics 3.2

1. $y = mx + b$ **2.** Parallel **3.** Perpendicular **4.** Vertical; (5, 0) **5.** Horizontal; (0, −5) **6.** $Ax + By = C$ **7.** 1 **8.** Point; slope; point; slope

Exercises 3.2

1. a. $m = 2$; y-intercept: (0, 5) **b.** $m = -\frac{3}{11}$; y-intercept: $\left(0, -\frac{4}{5}\right)$ **c.** $m = 6$; y-intercept: (0, 0) **3. a.** $m = 0$; y-intercept: (0, 2) **b.** m is undefined; no y-intercept **c.** $m = -1$; y-intercept: (0, 0) **5.** $y = 4x + 7$ **7.** $y = -\frac{2}{11}x + 5$ **9.** $y = -6$

11.

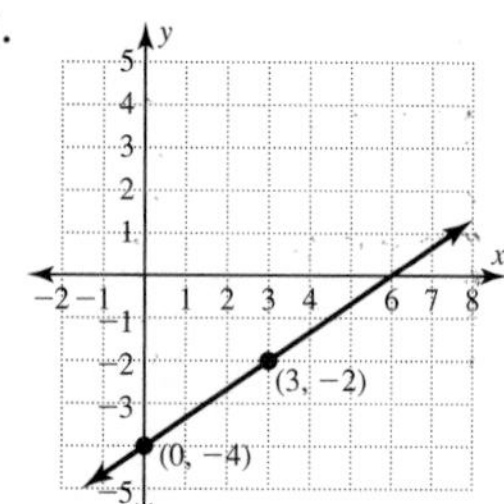

13.

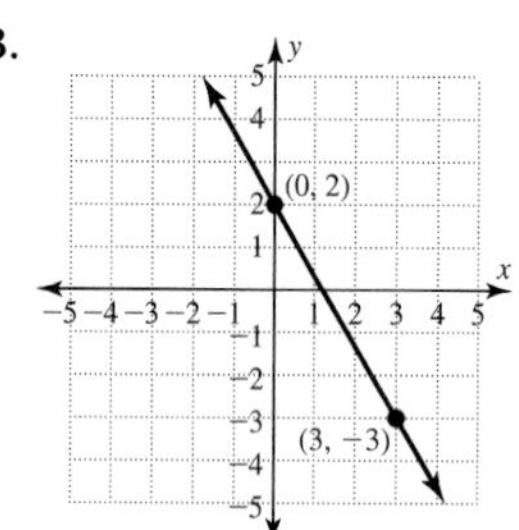

15.

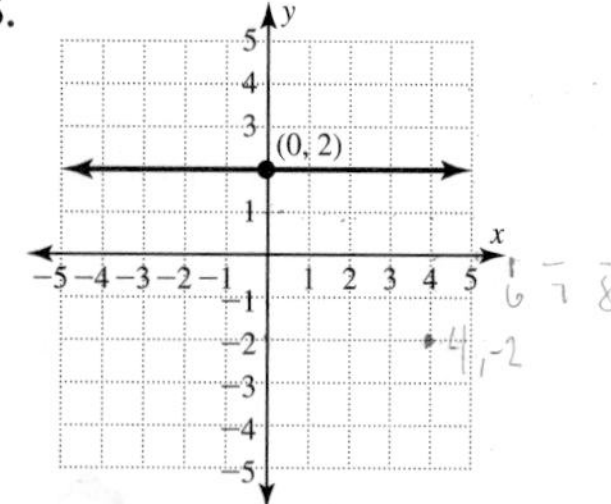

17. $f(x) = \frac{2}{5}x + 1$ **19.** $f(x) = -\frac{1}{4}x - 1$ **21.** $f(x) = -50x + 500$; the slope gives us the speed of the truck (50 mi/h). The y-intercept represents the initial distance from the dispatch center (500 mi). **23.** −2, 1, 4, 7, 10 **25. a.** 1 **b.** 7 **c.** 7 **d.** (0, 1) **e.** $y = 7x + 1$ **27. a.** 2 **b.** −1 **c.** $-\frac{1}{2}$ **d.** (0, 8) **e.** $y = -\frac{1}{2}x + 8$ **29. a.** −12 **b.** The jet is using 12 gal of fuel per min from the main tank. **c.** $f(x) = 3{,}500 - 12x$ **d.** No, there is not enough fuel in the tank for 300 min. **31.** $f(x) = 50x + 75$; the cost is \$50 per hour plus a flat fee of \$75 per visit **33.** $y = 5$ **35.** $x = -4$ **37.** Neither **39.** Perpendicular **41.** Parallel **43.** Neither

45. a. $m = 5$; point: (3, 4) **b.** $m = -7$; point: (2, −5) **c.** $m = \frac{1}{2}$; point: (−6, 0)

47.

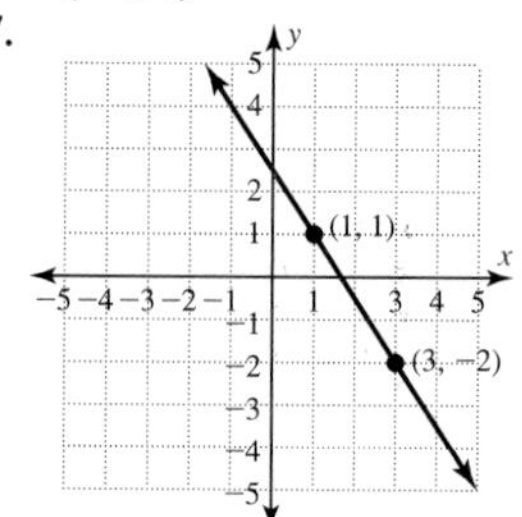

49.

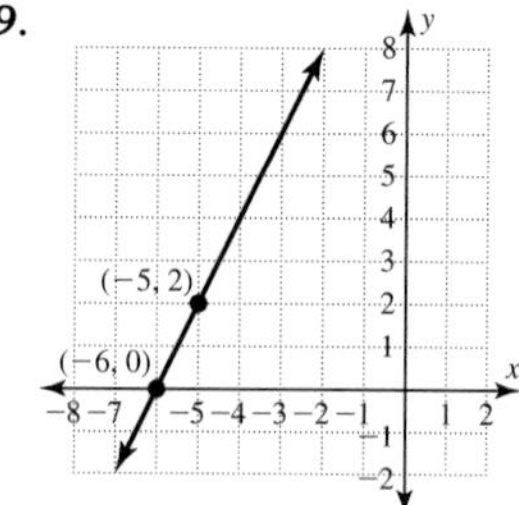

51.

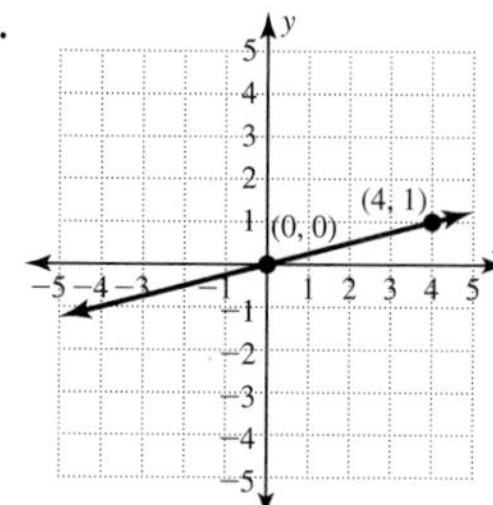

53.

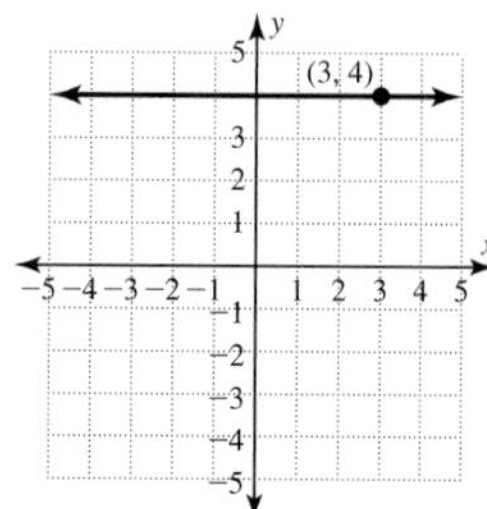

55. $y = -4x + 11$ **57.** $y - 3 = \frac{1}{3}(x - 1)$ **59.** $y - 1 = -\frac{2}{7}(x - 3)$ **61.** $y = \frac{1}{4}x + 3$ **63.** $y = 2x + 6$ **65. a.** $x = 2$ **b.** $y = 3$ **67.** $y = \frac{3}{7}x + \frac{15}{7}$ **69.** $y = \frac{3}{4}x + \frac{19}{4}$ **71.** $y = -8$ **73.** $f(x) = 0.8x + 3.25$; the cost is \$0.80 per mile plus a flat fee of \$3.25 per ride **75.** $y = 4x + 7$; $4x - y = -7$ **77.** $y = -\frac{2}{3}x - \frac{2}{3}$; $2x + 3y = -2$ **79.** $y = 2x + 8$; the spring stretches 2 cm for each kilogram attached; the length of the spring is 8 cm when no weight is attached **81. a.** $y = \frac{7}{300}x + \frac{79}{75}$ **b.** \$1.22

Using the Language and Symbolism of Mathematics 3.3

1. Solution **2.** Intersection **3.** Inconsistent; Parallel **4.** One; one **5.** Infinite; coincide **6. a.** Infinitely many **b.** Are; zero **c.** Are not; one

Exercises 3.3

1. (−2, 3) **3.** (3, 2) **5.** Solution **7.** Solution **9.** Solution **11.** Not a solution **13.** (2, 2) **15.** (2, −6) **17.** (5, −8) **19.** (−0.5, 4) **21.** (300, 90); each company charges \$90 when 300 mi are driven **23.** (−6, 4) **25.** (−3, 2) **27.** (6, −2) **29.** (4, 1) **31.** $\left(-\frac{9}{2}, -5\right)$ **33.** No solution **35.** Infinite number of solutions **37.** (1, −3) **39.** (2, 3) **41.** $\left(\frac{2}{7}, \frac{9}{7}\right)$ **43.** $\left(\frac{13}{15}, -\frac{9}{55}\right)$ **45.** (300, 700) **47.** (0.3, 0.7) **49.** (1, −3) **51.** (0, 3) **53.** (30, 40) **55.** One solution; consistent system of independent equations **57.** No solution; inconsistent system **59.** Infinite number of solutions; consistent system of dependent equations **61.** One solution; consistent system of independent equations **63.** Consistent system of independent equations **65.** Inconsistent system **67.** The cost for each option is \$11 when 4 units are produced. **69.** (1, 450); At 3 p.m. each plane is 450 mi from O'Hare Airport. **71.** 43.75°; 136.25° **73.** The daily costs will each be \$105 when 10 items are produced.

75.

x	$y_1 = 2x + 7$	$y_2 = 13 - x$
-3	1	16
-2	3	15
-1	5	14
0	7	13
1	9	12
2	11	11
3	13	10

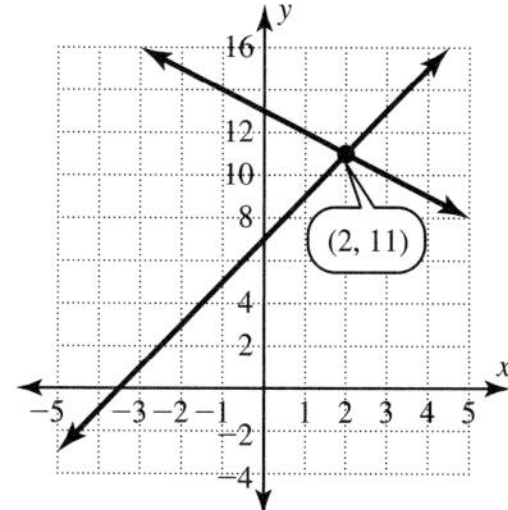

Answer: (2, 11); The numbers are 2 and 11.

77. $x = 3$; $y = 5$ **79.** $y = \frac{1}{2}x + 3$; $y = -4x + 3$

Using the Language and Symbolism of Mathematics 3.4

1. Equal **2.** Ordered pair **3.** False **4.** True **5.** A conditional equation **6.** An identity **7.** A contradiction **8.** Consistent; independent; (3, 4) **9.** Inconsistent; independent; no **10.** Consistent; dependent; infinite

Exercises 3.4

1. (4, 5) **3.** (2, 0) **5.** $(-34, 8)$ **7.** $\left(\frac{4}{7}, \frac{5}{7}\right)$ **9.** $(0, -6)$ **11.** $(-2, -1)$ **13.** $\left(3, -\frac{3}{5}\right)$ **15.** $\left(-\frac{13}{3}, -7\right)$ **17.** $(-1, 2)$ **19.** $(-2, -3)$ **21.** (12, 9) **23.** $(0, -6)$ **25.** (3, 2) **27.** $(-4, -2)$ **29.** (12, 15) **31.** Infinite number of solutions **33.** No solution **35.** (0, 0) **37.** (3, 4) **39.** $(2, -1)$ **41.** $\left(\frac{1}{2}, -\frac{1}{3}\right)$ **43.** B; the numbers are 75 and 45. **45.** C; the numbers are 30 and 0.

47.

x	$y_1 = 8 - x$	$y_2 = x - 2$
2	6	0
3	5	1
4	4	2
5	3	3
6	2	4
7	1	5
8	0	6

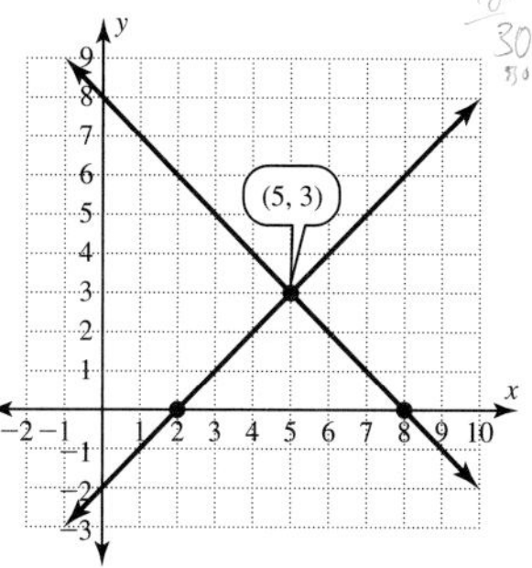

Answer: (5, 3); The two numbers are 5 and 3.

49. 36° and 54° **51.** $\frac{x + y}{2} = 76$; $x - y = 28$; the test scores are 90 and 62. **53.** $\frac{3x + y}{4} = 66$; $x - y = 8$; Tiger scored a 68 in each of his first three rounds, and he scored a 60 in the last round. **55. a.** Yes **b.** ≈ 1.1 years **57. a.** $y = 0.5x + 1$ **b.** $y = 0.2x + 2$ **c.** $\left(\frac{10}{3}, \frac{8}{3}\right)$ **d.** Yes **e.** Each company charges \$2.67 when 3.3 mi is driven. **59. a.** $y = 180x + 1{,}500$ **b.** $y = 140x + 2{,}000$ **c.** (12.5, 3,750) **d.** Yes **e.** After 12.5 years each plan is worth \$3,750.00. **61.** 1; 2; $\left(\frac{5}{3}, \frac{11}{3}\right)$ **63.** $B = -6$ **65.** $C = 6$ **67.** (4.65, 2.75) **69.** $(-11.3, -1.1)$

Using the Language and Symbolism of Mathematics 3.5

1. Equal **2.** Elimination; eliminate **3.** Opposites **4.** Inconsistent; no **5.** Consistent; dependent; infinite

Exercises 3.5

1. (2, 2) **3.** $(-8, 7)$ **5.** $\left(\frac{1}{6}, -1\right)$ **7.** $\left(-2, \frac{3}{2}\right)$ **9.** $(3, -5)$ **11.** (4, 5) **13.** $(1, -1)$ **15.** $(-9, 11)$ **17.** (19, 0) **19.** (0, 0) **21.** $(-3, 5)$ **23.** (6, 6) **25.** No solution **27.** Infinite number of solutions **29.** $\left(\frac{4}{3}, -\frac{1}{6}\right)$ **31.** (5, 9) **33.** (2, 2) **35.** $\left(\frac{1}{2}, -\frac{1}{2}\right)$ **37.** No solution **39.** (0, 3) **41.** C; the numbers are 2 and -5 **43.** D; the numbers are -2 and 6 **45.** $x + y = 88$; $x - y = 28$; the numbers are 58 and 30 **47.** $x + y = 102$; $y = 2x$; the numbers are 34 and 68 **49.** $\frac{4x + y}{5} = 25$; $y - x = 15$; He scored 22 points in each of the first 4 games and 37 points in the fifth game. **51. a.** $y = -0.5x + 1{,}030.5$ **b.** $y = 1.5x - 2{,}959.5$ **c.** (1995, 33) **d.** In 1995, 33% attended 4-year colleges and 33% attended community colleges. **53.** 0; 1; $\left(\frac{5}{11}, 2\right)$ **55.** (10, 20) **57.** (4.4, 5.5)

Using the Language and Symbolism of Mathematics 3.6

1. Find **2.** Variable **3.** Algebraic **4.** Solve **5.** Reasonable **6.** Mixture **7.** Rate **8.** Rate; time **9.** Principal; rate; time **10.** Work; time

Exercises 3.6

1. a. Let x = width in centimeters and y = length in centimeters. **b.** The perimeter is 204 cm; the length is 6 more than the width. **c.** $2x + 2y = 204$; $y = x + 6$ **d.** (48, 54) **e.** Yes **f.** The dimensions are 48 by 54 cm. **3.** The numbers are 28 and 72. **5.** The base is 32 cm, and the two equal sides measure 20 cm each. **7.** 22 and 35 **9.** 81, 81, 81, 81, and 91 **11. a.** 30 **b.** 2.31 **c.** 9 **13.** 6; 99 **15.** The fixed cost is \$30.00, and the pay-per-view movies are \$3.50 each. **17.** 47 union employees and 3 union stewards will be needed. **19.** 332 adult tickets and 456 student tickets were sold. **21.** 31° and 59° **23.** 35° and 145° **25.** 540 right-handed desks and 60 left-handed desks should be ordered. **27.** Each bulb will cost \$2.90 to purchase and use for 20 days. **29.** \$9,500 at 7% and \$2,500 at 9% **31.** \$6,500 in bonds and \$3,500 in stocks **33.** 84 and 94 km/h **35.** 85 and 100 km/h **37.** The rate of the boat is 15 km/h and the rate of the current is 9 km/h. **39.** Mix 60 L of the 5% solution and 20 L of the 33% solution. **41.** Use 220 g of 80% pure gold alloy. **43.** Add 8.4 L of pure water. **45.** Older bricklayer: 280 bricks/h; younger bricklayer: 220 bricks/h **47. a.** (5.88, 200) **b.** In approximately 1996, the United States and Europe were producing 200,000 publications. **49. a.** $y = 200x + 500$ **b.** $y = 100x + 1{,}000$ **c.** (5, 1,500) **d.** Yes **e.** In each option, \$1,500 is paid after 5 months. **51. a.** Fixed cost: \$2,500; variable cost: \$8 per pair of shoes **b.** The proposed answer does not meet the restrictions on the variable because the fixed cost cannot be negative. **53. a.** It will take the helicopter 2 h 20 min to fly to Pearl Harbor. **b.** From the point of no return, it will take 1 h to fly to Pearl Harbor. **c.** It will take the helicopter 1 h 20 min to reach the point of no return.

Review Exercises for Chapter 3

1. 2 **2.** $-\frac{3}{5}$ **3.** $\frac{4}{7}$ **4.** -1 **5.** 0 **6.** $\frac{2}{5}$ **7.** Undefined **8.** 0 **9.** Undefined **10.** 0.5 **11. a.** 4 **b.** 5 **c.** -4 **d.** $\frac{4}{5}$ **e.** $-\frac{4}{5}$

12. a.

x	y
0	6
5	9
10	12
15	15
20	18

b.

x	y
0	6
10	12
20	18
30	24
40	30

c.

x	y
0	6
-5	3
-10	0
-15	-3
-20	-6

13. a. 3,800, 4,600, 5,400, 6,200
b.

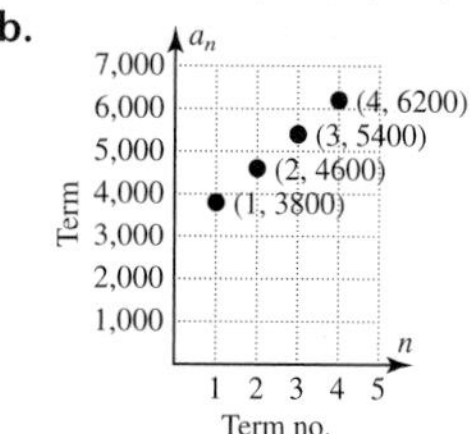

c. 800; monthly payment is \$800 **d.** $a_0 = 3{,}000$; down payment is \$3,000
14. $\frac{1}{6}$ **15. a.** Parallel **b.** Perpendicular **c.** Neither **16.** $y = -\frac{1}{2}x + 6$
17. $y = \frac{2}{3}x - 2$ **18.** $y = -3x + 11$ **19.** $y = \frac{3}{2}x - 4$ **20.** $y = 5x$
21. $y = \frac{1}{2}x + 4$ **22.** $y = 9$ **23.** $x = 4$

24.

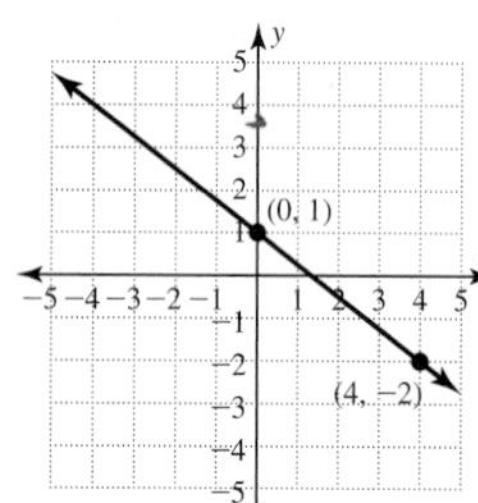

25.

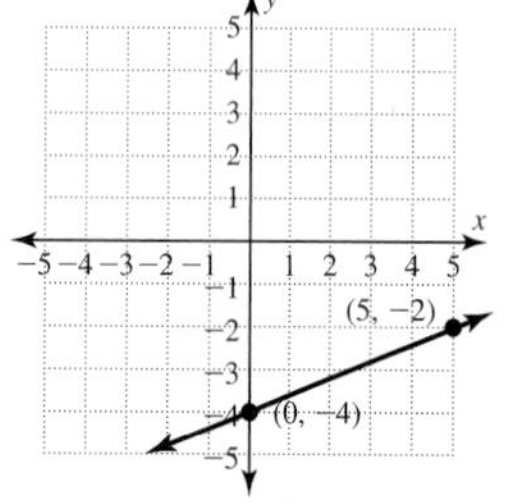

26. $m \approx 1$; $m = 0.9$ **27.** $m \approx -1$; $m = -1.1$ **28. a.** $(0, -4)$ **b.** 2 **c.** 3 **d.** $\frac{3}{2}$ **e.** $y = \frac{3}{2}x - 4$ **29.** a, b, d **30.** d **31.** $(3, -1)$ **32.** $(-5, 15)$
33. Consistent system of dependent equations **34.** Inconsistent system **35.** (6, 5) **36.** Infinitely many solutions **37.** No solution
38. $(-0.5, 0.25)$ **39.** $(11, -7)$ **40.** No solution **41.** $\left(0, \frac{4}{3}\right)$
42. $(-1, -1)$ **43.** (5, 11) **44.** $\left(\frac{18}{11}, \frac{1}{22}\right)$ **45.** $(-148, 225)$ **46.** (25, 60)
47. No solution **48.** $\left(-\frac{2}{7}, \frac{3}{5}\right)$ **49.** One solution; consistent system of independent equations **50.** No solution; inconsistent system **51.** An infinite number of solutions; consistent system of dependent equations **52.** No solution; inconsistent system **53.** B; (60, 40); the numbers are 60 and 40. **54.** D; $(60, -40)$; the first number is 60 and the second is -40. **55.** C; (20, 0); the numbers are 20 and 0. **56.** A; (70, 50); The numbers are 70 and 50. **57.** 9; 162 **58.** Both options cost \$3,500 when 10,000 pages are printed. **59.** 46° and 134° **60.** The first nine measure 24.975 cm; the tenth measures 25.725 cm. **61.** Car loan: \$3,000, educational loan: \$6,000 **62.** 9 cm **63.** 96 and 104 km/h **64.** Older bricklayer: 26 bricks/h; younger bricklayer: 22 bricks/h **65.** 45 L of the \$6 chemical; 55 L of the \$9 chemical **66. a.** $y = 0.95x + 6$ **b.** The slope $m = 0.95$ represents the cost per CD. The y-intercept (0, 6) represents the initiation fee \$6 **c.** $y = 1.95x$ **d.** The slope $m = 1.95$ represents the cost per CD; the y-intercept (0, 0) represents the initiation fee, \$0. **e.** (6, 11.7) **f.** When 6 CDs are purchased, each plan costs \$11.70. **67. a.** $y = 30x + 20$ **b.** The slope $m = 30$ represents the charge per hour; the y-intercept represents a \$20 flat fee. **c.** $y = 26x + 30$ **d.** The slope $m = 26$ represents the charge per hour; the y-intercept represents a \$30 flat fee. **e.** (2.5, 95) **f.** The total charge for each shop is \$95 when the repairs require 2.5 hours.

Mastery Test for Chapter 3

1. a. 2 **b.** -3 **c.** $-\frac{3}{2}$ **d.** $\frac{2}{5}$ **e.** -3 **f.** 7 **2. a.** 5 **b.** 20 **c.** 4 **d.** The altitude is increasing 4 ft/s. **e.** 5 minutes **3. a.** Perpendicular **b.** Parallel **c.** Perpendicular **d.** Neither **4. a.** $y = -2x + 6$ **b.** $y = 4x + 7$ **c.** $y = \frac{2}{3}x + 5$ **d.** $y = -\frac{5}{3}x - 2$ **5. a.** $y = 3$ **b.** $x = -4$ **c.** $x = -4$ **d.** $y = 3$

6. a.

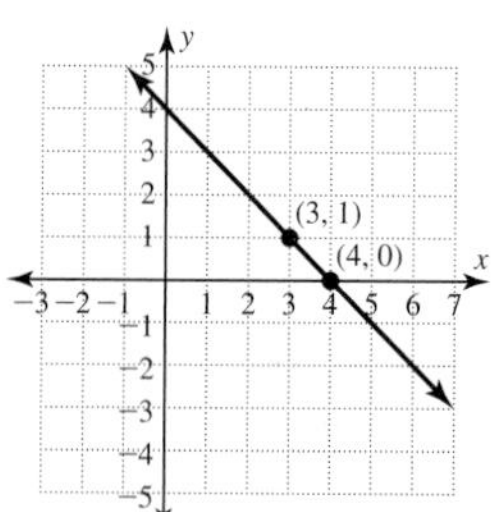

b.

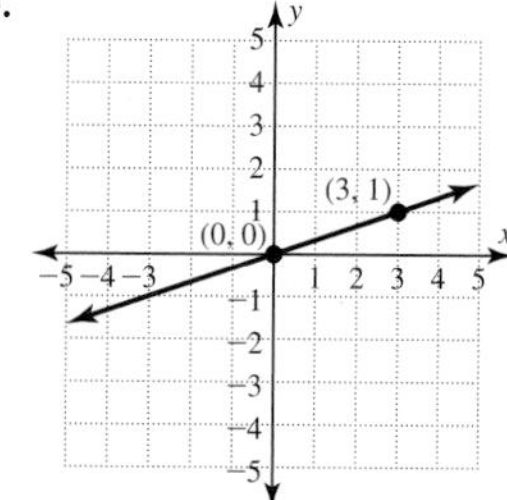

c.

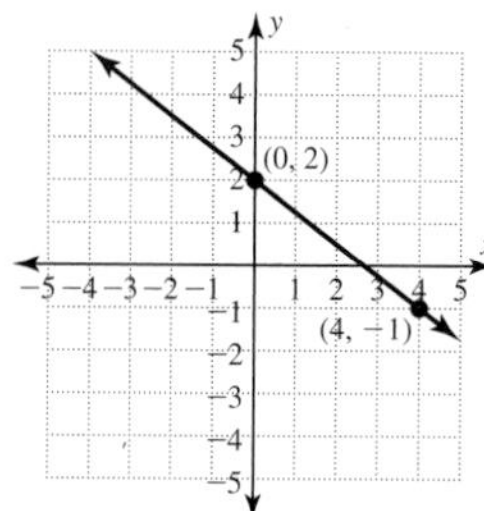

d.

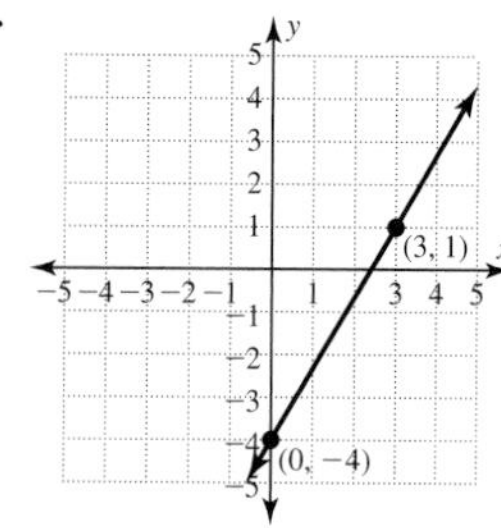

7. a. (0, 2) **b.** (2, 5) **c.** $(-2, -1)$ **d.** (12, 5) **8. a.** B **b.** C **c.** A
9. a. $(-2, 7)$ **b.** $\left(\frac{1}{14}, -\frac{5}{7}\right)$ **c.** $\left(2, -\frac{1}{3}\right)$ **d.** $\left(\frac{37}{10}, -\frac{13}{20}\right)$ **10. a.** $(-2, 1)$ **b.** $(-1, 1)$ **c.** Infinitely many solutions **d.** No solution **11. a.** 13°, 77° **b.** \$7,500 in bonds, \$2,500 in the savings account **c.** 15 mL 30% solution, 10 mL 80% solution **d.** 475 mi/h, 225 mi/h

Reality Check for End of Chapter 3

1. -1.5; the cost per kilowatt-hour is dropping by \$0.015/yr.
2. \$0.56/kWh **3.** \$0.05/kWh

Chapter 4

Reality Check for Chapter 4 Opener

30,000 lb.

Using the Language and Symbolism of Mathematics 4.1

1. Linear **2.** 1 **3.** Equivalent **4.** Conditional **5.** Solution **6.** Addition **7.** Subtraction **8. a.** $>$ **b.** $>$ **c.** $>$ **9.** Distributive **10.** $\geq$ **11.** $\leq$ **12.** $>$ **13.** $\leq$ **14.** $\geq$

Exercises 4.1

1. b **3. a.** $(-\infty, -4]$ **b.** $(5, \infty)$ **c.** $(-\infty, 3)$ **d.** $[-1, \infty)$ **5. a.** Not a solution **b.** Solution **c.** Not a solution **d.** Solution **7. a.** Solution **b.** Not a solution **c.** Solution **d.** Not a solution **9. a.** $x = 1$ **b.** $(-\infty, 1)$ **c.** $(1, \infty)$ **11. a.** $x = -3$ **b.** $(-\infty, -3]$ **c.** $[-3, \infty)$ **13. a.** $x = 500$ **b.** $(500, 800]$ **c.** $[0, 500)$ **d.** When 500 hours of overtime is worked, the spending limit is met. When more than 500 hours of overtime is worked, the spending limit is exceeded. When less than 500 hours of overtime is worked, the spending limit is not met. **15. a.** $x = 5$ **b.** $(-\infty, 5]$ **c.** $[5, \infty)$ **17. a.** $x = -5$ **b.** $(-5, \infty)$ **c.** $(-\infty, -5)$ **19. a.** $x = 8$ **b.** $[0, 8)$ **c.** $(8, 24]$ **d.** Each firm charges the same amount when 8 boxes of paper are ordered. Ace Office Supply charges less for orders of fewer than 8 boxes of paper and more for orders of more than 8 boxes of paper. **21.** $(4, \infty)$ **23.** $(-\infty, 1]$ **25.** $[3, \infty)$; all values of x greater than or equal to 3 **27.** $(-\infty, 2)$; all values of x less than 2 **29.** $[4, \infty)$ **31.** $(-\infty, 0)$ **33.** $[7, \infty)$ **35.** $(-\infty, -1)$ **37.** $(-\infty, 1)$ **39.** $[2, \infty)$ **41.** $(-\infty, -2)$ **43.** $[40, \infty)$ **45.** $(1, \infty)$; $(1.112, \infty)$ **47.** $(-\infty, -22)$; $(-\infty, -21.93)$ **49.** $(8, \infty)$ **51.** $(-\infty, -1]$ **53.** $[-20, \infty)$ **55.** $(1.5, \infty)$ **57.** $(2, \infty)$ **59.** $[2, \infty)$ **61.** $5 - 2x \leq 7 - 3x$; $(-\infty, 2]$ **63.** $2w + 3 \geq w + 11$; $[8, \infty)$ **65.** $3(m + 5) > 2m + 12$; $(-3, \infty)$ **67.** A maximum of 8 examiners **69.** From 0 to 60 ft **71.** The company will lose money when less than 450 mice are produced; the company will have a profit if more than 450 mice are produced.

Using the Language and Symbolism of Mathematics 4.2

1. Multiplication **2.** Multiplication **3.** $>$ **4.** $<$ **5.** Least common denominator **6.** Distributive

Exercises 4.2

1. a. $>$ **b.** $>$ **c.** $>$ **d.** $<$ **3. a.** $x = -1$ **b.** $(-\infty, -1)$ **c.** $(-1, \infty)$ **5. a.** $x = -2$ **b.** $[-2, \infty)$ **c.** $(-\infty, -2]$ **7. a.** $x = 2$; each company charges the same amount when the machines are rented for 2 hours **b.** $(2, \infty)$; the Dependable Rental Company charges less when the machines are rented for more than 2 hours **c.** $(0, 2)$; the Dependable Rental Company charges more when the machines are rented for less than 2 hours **9. a.** $x = 3$ **b.** $(-\infty, 3)$ **c.** $(3, \infty)$ **11. a.** $x = -5$ **b.** $[-5, \infty)$ **c.** $(-\infty, -5]$ **13. a.** $x = 30$; each store charges the same amount when 30 shirts are ordered **b.** $(0, 30)$; Mel's Sports charges less for orders of less than 30 shirts **c.** $(30, 60]$; Mel's Sports charges more for orders of more than 30 shirts **15.** $-2x < 4$; $\frac{-2x}{-2} > \frac{4}{-2}$; $x > -2$; $(-2, \infty)$ **17.** $-3x + 4 \geq 7$; $-3x + 4 - 4 \geq 7 - 4$; $-3x \geq 3$; $\frac{-3x}{-3} \leq \frac{3}{-3}$; $x \leq -1$; $(-\infty, -1]$ **19.** $(-2, \infty)$; all values of x greater than -2 satisfy the inequality **21.** $(-\infty, 4]$; all values of x less than or equal to 4 satisfy the inequality **23.** $[2, \infty)$ **25.** $(-\infty, -8)$ **27.** $(16, \infty)$ **29.** $(-\infty, 0]$ **31.** $(-\infty, 2.3)$ **33.** $(-\infty, -2)$ **35.** $(-\infty, -5]$ **37.** $(-\infty, 0]$ **39.** $(2, \infty)$ **41.** $(-\infty, 3]$ **43.** $(-5, \infty)$ **45.** $(-\infty, -5)$ **47.** $[-2, \infty)$ **49.** $(1, \infty)$ **51.** $\left[\frac{2}{3}, \infty\right)$ **53.** $[-3, \infty)$ **55.** $\left(-\frac{31}{3}, \infty\right)$ **57.** $(1, \infty)$ **59.** $[-8, \infty)$ **61.** $(-\infty, 7]$; $(-\infty, 7.1]$ **63.** $(-\infty, 8)$; $(-\infty, 8.9)$ **65.** $2(x + 5) < 6$; $(-\infty, -2)$ **67.** $4(y - 7) \geq 6(y + 3)$; $(-\infty, -23]$ **69.** At least 5°C **71.** At most \$60,000 **73. a.** $[0, 200)$ The company loses money when less than 200 bricks are produced. **b.** $(200, \infty)$; The company earns a profit when more than 200 bricks are produced.

Using the Language and Symbolism of Mathematics 4.3

1. Conditional **2.** Unconditional **3.** Contradiction **4.** Compound; and **5.** Intersection **6.** Union

Exercises 4.3

1. C **3.** B **5. a.** $-5 < x \leq 1$ **b.** $x < -5$ or $x > 1$ **7. a.** $0 < x < 6$ **b.** $6 < x < 7$ **9. a.** $150 \leq x \leq 450$ **b.** $5 \leq x < 9$ **11. a.** $x > 2$ and $x \leq 8$ **b.** $x \geq -5$ and $x \leq 22$ **13. a.** $5x + 2 > -13$ and $5x + 2 < 7$ **b.** $5x - 1 \geq -6$ and $5x - 1 < 14$ **15. a.** $x < -2$ or $x > 3$ **b.** $-4 \leq x \leq 2$ or $5 < x \leq 9$ **17.** $A \cap B = [5, 7]$; $A \cup B = (-4, 11)$ **19.** $A \cap B = [5, 11)$; $A \cup B = (-4, \infty)$ **21.** $A \cap B = [-4, -2]$; $A \cup B = \mathbb{R}$ **23.** $[-1, 6)$;

[number line: −1 to 6]; x is greater than or equal to -1 and less than 6 **25. a.** $-2 < x \leq 5$ **b.** $0 \leq x \leq 4$ **27.** $[-1, 2)$ **29.** $(-1, 4)$ **31.** $155 < x < 190$ **33.** $[0, 150] \cup [220, 270]$ **35. a.** $4{,}000 \leq 250x \leq 6{,}000$ **b.** $[16, 24]$; For a safety factor from 2 to 3 use 16 to 24 steel wires in the cable. **37.** $[50, 100]$; The strength requirements are met when 50 to 100 fiber strands are used. **39.** $(1, 6)$ **41.** $(-2, 3]$ **43.** Contradiction; no solution **45.** Unconditional; $\mathbb{R}$ **47.** Conditional; $(-\infty, 2]$ **49.** $[-5, 1)$; all real numbers greater than or equal to -5 and less than 1 **51.** $[1, 2]$; all real numbers greater than or equal to 1 and less than or equal to 2 **53.** $(-6, -4)$ **55.** $[-2, 1)$ **57.** $[0, 4]$ **59.** $[0, 4)$ **61.** $(2, 8]$ **63.** $(-1, 2)$ **65.** $[2, 4]$ **67.** $(-1, 1]$ **69.** $\left(-\infty, \frac{11}{7}\right]$ **71.** $(-1, 2)$ **73.** $[-5, 1]$ **75.** $(2, 3)$; $(1.9, 3.1)$ **77.** $(-3, 2]$; $(-2.9, 2.1]$ **79.** No solution **81.** $\left(\frac{1}{3}, \frac{8}{5}\right)$ **83.** $(-\infty, -3) \cup (-2, \infty)$ **85.** The acceptable range of temperatures is between 5 and 35°C. **87.** The third side must be between 4 and 20 m.

Using the Language and Symbolism of Mathematics 4.4

1. Distance **2. a.** $<$ **b.** $=$ **c.** $>$ **3.** $|x - a| = d$ **4.** Tolerance **5.** Isolate

Exercises 4.4

1. a. $|x| < 5$ **b.** $|x| \geq 2$ **c.** $|x - 1| = 3$ **d.** $|x| < 4$ **3. a.** $|x - 4| \leq 2$ **b.** $|x - 1| \geq 2$ **c.** $|x + 7| < 3$ **d.** $|x| \leq 5$ **5. a.** $|x| \leq 3$ **b.** $|x| > 3$ **c.** $|x| < 7$ **7. a.** $x = -6$ or $x = 6$ **b.** $x = 0$ **c.** $x = -3$ or $x = 7$ **d.** $x = -7$ or $x = 3$ **9. a.** $x = 6$ or $x = -2$ **b.** $(-2, 6)$ **c.** $(-\infty, -2) \cup (6, \infty)$ **11. a.** $x = 6$ or $x = 2$ **b.** $(2, 6)$ **c.** $(-\infty, 2] \cup [6, \infty)$ **13. a.** $x = -4$ or $x = 0$ **b.** $[-4, 0]$ **c.** $(-\infty, -4] \cup [0, \infty)$ **15. a.** $x = -5$ or $x = 15$ **b.** $(-5, 15)$ **c.** $(-\infty, -5] \cup [15, \infty)$ **17.** $(-2, 0)$; values greater than -2 and less than 0 **19.** $(-1, 2)$; values greater than -1 and less than 2 **21.** $(-\infty, -2] \cup \left[-\frac{2}{3}, \infty\right)$ **23.** $[-4, 4]$ **25.** $(-4, 1)$ **27.** $\left(-\frac{3}{2}, -\frac{1}{2}\right)$ **29.** $(-\infty, -4) \cup (3, \infty)$ **31.** $(-\infty, -97] \cup [99, \infty)$ **33.** $\left(-4, \frac{10}{3}\right)$ **35.** $x = \frac{1}{3}$ or $x = -1$ **37.** $x = 3$ or $x = \frac{1}{3}$ **39.** $x = 4$ or $x = -\frac{2}{5}$ **41.** $x \approx \pm 5$; $x = \pm 4.9$ **43.** $x \approx 7$ or $x \approx -23$; $x = 7.1$ or $x = -22.96$ **45.** $|x + 3| < 9$ **47.** $|x - 11| \leq 15$ **49.** $|x - 2| > 4$ **51.** $(16, 28)$ **53.** $|x - 15| \leq 0.001$; $[14.999, 15.001]$ **55.** $|x - 16| \leq 0.25$; $[15.75, 16.25]$ **57.** $|x - 25| \leq 2$; $[23, 27]$ **59. a.** $3000 \leq 500x + 250 \leq 9000$ where x is the number of synthetic strands. **b.** Between 6 and 17 strands.

Using the Language and Symbolism of Mathematics 4.5

1. Ordered pair; true **2.** Half-planes **3.** Solid **4.** Dashed **5.** Contains **6.** Not contain **7.** Upper **8.** Lower

Exercises 4.5

1. a. Solution **b.** Solution **c.** Not a solution **d.** Not a solution **3. a.** Not a solution **b.** Solution **c.** Not a solution **d.** Solution **5.** A and B are solutions **7.** B and C are solutions **9.** a **11.** A and C

13. a.

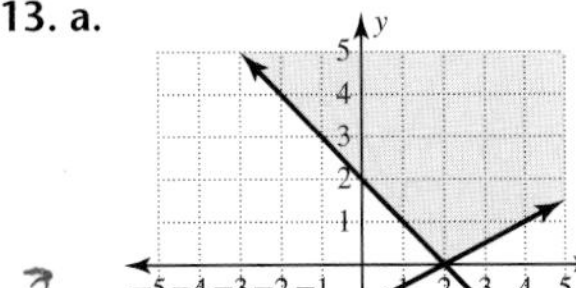

b.

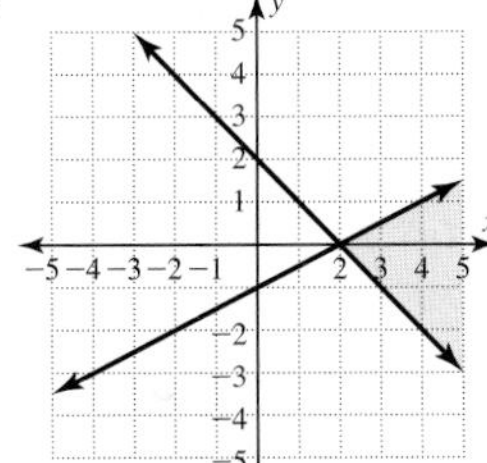

c.

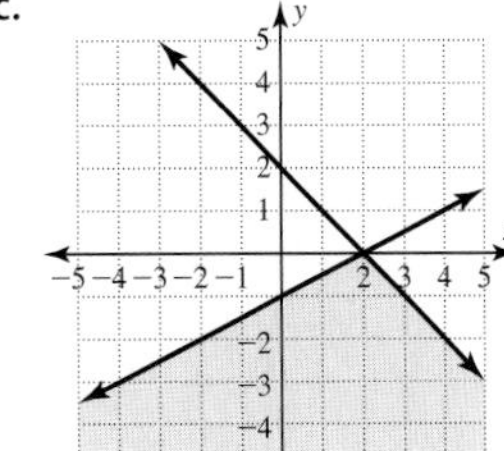

d.

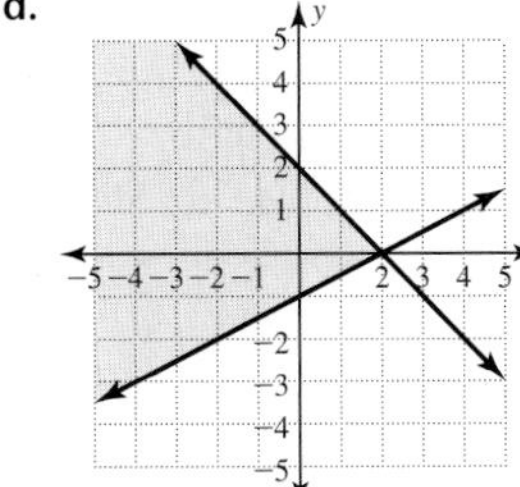

15. a.

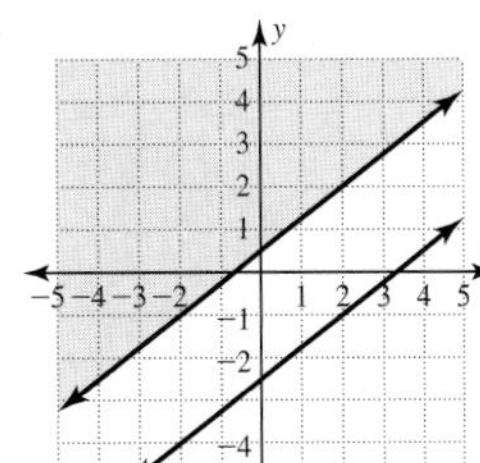

b. No solution

c.

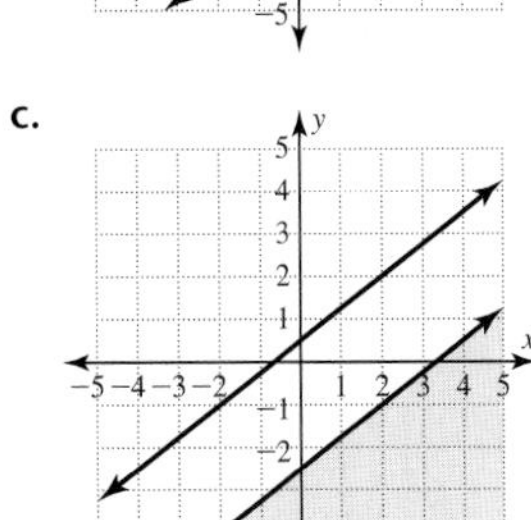

d.

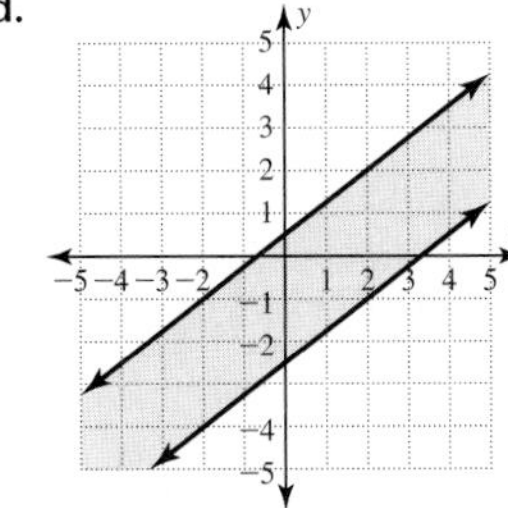

17.
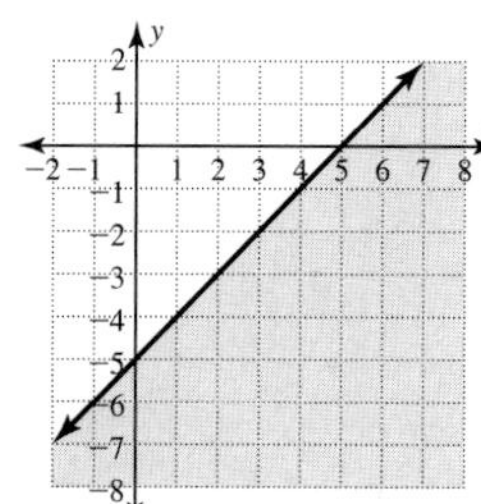

19.
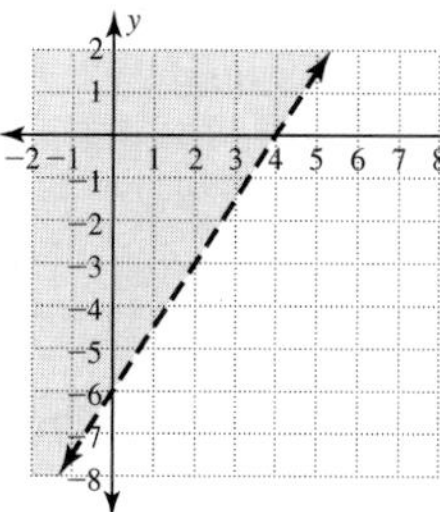

21.
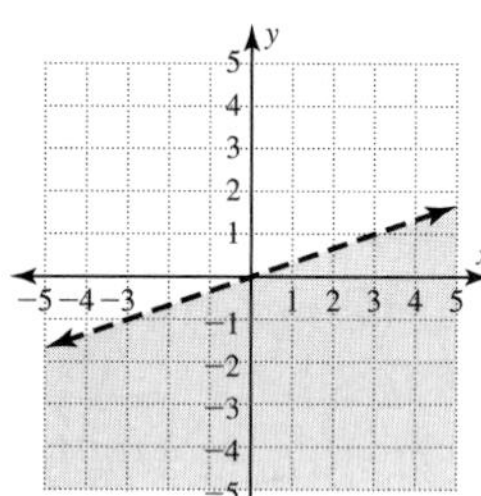

23.
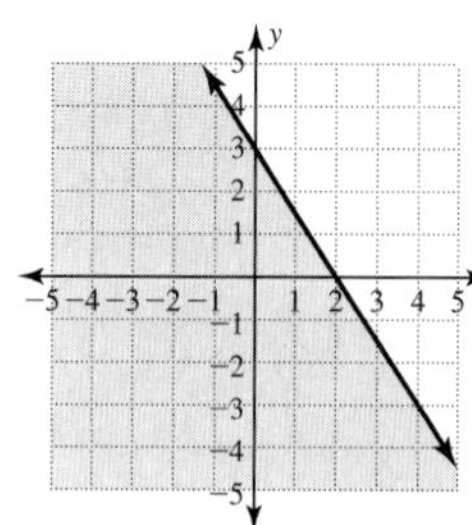

25.
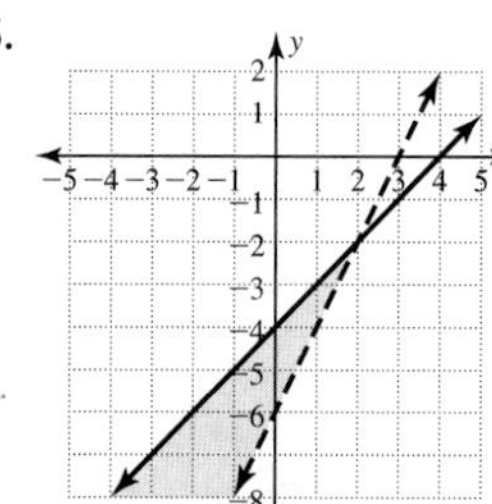

27.
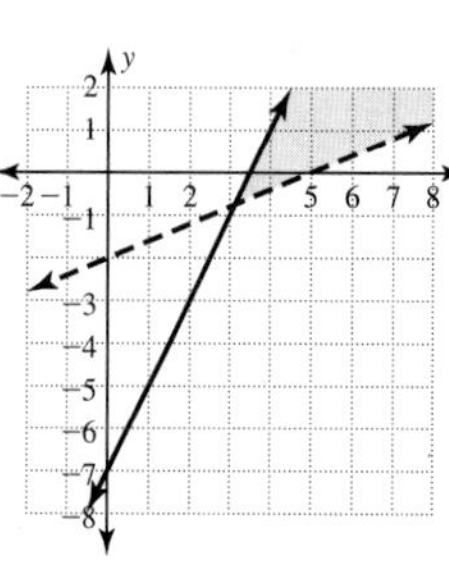

29.
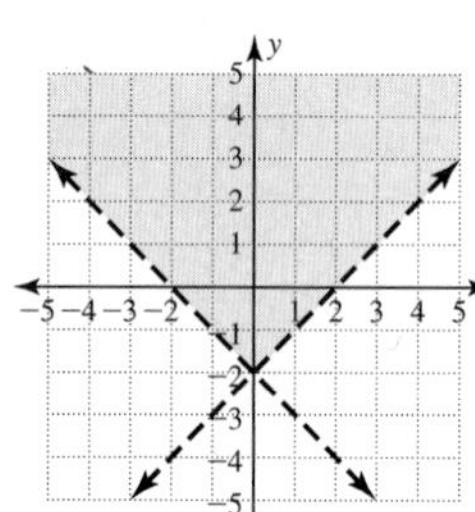

31.
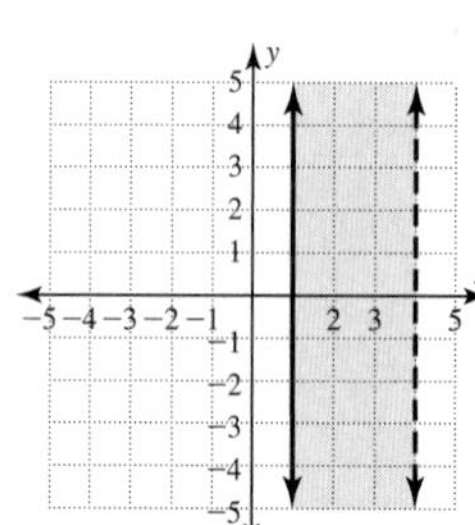

33.
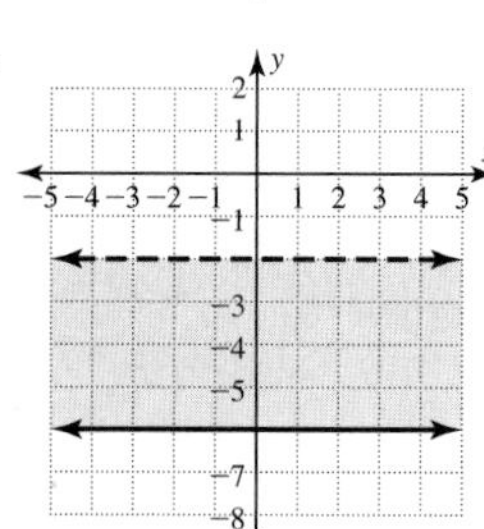

35.
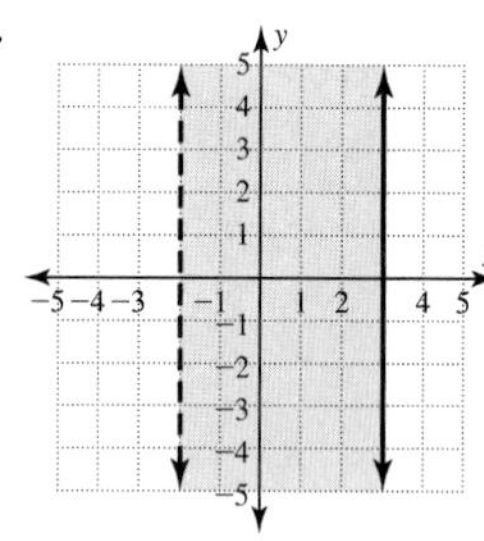

37. No solution

39.
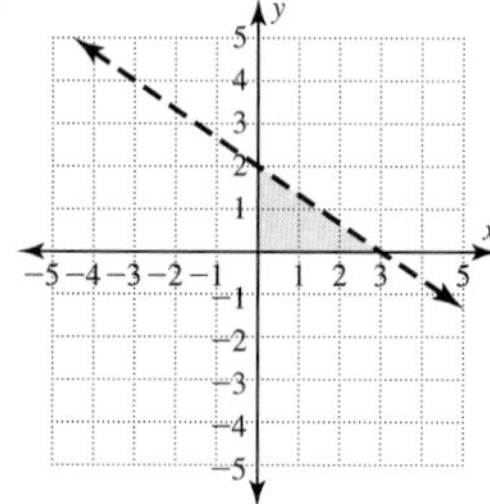

41.
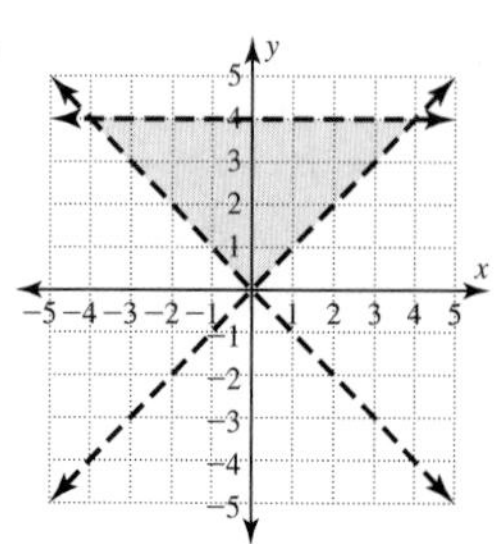

43.
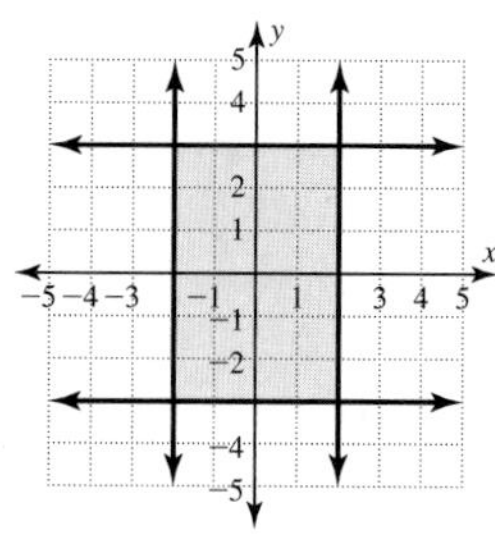

45.
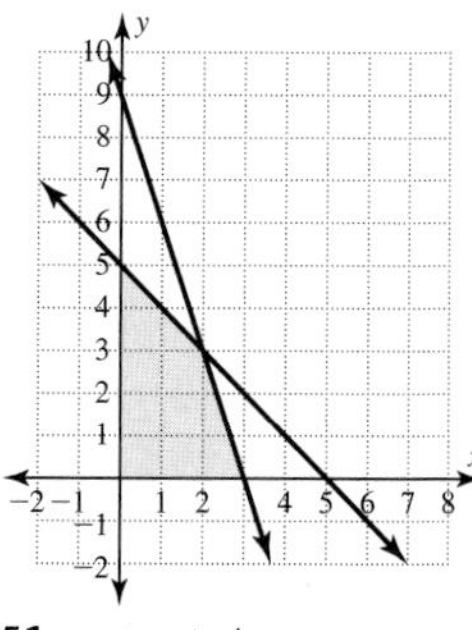

47. $y \le x - 4$; $y \le \frac{x}{2} - 3$ **49.** $x \ge y + 2$ **51.** $y \le x + 4$
53. $x > 0$; $y > 0$; $x + y \le 10$ **55.** $x \ge 0$; $y \ge 0$; $x + 2y \le 5$
57. $x > 0$; $y > 0$; $x + y \le 150$ **59.** $x \ge 0$; $y \ge 0$; $40x + 50y \le 4{,}400$

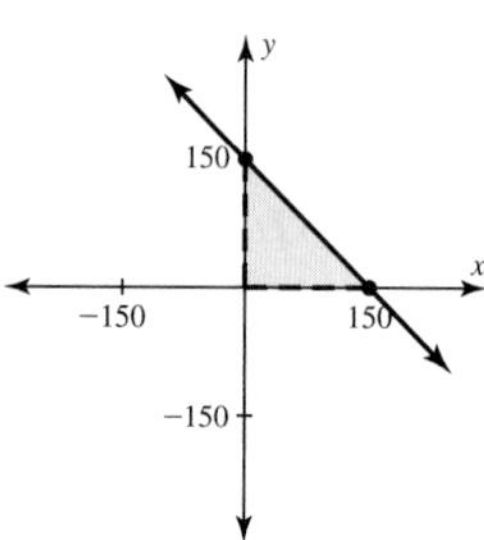

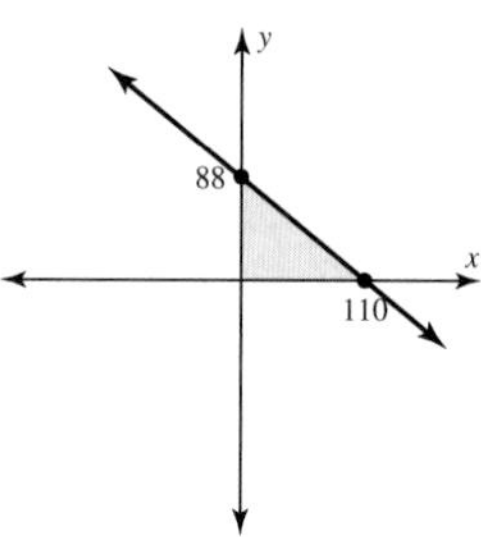

Review Exercises for Chapter 4

1. a. Not a solution **b.** Solution **c.** Solution **d.** Not a solution **2.** C and D
3. a. Not a solution **b.** Solution **c.** Not a solution **d.** Solution
4. a. $x = 3$ **b.** $(3, \infty)$ **c.** $(-\infty, 3)$ **5. a.** $x = -1$ **b.** $(-\infty, -1)$
c. $(-1, \infty)$ **6. a.** $x = 1998$ **b.** $[1990, 1998)$ **c.** $(1998, 2000]$ **d.** Before 1998 the community used less electricity generated by wind than by coal. This changed in 1998, and after 1998 more was generated by wind.
7. a. $x = 2$ **b.** $(0, 2)$ **c.** $(2, 4]$ **d.** If 2 hours are required, both shops charge \$85. For less than 2 hours, shop A charges less. For more than 2 hours, shop A charges more. **8.** (Verbal answers can vary.)

a. $x > 3$; $(3, \infty)$;
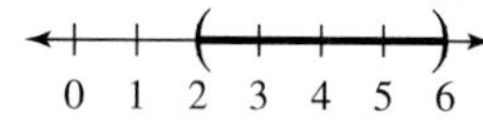

b. x is at most 4; $(-\infty, 4]$;
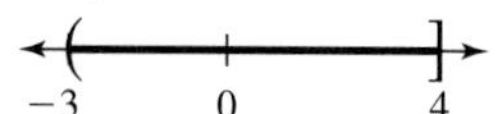

c. x is at least 2; $x \ge 2$;
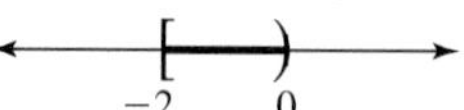

d. x is less than 5; $x < 5$; $(-\infty, 5)$

9. (Verbal answers can vary.) **a.** $2 < x < 6$; $(2, 6)$;

b. x is greater than -3 and less than or equal to 4; $(-3, 4]$;

c. x is greater than or equal to -2 and less than 0; $-2 \le x < 0$;

d. x is greater than or equal to -1 and less than or equal to 3; $-1 \le x \le 3$; $[-1, 3]$ **10.** $[-1, 3)$ **11.** $[-1, 5]$ **12. a.** $x + 7 \le 11$
b. $2(x - 1) > 13$ **c.** $3x - 5 \ge 7$ **d.** $4x + 9 \le 21$ **13.** $(-2, \infty)$; all values greater than -2 **14.** $(-\infty, 4]$; all values less than or equal to 4
15. $[-3, \infty)$; all values greater than or equal to -3 **16.** $(-\infty, 3)$; all values less than 3 **17.** $[1, \infty)$; all values greater than or equal to 1
18. $(0, 4)$; all values between 0 and 4 **19.** $(-\infty, -2)$ **20.** $(1, \infty)$
21. $(-\infty, -5]$ **22.** $[-4, \infty)$ **23.** $(-\infty, -3)$ **24.** $[0, \infty)$ **25.** $\left(-\frac{1}{2}, \infty\right)$

26. $(-\infty, 3)$ **27.** $[2, \infty)$ **28.** $(-\infty, -3]$ **29.** $(-3, \infty)$ **30.** $\left(-\infty, -\frac{4}{7}\right)$
31. $(-\infty, -3)$ **32.** $(-\infty, -13]$ **33.** $\mathbb{R}$ **34.** $\left(-\infty, \frac{25}{6}\right]$ **35.** $(-\infty, -22)$
36. $(-24, \infty)$ **37.** $(2, 6]$ **38.** $[0, 25)$ **39.** $(-140, -98]$ **40.** $[4, 14]$
41. $[-2, 2]$ **42.** $(-1, 5)$ **43.** $x = 7$ or $x = -1$ **44.** $x = 4$ or $x = 1$
45. $\left(-\infty, -\frac{6}{5}\right) \cup (2, \infty)$ **46.** $(2, 8)$ **47.** Conditional; $(0, \infty)$
48. Conditional; $(-\infty, 0]$ **49.** Unconditional; $\mathbb{R}$ **50.** Contradiction; no solution **51.** $A \cap B = (2, 4); A \cup B = (1, 7)$
52. $A \cap B = [3, 5); A \cup B = [2, 6]$ **53.** $A \cap B = [-3, 2]; A \cup B = \mathbb{R}$
54. $A \cap B = [-2, 5); A \cup B = (-\infty, 6)$ **55.** $-3 < x \le 4$
56. $0 \le x < \pi$ **57.** $(-\infty, 1)$ **58.** $(-\infty, -3) \cup \left(-\frac{4}{3}, \infty\right)$

59.
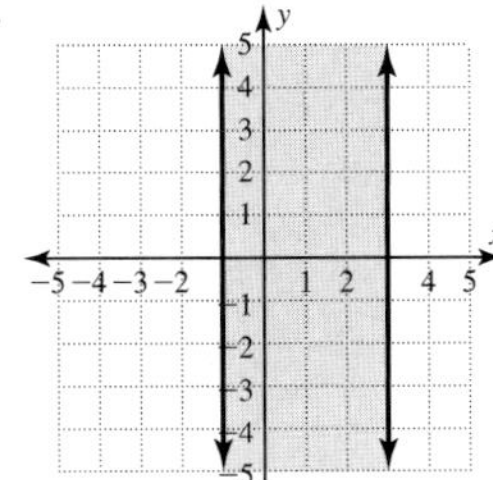

60.
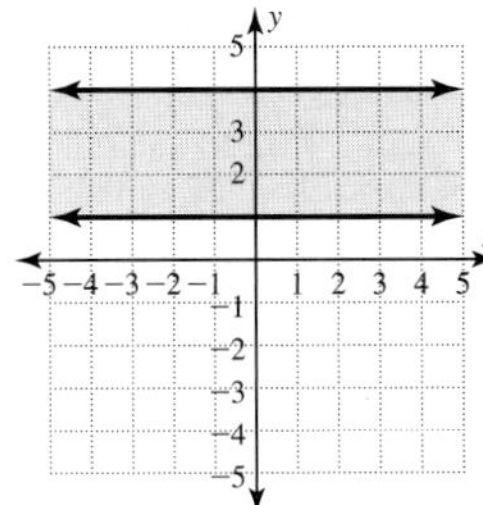

61.
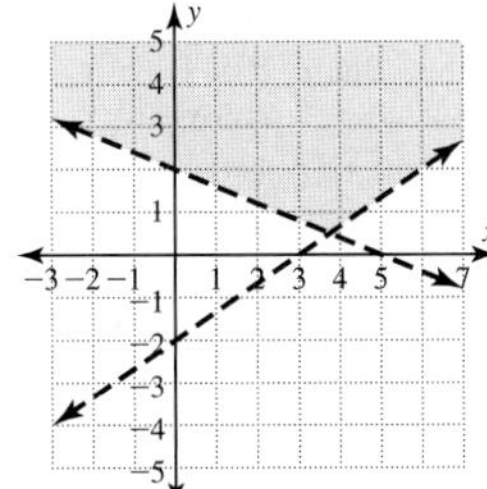

62.
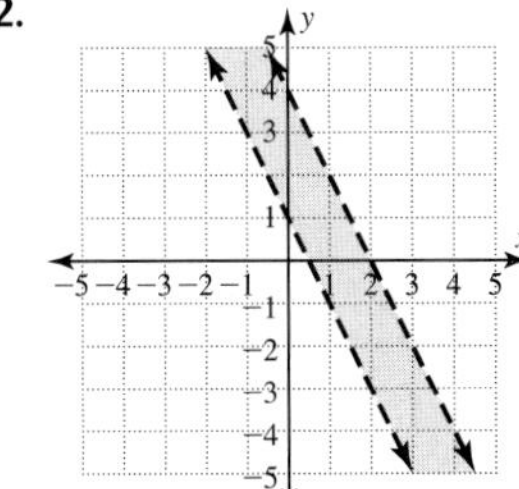

63.
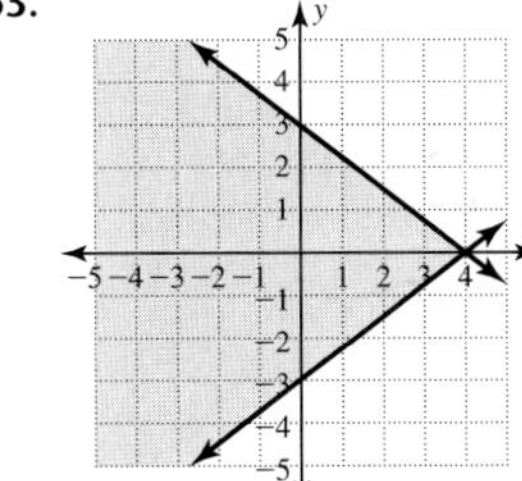

64.
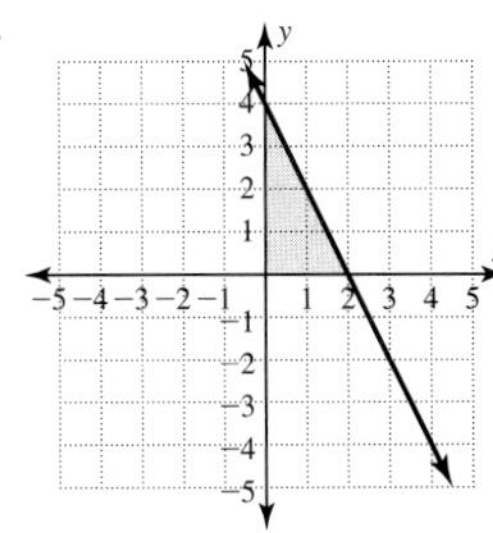

65. $|x| < 5$ **66.** $|x| \le 4$ **67.** $|x - 4| < 6$ **68.** $|x - 8| \le 12$
69. $4(x + 3) > 8; (-1, \infty)$ **70.** $3(x - 17) \le 2(x + 11); (-\infty, 73]$
71. $7 \le 3x - 2 < 19; [3, 7)$ **72.** She must score at least 18 points.
73. She must make at least \$510. **74.** The length must be between 10 and 20 cm. **75.** $160 < x < 200$ **76.** $|x - 7.2| \le 0.25; [6.95, 7.45]$
77. There is a loss if less than 50 posters are ordered; there is a profit if more than 50 posters are ordered. **78.** $(-\infty, 10); (-\infty, 10.05)$
79. $(-\infty, -2]; (-\infty, -2.01]$ **80.** $(1, 3]; (1.1, 2.99]$

Mastery Test for Chapter 4

1. a. Solution **b.** Solution **c.** Not a solution **d.** Not a solution
2. a. $(-\infty, 3)$ **b.** $[3, \infty)$ **c.** $(-\infty, 4]$ **d.** $(4, \infty)$ **3. a.** $(-\infty, 5]$
b. $(-\infty, -2)$ **c.** $(-11, \infty)$ **d.** $[4, \infty)$ **4. a.** $[-15, \infty)$ **b.** $(245, \infty)$
c. $(-\infty, -9)$ **d.** $\left[-\frac{3}{2}, \infty\right)$ **5. a.** Conditional; $(-\infty, 0)$
b. Unconditional; $\mathbb{R}$ **c.** Contradiction; no solution **d.** Conditional; $[0, \infty)$
6. a. $[-8, 5)$ **b.** $[-24, 54)$ **c.** $[-1, 4]$ **d.** $(-\infty, -4] \cup [3, \infty)$
7. a. $x = 8$ or $x = -8$ **b.** $x = 22$ or $x = -27$ **c.** $(-\infty, 1) \cup (9, \infty)$
d. $\left[-1, \frac{11}{5}\right]$ **8. a.** $(0, 3)$ **b.** $(-\infty, 0) \cup (3, \infty)$ **c.** $[-4, 0]$
d. $(-\infty, -4] \cup [0, \infty)$

9. a.
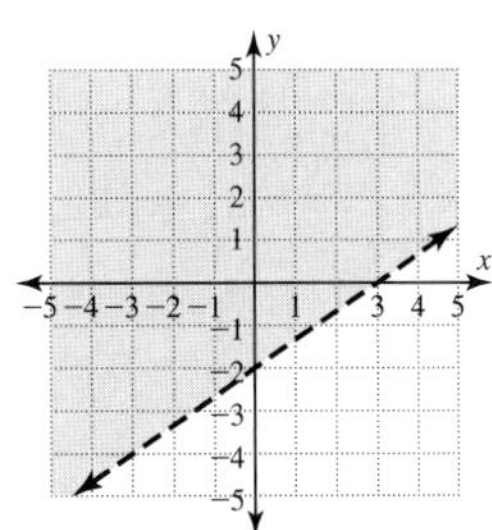

b.
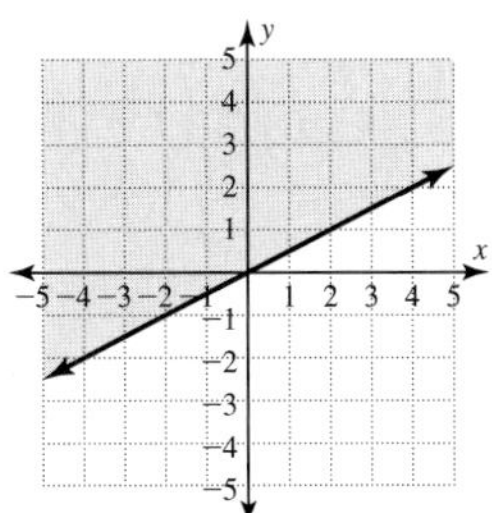

c.
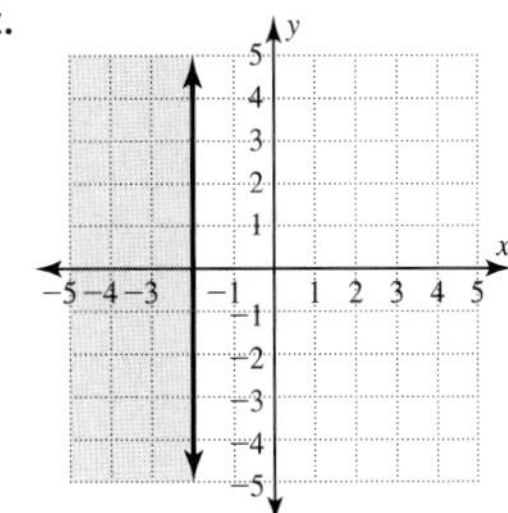

d.
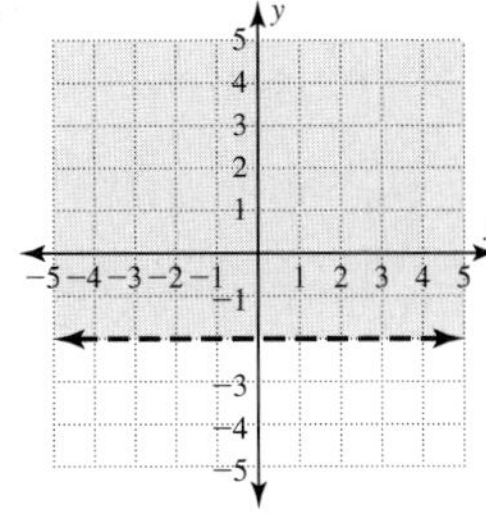

10. a.
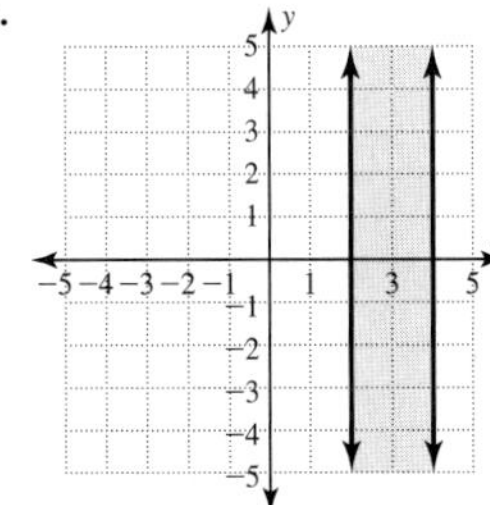

b.
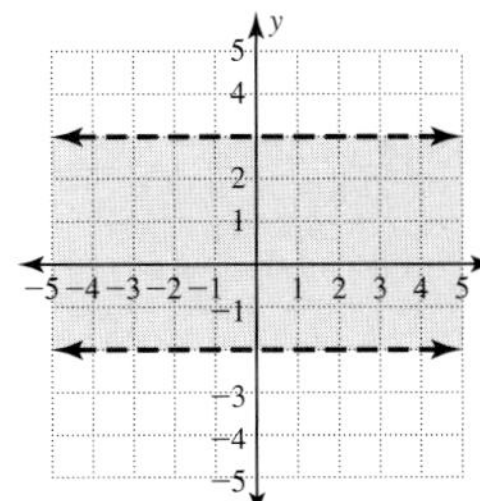

c.
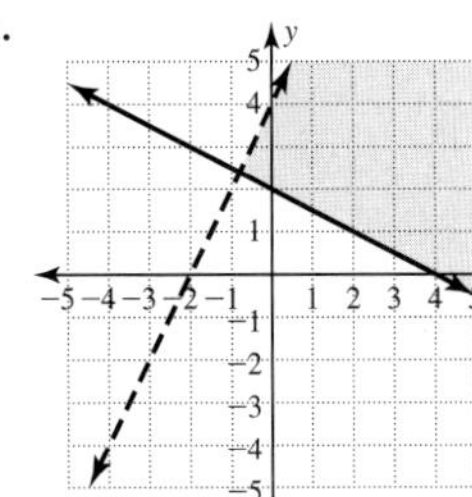

d.
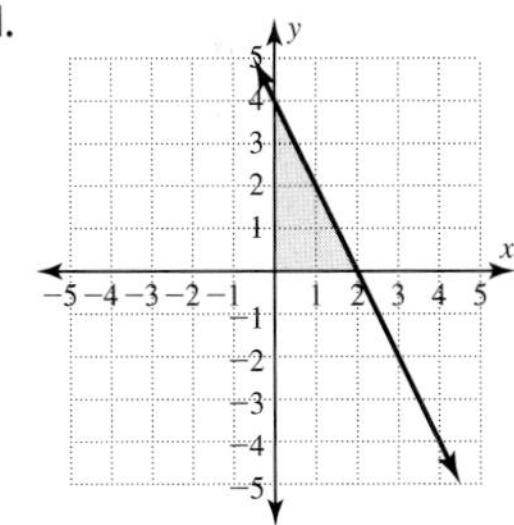

Reality Check for End of Chapter 4

a. $100 < 250x - 150 < 250$ **b.** $1.0 < x < 1.6$; Use a strand diameter between 1.0 and 1.6 inches to produce a cable that can support a load with a safety factor between 2 and 5 (from 100,000 to 250,000 lb) for this bridge.

Chapter 5

Reality Check for Chapter 5 Opener

B

Using the Language and Symbolism of Mathematics 5.1

1. Base; exponent **2.** Factors **3.** Terms; addends **4.** x **5.** $-x$ **6.** xy **7.** y
8. Exponential; expanded **9.** x^{m+n} **10.** $x^{m \cdot n}$ **11.** $x^m y^m$ **12.** $\frac{x^m}{y^m}$

Exercises 5.1

1. a. x^5 **b.** xy^4 **c.** $(xy)^3$ **3. a.** $(a + b)^3$ **b.** $a^3 + b^3$ **c.** $a + b^3$
5. a. $m \cdot m \cdot m \cdot m \cdot m \cdot m \cdot m$ **b.** $(-m)(-m)(-m)(-m)(-m)(-m)$
c. $-m \cdot m \cdot m \cdot m \cdot m \cdot m \cdot m$ **7. a.** $(m + n)(m + n)$
b. $m \cdot m + n \cdot n$ **c.** $m + n \cdot n$ **9. a.** $\frac{a \cdot a}{b \cdot b \cdot b}$ **b.** $\left(\frac{a}{b}\right)\left(\frac{a}{b}\right)\left(\frac{a}{b}\right)\left(\frac{a}{b}\right)$

c. $\frac{a}{b \cdot b \cdot b \cdot b}$ **11. a.** 36 **b.** 36 **c.** -36 **d.** -1 **13. a.** -1 **b.** 1 **c.** -1 **d.** 1,000 **15. a.** 169 **b.** 289 **17.** $x^2 \cdot x^3 = x^5$ **19.** $(x^3)^2 = x^6$ **21. a.** x^{20} **b.** x^{99} **23. a.** $15m^{10}$ **b.** $24m^8$ **25. a.** $-6x^6$ **b.** $-6x^6$ **27. a.** $9x^2y^2$ **b.** $8x^3y^3$ **29. a.** n^{12} **b.** $-n^{12}$ **31. a.** $5a^2b^2c^2$ **b.** $25a^2b^2c^2$ **33. a.** x^3y^6 **b.** x^2y^6 **35. a.** $\frac{9}{16}$ **b.** $\frac{16}{81}$ **37. a.** $\frac{64}{w^3}$ **b.** $\frac{81}{w^4}$ **39. a.** $\frac{x^{22}}{y^{33}}$ **b.** $\frac{x^6}{y^{33}}$ **41. a.** $\frac{16x^8}{81y^{12}}$ **b.** $-\frac{x^6}{8y^{12}}$ **43. a.** -4 **b.** 4 **45. a.** 36 **b.** 18 **47. a.** 35 **b.** 125 **49. a.** 180 **b.** 900 **51. a.** -19 **b.** -1 **53. a.** 54 **b.** 40 **55. a.** $3v$ **b.** v^3 **57. a.** $10m^3$ **b.** $24m^6$ **59.** -49; -48.7204 **61.** 4; 4.0804 **63.** 28; 29.0301 **65.** \$2.16 **67. a.** $f(20) = 80$; a wind speed of 20 mi/h generates 80 kW of electricity. **b.** $f(27) = 196.83$; a wind speed of 27 mi/h generates 196.83 kW of electricity. **69.** Doubling the wind speed increases the power by a factor of 8. **71.** $-125x^6y^{12}$ **73.** $6x^3y^6$ **75.** $200x^{13}$ **77.** $\frac{27x^3}{8y^6}$ **79. a.** x^2 **b.** y^2 **c.** $x^2 - y^2$ **81.** The volume of the larger cube is 8 times the volume of the smaller cube.

Using the Language and Symbolism of Mathematics 5.2

1. Terms; addends **2.** Factors **3.** Quotient **4.** x^{m-n}; 0 **5.** $\frac{1}{x^{n-m}}$ **6.** Subtract **7.** 1; 0 **8.** Undefined

Exercises 5.2

1. a. -5 **b.** 0 **c.** 1 **d.** 0 **e.** Undefined **3. a.** 1,000 **b.** 1 **c.** $\frac{1}{1{,}000}$ **5. a.** 1 **b.** -1 **c.** 1 **7. a.** 2 **b.** 1 **c.** 0 **9. a.** 1 **b.** 0 **c.** -1 **11. a.** 1 **b.** Undefined **13. a.** -3 **b.** Undefined **15.** $\frac{x^6}{x^4} = x^2$ for all $x \neq 0$ **17.** $\frac{x^4}{x^5} = \frac{1}{x}$ for all $x \neq 0$ **19.** x^{10} **21.** m^{24} **23.** 1 **25.** $-x^4$ **27.** $\frac{1}{t^4}$ **29.** $\frac{1}{v^5}$ **31.** $-\frac{1}{a^3}$ **33.** $\frac{a^3}{b^4}$ **35.** $\frac{2m^3}{3}$ **37.** $\frac{6}{11v^3}$ **39.** $\frac{3b}{4}$ **41.** $\frac{4n^5}{3m^2}$ **43.** $-\frac{3w^3}{5v^2}$ **45.** x^6 **47.** $\frac{1}{m^{12}}$ **49.** $36x^{18}$ **51. a.** $36x$ **b.** 19 **53. a.** $3x^2 - 3x$ **b.** x **55.** 8; 8.04 **57.** 200; 198.59 **59. a.** 1 **b.** 4 **c.** 5 **61. a.** 1 **b.** 0 **c.** 1 **63. a.** 4 **b.** $\frac{1}{4}$ **c.** 1 **65.** 11 **67.** $36x^{10}$ **69.** $32x^{15}$ **71.** $144m^{18}$ **73.** $200x^{21}$ **75.** $6x^6$ **77.** $3a^4b^5$ **79.** $\frac{x^4}{y}$ **81.** \$7,518.30 **83.** \$5,105.13

Using the Language and Symbolism of Mathematics 5.3

1. 0; $\frac{1}{x^n}$ **2.** x^{m+n} **3.** x^{mn} **4.** x^{m-n} **5.** 10 **6.** 10

Exercises 5.3

1. $x^{-2} = \frac{1}{x^2}$ **3.** $\frac{x^4}{x^7} = x^{-3}$ **5. a.** $\frac{1}{32}$ **b.** -32 **c.** -25 **7. a.** $\frac{5}{6}$ **b.** $\frac{25}{36}$ **c.** 1 **9. a.** $\frac{1}{100}$ **b.** $\frac{1}{1{,}000}$ **c.** $-\frac{1}{100}$ **11. a.** $\frac{1}{7}$ **b.** $\frac{7}{10}$ **c.** $-\frac{9}{5}$ **13. a.** $\frac{10}{7}$ **b.** 7 **c.** 2 **15. a.** $\frac{y}{x}$ **b.** $\frac{y^2}{x^2}$ **c.** $\frac{1}{x^2y}$ **17. a.** $\frac{1}{9x^2}$ **b.** $\frac{3}{x^2}$ **c.** $9x^2$ **19. a.** $\frac{1}{m+n}$ **b.** $\frac{1}{m} + \frac{1}{n}$ **c.** $m + \frac{1}{n}$ **21. a.** 45,800 **b.** 0.000458 **c.** 4,580,000 **d.** 0.00000458 **23. a.** 8,100 **b.** 0.0081 **c.** $-8{,}100$ **d.** -0.00000081 **25. a.** 9.7×10^3 **b.** 9.7×10^7 **c.** 9.7×10^{-1} **d.** 9.7×10^{-4} **27. a.** 3.5×10^{10} **b.** 3.5×10^{-9} **c.** -3.5×10^3 **d.** -3.5×10^{-2} **29.** 4,493,000,000 **31.** 0.000000000001 **33.** 1.4×10^7 **35.** 5.68×10^{-5} **37. a.** 2,300 **b.** 220,000,000 **c.** The claim is accurate. **39.** ≈21 minutes **41.** 0.8; 0.7953 **43.** 0.05; 0.05088 **45.** 0.000027; 0.00002628 **47.** 0.0001764 **49. a.** 0.367 **b.** 51,300,000 **51. a.** v^9 **b.** v^{15} **c.** $\frac{1}{v^{15}}$ **53. a.** $\frac{1}{x^4}$ **b.** 1 **c.** $\frac{1}{x^{10}}$ **55. a.** $\frac{3}{x}$ **b.** $12x$ **c.** $\frac{12}{x^7}$ **57. a.** $\frac{9x^2}{25y^2}$ **b.** $\frac{25y^2}{9x^2}$ **c.** $225x^2y^2$ **59.** $-\frac{1}{20x}$ **61.** $\frac{72y^8}{x}$ **63.** $-\frac{3x^{10}}{2}$ **65.** $-\frac{4x^{15}}{y^{32}}$ **67.** $4x^{10}$ **69.** $\frac{m^{12}}{n^{12}}$ **71.** $\frac{3y^6}{2x^7}$ **73.** 13; 25; $\frac{5}{6}$; $\frac{1}{5}$ **75.** 25; 49; $-\frac{7}{12}$; $-\frac{1}{7}$

Using the Language and Symbolism of Mathematics 5.4

1. Factors **2.** Plus; minus **3.** 0, 1, 2, 3, . . . **4.** Monomial **5.** Polynomial **6.** Coefficient **7.** Monomial **8.** Binomial **9.** Trinomial **10.** Sum **11.** Highest **12.** Zero **13.** 0 **14.** Alphabetical; decreasing **15.** 1 **16.** -1 **17.** 1 **18.** Identical; identical

Exercises 5.4

1. a. Monomial; coefficient: -12; degree: 0 **b.** Monomial; coefficient: -12; degree: 1 **c.** Monomial; coefficient: -1; degree: 12 **d.** Not a monomial **3. a.** Monomial; coefficient: $\frac{2}{5}$; degree: 4 **b.** Monomial; coefficient: $\frac{5}{2}$; degree: 4 **c.** Monomial; coefficient: $-\frac{5}{2}$; degree: 4 **d.** Not a monomial **5. a.** Monomial; coefficient: 5; degree: 4 **b.** Monomial; coefficient: -5; degree: 4 **c.** Not a monomial **d.** Not a monomial **7. a.** Second-degree binomial **b.** Third-degree trinomial **c.** Fourth-degree monomial **d.** Monomial of degree 0 **9. a.** Second-degree monomial **b.** First-degree binomial **c.** Not a polynomial **d.** Third-degree trinomial **11. a.** $8x^2yz^3$ **b.** $x^2 - 5x + 9$ **c.** $-4x^3 - 2x^2 + 9x + 7$ **13.** $Y_1 = Y_2$ **15.** $Y_1 \neq Y_2$ **17. a.** Like **b.** Unlike **c.** Unlike **19. a.** $-7xy$ **b.** $-7x + y$ **c.** $2x^2 - 3xy + 7y^2$ **21. a.** $21x$ **b.** $-7x$ **23. a.** $7x^2 - 12x$ **b.** $-x^2 - 2x$ **25. a.** $9x^2 - x + 6$ **b.** $x^2 - 13x + 12$ **27.** $14x - 2$ **29.** $-9v + 14$ **31.** $9a^2 - a + 6$ **33.** $6x^2 - x + 5$ **35.** $3w^2 - 7w + 10$ **37.** $-m^4 + 3m^3 - 2m + 4$ **39.** $-4n^4 + 9n^3 + n^2 + n - 12$ **41.** $-4y^6 + y^5 - 6y^4 + 11y^2 + 4y - 18$ **43.** $-5x^2 - 20x - 3$ **45.** $-13a + 13$ **47.** $12x - 12$ **49.** $6x + 1$ **51.** $11x^2 + 4xy - 8y^2$ **53.** $-20x^2 + 7xy + 11y^2$ **55.** $9x^3 + 6x^2 - 12x + 30$ **57.** $-8x^4 - 3x^3y + 3x^2y^2 + xy^3 + 2y^4$ **59. a.** -2 **b.** 24 **c.** 528 **d.** 342 **61.** $P(0) = -40$; \$40 is lost if no units are sold. **63.** $P(2) = 0$; a break-even point occurs when 2 units are sold. **65.** $P(20) = 0$; a break-even point occurs when 20 units are sold.

67.

x	Y_1
0	-1.1
0.2	-2.672
0.4	-3.668
0.6	-4.088
0.8	-3.932
1	-3.2
1.2	-1.892

69. $9.02x^2 + 4.02x$ **71.** $11.18x^2 - 1.97x - 5.98$ **73.** $2x^2 + 4x + 8$ **75.** $3x^2 + 10x + 21$ **77.** $2x^3 - 5x + 11$ **79.** $3w^6 + 5w^4$ **81.** $x^2 - y^2$ **83.** $2w + 7$ **85.** 350,000 tons

Using the Language and Symbolism of Mathematics 5.5

1. 2; 5 **2.** x^{m+n} **3.** Multiplied; added **4.** Distributive **5.** Each **6.** Binomial **7.** Trinomial **8.** First, outer, inner, last **9. a.** 2 **b.** 2 **c.** 3 **10. a.** 1 **b.** 1 **c.** 2 **11.** x

Exercises 5.5

1. $63v^5$ **3.** $4a^9b^5$ **5.** $24x^3 - 20x^2$ **7.** $-32v^3 - 12v^2$ **9.** $14x^3 - 18x^2 - 8x$ **11.** $-44x^4 + 55x^3 + 88x^2$ **13.** $10x^3y - 5x^2y^2 - 5xy^3$ **15.** $6a^4b + 12a^3b^2 - 3a^2b^3$ **17.** $y^2 + 5y + 6$ **19.** $6v^2 + 7v - 20$ **21.** $54x^2 + 63x + 15$ **23.** $2x^2 + xy - y^2$ **25.** $10x^2 + 6xy - 28y^2$ **27.** $x^3 - x^2 - 10x - 8$ **29.** $2m^3 + m^2 - 11m - 10$ **31.** $x^3 + y^3$ **33.** $6x^4 - 7x^3 - 5x^2 + 89x - 99$ **35.** $5x^3 + 5x^2 - 30x$ **37.** $2x^3 + 13x^2 + 17x - 12$ **39.** $m^2 + 7m + 12$ **41.** $n^2 + 3n - 40$ **43.** $20x^2 + 13x - 21$ **45.** $9y^2 - 64y + 7$ **47.** $12a^2 + 13ab - 14b^2$ **49.** $90x^2 - 43xy - 21y^2$ **51.** $(8x^2 + 10x - 3)$ cm^2 **53.** $(x^2 + x + 1)$ cm^2 **55.** $(4x^2 - 4x + 1)$ cm^2 **57.** $\left(\frac{3}{2}x^2 + \frac{5}{2}x + 1\right)$ cm^2 **59.** $(4x^2 + 4x - 3)$ cm^2 **61.** $(x^3 + 6x^2 + 5x)$ cm^3 **63.** $Y_1 \neq Y_2$ **65.** $Y_1 = Y_2$ **67. a.** $40v^2 - 35v$ **b.** $5v(8v - 7)$

69. a. $2m^3n - 2m^2n^2 + 10mn^3$ **b.** $2mn(m^2 - mn + 5n^2)$ **71. a.** $x^2 - x - 6$ **b.** $(x + 2)(x - 3)$ **73. a.** $15x^2 - 26x + 8$ **b.** $(3x - 4)(5x - 2)$ **75.** $-6x^4 + 7x^2 + 20$ **77.** $-x^4 + 2x^3 - 2x^2 + x + 20$ **79.** $v^2 + 10v + 25$ **81.** $x^3 + 6x^2 + 12x + 8$ **83.** $V = 4x^3 - 72x^2 + 288x$ **85.** $A = \pi x^2 + 30\pi x$ **87. a.** $R(x) = 280x - 2x^2$ **b.** $R(50) = 9{,}000$

Using the Language and Symbolism of Mathematics 5.6

1. a. C **b.** D **c.** A **d.** B **2.** 2 **3.** 3 **4.** Factoring **5.** Distributive **6.** $x^2 - 4$ **7.** $(x + 2)(x - 2)$

Exercises 5.6

1. $49a^2 - 1$ **3.** $z^2 - 100$ **5.** $4w^2 - 9$ **7.** $81a^2 - 16b^2$ **9.** $16x^2 - 121y^2$ **11.** $x^4 - 4$ **13.** $16m^2 + 8m + 1$ **15.** $n^2 - 18n + 81$ **17.** $9t^2 + 12t + 4$ **19.** $25v^2 - 80v + 64$ **21.** $16x^2 + 40xy + 25y^2$ **23.** $36a^2 - 132ab + 121b^2$ **25.** $a^2 + 2abc + b^2c^2$ **27.** $x^4 - 6x^2 + 9$ **29.** $x^4 + 2x^2y + y^2$ **31. a.** $-25x - 36$ **b.** $25x^2 - 36$ **33. a.** $16x^2 - 72x + 81$ **b.** $16x^2 - 81$ **35.** $14v + 98$ **37.** $4xy$ **39.** $2vw - 2w^2$ **41.** 0 **43.** $9x^2 - 36xy + 4y^2$ **45.** $(3x + 7y)^2 = 9x^2 + 42xy + 49y^2$ **47.** $(x + 8)^2 - (x - 6)^2 = 28x + 28$ **49.** $(a^2 + 1)(a + 1)(a - 1) = a^4 - 1$ **51.** $Y_1 = Y_2$ **53.** $Y_1 = Y_2$ **55.** $81x^2 - 1; (9x - 1)(9x + 1)$ **57.** $100w^2 - 60wx + 9x^2; (10w - 3x)^2$

Using the Language and Symbolism of Mathematics 5.7

1. Quotient **2.** $x - 3$ **3.** Product **4.** $(x - 3); (x + 2); x^2 - x - 6$ **5.** Standard **6.** Zero; divisor **7.** 0 **8.** Average

Exercises 5.7

1. $3x^2$ **3.** $-5a^2b$ **5.** $3a - 4$ **7.** $3a - 5 + \frac{2}{a}$ **9.** $8m^3 - 4m^2 + 5m + 3$ **11.** $5x^3 - 6x^2y^3 + 11y^2$ **13.** $6v^3 + 3v^2 - 9v - 2$ **15.** $5v^2 - v - 4w$ **17.** $10x^{11} - 5x^3 + 3$ **19.** $x + 7$ **21.** $v + 6$ **23.** $4m^2 + 3m + 19 + \frac{64}{m - 3}$ **25.** $3w - 4$ **27.** $4m - 7$ **29.** $9y - 7$ **31.** $7x^3 - 9x^2 + 6x + 2$ **33.** $x - 7$ **35.** $3a + 2$ **37.** $3v^2 + 2v + 4$ **39.** $x^2 + 2x + 4$ **41.** $4y^2 + 8y + 9$ **43.** $7x^2 - 9$ **45.** $(2x - 1)$ cm **47.** $(6x + 2)$ cm **49.** $(2x + 3)$ cm **51.** $f_1(x) = f_2(x)$ for $x \neq 2.5$ **53.** $5k + 4$ **55.** $x^3 - 14x - 15$ **57.** $3x + 5$ **59. a.** 4 **b.** 6 **c.** 2 **61.** $2x^2 + 7x + 10$ **63.** $4x + 5$ **65. a.** $3x^2 - 19x + 20$ **b.** $x - 5$ **67. a.** $A(x) = \frac{125x + 960}{x}$ **b.** $A(50) = 144.2$; when 50 units are produced, the average cost per unit is $144.20. **69. a.** $A(x) = \frac{8x + 1{,}200}{x}$ **b.** $C(200) = 2{,}800$; the total cost of producing 200 chairs is $2,800. **c.** $A(200) = 14$; the average cost of producing 200 chairs is $14.00 per chair. **d.** $C(250) = 3{,}200$; the total cost of producing 250 chairs is $3,200. **e.** $A(250) = 12.8$; the average cost of producing 250 chairs is $12.80 per chair. **f.** The cost increases by $400. **g.** The average cost decreases by $1.20 per chair.

Review Exercises for Chapter 5

1. a. xy^3 **b.** $(xy)^3$ **c.** $x^2 + y^2$ **d.** $(x + y)^2$ **2. a.** $x \cdot x + y \cdot y$ **b.** $(x + y)(x + y)$ **c.** $-x \cdot x \cdot x \cdot x$ **d.** $(-x)(-x)(-x)(-x)$ **3. a.** 9 **b.** $\frac{1}{9}$ **c.** -9 **d.** $\frac{1}{9}$ **4. a.** 1 **b.** 0 **c.** 2 **d.** 1 **5. a.** 5 **b.** $\frac{6}{5}$ **c.** $\frac{3}{2}$ **d.** $\frac{1}{6}$ **6. a.** -1 **b.** 1 **c.** $\frac{1}{6}$ **d.** $-\frac{1}{6}$ **7. a.** $2x^3$ **b.** x^6 **c.** 1 **d.** 0 **8. a.** 100 **b.** 100 **c.** $\frac{1}{100}$ **d.** $-\frac{1}{100}$ **9. a.** 3 **b.** 2 **c.** 1 **d.** 0 **10. a.** 9 **b.** 1 **c.** 3 **d.** 3 **11. a.** x^7 **b.** x^3 **c.** x^{10} **d.** x^8 **12. a.** $48x^8y^6$ **b.** $3x^2y^2$ **c.** $16x^6y^4$ **d.** $27x^6y^6$ **13. a.** -9 **b.** 9 **14. a.** 48 **b.** 144 **15. a.** 25 **b.** 49 **16. a.** $\frac{7}{12}$ **b.** $\frac{1}{7}$ **17. a.** 2 **b.** 1 **18. a.** $\frac{3}{16}$ **b.** $\frac{1}{144}$ **19.** $30x^3y^7z^{11}$ **20.** $25x^4y^6z^8$ **21.** $\frac{3ac^2}{5b^3}$ **22.** $\frac{n^4}{9m^6}$ **23.** $60x^{16}$ **24.** $9x^3$ **25.** $-200x^{21}$ **26.** $56y^2$ **27.** $2w^2$ **28.** $\frac{m^{24}}{625}$ **29.** $\frac{6}{x^7}$ **30.** 1 **31.** a^3b^4 **32.** $\frac{y^7z^{10}}{x^8}$ **33.** 0.0000000000001 W **34.** 299,000,000 m/s **35.** 5,230,000 **36.** 0.00000523 **37.** 4 h 11 min **38.** b **39.** Monomial of degree 0 **40.** Sixth-degree binomial **41.** Third-degree polynomial with four terms **42.** Sixth-degree trinomial **43.** $7x^5 + 11x^4 + 9x^3 - 3x^2 - 8x + 4$ **44.** $-7x^5$ **45.** $x^2 - 3$ **46.** $11x^2 - 3x + 2$ **47.** $5x^3 - 5x^2 - 8x + 4$ **48.** $5x^4 - 8x^3 + 13x^2 + 9x + 5$ **49.** $3x^5 + 3x^4 + 16x^3 + x^2 + 7x - 11$ **50.** $2x^2 - 11x - 12$ **51.** $35x^5 - 45x^4 + 15x^3 + 5x^2$ **52.** $35v^2 + 2v - 1$ **53.** $25y^2 - 70y + 49$ **54.** $81y^2 + 90y + 25$ **55.** $9a^2 - 25b^2$ **56.** $6m^3 - 8m^2 - 9m + 35$ **57.** $-\frac{4b^3}{a}$ **58.** $5m^3n - 7m^2n^2 - mn^3$ **59.** $v - 2$ **60.** $7w + 3$ **61.** $a + b$ **62.** $24xy + 32y^2$ **63.** $24xy$ **64.** $x - 5$ **65.** $(x + 4)(x^2 - 4x + 16) = x^3 + 64$ **66.** $\left(x + 6 + \frac{7}{x - 2}\right)(x - 2) = (x + 6)(x - 2) + 7 = x^2 + 4x - 12 + 7 = x^2 + 4x - 5$ **67. a.** $3x^2 - 12x$ **b.** $3x(x - 4)$ **68. a.** $5x^2 + 8x - 21$ **b.** $(5x - 7)(x + 3)$ **69. a.** $16x^2 - 9$ **b.** $(4x + 3)(4x - 3)$ **70. a.** $36x^2 - 84x + 49$ **b.** $(6x - 7)^2$ **71.** $45x^2 + 17x - 6$ **72.** $20x^2 + 9xy - 18y^2$ **73.** $25x^2 - 36y^2$ **74.** $25x^2 - 60xy + 36y^2$ **75.** $49x^2 + 28xy + 4y^2$ **76.** $x^4 - 81$ **77. a.** $-25x - 7$ **b.** $12x^2 - 25x - 7$ **78. a.** $64x^2 - 25y^2$ **b.** $64x^2 - 80xy + 25y^2$ **79.** $98a^2 + 8b^2$ **80.** $9a^2 + 6ab + 90b^2$ **81.** $4x - 1$ **82.** $27x^3 + 18x^2 + 12x + 8$ **83.** $(6x^2 - 7x - 3)$ ft^2 **84.** $(x + 1)$ cm **85. a.** $R(x) = x(350 - 4x) = -4x^2 + 350x$ **b.** $R(0) = 0$; when the price is $0 per unit, the revenue is $0 **c.** $R(40) = 7{,}600$; when the price is $40 per unit, the revenue is $7,600 **d.** $R(80) = 2{,}400$; when the price is $80 per unit, the revenue is $2,400 **86.** $17,834.78 **87. a.** $A(x) = \frac{15x + 1{,}000}{x}$ **b.** $C(100) = 2{,}500$; the total cost of producing 100 range finders is $2,500. **c.** $A(100) = 25$; the average cost of producing 100 range finders is $25 per range finder. **d.** $C(200) = 4{,}000$; the total cost of producing 200 range finders is $4,000. **e.** $A(200) = 20$; the average cost of producing 200 range finders is $20 per range finder. **f.** The cost increases by $1,500. **g.** The average cost decreases by $5 per range finder. **88.** $x^3 - 1 = (x - 1)(x^2 + x + 1)$; $x^3 + 1 = (x + 1)(x^2 - x + 1)$

Mastery Test for Chapter 5

1. a. x^{13} **b.** $40y^7$ **c.** $-v^{15}$ **d.** m^4n^3 **2. a.** w^{40} **b.** $32y^{20}$ **c.** v^8w^{12} **d.** $-\frac{27m^3}{8n^3}$ **3. a.** z^6 **b.** $2x^6$ **c.** -1 **d.** $-2a^2b^4$ **4. a.** 1 **b.** 1 **c.** 1 **d.** 8 **5. a.** x^5y^{11} **b.** x^3y^6 **c.** $\frac{4}{3x^5}$ **d.** $36x^{12}$ **6. a.** $\frac{1}{3}$ **b.** $\frac{1}{49}$ **c.** $\frac{9}{4}$ **d.** 9 **7. a.** 357,000 **b.** 0.0000735 **c.** 5.09×10^{-4} **d.** 9.305×10^7 **8. a.** Third-degree binomial **b.** Monomial of degree 0 **c.** Second-degree trinomial **d.** Fifth-degree polynomial with four terms **9. a.** $7x - y$ **b.** $-3x^2 + 6x + 18$ **c.** $9x^4 - 9x^3 + 19x^2 + 9x + 8$ **d.** $3x^3 - 12x^2 - 16x + 6$ **10. a.** $55x^6y^{10}$ **b.** $-15x^5 + 6x^4 - 21x^3 + 27x^2$ **c.** $x^3 + 8x^2 + 16x + 5$ **d.** $x^3 - 2x^2y - 2xy^2 - 3y^3$ **11. a.** $x^2 + 19x + 88$ **b.** $x^2 - 14x + 45$ **c.** $6x^2 + 7xy - 20y^2$ **d.** $10x^2 + 29xy - 21y^2$ **12. a.** $36x^2 - 1$ **b.** $x^2 - 81y^2$ **c.** $25x^2 - 121$ **d.** $9x^2 - 64y^2$ **13. a.** $36x^2 + 12x + 1$ **b.** $x^2 - 18xy + 81y^2$ **c.** $25x^2 - 110x + 121$ **d.** $9x^2 + 48xy + 64y^2$ **14. a.** $3a^2$ **b.** $3m^2n - 4n^3$ **c.** $x + 11$ **d.** $2x^2 + 6x + 8$

Reality Check for End of Chapter 5

The graph of $y = -0.0657x^2 + 0.394x$ models the shape of the cornea on eye B, and the graph of $y = -0.0867x^2 + 0.52x - 0.1$ models the shape of the cornea on eye A.

Chapter 6

Reality Check for Chapter 6 Opener

A trapezoidal cross section has greater stability.

Using the Language and Symbolism of Mathematics 6.1

1. Greatest common **2.** Coefficient; exponent **3.** Distributive **4.** Grouping **5.** 0 **6.** Zero **7.** $(-4, 0)$ **8.** -2 **9.** $x - 8$ **10.** Real **11.** n

Quick Review 6.1

1. $18x^5y^9$ **2.** $6x^3$ **3.** $-12x^4y^3 + 4x^3y^4 - 20x^2y^5$ **4.** $15x^3y^2 - 6x^2y^3 + 12xy^4$ **5.** $\frac{5}{2}x + \frac{15}{2}xy - 5y^2$

Exercises 6.1

1. $2x - 3$ **3.** $5a + 3b$ **5.** $2x^2 - 3xy + 4y^2$ **7.** $2x + y$ **9.** $7x - 9$ **11.** $7x(2x + 11)$ **13.** $4xy(2x - 5y)$ **15.** $2x(3x^2 - 4x + 5)$ **17.** $(5x - 2)(x + 3)$ **19.** $(x + 2y)(6x - 7y)$ **21.** $(5x - 2)(a + 3)$ **23.** $(a - b)(2x - 7)$ **25.** $(3x - 5)(x + 4)$ **27.** $(3x + 1)(4x - 5)$ **29.** $(x + 2)(x + 4)$; $x^2 + 6x + 8$ **31.** $(2x - 7)(5x - 2)$; $10x^2 - 39x + 14$ **33. a.** $(2, 0), (5, 0)$ **b.** 2, 5 **c.** $(x - 2)(x - 5)$ **35. a.** $(-15, 0), (10, 0)$ **b.** -15, 10 **c.** $(x + 15)(x - 10)$ **37. a.** 5, 25 **b.** $(5, 0), (25, 0)$ **c.** $(x - 5)(x - 25)$ **39. a.** $-4, 2, 4$ **b.** $(-4, 0), (2, 0), (4, 0)$ **c.** $(x + 4)(x - 2)(x - 4)$ **41. a.** $(2, 0)$; $(5, 0)$ **b.** $(x - 2)(x - 5)$ **43. a.** $(-2, 0)$; $(0, 0)$; $(3, 0)$ **b.** $x(x + 2)(x - 3)$ **45. a.** $-5, 9$ **b.** $(-5, 0), (9, 0)$ **47. a.** $-3, 17$ **b.** $(-3, 0), (17, 0)$ **49. a.** $-3, -1, 8$ **b.** $(-3, 0), (-1, 0), (8, 0)$

51. a.

x	y_1
0	300
5	150
10	50
15	0
20	0
25	50
30	150

b. $(x - 15)(x - 20)$

53. a.

x	y_1
0	0
1	15
2	16
3	9
4	0
5	-5
6	0

b. $x(x - 4)(x - 6)$

55. -21, 20; $(x + 21)(x - 20)$ **57.** -200, 210; $(x + 200)(x - 210)$ **59.** $(x - 5)(x + 25)$ **61.** $(x + 9)(x - 1)(x - 5)$ **63.** $(x + 2)(x + 1)(x - 1)(x - 2)$ **65.** 7 and -9; $(7, 0), (-9, 0)$ **67.** $(x + 8)(x - 10)$; -8 and 10 **69.** $(x + 8y)(x - 9y)$ **71.** $(x - 11y)(x - 13y)$ **73.** $A = P(1 + r)$ **75.** $A = 2\pi r(r + h)$

Using the Language and Symbolism of Mathematics 6.2

1. Prime **2.** Quadratic **3. a.** Quadratic **b.** Linear **c.** Constant **4.** Distributive **5.** Product; sum **6.** c; b **7.** Same **8.** Opposite **9.** Inspection

Quick Review 6.2

1. $x^2 - 2x - 15$ **2.** $x^2 + 11xy + 28y^2$ **3.** $x^3 - 7x^2 + 10x$ **4.** $-2x^3y^2 + 8x^2y^3 + 10xy^4$ **5.** $x^3 + x^2 - 17x + 15$

Exercises 6.2

1. B **3.** A **5.** $(x - 4)(x - 2)$ **7.** $(x + 1)(x - 12)$ **9.** $(x + 2)(x + 1)$ **11.** $(x - 2)(x + 1)$ **13.** $(x + 5)(x - 4)$ **15.** $(x - 4)(x - 5)$ **17.** $(x - 9)(x + 2)$ **19.** $(x + 6)(x - 4)$ **21.** $x - 5$; $x - 5$; $x - 5$ **23.** $10x$; $x - 8$; $x - 8$; $x - 8$ **25.** $x + 4y$; $x + 4y$; $x + 4y$ **27. a.** $(x + 3)(x + 4)$ **b.** $(x + 6)(x + 2)$ **c.** Prime **d.** $(x + 12)(x + 1)$ **29. a.** $(x + 4)(x - 7)$ **b.** $(x + 2)(x - 14)$ **c.** Prime **d.** $(x + 1)(x - 28)$ **31. a.** $(x - 6)(x + 6)$ **b.** $(x + 12)(x - 3)$ **c.** $(x - 2)(x + 18)$ **d.** Prime **33.** $(a - 11)(a + 1)$ **35.** $(y - 17)(y - 1)$ **37.** $(x - 7)(x - 1)$ **39.** $(y - 11)(y + 9)$ **41.** $(p - 25)(p - 4)$ **43.** $(t + 6)(t + 8)$ **45.** $(n + 9)(n - 4)$ **47.** $(x + 2y)(x + 3y)$ **49.** $(x + 3y)(x - 2y)$ **51.** $(x - 5y)(x - 5y)$ **53.** $(a - 18b)(a + 2b)$ **55.** $(x + 30)(x + 3)$ **57.** Prime **59.** $(m - 14n)(m - 7n)$ **61.** Prime **63.** Prime **65.** $5x(x + 1)(x + 2)$ **67.** $4ax(x - 3)(x - 2)$ **69.** $10a(z + 4)(z + 25)$ **71.** $2ab(x - 6)(x + 8)$ **73.** $(a + b)(x - 2)(x - 4)$ **75.** $(x - 20)(x - 4)$ **77.** $(x - 24)(x + 6)$ **79.** $-(x - 7)(x + 5)$ **81.** $-(m - 3)(m + 6)$ **83.** $-(x - 15y)(x + 3y)$ **85.** $(x - 2)(x - 18)$ **87.** $x^2 - 5x + 6$; $(x - 2)(x - 3)$ **89.** $(x + 5)(x - 3)$; $-5, 3$; $(-5, 0), (3, 0)$

Using the Language and Symbolism of Mathematics 6.3

1. G C F **2.** b **3.** Same **4.** Opposite **5.** 1; -1

Quick Review 6.3

1. $6x^2 + 19x - 7$ **2.** $10x^2 + 43xy + 28y^2$ **3.** $6x^2 + 7x - 20$ **4.** $-54x^3y + 27x^2y^2 + 15xy^3$ **5.** $6x^3 + 29x^2 - 7x - 10$

Exercises 6.3

1. B **3.** A **5. a.** $(3x + 1)(x + 1)$ **b.** $(3x + 1)(x - 1)$ **c.** $(3x - 1)(x - 1)$ **d.** $(3x - 1)(x + 1)$ **7. a.** $(2x - 1)(x + 1)$ **b.** $(2x - 1)(x - 1)$ **c.** $(2x + 1)(x - 1)$ **d.** $(2x + 1)(x + 1)$ **9.** $(5x - 1)(x + 2)$ **11.** $(5x - 2)(x - 1)$ **13.** $(3v + 2)(2v - 3)$ **15.** $(6v - 1)(v - 6)$ **17.** $(5x + 1)(x + 1)$ **19.** $(7x - 1)(x - 1)$ **21.** $(7x + 1)(x - 1)$ **23.** $(11x - 1)(x + 1)$ **25.** $(2x + 3)(x + 1)$ **27.** $a + 5$; $a + 5$; $a + 5$ **29.** $2m$; $m - 4$; $m - 4$; $m - 4$ **31. a.** $(m + 2)(3m + 4)$ **b.** $(3m + 8)(m + 1)$ **c.** $(m + 4)(3m + 2)$ **d.** $(m + 8)(3m + 1)$ **33. a.** $(n + 10)(4n - 1)$ **b.** $(n + 4)(2n - 5)$ **c.** $(n + 2)(5n - 4)$ **d.** Prime **35. a.** Prime **b.** $(4x - 3y)(x - 3y)$ **c.** $(x - 2y)(2x - 9y)$ **d.** $(6x - y)(x - 6y)$ **37.** Prime **39.** $(w - 1)(4w - 5)$ **41.** $(3z + 2)(3z - 5)$ **43.** $(2b + 5)(5b + 2)$ **45.** $(3m + 5n)(4m - 7n)$ **47.** $(2y + 7)(9y - 4)$ **49.** $(11y + 3)(5y - 4)$ **51.** $(4x + 5y)(5x + 3y)$ **53.** $(7x - 2y)(2x - 5y)$ **55.** $(3m + 2n)(11m - 3n)$ **57.** $(2v + 3w)(3v + 2w)$ **59.** $-(6x - 1)(2x - 3)$ **61.** $-(7y + 2)(5y - 2)$ **63.** $-(3z + 8)(4z + 5)$ **65.** $-2a(19x + 3)(x - 1)$ **67.** $5x^3(4x - 3)(3x - 5)$ **69.** $3mn(m + n)(5m - 7n)$ **71.** $7av(5v + 6w)(6v - 5w)$ **73.** $(a + b)(5x - 2)(7x - 3)$

Using the Language and Symbolism of Mathematics 6.4

1. D **2.** E **3.** A **4.** C **5.** B

Quick Review 6.4

1. $x^2 - 49$ **2.** $25x^2 + 20xy + 4y^2$ **3.** $36x^2 - 60x + 25$ **4.** $27x^3 - 8$ **5.** $42x + 98$

Exercises 6.4

1. 3, 3, 3 **3.** $2w$, $2w$, $2w$ **5.** $(m - 1)^2$ **7.** $(6v + 1)^2$ **9.** $(5x - 2y)^2$ **11.** $(11m + 4n)^2$ **13.** $-(x - 8y)^2$ **15.** $(4x - 3)^2$ **17.** $2x(x + 4)^2$ **19.** $(w + 7)(w - 7)$ **21.** $(3v + 1)(3v - 1)$ **23.** $(9m + 5)(9m - 5)$ **25.** $(2a + 3b)(2a - 3b)$ **27.** $(4v + 11w)(4v - 11w)$ **29.** $(6 + m)(6 - m) = -(m + 6)(m - 6)$ **31.** $5(2x + 3y)(2x - 3y)$ **33. a.** $(x - 3)^2$ **b.** Prime **c.** $(x - 9)(x - 1)$ **35. a.** $(x - 6y)^2$ **b.** $(x - 6y)(x + 6y)$ **c.** $(x - 9y)(x - 4y)$ **37. a.** Prime **b.** $(2x - 3y)^2$ **c.** $(x + 3y)(4x + 3y)$ **39. a.** $(5x + 8y)(5x - 8y)$ **b.** $(5x - 8y)^2$ **c.** $(5x - 16y)(5x - 4y)$ **41. a.** $12x$ **b.** 36 **43. a.** $20x$ **b.** 100 **45. a.** $(-2, 0), (0, 0), (2, 0)$ **b.** $-2, 0, 2$ **c.** $x(x + 2)(x - 2)$ **47.** $(x + 3)(x^2 - 3x + 9)$ **49.** $(m - 5)(m^2 + 5m + 25)$ **51.** $(4a - b)(16a^2 + 4ab + b^2)$ **53.** $(5x + 2y)(25x^2 - 10xy + 4y^2)$ **55.** $(x_1 + x_2)(x_1 - x_2)$ **57.** $\pi h(r_1 + r_2)(r_1 - r_2)$ **59. a.** $V(x) = x(7 - 2x)(9 - 2x)$ **b.** The height is x, the width is $7 - 2x$, and the length is $9 - 2x$. **c.** It depends on what you plan to do with the function. The factored form displays the role of the length, width, and height. **61.** $(x^2 + 3)(x^2 - 3)$ **63.** $(5y^2 + 1)(5y^2 - 1)$ **65.** $(9y^2 + 1)(3y + 1)(3y - 1)$ **67.** $-3(a + b)(a - b)$ **69.** $(a + b)(x + 3)^2$ **71.** $(x + y)(x - y)(a + b)(a - b)$ **73.** $(z - 10)(z^2 + 10z + 100)$ **75.** $(7v + w)(49v^2 - 7vw + w^2)$ **77.** $x^2 - 12x + 36$; $x - 6$ **79.** $v^2 - 25$; $v + 5$, $v - 5$

Using the Language and Symbolism of Mathematics 6.5

1. Grouping **2.** Greatest common factor **3.** Prime **4.** Sum **5.** Difference **6.** Cubes **7.** Difference

Quick Review 6.5

1. $a(x - y - 3)$ **2.** $2b(x - 3y + 3)$ **3.** $(3a - b)(x^2 - 3x + 9)$ **4.** $(a + 2b + 3y)(a + 2b - 3y)$ **5.** $(5x + a - 3b)(5x - a + 3b)$

Exercises 6.5

1. $x - y$; $x - y$; $x - y$ **3.** $y^2 - 2y + 1$; $y - 1$; $y - 1$; $y - 1$; $x - y + 1$ **5.** $(a + b)(c + d)$ **7.** $(5c + 3)(a - 2b)$ **9.** $(a + c)(b - d)$ **11.** $(v - 7)(v - w)$ **13.** $(2a + 4b + 3)(2a - 4b + 3)$ **15.** $(n + 5)(3m - k)$ **17.** $(z^3 + w^2)(a + b)$

19. $(3b + a - 4)(3b - a - 4)$ **21.** $(y + 1)(ay + a - 1)$ **23.** $(8y + 3z)(8y - 3z)$ **25.** Prime **27.** $3(x - 1)(4x - 5)$ **29.** $(7a - 2)^2$ **31.** $(a - b)(x + y)$ **33.** Prime **35.** Prime **37.** $(2x^5 + 3y^3)^2$ **39.** $12xy(x - y)(x + y)$ **41.** $(c + d)(x + y)$ **43.** $(x + 1)(ax + by)$ **45.** $(x^3 + 2y)^2$ **47.** $8a(x - 1)(8x - 5)$ **49.** $5(x^2 - 11)$ **51.** $7ab(3a + 5)(3a - 5)$ **53.** $-3(2x - y)^2$ **55.** $9(16x^2 + 9)$ **57.** $(3x - 5y - 1)(3x + 5y - 1)$ **59.** $2a(2x - 3y)(3x + 2y)$ **61.** $7st(s^2 + t^2)(s + t)(s - t)$ **63.** $3a(x + y - 1)(x - y + 1)$ **65.** $2ax(x - 2)(x + 7)$ **67.** $(a + 1)(a + b + 1)$ **69.** $(a - 2b)(x + 4)(x - 6)$ **71.** $(2x - 3y + a)(2x - 3y - 25a)$ **73.** $(-3, 0)$; $(2, 0)$, $(4, 0)$; $-3, 2, 4$; $(x + 3)(x - 2)(x - 4)$ **75.** $5a(b - 1)(b^2 + b + 1)$ **77.** $(x + y)(x^2 + x - xy - y + y^2)$

Using the Language and Symbolism of Mathematics 6.6

1. Quadratic **2.** Quadratic; linear; constant **3.** $x - 3$; $x + 6$ **4.** LCD **5.** $(c, 0)$ **6.** $x - c$ **7.** $=$ **8.** $>$ **9.** $<$

Quick Review 6.6

1. $x = 15$ **2.** $x = -7$ **3.** $x = \frac{7}{5}$ **4.** $x = \frac{2}{3}$ is a solution. **5.** $x = 3$ is not a solution.

Exercises 6.6

1. a. D **b.** C **c.** A **d.** B **3. a.** $2x^2 - 7x + 3 = 0$; $a = 2, b = -7, c = 3$ **b.** $8x^2 - 3x = 0$; $a = 8, b = -3, c = 0$ **c.** $7x^2 + 5 = 0$; $a = 7, b = 0, c = 5$ **d.** $2x^2 - x - 4 = 0$; $a = 2, b = -1, c = -4$ **5.** $m = 8, m = -17$ **7.** $n = \frac{5}{2}, n = -\frac{1}{3}$ **9.** $z = 1, z = -2, z = \frac{7}{2}$ **11.** $v = -11, v = 11$ **13.** $x = -2, x = -1$ **15.** $y = -3, y = 6$ **17.** $v = 0, v = -\frac{1}{3}$ **19.** $w = -\frac{3}{2}, w = 5$ **21.** $x = -\frac{5}{2}, x = -\frac{2}{3}$ **23.** $x = 3, x = 8$ **25.** $z = -\frac{3}{7}, z = \frac{1}{2}$ **27.** $x = -\frac{5}{3}, x = \frac{5}{3}$ **29.** $r = -5, r = 2$ **31.** $m = -3, m = 6$ **33.** $x = -2, x = 6$ **35.** $v = 4, v = 7$ **37.** $w = -1, w = 4$ **39.** $x = -1, x = \frac{5}{2}$ **41.** $x = -\frac{1}{3}$ **43.** $v = 0, v = -3, v = 8$ **45.** $w = 0, w = -\frac{3}{2}, w = \frac{2}{3}$ **47.** $y = 0, y = -\frac{3}{2}, y = \frac{1}{7}$ **49.** $m = \frac{7}{3}$ **51. a.** $(-13, 0)$, $(7, 0)$ **b.** $(x + 13)(x - 7)$ **c.** $-13, 7$ **d.** $x = -13, x = 7$ **53. a.** $(-1, 0)$, $(0, 0)$, $(2, 0)$ **b.** $x(x + 1)(x - 2)$ **c.** $-1, 0, 2$ **d.** $x = -1, x = 0, x = 2$ **55. a.** $m = \frac{3}{5}, m = 2$ **b.** $5m^2 - 13m + 6$ **57. a.** $x = 3, x = -\frac{2}{5}$ **b.** $20x^2 - 52x - 24$ **59. a.** $x = -3, x = 5$ **b.** $(-3, 5)$ **c.** $(-\infty, -3) \cup (5, \infty)$ **61. a.** $x = -5, x = 2$ **b.** $(-\infty, -5] \cup [2, \infty)$ **c.** $[-5, 2]$ **63.** $x = 2, x = -7$; $(-7, 2)$; $(-\infty, -7) \cup (2, \infty)$ **65.** $x = -7, x = \frac{2}{3}$; $(-\infty, -7) \cup \left(\frac{2}{3}, \infty\right)$; $\left(-7, \frac{2}{3}\right)$ **67.** 3 cm × 11 cm **69.** 8 m **71.** 15 cm

Review Exercises for Chapter 6

1. $4(x - 9)$ **2.** $2x(x - 5)$ **3.** $12ax(x - 2)$ **4.** $5x(x^2 - 3)$ **5.** $(a + b)(x + y)$ **6.** $(a - 3)(x + 2y)$ **7.** $(2x + 7)(3x - 5)$ **8.** $(4x + 3)(6x - 5)$ **9.** $(x + 2)(x - 2)$ **10.** $(2x + 1)(2x - 1)$ **11.** $7(x + 2)(x - 2)$ **12.** $2(m + 8)(m - 8)$ **13.** $11(x + 2y)(x - 2y)$ **14.** $a(2x + 3y)(2x - 3y)$ **15.** $(x - 9)(x - 2)$ **16.** $(m + 3)(m + 7)$ **17.** $(x + 2)^2$ **18.** $(m + 5)^2$ **19.** $(x - 7y)(x + 6y)$ **20.** $(x + 25y)(x + y)$ **21.** $(2x - 3y)^2$ **22.** $(3y - 5)^2$ **23.** $(a - 7)(x + 2y)$ **24.** $xy(x - 4)(x - 1)$ **25.** $(a + b)(x + 5)(x - 5)$ **26.** $11(x + y)(x - y + 3)$ **27.** $(v + w + 1)(v - w + 1)$ **28.** $(5a - 3b)(2x - 3y)$ **29.** Prime **30.** $(v + 3w)(v - 3w)$ **31.** $(10x + 7y)(10x - 7y)$ **32.** Prime **33.** $2(x + 3y)(2x - 5y)$ **34.** $(2x - 5y)(3x + 4y)$ **35.** $10(x + 3)(2x - 1)$ **36.** $(x + 10y)(6x + y)$ **37.** $(v^2 + 1)(v + 1)(v - 1)$ **38.** $(9w^2 + 4)(3w + 2)(3w - 2)$ **39.** $(x + 5y)(a + b)^2$ **40.** $(3x - y)(3x + y + 1)$ **41. a.** $(x - 50)(x + 2)$ **b.** $(x - 20)(x + 5)$ **c.** $(x + 10)(x - 10)$ **d.** Prime **42. a.** $(x - 2)(x + 24)$ **b.** $(x - 3)(x + 16)$ **c.** $(x - 6)(x + 8)$ **d.** Prime **43. a.** $(2x + 1)(50x + 1)$ **b.** $(2x + 1)(2x + 25)$ **c.** $(x + 5y)(5x + 4y)$ **d.** $(5x + 2y)(2x + 5y)$ **44. a.** $(3x - 1)(16x - 1)$ **b.** $(x - 8)(3x - 2)$ **c.** $(x - 24y)(2x - y)$ **d.** $(4x - 3y)(4x - y)$ **45.** $(x + 5)(x - 7)$ **46.** $(x + 3)(x + 1)(x - 2)$ **47.** $(2x - y)(4x^2 + 2xy + y^2)$ **48.** $(x + 4)(x^2 - 4x + 16)$ **49.** $2x(x - 2)(x^2 + 2x + 4)$ **50.** $(v - 1)^2(v^2 + v + 1)$ **51.** $x = 5, x = -1$ **52.** $x = \frac{3}{2}, x = -\frac{2}{3}$ **53.** $x = 0, x = 7, x = \frac{2}{7}$ **54.** $v = 7, v = -3$ **55.** $y = \frac{1}{5}, y = -\frac{3}{2}$ **56.** $w = 3, w = -\frac{7}{6}$ **57.** $v = 5, v = -3$ **58.** $x = 3, x = -7$ **59.** $x = 0, x = 6, x = -6$ **60.** $x = 0, x = 5$ **61.** $z = -2, z = 6$ **62.** $x = -\frac{9}{2}$ **63.** $w = -4, w = -2$ **64.** $x = -15, x = 5$ **65. a.** $x = 3, x = -12$ **b.** $x^2 + 9x - 36$ **66. a.** $x = \frac{1}{3}, x = -\frac{5}{2}$ **b.** $6x^2 + 13x - 5$ **67. a.** $(x + 6)(x - 4)$ **b.** $-6, 4$ **c.** $x = -6, x = 4$ **d.** $(-6, 4)$ **e.** $(-\infty, -6) \cup (4, \infty)$ **68. a.** $(x + 3)(x - 12)$ **b.** $-3, 12$ **c.** $x = -3, x = 12$ **d.** $[-3, 12]$ **e.** $(-\infty, -3] \cup [12, \infty)$ **69. a.** $(x + 3)(x + 1)(x - 1)$ **b.** $-3, -1, 1$ **c.** $x = -3, x = -1, x = 1$ **d.** $(-\infty, -3] \cup [-1, 1]$ **e.** $[-3, -1] \cup [1, \infty)$ **70. a.** $(x + 6)(x + 4)(x - 2)$ **b.** $-6, -4, 2$ **c.** $x = -6, x = -4, x = 2$ **d.** $(-\infty, -6) \cup (-4, 2)$ **e.** $(-6, -4) \cup (2, \infty)$ **71.** $(-11, 0)$, $(7, 0)$; $(x + 11)(x - 7)$; $-11, 7$; $x = -11, x = 7$ **72.** $(-15, 0)$, $(35, 0)$; $(x + 15)(x - 35)$; $-15, 35$; $x = -15, x = 35$ **73.** $2\pi(r_1 - r_2)$ **74.** $\pi(L_1 + L_2)(L_1 - L_2)$ **75.** $\frac{4}{3}\pi(r_1 - r_2)(r_1^2 + r_1r_2 + r_2^2)$ **76.** $h = 4$ cm **77.** Either 10 cm or 20 cm

Mastery Test for Chapter 6

1. a. $9x^2(2x - 11)$ **b.** $2xy^2(2x^2 - 3xy + 5y^2)$ **c.** $(7x - 2)(3x + 4)$ **d.** $(2x + 3y)(x - 5y)$ **2. a.** $(a + 3)(7x + 1)$ **b.** $(3x - 2)(4a - 5b)$ **c.** $2(x + 2)(x - 3)$ **d.** $(5x - 6)(2x - 3)$ **3. a.** $(-8, 0)$, $(-2, 0)$, $(5, 0)$ **b.** $-8, -2, 5$ **c.** $x = -8, x = -2, x = 5$ **d.** $(x + 8)(x + 2)(x - 5)$ **4. a.** $(w - 9)(w + 5)$ **b.** $(w + 9)(w + 5)$ **c.** $(v - 6)(v - 4)$ **d.** $(v + 9)(v - 4)$ **5. a.** $(x - 4y)(x + 3y)$ **b.** $(x - 12y)(x - y)$ **c.** $(a + 6b)(a + 24b)$ **d.** $(a + 8b)(a - 6b)$ **6. a.** $(2x - 7)(3x + 2)$ **b.** $(2x - 1)(3x - 14)$ **c.** $(x + 14)(6x + 1)$ **d.** $(x + 2)(10x - 3)$ **7. a.** $(3x + 4y)(4x + 3y)$ **b.** $(3x - 4y)(4x + 3y)$ **c.** $(x - 12y)(12x - y)$ **d.** $(5x - y)(8x + 3y)$ **8. a.** $(x + 7)^2$ **b.** $(x - 8)^2$ **c.** $(3x + 10)^2$ **d.** $(5x - 11)^2$ **9. a.** $(x + 2y)(x - 2y)$ **b.** $(20a + b)(20a - b)$ **c.** $(4v + 7w)(4v - 7w)$ **d.** $(6x + 5y)(6x - 5y)$ **10. a.** $(4v - 1)(16v^2 + 4v + 1)$ **b.** $(v + 5)(v^2 - 5v + 25)$ **c.** $(2x + 5y)(4x^2 - 10xy + 25y^2)$ **d.** $(3a - 10b)(9a^2 + 30ab + 100b^2)$ **11. a.** $(2x + 3)(a + b)$ **b.** $(2x - 5y)(7a - 3b)$ **c.** $(a - 2b)(a + 2b + 1)$ **d.** $(x + 5y + 2)(x + 5y - 2)$ **12. a.** $5(x + 7)(x - 7)$ **b.** $2a(x + 5)^2$ **c.** $3x(x - 7)(x + 2)$ **d.** $5a(a - 3b)(x + y)$ **13. a.** $x = -\frac{3}{2}, x = \frac{4}{3}$ **b.** $x = \frac{5}{2}$ **c.** $x = 5, x = -4$ **d.** $x = 0, x = \frac{1}{2}, x = -3$ **14. a.** $(-1, 0)$, $(6, 0)$ **b.** $x = -1, x = 6$ **c.** $(-1, 6)$ **d.** $(-\infty, -1) \cup (6, \infty)$

Reality Check for End of Chapter 6

$A(x) = 4(x - 1)(x + 8)$ or $A(x) = 4x^2 + 28x - 32$; $A(5) = 208$, 208 ft^2; $m = -\frac{1}{5}$

Chapter 7

Reality Check for Chapter 7 Opener

B

Using the Language and Symbolism of Mathematics 7.1

1. One: output **2.** Domain **3.** Range **4.** x **5.** y **6.** x **7.** y **8.** x **9.** y **10.** Vertical **11.** Function **12.** f; x; x; output; f

Quick Review 7.1

1. a. -1 **b.** 249 **c.** 349

2.

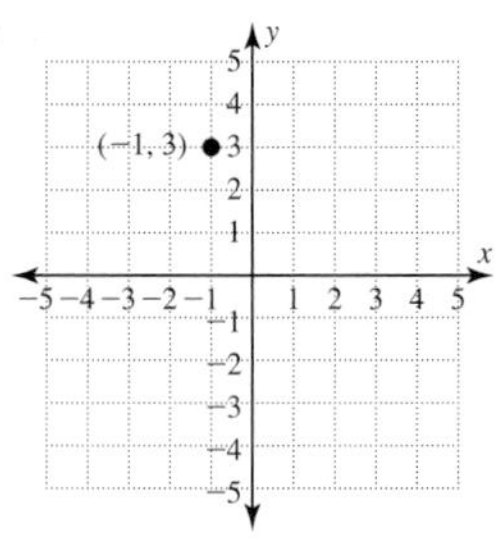

3.

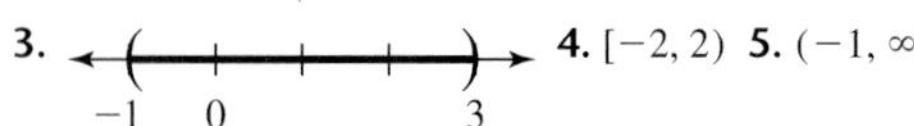

4. $[-2, 2)$ **5.** $(-1, \infty)$

Exercises 7.1

1. a. Not a function **b.** Function; $D = \{-3, -1, 1, 2\}$; $R = \{1, 2, 3\}$ **c.** Function; $D = \{0, 1, 2, 3\}$; $R = \{-1, 0, 1, 3\}$ **3. a.** Function; $D = \{-1, 1, 2\}$; $R = \{0, 1\}$ **b.** Not a function **c.** Function; $D = \{-3, -2, -1, 0, 1\}$; $R = \{-1, 0, 1, 2, 3\}$ **5. a.** Function **b.** Not a function **c.** Function **7. a.** Function **b.** Not a function **c.** Not a function **9. a.** Function **b.** Function **c.** Not a function **11. a.** -27 **b.** -7 **c.** 3 **13. a.** 0 **b.** 3 **c.** $\frac{1}{2}$ **15. a.** Output value **b.** Input value **c.** 3 **d.** 8 **17. a.** Input value **b.** Output value **c.** 0 **d.** 3 **19. a.** 8 **b.** -2 **c.** -5 **21. a.** 0 **b.** -1 **c.** -3 **23. a.** 4 **b.** 0 **c.** -3 **25. a.** -2 **b.** 0 **c.** 2 **27. a.** $D = \{-3, -2, -1, 0, 1, 2, 3\}$; $R = \{2\}$ **b.** $D = (-\pi, \pi]$; $R = [-1, 1]$ **c.** $D = [-2, \infty)$; $R = (-\infty, 2]$ **29. a.** $D = [-2, 2]$; $R = [0, 2]$ **b.** $D = [-2, 4)$; $R = [-1, 4]$ **c.** $D = \mathbb{R}$; $R = [-2, 2]$

31. a.

D	R
$-5 \to$	4
$-3 \to$	2
$-2 \to$	0
$0 \to$	-2
$1 \to$	-3
$4 \to$	-4

b. $\{(-5, 4), (-3, 2), (-2, 0), (0, -2), (1, -3), (4, -4)\}$

c.

33. a.

D	R
$-3 \to$	1
$-2 \to$	-1
$-1 \to$	3
$1 \to$	-2
$3 \to$	2

b. $\{(-3, 1), (-2, -1), (-1, 3), (1, -2), (3, 2)\}$

c.

x	y
-3	1
-2	-1
-1	3
1	-2
3	2

35. a.

x	y
-1	1
1	3
2	-1
4	-2

b. $\{(-1, 1), (1, 3), (2, -1), (4, -2)\}$

c.

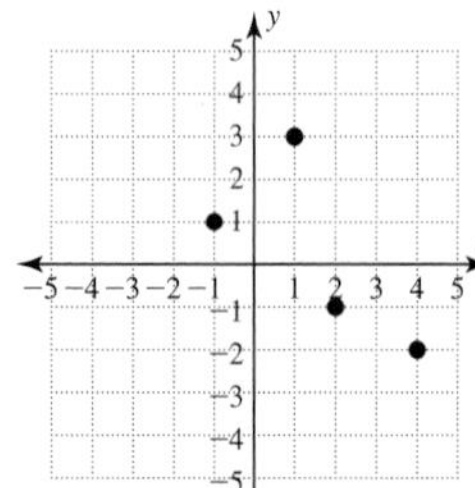

37. a.

x	y
-5	4
-3	4
1	4
2	4
3	4

b.

D	R
$-5 \to$	4
$-3 \to$	4
$1 \to$	4
$2 \to$	4
$3 \to$	4

c.

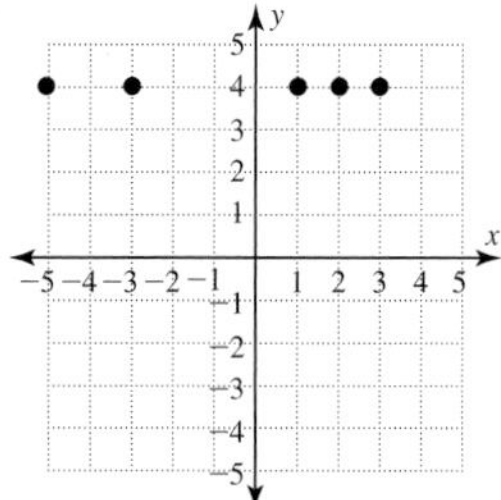

39. a. $f(x) = 2x$

b.

D	R
$25,000 →	$ 50,000
40,000 →	80,000
50,000 →	100,000
65,000 →	130,000

c.

x	f(x)
$25,000	$ 50,000
40,000	80,000
50,000	100,000
65,000	130,000

d. {(25,000, 50,000), (40,000, 80,000), (50,000, 100,000), (65,000, 130,000)}

41. a. Pressure **b.** 16 in^3 **c.** 15 lb/in^2

43. a.

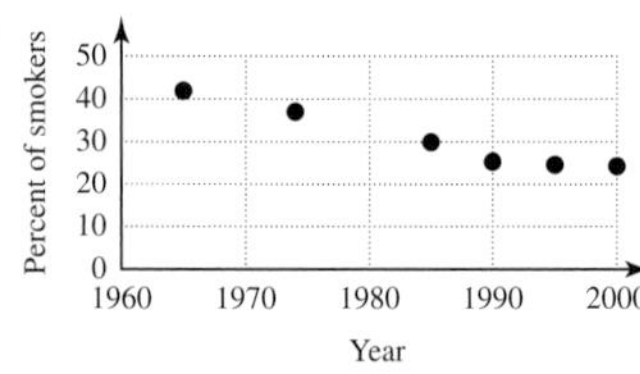

b. 29.9% **c.** 1974

45. a. $T(x) = 0.0825x$

b.

x	Y_1
0	0
100	8.25
200	16.5
300	24.75
400	33
500	41.25
600	49.5

47. a.

x	y
1	1,800
2	2,100
3	2,400
4	2,700
5	3,000
6	3,300

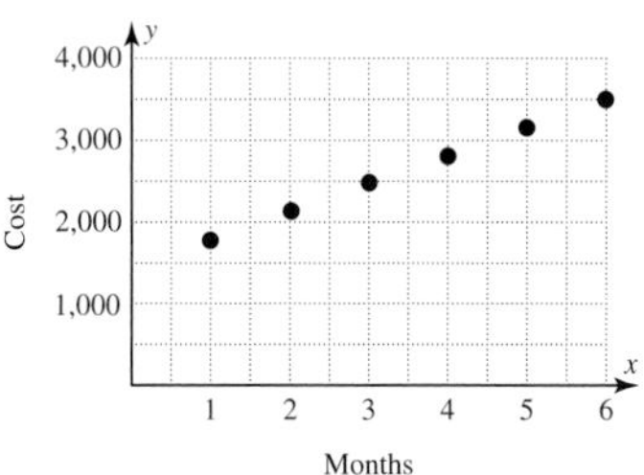

b. $8,700 **c.** 11

49. $D = [0, \infty)$; $R = [0, \infty)$ **51.** $D = \mathbb{R}$; $R = (-\infty, 1]$ **53.** $D = (-\infty, 0]$; $R = [0, \infty)$ **55.** D **57.** B

59.

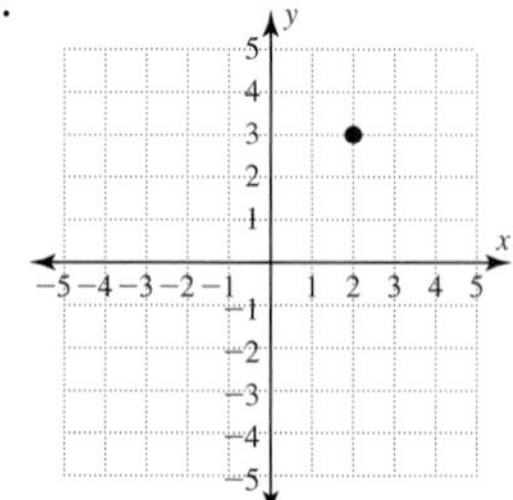

61.

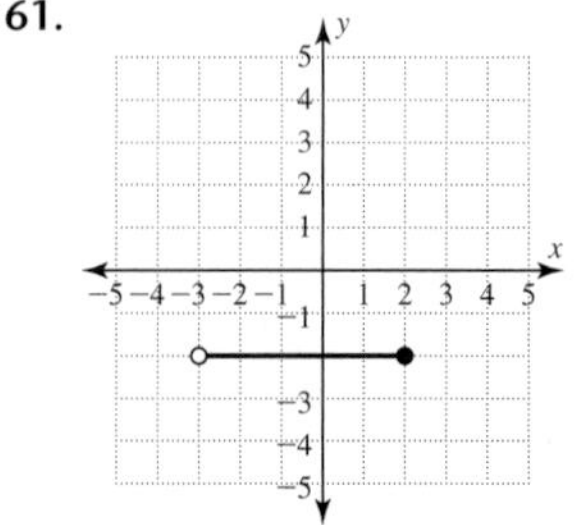

63.

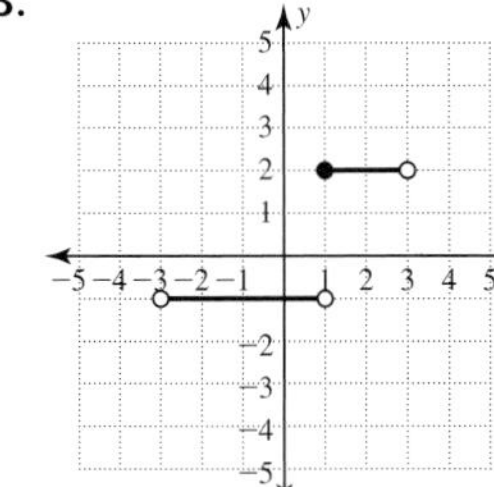

65.

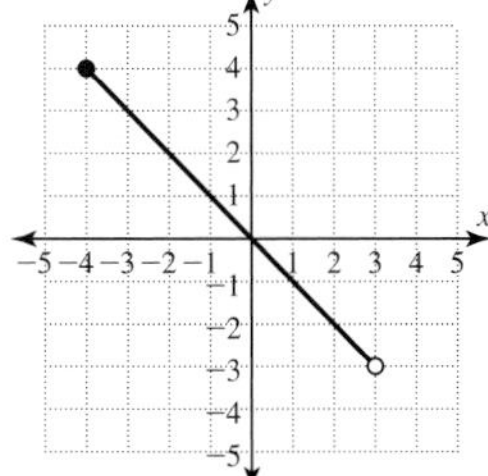

67.

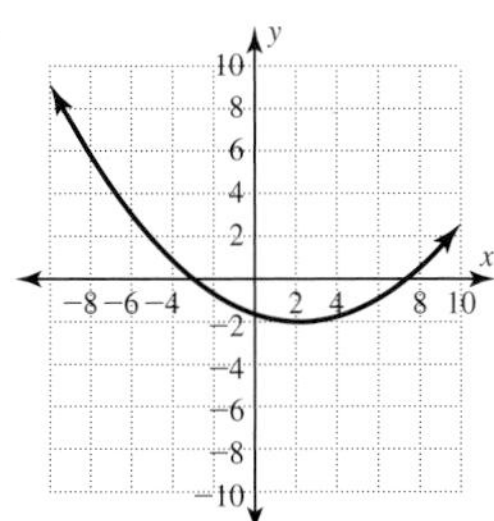

69. a.

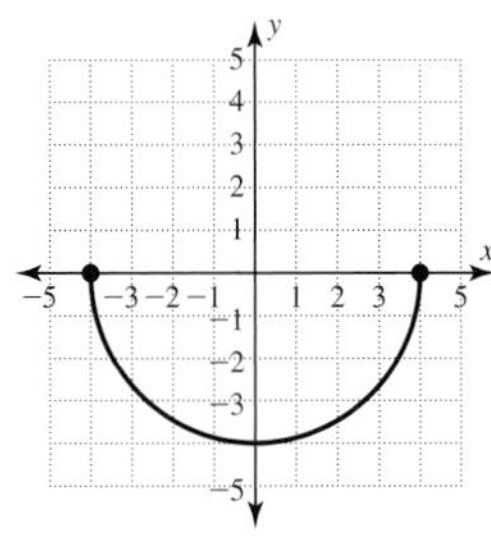

b.

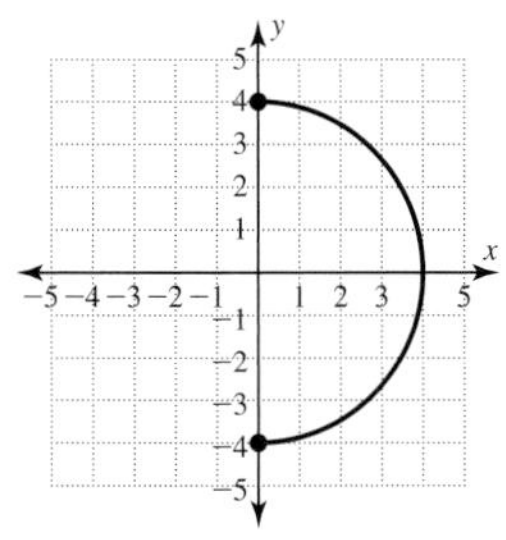

Using the Language and Symbolism of Mathematics 7.2

1. Linear **2.** 2; (0, 5) **3.** Increasing **4.** Decreasing **5.** Increases **6.** Decreases **7.** Positive **8.** Negative **9.** Highest **10.** Lowest **11.** V **12.** Vertex **13.** Above **14.** Below **15.** Loss interval **16.** Profit interval

Quick Review 7.2

1. Run **2.** $m = \frac{2}{5}$ **3.** 8, 11, 14, 17 **4.** 34 **5.** $(-\infty, -4)$

Exercises 7.2

1.

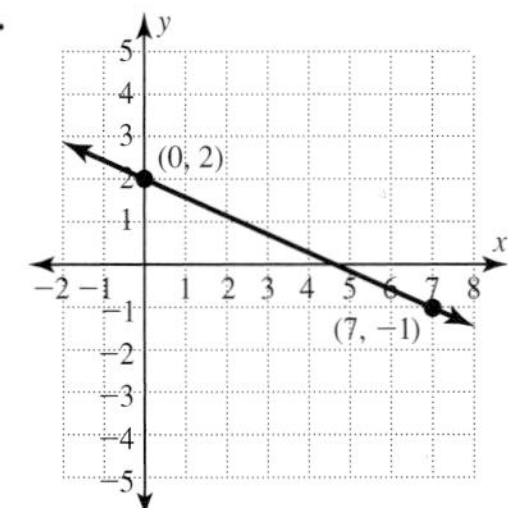

3. a. $(0, -2)$ **b.** $m = \frac{4}{3}$ **c.** $y = \frac{4}{3}x - 2$ **5.** Decreasing **7.** Increasing **9.** Decreasing **11.** Increasing **13.** Neither **15. a.** $m = \frac{1}{2}$ **b.** Increasing **c.** $y = \frac{1}{2}x + 3$ **17. a.** $m = -\frac{5}{4}$ **b.** Decreasing **c.** $y = -\frac{5}{4}x + 5$

19.

x	−3	−2	−1	0	1	2	3
y	1	0	1	2	3	4	5

$D = \mathbb{R}$; $R = [0, \infty)$; vertex: $(-2, 0)$

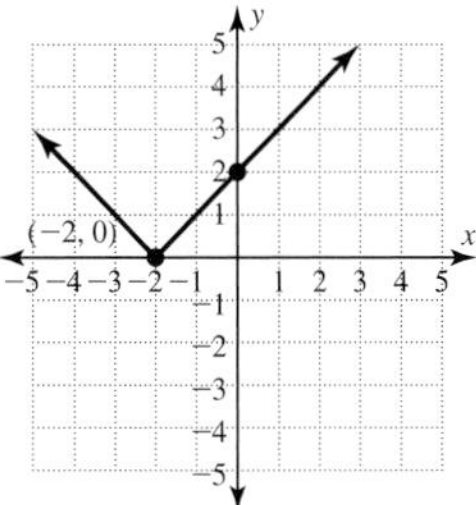

21.

x	−3	−2	−1	0	1	2	3
y	0	−1	−2	−3	−2	−1	0

$D = \mathbb{R}$; $R = [-3, \infty)$; vertex: $(0, -3)$

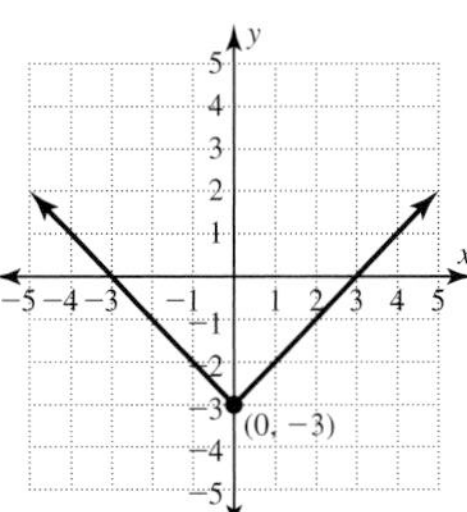

23. C **25.** D **27.** $D = \mathbb{R}$; $R = \mathbb{R}$ **29.** $D = \mathbb{R}$; $R = \{3\}$ **31.** $D = \mathbb{R}$; $R = [0, \infty)$; vertex: (3, 0) **33.** $D = \mathbb{R}$; $R = (-\infty, 2]$; vertex: (1, 2) **35.** Increasing: $(-\infty, -2)$; decreasing: $(-2, \infty)$ **37.** Increasing: $(1, \infty)$; decreasing: $(-\infty, 1)$ **39.** Positive: $(-\infty, -2)$; negative: $(-2, \infty)$ **41.** Positive: $(-\infty, -4) \cup (2, \infty)$; negative: $(-4, 2)$ **43. a.** $(-3, 4)$ **b.** 4 **c.** −3

45. a.

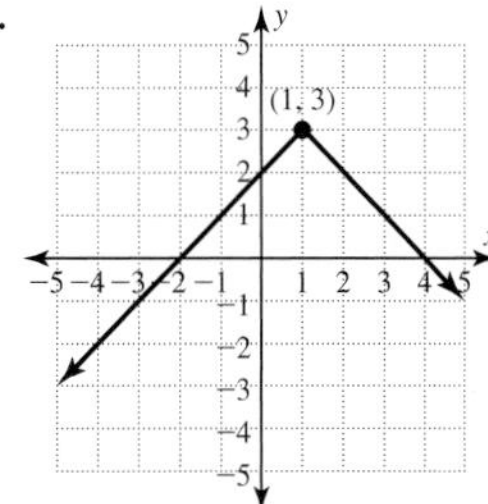

b. (1, 3) **c.** Maximum 3 **d.** $\mathbb{R}$ **e.** $(-\infty, 3]$ **f.** $(1, \infty)$ **g.** $(-\infty, 1)$ **h.** $(-2, 0)$, $(4, 0)$ **i.** $(0, 2)$ **j.** $(-2, 4)$ **k.** $(-\infty, -2) \cup (4, \infty)$

47. Profit interval: (50, 150]; loss interval: [0, 50) **49. a.** $x = -\frac{4}{3}$; $x = 2$ **b.** $\left(-\frac{4}{3}, 2\right)$ **c.** $\left(-\infty, -\frac{4}{3}\right) \cup (2, \infty)$ **51. a.** $T(0) = -21.6$; a 10-mi/h wind makes 0° feel like −21.6° **b.** $T(5) = -15.6$; a 10-mi/h wind makes 5° feel like −15.6° **c.** $T(20) = 2.4$; a 10-mi/h wind makes 20° feel like 2.4° **d.** Increasing **e.** Increasing **53. a.** $P(2) = \frac{1}{36}$; the probability of rolling a sum of 2 on a pair of dice is $\frac{1}{36}$. **b.** $P(7) = \frac{1}{6}$; the probability of rolling a sum of 7 on a pair of dice is $\frac{1}{6}$. **c.** $P(12) = \frac{1}{36}$; the probability of rolling a sum of 12 on a pair of dice is $\frac{1}{36}$. **d.** 2, 3, 4, 5, 6, 7, 8, 9, 10, 11, 12 **e.** 7

55. a. $C(x) = 0.2x$

b.

x	Y_1
0	0
500	100
1,000	200
1,500	300
2,000	400
2,500	500
3,000	600

c. \$200 **d.** \$2,125

57. a. $V(x) = -50x + 500$ **b.** \$150 **c.** 9 years

d.

x	0	1	2	3	4	5	6
$V(x)$	500	450	400	350	300	250	200

e. Decreasing **59. a.** $C(x) = 0.60x + 2.00$ **b.** There is a charge of \$0.60 per minute. **c.** Increasing **d.** There is a flat fee of \$2 per call.

Using the Language and Symbolism of Mathematics 7.3

1. Parabola **2.** Quadratic **3.** Vertex **4.** Upward **5.** Axis; symmetry **6.** Quadratic; linear; constant **7.** Linear **8.** Second **9.** Upward **10.** Upward **11.** Downward **12.** Curve **13.** Scatter **14.** Linear **15.** Parabola

Quick Review 7.3

1. Polynomial **2.** Point; slope **3.** Overhead **4.** Negative **5.** III

Exercises 7.3

1. D **3.** A **5.** Line; negative slope **7.** Line; positive slope **9.** Parabola; opens upward **11.** Parabola; opens downward **13.** Line; zero slope

15. a.

x	y
−3	7
−2	2
−1	−1
0	−2
1	−1
2	2
3	7

b, c.

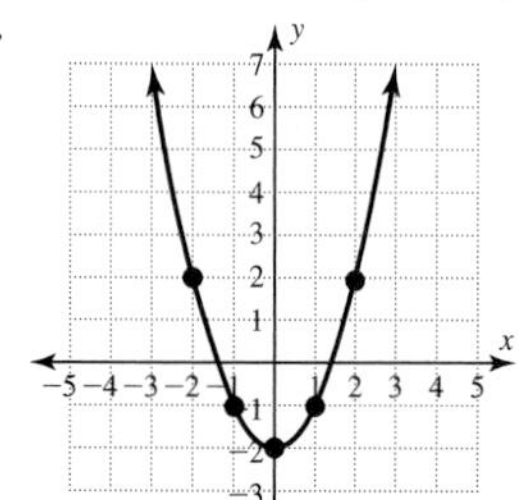

17. a.

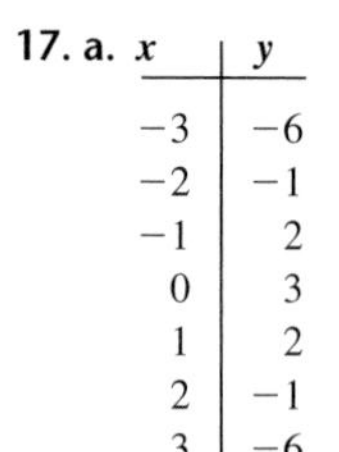

x	y
−3	−6
−2	−1
−1	2
0	3
1	2
2	−1
3	−6

b, c.

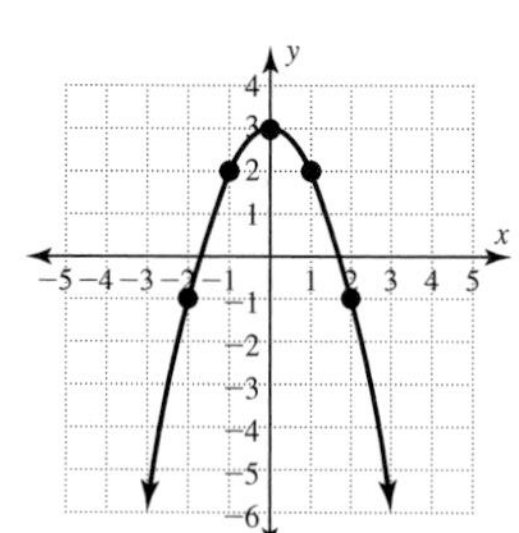

19. a.

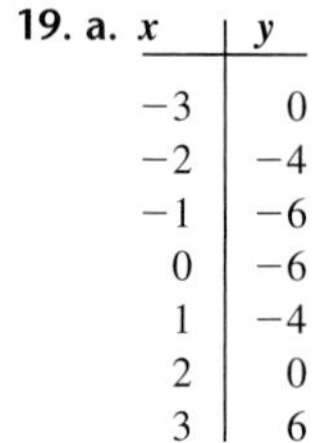

x	y
−3	0
−2	−4
−1	−6
0	−6
1	−4
2	0
3	6

b, c.

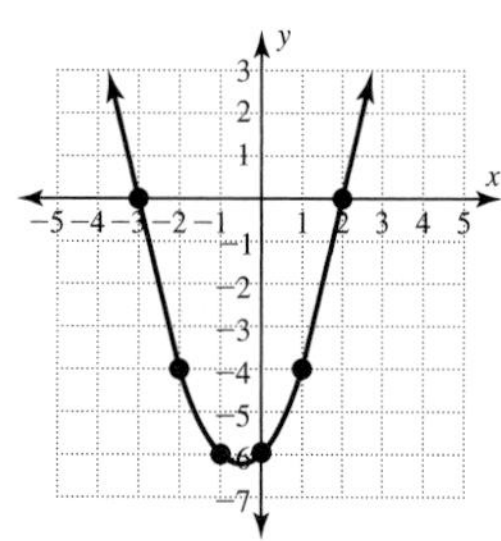

21. a. Upward **b.** $(1, -4)$ **c.** $(0, -3)$ **d.** $(-1, 0)$, $(3, 0)$ **e.** $\mathbb{R}$ **f.** $[-4, \infty)$ **g.** $(-\infty, 1)$ **h.** $(1, \infty)$ **23. a.** Downward **b.** $(0, 4)$ **c.** $(0, 4)$ **d.** $(-2, 0)$, $(2, 0)$ **e.** $\mathbb{R}$ **f.** $(-\infty, 4]$ **g.** $(0, \infty)$ **h.** $(-\infty, 0)$ **25. a.** Upward **b.** $(1, 0)$ **c.** $(0, 1)$ **d.** $(1, 0)$ **e.** $\mathbb{R}$ **f.** $[0, \infty)$ **g.** $(-\infty, 1)$ **h.** $(1, \infty)$ **27. a.** -16 **b.** $(0, -16)$ **c.** $(-4, 0)$, $(4, 0)$ **d.** $x = -4, x = 4$ **29. a.** -5 **b.** $(0, -5)$ **c.** $(-5, 0)$, $(-1, 0)$ **d.** $x = -5, x = -1$ **31.** C **33.** D **35. a.** $(-1, 4)$ **b.** $(0, 3)$ **c.** $(-3, 0)$, $(1, 0)$ **37. a.** $(-2, -1)$ **b.** $(0, 3)$ **c.** $(-3, 0)$, $(-1, 0)$ **39. a.** \$15,000 **b.** 20 and 80 units **c.** \$9,000; 50 units **41. a.** \$1,000 **b.** 2 and 22 units **c.** \$2,000; 12 units **43.** C **45.** D **47.** Negative slope; a decreasing function **49.** Positive slope; an increasing function **51.** $y = \frac{4}{7}x + 3$ **53.** $y = -3x - 6$ **55.** $y = -\frac{2}{3}x - \frac{8}{3}$ **57.** $-\frac{4}{7}x - \frac{13}{7}$

59. a.

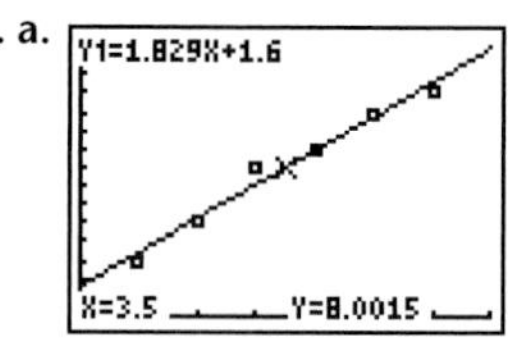

[0, 7, 1] by [0, 15, 1]

b. $y \approx 1.829x + 1.600$ **c.** $y \approx 8.002$

61. a.

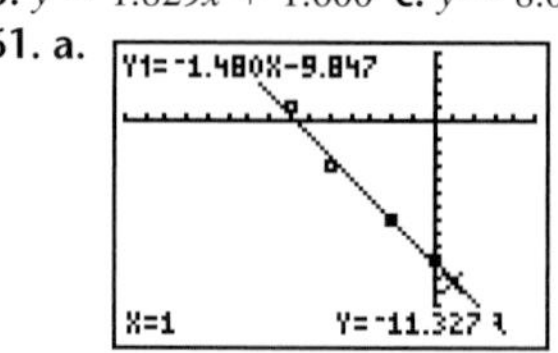

[−15, 5, 1] by [−15, 5, 1]

b. $y \approx -1.480x - 9.847$ **c.** $y \approx -11.327$

63. a.

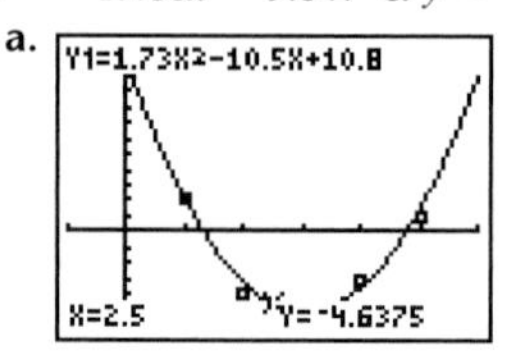

[−1, 6, 1] by [−6, 12, 1]

b. $y \approx 1.73x^2 - 10.5x + 10.8$ **c.** $y \approx -4.64$

65. a.

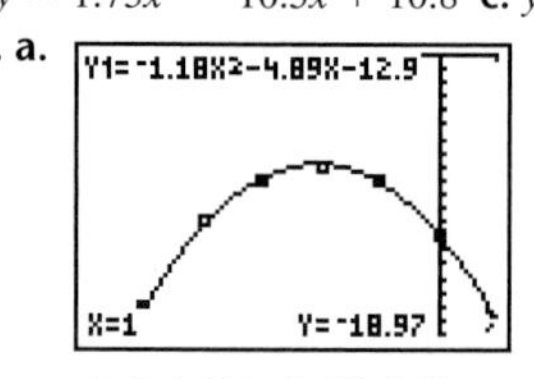

[−6, 1, 1] by [−20, 0, 1]

b. $y \approx -1.18x^2 - 4.89x - 12.9$ **c.** $y \approx -19.0$

67. a. $y = -0.005x + 3.201$ **b.** For each cigarette per day the birth weight decreases by 5 g. **c.** 3.11 kg **d.** 2.98 kg **69. a.** $y = 0.015x + 7.179$ **b.** The length of the winning long-jump record is increasing by 6 cm per Olympics (that is, per 4 years). **c.** 8.44 m **d.** 2024

71. a. $y = 2.177x - 159.985$ **b.** The number of women in the U.S. House has been increasing by a little over 2 per year. **c.** Approximately 5, the estimate was too low by approximately 13. **d.** 75

73. a. $y = 48.8x^2 + 25.0x - 181$ **b.** The load the beam can support is approximately 11,000 kg.

Using the Language and Symbolism of Mathematics 7.4

1. Plus or minus **2.** Principal **3.** Extraction of roots **4.** Completing the square **5.** Quadratic formula **6.** Standard **7.** $x^2 + 2ax + a^2$ **8.** $3x^2 + 5x - 7 = 0$ **9.** Discriminant

Quick Review 7.4

1. Integers **2.** Irrational **3.** Terminating **4.** Nonrepeating **5.** 6 **6.** $(x + 5)^2$ **7.** $(x - 8)^2$

Exercises 7.4

1. $v = \pm 9$ **3.** $x = \pm\sqrt{7}$ **5.** $z = \pm 3\sqrt{2}$ **7.** $x = 3, x = -1$ **9.** $x = 1, x = -2$ **11.** $x = \frac{2 \pm \sqrt{3}}{3}$ **13.** $t = \frac{1}{2}, t = -\frac{1}{4}$ **15.** $m = \frac{6}{5}, m = 0$ **17.** $w = \pm 3$ **19.** $v = -\frac{2}{3}$ **21.** $y = \pm\sqrt{5}$; $y \approx \pm 2.24$ **23.** $w = \frac{1 \pm \sqrt{6}}{5}$; $w \approx -0.29, w \approx 0.69$

25. y-intercept: $(0, -6)$; x-intercepts: $(\sqrt{6}, 0)$, $(-\sqrt{6}, 0)$ or $(2.45, 0)$, $(-2.45, 0)$ **27.** y-intercept: $(0, -1)$; x-intercepts: $(2 - \sqrt{5}, 0)$, $(2 + \sqrt{5}, 0)$ or $(-0.24, 0)$, $(4.24, 0)$ **29.** $-3 + \sqrt{17}$ and $3 + \sqrt{17}$ or $-3 - \sqrt{17}$ and $3 - \sqrt{17}$ **31.** $(-2 + 2\sqrt{3})$ cm by $(2 + 2\sqrt{3})$ cm **33. a.** \$61,200 **b.** 15 or 51 units **35. a.** 0.22 seconds, 5.78 seconds

b. (0.22, 5.78) **37.** C **39.** A **41.** 24; two distinct real solutions **43.** $(x + 5)^2 = 3$ **45.** $(x - 7)^2 = 1$ **47.** $(x - 9)^2 = 90$ **49. a.** 4 **b.** 1 **c.** 1 **d.** 1 **e.** −1 $\pm\sqrt{5}$ **51.** $z = 0, z = 4$ **53.** $x = 1, x = -5$ **55.** $z = -2 - \sqrt{2}, z = -2 + \sqrt{2}$ **57. a.** $y = -0.311x^2 + 30.8x - 358$ **b.** 13 or 86 units **59.** $x = 0, x = 4, x = -7$ **61.** $x = 0, x = 4, x = 1$ **63.** $x = 0, x = \pm\sqrt{10}$ **65. a.** $x = -1, x = 5$ **b.** $(-1, 5)$ **c.** $(-\infty, -1) \cup (5, \infty)$ **67. a.** $x = 1 \pm \sqrt{7}$ **b.** $(-\infty, 1 - \sqrt{7}] \cup [1 + \sqrt{7}, \infty)$ **c.** $[1 - \sqrt{7}, 1 + \sqrt{7}]$ **69. a.** $x = -5, x = \frac{1}{2}$ **b.** $\left(-5, \frac{1}{2}\right)$ **c.** $(-\infty, -5) \cup \left(\frac{1}{2}, \infty\right)$ **71. a.** $x = -\sqrt{7}, x = \sqrt{7}$ **b.** $(-\sqrt{7}, \sqrt{7})$ **c.** $(-\infty, -\sqrt{7}) \cup (\sqrt{7}, \infty)$

Using the Language and Symbolism of Mathematics 7.5

1. Vertex **2.** Symmetry **3.** $-\frac{b}{2a}$ **4.** $f\left(-\frac{b}{2a}\right)$ **5.** Second **6.** Third

Quick Review 7.5

1. 123.46 **2.** 100 **3.** 120 **4.** Zero **5.** 0

Exercises 7.5

1. a. (0, −8) **b.** (−2, 0) (4, 0) **c.** (1, −9) **3. a.** (0, −3) **b.** No x-intercepts **c.** (0, −3) **5.** 2 **7.** (3, 2) **9.** (1, 11) **11. a.** (0, −2) **b.** $\left(-\frac{1}{2}, 0\right)$; (2, 0) **c.** $\left(\frac{3}{4}, -\frac{25}{8}\right)$ **13. a.** (0, −12) **b.** $\left(\frac{3}{2}, 0\right)$; (−4, 0) **c.** $\left(-\frac{5}{4}, -\frac{121}{8}\right)$ **15.** $x^2 - 9 = 0$ **17.** $x^2 - 6x = 0$ **19.** $5x^2 - 3x - 14 = 0$ **21.** $10x^2 - 21x - 10 = 0$ **23.** $x^2 - 5 = 0$ **25.** $x^2 - 4x - 1 = 0$ **27.** $x^3 - 7x^2 + 10x = 0$ **29.** $x^3 - 6x^2 - x + 6 = 0$ **31.** $y = x^2 + 2x - 8$ **33.** $y = -x^2 - 2x + 8$ **35. a.** \$240,000 **b.** 20 and 120 tons **c.** \$250,000; 70 tons **37. a.** \$28,665 **b.** 4.5 and 45.5 tons **c.** \$58,835; 25 tons **39.** Maximum height: 144 ft after 3 seconds **41.** 100 m^2 **43.** 2,100 rev/min; maximum horsepower: 850 **45. a.** $y = -0.0000716x^2 + 0.299x - 5.07$ **b.** (2,100, 310) **c.** 310 maximum horsepower at 2,100 rev/min **47. a.** $y = -0.185x^2 + 14.2x - 62.5$ **b.** 5 or 72 units **c.** \$210; 38 units **49. a.**

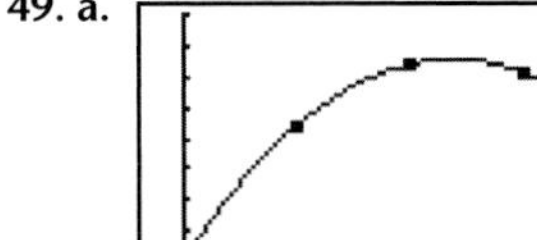

[−30, 330, 50] by [−25, 200, 25]

b. $f(x) = -0.00285x^2 + 1.36x + 3.00$ **c.** $f(325) = 144$ **d.** The ball has a height of 144 ft when it has traveled 325 ft horizontally. **e.** (−2.2, 0), (479.4, 0) **f.** 479.4 is the horizontal distance the ball would travel before reaching ground level. **g.** (238.6, 165.2) **h.** The ball has a maximum height of 165.2 ft when it is 238.6 ft from home plate.

Using the Language and Symbolism of Mathematics 7.6

1. Find **2.** Variable **3.** Algebraic **4.** Solve **5.** Reasonable **6.** Rate; time **7.** Legs; hypotenuse **8.** $a^2 + b^2 = c^2$ **9.** Practical

Quick Review 7.6

1. $n + 1$; $n + 2$ **2.** $n + 2$; $n + 4$ **3.** $x = -8, x = 8$ **4.** [0, 20] **5.** {0, 1, 2, 3, . . . , 20}

Exercises 7.6

1. −13 and −12 or 12 and 13 **3.** −18 and −16 or 16 and 18 **5.** −8 and −7 or 7 and 8 **7.** $-\frac{25}{3}$ and −21 or 7 and 25 **9.** $\frac{7}{3}$ cm by 9 cm **11.** 14 m **13.** 7.2 cm **15.** Between 0.6 and 5.4 seconds **17.** More than 10 and less than 60 windmills **19.** 5 cm **21.** 15 ft **23.** 18.0 ft **25.** 10.96 in **27.** 10.95 in **29.** 13.6 in by 10.2 in; 139 in^2 **31.** 14 in **33.** 7 mi **35.** 90 mi/h and 120 mi/h **37.** 28 ft **38.** 8 m **39.** 44 m **41.** 37.3 ft **43.** 9.28% **45.** 180 mi **47.** 18 cm **49. a.** $A(h) = h(60 - 2h)$ **b.** [0, 30] **c.** $A(5) = 250$; when 5 cm is folded up, the resulting trough has a cross-sectional area of 250 cm^2 **d.** 10 cm or 20 cm **e.** 450 cm^2; 15 cm **51. a.**

[−10, 200, 20] by [−5, 25, 5]

b. $f(x) = -0.00188x^2 + 0.375x$ **c.** $f(50) = 14.1$ **d.** When the horizontal distance is 50 ft, the height of the ball is 14.1 ft. **e.** (0, 0); (200, 0) **f.** The initial height of the ball is 0; the ball travels 200 yd **g.** (100, 18.7) **h.** The maximum height is 18.7 yd when the ball is 100 yd from the tee. **53. a.** 80 ft **b.** 5 seconds **c.** (0, 2) **d.** (2, 5) **e.** 2 seconds **f.** 140 ft **55. a.** 5.7 seconds **b.** From 1.3 to 4.3 seconds **c.** From 1.9 to 3.8 seconds **d.** From 2.3 to 3.3 seconds **e.** 134.6 ft **f.** 2.8 seconds

Using the Language and Symbolism of Mathematics 7.7

1. $\sqrt{-1}$ **2.** a; bi **3.** Sometimes **4.** Always **5.** Always **6.** Sometimes **7.** Never **8.** $a + bi$ **9.** Four **10.** Conjugate **11.** 2 **12.** $7 - 2i$

Quick Review 7.7

1. 3.746921×10^4 **2.** 0.0000406 **3.** $10x^2 - 7x - 12$ **4.** $(5x + 4)(2x - 3)$ **5.** $25x^2 - 9y^2$

Exercises 7.7

1. a. −6 **b.** $6i$ **c.** $-6i$ **d.** 6 **3. a.** $4 + 3i$ **b.** $3 + 4i$ **c.** $-3 - 4i$ **d.** $-4 + 3i$ **5. a.** $5i$ **b.** $-5i$ **c.** −6 **d.** 6 **7. a.** −10 **b.** 10 **c.** $-10i$ **d.** 10 **9. a.** $\frac{5}{3}i$ **b.** $\frac{5}{3}i$ **c.** $-\frac{5}{3}i$ **d.** $\frac{5}{3}$ **11.** $9 - i$ **13.** $3 + i$ **15.** $60 + 24i$ **17.** 7 **19.** −6 **21.** $10 + 6i$ **23.** $8 - 3i$ **25.** 53 **27.** $34 + 27i$ **29.** $24 + 10i$ **31.** $7 - 24i$ **33.** $12i$ **35.** $-1 - 2i$ **37.** $2 - 5i$, 29 **39.** $-13i$, 169 **41.** $2 - 2i$ **43.** $\frac{15}{17} - \frac{8}{17}i$ **45.** $7 + 6i$ **47.** i **49.** −1 **51.** 0; a double real solution **53.** −4; complex solutions with imaginary parts **55.** −8; complex solutions with imaginary parts **57.** $-30 - 29i$ **59.** −4 **61.** $4 + 10i$ **63.** $i^{46} = -1$ **65.** $z = \pm 10i$ **67.** $x = -\frac{3}{2} \pm \frac{3}{2}i$ **69.** $v = 11, v = -\frac{3}{2}$ **71.** $x = 2 \pm 5i$ **73.** $x = 3 \pm 2i$ **75.** $w = \frac{3}{2} \pm \frac{1}{2}i$ **77. a.** $x = 1, x = -3$ **b.** $x = -1$ **c.** $x = -1 \pm i$ **79.** $x^2 + 16 = 0$ **81.** $x^2 - 6x + 13 = 0$ **83.** $x = 2, x = -\frac{1}{2} \pm \frac{\sqrt{3}}{2}i$

Review Exercises for Chapter 7

1. a. Function **b.** Function **c.** Not a function **d.** Function **2. a.** Not a function **b.** Function **c.** Function **d.** Not a function **3. a.** Domain: {5, 6, 7}; range: {8} **b.** Domain: (−1, 3]; range: [0, 2] **c.** Domain {−3, −2, −1, 0, 1}; range: {1, 4, 6, 9, 11} **d.** Domain: ℝ ; range: [2, ∞) **4. a.** 3 **b.** 7 **c.** 12 **d.** 103 **5. a.** 4 **b.** 2 **c.** 2 **d.** −1 **6. a.** −2 **b.** 2 **c.** −1 **d.** −2

7. a.

D		*R*
−10	→	4
−5	→	3
0	→	2
5	→	1
10	→	0
15	→	−1
20	→	−2

b. {(−10, 4), (−5, 3), (0, 2), (5, 1), (10, 0), (15, −1), (20, −2)}

c.

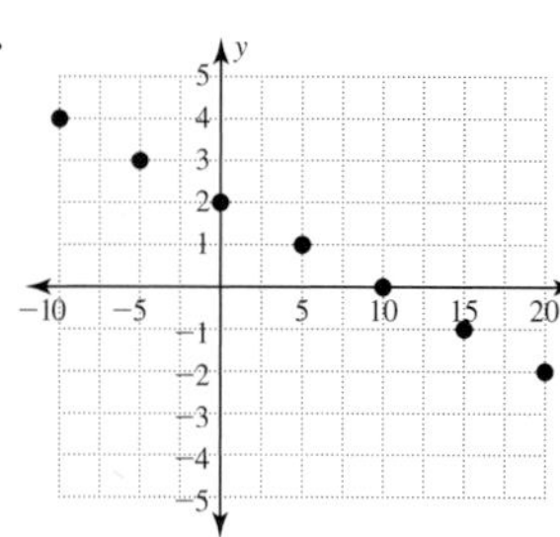

d. $y = -\frac{1}{5}x + 2$

8. a. $\{(-4, -1), (-2, 0), (0, 1), (2, 2), (4, 3)\}$

b.

x	y
-4	-1
-2	0
0	1
2	2
4	3

c.

D		R
-4	$\rightarrow$	-1
-2	$\rightarrow$	0
0	$\rightarrow$	1
2	$\rightarrow$	2
4	$\rightarrow$	3

d. $y = \frac{1}{2}x + 1$

9. $y = -\frac{1}{2}x + 4$ **10.** $y = \frac{2}{3}x - 4$ **11.** $y = 2x$

12.

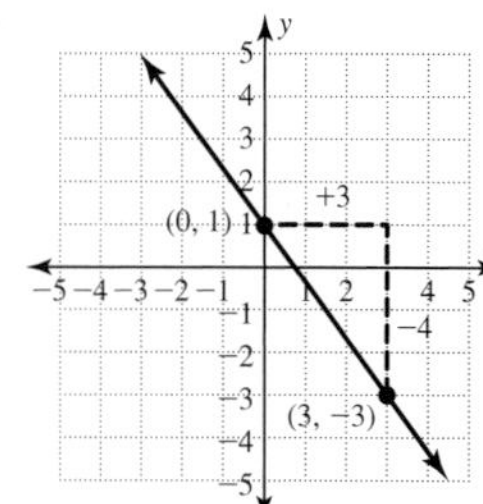

13.

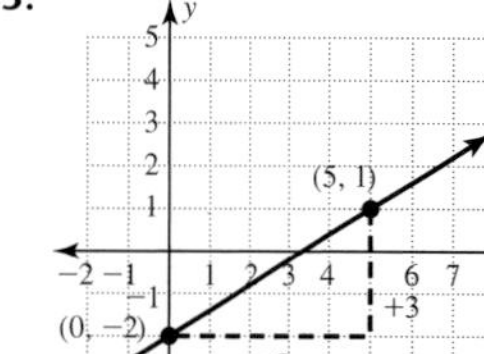

14. a. $(0, -6)$ **b.** 5 **c.** 4 **d.** $\frac{4}{5}$ **e.** $y = \frac{4}{5}x - 6$ **15.** Increasing **16.** Decreasing **17.** Decreasing **18.** Increasing **19.** Increasing: $(-\infty, -2)$; decreasing: $(-2, \infty)$ **20.** Increasing: $(-\infty, -1)$; decreasing: $(-1, \infty)$ **21.** Positive: $(3, \infty)$; negative: $(-\infty, 3)$ **22.** Positive: $(-6, 2)$; negative: $(-\infty, -6)\cup(2, \infty)$

23.

x	$f(x) = -\lvert x + 3\rvert + 4$
-8	-1
-7	0
-3	4
-1	2
0	1
1	0
3	-2

24.

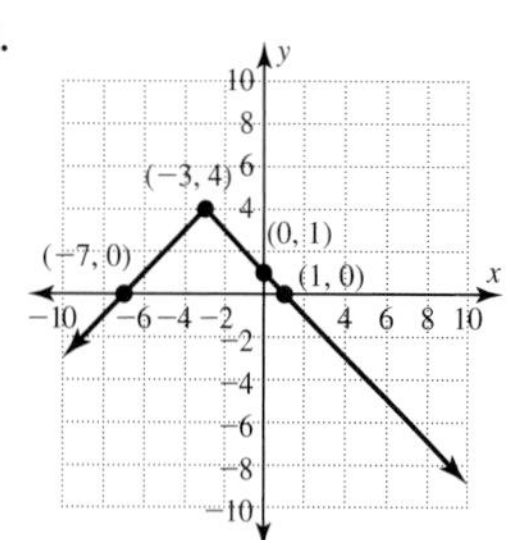

25. $(-7, 0);(1, 0)$ **26.** $(0, 1)$ **27.** $(-3, 4)$ **28.** 4 **29.** $\mathbb{R}$ **30.** $(-\infty, 4]$ **31.** $(-\infty, -3)$ **32.** $(-3, \infty)$ **33.** $(-7, 1)$ **34.** $(-\infty, -7)\cup(1, \infty)$ **35.** $y = 2.50x + 3.86$ **36.** $y = -3.19x + 4.49$ **37. a.** $y = 0.080x + 0.057$ **b.** For each 1 kg of mass added the spring stretches 0.08 cm. **c.** 1.0 cm **d.** 29.3 kg **38. a.** $(-2, 9)$ **b.** $(0, 5)$ **c.** $(-5, 0)(1, 0)$ **39. a.** $(2, -8)$ **b.** $(0, -6)$ **c.** $(-2, 0), (6, 0)$ **40. a.** $(4, 0)$ **b.** $(0, 4)$ **c.** $(4, 0)$ **41.** D **42.** A **43.** C **44.** B **45. a.** Downward **b.** $(2, 8)$ **c.** $(0, 6)$ **d.** $(-2, 0), (6, 0)$ **e.** $(-\infty, 2)$ **f.** $(2, \infty)$ **46. a.** Upward **b.** $(4, -6.25)$ **c.** $(0, -2.25)$ **d.** $(-1, 0), (9, 0)$ **e.** $(4, \infty)$ **f.** $(-\infty, 4)$ **47.** $x = 4, x = 3$ **48.** $x = \frac{5}{2}, x = -\frac{3}{4}$ **49.** $x = \pm 8$ **50.** $y = 2, y = 1$ **51.** $v = 5, v = 4$ **52.** $w = 2 \pm \sqrt{2}$ **53.** $m = \frac{5}{2}, m = -\frac{2}{5}$ **54.** $x = \frac{3 \pm 3\sqrt{3}}{2}$ **55.** $x = -10, x = 15$ **56.** $x = -11, x = -7$ **57. a.** $x = -5, x = 2$ **b.** $(-\infty, -5)\cup(2, \infty)$ **c.** $(-5, 2)$

58. $\left(-\frac{1}{3}, 0\right), (3, 0); -(3x + 1)(x - 3); -\frac{1}{3}, 3; x = -\frac{1}{3}, x = 3$ **59.** $(0, 0), (-2, 0), (1, 0); x(x + 2)(x - 1); 0, -2, 1; x = 0, x = -2, x = 1$ **60.** 8 cm, 15 cm, 17 cm **61.** 9.8 in **62. a.** 17.9 ft **b.** 18 ft **63.** 308 mi/h and 328 mi/h **64.** 4.24-cm diameter **65. a.** $(5, 40)$ **b.** \$306 at 22 units or 23 units **66. a.** 5.04 seconds **b.** 103 ft **c.** After 2.5 seconds **d.** Between 1 and 4 seconds **67.** 4.9% **68.** $y = 1.149x^2 + 2.339x + 4.548$ **69. a.** $y = 7.92x^2 + 45.3x - 28.0$ **b.** 94.3 cm **70. a.** $y = -0.224x^2 + 52.5x - 144$ **b.** \$260,000 **c.** 3 units and 231 units **d.** \$290,000 at 117 units; actual: \$300,000 at 100 units **71.** $5x^2 + 2x - 3 = 0$ **72.** $x^2 - 3 = 0$ **73.** $49x^2 - 70x + 25 = 0$ **74.** $10i$ **75.** $8 - 8i$ **76.** $2 - 4i$ **77.** $-2 - 36i$ **78.** $12 + 10i$ **79.** $51 - 27i$ **80.** $40 - 42i$ **81.** 29 **82.** i **83.** $-1 - i$ **84.** i **85.** $4 - 10i$ **86.** $2 - 3i$ **87. a.** i **b.** i **c.** 1 **88. a.** 0; a double real solution **b.** Positive; two distinct real solutions **c.** Negative; complex solutions with imaginary parts **89.** 0; a double real solution **90.** -12; complex solutions with imaginary parts **91.** 19; two distinct real solutions **92.** $z = \pm 7i$ **93.** $z = -\frac{1}{2} \pm \frac{3}{2}i$ **94.** $x = 3 \pm i$ **95.** $x = 3 \pm i\sqrt{2}$ **96.** $x^2 + 121 = 0$ **97.** $x^2 - 10x + 28 = 0$ **98.** $x = -1, x = -5$

Mastery Test for Chapter 7

1. a. Function **b.** Not a function **c.** Function **d.** Function **e.** Not a function **f.** Function **2. a.** Domain: $\{1, 2, 7, 8\}$; range: $\{4, 8, 11, 13\}$ **b.** Domain: $\mathbb{R}$; range: $\{7\}$ **c.** Domain: $\{-3, -2, -1, 0, 1\}$; range: $\{-2, -1, 0, 1, 2\}$ **d.** Domain: $[-2, \infty)$; range: $(-\infty, 2]$ **e.** Domain: $\{-3, -2, -1, 0\}$; range: $\{0, 1, 2, 7\}$ **f.** Domain: $[3, \infty)$; range: $[0, \infty)$ **3. a.** 2 **b.** -2 **c.** -3 **d.** 2 **e.** -18 **f.** 10

4. a.

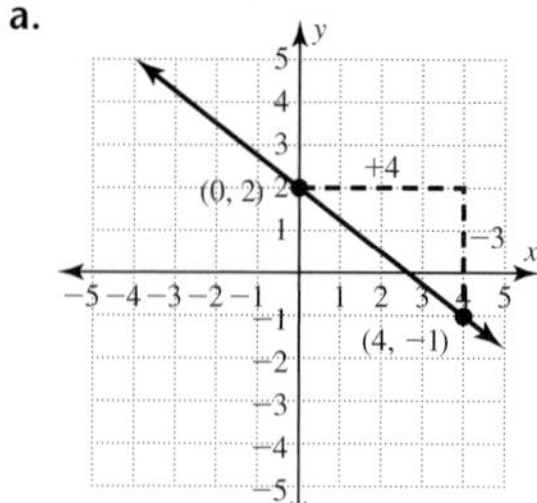

b.

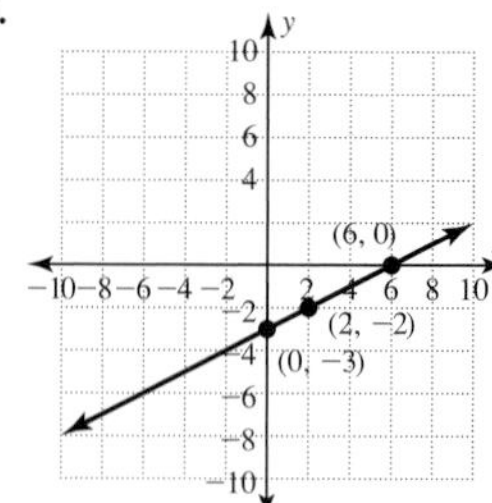

5. a. Decreasing **b.** Increasing **c.** Increasing **d.** Decreasing **6. a.** Positive: $\left(-\infty, \frac{4}{3}\right)$; negative: $\left(\frac{4}{3}, \infty\right)$ **b.** Positive: $(-2, 4)$; negative: $(-\infty, -2)\cup(4, \infty)$

7. a.

x	$f(x) = \lvert x + 1\rvert - 3$
-5	1
-3	-1
-1	-3
1	-1
3	1
5	3
7	5

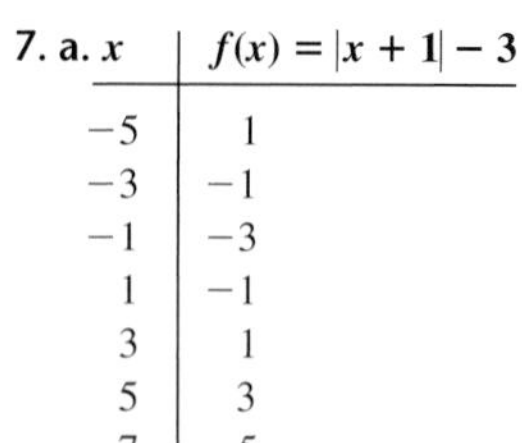

b.

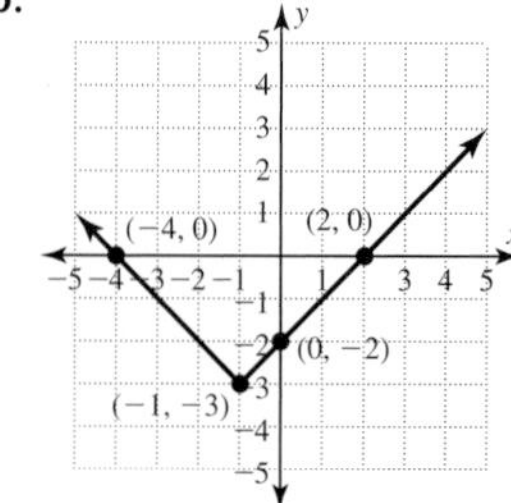

c. $(-4, 0); (2, 0)$ **d.** $(0, -2)$ **e.** $(-1, -3)$ **f.** -3 **g.** $\mathbb{R}$ **h.** $[-3, \infty)$ **i.** $(-\infty, -1)$ **j.** $(-1, \infty)$ **k.** $(-\infty, -4)\cup(2, \infty)$ **l.** $(-4, 2)$ **8. a.** Vertex: $(-1, 9)$; y-intercept: $(0, 8)$; x-intercepts: $(-4, 0), (2, 0)$ **b.** Vertex: $(2, 4)$; y-intercept: $(0, 3)$; x-intercepts: $(-2, 0), (6, 0)$ **c.** Opens upward; vertex: $(-2, -25)$; y-intercept: $(0, -21)$; x-intercepts: $(-7, 0), (3, 0)$ **d.** Opens downward; vertex: $(1, 16)$; y-intercept: $(0, 15)$; x-intercepts: $(-3, 0), (5, 0)$ **9. a.** $y = 0.007x + 1.364$ **b.** The women's high jump record is increasing by approximately 0.7 cm/yr. (This is approximately 2.8 cm every 4 years or every Olympics.) **c.** 1.87 m, this is 5 cm less than the actual record in 1972. **d.** 2.09 m; this is 3 cm more than the actual record in 2004. **10.** $f(x) \approx 0.210x^2 - 0.497x - 3.49$; $f(7) \approx 3.32$ **11. a.** $x = \frac{5}{2}, x = \frac{2}{3}$

b. $w = -1 \pm \sqrt{3}$ **c.** $v = \frac{7}{2}$ **d.** $x = 2 \pm \sqrt{5}$ **12. a.** 5, distinct real solutions **b.** 0, a double real solution **c.** 16, distinct real solutions **d.** -3, complex solutions with imaginary parts **13. a.** $(2, -25)$ **b.** $(-0.25, -15.125)$ **c.** $(2, 4)$ **d.** $(-2.6, -57.8)$ **14.** 40 ft; 1.5 seconds **15. a.** $9x^2 - 61x - 14 = 0$ **b.** $x^2 - 7 = 0$ **c.** $x^2 - 4x + 13 = 0$ **d.** $x^3 - 2x^2 + 9x - 18 = 0$ **16. a.** 10 in (diameter) **b.** 7 cm and 12 cm **c.** (5, 185); \$8,100 **d.** 156 mi/h and 256 mi/h **17. a.** $9i$ **b.** $3i$ **c.** $-5 + 4i$ **d.** $-1 - i$ **18. a.** $-1 + 2i$ **b.** $-12 - 26i$ **c.** $\frac{7}{5} + \frac{3}{10}i$ **d.** $8 - 6i$ **19. a.** $y = 7, y = -5$ **b.** $y = 1 \pm 6i$ **c.** $x = 5 \pm 2i$ **d.** $x = 3 \pm 5i$

Reality Check for End of Chapter 7

Without changing the water pressure, aiming the hose at different angles will cause the water to travel different horizontal distances. Aiming the nozzle at a 45° angle will allow the water to travel the greatest horizontal distance.

Chapter 8

Reality Check for Chapter 8 Opener

The gravitational attraction for objects on the Moon is about $\frac{1}{6}$ that on Earth. Thus less energy would be required to overcome the gravitational attraction for a lunar launch.

Using the Language and Symbolism of Mathematics 8.1

1. Ratio **2.** Undefined **3.** Rational **4.** Denominator; 0 **5.** Asymptote **6.** Asymptote **7.** Asymptote **8.** Lowest terms **9.** $\frac{A}{B}$ **10.** -1

Quick Review 8.1

1. $2^3 \cdot 7$ **2.** $\frac{2}{5}$ **3.** $-x^2(3x - 4)$ **4.** $x(x - 10)(x + 4)$ **5.** $(2x + 5y)(3x - 7y)$

Exercises 8.1

1. a. -1 **b.** 3 **c.** Undefined **3. a.** 4 **b.** $\mathbb{R} \sim \{4\}$ **c.** $x = 4$ **5.** C **7.** D **9.** D **11.** C **13. a.** $\mathbb{R} \sim \left\{\frac{1}{2}\right\}$ **b.** $\mathbb{R} \sim \{-9, 9\}$ **15.** $\mathbb{R} \sim \{1.5\}$ **17.** $\mathbb{R} \sim \{-3, 2\}$ **19.** D **21.** B **23.** C **25.** D **27.** $\frac{2b^2}{3a}$ **29.** $6x - 9$ **31.** $\frac{1}{2x - 3}$ **33.** $-\frac{a}{b}$ **35.** -1 **37.** $-\frac{a}{b}$ **39.** $\frac{x - 2y}{x + 2y}$ **41.** $\frac{x - y}{3}$ **43.** $-\frac{5x + 2}{7}$ **45.** $\frac{x}{x - y}$ **47.** $\frac{2x + 3y}{7}$ **49.** $\frac{2x + 3}{5y + 6}$ **51.** -1 **53.** $\frac{2a + b}{a + b}$ **55.** $\frac{x + y}{v + w}$ **57.** $\frac{x - 5}{3x - 1}$ **59.** $\frac{a + b}{a - b}$ **61.** $\frac{b - a - 1}{5a}$ **63.** -1 **65.** $\frac{3(x + y)}{4(x - y)}$ **67.** $-\frac{2x - y}{y}$ **69.** $\frac{(b^2 + 1)(b - 1)}{5}$ **71. a.** $T(6) = 3$; when paddling at 6 mi/h, it takes 3 h to go 12 mi upstream. **b.** $T(8) = 2$; when paddling at 8 mi/h, it takes 2 h to go 12 mi upstream. **c.** $T(2)$ is undefined; paddling at 2 mi/h would only match the speed of the current and would not make progress upstream. **73. a.** The average cost increases dramatically without bound. **b.** The average cost decreases and approaches \$4 per meter. **75.** 14 **77.** $14x$ **79.** $16y - 14x$ **81.** $10a - 10b$ **83.** $5x - 5$ **85.** $x^2 + 4xy + 3y^2$

Using the Language and Symbolism of Mathematics 8.2

1. $\frac{AC}{BD}$ **2. a.** Fraction **b.** Nonzero **3.** $\frac{AD}{BC}$ **4.** Reciprocal **5.** Both

Quick Review 8.2

1. $\frac{16}{15}$ **2.** $\frac{20}{3}$ **3.** $\frac{125}{48}$ **4.** $\frac{2}{2}$ **5.** $\frac{7}{7}$

Exercises 8.2

1. $\frac{11}{9}$ **3.** $\frac{11}{21}$ **5.** $\frac{1}{3m}$ **7.** $\frac{9y^2}{7x^6}$ **9.** $-\frac{1}{2}$ **11.** $\frac{x - 1}{x(x + 1)}$ **13.** $\frac{x - 3}{2(x + 3)}$ **15.** $\frac{7(x - 3)}{3(x - y)}$ **17.** $\frac{3(x + 2)}{x - 1}$ **19.** $\frac{a - 7}{y - c}$ **21.** $-\frac{2(x - 1)}{3}$ **23.** $\frac{6(x - y)}{5}$ **25.** $-\frac{4x}{3}$ **27.** $(x + 5)(x - 5)$ **29.** $-\frac{1}{7x(3x - 5)}$ **31.** 2 **33.** $\frac{11x}{9a}$ **35.** $\frac{7xy}{15}$ **37.** $\frac{5}{x - 3}$ **39.** $-\frac{2}{ab^2}$ **41.** $\frac{3}{11}$ **43.** $\frac{7y^2}{2a}$ **45.** $\frac{x + y}{12y}$ **47.** 4; 3.99 **49.** 3; 2.99 **51.** D **53.** B **55.** 5 **57.** xy^2 **59.** $x - 8$ **61.** $\frac{x - 3}{x - 3}$ **63.** $\frac{x - 2}{x - 1}$ **65. a.** $A(t) = \frac{100(t + 1)}{t(t - 1)}$

b.

t	2	6	10	14	18	22
$A(t)$	150	23.33	12.22	8.24	6.21	4.98

c. The average cost increases without bound. **d.** The average cost decreases to approximately \$4.53 per lamp.

Using the Language and Symbolism of Mathematics 8.3

1. $\frac{A + B}{C}$; $\frac{A - B}{C}$ **2.** Add **3.** Subtract **4. a.** Denominator; exponential **b.** Highest **c.** Multiplying **5. a.** LCD **b.** Denominator **c.** Numerators **d.** Lowest terms

Quick Review 8.3

1. $\frac{5}{6}$ **2.** $\frac{1}{2}$ **3.** $\frac{7}{12}$ **4.** $\frac{7}{16}$ **5.** $\frac{7}{12}$

Exercises 8.3

1. $\frac{2b + 3}{b^2}$ **3.** -1 **5.** $\frac{4}{s - 3}$ **7.** $\frac{1}{x + y}$

9. $\frac{1}{15xy^2}\left(\frac{9x}{9x}\right) + \frac{2}{27x^2y}\left(\frac{5y}{5y}\right) = \frac{9x}{135x^2y^2} + \frac{10y}{135x^2y^2} = \frac{9x + 10y}{135x^2y^2}$

11. $\frac{a + b}{21(3a - b)}\left(\frac{2}{2}\right) - \frac{a - b}{14(3a - b)}\left(\frac{3}{3}\right) = \frac{2a + 2b}{42(3a - b)} - \frac{3a - 3b}{42(3a - b)} = \frac{-a + 5b}{42(3a - b)}$

13. $\frac{2}{(x - 3)(x - 15)}\left(\frac{x + 3}{x + 3}\right) + \frac{1}{(x - 3)(x + 3)}\left(\frac{x - 15}{x - 15}\right) = \frac{2x + 6}{(x - 3)(x - 15)(x + 3)} + \frac{x - 15}{(x - 3)(x + 3)(x - 15)} = \frac{3x - 9}{(x - 3)(x - 15)(x + 3)} = \frac{3}{(x - 15)(x + 3)}$

15. $-\frac{13}{18w}$ **17.** $\frac{5}{14}$ **19.** $\frac{b^2 + 4b + 16}{b(b + 4)}$ **21.** $\frac{5x - 1}{x}$ **23.** $\frac{x^2 - 2x + 3}{x^3}$ **25.** $\frac{x + 13}{(x - 2)(x + 3)}$ **27.** $\frac{2x}{(x - 1)(x + 1)}$ **29.** $\frac{8x + 22}{(x - 4)(x + 2)}$ **31.** $\frac{2x^2 + 2x + 5}{(x + 2)(x - 1)}$ **33.** $\frac{1}{77}$ **35.** $\frac{5m + 9}{(m + 1)(m - 2)(m + 3)}$ **37.** $\frac{4m - 13}{(m - 3)(m - 4)}$ **39.** $\frac{2}{a - 4b}$ **41.** $\frac{3x + 2}{2(x - 1)}$ cm^2 **43. a.** $\frac{8(2t + 5)}{t(t + 5)}$ **b.** If the time required to paint the house becomes very large, the fractional portion of the job completed in 8 h approaches 0. **45. a.** $\frac{255r + 900}{r(r + 10)}$

b. 5 h **47.** A **49.** D **51.** $\dfrac{-x^2+6x-3}{x^2-3x+4}$ **53.** $-\dfrac{3}{(x-1)(x+1)}$

55. $\dfrac{2}{x+5}$ **57.** $\dfrac{10}{s-2t}$ **59.** $\dfrac{1}{z+3}$ **61.** $\dfrac{1}{v-w}$ **63.** $\dfrac{3w-4}{(w-1)(w-2)}$

65. $\dfrac{1}{w-3}$ **67.** $-\dfrac{2m}{m-1}$ **69.** $\dfrac{2x+5}{(x+5)(2x-5)}$

Using the Language and Symbolism of Mathematics 8.4

1. a. Grouping symbols **b.** Exponentiations **c.** Multiplications; divisions **d.** Additions; subtractions **2.** Numerator; denominator **3.** Reciprocal **4.** Addition; subtraction

Quick Review 8.4

1. 180 **2.** $(x+y)(x-y)$ **3.** -11 **4.** 43 **5.** $\dfrac{1}{x}+\dfrac{3}{x^2}$

Exercises 8.4

1. a. x **b.** $\dfrac{x+10}{3}$ **3. a.** $\dfrac{5y}{6}$ **b.** $\dfrac{3(y-4)}{2}$ **5. a.** $\dfrac{v^2-6}{v(v-6)}$ **b.** $\dfrac{2(v^2-v+18)}{v(v-6)^2}$ **7. a.** $\dfrac{16}{15}$ **b.** $\dfrac{10}{3}$ **9.** $\dfrac{3}{2}$ **11.** -2 **13.** -2 **15.** v^2-1 **17.** $x-3$ **19.** $-\dfrac{4}{x^2}$ **21.** $\dfrac{12}{5x}$ **23.** $\dfrac{9y}{4}$ **25.** $\dfrac{2}{xy(w+z)}$ **27.** $-\dfrac{x(x-7)}{10(x+2)}$ **29.** $\dfrac{x}{2x+1}$ **31.** $\dfrac{v^2w^2}{v+w}$ **33.** $\dfrac{x-1}{6(x+1)}$ **35.** $\dfrac{3x^2}{x+5}$ **37.** $-\dfrac{4aw}{w^2+a^2}$ **39.** $\dfrac{v+3}{v+6}$ **41.** $\dfrac{3}{2}=1.5$; 1.494 **43.** $\dfrac{4}{5}=0.8$; 0.820 **45.** $\dfrac{4a-4b+1}{4ab}$ **47.** $\dfrac{y^2}{x^2+y^2}$ **49.** $\dfrac{w+v}{w-v}$ **51.** $-m^2n-mn^2$ **53.** $\dfrac{a+b}{ab}$ **55.** $\dfrac{1}{x-y}$ **57.** $\dfrac{a+b}{a^2b^2}$ **59.** $-\dfrac{x+y}{x}$ **61.** B **63.** D **65.** $x(x+5)$ cm^2

67. a. $A(t)=\dfrac{1{,}680}{t}$ for $t\neq 2$ **b.** $(0,2)\cup(2,168]$ **c.** The average cost decreases to \$10 per unit when the factory operates 168 h/wk.

69. a. $f=\dfrac{d_od_i}{d_o+d_i}$ **b.** $f\approx 0.488$ ft

Using the Language and Symbolism of Mathematics 8.5

1. Same **2.** Zero **3. a.** LCD **b.** Solve **c.** Extraneous **4.** Zero

Quick Review 8.5

1. $x=6$ **2.** $x=3$ **3.** $x=37$ **4.** $x=-4$ **5.** There is no solution.

Exercises 8.5

1. a. 2 and 3 **b.** $-\dfrac{3}{2}$ and $\dfrac{2}{3}$ **c.** -3, $-\dfrac{1}{3}$, and $\dfrac{1}{2}$ **3.** $x=3$ **5.** $x=0.5$ **7.** $z=2$ **9.** There is no solution. **11.** $p=-5$ **13.** $n=-16$ **15.** $k=1$ **17.** There is no solution. **19.** $y=5$ **21.** $x=3, x=23$ **23.** $t=\dfrac{2}{3}, t=2$ **25.** $z=0, z=\dfrac{2}{3}$ **27.** $v=\dfrac{13}{3}$ **29.** $m=2$ **31.** $x=-2, x=3$ **33.** $m=-\dfrac{2}{3}$ **35.** $z=\dfrac{11}{4}$ **37.** $n=\dfrac{2}{3}$ **39.** $x=-2, x=0, x=1$ **41. a.** $-\dfrac{2(p-2)}{(p-1)(p+1)}$ **b.** $p=2$ **43. a.** $-\dfrac{(x-5)(x+2)}{(x+1)(x-2)}$ **b.** $x=-2, x=5$ **45. a.** 0 **b.** $\mathbb{R}\sim\left\{-\dfrac{2}{3},\dfrac{1}{2}\right\}$ **47.** 4.12 ft **49.** $r_1=6\ \Omega$; $r_2=30\ \Omega$ **51.** $b=\dfrac{ad}{c}$ **53.** $b=\dfrac{ad+a+c}{c}$ **55.** $d=\dfrac{k}{I}$ **57.** $R=\dfrac{r_1r_2}{r_1+r_2}$ **59.** $B=\dfrac{2A-bh}{h}$ **61.** $z=\dfrac{xy}{x-y}$ **63. a.** $\mathbb{R}\sim\left\{-3,\dfrac{5}{2}\right\}$ **b.** There is no solution. **65.** $x=-3, x=0, x=2$ **67.** $z=2$ **69.** $z=2$

Using the Language and Symbolism of Mathematics 8.6

1. kb **2.** $\dfrac{k}{b}$ **3.** kbc **4.** Increase **5.** Decrease **6.** Decrease **7.** Increase **8.** Rate; time **9.** Principal; rate; time **10.** Rate; time

Quick Review 8.6

1. 4 hours **2.** 4 h 48 min **3.** 480 mi/h **4.** \$300 **5.** 8%

Exercises 8.6

1. a. $m=kn$ **b.** $m=\dfrac{k}{p}$ **c.** $m=\dfrac{kn}{p}$ **d.** $m=knp$ **3. a.** $v=k\sqrt{w}$ **b.** $v=\dfrac{k}{x^2}$ **c.** $v=\dfrac{k\sqrt{w}}{x^2}$ **d.** $v=kx^2\sqrt{w}$ **5.** $I=kV$ **7.** $w=\dfrac{k}{d^2}$ **9.** $R=kv^2$ **11.** $V=kr^2h$; $\left(k=\dfrac{1}{3}\pi\right)$ **13. a.** $y=30$ **b.** $y=19.2$ **15.** $a=10$ **17.** $a=22.5$ **19.** D **21.** A **23.** F **25.** C **27.** D **29. a.** 34.8 **b.** $N=34.8P$

c.

Population	Kilograms of newsprint
700,000	24,360,000
710,000	24,708,000
720,000	25,056,000
730,000	25,404,000

d. 25,056,000 kg **31. a.** 40 **b.** $t=\dfrac{40}{w}$

c.

Workers	Time (hours)
5	8.00
10	4.00
15	2.67
20	2.00

d. 2 hours **33.** 9 min **35.** 2 Ω **37.** 270 lb **39.** 5, 6 **41.** 7, 9 **43.** 6 **45.** $\dfrac{2}{3}$ or $\dfrac{3}{2}$ **47.** 5 **49.** 12 m and 4 m **51.** 60 Ω and 120 Ω **53.** 54° and 36° **55.** \$8,400 **57.** 20 vehicles **59.** 11 km/h **61.** 480 mi/h and 520 mi/h **63.** 6 hours **65.** 6 hours **67.** 75 hours **69.** 3% and 5% **71.** 8.5% and 7%

Review Exercises for Chapter 8

1. 9 **2.** Undefined **3.** $-\dfrac{1}{15}$ **4.** $\dfrac{1}{9}$ **5.** $\mathbb{R}\sim\{-2,3\}$ **6.** $\mathbb{R}\sim\{-5,3\}$ **7.** $\mathbb{R}\sim\left\{-5,\dfrac{1}{3}\right\}$ **8.** $\mathbb{R}$ **9.** $x=\dfrac{1}{2}$ **10.** $x=-6, x=6$ **11.** $x=0, x=9$ **12.** $\dfrac{3x}{y^2}$ **13.** $-\dfrac{3}{2}$ **14.** $\dfrac{3(x-1)}{5}$ **15.** $\dfrac{x-5}{2x-1}$ **16.** $\dfrac{c}{a+b}$ **17.** $-\dfrac{1}{2m}$ **18.** $\dfrac{5x+2y}{3x-y}$ **19.** $\dfrac{3x-4y}{4x-3y}$ **20.** $\dfrac{x+2y}{x-2y}$ **21.** $2x^2-5x-3$ **22.** $2y^2-5y-3$ **23.** D **24.** C **25.** A **26.** B **27.** $6x-4$ **28.** $\dfrac{4xy+2y^2}{x}$ **29.** $\dfrac{6t^2-t-2}{2}$ **30.** $\dfrac{v+6w}{v-w}$ **31.** $\dfrac{1}{2x+1}$ **32.** 0 **33.** $\dfrac{7v}{(v-1)(2v+1)(3v-2)}$ **34.** 3 **35.** $\dfrac{3w^2+17w+66}{(w-3)(w-2)(w+6)}$ **36.** $\dfrac{6v-1}{3v-2}$ **37.** $y-4$ **38.** $\dfrac{2}{3z+2}$ **39.** $\dfrac{a(2a+1)}{(a+1)^2}$ **40.** $\dfrac{x+1}{x+2}$ **41.** $\dfrac{v+3}{v-3}$ **42.** $-\dfrac{x-1}{x+1}$ **43.** $\dfrac{xy+1}{xy-1}$ **44.** $-\dfrac{w-4}{3w-4}$ **45.** $\dfrac{m+5}{m+4}$ **46.** 2 **47.** $\dfrac{1}{5y+1}$ **48.** 2 **49. a.** $-\dfrac{2(m+1)}{(m+2)(m+4)}$ **b.** $m=-1$ **50. a.** $\dfrac{-5x+16}{(x+1)(x-6)}$ **b.** $x=\dfrac{16}{5}$ **51. a.** $\dfrac{2x^2-x-12}{(x-3)(x-2)}$ **b.** $x=4$

52. a. $\frac{2}{x-1}$ **b.** There is no solution. **53.** $x = -1$ **54.** There is no solution.
55. $y = -\frac{8}{3}$ **56.** $y = 5$ **57.** $w = 1, w = 2$ **58.** $v = -1$
59. $\mathbb{R} \sim \left\{-4, \frac{1}{2}\right\}$ **60.** $\mathbb{R} \sim \left\{-2, -\frac{1}{2}, 5\right\}$ **61.** $b = -a - 1$
62. $r_2 = \frac{E - r_1 I}{I}$ **63.** 14; 14.10 **64.** 0; -0.01
65. a. $\frac{4(13x-1)(x+2)}{5x(x-2)}$ cm **b.** $\frac{x+2}{x}$ cm^2 **66.** 45 cm^2
67. $\frac{100(x+2)}{x(x-2)}$ cm^3 **68. a.** (0, 24] **b.** $A(t) = \frac{200(t+5)}{t(t+20)}$
c. $N(10) = 150$; the plant can produce 150 units in 10 hours.
d. $C(10) = 1{,}500$; it costs the plant \$1,500 to operate for 10 hours.
e. $A(10) = 10$; it costs an average of \$10 per unit when the plant operates for 10 hours. **f.** The average cost increases without bound. **g.** The average cost decreases to approximately \$5.49 per unit as the time approaches 24 hours. **69.** 6 hours; the constant of variation $k = 300$ is the distance in miles between St. Louis and Chicago. **70.** 4.5 in **71.** 180 lb **72.** -3 **73.** 25 and 10 **74.** 5 and 7 **75.** 4 Ω and 12 Ω **76.** 21 min **77.** 1,000 desks **78.** 8 cm × 14 cm **79.** 21 mi/h **80. a.** Directly **b.** Inversely **c.** Inversely **d.** Directly

Mastery Test for Chapter 8

1. a. $\mathbb{R} \sim \{6\}$ **b.** $\mathbb{R} \sim \{-6, 1\}$ **c.** $\mathbb{R} \sim \{-8, 8\}$ **d.** $\mathbb{R}$ **2. a.** $x = 6$
b. $x = 1, x = -6$ **c.** $x = -2$ **d.** $x = 3$ **3. a.** $\frac{1}{2x-3}$ **b.** $\frac{2}{x+3}$
c. $\frac{2x+3}{x+5}$ **d.** $-\frac{2x-3y}{a-2b}$ **4. a.** $\frac{5}{4}$ **b.** 1 **c.** $\frac{2}{3}$ **d.** $\frac{(x+2)(x+5)}{(x+3)(x-5)}$
5. a. 2 **b.** $\frac{1}{x-3}$ **c.** $\frac{9}{(w-5)(w+4)}$ **d.** $\frac{4}{x-y}$ **6. a.** x **b.** $\frac{x+12}{5}$
c. 1 **d.** $\frac{w+1}{w-3}$ **7. a.** $\frac{24}{5}$ **b.** $\frac{x-3}{3}$ **c.** $\frac{3(2x+3)}{4x-3}$ **d.** $-\frac{2(v-11)}{(v-4)^2}$
8. a. $z = 5$ **b.** $z = 2, z = 4$ **c.** There is no solution. **d.** $x = \frac{5y+4}{y-1}$
9. a. $y = 18$ **b.** 8 N/cm^2 **c.** $z = 40$ **d.** 1,215 lb **11. a.** 10 **b.** 10 hours **c.** 8.5%, 6% **d.** 475 mi/h, 525 mi/h

Reality Check for End of Chapter 8

784 N on the Earth
130 N on the Moon

Chapter 9

Reality Check for Chapter 9 Opener

A geosynchronous orbit keeps the satellite in constant communication with a fixed region on Earth.

Using the Language and Symbolism of Mathematics 9.1

1. Radicand **2.** Radical **3.** Index; order **4.** $\sqrt[n]{x}$ **5.** Cube **6.** Fourth **7.** Even; negative **8.** Plus; minus **9.** Real **10.** Zero **11.** Negative

Quick Review 9.1

1. 32 **2.** -18 **3.** $(-\infty, 3) \cup (3, \infty)$ or $\mathbb{R} \sim \{3\}$ **4.** $(-3, 4]$ **5.** $[-1, 3]$

Exercises 9.1

1. a. $\sqrt[3]{w}$ **b.** $\sqrt[4]{x}$ **c.** $\sqrt[7]{v}$ **3. a.** $\sqrt{4} = 2$ **b.** $\sqrt[3]{1{,}000} = 10$ **c.** $\sqrt[5]{32} = 2$ **5. a.** 4 **b.** 2 **c.** 0 **7. a.** 6 **b.** -6 **c.** 1 **9. a.** 2 **b.** -2 **c.** 0.8 **11. a.** 10 **b.** 100 **c.** 10 **13. a.** $\frac{1}{6}$ **b.** $\frac{4}{7}$ **c.** $-\frac{1}{2}$ **15. a.** 17 **b.** 13 **c.** 1 **17. a.** 19 **b.** 13 **c.** 2 **19. a.** 5 **b.** -25 **c.** -25 **21.** 8; 8.124 **23.** -1; -0.746 **25.** 13; 13.080 **27. a.** 5 **b.** 3 **c.** 2 **d.** 0 **29. a.** 2 **b.** 1 **c.** -2 **d.** 3 **31.** $x = 5, x = -7$ **33.** 13 cm **35.** 6 cm **37.** 0.5 sec **39.** 3.7 cm **41.** E **43.** B **45.** G **47.** A **49.** H; $\mathbb{R}$ **51.** B; $(-\infty, 2]$ **53.** C; $\mathbb{R}$ **55.** F; $\mathbb{R}$ **57.** D; $\mathbb{R}$ **59.** $\mathbb{R}$ **61.** $\mathbb{R}$ **63.** $[3, \infty)$ **65. a.** $C(0) = \$1{,}200$ **b.** $C(2) = \$2{,}683.28$ **c.** $C(5) = \$6{,}118.82$

Using the Language and Symbolism of Mathematics 9.2

1. Radicand; index **2.** $\sqrt[n]{x}\sqrt[n]{y}$ **3.** $\frac{\sqrt[n]{x}}{\sqrt[n]{y}}$ **4.** $|x|$ **5.** x

Quick Review 9.2

1. $9x - 6y$ **2.** $2x + 6y$ **3.** 17 **4.** 22 **5.** 5

Exercises 9.2

1. d **3.** 12 **5.** -6 **7.** $30\sqrt{2}$ **9.** $-4\sqrt{7}$ **11.** $-5\sqrt[3]{6}$ **13.** $10\sqrt[4]{13}$ **15.** $6\sqrt{5} - 4\sqrt{7}$ **17.** $-3\sqrt{7x}$ **19.** 0 **21.** $5\sqrt{3}$ **23.** $3\sqrt{7}$ **25.** $2\sqrt[3]{3}$ **27.** $2\sqrt[4]{3}$ **29.** $5\sqrt{7}$ **31.** $\sqrt{3}$ **33.** $-13\sqrt{2v}$ **35.** $-2\sqrt{7w}$ **37.** $-3\sqrt[3]{3}$ **39.** $-74t\sqrt[3]{3}$ **41.** $-2\sqrt[3]{5z^2}$ **43.** $\frac{5\sqrt{5}}{6}$ **45.** $\frac{9\sqrt{11x}}{35}$ **47.** $7.9\sqrt{2} + 0.8\sqrt{3}$ **49.** 0 **51. a.** $5x$ **b.** $5|x|$ **53. a.** $2x$ **b.** $2x$ **55. a.** $1{,}000x^3$ **b.** $1{,}000|x|^3$ **57. a.** x^6 **b.** x^6 **59.** D **61.** A **63. a.** 4 **b.** $4 > \sqrt[3]{63}$ **c.** 3.979 **65. a.** 2 **b.** $2 < \frac{1 + \sqrt{9.05}}{2}$ **c.** 2.004 **67. a.** -5 **b.** $-5 < \frac{-4 - \sqrt{35.97}}{2}$ **c.** -4.999 **69.** c **71.** 46.3 cm

Using the Language and Symbolism of Mathematics 9.3

1. a. $\sqrt[n]{xy}$ **b.** $\sqrt[n]{\frac{x}{y}}$ **2.** Order **3.** Even **4.** Conjugates **5.** Pair **6.** Rationalizing

Quick Review 9.3

1. $15x^2y - 10xy^2$ **2.** $14x^2 + 29xy - 15y^2$ **3.** $16x^2 - 25y^2$ **4.** $16x^2 + 40xy + 25y^2$ **5.** $5 + 7i$

Exercises 9.3

1. $2\sqrt{3}$ **3.** $10 - \sqrt{2}$ **5.** $8\sqrt{15}$ **7.** $84\sqrt{35}$ **9.** $\sqrt{6} + \sqrt{15}$ **11.** $30\sqrt{3} - 105\sqrt{7}$ **13.** 240 **15.** $4w$ **17.** $24z\sqrt{2}$ **19.** $6x - 15\sqrt{x}$ **21.** $2x^3\sqrt{3x}$ **23.** $x^2y\sqrt[3]{xy^2}$ **25.** $3x + 3\sqrt{3xy} - 4y$ **27.** $14 - 4\sqrt{6}$ **29.** $a + 10\sqrt{3ab} + 75b$ **31.** $v - 11$ **33.** 7 **35.** $-3v$ **37.** $-4 - \sqrt[3]{20}$ **39.** $y\sqrt[5]{x^2y}$ **41.** $10\sqrt{3} + 7\sqrt{2} - 6$ **43.** 2 **45.** $x^2 - 3y$ **47.** $v^2 - 3v + 1$ **49.** $\frac{\sqrt{3}}{2}$ **51.** $\frac{\sqrt{6}}{3}$ **53.** $\frac{\sqrt{10}}{4}$ **55.** $\sqrt{5}$ **57.** $3\sqrt{6}$ **59.** $\frac{13\sqrt{10}}{5}$ **61.** $\frac{5\sqrt{3x}}{x}$ **63.** $\frac{\sqrt{7} - 1}{2}$ **65.** $-15 - 3\sqrt{13}$ **67.** $-3\sqrt{7} - 3\sqrt{2}$ **69.** $\frac{a + \sqrt{ab}}{a - b}$ **71.** $4\sqrt[3]{9}$ **73.** $\frac{\sqrt[3]{6}}{2}$ **75.** $\frac{\sqrt[3]{3vw}}{3w}$ **77.** $\frac{5(\sqrt{3x} - \sqrt{2y})}{3x - 2y}$ **79.** Solution

81. $\left(-\frac{1}{2} + \frac{\sqrt{3}}{2}i\right)^3 = \left(-\frac{1}{2} + \frac{\sqrt{3}}{2}i\right)\left(-\frac{1}{2} + \frac{\sqrt{3}}{2}i\right)\left(-\frac{1}{2} + \frac{\sqrt{3}}{2}i\right)$
$= \left(\frac{1}{4} - \frac{\sqrt{3}}{2}i + \frac{3}{4}i^2\right)\left(-\frac{1}{2} + \frac{\sqrt{3}}{2}i\right) = \left(-\frac{1}{2} - \frac{\sqrt{3}}{2}i\right)\left(-\frac{1}{2} + \frac{\sqrt{3}}{2}i\right)$
$= \frac{1}{4} - \frac{3}{4}i^2 = \frac{1}{4} + \frac{3}{4} = 1$

83. 20; $20 < \frac{100}{\sqrt{24}}$; 20.412 **85.** 10; $10 > \frac{50}{\sqrt{37} - \sqrt{0.98}}$; 9.818

Using the Language and Symbolism of Mathematics 9.4

1. Radicand **2.** $x^n = y^n$ **3.** Extraneous **4.** Legs, hypotenuse **5.** $a^2 + b^2 = c^2$ **6.** $\sqrt{(x_2 - x_1)^2 + (y_2 - y_1)^2}$

Quick Review 9.4

1. $x = 13$ **2.** $x = \frac{9}{5}$; $x = -\frac{7}{5}$ **3.** Yes **4.** No **5.** Yes

Exercises 9.4

1. $x = 5$ **3.** $x = -1$ **5.** $x = 10$ **7.** $x = 6$ **9.** $t = 13$ **11.** $c = 393$ **13.** No solution **15.** $w = -\frac{9}{2}$ **17.** $v = 3$ **19.** $w = -4, w = 8$

21. $y = 4$ **23.** $t = 2$ **25.** $w = \frac{1}{3}$ **27.** $x = 4$ **29.** $u = 0$ **31.** x-intercept: $(-5, 0)$ y-intercept: $(0, 1)$ **33.** x-intercept: $(-9, 0)$ y-intercept: $(0, 3)$ **35.** $c = 13$ cm **37.** $b = 9$ cm **39.** $c = \sqrt{113}$ cm **41.** 5 **43.** 17 **45.** $\sqrt{2}$ **47.** 3 **49.** $7\sqrt{2} + \sqrt{58}$; right triangle **51.** $2\sqrt{10} + 5\sqrt{2} + \sqrt{106}$; not a right triangle **53.** 42.5 m **55.** $(5, -4)$ and $(5, 8)$ **57.** $x = 4$ **59.** $x = 9$ **61.** $x = 1 + \sqrt{7}$ **63.** $x = \frac{-1 + \sqrt{5}}{2}$ **65.** $x = 0, x = 2$ **67.** No solution **69.** \$450 **71.** 64 **73.** $S(0) = 0$; the strength of a beam with no volume is 0 N. **75.** $V = 1$; a square box beam with a strength of 750 N has a volume of 1 cm^3. **77. a.** $f(x) = \sqrt{289 - x^2}$ **b.** 15 ft **79. a.** $L(x) = \sqrt{x^2 + 121}$ **b.** $L(30) \approx 32$, when the horse is 30 ft from the pole, the length of the rope is 32 ft. **c.** 3,468 ft^2 **81.** $x = 3$ mi

Using the Language and Symbolism of Mathematics 9.5

1. $\sqrt[5]{x}$ **2.** $x^{1/7}$ **3.** $\sqrt[5]{x^2}$ **4.** $x^{3/4}$ **5.** $x^{1/n}$ **6.** Negative **7.** Radicand

Quick Review 9.5

1. $\frac{25}{16}$ **2.** x^{18} **3.** x^{77} **4.** x^4 **5.** x^{16}

Exercises 9.5

1. a. $w^{1/3}$ **b.** $x^{1/4}$ **c.** $v^{1/7}$ **3. a.** $\sqrt{100} = 100^{1/2} = 10$ **b.** $\sqrt[3]{125} = 125^{1/3} = 5$ **c.** $\sqrt[5]{-32} = (-32)^{1/5} = -2$ **5. a.** $16^{1/2} = 4$ **b.** $16^{1/4} = 2$ **c.** $0^{1/5} = 0$ **7. a.** $(-8)^{1/3} = -2$ **b.** $\left(-\frac{1}{32}\right)^{1/5} = -\frac{1}{2}$ **c.** $0.01^{1/2} = 0.1$ **9. a.** $\sqrt{36} = 6$ **b.** $\frac{1}{\sqrt{36}} = \frac{1}{6}$ **c.** $-\sqrt{36} = -6$ **11. a.** $\sqrt[3]{27} = 3$ **b.** $\sqrt[3]{-27} = -3$ **c.** $\frac{1}{\sqrt[3]{27}} = \frac{1}{3}$ **13. a.** $\sqrt{0.09} = 0.3$ **b.** $-\sqrt{0.09} = -0.3$ **c.** $\frac{1}{\sqrt{400}} = \frac{1}{20}$ **15. a.** $\sqrt[3]{0.008} = 0.2$ **b.** $\sqrt[3]{-0.008} = -0.2$ **c.** $\frac{1}{\sqrt[3]{0.008}} = \frac{1}{0.2} = 5$ **17. a.** $\sqrt[3]{64} = 4$ **b.** $\frac{1}{\sqrt[3]{64}} = \frac{1}{4}$ **c.** $-\sqrt[3]{64} = -4$ **19. a.** $\left(\sqrt[3]{\frac{8}{125}}\right)^2 = \frac{4}{25}$ **b.** $\left(\sqrt{\frac{9}{4}}\right)^3 = \frac{27}{8}$ **c.** $\left(\sqrt[3]{\frac{125}{8}}\right)^2 = \frac{25}{4}$ **21. a.** $\sqrt{25} + \sqrt{144} = 17$ **b.** $\sqrt{25 + 144} = 13$ **c.** $\sqrt{25 - 16} = 3$ **23. a.** $\sqrt[3]{8} + \sqrt[3]{1} = 3$ **b.** $\sqrt[3]{26 + 1} = 3$ **c.** $\sqrt[3]{1001 - 1} = 10$ **25. a.** $\sqrt{0.000001} = 0.001$ **b.** $\sqrt[3]{0.000001} = 0.01$ **c.** $\sqrt[6]{0.000001} = 0.1$ **27. a.** 2.943 **b.** 3.267 **29. a.** 2.668 **b.** 4.601 **31.** 3; 3.107 **33.** 25 **35.** 4 **37.** 11 **39.** 9 **41.** $x^{5/6}$ **43.** $x^{1/6}$ **45.** $z^{3/14}$ **47.** w **49.** v^2w^3 **51.** $\frac{64}{v^{3/5}}$ **53.** $\frac{27}{8n}$ **55.** $\frac{9x}{5y}$ **57.** $x - 1$ **59.** $2y - 3$ **60.** $5y^2 - 6y$ **61.** $6w^2 - 15w - 27$ **63.** $a - 9$ **65.** $b^{6/5} - c^{10/3}$ **67.** $b^{6/5} - 2b^{3/5}c^{5/3} + c^{10/3}$ **69.** $\frac{1}{x} + 2 + x$ **71.** $S(1{,}000) = 75{,}000$; a square box beam with a volume of 1,000 cm^3 has a load strength of 75,000 N. **73.** 4

Review Exercises for Chapter 9

1. a. 7 **b.** 1 **c.** $\frac{24}{25}$ **2. a.** -2 **b.** -1 **c.** 0 **3. a.** 7 **b.** -7 **c.** 5

4.

$\sqrt[3]{-27}$ $\sqrt[3]{27}$ $\sqrt{27}$

-5 0 5

($\sqrt{-27}$ is not real and cannot be located on a number line.)

5. a. 8.37 **b.** 4.12 **c.** -2.34 **6.** $20\sqrt{3}$ **7.** $4\sqrt{3}$ **8.** $4\sqrt{2} - 3\sqrt{3}$ **9.** $4\sqrt[3]{5}$ **10.** 3 **11.** $12\sqrt[3]{25}$ **12.** $2\sqrt{5}$ **13.** $2\sqrt[3]{3}$ **14.** $4\sqrt{2}$ **15.** $14\sqrt{3x}$ **16.** $4\sqrt{2v}$ **17.** $-\sqrt[3]{v}$ **18.** 5 **19.** $\frac{3x}{y^2}$ **20.** $\frac{\sqrt[3]{25}}{3}$ **21.** $-29\sqrt{2}$ **22.** -369 **23.** -67 **24.** $90\sqrt{7}$ **25.** $\frac{\sqrt{7}}{5}$ **26.** $\frac{5\sqrt{7}}{7}$ **27.** $5 - 2\sqrt{6}$ **28.** $2\sqrt{6} - 5$ **29.** $3\sqrt{7} + 3\sqrt{3}$ **30.** 180 **31.** 3 **32.** $\sqrt{6}$ **33.** $4x^3\sqrt{2x}$ **34.** $2x^2\sqrt[3]{4x}$ **35.** $2xy^5\sqrt{3xy}$ **36. a.** C **b.** G **c.** D **d.** E **e.** B **f.** A **g.** F **37. a.** E **b.** D **c.** A **d.** C **e.** B **38.** $\mathbb{R}$; $(-\infty, 3]$ **39.** $\mathbb{R}$; $\mathbb{R}$ **40.** $(-\infty, 4]$; $[0, \infty)$ **41.** $\mathbb{R}$; $\mathbb{R}$ **42.** $\mathbb{R}$; $[-3, \infty)$ **43.** $\mathbb{R}$; $\{1\}$ **44.** $\mathbb{R}$ **45.** $\mathbb{R}$ **46.** $\mathbb{R}$ **47.** $\mathbb{R}$ **48.** $[5, \infty)$ **49.** $\mathbb{R}$ **50.** $\mathbb{R} \sim \{5\}$ **51.** $\mathbb{R} \sim \{-3, 3\}$ **52.** $x = 21$ **53.** $x = 4$ **54.** $v = 5$ **55.** $w = -8$ **56.** x-intercept: $(24, 0)$ y-intercept: $(0, -2)$ **57.** 13 **58.** $4\sqrt{5} + 2\sqrt{10}$; a right triangle **59.** 8 cm, 15 cm, 17 cm **60.** $(8, 7)$, $(8, 1)$ **61. a.** $x^{1/2}$ **b.** $x^{2/3}$ **c.** $x^{-1/3}$ **62. a.** $\sqrt[4]{x}$ **b.** $\frac{1}{\sqrt[4]{x}}$ **c.** $\sqrt[4]{x^3}$ **63. a.** $2\sqrt{w} = 2w^{1/2}$ **b.** $\sqrt{2w} = (2w)^{1/2}$ **c.** $\sqrt[3]{x + 4} = (x + 4)^{1/3}$ **64.** 7 **65.** -7 **66.** $\frac{1}{7}$ **67.** $\frac{9}{25}$ **68.** $\frac{9}{25}$ **69.** $\frac{25}{9}$ **70.** 1,000 **71.** 100 **72.** 10 **73.** 16 **74.** 4 **75.** 2 **76.** $\frac{4y^6}{x^4}$ **77.** $4x^{2/3}y^{3/5}$ **78.** $x^2y^2z^2$ **79.** $6x - 10$ **80.** $9x - 25$ **81.** $x + y$

Mastery Test for Chapter 9

1. a. $[2, \infty)$ **b.** $(-\infty, 3]$ **c.** $\mathbb{R}$ **d.** B **e.** E **f.** A **g.** C **h.** D **2. a.** $\mathbb{R}$ **b.** $\mathbb{R} \sim \{6\}$ **c.** $\mathbb{R}$ **d.** $[6, \infty)$ **3. a.** 9 **b.** 3 **c.** -0.2 **d.** -2 **e.** 17 **f.** 13 **g.** 6; 5.74 **h.** 2; 2.01 **i.** 4; 4.12 **j.** 2; 2.03 **4. a.** $5\sqrt{7}$ **b.** $3\sqrt{2} + 4\sqrt{5}$ **c.** $-\sqrt[3]{7}$ **d.** $8\sqrt{2x}$ **5. a.** $2\sqrt{10}$ **b.** $2\sqrt[3]{3}$ **c.** $-9\sqrt{7}$ **d.** $\frac{10}{3}$ **6. a.** 30 **b.** $30 - 20\sqrt{2}$ **c.** $-4x^2$ **d.** $12x - 5$ **7. a.** 3 **b.** $3\sqrt{6}$ **c.** $8\sqrt{7} + 8\sqrt{2}$ **d.** $-8 - 4\sqrt{7}$ **8. a.** $x = 124$ **b.** $x = -5$ **c.** No real solution **d.** $x = 8$ **9. a.** 8 **b.** 12 **c.** 10 **d.** $\sqrt{29}$ **10. a.** 9 **b.** -10 **c.** 27 **d.** $x^{1/7}$ **e.** 2 **f.** 6

Reality Check for End of Chapter 9

$r \approx 42{,}242{,}900$ m; distance above the Earth $\approx 35{,}864{,}800$ m

Chapter 10

Reality Check for Chapter 10 Opener

It would take 30 years for the investment to be worth \$4,000. It would take 40 years for the investment to be worth \$8,000. It takes 10 years for the investment to double in value.

Using the Language and Symbolism of Mathematics 10.1

1. Difference **2.** Ratio **3.** Growth **4.** Decay **5.** b^x; base; exponent **6.** Growth **7.** Decay **8.** x **9.** 2.718 **10.** x; y **11.** a; b

Quick Review 10.1

1. Sub **2.** $(-1, 1)$ **3.** $(1, 2)$ **4.** $(-1, 2)$ **5.** $(1, 3]$

Exercises 10.1

1. a. Geometric; $r = 5$ **b.** Geometric; $r = \frac{1}{5}$ **c.** Geometric; $r = -5$ **d.** Geometric; $r = -\frac{1}{5}$ **3. a.** Geometric; $r = \frac{2}{3}$ **b.** Not geometric **c.** Not geometric **d.** Geometric; $r = 2$ **5. a.** 10, 12, 14, 16, 18 **b.** 10, 8, 6, 4, 2 **c.** 10, 20, 40, 80, 160 **d.** 10, -20, 40, -80, 160 **7.** B **9.** D **11.** E **13. a.** 3, 6, 9, 12, 15 **b.** 3, 6, 12, 24, 48 **15. a.** 2, 2, 2, 2, 2 **b.** 2, 2, 2, 2, 2 **17. a.** 4 **b.** 64 **c.** 1 **d.** $\frac{1}{16}$ **e.** 2 **19.** 16; 15.21 **21.** $\frac{1}{4} = 0.25$; 0.26 **23. a.** 2,824.295 **b.** 0.102 **c.** 22.655 **25. a.** 16.919 **b.** 0.202 **c.** 23.141

27.

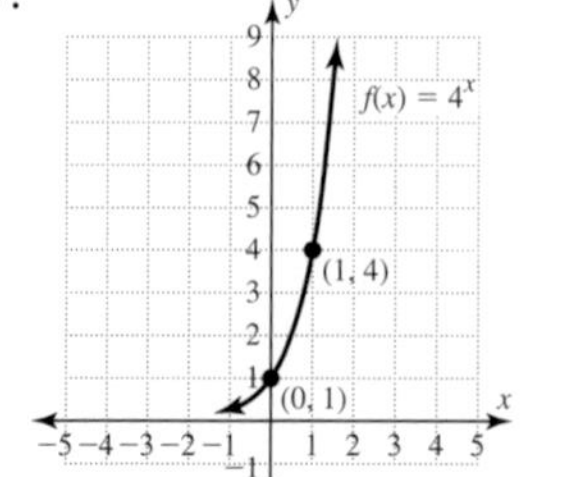

29.

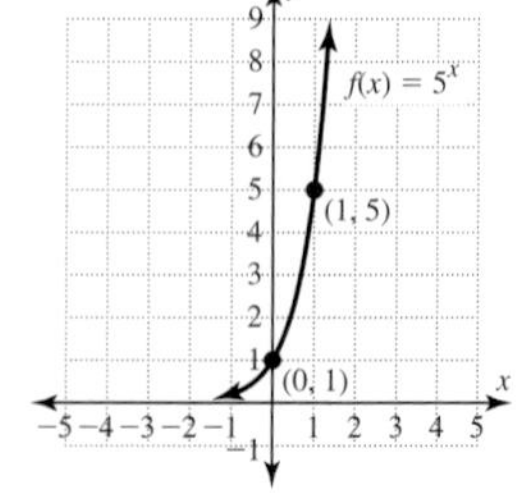

31. a. $x = 32$ **b.** $x = -8, x = 8$ **c.** $x = 6$ **33.** $x = 2$ **35.** $v = 2$ **37.** $w = 3$ **39.** $m = -1$ **41.** $n = 0$ **43.** $x = \frac{1}{2}$ **45.** $x = \frac{1}{2}$ **47.** $x = -2$ **49.** $w = -3$ **51.** $n = \frac{26}{15}$ **53.** $v = \frac{1}{2}$ **55.** $x = 4$ **57.** $y = -3$ **59.** $x = -5$ **61.** $x = \frac{1}{6}$ **63.** $x = -2$ **65. a.** \$110, \$120, \$130, \$140 **b.** \$90, \$80, \$70, \$60 **c.** \$110, \$121, \$133.10, \$146.41 **d.** \$90, \$81, \$72.90, \$65.61 **67.** 18 m, 9 m, 4.5 m, 2.25 m **69.** $A(t) = 1{,}000(1.06)^t$; \$1,503.63 **71.** 78.0 mg

Using the Language and Symbolism of Mathematics 10.2

1. y; x **2.** Different **3.** One-to-one **4.** f^{-1} **5.** $x = y$

Quick Review 10.2

1. Function **2.** Not a function **3.** Function

4.

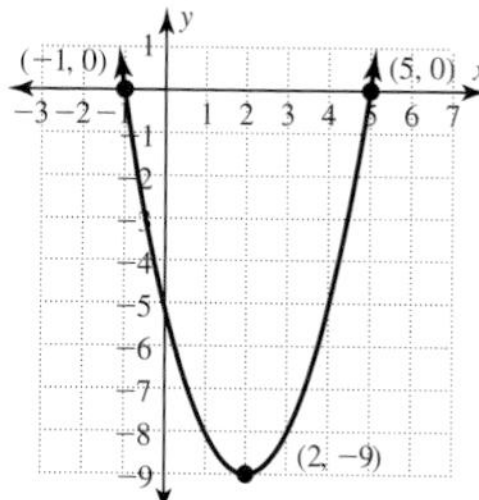

5.

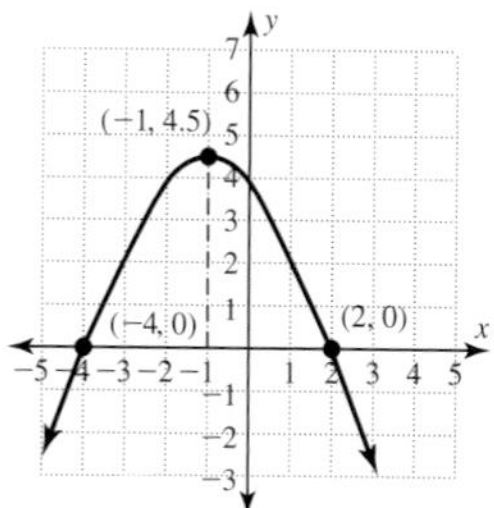

Exercises 10.2

1. $\{(4, 1), (11, 3), (2, 8)\}$ **3.** $\{(2, -3), (2, -1), (2, 0), (2, 2)\}$ **5.** $\{(b, a), (d, c)\}$ **7.** $\{(7, 4), (9, e), (-2, -3)\}$ **9.** $\{(0, 9), (3, 4), (-2, 6), (5, -4)\}$ **11.** $f = \{(-3, 2), (-1, 1), (1, 0), (3, -1)\}$; $f^{-1} = \{(2, -3), (1, -1), (0, 1), (-1, 3)\}$ **13.** Domain of f: $\{0, 1, 2, 3, 4\}$; range of f: $\{-3, -2, -1, 0, 1\}$; domain of f^{-1}: $\{-3, -2, -1, 0, 1\}$; range of f^{-1}: $\{0, 1, 2, 3, 4\}$ **15.** Domain of f: $\{-2, 0, 2\}$; range of f: $\{0, 3, 1\}$; domain of f^{-1}: $\{0, 3, 1\}$; range of f^{-1}: $\{-2, 0, 2\}$ **17.** One-to-one **19.** Not one-to-one **21.** One-to-one **23.** Not one-to-one **25.** Not one-to-one **27.** Not one-to-one **29.** Not one-to-one

31.

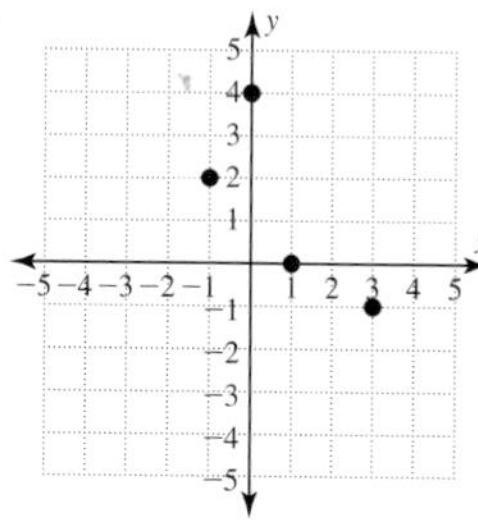

33.

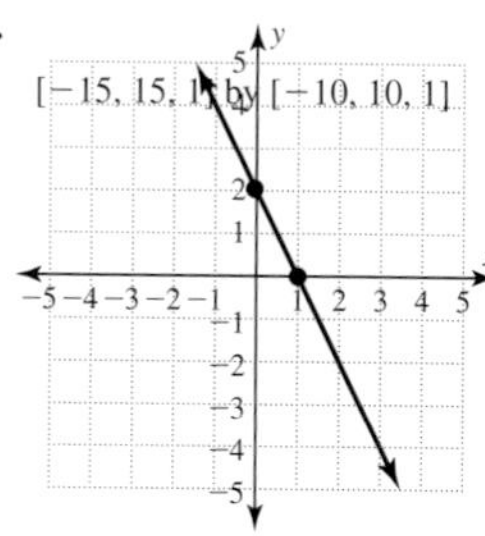

35.

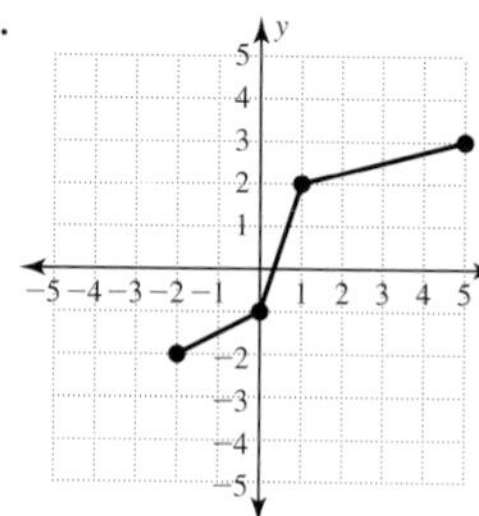

37.

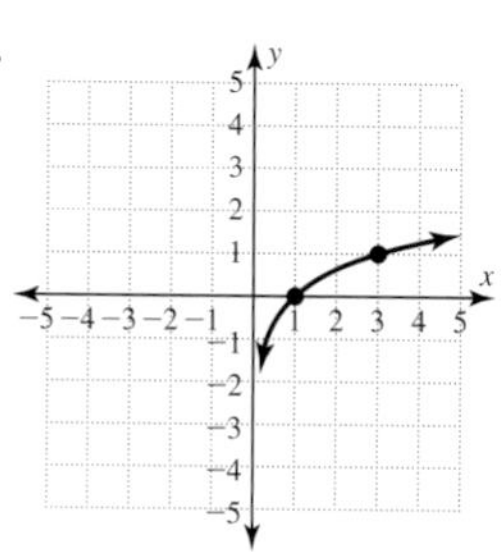

39.

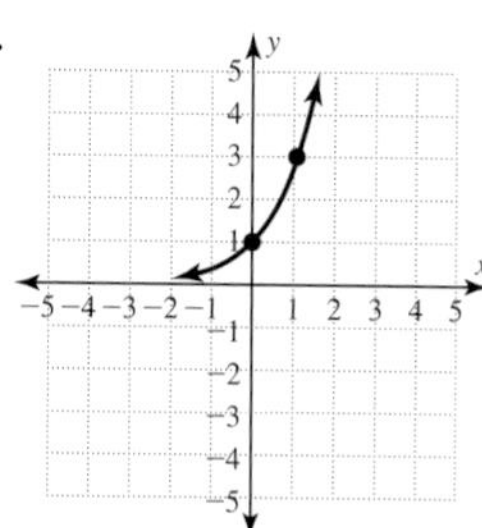

41.

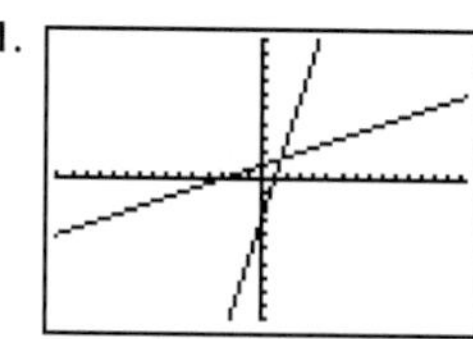

43.

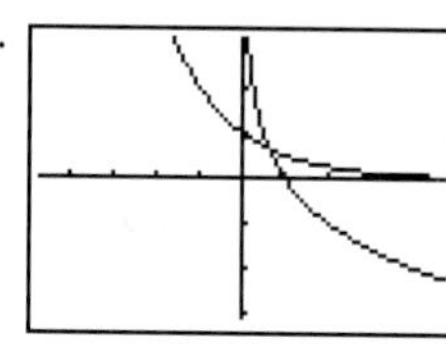

[−4.7, 4.7, 1] by [−3.1, 3.1, 1]

45. $f^{-1}(x)$ increases x by 2; $f^{-1}(x) = x + 2$ **47.** $f^{-1}(x)$ takes one-fourth of x; $f^{-1}(x) = \frac{x}{4}$ **49.** $f^{-1}(x)$ increases x by 3 and then divides this quantity by 2; $f^{-1}(x) = \frac{x + 3}{2}$ **51.** $f^{-1}(x) = \frac{x - 2}{5}$ **53.** $f^{-1}(x) = 3(x + 7)$ **55.** $f^{-1}(x) = (x - 2)^3$ **57.** $f^{-1}(x) = -x$ **59. a.** 17 **b.** 5 **61. a.** 5 **b.** 2 **63. a.** 1 **b.** 1

65.

European Size	U.S. Size
39	7
40.5	8
42	9
43	10
44	11

The inverse function is useful in converting a European shoe size to the equivalent U.S. size.

67.

Monthly Payment (\$)	Loan Amount (\$)
120.89	5,000
241.79	10,000
362.68	15,000
483.58	20,000
604.47	25,000

The inverse function is useful in determining the cost of the car one can afford on a given monthy budget.

69.

Celsius	Fahrenheit
−17.8°	0°
−6.7°	20°
4.4°	40°
15.6°	60°
26.7°	80°
37.8°	100°

The inverse function is useful in converting Celsius to Fahrenheit.

71. a. The number of units produced **b.** The cost of producing x units **c.** $C^{-1}(x) = \frac{x - 350}{12}$ **d.** The cost **e.** The number of units produced **f.** $C(100) = \$1{,}550$ **g.** $C^{-1}(1934) = 132$ units

Using the Language and Symbolism of Mathematics 10.3

1. $\log_b x$ **2.** b; x; y **3.** Undefined **4.** 0 **5.** 1 **6.** -1 **7.** x **8.** y

Quick Review 10.3

1. 1 **2.** Undefined **3.** $\frac{3}{2}$ **4.** $x = \frac{9}{2}$ **5.** $x = 3, x = -3$

Exercises 10.3

1. The log base 5 of 125 is 3; $5^3 = 125$ **3.** The log base 3 of $\sqrt{3}$ is $\frac{1}{2}$; $3^{1/2} = \sqrt{3}$ **5.** The log base 5 of $\frac{1}{5}$ is -1; $5^{-1} = \frac{1}{5}$ **7.** $\log_{16} 4 = \frac{1}{2}$; $16^{1/2} = 4$ **9.** $\log_8 4 = \frac{2}{3}$; the log base 8 of 4 is $\frac{2}{3}$ **11.** $\log_3 \frac{1}{9} = -2$; the log base 3 of $\frac{1}{9}$ is -2 **13.** $\log_m n = p$; the log base m of n is p **15.** $\log_n m = k$; $n^k = m$ **17.** 2 **19.** -3 **21.** $\frac{1}{3}$ **23.** 5 **25.** -1 **27.** 0 **29.** -1 **31.** 4 **33.** 6 **35.** 3 **37.** $\frac{1}{4}$ **39.** $\frac{3}{2}$ **41.** 1 **43.** 2.7 **45.** -2 **47. a.** Defined **b.** Undefined **49. a.** Undefined **b.** Defined **51.** $x = 2$ **53.** $x = 6$ **55.** $x = \frac{1}{6}$ **57.** $x = \sqrt{6}$ **59.** $x = 2$ **61.** $x = 5$ **63.** $x = \frac{1}{5}$ **65.** $x = -\frac{1}{2}$ **67.** $x = 4$ **69.** C **71.** B **73.** D: $(0, \infty)$; R: $\mathbb{R}$

75.

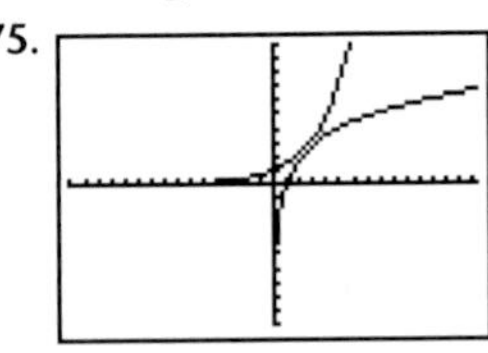

[−15, 15, 1] by [−10, 10, 1]

Using the Language and Symbolism of Mathematics 10.4

1. 2.718 **2.** $\log_{10} x$ **3.** $\log_e x$

Quick Review 10.4

1. 0 **2.** Undefined **3.** $\frac{1}{10{,}000}$ **4.** 7.5345247×10^7 **5.** 8.7×10^{-5}

Exercises 10.4

1. a. 2 **b.** 1.556 **c.** 3.584 **3. a.** -2 **b.** -0.954 **c.** -2.197 **5. a.** 0 **b.** 0 **c.** 0 **7. a.** 3 **b.** 1.303 **c.** 6.908 **9. a.** 5 **b.** -4 **c.** 9 **11. a.** 5 **b.** -5 **c.** $\frac{1}{2}$ **13. a.** 1.672 **b.** 46.989 **15. a.** 3.850 **b.** 46.993 **17. a.** 1 **b.** 2.303 **19. a.** 0.434 **b.** 1 **21. a.** 2.053 **b.** 112.980 **23. a.** -5.082 **b.** 0.006 **25. a.** 0.497 **b.** 1.145 **27. a.** -4.046 **b.** -9.316 **29. a.** -11.227 **b.** -25.851 **31. a.** 11.841 **b.** 27.264 **33. a.** 0.409 **b.** 0.108 **35. a.** 0.030 **b.** -0.349 **37. a.** 0.531 **b.** 1.760 **39. a.** 4.560 **b.** 4.443 **41. a.** 0.439 **b.** 0.845 **43. a.** -2.079 **b.** Undefined **45. a.** 0 **b.** Undefined **47. a.** -0.111 **b.** Undefined **49.** $x = 299.916$ **51.** $x = 0.500$ **53.** $x = 7.996$ **55.** $x = 0.050$ **57.** 2 **59.** 81

Using the Language and Symbolism of Mathematics 10.5

1. b^{x+y} **2.** $\log_b x + \log_b y$ **3.** b^{x-y} **4.** $\log_b x - \log_b y$ **5.** b^{xp} **6.** $p \log_b x$ **7.** $\frac{\log_b x}{\log_b a}$ **8.** $b^{x \log_b a}$ **9.** 7^x **10.** x **11.** 2; $5x - 1$

Quick Review 10.5

1. $15x^7y^9$ **2.** $\frac{3x}{y^5}$ **3.** $25x^6y^{14}$ **4.** $xy^{1/2}$ **5.** $x^{1/3}y^{2/3}$

Exercises 10.5

1. $\log a + \log b + \log c$ **3.** $\ln x - \ln 11$ **5.** $\log x + 5 \log y$ **7.** $2 \ln x + 3 \ln y + 4 \ln z$ **9.** $\ln(2x + 3) - \ln(x + 7)$ **11.** $\frac{1}{2}\log(4x + 7)$ **13.** $\frac{1}{2}\ln(x + 4) - 2\ln(y + 5)$ **15.** $\frac{1}{2}[\log x + \log y - \log(z - 8)]$ **17.** $2 \log x + 3 \log(2y + 3) - 4 \log z$ **19.** $\ln x + 2 \ln y - 3 \ln z$ **21.** $\log(x^2y^5)$ **23.** $\ln\left(\frac{x^3y^7}{z}\right)$ **25.** $\log\left(\frac{\sqrt{x+1}}{2x+3}\right)$ **27.** $\ln\sqrt[3]{(2x + 7)(7x + 1)}$ **29.** $\log_5 x^2\sqrt[3]{y^2}$ **31.** 34 **33.** 0.53 **35.** 37 **37.** 0.045 **39.** x **41.** 1.1833 **43.** 1.6542 **45.** 0.1187 **47.** 1.3562 **49.** -0.4709 **51.** 0.6438 **53.** C **55.** B **57.** F **59.** E **61.** A **63.** 2.2453 **65.** 0.7631 **67.** -1.2860 **69.** -1.8934 **71.** 3; 2.989 **73.** 32; 31.640 **75. a.** $f(x) = e^{x \ln 5}$ **b.** $f(x) = e^{-x \ln 5}$ **77.** $\ln\frac{1}{2} = \ln 1 - \ln 2 = 0 - \ln 2 = -\ln 2$ **79.** $\ln y = kt + c_1$ is equivalent to $y = e^{kt+c_1} = e^{kt}e^{c_1} = c_2e^{kt}$ (where $c_2 = e^{c_1}$)

Using the Language and Symbolism of Mathematics 10.6

1. Exponential **2.** Logarithmic **3.** x; y **4.** Extraneous

Quick Review 10.6

1. 0.982 **2.** 0.336 **3.** 29.824 **4.** $(5, -2)$ **5.** $(1, 2.5)$

Exercises 10.6

1. $w = 8$ **3.** $x = -2, x = 2$ **5.** $y = 15$ **7.** $n = 35$ **9.** $m = 16$ **11.** $w = -\frac{1}{2}$ **13.** $x = -2$ **15.** $t = 2$ **17.** $v = 10$ **19.** No solution **21.** $y = -7$ **23.** No solution **25.** $x = 2.5$ **27.** $x = 2$ **29.** $v = \frac{\ln 15}{\ln 4} \approx 1.953$ **31.** $w = 7 - \frac{\ln 22}{\ln 3} \approx 4.186$ **33.** $t = \frac{\ln 9.2}{\ln 11.3 - 2 \ln 9.2} \approx -1.102$ **35.** $z = \frac{\ln 5.3}{2 \ln 5.3 + 2 \ln 7.6} \approx 0.226$ **37.** $x = \frac{\ln 78.9}{3} \approx 1.456$ **39.** $y = \frac{\log 51.3 - 1}{2} \approx 0.355$ **41.** $v = \pm\sqrt{\frac{\ln 0.68}{\ln 0.83}} \approx \pm 1.439$ **43.** $x = \pm\sqrt{\ln\left(\frac{689.7}{3.7}\right) - 1} \approx \pm 2.056$ **45.** $x = 2$ **47.** $x = -1, x = 1$ **49.** $x = \frac{10^{0.83452} + 17}{5} \approx 4.766$ **51.** $y = e^e \approx 15.154$ **53.** $v = 3 + \sqrt{2} \approx 4.414$ **55.** No solution **57.** $x = 1, e^2 \approx 7.389$ **59.** $x = -0.005, x = 2.500$ **61.** $x = 2.998$ **63.** $10^{-\log x} = 10^{\log x^{-1}} = x^{-1} = \frac{1}{x}$ **65.** $e^{-x \ln 3} = e^{\ln 3^{-x}} = 3^{-x} = \frac{1}{3^x} = \left(\frac{1}{3}\right)^x$ **67.** $\log 60^x - \log 6^x = \log \frac{60^x}{6^x} = \log\left(\frac{60}{6}\right)^x = \log 10^x = x$ **69.** $\ln\left(\frac{4}{5}\right)^x + \ln\left(\frac{5}{3}\right)^x + \ln\left(\frac{3}{4}\right)^x = x \ln\frac{4}{5} + x \ln\frac{5}{3} + x \ln\frac{3}{4} = x\left(\ln\frac{4}{5} + \ln\frac{5}{3} + \ln\frac{3}{4}\right) = x \ln\left(\frac{4}{5}\right)\left(\frac{5}{3}\right)\left(\frac{3}{4}\right) = x \ln 1 = x(0) = 0$ **71.** 6.8 years

Using the Language and Symbolism of Mathematics 10.7

1. Periodic **2.** Continuous **3.** $P\left(1 + \frac{r}{n}\right)^{nt}$ **4.** Pe^{rt} **5.** Earthquakes **6.** Sound

Quick Review 10.7

1. $\log 10{,}000 = 4$ **2.** $10^{-3} = 0.001$ **3.** $e^{4.357} \approx 78$ **4.** 4.41 **5.** -3.59

Exercises 10.7

1. \$234.08 **3.** 9 years **5.** 18.4 years **7.** 8.7% **9.** 6.9% **11.** 9.9 years **13.** 24,000 years **15. a.** $B(0) = 4$; there are 4 units of bacteria present initially. **b.** $B(5) \approx 13.3$; after 5 days there are approximately 13.3 units of bacteria present. **c.** $B(10) \approx 44$; after 10 days there are approximately 44 units of bacteria present. **d.** $B(13.4) \approx 100$; the bacteria level reaches 100 units after approximately 13 days. **17.** 9,200 days **19.** There will be approximately 10,000 whales left in 10 years; it will take approximately 27.5 years for the whale population to reach 5,000. **21.** 13.5% **23.** 7.8 **25.** 7 times **27.** 24.8 decibels **29.** 3.2×10^{-8} W/cm^2 **31.** 3.96 **33.** 7.41×10^{-10} mol/L **35.** \$411.60 **37.** 36 payments

39. a.

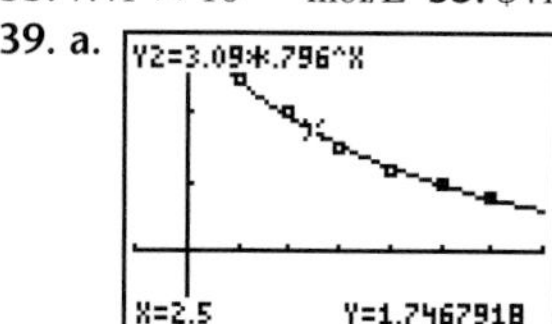

[−1, 7, 1] by [−1, 3, 1]

b. $y = 3.09(0.796^x)$ **c.** Exponential decay **d.** $y \approx 1.7$

41. a.

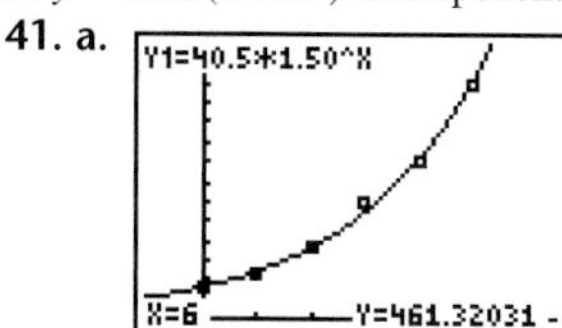

[−1, 6.5, 1] by [−5, 350, 25]

b. $f(x) = 40.5(1.50^x)$ **c.** In 2006 the storage requirements are estimated to be 460 GB.

Review Exercises for Chapter 10

1. a. Arithmetic; $d = 2$ **b.** Geometric; $r = 2$ **c.** Neither **2. a.** Geometric; $r = 1$; Arithmetic; $d = 0$ **b.** Arithmetic; $d = -4$ **c.** Geometric; $r = -1$

3. 100, 90, 81, 72.9, 65.61, and 59.049 gal **4. a.** 1 **b.** $\frac{1}{9}$ **c.** 81 **d.** 3

5. a. 0.318 **b.** 9.870 **c.** 36.462 **d.** 22.459 **6.** B **7.** C **8.** A

9. a.

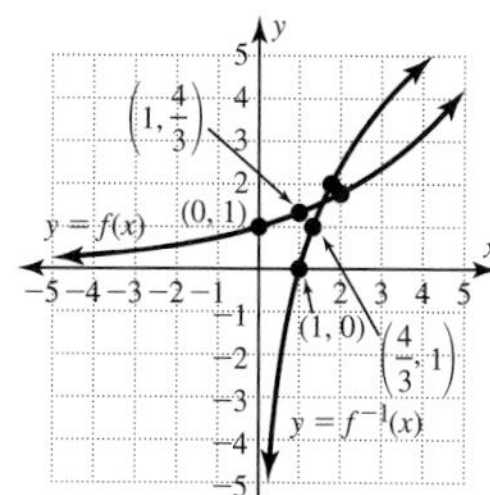

b.

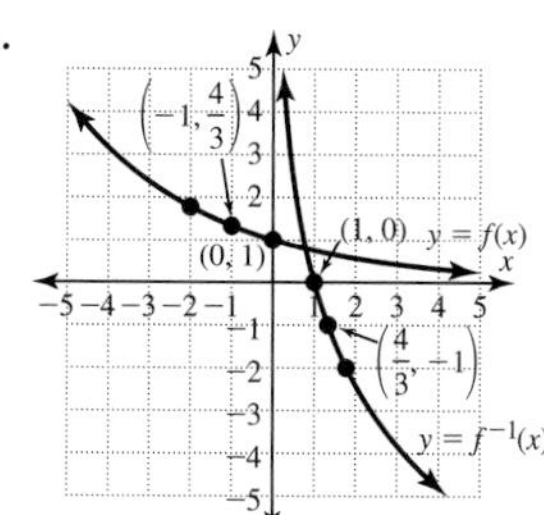

c.

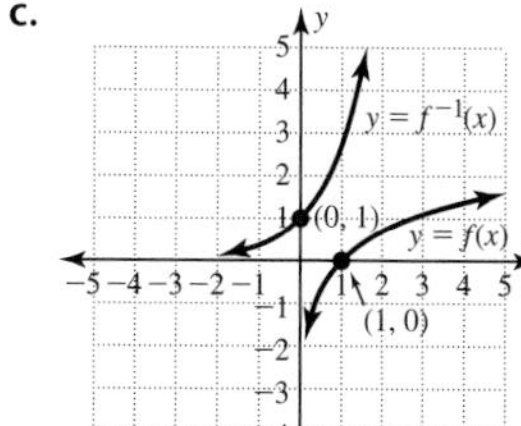

10. f^{-1} adds 2 and then divides the quantity by 3; $f^{-1}(x) = \dfrac{x+2}{3}$

11. f^{-1} subtracts 6 from x and then multiplies the quantity by 4; $f^{-1}(x) = 4(x - 6)$ **12.** $f^{-1} = \{(-5, -4), (-3, -3), (3, 0), (7, 2), (9, 3)\}$ **13.** $f^{-1} = \{(0, -2), (2, -1), (4, 0), (6, 1), (8, 2), (10, 3)\}$ **14.** $f^{-1}(x) = 3(x + 4)$ **15.** $f^{-1}(x) = \log_3 x$ **16. a.** x represents the number of pizzas. **b.** $C^{-1}(x) = \dfrac{x - 250}{2}$ **c.** The cost of producing the pizzas **d.** The number of pizzas **e.** \$450 **f.** 74 pizzas

17.

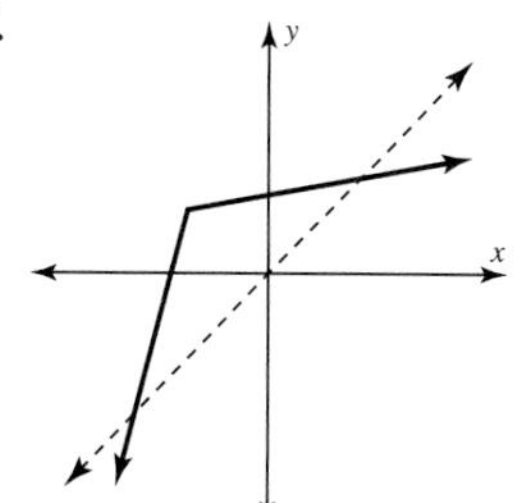

18.

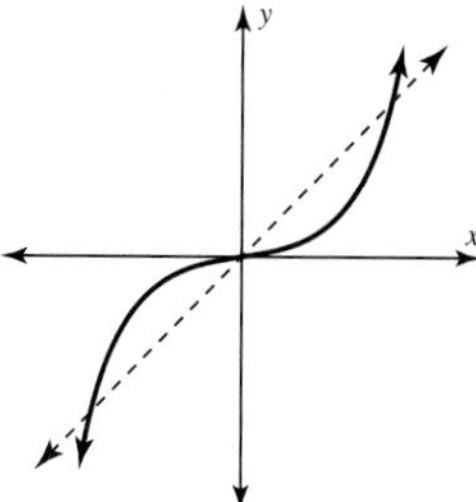

19. a. $6^{1/2} = \sqrt{6}$ **b.** $17^0 = 1$ **c.** $8^{-2} = \frac{1}{64}$ **20. a.** $a = b^c$ **b.** $a = e^c$ **c.** $c = 10^d$ **21. a.** $\log_7 343 = 3$ **b.** $\log_{19} \sqrt[3]{19} = \frac{1}{3}$ **c.** $\log_{4/7} \frac{49}{16} = -2$ **22. a.** $\ln \frac{1}{e} = -1$ **b.** $\log 0.0001 = -4$ **c.** $\log_8 y = x$ **23. a.** 2 **b.** $\frac{1}{2}$ **c.** -1 **24. a.** 0 **b.** 0.23 **c.** $-\frac{1}{3}$ **25. a.** 21.256 **b.** 2514.929 **c.** 0.155 **26. a.** 20.670 **b.** 0.083 **c.** 67,608.298 **27. a.** 2.055 **b.** 4.733 **c.** 2.940 **28. a.** -3.092 **b.** -7.118 **c.** -4.423 **29. a.** 1.065 **b.** 0.191 **c.** 0.165 **30. a.** 2.398 **b.** 3.401 **c.** 3.401 **31. a.** 2.010 **b.** 1.057 **c.** 3.170 **32. a.** 2.989 **b.** 2.016 **c.** 1.893 **33.** $x = -2$ **34.** $x = \frac{2}{3}$ **35.** $y = \frac{2}{3}$ **36.** $y = -\frac{1}{2}$ **37.** $w = 1$ **38.** $v = \frac{1}{4}$ **39.** $x = 2, x = -2$ **40.** $x = -\frac{1}{2}, x = 1$ **41.** $z = 2$ **42.** $x = \frac{3}{2}$ **43.** $x = 4$ **44.** $x = \sqrt{13}$ **45.** $w = \frac{1}{9}$ **46.** No solution **47.** No solution **48.** $t = 13$ **49.** $t = \frac{1}{2}$ **50.** $x = 11$ **51.** $x = 17$ **52.** $x = 7$ **53.** No solution **54.** $n = 3$ **55.** $y = 140$ **56.** $x = 33$ **57.** $x = 10, x = -10$ **58.** $w = 1$ **59.** $w = -2$ **60.** $v = -\frac{1}{2}$ **61.** $v = 3$ **62.** $y = \frac{27}{2}$ **63.** $v = -3, v = 0$ **64.** $x = 3$ **65.** $x = 10$ **66.** $x = 4$ **67.** $3 \log x + 5 \log y$ **68.** $\ln(7x - 9) - \ln(2x + 3)$ **69.** $\frac{1}{2}\ln(2x + 1) - \ln(5x + 9)$ **70.** $\frac{1}{2}(2 \log x + 3 \log y - \log z)$ **71.** $\ln x^2y^3$ **72.** $\ln \frac{x^5}{y^4}$ **73.** $\ln(x + 1)$ for $x > 4$ **74.** $\ln \sqrt{\frac{x}{y}}$ **75.** $x = 2.146$

76. $w = -0.293$ **77.** $y = e^{x \ln 5}$ **78.** $\log 50^x + \log 6^x - \log 3^x = x \log 50 + x \log 6 - x \log 3 = x(\log 50 + \log 6 - \log 3) = x\left(\log \frac{50 \cdot 6}{3}\right) = x \log 100 = x(2) = 2x$ **79.** $1000^{\log x} = (10^3)^{\log x} = 10^{3 \log x} = 10^{\log x^3} = x^3$ **80.** 3; 2.996 **81.** 1; 1.099 **82.** 3; 3.096 **83.** 3; 2.975 **84. a.** 13 **b.** 3 **c.** $\frac{1}{13}$ **85.** D **86.** E **87.** C **88.** B **89.** A **90. a.** \$2,938.66 **b.** \$2,979.69 **c.** \$2,983.65 **91.** 10,200 days **92.** 9.6 years **93.** 8.7% **94.** 6.8 **95. a.** $B(0) = 3$; the initial amount of bacteria is 3 units. **b.** $B(5) \approx 12.8$; after 5 days there are approximately 12.8 units of bacteria. **c.** $B(10) \approx 54.5$; after 10 days there are approximately 54.5 units of bacteria. **d.** $t \approx 12.1$ days; the amount of bacteria reaches 100 units after approximately 12.1 days.

Mastery Test for Chapter 10

1. a. Geometric; $r = 5$ **b.** Not geometric **c.** Geometric; $r = \frac{1}{2}$ **d.** Geometric; $r = 0.1$

2. a.

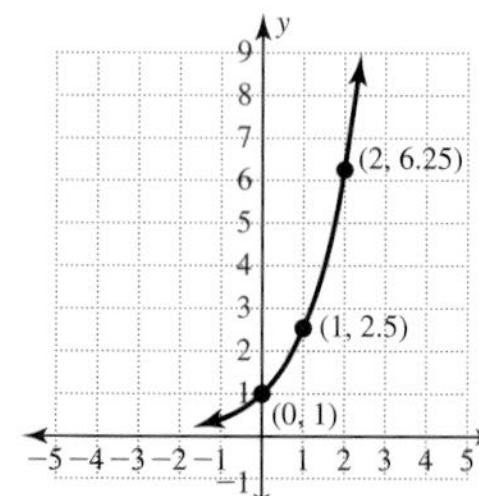

b.

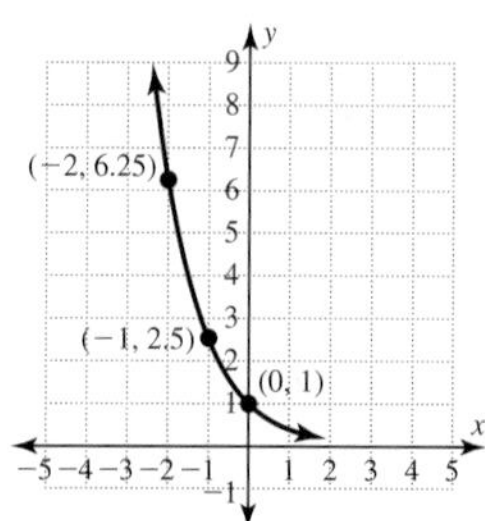

c. $\frac{1}{16}$ **d.** 4 **3. a.** $\{(4, -1), (9, 8), (11, -7)\}$ **b.** $f^{-1}(x) = \frac{x + 6}{3}$ **c.** $\{(-1, -2), (0, -1), (1, 0), (2, 1), (-2, 3)\}$ **d.** $\left\{\left(2, \frac{1}{2}\right), \left(3, \frac{1}{3}\right), \left(\frac{1}{6}, 6\right), (1, 1)\right\}$

4. a.

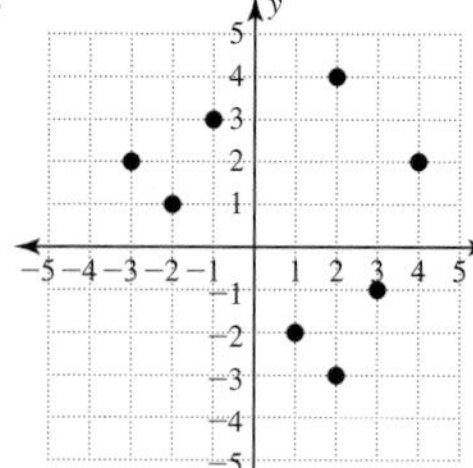

b.

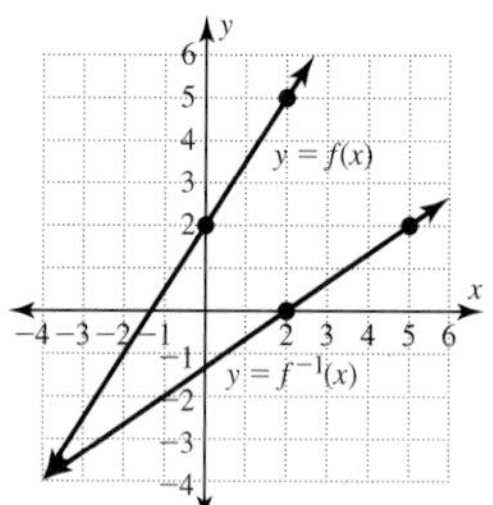

c.

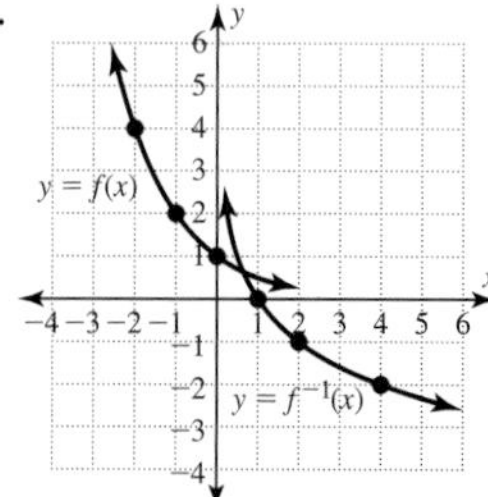

d.

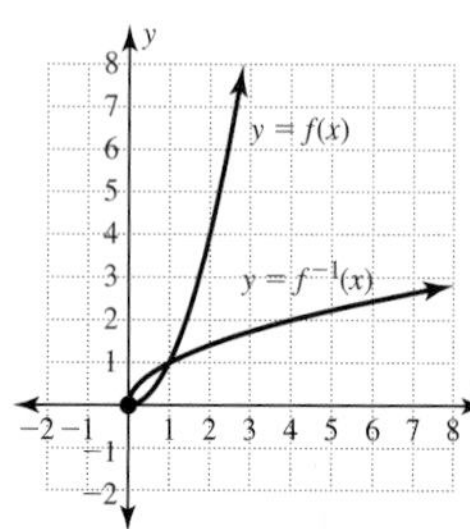

5. a. $5^{2/3} = \sqrt[3]{25}$ **b.** $5^{-3} = \frac{1}{125}$ **c.** $b^x = y + 1$ **d.** $b^{y+1} = x$ **6. a.** 0 **b.** 1 **c.** -1 **d.** 17 **7. a.** 1.2810 **b.** 2.9497 **c.** 12.6881 **d.** -10.4011 **8. a.** $4 \log x + 5 \log y$ **b.** $3 \ln x - \frac{1}{2} \ln y$ **c.** $\ln \frac{(7x + 9)^2}{x}$ **d.** $\log x\sqrt{x + 3}$ **9. a.** 1.730 **b.** 0.850 **c.** 6.644 **d.** 2.406 **10. a.** $x = -4$ **b.** $w = \frac{3}{2}$ **c.** $x = 16$ **d.** $y = \frac{1}{3}$ **e.** $t = -2$ **f.** $z = -3$ **g.** $y = 0.406$ **h.** $x = 1.549$ **11. a.** 35.8 years **b.** 8.4 years

Reality Check for End of Chapter 10

Approximately $60,810

Chapter 11

Reality Check for Chapter 11 Opener

The right wheel whose radius is r_2 is the outside wheel. Because $r_2 > r_1$, it will travel farther during one turn than the inside wheel.

Using the Language and Symbolism of Mathematics 11.1

1. Rectangular **2.** Coefficients; constants **3.** Equivalent **4.** Interchanged **5.** One; one; $\frac{1}{2}$ **6.** $r_2' = r_2 + 2r_1$ **7.** General **8.** Particular

Quick Review 11.1

1. Not a solution **2.** (4, 1) **3.** (3, −1) **4.** (2, 1) **5.** (2, 6)

Exercises 11.1

1. $\left[\begin{array}{cc|c} 3 & 1 & 0 \\ 2 & -1 & -5 \end{array}\right]$ **3.** $\left[\begin{array}{cc|c} 4 & 0 & 12 \\ 3 & 2 & 1 \end{array}\right]$ **5.** $\left[\begin{array}{cc|c} 1 & 0 & 5 \\ 0 & 1 & -6 \end{array}\right]$ **7.** $2x + 3y = 2$; $4x - 3y = 1$ **9.** $2x + y = 1$; $x + 3y = 0$ **11.** $x = 7$; $y = -8$ **13.** $\left[\begin{array}{cc|c} 1 & 3 & 0 \\ 2 & 1 & 1 \end{array}\right]$ **15.** $\left[\begin{array}{cc|c} 1 & -3 & -2 \\ 0 & 1 & 8 \end{array}\right]$ **17.** $\left[\begin{array}{cc|c} 1 & -1/3 & 1 \\ 6 & 4 & -6 \end{array}\right]$ **19.** $\left[\begin{array}{cc|c} 1 & 0 & 1 \\ 0 & 1 & 3 \end{array}\right]$ **21.** $(-5, 9)$ **23.** No solution **25.** $(5 - 3y, y)$; $(5, 0)$, $(2, 1)$, $(8, -1)$ **27.** $r_1' = \frac{1}{2}r_1$ **29.** $r_2' = r_2 - 3r_1$ **31.** $r_2' = \frac{1}{3}r_2$ **33.** $r_1' = r_1 - 2r_2$ **35.** $(-4, 3)$ **37.** $(7, -2)$ **39.** $(3, -2)$ **41.** $\left(2, \frac{1}{3}\right)$ **43.** $\left(\frac{8}{5}, \frac{14}{5}\right)$ **45.** $\left(\frac{1}{2}, 2\right)$ **47.** No solution **49.** $\left(\frac{5}{2} + \frac{1}{2}y, y\right)$ **51.** $x + y = 160$; $x - y = 4$; 82 and 78 **53.** $x + y = 90$; $x - y = 32$; 61° and 29° **55.** $20x + y = 3{,}200$; $30x + y = 4{,}300$; the fixed cost is $1,000 and the variable cost per costume is $110. **57.** $x + y = 30$; $x - y = 14$; the rate of the boat is 22 km/h, and the rate of the current is 8 km/h. **59.** $x + y = 100$; $0.15x + y = 83$; use 20 L of the concentrate and 80 L of water.

Using the Language and Symbolism of Mathematics 11.2

1. Plane **2.** Solution **3.** Three; three **4.** Echelon **5.** Inconsistent

Quick Review 11.2

1. Solution **2.** (4, 3) **3.** C **4.** B **5.** A

Exercises 11.2

1. $(2, 3, 3)$ **3.** $(1, 1, 1)$ **5.** $(1, 0, -2)$ **7.** $\left(1, -\frac{3}{5}, -4\right)$ **9.** $\left(-\frac{1}{2}, -\frac{3}{2}, \frac{1}{2}\right)$ **11.** $(2, 1, 3)$ **13.** No solution **15.** $\left[\begin{array}{ccc|c} 1 & 1 & 1 & 2 \\ -1 & 1 & -2 & 1 \\ 1 & 1 & -1 & 0 \end{array}\right]$ **17.** $\left[\begin{array}{ccc|c} 1 & 2 & -1 & 19 \\ 2 & -1 & 0 & -1 \\ 3 & -2 & 4 & -32 \end{array}\right]$ **19.** $x - 2y + 5z = 0$; $2x + 4y - 3z = 8$; $3x + 5y + 7z = 11$ **21.** $x + 2z = 5$; $x + y = 0$; $y + z = 4$ **23.** $x = 7$; $y = -5$; $z = 8$ **25.** $\left[\begin{array}{ccc|c} 1 & 2 & 3 & 16 \\ 2 & 1 & -2 & -11 \\ 3 & 2 & 1 & 3 \end{array}\right]$ **27.** $\left[\begin{array}{ccc|c} 1 & 2 & 4 & 3 \\ 3 & 5 & 7 & 1 \\ 4 & 9 & 2 & 8 \end{array}\right]$ **29.** $\left[\begin{array}{ccc|c} 1 & 3 & 5 & 11 \\ 0 & 1 & -1 & -9 \\ 4 & 8 & 3 & 7 \end{array}\right]$ **31.** $\left[\begin{array}{ccc|c} 1 & 2 & 2 & 3 \\ 0 & -3 & -1 & -7 \\ 0 & -7 & -2 & -17 \end{array}\right]$ **33.** $\left[\begin{array}{ccc|c} 1 & 3 & -2 & 19 \\ 0 & 1 & 2 & -5 \\ 0 & 4 & 1 & 8 \end{array}\right]$ **35.** $\left[\begin{array}{ccc|c} 1 & 0 & 0 & 3 \\ 0 & 1 & 2 & 3 \\ 0 & 0 & 11 & 22 \end{array}\right]$ **37.** $(-2, 7, 3)$ **39.** No solution **41.** $(2 - 3z, -5 - 2z, z)$, $(2, -5, 0)$, $(-1, -7, 1)$, $(-4, -9, 2)$ **43.** $r_1' = \frac{1}{2}r_1$ **45.** $r_2' = r_2 + 3r_1$ **47.** $r_1' = r_1 + 3r_2$ **49.** $(3, 1, -1)$ **51.** $(2, 3, 3)$ **53.** $(1 - 4x_3, -4 + 3x_3, x_3)$ **55.** $\left(\frac{1}{2}, -\frac{3}{2}, 2\right)$

57. No solution **59.** $(-3 + c, -1 - c, c)$; $(-3, -1, 0)$, $(-1, -3, 2)$
61. 26, 36, 46 **63.** 72 cm, 60 cm, 36 cm **65.** 94.5°, 28.5°, 57°
67. Use 600 g of A, 800 g of B, and 1,300 g of C. **69.** B
71.

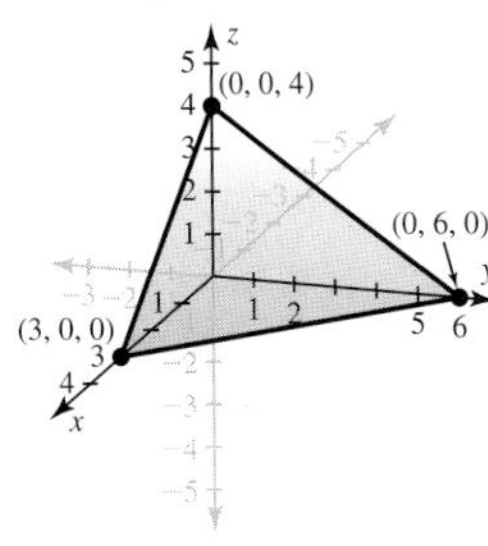

Using the Language and Symbolism of Mathematics 11.3

1. Shape **2.** Vertical; vertical **3.** Horizontal; horizontal **4. a.** Vertical; up **b.** Vertical; down **c.** Horizontal; left **d.** Horizontal; right
5. a. Horizontal; left; vertical; up **b.** Horizontal; right; vertical; down
6. Parabola; vertex

Quick Review 11.3

1. Domain: $\mathbb{R}$
Range: $[0, \infty)$
2. Domain: $\mathbb{R}$
Range: $[-2, \infty)$
3. Domain: $\mathbb{R}$
Range: $[-4, \infty)$
4. Domain: $[0, \infty)$
Range: $[0, \infty)$
5. Domain: $[-4, \infty)$
Range: $[0, \infty)$

Exercises 11.3

1. C **3.** D **5.** D **7.** B **9.** C **11.** D **13.** B **15.** D
17.

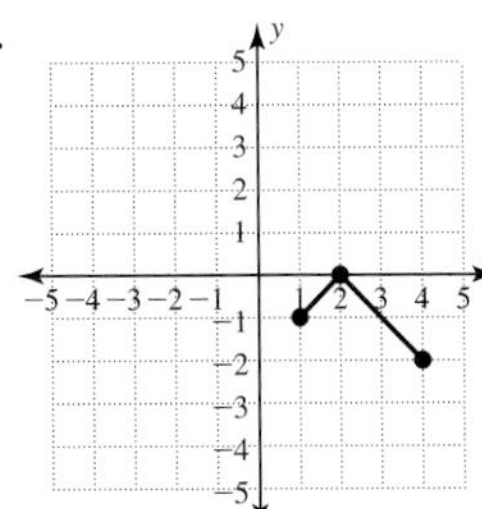

19.

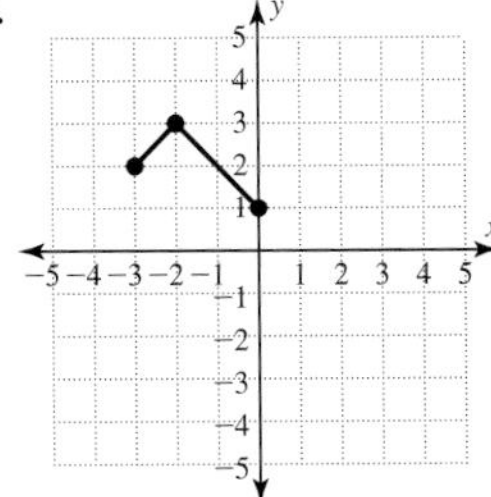

21.

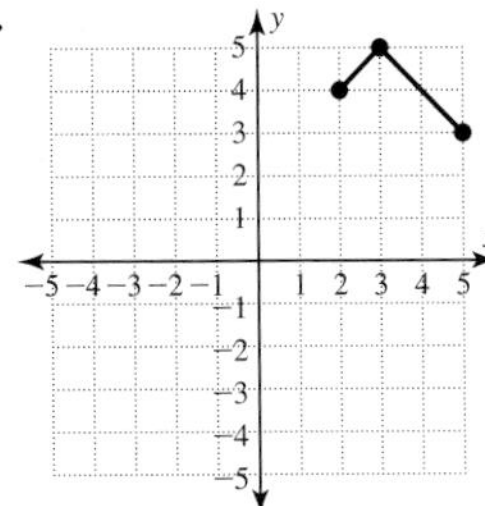

23. $y = \frac{x}{2} + 3$; $(0, 3)$

25.

x	$y = f(x) + 10$
-3	10
-2	18
-1	16
0	10
1	6
2	10

27. $y_2 = f(x) - 7$

29.

x	$y_2 = f(x - 4)$
4	-3
5	-1
6	1
7	3
8	5
9	7

31. $y_2 = f(x - 3)$ **33.** $(6, 0)$ **35.** $(0, 11)$ **37.** $(-5, -8)$ **39.** $(0, -13)$
41. $(-15, 0)$ **43.** $(7, -6)$ **45.** E **47.** D **49.** A **51.** $D = [3, 8)$; $R = [2, 4)$
53. $D = [0, 5)$; $R = [-5, -3)$ **55.** $D = [-4, 1)$; $R = [7, 9)$ **57.** D **59.** A
61. E **63.** H **65. a.** 2, 4, 6, 8, 10 **b.** 5, 7, 9, 11, 13 **c.** 8, 10, 12, 14, 16
d. $-2, -4, -6, -8, -10$

67.

$t°$ K	$V = 4(t - 273)$
473	800
423	600
373	400
323	200
273	0

69.

Item no. x	New Price ($)
1	25.95
2	34.79
3	13.98
4	27.78
5	20.95

Using the Language and Symbolism of Mathematics 11.4

1. Same; same **2.** $(x, -y)$ **3.** Stretching **4.** Shrinking **5.** Reflection; x
6. Same; stretching; 2

Quick Review 11.4

1. 8 **2.** 8
3. Domain: $\mathbb{R}$
Range: $[0, \infty)$
4. Domain: $\mathbb{R}$
Range: $(-\infty, 0]$
5. Domain: $\mathbb{R}$
Range: $[0, \infty)$

Exercises 11.4

1. G **3.** B **5.** E **7.** D **9.** B **11.** A **13.** D **15.** A **17.** D **19.** A **21.** B

23. A
25.

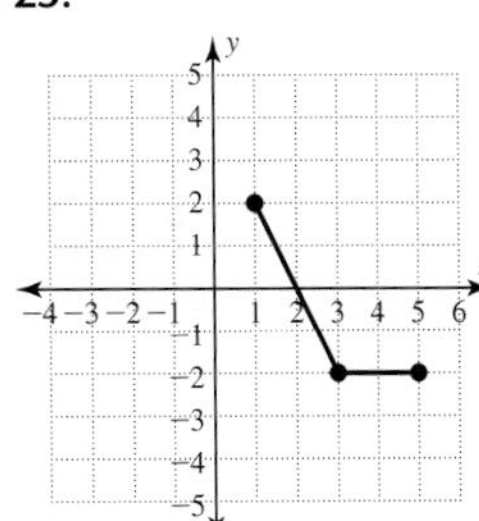

27.

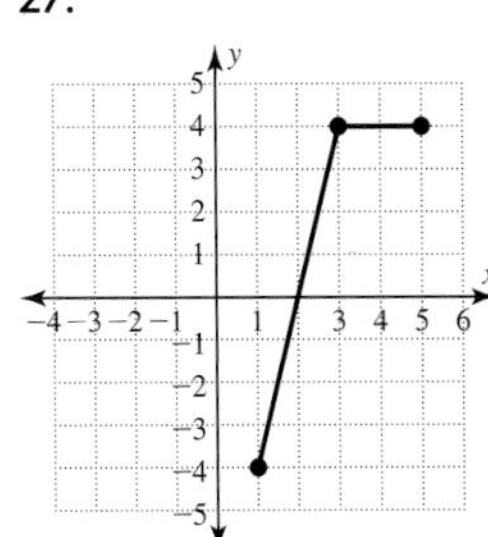

29.

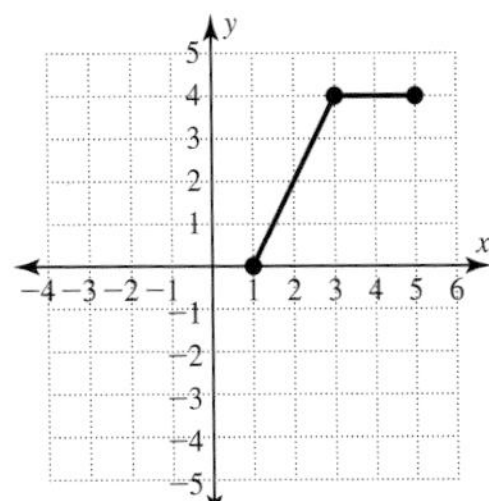

31.

x	$y = -f(x)$
-2	-18
-1	-9
0	-3
1	-1
2	-3

33.

x	$y = 2f(x)$
-2	36
-1	18
0	6
1	2
2	6

35. $y_2 = \frac{1}{2}f(x)$ **37.** $y_2 = -\frac{1}{4}f(x)$ **39.** $y_2 = f(x) - 8$ **41.** $y_2 = f(x + 3)$
43. (0, 0) **45.** (0, −7) **47.** (8, −7) **49.** (8, 7)

51. a.

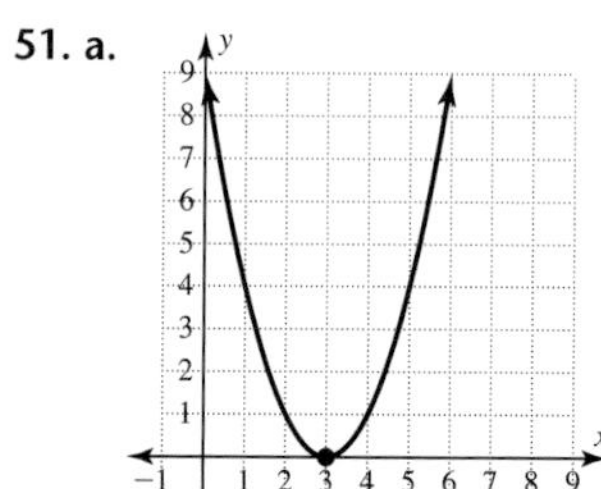

The graph of $f(x) = (x - 3)^2$ is obtained by shifting $f(x) = x^2$ three units to the right.

b.

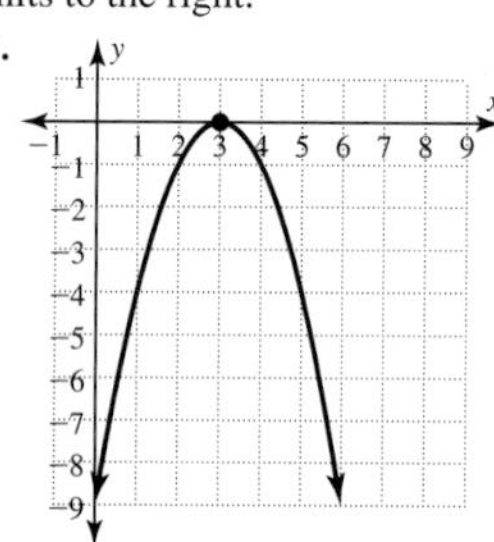

The graph of $f(x) = -(x - 3)^2$ is obtained by reflecting the graph of $f(x) = (x - 3)^2$ about the x-axis.

c.

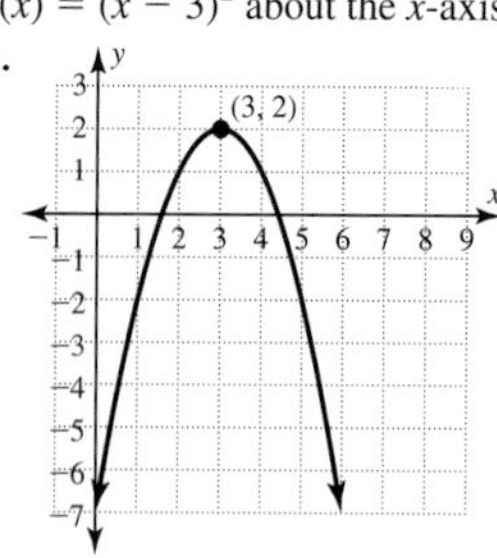

The graph of $f(x) = -(x - 3)^2 + 2$ is obtained by shifting the graph of $f(x) = -(x - 3)^2$ up 2 units.

53. B **55.** D **57.** F **59.** [4, 8) **61.** [4, 12) **63.** (−6, −2] **65. a.** 1, 8, 27, 64, 125 **b.** −1, −8, −27, −64, −125 **c.** 2, 16, 54, 128, 250 **d.** −1, 6, 25, 62, 123 **67. a.** $A_n = n$ **b.** $V_n = 50n$

69.

Years T	Interest I
5	3,000
10	6,000
15	9,000
20	12,000

Using the Language and Symbolism of Mathematics 11.5

1. $f(x) + g(x)$ **2.** $f(x) - g(x)$ **3.** $f(x) \cdot g(x)$ **4.** $\frac{f(x)}{g(x)}$; 0 **5.** $f[g(x)]$
6. $g(x)$; f **7.** x; f^{-1}; x; f

Quick Review 11.5

1. $2x^2 - 5x - 12$ **2.** $2x^2 - 9x - 18$ **3.** $4x^3 - 8x^2 - 51x - 45$ **4.** $x - 5$
5. $f^{-1}(x) = \frac{x - 3}{2}$

Exercises 11.5

1. a. 12 **b.** −6 **c.** 27 **d.** $\frac{1}{3}$ **3. a.** 80 **b.** 11 **c.** 8 **d.** 23

5.

x	$(f + g)(x)$
0	-3
3	5
8	16

7.

x	$(g - f)(x)$
0	1
3	5
8	2

9.

x	$(g \cdot f)(x)$
0	2
3	0
8	63

11. $f - g = \{(-2, -1), (1, 6), (4, 1)\}$ **13.** $\frac{f}{g} = \left\{\left(-2, \frac{3}{4}\right), (1, -5), \left(4, \frac{7}{6}\right)\right\}$

15. **17.**

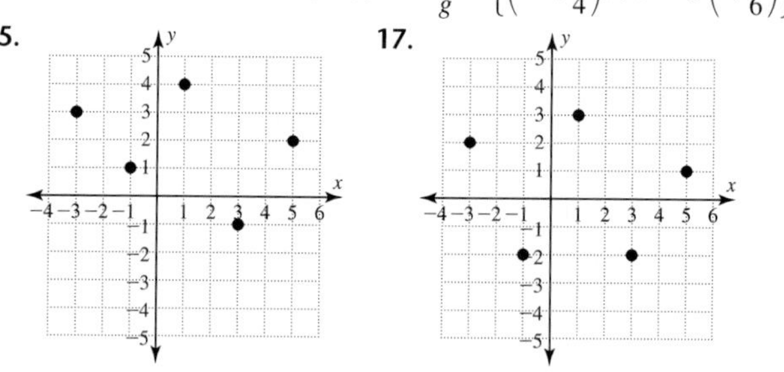

19. $(f + g)(x) = 2x^2 + x - 6, D = \mathbb{R}$; $(f - g)(x) = 2x^2 - 3x$, $D = \mathbb{R}$; $(f \cdot g)(x) = 4x^3 - 8x^2 - 3x + 9, D = \mathbb{R}$; $\left(\frac{f}{g}\right)(x) = x + 1$, $D = \mathbb{R} \sim \left\{\frac{3}{2}\right\}$; $(f \circ g)(x) = 8x^2 - 26x + 18, D = \mathbb{R}$

21. $(f + g)(x) = \frac{x^2 + x + 1}{x^2 + x}, D = \mathbb{R} \sim \{-1, 0\}$

$(f - g)(x) = \frac{x^2 - x - 1}{x^2 + x}, D = \mathbb{R} \sim \{-1, 0\}$

$(f \cdot g)(x) = \frac{1}{x + 1}, D = \mathbb{R} \sim \{-1, 0\}$

$\left(\frac{f}{g}\right)(x) = \frac{x^2}{x + 1}, D = \mathbb{R} \sim \{-1, 0\}$

$(f \circ g)(x) = \frac{1}{x + 1}, D = \mathbb{R} \sim \{-1, 0\}$

23. Not equal; $f(2)$ is defined and $g(2)$ is not defined. **25.** Not equal; the two functions have different domains. **27.** $f(x) = g(x)$

29.

$x \rightarrow g(x) \rightarrow f[g(x)]$		
$3 \rightarrow$	$5 \rightarrow$	9
$4 \rightarrow$	$2 \rightarrow$	8
$7 \rightarrow$	$1 \rightarrow$	0
$8 \rightarrow$	$4 \rightarrow$	6

$x \rightarrow f[g(x)]$	
$3 \rightarrow$	9
$4 \rightarrow$	8
$7 \rightarrow$	0
$8 \rightarrow$	6

31.

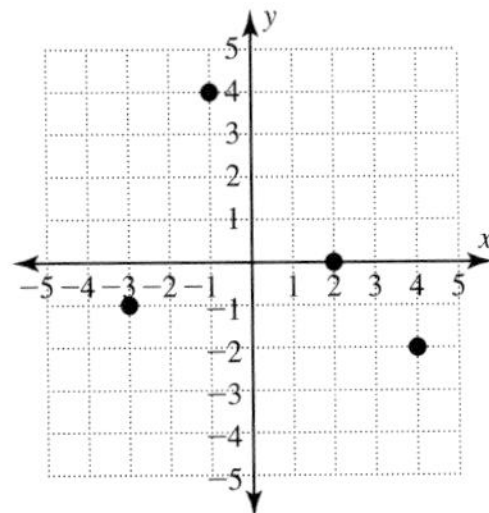

33. $(f \circ g)(x) = 16x^2 - 36x + 17$; $(g \circ f)(x) = 4x^2 - 20x + 10$ **35.** $(f \circ g)(x) = \sqrt{x + 5}$; $(g \circ f)(x) = \sqrt{x} + 5$ **37.** $(f \circ g)(x) = \frac{1}{x^2 - 2}$; $(g \circ f)(x) = \frac{1}{(x + 2)^2} - 4 = -\frac{(2x + 3)(2x + 5)}{(x + 2)^2}$

39.

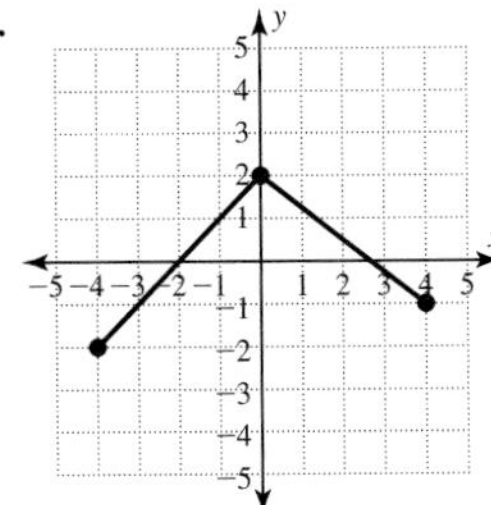

41. a. \$5,000 **b.** \$252,500 **c.** \$257,500 **d.** \$515 per unit **e.** $C(u) = u^2 + 5u + 5{,}000$ **f.** $A(u) = \frac{u^2 + 5u + 5{,}000}{u}$ **43. a.** 203 items **b.** \$80 **c.** \$16,240 **d.** $R(m) = -5m^3 + 135m^2 + 1620m$ **45. a.** $C(5{,}000) = 1{,}900$; the cost of making 5,000 doses is \$1,900. **b.** $R(1{,}900) = 2{,}850$; the revenue derived from vaccine costing \$1,900 is \$2,850. **c.** $(R \circ C)(5{,}000) = 2{,}850$; the revenue derived from making 5,000 doses of the vaccine is \$2,850. **d.** $(R \circ C)(d) = 0.45d + 600$; the revenue derived from making d doses of vaccine. **47. a.** $A(w) = 576 - w^2$ **b.** $w(x) = 24 - 2x$ **c.** $(A \circ w)(x) = 96x - 4x^2$; $(A \circ w)(x)$ is the area of the board when x cm is trimmed from all sides.

49. $(f \circ f^{-1})(x) = x$; $(f^{-1} \circ f)(x) = x$ **51. a.** $f^{-1}(x) = \frac{x - 5}{4}$ **b.** $(f \circ f^{-1})(x) = x$ **c.** $(f^{-1} \circ f)(x) = x$ **53.** $h(x) = (f \circ g)(x)$ **55.** $h(x) = \frac{g(x)}{f(x)}$ **57.** $f(x) = \frac{1}{x}$; $g(x) = 3x^2 - 7x + 9$ **59.** $f(x) = x + \frac{1}{x}$; $g(x) = x^3 - 2$

61.

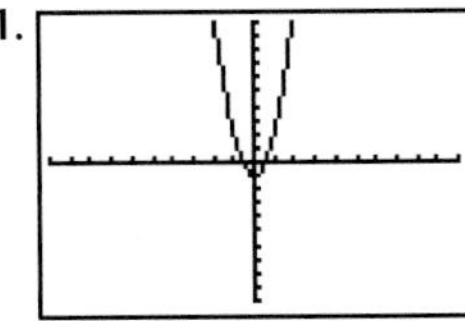

[−10, 10, 1] by [−10, 10, 1]

Using the Language and Symbolism of Mathematics 11.6

1. Natural **2.** Finite **3.** Infinite **4.** Arithmetic **5.** Geometric **6.** Fibonacci **7.** General **8.** Recursively **9.** $a_1 + (n - 1)d$ **10.** a_1r^{n-1} **11.** Summation; i; 1; n **12.** $\frac{n}{2}(a_1 + a_n)$ or $\frac{n}{2}[2a_1 + (n - 1)d]$ **13.** $\frac{a_1(1 - r^n)}{1 - r}$ or $\frac{a_1 - ra_n}{1 - r}$ **14.** $\frac{a_1}{1 - r}$

Quick Review 11.6

1. $a_5 = 23$ **2.** $a_n = 4n + 8$ **3.** 8, 11, 14, 17, 20 **4.** 4, 2, 2, 2, 4, 8 **5.** 5, 10, 20, 40, 80

Exercises 11.6

1. a. Geometric; $r = 6$ **b.** Arithmetic; $d = 5$ **c.** Neither **d.** Geometric; $r = 1$; arithmetic; $d = 0$ **3. a.** Arithmetic; $d = 3$ **b.** Neither **c.** Geometric; $r = 4$ **d.** Geometric; $r = 1$; arithmetic; $d = 0$ **5. a.** 8, 13, 18, 23, 28, 33 **b.** 1, 3, 7, 13, 21, 31 **c.** 8, 4, 2, 1, $\frac{1}{2}$, $\frac{1}{4}$ **d.** −10, 20, −40, 80, −160, 320 **7. a.** 18, 14, 10, 6, 2, −2 **b.** 18, 16, 14, 12, 10, 8 **c.** 7, 5, 3, 1, −1, −3 **d.** −9, −6, −3, 0, 3, 6 **9. a.** 1, 4, 16, 64, 256 **b.** 1, 2, 4, 8, 16 or 1, −2, 4, −8, 16 **c.** 16, 8, 4, 2, 1 **d.** 5, −15, 45, −135, 405 **11. a.** 416 **b.** −83 **c.** 41 **d.** 502 **13.** 21 **15.** −300 **17.** $\frac{1}{2}$ **19.** 46 **21.** 7 **23.** $\frac{3}{2}$ **25.** −3 or 3 **27.** 125 **29. a.** $5 + 7 + 9 + 11 + 13 + 15 = 60$ **b.** $0 + 3 + 8 + 15 + 24 = 50$ **c.** $2 + 4 + 8 + 16 + 32 + 64 = 126$ **d.** $5 + 5 + 5 + 5 + 5 + 5 + 5 + 5 + 5 + 5 = 50$ **31.** 1,640 **33.** 5 **35.** 9,240 **37.** 3,965 **39.** $\frac{372}{5}$ **41.** 381 **43.** 0.22222 **45.** $\frac{255}{8}$ **47.** 12.75603 **49.** 0.8888888 **51.** $\frac{3{,}1248}{625}$ **53.** 15 **55.** 8 **57.** $\frac{4}{33}$ **59.** $\frac{4}{5}$ **61.** $\frac{18}{5}$ **63.** $\frac{192}{5}$ **65. a.** $\frac{4}{9}$ **b.** $\frac{7}{33}$ **c.** $\frac{409}{999}$ **d.** $\frac{23}{9}$ **67.** 33,649 **69.** 11 **71.** $\frac{4}{13}$ **73.** 6 **75.** 4 **77.** $\frac{49}{3}$ **79.** 272 logs **81.** \$81,000 **83.** 21,845 people **85.** 96% **87.** \$1,800,000

Using the Language and Symbolism of Mathematics 11.7

1. Conic sections **2.** Circle **3.** Ellipse **4.** Hyperbola **5.** Parabola **6.** Radius **7.** Diameter **8.** Vertices **9.** Covertices **10.** Asymptotic

Quick Review 11.7

1. $-\frac{b}{2a}$ **2.** $(-1, 1)$ **3.** $9x^2 + 4y^2 - 36x + 8y + 4$ **4.** $x = -\frac{5}{4} \pm \frac{\sqrt{31}}{4}i$ **5.** $f(x) = (x + 6)^2 - 66$

Exercises 11.7

1. 5; (−1, 0.5) **3.** $\sqrt{2}$; (0.5, 1.5) **5.** 25 **7.** (2, 1) **9.** B **11.** D **13.** $x^2 + y^2 = 100$ **15.** $(x - 2)^2 + (y - 6)^2 = 2$ **17.** $x^2 + \left(y - \frac{1}{2}\right)^2 = \frac{1}{4}$ **19.** (−5, 4); 8 **21.** (3, 0); 3 **23.** (1, −5); 2 **25.** (0, 0); length of major axis: 12; length of minor axis: 8

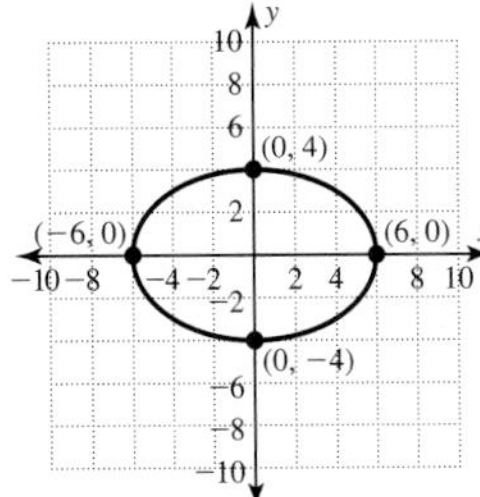

27. (−4, 3); length of major axis: 10; length of minor axis: 8

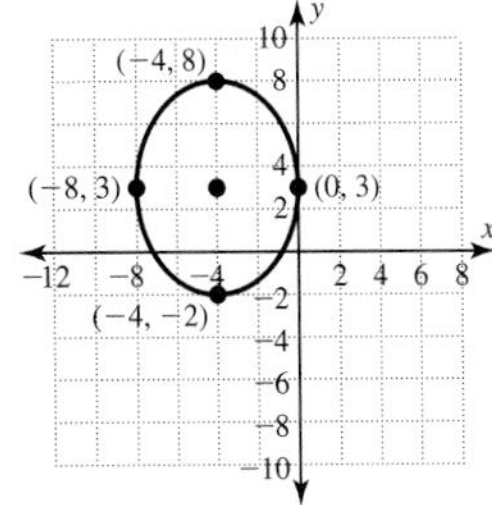

29. $\frac{x^2}{81} + \frac{y^2}{25} = 1$ **31.** $\frac{(x-3)^2}{4} + \frac{(y-4)^2}{1} = 1$

33. $\frac{(x-6)^2}{9} + \frac{(y+2)^2}{25} = 1$ **35.** $\frac{(x+3)^2}{4} + \frac{(y+4)^2}{36} = 1$

37. Center: (0, 0); $a = 5$, $b = 9$; opens horizontally

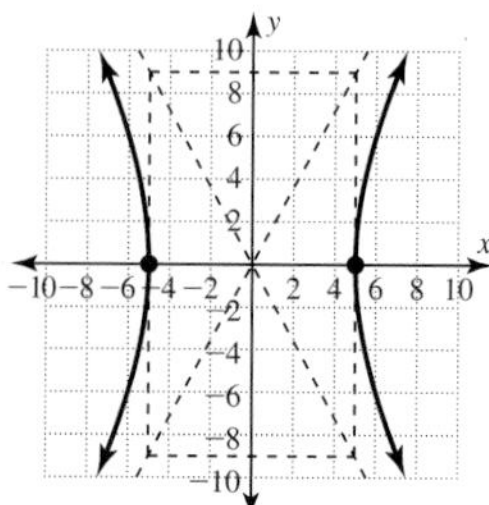

39. Center: (0, 0); $a = 2$, $b = 1$; opens horizontally

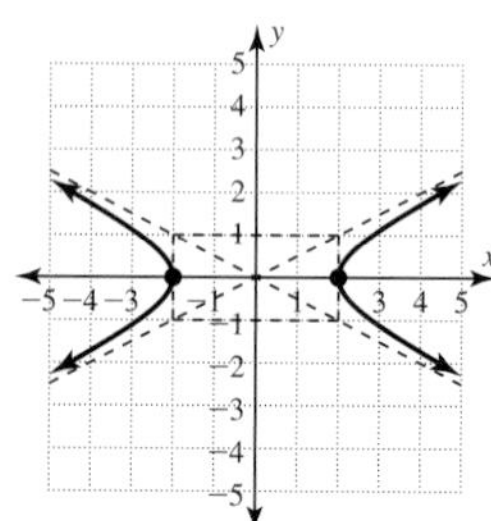

41. Center: (4, 6); $a = 3$, $b = 7$; opens vertically

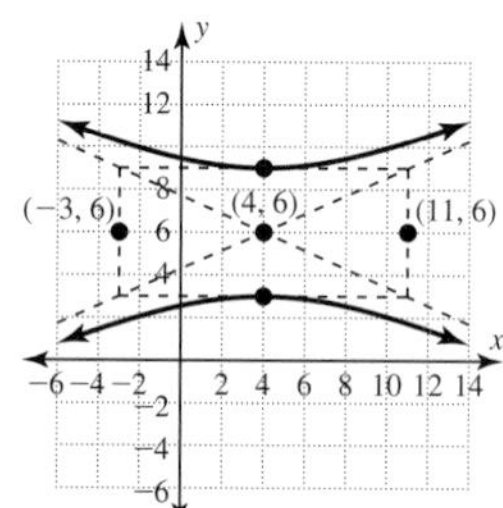

43. $\frac{y^2}{25} - \frac{x^2}{16} = 1$ **45.** $\frac{x^2}{9} - \frac{y^2}{49} = 1$

47. a.

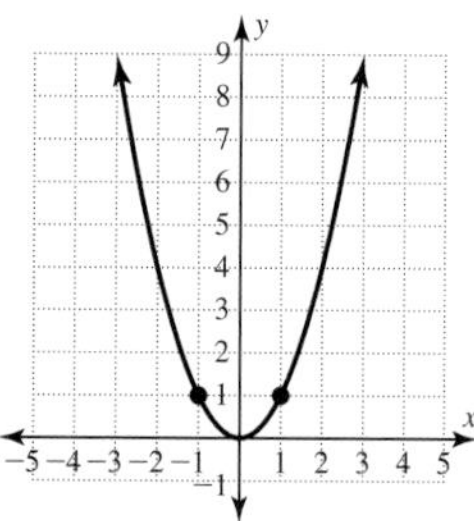

b.

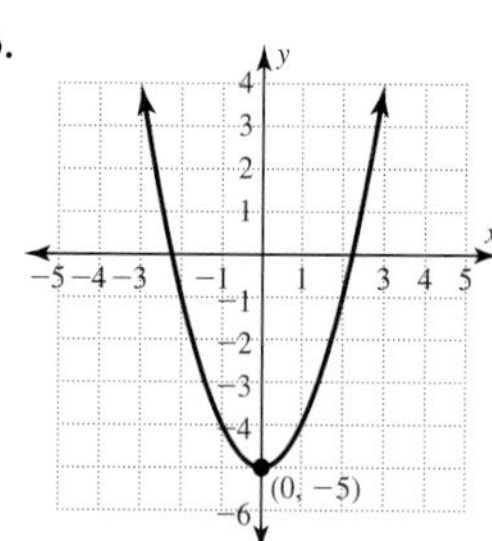

c.

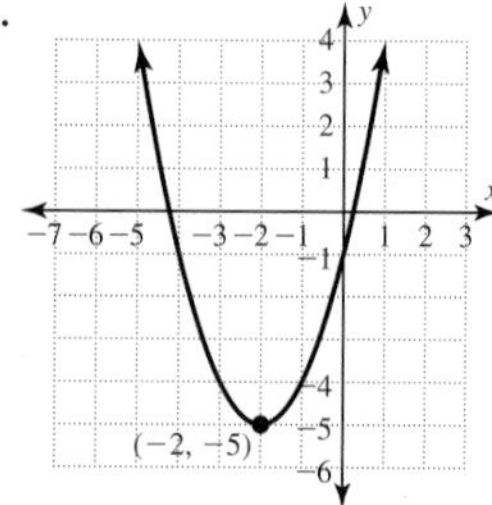

49. a.

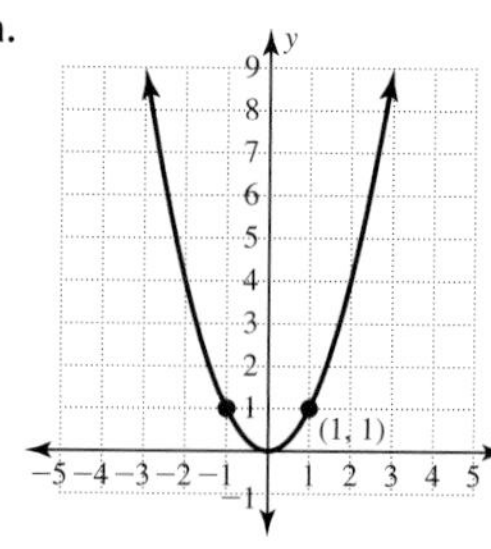

b.

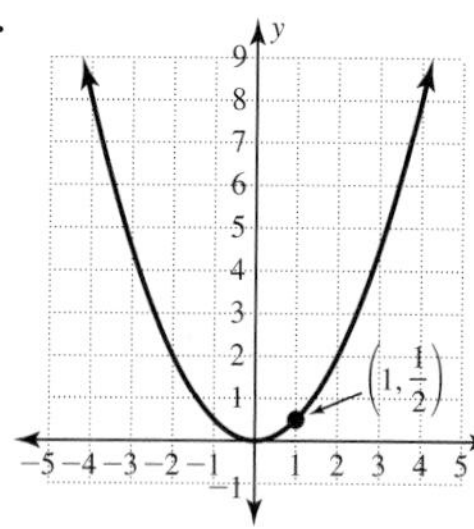

c.

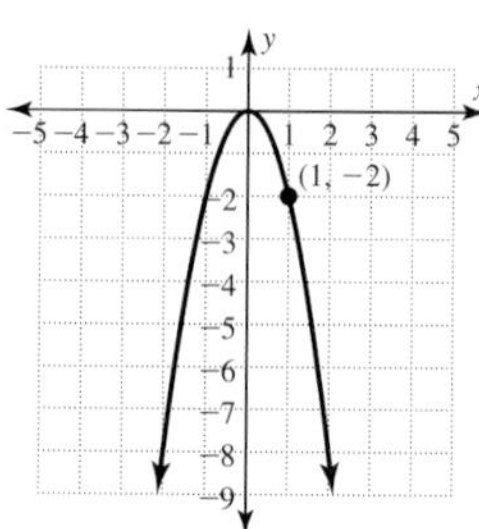

51.

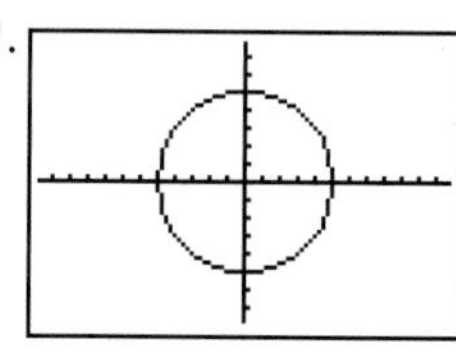

[−11.8, 11.8, 1] by [−7.8, 7.8, 1]

53.

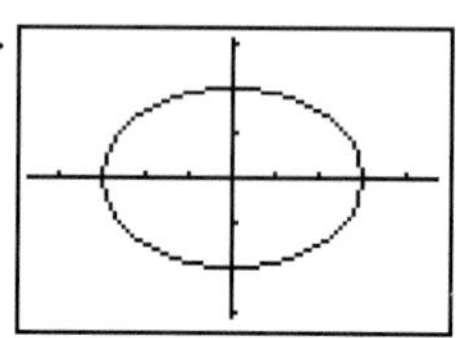

[−4.7, 4.7, 1] by [−3.1, 3.1, 1]

55. $Y_1 = \sqrt{36 - x^2}$; $Y_2 = -\sqrt{36 - x^2}$

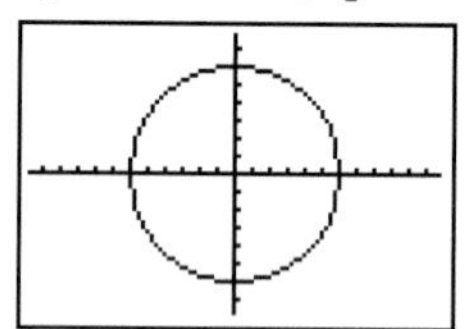

[−11.8, 11.8, 1] by [−7.8, 7.8, 1]

57. a. (−2, 7), (2, −1) **b.**

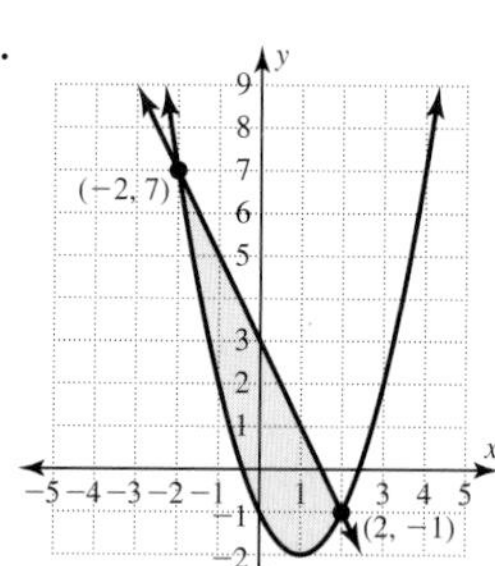

59. a. (2, 2), (−2, −2) **b.**

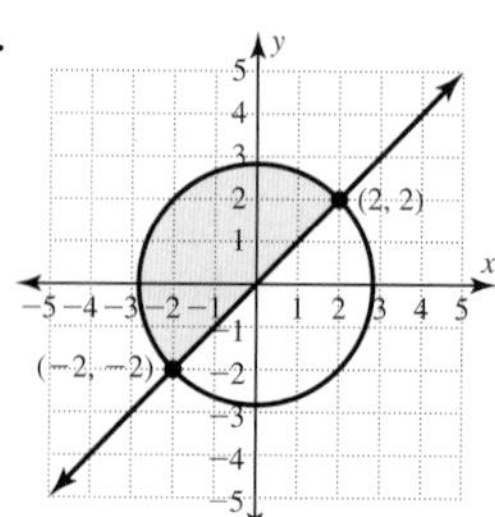

Review Exercises for Chapter 11

1. a. $\left[\begin{array}{rr|r} 2 & -5 & 17 \\ 3 & 4 & 14 \end{array}\right]$ **b.** $\left[\begin{array}{rrr|r} 3 & -4 & 2 & -11 \\ 2 & 2 & 3 & -3 \\ 4 & -1 & 5 & -13 \end{array}\right]$ **2. a.** $(-2, 6)$ **b.** $(4, 5, 8)$ **3.** $(5 - 3z, 4 + 2z, z)$; $(5, 4, 0)$; $(2, 6, 1)$, $(-10, 14, 5)$

4. $\left[\begin{array}{rr|r} 1 & 4 & 2 \\ 3 & 5 & 13 \end{array}\right]$ **5.** $\left[\begin{array}{rr|r} 1 & 2 & -5 \\ 3 & 2 & -3 \end{array}\right]$ **6.** $\left[\begin{array}{rr|r} 1 & 2 & 14 \\ 0 & -7 & -29 \end{array}\right]$

7. $\left[\begin{array}{rrr|r} 1 & 0 & -8 & -15 \\ 0 & 1 & 2 & 3 \\ 0 & 0 & -1 & -2 \end{array}\right]$ **8.** $(3, 7)$ **9.** $(-2, 5)$ **10.** $(1, -1)$

11. No solution **12.** $(1, -1, 1)$ **13.** $(1, 4, -3)$ **14.** $(2, 3, 4)$

15.

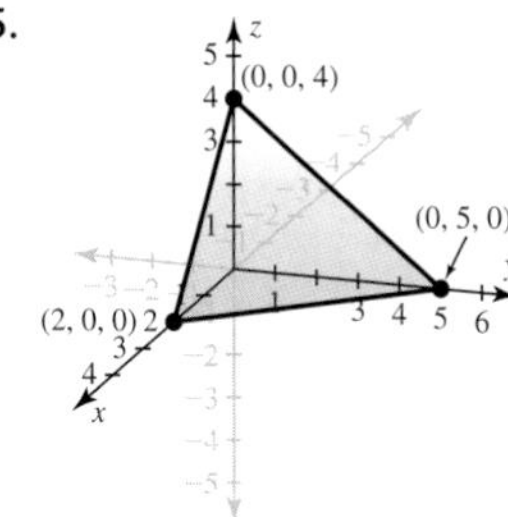

16. 4080 lb; 3280 lb **17.** \$61 for A, \$25 for B, \$9 for C

18. a. B **b.** A **c.** C **19. a.** A **b.** C **c.** B **20. a.** B **b.** A **c.** C

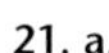

21. a.

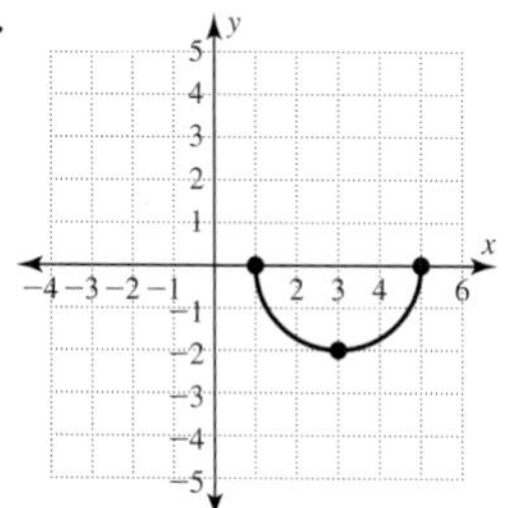

b.

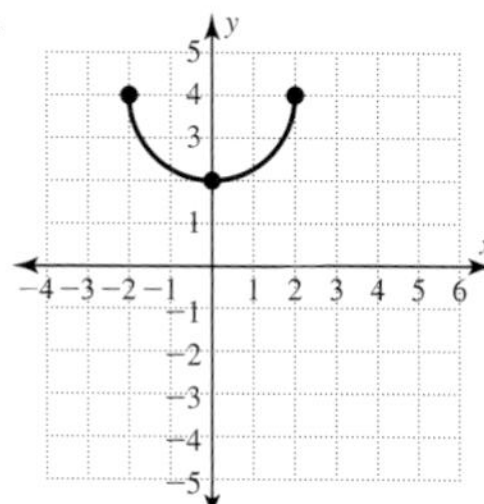

c.

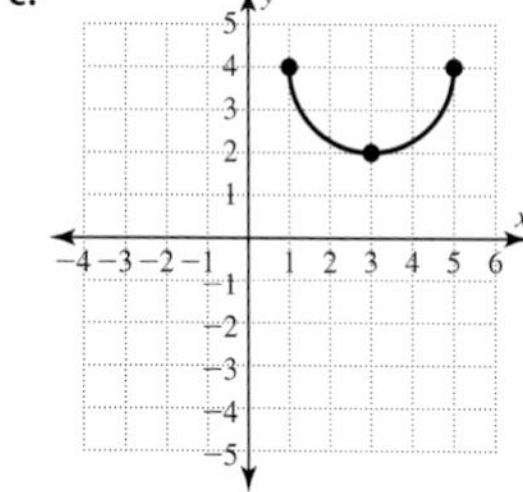

22. a. $(0, 7)$ **b.** $(9, 0)$ **c.** $(-11, -12)$

23. a.

x	$y = f(x) + 5$
−2	10
−1	14
0	8
1	1
2	5

b.

x	$y = f(x + 5)$
−7	5
−6	9
−5	3
−4	−4
−3	0

24. a. D **b.** A **c.** B **d.** C **25. a.** D **b.** B **c.** C **d.** A **26. a.** B **b.** C **c.** D **d.** A **27. a.** C **b.** D **c.** A **d.** B

28. a.

x	$y = -f(x)$
0	0
1	−8
2	−4
3	−12
4	−24

b.

x	$y = \frac{1}{4}f(x)$
0	0
1	2
2	1
3	3
4	6

c.

x	$y = 4f(x)$
0	0
1	32
2	16
3	48
4	96

d.

x	$y = -3f(x)$
0	0
1	−24
2	−12
3	−36
4	−72

29. a. D **b.** A **c.** B **d.** C **30. a.** 23 **b.** 201 **c.** 224 **d.** −178 **31. a.** −9 **b.** $-\frac{1}{9}$ **c.** −9 **d.** 10 **32. a.** 20 **b.** 3 **c.** −10 **d.** 163

33. a.

x	$f + g$
0	−6
1	−1
2	8
3	21

b.

x	$f - g$
0	−8
1	−7
2	−10
3	−17

c.

x	$f \cdot g$
0	−7
1	−12
2	−9
3	38

d.

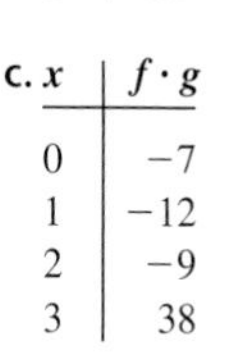

x	$\frac{f}{g}$
0	−7
1	$-\frac{4}{3}$
2	$-\frac{1}{9}$
3	$\frac{2}{19}$

34. a. $f + g = \{(-3, -4), (-1, -2), (1, 0), (2, 1), (3, 2)\}$
b. $f - g = \{(-3, -2), (-1, 0), (1, 2), (2, 3), (3, 4)\}$
c. $f \cdot g = \{(-3, 3), (-1, 1), (1, -1), (2, -2), (3, -3)\}$
d. $\frac{f}{g} = \{(-3, 3), (-1, 1), (1, -1), (2, -2), (3, -3)\}$

35.

x	$f \circ g$
−1	5
2	11
3	7
5	9

36.

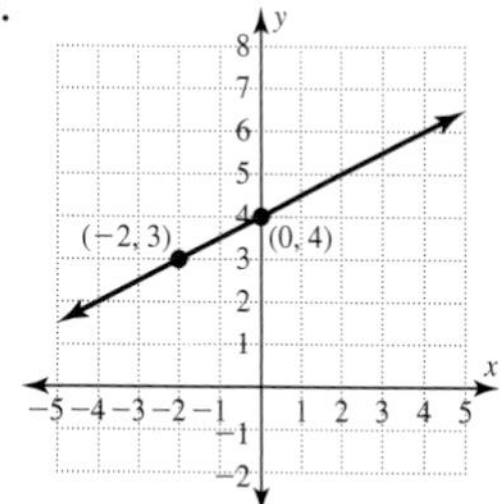

37. a. $(f + g)(x) = 2x - 1$; $D = \mathbb{R}$ **b.** $(f \cdot g)(x) = -8x^2 + 8x - 2$; $D = \mathbb{R}$ **c.** $(f \circ g)(x) = 2 - 8x$; $D = \mathbb{R}$ **38. a.** $(f - g)(x) = 6x - 3$; $D = \mathbb{R}$ **b.** $\left(\frac{f}{g}\right)(x) = -2$; $D = \mathbb{R} \sim \left\{\frac{1}{2}\right\}$ **c.** $(g \circ f)(x) = 5 - 8x$; $D = \mathbb{R}$ **39.** $f(x)$ is defined when $x = 3$, and $g(x)$ is undefined when $x = 3$.

40. $f^{-1}(x) = \frac{x - 3}{4}$; $(f \circ f^{-1})(x) = x$ **41. a.** $F(100) = 400$ **b.** $V(100) = 150$ **c.** $C(x) = 1.5x + 400$ **d.** $A(x) = \frac{1.5x + 400}{x}$

42. a. $N(40) = 120$; the factory can produce 120 desks in 40 h.
b. $C(120) = 14{,}400$; the cost of producing 120 desks is \$14,400.
c. $(C \circ N)(40) = 14{,}400$; the cost of operating the factory for 40 h is \$14,400. **d.** $(C \circ N)(t) = \left(150 - \frac{3t}{4}\right)(3t)$ **e.** There are only 168 hours in a

week. **43. a.** 5, 9, 13, 17, 21, 25 **b.** 5, 1, −3, −7, −11, −15 **c.** 5, 8, 11, 14, 17, 20 **d.** 5, 8, 11, 14, 17, 20 **44. a.** 3, 6, 12, 24, 48, 96 **b.** 3, −6, 12, −24, 48, −96 **c.** 3, 12, 48, 192, 768, 3072 **d.** 1, 3, 9, 27, 81, 243 **45. a.** 9, 11, 13, 15, 17 **b.** 4, 7, 12, 19, 28 **c.** 32, 16, 8, 4, 2 **d.** 3, 10, 17, 24, 31 **46.** 311 **47.** 16 **48.** 41 **49.** 14 **50. a.** $1 + 4 + 7 + 10 + 13 + 16 = 51$ **b.** $3 + 8 + 15 + 24 + 35 = 85$ **c.** $1 + 8 + 27 + 64 + 125 = 225$ **d.** $10 + 10 + 10 + 10 + 10 + 10 = 60$ **51.** 9,725 **52.** $\frac{4}{33}$ **53.** $\frac{3}{2}$ **54.** 1,071 cm **55.** 40 cm; $(160 + 80\sqrt{2})$ cm **56.** 13 **57.** (1, 7) **58.** (3, −3) **59.** 2 **60.** $(x - 3)^2 + (y + 3)^2 = 4$ **61.** C **62.** D **63.** A **64.** F **65.** E **66.** B

67.

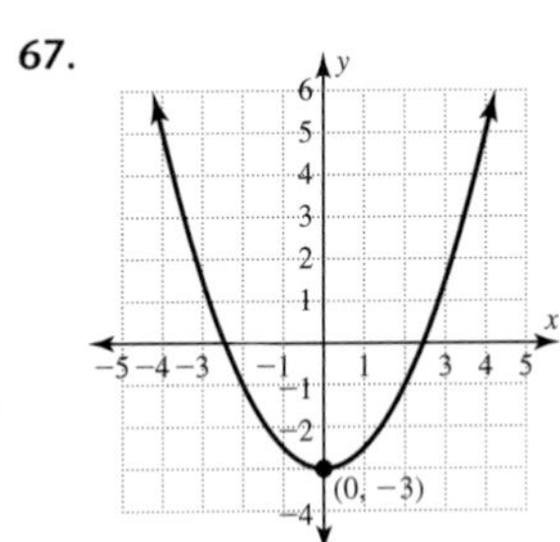

68.

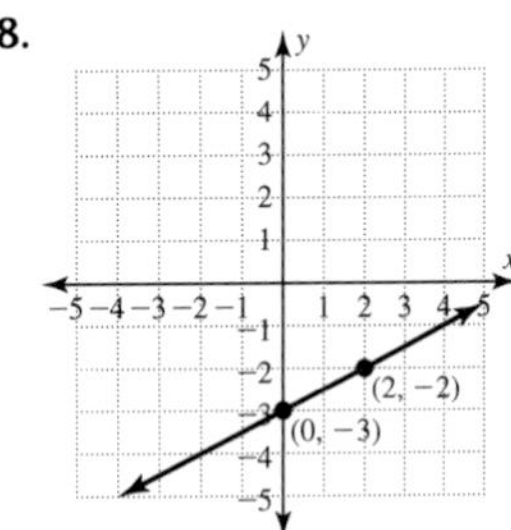

69.

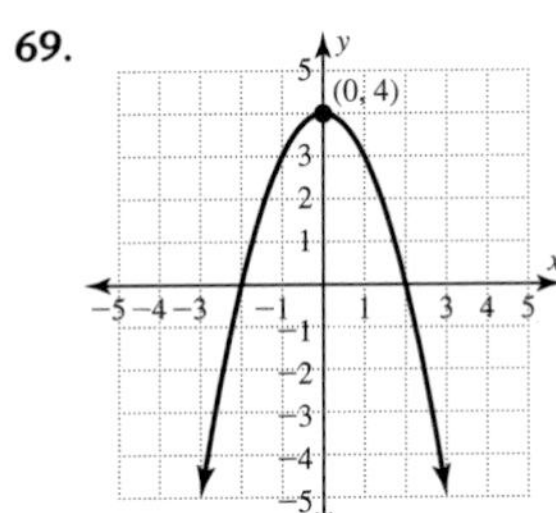

70.

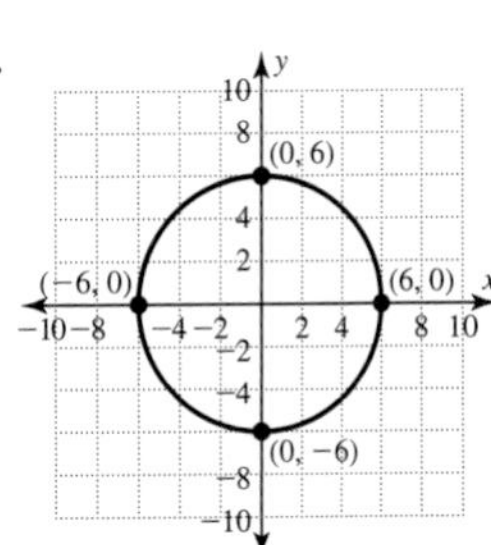

71.

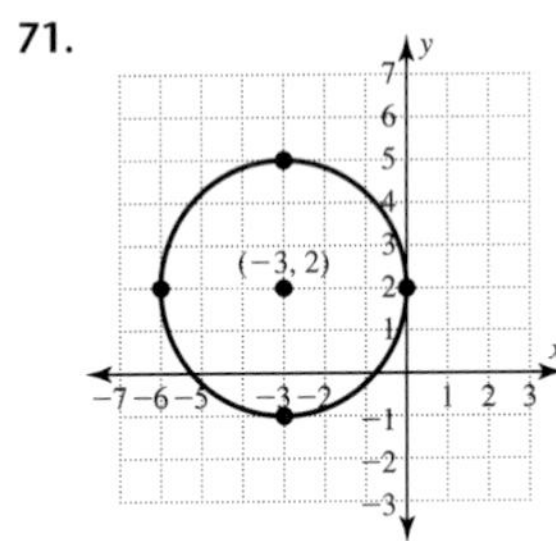

72.

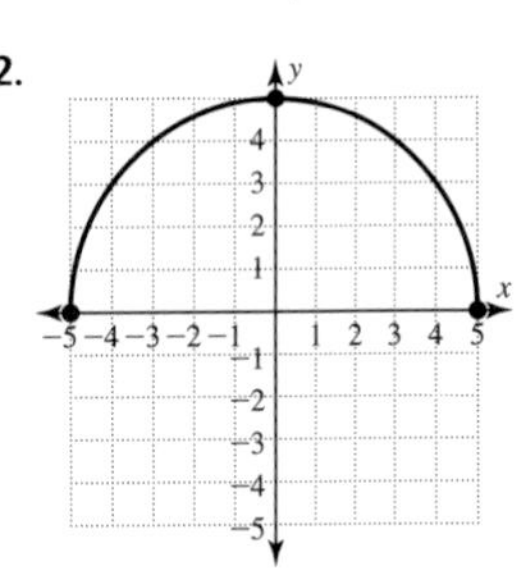

73.

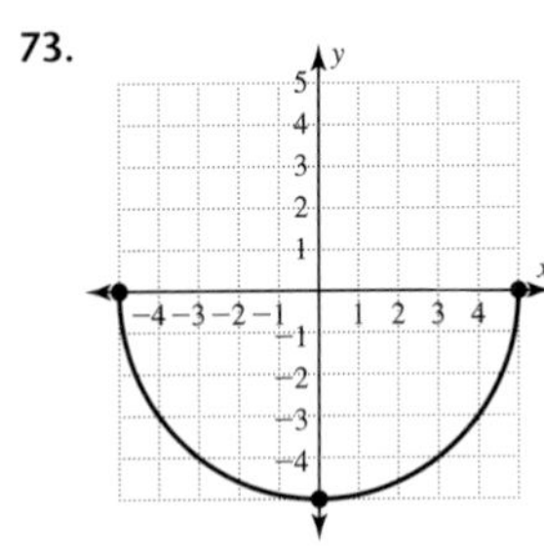

74.

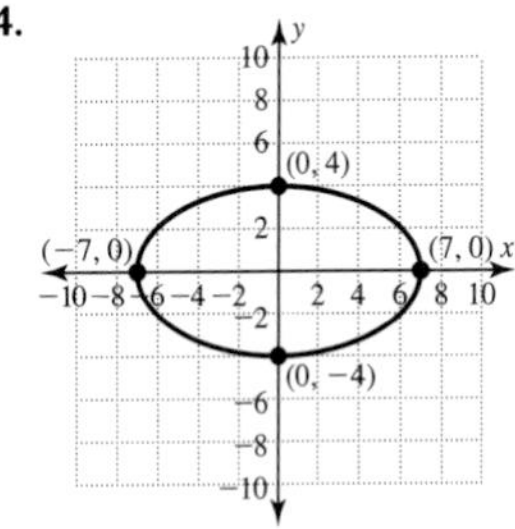

75.

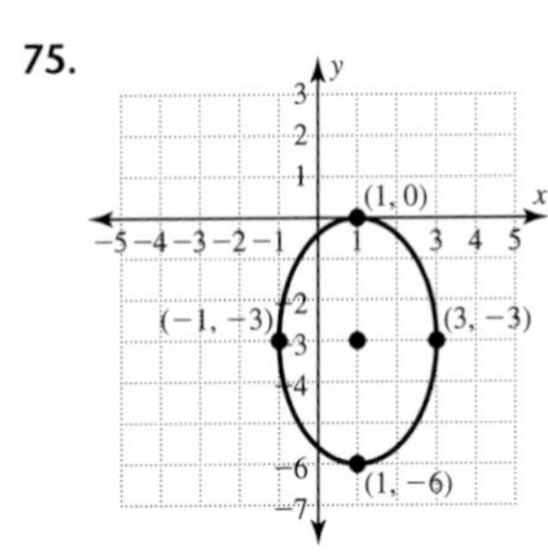

76.

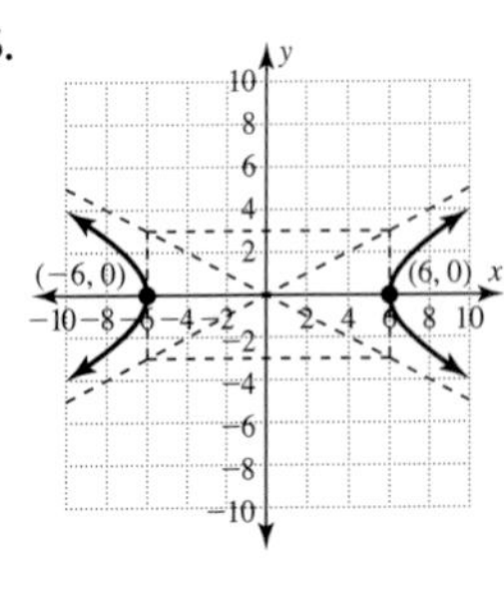

77. (−2, 3), (2, 3)

78.

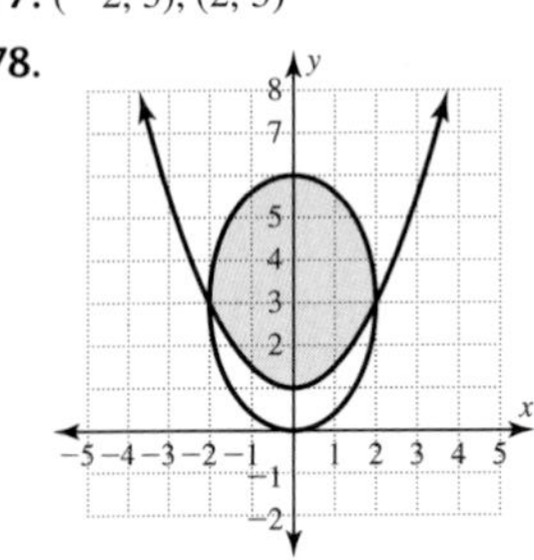

Mastery Test for Chapter 11

1. a. (−2, 7) **b.** No solution **c.** $(6 - 2y, y)$; (6, 0), (0, 3), (4, 1) **d.** (11, −7) **2. a.** (2, 3, 1) **b.** $\left(-\frac{1}{7} - \frac{1}{7}z, \frac{5}{7} + \frac{5}{7}z, z\right)$ **c.** No solution

3. a. A **b.** B

c.

x	$f(x) - 2$
0	−7
1	−6
2	1
3	20
4	57

d.

x	$f(x) + 2$
0	−3
1	−2
2	5
3	24
4	61

4. a. A **b.** B

c.

x	$f(x - 3)$
3	−5
4	−4
5	3
6	22
7	59

d.

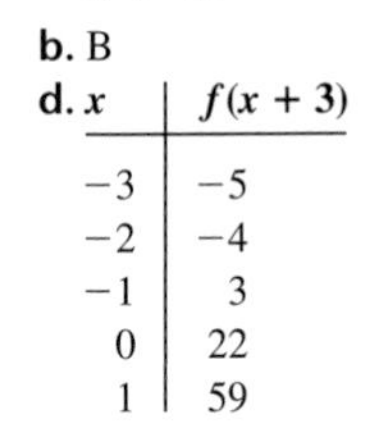

x	$f(x + 3)$
−3	−5
−2	−4
−1	3
0	22
1	59

5. a.

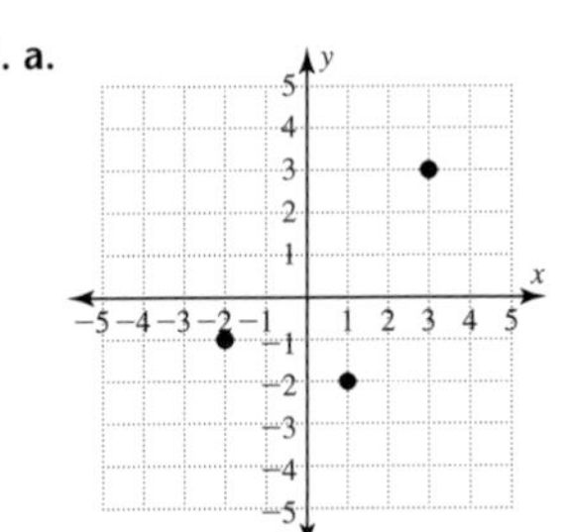

b.

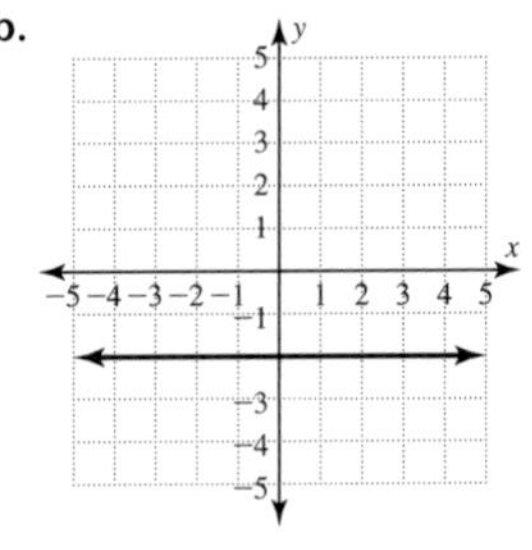

c.

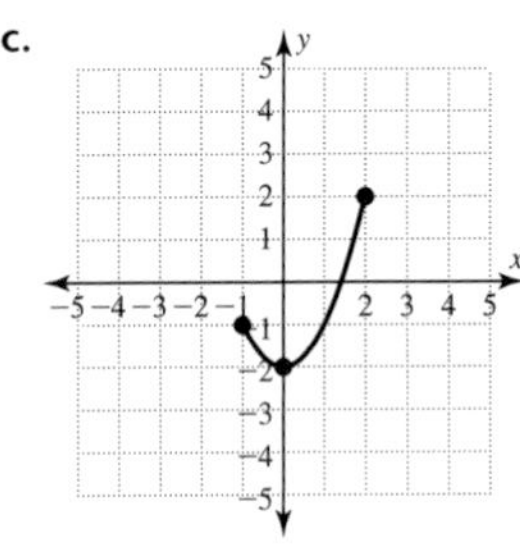

d.

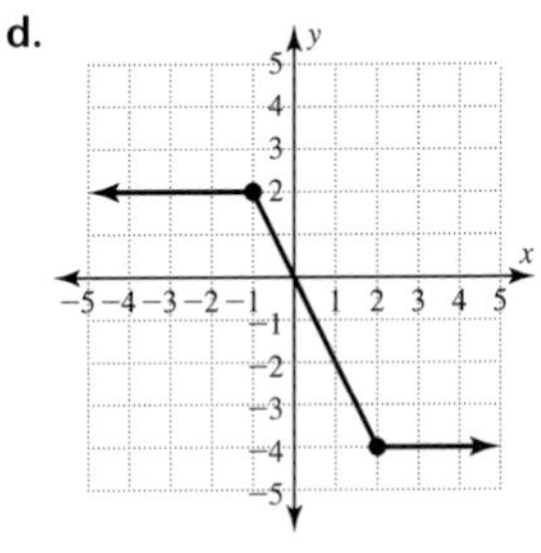

6. a. C **b.** D **c.** A **d.** B **7. a.** $(f + g)(x) = 4x - 3$ **b.** $(f - g)(x) = 2x - 7$ **c.** $(f \cdot g)(x) = 3x^2 + x - 10$ **d.** $\left(\frac{f}{g}\right)(x) = \frac{3x - 5}{x + 2}$ for $x \neq -2$ **8. a.** 82 **b.** 51 **c.** $4x^2 - 4x + 2$ **d.** $2x^2 + 1$ **9. a.** 4, 7, 10, 13, 16, 19 **b.** 4, 12, 36, 108, 324, 972 **c.** 2, 8, 14, 20, 26, 32 **d.** 2, 8, 32, 128, 512, 2048 **10. a.** 60 **b.** 1,562 **c.** 1,550 **d.** 4,194,306 **e.** 294 **11. a.** $\frac{1{,}023}{512}$ **b.** 2 **c.** $\frac{8}{9}$ **d.** $\frac{6}{11}$

12. a.

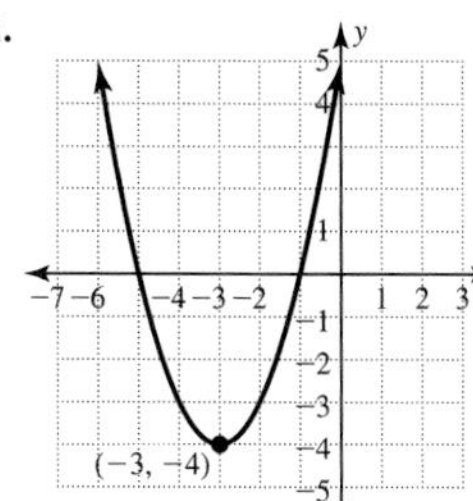

b.

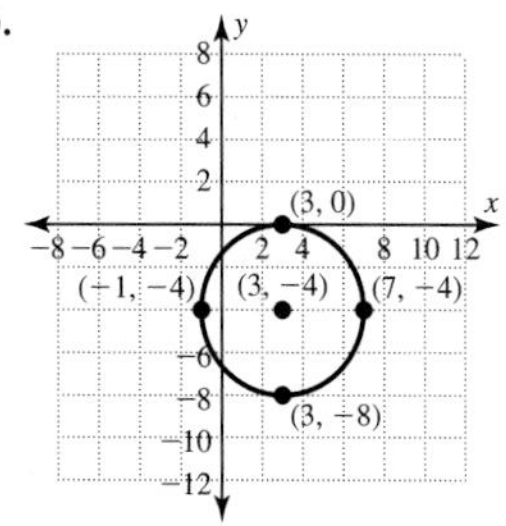

c.

(3, 0) (−2, −4) (3, −4) (8, −4) (3, −8)

d.

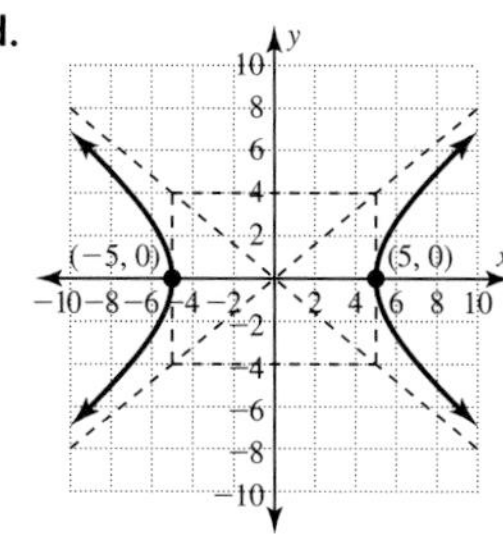

13. a. $y = -x^2 + 4$ **b.** $x^2 + (y - 5)^2 = 25$ **c.** $\dfrac{x^2}{49} + \dfrac{y^2}{25} = 1$

d. $\dfrac{x^2}{49} - \dfrac{y^2}{25} = 1$

Reality Check for End of Chapter 11

$\dfrac{\pi}{16}\text{ in} \approx 0.2\text{ in}$

Credits

Chapter 1

Opener: © PhotoDisc/Vol. 68; 32(both): © PhotoDisc/Vol. 1; 39(top): © Royalty-Free/Corbis; 39(bottom): © Digital Vision/PunchStock; 43: © PhotoDisc/Vol. 1; 44: © Corbis/Vol. 87; 57 (refrigerator): Courtesy G.E. Appliances; 57 (box): © McGraw-Hill, photo by C.P. Hammond; 58: © PhotoDisc/Vol. 1; 69: © Royalty-Free/Corbis; 83: Steve Cole/Photodisc/Getty Images

Chapter 2

Opener: Stone/Getty Images; 91: © Reuters/Corbis; 100: Photo by Jamie Squire/Getty Images; 112: © PhotoDisc Website; 138: © Royalty-Free/Corbis; 139: Courtesy Resilite; 147: Taxi/Getty Images; 148: © Royalty-Free/Corbis; 152: © Royalty-Free/Corbis; 159: © James Noble/Corbis; 167: © Royalty-Free/Corbis; 169: Transparencies, Inc.; 172: © Lester Lefkowitz/Corbis

Chapter 3

Opener: © PhotoDisc/Vol. 1; 178: © Manette Hall; 187: James W. Hall; 193: © Henry Diltz/Corbis; 208: Cole Group/Getty Images; 211: © Richard Pasley/Stock Boston; 231: © Duomo/Corbis; 252: © Spencer Grant/PhotoEdit; 253: © SuperStock; 261: © Stair Parts, Inc. George, USA/Schilling Photography

Chapter 4

Opener: © 1999 EyeWire, Inc.; 281: © Corbis/Vol. 1; 283 & 287: Brian Mercer; 290: James W. Hall; 299: © Michael S. Yamashita/Corbis; 300(top): © Dave Bartruff/Corbis; 300(bottom): © Royalty-Free/Corbis; 310(truck): © Royalty-Free/Corbis; 310(bottles): Photo by Robert Nickelsberg/Liaison/Getty Images; 320: © Corbis/Vol. 75; 323 & 325: AP Wide World Photos

Chapter 5

Opener: © Royalty-Free/Corbis; 337: James W. Hall; 347 & 356: NASA; 358(left): Photo by Geoffrey Wheeler/National Institute of Standards & Technology; 358(right): Brian Mercer; 368(left): Creatas/PunchStock; 368(right) & 395: James W. Hall

Chapter 6

Opener: Brian Mercer; 422: © PhotoDisc/Vol. 90; 453: © Corbis/Vol. 97; 466: © Corbis/Vol. 49

Chapter 7

Opener: James W. Hall; 490: James W. Hall; 498: © Royalty-Free/Corbis; 500(left): © PhotoDisc/Vol. 2; 500(right): © Royalty-Free/Corbis; 516: © David Gray/Reuters/Corbis; 530: © Jeff Greenberg/PhotoEdit; 532: © Royalty-Free/Corbis; 536: James W. Hall; 543(left): © Royalty-Free/Corbis; 543(right): © PhotoDisc/Vol. 27; 551: Brian Mercer; 554(left): © Manette Hall; 554(right): © Corbis/Vol. 159; 556(top): James W. Hall; 556(bottom): © PhotoDisc/Vol. 27; 557: © Patrik Giardino/Corbis; 578: © PhotoDisc/Vol. 20

Chapter 8

Opener: © NASA/Roger Ressmeyer/Corbis; 582: © Brent Smith/Reuters/Corbis; 601: © age fotostock/SuperStock; 630: © NASA; 641(left): AP Wide World Photos; 641(right): PhotoDisc/Getty RF; 640: © NASA/Roger Ressmeyer

Chapter 9

Opener: NASA Glenn Research Center (NASA-GRC); 664: © William James Warren/Corbis; 682: © PhotoDisc/Vol. 20; 701(solar): © PhotoDisc/Vol. 31; 701(bridge): © Royalty-Free/Corbis; 701(chimney): © Manette Hall; 712: © Royalty-Free/Corbis

Chapter 10

Opener: © Reuters/Corbis; 718: Bernhard Edmaier/Photo Researchers, Inc.; 770(top): © Corbis/Vol. 102; 770(bottom): Department of Defense photo by Master Sgt. Keith Reed, U.S. Air Force; 773: © Corbis/Vol. 154; 777: © PhotoDisc/Vol. 19; 778(left): Paul Hermansen/Photo Researchers, Inc.; 778(right): Department of Defense photo by Master Sgt. Keith Reed, U.S. Air Force; 783: © NOAA; 784: © NASA

Chapter 11

Opener: © Royalty-Free/Corbis; 818: © PhotoDisc/Website; 851: © Corbis/Vol. 13; 865: © PhotoDisc/Vol. 45; 882: Courtesy Jeff Hammond; 884: Courtesy of the Gunlocke Company, LLC www.gunlocke.com

Index of Features

Index of Spreadsheet References

Title and Page

Index of Features

Index of Calculator Perspectives

Index of Features

Index of Mathematical Notes

Index

A

B

E

H

I

M

N

O

Q

R

S

Y

Z

Strategy for Solving Word Problems

Step 1. Read the problem carefully to determine what you are being asked to find.
Step 2. Select a variable to represent each unknown quantity. Specify precisely what each variable represents and note any restrictions on each variable.
Step 3. If necessary, make a sketch and translate the problem into a word equation or a system of word equations. Then translate each word equation into an algebraic equation.
Step 4. Solve the equation or the system of equations, and answer the question asked by the problem.
Step 5. Check the reasonableness of your answer.

Statements of Variation

For $k \neq 0$,

Direct variation: $y = kx$, y varies directly as x

Inverse variation: $y = \frac{k}{x}$, y varies inversely as x

Joint variation: $z = kxy$, z varies jointly as x and y

Absolute Value Equations and Inequalities

For $a > 0$,

$|x| = a$ means $x = -a$ or $x = a$

$-a$ 0 a

$|x| \leq a$ means $-a \leq x \leq a$

$-a$ 0 a

$|x| \geq a$ means $x \leq -a$ or $x \geq a$

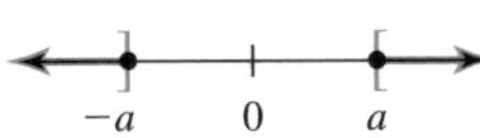

Properties of Radicals

If $\sqrt[n]{x}$ and $\sqrt[n]{y}$ are both real numbers, then:

$\sqrt[n]{xy} = \sqrt[n]{x}\sqrt[n]{y}$ and

$\sqrt[n]{\frac{x}{y}} = \frac{\sqrt[n]{x}}{\sqrt[n]{y}}$ for $y \neq 0$

$\sqrt[n]{x^m} = (\sqrt[n]{x})^m = x^{m/n}$

For any real number x and natural number n:

$\sqrt[n]{x^n} = |x|$ if n is even

$\sqrt[n]{x^n} = x$ if n is odd

Properties of Logarithms

For $x > 0$, $y > 0$, $b > 0$, and $b \neq 1$, $\log_b x = m$ is equivalent to $b^m = x$.

Product rule: $\log_b xy = \log_b x + \log_b y$

Quotient rule: $\log_b \frac{x}{y} = \log_b x - \log_b y$

Power rule: $\log_b x^p = p \log_b x$

Special identities:
$\log_b 1 = 0$
$\log_b b = 1$
$\log_b \frac{1}{b} = -1$
$\log_b b^x = x$
$b^{\log_b x} = x$

Sequences and Series

a_n is read as "a sub n."

Arithmetic sequences: $d = a_n - a_{n-1}$

$a_n = a_1 + (n - 1)d$

$S_n = \frac{n}{2}(a_1 + a_n)$

Geometric sequences: $r = \frac{a_n}{a_{n-1}}$

$a_n = a_1 r^{n-1}$

$S_n = \frac{a_1(1 - r^n)}{1 - r}$

Infinite geometric sequences: $S = \frac{a_1}{1 - r}$, $|r| < 1$.

Summation notation:

$\sum_{i=1}^{n} a_i = a_1 + a_2 + \ldots + a_{n-1} + a_n$

Methods of Solving Quadratic Equations

$ax^2 + bx + c = 0$ with $a \neq 0$.
Graphically
Numerically
Factoring
Extraction of roots
Completing the square
The quadratic formula: $x = \frac{-b \pm \sqrt{b^2 - 4ac}}{2a}$

Slope

The slope of a line through (x_1, y_1) and (x_2, y_2) is given by

$$m = \frac{\text{Change in } y}{\text{Change in } x} = \frac{\Delta y}{\Delta x} = \frac{y_2 - y_1}{x_2 - x_1} \text{ for } x_1 \neq x_2.$$

Positive slope: The line goes upward to the right.
Negative slope: The line goes downward to the right.
Zero slope: The line is horizontal.
Undefined slope: The line is vertical.

Forms of Linear Equations

Slope-intercept form: $y = mx + b$ or $f(x) = mx + b$ with slope m and y-intercept $(0, b)$
Vertical line: $x = a$ for a real constant a
Horizontal line: $y = b$ for a real constant b
General form: $Ax + By = C$
Point-slope form: $y - y_1 = m(x - x_1)$ through (x_1, y_1) with slope m

Properties of Exponents

Let $m, n, x, x^m, x^n, y, y^m$, and y^n be real numbers.
Product rule: $x^m \cdot x^n = x^{m+n}$
Power rule: $(x^m)^n = x^{mn}$
Product to a power: $(xy)^m = x^m y^m$
Quotient to a power: $\left(\frac{x}{y}\right)^m = \frac{x^m}{y^m}$ for $y \neq 0$
Quotient rule: $\frac{x^m}{x^n} = x^{m-n}$ for $x \neq 0$
Negative power: $\left(\frac{x}{y}\right)^{-n} = \left(\frac{y}{x}\right)^n$ for $x \neq 0, y \neq 0$
Special identities:
$x^0 = 1$ for $x \neq 0$ $\quad x^1 = x \quad x^{-1} = \frac{1}{x}$ for $x \neq 0$

Factoring Special Forms

Difference of two squares: $x^2 - y^2 = (x + y)(x - y)$
Perfect square trinomial: $x^2 + 2xy + y^2 = (x + y)^2$
Perfect square trinomial: $x^2 - 2xy + y^2 = (x - y)^2$
Difference of two cubes:
$x^3 - y^3 = (x - y)(x^2 + xy + y^2)$
Sum of two cubes:
$x^3 + y^3 = (x + y)(x^2 - xy + y^2)$

Systems of Two Linear Equations

One Solution

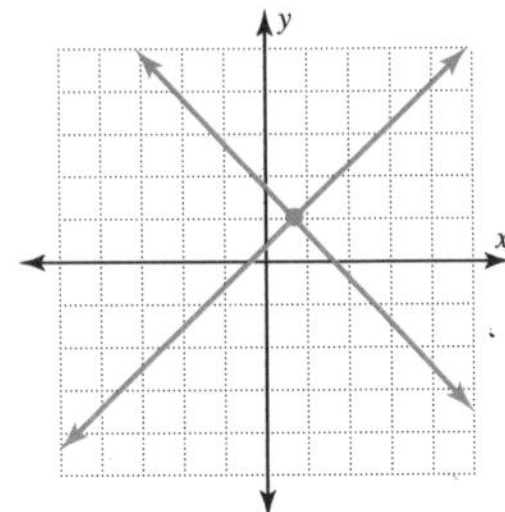

A consistent system of independent equations: The solution process will produce unique x- and y-coordinates.

No Solution

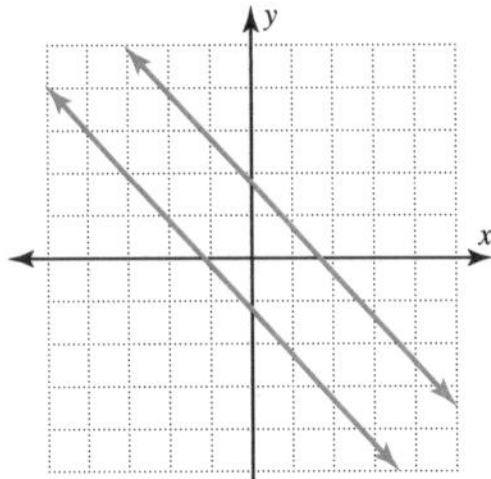

An inconsistent system: The solution process will produce a contradiction.

An Infinite Number of Solutions

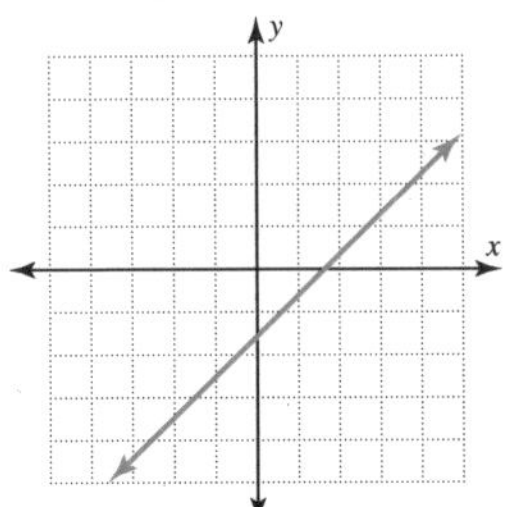

A consistent system of dependent equations: The solution process will produce an identity.

Equivalent Statements about Linear Factors of a Polynomial

For a real constant c and a real polynomial $P(x)$, the following statements are equivalent:

Algebraically

$x - c$ is a factor of $P(x)$.
$x = c$ is a solution of $P(x) = 0$.

Numerically

$P(c) = 0$; that is,
c is a zero of $P(x)$.

Graphically

$(c, 0)$ is an x-intercept
of the graph of $y = P(x)$.